International Table of Atomic Masses*†

Name	Symbol	Atomic Number	Atomic Mass	Name	Symbol	Atomic Number	Atomic Mass
Actinium	Ac	89	(227)	Neodymium	Nd	60	144.242
Aluminum	Al	13	26.9815386	Neon	Ne	10	20.1797
Americium	Am	95	(243)	Neptunium	Np	93	(237)
Antimony	Sb	51	121.760	Nickel	Ni	28	58.6934
Argon	Ar	18	39.948	Niobium	Nb	41	92.90638
Arsenic	As	33	74.92160	Nitrogen	N	7	14.0067
Astatine	At	85	(210)	Nobelium	No	102	(259)
Barium	Ba	56	137.327	Osmium	Os	76	190.23
Berkelium	Bk	97	(247)	Oxygen	O	8	15.9994
Beryllium	Be	4	9.012182	Palladium	Pd	46	106.42
Bismuth	Bi	83	208.98040	Phosphorus	P	15	30.973763
Bohrium	Bh	107	(272)	Platinum	Pt	78	195.084
Boron	B	5	10.811	Plutonium	Pu	94	(244)
Bromine	Br	35	79.904	Polonium	Po	84	(209)
Cadmium	Cd	48	112.411	Potassium	K	19	39.0983
Calcium	Ca	20	40.078	Praseodymium	Pr	59	140.90765
Californium	Cf	98	(251)	Promethium	Pm	61	(145)
Carbon	C	6	12.0107	Protactinium	Pa	91	231.03588
Cerium	Ce	58	140.116	Radium	Ra	88	(226)
Cesium	Cs	55	132.9054519	Radon	Rn	86	(222)
Chlorine	Cl	17	35.453	Rhenium	Re	75	186.207
Chromium	Cr	24	51.9961	Rhodium	Rh	45	102.90550
Cobalt	Co	27	58.933195	Roentgenium	Rg	111	(280)
Copper	Cu	29	63.546	Rubidium	Rb	37	85.4678
Curium	Cm	96	(247)	Ruthenium	Ru	44	101.07
Darmstadtium	Ds	110	(281)	Rutherfordium	Rf	104	(267)
Dubnium	Db	105	(268)	Samarium	Sm	62	150.36
Dysprosium	Dy	66	162.500	Scandium	Sc	21	44.955912
Einsteinium	Es	99	(252)	Seaborgium	Sg	106	(271)
Erbium	Er	68	167.259	Selenium	Se	34	78.96
Europium	Eu	63	151.964	Silicon	Si	14	28.0855
Fermium	Fm	100	(257)	Silver	Ag	47	107.8682
Fluorine	F	9	18.9984032	Sodium	Na	11	22.98976928
Francium	Fr	87	(223)	Strontium	Sr	38	87.62
Gadolinium	Gd	64	157.25	Sulfur	S	16	32.065
Gallium	Ga	31	69.723	Tantalum	Ta	73	180.94788
Germanium	Ge	32	72.64	Technetium	Tc	43	(98)
Gold	Au	79	196.966569	Tellurium	Te	52	127.60
Hafnium	Hf	72	178.49	Terbium	Tb	65	158.92535
Hassium	Hs	108	(270)	Thallium	Tl	81	204.3833
Helium	He	2	4.002602	Thorium	Th	90	232.03806
Holmium	Ho	67	164.93032	Thulium	Tm	69	168.93421
Hydrogen	H	1	1.00794	Tin	Sn	50	118.710
Indium	In	49	114.818	Titanium	Ti	22	47.867
Iodine	I	53	126.90447	Tungsten	W	74	183.84
Iridium	Ir	77	192.217	Ununbium	Uub	112	(285)
Iron	Fe	26	55.845	Ununhexium	Uuh	116	(292)
Krypton	Kr	36	83.798	Ununoctium	Uuo	118	(294)
Lanthanum	La	57	138.90547	Ununpentium	Uup	115	(228)
Lawrencium	Lr	103	(262)	Ununquadium	Uuq	114	(289)
Lead	Pb	82	207.2	Ununtrium	Uut	113	(284)
Lithium	Li	3	6.941	Uranium	U	92	238.02891
Lutetium	Lu	71	174.9668	Vanadium	V	23	50.9415
Magnesium	Mg	12	24.3050	Xenon	Xe	54	131.293
Manganese	Mn	25	54.938045	Ytterbium	Yb	70	173.054
Meitnerium	Mt	109	(276)	Yttrium	Y	39	88.90585
Mendelevium	Md	101	(258)	Zinc	Zn	30	65.38
Mercury	Hg	80	200.59	Zirconium	Zr	40	91.224
Molybdenum	Mo	42	95.96				

*Based on relative atomic mass of $^{12}C = 12$.

†The values given in the table apply to elements as they exist in materials of terrestrial origin and to certain artificial elements. Values in parentheses are the mass number of the isotope of the longest half-life.

OWL—Online Web-based Learning

OWL Instant Access (two semesters) ISBN-10: 0-495-39165-4 • ISBN-13: 978-0-495-39165-4
OWL e-Book Instant Access (two semesters) ISBN-10: 0-495-39166-2 • ISBN-13: 978-0-495-39166-1

Developed at the University of Massachusetts, Amherst, and class tested by tens of thousands of chemistry students, **OWL** is a fully customizable and flexible web-based learning system. **OWL** supports mastery learning and offers numerical, chemical, and contextual parameterization to produce thousands of problems correlated to this text. The **OWL** system also features a database of simulations, tutorials, and exercises, as well as end-of-chapter problems from the text. With **OWL**, you get the most widely used online learning system available for chemistry with unsurpassed reliability and dedicated training and support.

With thousands of problems correlated to this text and a database of simulations, tutorials, and exercises, as well as end-of-chapter problems from the text, you get the most widely used online learning system available for chemistry with unsurpassed reliability and dedicated training and support.

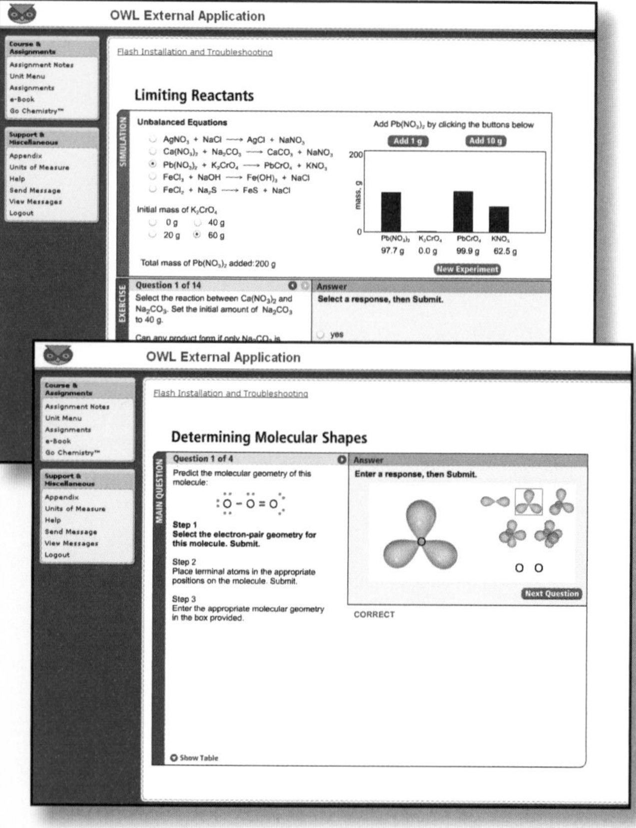

Features

- Interactive simulations of chemical systems are accompanied by guiding questions that lead you through an exploration of the simulation. These concept-building tools guide you to your own discovery of chemical concepts and relationships.
- Interactive problem-solving tutors ask questions and then give feedback that helps you solve the problem.
- Explorations of animations, movies, and graphic images help you examine the chemical principles behind multimedia presentations of chemical events.
- Re-try questions over and over, covering the same concept, but using different numerical values and chemical systems until you get it right.

A Complete e-Book!

Would you rather study online? If so, the **e-Book in OWL** is for you, with all chapters and sections in the textbook fully correlated to **OWL** homework content.

And now **OWL for Introductory and General Chemistry** includes **Go Chemistry**™—27 mini video lectures covering key chemistry concepts that you can view onscreen or download to your portable video player for study on the go!

Go Chemistry's mini video lectures include animations and problems for a quick summary of key concepts and for exam prep. The program's e-flashcards, which briefly introduce a key concept and then test your understanding of the basics with a series of questions, make study efficient and easy.

These 5-8 minute movies play on video iPods®, iPhones®, and other personal video players, and can be viewed on the desktop in QuickTime, Windows Media Player, and iTunes®.

ichapters.com

If an access card came packaged with your text, you may be able to use these assets immediately. If not, visit www.ichapters.com, our preferred online store, to purchase access to Go Chemistry and other exciting products from Cengage Learning.

THIRD EDITION

CHEMISTRY:
Principles and Practice

Daniel L. Reger
University of South Carolina

Scott R. Goode
University of South Carolina

David W. Ball
Cleveland State University

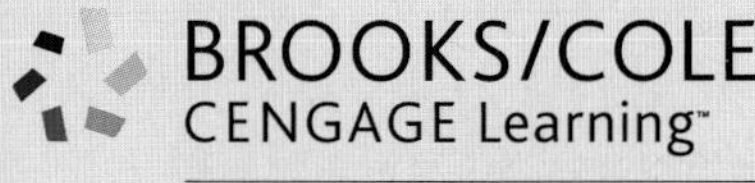

Australia • Brazil • Japan • Korea • Mexico • Singapore • Spain • United Kingdom • United States

***Chemistry: Principles and Practice,* Third Edition**
Daniel L. Reger, Scott R. Goode, and David W. Ball

Publisher: Mary Finch
Senior Acquisitions Editor: Lisa Lockwood
Senior Development Editor: Jay Campbell
Assistant Editor: Ashley Summers
Editorial Assistant: Elizabeth Woods
Senior Media Editor: Lisa Weber
Marketing Manager: Nicole Hamm
Marketing Assistant: Kevin Carroll
Marketing Communications Manager: Linda Yip
Project Manager, Editorial Production: Teresa L. Trego
Creative Director: Rob Hugel
Art Director: John Walker
Print Buyer: Karen Hunt
Permissions Editor: Roberta Broyer
Production Service: Graphic World, Inc.
Production Editor: Dan Fitzgerald, Graphic World, Inc.
Text Designer: tani hasegawa
Photo Researcher: Sue C. Howard
Copy Editor: Graphic World, Inc.
Illustrator: Greg Gambino/2064 Design
OWL Producers: Stephen Battisti, Cindy Stein, and David Hart, Center for Educational Software Development at the University of Massachusetts, Amherst, and Cow Town Productions
Cover Designer: John Walker; Bartay
Cover Image: Main image: Fundamental Photographs; Lake Nyos: Louise Gubb/Corbis
Compositor: Graphic World, Inc.

On the Cover:

The photographs show carbon dioxide escaping from a bottle of soda and carbon dioxide being released from Lake Nyos in Cameroon. In 1986, carbon dioxide erupted from Lake Nyos with tragic consequences. Scientists and engineers, using laboratory measurement and theoretical models of the solubility of carbon dioxide, developed a release system for the gas to make the area safe for the inhabitants. These topics are discussed in more detail in Chapter 12.

Library of Congress Control Number: 2008942748

ISBN-13: 978-0-534-42012-3
ISBN-10: 0-534-42012-5

Brooks/Cole
10 Davis Drive
Belmont, CA 94002-3098
USA

Cengage Learning is a leading provider of customized learning solutions with office locations around the globe, including Singapore, the United Kingdom, Australia, Mexico, Brazil, and Japan. Locate your local office at **international.cengage.com/region.**

Cengage Learning products are represented in Canada by Nelson Education, Ltd.

For your course and learning solutions, visit **academic.cengage.com.**

Purchase any of our products at your local college store or at our preferred online store **www.ichapters.com.**

Printed in China
2 3 4 5 6 7 14 13 12 11

We dedicate this book to our families,
our colleagues, and our students.
They all conspire to keep us on our toes.

Contents Overview

Table of Contents

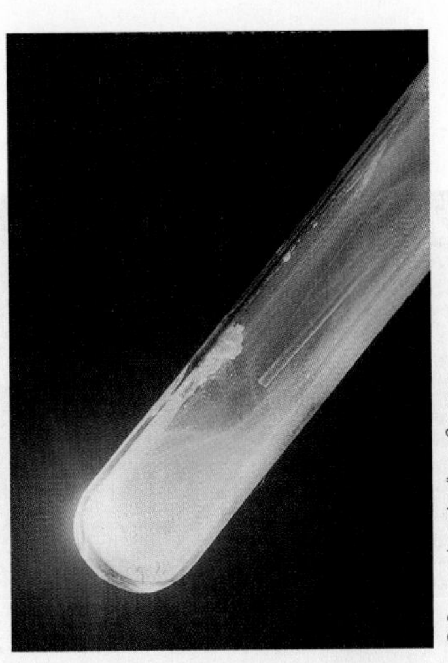

© Cengage Learning/Larry Cameron

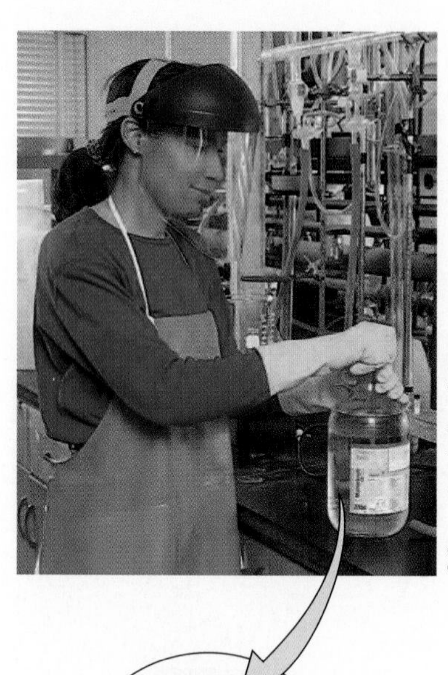
© Cengage Learning/Larry Cameron

© Digital Vision/Photolibrary

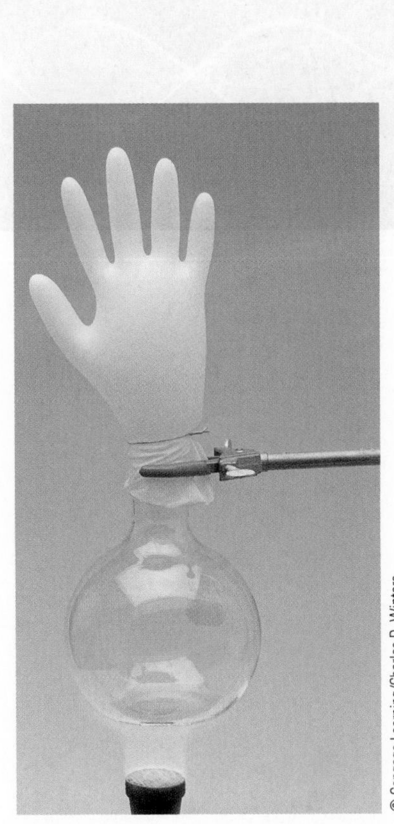

CHAPTER 7

Electronic Structure 248

FORENSIC ANALYSIS OF BULLETS

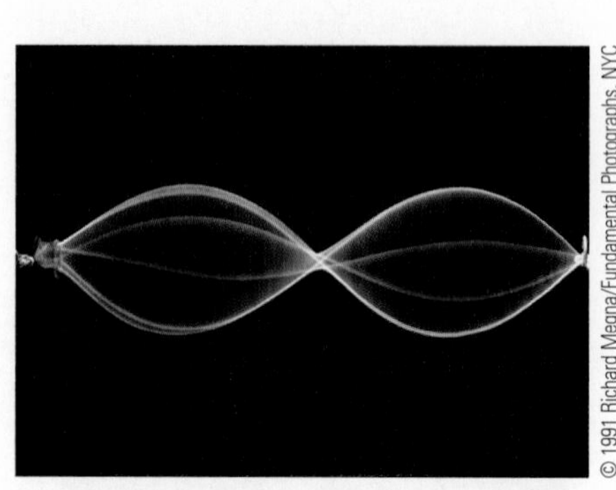

© 1991 Richard Megna/Fundamental Photographs, NYC

CHAPTER 8
The Periodic Table: Structure and Trends 290

THE STORY OF FLUORINE

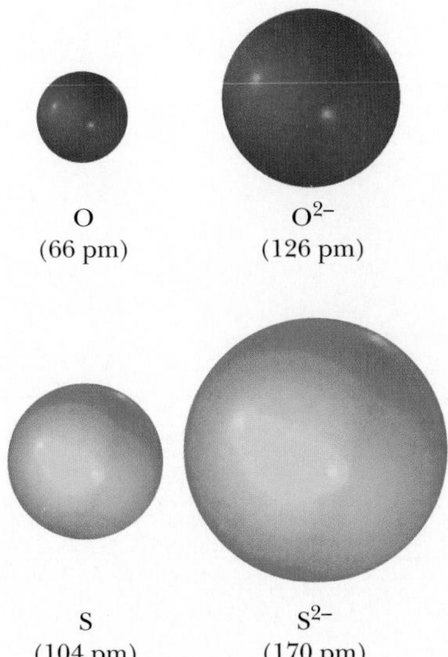

CHAPTER 9
Chemical Bonds 324

NITRIC OXIDE: A SMALL MOLECULE WITH IMPORTANT FUNCTIONS

Keith Kent/Photo Researchers, Inc.

CHAPTER 10
Molecular Structure and Bonding Theories 370

MOLECULES AND THE WAR ON TERROR

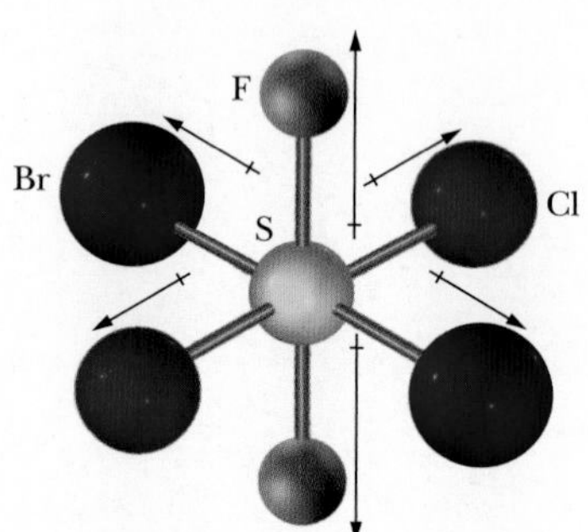

No dipole moment

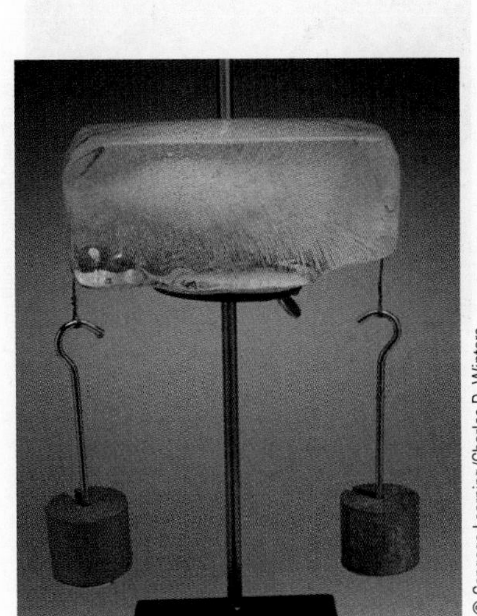

© Cengage Learning/Charles D. Winters

CHAPTER 13

Chemical Kinetics 510

THE ICE MAN

Doug Martin/Photo Researchers, Inc.

CHAPTER 14

Chemical Equilibrium 572

TRAGEDY IN BHOPAL

CHAPTER 15 Solutions of Acids and Bases 628

HYDROFLUORIC ACID

© Cengage Learning/Charles D. Winters

CHAPTER 16

Reactions between Acids and Bases 680

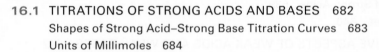

MODERN CHEMISTRY SOLVES CIVIL WAR MYSTERY

NASA Kennedy Space Center (NASA-KSC)

© VStoc/Alamy

CHAPTER 18

Electrochemistry 774

CORROSION IN THE BODY

CHAPTER 19
Transition Metals, Coordination Chemistry, and Metallurgy 826

CISPLATIN: UNUSUAL CANCER-FIGHTING MOLECULE

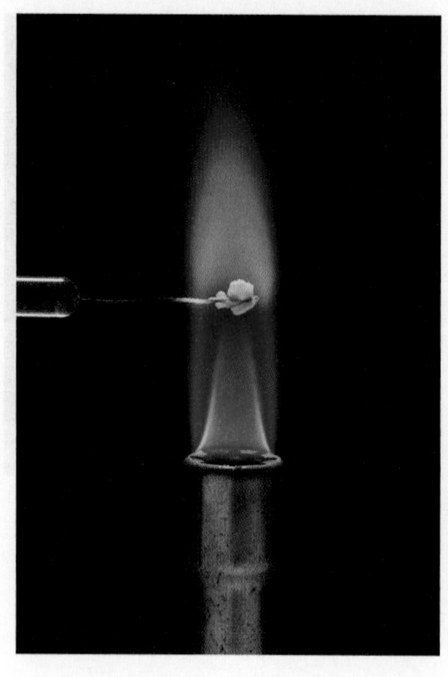
© Cengage Learning/Charles D. Winters

CHAPTER 20

Chemistry of Hydrogen, Elements in Groups 3A through 6A, and the Noble Gases 864

PROPERTIES OF ELEMENTS AND COMPOUNDS

CHAPTER 21

Nuclear Chemistry 894

SMOKE DETECTORS: LIFE-SAVING RADIOACTIVITY

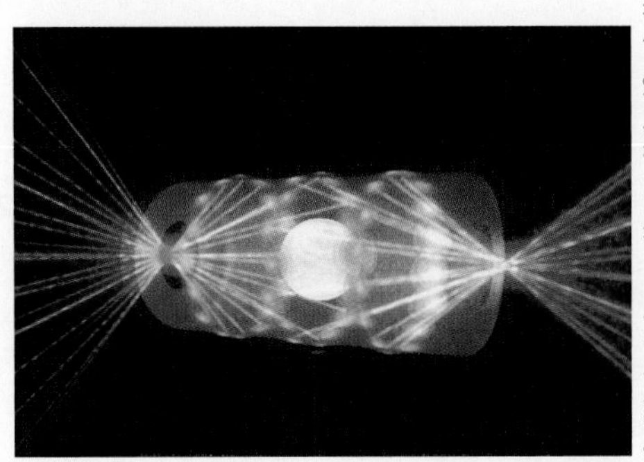

Credit is given to Lawrence Livermore National Security, LLC, Lawrence Livermore National Laboratory, and the Department of Energy under whose auspices this work was performed

CHAPTER 22

Organic Chemistry and Biochemistry 936

THE STRUCTURE OF DNA

Preface

Why Another General Chemistry Book?

Many books are available for the general chemistry course. Many have been published in various editions for years. So with the number of books on the market, why should you consider our book, *Chemistry: Principles and Practice* by Reger, Goode, and Ball? What makes this book special and different from other general chemistry books?

The Utility of Chemistry

Few students appreciate that chemistry is a living, evolving science in which people frequently discover new facts, develop new concepts, and solve problems both big and small. Students often see the discipline as simply a static set of facts and equations, and fail to grasp the relevance and sheer power of chemistry. *Chemistry: Principles and Practice* truly embodies its title by connecting the chemistry taught in the classroom *(principles)* with its real-world uses *(practice)*. We draw our applications from various fields, including forensics, organic chemistry, biochemistry, and industry.

Chapter Introductions – Each chapter opens with an application to entice students to read the chapter and show them how chemistry explains what they see in the real world. These openers are referenced throughout the chapter and often emphasize the experimental nature of chemistry.

Specialty Essays – The text also features *Principles of Chemistry* and *Practice of Chemistry* boxes. These are real-world applications of chemistry that show why and how chemists and other professionals actually use chemistry in their jobs and daily lives.

Case Studies and Summary Problems – These features are multipronged, multistep problems that examine real-world uses of chemistry. These appear at the end of the chapters.

Narrative – The presentation of chemistry in this text is extremely readable and concise. The scope and sequence presents topics as logical extensions of material previously covered. The material is presented with numerous concrete examples that stress logical, problem-solving approaches rather than rote learning. The text narrative uses analogies to which students can relate in their daily lives. For instance, when a compound or element is used in an exercise or an example problem, we often also briefly explain the real-world significance of that compound or element by mentioning its use in an important application.

Emphasis on Experiment

Chemistry is first and foremost an experimental science, and the observations and explanations that are its foundation have come from many years of experimentation. We emphasize the role of experiment and observation in the formulation of chemical theories, and we present the principles of chemistry in this context to show that chemistry comes from experiments and not from textbooks. Margin icons have been placed throughout the chapters as appropriate to emphasize this aspect of the text.

Developing Problem-Solving Skills

Too often, students come to us and say, "I knew how to do the problem—I just didn't know where to start." We use a focused approach, by teaching methods to solve a generic problem, then the skills to apply this method to new and different situations. We work hard to simplify equilibrium problems, often a difficult topic for students. We utilize a consistent five-step approach that works for each problem, regardless of the starting point—we feel our technique makes all these problems look alike. Our goal is to teach students to rearrange problems to look like something they know how to solve rather than looking for a different equation for each new problem.

One way this text organizes problem solving is to color-code the important given material, intermediate results, and final answer in most examples. The green–yellow–red sequence is familiar to most people. This system is exceptionally valuable in problems in which the given data are used in the middle of the problem. Many times, students see "new" data in the problem and are frustrated because they do not know where it came from, and the color code helps the student determine the source of the information.

Text Features

This textbook is designed to be used by students who are interested in further study in chemistry and related areas, such as biology, engineering, geology, and the medical professions. We have tailored the presentation of information by carefully considering the scope and sequence of the material. We introduce a new topic after consideration of why it is important and only when the students' current knowledge base will allow them to understand the principles on a conceptual basis. We believe that students learn new concepts more readily when they know why the material is important. Topics are raised when they can be explained clearly and completely. We carefully develop the language of chemistry; new terms are defined when they are first introduced.

Although maintaining the hallmark features of the second edition (readability, emphasis on experiment, problem solving), this third edition has been revamped with new features to meet the needs of today's students. Previous editions of Reger/Goode/Mercer were known for an emphasis on the experimental nature of chemistry, a focused approach to problem solving, lucid explanations, and intriguing applications. This third edition not only builds on these strengths, it also increases the emphasis on conceptual understanding and relevance, and has a completely new design and updated art program. The following list specifically notates which features are new to this edition.

- NEW **Co-author**—Accomplished teacher and physical chemist David W. Ball of Cleveland State University joins the author team. David, author of *Physical Chemistry,* also published by Brooks/Cole, has received numerous teaching awards and is active in the American Chemical Society. In addition to textbook writing, David has made other valuable contributions to the chemical education world.

IN-TEXT FEATURES

- NEW **Introductions**—Unique to the market, each chapter begins with an opening application drawn from various chemical fields, which then is revisited throughout the chapter. The applications are revisited in the text, in specialty features, and in the problem sets.
- **Learning Objectives**—Each section begins with a set of Learning Objectives that clearly indicate the important concepts and ends with an end-of-section Objectives Review. The exercises also are grouped by objectives.
- **Margin Notes**—As the material for each objective is covered, highly focused margin notes address each objective; the margin notes are reserved solely for this purpose. Students can use the objectives and margin notes to identify and learn the key concepts of each section.
- NEW **Enhanced Problem Solving**—The problem-solving pedagogy utilizes logical "thinking" strategies and "visual" road maps.
  - *Color and Flow Diagrams:* Flow diagrams are used in many problems that involve mathematical calculations. The flow diagrams show the starting point in the problem and the operations needed to get to the solution. In our experience, students often need aid in following the initial data through the intermediate steps to the solution. Therefore, we make unique pedagogical use of color by color coding various information in the *Example problems.* Initial data are marked in green, significant intermediate steps are marked in yellow, and the solution is highlighted in red. The flow diagrams are color coded to match

the solutions and are carefully formulated to aid the student in developing problem-solving thought processes and strategies.

- *Example Problems:* Example problems are numerous and worked in a straightforward and logical fashion. The strategy is presented, the problem solved, and warnings of potential errors and pitfalls provided. As mentioned earlier, example problems include color-coded flow diagrams that map out the problem-solving process in a graphical manner, thus aiding the student in developing problem-solving thought processes and strategies. Each worked example problem is followed by a similar Understanding problem and answer, so that the students can test their comprehension of the topic. Practical descriptive chemistry is incorporated into the example problems.

- **Specialty Essays**—*Principles of Chemistry* essays expand and/or reinforce important topics discussed in the book. *Practice of Chemistry* essays are real-world applications of chemistry, that is, why and how chemists and other professionals actually use chemistry in their jobs and daily lives. Many of the essays are new or updated.

END-OF-CHAPTER FEATURES

- NEW **Case Studies and Summary Problems**—*Case Studies* are multipronged, multistep problems that examine real-world uses of chemistry. *Summary Problems* focus on problems of chemical interest and draw on material from the entire chapter for their solutions. Either a Case Study or a Summary Problem ends each chapter.
- NEW **Ethics in Chemistry**—Unique to general chemistry texts, *Ethics in Chemistry* sections, located at the end of each chapter before the problem sets, emphasize the human side of chemistry and remind students that chemistry is not just a set of facts. These questions discuss the ethical issues and dilemmas scientists face in practicing their profession. They also are good exercises for schools that have a writing-across-the-curriculum requirement.
- NEW **Visual Chapter Summaries**—This large flow chart shows connections among the various concepts in the chapter.
- **Chapter Summary**—This summarizes the main points of the chapter.
- **Chapter Terms**—This contains the important terms of the chapter, separated by section. A comprehensive Glossary is also available in the appendix.
- NEW **Key Equations**—This recaps the important equations within the chapter.

END-OF-CHAPTER PROBLEM SETS

- *Questions* and *Exercises*—The text includes approximately 2000 *questions* and *exercises.* Questions are qualitative in nature, often conceptual, and include problem-solving skills. Questions that are suitable for a brief writing exercise are designated with the symbol of a pencil. The more challenging items are designated with a triangle. Selected exercises are marked with ■ to indicate that they are available in interactive form in OWL, Brooks/Cole's online learning system. Exercises are paired, with the odd-numbered ones having answers in the appendix and a similar even-numbered problem immediately following. Although most exercises appear in the order in which they are discussed, some Chapter Exercises are uncategorized, and Cumulative Exercises integrate concepts and methods introduced in earlier chapters. Cumulative Exercises often contain multiple parts, multiple steps, or both.

ELECTRONIC ANCILLARY MATERIALS

- NEW **Technology**—This edition fully integrates OWL (Online Web-based Learning), the online learning system trusted by tens of thousands of students. Integrated end-of-chapter questions correlate to OWL. An optional e-book of this edition is also available in OWL. In addition, Go Chemistry learning modules developed by award-winning chemists offer minilectures and learning tools that play on video iPods, personal video players, Windows Media View, and iTunes.

Organization

The overall organization of the material in this text follows a general order that has been established over the years. We have refined the general presentation of the key topics. The first two chapters introduce the student to the basic concepts and language of chemistry. For programs with well-prepared students, these chapters are designed so they can be made assigned reading. A new introduction to organic chemistry appears in Chapter 2. Stoichiometry in Chapter 3 is the first main topic, and we develop a general method for executing calculations based on chemical equations. The method is not "plug into this formula," but rather a sequential reasoning process that applies to a whole series of calculations ranging from mass-mass conversions to reactions in solution, to enthalpy changes in chemical equations, and to reactions that involve gases. Students using the first two editions have found that the material in these four consecutive chapters is interrelated—that is, each chapter's material does not require a new learning event. Example problems in the text are complemented with flow diagrams to help students organize the problem-solving process. Our approach fosters critical thinking skills by helping students develop a strategy rather than relying on rote memorization operations.

Chapter 3 begins with chemical equations so that students see the "chemistry" behind stoichiometry calculations. Empirical and molecular formulas, balancing equations, and the use of chemical equations in stoichiometry calculations starting with mass data are also presented in the first stoichiometry chapter. Students using the text have found the coverage of limiting reactant problems particularly successful, and they have demonstrated an ability to apply this knowledge to similar problems in the next three chapters. Chapter 4 covers solution stoichiometry, and Chapter 5 discusses thermochemistry. Chapter 4 emphasizes first the experimental approach of how ionic compounds behave in solution, as an introduction to quantitative solution stoichiometry calculations. The calculations are presented so that the students can combine what they have learned in the first stoichiometry chapter with the new information. In Chapter 5, enthalpy in chemical reactions is also introduced as a natural progression of stoichiometry; enthalpy is introduced as part of the chemical equation. Gases are covered next in Chapter 6 because we believe that early placement of this material is helpful for the first-semester laboratory, although the chapter can be taught after structure and bonding. Again, reaction stoichiometry is emphasized. In all of these chapters, the concepts are illustrated with important, real-life chemical reactions in the many example problems. We believe our integration of descriptive chemistry throughout the text, in worked examples, in featured topics, and in exercises helps solidify the concepts of chemical reactivity.

Chapters 7 through 10 develop the models for atomic and molecular structure. The models and theories are developed as a natural progression from experimental observations. We emphasize the periodic table as a tool to help learn electron configurations, as well as trends in ionization energies and the sizes of atoms and ions. The presentation of bonding and shapes of molecules is supported by high-quality drawings that picture atoms and orbitals in proper perspective. Users of the first two editions have found that their students developed a "visual" understanding of bonding and shapes of molecules. The organization of the molecular orbital section allows the instructor to omit, teach a basic introduction, or defer molecular orbital theory to a later time in the course. Chapter 11 has been reorganized to emphasize the experimental results that have led to the development of the models that explain the physical properties of different types of materials. Chapter 12, which examines the properties of solutions, gives a qualitative treatment of disorder as a driving force in the solution process. We emphasize the common features of the different colligative properties in a way to reduce rote memorization in the learning process.

Chapter 13 introduces kinetics. This chapter precedes the equilibrium chapters but can be deferred until later in the course. Throughout the chapter, realistic laboratory data help explain the concepts and give the students a feel for rates of reactions. The microscopic models of reaction rates stress that chemical reactions occur as a result of collisions between reacting species. The chapter includes a section on catalysis and concludes with a discussion of mechanisms. Because many programs defer study of reaction mechanisms to later courses in chemistry, the section on catalysis is an important topic for all students.

A systematic approach to equilibria is presented in Chapters 14 through 16. Many students of general chemistry find this topic difficult, but we clarify the material by introducing a strategy that works for all equilibrium systems. The introduction to equilibria uses simple gas-phase reactions. Solubility equilibria and the common ion effect are introduced at this point so that relevant and descriptive chemical problems can be treated early. Chapter 15 extends the concepts of solution equilibria to acid–base reactions. We present strong and weak electrolytes in this equilibrium chapter. In Chapter 16, the systematic treatment continues through acid–base titration curves. These three chapters can be taught in the first semester or may be moved to later in the second semester, depending on the needs of individual courses.

The material on equilibria is followed by thermodynamics. In Chapter 17, experimental data are used to introduce the concepts. This chapter integrates stoichiometry and concepts such as Le Chatelier's principle.

A comprehensive discussion on oxidation-reduction reactions and electrochemistry follows in Chapter 18. Oxidation numbers and redox equations briefly introduced in the equation section of Chapter 3 are fully developed in Chapter 18. Rather than confuse students with two different ways to balance complex redox reactions, as some texts do, the half-reaction method is used exclusively. The text is completed with comprehensive chapters on metallurgy and transition-metal chemistry, main-group chemistry, nuclear chemistry, and a combined organic chemistry and biochemistry chapter. The scope and sequence of this material allows the individual instructors to select the portions that are most appropriate for their course goals.

Overall, the design of the text enables students with different backgrounds and different methods of learning to master the wide-ranging mixture of material that constitutes a general chemistry course. The material is presented within the context that chemistry is based on experimental results. Importantly, students will leave the course with an appreciation for chemistry, its principles, and its practices.

Supporting Materials for the Instructor

OWL (Online Web-based Learning) for General Chemistry

Instant Access to OWL (two semesters): ISBN-10: 0-495-05099-7; ISBN-13: 978-0-495-05099-5

Instant Access to OWL e-Book (two semesters): ISBN-10: 0-495-55988-1; ISBN-13: 978-0-495-55988-7

Authored by Roberta Day and Beatrice Botch, University of Massachusetts, Amherst, and William Vining, State University of New York at Oneonta

Developed at the University of Massachusetts, Amherst, and class tested by more than a million chemistry students, OWL is a fully customizable and flexible Web-based learning system. OWL supports mastery learning and offers numerical, chemical, and contextual parameterization to produce thousands of problems correlated to this text. The OWL system also features a database of simulations, tutorials, and exercises, as well as end-of-chapter problems from the text. With OWL, you get the most widely used online learning system available for chemistry with unsurpassed reliability and dedicated training and support. Now OWL for General Chemistry includes Go

Chemistry—27 mini video lectures covering key chemistry concepts that students can view onscreen or download to their portable video player to study on the go! For this third edition, OWL includes parameterized end-of-chapter questions from the text (marked in the text with ■).

The optional **e-Book in OWL** includes the complete electronic version of the text, fully integrated and linked to OWL homework problems. Most e-Books in OWL are interactive and offer highlighting, note-taking, and bookmarking features that can all be saved. To view an OWL demo and for more information, visit **www.cengage.com/owl** or contact your Cengage Learning Brooks/Cole representative.

Online Test Bank

by James Collins, East Carolina University

The Online Test Bank contains more than 1200 multiple-choice questions of varying difficulty. Instructors can customize tests using the Test Bank files on the PowerLecture CD. Blackboard and WebCT versions of the Test Bank files are also available on the Faculty Companion Web site, accessible from **www.cengage.com/chemistry/reger.**

Online Instructor's Manual

by Christopher Dockery and John Cody, Kennesaw State University
ISBN-10: 0-495-55977-6; ISBN-13: 978-0-495-55977-1

The online Instructor's Manual offers suggestions for organization of the course. This manual presents detailed solutions of all even-numbered end-of-chapter exercises and problems in the text for the convenience of instructors and staff involved in teaching the course. Download the manual from the book's companion Web site, which is accessible from **www.cengage.com/chemistry/reger.**

PowerLecture™

PowerLecture with ExamView® and JoinIn Instructor's CD-ROM

ISBN-10: 0-495-55984-9; ISBN-13: 978-0-495-55984-9

PowerLecture is a one-stop digital library and presentation tool that includes:

- Prepared Microsoft® PowerPoint® Lecture Slides authored by the textbook authors that cover all key points from the text in a convenient format that you can enhance with your own materials or with additional interactive video and animations on the CD-ROM for personalized, media-enhanced lectures.
- Image libraries in PowerPoint and JPEG formats that contain electronic files for all text art, most photographs, and all numbered tables in the text. These files can be used to create your own transparencies or PowerPoint lectures.
- Electronic files for the complete Instructor's Manual and Test Bank.
- Sample chapters from the Student Solutions Manual and Study Guide.
- ExamView testing software, with all the test items from the printed Test Bank in electronic format, enables you to create customized tests of up to 250 items in print or online.
- JoinIn clicker questions for this text, for use with the classroom response system of your choice. Assess student progress with instant quizzes and polls, and display student answers seamlessly within the Microsoft PowerPoint slides of your own lecture. Please consult your Brooks/Cole representative for more details.

Faculty Companion Web Site

This site contains the Online Instructor's Manual, as well as WebCT and Blackboard versions of the Test Bank. Access the site from **www.cengage.com/chemistry/reger.**

Cengage Learning Custom Solutions

Cengage Learning Custom Solutions develops personalized text solutions to meet your course needs. Match your learning materials to your syllabus and create the perfect learning solution—your customized text will contain the same thought-provoking, scientifically sound content, superior authorship, and stunning art that you've come to expect from Cengage Learning, Brooks/Cole texts, yet in a more flexible format. Visit **www.cengage.com/custom** to start building your book today.

Laboratory Manual

Customized laboratory manuals of tested experiments will be produced as desired by individual colleges and universities.

Cengage Learning, Brooks/Cole Lab Manuals

We offer a variety of printed manuals to meet all your general chemistry laboratory needs. Instructors can visit the chemistry site at **www.cengage.com/chemistry** for a full listing and description of these laboratory manuals and laboratory notebooks. All Cengage Learning laboratory manuals can be customized for your specific needs. For more details, contact your Cengage Learning, Brooks/Cole representative.

Signature Labs. . . for the Customized Laboratory

Signature Labs combines the resources of Brooks/Cole, CER, and OuterNet Publishing to provide you unparalleled service in creating your ideal customized laboratory program. Select the experiments and artwork you need from our collection of content and imagery to find the perfect laboratories to match your course. Visit **www.signaturelabs.com** or contact your Cengage Learning representative for more information.

Supporting Materials for the Student

OWL for General Chemistry

OWL

See the above description in the instructor support materials section.

Go Chemistry for General Chemistry

ISBN-10: 0-495-38228-0; ISBN-13: 978-0-495-38228-7

Go Chemistry is a set of easy-to-use essential videos that can be downloaded to your video iPod, iPhone, or portable video player—ideal for the student on the go! Developed by award-winning chemists, these new electronic tools are designed to help students quickly review essential chemistry topics. Mini video lectures include animations and problems for a quick summary of key concepts. Selected Go Chemistry modules have e-flashcards to briefly introduce a key concept and then test student understanding of the basics with a series of questions. Go Chemistry also plays on QuickTime, iTunes, and Windows Media Player. OWL contains five Go Chemistry modules. To purchase modules, enter ISBN 0-495-38228-0 at **www.ichapters.com.**

Student Solutions Manual

by William Quintana, New Mexico State University

ISBN-10: 0-495-55980-6; ISBN-13: 978-0-495-55980-1

With an emphasis on accuracy and clarity, this meticulously prepared manual presents fully worked-out solutions to all of the odd-numbered end-of-chapter exercises and problems (numbers printed in blue). Informative and helpful, the manual refers students to any pertinent text, tables, and art in the book that would enhance understanding of the problem to be solved, and where appropriate, also briefly notes information to clarify the problem solving.

Study Guide

by Simon Bott, University of Houston, Calhoun

ISBN-10: 0-495-55979-2; ISBN-13: 978-0-495-55979-5

Developed to complement the approach of the textbook, the Study Guide is an interactive way for the student to review objectives by section, terminology of the chapter, and the math used in the chapter. Opening with a Self Test and closing with a Chapter Test, each chapter of the Study Guide gives the student ample opportunity to practice taking examinations. Numerous exercises are provided for problem-solving mastery. Answers to the Self Test, Chapter Test, and Practice Exercises are given at the end of each Study Guide chapter.

Student Companion Web Site

Accessible at **www.cengage.com/chemistry/reger,** this Web site provides an online glossary from the text, glossary flashcards, a crossword puzzle for each chapter based on key terms, as well as an interactive Periodic Table.

Acknowledgments

A book is not simply written by authors; it is very much a team project, with players from all quarters. We are truly indebted to our colleagues and our reviewers, who have patiently explained chemistry, worked problems, provided their best examples, discussed strategies, and looked for errors. Among the many people who have helped was John Holdcroft, who got this whole project started. Special thanks to Jeff Appling for helping crystallize many ideas early in the project, and David Shinn at University of Hawaii at Manoa, David Garza at Samford University, Amy Taylor at University of South Carolina, Scott Mason at Mount Union College, and Ed Mercer for careful reviews and attention to detail toward the end of the project. Regis Goode at Ridge View High School and Bob Conley of the New Jersey Institute of Technology have used every edition of the book, and have provided excellent reviews and discussion of new ideas. Andrea Thomas at Wilkes College is a long-time user who kept records of conversations with her students and their conceptions and misconceptions of the presentation. Don Neu at St. Cloud State University read the entire manuscript, and helped refine and bring consistency to the presentation of our material.

The members of the team at Cengage Learning were not only helpful and competent, they provided support, guidance, and reason, as needed. Most important to the project were our editors, Lisa Lockwood and Jay Campbell. Their knowledge and expertise, in concert with unflappable demeanors, therapeutic conference calls, and all-too-modest business lunches, kept the project under control, and the importance of their sincere belief in the author team cannot be underestimated. Senior Media Editor Lisa Weber handled the media products that accompany the book, PowerLecture and OWL in particular. Teresa Trego, Senior Content Project Manager, oversaw the production of the book and kept the book on schedule. Assistant Editor Ashley Summers coordinated the production of the print ancillary materials.

Dan Fitzgerald, our production editor at Graphic World Publishing Services, was able to marshal resources and throttle the flow of manuscript, art, photographs, and page proofs in a manner that accommodated the academic, professional, and personal schedules. Copyeditor Sheila Higgins was extremely helpful in polishing our writing. Greg Gambino of 2064 Design skillfully overhauled our art program with great success.

Our photographic team, Larry Cameron, Bob Philp, Charles Winters, and Richard Megna, brought a wonderful sense of design, photography, and chemistry to our book. And our photo researcher, Sue Howard, applied her unique skills to obtain photographs that exactly matched our needs.

We also acknowledge the reviewers of the book. They provided knowledge, insight, and plain common sense to help guide us during a sometimes arduous development path.

Reviewers of the Third Edition

Jeffrey R. Appling, *Clemson University*
Robert J. Balahura, *University of Guelph*
David Ballantine, *Northern Illinois University*
Mufeed Basti, *North Carolina A&T University*
Mark Benvenuto, *University of Detroit-Mercy*
Silas Blackstock, *University of Alabama*
Chris Bowers, *Ohio Northern University*
Fitzgerald B. Bramwell, *University of Kentucky*

Kristine Butcher, *California Lutheran University*
James Collins, *East Carolina University*
Robert Conley, *New Jersey Institute of Technology*
Allison Dobson, *Georgia Southern University*
Bill Donovan, *University of Akron*
Kenneth Dorris, *Lamar University*
Randall S. Dumont, *McMaster University*
Cassandra Eagle, *Appalachian State University*
Barb Edgar, *University of Minnesota*
George Evans, *East Carolina University*
Nancy Faulk, *Blinn College-Bryan Campus*
Galen George, *Santa Rosa Junior College*
Graeme Gerrans, *University of Virginia*
Y. C. Jean, *University of Missouri-Kansas City*
Eric Johnson, *Ball State University*
David Katz, *Pima Community College*
Jim Konzelman, *Gainesville State College*
Richard Kopp, *East Tennessee State University*
Craig McLauchlin, *Illinois State University*
Dave Metcalf, *University of Virginia*
Don Neu, *St. Cloud State University*
Daphne Norton, *Emory University*
Mark Ott, *Jackson Community College*
Preetha Ram, *Emory University*
Steve Rathbone, *Blinn College-Bryan Campus*
Kevin Redig, *Pima Community College*
Tracey Simmons-Willis, *Texas Southern University*
Cheryl Snyder, *Schoolcraft College*
Michael Starzak, *Binghamton University*
Bruce Storhoff, *Ball State University*
Andrea Thomas, *Wilkes Community College*
John Thompson, *Lane Community College*
Petr Vanýsek, *Northern Illinois University*
Rashmi Venkateswaran, *University of Ottawa*
Kristine Wammer, *St. Thomas University*
Thomas Webb, *Auburn University*
Marcy Whitney, *University of Alabama*

Reviewers of the Second Edition

Robert D. Allendoerfer, *State University of New York at Buffalo*
Jeffrey R. Appling, *Clemson University*
Robert J. Balahura, *University of Guelph*
Kristine Butcher, *California Lutheran University*
Robert Conley, *New Jersey Institute of Technology*
Geoffrey Davies, *Northeastern University*
John DeKorte, *Glendale Community College*
Raymond G. Fort, Jr., *University of Maine*
Donald G. Hicks, *Georgia State University*
Stuart Nowinski, *Glendale Community College*
Barbara N. O'Keeffe, *GMI Engineering & Management Institute*
Joseph M. Prokipcak, *University of Guelph*
David F. Rieck, *Salisbury State University*
Patricia Rogers, *University of California, Irvine*
Gary W. Simmons, *Lehigh University*
Bruce Storhoff, *Ball State University*
Edward Witten, *Northeastern University*
Orville Ziebarth, *Mankato State University*

Reviewers of the First Edition

Toby Block, *Georgia Institute of Technology*
Robert S. Bly, *University of South Carolina*
Lawrence Brown, *Appalachian State University*
Juliette Bryson, *Chabot College*
Allan Colter, *University of Guelph*
Ernest Davidson, *Indiana University, Bloomington*
Geoffrey Davies, *Northeastern University*
John DeKorte, *Glendale Community College*
Grover Everett, *University of Kansas, Lawrence*
David Garza, *Cumberland College*
Michael Golde, *University of Pittsburgh*
Frank Gomba, *United States Naval Academy*
Robert Gordon, *Queen's University*
Henry Heikkinen, *University of Northern Colorado*
James Holler, *University of Kentucky*
Thomas Huang, *Eastern Tennessee University*
Colin Hubbard, *University of New Hampshire*
Wilbert Hutton, *Iowa State University*
Philip Lamprey, *University of Massachusetts, Lowell*
Bruce Mattson, *Creighton University*
Hector McDonald, *University of Missouri, Rolla*
Jack McKenna, *St. Cloud State University*

Jennifer Merlic, *Santa Monica College*
Stephen L. Morgan, *University of South Carolina*
Gardiner Myers, *University of Florida*
George Pfeffer, *University of Nebraska*
Robert H. Philp, Jr., *University of South Carolina*
David Pringle, *University of Northern Colorado*
Joseph M. Prokipcak, *University of Guelph*
Ronald Ragsdale, *University of Utah*
Robert Richman, *Mt. St. Mary's College*
Eugene Rochow, *Harvard University*
Dennis Rushforth, *University of Texas, San Antonio*
James Sodetz, *University of South Carolina*
Helen Stone, *Ben L. Smith High School*
Ronald Strange, *Fairleigh Dickinson University*
Raymond Trautman, *San Francisco State University*
Eugene R. Weiner, *University of Denver*
Edward Wong, *University of New Hampshire*

Finally, we would be remiss if we did not express our appreciation to our spouses, Cheryl, Regis, and Gail. This textbook would not exist without their steadfast support.

Daniel L. Reger **Scott R. Goode** **David W. Ball** *October 2008*

About the Authors

Daniel L. Reger is a decorated inorganic chemist from the University of South Carolina. He is Carolina Distinguished Professor. He received his B.S. in 1967 from Dickinson College and his Ph.D. in 1972 from the Massachusetts Institute of Technology. In 1985 and 1994, he was a Visiting Fellow at Australian National University. In his 30+ years of teaching at South Carolina, he has received numerous university awards, including the Educational Foundation Research Award for Science, Mathematics, and Engineering in 1995; the Michael J. Mungo Award for Excellence in Undergraduate Teaching in 1995 and for Graduate Teaching in 2003; the Amoco Foundation Outstanding Teaching Award in 1996; the Carolina Trustee Professorship in 2000; and the Educational Foundation Outstanding Service Award in 2008. In 2007, he was awarded the South Carolina Governor's Award for Excellence in Scientific Research, and in 2008, he was the American Chemical Society's Outstanding South Carolina Chemist of the Year. Dr. Reger's research interests are in synthetic inorganic chemistry, and he has directed 28 Ph.D. students. He has authored more than 190 published research articles and has made more than 100 presentations at professional meetings.

Scott R. Goode is a distinguished analytical chemist also from the University of South Carolina. He received his B.S. in 1969 from University of Illinois at Urbana-Champaign and his Ph.D. from Michigan State University in 1973. Scott is an equally decorated teacher, having received numerous awards such as the Amoco Teaching Award in 1991, the Mungo Teaching Award in 1999, and the Ada Thomas Advising Award 2000. He twice received the Distinguished Honors Professor Award for his innovative course in General Chemistry. Dr. Goode's research interests include chemical education, forensics, and environmental chemistry, and he has directed 19 Ph.D. dissertations, 6 M.S. theses, and the programs of 19 M.A.T. students. His publishing achievements include more than 55 research articles and more than 150 presentations at professional meetings. He is highly active in the American Chemical Society and the Society for Applied Spectroscopy.

David W. Ball is a Professor of Chemistry at Cleveland State University. His research interests include computational chemistry of new high-energy materials, matrix isolation spectroscopy, and various topics in chemical education. He has authored more than 160 publications, equally split between research articles and educational articles, including five books currently in print. He has won recognition for the quality of his teaching, receiving several departmental and college teaching awards, as well as his university's Distinguished Faculty Teaching Award in 2002. He has been a contributing editor to *Spectroscopy* magazine since 1994, where he writes "The Baseline" column on fundamental topics in spectroscopy. He is also active in professional service, serving on the Board of Trustees for the Northeastern Ohio Science and Engineering Fair and the Board of Governors of the Cleveland Technical Societies Council. He is also active in the American Chemical Society, serving the Cleveland Section as chair twice (in 1998 and 2009) and Councilor from 2001 to the present.

Photo by Hulton Archive/Getty Images

Mary Blandy.

Forensic chemistry, the application of chemistry to criminal investigation, dates back to 1752 in England when Dr. Anthony Addington, a noted British physician of the time, used his skills as a chemist to unravel the mysterious death of prosperous English lawyer Francis Blandy. The story began when Mr. Blandy unwisely advertised a dowry of £10,000—a huge sum for those days and equivalent to more than $2,000,000 today—to the man who would marry his daughter, Mary. The sizable dowry attracted many suitors, all of whom were promptly rejected, save one. Captain William Henry Cranstoun was the son of a Scottish nobleman, and though not a handsome man, his rank and social status made him a suitable husband for Mary. By all accounts, Mary fell in love with him, and shortly thereafter Cranstoun moved into the Blandy household.

All went well for the first year, but then it was discovered that Cranstoun already had a wife back in Scotland. Mary's father became furious with Cranstoun and began to see him as a devious scoundrel who was interested only in the dowry.

To calm Mr. Blandy, Cranstoun persuaded Mary to secretly give her father a white powder. Cranstoun described this powder as an ancient formula that would make Mr. Blandy like him. Mary, wanting to keep Cranstoun's affections, began regularly administering the powder in her father's tea and gruel. As time went on, Mr. Blandy became progressively ill. Several servants also had become ill from eating some of the leftover food, though the servants eventually recovered after they stopped eating the food. Although the servants were suspicious, even to the point of preserving some of the tainted food, these incidents did not register with Mary. She never thought that the powder might be the cause of her father's deteriorating health.

When her father neared death, Mary's uncle visited and was told by the servants that Mr. Blandy might have been poisoned. Mary's uncle sent for Dr. Addington, a famous physician. After examining Mr. Blandy, Dr. Addington told Mary that the powder might be a poison. Though Mary immediately stopped feeding the powder to her father and quickly disposed of the remaining supply, by that time it was too late. Francis Blandy finally died on August 14, 1751.

Introduction to Chemistry

1

CHAPTER CONTENTS

OWL Online homework for this chapter may be assigned in OWL.

Look for the green colored vertical bar throughout this chapter, for integrated references to this chapter introduction.

Later, the powder was identified as arsenic, which is a cumulative poison that is lethal only when sufficient levels have built up in the body. This information helps explain why Mr. Blandy eventually succumbed to the poison but the servants did not. Despite the suspicious circumstances surrounding Mr. Blandy's death, it was some time before Mary was arrested. Cranstoun heard of her likely arrest and deserted her; he escaped to France where he died penniless in late 1752.

Mary Blandy came to trial on March 3, 1752. The trial was of particular interest because it was the first time detailed chemical evidence had been presented in court on a charge of murder by poisoning. Dr. Addington was brought in by the Crown to prove by scientific means that Mr. Blandy was poisoned. Although Dr. Addington could not analyze Francis Blandy's organs for traces of arsenic, because the technology did not exist at the time, he was able to convince the court on the basis of his tests that the powder Mary had put in her father's food was indeed arsenic. The servants also testified that they had seen Mary administering the powders to her father's food, and that she had tried to destroy the evidence.

Mary's counsel defended her vigorously, and Mary herself made an impassioned speech for her own defense. Although she admitted placing a powder in her father's food, she did state that the powder "had been given me with another intent" (*The Life and Trial of Mary Blandy,* by Gerald Firth). Unfortunately, the jury felt Cranstoun's actions did not mitigate her own, and at the end of the 13-hour trial, the jury swiftly convicted her of murder. She received the mandatory death sentence, and on April 6, 1752, Mary was publicly hanged in front of Oxford Castle.

Dr. Addington's chemical analysis involved many of the key features of this chapter, including the scientific method of investigation, measuring chemical and physical properties of matter, and separating the components of a complex mixture. His findings are highlighted throughout this chapter. ▮

Dr. Anthony Addington, one of the first forensic pathologists.

Chemistry is the study of matter and its interactions with other matter and with energy. Everything that we see, touch, and feel is matter. Everyone, not just the scientist, uses chemistry, because it describes everyday occurrences, as well as those in test tubes. No definition of chemistry, however, conveys the wide variety of projects that chemists work on, the urgency of many chemical problems, and the excitement of the search for solutions.

In this book, a description of the experiments that guided scientists toward their conclusions introduces most topics. This "experimental" approach conveys the crucial role of experiments in the development of science because experiments are the foundation that supports all science, including chemistry. Experiments may involve many people, may take months to design, and may require sophisticated equipment for analysis of data. However, in the end, they provide the same kind of information as do observations of the results of a simple chemical reaction. Chemistry is first and foremost an experimental science, and we derive our knowledge from carefully planned and performed experiments. The icon in the margin helps to emphasize the experimental nature of the topics.

Many students take chemistry because it is a prerequisite for other courses in their college careers. One important reason for this requirement is that chemistry provides a balance of experimental observations, mathematical models, and theoretical concepts. It teaches many important aspects of problem solving that are applicable to all areas of study.

Chemists investigate many different aspects of chemistry as they do their jobs, which may include the following diverse tasks, as well as many others:

- Develop methods to identify illicit narcotics (Chapter 2)
- Prevent, neutralize, and reverse the effects of acid rain (Chapter 3)
- Create the systems needed for exploration of our solar system (Chapter 5)
- Design new light sources that utilize energy more efficiently and minimize environmental harm (Chapter 7)

Chemistry is at the center of our knowledge of the physical world around us. Chemistry explores the fundamental properties of materials and their interactions with each other and with energy. Figure 1.1 illustrates the relationships between chemistry and other natural sciences.

Each of us feels the impact of chemistry every day of our lives. It is difficult to name an issue that affects society that does not involve chemistry in some way. The need for abundant pure water, the uses of petroleum, the fight against disease, and trips to the boundaries of our solar system all involve chemistry. Chemists have studied many aspects of our daily lives, from the compositions of the stars to the development of nonstick cookware. Many aspects of chemistry are not completely understood yet, but the field is always moving forward. Chemistry is an evolving experimental science, not a static body of knowledge mastered by long-dead scholars. New advances and discoveries occur every day.

1.1 The Nature of Science and Chemistry

OBJECTIVES

- ☐ Define science and chemistry
- ☐ Describe the scientific method of investigation
- ☐ Compare and contrast hypothesis, law, and theory

Science is derived from the Latin *scientia,* translated as "knowledge." For many, the term *science* refers to the systematic knowledge of the world around us, but an inclusive definition would also have to include the process through which this body of knowledge is formed. Science is both a particular kind of activity and also the result of that activity. The process of science involves *observation* and *experiment,* and the results are a knowledge that is based on experience. Science is the study of the *natural universe*—that is,

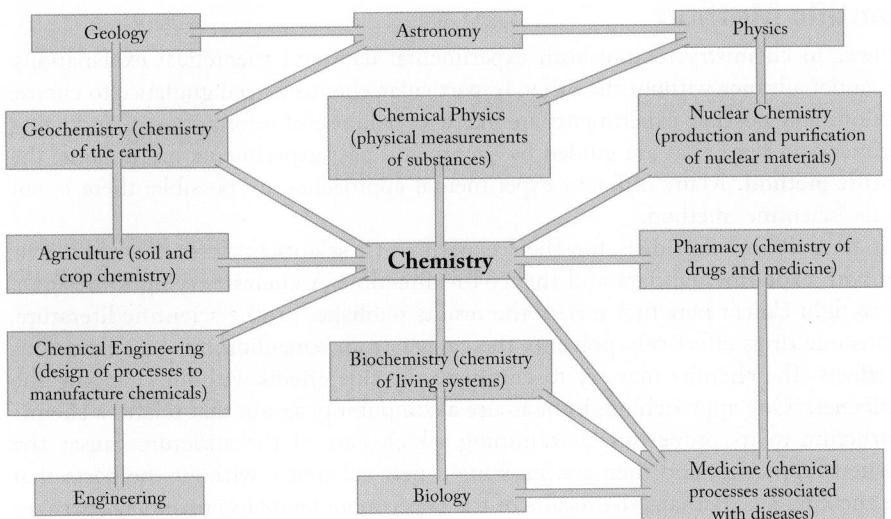

Figure 1.1 Chemistry and the natural sciences.

things that exist and are happening around us. Science is done by human beings who observe, experiment, and test their ideas. It is only by observing and experimenting that we can learn how the natural universe actually works.

Science is a broad field, and for a long time science has been broken into smaller, more specialized areas. **Chemistry** is the study of matter and its interactions. All chemists make observations of the behavior of matter and try to explain the results with principles that they hope will help predict the results of new experiments. If the results of the first experiments are consistent with the predictions, the principles are tested using more extensive experiments. If the predictions disagree with the observed results, the principles are modified to include the new results.

Chemistry is based on observations of the changes that occur during the experiments with matter and the understanding of these changes.

There are some practical reasons to learn the principles of chemistry. Millions of chemicals are known, and billions of reactions occur. Rather than record every individual action, it is more efficient to develop a few models that enable scientists to predict the products of related reactions. The term *model* applies to both a qualitative or nonmathematical picture (e.g., "Heating a reaction causes it to proceed faster.") and a quantitative or mathematical relationship (e.g., "The velocity increases in proportion to the square root of temperature."). Some models are quite mature and widely accepted, whereas others are still tentative.

This section describes some methods that chemists use to perform their investigations, together with some of the corresponding vocabulary.

Corbis/Photolibrary

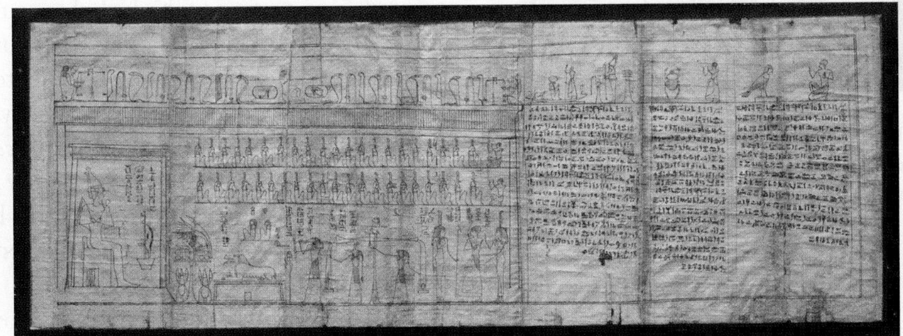

Scala/Art Resource, NY

Modern scientists record the results of their observations, as did scientists thousands of years ago.

Scientific Method

Advances in chemistry require both experimental data and theoretical explanations. One cannot advance without the other. In particular, chemists need guidance to choose which of many possible experiments are likely to yield useful information. Conducting experiments in ways that are guided by theory and past experiments has a name: the **scientific method.** Many different experimental approaches are possible; there is not just one "scientific" method.

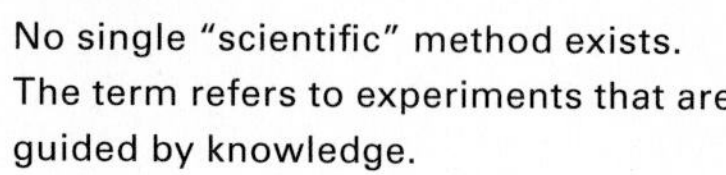

No single "scientific" method exists. The term refers to experiments that are guided by knowledge.

In formulating the ideas for their experiments, scientists draw on experience, using both experimental data and theory for direction. A chemist trying to design a drug to fight cancer may first review the results published in the scientific literature. Perhaps one drug effectively prevents the cancer from spreading but has dangerous side effects. The chemist may try to eliminate the side effects without changing the effectiveness. One approach might be to use a computer program that relates a chemical structure to its properties, determining which part of the structure causes the undesired properties, and then synthesizing a new substance without the parts that cause these effects. Perhaps the results of the experiment show improvement, perhaps not, but more research can be performed to achieve the final goal. Scientific investigations seldom proceed along a straight line but more often are cyclic. The improving, modifying, refining, and extending of our knowledge are all components of the scientific method.

Dr. Addington, the forensic pathologist mentioned at the beginning of this chapter, needed to identify the white powder found in Mr. Blandy's food. Dr. Addington's methods were a model for scientists to study. He took the powder obtained from the Blandy residence, weighed an exact amount, and added it to water, which he then boiled and filtered to obtain a liquid that was then called a "decoction." He performed five chemical tests on this material. He repeated the tests on a decoction of pure white arsenic that he bought from the pharmacist. Arsenic was widely available during this time because low concentrations of arsenic were prescribed as a "tonic."

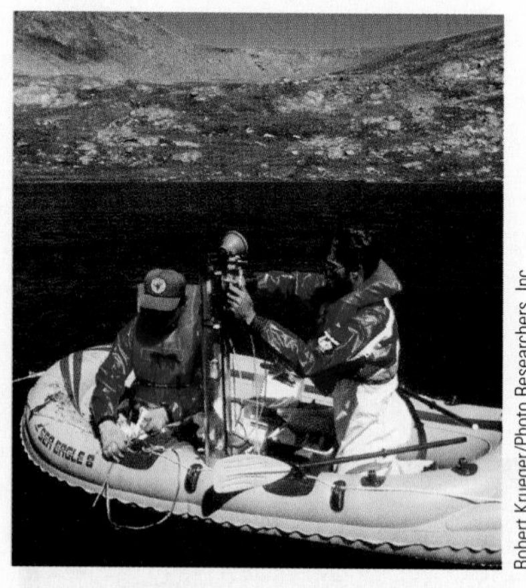

Robert Krueger/Photo Researchers, Inc.

U.S. Forest Service researchers test for evidence of acid rain in Wilderness Lake near Aspen, CO.

Over the years, scientists have developed a systematic language to describe their investigations. It is important that you master some of this language to be able to understand science and chemistry. If a statement (or equation) can summarize a large number of observations, the statement is called a **law.** For example, the English scientist Robert Boyle made careful measurements of the volume of a gas as it varied with pressure. He observed that the volume of a gas changed in opposite direction as did the pressure. This observation, now called *Boyle's law,* is discussed in detail in Chapter 6.

A law summarizes observations but provides no explanation. The word **hypothesis** describes a possible explanation for an observation. A hypothesis often starts as an untested assumption, which helps guide further investigation. A confirmed and accepted explanation of the laws of nature is called a **theory.** For example, scientists know that a gas expands when it is heated. But even more useful is the fact that a relatively simple theory, the kinetic theory of gases, explains these observations. Please note that scientists reserve the word *theory* for an *explanation* of the laws of nature—a narrow meaning that contrasts sharply with everyday usage ("I have a theory about why the basketball team lost last night. I think…"). Many people equate a theory with a hunch or an educated guess, but a scientist must carefully distinguish between theory and hypothesis.

"Theory guides, experiment decides."—*Chemist and educator I. M. Kolthoff (1899-1993)*

Chemistry is first and foremost an experimental science. A theory is only the best understanding available at a given time, so scientists are prepared to modify, extend, and even reject accepted theories as new data become available. When theory suggests that the results of an experiment were incorrect, the experiment may be repeated, usually under more carefully controlled conditions. The best experiments are designed to subject current theories to rigorous tests to obtain the best descriptions of nature.

Ethics and Integrity in Science

Honesty and integrity are perhaps among the most important traits of scientists. Scientists often disagree, and sometimes the same experiment, when repeated, appears to give different results. But scientists strive for accurate data and seek an explanation for differences from one laboratory to another.

If experimental data do not agree with theory, a scientist first repeats the experiment and looks for potential errors. If the experiment is found to be sound and properly executed, the scientist could change the data to match the theory, or modify the theory to explain the results. Changing data is completely unethical; cases of scientific fraud are known, but fortunately are rare. Generations of careful and accurate measurements, often repeated many times, provide the data that help science evolve.

OBJECTIVES REVIEW *Can you:*

- ☑ define science and chemistry?
- ☑ describe the scientific method of investigation?
- ☑ compare and contrast hypothesis, law, and theory?

1.2 Matter

OBJECTIVES

- ☐ Define matter and its properties
- ☐ Identify the properties of matter as intensive or extensive
- ☐ Differentiate between chemical and physical properties and changes
- ☐ Classify matter by its properties and composition
- ☐ Distinguish elements from compounds

Everything we see around us is composed of matter. **Matter** is defined as anything that has mass and occupies space. The food we eat, the air we breathe, and the books we read are all examples of matter. Few subjects in chemistry are as fundamental as matter and its properties.

Matter has mass and occupies space.

The definition of matter includes the term mass. **Mass** measures the quantity of matter in an object. **Weight,** a force of attraction between a particular object and Earth, is the most familiar property of matter. The weight of an object varies from one location to another, but the mass of that object is always the same. We can measure mass with a *balance* such as those shown in Figure 1.2. A balance compares the mass of an object

(*a*)

(*b*)

Figure 1.2 Laboratory balances. Both the older double-pan balance *(a)* and the modern single-pan balance *(b)* measure mass.

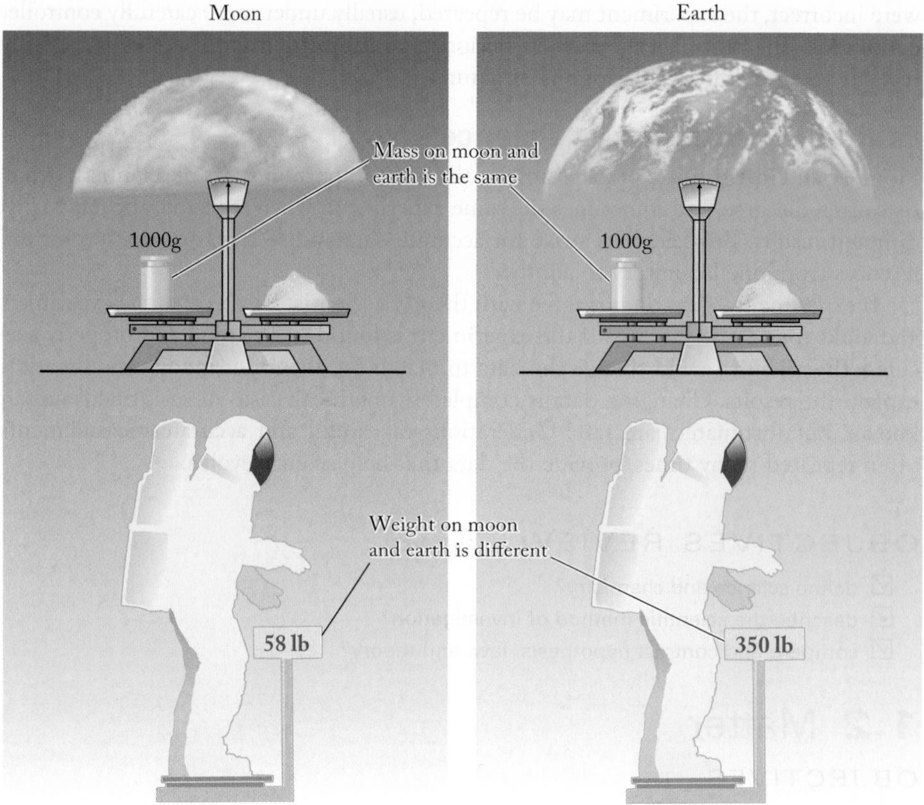

Figure 1.3 Mass and weight. A balance determines mass; a scale measures the weight of an object. On the moon, an object has the same mass as it has on the Earth, but it has less weight.

with objects of known mass. The balance determines the mass rather than the weight because the object and the standard mass are in the same gravitational environment. A device such as a spring scale measures the weight of an object, not the mass, so the reading depends on gravitational attractions (Figure 1.3). The weights of the famous moon rocks increased sixfold when they were brought to Earth because of Earth's greater gravitational attraction. The masses of the rocks, however, did not change.

Properties of Matter

Anything we observe or measure about a sample of matter is called a **property.** We all strive to understand matter and its properties; whether we speak about chemicals in a beaker or the food we eat, we are talking about matter. As a result of observations made through the centuries, scientists have developed several ways of classifying properties.

Physical properties can be measured without changing the composition of the sample. Chemical properties describe the tendency of a material to react, forming new and different substances.

One way to classify properties is to divide them into physical and chemical properties. **Physical properties** can be measured without changing the composition of the sample. The mass of a sample, the volume it occupies, and its color can be observed without changing the composition of the sample. The phase of the sample as solid, liquid, or gas can also be described. Mass, volume, color, and phase are all physical properties. **Chemical properties** describe the reactivity of a material. When chemical properties are measured, new and different substances form. Explosiveness and flammability are examples of chemical properties, because both of them relate how a substance can react chemically. The failure of a sample of matter to undergo chemical change is also considered to be a chemical property. The fact that gold does not react with water is a chemical property of gold.

When matter undergoes a physical change, the chemical composition does not change. In a chemical change, some matter is converted to a different kind of matter.

Changes in the properties of a substance can be classified as either physical or chemical changes. A **physical change** occurs without a change in the composition of the substance. Freezing, for example, is a physical change, as a substance goes from liquid phase to solid phase without changing its chemical composition. When a substance undergoes a **chemical change,** the substance is converted to a different kind (or

kinds) of matter. The rusting of iron and the burning of wood both produce new kinds of matter with properties quite different from those of the initial sample; these are examples of chemical changes. Most chemical changes are accompanied by physical changes because different materials with different physical properties are produced by chemical changes.

Dr. Addington kept careful notes and determined that the unknown powder behaved in exactly the same manner as the pure white arsenic. Comparing the behavior of an unknown with that of a known, often called a *control sample* or just a *control,* is an important part of most chemical analyses. Today, when chemists try to detect harmful compounds, the entire analytical procedure, including the number of samples, number of repetitions, and number of control samples, is often specified. No such information was available in 1752, but Dr. Addington developed a procedure based on his skills as a chemist.

Dr. Addington testified, "There was an exact similitude between the experiments made on the two decoctions. They corresponded so nicely in each trial that I declare I never saw any two things in Nature more alike than the decoction made from the powder found in Mr. Blandy's gruel and that made with white arsenic. From these experiments, and others which I am ready to produce if desired, I believe the powder to be white arsenic" *(The Life and Trial of Mary Blandy,* by Gerald Firth).

Finally, properties can be divided into extensive and intensive types. **Extensive properties** are those that depend on the size of the sample; they measure *how much* matter is in a particular sample. Mass and volume are typical extensive properties. **Intensive properties** are those that are independent of the size of the sample; they depend on *what* the sample is, not how much of it is present. Colors, melting points, and densities are all examples of intensive properties; none depends on the size of the sample. If all the intensive properties of two samples are identical, then it is reasonable to assume that the samples are the same material.

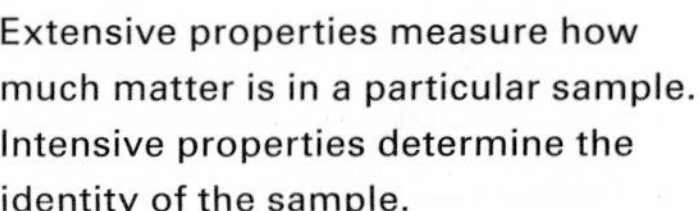

Extensive properties measure how much matter is in a particular sample. Intensive properties determine the identity of the sample.

BARBARA SAX/AFP/Getty Images

(*a*)

© ALEKSANDR S. KHACHUNTS, 2008/Used under license from Shutterstock.com

(*b*)

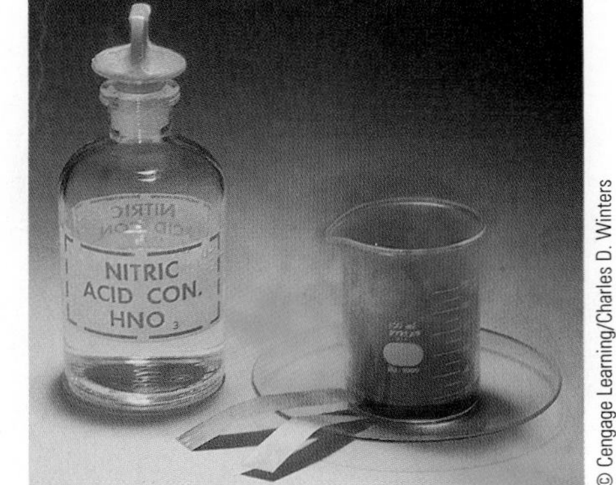

© Cengage Learning/Charles D. Winters

(*c*)

Chemical and physical changes. *(a)* Melting a metal to make a figurine *(b)* is a physical change. *(c)* Dissolving a metal by adding acid is a chemical change.

EXAMPLE 1.1 Properties of Matter

Classify each underlined property or change as either intensive or extensive, and either chemical or physical.

(a) The color of mercury is silvery.
(b) The sample of iron rusts by reaction with oxygen.
(c) The heat released by burning coal can power a city.
(d) Water boils at 100 °C.
(e) A new pencil is 10 inches long.

Strategy Look at the property or change to determine whether it depends on the amount of matter (intensive as opposed to extensive), and notice whether new and different matter forms (chemical as opposed to physical).

Solution

(a) Intensive, physical
(b) Intensive, chemical
(c) Extensive, chemical
(d) Intensive, physical
(e) Extensive, physical

Classifications of Matter

Classifying matter is the first step toward understanding matter and its properties. One way to classify matter is by color; another is by physical state—solid, liquid, or gas. From the point of view of the chemist, the most useful classification of matter is one that broadly divides matter into substances and mixtures, as shown in Figure 1.4.

Substances

A material that is chemically the same throughout is called a ***substance.*** By definition, any single substance is pure—if it were impure, there would be more than one substance present. The chemist's precise definition of a substance differs from that of the general public for whom "substance" and "matter" are synonymous. A detective might say, "We found a white substance that proved to be a mixture of painkillers," but a chemist would not. Millions of substances are known, and more are discovered every day, but there are only two types of substances: elements and compounds. If the substance cannot be broken down into simpler substances by chemical means, then the substance is an **element.** Currently, only 117 elements are known. If the substance can be broken down chemically into simpler substances, then the original substance is a **compound.** More than 30 million compounds are known, each with a unique set of physical and chemical properties.

Compounds can be broken down into simpler substances by chemical methods. Elements cannot.

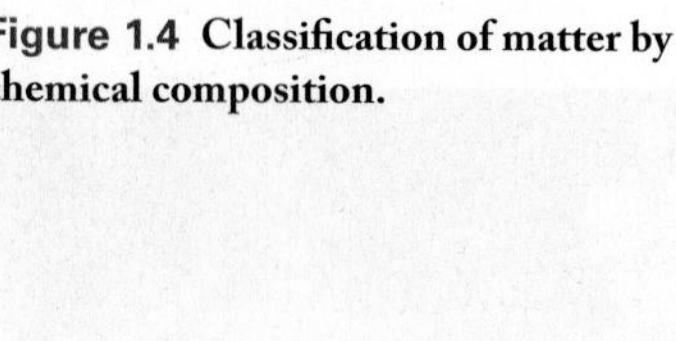

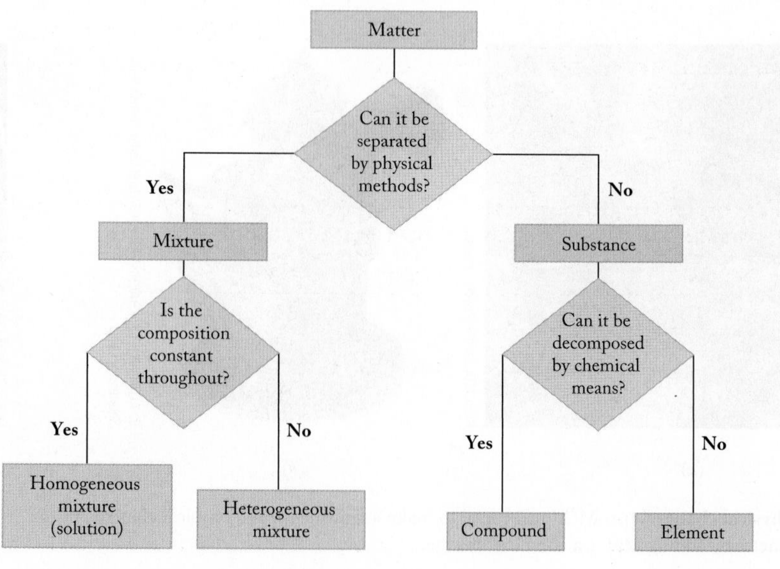

Figure 1.4 **Classification of matter by chemical composition.**

Elements and Their Symbols

Elements are the fundamental building blocks of all matter. Each element has a unique name. Many elements, such as the metals silver, gold, and copper, have been known since ancient times. Other naturally occurring elements have been isolated and purified only during the past 200 years. The most recently discovered elements do not occur in nature but have been produced using the techniques of high-energy nuclear chemistry.

Every element is represented by a characteristic **symbol.** The symbols of the elements are abbreviations of their names. Each consists of one or two letters, with the first letter always capitalized and the second lowercase. The symbols of many of the elements are obvious abbreviations of their names; for example, C is the symbol for carbon, N for nitrogen, Ca for calcium, Ar for argon, and As for arsenic. In other cases, particularly those of elements known since antiquity, the symbol is an abbreviation of the ancient name of the element, often in Latin. Examples include Na for sodium *(natrium),* Pb for lead *(plumbum),* Au for gold *(aurum),* and Sn for tin *(stannum).* One symbol derives from the German form of the element's name—tungsten, W (from *wolfram*). An alphabetical list of the elements, together with their symbols and some other important information, appears on the inside front cover. Table 1.1 lists the names and symbols of several common elements that appear frequently in this text. You should become familiar with these elements and their symbols.

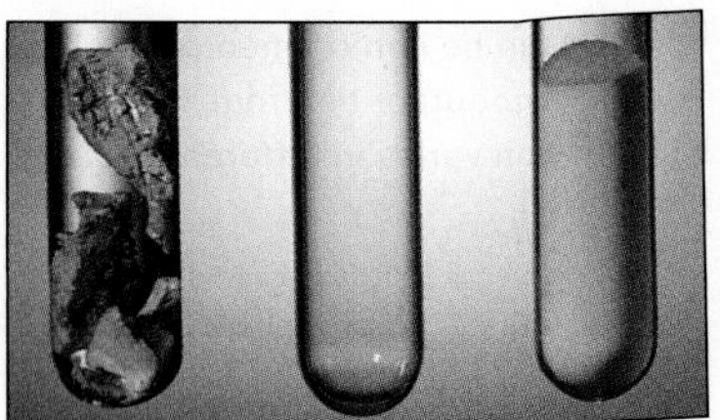

Sodium, chlorine, and sodium chloride.

Compounds

Most substances are compounds. A compound can be decomposed into simpler substances, and eventually into its constituent elements, by chemical methods. Because every compound is composed of two or more elements, the systematic names of most simple compounds are based on the names of their constituent elements. Examples include potassium chloride and aluminum fluoride. This topic is discussed in more detail later. In this section, let us consider some important characteristics of compounds; that is, those used to distinguish them from other matter.

All samples of a compound have the same composition and intensive properties.

A compound always contains the same elements in the same proportions. In any sample of sodium chloride, 39.3% of the mass is the element sodium and 60.7% is chlorine. Water consists of 11.2% hydrogen and 88.8% oxygen. Carbon dioxide contains 27.3% carbon and 72.7% oxygen. Not only are the compositions of all samples of a given compound identical, but all of the chemical and physical properties of the samples are also the same. For example, all samples of pure water have a freezing point of 0 °C, whether the water is obtained from the ocean (and purified) or from the kitchen faucet.

Mixtures

A **mixture** (Figure 1.5) is a combination of two or more substances that can be separated by differences in the physical properties of the substances. Mixtures can be separated by various means, such as filtering a solid out of a liquid or evaporating a liquid away from a dissolved solid.

Figure 1.5 Mixtures and pure substances. The beaker on the left contains a heterogeneous mixture of iron and sand. The beaker in the center contains a pure substance, copper sulfate. The beaker on the right contains a mixture of sugar and ground glass. As mixtures, the substances in the left and right beakers can be separated by physical processes. The sand-iron mixture can be separated using a magnet (to remove the iron). The sugar-glass mixture can also be separated by physical means because sugar dissolves in water but glass does not.

TABLE 1.1 Names and Symbols of Several Common Elements

Name	Symbol	Name	Symbol	Name	Symbol
Aluminum	Al	Fluorine	F	Nitrogen	N
Arsenic	As	Gold	Au	Oxygen	O
Barium	Ba	Hydrogen	H	Phosphorus	P
Bromine	Br	Iodine	I	Potassium	K
Calcium	Ca	Iron	Fe	Silicon	Si
Carbon	C	Lead	Pb	Silver	Ag
Chlorine	Cl	Magnesium	Mg	Sodium	Na
Chromium	Cr	Mercury	Hg	Sulfur	S
Copper	Cu	Nickel	Ni	Tin	Sn

Mixtures can be homogeneous (uniform throughout) or heterogeneous (composition varies in different parts of the sample).

We can further classify mixtures by determining whether the matter is uniform throughout. If the composition changes from one part to another, then the sample is a **heterogeneous mixture.** An example of a heterogeneous mixture is a combination of salt and pepper. If the composition is uniform throughout the mixture, then it is a **homogeneous mixture,** also called a **solution**. A common example of a homogeneous mixture is sugar dissolved in water. Because a homogeneous mixture, such as a sugar solution, is uniform throughout, it may be difficult to distinguish it from a pure substance.

One important difference between mixtures and substances is that a mixture can exhibit variable composition, whereas a substance cannot. A solution of sugar in water could consist of a teaspoon, a tablespoon, or even five tablespoons of sugar in a cup of water. In contrast, all samples of a substance such as sodium chloride are the same, whether made in the laboratory by the combination of sodium and chlorine, mined from the ground, or separated from seawater.

TABLE 1.2 Composition of Dry Air*

Substance	Concentration (% by volume)
Nitrogen	78.084
Oxygen	20.946
Argon	0.934
Carbon dioxide	0.033
Other	0.003

*"Dry" air has the water (humidity) removed.

Mixtures may have any phase: solid, liquid, or gas. Air is one example of a gaseous solution. Table 1.2 provides the composition of dry air. A common type of solid solution is glass. The glass factory adds different substances to change the tint, melting point, and other properties of the glass. Another solid solution, called an **alloy,** consists of a metal and another substance (usually another metal). Bronze, a homogeneous mixture of copper and tin, is a common alloy that is used to make statues because it is easy to cast and resists weathering well.

Solutions are homogeneous mixtures. Solutions can be solids, liquids, or gases.

Dr. Addington was not provided a sample of the unknown white powder, but he was provided a sample of some of the food. One of the servants had eaten leftover gruel and became violently ill. A maid also ate some of it and fell sick. Because of this peculiar chain of events, the servants became suspicious, and examined the pan used to prepare the gruel and discovered a white sediment at the bottom. The servants locked up the pan and gave it to the doctor when he arrived.

To determine the nature of the white powder, Dr. Addington needed to perform a series of tests on the powder from the pan and the control sample of white arsenic. However, the presence of any other matter from the pan would likely confuse his results. Dr. Addington performed a physical separation to extract the powder from the complex mixture of gruel in the pan. Dr. Addington used a powerful magnifying glass and fine tweezers to carry out the separation.

Scala/Ministero per i Beni e le Attività culturali/Art Resource, NY. Museo Nazionale (Terme di Diocleziano), Rome, Italy.

Bronze statues. The Greeks cast statues in bronze, a copper-tin alloy. Bronze resists weathering; this statue was made about 2100 years ago.

EXAMPLE 1.2 Classifications of Matter

Identify the following types of matter as elements, compounds, heterogeneous mixtures, or homogeneous mixtures.

(a) sodium chloride
(b) stainless steel (an alloy of iron, carbon, and other elements)
(c) chlorine
(d) soil

Strategy Figure 1.4 outlines the steps needed to classify matter as substances or mixtures.

Solution

(a) Sodium chloride is a compound made from the elements sodium and chlorine. (How to name compounds is explained in Chapter 2.)
(b) As an alloy, stainless steel is a homogeneous mixture, or a solution.
(c) Chlorine is an element (as seen by the list of the elements on the inside cover).
(d) Soil is a heterogeneous mixture.

OBJECTIVES REVIEW *Can you:*

- ☑ define matter and its properties?
- ☑ identify the properties of matter as intensive or extensive?
- ☑ differentiate between chemical and physical properties and changes?
- ☑ classify matter by its properties and composition?
- ☑ distinguish elements from compounds?

1.3 Measurements and Uncertainty

OBJECTIVES

- ☐ Distinguish between accuracy and precision
- ☐ Use the convention of significant figures to express the uncertainty of measurements
- ☐ Express the results of calculations to the correct numbers of significant figures

Chemistry, like most science, involves the interpretation of quantitative measurements, usually made as part of an experiment. It is important to realize that each measurement has four aspects: the object of the measurement, the value, its units, and the reliability of the measurement. When the results of a measurement are communicated (e.g., "The mass of iron was 4.0501 grams."), the object (iron), the value (4.0501), and the units (grams) are apparent, but the reliability of the measurement is not obvious. This section focuses on the value and its reliability; the next section considers the units.

It is often crucial to know the reliability of a particular value, as well as the value itself. Although exacting laboratory measurements and quick estimates both have their places in science, usually they are not comparable. When scientists analyze laboratory data, the interpretation generally places greater weight on more reliable measurements. This section introduces ways to assess the reliability of measurements, together with the methods used to determine the reliability of calculations based on those measurements.

Accuracy and Precision

Seldom is an experimental measurement taken just once. Why? Because a single measurement may be subject to error. Thus, it is common in science to measure the same quantity more than once; in some cases, scientists repeat their measurements many times. In normal practice, each measurement may result in a slightly different answer. In addition, a given parameter may vary for a given object. For example, the width of a piece of lumber may vary slightly down its length. Because of this, when a certain quantity is expressed, there must be some way of understanding the *reliability* of the quantity.

The concept of reliability has two components: accuracy and precision. **Accuracy** is the term used to express the agreement of the measured value with the true or accepted value. **Precision** expresses the agreement among repeated measurements; a

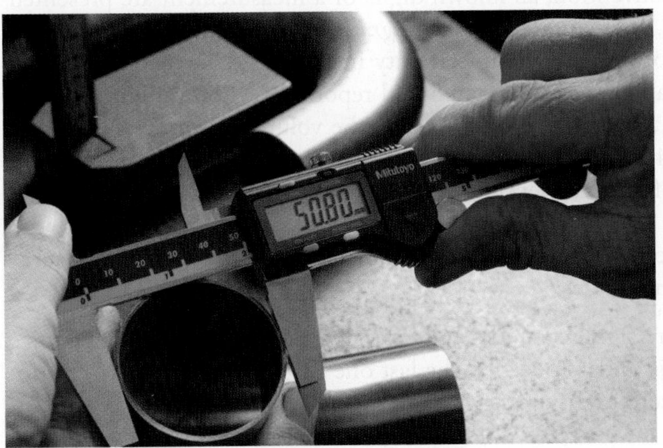

High-precision, high-accuracy measurements. A digital micrometer can measure dimensions of items several centimeters long with high accuracy and precision.

TABLE 1.3 Accuracy and Precision: Repetitive Weighing of an Object (True Mass = 5.11 g) on Several Balances

	Measured Mass (g)			
	Balance 1	Balance 2	Balance 3	Balance 4
	5.10	5.02	5.23	5.35
	5.13	5.20	5.21	5.10
	5.11	5.25	5.21	5.40
	5.11	4.97	5.20	5.15
	5.10	5.08	5.20	5.21
Average	5.11	5.10	5.21	5.24
Range	0.03	0.28	0.03	0.30

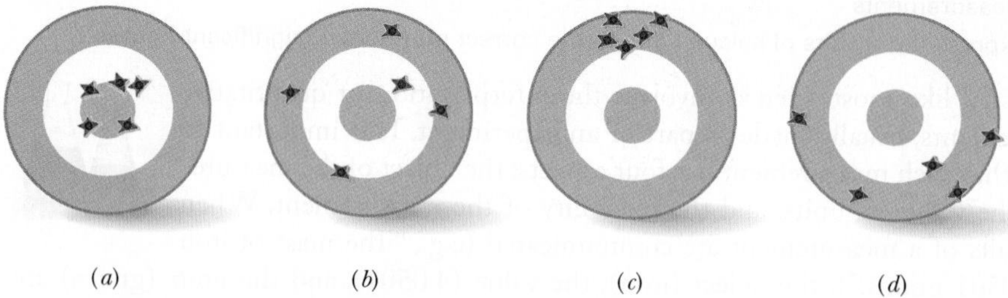

Accuracy and precision. Accuracy and precision are not the same thing, as the bullet holes in these targets illustrate. One can be *(a)* accurate and precise, *(b)* accurate but not precise, *(c)* precise but not accurate, or *(d)* neither accurate nor precise.

high-precision measurement is one that produces nearly the same value time after time. An accurate number has a small **error,** whereas a precise number has a small **uncertainty.**

Accuracy expresses how close a measurement is to the true value. Precision expresses how closely repeated measurements agree with each other.

Table 1.3 lists the results of measurements of the mass of a coin whose true mass is 5.11 g; the data in the table illustrate both precision and accuracy. The same coin was measured five times on each of four different balances. The average value for each set is taken as the best value, and the range of values (**range** is defined as the difference between the largest and smallest values) is a measure of the agreement among the individual determinations. The determination of mass on balance 1 is both precise and accurate, because the range of values is small and the average agrees with the true value. Balance 2 provides an accurate value but not a precise one, because the range of individual measurements is relatively large. When the precision is poor, scientists typically average many individual measurements. Balance 3 is precise but not accurate, perhaps because the balance was not properly zeroed. Balance 4 is neither precise nor accurate. It is important to understand the difference between "precise" and "accurate."

Significant Figures

Although many scientists use statistical methods to analyze their data, a relatively simple method is frequently used to estimate the uncertainty of the results of a computation or measurement. By convention, all known digits of a measurement are presented plus a final, estimated digit. These digits are called *significant figures* (or *significant digits*) of the measurement. By convention, the uncertainty in the last digit reported is presumed to be ±1. Thus, if a volume is measured and reported as 12.3 milliliters (mL), the implied uncertainty is ±0.1 mL; in other words, the volume might be as small as 12.2 mL or as large as 12.4 mL. It is the responsibility of the scientist who reports the data to use the significant-figure convention correctly to express the uncertainty in the measurement.

Numbers are presumed to have an uncertainty of ±1 in the last digit.

How do you determine the **number of significant figures** in a reported measurement? There are several simple rules:

1. All nonzero digits are significant. Thus, in the value 123.4, there are four significant figures, and it is understood that the last one (4) is estimated.
2. Zeros between nonzero digits are significant. In the value 102, there are three significant figures.

3. In a number with no decimal point, zeros at the end of the number ("trailing zeros") are not necessarily significant. Thus, in the value 602,000, there are at least three significant figures (the 6, the first 0 [see rule 2], and the 2). The three trailing zeros may or may not be significant because their primary purpose is to put the 6, the 0, and the 2 in the correct positions. A method to avoid this ambiguity, by expressing the value in scientific notation, is presented shortly.
4. If a number contains a decimal point, zeros at the beginning ("leading zeros") are not significant, but zeros at the *end* of the number are significant. For example, the value 0.0044 has two significant figures, whereas the number 0.0000340 has three significant figures (the last zero is significant).

Zeros may or may not be significant in a number. Follow the rules to determine whether they are significant.

EXAMPLE 1.3 Significant Figures

How many significant figures are there in the following values?

(a) 57.8
(b) 57.80
(c) 0.00271
(d) 96,500

Solution

(a) All nonzero digits are significant (rule 1). There are three significant figures in 57.8.
(b) The final zero is significant (rule 4), so there are four significant figures in this value.
(c) Leading zeros are not significant (rule 4), so there are only three significant figures in this value.
(d) The trailing zeros may or may not be significant (rule 3), so there are at least three significant figures in this value.

Again, it is the convention that the final reported digit gives the indication of how uncertain the value is. The following example demonstrates this.

EXAMPLE 1.4 Determining the Number of Significant Figures and Reliability

How many significant figures are present in each of the following measured quantities, and what is the uncertainty based on the convention of significant figures?

(a) A package of candy has a mass of 103.42 g.
(b) The mass of a milliliter of a gas is 0.003 g.
(c) The volume of a solution is 0.2500 L.
(d) The circumference of Earth is 24,900 miles.

Solution

(a) In the value 103.42, there are five significant figures. The uncertainty is ± 0.01 g.
(b) None of the zeros in 0.003 are significant; they show only the location of the decimal point with respect to the 3. There is only one significant digit, so the uncertainty is ± 0.001 g.
(c) The first 0 in 0.2500 is not significant, but the trailing zeros after the decimal point are. There are four significant figures in this value. The uncertainty is ± 0.0001 L.
(d) There are at least three significant figures, and maybe as many as five. Using scientific notation would clarify the number of significant figures.

Understanding

How many significant figures are in the number 0.01020? What is the uncertainty?

Answer Four significant figures; uncertainty is ± 0.00001

In the absence of a decimal point, trailing zeros are ambiguous (rule 3). One way to remove any ambiguity is to express the measurement in *scientific notation.* This method expresses a quantity as the product of two numbers. The first is a number between 1 and 10, and the second is 10 raised to some whole-number power. When scientific notation is used, the uncertainty is expressed more clearly. If we write 2000 as 2.00×10^3, it has three significant figures and the uncertainty is understood to be $\pm 0.01 \times 10^3$, which equals 10. Scientific notation is also a space-saving way to represent very small and very large numbers. For example, the number 0.00000000431 is much more compact when expressed as 4.31×10^{-9}; the number of significant figures (three) and the uncertainty, $\pm 0.01 \times 10^{-9}$, are apparent. Appendix A reviews scientific notation.

(*a*)

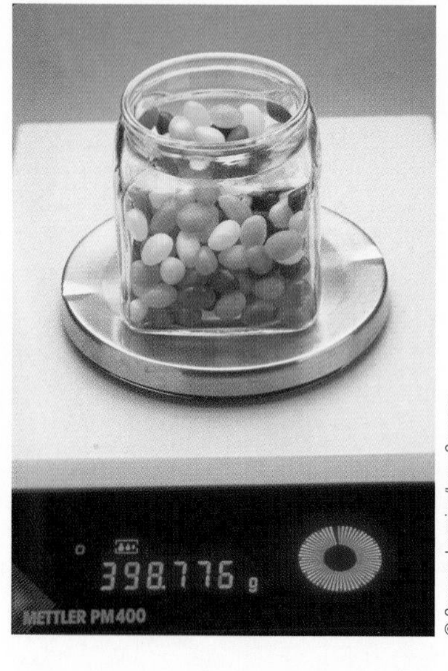

(*b*)

Measuring the mass of a sample. The mass of the sample is determined by subtracting the mass of the empty container from the mass of the container plus its contents.

Significant Figures in Calculations

In many experiments, the quantity of interest is not measured directly but is calculated from several measured values. For example, we must often determine the mass of a sample from the mass of an empty container and the total mass of the sample plus the container.

$$\text{mass of sample} = \text{total mass (sample + container)} - \text{mass of container}$$

The uncertainty in the mass of the sample depends on the uncertainties in the two measurements from which it is calculated.

The number of significant figures in a calculated value depends on the uncertainties of the measurements and the type of mathematical operations used. Electronic calculators typically do not follow the significant-figure conventions; they generally display as many digits as can fit across the calculator face. It is your responsibility to determine the number of significant figures in the result of any calculation because the significant figures represent the uncertainty in the measurement.

As with determining the number of significant figures in a given value, determining the number of significant figures after a calculation follows some simple rules.

Addition and Subtraction

In addition and subtraction, we look at the number of decimal places. The result is expressed to the smallest number of decimal places of the numbers involved. Consider the difference of two numbers using a calculator:

$$\begin{array}{r} 25.34 \\ -\ 24.0 \\ \hline 1.34 \end{array}$$

The calculator would show 1.34 as the "answer." The first number has two decimal places, but the second number does not; its least significant digit is in the tenths' place. Therefore, the rule is that we should limit our final answer to one decimal place: 1.3.

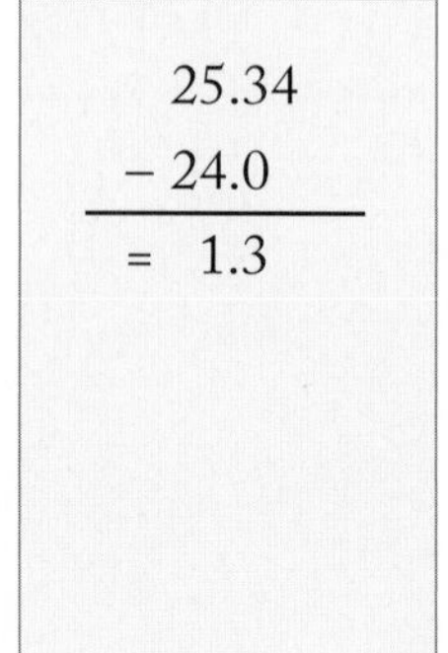

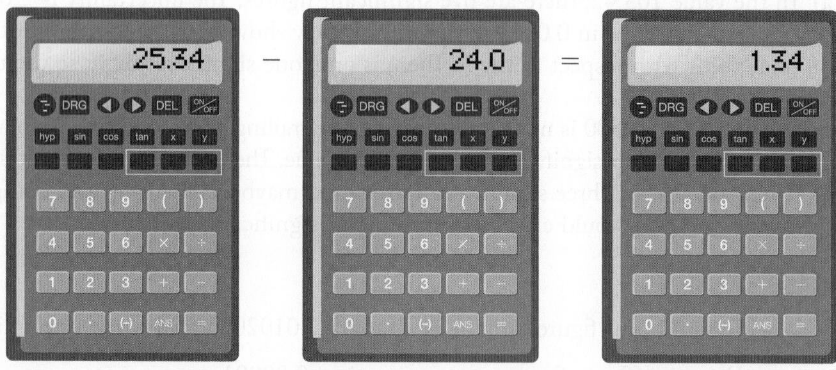

Significant figures. The operator must report the proper number of significant figures; the calculator will not.

The uncertainty in the final answer is dictated by the place of the last significant figure. In this case, the uncertainty would be ±0.1.

A calculator may also display fewer digits than are significant.

28.39	28.39
− 6.39	− 6.39
22	22.00
(calculator)	(correct)

Because the answer is derived from numbers with two decimal places, the answer also should be expressed to two decimal places. In such a case, many calculators display only two figures instead of the four that are significant.

When the result of a calculation has too many digits, round the number up or down to reflect the proper number of significant digits.[1] If the digit after the least significant figure is less than 5, round down; if the digit is 5, round to even; if the digit is greater than 5, round up. For example, if a calculation of a quantity has three significant figures and your calculator displays 12.35, then report 12.4. If the display is 12.25, then report 12.2.

In addition or subtraction, the number with the fewest decimal places determines the number of decimal places in the result.

If the first digit after the least significant figure is 0 to 4, round down; if 5, round to even; if 6 to 9, round up.

Multiplication and Division

In multiplication and division, the number of significant figures in the final answer is based on the number of significant figures of the values being multiplied and/or divided. The result has the same number of significant figures as the multiplier or divisor with the *fewest* significant figures. For example, on a calculator, the division of 227 by 365 would yield

$$227 \div 365 = 0.621917808\ldots$$

Because the numbers 227 and 365 each have only three significant figures, the final answer should be limited to three significant figures (and is rounded up, because the first digit being dropped is a 9):

$$227 \div 365 = 0.622$$

If the initial values have different numbers of significant figures, then the value with the fewest number of significant figures is the deciding value. Hence,

$$6.7 \times 0.345 = 2.3, \textit{ not } 2.31 \textit{ or } 2.3115$$

In multiplication or division, the number with the fewest significant figures determines the number of significant figures in the result.

One common calculation involving division is the determination of density from a measured mass and a measured volume. Density, d, is the mass of an object (m) divided by its volume (V):

$$d = \frac{m}{V}$$

For a sample with a mass of 7.311 g and a volume of 7.7 cubic centimeters (cm^3), the density is

$$d = \frac{7.311\text{ g}}{7.7\text{ cm}^3} = \underset{\text{(calculator)}}{0.9494805\text{ g/cm}^3} = \underset{\text{(correct)}}{0.95\text{ g/cm}^3}$$

The density is expressed to two significant digits, the same as the measured volume. Note that, although the implied uncertainty in the volume is ±0.1, the implied uncertainty in the density is correctly expressed as ±0.01.

As with addition and subtraction, calculators may show fewer significant figures than are needed when multiplying or dividing. If the product 0.5000 × 6.0000 is evaluated on a calculator, the display shows 3 as the result. The component numbers have four and five significant figures, so the result must have four significant figures, and 3.000 is the correct representation.

Table 1.4 gives a summary of how significant figures are treated in calculations.

[1]There are two common methods of rounding. The one presented is called *unbiased rounding*. Another method, called *symmetric rounding*, would round 0 to 4 down and 5 to 9 up.

TABLE 1.4 Determination of Significant Figures in Computed Results

Operation	Procedure	Example
Addition or subtraction	The answer has the same number of decimal places as the component with the fewest number of decimal places.	$12.314 + 2.\mathbf{32} = 14.\mathbf{63}$
Multiplication or division	The answer has the same number of significant digits as the component with the fewest number of significant digits.	$\mathbf{12.31} \times 9.1416 = \mathbf{112.5}$

EXAMPLE 1.5 Significant Figures in Calculated Quantities

Express the result of each calculation to the correct number of significant figures. In some, you may have to remember the correct order of operations.

(a) $(0.082 \times 25.32)/27.41$
(b) $55.8752 - 56.533$
(c) $0.198 \times 10.012937 + 0.8021 \times 11.009305$
(d) $2.334 \times 10^{-2} - 3.1 \times 10^{-3}$
(e) $(25.7 - 25.2) \times 0.4184$

Strategy In addition or subtraction, the final result has the same number of decimal places as the number with the fewest number of decimal places. In multiplication or division, the final result has the same number of significant figures as the number with the fewest number of significant figures.

Solution

(a) This calculation involves only multiplication and division. Two of the three numbers have four significant figures, but the third has only two. The result of the calculation will have only two significant figures.

$$\frac{0.082 \times 25.32}{27.41} = \underset{\text{(calculator)}}{0.075747537} = \underset{\text{(correct)}}{0.076}$$

(b) Only subtraction is involved in this calculation. We can identify the last significant digit more easily by writing the numbers in a column:

$$\begin{array}{r} 55.8752 \\ -56.533 \\ \hline -0.06578 \text{ rounds to } -0.0658 \end{array}$$

The last decimal place that both numbers have in common is the third, or thousandths' place. The final answer is limited to that place (and we have rounded up). Even though the original values have four and five significant figures, the result has only three.

(c) With regard to the order of operations, first evaluate the two products, then add the results. The first product has three significant digits, and the second product has four.

$$\begin{array}{rl} 0.198 \times 10.012937 = & 1.98256 \text{ round to } 1.98 \\ +0.8021 \times 11.009305 = & 8.83056 \text{ round to } 8.830 \\ \hline = & 10.81312 \text{ round to } 10.81 \end{array}$$

We limit the final answer to two decimal places, as indicated by the first product, 1.98. The numbers are rounded in the last column so that the number of significant digits is clear.

(d) The problem is simplified if both numbers are expressed in scientific notation with the same power of 10. Choosing the second number to change,

$$3.1 \times 10^{-3} = 0.31 \times 10^{-2}$$

Now the uncertainty in the addition operation is easy to interpret.

$$\begin{array}{r} 2.334 \times 10^{-2} \\ -0.31\ \ \times 10^{-2} \\ \hline 2.024 \times 10^{-2} \\ 2.02\ \ \times 10^{-2} \end{array}$$

In the last step, we are limiting the final answer to two decimal places, which is the limit imposed from the numbers being subtracted.

(e) Perform the operation inside the parentheses first, and note the number of significant figures.

$$\begin{array}{ccccc} (25.7 - 25.2) & \times & 0.4184 & & \\ 0.5 & \times & 0.4184 & = & 0.2092 \end{array}$$

1 significant figure $\times$ 4 significant figures = 1 significant figure = 0.2

The correct expression of the answer has only one significant figure.

Understanding

Express the results of the calculation to the correct number of significant figures: 1.33/55.494 + 10.00.

Answer 10.02

Sequential Calculations and Roundoff Error

When you perform several calculations in sequence, be careful not to introduce an error by rounding intermediate results. Consider multiplying three numbers:

$2.5 \times 4.50 \times 3.000 = ?$

(a) Solve in two separate steps, rounding at each place:

$2.5 \times 4.50 = 11.25$, which is rounded to two significant digits

$11 \times 3.000 = 33$

(b) Solve, but do not round, intermediate calculations:

$2.5 \times 4.50 = 11.25 \times 3.000 = 33.75 = 34$

If you need to write down intermediate results, it is a good practice to write them with one more digit than suggested by the significant figures, and round off the final result appropriately. This practice minimizes roundoff error.

Write intermediate results with one more digit than needed and round off the final result appropriately to minimize roundoff error.

Quantities That Are Not Limited by Significant Figures

The concept of significant figures applies only to measured numbers, or to quantities calculated from measured numbers. Three kinds of numbers never limit significant figures:

1. *Counted numbers.* There are exactly five fingers on a hand or 24 students in a class; there is no uncertainty in the numbers 5 and 24.
2. *Defined numbers.* There are exactly 12 inches in a foot, so there is no uncertainty at all in this number.
3. *The power of 10.* The power of 10, when exponential notation is used, is an exact number and never limits the number of significant figures.

OBJECTIVES REVIEW *Can you:*

- ☑ distinguish between accuracy and precision?
- ☑ use the convention of significant figures to express the uncertainty of measurements?
- ☑ express the results of calculations to the correct numbers of significant figures?

PRINCIPLES OF CHEMISTRY
Accuracy and Precision

The issue of accuracy versus precision can lead to some dramatic consequences. Consider the story of the Hubble Space Telescope (HST).

In 1946, astronomer Lyman Spitzer wrote that an orbiting space telescope would have two enormous advantages over a ground-based system. First, the space telescope could observe regions of the electromagnetic spectrum (principally ultraviolet and infrared) outside of the visible light range because it would be in orbit above the ultraviolet- and infrared-absorbing atmosphere of Earth. Second, the resolution would be limited only by the imperfections in the optics rather than by atmospheric turbulence. This advantage would increase the resolution by a factor of 10 over ground-based telescopes.

After almost a decade of debate, the U.S. Congress approved startup expenses in 1978 with a launch date of 1983. The project was named the Hubble Space Telescope in honor of the late astronomer Edwin Hubble, whose careful measurements and calculations showed that the universe is expanding.

In 1979, an optical company began to polish the telescope's primary mirror, 2.4 m in diameter. The design required the surface irregularities reduced to dimensions smaller than 20 nm. This level of smoothness required precise and accurate grinding—had the mirror been expanded to the diameter of Earth, the 20-nm roughness would correspond to a height of about 4 inches.

Delays in the manufacture of the mirror and other components to be sent aloft, as well as the loss of the space shuttle *Challenger* in 1986, set the schedule back. The telescope and package of five measurement instruments were finally launched on April 24, 1990.

The first images returned by the HST were disappointing. The scientists tweaked and focused, but their best efforts were still well below expectations. In fact, they were little better than could be achieved by a terrestrial telescope. The scientists analyzed the images and hypothesized that the mirror had been

Denise Applewhite/Princeton University

Lyman Spitzer first appreciated the improvement in image quality from a space telescope.

NASA Marshall Space Flight Center (NASA-MSFC)

Polishing the mirror begins in 1979.

NASA

Space shuttle *Atlantis* carries the Hubble Space Telescope into orbit.

1.4 Measurements and Units

OBJECTIVES

- ☐ List the SI base units
- ☐ Derive unit conversion factors
- ☐ Convert measurements from one set of units to another
- ☐ Derive conversion factors from equivalent quantities

Scientific progress is based on gathering and interpreting careful observations. Many measurements use *quantities* to describe properties and communicate information with known precision. A quantity has two parts: a value and a unit. The previous section discussed the treatment of values. This section discusses the units of measurements.

Accepted standards of comparison are necessary for meaningful measurements. It is important that quantities reported, such as distance, time, volume, and mass, have the same meanings for everyone. When we read that a pen is 6 inches long, we understand that the pen is six times as long as the length that has been defined as 1 inch. **Units** are standards that are used to compare measurements. The scientific community has ground precisely but to the wrong shape—the images suggested that the mirror was too flat near the edges by about 2 μm, about the diameter of a bacterial cell.

An investigative team went to the manufacturer and reviewed the procedures. They found a manufacturing alignment tool had been misadjusted when a technician centered a crosshair not on the target, but on a scratch exactly 1.3 mm away. Consequently, the mirror was polished very precisely, but not very accurately! The scientists quickly determined that the magnitude of the error agreed exactly with the error calculated from the analysis of the images. Because grinding was inaccurate, but precise, the scientists proposed sending a pair of small correcting mirrors, ground to compensate for the error in the grinding of the main mirror. These mirrors were launched in December 1993, and seven astronauts, who had trained for months with the highly specialized tools, corrected the mirror and replaced, upgraded, or repaired several other components. On January 10, 1994, NASA declared the HST a success.

The telescope is now highly precise and highly accurate. The HST has produced some of the most remarkable images seen by humans. It has provided data for scientists to determine how fast the universe is expanding, to refine estimates of the age of the universe, and has allowed astronomers to find the first planets outside of our solar system. ∎

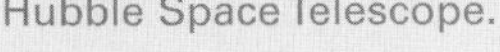

NASA Johnson Space Center

Hubble Space Telescope.

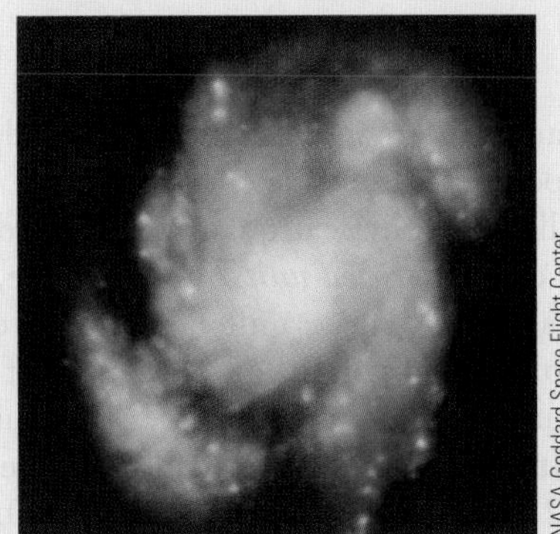

NASA Goddard Space Flight Center

NASA Goddard Space Flight Center

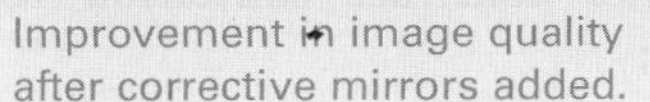

Improvement in image quality after corrective mirrors added.

NASA, ESA, STScI, J. Hester and P. Scowen (Arizona State University)

Star-forming pillars in the Eagle Nebula.

adopted the SI *(Le Système International d'Unités)* units to express measurements. These units are an outgrowth of the metric system that was created during the French Revolution, when the French rejected anything related to the deposed monarchy.

Although most SI units have been accepted by the scientific community, a few have not; some nonsystematic units are still commonly used for certain measurements, such as the atmosphere for pressure and the liter for volume.

Base Units

Any measurement can be expressed in terms of one or a combination of the seven fundamental quantities: length, mass, time, temperature, amount of substance, electrical current, and luminous intensity. The SI defines a **base unit** for each of these. Table 1.5 lists all the base units, together with the abbreviation used to represent that unit. For example, the base unit of time is the second (abbreviated s), whereas the base unit of length is the meter (abbreviated m). All base units except one are defined in terms of experiments that can be reproduced in laboratories around the world. The only unit that is based on a physical

AFP/Getty Images

The kilogram standard. The standard for the unit kilogram is a metal cylinder kept in a special vault outside of Paris, France. Anything that has the same mass has a mass of exactly 1 kg.

TABLE 1.5 The SI Base Units

Quantity	Unit	Abbreviation
Length	meter	m
Mass	kilogram	kg
Time	second	s
Temperature	kelvin	K
Amount	mole	mol
Electric current	ampere	A
Luminous intensity	candela	cd

standard is the kilogram, which is defined as the mass of a platinum/iridium cylinder kept in a special vault outside of Paris, France.

All other physical quantities can be expressed as algebraic combinations of base units, called ***derived units.*** For example, area has units of length squared (m^2, square meters in the SI base units); volume is expressed in cubic meters (m^3). Density is the ratio of mass to volume and has units of kilogram per cubic meter (kg/m^3).

There are only seven base units, but there are a large number of derived units.

Quantities are usually expressed in units that avoid very large and very small numbers. It is more convenient and more meaningful to express the width of a human hair as 122 micrometers (μm) rather than 0.000122 m. Fortunately, conversion among SI units is generally simple, especially if you adopt a consistent algebraic methodology.

The prefix of an SI unit indicates the power of 10 by which the base unit is multiplied.

The SI creates units of different sizes by attaching prefixes that move the decimal point. Table 1.6 gives these prefixes and their meanings and abbreviations. The prefix *kilo-* (used in the base unit for mass) means 1000, or 10^3. An object that has a mass of 1 kg has a mass of *exactly* 1000 g. The prefixes used most frequently in this book appear in bold type in the table. Abbreviations for the prefix/base unit combinations are made by simply placing the abbreviations next to each other, first the prefix abbreviation, then the base unit abbreviation. Thus, 1 kg is exactly 1000 g, and a human hair ranges in width between 20 and 200 μm.

Conversion Factors

It is easy to convert from one SI unit to another, because the meanings of the prefixes allow us to construct conversion factors. For example, consider the following statement:

$$1 \text{ kg} = 1000 \text{ g}$$

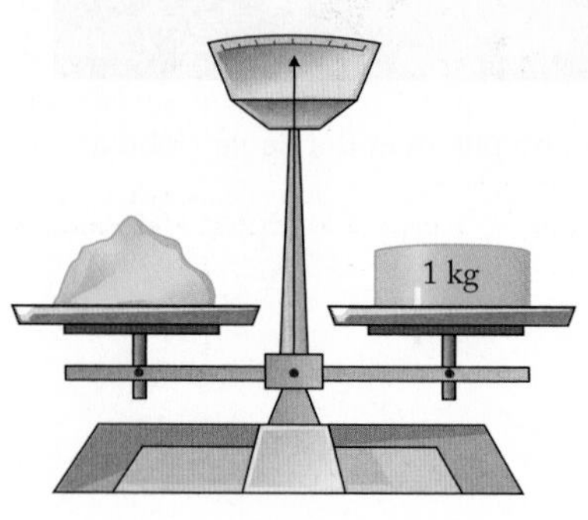

Because 1 kg equals 1000 g (because of the definition of the prefix *kilo-*), the above statement is an algebraic equality. Suppose we divide both sides of the equation by the same quantity, in this case, 1 kg:

$$\frac{1 \text{ kg}}{1 \text{ kg}} = \frac{1000 \text{ g}}{1 \text{ kg}}$$

The expression is an equality because when you divide both sides of an equality by the same thing, the new expression is still an equality. Note, now, that the left side has the same quantity in the numerator and denominator of the fraction: 1 kg. If the same thing appears in the numerator and denominator of a fraction, they cancel out, and in this case, what is leftover is simply 1:

$$1 = \left(\frac{1000 \text{ g}}{1 \text{ kg}}\right)$$

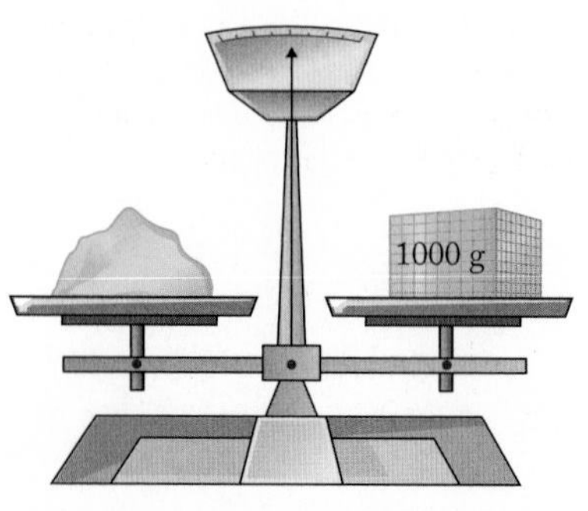

Kilo- means 1000. One kilogram (kg) equals 1000 g and 1 km equals 1000 m.

This should make sense, because 1000 g equals 1 kg, so the fraction on the right still has the same quantity in the numerator and denominator of the fraction, but this time expressed in different units. The expression on the right is known as a **conversion factor** (or *unit conversion factor*).

When we multiply a quantity by 1, the quantity does not change. However, when that "1" is a conversion factor, what happens is that we change the units of the quantity (and usually the numeric value associated with that quantity).

As an example, suppose we convert 2.45 kg into gram units. Setting up a conversion usually means starting with the quantity given, then multiplying it by 1 in terms of the conversion factor. Thus, we have

$$2.45 \text{ kg} \times \left(\frac{1000 \text{ g}}{1 \text{ kg}}\right)$$

Note that we have the unit kilogram both in the numerator (of the first term, which is assumed to be a fraction with 1 in the denominator) and the denominator (of the second term). Algebraically, the kilogram units cancel, leaving units of grams in the numerator.

$$2.45 \cancel{\text{kg}} \times \left(\frac{1000 \text{ g}}{1 \cancel{\text{kg}}}\right)$$

Now complete the numerical multiplication and divisions:

$$2.45 \cancel{\text{kg}} \times \left(\frac{1000 \text{ g}}{1 \cancel{\text{kg}}}\right) = 2450 \text{ g} = 2.45 \times 10^3 \text{ g}$$

The final answer is expressed using scientific notation to emphasize that there are three significant figures. Because the relationship between kilograms and grams is a defined number, the 1000 and the 1 do not affect the determination of significant figures in the final answer.

A second conversion factor can be derived from our relationship "1 kg = 1000 g":

$$1 \text{ kg} = 1000 \text{ g}$$

$$\left(\frac{1 \text{ kg}}{1000 \text{ g}}\right) = 1$$

How did we know to use the first conversion factor and not this one? The key is in noting which units need to be eliminated and which units need to be introduced. In converting from kilograms to grams, we need to eliminate the kilogram unit. Since the given quantity, 2.45 kg, has the kilogram unit in the numerator, we use a conversion factor that has kilograms in the denominator and grams in the numerator.

A second way to determine which conversion factor is correct is to estimate the answer. When a mass of 2.45 kg is expressed in grams, the number of grams will be larger because grams is a smaller unit.

In going from one prefixed unit to another prefixed unit (e.g., from kilometer to millimeter), it may be convenient to first convert to the base unit (meter) and then to the final desired prefixed unit (millimeter). The following example demonstrates this conversion.

TABLE 1.6 Prefixes Used with SI Units

Prefix	Abbreviation	Meaning
yotta-	Y	10^{24}
zetta-	Z	10^{21}
exa-	E	10^{18}
peta-	P	10^{15}
tera-	T	10^{12}
giga-	G	10^{9}
mega-	M	10^{6}
kilo-	k	10^{3}
hecto-	h	10^{2}
deka-	da	10^{1}
deci-	d	10^{-1}
centi-	c	10^{-2}
milli-	m	10^{-3}
micro-	μ	10^{-6}
nano-	n	10^{-9}
pico-	p	10^{-12}
femto-	f	10^{-15}
atto-	a	10^{-18}
zepto-	z	10^{-21}
yocto-	y	10^{-24}

Students unfamiliar with exponential notation should refer to Appendix A for an explanation.

Multiplying by the conversion factor does not change the quantity, just the units in which it is expressed.

Choose a conversion factor that cancels the unwanted units and leaves the desired units.

EXAMPLE 1.6 Converting Units

How many millimeters are there in 17.43 km?

Strategy Conversions often take two steps. Convert from the given unit to the base unit, then from the base unit to the wanted unit. You can estimate the result to confirm your calculation.

Solution

We can estimate that the length of 17.43 km will be a very large number of millimeters. The first step in the conversion is to convert from the given unit to the base unit, meter:

$$17.43 \cancel{\text{km}} \times \left(\frac{1000 \text{ m}}{1 \cancel{\text{km}}}\right) = 17{,}430 \text{ m}$$

The green shading indicates data that is given with the problem, the yellow indicates intermediate results, and the red is the final answer.

Next, we take this quantity and convert it to millimeter units:

$$17{,}430\ \cancel{\text{m}} \times \left(\frac{1000\ \text{mm}}{1\ \cancel{\text{m}}}\right) = 17{,}430{,}000\ \text{mm} = 1.743 \times 10^7\ \text{mm}$$

Note how in both conversions how the units cancel algebraically. The final quantity is expressed in scientific notation, showing the proper number of significant figures.

We went from kilometers, which are close in length to miles, to millimeters, which are fractions of an inch, so we are not surprised that the answer is a million times larger.

It is sometimes convenient to combine the two conversion factors on one line, cancel out all appropriate units, and perform the final multiplications and divisions in one longer series. The conversion can also be performed as

$$17.43\ \cancel{\text{km}} \times \left(\frac{1000\ \cancel{\text{m}}}{1\ \cancel{\text{km}}}\right) \times \left(\frac{1000\ \text{mm}}{1\ \cancel{\text{m}}}\right) = 17{,}430{,}000\ \text{mm} = 1.743 \times 10^7\ \text{mm}$$

Effectively, we are performing the same conversion as we did earlier, except in one longer step as opposed to two separate steps. The one-step process minimizes roundoff error because there is no opportunity to drop any digits in intermediate steps.

Understanding

How many kilograms are there in 165 μg?

Answer 1.65×10^{-7} kg

The conversion factor method is also called the *factor-label method,* or *dimensional analysis.*

Conversion among Derived Units

Conversion among derived units is not significantly different from conversion among base units. We form the conversion factors from the relations between units, but sometimes operations such as squaring and cubing are required to make the unit conversion factors contain the desired units.

Volume

The volume of a rectangular box is the product of its length times its width times its height. Because volume is a product of three lengths, the standard unit of volume is a cube with dimensions equal to the base unit of length: $1\ \text{m} \times 1\ \text{m} \times 1\ \text{m} = 1\ \text{m}^3$. A cubic meter is an inconveniently large volume for laboratory-scale experiments—a cubic meter of water weighs 1000 kg, or about 2200 pounds. Volume measurements in cubic centimeters are much more common. Conversions between these volume units are similar to those of length or mass except that volume is length cubed. The relationship between the lengths (meters and centimeters) can be written first; then the relationship between the volumes can be determined by cubing the equivalent lengths.

Identical lengths: $100\ \text{cm} = 1\ \text{m}$

Identical volumes: $(100\ \text{cm})^3 = (1\ \text{m})^3$

$10^6\ \text{cm}^3 = 1\ \text{m}^3$

$$\text{conversion factor} = \left(\frac{10^6\ \text{cm}^3}{1\ \text{m}^3}\right)$$

A nonsystematic unit widely used by chemists to express volume is the liter, abbreviated L. A liter is defined as 0.001 m³. There are 1000 L in 1 m³. The milliliter

(10^{-3} L) is also commonly used for small volumes. It can be shown that 1 mL is identical to 1 cm^3.

$$
\begin{aligned}
1\ \text{cm}^3 = 1\ \text{mL} &= 0.001\ \text{L} &&= 10^{-6}\ \text{m}^3 \\
1000\ \text{mL} &= 1\ \text{L} &&= 0.001\ \text{m}^3 \\
10^6\ \text{mL} &= 1000\ \text{L} &&= 1\ \text{m}^3
\end{aligned}
$$

Example 1.7 illustrates conversions among these units of volume.

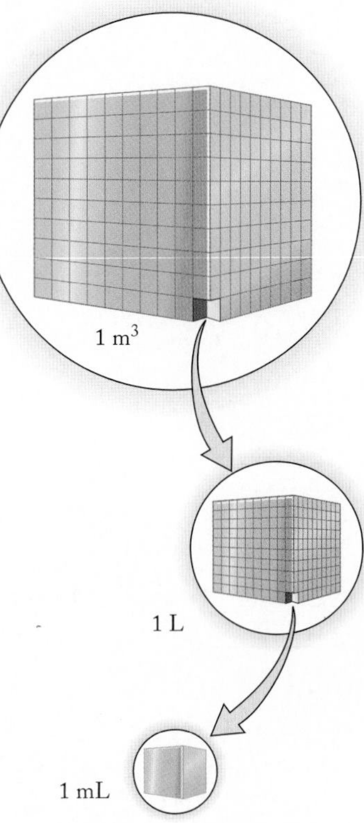

Volume. Volumes can be expressed in different units depending on the size of the object.

EXAMPLE 1.7 Conversions among Volume Units

Express a volume of 322 mL in units of:

(a) liters (b) cm^3 (c) m^3

Strategy A flow diagram helps explain the process. Quantities in the flow diagrams are in colored boxes, and processes appear above arrows. The process in this problem is a unit conversion, so the appropriate unit conversion factor is defined before the quantities are converted. The first part uses a milliliter-to-liter conversion factor.

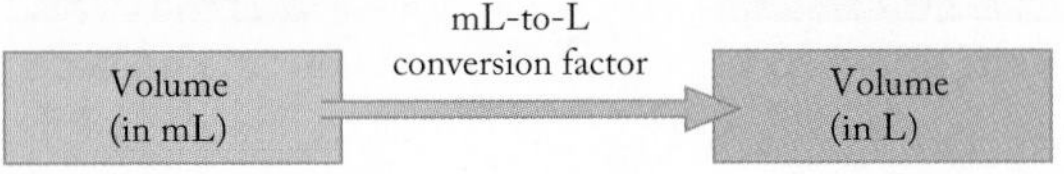

Solution

(a) We can derive the conversion factor from the meaning of the prefix *milli-*. One milliliter is equal to 0.001 L, so there are 1000 mL in 1 L:

$$\text{conversion factor} = \left(\frac{1\ \text{L}}{1000\ \text{mL}}\right)$$

$$\text{volume} = 332\ \cancel{\text{mL}} \times \left(\frac{1\ \text{L}}{1000\ \cancel{\text{mL}}}\right) = 0.322\ \text{L}$$

The amount of substance and its volume remain the same, and only the units change; 0.322 L is identical to 322 mL. The final answer has three significant digits, from the three significant digits in 322 mL. The conversion factor is exact.

(b) The milliliter and the cubic centimeter represent the same volume, so the conversion factor is

$$\text{conversion factor} = \left(\frac{1\ \text{cm}^3}{1\ \text{mL}}\right)$$

Note that the numerical value of the volume will not change, just the units:

$$\text{volume} = 332\ \cancel{\text{mL}} \times \left(\frac{1\ \text{cm}^3}{1\ \cancel{\text{mL}}}\right) = 322\ \text{cm}^3$$

(c) From the answer to part a and the knowledge that 1 m^3 is the same as 1000 L:

$$\text{volume} = 0.322\ \cancel{\text{L}} \times \left(\frac{1\ \text{m}^3}{1000\ \cancel{\text{L}}}\right) = 3.22 \times 10^{-4}\ \text{m}^3$$

Density

Density, defined as mass per unit volume, is an intensive property that helps identify substances. Density always relates the mass of a substance to its volume. Each substance has its own characteristic density, so densities cannot be used to convert between masses and volumes of different substances. When density is expressed in the base units of the

International System, the units are kg/m^3, or kg · m^{-3}, units that are inconvenient for densities of most samples of matter. Common practice is to express the densities of solids and liquids in g/cm^3 (which is the same as g/mL); densities of gases are generally expressed in g/L. A unit conversion in which *two* units change (kg to g *and* m^3 to cm^3) requires *two* conversion factors. The factors can be applied separately or together; the next example presents both processes.

EXAMPLE 1.8 Conversions between Density Units

Express a density of 8.4 g/cm^3 in terms of the SI base units of kg/m^3.

Strategy The factors needed to convert g to kg and cm^3 to m^3 have already been derived. In separate steps, our flow diagram is:

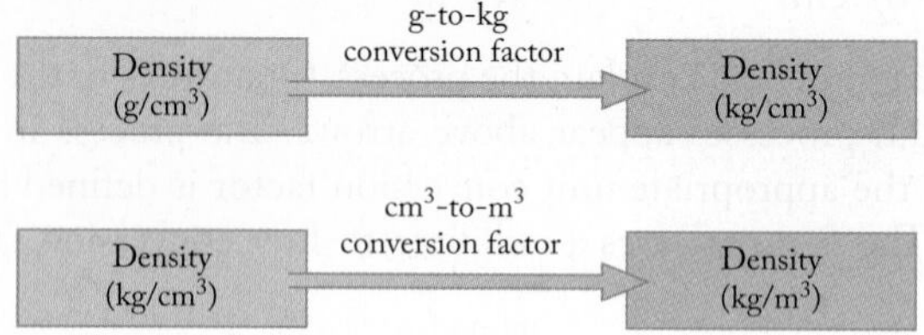

Solution

$$\text{Density} = 8.4\,\frac{\cancel{\text{g}}}{\text{cm}^3} \times \left(\frac{1\ \text{kg}}{1000\ \cancel{\text{g}}}\right) = 8.4 \times 10^{-3}\,\frac{\text{kg}}{\text{cm}^3}$$

$$8.4 \times 10^{-3}\,\frac{\text{kg}}{\cancel{\text{cm}}^3} \times \left(\frac{100\ \cancel{\text{cm}}}{1\ \text{m}}\right)^3 = 8.4 \times 10^3\,\frac{\text{kg}}{\text{m}^3}$$

Note that we cube the conversion factor so we get cubic centimeters in the numerator to cancel the cubic centimeters initially present in the denominator.

We can also do this conversion in a single, multistep calculation. The flow diagram is:

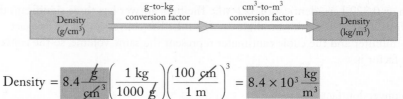

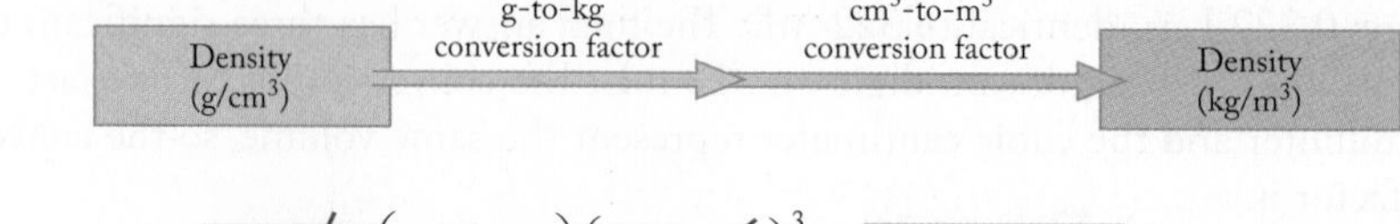

$$\text{Density} = 8.4\,\frac{\cancel{\text{g}}}{\cancel{\text{cm}}^3}\left(\frac{1\ \text{kg}}{1000\ \cancel{\text{g}}}\right)\left(\frac{100\ \cancel{\text{cm}}}{1\ \text{m}}\right)^3 = 8.4 \times 10^3\,\frac{\text{kg}}{\text{m}^3}$$

It is sometimes clearer to leave out the explicit multiplication symbols, especially when more than one conversion factor is needed.

Understanding

The density of a gas is 1.05 g/L. Express this quantity in terms of SI base units.

Answer The density of the gas is 1.05 kg/m^3.

English System

Most people in the United States are more familiar with the English system of measurements than with the International System. Table 1.7 summarizes the relationships between the SI and English systems. This table, taken from a more complete presentation in Appendix C, demonstrates the simplicity of the SI units and prefixes, and provides the information needed to make conversions from one system to the other. Note that, with one exception, none of the relationships between English and SI units is

TABLE 1.7 Relationships in the SI and English Systems

SI Units	English Units	SI-English Equivalents
Length		
1 km = 10^3 m	1 mi = 5280 ft	1 mi = 1.609 km
1 cm = 10^{-2} m	1 yd = 3 ft	1 m = 39.37 in
1 mm = 10^{-3} m	1 ft = 12 in	1 in = 2.54 cm*
1 nm = 10^{-9} m		
Volume		
1 m^3 = 10^6 cm^3	1 gal = 4 qt	1 L = 1.057 qt
1 cm^3 = 1 mL	1 qt = 57.75 in^3	1 qt = 0.946 L
	1 qt = 32 fluid ounces	
Mass		
1 kg = 10^3 g	1 lb = 16 ounces avdp†	1 lb = 453.6 g
1 mg = 10^{-3} g	1 ton = 2000 lb	1 avdp ounce = 28.35 g†
		1 troy ounce = 31.10 g†

*The inch-to-centimeter conversion is exact; other SI-English conversions are approximate.

†Ounces avoirdupois are used to express the weights of most items of commerce other than gems, precious metals, and drugs. Jewelers and pharmacists use troy ounces.

exact, so the numbers in these conversion factors *are* considered when evaluating significant figures.

The green shading indicates data that is given with the problem, the yellow indicates intermediate results, and the red is the final answer.

EXAMPLE 1.9 Conversions between the SI and English Systems

Perform the following conversions.

(a) Express the mass of 12.2 ounces of ground beef in kilograms.
(b) A soft drink contains 355 mL. What is this volume in quarts?

Strategy Look at Table 1.7 for the relationships between English and metric quantities; then convert the metric quantities to the desired units.

Solution

(a) The relationship between the SI and English systems for units of mass is 1 ounce (avoirdupois) = 28.35 g. We can derive the relevant conversion factor from this relationship.

$$\text{Mass of ground beef} = 12.2\ \cancel{\text{oz}} \times \left(\frac{28.35\ \text{g}}{1\ \cancel{\text{oz}}}\right) = 346\ \text{g}$$

To obtain the mass in the desired units of kilograms requires a second conversion.

$$\text{Mass of ground beef} = 346\ \cancel{\text{g}} \times \left(\frac{1\ \text{kg}}{1000\ \cancel{\text{g}}}\right) = 0.346\ \text{kg}$$

The final answer has three significant figures, limited by the three significant figures in the mass of the ground beef, 12.2 oz. The calculations can be chained together, as shown in part b.

(b) $$\text{Volume of soft drink} = 355\ \cancel{\text{mL}}\left(\frac{1\ \cancel{\text{L}}}{1000\ \cancel{\text{mL}}}\right)\left(\frac{1\ \text{qt}}{0.946\ \cancel{\text{L}}}\right) = 0.375\ \text{qt}$$

You could also use 1 L = 1.057 qt and get the same answer.

Understanding

How many yards does a runner travel when running the 100.0-m dash?

Answer 109.4 yards

Rob Walls/Alamy

English and metric units of length.

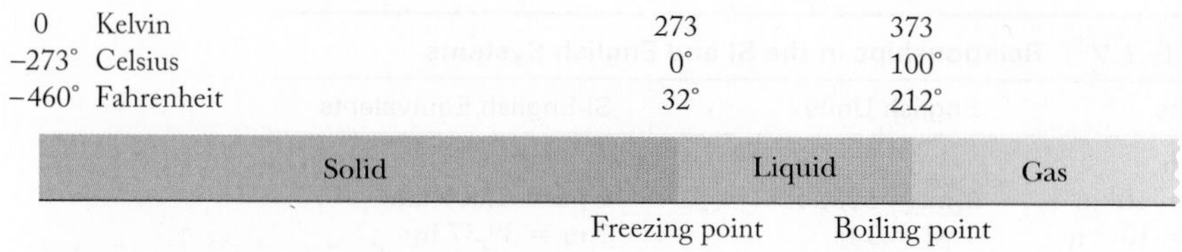

0	Kelvin	273	373
−273°	Celsius	0°	100°
−460°	Fahrenheit	32°	212°

Solid | Liquid | Gas

Freezing point | Boiling point

Figure 1.6 Freezing and boiling points of water on the Kelvin, Celsius, and Fahrenheit scales.

Temperature Conversion Factors

Temperature is a familiar quantity to most of us. In the scientific community and in much of the world, temperatures are measured in units of degrees Celsius, abbreviated °C. Water has a freezing point of 0 °C, and its boiling point is 100 °C. The scale was formerly called the *centigrade scale* because the interval between freezing and boiling is divided into 100 equal units.

In the United States, we often use the Fahrenheit scale (°F) to express temperature. This scale fixes the freezing and boiling points of water at 32 °F and 212 °F, so the difference between these two temperatures is 180 °F. Figure 1.6 shows the relation between the Celsius and Fahrenheit scales.

The conversion factor between the Celsius temperature (T_C) and Fahrenheit temperature (T_F) is.

$$T_F = T_C \times \left(\frac{1.8\ °F}{1.0\ °C}\right) + 32\ °F$$

Unit conversions may require more than one mathematical operation.

This formula takes into account the relative sizes of the two degrees, as well as the offset (by 32 °F) from a zero value. This formula can be easily rearranged to solve for the Celsius temperature:

$$T_C = (T_F - 32\ °F) \times \left(\frac{1.0\ °C}{1.8\ °F}\right)$$

Many years after the definition of the Celsius temperature scale, laboratory scientists discovered that they could obtain no temperature below −273.15 °C. This lowest possible temperature is referred to as absolute zero. The SI temperature scale, named after Lord Kelvin, uses the same size unit as the Celsius scale but starts at absolute zero, so the relationship between Celsius and Kelvin temperatures is

$$T_K = T_C + 273.15$$

T_K is the temperature on the Kelvin scale. The SI unit of temperature is the kelvin, abbreviated K; the International System does not use degrees (so the proper way to state 298 K is "two hundred and ninety-eight kelvin[s]," not "two hundred and ninety-eight *degrees* kelvin"). Because the units on the Kelvin and Celsius temperature scales are the same size, a change in temperature of 12 °C is also a change of 12 K.

EXAMPLE 1.10 Conversions among Temperature Scales

We usually measure the densities of solids and liquids at a temperature of 25 °C. Express this temperature on the Fahrenheit and Kelvin scales.

Strategy There are 1.8 °F for every 1.0 °C. The scales do not begin at the same point, however, so you will have to add or subtract the starting temperature as appropriate.

Solution

The relationship needed to convert between the Fahrenheit and Celsius scales is

$$T_F = T_C \times \left(\frac{1.8\ °F}{1.0\ °C}\right) + 32\ °F$$

$$T_F = 25\ °\cancel{C} \times \left(\frac{1.8\ °F}{1.0\ °\cancel{C}}\right) + 32\ °F = 77\ °F$$

To find the temperature on the Kelvin scale, add 273.15:

$$T_K = 25 + 273.15 = 298.15 = 298\ K$$

The convention of significant figures indicates that the final digit to be reported in our answer is in the units ("ones") place.

A temperature of 25 °C or 298 K is convenient for experiments because it is a little warmer than room temperature.

Understanding

The boiling point of benzene, an important industrial compound found in crude oil, is 80 °C. Express this temperature in degrees Fahrenheit and in kelvins.

Answer 176 °F, 353 K

Conversions between Unit Types

We can extend the conversion factor method to calculations in which we change from one type of measurement to another, as well as between types of units. For example, if we know the mass of a sample and its density, we can use the density to find the volume occupied by that sample. The density of copper is 8.92 g/cm^3, so a volume of 1 cm^3 is equivalent to a mass of 8.92 g; that is, each 1 cm^3 of the sample has a mass of 8.92 g.

$$1\ cm^3\ Cu = 8.92\ g\ Cu$$

The two conversion factors derived from the density of copper are

$$\left(\frac{8.92\ g\ Cu}{1\ cm^3\ Cu}\right) \text{ and } \left(\frac{1\ cm^3\ Cu}{8.92\ g\ Cu}\right)$$

Example 1.11 illustrates this type of conversion factor.

Some conversion factors allow the change from one type of unit to another.

© Cengage Learning/Larry Cameron

The density of copper. A cube of copper that is 1.00 cm on a side (giving it a volume of 1 cm^3) has a mass of 8.92 g.

EXAMPLE 1.11 Conversions between Unit Types

What is the volume occupied by 25.0 g aluminum? The density of aluminum is 2.70 g/cm^3.

Strategy Because the mass of this sample is known and we want to find the volume, we multiply the mass by a conversion factor that has units of grams in the denominator and volume units in the numerator.

Solution

We derive the conversion factor from the density of aluminum:

$$1\ cm^3\ Al = 2.70\ g\ Al$$

$$\text{conversion factor} = \left(\frac{1\ cm^3\ Al}{2.70\ g\ Al}\right)$$

Applying this conversion factor, we obtain the desired volume:

$$\text{Volume of aluminum} = 25.0\ g\ Al \times \left(\frac{1\ cm^3\ Al}{2.70\ g\ Al}\right) = 9.26\ cm^3\ Al$$

Understanding

A jeweler must estimate the mass of a diamond without removing it from its setting. The jeweler determines that the diamond has a volume of 0.0569 cm^3. If the density of diamond is 3.513 g/cm^3, what is the mass of this diamond?

Answer 0.200 g, which the jeweler would label as 1.00 carats

In Example 1.11, we used a conversion factor to change one type of quantity into an equivalent quantity having a different unit. Conversion factors based on known *chemical* relationships (first presented in Chapter 3) are used frequently throughout this text. These relationships will enable us to predict, among other things, the amount of material formed in a laboratory-scale reaction, how much gasoline can be refined from a barrel of oil, how much limestone acid rain consumes, and how much heat a cubic foot of natural gas can produce. Unit conversion truly is a powerful calculation technique.

OBJECTIVES REVIEW *Can you:*

- ☑ list the SI base units?
- ☑ derive unit conversion factors?
- ☑ convert measurements from one set of units to another?
- ☑ derive conversion factors from equivalent quantities?

CASE STUDY **Unit Conversions**

The Mars Climate Orbiter was launched from Kennedy Space Center in Cape Canaveral, Florida, in December of 1998. The Climate Orbiter and a companion mission, the Mars Polar Lander, were designed to investigate the planet's geological history and to search for historical evidence of previous life on Mars. There is strong evidence that Mars once contained abundant water, but scientists don't know what happened to the water or what forces drove it away. The Polar Lander was designed to search for water, a critical component of life, at the edge of the Martian South Pole and to relay its findings back to Earth via the Climate Observer. All of NASA's previous Mars missions had landed at the equator, where the evidence of water is less convincing. On the other hand, the poles are capped with frozen carbon dioxide and are more likely to retain water, frozen as ice.

On September 23, 1999, almost ten months after launch, the Mars Climate Orbiter engine ignited as expected, but engineers never received a signal confirming it achieved orbit around Mars. They soon determined that a navigation error placed the Climate Observer much too close to the planet before the rockets fired. The spacecraft came within 60 km (37 miles) of the planet, much closer than the planned 100 km. The error in distance caused the spacecraft to crash into Mars.

A review panel was appointed to determine the root cause of the crash. They gathered the facts and published their results a few days later. The spacecraft crashed because the engineering team that built the Climate Observer used English units (pounds, inches, and so forth) while the team operating the spacecraft used metric units (meters, kilograms, etc.).

The real problem, according to the final report, was a failure to recognize and correct the fact that the two teams were using different units. Unknowingly, neither team converted their units to the other system. NASA had a system in place designed to look for problems like different measurement units, but the system failed, ultimately dooming the Mars mission.

The units of a quantity are just as important as its number. In most cases, an interplanetary visit is not at stake, but proper communication of a quantity requires not only the correct expression of the amount, but the correct expression of the unit involved.

sci-tech> space> story page

exploringmars in-depth specials

NASA's metric confusion caused Mars orbiter loss

September 30, 1999
Web posted at: 1:46 p.m. EDT (1746 GMT)

(CNN) -- NASA lost a $125 million Mars orbiter because one engineering team used metric units while another used English units for a key spacecraft operation, according to a review finding released Thursday.

NASA

NASA's Climate Orbiter was lost September 23, 1999.

ETHICS IN CHEMISTRY

These questions can be done as a group, perhaps as a classroom activity, or assigned as individual writing exercises.

1. Read the three case studies and select the one in which the actions are the least defensible. Write a paragraph identifying the ethical issue and explain why you chose this point. (These studies were taken, in part, from Paul Treichel, Jr.: Ethical conduct in science-the joys of teaching and the joys of learning. *Journal of Chemical Education* 1999, vol 76, p. 1327).

Case 1 Able and Baker were assigned an experiment asking them to confirm Boyle's law, which involved measuring the volume of a gas sample at various pressures. Boyle's law is discussed in Chapter 6. In all, they collected 12 different sets of data. When they met after class to graph the data, however, they discovered that two measurements differed greatly from the others. After deliberation , they concluded that they must have made inadvertent errors in these measurements, perhaps by misreading their ruler or by writing the numbers down incorrectly in their notebook. Able and Baker reconciled the discrepancies by simply dropping the two sets of "erroneous" data and recopying their remaining data onto a new sheet to turn in. Thus, their written laboratory report contained a neat table of the satisfactory data (10 sets of pressure-volume measurements), with their graph showing all points lying on the line. They did not mention the omitted data in their report.

Case 2 Later in the semester, Able and Baker performed a heat-of-reaction experiment similar to those mentioned in Chapter 5. Here, they measured the increase in temperature generated by a reaction between an acid and a base. By this time, they had become rather capable in the laboratory, so preparing solutions took little time and they quickly were able to carry out the reactions in triplicate. Later that evening, they calculated the results of their three experiments. Two of the three determinations gave almost identical results, but the third differed by about 20%. Able and Baker considered dropping the third value and showing only the first two results, but they thought that reporting three determinations would look better than reporting only two. Plus, the grader might see from their data sheets that they had done the experiment a third time and question the omission. So instead, they decided simply to change the data. They scratched out the final temperature in the errant data set and wrote in a value that was 20% higher. Using this number, they recalculated the result and the answer was close enough to the first two results to pass any reasonable inspection.

Case 3 Able and Baker passed Chemistry 1 and continued on to Chemistry 2. In their second experiment in the laboratory, they determined the rate at which a product formed by measuring the absorption of light by a colored product in one of the instruments in the laboratory. Two nights later, while Able and Baker were in the chemistry computer laboratory working up their data, they ran into a problem. The graph of the first four measurements gave a straight line, but the next four points were off the line. For a while, Able and Baker were puzzled as to which data to use, but then they remembered that midway through the experiment, the original instrument stopped working and they switched to a different one. (In fact, they had even made a note of this in their notebook.) Clearly, the problem must have been with the instrument. They decided that the logical way to deal with this problem was to impose a correction factor, so they multiplied each of the values obtained using the first instrument by a factor of 1.04. Both sets of data were then used to plot a nice, straight line. However, they decided not to mention the correction factor in their report because it was just too complicated for them to explain.

2. The General Chemistry grader approaches the professor with a quizzical look. She is grading a laboratory report submitted by Ann and Bob who did the experiment in the Wednesday section. They shared a laboratory station, so they submitted a single

report with both of their names. The report has a word badly misspelled. The oddity is that the grader saw the same badly misspelled word in Charlie's report in the Monday section. The professor asks Charlie, who did the experiment by himself, to hand his report back for regrading.

Charlie's report is word-for-word identical to Ann and Bob's. Even the data are identical. Ann, Bob, and Charlie are summoned into the professor's office and the laboratory reports are read by all. Bob is stunned. It seems that Ann and Charlie are dating, and that Charlie did the report first and gave his report on disk to Ann. Ann said that she meant only to look at the report as a template, but she accidentally pasted it into her report, overwriting all her data. The professor points out that the only change he can see is that the Charlie's name was changed to Ann and Bob—everything else is the same.

Bob is getting physically ill and protests that he knew nothing about any plagiarism, that he submitted his part to Ann and that he should not be punished by her transgressions. Charlie and Ann say it was an innocent mistake, and that the professor strongly encourages students to collaborate and, therefore, should be understanding and lenient. The professor asks them to recommend an action.

What do you recommend? First, decide whether you will recommend the same action for each, then what the action is and why you would recommend it.

3. By the standards of justice prevailing in 1752, Mary Blandy had a fair trial and a fair sentence. Ironically, modern forensic science might have made it easier to convict her, but her defense lawyer would raise doubt of her intent. She loved her father and never meant to kill him but rather wanted to believe what Cranstoun had told her, that the powder would make her father accept him.

Write a brief paragraph stating how you would vote on the Blandy case and justify your vote.

Chapter 1 Visual Summary

The chart shows the connections between the major topics discussed in this chapter.

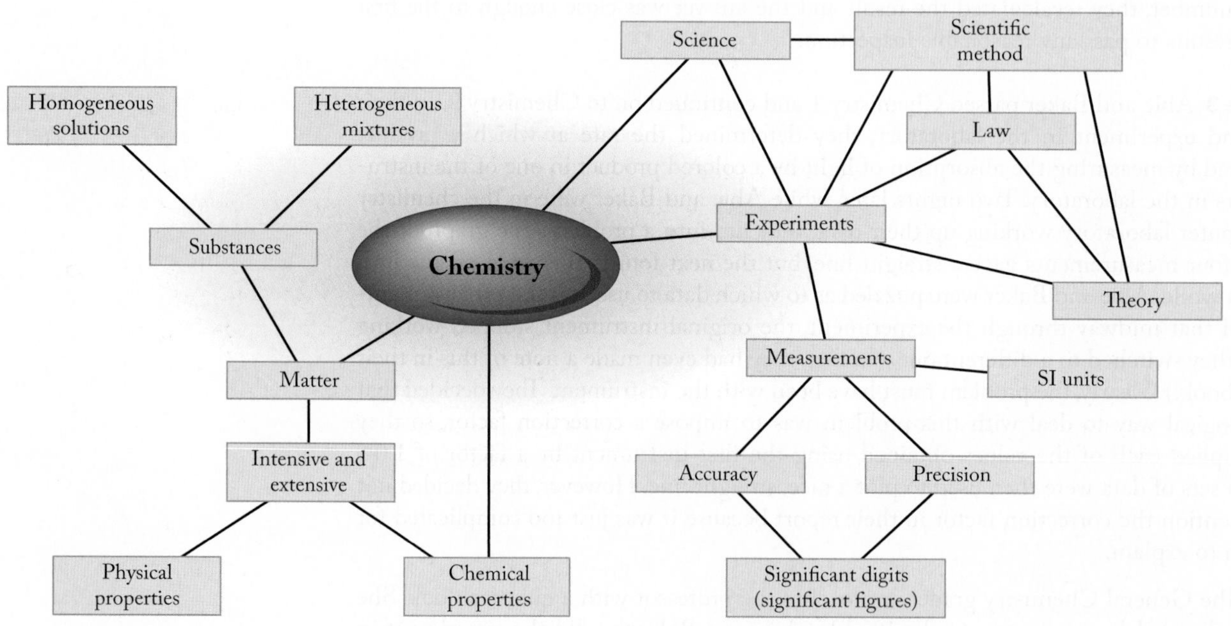

Summary

1.1 The Nature of Science and Chemistry

Chemistry is the science that explores interaction of matter with other matter and with energy. Advances in chemistry occur through careful experiments guided by the *scientific method. Laws* summarize the results of numerous experimental results. Scientists offer a *hypothesis* to explain the experiments. The hypothesis can evolve into a *theory* after additional experiments are performed and scientists widely accept the explanation.

1.2 Matter

Matter is anything that has *mass* and occupies space. The *properties* of matter can be divided into *physical properties* that are observed without changing the composition of the sample, whereas observation of *chemical properties* requires that the sample undergo *chemical change. Extensive properties* are related to *how much* matter is present in a sample, and *intensive properties* are characteristic of the *type* of matter. *Density,* the ratio of mass to volume, is an important intensive property derived from the ratio of two extensive properties.

The broadest division of matter is into *substances* and *mixtures.* Pure substances are classified as *compounds* or *elements.* Compounds can be decomposed into elements by chemical means; elements cannot. *Mixtures* can be separated into components by physical processes and are called *homogeneous* when the mixture has the same composition throughout or *heterogeneous* when different parts of the mixture have different properties. Another name for a homogeneous mixture is a *solution.*

1.3 Measurements and Uncertainty

Every measurement has an uncertainty associated with it, indicated by the number of *significant figures* or *significant digits.* The uncertainties in the measurements and how they are combined determine the proper *number of significant figures* in a reported value. In addition or subtraction, the number with the fewest number of decimal places determines the number of decimal places in the result. In multiplication or division, the number of significant figures in the answer is the same as that in the quantity with the fewest number of significant figures.

1.4 Measurements and Units

Scientists express measurements using the *SI* units. In this system, seven *base units* are defined, and all other units of measure are derived from them. *Conversion factors,* based on equalities or equivalencies, are useful in changes from one unit type to another. Chemists use both the Celsius and Kelvin scales to express temperature. These scales have units of measure that are the same size, but the Kelvin scale is based on an absolute zero.

Chapter Terms

The following terms are defined in the Glossary, Appendix I.

Section 1.1
Chemistry
Hypothesis
Law
Science
Scientific method
Theory

Section 1.2
Alloy
Chemical change
Chemical property
Compound
Element
Extensive property
Heterogeneous mixture
Homogeneous mixture
Intensive property
Mass
Matter
Mixture
Physical change
Physical property
Property
Solution
Substance
Symbol
Weight

Section 1.3
Accuracy
Error
Precision
Number of significant figures
Range
Scientific notation (exponential notation)
Significant figure (significant digit)
Uncertainty

Section 1.4
Base unit
Conversion factor
Density
Derived unit
Unit

Questions and Exercises

OWL Selected end of chapter Questions and Exercises may be assigned in OWL.

Blue-numbered Questions and Exercises are answered in Appendix J; questions are qualitative, are often conceptual, and include problem-solving skills.

■ Questions assignable in OWL

✎ Questions suitable for brief writing exercises

▲ More challenging questions

Questions

1.1 Define *science* in your own words. List three fields that are science and three fields that are not science.

1.2 ✎ Compare the uses of the words *theory* and *hypothesis* by scientists and by the general public.

1.3 ✎ Explain how the coach of an athletic team might use scientific methods to enhance the team's performance.

1.4 ✎ Draw a diagram similar to Figure 1.1 that places the following words in the proper relationships: *theory, hypothesis, model, data, guess,* and *law.*

1.5 ✎ Some scientists think the extinction of the dinosaurs was due to a collision with a large comet or meteor. Is this statement a hypothesis or a theory? Justify your answer.

1.6 ✎ List three intensive and three extensive properties of air.

1.7 Define *matter, mass,* and *weight.*

1.8 Matter occupies space and has mass. Are the astronauts in a space shuttle composed of matter while they are weightless? Explain your answer.

1.9 Give three examples of homogeneous and heterogeneous mixtures.

1.10 ✎ Do you think it is easier to separate a homogeneous mixture or a heterogeneous mixture, or would both be equally difficult? Explain your answer.

1.11 ✎ A solution made by dissolving sugar in water is homogeneous because the composition is the same everywhere. But if you could look with very high magnification, you would see locations with water particles and other locations with particles of sugar. How can we say that a sugar solution is homogenous?

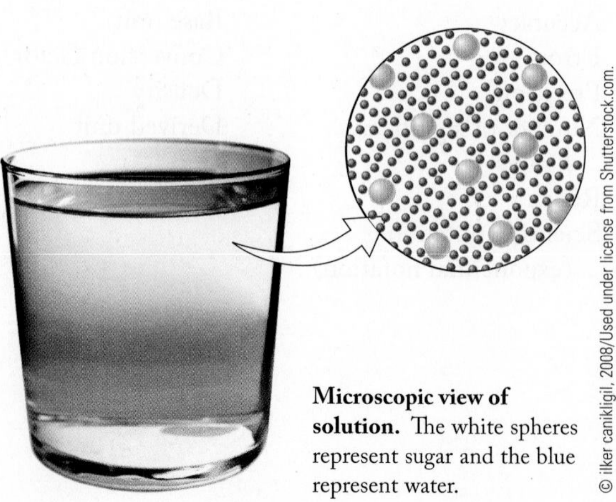

Microscopic view of solution. The white spheres represent sugar and the blue represent water.

1.12 ▲ Is the light from an electric bulb an intensive or extensive property?

1.13 ▲ Are all alloys homogeneous solutions? Explain your answer.

1.14 Explain the differences between substances, compounds, and elements.

1.15 Football referees mark the ball from its position when the player is down or steps out of bounds. A cumbersome but accurate chain is used to determine whether the ball has advanced 10 yards. Given the high accuracy of the measurement chain, why do many fans and players question the officials when they make these measurements?

1.16 Explain the differences between *accuracy* and *precision,* and the relationship between the number of significant digits and the precision of a measurement.

1.17 Describe a computation in which *your* calculator does not display the correct number of significant digits.

1.18 Draw a block diagram (see Example 1.8) that illustrates the processes used to convert km/hr to m/s.

1.19 Give examples of two numbers, one that is exact (no uncertainty) and one that is not, by using them in a sentence.

1.20 ✎ If you repeat the same measurement many times, will you always obtain exactly the same result? Why or why not? What factors influence the repeatability of a measurement?

1.21 Propose the appropriate SI units and prefixes to express the following values:

(a) Diameter of a human hair
(b) Distance between New York City and Auckland, New Zealand
(c) Mass of water in Lake Michigan
(d) Volume of 5 lb table salt
(e) Mass of the average house

1.22 ✎ For centuries, a foot was designated as literally a *foot*—the length of the king's foot. What are the disadvantages of such a measurement system? Are there any advantages?

1.23 Give an example of a conversion factor that (a) can convert between SI units, and (b) can convert between units of the SI and English system.

1.24 ✎ ▲ With some simple research, determine what experimental phenomena provide the basis for the standards for six base units. Is there a commonality between any of these phenomena?

Exercises

In this section, similar exercises are arranged in pairs.

OBJECTIVES Identify properties of matter as intensive and extensive. Differentiate between chemical and physical properties and changes.

1.25 Each of the following parts contains an underlined property. Classify the property as intensive or extensive and as chemical or physical.
(a) Bromine is a reddish liquid.
(b) A ball is a spherical object.
(c) Sodium and chlorine react to form table salt.
(d) A sample of water has a mass of 45 g.
(e) The density of aluminum is 2.70 g/cm^3.

1.26 Each of the following parts contains an underlined property. Classify the property as intensive or extensive and as chemical or physical.
(a) A lemon is yellow.
(b) Sulfuric acid converts sugar to carbon and steam.
(c) The sample has a mass of 1 kg.
(d) Sand is insoluble in water.
(e) Wood burns in air, forming carbon dioxide and water.

1.27 Classify each of the following processes as a chemical change or a physical change.
(a) Water boiling
(b) Glass breaking
(c) Leaves changing color
(d) Iron rusting

© Cengage Learning/George Semple

Leaves changing color.

1.28 Classify each of the following processes as a chemical change or a physical change.
(a) Tea leaves soaking in warm water
(b) A firecracker exploding
(c) Magnetization of an iron nail
(d) A cake baking

1.29 Which of the following processes describe physical changes, and which describe chemical changes?
(a) Milk souring
(b) Water evaporating
(c) The forming of copper wire from a bar of copper
(d) An egg frying

1.30 Which of the following processes describe physical changes, and which describe chemical changes?
(a) A seed growing into a plant
(b) Distillation of alcohol
(c) Mixing an Alka-Seltzer tablet with water
(d) Hammering iron into a horseshoe

1.31 Which of the following processes describe physical changes, and which describe chemical changes?
(a) Alcohol burns
(b) Sugar crystallizes
(c) Gas bubbles rise out of a glass of soda
(d) A tomato ripens

1.32 ■ Which of the following processes describe physical changes, and which describe chemical changes?
(a) Meat cooks
(b) A candle burns
(c) Wood is attached with nails
(d) Newspaper yellows with age

1.33 In the following description of the element fluorine, identify which of the properties are chemical and which are physical. "Fluorine is a pale-yellow corrosive gas that reacts with practically all substances. Finely divided metals, glass, ceramics, carbon, and even water burn in fluorine with a bright flame. Small amounts of compounds of this element in drinking water and toothpaste prevent dental cavities. The free element has a melting point of −219.6 °C and boils at −188.1 °C. Fluorine is one of the few elements that forms compounds with the element xenon."

1.34 In the following description of the element iron, identify which of the properties are chemical and which are physical. "Iron is rarely found as the free element in nature. Mostly it is found combined with oxygen in an ore. The metal itself can be obtained by reacting the ore with carbon, producing iron and carbon dioxide. Iron is a silver-colored metal that conducts heat and electricity well. It is one of the most structurally important metals because of its hardness and mechanical strength, and it makes alloys with many other metals. Stainless steel is one useful alloy of iron that does not corrode in the presence of water and oxygen, like pure iron does."

1.35 In the following description of the element sodium, identify which of the properties are chemical and which are physical. "Sodium is a soft, silver-colored metal that reacts with water to form sodium hydroxide and hydrogen gas. It is stored under oil because it reacts with air. Sodium melts at 98 °C, which is relatively low for a metal."

1.36 In the following description of the element bromine, identify which of the properties are chemical and which are physical. "Bromine is one of the few elements that is a liquid at room temperature. It is an acrid-smelling substance that reacts readily with most metals. It evaporates easily, so most containers of bromine are filled with visible amounts of red fumes. Most bromine is obtained from sodium bromide, a compound found in salt beds."

OBJECTIVES Classify matter by its properties and composition. Distinguish elements from compounds.

1.37 Classify each of the following as an element, a compound, or a mixture. Identify mixtures as homogeneous or heterogeneous.
(a) Air (b) Sugar
(c) Cough syrup (d) Cadmium

1.38 Classify each of the following as an element, a compound, or a mixture. Identify mixtures as homogeneous or heterogeneous.
(a) Water (b) Window cleaner
(c) 14-karat gold (d) Copper

1.39 Classify each of the following as an element, a compound, or a mixture. Identify mixtures as homogeneous or heterogeneous.
(a) Helium (b) A muddy river
(c) Window glass (d) Paint

1.40 ■ Classify each of the following as an element, a compound, or a mixture.
(a) Gold (b) Milk
(c) Sugar (d) Vinaigrette dressing with herbs

1.41. Which of the following mixtures is a solution?
(a) Air (b) A printed page
(c) Milk of magnesia (d) Clear tea

1.42 Which of the following mixtures is a solution?
(a) Wood (b) Champagne
(c) Salt water (d) Cloudy tea

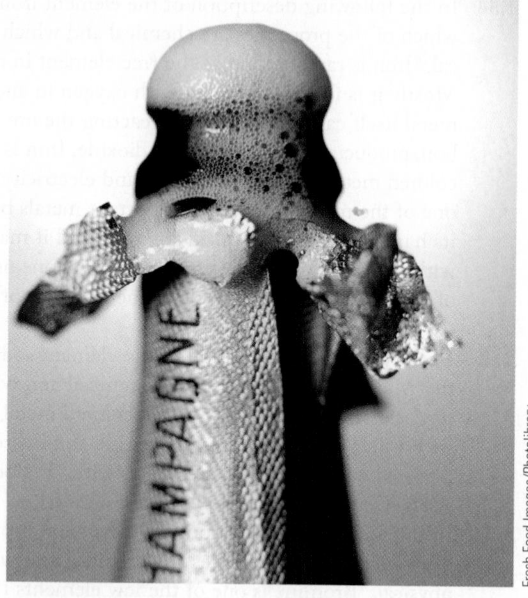

Champagne.

OBJECTIVE Distinguish between accuracy and precision; express the uncertainty of a measurement or calculation to the correct number of significant figures.

1.43 A sample's true mass is 2.54 g. For each set of measurements, characterize the set as accurate, precise, both, or neither.
(a) 2.50, 2.55, 2.59, 2.60
(b) 2.53, 2.54, 2.54, 2.55
(c) 2.49, 2.51, 2.53, 2.63
(d) 2.44, 2.44, 2.45, 2.47

1.44 ■ A measurement's true value is 17.3 g. For each set of measurements, characterize the set as accurate, precise, both, or neither.
(a) 17.2, 17.2, 17.3, 17.3 g
(b) 16.9, 17.3, 17.5, 17.9 g
(c) 16.9, 17.2, 17.9, 18.8 g
(d) 17.8, 17.8, 17.9, 18.0 g

1.45 How many significant figures are in each value?
(a) 1.5003 (b) 0.007
(c) 5.70 (d) 2.00×10^7

1.46 ■ How many significant figures are there in each of the following?
(a) 0.136 m (b) 0.0001050 g
(c) 2.700×10^3 nm (d) 6×10^{-4} L
(e) 56003 cm^3

1.47 How many significant figures are in each measurement?
(a) 5×10^3 m (b) 5.0005 g/mL
(c) 22.9898 g (d) 0.0040 V

1.48 How many significant figures are in each measurement?
(a) 3.1416 degrees (b) 0.00314 K
(c) 1.0079 s (d) 6.022×10^{23} particles

1.49 Express the measurements to the requested number of significant figures.
(a) 96,485 J/C to three significant figures
(b) 2.9979 g/cm^3 to three significant figures
(c) 0.0597 mL to one significant figure
(d) 6.626×10^{-34} kg to two significant figures

1.50 Express the measurements to the requested number of significant figures.
(a) 0.08205 kg to three significant figures
(b) 1.00795 m to three significant figures
(c) 18.9984032 g to five significant figures
(d) 18.9984032 g to four significant figures

1.51 Look at the photographs and measure the volume of solution shown in Figure 1.7*a*, the temperature in Figure 1.7*b*, and the pressure shown in Figure 1.7*c*. Estimate the amounts as accurately as possible and express them to the appropriate number of significant digits.

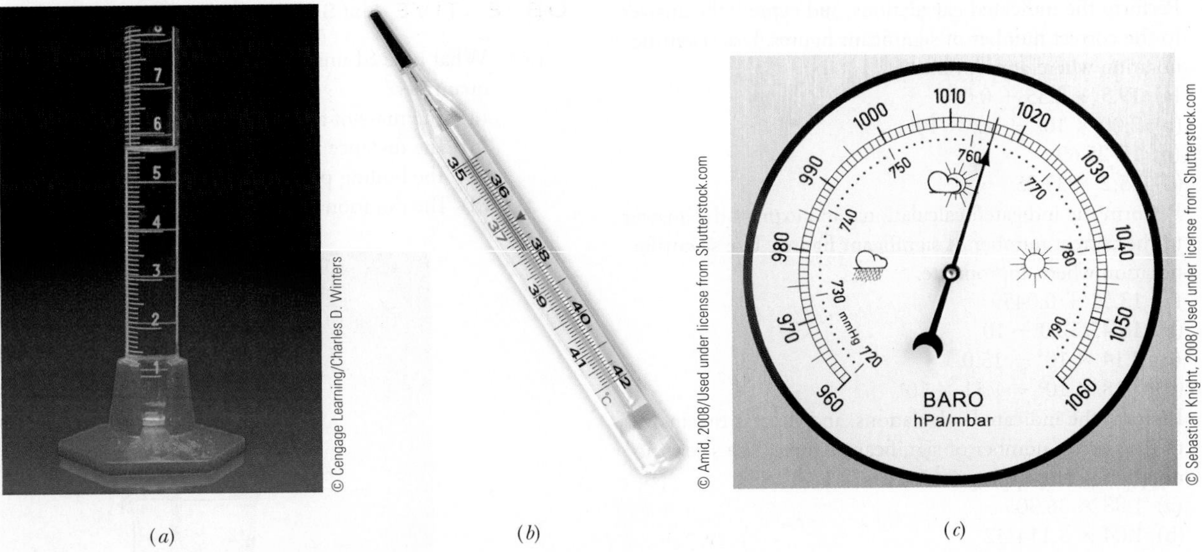

(*a*) (*b*) (*c*)

Figure 1.7 Measurements. *(a)* Graduated cylinder. *(b)* Thermometer. *(c)* Barometer.

1.52 Look at the photographs and measure the temperature shown in Figure 1.8*a*, the volume of solution in Figure 1.8*b*, and the elapsed time shown in Figure 1.8*c*. Estimate the amounts as accurately as possible and express them to the appropriate number of significant digits.

1.53 Perform the indicated calculations, and express the answer to the correct number of significant figures. Use scientific notation where appropriate.
(a) 17.2 × 12.55
(b) 1.4 × 1.11/42.33
(c) 18.33 × 0.0122
(d) 25.7 − 25.25

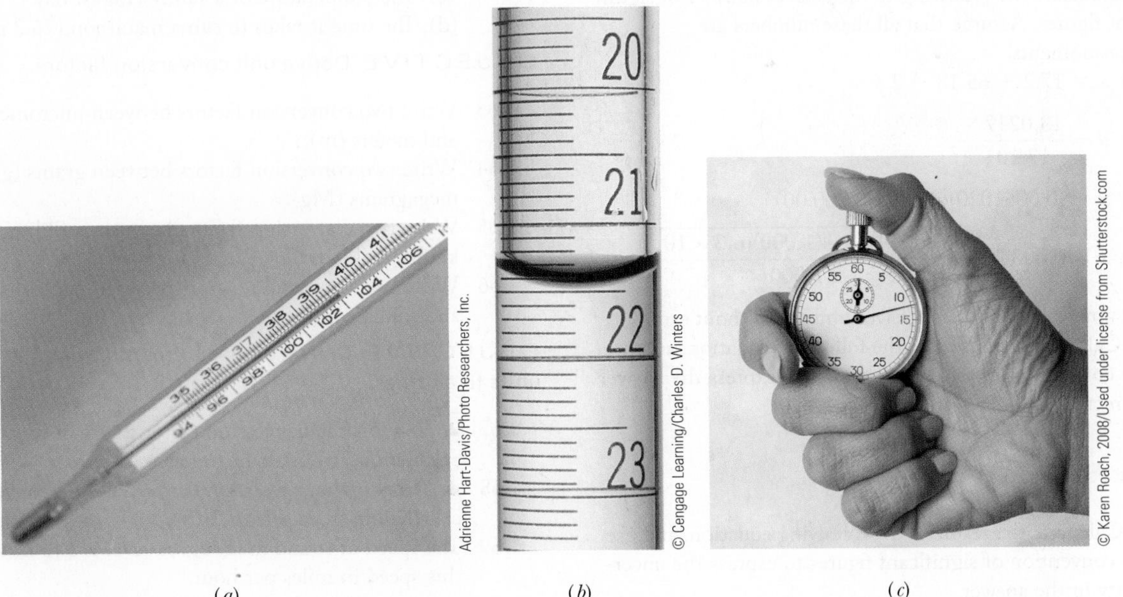

(*a*) (*b*) (*c*)

Figure 1.8 Measurements. *(a)* Thermometer. *(b)* Burette. *(c)* Stopwatch.

1.54 Perform the indicated calculations, and express the answer to the correct number of significant figures. Use scientific notation where appropriate.
(a) $19.5 + 2.35 + 0.037$
(b) $2.00 \times 10^3 - 1.7 \times 10^1$
(c) $15/25.69$
(d) $45.2 - 37.25$

1.55 Perform the indicated calculations, and express the answer to the correct number of significant figures. Use scientific notation where appropriate.
(a) $13.51 + 0.0459$
(b) $16.45/32.0 + 10$
(c) $3.14 \times 10^4 - 15.0$
(d) $7.18 \times 10^3 \div 1.51 \times 10^5$

1.56 Perform the indicated calculations, and express the answer to the correct number of significant figures. Use scientific notation where appropriate.
(a) 1.88×36.305
(b) $1.04 \times 3.114/42$
(c) $28.5 + 4.43 + 0.073$
(d) $3.10 \times 10^2 - 5.1 \times 10^1$

1.57 The following expressions involve multiplication/division and addition/subtraction operations of measured values in the same problem. Evaluate each, and express the answer to the correct number of significant figures.

(a) $\dfrac{(25.12 - 1.75) \times 0.01920}{(24.339 - 23.15)}$

(b) $\dfrac{55.4}{(26.3 - 18.904)}$

(c) $(0.921 \times 27.977) + (0.470 \times 28.976) + (3.09 \times 29.974)$

1.58 ■ Calculate the following to the correct number of significant figures. Assume that all these numbers are measurements.
(a) $x = 17.2 + 65.18 - 2.4$

(b) $x = \dfrac{13.0217}{17.10}$

(c) $x = (0.0061020)(2.0092)(1200.00)$

(d) $x = 0.0034 + \dfrac{\sqrt{(0.0034)^2 + 4(1.000)(6.3 \times 10^{-4})}}{2(1.000)}$

Assume the 4 and 2 are exact numbers, without error.

1.59 ▲ Calculate the result of the following equation, and use the convention of significant figures to express the answer correctly.

$$x = \frac{10^{121}}{10^{-121}} \times 1.01$$

1.60 ▲ Calculate the result of the following equation, and use the convention of significant figures to express the uncertainty in the answer.

$$x = \frac{2.05 \times 10^{-65}}{3.4 \times 10^{51}} + 1.9 \times 10^{-3}$$

OBJECTIVE List SI base units.

1.61 What base SI unit is used to express each of the following quantities?
(a) The mass of a person
(b) The distance from London to New York City
(c) The boiling point of water
(d) The duration of a movie

Charles D. Winters/Photo Researchers, Inc.

Water boiling.

1.62 ■ What base SI unit is used to express each of the following quantities?
(a) The mass of a bag of flour
(b) The distance from the Earth to the Sun
(c) The temperature of a sunny August day
(d) The time it takes to run a marathon (26.2 miles)

OBJECTIVE Derive unit conversion factors.

1.63 Write two conversion factors between micrometers (μm) and meters (m).

1.64 Write two conversion factors between grams (g) and megagrams (Mg).

1.65 Write two conversion factors between milliliters (mL) and kiloliters (kL).

1.66 Write two conversion factors between nanoseconds (ns) and milliseconds (ms).

OBJECTIVE Convert measurements from one set of units to another.

1.67 ▲ What is the conversion factor that will convert, in one calculation, from km/hr to ft/s.

1.68 ▲ What is the conversion factor that will convert, in one calculation, from g/L to lb/ft^3.

1.69 The speed of sound in air at sea level is 340 m/s. Express this speed in miles per hour.

1.70 ■ The area of the 48 contiguous states is 3.02×10^6 mi^2. Assume that these states are completely flat (no mountains and no valleys). What volume of water, in liters, would cover these states with a rainfall of two inches?

1.71 (a) A light-year, the distance light travels in 1 year, is a unit used by astronomers to measure the great distances between stars. Calculate the distance, in miles, represented by 1 light-year. Assume that the length of a year is 365.25 days, and that light travels at a rate of 3.00×10^8 m/s.
(b) The distance to the nearest star (other than the Sun) is 4.36 light-years. How many meters is this? Express the result in scientific notation and with all the zeros.

1.72 ■ Carry out each of the following conversions:
(a) 25.5 m to km (b) 36.3 km to m
(c) 487 kg to g (d) 1.32 L to mL
(e) 55.9 dL to L (f) 6251 L to cm^3

1.73 Perform the conversions needed to fill in the blanks. Use scientific notation where appropriate. Do the operations first without a calculator or spreadsheet, to check your understanding of SI prefixes.
(a) 6.39 cm = _____ m = _____ mm = _____ nm
(b) 55.0 cm^3 = ____ dm^3 = ____ mL = ____ L = ____ m^3
(c) 23.1 g = _____ mg = _____ kg
(d) 98.6 °F = _____ °C = _____ K

1.74 ■ Perform the conversions needed to fill in the blanks. Use scientific notation where appropriate. Do the operations first without a calculator or spreadsheet, to check your understanding of SI prefixes.
(a) 45 s = _____ ms = _____ minutes
(b) 550 nm = _____ cm = _____ m
(c) 4 °C = _____ K = _____ °F
(d) 2.00 L = _____ cm^3 = _____ m^3 = _____ qt

1.75 The 1500-m race is sometimes called the *metric mile.* Express this distance in miles.

1.76 A standard sheet of paper in the United States is 8.5 × 11 inches. Express the area of this sheet of paper in square centimeters.

1.77 Wine is sold in 750-mL bottles. How many quarts of wine are in a case of 12 bottles?

© Cephas Picture Library/Alamy

A case of wine.

1.78 The speed limit on limited-access roads in Canada is 100 km/h. How fast is this in miles per hour? In meters per second?

1.79 Wine sold in Europe has its volume labeled in centiliters (cL). If wine is sold in 750-mL bottles, how many centiliters is this?

1.80 Many soft drinks are sold in 2.00-L containers. How many fluid ounces is this?

1.81 Derive an equation, including units, to make conversions from kelvins to degrees Fahrenheit.

1.82 Derive an equation, including units, to make conversions from degrees Fahrenheit to kelvins.

1.83 (a) Helium has the lowest boiling point of any substance; it boils at 4.21 K. Express this temperature in degrees Celsius and degrees Fahrenheit.
(b) The oven temperature for a roast is 400 °F. Convert this temperature to degrees Celsius.

1.84 (a) The boiling point of octane is 126 °C. What is this temperature in degrees Fahrenheit and in kelvins?
(b) Potatoes are cooked in oil at a temperature of 350 °F. Convert this temperature to degrees Celsius.

1.85 The melting point of sodium chloride, table salt, is 801 °C. What is this temperature in degrees Fahrenheit and in kelvins?

1.86 At what temperature does a Celsius thermometer give the same numerical reading as a Fahrenheit thermometer?

OBJECTIVE **Derive conversion factors from equivalent quantities.**

1.87 The density of benzene at 25.0 °C is 0.879 g/cm^3. What is the volume, in liters, of 2.50 kg benzene?

1.88 Ethyl acetate, one of the compounds in nail polish remover, has a density of 0.9006 g/cm^3. Calculate the volume of 25.0 g ethyl acetate.

1.89 Lead has a density of 11.4 g/cm^3. What is the mass, in kilograms, of a lead brick measuring 8.50 × 5.10 × 3.20 cm?

1.90 What is the radius, r, of a copper sphere (density = 8.92 g/cm^3) whose mass is 3.75×10^3 g? The volume, V, of a sphere is given by the equation $V = (4/3)\pi r^3$.

1.91 An irregularly shaped piece of metal with a mass of 147.8 g is placed in a graduated cylinder containing 30.0 mL water. The water level rises to 48.5 mL. What is the density of the metal in g/cm^3?

1.92 ■ A solid with an irregular shape and a mass of 11.33 g is added to a graduated cylinder filled with water (d = 1.00 g/mL) to the 35.0-mL mark. After the solid sinks to the bottom, the water level is read to be at the 42.3-mL mark. What is the density of the solid?

1.93 ▲ How many square meters will 4.0 L (about 1 gal) of paint cover if it is applied to a uniform thickness of 8.00×10^{-2} mm (volume = thickness × area)?

1.94 ▲ A package of aluminum foil with an area of 75 ft^2 weighs 12 ounces avdp. Use the density of aluminum, 2.70 g/cm^3, to find the average thickness of this foil, in nanometers (volume = thickness × area).

Chapter Exercises

1.95 In describing the phase of a substance, is it possible that a substance can have two phases at the same time, say, solid and liquid phase? Give examples or circumstances to support your answer.

1.96 ■ To determine the density of a material, a scientist first weighs it (Figure 1.9*a*). She would then add it to a graduated cylinder (see Figure 1.9*b*) that contains some water (see Figure 1.9*c*) and record the mass, volume of water, and volume of water plus the metal in her notebook. Use the photographs as the source of the data to calculate the density. Make sure you express the density to the correct number of significant digits.

1.97 ▲ Gold leaf, which is used for many decorative purposes, is made by hammering pure gold into very thin sheets. Assuming that a sheet of gold leaf is 1.27×10^{-5} cm thick, how many square feet of gold leaf could be obtained from 28.35 g gold? The density of gold is 19.3 g/cm^3.

1.98 The speed of light is 3.00×10^8 m/s. Assuming that the distance from the Earth to the Sun is 93,000,000 miles, (a) how many light-years is this (see question 1.71)? (b) How many minutes does it take for light to reach the Earth from the Sun?

1.99 The mass of a piece of metal is 134.412 g. It is placed in a graduated cylinder that contains 12.35 mL water. The volume of the metal and water in the cylinder is found to be 19.40 mL. Calculate the density of the metal.

1.100 ▲ Consider two liquids: liquid A, with a density of 0.98 g/mL, and liquid B, with a density of 1.03 g/mL. Notice that one density is known to have two significant figures and the other to have three. Calculate the volume of liquid A in a sample that weighs 9.9132 g; be sure to express your result to the proper number of significant digits. Calculate the volume of the same mass of liquid B, again making sure that you have the appropriate number of significant figures.

Recording the number of significant figures is only one way to estimate the uncertainty. Repeat the calculations of volume by using the minimum and maximum values of density to calculate maximum and minimum volumes. The range between the two is also a measure of uncertainty.

Compare the estimated uncertainties in the two liquids as measured by the two techniques. Do all estimates give the same answer? Should they? Explain any disagreements.

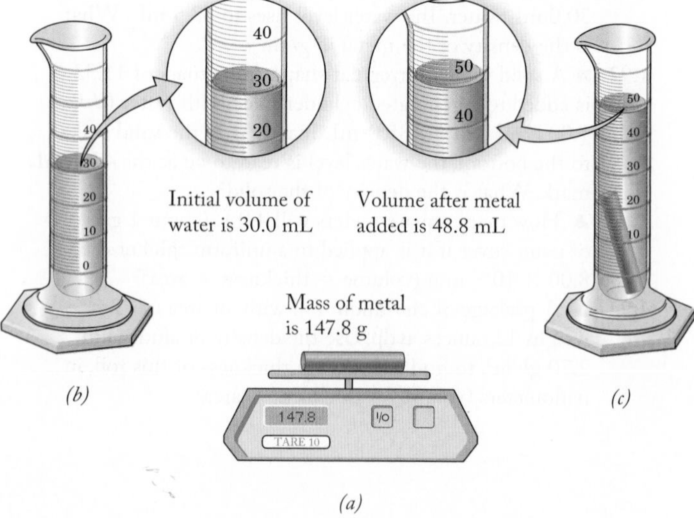

Figure 1.9 Measuring density. *(a)* Mass of metal. *(b)* Volume of water. *(c)* Volume of water plus metal.

Cumulative Exercises

1.101 ▲ A student puts a pulsed laser and detector in the hall of the chemistry building. She places a mirror at the other end of the building and measures the roundtrip distance at 312 ft 6 in. She calibrates a time-measuring device called an *oscilloscope,* which records the amplitude of the laser signal as a function of time.

The results of two experiments, in which the vertical axis shows the magnitude of the signal and the horizontal axis shows the time, are shown. The first peak indicates the moment at which the laser fired, and the second peak is due to the pulse returning from the mirror.

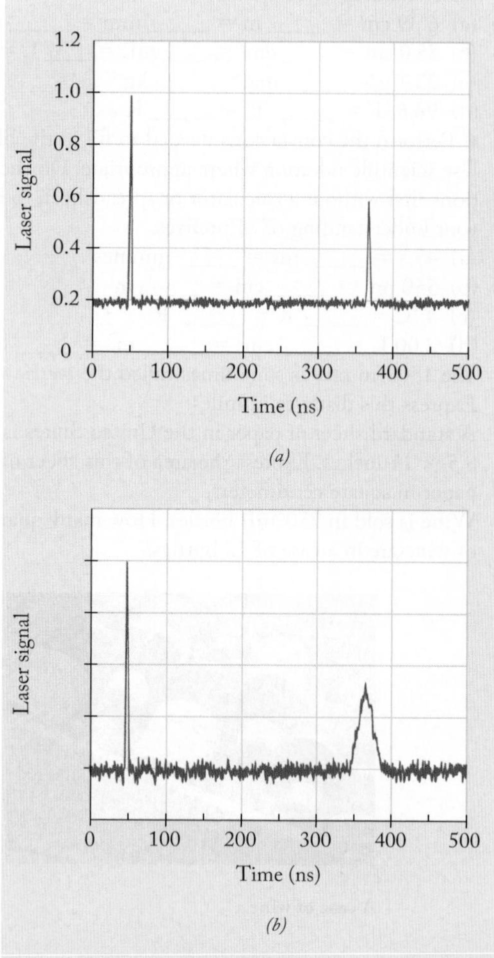

Measuring the velocity of light.

(a) Calculate the velocity of light in both experiments. Use the convention of significant digits to express the uncertainty.

(b) Explain what factors limit the uncertainty. In other words, how can the student improve the experiment?

1.102 A scientific oven is programmed to change temperature from 80.0 °F to 215.0 °F in 1 minute. Express the rate of change in degrees Celsius per second, and use the convention of significant digits to express the uncertainty in the rate.

1.103 The average body temperature of a cow is about 101.5 °F. Express this in degrees Celsius and in kelvins, using the correct number of significant figures.

1.104 "No two substances can have the same complete set of physical and chemical properties." Present arguments for and against this statement.

1.105 ▲ The main weapon on a military tank is a cannon that fires a blunt projectile specially designed to cause a shock wave when it hits another tank. The projectile fits into a finned casing that improves its accuracy. Calculate the mass of the projectile, assuming it is a cylinder of uranium (density = 19.05 g/cm^3) that is 105 mm in diameter and 30 cm in height. The volume of a cylinder is given by the equation $V = \pi r^2 h$.

1.106 The U.S. debt in 2008 was $9.2 trillion.

(a) Estimate the height, in kilometers, of a stack of 9.2 trillion $1 bills. Assume that a $1 bill has a thickness of 0.166 mm.

(b) Estimate the mass of this stack if a $1 bill has a mass of 1.01 g.

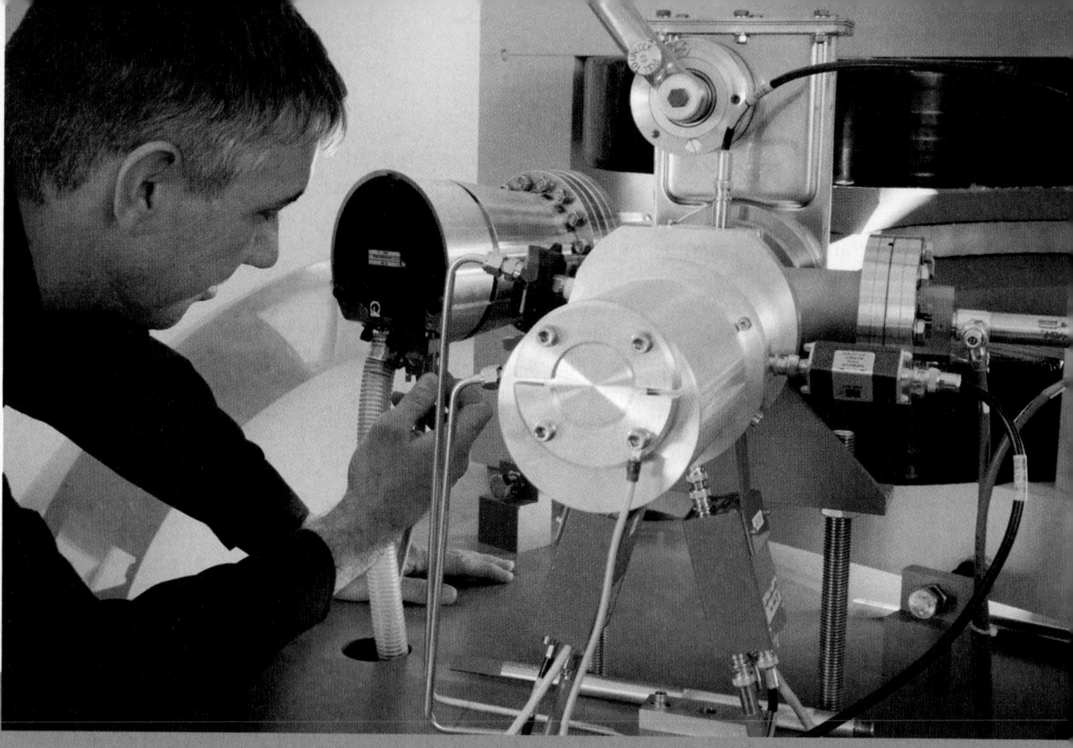

Mass spectrometer.

James King-Holmes/Photo Researchers, Inc.

According to the U.S. Bureau of Justice Statistics, between October 1, 2003, and September 30, 2004, there were 12,166 cocaine-related arrests at the federal level. That equals more than 33 arrests each day. The numbers are just as high at the state and local levels. To carry out these arrests, law-enforcement officers need to be able to identify quickly and accurately any seized cocaine. For attorneys to prosecute these arrests properly, crime laboratories need equipment that can verify the identity of the seized substance. Cocaine, particularly crack cocaine, is the drug most commonly associated with violent crime nationally, so the stakes are extremely high when it comes to prosecuting these cases.

Police officers often carry a test kit to determine whether a confiscated white powder is cocaine. In the test, the officer puts a small amount of the suspected material in a clear plastic envelope that contains two glass vials of chemicals. The officer seals the envelope with a clip and breaks the vials to allow the contents to mix. If cocaine is present, a blue solid forms.

Atoms, Molecules, and Ions

2

CHAPTER CONTENTS

OWL Online homework for this chapter may be assigned in OWL.

Look for the green colored vertical bar throughout this chapter, for integrated references to this chapter introduction.

The chemical test works well to screen samples, but the results are not necessarily conclusive. Although all samples of cocaine cause a blue precipitate to form, a few other compounds also would form a blue precipitate. This result, termed a *false positive,* means that the identification must be confirmed by another method.

The main tool used to definitively confirm the presence of cocaine is a device called a *mass spectrometer.* The "mass spec," as it is sometimes called, can analyze a compound and provide data that are an unambiguous fingerprint of cocaine, clearly distinguishing it from other substances. It can also quantify masses as small as a picogram (10^{-12} g). Chemists are developing sensitive methods to identify and quantify cocaine in a person's breath, urine, saliva, blood, and hair. ▮

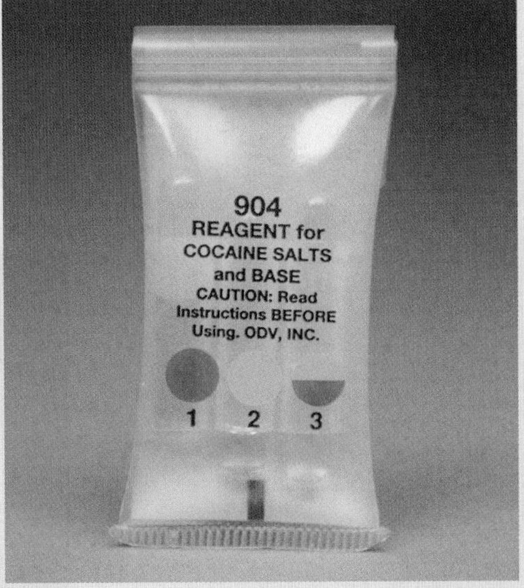

Scott R. Goode

For hundreds of years, scientists have conducted experiments to determine why one material differs from another. For example, at room temperature and pressure, chlorine is a greenish yellow gas, sulfur is a yellow solid, and mercury is a silver-gray liquid (Figure 2.1).

The physical properties of these elements differ, as do their chemical properties. This chapter lays the foundation for the development of concepts and models that can explain these observations about the properties of the elements. It is important to recognize that the models developed in this chapter involve experiments performed by thousands of individuals over a period of more than two centuries.

2.1 Dalton's Atomic Theory

OBJECTIVES

- ☐ Describe the four postulates of Dalton's atomic theory
- ☐ Relate the laws of constant composition, multiple proportions, and conservation of mass to Dalton's atomic theory

More than 2300 years ago, Greek philosophers first asked whether a sample of matter divided into smaller and smaller pieces would retain the properties of the substance. In other words, is matter "continuous," or is it "discontinuous"—that is, composed of some smallest indivisible particle that does not retain the properties of the sample when further subdivided? Scientists debated this idea widely for centuries but reached no conclusion until they performed experiments that could differentiate continuous from discontinuous matter.

An understanding of how matter is composed was developed by careful quantitative experiments. One important experiment was conducted by Antoine Lavoisier (1743–1794). He demonstrated that when a reaction was conducted in a closed container, the mass of the products was equal to the mass of the starting reactants. This result is summarized in the **law of conservation of mass:** There is no detectable change in mass when a chemical reaction occurs.

Other experiments showed that each compound is always formed by the same elements in the same mass ratios. For example, scientists determined that water always contained 1 g hydrogen for every 8 g oxygen. These results are summarized by the **law of constant composition:** All samples of a pure substance contain the same elements in the same proportions by mass.

It had also been observed experimentally that, in certain cases, more than one compound can form from the same elements. The compositions of such compounds reveal

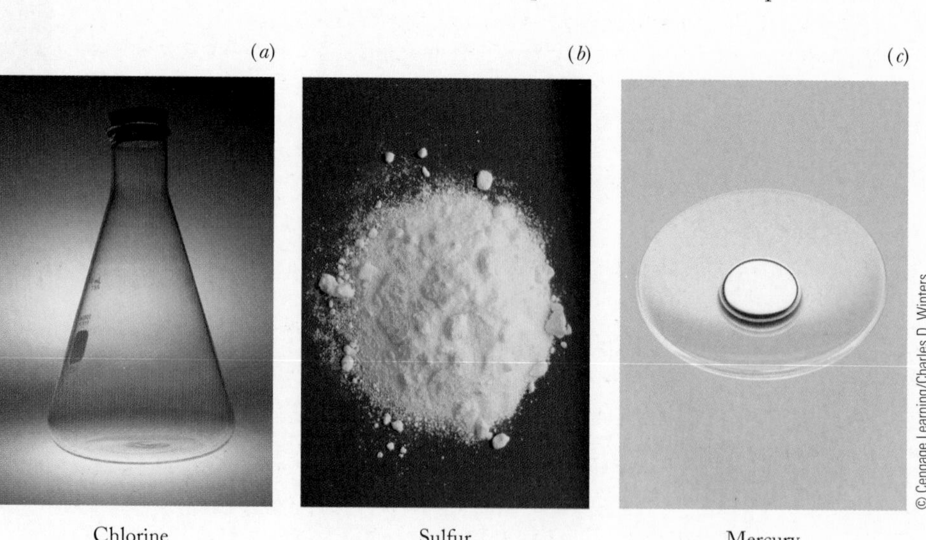

Figure 2.1 **Several elements.** (*a*) Chlorine (*b*) Sulfur (*c*) Mercury

an important relationship called the **law of multiple proportions:** In each pair of compounds formed by the same elements, the masses of one element that combine with a fixed mass of a second element are always in a ratio of small whole numbers. For example, two common compounds contain only carbon and oxygen: carbon monoxide and carbon dioxide. In carbon monoxide, 1.33 g oxygen combine with 1.00 g carbon; in carbon dioxide, 2.66 g oxygen combine with 1.00 g carbon. Thus, the ratio of the masses of oxygen that combine with 1.00 g carbon is 2.66:1.33, or 2:1.

Building on these and other experimental results, John Dalton (1766–1844) proposed a model that explained many of the properties of matter. At the core of his model is the assumption that matter is discontinuous. In modern terms, the four postulates of **Dalton's atomic theory** are as follows:

1. Matter is composed of small indivisible particles called *atoms*. The **atom** is the smallest unit of an element that has all the properties of that element.
2. An *element* is composed entirely of one type of atom. The chemical properties of all atoms of any element are the same.
3. A *compound* contains atoms of two or more different elements. The relative number of atoms of each element in a particular compound is always the same.
4. Atoms do not change their identities in chemical reactions. Chemical reactions rearrange only how atoms are joined together.

John Dalton.

Atoms are the building blocks of elements and compounds.

Each element is assigned a unique name and symbol, with the symbol being generally the first or first two letters of the name (see page opposite the periodic table on the inside cover of this textbook). Dalton's theory explains the experimental results known at that time. For example, the law of constant composition is explained by the premise that a given compound is always made up of the same types of atoms in the same ratios.

The explanation for the law of constant composition is that the relative numbers of atoms of each element in a given compound are always the same.

In a similar manner, atomic theory provides an explanation for the law of multiple proportions. Although the relative number of atoms of each element in a particular compound is always the same, there is no reason that two compounds cannot be made from the same elements *in different ratios* (Figure 2.2). When this happens, the ratio of the atoms will be in whole number ratios.

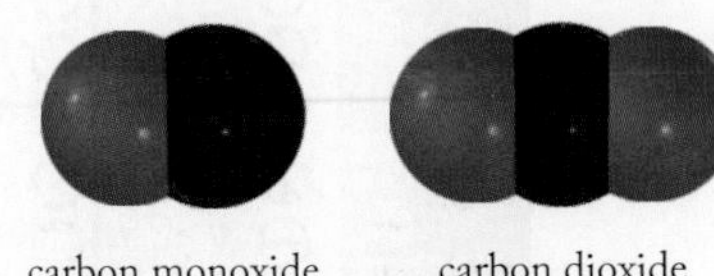

Figure 2.2 **Two compounds formed from carbon and oxygen.**

Dalton's postulates also explain the law of conservation of mass. In a chemical reaction, the combinations of atoms change, but neither the number of atoms nor the types of atoms change. Because the number and types of atoms do not change, the mass cannot change.

OBJECTIVES REVIEW *Can you:*

- ☑ explain the four postulates of Dalton's atomic theory?
- ☑ relate the laws of constant composition, multiple proportions, and conservation of mass to Dalton's atomic theory?

2.2 Atomic Composition and Structure

OBJECTIVES

- ☐ Describe the three subatomic particles that make up an atom, including their relative charges and masses
- ☐ Specify the locations of protons, neutrons, and electrons in the atom

Although atoms were initially viewed as indivisible, experiments performed or interpreted long after Dalton proposed his theory have shown that atoms are composed of three types of particles. The way in which atoms combine depends on how these subatomic particles are arranged in each atom. This section presents the developments that contributed to the discovery of these subatomic particles and other key discoveries that led to the modern description of the atom.

PRINCIPLES OF CHEMISTRY
The Existence of Atoms

Today, John Dalton is regarded as the father of modern atomic theory. However, he was not the first person to propose that all matter is composed of particles called *atoms.* For instance, Isaac Newton published statements that indicated his belief in the concept of atoms. What made Dalton's theory so significant to the history of chemistry was that his theory was firmly founded on the results of scientific experiments. Earlier atomic theories were based more on philosophical arguments and speculations than on physical evidence. Dalton, in addition to performing his own experiments, combed through at least 15 years of published data to derive and support his theory. Although he was a rather poor experimentalist, Dalton's ability to recognize and interpret relationships among experimental data was one of his greatest assets, and his atomic theory helped to explain many earlier experimental results.

Probably the most important factor in the acceptance of Dalton's work was his use of quantitative measurements of mass to describe chemical reactions. Dalton believed that all of the atoms of a particular element were of the same size and mass, whereas atoms of different elements had different masses. Although direct measurement was impossible, he believed that it was extremely important to determine the atomic mass of each element by experiment. In his attempts to determine the masses, he had to make certain assumptions about the formulas of substances, some of which were incorrect, which led to errors. Unfortunately, when others demonstrated these errors, Dalton was unwilling to correct his mistakes, causing inaccuracies and uncertainties about molecular formulas and relative masses of atoms to persist for almost 50 years after his theory was published. However, despite its flaws, Dalton's work was an important milestone in the development of chemistry as a quantitative science. It introduced the importance of mass as a characteristic of an element, and perhaps most importantly, it encouraged other scientists to perform quantitative experiments to determine accurate values for the masses of atoms.

Ever since Dalton's atomic theory was proposed, chemists have accumulated a vast amount of data that support the existence of atoms. Through the 19th and most of the 20th century, all of the experimental evidence for these very tiny particles was indirect. Only in recent years has the development of a technique called *scanning tunneling microscopy* allowed scientists to obtain pictures of individual atoms. ∎

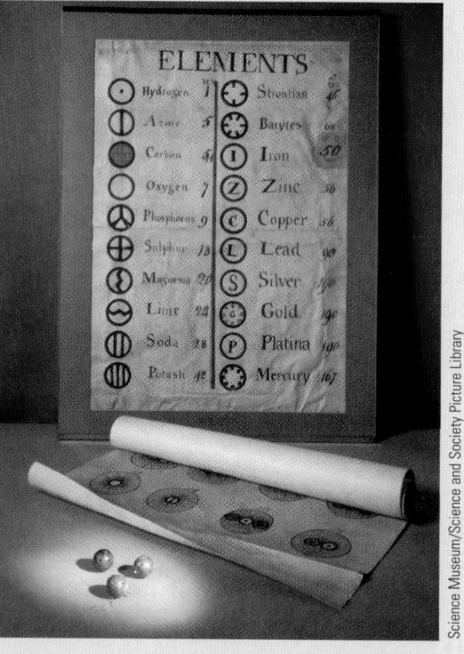

Science Museum/Science and Society Picture Library

Dalton's scale of relative atomic weights.

© P. Plailly/Look at Sciences/Phototake

Scanning tunneling microscope view of palladium atoms. Each sphere represents palladium (Pd) atoms deposited under ultrahigh vacuum on a graphite substrate.

The Electron

Toward the end of the 1800s, scientists began to investigate the flow of electricity in a device called a *gas discharge tube*—a glass tube with a metal electrode at each end and containing a small amount of gas. When high voltages are applied across the electrodes, an electrical discharge—a flow of electricity—occurs and the gas begins to glow. (Modern neon signs and fluorescent lights are examples of gas discharge tubes.) The late 19th-century experi-

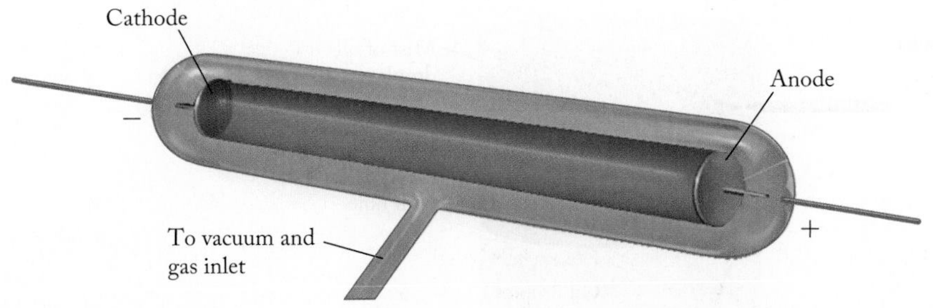

A gas discharge tube. When a high voltage is applied across a partially evacuated tube, the gas begins to glow.

ments showed that the negative electrode was the source of some unusual emission quite different from the light emitted by the gas. Because such emissions came from the negative electrode, called the *cathode,* they were named *cathode rays.* In further experiments, scientists found that cathode rays traveled in straight lines, heated a metal foil placed in their path, and could be deflected by electric and magnetic fields. One group of scientists believed cathode rays to be some sort of light or energy, whereas another group believed that cathode rays were electrically charged particles.

In 1897, British physicist J. J. Thomson (1856–1940) settled the controversy with a series of experiments using specially prepared gas discharge tubes. Thomson found that, by carefully applying controlled magnetic and electric fields to the cathode rays, he could establish that cathode rays were electrically charged particles, and the direction of their deflection by electric and magnetic fields indicated that they were negatively charged. The negative particle is called the **electron,** a name that was suggested years earlier for the particle that theoretically carried electricity. Thomson correctly concluded that electrons were constituents of all atoms, and in 1906, he received the Nobel prize in physics for his work on the electron.

Thomson and his coworkers, as well as many other research groups, then launched experiments designed to determine the charge of the electron. Robert A. Millikan (1868–1953) was the first to measure accurately the charge of the electron. Millikan injected tiny droplets of oil into a chamber and exposed the chamber contents to high-energy radiation, causing each oil drop to acquire an electrical charge. Millikan then measured the rate at which the drop fell in the absence and in the presence of an electric field. Depending on the charge on the drop and the strength of the electric field, the drop fell, rose, or remained stationary. From the data, he calculated the charge on the oil drop.

Millikan observed that the charge on any particular drop was always an integral multiple of a single quantity, which he assumed was the charge carried by a single electron (given by the symbol e). In 1913, Millikan published a value for e equal to

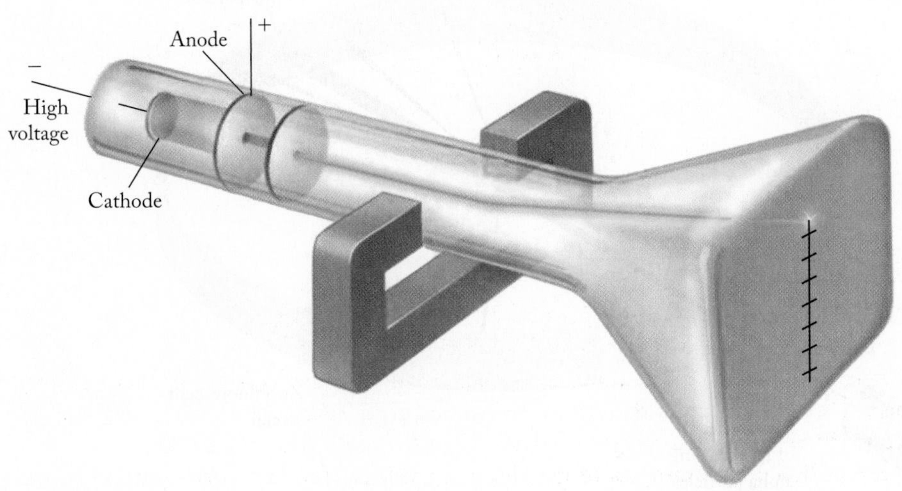

A magnetic field deflects an electron beam.

Millikan oil drop experiment. Oil drops are formed by the injector and are charged by capturing electrons produced by the interaction of high-energy radiation with a gas. From the rate at which the drop moved in the presence and absence of the electric field, Millikan calculated the charges on the oil drops.

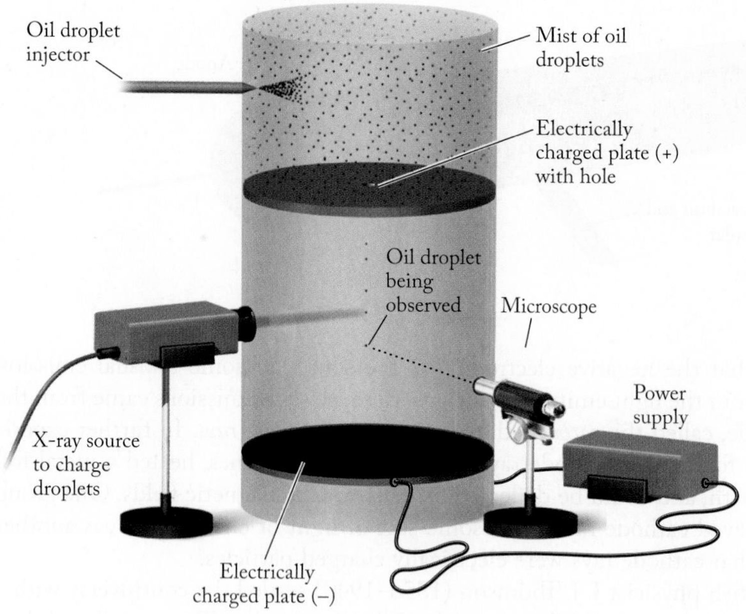

-1.60×10^{-19} coulombs (C), which is exceptionally close to the currently accepted value, $-1.602177 \times 10^{-19}$ C. Using Millikan's value for *e* and work published earlier by Thomson, scientists calculated the value for *m*, the mass of the electron. To three significant figures, the modern value for the mass of the electron is 9.11×10^{-31} kg. For his achievements, Millikan was awarded the Nobel prize in physics in 1923.

The electron is a subatomic particle with a mass of 9.11×10^{-31} kg and a charge of -1.602×10^{-19} coulombs.

The Nuclear Model of the Atom

Atoms are electrically neutral, so they must contain positive charges, as well as the negative electrons. In addition, the mass of an atom is much greater than that of an electron. An important experiment performed in the laboratory of Ernest Rutherford (Figure 2.3) showed how that positive charge and mass were arranged. In 1899, Rutherford and his coworkers had discovered that uranium emitted a particle they called the *alpha particle* (symbolized by α). Rutherford was able to characterize the alpha particle as having a charge of 2+ and a mass four times that of a hydrogen atom. Rutherford and his coworkers built an apparatus to study the deflection of alpha particles as they passed

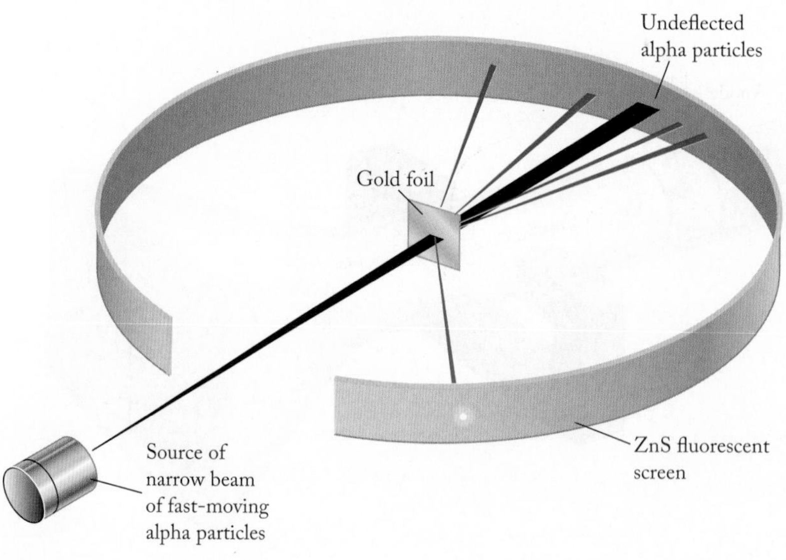

Figure 2.3 The Rutherford experiment. Alpha particles are directed toward a thin piece of metal foil inside a vacuum chamber. Detectors indicate that although most of the particles go through the foil (*black*), some are partially deflected (*red*) and a few are deflected back (*blue*) toward the direction from which they came.

through thin metal targets (like gold and platinum). In the experiment, most of the alpha particles went through the metal foil with no deflection, but the researchers were shocked to find that a few of the alpha particles were deflected through large angles; in fact, some were deflected back in the direction from which they came. To obtain this result, the atoms needed to contain areas of mass that were much greater than the alpha particles.

To explain these experimental results, Rutherford proposed the *nuclear model* of the atom, in which the positive charge and nearly all of the mass of the atom are in a central core, with the electrons at a relatively large distance from this core. Calculations based on the experimental data showed that the central core, which Rutherford called the **nucleus,** is extremely small, even in comparison with the size of the atom. The electrons, which are outside the nucleus, occupy most of the volume of the atom.

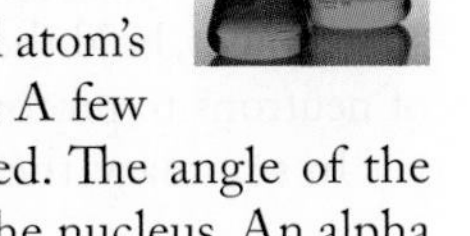

The nuclear model of the atom explains Rutherford's experimental results. Most of the volume of an atom is occupied by the electrons, which have low masses relative to alpha particles and do not measurably deflect them. Thus, most of the alpha particles do not come close to a metal atom's nucleus and travel through the metal foil without being deflected. A few come close to a massive, highly charged nucleus and are deflected. The angle of the deflection is determined by how close the alpha particle comes to the nucleus. An alpha particle that hits the nucleus rebounds significantly (Figure 2.4).

The Proton

Later experiments in Rutherford's laboratory proved that each element has a different positive charge on the nucleus, and the lightest element, hydrogen, has a positive nuclear charge equal in magnitude to that of the electron, or 1+. Rutherford proposed that the hydrogen nucleus was a fundamental particle, and he called it the **proton.** The mass of the proton is 1836 times the mass of the electron, or 1.673×10^{-27} kg. The nuclear

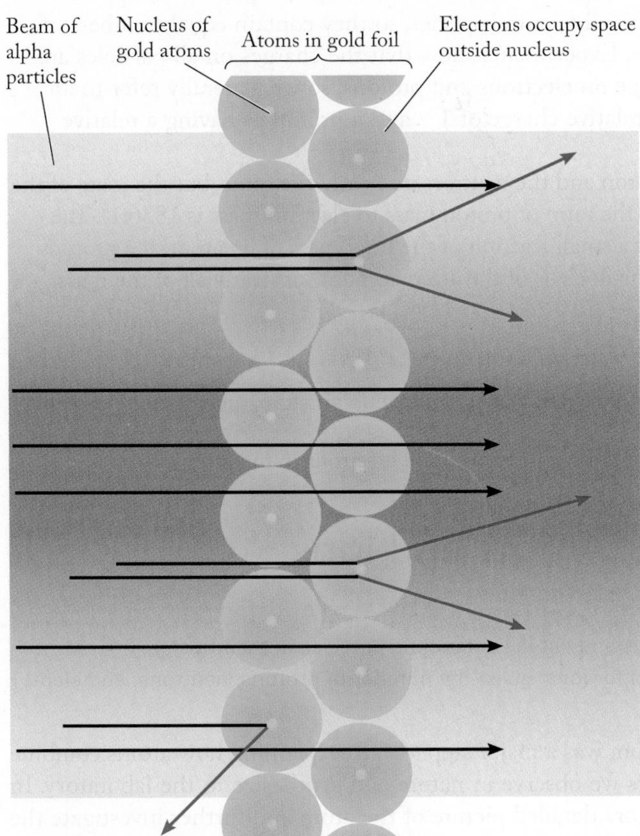

Figure 2.4 Deflection of alpha particles by the nucleus. The nuclear model of the atom predicts that most alpha particles pass through the thin metal foil, but a few will be deflected, some considerably, by the massive, highly charged nuclei.

charge for any element is a result of the protons in the nucleus. Thus, the nucleus of a helium atom, which has a charge of 2+, contains two protons; the nucleus of an oxygen atom, which has a charge of 8+, contains eight protons; and so on. Although he could not explain why the protons were close together in the nucleus—an unlikely situation in view of the repulsion between charges of the same sign—Rutherford knew that he could not explain the experimental results without proposing that all of the protons in the atom were contained in this dense, positively-charged nucleus.

Protons are subatomic particles with a mass of 1.673×10^{-27} kg and a charge of $+1.602 \times 10^{-19}$ coulombs.

The Neutron

For most elements, the mass of the protons in the nucleus accounted for less than half of the nuclear mass, so scientists inferred that a neutral particle must be present in atoms to account for the remaining mass. In 1932, James Chadwick (1891–1974) first observed the effect of this electrically neutral particle. This particle is now known as a **neutron** and has a mass almost the same as that of the proton. The nucleus contains both the protons and neutrons. Forces called *strong nuclear binding forces,* which are stronger than electrostatic forces, hold the neutrons and protons together in the nucleus. In general, the ratios of neutrons to protons in the nuclei of atoms range from 1.0 to 1.6.

The neutron has a mass of 1.675×10^{-27} kg and has no electrical charge.

In summary, the particles that are found within all atoms are as follows:

Particle	Charge (C)	Mass (kg)	Relative Charge	Relative Mass	Location
Electron	-1.602×10^{-19}	9.109×10^{-31}	1−	0	Outside nucleus
Proton	$+1.602 \times 10^{-19}$	1.673×10^{-27}	1+	1	In nucleus
Neutron	0	1.675×10^{-27}	0	1	In nucleus

Notice two important properties of these particles:

1. The charges of the electron and the proton are exactly equal but of opposite sign. Atoms are electrically neutral species, so they contain equal numbers of electrons and protons. Experiments show that the charges on all particles are multiples of the charge on electrons and protons, so we generally refer to an electron as having a relative charge of 1− and a proton as having a relative charge of 1+.
2. The masses of the proton and the neutron are nearly the same, but the mass of the electron is much less (the ratio of proton mass to electron mass is 1836:1). The electrons provide only a small fraction of the total mass of an atom; the protons and the neutrons in the nucleus of the atom account for nearly all of the mass.

Atoms contain protons and neutrons in a central core, the nucleus, which is surrounded by electrons.

OBJECTIVES REVIEW *Can you:*

☑ describe the three subatomic particles that make up an atom, including their relative charges and masses?

☑ specify the locations of protons, neutrons, and electrons in the atom?

2.3 Describing Atoms and Ions

OBJECTIVES

☐ Define isotopes of atoms and list the subatomic particles in their nuclei

☐ Write complete symbols for ions, given the number of protons, neutrons, and electrons that are present

The nuclear model of the atom was a major step toward explaining how atoms combine to form the many substances we observe in nature and synthesize in the laboratory. In this section, we develop a more detailed picture of the atom and further investigate the role of the subatomic particles in determining the chemical properties of an element.

Atoms

The **atomic number** (represented by the letter Z), the number of protons in the nucleus, determines the identity of an element. The atom with one proton in its nucleus is hydrogen; its atomic number is 1. Every helium atom contains two protons in the nucleus, so the atomic number of helium is 2. The atomic number of lithium is 3, and so on, as shown on the periodic table.

Because atoms have a neutral charge, the number of electrons in any element is always equal to the number of protons, but experiments show that the number of neutrons in atoms of the same element can vary. The mass of an atom is determined by the numbers of protons and neutrons. The **mass number** (represented by the letter A) is the sum of the numbers of protons and neutrons in an atom. The mass number does not identify a specific element; the atomic number does that. In fact, different atoms of the same element can differ in numbers of neutrons. **Isotopes** are atoms of one element whose nuclei contain different numbers of neutrons. That is, isotopes have the same atomic number but different mass numbers. For example, there are three isotopes of hydrogen. Most atoms of hydrogen have no neutrons in the nucleus, a few have one neutron, and even fewer have two neutrons (Figure 2.5). All hydrogen atoms have one proton and one electron, but they can have different mass numbers (1, 2, or 3).

Atoms of the same element have the same number of protons and electrons but can have different numbers of neutrons in their nuclei; each is a different isotope of the same element.

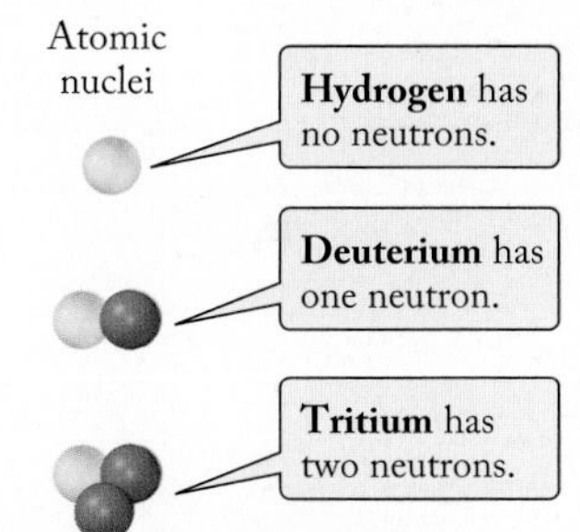

Figure 2.5 Isotopes of hydrogen. Hydrogen has three isotopes. Each isotope has one proton and one electron, but the number of neutrons ranges from zero to two.

The existence of isotopes is an interesting phenomenon. Although some isotopes are unstable and spontaneously decompose to other species, most elements occur in nature as mixtures of stable isotopes. About 75% of the naturally occurring elements have two or more stable isotopes. Titanium and nickel, for example, each have five stable isotopes, whereas copper and chlorine each have only two. Fluorine (mass number = 19) and phosphorus (mass number = 31) are examples of elements with just one stable isotope. Chemists use the different isotopes of elements for important experiments such as the dating of fossils and other geologic samples. An example of this is presented in the chapter opener of Chapter 13.

Note also that the existence of isotopes requires a modification of one of Dalton's postulates for the modern atomic theory. Dalton was unaware of isotopes, so he did not realize that atoms of the same element can, in fact, be different. An element *can* be composed of more than one type of atom, because isotopes of an element have different numbers of neutrons in the nucleus. We understand now that the defining characteristic of an element is the number of protons in the nucleus, and that the number of neutrons in nuclei of the same element can be different. This illustrates how science adapts itself as our understanding of nature improves.

To designate specific isotopes of an element a shorthand notation is used of the form

$${}^{A}_{Z}\mathrm{X}$$

where X is the symbol of the element, A is the mass number, and Z is the atomic number. The three isotopes of hydrogen are represented as

$${}^{1}_{1}\mathrm{H} \quad {}^{2}_{1}\mathrm{H} \quad {}^{3}_{1}\mathrm{H}$$

Notice that the atomic number is the same for all three of these isotopes. *If the atomic number is 1, the atom is hydrogen.* Oxygen has three naturally occurring isotopes:

$${}^{16}_{8}\mathrm{O} \quad {}^{17}_{8}\mathrm{O} \quad {}^{18}_{8}\mathrm{O}$$

Inclusion of the atomic number with the symbol is optional, because either one is sufficient to identify the particular element. The three oxygen isotopes are often written more simply as

$${}^{16}\mathrm{O} \quad {}^{17}\mathrm{O} \quad {}^{18}\mathrm{O}$$

Remember that atoms are electrically neutral, so the atoms of all isotopes of any element always have the same number of electrons as the number of protons in the nucleus. In the case of oxygen, the atoms of all three isotopes contain eight electrons.

EXAMPLE 2.1 Symbols of Atoms

Write the symbol for the atom with:

(a) 6 protons and 6 neutrons
(b) 13 protons and 14 neutrons

Strategy Use the atomic number (equal to the number of protons), to locate the symbol for the element in the periodic table on the inside front cover of this textbook; and sum the number of protons and neutrons to get the mass number.

Solution

(a) The element with six protons is carbon. The mass number is the sum of the numbers of protons and neutrons ($6 + 6 = 12$).

$$^{12}_{6}\mathrm{C} \quad \text{or} \quad ^{12}\mathrm{C}$$

(b) Aluminum is the element that has 13 protons; the mass number is 27.

$$^{27}_{13}\mathrm{Al} \quad \text{or} \quad ^{27}\mathrm{Al}$$

Understanding

Write the symbol for the atom with 19 protons and 20 neutrons.

Answer $^{39}_{19}\mathrm{K}$ or $^{39}\mathrm{K}$

Ions

Atoms can gain or lose electrons and become charged particles called *ions.*

In the courses of many chemical reactions, atoms lose or gain electrons and become charged particles called **ions.** A **cation** is an ion that has a positive charge; an **anion** has a negative charge. Cations have fewer electrons than protons, whereas anions have more electrons than protons. Ions are formed by the loss or gain of electrons by neutral atoms; the number of protons in the nucleus never changes in a chemical process. The charge of an ion = the number of protons − the number of electrons. The charge is positive if there are more protons and negative if there are more electrons.

Ion	Formed by	Composition	Charge
Cation	Loss of electrons by neutral atom	More protons than electrons	+ (positive)
Anion	Gain of electrons by neutral atom	More electrons than protons	− (negative)

The number of protons in the nucleus determines the symbol for an ion. A right superscript number and sign after the symbol indicate its charge. A sodium cation with a charge of 1+ is written as Na^+ (when the charge is 1, the number is omitted); an anion of oxygen with a 2− charge is written as O^{2-}. We can combine this notation with that for isotopes to indicate specific-charged isotopes of elements. The symbol $^{37}Cl^-$ represents the anion of chlorine that contains 17 protons (all isotopes of chlorine contain 17 protons), 20 neutrons, and 18 electrons (1 more electron than protons to give the ion the 1− charge). The cation of magnesium that contains 12 protons, 13 neutrons, and 10 electrons is written as $^{25}Mg^{2+}$.

EXAMPLE 2.2 Symbols of Ions

(a) Write the symbol for an ion with 8 protons, 9 neutrons, and 10 electrons.
(b) Write the symbol for an ion with 20 protons, 20 neutrons, and 18 electrons.

Strategy Determine the identity of the atom from the number of protons and sum the number of protons and neutrons to get the mass number. The charge is determined by the numbers of protons and electrons: charge = the number of protons − the number of electrons. If more protons are present, the charge will be positive, and if more electrons are present, the charge will be negative.

Solution

(a) The eight protons define the atomic number as 8, so the element is oxygen. The sum of the numbers of protons and neutrons is 17, the mass number. The charge = the number of protons − the number of electrons = 8 − 10 = 2−. The symbol is

$${}^{17}_{8}O^{2-} \qquad \text{or} \qquad {}^{17}O^{2-}$$

(b) This ion has an atomic number of 20, so it is the element calcium. The mass number is 40, the sum of the numbers of protons and neutrons. The charge = 20 − 18 = 2+. The symbol is

$${}^{40}_{20}Ca^{2} \qquad \text{or} \qquad {}^{40}Ca^{2}$$

Understanding

Write the symbol for an ion that contains 23 protons, 28 neutrons, and 20 electrons.

Answer ${}^{51}_{23}V^{3}$ or ${}^{51}V^{3}$

EXAMPLE 2.3 Particles in Ions

State the number of protons, neutrons, and electrons present, and identify each of the following ions as a cation or anion:

(a) ${}^{23}_{11}Na$ (b) ${}^{81}_{35}Br^{-}$

Strategy The element symbol and the atomic number located in the bottom left of each give the number of protons. The mass number located in the top left is the sum of the protons and neutrons, so the number of neutrons = mass number − atomic number. The number of electrons is calculated from this equation: charge = the number of protons − the number of electrons. If the charge is positive, the ion is a cation, and if the charge is negative, it is an anion.

Solution

(a) This sodium ion has a positive charge, so it is a cation. All sodium atoms contain 11 protons, making the number of neutrons = mass number − atomic number = 23 − 11 = 12. The number of electrons is calculated from the following equations:

$$\text{Charge} = \text{number of protons} - \text{number of electrons}$$

$$\text{Number of electrons} = \text{number of protons} - \text{charge}$$

$$\text{Number of electrons} = 11 - 1$$

$$\text{Number of electrons} = 10$$

(b) Because the charge of the ion is negative, it is an anion of bromine. All bromine atoms contain 35 protons, making the number of neutrons = 81 − 35 = 46. The number of electrons is calculated from the following equations:

$$\text{Number of electrons} = \text{number of protons} - \text{charge}$$

$$\text{Number of electrons} = 35 - (-1)$$

$$\text{Number of electrons} = 36$$

Understanding

How many protons, neutrons, and electrons are in $^{39}K^+$?

Answer 19 protons, 20 neutrons, and 18 electrons

EXAMPLE 2.4 Components of Ions

Fill in the blanks in the following table.

	Complete Symbol	Atomic Number	Mass Number	Charge	Number of Protons	Number of Electrons	Number of Neutrons
a	$^{15}N^{3-}$						
b			24	2+	12		

Strategy Use the information given in the table to determine the numbers of protons, neutrons, and electrons, and use that data to complete the table.

Solution

(a) The symbol given includes the mass number and the charge of the ion. The atomic number subscript has been omitted, but from the periodic table, we know the element nitrogen has an atomic number of 7. The mass number is the superscript 15 before the symbol, and the charge is the superscript 3− that follows the symbol. The number of protons is 7, the atomic number. The number of electrons is calculated from: number of electrons = number of protons (7) − charge (−3) = 10, and the number of neutrons is the mass number (15) − the atomic number (7) = 8.

(b) The presence of 12 protons in this ion means that its atomic number is 12, so the symbol used is Mg. The mass number is 24 and the charge 2+, making the correct symbol $^{24}Mg^{2+}$. The number of electrons is number of protons (12) − charge (2) = 10, and the number of neutrons is the mass number − the atomic number $(A - Z) = 12$. The completed table is as follows:

	Complete Symbol	Atomic Number	Mass Number	Charge	Number of Protons	Number of Electrons	Number of Neutrons
a	$^{15}N^{3-}$	7	15	3−	7	10	8
b	$^{24}Mg^{2+}$	12	24	2+	12	10	12

Understanding

Write the symbol for the ion that has a mass number of 79, an atomic number of 34, and contains 36 electrons.

Answer $^{79}Se^{2-}$

OBJECTIVES REVIEW *Can you:*

☑ define isotopes of atoms and list the subatomic particles in their nuclei?

☑ write complete symbols for ions, given the number of protons, neutrons, and electrons that are present?

2.4 Atomic Masses

OBJECTIVES

☐ Define the atomic mass unit

☐ Determine the atomic mass of an element from isotopic masses and their natural abundances

PRACTICE OF CHEMISTRY
Isotopes of Hydrogen

Although many elements occur as mixtures of isotopes, hydrogen, as the lightest element, is the only element with isotopes that differ in mass by factors of 2 and 3. This fact has significant effects on the physical properties of the compounds formed by the isotopes. For example, as shown in the table, D_2O has a melting point almost 4 °C higher than H_2O. The isotope with a mass number of 1 is by far the most abundant. The isotope with a mass number of 2 is called *deuterium;* 1 atom of deuterium is present for every 7000 hydrogen atoms. The isotope with a mass number of 3 is called *tritium;* 1 atom of tritium is present for every 10^{18} hydrogen atoms. Tritium is unstable (radioactive, which is discussed in Chapter 21), whereas the other two isotopes are stable. Tritium has a wide variety of applications, ranging from being part of the triggering devices in nuclear weapons to being a component in various self-luminescent devices, such as exit signs in buildings, aircraft dials and gauges, luminous paints, and wristwatches. However, because tritium occurs naturally in such minute amounts, it also must be produced artificially to meet scientists' demand. Tritium can be created by a variety of nuclear reactions in nuclear power plants or specially designed nuclear reactors.

Deuterium can be obtained from natural sources. The usual source is one of its compounds, deuterium oxide, D_2O, also known as *heavy water.* The following table compares the properties of heavy water with those of H_2O. Because of the difference in their boiling points, H_2O and D_2O can be separated by distillation. Pure deuterium, D_2, can then be obtained by electrolysis (a process in which electricity is used to split water into oxygen and hydrogen) from the heavy water.

Properties of H_2O and D_2O

	H_2O	D_2O
Melting point	0.0 °C	3.8 °C
Boiling point	100.0 °C	101.4 °C
Density at 4 °C	1.000 g/cm^3	1.108 g/cm^3

Deuterium and tritium have become valuable tools for studying the reactions of compounds that contain hydrogen. Chemists can "label" a compound by replacing one or more of its ordinary hydrogen atoms with deuterium or tritium atoms. The resulting compound is chemically nearly identical to the original compound. As the compound reacts, the path taken by the heavier isotopes can be monitored: deuterium by mass spectroscopic analysis, and tritium by counting its radioactive decay. (Isotopes used in this manner are also called *tracers.*) Scientists use this technique to study many important reactions, including digestion and body metabolism. It may sound a little scary that a scientist would inject a radioactive substance (one used in nuclear weapons, no less!) in a person to study certain bodily functions. However, tritium is one of the *least* dangerous radioactive substances known to humans and in low enough concentrations has little to no effect on the human body. ▮

Chemists need to know the masses of the atoms in elements and compounds to obtain a quantitative understanding of chemical reactions. The mass number of an isotope tells us the number of protons plus neutrons, and we know that nearly all the mass of an atom comes from these particles, but the absolute mass of these particles is small. In addition, many elements occur as mixtures of isotopes that have different masses. This section introduces the units used for measuring the masses of atoms and describes the experiments used to measure atomic masses.

Atomic Mass Unit

Long before scientists had the ability to measure the masses of individual atoms, they established a relative scale to compare the mass of an atom of one element with that of another. A mass of 1 was assigned to the lightest element, hydrogen, and the masses of all other elements were relative to it. Several different atomic mass scales have been used since early in the 1800s, each based on assigning a mass to an atom of one isotope or element and comparing the masses of all other atoms with it. Today, all scientists have agreed to use a single scale, in which the **atomic mass unit (u)**[1] is exactly $\frac{1}{12}$ the mass of

[1]Many other general chemistry textbooks use the abbreviation amu for the atomic mass unit. We have chosen to use the International Union of Pure and Applied Chemistry (IUPAC) recommended abbreviation of u throughout this text.

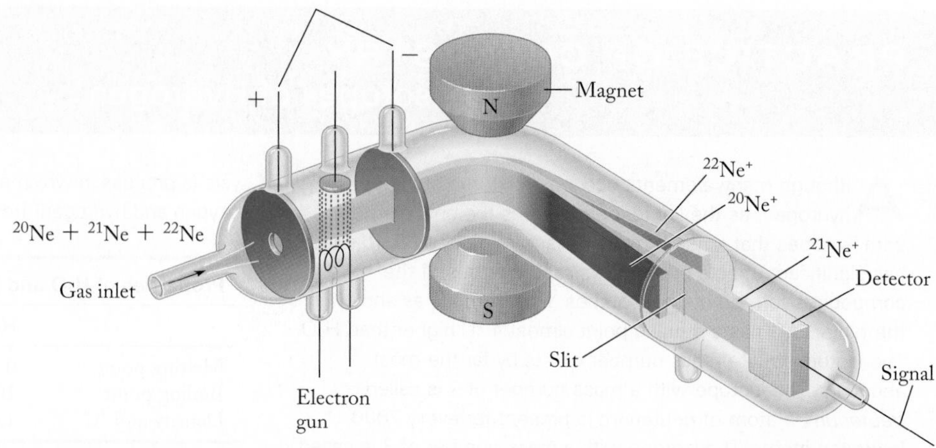

Figure 2.6 Diagram of a mass spectrometer. Ions are formed from a gaseous sample (neon in this case) by bombardment with high energy electrons and accelerated by an electric field. The amount of deflection of the ions by the magnetic field depends on the mass-to-charge ratios, which are different for each isotope, with the heavier isotopes deflected less. Changing the electric or magnetic field allows all ions to hit the slit and be detected.

a ^{12}C atom. The mass of an atom of ^{12}C is defined as exactly 12 u, and all other atoms are compared with this standard. The atomic mass unit has been measured.

$$1 \text{ u} = 1.66054 \times 10^{-27} \text{ kg}$$

The choice of carbon for the standard is somewhat arbitrary, but notice that 12 u is numerically equal to the mass number of ^{12}C. Thus, the masses of both the proton and the neutron are about 1 u. On this scale, ^{24}Mg has an atomic mass of 24 u; that is, one atom of ^{24}Mg has about twice the mass of one atom of ^{12}C. ^{4}He has an atomic mass of 4 u, so three ^{4}He atoms have approximately the same mass as one ^{12}C atom. Rounded to whole numbers, these values are the same as the mass numbers of the atoms, but when expressed more precisely they differ slightly from whole numbers (^{24}Mg = 23.98504 u, ^{4}He = 4.00260 u) because other factors in addition to the masses of the subatomic particles determine the mass of an atom (see Chapter 21).

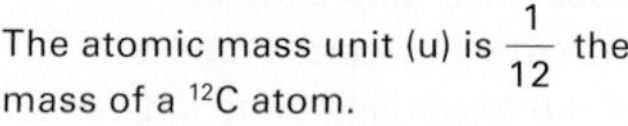

The atomic mass unit (u) is $\frac{1}{12}$ the mass of a ^{12}C atom.

The Mass Spectrometer

In the 19th and early 20th centuries, scientists determined the atomic masses of the elements by careful analysis of the mass compositions of compounds with known formulas. Today, the atomic masses of all the elements have been determined experimentally with mass spectrometers.

The accurate isotopic masses of stable atoms that are available today have all been measured with a mass spectrometer.

A mass spectrometer measures the masses and relative abundances of the isotopes present in a sample of an element. Figure 2.6 shows one type of mass spectrometer. A curved tube is evacuated with a vacuum pump, and a sample of the element is then introduced as a gas into one end. The gas is exposed to a beam of high-energy electrons that convert the atoms of the element to cations. The high voltage between the plates accelerates these cations through a slit so that they travel down the tube.

The curved section of the tube has a magnetic field perpendicular to the direction of the ions. This magnetic field deflects the ions into a curved path. The degree of curvature of the path depends on the mass and charge of the ion, the magnitude of the accelerating voltage, and the magnetic field strength. The mass-to-charge ratio of the ions that reach the detector can be determined from the known voltage and magnetic field strength, which is varied to bring each set of ions to the detector. Figure 2.7 shows the output for a sample of neon. The position of the peaks indicates the mass of each isotope, and the strength of the signal (represented by signal height in the drawing) gives the relative abundance. For neon, the major isotope is ^{20}Ne (~90%), and minor isotopes are ^{21}Ne (~0.3%) and ^{22}Ne (~10%).

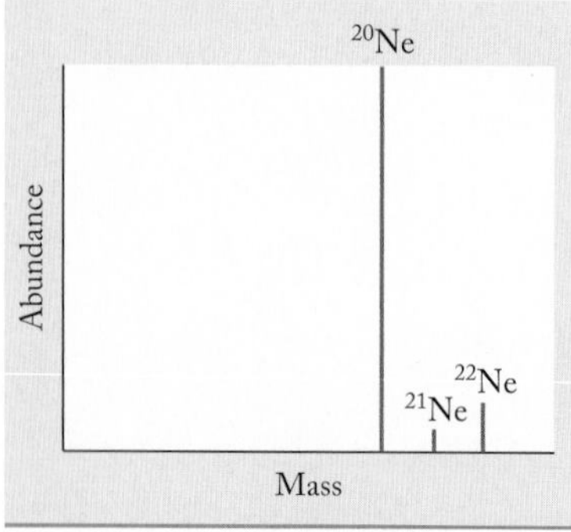

Figure 2.7 Mass spectrum of neon. The mass spectrum of neon shows three isotopes. The most abundant is ^{20}Ne, but some ^{21}Ne (not to scale) and ^{22}Ne atoms are also observed.

Isotopic Distributions and Atomic Mass

Most elements have several isotopes, but the isotopic compositions of most naturally occurring elements are generally constant and independent of the origin of the sample. Because the isotopes of a given element have different numbers of neutrons, they have different masses, referred to as **isotopic masses.** For example, naturally occurring lithium is a mixture of two isotopes: 7.42% of the atoms are ^{6}Li (isotopic mass = 6.015 u) and 92.58% are ^{7}Li (isotopic mass = 7.016 u). If natural lithium were 50% ^{6}Li and

50% ^{7}Li, then the average mass would be about 6.5 u, but natural lithium is mainly ^{7}Li, so the average is much closer to 7 u. A weighted average mass can be calculated that takes into account the natural abundance of each isotope. In this calculation, 7.42% and 92.58% are expressed as the decimal fractions 0.0742 and 0.9258, respectively.

$$\text{Average mass of Li} = 0.0742 \times 6.015\ \text{u} + 0.9258 \times 7.016\ \text{u}$$
$$= 6.941\ \text{u}$$

This result is the **atomic mass,** which is the *weighted average mass,* in atomic mass units, of the naturally occurring element. The term *atomic weight* has frequently been used to refer to this average mass, but we will use the technically correct term *atomic mass.* Remember that atomic mass is an *average* that reflects the natural isotopic distribution of the element, and only in the case of an element that has only one naturally occurring isotope do any of the atoms have this mass.

The atomic mass of an element is the average mass of the atoms in a natural sample of the element.

EXAMPLE 2.5 Calculating the Atomic Mass

The green shading indicates data that is given with the problem, the yellow indicates intermediate results, and the red is the final answer.

Chlorine has two stable isotopes: ^{35}Cl, with a natural abundance of 75.77% and a mass of 34.97 u, and ^{37}Cl, with a natural abundance of 24.23% and an atomic mass of 36.97 u. Calculate the atomic mass of chlorine.

Strategy Atomic mass is a weighted average of each isotope of the element.

Solution

Calculate the weighted average of the two isotopes:

$$\text{Atomic mass Cl} = 0.7577 \times 34.97\ \text{u} + 0.2423 \times 36.97\ \text{u}$$
$$= 35.45\ \text{u}$$

Understanding

Boron has two stable isotopes: ^{10}B, with a natural abundance of 19.9% and an atomic mass of 10.01 u, and ^{11}B, with a natural abundance of 80.1% and an atomic mass of 11.01 u. Calculate the atomic mass of boron.

Answer 10.81 u

OBJECTIVES REVIEW *Can you:*

- ☑ define the atomic mass unit?
- ☑ determine the atomic mass of an element from isotopic masses and their natural abundances?

2.5 The Periodic Table

OBJECTIVES

- ☐ Define groups and periods
- ☐ Use the periodic table as a guide to classify elements as metals, nonmetals, or metalloids
- ☐ Classify elements as representative, transition, lanthanide, or actinide
- ☐ Describe the properties of elements in the alkali metal, alkaline earth metal, halogen, and noble-gas groups

By the middle of the 19th century, chemists had isolated many of the elements and begun a systematic investigation of their properties. As the chemical and physical properties of the elements were determined, scientists noted that some elements were quite similar to others. For example, lithium, sodium, and potassium have similar chemical properties. The same is true of the elements chlorine, bromine, and iodine. These two groups of elements, however, have different properties. The grouping and classifying of elements is the first step toward understanding the properties of those elements. The most widely used classification scheme is the periodic table of the elements.

Working independently, in the 1860s, Russian chemist Dimitri Mendeleev (1834–1907) and German physicist Julius Lothar Meyer (1830–1895) proposed to arrange the elements in a table, the modern version of which is shown in Figure 2.8 and on the inside cover of this textbook. This **periodic table** arranges elements (represented by their symbols) into rows and places elements with similar chemical properties in the same columns. The lighter elements are at the top of the column, and the heavier elements are at the bottom.

In his original table, Mendeleev arranged elements by increasing atomic mass, but this order placed several elements in locations that did not fit for their observed chemical properties. He changed their orders so that the chemical properties of the elements took precedence, not the atomic mass. Mendeleev also could not fill all the spaces with known elements. He inferred the existence of several elements that would occupy these spaces and predicted their properties. The value of this classification system was demonstrated when one of the missing elements, gallium, was discovered 4 years later. Gallium had properties close to those that Mendeleev predicted.

In 1914, English physicist Henry Gwyn-Jeffries Moseley sought to reconcile the discrepancies in Mendeleev's table. Moseley ordered the elements based on their atomic number, not their atomic mass, and thus the modern version of the periodic table was born. Each horizontal row of the table is called a **period** and is numbered. The properties of the elements change regularly across a period. The elements in each column, called a **group,** have similar properties. The groups are numbered across the top. In the traditional numbering method (in North America), each group is labeled with a combination of a number and the letter A or B. Recently, another scheme, also shown in Figure 2.8, has been adopted. In this newer method, the groups are numbered 1 through 18. We generally use the older labeling scheme in this book.

The elements in a column make up a group and have similar chemical properties.

One important way to classify elements is to divide them into metals, nonmetals, and metalloids. A **metal** is a material that is shiny and is a good electrical conductor. Most of the elements, those in the center and on the left side of the table, are metals. **Nonmetals,** elements that typically do not conduct an electrical current, include the elements in the top right part of the table. Periodic tables such as the one in Figure 2.8 have a line dividing the metals from the nonmetals. The elements along the line have some properties of both metals and nonmetals and are called **metalloids.** A particularly

Figure 2.8 Periodic table of the elements. Metals are shown in blue, metalloids in green, and nonmetals in yellow.

	1 1A	2 2A	3 3B	4 4B	5 5B	6 6B	7 7B	8	9 8B	10	11 1B	12 2B	13 3A	14 4A	15 5A	16 6A	17 7A	18 8A
1	1 H																	2 He
2	3 Li	4 Be											5 B	6 C	7 N	8 O	9 F	10 Ne
3	11 Na	12 Mg											13 Al	14 Si	15 P	16 S	17 Cl	18 Ar
4	19 K	20 Ca	21 Sc	22 Ti	23 V	24 Cr	25 Mn	26 Fe	27 Co	28 Ni	29 Cu	30 Zn	31 Ga	32 Ge	33 As	34 Se	35 Br	36 Kr
5	37 Rb	38 Sr	39 Y	40 Zr	41 Nb	42 Mo	43 Tc	44 Ru	45 Rh	46 Pd	47 Ag	48 Cd	49 In	50 Sn	51 Sb	52 Te	53 I	54 Xe
6	55 Cs	56 Ba	57 La*	72 Hf	73 Ta	74 W	75 Re	76 Os	77 Ir	78 Pt	79 Au	80 Hg	81 Tl	82 Pb	83 Bi	84 Po	85 At	86 Rn
7	87 Fr	88 Ra	89 Ac†	104 Rf	105 Db	106 Sg	107 Bh	108 Hs	109 Mt	110 Ds	111 Rg							

*Lanthanides	58 Ce	59 Pr	60 Nd	61 Pm	62 Sm	63 Eu	64 Gd	65 Tb	66 Dy	67 Ho	68 Er	69 Tm	70 Yb	71 Lu
†Actinides	90 Th	91 Pa	92 U	93 Np	94 Pu	95 Am	96 Cm	97 Bk	98 Cf	99 Es	100 Fm	101 Md	102 No	103 Lr

interesting feature of metalloids is that they are **semiconductors,** that is, weak conductors of electricity. This property makes them extremely useful in solid-state electronics such as MP3 players and cell phones. Hydrogen, the lightest element, is generally listed in Group 1A, but it is a nonmetal and is sometimes also shown in Group 7A.

The elements are divided into metals, metalloids, and nonmetals.

The elements in groups labeled A are historically called the **representative elements** or the **main-group elements.** The metals in the center part of the table, the B groups, are called the **transition metals.** Two series of heavier elements are set off at the bottom of the table to save space. The **lanthanides** (cerium [Ce] through lutetium [Lu]) are the elements that follow lanthanum (La) in period 6. The **actinides** (thorium [Th] through lawrencium [Lr]) follow actinium (Ac) in period 7. These two series of elements are also known as the **inner transition metals.** Most of the actinide elements do not occur in nature but have been made in laboratories via nuclear reactions, a subject that is discussed in Chapter 21.

The transition metals are located in the center part of the periodic table, labeled as B groups.

Important Groups of Elements

Several important groups of elements have specific names and characteristic properties (Figure 2.9).

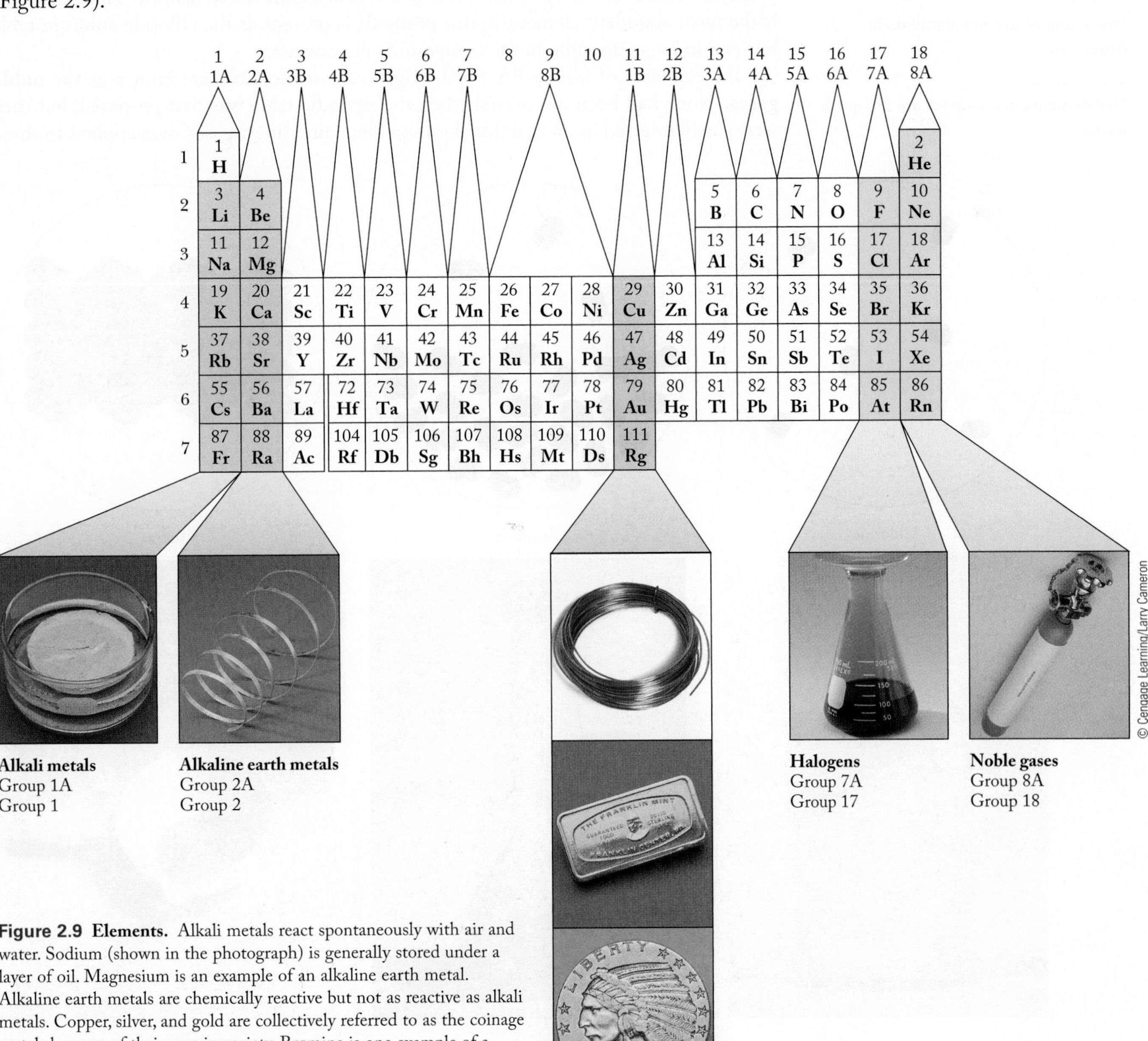

Figure 2.9 Elements. Alkali metals react spontaneously with air and water. Sodium (shown in the photograph) is generally stored under a layer of oil. Magnesium is an example of an alkaline earth metal. Alkaline earth metals are chemically reactive but not as reactive as alkali metals. Copper, silver, and gold are collectively referred to as the coinage metals because of their uses in society. Bromine is one example of a halogen and is a liquid at room temperatures; other halogens are gases (fluorine and chlorine) or solid (iodine). The noble gases exhibit little chemical reactivity.

The elements in Group 1A are known as the **alkali metals.** They are soft, low-melting metals that are quite reactive, with their reactivity increasing down the group. Their high reactivity toward water and many other substances requires that they be handled with extreme care in the laboratory. Figure 2.10 shows a picture of the reaction that occurs when sodium contacts water. The elements sodium (Na) and potassium (K) are abundant in the earth's crust and occur in compounds such as sodium chloride rather than as free elements.

The elements in Group 1A are the alkali metals.

Elements in Group 2A are known as the **alkaline earth metals.** They are less reactive than the alkali metals. Magnesium (Mg) and calcium (Ca) are also abundant in the earth's crust, and calcium is an important constituent of bones, seashells, and coral reefs.

The elements in Group 2A are the alkaline earth metals.

Group 7A elements are called the **halogens,** a word that means "salt-formers." The halogens are among the most reactive of the nonmetals, with their reactivity decreasing down the column. Fluorine is the most reactive of all the elements, readily forming compounds with most other elements. At room temperature and pressure, fluorine is a yellow gas, chlorine is a greenish yellow gas, bromine is a red liquid (some is in the gas phase, as can be seen in the picture), and iodine is a dark violet, lustrous solid. Chlorine is the most abundant element in this group. It is present as the chloride anion in table salt and in large amounts in the compounds in seawater.

The halogens are the elements in Group 7A.

The elements of Group 8A, on the right side of the table, are known as the **noble gases.** None had been discovered when the periodic table was first proposed, but they were easily inserted as an additional group. The name "inert gases" was applied to these

The elements in Group 8A are the noble gases.

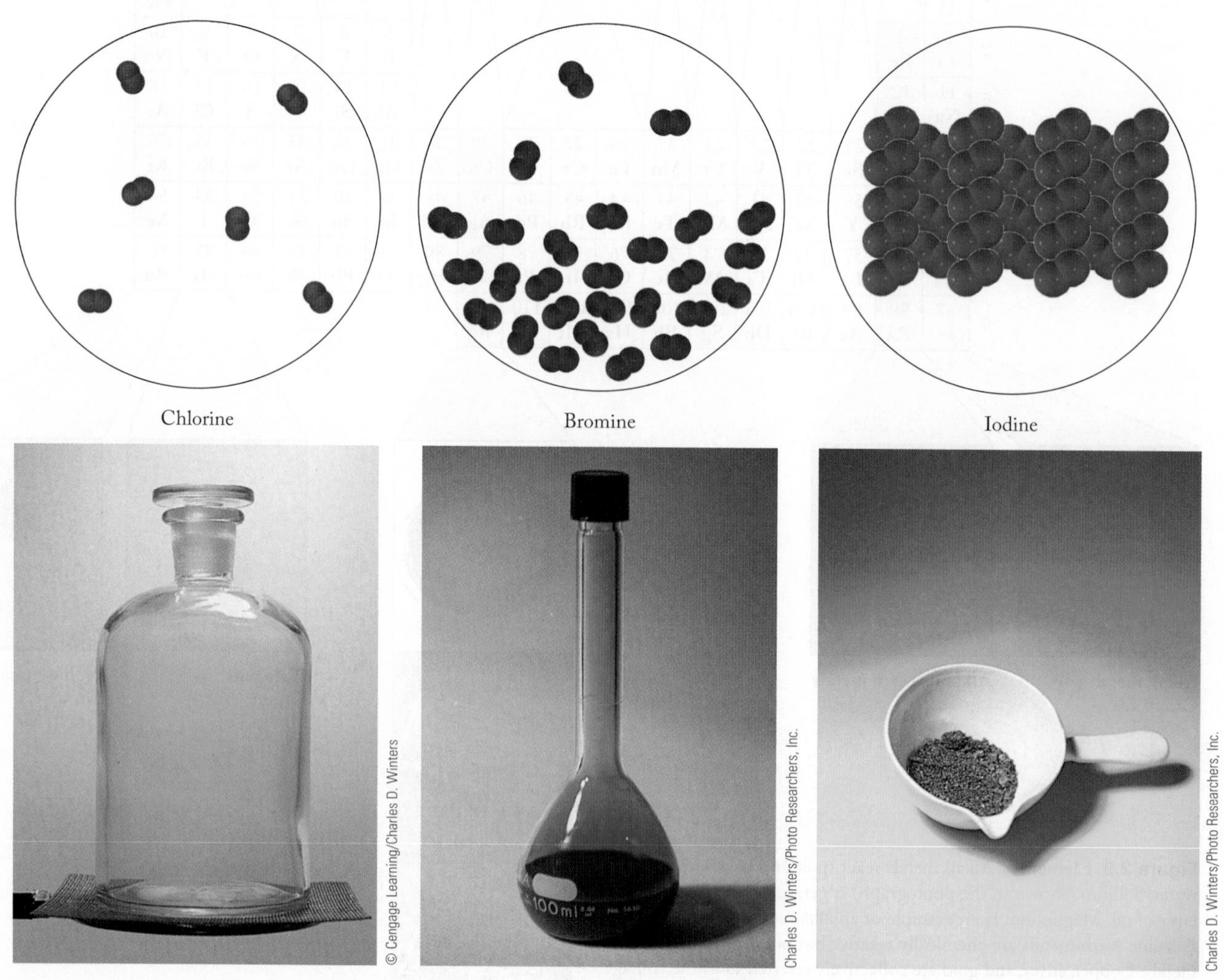

The halogens. At room temperature, chlorine is as gas, bromine is a liquid, and iodine is a solid.

elements because they are all gases at room temperature and, before 1962, were thought to be completely nonreactive. Now, several compounds of xenon (Xe), krypton (Kr), radon (Rn), and even a low-temperature-stable compound of argon (Ar) have been made, so the word *inert* has been replaced by *noble*.

Charles D. Winters/Photo Researchers, Inc.

Figure 2.10 Mixing sodium with water. Sodium metal reacts violently with water. A flammable gas (hydrogen) is produced.

EXAMPLE 2.6 The Periodic Table

Identify an element that fits the following criteria:

(a) What element in the fourth period is an alkaline earth metal?
(b) What element in the second period is a halogen?

Strategy Use the periodic table to locate the element using the group name to find the column and the period number to locate the row.

Solution

(a) The fourth period starts with potassium, K, and ends with krypton, Kr. The alkaline earth metal in the fourth period is calcium, Ca.
(b) The second period goes from lithium, Li, to neon, Ne. The halogen in that period is fluorine, F. (Don't forget that the first period of the periodic table contains only two elements, hydrogen and helium.)

Understanding

Identify the alkali metal in the fifth period.

Answer Rubidium, Rb

OBJECTIVES REVIEW *Can you:*

- ☑ define groups and periods?
- ☑ use the periodic table as a guide to classify elements as metals, nonmetals, or metalloids?
- ☑ classify elements as representative, transition, lanthanide, or actinide?
- ☑ describe the properties of elements in the alkali metal, alkaline earth metal, halogen, and noble-gas groups?

2.6 Molecules and Molecular Masses

OBJECTIVES

- ☐ Interpret the molecular formula of a substance
- ☐ Determine molecular mass from the formula of a compound

Atoms are the basic building blocks in all matter, but the millions of known substances are the results of combinations of atoms. This section presents how chemists represent these substances using formulas.

Molecules

In many pure substances, both elements and compounds, the atoms are grouped into small clusters called *molecules.* A **molecule** is a combination of atoms joined so strongly that they behave as a single particle. Molecules, like atoms, are electrically neutral. If all of the atoms in the molecule are the same, the substance is an element. If atoms of two or more elements form a molecule, the substance is a **molecular compound.** The simplest molecules are **diatomic**—that is, composed of two atoms. The stable forms of the elements hydrogen, oxygen, nitrogen, and the halogens are diatomic molecules. Many compounds (e.g., carbon monoxide and hydrogen chloride) also exist as diatomic molecules. Most molecules are more complicated. For example, although hydrogen and

Diatomic Elements

Element	Formula
Hydrogen	H_2
Oxygen	O_2
Nitrogen	N_2
Fluorine	F_2
Chlorine	Cl_2
Bromine	Br_2
Iodine	I_2

A molecular formula gives the symbol and number of atoms of each element present in one molecule of the substance.

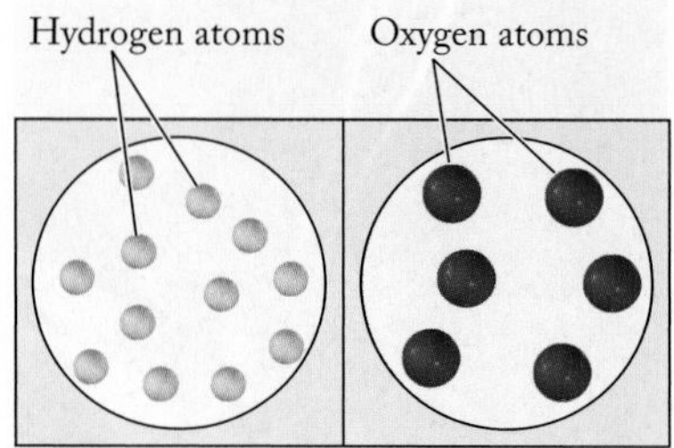

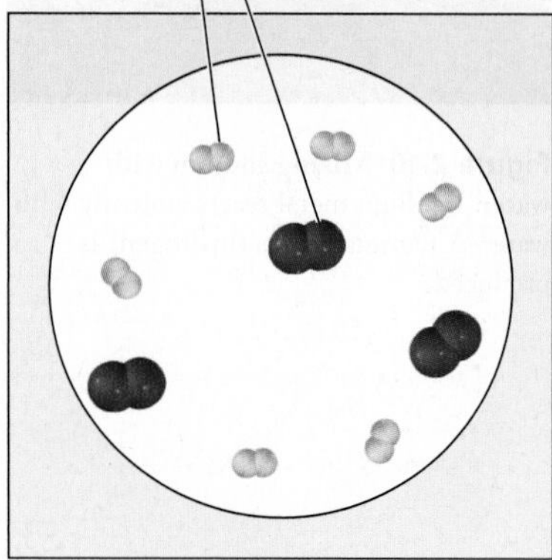

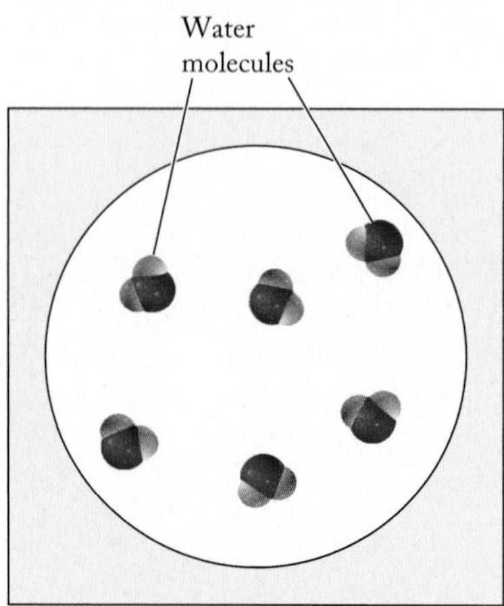

Figure 2.11 Atoms and molecules of hydrogen and oxygen, and molecules of water.

oxygen both exist as diatomic molecules, one atom of oxygen combines with two atoms of hydrogen to form a molecule of water (Figure 2.11). *Molecular compounds typically consist of a combination of nonmetallic elements.*

A **molecular formula** is an abbreviated description of the composition of a molecule and gives the number of every type of atom in a molecule. In the formula, chemical symbols identify the elements present, with each symbol followed by a numerical *subscript* indicating the number of atoms of that element that occur in the molecule. The absence of a subscript means that one atom of that element is present. Molecular hydrogen is written as H_2, and water is written as H_2O. In subsequent chapters, you will learn how to use experimental data to determine the formula of a compound.

Figure 2.12 shows different ways of writing the formulas of substances. The first line shows the molecular formula. The second line shows the **structural formula,** which indicates how the atoms are connected (indicated by lines between the atom symbols) in the molecule. Figure 2.12 also shows models that help us visualize the shapes of molecules.

EXAMPLE 2.7 Writing Molecular Formulas

Write the molecular formulas of the substances described or pictured.

(a) The element nitrogen exists as diatomic molecules.
(b) A sulfur dioxide molecule contains one sulfur and two oxygen atoms (symbol for sulfur is written first).
(c)

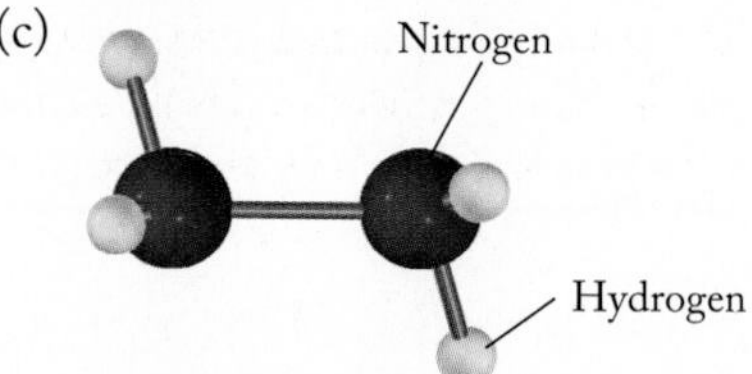

Strategy Write the symbol for each element in the molecule with a subscript after the symbol indicating the number of atoms of that type present.

Solution

(a) The symbol for nitrogen is followed by a subscript 2: N_2.
(b) When only one atom of an element is present in a substance, no subscript is given, so the formula of sulfur dioxide is SO_2.

Molecular formula	Cl_2	CH_4	C_2H_6
Structural formula	Cl—Cl	H—C—H (with H above and below C)	H—C—C—H (with H above and below each C)
Ball-and-stick model of molecule			
Space-filling model of molecule			

Figure 2.12 Molecular and structural formulas.

(c) The picture shows four hydrogen atoms and two nitrogen atoms combined into a single molecule. The formula is N_2H_4.

Understanding

What is the formula of the molecule pictured below?

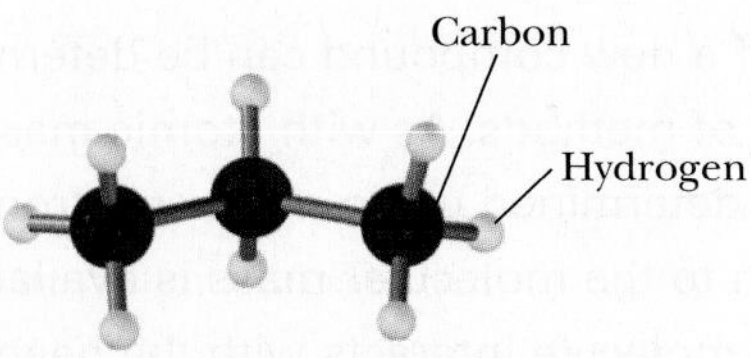

Answer C_3H_8

Molecular Mass

Because a molecule is a combination of atoms, the mass of a molecule is the sum of the masses of the atoms present. The **molecular mass** is the sum of the atomic masses of all atoms present in the molecular formula, expressed in atomic mass units (u). We can calculate the molecular mass of CO_2 from the atomic masses given on the periodic table, taking into account the subscripts in the molecular formula. The formula indicates that one molecule of CO_2 contains one atom of carbon and two atoms of oxygen. Look up the atomic masses of carbon and oxygen on the periodic table and then take the appropriate multiple.

1(C) $1 \times 12.01\ \text{u} = 12.01\ \text{u}$

2(O) $2 \times 16.00\ \text{u} = 32.00\ \text{u}$

Molecular mass $CO_2 = 44.01\ \text{u}$

The number of significant figures to use in this type of calculation is sometimes arbitrary. If you are asked for the molecular mass of a compound, use either one or two digits after the decimal point, depending on the desired precision. The molecular mass of carbon dioxide can be properly used as either 44.0 or 44.01 u depending on the number of significant figures needed.

The molecular mass is the sum of the atomic masses of all atoms present in the compound.

EXAMPLE 2.8 Calculating Molecular Mass

Hydrazine, N_2H_4, is a fuel that has been used as a rocket propellant. What is the molecular mass of hydrazine?

Strategy Calculate the molecular mass using the atomic masses given on the periodic table, taking into account the subscripts in the molecular formula. The flow diagram is:

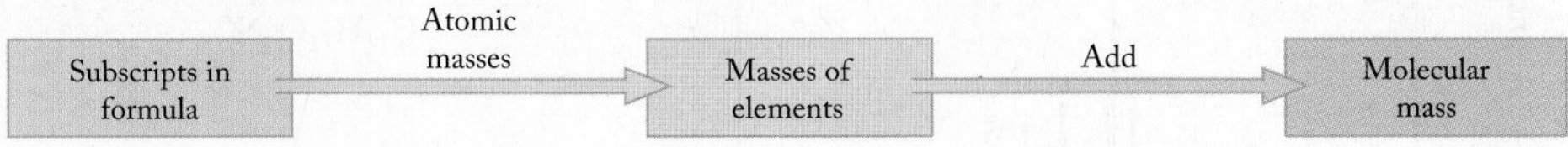

Solution

Work the problem systematically, using the data in the periodic table.

2(N) $2 \times 14.01\ \text{u} = 28.02\ \text{u}$

4(H) $4 \times 1.01\ \text{u} = 4.04\ \text{u}$

Molecular mass $N_2H_4 = 32.06\ \text{u}$

Understanding

What is the molecular mass of carbon tetrachloride, CCl_4?

Answer 153.81 u

The molecular mass of a new compound can be determined experimentally by a number of methods. As with atomic masses, most molecular masses are determined using mass spectrometers, but information in addition to the molecular mass is available in the experiment. When the molecule interacts with the beam of high-energy electrons inside the mass spectrometer, a number of processes occur. The molecular ion (molecule with a 1+ charge) forms, but in addition, the molecular ion fragments and the new ions of lower molecular mass that are produced are also detected by the mass spectrometer. Figure 2.13 shows the mass spectrum of cocaine. Cocaine has the molecular formula $C_{17}H_{21}O_4N$. Its molecular mass is

Molecular model of cocaine.

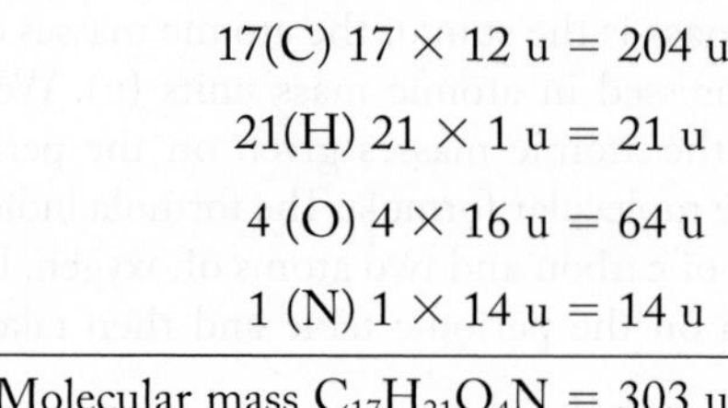
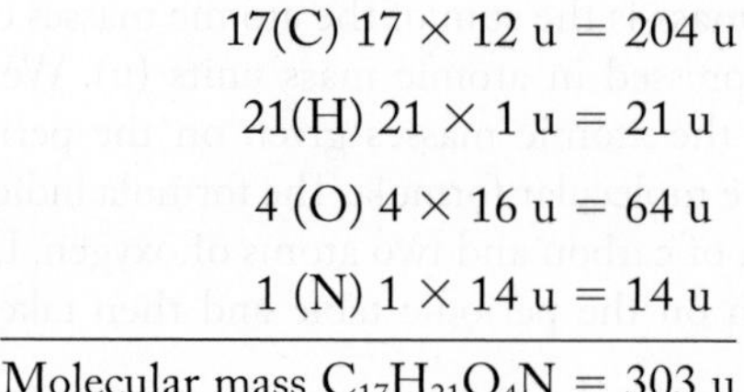

$$17(\text{C})\ 17 \times 12\ \text{u} = 204\ \text{u}$$
$$21(\text{H})\ 21 \times 1\ \text{u} = 21\ \text{u}$$
$$4\ (\text{O})\ 4 \times 16\ \text{u} = 64\ \text{u}$$
$$1\ (\text{N})\ 1 \times 14\ \text{u} = 14\ \text{u}$$
$$\text{Molecular mass } C_{17}H_{21}O_4N = 303\ \text{u}$$

As shown in Figure 2.13, the mass spectrum of cocaine shows a peak for the molecular mass (M^+) at 303, but it shows additional "fingerprint" peaks that represent lighter pieces of the molecule that form in the mass spectrometer. As each individual molecule breaks into lighter pieces in the mass spectrometer in its own characteristic way, the mass spectrum clearly identifies the substance, and is thus important in law enforcement for definitively characterizing illegal substances.

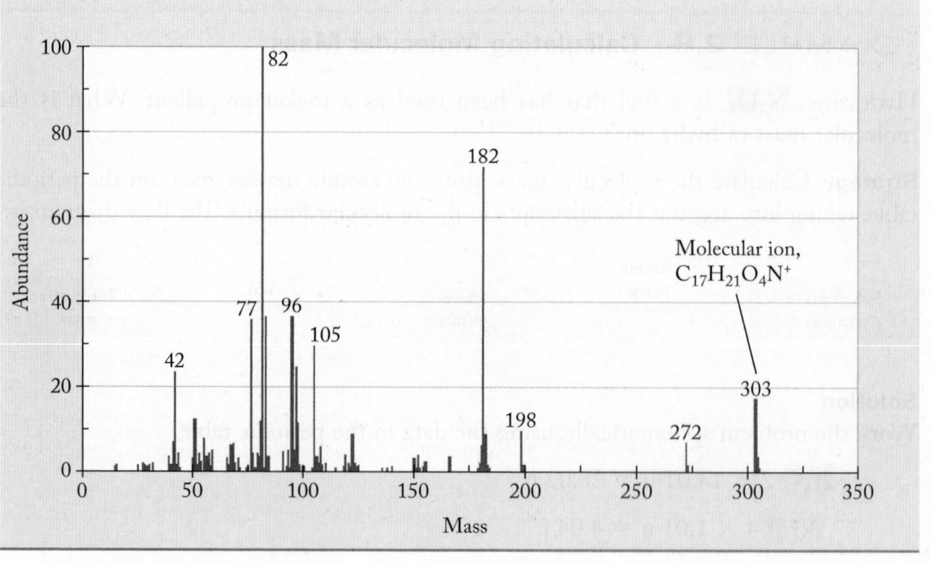

Figure 2.13 Mass spectrum of cocaine. Vertical axis shows the abundance of the ions formed when cocaine interacts with a beam of electrons. The molecular ion is the ion formed from the whole molecule.

OBJECTIVES REVIEW *Can you:*

- ☑ interpret the molecular formula of a substance?
- ☑ determine molecular mass from the formula of a compound?

2.7 Ionic Compounds

OBJECTIVES

- ☐ Define ionic compound, and compare and contrast ionic compounds and molecular compounds
- ☐ Predict ionic charges expected for cations of elements in Groups 1A, 2A, and 3B, and aluminum, and for anions of elements in Groups 6A, 7A, and nitrogen
- ☐ Write formulas for ionic compounds
- ☐ List the names, formulas, and charges of the important polyatomic ions
- ☐ Calculate the formula masses for ionic compounds

Many compounds exist in which the elements are present as *ions,* atoms that have gained or lost electrons. Most compounds that contain a metal and a single nonmetallic element consist of ions. An **ionic compound** is composed of cations and anions joined together. Such compounds are held together by electrostatic forces, and adopt structures that maximize the attraction of oppositely charged species and minimize the repulsion between charged species with the same sign. This section describes ionic compounds and how to write their formulas and distinguish them from molecular compounds, which are combinations of *atoms* held together by forces as outlined in Chapters 9 and 10.

Ionic compounds generally consist of a combination of metals with nonmetals. An example of an ionic compound is sodium chloride. It is made up of equal numbers of sodium cations (Na^+) and chloride anions (Cl^-). Figure 2.14 shows its structure. Each Na^+ ion is surrounded by six Cl^- ions, and in turn, each Cl^- ion is surrounded by six Na^+ ions. This arrangement forms an extended three-dimensional array.

Figure 2.14 **Structure of sodium chloride (NaCl).** Three-dimensional array of ions in solid NaCl.

The formula of sodium chloride is NaCl (it is customary to write the cation first). Each grain of table salt contains a large number of sodium cations and chloride anions, but they are always present in a ratio of 1:1. *All ionic compounds are overall electrically neutral;* the sum of the charges contributed by the cations and anions in the formula of an ionic compound must sum to zero.

Because no one sodium ion is uniquely combined with a single chloride ion, the subscripts in the formulas of ionic compounds have a slightly different meaning from those in the formulas of molecules. A molecular formula gives the actual numbers and types of atoms in a molecule, but the exact number of ions in the three-dimensional array of an ionic compound depends on the size of the sample. This type of chemical formula is an **empirical formula,** one that gives the relative numbers of atoms of each element in a substance with the smallest possible whole-number subscripts. The empirical formula of ionic compounds leads to electrical neutrality.

Formulas of Ionic Compounds

The periodic table helps predict the expected charges on many ions. In general, the metallic elements form cations, and the nonmetallic elements, especially those closest to the right side of the periodic table (excluding the noble gases), form anions. Experiment shows that the metals in Groups 1A, 2A, and 3B form cations with charges equal to their group numbers. Group 1A elements form cations with 1+ charges, Group 2A elements form cations with 2+ charges, and Group 3B metals and aluminum in Group 3A form cations with 3+ charges. Main-group nonmetals form anions, whose charges depend on how far to the right they are in the periodic table. That is, Group 7A elements form anions with 1− charges, Group 6A elements form anions with 2− charges, and nitrogen from Group 5A forms anions with 3− charges. Table 2.1 lists these common ions.

The charges on many monatomic ions can be determined from their group numbers found on the periodic table.

If the charges on the ions in a compound are known, we write the formula of the ionic compound by adjusting the subscripts so that the sum of the charges is zero. For

The empirical formulas of ionic compounds balance the charges of the ions.

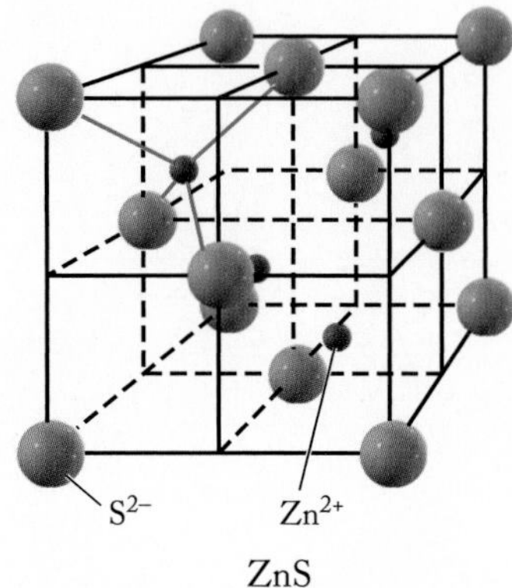

ZnS

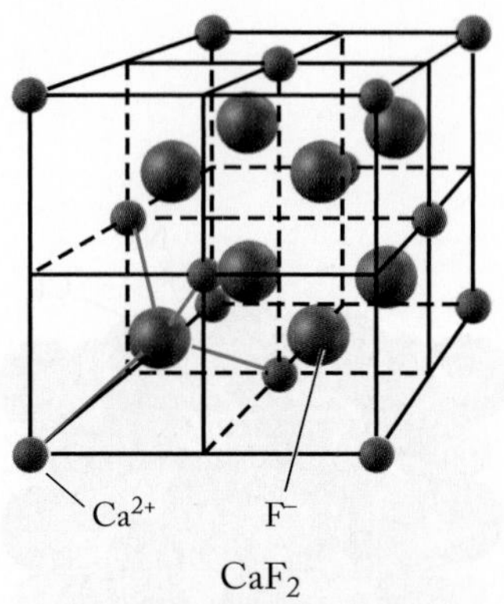

CaF_2

Structures of ionic compounds. Different arrangement of the ions in ionic compounds is possible and depends on the relative sizes and charges of the ions.

TABLE 2.1 **Charges on Common Ions**

1A	2A	3B	3A	5A	6A	7A
Li^+	Be^{2+}			N^{3-}	O^{2-}	F^-
Na^+	Mg^{2+}		Al^{3+}		S^{2-}	Cl^-
K^+	Ca^{2+}	Sc^{3+}			Se^{2-}	Br^-
Rb^+	Sr^{2+}	Y^{3+}			Te^{2-}	I^-
Cs^+	Ba^{2+}	La^{3+}				

cases in which the ions have equal but opposite charges, the subscripts are always 1 because an empirical formula is expressed as the smallest whole-number ratio. The formula of the compound formed by Zn^{2+} and S^{2-} is ZnS (remember that the subscript 1 is not written). An example of a case in which the charges are not the same is the ionic compound formed by Ca^{2+} and F^-. It takes two F^- anions to balance one Ca^{2+} cation, so the formula is CaF_2.

EXAMPLE 2.9 Empirical Formulas of Ionic Compounds

Write the empirical formula of the compound made from

(a) calcium cations and Br anions.
(b) magnesium cations and S anions.
(c) potassium cations and O anions.

Strategy Determine the charge of the species from their group number on the periodic table. Balance the overall charge of the compound with the appropriate subscripts by taking into account the charges on the anion and cation.

Solution

(a) The Ca cation has a 2+ charge because it is in Group 2A. The Br anion has a 1− charge because it is in Group 7A. Two Br^- anions are needed to balance the charge of one Ca^{2+} cation. Thus, the empirical formula is $CaBr_2$.
(b) These two ions have the same number for the charge: The Mg cation has a 2+ charge because it is in Group 2A, and the S anion has a 2− charge because it is in group 6A. The formula MgS balances the charges to zero.
(c) The potassium cation has a 1+ charge because it is in Group 1A, and the O anion has a 2− charge because it is in Group 6A. Two K^+ cations balance the charge of one O^{2-} anion. The formula is K_2O.

Understanding

What is the empirical formula of the compound made from sodium cations and Se anions?

Answer Na_2Se

Polyatomic Ions

So far, we have considered only **monatomic ions,** ions formed from single atoms by the loss or gain of electrons. Ions can also be formed by *groups* of atoms joined together by the same kinds of forces that hold atoms together in molecules. In such ions—e.g., NH_4 (ammonium ion) and OH^- (hydroxide ion)—the total number of protons and electrons in the entire group of atoms are not equal. A **polyatomic ion** is a group of atoms with a net charge that behave as a single particle. In ionic solids, ammonium, NH_4, is the most important polyatomic cation, but there are many important polyatomic anions. Table 2.2 is a short list of polyatomic anions that we will use in the next

A polyatomic ion is a group of atoms with a net charge that behaves as a single particle.

TABLE 2.2 Polyatomic Anions

Name	Formula	Name	Formula
Acetate	$CH_3CO_2^-$ (or CH_3COO^-)	Nitrate	NO_3^-
Carbonate	CO_3^{2-}	Nitrite	NO_2^-
Hydrogen carbonate (bicarbonate)	HCO_3^-	Permanganate	MnO_4^-
Chlorate	ClO_3^-	Phosphate	PO_4^{3-}
Perchlorate	ClO_4^-	Hydrogen phosphate	HPO_4^{2-}
Chromate	CrO_4^{2-}	Dihydrogen phosphate	$H_2PO_4^-$
Cyanide	CN^-	Sulfate	SO_4^{2-}
Dichromate	$Cr_2O_7^{2-}$	Hydrogen sulfate (bisulfate)	HSO_4^-
Hydroxide	OH^-	Sulfite	SO_3^{2-}

several chapters. Familiarize yourself with the formulas, names, and charges of the ions in Table 2.2. Appendix D provides a more extensive list of polyatomic ions.

A polyatomic ion behaves as a single particle because its atoms are held together strongly. The four atoms in a carbonate ion, CO_3^{2-}, behave as a single particle with a 2− charge. The ion does not break up whether in the solid phase, dissolved in solution, or even in the gas phase. The cyanide ion, CN^-, remains intact in many of its reactions, behaving as a single particle with a 1− charge, and has chemistry similar to the monatomic halide ions (Cl^-, Br^-, and I^-). The empirical formula of an ionic compound containing polyatomic ions is also deduced from the ionic charges. Treat each polyatomic ion as an inseparable group of atoms with the total charge given in Table 2.2, and write the empirical formula that yields a neutral compound. For a compound in which the subscript of the polyatomic ion is greater than 1, place parentheses around the entire polyatomic group to show that it acts as a single particle. For example, the formula of the compound containing ammonium and carbonate ions is written as $(NH_4)_2CO_3$.

EXAMPLE 2.10 Empirical Formulas of Ionic Compounds

Write the formula for the compound made up of

(a) barium cations and nitrate anions.
(b) sodium cations and hydroxide anions.
(c) potassium cations and dichromate anions.

Strategy Treat the polyatomic anions as a single-charged group and balance the overall charge of the compound with the appropriate subscripts by taking into account the charges on the anion and cation.

Solution

(a) Barium is in Group 2A and has a 2+ charge and nitrate has the formula NO_3^-. Two NO_3^- anions are needed to balance the charge of one Ba^{2+} cation. The formula is $Ba(NO_3)_2$. Note that the parentheses around the NO_3^- group mean that

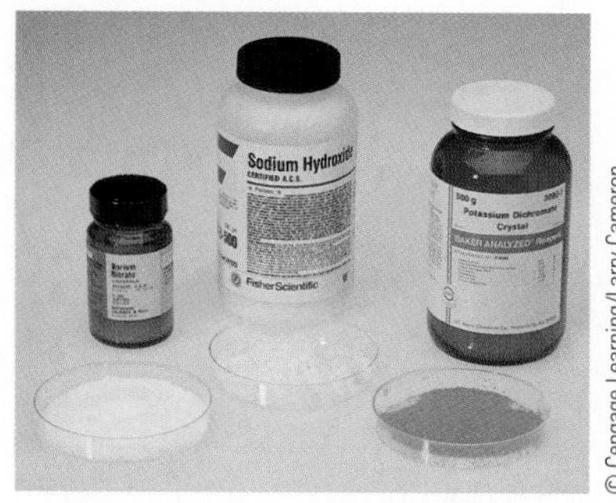

© Cengage Learning/Larry Cameron

(From left to right) Barium nitrate, sodium hydroxide, and potassium dichromate.

there are two complete NO_3^- groups for every Ba^{2+}. It is incorrect to write the formula as BaN_2O_6 because the NO_3^- polyatomic anion is a single group that does not change its formula or charge.

(b) The sodium cation has a charge of 1+ because it is in Group 1A, and the hydroxide anion is OH^-. The formula is NaOH.

(c) The potassium cation has a 1+ charge because it is in Group 1A, and the dichromate anion is $Cr_2O_7^{2-}$. The formula is $K_2Cr_2O_7$.

Understanding

Write the formula for the compound made from ammonium cations and sulfate anions.

Answer $(NH_4)_2SO_4$

Formula Masses of Ionic Compounds

The formula mass (in u) is the sum of the atomic masses of all atom types in the empirical formula of an ionic compound.

A quantity analogous to the molecular mass, called the **formula mass,** is the sum of the atomic masses of all the atoms in the empirical formula of an ionic compound. The term *molecular mass* should not be applied to ionic compounds. The formula of an ionic substance gives only the *relative* numbers of cations and anions. Example 2.11 illustrates the calculation of the formula mass for an ionic compound.

EXAMPLE 2.11 Formula Mass

Calculate the formula mass of the ionic compound barium nitrate, $Ba(NO_3)_2$.

Strategy The formula mass is calculated the same way as the molecular mass using the atomic masses given on the periodic table, taking into account the subscripts in the empirical formula.

Solution

The formula of barium nitrate indicates that two nitrate ions are present for each barium ion. Thus, a single formula unit of this compound contains one barium, two nitrogen, and six oxygen atoms. The calculation of the formula mass is

$$1(\text{Ba})\ 1 \times 137.33\text{ u} = 137.33\text{ u}$$
$$2(\text{N})\ 2 \times 14.01\text{ u} = 28.02\text{ u}$$
$$6(\text{O})\ 6 \times 16.00\text{ u} = 96.00\text{ u}$$
$$\text{Formula mass Ba(NO}_3)_2 = 261.35\text{ u}$$

Understanding

Calculate the formula mass of Na_2O.

Answer 61.98 u

OBJECTIVES REVIEW *Can you:*

☑ define ionic compound, and compare and contrast ionic compounds and molecular compounds?
☑ predict ionic charges expected for cations of elements in Groups 1A, 2A, and 3B, and aluminum, and for anions of elements in Groups 6A, 7A, and nitrogen?
☑ write formulas for ionic compounds?
☑ list the names, formulas, and charges of the important polyatomic ions?
☑ calculate the formula masses for ionic compounds?

2.8 Chemical Nomenclature

OBJECTIVES

- ☐ Name simple ionic and transition-metal compounds, and acids
- ☐ Name simple molecular compounds
- ☐ Name simple organic compounds

Chemists use names and formulas to identify compounds. In the early development of chemistry, many different methods were used to name substances. Scientists have isolated millions of different compounds and are preparing more daily. A unique name describes each compound. **Chemical nomenclature** is the organized system for the naming of substances. This section outlines the methods used to name ionic compounds, acids, and some simple molecular compounds.

TABLE 2.3 Common Monatomic Anions

Anion	Name	Anion	Name
H^-	Hydride	F^-	Fluoride
N^{3-}	Nitride	Cl^-	Chloride
O^{2-}	Oxide	Br^-	Bromide
S^{2-}	Sulfide	I^-	Iodide

Ionic Compounds

Chemists name ionic compounds composed of monatomic ions by using the name of the element that is present as the cation (generally a metal), followed by the name of the anion; the latter consists of the first part of the name (the root) of the element (generally a nonmetal) with the suffix *-ide* added. Table 2.3 gives the names of several important monatomic anions.

Binary compounds, compounds composed of only two elements, are easy to name. Table salt, NaCl, is sodium chloride. Magnesium bromide is the name for $MgBr_2$. Note that the numbers of ions in the empirical formula are inferred from the known charges of the ions; numerical prefixes are not included in the name.

The name of a binary ionic compound consists of the cation name first, followed by the root of the name of the element in the anion, with an *-ide* ending.

EXAMPLE 2.12 Naming Binary Ionic Compounds

(a) Name the compounds.
1. BaI_2 2. MgO

(b) Write the formula of the compounds
1. Sodium sulfide 2. Potassium fluoride

Strategy For part a, write the name of the cation followed by the name of the anion as it appears in Table 2.3; then reverse that strategy for part b, being careful to balance the charges to zero with appropriate subscripts.

Solution

(a) 1. Barium iodide 2. Magnesium oxide

(b) 1. Na_2S 2. KF

Understanding

Name the compound $CaCl_2$ and write the formula of lithium oxide.

Answer Calcium chloride, Li_2O

Table 2.2 contains the names of polyatomic ions. These names are used directly in the compound name, and the formula is determined by the overall charge on the polyatomic ion. Ammonium sulfide is the name for $(NH_4)_2S$. The formula of magnesium nitrate is $Mg(NO_3)_2$.

EXAMPLE 2.13 Formulas of Ionic Compounds That Contain Polyatomic Ions

Write the formula for each of the following compounds.

(a) ammonium chromate
(b) barium perchlorate
(c) sodium hydrogen sulfate

Strategy The polyatomic ions are treated as a single group with the formulas and charges listed in Table 2.2. Subscripts of the anions and cations are adjusted so that the compound will be neutral, overall.

Solution

(a) An ammonium ion, NH_4, has a 1+ charge, and chromate, CrO_4^{2-}, has a 2− charge, so two ammonium cations are needed to balance the charge. Place the formula of the ammonium ion in parentheses so that the 2 subscript clearly indicates two ammonium ions: $(NH_4)_2CrO_4$.

(b) Two polyatomic perchlorate anions, ClO_4^-, each with a 1− charge, are needed to balance the 2+ barium ion, so the formula is $Ba(ClO_4)_2$. Again, parentheses are used around the perchlorate anion to indicate it acts as a single unit with a fixed formula and charge.

(c) One sodium 1+ cation balances the charge of the 1− hydrogen sulfate anion, HSO_4^-, giving the formula $NaHSO_4$. Parentheses are not needed for the hydrogen sulfate ion because the subscript 1 is not written.

Understanding

Write the formulas for sodium carbonate and strontium phosphate.

Answer Na_2CO_3 and $Sr_3(PO_4)_2$

Charges on Transition Metal Ions

The charges of metal ions in Groups 1A, 2A, and 3B always equal the group number. Metals from other groups, most notably the transition metals, can form more than one cation. For example, iron combines with chlorine to form two different ionic compounds, with the formulas $FeCl_2$ and $FeCl_3$. Because the charge on a chloride ion is 1−, the charge of the iron ion is 2+ in $FeCl_2$ and 3+ in $FeCl_3$. The modern system of nomenclature for these compounds uses a Roman numeral in parentheses after the name of the metal to specify the charge. The compound $FeCl_2$ is iron(II) chloride, spoken as "iron two chloride." The compound $FeCl_3$ is iron(III) chloride ("iron three chloride").

A Roman numeral in parentheses represents the positive charge on the metal ion.

For some of the more common ions, an older system of nomenclature also exists; it uses the suffixes *-ous* and *-ic* to designate the lower and higher charged cations, respectively. With some metals, the Latin name for the element is used as the root. For example, *ferrous* and *ferric* are the names of Fe^{2+} and Fe^{3+}, and *cuprous* and *cupric* are the names of Cu^+ and Cu^{2+}, respectively. You may encounter this system in the older chemical literature. Table 2.4 contains examples of names of metal compounds.

EXAMPLE 2.14 Naming Transition-Metal Compounds

Write the modern name of each of the following compounds.

(a) $CoBr_2$ (b) $Cr_2(SO_4)_3$ (c) $Fe(OH)_3$

Strategy Determine the charge on the transition metal from the charges of the anions and subscripts, and indicate that charge with a Roman numeral after the name of the metal.

TABLE 2.4 Naming Metal Compounds

Compound	Modern Name	Older Name
$FeCl_2$	Iron(II) chloride	Ferrous chloride
$FeCl_3$	Iron(III) chloride	Ferric chloride
Cu_2O	Copper(I) oxide	Cuprous oxide
CuO	Copper(II) oxide	Cupric oxide
$CrCl_2$	Chromium(II) chloride	Chromous chloride
Cr_2S_3	Chromium(III) sulfide	Chromic sulfide
$TlBr$	Thallium(I) bromide	Thallous bromide
$TlCl_3$	Thallium(III) chloride	Thallic chloride
$SnCl_2$	Tin(II) chloride	Stannous chloride
$SnCl_4$	Tin(IV) chloride	Stannic chloride

Solution

(a) The two 1− charged bromide anions require a 2+ charge on cobalt to produce a neutral compound, so the compound is cobalt(II) bromide.

(b) Each of the three polyatomic sulfate anions has a 2− charge. These three 2− charged anions present in the formula of each unit of the compound generate a total charge of 6−. Each of the two chromium cations must have a 3+ charge to balance the charges to zero. The compound is chromium(III) sulfate.

(c) Each hydroxide polyatomic anions has a single negative charge, so the compound is iron(III) hydroxide.

Understanding

Write the modern name of $Co(CH_3CO_2)_2$.

Answer Cobalt(II) acetate

Acids

Acids are an important class of compounds that are named in a special way. We discuss acids and their reactions in Section 3.1 and cover them more extensively in Chapters 15 and 16. An acid may be defined as a compound that produces hydrogen ions in *aqueous* solution, that is, when it is dissolved in water. The following discussion of the naming of acids is limited to those related to the common anions presented in this chapter. Each of the anions, combined with a sufficient number of hydrogen ions (H^+) to give electrical neutrality, forms an acid. The acid related to the Cl^- ion is HCl; PO_4^{3-} forms the acid H_3PO_4.

When the name of the anion ends in *-ide* (except for hydroxide), we obtain the name of the acid by adding the prefix *hydro-* and changing the ending to *-ic,* followed by the word *acid. These names refer to water solutions of the compounds.* The molecular compound HCl(g) is named as a small molecule, hydrogen chloride, as described in the next section. When dissolved in water it forms a solution called *hydrochloric acid.* Some examples follow.

Anion	Anion Name	Formula	Aqueous Solution
F^-	Fluor*ide*	HF	*hydro*fluor*ic acid*
Cl^-	chlor*ide*	HCl	*hydro*chlor*ic acid*
Br^-	brom*ide*	HBr	*hydro*brom*ic acid*
I^-	iod*ide*	HI	*hydro*iod*ic acid*
CN^-	cyan*ide*	HCN	*hydro*cyan*ic acid*

Other polyatomic anions, most of which contain oxygen, also form acids. If the polyatomic anion name ends in *-ate,* form the name of the corresponding acid by changing the ending to *-ic,* followed by the word *acid.* For anions with the ending *-ite,* change the ending to *-ous,* followed by the word *acid.* The prefix in the name of the anion is retained. Some examples follow.

Anion		Acid	
PO_4^{3-}	phosph*ate*	H_3PO_4	phosphor*ic acid*
ClO_4^-	perchlor*ate*	$HClO_4$	perchlor*ic acid*
NO_3^-	nitr*ate*	HNO_3	nitr*ic acid*
NO_2^-	nitr*ite*	HNO_2	nitr*ous acid*
SO_4^{2-}	sulf*ate*	H_2SO_4	sulfur*ic acid*
SO_3^{2-}	sulf*ite*	H_2SO_3	sulfur*ous acid*

Molecular Compounds

Many important molecular compounds have nonsystematic common names. For example, H_2O is called water, NH_3 is ammonia, and CH_4 is methane; these names are not related to the formulas.

water, H_2O ammonia, NH_3 methane, CH_4

Models of common molecular compounds. Many common molecular compounds have historical rather than systematic names.

However, most binary molecular compounds have systematic names that are determined by methods similar to those used in naming ionic compounds. With ionic compounds, the cation is named before the anion. This distinction is not possible with molecular compounds because they form from two or more nonmetals, so we need general rules to decide which element should appear first in the name and in the formula.

1. The element farther to the left in the periodic table appears first.
2. The element closer to the bottom within any group appears first.

Hydrogen, which has chemical properties of elements in both Groups 1A and 7A on the periodic table, has its own rules. Hydrogen is the second element named in the compounds it forms with elements in Groups 1A through 5A, and the first element in its compounds with Group 6A and 7A elements. Oxygen is also special and always appears last except when it is combined with fluorine. These rules create the following order in which the elements are named: B, Si, C, As, P, N, **H**, Se, S, I, Br, Cl, **O**, F. Generally, this is also the order in which the elements appear in the formula of the compound. Figure 2.15 is a number-line representation of this order.

We name a binary compound by using the name of the first element followed by that of the second element with its ending changed to *-ide.* Hydrogen bromide is written as HBr (hydrogen comes before bromine on the list).

In many cases, more than one compound can be formed from the same elements. Carbon and oxygen form two stable compounds: CO and CO_2. When naming molecular compounds, we generally use a prefix to indicate the number of atoms of each element in the molecule. Table 2.5 lists several prefixes together with an example of each. The prefix *mono-* (for one) is used only with the second element. It is common practice to drop the last letter of a prefix that ends in *a* or *o* before elements that begin with a vowel, especially "oxide" (e.g., CO is carbon monoxide). Numerical prefixes occur only in the names of molecular compounds. *It is incorrect to use these prefixes in the names of ionic compounds.*

A prefix indicates the number of atoms of each element present in one molecule of a compound.

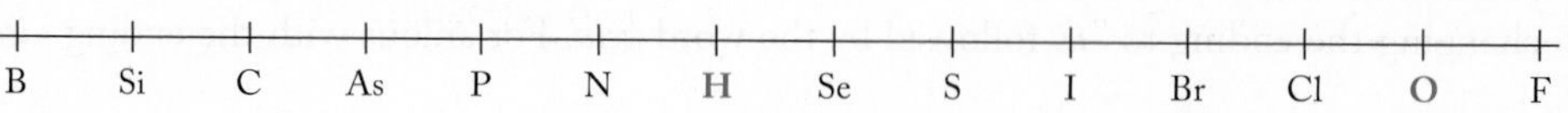

Figure 2.15 Number line representation of order. The element to the left is named first and appears first in the formula of binary molecular compounds.

TABLE 2.5 Prefixes Used to Name Molecular Compounds

Number	Prefix	Example	Name
One	mono-	CO	Carbon monoxide
Two	di-	CO_2	Carbon dioxide
Three	tri-	SO_3	Sulfur trioxide
Four	tetra-	CBr_4	Carbon tetrabromide
Five	penta-	PCl_5	Phosphorus pentachloride
Six	hexa-	SF_6	Sulfur hexafluoride
Seven	hepta-	IF_7	Iodine heptafluoride

This nomenclature is particularly useful for the oxides of nitrogen because there are six of them. Two are dinitrogen monoxide, N_2O, and nitrogen dioxide, NO_2. It could be a terrible mistake to confuse them because N_2O is a relatively nontoxic material sometimes used as an anesthetic ("laughing gas") and NO_2 is extremely toxic. Correct nomenclature is very important!

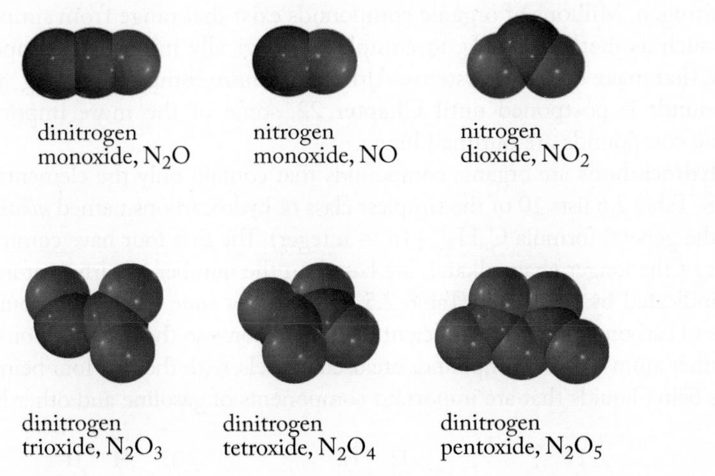

Six oxides of nitrogen.

EXAMPLE 2.15 Formulas of Molecular Compounds

Write the formula for each of the following compounds.

(a) selenium trioxide
(b) dinitrogen tetroxide

Strategy Use the prefixes to determine the number of each atom type.

Solution

(a) The prefix *tri-* means three: SeO_3
(b) The prefixes *di-* and *tetra-* stand for two and four: N_2O_4. Note that in the name of this compound, the *a* of tetra- is dropped.

Understanding

Write the formula for sulfur tetrachloride.

Answer SCl_4

EXAMPLE 2.16 Naming Molecular Compounds

Give the systematic names for each of the following compounds.

(a) N_2O_5 (b) AsI_3 (c) XeF_6

Strategy Use the appropriate prefixes to indicate the number of each atom type, remembering to leave off *mono-* for the first element in the name.

Solution

(a) This name is another example that drops the last letter of the prefix: dinitrogen pentoxide.
(b) This compound is arsenic triiodide.
(c) This is xenon hexafluoride (an interesting example of a noble gas compound).

Understanding

Give the name for the compound IF_3.

Answer Iodine trifluoride

Organic Compounds

Organic compounds are compounds that contain carbon atoms. In most organic compounds, carbon is found in combination with other elements such as hydrogen, oxygen, and nitrogen. Millions of organic compounds exist that range from simple, small molecules such as methane, CH_4, to complex, biologically important compounds, such as DNA, that make up living systems. Although a more complete naming system for these compounds is postponed until Chapter 22, some of the more important classes of organic compounds are outlined here.

Hydrocarbons are organic compounds that contain only the elements hydrogen and carbon. Table 2.6 lists 10 of the simplest class of hydrocarbons named *alkanes,* all of which have the general formula C_nH_{2n+2} (n = integer). The first four have common names. The names of the longer chain alkanes are based on the number of carbon atoms in the molecule, indicated by prefixes in Table 2.5, followed by *-ane.* These compounds have linear chains of carbon atoms with sufficient hydrogen atoms so that each carbon is connected to four other atoms. These compounds are used as fuels, with the first four being gases and the others being liquids that are important components of gasoline and other liquid fuels.

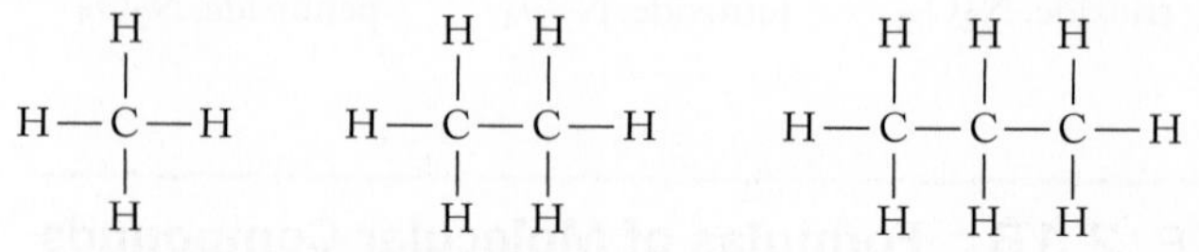

```
    H             H   H               H   H   H
    |             |   |               |   |   |
 H—C—H        H—C—C—H         H—C—C—C—H
    |             |   |               |   |   |
    H             H   H               H   H   H
 methane         ethane              propane
```

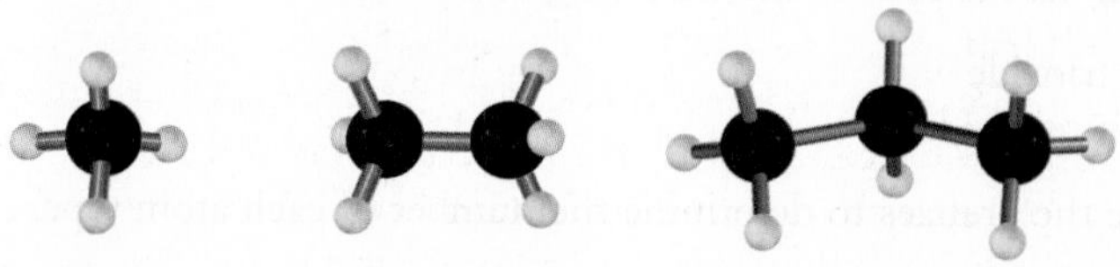

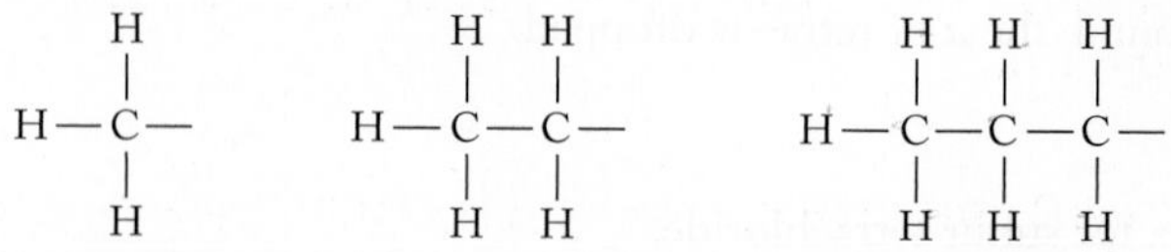

```
    H             H   H               H   H   H
    |             |   |               |   |   |
 H—C—         H—C—C—          H—C—C—C—
    |             |   |               |   |   |
    H             H   H               H   H   H
 methyl          ethyl               propyl
```

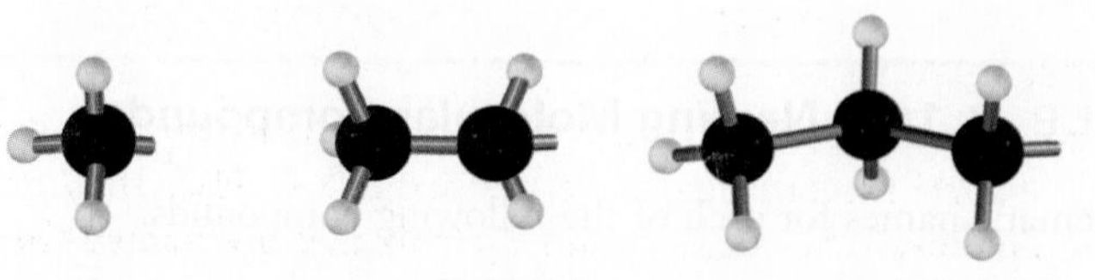

TABLE 2.6 Hydrocarbons

Name	Formula	Alkyl Group	Formula
Methane	CH_4	Methyl	CH_3-
Ethane	C_2H_6	Ethyl	C_2H_5-
Propane	C_3H_8	Propyl	C_3H_7-
n-Butane	C_4H_{10}	Butyl	C_4H_9-
n-Pentane	C_5H_{12}	Pentyl	C_5H_{11}-
n-Hexane	C_6H_{14}	Hexyl	C_6H_{13}-
n-Heptane	C_7H_{16}	Heptyl	C_7H_{15}-
n-Octane	C_8H_{18}	Octyl	C_8H_{17}-
n-Nonane	C_9H_{20}	Nonyl	C_9H_{19}-
n-Decane	$C_{10}H_{22}$	Decyl	$C_{10}H_{21}$-

Alkanes that contain more than three carbon atoms, such as butane with the formula C_4H_{10}, can have different arrangements of the carbon chain that show *branching*. Alkanes without branching are given an *n*- prefix.

```
 H  H  H  H                  H
 |  |  |  |                  |
H—C—C—C—C—H               H—C—H
 |  |  |  |              H   |   H
 H  H  H  H              |   |   |
                       H—C———C———C—H
                         |   |   |
                         H   H   H
```

n-butane

methylpropane

These alkanes are named using the longest chain as the base name. For the branched compound, the additional *substituent* attached to the longest chain is named as an *alkyl* group. As shown in Table 2.6, each of the hydrocarbons can become an alkyl group by removing a hydrogen atom from a terminal carbon atom. In this case, the -CH_3 substituent is named a methyl group—the base name of the hydrocarbon with a *-yl* ending. The name of the compound with the branched chain is methylpropane; the longest carbon chain has three carbon atoms and the substituent, named first, on that chain is a methyl group. For molecules that have longer chains, the chain is numbered, and these numbers are used to indicate the location of the substituent, with the numbers starting at the end of the chain that minimizes the number of the substituent. The position of the substituent is indicated by a number before its name followed by a dash. Other types of groups can also be substituents, such as the halogens shown in Table 2.7.

Alkanes are named based on the longest chain, with the position of the substituents attached to the longest chain group indicated by the number of the carbon atom to which it is attached in the chain.

```
1   2   3   4   5            1   2   3   4   5
CH3CHCH2CH2CH3               CH3CH2CHCH2CH3
     |                               |
    CH3                             CH3
```

2-methylpentane

3-methylpentane

TABLE 2.7 Substituents

Name	Formula
Fluoro	-F
Chloro	-Cl
Bromo	-Br
Iodo	-I

Hydrocarbons closely related to the alkanes are *cycloalkanes,* hydrocarbons that contain a ring of carbon atoms and have the formula C_nH_{2n}. The first three simple cycloalkanes are pictured. Substituents on cycloalkanes are named the same as with alkanes adding the prefix *cyclo-*, with the numbers starting at the location of the substituent.

cyclopropane　　　cyclobutane 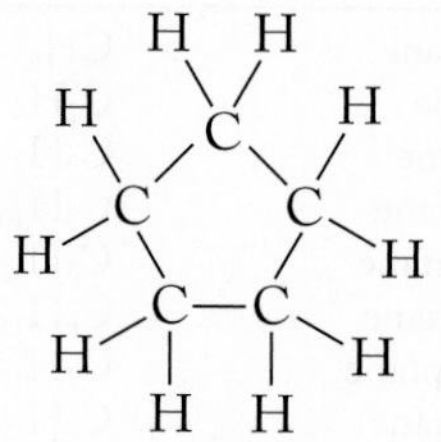　　　cyclopentane

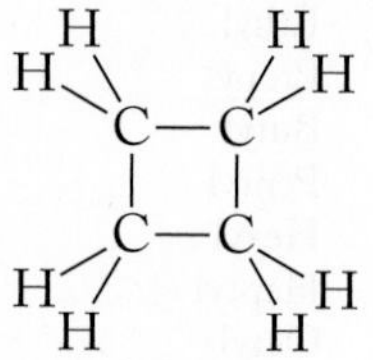

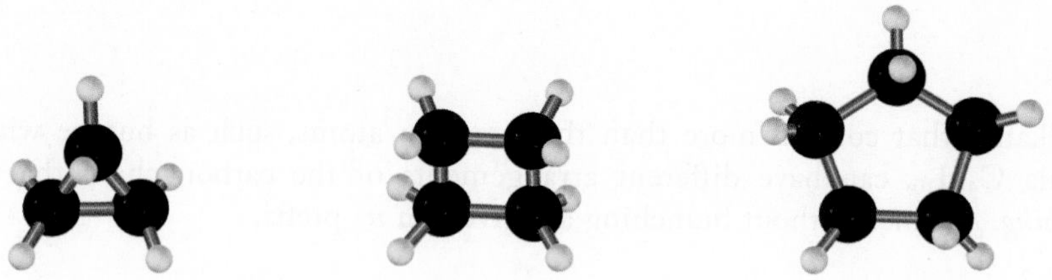

EXAMPLE 2.17 Naming Organic Compounds

Name the two compounds pictured below.

(a) $CH_3CH_2CHCH_2CH_3$ with CH_2CH_3 attached to the third carbon

(b) $CH_3CHCH_2CH_2CH_2CH_3$ with Cl attached to the second carbon

Strategy Locate the longest chain or biggest ring. Locate any substituents and number the chain to minimize the number of the substituent. Name the substituent, properly located on the chain by a number followed by a dash; then add the base alkane name of the longest carbon chain.

Solution

(a) The longest chain contains five carbon atoms, the base name is pentane. An ethyl group is located at the 3-position. Numbering the chain from either direction places it at the 3-position. The name is 3-ethylpentane.

(b) The longest chain contains six carbon atoms, so the base name is hexane. A chloro group located at the 2-position, so the complete name is 2-chlorohexane.

Understanding

Name the compound pictured below.

$CH_3CH_2CHCH_2CH_2CH_2CH_2CH_3$ with CH_3 attached to the third carbon

Answer 3-methyloctane

More complex organic compounds contain *functional groups,* atoms or small groups of atoms that undergo characteristic reactions. Although the naming of these compounds occurs later in Chapter 22, the halogens in Table 2.7 are functional groups.

TABLE 2.8 Functional Groups

Functional Group	Name	Example
–OH	Alcohol	CH_3OH (methanol) CH_3CH_2OH (ethanol)
C–O–C	Ether	$CH_3CH_2OCH_2CH_3$ (diethyl ether)

Table 2.8 lists two other functional groups together with their names. The names of some of these compounds, such as methanol (CH_3OH), are used in many of the examples in the text.

```
   H            H   H            H   H       H   H
   |            |   |            |   |       |   |
H—C—OH       H—C—C—OH        H—C—C—O—C—C—H
   |            |   |            |   |       |   |
   H            H   H            H   H       H   H
 methanol       ethanol            diethyl ether
```

OBJECTIVES REVIEW *Can you:*

- ☑ name simple ionic and transition-metal compounds, and acids?
- ☑ name simple molecular compounds?
- ☑ name simple organic compounds?

2.9 Physical Properties of Ionic and Molecular Compounds

OBJECTIVES

- ☐ Compare and contrast the physical properties of ionic compounds with those of molecular compounds
- ☐ Describe the process of dissociation, and relate the terms *electrolyte* and *nonelectrolyte* to the electrical conductivity of solutions

We introduced two general categories of compounds, ionic and molecular, in the earlier sections of this chapter. It is usually possible to classify a simple compound as either ionic or molecular from the elements in a compound. Generally, a compound is ionic if at least one metal is combined with one or more nonmetallic elements. A molecular compound typically results from the combination of two or more nonmetals.

Ionic compounds generally contain a metallic and a nonmetallic element, whereas molecular compounds generally contain two or more nonmetals.

Ionic and molecular substances usually have significantly different physical properties. An ionic solid has a three-dimensional structure that is held together by strong electrostatic forces because each ion is surrounded by several ions of the opposite charge. In general, ionic materials form hard, but brittle crystalline solids that must be heated to high temperatures before they melt and to extremely high temperatures before they vaporize.

An important property of ionic substances is that when the hard crystalline solids dissolve in water, they break up into separate individual cations and anions (each surrounded by water molecules). **Dissociation** is the separation of a compound into smaller units, in this case, individual cations and anions as the substance dissolves in water. For example, solid sodium chloride, NaCl, dissociates into Na^+ cations and Cl^-

PRINCIPLES OF CHEMISTRY
Physical Properties of Cocaine

The identification of cocaine has been discussed earlier in the chapter. Interestingly, cocaine has both a molecular and an ionic form. The properties are listed below and are quite consistent with all the other compounds that have been discussed.

	Cocaine	Cocaine hydrochloride
Formula	$C_{17}H_{21}O_4N$	$[C_{17}H_{21}O_4NH]^+Cl^-$
Structure	Molecular	Ionic
Temperature stability	Vaporizes at 98 °C	Decomposes at 196 °C
Water solubility	Insoluble	Soluble
Method of abuse	Inhalation	Insufflation
Length of jail sentence for possession of 5 g	120 months	18 months

The molecular form of cocaine is known by the street names of "crack" and "freebase." As with most molecular compounds, it changes when heated from a yellowish waxy solid ("rock") to a vapor. Crack is commonly abused by smoking it. The ionic form, generally called "coke" or "snow," is a white crystalline material. Like most ionic compounds, cocaine hydrochloride must be heated to high temperatures to melt it, and it actually decomposes and burns before it forms a vapor. The ionic form is quite soluble in water, however, and it is commonly abused by insufflation, known as "snorting" or "sniffing." The fine powder is not inhaled but deposited on the nasal membranes and absorbed because it is water soluble.

The penalties for possession of cocaine and for cocaine hydrochloride are quite different. In U.S. Federal court, possession of 5 g cocaine hydrochloride results in a mandatory jail sentence of 18 months whereas the same mass of cocaine results in a mandatory sentence of 120 months. The disparity in the sentences is due to information supplied to Congress that the hydrochloride form is much less addictive, but current information indicates the difference is much smaller than originally stated. In 2008, a commission recommended changing the Federal sentencing guidelines to eliminate the very large sentencing disparity between possession of the two forms of the same drug. ∎

When ionic substances dissolve in water, they **dissociate** into individual cations and anions.

anions when dissolved in water (Figure 2.16). As shown, the water molecules interact with the dissolved ions. When ionic substances that contain polyatomic ions dissolve in water, the individual polyatomic ions act as a single group in solution. For example, ammonium nitrate, NH_4NO_3, dissociates into NH_4 cations and NO_3^- anions.

Because most ionic compounds dissociate in water, measuring electrical conductivity is another way to distinguish ionic from most molecular compounds. A sample must contain mobile charges to conduct an electrical current. A solid ionic compound does not conduct electricity, because the charged particles (ions) are held tightly together and cannot move about (Figure 2.17*a*). An ionic compound in the molten state is a good conductor of electricity because the ions present can move and carry the electric current

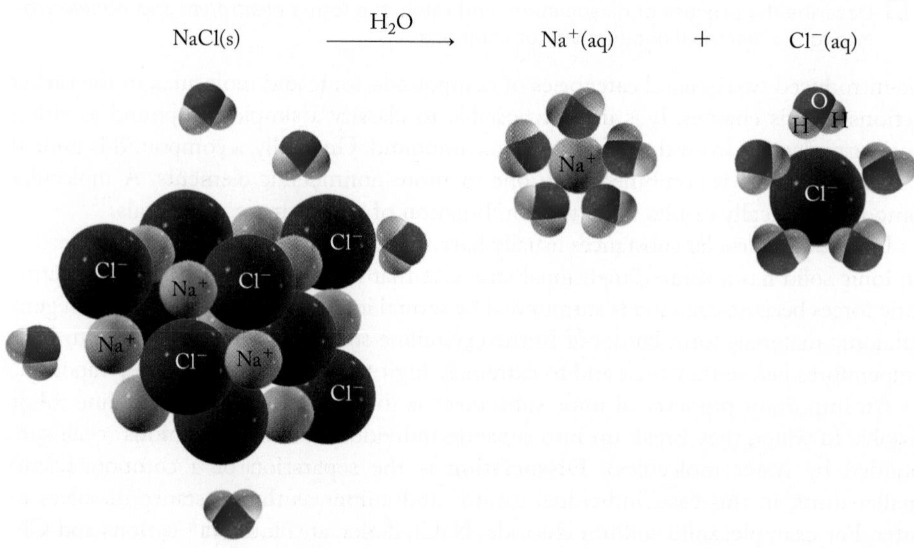

Figure 2.16 Dissociation of sodium chloride (NaCl) in water. Solid NaCl dissociates in water into Na^+ cations and Cl^- anions.

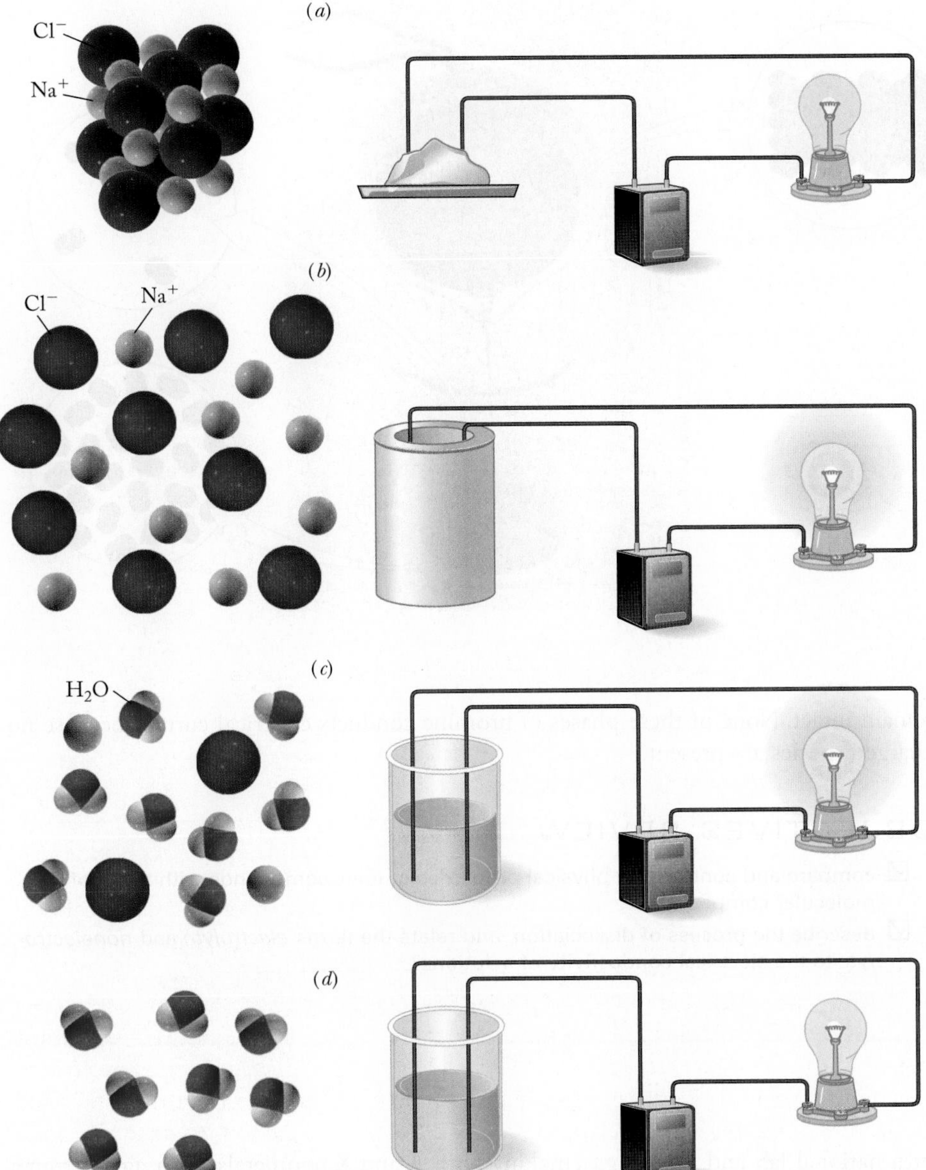

Figure 2.17 Electrical conductivity. Ionic solids do not conduct electrical current *(a)* but are good conductors when melted *(b)* or when dissolved in water *(c)*. Pure water *(d)* does not conduct electrical current. (Normal tap water does have a small amount of dissolved ions in it, so it is a weak electrical conductor.)

(see Figure 2.17*b*). Most aqueous solutions of ionic compound are also good conductors of electricity because dissociation into ions in solution allows them to move independently of each other (see Figure 2.17*c*). The term **electrolyte** refers to a substance that produces ions when dissolved in water because these solutions conduct electricity. Pure water and solutions of most other molecular compounds, such as sugar, are poor electrical conductors because almost no charged particles are present (see Figure 2.17*d*). Water and compounds that dissolve in water and remain as neutral molecules are called **nonelectrolytes** because these solutions do not conduct electricity.

Ionic compounds dissolved in water are good conductors of electricity and are termed electrolytes.

The physical properties of small molecular compounds are different from those of ionic compounds. Small molecular compounds at room temperature generally exist as gases, liquids, or low-melting solids. Strong forces hold individual molecules together, even in the gas phase, but the forces that hold one molecule to another are quite weak (see Chapter 11 for a discussion of these forces). Figure 2.18 depicts bromine molecules in all three phases. In the solid phase, the molecules are in fixed positions. The solid is easy to melt: It becomes a liquid at −7 °C and the liquid boils at 59 °C. In the liquid phase, the molecules are still in close contact but are free to move. In the gas phase, the molecules are in motion and are well separated. In all phases, the individual Br_2 molecules

Figure 2.18 Three phases of bromine. Bromine solid melts below room temperature; some red bromine gas can also be seen above the liquid in this photograph. Bromine is typical of molecular substances.

Ionic compounds are generally hard crystalline solids, whereas small molecular compounds are generally gases, liquids, or low-melting solids at room temperature.

remain intact. None of these phases of bromine conducts electrical current because no charged species are present.

OBJECTIVES REVIEW *Can you:*

- ☑ compare and contrast the physical properties of ionic compounds with those of molecular compounds?
- ☑ describe the process of dissociation, and relate the terms *electrolyte* and *nonelectrolyte* to the electrical conductivity of solutions?

Summary Problem

You have just been hired as an intern at a national lab and they ask you to look for safety violations. On your first day on the job, you are surprised to find some chemicals in a storage room that are not properly labeled. Although it is difficult to read the labels, you determine that two of the compounds have the formula A_2X (where A and X are element symbols). One is a hard, white solid that dissolves in water; the other is a gas in a metal cylinder. From notes on the label of the solid, you can figure out that A is a metal from Group 1A or 2A and X is a nonmetallic element. Also, there is an indication that in the solid A has 20 neutrons and 18 electrons, whereas X has 16 neutrons and 18 electrons. For the second compound, you assume that the material is molecular because it is a gas, making both A and X nonmetals. Also, again from a note on the cylinder, you think A has 7 neutrons and 7 electrons and X has 8 neutrons and 8 electrons. What is the formula and name of each compound?

In the case where A is a metal and the compound is water soluble, it is likely that the compound is ionic. From the formula, a 1+ charge on A and a 2− charge on X is likely because overall the compound has to be neutral. We can eliminate a 2+ charge on A because then X would have a 4− charge, an unlikely charge on a monatomic ion. Elements in Group 1A have a 1+ charge, and given that most elements have about the same number of protons as neutrons, A must be K^+. Potassium has 19 protons in its nucleus, so with 18 electrons K^+ would have the correct charge. Using similar reasoning, one can determine that X must be from Group 6A to have a 2− charge and must be S^{2-} given the number of neutrons and electrons. The empirical formula is K_2S, and the name is potassium sulfide.

For the case of the gas, the compound that is molecular, both A and X are not charged, so the number of protons must be equal to the number of electrons. Thus, A has 7 protons and is nitrogen, and X has 8 protons and is oxygen. The formula is N_2O, and the name is dinitrogen monoxide.

The ionic compound potassium sulfide is expected to dissolve in water (more on this issue is presented in Chapter 4). One would predict that it is a hard crystalline solid that melts only at high temperatures. Although the solid will not conduct electrical current, the molten liquid will. When it dissolves in water, it will dissociate into K^+ and S^{2-} ions, this solution will also conduct electrical current. Dinitrogen monoxide is molecular and is expected to be a gas or liquid at room temperature and pressure.

Question

1. What is the formula of a compound with the general formula AX_2 if A is from Group 2A with an atomic mass of 40 and X is from Group 7A and contains 36 electrons?

ETHICS IN CHEMISTRY

The publication of new scientific results in research articles and books is important to the careers of many scientists; their very jobs and salary levels require them to conduct experiments and publish the results. Most scientific articles and books are *peer reviewed* before they are published. The new proposed manuscripts and books are submitted to an editor and the editor sends it to other scientists working in the same field for their opinion of the work.

A scientist working in the 19th century might have been asked to review John Dalton's book *A New System of Chemical Philosophy, Part I,* in which he first proposed the existence of atoms (see Section 2.1). Although the book had convincing arguments that the atom proposal was correct, it still needed to gain acceptance from others in the scientific community. The book was published in 1808, and chemists subsequently continued to accumulate vast amounts of data that supported the existence of atoms.

The peer review process gives the paper or book, if published, significantly more credibility than one published without review. Although modern scientific journals have ethical guidelines for reviewers, in Dalton's time that was not the case.

1. What would you do if at the time you received Dalton's book describing the existence of atoms, you had just finished writing a similar book with similar experimental results and conclusions?

2. After reading the book and returning your review, you thought of a new experiment that would prove the existence of atoms more conclusively than any point made by Dalton in the book. Is it ethical to do the experiment on your own and publish the results, or should you contact Dalton, either before or after you do the experiment?

3. What would you do if at the time you received Dalton's book you had just done an experiment that showed one of the points made in the book was incorrect? Would those facts be enough for you to suggest the book not be published, or is some other course of action more appropriate?

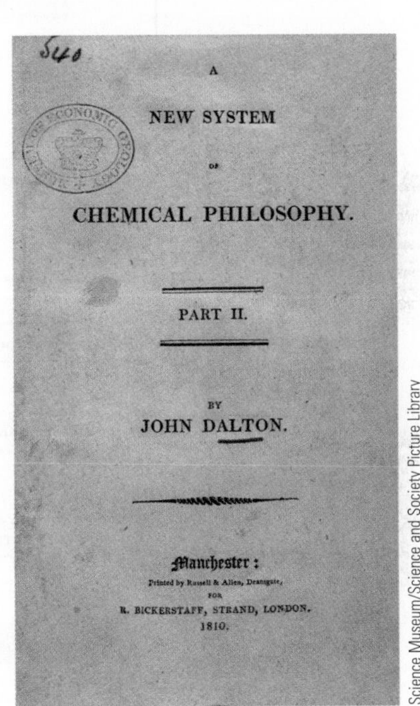
A

NEW SYSTEM

OF

CHEMICAL PHILOSOPHY.

PART II.

BY

JOHN DALTON.

Manchester:

Printed by Russell & Allen, Deansgate,

FOR

R. BICKERSTAFF, STRAND, LONDON.

1810.

Cover of Dalton's second book.

Chapter 2 Visual Summary

The chart shows the connections between the major topics discussed in this chapter.

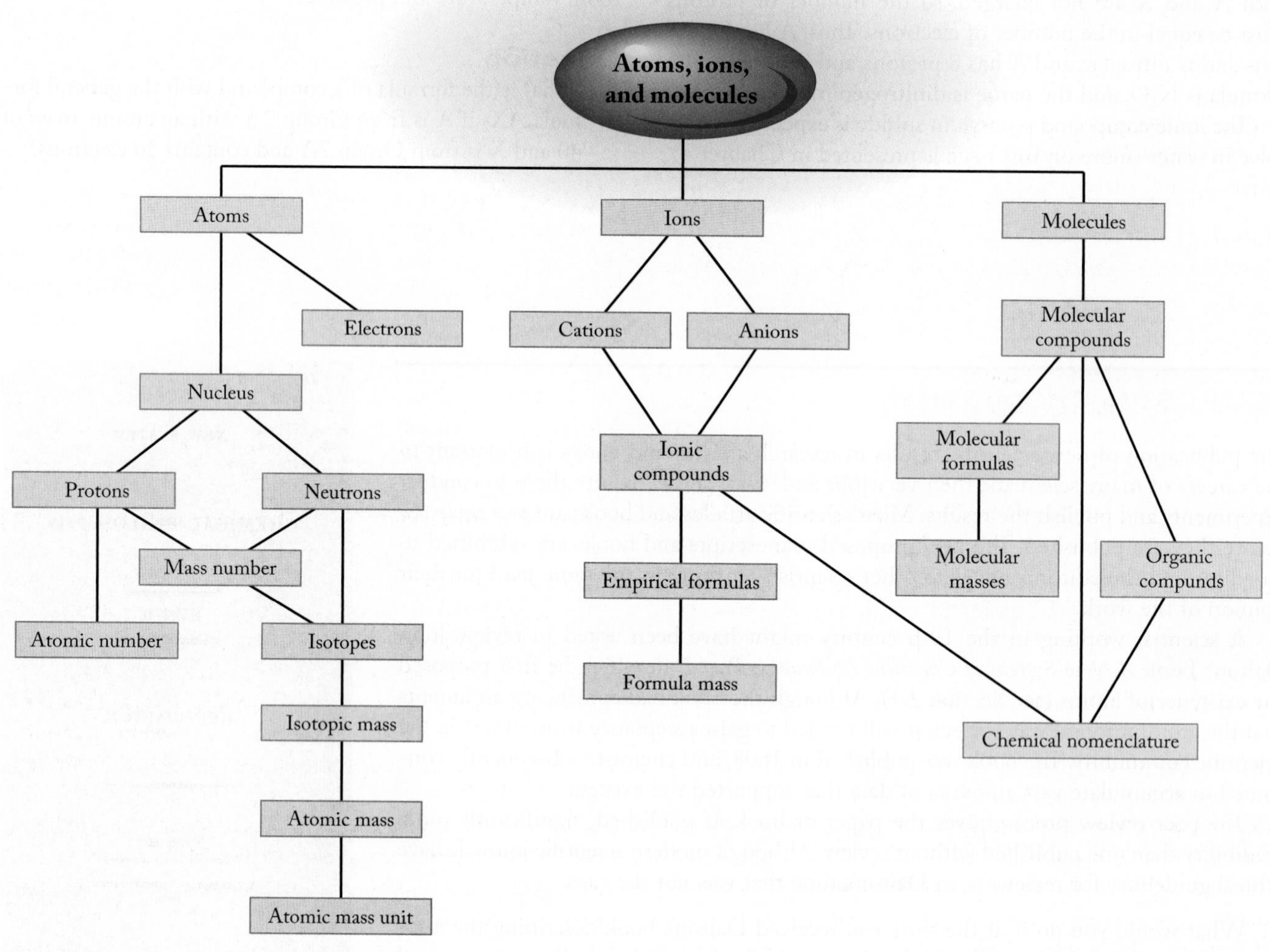

Summary

2.1 Dalton's Atomic Theory

In *Dalton's atomic theory,* all matter is composed of small individual particles called *atoms.* The existence of atoms explains the *law of constant composition* (all samples of a pure substance contain the same elements in the same proportions), the *law of multiple proportions* (for different compounds formed from the same elements, the masses of one element that combine with a fixed mass of the other are in a ratio of small whole numbers), and the *law of conservation of mass* (there is no loss or gain in mass when a chemical reaction takes place).

2.2 Atomic Composition and Structure and 2.3 Describing Atoms and Ions

Atoms contain three different kinds of particles: (1) *protons,* which have a relative charge of 1+ and a relative mass of 1;

(2) *neutrons,* which have no charge and a relative mass of 1; and (3) *electrons,* which have a relative charge of 1− and a relative mass of nearly 0. The protons and neutrons are closely packed in a central core of the atom called the *nucleus,* and the electrons are found at a relatively large distance from this core. The *atomic number* is the number of protons in the nucleus, and it defines the type of atom (i.e., the element). The *mass number* is the sum of the numbers of protons and neutrons in the atomic nucleus. Atoms that have the same atomic number but different mass numbers are known as *isotopes.* In many chemical reactions, atoms lose or gain electrons to form *ions. Cations* are positively charged ions, and *anions* are negatively charged ions.

2.4 Atomic Masses

The *atomic mass unit* (u) is defined as exactly $\frac{1}{12}$ of the mass of a ^{12}C atom. One measures the mass of an individual atom, called the *isotopic mass* and expressed in atomic mass units, by comparing it with that of the ^{12}C atom. The *atomic mass* for a naturally occurring element is a weighted average of the masses of its stable isotopes, taking into account the relative abundance of the isotopes.

2.5 The Periodic Table

The *periodic table* arranges the elements of similar chemical properties into rows and columns. Ultimately, the elements are listed in order of increasing atomic number. The rows of the periodic table are called *periods,* whereas the columns of chemically similar elements are called *groups.* Most periodic tables include the atomic number and atomic mass of each element for easy reference. Most chemists use a system of labeling groups that has a number and either the letter A or B. Groups that are labeled with an A are called the *representative* or *main group elements,* whereas groups labeled with a B are called the *transition metals.*

Several groups of elements have common names. The first column, labeled 1A, is the *alkali metal group.* The second column, labeled 2A, is the *alkaline earth group.* The last column, 8A, has the *noble gases,* so called because they are generally chemically unreactive. Next to this group are the *halogens,* in column 7A. Metallic elements are on the left of the periodic table, whereas nonmetals are located in the upper right side. Bordering the metals and nonmetals are elements that have intermediate properties. They are called *metalloids.*

2.6 Molecules and Molecular Masses

Atoms are nature's building blocks, and *molecules* are combinations of atoms—generally atoms of the nonmetallic elements—joined strongly together. If the atoms of the molecule are all the same, the substance is an element. *Molecular compounds* contain atoms of two or more elements. The *molecular formula* includes the number of each type of atom in the molecule, as a subscript following the symbol for the element. A *structural formula* indicates how the atoms are connected in the molecule. Chemists use atomic masses and the formula of a compound to find the *molecular mass* of the compound.

2.7 Ionic Compounds

Cations and anions form neutral species known as *ionic compounds.* Electrostatic attractive forces hold the ions in an ionic compound together. An ionic compound is generally formed from a metal cation and a nonmetal anion. The *empirical formula* of an ionic compound gives the relative numbers of ions, using the smallest possible whole numbers. The formula of an ionic compound can be written by balancing the charges of its ions. A *polyatomic ion* is a group of atoms that have a net charge. The empirical formula of an ionic compound is used to calculate its *formula mass,* the sum of the atomic masses of all the atoms in the empirical formula.

2.8 Chemical Nomenclature

Chemical nomenclature is a method of systematically naming compounds. Chemists name a *binary* ionic compound by first naming the cation, then the anion. The name of a monatomic anion consists of the first part of the element name plus an *-ide* suffix. The charge of a monatomic anion is related to the group number: 1− for Group 7A elements, 2− for Group 6A elements, and 3− for nitrogen in Group 5A. A metal in Group 1A, 2A, or 3B always forms an ion with the charge equal to the group number. The charge on cations of other metals can differ in different compounds and is indicated in the name by a Roman numeral in parentheses. The name of an acid is related to the ending used in the name of the corresponding anion.

Molecular compounds are named similarly to ionic compounds, with the element farther to the left on the periodic table generally named first. A prefix indicates the number of atoms of each element present in the molecules and must be used when the same two elements form more than one compound. *Organic compounds* are named using the longest carbon chain as the base name, with substituents named before the base name, preceded by a number that indicates its position in the chain.

2.9 Physical Properties of Ionic and Molecular Compounds

Ionic compounds are generally hard, brittle crystalline solids. They *dissociate* when dissolved in water into individual cations and anions, and are *electrolytes* because these solutions conduct electricity. When polyatomic ions are present in ionic compounds, the ions retain their identity when the solid dissociates on dissolving in water. Most substances consisting of small molecules form gases, liquids, or low-melting solids; molecular compounds are generally *nonelectrolytes* when dissolved in water.

Chapter Terms

The following terms are defined in the Glossary, Appendix I.

Section 2.1
Atom
Law of conservation of mass
Law of constant composition
Law of multiple proportions

Section 2.2
Electron
Neutron
Nucleus
Proton

Section 2.3
Anion
Atomic number
Cation
Ion
Isotopes
Mass number

Section 2.4
Atomic mass (atomic weight)
Atomic mass unit (u)
Isotopic mass

Section 2.5
Actinides
Alkali metals
Alkaline earth metals
Group
Halogens
Inner transition metals
Lanthanides
Metal
Metalloid
Noble gases
Nonmetal
Period
Periodic table
Representative elements (main group elements)
Semiconductor
Transition metals

Section 2.6
Diatomic molecule
Molecular compound
Molecular formula
Molecular mass
Molecule
Structural formula

Section 2.7
Empirical formula
Formula mass
Ionic compound
Monatomic ion
Polyatomic ion

Section 2.8
Binary compound
Chemical nomenclature
Hydrocarbon
Organic compound

Section 2.9
Dissociation
Electrolyte
Nonelectrolyte

Questions and Exercises

OWL Selected end of chapter Questions and Exercises may be assigned in OWL.

Blue-numbered Questions and Exercises are answered in Appendix J; questions are qualitative, are often conceptual, and include problem-solving skills.

■ Questions assignable in OWL

✎ Questions suitable for brief writing exercises

▲ More challenging questions

Questions

2.1 How does Dalton's atomic theory explain each of the following facts?
(a) A sample of pure NaCl (table salt) obtained from a mine in the United States contains sodium and chlorine in the same ratio as NaCl obtained from a mine in France.
(b) The mass of the hydrogen peroxide molecule, H_2O_2, equals the sum of the masses of the hydrogen, H_2, and oxygen, O_2, molecules from which it is formed.

2.2 State how Dalton's atomic theory explains
(a) the law of conservation of mass.
(b) the law of constant composition.

2.3 Compare and contrast the terms *atom, element, molecule,* and *compound.* Give an example of each. Some of your examples will fit more than one term, so clarify which term fits each example.

2.4 Compare the masses and charges of the three major particles that make up atoms.

2.5 ✎ Describe the experimental setup and results of the Rutherford experiment.

2.6 How does the nuclear model of the atom explain the results of the Rutherford experiment?

2.7 If aluminum foil had been used in the Rutherford experiment in place of gold foil, how might the outcome have differed?

2.8 Describe the arrangement of protons, neutrons, and electrons in an atom.

2.9 Define the following terms.
(a) atomic number (b) mass number
(c) isotope

2.10 What is the relationship between each of the following quantities and the numbers of the subatomic particles found in an atom?
(a) atomic number (b) mass number
(c) symbol for element

2.11 ▲ A mass spectrometer determines isotopic masses to eight or nine significant digits. What limits the atomic mass of carbon to only five significant digits?

2.12 Explain the difference in the meanings of 4 P and P_4.

2.13 Explain the difference in the meanings of 8 S and S_8.

Farrell Grehan/Photo Researchers, Inc.

Sulfur is produced on a large scale.

2.14 Methane, CH_4, is the principal component of natural gas. Interpret the molecular formula of this compound in words.

2.15 Dinitrogen tetroxide is a component of smog. Give the molecular formula of this gaseous compound, and interpret the formula in words.

2.16 ✎ Carbon monoxide, CO, is a molecular compound, whereas cesium bromide, CsBr, is ionic. Explain the difference in the meanings of these two formulas.

2.17 Sulfur dioxide, SO_2, is a molecular compound that contributes to acid rain, and $CaCO_3$ is an ionic compound that can neutralize acid rain. Explain the difference in the meanings of these two formulas.

2.18 The names of acids formed from oxygen containing polyatomic anions are related to the name of the anion. How are the names of the anions modified to obtain the names of the acids?

2.19 Does the name *nitrogen oxide* correctly apply to the compound NO? Explain why or why not.

2.20 What is missing from the name *chromium chloride* for the compound $CrCl_3$?

2.21 Describe the types of elements that generally combine to form ionic compounds and the types that combine to form molecular compounds.

2.22 ✎ How do the properties of ionic compounds differ from those of molecular compounds?

2.23 Explain why most ionic compounds are hard solids at room temperature, whereas most small molecular substances, such as H_2O and O_2, are liquids or gases.

2.24 ✎ A chemist received a white crystalline solid to identify. When she heated the solid to 350 °C, it did not melt. The solid dissolved in water to give a solution that conducted electricity. Based on this information, what might the chemist conclude about the solid? Explain why.

2.25 NaCl is said to *dissociate* in water. Draw a picture of this process.

2.26 Explain on an atomic level why molten ionic compounds conduct electricity, whereas molten molecular compounds do not.

2.27 Define *group* and *period.*

2.28 ■ Name and give the symbols for two elements that
(a) are metals.
(b) are nonmetals.
(c) are metalloids.
(d) consist of diatomic molecules.

Exercises

OBJECTIVES Define isotopes of atoms and list the subatomic particles in their nuclei.

2.29 Give the complete symbol (^A_ZX), including atomic number and mass number, of (a) a chlorine atom with 20 neutrons, and (b) a calcium atom with 20 neutrons.

2.30 ■ Give the complete symbol (^A_ZX), including atomic number and mass number, of (a) a nickel atom with 31 neutrons, and (b) a tungsten atom with 110 neutrons.

2.31 Write the symbol that describes each of the following isotopes.
(a) an atom that contains 7 protons and 8 neutrons
(b) an atom that contains 31 protons and 39 neutrons
(c) an atom that contains 18 protons and 22 neutrons

2.32 Write the symbol that describes each of the following isotopes.
(a) an atom that contains 5 protons and 6 neutrons
(b) an atom that contains 25 protons and 30 neutrons
(c) an atom that contains 14 protons and 14 neutrons

2.33 Give the numbers of protons and neutrons in
(a) $^{79}_{33}As$ (b) $^{51}_{23}V$ (c) $^{128}_{52}Te$

2.34 Give the numbers of protons and neutrons in
(a) $^{32}_{16}S$ (b) $^{24}_{12}Mg$ (c) $^{37}_{17}Cl$

OBJECTIVE Write complete symbols for ions, given the number of protons, neutrons, and electrons that are present.

2.35 Write the atomic symbol for the element whose monatomic ion has a 2+ charge, has 14 more neutrons than electrons, and has a mass number of 88.

2.36 ■ Write the atomic symbol for the element whose monatomic ion has a 2− charge, has 20 more neutrons than electrons, and has a mass number of 126.

2.37 Write the symbol for the ion with
(a) 8 protons, 10 electrons, and 8 neutrons.
(b) 34 protons, 36 electrons, and 45 neutrons.
(c) 28 protons, 26 electrons, and 31 neutrons.

2.38 Write the symbol for the ion with
(a) 4 protons, 2 electrons, and 5 neutrons.
(b) 32 protons, 30 electrons, and 40 neutrons.
(c) 35 protons, 36 electrons, and 44 neutrons.

2.39 Write the symbol for the atom or ion of the species that contains
(a) 12 protons, 13 neutrons, and 10 electrons.
(b) 13 protons, 14 neutrons, and 10 electrons.
(c) 14 protons, 15 neutrons, and 14 electrons.
(d) 35 protons, 44 neutrons, and 36 electrons.

2.40 Write the symbol for the atom or ion of the species that contains
(a) 23 protons, 28 neutrons, and 20 electrons.
(b) 53 protons, 74 neutrons, and 54 electrons.
(c) 44 protons, 58 neutrons, and 41 electrons.
(d) 15 protons, 16 neutrons, and 15 electrons.

2.41 Given the partial information in each column of the following table, fill in the blanks.

Symbol	—	$^{40}Ca^{2+}$	—	—
Atomic number	11	—	—	—
Mass number	—	—	81	—
Charge	—	—	1−	2−
Number of protons	—	—	35	52
Number of electrons	10	—	—	—
Number of neutrons	12	—	—	76

2.42 Complete the table below. If necessary, use the periodic table.

Symbol	Charge	Number of Protons	Number of Neutrons	Number of Electrons
—	0	9	10	—
^{31}P	0	—	16	—
—	3+	27	30	—
—	—	16	16	18

OBJECTIVE Determine the atomic mass of an element from isotopic masses and their natural abundances.

2.43 Data obtained with a mass spectrometer show that, in a sample of an element, 60.11% of the atoms have masses of 68.926 u, whereas the remaining 39.89% of the atoms have masses of 70.926 u. Calculate the atomic mass of this element and give its name and symbol.

2.44 ■ An element has two isotopes with masses of 62.9396 u and 64.9278 u, and 30.83% of the atoms are the heavier isotope. Calculate the atomic mass of this element and give its name and symbol.

2.45 Naturally occurring rubidium is 72.17% ^{85}Rb (atomic mass = 84.912 u). The remaining atoms are ^{87}Rb (atomic mass = 86.909 u). Calculate the atomic mass of Rb.

2.46 Naturally occurring indium is 95.7% ^{115}In (atomic mass = 114.904 u). The remaining atoms are ^{113}In (atomic mass = 112.904 u). Calculate the atomic mass of In.

2.47 ▲ The mass spectrum of an element shows that 78.99% of the atoms have a mass of 23.985 u, 10.00% have a mass of 24.986 u, and the remaining 11.01% have a mass of 25.982 u.
(a) Calculate the atomic mass of this element.
(b) Give the symbol for each of the isotopes present.

2.48 ▲ The mass spectrum of an element shows that 92.2% of the atoms have a mass of 27.977 u, 4.67% have a mass of 28.976 u, and the remaining 3.10% have a mass of 29.974 u.
(a) Calculate the atomic mass of this element.
(b) Give the symbol for each of the isotopes present.

2.49 ▲ The most intense peak in a mass spectrum is assigned a height of 100 units. The following spectrum was obtained from a sample of an element. Use the data to calculate the atomic mass of the element. Identify the element.

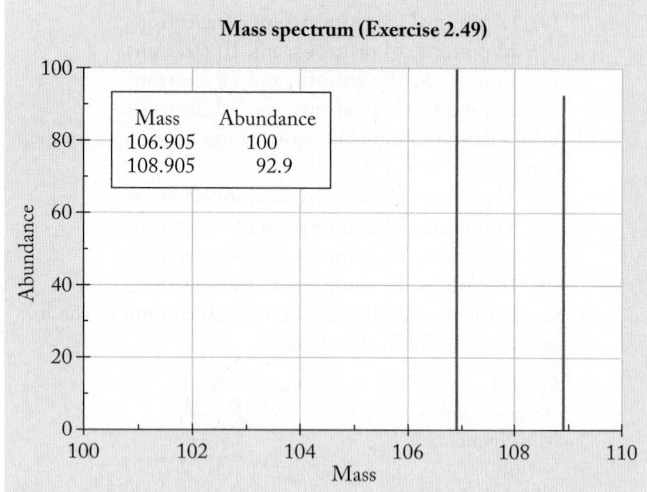

2.50 ▲ The most intense peak in a mass spectrum is assigned a height of 100 units. The following spectrum was obtained from a sample of an element. Use the data to calculate the atomic mass of the element. Identify the element.

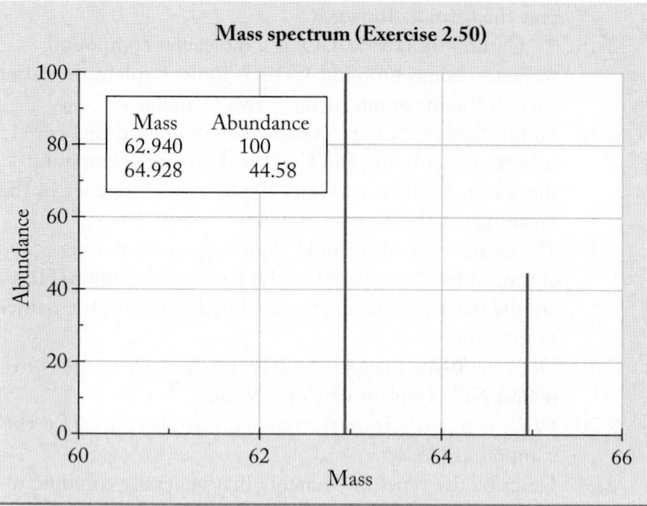

2.51 ▲ Antimony occurs naturally as two isotopes, one with a mass of 120.904 u and the other with a mass of 122.904 u.
(a) Give the symbol that identifies each of these isotopes of antimony.
(b) Get the atomic mass of antimony from the periodic table and use it to calculate the natural abundance of each of these isotopes.

Antimony triiodide.

2.52 ▲ Bromine occurs naturally as two isotopes, one with a mass of 78.918 u and the other with a mass of 80.916 u.
(a) Give the symbol that identifies each of these isotopes of bromine.
(b) Get the atomic mass of bromine from the periodic table and use it to calculate the natural abundance of each of these isotopes.

OBJECTIVE **Define groups and periods.**

2.53 Give a name and symbol for an element in the fifth period that is in the same group with
(a) sodium. (b) Fe.
(c) bromine. (d) Ne.

2.54 Give a name and symbol for an element in the sixth period that is in the same group with
(a) Ge. (b) magnesium.
(c) Y. (d) arsenic.

2.55 Give a name and symbol for an element that is in the same group with
(a) Ti. (b) oxygen.
(c) fluorine. (d) Ba.

2.56 Give a name and symbol for an element that is in the same group with
(a) argon. (b) N.
(c) Os. (d) tungsten.

2.57 Identify each of the following elements as a representative, a transition, or an inner transition element from its position in the periodic table.
(a) silicon (b) Cr
(c) magnesium (d) Np

2.58 Identify each of the following elements as a representative, a transition, or an inner transition element from its position in the periodic table.
(a) barium (b) Mo
(c) F (d) hafnium

2.59 Identify each of the following elements as a representative, a transition, or an inner transition element from its position in the periodic table.
(a) Xe (b) iron
(c) K (d) europium

2.60 Identify each of the following elements as a representative, a transition, or an inner transition element from its position in the periodic table.
(a) Br (b) platinum
(c) rubidium (d) U

2.61 Give the symbol and name for
(a) the alkali metal in the same period as chlorine.
(b) a halogen in the same period as magnesium.
(c) the heaviest alkaline earth metal.
(d) a noble gas in the same period as carbon.

2.62 Give the symbol and name for
(a) the alkaline earth element in the same period as sulfur.
(b) a noble gas in the same period as potassium.
(c) the heaviest alkali metal.
(d) a halogen in the same period as tin (Sn).

2.63 How many elements are in each of the following?
(a) the alkali metals
(b) the halogens
(c) the lanthanides
(d) the sixth period
(e) Group 2B

2.64 How many elements are there in Group 4A of the periodic table? Give the name and symbol of each of these elements. Tell whether each is a metal, nonmetal, or metalloid.

2.65 Which two elements would you expect to exhibit the greatest similarity in physical and chemical properties: Na, Kr, P, Ra, Sr, Te? Explain your choice.

2.66 ■ Of the following elements, which two elements would you expect to exhibit the greatest similarity in physical and chemical properties: Cl, P, S, Se, Ti? Explain your choice.

2.67 Which two elements would you expect to exhibit the greatest similarity in physical and chemical properties: B, C, Hf, Pb, Pr, Sn? Explain your choice.

2.68 Which two elements would you expect to exhibit the greatest similarity in physical and chemical properties: H, Cl, I, Te, W, U? Explain your choice.

OBJECTIVE **Interpret the molecular formula of a substance.**

2.69 Write the molecular formula of the molecules pictured below.

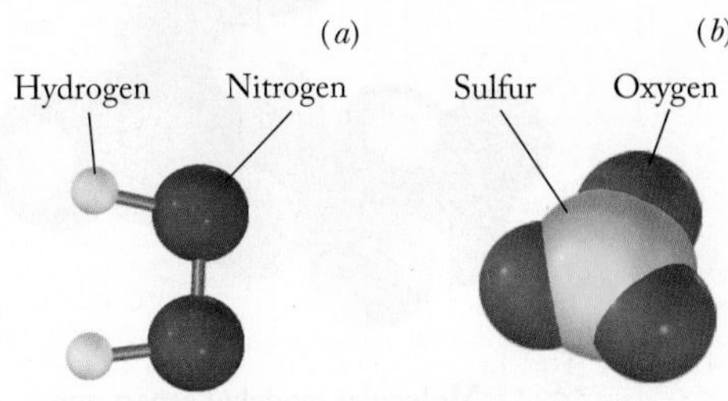

2.70 Write the molecular formula of the molecules pictured below.

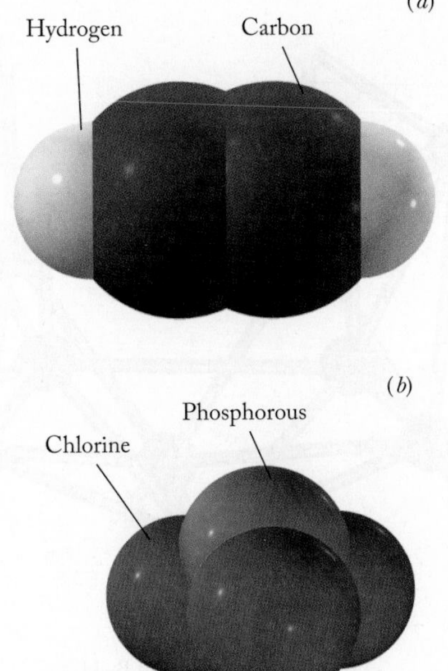

2.71 Draw a ball-and-stick picture of SF_2 (sulfur is located between the two fluorine atoms).

2.72 Draw a ball-and-stick picture of SO_2 (sulfur is located between the two oxygen atoms).

OBJECTIVE Determine molecular mass from the formula of a compound.

2.73 Calculate the molecular mass of each of the following molecules.
(a) C_4H_6O (b) $NOCl_2$ (c) N_2O_3

2.74 ■ Calculate the molecular mass of each of the following molecules.
(a) P_4O_{10} (b) C_6H_7N (c) H_3PO_4

2.75 Aspartame is an artificial sweetener that has the formula $C_{14}H_{18}N_2O_5$. What is the molecular mass of aspartame?

Molecular model of aspartame.

2.76 The compound $B_{10}H_{14}$ has an unusual structure, with some of the hydrogen atoms bridging between two of the boron atoms. What is the molecular mass of $B_{10}H_{14}$?

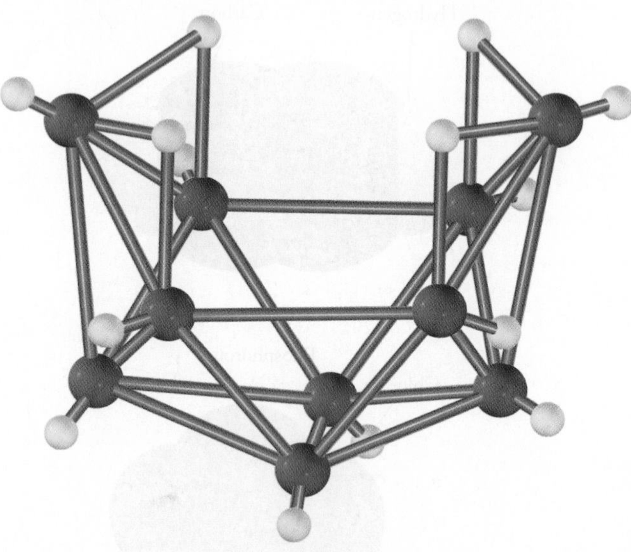

Molecular model of $B_{10}H_{14}$.

OBJECTIVES Predict ionic charges expected for cations of elements in Groups 1A, 2A, and 3B, and aluminum, and for anions of elements in Groups 6A, 7A, and nitrogen.

2.77 Write the symbol for the monatomic ion that is expected for each of the following elements.
(a) iodine (b) magnesium
(c) oxygen (d) sodium

2.78 Write the symbol for the monatomic ion that is expected for each of the following elements.
(a) potassium (b) bromine
(c) barium (d) sulfur

OBJECTIVES Write formulas for ionic compounds.

2.79 What is the empirical formula for the compound made from each of the following pairs of ions?
(a) Ca^{2+} and S^{2-} (b) Mg^{2+} and N^{3-}
(c) Fe^{2+} and F^-

2.80 What is the empirical formula for the compound made from each of the following pairs of ions?
(a) Li^+ and I^- (b) Cs^+ and O^{2-}
(c) Y^{3+} and Cl^-

2.81 Write the empirical formula for the ionic compound made from each of the following pairs of elements.
(a) calcium and chlorine
(b) rubidium and sulfur
(c) lithium and nitrogen
(d) yttrium and selenium

2.82 Write the empirical formula for the ionic compound made from each of the following pairs of elements.
(a) magnesium and fluorine
(b) sodium and oxygen
(c) scandium and selenium
(d) barium and nitrogen

OBJECTIVES List the names, formulas, and charges of the important polyatomic ions.

2.83 Write the formula and charge of
(a) the hydroxide ion.
(b) the chlorate ion.
(c) the permanganate ion.

2.84 Write the formula and charge of
(a) the chromate ion.
(b) the carbonate ion.
(c) the sulfate ion.

2.85 Write the formula and charge of
(a) the hydrogen sulfate ion.
(b) the cyanide ion.
(c) the dihydrogen phosphate ion.

2.86 Write the formula and charge of
(a) the perchlorate ion.
(b) the sulfite ion.
(c) the hydrogen carbonate ion.

2.87 Write the formula of
(a) magnesium nitrite.
(b) lithium phosphate.
(c) barium cyanide.
(d) ammonium sulfate.

2.88 Write the formula of
(a) sodium nitrate.
(b) beryllium hydroxide.
(c) ammonium acetate.
(d) potassium sulfite.

2.89 Write the formula of
(a) strontium nitrate.
(b) sodium dihydrogen phosphate.
(c) potassium perchlorate.
(d) lithium hydrogen sulfate.

2.90 ■ Give the symbol, including the correct charge, for each of the following ions.
(a) barium ion
(b) perchlorate ion
(c) cobalt(II) ion
(d) sulfate ion

OBJECTIVE Calculate the formula masses for ionic compounds.

2.91 Calculate the formula mass for each of the following compounds.
(a) K_2SO_4 (b) $AgNO_3$ (c) NH_4Cl

2.92 Calculate the formula mass for each of the following compounds.
(a) NaOH (b) K_2CO_3 (c) $Ca_3(PO_4)_2$

OBJECTIVES Name simple ionic and transition-metal compounds, and acids.

2.93 Write the name of each of the following ionic compounds.
(a) LiI (b) Mg_3N_2
(c) Na_3PO_4 (d) $Ba(ClO_4)_2$

2.94 Write the name of each of the following ionic compounds.
(a) NH_4Br (b) $BaCl_2$
(c) K_2O (d) $Sr(NO_3)_2$

2.95 Write the modern name of each of the following transition-metal compounds.
(a) $CoCl_3$ (b) $FeSO_4$ (c) CuO

2.96 Write the modern name of each of the following transition-metal compounds.
(a) $RhBr_2$ (b) CuCN (c) $V(NO_3)_3$

2.97 Write the formula of
(a) manganese(III) sulfide.
(b) iron(II) cyanide.
(c) potassium sulfide.
(d) mercury(II) chloride.

2.98 Write the formula of
(a) calcium nitride.
(b) chromium(III) perchlorate.
(c) tin(II) fluoride.
(d) potassium permanganate.

Potassium permanganate.

2.99 Write the formula and name of the acid related to the following ions.
(a) chloride (b) nitrite (c) perchlorate

2.100 Write the formula and name of the acid related to the following ions.
(a) cyanide (b) nitrate (c) phosphate

2.101 What is the name of each of the following acids?
(a) H_3PO_4 (b) H_2SO_3 (c) H_2Te

2.102 What is the name of each of the following acids?
(a) H_2CO_3 (b) HBr (c) HNO_2

2.103 Some people who have hypertension or heart problems use potassium chloride as a substitute for sodium chloride. What is the formula of potassium chloride?

2.104 ■ The compound MnO is added to glass during manufacture to improve its clarity. Write the name of MnO.

OBJECTIVES Name simple molecular compounds.

2.105 Write the formula for each of the following molecular compounds.
(a) sulfur tetrafluoride
(b) nitrogen trichloride
(c) dinitrogen pentoxide
(d) chlorine trifluoride

2.106 Write the formula for each of the following molecular compounds.
(a) sulfur difluoride
(b) silicon tetrachloride
(c) gallium trichloride
(d) dinitrogen trioxide

2.107 Write the name of each of the following molecular compounds.
(a) PBr_5 (b) SeO_2
(c) B_2Cl_4 (d) S_2Cl_2

2.108 ■ Write the name of each of the following molecular compounds.
(a) HI (b) NF_3
(c) SO_2 (d) N_2Cl_4

OBJECTIVE Name simple organic compounds.

2.109 Write the name of the organic compounds pictured below.
(a) $CH_3CH_2CH_2CH_2CH_3$
(b)
```
CH3CH2CHCH2CH2CH3
       |
      CH2CH3
```

2.110 Write the name of the organic compounds pictured below.
(a)
```
        H  H
     H   \ /   H
      \   C   /
       C/   \C
   H—C        C—H
      |        |
   H—C        C—H
       \C   C/
      /  / \  \
     H  /   \  H
       H     H
```
(b)
```
CH3CHCH2CH2CH3
   |
   Br
```

2.111 Write the name of the organic compounds pictured below.
(a) $CH_3CH_2CH_2CH_2CH_2CH_3$
(b)
```
CH3CH2CHCH2CH2CH2CH3
       |
       Cl
```

2.112 Write the name of the organic compounds pictured below.
(a)
```
CH3CHCH2CH2CH3
   |
   Br
```
(b)
```
CH3CHCH2CH3
   |
   I
```

OBJECTIVE Compare and contrast the physical properties of ionic compounds with those of molecular compounds.

2.113 Of the two compounds, LiCl and CO_2, which one do you predict will dissolve in water and which one will be a gas? Explain your answer.

2.114 Of the two compounds, Na_2CO_3 and Cl_2, which one do you predict will dissolve in water and which one will be a gas? Explain your answer.

OBJECTIVES Describe the process of dissociation, and relate the terms *electrolyte* and *nonelectrolyte* to the electrical conductivity of solutions.

2.115 Which beaker below best pictures sodium sulfite in water solution? Explain your answer.

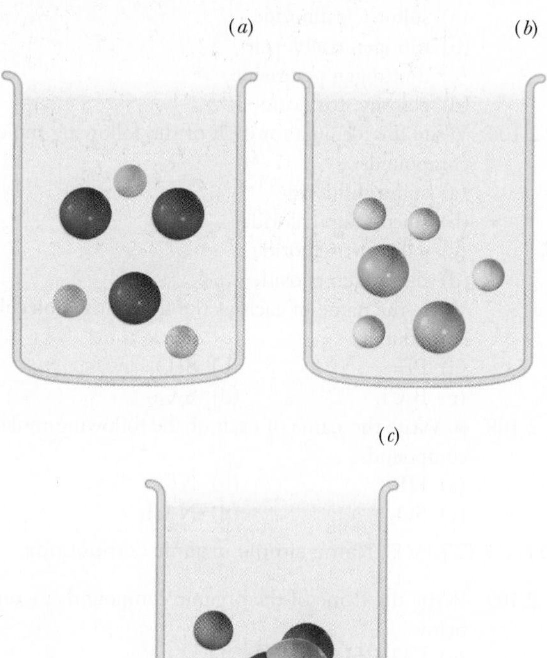

2.116 Which beaker in problem 2.115 best pictures lithium sulfide in water solution? Explain your answer.

2.117 Which of the following substances conducts an electrical current when dissolved in water? Identify the formulas and charges of the ions present in the conducting solutions.
(a) $FeCl_3$ (b) $CO(NH_2)_2$ (urea)
(c) NH_4Br (d) $NaClO_4$
(e) C_2H_5OH

2.118 Which of the following substances conducts an electrical current when dissolved in water? Identify the formulas and charges of the ions present in the conducting solutions.
(a) $AlBr_3$ (b) $C_2H_4(OH)_2$ (ethylene glycol)
(c) $Ca(NO_3)_2$ (d) $(NH_4)_2SO_4$
(e) $K_2Cr_2O_7$

Chapter Exercises

2.119 Predict the formula of an ionic compound formed from calcium and nitrogen.

2.120 Write the formula of iron(III) sulfate.

2.121 The common name for a slurry of $Mg(OH)_2$ in water is Milk of Magnesia. Give the proper name of this compound.

2.122 ■ Write the formula of potassium nitrate and ammonium carbonate.

2.123 Write the symbol, including atomic number, mass number, and charge, for each of the following species.
(a) a halogen with a mass number of 35 and a 1− charge
(b) an alkali metal with 18 electrons, 20 neutrons, and a 1+ charge

2.124 Write the symbol, including atomic number, mass number, and charge, for each of the following species.
(a) a neutral noble-gas element with 21 neutrons in its nucleus
(b) an alkaline earth metal with a mass number of 40 and a 2+ charge

2.125 Name each of the following compounds, and indicate whether each is ionic or molecular.
(a) NO (b) $Y_2(SO_4)_3$
(c) Na_2O (d) NBr_3

2.126 Write the formula of each of the following compounds, and indicate whether each is ionic or molecular.
(a) calcium phosphate
(b) germanium dioxide
(c) iron(III) sulfate
(d) phosphorus tribromide

2.127 Partial information is given in each column in the following table. Fill in the blank spaces.

Symbol	—	—	—	$^{28}Si^{2-}$
Atomic number	—	—	49	—
Mass number	70	103	—	—
Charge	—	3+	1+	—
Number of protons	31	—	—	—
Number of electrons	28	42	—	—
Number of neutrons	—	—	65	—

2.128 Plutonium was first isolated by Glenn Seaborg and coworkers in the early 1940s as the ^{239}Pu isotope; they made it by a nuclear reaction of deuterium with uranium. Give the numbers of protons, neutrons, and electrons in an atom of this isotope of plutonium.

2.129 ■ From the list of elements Li, Ca, Fe, Al, Cl, O, C, and N, write the formula and name of a compound that fits each of the following descriptions.
(a) an ionic compound with the formula MX_2, where M is an alkaline earth metal and X is a nonmetal
(b) a molecular substance with the formula AB_2, where A is a Group 4A element and B is a Group 6A element
(c) a compound with the formula M_2X_3, where M is a transition metal and X is a nonmetal

2.130 Describe the compositions of the three isotopes of hydrogen. Write the symbol and give the name of each isotope.

2.131 Write the symbol for each of the following species.
(a) a cation with a mass number of 23, an atomic number of 11, and a charge of 1+
(b) a member of the nitrogen group (Group 5A) that has a 3+ charge, 48 electrons, and 70 neutrons
(c) a noble gas with no charge and 48 neutrons

2.132 Write the formula for each of the following compounds.
(a) sodium selenide
(b) nickel(II) bromide
(c) dinitrogen pentoxide
(d) copper(II) sulfate
(e) ammonium sulfite

Cumulative Exercises

2.133 The relative abundance of ^{6}Li is known to only three significant figures (7.42%). How can the atomic mass of lithium have four significant figures?

2.134 An atom contains 38 protons and 40 neutrons.
(a) Write the symbol for this atom.
(b) In which group of the periodic table is this element located?
(c) What is the charge of the monatomic ion this element forms?
(d) What is the symbol of an atom in the same group that contains 12 neutrons?

2.135 The accepted atomic mass of nitrogen is 14.0067 u. Approximately 99.632% of natural nitrogen is ^{14}N, which has an isotopic mass of 14.0031 u. The remaining nitrogen is ^{15}N. What is the isotopic mass of ^{15}N in atomic mass units?

2.136 There are two stable isotopes of carbon. If natural carbon consists of 98.938% ^{12}C and the accepted atomic mass of carbon is 12.0107 u, what is the isotopic mass of a ^{13}C atom? Diamond is made of pure carbon and has excellent mechanical, electrical, and light transmission properties. Recently, scientists made some diamonds out of ^{13}C and found that they had even better physical properties than "normal" diamonds do.

Diamonds.

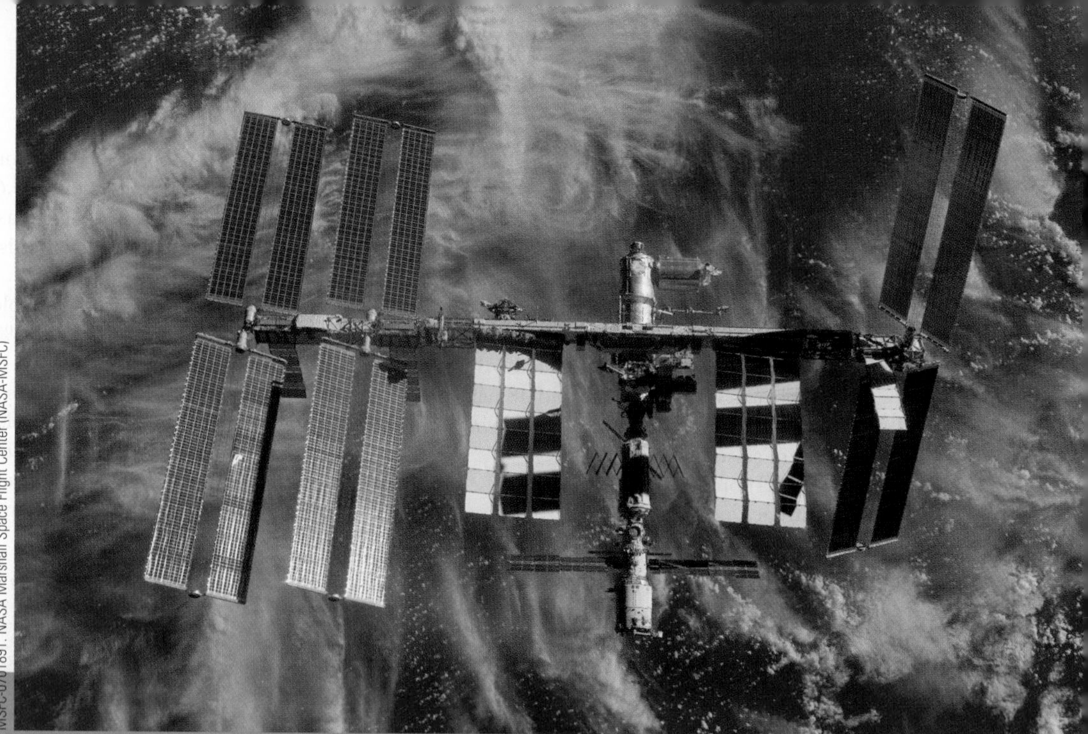

International Space Station.

MSFC-0701891. NASA Marshall Space Flight Center (NASA-MSFC)

Astronauts in space require a carefully designed life support system. This system must function in a self-contained environment and provide the astronauts with all of their necessities, such as electrical power, breathable air, and drinkable water. However, space vehicles do not have adequate room to store every possible supply needed. Therefore, astronauts bring various resources to produce what they need and to recycle and reuse most waste that forms. Chemical reactions, such as those of the life support system, and the amounts of materials consumed and formed in the reactions are the subject of this chapter.

The specific reactions in the life support systems are chosen for optimum safety given the mission and its duration. Astronauts spend only a week or two on the space shuttle, whereas the crews on the International Space Station (ISS) change every 6 months. Different approaches are used to supply the air, water, and power for the two orbiting platforms.

NASA

Astronaut Daniel W. Bursch, *Expedition Four* flight engineer, works on the Elektron Oxygen Generator in the Zvezda Service Module on the International Space Station.

The astronauts on the long-duration ISS and short-duration space shuttle have similar environmental requirements. The environmental system must supply oxygen, control water, and remove carbon dioxide and gases such as ammonia and acetone, which people emit in small amounts. The chemistry of the life support systems is not complex; in fact, simplicity is quite important in this application. Wherever possible, the chemicals are reused, rather than consumed or discarded, and the systems have several backups in case of malfunction.

The primary source of oxygen on the ISS comes from water in a process called *electrolysis.* Electricity

3

Equations, the Mole, and Chemical Formulas

CHAPTER CONTENTS

OWL Online homework for this chapter may be assigned in OWL.

Look for the green colored vertical bar throughout this chapter, for integrated references to this chapter introduction.

generated by solar panels is directed to the Russian-made Elektron Oxygen Generator that splits water into oxygen for breathing and hydrogen, which is vented to space. The main backup oxygen supply is the Solid Fuel Oxygen Generator, and several other tanks of oxygen serve as additional backups. The two backup systems together provide 100 days of oxygen. In contrast, the space shuttle carries tanks of liquid oxygen that supply the astronauts with breathing oxygen and supply oxygen to electrical generators called *fuel cells.* These devices use the oxygen and hydrogen to produce electricity and water. Interestingly, this formation of water is the reverse of the electrolysis reaction on the ISS.

The astronauts produce carbon dioxide when they exhale, which must be removed from the air for a safe breathing environment. To remove the carbon dioxide, the ISS uses the Regenerative Carbon Dioxide Removal System. This system contains compounds with large cavities, called molecular sieves, that selectively absorb the carbon dioxide. After they have absorbed carbon dioxide, the molecular sieves can be recycled by heating to drive off the absorbed carbon dioxide. The backup system on the ISS is the same as the primary system on the shuttle, a number of canisters that contain chemicals that react with the carbon dioxide. Trace gases such as ammonia and acetone are removed by activated charcoal.

NASA

Space Shuttle approaching International Space Station for resupply.

Water is supplied to the ISS by the unmanned *Progress* supply spacecraft from Russia or the U.S. space shuttle. In addition, the ISS has a water recovery and management system. In contrast, the space shuttle generates water as a by-product of the electrical power generated in the fuel cells. ■

Chemistry progressed from an art to a science when, in the course of performing experiments, scientists began to measure the quantities of each substance consumed in a chemical reaction and the amounts of the resulting substances produced. This chapter and Chapter 4 discuss **stoichiometry,** the study of *quantitative* relationships involving the substances in chemical reactions, and present a method for finding the formulas of the new substances produced in these reactions.

3.1 Chemical Equations

OBJECTIVES

- ☐ Write balanced equations for chemical reactions, given chemical formulas of reactants and products
- ☐ Identify an acid and a base
- ☐ Identify and balance chemical equations for neutralization reactions, combustion reactions, and oxidation–reduction reactions
- ☐ Assign oxidation numbers to elements in simple compounds

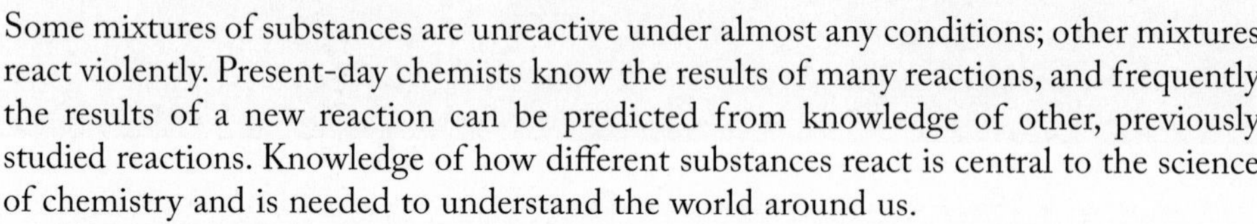

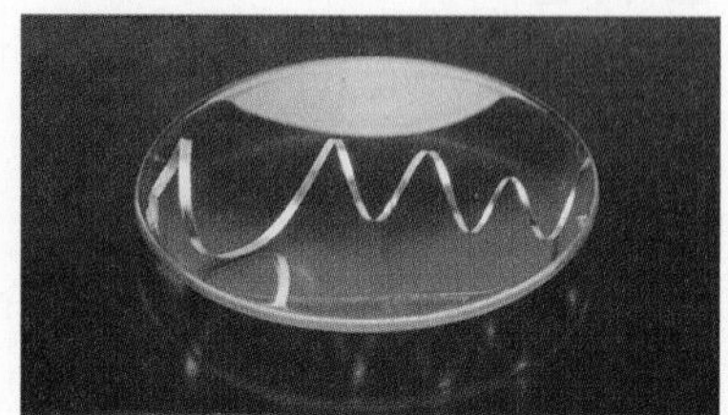

(*a*)

(*b*)

(*c*)

Magnesium reacting with oxygen.
(a) Magnesium. *(b)* Magnesium reacts with oxygen. The reaction is accompanied by a bright light and *(c)* magnesium oxide is formed as the product.

Some mixtures of substances are unreactive under almost any conditions; other mixtures react violently. Present-day chemists know the results of many reactions, and frequently the results of a new reaction can be predicted from knowledge of other, previously studied reactions. Knowledge of how different substances react is central to the science of chemistry and is needed to understand the world around us.

Equations compactly describe chemical changes. Rather than using an equal sign when writing equations, chemists generally use an arrow that means "yields." For example, the equation that describes the reaction of magnesium and oxygen to yield magnesium oxide is written as

$$2Mg + O_2 \rightarrow 2MgO$$

Magnesium and oxygen are the **reactants,** the substances that are consumed, and the magnesium oxide is the **product,** the substance that is formed. The arrow points from the reactants to the products. Such a **chemical equation** describes the identities and relative amounts of reactants and products in a chemical reaction.

Each side of the preceding chemical equation contains two magnesium atoms and two oxygen atoms, consistent with Dalton's atomic theory that atoms are conserved in chemical reactions, and is frequently called a *balanced* equation.

In general, the identities (i.e., chemical formulas) of the reactants and products in the chemical reaction will be given or can be determined by the information given. Constructing a chemical equation from this information involves two steps. First, write the formulas of the reactants and products on the appropriate sides of an arrow. Next, balance the numbers of each type of atom on both sides of the arrow. For example, the reaction of molecules of hydrogen, H_2, and chlorine, Cl_2, produces molecules of hydrogen chloride, HCl. For step one:

$$H_2 + Cl_2 \rightarrow HCl \qquad \textit{(unbalanced)}$$

This is *not* a balanced equation because there are different numbers of hydrogen atoms and chlorine atoms on the left and right sides of the arrow. The second step in writing an equation is to balance the numbers of each type of atom on both sides of the arrow. This balance is accomplished by adjusting the number that precedes each chemical formula. This number is the **coefficient,** the number of units of each substance involved in the equation. Balance the current example by placing a coefficient of 2 in front of the HCl.

$$H_2 + Cl_2 \rightarrow 2HCl \qquad \textit{(balanced)}$$

Now there are two atoms of hydrogen and two atoms of chlorine on each side of the arrow as pictured in Figure 3.1; the equation is balanced.

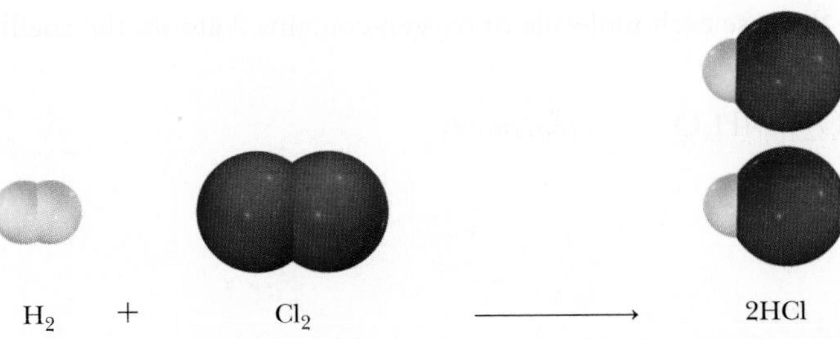

Figure 3.1 Reaction of hydrogen and chlorine. The reaction of one molecule of hydrogen with one molecule of chlorine forms two molecules of hydrogen chloride. Both sides of the equation contain the same number of hydrogen and chlorine atoms.

The accepted convention is to use the lowest possible whole numbers to write a balanced chemical equation. Although the chemical equation

$$2H_2 + 2Cl_2 \rightarrow 4HCl$$

satisfies the law of conservation of mass because there is the same number of H and Cl atoms on both sides of the equation, all coefficients are divisible by 2. Because the convention is to use the lowest ratio of whole numbers, the equation

$$H_2 + Cl_2 \rightarrow 2HCl$$

is the proper way to express this chemical change. Finally, it is incorrect to write balanced equations by changing the formula of any of the substances. The correct formulas of the substances in a chemical reaction are found experimentally using methods described in Section 3.3. *Do not alter the subscripts in any of the substances when writing balanced equations.* Changing the subscripts of hydrogen chloride is incorrect because experiments show that the formula is HCl and not H_2Cl_2.

Balanced equations have the same number of atoms of each element on both sides of the equation.

Writing Balanced Equations

The most common way to write balanced equations is to adjust the coefficients of the reactants and products of the reaction until the same number of atoms of each element are on both sides. Remember, the properly written formula of a substance cannot be changed to balance a chemical equation. The best way to learn to write balanced equations is by practice.

Equations are balanced by adjusting the coefficients of the reactants and products.

A good way to begin writing balanced equations is to assume that the equation contains one formula unit of the most complicated substance. Bring the atoms of this substance into balance by adjusting the coefficients of the substances on the other side of the equation. Last, balance the elements of the other substances on the same side of the equation as the most complicated substance. Consider, for example, the reaction of molecular oxygen, O_2, and propane, C_3H_8 (a fuel that is sometimes used for home heating and cooking). Experiment shows that the products of this reaction are carbon dioxide, CO_2, and water, H_2O.

$$C_3H_8 + O_2 \rightarrow CO_2 + H_2O \qquad \textit{(unbalanced)}$$

Assume a coefficient of 1 for C_3H_8, the most complicated substance, and proceed to balance its atoms on the opposite side, remembering to leave the O_2 for last. Place a coefficient of 3 in front of the CO_2 to balance the carbon atoms in C_3H_8. Each water molecule contains two hydrogen atoms, so four water molecules will contain eight hydrogen atoms. Placing a coefficient of 4 before the H_2O balances the hydrogen atoms.

$$C_3H_8 + O_2 \rightarrow 3CO_2 + 4H_2O \qquad \textit{(unbalanced)}$$

The next task is to make the number of oxygen atoms on the left side of the equation equal to the number on the right. There are 6 oxygen atoms in 3 CO_2 molecules (3 molecules with 2 oxygen atoms per molecule) and 4 oxygen atoms in 4 molecules of H_2O (4 molecules with 1 oxygen atom per molecule), giving a total of 10 atoms of oxygen on the product side. Because each molecule of oxygen contains 2 atoms, the coefficient of O_2 is 10/2, or 5.

$$C_3H_8 + 5O_2 \rightarrow 3CO_2 + 4H_2O \qquad \textit{(balanced)}$$

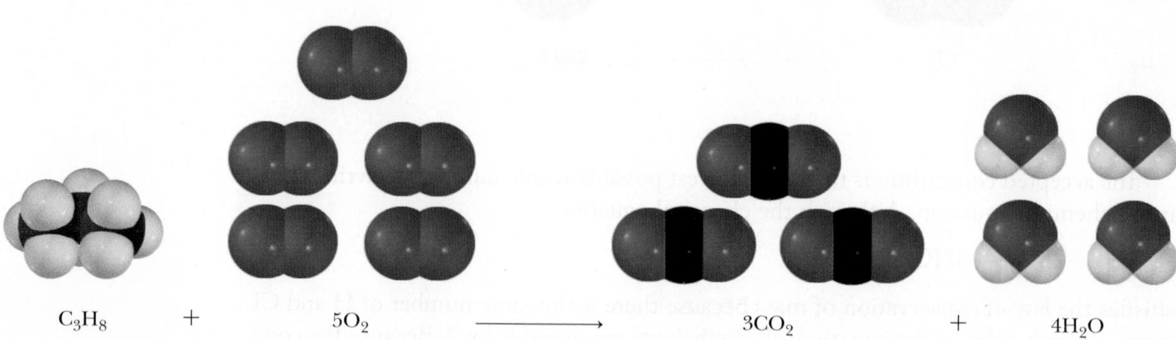

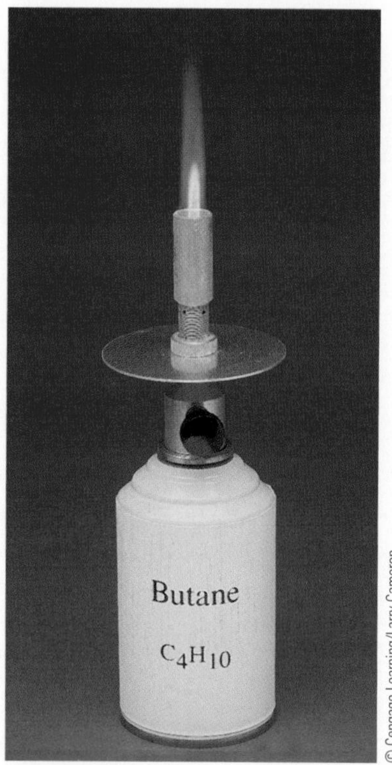

© Cengage Learning/Larry Cameron

Burning butane gas. Butane burns in air to produce carbon dioxide and water. It is a good fuel for a portable burner.

The final step is to check that your answer is correct. In this case, each side of the equation contains 3 carbon, 8 hydrogen, and 10 oxygen atoms.

Occasionally, another step needs to be added, illustrated by balancing the reaction of oxygen and butane (C_4H_{10}, another portable fuel) to also yield H_2O and CO_2.

$$C_4H_{10} + O_2 \rightarrow CO_2 + H_2O \qquad \textit{(unbalanced)}$$

Starting as detailed earlier, assume that the reaction involves one molecule of butane, and place a 4 in front of the CO_2 and a 5 in front of the H_2O to balance the carbon and hydrogen atoms. The product side has 13 oxygen atoms present $[(4 \times 2) + (5 \times 1) = 13]$, requiring a fraction, 13/2, as the coefficient of O_2 to yield a balanced equation.

$$C_4H_{10} + 13/2O_2 \rightarrow 4CO_2 + 5H_2O \qquad \textit{(balanced)}$$

Although these coefficients provide a balanced chemical equation, recall the convention that chemical equations are written with the smallest correct set of *whole number* coefficients. Fractions are generally avoided because a fraction of a molecule cannot exist. To eliminate the fraction, multiply the coefficients of *all* substances in the equation by 2.

$$2C_4H_{10} + 13O_2 \rightarrow 8CO_2 + 10H_2O \qquad \textit{(balanced)}$$

Check the final answer: we find 8 carbon, 20 hydrogen, and 26 oxygen atoms on each side of the equation. The coefficients are the smallest possible set of whole numbers.

EXAMPLE 3.1 Writing Balanced Equations

Balance the following reaction.

$$NaOH + Al + H_2O \rightarrow H_2 + NaAlO_2$$

Strategy Choose the most complicated species and balance the elements in it on the other side of the arrow. Then adjust the coefficients of the other species in the reaction to bring the equation into balance.

Solution

In this case, $NaAlO_2$ is the most complicated species. Its atoms are balanced without adding any coefficients to the reactant side. Only the hydrogen atoms, an atom type present in two of the reactants, are not balanced. They can be balanced with a 3/2 coefficient for H_2.

$$NaOH + Al + H_2O \rightarrow 3/2H_2 + NaAlO_2$$

The equation is balanced, but both sides of the equation should be multiplied by 2 to remove the fraction.

$$2NaOH + 2Al + 2H_2O \rightarrow 3H_2 + 2NaAlO_2$$

Check the final answer: 2 Na, 2 Al, 4 O, and 6 H atoms are on each side.

Understanding

The reaction of acetylene, C_2H_2, with oxygen, O_2, yields carbon dioxide, CO_2, and water. Write the balanced chemical equation for this reaction. As shown, this reaction produces a very hot flame that is used to weld metals.

Answer $2C_2H_2 + 5O_2 \rightarrow 4CO_2 + 2H_2O$

Acetylene torch.

EXAMPLE 3.2 Writing Balanced Equations

Write a balanced equation for the reaction pictured below. In the diagrams, the red spheres represent oxygen atoms and blue spheres represent nitrogen atoms.

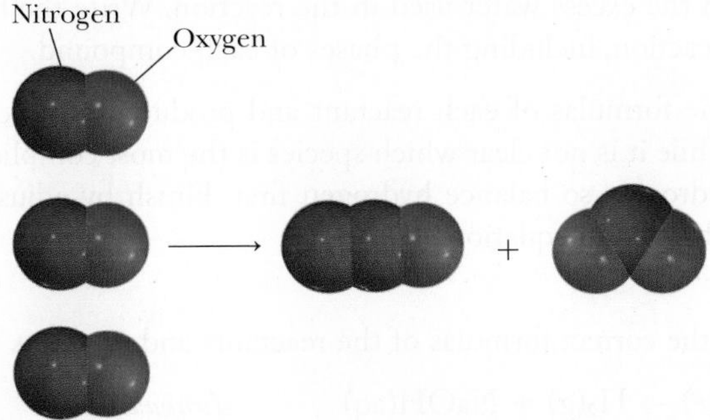

Strategy Examine the diagram to determine the formula of each molecule; then count the number of molecules of each type in the picture and use those numbers as the coefficients of the equation.

Solution

The left, reactant side, contains three NO molecules. The right, product side, contains one N_2O and one NO_2. The equation is

$$3NO \rightarrow N_2O + NO_2$$

Understanding

Write a balanced equation for the reaction pictured below.

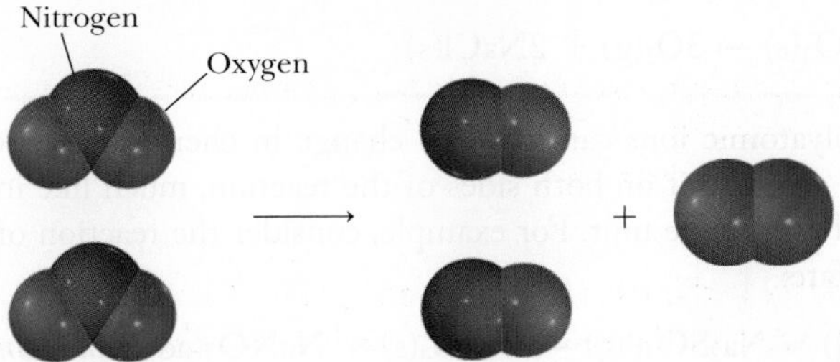

Answer $2NO_2 \rightarrow 2NO + O_2$

© Cengage Learning/Larry Cameron

Figure 3.2 Drain cleaners. Drano is a solid that contains sodium hydroxide (lye) and a small quantity of aluminum. The cleaning is provided by the reaction of sodium hydroxide with materials such as food and hair. The aluminum reacts with the sodium hydroxide to form hydrogen gas (making the liquid in the flask look milky), which stirs the mixture.

Usually, the physical state of each substance in the equation must be specified. The symbols used are: (s) for solid, (ℓ) for liquid, (g) for gas, and (aq) for substances dissolved in water (aqueous solution). Consider, for example, the reaction of Example 3.1 taking place in water. As shown in Figure 3.2, experiment shows that NaOH and $NaAlO_2$ dissolve in water, the Al is a solid, the H_2O is a liquid, and the H_2 is a gas at the temperature and pressure of the reaction. The equation that shows the physical states is

$$2NaOH(aq) + 2Al(s) + 2H_2O(\ell) \rightarrow 3H_2(g) + 2NaAlO_2(aq)$$

Several commercial drain cleaners contain a mixture of sodium hydroxide (NaOH) and aluminum. Although the major effect of the cleaner comes from the reaction of the sodium hydroxide (also called lye) with grease, hair, and other materials in the drain, the gaseous hydrogen that forms in the reaction stirs the mixture and helps unclog the drain.

EXAMPLE 3.3 Writing Balanced Equations

The reaction of sodium metal with water produces hydrogen gas and sodium hydroxide, which is soluble in the excess water used in the reaction. Write the balanced chemical equation for this reaction, including the phases of each compound.

Strategy Write the formulas of each reactant and product on the correct side of the equation arrow. While it is not clear which species is the most complicated, three of the species contain hydrogen, so balance hydrogen first. Finish by adjusting other coefficients to balance the overall equation.

Solution

The reaction with the correct formulas of the reactants and products is

$$Na(s) + H_2O(\ell) \rightarrow H_2(g) + NaOH(aq) \qquad \textit{(unbalanced)}$$

A coefficient of 2 in front of both $H_2O(\ell)$ and NaOH(aq) balances both hydrogen and oxygen.

$$Na(s) + 2H_2O(\ell) \rightarrow H_2(g) + 2NaOH(aq) \qquad \textit{(unbalanced)}$$

A balanced equation is created by placing a coefficient of 2 in front of Na(s):

$$2Na(s) + 2H_2O(\ell) \rightarrow H_2(g) + 2NaOH(aq) \qquad \textit{(balanced)}$$

Check the final answer: There are 2 Na, 4 H, and 2 O atoms on each side.

Understanding

The chapter introduction indicates that a backup oxygen source for the ISS is a Solid Fuel Oxygen Generator. The main compound in the generator is solid sodium chlorate, $NaClO_3$. Heating this compound produces oxygen gas and solid sodium chloride, NaCl. Write the balanced equation for this reaction.

Answer $2NaClO_3(s) \rightarrow 3O_2(g) + 2NaCl(s)$

Although polyatomic ions can undergo change in chemical reactions, most often they behave as a single unit on both sides of the reaction, much like individual atoms, and are balanced as a single unit. For example, consider the reaction of barium nitrate with sodium sulfate:

$$Ba(NO_3)_2(aq) + Na_2SO_4(aq) \rightarrow BaSO_4(s) + NaNO_3(aq) \; \textit{(unbalanced)}$$

Rather than balancing nitrogen, sulfur, and oxygen atoms individually, the nitrate ion can be balanced as a unit and the sulfate ion as another unit. Each side has one sulfate ion, so the sulfate is balanced. However, the reactant side has two nitrate ions, so two nitrate ions are needed as products. We balance the nitrate by placing a coefficient of 2 in front of the sodium nitrate:

$$Ba(NO_3)_2(aq) + Na_2SO_4(aq) \rightarrow BaSO_4(s) + 2NaNO_3(aq)$$

Now the nitrate ions are balanced, and we have the added benefit of balancing sodium as well, producing the balanced equation.

Types of Chemical Reactions

Chemists have recognized and classified many types of chemical reactions. Knowing a particular reaction type is often useful in writing chemical equations. Three types of reactions—*neutralization, combustion* of organic compounds, and *oxidation–reduction*—are particularly important to learn at this point because they are frequently encountered. Chapter 4 presents a fourth common type of reaction: *precipitation.*

Acids and Bases

In Chapter 2, we learned that ionic compounds that dissolve in water dissociate into individual cations and anions forming solutions that will conduct electrical current. For example, $CaCl_2$ *dissociates* into Ca^{2+} and Cl^- ions when it dissolves in water, as shown in the following equation:

$$CaCl_2(s) \xrightarrow{H_2O} Ca^{2+}(aq) + 2Cl^-(aq)$$

The (aq) label after the ions mean that the species are dissolved in water. A coefficient of 2 before the Cl^- is needed so that the equation is balanced.

Two important classes of compounds that form solutions containing ions are acids and bases. The simplest definition of an **acid** is any substance that dissolves in water and yields the hydrogen cation (H^+). An example is nitric acid, HNO_3.

$$HNO_3(\ell) \xrightarrow{H_2O} H^+(aq) + NO_3^-(aq)$$

Acids are generally molecular compounds, but unlike many molecular compounds such as sugar that do not change in solution, when HNO_3 dissolves in water it forms ions in a process known as *ionization.*[1]

Another way to write H^+(aq) ion is H_3O^+ (aq), which is the *hydronium ion.* Writing H_3O^+ indicates that the hydrogen cation is associated with a water molecule, and that bare H^+ ions are not present in solution. The H_3O^+ representation is particularly useful in displaying drawings of individual ions and in certain problems considered in later chapters. In reaction stoichiometry, the H^+(aq) representation is preferred because it simplifies equations.

Table 3.1 lists several compounds that are acids in water solution. Appendix D contains a more complete list. Note that the names of some compounds change when the pure substances are dissolved in water. For example, at room temperature and pressure, pure HCl exists as a gas and is named hydrogen chloride. When dissolved in water, it ionizes into H^+(aq) and Cl^-(aq), and becomes an acid called *hydrochloric acid.*

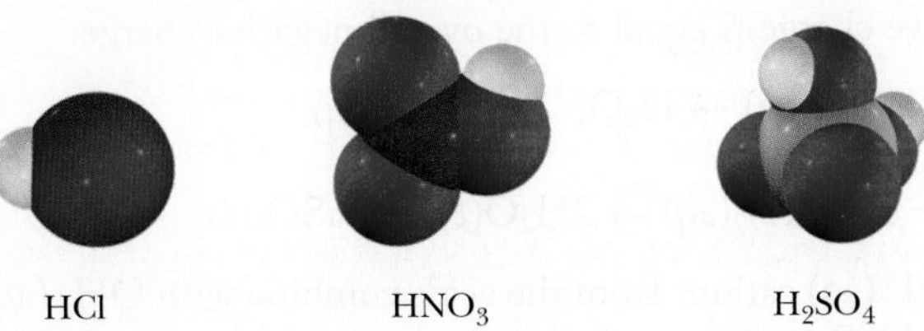

Molecular models of common acids.

TABLE 3.1 Common Acids

Acid	Name
HF	Hydrofluoric acid
HCl	Hydrochloric acid
HBr	Hydrobromic acid
HCN	Hydrocyanic acid
HNO_2	Nitrous acid
HNO_3	Nitric acid
H_2SO_3	Sulfurous acid
H_2SO_4	Sulfuric acid
$HClO_4$	Perchloric acid
H_3PO_4	Phosphoric acid

[1]Although general agreement does not exist on the issue, *dissociation* often is used for the separation of ionic compounds in solution, whereas *ionization* is used to describe those cases where molecular compounds separate into ions in solution.

The simplest definition of a **base** is any substance that produces hydroxide anion (OH^-) in water. The most common bases are the hydroxides of elements in Group 1A and the heavier Group 2A elements. Two examples of bases are NaOH and $Sr(OH)_2$.

$$NaOH(s) \xrightarrow{H_2O} Na^+(aq) + OH^-(aq)$$
$$Sr(OH)_2(s) \xrightarrow{H_2O} Sr^{2+}(aq) + 2OH^-(aq)$$

The equations for acids and bases dissolving in water show charged species. When you write an equation that contains charged species, *the sum of the charges on each side of the equation must be the same,* as well as the number of atoms of each element. This process gives us an additional check to determine whether a chemical equation is properly balanced.

Acids dissolve in water to produce $H^+(aq)$, and bases dissolve in water to produce $OH^-(aq)$.

EXAMPLE 3.4 Equations

Write the equations for hydrogen chloride gas dissolving in water.

Strategy HCl(g) forms $H^+(aq)$ and $Cl^-(aq)$ when it dissolves in water.

Solution

$$HCl(g) \xrightarrow{H_2O} H^+(aq) + Cl^-(aq)$$

An alternative solution uses the hydronium ion for $H^+(aq)$ (both are correct); the equation is

$$HCl(g) + H_2O \longrightarrow H_3O^+(aq) + Cl^-(aq)$$

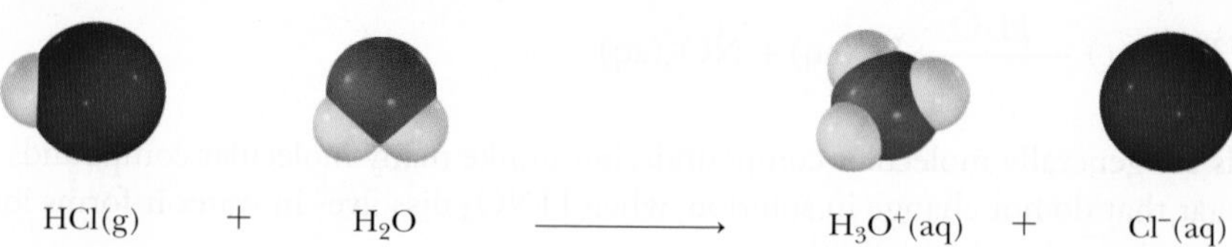

Write the equation for sodium hydroxide dissolving in water.

Answer $NaOH(s) \xrightarrow{H_2O} Na^+(aq) + OH^-(aq)$

Acid–Base Reactions: Neutralization

The reaction of an acid with a base yields water and the respective salt. This reaction is called **neutralization.** A **salt** is an ionic compound composed of a cation from a base and an anion from an acid. Be careful to write the formula of the salt correctly, so that the overall positive charge is equal to the overall negative charge.

$$HCl(aq) + NaOH(aq) \rightarrow H_2O(\ell) + NaCl(aq)$$

$$H_2SO_4(aq) + Ba(OH)_2(aq) \rightarrow 2H_2O(\ell) + BaSO_4(s)$$

Notice how $H^+(aq)$ cations from the acids combine with $OH^-(aq)$ anions from the bases to produce H_2O.

$$H^+(aq) + OH^-(aq) \rightarrow H_2O(\ell)$$

A relationship that is useful in balancing any acid–base reaction is that the number of hydrogen ions contributed by the acid and the number of hydroxide ions contributed by the base are equal to each other and to the number of water molecules formed.

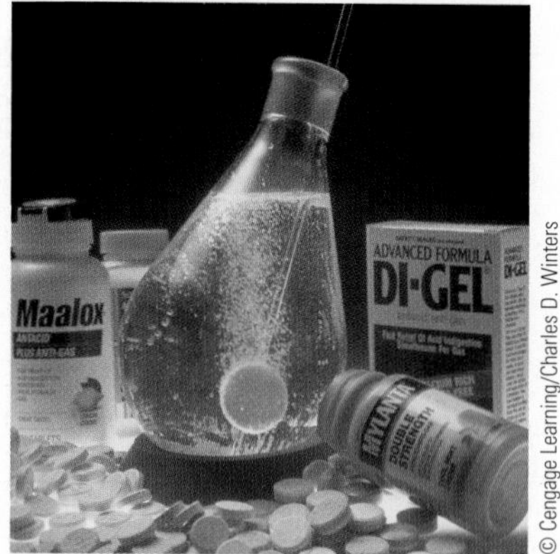

Antacids. The feeling of "heartburn" is caused when acid in the stomach fluxes into the esophagus. Commercial antacids are bases that neutralize the acid.

The reaction of an acid and a base yields water and a salt. This reaction is a neutralization reaction.

EXAMPLE 3.5 Neutralization Reactions

Write the balanced chemical equation for the reaction of aqueous nitric acid with solid magnesium hydroxide.

Strategy This reaction is an acid–base reaction making the products water and the respective salt. The formulas of the reactants, to be written to the left of the arrow, must be written correctly: HNO_3 and $Mg(OH)_2$. It is important to remember the charge on magnesium, a Group 2A element, is 2+, so two OH^- polyatomic anions are needed to balance the charges in the formula of magnesium hydroxide. The products in this acid–base reaction, placed to the right of the arrow, are H_2O and the salt $Mg(NO_3)_2$. After the reactants and products are written, adjust the coefficients to produce the equation.

Solution

The products of a neutralization reaction are water and the appropriate salt.

$$HNO_3(aq) + Mg(OH)_2(s) \rightarrow H_2O(\ell) + Mg(NO_3)_2(aq) \qquad \textit{(unbalanced)}$$

Each $Mg(OH)_2$ provides 2 OH^-, so 2 HNO_3 are needed to provide 2 H^+ ions and two water molecules are produced.

$$2HNO_3 + Mg(OH)_2 \rightarrow 2H_2O + Mg(NO_3)_2 \qquad \textit{(balanced)}$$

These coefficients also balance the Mg^{2+} and NO_3^- ions.

Understanding

Write the balanced chemical equation for the reaction of hydrochloric acid with calcium hydroxide.

Answer $2HCl(aq) + Ca(OH)_2(s) \rightarrow 2H_2O(\ell) + CaCl_2(aq)$

In a neutralization reaction, each $H^+(aq)$ provided by the acid is neutralized by one $OH^-(aq)$ from the base, forming one molecule of H_2O and a salt.

Combustion Reactions

A **combustion reaction** is the process of burning, and most combustion involves reaction with oxygen. Most organic compounds react with oxygen in combustion reactions and give off large amounts of heat. The burning of wood and the burning of natural gas are examples of the combustion of organic compounds. Unless told differently, assume that the products of the combustion of organic compounds that contain only carbon, hydrogen, and oxygen are always CO_2 and H_2O. Depending on how the reaction is carried out, the water molecules could be in either the gas (if the temperature of the reaction is above 100 °C) or the liquid state.

Organic compounds react with oxygen to produce CO_2 and H_2O—a combustion reaction.

AP Photo

Combustion reaction. An oil well burning out of control is an example of a combustion reaction.

EXAMPLE 3.6 Combustion Reactions

Write the equation for the combustion of liquid ethanol, also called ethyl alcohol, C_2H_5OH.

Strategy The combustion of an organic compound such as ethanol produces CO_2 and H_2O. In balancing this reaction, change the coefficients of CO_2 and H_2O to balance the carbon and hydrogen atoms in the ethyl alcohol, and finish by balancing oxygen, remembering to count the oxygen that is present in the ethyl alcohol.

Solution

The reaction is

$$C_2H_5OH(\ell) + O_2(g) \rightarrow CO_2(g) + H_2O(g)$$

First, assume a coefficient of 1 for the most complicated species, ethanol. To balance carbon and hydrogen, place a coefficient of 2 before CO_2 and a 3 before H_2O.

$$C_2H_5OH(\ell) + O_2(g) \rightarrow 2CO_2(g) + 3H_2O(g)$$

Seven oxygen atoms are on the product side. Because the reactant side has one oxygen atom in the ethanol, a coefficient of 3 for O_2 will yield the balanced equation.

$$C_2H_5OH(\ell) + 3O_2(g) \rightarrow 2CO_2(g) + 3H_2O(g)$$

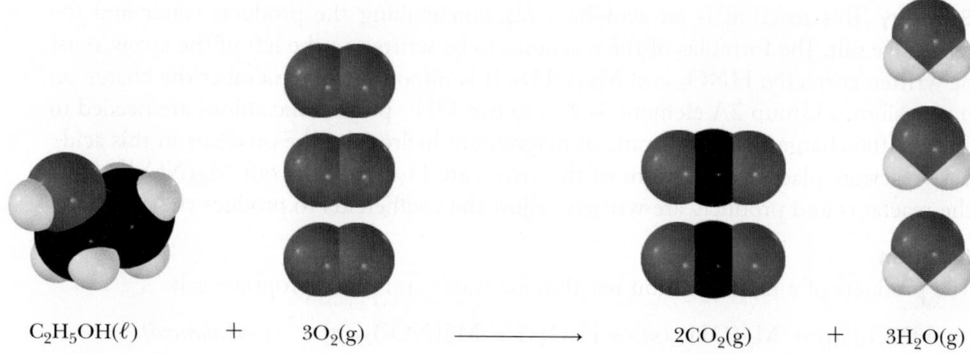

Understanding

Write the equation for the combustion of liquid methanol, CH_3OH.

Answer $2CH_3OH(\ell) + 3O_2(g) \rightarrow 2CO_2(g) + 4H_2O(g)$

Combustion reactions of alcohols. *(a)* Small ethanol burner used by a jeweler. *(b)* Sterno is a brand of "jellied" methanol used in cooking. *(c)* Some racing cars use alcohol fuel because it is less likely than gasoline to explode in a collision.

(*a*)

(*b*)

(*c*)

Oxidation–Reduction Reactions

Combustion reactions are a special class of chemical reactions known as **oxidation–reduction reactions.** This topic is covered in detail in Chapter 18, but a brief overview of the topic is appropriate here.

An example of an oxidation–reduction reaction is the combustion of lithium to yield lithium oxide.

$$4Li(s) + O_2(g) \rightarrow 2Li_2O$$

In this reaction, Li(s) is converted into Li^+, changing its charge from zero, as the element, to 1+, as an ion. The Li(s) has *lost* an electron during the course of the reaction. The loss of electrons by a substance is known as **oxidation.** The term *oxidation* originally meant reaction with oxygen but has been expanded to apply to any element that loses electrons in a chemical reaction. The charge of the oxygen in the reactant O_2 is also zero. In the reaction, each oxygen atom changes its charge from zero to 2−. The oxygen atoms *gain* electrons in this reaction. **Reduction** is the gain of electrons by a substance.

An **oxidation–reduction reaction** is one in which electrons are transferred from one species to another. In all oxidation–reduction reactions, some atoms are oxidized and some are reduced. The term *redox* is often used in place of *oxidation–reduction.*

Oxidation is the loss of electrons; reduction is the gain of electrons.

Another example of a redox reaction is the reaction of $H_2(g)$ and $Cl_2(g)$ to produce $HCl(g)$:

$$H_2(g) + Cl_2(g) \rightarrow 2HCl(g)$$

As in the first example, the charges of the atoms in the elements $H_2(g)$ and $Cl_2(g)$ are zero. But what are the "charges" on the atoms in the HCl molecule? Remember that when HCl(g) is dissolved in water, it ionizes into $H^+(aq)$ and $Cl^-(aq)$, so we expect that the hydrogen atoms in $H_2(g)$ lose electrons and are thus oxidized, and the chlorine atoms in $Cl_2(g)$ gain electrons and are reduced. We often need to identify which compounds are oxidized and which are reduced to understand the chemistry of the reaction and, in many cases, to help balance complicated reactions. Thus, we need some system to keep track of electrons as atoms undergo chemical reactions.

We use a bookkeeping method to keep track of the electrons in molecular compounds such as HCl(g). **Oxidation numbers** (frequently referred to as *oxidation states*) are integer numbers assigned to atoms in molecules or ions based on a set of rules. Use the following rules to assign an oxidation number to each element in a compound.

Oxidation numbers are assigned by following a series of rules.

Rules for Assigning Oxidation Numbers

1. An atom in its elemental state has an oxidation number of zero. For example, the oxidation numbers of atoms in $H_2(g)$ and Li(s) are zero.
2. Monatomic ions in ionic compounds have an oxidation number equal to the charge of the ion. For example, the oxidation number of each lithium ion (Li^+) in Li_2CO_3 is $+1$, and the oxidation number of the chloride (Cl^-) in NaCl is -1.
3. In compounds, fluorine has the oxidation number of -1; oxygen is generally -2; hydrogen combined with a nonmetal is generally $+1$, and -1 when combined with metals; and the other halogens are generally -1. (For oxygen, hydrogen, and the halogens, there are some exceptions that are considered in later sections.)
4. All other atoms are assigned oxidation numbers so that the sum of the oxidation numbers for all of the atoms in a species is equal to the charge of the species. For example, the oxidation numbers in neutral compounds, such as CO_2, must sum to zero (where carbon must be assigned a $+4$ oxidation number to balance the *two* -2 oxygens). The oxidation numbers in charged species, such as the nitrate ion (NO_3^-), sum to the charge of the species (where nitrogen must be assigned a $+5$ oxidation number to yield the overall 1− charge when summed with the *three* -2 oxygens).

Charges on atoms are written with the sign after the number; oxidation numbers are written with the sign before the number.

Consider assigning oxidation numbers in the combustion of elemental sulfur, S_8, to yield SO_2.

$$S_8(s) + 8O_2(g) \rightarrow 8SO_2(g)$$

Both sulfur and oxygen are elements, so the oxidation numbers of each of those atoms are 0 (rule 1). Because SO_2 contains two oxygen atoms, each with an oxidation number of -2, for the oxidation numbers to sum to the zero charge of this molecule (rule 4) the sulfur must have an oxidation number of $+4$.

$$\underset{0}{S_8(s)} + \underset{0}{8O_2(g)} \rightarrow 8\underset{+4}{S}\underset{-2}{O_2}(g)$$

In going from an oxidation number of 0 to $+4$, each sulfur atom loses 4 electrons, so S is oxidized. In going from an oxidation number of 0 to -2, each oxygen atom gains 2 electrons, so O is reduced

© Cengage Learning/Leon Lewandowski

Reaction of sulfur with oxygen. Sulfur burns in oxygen with a blue flame. This reaction is both a combustion and oxidation–reduction reaction.

An interesting example of a redox reaction is the decomposition of $NaClO_3$ to yield O_2 and NaCl, a reaction that takes place at an elevated temperature. A device called an "oxygen candle" uses this reaction to generate oxygen in emergency situations and is similar to the backup oxygen generator on the space station. Iron powder in the "oxygen candle" reacts, heating the sodium chlorate to a temperature at which oxygen is released at the desired rate. Some of the original "candles" used on the space station did not work and had to be replaced. The equation for this decomposition is:

$$2NaClO_3(s) \rightarrow 3O_2(g) + 2NaCl(s)$$

To determine which species are oxidized and reduced, we start by assigning oxidation numbers. For $NaClO_3$, the guidelines state that the Na^+ in this ionic compound has a +1 oxidation number. In ClO_3^-, the rules state that each of the three oxygens will have an oxidation number of −2. For the overall charge of ClO_3^- to come out to 1−, the Cl must have an oxidation number of +5. In NaCl, the Na^+ is +1 and the Cl^- is −1.

$$\underset{+1\ +5\ -2}{2NaClO_3(s)} \rightarrow \underset{0}{3O_2(g)} + \underset{+1\ -1}{2NaCl(s)}$$

In this reaction, the oxygen in $NaClO_3$ is oxidized, changing in oxidation number from −2 to 0. The chlorine is reduced, changing in oxidation number from +5 to −1; the sodium does not change its oxidation number.

EXAMPLE 3.7 Oxidation–Reduction Reactions

Lead(II) oxide can be converted to metallic lead by reaction with carbon monoxide. The other product of the reaction is carbon dioxide. (a) Write the balanced equation for this reaction. (b) Assign oxidation numbers to each element in the reactants and products, and indicate which element is oxidized and which is reduced.

Strategy First, write a properly balanced chemical equation. Then use the rules to assign the oxidation numbers, and identify the elements that gain and lose electrons in the reaction.

Solution

(a) First, write the reactants and products.

$$PbO(s) + CO(g) \rightarrow Pb(s) + CO_2(g)$$

The coefficient of each compound is 1 in the balanced equation.

(b) All of the oxygen atoms have an oxidation number of −2. This makes +2 the oxidation number for lead in PbO and the carbon atom in CO. The elemental lead has an oxidation number of zero and the carbon atom in CO_2 has an oxidation number of +4 to balance the −2 assigned to each of the two oxygen atoms.

$$\underset{+2\ -2}{PbO(s)} + \underset{+2\ -2}{CO(g)} \rightarrow \underset{0}{Pb(s)} + \underset{+4\ -2}{CO_2(g)}$$

In the reaction, the carbon atom in CO loses two electrons and is oxidized and the lead atom in PbO gains two electrons and is reduced.

Understanding

The reaction of magnesium metal, Mg(s), with oxygen, $O_2(g)$, yields magnesium oxide, MgO(s). Write the balanced equation for this reaction, assign an oxidation number to

PRACTICE OF CHEMISTRY
Nitric and Sulfuric Acids Are Culprits in Acid Rain: No Easy Answers

The harmful effects of acid rain have been highly publicized—lakes unable to support aquatic life, dying forests, and buildings and monuments that are literally dissolving away with every rainfall. The problem is not restricted to highly industrialized nations—it is a worldwide problem. The causes are fairly well understood, but there is no simple solution to this multifaceted problem.

The increased acidity of rain is mainly caused by oxides of nitrogen and sulfur, which are present in the atmosphere from a variety of sources. Nitrogen oxides are formed in the atmosphere by electrical storms, as well as in combustion processes, particularly in automobile engines. Sulfur oxides also are major contributors to acid rain. Volcanic eruptions may spew thousands of tons of sulfur dioxide, SO_2, into the atmosphere; other natural sources are forest fires and the bacterial decay of organic matter. The main man-made sources are the burning of sulfur-containing coal and other fossil fuels, and the roasting of metal sulfides in the production of metals such as zinc and copper.

What can be done to solve the problem of acid rain? An obvious answer is to stop releasing nitrogen and sulfur oxides into the atmosphere. Some states, led by California, have enacted strict emission control standards for cars sold and licensed within their borders. Removing sulfur from fossil fuels is possible, but it is extremely expensive and technically difficult to accomplish. A cheaper but less efficient method, called *wet scrubbing,* removes SO_2 after it has been formed in fuel combustion by using mixtures of limestone to neutralize the acid formed when the gas dissolves in water. The technology for this process is fairly simple, but the installation costs and the disposal of the resulting solid waste present other problems. A number of new technologies are being developed that remove the SO_2 effectively and produce useful products rather than solid waste. Both government and industry are testing these new procedures.

Alternate energy sources would help to solve the problem, but these alternatives all have advantages and disadvantages. Solar power is one option and is being used extensively, especially at isolated locations in sunny areas. Many scientists believe that solar power is the most important energy source for the future. Nuclear power is another option that does not produce acid rain, but it poses a number of risks such as how to deal with the radioactive waste. Wind-generated power is clean from a "chemical" viewpoint, but many are resisting the "visual pollution" caused by large fields of windmills. Of course, all of the above solutions to acid rain also have important relevance to reducing the production of the greenhouse gas CO_2. Difficult choices to reduce acid rain need to be made; the problem will not just "wash" away. ▮

Volcanic eruption. The eruption of a volcano can spew large amounts of $SO_2(g)$ into the atmosphere.

each element in the reactants and products, and indicate which element is oxidized and which is reduced.

Answer

$$\underset{0}{2\text{Mg}(s)} + \underset{0}{\text{O}_2(g)} \rightarrow 2\underset{+2}{\text{Mg}}\underset{-2}{\text{O}}(s)$$

The magnesium metal is oxidized and the oxygen atoms in the O_2 are reduced.

OBJECTIVES REVIEW *Can you:*

☑ write balanced equations for chemical reactions, given chemical formulas of reactants and products?

☑ identify an acid and a base?

- ☑ identify and balance chemical equations for neutralization reactions, combustion reactions, and oxidation–reduction reactions?
- ☑ assign oxidation numbers to elements in simple compounds?

3.2 The Mole and Molar Mass

OBJECTIVES

- ☐ Express the amounts of substances using moles
- ☐ Determine the molar mass of any element or compound from its formula
- ☐ Use molar mass and Avogadro's number to interconvert between mass, moles, and numbers of atoms, ions, or molecules

The fact that individual atoms and molecules are so small and have so little mass makes it impossible to count them in the laboratory; therefore, a convenient, larger unit is needed for counting atoms and molecules. If we were selling cans of soda, we might use the unit of the six-pack; if we were selling eggs, we might use the unit of the dozen. To count atoms and molecules conveniently, we need a unit that contains many more entities than a six-pack or a dozen.

The SI unit of *amount* of substance is the *mole* (abbreviated mol). One **mole** is equal to the number of atoms in exactly 12 g of the ^{12}C isotope of carbon. The mole is the *unit* of the quantity "amount of a substance," as the meter is the unit of the quantity "length." The number of atoms in 12 g ^{12}C has been experimentally measured and found to be 6.022×10^{23} atoms (when expressed to four significant figures) and is known as **Avogadro's number,** after the 19th century physicist Amedeo Avogadro. Thus, 1 mol of anything has 6.022×10^{23} of those things, no matter what those things are. Figure 3.3 shows photographs of 1-mol quantities for several elements. Each of these samples has 6.022×10^{23} atoms of the respective element.

The mole is the SI unit for amount and contains 6.022×10^{23} atoms, ions, or molecules.

The definition of the mole and the measurement of Avogadro's number generate relationships that allow the interconversion of moles and number of atoms or molecules. For example:

$$1 \text{ mol O} = 6.022 \times 10^{23} \text{ atoms of oxygen}$$

$$1 \text{ mol } H_2O = 6.022 \times 10^{23} \text{ molecules of } H_2O$$

Either of these relationships can be used to construct conversion factors, such as was done in Chapter 1. Example 3.8 demonstrates conversion between moles to number of molecules.

Figure 3.3 One mole of elements. One mole of each of several elements: iron as a rod, liquid mercury in the cylinder, copper wire, sodium metal pictured under oil to protect it from the air, granular aluminum, and argon gas in the balloons.

The green shading indicates data that is given with the problem, the yellow indicates intermediate results, and the red is the final answer.

EXAMPLE 3.8 Conversion of Moles and Number of Atoms and Molecules

(a) How many atoms are present in 0.11 mol argon (symbol Ar)?
(b) How many moles are present in 2.67×10^{23} molecules of N_2H_4?

Strategy In both parts, the unit conversions come from the definition of the mole: 1 mol Ar = 6.022×10^{23} atoms Ar, and 1 mol N_2H_4 = 6.022×10^{23} molecules of N_2H_4.

Solution

(a) We know that 1 mol Ar = 6.022×10^{23} atoms argon; use this equality to calculate the number of argon atoms in 0.11 mol argon.

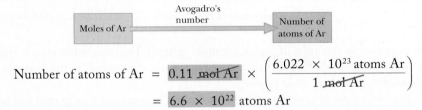

Moles of Ar → (Avogadro's number) → Number of atoms of Ar

$$\text{Number of atoms of Ar} = 0.11 \text{ mol Ar} \times \left(\frac{6.022 \times 10^{23} \text{ atoms Ar}}{1 \text{ mol Ar}}\right)$$

$$= 6.6 \times 10^{22} \text{ atoms Ar}$$

(b) From Avogadro's number, we know that 1 mol N_2H_4 = 6.022×10^{23} molecules of N_2H_4.

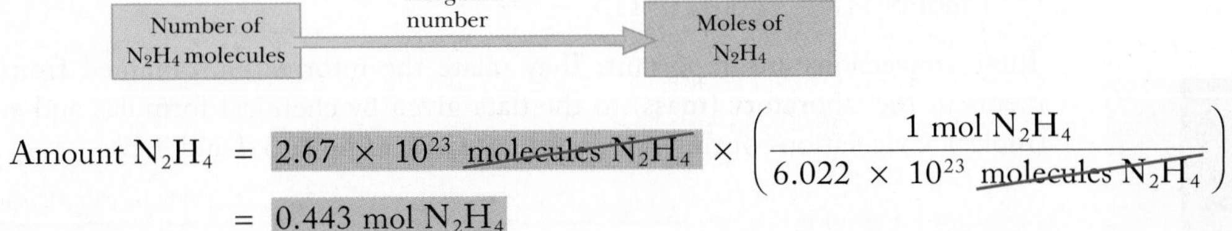

$$\text{Amount } N_2H_4 = 2.67 \times 10^{23} \text{ molecules } N_2H_4 \times \left(\frac{1 \text{ mol } N_2H_4}{6.022 \times 10^{23} \text{ molecules } N_2H_4}\right)$$

$$= 0.443 \text{ mol } N_2H_4$$

This answer, about half a mole of N_2H_4, is reasonable because about half of Avogadro's number of molecules are present.

Understanding

How many molecules are present in 0.241 mol N_2O?

Answer 1.45×10^{23} molecules of N_2O

Balanced chemical equations are balanced in terms of moles, as well as molecules. For example, the balanced chemical equation

$$H_2(g) + Cl_2(g) \rightarrow 2HCl(g)$$

implies that one molecule of hydrogen gas reacts with one molecule of chlorine gas to make two molecules of hydrogen chloride gas. It also implies that one *mole* of hydrogen gas reacts with one *mole* of chlorine gas to make two *moles* of hydrogen chloride gas. Thus, the coefficients in a balanced chemical reaction stand for molar amounts in addition to molecular amounts.

The coefficients of the balanced equation represent molecular or molar amounts.

Molar Mass

A chemical equation gives the number of molecules or moles of each reactant and product, but people working in the laboratory usually measure the *masses* of each reactant and product. The **molar mass** *(M)* of any atom, molecule, or compound is the mass (in grams) of one mole of that substance (Table 3.2). The molar mass of an element is numerically equal to its atomic mass, but in units of grams per mole (g/mol) rather than u. The molar mass of a molecular substance is numerically equal to its molecular mass,

TABLE 3.2 Molar Mass of Ar, C_2H_6, and NaF

	Atomic Scale		Laboratory Scale
Substance	Name	Mass	Molar Mass
Ar (atom)	Atomic mass	39.95 u	39.95 g/mol
C_2H_6 (molecule)	Molecular mass	30.07 u	30.07 g/mol
NaF (ionic)	Formula mass	41.98 u	41.98 g/mol

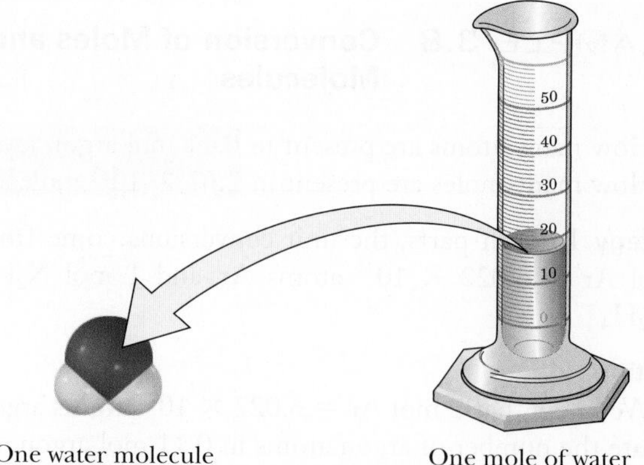

Molecular mass of H_2O = 18.0 u   Molar mass of H_2O = 18.0 g/mol

The molar mass of an atom, molecule, or ionic compound is the atomic, molecular, or formula mass, respectively, expressed in grams per mole (g/mol).

also expressed in the units of grams per mole (g/mol). The term *molar mass* is also used for ionic compounds; the molar mass of an ionic compound is the formula mass expressed in grams per mole (g/mol).

The molar mass of a substance is used to convert between mass (in grams) and amount (in moles). For example, in Chapter 2, we calculated the molecular mass of hydrazine, N_2H_4, to be 32.06 u. The mass of 1 mol hydrazine, the molar mass, is 32.06 g/mol. The molar mass provides the conversion factor that relates the mass to 1 mol of a substance:

The molar mass of an element or compound is the basis of the conversion factors used to relate mass and number of moles.

$$1 \text{ mol } N_2H_4 = 32.06 \text{ g } N_2H_4$$

These conversions are important. They relate the information obtained from experiments in the laboratory (mass) to the data given by chemical formulas and equations (moles). Calculations of this type are shown in Examples 3.9 and 3.10.

(*a*)

(*b*)

Dr. Marita Cantwell

Scientists have discovered that plants produce ethylene (C_2H_4) as part of the ripening process. *(a)* Boxes of green tomatoes are placed in ventilated chambers where a small amount of ethylene gas is added to accelerate the ripening process. *(b)* The green tomatoes on the left were picked at the same time as the red ones on the right, but the ones on the right have been in the ethylene-filled chamber for a few days.

EXAMPLE 3.9 Converting Moles to Mass

Ethylene, C_2H_4, is used in such diverse applications as the ripening of tomatoes and the preparation of plastics.

(a) Calculate the molar mass of ethylene.
(b) Calculate the mass of 3.22 mol ethylene.

Strategy Use the formula of ethylene, C_2H_4, and the atomic masses of its two elements to calculate its molar mass, and use the molar mass as the conversion between mass and moles.

Solution

(a) The molecular mass of C_2H_4 is the sum of the atomic mass of each of its elements multiplied by the numbers of each type of atoms present:

$$\begin{array}{ll} 2(\text{C}) & 2 \times 12.01 \text{ u} = 24.02 \text{ u} \\ 4(\text{H}) & 4 \times 1.008 \text{ u} = \ \ 4.03 \text{ u} \\ \hline \multicolumn{2}{l}{\text{Molecular mass } C_2H_4 = 28.05 \text{ u}} \end{array}$$

Given that the molar mass of a molecular substance is numerically equal to its molecular mass, the molar mass of C_2H_4 is 28.05 g/mol.

(b) The molar mass of C_2H_4 is 28.05 g/mol, and this value is used as the conversion factor to calculate the mass of 3.22 mol C_2H_4.

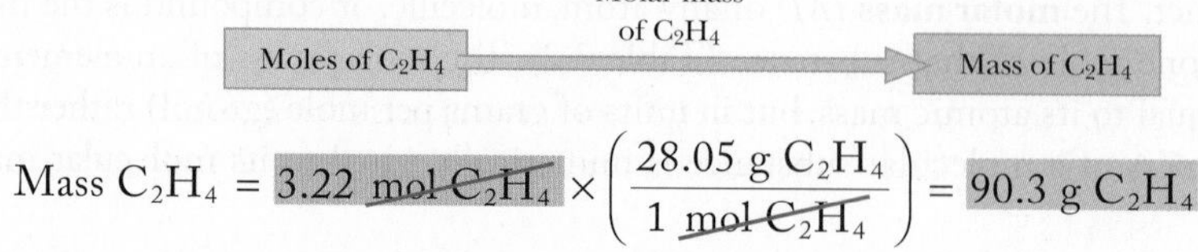

$$\text{Mass } C_2H_4 = 3.22 \text{ mol } C_2H_4 \times \left(\frac{28.05 \text{ g } C_2H_4}{1 \text{ mol } C_2H_4}\right) = 90.3 \text{ g } C_2H_4$$

Understanding

What is the mass of 43.1 mol phosphorus pentachloride?

Answer 8.97×10^3 g, or 8.97 kg

EXAMPLE 3.10 Converting Mass to Moles

How many moles are present in 14.2 g hydrazine, N_2H_4?

Strategy Use the molar mass of N_2H_4 as the conversion factor for the conversion of grams N_2H_4 into moles N_2H_4.

Solution

The molar mass of N_2H_4 is 32.06 g/mol. Because 14.2 g is about half the mass of 1 mol, we can estimate even before doing any calculation that our final answer will be about half a mole of N_2H_4. The molar mass (32.06 g/mol N_2H_4) is used for the exact conversion of grams into moles.

Mass of N_2H_4 → (Molar mass of N_2H_4) → Moles of N_2H_4

$$\text{Amount } N_2H_4 = 14.2 \text{ g } N_2H_4 \times \left(\frac{1 \text{ mol } N_2H_4}{32.06 \text{ g } N_2H_4}\right) = 0.443 \text{ mol } N_2H_4$$

Understanding

How many moles are present in a 12.7-g sample of NO?

Answer 0.423 mol

We frequently need to know the number of moles of each element present in a compound. The formula N_2H_4 indicates that there are two moles of nitrogen atoms and four moles of hydrogen atoms in every mole of N_2H_4. For N_2H_4,

1 mol N_2H_4 contains 2 mol N, and 4 mol H

This information in the formulas of compounds generates conversion factors such as

$$\frac{2 \text{ mol N}}{1 \text{ mol } N_2H_4} \quad \text{or} \quad \frac{4 \text{ mol H}}{1 \text{ mol } N_2H_4}$$

The reciprocals of these expressions can also be used as conversion factors. These conversion factors can be used to calculate the number of moles of *atoms* present and are specific to the compound being studied. For example, calculate the number of moles of nitrogen and hydrogen atoms in 0.443 mol hydrazine:

$$\text{Amount N} = 0.443 \text{ mol } N_2H_4 \times \frac{2 \text{ mol N}}{1 \text{ mol } N_2H_4} = 0.886 \text{ mol N}$$

$$\text{Amount H} = 0.443 \text{ mol } N_2H_4 \times \frac{4 \text{ mol H}}{1 \text{ mol } N_2H_4} = 1.77 \text{ mol H}$$

Moles of N_2H_4 → (Formula of N_2H_4) → Moles of N and H atoms

EXAMPLE 3.11 Moles of Atoms

Shown in the margin is one molecule of methylamine; the NH_2 group is an example of an amine functional group, a group present in many important chemicals. Write the formula of methylamine and use it to calculate the number of moles of hydrogen atoms in 0.22 mol methylamine.

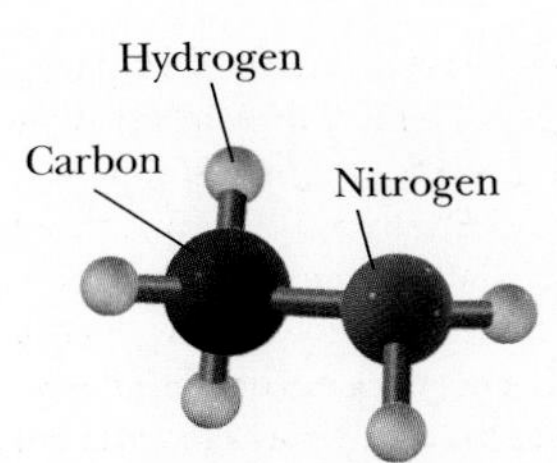

Methylamine.

Strategy From the figure, count the number of each type of atom. Then use the formula to generate the conversion factor to calculate the moles of hydrogen.

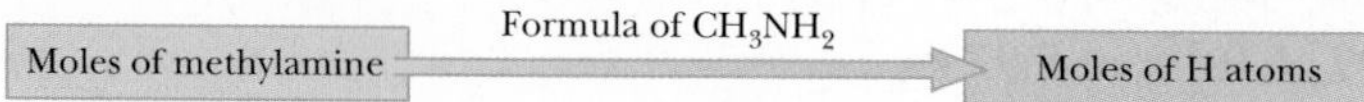

Solution

There are five hydrogen atoms (three attached to C and two attached to N) and one carbon and one nitrogen atom in the figure—the formula is generally written as CH_3NH_2. For CH_3NH_2,

1 mol CH_3NH_2 contains 5 mol H

The conversion factor needed to calculate the moles of hydrogen atoms is thus

$$\frac{5 \text{ mol H}}{1 \text{ mol } CH_3NH_2}$$

$$\text{Amount H} = 0.22 \text{ mol } CH_3NH_2 \times \left(\frac{5 \text{ mol H}}{1 \text{ mol } CH_3NH_2}\right) = 1.1 \text{ mol H}$$

Understanding

How many moles of C are present in 0.22 mol CH_3NH_2?

Answer 0.22 mol C

OBJECTIVES REVIEW *Can you:*

- ☑ express the amounts of substances using moles?
- ☑ determine the molar mass of any element or compound from its formula?
- ☑ use molar mass and Avogadro's number to interconvert between mass, moles, and numbers of atoms, ions, or molecules?

3.3 Chemical Formulas

OBJECTIVES

- ☐ Calculate the mass percentage of each element in a compound (percentage composition) from the chemical formula of that compound
- ☐ Calculate the mass of each element present in a sample from elemental analysis data such as that produced by combustion analysis
- ☐ Determine the empirical formula of a compound from mass or mass percentage data
- ☐ Use molar mass to determine a molecular formula from an empirical formula

Percentage Composition of Compounds

When a new compound is prepared or isolated, one of the first experimental tasks is to determine its molecular formula. A molecular formula represents, in part, numerical information. For example, from the formula of benzene, C_6H_6, we know that one molecule of benzene consists of six carbon and six hydrogen atoms. We also know that one mole of benzene contains six moles of carbon and six moles of hydrogen atoms. This numerical information can be used to determine the molecular mass of benzene:

$$\begin{array}{lr} 6(\text{C})\ 6 \times 12.01 \text{ u} = & 72.06 \text{ u} \\ 6(\text{H})\ 6 \times 1.01 \text{ u} = & 6.06 \text{ u} \\ \hline \text{Molecular mass } C_6H_6 = & 78.12 \text{ u} \end{array}$$

From this molecular mass calculation, we know that 1 mol benzene has a total mass of 78.12 g from 72.06 g carbon and 6.06 g hydrogen. This information can be used to calculate the mass percentage of each element.

$$\text{Mass percentage C} = \frac{72.06 \text{ g C}}{78.12 \text{ g compound}} \times 100\% = 92.24\% \text{ C}$$

$$\text{Mass percentage H} = \frac{6.06 \text{ g H}}{78.12 \text{ g compound}} \times 100\% = 7.76\% \text{ H}$$

The mass percentage of each element in a compound is calculated from the chemical formula and the atomic masses of each element.

An important observation is we obtain the same answer for any compound having the general formula C_nH_n. For example, the compound acetylene, C_2H_2, also is 92.24% carbon and 7.76% hydrogen by mass. Thus, the percentage composition of a compound can be based on its *empirical formula* (the *relative* numbers of atoms of the elements in a compound expressed as the smallest whole-number ratio), as well as on its molecular formula. The percentage composition calculated from the empirical formula of both benzene and acetylene, CH, is the same as that calculated from the molecular formulas.

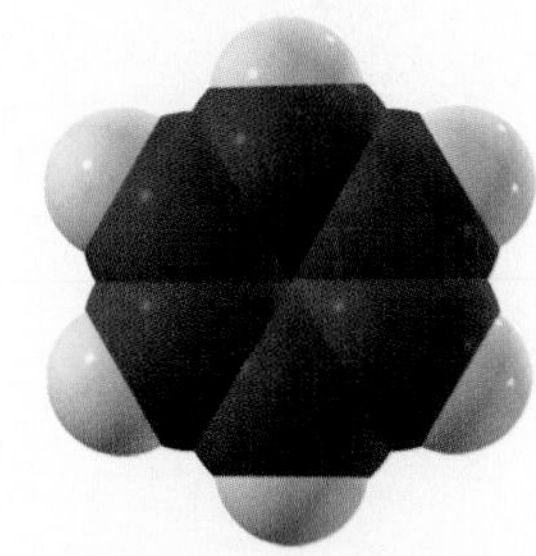

Molecular model of benzene.

EXAMPLE 3.12 Mass Composition

Aspirin is a remarkable analgesic (painkiller) that also appears to prevent certain heart conditions. From its chemical formula, $C_9H_8O_4$, calculate the percentage by mass of each element in aspirin.

Strategy A flow diagram outlines the strategy for this problem. The formula is used to calculate the masses of each element in the compound, and those numbers are used to calculate the percentage composition.

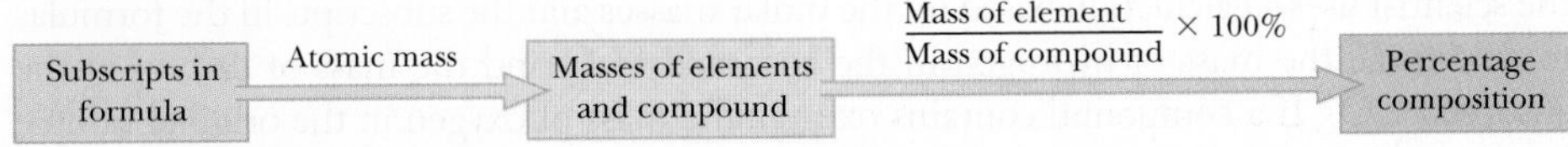

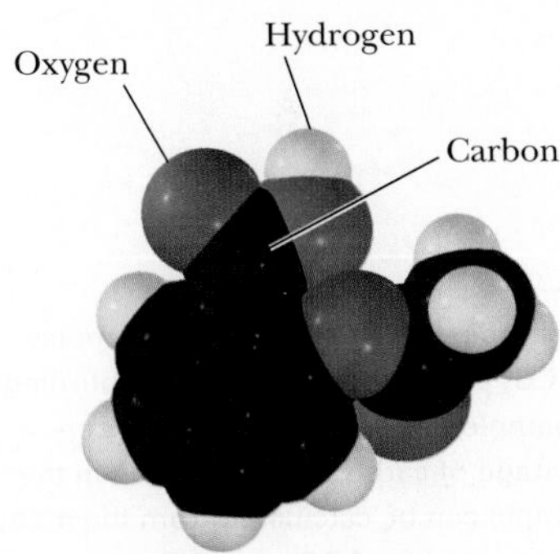

Molecular model of aspirin.

Solution

First, calculate the molecular mass of aspirin.

$$\begin{array}{ll} 9(\text{C}) & 9 \times 12.01 \text{ u} = 108.09 \text{ u} \\ 8(\text{H}) & 8 \times 1.01 \text{ u} = 8.08 \text{ u} \\ 4(\text{O}) & 4 \times 16.00 \text{ u} = 64.00 \text{ u} \\ \hline \text{Molecular mass } C_9H_8O_4 & = 180.17 \text{ u} \end{array}$$

Thus, 1 mol aspirin has a mass of 180.17 g that comes from 108.09 g carbon, 8.08 g hydrogen, and 64.00 g oxygen. Use these values to express the mass composition of the compound as percentages.

$$\% \text{ C:} \quad \frac{108.09 \text{ g C}}{180.17 \text{ g } C_9H_8O_4} \times 100\% = 59.99\% \text{ C}$$

$$\% \text{ H:} \quad \frac{8.08 \text{ g H}}{180.17 \text{ g } C_9H_8O_4} \times 100\% = 4.48\% \text{ H}$$

$$\% \text{ O:} \quad \frac{64.00 \text{ g O}}{180.17 \text{ g } C_9H_8O_4} \times 100\% = 35.52\% \text{ O}$$

Understanding

Calculate the percentage by mass of each element in $C_2H_2F_4$.

Answer 23.54% C, 1.98% H, 74.48% F

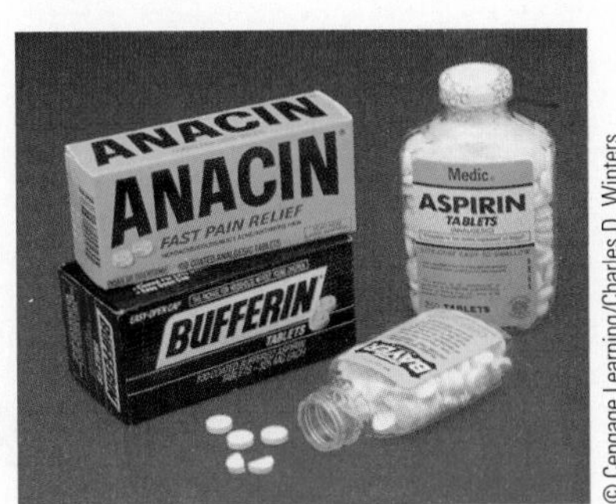

Aspirin is an important analgesic and, taken in low doses, appears to reduce chances of fatal heart attacks.

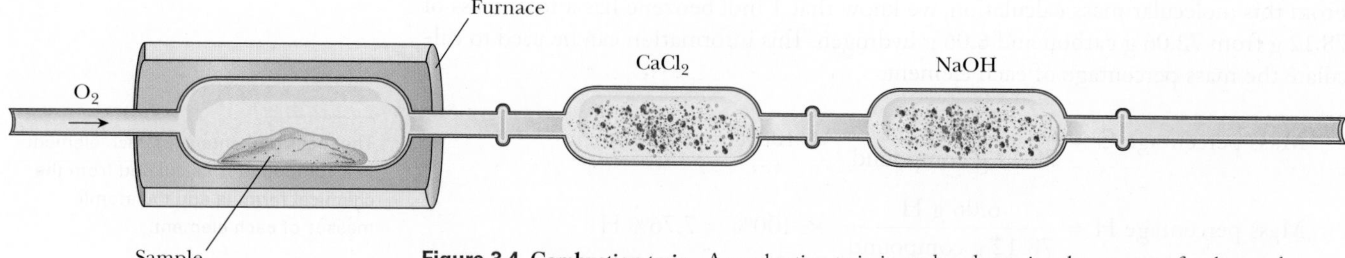

Figure 3.4 Combustion train. A combustion train is used to determine the amount of carbon and hydrogen in a compound. $CaCl_2$ is frequently used in the first trap because it absorbs the H_2O but not the CO_2. Sodium hydroxide is used in the second trap. It cannot be used in the first trap because NaOH absorbs both H_2O and CO_2.

Combustion Analysis

By reversing the mass percentage calculation, chemists can calculate the empirical formula of a newly prepared compound. Chemists have developed a number of experimental methods to determine mass percentages. One important experiment is **combustion analysis,** which determines the quantity of carbon and hydrogen in a sample of an organic compound. Figure 3.4 is a schematic diagram of an apparatus that can be used in this experiment.

In this analysis, a small, carefully weighed sample of a compound is completely burned in a stream of O_2. Oxygen is added to ensure complete conversion of all the carbon into CO_2 and all the hydrogen into H_2O. The H_2O is collected in the first trap, and the CO_2 is collected in the second trap. The two traps are weighed before and after the combustion of the sample to determine the masses of the H_2O and CO_2 absorbed. In this type of experiment, the mass of each of the *elements* present, the desired information, is not determined directly. Instead, the scientist uses a calculation based on the molar masses and the subscripts in the formulas to determine the mass of hydrogen in the absorbed H_2O and the mass of carbon in the absorbed CO_2. If a compound contains oxygen, the mass of oxygen in the original sample has to be determined by subtraction, after the masses of the other elements have been determined.

A combustion analysis gives the mass of CO_2 and H_2O produced when burning a sample in excess oxygen. The percentage of carbon and hydrogen in the sample can be calculated from the measured masses of CO_2 and H_2O.

EXAMPLE 3.13 Combustion Analysis

A chemist uses the apparatus shown in Figure 3.4 to determine the composition of a compound made up of only carbon, hydrogen, and oxygen. During combustion of a 0.1000-g sample, the mass of the first trap (collecting H_2O) increased by 0.0928 g H_2O and the mass of the second trap (collecting CO_2) increased by 0.228 g CO_2. (Note that it is possible for the sum of the masses of the H_2O and CO_2 to be greater than the mass of the starting sample because some of the oxygen in the products comes from the added O_2.) Calculate the mass and mass percentage of each element in the 0.1000-g sample.

Strategy The formulas of water and carbon dioxide, coupled with the periodic table, allow the conversion of mass of each compound to mass of the two respective elements by using the following sequence: mass compound → moles compound; moles compound → moles of element; moles element → grams element. Oxygen is calculated by difference from the mass of C and H, and the total mass of the sample.

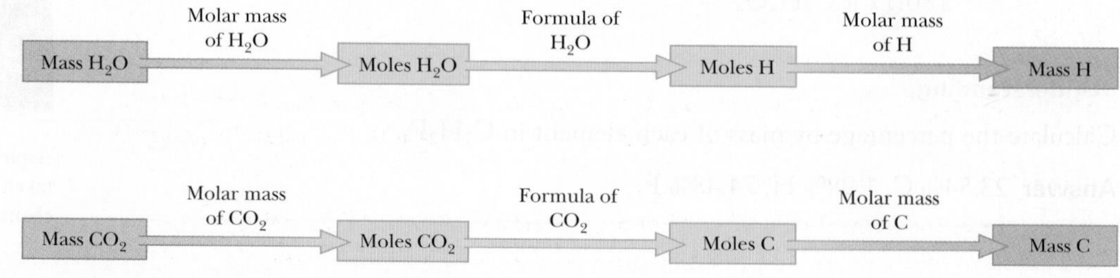

Solution

From the formulas, 1 mol H_2O has a mass of 18.02 g, and 1 mol CO_2 has a mass of 44.01 g. Use these values to convert the masses of CO_2 and H_2O determined in the combustion analysis into moles.

$$\text{Amount } H_2O = 0.0928 \text{ g } H_2O \times \left(\frac{1 \text{ mol } H_2O}{18.02 \text{ g } H_2O}\right) = 0.00515 \text{ mol } H_2O$$

$$\text{Amount } CO_2 = 0.228 \text{ g } CO_2 \times \left(\frac{1 \text{ mol } CO_2}{44.01 \text{ g } CO_2}\right) = 0.00518 \text{ mol } CO_2$$

From the formulas of water and carbon dioxide, we know that 1 mol H_2O has 2 mol H, and 1 mol CO_2 has 1 mol C. These relationships are used to convert moles of each compound into moles of each element:

$$\text{Amount H} = 0.0515 \text{ mol } H_2O \times \left(\frac{2 \text{ mol H}}{1 \text{ mol } H_2O}\right) = 0.0103 \text{ mol H}$$

$$\text{Amount C} = 0.00518 \text{ mol } CO_2 \times \left(\frac{1 \text{ mol C}}{1 \text{ mol } CO_2}\right) = 0.00518 \text{ mol C}$$

Use the molar mass of hydrogen and carbon to calculate the mass of each present in the sample.

$$\text{Mass H} = 0.0103 \text{ mol H} \times \left(\frac{1.01 \text{ g H}}{1 \text{ mol H}}\right) = 0.0104 \text{ g H}$$

$$\text{Mass C} = 0.00518 \text{ mol C} \times \left(\frac{12.01 \text{ g C}}{1 \text{ mol C}}\right) = 0.0622 \text{ g C}$$

Note that it is possible to combine these into a chain calculation.

$$\text{Mass of H} = 0.0928 \text{ g } H_2O \times \left(\frac{1 \text{ mol } H_2O}{18.02 \text{ g } H_2O}\right)\left(\frac{2 \text{ mol H}}{1 \text{ mol } H_2O}\right)\left(\frac{1.01 \text{ g H}}{1 \text{ mol H}}\right) = 0.0104 \text{ g H}$$

$$\text{Mass of C} = 0.228 \text{ g } CO_2 \times \left(\frac{1 \text{ mol } CO_2}{44.01 \text{ g } CO_2}\right)\left(\frac{1 \text{ mol C}}{1 \text{ mol } CO_2}\right)\left(\frac{12.01 \text{ g C}}{1 \text{ mol C}}\right) = 0.0622 \text{ g C}$$

The mass of oxygen in the sample cannot be determined directly in this experiment. Because the sample contains only carbon, hydrogen, and oxygen, the mass of oxygen is determined by subtracting the mass of carbon and hydrogen from the total mass, 0.1000 g, of the sample.

$$\text{Mass}_{\text{sample}} = \text{mass}_C + \text{mass}_H + \text{mass}_O$$

$$\text{Mass}_O = \text{mass}_{\text{sample}} - (\text{mass}_C + \text{mass}_H)$$

$$\text{Mass}_O = 0.1000 \text{ g total} - (0.0622 \text{ g C} + 0.0104 \text{ g H}) = 0.0274 \text{ g O}$$

In terms of mass percentages:

$$\text{Mass percentage C} = \frac{0.0622 \text{ g C}}{0.1000 \text{ g total}} \times 100\% = 62.2\% \text{ C}$$

$$\text{Mass percentage H} = \frac{0.0104 \text{ g H}}{0.1000 \text{ g total}} \times 100\% = 10.4\% \text{ H}$$

$$\text{Mass percentage O} = \frac{0.0274 \text{ g O}}{0.1000 \text{ g total}} \times 100\% = 27.4\% \text{ O}$$

Understanding

Combustion of a 0.2000-g sample of a compound made up of only carbon, hydrogen, and oxygen yields 0.200 g H_2O and 0.4880 g CO_2. Calculate the mass and mass percentage of each element present in the 0.2000-g sample.

Answer 0.1332 g C, 0.0224 g H, 0.0444 g O; 66.6% C, 11.2% H, 22.2% O

Empirical Formulas

New compounds are isolated every day. Each of these new compounds needs to be characterized, and an important part of that is to determine the molecular formula. The first step in this process is to experimentally determine the empirical formula. The empirical formula of a compound can be determined from either the masses or mass percentages of the elements in a sample. This calculation yields *only* the empirical formula, because the composition by mass is based only on the relative number of atoms of each element in the compound. The empirical formula is usually all you need to describe the composition of an *ionic* compound. Additional experimental information (such as the molar mass of the compound) is needed to determine the correct formula of a *molecular* compound.

Consider an example of an experiment designed to determine the empirical formula of a compound. A combustion analysis experiment similar to that described previously shows that a 2.000-g sample of a particular compound consists of 1.714 g carbon, 0.286 g hydrogen, and no other elements. If we can calculate the relative number of moles of each element, we will also know the relative number of atoms, and thus the empirical formula. To calculate the relative number of moles, we need to convert the mass of each element into moles of *atoms* of that element. The molar mass of each element is used for this conversion.

$$1 \text{ mol C} = 12.011 \text{ g C}$$

$$1 \text{ mol H} = 1.008 \text{ g H}$$

From the masses of the elements (as determined in the experiment), we calculate the number of moles of each element in this sample:

$$\text{Amount C} = 1.714 \text{ g C} \times \left(\frac{1 \text{ mol C}}{12.011 \text{ g C}}\right) = 0.1427 \text{ mol C}$$

$$\text{Amount H} = 0.286 \text{ g H} \times \left(\frac{1 \text{ mol H}}{1.008 \text{ g H}}\right) = 0.284 \text{ mol H}$$

Thus, the empirical formula of this compound is $C_{0.1427}H_{0.284}$, but this is not how we typically express chemical formulas. In formulas, the relative number of atoms are expressed as *whole numbers*. The molar values must to be adjusted to whole numbers. A method of making this adjustment that is usually successful is to divide the number of moles of each element by the *smallest* number of moles found. This procedure will convert the smallest number to 1 and all other values to a number greater than 1, without changing their relative values.

$$\text{Amount C} = 0.1427 \text{ mol C} \div 0.1427 = 1.000 \text{ mol C}$$

$$\text{Amount H} = 0.284 \text{ mol H} \div 0.1427 = 1.99 \text{ mol H}$$

The molar masses of elements are used to calculate empirical formulas from mass composition.

The proper empirical formula is CH_2—the 1.99 is within experimental error of a whole number.

The following steps represent the process for determining the empirical formula.

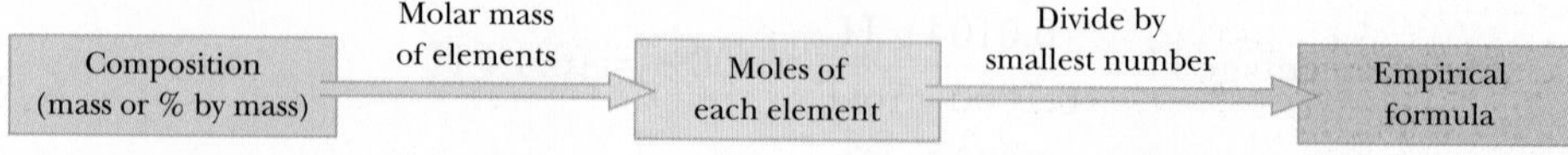

The method of dividing through by the smallest number does not always yield whole numbers; it just makes the smallest subscript equal to 1. Many empirical formulas, such as $C_5H_{10}O_3$, do not have any of the elements with a subscript of 1. In these cases, dividing through by the smallest number yields numbers that end in decimals that are simple fractions, such as 0.25 (1/4), 0.33 (1/3), 0.5 (1/2), and 0.67 (2/3). These numbers cannot be rounded to the nearest whole number; instead, multiply all the subscripts by the number that converts them to integers (generally the denominator of the fraction). For example, an experiment to determine the empirical formula of $C_5H_{10}O_3$ might yield numbers such as C = 0.0924 mol, H = 0.186 mol, and O = 0.0556 mol. Dividing by the smallest value, 0.0556, yields the values C = 1.66 mol, H = 3.34 mol, and O = 1.00 mol. To find the correct formula, we must multiply each of these numbers by the *same* smallest whole number (so the ratio is not changed) that will convert them all to integers. In this case, both decimals are fractions with 3 in the denominator, so multiplying each number by 3 converts all of them to whole numbers: C = 4.98, H = 10.0, and O = 3.00. This method yields the correct empirical formula of $C_5H_{10}O_3$ (because 4.98 is within experimental error of 5). The overall strategy follows.

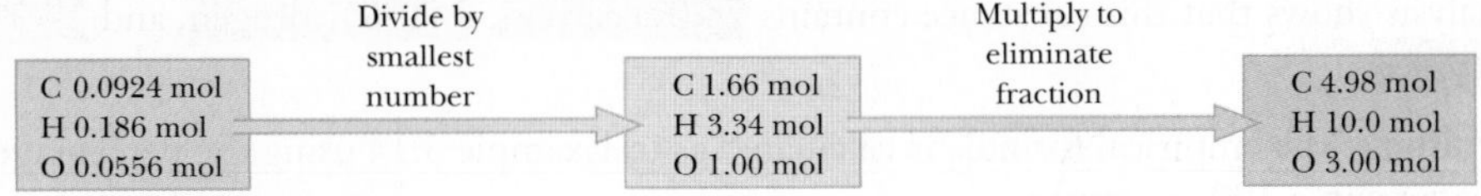

EXAMPLE 3.14 Empirical Formulas

Analysis of a 0.330-g sample of a compound shows that it contains 0.226 g chromium and 0.104 g oxygen. What is the empirical formula?

Strategy The strategy is outlined in the following diagram.

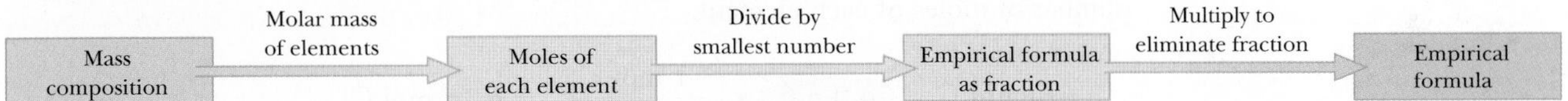

The mass composition values are converted to moles using periodic table information and the molar values converted to whole numbers.

Solution

First, determine the number of moles of each element present in the sample.

$$\text{Amount Cr} = 0.226\ \text{g Cr} \times \left(\frac{1\ \text{mol Cr}}{52.00\ \text{g Cr}}\right) = 0.00435\ \text{mol Cr}$$

$$\text{Amount O} = 0.104\ \text{g O} \times \left(\frac{1\ \text{mol O}}{16.00\ \text{g O}}\right) = 0.00650\ \text{mol O}$$

Divide by the smallest number of moles found.

$$\text{Amount Cr} = 0.00435\ \text{mol Cr} \div 0.00435 = 1.00\ \text{mol Cr}$$
$$\text{Amount O} = 0.00650\ \text{mol O} \div 0.00435 = 1.49\ \text{mol O}$$

The data given in this problem are known to three significant figures. Because 1.49 cannot be rounded to the nearest integer, we need to multiply each mole value by the same smallest number that will convert them into integers. In this case, *multiply all of the molar quantities by 2.*

$$\text{Amount Cr} = 1.00\ \text{mol Cr} \times 2 = 2.00\ \text{mol Cr}$$
$$\text{Amount O} = 1.49\ \text{mol O} \times 2 = 2.98\ \text{mol O}$$

These coefficients are now integers within the accuracy of the measurement, and the empirical formula is Cr_2O_3.

This green solid is Cr_2O_3.

Understanding

Analysis of a substance shows that a 0.902-g sample contains 0.801 g carbon and 0.101 g hydrogen. What is the empirical formula of the substance?

Answer C_2H_3

The mass percentage of each element in a compound is equal to the number of grams of each element present in a 100.00-g sample.

Empirical formulas are often determined from the results of experiments that provide mass percentage composition. If the composition is given as percentages, assume that a 100.00-g sample has been analyzed. The percentage of each element is then equal to the number of grams of that element in the 100.00-g sample. The next example illustrates determining an empirical formula from mass percentage composition.

EXAMPLE 3.15 Empirical Formulas from Percentage Data

Calculate the empirical formula of a compound extracted from tobacco. Chemical analysis shows that this substance contains 74.0% carbon, 8.70% hydrogen, and 17.3% nitrogen.

Strategy The empirical formula is calculated as in Example 3.14 using the percentages as grams in a 100-g sample.

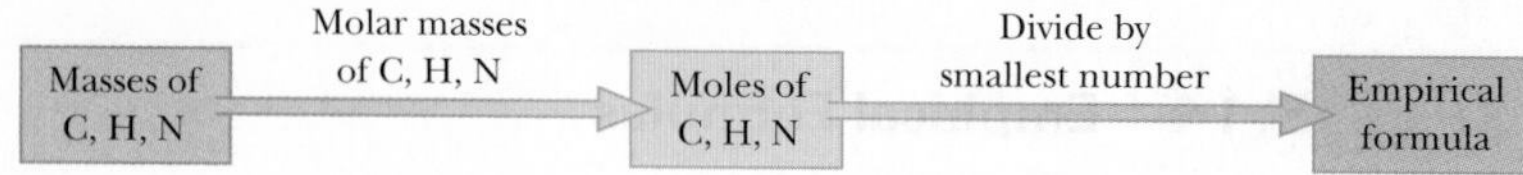

Solution

Assume the sample has a mass of exactly 100 g. This sample contains 74.0 g carbon, 8.70 g hydrogen, and 17.3 g nitrogen. Use the molar masses of the elements to calculate the number of moles of each element.

$$\text{Amount C} = 74.0\ \cancel{\text{g C}} \times \left(\frac{1\ \text{mol C}}{12.01\ \cancel{\text{g C}}}\right) = 6.16\ \text{mol C}$$

$$\text{Amount H} = 8.70\ \cancel{\text{g H}} \times \left(\frac{1\ \text{mol H}}{1.008\ \cancel{\text{g H}}}\right) = 8.63\ \text{mol H}$$

$$\text{Amount N} = 17.3\ \cancel{\text{g N}} \times \left(\frac{1\ \text{mol N}}{14.01\ \cancel{\text{g N}}}\right) = 1.23\ \text{mol N}$$

This calculation yields the relative number of moles of each element. To convert to integers, divide each by the smallest number, 1.23:

$$\text{Amount C} = 6.16\ \text{mol C} \div 1.23 = 5.01\ \text{mol C}$$

$$\text{Amount H} = 8.63\ \text{mol H} \div 1.23 = 7.02\ \text{mol H}$$

$$\text{Amount N} = 1.23\ \text{mol N} \div 1.23 = 1.00\ \text{mol N}$$

The empirical formula is C_5H_7N.

Understanding

Analysis of a substance shows that its composition is 30.4% nitrogen and 69.6% oxygen. What is the empirical formula of the substance?

Answer NO_2

Molecular Formulas

The methods listed earlier can yield only empirical formulas. The molecular formula must be known to properly characterize a new molecular compound. For example, both C_2H_2 and C_6H_6 will yield the same results in a combustion analysis, but they have different properties. To calculate the molecular formula from the empirical formula, we must know the molar mass of the compound from experiment. For example, a compound with an empirical formula of CH_2 is found experimentally (such experiments are described later) to have a molar mass of 42 g/mol. The molar mass of the empirical formula, CH_2, is 14 g/mol. The molecular formula must be a whole-number multiple of the empirical formula, CH_2, C_2H_4, C_3H_6. . . . $(CH_2)_n$, where n is the number of times the empirical formula occurs in the molecular formula. The value of n is calculated as follows:

$$n = \text{molar mass of compound/molar mass of empirical formula}$$

In this case,

$$n = \frac{42 \text{ g/mol}}{14 \text{ g/mol}} = 3$$

The molecular formula is $(CH_2)_3$ or C_3H_6, three times the empirical formula.

The molar mass of the compound must be known to determine a molecular formula from the empirical formula.

EXAMPLE 3.16 Molecular Formulas

In Example 3.15, the empirical formula of a compound extracted from tobacco was calculated to be C_5H_7N. In a separate experiment, the molar mass of this compound is found to be 162 g/mol. Calculate the molecular formula.

Strategy The molecular formula is determined from the empirical formula by dividing the molar mass found in the experiment by the molar mass of the empirical formula and multiplying the subscripts of the empirical formula by the resultant whole number.

Solution

To determine the molecular formula, we need to calculate n for the formula $(C_5H_7N)_n$. The empirical formula C_5H_7N has a molar mass of 81 g/mol, and the molar mass of the compound was found to be 162 g/mol.

$$n = \frac{162 \text{ g/mol}}{81 \text{ g/mol}} = 2$$

The molecular formula is therefore $(C_5H_7N)_2$, or $C_{10}H_{14}N_2$. The compound is nicotine, which is found in tobacco leaves and is widely used as an agricultural insecticide. Figure 3.5 is a computer drawing of the structure of nicotine.

Understanding

Determine the molecular formula of a compound that has the empirical formula of C_4H_8O and has a molar mass of 144 g/mol.

Answer $C_8H_{16}O_2$

Figure 3.5 Structure of nicotine. Nicotine, $C_{10}H_{14}N_2$, has an interesting structure containing both a five-member and a six-member ring of atoms.

OBJECTIVES REVIEW *Can you:*

- ☑ calculate the mass percentage of each element in a compound (percentage composition) from the chemical formula of that compound?
- ☑ calculate the mass of each element present in a sample from elemental analysis data such as that produced by combustion analysis?
- ☑ determine the empirical formula of a compound from mass or mass percentage data?
- ☑ use molar mass to determine a molecular formula from an empirical formula?

3.4 Mass Relationships in Chemical Equations

OBJECTIVES

- ☐ Calculate the mass of a product formed or a reactant consumed in a chemical reaction
- ☐ Determine theoretical yields from reaction data

Chemists often need to calculate the masses of reactants needed to produce a given amount of a product. The chemical equation, most importantly the coefficients, provides the starting point for these calculations. This section presents the quantitative methods used to predict the relationships between masses of reactants and products.

The chemical equation is a convenient and quantitative way to describe any chemical reaction (see Section 3.1). The equation not only tells us what happens, it also expresses stoichiometry, the quantitative relationships among the species involved, in molecular or molar amounts. For example, the following information can be interpreted in terms of either molecules or moles.

The coefficients of the balanced chemical equation relate amounts of each substance in the equation to any other substance in the equation. The equation for the formation of water from hydrogen and oxygen generates a series of conversion factors that allow us to calculate the number of moles of one substance in the equation if we know the number of moles of another substance in the equation.

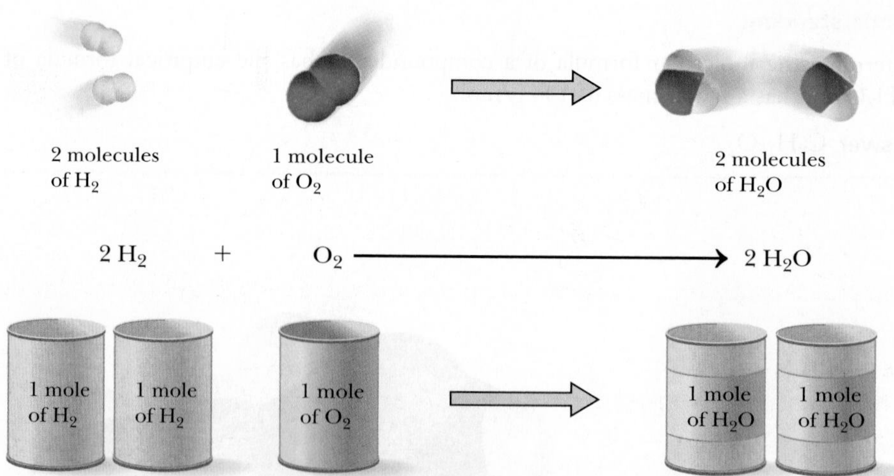

Because we know how to calculate moles of any substance from the mass of the substance and vice versa, this knowledge can be combined with the stoichiometry of the chemical equation to answer questions such as: "What mass of water is produced when 5.0 g hydrogen is burned with excess oxygen?" A balanced chemical equation is necessary if we are to successfully relate the amounts of two or more substances involved in a reaction. The key to these conversions is that chemical equations quantitatively express stochiometric relationships in both numbers of molecules and in moles.

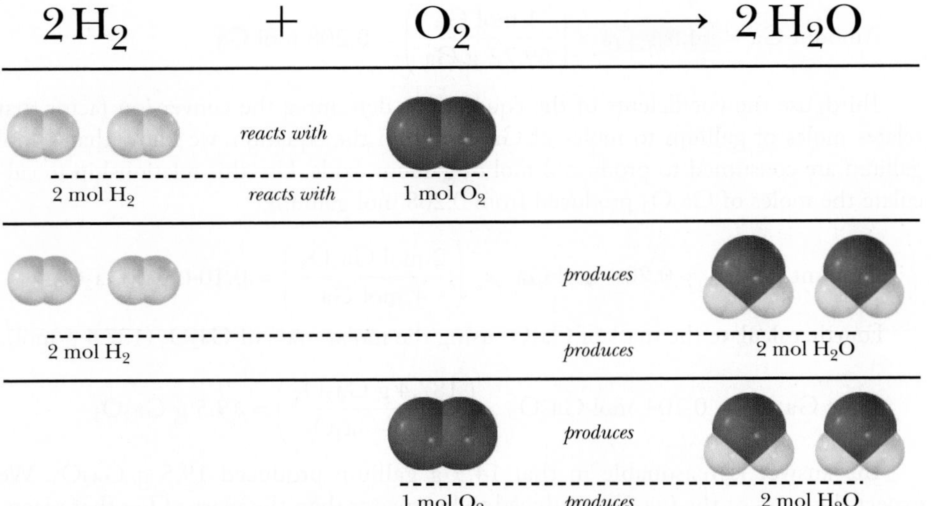

Several different ways of expressing the same stoichiometric relationship.

The complete procedure for using an equation to calculate the mass of a product formed or a reactant consumed in a chemical reaction is as follows:

1. Write the balanced chemical equation.
2. Start with the (given) mass of one substance and calculate the number of moles of this substance, using the appropriate mass–mole conversion factor.
3. Use the coefficients of the balanced equation to calculate the moles of the desired substance from the moles of the given substance.
4. Calculate the mass of the desired substance, using the appropriate mole–mass conversion factor.

Steps 2 and 4 involve unit conversions based on the molar masses calculated from the atomic mass given on the periodic table. Step 3 involves a unit conversion based on the coefficients in the balanced equation written in step 1. The following diagram summarizes this procedure.

The coefficients in a chemical equation are used to calculate moles of one substance in the equation from the known number of moles of a second substance.

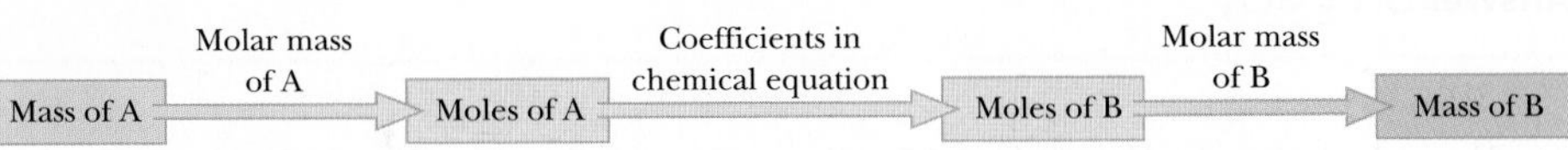

EXAMPLE 3.17 Stoichiometry Calculations

Determine the mass of Ga_2O_3 formed from the reaction of 14.5 g gallium metal with excess O_2.

Strategy Solve the problem by using the four-step procedure just outlined: (1) Write the balanced equation; (2) calculate the moles of the given substance, gallium, from the mass; (3) use the chemical equation to calculate the moles of desired species, Ga_2O_3, from the moles of gallium; and (4) finish the problem by calculating the mass of Ga_2O_3 from the moles of Ga_2O_3. The information needed for steps 2 and 4 comes from the periodic table and the chemical formulas, and step 3 from the coefficients in the equation. Because the oxygen is in excess, the amount of product formed will depend only on the amount of the gallium.

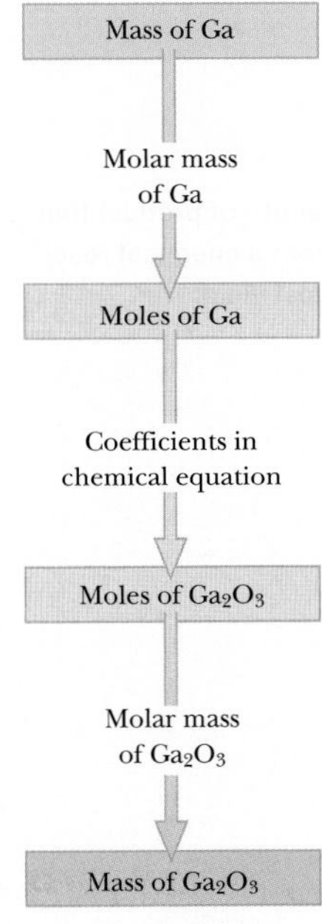

Solution

First, write the balanced equation.

$4Ga(s) + 3O_2(g) \rightarrow 2Ga_2O_3(s)$

Second, calculate the number of moles of the given substance, gallium. The molar mass of gallium is 69.72 g/mol.

$$\text{Amount Ga} = 14.5\ \cancel{\text{g Ga}} \times \left(\frac{1\ \text{mol Ga}}{69.72\ \cancel{\text{g Ga}}}\right) = 0.208\ \text{mol Ga}$$

Third, use the coefficients of the equation to determine the conversion factor that relates moles of gallium to moles of Ga_2O_3. From the equation, we know that 4 mol gallium are consumed to produce 2 mol of gallium oxide. Use this relationship to calculate the moles of Ga_2O_3 produced from 0.208 mol gallium:

$$\text{Amount Ga}_2\text{O}_3 = 0.208\ \cancel{\text{mol Ga}} \times \left(\frac{2\ \text{mol Ga}_2\text{O}_3}{4\ \cancel{\text{mol Ga}}}\right) = 0.104\ \text{mol Ga}_2\text{O}_3$$

Fourth, calculate the mass of Ga_2O_3 using the molar mass of Ga_2O_3 (187.4 g/mol).

$$\text{Mass Ga}_2\text{O}_3 = 0.104\ \cancel{\text{mol Ga}_2\text{O}_3} \times \left(\frac{187.4\ \text{g Ga}_2\text{O}_3}{1\ \cancel{\text{mol Ga}_2\text{O}_3}}\right) = 19.5\ \text{g Ga}_2\text{O}_3$$

The answer is reasonable in that 14.5 g gallium produced 19.5 g Ga_2O_3. We expect the mass of the Ga_2O_3 produced to be greater than the mass of Ga that reacts, because it also includes some oxygen. The gallium seems to magically gain mass as it reacts because the final product, Ga_2O_3, adds oxygen atoms to it from the O_2 in the chemical reaction.

Note that it is possible to combine the unit conversion steps in this problem into a single, longer calculation.

$$\text{Mass Ga}_2\text{O}_3 = 14.5\ \cancel{\text{g Ga}} \times \left(\frac{1\ \cancel{\text{mol Ga}}}{69.72\ \cancel{\text{g Ga}}}\right)\left(\frac{2\ \cancel{\text{mol Ga}_2\text{O}_3}}{4\ \cancel{\text{mol Ga}}}\right)\left(\frac{187.4\ \text{g Ga}_2\text{O}_3}{1\ \cancel{\text{mol Ga}_2\text{O}_3}}\right) = 19.5\ \text{g Ga}_2\text{O}_3$$

This combined procedure may be simpler, but be careful to include units in each of the conversion factors, and check that all units cancel except those desired for the answer.

Understanding

Calculate the mass of sulfur trioxide that will form by the reaction of 4.1 g sulfur dioxide with excess O_2.

Answer 5.1 g SO_3

The calculation in Example 3.17 tells us that 19.5 g Ga_2O_3 can be formed when 14.5 g gallium burns in excess of O_2. The 19.5 g Ga_2O_3 is the maximum mass that can be produced because when all of the gallium is consumed, the reaction stops. This calculated mass of product is the **theoretical yield,** the maximum quantity of product that can be obtained from a chemical reaction, based on the amounts of starting materials. Note that to use the Ga to calculate the theoretical yield, it was important to know that an excess of O_2 was present in the reaction.

The maximum quantity of product that can be obtained from a chemical reaction is the theoretical yield.

EXAMPLE 3.18 Theoretical Yield

Given the following equation, answer the questions that follow.

$$2PbS + 3O_2 \rightarrow 2PbO + 2SO_2$$

(a) What mass of O_2 will react with 4.10 g PbS?
(b) What is the theoretical yield of PbO?

Strategy The same series of steps used in Example 3.17 is used for this problem. The chemical equation can be used to determine the amounts of reactants consumed (part a), as well as the amounts of products formed in the reaction (needed in part b). The flow diagram that outlines the strategy for part a is

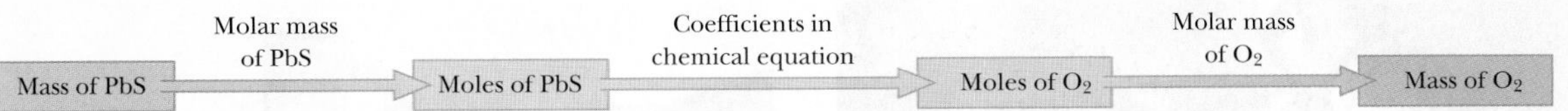

Solution

(a) The balanced equation was given, so the first step in our procedure is already done. The second step is to calculate the number of moles of PbS that react. (The molar mass of PbS is 239.3 g/mol.)

$$\text{Amount PbS} = 4.10\ \text{g PbS} \times \left(\frac{1\ \text{mol PbS}}{239.3\ \text{g PbS}}\right) = 0.0171\ \text{mol PbS}$$

From the chemical equation, we know that 3 mol O_2 reacts with 2 mol PbS. Use this conversion in the third step to calculate the number of moles of O_2 that react with 0.0171 mol PbS.

$$\text{Amount } O_2 = 0.0171\ \text{mol PbS} \times \left(\frac{3\ \text{mol } O_2}{2\ \text{mol PbS}}\right) = 0.0257\ \text{mol } O_2$$

Finish the problem (step 4) by calculating the mass of O_2 consumed, using the molar mass of O_2.

$$\text{Mass } O_2 = 0.0257\ \text{mol } O_2 \times \left(\frac{32.00\ \text{g } O_2}{1\ \text{mol } O_2}\right) = 0.822\ \text{g } O_2$$

(b) The strategy is the same as in part a, except that we use the equation to determine the amount of PbO rather than O_2.

Moles of PbS → (Coefficients in chemical equation) → Moles of PbO → (Molar mass of PbO) → Mass of PbO

The equation and the amount of PbS, 0.0171 mol, are known from part a. Calculate the number of moles of PbO that are made.

$$\text{Amount PbO} = 0.0171\ \text{mol PbS} \times \left(\frac{2\ \text{mol PbO}}{2\ \text{mol PbS}}\right) = 0.0171\ \text{mol PbO}$$

Calculate the theoretical yield by converting moles of PbO into grams, using the formula mass of PbO (223.2 g/mol).

$$\text{Mass PbO} = 0.0171\ \text{mol PbO} \times \left(\frac{223.2\ \text{g PbO}}{1\ \text{mol PbO}}\right) = 3.82\ \text{g PbO}$$

The reaction in this example is commercially important in the mining and metals industry. Many metal ores are mined as mixtures of sulfides and oxides. The first step in the refining process is to convert all of the ore to metal oxides by treating it with oxygen at high temperatures (Figure 3.6). This process is called *roasting*. The oxide is then used as a reactant in additional processes that ultimately produce the pure metal. The roasting process produces large amounts of the toxic gas SO_2. In the past, this SO_2 was a serious pollution problem. Today, most of the SO_2 is collected and used in the synthesis of sulfuric acid, H_2SO_4.

Understanding

Given the following equation, calculate the mass of O_2 needed to react completely with 7.4 g NO.

$$2NO + O_2 \rightarrow 2NO_2$$

Answer 3.9 g O_2

(a)

(b)

Photos compliments of The Doe Run Company North America, Herculaneum, Missouri Smelting Division

Figure 3.6 Roasting of PbS. *(a)* Lead(II) sulfide is converted into lead(II) oxide in the high temperature "roasting" process. The equation is $2PbS + 3O_2 \rightarrow 2PbO + 2SO_2$. *(b)* The SO_2 gas is collected from the gas stream and used in the synthesis of sulfuric acid.

OBJECTIVES REVIEW *Can you:*

- ☑ calculate the mass of a product formed or a reactant consumed in a chemical reaction?
- ☑ determine theoretical yields from reaction data?

3.5 Limiting Reactants

OBJECTIVES

- ☐ Identify the limiting reactant in a chemical reaction and use it to determine theoretical yields of products that form in a chemical reaction
- ☐ Determine percent yields in chemical reactions

Each stoichiometry problem that we have solved so far has had one clearly identified reactant on which the calculation was based, and it was assumed that all other reactants were present in excess. A more normal situation is that we have information on two or more of the reactants and initially do not know which reactant should be used for the calculation of how much product will form. The situation is analogous to that of a hot-dog vendor who has three hot dogs but only two buns. The vendor can sell only two hot dogs in buns; the number of buns *limits* the sale to two. After the two hot dogs in buns are sold, one hot dog remains.

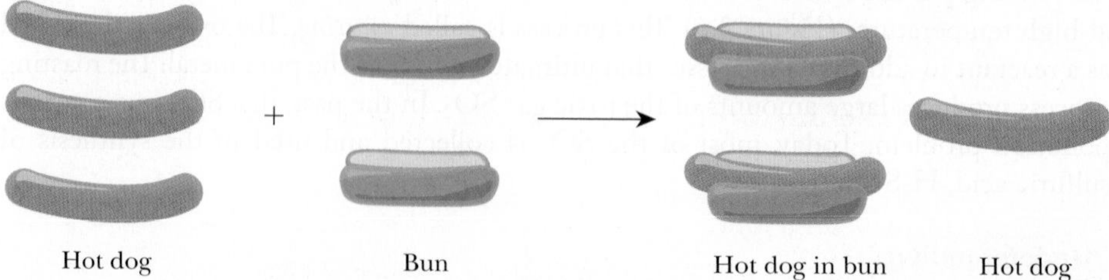

As an example of a chemical reaction that is limited by one reactant, consider a chemical reaction taking place between 3 mol sodium and 1 mol chlorine. From the coefficients of the below balanced equation, we know that 2 mol Na react with every 1 mol Cl_2.

$$2Na + Cl_2 \rightarrow 2NaCl$$

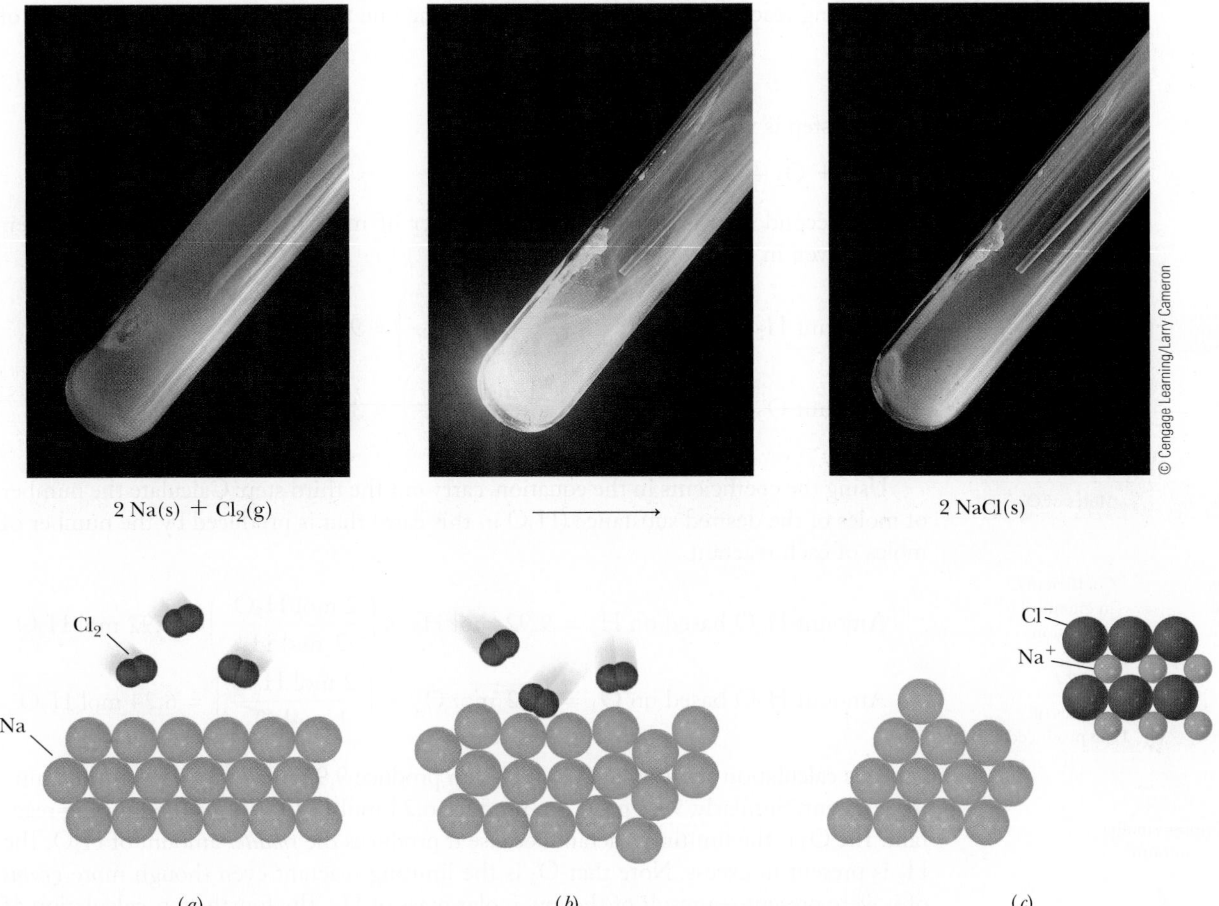

Figure 3.7 Limiting reactant. *(a)* Sodium is placed in a test tube containing a gas inlet tube connected to a source of chlorine gas. *(b)* Chlorine is added through the inlet tube and the reaction $2Na + Cl_2 \rightarrow 2NaCl$ takes place. *(c)* The quantity of chlorine introduced was not sufficient to react with all of the sodium. When the reaction ends, some of the sodium remains together with the sodium chloride product. Chlorine is the limiting reactant. The atoms in Na(s) and Cl_2(g) change size when they react to form ionic NaCl(s).

In a reaction between 3 mol sodium and 1 mol chlorine, all of the chlorine is consumed, but some of the sodium remains unreacted (Figure 3.7). When the last molecule of Cl_2 is consumed, the reaction stops. In this case, chlorine is the **limiting reactant,** the reactant that is completely consumed when the chemical reaction occurs. When we calculate the amount of product formed, the calculation must be based on the limiting reactant, not the reactants that are present in excess.

The limiting reactant determines the amount of product formed in the reaction.

How do we determine which substance is the limiting reactant in a case where we are given grams of reactants? There is a simple test: *The limiting reactant is the one that yields the smallest amount (either in moles or grams) of any one product.* If the mass of more than one reactant is given, approach the stoichiometry problem by calculating the number of moles of product formed from the given quantity of *each* reactant. The reactant that yields the *smallest amount* of product is limiting; use it for the stoichiometry calculation. Example 3.19 illustrates this process.

The limiting reactant is determined by calculating the number of moles of product formed from the given quantity of *each* reactant. The reactant that yields the least amount of product is the limiting reactant.

EXAMPLE 3.19 Limiting Reactant Calculations

A reaction is conducted with 20.0 g H_2 and 99.8 g O_2. (a) State the limiting reactant. (b) Calculate the mass, in grams, of H_2O that can be produced from this reaction.

Strategy We use the same four-step procedure outlined in Section 3.4, but we must calculate the quantities of product that could *form from each of the reactants* to determine the limiting reactant. The limiting reactant is the one that yields the *smallest* amount of product.

Solution

The first step is writing the equation.

$$2H_2 + O_2 \rightarrow 2H_2O$$

The second step is to calculate the number of moles of the reactants from their masses given in the problem.

Mass of H_2 → Molar mass of H_2 → Moles of H_2 → Coefficients in chemical equation → Moles of H_2O produced

Mass of O_2 → Molar mass of O_2 → Moles of O_2 → Coefficients in chemical equation → Moles of H_2O produced

Choose smaller amount → Molar mass of H_2O → Mass of H_2O

$$\text{Amount } H_2 = 20.0 \text{ g } H_2 \times \left(\frac{1 \text{ mol } H_2}{2.016 \text{ g } H_2}\right) = 9.92 \text{ mol } H_2$$

$$\text{Amount } O_2 = 99.8 \text{ g } O_2 \times \left(\frac{1 \text{ mol } O_2}{32.00 \text{ g } O_2}\right) = 3.12 \text{ mol } O_2$$

Using the coefficients in the equation, carry out the third step: Calculate the number of moles of the desired substance (H_2O in this case) that is produced by the number of moles of each reactant.

$$\text{Amount } H_2O \text{ based on } H_2 = 9.92 \text{ mol } H_2 \times \left(\frac{2 \text{ mol } H_2O}{2 \text{ mol } H_2}\right) = 9.92 \text{ mol } H_2O$$

$$\text{Amount } H_2O \text{ based on } O_2 = 3.12 \text{ mol } O_2 \times \left(\frac{2 \text{ mol } H_2O}{1 \text{ mol } O_2}\right) = 6.24 \text{ mol } H_2O$$

This calculation shows that 20.0 g H_2 can produce 9.92 mol H_2O if H_2 is the limiting reactant. Similarly, 99.8 g O_2 can produce 6.24 mol H_2O if O_2 is the limiting reactant. The O_2 is the limiting reactant because it produces the *smaller* amount of H_2O. The H_2 is present in excess. Note that O_2 is the limiting reactant even though more *grams* of it were present—a result of the low molar mass of H_2. The fourth step, calculation of the mass of H_2O formed in the reaction, is based on the amount of H_2O that can be formed by the limiting reactant.

$$\text{Mass } H_2O = 6.24 \text{ mol } H_2O \times \left(\frac{18.02 \text{ g } H_2O}{1 \text{ mol } H_2O}\right) = 112 \text{ g } H_2O$$

The theoretical yield of water produced in the reaction of 20.0 g H_2 and 99.8 g O_2 is 112 g; at the end of the reaction, no O_2, the limiting reactant, remains, but some H_2 does (an amount that can be calculated).

Understanding

Given the following equation, calculate the mass (in grams) of $AlCl_3$ that can be produced from 4.40 g Al and 12.0 g Cl_2?

$$2Al + 3Cl_2 \rightarrow 2AlCl_3$$

Answer 15.0 g $AlCl_3$

Actual and Percent Yield

The previous example illustrated the calculation of the *theoretical yield.* However, it is often difficult to achieve the theoretical yield in the laboratory or an industrial process. For example, a reaction might produce a gas that is difficult to collect. If a solid forms, some of it might stick to the walls of the reaction vessel and remain uncollected (Figure 3.8). Sometimes reactions other than the one described by the equation, called side reactions, occur and consume some starting material without forming the expected product. Many times a reaction simply does not go completely to products. Because of these problems, not all of the product predicted by the stoichiometry calculation is isolated.

Figure 3.8 Actual yield. In collecting a solid product from a chemical reaction, some of the solid cannot be recovered from the reaction container.

The mass of product isolated from a reaction is known as the **actual yield,** a mass that is always less than the theoretical yield. Chemists try to come as close to the theoretical yield as possible, and their ability to do so is expressed as a **percent yield:**

$$\text{Percent yield} = \frac{\text{actual yield}}{\text{theoretical yield}} \times 100\%$$

The actual yield of a product, the result of a laboratory experiment, is less than the theoretical yield and is expressed as the percent yield.

EXAMPLE 3.20 Calculating Percent Yield

Earlier, the reaction of $NaClO_3(s)$ shown below was presented as a portable source of oxygen that is similar to the chemistry used as a backup source for oxygen on the space station. If this reaction is needed to produce oxygen, it is important that the actual yield be close to the theoretical yield—the oxygen is needed for the astronauts to breathe!

$$2NaClO_3(s) \rightarrow 3O_2(g) + 2NaCl(s)$$

(a) Calculate the theoretical yield of O_2 expected for the reaction of 45.4 g $NaClO_3$.
(b) What is the percent yield if 20.0 g O_2 is isolated in this experiment?

Strategy We again use the same series of four steps to calculate the theoretical yield. In part b, this theoretical yield is divided into the actual yield of product isolated in the experiment times 100% to convert this number to a percentage.

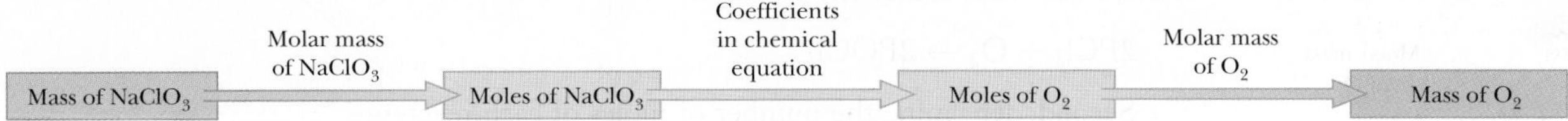

Solution

(a) The first step, the balanced equation, is given. Next, calculate the number of moles of $NaClO_3$ from the mass given in the problem and the molar mass of $NaClO_3$ (106.4 g/mol)

$$\text{Amount } NaClO_3 = 45.4 \text{ g } NaClO_3 \times \left(\frac{1 \text{ mol } NaClO_3}{106.4 \text{ g } NaClO_3}\right) = 0.427 \text{ mol } NaClO_3$$

Third, use the coefficients of the balanced equation to calculate the number of moles of O_2 that can form from 0.426 mol $NaClO_3$.

$$\text{Amount } O_2 = 0.427 \text{ mol } NaClO_3 \times \left(\frac{3 \text{ mol } O_2}{2 \text{ mol } NaClO_3}\right) = 0.640 \text{ mol } O_2$$

Fourth, calculate the mass of O_2 produced.

$$\text{Mass } O_2 = 0.640 \text{ mol } O_2 \times \left(\frac{32.00 \text{ g } O_2}{1 \text{ mol } O_2}\right) = 20.5 \text{ g } O_2$$

(b) Calculate the percent yield by dividing the mass of oxygen actually isolated in the reaction, 20.0 g, by the theoretical yield (times 100%).

$$\text{Percent yield} = \frac{20.0 \text{ g isolated}}{20.5 \text{ g calculated}} \times 100\% = 97.6\%$$

The calculation shows that 45.4 g $NaClO_3$ react to produce a maximum of 20.5 g O_2. In this example, the chemist collected only 20.0 g of the product, for a percent yield of 97.6%.

Understanding

What is the percent yield if 2.4 g NH_3 is obtained from the reaction of 0.64 g H_2 with excess N_2?

Answer 67%

Laboratory workers occasionally observe an actual yield that is *greater* than the theoretical yield because the desired substance may be contaminated by other products or by excess reactants. In these cases, a purification procedure is needed. Any time the actual yield exceeds the theoretical yield, further investigation must be done to determine the source of the error.

A number of the calculations necessary to study chemical reactions in a quantitative manner have now been presented. The following problem integrates many of the concepts presented up to this point.

EXAMPLE 3.21 Stoichiometry Calculations

Phosphorus trichloride reacts with oxygen to yield $POCl_3$. In an experiment performed in the laboratory, 11.0 g PCl_3 and 1.34 g O_2 are mixed, and 11.2 g $POCl_3$ is isolated. What is the percent yield?

Strategy To determine the percent yield, we must calculate the theoretical yield based on the limiting reactant. The strategy for calculating the theoretical yield is the same as in Example 3.19. Use this calculated theoretical yield and the actual yield given in the problem to calculate the percent yield.

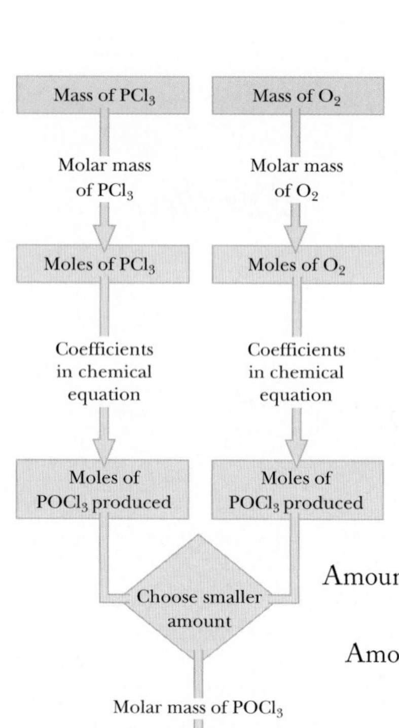

Solution

First, write and balance the equation.

$$2PCl_3 + O_2 \rightarrow 2POCl_3$$

Second, determine the number of moles of each reactant.

$$\text{Amount } PCl_3 = 11.0 \text{ g } PCl_3 \times \left(\frac{1 \text{ mol } PCl_3}{137.3 \text{ g } PCl_3}\right) = 0.0801 \text{ mol } PCl_3$$

$$\text{Amount } O_2 = 1.34 \text{ g } O_2 \times \left(\frac{1 \text{ mol } O_2}{32.00 \text{ g } O_2}\right) = 0.0419 \text{ mol } O_2$$

Third, calculate the equivalent amount of $POCl_3$ produced from the moles of each reactant. The reactant that yields the smaller number of moles of $POCl_3$ is the limiting reactant.

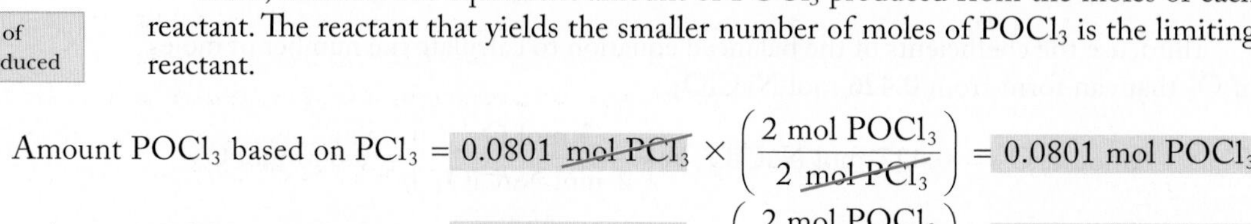

$$\text{Amount } POCl_3 \text{ based on } PCl_3 = 0.0801 \text{ mol } PCl_3 \times \left(\frac{2 \text{ mol } POCl_3}{2 \text{ mol } PCl_3}\right) = 0.0801 \text{ mol } POCl_3$$

$$\text{Amount } POCl_3 \text{ based on } O_2 = 0.0419 \text{ mol } O_2 \times \left(\frac{2 \text{ mol } POCl_3}{1 \text{ mol } O_2}\right) = 0.0838 \text{ mol } POCl_3$$

The PCl_3 is the limiting reactant; less $POCl_3$ is produced from 11.9 g of it than 1.34 g O_2. In the fourth step, use the number of moles of the limiting reactant PCl_3 to calculate the theoretical yield:

$$\text{Mass } POCl_3 = 0.0801 \text{ mol } POCl_3 \times \left(\frac{153.3 \text{ g } POCl_3}{1 \text{ mol } POCl_3}\right) = 12.3 \text{ g } POCl_3$$

The percent yield is the actual yield of the reaction divided by the theoretical yield, times 100%.

$$\text{Percent yield} = \frac{11.2 \text{ g } POCl_3 \text{ actual}}{12.3 \text{ g } POCl_3 \text{ theoretical}} \times 100\% = 91.1\% \text{ yield}$$

In summary, when 1.34 g O_2 reacts with 11.0 g PCl_3, the theoretical yield is 12.3 g $POCl_3$. Experimentally, 11.2 g $POCl_3$ was collected; therefore, the percent yield is 91.1%.

Understanding

In an experiment performed in the laboratory, 44 g NH_3 is mixed with 120 g O_2, and 73 g NO is isolated. Given the following equation, what is the percent yield?

$$4NH_3 + 5O_2 \rightarrow 4NO + 6H_2O$$

Answer 94%

Chemists have practical reasons for performing reactions with some reactants in excess rather than always using the exact amount required. Sometimes the result of a chemical reaction depends on an excess of one or more reactants. For example, the combustion of many organic compounds actually produces a mixture of CO, CO_2, and H_2O if oxygen is not present in excess. An excess of one or more reactants can usually avoid undesirable side products. In other cases, an excess of certain reactants may be needed to increase the yield or shorten the length of time it takes for the reaction to occur.

OBJECTIVES REVIEW *Can you:*

- ☑ identify the limiting reactant in a chemical reaction and use it to determine theoretical yields of products that form in a chemical reaction?
- ☑ determine percent yields in chemical reactions?

Summary Problem

A chemist working for the Bayer Company in Germany, Felix Hoffmann, first synthesized aspirin to treat his father's arthritis. Aspirin reduces pain and fever by reducing the production of prostaglandins, inflammatory compounds released when cells are damaged. Felix Hoffmann's father was using salicylic acid, a precursor to aspirin, for his condition in the late 19th century. Salicylic acid was probably first discovered when humans learned that chewing willow bark tended to reduce fever. The father of modern medicine, Hippocrates (460–377 BC) left historical records of pain relief treatments, including the use of powder made from the bark and leaves of the willow tree to help heal headaches, pain, and fever.

Later, salicylic acid was isolated as the substance in willow bark that had the analgesic effect. Unfortunately, the drug was terribly irritating to the stomach and was associated with other ill effects—an unpleasant, sometimes nauseating taste, and digestive problems, among others—and it was believed that salicylic acid weakened the heart. Felix Hoffmann decided to determine whether he could modify the substance to reduce the side effects without sacrificing its ability to reduce fever and inflammation. He synthesized a related compound, acetylsalicylic acid, and his father reported good results. Felix Hoffmann convinced his employer, the Bayer company, to market the new wonder drug, which was patented with the name "aspirin" on March 6, 1889. As is frequently the case with these types of stories, a number of other versions of how aspirin was introduced exist.

Aspirin can be synthesized from the reaction of salicylic acid, $C_7H_6O_3$, and acetic anhydride, $C_4H_6O_3$. The products of the reaction are aspirin, isolated as a white solid, and acetic acid, $C_2H_4O_2$, a liquid. In the reaction of 2.01 g salicylic acid and 2.04 g acetic anhydride, 1.81 g aspirin is isolated. When a 0.1134-g sample of the aspirin was analyzed by combustion analyses, it yielded 0.04537 g water and 0.2492 g carbon dioxide. Mass spectroscopy indicates that the molar mass of aspirin is 180 g/mol. Use these data to determine the molecular formula of aspirin, write a balanced equation for the reaction, and determine the percent yield of aspirin.

The first stage of the problem is to use the combustion analysis data to determine the empirical formula of aspirin. The data for water and carbon dioxide are used to calculate the moles and mass of C and H. We will calculate the mass of O by difference; given that the starting materials contain only C, H, and O, the same is true of the products.

$$\text{Amount } H_2O = 0.04537 \ \cancel{\text{g } H_2O} \times \left(\frac{1 \text{ mol } H_2O}{18.02 \ \cancel{\text{g } H_2O}}\right) = 0.002518 \text{ mol } H_2O$$

$$\text{Amount } CO_2 = 0.2492 \ \cancel{\text{g } CO_2} \times \left(\frac{1 \text{ mol } CO_2}{44.01 \ \cancel{\text{g } CO_2}}\right) = 0.005662 \text{ mol } CO_2$$

From the formulas of water and carbon dioxide, we know that 1 mol H_2O has 2 mol H, and 1 mol CO_2 has 1 mol C.

$$\text{Amount H} = 0.002518 \ \cancel{\text{mol } H_2O} \times \left(\frac{2 \text{ mol H}}{1 \ \cancel{\text{mol } H_2O}}\right) = 0.005036 \text{ mol H}$$

$$\text{Amount C} = 0.005662 \ \cancel{\text{mol } CO_2} \times \left(\frac{1 \text{ mol C}}{1 \ \cancel{\text{mol } CO_2}}\right) = 0.005662 \text{ mol C}$$

The molar amounts need to be converted to mass to calculate the mass of oxygen by difference.

$$\text{Mass H} = 0.005036 \ \cancel{\text{mol H}} \times \left(\frac{1.01 \text{ g H}}{1 \ \cancel{\text{mol H}}}\right) = 0.005086 \text{ g H}$$

$$\text{Mass C} = 0.005662 \ \cancel{\text{mol C}} \times \left(\frac{12.01 \text{ g C}}{1 \ \cancel{\text{mol C}}}\right) = 0.06800 \text{ g C}$$

$$\text{Mass}_O = \text{mass}_{sample} - (\text{mass}_C + \text{mass}_H) = 0.1134 \text{ g} - (0.005086 \text{ g} + 0.06800 \text{ g}) = 0.04031\text{g}$$

$$\text{Amount O} = 0.04031 \ \cancel{\text{g O}} \times \left(\frac{1 \text{ mol O}}{16.00 \ \cancel{\text{g O}}}\right) = 0.002519 \text{ mol O}$$

Divide each of the molar amounts by the smallest number to try to convert these calculated molar amounts to whole numbers.

$$\text{Amount H} = 0.005036 \text{ mol H} \div 0.002519 = 1.999 \text{ mol H}$$

$$\text{Amount C} = 0.005662 \text{ mol C} \div 0.002519 = 2.248 \text{ mol C}$$

$$\text{Amount O} = 0.002519 \text{ mol O} \div 0.002519 = 1.000 \text{ mol O}$$

We still need to multiply each number by 4 to convert them to whole numbers.

$$\text{Amount H} = 1.999 \text{ mol H} \times 4 = 7.996 \text{ mol H}$$

$$\text{Amount C} = 2.248 \text{ mol C} \times 4 = 8.992 \text{ mol C}$$

$$\text{Amount O} = 1.000 \text{ mol O} \times 4 = 4.000 \text{ mol O}$$

The empirical formula is $C_9H_8O_4$. The molar mass of this formula is 180 g/mol, the same as the molar mass determined for aspirin in the experiments, so this is the molecular formula as well. The coefficients are all one in the balanced equation.

$$C_7H_6O_3 + C_4H_6O_3 \rightarrow C_9H_8O_4 + C_2H_4O_2$$

salicylic acid + acetic anhydride → aspirin + acetic acid

With the balanced equation, the limiting reactant must be determined. To do this, convert the mass of both reactants to moles of aspirin, the limiting reactant will be the one that produces the smaller amount.

$$\text{Amount } C_7H_6O_3 = 2.01 \cancel{\text{ g } C_7H_6O_3} \times \left(\frac{1 \text{ mol } C_7H_6O_3}{138.1 \cancel{\text{ g } C_7H_6O_3}}\right) = 0.0146 \text{ mol } C_7H_6O_3$$

$$\text{Amount } C_4H_6O_3 = 2.04 \cancel{\text{ g } C_4H_6O_3} \times \left(\frac{1 \text{ mol } C_4H_6O_3}{102.1 \cancel{\text{ g } C_4H_6O_3}}\right) = 0.0200 \text{ mol } C_4H_6O_3$$

The molar amounts of aspirin produced from each are numerically the same as those calculated earlier because the coefficients are all one. The salicylic acid is the limiting reactant. The theoretical yield is

$$\text{Mass } C_9H_8O_4 = 0.0146 \cancel{\text{ mol } C_9H_8O_4} \times \left(\frac{180.2 \text{ g } C_9H_8O_4}{1 \cancel{\text{ mol } C_9H_8O_4}}\right) = 2.63 \text{ g } C_9H_8O_4$$

In the actual experiment, 1.81 g aspirin is isolated. The percent yield is

$$\text{Percent yield} = \frac{1.81 \text{ g isolated}}{2.63 \text{ g calculated}} \times 100\% = 68.8\ \%$$

ETHICS IN CHEMISTRY

Many scientists believe that if aspirin had been invented in recent times that it would be available only by prescription, and there would be numerous warnings about adverse effects such as gastrointestinal bleeding.

1. Aspirin can cause problems to stomach lining in some people and cause internal bleeding. Based on this well-known information, discuss if the government should stop over-the-counter sale of aspirin and restrict it to prescription only?

2. You have just carried out two reactions to prepare aspirin. The yield of the two reactions was 56% and 78%. How should you report this in your write-up of the experiment for your grade? Should you emphasize the higher-yielding reaction?

3. The acid rain feature pointed out that "wet scrubbers" could remove most of the acids produced in the burning of coal, but the introduction of these devices reduces the efficiency of the plant, causing it to produce less electricity per ton of coal. Thus, more coal is burned to produce the same amount of electricity as before the scrubbers were installed, which increases the amount of CO_2, an important greenhouse gas. Discuss the ethics of this trade-off of acid rain for possible problems with global warming.

Chapter 3 Visual Summary

The chart shows the connections between the major topics discussed in this chapter.

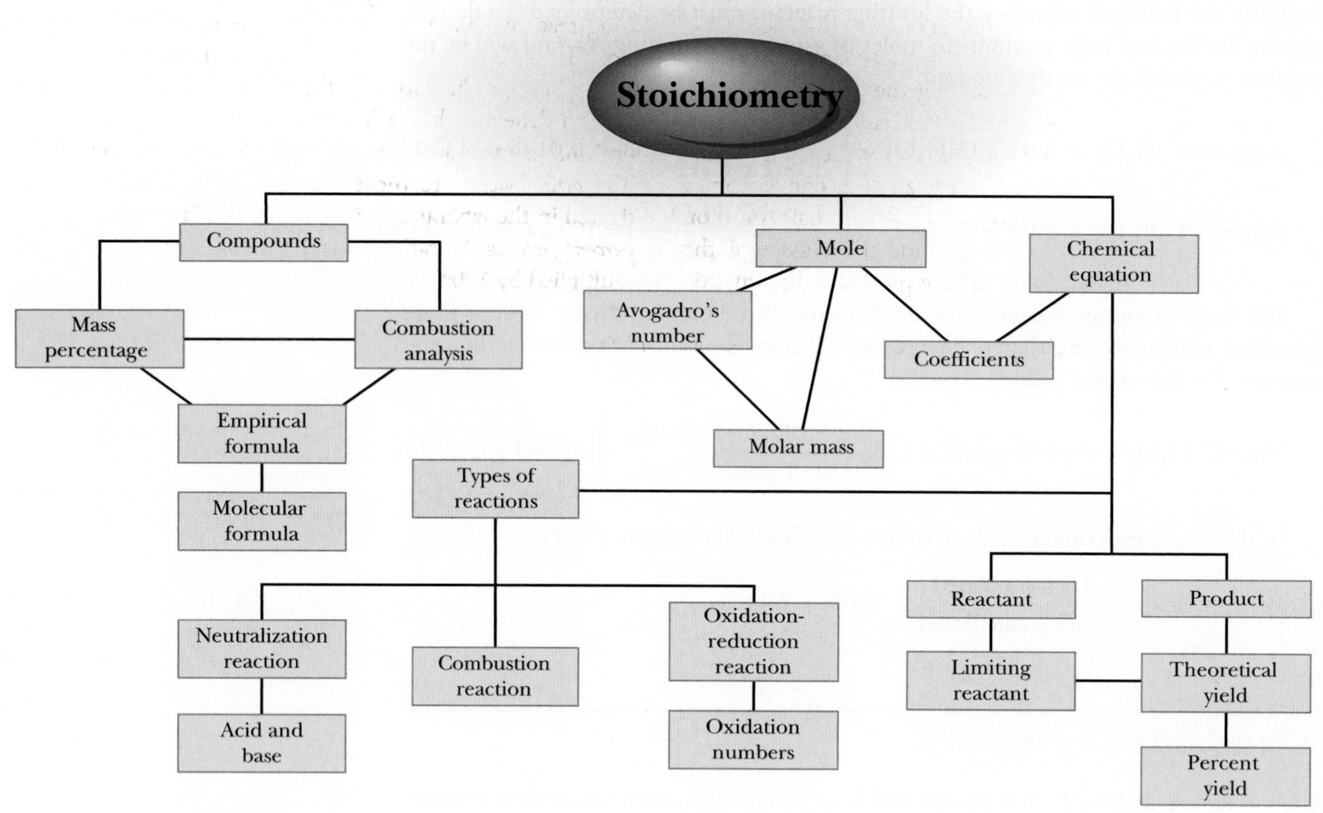

Summary

3.1 Chemical Equations

Stoichiometry is the study of quantitative relationships involving substances and their reactions. Many stoichiometry calculations are based on the *chemical equation,* a quantitative description of a chemical reaction. The first step in writing a chemical equation is to write the correct formulas of the *reactants* and *products* of the reaction. The second step is adjusting the *coefficients* of the substances so that the number of atoms of each element is the same on both the product and reactant sides. One important type of chemical reaction takes place between an *acid,* a substance that dissolves in water to yield the hydrogen cation (H^+), and a *base,* a substance that produces the hydroxide anion (OH^-) in water, is known as a *neutralization reaction.* The products of this reaction are water and a *salt.* A *combustion reaction* is the process of burning; most combustions involve reaction with oxygen. Combustion reactions are a special class of *oxidation–reduction* reactions.

3.2 The Mole and Molar Mass

The *mole* is the SI unit of *amount.* The mole is defined as the number of atoms in 12 g of carbon-12. This number, called *Avogadro's number,* has a value of 6.022×10^{23} units/mol. The mass, in grams, of 1 mol of any substance is its *molar mass.* Molecular mass, molar mass, and Avogadro's number are the key quantities in the important stoichiometric relationships in this chapter. These relationships provide conversion factors from the molecular to the molar scale and from mass to number of moles.

3.3 Chemical Formulas

The percentage composition by mass of each element in a compound is determined from either the empirical or the molecular formula of that compound. *Combustion analysis* is used to determine the compositions of organic compounds. The *empirical formula* of a substance is calculated from the mass or the percentage composition. Determination of a molecular formula of a compound requires a value for the molar mass, in addition to its empirical formula.

3.4 Mass Relationships in Chemical Equations

The coefficients in the chemical equation express not only the relative number of molecules involved in the chemical change, but also the relative number of moles of the substances consumed and produced by the reaction. Reaction stoichiometry problems are solved by first calculating the number of moles from the given masses. Then the coefficients in the equation are used to calculate the number of moles of the desired substances. The final solution may require another conversion of units; for example, moles into grams. The most common reaction stoichiometry problems provide the masses of the reactants and ask for the masses of the products. The quantities of products that are calculated from the chemical equation are *theoretical yields*.

3.5 Limiting Reactants

When the quantity of more than one reactant is known, it is necessary to determine which is the *limiting reactant*—the reactant that is completely consumed in the reaction. The limiting reactant is used to complete the stoichiometry calculation. In a stoichiometry problem, calculate the number of moles of a product formed from the given quantity of *each* reactant. The one that yields the least amount of product is the limiting reactant and is used to calculate the theoretical yield. The *actual yields* are those obtained from experiments conducted in the laboratory or factory. Chemists often record the *percent yield* as the ratio of the actual yield to theoretical yield multiplied by 100%.

Download Go Chemistry concept review videos from OWL or purchase them from **www.ichapters.com**

Chapter Terms

The following terms are defined in the Glossary, Appendix I.

Section 3.1
Acid
Base
Chemical equation
Coefficient
Combustion reaction
Neutralization
Organic compound
Oxidation
Oxidation numbers
Oxidation–reduction reactions
Product
Reactant
Reduction
Salt
Stoichiometry

Section 3.2
Avogadro's number
Molar mass
Mole

Section 3.3
Combustion analysis

Section 3.4
Theoretical yield

Section 3.5
Actual yield
Limiting reactant
Percent yield

Questions and Exercises

OWL Selected end of chapter Questions and Exercises may be assigned in OWL.

Blue-numbered Questions and Exercises are answered in Appendix J; questions are qualitative, are often conceptual, and include problem-solving skills.

■ Questions assignable in OWL

✎ Questions suitable for brief writing exercises

▲ More challenging questions

Questions

3.1 What is the difference between writing the names of the reactants and products of the reaction, and writing the chemical equation?

3.2 Describe the steps needed to write balanced equations.

3.3 Using solid circles for H atoms and open circles for O atoms, make a drawing that shows the molecular level representation for the balanced equation of H_2 and O_2 reacting to form H_2O.

3.4 Using solid circles for H atoms and open circles for O atoms, make a drawing that shows the molecular level representation for the balanced equation of H_2 and O_2 reacting to form H_2O_2. The two oxygen atoms are bonded to each other, and a hydrogen atom is bonded to each oxygen.

3.5 Give the name and definition of the SI unit for amount of substance.

3.6 How many objects are in 1 mol? What is the common name for this number of objects?

3.7 ✎ When writing and balancing a chemical equation, we generally count the number of molecules that are present, but when carrying out reactions in the laboratory, we think of the equation in terms of moles. Why is the unit of moles more convenient on the laboratory scale?

3.8 What are the units for molecular mass, formula mass, and molar mass?

3.9 Draw a diagram that outlines the conversion of number of atoms into moles.

3.10 Draw a flow diagram that outlines the conversion of moles to number of atoms.

3.11 Describe an experiment that would enable someone to determine the percentages of carbon and hydrogen in a sample of a newly prepared hydrocarbon.

3.12 Explain how a combustion analysis is used to determine the percentage of oxygen in a new compound that contains only carbon, hydrogen, and oxygen.

3.13 Only the empirical formula can be calculated from percentage composition data. What additional information is needed to calculate the molecular formula from the empirical formula, and if given this information, how is the molecular formula determined?

3.14 Interpret the following equation in terms of number of moles.

$$N_2 + 2H_2 \rightarrow N_2H_4$$

3.15 Draw a flow diagram used to answer the following question, "How many grams of N_2 are needed to exactly react with 2.44 g H_2 given the equation $N_2 + 2H_2 \rightarrow N_2H_4$?"

3.16 ✎ Describe what is meant by the expression, "The reaction was carried out with the reactants present in stoichiometric amounts."

3.17 Describe a method of determining the limiting reactant in a calculation based on the individual masses of each of the reactants.

3.18 Describe what is meant by the statement, "In a combustion reaction, C_2H_4 is the limiting reactant and oxygen is present in excess."

Exercises

OBJECTIVE **Write balanced equations for chemical reactions.**

3.19 A mixture of carbon monoxide and oxygen gas reacts as shown below.

(a) Write the balanced equation (remember to express the coefficients as the lowest set of whole numbers).

(b) Name the product.

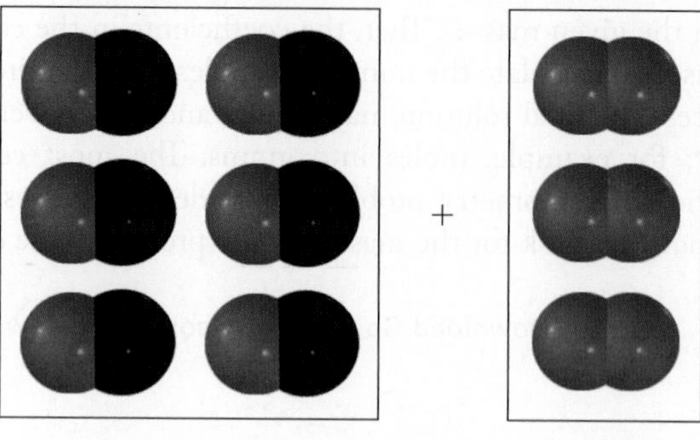

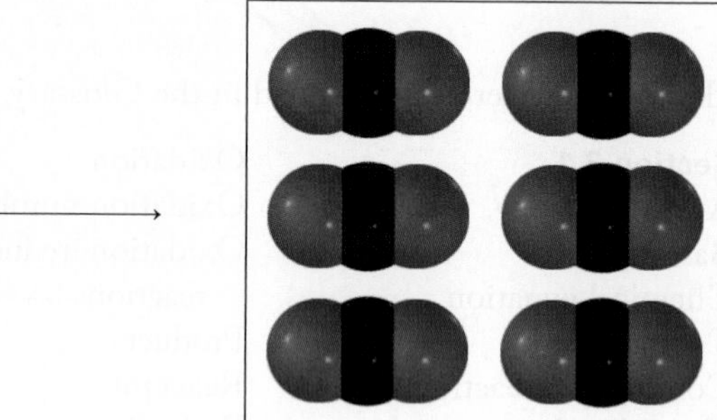

3.20 A mixture of sulfur dioxide and oxygen gas reacts as shown below.
(a) Write the balanced equation (remember to express the coefficients as the lowest set of whole numbers).
(b) Name the product.

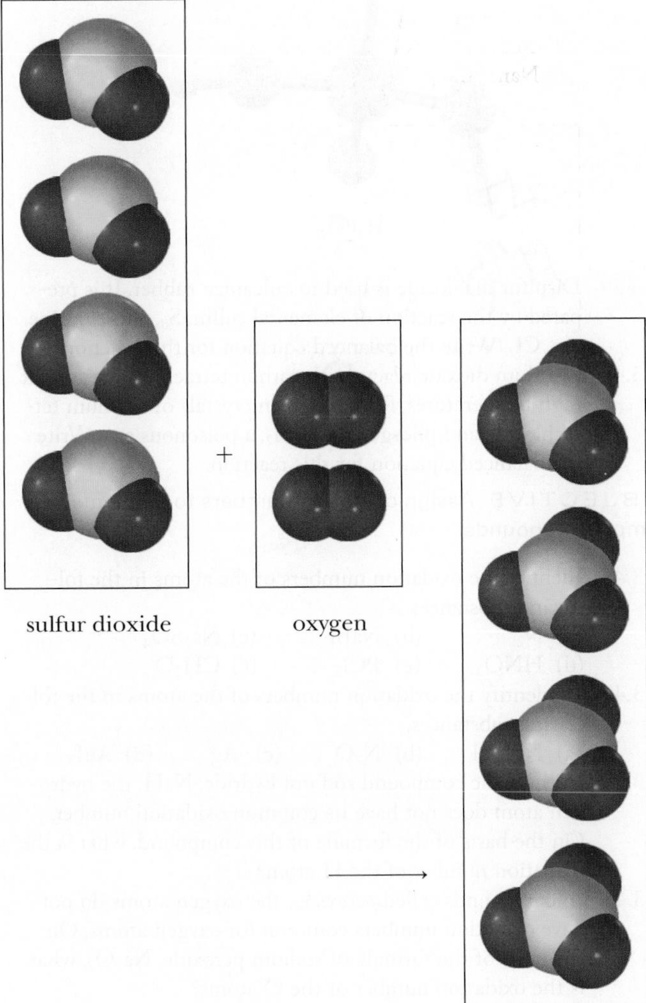

3.21 Write balanced equations for the following reactions.
(a) $C_5H_{12} + O_2 \rightarrow CO_2 + H_2O$
(b) $NH_3 + O_2 \rightarrow N_2 + H_2O$
(c) $KOH + H_2SO_4 \rightarrow K_2SO_4 + H_2O$

3.22 Write balanced equations for the following reactions.
(a) $Mg_3N_2 + H_2O \rightarrow NH_3 + Mg(OH)_2$
(b) $Fe + O_2 \rightarrow Fe_2O_3$

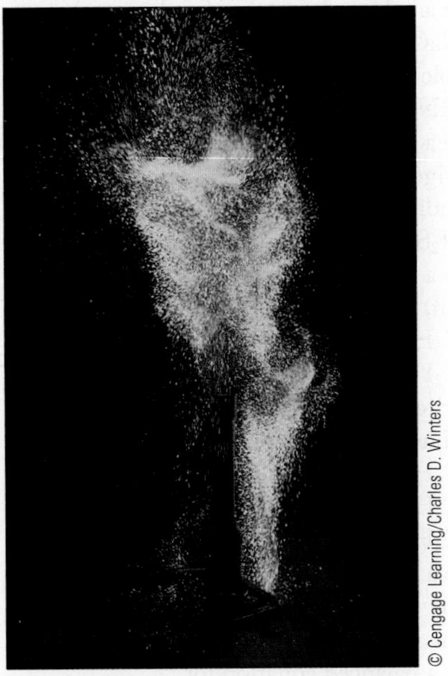

Iron powder burns in oxygen to form Fe_2O_3.

(c) $Zn + H_3PO_4 \rightarrow H_2 + Zn_3(PO_4)_2$

3.23 Write balanced equations for the following reactions.
(a) $N_2H_4 + N_2O_4 \rightarrow N_2 + H_2O$
(b) $F_2 + H_2O \rightarrow HF + O_2$
(c) $Na_2O + H_2O \rightarrow NaOH$

3.24 ■ Balance these reactions.
(a) $Al(s) + O_2(g) \rightarrow Al_2O_3(s)$
(b) $N_2(g) + H_2(g) \rightarrow NH_3(g)$
(c) $C_6H_6(\ell) + O_2(g) \rightarrow H_2O(\ell) + CO_2(g)$

3.25 (a) Write the equation for perchloric acid ($HClO_4$) dissolving in water.
(b) Write the equation for sodium nitrate dissolving in water.

3.26 (a) Write the equation for hydrofluoric acid (HF) dissolving in water.
(b) Write the equation for lithium sulfate dissolving in water.

3.27 (a) Write the equation for nitric acid (HNO_3) dissolving in water.
(b) Write the equation for potassium carbonate dissolving in water.

3.28 (a) Write the equation for hydrochloric acid dissolving in water.
(b) Write the equation for barium nitrate dissolving in water.

OBJECTIVES Identify and balance chemical equations for neutralization reactions, combustion reactions, and oxidation–reduction reactions.

3.29 Write a balanced equation for the reaction of
(a) NaOH and H_2SO_4.
(b) calcium hydroxide and HCl.
(c) HNO_3 and lithium hydroxide.

3.30 Write a balanced equation for the reaction of
(a) $Mg(OH)_2$ and HF.
(b) sodium hydroxide and HCl.
(c) H_2SO_4 and strontium hydroxide.

3.31 Write a balanced equation for the combustion (in excess oxygen) of each of the following compounds.
(a) C_6H_{12} (b) C_4H_8
(c) C_2H_4O (d) $C_4H_6O_2$

3.32 ■ Write a balanced equation for each of these combustion reactions.
(a) $C_4H_{10}(g) + O_2(g) \rightarrow$
(b) $C_6H_{12}O_6(s) + O_2(g) \rightarrow$
(c) $C_4H_8O(\ell) + O_2(g) \rightarrow$

3.33 Write a balanced equation for
(a) the combustion of C_6H_{10} and O_2.
(b) the reaction of $Be(OH)_2$ and nitric acid.

3.34 Write a balanced equation for
(a) the combustion of C_8H_8 and O_2.
(b) the reaction of potassium hydroxide and HCl.

3.35 The reaction of carbon disulfide and oxygen yields sulfur dioxide and carbon dioxide. Write the balanced equation for this reaction.

3.36 Methyl tertiarybutyl ether, $CH_3OC(CH_3)_3$, MTBE, is a compound that is added to gasoline to increase the octane rating, replacing the toxic compound tetraethyl lead that was used previously. Unfortunately, methyl tertiarybutyl ether has contaminated groundwater in certain locations and is being phased out. Write the equation for the combustion of MTBE in excess oxygen.

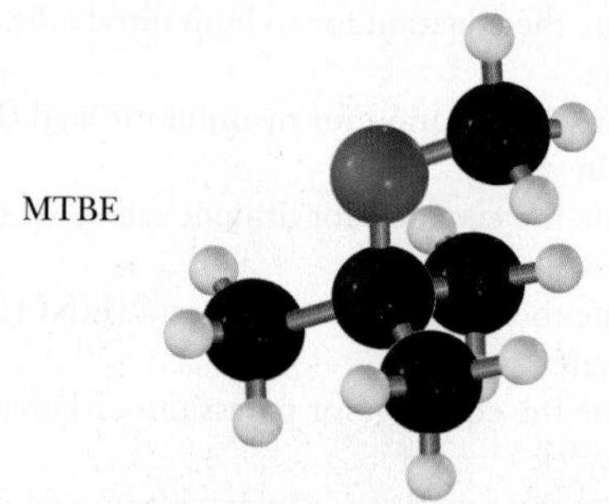
MTBE

3.37 Acetone, $(CH_3)_2CO$, is an important industrial compound. Although its toxicity is relatively low, workers using it must be careful to avoid flames and sparks because this compound burns readily in air. Write the balanced equation for the combustion of acetone.

3.38 The substance H_3PO_3 can be converted into H_3PO_4 and PH_3 by heating. Write the balanced equation for this reaction.

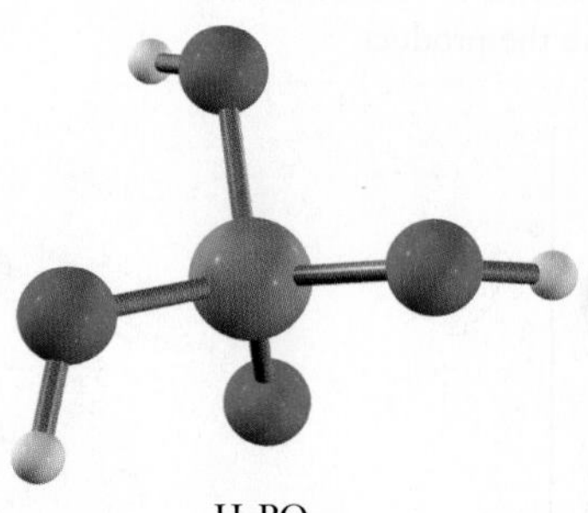
H_3PO_4

3.39 Disulfur dichloride is used to vulcanize rubber. It is prepared by the reaction of elemental sulfur, S_8, and chlorine gas, Cl_2. Write the balanced equation for this reaction.

3.40 Uranium dioxide reacts with carbon tetrachloride vapor at high temperatures, forming green crystals of uranium tetrachloride and phosgene, $COCl_2$, a poisonous gas. Write the balanced equation for this reaction.

OBJECTIVE Assign oxidation numbers to elements in simple compounds.

3.41 Identify the oxidation numbers of the atoms in the following substances.
(a) N_2 (b) NaBr (c) Na_2SO_4
(d) HNO_3 (e) PCl_5 (f) CH_2O

3.42 ■ Identify the oxidation numbers of the atoms in the following substances.
(a) NH_4Cl (b) N_2O (c) Ag (d) AuI_3

3.43 In the ionic compound sodium hydride, NaH, the hydrogen atom does not have its common oxidation number. On the basis of the formula of this compound, what is the oxidation number of the H atom?

3.44 In compounds called *peroxides,* the oxygen atoms do not have oxidation numbers common for oxygen atoms. On the basis of the formula of sodium peroxide, Na_2O_2, what is the oxidation number of the O atoms?

3.45 ▲ The reaction of hydrazine, N_2H_4, with molecular oxygen is violent because it rapidly produces large quantities of gases and heat. For this reason, hydrazine has been used as a rocket fuel. The products of the reaction are NO_2 and water. Write the balanced equation for this reaction. Assign oxidation numbers to each element in the reactants and products, and indicate which element is oxidized and which is reduced.

3.46 The reaction of iron metal with oxygen gas at increased temperatures yields iron(III) oxide. Write the balanced equation for this reaction. Assign an oxidation number to each element in the reactants and products, and indicate which element is oxidized and which is reduced.

3.47 Zinc metal and HCl react to yield zinc(II) chloride and hydrogen gas. Write the balanced equation for this reaction. Assign oxidation numbers to each element in the reactants and products, and indicate which element is oxidized and which is reduced.

3.48 White phosphorus, P_4, is a solid at room temperature. It reacts with molecular oxygen to yield solid P_4O_{10}. Write the balanced equation for this reaction, including the physical states. Assign an oxidation number to each element in the reactants and products, and indicate which element is oxidized and which is reduced.

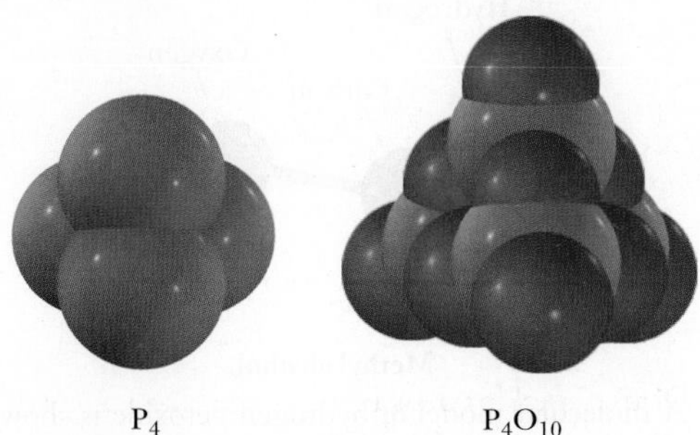

3.49 One of the ways to remove nitrogen monoxide gas, a serious source of air pollution, from smokestack emissions is by reaction with ammonia gas, NH_3. The products of the reaction, N_2 and H_2O, are not toxic. Write the balanced equation for this reaction. Assign an oxidation number to each element in the reactants and products, and indicate which element is oxidized and which is reduced.

3.50 The reaction of MnO_2 and HCl yields $MnCl_2$, Cl_2, and water. Write the balanced equation for this reaction. Assign an oxidation number to each element in the reactants and products, and indicate which element is oxidized and which is reduced.

OBJECTIVE **Express the amounts of substances using moles.**

3.51 State how many atoms are present in the following samples.
(a) 1.44 mol Mg
(b) 9.77 mol Ne
(c) 0.099 mol Fe

3.52 State how many atoms are present in the following samples.
(a) 0.0778 mol Xe
(b) 1.45 mol K
(c) 55.8 mol Ti

3.53 State how many molecules are present in the following samples.
(a) 99.2 mol H_2O
(b) 1.22 mol N_2
(c) 22.9 mol C_3H_6
(d) 0.0022 mol N_2O

3.54 State how many molecules are present in the following samples.
(a) 0.223 mol Cl_2
(b) 14.7 mol N_2H_4
(c) 0.334 mol C_9H_{18}
(d) 1.22 mol CO_2

3.55 State how many moles are present in the following samples.
(a) 3.44×10^{24} molecules of O_2
(b) 1.11×10^{22} atoms of Na
(c) 5.57×10^{30} molecules of C_2H_6
(d) 1.66×10^{24} molecules of CO

3.56 State how many moles are present in the following samples.
(a) 1.33×10^{26} molecules of Br_2
(b) 7.71×10^{26} molecules of C_5H_{12}
(c) 2.34×10^{23} molecules of B_2H_6
(d) 7.76×10^{23} atoms of Ne

OBJECTIVE **Determine the molar mass of any element or compound from its formula.**

3.57 Give the molar mass of the following substances.
(a) NaOH
(b) C_2H_4
(c) $Mg(OH)_2$

3.58 Give the molar mass of the following substances.
(a) N_2O_4
(b) Na_2SO_4
(c) $C_6H_{10}O_2$

3.59 Give the molar mass of the following substances.
(a) $ZnBr_2$
(b) K_2CrO_4
(c) BaS

3.60 Give the molar mass of the following substances.
(a) N_2O_2
(b) $(NH_4)_2CO_3$
(c) $C_8H_{15}N$

OBJECTIVES **Use molar mass and Avogadro's number to interconvert between mass, moles, and numbers of atoms, ions, or molecules.**

3.61 (a) Calculate the number of moles in 9.40 g SO_2.
(b) Calculate the mass in 3.30 mol $AlCl_3$.
(c) Calculate the number of moles in 1.12×10^{23} molecules of H_2SO_4.

3.62 ■ How many moles of compound are in
(a) 39.2 g H_2SO_4?
(b) 8.00 g O_2?
(c) 10.7 g NH_3?

3.63 (a) Calculate the number of moles in 14.3 g C_6H_6.
(b) Calculate the mass of 0.0535 mol SiH_4.
(c) Calculate the number of molecules in 1.11 g H_2O.

3.64 (a) Calculate the mass of 78.4 mol CO_2.
(b) Calculate the number of moles in 192 g $AgNO_3$.
(c) Calculate the number of molecules in 9.22 g CH_4.

3.65 Calculate the number of moles in the following samples.
(a) 2.2 g K_2SO_4
(b) 6.4 g $C_8H_{12}N_4$
(c) 7.13 g $Fe(C_5H_5)_2$

3.66 Calculate the mass, in grams, of the following samples.
(a) 7.55 mol N_2O_4
(b) 9.2 mol $CaCl_2$
(c) 0.44 mol CO

3.67 (a) Calculate the number of moles in 48.0 g H_2O_2.
(b) Calculate the number of oxygen atoms in this sample.

3.68 (a) Calculate the number of molecules in 3.4 g H_2.
(b) Calculate the number of hydrogen atoms in this sample.

3.69 (a) Calculate the mass, in grams, of 3.50 mol NO_2.
(b) Calculate the number of molecules in this sample.
(c) Calculate the number of nitrogen and oxygen atoms in the sample.

3.70 (a) Calculate the number of moles in 33.1 g SO_3.
(b) Calculate the number of molecules in this sample.
(c) Calculate the number of sulfur and oxygen atoms in the sample.

3.71 Possession of 5.0 g "crack" cocaine, $C_{17}H_{21}NO_4$, is a felony in most states, the conviction of which carries mandatory jail time. How many moles of cocaine is this quantity?

3.72 A standard serving of alcohol is 0.9 fluid ounce pure ethanol, C_2H_5OH. If there is 29.56 mL in one fluid ounce and the density of ethanol is 0.7894 g/mL, how many moles of ethanol are in a standard serving?

3.73 Colchicine, $C_{22}H_{25}NO_6$, is a naturally occurring compound that has been used as a medicine since the time of the pharaohs in ancient Egypt. Although the reasons for its effectiveness are not yet clearly understood, it is still used to treat the inflammation in joints caused by a gout attack.
(a) What is the molar mass of colchicine?
(b) What is the mass, in grams, of 3.2×10^{22} molecules of colchicine?
(c) How many moles of colchicine are in a 326-g sample?
(d) How many carbon atoms are present in 50 molecules of colchicine?

3.74 Nickel tetracarbonyl, $Ni(CO)_4$, is a volatile (easily converted to the gas phase), extremely toxic compound that forms when carbon monoxide gas is passed over finely divided nickel. Despite this toxicity, it has been used for more than a century in a method of making highly purified nickel.
(a) What is the mass of 1.00 mol $Ni(CO)_4$?
(b) How many moles of $Ni(CO)_4$ are in a 3.22-g sample?
(c) How many molecules of $Ni(CO)_4$ are in a 5.67-g sample?
(d) How many atoms of carbon are present in 34 g $Ni(CO)_4$?

3.75 A molecular model of methyl alcohol is shown below; the OH group is an alcohol functional group, a group present in many important chemicals. Write the formula of methyl alcohol and use it to calculate the number of moles of hydrogen atoms in 0.33 mol methyl alcohol.

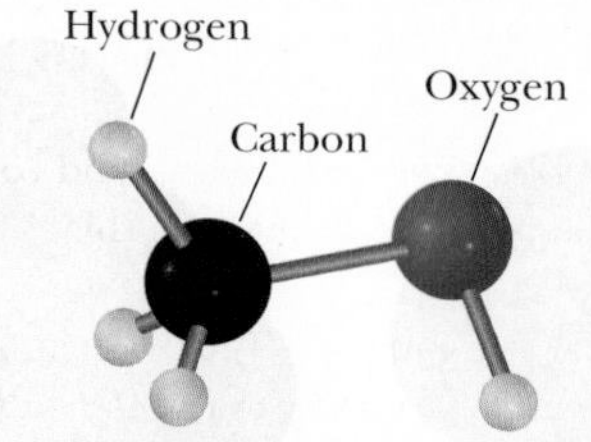

Methyl alcohol.

3.76 A molecular model of hydrogen peroxide is shown below. Write the formula of hydrogen peroxide and use it to calculate the number of moles of hydrogen atoms in 0.011 mol hydrogen peroxide.

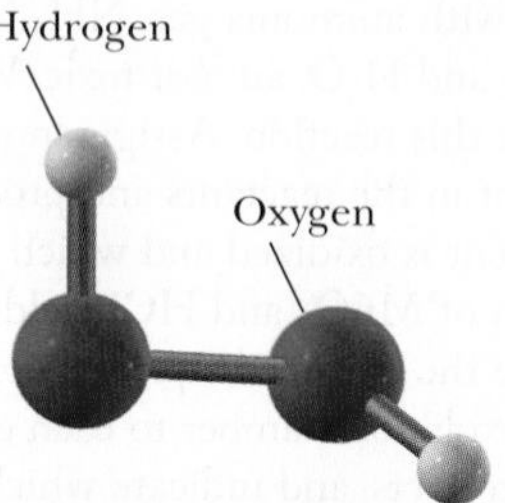

Hydrogen peroxide.

OBJECTIVE Calculate the mass percentage of each element in a compound (percentage composition) from the chemical formula of that compound.

3.77 What is the percentage, by mass, of each element in the following substances?
(a) C_4H_8
(b) $C_3H_4N_2$
(c) Fe_2O_3

3.78 What is the percentage, by mass, of each element in the following substances?
(a) C_6H_{12}
(b) $C_5H_{12}O$
(c) $NiCl_2$

3.79 What is the mass percentage of each element in acetone, C_3H_6O?

3.80 ■ Calculate the mass percentage of copper in CuS, copper(II) sulfide.

3.81 What is the mass percentage of each element in sodium sulfate?

3.82 What is the mass percentage of each element in magnesium carbonate?

3.83 The compound sodium borohydride, $NaBH_4$, is used in the preparation of many organic compounds.
(a) What is the molar mass of sodium borohydride?
(b) What is the mass percentage of each element in this compound?

3.84 Calcium carbonate is popular as an antacid because, in addition to neutralizing stomach acid, it provides calcium, a necessary mineral to the body.
(a) What is the formula mass of calcium carbonate?
(b) What is the mass percentage of each element in this compound?

3.85 A chemist prepared a compound that she thought had the formula FeI_3. When the compound was analyzed, it contained 18.0% Fe and 82.0% I. Calculate the mass percentage composition expected for FeI_3 and compare the result with that found in the analysis. Is this the correct formula of the compound?

3.86 A compound was prepared and analyzed. It was 56.0% C, 3.92% H, and 27.6% Cl by mass. The compound was thought to have the formula $C_6H_4(OH)Cl$. Calculate the mass percentage of each element in this formula. Is the analysis consistent with this formula?

3.87 Calculate the mass of carbon in the following compounds.
(a) 4.9 g CO
(b) 2.2 g C_3H_6
(c) 9.33 g C_2H_6O

3.88 Calculate the mass of carbon in the following compounds.
(a) 1.80 g $C_4H_{10}O$
(b) 0.00223 g Na_2CO_3
(c) 22.1 g $C_5H_{11}N$

3.89 Calculate the mass of carbon in the following compounds.
(a) 4.32 g CO_2
(b) 2.21 g C_2H_4
(c) 0.0443 g CS_2

3.90 ■ Calculate the mass of hydrogen in the following compounds.
(a) 4.33 g H_2O
(b) 1.22 g C_2H_2
(c) 4.44 g N_2H_4

OBJECTIVE Calculate the mass of each element present in a sample from elemental analysis data such as that produced by combustion analysis.

3.91 A 1.070-g sample of a compound containing only carbon, hydrogen, and oxygen burns in excess O_2 to produce 1.80 g CO_2 and 1.02 g H_2O. Calculate the mass of each element in the sample and the mass percentage of each element in the compound.

3.92 A 2.770-g sample containing only carbon, hydrogen, and oxygen burns in excess O_2 to produce 4.06 g CO_2 and 1.66 g H_2O. Calculate the mass of each element in the sample and the mass percentage of each element in the compound.

3.93 A 3.11-g sample containing only carbon, hydrogen, and nitrogen burns in excess O_2 to produce 5.06 g CO_2 and 2.07 g H_2O. Calculate the mass of each element in the sample and the mass percentage of each element in the compound.

3.94 A 0.513-g sample containing only carbon, hydrogen, and nitrogen burns in excess O_2 to produce 1.04 g CO_2 and 0.704 g H_2O. Calculate the mass of each element in the sample and the mass percentage of each element in the compound.

OBJECTIVE Determine the empirical formula of a compound from mass or mass percentage data.

3.95 What is the empirical formula of a compound that contains 0.139 g hydrogen and 0.831 g carbon?

3.96 What is the empirical formula of a substance that contains 0.80 g carbon and 0.20 g hydrogen?

3.97 A sample contains 0.571 g carbon, 0.072 g hydrogen, and 0.333 g nitrogen. What is the empirical formula of this substance?

3.98 A sample contains 0.152 g nitrogen and 0.348 g oxygen. What is the empirical formula of this substance?

3.99 What is the empirical formula of a substance that contains only iron and chlorine, and is 44.06% by mass iron?

3.100 What is the empirical formula of a substance containing only selenium and chlorine, and is 52.7% selenium by mass?

3.101 A 1.000-g sample of a compound contains 0.252 g titanium and 0.748 g chlorine. Determine the empirical formula of this compound.

3.102 A sample is shown to contain 0.173 g chromium and 0.160 g oxygen. What is the empirical formula of this substance?

3.103 A compound contains only carbon, hydrogen, and oxygen, and is 66.6% carbon and 11.2% hydrogen. What is the empirical formula of this substance?

3.104 ■ A 1.20-g sample of a compound gave 2.92 g of CO_2 and 1.22 g of H_2O on combustion in oxygen. The compound is known to contain only C, H, and O. What is its empirical formula?

3.105 A platinum compound named cisplatin is effective in the treatment of certain types of cancer. Analysis shows that it contains 65.02% platinum, 2.02% hydrogen, 9.34% nitrogen, and 23.63% chlorine. What is its empirical formula?

3.106 Carvone is an oil isolated from caraway seeds that is used in perfumes and soaps. This compound contains 79.95% carbon, 9.40% hydrogen, and 10.65% oxygen. What is its empirical formula?

3.107 When a 2.074-g sample that contains only carbon, hydrogen, and oxygen burns in excess O_2, the products are 3.80 g CO_2 and 1.04 g H_2O. What is the empirical formula of this compound?

3.108 ▲ A 0.459-g sample that contains only carbon, hydrogen, and oxygen reacts with an excess of O_2 to produce 0.170 g CO_2 and 0.0348 g H_2O. What is the percentage of C and H in the starting material?

3.109 ▲ A compound contains only C, H, N, and O. Combustion of a 1.48-g sample in excess O_2 yields 2.60 g CO_2 and 0.799 g H_2O. A separate experiment shows that a 2.43-g sample contains 0.340 g N. What is the empirical formula of the compound?

3.110 ▲ A compound contains only C, H, N, and O. Combustion of a 2.18-g sample in excess O_2 yields 3.94 g CO_2 and 1.89 g H_2O. A separate experiment shows that a 1.23-g sample contains 0.235 g N. What is the empirical formula of the compound?

OBJECTIVE **Use molar mass to determine a molecular formula from an empirical formula.**

3.111 What is the molecular formula of a compound with an empirical formula of CH_2O and a molar mass of 90 g/mol?

3.112 What is the molecular formula of a compound with an empirical formula of HO and a molar mass of 34 g/mol?

3.113 What is the molecular formula of each of the following compounds?
(a) empirical formula of C_2H_4O and molar mass of 132 g/mol
(b) empirical formula of $C_3H_4NO_3$ and molar mass of 408 g/mol

3.114 ■ What is the molecular formula of each of the following compounds?
(a) empirical formula $C_5H_{10}O$ and molar mass of 258 g/mol
(b) empirical formula PCl_3 and molar mass of 137.3 g/mol?

3.115 A compound contains 62.0% carbon, 10.4% hydrogen, and 27.5% oxygen by mass, and has a molar mass of 174 g/mol. What is the molecular formula of the compound?

3.116 ■ Mandelic acid is an organic acid composed of carbon (63.15%), hydrogen (5.30%), and oxygen (31.55%). Its molar mass is 152.14 g/mol. Determine the empirical and molecular formulas of the acid.

3.117 Acetic acid gives vinegar its sour taste. Analysis of acetic acid shows it is 40.0% carbon, 6.71% hydrogen, and 53.3% oxygen. Its molar mass is 60 g/mol. What is its molecular formula?

Vinegar.

3.118 Fructose, an important sugar, is made up of 40.0% carbon, 6.71% hydrogen, and 53.3% oxygen. Its molar mass is 180 g/mol. What is its molecular formula?

OBJECTIVE **Calculate the mass of a product formed (theoretical yield) or a reactant consumed in a chemical reaction.**

3.119 (a) Write the equation for the combustion of propylene, C_3H_6.
(b) Calculate the mass of CO_2 produced when 2.45 g C_3H_6 burns in excess oxygen.

3.120 (a) Write the equation for the combustion of C_4H_8O.
(b) Calculate the mass of O_2 consumed in the combustion of a 5.33-g sample of C_4H_8O.

3.121 The reaction of P_4, a common elemental form of phosphorus, with Cl_2 yields PCl_5. Calculate the mass of Cl_2 needed to react completely with 0.567 g P_4.

3.122 What mass of NH_3 forms from the reaction of 5.33 g N_2 with excess H_2?

3.123 Aluminum metal reacts with sulfuric acid, H_2SO_4, to yield aluminum sulfate and hydrogen gas. Calculate the mass of aluminum metal needed to produce 13.2 g hydrogen.

3.124 ■ Chlorine can be produced in the laboratory by the reaction of hydrochloric acid with excess manganese(IV) oxide.

$$4HCl(aq) + MnO_2(s) \rightarrow Cl_2(g) + 2H_2O(\ell) + MnCl_2(aq)$$

How many moles of HCl are needed to form 12.5 mol Cl_2?

3.125 Lithium metal reacts with O_2 to form lithium oxide. What is the theoretical yield of lithium oxide when 0.45 g lithium reacts with excess O_2?

3.126 In a reaction of HCl and NaOH, the theoretical yield of H_2O is 78.2 g. What is the theoretical yield of NaCl?

OBJECTIVES **Identify the limiting reactant in a chemical reaction and use it to determine theoretical yield of products that form in a chemical reaction.**

3.127 A mixture of hydrogen and nitrogen gas reacts as shown in the drawing below.
(a) Write the balanced equation.
(b) Which reactant is the limiting reactant?

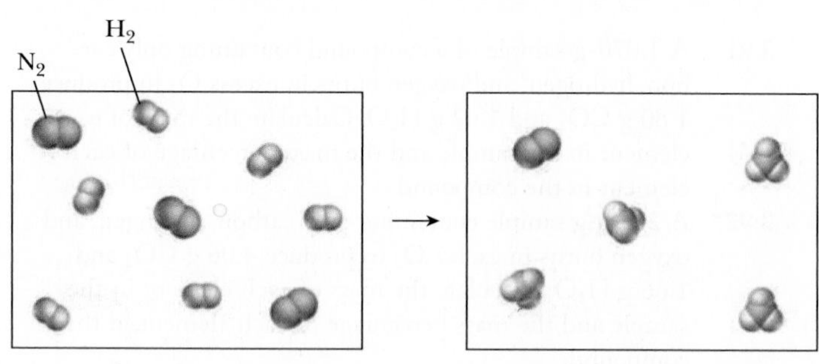

3.128 A mixture of antimony atoms and Cl_2 in the gas phase reacts as shown in the drawing below.
(a) Write the balanced equation.
(b) Which reactant is the limiting reactant?

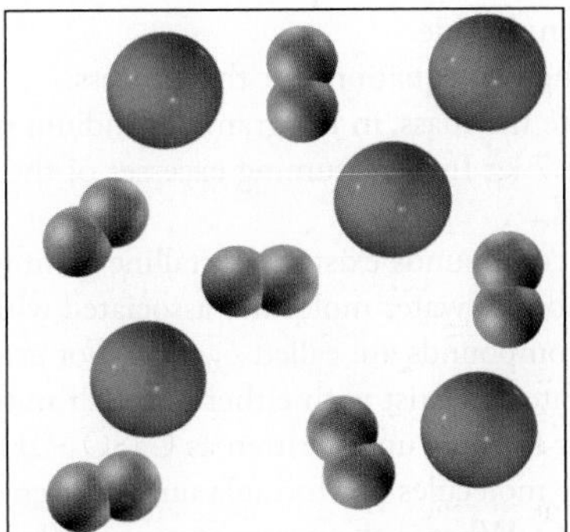

→

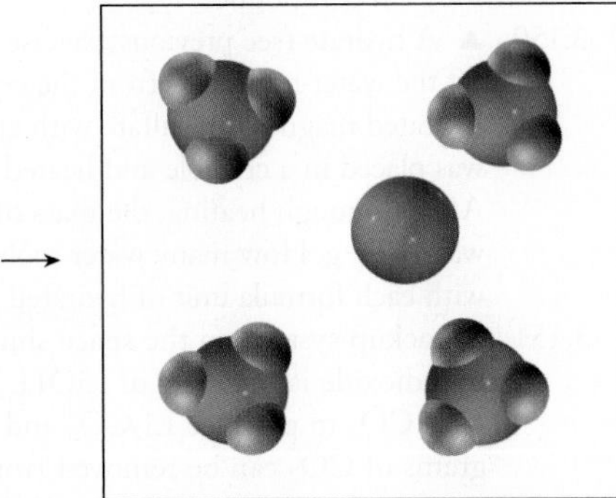

3.129 Hydrogen and nitrogen react to form ammonia, NH_3. Calculate the mass of NH_3 produced from the reaction of 14 g N_2 and 1.0 g H_2.

3.130 ■ Calculate the mass of silver produced if 3.22 g zinc metal and 4.35 g $AgNO_3$ react according to the following equation.

$$Zn(s) + 2AgNO_3(aq) \rightarrow 2Ag(s) + Zn(NO_3)_2(aq)$$

3.131 What is the theoretical yield, in grams, of CO_2 formed from the reaction of 3.12 g CS_2 and 1.88 g O_2? The second product is SO_2.

3.132 What is the theoretical yield, in grams, of P_4O_{10} formed from the reaction of 2.2 g P_4 with 4.2 g O_2?

OBJECTIVE Determine percent yields in chemical reactions.

3.133 In a reaction of 3.3 g Al with excess HCl, 3.5 g $AlCl_3$ is isolated (hydrogen gas also forms in this reaction). What is the percent yield of the aluminum compound?

3.134 A reaction of 43.1 g CS_2 with excess Cl_2 yields 45.2 g CCl_4 and 41.3 g S_2Cl_2. What is the percent yield of each product?

3.135 The reaction of 9.66 g O_2 with 9.33 g NO produces 10.1 g NO_2. What is the percent yield?

3.136 The reaction of 7.0 g Cl_2 with 2.3 g P_4 produces 7.1 g PCl_5. What is the percent yield?

3.137 The combustion of 33.5 g C_3H_6 with 127 g O_2 yields 16.1 g H_2O. What is the percent yield?

3.138 ■ Methanol, CH_3OH, is used in racing cars because it is a clean-burning fuel. It can be made by this reaction:

$$CO(g) + 2H_2(g) \rightarrow CH_3OH(\ell)$$

What is the percent yield if 5.0×10^3 g H_2 reacts with excess CO to form 3.5×10^3 g CH_3OH?

3.139 The reaction of 23.1 g NaOH with 21.2 g HNO_3 yields a 12.9-g sample of $NaNO_3$.
(a) What is the percent yield?
(b) Identify the reactant that is present in excess and calculate the mass of it that remains at the end of the reaction.

3.140 ■ When heated, potassium chlorate, $KClO_3$, melts and decomposes to potassium chloride and diatomic oxygen.
(a) What is the theoretical yield of O_2 from 3.75 g $KClO_3$?
(b) If 1.05 g of O_2 is obtained, what is the percent yield?

3.141 The Ostwald process is used to make nitric acid from ammonia. The first step of the process is the oxidation of ammonia as pictured below:

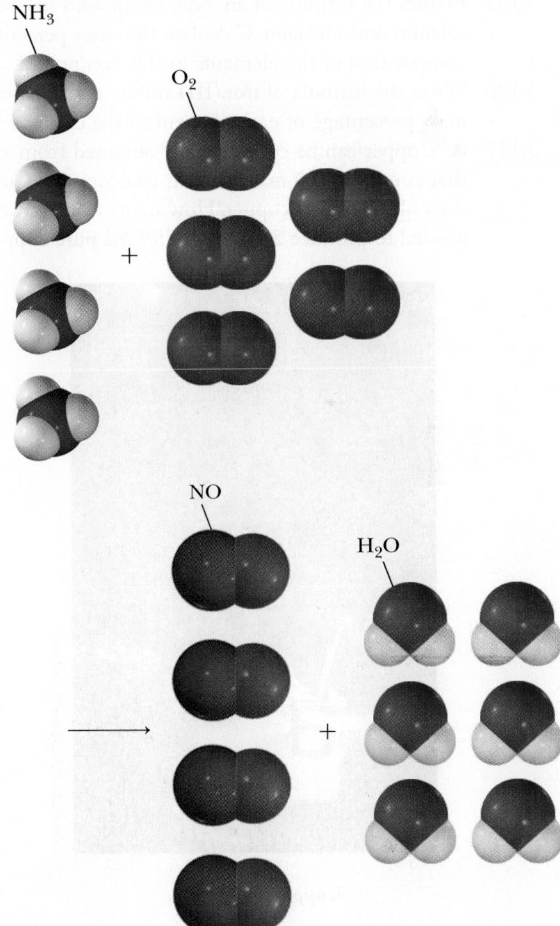

In an experiment, 50.0 g of each reactant is sealed in a container and heated so the reaction goes to completion.
(a) What is the limiting reactant?
(b) How much of the nonlimiting reactant remains after the reaction is completed? Assume that all of the limiting reactant is consumed.

3.142 In the second step of the Ostwald process (see previous exercise), the nitrogen monoxide reacts with more O_2 to yield nitrogen dioxide.

In an experiment, 75.0 g NO and 45.0 g O_2 are sealed in a container and heated so the reaction goes to completion.

(a) What is the limiting reactant?

(b) How much of the nonlimiting reactant remains after the reaction is complete? Assume that all of the limiting reactant is consumed.

3.143 ▲ A 2.24-g sample of an unknown metal reacts with HCl to produce 0.0808 g H_2 gas. The reaction is $M + 2HCl \rightarrow MCl_2 + H_2$. Assuming that the percent yield of product is 100%, identify the metal.

3.144 ▲ A 3.11-g sample of one of the halogens, X_2, is shown to react with NaOH to produce 2.00 g NaX. The equation is $2NaOH + X_2 \rightarrow NaX + NaXO + H_2O$. Assuming that the percent yield of product is 100%, identify the halogen.

Chapter Exercises

3.145 Predict the formula of an ionic compound formed from calcium and nitrogen. Calculate the mass percentage composition of the elements in this compound.

3.146 Write the formula of iron(III) sulfate and calculate the mass percentage of each element in the compound.

3.147 ▲ Copper can be commercially obtained from an ore that contains 10.0 mass percent chalcopyrite, $CuFeS_2$, as the only source of copper. How many tons of the ore are needed to produce 20.0 tons of 99.0% pure copper?

Copper production.

3.148 In_2S_3 can be converted into metallic indium by a two-step process. First, it is converted into In_2O_3 by reaction with oxygen. The other product of the reaction is SO_2. Indium metal is obtained by reaction of In_2O_3 with carbon. Assume that the other product of the second reaction is carbon dioxide.

(a) Write the two equations for this process.

(b) Calculate the mass, in kilograms, of indium produced from 35.7 kg In_2S_3, assuming excesses of the other reactants.

3.149 Some ionic compounds exist in crystalline form with a certain number of water molecules associated with the ions. Such compounds are called *hydrates.* For example, calcium sulfate can exist with either one-half water molecule per formula unit, written as $CaSO_4 \cdot \frac{1}{2}H_2O$, or two water molecules per formula unit, written as $CaSO_4 \cdot 2H_2O$. What is the percentage water (by mass) for each compound?

3.150 ▲ A hydrate (see previous exercise) can be heated to drive off the water molecules from the crystal. A sample of hydrated magnesium sulfate with an initial mass of 3.650 g was placed in a crucible and heated with a Bunsen burner. After thorough heating, the mass of the solid remaining was 1.782 g. How many water molecules are associated with each formula unit of hydrated magnesium sulfate?

3.151 A backup system on the space shuttle that removes carbon dioxide is canisters of LiOH. This compound reacts with CO_2 to produce Li_2CO_3 and water. How many grams of CO_2 can be removed from the atmosphere by a canister that contains 83 g LiOH?

3.152 ▲ Copper sulfate is generally isolated as its hydrate, $CuSO_4 \cdot xH_2O$. If a sample contains 25.5% Cu, 12.8% S, 57.7% O, and 4.04% H, what is the value of *x*?

3.153 The compound dinitrogen monoxide, N_2O, is a nontoxic gas that is used as the propellant in cans of whipped cream. How many nitrogen *atoms* are in a 34.7-g sample of N_2O?

3.154 Morphine is a narcotic substance that has been used medically as a painkiller. Its use has been highly restricted because of its addictive nature. Morphine is 71.56% C, 6.71% H, 4.91% N, and 16.82% O, and its molar mass is 285 g/mol. What is its molecular formula?

3.155 Fill in the blanks in the following table.

Name	Empirical Formula	Molar Mass (g/mol)	Molecular Formula
Dimethyl sulfoxide	C_2H_6SO	78	—
Cyclopropane	—	—	C_3H_6
Tryptamine	C_5H_6N	160	—
Lactose	—	—	$C_{12}H_{22}O_{11}$

3.156 ■ Heating $NaWCl_6$ at 300° C converts it into Na_2WCl_6 and WCl_6. If the reaction of 5.64 g $NaWCl_6$ produces 1.52 g WCl_6, what is the percentage yield?

3.157 The compound $K[PtCl_3(C_2H_4)]$ was first prepared by a Danish pharmacist around 1830. Scientists have only recently determined its structure. This complex is now known to be the first example of an important class of compounds known as organometallics. It can be prepared by the reaction of $K_2[PtCl_4]$ with ethylene, C_2H_4. The other product is potassium chloride.

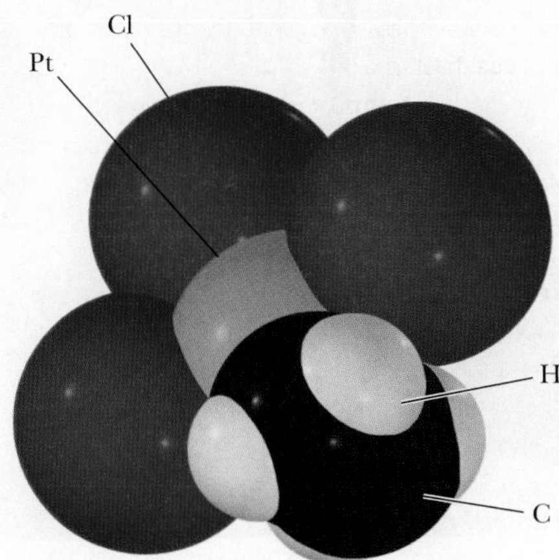

Model of $[PtCl_3(C_2H_4)]^-$.

(a) What mass of $K[PtCl_3(C_2H_4)]$ can be prepared from 45.8 g $K_2[PtCl_4]$ and 12.5 g ethylene?

(b) Identify the reactant present in excess and calculate the mass of it that remains at the end of the reaction.

3.158 ▲ Many important chemical processes require two (or more) steps. One example is a process used to determine the amount copper in a sample. Many copper compounds, dissolved in water, will react with zinc metal to yield copper metal and a water-soluble zinc compound.

$$CuSO_4(aq) + Zn(s) \rightarrow Cu(s) + ZnSO_4(aq)$$

The metallic copper is collected and weighed, whereas the $ZnSO_4$ stays in the water in which the reaction takes place. To ensure that all of the copper is converted to the metallic form, an excess of zinc is added to the reaction. The excess zinc must be removed before the weight of the copper is determined because it is also a solid. An excess of H_2SO_4 is added to remove the zinc. This acid will react with the zinc metal but *not* the copper metal.

$$Zn(s) + H_2SO_4(aq) \rightarrow ZnSO_4(aq) + H_2(g)$$

With the solid zinc thus removed, the copper metal is weighed and the mass used to find the percentage copper. What is the percent of copper in a hydrated sample of the formula $CuSO_4 \cdot xH_2O$ (x = unknown number of water molecules), if a 1.20-g sample reacts with excess zinc followed by addition of excess H_2SO_4 to yield 0.306 g copper metal? What is the value of x?

3.159 ▲ Molecular nitrogen can be converted to NO in two steps. The equations follow. These two reactions are the first steps in the industrially important conversion of nitrogen into nitric acid. Calculate the mass of NO formed from 100 g N_2 and excess H_2 and O_2.

$$N_2 + 3H_2 \rightarrow 2NH_3$$

$$4NH_3 + 5O_2 \rightarrow 4NO + 6H_2O$$

Cumulative Exercises

3.160 The reaction of sulfur dichloride and sodium fluoride yields sulfur tetrafluoride, disulfur dichloride, and sodium chloride. Write the balanced equation. What mass of sulfur tetrafluoride is formed by the reaction of 12.44 g sulfur dichloride and 10.11 g sodium fluoride?

3.161 The reaction of equal molar amounts of benzene, C_6H_6, and chlorine, Cl_2, carried out under special conditions completely consumes the reactants and yields a gas and a clear liquid. Analysis of the liquid shows that it contains 64.03% carbon, 4.48% hydrogen, and 31.49% chlorine, and has a molar mass of 112.5 g/mol. Write the balanced equation for this reaction.

3.162 ▲ Although copper does not usually react with acids, it does react with concentrated nitric acid. The reaction is complicated, but one outcome is

$$Cu(s) + HNO_3(conc) \rightarrow Cu(NO_3)_2(aq) + NO_2(g) + H_2O(\ell)$$

(a) Name all of the reactants and products.

(b) Balance the reaction.

(c) Assign oxidation numbers to the atoms. Is this a redox reaction?

(d) Pre-1983 pennies were made of pure copper. If such a penny had a mass of 3.10 g, how many moles of Cu are in one penny? How many atoms of copper are in one penny?

(e) What mass of HNO_3 would be needed to completely react with a pre-1983 penny?

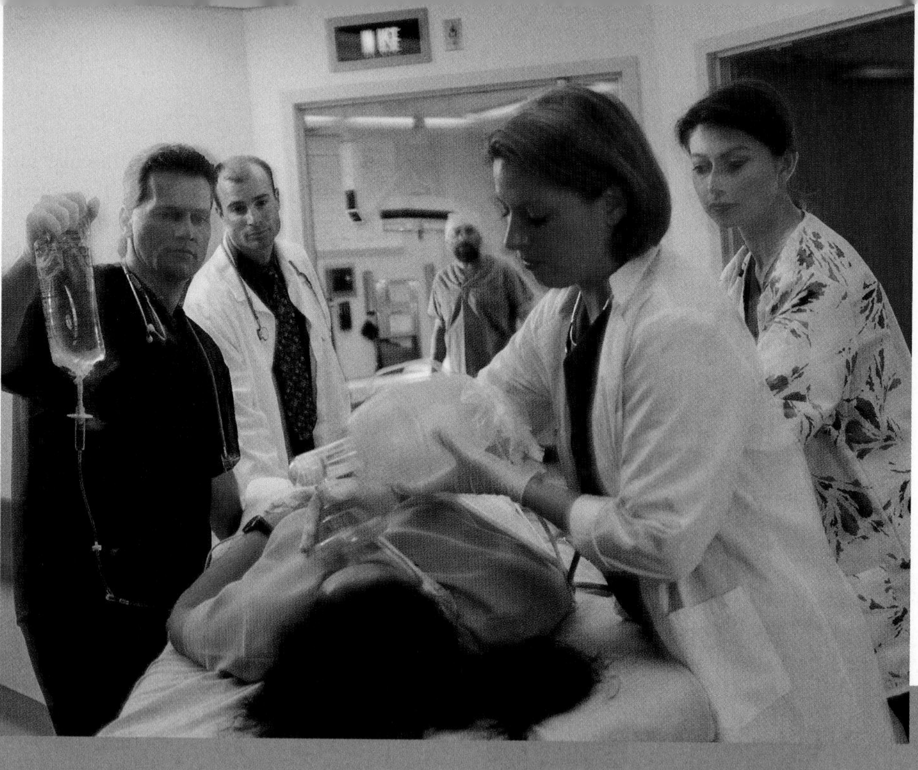

Stockbyte/Photolibrary

People do not often think of themselves as aqueous solutions, but water makes up the bulk of your body mass. When we are sick, the amounts of water and dissolved species in our blood are often out of balance. Determining concentrations of compounds dissolved in water and the reactions that occur in aqueous solutions are important goals of this chapter.

Television medical dramas often have hectic scenes where an unconscious patient is being wheeled into the emergency department and the staff is feverishly barking orders. An order you frequently hear is what sounds like "Lights." Actually, the staff member is saying "Lytes," shorthand for a request for the chemical analysis of the electrolytes in the patient's blood. Remember from Chapter 2 that electrolytes are compounds that form ions when dissolved in water. Electrolytes in the body help regulate many of the body's functions, such as the flow of nutrients into and waste products out of cells. An abnormality in electrolyte function is a primary marker for disease or bodily injury.

About 60% of your body weight is water. Approximately two thirds of this water is found inside the cells, referred to as intracellular fluid (ICF). ICF generally is relatively high in potassium and low in sodium, though the actual composition of ICF depends on the specific cell.

The remaining one third of your body weight that is water surrounds the cells and is called extracellular fluid (ECF). Most of the body's ECF is blood. The ECF is relatively high in sodium and low in potassium, exactly opposite of the ICF. Proper body function requires a subtle yet complex electrolyte differential between ICF and ECF.

The body works to keep the total amount of water and the concentrations of electrolytes within a certain range. One way to do this is to increase the amount of water brought into or excreted from the body. For example, if the sodium concentration is too high, the body produces a substance that acts to make you thirsty and drink more fluids. In addition, the adrenal gland (located in your

Chemical Reactions in Solution

4

CHAPTER CONTENTS

OWL Online homework for this chapter may be assigned in OWL.

Look for the green colored vertical bar throughout this chapter, for integrated references to this chapter introduction.

abdomen) makes another substance called aldosterone that directs the kidney to increase the concentration of sodium in the urine. A substance such as aldosterone that is produced by one tissue to cause an activity in another tissue is called a hormone. So overall, when you have too much sodium, your body tells you to drink more water, and it also acts to eliminate more sodium in the urine. A low sodium concentration in the body is a rarer condition and is often caused by drinking too much water in a short period (as might happen to joggers or other long-term exercisers who mistakenly drink too much water in an attempt to remain hydrated). Clearly, good hydration is an important way to stay healthy, but drinking too much water in a short period is also unhealthy.

Back in the emergency department, the hospital laboratory swiftly returns the results of the electrolyte analysis. After reviewing the data, showing the concentrations of sodium and potassium and other information, the physician quickly determines that the patient has high levels of potassium in the blood and suspects some kind of infection—perhaps bacteria have destroyed some of the cell walls, allowing potassium into the blood and affecting the ICF-ECF differential. The physician prescribes that the patient be given insulin to help reduce potassium levels and orders an immediate electrocardiogram because heart cells are sensitive to increased levels of potassium. The patient is then monitored closely to determine the source of the infection. ∎

Kim Truett/University South Carolina Publications

We encounter liquid solutions every day of our lives. Mornings frequently start with a cup of coffee or tea. The water in the shower is a solution, because common tap water contains many dissolved substances (such as magnesium sulfate) in addition to water molecules.

A majority of chemical reactions take place in solution because in solution the reacting species are in constant motion and can readily collide, a necessary requirement for reaction to occur. The most common solutions are made by dissolving substances in water. Water is inexpensive and nontoxic, and it dissolves many substances.

For a reaction to occur, species must collide.

4.1 Ionic Compounds in Aqueous Solution

OBJECTIVES

- ☐ Define solvent and solute
- ☐ Describe the behavior in water of strong electrolytes, weak electrolytes, and nonelectrolytes
- ☐ Predict the solubility of common ionic substances in water
- ☐ Predict products of chemical reactions in solutions of ionic compounds
- ☐ Write net ionic equations

© Cengage Learning/Charles D. Winters

Solid copper(II) chloride ($CuCl_2$) dissolves in water producing individual Cu^{2+} and Cl^- ions.

Chemists commonly prepare a solution by dissolving a solid in a liquid. The liquid is the **solvent,** the component that has the same physical state as the solution. The substance being dissolved is called the **solute.** It is often a solid but can also be a gas or a liquid. When a solution is formed from two liquids, the solvent is generally assumed to be the liquid that is present in greater quantity. The most common solvent is water; solutions with water as the solvent are called **aqueous solutions.**

Experiments show that most ionic compounds are electrolytes because they dissociate into ions when dissolved in water. The experiment is straightforward: As detailed in Chapter 2, ions in solution conduct electricity; neutral molecules do not. Many electrolytes separate completely into ions and are referred to as **strong electrolytes.** Notations such as NaCl(aq) and $CuCl_2$(aq) properly represent strong electrolytes in aqueous solution, but more accurate representations of the solutes in these solutions would be Na^+(aq) and Cl^-(aq), or Cu^{2+}(aq) and $2Cl^-$(aq). We can write a chemical equation to describe the dissociation process that occurs when ionic compounds dissolve in water.

Strong electrolytes dissociate completely into ions as they dissolve. Solutions of strong electrolytes are good conductors of electricity.

$$\mathrm{NaCl(s)} \xrightarrow{\mathrm{H_2O}} \mathrm{Na^+(aq) + Cl^-(aq)}$$

$$\mathrm{CuCl_2(s)} \xrightarrow{\mathrm{H_2O}} \mathrm{Cu^{2+}(aq) + 2Cl^-(aq)}$$

Nonelectrolytes do not conduct electricity. Many molecular compounds can show this type of behavior when dissolved in water. For example, sucrose, $C_{12}H_{22}O_{11}$, when dissolved in water does not conduct an electrical current because it does not form any ions when it dissolves.

Compounds that dissolve but do not conduct electricity are nonelectrolytes. They do not dissociate into ions when dissolved.

$$\mathrm{C_{12}H_{22}O_{11}(s)} \xrightarrow{\mathrm{H_2O}} \mathrm{C_{12}H_{22}O_{11}(aq)}$$

In contrast, as outlined in Chapter 3, a few molecular compounds, such as the acids HCl, HI, and HNO_3, completely ionize when dissolved in water and form acidic solutions; they are strong electrolytes. A properly balanced chemical equation for gaseous HCl dissolving in water is

$$\mathrm{HCl(g)} \xrightarrow{\mathrm{H_2O}} \mathrm{H^+(aq) + Cl^-(aq)}$$

Molecular compounds that only partially ionize in solution are called **weak electrolytes.** Water solutions of acetic acid (CH_3COOH) only weakly conduct electrical current, indicating that acetic acid only partially ionizes in solution. In water solution, most of the compound is present as molecules, only a small fraction of the molecules ionize. From conductivity measurements, scientists calculate that acetic acid is about 4% ionized. The chemical equation that describes dissolving acetic acid in water is

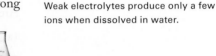

$$\underset{96\%}{CH_3COOH(\ell)} + H_2O(\ell) \rightleftharpoons \underset{4\%}{CH_3COO^-(aq)} + H_3O^+(aq)$$

In writing the equation with a double arrow, we indicate that the reaction does proceed in both directions. At any given point in time, some of the acetic acid molecules ionize, but an equal number of the acetate anions recombine with hydrogen ions to remake acetic acid; the reaction is said to be at *equilibrium.*

Acetic acid only partially ionizes in water and is an example of a *weak acid.* Other acids, like HI, HCl, and HNO_3, ionize completely in water; acids that ionize completely are known as *strong acids.* As their names imply, strong and weak acids are also strong and weak electrolytes, respectively.

Weak electrolytes produce only a few ions when dissolved in water.

It is important to note that chemists determine which compounds are strong electrolytes, weak electrolytes, or nonelectrolytes by evaluating the results of experiments, such as the conductivity experiment pictured in Chapter 2. Later (see Chapter 15) we discuss weak electrolytes, but in this chapter, we will only present strong acids and soluble ionic compounds that dissociate completely into ions in aqueous solution.

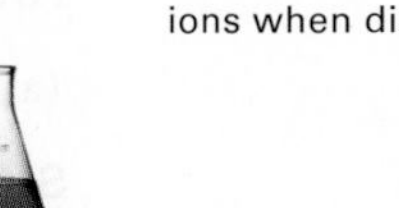

Solubility of Ionic Compounds

The best way to determine which ionic compounds will dissolve in water is by experiment. One way is to place some of the solid in water and observe whether it dissolves (Figure 4.1). The amount that dissolves is referred to as its **solubility,** the concentration of solute that exists in equilibrium with an excess of that substance.

For example, all of the nitrates (see Figure 4.1*b*) tested in Figure 4.1 dissolved in water, but two of the hydroxides tested (see Figure 4.1*d*) were insoluble. Chemists have used the results of such experiments to develop a series of rules that help predict the solubility of ionic compounds. Table 4.1 lists some of these solubility rules.

The solubility of a substance is determined by experiment.

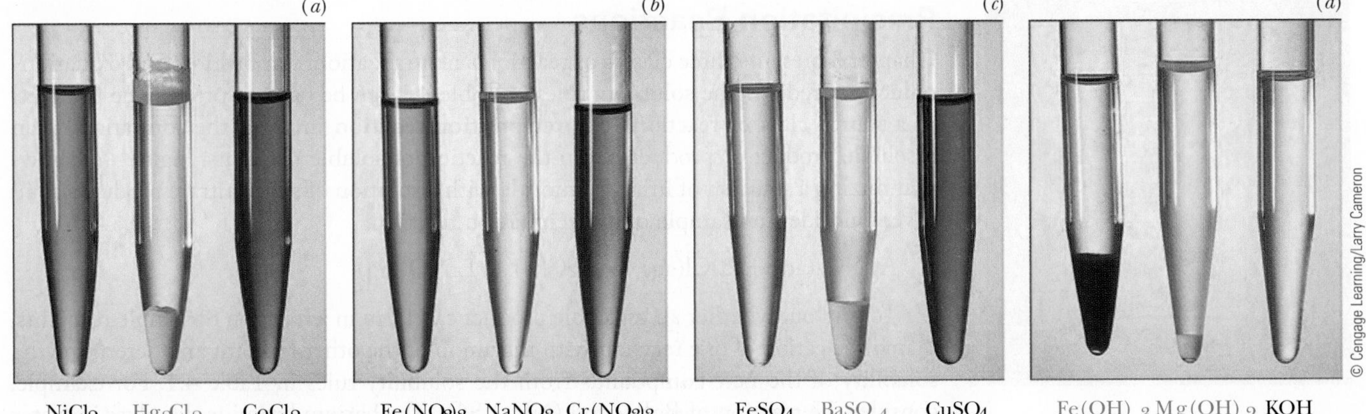

Figure 4.1 Determining solubility. These tubes show the results of experiments in which ionic compounds are added to water. The solubility rules in Table 4.1 are based on the results of experiments of this type. Insoluble compounds are identified by the blue labels.

TABLE 4.1 Solubility Rules for Ionic Compounds in Water

Compounds	Exceptions
Soluble Ionic Compounds	
Group 1A cations and NH_4^+	
Nitrates (NO_3^-)	
Perchlorates (ClO_4^-)	
Acetates (CH_3COO^-)	
Chlorides, bromides, iodides (Cl^-, Br^-, I^-)	Ag^+, Hg_2^{2+}, Pb^{2+}
Sulfates (SO_4^{2-})	Hg_2^{2+}, Pb^{2+}, Sr^{2+}, Ba^{2+}
Insoluble Ionic Compounds	
Carbonates (CO_3^{2-})	Group 1A cations, NH_4^+
Phosphates (PO_4^{3-})	Group 1A cations, NH_4^+
Hydroxides (OH^-)	Group 1A cations, NH_4^+, Sr^{2+}, Ba^{2+}

EXAMPLE 4.1 Solubility of Ionic Compounds

Are the compounds listed below soluble or insoluble in water?

(a) $Ba(NO_3)_2$ (b) $PbSO_4$ (c) LiOH (d) AgCl

Strategy Use the solubility rules in Table 4.1 to predict whether each compound is soluble or insoluble in water.

Solution

(a) Table 4.1 indicates that all ionic compounds of the nitrate ion are soluble; therefore, $Ba(NO_3)_2$ is soluble in water.

(b) Table 4.1 also indicates that most ionic compounds of the sulfate ion are soluble, but one of the exceptions is Pb^{2+}. $PbSO_4$ is insoluble in water.

(c) Even though most ionic compounds of the hydroxide ion are insoluble, ionic compounds of the Group 1A elements are soluble. LiOH is soluble in water.

(d) Most ionic compounds of the chloride ion are soluble, but one of the exceptions is Ag^+. AgCl is insoluble in water.

Understanding

Use the solubility rules in Table 4.1 to predict whether $BaSO_4$ is soluble or insoluble in water.

Answer $BaSO_4$ is insoluble.

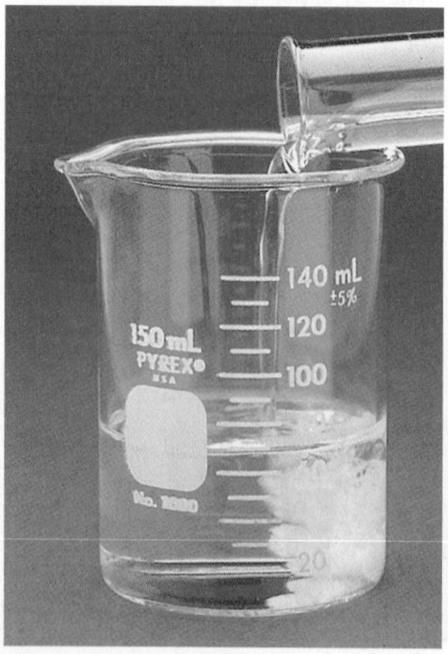

Precipitation Reactions

Chapter 3 presents three classes of reactions: neutralization, combustion, and oxidation-reduction (redox). The solubility rules of Table 4.1 can be used to predict the products of a fourth class of reactions. A **precipitation reaction** involves the formation of an insoluble product or products from the reaction of soluble reactants. Figure 4.2 shows that mixing a solution of lithium chloride with a solution of silver nitrate produces solid silver chloride, an example of a precipitation reaction.

$$AgNO_3(aq) + LiCl(aq) \rightarrow AgCl(s) + LiNO_3(aq)$$

To find out whether an insoluble product can form in a reaction of soluble reactants, match the cation of one reactant with the anion of the other reactant and determine the solubility of the new compounds from the solubility rules in Table 4.1. For example, consider the reaction of $BaBr_2$ and $(NH_4)_2SO_4$. The barium bromide dissolves in water to give the $Ba^{2+}(aq)$ and $Br^-(aq)$ ions, and ammonium sulfate dissolves to give NH_4^+ (aq) and $SO_4^{2-}(aq)$ ions. To determine whether an insoluble product forms, make a table with the cations listed along the top and the anions listed down the left side. Write all

Figure 4.2 Precipitation of silver chloride. Mixing solutions of lithium chloride and silver nitrate yields the white solid silver chloride.

of the possible combinations, and use the solubility rules to determine whether the reactants dissolve to form ions and whether insoluble products form.

	Cations	
Anions	Ba^{2+}	NH_4^+
Br^-	$BaBr_2$, soluble reactant	NH_4Br, soluble product
SO_4^{2-}	$BaSO_4$, insoluble product	$(NH_4)_2SO_4$, soluble reactant

One of the products, NH_4Br, is soluble, but the other product, $BaSO_4$, is insoluble and precipitates when the reactant solutions are mixed. Therefore, a chemical reaction occurs.

$$BaBr_2(aq) + (NH_4)_2SO_4(aq) \rightarrow BaSO_4(s) + 2NH_4Br(aq)$$

The products, if any, of many reactions of ionic compounds can be predicted from the solubility rules.

EXAMPLE 4.2 Precipitation Reactions

Using the solubility rules in Table 4.1 as a guide, predict whether an insoluble product forms when each of the following pairs of solutions is mixed. Write the balanced chemical equation if a precipitation reaction does occur.

(a) $Pb(NO_3)_2$ and sodium carbonate
(b) ammonium bromide and $AgClO_4$
(c) potassium hydroxide and copper(II) chloride
(d) ammonium bromide and cobalt(II) sulfate

Strategy Write the formulas of the new potential ionic compounds by making a table of cations and anions and their combinations. Use the solubility rules (see Table 4.1) to determine whether any of the combinations are insoluble.

Solution

(a) Write the table, showing the ions and all of their possible combinations.

	Cations	
Anions	Pb^{2+}	Na^+
NO_3^-	$Pb(NO_3)_2$, soluble reactant	$NaNO_3$, soluble product
CO_3^{2-}	$PbCO_3$, insoluble product	Na_2CO_3, soluble reactant

The two products are $NaNO_3$ and $PbCO_3$. Sodium nitrate is soluble, but lead carbonate is not. The equation is

$$Na_2CO_3(aq) + Pb(NO_3)_2(aq) \rightarrow PbCO_3(s) + 2NaNO_3(aq)$$

(b) Write the table, showing the ions and all of their possible combinations.

	Cations	
Anions	NH_4^+	Ag^+
Br^-	NH_4Br, soluble reactant	$AgBr$, insoluble product
ClO_4^-	NH_4ClO_4, soluble product	$AgClO_4$, soluble reactant

The two products are NH_4ClO_4 and AgBr. The NH_4ClO_4 is soluble, but the AgBr is one of the few insoluble halides. The balanced chemical equation is

$$AgClO_4(aq) + NH_4Br(aq) \rightarrow AgBr(s) + NH_4ClO_4(aq)$$

(c) One of the two possible products, KCl, is soluble, but the other, $Cu(OH)_2$, is insoluble. The balanced chemical equation is

$$2KOH(aq) + CuCl_2(aq) \rightarrow 2KCl(aq) + Cu(OH)_2(s)$$

(d) Both products, $(NH_4)_2SO_4$ and $CoBr_2$, are soluble; thus, no insoluble product forms. No chemical reaction occurs when the two reactant solutions are combined.

Understanding

Predict whether an insoluble product forms when solutions of strontium nitrate, $Sr(NO_3)_2$, and sodium sulfate are mixed.

Answer Insoluble $SrSO_4(s)$ forms

Net Ionic Equations

In many chemical reactions involving ionic compounds, some of the ions remain in solution and do not undergo change. Figure 4.2 showed that mixing solutions of $AgNO_3(aq)$ and $LiCl(aq)$ yields an insoluble white solid that we can identify as $AgCl(s)$. We can write this reaction as the **overall equation,** which shows all of the reactants and products in undissociated form.

$$AgNO_3(aq) + LiCl(aq) \rightarrow AgCl(s) + LiNO_3(aq)$$

The solubility rules indicate that all of the compounds except silver chloride are soluble in water and exist in solution as ions. A more accurate description of this reaction is the **complete ionic equation,** an equation in which strong electrolytes are shown as ions in the solution.

$$Ag^+(aq) + NO_3^-(aq) + Li^+(aq) + Cl^-(aq) \rightarrow AgCl(s) + Li^+(aq) + NO_3^-(aq)$$

This equation represents the species as they exist in solution. Notice that the lithium ions and nitrate ions are present in the same forms on both sides of the equation. Because they undergo no change, we can omit them from the equation; they are referred to as **spectator ions** because they do not participate in any chemical change. The only chemical change is represented by the **net ionic equation,** which shows only those species in the solution that actually undergo a chemical change:

$$Ag^+(aq) + \cancel{NO_3^-(aq)} + \cancel{Li^+(aq)} + Cl^-(aq) \rightarrow AgCl(s) + \cancel{Li^+(aq)} + \cancel{NO_3^-(aq)}$$

$$Ag^+(aq) + Cl^-(aq) \rightarrow AgCl(s)$$

The net ionic equation shows only those species in a chemical reaction that undergo change.

The net ionic equation is a simpler, and often a more useful description of the chemical reaction. Remember that the spectator ions are still present in the solution even though they do not participate in the reaction.

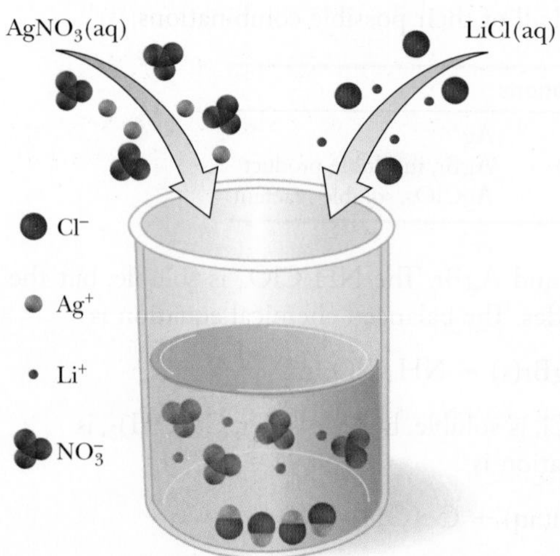

The compounds silver nitrate and lithium chloride dissociate into ions in water solution. Mixing these solutions produces insoluble silver chloride.

EXAMPLE 4.3 Writing Net Ionic Equations

Mixing aqueous solutions of magnesium nitrate with potassium hydroxide produces insoluble magnesium hydroxide. Write the overall equation, the complete ionic equation, and the net ionic equation for this reaction.

Strategy Write the overall equation that shows all of the compounds in the reaction. Then write the complete ionic equation showing all of the soluble compounds dissociated into ions. To write the net ionic equation, cancel the spectator ions that appear in equal amounts on both sides of the equation leaving only the species that undergo change.

Solution

The overall equation for this reaction is

$$Mg(NO_3)_2(aq) + 2KOH(aq) \rightarrow Mg(OH)_2(s) + 2KNO_3(aq)$$

Both of the reactants and KNO_3 are soluble and exist as ions in water, but $Mg(OH)_2$ is an insoluble solid. The complete ionic equation is

$$Mg^{2+}(aq) + \cancel{2NO_3^-(aq)} + \cancel{2K^+(aq)} + 2OH^-(aq) \rightarrow Mg(OH)_2(s) + \cancel{2K^+(aq)} + \cancel{2NO_3^-(aq)}$$

The $2K^+(aq)$ and $2NO_3^-(aq)$ are present on both sides of the equation and do not undergo change. We remove these spectator ions to obtain the net ionic equation.

$$Mg^{2+}(aq) + 2OH^-(aq) \rightarrow Mg(OH)_2(s)$$

Understanding

Mixing aqueous solutions of lead(II) nitrate and potassium sulfate produces insoluble lead sulfate. Write the net ionic equation for this reaction.

Answer $Pb^{2+}(aq) + SO_4^{2-}(aq) \rightarrow PbSO_4(s)$

The net ionic equation in Example 4.3 suggests that any soluble magnesium salt and any soluble hydroxide can form $Mg(OH)_2$. Someone who needed $Mg(OH)_2(s)$ but had no $Mg(NO_3)_2$ could substitute $MgCl_2$ (also soluble in water) in the preparation. The net ionic equation is the same regardless of the source of Mg^{2+}.

Net ionic equations are useful in writing acid–base reactions. Consider the acid–base reaction that occurs when a solution of potassium hydroxide is mixed with a solution of hydrochloric acid. The overall equation is

$$HCl(aq) + KOH(aq) \rightarrow KCl(aq) + H_2O(\ell)$$

However, all of the compounds except water exist in solution as ions (recall that HCl ionizes in water to produce $H^+(aq)$ and $Cl^-(aq)$), so the complete ionic equation is

$$H^+(aq) + \cancel{Cl^-(aq)} + \cancel{K^+(aq)} + OH^-(aq) \rightarrow \cancel{K^+(aq)} + \cancel{Cl^-(aq)} + H_2O(\ell)$$

The K^+ and Cl^- ions are present in the same forms on both sides of the equation. Because they undergo no change, they can be omitted from the equation (Figure 4.3). The net ionic equation is

$$H^+(aq) + OH^-(aq) \rightarrow H_2O(\ell)$$

OBJECTIVES REVIEW *Can you:*

- ☑ define solvent and solute?
- ☑ describe the behavior in water of strong electrolytes, weak electrolytes, and nonelectrolytes?
- ☑ predict the solubility of common ionic substances in water?
- ☑ predict products of chemical reactions in solutions of ionic compounds?
- ☑ write net ionic equations?

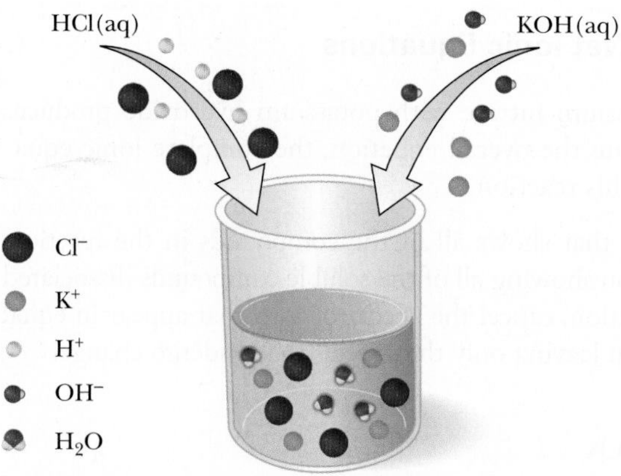

Figure 4.3 A net ionic equation. The acid HCl and the base KOH in separate solutions are present as ions. When mixed, the $K^+(aq)$ and $Cl^-(aq)$ ions undergo no change, but the $H^+(aq)$ and $OH^-(aq)$ ions react to form water.

4.2 Molarity

OBJECTIVES

- ☐ Define concentration and molarity
- ☐ Describe how to prepare solutions of known molarity from weighed samples of solute
- ☐ Describe how to prepare solutions of known molarity from concentrated solutions
- ☐ Calculate the amount of solute, the volume of solution, or the molar concentration of solution, given the other two quantities

In the laboratory, it is generally much simpler to measure volumes of liquid solutions than to weigh them to determine their masses. For quantitative calculations, we need to know the **concentration** of a solution—that is, the amount of solute in a given quantity of that solution. Scientists use several different units to express concentration, but for calculations involving stoichiometry, the most useful unit of concentration is **molarity** (symbolized by *M*), the number of moles of solute per liter of solution.

Concentration is the ratio of the quantity of solute divided by the quantity of solution. Molarity expresses concentration as moles of solute per liter of solution.

$$\text{Molarity} = \frac{\text{moles of solute}}{\text{liter of solution}}$$

Recall from the chapter introduction that your body controls excess sodium in two ways. One hormone decreases the amount of sodium in blood by increasing the excretion of sodium in urine. The other hormone makes you thirsty, increasing the volume of blood.

$$\begin{array}{l} \text{Decrease sodium} \rightarrow \\ \text{Increase blood volume} \rightarrow \end{array} \frac{\text{fewer moles of solute}}{\text{larger volume of solution}} \Rightarrow \text{decreased sodium concentration}$$

One hormone acts on the numerator and one the denominator, but they both act to decrease the sodium concentration to bring it back into the normal range.

Figure 4.4 illustrates one way to prepare a solution of known molar concentration. The solute is weighed and placed in a **volumetric flask** that has been calibrated to contain a known volume of liquid. Next, solvent is added to the flask to dissolve the solute. The flask is then filled to the calibration mark with more solvent; then the solution is thoroughly mixed. Note that the molarity of a solution is based on the *total volume of solution* and not on the volume of added solvent. Also, note that the definition of molarity requires a volume of solution in *liters*. If the stated volume is not in units of liters, it must be converted to liters before a concentration in molarity can be determined.

Solutions of known molarity can be prepared from a weighed sample dissolved in a solvent, then diluted to a known volume of solution.

(a)

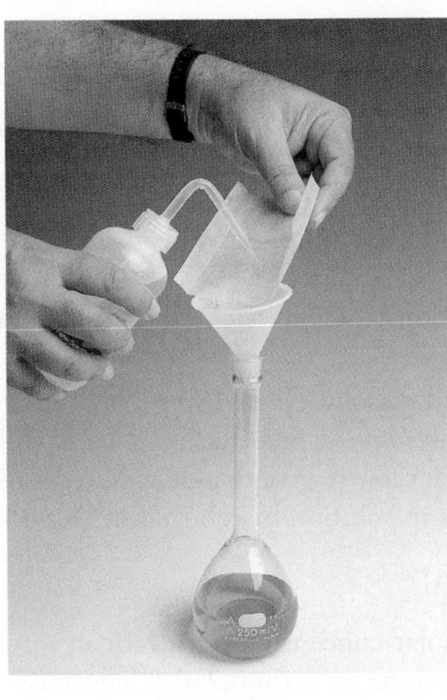
(b)

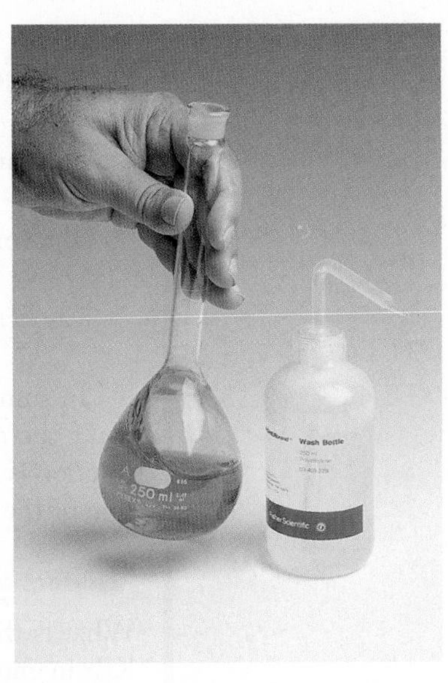
(c)

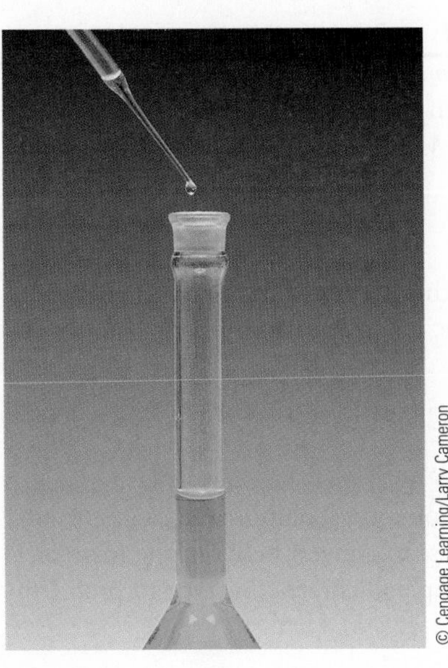
(d)

Figure 4.4 Preparation of a solution of known molarity. *(a)* Add the carefully weighed sample to the volumetric flask. *(b)* Use the solvent to wash the residue of the solid on the weighing paper into the flask, and *(c)* swirl the flask to dissolve the solute. *(d)* Add more solvent until the level of solution is at the calibration mark on the neck of the flask.

EXAMPLE 4.4 Calculating Molar Concentration of a Solution

A physician attending to a dehydrated patient ordered that the patient be given intravenous normal saline. The recipe for normal saline solution is to weigh 180 g of very pure NaCl into a container and add enough water to produce 20.0 L of solution. What is the molar concentration of NaCl?

Strategy Because molarity is defined as moles of solute per liter of solution, convert the mass of solute to moles. The flow diagram follows:

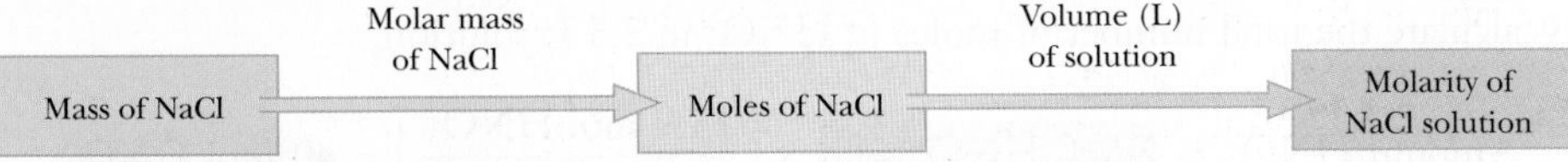

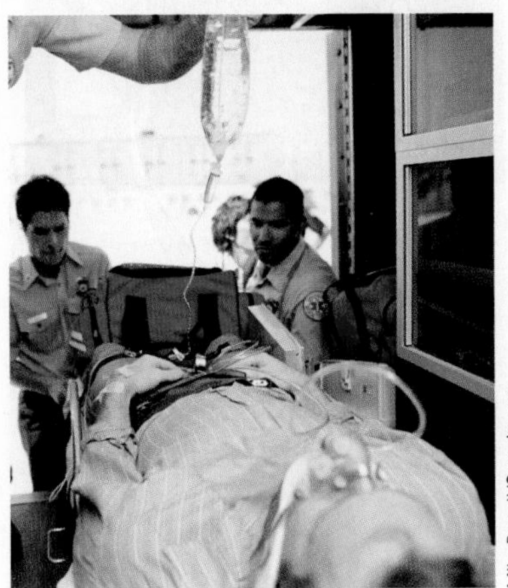

A saline solution is administered to a dehydrated patient.

Solution

Use the molar mass of NaCl (58.44 g/mol) to calculate the number of moles of NaCl.

$$\text{Amount NaCl} = 180\,\text{g NaCl} \times \left(\frac{1\ \text{mol NaCl}}{58.44\,\text{g NaCl}}\right) = 3.08\ \text{mol NaCl}$$

The volume of solution is already given in liters, so the molarity is

$$\text{Concentration NaCl} = \frac{3.08\ \text{mol NaCl}}{20.0\ \text{L soln}} = 0.154\ M$$

In summary, dissolution of 180 g NaCl in water and addition of enough water to make the *total volume* of the solution 20.0 L will yield a 0.154 *M* NaCl solution. This solution is much safer to give to a dehydrated patient than pure water as the concentration of the ions in solution match that in blood fairly closely.

Understanding

What is the molar concentration of KBr in a solution prepared by dissolving 0.321 g KBr in enough water to form 0.250 L solution?

Answer 0.0108 *M*

Calculation of Moles from Molarity

The molarity of a solution provides the relation that converts between volume of solution and moles of solute.

The molar concentration *relates the volume of solution (expressed in liters) to the number of moles of solute present.* It allows us to convert between volume of a solution and number of moles of solute. If, for example, we have 0.154 *M sodium chloride,* each liter of the solution contains 0.154 mol NaCl.

1 L NaCl soln contains 0.154 mol NaCl

For calculations involving this solution, we can use this relationship to convert between volume of solution and number of moles of solute, using the appropriate conversion factor:

$$\left(\frac{1\ \text{L NaCl soln}}{0.154\ \text{mol NaCl}}\right) \quad \text{or} \quad \left(\frac{0.154\ \text{mol NaCl}}{1\ \text{L NaCl soln}}\right)$$

Using the molarity of a solution to convert between volume of solution and moles of solvent is the same type of procedure as using molar mass to convert between the mass of a sample and the number of moles. The following example illustrates this type of problem.

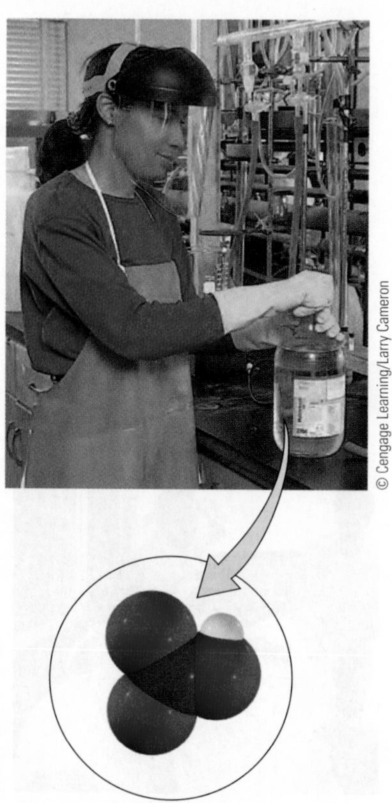

Figure 4.5 Concentrated HNO_3. Wear proper protective equipment when handling concentrated acids.

EXAMPLE 4.5 Moles of Solute

Concentrated nitric acid, HNO_3, is 15.9 *M* and is frequently sold in 2.5-L containers (Figure 4.5). How many moles of HNO_3 are present in each container?

Strategy Molarity, given in the problem, is the conversion between volume and moles present in a given volume of solution.

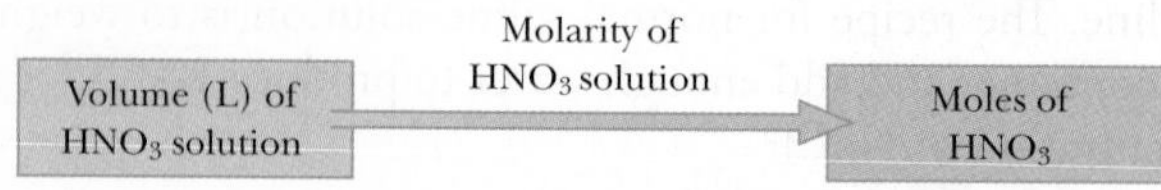

Solution

One liter of a 15.9 *M* HNO_3 solution contains 15.9 mol HNO_3. Use this relationship to calculate the total number of moles of HNO_3 in 2.5 L solution.

$$\text{Amount HNO}_3 = 2.5\ \text{L HNO}_3\ \text{soln} \times \left(\frac{15.9\ \text{mol HNO}_3}{1\ \text{L HNO}_3\ \text{soln}}\right) = 40\ \text{mol HNO}_3$$

Understanding

Frequently physicians prescribe half-saline solutions rather than normal saline solution for their patients to reduce the amount of sodium the patient receives. The concentration of sodium chloride in half-saline solutions is 0.0770 *M*. How many moles of sodium chloride are in a 500-mL bag of the half-saline solution?

Answer 0.0385 mol NaCl

EXAMPLE 4.6 Preparing Solutions of Known Molarity

What mass of potassium sulfate, K_2SO_4, is needed to prepare 500 mL of a 0.200 *M* K_2SO_4 solution?

Strategy Use the molarity and volume of solution to determine the amount (moles) of K_2SO_4 that is needed to prepare the solution; then use the molar mass to calculate the mass of K_2SO_4.

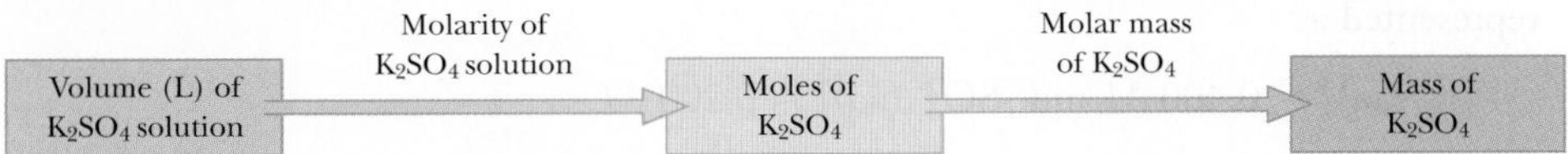

Solution
First, determine the amount of K_2SO_4 needed. Remember to convert the volume to liters.

$$\text{Amount } K_2SO_4 = 0.500 \text{ L } \cancel{K_2SO_4 \text{ soln}} \times \left(\frac{0.200 \text{ mol } K_2SO_4}{1 \text{ L } \cancel{K_2SO_4 \text{ soln}}}\right) = 0.100 \text{ mol } K_2SO_4$$

Second, use the molar mass of K_2SO_4 (174.3 g/mol) to calculate the mass.

$$\text{Mass } K_2SO_4 = 0.100 \text{ } \cancel{\text{mol } K_2SO_4} \times \left(\frac{174.3 \text{ g } K_2SO_4}{1 \text{ } \cancel{\text{mol } K_2SO_4}}\right) = 17.4 \text{ g } K_2SO_4$$

Understanding

Calculate the mass of $AgNO_3$ needed to prepare 1.00 L of a 0.150 *M* $AgNO_3$ solution?

Answer 25.5 g $AgNO_3$

Calculating the Molar Concentration of Ions

When the potassium sulfate in Example 4.6 dissolves in water, it dissociates into K^+ and SO_4^{2-} ions, as pictured in Figure 4.6. Measurements of electrical conductivity show that potassium sulfate is a strong electrolyte and that two K^+ and one SO_4^{2-} ions are produced in solution for every one K_2SO_4 that dissolves.

$$K_2SO_4 \xrightarrow{H_2O} 2K^+(aq) + SO_4^{2-}(aq)$$

Because there are 2 mol potassium ions in every 1 mol potassium sulfate, the molar concentration of the K^+ ions in the solution is twice the molar concentration of the K_2SO_4. In the 0.200 *M* K_2SO_4 solution described in Example 4.6, the concentration of K^+ *ions* is 0.400 *M*. When dealing with solutions of ionic materials, it is important to carefully specify the species to which the molarity refers.

It is common to use square brackets around a species to imply "concentration of this species in units of molarity." Thus, in the above K_2SO_4 solution, rather than say

$$\text{Concentration of } K_2SO_4 = 0.200\ M$$

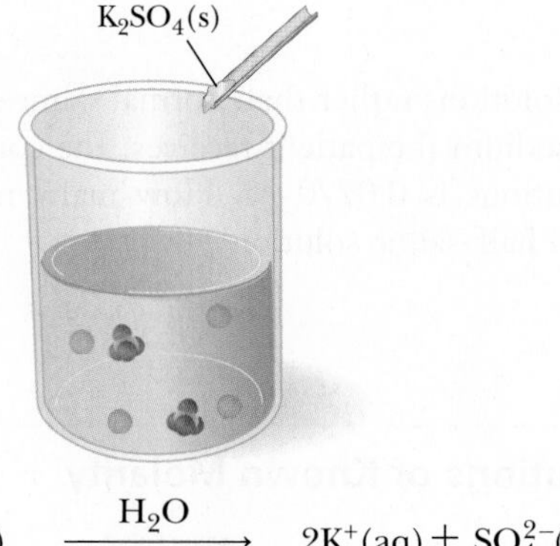

Figure 4.6 Dissolving ionic compounds. When K_2SO_4 dissolves in water, two moles of $K^+(aq)$ and one mole of the polyatomic anion $SO_4^{2-}(aq)$ form in solution for every one mole of K_2SO_4.

$$K_2SO_4(s) \xrightarrow{H_2O} 2K^+(aq) + SO_4^{2-}(aq)$$

we would use, more succinctly,

$$[K_2SO_4(aq)] = 0.200\ M$$

Also, based on the paragraph above, the concentrations of the individual ions can be represented as

$$[K^+(aq)] = 0.400\ M \text{ and } [SO_4^{2-}(aq)] = 0.200\ M$$

EXAMPLE 4.7 Calculating Molar Concentrations of Ions

What are the molar concentrations of the ions in a 0.20 M calcium nitrate, $Ca(NO_3)_2$, solution?

Strategy In aqueous solution, $Ca(NO_3)_2$ dissociates into one $Ca^{2+}(aq)$ and two $NO_3^-(aq)$ ions.

$$Ca(NO_3)_2(s) \xrightarrow{H_2O} Ca^{2+}(aq) + 2NO_3^-(aq)$$

Use the coefficients for $Ca(NO_3)_2$, $Ca^{2+}(aq)$, and $NO_3^-(aq)$ to calculate the $Ca^{2+}(aq)$ and $NO_3^-(aq)$ concentration.

Solution

Based on the relationships in the chemical equation

1 mol $Ca(NO_3)_2$ yields 1 mol $Ca^{2+}(aq)$ and 2 mol $NO_3^-(aq)$

Because 1 mol $Ca(NO_3)_2$ produces 1 mol $Ca^{2+}(aq)$, the concentrations are the same. Thus, $[Ca^{2+}(aq)] = 0.20\ M$. To calculate the concentration of the nitrate ion

$$[NO_3^-(aq)] = \frac{0.20 \text{ mol } Ca(NO_3)_2}{\text{L soln}} \times \left(\frac{2 \text{ mol } NO_3^-(aq)}{1 \text{ mol } Ca(NO_3)_2}\right) = 0.40\ M$$

Understanding

What are the molar concentrations of the ions in a 1.10 M Li_2CO_3 solution?

Answer 2.20 M $Li^+(aq)$; 1.10 M $CO_3^{2-}(aq)$

Dilution

Chemists frequently need to prepare dilute (low-concentration) solutions from the more concentrated solutions that allow for more convenient storage of larger amounts of solute. To prepare a dilute solution, they mix pure solvent with a certain volume of the concentrated solution. This procedure, illustrated in Figure 4.7, is conceptually similar to the preparation of a solution directly from a solid. The difference is that the con-

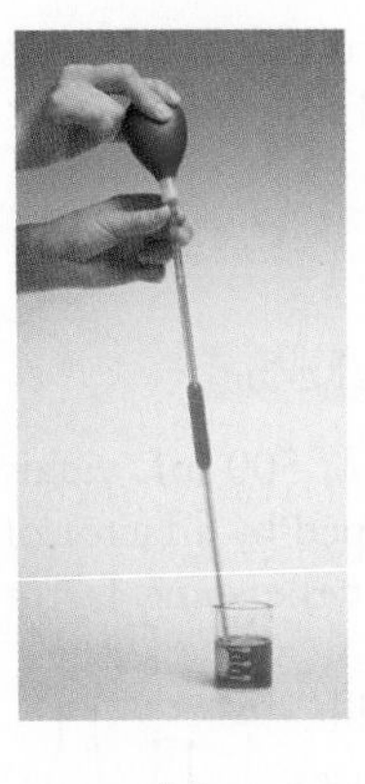
(a)

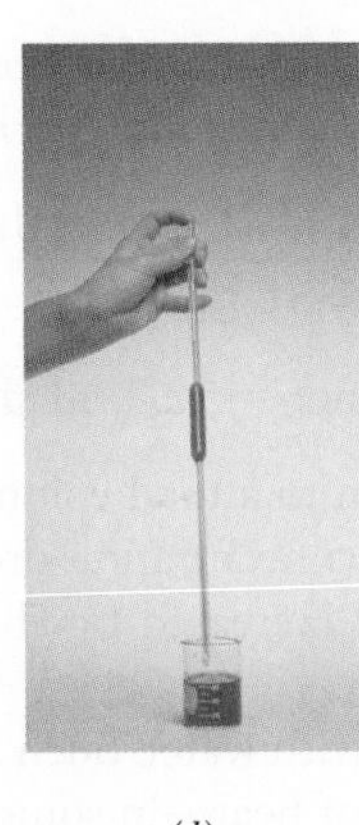
(b)

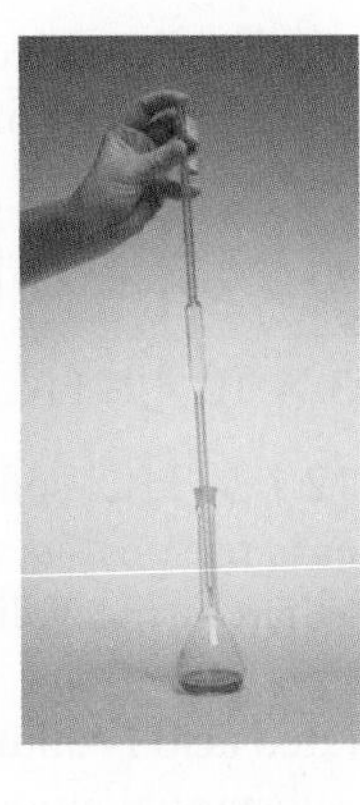
(c)

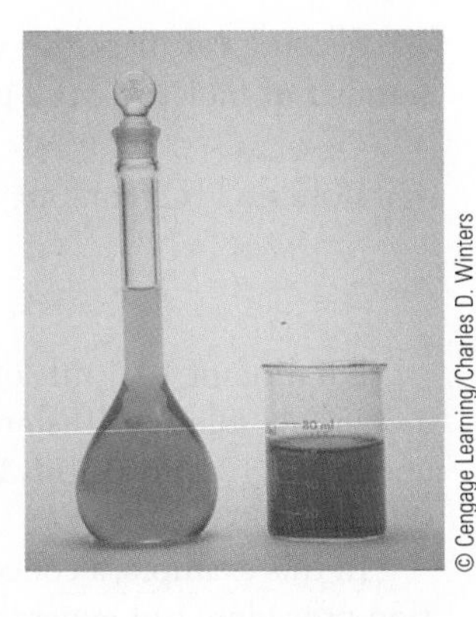
(d)

Figure 4.7 Preparing a dilute solution from a concentrated solution. *(a)* Draw the concentrated solution into a pipet to a level just above the calibration mark. *(b)* Allow the liquid to settle down to the calibration line. Touch the tip of the pipet to the side of the container to remove any extra liquid. *(c)* Transfer this solution to a volumetric flask. Again touch the pipet to the wall of the flask to ensure complete transfer, but do not blow out the pipet because it is calibrated for a small amount of liquid to remain in the tip. *(d)* Dilute to the mark with solvent. Frequently, especially with concentrated acids, it is best to have some solvent present in the volumetric flask before you add the concentrated sample.

centration of the new solution is based on a known *volume* of the concentrated solution rather than on the mass of the solute.

Solutions of known molarity are often prepared by diluting more concentrated solutions.

One device used to measure the volume of a solution accurately is a **pipet,** a calibrated device designed to deliver an accurately known volume of liquid with high precision (Figure 4.7). The liquid is drawn into the pipet by means of suction from a rubber bulb. When 10 mL of a concentrated solution (often called the *stock solution*) are diluted to 100 mL with pure solvent, as shown in Figure 4.7, the concentration of the dilute solution is determined from the volume and concentration of the stock solution and the total volume of the dilute solution.

Dilution of a given amount of a concentrated solution *does not change the number of moles of solute;* the moles of solute in the concentrated solution are the same as in the dilute solution. The only difference between the two solutions is that more solvent is present (i.e., there is a larger volume) in the dilute solution.

EXAMPLE 4.8 Dilution

Concentrated hydrochloric acid is sold as a 12.1 *M* solution. What volume of this solution of concentrated HCl is needed to prepare 0.500 L of 0.250 *M* HCl?

Strategy We know the volume and the concentration of the dilute solution; this information is used to calculate the number of moles of HCl needed to prepare the dilute solution. Because the source of the HCl is the concentrated solution (conc), we can calculate the volume required to add this needed amount of HCl using the molarity of the concentrated solution.

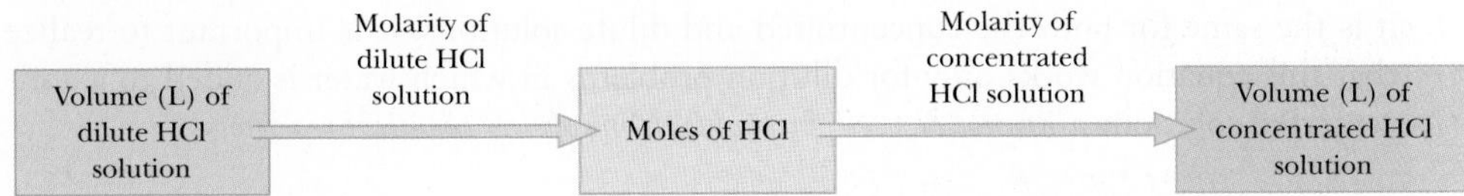

Solution

First, calculate the number of moles of HCl required in the dilute solution (dil) from its volume and molarity. The conversion factor comes from the given molarity of the dilute HCl.

$$\text{Amount HCl} = 0.500\ \cancel{\text{L HCl(dil) soln}} \times \left(\frac{0.250\ \text{mol HCl}}{1\ \cancel{\text{L HCl(dil) soln}}}\right) = 0.125\ \text{mol HCl}$$

Second, calculate the volume of the concentrated solution needed to contain this number of moles of HCl using the molarity of the concentrated HCl solution.

$$\text{Volume HCl(conc)} = 0.125 \cancel{\text{mol HCl}} \times \left(\frac{1 \text{ L HCl(conc) soln}}{12.1 \cancel{\text{mol HCl}}}\right)$$

$$= 1.03 \times 10^{-2} \text{ L HCl(conc)} = 10.3 \text{ mL HCl(conc)}$$

Dilution of 10.3 mL of a 12.1 *M* HCl solution to a total volume of 500 mL yields a 0.250 *M* solution. The answer is reasonable; the concentration dropped by a factor of about 50 (12.1 to 0.250 *M*), whereas the volume increased by about the same factor (10.3 to 500 mL).

In this example, a concentrated acid is diluted with water. Such a dilution procedure can generate a substantial amount of heat—in some cases, enough to boil the water, causing it to spatter out of the container. For this reason, an important rule for chemists to remember is *Add acid to water.*

Understanding

Sodium hydroxide is sold as a 1.00 *M* solution. What volume of this solution of NaOH is needed to prepare 250 mL of 0.110 *M* NaOH?

Answer 27.5 mL

In Example 4.8, we used a conversion factor to calculate the volume of a concentrated solution needed to prepare a dilute solution. The key to the calculation is that the numbers of moles of the solute in the 10.3-mL sample of concentrated HCl is the same as in the 250 mL of dilute solution. The problem can also be solved by a second method, an algebraic method. Because the product molarity (moles/liter) × liters yields moles, and the numbers of moles in the two solutions are equal, we can write the following equation:

$$\text{molarity(conc)} \times \text{liters(conc)} = \text{molarity(dil)} \times \text{liters(dil)}$$

$$M\text{(conc)} \times \text{L(conc)} = M\text{(dil)} \times \text{L(dil)}$$

In general terms, the dilution relationship is written as M(conc) × V(conc) = M(dil) × V(dil), where V is the volume, expressed in any consistent unit. In using this equation for the problem in Example 4.8, solve the equation for the unknown quantity—the volume of the concentrated solution—and substitute the three values given in the problem:

$$V\text{(conc)} = \frac{M\text{(dil)} \times V\text{(dil)}}{M\text{(conc)}}$$

$$V\text{(conc)} = \frac{0.250 \cancel{M \text{ HCl}} \times 500 \text{ mL HCl}}{12.1 \cancel{M \text{ HCl}}} = 10.3 \text{ mL HCl}$$

With the algebraic method, any volume unit may be used (mL, for example), as long as it is the same for both the concentrated and dilute solutions. It is important to realize that this equation works only for dilution problems in which water is added to a concentrated solution; it *cannot be used for problems involving chemical reactions.*

EXAMPLE 4.9 Dilution

Calculate the molar concentration of a solution prepared by diluting 50 mL of 5.23 *M* NaOH to 2.0 L.

Strategy Using the algebraic method described above, substitute quantities into the formula and solve for the desired quantity, the concentration of the prepared (i.e., diluted) solution.

Solution

Starting with the original expression:

$$M(\text{conc}) \times V(\text{conc}) = M(\text{dil}) \times V(\text{dil})$$

Rearrange this to solve for the quantity we are looking for, which is M(dil):

$$M(\text{dil}) = \frac{M(\text{conc}) \times V(\text{conc})}{V(\text{dil})}$$

Now substitute the given quantities:

$$M(\text{dil}) = \frac{5.23\ M\ \text{NaOH} \times 50\ \cancel{\text{mL NaOH}}}{2000\ \cancel{\text{mL NaOH}}}$$

$$M(\text{dil}) = 0.13\ M\ \text{NaOH}$$

Understanding

What is the molar concentration of a solution prepared by diluting 20 mL of 5.2 M HNO_3 to 0.50 L?

Answer 0.21 M

OBJECTIVES REVIEW *Can you:*

- ☑ define concentration and molarity?
- ☑ describe how to prepare solutions of known molarity from weighed samples of solute?
- ☑ describe how to prepare solutions of known molarity from concentrated solutions?
- ☑ calculate the amount of solute, the volume of solution, or the molar concentration of solution, given the other two quantities?

4.3 Stoichiometry Calculations for Reactions in Solution

OBJECTIVE

- ☐ Perform stoichiometric calculations with chemical reactions, given the molar concentrations and volumes of solutions of reactants or products

Stoichiometry calculations for reactions in solution are similar to those already illustrated in Chapter 3, but the amounts are calculated from the volumes of solutions of known concentrations rather than from masses. The important similarity is that the chemical equation relates the number of moles of one substance to the number of moles of another, regardless of whether we measure the mass of solid or the volume and concentration of a solution.

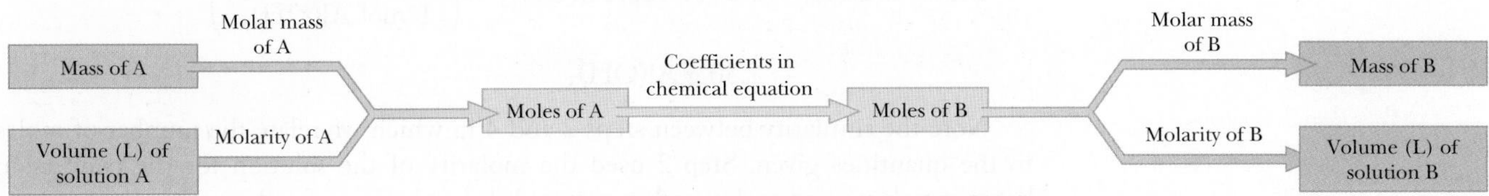

We now have two methods for converting between quantity and number of moles. If the mass of a compound is given or needed, we use the molar mass of the compound. If the volume of solution of a compound is given or needed, we use the molarity of the solution. In both cases, the stoichiometric relationships of the chemical equation enable us to calculate the number of moles of any other substance in the equation.

The number of moles of a reactant or product in solution is calculated from the molarity and volume of solution.

The following two example problems demonstrate stoichiometry calculations in which the reactants or products are in solution.

EXAMPLE 4.10 Solution Stoichiometry

Calculate the mass, in grams, of $Al(OH)_3$ (molar mass = 78.00 g/mol) formed by the reaction of exactly 0.500 L of 0.100 *M* NaOH with excess $Al(NO_3)_3$.

Strategy This problem is identical in strategy to the stoichiometry problems in Chapter 3. We write the balanced equation, then calculate the number of moles of the given species. Then we use the coefficients of the equation to calculate the number of moles of the desired species. Finally, we convert from moles of the desired species to mass, using the molar mass of $Al(OH)_3$. The only change in the solution when compared to Chapter 3 is that we will use volume and concentration to compute the number of moles of the given species.

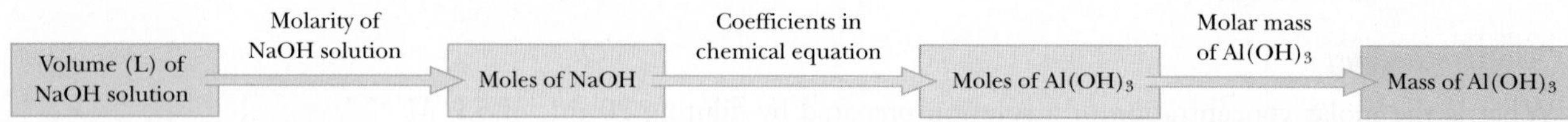

Precipitation of aluminum hydroxide. The reaction of sodium hydroxide and aluminum nitrate yields insoluble aluminum hydroxide. Aluminum hydroxide is a gelatinous solid used in water purification.

Solution

First, write the chemical equation.

$$Al(NO_3)_3(aq) + 3NaOH(aq) \rightarrow 3NaNO_3(aq) + Al(OH)_3(s)$$

Second, calculate the number of moles of NaOH from the volume and concentration of the NaOH solution. NaOH is the limiting reactant because $Al(NO_3)_3$ is in excess. The important relationship is

1 L NaOH soln contains 0.100 mol NaOH

$$\text{Amount NaOH} = 0.500 \text{ L NaOH soln} \times \left(\frac{0.100 \text{ mol NaOH}}{1 \text{ L NaOH soln}}\right)$$

$$= 0.0500 \text{ mol NaOH}$$

Third, use the stoichiometric relationships from the chemical equation to calculate the number of moles of $Al(OH)_3$ formed in the reaction.

$$\text{Amount Al(OH)}_3 = 0.0500 \text{ mol NaOH} \times \left(\frac{1 \text{ mol Al(OH)}_3}{3 \text{ mol NaOH}}\right)$$

$$= 0.0167 \text{ mol Al(OH)}_3$$

Fourth, calculate the mass of $Al(OH)_3$, using its molar mass.

$$\text{Mass Al(OH)}_3 = 0.0167 \text{ mol Al(OH)}_3 \times \left(\frac{78.00 \text{ g Al(OH)}_3}{1 \text{ mol Al(OH)}_3}\right)$$

$$= 1.30 \text{ g Al(OH)}_3$$

Note the similarity between steps 2 and 4 in which we relate the number of moles to the quantities given. Step 2 used the molarity of the solution for the conversion between volume and moles, and step 4 used the molar mass for the conversion between moles and mass. The coefficients of the equation are used in a separate step, to convert moles of one substance to moles of another.

It is possible to combine the steps in this problem into a single, multistep calculation.

$$\text{Mass Al(OH)}_3 = 0.500 \text{ L NaOH soln} \times \left(\frac{0.100 \text{ mol NaOH}}{1 \text{ L NaOH soln}}\right)\left(\frac{1 \text{ mol Al(OH)}_3}{3 \text{ mol NaOH}}\right)\left(\frac{78.00 \text{ g Al(OH)}_3}{1 \text{ mol Al(OH)}_3}\right) = 1.30 \text{ g Al(OH)}_3$$

Understanding

Calculate the mass of AgCl that forms in the reaction of 0.500 L of 1.30 *M* $CaCl_2$ with excess $AgNO_3$.

Answer 186 g AgCl

In Example 4.10, we calculated the amount of product from the concentration and volume of NaOH, and the coefficients in the balanced equation. We then calculated the mass of product from the molar mass of $Al(OH)_3$. It is often necessary to calculate the concentration or volume of a reactant or product from a given mass, as illustrated in Example 4.11.

EXAMPLE 4.11 Solution Stoichiometry

What volume of 0.20 *M* HNO_3 is needed to react completely with 37 g $Ca(OH)_2$?

Strategy We use the same four steps as for all stoichiometry problems. Write the balanced equation, calculate the amount of the given substance from the data supplied, calculate the amount of the desired substance from the coefficients of the chemical equation, then convert to the desired units using the molarity of HNO_3.

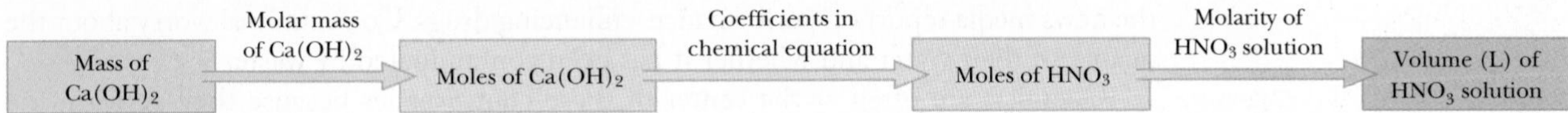

Solution

The first step is to write the balanced equation for the acid–base reaction.

$$2HNO_3(aq) + Ca(OH)_2(s) \rightarrow 2H_2O(\ell) + Ca(NO_3)_2(aq)$$

Second, calculate the number of moles of the given substance, $Ca(OH)_2$ (molar mass = 74.1 g/mol), from the given mass.

$$Ca(OH)_2 = 37\ g\ Ca(OH)_2 \times \left(\frac{1\ mol\ Ca(OH)_2}{74.1\ g\ Ca(OH)_2}\right) = 0.50\ mol\ Ca(OH)_2$$

Third, use the coefficients of the chemical equation to calculate the equivalent number of moles of HNO_3.

$$\text{Amount } HNO_3 = 0.50\ mol\ Ca(OH)_2 \times \left(\frac{2\ mol\ HNO_3}{1\ mol\ Ca(OH)_2}\right) = 1.0\ mol\ HNO_3$$

Fourth, finish the problem by calculating the volume of 0.20 *M* HNO_3 that contains 1.0 mol HNO_3.

$$\text{Volume } HNO_3 \text{ soln} = 1.0\ mol\ HNO_3 \times \left(\frac{1\ L\ HNO_3\ soln}{0.20\ mol\ HNO_3}\right)$$

$$= 5.0\ L\ HNO_3\ soln$$

It is possible to combine the steps in this problem into a single, multistep calculation.

$$\text{Volume } HNO_3 \text{ soln} = 37\ g\ Ca(OH)_2 \times \left(\frac{1\ mol\ Ca(OH)_2}{74.1\ g\ Ca(OH)_2}\right)\left(\frac{2\ mol\ HNO_3}{1\ mol\ Ca(OH)_2}\right)\left(\frac{1\ L\ HNO_3\ soln}{0.20\ mol\ HNO_3}\right) = 5.0\ L\ HNO_3\ soln$$

Understanding

What volume of 1.50 *M* hydrochloric acid is needed to react completely with 4.50 g magnesium hydroxide?

Answer 103 mL

OBJECTIVE REVIEW *Can you:*

☑ perform stoichiometric calculations with chemical reactions, given the molar concentrations and volumes of solutions of reactants or products?

4.4 Chemical Analysis

OBJECTIVES

☐ Determine the concentration of a solution from data obtained in a titration experiment (volumetric analysis)

☐ Calculate solution concentration from the mass of product formed in a precipitation reaction (gravimetric analysis)

The identification of chemical species *(qualitative analysis)* and the determination of amounts or concentrations *(quantitative analysis)* are important not only in chemistry but in fields such as medicine, agriculture, and law. Chemical analyses also influence many economic and political decisions. Newspapers are filled with stories about the impacts of chemicals on our lives. At the Olympic Games and other sporting events, the news media report on performance-enhancing drugs. Communities worry about the quality of their water and whether it has been contaminated by waste.

Chemists are often at the center of these controversies because they perform the analyses that are necessary for rational action on many important issues. They have designed many techniques to analyze substances and are developing new methods daily. This section outlines two of the most widely used methods.

Acid–Base Titrations

A truck carrying several containers of acidic materials has overturned into an irrigation canal. The canal water must be tested to ascertain its level of acidity. If it is too acidic, it must be neutralized or it may kill the species living in the canal and the crops to which the water is added.

To determine whether an irrigation canal along a road has been damaged by an acid spill, scientists must analyze the water. One way to perform this analysis is to measure the amount of base needed to neutralize the acid present in a sample of the canal water. The scientists can make this measurement using a **titration,** a procedure to determine the quantity of one substance by adding a measured amount of a second substance. The point at which the stoichiometrically equivalent amounts of the two reactants are present is called the **equivalence point.** The reaction stoichiometry is used to calculate the amount of acid in the sample from the measured amount of base added to reach the equivalence point.

A common way to detect the equivalence point in an acid–base titration is to add an **indicator,** a compound that changes color as an acidic solution becomes basic, or vice versa. The point at which the indicator changes color is called the **end point** of the titration. For the analysis to be accurate, the analyst must select an indicator that changes color close to the equivalence point.

An analysis of the acidity of the canal water involves several steps. A measured volume of the canal water is placed in a flask, and a few drops of indicator solution are added. A **standard solution** (a solution with an accurately known concentration) of base is added from a buret. The indicator changes color when the end point is reached (Figure 4.8). The addition of the base is stopped when the end point is reached, and the volume of solution delivered is read from the buret. The concentration of the acid in the sample is calculated from the data. A chemical analysis like a titration that involves measurement of the volume of a solution or substance is called a **volumetric analysis.**

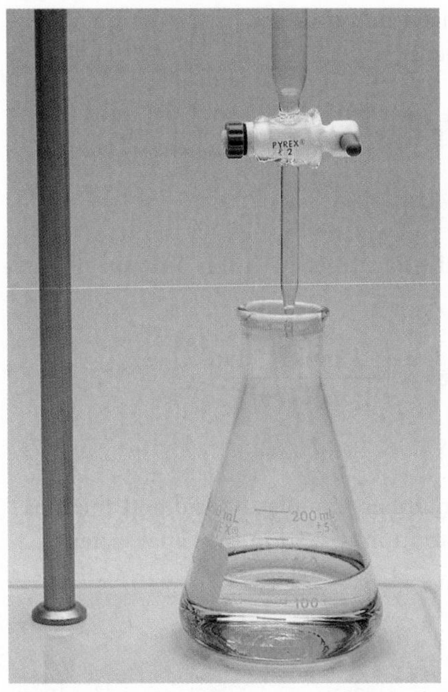

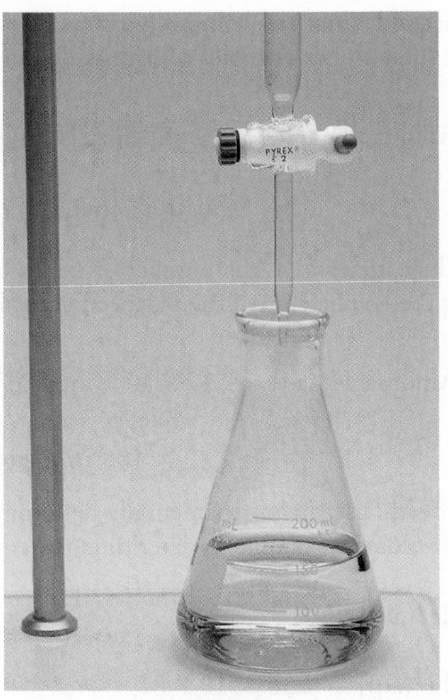

(*a*) (*b*) (*c*)

Figure 4.8 An acid–base titration using a phenolphthalein indicator. We titrate an acidic solution (in the flask) by adding standard sodium hydroxide solution from the buret. *(a)* Acid solution containing phenolphthalein indicator, before the titration is started. *(b)* Acid solution containing phenolphthalein, after the addition of exactly the correct volume of base to reach the end point. *(c)* Excess base has been added to the solution.

EXAMPLE 4.12 Acid–Base Titration

In a titration, volume and known concentration of one solution are used to determine the unknown concentration of a second solution.

A tank car carrying concentrated hydrochloric acid overturns and spills into a small irrigation canal. An analysis of the canal water must be performed before chemists will decide how to remediate the spill. A chemist titrates a 200-mL sample of water from the canal with a 0.00100 *M* NaOH solution. What is the molar concentration of the acid in the canal if 23.20 mL of the base solution are needed to reach the equivalence point?

Strategy As in other equation stoichiometry problems, we must perform the following steps: (1) write the balanced equation; (2) use the information in the problem to determine the number of moles of the given substance (NaOH in this problem); (3) use the coefficients in the equation to convert from moles of NaOH to moles of the desired substance, hydrochloric acid; and (4) use the number of moles of the acid calculated in step 3 and the measured volume of sample to determine the concentration of acid.

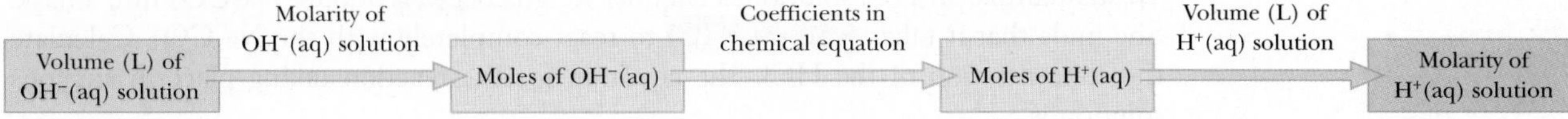

Solution

First, write the equation. The balanced chemical equation is

$$HCl(aq) + NaOH(aq) \rightarrow H_2O(\ell) + NaCl(aq)$$

Both HCl and NaOH dissociate into ions in water producing H^+(aq) and Cl^-(aq), Na^+(aq) and OH^-(aq). The sodium cations and chloride anions are spectator ions, so we need only the net ionic equation of this acid–base reaction.

$$H^+(aq) + OH^-(aq) \rightarrow H_2O(\ell)$$

Second, from the volume (in liters) and concentration of the base solution, determine the number of moles of hydroxide ion added.

$$\text{Amount OH}^-(\text{aq}) = 0.02320 \text{ L OH}^-(\text{aq}) \text{ soln} \times \left(\frac{0.00100 \text{ mol OH}^- (\text{aq})}{1 \text{ L OH}^-(\text{aq}) \text{ soln}}\right)$$

$$= 2.32 \times 10^{-5} \text{ mol OH}^-(\text{aq})$$

Third, determine the number of moles of acid that react with this amount of base using the coefficients of the balanced equation.

$$\text{Amount H}^+(\text{aq}) = 2.32 \times 10^{-5} \text{ mol OH}^- (\text{aq}) \times \left(\frac{1 \text{ mol H}^+ (\text{aq})}{1 \text{ mol OH}^- (\text{aq})}\right)$$

$$= 2.32 \times 10^{-5} \text{ mol H}^+(\text{aq})$$

Fourth, use this experimentally determined number of moles of acid and the measured volume of sample to determine the concentration of acid in the canal water.

$$\left[\text{H}^+(\text{aq})\right] = \frac{2.32 \times 10^{-5} \text{ mol H}^+ (\text{aq})}{0.200 \text{ L pond water}}$$

$$[\text{H}^+(\text{aq})] = 1.16 \times 10^{-4}\ M$$

This information is communicated to crop scientists who conclude that this concentration of acid threatens species living in the canal and is too high for the water to be used on crops. The acid in the water must be neutralized in a manner that will not produce additional contamination.

Understanding

What is the molarity of NaOH in a 300-mL sample that is neutralized by 55.00 mL of 1.33 *M* HCl solution?

Answer 0.244 *M* NaOH

The titration in Example 4.12 used a standard solution of sodium hydroxide. Chemists often need standard solutions of acid to measure the concentrations of unknown bases. A high-purity base is chosen to standardize an acid. The next example shows how hydrochloric acid can be standardized by using a carefully weighed quantity of sodium carbonate. Sodium carbonate is available in high purity, at low cost, and is frequently used to standardize acids.

EXAMPLE 4.13 Standardization of a Solution of HCl

To standardize an HCl solution, a chemist weighs 0.210 g of pure Na_2CO_3 into a flask. She finds that it takes 5.50 mL HCl to react completely with the Na_2CO_3. Calculate the concentration of the HCl. The equation for the reaction taking place in aqueous solution is

$$2\text{HCl(aq)} + \text{Na}_2\text{CO}_3\text{(aq)} \rightarrow \text{H}_2\text{O}(\ell) + 2\text{NaCl(aq)} + \text{CO}_2\text{(g)}$$

Strategy Use the strategy outlined in the following flow diagram to solve this example.

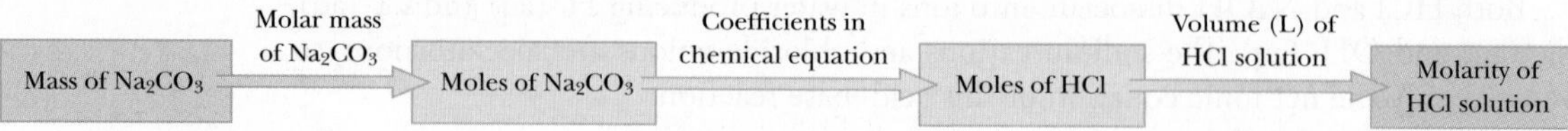

PRACTICE OF CHEMISTRY
Titrations in the Emergency Department

"The patient is septic. Titrate the BP with Levophed to a systolic of 90," the physician says calmly and distinctly. The physician realized that the patient is suffering from a bacterial infection. One symptom is dilation of blood vessels and a corresponding decline in blood pressure, which is measured by two numbers—a maximum pressure called the *systolic* and a minimum called the *diastolic.* Normal blood pressures are approximately 120 mm Hg for the systolic and 80 mm Hg for the diastolic, written as 120/80. The pressure units are millimeters of mercury (mm Hg), or torr, which are common units for measuring pressures (see Chapter 6). The physician orders a vasoconstrictor, Levophed, given to the patient to constrict the blood vessels until the systolic blood pressure reaches 90 mm Hg.

The medicine is given in small quantities, just like a titration, with the blood pressure monitored constantly. The physician asked that the medicine delivery stop when the systolic reaches 90 mm Hg. In this titration, the systolic blood pressure is equivalent to the indicator that changes color when the reaction is complete in an acid-base titration in the laboratory. ∎

Solution

All problems of this type start with the chemical equation, which is given in this case. Next, determine the number of moles of the given substance, Na_2CO_3, from the given mass.

$$\text{Amount } Na_2CO_3 = 0.210 \text{ g } Na_2CO_3 \times \left(\frac{1 \text{ mol } Na_2CO_3}{106.0 \text{ g } Na_2CO_3}\right) = 1.98 \times 10^{-3} \text{ mol } Na_2CO_3$$

Use the coefficients of the equation to calculate the number of moles of the desired substance, HCl.

$$\text{Amount HCl} = 1.98 \times 10^{-3} \text{ mol } Na_2CO_3 \times \left(\frac{2 \text{ mol HCl}}{1 \text{ mol } Na_2CO_3}\right) = 3.96 \times 10^{-3} \text{ mol HCl}$$

Now compute the concentration of the HCl solution from the amount of HCl and the given volume of HCl solution (converted to liters).

$$\text{Concentration HCl} = \frac{3.96 \times 10^{-3} \text{ mol HCl}}{0.00550 \text{ L HCl soln}} = 0.720\ M \text{ HCl}$$

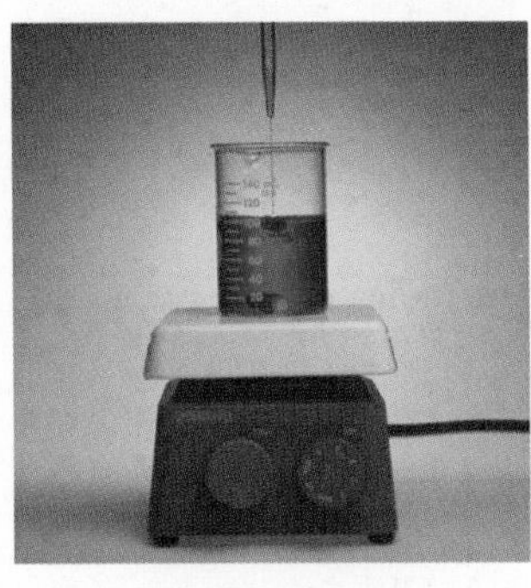

(*a*)

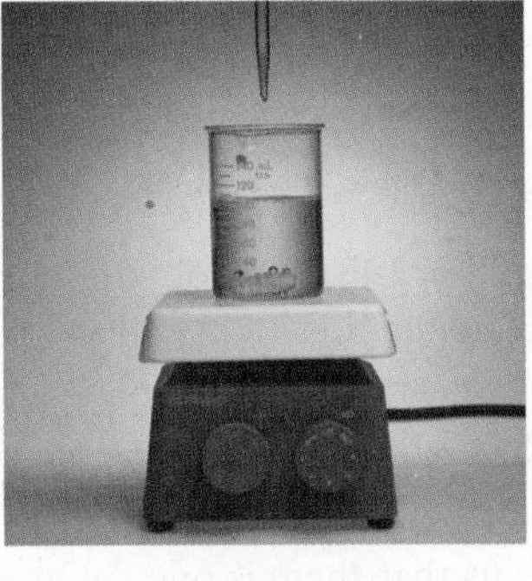

(*b*)

(*c*)

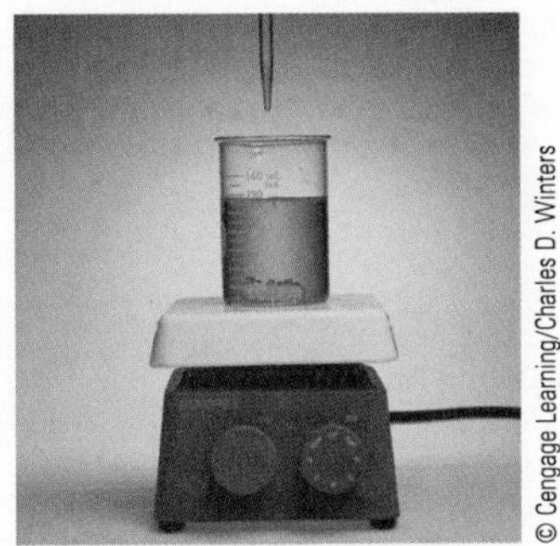

(*d*)

Standardization of HCl with Na_2CO_3. Solid Na_2CO_3 is dried, weighed, and dissolved in water containing an indicator that is blue in basic solutions. *(a)* HCl solution is added to the Na_2CO_3 solution. The bubbles that form are the gaseous CO_2. *(b)* Enough HCl is added to change the indicator color to light green. *(c)* Because the CO_2 that remains dissolved in water acts as an acid, the solution is heated to remove the CO_2 gas. The solution turns back to blue. *(d)* HCl is again added until the green end point is reached. In many titrations, as in this example, care must be taken to ensure accurate results.

Understanding

What is the molarity of an HCl solution if a 50.0-mL sample is neutralized by 10.00 mL of 0.23 *M* sodium hydroxide?

Answer 0.046 *M*

Gravimetric Analysis

A precipitation reaction can be used to isolate a desired ion from a solution.

Chemists use the limited solubilities of certain compounds in many chemical applications. Adding an appropriate compound to a solution can precipitate a particular ion as a solid. Using the solubility rules in Table 4.1, we can choose a reactant that will cause the desired ion to precipitate as an insoluble product while leaving the other ions in solution.

Chemists use precipitation reactions of this type for chemical analyses. If one component of a solution is precipitated selectively, it can then be separated from solution, dried, and weighed. This procedure, analysis by mass, is known as **gravimetric analysis.** One of the most widely used gravimetric procedures is the determination of halides by the addition of silver nitrate to precipitate the silver halides. Another important gravimetric analysis is the determination of sulfate ion, SO_4^{2-}, by the addition of $BaCl_2$ to form insoluble $BaSO_4$. An example of this type of experiment is explained in Example 4.14 and the Case Study.

EXAMPLE 4.14 Gravimetric Analysis

A recycling company has purchased several tons of scrap wire that is known to contain silver. A sample of wire with a mass of 2.0764 g is completely dissolved in nitric acid. Then dilute hydrochloric acid is added until precipitation stops. The precipitate is filtered, dried thoroughly, and weighed. The precipitate has a mass of 0.1656 g. Assuming that the precipitate is pure AgCl, what is the percentage by mass of silver in the wire sample?

Strategy Determine the amount of silver in the AgCl formed in the precipitation reaction. To do this, calculate the number of moles of given species (silver chloride in the precipitate) from the given data. Use the stoichiometry in the chemical equation to determine the number of moles of desired species, Ag, present in the sample.

Convert this number of moles to grams, and determine the mass percentage of Ag in the original sample.

Solution

The precipitation reaction between silver ions and chloride ions is

$$Ag^+(aq) + Cl^-(aq) \rightarrow AgCl(s)$$

The molar mass of AgCl is 143.4 g/mol. The number of moles of the given substance, AgCl, that were precipitated is

$$\text{Amount AgCl} = 0.1656\ \cancel{\text{g AgCl}} \times \left(\frac{1\ \text{mol AgCl}}{143.32\ \cancel{\text{g AgCl}}}\right) = 1.155 \times 10^{-3}\ \text{mol AgCl}$$

The stoichiometry tells us that there is one Ag atom per AgCl unit,

$$\text{Amount Ag} = 1.155 \times 10^{-3}\ \cancel{\text{mol AgCl}} \times \left(\frac{1\ \text{mol Ag}}{1\ \cancel{\text{mol AgCl}}}\right) = 1.155 \times 10^{-3}\ \text{mol Ag}$$

We determine the mass of this amount of Ag using its atomic mass, 107.87 g/mol:

$$\text{Mass Ag} = 1.155 \times 10^{-3}\ \cancel{\text{mol Ag}} \times \left(\frac{107.87\ \text{g Ag}}{1\ \cancel{\text{mol Ag}}}\right) = 0.1246\ \text{g Ag}$$

Determine the percentage Ag by dividing this mass by the mass of the entire sample and multiply by 100%:

$$\% \text{ Ag} = \frac{0.1246 \text{ g Ag}}{2.0764 \text{ g sample}} \times 100\% = 6.001\% \text{ Ag}$$

The sample is slightly more than 6% Ag.

Understanding

An unknown sample of a carbonate salt has a mass of 3.775 g. The sample is dissolved in water, and aqueous barium nitrate is added until precipitation is complete. The precipitate is filtered, dried, and weighed. Its mass is 2.006 g. Assuming the precipitate is pure $BaCO_3$, what is the percentage carbonate in the sample?

Answer 16.16% carbonate

OBJECTIVES REVIEW *Can you:*

- ☑ determine the concentration of a solution from data obtained in a titration experiment (volumetric analysis)?
- ☑ calculate solution concentration from the mass of product formed in a precipitation reaction (gravimetric analysis)?

CASE STUDY Determination of Sulfur Content in Fuel Oil

A sample of fuel oil had to be analyzed for sulfur content, to determine whether it would meet pollution standards if it was used. As mentioned in the Practice of Chemistry essay in Chapter 3, burning fuels with high sulfur content contributes to acid rain and is against the law. To carry out the assay, the chemist uses a classic gravimetric analysis generally called *Eschka's method* after the scientist who first developed the technique. A brief summary of the method is as follows: A sample is heated in air in the presence of a mixture of magnesium oxide and calcium carbonate, called *Eschka's mixture.* This process converts all of the sulfur in the sample to SO_2 or SO_3, which in the presence of the Eschka's mixture forms mostly calcium and magnesium sulfate and sulfite. All the sulfur-containing species are then converted to sulfate ions by reaction with Br_2, and these ions precipitated as insoluble $BaSO_4$ by adding a barium chloride solution. The precipitate is separated, dried, and weighed, and the mass is used to calculate the percent sulfur in the oil sample.

In the actual experiment, 10 g of the Eschka's mixture and 1.8939 g of the oil to be analyzed for percentage sulfur is added to a nickel crucible. The crucible is placed in a furnace at 800 °C for 4 hours and then allowed to cool. The high temperature reaction converts all of the sulfur to $SO_2 + SO_3$, and these compounds react with the Group 2 metal compounds in the Eschka's mixture as follows:

$$MgO + SO_2 \rightarrow MgSO_3$$

$$CaCO_3 + SO_3 \rightarrow CaSO_4 + CO_2$$

The contents of the crucible are transferred to a beaker containing water, and about 75 mL of 6 *M* HCl is added to neutralize all the oxides and carbonates.

$$MgO + 2HCl \rightarrow MgCl_2 + H_2O$$

$$CaCO_3 + 2\,HCl \rightarrow CaCl_2 + H_2O + CO_2$$

One milliliter of bromine water (bromine liquid dissolved in water) is added to this mixture to convert all of the sulfur compounds into sulfate:

$$SO_3^{2-} + Br_2 + H_2O \rightarrow SO_4^{2-} + 2HBr$$

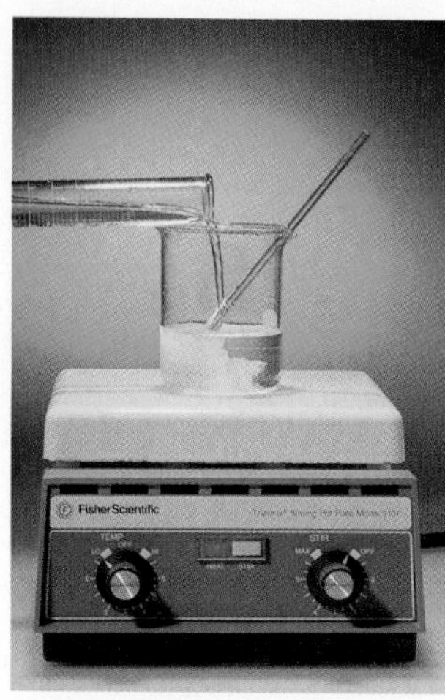
(a)

(b)

(c)

(d)

Figure 4.9 Analysis for sulfate. *(a)* $BaSO_4$ precipitates on addition of excess $BaCl_2$ to the solution containing SO_4^{2-}. *(b)* The solution is filtered to collect the $BaSO_4$. *(c)* The solid $BaSO_4$ is heated to remove all volatile (easily converted to the gas phase) impurities, then cooled. *(d)* The $BaSO_4$ is weighed to determine its mass.

The solution is then heated to drive off any excess bromine:

$$Br_2(aq) \rightarrow Br_2(g)$$

Finally, 10 mL of 0.1 *M* $BaCl_2$ is added to precipitate the sulfate as barium sulfate:

$$SO_4^{2-} + Ba^{2+}(aq) \rightarrow BaSO_4(s)$$

A clean porcelain crucible is heated for an hour at 800 °C to remove any water. It is cooled in a desiccator (dry closed container) and weighed (mass = 14.5821 g). The barium sulfate that was precipitated by the barium chloride is collected by passing the solution through very fine filter paper. The mass of the filter paper is not measured because it will be burned. The filter paper and barium sulfate precipitate that was collected is placed in the crucible and the crucible slowly heated so that the filter paper does not catch on fire and spatter the barium sulfate out of the crucible. The crucible was heated at 800 °C for an hour, cooled in the desiccator, and weighed. It was found to have a mass of 14.5886 g. These steps in the analysis are shown in Figure 4.9.

We now can calculate the percentage of sulfur in the oil sample. The strategy involves working backward. We know the mass of barium sulfate collected in the last step, and we can compute the mass of sulfur in that material. Because the procedure converted all of the sulfur in the oil sample into sulfate, this mass will represent the mass of the sulfur in the sample.

The mass of $BaSO_4$ is the mass of the crucible and barium sulfate weighed after being heated minus the mass of the empty crucible weighed before the barium sulfate was added.

$$\text{Mass of } BaSO_4 = 14.5886 \text{ g} - 14.5821 \text{ g} = 0.0065 \text{ g}$$

We compute the mass of sulfur from the molar masses of barium sulfate (233.4 g/mol) and sulfur (32.07 g/mol), and the formula that indicates that 1 mol barium sulfate contains 1 mol sulfur.

$$\text{Mass of S} = 0.0065 \cancel{\text{g BaSO}_4} \times \left(\frac{1 \cancel{\text{mol BaSO}_4}}{233.4 \cancel{\text{g BaSO}_4}}\right)\left(\frac{1 \cancel{\text{mol S}}}{1 \cancel{\text{mol BaSO}_4}}\right)\left(\frac{32.07 \text{ g S}}{1 \cancel{\text{mol S}}}\right)$$

$$= 8.9 \times 10^{-4} \text{ g S}$$

The mass of the original oil sample is 1.8939 g.

$$\text{Percentage S} = \left(\frac{8.9 \times 10^{-4}\ \text{g S}}{1.8939\ \text{g sample}}\right) \times 100\% = 0.047\%$$

The chemist is suspicious of these results for two reasons. First, the mass of the barium sulfate precipitate, 0.0065 g, is too small to weigh without sizable error. In general, chemists want a precipitate that weighs approximately 0.1 g so that small losses in transferring the compound from beaker to beaker are insignificant. Second, the chemist knows that home heating fuels are about 0.2% sulfur, substantially higher than she calculated.

She asks her manager about these problems and is told that the sample is actually diesel fuel, which has an allowed maximum sulfur concentration of 0.05%. The chemist points out that she collected only 0.0065 g of material and does not have a lot of confidence when dealing with such small quantities, so her manager tells her to perform a slightly different analysis. The combustion in Eschka's mixture is the same, but the determination is by titration with barium perchlorate. The titration uses a compound called *thorin* as an indicator, and is done in 20% water/80% isopropyl alcohol.

In the experiment, the combustion was performed on a sample of mass 2.0023 g, as described earlier, producing a bromine-free aqueous solution in which all the sulfur in the sample has been converted to sulfate. This sample was added to 80% isopropyl alcohol and titrated with 0.0250 *M* barium perchlorate in a small (5-mL) accurate burette.

$$SO_4^{2-} + Ba(ClO_4)_2 \rightarrow BaSO_4(s) + 2ClO_4^-$$

The thorin indicator is yellow, but it turns pink in the presence of barium ions. The titration ends at the first indication of pink, the point at which the barium ions are no longer precipitated by the sulfate because all of it has reacted. The initial reading of the burette was 4.88 mL, and the final reading was 3.40 mL.

The concentration of sulfate in the unknown solution is calculated from the concentration and volume of barium perchlorate used in the titration.

$$\text{Volume } Ba(ClO_4)_2 = 4.88 - 3.40 = 1.48\ \text{mL}$$

$$\text{Amount } Ba(ClO_4)_2 = 1.48\ \cancel{\text{mL } Ba(ClO_4)_2} \times \left(\frac{1\ \cancel{\text{L}}}{1000\ \cancel{\text{mL}}}\right)\left(\frac{0.0250\ \text{mol } Ba(ClO_4)_2}{1\ \cancel{\text{L } Ba(ClO_4)_2}}\right)$$

$$= 3.70 \times 10^{-5}\ \text{mol } Ba(ClO_4)_2$$

Next, we calculate the amount of sulfate:

$$\text{Amount } SO_4^{2-} = 3.70 \times 10^{-5}\ \cancel{\text{mol } Ba(ClO_4)_2} \times \left(\frac{1\ \text{mol } SO_4^{2-}}{1\ \cancel{\text{mol } Ba(ClO_4)_2}}\right) = 3.70 \times 10^{-5}\ \text{mol } SO_4^{2-}$$

The mass of sulfur in the original sample comes from the molar masses of sulfur and sulfate ion:

$$\text{Mass of S} = 3.70 \times 10^{-5}\ \cancel{\text{mol } SO_4^{2-}} \times \left(\frac{32.07\ \text{g S}}{1\ \cancel{\text{mol } SO_4^{2-}}}\right) = 1.19 \times 10^{-3}\ \text{g S}$$

The mass of the original sample that was used for the analysis was 2.0023 g, this value is used to calculate the percentage sulfur in the sample.

$$\%\ \text{sulfur} = \left(\frac{1.19 \times 10^{-3}\ \text{g S}}{2.0023\ \text{g sample}}\right) \times 100\% = 0.0583\ \%\ \text{S}$$

The results from the titration are more accurate and have an additional significant figure. The additional accuracy is needed to support the conclusion that the fuel contains too much sulfur to be sold as diesel fuel.

Questions

1. What is the percentage of sulfur in a sample of diesel fuel if a 3.44-g sample produces 0.0073 g $BaSO_4$ using the method described earlier?
2. The company must mix some sulfur-free diesel fuel with the 0.0583% sulfur fuel to bring the sulfur down to 0.050%. How many grams of sulfur-free fuel are needed to bring a kilogram of the diesel fuel down to 0.050% S?

ETHICS IN CHEMISTRY

1. You just spent nearly a whole working day analyzing a sample of fuel oil that possibly contains too much sulfur to be sold and came up with a result that indicated the fuel oil contained 0.048% sulfur. The maximum concentration that can be sold is 0.05%. What is your recommendation as to whether the fuel can be sold?

2. You analyze the fuel oil by two different methods and come up with two results that are fairly different. One analyses would allow the fuel oil to be sold, the other would not let it be sold. What action do you take next?

3. To be certain that your analysis for percentage of sulfur in fuel oil is correct, you conduct the same procedure three times on three portions of the same sample and get the following results: 0.051%, 0.052%, and 0.046%. What do you report as the percentage of sulfur in the oil? Can you reliably conclude that the sulfur concentration is less than 0.050%? What steps might you take to improve the reliability of your conclusions?

Chapter 4 Visual Summary

The chart shows the connections between the major topics discussed in this chapter.

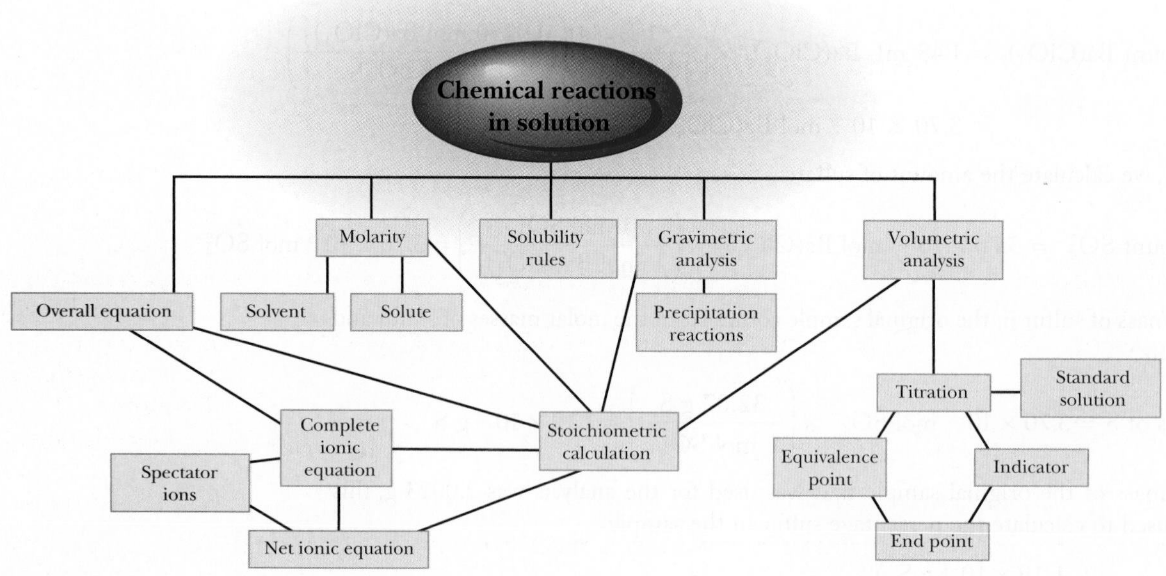

Summary

4.1 Ionic Compounds in Aqueous Solution

Many chemical reactions are conducted with one or more of the reactants dissolved in a *solvent.* Most ionic compounds and a few molecular compounds dissociate completely into ions in water and are known as *strong electrolytes.* Most molecular compounds produce no ions in solution and are known as *nonelectrolytes,* whereas a few partially dissociate into ions and are known as *weak electrolytes.* A series of rules for ionic compounds, based on experimental observations, help chemists predict which substances dissolve in water—that is, determine their *solubility.* The solubility rules allow the prediction of which substances (if any) precipitate during reactions that occur in water solution. In a *precipitation reaction,* soluble compounds react to form an insoluble product. For this type of reaction, chemists frequently write the *net ionic equation,* one that shows only those species in the reaction that undergo change.

4.2 Molarity

To perform stoichiometric calculations, you must know the concentration of the *solute* in the solution. *Molarity,* defined as the number of moles of solute per liter of solution, is the most convenient unit of concentration for stoichiometry calculations. The chemist can prepare a solution of known concentration by adding solvent to a weighed sample of solute and measuring the volume of the solution or by diluting a solution of known (higher) concentration. In the latter case, a *pipet* is often used to deliver a fixed volume of the concentrated solution, which is then diluted with pure solvent to form a known volume of dilute solution. Molarity is the conversion factor for volume of solution and moles of solute.

4.3 Stoichiometry Calculations for Reactions in Solution

Reactions performed in solution generally require the calculation of number of moles from the molar concentration and volume of solution. As was presented in Chapter 3, equations are used to convert moles of one compound to moles of another.

4.4 Chemical Analysis

Chemical analysis in which the analyst measures the volume of solution is known as *volumetric analysis.* An acid–base titration is a common example of volumetric analysis. To determine the concentration of a solution, the chemist titrates it by adding an equivalent amount of a reactant in a solution of known concentration. A buret is used to measure accurately the volume of an added liquid. *Indicators* that change color at or near the *equivalence point* are used to determine when an equivalent amount of the solution of known concentration has been added. Precipitation reactions are also used for chemical analyses. One component of a solution is precipitated selectively, then separated from solution, dried, and weighed. This procedure, analysis by mass, is known as *gravimetric analysis.*

Download Go Chemistry concept review videos from OWL or purchase them from **www.ichapters.com**

Chapter Terms

The following terms are defined in the Glossary, Appendix I.

Section 4.1
Aqueous solution
Complete ionic equation
Net ionic equation
Overall equation
Precipitation reaction
Solubility
Solute
Solvent
Spectator ions
Strong electrolytes
Weak electrolytes

Section 4.2
Concentration
Molarity
Pipet
Volumetric flask

Section 4.4
End point
Equivalence point
Gravimetric analysis
Indicator
Standard solution
Titration
Volumetric analysis

Key Equations

$$\text{Molarity} = \frac{\text{moles of solute}}{\text{liter of solution}} \quad (4.2)$$

$$\text{Molarity(conc)} \times \text{volume(conc)} = \text{molarity(dil)} \times \text{volume(dil)} \quad (4.2)$$

Questions and Exercises

OWL Selected end of chapter Questions and Exercises may be assigned in OWL.

Blue-numbered Questions and Exercises are answered in Appendix J; questions are qualitative, are often conceptual, and include problem-solving skills.

■ Questions assignable in OWL

✎ Questions suitable for brief writing exercises

▲ More challenging questions

Questions

4.1 A solution is formed by dissolving 3 g sugar in 100 mL water. Identify the solvent and the solute.

4.2 A solution is formed by mixing 1 gal ethanol with 10 gal gasoline. Identify the solvent and the solute.

4.3 An aqueous sample is known to contain either Sr^{2+} or Hg_2^{2+} ions. Use the solubility rules (see Table 4.1) to propose an experiment that will determine which ion is present.

4.4 Ammonium chloride is a strong electrolyte. Draw a molecular-level picture of this substance after it dissolves in water.

4.5 Experiments show that propionic acid (CH_3CH_2COOH) is a weak acid. Write the chemical equation.

4.6 Describe the procedure used to make 1.250 L of 0.154 *M* sodium chloride from solid NaCl and water.

4.7 If enough Li_2SO_4 dissolves in water to make a 0.33 *M* solution, explain why the molar concentration of Li^+ is different from the molar concentration of $Li_2SO_4(aq)$.

4.8 Describe how 500 mL of a 1.5 *M* solution of HCl can be prepared from 12.1 *M* HCl and pure solvent.

4.9 ✎ Addition of water to concentrated sulfuric acid is dangerous because it generates enough heat to boil the water, causing it to spatter out of the container. For this reason, chemists remember to add acid to water. In a dilution experiment, we calculate the amount of the more concentrated solution that must be measured out. If we place this concentrated solution (Figure 4.7) in the volumetric flask first, then dilute with water, we violate the caution "Add acid to water." Describe a safer variation on the method shown in Figure 4.7 that allows the quantitative dilution of concentrated sulfuric acid.

4.10 Draw the flow diagram for a calculation that illustrates how to use a titration to determine the concentration of a solution of HNO_3, by reaction with 1.00 g Na_2CO_3.

4.11 Explain why the algebraic expression

$$V(\text{conc}) = \frac{M(\text{dil}) \times V(\text{dil})}{M(\text{conc})}$$

can be used for dilution problems but not for titration calculations.

4.12 ✎ Describe in words the titration of an acid with a base. Be sure to use the terms *equivalence point, indicator,* and *end point* correctly.

4.13 ✎ Describe the use of gravimetric analysis to determine the percentage of chlorine in a water-soluble unknown solid.

4.14 Draw the contents of a beaker of water that contains dissolved forms of the following (draw only the substances added to the water):
(a) potassium chloride
(b) barium hydroxide
(c) molecular oxygen, O_2

Exercises

OBJECTIVE **Predict the solubility of common ionic substances in water.**

4.15 Which of the following compounds dissolves in water?
(a) BaI_2
(b) lead (II) chloride
(c) Na_2CO_3
(d) ammonium sulfate

4.16 Which of the following compounds dissolves in water?
(a) Hg_2Cl_2
(b) calcium bromide
(c) KNO_3
(d) silver perchlorate

4.17 Which of the following compounds dissolves in water?
(a) $CaCl_2$
(b) barium hydroxide
(c) $AgNO_3$
(d) calcium carbonate

4.18 Which of the following compounds dissolves in water?
(a) Na_3PO_4
(b) ammonium carbonate
(c) NH_4Cl
(d) strontium sulfate

OBJECTIVES Predict products of chemical reactions in solutions of ionic compounds and use the overall chemical equation to write net ionic equations.

4.19 Write the net ionic equation for the reaction, if any, that occurs on mixing
(a) solutions of sodium hydroxide and magnesium chloride.
(b) solutions of sodium nitrate and magnesium bromide.
(c) magnesium metal and a solution of hydrochloric acid to produce magnesium chloride and hydrogen.

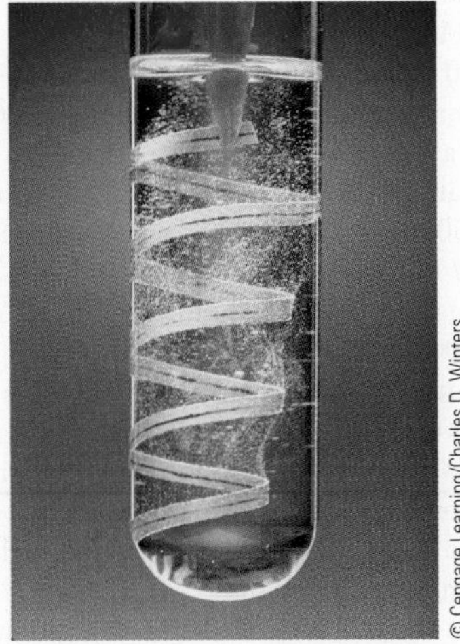

Magnesium metal reacting with HCl.

4.20 Write the net ionic equation for the reaction, if any, that occurs on mixing
(a) solutions of ammonium carbonate and magnesium chloride.
(b) solutions of nitric acid and sodium hydroxide.
(c) solutions of beryllium sulfate and sodium hydroxide.

4.21 Write the net ionic equation for the reaction, if any, that occurs on mixing
(a) solutions of hydrochloric acid and calcium hydroxide.
(b) solutions of ammonium chloride and $AgClO_4$.
(c) solutions of $Ba(ClO_4)_2$ and sodium carbonate.

4.22 Write the net ionic equation for the reaction, if any, that occurs on mixing
(a) solutions of potassium bromide and silver nitrate.
(b) a solution of nitric acid and calcium metal to produce calcium nitrate and hydrogen gas.
(c) solutions of lithium hydroxide and iron(III) chloride.

4.23 Write the overall equation (including the physical states), the complete ionic equation, and the net ionic equation for the reaction that occurs when aqueous solutions of silver nitrate and calcium chloride are mixed.

4.24 Write the overall equation (including the physical states), the complete ionic equation, and the net ionic equation for the reaction that occurs when aqueous solutions of cobalt(II) bromide and sodium hydroxide are mixed.

4.25 Write the overall equation (including the physical states), the complete ionic equation, and the net ionic equation for the reaction that occurs when aqueous solutions of ammonium phosphate and silver nitrate are mixed.

4.26 Write the overall equation (including the physical states), the complete ionic equation, and the net ionic equation for the reaction that occurs when aqueous solutions of lead (II) acetate and barium bromide are mixed.

4.27 An aqueous sample is known to contain either Pb^{2+} or Ba^{2+}. Treatment of the sample with NaCl produces a precipitate. Use the solubility rules (see Table 4.1) to determine which cation is present.

4.28 An aqueous sample is known to contain either Ag^{+} or Mg^{2+} ions. Treatment of the sample with NaOH produces a precipitate, but treatment with KBr does not. Use the solubility rules (see Table 4.1) to determine which cation is present.

4.29 An aqueous sample is known to contain either Mg^{2+} or Ba^{2+} ions. Treatment of the sample with Na_2CO_3 produces a precipitate, but treatment with ammonium sulfate does not. Use the solubility rules (see Table 4.1) to determine which cation is present.

4.30 An aqueous sample is known to contain either Pb^{2+} or Fe^{3+} ions. Treatment of the sample with Na_2SO_4 produces a precipitate. Use the solubility rules (see Table 4.1) to determine which cation is present.

4.31 In the beakers shown below, the colored spheres represent a particular ion, with the dark gray balls representing Pb^{2+}. In one reactant beaker is $Pb(NO_3)_2$ and in the other is NaCl. In the product beaker, the organized solid represents an insoluble compound. Write the overall equation, the complete ionic equation, and the net ionic equation.

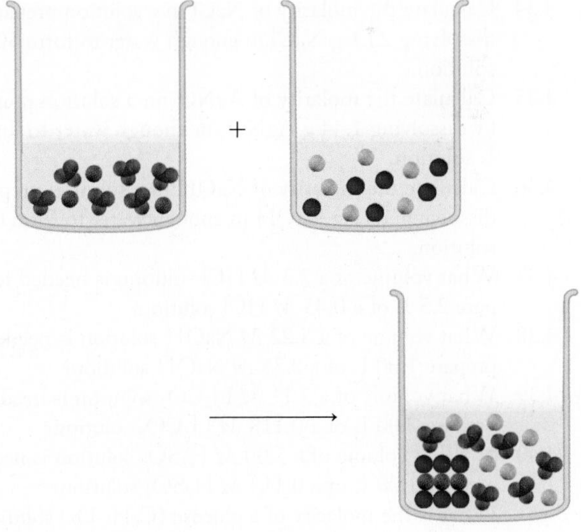

4.32 In the beakers shown below, the colored spheres represent a particular ion, with the dark gray balls representing Ag^+. In one reactant beaker is $AgNO_3$ and in the other is NaBr. In the product beaker, the organized solid represents an insoluble compound. Write the overall equation, the complete ionic equation, and the net ionic equation.

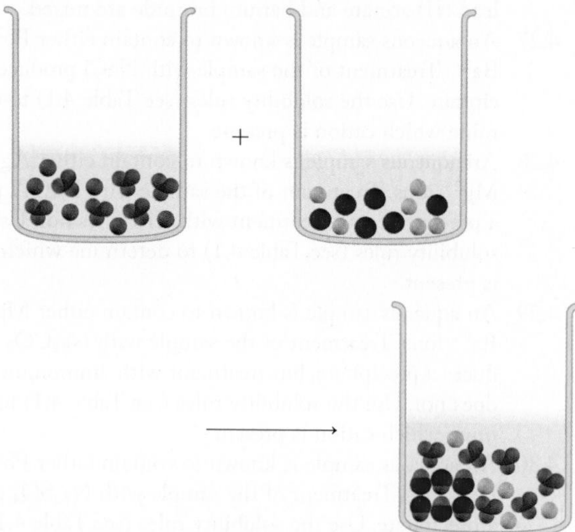

OBJECTIVES Describe how to prepare solutions of known molarity from weighed samples of solute or from concentrated solutions.

4.33 Calculate the molarity of KOH in a solution prepared by dissolving 8.23 g KOH in enough water to form 250 mL solution.

4.34 Calculate the molarity of NaCl in a solution prepared by dissolving 23.1 g NaCl in enough water to form 500 mL solution.

4.35 Calculate the molarity of $AgNO_3$ in a solution prepared by dissolving 1.44 g $AgNO_3$ in enough water to form 1.00 L solution.

4.36 Calculate the molarity of NaOH in a solution prepared by dissolving 1.11 g NaOH in enough water to form 0.250 L solution.

4.37 What volume of a 2.3 *M* HCl solution is needed to prepare 2.5 L of a 0.45 *M* HCl solution?

4.38 What volume of a 5.22 *M* NaOH solution is needed to prepare 1.00 L of a 2.35 *M* NaOH solution?

4.39 What volume of a 2.11 *M* Li_2CO_3 solution is needed to prepare 2.00 L of a 0.118 *M* Li_2CO_3 solution?

4.40 ■ What volume of a 5.00 *M* H_2SO_4 solution is needed to prepare 1.00 L of a 0.113 *M* H_2SO_4 solution?

4.41 What is the molarity of a glucose ($C_6H_{12}O_6$) solution prepared from 55.0 mL of a 1.0 *M* solution that is diluted with water to a final volume of 2.0 L?

4.42 ■ If you dilute 25.0 mL of 1.50 *M* hydrochloric acid to 500 mL, what is the molar concentration of the dilute acid?

4.43 Calculate the molarity of 2.0 L solution prepared by dilution with water of
(a) 3.56 g NaOH.
(b) 25 mL of a 1.4 *M* NaOH solution.

4.44 Calculate the molarity of 250 mL solution prepared by dilution with water of
(a) 0.12 g sodium nitrate.
(b) 0.75 mL of a 0.42 *M* NaOH solution.

OBJECTIVES Calculate the amount of solute, the volume of solution, or the molar concentration of solution, given the other two quantities.

4.45 Calculate the mass of solute in
(a) 3.13 L of a 2.21 *M* HCl solution.
(b) 1.5 L of a 1.2 *M* KCl solution.

4.46 Calculate the mass of solute in
(a) 0.113 L of a 1.00 *M* KBr solution.
(b) 120 mL of a 2.11 *M* KNO_3 solution.

4.47 How many grams of $AgNO_3$ are needed to prepare 300 mL of a 1.00 *M* solution?

4.48 ■ What mass of oxalic acid, $H_2C_2O_4$, is required to prepare 250 mL of a solution that has a concentration of 0.15 *M* $H_2C_2O_4$?

4.49 How many grams of barium chloride are needed to prepare 1.00 L of a 0.100 *M* solution?

4.50 What mass of sodium sulfate, in grams, is needed to prepare 400 mL of a 2.50 *M* solution?

4.51 What is the molarity of a solution of strontium chloride that is prepared by dissolving 4.11 g $SrCl_2$ in enough water to form 1.00-L solution? What is the molarity of each ion in the solution?

4.52 What is the molarity of a solution of sodium hydrogen sulfate that is prepared by dissolving 9.21 g $NaHSO_4$ in enough water to form 2.00-L solution? What is the molarity of each ion in the solution?

4.53 What is the molarity of a solution of magnesium nitrate that is prepared by dissolving 21.5 g $Mg(NO_3)_2$ in enough water to form 5.00 L solution? What is the molarity of each ion in the solution?

4.54 ■ If 6.73 g of Na_2CO_3 is dissolved in enough water to make 250 mL of solution, what is the molar concentration of the sodium carbonate? What are the molar concentrations of the Na^+ and CO_3^{2-} ions?

4.55 The substance KSCN is frequently used to test for iron in solution, because a distinctive red color forms when it is added to a solution of the Fe^{3+} cation. As a laboratory assistant, you are supposed to prepare 1.00 L of a 0.200 *M* KSCN solution. What mass, in grams, of KSCN do you need?

A solution containing Fe^{3+} turns red when potassium thiocyanate (KSCN) is added.

4.56 Potassium permanganate ($KMnO_4$) solutions are used for the determination of Fe^{2+} in samples of unknown concentration. As a laboratory assistant, you are supposed to prepare 500 mL of a 0.1000 *M* $KMnO_4$ solution. What mass of $KMnO_4$, in grams, do you need?

4.57 Two liters of a 1.5 *M* solution of sodium hydroxide are needed for a laboratory experiment. A stock solution of 5.0 *M* NaOH is available. How is the desired solution prepared?

4.58 ▲ A 6.00-g sample of sodium hydroxide is added to a 1.00-L volumetric flask, and water is added to dissolve the solid and fill the flask to the mark. A 100-mL portion of this solution is added to a 5.00-L volumetric flask, and water is added to fill the flask to the mark. What is the concentration of NaOH in the second flask?

4.59 Calculate the number of moles of solute in
(a) 33 mL of a 3.11 *M* HNO_3 solution.
(b) 1.0 L of a 3.2 *M* HNO_3 solution.

4.60 Calculate the number of moles of solute in
(a) 0.22 L of a 1.2 *M* NaCl solution.
(b) 500 mL of a 0.22 *M* solution of $AgNO_3$.

4.61 Calculate the number of moles of solute in
(a) 1.33 L of a 0.211 *M* $AgNO_3$ solution.
(b) 1000 mL of a 0.00113 *M* solution of calcium chloride.

4.62 Calculate the number of moles of solute in
(a) 238 mL of a 0.211 *M* NaBr solution.
(b) 1.2 L of a 0.077 *M* solution of ammonium chloride.

4.63 Calculate the number of moles of solute in
(a) 34 mL of a 0.11 *M* potassium sulfate solution.
(b) 10 mL of an 8.3 *M* solution of sodium chloride.

4.64 Calculate the number of moles of solute in
(a) 12.4 mL of a 1.2 *M* NaCl solution.
(b) 22 L of a 2.2 *M* solution of calcium nitrate.

4.65 What volume of 2.4 *M* HCl is needed to obtain 1.3 mol HCl?

4.66 What volume of 0.022 *M* $CaCl_2$ is needed to obtain 0.13 mol $CaCl_2$?

OBJECTIVE Perform stoichiometric calculations with chemical reactions, given the molar concentrations and volumes of solutions of reactant or products.

4.67 What mass of AgCl, in grams, forms in the reaction of 3.11 mL of 0.11 *M* $AgNO_3$ with excess $CaCl_2$?

4.68 ■ What mass of barium sulfate, in grams, forms in the reaction of 25.0 mL of 0.11 *M* $Ba(OH)_2$ with excess H_2SO_4?

4.69 What mass of sodium hydroxide, in grams, is needed to react with 100.0 mL of 3.13 *M* H_2SO_4?

4.70 What mass of calcium hydroxide, in grams, is needed to react with 100.0 mL of 0.0922 *M* HCl?

4.71 What volume of 0.66 *M* HNO_3 is needed to react completely with 22 g of strontium hydroxide?

4.72 What volume of 0.22 *M* hydrochloric acid is needed to react completely with 2.5 g magnesium hydroxide?

4.73 What is the molar concentration of a solution of HCl if 135 mL react completely with 2.55 g $Ba(OH)_2$?

4.74 What is the molar concentration of a solution of H_2SO_4 if 5.11 mL react completely with 0.155 g NaOH?

4.75 What mass, in grams, of $BaSO_4$ forms in the reaction of 355 mL of 0.032 *M* H_2SO_4 with 266 mL of 0.015 *M* $Ba(OH)_2$?

4.76 Calculate the mass of magnesium hydroxide formed in the reaction of 1.2 L of a 5.5 *M* solution of sodium hydroxide and excess magnesium nitrate.

4.77 What mass of lead(II) sulfate precipitates on mixing 20.0 mL of a 1.11 *M* solution of lead(II) acetate with an excess of sodium sulfate solution?

4.78 What mass of iron (III) hydroxide precipitates on mixing 100.0 mL of a 1.545 *M* solution of iron (III) nitrate with an excess of sodium hydroxide solution?

4.79 What is the solid that precipitates, and how much of it forms, when an excess of sodium sulfate solution is mixed with 10.0 mL of a 2.10 *M* barium bromide solution?

4.80 ■ What is the solid that precipitates, and how much of it forms, when an excess of sodium chloride solution is mixed with 10.0 mL of a 2.10 *M* silver nitrate solution?

4.81 What volume of 1.212 *M* silver nitrate is needed to precipitate all of the iodide ions in 120.0 mL of a 1.200 *M* solution of sodium iodide?

4.82 What volume of 0.112 *M* potassium carbonate is needed to precipitate all of the calcium ions in 50.0 mL of a 0.100 *M* solution of calcium chloride?

4.83 A solid forms when excess barium chloride is added to 21 mL of 3.5 *M* ammonium sulfate. Write the overall equation, and calculate the mass of the precipitate.

4.84 A solid forms when excess iron(II) chloride is added to 220 mL of 1.22 *M* sodium hydroxide. Write the overall equation, and calculate the mass of the precipitate.

4.85 Write the overall equation (including the physical states), the complete ionic equation, and the net ionic equation for the reaction that occurs on mixing aqueous solutions of silver nitrate and sodium bromide. What mass of solid precipitates if 345 mL of a 0.330 *M* silver nitrate solution mixes with 100.0 mL of a 1.30 *M* sodium bromide solution?

4.86 Write the overall equation (including the physical states), the complete ionic equation, and the net ionic equation for the reaction of aqueous solutions of sodium hydroxide and magnesium chloride. What mass of solid forms on mixing 50.0 mL of 3.30 *M* sodium hydroxide with 35.0 mL of 1.00 *M* magnesium chloride?

OBJECTIVE Determine the concentration of a solution from data obtained in a titration experiment.

4.87 What is the molar concentration of a solution of HNO_3 if 50.00 mL react completely with 22.40 mL of a 0.0229 *M* solution of $Sr(OH)_2$?

4.88 ■ If a volume of 32.45 mL HCl is used to completely neutralize 2.050 g Na_2CO_3 according to the following equation, what is the molarity of the HCl?

$$Na_2CO_3(aq) + 2HCl(aq) \rightarrow 2NaCl(aq) + CO_2(g) + H_2O(\ell)$$

4.89 What is the molar concentration of an HCl solution if a 100.0-mL sample requires 33.40 mL of a 2.20 *M* solution of KOH to reach the equivalence point?

4.90 What is the molar concentration of an H_2SO_4 solution if a 50.0-mL sample requires 9.65 mL of a 1.33 *M* solution of NaOH to reach the equivalence point?

4.91 (a) What volume of 0.223 M HNO_3 is required to neutralize 50.00 mL of 0.033 M barium hydroxide?
(b) What volume of 1.13 M $AgNO_3$ is required to precipitate all of the chloride ions in 10.00 mL of 2.43 M calcium chloride?

4.92 ■ What volume, in milliliters, of 0.512 M NaOH is required to react completely with 25.0 mL 0.234 M H_2SO_4?

4.93 The pungent odor of vinegar is a result of the presence of acetic acid, CH_3COOH. Only one hydrogen atom of the CH_3COOH reacts with a base in a neutralization reaction. What is the concentration of acetic acid if a 10.00-mL sample is neutralized by 3.32 mL of 0.0100 M strontium hydroxide?

Acetic acid.

4.94 ■ What volume of 0.109 M HNO_3, in milliliters, is required to react completely with 2.50 g of $Ba(OH)_2$?

$$2HNO_3(aq) + Ba(OH)_2(aq) \rightarrow 2H_2O(\ell) + Ba(NO_3)_2(aq)$$

4.95 ▲ Oranges and grapefruits are known as citrus fruits because their acidity comes mainly from citric acid, $H_3C_6H_5O_7$. Calculate the concentration of citric acid in a solution if a 30.00-mL sample is neutralized by 15.10 mL of 0.0100 M KOH. Assume that three acidic hydrogens of each citric acid molecule are neutralized in the reaction.

4.96 Oxalic acid, $H_2C_2O_4$, is an acid in which both of the hydrogens react with base in a neutralization reaction. What is the concentration of an oxalic acid solution if 10.00 mL of the solution is neutralized by 22.05 mL of 0.100 M sodium hydroxide?

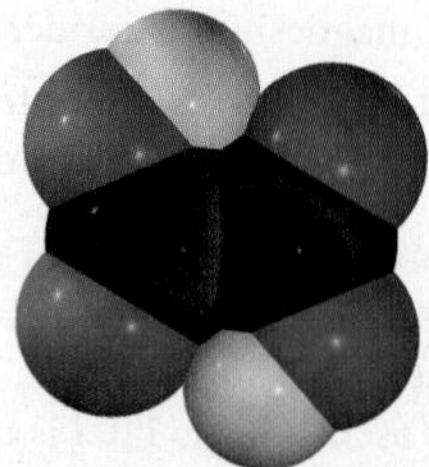

Oxalic acid.

4.97 ▲ A 125-mL sample of a $Ba(OH)_2$ solution is mixed with 75 mL of 0.10 M HCl. The resulting solution is still basic. An additional 35 mL of 0.012 M HCl is needed to neutralize the base. What is the molarity of the $Ba(OH)_2$ solution?

4.98 ▲ A solution is prepared by placing 14.2 g KCl in a 1.00-L volumetric flask and adding water to dissolve the solid, then filling the flask to the mark. What is the molarity of an $AgNO_3$ solution if 25.0 mL of the KCl solution reacts with exactly 33.2 mL of the $AgNO_3$ solution?

OBJECTIVE **Calculate solution concentration from the mass of product formed in a precipitation reaction.**

4.99 Mixing excess potassium carbonate with 300 mL of a calcium chloride solution of unknown concentration yields 4.50 g of a solid. Give the formula of the solid, and calculate the molar concentration of the calcium chloride solution.

4.100 ■ Mixing excess silver nitrate with 246 mL of a magnesium chloride solution of unknown concentration yields 2.21 g of a solid. Give the formula of the solid, and calculate the molar concentration of the magnesium chloride solution.

4.101 What is the percentage of barium in an ionic compound of unknown composition if a 2.11-g sample of the compound is completely dissolved in water and produces 1.22 g barium sulfate on addition of an excess of a sodium sulfate solution?

4.102 What is the percentage of silver in an ionic compound of unknown composition if a 3.13-g sample of the compound is completely dissolved in water and produces 2.02 g silver chloride on addition of an excess of a sodium chloride solution?

4.103 Sterling silver is a mixture of silver and copper. It dissolves in nitric acid to form the Ag^+ and Cu^{2+} ions. A 0.360-g sample of sterling silver is dissolved in nitric acid, and the Ag^+ precipitates with excess NaCl as AgCl. The mass of the AgCl produced is 0.435 g. What is the mass percentage of silver in the sterling silver?

4.104 The percentage of copper ions in a sample can be determined by reaction with zinc metal to produce copper metal.

$$Cu^{2+} + Zn(s) \rightarrow Cu(s) + Zn^{2+}$$

The copper produced is collected and weighed. If a 15.5-g ore sample containing copper ions is dissolved and produces 4.33 g copper metal when it reacts with zinc, what is the percentage of copper in the ore sample?

4.105 What is the molarity of a sodium chloride solution if addition of excess $AgNO_3$ to a 20-mL sample yields 0.0112 g precipitate?

4.106 What is the molarity of a potassium sulfate solution if addition of excess $BaCl_2$ to a 100-mL sample yields 0.233 g precipitate?

Chapter Exercises

4.107 A solution contains Be^{2+}, Ca^{2+}, and Ba^{2+}. Predict what happens if NaOH is added to the solution.

4.108 An environmental laboratory wants to remove Hg_2^{2+} from a water solution. Suggest a method of removal.

4.109 ▲ A 5.30-g sample of NaOH is placed in a 1.00-L volumetric flask and water is added to the mark. A 100.0-mL sample of the resulting solution is placed in a 500.0-mL volumetric and diluted to the mark with water. What volume of the second sample is needed to neutralize 33.0 mL of 0.0220 M H_2SO_4?

4.110 A 10.0-mL sample of solution that is 0.332 M NaCl and 0.222 M KBr is evaporated to dryness. What mass of solid remains?

4.111 What is the concentration of hydroxide ion in a solution made by mixing 200.0 mL of 0.0123 M NaOH with 200.0 mL of 0.0154 M $Ba(OH)_2$, followed by dilution of the mixture to 500.0 mL?

4.112 What is the molar concentration of chloride ion in a solution formed by mixing 150.0 mL of 1.54 M sodium chloride with 200.0 mL of 2.00 M calcium chloride, followed by dilution of the mixture to 500.0 mL?

4.113 What mass of NaOH is needed to prepare 1.00 L of an NaOH solution with the correct concentration such that 50.0 mL of it will exactly neutralize 10.0 mL of 3.11 M H_2SO_4?

4.114 Sodium thiosulfate, $Na_2S_2O_3$, is used in photographic film developing. The amount of $Na_2S_2O_3$ in a solution can be determined by a titration with I_2, according to the following equation:

$$2Na_2S_2O_3(aq) + I_2(aq) \rightarrow Na_2S_4O_6(aq) + 2NaI(aq)$$

Calculate the concentration of the $Na_2S_2O_3$ solution if 30.30 mL of a 0.1120 M I_2 solution reacts completely with a 100.0-mL sample of the $Na_2S_2O_3$ solution. In the actual experiment, excess KI is added to solubilize the I_2, but it is not part of the chemical change.

4.115 Toxic nitrogen monoxide gas can be prepared in the laboratory by carefully mixing a dilute sulfuric acid with an aqueous solution of sodium nitrite, as the following equation shows. What volume of 1.22 M sulfuric acid (assume excess sodium nitrite) is needed to prepare 2.44 g NO?

$$3H_2SO_4(aq) + 3NaNO_2(aq) \rightarrow 2NO(g) + HNO_3(aq) + 3NaHSO_4(aq) + H_2O(\ell)$$

4.116 Although silver chloride is insoluble in water, adding ammonia to a mixture of water and silver chloride causes the silver ions to dissolve because of the formation of $[Ag(NH_3)_2]^+$ ions. What is the concentration of $[Ag(NH_3)_2]^+$ ions that results from the addition of excess ammonia to a mixture of water and 0.022 g silver chloride if the final volume of the solution is 150 mL?

Cumulative Exercises

4.117 ▲ An 83.5-g sample contains NaCl contaminated with a substance that is not water soluble. The sample is added to water, which is then filtered to remove the contaminant and diluted to form 250.0 mL of a homogeneous solution. That solution is analyzed and the concentration of NaCl is 1.23 M. What is the percentage of NaCl in the original sample?

4.118 A 0.3120-g sample of a soluble compound made up of aluminum and chlorine yields 1.006 g AgCl when mixed with enough $AgNO_3$ to react completely with all of the chloride ions. What is the empirical formula of the compound?

4.119 A 2.64-g sample of $Ba(OH)_2$ is dissolved in water to form 250.0 mL solution. This solution is titrated with 0.0554 M H_2SO_4. It takes 33.4 mL of the acid solution to neutralize 30.0 mL of the base solution. After the titration is complete, the water is evaporated from the flask, leaving a solid. What is the solid, and what mass of it is present?

4.120 ■ An aqueous solution of hydrazine, N_2H_4, can be prepared by the reaction of ammonia and sodium hypochlorite.

$$2NH_3(aq) + NaOCl(aq) \rightarrow N_2H_4(aq) + NaCl(aq) + H_2O(\ell)$$

What is the theoretical yield of hydrazine, in grams, prepared from the reaction of 50.0 mL of 1.22 M $NH_3(aq)$ with 100.0 mL of 0.440 M NaOCl(aq)?

4.121 ▲ Tin(II) fluoride (stannous fluoride) is added to toothpaste as a convenient source of fluoride ion, which is known to help minimize tooth decay. The concentration of stannous fluoride in a particular toothpaste can be determined by precipitating the fluoride as the mixed salt PbClF.

$$SnF_2(aq) + 2Pb^{2+}(aq) + 2Cl^-(aq) \rightarrow 2PbClF(s) + Sn^{2+}(aq)$$

The concentrations of Pb^{2+} and Cl^- are controlled so that $PbCl_2$ does not precipitate. If a sample of toothpaste that weighs 10.50 g produces a PbClF precipitate that weighs 0.105 g, what is the mass percentage of SnF_2 in the toothpaste?

4.122 What is the percentage of barium in an unknown if a 2.3-g sample of the compound dissolved in water produces 2.2 g barium sulfate on addition of an excess of sodium sulfate?

4.123 Most photographic films, both colored and black and white, contain silver compounds. Some of the silver is left in the film as part of the image, but much of it dissolves as $Ag^+(aq)$ ions in solution during the developing of the film. The silver is valuable and is generally recovered. One method of recovering the $Ag^+(aq)$ is to add enough NaCl to precipitate all of the $Ag^+(aq)$ ions as AgCl. What mass of silver chloride is produced when excess NaCl is added to 4.00 L of a solution that is 0.0438 M in $Ag^+(aq)$?

Developing a photograph.

NASA Marshall Space Flight Center (NASA-MSFC)

NASA Kennedy Space Center (NASA-KSC)

A lot of energy is needed to launch a space vehicle into space. In all cases, this energy is provided by chemical reactions.

A lot of energy is needed to launch a space vehicle into space—a minimum of 60,500 kJ for every kilogram of mass, which is enough energy to melt an ice cube 57 cm on each side! This energy is provided by chemical reactions. The energy evolved in chemical reactions is the subject of this chapter.

In the United States, the National Aeronautics and Space Administration (NASA) has utilized several different chemical reactions to launch vehicles into space. Several vehicles have been used to take humans to the moon and back. For the powerful, three-stage Saturn V rocket, a highly refined petroleum similar to kerosene, called *RP-1,* was used in the first stage. Liquid oxygen was used as the other reactant:

$$\text{RP-1}(\ell) + O_2(\ell) \rightarrow CO_2(g) + H_2O(g) + \text{energy} \quad \textit{(unbalanced)}$$

For the second and third stages, the Saturn V rocket used the reaction between liquid hydrogen and liquid oxygen:

$$2H_2(\ell) + O_2(\ell) \rightarrow 2H_2O(g) + \text{energy}$$

A smaller craft called the *Lunar Module* took astronauts down to the surface of the moon and returned them into lunar orbit. This craft used the reaction between dinitrogen tetroxide, N_2O_4, and unsymmetrical dimethylhydrazine (UDMH):

$$(CH_3)_2NNH_2 + 2N_2O_4 \rightarrow 3N_2(g) + 2CO_2(g) + 4H_2O(g) + \text{energy}$$

On some lunar visits, a carlike vehicle called a *Lunar Rover* was used. It used batteries for power (Chapter 18 explains how all batteries are based on chemical reactions).

The space shuttle uses two types of boosters to lift off from the ground. The first one, using liquid fuels, is also based on the reaction between liquid

Thermochemistry

5

CHAPTER CONTENTS

OWL Online homework for this chapter may be assigned in OWL.

Look for the green colored vertical bar throughout this chapter, for integrated references to this chapter introduction.

hydrogen and liquid oxygen to make water. The second type of booster is the solid rocket booster, which depends on a reaction between ammonium perchlorate and aluminum:

$$10Al(s) + 6NH_4ClO_4(s) \rightarrow 5Al_2O_3(s) + 9H_2O(g) + 3N_2(g) + 6HCl(g) + \text{energy}$$

In the course of both chemical reactions, enough energy is given off to provide thrust to boost the space shuttle into orbit.

Newer spacecraft are using a new type of engine called an *ion thruster* for in-space propulsion, although not for liftoff. The ions created in the drive are accelerated by a magnetic field, and the spacecraft is accelerated in the opposite direction of the ion stream. Xenon is a common "fuel" because it is easy to ionize and is a relatively massive atom. It takes energy to ionize the xenon atom:

$$Xe(g) + \text{energy} \rightarrow Xe^+(g) + e^-$$

The space probe Deep Space 1, launched by NASA in 1998, uses such a drive (pictured at right). Although it generates a force of only 0.092 newton (N; a force equivalent to one third of an ounce), an ion thruster can operate for hundreds of days at a time, ultimately generating a substantial velocity.

NASA Headquarters-Greatest Images of NASA (NASA-HQ-GRIN)

The energy changes that accompany chemical reactions are of fundamental interest, so we can understand how to use chemical reactions to provide energy for useful purposes, such as hot packs to keep us warm or cold packs that soothe sprained ankles. This chapter introduces the topic of energy and chemical reactions. ∎

Since humans first discovered fire and used it to heat their caves and cook their food, one of the principal applications of chemical reactions has been to supply energy. Our society consumes large quantities of energy to provide itself with heat, light, and transportation, as well as for manufacturing material goods. Most of this energy comes from chemical reactions, mainly by burning fossil fuels. This chapter presents the relationship between chemical reactions and energy, and introduces concepts to be used in the next several chapters.

5.1 Energy, Heat, and Work

OBJECTIVES

- ☐ Distinguish between system, surroundings, kinetic energy, potential energy, heat, work, and chemical energy
- ☐ Identify processes as exothermic or endothermic based on the heat of that process

(*a*)

We have used chemical equations to calculate the quantities of substances consumed or produced in chemical reactions. Nearly all chemical reactions occur with a simultaneous change in energy. For example, the burning of wood or natural gas is a chemical reaction that releases energy in the form of heat. Our everyday experience tells us that the quantity of heat produced by a fire depends on the amount and type of fuel that burns. Similarly, a complete chemical equation includes a quantitative measure of the energy produced or consumed. This section presents the ideas needed to calculate the energy changes associated with chemical reactions and to treat the energy changes as stoichiometric quantities. **Thermochemistry** is the study of the relationship between heat and chemical reactions.

Energy

Energy can take many forms; mechanical, electrical, and chemical energy are just a few examples. All forms of energy fall into two categories: kinetic energy and potential energy. **Kinetic energy** is energy possessed by matter because it is in motion. The kinetic energy of an object depends on both its mass *(m)* and its velocity *(v)*, and is given by the equation

$$\text{Kinetic energy} = \frac{1}{2}mv^2$$

(*b*)

The SI unit for energy is the **joule (J),** which is defined in terms of three of the base SI units for mass, length, and time:

$$\text{Joule} = \frac{(\text{kilogram})(\text{meter})^2}{(\text{second})^2}$$

$$\text{J} = \frac{\text{kg} \cdot \text{m}^2}{\text{s}^2}$$

A moving baseball is an example of an object that possesses kinetic energy. For example, a baseball having a mass of 145 g (0.145 kg) and a velocity of 40.0 m/s has a kinetic energy of

$$\frac{1}{2}(0.145\ \text{kg})(40.0\ \text{m/s})^2 = 116\ \frac{\text{kg} \cdot \text{m}^2}{\text{s}^2} = 116\ \text{J}$$

(*c*)

These things convert chemical energy [*(a)* gasoline fuel, *(b)* a battery, *(c)* food] a form of potential energy, into kinetic energy.

Thermal energy is kinetic energy in the form of random motion of the particles in a sample of matter. The greater the temperature of the matter, the faster its particles move and the higher its thermal energy. **Heat** is the flow of energy from one object to another that causes a change in the temperature of the object. When heat is added to or removed from a sample, it causes a change in the temperature of that sample.

Work is the application of a force across some distance. It takes energy to perform work, so like heat, quantities of work are expressed in units of joules. Work can take many forms, including mechanical work, chemical work, gravitational work, pressure-volume work, and electrical work. We consider some of these forms of work more explicitly later in this chapter.

Potential energy is energy possessed by matter because of its position or condition. A brick on top of a building has more potential energy than one lying on the ground, because the potential energy depends on the vertical position of the brick. If the brick were dropped from the top of the building, its potential energy would be converted to kinetic energy as it fell.

Compounds also possess potential energy as a result of the forces that hold the atoms together. This form of potential energy is called **chemical energy.** In a chemical reaction, because the chemical energy of the reactants is not the same as that of the products, energy is either absorbed or released during the reaction, usually in the form of heat.

Basic Definitions

Certain terms are used in special ways in thermochemistry, and their precise definitions are important. In thermochemistry, attention is focused on a sample of matter that is called the **system.** In this chapter, the chemical systems consist of the atoms that react. The **surroundings** are all other matter, including the reaction container, the laboratory bench, and the person observing the reaction (Figure 5.1). The **law of conservation of energy** states that the total energy of the universe—the system plus the surroundings—is constant during a chemical or physical change. Energy is often transferred between the system and the surroundings, but the total energy of the universe before and after a change is constant. The law of conservation of energy is also referred to as the **first law of thermodynamics.**

Figure 5.1 The system. The system is the matter of interest. The yellow liquid inside the flask is our system.

If energy does transfer between the system and the surroundings, then the total amount of energy contained in the system has changed. Experimental evidence has shown that if the energy of a system changes, that energy change manifests itself as either heat or work. Thus, we can construct the expression

Energy change = work + heat

for the energy change of a system. This expression is another way to state the first law of thermodynamics.

When a chemical reaction takes place, energy is either transferred to or absorbed from the surroundings. In most reactions, much of the energy is transferred as heat. A chemical reaction is called **exothermic** if it releases heat to the surroundings. The combustion of natural gas to produce carbon dioxide and water is an example of an exothermic reaction. Thus, in the chemical equation, we can write heat as a *product* of the reaction:

$$CH_4(g) + 2O_2(g) \longrightarrow CO_2(g) + 2H_2O(\ell) + \text{heat}$$

A reaction that absorbs heat is called **endothermic.** The formation of nitric oxide (NO) from the elements is an example of an endothermic reaction. Because the reaction system absorbs heat in an endothermic reaction, energy is a *reactant* in the equation.

$$N_2(g) + O_2(g) + \text{heat} \rightarrow 2NO(g)$$

As with all forms of energy, the SI unit of heat energy is the joule. Most people are familiar with the calorie as a measure of heat. A *calorie* (cal) was originally defined as the amount of heat needed to increase the temperature of 1 g water by 1 °C, from 14.5 °C to 15.5 °C. A calorie is now defined as

1 cal = 4.184 J

Thus, it takes 4.184 J to increase the temperature of 1 g water from 14.5 °C to 15.5 °C. Energy content of foods is listed as Calories (with a capital *C*), which are actually kilocalories—that is, a 400-Cal muffin contains 400,000 calories, not 400.

© Cengage Learning/Larry Cameron

Figure 5.2 Chemical reactions. Most chemical reactions in the laboratory, such as this reaction of K_2CrO_4 and $Pb(NO_3)_2$ to yield solid $PbCrO_4$, are carried out under conditions of constant pressure.

OBJECTIVES REVIEW *Can you:*

- ☑ distinguish between system, surroundings, kinetic energy, potential energy, heat, work, and chemical energy?
- ☑ identify processes as exothermic or endothermic based on the heat of that process?

5.2 Enthalpy and Thermochemical Equations

OBJECTIVES

- ☐ Define enthalpy
- ☐ Express energy changes in chemical reactions
- ☐ Calculate enthalpy changes from stoichiometric relationships

In the laboratory, most chemical reactions occur in open containers, where the pressure is essentially constant (Figure 5.2). The **enthalpy,** H, of a system is a measure of the total energy of the system at a given pressure and temperature. Although the value of the total enthalpy of any system cannot be known, the *change* in enthalpy that accompanies a change in the system can be measured. Under conditions of constant pressure and temperature, the quantity of heat *absorbed or given off by the system* at constant pressure and temperature is called the **change in enthalpy,** and is represented by the symbol ΔH. When the symbol Δ (delta) precedes another symbol, it means "final value − initial value," so ΔH means $(H_{final} - H_{initial})$. [Similarly, $\Delta T = (T_{final} - T_{initial})$ represents a change in temperature, and so forth.] The symbol ΔH is spoken as "delta H."

The change in enthalpy, ΔH, expresses the energy change caused by a chemical reaction that takes place at constant pressure and temperature.

The direction in which heat is transferred determines the *sign* of ΔH. If the chemical reaction gives off heat (is exothermic), the system has lost energy, which means that its enthalpy decreases and ΔH is negative. If the chemical reaction absorbs heat (is endothermic), the energy of the system increases and the sign of ΔH is positive (Figure 5.3).

Rather than showing energy as a reactant or product in an equation, as was done earlier, it is more common to write the value of ΔH of the reaction. A **thermochemical equation** is a chemical equation for which the value of ΔH is given. The chemical reaction is assumed to occur at constant pressure and temperature, because ΔH is used in the thermochemical equation. The enthalpy change is determined by experiment (see Section 5.3). For example, Equation 5.1 is the thermochemical equation for an exothermic reaction, the combustion of methane.

Exothermic processes transfer heat to the surroundings, and the sign of ΔH is negative.

$$CH_4(g) + 2O_2(g) \rightarrow CO_2(g) + 2H_2O(\ell) \qquad \Delta H = -890 \text{ kJ} \qquad [5.1]$$

Equation 5.2 is the thermochemical equation for an endothermic reaction, the formation of nitrogen monoxide from the elements.

$$N_2(g) + O_2(g) \rightarrow 2NO(g) \qquad \Delta H = +181.8 \text{ kJ} \qquad [5.2]$$

The value of ΔH in a thermochemical equation refers to coefficients that stand for moles, not molecules. The enthalpy change of −890 kJ in Equation 5.1 is observed when *one mole* of CH_4 and *two moles* of O_2 react to produce *one mole* of CO_2 and *two moles* of H_2O. Remember that a negative enthalpy change means that the *system* gives off heat.

In a thermochemical equation, ΔH assumes that the coefficients refer to molar quantities.

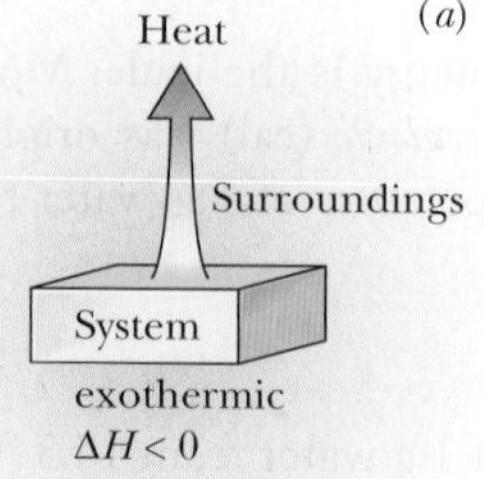

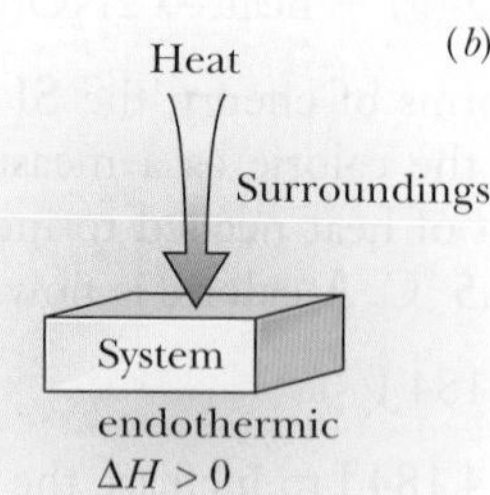

Figure 5.3 Heat and enthalpy change. *(a)* In exothermic reactions, heat is transferred from the system to the surroundings, and the enthalpy of the system decreases. *(b)* In endothermic reactions, heat is transferred from the surroundings to the system, increasing the enthalpy of the system.

PRACTICE OF CHEMISTRY
Hot and Cold Packs

One characteristic of exothermic reactions is that they feel hot to the touch, whereas endothermic reactions feel cold. Several consumer products take advantage of these characteristics.

Hot packs use the heat generated when a soluble salt dissolves in water. Typically, a plastic bag has two compartments with water and an ionic compound. When the barrier between the two compartments is broken and the contents mixed, the ionic compound gives off energy as it dissolves. Calcium chloride is commonly used:

$$CaCl_2(s) \xrightarrow{H_2O} Ca^{2+}(aq) + 2Cl^-(aq) \qquad \Delta H = -82.8 \text{ kJ}$$

The 82.8 kJ of energy given off by the dissolution of the solid calcium chloride is enough to increase the temperature of the hot pack to up to 90 °C (194 °F)—therefore, caution is advised when working with these products! Other hot packs have compartments of finely divided metal (such as iron or magnesium) that will react with water to generate heat. Typically, the hot pack stays hot for about 20 minutes or more. Uses of hot packs include thermal pain therapy and hand warming in cold weather.

Some compounds absorb heat when they dissolve; they form the basis of cold packs. Ammonium nitrate is an example:

$$NH_4NO_3(s) \xrightarrow{H_2O} NH_4^+(aq) + NO_3^-(aq) \qquad \Delta H = +25.5 \text{ kJ}$$

Because ammonium nitrate absorbs heat to dissolve, the solution feels cold, and when confined to a plastic bag, it can serve as a cold pack. Cold packs can get as cold as 0 °C (32 °F). They are also used for pain therapy, as well as keeping food cool so it does not spoil. ■

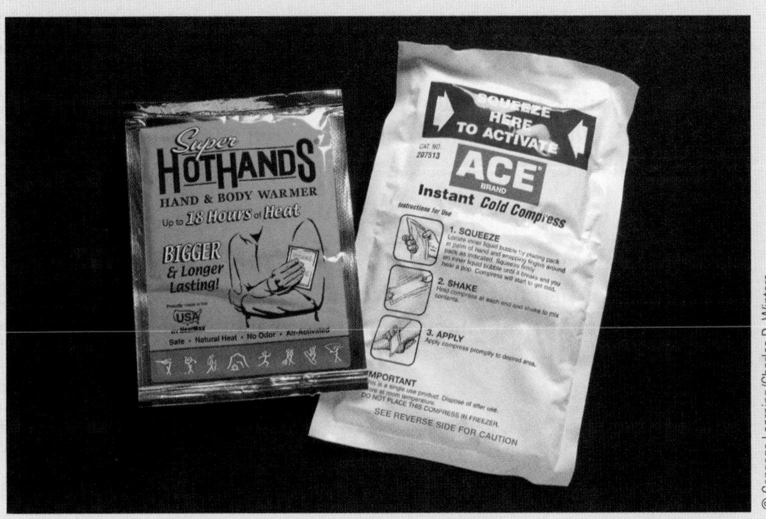

Chemical reactions are used to generate both heat and cold.

It is important to *include the physical state of every substance in any thermochemical equation.* Although it is good practice to include the physical states of the substances involved in any chemical equation, it is absolutely necessary to include them in a thermochemical equation, because the energy of a substance depends on its physical state. For example, in making liquid water from hydrogen and oxygen, the thermochemical equation is

$$2H_2(g) + O_2(g) \rightarrow 2H_2O(\ell) \qquad \Delta H = -571.7 \text{ kJ}$$

However, if the product is *gaseous* water, the thermochemical equation is

$$2H_2(g) + O_2(g) \rightarrow 2H_2O(g) \qquad \Delta H = -483.6 \text{ kJ}$$

The difference in enthalpy change is substantial and is due solely to the different phase of the product.

Stoichiometry of Enthalpy Change in Chemical Reactions

The enthalpy change is part of a thermochemical equation; ΔH is simply another stoichiometric quantity of the reaction. Using a thermochemical equation, we can calculate the quantity of heat produced or absorbed by a reaction just as we have calculated the masses of products formed in a reaction. The thermochemical equation expresses the stoichiometric relationship between the number of moles of any substance in the equation and the quantity of heat produced or absorbed in the reaction. For example, the thermochemical equation for the burning of ethane is

$$2C_2H_6(g) + 7O_2(g) \rightarrow 4CO_2(g) + 6H_2O(\ell) \qquad \Delta H = -3120 \text{ kJ} \qquad [5.3]$$

Some thermochemical relationships are:

Enthalpy changes are part of the stoichiometry of a thermochemical equation.

2 mol $C_2H_6(g)$ reacts and 3120 kJ is given off

7 mol $O_2(g)$ reacts and 3120 kJ is given off

4 mol $CO_2(g)$ is produced and 3120 kJ is given off

6 mol $H_2O(\ell)$ is produced and 3120 kJ is given off

These relations are used in the same way as are the mole-to-mole relations introduced in Chapter 3. The following examples illustrate the process.

EXAMPLE 5.1 Enthalpy as a Stoichiometric Quantity

Calculate the enthalpy change if 5.00 mol $N_2(g)$ reacts with $O_2(g)$ to make NO, a toxic air pollutant and important industrial compound, using the following thermochemical equation:

$$N_2(g) + O_2(g) \rightarrow 2NO(g) \qquad \Delta H = +181.8 \text{ kJ}$$

Strategy We will use the same approach as used in previous stoichiometry problems, except that this time a relationship exists between the number of moles of reactant and the quantity of energy consumed: 1 mol N_2 reacts and 181.8 kJ energy is consumed.

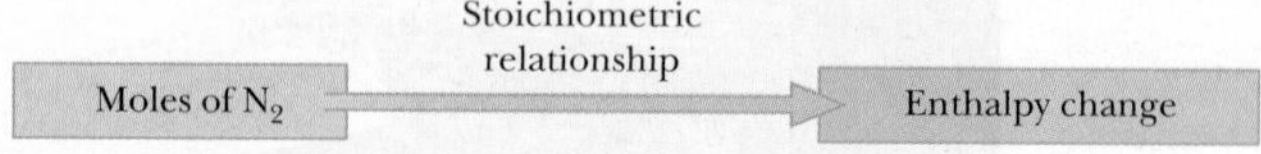

Solution

The equation is

$$N_2(g) + O_2(g) \rightarrow 2NO(g) \qquad \Delta H = +181.8 \text{ kJ}$$

We derive the conversion factor from the fact that when 1 mol nitrogen reacts, 181.8 kJ energy is consumed. Using the amount given, we have

$$\Delta H = 5.00 \text{ mol } N_2 \times \left(\frac{+181.8 \text{ kJ}}{1 \text{ mol } N_2}\right) = +909 \text{ kJ}$$

(Refer to Section 3.4 for a reminder on how to solve stoichiometry problems, if needed.) Because the enthalpy change is positive, the reaction absorbs energy as it proceeds.

Understanding

Use Equation 5.3 to determine the enthalpy change when 6.00 mol C_2H_6 burns in excess oxygen.

Answer −9360 kJ

EXAMPLE 5.2 Enthalpy as a Stoichiometric Quantity

Calculate the enthalpy change observed in the combustion reaction of 1.00 g ethane, using the thermochemical equation (see Equation 5.3).

Strategy We will use the same approach as in previous stoichiometry calculations. Convert the mass of ethane to moles; then use stoichiometric relations in the thermochemical equation to calculate ΔH.

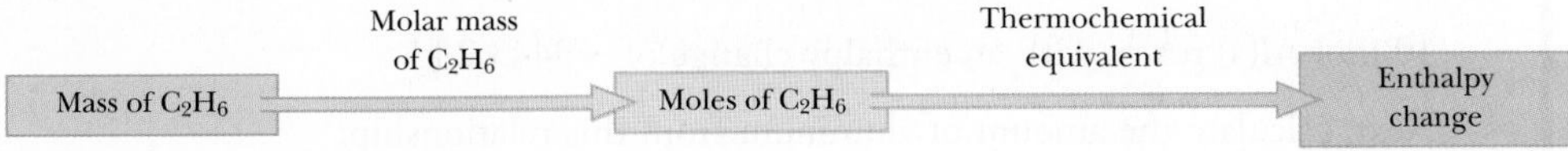

Solution

The equation, given earlier in this section, is

$$2C_2H_6(g) + 7O_2(g) \rightarrow 4CO_2(g) + 6H_2O(\ell) \qquad \Delta H = -3120 \text{ kJ}$$

First, we convert the mass of C_2H_6 to moles, using its molar mass (30.07 g/mol).

$$\text{mol } C_2H_6 = 1.00 \text{ g } C_2H_6 \times \left(\frac{1 \text{ mol } C_2H_6}{30.07 \text{ g } C_2H_6}\right)$$

$$= 3.33 \times 10^{-2} \text{ mol } C_2H_6$$

The change in enthalpy provided in the thermochemical equation gives us the relation between the moles of C_2H_6 and ΔH:

2 mol $C_2H_6(g)$ reacts and 3120 kJ is given off

Note that the relationship contains the coefficient of ethane, 2, that appears in the thermochemical equation, which must be included in the conversion factor to determine the enthalpy change. Because the process is exothermic, the enthalpy change is negative.

$$\Delta H = 3.33 \times 10^{-2} \text{ mol } C_2H_6(g) \times \left(\frac{-3120 \text{ kJ}}{2 \text{ mol } C_2H_6(g)}\right)$$

$$= -51.9 \text{ kJ}$$

Remember that a negative sign for the change in enthalpy means that heat is given off to the surroundings. We find that the system (the reaction) gives off 51.8 kJ of heat for each gram of ethane reacted.

Understanding

Use Equation 5.2 to calculate the enthalpy change when 5.00 g O_2 is consumed by reaction with N_2, forming NO.

Answer +28.4 kJ

EXAMPLE 5.3 Enthalpy as a Stoichiometric Quantity

The chapter introduction introduced the following reaction as one chemical reaction used to launch the space shuttle. Calculate the mass of aluminum required to generate 60,500 kJ energy (enough to launch 1 kg of matter into space). The thermochemical equation is

$$10Al(s) + 6NH_4ClO_4(s) \rightarrow 5Al_2O_3(s) + 9H_2O(g) + 3N_2(g) + 6HCl(g)$$
$$\Delta H = -9443.2 \text{ kJ}$$

Strategy The approach is the reverse of that used in Examples 5.1 and 5.2.

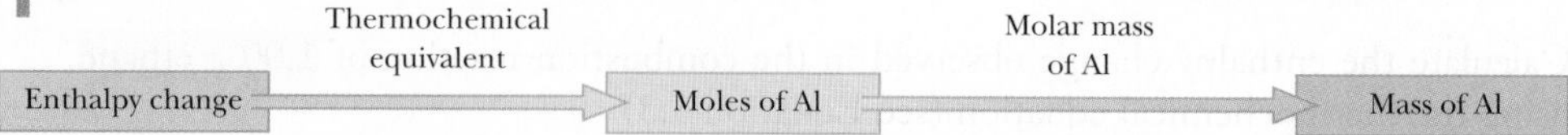

Solution

First, we use the thermochemical equation to calculate the number of moles of aluminum that is equivalent to the desired quantity of heat.

10 mol Al(s) reacts with an enthalpy change of −9443.2 kJ

Next, calculate the amount of aluminum from this relationship:

$$\text{Amount Al} = -60{,}500\ \cancel{\text{kJ}} \times \left(\frac{10\ \text{mol Al}}{-9443.2\ \cancel{\text{kJ}}}\right) = 64.1\ \text{mol Al}$$

Finally, we use the molar mass of aluminum (26.98 g/mol) to convert the moles of aluminum to the desired unit, grams.

$$\text{Mass Al} = 64.1\ \cancel{\text{mol Al}} \times \left(\frac{26.98\ \text{g Al}}{1\ \cancel{\text{mol Al}}}\right) = 1.73 \times 10^3\ \text{g Al}$$

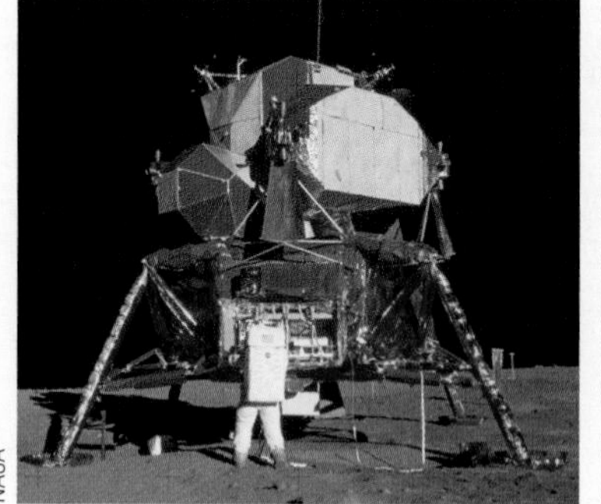
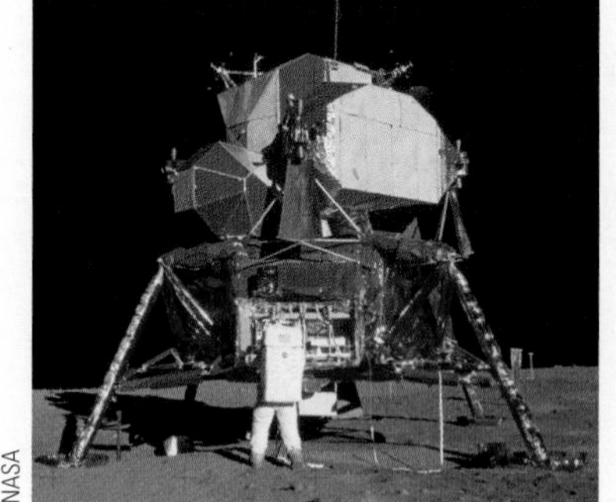

NASA

The lunar module used chemical reactions for propulsion.

Understanding

This chapter's introduction mentioned the following reaction as one chemical reaction used to power the lunar module in space. Calculate the mass of dinitrogen tetroxide, N_2O_4, required to generate 7632 J of energy (the approximate kinetic energy of the lunar module moving at a velocity of 1 m/s). The thermochemical equation is

$$(CH_3)_2NNH_2 + 2N_2O_4 \rightarrow 3N_2(g) + 2CO_2(g) + 4H_2O(g) \qquad \Delta H = -1763.5\ \text{kJ}$$

Answer 0.7963 g

OBJECTIVES REVIEW *Can you:*

- ☑ define enthalpy?
- ☑ express energy changes in chemical reactions?
- ☑ calculate enthalpy changes from stoichiometric relationships?

5.3 Calorimetry

OBJECTIVES

- ☐ Relate heat flow to temperature change
- ☐ Determine changes in enthalpy from calorimetry experiments

Scientists and engineers need to know the enthalpy changes that accompany chemical reactions to assess the value of fuels and to design chemical factories, among other things. When a reaction is highly exothermic, heat must be removed to avoid potential explosions. On the other hand, heat must be provided if reactions are endothermic. For example, the recovery of metals from their ores generally involves endothermic reactions, so fuels must be burned to maintain the reaction. This section presents one of the most common ways of measuring enthalpies of reaction.

Heat is determined by measuring the temperature change of the contents of the calorimeter.

Chemists determine all enthalpy changes for chemical reactions experimentally. In many cases, these experiments involve measurement of the heat released or absorbed when the chemical change occurs, a process called **calorimetry.** The device in which the reaction takes place and the heat is measured is known as a **calorimeter.** Calorimeters differ in the ways in which they measure heat and the conditions under which the reaction occurs. The

TABLE 5.1 Specific Heats of Some Common Substances

Substance	Formula	Specific Heat (J/g · K)
Water	$H_2O(\ell)$	4.184
Ethyl alcohol	$C_2H_5OH(\ell)$	2.419
Diethyl ether	$CH_3CH_2OCH_2CH_3(\ell)$	2.320
Aluminum	Al(s)	0.900
Gold	Au(s)	0.129
Mercury	$Hg(\ell)$	0.139
Graphite	C(s, graphite)	0.720
Magnesium oxide	MgO(s)	0.92

calorimeter considered here is an insulated vessel containing a solution in which the reaction occurs. The quantity of heat released or absorbed by the reaction (the system) causes a change in the temperature of the solution (the surroundings), which is measured with a thermometer. For the heat to correspond to the enthalpy change of the system, the calorimeter must be operated at constant pressure. A convenient calorimeter can be constructed from some nested disposable foam coffee cups (Figure 5.4). The reaction in the calorimeter proceeds at constant pressure because atmospheric pressure changes little during the course of the experiment.

Thermometer
Polystyrene
Polystyrene cups
Water
Glass stirring rod
Beaker

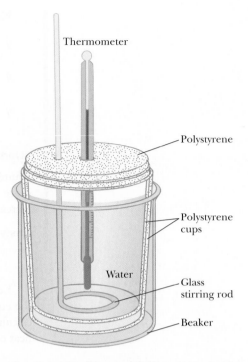

Figure 5.4 A coffee-cup calorimeter. Nested coffee cups can be used as a calorimeter to determine ΔH for reactions carried out in solution.

Because the calorimeter is an insulated vessel, its contents are the only part of the universe we must consider. The insulation prevents any transfer of heat into or out of the calorimeter. In this type of experiment, the amount of solution must be known because the observed temperature change depends on the amount of solution present.

Heat Capacity and Specific Heat

How is the heat related to the observed change in temperature? Experiments show that for different substances, the same amount of heat causes a different temperature change. We can define the **heat capacity** of a sample (such as the solution in a calorimeter) as the quantity of heat required to increase the temperature of that object by 1 K (or 1 °C). Heat capacity has units of J/K (or J/°C) and is nearly constant for a given substance over small ranges of temperature. The **specific heat,** C_s, is the heat needed to increase the temperature of a 1-g sample of the material by 1 K, and it has the units J/g · K (or J/g · °C). Table 5.1 lists the specific heats of several common substances. Note that water has a large specific heat; it takes more energy to increase the temperature of 1 g water by 1 K than 1 g of any of the other substances listed.

If the mass of a sample and its specific heat are known, the relationship between heat *(q)* and change in temperature (ΔT) is given by

$$q = mC_s\Delta T \quad [5.4]$$

where q is the heat in joules; m is the mass, in grams, of the sample; C_s is the specific heat of the sample; and ΔT is $T_{final} - T_{initial}$.

(*a*)

(*b*)

The relatively high specific heat of water requires the transfer of a large amount of heat to warm or cool the water. *(a)* Large bodies of water store thermal energy and can have significant impact on weather. *(b)* Hurricanes draw energy from the warmth of oceans.

EXAMPLE 5.4 Determining the Heat of a Process

What quantity of heat must be added to a 120-g sample of aluminum to change its temperature from 23.0 °C to 34.0 °C?

Strategy Use Equation 5.4 and the value of the specific heat of aluminum from Table 5.1 to determine the heat.

Solution

First, we need the temperature change, ΔT:

$$\Delta T = T_{final} - T_{initial} = 34.0\ °C - 23.0\ °C = 11.0°C$$

Now, using Equation 5.4 and substituting for the appropriate quantities:

$$q = mC_s\Delta T$$

$$q = 120\ \text{g} \times 0.900 \frac{\text{J}}{\text{g} \cdot {}^\circ\text{C}} \times 11.0\ {}^\circ\text{C}$$

$$q = 1190\ \text{J} = 1.19 \times 10^3\ \text{kJ}$$

Understanding

Aluminum oxide is one material used to construct the nozzles in the engines of the space shuttle (see this chapter's introduction). What quantity of heat is needed to increase the temperature of 1.20×10^2 g Al_2O_3 ($C_s = 0.773$ J/g · °C) by 11.0 °C?

Answer 1.02×10^3 J

Equation 5.4 can be algebraically rearranged to solve for any of the four variables in the equation. As long as you know three of the four quantities, the fourth one can be calculated. The following example illustrates a two-step determination of the specific heat of a metal.

EXAMPLE 5.5 Measuring Specific Heat

When a 60.0-g sample of metal at 100.0 °C is added to 45.0 g water at 22.60 °C, the final temperature of both the metal and the water is 32.81 °C. The specific heat of water is 4.184 J/g · °C. Calculate the specific heat of the metal.

Strategy Because of the law of conservation of energy, the energy lost by the metal will be gained by the water. We will determine the amount of heat *(q)* gained by the water and assume that this is the amount of heat lost by the metal. Knowing that quantity, we can determine C_s for the metal, because we also know ΔT for the metal. Thus, this will be a two-step calculation.

Solution

In the first step, we calculate the heat gained by the water, using Equation 5.4 as written:

$$q = mC_s\Delta T$$

$$q = 45.0\ \text{g} \times 4.184 \frac{\text{J}}{\text{g} \cdot {}^\circ\text{C}} \times (32.81 - 22.60)\ {}^\circ\text{C} = 1.92 \times 10^3\ \text{J}$$

Note how all units cancel except the unit of heat. The heat given up by the metal is therefore 1.92×10^3 J. Let us rearrange Equation 5.4 to algebraically solve for the specific heat:

$$q = mC_s\Delta T$$

$$C_s = \frac{q}{m\Delta T}$$

For our second step, we substitute our known quantities ($q = -1.92 \times 10^3$ J, $m =$ 60.0 g, $\Delta T = 32.81 - 100.0$ °C $= -67.2$ °C) for the metal, and solve for the specific heat of the metal:

$$C_s = \frac{-1.92 \times 10^3\ \text{J}}{(60.0\ \text{g}) \cdot (-67.2\ {}^\circ\text{C})} = 0.476 \frac{\text{J}}{\text{g} \cdot {}^\circ\text{C}}$$

Understanding

When a 43.0-g sample of metal at 100.0 °C is added to 38.0 g water at 23.72 °C, the final temperature of both the metal and the water is 29.33 °C. The specific heat of water is 4.184 J/g · °C. What is the specific heat of the metal?

Answer 0.293 J/g · °C

Calorimetry Calculations

Equation 5.4, relating heat, mass, specific heat, and temperature change, has a central role in calorimetry. To simplify the calculations in example problems, we make several assumptions.

1. The heat, q, is evaluated from the mass, temperature change, and specific heat of the solution.
2. The heat required to change the temperature of the vessel, stirrer, and thermometer is sufficiently small to be ignored.
3. The specific heat of the solution, as long as it is dilute, is the same as that of water, 4.184 J/g · °C.

EXAMPLE 5.6 Calorimetry

A 50.0-g sample of a dilute acid solution is added to 50.0 g of a base solution in a coffee-cup calorimeter. The temperature of the liquid increases from 18.20 °C to 21.30 °C. Calculate q for the neutralization reaction, assuming that the specific heat of the solution is the same as that of water (4.184 J/g · °C).

Strategy Use the temperature change and Equation 5.4 to determine the heat released by the reaction. Be careful to correctly determine the mass of the solution.

Solution

The total mass of solution in the calorimeter is 50.0 g + 50.0 g = 100.0 g. The change in temperature is

$$\Delta T = T_{\text{final}} - T_{\text{initial}} = 21.30\ ^\circ\text{C} - 18.20\ ^\circ\text{C} = 3.10\ ^\circ\text{C}$$

We can use Equation 5.4 directly to determine the heat of the reaction.

$$q = mC_s\Delta T$$

$$q = (100.0\ \cancel{\text{g}}) \times \left(4.184 \frac{\text{J}}{\cancel{\text{g}} \cdot \cancel{^\circ\text{C}}}\right) \times (3.10\ \cancel{^\circ\text{C}})$$

$$q = 1.30 \times 10^3\ \text{J} = 1.30\ \text{kJ}$$

Again, all units cancel except the unit of heat.

Understanding

A chemical reaction releases enough heat to increase the temperature of 49.9 g water from 17.82 °C to 19.72 °C. Calculate q for the reaction.

Answer 397 J

Many times, energy changes on a "per-mole" basis are needed. The following example illustrates how to determine a molar enthalpy change.

EXAMPLE 5.7 Enthalpy Change from Calorimetry

A 50.0 g-sample of acid takes 46.4 mL of 0.500 *M* NaOH solution to neutralize it. Assume the same amount of heat is given off as in Example 5.6.

(a) Calculate the enthalpy change for the neutralization per mole of hydrogen ions, described by the equation

$$H^+(aq) + OH^-(aq) \rightarrow H_2O(\ell) \qquad \Delta H = ?$$

(b) Is the neutralization reaction an endothermic or exothermic process?

Strategy The thermochemical equation in (a) is for one mole of each reactant. We need to find the number of moles of acid reacted in the titration information given, using a mole-mole calculation. Dividing the total heat by the total moles of acid will give the heat energy per mole of acid. The direction of temperature change (up or down) will indicate whether the reaction is exothermic or endothermic.

Solution

(a) We determined a heat of 1.30 kJ in Example 5.6:

$$q = 1.30 \text{ kJ}$$

First, we determine the number of moles of $H^+(aq)$ neutralized in that experiment from the titration data:

$$\text{Moles } H^+ = 0.0464 \text{ L NaOH} \times \left(\frac{0.500 \text{ mol } OH^-}{1 \text{ L NaOH}}\right) \times \left(\frac{1 \text{ mol } H^+}{1 \text{ mol } OH^-}\right) = 0.0232 \text{ mol } H^+$$

Thus, 1.30 kJ of heat results from the reaction of 0.0232 mol H^+ with base.

1.30 kJ is given off by 0.0232 mol H^+

In the thermochemical equation, 1 mol $H^+(aq)$ reacts, so the heat *per mole* of $H^+(aq)$ is

$$q = \left(\frac{1.30 \text{ kJ}}{0.0232 \text{ mol } H^+(aq)}\right) = 56.0 \text{ kJ/mol } H^+(aq)$$

When 1 mol $H^+(aq)$ reacts, 56.0 kJ of heat is generated.

(b) Because the reaction released heat, the reaction is exothermic. Because exothermic reactions are associated with negative ΔH's, if we were to write the ΔH of this process, we would write it as

$$\Delta H = -56.0 \text{ kJ/mol } H^+(aq)$$

Understanding

The reaction of 0.440 g magnesium with 400 g (excess) hydrochloric acid solution causes the temperature of the solution to increase by 5.04 °C. Assume that the specific heat of the solution is the same as that of water, and Mg is the limiting reactant. Calculate the ΔH for the reaction as written:

$$Mg(s) + 2HCl(aq) \rightarrow MgCl_2(aq) + H_2(g)$$

Answer −466 kJ/mol Mg

The coffee-cup calorimeter is not adequate for high-accuracy measurements. Scientists need to account for the heat needed to change the temperature of the calorimeter, stirrer, and thermometer, as well as its contents. In addition, the specific heats of dilute solutions are not exactly the same as that of water. Chemists can solve these problems and can measure temperature changes as small as 10 microdegrees (1 microdegree =

10^{-6} degrees). They can determine enthalpy changes to a precision of four or five significant digits when they take these details into account.

OBJECTIVES REVIEW *Can you:*

- ☑ relate heat flow to temperature change?
- ☑ determine changes in enthalpy from calorimetry experiments?

5.4 Hess's Law

OBJECTIVES

- ☐ Define a state function
- ☐ Draw and interpret enthalpy diagrams to illustrate energy changes in a reaction
- ☐ Use Hess's law to combine thermochemical equations to find an unknown ΔH

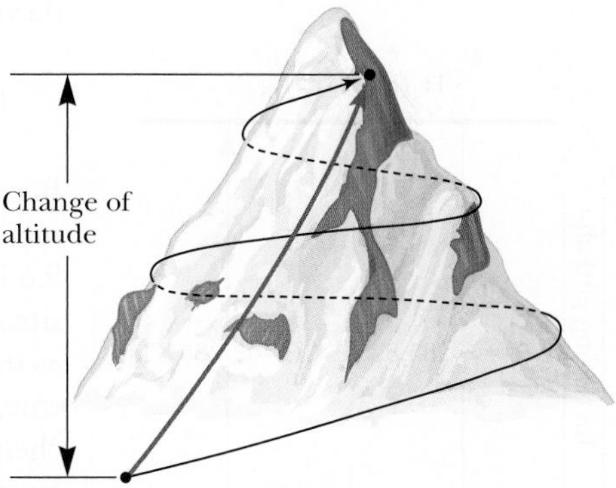

Figure 5.5 State function. The altitude of a mountain climber is analogous to a state function. The final altitude is the same whether the climber takes a direct, short path or a longer, more circuitous route.

The enthalpy change that accompanies a chemical reaction is an important and often necessary piece of information. However, it is difficult—and sometimes impossible—to determine experimentally the enthalpy changes for some chemical reactions. Fortunately, the enthalpy change for a reaction can be calculated from experimentally determined enthalpy changes for other reactions.

State Functions

A **state function** is any property of a system that is determined by the present conditions of the system. It is independent of how the system got to that set of conditions. For example, consider a mountain climber ascending a mountain. The climber can go straight up the mountain, or take a more circuitous route around and around the mountain (Figure 5.5). Whether the climber takes a direct route or a cyclic path, the individual's altitude at the end is the same. The altitude is analogous to a state function, in that it depends only on the final location, not how the climber got to that altitude. On the other hand, the distance traveled is not a state function. The direct path is shorter (although it may be more arduous!), whereas the circular path is longer. Because the distance traveled depends on path, it is not a state function.

The value of a state function does not depend on how the state was achieved, but rather only on the actual conditions of the state.

The enthalpy of a chemical system is a state function, as is the change in enthalpy. The value of ΔH for a process does not depend on how the process occurred. It depends only on the initial state of the system and the final state of the system. This fact will become important to us as we learn more about the enthalpy changes of chemical reactions.

Thermochemical Energy-Level Diagrams

A diagram is a convenient means of showing the enthalpy change in a chemical reaction. The enthalpy change that occurs when 1 mol of liquid water forms from the elements at 25 °C has been measured experimentally.

$$H_2(g) + \frac{1}{2}O_2(g) \rightarrow H_2O(\ell) \qquad \Delta H = -285.8 \text{ kJ} \qquad [5.5]$$

(You may remember this reaction from the introduction as one reaction used to launch spacecraft into space.) Because enthalpy is a state function, Equation 5.5 tells us that the enthalpy of 1 mol of liquid water is 285.8 kJ *less* than the enthalpy of 1 mol $H_2(g)$ plus one-half mole of $O_2(g)$. Because thermochemical equations are written in terms of moles, fractions are commonly encountered as coefficients in thermochemical equations in which 1 mol of product forms.

In thermochemical equations, the coefficients refer to molar amounts, so fractional coefficients can be used.

Figure 5.6 is an **energy-level diagram** representing the enthalpy change for the formation of water from hydrogen and oxygen. Energy-level diagrams are really one-dimensional graphs. This graph shows the relative enthalpies of the water and molecular

An energy-level diagram is a representation of the relative enthalpies of the reactants and products of a reaction.

hydrogen and oxygen. The energies on the vertical axis are not absolute numbers because we do not know the absolute enthalpy of any given substance. Experimental measurements give only the *difference* in enthalpy between the reactants and the products.

Under certain conditions, it is possible to reverse the direction of Equation 5.5 and decompose liquid water into the free elements:

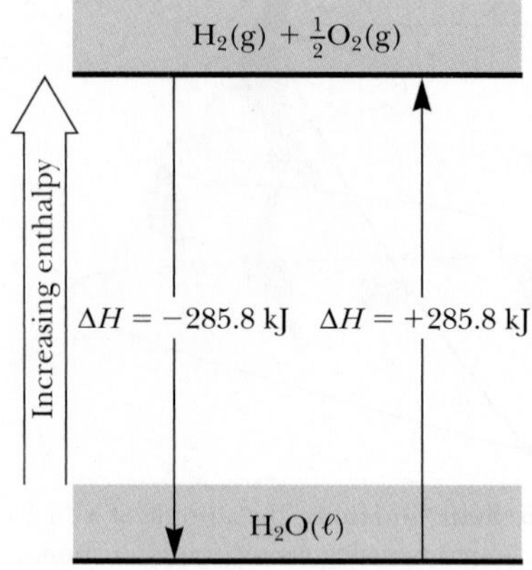

Figure 5.6 An energy-level diagram. One mole of $H_2O(\ell)$ has an enthalpy that is 285.8 kJ lower than that of one mole of $H_2(g)$ and one-half mole of $O_2(g)$.

$$H_2O(\ell) \rightarrow H_2(g) + \frac{1}{2}O_2(g) \qquad \Delta H = +285.8 \text{ kJ} \qquad [5.6]$$

Because energy is *released* (ΔH is negative) when water forms from the diatomic elements, energy is *absorbed* when the reverse reaction takes place. The ΔH for Equation 5.6 is numerically the same as the enthalpy change for Equation 5.5 but has the opposite sign. That is, the reverse reaction *absorbs* exactly as much heat from the surroundings as the forward reaction *releases* to the surroundings at the same pressure. If this were not true, the law of conservation of energy would be violated. Thus, when we reverse a chemical reaction, we change the sign on the original ΔH to get the ΔH of the new reaction.

Reversing the direction of a chemical reaction changes the sign on the reaction's ΔH.

For certain reactions, it is difficult or impossible to measure the change in enthalpy directly. If we try to measure ΔH of the reaction

$$C(s) + \frac{1}{2}O_2(g) \longrightarrow CO(g) \qquad [5.7]$$

by burning carbon, we find that a mixture of CO and CO_2 is produced. However, because ΔH is a state function, we can use an indirect approach instead. We can measure the changes in enthalpy for carbon and carbon monoxide separately reacting with a large excess of oxygen to form CO_2. The thermochemical equations that result from these experiments are

$$C(s) + O_2(g) \rightarrow CO_2(g) \qquad \Delta H_1 = -393.5 \text{ kJ} \qquad [5.8]$$

$$CO(g) + \frac{1}{2}O_2(g) \rightarrow CO_2(g) \qquad \Delta H_2 = -283.0 \text{ kJ} \qquad [5.9]$$

These two equations can be "added" algebraically in such a way that the desired Equation 5.7 is obtained. If we reverse Equation 5.9 (changing the sign of ΔH) and add it to Equation 5.8 (in such a way that the arrow is equivalent to the equal sign of an algebraic expression), the sum is Equation 5.7:

$$C(s) + O_2(g) \rightarrow \cancel{CO_2(g)} \qquad \Delta H = -393.5 \text{ kJ} \qquad [5.8]$$

$$\cancel{CO_2(g)} \rightarrow CO(s) + \cancel{\tfrac{1}{2}O_2(g)} \qquad \Delta H = +283.0 \text{ kJ} \qquad [5.9 \text{ reversed}]$$

$$C(s) + \frac{1}{2}O_2(g) \rightarrow CO(g) \qquad \Delta H = -110.5 \text{ kJ} \qquad [5.7 \text{ as desired}]$$

Note that one-half mole oxygen cancels from both sides of the combined equation. In the first equation, one mole of $O_2(g)$ reacts, whereas the second equation shows the formation of one-half mole of $O_2(g)$. Thus, the net amount of oxygen in the summed equation is one-half mole, as a reactant. Figure 5.7 is the energy-level diagram for this process. First, one mole of carbon and one mole of oxygen react to form one mole of CO_2 in an exothermic step. The second reaction shows the endothermic step in which this CO_2 decomposes to form one mole of CO and one-half mole of O_2. Although we can directly measure the enthalpy change for the first reaction, experimentally, we can measure only the reverse of the second reaction. As shown on the diagram, ΔH for the reverse reaction has the same magnitude as the reaction needed for the second step but has the opposite sign. Finally, we calculate ΔH for the desired reaction from the sum of the enthalpy changes for the two individual steps.

Calculating the enthalpy change in an overall chemical reaction by summing the enthalpy changes of each step is called **Hess's law.**

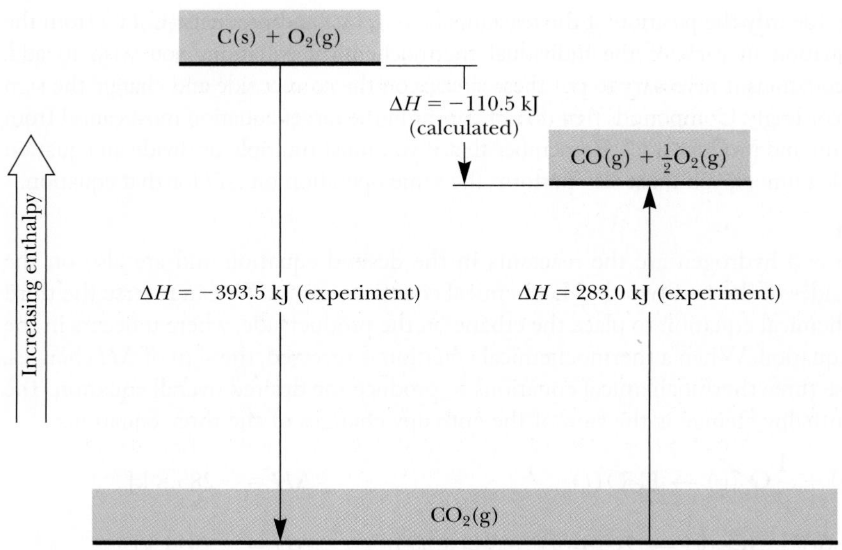

C(s) + O_2(g)

$\Delta H = -110.5$ kJ (calculated)

CO(g) + $\frac{1}{2}O_2$(g)

Increasing enthalpy

$\Delta H = -393.5$ kJ (experiment)

$\Delta H = 283.0$ kJ (experiment)

CO_2(g)

Figure 5.7 The enthalpy change for creation of 1 mol CO(g). Hess's law allows combination of experimentally measured enthalpy changes to calculate the desired enthalpy change.

The properties that govern the combination of thermochemical equations are natural consequences of the law of conservation of energy and the fact that enthalpy is a state function:

1. The change in enthalpy for an equation obtained by adding two or more thermochemical equations is the sum of the enthalpy changes of the added equations (as illustrated by Figure 5.7).
2. When a thermochemical equation is written in the reverse direction, the enthalpy change is numerically the same but has the opposite sign (as illustrated by Figure 5.6).
3. The enthalpy change is an extensive property that depends on the amounts of the substances that react. For example, when the coefficients in a thermochemical equation are doubled, the enthalpy change also doubles. (We assumed this fact in the solutions of Examples 5.1 and 5.2.) Whenever the coefficients in an equation are multiplied by a factor, the enthalpy change must be multiplied by the same factor.

Hess's law is a powerful tool for determining the enthalpy change that accompanies a reaction. It is not necessary to measure the enthalpy change for every reaction; Hess's law lets us calculate the enthalpy change for one reaction from thermochemical equations for others.

Hess's law allows the calculation of ΔH of a reaction from the ΔH values of other reactions.

Examples 5.8 and 5.9 show problems that use Hess's law and the other properties of thermochemical equations.

EXAMPLE 5.8 Hess's Law

Hydrogenation of hydrocarbons is an important reaction in the chemical industry. A simple example is the hydrogenation of ethylene to form ethane. Calculate the enthalpy change for

$$\underset{\text{ethylene}}{C_2H_4(g)} + H_2(g) \rightarrow \underset{\text{ethane}}{C_2H_6(g)} \qquad \Delta H = ?$$

Use the following thermochemical equations to determine the overall enthalpy change.

$$H_2(g) + \frac{1}{2}O_2(g) \rightarrow H_2O(\ell) \qquad \Delta H = -285.8 \text{ kJ}$$

$$C_2H_4(g) + 3O_2(g) \rightarrow 2CO_2(g) + 2H_2O(\ell) \qquad \Delta H = -1411 \text{ kJ}$$

$$C_2H_6(g) + \frac{7}{2}O_2(g) \rightarrow 2CO_2(g) + 3H_2O(\ell) \qquad \Delta H = -1560 \text{ kJ}$$

Strategy Identify the position of the reactants (C_2H_4, H_2) and products (C_2H_6) from the target equation in each of the individual thermochemical equations you wish to add. Reverse equations if necessary to put these species on the correct side and change the sign of ΔH accordingly. Compounds that do not appear in the target equation must cancel from the reactant and product sides. Remember that if you must multiply or divide an equation by a whole number, you must also perform the same operation on ΔH for that equation.

Solution

Ethylene and hydrogen are the reactants in the desired equation and are also on the reactant sides of the first two thermochemical equations given. We can reverse the third thermochemical equation to place the ethane on the product side, where it occurs in the desired equation. When a thermochemical equation is reversed, the sign of ΔH changes. Add these three thermochemical equations to produce the desired overall equation. The overall enthalpy change is the sum of the enthalpy changes of the three equations.

$$H_2(g) + \cancel{\frac{1}{2}O_2(g)} \rightarrow \cancel{H_2O(\ell)} \qquad \Delta H = -285.8 \text{ kJ}$$

$$C_2H_4(g) + \cancel{3O_2(g)} \rightarrow \cancel{2CO_2(g)} + \cancel{2H_2O(\ell)} \qquad \Delta H = -1411 \text{ kJ}$$

$$\cancel{2CO_2(g)} + \cancel{3H_2O(\ell)} \rightarrow C_2H_6(g) + \cancel{\frac{7}{2}O_2(g)} \qquad \Delta H = +1560 \text{ kJ}$$

$$C_2H_4(g) + H_2(g) \rightarrow C_2H_6(g) \qquad \Delta H = -137 \text{ kJ}$$

Note that many of the reactants and products are canceled out, because they appear on the reactant side and the product side. The net equation does not include these substances; in this example, the $\frac{7}{2}$ mol oxygen, 2 mol carbon dioxide, and 3 mol water are absent from the final equation.

Understanding

Calculate the enthalpy change for

$$C_2H_4(g) + H_2O(\ell) \rightarrow C_2H_5OH(\ell) \qquad \Delta H = ?$$

using the thermochemical equations

$$C_2H_5OH(\ell) + 3O_2(g) \rightarrow 2CO_2(g) + 3H_2O(\ell) \qquad \Delta H = -1367 \text{ kJ}$$

$$C_2H_4(g) + 3O_2(g) \rightarrow 2CO_2(g) + 2H_2O(\ell) \qquad \Delta H = -1411 \text{ kJ}$$

Answer −44 kJ

EXAMPLE 5.9 Hess's Law

The chemical industry converts hydrocarbons of low molecular mass to larger and more useful compounds. Calculate the change in enthalpy for the synthesis of cyclohexane (C_6H_{12}), a compound used in the production of nylon, from ethylene.

$$3C_2H_4(g) \rightarrow C_6H_{12}(\ell) \qquad \Delta H = ?$$

Use the information in Example 5.8 and the thermochemical equation for the combustion of cyclohexane:

$$C_6H_{12}(\ell) + 9O_2(g) \rightarrow 6CO_2(g) + 6H_2O(\ell) \qquad \Delta H = -3920 \text{ kJ}$$

Strategy Arrange the thermochemical equations so that the reactant, C_2H_4, is on the left and the product, C_6H_{12}, is on the right. The other products and reactants should all cancel so that the desired reaction is all that remains.

Solution

Because $C_6H_{12}(\ell)$ appears as a product in the desired equation, we reverse the direction of the given thermochemical equation and change the sign of ΔH.

$$6CO_2(g) + 6H_2O(\ell) \rightarrow C_6H_{12}(\ell) + 9O_2(g) \qquad \Delta H = +3920 \text{ kJ}$$

In the desired equation, 3 mol C_2H_4 is on the reactant side. We use the thermochemical equation for the reaction of 1 mol ethylene with oxygen, and multiply all of the coefficients and the enthalpy change by 3.

$$3C_2H_4(g) + 9O_2(g) \rightarrow 6CO_2(g) + 6H_2O(\ell) \qquad \Delta H = 3 \times (-1411) \text{ kJ}$$

We add these last two equations to obtain the desired reaction and the enthalpy change.

$$\cancel{6CO_2(g)} + \cancel{6H_2O(\ell)} \rightarrow C_6H_{12}(\ell) + \cancel{9O_2(g)} \qquad \Delta H = +3920 \text{ kJ}$$

$$3C_2H_4(g) + \cancel{9O_2(g)} \rightarrow \cancel{6CO_2(g)} + \cancel{6H_2O(\ell)} \qquad \Delta H = 3 \times (-1411) \text{ kJ}$$

$$3C_2H_4(g) \rightarrow C_6H_{12}(\ell) \qquad \Delta H = -313 \text{ kJ}$$

Again, we struck through those substances that appear on both the reactant side and the product side; these substances do not appear in the final, overall reaction.

Understanding

Given the thermochemical equations

$$Sn(s) + Cl_2(g) \rightarrow SnCl_2(s) \qquad \Delta H = -325 \text{ kJ}$$

$$SnCl_2(s) + Cl_2(g) \rightarrow SnCl_4(\ell) \qquad \Delta H = -186 \text{ kJ}$$

determine ΔH for

$$2SnCl_2(s) \rightarrow Sn(s) + SnCl_4(\ell) \qquad \Delta H = ?$$

Answer $+139$ kJ

OBJECTIVES REVIEW *Can you:*

- ☑ define a state function?
- ☑ draw and interpret enthalpy diagrams to illustrate energy changes in a reaction?
- ☑ use Hess's law to combine thermochemical equations to find an unknown ΔH?

5.5 Standard Enthalpy of Formation

OBJECTIVES

- ☐ Identify formation reactions and their enthalpy changes
- ☐ Calculate the enthalpy change of a reaction from standard enthalpies of formation

The enthalpy changes for many thousands of chemical reactions have been measured, and many more can be calculated using Hess's law. We need a convenient way of reducing this large data set to a more manageable size. If we take advantage of Hess's law, we need only a small number of enthalpy changes to deal with any chemical reaction.

Careful measurements show that the enthalpy change for any reaction is influenced by the pressure and the temperature. Although these effects are quite small compared with typical enthalpy changes for chemical reactions, they cannot be ignored. Therefore, in tabulations of enthalpy changes, all of the data must be measured at the same temperature and pressure. Scientists define the **standard state** of a substance at a specified temperature as its pure form at 1 atm pressure. Although there is no defined standard temperature, this book uses the reference temperature of 298.15 K (25 °C) that is conventional for nearly all thermochemical data. An enthalpy change in which all reactants and products are in their standard states is called a *standard enthalpy change* and is designated by the symbol ΔH°. The superscript ° symbol means that all reactants and products are in the standard state of 1 atm pressure and 298.15 K.

The standard state of a substance is the pure solid, liquid, or gas at one atmosphere pressure and the designated temperature (usually 298.15 K or 25.0 °C).

The enthalpy changes we focus on are enthalpy changes for formation reactions. A **formation reaction** is a chemical reaction that makes one mole of a substance from its constituent elements in their standard states. The enthalpy change for a formation reaction is symbolized by ΔH_f°, with the f subscript standing for "formation." ΔH_f° is referred to as the **standard enthalpy of formation.**

The standard enthalpy of formation is the ΔH of a reaction with 1 mol of product created from elements in their standard states.

One example is the formation reaction for $H_2O(\ell)$:

$$H_2(g) + \frac{1}{2}O_2(g) \rightarrow H_2O(\ell)$$

The coefficient on the product, $H_2O(\ell)$, is understood to be 1, so this reaction shows the formation of 1 mol $H_2O(\ell)$. The coefficients of the reactants balance the overall reaction and lead to a fractional coefficient for oxygen gas.

The following reaction is not a formation reaction:

$$C(g) + CO_2(g) \rightarrow 2CO(g)$$

It is not a formation reaction for the following reasons:

- One mole of product is not being made, two moles are.
- The standard state of carbon is not the gas phase.
- $CO_2(g)$ is not an element; all of the reactants in a formation reaction must be elements.

Thus, formation reactions have specific requirements. You should be able to recognize and write a formation reaction for any substance.

EXAMPLE 5.10 Formation Reactions

Which of the following reactions are formation reactions?

(a) $2H_2(g) + O_2(g) \rightarrow 2H_2O(\ell)$
(b) $Fe_2O_3(s) + 3SO_3(g) \rightarrow Fe_2(SO_4)_3(s)$
(c) $\frac{1}{2}N_2(g) + \frac{3}{2}H_2(g) \rightarrow NH_3(g)$

Strategy Look for reactions that have all elements in their standard states as reactants and one mole of a compound as a product.

Solution

(a) No, this is not a proper formation reaction because two moles of product, not one mole, are being formed.
(b) No, this is not a formation reaction, because the reactants are not elements in their standard states. Instead, the reactants are compounds.
(c) Yes, this is a formation reaction, for $NH_3(g)$.

Understanding

Which of the following reactions are formation reactions?

(a) $H_2(g) + O(g) \rightarrow H_2O(g)$
(b) $Fe(s) + N_2(g) + 3O_2(g) \rightarrow Fe(NO_3)_2(s)$
(c) $NH_3(g) \rightarrow \frac{1}{2}N_2(g) + \frac{3}{2}H_2(g)$

Answer (a) No (b) Yes (c) No

EXAMPLE 5.11 Writing Formation Reactions

Write the correct formation reactions for the following substances. Consult a periodic table for the proper phases of the elements involved.

(a) $NCl_3(g)$ (b) $Ca(NO_3)_2(s)$ (c) $O_3(g)$

Strategy Write chemical reactions for the formation of one mole of the given substance, with the reactants being the constituent elements of the substance in their standard states, not forgetting diatomic elements and the proper phases at 25.0 °C.

Solution

(a) $\frac{1}{2}N_2(g) + \frac{3}{2}Cl_2(g) \rightarrow NCl_3(g)$

(b) $Ca(s) + N_2(g) + 3O_2(g) \rightarrow Ca(NO_3)_2(s)$

(c) $\frac{3}{2}O_2(g) \rightarrow O_3(g)$

You should verify that each reaction is properly balanced.

Understanding

Write the correct formation reactions for the following substances. Consult a periodic table for the proper phases of the elements involved.

(a) $C_6H_6(\ell)$ (b) $C_6H_{12}O_6(s)$ (c) $BaCO_3(s)$

Answer

(a) $6C(\text{graphite}) + 3H_2(g) \rightarrow C_6H_6(\ell)$
(b) $6C(\text{graphite}) + 6H_2(g) + 3O_2(g) \rightarrow C_6H_{12}O_6(s)$
(c) $Ba(s) + C(\text{graphite}) + \frac{3}{2}O_2(g) \rightarrow BaCO_3(s)$

For every formation reaction, there is a corresponding and characteristic standard enthalpy of formation, labeled ΔH_f°. As examples, the equations and values for some standard enthalpies of formations are:

$$C(\text{graphite}) + O_2(g) \rightarrow CO_2(g) \qquad \Delta H_f^\circ[CO_2(g)] = -393.51 \text{ kJ/mol}$$

$$H_2(g) + \frac{1}{2}O_2(g) \rightarrow H_2O(\ell) \qquad \Delta H_f^\circ[H_2O(\ell)] = -285.83 \text{ kJ/mol}$$

$$Na(s) + \frac{1}{2}N_2(g) + \frac{3}{2}O_2(g) \rightarrow NaNO_3(s) \qquad \Delta H_f^\circ[NaNO_3(s)] = -467.9 \text{ kJ/mol}$$

$$O_2(g) \rightarrow O_2(g) \qquad \Delta H_f^\circ[O_2(g)] = 0 \text{ kJ/mol}$$

The enthalpy change for each of these reactions is the standard enthalpy of formation of the substance that appears as the product of the reaction. Note several points about these equations.

1. Only 1 mol of a single substance appears on the product side of each reaction.
2. Even though some reactions are impractical, such as the production of sodium nitrate by reaction of the elements at 25.0 °C (298.15 K), it is still possible to calculate the enthalpy change (the standard enthalpy of formation) for the reaction.
3. The enthalpy of formation of $O_2(g)$ in its standard state is exactly zero, because the equation defining the enthalpy of formation involves no net change (i.e., an element "reacting" to make an element). In fact, the ΔH_f° of all elements in their standard states is zero.

Why do we focus on formation reactions and enthalpies of formation? Simply this: every chemical reaction can be broken down into formation reactions of the products and reactants and recombined, using Hess's law, into the overall reaction. The enthalpy of the overall reaction is the algebraic sum of the enthalpies of formation of the reactants and products. Analysis of chemical reactions leads us to the following rule: For any chemical reaction, the standard enthalpy change of that reaction, ΔH_{rxn}°, is given by

$$\Delta H_{rxn}^\circ = \Sigma m \Delta H_f^\circ \text{ [products]} - \Sigma n \Delta H_f^\circ \text{ [reactants]} \qquad [5.10]$$

The standard enthalpy of any chemical reaction is the sum of the enthalpies of formation of the products minus the sum of the enthalpies of formation of the reactants.

where m is the number of moles of each product, and n is the number of moles of each reactant in the chemical equation. Because ΔH_f°s are typically expressed in units of kilojoules per mole (kJ/mol), after multiplying them by an amount in moles, the resulting ΔH°_{rxn} has units of kilojoules and is assumed to correspond to the chemical reaction as it occurs in molar amounts. Note also that because we are considering standard enthalpies, conditions are 1 atm and (usually) 298.15 K.

Example 5.12 illustrates the use of standard enthalpies of formation to calculate the enthalpy change of a reaction.

EXAMPLE 5.12 Calculating Enthalpy of Reaction from Enthalpies of Formation

One step in the production of nitric acid, a powerful acid used in the production of fertilizers and explosives, is the combustion of ammonia.

$$4NH_3(g) + 5O_2(g) \rightarrow 4NO(g) + 6H_2O(g)$$

Use Equation 5.10, with the enthalpies of formation of these substances in Table 5.2, to find ΔH°_{rxn} of this reaction.

Strategy Look up the standard enthalpy of formation of each substance in Table 5.2, recalling that the value of ΔH°_f for any element in its standard state is zero. Multiply each value from Table 5.2 by the coefficients from the balanced chemical equation. Sum the resulting values for the products and subtract the resulting values of the reactants. When collecting terms together, watch the signs of each ΔH°_f value.

Solution
We will substitute the coefficients and the standard enthalpies of formation for the substances into Equation 5.10, and solve for the enthalpy change of the reaction.

$$\Delta H^\circ_{rxn} = \Sigma\, m\, \Delta H^\circ_f[\text{products}] - \Sigma\, n\, \Delta H^\circ_f[\text{reactants}]$$

$$\Delta H^\circ_{rxn} = (4\Delta H^\circ_f[NO(g)] + 6\Delta H^\circ_f[H_2O(g)]) - (4\Delta H^\circ_f[NH_3(g)] + 5\Delta H^\circ_f[O_2(g)])$$

$$\Delta H^\circ_{rxn} = ([4\text{ mol}][90.25\text{ kJ/mol}] + [6\text{ mol}][-241.82\text{ kJ/mol}]) - ([4\text{ mol}][-46.11\text{ kJ/mol}] + [5\text{ mol}][0\text{ kJ/mol}])$$

$$\Delta H^\circ_{rxn} = -905.48\text{ kJ}$$

Note how the units of moles cancel out of each term, leaving kilojoules as the final unit. The following energy-level diagram represents this calculation.

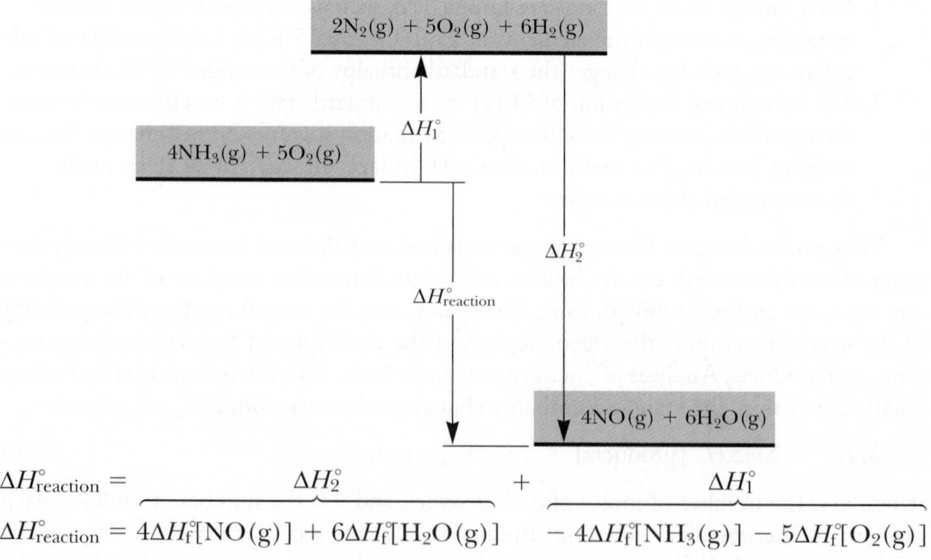

$$\Delta H^\circ_{reaction} = \overbrace{4\Delta H^\circ_f[NO(g)] + 6\Delta H^\circ_f[H_2O(g)]}^{\Delta H^\circ_2} + \overbrace{-4\Delta H^\circ_f[NH_3(g)] - 5\Delta H^\circ_f[O_2(g)]}^{\Delta H^\circ_1}$$

PRINCIPLES OF CHEMISTRY
Using Enthalpies of Formation to Determine ΔH_{rxn}

It is easy to establish the validity of Equation 5.10, sometimes referred to as the "products-minus-reactants" approach for determining the enthalpy of reaction, using a simple example. Consider the balanced chemical reaction

$$Fe_2O_3(s) + 3SO_3(g) \rightarrow Fe_2(SO_4)_3(s) \qquad \Delta H_{rxn}$$

This reaction can be rewritten as the combination of three reactions, all based on formation reactions of the products and reactants:

$$Fe_2O_3(s) \rightarrow 2Fe(s) + \frac{3}{2}O_2(g) \qquad -\Delta H_f[Fe_2O_3(s)]$$

$$3 \times [SO_3(g) \rightarrow \frac{1}{8}S_8(s) + \frac{3}{2}O_2(g)] \qquad 3 \times -\Delta H_f[SO_3(g)]$$

$$2Fe(s) + \frac{3}{8}S_8(s) + 6O_2(g) \rightarrow Fe_2(SO_4)_3(s) \qquad \Delta H_f[Fe_2(SO_4)_3(s)]$$

$$Fe_2O_3(s) + 3SO_3(g) \rightarrow Fe_2(SO_4)_3(s) \qquad \Delta H_{rxn} = \Delta H_f[Fe_2(SO_4)_3(s)] - \Delta H_f[Fe_2O_3(s)] - 3\Delta H_f[SO_3(g)]$$ ∎

TABLE 5.2 Standard Enthalpies of Formation

Substance	Name	ΔH_f° (kJ/mol)
$Br_2(\ell)$	Bromine	0
C(s, diamond)	Diamond	+1.895
C(s, graphite)	Graphite	0
$CH_4(g)$	Methane	−74.81
$C_2H_6(g)$	Ethane	−84.68
$C_3H_8(g)$	Propane	−103.85
$C_4H_{10}(g)$	*n*-Butane	−124.73
$CH_3OH(\ell)$	Methyl alcohol	−238.66
$C_2H_5OH(\ell)$	Ethyl alcohol	−277.69
$CO(g)$	Carbon monoxide	−110.52
$CO_2(g)$	Carbon dioxide	−393.51
$H_2(g)$	Hydrogen	0
$H_2O(g)$	Water	−241.82
$H_2O(\ell)$	Water	−285.83
$N_2(g)$	Nitrogen	0
$NH_3(g)$	Ammonia	−46.11
$NO(g)$	Nitrogen monoxide	+90.25
$O_2(g)$	Oxygen	0

Understanding

Another step in the production of nitric acid is the conversion of nitrogen monoxide to nitrogen dioxide. Using enthalpies of formation from Appendix G, calculate the enthalpy change that accompanies the reaction

$$2NO(g) + O_2(g) \longrightarrow 2NO_2(g)$$

Answer −114.14 kJ

Notice that the reactants contribute the *negative* of their enthalpies of formation to the overall combination, whereas products contribute the *positive* of their enthalpies of formation. Also, one of the reactions is taken three times, so its enthalpy of formation is also taken three times. The algebraic combination of the enthalpies of reaction shows that the enthalpy of reaction is found by taking the ΔH_f of the products and subtracting the ΔH_f of the reactants.

Chemists conveniently measure the **enthalpy of combustion,** the energy change for a combustion reaction, for organic compounds in the laboratory by performing calorimetry, and often use these data to determine the enthalpy of formation of the compound. Example 5.13 illustrates this process.

EXAMPLE 5.13 Calculating Enthalpy of Formation from Combustion Information

Calculate the standard enthalpy of formation for glucose, $C_6H_{12}O_6(s)$, from the following information. A calorimetry experiment shows that the enthalpy of combustion of 1 mol glucose to form carbon dioxide and water at 298.15 K is −2807.8 kJ. Use the data in Table 5.2 for the standard enthalpies of formation of carbon dioxide and water.

Strategy ΔH for the reaction was measured and ΔH_f° for the products are in Table 5.2. The enthalpy of formation of oxygen gas is zero, so we can calculate ΔH_f° for glucose.

Solution

The balanced combustion reaction is

$$C_6H_{12}O_6(s) + 6O_2(g) \rightarrow 6CO_2(g) + 6H_2O(\ell) \qquad \Delta H^\circ_{rxn} = -2807.8 \text{ kJ}$$

We know ΔH°_{rxn} but not the ΔH_f° of glucose. We will omit the units for clarity, recognizing that the proper units for an enthalpy of formation are kilojoules per mole (kJ/mol) and that our final answer will have units of kilojoules:

$$\Delta H^\circ_{rxn} = \{6\Delta H_f^\circ[CO_2(g)] + 6\Delta H_f^\circ[H_2O(\ell)]\} - \{\Delta H_f^\circ[C_6H_{12}O_6(s)] + 6\Delta H_f^\circ[O_2(g)]\}$$

$$-2807.8 = \{6(-393.51) + 6(-285.83)\} - \{\Delta H_f^\circ[C_6H_{12}O_6] + 6(0)\}$$

Solve for the unknown enthalpy of formation.

$$\Delta H_f^\circ[C_6H_{12}O_6(s)] = -1268.2 \text{ kJ/mol}$$

Understanding

The standard enthalpy change when 1 mol rubbing alcohol, isopropanol, $C_3H_7OH(\ell)$, burns to form carbon dioxide and liquid water at 298.15 K is −2005.8 kJ. Calculate the standard enthalpy of formation of rubbing alcohol.

Answer $\Delta H_f^\circ[C_3H_7OH] = -318.0$ kJ/mol

OBJECTIVES REVIEW *Can you:*

☑ identify formation reactions and their enthalpy changes?

☑ calculate the enthalpy change of a reaction from standard enthalpies of formation?

CASE STUDY Refining versus Recycling Aluminum

Our society has two methods for generating useful aluminum products such as cans and automobile parts: aluminum can be refined from its ores, or used aluminum products can be melted down and recycled into new products. Which process requires more energy?

The amount of energy needed to heat 1 mol aluminum to its melting point (660 °C) from room temperature (assumed to be 22 °C) to recycle the metal can be calculated from its specific heat, assuming that the specific heat does not vary between room temperature and its melting point. One mole of aluminum has a mass of 27.0 g, and we will use 0.900 J/g · °C for the specific heat:

$$q = 27.0 \text{ g} \times (0.900 \text{ J/g} \cdot {}^\circ\text{C})(660 \,{}^\circ\text{C} - 22 \,{}^\circ\text{C})$$

$$q = 15{,}500 \text{ J}$$

Energy also is needed to melt the aluminum, a quantity of energy known as the enthalpy of fusion. For aluminum, the enthalpy of fusion is 399.9 J/g, so the amount of energy needed to melt 1 mol aluminum is

$$q_{\text{melt}} = 27.0 \text{ g} \times 399.9 \text{ J/g}$$

$$q_{\text{melt}} = 10{,}800 \text{ J}$$

The total amount of energy needed to melt aluminum for recycling is thus

$$q_{\text{tot}} = 15{,}500 + 10{,}800 \text{ J}$$

$$q_{\text{tot}} = 26{,}300 \text{ J} = 26.3 \text{ kJ}$$

To determine how much energy we need to obtain 1 mol aluminum from refining, we need to know the chemical reactions involved and the enthalpy changes of those reactions. The main aluminum ore is bauxite, which is a mixture of minerals including hydrated aluminum hydroxide. In refining bauxite, it is separated, washed, and finally heated to a high temperature to drive off excess water. What remains is largely aluminum oxide, Al_2O_3. This aluminum oxide is dissolved in molten cryolite, Na_3AlF_6, at about 1100 °C. Using carbon electrodes, an electric current is passed through the solution to generate liquid aluminum according to the following reaction:

$$2Al_2O_3(\text{solv}) + 3C(s) \rightarrow 4Al(\ell) + 3CO_2(g)$$

The production of 1 g aluminum requires 71.6 kJ energy, so the production of one mole (27.0 g) of aluminum requires

$$q = 27.0 \text{ g} \times 71.6 \text{ kJ/g}$$

$$q = 1930 \text{ kJ}$$

of energy. Thus, it requires more than *70 times* more energy to generate aluminum from ore as it does to melt down scrap aluminum for recycling. Granted, this analysis does not include the other factors required for both processes, such as collection of raw materials, transport, facilities costs, and manpower. But a simple analysis using basic chemical principles illustrates how recycling aluminum is much less energy intensive than production of aluminum from its ores.

Questions

1. In the last chemical reaction for the isolation of Al from Al_2O_3, what does the "solv" label on aluminum oxide mean? Why can the "aq" label not be used?
2. A single aluminum can has a mass of 15.0 g. Calculate how much energy is required to melt one can and how much energy is needed to isolate one can's worth of aluminum from its ore.

Simple chemical principles can demonstrate that it usually requires less energy to recycle metals than it does to refine them from their ores.

ETHICS IN CHEMISTRY

1. Power plants that generate electricity by burning natural gas produce substantial CO_2, a "greenhouse gas" that most scientists believe is contributing to global warming. However, such plants produce much less pollution that causes acid rain than a power plant that burns coal. In the United States, the construction of new nuclear power plants, which produce neither CO_2 nor acid rain problems, have been halted for many years because of fears of a runaway nuclear reaction and problems with storage of radioactive waste. Thus, in recent years, many new power plants in the United States burn natural gas; but as the cost of natural gas has increased recently, so has the cost of electricity. What factors would you consider most important if your electric company proposed building a new power plant to produce electricity for your town?

2. The hills outside your university contain an enormous amount of energy in the form of hydrocarbons that are present in shale deposits. The increase in the price of petroleum has made the extraction of shale oil economically feasible, but only if done on a large scale. The proposed extraction plan will involve building a plant, pulverizing the shale deposit, and extracting the oil. The estimates are that a huge hole about 0.5 mile × 2 miles × 500 feet deep will be excavated. Proponents of the plan state the excavation will fill with water and become an attractive lake. The opponents say that the excavation will leave an environmental scar that will take thousands of years and billions of dollars to heal. Propose arguments in support and in opposition of the oil shale extraction. Use current data from the Internet in support of your argument.

3. The introduction to this chapter discusses a number of chemical fuels that are used to launch rockets. A switch from chemical fuels to nuclear fuels has been considered. If it could be shown that nuclear fuels are an efficient method to launch rockets, would you support such a decision?

4. Consider the Case Study on p. 197 that discusses the energy requirements of refining versus recycling aluminum. Are there any ethical considerations in the decision to refine, reuse, or recycle aluminum (or any other substances, for that matter)?

Chapter 5 Visual Summary

The chart shows the connections between the major topics discussed in this chapter.

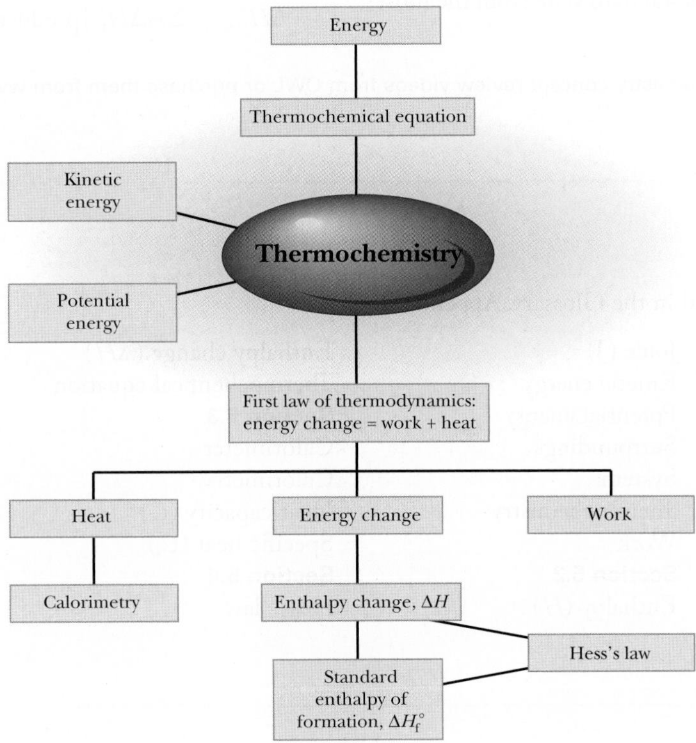

Summary

5.1 Energy, Heat, and Work

Thermochemistry is the study of the energy changes that accompany virtually all chemical reactions. *Kinetic energy* is the energy of motion, and *potential energy* is the energy of condition or position. *Chemical energy* is a form of potential energy arising from the forces that hold atoms together. In thermochemistry, the *system*—usually the atoms that are undergoing some change—is the matter of interest. The *surroundings* are the rest of the matter in the universe. According to the *law of conservation of energy*, all the energy lost or gained by the system is transferred to or from the surroundings.

5.2 Enthalpy and Thermochemical Equations

A *thermochemical equation* includes information about the energy changes that accompany the reaction. The *change in enthalpy*, ΔH, is equal to the heat change of the system if the change occurs at constant pressure. Reactions that give off heat to the surroundings are *exothermic* and have negative ΔH values, whereas reactions that absorb heat are *endothermic* and have positive ΔH values. Chemists perform stoichiometric calculations that determine the enthalpy changes in chemical reactions by using the relationships derived from the thermochemical equation.

5.3 Calorimetry

Determinations of enthalpy changes that accompany chemical reactions are based on *calorimetry*. The heat released or absorbed when a chemical reaction occurs produces an increase or decrease in the temperature of the surroundings, which are the contents of the calorimeter. The enthalpy change is calculated from the *heat capacity* of the calorimeter system and the change in the temperature.

5.4 Hess's Law

Because enthalpy is a *state function*, changes in enthalpy can be expressed in *energy-level diagrams*. These diagrams are used to demonstrate *Hess's law*, which states that the change in enthalpy for an equation obtained by adding two or more thermochemical equations is the sum of the enthalpy changes of the equations that have been added. Chemists can use Hess's law to determine enthalpy changes of reactions that cannot be obtained by direct experimental methods.

5.5 Standard Enthalpy of Formation

A convenient way of tabulating enthalpy data is as standard enthalpies of formation. The *standard enthalpy of formation,* ΔH_f°, is the enthalpy change that accompanies the formation of one mole of a substance in its standard state from the most stable forms of the elements in their standard states. The enthalpy change for any reaction can be calculated from the standard enthalpies of formation of the substances involved, using the equation

$$\Delta H_{rxn}^\circ = \Sigma m\Delta H_f^\circ\,[\text{products}] - \Sigma n\Delta H_f^\circ\,[\text{reactants}]$$

Download Go Chemistry concept review videos from OWL or purchase them from **www.ichapters.com**

Chapter Terms

The following terms are defined in the Glossary, Appendix I.

Section 5.1
Chemical energy
Conservation of energy (law)
Endothermic processes
Exothermic processes
First law of thermodynamics
Heat (q)
Joule (J)
Kinetic energy
Potential energy
Surroundings
System
Thermochemistry
Work

Section 5.2
Enthalpy (H)
Enthalpy change (ΔH)
Thermochemical equation

Section 5.3
Calorimeter
Calorimetry
Heat capacity (C)
Specific heat (C_s)

Section 5.4
Hess's law
State function
Thermochemical energy-level diagram

Section 5.5
Enthalpy of combustion
Formation reaction
Standard enthalpy of formation (ΔH_f°)
Standard state

Key Equations

The first law of thermodynamics (5.1)

Energy change = heat + work

Relationship between heat and temperature change (5.3)

$$q = mC_s\Delta T$$

Calculation of enthalpy change of a chemical reaction from the enthalpies of formation (5.5)

$$\Delta H_{rxn}^\circ = \Sigma m\Delta H_f^\circ\,[\text{products}] - \Sigma n\Delta H_f^\circ\,[\text{reactants}]$$

Questions and Exercises

OWL Selected end of chapter Questions and Exercises may be assigned in OWL.

Blue-numbered Questions and Exercises are answered in Appendix J; questions are qualitative, are often conceptual, and include problem-solving skills.

■ Questions assignable in OWL

✎ Questions suitable for brief writing exercises

▲ More challenging questions

Questions

5.1 Why must the physical states of all reactants and products be specified in a thermochemical equation?

5.2 Why is chemical energy classified as a form of potential energy?

5.3 What is the difference between the enthalpy of reaction and the enthalpy of formation? For what chemical reaction(s) are the two quantities the same?

5.4 Classify each process as exothermic or endothermic.
(a) ice melts
(b) gasoline burns
(c) steam condenses
(d) reactants → products, $\Delta H = -50$ kJ

5.5 Explain why the specific heat of the contents of the calorimeter must be known in a calorimetry experiment.

5.6 Define energy. What are its units?

5.7 Define heat. What are its units? How does it differ from energy?

5.8 Differentiate between kinetic energy and potential energy.

5.9 Describe the difference between the system and the surroundings.

5.10 What characteristic does every exothermic reaction have?

5.11 What characteristic does every endothermic reaction have?

5.12 Is the Sun exothermic or endothermic? Is it any less exothermic or endothermic in the winter, as opposed to the summer?

5.13 Under what circumstances is the heat of a process equal to the enthalpy change for the process?

5.14 Cheryl walks upstairs from the lobby of her residence hall to the roof, where she studies chemistry in the open air. She is joined by Carol, who rode the elevator from the lobby. Consider the two students' journeys, and identify which of the following are state functions and which are path functions.
(a) energy expended (b) time expended
(c) change in altitude (d) change in potential energy

5.15 State the first law of thermodynamics.

5.16 State in words the meaning of the following thermochemical equation:

$$C_2H_4(g) + 3O_2(g) \rightarrow 2CO_2(g) + 2H_2O(\ell) \qquad \Delta H = -1411 \text{ kJ}$$

5.17 Draw an energy-level diagram for an exothermic reaction of the following type:

reactants → products

5.18 Draw an energy-level diagram for an endothermic reaction of the following type:

reactants → products

5.19 Draw an enthalpy diagram for

reactants → products

that illustrates the use of enthalpies of formation in the calculation of the enthalpy change for the reaction. The diagram should have three levels—one for reactants, one for products, and one for the free elements. Draw arrows between the levels labeled in terms of the ΔH_f° of products and reactants and the ΔH_{rxn}.

5.20 ✎ Explain why absolute enthalpies cannot be measured and only changes can be determined.

5.21 Methane, $CH_4(g)$, and octane, $C_8H_{18}(\ell)$, are important components of the widely used fossil fuels. The enthalpy change for combustion of 1 mol methane is −890 kJ, and that for 1 mol octane is −5466 kJ. Which of these fuels produces more energy per gram of compound burned? What is the difference in energy produced per gram of compound?

5.22 The formation of hydrogen chloride is exothermic:

$$\frac{1}{2}H_2(g) + \frac{1}{2}Cl_2(g) \rightarrow HCl(g) \qquad \Delta H = -92.3 \text{ kJ}$$

What are the values of ΔH_{rxn} for

(a) $HCl(g) \rightarrow \frac{1}{2}H_2(g) + \frac{1}{2}Cl_2(g)$

(b) $H_2(g) + Cl_2(g) \rightarrow 2HCl(g)$

5.23 Explain why the calorimeter and its contents are the only part of the surroundings that are used to calculate the ΔH of reaction.

5.24 Addition of solid ammonium nitrate to water in a coffee-cup calorimeter results in a solution with a temperature lower than the original temperature of the water. The $NH_4NO_3(s)$ absorbs heat in the process of dissolving,

$$NH_4NO_3(s) + \text{heat} \rightarrow NH_4^+(aq) + NO_3^-(aq)$$

If the calorimeter is perfectly insulating (no heat can enter or leave), what provides the heat?

5.25 ✎ Describe how Hess's law leads to Equation 5.10. Use the reaction

$$2NaHCO_3(s) \rightarrow Na_2CO_3(s) + H_2O(\ell) + CO_2(g)$$

to justify your description.

5.26 Under what conditions can the value of ΔH for a reaction be denoted by the symbol ΔH°?

5.27 Why is it unnecessary to include the enthalpies of formation of elements, such as $P_4(s)$, $H_2(g)$, or C(graphite), in a table of standard enthalpies of formation?

5.28 A toaster toasts some bread at high temperature, then cools. After it has cooled down, the kitchen is found to have warmed up by 0.024 °C. Identify the system, the surroundings, and indicate whether the process that has occurred was exothermic or endothermic.

5.29 What are the two factors about a system that relate the heat of a process and the temperature change that the process causes the system?

5.30. ✎ A perpetual motion machine of the first kind generates more energy than it uses. Explain why this violates the first law of thermodynamics.

Exercises

In this section, similar exercises are arranged in pairs.

OBJECTIVES Distinguish between kinetic energy, potential energy, heat, work, and chemical energy. Identify processes as exothermic or endothermic based on the heat of a process.

5.31 A chemical reaction occurs and gives off 32,500 J. How many calories is this? Is the reaction endothermic or exothermic?

5.32 ■ A chemical reaction occurs and absorbs 64.7 cal. How many joules is this? Is the reaction endothermic or exothermic?

OBJECTIVES Define enthalpy. Express energy changes in chemical reactions. Calculate enthalpy changes from stoichiometric relationships.

5.33 The enthalpy change for the following reaction is −393.5 kJ.

$$C(s, \text{graphite}) + O_2(g) \rightarrow CO_2(g)$$

(a) Is energy released from or absorbed by the system in this reaction?
(b) What quantities of reactants and products are assumed?
(c) Predict the enthalpy change observed when 3.00 g carbon burns in an excess of oxygen.

5.34 The enthalpy change for the following reaction is +131.3 kJ.

$$C(s, graphite) + H_2O(g) \rightarrow CO(g) + H_2(g)$$

(a) Is energy released from or absorbed by the system in this reaction?
(b) What quantities of reactants and products are assumed if $\Delta H = +131.3$ kJ?
(c) What is the enthalpy change when 6.00 g carbon is reacted with excess $H_2O(g)$?

5.35 The thermochemical equation for the burning of methane, the main component of natural gas, is

$$CH_4(g) + 2O_2(g) \rightarrow CO_2(g) + 2H_2O(\ell) \qquad \Delta H = -890 \text{ kJ}$$

(a) Is this reaction endothermic or exothermic?
(b) What quantities of reactants and products are assumed if $\Delta H = -890$ kJ?
(c) What is the enthalpy change when 1.00 g methane burns in an excess of oxygen?

5.36 When lightning strikes, the energy can force atmospheric nitrogen and oxygen to react to make NO:

$$N_2(g) + O_2(g) \rightarrow 2\,NO(g) \qquad \Delta H = +181.8 \text{ kJ}$$

(a) Is this reaction endothermic or exothermic?
(b) What quantities of reactants and products are assumed if $\Delta H = +181.8$ kJ?
(c) What is the enthalpy change when 3.50 g nitrogen is reacted with excess $O_2(g)$?

5.37 One step in the manufacturing of sulfuric acid is the conversion of $SO_2(g)$ to $SO_3(g)$. The thermochemical equation for this process is

$$SO_2(g) + \frac{1}{2}O_2(g) \rightarrow SO_3(g) \qquad \Delta H = -98.9 \text{ kJ}$$

The second step combines the SO_3 with H_2O to make H_2SO_4.

(a) Calculate the enthalpy change that accompanies the reaction to make 1.00 kg $SO_3(g)$.
(b) Is heat absorbed or released in this process?

5.38 If nitric acid were sufficiently heated, it can be decomposed into dinitrogen pentoxide and water vapor:

$$2HNO_3(\ell) \rightarrow N_2O_5(g) + H_2O(g) \qquad \Delta H_{rxn} = +176 \text{ kJ}$$

(a) Calculate the enthalpy change that accompanies the reaction of 1.00 kg $HNO_3(\ell)$.
(b) Is heat absorbed or released during the course of the reaction?

5.39 The thermite reaction produces a large quantity of heat, enough to melt the iron metal that is a product of the reaction:

$$2Al(s) + Fe_2O_3(s) \rightarrow Al_2O_3(s) + 2Fe(s) \qquad \Delta H_{rxn} = -852 \text{ kJ}$$

What is the enthalpy change if 50.0 g Al reacts with excess iron(III) oxide?

5.40 ■ Hydrazine, N_2H_4, is used as a fuel in some rockets:

$$N_2H_4(\ell) + O_2(g) \rightarrow N_2(g) + 2H_2O(\ell) \qquad \Delta H = -622 \text{ kJ}$$

What is the enthalpy change if 110.0 g N_2H_4 reacts with excess oxygen?

5.41 The combustion of 1.00 mol liquid octane (C_8H_{18}), a component of gasoline, in excess oxygen is exothermic, producing 5.46×10^3 kJ of heat.

(a) Write the thermochemical equation for this reaction.
(b) Calculate the enthalpy change that accompanies the burning of 10.0 g octane.

5.42 The combustion of 1.00 mol liquid methyl alcohol (CH_3OH) in excess oxygen is exothermic, giving 727 kJ of heat.

(a) Write the thermochemical equation for this reaction.
(b) Calculate the enthalpy change that accompanies the burning 10.0 g methanol.
(c) Compare this with the amount of heat produced by 10.0 g octane, C_8H_{18}, a component of gasoline (see Exercise 5.41).

5.43 Another reaction that is used to propel rockets is

$$N_2O_4(\ell) + 2N_2H_4(\ell) \longrightarrow 3N_2(g) + 4H_2O(g)$$

This reaction has the advantage that neither product is toxic, so no dangerous pollution is released. When the reaction consumes 10.0 g liquid N_2O_4, it releases 124 kJ of heat.

(a) Is the sign of the enthalpy change positive or negative?
(b) What is the value of ΔH for the chemical equation if it is understood to be written in molar quantities?

5.44 Ammonia is produced commercially by the direct reaction of the elements. The formation of 5.00 g gaseous NH_3 by this reaction releases 13.56 kJ of heat.

$$N_2(g) + 3H_2(g) \rightarrow 2NH_3(g)$$

(a) What is the sign of the enthalpy change for this reaction?
(b) Calculate ΔH for the reaction, assuming molar amounts of reactants and products.

5.45 The reaction of 1 mol $O_2(g)$ and 1 mol $N_2(g)$ to yield 2 mol NO(g) is endothermic, with $\Delta H = +181.8$ kJ. Calculate the enthalpy change observed when 2.20 g $N_2(g)$ reacts with an excess of oxygen.

5.46 The reaction of 1 mol C(s, graphite) with 0.5 mol $O_2(g)$ to yield 1 mol CO(g) gives off 110.5 kJ. Calculate the enthalpy change when 52.0 g CO(g) is formed.

5.47 The reaction of 2 mol Fe(s) with 1 mol $O_2(g)$ to make 2 mol FeO(s) gives off 544 kJ. Calculate the enthalpy change that accompanies the formation of 100.0 g FeO.

5.48 The reaction of 2 mol $H_2(g)$ with 1 mol $O_2(g)$ to yield 2 mol $H_2O(\ell)$ is exothermic, with $\Delta H = -572$ kJ. Calculate the enthalpy change observed when 10.0 g $O_2(g)$ reacts with an excess of hydrogen.

5.49 Gasohol, a mixture of ethyl alcohol and gasoline, has been proposed as a fuel to help conserve our petroleum resources. It is available on a limited basis. The thermochemical equation for the burning of ethyl alcohol is

$$C_2H_5OH(\ell) + 3O_2(g) \rightarrow 2CO_2(g) + 3H_2O(\ell) \quad \Delta H = -1366.8 \text{ kJ}$$

Calculate the enthalpy change observed when burning 2.00 g ethyl alcohol.

5.50 ■ Isooctane (2,2,4-trimethylpentane), one of the many hydrocarbons that makes up gasoline, burns in air to give water and carbon dioxide.

$$2C_8H_{18}(\ell) + 25O_2(g) \rightarrow 16CO_2(g) + 18H_2O(\ell) \quad \Delta H^\circ = -10{,}922 \text{ kJ}$$

What is the enthalpy change if you burn 1.00 L of isooctane (density = 0.69 g/mL)?

5.51 ▲ The enthalpy change when 1 mol methane (CH_4) is burned is −890 kJ. It takes 44.0 kJ to vaporize 1 mol water. What mass of methane must be burned to provide the heat needed to vaporize 1.00 g water?

5.52 ▲ It takes 6.01 kJ to melt 1 mol of ice at 0 °C. Based on the data given in Exercise 5.51, how many grams of CH_4 must be burned to melt an ice cube having a mass of 35.0 g?

©Sebastian Duda, 2008/Used under license from Shutterstock.com

OBJECTIVES Relate heat flow to temperature change. Determine changes in enthalpy from calorimetry data.

5.53 How much heat, in kilojoules, must be added to increase the temperature of 500 g water from 22.5 °C to 39.1 °C? (See Table 5.1 for the specific heat of water.)

5.54 ■ How much energy is required to raise the temperature of 50.00 mL of water from 25.52 °C to 28.75 °C? (The density of water at this temperature is 0.997 g/mL.)

5.55 How much heat, in kilojoules, must be removed to decrease the temperature of a 20.0-g bar of aluminum from 34.2 °C to 22.5 °C? (See Table 5.1 for the specific heat of aluminum.)

5.56 How much heat, in kilojoules, must be removed to reduce the temperature of a 300-g bar of gold from 800 °C to 24.5 °C? (See Table 5.1 for the specific heat of gold.)

5.57 A 50.0-g sample of metal at 100.00 °C is added to 40.0 g water that is initially 23.50 °C. The final temperature of both the water and the metal is 28.46 °C.

(a) Use the specific heat of water to find the heat absorbed by the water.
(b) How much heat did the metal sample lose?
(c) Calculate the specific heat of the metal.

5.58 A 50.0-g sample of metal at 100.00 °C is added to 60.0 g water that is initially 25.00 °C. The final temperature of both the water and the metal is 31.51 °C.

(a) Use the specific heat of water to find the heat absorbed by the water.
(b) How much heat did the metal sample lose?
(c) Calculate the specific heat of the metal.

5.59 A 59.9-g sample of ethyl alcohol at 70.30 °C is mixed with 40.1 g water that is initially 22.00 °C. The specific heat of ethyl alcohol is 2.419 J/g · °C. What is the final temperature of the resulting solution?

5.60 ■ A 40.0-g sample of gold powder at 91.50 °C is dissolved into 51.2 g mercury that is initially 22.00 °C. Using the specific heats given in Table 5.1, calculate the final temperature of the resulting solution, called an *amalgam*.

5.61 When 7.11 g NH_4NO_3 is added to 100 mL water, the temperature of the calorimeter contents decreases from 22.1 °C to 17.1 °C. Assuming that the mixture has the same specific heat as water and a mass of 107 g, calculate the heat q. Is the dissolution of ammonium nitrate exothermic or endothermic?

5.62 A 50-mL solution of a dilute $AgNO_3$ solution is added to 100 mL of a base solution in a coffee-cup calorimeter. As $Ag_2O(s)$ precipitates, the temperature of the solution increases from 23.78 °C to 25.19 °C. Assuming that the mixture has the same specific heat as water and a mass of 150 g, calculate the heat q. Is the precipitation reaction exothermic or endothermic?

5.63 A 0.470-g sample of magnesium reacts with 200 g dilute HCl in a coffee-cup calorimeter to form $MgCl_2(aq)$ and $H_2(g)$. The temperature increases by 10.9 °C as the magnesium reacts. Assume that the mixture has the same specific heat as water and a mass of 200 g.

(a) Calculate the enthalpy change for the reaction. Is the process exothermic or endothermic?
(b) Write the chemical equation and evaluate ΔH.

5.64 Dissolving 6.00 g $CaCl_2$ in 300 mL of water causes the temperature of the solution to increase by 3.43 °C. Assume that the specific heat of the solution is 4.18 J/g · K and its mass is 306 g.

(a) Calculate the enthalpy change when the $CaCl_2$ dissolves. Is the process exothermic or endothermic?
(b) Determine ΔH on a molar basis for

$$CaCl_2(s) \xrightarrow{H_2O} Ca^{2+}(aq) + 2Cl^-(aq)$$

OBJECTIVES Define a state function. Draw and interpret enthalpy diagrams to illustrate energy changes in a reaction. Use Hess's law to combine thermochemical equations to find an unknown ΔH.

5.65 Draw an energy-level diagram (e.g., see Figure 5.6) based on each of the following thermochemical equations. Label each level with the amounts of substances present, and use an arrow between levels for the given enthalpy change. (Do not show the reverse process on the diagram.)
(a) $C(s, graphite) + H_2O(g) \rightarrow CO(g) + H_2(g)$ $\Delta H = +131.3$ kJ
(b) $CO(g) + H_2O(g) \rightarrow CO_2(g) + H_2(g)$ $\Delta H = -41.2$ kJ
(c) $2SO_2(g) + O_2(g) \rightarrow 2SO_3(g)$ $\Delta H = -197.8$ kJ

5.66 Draw an energy-level diagram (e.g., see Figure 5.6) based on each of the following thermochemical equations. Label each level with the amounts of substances present, and use an arrow between levels for the given enthalpy change. (Do not show the reverse process on the diagram.)
(a) $Zn(s) + 2HCl(aq) \rightarrow ZnCl_2(aq) + H_2(g)$ $\Delta H = -152.4$ kJ
(b) $N_2(g) + 2O_2(g) \rightarrow 2NO_2(g)$ $\Delta H = +66.36$ kJ
(c) $2C_2H_6(g) + 7O_2(g) \rightarrow 4CO_2(g) + 6H_2O(\ell)$ $\Delta H = -3120$ kJ

5.67 Using the following thermochemical equations

$$C_2H_6(g) + \frac{7}{2}O_2(g) \rightarrow 2CO_2(g) + 3H_2O(\ell) \quad \Delta H = -1560 \text{ kJ}$$

$$2C_2H_2(g) + 5O_2(g) \rightarrow 4CO_2(g) + 2H_2O(\ell) \quad \Delta H = -2599 \text{ kJ}$$

$$H_2(g) + \frac{1}{2}O_2(g) \rightarrow H_2O(\ell) \quad \Delta H = -286 \text{ kJ}$$

calculate ΔH for

$$C_2H_2(g) + 2H_2(g) \rightarrow C_2H_6(g) \quad \Delta H = ?$$

5.68 Using the thermochemical equations in Exercise 5.67 as needed and in addition

$$CH_4(g) + 2O_2(g) \rightarrow CO_2(g) + 2H_2O(\ell) \quad \Delta H = -890 \text{ kJ}$$

$$C_2H_4(g) + 3O_2(g) \rightarrow 2CO_2(g) + 2H_2O(\ell) \quad \Delta H = -1411 \text{ kJ}$$

calculate ΔH for

$$C_2H_4(g) + 2H_2(g) \rightarrow 2CH_4(g) \quad \Delta H = ?$$

5.69 Calculate ΔH for the reaction

$$Zn(s) + \frac{1}{2}O_2(g) \rightarrow ZnO(s) \quad \Delta H = ?$$

given the equations

$$Zn(s) + 2HCl(aq) \rightarrow ZnCl_2(aq) + H_2(g) \quad \Delta H = -152.4 \text{ kJ}$$

$$ZnO(s) + 2HCl(aq) \rightarrow ZnCl_2(aq) + H_2O(\ell) \quad \Delta H = -90.2 \text{ kJ}$$

$$2H_2(g) + O_2(g) \rightarrow 2H_2O(\ell) \quad \Delta H = -571.6 \text{ kJ}$$

5.70 Calculate ΔH for

$$Mg(s) + \frac{1}{2}O_2(g) \rightarrow MgO(s) \quad \Delta H = ?$$

given the equations

$$Mg(s) + 2HCl(aq) \rightarrow MgCl_2(aq) + H_2(g) \quad \Delta H = -462 \text{ kJ}$$

$$MgO(s) + 2HCl(aq) \rightarrow MgCl_2(aq) + H_2O(\ell) \quad \Delta H = -146 \text{ kJ}$$

$$2H_2(g) + O_2(g) \rightarrow 2H_2O(\ell) \quad \Delta H = -571.6 \text{ kJ}$$

5.71 Given the thermochemical equations

$$2Cu(s) + Cl_2(g) \rightarrow 2CuCl(s) \quad \Delta H = -274.4 \text{ kJ}$$

$$2CuCl(s) + Cl_2(g) \rightarrow 2CuCl_2(s) \quad \Delta H = -165.8 \text{ kJ}$$

find the enthalpy change for

$$Cu(s) + Cl_2(g) \rightarrow CuCl_2(s) \quad \Delta H = ?$$

5.72 In the process of isolating iron from its ores, carbon monoxide reacts with iron(III) oxide, as described by the following equation:

$$Fe_2O_3(s) + 3CO(g) \rightarrow 2Fe(s) + 3CO_2(g) \quad \Delta H = -24.8 \text{ kJ}$$

The enthalpy change for the combustion of carbon monoxide is

$$2CO(g) + O_2(g) \rightarrow 2CO_2(g) \quad \Delta H = -566 \text{ kJ}$$

Use this information to calculate the enthalpy change for the equation

$$4Fe(s) + 3O_2(g) \rightarrow 2Fe_2O_3(s) \quad \Delta H = ?$$

5.73 Draw an energy-level diagram that represents the Hess's law calculation in Exercise 5.71.

5.74 Draw an energy-level diagram that represents the Hess's law calculation in Exercise 5.72.

5.75 What does an energy-level diagram for the *reverse* reaction from Exercise 5.71 look like?

5.76 What does an energy-level diagram for the *reverse* reaction from Exercise 5.72 look like?

OBJECTIVES Identify formation reactions and their enthalpy changes. Calculate a reaction enthalpy change from standard enthalpies of formation.

5.77 Write the formation reaction for each of the following substances.
(a) $HBr(g)$
(b) $H_2SO_4(\ell)$
(c) $O_3(g)$
(d) $NaHSO_4(s)$

5.78 Write the chemical equation for the reaction whose energy change is the standard enthalpy of formation of each of the following substances.
(a) $CH_3COOH(\ell)$
(b) $H_3PO_4(\ell)$
(c) $CaSO_4 \cdot 2H_2O(s)$
(d) C(s, diamond)

© Cengage Learning/Charles D. Winters

5.79 Use standard enthalpies of formation to calculate the enthalpy change for each of the following reactions at 298.15 K and 1 atm. Label each as endothermic or exothermic.
(a) The fermentation of glucose to ethyl alcohol and carbon dioxide:

$$C_6H_{12}O_6(s) \rightarrow 2C_2H_5OH(\ell) + 2CO_2(g)$$

(b) The combustion of normal (straight-chain) butane:

$$n\text{-}C_4H_{10}(g) + \frac{13}{2}O_2(g) \rightarrow 4CO_2(g) + 5H_2O(\ell)$$

5.80 ■ Use the standard enthalpies of formation from Appendix G to calculate the enthalpy change for each of the following reactions at 298.15 K and 1 atm. Label each as endothermic or exothermic.
(a) The photosynthesis of glucose:

$$6CO_2(g) + 6H_2O(\ell) \rightarrow C_6H_{12}O_6(s) + 6O_2(g)$$

(b) The reduction of iron(III) oxide with carbon:

$$2Fe_2O_3(s) + 3C(s) \rightarrow 4Fe(s) + 3CO_2(g)$$

5.81 Use the standard enthalpies of formation from Appendix G to calculate the enthalpy change for each of the following reactions at 298.15 K and 1 atm. Label each as endothermic or exothermic.
(a) $NaHCO_3(s) \rightarrow NaOH(s) + CO_2(g)$
(b) $H_2O(\ell) + SO_3(g) \rightarrow H_2SO_4(\ell)$
(c) $H_2O(g) + SO_3(g) \rightarrow H_2SO_4(\ell)$

5.82 ■ Use data in Appendix G to find the enthalpy of reaction for
(a) $CaCO_3(s) \rightarrow CaO(s) + CO_2(g)$
(b) $2HI(g) + F_2(g) \rightarrow 2HF(g) + I_2(s)$
(c) $SF_6(g) + 3H_2O(\ell) \rightarrow 6HF(g) + SO_3(g)$

5.83 Calculate $\Delta H°$ when a 38-g sample of glucose, $C_6H_{12}O_6(s)$, burns in excess $O_2(g)$ to form $CO_2(g)$ and $H_2O(\ell)$ in a reaction at constant pressure and 298.15 K.

5.84 Calculate the amount of heat evolved or absorbed when a 0.2045-g sample of acetylene, $C_2H_2(g)$, burns in excess oxygen to form $CO_2(g)$ and $H_2O(\ell)$ in a reaction at constant pressure and 298.15 K.

5.85 The octane number of gasoline is based on a comparison of the gasoline's behavior with that of 2,2,4-trimethylpentane, $C_8H_{18}(\ell)$, which is arbitrarily assigned an octane number of 100. The standard enthalpy of combustion of this compound is −5456.6 kJ/mol.
(a) Write the thermochemical equation for the combustion of 2,2,4-trimethylpentane.
(b) Use the standard enthalpies of formation in Appendix G to calculate the standard enthalpy of formation of 2,2,4-trimethylpentane.

5.86 One of the components of jet engine fuel is *n*-dodecane, $C_{12}H_{26}(\ell)$, which has a standard enthalpy of combustion of −8080.1 kJ/mol.
(a) Write the thermochemical equation for the combustion of *n*-dodecane.
(b) Use the standard enthalpies of formation in Appendix G to calculate the standard enthalpy of formation of *n*-dodecane.

Chapter Exercises

5.87 A fission nuclear reactor produces about 8.1×10^7 kJ of energy for each gram of uranium consumed. One kilogram of high-grade coal produces about 2.8×10^4 kJ of energy when it is burned.
(a) How many metric tons (1 metric ton = 1000 kg) of coal must be burned to produce the same energy as produced by the fission of 1 g uranium?
(b) How many kilograms of sulfur dioxide are produced from the burning of the coal in part (a), if the coal is 0.90% by mass sulfur?
(c) ✎ Compare the environmental hazards of approximately 1 g radioactive waste with those of the sulfur dioxide produced by the burning coal to produce the same amount of energy.

5.88 Propane, $C_3H_8(g)$, and *n*-octane, $C_8H_{18}(\ell)$, are important components of the widely used fossil fuels. The enthalpy change for combustion of 1 mol propane is −2219 kJ, and that for 1 mol octane is −5466 kJ. Calculate the enthalpy change per gram for each compound.

5.89 When a 2.30-g sample of magnesium dissolves in dilute hydrochloric acid, 16.25 kJ of heat is released. Determine the enthalpy change for the thermochemical equation

$$Mg(s) + 2HCl(aq) \rightarrow MgCl_2(aq) + H_2(g) \qquad \Delta H = ?$$

5.90 ▲ A 1:1 mole ratio of CO(g) and H_2(g) is called *water gas.* It is used as a fuel because it can be burned in air:

$$2CO(g) + O_2(g) \rightarrow 2CO_2(g) \qquad \Delta H = -566 \text{ kJ}$$

$$2H_2(g) + O_2(g) \rightarrow 2H_2O(\ell) \qquad \Delta H = -571.7 \text{ kJ}$$

(a) Find the number of moles of CO(g) and H_2(g) present in 10.0 g water gas. (Remember that they are present in a 1:1 mole ratio.)
(b) Use the preceding thermochemical equations to find the enthalpy change when 10.0 g water gas is burned in air.

5.91 What mass of ethylene, C_2H_4(g), must be burned to produce 3420 kJ of heat, given that its enthalpy of combustion is −1410.1 kJ/mol?

5.92 What mass of acetylene, C_2H_2(g), must be burned to produce 3420 kJ of heat, given that its enthalpy of combustion is −1301 kJ/mol? Compare this with the answer to Exercise 5.91 and determine which substance produces more heat per gram.

5.93 It takes 677 J of heat to increase the temperature of 25.0 g liquid ethanol (C_2H_5OH) from 23.5 °C to 34.7 °C. What is the specific heat of this substance?

5.94 100.0 J of heat is added to a 3.45-g sample of an unknown metal. The temperature of the metal increases from 22.37 °C to 54.58 °C. Use the date in Table 5.1 to identify the metal.

5.95 ▲ When 50.0 g water at 41.6 °C was added to 50.0 g water at 24.3 °C in a calorimeter, the temperature increased to 32.7 °C. When 4.82 g $KClO_3$(s) was added to 100.0 g water in the calorimeter (at 24.3 °C), the temperature decreased to 20.6 °C.
(a) What is the heat capacity of this calorimeter?
(b) What is the enthalpy of solution of $KClO_3$(s) in kJ/mol?

5.96 A typical waterbed measures 84 in. × 60 in. × 9 in. How many kilocalories are required to heat the water in the waterbed from 55 °F (cold water from the faucet) to 85 °F, the operating temperature of the waterbed?

5.97 The enthalpy of combustion of liquid *n*-hexane, C_6H_{14}, is −4159.5 kJ/mol, and that of gaseous *n*-hexane is −4191.1 kJ/mol. Use Hess's law to determine ΔH for the vaporization of 1 mol of *n*-hexane:

$$C_6H_{14}(\ell) \rightarrow C_6H_{14}(g)$$

5.98 What is ΔH_{rxn} for reaction of iron(III) oxide and carbon monoxide to give iron metal and carbon dioxide gas? Use the following reactions:

$$4Fe(s) + 3O_2(g) \rightarrow 2Fe_2O_3(s) \qquad \Delta H = -1648.4 \text{ kJ}$$

$$2CO(g) + O_2(g) \rightarrow 2CO_2(g) \qquad \Delta H = -565.98 \text{ kJ}$$

5.99 Cyclopropane, C_3H_6(g), is a flammable compound that has been used in the past as an anesthetic. It has an enthalpy of combustion of −2091 kJ/mol.
(a) Write the thermochemical equation for the combustion of cyclopropane.
(b) Use the standard enthalpies of formation in Appendix G to calculate the standard enthalpy of formation of cyclopropane.

Cumulative Exercises

5.100 Ammonium nitrate, a common fertilizer, has been used by terrorists to construct car bombs. The products of the explosion of ammonium nitrate are nitrogen gas, oxygen gas, and water vapor. ΔH_f° for ammonium nitrate is −87.37 kcal/mol.
(a) Write the balanced chemical reaction for the decomposition of ammonium nitrate.
(b) How many moles of gas are produced if 1.000 kg NH_4NO_3 is reacted?
(c) How many kilojoules of energy are released per pound (453.6 g) of ammonium nitrate?

5.101 ▲ In the 1880s, Frederick Trouton noted that the enthalpy of vaporization of 1 mol pure liquid is approximately 88 times the boiling point, T_b, of the liquid on the Kelvin scale. This relationship is called **Trouton's rule** and is represented by the thermochemical equation

$$\text{liquid} \rightarrow \text{gas} \qquad \Delta H = 88 \cdot T_b \text{ joules}$$

Combined with an empirical formula from chemical analysis, Trouton's rule can be used to find the molecular formula of a compound, as illustrated here. A compound that contains only carbon and hydrogen is 85.6% C and 14.4% H. Its enthalpy of vaporization is 389 J/g, and it boils at a temperature of 322 K.
(a) What is the empirical formula of this compound?
(b) Use Trouton's rule to calculate the approximate enthalpy of vaporization of one mole of the compound. Combine the enthalpy of vaporization per mole with that same quantity per gram to obtain an approximate molar mass of the compound.
(c) Use the results of parts (a) and (b) to find the molecular formula of this compound. Remember that the molecular mass must be exactly a whole-number multiple of the empirical formula mass, so considerable rounding may be needed.

5.102 ▲ (See Exercise 5.101 for an explanation of Trouton's rule.) A compound that contains only carbon, hydrogen, and oxygen is 54.5% C and 9.15% H. Its enthalpy of vaporization is 388 J/g, and it boils at a temperature of 374 K.
(a) What is the empirical formula of this compound?
(b) Use Trouton's rule to calculate the approximate enthalpy of vaporization of one mole of the compound. Combine the enthalpy of vaporization per mole with that same quantity per gram to obtain an approximate molar mass of the compound.
(c) Use the information in parts (a) and (b) to find the molecular formula of this compound. Remember that the molecular mass must be exactly a whole-number multiple of the empirical formula mass, so considerable rounding may be needed.

5.103 The price of silver is \$16.74 per troy ounce at this writing (1 troy oz = 31.10 g).
(a) Calculate the cost of 1 mol silver.
(b) How much heat is needed to increase the temperature of \$1000.00 worth of silver from 15.0 °C to 99.0 °C? $C_s(Ag) = 0.235$ J/g · °C.

PhotoSpin, Inc/Alamy

5.104 ▲ The law of Dulong and Petit states that the heat capacity of metallic elements is approximately 25 J/mol · °C at 25 °C. In the 19th century, scientists used this relationship to obtain approximate atomic masses of metals, from which they determined the formulas of compounds. Once the formula of a compound of the metal with an element of known atomic mass is known, the mass percentage composition of the compound is used to find the atomic mass of the metal. The following example shows the calculations involved.
(a) Experimentally, the specific heat of a metal is found to be 0.24 J/g · °C. Use the law of Dulong and Petit to calculate the approximate atomic mass of the metal.
(b) An oxide of this element is 6.90% oxygen by mass. Use the molar mass of 16.00 g/mol for oxygen and the approximate atomic mass found in part (a) to determine the subscripts x and y in the formula of the oxide, M_xO_y. (The mole ratio of the elements you find will not be exactly whole numbers, so considerable rounding is needed to obtain whole numbers in the formula.)
(c) From the formula established in part (b), x mol M are combined with y mol O. Calculate the mass of the metal that is combined with y mol O, using the percent composition of the oxide, and find the atomic mass of the metal. What is the element M?

5.105 ▲ See Exercise 5.104 for a description of the law of Dulong and Petit.
(a) Experimentally, the specific heat of a metal is found to be 0.460 J/g · °C. Use the law of Dulong and Petit to calculate the approximate atomic mass of the metal.
(b) A chloride of this element is 67.2% chlorine by mass. Use the molar mass of 35.45 g/mol for chlorine and the approximate atomic mass found in part (a) to determine the subscripts x and y in the formula of the chloride, M_xCl_y. (The mole ratio of the elements you find will not be exactly whole numbers, so considerable rounding may be needed to obtain whole numbers in the formula.)
(c) From the formula established in part (b), x mol M is combined with y mol Cl. Calculate the mass of the metal that is combined with y mol chlorine, using the percent composition of the chloride, and find the atomic mass of the metal. What is the element M?

5.106 A compound is 82.7% carbon and 17.3% hydrogen, and has a molar mass of approximately 60 g/mol. When 1.000 g of this compound burns in excess oxygen, the enthalpy change is −49.53 kJ.
(a) What is the empirical formula of this compound?
(b) What is the molecular formula of this compound?
(c) What is the standard enthalpy of formation of this compound?
(d) Two compounds that have this molecular formula appear in Appendix G. Which one was used in this exercise?

5.107 ■ When wood is burned we may assume that the reaction is the combustion of cellulose (empirical formula, CH_2O).

$$CH_2O(s) + O_2(g) \rightarrow CO_2(g) + H_2O(g) \qquad \Delta H^\circ = -425 \text{ kJ}$$

How much energy is released when a 10-lb wood log burns completely? (Assume the wood is 100% dry and burns via the reaction above.)

5.108 ■ You want to heat the air in your house with natural gas (CH_4). Assume your house has 275 m^2 (about 2960 ft^2) of floor area and that the ceilings are 2.50 m from the floors. The air in the house has a molar heat capacity of 29.1 J/mol K. (The number of moles of air in the house can be found by assuming that the average molar mass of air is 28.9 g/mol and that the density of air at these temperatures is 1.22 g/L.) What mass of methane do you have to burn to heat the air from 15.0 °C to 22.0 °C?

SCUBA Diving. Underwater divers use pressurized air tanks and breathing masks.

The interaction between gases and human organs and tissues has profound impact on health and well-being. One extreme example results from the fact that divers in lakes and oceans must take some atmosphere with them so they can breathe underwater; otherwise, they would be tied to the water's surface. Divers use SCUBA gear (*s*elf-*c*ontained *u*nderwater *b*reathing *a*pparatus) to carry air with them.

"Normal" air is a mixture of various gases, mostly nitrogen (approximately 78% by volume) and oxygen (approximately 21% by volume). Unfortunately, air with this concentration of nitrogen and oxygen can be used only for diving to depths up to ~50 m (~150 ft). The high pressures caused by water at depths greater than 50 m starts to force more nitrogen gas to dissolve into the bloodstream and other tissues. This leads to a state of motor function loss, decision-making inability, and impairment in judgment known as *nitrogen narcosis.* There is also danger from the *bends,* a condition in which, as a diver ascends toward the surface and the surrounding water pressure lessens, nitrogen bubbles come out of the body tissues. These bubbles collect in the joints, causing

The Gaseous State

6

CHAPTER CONTENTS

OWL Online homework for this chapter may be assigned in OWL.

Look for the green colored vertical bar throughout this chapter, for integrated references to this chapter introduction.

extreme pain and the body to curl up (hence the name). You might think that divers could avoid these afflictions altogether by diving with air that has a greater concentration of oxygen. Ironically, diving with pure oxygen is dangerous at depths more than ~10 feet because oxygen actually becomes toxic at high pressures.

Deep-sea divers use an air mix that contains a substantial amount of helium gas, because helium does not dissolve in body tissues to a large extent. For depths between about 60 and 100 m (200–300 ft), *heliox* can be used. Heliox is a mixture of 30% O_2 and 70% He. At depths deeper than 100 m, heliox may cause *high-pressure nervous syndrome,* which can cause uncontrollable shaking. The root cause of high-pressure nervous syndrome remains unclear; however, scientists have found that adding nitrogen to heliox allows divers to dive deeper than 100 m. This combination of O_2, He, and N_2 is called *trimix.* Trimix is a mixture of about 10% O_2, 20% N_2, and 70% He. The presence of both nitrogen and helium seems to counteract each other's effects on the body, and depths of more than 130 m (400 ft) can be attained. ▮

Figure 6.1 Steel production. The initial process in the production of steel results in an impure substance called *pig iron.* Oxygen is injected into the molten pig iron to remove impurities, particularly the carbon.

This chapter discusses the behavior and properties of gases. Matter commonly exists on the earth in three physical states: solid, liquid, and gas. We come into contact with gases every day. The atmosphere is a sea of gas, consisting mainly of nitrogen and oxygen. Other gases are present in the atmosphere at low concentrations, and some of them are important to life. Carbon dioxide, CO_2, is necessary for the survival of plants, but it has become evident to most people that an increase in the concentration of CO_2 in the air is contributing to global warming. Another important atmospheric gas is ozone, O_3. This highly toxic gas is a source of air pollution at ground level. However, at high altitudes, where it acts to absorb dangerous radiation from the sun, O_3 is beneficial to our health.

Many gases are important in industrial processes. More than 60 billion pounds of nitrogen and 45 billion pounds of oxygen are produced for sale in the United States each year. Most of the nitrogen is converted to another important gas, ammonia (NH_3), for use in the production of fertilizers and plastics. Oxygen is used in hospitals, in the production of steel (Figure 6.1) and other metals, and in the propulsion of NASA space shuttles.

Natural gas, which is mainly methane (CH_4) formed by the decay of plants, is trapped underground. People use it to heat homes and water, for cooking, and to manufacture hydrogen gas. Hydrogen is used widely in industry, and in combination with oxygen provides propulsion for the NASA space shuttles. Other gases are manufactured or separated from crude oil. An example is ethylene, C_2H_4, which has many applications, including the production of the plastic polyethylene.

Because gases play such important roles in industry and in our everyday lives, it is important to know how they behave when their conditions, such as temperature or pressure, are modified. Gases have similar physical behaviors, which allows us to develop models to predict their properties. We also present a model that explains the behavior of gases on the molecular level.

6.1 Properties and Measurements of Gases

OBJECTIVES

- ☐ Describe the characteristics of the three states of matter: solid, liquid, and gas
- ☐ Define the pressure of a gas and know the units in which it is measured

The distinctions among the gas, liquid, and solid phases are readily apparent when physical properties are observed. A **gas** is a fluid with no definite shape or fixed volume; it fills the total volume of its container. When a gas expands, the volume of the empty space between gas particles changes. Because a gas is mostly empty space, a gas is also *compressible;* the volume of a gas sample decreases when an external force is applied. A **liquid** is a fluid with a fixed volume but no definite shape. Like a gas, a liquid takes the shape of its container, but a liquid has definite volume and does not expand to fill the container. A **solid** has both fixed shape and fixed volume. Liquids and solids are **condensed phases**—that is, phases that are resistant to volume changes because the spaces between the particles are small and cannot readily change. Figure 6.2 shows how diatomic molecules of bromine are arranged in each of the three states.

Gases expand to fill a container, but samples of liquids and solids have fixed volumes.

Because the individual particles in both the liquid and solid phases are closely packed, but in the gas phase are separated, the density of the gas phase is much lower than the density of either of the condensed phases. Density is generally expressed in grams per liter (g/L) for a gas, but the densities of liquids and solids are expressed in grams per milliliter (g/mL). When a gas under atmospheric conditions condenses to a solid or a liquid, the density increases by a factor of about 1000.

Pressure of a Gas

Pressure is defined as the force exerted on a surface divided by the area of the surface. The atmosphere, a sea of gas more than 10 miles high, exerts a pressure because of the weight of the gas molecules in the air. We generally do not notice this pressure because it surrounds everything equally, but if you change altitude rapidly, you can feel your ears "pop" because the pressure on the inner side of the eardrum changes more slowly than

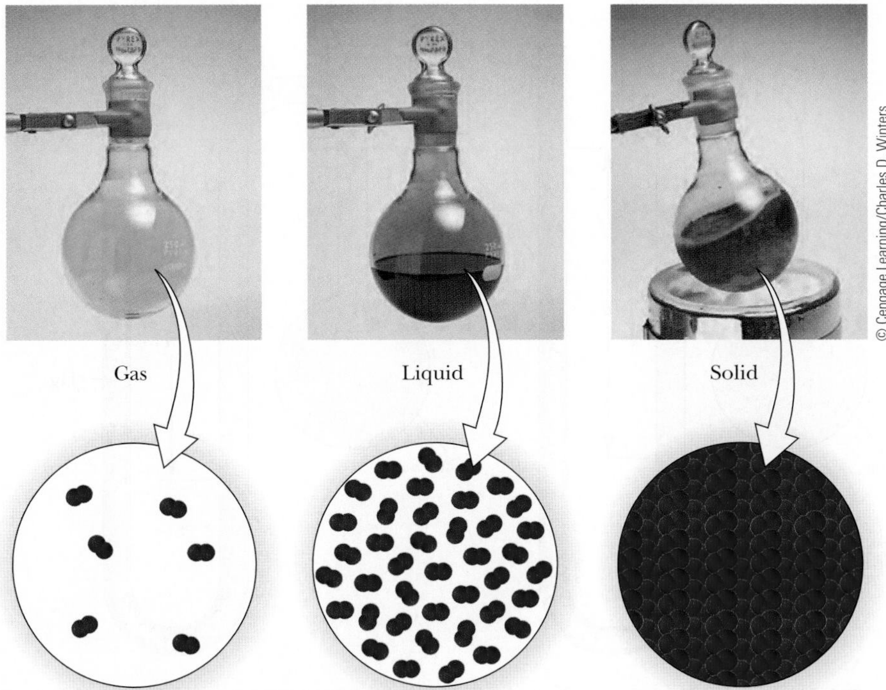

Figure 6.2 Gas, liquid, and solid phases of bromine. The gas phase has neither definite volume nor fixed shape. The liquid phase has a definite volume but not a fixed shape. The solid phase has fixed shape and volume.

the outer pressure. At high altitudes, on a mountain, for example, the pressure of the atmosphere is lower than at sea level because up high there are fewer gas molecules above you.

A barometer (Figure 6.3) measures the pressure of the atmosphere. A long glass tube, sealed at one end, is filled completely with mercury and inverted into an open dish of mercury. Gravitational attraction pulls down the column of mercury, leaving a vacuum above it in the tube. The column of liquid stops falling when the pressure caused by the weight of the mercury in the column is equal to the pressure exerted by the atmosphere on the surface of the mercury in the dish. Measuring the height of the mercury column is a method to determine the atmospheric pressure. At sea level, the mercury column is about 760 mm high on an average day. If the mercury level in the barometer rises, the weather forecaster reports high pressure; if the atmospheric pressure is low, the mercury level in the tube decreases. Mercury is used in barometers because it is a liquid with a high density, 13.6 g/mL. When water is used in a barometer, the column of water is more than 10 m high.

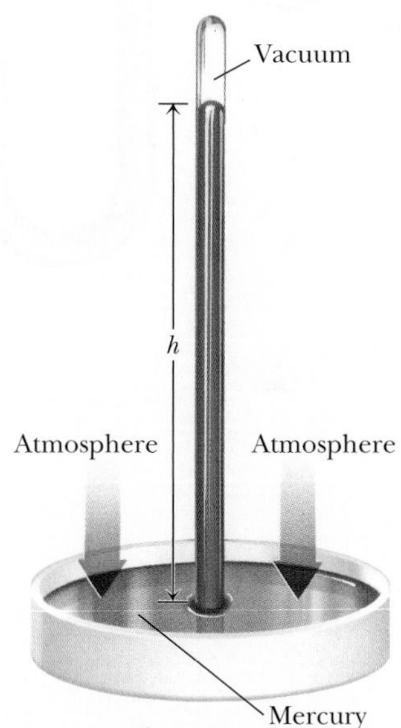

Figure 6.3 Barometer. The pressure exerted by the atmosphere supports a column of mercury. The height of the column is used to measure the pressure of the atmosphere.

A manometer measures pressure *differences.* Figures 6.4*a* and 6.4*b* show open-end manometers. A U-shaped tube containing mercury is connected to the container of a gas sample. The atmosphere exerts a pressure on the mercury surface at the open end of the tube, and the gas within the container exerts pressure on the other surface of the mercury. The difference between the heights of the two mercury surfaces corresponds to the difference between the gas pressure in the container and the *atmospheric pressure.* The mercury column is lower on the end of the U-tube that experiences the greater pressure.

The pressure exerted by a gas is measured with a barometer or a manometer.

Figure 6.4*c* shows a closed-end manometer, generally used to measure low gas pressures. In this case, one end of the U-tube is evacuated and sealed. The pressure of the gas is *equal* to the difference between the heights of the two mercury surfaces.

Units of Pressure Measurement

The SI unit of pressure is the pascal (Pa), named for the French scientist Blaise Pascal (1623–1662):

$$1 \text{ Pa} = 1 \text{ N/m}^2 = \frac{1 \text{ kg}}{\text{m} \cdot \text{s}^2}$$

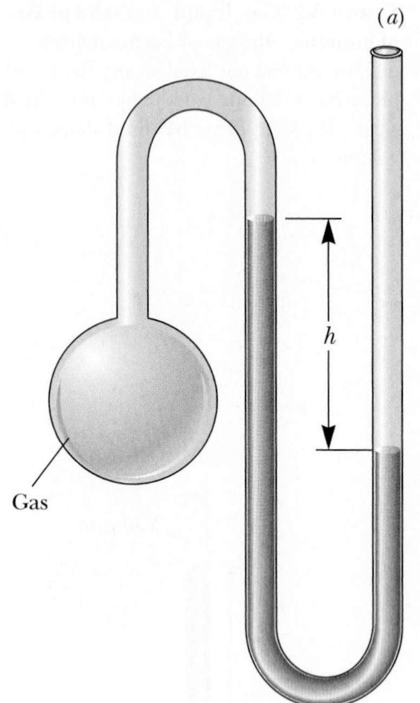

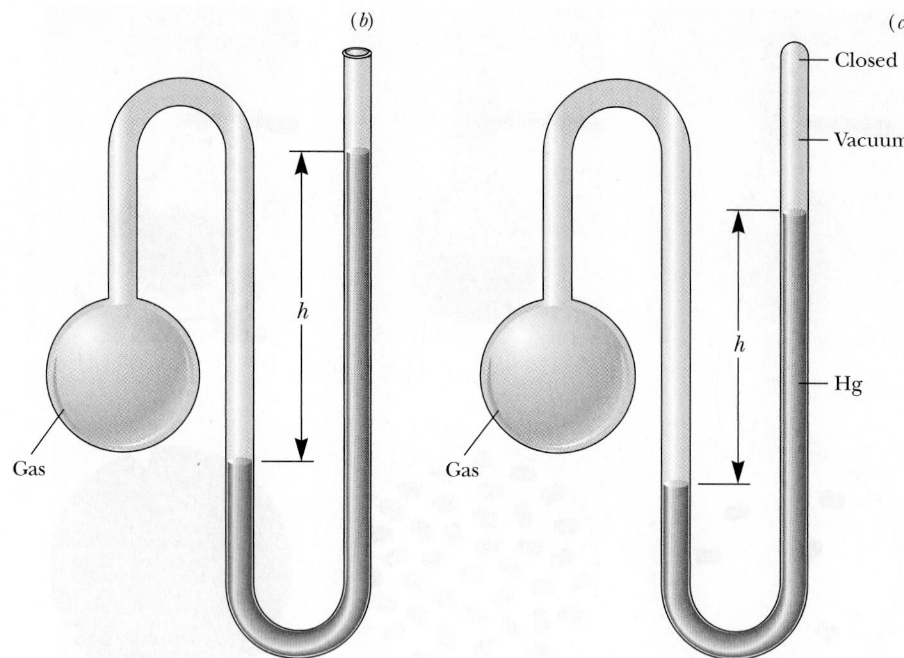

Figure 6.4 Manometer. In an open-end manometer *(a, b)*, the difference between the heights of the mercury surfaces *(h)* in a U-tube measures the *difference* between the pressure of a gas sample and *atmospheric pressure. (a)* The pressure of the gas in the container is less than atmospheric. *(b)* The pressure of the gas in the container is greater than atmospheric. *(c)* In a closed-end manometer, the pressure of the gas is *equal* to the difference between the heights of the two mercury surfaces.

where N is the newton, the SI unit for force ($1\ N = 1\ kg \cdot m/s^2$), m is the meter, and s is the second. This unit of pressure is quite small for experiments typically conducted by chemists. A related unit is the bar:

$$1\ \text{bar} = 10^5\ \text{Pa}$$

Whereas the pascal and bar may be the SI-defined units of pressure, other units are commonly used, two of which are based on the mercury barometer and the manometer. One atmosphere (1 atm) of pressure is the average pressure of the atmosphere at sea level, and is now defined as the pressure exerted by a column of mercury exactly 760 mm high:

$$1\ \text{atm} = 760\ \text{mm Hg} = 101.325\ \text{kPa}$$

Another name for the unit millimeters of mercury (mm Hg) is the *torr,* so

$$1\ \text{torr} = 1\ \text{mm Hg}$$

The torr is named for the inventor of the barometer, Evangelista Torricelli (1608–1647), an Italian scientist who studied under Galileo.

The pressure unit in the English system of measurement, pounds per square inch (psi), is used in many engineering applications. This text generally uses torr and atmosphere to express pressure. Table 6.1 shows important relationships needed to convert between various pressure units.

TABLE 6.1 Relationships between Pressure Units

Relationship
1 atm = 760 mm Hg
1 torr = 133.3 Pa
1 atm = 760 torr
1 atm = 14.7 psi
1 atm = 101.325 kPa
1 atm = 1.01325 bar
1 atm = 29.92 in. Hg

EXAMPLE 6.1 Converting among Pressure Units

Express a pressure of 0.450 atm in the following units:

(a) torr (b) kPa

Strategy Use the relationship 1 atm = 760 torr to determine the pressure in torr and the equality 1 atm = 101.3 kPa to determine the pressure in kilopascals (kPa).

Solution

(a) $\text{Pressure} = 0.450 \cancel{\text{atm}} \times \left(\frac{760 \text{ torr}}{1 \cancel{\text{atm}}}\right) = 342 \text{ torr}$

(b) $\text{Pressure} = 0.450 \cancel{\text{atm}} \times \left(\frac{101.325 \text{ kPa}}{1 \cancel{\text{atm}}}\right) = 45.6 \text{ kPa}$

Understanding

Express a pressure of 433 torr in atmospheres.

Answer 0.570 atm

OBJECTIVES REVIEW *Can you:*

- ☑ describe the characteristics of the three states of matter?
- ☑ define the pressure of a gas and the units in which it is measured?

6.2 Gas Laws

OBJECTIVE

- ☐ Determine how a gas sample responds to changes in volume, pressure, moles, and temperature

The results of experiments performed over centuries led to a remarkable conclusion: *The physical properties of all gases behave in the same general manner, regardless of the identity of the gas.* Careful analysis demonstrates that four independent properties define the physical state of a gas: pressure *(P)*, volume *(V)*, temperature *(T)*, and number of moles *(n)*. A change in any one of these properties influences the others. To illustrate these interrelationships, we will examine the results of experiments in which the change in the volume of a gas will be measured as any one of the other three properties is varied and the remaining two are held constant. Remember that these relationships, known as **gas laws,** apply to the gas phase only.

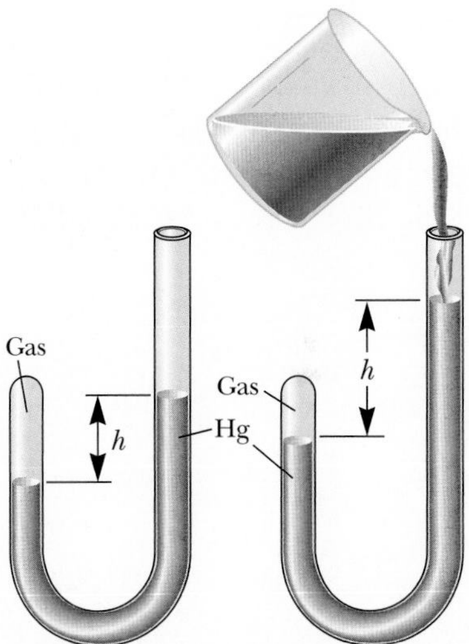

Figure 6.5 Change in volume of a gas with a change in pressure. Increasing the pressure on a sample of gas caused by the addition of mercury to the right side of the tube decreases the volume of the gas occupied by the sample.

Volume and Pressure: Boyle's Law

Figure 6.5 shows an experiment that determines how changes in pressure, measured with a manometer, influence the volume of a gas sample. The experimenter increases the pressure on the sample of gas by adding mercury to the open end of the manometer while temperature is constant. The pressure of the gas sample in the closed end of the tube is equal to atmospheric pressure plus the difference in height (*h*) of the mercury surfaces. The experiment shows that the volume of the gas decreases as the pressure increases.

A plot of the volume measured in this experiment, as a function of the inverse of the pressure, is a straight line (Figure 6.6). An Irish chemist, Robert Boyle (1627–1691), was the first to note the mathematical description of this relationship. **Boyle's law** states that at constant temperature, the volume of a sample of gas is inversely proportional to the pressure. In equation form,

$$V = \text{constant} \times \frac{1}{P}$$

where the constant is the slope of the line in Figure 6.6, which is dependent on the temperature and the amount of matter in the gas sample—the values of the two properties that were held constant in this experiment.

Boyle's law can be rewritten as

$$PV = \text{constant}$$

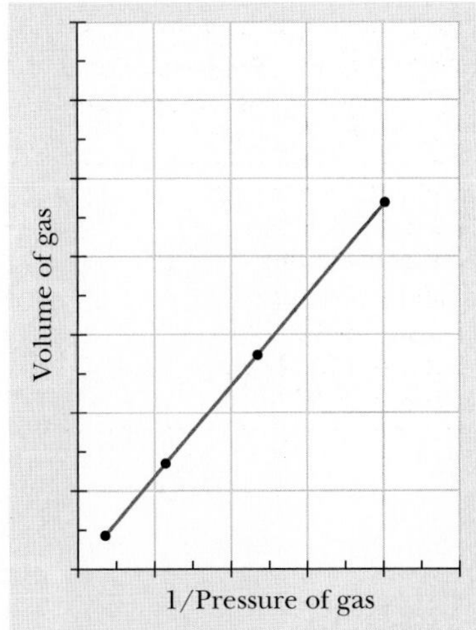

Figure 6.6 A plot of volume versus the inverse of pressure. The volume of a gas is proportional to the reciprocal of the pressure.

which states that the product of the pressure and volume is constant (as long as temperature and amount remain constant as well). If pressure or volume were to change, the other variable would have to change in concert so that the product PV remains constant. If we use the subscripts 1 and 2 to indicate the initial and changed pressure and volume, then

$$P_1V_1 = \text{constant} = P_2V_2$$

A more concise way of writing this is:

$$P_1V_1 = P_2V_2 \qquad [6.1]$$

Pressure × volume = constant, if the temperature and amount are held constant.

This equation is another form of Boyle's law. Using this expression, you can predict what will happen to the pressure or the volume of a gas if its volume or pressure changes. Note that this law applies to any substance that is in the gas phase, as well as mixtures of gases.

EXAMPLE 6.2 Using Boyle's Law

A sample of argon gas at an initial pressure of 1.35 atm and an initial volume of 18.5 L is compressed to a final pressure of 3.89 atm. What is the final volume of the argon? Assume temperature and amount remain constant.

Strategy Use Equation 6.1 and solve for V_2.

Solution

First, list the information given in the problem.

	Pressure	Volume
Initial	$P_1 = 1.35$ atm	$V_1 = 18.5$ L
Final	$P_2 = 3.89$ atm	$V_2 = ?$

Solve Boyle's law for the unknown variable V_2; then substitute the values from the above table:

$$V_2 = \frac{V_1P_1}{P_2}$$

$$V_2 = \frac{(1.35\ \cancel{\text{atm}})(18.5\ \text{L})}{3.89\ \cancel{\text{atm}}}$$

$$V_2 = 6.42\ \text{L}$$

Note that the units of atmosphere cancel out, leaving the volume unit of liter for the correct answer.

Understanding

A balloon containing 575 mL nitrogen gas at a pressure of 1.03 atm is compressed to a final volume of 355 mL. What is the resulting pressure of the nitrogen?

Answer 1.67 atm

It is crucial to express both pressures or both volumes in the same units when applying Boyle's law. In most cases, it does not matter which unit is used to express pressure or volume, as long as the same units are used for initial and final conditions. The following example illustrates this application.

EXAMPLE 6.3 Pressure and Volume Changes

In the lungs of a deep-sea diver ($V = 6.0$ L) at a depth of 100 m, the pressure of the air is 7400 torr. At a constant temperature of 37 °C, to what volume would the air expand if the diver were immediately brought to the surface (1.0 atm)?

Strategy After making sure the units are consistent, use Boyle's law to derive the algebraic equation that relates pressure and volume.

Solution

List the information given in the problem.

	Pressure	Volume
Initial	$P_1 = 7400$ torr	$V_1 = 6.0$ L
Final	$P_2 = 1.0$ atm	$V_2 = ?$

Here, pressure values are given in two different units. We need to convert one quantity to a different unit. Let us convert the P_1 to units of atm (we could have just as easily converted P_2 to torr):

$$7400 \text{ torr} \times \left(\frac{1 \text{ atm}}{760 \text{ torr}}\right) = 9.74 \text{ atm}$$

Solve Boyle's law for the unknown variable V_2; then substitute the values from the above table, using the converted value for P_1:

$$V_2 = \frac{P_1 \times V_1}{P_2} = \frac{(9.74 \text{ atm})(6.0 \text{ L})}{1.0 \text{ atm}} = 58 \text{ L}$$

Clearly, the diver needs to expel gas when rising to the surface—58 L is a much larger volume than the lungs can hold.

Understanding

At a pressure of 740 torr, a sample of gas occupies 5.00 L. Calculate the volume of the sample if the pressure is changed to 1.00 atm at constant temperature.

Answer 4.87 L

Zac Macaulay/Getty Images

A deep-sea diver. A diver must rise from the bottom very slowly while breathing normally, to allow time to expel the excess air from the lungs. (This gas expansion is a separate situation from the better-known problem of the bends, which involves gases dissolved in blood.)

Volume and Temperature: Charles's Law

Figure 6.7 shows the effect of a change in temperature on the volume of a gas, with the pressure and amount of gas in the sample held constant. Heating the gas increases the volume.

Figure 6.8 is a plot of the experimentally determined volumes of three different samples of gas as the temperature varies. When the Kelvin scale is used to measure temperature, doubling the temperature causes the volume of the gas to double. A French chemist and balloonist, Jacques Charles (1746–1823), determined this relationship. **Charles's law** states that at constant pressure, the volume of a fixed amount of gas is proportional to the absolute temperature, or

$$V = \text{constant} \times T$$

The graphs in Figure 6.8 give the experimental basis for the development of the Kelvin temperature scale and describe one of the first measurements to suggest the existence of an absolute zero of temperature—a temperature that is the lowest possible that can be obtained. Charles's law indicates that at absolute zero the volume of the gas must be zero. Does matter disappear at absolute zero? No, all gases condense to the liquid or solid phase before they reach this temperature. Because the basis for the graph is the measurement of the volume of a *gas,* Charles's law no longer applies once the sam-

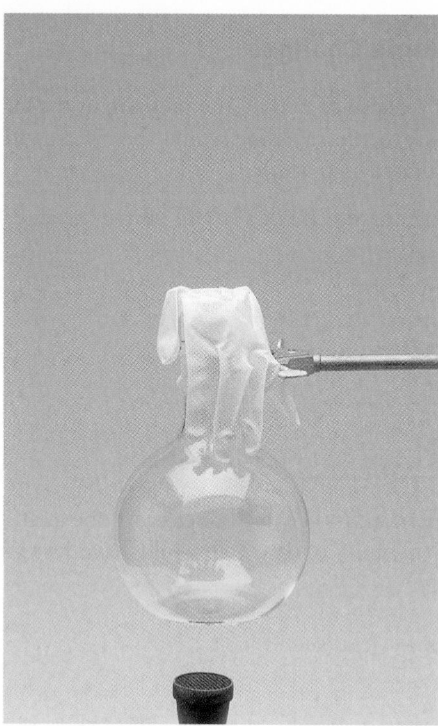

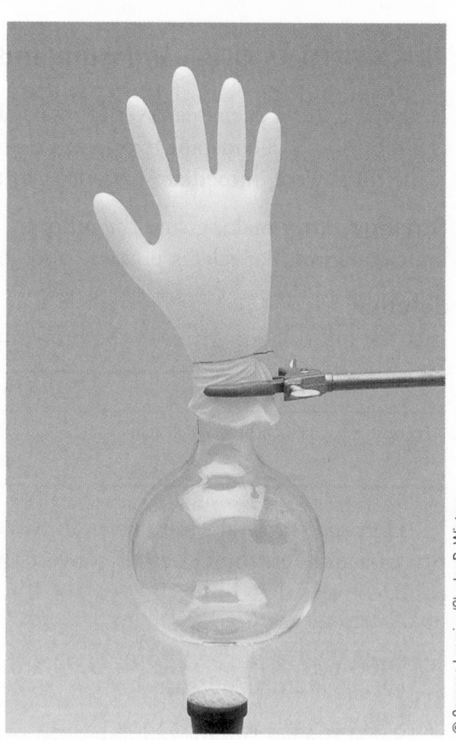

Figure 6.7 Heating a gas. Heating a sample of a gas causes the volume of the gas to increase when the pressure remains constant.

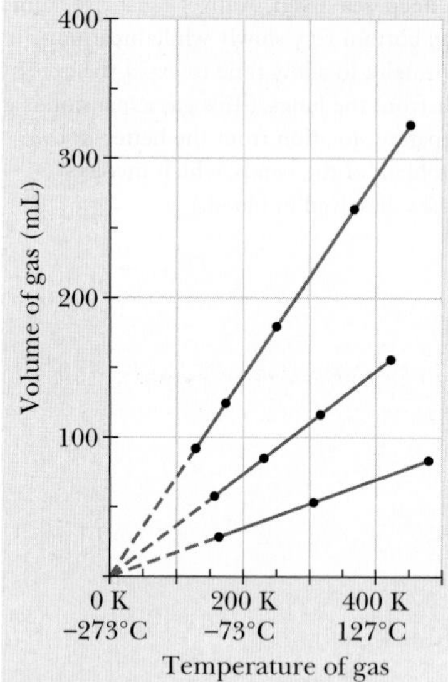

Figure 6.8 Plot of volume versus temperature. *Solid lines* connect experimentally determined volumes of three gas samples as temperature changes. *Dotted lines* are extensions of the experimental straight lines, taken to lower temperatures. These extensions all reach zero volume at −273 °C.

Volume = constant × temperature, if the pressure and amount are held constant.

ple becomes a liquid or a solid. Nevertheless, the graph can be extrapolated to zero volume, allowing the determination of the zero on the temperature scale. All three samples reach a volume of zero at the same temperature. This temperature, absolute zero, has the value −273.15 °C, which is the zero point of the Kelvin scale, as outlined in Chapter 1.

As with Boyle's law, Charles's law can be rewritten into a form that allows us to predict changes in the properties of a given sample of gas. The form of Charles's law above can be rewritten as

$$\frac{V}{T} = \text{constant}$$

If the volume or temperature of a given sample of a gas at constant pressure changes, the two sets of volume/temperature values can be related by the expression

$$\frac{V_1}{T_1} = \frac{V_2}{T_2} \qquad [6.2]$$

Equation 6.2 is another form of Charles's law. Remember, the temperature *must* be expressed in units of kelvins.

EXAMPLE 6.4 Temperature and Volume Changes

A balloon filled with oxygen gas at 25 °C occupies a volume of 2.1 L. Assuming that the pressure remains constant, what is the volume at 100 °C?

Strategy Convert the temperatures to kelvins and use Equation 6.2.

Solution

List the data given by the problem.

	Volume	Temperature
Initial	$V_1 = 2.1$ L	$T_1 = 25 + 273 = 298$ K
Final	$V_2 = ?$	$T_2 = 100 + 273 = 373$ K

Rearrange the equation to place only the unknown property on the left, and solve the problem by substituting the known values.

$$V_2 = \frac{V_1 \times T_2}{T_1} = \frac{(2.1\ \text{L})(373\ \cancel{\text{K}})}{298\ \cancel{\text{K}}} = 2.6\ \text{L}$$

As predicted by Charles's law, the volume increases as the temperature increases.

Understanding

The volume of a sample of nitrogen gas increases from 0.440 L at 27 °C to 1.01 L as it is heated to a new temperature. Calculate the new temperature of the nitrogen.

Answer 416 °C

Figure 6.9 Masses and moles of equal volumes of two gases. Identical flasks of hydrogen and nitrogen gas at the same temperature and pressure contain the same number of moles and molecules but have different masses.

Avogadro's Law and the Combined Gas Law

In 1811, Amedeo Avogadro proposed that at the same temperature and pressure, equal volumes of gases contain the same number of particles. Over several decades, scientists tested Avogadro's hypothesis and found it to be true within experimental error. The flasks in Figure 6.9 illustrate Avogadro's hypothesis for samples of hydrogen and nitrogen at normal temperature and pressure.

Avogadro's law states that at constant pressure and temperature, the volume of a gas sample is proportional to the number of moles of gas present.

$$V = \text{constant} \times n$$

Volume = constant × amount, if the pressure and temperature are held constant.

As with all three of the laws presented earlier, Avogadro's law applies to all gas samples. Figure 6.10 presents a graphic representation of Avogadro's law.

As with the previous gas laws, Avogadro's law can be written in a way that allows us to predict changes in the conditions of a gas. This form is

$$\frac{V_1}{n_1} = \frac{V_2}{n_2} \qquad [6.3]$$

Finally, for a given amount of gas (i.e., n is constant), the three remaining properties of a gas can be related by an expression called the **combined gas law:**

$$\frac{P_1V_1}{T_1} = \frac{P_2V_2}{T_2} \qquad [6.4]$$

This gas law can be used for a fixed amount of gas if the change in conditions involves more than one of the properties. Again, temperature must be expressed in kelvins, and the units of the two pressure quantities and the two volume quantities must be the same.

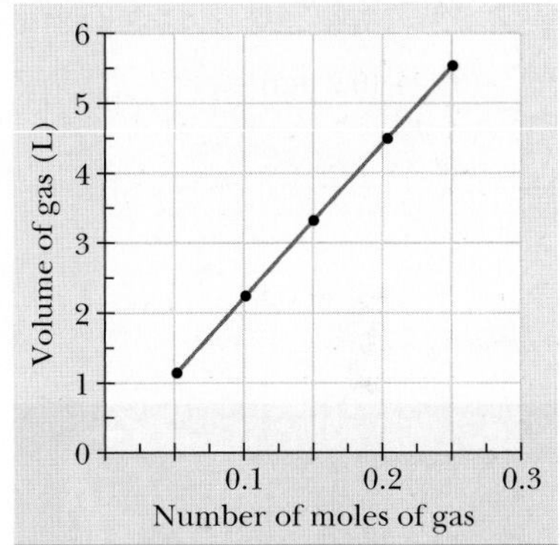

Figure 6.10 Plot of volume versus amount. The volume of a gas at constant pressure and temperature is directly proportional to the amount (number of moles of gas present).

EXAMPLE 6.5 Pressure, Volume, and Temperature Changes

A helium weather balloon is filled to a volume of 219 m³ on the ground, where the pressure is 754 torr and the temperature is 25 °C. As the balloon rises, the pressure and temperature decrease, so it is important to know how much the gas will expand to ensure that the balloon can withstand the expansion. What is the volume at an altitude of 10,000 m, where the atmospheric pressure is 210 torr and the temperature is −43 °C?

Strategy Verify that the units are appropriate (temperature must be in kelvins), and use the combined gas law to solve for the final volume.

Solution

List the values given in the problem. The temperatures must be converted to kelvins.

	Volume	Pressure	Temperature
Initial	$V_1 = 219\ \text{m}^3$	$P_1 = 754$ torr	$T_1 = 25 + 273 = 298$ K
Final	$V_2 = ?$	$P_2 = 210$ torr	$T_2 = -43 + 273 = 230$ K

PRACTICE OF CHEMISTRY
Internal Combustion Engine Cylinders

Internal combustion engines ultimately derive their power from gas pressure. Engines have cylinders with pistons inside them that can go up and down, and are connected to a crankshaft. Gasoline vapors and air are brought into the cylinder chamber and compressed. A spark plug ignites the flammable mixture, and the formation of gaseous products (mostly CO_2 and H_2O) at an elevated temperature creates a high pressure inside the cylinder, pushing the piston away. This motion ultimately provides the power to turn wheels and power the other systems of the automobile (or other vehicle).

An important application of gas laws is to calculate how much force is generated inside an engine cylinder. For example, suppose 300 cm³ of a gasoline/air mixture at 70 °C and a pressure of 0.967 atm is drawn into a cylinder and compressed to 31.5 cm³. What is the resulting pressure if the temperature of the gases after ignition and combustion is 350 °C?

First, list all of our data:

$V_1 = 300\ \text{cm}^3 \quad T_1 = 70 + 273 = 343\ \text{K} \quad P_1 = 0.967\ \text{atm}$

$V_2 = 31.5\ \text{cm}^3 \quad T_2 = 350 + 273 = 623\ \text{K} \quad P_2 = ?$

Now, use the combined gas law and solve for P_2:

$$P_2 = \frac{P_1V_1T_2}{T_1V_2} = \frac{(0.967\ \text{atm})(300\ \cancel{\text{cm}^3})(623\ \cancel{\text{K}})}{(343\ \cancel{\text{K}})(31.5\ \cancel{\text{cm}^3})}$$

$P_2 = 16.7$ atm

Pressure is force divided by area. If we know that a cylinder has a diameter of 2.80 inches, it can be calculated that the force generated inside each piston is equivalent to nearly a ton! Lest you be skeptical, be assured that realistic numbers were used in this example. This example demonstrates an important application of gas laws.

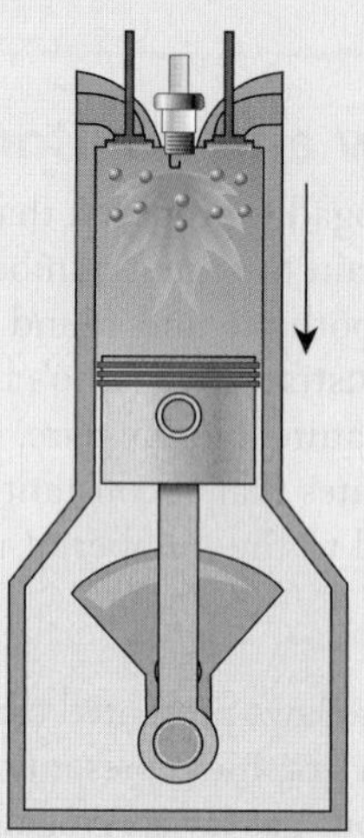

The high pressure of the gas in the internal combustion engine is used to convert the heat generated by burning the fuel into mechanical energy that propels an auto.

© Phil Dauber/Photo Researchers, Inc.

A weather balloon. Meteorologists use weather balloons to sample conditions in the upper atmosphere. They do not completely fill the balloons at launch because the helium expands as a balloon rises because of the decrease in pressure.

Rearrange the combined gas law (see Equation 6.4) to place only V_2 on the left side, and solve the problem by substituting the known values.

$$\frac{P_1V_1}{T_1} = \frac{P_2V_2}{T_2}$$

$$V_2 = \frac{P_1V_1T_2}{T_1P_2} = \frac{(754\ \cancel{\text{torr}})(219\ \text{m}^3)(230\ \cancel{\text{K}})}{(298\ \cancel{\text{K}})(210\ \cancel{\text{torr}})} = \boxed{607\ \text{m}^3}$$

The volume of the balloon nearly triples as it rises.

Understanding

The pressure of a sample of gas is 2.60 atm in a 1.54-L container at a temperature of 0 °C. Calculate the pressure exerted by this sample if the volume changes to 1.00 L and the temperature changes to 27 °C.

Answer 4.40 atm

OBJECTIVE REVIEW *Can you:*

☑ determine how a gas sample responds to changes in volume, pressure, moles, and temperature?

6.3 The Ideal Gas Law

OBJECTIVES

- ☐ Write the ideal gas law
- ☐ Calculate the pressure, volume, amount, or temperature of a gas, given values of the other three properties
- ☐ Calculate the molar mass and the density of gas samples by using the ideal gas law

Boyle's, Charles's, and Avogadro's laws—laws that apply to all gaseous samples—state how volume changes with changes in pressure, temperature, and number of moles, respectively:

$V = \text{constant} \times \frac{1}{P}$ or $P_1V_1 = P_2V_2$	Boyle's law
$V = \text{constant} \times T$ or $\frac{V_1}{T_1} = \frac{V_2}{T_2}$	Charles's law
$V = \text{constant} \times n$ or $\frac{V_1}{n_1} = \frac{V_2}{n_2}$	Avogadro's law

The volume of a gas sample is inversely proportional to the pressure and directly proportional to both the number of moles and the temperature (in kelvins).

We can combine these three laws into a single equation known as the **ideal gas law:**

$$PV = nRT \qquad [6.5]$$

where R is known as the ideal gas law constant. The value of the constant R is determined experimentally.

Measurements show that the volume of *1 mol* of an ideal gas at 273.15 K (0 °C) and 1.000 atm is 22.41 L. The conditions of 0 °C and 1 atm are known as **standard temperature and pressure (STP).** By substituting these values in the ideal gas equation, we calculate the value of R:

The ideal gas law expresses the interrelationships of volume, pressure, amount, and temperature.

$$R = \frac{PV}{nT} = \frac{(1.000 \text{ atm})(22.41 \text{ L})}{(1 \text{ mol})(273.1 \text{ K})} = 0.08206 \frac{\text{L} \cdot \text{atm}}{\text{mol} \cdot \text{K}}$$

As shown in Table 6.2, the numeric value of R depends on the units used to measure pressure and volume. All gases, such as H_2, O_2, and N_2, and mixtures of gases follow the ideal gas law at normal temperatures and pressures. We use the term *ideal* in the name because, as is outlined later, under certain conditions, the behavior of gases deviates from that predicted by the ideal gas law.

The ideal gas law relates the four independent properties of a gas (P, V, n, and T) as they exist at any point in time. The other gas laws introduced in the previous section require that one of the properties of a gas sample change: As volume changes, we can follow changes in pressure (at constant T) or temperature (at constant P). The ideal gas law does not require a change. It relates the properties of a gas at any instant, not over some change in conditions. These calculations require a value for R, so it is necessary to match the units used in R with the units used for pressure and volume, generally atmospheres and liters.

The ideal gas law is used to determine the value of any of the four properties—pressure, volume, amount, and temperature of a gas, given values of the other three.

TABLE 6.2 Values for the Ideal Gas Constant

R	Units
0.08206	$\frac{\text{L} \cdot \text{atm}}{\text{mol} \cdot \text{K}}$
8.314	$\frac{\text{kg} \cdot \text{m}^2}{\text{s}^2 \cdot \text{mol} \cdot \text{K}}$
8.314	$\frac{\text{J}}{\text{mol} \cdot \text{K}}$
1.987	$\frac{\text{cal}}{\text{mol} \cdot \text{K}}$

We can illustrate the procedure by calculating the number of moles in a sample of argon gas that occupies a volume of 298 mL at a pressure of 351 torr and a temperature of 25 °C. First, the known values must be converted to match the units used in R.

For volume, 298 mL = 0.298 L. Because 1 atm = 760 torr, the conversion of pressure to atmospheres is

$$\text{Pressure in atm} = 351 \cancel{\text{torr}} \times \left(\frac{1 \text{ atm}}{760 \cancel{\text{torr}}}\right) = 0.462 \text{ atm}$$

For temperature,

$$T_K = T_C + 273 = 25 + 273 = 298 \text{ K}$$

Rearrange the ideal gas law to place the unknown, the number of moles, on the left, and solve the equation by substituting the known quantities in the appropriate units.

$$PV = nRT$$

$$n = \frac{PV}{RT} = \frac{(0.462 \text{ atm})(0.298 \text{ L})}{(0.08206 \text{ L} \cdot \text{atm/mol} \cdot \text{K})(298 \text{ K})} = 5.63 \times 10^{-3} \text{ mol}$$

Note that most of the units cancel out, leaving moles, the correct unit for the answer. Always write and cancel units to ensure that you have used the proper ones and combined the properties correctly.

EXAMPLE 6.6 Pressure of a Gas

Calculate the pressure of a 1.2-mol sample of methane gas in a 3.3-L container at 25 °C.

Strategy Substitute the given values into the ideal gas law, being sure to match the units given in the ideal gas constant, R.

Solution

List the given values with the appropriate units.

$$V = 3.3 \text{ L} \qquad n = 1.2 \text{ mol} \qquad T = 25 + 273 = 298 \text{ K}$$

Rearrange the ideal gas law with pressure on the left, and solve the problem by substituting the known values.

$$PV = nRT$$

$$P = \frac{nRT}{V} = \frac{(1.2 \text{ mol})(0.0821 \text{ L} \cdot \text{atm/mol} \cdot \text{K})(298 \text{ K})}{3.3 \text{ L}} = 8.9 \text{ atm}$$

Understanding

Calculate the temperature of a 350-mL container that holds 0.620 mol of an ideal gas at a pressure of 42.0 atm.

Answer 289 K

Molar Mass and Density

Determination of molar mass is an important step in the identification of a new substance because, together with percentage composition, the molar mass is needed to establish the molecular formula. Before the development of mass spectrometry, the molar masses of many substances were determined by using the ideal gas law. When the number of moles (n) in a gas sample of known mass (m) is calculated with the ideal gas law, then the molar mass is found by dividing m grams by n moles, as shown in Example 6.7.

The molar mass of a gas can be determined by measuring temperature, pressure, and volume of a known mass of the gas.

$$\text{Molar mass} = \frac{m}{n}$$

EXAMPLE 6.7 Molar Mass

An experiment shows that a 0.495-g sample of an unknown gas occupies 127 mL at 98 °C and 754 torr pressure. Calculate the molar mass of the gas.

Strategy Use the data given in the problem to calculate moles, using the ideal gas law; then combine this result with the measured mass of the sample to calculate the molar mass.

Solution
List the measured values of temperature, pressure, and volume of the gas, with the correct units.

$T = 371$ K; $P = 0.992$ atm; $V = 0.127$ L

Use the ideal gas law to calculate the number of moles, n, of gas.

$$n = \frac{PV}{RT} = \frac{(0.992 \text{ atm})(0.127 \text{ L})}{(0.08206 \text{ L} \cdot \text{atm/mol} \cdot \text{K})(371 \text{ K})}$$

$$n = 4.14 \times 10^{-3} \text{ mol}$$

Use this number of moles and the mass of sample measured in the experiment (0.495 g) to calculate the molar mass.

$$\text{Molar mass} = \frac{m}{n} = \frac{0.495 \text{ g}}{4.14 \times 10^{-3} \text{ mol}}$$

$$\text{Molar mass} = 1.20 \times 10^{2} \ \frac{\text{g}}{\text{mol}}$$

Understanding

Calculate the molar mass of a gas if a 9.21-g sample occupies 4.30 L at 127 °C and a pressure of 342 torr.

Answer 156 g/mol

The density of any given gas under a fixed set of conditions is also calculated from the ideal gas law, as shown in Example 6.8. The density is important information related to properties such as the speed of sound and the thermal conductivity of a sample of gas.

EXAMPLE 6.8 Density of a Gas

What is the density of N_2 gas at 1.00 atm and 100 °C?

Strategy Density is mass per unit volume. The mass of 1 mol nitrogen is 28.0 g. Use the ideal gas law to calculate the volume of 1 mol of a gas under the given conditions.

Solution
First, calculate the volume of 1 mol N_2 under the given conditions using the ideal gas law.

$$V = \frac{nRT}{P} = \frac{(1 \text{ mol})(0.08206 \text{ L} \cdot \text{atm/mol} \cdot \text{K})(373 \text{ K})}{1.00 \text{ atm}}$$

$$V = 30.6 \text{ L}$$

Calculate the density from this value of the volume and the mass of 1 mol N_2.

$$d = \frac{\text{mass}}{\text{volume}} = \frac{28.0 \text{ g}}{30.6 \text{ L}} = 0.915 \ \frac{\text{g}}{\text{L}}$$

Understanding

Calculate the density of H_2 gas at 1.00 atm and 100 °C.

Answer 0.0659 g/L

The two calculations in Example 6.8 show that at constant pressure and temperature, the density of a gas is directly related to its molar mass. The density of H_2 is much lower than the density of N_2, because the volume occupied by 1 mol of each under fixed conditions is the same, but the masses of 1-mol samples of the two gases are quite different.

OBJECTIVES REVIEW *Can you:*

- ☑ write the ideal gas law?
- ☑ calculate the pressure, volume, amount, or temperature of a gas, given values of the other three properties?
- ☑ calculate the molar mass and the density of gas samples by using the ideal gas law?

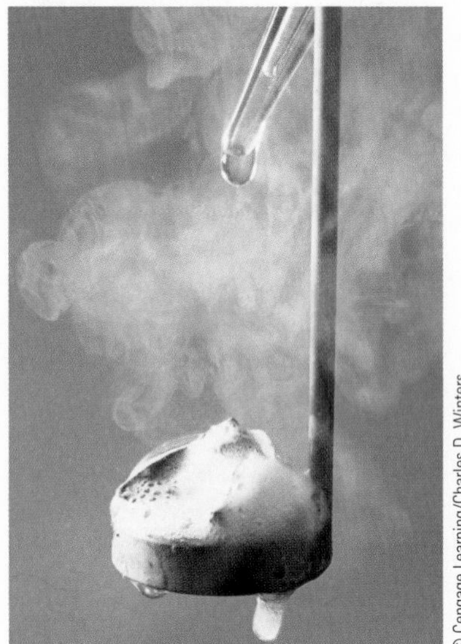

Reaction of Li with water produces hydrogen gas and LiOH.

6.4 Stoichiometry Calculations Involving Gases

OBJECTIVES

- ☐ Perform stoichiometric calculations for reactions in which some or all of the reactants or products are gases
- ☐ Use relative volumes of gases directly in equation stoichiometry problems

The reactants and products in chemical reactions are frequently gases. Just as in solution, reacting species in the gas phase can readily collide, a necessary requirement for reaction to occur. We can use the ideal gas law to determine the number of moles, *n*, for use in problems involving reactions in much the same way that we use molar mass for solids and molarity for compounds in solution. From the coefficients in the chemical equation (as in Chapters 3 and 4), we determine the conversion factors that relate moles of one substance to moles of another.

For example, we can determine the volume of hydrogen gas produced in a reaction of 4.40 g lithium with excess water. The temperature, 27 °C, and the pressure, 0.993 atm, at which the reaction occurs must also be known for this calculation. The strategy for the problem is similar to those for the stoichiometric calculations conducted in Chapters 3 and 4.

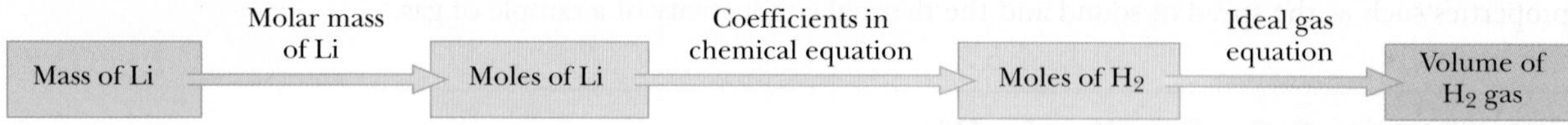

The first step, as always in stoichiometry calculations, is to write the chemical equation.

$$2Li(s) + 2H_2O(\ell) \rightarrow 2LiOH(aq) + H_2(g)$$

Second, convert grams of lithium to moles.

$$\text{Amount Li} = 4.40\ \cancel{\text{g Li}} \times \left(\frac{1\ \text{mol Li}}{6.941\ \cancel{\text{g Li}}}\right) = 0.634\ \text{mol Li}$$

Third, use the coefficients in the equation to calculate the number of moles of hydrogen gas that is equivalent to 0.634 mol lithium.

Use the ideal gas law to convert the moles of a gas sample to its equivalent volume.

$$\text{Amount } H_2 = 0.634\ \cancel{\text{mol Li}} \times \left(\frac{1\ \text{mol } H_2}{2\ \cancel{\text{mol Li}}}\right) = 0.317\ \text{mol } H_2$$

Fourth, use the moles of hydrogen gas and the ideal gas law to calculate the volume of hydrogen gas produced. The known values are

$P = 0.993$ atm $\quad V = ?$

$n = 0.317$ mol $H_2 \quad T = 300$ K

Solve the ideal gas law for volume.

$$V = \frac{nRT}{P} = \frac{(0.317\ \cancel{\text{mol}}\ H_2)(0.08206\ \text{L} \cdot \cancel{\text{atm}}/\cancel{\text{mol}} \cdot \cancel{\text{K}})(300\ \cancel{\text{K}})}{0.993\ \cancel{\text{atm}}}$$

$$= 7.86\ \text{L } H_2$$

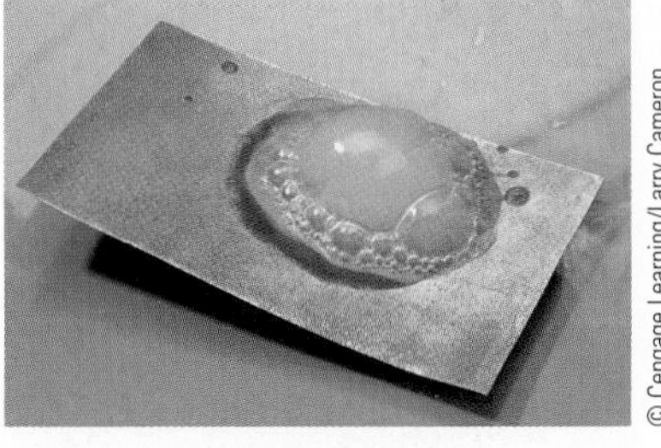

Zinc reacts with hydrochloric acid to give off bubbles of hydrogen gas.

EXAMPLE 6.9 Using Volumes of Gases in Equations

Chemists frequently prepare hydrogen gas in the laboratory by the reaction of zinc and hydrochloric acid. The other product is $ZnCl_2(aq)$. Calculate the volume of hydrogen produced at 744 torr pressure and 27 °C by the reaction of 32.2 g zinc and 500 mL of 2.20 *M* HCl.

Strategy The strategy of this example is interesting, because we use three different methods to calculate the number of moles of the three different substances: (1) Use the molar mass to calculate the number of moles from the mass of zinc. (2) Use the molarity and the volume of solution to calculate the number of moles of HCl. (3) Use the ideal gas law to convert the number of moles of hydrogen gas to volume of hydrogen gas. As always, use the chemical equation to relate the number of moles of one substance to moles of another.

Solution

First, write the chemical equation.

$$Zn(s) + 2HCl(aq) \rightarrow ZnCl_2(aq) + H_2(g)$$

Second, use the information given in the problem to calculate the number of moles of zinc and hydrochloric acid. Because the amounts of both reactants are given, this is a limiting-reactant problem. We need to calculate the number of moles of hydrogen gas each reactant would produce if it were consumed completely.

$$\text{Amount Zn} = 32.2\ \cancel{\text{g Zn}} \times \left(\frac{1\ \text{mol Zn}}{65.39\ \cancel{\text{g Zn}}}\right) = 0.492\ \text{mol Zn}$$

$$\text{Amount HCl} = 0.500\ \cancel{\text{L HCl soln}} \times \left(\frac{2.20\ \text{mol HCl}}{1\ \cancel{\text{L HCl soln}}}\right) = 1.10\ \text{mol HCl}$$

Use the coefficients in the equation to calculate the amount of hydrogen one could obtain from each of the reactants.

$$\text{Amount } H_2 \text{ based on Zn} = 0.492\ \cancel{\text{mol Zn}} \times \left(\frac{1\ \text{mol } H_2}{1\ \cancel{\text{mol Zn}}}\right) = 0.492\ \text{mol } H_2$$

$$\text{Amount } H_2 \text{ based on HCl} = 1.10\ \cancel{\text{mol HCl}} \times \left(\frac{1\ \text{mol } H_2}{2\ \cancel{\text{mol HCl}}}\right) = 0.550\ \text{mol } H_2$$

The zinc yields the smaller amount of hydrogen and, therefore, is the limiting reactant. Complete the problem by using the ideal gas law.

$$P = \frac{744}{760}\ \text{atm} = 0.979\ \text{atm} \qquad V = ?$$

$$n = 0.492\ \text{mol } H_2 \qquad T = 300\ \text{K}$$

$$V = \frac{nRT}{P} = \frac{(0.492\ \cancel{\text{mol}}\ H_2)(0.08206\ \text{L}\cdot\cancel{\text{atm}}/\cancel{\text{mol}}\cdot\cancel{\text{K}})(300\ \cancel{\text{K}})}{0.979\ \cancel{\text{atm}}}$$

$$= 12.4\ \text{L } H_2$$

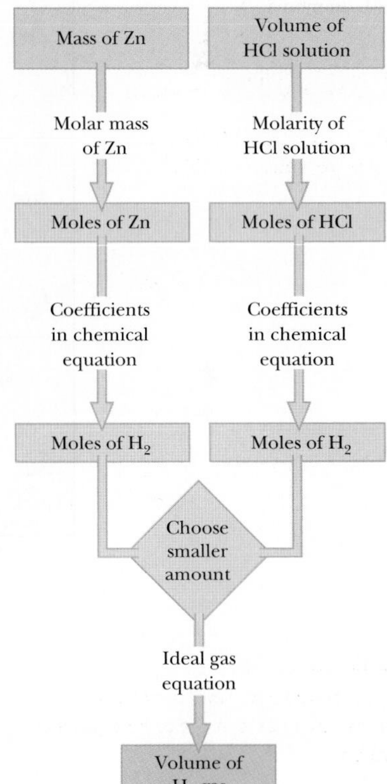

Understanding

Many scientists believe that when Earth's atmosphere evolved, some of the oxygen gas came from the decomposition of water induced by solar radiation.

$$2H_2O(\ell) \xrightarrow{\text{light}} 2H_2(g) + O_2(g)$$

What volume of oxygen at 754 torr and 40 °C does the decomposition of 2.33 g of H_2O produce?

Answer 1.68 L O_2

Volumes of Gases in Chemical Reactions

We have already seen that equal volumes of gases at the same temperature and pressure contain the same number of moles of each gas. *In chemical reactions under these conditions, the volumes of gases combine in the same proportions as the coefficients of the equation.* This statement is a direct consequence of Avogadro's law. We can thus directly calculate the volume (rather than number of moles) of a gas produced by a reaction of gases, as long as the pressure and temperature of the gases are the same. For example, chemists prepare ammonia gas by the reaction of nitrogen gas and hydrogen gas.

$$3H_2(g) + N_2(g) \rightarrow 2NH_3(g)$$

The equation states that 3 mol hydrogen reacts with 1 mol nitrogen to yield 2 mol ammonia. It also states that 3 L hydrogen gas reacts with 1 L nitrogen gas to produce 2 L ammonia gas (Figure 6.11).

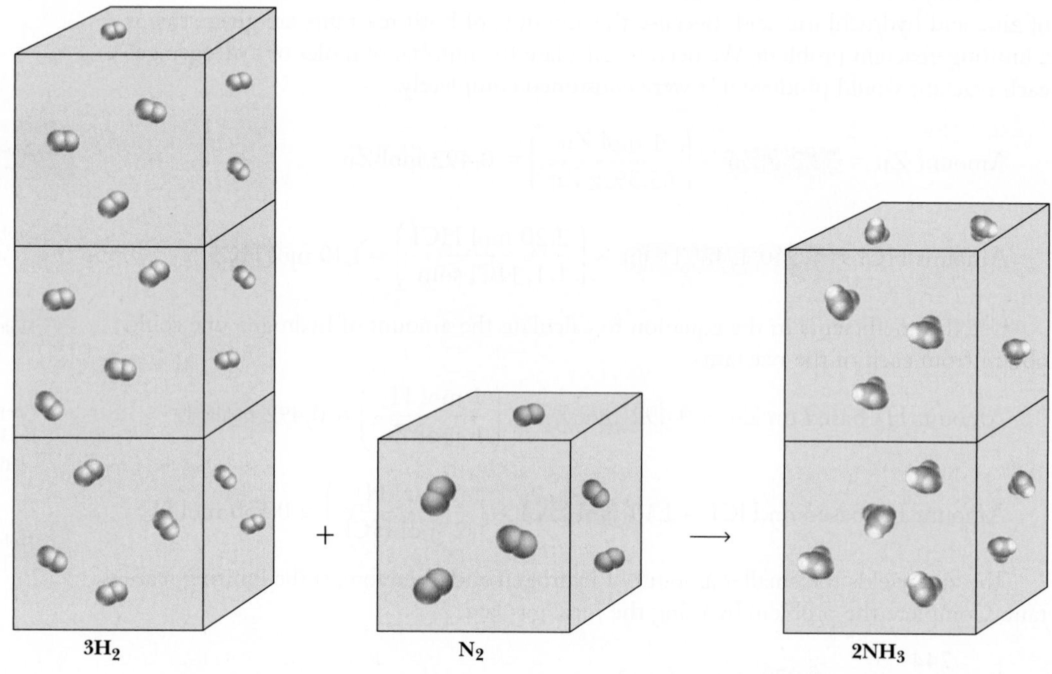

Figure 6.11 Volumes of gases in chemical reactions. The reaction of 3 L hydrogen gas with 1 L nitrogen gas yields 2 L ammonia.

EXAMPLE 6.10 Volumes of Gases in Chemical Reactions

Nitrogen monoxide, NO, is a pollutant formed in running automobile engines. It reacts with oxygen in the air to produce nitrogen dioxide, NO_2. Calculate the volume of NO_2 gas produced and the volume of O_2 gas consumed when 2.34 L NO gas reacts with excess O_2. Assume that all volumes are measured at the same pressure and temperature.

Strategy Volumes of gases combine in the same proportions as the coefficients in the equation.

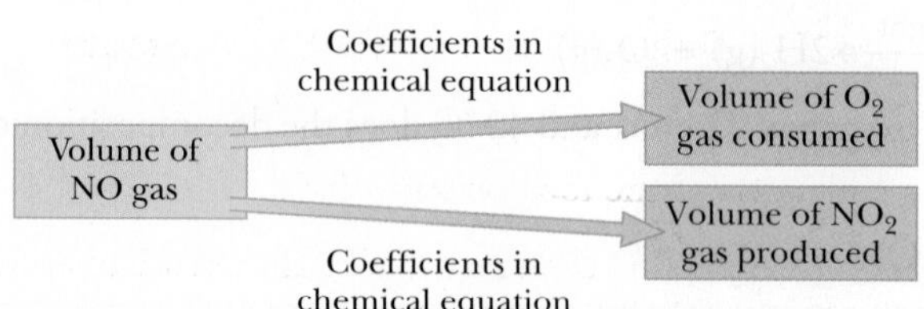

Solution

The equation is

$$2NO(g) + O_2(g) \rightarrow 2NO_2(g)$$

and states that two volumes of NO are needed to react with one volume of O_2, producing two volumes of NO_2. Using liters as the measure of volume,

2 L NO reacts with 1 L O_2

2 L NO produces 2 L NO_2

Use these equivalencies to calculate the volume of O_2 needed in the reaction and the volume of NO_2 produced.

$$\text{Volume of } O_2 = 2.34 \text{ L NO} \times \left(\frac{1 \text{ L } O_2}{2 \text{ L NO}}\right) = 1.17 \text{ L } O_2$$

$$\text{Volume of } NO_2 = 2.34 \text{ L NO} \times \left(\frac{2 \text{ L } NO_2}{2 \text{ L NO}}\right) = 2.34 \text{ L } NO_2$$

Understanding

Hydrogen, H_2, and chlorine, Cl_2, react to form hydrogen chloride, HCl. Calculate the volume of HCl formed by the reaction of 2.34 L H_2 and 3.22 L Cl_2.

Answer 4.68 L HCl

© Jon Arnold Images Ltd/Alamy

Air pollution. Nitrogen oxides formed in combustion reactions in the engines of automobiles contribute to smog.

OBJECTIVES REVIEW *Can you:*

- ☑ perform equation stoichiometric calculations for reactions in which some or all of the reactants or products are gases?
- ☑ use relative volumes of gases directly in stoichiometry problems?

6.5 Dalton's Law of Partial Pressure

OBJECTIVES

- ☐ Use Dalton's law of partial pressure in calculations involving mixtures of gases
- ☐ Calculate the partial pressure of a gas in a mixture from its mole fractions

In many of the examples in the preceding sections, the identities of the gases were not needed to solve the problems because all gases follow the ideal gas law at modest temperatures and pressures. In fact, we do not even need a pure sample of gas to use the ideal gas law. Many of the early experiments that led to the formulation of the gas laws were performed with samples of air rather than pure substances.

In 1801, English scientist John Dalton realized that each gas in a mixture of gases exerts a pressure, called a **partial pressure,** which is the same as if the gas occupied the container by itself. **Dalton's law of partial pressure** summarizes his observations: The total pressure of a mixture of gases is the sum of the partial pressures of all the components of the mixture. For a mixture of two gases, A and B, the total pressure, P_T, is

$$P_T = P_A + P_B$$

where P_A and P_B are the partial pressures of gases A and B (Figure 6.12).

The total pressure of a mixture of gases is the sum of the partial pressure each component exerts.

EXAMPLE 6.11 Dalton's Law of Partial Pressure

A gas sample in a 1.2-L container holds 0.22 mol N_2 and 0.13 mol O_2. Calculate the partial pressure of each gas and the total pressure at 50 °C.

Strategy Use the ideal gas law to calculate the partial pressure of each gas in the container, and sum these two numbers to obtain the total pressure.

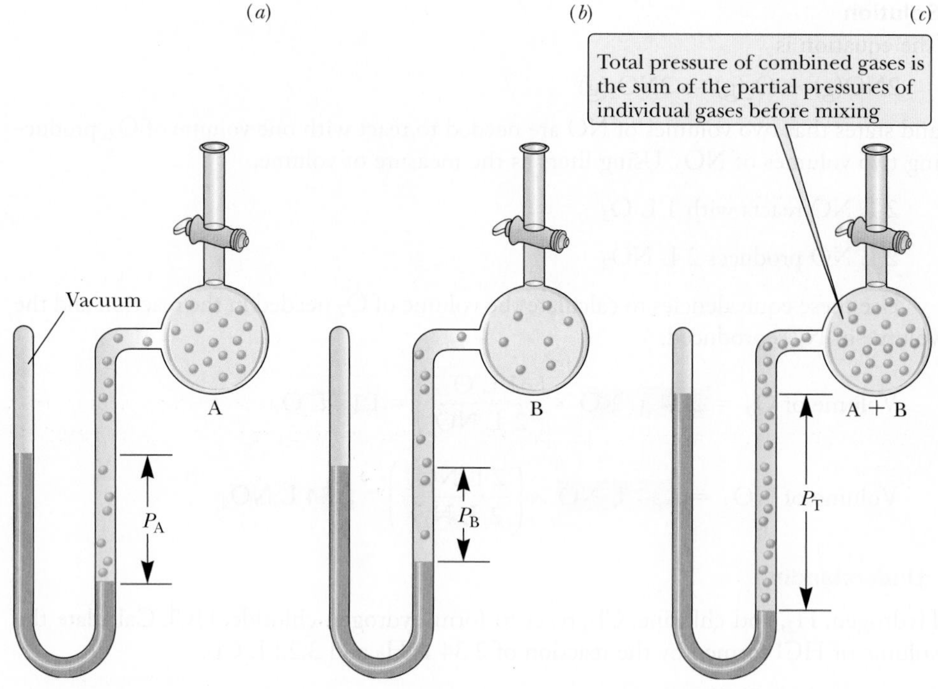

Figure 6.12 **Pressure of a mixture of gases.**

Solution

Make a table of the information given.

$P_{N_2} = ?$	$V_{N_2} = 1.2$ L	$n_{N_2} = 0.22$ mol	$T_{N_2} = 323$ K
$P_{O_2} = ?$	$V_{O_2} = 1.2$ L	$n_{O_2} = 0.13$ mol	$T_{O_2} = 323$ K

$$P_{N_2} = \frac{(n_{N_2})RT}{V_{N_2}} = \frac{(0.22 \text{ mol N}_2)(0.08206 \text{ L} \cdot \text{atm/mol} \cdot \text{K})(323 \text{ K})}{1.2 \text{ L}}$$

$$= 4.9 \text{ atm N}_2$$

$$P_{O_2} = \frac{(n_{O_2})RT}{V_{O_2}} = \frac{(0.13 \text{ mol O}_2)(0.08206 \text{ L} \cdot \text{atm/mol} \cdot \text{K})(323 \text{K})}{1.2 \text{ L}}$$

$$= 2.9 \text{ atm O}_2$$

The total pressure is the sum of the partial pressures of the oxygen and nitrogen.

$$P_T = P_{N_2} + P_{O_2} = 4.9 \text{ atm} + 2.9 \text{ atm} = 7.8 \text{ atm}$$

Understanding

Calculate the partial pressure of each gas and the total pressure in a 4.6-L container at 27 °C that contains 3.22 g Ar and 4.11 g Ne.

Answer $P_{Ar} = 0.43$ atm; $P_{Ne} = 1.1$ atm; $P_T = 1.5$ atm

Partial Pressures and Mole Fractions

A mixture of gases is a solution. A convenient concentration unit to describe this gaseous mixture is the **mole fraction**—the number of moles of one component of a mixture divided by the total number of moles of all substances present in the mixture. The symbol χ (the Greek letter chi) represents mole fraction:

$$\chi_A = \frac{\text{moles of component A}}{\text{total moles of all substances}} = \frac{n_A}{n_{total}}$$

If the container shown in Figure 6.13 holds 0.030 mol argon and 0.090 mol neon, the mole fractions of the gases are

$$\chi_{Ar} = \frac{0.030 \text{ mol Ar}}{0.120 \text{ mol total}} = 0.25$$

$$\chi_{Ne} = \frac{0.090 \text{ mol Ne}}{0.120 \text{ mol total}} = 0.75$$

Note that mole fraction is a unitless quantity. The sum of the mole fractions of all components in the mixture is always exactly 1.

$$\chi_A + \chi_B + \chi_C + \cdots + \chi_n = 1$$

Mole fraction is a convenient concentration unit for partial-pressure calculations, because at constant volume and temperature, the partial pressure of any gas in a mixture is given by

$$P_A = \chi_A \times P_T$$

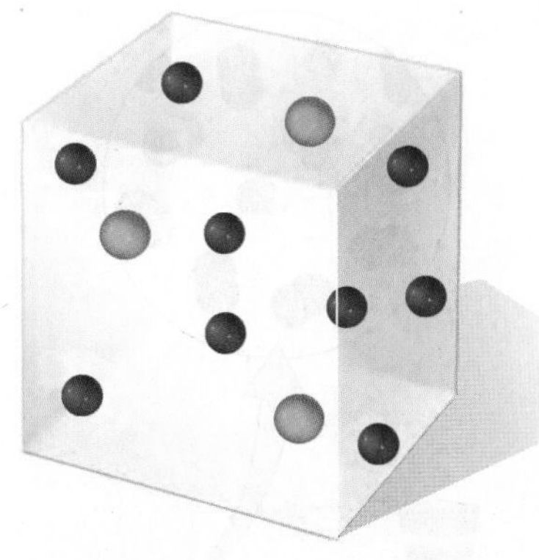

Figure 6.13 Mole fraction. The mole fraction expresses the concentration of each gas in a mixture of argon *(yellow spheres)* and neon *(red spheres)*.

Mole fraction is a convenient concentration unit for mixtures of gases.

EXAMPLE 6.12 Partial Pressure of a Gas

Trimix, as outlined in the introduction to this chapter, is a mixture of O_2, N_2, and He that is used for very deep SCUBA dives. What is the partial pressure of oxygen if 0.10 mol oxygen is mixed with 0.20 mol nitrogen and 0.70 mol helium? The total pressure of gas is 4.2 atm.

Strategy The partial pressure of oxygen is its mole fraction times the total pressure.

Solution

First, calculate the total number of moles.

$$n_T = n_{O_2} + n_{N_2} + n_{He}$$

$$n_T = 0.10 \text{ mol } O_2 + 0.20 \text{ mol } N_2 + 0.70 \text{ mol He}$$

$$= 1.00 \text{ mol}$$

The mole fraction of oxygen is the number of moles of oxygen divided by the total number of moles of all three gases in the mixture.

$$\chi = \frac{0.10 \text{ mol } O_2}{1.00 \text{ mol total}} = 0.10$$

The partial pressure of oxygen is its mole fraction times the total pressure of the gas.

$$P_{O_2} = \chi_{O_2} \times P_T = 0.10 \times 4.2 \text{ atm} = 0.42 \text{ atm oxygen}$$

Understanding

What is the partial pressure of helium in a flask at a total pressure of 700 torr, if the sample contains 10.2 mol argon and 10.4 mol helium?

Answer 353 torr

Collecting Gases by Water Displacement

Chemists frequently use an apparatus such as that shown in Figure 6.14 to collect the gases produced in chemical reactions. They measure the volume of gas generated in a reaction by determining the volume of water displaced.

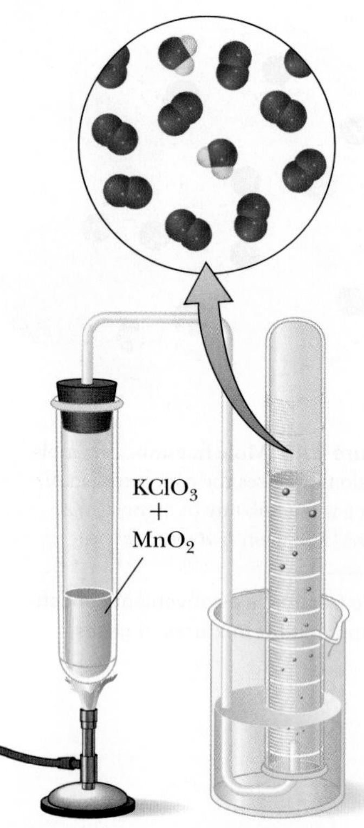

Figure 6.14 Collecting a gas by water displacement. The volume of gas produced in a chemical reaction can be measured by the displacement of water. The reaction shown is the thermal decomposition of $KClO_3$ (with MnO_2 added to speed up the reaction) to yield O_2 gas: $2KClO_3(s) \xrightarrow{MnO_2} 2KCl(s) + 3O_2(g)$.

TABLE 6.3 Pressure of Water Vapor at Selected Temperatures

Temperature (°C)	Pressure of Water Vapor (torr)
5	6.54
10	9.21
15	12.79
20	17.54
21	18.66
22	19.84
23	21.08
24	22.39
25	23.77
26	25.21
27	26.76
28	28.37
29	30.06
30	31.84
35	42.20
40	55.36
50	92.59
60	149.5
70	233.8
80	355.3
90	525.9
100	760.0

The gas sample collected by displacement of water is not pure, because some water molecules are also present in the gas phase. Thus, the total pressure of the gas collected in the apparatus shown in Figure 6.14 is due to both the collected O_2 gas and the water vapor. As shown in Table 6.3, the partial pressure of water present in the gas phase depends on the temperature of the water. To determine the partial pressure of the gas collected, you must subtract the partial pressure of the water vapor from the total pressure, as required by Dalton's law.

EXAMPLE 6.13 Collecting Gases by Water Displacement

A sample of $KClO_3$ is heated and decomposes to produce O_2 gas. The gas is collected by water displacement at 26 °C. The total volume of the collected gas is 229 mL at a pressure equal to the measured atmospheric pressure, 754 torr. How many moles of O_2 form?

Strategy The collected gas is a mixture of O_2 and H_2O. Because we are interested in the amount of O_2, we must first determine the partial pressure of the O_2 gas, then use the ideal gas law to find the amount of O_2.

Solution

First, determine the partial pressure of the pure O_2 gas in the sample. The partial pressure of water vapor at 26 °C is 25 torr (Table 6.3). From Dalton's law of partial pressure,

$$P_T = P_{O_2} + P_{H_2O}$$

$$P_{O_2} = P_T - P_{H_2O}$$

$$P_{O_2} = 754 \text{ torr} - 25 \text{ torr} = 729 \text{ torr } O_2$$

Calculate the amount of O_2 from the ideal gas law.

$$P_{O_2} = 729 \text{ torr} \times \left(\frac{1 \text{ atm}}{760 \text{ torr}}\right) = 0.959 \text{ atm}$$

$$V = 0.229 \text{ L};\ T = 26 + 273 = 299 \text{ K};\quad n = ?$$

$$n = \frac{PV}{RT} = \frac{(0.959 \text{ atm})(0.229 \text{ L})}{(0.08206 \text{ L} \cdot \text{atm/mol} \cdot \text{K})(299 \text{ K})}$$

$$= 8.95 \times 10^{-3} \text{ mol } O_2$$

Understanding

Calculate the number of moles of hydrogen produced by the reaction of sodium with water. In the reaction, 1.3 L gas is collected by water displacement at 26 °C. The atmospheric pressure is 756 torr.

Answer 0.051 mol H_2

OBJECTIVES REVIEW *Can you:*

- ☑ use Dalton's law of partial pressure in calculations involving mixtures of gases?
- ☑ calculate the partial pressure of a gas in a mixture from its mole fractions?

6.6 Kinetic Molecular Theory of Gases

OBJECTIVES

- ☐ Show that the predictions of the kinetic molecular theory are consistent with experimental observations
- ☐ Sketch a Maxwell–Boltzmann distribution curve for the distribution of speeds of gas molecules
- ☐ Perform calculations using the relationships among molecular speed and the temperature and molar mass of a gas

Like all laws of nature, all of the gas laws were discovered experimentally. For example, scientists made measurements of volumes and temperatures of gases to show that the volume of a gas at constant pressure is proportional to its temperature in kelvins. Chemists sought to understand why a single law can describe the physical behavior of all gases, regardless of the nature or size of the gas particles. The **kinetic molecular theory** describes the behavior of gas particles at the molecular level. The theory is built on four postulates:

1. A gas consists of small particles that are in constant and random motion. No forces of attraction or repulsion exist between any two gas particles.
2. Gas particles are very small compared with the average distance that separates them.
3. Collisions of gas particles with each other and with the walls of the container are *elastic;* that is, no loss in total kinetic energy occurs when the particles collide.
4. The average kinetic energy of gas particles is proportional to the temperature on the Kelvin scale.

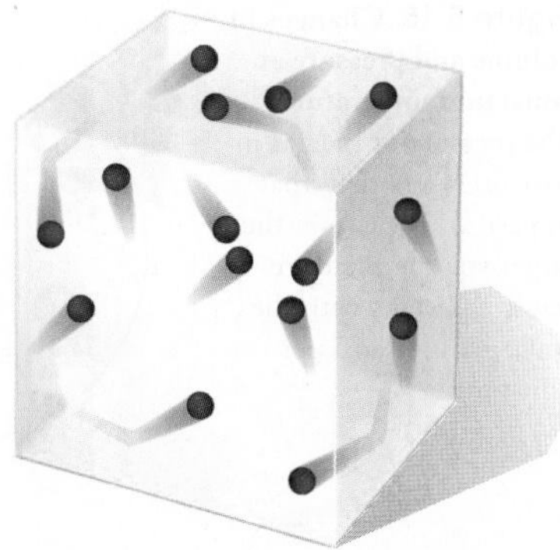

Figure 6.15 **Kinetic molecular theory of gases.** Although in rapid motion, gas particles occupy only a small percentage of the total volume of the container. The collisions with the walls exert pressure.

Figure 6.15 describes the behavior of particles in the gas phase. The particles occupy only a small part of the volume of the container; most of the volume is empty space. The gas particles are in constant motion; they collide with each other and with the walls of the box. The direction and speed of the particles change when they collide, but the total energy of the gas does not change. The energy of the gas changes only if the temperature changes.

Recall that pressure is the force per unit area. The kinetic molecular theory assumes that the pressure exerted by a gas comes from the collisions of the individual gas particles with the walls of the container. Pressure increases if the energy of the collisions or the number of wall collisions per second increases, because both will increase the force on the wall. The pressure of a gas is the same on all walls of its container.

Comparison of Kinetic Molecular Theory and the Ideal Gas Law

For a theory to be useful, it must be able to account for experimental observations. It is important, therefore, to compare the predictions of kinetic molecular theory with experimental observations of relationships among volume, pressure, temperature, and amount. All the comparisons shown here are qualitative, but calculations show that the quantitative relations from kinetic molecular theory are also correct.

The kinetic molecular theory is consistent with the ideal gas law.

Volume and Pressure: Compression of Gases

Kinetic molecular theory assumes that gas particles are small compared with the distances that separate them. Gases can expand to fill a larger container or compress to fit into a smaller container, because most of the volume of a gas is empty space. Solids and liquids are different from gases; they do not readily compress because the particles are in close contact.

Boyle observed that the pressure of a gas increases when the volume decreases as long as the temperature is kept constant. The kinetic molecular theory explains this observation: As the size of a container decreases (at constant temperature), the number of collisions of the gas particles with the walls per unit area during any time interval increases, because the particles have less distance to travel between collisions with the walls. At constant temperature, the average force of each collision does not change, but in a smaller volume, the same number of particles strike a given area of the wall more often, so the pressure of gas in the container increases as volume decreases (Figure 6.16).

Volume and Temperature

The kinetic molecular theory states that the average kinetic energy of gas particles is proportional to the temperature. Two consequences of this increased kinetic energy are that each collision exerts a greater force on the walls, and that the number of collisions per unit area per unit time increases. If pressure is to remain constant, the size of the container must increase, reducing the number of these more energetic collisions per unit

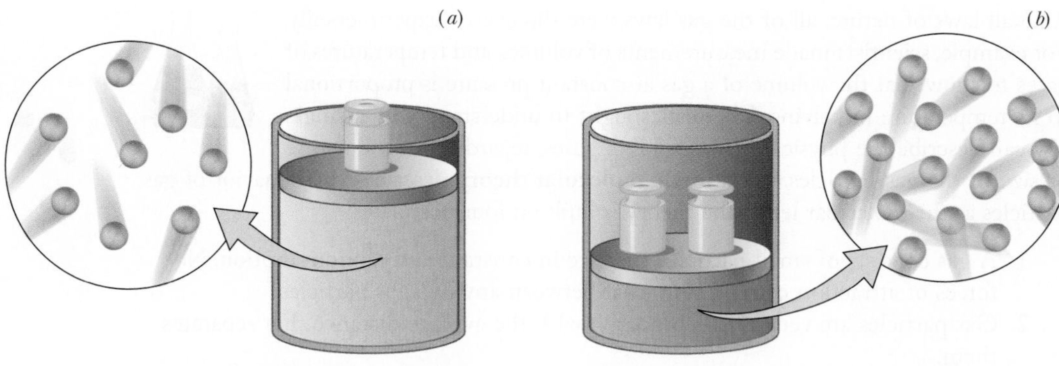

Figure 6.16 Changes in volume and pressure at constant temperature. The pressure of the gas in part *(a)* is less than that in part *(b)* because in the larger volume there are fewer collisions with the walls per unit area per unit time.

area. The kinetic molecular theory thus predicts an increase in volume with an increase in temperature at constant pressure—Charles's law.

Volume and Amount

Increasing the number of gas particles in a container increases the number of collisions with the walls per unit area per unit time. If the pressure were to remain constant, the volume of the container must increase as predicted by Avogadro's law.

The kinetic molecular theory of gases interprets the observed behavior of gases on a molecular scale.

Average Speed of Gas Particles

Kinetic molecular theory assumes that the average kinetic energy of the gas particles is directly proportional to the temperature in kelvins. Not all of the gas particles will move at the same speed, so we refer to the average kinetic energy. The relationship of the average kinetic energy of the gas particles to the speed *(u)* of the particles is

$$\overline{KE} = \frac{1}{2}m\overline{u^2}$$

where the bars over kinetic energy *(KE)* and the squared speed (u^2) indicate average values, and m is the mass of the particles. The square root of $\overline{u^2}$ is called the **root-mean-square (rms) speed,** labeled u_{rms}, and is used to indicate the average speed of a gas.

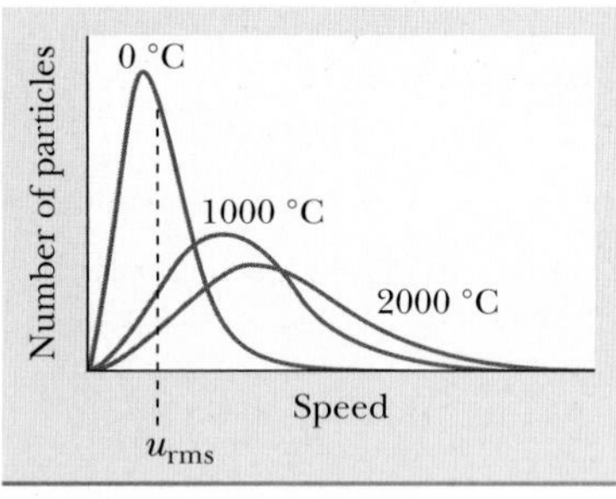

Figure 6.17 Maxwell–Boltzmann distribution. Graph shows the number of particles that have a given speed versus the speed. The curve broadens and the maximum shifts to higher speeds as the temperature increases. The root-mean-square speed (u_{rms}) at 0 °C is shown.

From the mathematical treatment of the kinetic theory of gases, we can determine the relative number of gas particles that have any particular speed. Figure 6.17 is a plot of the number of gas particles with a given speed versus the speed. Some of the particles have low speeds, whereas others move rapidly. The root-mean-square speed for 0 °C is indicated in the plot. Notice that it is not the same as the most probable speed, which would be at the maximum of the plot.

Plots of this type are known as *Maxwell–Boltzmann distribution curves.* The u_{rms} speed is a little higher than the most probable speed because the graphs are not symmetric. If the temperature increases, the average speed increases, the curve broadens, and both the most probable speed and u_{rms} shift to greater values.

A Maxwell–Boltzmann distribution describes the speed of gas molecules.

The average kinetic energy of the gas particles is proportional to the temperature, and kinetic molecular theory predicts that the rms speed is related to temperature and molar mass by the equation

$$u_{rms} = \sqrt{\frac{3RT}{M}} \qquad [6.6]$$

Increasing the temperature of a gas increases its average speed.

To obtain the rms speed of gas molecules using Equation 6.6, we must express R as 8.314 J/mol · K and the molar mass, M, in *kilograms* per mole.

EXAMPLE 6.14 Root-Mean-Square Speed of Gas Particles

Calculate the rms speed in meters per second of argon atoms at 27 °C.

Strategy Use Equation 6.6, remembering to use the proper values and units for R, to convert the molar mass into units of kilograms per mole (kg/mol).

Solution

The molar mass of argon, 39.95 g/mol, in the proper units, is 0.03995 kg/mol. So that units can be canceled out in the calculation, expand the joule into its base units, kg · m²/s², when you substitute the value of R into Equation 6.6.

$$u_{\mathrm{rms}} = \sqrt{\frac{3RT}{M}} = \sqrt{\frac{3\left(8.314\dfrac{\cancel{\mathrm{kg}}\cdot \mathrm{m}^2}{\mathrm{s}^2\cdot \cancel{\mathrm{mol}}\cdot \cancel{\mathrm{K}}}\right)(300\ \cancel{\mathrm{K}})}{0.03995\dfrac{\cancel{\mathrm{kg}}}{\cancel{\mathrm{mol}}}}} = 433\frac{\mathrm{m}}{\mathrm{s}}$$

Understanding

Calculate the rms speed of neon atoms at 27 °C.

Answer 609 m/s

Equation 6.6 shows that the rms speed of a gas sample is proportional to the square root of temperature and inversely proportional to the square root of molar mass. Figure 6.17 shows how the distribution of speed changes with temperature. Figure 6.18 is a similar plot showing the speed distributions for three different gases at the same temperature. At constant temperature, gases with greater molar masses have lower rms speeds. You can see this trend by comparing the result in Example 6.14 with the answer in the Understanding section.

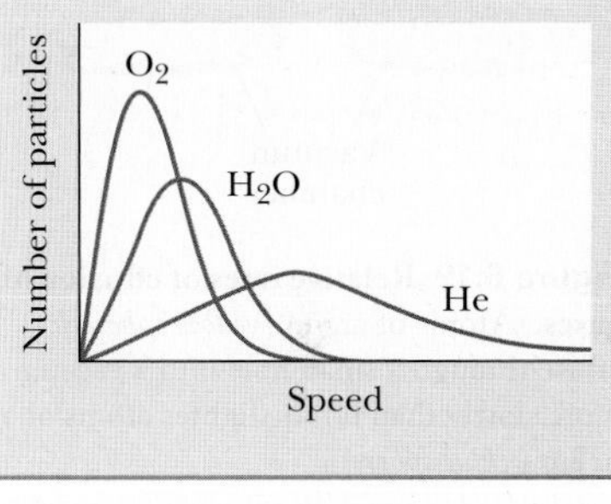

Figure 6.18 Distribution of speeds for particles of different masses. At constant temperature, the root-mean-square speed of a gas increases as the molar mass decreases.

Lighter gases move faster than heavier gases at the same temperature.

The observation that heavier particles have lower rms speeds is expected, because the average kinetic energies of all gases are the same at a given temperature. Thus, the molecules in a sample made up of heavier particles must be moving more slowly, on average, than the molecules in a sample made up of lighter particles, because the two samples have the same average kinetic energy.

OBJECTIVES REVIEW *Can you:*

- ☑ show that the predictions of the kinetic molecular theory are consistent with experimental observations?
- ☑ sketch a Maxwell–Boltzmann distribution curve for the distribution of speeds of gas molecules?
- ☑ perform calculations using the relationships among molecular speed and the temperature and molar mass of a gas?

6.7 Diffusion and Effusion

OBJECTIVE

- ☐ Calculate the molar mass of a gas from the relative rates of effusion of two gases

Any theory must undergo tests of its ability to predict the results of new observations. The kinetic molecular theory correctly describes the mixing of gases, a process called *diffusion.* **Diffusion** is the mixing of particles caused by motion. The faster the molecular motion, the faster a gas diffuses. However, the rate of diffusion is always less than the rms speed of the gas, because collisions prevent the particles from moving in a straight line.

Closely related to diffusion is **effusion,** the passage of a gas through a small hole into an evacuated space. Thomas Graham (1805–1869) carefully measured the rates of effusion of several gases. **Graham's law** states that the rate of effusion of a gas is inversely proportional to the square root of its molar mass. The kinetic molecular theory explains Graham's law because the rms speed of the gas particles is inversely proportional to the square root of their molar mass.

The rates of effusion are inversely proportional to the square root of the molar mass.

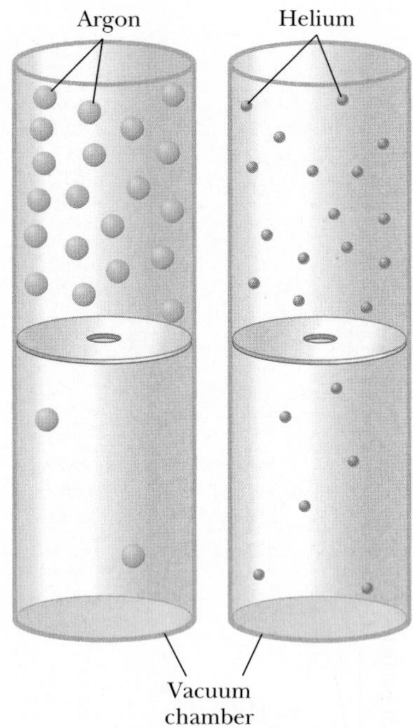

Figure 6.19 Relative rates of effusion of gases. Atoms of argon *(yellow spheres)* effuse through a small hole into a vacuum more slowly than do the lighter atoms of helium *(blue spheres)*.

The molar mass of a gas can be determined from relative rates of effusion.

Graham's law is frequently used to compare the rates of effusion of two gases, written as

$$\frac{\text{rate of effusion of gas A}}{\text{rate of effusion of gas B}} = \sqrt{\frac{\text{molar mass of B}}{\text{molar mass of A}}} \qquad [6.7]$$

This is a useful equation because scientists frequently find relative measurements easier to carry out than absolute measurements. We can use this expression to compare the rates at which two gases effuse through a small hole. For example, the relative rates of effusion of helium and argon are

$$\frac{\text{rate of effusion of helium}}{\text{rate of effusion of argon}} = \sqrt{\frac{\text{molar mass of Ar}}{\text{molar mass of He}}} = \sqrt{\frac{40\text{ g/mol}}{4.0\text{ g/mol}}} = 3.2$$

In the same amount of time, the helium atoms are a little more than three times faster at effusing through the hole (Figure 6.19) because of their smaller molar mass. Graham's law can also be applied to the relative rates of diffusion of gases.

Molar Mass Determinations by Graham's Law

We can use Graham's law to determine the molar mass of an unknown gas by measuring the times needed for equal volumes of a known gas and an unknown gas to effuse through the same small hole at constant pressure and temperature. Equation 6.7 relates the rate of effusion to molar mass. Gases with greater rates of effusion escape through the hole in shorter lengths of time; the time it takes for a gas to effuse, t, is inversely proportional to the rate of effusion. Thus, Equation 6.7 becomes

$$\frac{\text{rate of effusion of gas A}}{\text{rate of effusion of gas B}} = \frac{t_B}{t_A} = \sqrt{\frac{\text{molar mass of B}}{\text{molar mass of A}}} \qquad [6.8]$$

EXAMPLE 6.15 Determination of Molar Mass by Effusion

Calculate the molar mass of a gas if equal volumes of nitrogen and the unknown gas take 2.2 and 4.1 minutes, respectively, to effuse through the same small hole under conditions of constant pressure and temperature.

Strategy Because the rates of effusion are inversely proportional to the square root of the molar masses, we can use the relative rate of effusion of the two gases and the molar mass of one (nitrogen) to calculate the molar mass of the unknown gas.

Solution

Solve Equation 6.8 for the molar mass of the unknown gas *(x)* by squaring both sides and rearranging.

$$\frac{t_x}{t_{N_2}} = \sqrt{\frac{\text{molar mass of } x}{\text{molar mass of } N_2}}$$

$$\frac{(t_x)^2}{(t_{N_2})^2} = \frac{\text{molar mass of } x}{\text{molar mass of } N_2}$$

$$\text{Molar mass } x = \text{molar mass } N_2 \times \frac{(t_x)^2}{(t_{N_2})^2} = 28\frac{\text{g}}{\text{mol}} \times \frac{(4.1\ \cancel{\text{min}})^2}{(2.2\ \cancel{\text{min}})^2} = 97\frac{\text{g}}{\text{mol}}$$

Understanding

Calculate the molar mass of a gas if equal volumes of oxygen gas and the unknown gas take 3.25 and 8.41 minutes, respectively, to effuse through a small hole under conditions of constant pressure and temperature.

Answer 214 g/mol

OBJECTIVE REVIEW *Can you:*

☑ calculate molar mass of a gas from the relative rates of effusion of two gases?

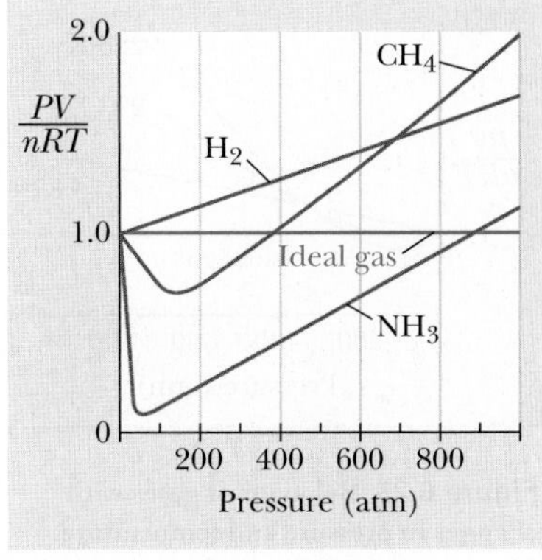

Figure 6.20 Influence of high pressure on gases. A plot of PV/nRT versus P for three gases. An ideal gas has a value of 1 for PV/nRT at all pressures.

6.8 Deviations from Ideal Behavior

OBJECTIVES

- ☐ Explain why gases deviate from the ideal gas law under certain conditions
- ☐ Use the van der Waals equation to account for deviations from the ideal gas law

The kinetic molecular theory is a model that explains the ideal gas law on the molecular level. The ideal gas law was discovered through careful experimental observations and applies to an *ideal* gas—that is, any gas that follows the assumptions of the kinetic molecular theory. Most gases obey the ideal gas law quite closely at a pressure of about 1 atm and a temperature well above the boiling point of the substance.

Figure 6.20 is a plot of measured values of PV/nRT versus P for three gases. For a gas that follows the ideal gas law, the measured values of PV/nRT follow the blue line in the figure. At low pressures, less than a few atmospheres, all of the gases follow the ideal gas law, but as the pressure increases to high values (100 atm is a substantial pressure), deviations occur.

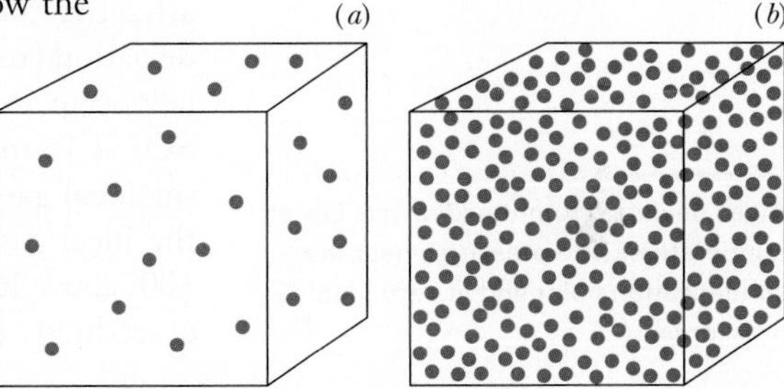

Figure 6.21 Gases at low and high pressures. *(a)* At a low pressure, the size of the gas particles in the container is small in comparison with the volume occupied by the gas. *(b)* At a high pressure, the particles occupy a significant percentage of the volume of the gas.

Figure 6.20 shows that gases at high pressures do not behave as predicted by the ideal gas law. Does that mean that we should discard the law? No, it is useful at the pressures at which chemists generally work. When experimental observations are inconsistent with a theory, scientists reevaluate both the theory and the experiments. Deviations from the ideal gas law occur under extreme conditions because two of the assumptions of the kinetic molecular theory simply are not correct when gas particles are close together. These assumptions are: (1) that gas particles are small compared with the distances separating them, and (2) that no attractive forces exist between gas particles.

Deviations Due to the Volume Occupied by Gas Particles

Kinetic molecular theory assumed that the volume of the gas particles is negligible with respect to the space occupied by the gas. At high pressure, the volume occupied by the individual particles is no longer negligible compared with the volume of the gas sample (Figure 6.21). When the particle's size is no longer negligible, the actual volume available for the gas particles to move is reduced. Because of the inverse relationship of volume and pressure, this effect at very high pressures causes the measured pressure to be greater for all gases than predicted by the ideal gas law, causing the deviations above the line in Figure 6.20.

Deviations Due to Attractive Forces

Ammonia shows a deviation *below* the line at moderate pressures; methane does as well, but less so. These deviations arise from forces of attraction between gas particles that are close together (similar to the forces discussed in Chapter 11 that hold molecules together in liquids). Gas particles that are attracted to each other do not strike the wall as hard as predicted (Figure 6.22), reducing the pressure below that predicted by the ideal gas law. As the pressure increases, the particles are forced closer together, making this attractive interaction more important because more gas particles are close to the one about to hit the wall. Ammonia dips most significantly below the line because, among the gases shown, it has the strongest attractive forces between its molecules. Hydrogen is not observed to go below the line because its attractive forces are very small. At very high pressures, the effect of molecular volume is greater than that of the attractive forces, so all three gases deviate above the ideal line.

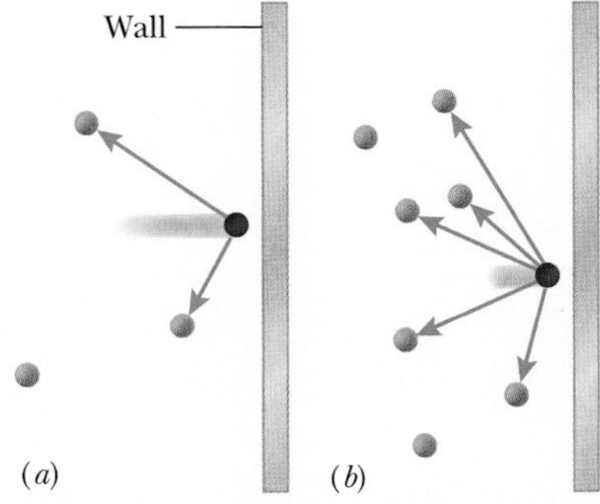

Figure 6.22 Forces of attraction in gases. *(a)* At low pressures, only a few particles are close to a particle that is about to hit the wall (colored green). *(b)* At higher pressures, many particles are close to the particle that is about to hit the wall. The attractive forces between the closely packed gas particles reduce the net force of the collision with the wall, reducing the pressure of the gas to less than that predicted by the ideal gas law.

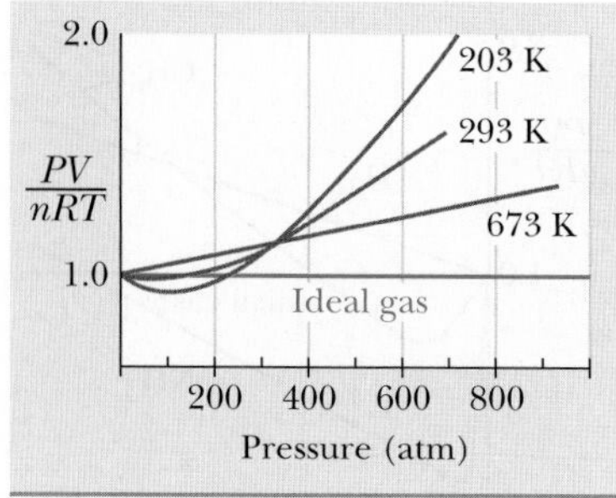

Figure 6.23 Behavior of gases with changes in pressure and temperature. The deviations of O_2 gas from the ideal gas law change with changes in temperature.

Figure 6.23 is a plot of PV/nRT versus pressure for oxygen at three different temperatures. At the lowest temperature, 203 K, the deviation is initially below the line because of attractive forces, but at greater pressures, the size factor dominates, causing significant deviations above the line. At 293 K, the average speed of the molecules is greater, reducing the effects of attractive forces at moderate pressures and the effects of size at greater pressures when compared with the behavior at 203 K. At the highest temperature, the greater rms speeds of gas molecules reduce the effect of attractive forces such that the deviation below the line that was observed at moderate pressures is not observed and cause the deviations at high pressures to be less important. Although the behavior of the gases at each temperature and pressure is complex because of the competing nature of attractive forces and size effects, in general, *gases behave most ideally at low pressures and high temperatures.*

Each type of molecule or atom behaves differently, but for gases to follow the ideal gas law, conditions must be far from the temperature and pressure under which they would condense to a liquid. Consider SO_2, with a boiling point of −10 °C at 1 atm. At less than −10 °C, the attractive forces between SO_2 molecules hold them close together, so it is a liquid. When the temperature is barely above the boiling point, the attractive forces between the molecules are sufficiently strong to cause considerable deviation from the ideal gas law. Measurable deviations of SO_2 gas from the ideal gas law occur even at room temperature, about 30 °C greater than the boiling point of SO_2 at 1 atm. In comparison, N_2, with a boiling point of −196 °C at 1 atm, follows the ideal gas law closely at normal temperatures and pressures. A gas usually follows the ideal gas law under conditions of temperature and pressure that are more than 100° above its condensation temperature (boiling point), as long as the pressure is not exceedingly high.

A gas deviates from the ideal gas law at temperatures and pressures near the condensation point and at very high pressures.

EXAMPLE 6.16 Deviations from the Ideal Gas Law

In each part, predict which gas sample is likely to follow the ideal gas law more closely:

(a) SO_2 gas at 0 °C or SO_2 at 100 °C, both at 1 atm
(b) N_2 gas at 1 atm or N_2 at 100 atm, both at 25 °C
(c) O_2 gas or NH_3 gas, both at −20 °C and 1 atm

Strategy In each case, choose the gas that is farther away from its condensation point or at higher temperature; gases follow the ideal gas law at high temperatures and low pressures.

Solution

(a) At 0 °C and 1 atm, the SO_2 is close to its condensation point of −10 °C and does not behave ideally. At 100 °C, it is well above its condensation point and follows the ideal gas law.
(b) Gases at a given temperature follow the ideal gas law better at lower pressures; therefore, N_2 at 1 atm follows the law better than N_2 at 100 atm.
(c) At 1 atm, O_2 boils at −183 °C and NH_3 boils at −33 °C. Oxygen follows the ideal gas law more closely because it is farther from the temperature at which it would condense to a liquid.

Understanding

Which gas and set of conditions best follows the ideal gas law: (a) N_2 at 25 °C and 1 atm, (b) SO_2 at 25 °C and 1 atm, or (c) N_2 at 25 °C and 100 atm?

Answer (a)

van der Waals Equation

The ideal gas law can be modified to include the effects of attractive forces and the volume occupied by the particles. To correct for the volume occupied by the gas particles, we subtract the term nb from the volume, where n is the number of moles of gas and b is a constant that depends on the size of the gas particles. The volume term in the gas law then becomes $(V - nb)$. This corrected volume is the empty space, which is the only part of the sample that can be compressed.

We can also modify the pressure term to correct for attractive forces by adding the term an^2/V^2 to the pressure, where a is a constant related to the strength of the attractive forces, and n and V are the number of moles and the volume of the gas. The pressure term in the gas law then becomes $(P + an^2/V^2)$. Substituting the new pressure and volume terms into the ideal gas law, we get the **van der Waals equation:**

$$\left(P + \frac{an^2}{V^2}\right)(V - nb) = nRT \qquad [6.9]$$

The experimentally determined van der Waals constants are different for each gas; Table 6.4 provides a few values.

TABLE 6.4 Van der Waals Constants

Gas	a (atm L^2/mol^2)	b (L/mol)
H_2	0.244	0.0266
He	0.0341	0.0237
Ne	0.211	0.0171
H_2O	5.46	0.0305
NH_3	4.17	0.0371
CH_4	2.25	0.0428
N_2	1.39	0.0391
O_2	1.36	0.0318
Ar	1.34	0.0322
CO_2	3.59	0.0427

The van der Waals equation corrects the ideal gas law for the effects of attractive forces and the volume occupied by the particles.

EXAMPLE 6.17 Van der Waals Equation

Calculate the pressure in atmospheres of 2.01 mol gaseous H_2O at 400 °C in a 2.55-L container, using the ideal gas law and van der Waals equation. Compare the two answers.

Strategy Use both the ideal gas law and the van der Waals equation to calculate the pressure under the conditions given.

Solution

For the ideal gas law,

$$P = \frac{nRT}{V} = \frac{(2.01\ \cancel{\text{mol}})(0.08206\ \cancel{\text{L}}\cdot\text{atm}/\cancel{\text{mol}}\cdot\cancel{\text{K}})(673\ \cancel{\text{K}})}{2.55\ \cancel{\text{L}}}$$

$$= 43.5\ \text{atm}$$

Rearranging the van der Waals equation to solve for pressure yields

$$P = \frac{nRT}{V - nb} - \frac{an^2}{V^2}$$

Substitute the measured values and the constants from Table 6.4 into this equation.

$$P = \frac{(2.01\ \cancel{\text{mol}})(0.08206\ \cancel{\text{L}}\cdot\text{atm}/\cancel{\text{mol}}\cdot\cancel{\text{K}})(673\ \cancel{\text{K}})}{2.55\ \cancel{\text{L}} - (2.01\ \cancel{\text{mol}})(0.0305\ \cancel{\text{L}}/\cancel{\text{mol}})} - \frac{(5.46\ \text{atm}\ \cancel{\text{L}^2}/\cancel{\text{mol}^2})(2.01\ \cancel{\text{mol}})^2}{(2.55\ \cancel{\text{L}})^2}$$

$$= 44.6\ \text{atm} - 3.39\ \text{atm} = 41.2\ \text{atm}$$

Under these conditions, the ideal gas law and van der Waals equation yield values that differ by about 5%.

Understanding

Calculate the pressure in atmospheres of 0.223 mol ammonia gas at 30.0 °C in a 3.23-L container, using the ideal gas law and van der Waals equation.

Answer Ideal gas law = 1.72 atm, van der Waals = 1.70 atm. Under these conditions, the correction is small.

OBJECTIVES REVIEW *Can you:*

- ☑ explain why gases deviate from the ideal gas law under certain conditions?
- ☑ use the van der Waals equation to account for deviations from the ideal gas law?

Summary Problem

The Practice of Chemistry box on the internal combustion engine made an assumption that is not exactly true. In the problem, 300 cm^3 of a mixture of gasoline and air at 70 °C and a pressure of 0.967 atm were compressed to 31.5 cm^3, and after ignition the temperature of the gases was 350 °C. We assumed then that the number of moles in the cylinder did not change, but they do. In other problems worked in the text, the stoichiometry was always carefully worked into the solution—we should use it here. If we use the formula of octane, C_8H_{18}, as a representative molecule for the complex mixture of compounds in gasoline, we can write the equation for the combustion reaction.

$$2C_8H_{18}(g) + 25O_2(g) \rightarrow 16CO_2(g) + 18H_2O(g)$$

If the mole fraction of the octane in the cylinder is 0.0100 and that of oxygen and nitrogen are 0.210 and 0.780, calculate the pressure in the cylinder after combustion, taking the change in moles as well as temperature and volume into account.

To solve for the final pressure, we know the initial pressure (0.967 atm), initial and final temperatures (70 °C = 343 K and 350 °C = 623 K), and initial and final volumes (300 cm^3 and 31.5 cm^3). We need to determine the number of moles of gas present under both conditions. The total number of moles in the cylinder at 343 K can be calculated using the ideal gas law.

$$n = \frac{PV}{RT} = \frac{(0.967\ \cancel{\text{atm}})(0.300\ \cancel{\text{L}})}{(0.08206\ \cancel{\text{L}} \cdot \cancel{\text{atm}}/\text{mol} \cdot \cancel{\text{K}})(343\ \cancel{\text{K}})} = 0.0103\ \text{mol}$$

The moles of each of the three gases present at the start of the problem can be calculated from this total number of moles and the mole fraction of each.

$$\text{Amount } C_8H_{18} = \chi_{\text{octane}} \times \text{mol}_{\text{total}} = 0.0100 \times 0.0103\ \text{mol}$$
$$= 1.03 \times 10^{-4}\ \text{mol } C_8H_{18}$$

$$\text{Amount } O_2 = \chi_{\text{oxygen}} \times \text{mol}_{\text{total}} = 0.210 \times 0.0103\ \text{mol}$$
$$= 2.16 \times 10^{-3}\ \text{mol } O_2$$

$$\text{Amount } N_2 = \chi_{\text{nitrogen}} \times \text{mol}_{\text{total}} = 0.780 \times 0.0103\ \text{mol}$$
$$= 8.03 \times 10^{-3}\ \text{mol } N_2$$

We assume that the moles of nitrogen will not change (see the Ethics questions), but we need to calculate the limiting reactant, C_8H_{18} or O_2, in the combustion reaction. We will use water as the product in the calculation.

$$\text{Amount } H_2O \text{ based on } C_8H_{18} = 1.03 \times 10^{-4}\ \cancel{\text{mol } C_8H_{18}} \times \left(\frac{18\ \text{mol } H_2O}{2\ \cancel{\text{mol } C_8H_{18}}}\right)$$
$$= 9.27 \times 10^{-4}\ \text{mol } H_2O$$

$$\text{Amount } H_2O \text{ based on } O_2 = 2.16 \times 10^{-3}\ \cancel{\text{mol } O_2} \times \left(\frac{18\ \text{mol } H_2O}{25\ \cancel{\text{mol } O_2}}\right)$$
$$= 1.56 \times 10^{-3}\ \text{mol } H_2O$$

Octane is the limiting reactant and is assumed to be completely consumed in the reaction producing 9.27×10^{-4} mol H_2O. We need to calculate the amount of the original O_2 that is consumed and the CO_2 that is produced in the reactions based on the consumption of all of the octane.

$$\text{Amount } CO_2 \text{ produced} = 1.03 \times 10^{-4} \cancel{\text{mol } C_8H_{18}} \times \left(\frac{16 \text{ mol } CO_2}{2 \cancel{\text{mol } C_8H_{18}}}\right)$$

$$= 8.24 \times 10^{-4} \text{ mol } CO_2$$

$$\text{Amount } O_2 \text{ consumed} = 1.03 \times 10^{-4} \cancel{\text{mol } C_8H_{18}} \times \left(\frac{25 \text{ mol } O_2}{2 \cancel{\text{mol } C_8H_{18}}}\right)$$

$$= 1.29 \times 10^{-3} \text{ mol } O_2$$

The final number of moles present is the CO_2 (8.24×10^{-4}) and H_2O (9.27×10^{-4}) that is produced in the reaction plus the O_2 that was not consumed ($[2.16 - 1.29] \times 10^{-3}$) and the N_2 of which none was consumed (8.03×10^{-3}) $= 1.06 \times 10^{-2}$ mol total. Now use the combined gas law to calculate the final pressure. First, we list all of our data:

$V_1 = 300 \text{ cm}^3$ $T_1 = 70 + 273 = 343 \text{ K}$ $P_1 = 0.967 \text{ atm}$ $n_1 = 0.0103 \text{ mol}$

$V_2 = 31.5 \text{ cm}^3$ $T_2 = 350 + 273 = 623 \text{ K}$ $P_2 = ?$ $n_2 = 0.0106 \text{ mol}$

Rearrange the combined gas law, solving for P_2:

$$P_2 = \frac{P_1V_1T_2n_2}{T_1V_2n_1} = \frac{(0.967 \text{ atm})(300 \cancel{\text{cm}^3})(623 \cancel{\text{K}})(0.0106 \cancel{\text{mol}})}{(343 \cancel{\text{K}})(31.5 \cancel{\text{cm}^3})(0.0103 \cancel{\text{mol}})} = 17.2 \text{ atm}$$

In the Practice of Chemistry box where we assumed no change in the number of moles, the final pressure was calculated to be 16.7 atm. Our more accurate number of 17.2 atm is just slightly larger.

ETHICS IN CHEMISTRY

1. In the Summary Problem, many assumptions were made. We assumed that the formula for octane could be substituted for a complex mixture of gasoline, the mixture of which can change dramatically in different regions of the country. Another major assumption, that the nitrogen does not react, is also untrue. Under the conditions of the reaction, some of the nitrogen reacts with oxygen to produce several different nitrogen oxides, mainly NO and NO_2, which are described collectively as NO_x, causing air pollution. Automobile manufacturers can adjust the conditions in the engine to reduce NO_x production, but those changes can reduce the efficiency of the engine. Is it ethical of the manufacturers to adjust the engine for maximum power, thus producing more NO_x, even if the adjustment consumes less gasoline and saves the owner of the vehicle money?

2. Many states in the Midwest of the United States use local coal for their power generation. The Midwestern coal produces sulfur oxides; most are removed at the plant by scrubbers, and the emissions are within legal limits. The prevailing weather moves the exhaust plume, and it tends to concentrate and precipitate in the Northeast where it can produce acid rain, which has defoliated several important forest areas. The management of the power company could buy low-sulfur coal from distant sources, or they could improve their scrubbers. Both proposed solutions decrease efficiency, increase energy costs to the customer, produce more carbon dioxide, and decrease profits. Should the power companies change their operations, even though the effect of the pollution is distant? What criteria should be used to make this decision?

Chapter 6 Visual Summary

The chart shows the connections between the major topics discussed in this chapter.

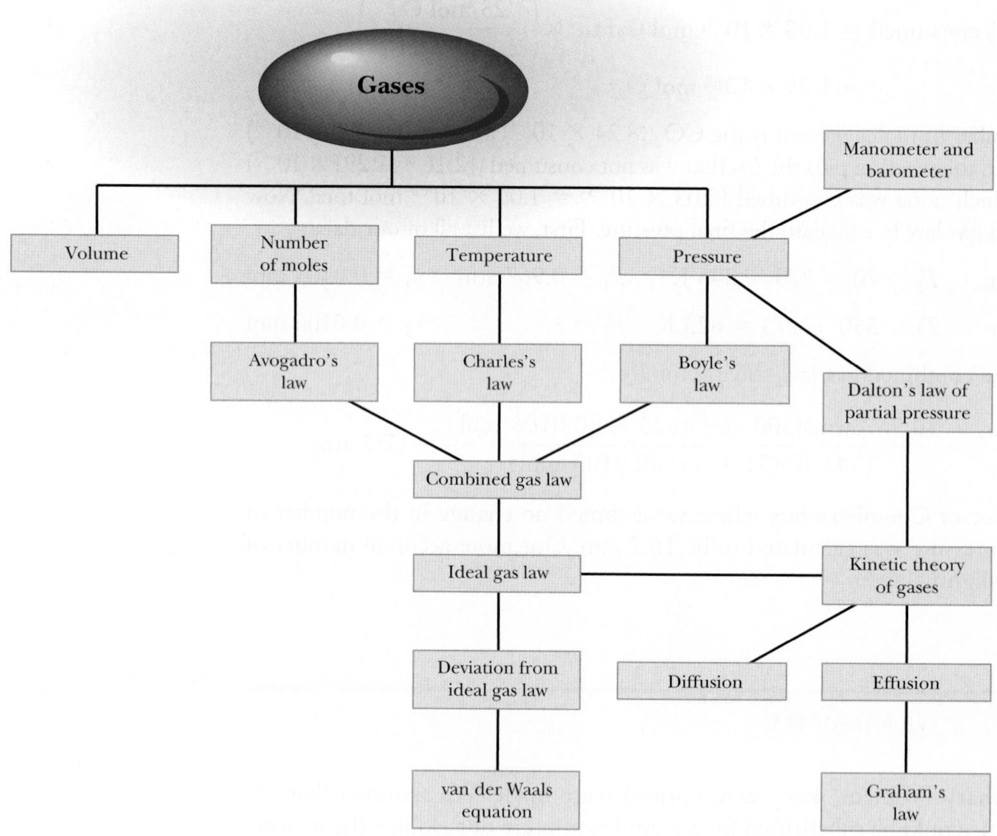

Summary

6.1 Properties and Measurements of Gases

A *gas* is a fluid without definite volume or shape. The *pressure* of a gas is the force per unit area exerted by it. Pressure is expressed in a number of different units; 1 atm is the pressure of the atmosphere at sea level, and 1 atm is 760 torr. The volume, pressure, temperature, and amount of gas in a sample describe the state of that sample.

6.2 Gas Laws and 6.3 The Ideal Gas Law

Experiments have shown that the volume of a gas sample is inversely proportional to pressure (Boyle's law), and directly proportional to temperature (Charles's law) and amount of sample (Avogadro's law). These laws allow calculations of changes in the state of a gas sample when one or more of the properties are changed. These relationships can also be combined into a single law, *the ideal gas law:*

$$PV = nRT$$

where R is the ideal gas law constant, determined experimentally to be 0.08206 L · atm/mol · K. The ideal gas law can be used to calculate any one of the four variables, if three have been measured, or to determine the molar mass, given the density or volume and mass of a sample.

6.4 Stoichiometry Calculations Involving Gases

The ideal gas law can be used to determine n, the number of moles of gas, and this value is used in standard stoichiometry calculations involving chemical equations. In a reaction that involves two or more gases at the same temperature and pressure, the coefficients in the equation can be interpreted as volumes.

6.5 Dalton's Law of Partial Pressure

Dalton's law of partial pressure states that the total pressure of a mixture of gases is the sum of the partial pressures of the component gases. The *partial pressure* of a gas is the pressure it would exert if it alone occupied the container at the same temperature. When a gas is collected over water, the total pressure is the sum of the pressure of the gas and the pressure of the water vapor that is present. The partial pressures of the gases in a mixture are proportional to their *mole fractions.*

6.6 Kinetic Molecular Theory of Gases

The *kinetic molecular theory* explains the behavior of gas particles at the molecular level. It makes the following assumptions: (1) gases are small particles in constant and random motion, and there are no forces of attraction or repulsion between any two gas particles; (2) gas particles are very small compared with sample volumes; (3) collisions of gas particles with each other and with the walls of the container are *elastic;* and (4) the average kinetic energy of the gas particles is proportional to the temperature on the Kelvin scale. The *pressure* of a gas comes from the particles rebounding from the walls of the container. The gas particles do not all move at the same speed but have speeds given by the Maxwell–Boltzmann distribution. The root-mean-square speed, u_{rms}, of a gas is proportional to the square root of temperature and inversely proportional to the square root of the molar mass. The assumptions of the kinetic molecular theory can be used to explain the ideal gas law.

6.7 Diffusion and Effusion

Diffusion and *effusion* are related to the speed of the gas molecules. *Graham's law* states that the rate of effusion is inversely proportional to the square root of the molar mass, and it can be used to determine the molar mass.

6.8 Deviations from Ideal Behavior

The ideal gas law predicts the behavior of most gases at typical laboratory pressures and temperatures. Deviations from the law are observed at high pressures and low temperatures because of attractive forces between molecules and the actual volume of the particles. The van der Waals equation describes the behavior of gases at high pressures more accurately than does the ideal gas law; it is an extension of the ideal gas law that contains terms that account for attractive forces and molecular size.

Chapter Terms

The following terms are defined in the Glossary, Appendix I.

Section 6.1
Condensed phase
Gas
Liquid
Pressure
Solid

Section 6.2
Avogadro's law
Boyle's law
Charles's law
Combined gas law

Section 6.3
Ideal gas law
Standard temperature and pressure (STP)

Section 6.5
Dalton's law of partial pressure
Mole fraction
Partial pressure

Section 6.6
Kinetic molecular theory
Root-mean-square (rms) speed, u_{rms}

Section 6.7
Diffusion
Effusion
Graham's law

Section 6.8
van der Waals equation

Key Equations

Boyle's law (6.2)

$$V = \text{constant} \times \frac{1}{P} \text{ and } P_1V_1 = P_2V_2$$

Charles's law (6.2)

$$V = \text{constant} \times T \text{ and } \frac{V_1}{T_1} = \frac{V_2}{T_2}$$

Avogadro's law (6.2)

$$V = \text{constant} \times n \text{ and } \frac{V_1}{n_1} = \frac{V_2}{n_2}$$

Combined gas law (6.2)

$$\frac{P_1V_1}{T_1} = \frac{P_2V_2}{T_2}$$

Ideal gas law (6.3)

$$PV = nRT$$

Dalton's law of partial pressure (6.5)

$$P_T = P_A + P_B$$

Mole fraction, χ (6.5)

$$\chi_A = \frac{\text{moles of component A}}{\text{total moles of all substances}} = \frac{n_A}{n_{total}}$$

$$P_A = \chi_A \times P_T$$

Average kinetic energy (expressed as root-mean-square, u_{rms}, speed) of a gas (6.6)

$$u_{rms} = \sqrt{\frac{3RT}{M}}$$

Graham's law (rate of effusion) (6.7)

$$\frac{\text{rate of effusion of gas A}}{\text{rate of effusion of gas B}} = \sqrt{\frac{\text{molar mass of B}}{\text{molar mass of A}}} \text{ and}$$

$$\frac{\text{rate of effusion of gas A}}{\text{rate of effusion of gas B}} = \frac{t_B}{t_A} = \sqrt{\frac{\text{molar mass of B}}{\text{molar mass of A}}}$$

van der Waals equation (for a gas not following the ideal gas law, 6.8)

$$\left(P + \frac{an^2}{V^2}\right)(V - nb) = nRT$$

where a is a constant related to the strength of the attractive forces and b is a constant that depends on the size of the gas particles

Questions and Exercises

OWL Selected end of chapter Questions and Exercises may be assigned in OWL.

Blue-numbered Questions and Exercises are answered in Appendix J; questions are qualitative, are often conceptual, and include problem-solving skills.

■ Questions assignable in OWL

✎ Questions suitable for brief writing exercises

▲ More challenging questions

Questions

6.1 Describe the similarities and differences between the ways in which a gas and a liquid occupy a container.

6.2 Compare the densities of a single substance as a solid, a liquid, and a gas.

6.3 ✎ Describe how atmospheric pressure is measured with a barometer and how pressure differences are measured with an open-end manometer.

6.4 Make a drawing of an open-end manometer measuring the pressure of a sample of a gas that is at 200 torr. Assume the pressure on the open end is 1.00 atm.

6.5 Define three units that are used to express pressure.

6.6 Describe the change in the volume of a gas sample that occurs when each of the following three properties is increased with the other two held constant:
(a) pressure (b) temperature (c) amount

6.7 ✎ Describe an experiment with a gas that allows the determination of absolute zero on the temperature scale. In this experiment, is the temperature of absolute zero measured directly?

6.8 ▲ Demonstrate how Boyle's, Charles's, and Avogadro's laws can be obtained from the ideal gas law.

6.9 ▲ Derive an equation for density of a gas from the ideal gas law.

6.10 ✎ Why do 1 mol N_2 and 1 mol O_2 both exert the same pressure if placed in the same 20-L container? Is the mass of the gas sample the same in both cases? Explain why it is the same or different, and if it is different, predict which gas sample weighs more.

6.11 Explain why most gases deviate from ideal behavior at low temperatures but not at high temperatures.

6.12 Explain the differences between the concentration unit's molarity and mole fraction. Can molarity be used to describe the concentration of a mixture of gases?

6.13 List the four assumptions of the kinetic molecular theory.

6.14 ✎ Discuss the origin of gas pressure in terms of the kinetic molecular theory.

6.15 Draw an approximate Maxwell–Boltzmann distribution curve for the distribution of speeds of gas molecules. What are the units on each axis? About where is the root-mean-square speed on the curve?

6.16 Define the terms *diffusion* and *effusion.*

6.17 ✎ Describe why gases at high pressures do not follow the ideal gas law.

6.18 In each of the following cases, does the ratio PV/nRT for a real gas have a value greater than or less than 1?
(a) Attractive forces between particles are strong.
(b) The volume of the gas particle becomes important relative to the total volume of the gas.

Exercises

OBJECTIVE Determine how a gas sample responds to changes in volume, pressure, moles, and temperature.

6.19 Express a pressure of
(a) 334 torr in atm.
(b) 3944 Pa in atm.
(c) 2.4 atm in torr.

6.20 Express a pressure of
(a) 3.2 atm in torr.
(b) 54.9 atm in kPa.
(c) 356 torr in atm.

6.21 The temperature terms for gas law problems must always be expressed in kelvins. Convert the following temperatures to kelvins.
(a) 45 °C (b) −28 °C (c) 230 °C

6.22 Convert the following kelvin temperatures to degrees Celsius.
(a) 344 K (b) 122 K (c) 1537 K

6.23 A sample of gas at 1.02 atm of pressure and 39 °C is heated to 499 °C at constant volume. What is the new pressure in atmospheres?

6.24 ■ A 256-mL sample of a gas exerts a pressure of 2.75 atm at 16.0 °C. What volume would it occupy at 1.00 atm and 100 °C?

6.25 A 39.6-mL sample of gas is trapped in a syringe and heated from 27 °C to 127 °C. What is the new volume (in mL) in the syringe if the pressure is constant?

6.26 The quantity of gas in a 34-L balloon is increased from 3.2 to 5.3 mol at constant pressure. What is the new volume of the balloon at constant temperature?

6.27 The pressure on a balloon holding 166 mL of gas is increased from 399 torr to 1.00 atm. What is the new volume of the balloon (in mL) at constant temperature?

6.28 A sample of hydrogen gas is in a 2.33-L container at 745 torr and 27 °C. Express the pressure of hydrogen (in atm) after the volume is changed to 1.22 L and the temperature is increased to 100 °C.

6.29 The pressure of a 900-mL sample of helium is increased from 2.11 to 4.33 atm, and the temperature is also increased from 0 °C to 22 °C. What is the new volume (in mL) of the sample?

6.30 ■ A balloon is filled to the volume of 135 L on a day when the temperature is 21 °C. If no gases escaped, what would be the volume of the balloon after its temperature has changed to −8 °C?

6.31 Natural gas has been stored in an expandable tank that keeps a constant pressure as gas is added or removed. The tank has a volume of 4.50×10^4 ft^3 when it contains 77.4 million mol natural gas at −5 °C. What is the new volume of the tank if consumers use up 5.3 million mol and the temperature increases to 7 °C?

6.32 ■ A sample of gas occupies 135 mL at 22.5 °C; the pressure is 165 torr. What is the pressure of the gas sample when it is placed in a 252-mL flask at a temperature of 0.0 °C?

6.33 ▲ A 10-L cylinder contains helium gas at a pressure of 3.3 atm. The hosts of a party use the gas to fill balloons. How many 4-L balloons can be filled if the ambient pressure is 1.03 atm and the temperature remains constant? The final pressure in the tank will be 1.03 atm.

6.34 ▲ A 40-L cylinder contains helium gas at a pressure of 20.3 atm. Meteorologists fill a balloon with the gas to lift weather equipment into the stratosphere. What is the final pressure in the cylinder after a 105-L balloon is filled to a pressure of 1.03 atm?

6.35 The container below contains a gas and has a piston that can move without changing the pressure in the container. Redraw this container; then draw the container again after the temperature of the container has doubled on the Kelvin scale.

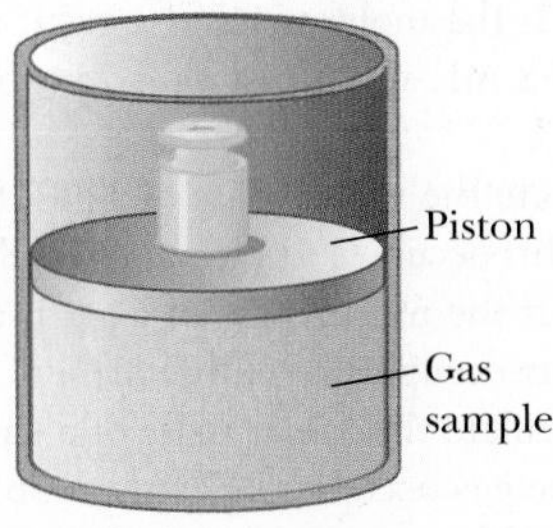

6.36 The container above contains a gas and has a piston that can move. Redraw this container; then draw the container again after the pressure on top of the piston has doubled.

OBJECTIVE Calculate the pressure, volume, amount, or temperature of a gas, given values of the other three properties.

6.37 A sample of argon occupies 3.22 L at 33 °C and 230 torr. How many moles of argon are present in the sample?

6.38 What is the temperature of a gas, in °C, if a 2.49-mol sample in a 24.0-L container is under a pressure of 2.44 atm?

6.39 A 3.00-L container is rated to hold a gas at a pressure no greater than 100 atm. Assuming that the gas behaves ideally, what is the maximum number of moles of gas that this vessel can hold at 27 °C?

6.40 What is the pressure, in atm, of 0.322 g N_2 gas in a 300-mL container at 24 °C?

6.41 What is the volume, in liters, of a balloon that contains 82.3 mol H_2 gas at 25 °C and 1.01×10^5 Pa?

6.42 What is the temperature of an ideal gas if 1.33 mol occupies 22.1 L at a pressure of 1.21 atm?

6.43 What is the pressure in a 2.33-L container holding 1.44 g CO_2 at 211 °C?

6.44 ■ What is the pressure exerted by 1.55 g Xe gas at 20 °C in a 560-mL flask?

6.45 How many N_2 molecules are in a 33.2-L container that is at 1.13 atm of pressure and 122 °C?

6.46 Calculate the volume of a gas sample containing 2.35×10^{25} water molecules at 0.173 atm of pressure and 229 °C.

6.47 In the cubical container below, each dot represents 0.10 of a mole of gas. If the container volume is 2.3 L and is at 27 °C, calculate the pressure in the container.

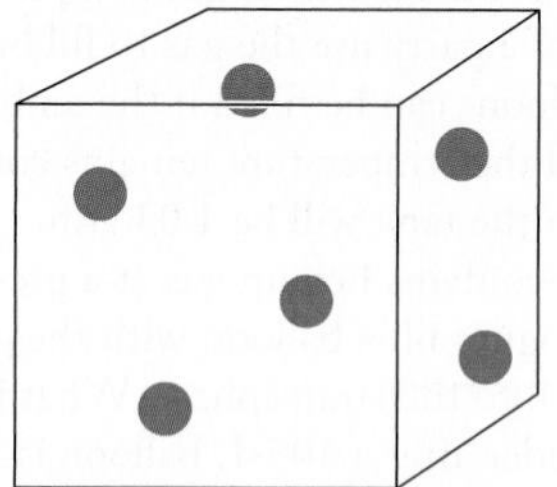

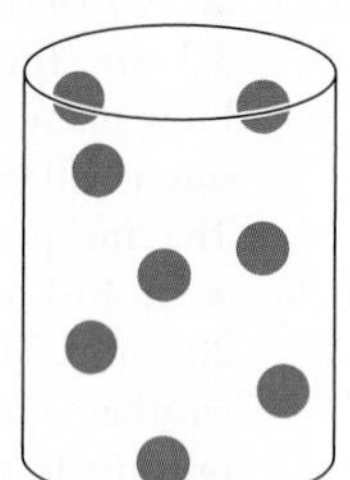

6.48 In the cylindrical container above, each dot represents 0.22 of a mole of gas. If the container pressure is 2.3 atm and is at 127 °C, calculate the volume of the container.

OBJECTIVE Calculate the molar mass and the density of gas samples by using the ideal gas law.

6.49 What is the molar mass of a gas if a 0.550-g sample occupies 258 mL at a pressure of 744 torr and a temperature of 22 °C?

6.50 Calculate the molar mass of a gas if a 0.165-g sample at 1.22 atm occupies a volume of 34.8 mL at 50 °C.

6.51 What is the molar mass of a gas if a 0.121-g sample at 740 torr occupies a volume of 21.0 mL at 29 °C?

6.52 ■ Calculate the molar mass of a gaseous element if 0.480 g of the gas occupies 367 mL at 365 torr and 45 °C. Suggest the identity of the element.

6.53 What is the density of He gas at 10.00 atm and 0 °C?

6.54 ■ Diethyl ether, $(C_2H_5)_2O$, vaporizes easily at room temperature. If the vapor exerts a pressure of 233 mm Hg in a flask at 25 °C, what is the density of the vapor?

6.55 What is the density of CO_2 gas at 1.00 atm and 27 °C?

6.56 What is the density of C_2H_6 gas at 0.55 atm and 100 °C?

6.57 Assuming the ideal gas law holds, what is the density of the atmosphere on the planet Venus if it is composed of $CO_2(g)$ at 730 K and 91.2 atm?

6.58 Assuming the ideal gas law holds, what is the density of the atmosphere on the planet Mars if it is composed of $CO_2(g)$ at −55 °C and 700 Pa?

OBJECTIVE Perform stoichiometric calculations for reactions in which some or all of the reactants or products are gases.

6.59 What volume, in milliliters, of hydrogen gas at 1.33 atm and 33 °C is produced by the reaction of 0.0223 g lithium metal with excess water? The other product is LiOH.

6.60 ■ Calculate the volume of methane, CH_4, measured at 300 K and 825 torr, that can be produced by the bacterial breakdown of 1.25 kg of a simple sugar.

$$C_6H_{12}O_6 \rightarrow 3CH_4 + 3CO_2$$

6.61 Heating potassium chlorate, $KClO_3$, yields oxygen gas and potassium chloride. What volume, in liters, of oxygen at 23 °C and 760 torr is produced by the decomposition of 4.42 g potassium chlorate?

6.62 What volume of oxygen gas, in liters, at 30 °C and 0.993 atm reacts with excess hydrogen to produce 4.22 g water?

6.63 What volume of hydrogen gas, in liters, is produced by the reaction of 1.33 g zinc metal with 300 mL of 2.33 *M* H_2SO_4? The gas is collected at 1.12 atm of pressure and 25 °C. The other product is $ZnSO_4(aq)$.

6.64 ■ What volume of hydrogen gas, in liters, is produced by the reaction of 3.43 g of iron metal with 40.0 mL of 2.43 *M* HCl? The gas is collected at 2.25 atm of pressure and 23 °C. The other product is $FeCl_2(aq)$.

6.65 The "air" that fills the air bags installed in automobiles is generally nitrogen produced by a complicated process involving sodium azide, NaN_3, and KNO_3. Assuming that one mole of NaN_3 produces one mol of N_2, what volume, in liters, of nitrogen gas is released from the decomposition of 1.88 g sodium azide? The pressure is 755 torr and the temperature is 24 °C.

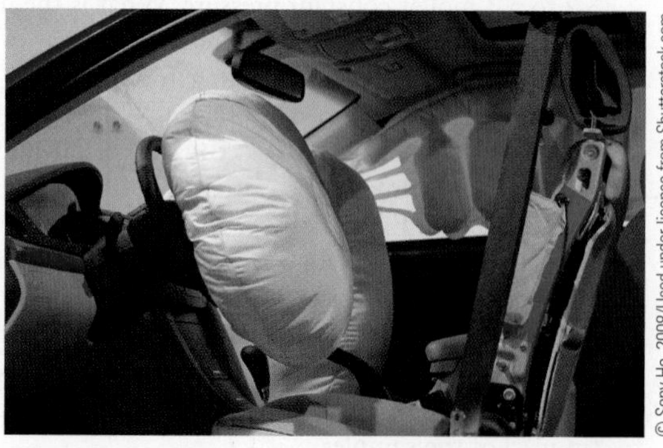

6.66 ■ Ammonia gas is synthesized from hydrogen and nitrogen:

$$3H_2(g) + N_2(g) \rightarrow 2NH_3(g)$$

If you want to produce 562 g of NH_3, what volume of H_2 gas, at 56 °C and 745 torr, is required?

OBJECTIVE Use volumes of gases directly in stoichiometry problems.

6.67 What volume of hydrogen gas is needed to exactly react with 4.2 L nitrogen gas to produce ammonia?

6.68 ■ Assuming the volumes of all gases in the reaction are measured at the same temperature and pressure, calculate the volume of water vapor obtainable by the explosive reaction of a mixture of 725 mL of hydrogen gas and 325 mL of oxygen gas.

6.69 The gas hydrogen sulfide, H_2S, has the offensive smell associated with rotten eggs. It reacts slowly with the oxygen in the atmosphere to form sulfur dioxide and water. What volume of sulfur dioxide gas, in liters, forms at constant pressure and temperature from 2.44 L hydrogen sulfide, and what volume of oxygen gas is consumed?

6.70 Considerable concern exists that an increase in the concentration of CO_2 in the atmosphere will lead to global warming. This gas is the product of the combustion of hydrocarbons used as energy sources. What volume of CO_2 gas, at constant temperature and pressure, is produced by the combustion of 2.00×10^3 L CH_4 gas? What volume of oxygen gas is consumed?

6.71 What volume of ammonia, NH_3, is produced from the reaction of 3 L hydrogen gas with 3 L nitrogen gas? What volume, if any, of the reactants will remain after the reaction ends. Assume all volumes are measured at the same pressure and temperature.

6.72 Nitrogen monoxide gas reacts with oxygen gas to produce nitrogen dioxide gas. What volume of nitrogen dioxide is produced from the reaction of 1 L nitrogen monoxide gas with 3 L oxygen gas? What volume, if any, of the reactants will remain after the reaction ends? Assume all volumes are measured at the same pressure and temperature.

6.73 Nitrogen dioxide can form in the reaction of oxygen gas and nitrogen gas. In the containers below, the red molecules represent oxygen gas and the blue molecules represent nitrogen gas. Redraw these two containers; then draw a container of the products of the reaction along with any unreacted nitrogen or oxygen after the two gases in the two containers have been mixed of appropriate size with the appropriate number of molecules, assuming the pressure and temperature of the gases has not changed.

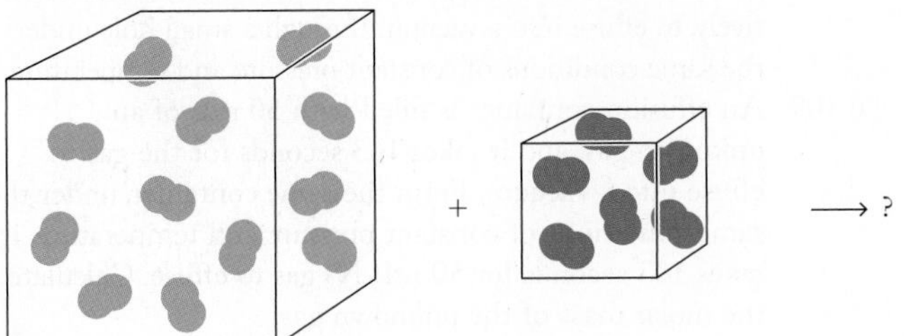

6.74 What mass of water forms when oxygen gas in the container below, where each red molecule represents 0.10 of a mole, reacts with hydrogen gas in the other container, where each white molecule represents 0.10 of a mole.

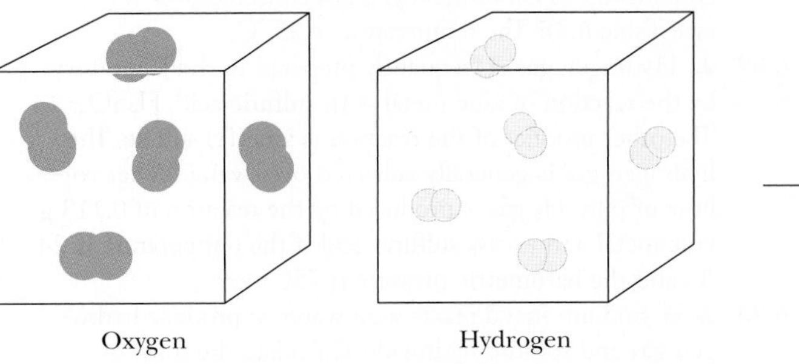

OBJECTIVE Use Dalton's law of partial pressure in calculations involving pressures with mixtures of gases.

6.75 What is the total pressure, in atm, in a container that holds 1.22 atm of hydrogen gas and 4.33 atm of argon gas?

6.76 What is the partial pressure of argon, in torr, in a container that also contains neon at 235 torr and is at a total pressure of 500 torr?

6.77 ▲ The pressure in a 3.11-L container is 4.33 atm. What is the new pressure in the tank when 2.11 L gas at 2.55 atm is added to the container? All the gases are at 27 °C.

6.78 ▲ A 4.53-L sample of neon at 3.22 atm of pressure is added to a 10.0-L cylinder that contains argon. If the pressure in the cylinder is 5.32 atm after the neon is added, what was the original pressure of argon in the cylinder?

6.79 What is the pressure, in atm, in a 3.22-L container that holds 0.322 mol oxygen and 1.53 mol nitrogen? The temperature of the gases is 100 °C.

6.80 Calculate the partial pressure of oxygen, in atm, in a container that holds 3.22 mol oxygen and 4.53 mol nitrogen. The total pressure in the container is 7.32 atm.

6.81 A 10.5-g sample of hydrogen is added to a 30-L container that also holds argon gas at 1.53 atm. The gases are at 120 °C. What is the partial pressure of hydrogen gas in the mixture, and what is the total pressure in the container?

6.82 ■ What is the total pressure exerted by a mixture of 1.50 g H_2 and 5.00 g N_2 in a 5.00-L vessel at 25 °C?

OBJECTIVE Calculate the partial pressure of a gas in a mixture from its mole fractions.

6.83 Calculate the partial pressure of hydrogen gas, in atm, in a container that holds 0.220 mol hydrogen and 0.432 mol nitrogen. The total pressure is 5.22 atm.

6.84 ■ What is the partial pressure of neon, in torr, in a flask that contains 3.11 mol of neon and 1.02 mol of argon under a total pressure of 209 torr?

6.85 What is the partial pressure of each gas in a flask that contains 0.22 mol neon, 0.33 mol nitrogen, and 0.22 mol oxygen if the total pressure in the flask is 2.6 atm?

6.86 What is the partial pressure of each gas in a flask that contains 2.3 g neon, 0.33 g xenon, and 1.1 g argon if the total pressure in the flask is 2.6 atm?

6.87 What is the partial pressure of oxygen gas, in torr, collected over water at 26 °C if the total pressure is 755 torr (see Table 6.3)?

6.88 What is the total pressure, in torr, in a 1.00-L flask that contains 0.0311 mol hydrogen gas collected over water (see Table 6.3)? The temperature is 25 °C.

6.89 ▲ Hydrogen gas is frequently prepared in the laboratory by the reaction of zinc metal with sulfuric acid, H_2SO_4. The other product of the reaction is zinc(II) sulfate. The hydrogen gas is generally collected over water. What volume of pure H_2 gas is produced by the reaction of 0.113 g zinc metal and excess sulfuric acid if the temperature is 24 °C and the barometric pressure is 750 torr?

6.90 ▲ ■ Sodium metal reacts with water to produce hydrogen gas and sodium hydroxide. Calculate the mass of sodium used in a reaction if 499 mL of wet hydrogen gas are collected over water at 22 °C and the barometric pressure is 755 torr. The vapor pressure of the water at 22 °C is 22 torr.

6.91 Two 1-L containers at 27 °C are connected by a stopcock as pictured below. If each dot in the containers represents 0.0050 mol of a nonreactive gas, what is the pressure in each container before and after the stopcock is opened?

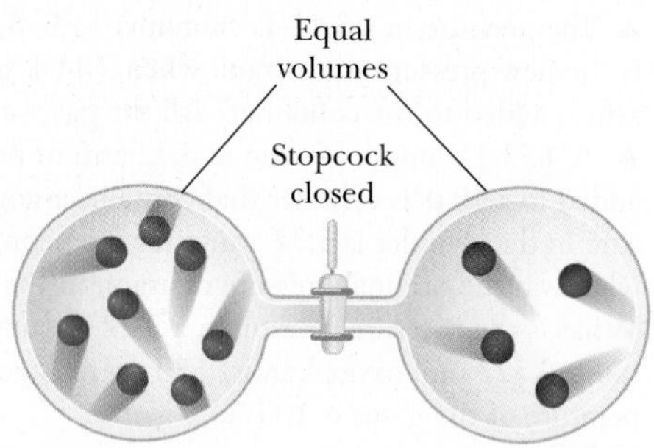

6.92 Two 1-L containers at 27 °C are connected by a stopcock as pictured below. If each dot in the containers represents 0.0020 mol of a nonreactive gas, what is the pressure in each container before and after the stopcock is opened? Draw the container after the stopcock is opened, indicating the number of red and blue dots on each side.

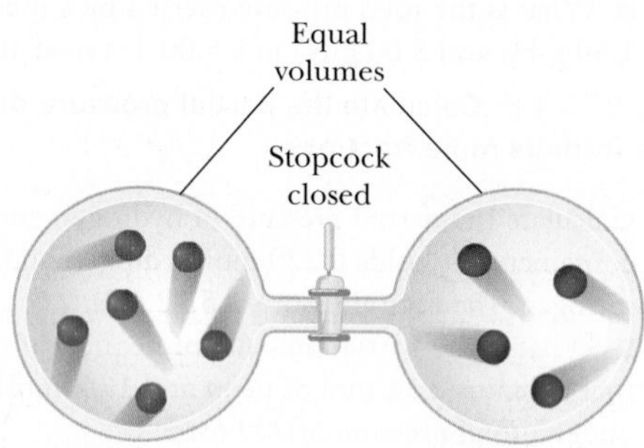

6.93 A robotic analysis of the atmosphere of the planet Venus shows that it has $\chi_{CO_2} = 0.964$, $\chi_{N_2} = 0.034$, and $\chi_{H_2O} = 0.0020$. If the total atmospheric pressure on Venus is 91.2 atm, what are the partial pressures (in atm) of each gas?

6.94 A robotic analysis of the atmosphere of the planet Mars shows that it has $\chi_{CO_2} = 0.9532$, $\chi_{N_2} = 0.027$, $\chi_{Ar} = 0.016$, and $\chi_{O_2} = 0.0013$. If the total atmospheric pressure on Mars is 7.00×10^2 Pa, what are the partial pressures (in Pa) of each gas?

OBJECTIVE Predict relative speeds of gases and perform calculations using the relationships among molecular speed and the temperature and molar mass of a gas.

6.95 Arrange the following gases in order of increasing rms speed of the particles at the same temperature: N_2, O_2, Ne.

6.96 ■ Place the following gases in order of increasing average molecular speed at 25 °C: Ar, CH_4, N_2, CH_2F_2.

6.97 Arrange the following gases, at the temperatures indicated, in order of increasing rms speed of the particles: neon at 25 °C, neon at 100 °C, argon at 25 °C.

6.98 Arrange the following gases, at the temperatures indicated, in order of increasing rms speed of the particles: helium at 100 °C, neon at 50 °C, argon at 0 °C.

6.99 Calculate the rms speed of neon atoms at 100 °C.

6.100 Calculate the rms speed of SO_2 molecules at 127 °C. What is the rms speed if the temperature is doubled on the Kelvin scale?

6.101 Calculate the molar mass of a gas that has an rms speed of 518 m/s at 28 °C.

6.102 What is the temperature, in kelvins, of neon atoms that have an rms speed of 700 m/s?

OBJECTIVE Use the relationship that relative rates of effusion of two gases are inversely proportional to the square root of its molar mass to calculate molar mass.

6.103 Calculate the ratio of the rate of effusion of helium to that of neon gas under the same conditions.

6.104 Calculate the ratio of the rate of effusion of CO_2 to that of CH_4 gas under the same conditions.

6.105 Calculate the ratio of the rate of effusion of helium to that of argon under the same conditions.

6.106 A container is filled with equal molar amounts of N_2 and SO_2 gas. Calculate the ratio of the rates of effusion of the two gases.

6.107 Calculate the molar mass of a gas if equal volumes of it and hydrogen take 9.12 and 1.20 minutes, respectively, to effuse into a vacuum through a small hole under the same conditions of constant pressure and temperature.

6.108 Calculate the molar mass of a gas if equal volumes of oxygen and the unknown gas take 5.2 and 8.3 minutes, respectively, to effuse into a vacuum through a small hole under the same conditions of constant pressure and temperature.

6.109 An effusion container is filled with 50 mL of an unknown gas, and it takes 163 seconds for the gas to effuse into a vacuum. From the same container, under the same conditions of constant pressure and temperature, it takes 103 seconds for 50 mL N_2 gas to effuse. Calculate the molar mass of the unknown gas.

6.110 ■ A gas effuses 1.55 times faster than propane (C_3H_8) at the same temperature and pressure.
(a) Is the gas heavier or lighter than propane?
(b) What is the molar mass of the gas?

OBJECTIVE Explain why gases deviate from the ideal gas law under certain conditions.

6.111 For each of the following pairs of gases at the given conditions, predict which one would more closely follow the ideal gas law. Explain your choice.
(a) oxygen (boiling point = −183 °C) gas at −150 °C or at 30 °C, both measured at 1.0 atm
(b) nitrogen (boiling point = −196 °C) or xenon (boiling point = −107 °C) gas at −100 °C, both measured at 1.0 atm
(c) argon gas at 1 atm or at 50 atm of pressure, both measured at 25 °C

6.112 ■ For each of the following pairs of gases at the given conditions, predict which one would more closely follow the ideal gas law. Explain your choice.
(a) Oxygen (boiling point = −183 °C) or sulfur dioxide (boiling point = −10 °C), both measured at 25 °C and 1 atm
(b) Nitrogen (boiling point = −196 °C) at −150 °C or at 100 °C, both measured at 1 atm
(c) Argon gas at 1 atm or at 200 atm, both measured at 200 °C

6.113 For the following pairs of gases at the given conditions, predict which one would more closely follow the ideal gas law. Explain your choice.
(a) CO_2 gas at 0.05 atm or at 10 atm of pressure
(b) Propane (boiling point = −45 °C) or neon (boiling point = −246 °C) gas at −20 °C and 1 atm
(c) Sulfur dioxide at 0 °C or at 50 °C, both measured at 1 atm

6.114 For the following pairs of gases at the given conditions, predict which one would more closely follow the ideal gas law. Explain your choice.
(a) nitrogen (boiling point = −196 °C) or butane (boiling point = −1 °C), both measured at 25 °C and 1 atm
(b) Oxygen gas at 0.50 atm or at 150 atm, both measured at 200 °C
(c) Argon gas (boiling point = −186 °C) at −160 °C or at 10 °C, both measured at 1.0 atm

OBJECTIVE Use the van der Waals equation to correct deviations observed for the ideal gas law.

6.115 Calculate the pressure, in atm, of 10.2 mol argon at 530 °C in a 3.23-L container, using both the ideal gas law and the van der Waals equation.

6.116 Calculate the pressure, in atm, of 1.55 mol nitrogen at 530 °C in a 3.23-L container, using both the ideal gas law and the van der Waals equation.

6.117 Calculate the pressure, in atm, of 13.9 mol neon at 420 °C in a 4.73-L container, using both the ideal gas law and the van der Waals equation.

6.118 Calculate the pressure, in atm, of 5.75 mol methane (CH_4) at 440 °C in a 4.93-L container, using both the ideal gas law and the van der Waals equation.

Chapter Exercises

6.119 It is important to check the pressure in car tires at the start of the winter, because the large temperature change will cause the pressure to drop. Calculate the pressure change in a tire inflated to 32 pounds per square inch (psi) at 90 °F if the temperature declines to 32 °F. Assume that atmospheric pressure is 15 psi; this information is important because the tire pressure is measured as that above atmospheric pressure.

6.120 Workers at a research station in the Antarctic collected a sample of air to test for airborne pollutants. They collected the sample in a 1.00-L container at 764 torr and −20 °C. Calculate the pressure in the container when it was opened for analysis in a particulate-free clean room in a laboratory in South Carolina, at a temperature of 22 °C.

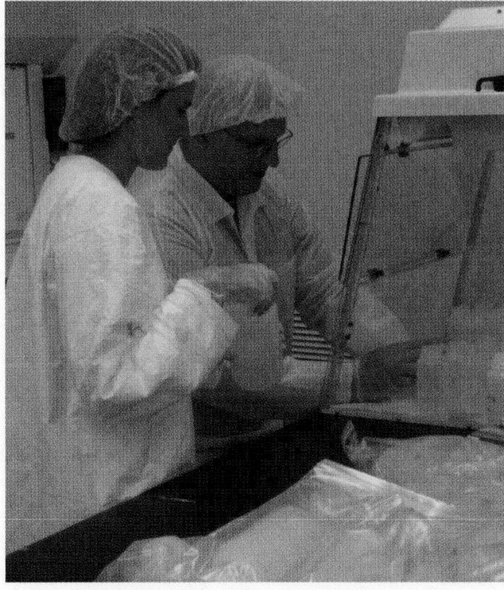

Courtesy of Scott Goode

6.121 A 2.8-L tank is filled with 0.24 kg oxygen. What is the pressure in the tank at 20 °C? Assume ideal behavior.

6.122 A 1.26-g sample of a gas occupies a volume of 544 mL at 27 °C and 744 torr? What is the molecular formula and name of the gas if its empirical formula is C_2H_5.

6.123 ▲ To lose weight, we are told to exercise to "burn off the fat." Although fat is a complicated mixture, it has approximately the formula $C_{56}H_{108}O_6$. Calculate the volume of oxygen that must be consumed at 22 °C and 1.00 atm of pressure to "burn off" 5.0 pounds of fat. (*Hint:* Start by writing the equation for the combustion of the fat.)

6.124 ▲ Three bulbs are connected by tubing, and the tubing is evacuated. The volume of the tubing is 22.0 mL. The first bulb has a volume of 50.0 mL and contains 2.00 atm argon, the second bulb has a volume of 250 mL and contains 1.00 atm neon, and the third bulb has a volume of 25.0 mL and contains 5.00 atm hydrogen. If the stopcocks (valves) that isolate all three bulbs are opened, what is the final pressure of the whole system?

6.125 Calculate the mass of water produced in the reaction of 4.33 L oxygen and 6.77 L hydrogen gas. Both gases are at a pressure of 1.22 atm and a temperature of 27 °C.

6.126 ▲ Calculate (a) the rms speed (in m/s) of samples of hydrogen and nitrogen at STP; and (b) the average kinetic energies per molecule (in kg m^2/s^2) of the two gases under these conditions.

6.127 Lithium hydroxide is used to remove the CO_2 produced by the respiration of astronauts. An astronaut produces about 400 L CO_2 at 24 °C and 1.00 atm of pressure every 24 hours. What mass of lithium hydroxide, in grams, is needed to remove the CO_2 produced by the astronaut in 24 hours? The equation is

$$2LiOH(s) + CO_2(g) \rightarrow Li_2CO_3(s) + H_2O(\ell)$$

6.128 (a) Use the van der Waals equation to calculate the pressure, in atm, of 30.33 mol hydrogen at 240 °C in a 2.44-L container.
(b) Do the same calculation for methane under the same conditions.
(c) What difference between the two gases causes the pressure in the containers to be different?

6.129 The graphs below represent two plots of average speed of a gas versus the number of particles with that speed.
(a) If one plot is for argon and the other neon, which plot would be for neon?
(b) If both gases are the same, which plot represents the gas at a greater temperature?

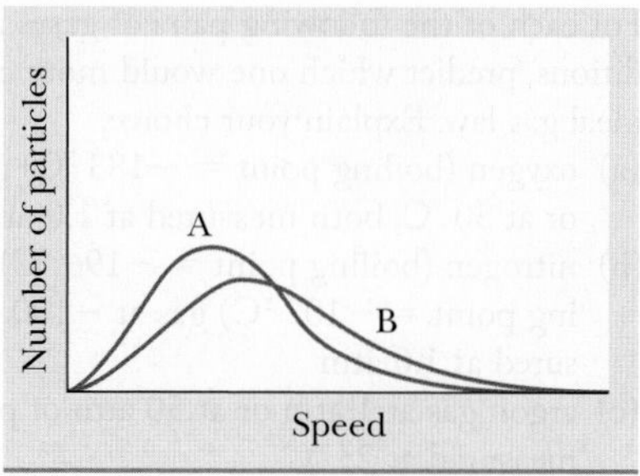

6.130 Draw a plot similar to that shown above for the following:
(a) Ar at 0 °C (b) Ar at 100 °C

Cumulative Exercises

6.131 An enzyme in yeast can convert pyruvic acid, $C_3H_4O_3$, to CO_2 and C_2H_4O. What volume of CO_2 gas is produced from 0.113 g pyruvic acid if the gas is collected at 755 torr and 25 °C?

6.132 Diborane, B_2H_6, is a gas at 744.0 torr and 120.0 °C. It reacts violently with $O_2(g)$, yielding $B_2O_3(s)$ and water vapor. The reaction is so energetic that it once was considered as a possible rocket fuel. What is ΔH for the reaction of 1 mol B_2H_6 if the reaction of 2.329 L of B_2H_6 under the above conditions yields 143.9 kJ heat at constant pressure?

6.133 A 10.0-L container is filled with 2.66 g H_2 and 4.88 g Cl_2, and heated to 111 °C. After a few days, the $H_2(g)$ and $Cl_2(g)$ had reacted to form HCl(g), but some of one of the reactant gases remains because it is present in excess. What is the pressure in the container?

6.134 A compound that contains only hydrogen and carbon is burned in oxygen gas at 180 °C and 755 torr of pressure to produce 1.23 L $H_2O(g)$ and 0.984 L $CO_2(g)$. What is the empirical formula of the compound?

6.135 Combustion of a 4.33-g sample of a compound yielded 2.20 L CO_2 gas at a temperature of 27 °C and a pressure of 0.99 atm. What is the percentage of carbon in the sample?

6.136 If a 100.0-mL sample of 0.88 M H_2O_2 (hydrogen peroxide) solution decomposes into oxygen gas and water, what volume of oxygen is produced at a temperature of 22 °C and a pressure of 0.971 atm?

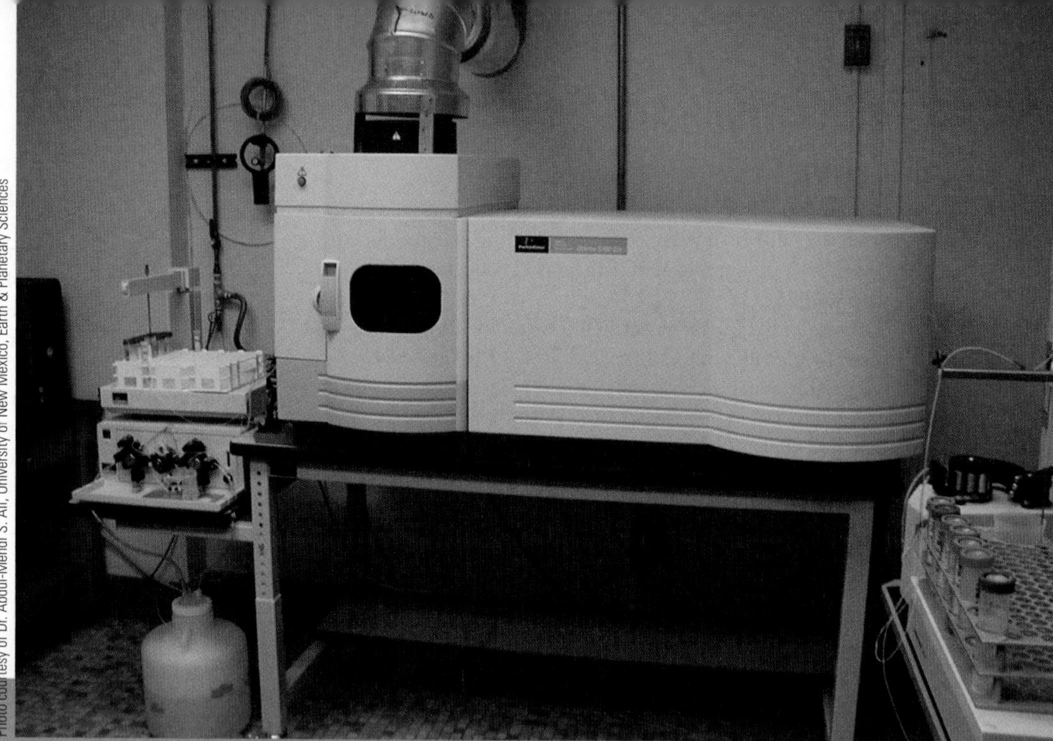

Photo courtesy of Dr. Abdul-Mehdi S. Ali, University of New Mexico, Earth & Planetary Sciences

Modern emission spectrometers can measure concentrations of up to 40 elements in less than a minute.

Chemists have studied the interaction of matter and energy for hundreds of years. For example, when an element encounters a high-temperature environment, that element can emit light. The heat elevates some of the element's electrons from the normal lowest energy state of the atom (the "ground state") to a higher energy state (an "excited state"). The atoms then emit this excess energy as light when the electron moves back to the lowest energy state. Studying the light emitted provides insight into how electrons are arranged in atoms, which is the subject of this chapter.

The emission of light from the high-temperature atoms also provides information about the composition of a sample of matter. The properties of the light identify the energy level of the excited state of the atom, and these levels are unique. All elements have a different set of energy levels, and elements can be distinguished from one another based on these levels. In addition, the amount of light provides information about the quantity of the element present. A technique called *atomic emission spectroscopy* is one of the most widely used methods for determining the elemental composition of a sample of matter. Emission spectroscopy has been used for decades in areas such as metallurgy, water analysis, environmental samples, and forensic science.

When investigating and prosecuting crimes involving firearms, forensic scientists often need to analyze bullets, both their physical markings (see photo on page 249) and their chemical composition. The Federal Bureau of Investigation (FBI) has been using the compositional analysis of bullet lead since the 1970s to help determine whether a bullet found at a crime scene can be matched to one found in the possession of a suspect. The FBI argues that although thousands of bullets are produced with essentially identical concentrations, the bullets that match most closely are packaged in the same box. Chemical analysis can be used to determine whether bullets come from the same or different sources.

Using atomic emission spectroscopy, the FBI analyzed samples from four major manufacturers and found a number of elements in the bullet lead.

Electronic Structure 7

CHAPTER CONTENTS

OWL Online homework for this chapter may be assigned in OWL.

Look for the green colored vertical bar throughout this chapter, for integrated references to this chapter introduction.

The FBI concluded that the wide ranges in concentrations of all of these elements allowed them to distinguish among thousands of different packages of bullets.

Concentrations of Elements in Bullet Lead as Determined by Atomic Emission Spectroscopy

Brand	As	Sb	Sn	Cu	Bi	Ag
CCI	—	23,800–29,900	—	97–381	56–180	18–69
Federal	1127–1645	25,700–29,000	1100–2880	233–329	30–91	14–19
Remington	—	5670–9620	—	62–962	67–365	21–118
Winchester	—	2360–6650	—	54–470	35–208	14–61

Concentrations are measured in units of microgram (μg) of element per gram (g) of bullet. Dashes indicate that none of that particular element was detected.

Data from Peters CA. *Comparative Elemental Analysis of Firearms Projectile Lead by ICP-OES.* Washington, DC: FBI Laboratory Chemistry Unit. October 11, 2002.

In the past decade, a growing body of research has revealed that the practice of chemically matching bullets is seriously flawed. In February 2005, a select committee of the Board of Chemical Sciences and Technology of the National Academy of Science issued a report that asked the FBI to limit how its examiners present their data in the courtroom. The report suggests that when two bullets have matching compositions, instead of stating they came from the same box of ammunition, an FBI expert should be instructed to testify that there is an increased probability that the two bullets came from a "compositionally indistinguishable volume of lead." The experts were asked to explain to jurors that the same composition is found in as few as 12,000 bullets or as many as 35 million bullets. Currently, elemental analysis of this sort can be used as strong evidence that a bullet is *not* from a particular lot rather than as evidence that a bullet must be included within a small batch, such as a box, of bullets. ▮

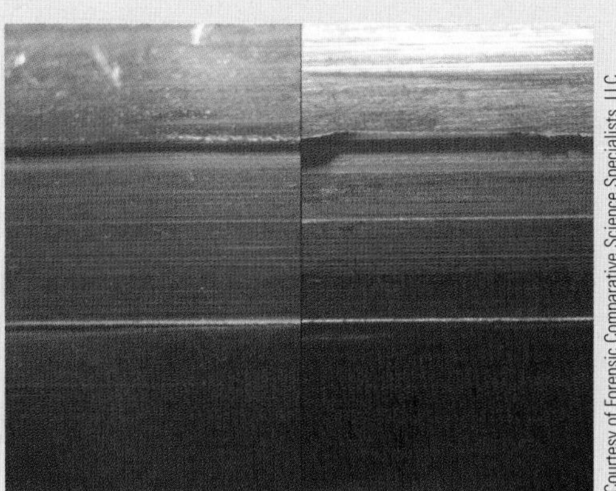

Courtesy of Forensic Comparative Science Specialists, LLC

Gun barrels are grooved to increase a bullet's accuracy. These grooves are unique to a gun's make and model, and the marks they make on bullets are visible under a microscope. The continuity of the scratches shows the two bullets were fired from the same gun.

Chapter 2 presented the structure of an atom as initially proposed by Rutherford: massive protons and neutrons in a central nucleus, the lighter electrons occupying the space around the nucleus. Rutherford's model did not, however, address exactly *how* the electrons occupied the space around the nucleus.

Initially, it was assumed that electrons held fixed orbits about the nucleus, leading to the so-called planetary model of the atom. Experimental data soon indicated that the true situation is more complex. In particular, evidence that small pieces of matter, such as electrons, exhibited wave behavior required that the behavior of electrons be understood in different terms. Advances in the 1920s and 1930s helped scientists develop a better understanding of how electrons participate in atomic structure.

Although the planetary model is useful, it is too simplistic. The arrangement of electrons in atoms is more complicated than that model. We now use a model of electronic structure that agrees with experimental evidence. This model also helps us understand some of the properties of atoms (see discussion in Chapter 8), how the atoms can make positively and negatively charged ions (see Chapter 9), and how atoms can combine to make molecules (see Chapter 10). This chapter presents the current model of how electrons are arranged in atoms.

Because much of the knowledge of the arrangement of electrons in atoms is based on observations of their interaction with light, we must first consider the nature of electromagnetic radiation. Keep in mind that the main goal in this chapter is to understand the properties of electrons in atoms.

7.1 The Nature of Light

OBJECTIVES

- ☐ Describe the relationships among the wavelength, frequency, and energy of electromagnetic radiation
- ☐ Describe the models that are used to explain the behavior of light
- ☐ Calculate the quantized energy of light

Under certain conditions, atoms and molecules emit and absorb energy in the form of light. The nature of light is key to the modern description of the atom.

The Wave Nature of Light

In the late 19th century, physicists knew that light could be described as **waves** similar to the waves that move through water. In water, a disturbance produces an up-and-down motion of the surface. Although the crests of the waves move horizontally with time, both the liquid and an object floating on it simply move up and down, as shown in Figure 7.1. Waves are periodic in nature: They repeat at regular intervals of both time and distance.

Any wave is described by its wavelength, frequency, and amplitude, some of which are shown in Figure 7.2. The **wavelength** (λ, lambda) is the distance between one peak and the next. In the SI system, wavelength is measured in meters, although other units

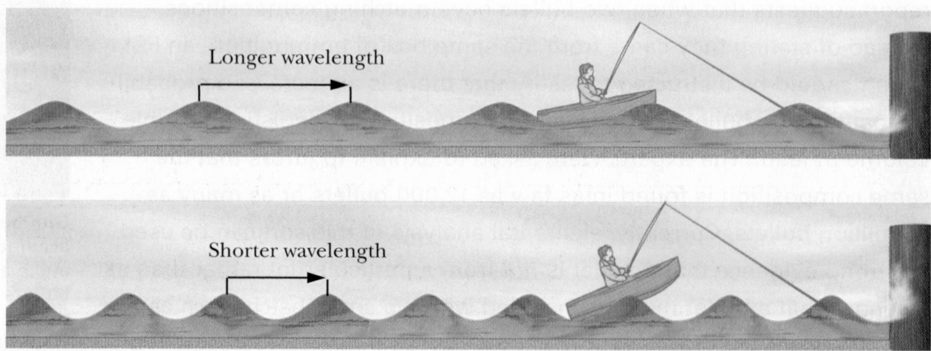

Figure 7.1 Water waves. The distance between two neighboring peaks is called the *wavelength*.

Waves. The periodic nature of wave motion is not always easily seen.

of length are common. The **frequency** (ν, nu) of a wave is the number of waves that pass a fixed point in 1 second. The SI unit for frequency is s^{-1} (standing for 1/s, and spoken of as "per second") and is called **hertz (Hz).** The maximum height of a wave is called its **amplitude;** the height of a wave varies between $+A_{max}$ and $-A_{max}$.

Light waves are called **electromagnetic radiation** because they consist of oscillating electric and magnetic fields, which are perpendicular to each other and perpendicular to the direction of propagation, as shown in Figure 7.3. The periodic variations of the electric and magnetic fields of light are analogous to the motion of the water in Figure 7.1.

The speed at which a wave travels is the product of its wavelength and frequency. *The experimentally measured speed of light shows that all electromagnetic radiation travels at the same speed in a vacuum, no matter what its wavelength.* The speed of light in a vacuum, 3.00×10^8 m/s (rounded to three significant digits), is one of the fundamental constants of nature.

$$c = \lambda\nu = 3.00 \times 10^8 \text{ m/s} \qquad [7.1]$$

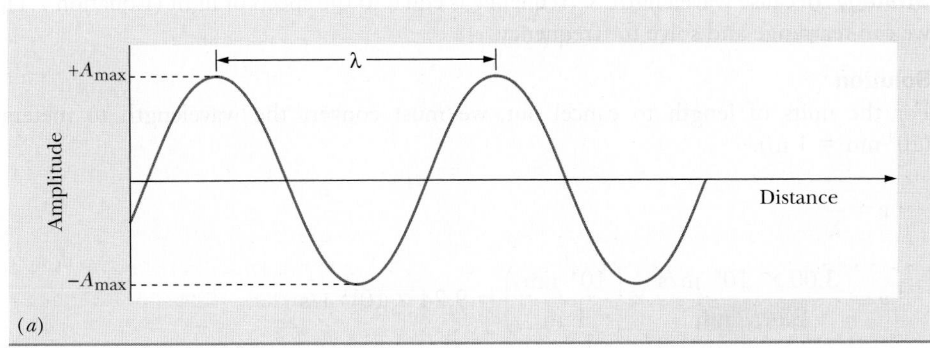

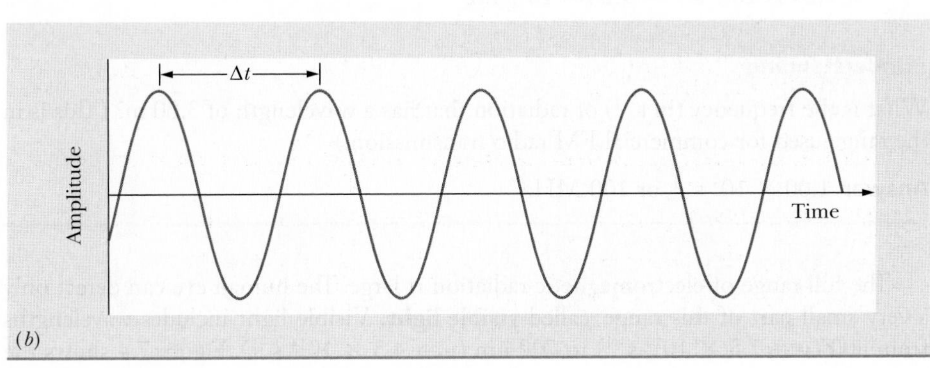

Figure 7.2 Typical waves. *(a)* The wavelength, λ, is the distance between two successive peaks of the wave, and the amplitude is the vertical displacement from the undisturbed medium. The amplitude, A, varies from $-A_{max}$ to $+A_{max}$. *(b)* The length of time it takes for one complete wave to pass a point is Δt. The frequency of the wave, ν, is the number of waves that pass a point in each second.

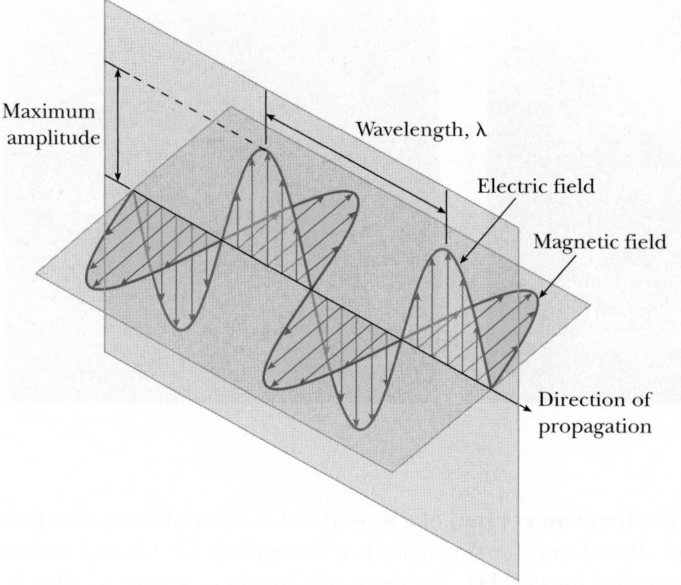

Figure 7.3 Electromagnetic radiation. Light, or electromagnetic radiation, consists of oscillating electric and magnetic fields that have the same frequency and wavelength but are perpendicular to each other and to the direction of motion of the wave.

If either the wavelength or the frequency of electromagnetic radiation is known, the other can be calculated from the equation $c = \lambda\nu$.

Note from Equation 7.1 that, as the wavelength of the electromagnetic radiation increases, the frequency decreases, and vice versa. Because the speed of light is a constant, a known wavelength or frequency allows us to calculate the other, as shown in Example 7.1.

EXAMPLE 7.1 Frequency and Wavelength

The table in the introduction to this chapter shows that copper is a common component of bullets. A characteristic light emission of excited copper atoms occurs at 324.7 nm. What is the frequency of this light?

Strategy Because wavelength × frequency is equal to the speed of light (Equation 7.1), we can rearrange and solve for frequency.

Solution

For the units of length to cancel out, we must convert the wavelength to meters (10^9 nm = 1 m).

$$\nu = \frac{c}{\lambda}$$

$$\nu = \frac{3.00 \times 10^8 \text{ m/s}}{324.7 \text{ nm}} \times \left(\frac{10^9 \text{ nm}}{1 \text{ m}}\right) = 9.24 \times 10^{14} \text{ 1/s}$$

$$= 9.24 \times 10^{14} \text{ s}^{-1} = 9.24 \times 10^{14} \text{ Hz}$$

Understanding

What is the frequency (in s^{-1}) of radiation that has a wavelength of 3.00 m? (This is in the range used for commercial FM radio transmission.)

Answer 1.00×10^8 s^{-1}, or 100 MHz

The full range of electromagnetic radiation is large. The human eye can detect only a very small part of this range, called **visible light.** Visible light includes wavelengths from 400 ($\nu = 7.5 \times 10^{14}$ s^{-1}) to 700 nm ($\nu = 4.3 \times 10^{14}$ s^{-1}). Figure 7.4 shows the

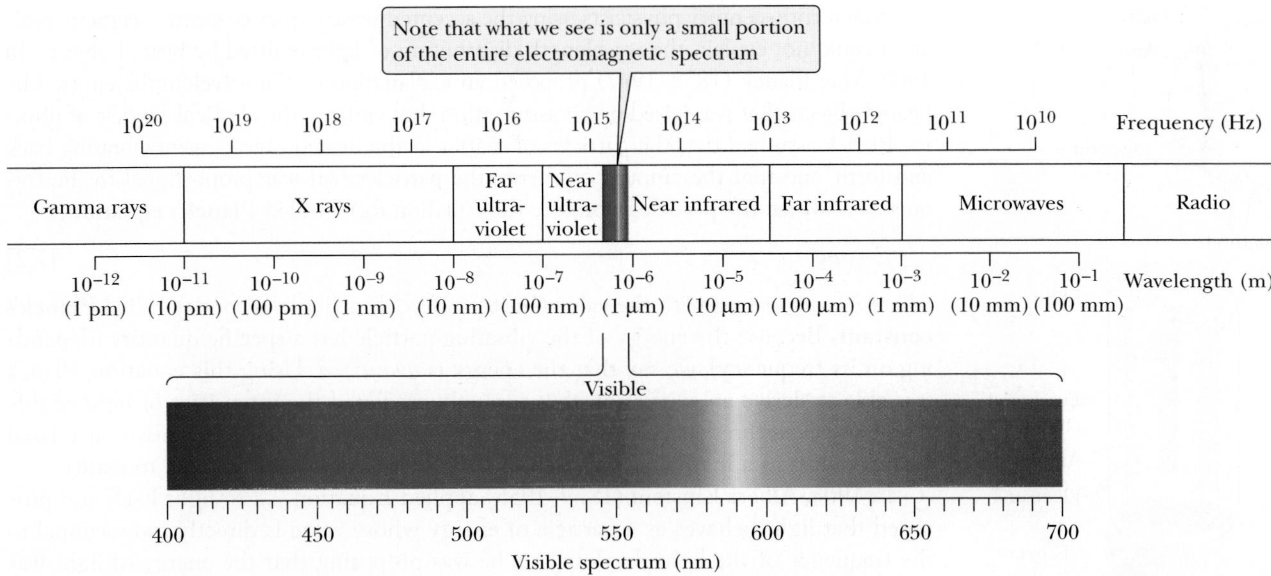

Figure 7.4 Electromagnetic spectrum. The range of electromagnetic radiation is shown, and names commonly used to refer to different regions are identified. Both frequencies and wavelengths are shown. Divisions between the regions are not defined precisely.

full range of electromagnetic radiation together with the common names used to identify different ranges of wavelengths. We encounter many of these names in everyday conversation, such as x rays used for medical diagnosis, microwaves used to heat food, and radio waves used in communication.

Quantization of Energy

At temperatures greater than absolute zero (0 K), matter emits electromagnetic radiation of all wavelengths, and the emission is referred to as a continuum. Not all wavelengths of light are emitted with equal intensity, however. The distribution of the intensity of the different wavelengths changes with temperature. A dull red glow might be emitted from an electric stove's heating element, whereas the white light from a common light bulb is produced by the electrical heating of a small tungsten wire to a much greater temperature.

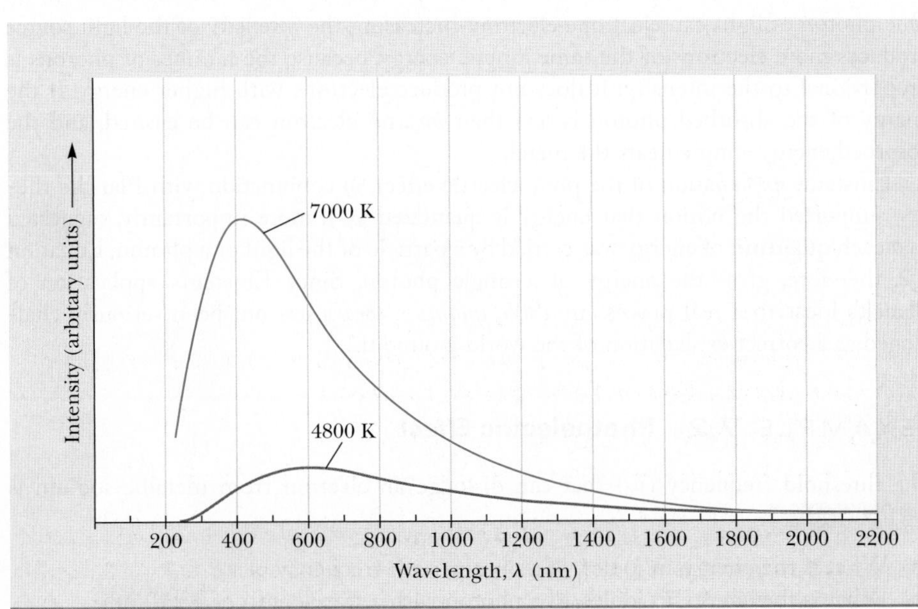

Intensity of light. The intensity of light emitted by an object varies by wavelength and by temperature, as illustrated by these curves.

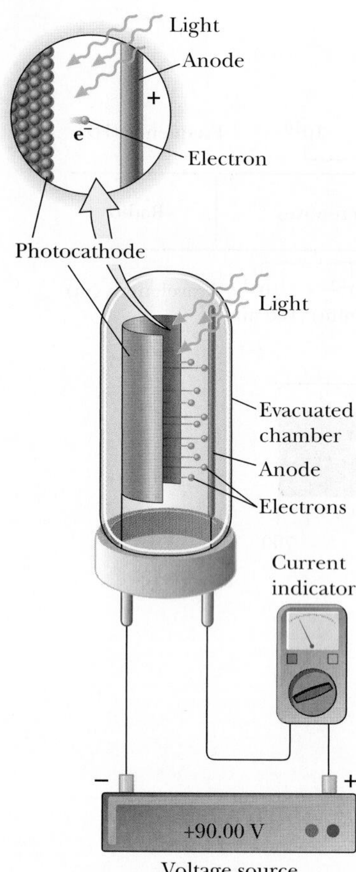

Photoelectric cell. When light of high enough frequency strikes the metal surface in the tube, electrons are ejected. The electrons are attracted to the other electrode in the cell, producing an electric current in the external circuit. Some automatic door openers are activated by the electric current from a photoelectric cell.

Nineteenth-century physicists using the accepted wave theory of electromagnetic radiation could not explain the wavelength distribution of light emitted by heated objects. In 1900, Max Planck (1858–1947) proposed an explanation of the wavelengths emitted by heated objects that was based on an assumption that violated the classical models of physics. Planck assumed that the particles of matter in the heated objects were vibrating back and forth, and that the amount of energy the particles had was proportional to the frequency at which the particles vibrated. The equation form, called **Planck's equation,** is

$$E = h\nu \qquad [7.2]$$

where h is a constant with the value 6.626×10^{-34} J · s (joule-seconds), called **Planck's constant.** Because the energy of the vibrating particle has a specific quantity (depending on its frequency), we say that the energy is *quantized.* Using this equation, Planck was able to derive an expression that correctly predicted the intensities of light of different wavelengths that are given off by objects. However, many scientists dismissed Planck's ideas as a mathematical trick that did work but was not related to reality.

In 1905, Albert Einstein (1879–1955) applied Equation 7.2 to light itself and proposed that light behaves as a particle of energy whose value is directly proportional to the frequency of the light. In doing so, he was proposing that the energy of light was quantized—that is, it could have only a certain amount of energy. Einstein used this idea to explain the **photoelectric effect,** the process in which electrons are ejected from a solid metal when it is exposed to light. Each metal has a characteristic minimum frequency of light, ν_0, that is necessary before any electrons are emitted. As the frequency of light increases from ν_0, the kinetic energy of the ejected electrons also increases. Light of lower frequency than this threshold, no matter how intense, does not eject any electrons. More intense light does not increase the kinetic energy of the electrons, but it does increase the number of electrons emitted. These observations contradicted the predictions of classical physics. In the classical wave picture of light, any frequency of light, as long as it was bright enough, could eject electrons.

Einstein interpreted these results by applying Planck's theory. He suggested that light, in addition to having the properties of waves, could also be viewed as a stream of tiny particles, now referred to as **photons.** A *single* photon with an energy of $h\nu$ must provide enough energy to dislodge an electron from the solid. Some of the energy, $h\nu_0$, must be used to overcome the attraction the solid has for the electrons, and the rest appears as the kinetic energy (KE) of the electron.

$$h\nu = h\nu_0 + \text{KE}$$

One photon of light can eject one electron. Increasing the intensity of the light source produces more electrons of the same kinetic energy, because the number of photons is proportional to the intensity; it does not produce electrons with higher energy. If the energy of the absorbed photon is less than $h\nu_0$, no electron can be ejected, and the absorbed energy simply heats the metal.

Einstein's explanation of the photoelectric effect, in conjunction with Planck's theory, supported the notion that energy is quantized and, more importantly, suggested that each **quantum** of energy was carried by a particle of the light or a photon. Equation 7.2, therefore, gives the energy of a single photon. Since Einstein's application of Planck's ideas to a real process in 1905, *quantum theory* has not been seriously challenged as a correct explanation of the world around us.

In the interpretation of the photoelectric effect, electromagnetic radiation is treated as particles of light (photons) instead of waves.

EXAMPLE 7.2 Photoelectric Effect

The threshold frequency (ν_0) that can dislodge an electron from metallic sodium is $5.51 \times 10^{14}\ \text{s}^{-1}$.

(a) What is the energy, in joules, of a photon with frequency of ν_0?
(b) What is the energy, in joules, of a photon with a wavelength of 430.0 nm?

(c) What is the kinetic energy, in joules, of an electron that is ejected from sodium by light with a wavelength of 430.0 nm?
(d) What is the energy (in kJ/mol) of a mole of photons with frequency of ν_0?

Strategy (a) Use Planck's relationship between energy and frequency to determine the energy of the photon. (b) Use the relationship between c, λ, and ν to first determine the frequency of the photon, then Planck's relationship to determine its energy. (c) Use Einstein's relationship between E and the threshold frequency ν_0 to determine the kinetic energy of the ejected electron. (d) Multiply the energy of one photon by Avogadro's number and convert to kJ to determine the energy of a mole of photons having a frequency of ν_0.

Solution

(a) Equation 7.2 gives the energy of a photon:

$$E_0 = 6.626 \times 10^{-34}\ \text{J} \cdot \cancel{\text{s}} \times 5.51 \times 10^{14}\ \cancel{\text{s}}^{-1} = 3.65 \times 10^{-19}\ \text{J}$$

(b) First, calculate the frequency of the photon from its wavelength:

$$\nu = \frac{c}{\lambda} = \frac{3.00 \times 10^{8}\ \cancel{\text{m}}/\text{s}}{430.0\ \cancel{\text{nm}}} \times \left(\frac{10^{9}\ \cancel{\text{nm}}}{1\ \cancel{\text{m}}}\right) = 6.98 \times 10^{14}\ \text{s}^{-1}$$

Second, calculate the energy of the photon from Planck's equation.

$$E_{\text{photon}} = h\nu = 6.626 \times 10^{-34}\ \text{J} \cdot \cancel{\text{s}} \times 6.98 \times 10^{14}\ \cancel{\text{s}}^{-1} = 4.63 \times 10^{-19}\ \text{J}$$

(c) Using the energies found in parts (a) and (b), calculate the kinetic energy of the electron.

$$h\nu = h\nu_0 + \text{KE}$$

The quantities $h\nu_0$ and $h\nu$ were calculated in parts (a) and (b), respectively. Substituting:

$$4.63 \times 10^{-19}\ \text{J} = 3.65 \times 10^{-19}\ \text{J} + \text{KE}$$

$$\text{KE} = 9.8 \times 10^{-20}\ \text{J}$$

(d) Using the energy from (a) and multiplying by Avogadro's number:

$$3.65 \times 10^{-19}\ \cancel{\text{J}} \times 6.022 \times 10^{23}/\text{mol} \times \left(\frac{1\ \text{kJ}}{1000\ \cancel{\text{J}}}\right) = 220.\ \text{kJ/mol}$$

To put this molar energy into comparison, the bond energy of a C-H bond is about 400 kJ/mol. A photon having this frequency has only about half the energy needed to break a C-H bond.

Understanding

Light with a wavelength of 450.0 nm strikes metallic cesium and ejects electrons with a kinetic energy of 1.22×10^{-19} J. What is the photoelectric threshold frequency (in s^{-1}) for cesium?

Answer $4.83 \times 10^{14}\ \text{s}^{-1}$

The Dual Nature of Light?

Is light a particle, or is light a wave? It depends on what property of light you are measuring. Light refracts, reflects, interferes, and can be described with a wavelength and frequency. In these regards, light behaves as a wave. Light also behaves as a "particle" of energy. Some people speak of a "dual nature of light," going so far as to use the word *wavicle* to describe light. But perhaps the issue is more our own prejudices, rather than the nature of light. We presume that a phenomenon must be *either* a particle *or* a wave, and that the two are mutually exclusive. Up to 1900, such a dichotomy was valid based on observations of the world around us, but not now. Light has *both* wave and particle properties, depending on the property. Realizing this is an important step forward, for later in this chapter, we explain that matter, typically viewed as particulate in nature, has wave properties.

OBJECTIVES REVIEW *Can you:*

- ☑ describe the relationships among the wavelength, frequency, and energy of electromagnetic radiation?
- ☑ describe the models that are used to explain the behavior of light?
- ☑ calculate the quantized energy of light?

7.2 Line Spectra and the Bohr Atom

OBJECTIVES

- ☐ Describe the origin of atomic line spectra
- ☐ Calculate the observed lines in the emission and absorption spectra of the hydrogen atom
- ☐ Relate the electron energy levels in the hydrogen atom to the observed line spectrum and the Bohr model

When energy in the form of heat or an electric discharge is added to a sample of gaseous atoms in a process called *excitation,* the atoms can emit some of the added energy as light. Examination of the **spectrum** (the intensity of the light as a function of wavelength) reveals that the light from excited atoms is quite different from the light emitted by a heated solid. The heated solid produces a **continuous spectrum,**[1] one in which all wavelengths of light are present (next page, top). The light emitted by excited atoms (the atomic emission spectrum) is very different and is called a **line spectrum** because it contains light only at specific wavelengths. Figure 7.5 is a schematic representation of the experimental observation of a line spectrum. Each element produces a line spectrum that is characteristic of that element and different from the spectrum of any other element. Long before scientists understood the reason for this behavior, they used line spectra to identify the elements present in samples of matter. In fact, in the 1860s, the presence of unexpected emission lines observed in some samples of sodium and potassium led to the discovery of the elements cesium and rubidium. Figure 7.6 shows the line spectra of several elements.

The spectrum of the hydrogen atom was particularly simple: Four lines in the visible region of the spectrum, getting progressively closer together (Figure 7.5). Examination of other regions of the spectrum showed that other series of lines existed as well. A study

[1]Some references refer to it as a "continuum spectrum."

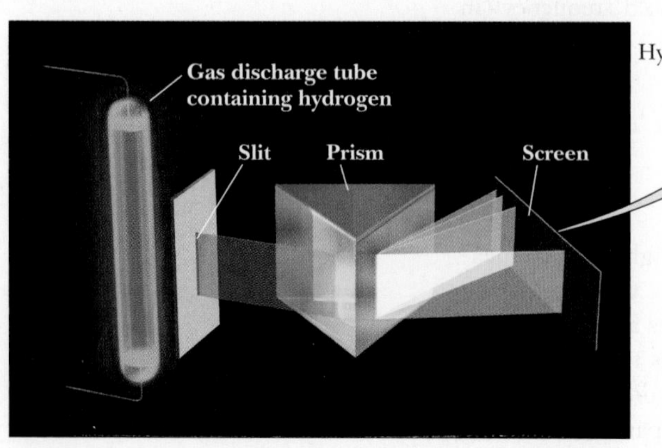

Figure 7.5 Experimental observation of a spectrum. First light from a source passes through a slit; then a prism separates it by wavelength. The separated light produces an image on a detector. The spectrum shown is that of hydrogen.

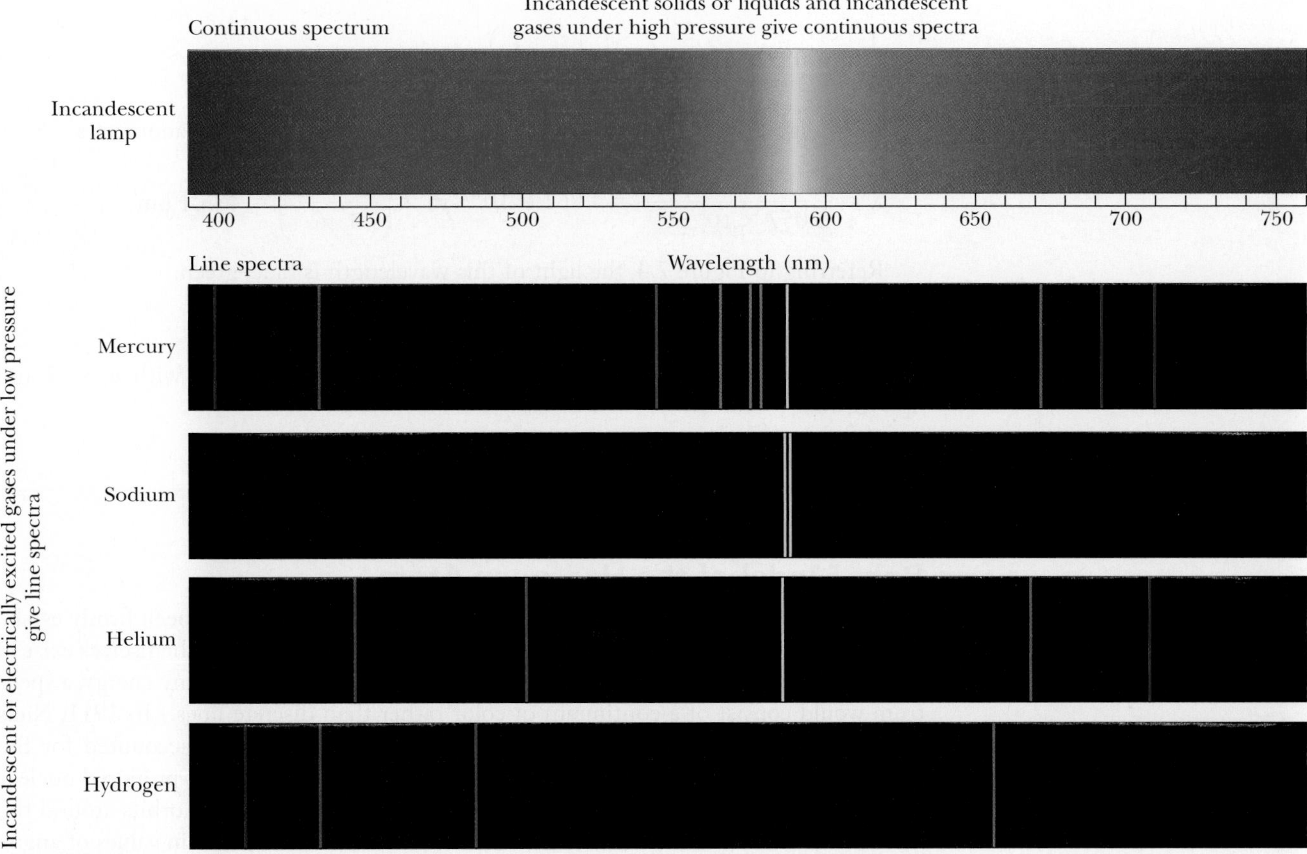

Figure 7.6 Emission spectra. The emission spectra in the visible region for an incandescent light (top) and several elements.

of the wavelengths of the lines by J. R. Rydberg in 1890 revealed that the wavelengths of all the lines could be predicted by a simple formula called the **Rydberg equation:**

$$\frac{1}{\lambda} = R_H\left(\frac{1}{n_1^2} - \frac{1}{n_2^2}\right) \qquad [7.3]$$

Here, n_1 and n_2 are positive integers with $n_1 < n_2$, and R_H is a constant, called the **Rydberg constant,** with a value of 1.097×10^7 m^{-1}. Notably, this equation was determined empirically, based solely on the experimentally observed wavelengths of lines in the spectrum of the hydrogen atom. It correctly gives the wavelengths of the light emitted by the H atom, but at the time, there was no theoretical explanation for this correlation. The hydrogen atom spectrum consists of series of lines that are named after the individuals who discovered them: The Lyman ($n_1 = 1$), Balmer ($n_1 = 2$), Paschen ($n_1 = 3$), Brackett ($n_1 = 4$), and Pfund ($n_1 = 5$) series. All lines in any series have the same value of n_1, with each line having a different value for n_2.

The Rydberg equation accurately predicts the wavelengths of all the observed lines in the spectrum of hydrogen atoms.

EXAMPLE 7.3 Calculating Wavelengths from the Rydberg Equation

Calculate the wavelength (in nm) of the line in the hydrogen spectrum for $n_1 = 2$ and $n_2 = 4$ (the second line of the Balmer series).

Strategy The Rydberg equation is used to relate the wavelength of light to n_1 and n_2.

Solution

Substitute the values for n_1 and n_2 into Equation 7.3:

$$\frac{1}{\lambda} = 1.097 \times 10^7 \text{ m}^{-1} \times \left(\frac{1}{2^2} - \frac{1}{4^2}\right) = 2.057 \times 10^6 \text{ m}^{-1}$$

Rearrange to solve for the wavelength, and convert the units to nanometers.

$$\lambda = \frac{1}{2.057 \times 10^6 \text{ m}^{-1}} = 4.861 \times 10^{-7} \text{ m} \times \left(\frac{10^9 \text{ nm}}{1 \text{ m}}\right) = 486.1 \text{ nm}$$

Referring to Figure 7.4, the light of this wavelength is blue–green.

Understanding

Find the wavelength (in nm) of the next line in the Balmer series, with $n_1 = 2$ and $n_2 = 5$.

Answer 434.0 nm

Bohr Model of the Hydrogen Atom

Once the relation between the energy of light and its frequency had been firmly established, the discrete line spectra of atoms suggested that the electrons themselves exist in only certain allowed energy levels. (After all, if electrons could have any energy, a spectrum would consist of a continuum of color rather than discrete lines.) In 1911, Niels Bohr (1885–1962) proposed a model for the hydrogen atom that accounted for the observed spectrum of hydrogen. Bohr began with Ernest Rutherford's proposed nuclear model for the atom, and assumed that the electron moved in circular orbits around the nucleus. Bohr further assumed that the electron could have only certain values of angular momentum (i.e., momentum of a mass moving in a circle). From these assumptions, Bohr found that the allowed radii and energies are also quantized, and the allowed energies are given by

$$E_n = -\frac{2\pi^2 me^4}{h^2}\left(\frac{1}{n^2}\right) = \frac{-B}{n^2} \qquad [7.4]$$

where m is the mass of the electron, e is charge of the electron, h is Planck's constant, and n is a positive integer that indicates the electron's energy level. Substitution of the values for π, m, e, and h, after some unit conversions, gives a value of $B = 2.18 \times 10^{-18}$ J. The allowed energies, E_n, are found by using any positive integer for n (1, 2, 3, …) in Equation 7.4; therefore, many different energy levels are possible. The lowest energy level of the hydrogen atom with $n = 1$ is, therefore, -2.18×10^{-18} J. Bohr concluded that the energy levels of the electron in a hydrogen atom are *quantized.*

The existence of line spectra for the elements suggests that the energies of atoms are quantized.

Bohr realized that the light emitted by the atom must have energy ($h\nu$) that is exactly equal to the difference between the energies of two of its allowed levels:

$$E_{\text{light}} = E_2 - E_1 = -\frac{B}{n_2{}^2} - \left(-\frac{B}{n_1{}^2}\right) = \frac{B}{n_1{}^2} - \frac{B}{n_2{}^2}$$

$$h\nu = B\left(\frac{1}{n_1^2} - \frac{1}{n_2^2}\right) = 2.18 \times 10^{-18}\left(\frac{1}{n_1^2} - \frac{1}{n_2^2}\right) \qquad [7.5]$$

Equation 7.5 is equivalent to the Rydberg equation. Comparison of Equations 7.3 and 7.5 shows that the value of R_H, the Rydberg constant, is B/hc. The value of the Rydberg constant calculated from the Bohr model is nearly identical to that found experimentally. The ability to calculate the experimental value of the Rydberg constant in terms of other physical constants was a major triumph of the Bohr model.

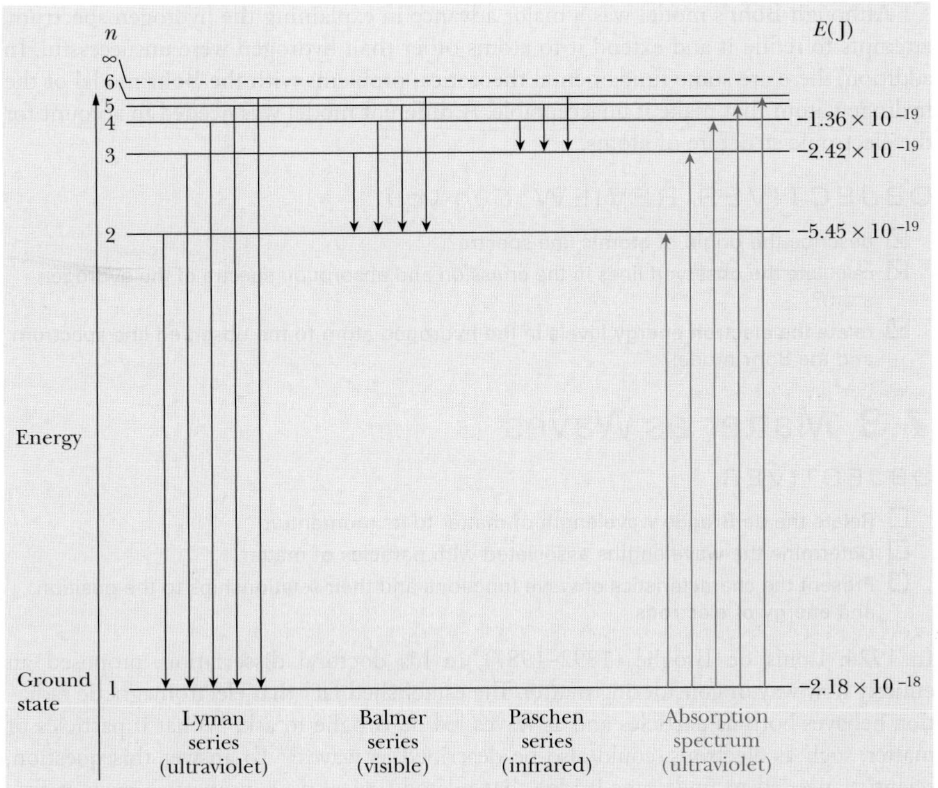

Figure 7.7 Transitions in the hydrogen atom. *(Left)* Electron transitions that produce the lines in the Lyman, Balmer, and Paschen series in the emission spectrum of hydrogen. *(Right)* The absorption spectrum contains only the lines in the Lyman series, because in a sample of hydrogen, nearly all the atoms are in the ground state, which has $n = 1$.

Figure 7.7 shows the energy-level diagram for the hydrogen atom. When the electron has been completely removed from the atom ($n = \infty$), the energy is zero. As the electron and the H^+ nucleus move closer together, the atom becomes more stable (lower in energy), so the energies of all the allowed states have a negative sign. Because the energy of an allowed state is proportional to $1/n^2$, the energies of the allowed states get closer together as n increases.

The vertical arrows in Figure 7.7 show the transitions of the electron between the quantized energy states of the atom. When an electron goes from one quantized energy state to a lower one, the difference in energy is released as a single photon. A hydrogen atom with its electron in the $n = 4$ state ($E_4 = -1.36 \times 10^{-19}$ J) may return to the lowest energy state ($n = 1, E_1 = -2.18 \times 10^{-18}$ J) by emitting light in several ways. The electron can return to the $n = 1$ state in one step, by emitting a single photon with an energy equal to the energy difference between the $n = 4$ and $n = 1$ states, or 2.04×10^{-18} J. Alternatively, the same energy change can occur by emission of as many as three photons, corresponding to the energies of the transitions from $n = 4$ to $n = 3$, then from $n = 3$ to $n = 2$, and finally from $n = 2$ to $n = 1$. For each transition, however, the energy must be emitted as a single photon. Each of the spectral series mentioned earlier corresponds to a set of transitions in which the final energy states of the atom are identical. For example, all transitions that end with the hydrogen atom having its electron in the $n = 2$ state belong to the Balmer series. These transitions would be visible if hydrogen were being studied in an emission spectrometer (see discussion in this chapter's introduction).

All of the energy released when an atom goes from one allowed energy state to a lower one is contained in a single photon of light.

The light emitted by the hydrogen atom produces lines in all of the series shown in Figure 7.7. The electron in the hydrogen atom can also be excited to higher levels by the *absorption* of a photon. The only photons absorbed are those with energy identical to the energy difference between two allowed states of the atom. The **ground state** of an atom is its lowest quantized energy state. At normal temperatures, nearly all hydrogen atoms are present in the ground state, so the observed absorption lines arise from transitions from the ground state ($n = 1$) to the excited states ($n > 1$). Thus, the only lines observed in the absorption spectrum of hydrogen atoms are those in the Lyman series.

Although Bohr's model was a major advance in explaining the hydrogen spectrum, attempts to refine it and extend it to atoms other than hydrogen were unsuccessful. In addition, there are some fundamental theoretical problems with the Bohr model of the hydrogen atom that make it unacceptable. A different model was needed to account for the electronic structure of atoms.

OBJECTIVES REVIEW *Can you:*

- ☑ describe the origin of atomic line spectra?
- ☑ calculate the observed lines in the emission and absorption spectra of the hydrogen atom?
- ☑ relate the electron energy levels in the hydrogen atom to the observed line spectrum and the Bohr model?

7.3 Matter as Waves

OBJECTIVES

- ☐ Relate the de Broglie wavelength of matter to its momentum
- ☐ Determine the wavelengths associated with particles of matter
- ☐ Present the characteristics of wave functions and their relationships to the position and energy of electrons

In 1924, Louis de Broglie (1892–1987), in his doctoral dissertation, proposed an entirely new way of considering matter. The established fact that electromagnetic radiation behaves both as particles and as waves led de Broglie to ask, "What if particles of matter, such as electrons, could also be described as waves?" To answer this question, scientists needed to find some bridge that related typical wave properties, such as frequency or wavelength, to properties usually associated with particles of matter. A few years earlier, Arthur Compton (1892–1962) had performed experiments that showed that the momentum of a photon is given by the expression

$$\text{Momentum} = p = h/\lambda$$

de Broglie suggested that the same relationship between wavelength and momentum of a photon might be used to relate the wave and particle properties of matter. The momentum of matter is the product of mass × velocity, so de Broglie proposed the use of the following equation to calculate the wavelength associated with an electron:

$$p = mv = h/\lambda$$

which rearranges to

$$\lambda = h/p = h/mv \qquad [7.6]$$

Thus, de Broglie predicted that a particle of matter would have a wavelength that is inversely proportional to its mass. The smaller the mass, the larger the associated wavelength. Because h is so small, de Broglie wavelengths of particles of matter are extremely small—unless the particle itself is tiny, such as an electron. Equation 7.6 is called the **de Broglie equation.**

Only a few years later, in 1927, American physicists Clinton Davisson and Lester Germer performed an experiment in which they observed the diffraction of electrons by a crystal of nickel metal. Diffraction, however, is a property of *waves.* The electron diffraction experiments confirmed that the de Broglie equation correctly calculated the wavelength of the electrons, and that small particles of matter do exhibit wave properties.

The Davisson–Germer experiment, which demonstrated that matter exhibits wave properties and particle properties, was a significant step forward in understanding the properties of the electron.

EXAMPLE 7.4 Calculating the Wavelength of an Electron

Find the wavelength of electrons that have a velocity of 3.00×10^6 m/s.

Strategy The relationship between the wavelength of electrons and their velocity is given in the de Broglie equation.

Solution

Substitute the known values into Equation 7.6, using the appropriate SI units. The mass of the electron must be expressed in kilograms ($m = 9.11 \times 10^{-31}$ kg) and the velocity in meters per second (m/s) when Planck's constant is expressed in J · s, because 1 J = 1 kg m^2/s^2. In this example, the base units are used for Planck's constant.

$$\lambda = \frac{h}{mv} = \frac{6.626 \times 10^{-34}\ \text{kg} \cdot \text{m}^2/\text{s}}{(9.11 \times 10^{-31}\ \text{kg})(3.00 \times 10^{6}\ \text{m/s})} = 2.42 \times 10^{-10}\ \text{m}$$

This wavelength is comparable with that of x rays (see Figure 7.4), which are also diffracted by crystalline solids.

Understanding

What is the velocity (in m/s) of neutrons that have a wavelength of 0.200 nm? The mass of a neutron is 1.67×10^{-27} kg.

Answer 1.98×10^3 m/s

---

de Broglie's equation offered an explanation for the assumption of quantized angular momentum of the electron in the hydrogen atom by suggesting that the electron "wave" in an atom must be a *standing wave,* which is a wave that stays in a constant position. The vibration of a violin string is a simple example of a standing wave. When a violin string is plucked, its vibration is restricted to certain wavelengths, because the ends of the string cannot move. The wavelength of the vibration times a whole number must equal twice the length of the string (Figure 7.8). de Broglie's equation suggested that the circumference of a Bohr orbit must be a whole-number multiple of the electron's wavelength so that a standing wave is produced. If the electron were not a standing wave, it would partially cancel itself on each successive orbit until its amplitude was zero, and the electron (the wave) would no longer exist!

de Broglie's restriction for a standing wave is expressed in the equation

$$2\pi r = n\lambda \qquad [7.7]$$

Note the similarity between the condition for the standing wave in a vibrating string (see Figure 7.8), $2L = n\lambda$, and Equation 7.7. Figure 7.9 shows a graphic representation of an allowed wave and a forbidden wave. de Broglie showed that treating the electron as a standing wave results in the quantization of angular momentum assumed by Bohr. Thus, treatment of the electron as a wave justified Bohr's assumption.

Although Bohr's treatment of the hydrogen atom explains the atom's spectrum, it cannot be extended to larger atoms, and it does not have a basis in theory beyond Bohr's own assumptions. The modern model of electrons as waves, however, not only explains

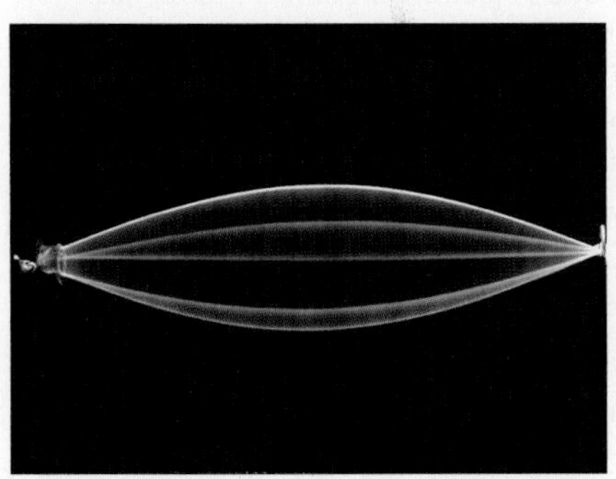
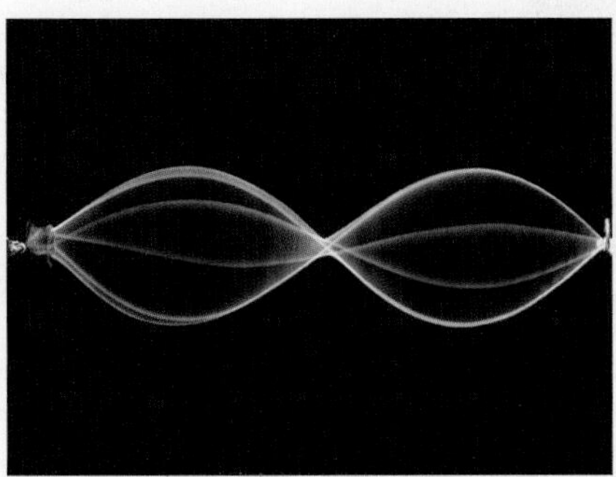

Figure 7.8 Standing waves. The wavelengths at which the stretched rubber tube (fixed at the ends) vibrates are those that satisfy the equation $2L = n\lambda$, where L is the length of the string, λ is the wavelength of the vibration, and n is any whole positive number. The waves with $n = 1, 2,$ and 3 are shown.

the energy levels of the H atom, it can be extended to describe other properties, as well as other atoms (and even molecules).

Several modern experimental techniques are based on the wave properties of matter. The diffraction of electrons and neutrons by molecules provides important information about their structures by allowing us to measure the distances between atoms accurately. The electron microscope, which is capable of higher magnifications than those achieved by a light microscope, is based on the wave properties of electrons. Thus, the behavior of matter as waves is firmly established by experiment.

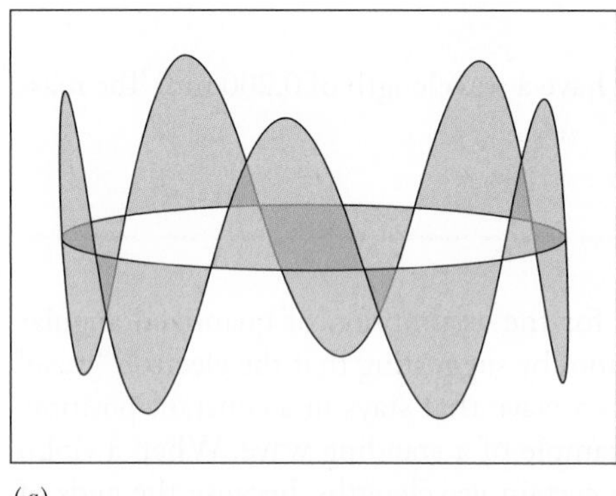

(*a*)

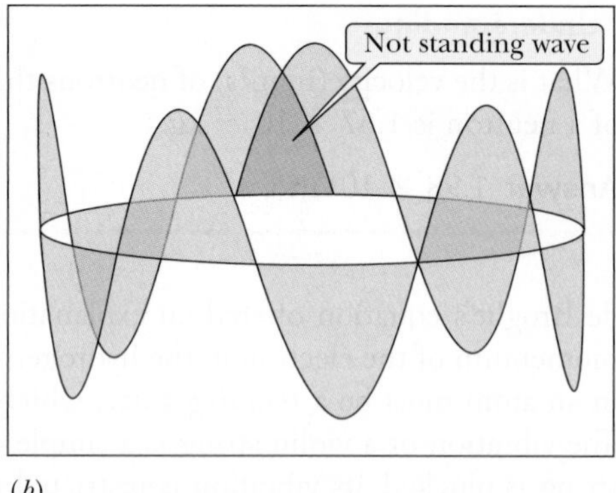

(*b*)

Figure 7.9 Circular standing waves. *(a)* The circumference of the circle is exactly five times the wavelength, so a stationary wave is produced. This is an allowed orbit. *(b)* The wave does not close on itself, because the circumference is 5.2 times the wavelength. This orbit is not allowed.

PRINCIPLES OF CHEMISTRY

Heisenberg's Uncertainty Principle Limits Bohr's Atomic Model

Werner Heisenberg (1901–1976) postulated an important principle of nature, one that limits the knowledge we may have about particles. This **Heisenberg uncertainty principle** states that it is not possible to know simultaneously both the precise position and the precise momentum of a particle. Expressed mathematically, the uncertainty principle is

$$\Delta x \cdot \Delta p \geq \frac{h}{4\pi} = 5.3 \times 10^{-35} \frac{\text{kg} \cdot \text{m}^2}{\text{s}}$$

Werner Heisenberg. Heisenberg (1901–1976), a German physicist, was one of the pioneers in the field of quantum theory and the discoverer of the uncertainty principle. He received the Nobel Prize in Physics in 1932.

where Δx and Δp are the uncertainties in position and momentum, respectively, and h is Planck's constant.

The uncertainty dictated by the Heisenberg principle is of no importance when we consider normal-size objects, such as baseballs and automobiles, because the product of uncertainties is so small. A baseball with a mass of about 142 g, traveling at 95 miles per hour (42 m/s), has an inherent uncertainty in its position of only about 1×10^{-33} m! Such a small distance is not measurable even in today's laboratories. Only for very small particles does the uncertainty principle become a significant limitation. The uncertainty principle makes it clear that the Bohr model of the atom is unacceptable, despite whatever support it might get from the de Broglie equation. Bohr's model predicts that an electron in the $n = 1$ orbit has a distance of 53 pm from the nucleus and a momentum of 1.99×10^{-24} kg · m/s. If we assume that the uncertainty of the momentum is 1% of its value, or 1.99×10^{-26} kg · m/s, then the uncertainty in its position is

$$\Delta x = \frac{5.3 \times 10^{-35}}{1.99 \times 10^{-26}} = 2.7 \times 10^{-9} \text{ m} = 2700 \text{ pm}$$

The uncertainty in the position of the electron is about *50 times* the radius of the Bohr orbit. The Bohr model thus calculates the position and the momentum of the electron more accurately than is possible within the limitations of the Heisenberg uncertainty principle. ∎

Schrödinger Wave Model

Shortly after de Broglie proposed that very small particles of matter might be described as waves, Erwin Schrödinger (1887–1961) devised a wave model to describe the behavior of the electrons in atoms. A complete description of the mathematics of his model is complicated and will not be given in this textbook. However, we present the results here because they are important to understanding the electronic structure of atoms.

1. The electron wave can be described by a mathematical function that gives the amplitude of the wave at any point in space. This function is called a **wave function** and is usually represented by the Greek letter ψ (psi).
2. The square of the wave function, ψ^2, gives the *probability* of finding the electron at any point in space. It is not possible to say exactly where the electron is located when we describe it as a wave. The wave model does not conflict with the Heisenberg uncertainty principle (see Principles of Chemistry), because it does not precisely define the location of the electron.
3. Many wave functions are acceptable descriptions of the electron wave in an atom. Each is characterized by a set of **quantum numbers.** The values of the quantum numbers are related to the shape and size of the electron wave and the location of the electron in three-dimensional space.
4. It is possible to calculate the energy of an electron having each possible wave function. When the wave model is applied to hydrogen, it predicts quantized energy levels identical to those predicted by Bohr and measured by experiment. The angular momentum of the electron is also quantized, but this is a natural consequence of the wave function, not an assumption of the wave model.
5. The wave function allows us to understand the properties of electrons in atoms other than hydrogen as well. This makes Schrödinger's wave model, a fundamental idea in the theory called **quantum mechanics,** superior to Bohr's theory, which is limited to the hydrogen atom.

No adequate physical analogy exists for the wave model of the atom as proposed by Schrödinger. Probably one of the best ways to visualize an electron in an atom is as a cloud of negative charge distributed about the nucleus of the atom, rather than as a rapidly moving particle. The cloud is spread out in proportion to the value of ψ^2 at each location. The following section discusses the electron-cloud interpretation of wave functions.

OBJECTIVES REVIEW *Can you:*

- ☑ relate the de Broglie wavelength of matter to its momentum?
- ☑ determine the wavelengths associated with particles of matter?
- ☑ present the characteristics of wave functions and their relationships to the position and energy of electrons?

7.4 Quantum Numbers in the Hydrogen Atom

OBJECTIVES

- ☐ List the quantum numbers in the hydrogen atom and their allowed values and combinations
- ☐ Relate the values of quantum numbers to the energy, shape, size, and orientation of the electron cloud in the hydrogen atom
- ☐ Draw contour surfaces and electron-density representations of the electron in the hydrogen atom
- ☐ Give the notations used to represent the shells, subshells, and orbitals in the hydrogen atom

The best current description of the electronic structure of the atom treats the electron as a wave. No wave has a precise position; rather, it is defined over a complete period, which is one wavelength in length. The wave model provides quantum numbers that describe the characteristics of the wave that represents the electron, instead of a specific location for the electron. These quantum numbers are analogous to the coordinates used to locate the position of a particle. For example, the location of an airplane in flight is given by three numbers: the longitude, latitude, and altitude. The wave model initially produces three kinds of quantum numbers that must be specified to define the wave function of an electron. They are represented by the symbols n, ℓ, and m_ℓ. The values of these quantum numbers give as much information about the location and the energy of the electron as is possible. The three-dimensional wave function of an electron, described by specific values of n, ℓ, and m_ℓ, is called an **atomic orbital.** Each of the quantum numbers is restricted to certain whole-number values. Furthermore, the value of n restricts the values of ℓ, which, in turn, places restrictions on the values that m_ℓ may have. We describe each of these quantum numbers in the following paragraphs.

An atomic orbital is a wave function described by specific allowed values of the n, ℓ, and m_ℓ quantum numbers.

The **principal quantum number** is represented by n. The allowed values for n are all positive whole numbers: $n = 1, 2, 3, \ldots$. The principal quantum number gives information about the *distance of the electron from the nucleus.* The larger the value of the principal quantum number, the greater the average distance of the electron from the nucleus and, therefore, the size of the orbital. Remember that the wave model does not provide a precise distance, and there is a small probability that any electron is very close to or very far from the nucleus, regardless of the value of n. As we shall see, several different wave functions can have the same value of n (except for $n = 1$). The term **principal shell** (or more simply, **shell**) refers to all atomic orbitals that have the same value of n, because they all have approximately the same average distance from the nucleus. The n quantum number is important in determining the energy of the atom, because the distance of the electron from the nucleus is related to the energy of the atom. The wave model gives the same energy for the hydrogen atom, -2.18×10^{-18} J/n^2, as Bohr found; therefore, the smaller the value of n, the lower the energy of the atom.

The principal quantum number is designated by n and provides information about the distance of the electron from the nucleus.

All orbitals that have the same value of n are in the same principal shell.

The **angular momentum quantum number** is represented by ℓ. The possible values of ℓ for a given n are all positive integers from zero up to $n - 1$: $\ell = 0, 1, 2, \ldots (n - 1)$. Thus, the ℓ quantum number must equal 0 for an orbital in the $n = 1$ shell. When the principal quantum number n equals 4, ℓ can have the value 0, 1, 2, or 3. The angular momentum quantum number, ℓ, can be associated with the shape that the atomic orbital may have (which we will consider shortly). Each value of the ℓ quantum number corresponds to a particular *shape* for the atomic orbital. A **subshell** is the set of all the possible orbitals that have the same values of both the n and ℓ quantum numbers. Just as in the case of a shell, each subshell may consist of more than one orbital.

The angular momentum quantum number is ℓ. It describes the shape of the orbital.

To identify a subshell, we use a notation that specifies values for both the principal and the angular momentum quantum numbers. The numerical value for n is used, but lowercase letters are used for different values of ℓ, as follows:

A subshell contains all orbitals that have the same values for n and ℓ. The notation for a subshell consists of a number, which is the value of the n quantum number, followed by a lowercase letter (s, p, d, or f) that identifies the value of the ℓ quantum number.

Angular momentum quantum number, ℓ	0	1	2	3	4	5	6
Letter used	*s*	*p*	*d*	*f*	*g*	*h*	*i*

The first four letters, *s*, *p*, *d*, and *f*, are related to the words used by early scientists to describe lines in atomic spectra: *sharp*, *principal*, *diffuse*, and *fundamental.* With the advent of quantum mechanics, this same terminology was applied. Thus, the subshell with $n = 3$ and $\ell = 1$ is called the 3p subshell.

EXAMPLE 7.5 Allowed Combinations of Quantum Numbers

Give the notation for each of the following subshells that is an allowed combination of quantum numbers. If it is not an allowed combination, explain why.

(a) $n = 2, \ell = 0$ (b) $n = 1, \ell = 1$ (c) $n = 4, \ell = 2$

TABLE 7.1 Allowed Combinations of the n, ℓ, and m_ℓ Quantum Numbers

Shell, n	Subshell, ℓ (label)	Orbital, m_ℓ	Number of Orbitals in Subshell
1	0 (1*s*)	0	1
2	0 (2*s*)	0	1
	1 (2*p*)	−1, 0, +1	3
3	0 (3*s*)	0	1
	1 (3*p*)	−1, 0, +1	3
	2 (3*d*)	−2, −1, 0, +1, +2	5
4	0 (4*s*)	0	1
	1 (4*p*)	−1, 0, +1	3
	2 (4*d*)	−2, −1, 0, +1, +2	5
	3 (4*f*)	−3, −2, −1, 0, +1, +2, +3	7

Strategy Apply the rules for the possible values of n and ℓ.

Solution

(a) $n = 2, \ell = 0$ is an allowed subshell. We use the letter *s* to express the value of $\ell = 0$, so the correct notation is 2*s*.
(b) Because ℓ must be less than n, a value of $\ell = 1$ is not possible when $n = 1$.
(c) The letter *d* means that $\ell = 2$, so this subshell is referred to as 4*d*.

Understanding

What is the notation for the subshell with $n = 3$ and $\ell = 1$?

Answer 3*p*

The **magnetic quantum number** is represented by m_ℓ. Allowed values for m_ℓ are all integers from $-\ell$ to $+\ell$. For example, if the ℓ quantum number is 2 (a *d* subshell), then m_ℓ may have the values −2, −1, 0, +1, and +2. The m_ℓ quantum number provides information about the *orientation in space* of the atomic orbital. Each subshell consists of one or more atomic orbitals. The number of orbitals in any given subshell is equal to $(2\ell + 1)$, corresponding to the $(2\ell + 1)$ allowed values of the m_ℓ quantum number. An *s* subshell has only one orbital $[2(0) + 1 = 1]$, a *p* subshell has three orbitals $[2(1) + 1 = 3]$, a *d* subshell consists of five orbitals $[2(2) + 1 = 5]$, and so on.

The magnetic quantum number, m_ℓ, tells about the orientation of the orbital.

Once values for these three quantum numbers are specified, most of the information that can be known about the location of the electron in three-dimensional space has been given. The n quantum number specifies the size of the orbital, the ℓ quantum number the shape of the orbital, and the m_ℓ quantum number the orientation of the orbital in space. Table 7.1 shows the allowed combinations of these three quantum numbers, through the fourth shell.

EXAMPLE 7.6 Allowed Combinations of Quantum Numbers

Give the notation for each of the following orbitals that is an allowed combination of quantum numbers. If it is not an allowed combination, explain why.

(a) $n = 3, \ell = 0, m_\ell = 0$
(b) $n = 3, \ell = 1, m_\ell = +2$
(c) $n = 2, \ell = 2, m_\ell = +1$
(d) $n = 4, \ell = 1, m_\ell = -1$
(e) $n = 3, \ell = -1, m_\ell = +1$

Strategy Remember that ℓ goes from 0 to $(n - 1)$ and m_ℓ ranges from $-\ell$ to ℓ.

Solution

(a) This set of quantum numbers is an allowed combination. The values of n and ℓ indicate that the electron is in a 3*s* orbital.
(b) This set of quantum numbers is not allowed, because m_ℓ must be between $-\ell$ and $+\ell$.

(c) This set of quantum numbers is not allowed, because the value of ℓ is greater than $(n - 1)$.
(d) This set of quantum numbers is allowed and, from the values of n and ℓ, is a $4p$ orbital.
(e) This set of quantum numbers is not allowed, because ℓ cannot be negative.

Understanding

Give the notation for each of the following orbitals that is an allowed combination. If it is not an allowed combination, explain why.

(a) $n = 2, \ell = 1, m_\ell = 0$ (b) $n = 5, \ell = 3, m_\ell = -3$

Answer (a) Allowed; $2p$ (b) Allowed; $5f$

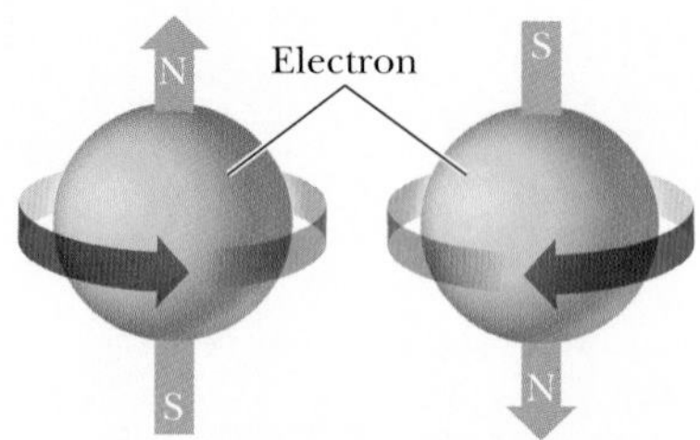

Figure 7.10 Electron spin. The electron is visualized as a sphere with the charge on its surface. When the charge of the electron spins counterclockwise or clockwise, magnetic fields are generated in opposite directions.

The electron spin quantum number, m_s, has only two allowed values, $+\frac{1}{2}$ and $-\frac{1}{2}$.

Electron Spin

A fourth quantum number does not come directly from the wave model but is necessary to account for an important property of electrons. Scientists have observed that electrons act as small magnets when placed in a magnetic field. For example, when a beam of hydrogen atoms passes through a magnetic field, half of the atoms are deflected in one direction, and the other half are deflected in the opposite direction. Visualize the electron in the atom as a sphere, with its charge on the surface, that may spin only in a clockwise or counterclockwise direction (Figure 7.10). The electric current produced by this spin causes the electron to behave as a magnet with its poles in one of two possible directions with respect to the external magnetic field.

In the wave model, the magnetic behavior of the electron is described by the **electron spin quantum number,** represented by the symbol m_s. The allowed values of m_s are $+\frac{1}{2}$ and $-\frac{1}{2}$, corresponding to the two possible spin states for an electron. The electron spin does not depend on the values of any of the other quantum numbers. Two electrons that have the same spin are said to be *parallel,* whereas electrons with different spins (one $+\frac{1}{2}$ and the other $-\frac{1}{2}$) are called *paired.*

We can now summarize the wave description of the electron in the hydrogen atom. Four quantum numbers (n, ℓ, m_ℓ, and m_s) are needed to describe the electron in any hydrogen atom. Each quantum number provides some information about the probable location in space or the magnetic behavior of the electron. Remember, we must be satisfied with a probability distribution for the electron because there is no exact location for a wave.

Quantum Number	Property
Principal quantum number n	Orbital size
Angular momentum quantum number ℓ	Orbital shape
Magnetic quantum number m_ℓ	Orbital orientation in space
Electron spin quantum number m_s	Electron spin direction

Representations of Orbitals

The wave function gives the shape, size, and orientation of an orbital. The *square of the wave function,* ψ^2, gives the *probability* that the electron will be found at any specific location in space. Plotting ψ^2 helps us visualize the orbitals, showing different spatial characteristics of each. There are several different ways to depict the location of the electron in an atom that emphasize that the plot is a probability and that the location has uncertainty.

One method is to use different densities of dots to represent the probability of finding the electron at a particular location. At places where the probability is high, the dots are highly concentrated. At locations where the probability is low, few dots are present.

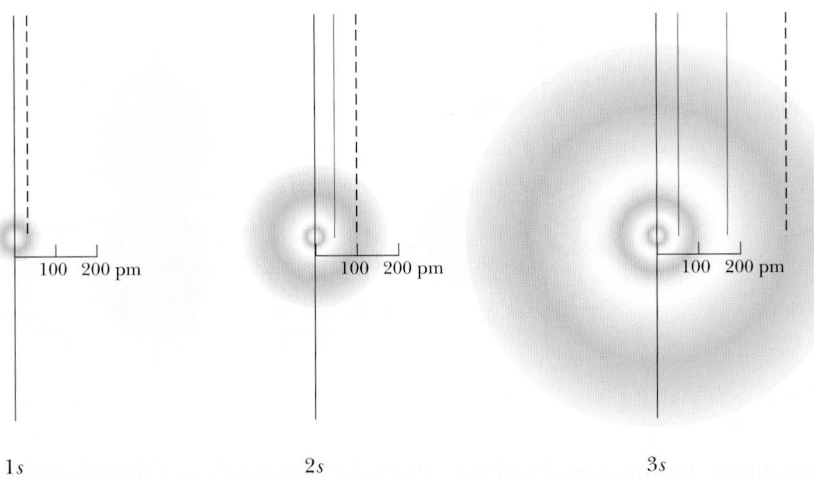

Figure 7.11 Hydrogen *s* orbitals. The nucleus is at the center of the sphere, and the concentration of color is proportional to the probability of the electron's location for the 1*s*, 2*s*, and 3*s* orbitals. The dotted vertical lines indicate the maximum probability of where the electron would be, while the solid blue vertical lines indicate regions where the probability that the electron is there is zero; these are the *nodes.*

There are even regions in space where the probability of finding an electron is exactly zero; these regions are called **nodes.** Figure 7.11 shows this representation of the 1*s*, 2*s*, and 3*s* orbitals for the hydrogen atom. Note that the electron probability extends farther from the nucleus as the value of the principal quantum number increases, but for all three of the wave functions, significant electron density occurs close to the nucleus. Drawings such as those in Figure 7.11 are often referred to as electron-cloud or electron-density representations, because the shading shows the electron as spread out over a region of space. The *s* orbitals are all spherical, because the electron probability depends only on the distance of the electron from the nucleus, not on the direction.

All *s* orbitals have a spherical shape.

A second and more common way of representing an electron orbital is to use contour diagrams. In a contour diagram, a surface is drawn that encloses some fraction of the electron probability, usually 90%. The value of ψ^2 is the same everywhere on the surface. Figure 7.12 presents 90% contour surfaces for the *s* orbitals ($\ell = 0$) with $n = 1, 2,$ and 3. As the principal quantum number increases in value, the average distance of the electron from the nucleus increases, and thus the size of the contour surface increases.

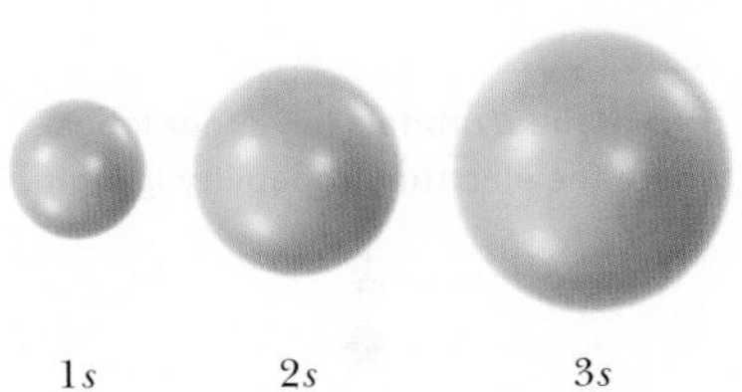

Figure 7.12 Contours for the *s* orbitals. Contours that enclose 90% of the electron's probability are given for the 1*s*, 2*s*, and 3*s* orbitals. The sizes are to scale.

The *p* orbitals ($\ell = 1$) have a different shape from the *s* orbitals. They have two lobes, one on each side of the nucleus. Figures 7.13*a* and 7.13*b* are graphs of ψ and ψ^2 for a 2*p* orbital. Although the wave function itself has different mathematical signs on opposite sides of the vertical axis, the square of ψ (the electron density) has the same pattern on both sides of the vertical axis. When we study the molecular wave functions (see Chapters 9 and 10), the signs of the atomic wave functions on each atom in the molecule become important. Figures 7.13*c* and 7.13*d* are the electron density diagram

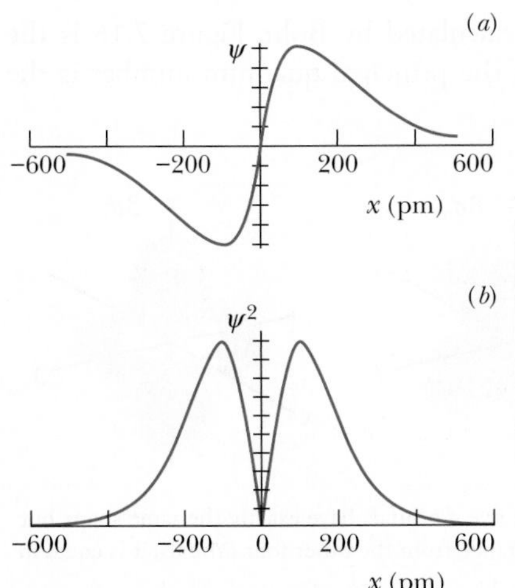

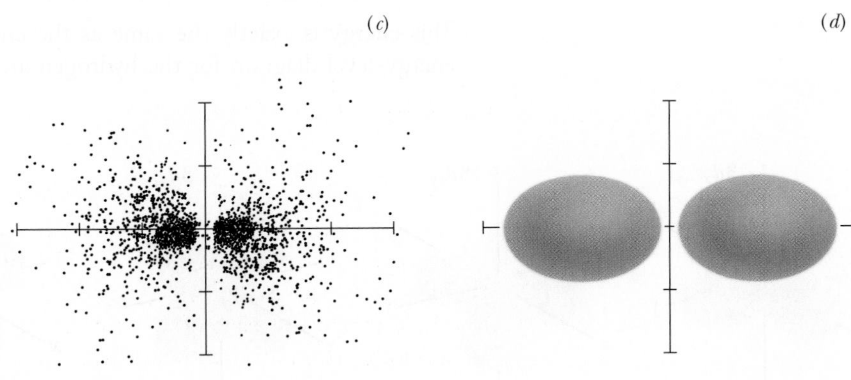

Figure 7.13 Representations of a 2*p* orbital. *(a)* The graph of ψ for a 2*p* orbital directed along the *x* axis. *(b)* The square of the wave function, ψ^2, is proportional to the electron probability. *(c)* The electron density is also represented by the distribution of dots. *(d)* The surface encloses 90% of the electron's distribution.

Figure 7.14 The three 2*p* orbitals. The contour surfaces for the three 2*p* orbitals are identical in size and shape, but each is directed along a different axis.

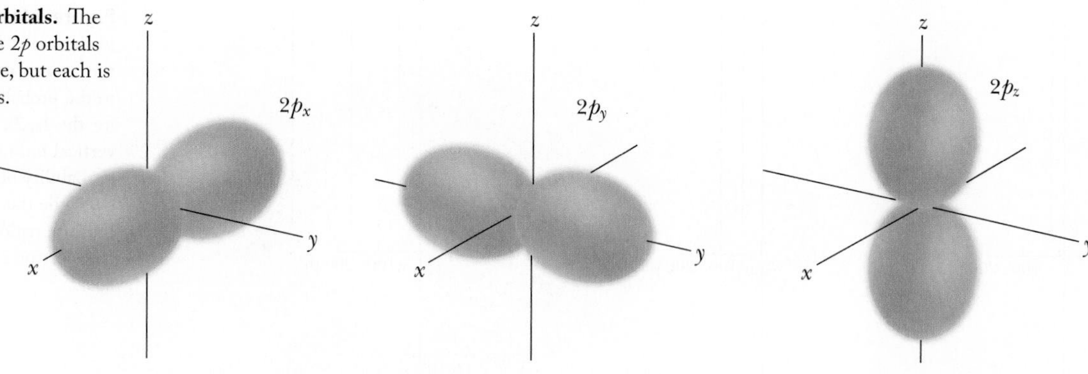

and contour surface for the same 2*p* orbital. All *p* orbitals, regardless of the value of the principal quantum number, have this same "dumbbell" shape, consisting of two lobes on opposite sides of the nucleus. Unlike *s* orbitals, the electron density for all *p* orbitals is zero at the nucleus. In fact, any orbital with ℓ greater than zero has a node at the nucleus. This concept is discussed further in Section 7.5.

There are two lobes in *p* orbitals, directed at 180 degrees.

When the angular momentum quantum number (ℓ) is equal to 1 (a *p* subshell), three values for the magnetic quantum number are allowed, so each *p* subshell must consist of three orbitals. In any principal shell, the three different *p* orbitals have exactly the same size and shape but different orientations. One *p* orbital is directed along each of the three Cartesian axes; these orbitals are referred to as p_x, p_y, and p_z. Figure 7.14 shows contours illustrating the relative orientations of the 2*p* orbitals. Each shell beyond the first has a subshell containing three *p* orbitals (2*p*, 3*p*, 4*p*, and so on). Just as in the case of the *s* orbitals, the contours for *p* orbitals increase in size as the value of the principal quantum number increases.

Figure 7.15 shows the contours for the five *d* orbitals ($\ell = 2$). Four of these have the same shape, with four identical lobes that point at the corners of a square; these are labeled d_{xy}, d_{xz}, d_{yz}, and $d_{x^2-y^2}$. The remaining *d* orbital (d_{z^2}) looks different but is mathematically equivalent to the other four.

Four of the *d* orbitals have four lobes where the electron probability is high.

The shapes of the seven *f* orbitals have also been calculated, but they are more complex than those already shown. We will not need them in later chapters, but we have included them here for reference.

Energies of the Hydrogen Atom

The energy of the hydrogen atom depends only on the value of the principal quantum number of the wave function of the electron,

$$E_n = -B/n^2 = -2.18 \times 10^{-18}\ \text{J}/n^2$$

This energy is exactly the same as the energy calculated by Bohr. Figure 7.16 is the energy-level diagram for the hydrogen atom. If the principal quantum number is the

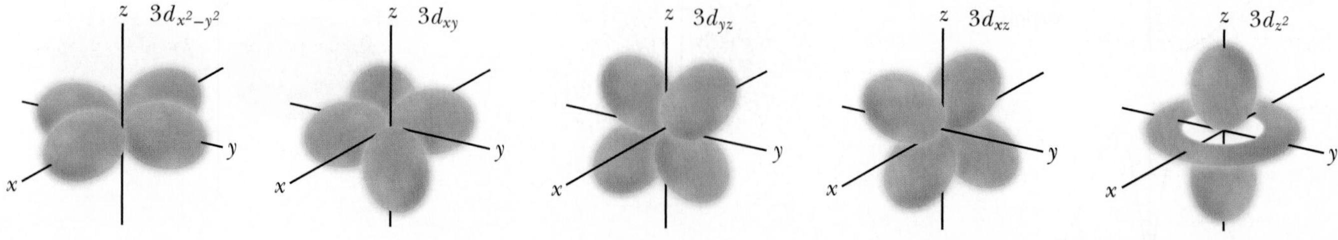

Figure 7.15 Contours for the five 3*d* orbitals. Four of the five *d* orbitals have exactly the same shape but differ in orientation. Although the d_{z^2} has a different appearance from the other four orbitals, it is equal in energy.

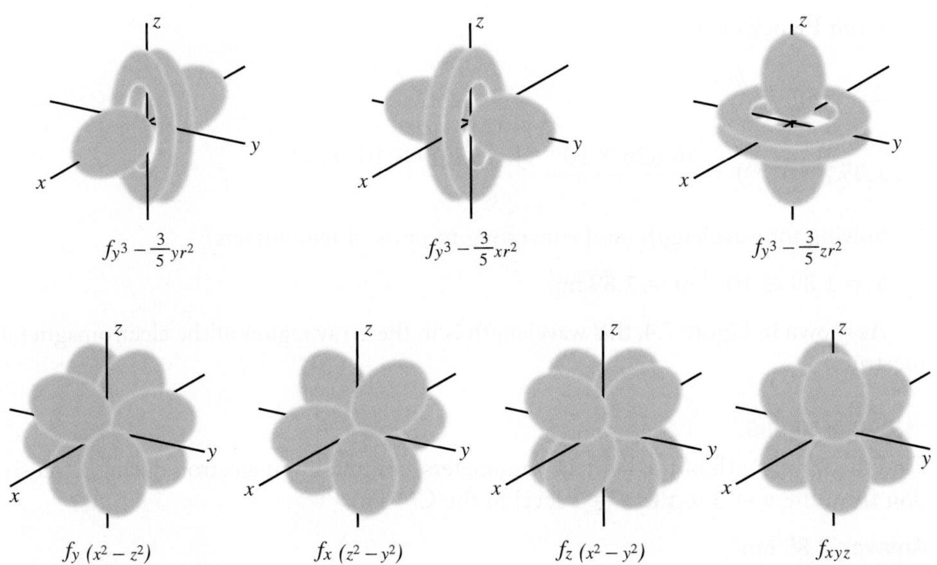

The *f* orbitals. The *f* orbitals have these shapes.

same, no matter which subshell or orbital the electron occupies, the hydrogen atom has exactly the same energy. This energy-level diagram is exactly the same as Bohr's (see Figure 7.7), except that in Figure 7.16, the different subshells that comprise each principal shell are identified as connected boxes.

The value of the principal quantum number determines the energy of any one-electron wave function.

The energy of each wave function for any atomic species containing only one electron is given by

$$E_n = \frac{-Z^2B}{n^2} = \frac{-2.18 \times 10^{-18} Z^2 \text{J}}{n^2} \qquad [7.8]$$

where Z is the nuclear charge (the number of protons in the nucleus), and the other symbols have their usual meanings. With this equation we can calculate the spectrum of any ion that contains one electron, for example, He^+ or Li^{2+}. The one-electron spectrum of the O^{7+} ion has been used to identify the presence of oxygen in the atmosphere of the Sun.

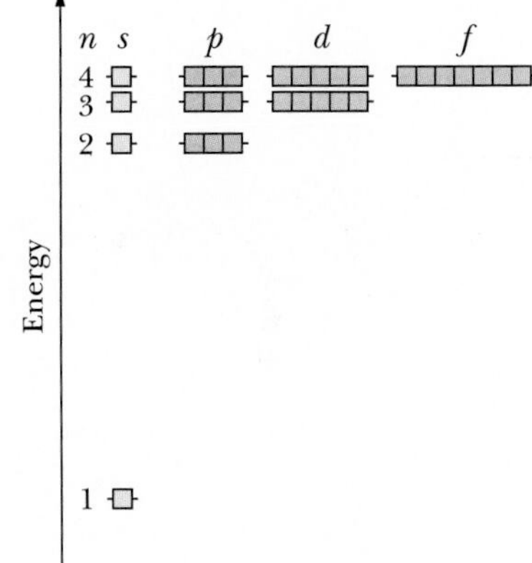

Figure 7.16 Energy-level diagram for the hydrogen atom. Each *box* represents one of the orbitals. The *short horizontal line* at the center of each box or set of connected boxes locates the energy of each subshell. In hydrogen or any other one-electron species (e.g., He^+ or O^{7+}), all of the orbitals having the same principal quantum number have identical energies.

EXAMPLE 7.7 Calculating Lines in the O^{7+} Spectrum

What is the wavelength of light, in nanometers, required to raise an electron in the O^{7+} ion from the $n = 1$ shell to the $n = 2$ shell?

Strategy Use Equation 7.8 and the knowledge that a photon must have the same energy as the difference in energies of two quantized energy levels. Note that for an oxygen atom, $Z = 8$.

Solution

For $n = 1$:

$$E(n = 1) = \frac{-2.18 \times 10^{-18}(8^2)\text{ J}}{1^2} = -1.40 \times 10^{-16}\text{ J}$$

For $n = 2$:

$$E(n = 2) = \frac{-2.18 \times 10^{-18}(8^2)\text{ J}}{2^2} = -3.49 \times 10^{-17}\text{ J}$$

The change in energy is thus

$$\begin{aligned}\Delta E &= E(n = 2) - E(n = 1) = -3.49 \times 10^{-17}\text{ J} - (-1.40 \times 10^{-16}\text{ J}) \\ &= 1.05 \times 10^{-16}\text{ J}\end{aligned}$$

From Planck's law:

$$\Delta E = h\nu = \frac{hc}{\lambda}$$

$$1.05 \times 10^{-16}\ \text{J} = \frac{(6.626 \times 10^{-34}\ \text{J} \cdot \text{s})(3.00 \times 10^{8}\ \text{m/s})}{\lambda}$$

Solving for wavelength (and converting to units of nanometers):

$$\lambda = 1.89 \times 10^{-9}\ \text{m} = 1.89\ \text{nm}$$

As shown in Figure 7.4, this wavelength is in the x-ray region of the electromagnetic spectrum.

Understanding

Find the wavelength of the light, in nanometers, emitted by an electron during a transition from the $n = 3$ to the $n = 1$ level in the C^{5+} ion.

Answer 2.85 nm

OBJECTIVES REVIEW *Can you:*

- ☑ list the quantum numbers in the hydrogen atom and their allowed values and combinations?
- ☑ relate the values of quantum numbers to the energy, shape, size, and orientation of the electron cloud in the hydrogen atom?
- ☑ draw contour surfaces and electron-density representations of the electron in the hydrogen atom?
- ☑ give the notations used to represent the shells, the subshells, and the orbitals in the hydrogen atom?

7.5 Energy Levels for Multielectron Atoms

OBJECTIVES

- ☐ Define screening and effective nuclear charge
- ☐ Relate penetration effects to the relative energies of subshells within the same shell

Although the wave model of the hydrogen atom gave new insight into the structure of matter, an important goal of our study is to understand the nature of all elements, not just hydrogen. For any atom or ion that contains more than one electron, exact mathematical expressions for the electron waves are not known, and approximate wave functions must be used. Despite this limitation, the wave properties of matter are extremely useful in interpreting the chemical properties of atoms.

The same four quantum numbers that are used for the hydrogen atom (n, ℓ, m_ℓ, and m_s) describe the electrons in multielectron atoms. Unlike the subshells in the hydrogen atom, *the different subshells within the same shell of a multielectron atom do not have the same energy.* The dependence of the energy on the angular momentum quantum number causes the line spectra for all the elements beyond hydrogen to be much more complex; in many cases, they contain thousands of lines in the visible region alone. In fact, the wavelengths of the emission lines are among the primary tools used to determine the energy-level diagrams of atoms. Figure 7.17 is the emission spectrum of chromium.

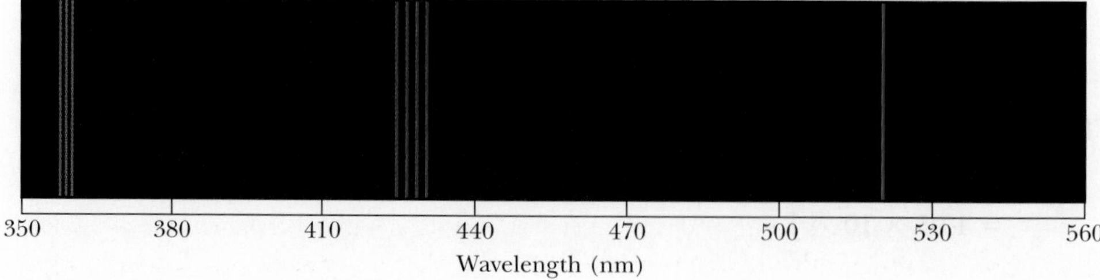

Figure 7.17 Emission spectrum of chromium. The presence of these lines is used to identify the presence of chromium in a sample. These are the most intense lines of the almost 600 lines in the visible spectrum of chromium. The green "line" is actually three lines that are not separated on this scale. Each element has a unique spectrum that may contain thousands of lines.

Effective Nuclear Charge

It is important to understand why subshells within the same shell differ in energy when an atom or ion contains more than one electron. Because charges of opposite sign attract each other, the energy of the atom decreases (the atom becomes more stable) as the electron gets closer to the nucleus. Thus, the energy of a 1*s* electron is less than that of a 2*s* electron because the 1*s* electron is, on average, closer to the nucleus. Experimental data show that, for any atom that contains more than one electron, the energy is also influenced by the ℓ quantum number; for example, the 2*s* subshell is lower in energy than the 2*p* subshell. A qualitative understanding of the dependence of the energy on the ℓ quantum number can be obtained by considering the electrostatic forces that act on the electrons in a multielectron atom.

Energies in multielectron atoms depend on the values of *both* the n and ℓ quantum numbers. In one-electron atoms and ions, the energy depends only on the value of the n quantum number.

The single electron in any one-electron species, regardless of its location or the orbital it occupies, is attracted by the nuclear charge. For example, the single electron in the Li^{2+} ion is attracted by the +3 charge on the nucleus (Figure 7.18*a*). The situation in the neutral lithium atom, with three electrons, is more complicated. Each electron is not only attracted by the +3 charge of the nucleus but is also repelled by the negative charges of the other two electrons. The electron-electron repulsions, known as **interelectronic repulsions,** reduce the effect of the positive charge of the nucleus on each electron, thus influencing its energy.

The net attraction of the nucleus for an electron at any distance r is reduced, or shielded, by the repulsive forces from the electrons between it and the nucleus. Figure 7.18*b* represents this situation schematically for the lithium atom. The lowest energy state of lithium has two electrons in the 1*s* subshell and one electron in the second shell. Because the two electrons in the 1*s* orbital are much closer to the nucleus than an electron in the second shell, most of the time the 1*s* electrons are between the nucleus and the third electron. The **effective nuclear charge,** Z_{eff}, is the weighted average of the nuclear charge that affects an electron in the atom, after correction for the shielding of nuclear charge by inner electrons and the interelectronic repulsions. The effective nuclear charge for the electron in the second shell is considerably less than +3, because both 1*s* electrons are usually much closer to the nucleus than an electron in the second principal shell. The result of the influence of inner electrons on the effective nuclear charge is frequently called **electron shielding.**

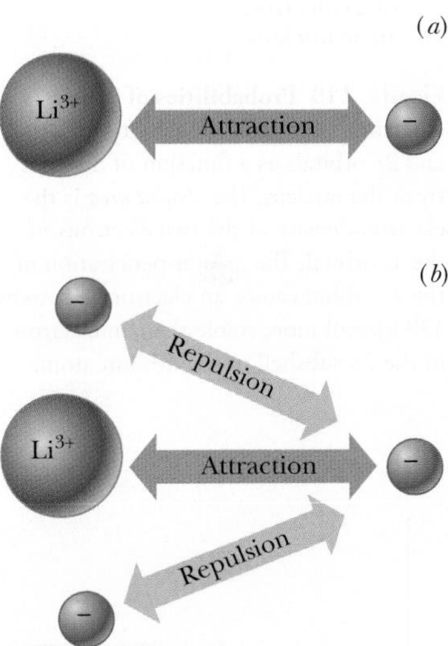

Figure 7.18 Effective nuclear charge. *(a)* A single electron in Li^{2+} is subject to the full +3 charge of the nucleus. *(b)* In the lithium atom, the +3 attractive force of the nucleus on the outer electron is reduced, or shielded, by the repulsive forces from the inner electrons.

To determine the effective nuclear charge for each electron, we need to know whether the other electrons in the atom are between it and the nucleus. In the lithium atom, the 1*s* electrons are very close to the nucleus, and experimental measurements show that the effective nuclear charge for them is close to +3. In the lithium atom's lowest energy state, the third electron is in the second shell and, on average, is farther from the nucleus than are the 1*s* electrons. However, the electron in the second shell has a smaller probability of being closer to the nucleus than a 1*s* electron. The extent of shielding of the third electron by the 1*s* electrons depends on the distance of the third electron from the nucleus. Close to the nucleus, where the electron has a small probability of being, it experiences nearly all of the +3 nuclear charge. At large distances, the shielding by the 1*s* electrons is nearly complete and the electron experiences a nuclear charge of essentially +1, the charge of the nucleus minus the charge of the two 1*s* electrons. Because the third electron spends most of its time farther from the nucleus than the 1*s* electrons, the effective nuclear charge is a good deal smaller than +3.

The *effective* nuclear charge is the total nuclear charge corrected for the effect of the charges of the inner electrons that are present in the atom or ion.

Figure 7.19 shows plots of the electron probabilities (ψ^2) for the 2*s* and 2*p* orbitals as a function of the distance from the nucleus. Although the average distances of the 2*s* and 2*p* electrons are about the same, the probability that the electron is close to the nucleus is greater for the 2*s* electron than for the 2*p* electron (see the electron density plots for the 2*s* and 2*p* electrons in Figures 7.11 and 7.13). Figure 7.19 shows that the 2*s* electron *penetrates* the electron density of the filled 1*s* shell more than does the 2*p* electron, so it is influenced by a greater effective nuclear charge. As such, the energy of an *s* electron is lower than the energy of a *p* electron in the same shell. Within *any* shell, the penetration of the *s* orbital is always greater than that of the *p* orbitals, which, in turn, is greater than that of the *d* orbitals. This means that within any shell, the subshells increase in energy

The greater the penetration of the inner shell by an electron, the greater the effective nuclear charge.

in the order of increasing value of the quantum number ℓ. In the fourth shell, the greater penetration of electrons with lower values of the ℓ quantum number is reflected in the increasing order of energy for the subshells of $4s < 4p < 4d < 4f$.

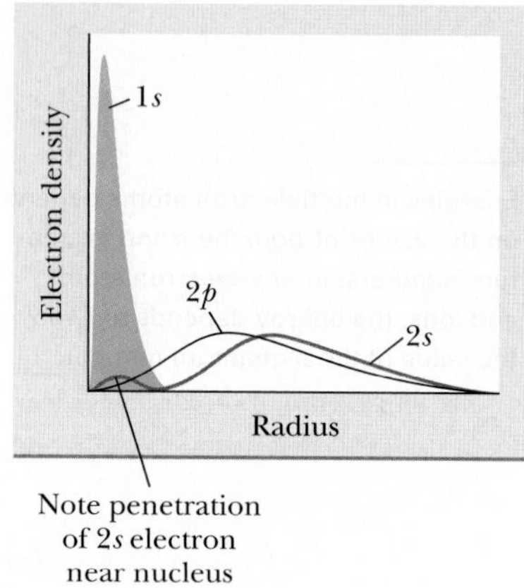

Figure 7.19 Probabilities of 2s and 2p electrons. Electron probability for the 2s and 2p orbitals as a function of distance from the nucleus. The *shaded area* is the electron density of the two electrons in the 1s orbital. The greater penetration of the 2s orbital causes an electron in it to be 179 kJ/mol more stable than an electron in the 2p subshell in the lithium atom.

Energy-Level Diagrams of Multielectron Atoms

Different interelectronic repulsive forces affect electrons in different subshells, so the energy of an atom depends on which subshells are occupied. Figure 7.20, which presents the energy-level diagram for the arsenic atom (another element in bullets; see the introduction to this chapter), is based on the interpretation of the line spectrum of that element. The order in which the subshells are occupied is typical for atoms up to radon.

Initially, each shell fills completely, starting with the lowest-energy orbital and filling in order of energy, before the next higher one is occupied. However, as seen in Figure 7.20, the energy of the 4s subshell is less than that of the 3d subshell because of the greater penetration by the 4s orbital of the electrons into the first and second shells. This overlap in the energy of different shells becomes more common as n increases. As can be seen in Figure 7.20, the energy separation between subshells gets quite small in the higher shells, so small changes in the shielding effects may cause the energy order to change from one element to the next. We examine these situations in more detail in Chapter 8. Based on experimental observations, the subshells are usually occupied by electrons in the following order: $1s < 2s < 2p < 3s < 3p < 4s < 3d < 4p < 5s < 4d < 5p < 6s < 4f < 5d < 6p < 7s < 5f < 6d$.

Figure 7.21 shows a chart to help remember the order of filling electron shells and subshells. In Chapter 8, we will see how other tools can help us remember the order of filling.

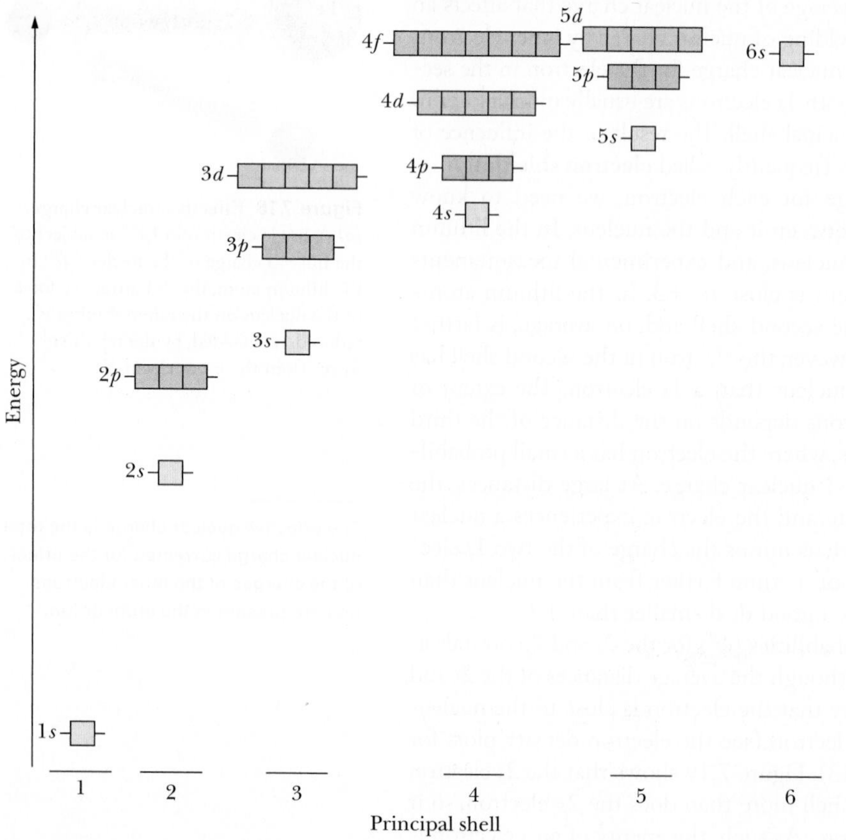

Figure 7.20 The energy levels for electrons in multielectron atoms are dependent on both the n and ℓ quantum numbers.

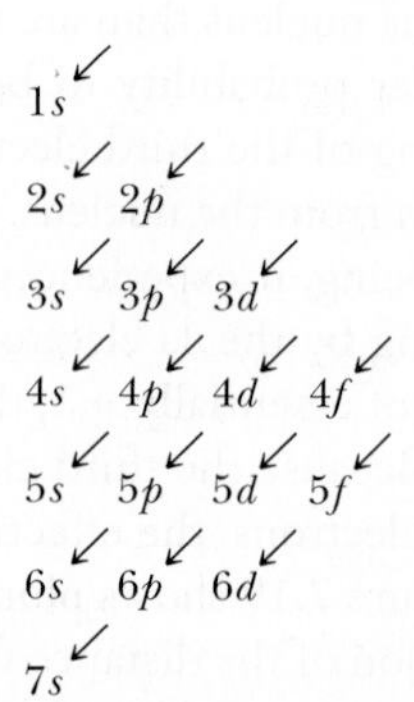

Figure 7.21 Diagonal mnemonic for remembering order of filling electron shells and subshells. By following each arrow along its backward diagonal, the proper order for filling electron shells and subshells in multielectron atoms can be reproduced easily.

Because the energies of different orbitals depend only on the values of the n and ℓ quantum numbers and not on the value of m_ℓ, all of the orbitals in a subshell (designated by different values of the m_ℓ quantum number) have exactly the same energy. When orbitals are of *exactly* the same energy—for example, the three different $2p$ orbitals—they are referred to as **degenerate orbitals.**

OBJECTIVES REVIEW *Can you:*

- ☑ define screening and effective nuclear charge?
- ☑ relate penetration effects to the relative energies of subshells within the same shell?

7.6 Electrons in Multielectron Atoms

OBJECTIVES

- ☐ Use the Pauli exclusion principle to determine the maximum number of electrons in an orbital, subshell, or shell
- ☐ Write the electron configuration of an atom
- ☐ Construct an orbital diagram and an energy-level diagram for a given atom
- ☐ Predict the number of unpaired electrons in an atom

Knowledge of the wave functions in atoms is extremely useful in determining the chemical properties of the element. This section describes ways of representing multiple electrons in atoms.

Pauli Exclusion Principle

One of the most important steps in the development of the description of the multielectron atom was the statement of the Pauli exclusion principle. In 1925, Wolfgang Pauli (1900–1958) summarized the results of many experimental observations with what is now known as the **Pauli exclusion principle:** *No two electrons in the same atom can have the same set of all four quantum numbers.* The Pauli exclusion principle is the quantum-mechanical equivalent of saying that two objects cannot occupy the same space at the same time. Using the Pauli exclusion principle, we find that any orbital (described by the three quantum numbers n, ℓ, and m_ℓ) can have a maximum of two electrons in it, one with a spin of $+\frac{1}{2}$ and the other with a spin of $-\frac{1}{2}$. Thus, *the maximum number of electrons that can share a single orbital in an atom is two.* Two electrons in the same orbital are referred to as an **electron pair** or **paired electrons,** because they must have different spin quantum numbers. When a single electron is in an orbital, it is called an **unpaired electron.** The Pauli exclusion principle explains why there are maxima on the number of electrons that can be present in each type of subshell and in each shell. Table 7.2 gives these maxima.

The restrictions on the quantum numbers and the Pauli exclusion principle determine the capacities of orbitals, subshells, and principal shells.

Aufbau Principle

We can now present the quantum-mechanical description of electrons in atoms. In a procedure called the **aufbau principle** (*aufbauen* is German for "building up"), electrons are added to the atom one at a time until the proper number is present. As each electron

TABLE 7.2 Maximum Number of Electrons in Shells and Subshells

Capacity of Subshells				
Subshell	s ($\ell = 0$)	p ($\ell = 1$)	d ($\ell = 2$)	f ($\ell = 3$)
Number of orbitals ($2\ell + 1$)	1	3	5	7
Number of electrons $2(2\ell + 1)$	2	6	10	14
Capacity of Shells				
Principal quantum number (n)	1	2	3	4
Number of orbitals (n^2)	1	4	9	16
Number of electrons ($2n^2$)	2	8	18	32

is added, it is assigned the quantum numbers of the lowest energy orbital available. The resulting list of occupied orbitals of the atom (called the **electron configuration** of the atom) is its lowest energy state, which is called the ground state (Section 7.2). Practically all of the atoms in a sample are in the ground state at normal temperatures.

In the hydrogen atom, there is only one electron, which occupies the 1*s* orbital in its ground state. The helium atom, with two electrons, has a ground state with both electrons in the 1*s* orbital ($n = 1, \ell = 0, m_\ell = 0$). According to the Pauli exclusion principle, these electrons must have opposite spins. The He atom contains one pair of electrons in the 1*s* subshell. On an energy diagram, electrons are designated by arrows that represent the electron spin quantum numbers. An arrow points up if it has one spin quantum number and down if it has the other. This notation is used in Figure 7.22*a* to show the ground state of the helium atom. If one or more of the electrons is in any other allowed orbital of the diagram (see Figure 7.22*b*), the atom is in an **excited state.** The excited state is of higher energy, and the atom tends to return to its ground state by losing energy, often by emitting a photon of light. Do not confuse an excited state with an *impossible state,* in which forbidden combinations of quantum numbers are present; for example, the state is impossible if both electrons in the 1*s* orbital have the same spin (see Figure 7.22*c*).

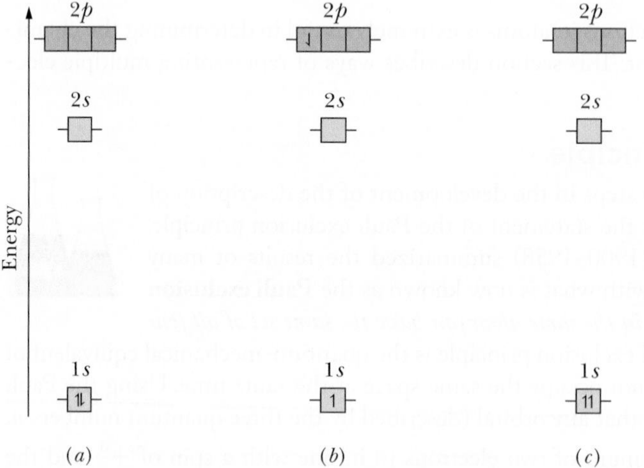

Figure 7.22 Energy-level diagram for the helium atom. *(a)* Ground state of the helium (He) atom. *(b)* An excited state of the helium atom in which one electron occupies the 2*p* subshell. An excited atom returns to the ground state by losing energy. *(c)* An impossible electronic state of He. As indicated, the electrons have the same spin in the 1*s* orbital, and thus would have the same set of four quantum numbers.

Although the energy-level diagram is the most complete way to show the arrangement of electrons in atoms, chemists have developed a number of shorthand descriptions. An **orbital diagram** is one way to show how the electrons are present in an atom. Each orbital is represented by a box, with orbitals in the same subshell shown as grouped boxes. The electrons in each orbital are represented by arrows pointing up or down to indicate one of the two allowed values of the spin quantum number. Just as in the energy-level diagrams, if an orbital contains two electrons (an electron pair), the directions of the two arrows must be opposite to be consistent with the Pauli exclusion principle. The orbital diagram for the hydrogen atom is

Two electrons in the same orbital must always have opposing spins, represented by "up" and "down" arrows.

$$\text{H} \quad \underset{\boxed{\uparrow}}{1s}$$

It would be equally correct to show the single electron as an arrow pointing down, but most chemists follow a convention of representing the first electron in an orbital with an "up" arrow. In orbital diagrams, the electrons are represented as they are in energy-level diagrams except that all of the orbitals are shown on a single line. The orbitals appear in order of increasing energy, with gaps between groups of boxes to indicate a difference in the energies of the orbitals.

Both energy-level diagrams and orbital diagrams are used to represent the electrons in atoms.

An electron configuration lists the occupied subshells, using the usual notations (e.g., 1*s* or 3*d*), with a superscript number indicating the number of electrons in the subshell. In this notation, the electron configuration of a ground state hydrogen atom is

$1s^1$; this is read as "one ess one" to indicate that the single electron in the ground state hydrogen atom is in the $1s$ subshell. The spins of the electrons are not explicitly given in an electron configuration as they are in an orbital diagram.

The electron configuration of an atom is compact; it does not contain the detailed information about electron spins that an orbital diagram provides.

Each atom of helium has two electrons. The energy-level diagram of this atom has already been shown in Figure 7.22. The electron configuration and orbital diagram for helium are

		$1s$
He	$1s^2$	[↑↓]

The lithium atom, Li, contains three electrons, and the first two enter the $1s$ subshell with opposite spins. The third electron must go into the subshell with the next higher energy ($2s$) so that the Pauli exclusion principle is not violated. The electron configuration and orbital diagram are

		$1s$	$2s$
Li	$1s^22s^1$	[↑↓]	[↑]

Beryllium, with four electrons, completes the filling of the $2s$ subshell.

		$1s$	$2s$
Be	$1s^22s^2$	[↑↓]	[↑↓]

Because the $1s$ and $2s$ orbitals are filled with four electrons, the fifth electron in the boron atom must occupy the $2p$ subshell, which consists of three orbitals. The electron configuration and orbital diagram for boron are

		$1s$	$2s$	$2p$
B	$1s^22s^22p^1$	[↑↓]	[↑↓]	[↑][ ][ ]

The three p orbitals are shown as connected boxes, to indicate that they form a degenerate set (all have the same energy). Any one of the three boxes could contain the electron, but by convention we usually proceed from left to right when we place electrons in the boxes.

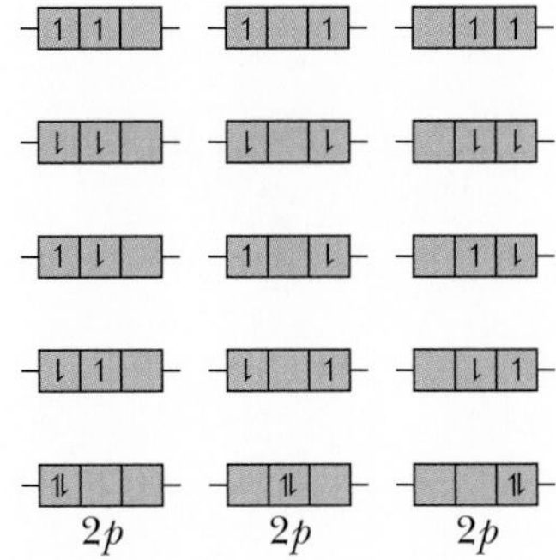

The $2p$ electrons of carbon. There are 15 possible ways electrons and their spins can exist in the $2p$ orbitals of the carbon atom.

The next element is carbon, which contains six electrons and must have two electrons in the $2p$ subshell. Fifteen ways exist in which to assign two electrons in the $2p$ subshell, but not all of these have the same energy. The experimentally determined magnetic properties of the carbon atom show that it contains two unpaired electrons of the same-direction spin. The second $2p$ electron must occupy a different orbital and have the same spin as the first electron to be consistent with the observed properties of the atom.

Whenever electrons are added to a subshell that contains more than one orbital, the electrons enter separate orbitals until there is one electron in each. These observations can be explained by the differences in interelectronic repulsions. Two electrons in the same orbital are closer together than they would be if they were in separate orbitals, and they therefore repel each other more strongly. Furthermore, experiments show that the spins of all the unpaired electrons are the same. This order is summarized by **Hund's rule:** In the filling of degenerate orbitals (orbitals with identical energies), one electron occupies each orbital, and all electrons have identical spins, before any two electrons are placed in the same orbital. Following Hund's rule, the electron configuration and orbital diagram for the carbon atom are

Hund's rule states that degenerate orbitals are filled with one electron in each before any electrons are paired.

		$1s$	$2s$	$2p$
C	$1s^22s^22p^2$	[↑↓]	[↑↓]	[↑][↑][ ]

We use Hund's rule to write the ground-state electron configurations and orbital diagrams for the elements with atomic numbers 7 through 10. Note that the added electrons must form pairs starting with oxygen because there are only three degenerate orbitals in the 2*p* subshell.

		1*s*	2*s*	2*p*
N	$1s^2 2s^2 2p^3$	↑↓	↑↓	↑ ↑ ↑
O	$1s^2 2s^2 2p^4$	↑↓	↑↓	↑↓ ↑ ↑
F	$1s^2 2s^2 2p^5$	↑↓	↑↓	↑↓ ↑↓ ↑
Ne	$1s^2 2s^2 2p^6$	↑↓	↑↓	↑↓ ↑↓ ↑↓

OBJECTIVES REVIEW *Can you:*

- ☑ use the Pauli exclusion principle to determine the maximum number of electrons in an orbital, subshell, or shell?
- ☑ write the electron configuration of an atom?
- ☑ construct an orbital diagram and an energy-level diagram for a given atom?
- ☑ predict the number of unpaired electrons in an atom?

7.7 Electron Configurations of Heavier Atoms

OBJECTIVES

- ☐ Write the ground-state electron configuration of heavier atoms
- ☐ Write abbreviated electron configurations
- ☐ Determine whether an electron configuration is anomalous

Through the element argon, electrons fill shells and subshells in expected order: first the 1*s*, then the 2*s* and 2*p*, then the 3*s* and 3*p*. The next subshell filled, for a potassium atom, is the 4*s*, not the 3*d*.

Why is this? Experiments show that in the ground state of potassium atoms, the final electron is in the 4*s* subshell, not the 3*d*. In Section 7.5, we argued that this was due to shielding and penetration effects. So the electron configuration of a ground-state potassium atom is not

K $1s^2\ 2s^2\ 2p^6\ 3s^2\ 3p^6\ 3d^1$ ← INCORRECT

This is a higher energy excited state of the potassium atom. The correct ground state electron configuration of a potassium atom is

K $1s^2\ 2s^2\ 2p^6\ 3s^2\ 3p^6\ 4s^1$ ← CORRECT

Andrew Lambert Photography/Photo Researchers, Inc.

Calcium metal. Pure calcium metal is silvery and soft, and reacts slowly with water.

The next larger atom, calcium, has its final electron in the 4*s* subshell also, so the electron configuration of a calcium atom is

Ca $1s^2\ 2s^2\ 2p^6\ 3s^2\ 3p^6\ 4s^2$

Now that the 4*s* subshell is filled, the next subshell to be filled is the 3*d* subshell. Five orbitals are in a *d* subshell, and as with the three orbitals in the *p* subshells, we must follow Hund's rule and put a single electron in each orbital, with the same spin, before pairing electrons. So for a manganese atom, whose electron configuration is $1s^2\ 2s^2\ 2p^6\ 3s^2\ 3p^6\ 4s^2\ 3d^5$, the orbital diagram would be

	$1s$	$2s$	$2p$	$3s$	$3p$	$4s$	$3d$
Mn	⇅	⇅	⇅ ⇅ ↑	⇅	⇅ ⇅ ⇅	⇅	↑ ↑ ↑ ↑ ↑

A manganese atom thus has five unpaired electrons in its ground state. With the next atom, iron, electrons in the *d* orbitals begin to pair until the *d* subshell is filled. For larger atoms, shells and subshells are filled in the order indicated by Figure 7.21.

EXAMPLE 7.8 Electron Configurations of Heavier Atoms

Bromine atoms have 35 electrons around the nucleus. What is the electron configuration of a bromine atom?

Strategy Use Figure 7.21 to determine the order of filling of subshells beyond the $3p$ subshell. Fill subshells until the total number of electrons is 35.

Solution
Let us construct a table so we can keep a running count of total electrons:

Shell/subshell	No. of Electrons	Total No. of Electrons
$1s$	2	2
$2s$	2	4
$2p$	6	10
$3s$	2	12
$3p$	6	18
$4s$	2	20
$3d$	10	30
$4p$	5	35

We need to go up to the $4p$ subshell to accommodate 35 electrons. The complete electron configuration of a Br atom is $1s^2\ 2s^2\ 2p^6\ 3s^2\ 3p^6\ 4s^2\ 3d^{10}\ 4p^5$.

Understanding

What is the electron configuration of Zr, whose atomic number is 40?

Answer $1s^2\ 2s^2\ 2p^6\ 3s^2\ 3p^6\ 4s^2\ 3d^{10}\ 4p^6\ 5s^2\ 4d^2$

Abbreviated Electron Configurations

Electron configurations can get long, especially for larger atoms. Chemists simplify long electron configurations to focus on the outer electrons that are involved in most chemical reactions.

One simplification, leading to an **abbreviated electron configuration,** is to use the noble gases to represent the partial electron configuration up to the number of electrons for that gas. For example, the electron configuration of lithium is $1s^2\ 2s^1$. Because the electron configuration of helium is $1s^2$, the electron configuration of lithium could be written as [He] $2s^1$, where [He] represents the electron configuration of helium, or $1s^2$. Granted, this is not much of a simplification, but now consider the electron configuration of sodium:

$$\underbrace{1s^2\ 2s^2\ 2p^6}_{\text{electron configuration of Ne}}\ 3s^1$$

We can abbreviate the electron configuration of Na as [Ne] $3s^1$, which is a significant simplification. Not only does this method simplify writing the electron configuration, it emphasizes the configuration of the outermost electrons, which are the ones that usually participate in chemical reactions.

Abbreviated electron configurations are more convenient for expressing the electron configurations of heavier atoms.

EXAMPLE 7.9 Abbreviated Electron Configurations

What is the abbreviated electron configuration for antimony, an element found in bullets (see the introduction to this chapter), whose atomic number is 51?

Strategy Use the periodic table to find the next lower noble gas and build on its electron configuration.

Solution

The closest noble gas with fewer electrons than antimony is krypton, whose atomic number is 36. The electron configuration of krypton is $1s^2\ 2s^2\ 2p^6\ 3s^2\ 3p^6\ 4s^2\ 3d^{10}\ 4p^6$. Using [Kr] to represent these electrons, we have as the abbreviated electron configuration for antimony:

[Kr] $5s^2\ 4d^{10}\ 5p^3$

This is a much more compact way to represent the electron configuration.

Understanding

What is the abbreviated electron configuration of barium?

Answer [Xe] $6s^2$

Table 7.3 lists the ground-state electron configurations of the atoms.

TABLE 7.3 Ground-State Electron Configurations of the Atoms

1	H	$1s^1$	38	Sr	[Kr]$5s^2$	75	Re	[Xe]$6s^24f^{14}5d^5$
2	He	$1s^2$	39	Y	[Kr]$5s^24d^1$	76	Os	[Xe]$6s^24f^{14}5d^6$
3	Li	[He]$2s^1$	40	Zr	[Kr]$5s^24d^2$	77	Ir	[Xe]$6s^24f^{14}5d^7$
4	Be	[He]$2s^2$	41	Nb	[Kr]$5s^14d^4$	78	Pt	[Xe]$6s^14f^{14}5d^9$
5	B	[He]$2s^22p^1$	42	Mo	[Kr]$5s^14d^5$	79	Au	[Xe]$6s^14f^{14}5d^{10}$
6	C	[He]$2s^22p^2$	43	Tc	[Kr]$5s^24d^5$	80	Hg	[Xe]$6s^24f^{14}5d^{10}$
7	N	[He]$2s^22p^3$	44	Ru	[Kr]$5s^14d^7$	81	Tl	[Xe]$6s^24f^{14}5d^{10}6p^1$
8	O	[He]$2s^22p^4$	45	Rh	[Kr]$5s^14d^8$	82	Pb	[Xe]$6s^24f^{14}5d^{10}6p^2$
9	F	[He]$2s^22p^5$	46	Pd	[Kr]$4d^{10}$	83	Bi	[Xe]$6s^24f^{14}5d^{10}6p^3$
10	Ne	[He]$2s^22p^6$	47	Ag	[Kr]$5s^14d^{10}$	84	Po	[Xe]$6s^24f^{14}5d^{10}6p^4$
11	Na	[Ne]$3s^1$	48	Cd	[Kr]$5s^24d^{10}$	85	At	[Xe]$6s^24f^{14}5d^{10}6p^5$
12	Mg	[Ne]$3s^2$	49	In	[Kr]$5s^24d^{10}5p^1$	86	Rn	[Xe]$6s^24f^{14}5d^{10}6p^6$
13	Al	[Ne]$3s^23p^1$	50	Sn	[Kr]$5s^24d^{10}5p^2$	87	Fr	[Rn]$7s^1$
14	Si	[Ne]$3s^23p^2$	51	Sb	[Kr]$5s^24d^{10}5p^3$	88	Ra	[Rn]$7s^2$
15	P	[Ne]$3s^23p^3$	52	Te	[Kr]$5s^24d^{10}5p^4$	89	Ac	[Rn]$7s^26d^1$
16	S	[Ne]$3s^23p^4$	53	I	[Kr]$5s^24d^{10}5p^5$	90	Th	[Rn]$7s^26d^2$
17	Cl	[Ne]$3s^23p^5$	54	Xe	[Kr]$5s^24d^{10}5p^6$	91	Pa	[Rn]$7s^25f^26d^1$
18	Ar	[Ne]$3s^23p^6$	55	Cs	[Xe]$6s^1$	92	U	[Rn]$7s^25f^36d^1$
19	K	[Ar]$4s^1$	56	Ba	[Xe]$6s^2$	93	Np	[Rn]$7s^25f^46d^1$
20	Ca	[Ar]$4s^2$	57	La	[Xe]$6s^25d^1$	94	Pu	[Rn]$7s^25f^6$
21	Sc	[Ar]$4s^23d^1$	58	Ce	[Xe]$6s^24f^15d^1$	95	Am	[Rn]$7s^25f^7$
22	Ti	[Ar]$4s^23d^2$	59	Pr	[Xe]$6s^24f^3$	96	Cm	[Rn]$7s^25f^76d^1$
23	V	[Ar]$4s^23d^3$	60	Nd	[Xe]$6s^24f^4$	97	Bk	[Rn]$7s^25f^9$
24	Cr	[Ar]$4s^13d^5$	61	Pm	[Xe]$6s^24f^5$	98	Cf	[Rn]$7s^25f^{10}$
25	Mn	[Ar]$4s^23d^5$	62	Sm	[Xe]$6s^24f^6$	99	Es	[Rn]$7s^25f^{11}$
26	Fe	[Ar]$4s^23d^6$	63	Eu	[Xe]$6s^24f^7$	100	Fm	[Rn]$7s^25f^{12}$
27	Co	[Ar]$4s^23d^7$	64	Gd	[Xe]$6s^24f^75d^1$	101	Md	[Rn]$7s^25f^{13}$
28	Ni	[Ar]$4s^23d^8$	65	Tb	[Xe]$6s^24f^9$	102	No	[Rn]$7s^25f^{14}$
29	Cu	[Ar]$4s^13d^{10}$	66	Dy	[Xe]$6s^24f^{10}$	103	Lr	[Rn]$7s^25f^{14}6d^1$
30	Zn	[Ar]$4s^23d^{10}$	67	Ho	[Xe]$6s^24f^{11}$	104	Rf	[Rn]$7s^25f^{14}6d^2$
31	Ga	[Ar]$4s^23d^{10}4p^1$	68	Er	[Xe]$6s^24f^{12}$	105	Db	[Rn]$7s^25f^{14}6d^3$
32	Ge	[Ar]$4s^23d^{10}4p^2$	69	Tm	[Xe]$6s^24f^{13}$	106	Sg	[Rn]$7s^25f^{14}6d^4$
33	As	[Ar]$4s^23d^{10}4p^3$	70	Yb	[Xe]$6s^24f^{14}$	107	Bh	[Rn]$7s^25f^{14}6d^5$
34	Se	[Ar]$4s^23d^{10}4p^4$	71	Lu	[Xe]$6s^24f^{14}5d^1$	108	Hs	[Rn]$7s^25f^{14}6d^6$
35	Br	[Ar]$4s^23d^{10}4p^5$	72	Hf	[Xe]$6s^24f^{14}5d^2$	109	Mt	[Rn]$7s^25f^{14}6d^7$
36	Kr	[Ar]$4s^23d^{10}4p^6$	73	Ta	[Xe]$6s^24f^{14}5d^3$	110	Ds	[Rn]$7s^25f^{14}6d^8$
37	Rb	[Kr]$5s^1$	74	W	[Xe]$6s^24f^{14}5d^4$			

PRACTICE OF CHEMISTRY Magnets

Magnetism is caused by moving charges, specifically electrons. If electricity is moving through a straight wire, a circular *magnetic field* is produced. If electricity is moving in a circle or loop, then a doughnut-shaped field is produced. This field is reinforced in the center of the loop, forming what is known as a *magnetic dipole. Electromagnets* are magnets formed by wires in such configurations and are used by society in various ways. One of the more exciting ways is in *magnetic resonance imaging* (MRI). MRI is a technique that uses radio waves in conjunction with magnetic fields produced by large magnets. Interactions among hydrogen atoms, the radio waves, and the magnetic field produce signals that differ with body tissue; these signals are collected by detectors and displayed by a computer as an image. Trained medical personnel can differentiate between the tissues and diagnose disease.

All matter reacts to the presence of an external magnetic field. Matter that has no unpaired electrons is slightly repelled by a magnetic field. Such matter is called *diamagnetic.* In matter that has unpaired electrons, the unpaired electrons act as tiny magnets themselves. In the presence of an external magnetic field, the matter is attracted to the field. Such matter is called *paramagnetic.* Precise measurements of the force of attraction between a sample of matter and an external magnetic field can experimentally determine how many unpaired electrons are in the atoms in a sample.

In solid substances where the magnetic fields of the individual atoms are highly aligned, strong magnetic behavior is seen, and the material is a permanent magnet. Such materials are called *ferromagnetic,* because this behavior is typified by certain samples of iron (L. *ferrum*). Though typified by iron, this effect is not exclusive to iron; other metallic elements and mixtures of metallic elements called *alloys* are also ferromagnetic. One of the strongest permanent magnets is an alloy of aluminum, nickel, and cobalt called *alnico.* Alnico magnets having a magnetic field 25,000 times that of Earth's magnetic field are readily manufactured.

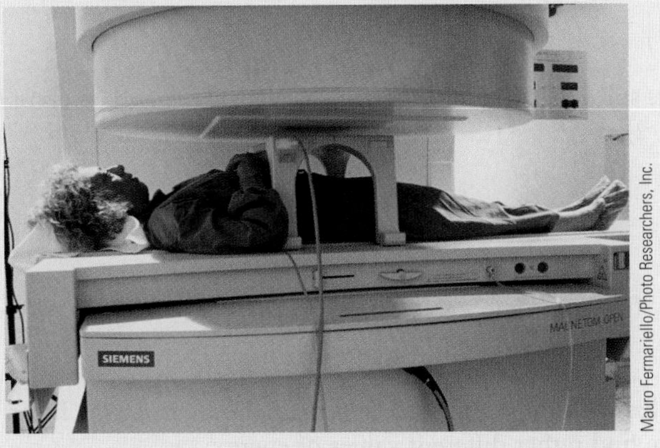

Mauro Fermariello/Photo Researchers, Inc.

Magnetic resonance imaging. A magnetic resonance imaging (MRI) system allows trained personnel to scan body tissues using a combination of radio waves and a magnetic field.

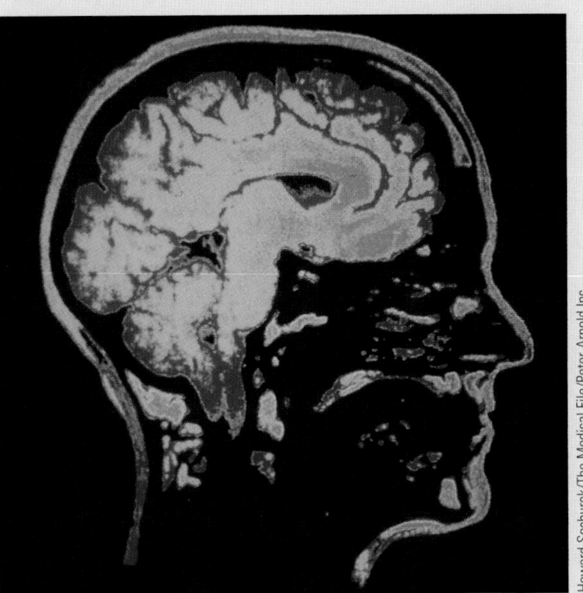

Howard Sochurek/The Medical File/Peter Arnold Inc.

Magnetic resonance imaging (MRI) scanning. MRI scans allow medical personnel to visualize body tissues to diagnose disease.

Anomalous Electron Configurations

A review of Table 7.3 shows that some elements, such as chromium and silver, have more than one unfilled subshell, or have a lower subshell less than completely filled with electrons. For example, the electron configuration of chromium is [Ar] $4s^1\ 3d^5$, not [Ar] $4s^2\ 3d^4$, as expected. Curiously, all of the exceptions are transition metals or inner transition metals (the lanthanides and actinides); none of the main group elements has such electron configurations. These electron configurations of the exceptions are **anomalous.**

Why do some atoms have anomalous electron configurations? In some, but not all, cases, the anomalous electron configuration leads to a combination of either two half-filled subshells or a half-filled and a completely filled subshell. For example, the electron configuration of chromium ([Kr] $4s^1\ 3d^5$) has two half-filled subshells, whereas the electron configuration of copper is [Kr] $4s^1\ 3d^{10}$. However, this does not happen with all possible cases (the electron configuration of tungsten is [Xe] $6s^2\ 4f^{14}\ 5d^4$, not [Xe] $6s^1\ 4f^{14}\ 5d^5$). Currently, we cannot predict which atoms will have anomalous electron configurations in advance, but we do know the reason why those atoms are anomalies: The total electronic energy of the atom is lower in an anomalous configuration in comparison with a "normal" electron configuration. The electron configuration [Xe] $6s^2\ 4f^{14}\ 5d^9$ might be the expected electron configuration of a gold atom, but experiment shows that gold atoms have the ground-state electron configuration [Xe] $6s^1\ 4f^{14}\ 5d^{10}$. This second electron configuration has a lower energy than the first; thus, it is the one found in Table 7.3.

OBJECTIVES REVIEW *Can you:*

- ☑ write the ground-state electron configuration of heavier atoms?
- ☑ write abbreviated electron configurations?
- ☑ determine whether an electron configuration is anomalous?

CASE STUDY Applications and Limits of Bohr's Theory

One of the problems with the Bohr theory of hydrogen (in addition to the fact that it did not treat electrons as waves) is that it applied only to hydrogen and other single-electron atoms (such as He^+, Li^{2+}, among others). To treat other one-electron systems, we would need to include a factor of Z^2 in the numerator of Equation 7.4, where Z represents the charge on the nucleus:

$$E_n = -\frac{Z^2 2\pi^2 me^4}{h^2}\left(\frac{1}{n^2}\right) = \frac{-Z^2B}{n^2}$$

Because the constant B is still 2.18×10^{-18} J, we can calculate the energy levels of the He^+ species (for which $Z = 2$):

$n =$	$E_n =$
1	-8.72×10^{-18} J
2	-2.18×10^{-18} J
3	-9.67×10^{-19} J
4	-5.45×10^{-19} J
etc.	

If we were to compare this with the energy levels of the helium *atom,* which can be measured experimentally, we find rather different values of energy:

$n =$	$E_n =$
1	-3.94×10^{-18} J
2	-7.63×10^{-19} J
3	-6.36×10^{-19} J
4	-5.81×10^{-19} J

Even the first four electronic energy levels of the helium atom defied any attempt to predict them mathematically. So although Bohr's model of the hydrogen atom was an important step in the development of the currently accepted theory of quantum mechanics (by extending the concept of quantization to measurable factors other than energy), it was powerless to predict the electronic energies of even the second-smallest atom.

Questions

1. What are the first three energy levels of Li^{2+}? (*Hint:* What is Z for Li^{2+}?)
2. What is the charge on a uranium ion that has a single electron?
3. Predict the wavelengths of light that cause the $n_1 = 2, 3,$ and $4 \rightarrow n_2 = 1$ electronic transition of the helium ion. These wavelengths of light would be detected in emission spectrometers (see the chapter introduction) when analyzing for helium.

ETHICS IN CHEMISTRY

1. Fluorescent light bulbs generate light by forcing atoms to go from one electronic state to another, whereas incandescent light bulbs generate light by heating a thin filament until it glows white-hot. Fluorescent lights require less energy to operate, but they contain small amounts of mercury, which can be hazardous to the environment. Incandescent lights do not contain any environmentally hazardous materials, but they do require more energy to operate and get very hot during operation. If you were the developer of a new housing subdivision, what factors would you consider when deciding to use incandescent versus fluorescent lights in various functions of the project, such as in street lights and outdoor house lighting?

2. X rays are a type of electromagnetic radiation called *ionizing radiation* because they have enough energy to remove electrons from atoms in matter, including our bodies. X rays are useful in medicine because they allow doctors to visualize bones and other tissues without surgery, yet they carry some risk because of their ionizing capabilities. Although use of x rays in a patient with an acute medical problem (such as a broken bone) is unquestionably worth the risk, use of x rays in healthy patients (as in annual dental x rays) carries a "risk versus benefit" question. In the past, x rays of much higher intensities were used, but with improvements in film and detector sensitivities, lower intensities—with proportionately lower risk—are currently used. Given your experience, under what circumstances do you think that the use of x rays is justified? (If you get the chance, discuss this with a physician or dentist and compare your answers.)

3. As a noninvasive technique to look for medical problems, the MRI method is rapidly expanding its scope. The method is typical of many of the advances that we are experiencing in medicine. New methods can improve health, but they are expensive tests to use for every problem. In your opinion, what sorts of medical conditions justify the expense of using MRI?

4. In the 1960s, investigations into the assassination of President John F. Kennedy included chemical analysis of the bullet fragments recovered from the crime scene. The chemists were trying to determine how many bullets were fired. They measured the silver and antimony concentrations; their measurements are plotted below.

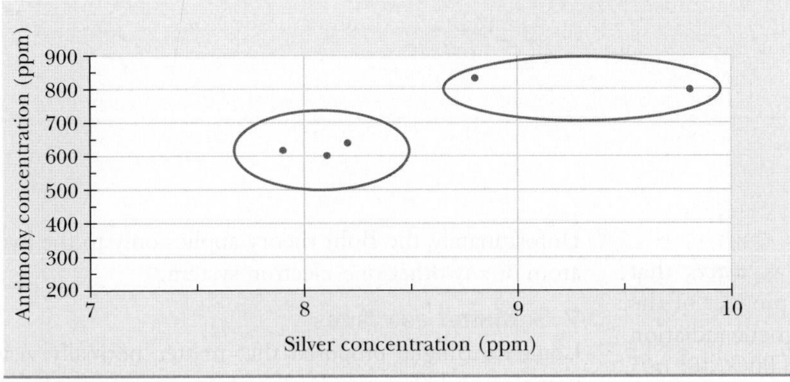

The chemists determined that three fragments (data from which are circled near the center of in the plot) came from one bullet and that two fragments (also circled, upper right) came from a second bullet, but both bullets came from the same box of ammunition. Do you think the evidence supports this conclusion? Justify your answer. (Refer to the chapter introduction for a discussion of forensic analysis of bullets.)

Chapter 7 Visual Summary

The chart shows the connections between the major topics discussed in this chapter.

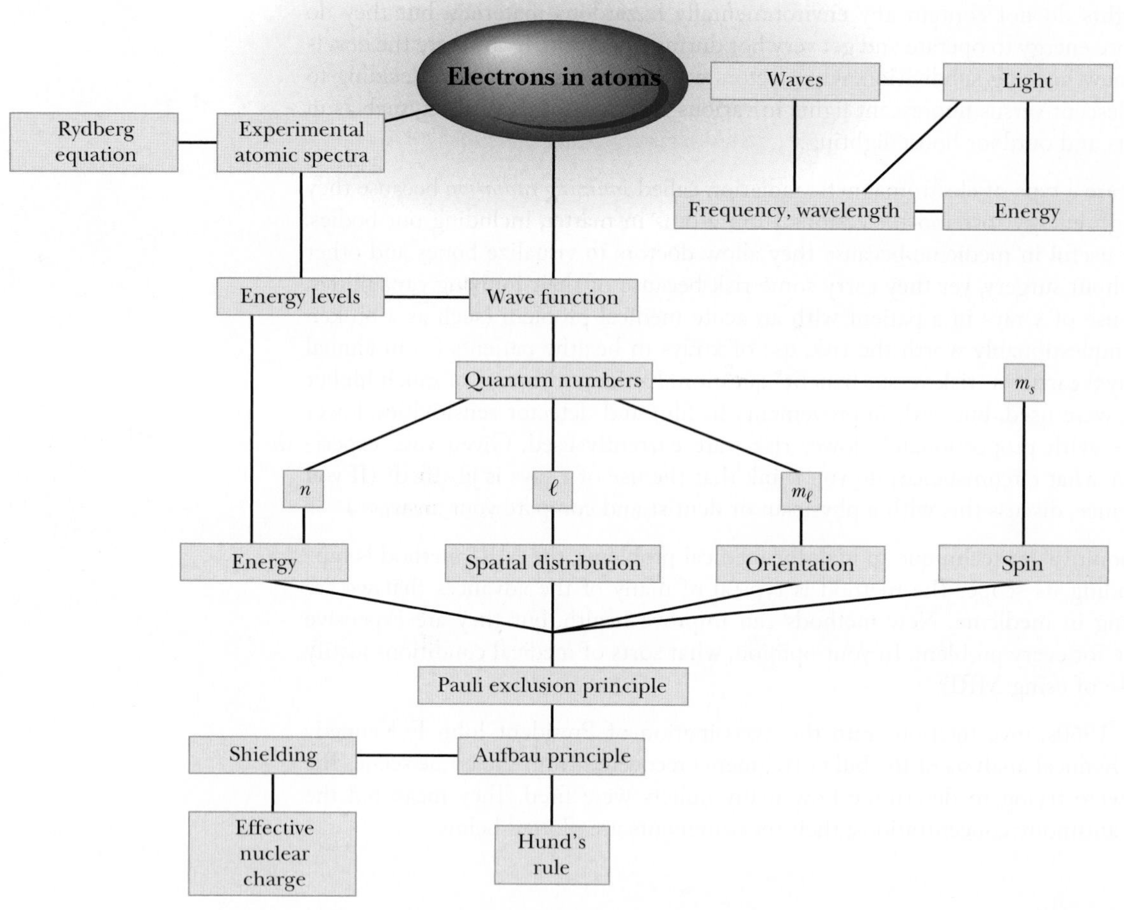

Summary

7.1 The Nature of Light

Electromagnetic radiation can be described as *waves* that travel at a constant speed in a vacuum. The product of the *wavelength* (λ) and *frequency (ν)* of electromagnetic radiation always equals the speed of light *(c)*, which is 3.00×10^8 m/s in a vacuum. Light also can be considered as a stream of *photons,* particles of light, each having an energy of $h\nu$, where h is Planck's constant and has the value 6.626×10^{-34} J · s. The particle nature of light explains the *photoelectric effect* and the *line spectra* of the elements.

7.2 Line Spectra and the Bohr Atom

When Niels Bohr assumed that the angular momentum of the electron was quantized, his model accurately predicted the positions of the lines in the hydrogen atom spectrum. Unfortunately, the Bohr theory applies only to the hydrogen atom or any other one-electron system.

7.3 Matter as Waves

Louis de Broglie proposed that matter, normally viewed as particles, could be considered as waves with a wavelength of $\lambda = h/mv$. de Broglie showed that Bohr's assumption of quantized angular momentum is a natural consequence resulting from considering the electron in the hydrogen atom as a wave. In developing quantum mechanics, Schrödinger found that the *wave function,* ψ, is described by three *quantum numbers:* the *principal quantum number n,* the *angular momentum quantum number* ℓ, and the *magnetic quantum number* m_ℓ. The square of the wave function, ψ^2, is the probability of finding the electron at any point in space. The value of each of the quantum numbers can be related to a characteristic of the wave function. The value of n determines the distance of the electron from the nucleus and the energy of the atom. The shape of the electron cloud is related to the quantum number ℓ, and its orientation in space is determined by the quantum number m_ℓ.

7.4 Quantum Numbers in the Hydrogen Atom

All wave functions with the same value of the quantum number n belong to the same *principal shell;* those in which both n and ℓ are specified constitute a *subshell.* When n, ℓ, and m_ℓ are all specified, the wave function is called an *atomic orbital.* Subshells and orbitals are identified by giving the value of the principal quantum number (1, 2, 3, ...) followed by the letter *s, p, d,* or *f,* representing the value 0, 1, 2, or 3, respectively, for the quantum number ℓ. Thus, 3*p* refers to an orbital or subshell for which $n = 3$ and $\ell = 1$. In addition to the n, ℓ, and m_ℓ quantum numbers, which describe the location of the electron, a fourth quantum number representing the *electron spin,* m_s, is needed to account for the magnetic properties of electrons and atoms.

Restrictions on the values of the quantum numbers mean that the following subshells are allowed:

1*s*
2*s*, 2*p*
3*s*, 3*p*, 3*d*
4*s*, 4*p*, 4*d*, 4*f*
.
.
.

The number of orbitals in the types of subshells are as follows: one for *s,* three for *p,* five for *d,* and seven for *f.*

7.5 Energy Levels for Multielectron Atoms

In an atom that contains two or more electrons, the energy of an electron depends on both the principal and angular momentum quantum numbers because of interelectronic repulsions. The effective nuclear charge is calculated by taking into account the effect of shielding by inner electrons and interelectronic repulsions.

7.6 Electrons in Multielectron Atoms

The *Pauli exclusion principle* restricts the number of electrons that can be placed in an orbital to two, because there are only two allowed values ($\pm\frac{1}{2}$) for the electron spin quantum number m_s. The Pauli exclusion principle determines the maximum number of electrons in any subshell as 2 for *s,* 6 for *p,* 10 for *d,* and 14 for *f.* Furthermore, *Hund's rule* states that electrons in *degenerate orbitals* (orbitals with identical energies) do not pair until there is one electron in each orbital of the set.

7.7 Electron Configurations of Heavier Atoms

We build up the predicted *ground-state* electron configuration of an atom by adding the appropriate number of electrons to the lowest energy subshells available, following the restrictions of the Pauli exclusion principle. The *electron configuration* ($1s^2 2s^2$, and so on) and the *orbital diagram* are two ways of representing electrons in atoms.

Download Go Chemistry concept review videos from OWL or purchase them from **www.ichapters.com**

Chapter Terms

The following terms are defined in the Glossary, Appendix I.

Section 7.1
Amplitude
Electromagnetic radiation
Frequency, ν
Hertz (Hz)
Photoelectric effect
Photon
Planck's constant, h
Planck's equation
Quantum
Visible light
Wave
Wavelength, λ

Section 7.2
Continuous spectrum
Ground state
Line spectrum
Rydberg constant
Rydberg equation
Spectrum

Section 7.3
de Broglie equation
Heisenberg uncertainty principle
Quantum mechanics
Quantum numbers
Wave function, ψ

Section 7.4
Angular momentum quantum number, ℓ
Atomic orbital
Electron spin quantum number, m_s
Principal quantum number, n
Principal shell
Magnetic quantum number, m_ℓ
Nodes
Shell
Subshell

Section 7.5
Degenerate orbitals
Effective nuclear charge
Electron shielding
Interelectronic repulsions

Section 7.6
Aufbau principle
Electron configuration
Electron pair (paired electrons)
Excited state
Ground state
Hund's rule
Orbital diagram
Paired electron
Pauli exclusion principle
Unpaired electron

Section 7.7
Electron configuration, abbreviated
Electron configuration, anomalous

Key Equations

Wave properties of light (7.1)

$$c = \lambda\nu$$

Energy of light (7.1)

$$E = h\nu$$

Predicted wavelengths of the hydrogen atom spectrum (7.2)

$$\frac{1}{\lambda} = R_H\left(\frac{1}{n_1^2} - \frac{1}{n_2^2}\right)$$

Energy of a hydrogen atom (7.2)

$$E = -\frac{2\pi^2 me^4}{h^2}\left(\frac{1}{n^2}\right) = -\frac{B}{n^2},\ B = 2.18 \times 10^{-18}\ \text{J}$$

de Broglie wavelength of matter (7.3)

$$\lambda = h/p = h/mv$$

Questions and Exercises

OWL Selected end of chapter Questions and Exercises may be assigned in OWL.

Blue-numbered Questions and Exercises are answered in Appendix J; questions are qualitative, are often conceptual, and include problem-solving skills.

■ Questions assignable in OWL

✎ Questions suitable for brief writing exercises

▲ More challenging questions

Questions

7.1 ✎ Two light sources have exactly the same color, but the second source has twice the brightness of the first. Which of the four characteristics of a light wave (amplitude, speed, frequency, wavelength) are the same and which are different for these two waves? Compare the energy of the photons and the total energy of the light from the two sources.

7.2 The most intense emissions of silver and bismuth, both elements found in bullets (see the chapter introduction), are 328.1 and 195.5 nm, respectively. Compare the energies and frequencies of light from these two elements.

7.3 Why are there many more lines observed in the emission spectra of hydrogen and other elements than are found in their absorption spectra?

7.4 What assumptions did Bohr make in explaining the hydrogen atom spectrum?

7.5 How did de Broglie justify Bohr's assumption that angular momentum was quantized?

7.6 How is the wave function, ψ, related to the location of the electron in space?

7.7 How does the electron spin quantum number affect the energy of the electron in the hydrogen atom?

7.8 (a) Which quantum number is related to the average distance of the electron from the nucleus?

(b) What name and symbol are used for the quantum number that determines the shape of the electron probability distribution?

(c) Which quantum number contains information about the orientation of the electron cloud?

7.9 In a carbon atom, do the $2s$ or $2p$ electrons experience a higher effective nuclear charge? Explain.

7.10 How is interelectronic repulsion related to shielding?

7.11 ✎ How is the effective nuclear charge different from the nuclear charge in a beryllium atom? Is the effective nuclear charge the same for all electrons in this atom? Explain.

7.12 Explain what is meant by "penetration" in the explanation of the dependence of electron energies on the ℓ quantum number.

7.13 Why is the Heisenberg uncertainty principle an important factor for electrons in atoms but unimportant for large objects such as a baseball?

7.14 How and why do the energy-level diagrams for the hydrogen atom and the many-electron atom differ?

7.15 Draw an energy-level diagram for a multielectron atom through $n = 3$.

7.16 State the Pauli exclusion principle; then use it to explain why an orbital can contain a maximum of two electrons.

7.17 ✎ State Hund's rule; then use it to explain why there are two unpaired electrons in both the carbon atom and the oxygen atom.

7.18 Use the Pauli exclusion principle to explain why the $3p$ subshell can contain a maximum of six electrons.

Exercises

OBJECTIVE Describe the relationships among wavelength, frequency, and energy of light.

7.19 Tin is a component of some bullets (see the chapter introduction). The most intense light that is emitted by excited tin atoms has a wavelength of 284.0 nm. What is the frequency of this light (measured in s^{-1})? Name the spectral region for this light (ultraviolet, x ray, and so on).

7.20 The introduction to this chapter indicates that lead is the main component of bullets. The strongest emission from excited-state Pb atoms has a wavelength of 405.8 nm. What is the frequency of this light (in s^{-1})? Name the spectral region for this light (ultraviolet, x ray, and so on).

7.21 The human eye is most sensitive to light that has a frequency of 5.41×10^{14} Hz. What is the wavelength of this light? What name is used for the spectral region of this radiation?

7.22 Microwave ovens operate using radiation that has a frequency of 2.45×10^{9} Hz. What is the wavelength of this radiation? What name is used for the spectral region of this radiation?

7.23 An AM radio station broadcasts at a frequency of 580 kHz. What is the wavelength, in meters and nanometers, of this signal?

7.24 An FM radio station broadcasts at a frequency of 101.3 MHz. What is the wavelength, in meters and nanometers, of this radiation?

7.25 Gamma rays are electromagnetic radiation of very short wavelength emitted by the nuclei of radioactive elements. Strontium-91 emits a gamma ray with a frequency of 2.47×10^{20} Hz. Express the wavelength of this radiation, in picometers.

7.26 Gamma rays are electromagnetic radiation of very short wavelength emitted by the nuclei of radioactive elements. Arsenic-74 emits a gamma ray with a frequency of 1.44×10^{20} Hz. Express the wavelength of this radiation, in picometers.

7.27 The wavelength used by citizen's band radio is 21 m. Calculate the frequency (in s^{-1}) of this electromagnetic radiation.

7.28 The electromagnetic radiation used by amateur radio operators has a wavelength of 10 m. Calculate the frequency (in s^{-1}) of this electromagnetic radiation.

OBJECTIVE Describe the models that are used to describe the behavior of light.

7.29 This laser emits green light with a wavelength of 533 nm.

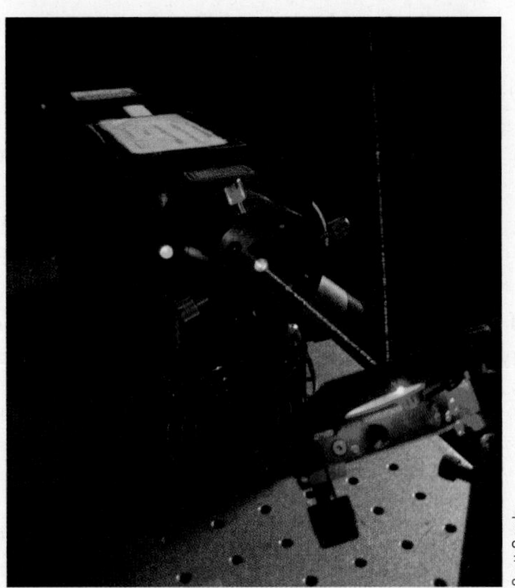

Scott Goode

(a) What is the energy, in joules, of one photon of light at this wavelength?
(b) If a particular laser produces 1.00 watt (W) of power (1 W = 1 J/s), how many photons are produced each second by the laser?

7.30 The photoelectric effect for cadmium has a threshold frequency of 9.83×10^{14} Hz. For light of this frequency, find the following characteristics:
(a) the wavelength
(b) the energy of one photon (in J)
(c) the energy of 1 mol of photons (in kJ)

7.31 What is the energy (in kJ) of 1 mol of photons with a frequency of 3.70×10^{15} Hz?

7.32 What is the energy (in kJ) of 1 mol of photons with a frequency of 2.50×10^{14} Hz?

7.33 The yellow light emitted by sodium vapor consists of photons with a wavelength of 589 nm. What is the energy change of a sodium atom that emits a photon with this wavelength?

7.34 The red color of neon signs is due to electromagnetic radiation with a wavelength of 640 nm. What is the change in the energy of a neon atom when it emits a photon of this wavelength?

7.35 The photoelectric threshold frequency for carbon is $1.16 \times 10^{15}\ s^{-1}$.
(a) What is the longest wavelength of light that will eject electrons from a sample of solid carbon? What is the name of the region of the electromagnetic spectrum with light of this wavelength?
(b) Are electrons ejected from carbon by any light in the visible region of the spectrum?

7.36 Electrons are ejected from sodium metal by any light that has a wavelength shorter than 544 nm. What is the photoelectric threshold frequency for sodium metal?

7.37 ▲ The charge of an electron is -1.602×10^{-19} C. How many electrons must be ejected from the metal each second to produce an electric current of 1.0 mA (1 A = 1 C/s)? How many photons must be absorbed each second to produce this number of photoelectrons, assuming that each photon causes an electron to be ejected?

7.38 ▲ The charge of an electron is -1.602×10^{-19} C. How many microamperes (1 A = 1 C/s) of electrical current are produced by a photoelectric cell that ejects 2.50×10^{13} electrons each second? How many photons must be absorbed each second to produce this number of photoelectrons, assuming that each photon causes an electron to be ejected?

OBJECTIVE Describe the origin of line spectra, especially that of hydrogen in light of Bohr's theory.

7.39 What is the wavelength (in nm) of the line in the spectrum of the hydrogen atom that arises from the transition of the electron from the Bohr orbit with $n = 3$ to the orbit with $n = 1$? In what region of the electromagnetic spectrum (ultraviolet, visible, and so on) is this radiation observed?

7.40 The spectrum below is of the hydrogen atom in the visible region. Label each emission line with the wavelength, the initial quantum number, and the final quantum number.

Hydrogen (H)

OBJECTIVE Relate the de Broglie wavelength of matter to its momentum.

7.41 What is the wavelength, in nanometers, of a neutron (mass = 1.67×10^{-27} kg) that is moving at a velocity of 1.7×10^2 m/s?

7.42 What is the wavelength, in nanometers, of an electron that is moving at a velocity of 2.9×10^5 m/s?

7.43 The velocity of an electron having $n = 1$ in Bohr's model is 2.19×10^6 m/s. What is the wavelength of this electron? The radius of its orbit is 52.9 pm. Compare the wavelength of the electron to the circumference of the orbit ($c = 2\pi r$).

7.44 Find the wavelength of an electron having $n = 2$ in Bohr's model of hydrogen if the velocity of the electron is 1.096×10^6 m/s. The radius of this orbit is 212 pm. Compare the wavelength of the electron with the circumference of the orbit ($c = 2\pi r$).

7.45 Find the de Broglie wavelength that is associated with each of the following objects:
(a) a ball with a mass of 0.100 kg traveling at 40.0 m/s
(b) a 753-kg car traveling at 24.6 m/s (55 mph)
(c) a neutron (mass = 1.67×10^{-27} kg) with a velocity of 2.70×10^3 m/s; this is the root-mean-square speed of a neutron at normal room temperature.

7.46 Find the de Broglie wavelength associated with each of the following objects:
(a) a 68-kg sprinter traveling at 10.0 m/s
(b) a 50.0-g ball traveling at 100 mph (44.7 m/s)
(c) an electron (mass = 9.11×10^{-31} kg) with a velocity of 1.2×10^5 m/s. This is the root-mean-square speed of an electron at normal room temperature.

7.47 Neutrons (mass = 1.67×10^{-27} kg) with a wavelength of 0.150 nm are needed for a diffraction experiment. What velocity must these neutrons have?

7.48 ■ What is the velocity of an electron that has a de Broglie wavelength of 1.00 nm?

OBJECTIVE List allowed combinations of quantum numbers in atoms.

7.49 Give the notation (1*s*, 2*s*, 2*p*, and so on) for each of the following subshells. If the combination is not allowed, state why.
(a) $n = 6, \ell = 1$ (b) $n = 3, \ell = 0$
(c) $n = 5, \ell = 2$ (d) $n = 4, \ell = 0$
(e) $n = 2, \ell = 3$

7.50 Give the notation (1*s*, 2*s*, 2*p*, and so on) for each of the following subshells. If the combination is not allowed, state why.
(a) $n = 5, \ell = 1$ (b) $n = 1, \ell = 1$
(c) $n = 3, \ell = 2$ (d) $n = 4, \ell = 3$
(e) $n = 7, \ell = 0$

7.51 Give the values of the n and ℓ quantum numbers for the subshells identified by the following designations.
(a) 3*p* (b) 5*d*
(c) 7*s* (d) 4*f*
(e) 2*s*

7.52 Give the values of the n and ℓ quantum numbers for the subshells identified by the following designations.
(a) 3*d* (b) 5*p*
(c) 6*s* (d) 5*f*
(e) 1*s*

7.53 In each part, a set of quantum numbers is given. If the set is an allowed combination of n, ℓ, m_ℓ, and m_s, give the subshell to which this wave function belongs (1*s*, 2*s*, 2*p*, and so on). If the combination of quantum numbers is not allowed, state why.
(a) $n = 2, \ell = 1, m_\ell = 0, m_s = -\frac{1}{2}$
(b) $n = 2, \ell = 2, m_\ell = -2, m_s = +\frac{1}{2}$
(c) $n = 3, \ell = 0, m_\ell = 0, m_s = +\frac{1}{2}$
(d) $n = 1, \ell = 0, m_\ell = 1, m_s = -\frac{1}{2}$
(e) $n = 3, \ell = 2, m_\ell = 2, m_s = +\frac{1}{2}$
(f) $n = 5, \ell = 0, m_\ell = 0, m_s = +\frac{1}{2}$

7.54 In each part, a set of quantum numbers is given. If the set is an allowed combination of n, ℓ, m_ℓ, and m_s, give the subshell to which this wave function belongs (1*s*, 2*s*, 2*p*, and so on). If the combination of quantum numbers is not allowed, state why.
(a) $n = 1, \ell = 1, m_\ell = 0, m_s = -\frac{1}{2}$
(b) $n = 4, \ell = 2, m_\ell = 2, m_s = +\frac{1}{2}$
(c) $n = 2, \ell = 0, m_\ell = 0, m_s = +\frac{1}{2}$
(d) $n = 3, \ell = 0, m_\ell = 1, m_s = -\frac{1}{2}$
(e) $n = 3, \ell = 2, m_\ell = 0, m_s = +\frac{1}{2}$
(f) $n = 6, \ell = 0, m_\ell = 0, m_s = +\frac{1}{2}$

7.55 (a) How many subshells are present in the $n = 4$ shell?
(b) How many orbitals are in the 3*d* subshell?
(c) What is the maximum value of ℓ that is allowed in the shell with $n = 3$?
(d) What are the values of n and ℓ for a 3*p* subshell? Give all allowed values of the m_ℓ quantum number for this subshell.

7.56 ■ (a) How many subshells are present in the $n = 3$ shell?
(b) How many orbitals are in the 4*p* subshell?
(c) What is the maximum value of ℓ that is allowed in the shell with $n = 4$?
(d) What are the values of n and ℓ for a 3*d* subshell? Give all allowed values of the m_ℓ quantum number for this subshell.

OBJECTIVE Relate values of quantum numbers to energy, shape, size, and orientation of electron cloud.

7.57 In each part, sketch the contour surface for the orbital described.
(a) $n = 2, \ell = 0$
(b) $3p_x$
(c) $4d_{xy}$
(d) $n = 2, \ell = 1$
(e) $n = 1, \ell = 0$

7.58 In each part, sketch the contour surface for the orbital described.
(a) $n = 3, \ell = 0$
(b) $3d_{xy}$
(c) $4p_y$
(d) $n = 3, \ell = 1$
(e) $n = 2, \ell = 0$

7.59 In what region of space is the probability of finding the p_y electron the greatest? Where is the probability of finding this electron smallest?

7.60 In what region of space is the probability of finding the p_z electron the greatest? Where is the probability of finding this electron smallest?

7.61 What is the designation for an orbital that has a spherical distribution about the nucleus?

7.62 Sketch the shape of the contour surface for an electron with the quantum numbers $n = 3$, $\ell = 0$, $m_\ell = 0$, and $m_s = -\frac{1}{2}$.

7.63 Sketch an orbital contour that is expected for an electron that has $n = 3$ and $\ell = 2$.

7.64 Show the shape of a contour for a p orbital.

7.65 Arrange the following orbitals for the hydrogen atom in order of increasing energy: $3p_x$, $2s$, $4d_{xy}$, $3s$, $4p_z$, $3p_y$, $4s$.

7.66 Arrange the following orbitals for the hydrogen atom in order of increasing energy: $1s$, $4d_{xy}$, $3s$, $4d_{yz}$, $3p_y$, $4s$, $4p_x$.

7.67 What is the wavelength (in nm) of the line in the spectrum of the Li^{2+} ion that comes from the transition of the electron from the Bohr orbit with $n = 3$ to the orbit with $n = 1$? In what region of the electromagnetic spectrum (ultraviolet, visible, and so on) is this radiation observed?

7.68 The absorption spectra of ions have been used to identify the presence of the elements in the atmospheres of the Sun and other stars. (In fact, the element helium was discovered in the spectrum of the Sun before it was identified on Earth, hence its name.) What is the wavelength of light (in nm) that is absorbed by He^+ ions when they are excited from the Bohr orbit with $n = 3$ to the $n = 4$ state?

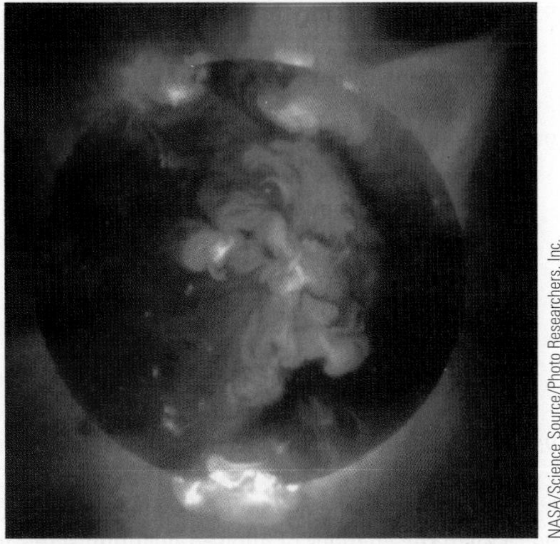
NASA/Science Source/Photo Researchers, Inc.

OBJECTIVE Represent the electronic structure of an atom by its orbital diagram or electron configuration.

7.69 In each part, arrange the subshells in order of increasing energy in a multielectron atom.
(a) $5p$, $2p$, $3d$, $2s$, $3p$
(b) $1s$, $2p$, $3d$, $2s$, $4d$, $3s$
(c) $1s$, $2s$, $3s$, $2p$, $3p$, $4p$, $3d$

7.70 ■ In each part, arrange the orbitals in order of increasing energy in a multielectron atom.
(a) $3p_x$, $2s$, $4d_{xy}$, $3s$, $4p_z$, $3p_y$, $4s$
(b) $1s$, $3p_x$, $3d_{xy}$, $4s$, $3p_y$
(c) $2s$, $4s$, $3p_x$, $3d_{xz}$, $5s$, $3d_{xy}$

7.71 For all elements with $Z \leq 10$, write the electron configuration for
(a) those that have two unpaired electrons.
(b) the element with the largest number of unpaired electrons.
(c) those that have only two occupied subshells.

7.72 For all elements with $Z \leq 10$, write the electron configuration for
(a) those that have a single unpaired electron.
(b) elements that have completely filled subshells.
(c) those that have two unpaired electrons.

7.73 Show the orbital diagram for each of the answers in Exercise 7.71.

7.74 Show the orbital diagram for each of the answers in Exercise 7.72.

7.75 Which of the following atoms have no unpaired electrons in the ground state: B, C, Ne, Pd?

7.76 Give the number of unpaired electrons present in the ground state of
(a) Li. (b) He.
(c) F. (d) V.

7.77 What is the highest occupied subshell in each of the following elements?
(a) helium (b) arsenic
(c) carbon (d) promethium

7.78 What is the highest occupied subshell in each of the following elements?
(a) hydrogen (b) iron
(c) nitrogen (d) uranium

7.79 What are the four quantum numbers of the highest energy electron in the ground state of a carbon atom? (You will have to make some arbitrary choices.)

7.80 ■ What are the four quantum numbers of the highest energy electron in the ground state of a nickel atom? (You will have to make some arbitrary choices.)

7.81 Assign the four quantum numbers (n, ℓ, m_ℓ, m_s) to the highest energy electron in the ground state of a lithium atom.

7.82 Assign the four quantum numbers (n, ℓ, m_ℓ, m_s) to the highest energy electron in the ground state of a scandium atom. (You will have to make some arbitrary choices.)

7.83 Give the maximum number of electrons that may occupy the following shells or subshells.
(a) the $3d$ subshell
(b) the $5s$ subshell
(c) the second principal shell
(d) the fifth principal shell

7.84 Give the maximum number of electrons that may occupy the following shells or subshells.
(a) the $3p$ subshell
(b) the $4d$ subshell
(c) the fourth principal shell
(d) the third principal shell

7.85 In each part, an orbital diagram for an atom is given. Identify the element and whether this is the ground state of the atom. For any excited states, show the orbital diagram for the ground state.

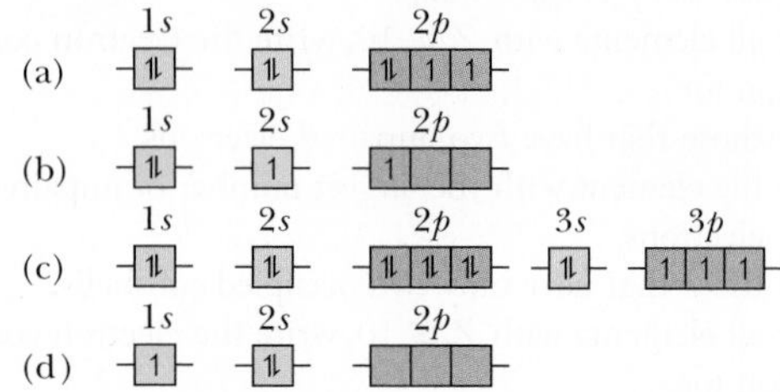

7.86 In each part, an orbital diagram for an atom is given. Identify the element and whether this is the ground state of the atom. For any excited states, show the orbital diagram for the ground state.

	1s	2s	2p	3s	3p	4s
(a)	↑↓	↑	↑ ↑ ↑			
(b)	↑↓	↑↓	(empty)			
(c)	↑	↑↓	↑↓			
(d)	↑↓	↑↓	↑↓ ↑↓ ↑↓	↑↓	↑↓ ↑↓ ↑	↑

Chapter Exercises

7.87 The speed of sound waves in air is 344 m/s, and the frequency of middle C is 512 Hz. What is the wavelength (in m) of this sound wave?

7.88 ■ ▲ In the photoelectric effect, the energy of the absorbed photon is equal to the sum of the energy for the threshold frequency ($h\nu_0$) of the metal and the kinetic energy of the ejected electron. When light with a wavelength of 400 nm strikes potassium metal, the ejected electrons have a kinetic energy of 1.38×10^{-19} J. What is the photoelectric threshold frequency (in s^{-1}) for potassium?

7.89 The flame color tests used to identify elements often depend on the emission spectrum of the atoms. Barium compounds impart a green color to a flame that is due to an emission line at 493 nm. What is the frequency (in s^{-1}) of this light and the energy (in J) of one photon?

Andrew Lambert Photography/Photo Researchers, Inc.

7.90 The flame color tests used to identify elements often depend on the emission spectrum of the atoms. Strontium compounds produce a bright red color in a flame that is due to an emission line at 641 nm. What is the frequency (in s^{-1}) of this light and the energy (in J) of one photon?

Andrew Lambert Photography/Photo Researchers, Inc.

7.91 The Paschen series of lines in the hydrogen atom spectrum arises from transitions to the $n = 3$ state. Use the Rydberg equation to calculate the wavelength (in nm) of the two lowest energy lines in the Paschen series.

7.92 The Lyman series of lines in the hydrogen atom spectrum arises from transition of the electron to the $n = 1$ state. Use the Rydberg equation to calculate the wavelength (in nm) of the two lowest energy lines in the Lyman series.

7.93 Find the uncertainty in the position (in m) of a 650-kg automobile that is moving at 55 mph if the speed is known to within 1 mile/hr. Is this uncertainty in position significant?

7.94 According to both the Bohr model and the quantum-mechanical model, the energy of the hydrogen atom can be calculated from the quantum number n with the following equation:

$$E_n = -\frac{B}{n^2} = -\frac{2.18 \times 10^{-18}\ \text{J}}{n^2}$$

Express the energy, in joules, of the three lowest energy states of the hydrogen atom.

7.95 The energy expression given for the allowed states in the hydrogen atom, -2.18×10^{-18} J/n^2, refers to a single atom. Express the energy of the allowed states (in kJ/mol).

7.96 When the energies of allowed quantum states in a single-electron ion are expressed in kilojoules per mole (kJ/mol), Equation 7.4 becomes

$$E_n = -\frac{1312 \cdot Z^2}{n^2}\ \text{kJ}$$

where Z is the charge on the nucleus. What are the energies of the three lowest energy states of Li^{2+} (expressed in kJ/mol)?

7.97 Use the aufbau procedure to obtain the electron configuration and orbital diagram for atoms of the following elements.

(a) Li (b) F (c) O (d) Ga

7.98 Use the aufbau procedure to obtain the electron configuration and orbital diagram for atoms of the following elements.
(a) Be (b) B (c) Ne (d) Rb

7.99 In extremely energetic systems such as the Sun, hydrogen emission lines can be seen from shells as high as $n = 40$. The spectrum emitted is quite striking because the energy levels become spaced quite closely.
(a) Calculate the difference in energy (in J) between the $n = 2$ and $n = 3$ levels, and compare it with the difference in energy between the $n = 32$ and $n = 33$ levels.
(b) Calculate the largest energy difference (in J) that can be observed in the hydrogen atom.
(c) Explain why there is a limit to the energy difference.
(d) Is there a physical phenomenon that corresponds to this difference? Explain your answer.

7.100 An experiment uses single-photon counting techniques to measure light levels. If the wavelength of light emitted in an experiment is 589.0 nm, and the detector counts 1004 photons over a 10.0-second period, what is the power, in watts, striking the detector (1 W = 1 J/s)?

7.101 In each part, identify the orbital diagram as the ground state, the excited state, or an impossible state. If it is an excited state, give the ground-state diagram, and if it is an impossible state, explain why.

	1s	2s	2p	3s	3p
(a)	[↑↓]	[↑↓]	[↑][ ][ ]		
(b)	[↑↓]	[↑↓]	[↑][↑][ ]		
(c)	[↑↓]	[↑↓]	[↑↓][↑↓][↑↓]	[↑↓]	[↑↓][↑↓][ ]
(d)	[↑↓]	[↑↓]	[↑↑↓][↑][ ]		

7.102 In each part, identify the orbital diagram as the ground state, the excited state, or an impossible state. If it is an excited state, give the ground-state diagram, and if it is an impossible state, explain why.

	1s	2s	2p
(a)	[↑]	[↑↓]	[↑][ ][ ]
(b)	[↑↓]	[↑↓]	[↑↓][ ][ ]
(c)	[↑↓]	[↑↑]	[↑][↑][ ]
(d)	[↑↓]	[↑↓]	[↑][↑][↑]

Cumulative Exercises

7.103 The distance between layers of atoms in a crystal is measured via diffraction of waves with a wavelength comparable with the distance separating the atoms.
(a) What velocity must an electron have if a wavelength of 100 pm is needed for an electron diffraction experiment?
(b) Calculate the velocity of a neutron that has a wavelength of 100 pm. Compare this with the root-mean-square speed of a neutron at 300 K.

7.104 A baseball weighs 142 g. A professional pitcher throws a fast ball at a speed of 100 mph and a curve ball at 80 mph. What wavelengths are associated with the motions of the baseball? If the uncertainty in the position of the ball is $\frac{1}{2}$ wavelength, which ball (fast ball or curve) has a more precisely known position? Can the uncertainty in the position of a curve ball be used to explain why batters frequently miss it?

7.105 A scientist uses atomic emission spectroscopy (see the chapter introduction) to analyze an unknown sample. A series of lines in the far UV are observed at the following wavelengths (all in nanometer units):

1.91, 1.94, 1.98, 2.09, 2.48
8.36, 8.85, 9.91, 13.37
22.3, 26.1, 38.2
53.5, 82.6

The scientist notes a similarity to the hydrogen atom spectrum. Assume the first set are transitions that end in the $n = 1$ state, the second end in the $n = 2$ state, and so on. Calculate R (in J) from the data. Calculate the number of positive charges on the species and identify the species.

7.106 The proton also has a spin quantum number, just like the electron. In an H atom, if the spins of the p^+ and e^- are in the same direction, they are referred to as parallel. If they are in opposite directions, they are referred to as antiparallel. Suppose an H atom with its two particles having parallel spins is labeled H_p, and one with antiparallel spins is labeled H_a. Experiment shows that H_p is more stable than H_a. The energy between the two states has the same energy as a light wave with a wavelength of 21.0 cm.
(a) What is the difference in energy between the two states (in J)?
(b) What is the difference in energy between the two states (in J/mol)?
(c) The energy change of the reaction

$$H_2 \rightarrow H_p + H_p$$

is 435.996 kJ/mol. What are the energy changes (in kJ/mol) of the following two reactions?

$$H_2 \rightarrow H_p + H_a$$

$$H_2 \rightarrow H_a + H_a$$

(d) Under the right conditions, the metal sodium can make an ionic compound with hydrogen, with the resulting formula NaH. What is the proper name of this compound?
(e) What is the oxidation number of H_a? H_p? Na and H in NaH?
(f) Are there two possible compounds NaH_a and NaH_p? Why or why not?
(g) ▲ Given that hydrogen is the most common element in the universe, explain why radioastronomers detect virtually no emissions from the sky of radiation that has the wavelength 21.0 cm.

An artist's rendition of Henri Moissan trying to isolate fluorine, the most chemically reactive element.

(*a*)

(*b*)

(a) Fluorspar, a common mineral, was the historical source of the most reactive element known. *(b)* Compounds of fluorine can etch glass and is used industrially for that purpose.

How do chemists isolate the most reactive element known? Wouldn't the element, once formed, react immediately to form compounds?

The most reactive element is fluorine, which is on the top of Group 7A in the periodic table. Fluorine makes compounds with every other known element except helium and neon. As such, it can be a unique chemical challenge to produce a container filled with elemental fluorine!

In 1529, German mineralogist George Agricola described a mineral that, when added to an ore, helped the ore melt at a lower temperature. Agricola gave the name *fluores* to the mineral (from the Latin *fluere,* meaning "to flow"), which was eventually renamed *fluorite.* Fluorite is still used today in the iron smelting industry; we know it as calcium fluoride.

Treating fluorite with acid released a gas so corrosive that it etched glass. However, it was not until 1780 that Swedish chemist Carl Wilhelm Scheele studied this gas in detail. Scheele believed that this gas was also an acid and dubbed it "fluoric acid."

Around 1810, English chemist Humphry Davy was performing experiments on a compound called *muriatic acid* (which we know as *hydrochloric acid*) and was able to isolate a green gas from this compound. Davy called this green gas *chlorine,* from the Greek word for "green." Davy's experiments with fluoric acid convinced him that fluoric acid was similar to hydrochloric acid, and thus must contain an element similar to chlorine. He proposed the name *fluorine* for this new, not yet isolated substance. Davy tried a variety of chemical and electrical means to generate this substance, but all his attempts either failed or formed fluorine that immediately reacted with some other substance to produce fluorine-containing compounds. It became clear that fluorine, if it were an element, would be very reactive.

One common method of generating elements is electrolysis (see Chapter 18), in which scientists use electricity to decompose a compound into its elements. Several chemically reactive elements, such as potassium, sodium, and magnesium, were originally isolated using this method. In 1885, French chemist Edmond Frémy tried to isolate fluorine by electrolyzing fluorite, but failed. Any fluorine produced immediately

The Periodic Table: Structure and Trends

8

CHAPTER CONTENTS

OWL Online homework for this chapter may be assigned in OWL.

Look for the green colored vertical bar throughout this chapter, for integrated references to this chapter introduction.

reacted to form a new fluorine-containing compound. Next, Frémy tried pure fluoric acid, which we now know as hydrogen fluoride (HF). HF could be liquefied at temperatures less than 20 °C, not much lower than room temperature. The electrolysis of HF(ℓ) failed, too, but for a different reason—the substance would not pass a current. Without electricity flowing, there is no reaction to produce elements. Ultimately, Frémy stopped his efforts.

Other scientists tried to isolate elemental fluorine but with no success and, in some cases, life-threatening or even lethal results. Irish chemists George and Thomas Knox were both poisoned; George Gore of England narrowly escaped from a hydrogen and fluorine explosion; and at least two chemists, Jerome Nickels of France and Paulin Louyet of Belgium, perished in their attempts.

A student of Frémy's, Ferdinand Frédéric Henri Moissan, restarted the project. After a series of unsuccessful attempts, Moissan reasoned that he might be able to get a current to flow through pure hydrogen fluoride if he found a fluorine-containing salt that would dissolve in pure HF. He settled on potassium fluoride, used a platinum-iridium container with stoppers carved out of fluorite, and cooled everything down to about −50 °C. On June 26, 1886, Moissan successfully generated a sample of pure fluorine gas. Moissan received the 1906 Nobel Prize in Chemistry for his feat. Although he was ultimately successful in isolating fluorine, Moissan also suffered health problems that were likely related to fluorine exposure and died in 1907 at the age of 54.

Work on uranium hexafluoride (UF_6) during World War II's Manhattan Project, which developed the atomic bomb, required large amounts of fluorine. Other industrial and household products have required large amounts of fluorine, including chlorofluorocarbons (used in pressurized spray cans) and Teflon (a waxy substance that acts as a nonstick surface on kitchen utensils). Today, fluorine gas can be safely generated using equipment that is either coated with an inert substance or has an adhering metal fluoride that protects the metal container from further reaction. It is one of the few substances that cannot be stored in glass containers because, as mentioned earlier, it will react with the glass (which is largely silicon dioxide):

$$SiO_2(s) + 2F_2(g) \rightarrow SiF_4(g) + O_2(g)$$

Fluorine is a highly reactive substance, and any work with it must be done with extreme caution. ▌

In the periodic table, elements that have similar chemical properties are grouped in vertical columns. For example, all the elements in Group 1A are soft metals that react violently with water and form ions that have a 1+ charge. Modern knowledge of the electronic structure of atoms allows an explanation of this grouping—knowledge that developers of the periodic table, for example, Julius Lothar Meyer and Dmitri Mendeleev, did not have. A key point is that the electron configurations of the outermost electrons in each group are similar. The experimental trends in group properties on which the periodic table was based can now be explained by the arrangements of electrons in atoms.

8.1 Electronic Structure and the Periodic Table

OBJECTIVES

- ☐ Correlate the location of an element in the periodic table with its electron configuration
- ☐ Write the ground-state electron configuration of any element from its location in the periodic table

The chemical similarities from which the periodic table was constructed correlate with the outermost electron configurations of the atoms. Because of this, we can predict the electron configurations of the elements from their locations in the periodic table. To aid in our discussion, we define the **valence electrons** as the electrons with the highest principal quantum number in an atom, and any electrons in an unfilled subshell from a lower shell. Electrons in filled subshells that have lower principal quantum numbers are called **core electrons.** For example, from the electron configuration of magnesium,

Mg: $1s^22s^22p^63s^2$

Valence electrons have the highest principal quantum number in an atom, or occupy an unfilled subshell from a lower shell.

we find that magnesium atoms have two valence electrons ($3s^2$) and 10 core electrons ($1s^22s^22p^6$). Small atoms may have more valence electrons than core electrons; for example, fluorine atoms, the topic of the chapter opener, have an electron configuration of $1s^22s^22p^5$, with seven valence electrons and only two core electrons.

The orbitals in which the valence electrons reside are called the **valence orbitals.** Valence orbitals include the orbitals of the highest principal quantum number and, if an atom has *d* or *f* electrons, the orbitals of any partially-filled subshells of lower principal quantum number.

Figure 8.1*a* shows how the periodic table can be divided into four blocks of elements: elements with highest energy electrons in *s, p, d,* or *f* subshells. Also shown in Figure 8.1*b* is the energy-level diagram introduced in Chapter 7, with the subshells marked by the same color-coding scheme.

The arrangement of the elements in the periodic table correlates with the subshells that hold the electrons in the atom. The first period contains only two elements, hydrogen ($1s^1$) and helium ($1s^2$). The restrictions on the quantum numbers limit the first shell to a $1s$ orbital that can hold only two electrons.

The third element, lithium ($1s^22s^1$), has its valence electron in the $2s$ subshell. The periodic table reflects this because lithium is the first element in the second period. The second period (labeled by the number on the left of the periodic table) contains the elements for which the outermost electrons are in the second shell; note that the period number is the same as the shell number of the outermost electron. The second period contains eight elements, two for the $2s$ subshell and six for the $2p$ subshell (the three $2p$ orbitals with two electrons each).

The period number is the same as the principal quantum number of the outermost electron.

The $2p$ subshell is completely filled at the element neon. The next element, sodium ($1s^22s^22p^63s^1$, or [Ne]$3s^1$), has an electron in the $3s$ subshell. Sodium is the first element of the third period. *Elements that have one electron in a new principal shell in the energy-level diagram start a new period on the periodic table.* Each period always starts with elements for which the valence electron is in an *s* subshell because, within any shell, the *s*

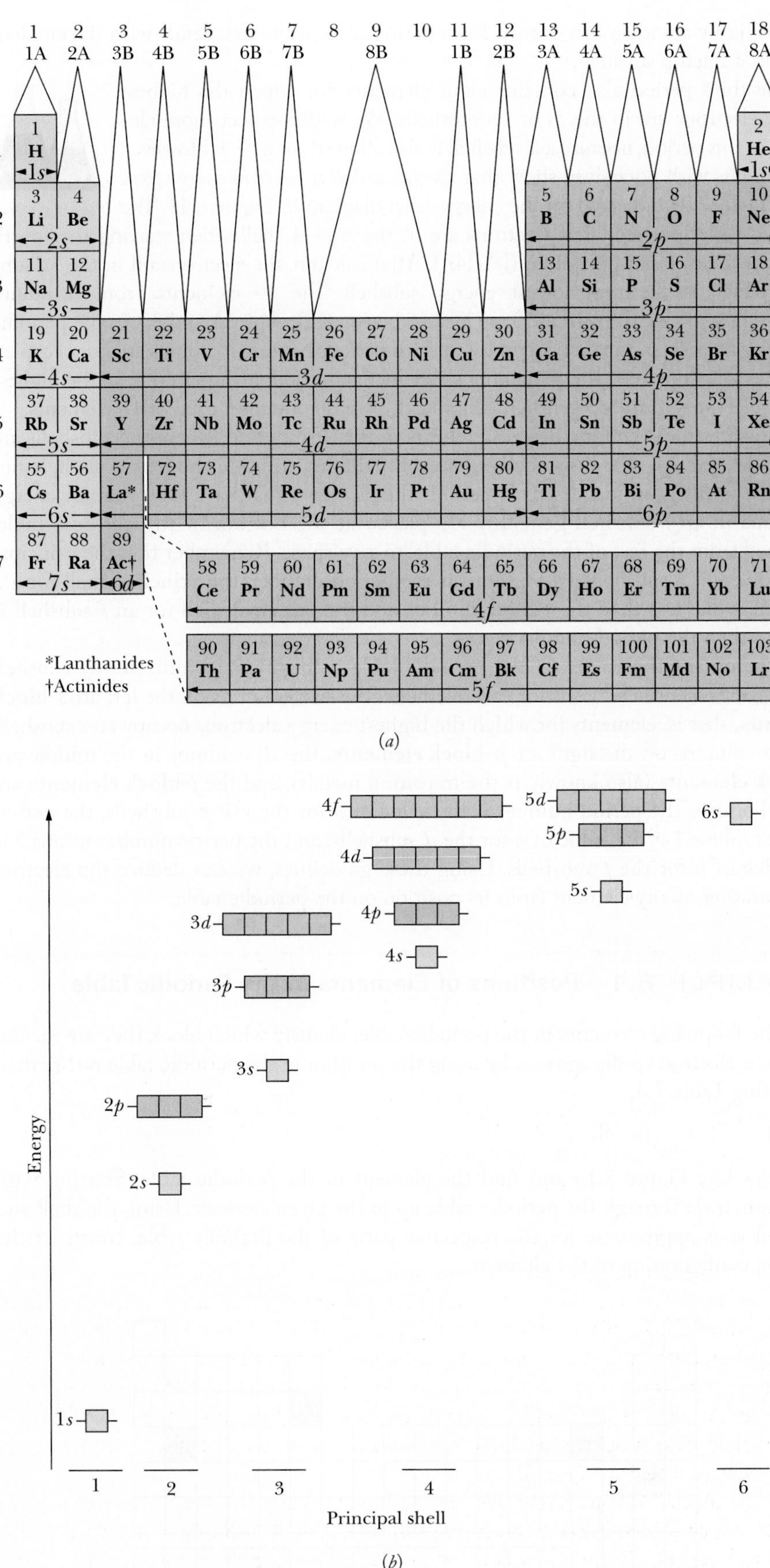

Figure 8.1 *(a)* Periodic table indicating blocks of elements. *(b)* Energy-level diagram appropriate for normal filling of orbitals. The arrangement of the elements in the periodic table, an arrangement based on results of chemical and physical properties, correlates with the energy-level diagram, as shown by the color coding in *(a)* and *(b)*.

subshell is always lowest in energy. Each period (except the first) ends with the element that has a filled *p* subshell.

The third period also contains eight elements, for which the highest energy electrons are in the 3*s* or 3*p* subshells. We might expect more elements in this period, because a *d* subshell is also allowed for $n = 3$. However, experiments with potassium show that *the 4*s *subshell is lower in energy than the 3*d *subshell,* as indicated on the energy-level diagram in Figure 8.1*b*. The periodic table indicates that electrons are in the $n = 4$ shell when starting the fourth period at the element potassium ([Ar]$4s^1$). After calcium, the electrons fill in the 3*d* subshell, which is the next lowest energy subshell. The 10 elements from scandium ([Ar]$4s^23d^1$) to zinc ([Ar]$4s^23d^{10}$) are located between Groups 2A and 3A in the periodic table. A gap exists in the first, second, and third periods because *d* orbitals are not occupied until the fourth period. The remaining six elements in the fourth period have the highest energy electrons in the 4*p* subshell. The fourth period contains a total of 18 elements.

The pattern of the fourth period repeats in the fifth period, where the subshells are the 5*s*, 4*d*, and 5*p*. The *f* subshell, allowed for the first time in the $n = 4$ shell, is not occupied until cerium in the sixth period. The elements that have their highest energy electrons in an *f* subshell generally are placed at the bottom of the periodic table, detached from the rest of the periodic table, to save space. Remember that the principal shell for *s* and *p* subshells is the same as the period number, the principal shell for a *d* subshell is one less than the period number, and the principal shell for an *f* subshell is two less than the period number.

In summary, the colors on the periodic table in Figure 8.1*a* indicate the subshell holding the valence electrons for each element. The two columns on the left are ***s*-block elements,** that is, elements for which the highest energy electrons occupy an *s* subshell. The six columns on the right are ***p*-block elements,** the 10 columns in the middle are ***d*-block elements** (also known as the transition metals), and the ***f*-block elements** are at the bottom. The period number is the value of *n* for the *s* or *p* subshells, the period number minus 1 is the value of *n* for the *d* subshells, and the period number minus 2 is the value of *n* for the *f* subshells. Using these guidelines, we can deduce the electron configuration of any element from its position on the periodic table.

The structure of the periodic table is directly related to the filling of shells and subshells with electrons.

EXAMPLE 8.1 Positions of Elements in the Periodic Table

Find the following elements in the periodic table, identify which block they are in, and give their electron configurations by using the position in the periodic table rather than consulting Table 7.3.

(a) Al (b) Sr

Strategy Use Figure 8.1*a* and find the element in the periodic table. Starting with hydrogen, trace through the periodic table up to the given element. Using the shell and subshell as is appropriate for the respective parts of the periodic table, construct the electron configuration of the element.

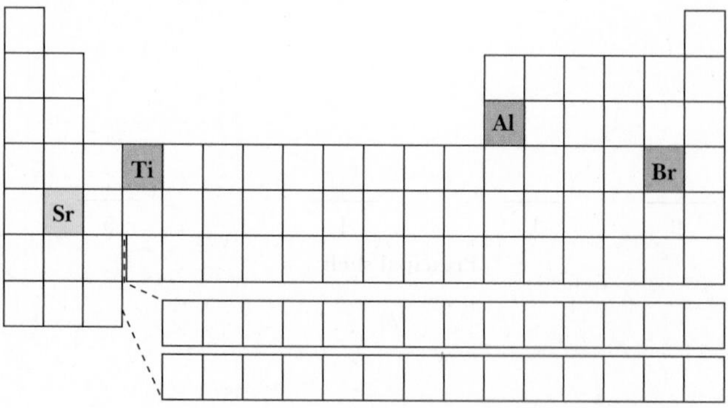

PRINCIPLES OF CHEMISTRY
The True Shape of the Periodic Table?

Now that we know the underlying reason for the structure of the periodic table, we can argue that it should really have the following outline:

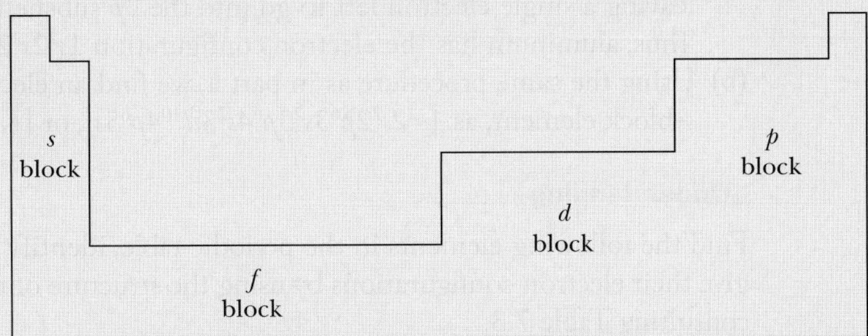

According to the aufbau principle, in the sixth and seventh periods, the *f* block comes before the *d* block; therefore, those elements should be positioned right next to the *s* block in those periods. However, if we look closely at the electron configurations of the elements right after the *s*-block elements barium and radium, we find that the next electron goes into a *d* subshell, not an *f* subshell. The electron configuration of lanthanum is [Xe]$6s^2 5d^1$ (rather than [Xe]$6s^2 4f^1$), and the electron configuration of actinium is [Rn]$7s^2 6d^1$ (rather than [Rn]$7s^2 5f^1$). Therefore, perhaps the correct shape of the periodic table should be

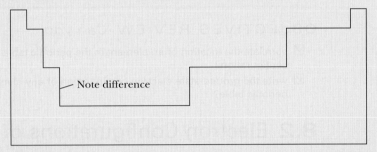

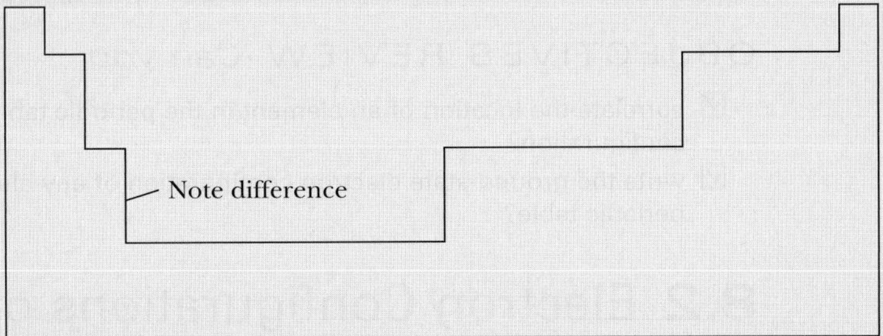

Now we have a periodic table with a split *d* block. We avoid these structures by splitting the *f* block off from the rest of the periodic table. This also has the advantage of being able to print a legible periodic table on a single page of paper.

The structure of the periodic table was last modified in 1944, when Glenn Seaborg took 14 elements out of the body of the periodic table and matched them with the trans-actinium elements that were just then being synthesized. Seaborg argued—correctly—that the chemistry of these elements was not entirely compatible with other elements in the *d* block, but rather belonged to a block of their own. Previous to Seaborg's action, the periodic table was depicted as follows:

Group 0	I	II	III	IV	V	VI	VII	VIII
	a b	a b	a b	a b	a b	a b	a b	
	H 1							
He 2	Li 3	Be 4	B 5	C 6	N 7	O 8	F 9	
Ne 10	Na 11	Mg 12	Al 13	Si 14	P 15	S 16	Cl 17	
Ar 18	K 19 / Cu 29	Ca 20 / Zn 30	Sc 21 / Ga 31	Ti 22 / Ge 32	V 23 / As 33	Cr 24 / Se 34	Mn 25 / Br 35	Fe 26, Co 27, Ni 28
Kr 36	Rb 37 / Ag 47	Sr 38 / Cd 48	Y 39 / In 49	Zr 40 / Sn 50	Nb 41 / Sb 51	Mo 42 / Te 52	/ I 53	Ru 44, Rh 45, Pd 46
Xe 54	Cs 55 / Au 79	Ba 56 / Hg 80	57-71* / Tl 81	Hf 72 / Pb 82	Ta 73 / Bi 83	W 74 / Po 84	Re 75 /	Os 76, Ir 77, Pt 78
Rn 86		Ra 88	Ac 89	Th 90	Pa 91	U 92		

This version of the periodic table was based on supposed similar chemical properties of the elements, just like the modern periodic table is based. However, because *d* and *f* electrons can have subtle effects on the exact chemistry of the elements, chemists of the early 20th century did not yet realize that adding a block with a width of 14 elements separated from the *d* block is the best way to represent these elements. ∎

Solution

(a) Aluminum has an atomic number of 13, and based on its position in the periodic table, it is in the *p* block. Starting at atomic number 1 and counting to higher atomic numbers, we put two electrons in the 1*s* subshell for the first period, two in the 2*s* subshell for the *s* block of the second period, and six in the 2*p* subshell for the *p* block of the second period. This gives us a total of 10 electrons so far. Two more electrons go into the 3*s* subshell for the *s* block of the third period, leaving a single electron left to go into the 3*p* subshell, taking us to aluminum. Thus, aluminum has the electron configuration $1s^22s^22p^63s^23p^1$, or $[Ne]3s^23p^1$.

(b) Using the same procedure as in part a, we find an electron configuration for Sr, an *s*-block element, as $1s^22s^22p^63s^23p^64s^23d^{10}4p^65s^2$, or $[Kr]5s^2$.

Understanding

Find the following elements in the periodic table, identify which block they are in, and give their electron configurations by using the structure of the periodic table rather than consulting Table 7.3.

(a) Br (b) Ti

Answer

(a) *p* block, $[Ar]4s^23d^{10}4p^5$
(b) *d* block, $[Ar]4s^23d^2$

OBJECTIVES REVIEW *Can you:*

☑ correlate the location of an element in the periodic table with its electron configuration?
☑ write the ground-state electron configuration of any element from its location in the periodic table?

8.2 Electron Configurations of Ions

OBJECTIVES

☐ Write the ground-state electron configurations of ions
☐ Recognize the order in which electrons are lost or gained to make ions
☐ Identify an isoelectronic series

Atoms that have gained or lost electrons are called *ions*. For example, an anion (a negatively charged ion) is formed by the addition of electrons to an atom. The additional electrons occupy empty orbitals, using the same rules that apply to atoms. The electron configuration of fluorine is $1s^22s^22p^5$, and that of the fluoride anion, F^-, is $1s^22s^22p^6$. The electron configuration of O^{2-} is also $1s^22s^22p^6$.

$$F + e^- \longrightarrow F^- \qquad O + 2e^- \longrightarrow O^{2-}$$

$$1s^22s^22p^5 \quad 1s^22s^22p^6 \qquad 1s^22s^22p^4 \quad 1s^22s^22p^6$$

As in these two examples, stable monatomic anions frequently have completely filled valence *p* orbitals. This gives them an electron configuration of the nearest noble gas. Both F^- and O^{2-}, for instance, have the electron configuration of neon. Other monatomic anions have electron configurations of other noble gases. (In part, it is the relative stability of F^- ions with respect to F_2 that makes elemental fluorine so chemically reactive, as described in the introduction to this chapter.)

Anions are formed by the addition of electrons to valence orbitals.

EXAMPLE 8.2 Electron Configurations of Anions

Write the electron configuration of the following anions, and state what noble-gas electron configuration they have.

(a) I^- (b) S^{2-}

Strategy Add electrons to the neutral atom electron configuration using the same rules as outlined for atoms. Use the periodic table to identify the appropriate noble gas.

Solution

(a) The iodine atom has the electron configuration $[Kr]5s^2 4d^{10} 5p^5$. The anion is formed by adding one electron to the $5p$ orbital. The electron configuration for I^- is $[Kr]5s^2 4d^{10} 5p^6$, which is the same as the next noble gas, xenon.

(b) The S^{2-} anion has 18 electrons, 2 more than the sulfur atom. Its electron configuration is $[Ne]3s^2 3p^6$, the same as that of the next noble gas, argon.

Understanding

Write the electron configuration of Cl^-.

Answer $[Ne]3s^2 3p^6$, which is the electron configuration of the noble gas argon.

A cation (a positively charged ion) is formed when an atom loses one or more electrons. Experiments show that *electrons of highest* n *value are removed first. For subshells of the same* n *level, electrons are removed from the subshell of highest* ℓ *value first.* The removal order may not be the same as the order in which electrons filled the subshells of the original atom, although for elements in the s and p blocks, the valence electrons that are lost are the ones that were added last. The electron configuration of beryllium is $1s^2 2s^2$, so the Be^{2+} ion has the electron configuration $1s^2$.

In contrast, for the d-block transition elements, *the* ns *electrons are lost before the* (n − 1)d *electrons.* For example, iron loses its $4s^2$ electrons before any of its $3d$ electrons. For the formation of the Fe^{2+} ion:

$$Fe \longrightarrow Fe^{2+} + 2e^-$$

$$[Ar]4s^2 3d^6 \qquad [Ar]3d^6$$

The electron configuration of Fe^{2+} is not $[Ar]4s^2 3d^4$. The $4s$ electrons are lost before the $3d$ electrons. For the formation of the Fe^{3+} ion from the Fe^{2+} ion:

$$Fe^{2+} \longrightarrow Fe^{3+} + e^-$$

$$[Ar]3d^6 \qquad [Ar]3d^5$$

How do we know that this is the electron configuration of Fe^{3+} ions? Experimental measurements, such as spectroscopic and magnetic experiments, tell us that this is correct.

Another example is copper(II).

$$Cu \longrightarrow Cu^{2+} + 2e^-$$

$$[Ar]4s^1 3d^{10} \quad [Ar]3d^9$$

In this case, the $4s$ and one of the $3d$ electrons are removed. Note that the electron configuration of Cu^{2+} would be $[Ar]3d^9$ even if the electron configuration of Cu atoms were not exceptions to the normal aufbau principle of filling $4s$ before $3d$.

2s ⟶ 2s
1s 1s
$1s^2 2s^2 \longrightarrow 1s^2$
$Be \longrightarrow Be^{2+} + 2e^-$

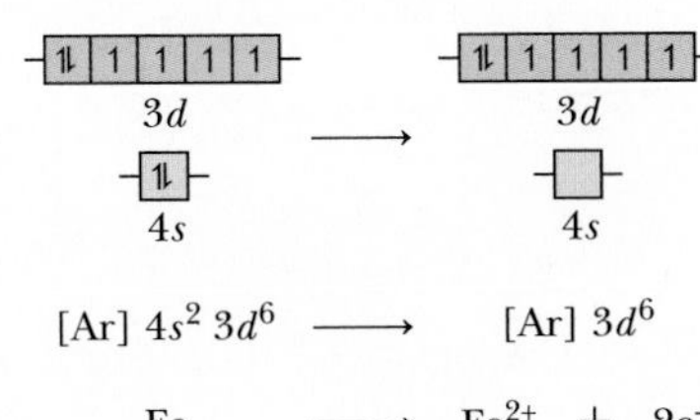

Cation formation for beryllium and iron.

Cations are formed by the loss of valence electrons from the orbitals with the greatest n values.

EXAMPLE 8.3 Electron Configurations of Cations

Write the electron configuration of the following cations.

(a) Na^+ (b) Ni^{2+} (c) Zr^{3+} (d) Ga^+ (e) Ga^{3+}

Strategy In each case, write the electron configuration of the neutral atom and *remove electrons of highest* n *value first,* until the proper charge is reached. The np and ns electrons are lost before the $(n-1)d$ electrons.

Solution

(a) The sodium atom has the electron configuration $1s^22s^22p^63s^1$. The ion is formed by the loss of one electron from the orbital with the highest n level, the $3s$ orbital. The electron configuration for the Na^+ cation is $1s^22s^22p^6$ or [Ne].

(b) The Ni atom has the electron configuration $[Ar]4s^23d^8$. The two $4s$ electrons (those in the orbital of highest n value) are lost before the $3d$ electrons. The electron configuration of Ni^{2+} is $[Ar]3d^8$.

(c) The Zr atom has the electron configuration $[Kr]5s^24d^2$. The two $5s$ electrons (those in the orbital of highest n value) and one of the $4d$ electrons are lost. The electron configuration of Zr^{3+} is $[Kr]4d^1$.

(d) The Ga atom has the electron configuration $[Ar]4s^23d^{10}4p^1$. Electrons of highest n value are removed first, and for subshells of the same n level, *electrons are removed from the subshell of highest ℓ value first.* The $4p^1$ is removed, making the electron configuration of Ga^+ $[Ar]4s^23d^{10}$.

(e) The Ga atom has the electron configuration $[Ar]4s^23d^{10}4p^1$. Electrons of highest n value are removed. The $4p^1$ and $4s^2$ are removed making the electron configuration of Ga^{3+} $[Ar]3d^{10}$.

Understanding

Write the electron configuration of Co^{3+}.

Answer $[Ar]3d^6$

Isoelectronic Series

An **isoelectronic series** is a group of atoms and ions that contain the same number of electrons. The species O^{2-}, F^-, Ne, Na^+, and Mg^{2+} are isoelectronic—they all have 10 electrons, and the electron configuration $1s^22s^22p^6$.

As outlined in Chapter 2, the charges on the ions formed by the s- and p-block elements (the A group elements) can be determined from the group number. The elements of Group 1A form cations with a 1+ charge, Group 2A elements form cations with a 2+ charge, Group 6A elements form anions with a 2− charge, and Group 7A elements form anions with a 1− charge. The electron configurations of the common ions of oxygen, fluorine, sodium, and magnesium are:

$$\underset{1s^22s^22p^4}{O} + 2e^- \longrightarrow \underset{1s^22s^22p^6}{O^{2-}} \qquad \underset{1s^22s^22p^5}{F} + e^- \longrightarrow \underset{1s^22s^22p^6}{F^-}$$

$$\underset{1s^22s^22p^63s^1}{Na} \longrightarrow \underset{1s^22s^22p^6}{Na^+} + e^- \qquad \underset{1s^22s^22p^63s^2}{Mg} \longrightarrow \underset{1s^22s^22p^6}{Mg^{2+}} + 2e^-$$

All four of these ions have the same electron configuration as neon. Atoms of neon are stable, so we expect that ions with the same electron configuration will also be stable.

The species in an isoelectronic series have the same electron configuration.

EXAMPLE 8.4 Electron Configurations of Isoelectronic Series

(a) Which atom or ions among Ar, S^{2-}, Si^-, and Cl^{3+} are isoelectronic with P^+?

(b) Which ions among Fe^{3+}, Ni^{3+}, and Co^{3+} are isoelectronic with Mn^{2+}?

Strategy Count the number of electrons in each species and see which have the same number.

Solution

(a) The P^+ cation has 14 electrons. Ar and S^{2-} have 18 electrons, Si^- has 15 electrons, and Cl^{3+} has 14 electrons. Only Cl^{3+} is isoelectronic with P^+.

(b) The Mn^{2+} cation has 23 electrons. The Fe^{3+} ion also has 23 electrons and the same electron configuration as Mn^{2+}; they are isoelectronic. The Ni^{3+} ion has 25 electrons, and Co^{3+} has 24 electrons. Neither is isoelectronic with Mn^{2+}.

Understanding

Which elements or ions among K, K^+, Ca^+, and Sc^{3+} are isoelectronic with Ar?

Answer K^+ and Sc^{3+}

OBJECTIVES REVIEW *Can you:*

- ☑ write the ground-state electron configurations of ions?
- ☑ recognize the order in which electrons are lost or gained to make ions?
- ☑ identify an isoelectronic series?

8.3 Sizes of Atoms and Ions

OBJECTIVES

- ☐ Arrange atoms and ions according to size
- ☐ Correlate the trends in sizes with the effective nuclear charge experienced by the outer electrons

A detailed knowledge of the electronic structure of atoms and ions leads to an understanding of the chemical and physical properties of the elements. The relative sizes of atoms and ions, which are determined by the sizes of their electron clouds, are important because they provide information about the structure and reactivity of molecules, metals, and ionic compounds.

Measurement of Sizes of Atoms and Ions

The size of an atom or ion is not easy to measure. We have already seen that the electron cloud does not have a fixed boundary; only a probability distribution can be determined. An **atomic radius** is half the distance between adjacent atoms of the same element in a molecule. The atomic radius of an element varies from one type of molecule to another, so a representative molecule must be chosen for the measurement. For example, chemists determine the atomic radii of chlorine and bromine atoms by measuring the distance between the nuclei in the diatomic molecules. The distances between the nuclei in Cl_2 and Br_2 are 198 and 228 pm, respectively. The atomic radius of each atom is simply half of this distance—99 pm for chlorine and 114 pm for bromine (Figure 8.2). These data enable us to predict that the distance between the chlorine and bromine nuclei in BrCl should be about 213 pm, which is close to the measured value, 214 pm. This method works well for the nonmetallic elements, but the determination of the sizes of metals is more difficult. Atomic radii of most metals have been assigned on the basis of a variety of experimental evidence. Using similar methods with ionic compounds, chemists have determined an approximate **ionic radius,** a measure of the size of an ion in an ionic solid.

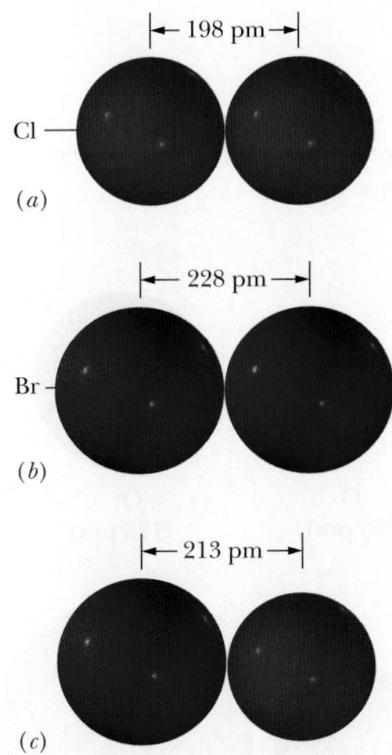

Figure 8.2 Atomic radii. The atomic radii of chlorine *(a)* and bromine *(b)* are half the distance between nuclei in the homonuclear diatomic molecules. These values can be used to predict the distance between the nuclei in BrCl *(c)*.

Comparative Sizes of Atoms and Their Ions

The size of an atom can readily be compared with the sizes of its own ions. Consider the sizes of the lithium atom and the lithium cation. The electron configuration of Li is $1s^22s^1$, and that of Li^+ is $1s^2$. In this case, the number of protons in the nucleus is constant, but the lithium atom has one more electron than the ion, and that electron is in the $n = 2$ level. The cation has no electrons in the $n = 2$ level. Because the size of an orbital in the $n = 2$ shell is larger than one in the $n = 1$ shell, the lithium atom is much larger than its cation. This trend is general: *An atom is always larger than any of its cations* (Figure 8.3).

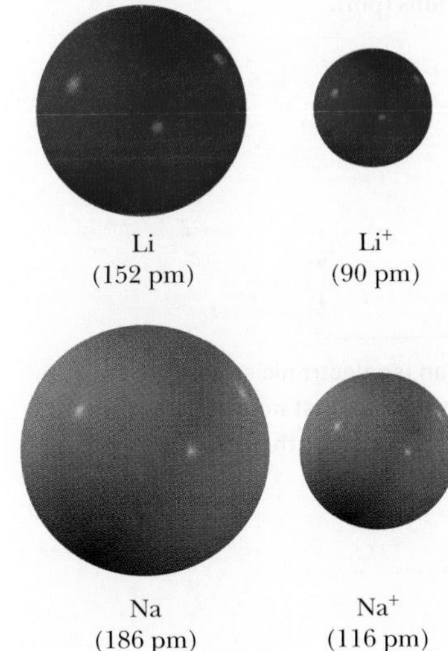

Figure 8.3 Radii of atoms and cations (pm).

An atom is larger than its cations.

Figure 8.4 shows the sizes of the iron atom, and its 2+ and 3+ ions. The iron atom ($[Ar]4s^23d^6$) has two 4*s* electrons, whereas the Fe^{2+} cation ($[Ar]3d^6$) has none. The cation is considerably smaller in size because the electrons in the $n = 4$ level have been removed from the iron atom. Removing another electron to form Fe^{3+} ($[Ar]3d^5$) produces a small but noticeable decrease in radius. The trend is as follows: *The greater the positive charge on the cation of the same element, the smaller the ionic radius.*

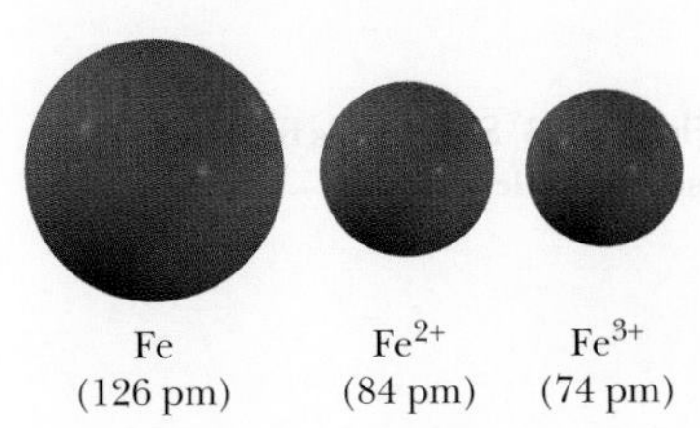

Figure 8.4 Radii of iron and two of its cations (pm).

Measurements of the sizes of anions show that they are larger than their parent atoms. Adding an electron to a fluorine atom ($1s^22s^22p^5$), forming a fluoride ion ($1s^22s^22p^6$), increases the amount of negative charge interacting with the positively charged nucleus, producing a larger electron cloud. Again, this trend is as follows: *Anions are always larger than the neutral atoms* (Figure 8.5).

Anions are larger than the neutral atoms.

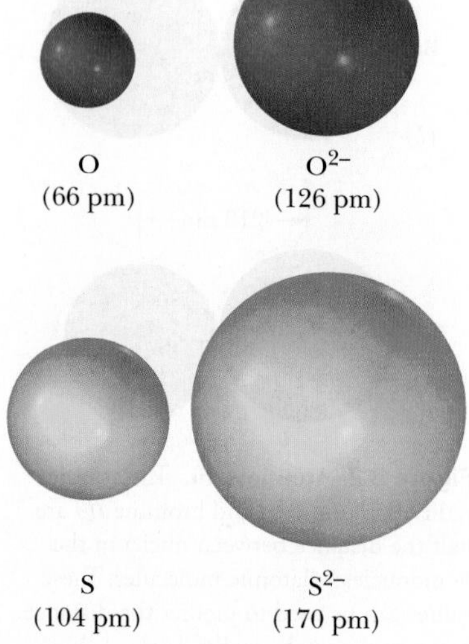

Figure 8.5 Radii of atoms and anions (pm).

Size Trends in Isoelectronic Series

Because each member of an isoelectronic series has the same number of electrons, the number of protons in the nucleus determines the size within the series. Figure 8.6 compares the experimentally measured sizes of four ions with the electron configuration $1s^22s^22p^6$.

The observed trend toward smaller radii observed as the atomic numbers increases in an isoelectronic series is explained by changes in the nuclear charge experienced by the electrons. Increasing the nuclear charge increases the electrostatic attractions that the nucleus exerts on the electrons, causing the radius of the species to decrease. For example, F^- and O^{2-} are isoelectronic, but the F^- ion has one more proton in its nucleus. The greater nuclear charge has a greater attractive force for the electron cloud, decreasing the size of F^- compared with O^{2-}.

EXAMPLE 8.5 Sizes of Atoms and Ions

Identify the larger species in each of the following pairs.

(a) K or K^+ (b) S^{2-} or Cl^- (c) Co^{2+} or Co^{3+}

Strategy For each pair, consider the shell the valence electrons are in. Electrons in higher shells cause the species to be larger. If they are in the same shell, the greater the nuclear charge, the smaller the species will be.

Solution

(a) K has the electron configuration $[Ar]4s^1$, and K^+ has the electron configuration [Ar]. The presence of the electron in the 4*s* orbital makes K much larger than K^+.

(b) These two species are isoelectronic and have the same electron configuration, [Ar], but the chloride ion ($Z = 17$) has one more proton in its nucleus than does the sulfide ion ($Z = 16$). The greater nuclear charge of Cl^- causes it to be smaller than S^{2-}.

(c) Both Co^{2+} ($[Ar]3d^7$) and Co^{3+} ($[Ar]3d^6$) have valence electrons in the same subshell, but Co^{2+} is slightly larger because it has an extra electron, expanding the electron cloud.

In an isoelectronic series, the species with the largest number of protons in its nucleus has the smallest radius.

Understanding

Which is larger: Se^{2-} or Br^-?

Answer Se^{2-}

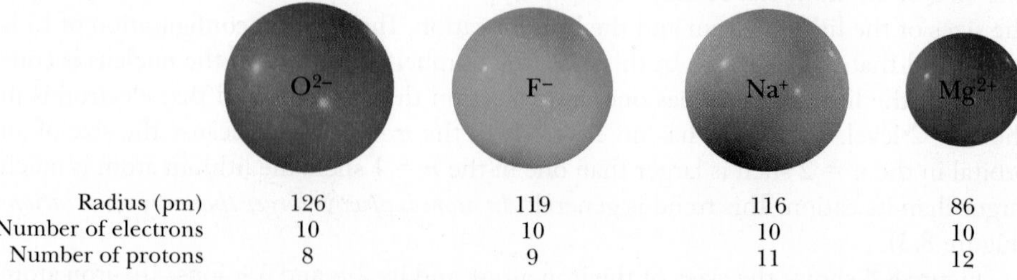

	O^{2-}	F^-	Na^+	Mg^{2+}
Radius (pm)	126	119	116	86
Number of electrons	10	10	10	10
Number of protons	8	9	11	12

Figure 8.6 Size trends for an isoelectronic series.

Trends in the Sizes of Atoms

The periodic tables in Figure 8.7 show the atomic radii for the *s*- and *p*-block elements. These radii are measured values based on the spacing of adjacent atoms in molecules. The values for helium, neon, and argon are estimated, because no

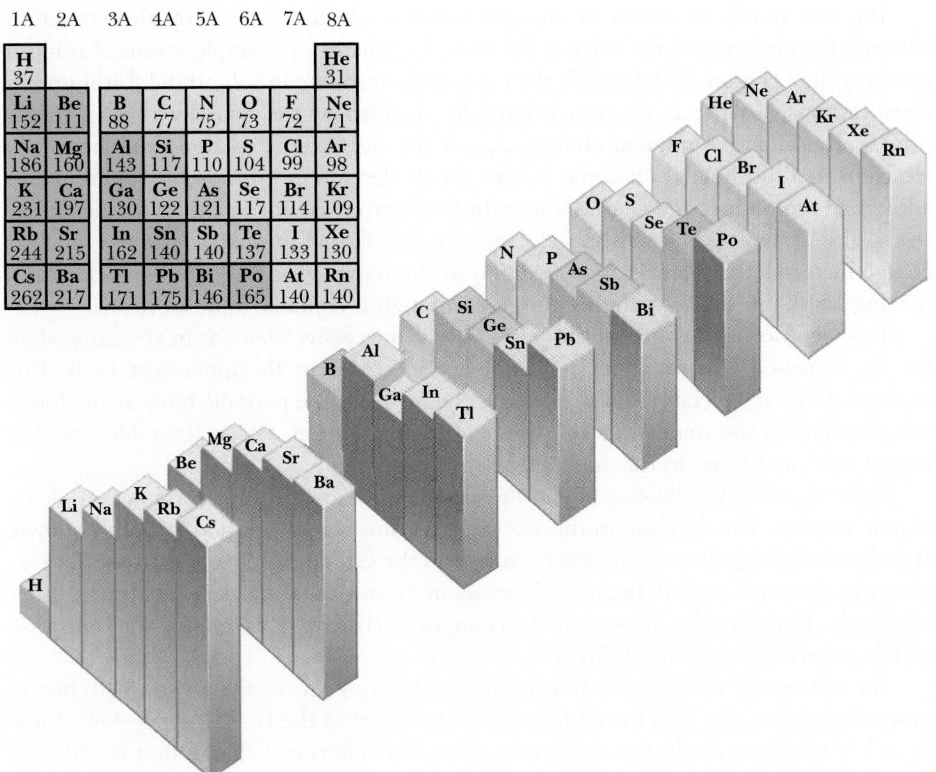

Figure 8.7 Atomic radii of the main-group elements (pm).

compounds have been isolated for these elements. Size trends in any group are easily predicted: the heavier the element in the group, the larger it will be. For example, silicon, $1s^22s^22p^63s^23p^2$, is larger than carbon, $1s^12s^22p^2$, because the 3*s* and 3*p* orbitals are larger than the 2*s* and 2*p* orbitals. Figure 8.8 includes plots of atomic radius versus period number for the alkali metals (Group 1A) and the halogens (Group 7A). These plots show the expected increases in size for the heavier members of these two groups.

The radii of atoms increase down any group because of the increase in the principal quantum number of the valence orbitals.

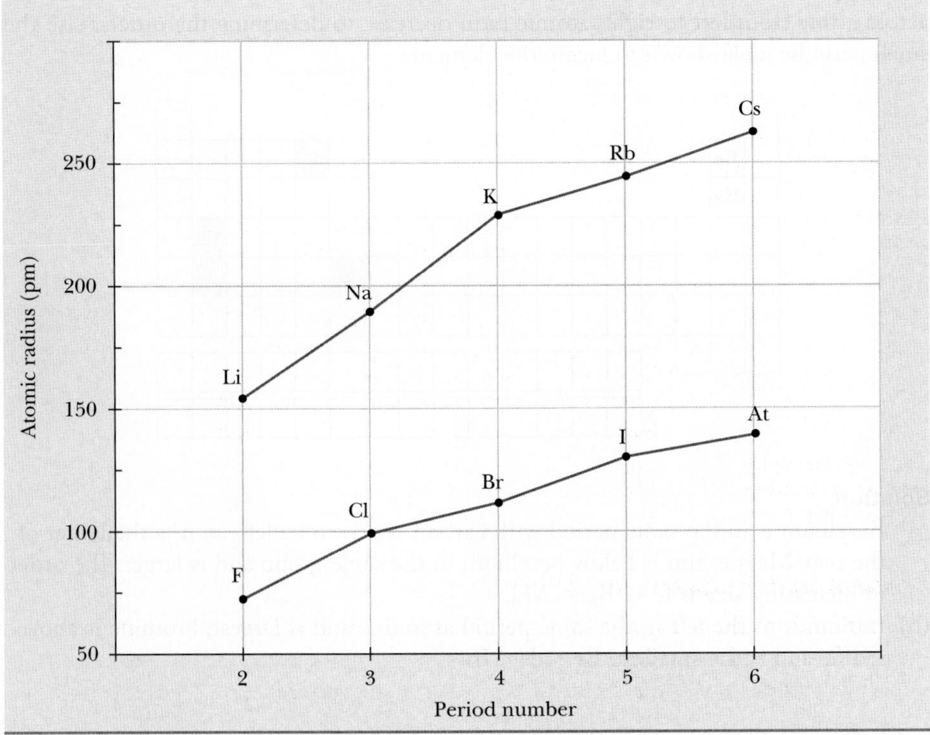

Figure 8.8 Atomic radius versus period number for Group 1A and 7A atoms.

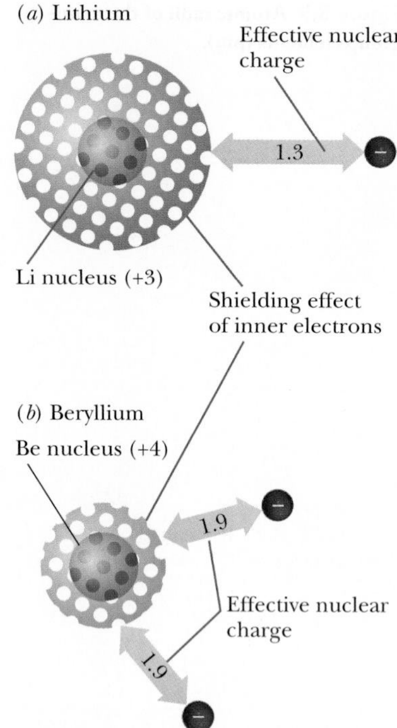

Figure 8.9 Effective nuclear charge for lithium and beryllium. *(a)* In lithium, two 1*s* electrons screen the outermost electron from the 3+ charged nucleus, reducing its effective nuclear charge to 1.3. *(b)* In beryllium, the outermost electrons are screened from the 4+ charge by the two 1*s* electrons, but the two 2*s* electrons do not shield each other very much. The effective nuclear charge of the outer electrons is about 1.9, greater than that for lithium and making beryllium smaller.

The size trends of atoms in any given *period* are determined by the attraction between the nucleus and the valence electrons. Compare, for example, atoms of lithium and beryllium (Figure 8.9). In each, the valence electrons are in a 2*s* orbital. Lithium has three protons, but the 2*s* electron is partially shielded by the two electrons in the 1*s* orbital. The **effective nuclear charge,** Z_{eff}, is the net positive charge experienced by electrons in a subshell. Z_{eff} is about 1.3 for the 2*s* electron in lithium. Beryllium has an additional proton in its nucleus. Because the 2*s* electrons do not shield each other effectively, Z_{eff} is about 1.9 for the 2*s* electrons in beryllium. The larger effective nuclear charge increases the electrostatic attraction of the nucleus for the outermost electrons, making beryllium *smaller* than lithium, even though it contains more electrons.

This trend continues across the period. Each successive element in the same shell has an increased effective nuclear charge for electrons in the outermost shell. This increase draws the electron cloud closer to the nucleus. The periodic table at the lower left summarizes the trends in sizes. Elements at the bottom left of the table have the largest radii, and those in the upper right are smallest.

Figure 8.10 shows these trends in the atomic radii graphically, as a function of the atomic number. The decrease in the size of the atoms across a period is clearly shown. A particularly large decrease in size occurs from the Group 1A elements to the 2A elements in the same period, because electrons in an *ns* orbital do not shield each other effectively. The plots also show that the changes in size are not large for the transition metals, especially after Group 5B.

The radii of atoms decrease across a period because of the increase in effective nuclear charge.

The outermost electrons in transition-metal atoms are in the *ns* subshell, but in proceeding across the row, the additional electrons are in the $(n - 1)d$ subshell. These $(n - 1)d$ electrons shield the *ns* electrons from the additional charge that is added to the nucleus with each additional electron, so the effective nuclear charge on the outer *ns* electrons increases only slowly.

EXAMPLE 8.6 Size Trends

List the following series of elements in order of increasing atomic radius.

(a) Be, C, Mg
(b) In, I, Br

Strategy Use the two size trends: (1) down a group, the atomic radii increase; and (2) across a row from left to right, atomic radii decrease to determine the orders. Use the small periodic table shown to locate the elements.

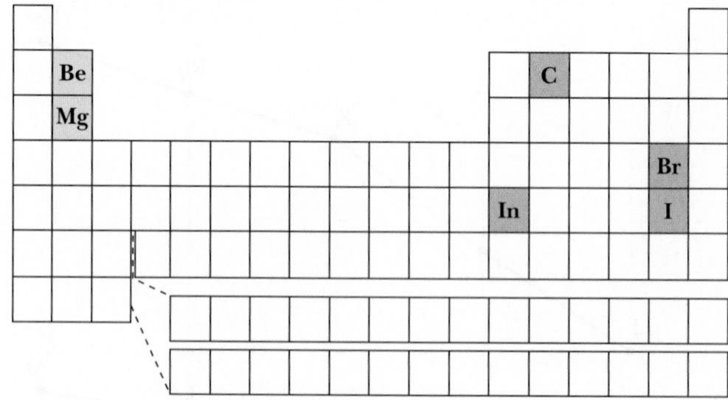

Solution

(a) Beryllium is in the same period with carbon and is to its left, so it is the larger of the two. Magnesium is below beryllium in the same group and is larger. The order of increasing size is C < Be < Mg.
(b) Indium is to the left in the same period as iodine and is largest; bromine is above iodine and is the smallest: Br < I < In.

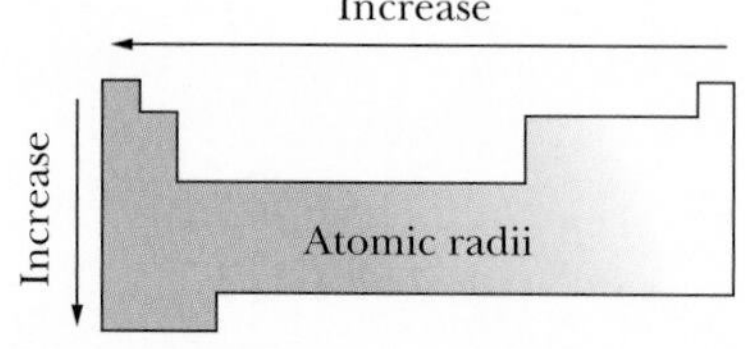

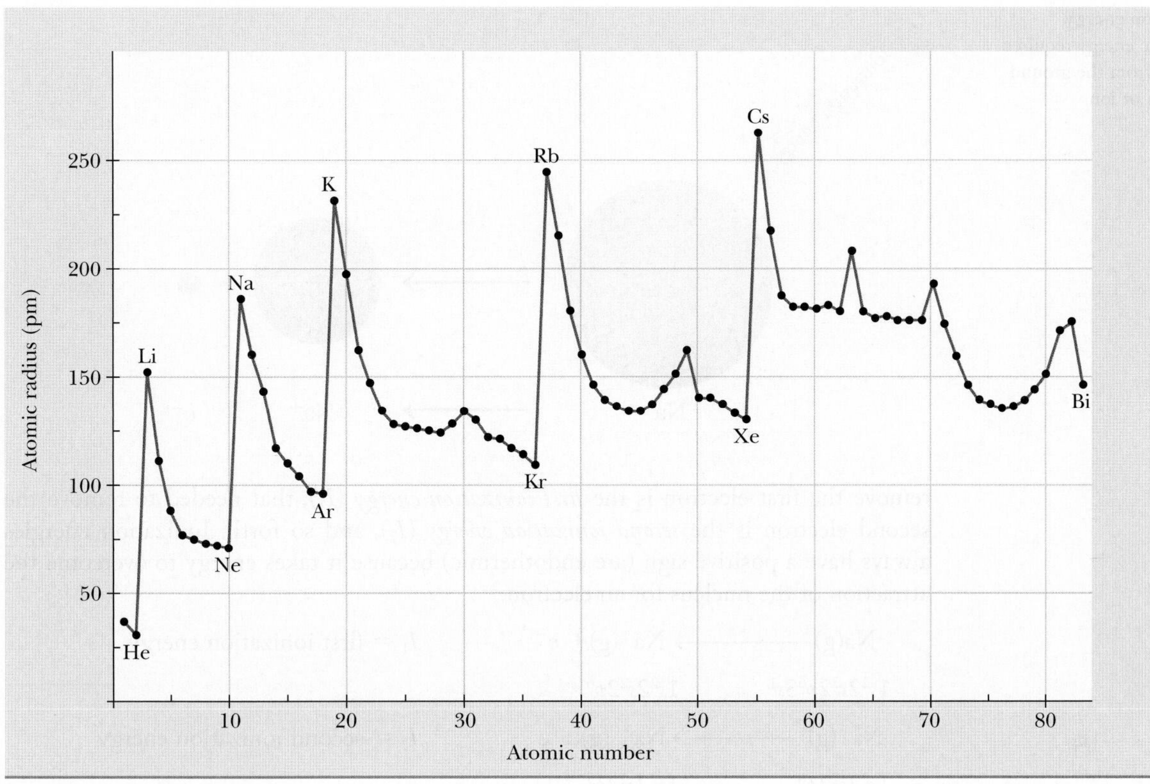

Figure 8.10 Atomic radii of atoms.

Understanding

List the elements O, S, and Al in order of increasing size.

Answer O < S < Al

OBJECTIVES REVIEW *Can you:*

- ☑ arrange atoms and ions according to size?
- ☑ correlate the trends in sizes with the effective nuclear charge experienced by the outer electrons?

8.4 Ionization Energy

OBJECTIVES

- ☐ Define ionization energy
- ☐ Arrange atoms according to ionization energies
- ☐ Predict the relative energies needed for successive ionizations

Atoms become ions by gaining or losing electrons. As with any chemical process, gaining or losing electrons is accompanied by a change in energy. In this section and the next section, we consider the energy change that occurs with the formation of ions. All of these energy changes are measured experimentally, and we can relate the energy changes to the organization of electrons in atoms.

Many metallic elements exist in ionic compounds as cations, in which atoms have lost electrons. The **ionization energy** is the energy required to remove an electron from a gaseous atom or ion in its electronic ground state (Figure 8.11). The energy needed to

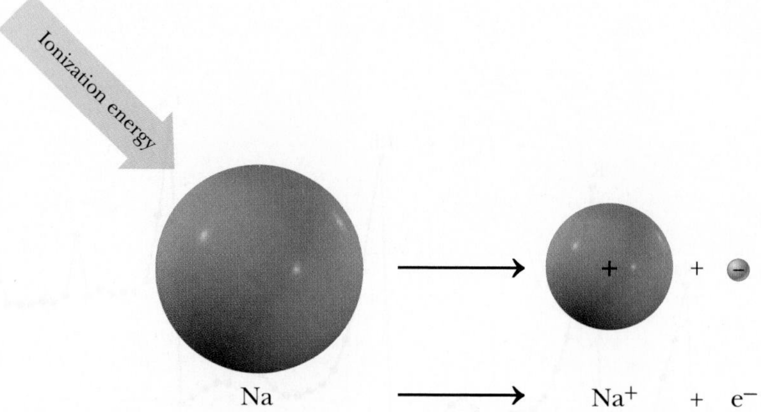

Figure 8.11 Ionization energy. Ionization energy is the energy required to remove an electron from the ground state of a gaseous atom or ion.

remove the first electron is the *first ionization energy* (I_1), that needed to remove the second electron is the *second ionization energy* (I_2), and so forth. Ionization energies always have a positive sign (are endothermic) because it takes energy to overcome the attraction of the nucleus for an electron.

$$Na(g) \longrightarrow Na^{+}(g) + e^{-} \qquad I_1 = \text{first ionization energy}$$
$$1s^22s^22p^63s^1 \qquad 1s^22s^22p^6$$

$$Na^{+}(g) \longrightarrow Na^{2+}(g) + e^{-} \qquad I_2 = \text{second ionization energy}$$
$$1s^22s^22p^6 \qquad 1s^22s^22p^5$$

Within the same shell *(n),* electrons are removed from the subshell that has the greatest angular momentum quantum number *(ℓ)*. For an aluminum atom, a $3p$ electron is removed first:

$$Al(g) \longrightarrow Al^{+}(g) + e^{-}$$
$$1s^22s^22p^63s^23p^1 \qquad 1s^22s^22p^63s^2$$

Trends in First Ionization Energies

Ionization energies are determined by the strength of the interaction between the nucleus and the valence electrons. Electrons that are attracted by a large effective nuclear charge, Z_{eff}, are difficult to remove, whereas weakly attracted electrons are easily removed. Figure 8.12 is a plot of first ionization energy versus atomic number. The periodic tables in Figure 8.13 show these trends and values for the representative elements.

An important trend in the first ionization energies of the elements is a general increase across rows in each period, which can be explained by the changes that occur in Z_{eff}. Consider the first ionization energy for the atoms in the second period. The valence electron for lithium, the $2s$ electron, is not tightly held by the nucleus because the two electrons in the $1s$ orbital effectively screen it ($Z_{eff} \approx 1.3$).

Next, consider beryllium. The screening of the $2s$ valence electrons of beryllium by the $1s$ electrons is about the same as that of lithium, but these electrons experience an effective nuclear charge of about 1.9, considerably greater than that for the $2s$ electron in lithium. Because of the increased effective nuclear charge, beryllium has a higher first ionization energy than lithium. This logic is similar to that used to explain size trends; a beryllium atom is smaller than a lithium atom because the valence electrons are attracted by a larger effective nuclear charge. The ionization energy increases for the same reason. This general trend, increasing first ionization energy from left to right as Z_{eff} increases, continues across the period.

Ionization energies increase across rows in a period because of the increase in effective nuclear charge.

Ionization energies generally increase across any period, but not as smoothly as the sizes change. The ionization energies are more sensitive than the radii to changes in the

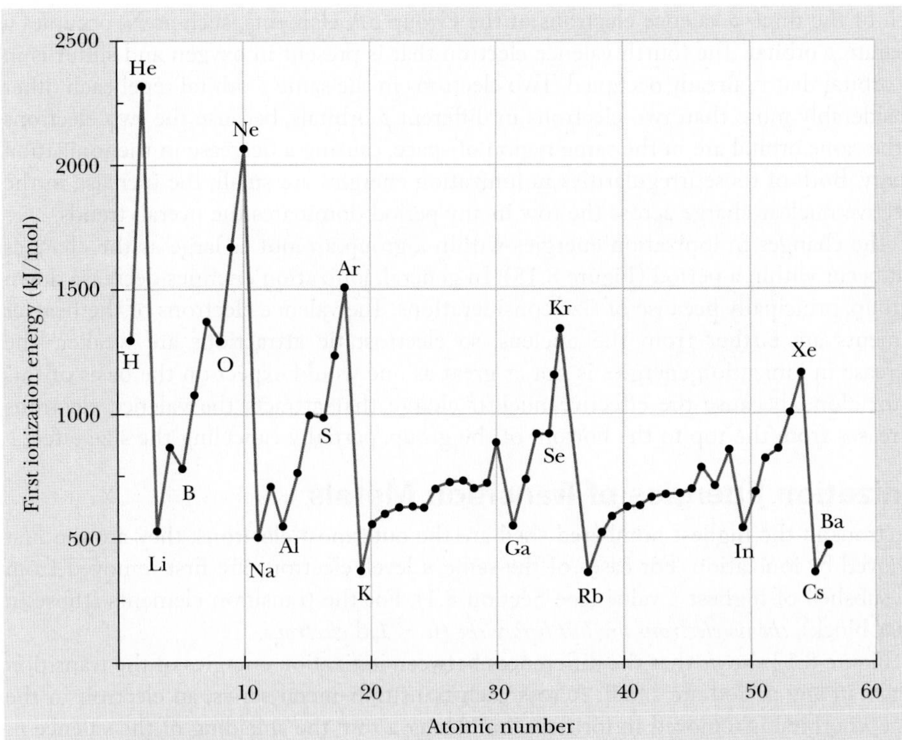

Figure 8.12 First ionization energies of atoms.

occupied subshells. Across the second and third periods are two irregularities in the general trend of increasing ionization energies. The first is a decrease in ionization energy between some Group 2A and 3A elements. The valence electron configurations for these groups change from ns^2 to ns^2np^1. The slight decrease in the ionization energy trend is explained by the smaller penetration of the inner electrons by an np^1 electron in comparison with the ns^2 electrons.

Slight breaks in the increasing ionization energies across a period are observed at Group 3A, where electrons first enter the *p* subshell, and at Group 6A, where electrons are first paired in the *p* subshell.

A second decline in ionization energy occurs for the Group 6A elements oxygen and sulfur. Figure 8.14 is the energy-level diagram of oxygen. Experiments indicate that

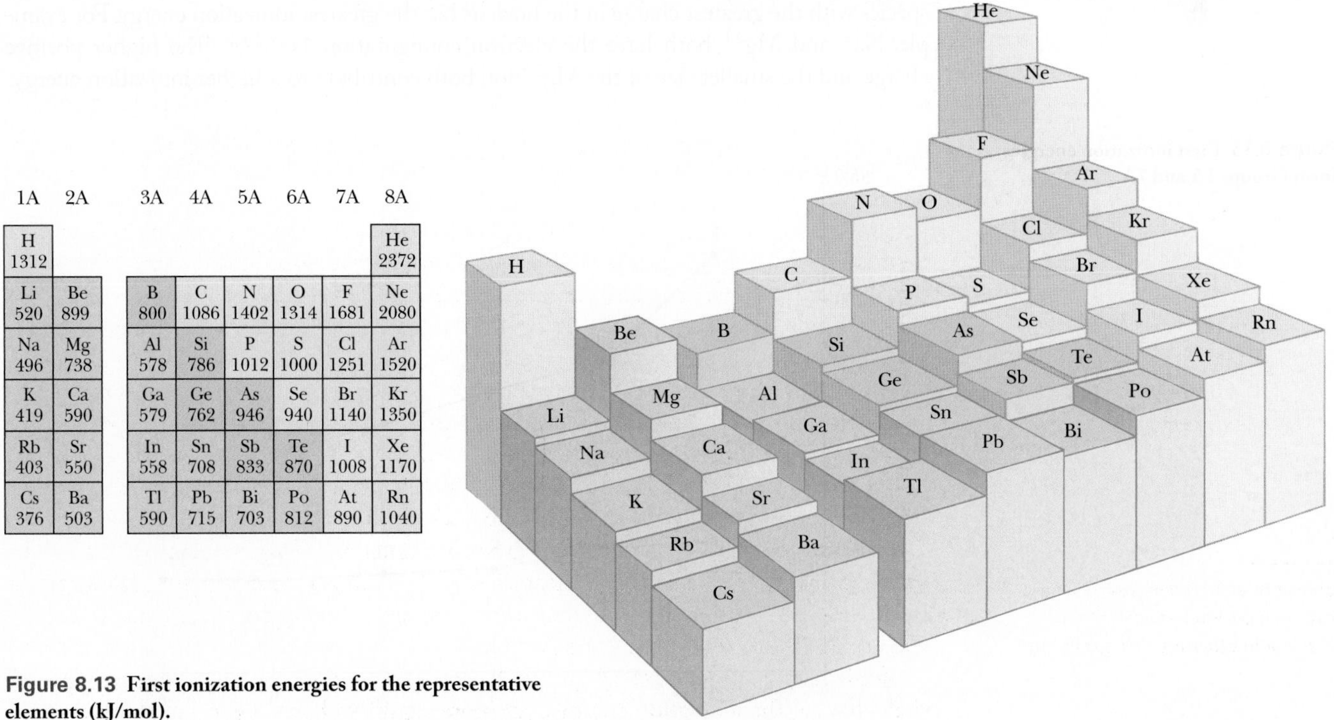

1A	2A	3A	4A	5A	6A	7A	8A
H 1312							He 2372
Li 520	Be 899	B 800	C 1086	N 1402	O 1314	F 1681	Ne 2080
Na 496	Mg 738	Al 578	Si 786	P 1012	S 1000	Cl 1251	Ar 1520
K 419	Ca 590	Ga 579	Ge 762	As 946	Se 940	Br 1140	Kr 1350
Rb 403	Sr 550	In 558	Sn 708	Sb 833	Te 870	I 1008	Xe 1170
Cs 376	Ba 503	Tl 590	Pb 715	Bi 703	Po 812	At 890	Rn 1040

Figure 8.13 First ionization energies for the representative elements (kJ/mol).

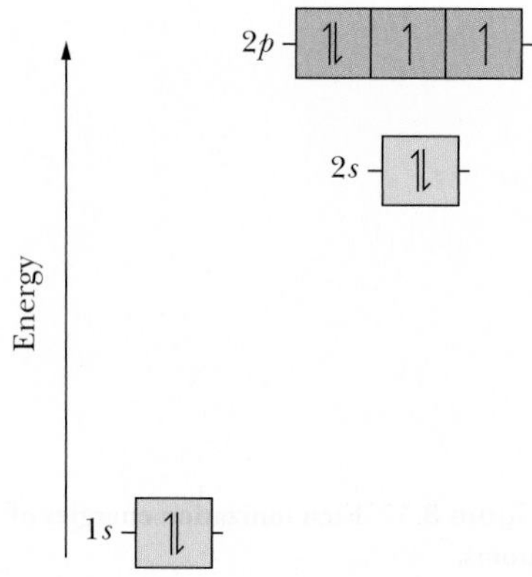

Figure 8.14 Energy-level diagram of oxygen. Electrons are paired in one of the $2p$ orbitals.

Ionization energies of representative elements decrease slightly down a group.

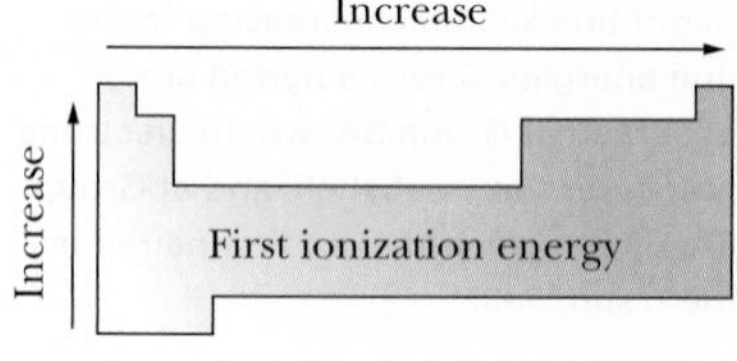

each of the three p valence electrons of the Group 5A elements, such as N, occupies a separate p orbital. The fourth valence electron that is present in oxygen and sulfur is in a p orbital that is already occupied. Two electrons in the same p orbital repel each other considerably more than two electrons in different p orbitals, because the two electrons in the same orbital are in the same region of space, causing a decrease in the ionization energy. Both of these irregularities in ionization energies are small; the increase in the effective nuclear charge across the row in any period dominates the overall trends.

The changes in ionization energies within a group are not as large as the changes that occur within a period (Figure 8.15). In general, ionization energies decrease down a group, principally because of size considerations. The valence electrons of the heavier elements are farther from the nucleus, so electrostatic attractions are weaker. The decrease in ionization energies is not as great as one would expect on the basis of this factor alone, because the effective nuclear charge that attracts the valence electrons increases from the top to the bottom of the group, partially canceling the size effect.

Ionization Energies of Transition Metals

Electrons in the highest numbered shell are the outermost electrons; they are the first removed by ionization. For cases of the same n level, electrons are first removed from the subshell of highest ℓ value (see Section 8.1). For the transition elements (those in the d block), *the* ns *electrons are lost before the (*n *−* *1)*d *electrons.*

Figure 8.12 shows that the differences between ionization energies of the transition metals in any period are small. Across each transition-metal series, an electron in the same ns orbital is removed to form an ion. Along a row, the shielding of the valence ns electrons by the added $(n - 1)d$ electrons almost cancels the increase in the nuclear charge. The effective nuclear charge that attracts the ns electrons increases quite slowly, as reflected by the small increases in the ionization energies.

The fact that ionization energies do not change much influences the chemistry of the transition metals. Frequently, transition-metal elements in the same period form similar compounds. For example, many transition metals between scandium and zinc form 2+ ions. The main-group elements do not exhibit such similarity.

Ionization Energy Trends in an Isoelectronic Series

The trends in ionization energies within an isoelectronic series are easy to determine: The species with the greatest charge in the nucleus has the greatest ionization energy. For example, Na^+ and Mg^{2+} both have the electron configuration $1s^22s^22p^6$. The higher positive charge and the smaller size of the Mg^{2+} ion both contribute to a higher ionization energy.

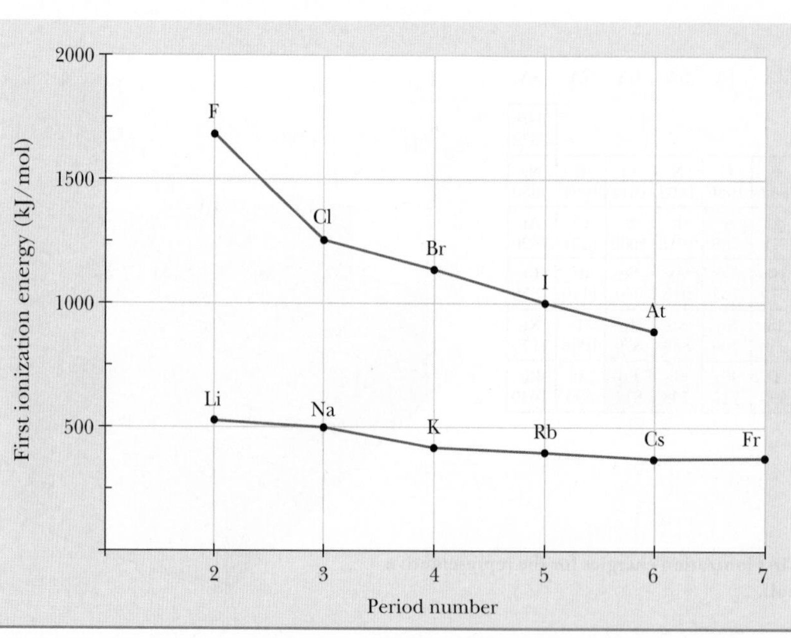

Figure 8.15 First ionization energies down Groups 1A and 7A.

EXAMPLE 8.7 Ionization Energy Trends

Predict which species in each of the following pairs has the higher first ionization energy.

(a) Mg or P (b) B or Cl (c) K^+ or Ca^{2+}

Strategy Use the ionization energy trends—(1) up a group, the ionization energies increase; and (2) across a row from left to right, ionization energies mainly increase—to determine which has the greater ionization energy. For two species that are isoelectronic, the species with the greater charge has the higher ionization energy. Use the small periodic table shown to locate the elements.

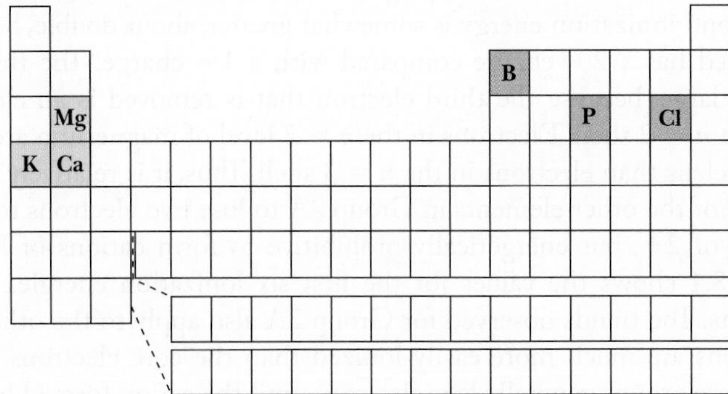

Solution

(a) Magnesium and phosphorus are in the same period. Phosphorus is to the right and has the higher first ionization energy.

(b) Boron and chlorine are in different periods, but chlorine is four groups to the right of boron. A small decrease in ionization energies occurs from the second period to the third period, but this change is small compared with the increase from Group 3A to Group 7A. Chlorine has the greater first ionization energy.

(c) The ions K^+ and Ca^{2+} are isoelectronic, but Ca^{2+} has the higher charge and thus has the greater first ionization energy.

Understanding

Predict which species has the greater first ionization energy: Al or Si.

Answer Si

Ionization Energies and Charges of Cations

In determining the formulas of many ionic compounds, we have already used the facts that elements in Group 1A form 1+ cations, those in Group 2A form 2+ cations, and those in Group 3A form 3+ cations. We can explain these observations by measuring the successive ionization energies of the atoms in the laboratory. Consider the first three ionization energies for magnesium:

$$\underset{[\text{Ne}]3s^2}{\text{Mg(g)}} \longrightarrow \underset{[\text{Ne}]3s^1}{\text{Mg}^+\text{(g)}} + e^- \qquad I_1 = 738 \text{ kJ/mol}$$

$$\underset{[\text{Ne}]3s^1}{\text{Mg}^+\text{(g)}} \longrightarrow \underset{[\text{Ne}]}{\text{Mg}^{2+}\text{(g)}} + e^- \qquad I_2 = 1450 \text{ kJ/mol}$$

$$\underset{[\text{Ne}]}{\text{Mg}^{2+}\text{(g)}} \longrightarrow \underset{[\text{He}]2s^22p^5}{\text{Mg}^{3+}\text{(g)}} + e^- \qquad I_3 = 7734 \text{ kJ/mol}$$

TABLE 8.1 Successive Ionization Energies for Some Third-Period Elements (kJ/mol)

Element	I_1	I_2	I_3	I_4	I_5	I_6
Sodium	496	4,562	6,912	9,540	13,360	16,620
Magnesium	738	1,450	7,734	10,550	13,640	18,030
Aluminum	578	1,817	2,745	11,600	14,840	18,390
Silicon	786	1,577	3,232	4,350	16,100	19,800
Phosphorus	1,012	1,903	2,912	4,940	6,270	21,300

Magnesium lies near the left side of the periodic table and has two valence electrons. Both the first and second ionization energies are low, because the Z_{eff} for both electrons is low. The second ionization energy is somewhat greater, about double, because the ion that is produced has a 2+ charge compared with a 1+ charge. The third ionization energy is very large, because the third electron that is removed is an electron from a lower shell, the $n = 2$ shell. Electrons in the $n = 2$ level of magnesium are more tightly held by the nucleus than electrons in the $n = 3$ shell. Thus, it is relatively easy for magnesium and all of the other elements in Group 2A to lose two electrons to form cations with a charge of 2+, but energetically prohibitive to form cations of larger positive charge. Table 8.1 shows the values for the first six ionization energies of the third-period elements. The trends observed for Group 2A also apply to the other groups; the valence electrons are much more easily ionized than the core electrons. In each case, elements in these groups generally lose electrons until the cation formed has an electron configuration of a noble gas.

Valence electrons are much easier to remove than core electrons.

The blue ionization energies represent the removal of core electrons. The steep increase in ionization energy on extracting an electron from a lower quantum level explains why only certain ions are formed: Na^+, Mg^{2+}, Al^{3+}, and so forth.

EXAMPLE 8.8 Ionization Energy Trends

Experiment shows that the first ionization energy of lithium is less than the first ionization energy of beryllium, but the second ionization energy of beryllium is less than the second ionization energy of lithium. Explain these observations.

Strategy The key issue in this problem is to consider whether each ionization process removes valence or core electrons. If valence electrons are removed, consider the Z_{eff} of the electrons. The ionization energy to remove core electrons is always high.

Solution
For both lithium and beryllium atoms, the first electron is removed from the 2*s* subshell. Beryllium has one more proton in its nucleus, and thus has a higher Z_{eff} for electrons in the 2*s* subshell—its first ionization energy is greater. For beryllium atoms, the second electron is also removed from the 2*s* subshell, but in lithium atoms, the second electron is removed from an inner 1*s* subshell. Removing a core electron requires more energy than removing a valence electron, so the second ionization energy for lithium is greater than the second ionization energy for beryllium.

Understanding

Which has a larger second ionization energy: beryllium or boron?

Answer Boron

Elements in Groups 4A through 8A generally do not form cations. These elements, at the right of the periodic table, have more valence electrons, and these electrons have a higher effective nuclear charge than the valence electrons of the elements to the left. The increased attraction makes it difficult to remove them at all.

Ionization of the heavier elements in Group 3A (gallium and below) does not lead to a noble-gas electron configuration. For indium atoms,

$$\text{In} \longrightarrow \text{In}^{3+} + 3e^-$$

$$[\text{Kr}]5s^2 4d^{10} 5p^1 \quad [\text{Kr}]4d^{10}$$

The 4*d* electrons are not valence electrons and are not easily removed, so the fourth ionization energy is much larger than the third one.

$$I_1 = 558 \text{ kJ/mol}; I_2 = 1820 \text{ kJ/mol}; I_3 = 2704 \text{ kJ/mol}; I_4 = 5210 \text{ kJ/mol}$$

The [noble gas]$(n - 1)d^{10}$ electron configuration is known as a **pseudo-noble-gas electron configuration** because several cations with this electron arrangement are stable. One other related trend for the heavier elements in Group 3A is that two differently charged cations of the same element are found in stable compounds. For example, both InCl and $InCl_3$ are stable compounds. To form In^+, only the $5p^1$ electron is removed, leaving the $[Kr]5s^2 4d^{10}$ electron configuration. The relatively large increase in energy between I_1 and I_2 is an important factor contributing to the stability of the In^+ ion. This electron configuration, as well as the pseudo-noble-gas electron configuration of In^{3+}, $[Kr]4d^{10}$, is common in compounds formed by the metals in this region of the periodic table.

Representative-element metals can lose electrons until the cations attain a noble-gas or a pseudo-noble-gas electron configuration.

The charges on transition-metal ions are not as predictable. In general, the lowest positive charge found on a transition metal occurs when the *ns* electrons are removed, leading to cations with a 2+ charge. A few other electron arrangements are particularly stable but depend on factors to be discussed later. The main point is that the transition metals form cations, but most form more than one stable cation.

EXAMPLE 8.9 Metal Cations

Tin forms two stable cations. Predict the charges and electron configurations of the two cations.

Strategy Use the position of tin in the periodic table and the expected relative magnitudes of the ionization energies to predict the charges of the ions.

Solution

Tin has the electron configuration $[Kr]5s^2 4d^{10} 5p^2$. The first two ionization energies are probably relatively low, so it is reasonable to presume that Sn^{2+} is one of the cations of tin. Sn^{2+} would have the electron configuration $[Kr]5s^2 4d^{10}$. The second cation would probably result from the loss of the two 5*s* valence electrons, giving the Sn^{4+} ion with a pseudo-noble-gas electron configuration of $[Kr]4d^{10}$.

Understanding

Predict the charges on the two cations of thallium.

Answer Tl^+ and Tl^{3+}

OBJECTIVES REVIEW *Can you:*

- ☑ define ionization energy?
- ☑ arrange atoms according to ionization energies?
- ☑ predict the relative energies needed for successive ionizations?

8.5 Electron Affinity

OBJECTIVES

- ☐ Define electron affinity
- ☐ Relate trends in electron affinity to the periodic table
- ☐ State how electron affinity affects the formation of anions

1A	2A	3A	4A	5A	6A	7A
H −73						
Li −60	Be 241	B −27	C −122	N ~0	O −141	F −328
Na −53	Mg 230	Al −43	Si −134	P −72	S −200	Cl −349
K −48	Ca 156	Ga −29	Ge −119	As −78	Se −195	Br −325
Rb −47	Sr 167	In −29	Sn −107	Sb −103	Te −190	I −295
Cs −46	Ba 52	Tl −19	Pb −35	Bi −91		

Figure 8.16 Electron affinities for selected main-group elements (kJ/mol).

Ionization energies provide information about how elements form cations from the atoms. Many atoms, especially those of the elements on the right side of the periodic table, accept electrons to form anions. The **electron affinity** of an element is the energy change when an electron is added to a gaseous atom to form an anion.

$$A(g) + e^- \longrightarrow A^-(g)$$

Figure 8.16 gives representative values of electron affinities.[1]

The elements with the highest (most favorable) electron affinities are the Group 7A elements. They have the highest effective nuclear charges in their period and a vacancy in the valence p orbitals to hold an additional electron and complete a valence shell octet.

$$F(g) + e^- \longrightarrow F^-(g)$$

$$[He]2s^22p^5 \qquad [He]2s^22p^6 = [Ne]$$

In fact, fluorine (the subject of the introduction to this chapter) has one of the largest electron affinities of all the elements, as shown in Figure 8.16.

The Group 6A elements have electron affinities that are quite exothermic as well. The elements of Groups 6A and 7A also have high ionization energies, so it is energetically favorable to gain an electron, but it is difficult to remove an electron. It is no surprise that elements in Groups 6A and 7A form anions.

The electron affinity of nitrogen is near zero, despite the fact that the electron affinities of the elements on both sides of it in the periodic table are exothermic. The change in electron configuration of nitrogen on addition of an electron is

$$N(g) + e^- \longrightarrow N^-(g)$$

$$[He]2s^22p^3 \qquad [He]2s^22p^4$$

The nitrogen atom has an electron in each of its three p orbitals. The additional electron must pair with one of these electrons, causing larger electron-electron repulsions. These unfavorable interactions make the formation of a nitrogen anion less favorable than expected. Nitrogen has a less favorable electron affinity for the same reason that oxygen has a lower ionization energy, in comparison with neighbors in the second period. A similar argument explains the endothermic electron affinity of beryllium. In this case, the added electron is the first one to enter the $2p$ subshell.

In contrast, the electron affinity of carbon is quite exothermic. In this case, the electron is added to an empty $2p$ orbital.

$$C(g) + e^- \longrightarrow C^-(g)$$

$$[He]2s^22p^2 \qquad [He]2s^22p^3$$

Electron affinities are generally most favorable for elements with high ionization energies.

Electron affinities do not change dramatically down the groups. As with ionization energy trends down a group, the changes in size and Z_{eff} affect the electron affinities in opposite directions and largely cancel each other.

EXAMPLE 8.10 Electron Affinity Trends

Which atom has a higher electron affinity: chlorine or sulfur?

Strategy Use the electron affinity trends: (1) down a group, the electron affinity does not change dramatically; and (2) across a row from left to right, electron affinity mainly increases to determine the orders.

[1]The formal International Union of Pure and Applied Chemistry (IUPAC) definition of electron affinity is that it is the energy *given off*, so technically it is understood that electron affinities are exothermic. Some versions of Figure 8.16, therefore, change the signs on all electron affinities, leading to negative values for those atoms whose electron affinities are endothermic. This creates potential confusion for students who are learning that exothermic processes have a negative energy change, so we—together with a majority of textbooks—adopt the convention shown in Figure 8.16.

Solution

Chlorine and sulfur are in the same row, but chlorine is to the right of sulfur and will have the higher electron affinity.

Understanding

Which atom has a higher electron affinity: oxygen or sulfur?

Answer Oxygen has a greater electron affinity.

OBJECTIVES REVIEW *Can you:*

- ☑ define electron affinity?
- ☑ relate trends in electron affinity to the periodic table?
- ☑ state how electron affinity affects the formation of anions?

8.6 Trends in the Chemistry of Elements in Groups 1A, 2A, and 7A

OBJECTIVES

- ☐ Write equations for the common reactions of the elements in Groups 1A, 2A, and 7A
- ☐ Identify reactivity trends within each of these groups

H							He
Li	Be	B	C	N	O	F	Ne
Na	Mg	Al	Si	P	S	Cl	Ar
K	Ca	Ga	Ge	As	Se	Br	Kr
Rb	Sr	In	Sn	Sb	Te	I	Xe
Cs	Ba	Tl	Pb	Bi	Po	At	Rn

The periodic table groups elements by chemical and physical properties. We now know that each group of elements has the same number of valence electrons. The chemical properties of the elements are strongly influenced by factors such as size, ionization energy, and electron affinity in ways that have already been discussed. For example, it is known that elements with low ionization energies are generally metals and form cations in their compounds, whereas those with high ionization energies are nonmetals and form anions. The chemistry of the Group 1A, 2A, and 7A elements is outlined in this section.

Group 1A: Alkali Metals

The Group 1A metals—lithium, sodium, potassium, rubidium, cesium, and francium—all have a single electron in the *s* valence orbital. Because this electron is easily removed, these elements are highly reactive, forming compounds that contain the metals as 1+ ions. In nature, these elements are found only in combination with other elements, because they are too reactive to exist in the environment as the uncombined elements. Hydrogen, which also has a $1s^1$ electron configuration, is not a member of this group (even though it appears on top of it in most periodic tables). Because it is the first element in the periodic table, hydrogen has unique chemical properties and is really in a group by itself.

Both sodium and potassium are abundant in nature, but lithium, rubidium, and cesium are relatively rare. Francium is exceedingly scarce, because all of its isotopes are unstable and decompose to other elements. The elements of this group are soft, silver-colored metals (Figure 8.17) (except cesium, which is golden), and they melt at lower temperatures than most other metals. Cesium, in fact, melts just above room temperature. The low melting point of sodium, 98 °C, is one of the reasons that sodium is a cooling liquid in nuclear reactors, particularly in submarines.

Each of the alkali metals emits light of a characteristic color when the metal or a compound containing the metal is placed in a flame, a procedure called a **flame test.** Figure 8.18 shows the colors observed when salts of these elements are heated in a flame. The colors of light emitted by these elements are related to their line spectra. Flame tests are used to determine which elements are present in samples of unknown composition (qualitative analysis), and the intensity of the colors is used to determine the amounts of the elements in the samples (quantitative analysis).

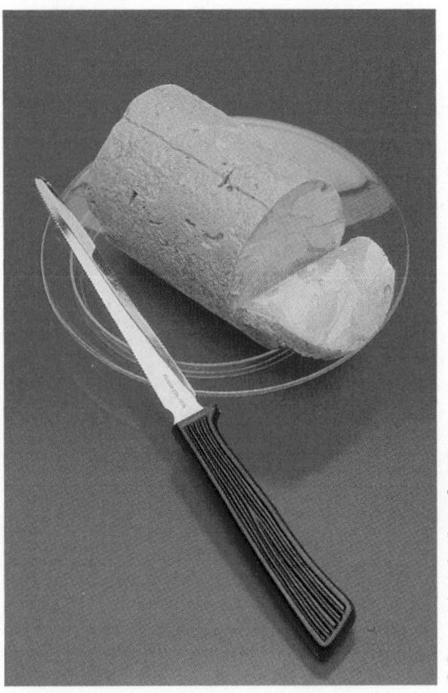

Figure 8.17 Sodium metal. Sodium metal is soft and is easily cut with a knife. It has a silvery color when first cut but rapidly loses its luster as it reacts with air.

Figure 8.18 Flame test for alkali-metal compounds. The alkali metals emit radiation of characteristic color when heated in a flame. A drop of a solution that contains a salt of the metal is placed on a wire loop, and the loop is placed in a flame.

All of the alkali metals are very reactive, with reactivity generally increasing from Li to Cs, as expected from the decrease in ionization energies. These metals react with hydrogen to make ionic compounds of the hydride ion, H^-. These compounds are unusual in that the hydrogen has a −1 oxidation number.

$$2M(s) + H_2(g) \rightarrow 2MH(s) \qquad M = Li, Na, K, Rb, Cs$$

The ion H^- is isoelectronic with helium, but in contrast with the noble gas helium, the hydride ion is very reactive. NaH and KH will spontaneously react with water to make hydrogen gas and the respective hydroxide.

$$MH(s) + H_2O(\ell) \rightarrow MOH(aq) + H_2(g) \qquad M = Na^+, K^+$$

The Group 1A metals react with molecular oxygen, but only lithium reacts to form the compound we expect, Li_2O.

$$4Li(s) + O_2(g) \rightarrow 2Li_2O(s)$$

Sodium reacts with excess O_2 to form mainly a compound that contains the polyatomic peroxide anion, O_2^{2-}.

$$2Na(s) + O_2(g) \rightarrow Na_2O_2(s)$$

This reaction is complicated by the fact that a second reaction occurs at the same time, producing sodium oxide, Na_2O. Potassium, rubidium, and cesium react to form mixtures of three compounds. In addition to forming the oxide and peroxide, these metals produce compounds that contain the polyatomic superoxide anion, O_2^-.

In peroxide, oxygen has an oxidation number of −1. In superoxide, oxygen has an oxidation number of −1/2.

$$K(s) + O_2(g) \rightarrow KO_2(s)$$

The following table summarizes the products of the reaction of the Group 1A metals with oxygen:

Reactions of Group 1A Elements with Oxygen

Element	Oxide	Peroxide	Superoxide
M = Li	M_2O	Not formed	Not formed
M = Na	M_2O	M_2O_2	Not formed
M = K, Cs, Rb	M_2O	M_2O_2	MO_2

All Group 1A elements react with water to give similar products, hydrogen gas and solutions of the metal hydroxide.

$$2M(s) + 2H_2O(\ell) \rightarrow 2MOH(aq) + H_2(g) \qquad M = Li, Na, K, Rb, Cs$$

These reactions can be dangerous because the hydrogen gas that is produced reacts with oxygen in air to form water, sometimes quite violently (Figure 8.19). Sodium, which is

(*a*)

(*b*)

(*c*)

Figure 8.19 Reaction of potassium and water. *(a)* Potassium reacts rapidly with water, giving off H_2 gas. *(b)* In air, flames are seen as the hydrogen gas produced reacts further with oxygen in the air, forming water. The red color is from excited potassium atoms in the vapor phase. *(c)* Under an atmosphere of argon (held in by the inverted funnel), the potassium still reacts rapidly with water to produce hydrogen, but there is no flame because there is no oxygen present.

commonly used in the laboratory, must be treated carefully and is usually stored under oil to protect it from water vapor and oxygen in the air.

These metals also react with the halogens. The typical reaction is

$$2M(s) + X_2 \rightarrow 2MX(s) \qquad X = F, Cl, Br, I$$

The heavier members of Group 1A are generally more reactive. For example, lithium and sodium react slowly with liquid bromine; the other Group 1A elements react violently.

The reactivity of the Group 1A metals increases down the group. Their chemistry is dominated by the formation of M^+ ions.

The small size of lithium causes it to have some unusual chemical properties. The most unusual is that it reacts with molecular nitrogen to form lithium nitride, a compound that contains the N^{3-} ion.

$$6Li(s) + N_2(g) \rightarrow 2Li_3N(s)$$

Nitrogen is not a very reactive gas. In fact, nitrogen gas is often used to protect materials that react with oxygen and water. This method cannot be used with lithium. Although lithium is generally the least reactive of the Group 1A metals, it is the only one of them that reacts directly with molecular nitrogen. Lithium fluoride, LiF, is also not very soluble in water because both ions are very small, making the ions tightly held together.

H							He
Li	Be	B	C	N	O	F	Ne
Na	Mg	Al	Si	P	S	Cl	Ar
K	Ca	Ga	Ge	As	Se	Br	Kr
Rb	Sr	In	Sn	Sb	Te	I	Xe
Cs	Ba	Tl	Pb	Bi	Po	At	Rn

Group 2A: Alkaline Earth Metals

The alkaline earth metals (Group 2A)—beryllium, magnesium, calcium, strontium, barium, and radium—have the valence electron configuration ns^2. These metals are reactive, but not as reactive as the Group 1A metals, because the Group 2A metals have greater ionization energies. With the exception of beryllium, most reactions of these elements lead to the formation of ionic compounds in which the two valence electrons are removed, yielding the metals as 2+ cations.

Both magnesium and calcium are abundant in nature. Magnesium is found in many minerals, including the magnesium carbonate-limestone combination known as dolomite, $MgCO_3 \cdot CaCO_3$. Large land masses, such as the Dolomite Mountains in Italy, consist of this mineral. Large deposits of $CaCO_3$ formed from the fossilized remains of ancient life are found in all parts of the world. Coral and seashells are also composed mainly of $CaCO_3$. Strontium and barium are moderately abundant, but beryllium and radium are rare. All of the isotopes of radium are unstable.

The famous White Cliffs of Dover in England are composed largely of calcium carbonate.

The alkaline earth metals have a silvery white appearance. They are softer than most other metals but are harder and have higher melting points than the alkali metals. Like the alkali metals, the heavier members of the alkaline earth group (calcium, strontium, and barium) show characteristic colors in a flame test (Figure 8.20). Beryllium and magnesium emit light when heated, but characteristic spectral lines are not in the visible region of the spectrum.

(a)

(b)

(c)

Andrew Lambert Photography/Photo Researchers, Inc.

Figure 8.20 Flame test for alkaline earth metals. The heavier alkaline earth metals emit radiation of characteristic color when heated in a flame. *(a)* calcium, *(b)* strontium, and *(c)* barium.

Magnesium has widespread industrial uses. Although it is now about twice as expensive as aluminum, its use might eventually increase because it is easily isolated from seawater and is even less dense than aluminum. It is particularly useful when mixed with other metals to form *alloys.* Magnesium alloys are useful in aeronautical applications, where low density and high strength are important. It is also added to aluminum to improve its mechanical properties. Magnesium-aluminum alloys can be machined much more easily than pure aluminum. In its ionized form of Mg^{2+}, magnesium is an important constituent of the chlorophylls, which are compounds of great importance in photosynthesis.

The reactivity of the Group 2A metals increases down the group, a trend that follows the decrease in ionization energies. Although all Group 2A metals form hydrides, BeH_2 and MgH_2 are not ionic but are better characterized as having repeating units, with metal atoms bridged by two hydrogen atoms. The rest are ionic in character.

$$M(s) + H_2(g) \rightarrow MH_2(s) \qquad M = Ca, Sr, Ba$$

Beryllium and magnesium react with oxygen at high temperatures to form their oxides, which are generally unreactive and act as coatings that protect the metal from further reaction. Calcium metal reacts with oxygen at room temperature.

$$2Ca(s) + O_2(g) \rightarrow 2CaO(s)$$

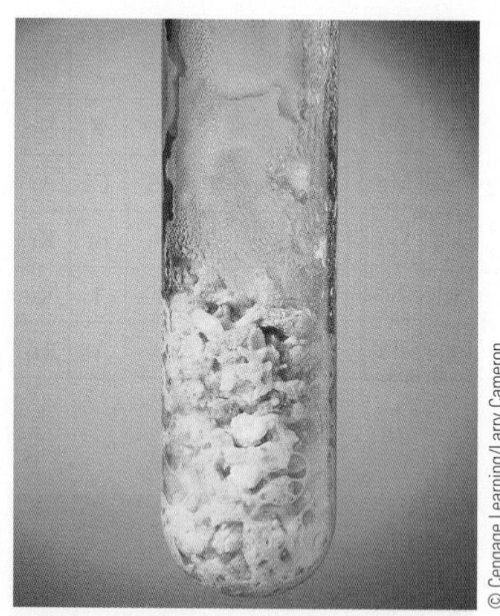
© Cengage Learning/Larry Cameron

Reaction of calcium and water. Calcium reacts with water to yield $Ca(OH)_2$ and H_2 gas.

The Group 2A metals are not as reactive as the Group 1A metals. They form M^{2+} ions.

Just as with the Group 1A metals, the heavier elements of Group 2A form salts of the peroxide ion, O_2^{2-}. This reactivity is used in the production of hydrogen peroxide on a commercial scale. The reaction of barium metal with oxygen forms barium peroxide, which is then treated with sulfuric acid to produce hydrogen peroxide.

$$Ba(s) + O_2(g) \rightarrow BaO_2(s)$$

$$BaO_2(s) + H_2SO_4(aq) \rightarrow BaSO_4(s) + H_2O_2(aq)$$

Beryllium does not react with water or steam. Magnesium reacts slowly with steam. Calcium, strontium, and barium react at room temperature with water to give the metal hydroxide and hydrogen gas.

$$M(s) + 2H_2O(\ell) \rightarrow M(OH)_2(aq) + H_2(g) \qquad M = Ca, Sr, Ba$$

Halides of the Group 2A metals are made by direct combination of the metal and halogen or, as is the case with fluorite in the chapter introduction, are mined.

PRACTICE OF CHEMISTRY
Fireworks

Fireworks have been known and enjoyed for more than a thousand years. It might be surprising to know that the basics of fireworks have changed little in hundreds of years. What has changed, however, is our understanding of the chemistry behind fireworks.

Although the specific chemicals depend on the type of firework, most fireworks are composed of a fuel, an oxidizing agent, color-producing chemicals, effects-producing chemicals (i.e., smoke generators), binders, and various other additives. In this discussion, we are particularly concerned with the additives that add colors to the lights that these pyrotechnic displays produce.

As shown in the flame tests of Group 1A and 2A elements in Figures 8.18 and 8.20, different elements produce different colors of light when their atoms are excited in a flame. Fireworks producers take advantage of that fact by including certain chemicals, typically simple salts, in the production of the fireworks. For example, barium salts such as barium chlorate or barium nitrate are used when green light is desired. A copper arsenite compound curiously named Paris Green is used for bright blue light; however, because of the toxicity of arsenic, basic copper carbonate is also used to produce blue light. Although it may be a metal ion that produces light in a fireworks display, the specific metal compound used depends on a variety of factors, including the chemical compatibility with other ingredients such as fuels, oxidizers, and other additives. Mixing the wrong compound with a fuel and oxidizer could lead to some disastrous premature explosions of a pyrotechnic recipe!

The following table lists the elements that produce specific colors in fireworks—perhaps the most enjoyable of "flame tests." ▮

Element	Color Produced
Antimony	White
Barium	Green
Carbon	Gold
Copper	Blue
Lithium	Red
Magnesium	White
Sodium	Yellow
Strontium	Red
Titanium	Silver

From Lancaster R. 1992. *Fireworks: Principles and practice.* New York: Chemical Publishing Co., Inc.

Group 7A: The Halogens

The chemistry of the halogens—fluorine, chlorine, bromine, iodine, and presumably astatine—is dominated by the gain of one electron to attain a noble-gas electron configuration. (The chapter introduction recounts the extreme reactivity of the most reactive halogen, fluorine.) Elemental halogens all exist as diatomic molecules, but they are very reactive and occur in nature combined with other elements. Under standard conditions, fluorine is a pale yellow gas and chlorine a deeper greenish-yellow gas, bromine is a deep red liquid (the only nonmetallic element that is a liquid at room temperature), and iodine is a shiny violet-black solid.

H							He
Li	Be	B	C	N	O	F	Ne
Na	Mg	Al	Si	P	S	Cl	Ar
K	Ca	Ga	Ge	As	Se	Br	Kr
Rb	Sr	In	Sn	Sb	Te	I	Xe
Cs	Ba	Tl	Pb	Bi	Po	At	Rn

The reactivity of the halogens decreases down the group. Their chemistry is dominated by the formation of X^- ions.

Fluorine and chlorine are both abundant in nature, occurring as anions. Chloride ion is one of the main components of seawater. Bromine is less abundant but is found in sizable quantities as bromide ions in the ocean and in certain inland seas such as the Dead Sea and the Great Salt Lake. Iodine is scarce but is found in certain subterranean wells and in seaweed, where it was first discovered. Fluorine and chlorine are prepared industrially by treating their ionic compounds with electricity. Chlorine is used to convert bromide and iodide to the respective elements. Astatine is rare because all of its isotopes are unstable. Little is known about the chemical behavior of this element.

Fluorine is the most reactive of all the nonmetals (see chapter introduction). Because of its reactivity, it was not isolated until 1886, 60 years after the other halogens were prepared. It forms compounds with all other elements except helium and neon.

PRINCIPLES OF CHEMISTRY
Salt

Chemistry has a specific definition of the word *salt*: an ionic compound derived from an acid and a base. But when most people use the word *salt,* they are referring to a specific substance: table salt, or sodium chloride. In this context, the compound salt has an illustrious history and unique position in human activities.

Both sodium and chlorine are essential elements for biological function, and salt supplies both. Archeological evidence exists for the intentional use of salt going back to 3000 BC. Around 200 BC, there were references to the use of salt as payment to Roman soldiers for services rendered (the word *salary* shares its root with *salt*), although some historians question this. What is not questioned is that even at this early date, salt was a valuable commodity. Salt has been (and still is) used as a flavor enhancer and preservative (either by packing in the solid or immersing in highly saline solutions called *brines*). In the past, excessive salt ingestion has been linked to high blood pressure, but a direct link is now in question.

More than 200 million tons of salt are produced every year globally, either by evaporation of brines or mining of salt formations in the ground. Despite its popularity as a seasoning agent, most salt is used as a de-icer, because salt water has a lower freezing point than pure water. The salt we use in cooking is not pure salt. The NaCl in the familiar cylindrical packages has magnesium carbonate added to keep the salt crystals from sticking together. Much salt also has a small amount of sodium iodide added, creating "iodized salt." This provides the necessary amount of dietary iodine so people do not suffer from iodine-deficiency diseases.

Although all table salt is sodium chloride, specialty salts have become widely available for the home cook. Kosher salt, sea salt, and *fleur de sal* (French for "flower of salt") are used in kitchens, but at prices sometimes exceeding $20 per pound! Ordinary table salt, at less than a dollar a pound, provides more NaCl for your money. ∎

Over 200 million tons of sodium chloride are produced every year.

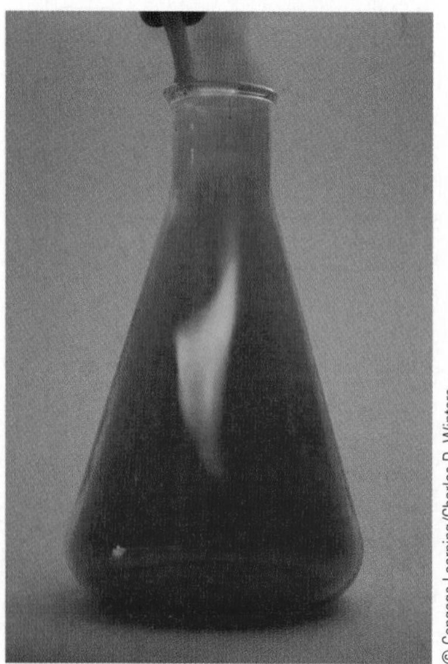

Elemental bromine has dark orange vapors.

It has an interesting chemistry with xenon, reacting under different conditions to form XeF_2, XeF_4, and XeF_6. It also forms a special group of compounds called the **interhalogens,** compounds formed from two different halogens. Most of the interhalogens have the general formula XF_n, where X is one of the heavier halogens and $n = 1, 3, 5$, and even 7 in the compound IF_7.

The reactivity of the halogens *decreases* down the group. For example, fluorine reacts explosively with hydrogen, but the reactions with hydrogen become less violent farther down the group, and that of iodine is slow. In each case, the hydrogen halide forms.

$$X_2(g) + H_2(g) \rightarrow 2HX(g) \qquad X = F, Cl, Br, I$$

The hydrogen halides are all molecular compounds and produce acidic solutions when they dissolve in water. These acid solutions, particularly HF(aq) and HCl(aq), are important industrially. HF is a very reactive, toxic material that can dissolve glass. It is used in the production of fluorocarbons, compounds that are used as refrigerants and aerosol propellants. Aqueous HCl is an inexpensive acid with a variety of applications in industry, such as the synthesis of metal chlorides and other important chemicals.

EXAMPLE 8.11 Reactivity Trends

Write the equations for the reactions of potassium and calcium with water. Which reaction do you predict will be more energetic?

Strategy Both metallic elements react with water to form an aqueous hydroxide compound and hydrogen gas. The metal that gives its electron(s) up more easily will produce the more energetic reaction.

Solution

Group 1A metals form 1+ ions, and Group 2A metals form 2+ ions. Writing the balanced reactions:

$$2K(s) + 2H_2O(\ell) \rightarrow 2KOH(aq) + H_2(g)$$

$$Ca(s) + 2H_2O(\ell) \rightarrow Ca(OH)_2(aq) + H_2(g)$$

For two elements from the same row, the Group 1A metal, potassium, is more reactive and will produce the more energetic reaction.

Understanding

Write the equation for the reactions of sodium and hydrogen.

Answer $2Na(s) + H_2(g) \rightarrow 2NaH(s)$

OBJECTIVES REVIEW *Can you:*

- ☑ write equations for the common reactions of the elements in Groups 1A, 2A, and 7A?
- ☑ identify reactivity trends within each of these groups?

CASE STUDY Cesium Fluoride

Cesium (spelled caesium outside of North America) is the most reactive metal that can be isolated in quantity, and fluorine is the most reactive nonmetal. The compound that forms when they react is cesium fluoride, CsF. Cesium fluoride has some useful properties that make it a desirable substance for certain applications. For example, solid CsF is transparent to infrared (IR) radiation, so it is used to make windows that must be IR transparent. Cesium fluoride is also a useful reactant in organic chemistry, where it is used to add fluorine atoms to organic compounds. Its usefulness comes from the fact that it is soluble in many organic solvents, as well as water. Care must be taken not to expose CsF to acid, because toxic HF(g) will be produced.

Because both cesium and fluorine are so reactive, the enthalpy change to make 1 mol CsF is fairly large:

$$Cs(s) + \frac{1}{2}F_2(g) \rightarrow CsF(s) \qquad \Delta H = -553.5 \text{ kJ}$$

With a molar mass of 152 g/mol, this corresponds to 3.6 kJ per gram of CsF formed; this value is substantial, but not as much as the 15.9 kJ per gram given off when hydrogen and oxygen react to make water. In the ionic form, the F^- ions have the electron configuration of neon, whereas the Cs^+ ions have the electron configuration of xenon. Because the cesium ion is so massive, CsF has a density of about 4.1 g/cm^3, almost twice the density of sodium chloride, which is 2.2 g/cm^3.

Curiously, although most elements exist as mixtures of isotopes, both cesium and fluorine exist naturally as 100% of a particular isotope. The fluorine isotope is fluorine-19 and has 10 neutrons in its nucleus. The cesium isotope is cesium-133, which has 78 neutrons in its nucleus.

Many of the properties of CsF are a direct result of the periodic trends discussed in this chapter: ionization energy, electron affinities, and ionic sizes. The fact that both cesium and fluorine exist as single isotopes in nature is one of the properties that *cannot* be predicted from periodic trends!

ETHICS IN CHEMISTRY

1. In 1962, chemist Neil Bartlett performed an experiment reacting PtF_6 (an extremely reactive chemical) with O_2, and subsequently isolated the compound $O_2^+PtF_6^-$ (charges are shown explicitly for clarity). Bartlett noticed that the O_2 molecule has approximately the same radius as the Xe atom, and wondered whether PtF_6 would react

similarly with xenon. If you were a research advisor, would you assign a student this research project solely on the basis of similar radii?

2. Fluoridation—the intentional addition of fluoride ion to drinking water to decrease the occurrence of dental caries (cavities)—is an emotional issue for some people. Proponents of fluoridation argue that it, together with better overall dental care, has reduced the occurrence of cavities by more than 60% and has improved the dental health of millions of people. Opponents argue that the 60%+ reduction is based on faulty data, that fluoridation of municipal drinking water is forced medication, and that fluoride treatment can be obtained by other, voluntary ways such as by using fluoride toothpaste or getting fluoride treatments from the dentist. Discuss the ethics of this issue. (Consider discussing it with your dentist as well.)

3. Fluorine (see the chapter introduction) is a chemically reactive substance in its elemental form. Contact of fluorine with the skin produces painful chemical burns, because the fluorine will react with calcium ions in intracellular fluid to form insoluble CaF_2 crystals. Yet, chemists need to know its chemical and physical properties. Would you ever consider a research project that included working with elemental fluorine? Why or why not?

Chapter 8 Visual Summary

The chart shows the connections between the major topics discussed in this chapter.

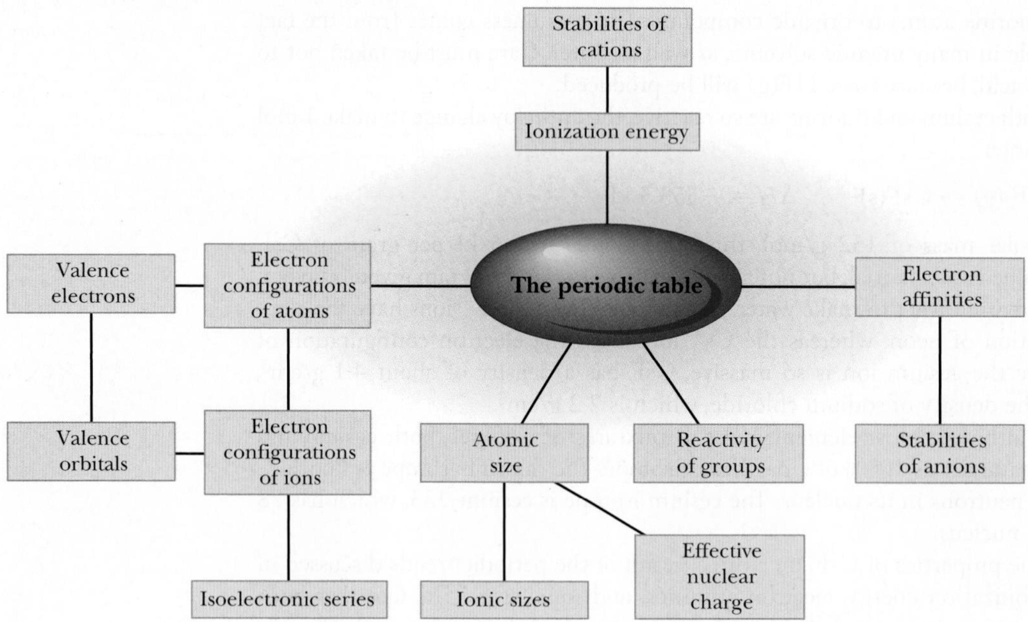

Summary

8.1 Electronic Structure and the Periodic Table

The structure of the periodic table follows the order of filling shells and subshells with electrons. Elements in a column share the same valence shell electron configurations. Each period marks the beginning of a new valence shell. We can divide the periodic table into an *s, p, d,* and *f* block, depending on which type of subshell is being filled with electrons. Knowing this, it is a straightforward task to determine the electron configuration of an element.

8.2 Electron Configurations of Ions

The electron configurations of ions are determined by starting with the electron configurations of the atoms and then adding or removing the correct number of electrons. Adding electrons to form anions follows the rules for predicting electron configurations of atoms. When cations form, electrons of highest principal shell are removed first, and for subshells of the same n level, electrons are removed from the subshell of highest ℓ value first. Thus, for the transition metals, the ns electrons are removed before the $(n - 1)d$ electrons. Species with the same number of electrons are *isoelectronic.*

8.3 Sizes of Atoms and Ions

The relative sizes of atoms and ions can be estimated from the electron configurations of the species. A cation is smaller than its neutral atom; an anion is larger. Within an isoelectronic series, the species with the most protons in the nucleus exerts the strongest attraction for the electron cloud and has the smallest radius. For the atoms in any given period, the valence electrons occupy the same shell, but the number of protons in the nucleus increases as the atomic number increases. Because electrons in the same shell do not shield each other effectively, the increasing effective nuclear charge, Z_{eff}, draws the outermost electrons closer to the nucleus and causes the atoms to decrease in size along the period.

8.4 Ionization Energy

The trends in *ionization energy,* the energy required to remove the highest energy electron from a gas-phase atom or ion, relate to the size trends. Across a period, the ionization energies increase because the effective nuclear charge is increasing and the radius is decreasing. There are two breaks in this general trend: at Group 3A, where the p level is first occupied, and at Group 6A, where electrons are first paired in one of the p orbitals. Ionization energies decrease slightly down a group and change only slowly along a transition-metal series. In an isoelectronic series, the ionization energy is greatest for the species with the most protons in the nucleus. The energy needed to remove the second electron is always greater than the first, and the ionization energies for inner electrons are very high.

8.5 Electron Affinity

The *electron affinity* is the energy change that accompanies the addition of an electron to a gas-phase atom or ion. In general, elements to the right side and at the top of the periodic table have exothermic (favorable) electron affinities. These are the elements that are generally observed to exist as anions in compounds with metals.

8.6 Trends in the Chemistry of Elements in Groups 1A, 2A, and 7A

The alkali metals, Group 1A, all have the outer electron configuration ns^1. Because of their low first ionization energies, they are very reactive, generally forming 1+ ions. They can be identified by the characteristic color given off when their salts are heated in a flame test. The alkaline earth metals, Group 2A, all have the outer electron configuration ns^2. These metals are reactive, forming 2+ cations, and the reactivities of elements in both groups increase down the group. The halogens, Group 7A, are also very reactive (generally forming 1− ions), with reactivity decreasing down the group.

Chapter Terms

The following terms are defined in the Glossary, Appendix I.

Section 8.1
Core electrons
d-block elements
f-block elements
p-block elements
s-block elements
Valence electrons
Valence orbitals

Section 8.2
Isoelectronic series

Section 8.3
Atomic radius
Effective nuclear charge
Ionic radius

Section 8.4
Ionization energy
Pseudo-noble-gas electron configuration

Section 8.5
Electron affinity

Section 8.6
Flame test
Interhalogens

Questions and Exercises

OWL Selected end of chapter Questions and Exercises may be assigned in OWL.

Blue-numbered Questions and Exercises are answered in Appendix J; questions are qualitative, are often conceptual, and include problem-solving skills.

■ Questions assignable in OWL

✎ Questions suitable for brief writing exercises

▲ More challenging questions

Questions

8.1 Define an isoelectronic series. Give the symbols for four species that are isoelectronic.

8.2 Discuss how measurements of the F–F bond length, 143 pm, and the Cl–Cl bond length, 198 pm, can be used to predict the Cl–F bond length in ClF. What is the predicted bond length in ClF?

8.3 Graph the atomic radii versus atomic number of the first 18 elements. Explain the trends in radii across the second period and down Group 1A.

8.4 Explain why carbon atoms are larger than oxygen atoms even though oxygen contains more electrons.

8.5 Why are sulfur atoms larger than oxygen atoms?

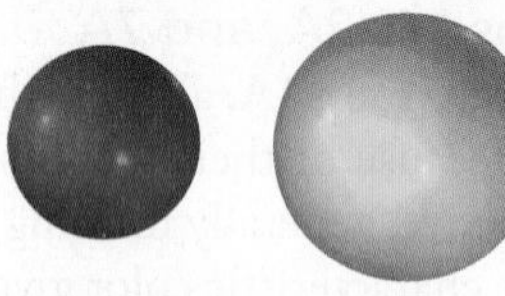

Oxygen Sulfur

8.6 Define ionization energy. Write an equation for the first ionization energy for lithium. Write the electron configuration of each species in the equation.

8.7 How many different values of ionization energy does an atom have? What determines this number?

8.8 Graph ionization energy versus atomic number for the second-period elements. Explain the trends and any discontinuities in the graph.

8.9 ✎ Even though ionization energies generally increase from left to right across the periodic table, the first ionization energy for aluminum is lower than that for magnesium. How can this observation be explained?

8.10 ✎ Explain why the first ionization energy of sodium is slightly lower than that of lithium.

8.11 ✎ Explain why the first ionization energies of manganese, iron, and cobalt increase very slightly, whereas in the series of gallium, germanium, and arsenic, the first ionization energy increases considerably.

8.12 ✎ Explain why the second ionization energy of magnesium is about twice the first, but the third ionization energy is more than four times the second.

8.13 ✎ The first ionization energy of boron is 800 kJ/mol. Qualitatively discuss the expected values of the next three ionization energies of boron. Discuss the reasons for any big differences between them.

8.14 ✎ Explain why the first ionization energy of magnesium is greater than the first ionization energies of both sodium and aluminum.

8.15 ✎ Aluminum atoms are larger than silicon atoms, and the first ionization energy of silicon is greater than that of aluminum. Explain these trends, using differences in the effective nuclear charges.

8.16 Define electron affinity.

8.17 In which group in the periodic table would you expect the elements to have strongly exothermic electron affinities? Explain your answer.

8.18 How do electron affinities vary down a group?

8.19 Explain why the electron affinity of lithium is slightly favorable (exothermic), whereas the electron affinity of beryllium is unfavorable (endothermic). Contrast these trends with the ionization energy trends of these two elements.

8.20 Describe the physical properties of the elements in Group 1A.

8.21 Describe the physical properties of the elements in Group 2A.

8.22 Describe the physical properties of the elements in Group 7A.

8.23 State the reactivity trend down the group for the elements in the following groups.
(a) 1A (b) 2A (c) 7A

8.24 Using Figure 8.18, suggest compounds from Group 1A to put in fireworks that would burn
(a) red. (b) yellow.

Exercises

OBJECTIVE Relate the electron configuration of an element to its location in the periodic table.

8.25 Write the electron configurations of the following elements after finding their locations in the periodic table.
(a) P (b) Sr (c) Sm (d) Ra

8.26 ■ Write the electron configurations of the following elements after finding their locations in the periodic table.
(a) B (b) Bi (c) Ba (d) Cd

8.27 Identify the block of the periodic table where each of the following elements is.
(a) Gd (b) Ir (c) As (d) Sc

8.28 Identify the block of the periodic table where each of the following elements is.
(a) Tl (b) Be (c) Xe (d) U

OBJECTIVE Write the electron configurations of ions.

8.29 Using the abbreviated notation, write the ground-state electron configuration of the following ions.
(a) S^{2-} (b) Mn^{2+} (c) Ge^{2+}

8.30 ■ Using the abbreviated notation, write the ground-state electron configuration of the following ions.
(a) Y^{3+} (b) Br^- (c) Rh^{2+}

8.31 Using the abbreviated notation, write the ground-state electron configuration of the following ions.
(a) P^{2-} (b) Fe^{2+} (c) Co^{3+}

8.32 Using the abbreviated notation, write the ground-state electron configuration of the following ions.
(a) Ga^{2+} (b) Se^{2-} (c) Ru^{2+}

8.33 Identify the cations with a 1+ charge that have the following electron configurations.
(a) $1s^22s^22p^63s^23p^1$
(b) $1s^22s^22p^63s^1$
(c) $1s^22s^22p^63s^23p^64s^23d^{10}$

8.34 ■ Identify the cations with a 2+ charge that have the following electron configurations.
(a) $[Ar]4s^23d^{10}4p^1$
(b) $1s^22s^22p^1$
(c) $1s^22s^22p^63s^23p^1$

8.35 Identify the anions with a 2− charge that have the following electron configurations.
(a) $[Ar]4s^23d^{10}4p^6$
(b) $1s^22s^22p^5$
(c) $1s^22s^22p^63s^23p^6$

8.36 Identify the anions with a 1− charge that have the following electron configurations.
(a) $1s^22s^22p^63s^23p^6$
(b) $1s^22s^22p^5$
(c) $1s^22s^22p^63s^23p^64s^23d^{10}4p^5$

8.37 Write the ground-state electron configurations for Fe^{3+} and Cr^{3+}.

8.38 Which transition-metal ion with a 3+ charge has the ground-state electron configuration $[Kr]4d^5$?

OBJECTIVE Identify an isoelectronic series.

8.39 Write the symbols for a cation and an anion that are isoelectronic with Se.

8.40 Write the symbols for two cations and two anions that are isoelectronic with Kr.

8.41 What neutral atoms are isoelectronic with the following ions?
(a) O^{2-} (b) Fe^{2+} (c) Fe^{3+} (d) In^+

8.42 What neutral atoms are isoelectronic with the following ions?
(a) Pb^{4+} (b) Br^- (c) S^{2-} (d) Ni^{3+}

8.43 Which of the following species are not isoelectronic with the rest: F^-, Ne, Na^+, Ca^{2+}?

8.44 ■ Which of the following species are not isoelectronic with the rest: Ar, K^+, Ca^{2+}, Y^{3+}?

OBJECTIVE Recognize size trends of atoms.

8.45 Which species in each of the following pairs is larger? Explain your answer.
(a) Na or Na^+ (b) O^{2-} or F^- (c) Ni^{2+} or Ni^{3+}

8.46 ■ Select the atom or ion in each pair that has the smaller radius.
(a) Cs or Rb (b) O^{2-} or O (c) Br or As

8.47 Which species in each of the following pairs is larger? Explain your answer.
(a) Na or Mg (b) B or O (c) Be^{2+} or Be^{3+}

8.48 Which species in each of the following pairs is larger? Explain your answer.
(a) Li or Na (b) O or P (c) Rb or Rb^+

8.49 Using only a periodic table as a guide, arrange each of the following series of atoms in order of increasing size.
(a) B, O, Li (b) C, N, Si (c) S, As, Sn

8.50 ■ Using only a periodic table as a guide, arrange each of the following series of atoms in order of increasing size.
(a) Na, Be, Li (b) P, N, F (c) I, O, Sn

8.51 Using only a periodic table as a guide, arrange each of the following series of species in order of increasing size.
(a) Li, Be^{2+}, Be (b) S, S^{2-}, Cl (c) O, S, Si

8.52 Using only a periodic table as a guide, arrange each of the following series of species in order of increasing size.
(a) F, F^-, O^{2-} (b) Al^{3+}, Mg, Na (c) N, P, Si

8.53 Which element in the second period has the largest atomic radius? Why?

8.54 ■ Which is larger, Na^+ or F^-? For each of these ions, draw a representation of the shape of the highest energy occupied orbital.

8.55 Of the atoms with the electron configurations $1s^22s^22p^63s^23p^5$ and $1s^22s^22p^63s^23p^3$, which is larger?

8.56 Of the atoms with the electron configurations $1s^22s^22p^4$ and $1s^22s^22p^2$, which is smaller?

OBJECTIVE Arrange atoms according to ionization energies.

8.57 Indicate which species in each pair has the higher first ionization energy. Explain your answer.
(a) Si or Cl (b) Na or Rb (c) O^{2-} or F^-

8.58 Indicate which species in each pair has the higher ionization energy. Explain your answer.
(a) N or F (b) Mg^{2+} or Na^+ (c) K or Si

8.59 Indicate which species in each pair has the higher ionization energy. Explain your answer.
(a) Ge or Cl (b) B or F (c) Al^{3+} or Na^+

8.60 Indicate which species in each pair has the higher ionization energy. Explain your answer.
(a) K or I (b) Al or Al^+ (c) Cl^- or Ar

8.61 Using only a periodic table as a guide, arrange each of the following series of species in order of increasing first ionization energy.
(a) O, O^{2-}, F (b) C, Si, N (c) Te, Ru, Sr

8.62 Using only a periodic table as a guide, arrange each of the following series of species in order of increasing first ionization energy.
(a) S, Se^{2-}, O (b) Fe, Br, F (c) Cl, Cl^-, F

8.63 Using only a periodic table as a guide, arrange each of the following series of species in order of increasing first ionization energy.
(a) N, N^{3-}, Ne (b) P, Si, Cl (c) Ga, O, Se

8.64 Using only a periodic table as a guide, arrange each of the following series of species in order of increasing first ionization energy.
(a) K, Se, S (b) Cr, As, O (c) O, O^-, F

8.65 Which second ionization energy is greater, that of aluminum or that of magnesium? Explain your answer.

8.66 ■ Predict which of these elements would have the greatest difference between the first and second ionization energies: Si, Na, P, Mg. Briefly explain your answer.

8.67 Which third ionization energy is greater, that of aluminum or that of magnesium? Explain your answer.

8.68 Which second ionization energy is greater, that of oxygen or that of fluorine? Explain your answer.

OBJECTIVE Predict relative energies needed for successive ionization.

8.69 Li^{2+} is not a common charge for a lithium ion. Use the ionization energies of lithium to explain why.

8.70 Suggest a reason for the lack of metal cations with a 4+ charge.

8.71 What is the electron configuration of the Ba^{3+} ion? Suggest a reason why this ion is not normally found in nature.

8.72 On which ionization (first, second, third, fourth, etc.) will the ionization energy of an Al atom increase dramatically? Explain why.

OBJECTIVE State how electron affinity affects formation of anions.

8.73 Indicate which species in each pair has the more negative electron affinity. Explain your answer.
(a) S or Cl (b) N or O (c) S or F

8.74 ■ Indicate which species in each pair has the more favorable (more negative) electron affinity. Explain your answer.
(a) Se or Br (b) S or P (c) Br or As

8.75 Indicate which species in each pair has the more favorable (more negative) electron affinity. Explain your answer.
(a) O or F (b) P or Cl (c) Se or Br

8.76 Indicate which species in each pair has the more favorable (more negative) electron affinity. Explain your answer.
(a) Br or Te (b) O or B (c) In or Se

OBJECTIVE Write equations for some chemistry of Groups 1A, 2A, and 7A.

8.77 Write the equation for the reaction, if any, of sodium with the following substances.
(a) oxygen (b) nitrogen
(c) chlorine (d) water

8.78 Write the equation for the reaction, if any, of lithium with the following substances.
(a) oxygen (b) nitrogen
(c) chlorine (d) water

8.79 Write the equation for the reaction, if any, of barium with the following substances.
(a) oxygen (b) water

8.80 Write the equation for the reaction, if any, of calcium with the following substances.
(a) oxygen (b) water

OBJECTIVE Identify reactivity trends in the elements.

8.81 What is the trend in reactivity as you go down from Li to Cs in the periodic table? Give a reason for this trend.

8.82 What is the trend in reactivity as you go down from F to I in the periodic table? Give a reason for this trend.

8.83 One way to generate hydrogen gas in the laboratory is to combine a reactive metal with water. The ensuing chemical reaction produces hydrogen gas as one product. Which element would generate hydrogen gas more vigorously when combined with water: potassium or calcium? Explain your answer. Write the balanced chemical reaction for both processes.

8.84 The presence of radioactive strontium in nuclear fallout (which is residue from atmospheric testing of nuclear bomb testing) is a major health concern because of strontium's periodic table relationship with calcium, a necessary nutrient. What is the reason for this concern?

Chapter Exercises

8.85 What are the changes in (a) size, (b) ionization energy, and (c) electron affinity from potassium to calcium?

8.86 Write the electron configuration for the 3+ cation of titanium.

8.87 Write the electron configuration for the 2+ cation of calcium.

8.88 One sphere below represents a boron atom, while the other sphere represents an oxygen atom. Select the one that represents the boron atom. Explain your choice.

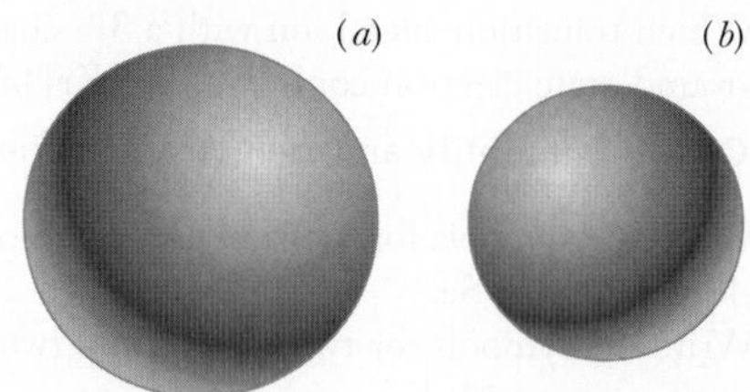

8.89 Three elements have the electron configurations $1s^22s^22p^63s^2$, $[Ar]4s^2$, and $1s^22s^22p^63s^23p^5$. The atomic radii of these elements (not necessarily in the same order) are 99, 160, and 231 pm. Identify the elements from the electron configurations, and match the sizes with these elements.

8.90 Write the electron configuration of the copper atom and the 2+ cation of copper. Remember that the electron configuration of copper atom is unusual. Does the fact that copper is an exception to the aufbau principle influence the electron configuration of Cu^{2+}?

8.91 Palladium, with an electron configuration of $[Kr]4d^{10}$, is an exception to the aufbau principle. Write the electron configuration of the 2+ cation of palladium. Does the fact that palladium is an exception influence the electron configuration of Pd^{2+}?

Johnson Matthey Platinum Today www.platinum.matthey.com

8.92 List the element for which the 2+ cation has the electron configuration $[Ar]3d^4$.

8.93 Determine the number of unpaired electrons for a Cr^{4+} cation.

8.94 Identify which member of Group 5A has the following characteristics:
(a) largest size (b) smallest ionization energy

8.95 Of the following electron configurations:
(1) $1s^22s^22p^1$
(2) $1s^22s^22p^4$
(3) $1s^22s^22p^5$
which represents the element with the
(a) largest size?
(b) smallest ionization energy?
(c) greatest electron affinity?

8.96 The mineral magnetite has the overall formula Fe_3O_4 (as $Fe_2O_3{\cdot}FeO$) and contains both Fe^{2+} and Fe^{3+}. Write the electron configurations of both cations of iron.

8.97 Which of the cations Ga^{4+} and Mn^{4+} is not a stable species? Why?

8.98 Rank the following ions in order of increasing sizes and increasing ionization energies: S^{2-}, K^+, Ca^{2+}.

8.99 Arrange the elements lithium, carbon, and oxygen in order of
(a) increasing size.
(b) increasing first ionization energy.
(c) increasing second ionization energy.
(d) number of unpaired electrons.

Cumulative Exercises

8.100 Chlorine gas can be prepared by passing electricity through a solution of NaCl. What volume of chlorine at standard temperature and pressure can be prepared from 2.44 g NaCl?

8.101 Write the electron configuration and orbital diagram for the noble gas that has a density of 1.62 g/L at 27 °C and 1.00 atm pressure.

8.102 ■ ▲ Write the electron configuration of the alkali metal (M) that reacts with oxygen to yield an oxide, M_2O, if 1.22 g of the metal reacts with 1.41 g of oxygen to form 2.63 g of the oxide.

8.103 ▲ Write an equation for the reaction with water of a metal that, in its ground state, has one electron in a 3*s* orbital. What mass of water is needed to react with 2.34 g of this metal?

8.104 What is the mass percent of a transition metal, M, in a compound of the formula MCl_3 if the electron configuration of the metal cation in this compound is $1s^22s^22p^63s^23p^63d^5$?

8.105 What masses of iron and Cl_2 are needed to prepare 7.88 g of the metal halide product if, under the conditions of the reaction, the electron configuration of the iron cation in the product is $1s^22s^22p^63s^23p^63d^6$?

8.106 Given the following two equations, where the metals (M and M′) are from the second period, what are the metals, and which metal cation in the product has a larger radius?

$$2M + F_2 \rightarrow 2MF$$

$$M' + F_2 \rightarrow M'F_2$$

8.107 Write the electron configuration and orbital diagram for an atom of a gaseous diatomic element that has a density of 1.14 g/L at 27 °C and 1.00 atm pressure.

8.108 ■ ▲ The Case Study in this chapter introduced cesium fluoride, CsF.
(a) A 10.76-g sample of impure CsF was dissolved in 100 mL water. The sample was mixed with excess aqueous calcium nitrate solution to precipitate insoluble calcium fluoride. The precipitate was filtered, dried, and weighed. A total of 2.35 g calcium fluoride was collected. What was the percentage of CsF in the original sample?
(b) Given the reaction

$$CsF(s) + HNO_3(aq) \rightarrow HF(g) + CsNO_3(aq)$$

what volume of HF(g) is produced at standard temperature and pressure (STP) if 100.0 g CsF reacts?
(c) CsF is not only colorless but transparent to infrared light up to a wavelength limit of 15.0 μm. What is the energy of a photon of light having that wavelength? What is the energy of a mole of photons of that wavelength? How many moles of photons are needed to supply the same amount of energy as produced in the reaction given in the Case Study?

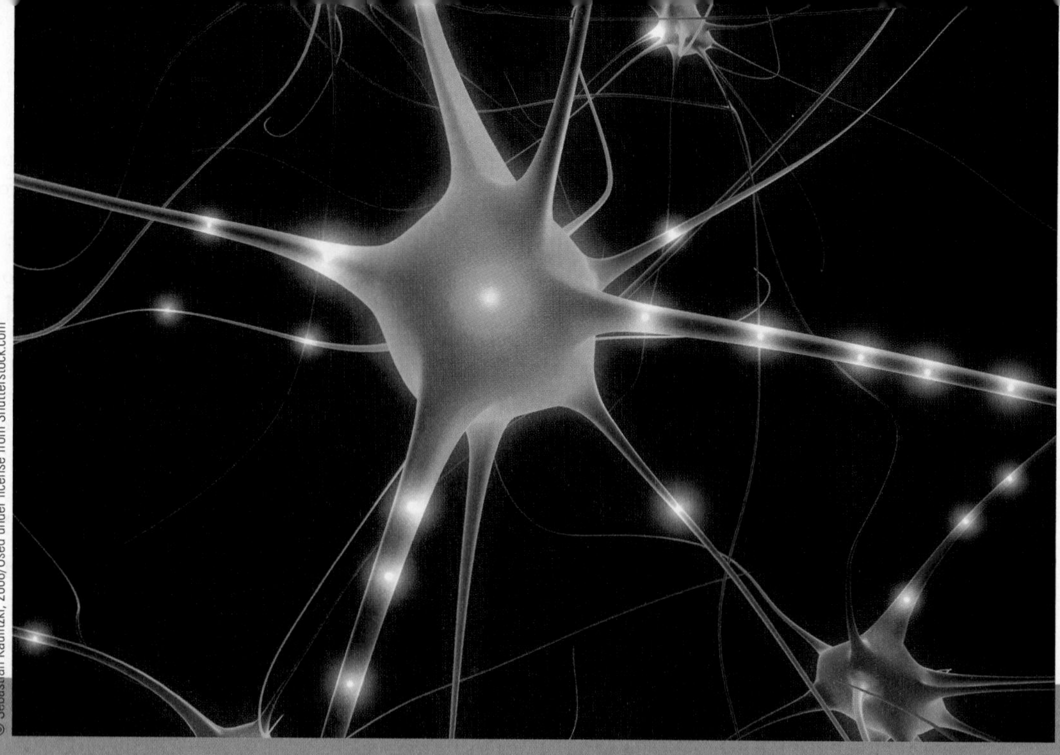

Nerve cell image. The NO molecule can affect nerve cells.

When we think of processes that go on in the human body, we generally consider huge molecules such as DNA or proteins. It was, therefore, a great surprise when, in the mid- to late-1980s, researchers conclusively established a fascinating notion: The molecule NO, a component of smog, actually has a strong physiologic effect in the human body. In fact, the range of biological functions of this simple molecule continues to astound researchers.

Nitrogen monoxide (NO; commonly called *nitric oxide*) is produced in the body by the reaction of the amino acid arginine ($NH_2CHCOOH[CH_2]_3NHCNHNH_2$) with oxygen. This molecule releases NO with the assistance of an enzyme called *nitric oxide synthase* (NOS). One of the major functions of NO is as a vasodilator; it causes the smooth muscle of blood vessels to relax, increasing blood flow. Because of its many important functions, a number of pharmaceuticals have been developed that also deliver NO. People experiencing chest pains caused by angina (a partial blocking of the blood vessels, reducing the flow of blood—and therefore oxygen—to the heart) benefit by taking nitroglycerin tablets. The nitroglycerin promotes the formation of NO in the bloodstream, thus increasing blood flow. Another pharmaceutical that promotes NO production is sildenafil (Viagra).

NO is also an effective neurotransmitter, in part because it is a small molecule (approximately 120 pm) that can diffuse rapidly in all directions, affecting nerve cells that are not even connected to the primary nerve cell synaptically. Neurochemists postulate that NO is important in establishing memory.

NO is important for the body's immune response. In our bodies, cells called *macrophages* release NO as a defense against invading bacteria. Understandably, one side reaction of massive infection is a decline in blood pressure because the larger influx of NO from macrophages works to relax blood vessel walls systemically. Also, physicians recently have begun using NO to help treat sickle cell anemia.

Chemical Bonds

9

CHAPTER CONTENTS

OWL Online homework for this chapter may be assigned in OWL.

Look for the green colored vertical bar throughout this chapter, for integrated references to this chapter introduction.

The excitement around the discovery of the activity of NO in the human body was based partially on the realization that a small molecule played such a crucial role in the body. Not only did *Science,* a major scientific journal, name NO as "Molecule of the Year" in 1992, but the 1998 Nobel Prize in Physiology or Medicine was awarded to a team of three scientists who discovered the signaling activity of NO in the body. Little molecules can do big things! ▮

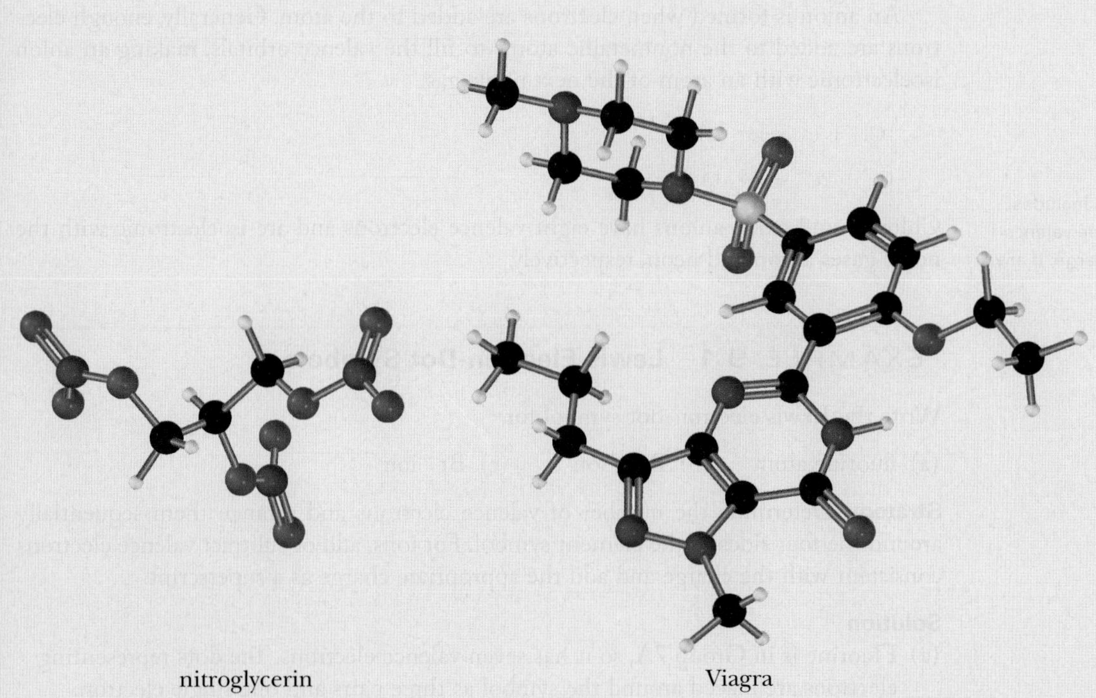

Chemists strive to understand how and why elements combine to form compounds. Now that we have developed a model that relates the electronic structures of atoms and ions to other properties, we will extend this model to include compounds. **Chemical bonds** are the forces that hold the atoms together in substances. We will discuss two main types of chemical bonds that describe the forces holding the atoms together in ionic and molecular compounds.

9.1 Lewis Symbols

OBJECTIVE

☐ Write Lewis electron-dot symbols for elements and ions

It is the valence electrons that form chemical bonds—the inner (core) electrons are held more tightly by the nucleus and are not usually involved in bonding. A convenient and useful way to represent the valence electrons is by Lewis electron-dot symbols, first proposed by the American chemist G. N. Lewis (1875–1946). A **Lewis electron-dot symbol** for an atom consists of the symbol for the element surrounded by dots, one for each valence electron. Figure 9.1 shows Lewis electron-dot symbols for the representative elements through radium. By convention, the first four electron dots are placed sequentially around the four sides of the element symbol, and additional valence electrons form pairs. It does not matter which sides are marked with a dot, as long as the first four go around the four sides of the elemental symbol. For the representative (A group) elements, the number of dots is equal to the group number, so that elements in the same column of the periodic table have similar Lewis electron-dot symbols.

1A	2A	3A	4A	5A	6A	7A	8A
·H							He:
·Li	·Be·	·B·	·C·	·N·	·O·	:F·	:Ne:
·Na	·Mg·	·Al·	·Si·	·P·	·S·	:Cl·	:Ar:
·K	·Ca·	·Ga·	·Ge·	·As·	·Se·	:Br·	:Kr:
·Rb	·Sr·	·In·	·Sn·	·Sb·	·Te·	:I·	:Xe:
·Cs	·Ba·	·Tl·	·Pb·	·Bi·	·Po·	:At·	:Rn:
·Fr	·Ra·						

Figure 9.1 **Lewis electron-dot symbols for the representative elements.**

Lewis electron-dot symbols can also be written for ions. The Lewis electron-dot symbols for many cations show no electrons at all, because the ionization process removes all of the original valence electrons from the atom.

$$\text{Na}\cdot \rightarrow \text{Na}^+ + e^-$$

$$\cdot\text{Ca}\cdot \rightarrow \text{Ca}^{2+} + 2e^-$$

The cations shown are isoelectronic (i.e., contain the same number of electrons) with noble gases—Na^+ with neon and Ca^{2+} with argon.

An anion is formed when electrons are added to the atom. Generally, enough electrons are added to the nonmetallic atoms to fill the valence orbitals, making an anion isoelectronic with an atom of the next noble gas.

$$:\ddot{\underset{\cdot\cdot}{\text{Cl}}}\cdot + e^- \longrightarrow :\ddot{\underset{\cdot\cdot}{\text{Cl}}}:^-$$

$$\cdot\ddot{\underset{\cdot\cdot}{\text{O}}}\cdot + 2e^- \longrightarrow :\ddot{\underset{\cdot\cdot}{\text{O}}}:^{2-}$$

A Lewis electron-dot symbol includes the symbol of the element, the valence electrons as dots, and the charge, if any.

Chloride and oxide anions have eight valence electrons and are isoelectronic with the noble gases argon and neon, respectively.

EXAMPLE 9.1 Lewis Electron-Dot Symbols

Write the Lewis electron-dot symbol for

(a) fluorine atom (b) Be^{2+} ion (c) Br^- ion

Strategy Determine the number of valence electrons and arrange them sequentially around the four sides of the element symbol. For ions, add or subtract valence electrons consistent with the charge and add the appropriate charge as a superscript.

Solution

(a) Fluorine is in Group 7A, so it has seven valence electrons. The dots representing electrons are placed around the symbol as three pairs and one single electron.

$$:\ddot{\underset{\cdot\cdot}{\text{F}}}\cdot$$

(b) The beryllium atom in Group 2A has two valence electrons, and both of them are lost in the formation of the 2+ ion. The Lewis electron-dot symbol is the same as the symbol of the ion

$$Be^{2+}$$

(c) Bromine atom is in Group 7A and has seven valence electrons. The 1− charge indicates an additional electron giving it eight and making it isoelectronic with Kr.

$$:\ddot{\underset{..}{Br}}:^{-}$$

Understanding

What are the Lewis electron-dot symbols for a tin atom and the Sn^{2+} ion?

Answer

$$\cdot\dot{\underset{\cdot}{Sn}}\cdot \qquad \cdot Sn\cdot^{2+}$$

OBJECTIVES REVIEW *Can you:*

☑ write Lewis electron-dot symbols for elements and ions?

9.2 Ionic Bonding

OBJECTIVES

- ☐ Represent the formation of ionic compounds through the use of Lewis symbols
- ☐ Define lattice energy
- ☐ Describe how the charges and the sizes of ions influence lattice energies
- ☐ Explain how the correlations between ionization energies and lattice energies predict the charges on Groups 1A, 2A, 6A, and 7A elements when they form ionic compounds.

As outlined in Section 8.6, metals in Groups 1A and 2A tend to react with the nonmetals in Groups 6A and 7A, and form ionic compounds. Electrons are easily removed from Groups 1A and 2A metals because they have low ionization energies; electrons are easily added to nonmetals because their electron affinities are generally favorable. **Ionic bonding** is the bonding that results from electrostatic attraction between positively charged cations and negatively charged anions. The formation of binary ionic compounds can be represented through the use of Lewis electron-dot symbols that show how electrons are lost from metals and gained by nonmetals.

$$\underset{[He]2s^1}{Li\cdot} + \underset{[Ne]3s^23p^5}{\cdot\ddot{\underset{..}{Cl}}:} \longrightarrow \underset{[He]}{Li^+} + \underset{[Ar]}{:\ddot{\underset{..}{Cl}}:^-} \quad \text{(or LiCl)}$$

$$\underset{[Ne]3s^1}{Na\cdot} + \underset{[He]2s^22p^4}{\cdot\ddot{\underset{..}{O}}\cdot} + \underset{[Ne]3s^1}{\cdot Na} \longrightarrow \underset{[Ne]}{2Na^+} + \underset{[He]\,2s^22p^6 \text{ or } [Ne]}{:\ddot{\underset{..}{O}}:^{2-}} \quad \text{(or } Na_2O\text{)}$$

Each of the lithium and sodium atoms loses one electron to form a cation with a 1+ charge. The chlorine atom has seven valence electrons and gains one electron, filling its valence orbitals. Oxygen fills its valence orbitals, gaining two electrons to form the oxide ion. In forming Na_2O, two sodium atoms must transfer one electron each to supply the two electrons to oxygen.

When an ionic compound forms, the number of electrons lost in forming the cation(s) must equal the number gained to form the anion(s).

EXAMPLE 9.2 Lewis Electron-Dot Symbols of Ionic Compounds

Use Lewis electron-dot symbols to show the formation of (a) magnesium oxide and (b) calcium fluoride from the atoms.

Strategy Move the valence electrons from the metal to the nonmetal, placing eight around each nonmetal, and indicate the charges as superscripts.

Solution

(a) The magnesium atom has two valence electrons that it readily loses. The oxygen atom has six valence electrons and can gain two more electrons to fill its valence orbitals.

$$\cdot\text{Mg}\cdot + \cdot\ddot{\underset{\cdot\cdot}{\text{O}}}\cdot \longrightarrow \text{Mg}^{2+} + :\ddot{\underset{\cdot\cdot}{\text{O}}}:^{2-} \quad (\text{or MgO})$$

(b) The calcium atom has two valence electrons to lose, but the fluorine atom with seven valence electrons can gain only one additional electron to fill its valence shell, so two fluorine atoms are needed and two fluoride ions are formed.

$$\cdot\text{Ca}\cdot + \cdot\ddot{\underset{\cdot\cdot}{\text{F}}}: + \cdot\ddot{\underset{\cdot\cdot}{\text{F}}}: \longrightarrow \text{Ca}^{2+} + 2:\ddot{\underset{\cdot\cdot}{\text{F}}}:^{-} \quad (\text{or } \text{CaF}_2)$$

Understanding

Use Lewis electron-dot symbols to show the formation of lithium sulfide from the atoms.

Answer

$$\text{Li}\cdot + \text{Li}\cdot + \cdot\ddot{\underset{\cdot\cdot}{\text{S}}}\cdot \longrightarrow 2\text{Li}^{+} + :\ddot{\underset{\cdot\cdot}{\text{S}}}:^{2-} \quad (\text{or } \text{Li}_2\text{S})$$

Lattice Energy

When sodium metal is added to a container of chlorine, an exothermic reaction takes place (Figure 9.2).

$$\text{Na(s)} + \tfrac{1}{2}\text{Cl}_2\text{(g)} \rightarrow \text{NaCl(s)} \qquad \Delta H = -411 \text{ kJ}$$

Many of the properties of ionic compounds can be explained by examining the factors that cause this reaction to be highly exothermic. The reaction can be viewed as the sum of five simpler reactions (i.e., we are applying Hess's law), a process known as a *Born–Haber cycle.*

1	$\text{Na(s)} \rightarrow \text{Na(g)}$	$\Delta H_1 = 107$ kJ
2	$\text{Na(g)} \rightarrow \text{Na}^+\text{(g)} + \text{e}^-$	$\Delta H_2 = 496$ kJ
3	$\tfrac{1}{2}\text{Cl}_2\text{(g)} \rightarrow \text{Cl(g)}$	$\Delta H_3 = 121$ kJ
4	$\text{Cl(g)} + \text{e}^- \rightarrow \text{Cl}^-\text{(g)}$	$\Delta H_4 = -349$ kJ
5	$\text{Na}^+\text{(g)} + \text{Cl}^-\text{(g)} \rightarrow \text{NaCl(s)}$	$\Delta H_5 = -786$ kJ
	$\text{Na(s)} + \tfrac{1}{2}\text{Cl}_2\text{(g)} \rightarrow \text{NaCl(s)}$	$\Delta H = \Delta H_1 + \Delta H_2 + \Delta H_3 + \Delta H_4 + \Delta H_5 = -411$ kJ

The first three steps are always endothermic, whereas the fourth step is exothermic in this case. Different ionic compounds may have different values for the energy changes of these steps, and their overall combination may be endothermic or exothermic. However, the final step—the bringing together of positive and negative ions to make the solid ionic compound—is *always* exothermic.

The reverse of equation 5 is used to define a quantity called the **lattice energy**—the energy required to separate one mole of an ionic crystalline solid into the isolated gaseous ions (Figure 9.3).

$$\text{MX(s)} \rightarrow \text{M}^+\text{(g)} + \text{X}^-\text{(g)} \qquad \Delta H = \text{lattice energy}$$

Lattice energies are difficult to measure and are generally calculated from the Born–Haber cycle, using the measured enthalpy of formation of the ionic solid.

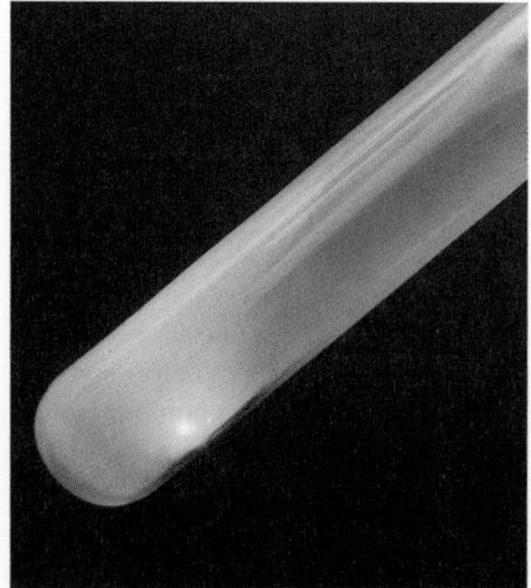

Figure 9.2 Reaction of sodium and chlorine. Sodium metal reacts violently with chlorine gas to form sodium chloride.

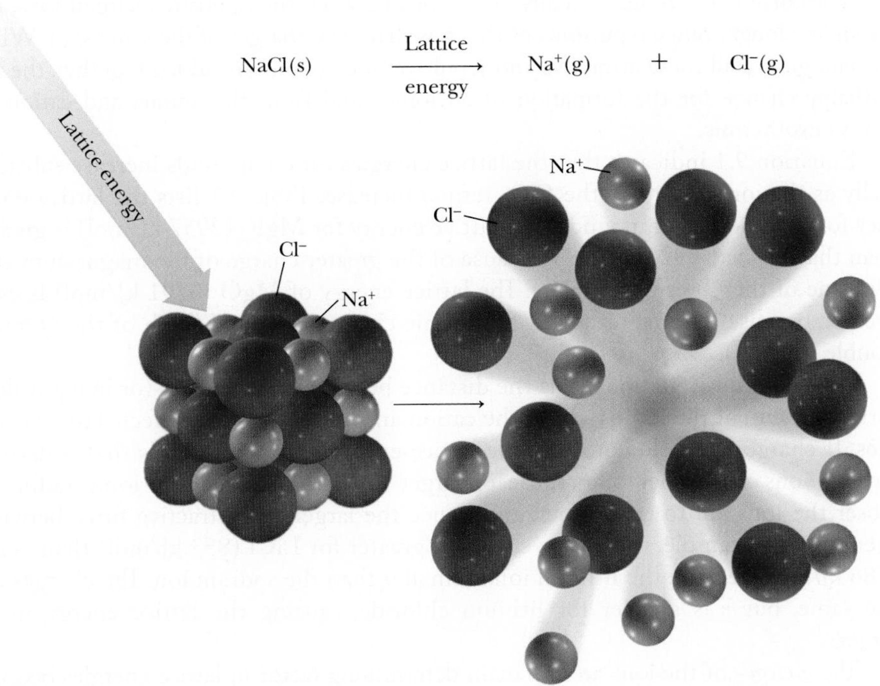

Figure 9.3 Lattice energy. Lattice energy is the energy required to separate one mole of an ionic crystalline solid into isolated gaseous ions.

The magnitude of the lattice energy for any given ionic solid is described in Equation 9.1:

$$E = \frac{kQ_1Q_2}{r} \qquad [9.1]$$

where k is a constant, Q_1 and Q_2 are the charges on the two particles, and r is their distance of separation in the compound. Energy is released if the particles come together and have opposite signs for their charges (an exothermic process), and is absorbed if the charges are of the same sign (an endothermic process).

The formation of an ionic solid involves many interactions, because many cations and anions come together and, as discussed in Chapter 2, form a three-dimensional arrangement of alternating positive and negative ions. The arrangement of the lattice depends on the relative sizes and charges of the ions that compose the lattice. Figure 9.4 shows the arrangement of the ions in sodium chloride (NaCl) and in cesium chloride (CsCl). In solid sodium chloride, each sodium cation is surrounded by six chloride anions, and each chloride is surrounded by six sodium cations. In the cesium chloride structure, each cation or anion is surrounded by eight of the oppositely charged ions.

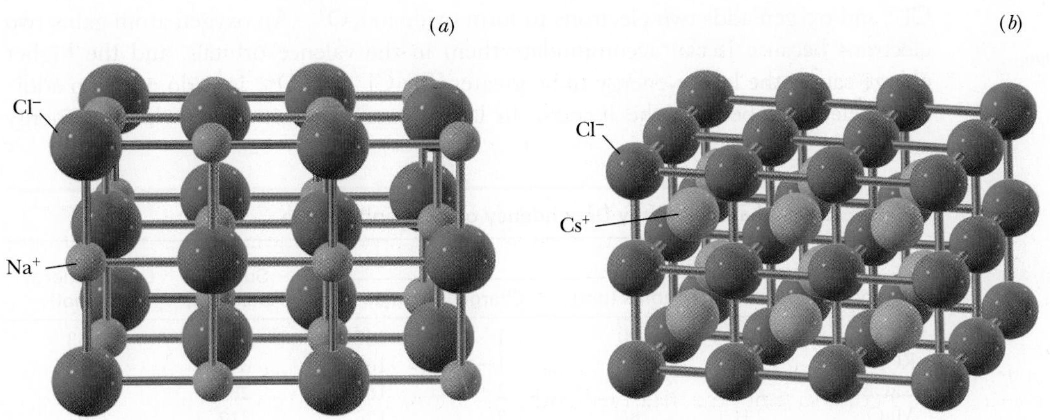

Figure 9.4 Ionic structures of sodium chloride (NaCl) and cesium chloride (CsCl). In NaCl, each cation is surrounded by six anions *(a)*, but in CsCl, each cation is surrounded by eight anions *(b)*.

Ionic solids are stable because of high lattice energies.

These arrangements result in many attractions between the opposite charged ions, but also many longer range repulsions of the ions that have charges of the same sign. When the energies of all these attractive and repulsive interactions are added together, the net enthalpy change for the formation of an ionic solid from the anions and cations is always exothermic.

Equation 9.1 indicates that the lattice energies for compounds increase substantially as the ionic charges (the Q_1Q_2 terms) increase. Table 9.1 lists the lattice energies for some ionic compounds. The lattice energy for MgF_2 (2957 kJ/mol) is greater than that for NaF (923 kJ/mol) because of the greater charge of the magnesium cation (one of the Q terms doubles). The lattice energy of MgO (3791 kJ/mol) is even greater, because the magnitude of both ionic charges is now 2 (both of the Q terms double vs. NaF).

TABLE 9.1 Lattice Energies

Compound	Lattice Energy (kJ/mol)
LiF	1036
LiCl	853
LiBr	807
LiI	757
NaF	923
NaCl	786
NaBr	747
NaI	704
KF	821
KCl	715
KBr	682
KI	649
CsCl	657
MgF_2	2957
$MgCl_2$	2526
$MgBr_2$	2440
MgI_2	2327
$CaCl_2$	2258
Na_2O	2481
MgO	3791
CaO	3401

The r term in Equation 9.1 is the distance between the two ions (or in equivalent terms the sum of the ionic radii of the cation and anion) and also affects lattice energies. If charges are held constant, the lattice energies for ionic solids that consist of smaller ions are higher than those of larger ions. The smaller the ionic radii, the closer the ions are to each other, and hence the larger the attractive force between the ions. For example, the lattice energy is greater for LiCl (853 kJ/mol) than NaCl (786 kJ/mol) because the lithium ion is smaller than the sodium ion. The charges are the same, but r is smaller for lithium chloride, causing the lattice energy to be larger.

The charges of the ions are the main determining factor in lattice energies because size differences are generally small, whereas the charges change by factors of two or greater. As seen in Table 9.2, the lattice energy change from LiCl to the larger cation in KCl is a reduction of only 138 kJ/mol. In contrast, from LiCl to a larger but more highly charged cation in $CaCl_2$, the difference is 1405 kJ/mol. When evaluating the lattice energy of MgO, where both ions have higher charges than LiCl, the difference is even greater. The lattice energy for MgO is the largest in Table 9.2 because of both the high charges of the ions and their small sizes.

The lattice energies of pairs of ions explain many factors of ionic bonding. For instance, why does a metal such as magnesium lose two electrons, rather than just one, when it forms an ionic compound? The first ionization energy of magnesium is 738 kJ/mol, and the second is 1450 kJ/mol. The answer is that the increase in lattice energy obtained from the more highly charged ion is more than the cost of the additional energy needed to form it. As shown in Table 9.1, the lattice energies of MgX_2 (X = halide) compounds are considerably larger than those of Group 1A halide compounds, because of the greater charge on magnesium. These attractive forces would increase again if magnesium was to lose a third electron, but this process does not occur because the third ionization energy, the energy needed to remove a core electron, is extremely high at 7734 kJ/mol. The increase in lattice energy is not nearly as great as the additional energy needed to remove a core electron.

Lattice energies are greatest for compounds made from higher charged, small ions.

Atoms of elements on the right side of the periodic table gain electrons until their valence orbitals are filled. For example, chlorine adds one electron to form chloride ion, Cl^-, and oxygen adds two electrons to form oxide ion, O^{2-}. An oxygen atom gains two electrons because it can accommodate them in the valence orbitals, and the higher charge causes the lattice energy to be greater. The Cl^- and O^{2-} ions do not gain additional electrons because the increase in lattice energy is not as great as the energy

Group 2A metals lose both valence electrons when forming ionic compounds.

TABLE 9.2 Lattice Energy Dependency on Size and Charge

	Cation		Anion			
Compound	Charge	Radius (pm)	Charge	Radius (pm)	Sum of Radii (pm)	Lattice Energy (kJ/mol)
LiCl	1+	90	1−	167	257	853
KCl	1+	152	1−	167	319	715
$CaCl_2$	2+	114	1−	167	281	2258
MgO	2+	86	2−	126	212	3791

needed to add an additional electron to the anions, which is large because their valence orbitals are full.

Elements in Groups 6A and 7A gain enough electrons to fill the valence orbitals.

EXAMPLE 9.3 Lattice Energies

Explain why the lattice energy of Na_2O is considerably greater than that of NaF.

Strategy Lattice energies are evaluated by using the relationship

$$E = \frac{kQ_1Q_2}{r}$$

In general, the changes that can take place with the Q terms are more important than small changes in r.

Solution

The lattice energy of Na_2O is greater because the greater charge on O^{2-} than on F^- leads to a Q term that is twice as large. Although the F^- ion is smaller than O^{2-}, a change that increases the lattice energy, doubling one of the Q terms, has a greater impact on the change in lattice energy when compared with small changes in the sizes of the ions.

Understanding

Explain why the lattice energy of KCl is greater than that of KI.

Answer The charges of the ions in these two compounds are the same, but Cl^- is smaller than I^-, decreasing r and thus increasing E for KCl.

OBJECTIVES REVIEW *Can you:*

- ☑ represent the formation of ionic compounds through the use of Lewis symbols?
- ☑ define lattice energy?
- ☑ describe how the charges and the sizes of ions influence lattice energies?
- ☑ explain how the correlations between ionization energies and lattice energies predict the charges on Groups 1A, 2A, 6A, and 7A elements when they form ionic compounds?

9.3 Covalent Bonding

OBJECTIVES

- ☐ Define covalent bonding
- ☐ Write Lewis structures of molecules and polyatomic ions

Compounds that contain only nonmetals usually consist of isolated molecules that are generally gases, liquids, or low melting solids. For example, at room temperature, benzene, C_6H_6, is a liquid, and nitrogen dioxide, NO_2, is a gas that will condense into a liquid only at low temperatures (Figure 9.5). Room temperature water can be changed from the liquid to either the solid or gas phase at atmospheric pressure simply by changing its temperature to less than 0 °C or to more than 100 °C. By comparison, ionic compounds are generally hard crystalline solids with high melting and boiling points. These dramatic differences in physical properties indicate that we need a second bonding model to explain the attractive forces in molecules.

The bonding in the simplest molecule, H_2, is described by the *sharing* of the two electrons. By sharing, both electrons are attracted by each nucleus. *Each* nucleus is considered to have *gained* a share of both electrons and now has the helium noble-gas electron configuration. Figure 9.6 depicts the sharing process graphically. No new electrons have been added, but the two shared electrons can be counted as being under the influ-

(*a*)

(*b*)

(*c*)

Figure 9.5 Ionic and molecular compounds. Ionic compounds, such as $Ni(NO_3)_2 \cdot 6H_2O$ *(a),* are generally brittle solids. Many molecular compounds formed from nonmetallic elements are liquids, such as benzene, C_6H_6 *(b),* or gases, such as nitrogen dioxide, NO_2 *(c).*

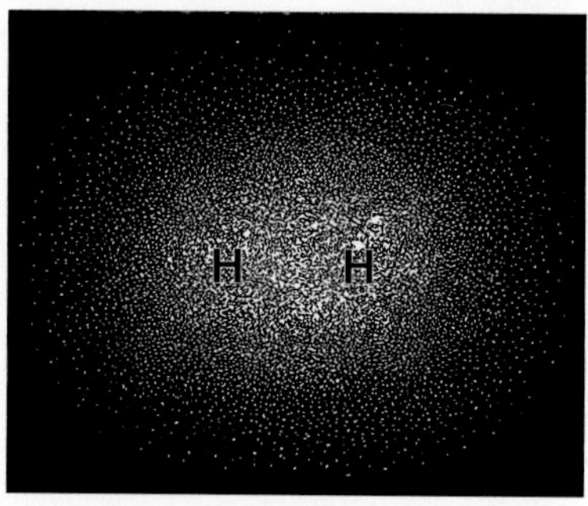

Figure 9.6 Sharing of electrons in H_2. The two nuclei in H_2 are held together by shared electrons.

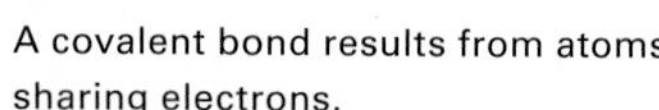

A covalent bond results from atoms sharing electrons.

ence of both nuclear charges. A **covalent bond** is a bond that results from atoms sharing electrons.

The diagram in Figure 9.7 shows how the energy of interaction of two hydrogen atoms changes as the distance between them varies. The energy of interaction is zero when the atoms are far apart. As the atoms approach each other, the electron clouds start to overlap. Both electrons are attracted by the positive charges of both nuclei. The overall energy of the atoms decreases as the two nuclei and their electrons approach each other; this sharing is thus an exothermic process. The energy continues to decrease as the overlap increases, until at very short distances the repulsion between the *positively* charged nuclei begins to dominate, increasing the energy. The distance that corresponds to the lowest energy is the distance at which the molecule is most stable. The **bond length** is the minimum energy distance between the nuclei of two bonded atoms in a molecule, as

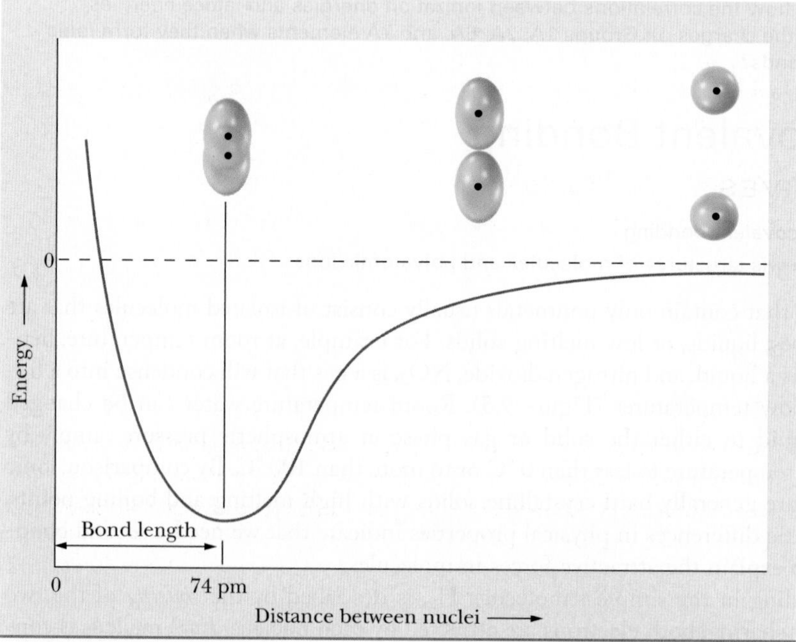

Figure 9.7 Potential energy curve for H_2. Two hydrogen atoms are most stable when their electron clouds overlap (shown on the left) to form a covalent bond. The isolated hydrogen atoms (shown on the right) are at higher energy. Completely separated atoms have zero energy because they are too far away to interact with each other.

PRACTICE OF CHEMISTRY
Chemical Bonding and Gilbert N. Lewis

Few people have had the impact on chemistry that Gilbert N. Lewis had. Not only did Lewis build the chemistry department at the University of California at Berkeley into an internationally known center of chemical discovery, but Lewis himself made several seminal contributions to our understanding of atoms, molecules, and electrons. Most importantly, it was Lewis who established the key role of electrons in chemical bonding.

In 1916, Lewis published an article titled "The Atom and the Molecule." In it, Lewis proposed that the chemical bond—named a "covalent bond" by contemporaneous researcher Irving Langmuir—is a pair of electrons shared by two atoms. Lewis fashioned two models to represent covalent bonds. In the first model, the potential octet was represented as electrons at the corners of a cube. Atoms whose valence shells were missing electrons could accept one or more electrons into its cube, ultimately getting all eight corners filled—a complete octet. In a second model, Lewis envisioned pairs of electrons occupying the corners of a tetrahedron. The sharing of one corner between two atoms constitutes a single bond; sharing two or three corners of two tetrahedral would make double or triple covalent bonds (discussed later).

Within about 20 years, Lewis's ideas would be superseded by new, wave-based descriptions of electrons and bonding. However, Lewis dot structures continue to have an enormous impact on our basic understanding of chemical bonding. This simple and elegant model is quite successful at predicting most of the major features of structure and bonding. ▮

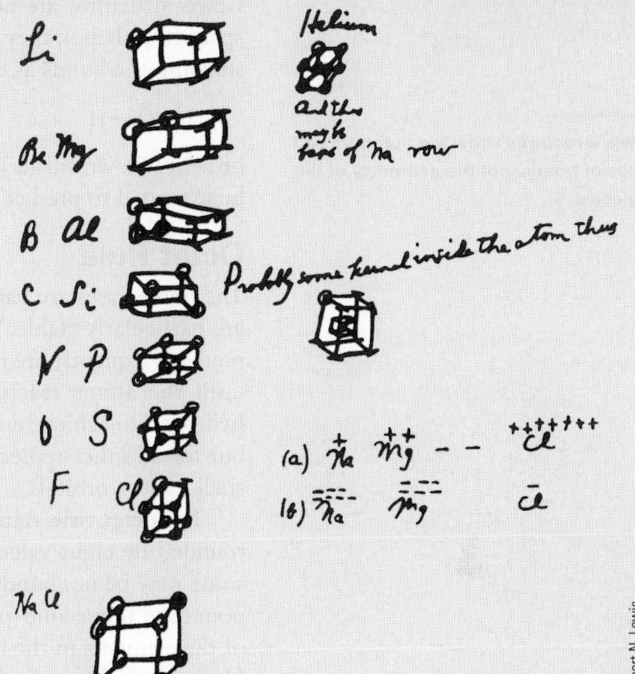

Gilbert N. Lewis

shown in Figure 9.7. In the H_2 molecule, this distance is 74 pm. For a covalent bond to form, nuclei must be close enough together for their electron clouds to overlap and concentrate electron density between them, minimizing the total energy.

A covalent bond forms because two atoms sharing electrons are lower in energy than the two isolated atoms.

Lewis Structures of Molecules

The Lewis electron-dot symbol notation can be used to describe the covalent bonds in molecules.

H· + ·C̤̈l: ⟶ H:C̤̈l:

:F̤̈· + ·F̤̈: ⟶ :F̤̈:F̤̈:

H· + ·Ö̤· + ·H ⟶ H:Ö̤:H

Two types of pairs of electrons appear in these structures. **Bonding pairs** of electrons are shared between two atoms. **Lone,** or **nonbonding, pairs** of electrons are entirely on one atom and are not shared. A molecule of water has both types of electron pairs.

H:Ö:
   H̤

Bonding pairs — Lone pairs

Frequently, a line is used instead of two dots to indicate a bonding pair.

H—Ö:
   |
   H

Drawings of this type are known as **Lewis structures,** representations of covalent bonding in which Lewis symbols show how the *valence electrons* are present in the molecule. Although Lewis structures are simple models of bonding, they show remarkable agreement with the experimentally determined structures. Chemists use Lewis structures to predict bonding and the chemical properties of a large number of compounds. Lewis structures are not intended to show the positions of the atoms or electrons in space (the shape); they show the numbers and types of bonds. Lewis structures of water showing the bonds as either linear

$$H-\ddot{\underset{..}{O}}-H$$

or bent (shown earlier) are both correct. Chapter 10 outlines how Lewis structures can be extended to predict the molecular shape.

Lewis structures show the numbers and types of bonds, not the geometry of the molecule.

Octet Rule

The noble gases are particularly unreactive, indicating that their electron arrangements are particularly stable. We have seen that the ions such as Na^+ and Cl^- in ionic compounds frequently are isoelectronic with the noble gases. Similarly, covalent bonds form until the atoms reach a noble-gas electron configuration by sharing electrons. For hydrogen to achieve a noble-gas electron configuration, only two electrons are needed, but for all other representative elements, eight electrons are needed to fill the valence shell *s* and *p* orbitals.

The **octet rule** states that each atom in a molecule shares electrons until it is surrounded by eight valence electrons. Some of the electrons may be bonding electrons; some may be nonbonding (lone-pair) electrons. The octet rule is most useful for compounds of the second-period elements. We will see later that there are many compounds of the elements in the third and higher periods for which the octet rule is not followed. Hydrogen, of course, shares only two electrons because the 1*s* orbital in the hydrogen atom can only hold two electrons.

The octet rule is followed in the Lewis structures of H_2O, NH_3, and CH_4.

```
H—Ö:          H—N̈—H          H
  |             |             |
  H             H          H—C—H
                              |
                              H
```

An oxygen atom has six valence electrons, and after sharing two more electrons with two hydrogen atoms, it has filled its octet. Nitrogen has five valence electrons and can share three more electrons, and carbon has four valence electrons and can share four more electrons to attain an octet. In most molecules that contain oxygen, each oxygen atom forms two covalent bonds, which complete the octet of valence electrons. In a similar manner, nitrogen usually forms three covalent bonds, and carbon usually forms four covalent bonds.

In most Lewis structures, oxygen makes two bonds and has two lone pairs; nitrogen makes three bonds and has one lone pair; carbon makes four bonds and has no lone pairs.

Two atoms can share more than one pair of electrons between them. For example, consider molecular nitrogen, N_2. For both nitrogen atoms to attain an octet, each atom must share three additional electrons, a total of six for both.

$$\cdot\dot{\underset{..}{N}}\cdot + \cdot\dot{\underset{..}{N}}\cdot \longrightarrow :N:::N: \text{ or } :N{\equiv}N:$$

The sharing of one pair of electrons is a **single bond,** the sharing of two pairs is a **double bond,** and the sharing of three pairs, as shown of N_2, is a **triple bond.** CO_2 is an example of a compound that contains double bonds. In this case, each oxygen atom shares two electron pairs with the carbon atom. These four bonds also place an octet of electrons around the carbon atom.

$$\cdot\ddot{\underset{..}{O}}\cdot + \cdot\dot{C}\cdot + \cdot\ddot{\underset{..}{O}}\cdot \longrightarrow \ddot{\underset{..}{O}}::C::\ddot{\underset{..}{O}} \text{ or } \ddot{\underset{..}{O}}{=}C{=}\ddot{\underset{..}{O}}$$

The **bond order** is the number of electron pairs that are shared between two atoms. The bond order in a double bond such as each of the carbon-oxygen bonds in CO_2 is two; a triple bond such as the one in N_2 has a bond order of three. As the bond order between two atoms increases, the strength of the bond increases and the bond distance generally decreases. As an example, the average bond distances measured in compounds with single and double bonds between two nitrogen atoms and the triple-bond distance in N_2 are

N−N	N=N	N≡N
147 pm	125 pm	110 pm

Writing Lewis Structures

The Lewis structure of a molecule is useful because this relatively simple model describes the bonding. To write a correct structure, we need to know the formula of the compound and which atoms are connected—that is, which atoms are sharing electrons. In H_2O, the two hydrogen atoms are bonded to the oxygen atom but not to each other. The bonding arrangement in this simple molecule is not difficult to figure out, because each hydrogen atom can make only one bond. In other molecules, the connectivity is a more difficult problem that must be determined by experiment. The **skeleton structure** shows which atoms are bonded to each other in a molecule; each connection represents *at least* a single bond. In this textbook, we generally give the skeleton structure, but after you gain some experience in writing Lewis structures and learn a few general rules, you can frequently write the skeleton structure yourself.

Most compounds consist of a **central atom,** an atom bonded to two or more other atoms, and several terminal atoms. Hydrogen and fluorine must always be *terminal* atoms because each of these atoms can make only one covalent bond. The same is generally true of the other halogens (exceptions are covered in Section 9.7). The formula often provides a hint because the central atom is generally named first, such as in NO_2 and SO_3. When in doubt, the central atom is generally the element that is located toward *the bottom and left side in the periodic table*—the element with the lowest ionization energy. For compounds with more than one central atom, the molecular formula is frequently written to indicate the connectivity and help in writing the skeleton structure. For example, ethanol is often written as CH_3CH_2OH to show that the skeleton structure is

```
    H   H
    |   |
H — C — C — O — H
    |   |
    H   H
```

The following steps can be followed to determine the Lewis structure for a molecule.

Steps for Writing Lewis Structures

1. Write the skeleton structure with single bonds between all bonded atoms.
2. Sum the valence electrons of the atoms in the molecule. For cations, subtract one electron for each positive charge; for anions, add one electron for each negative charge.
3. Subtract two electrons for each bond in the skeleton structure.
4. Count the number of electrons needed to satisfy the octet rule for each atom in the structure, remembering that hydrogen needs only two electrons.
   (a) If the number of electrons needed equals the number remaining, go to 5.
   (b) If the number of electrons needed is less than the number remaining, add *one* bond for every *two* additional electrons needed between atoms that have not obtained an octet.
5. After accounting for electrons used in step 4b, if it is necessary, place the remaining electrons as lone pairs on atoms that need them to satisfy the octet rule.

Lewis structures are written by arranging the valence electrons as bond pairs and lone pairs to place an octet of electrons around each atom (two for hydrogen).

This procedure is an organized method of arranging the available valence electrons so that there is an octet for each atom. After writing a Lewis structure, check to make sure that the structure shows the *correct number of valence electrons.* In deciding where to place multiple bonds, be careful to place no more than eight electrons around an atom from the second period and only two electrons on each hydrogen atom.

These rules can be applied to water to obtain the Lewis structure given earlier.

1. Only one skeleton structure is possible. Hydrogen atoms make only one bond apiece.

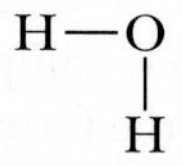

```
H—O
  |
  H
```

2. The total number of valence electrons is

1(O)	$1 \times 6 = 6$
2(H)	$2 \times 1 = 2$
	Total = 8

3. The skeleton structure shows two bonds. Each bond uses two electrons, for a total of four.

$$\begin{aligned} \text{total number of valence electrons} &= 8 \\ -\ \text{electrons used in skeleton structure} &= 4 \\ \hline \text{remaining valence electrons} &= 4 \end{aligned}$$

4. To obey the octet rule, the oxygen atom needs four unshared electrons.

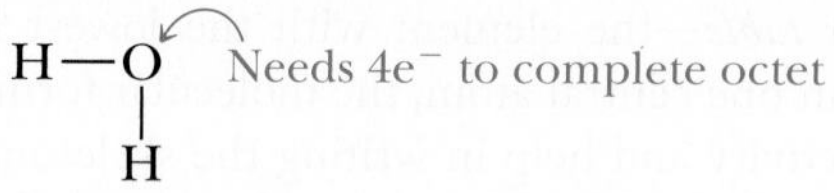

```
H—O  Needs 4e⁻ to complete octet
  |
  H
```

Four is the number of electrons remaining.

5. Satisfy the octet rule with lone pairs.

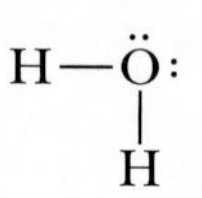

```
    ..
H—O:
  |
  H
```

This Lewis structure is the same as that shown earlier.

EXAMPLE 9.4 Lewis Structure of BF_4^-

Write the Lewis structure of BF_4^-.

Strategy Write the skeleton structure, sum the valence electrons, take into account the number of bonds in the skeleton structure, and determine whether the structure can be finished by lone pairs. For this polyatomic ion, one electron is added to the valence electron count to account for the negative charge.

Solution

1. Fluorine makes only one bond, like hydrogen, making B the central atom. The skeleton structure is

```
    F
    |
F—B—F
    |
    F
```

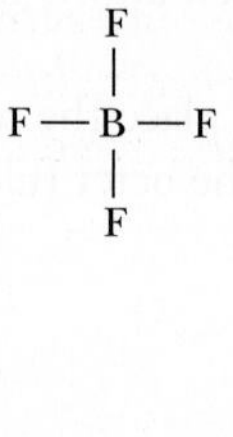

2. The total number of valence electrons is

1(B)	1 × 3 = 3
4(F)	4 × 7 = 28
negative charge	= 1
	32

3. The skeleton structure shows four B-F bonds. Each of these bonds uses two electrons, for a total of eight electrons.

total number of valence electrons	= 32
− electrons used in skeleton structure	= 8
remaining valence electrons	= 24

4. Calculate the number of electrons needed to satisfy the octet rule. The boron atom has four bonds, and thus already satisfies the octet rule. Each fluorine atom has one bond and requires six electrons to complete the octet for a total of 24, the number that are available.

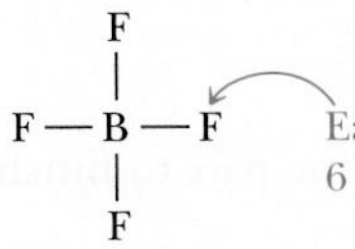

```
      F
      |
  F — B — F     Each fluorine needs
      |         6 e⁻ to complete octet
      F
```

5. The 24 electrons are added as lone pairs on the fluorine atoms to finish the Lewis structure. The whole structure is placed in brackets with a superscript minus sign to show the charge.

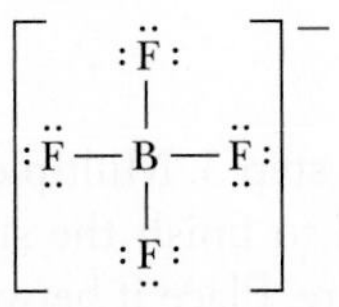

```
 ┌      ..      ┐ −
 |     :F:      |
 |  ..  |   ..  |
 | :F — B — F:  |
 |  ..  |   ..  |
 |     :F:      |
 └      ..      ┘
```

Always check that each atom has an octet and that the final Lewis structure has the correct number of valence electrons. For BF_4^-, there are four bonding pairs and 12 lone pairs of electrons for a total of 32 electrons, as calculated in step 2.

Understanding

Write the Lewis structure of dimethyl ether, CH_3OCH_3.

Answer

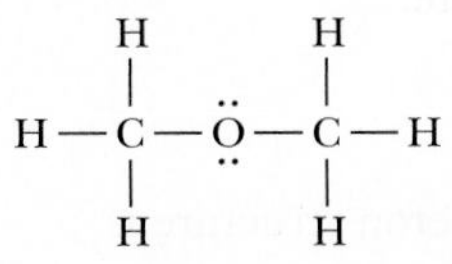

```
    H       H
    |  ..   |
H — C — O — C — H
    |  ..   |
    H       H
```

EXAMPLE 9.5 Lewis Structure of Ethylene

Write the Lewis structure of ethylene, CH_2CH_2.

Strategy Write the skeleton structure, sum the valence electrons, take into account the number of bonds in the skeleton structure, and determine whether the structure can be finished by lone pairs. If there are fewer electrons than needed, add one bond for every two additional electrons needed.

Solution

1. The skeleton structure is given by the formula and the fact that hydrogen atoms make only one bond.

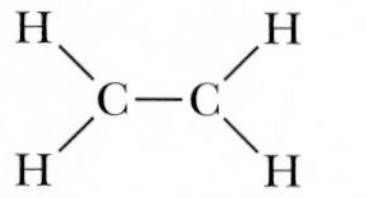

```
H       H
 \     /
  C—C
 /     \
H       H
```

2. The total number of valence electrons is

2(C)	$2 \times 4 = 8$
4(H)	$4 \times 1 = 4$
	12

3. The skeleton structure shows 5 bonds that use 2 electrons each, for a total of 10.

$$\begin{aligned} \text{total number of valence electrons} &= 12 \\ -\ \text{electrons used in skeleton structure} &= 10 \\ \hline \text{remaining valence electrons} &= 2 \end{aligned}$$

4. Each carbon atom has three bonds; thus, each needs one lone pair to finish the structure.

```
H       H
 \     /
  C—C
 /     \
H       H
```

Needs 2e⁻, 2e⁻ to complete octet

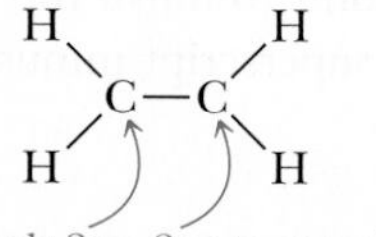

Four electrons are needed, but only two electrons remain in step 3. Multiple bonds are necessary. We have one pair of electrons less than we need to finish the structure with lone pairs. One double bond is needed to finish the structure. Place it between the carbon atoms, because the hydrogen atoms can make only one bond.

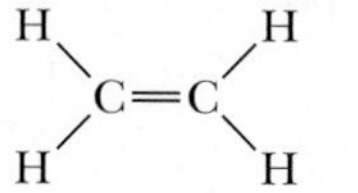

```
H       H
 \     /
  C=C
 /     \
H       H
```

Many molecules need multiple bonds in their Lewis structures to complete an octet of electrons around each atom with the available valence electrons.

The Lewis structure shows six bonds and accounts for all 12 valence electrons calculated in step 2. Each of the hydrogen atoms shares two electrons, and each of the carbon atoms shares eight electrons, so the structure is complete.

Understanding

Write the Lewis structure of acetone, $CH_3C(O)CH_3$. The skeleton structure is

```
    H   O   H
    |   |   |
H—C—C—C—H
    |       |
    H       H
```

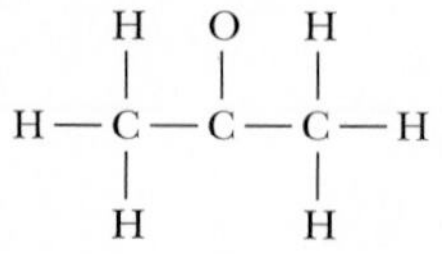

Answer

```
    H  :O:  H
    |   ‖   |
H—C—C—C—H
    |       |
    H       H
```

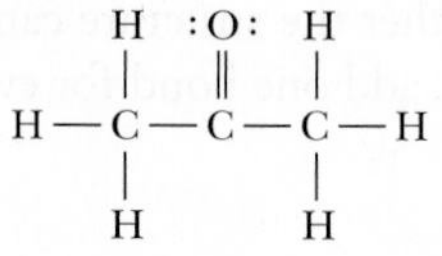

EXAMPLE 9.6 Lewis Structure of Acetonitrile

Write the Lewis structure of CH_3CN.

Solution

1. You can deduce the skeleton structure from the formula.

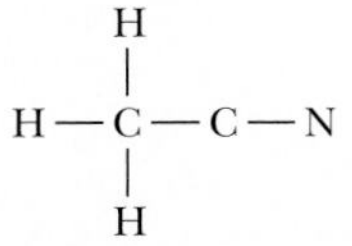

```
      H
      |
  H—C—C—N
      |
      H
```

2. The total number of valence electrons is

3(H)	$3 \times 1 = 3$
2(C)	$2 \times 4 = 8$
1(N)	$1 \times 5 = 5$
	16

3. The skeleton structure shows 5 bonds that use 2 electrons each, for a total of 10.

$$\begin{array}{r} \text{total number of valence electrons} = 16 \\ -\ \text{electrons used in skeleton structure} = 10 \\ \hline \text{remaining valence electrons} = \ \ 6 \end{array}$$

4. The first carbon atom has four bonds and needs no additional electrons for an octet. The second carbon atom has two bonds and would need two lone pairs, and the nitrogen atom has only one bond and would need three lone pairs to achieve an octet of electrons.

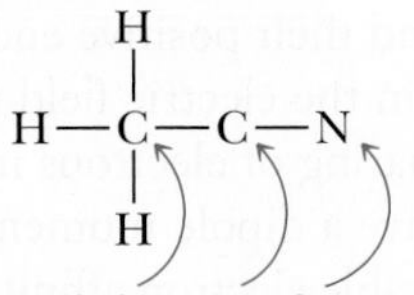

```
      H
      |
  H—C—C—N
      |
      H
```

Needs $0e^-$, $4e^-$, $6e^-$ to complete octet

Ten electrons (five pairs) are needed to finish the structure with lone pairs, but only six remain, so multiple bonds are necessary. We are two pairs of electrons short, so two additional bonds are needed to complete the structure. Place both additional bonds between the carbon and nitrogen atoms, because any other arrangement would place more than an octet of electrons around one of the atoms.

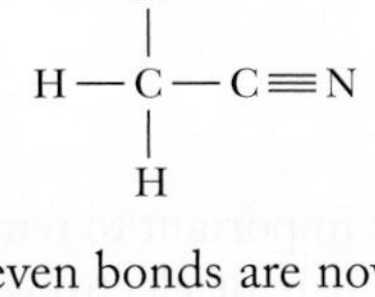

```
      H
      |
  H—C—C≡N
      |
      H
```

5. Seven bonds are now present in the structure, using 14 of the 16 valence electrons. The remaining two electrons are placed on the nitrogen atom to complete its octet.

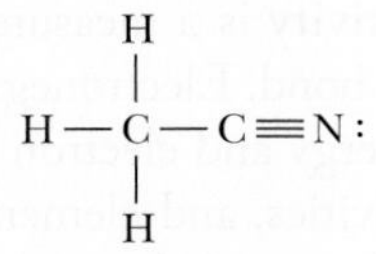

```
      H
      |
  H—C—C≡N:
      |
      H
```

Understanding

Write the Lewis structure of CH_2CCH_2.

Answer

$$\begin{array}{c} \text{H} \quad\quad\quad\quad\quad \text{H} \\ \diagdown \quad\quad\quad\quad\quad \diagup \\ \text{C}{=}\text{C}{=}\text{C} \\ \diagup \quad\quad\quad\quad\quad \diagdown \\ \text{H} \quad\quad\quad\quad\quad \text{H} \end{array}$$

OBJECTIVES REVIEW *Can you:*

- ☑ define covalent bonding?
- ☑ write Lewis structures of molecules and polyatomic ions?

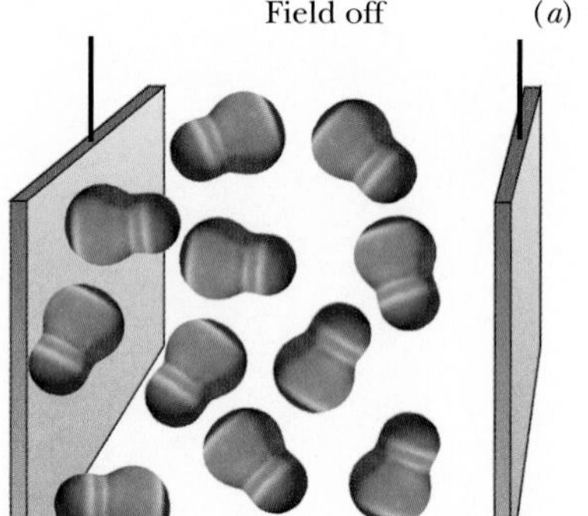

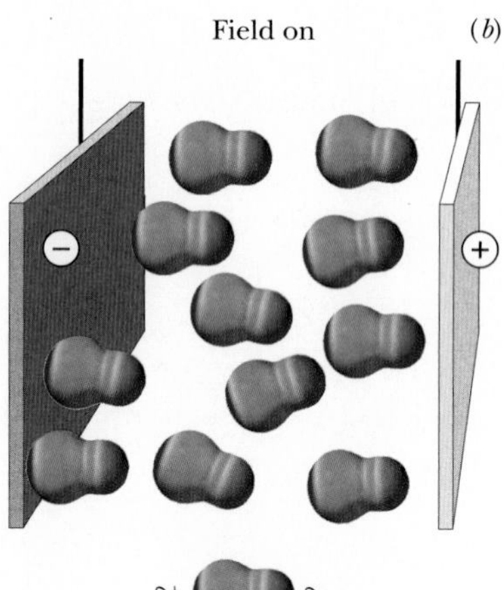

Figure 9.8 Measurement of dipole moment. *(a)* Molecules of ClF are randomly oriented when the field is off. *(b)* Because of the unequal electron distribution (red indicates areas of high-electron density, and blue indicates areas of low-electron density, with the green regions being closer to electron neutral) in ClF, the molecules align when the field is on such that the negatively charged fluorine atoms point toward the positively charged plate.

The dipole moment of a compound can be determined by placing the compound in an electric field.

9.4 Electronegativity

OBJECTIVES

- ☐ Define dipole moment and electronegativity
- ☐ Describe the periodic trends of electronegativity
- ☐ Predict bond polarities from electronegativity differences

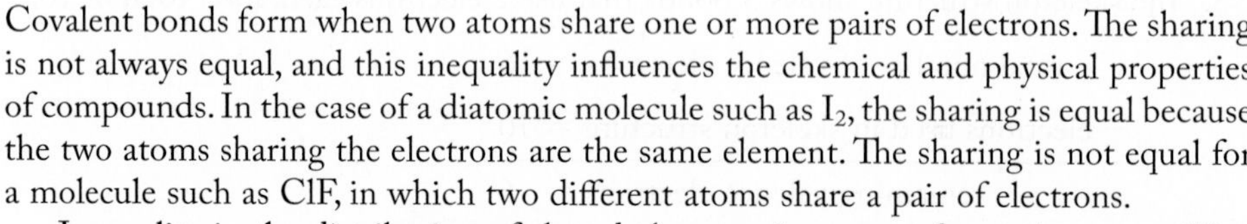

Covalent bonds form when two atoms share one or more pairs of electrons. The sharing is not always equal, and this inequality influences the chemical and physical properties of compounds. In the case of a diatomic molecule such as I_2, the sharing is equal because the two atoms sharing the electrons are the same element. The sharing is not equal for a molecule such as ClF, in which two different atoms share a pair of electrons.

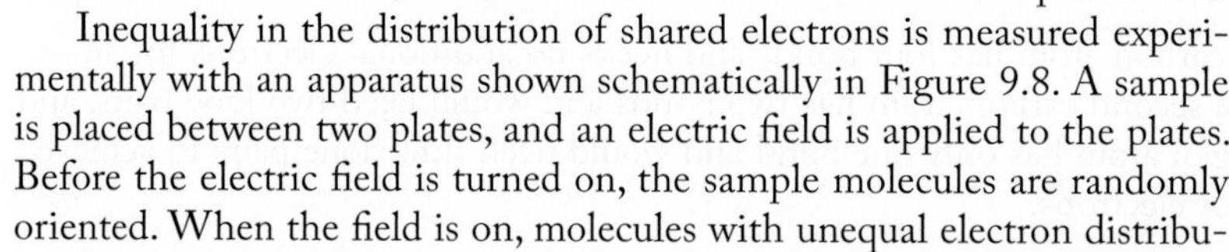

Inequality in the distribution of shared electrons is measured experimentally with an apparatus shown schematically in Figure 9.8. A sample is placed between two plates, and an electric field is applied to the plates. Before the electric field is turned on, the sample molecules are randomly oriented. When the field is on, molecules with unequal electron distributions orient their negative ends toward the positive plate and their positive ends toward the negative plate. The tendency of the molecules to align in the electric field is dependent on their **dipole moment,** a measure of the unequal sharing of electrons in a molecule. Molecules such as I_2 do not orient in the field and have a dipole moment of zero.

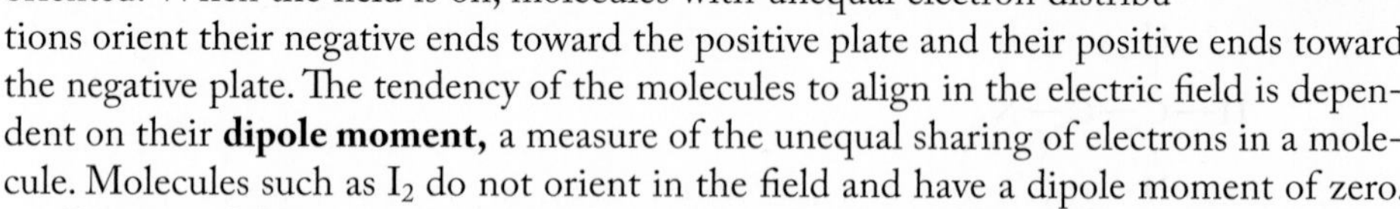

Because of the greater ionization energy and more favorable electron affinity of fluorine, fluorine attracts the shared pair of electrons in the covalent bond of ClF more strongly than does the chlorine. As shown, the ClF molecules are aligned with the fluorine end toward the positive plate and the chlorine end toward the negative plate. The fluorine end has a partial negative charge because it attracts the shared pair of electrons more strongly than does the chlorine. The chlorine end has a partial positive charge. These *partial* charges are less than the whole units of charge observed on ions. This partial charge separation is indicated as:

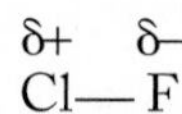

$$\overset{\delta+}{\text{Cl}}\text{—}\overset{\delta-}{\text{F}}$$

where δ (lower case Greek delta) indicates a partial charge. It is important to remember that the bond in ClF is still covalent, but the sharing of the pair of electrons in the covalent bond is not equal.

Dipole moments arise from unequal sharing of the electrons in covalent bonds and it is important to quantify this unequal sharing. **Electronegativity** is a measure of the ability of an atom to attract the shared electrons in a chemical bond. Electronegativities of atoms correlate with many properties, such as ionization energy and electron affinity. Elements with low ionization energies have low electronegativities, and elements with high ionization energies have high electronegativities. Although several electronegativity scales have been suggested, the scale developed by Linus Pauling, who first introduced the concept of electronegativity, is most widely accepted. This experimentally based scale

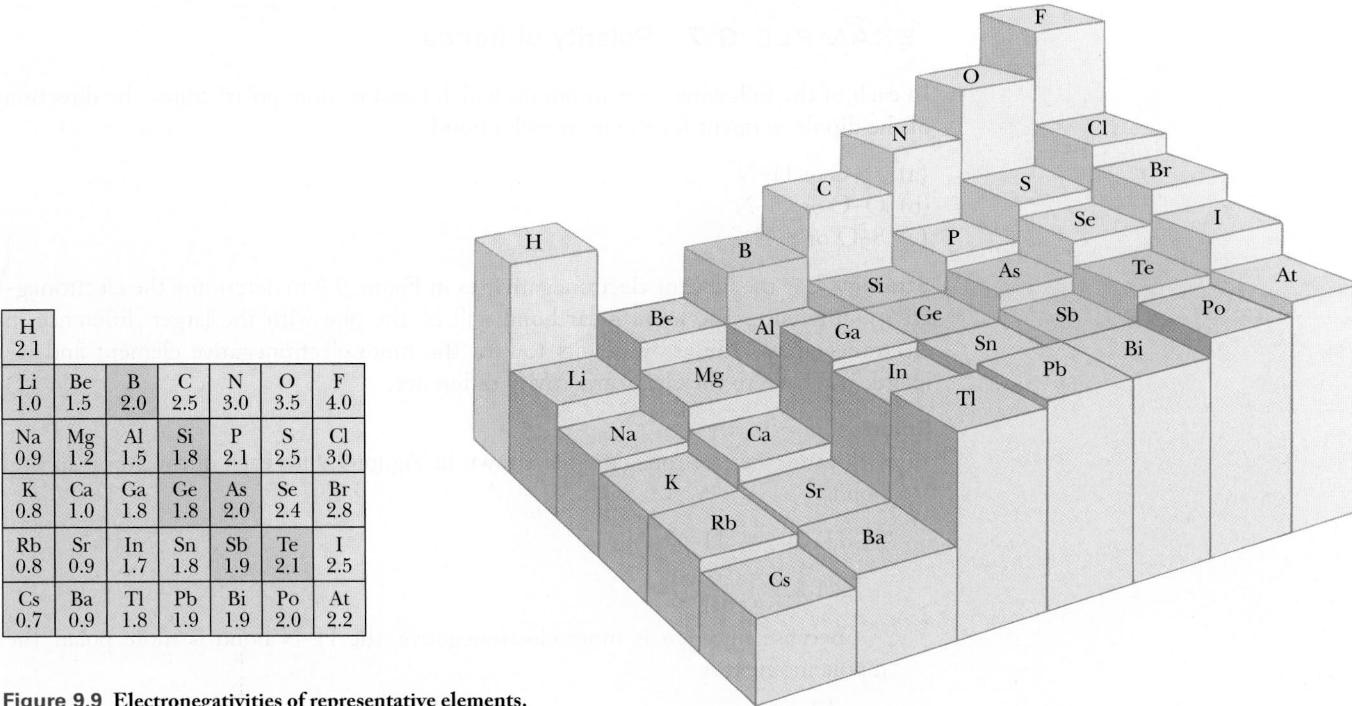

H 2.1						
Li 1.0	Be 1.5	B 2.0	C 2.5	N 3.0	O 3.5	F 4.0
Na 0.9	Mg 1.2	Al 1.5	Si 1.8	P 2.1	S 2.5	Cl 3.0
K 0.8	Ca 1.0	Ga 1.8	Ge 1.8	As 2.0	Se 2.4	Br 2.8
Rb 0.8	Sr 0.9	In 1.7	Sn 1.8	Sb 1.9	Te 2.1	I 2.5
Cs 0.7	Ba 0.9	Tl 1.8	Pb 1.9	Bi 1.9	Po 2.0	At 2.2

Figure 9.9 **Electronegativities of representative elements.**

gives a value of 4.0 to fluorine, the most electronegative element. Electronegativity values decrease down the periodic table and to the left across the table (Figure 9.9), with cesium (Cs) having the lowest at 0.7. Oxygen has the second-highest electronegativity at 3.5, and nitrogen and chlorine tie for third at 3.0. Carbon and hydrogen have intermediate electronegativities of 2.5 and 2.1, respectively.

Elements with high ionization energies, those in the top right of the periodic table, have the greatest electronegativities.

The dipole moment in diatomic molecules is proportional to the difference in electronegativity between two covalently bonded atoms and the distance between the two nuclei. An arrow, as shown below, pointing in the direction of the more electronegative atom represents the dipole moment. The tail of the arrow is crossed to emphasize that it is the positive end. A longer arrow shows a larger measured dipole moment is obtained in the experiment outlined in Figure 9.8.

$$\overset{+\!\!\longrightarrow}{\text{H—F}} \qquad \overset{+\!\!\rightarrow}{\text{H—Br}}$$

Diatomic molecules with dipole moments are said to be *polar,* the electron distribution in the bond is unsymmetric. The greater the length of the arrow representing the dipole moment, the greater the *polarity* of the bond. A molecule such as ClF is *polar,* whereas I_2 is *nonpolar;* in I_2, the electron distribution of the covalent bond is symmetric. The polarity of any individual bond in larger molecules is also proportional to the difference in the electronegativity of the two atoms that form the covalent bond. The dipole moment of molecules other than diatomics is considered in Chapter 10 after a model is developed to determine the *shape* of molecules, a factor that is also important in the dipole moments of larger molecules.

The polarity of a covalent bond is proportional to the difference in electronegativities of the two atoms involved in the bond.

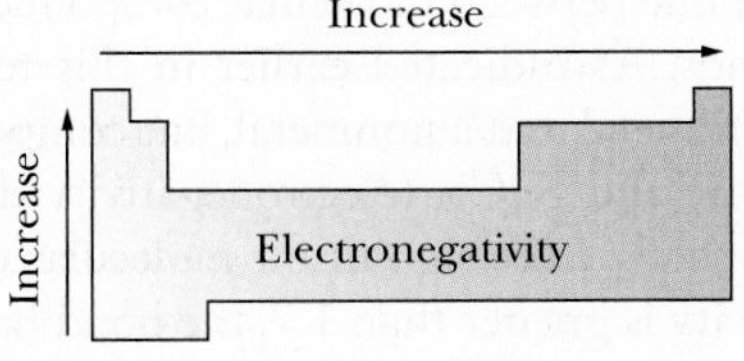

EXAMPLE 9.7 Polarity of Bonds

In each of the following pairs of bonds, which bond is more polar? Show the direction of the dipole moment for the more polar bond.

(a) H–C or H–N
(b) O–C or Cl–N
(c) S–O or S–F

Strategy Use the table of electronegativities in Figure 9.9 to determine the electronegativity difference. The more polar bond will be the one with the larger difference in electronegativity. The arrow points toward the more electronegative element and it's length is related to the electronegativity difference.

Solution

(a) Below are the electronegativities shown in Figure 9.9 for each of the atoms in the bonds.

H–C H–N

2.1 2.5 2.1 3.0

Because nitrogen is more electronegative, the H–N bond is more polar. The dipole moment is

$$\overset{\longleftrightarrow}{\text{H—N}}$$

(b) Below are the electronegativities shown in Figure 9.9 for each of the atoms in the bonds.

O–C Cl–N

3.5 2.5 3.0 3.0

The O–C bond is more polar with oxygen having the greater electronegativity.

$$\overset{\longleftrightarrow}{\text{C—O}}$$

(c) Fluorine is more electronegative than oxygen, and both are more electronegative than sulfur. The S–F bond is more polar because of the greater difference in electronegativities of the bonded atoms. The dipole moment is

$$\overset{\longleftrightarrow}{\text{S—F}}$$

Understanding

Which bond is more polar: P–S or Sb–S?

Answer Sb–S

Polar covalent bonds are intermediate between the two limiting cases of nonpolar covalent bonds (such as I_2) and ionic bonds (such as NaF, in which the valence electron on sodium transfers to the fluorine atom). Although large differences in the electronegativities of the bonded atoms produce more polar bonds, the electronegativity differences do not create a clean line between molecular compounds that have polar covalent bonds and ionic compounds. As indicated earlier in this textbook, ionic compounds generally are composed of a metal and a nonmetal, but compounds such as $AlCl_3$ (electronegativity difference, 1.5) and $SnCl_2$ (electronegativity difference, 1.2), containing the more electronegative metals, have significant molecular character. Generally, if the difference in electronegativity is greater than 1.7, the bond has significant ionic character. However, experimental measurements of properties are the only sure way to deter-

Compound	Cl_2	ICl	NaI	NaF
Difference in electronegativity	0	0.5	1.6	3.1
Polarity	nonpolar	polar	ionic	ionic
Phase (room temperature)	gas	solid	crystalline solid	crystalline solid
Melting point, °C	−101	27	661	995
Boiling point, °C	−35	97	1304	1695
Bonding	covalent	polar covalent	ionic	ionic

Figure 9.10 Effects of electronegativity differences on properties of compounds. The physical properties of substances must be measured to determine whether the bonding is ionic or covalent. The greater the difference in electronegativity, the more likely it is that the substance is a high-melting ionic compound.

mine whether the bonding in a compound is covalent or ionic. Figure 9.10 indicates how properties change for substances as the electronegativity difference of the bonded atoms increases.

OBJECTIVES REVIEW *Can you:*

- ☑ define dipole moment and electronegativity?
- ☑ describe the periodic trends of electronegativity?
- ☑ predict bond polarities from electronegativity differences?

9.5 Formal Charge

OBJECTIVES

- ☐ Assign formal charges on atoms in Lewis structures
- ☐ Predict the relative stabilities of Lewis structures by evaluating formal charges

In all of the Lewis structures we have written, the bonds have been formed by the sharing of one or more electron pairs, in which half of the electrons come from each bonding atom. Writing bonds in this manner has led to the generalization that Group 6A elements usually make two bonds to achieve an octet of electrons, Group 5A elements make three bonds, and Group 4A elements make four bonds. Not all Lewis structures can be completed by following this principle. An example is carbon monoxide, CO. Following the rules for writing the Lewis structure yields

:C≡O:

The structure has an octet of electrons about each atom and shows the correct number of valence electrons. This structure is different from those shown earlier, however, because the oxygen and carbon atoms each make three bonds instead of their normal numbers—two and four, respectively. This difference is evident if we separate the elements into their Lewis electron-dot symbols, assigning half of the shared electrons to each atom.

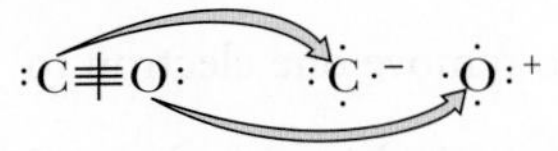

:C≡O: → :C·⁻ + ·O:⁺

Separating the shared electrons equally yields a Lewis symbol for carbon that has five valence electrons, two from the lone pair and three of the six electrons that formed the triple bond. Because the neutral carbon atom has four valence electrons, this species is an anion. The Lewis symbol obtained for oxygen by equally dividing the bonding electrons has only five valence electrons, compared with six for the atom, and thus is a cation. The Lewis structure is written with **formal charge,** a charge assigned to each atom in the structure by assuming that the shared electrons are divided equally between the bonded atoms, to indicate this bonding feature.

⊖ ⊕
:C≡O:

Formal charges are assigned by dividing bonding electrons equally between the two bonded atoms.

Formal charges are written next to the atoms and inside a circle. If Lewis structures have atoms with formal charges, only nonzero values are written. Assigning formal charges is one method of electron counting. Although they do not represent actual charges on atoms, the formal charges on neutral species must sum to zero. The sum of the formal charges on an ion must equal the charge on the ion.

Atoms in Lewis structures that have an unanticipated number of covalent bonds have nonzero formal charges.

Most Lewis structures yield values of zero for the formal charges of the atoms. Structures whose atoms have nonzero formal charges can be recognized by the number of bonds made by the atoms. Carbon atoms generally form four bonds, nitrogen atoms form three bonds, and oxygen atoms form two bonds. If the Lewis structure shows bonds that differ from these numbers, the atom will have a formal charge. Atoms with more bonds have a positive formal charge, and atoms with fewer bonds have a negative formal charge. For example, a nitrogen atom that forms four bonds will have a +1 formal charge, and one that forms only two bonds will have a −1 formal charge. Oxygen atoms with three bonds, as in CO, will have a +1 formal charge, and those with only one bond will have a −1 formal charge (Table 9.3).

TABLE 9.3 Formal Charge

	Formal Charge		
Atom	−1	0	+1
N	$-\underset{\cdot\cdot}{\ddot{\text{N}}}-$	$-\underset{\cdot\cdot}{\overset{\vert}{\text{N}}}-$	$-\underset{\vert}{\overset{\vert}{\text{N}}}-$
O	$-\underset{\cdot\cdot}{\ddot{\text{O}}}\!:$	$-\underset{\cdot\cdot}{\ddot{\text{O}}}-$	$-\underset{\cdot\cdot}{\overset{\vert}{\text{O}}}-$
C	$-\underset{\cdot\cdot}{\overset{\vert}{\text{C}}}-$	$-\underset{\vert}{\overset{\vert}{\text{C}}}-$	

The formal charge on an atom can also be calculated from the equation

$$\text{Formal charge} = (\text{number of valence electrons in atom}) - (\text{number of lone pair electrons}) - \left(\frac{1}{2}\ \text{number of shared electrons}\right)$$

We can apply this logic to determine the formal charges on the atoms in the Lewis structure of CO,

$$\text{Formal charge of C} = 4 - 2 - \frac{1}{2}(6) = -1$$

$$\text{Formal charge of O} = 6 - 2 - \frac{1}{2}(6) = +1$$

EXAMPLE 9.8 Formal Charge

Draw the Lewis structure of the ammonium cation, showing formal charges as needed.

Strategy After writing a correct Lewis structure, show nonzero formal charges on the atoms making an unusual number of bonds.

Solution

First, write the Lewis structures without showing any formal charge.

1. The skeleton structure of ammonium ion, NH_4^+, is

$$\begin{array}{c} \text{H} \\ \vert \\ \text{H}-\text{N}-\text{H} \\ \vert \\ \text{H} \end{array}$$

2. The total number of valence electrons, remembering to remove one electron to account for the charge on the cation, is

4(H)	4 × 1 =	4
1(N)	1 × 5 =	5
	1+ charge =	−1
		8

3. The skeleton structure shows four bonds, for a total of eight electrons, and uses all of the valence electrons. The nitrogen atom has eight electrons around it, and each hydrogen atom has the required two electrons. The Lewis structure is finished except for formal charges. In the Lewis structure, nitrogen makes four bonds

rather than the normal three, so it has a +1 formal charge. The final Lewis structure is

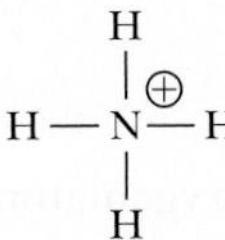

A final check shows the correct number of electrons around each atom and the correct number of valence electrons. Finally, the sum of the formal charges is the charge of the ammonium ion.

The formal charges could also have been calculated based on the formula method. Using the formula for the nitrogen atom

$$\text{Formal charge} = 5 - 0 - \frac{1}{2}(8) = +1$$

For the hydrogen atoms (which are all equivalent),

$$\text{Formal charge} = 1 - 0 - \frac{1}{2}(2) = 0$$

This method, of course, calculates the same result that the nitrogen has a +1 formal charge, whereas the hydrogen atoms have a formal charge of 0.

Understanding

Draw the Lewis structure of the NH_2^- anion showing all non-zero formal charges.

Answer

EXAMPLE 9.9 Formal Charge

Draw the Lewis structure of ozone, O_3, showing formal charges if needed.

Strategy The strategy is the same as for Example 9.8.

Solution

First, write the Lewis structure without showing any formal charge.

1. The skeleton structure of ozone is

   O–O–O (bent)

2. The total number of valence electrons is

   3(O) $3 \times 6 = 18$

3. The skeleton structure shows two bonds, for a total of four electrons.

$$\begin{array}{r} \text{total number of valence electrons} = 18 \\ -\ \text{electrons used in skeleton structure} = \ \ 4 \\ \hline \text{remaining valence electrons} = 14 \end{array}$$

4. The center oxygen atom requires 2 lone pairs and those on the outside require 3 pairs each to satisfy the octet rule, for a total of 16 additional electrons.

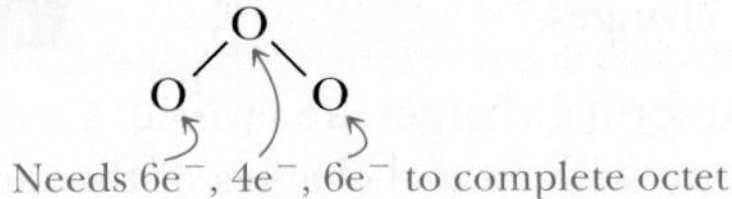

Because only 14 electrons remain at step 3, one additional bond is needed. You can place the double bond between either of the pairs of bonded oxygen atoms.

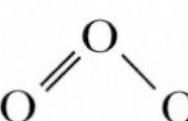

5. The three bonds use six electrons. Complete the octet for each oxygen atom by using the remaining 12 electrons as lone pairs.

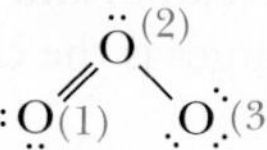

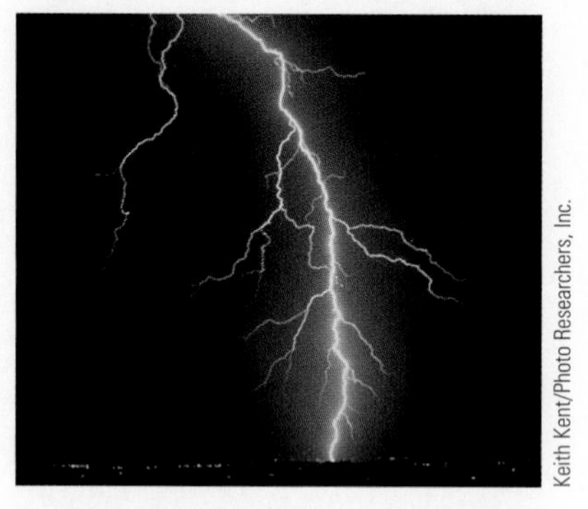

Keith Kent/Photo Researchers, Inc.

Ozone. Ozone, O_3, is produced from O_2 in the air during electrical storms causing the odor associated with thunderstorms.

Oxygen normally forms two bonds, and O(1) matches this description. However, O(2) forms three bonds and has a +1 formal charge, and O(3) forms one bond and has a −1 formal charge. Using the formula to calculate the formal charges:

$$\text{Formal charge O(1)} = 6 - 4 - \frac{1}{2}(4) = 0$$

$$\text{Formal charge O(2)} = 6 - 2 - \frac{1}{2}(6) = +1$$

$$\text{Formal charge O(3)} = 6 - 6 - \frac{1}{2}(2) = -1$$

The complete Lewis structure is:

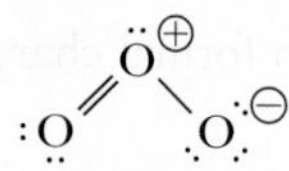

A final check shows that each oxygen atom has 8 electrons, and the total number of electrons is the correct number, 18. The sum of the formal charges equals the charge of the species, zero.

Understanding

Assign formal charges to the atoms in N_2O, given the Lewis structure:

N=N=O

Answer

⊖N=⊕N=O

Formal Charges and Structure Stability

Sometimes several different structures can be drawn for the same molecule. For example, in addition to the N—N—O skeleton structure for dinitrogen monoxide (commonly called *nitrous oxide*) shown in the Understanding section for Example 9.9, a second one with N—O—N connectivity can be written. The two Lewis structures are written below:

A: ⊖N=⊕N—O

B: ⊖N=(+2)O=N⊖

Scientists have used a number of different techniques to evaluate the actual structure. They found that structure A is correct. Comparing experimentally determined structures with those predicted by Lewis structures allows a number of conclusions about formal changes.

1. Lewis structures that show the smallest formal charges are favored.
2. Lewis structures that have adjacent atoms with formal charges of the same sign are much less favorable.

3. Lewis structures that place negative formal charges on the more electronegative atoms are favored.
4. Formal charges of opposite signs are usually on adjacent atoms.

The first rule can be used to "predict" that structure A shown above with the N—N—O connectivity for dinitrogen monoxide is correct, a prediction verified by experimental results. Example 9.10 demonstrates how these principles help predict which structures are more likely to be correct.

EXAMPLE 9.10 Structure of Hydrogen Cyanide

There are two possible ways to connect the atoms in hydrogen cyanide: HCN and HNC. Write the Lewis structures and use formal charges to predict which is the more likely structural arrangement.

H—C—N H—N—C

A B

Strategy Write the Lewis structure for each arrangement and evaluate the formal charges to predict the more likely structure.

Solution

The Lewis structure for arrangement A is

H—C≡N:

No formal charges are needed. The carbon atom forms four bonds, and the nitrogen atom forms three.

The Lewis structure of B is

$$\mathrm{H{-}\overset{\oplus}{N}{\equiv}\overset{\ominus}{C}:}$$

In this structure, a positive formal charge is assigned to the nitrogen atom because it forms four bonds, and a negative formal charge is on the carbon atom because it forms three bonds. Because Lewis structures that show the smallest formal charge are more stable, structure A is predicted to be favored and is the structure actually observed for this molecule.

Understanding

Use formal charge to predict which of the following skeleton structures is more likely to be correct for a molecule with the formula H_2CO.

```
H
 \
  C—O        H—C—O—H
 /
H
   A             B
```

Answer A

OBJECTIVES REVIEW *Can you:*

☑ assign formal charges on atoms in Lewis structures?
☑ predict the relative stabilities of Lewis structures by evaluating formal charges?

9.6 Resonance in Lewis Structures

OBJECTIVES

☐ Write all possible resonance structures for a given molecule or ion
☐ Predict the importance of different resonance structures

Some molecules and ions do not have one unique Lewis structure. An example is ozone, O_3, whose Lewis structure was shown in Example 9.9.

When this Lewis structure was written earlier, the placement of the double bond between atoms 1 and 2 was arbitrary. A second, equally correct structure places the double bond between atoms 2 and 3.

The two equally acceptable Lewis structures for O_3 are **resonance structures**—structures that differ only in the distribution of the valence electrons. The skeleton structure does *not* change, only the placement of the electons. Resonance structures are written with a double-headed arrow between them.

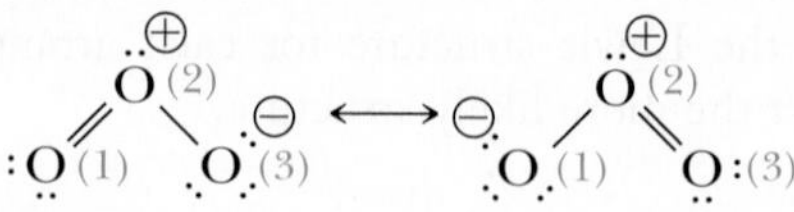

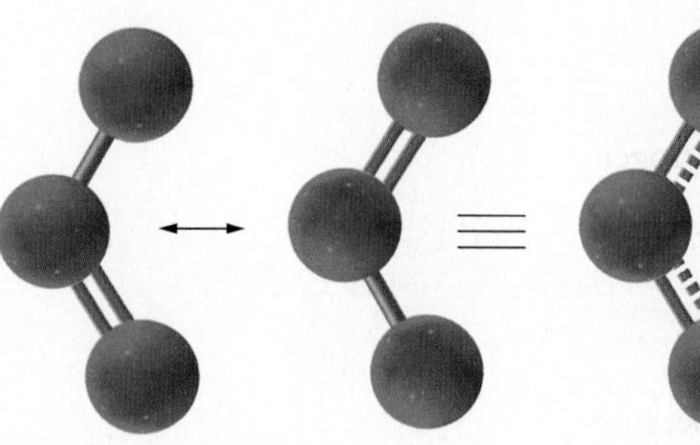

Figure 9.11 **Averaging of resonance structures.** The bonding in molecules with resonance structures is an average of the various resonance structures. In this case, the averaging of a single and a double bond leads to a predicted bond order of 1.5 for each.

The bonding in species with resonance structures is an average of all resonance forms.

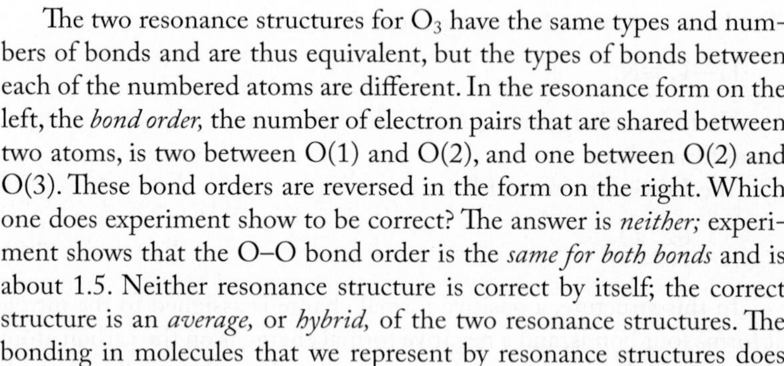

The two resonance structures for O_3 have the same types and numbers of bonds and are thus equivalent, but the types of bonds between each of the numbered atoms are different. In the resonance form on the left, the *bond order,* the number of electron pairs that are shared between two atoms, is two between O(1) and O(2), and one between O(2) and O(3). These bond orders are reversed in the form on the right. Which one does experiment show to be correct? The answer is *neither;* experiment shows that the O–O bond order is the *same for both bonds* and is about 1.5. Neither resonance structure is correct by itself; the correct structure is an *average,* or *hybrid,* of the two resonance structures. The bonding in molecules that we represent by resonance structures does not bounce back and forth between the various resonance structures but is the average of all forms, as illustrated in Figure 9.11. Chemists frequently write an averaged structure in which the bond order 1.5 is shown as a solid line and a dashed line.

One of the most important examples of resonance structures occurs in benzene, as shown in Example 9.11.

EXAMPLE 9.11 Resonance Structures of Benzene

Write the resonance structures of benzene, C_6H_6. The skeleton structure of this molecule is a six-member ring of carbon atoms, with one hydrogen atom bonded to each carbon atom.

Strategy Write all possible resonance forms, remembering not to change the skeleton structure, just the locations of bonds or lone pairs.

Solution

1. The skeleton structure is

2. The total number of valence electrons is

6(C)	$6 \times 4 = 24$
6(H)	$6 \times 1 = 6$
	30

3. The skeleton structure has 6 CC bonds and 6 CH bonds, using 24 of these electrons.

total number of valence electrons	= 30
− electrons used in the skeleton structure	= 24
remaining valence electrons	= 6

4. In the skeleton structure of benzene, each carbon atom forms three bonds, and thus shares six electrons. Each carbon atom would need an additional lone pair to satisfy the octet rule, for a total of 12 electrons. Only six electrons remain. Three additional bonds are needed. There are two ways to put these bonds in the cyclic structure that give each carbon atom four bonds:

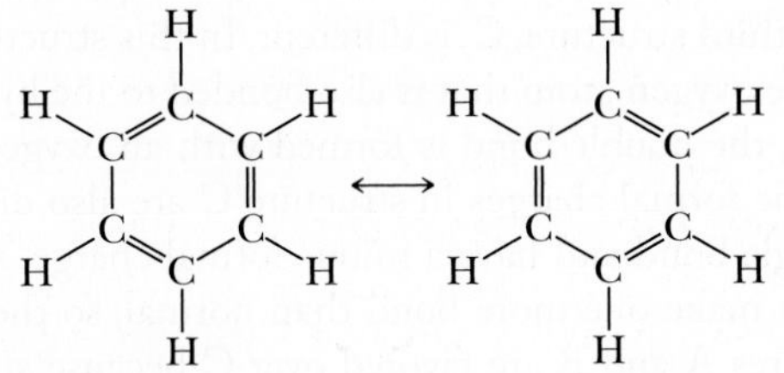

Both structures complete the octet around each carbon atom and show the correct number of valence electrons. They are both equally satisfactory Lewis structures. The actual structure of the molecule is the average of these two resonance structures.

Understanding

Write all the resonance structures for NO_2^-, an ion with O–N–O connectivity.

Answer

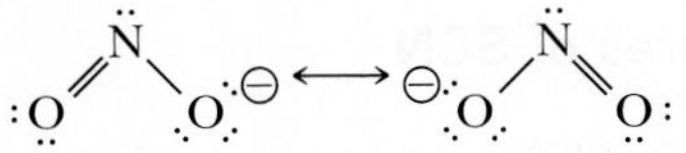

Note that although we will not learn to predict the geometry of molecules such as NO_2 and O_3 until Chapter 10, we generally draw them in a manner consistent with their experimentally measured geometry. In terms of a Lewis structure, it would be equally correct to write NO_2 as a linear molecule.

Example 9.11 shows that there are two resonance structures for benzene that have the same number and types of bonds but differ in the relative positions of the C–C single and double bonds. Because the two structures have the same types of bonds, they are equivalent. Remember that neither resonance structure is correct by itself; the correct structure is an average of the two. A single structure is frequently written for benzene, with a dashed circle to indicate that the true structure is an average of the two resonance structures.

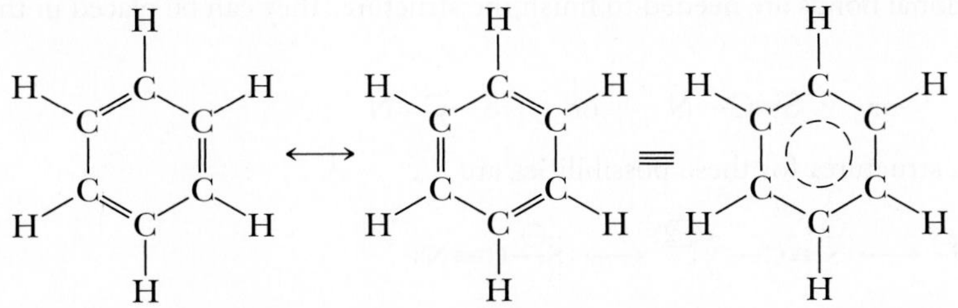

All of the C–C bond distances in benzene are found experimentally to be equal and about the length expected for a bond order of 1.5.

Species with Nonequivalent Resonance Structures

In the examples of resonance that have been shown so far in this chapter, the resonance structures have the same numbers and types of bonds, and are therefore equivalent. Sometimes resonance structures are not equivalent, and it is important to determine which one(s) best describe the actual bonding. Formal charge can be used to predict which resonance structures are favored. The resonance structures of nitric acid, HNO_3, are a good example. Three Lewis structures can be written for this molecule, given the known skeleton structure.

(1) (2) (3)

A B C

Structures A and B have the same number and types of bonds, lone pairs, and formal charges; they are equivalent. The third structure, C, is different. In this structure, the N=O double bond is formed with the oxygen atom that is also bonded to the hydrogen atom, whereas in structures A and B, the double bond is formed with an oxygen atom that is not bonded to other atoms. The formal charges in structure C are also different. Two of the oxygen atoms form a single bond and have a minus formal charge, and the nitrogen and remaining oxygen atom make one more bond than normal, so they both have positive formal charges. Structures A and B are favored over C because structure C shows more formal charges than A and B. Structure C also places formal charges of the same sign on adjacent atoms, an unfavorable situation. Resonance structure C does not contribute significantly to the bonding; an average of resonance structures A and B best describe the bonding in nitric acid. This conclusion is verified by experiment, which shows that the N–O bonds to O(1) and O(2) are equal in length and shorter than the bond to O(3), consistent with a higher bond order.

Formal charge can be used to determine which resonance structures are more important.

EXAMPLE 9.12 Resonance Structures of SCN^-

Write the resonance structures for thiocyanate anion, SCN^-, and indicate the relative contribution of each.

Solution

1. The connectivity is S–C–N.
2. The total number of valence electrons (remembering to add one for the overall negative charge) is 16.
3. The 2 bonds use 4 electrons, leaving 12 to finish the structure.
4. The nitrogen and sulfur atoms would each need 6 electrons and the carbon atom 4 electrons to satisfy the octet rule—a total of 16.

S—C—N

Needs $6e^-$, $4e^-$, $6e^-$ to complete octet

Two additional bonds are needed to finish the structure. They can be placed in three different ways.

S=C=N or S≡C—N or S—C≡N

The Lewis structures for these possibilities are

$$\ddot{S}=C=\ddot{N}^{\ominus} \longleftrightarrow :\overset{\oplus}{S}\equiv C-\ddot{N}:^{-2} \longleftrightarrow :\ddot{S}^{\ominus}-C\equiv N:$$

Each of these structures has 16 valence electrons and an octet around each atom. Also, in each structure, the sum of the formal charges is 1−, the charge on the thiocyanate anion. Each of these resonance structures is different. The center structure, with −2 and +1 formal charges, is the least favorable because it does not minimize the formal charges. We can conclude that this resonance structure does not make a significant contribution to the bonding in this anion. The two other structures are important resonance structures for thiocyanate. Of these, the first structure is favored because the negative formal charge is on the more electronegative element, nitrogen. The first structure contributes more to the overall bonding than the third, but both are important.

Understanding

Write all of the resonance structures for NCO^-, where the carbon atom is the central atom. Indicate the relative contribution of each.

Answer

$$:\!N\!\equiv\!C\!-\!\ddot{\underset{..}{O}}\!:^{\ominus} \longleftrightarrow {}^{\ominus}\ddot{\underset{..}{N}}\!=\!C\!=\!\ddot{\underset{..}{O}} \longleftrightarrow {}^{(-2)}\!:\!\ddot{\underset{..}{N}}\!-\!C\!\equiv\!O\!:^{\oplus}$$

A B C

Resonance structure A is most important, and B is second; C is unimportant. Experimental measurements of bond lengths indicate that the N–C bond is between a double and a triple bond, and the C–O bond is between a single and a double bond.

OBJECTIVES REVIEW *Can you:*

☑ write all possible resonance structures for a given molecule or ion?
☑ predict the importance of different resonance structures?

9.7 Molecules That Do Not Satisfy the Octet Rule

OBJECTIVES

☐ Be able to write the best Lewis structures for molecules that deviate from the octet rule
☐ Recognize and name the types of molecules for which the octet rule is not obeyed

The Lewis structures of most species have eight electrons around each atom (with only two around hydrogen). Three classes of molecules do not obey the octet rule: electron-deficient molecules, odd-electron molecules, and expanded valence shell molecules.

Electron-Deficient Molecules

Elements of Groups 2A and 3A have only two and three valence electrons, respectively, not enough to make four electron-pair bonds. Many compounds of these elements, especially those of beryllium, boron, and aluminum, form compounds that are called **electron deficient,** compounds for which the Lewis structures do not have eight electrons around the central atom. An example is BeH_2. The Lewis structure of BeH_2 is

H–Be–H

The two valence electrons in beryllium make electron-pair bonds with the two hydrogen atoms. Additional bonds cannot be made, because the two bonds in the skeleton structure use all four of the available valence electrons. The same is true for BH_3, an unstable molecule that has been observed in the gas phase. The three valence electrons in boron combine with the single electrons on the three hydrogen atoms, to form the three covalent bonds.

$$\mathrm{H{-}B(H)(H)}$$

Molecular compounds of elements in Groups 2A and 3A can be electron-deficient because of a shortage of valence electrons.

The beryllium atom in BeH_2 and the boron atom in BH_3 are electron deficient because there are only four and six electrons, respectively, around the central atom. The source of the electron deficiency is simply that elements in these groups do not have enough valence electrons to form four electron-pair bonds with other atoms.

EXAMPLE 9.13 Electron-Deficient Molecules

Draw the Lewis structure of AlH_3.

Strategy Draw the Lewis structure in the normal way but realize that, in certain cases, it is not possible to place eight electrons around the central atom.

Solution

1. The skeleton structure has to have the hydrogen atoms bonded to the central aluminum atom.
2. The total number of valence electrons is six.
3. The bonds of the skeleton structure use all six valence electrons, so the aluminum is electron deficient.

$$\mathrm{H{-}Al(H)(H)}$$

Understanding

Draw the Lewis structure of BeF_2.

Answer

$$:\ddot{\underset{..}{F}}-\mathrm{Be}-\ddot{\underset{..}{F}}:$$

The halides of beryllium, boron, and aluminum are also frequently given as examples of electron-deficient molecules. The Lewis structure of BF_3 is typical.

$$:\ddot{\underset{..}{F}}-\mathrm{B}(\ddot{\underset{..}{F}}:)(\ddot{\underset{..}{F}}:)$$

This case is different from BH_3 in that additional valence electrons (the fluorine lone pairs) are present. The chemical properties of BF_3 indicate an electron-deficient structure is correct.

Odd-Electron Molecules

Any molecule that has an odd number of valence electrons must violate the octet rule. An example is NO, an unusual molecule that the chapter introduction discusses because of its important physiologic properties. NO has 11 valence electrons. Two resonance structures for NO are

$$\cdot\ddot{N}=\ddot{O} \longleftrightarrow \ddot{N}^{\ominus}=\ddot{O}^{\oplus}\cdot$$

A B

PRACTICE OF CHEMISTRY
Inhaled Nitric Oxide May Help Sickle Cell Disease

As mentioned in the introduction to this chapter, NO has many important physiologic properties. Medical professionals have been doing extensive research into the use of NO to treat *sickle cell anemia,* an inherited disease that affects red blood cells. According to the U.S. Centers for Disease Control and Prevention, more than 70,000 people in the United States have the disease and millions of people are affected worldwide.

Red blood cells contain hundreds of millions of molecules of hemoglobin, a complex protein that carries oxygen throughout the body. In healthy red blood cells, hemoglobin molecules have a slightly negative charge, and hence repel each other and float freely within the cell. This property enables hemoglobin to move freely and creates the round, smooth shape of a healthy red blood cell. Hemoglobin molecules in sickle cells (called *hemoglobin S*) have a neutral charge and thus are not repelled from each other, producing twisted hemoglobin S chains. These rigid chains contort red blood cells into a sickle shape.

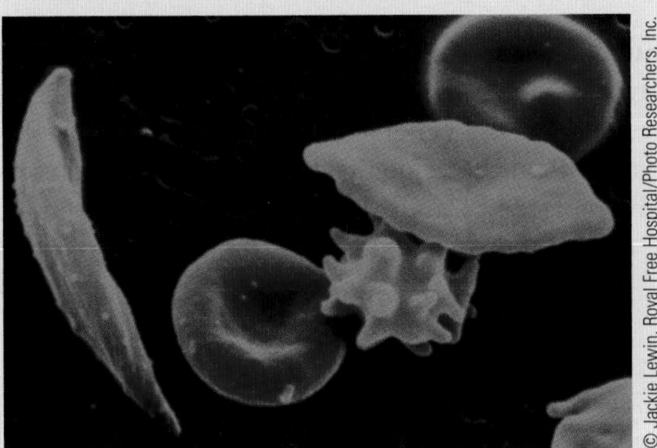

© Jackie Lewin, Royal Free Hospital/Photo Researchers, Inc.

Healthy red blood cells ("donut" shaped) and blood cells affected by sickle cell disease.

Because of their oblong shape, sickle cells have a difficult time flowing through the body's arteries. This lack of flow causes chronic and often severe pain for the patient. Until recently, little could be done to treat anyone with this disease.

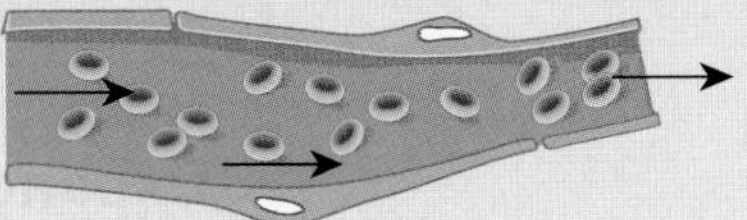

Normal cells move freely

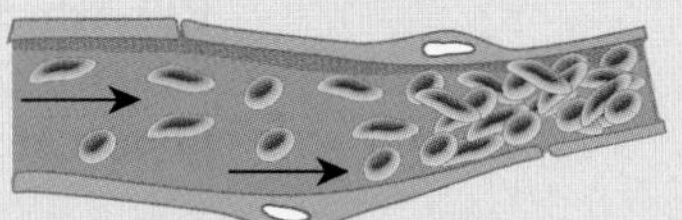

Sickled cells get stuck

Now, however, researchers have found that NO treatments help reduce pain and undo the sickling effect of the disease on red blood cells. The patient inhales the NO, which dissipates throughout the body. Though the actual chemistry of the effects of the treatment is complex, it is based in information presented in this chapter. Being a free radical and very reactive, NO is able to break down the hemoglobin S chains, allowing the red blood cells to assume a more circular shape. The NO then also reacts with the hemoglobin S itself and changes it back to regular hemoglobin, restoring its normal function. NO also can help prevent red blood cells from sickling by stopping the hemoglobin S chains from forming.

Currently, the treatment is limited in scope because even relatively small amounts of NO can be dangerous; therefore, doctors must take extreme care when administering the treatment. However, researchers are investigating methods to deliver NO in a more targeted fashion. ∎

Other resonance structures that place nine electrons about either atom are not important, because a second-period element has only four valence orbitals that can hold a maximum of eight electrons. The nitrogen atom in form A has only seven electrons, and the oxygen atom in form B has only seven electrons. Of the two resonance structures, A is the more important contributing structure because the atoms have zero formal charge.

In the Lewis structure of a molecule that contains an odd number of electrons, one atom has only seven valence electrons.

Odd-electron molecules are called *radicals.* Radicals are often reactive and form even-electron species that obey the octet rule. For example, when gaseous NO is frozen into a solid, it forms N_2O_2. The formation of N_2O_2 from two NO molecules can be viewed as making a new N–N bond, using the unpaired electrons on the nitrogen atoms shown in resonance structure A.

$$\ddot{\underset{..}{O}}=\ddot{N}\cdot + \cdot\ddot{N}=\ddot{\underset{..}{O}} \longrightarrow \ :\!\ddot{\underset{..}{O}}\!=\!\ddot{N}-\underset{..}{N}\!=\!\ddot{O}\!:$$

As mentioned in the chapter introduction, NO is a key molecule in the human body's fight against bacteria; the fact that NO is a radical is likely a main reason for this important physiologic property. When bacteria are discovered in the body, the immune system releases NO radicals. The highly reactive molecules enter the bacteria cells and attach themselves to proteins within the bacteria to form a compound called *SNO (S-nitrosothiol)*. When enough SNO gets absorbed into the bacteria, the protein becomes damaged and the bacteria become disabled. Macrophages, another part of the immune system's defenses mentioned in this chapter's introduction, then digest and dispose of the bacteria.

Expanded Valence Shell Molecules

Many compounds have more electrons than are needed to satisfy the octet rule. In these cases, the excess electrons are placed on the central atom, giving it more than an octet of electrons. For example, the reaction of phosphorus with fluorine yields the expected PF_3, which has a Lewis structure that follows the octet rule, but also forms PF_5. The 5 fluorine atoms are bonded to phosphorus, resulting in 10 electrons about the phosphorus atom in the Lewis structure.

```
      :F:  F:
       |  /
  :F — P
       |  \
      :F:  F:
```

The molecule has an **expanded valence shell,** having more than eight electrons about an atom in a Lewis structure. A large class of compounds in which the central atoms have an expanded valence shell has the general formula YF_n, where Y=P, S, Cl, As, Se, Br, Te, I, or Xe, and n is larger than the value needed to satisfy the octet rule. In some cases, a fluorine atom can be replaced by another halogen or an oxygen atom. To write Lewis structures for these compounds, we add an additional step, step 6, to the five steps given in Section 9.3.

6. When *more* electrons are available than are needed to satisfy the octet rule for all atoms present, place the extra electrons around the central atom (the central atom must be an element from the third or later row of the periodic table).

The octet rule can be exceeded for elements in the third and later periods, but not for elements in the second period. Orbital theory provides a good explanation. The octet rule is based on the idea that the valence s and p subshells can hold eight electrons. In the second period, there are no valence d orbitals (there are no $1d$ or $2d$ orbitals), and the octet rule is never exceeded in stable species. Atoms in the third and later periods have d orbitals that can hold additional electrons, exceeding an octet. Recent calculations have caused chemists to question the extent to which d orbitals are involved in expanded valence shell molecules and this issue remains an active area of research.

Elements from Groups 5A through 8A in the third and later periods can form compounds in which the central atom is surrounded by more than eight electrons.

EXAMPLE 9.14 Expanded Valence Shell Compounds

Write the Lewis structure for XeF_4.

Strategy Write the Lewis structure for this molecule in the normal way, but place any excess electrons available after each element satisfies the octet rule on the central atom.

Solution

Although xenon is a noble gas, it forms a number of compounds. In XeF_4, after using eight valence electrons to form the four Xe—F bonds and placing three lone pairs around each fluorine atom so they reach an octet, four electrons remain. Place them as lone pairs on the central xenon atom:

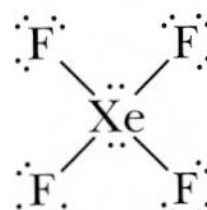

The Lewis structure places a total of 12 electrons, 4 bond pairs and 2 lone pairs, about the xenon atom.

Understanding

Write the Lewis structure for SF_6.

Answer

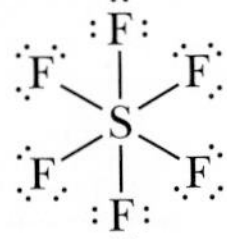

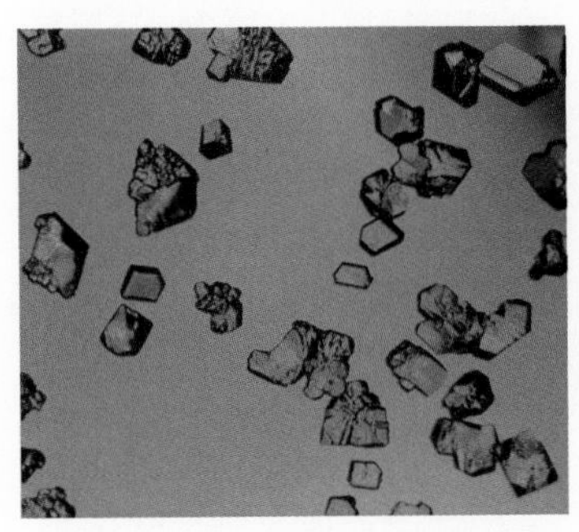

Crystals of XeF_4. XeF_4 was one of the first compounds of a noble gas to be prepared. It forms in the direct reaction of Xe and F_2.

The atoms in the Lewis structures of PF_5, XeF_4, and SF_6 have zero formal charge. In each case, dividing the bonding electrons equally leads to Lewis symbols for the neutral atoms. As shown in the next section, expanded valence shell Lewis structures frequently minimize or eliminate formal charges.

Oxides and Oxyacids of *p*-Block Elements from the Third and Later Periods

Many oxides contain a *p*-block central atom from the third or later periods. In these oxides, Lewis structures in which the central atom has an expanded valence shell are generally favored. An example is SO_2. Three resonance structures can be written:

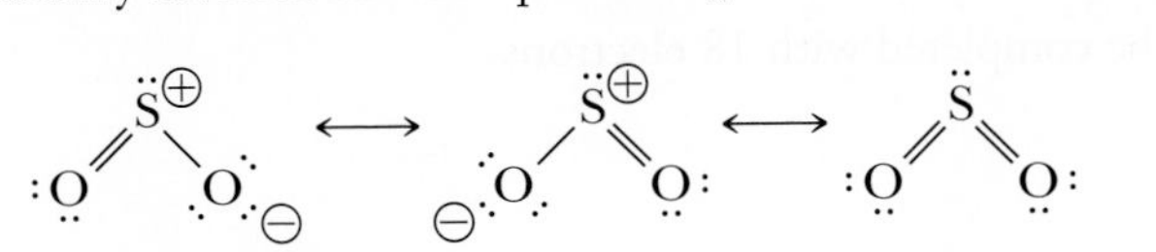

A ⟷ B ⟷ C

In structures A and B, each atom has eight electrons, but two atoms have nonzero formal charges. In C, sulfur is shown surrounded by 10 electrons, but all atoms in structure C have a zero formal charge. Structure C is favored because it minimizes formal charges. Ten and even 12 electrons about a central atom from the third row are perfectly acceptable. Experimental evidence also supports the importance of structure C. The S–O bond lengths are 143 pm, much shorter than the 157-pm bond distance typically observed for S–O single bonds (see the following data for H_2SO_4).

Another class of compounds in which the central atom has an expanded valence shell is the **oxyacids** with central atoms from the third or later periods. An oxyacid has at least one hydrogen atom attached to an oxygen atom and has the general formula $(HO)_mXO_n$. In writing the skeleton structures of oxyacids, connect the oxygen atoms to the central atom and the hydrogen atoms to the oxygen atoms. An example of an oxyacid is sulfuric acid, H_2SO_4. The oxygen atoms are all bonded to the sulfur atom, and the hydrogen atoms are bonded to two of the oxygen atoms. The two most representative Lewis structures are

```
         ..⊖                              :O:
        :O:                               ||
         |(+2)..                    ..    ||   ..
  H—O—S—O—H  ⟷  H—O—S—O—H
    ..   |   ..                     ..    ||   ..
        :O:                               ||
         ..⊖                              :O:
         A                                 B
```

In structure A, all atoms satisfy the octet rule, but a high formal charge of +2 is assigned to the sulfur atom. Lewis structure B has 12 electrons around the sulfur atom, but no formal charges are produced because the sulfur uses all six of its valence electrons to make six bonds. Resonance structure B is more important because formal charges should be minimized. Experimental evidence supports the importance of structure B. The S–OH bond lengths are 157 pm, as are typical for an S–O single bond. The other two S–O bond lengths are shorter at 142 pm, consistent with the length of a double bond. The relative importance of these different Lewis structures is still being debated in chemical literature.

Experimental evidence supports the importance of Lewis structures that contain atoms with expanded valence shells rather that those containing high formal charges.

EXAMPLE 9.15 Resonance Structures

Draw the important resonance structures for chloric acid, $HClO_3$.

Solution

1. The skeleton structure has the oxygen atoms bonded to chlorine and the hydrogen atom bonded to an oxygen atom.

```
         O
         |
  H—O—Cl
         |
         O
```

2. The total number of valence electrons is 26.
3. Eight electrons are used in the skeleton structure, leaving 18 to complete the structure with lone pairs.
4. The Lewis structure can be completed with 18 electrons.

```
          ..⊖
         :O:
    ..    |  (+2)
  H—O—Cl:
    ..    |
         :O:
          ..⊖
```

This structure is valid but has considerable formal charges. A second structure can be written that eliminates the formal charges.

```
         :O:
    ..    ||
  H—O—Cl:
    ..    ||
         :O:
```

This structure is favored because there are no formal charges.

Understanding

Write the important resonance structures for H_2SO_3.

Answer

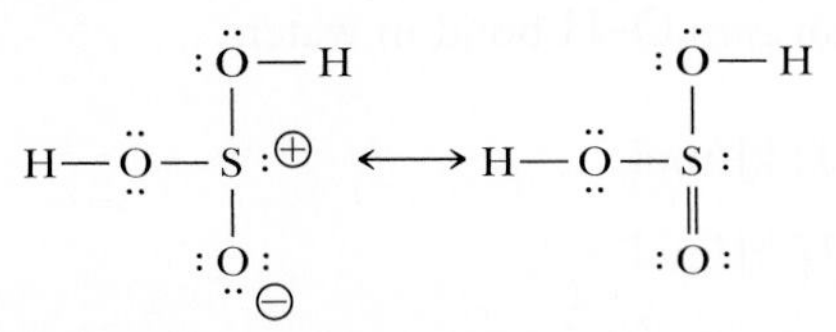

```
          ..                         ..
        : O — H                    : O — H
          ..                         ..
          |                          |
     ..                         ..
H — O — S :⊕   ⟷   H — O — S :
     ..                         ..
          |                          ||
        : O :                      : O :
          .. ⊖
```

In this case, the structure on the right is favored because there are no formal charges. Additional, less important Lewis structures can also be written.

OBJECTIVES REVIEW *Can you:*

- ☑ write the best Lewis structures for molecules that deviate from the octet rule?
- ☑ recognize and name the types of molecules for which the octet rule is not obeyed?

9.8 Bond Energies

OBJECTIVES

- ☐ Relate bond energies to bond strengths
- ☐ Calculate approximate enthalpies of reaction from bond energies

Covalent bonds form between many different elements. Not all types of covalent bonds are of equal strength; each involves different nuclei and electrons in different orbitals. Bond strengths are important, because *species with strong bonds are generally stable.*

The **bond dissociation energy** or **bond energy** *(D)* is the energy required to break one mole of bonds in a gaseous species. The thermochemical equation that describes the bond dissociation for H_2 is

$$H_2(g) \rightarrow H(g) + H(g) \qquad \Delta H = D(\text{H} - \text{H}) = 436 \text{ kJ/mol}$$

Bond energies are always endothermic and thus have a positive sign; it takes energy to break a bond.

Table 9.4 shows a number of important bond energies. For diatomic molecules, these numbers are measured exactly. A problem arises in measuring exact bond energies

TABLE 9.4 Bond Energies (kJ/mol)*

Single Bonds							
C–H	414	N–H	389	O–H	463	F–F	159
C–C	348	N–N	163	O–O	146	Cl–F	253
C–N	293	N–O	201	O–F	190	Cl–Cl	242
C–O	351	N–F	272	O–Cl	203	Br–F	237
C–F	439	N–Cl	200	O–I	234	Br–Cl	218
C–Cl	328	N–Br	243			Br–Br	193
C–Br	276			S–H	339	I–Cl	298
C–I	238	H–H	436	S–F	327	I–Br	180
C–S	259	H–F	569	S–Cl	251	I–I	151
		H–Cl	431	S–Br	218		
Si–H	293	H–Br	368	S–S	266		
Si–Si	226	H–I	297				
Si–C	301						
Si–O	368						
Multiple Bonds							
C=C	611	O=O (O_2)	498				
C≡C	837						
C=N	615	N=N	418				
C≡N	891	N≡N	946				
C=O	799						
C≡O	1072	S=O	523				
		S=S	418				

*The bond energies for the diatomic molecules can be measured directly; the other numbers are average bond energies.

TABLE 9.5 Selected Bond Lengths and Energies

Bond Type	Bond Length (pm)	Bond Energy (kJ/mol)
C–C	154	348
C=C	134	611
C≡C	120	837
C–N	147	293
C=N	131	615
C≡N	116	891
C–O	143	351
C=O	123	799
C≡O	113	1072
N–N	147	163
N=N	125	418
N≡N	110	946

in polyatomic molecules because the energy required to break a bond is influenced by the other atoms. Consider the bond energy for *each* O–H bond in water.

$$H_2O(g) \rightarrow OH(g) + H(g) \qquad \Delta H = 502 \text{ kJ/mol}$$

$$OH(g) \rightarrow O(g) + H(g) \qquad \Delta H = 427 \text{ kJ/mol}$$

The values are not the same because the species in which we are breaking the bonds are not the same. In a similar manner, the O–H bond energy in CH_3OH is different (435 kJ/mol) from both of these values. Because bond energies depend on the environment of the bonded atoms, the table gives *average* bond energies for all bond types other than those in diatomic molecules. Although not exact, these numbers are fairly accurate because most bonds between the same two atoms are of similar strengths.

The values of bond energies span a fairly wide range (Table 9.4). The C–H bond is more than twice the energy of the O–F bond. In a comparison of bond strengths between a single pair of atoms, double and triple bonds are stronger than single bonds. As mentioned earlier (see Section 9.3) and shown in Table 9.5, the sharing of multiple electron pairs also leads to shorter bond lengths.

Bond energies measure the strengths of chemical bonds.

Bond Energies and Enthalpies of Reaction

The data in Table 9.4 can be used with Hess's law to estimate enthalpies of reactions. Consider the reaction

$$H_2(g) + F_2(g) \rightarrow 2HF(g)$$

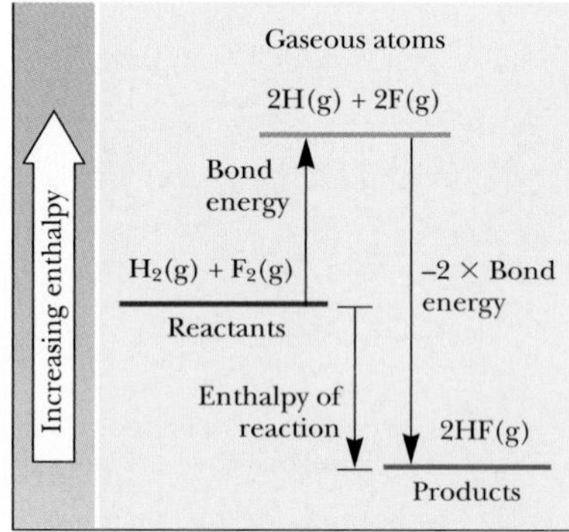

Figure 9.12 Calculation of enthalpy of reaction. An energy-level diagram showing the use of bond energies to calculate the enthalpy of reaction for $H_2(g) + F_2(g) \rightarrow 2HF(g)$.

The energy required to break the bonds in one mole of each reactant to form gaseous atoms is the respective bond energy *(D)*. As shown in Figure 9.12, these two processes are endothermic—it takes energy to break bonds. In the reaction, 2 mol of H–F bonds form in an exothermic process (see Figure 9.12). Combining the two processes yields the enthalpy of the reaction.

In equation form, the calculation of enthalpies of reaction from bond energies is

$$\Delta H_{\text{reaction}} = \Sigma(\text{moles of bonds broken} \times \text{bond energies of bonds broken}) - \Sigma(\text{moles of bonds formed} \times \text{bond energies of bonds formed}) \quad [9.2]$$

The negative sign in the equation indicates that we are making bonds in the products, an exothermic process, so the energy change is the negative of bond energy. Note in the following calculation that 1 mol H_2 and F_2 are consumed, and 2 mol HF are formed. The values given in Table 9.4 are for 1 mol of bonds.

$$\Delta H_{\text{reaction}} = [1 \text{ mol} \times D(\text{H}-\text{H}) + 1 \text{ mol} \times D(\text{F}-\text{F})] - [2 \text{ mol} \times D(\text{H}-\text{F})]$$

$$\Delta H_{\text{reaction}} = [1 \text{ mol} \times 436 \text{ kJ/mol} + 1 \text{ mol} \times 159 \text{ kJ/mol}] - [2 \text{ mol} \times 569 \text{ kJ/mol}]$$

$$\Delta H_{\text{reaction}} = -543 \text{ kJ}$$

The enthalpy of reaction can be estimated from the energy required to break all of the bonds in the reactants minus the bond energies of the bonds in the products.

The bonds formed are stronger than the bonds broken, so the reaction is exothermic. F_2 is very reactive, and the weak F–F bond is a major reason for this reactivity.

Calculations of this type yield accurate $\Delta H_{\text{reaction}}$ values for reactions of diatomic molecules in the gas phase. For most other reactions, the calculated enthalpies of reaction are only approximately correct, because *average* bond energies are used. The simplicity of determining ΔH values from a single small table, rather than measuring the enthalpy change of each reaction experimentally, is an advantage of this method.

EXAMPLE 9.16 Calculation of Enthalpies of Reaction

Use Table 9.4 to calculate an approximate enthalpy of reaction for

$$CH_4(g) + 2O_2(g) \rightarrow CO_2(g) + 2H_2O(g)$$

Strategy Use Equation 9.2, being careful to substitute the correct number of moles of *bonds* being formed or broken. Also, write the Lewis structure of each species to determine the *types* (single, double, or triple) of bonds being formed or broken.

Solution

Three important points are:

1. The Lewis structure of each compound in the reaction must be known, because bond energies depend on bond order. The Lewis structures for the three compounds in this equation are (the bond energy of O_2 is listed separately)

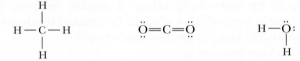

```
     H
     |                    ..  ..          ..
H—C—H               O=C=O          H—O:
     |                                     |
     H                                     H
```

   The Lewis structures show that the bonds in CO_2 are double bonds, whereas the bonds in CH_4 and H_2O are single bonds.
2. In the calculation, you must take into account the total number of moles of bonds being broken or made. Both the coefficients in the equation and the Lewis structures must be considered. One mole of CH_4 has four moles of C–H bonds, two moles of O_2 have two moles of O=O bonds, one mole of CO_2 has two moles of C=O double bonds, and two moles of H_2O contain four moles of O–H bonds.
3. Remember the minus sign in front of the bond energies of the *products,* because the energy of bonds formed in the products is the negative of bond energy.

Using Equation 9.2:

$$\Delta H_{\text{reaction}} = [4\text{ mol} \times D(\text{C}-\text{H}) + 2\text{ mol} \times D(\text{O}=\text{O})] - [2\text{ mol} \times D(\text{C}=\text{O}) + 4\text{ mol} \times D(\text{O}-\text{H})]$$

Substitute the bond energy values from Table 9.4.

$$\Delta H_{\text{reaction}} = [4\text{ mol} \times 414\text{ kJ/mol} + 2\text{ mol} \times 498\text{ kJ/mol}] - [2\text{ mol} \times 799\text{ kJ/mol} + 4\text{ mol} \times 463\text{ kJ/mol}]$$

$$\Delta H_{\text{reaction}} = [2652\text{ kJ}] - [3450\text{ kJ}] = -798\text{ kJ}$$

The reaction is exothermic. In this reaction, the strong C=O bonds in CO_2 contribute the main difference in bond strengths between the reactants and products. Remember that enthalpies of reaction calculated from bond energies are only approximate; the experimentally measured value for this reaction is −802 kJ.

Will & Deni McIntyre/Photo Researchers, Inc.

Combustion of methane. Methane gas burns in air in an exothermic reaction. Shown here is methane being flared in Saudi Arabia. The black smoke is evidence of incomplete combustion.

Understanding

Calculate the enthalpy of reaction for the reaction of N_2 and H_2 to form 2 mol NH_3, using data from Table 9.4.

Answer −80 kJ

OBJECTIVES REVIEW *Can you:*

- ☑ relate bond energies to bond strengths?
- ☑ calculate approximate enthalpies of reaction from bond energies?

Summary Problem

An *amino acid* contains a carboxyl group (–COOH, a group that contains both a C=O and –OH functional group) and an amine group ($–NH_2$) in a single molecule. The simplest amino acid is glycine, NH_2CH_2COOH. It is one of the 20 amino acids found in proteins, which are the building blocks of biological systems. Glycine occurs only in large amounts in proteins that are used for structural functions such as collagen, which is important in connective tissues in mammals. Interestingly, it is not an essential amino acid that humans need to consume because the body produces it from other foods we eat. By understanding the bonding in such a molecule (e.g., Does it have any polar bonds?), we can learn more about its physical properties, how it will react with other substances, and why it is observed to be extensively involved in some proteins and not others (this topic is covered in more detail in Chapter 22). From the formula as previously written, write the skeleton structure; then use the procedures in the chapter to write the Lewis structures of the molecule, including all possible resonance forms. Which resonance form is most important? Also, which of the covalent bonds would be most polar, and which bond should have the greatest bond energy?

From the formula, the skeleton structure is

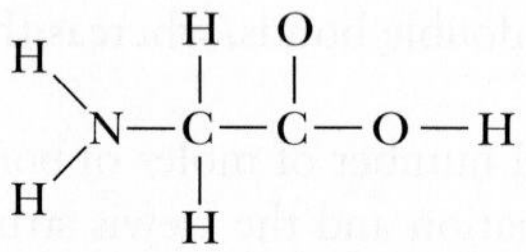

```
H     H  O
 \    |  |
  N—C—C—O—H
 /    |
H     H
```

The total number of valence electrons are:

2(C)	2 × 4 = 8
5(H)	5 × 1 = 5
1(N)	1 × 5 = 5
2(O)	2 × 6 = 12
	30

The skeleton structure has nine connections between the atoms in the molecule, each of which starts off with a single bond:

total number of valence electrons	= 30
−electrons used in the skeleton structure	= 18
remaining valence electrons	= 12

In the skeleton structure, one carbon atom is bonded to four other atoms and the second is bonded to three other atoms, the nitrogen atom is bonded to three other atoms, and one oxygen atom is bound to one atom and the other to two atoms. Thus, to reach an octet with lone pairs only, one carbon atom and the nitrogen atom need a lone pair, one oxygen atom needs three electron pairs, and the other one needs two lone pairs, making a total of 7 lone pairs or 14 electrons needed to finish the structure. Because only 12 electrons are available, 1 double bond is needed. It can be placed in two locations and the structures finished with the remaining lone pairs:

```
          H  :Ö:                          H  :O:
H          |   |               H           |   ||
  \        |   |                 \         |   ||  ..
   :N—C—C=O—H                     :N—C—C—O—H
  /        |       ..             /        |       ..
H          H                   H           H

           A                                B
```

One oxygen atom in structure A makes three bonds, so it needs a +1 formal charge, whereas the other oxygen atom makes one bond and needs a −1 formal charge. In structure B, all atoms have a formal charge of zero. The two resonance forms are:

```
          H  :Ö:⊖                           H  :O:
H          |   |                 H           |   ||
  \        |   |      ⊕            \         |   ||  ..
   :N—C—C=O—H        ⟷            :N—C—C—O—H
  /        |       ..               /        |       ..
H          H                     H           H

           A                                  B
```

Resonance form B is favored because it has no formal charges. From Figure 9.9, the electronegativities of the atoms are: H = 2.1, C = 2.5, N = 3.0, and O = 3.5. Using these values, the OH bond is most polar. There is a double bond in resonance form B, so we would predict that this C=O bond would have the highest bond energy.

ETHICS IN CHEMISTRY

1. You are conducting a study funded by a pharmaceutical firm on the effectiveness of a potential new medicine that delivers nitrogen monoxide to the blood to combat sickle cell anemia. The test group is small, consisting of about 100 patients, of which half receive the medicine and half receive a placebo. Early studies indicate that two of the patients experienced development of a skin rash that cleared up when the experimental drug was removed. As director of the study, what actions will you take because of this new development? Explain your logic.

Chapter 9 Visual Summary

The chart shows the connections between the major topics discussed in this chapter.

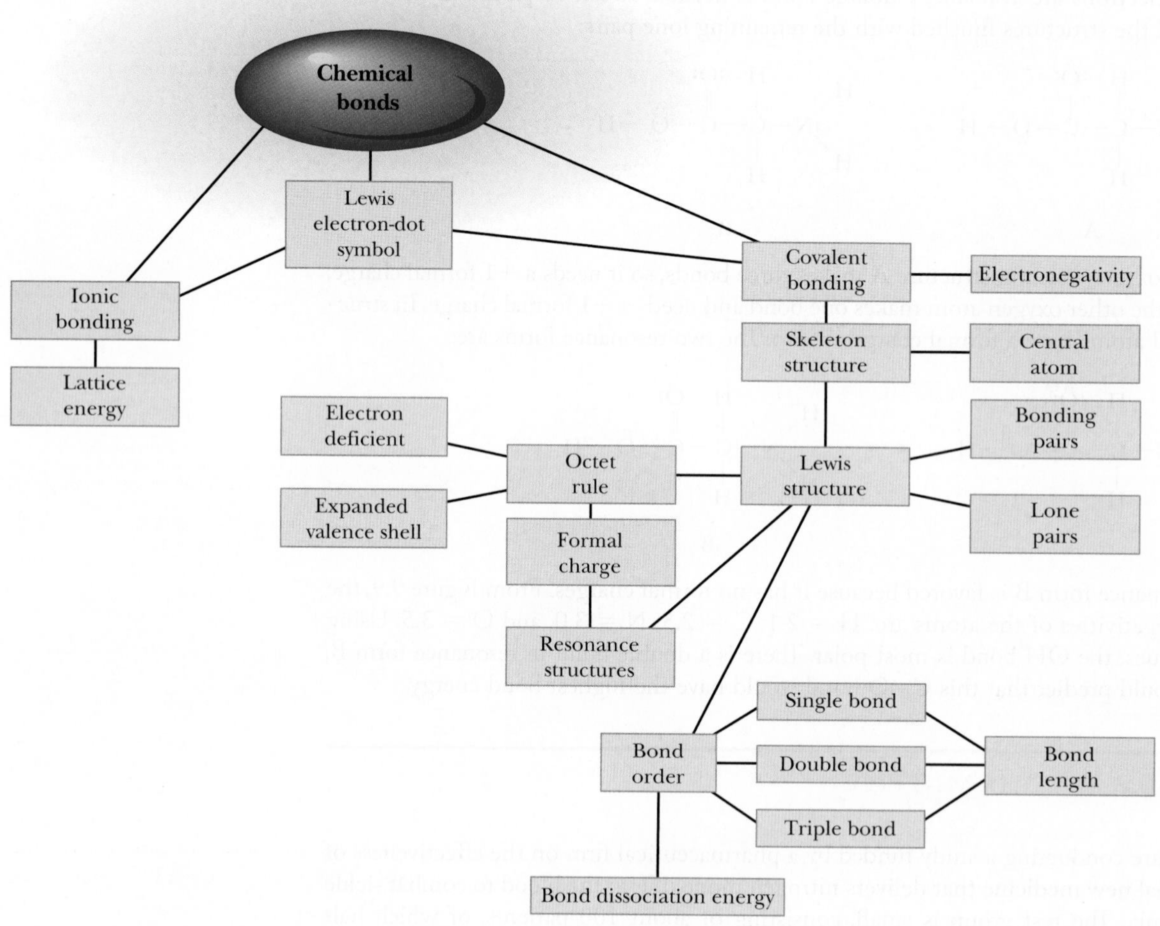

Summary

9.1 Lewis Symbols and 9.2 Ionic Bonding

Lewis electron-dot symbols represent the valence electrons (the outer electrons) with dots. In *ionic bonding,* some of the valence electrons are transferred from a metal to a nonmetal; this process produces cations and anions that, for the representative elements, generally have noble-gas electron configurations. The main driving force for the formation of ionic solids is the *lattice energy,* the energy required to separate one mole of an ionic crystalline solid into the gaseous ions. The lattice energy is greatest for small, higher charged ions.

9.3 Covalent Bonding

Covalent bonds are formed by the sharing of valence electrons between atoms. *Lewis structures,* showing *bonding pairs* and *lone pairs,* are used to represent covalent bonding. In Lewis structures, atoms (particularly those of the second period) share valence electrons to attain an *octet* of electrons (a hydrogen atom has only two). The Lewis structure shows the number of valence electrons around each atom of the molecule or ion. Atoms can share one pair of electrons to form a *single bond,* two pairs to form a *double bond,* or three pairs to form a *triple bond.*

9.4 Electronegativity

Atoms of different elements share electrons unequally, producing *polar bonds.* The *electronegativity* differences of the elements are used to determine the extent of this unequal sharing. Fluorine is the most electronegative element, and electronegativity decreases down a group and to the left across a period.

9.5 Formal Charge and 9.6 Resonance in Lewis Structures

Formal charges can be assigned to atoms in Lewis structures to indicate the charge that would result if the shared electrons were divided equally between the bonded atoms. Frequently, more than one Lewis structure can be written for a compound. Lewis structures that differ only in the placement of multiple bonds are called *resonance structures.* The actual bonding is an average of the bonding in the resonance structures, but not all resonance structures contribute equally in every case. Resonance structures that have high formal charges or that place charges of the same sign on adjacent atoms do not contribute significantly to the bonding.

9.7 Molecules That Do Not Satisfy the Octet Rule

In certain cases, it may not be possible to write Lewis structures that obey the octet rule. A molecule in which an atom does not attain an octet is called *electron deficient;* such a molecule can accept an electron pair from another molecule or ion. Species with *central atoms* from the third and later periods often exceed an octet around the central atom; such species are said to have an *expanded valence shell.*

9.8 Bond Energies

Bond dissociation energy is the energy needed to break one mole of a particular type of covalent bond. In most cases, it is an average number, because the energy needed to break a given type of bond may be different for different compounds. Bond energies can be used to calculate approximate values for the enthalpies of reactions by determining the energy required to break the bonds of the reactants and the energy released to form all the bonds in the products.

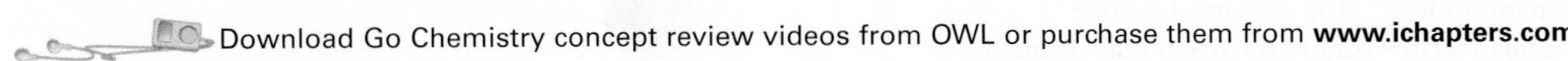
Download Go Chemistry concept review videos from OWL or purchase them from **www.ichapters.com**

Chapter Terms

The following terms are defined in the Glossary, Appendix I.

Chemical bond

Section 9.1

Lewis electron-dot symbol

Section 9.2

Ionic bonding

Lattice energy

Section 9.3

Bond length

Bond order

Bonding pairs

Central atom

Covalent bond

Double bond

Lewis structure

Lone or nonbonding pairs

Octet rule

Single bond

Skeleton structure

Triple bond

Section 9.4

Dipole moment

Electronegativity

Section 9.5

Formal charge

Section 9.6

Resonance structure

Section 9.7

Electron-deficient molecule

Expanded valence shell molecule

Odd electron molecule

Oxyacid

Section 9.8

Bond dissociation energy or bond energy

Key Equations

Formal charge

$$= (\text{number of valence electrons in atom}) - (\text{number of lone pair electrons}) - \left(\frac{1}{2}\text{ number of shared electrons}\right) \quad (9.5)$$

$\Delta H_{\text{reaction}}$

$$= \Sigma(\text{moles of bonds broken} \times \text{bond energies of bonds broken}) - \Sigma(\text{moles of bonds formed} \times \text{bond energies of bonds formed}) \quad (9.8)$$

Questions and Exercises

OWL Selected end of chapter Questions and Exercises may be assigned in OWL.

Blue-numbered Questions and Exercises are answered in Appendix J; questions are qualitative, are often conceptual, and include problem-solving skills.

■ Questions assignable in OWL

✎ Questions suitable for brief writing exercises

▲ More challenging questions

Questions

9.1 What is a Lewis electron-dot symbol?

9.2 Use Lewis electron-dot symbols to show the electron transfer during the formation of each compound from the appropriate atoms.
(a) barium bromide (b) potassium sulfide

9.3 Use Lewis electron-dot symbols to show the electron transfer during the formation of each compound from the appropriate atoms.
(a) beryllium oxide (b) yttrium chloride

9.4 ✎ What main factors control the magnitude of lattice energies? Give a specific example of a compound that should have a high lattice energy, and explain why its lattice energy is high.

9.5 Explain why
(a) the lattice energy of NaI is greater than that of KI.
(b) the lattice energy of $MgCl_2$ is greater than that of NaCl.

9.6 Explain why
(a) the lattice energy of LiCl is greater than that of LiBr.
(b) the lattice energy of Na_2O is greater than that of NaF.

9.7 ✎ Given that the lattice energy of an ionic solid increases with the charges of the anion and cation, discuss why the formula of NaF is not NaF_2.

9.8 ✎ Describe the main difference between covalent and ionic bonding.

9.9 Define the octet rule and rationalize why the rule applies to most compounds made up of representative elements.

9.10 Give the characteristics of a correct Lewis structure.

9.11 Explain the different behaviors expected from HCl and Cl_2 when each is placed between two oppositely charged plates.

9.12 What is a polar bond, and under what circumstances does it occur?

9.13 Outline the trends in the electronegativities of the representative elements. Which element has the highest electronegativity?

9.14 ✎ Compare the trends in electronegativity and ionization energy within (a) a group and (b) a period. Explain any differences.

9.15 Explain how formal charges are assigned to atoms in Lewis structures.

9.16 Describe what is meant by resonance structures.

9.17 ✎ Discuss factors that relate formal charges in Lewis structures to the relative stabilities of different connectivities of atoms in molecules and relative stabilities of different resonance structures.

9.18 What elements are most likely to form electron-deficient molecules?

9.19 What is a radical, and why do radicals violate the octet rule?

9.20 Which elements can form expanded valence shell molecules? Do the two xenon compounds shown below have an expanded valence shell? Explain your answer.

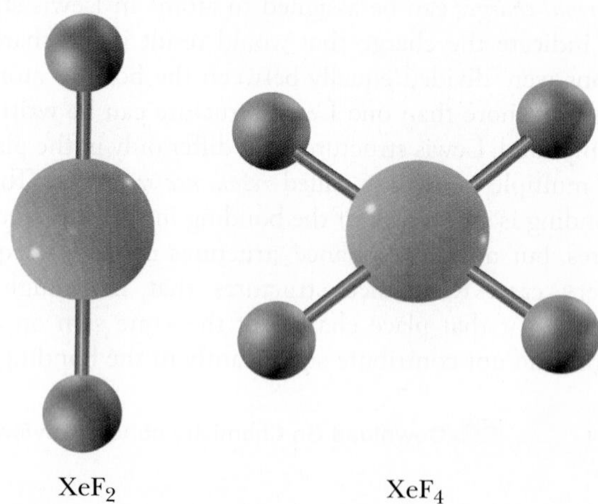

9.21 Define bond dissociation energy. Describe a method that uses bond dissociation energies to calculate approximate enthalpies of chemical reactions.

9.22 Explain why the bond dissociation energy for HCl is measured exactly, but that for a C–C bond is an average.

Exercises

OBJECTIVE Write Lewis electron-dot symbols for elements and ions.

9.23 Write the Lewis symbol for the following species.
(a) a sodium atom (b) a fluorine atom
(c) O^{2-} (d) Mg^{2+}

9.24 ■ Write the Lewis symbol for the following species.
(a) a sulfur atom (b) I^-
(c) a beryllium atom (d) Ga^{2+}

9.25 Write the Lewis symbol for the following species.
(a) a lithium atom (b) S^{2-}
(c) a magnesium atom (d) a bromine atom

9.26 Write the Lewis symbol for the following species.
(a) a potassium atom (b) a nitrogen atom
(c) a boron atom (d) F^-

OBJECTIVES Represent the formation of ionic compounds through the use of Lewis symbols and describe how the charges and the sizes of ions influence lattice energies.

9.27 Write the formulas of the ionic compounds that will form from the two pairs of elements given. Of each pair, which has the greater lattice energy? Explain your choice.
(a) lithium and oxygen; sodium and sulfur
(b) potassium and chlorine; magnesium and fluorine

9.28 Write the formulas of the ionic compounds that will form from the two pairs of elements given. Of each pair, which has the greater lattice energy? Explain your choice.
(a) potassium and sulfur; potassium and chlorine
(b) lithium and fluorine; rubidium and chlorine

9.29 Of the ionic solids LiCl and LiI, which has the greater lattice energy? Explain your choice.

9.30 Of the ionic solids CaO and BaO, which has the greater lattice energy? Explain your choice.

9.31 Arrange the following series of compounds in order of increasing lattice energies.
(a) NaBr, NaCl, KBr
(b) MgO, CaO, $CaCl_2$
(c) LiF, BeF_2, BeO

9.32 ■ Arrange the following series of compounds in order of increasing lattice energies.
(a) LiCl, NaCl, $BeCl_2$
(b) MgO, MgF_2, $MgBr_2$
(c) $MgCl_2$, BeO, $BeCl_2$

OBJECTIVE **Write Lewis structures of molecules and polyatomic ions.**

9.33 Write the formula of the simplest neutral compound made from the following:
(a) carbon and fluorine
(b) nitrogen and iodine
(c) oxygen and chlorine

9.34 Write the formula of the simplest neutral compound made from the following:
(a) silicon and hydrogen
(b) phosphorus and chlorine
(c) sulfur and fluorine

9.35 Write the Lewis structure for each of the following compounds. Label electrons as bonding pairs or lone pairs.
(a) H_2S (b) H_2CO (c) PF_3

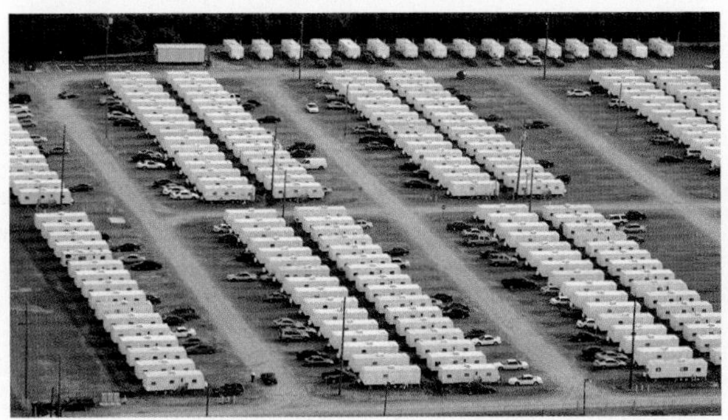

Photo by Mario Tama/Getty Images

Formaldehyde, H_2CO, is a toxic gas that was found to be at unacceptably high levels in some of the FEMA trailers supplied to homeless victims of hurricane Katrina.

9.36 ■ Draw Lewis structures for the following species. (The skeleton is indicated by the way the molecule is written.)
(a) Cl_2CO (b) H_3C–CN

9.37 What is the bond order between each pair of bonded atoms of the compounds in Exercise 9.35?

9.38 What is the bond order between each pair of bonded atoms of the compounds in Exercise 9.36?

9.39 Write the Lewis structure for the following compounds.
(a) AsH_3 (b) ClF (c) CF_3OH

9.40 ■ Write Lewis structures for these molecules or ions.
(a) CN^-
(b) tetrafluoroethylene, C_2F_4, the molecule from which Teflon is made
(c) acrylonitrile, CH_2CHCN, the molecule from which Orlon is made

9.41 Write the Lewis structure for the following species.
(a) CCl_4 (b) NO^+ (c) BCl_4^-

9.42 ■ Draw a Lewis structure for each of the following molecules or ions.
(a) CS_2
(b) BF_4^-
(c) HNO_2 (where the bonding is in the order HONO)
(d) $OSCl_2$ (where S is the central atom)

9.43 Write the Lewis structure for each compound, with the skeleton structure shown below.

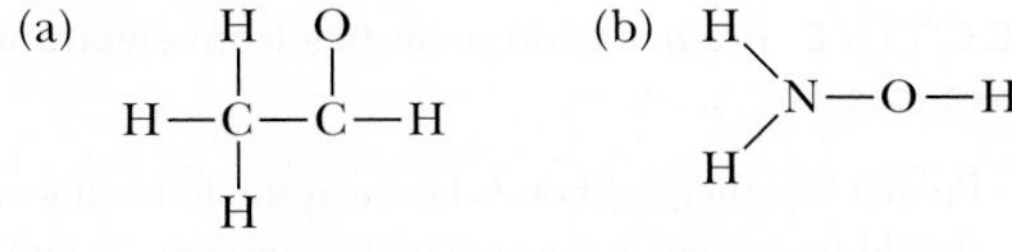

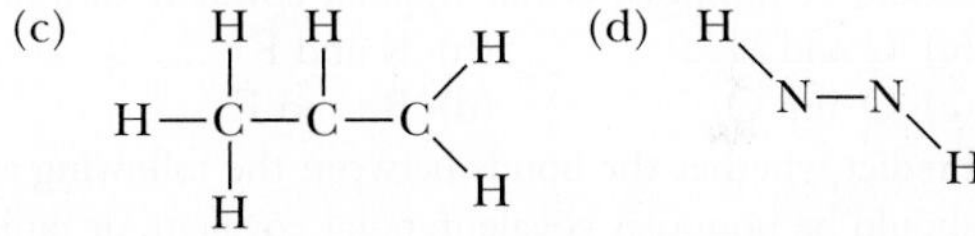

9.44 Write the Lewis structure for each compound, with the skeleton structure shown below.

(a) F—O—O—F (b) O above C in H—C—O—H

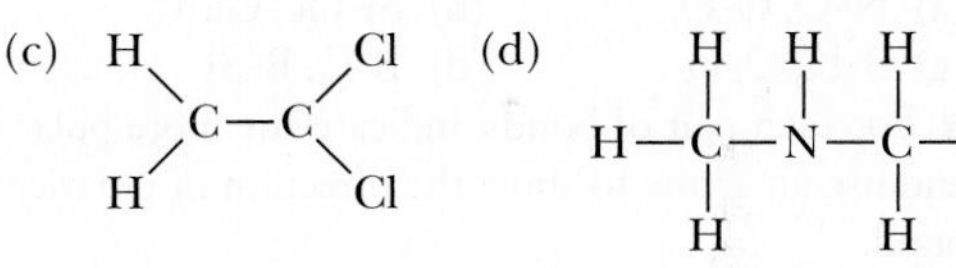

9.45 Write the Lewis structure for each compound, with the skeleton structure shown below.

(a) H—C—C—H (b) H—O—Cl
(c) H_2C—C(H)—C—N (d) H—O—O—H

9.46 Write the Lewis structure for each species, with the skeleton structure shown below.

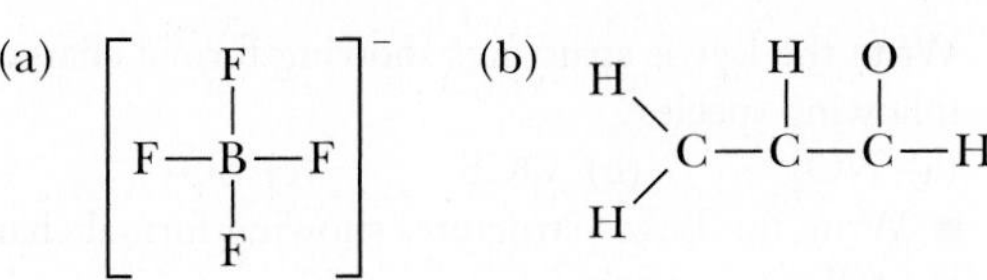

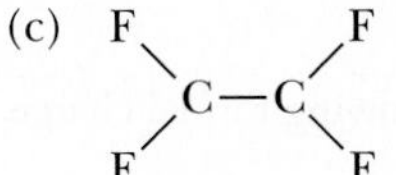

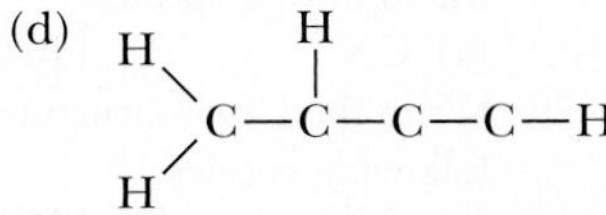

OBJECTIVE Describe the periodic trends of electronegativity.

9.47 Which atom in each of the following pairs has the greater electronegativity? Use only a periodic table to determine your answer.
(a) Br, I (b) S, Cl (c) C, N

9.48 Which atom in each of the following pairs has the greater electronegativity? Use only a periodic table to determine your answer.
(a) O, S (b) K, Ge (c) Br, Te

9.49 Arrange the members of each of the following sets of elements in order of increasing electronegativity (use only a periodic table to determine your answer).
(a) Cl, I, Br (b) Br, Ca, Ga (c) O, K, Ge

9.50 Arrange the members of each of the following sets of elements in order of increasing electronegativities.
(a) Pb, C, Sn, Ge (b) S, Na, Mg, Cl
(c) P, N, Sb, Bi (d) Se, Ba, F, Si, Sc

OBJECTIVE Predict bond polarities from electronegativity differences.

9.51 Predict whether the bonds between the following atoms should be nonpolar covalent, polar covalent, or ionic.
(a) C and H (b) K and F
(c) O and O (d) Be and F

9.52 Predict whether the bonds between the following atoms should be nonpolar covalent, polar covalent, or ionic.
(a) Li and O (b) F and Cl
(c) S and S (d) Na and I

9.53 For each pair of bonds, indicate which has the greater polarity, and show the direction of the dipole moment (use Figure 9.9 if necessary).
(a) N-O, C-O (b) Si-Ge, Ge-C
(c) S-H, O-H (d) B-C, B-Si

9.54 ■ For each pair of bonds, indicate the more polar bond, and use an arrow to show the direction of polarity in each bond.
(a) C—O and C—N (b) P—Br and P—Cl
(c) B—O and B—S (d) B—F and B—I

9.55 Which molecule has the most polar bond: N_2, BrF, or ClF? Use an arrow to show the direction of polarity in each bond.

9.56 ■ Given the bonds C—N, C—H, C—Br, and S—O,
(a) which atom in each is the more electronegative?
(b) which of these bonds is the most polar?

OBJECTIVE Assign formal charges on atoms in Lewis structures and use them to predict the relative stabilities.

9.57 Write the Lewis structures showing formal charge for the following species.
(a) NO_2^- (b) OCS (c) SO_3

9.58 ■ Write the Lewis structures showing formal charge for the following species.
(a) CN^- (b) HCO_2^-

9.59 Write the Lewis structures showing formal charge for the following species.
(a) FSO_3^- (b) HNC (c) SO_2Cl_2

9.60 Write the Lewis structures showing formal charge for the following species.
(a) ClO_3^- (b) NCCN (c) $SOCl_2$

9.61 The Understanding section in Example 9.10 showed two possible arrangements for the atoms in H_2CO. We concluded that structure A, repeated here, was better. A third possible structure follows as B. Use formal charge to predict which of these arrangements is the more favored.

A: H_2C—O (both H on C) B: H—C—O—H

9.62 ■ The connectivity of HNO could be either HNO or HON. Draw a Lewis structure for each and predict which connectivity is the more favorable arrangement.

9.63 The connectivity of H_3CN could be H_2CNH or $HCNH_2$ as well. Draw a Lewis structure for each of these two structures and predict which connectivity is the more favorable arrangement.

9.64 Predict which connectivity, SCN^- or CSN^-, is the more favorable arrangement.

OBJECTIVE Write all possible resonance forms for a given molecule or ion and predict the relative importance.

9.65 Write all possible resonance structures for the following species. Assign a formal charge to each atom. In each case, which resonance structure is the most important?
(a) NO_2^- (nitrogen is central)
(b) ClCN

9.66 ■ Show all possible resonance structures for each of the following molecules or ions:
(a) Nitrate ion, NO_3^-
(b) Nitrous oxide (laughing gas), N_2O (where the bonding is in the order N—N—O)

9.67 Write all possible resonance structures for the species with the skeleton structures shown below. In each case, which resonance structure is the most important?

(a) H_2C—N—N (both H on C) (b) $[O—C—N]^-$

9.68 Write all possible resonance structures for the species with the skeleton structures shown below. In each case, which resonance structure is the most important?

(a) H_3C—C(O)—NO_2 (skeleton: H—C(H)(H)—C(=O)—N(O)(O)) (b) $[O—N(O)—O]^-$

9.69 Write all possible resonance structures for the species with the skeleton structures shown below. In each case, which resonance structure is the most important?

(a) $[N—N—N]^-$ (b) $[O—C(O)(O)]^{2-}$

9.70 Write all possible resonance structures for the species with the skeleton structures shown below. In each case, which resonance structure is the most important?

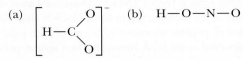

(a) $[H{-}C(O)(O)]^-$ (b) H—O—N—O

9.71 Write all resonance structures of toluene, $C_6H_5CH_3$, a molecule with the same cyclic structure as benzene. Which resonance structures are the most important?

9.72 ■ Write all resonance structures of chlorobenzene, C_6H_5Cl, a molecule with the same cyclic structure as benzene. In all structures, keep the C–Cl bond as a single bond. Which resonance structures are the most important?

9.73 Draw all resonance structures for methylisocyanate, CH_3NCO, a toxic gas used in the manufacturing of pesticides. Which resonance structures are the most important?

9.74 Write all possible resonance structures for the species with the skeleton structures shown below. In each case, which resonance structure is the most important?

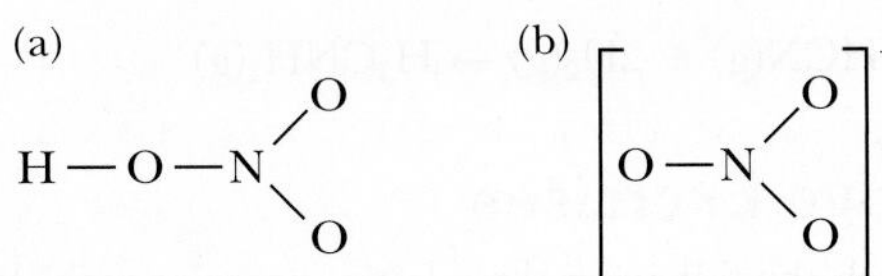

OBJECTIVE Write the Lewis structure and recognize the types of molecules for which the octet rule is not obeyed.

9.75 Write the Lewis structures of the following molecules, and indicate whether each is an odd-electron molecule, an electron-deficient molecule, or an expanded valence shell molecule.
(a) SeF_6 (b) BBr_3 (c) NO_2

9.76 ■ Write the Lewis structures for the following species, and indicate whether each is an odd-electron species, an electron-deficient species, or an expanded valence shell species.
(a) IF_3 (b) ICl_4^- (c) N_2^+

9.77 Write the Lewis structures for the following species, and indicate whether each is an odd-electron species, an electron-deficient species, or an expanded valence shell species. There is only one central atom in each.
(a) XeF_2
(b) $BeCl_2$
(c) XeO_2F_4 (both O and F are terminal atoms)

9.78 Write the Lewis structures for the following species, and indicate whether each is an odd-electron species, an electron-deficient species, or an expanded valence shell species.
(a) BI_3 (b) IF_5 (c) HN_2

9.79 For the following species, write a resonance structure with an octet about the central atom and a second resonance structure that minimizes formal charges.

(a) O—Se—O (b) O—S(—O)—O (S bonded to three O atoms)

9.80 For the following species, write a resonance structure with an octet about the central atom and a second resonance structure that minimizes formal charges.

(a) O—Cl(O)(O)—O—H (Cl bonded to O above, O below, O left, and O—H right) (b) H—O—P(O)(O—H)—O—H (P bonded to O above, O—H below)

9.81 For the following species, write a resonance structure with an octet about the central atom and a second resonance structure that minimizes formal charges.

(a) $[O{-}Cl{-}O]^-$ (b) $[O{-}S(O){-}O{-}H]^-$ (S bonded to O above)

9.82 For the following species, write a resonance structure with an octet about the central atom and a second resonance structure that minimizes formal charges.

(a) $[O{-}S(O){-}O]^{2-}$ (S bonded to O above) (b) $[O{-}Cl(O)(O){-}O]^-$ (Cl bonded to O above and O below)

OBJECTIVES Relate bond energies to bond strengths and calculate approximate enthalpies of reaction from bond energies.

9.83 Write the Lewis structures of H_2CNH and H_3CNH_2. Predict which molecule has the greater C–N bond energy.

9.84 ■ Write the Lewis structures of HNNH and H_2NNH_2. Predict which molecule has the greater N–N bond energy.

9.85 Using Table 9.4, calculate the energy required to break all of the bonds in one mole of the following compounds.
(a) NH_3 (b) CH_3OH

9.86 Using Table 9.4, calculate the energy required to break all of the bonds in one mole of the following compounds.
(a) CH_2CF_2 (b) N_2H_4

9.87 Using Table 9.4, calculate an approximate enthalpy change for
(a) the reaction of molecular hydrogen (H_2) and molecular oxygen (O_2) in the gas phase to produce 2 mol water vapor.
(b) the reaction of carbon monoxide and molecular oxygen to form 2 mol carbon dioxide.

9.88 ■ The equation for the combustion of gaseous methanol is

$$2CH_3OH(g) + 3O_2(g) \rightarrow 2CO_2(g) + 4H_2O(g)$$

Using the bond dissociation enthalpies, estimate the enthalpy change for this reaction.

9.89 Use Table 9.4 to calculate an approximate enthalpy change for
(a) the combustion of 1 mol C_2H_4 in excess molecular oxygen to form gaseous water and CO_2.
(b) the reaction of 1 mol formaldehyde, H_2CO, with molecular hydrogen to form gaseous methanol (CH_3OH).

9.90 Use Table 9.4 to calculate an approximate enthalpy change for
(a) the reaction of H_2 and C_2H_2 to form 1 mol C_2H_6.
(b) the reaction of molecular hydrogen and molecular nitrogen to form 1 mol ammonia.

© Cengage Learning/Charles D. Winters

Chapter Exercises

9.91 Acrolein has the formula $CH_2C(H)C(O)H$. Draw its skeleton structure and Lewis structure.

9.92 Use the octet rule to predict the element (E) from the second period that would be the central atom in the following ions.
(a) EF_4^- (b) EF_4^+

9.93 The compound disulfur dinitride, S_2N_2, has a cyclic structure with alternating sulfur and nitrogen atoms. Draw two Lewis structures for S_2N_2, one that places an octet about each atom and another that minimizes formal charges.

9.94 ■ Consider the nitrogen–oxygen bond lengths in NO_2^+, NO_2^-, and NO_3^-. In which ion is the bond predicted to be longest? In which is it predicted to be the shortest? Explain briefly.

9.95 The compound SF_4CH_2 has an unusually short SC distance of 155 pm. Use the Lewis structure to explain this short distance.

9.96 The compound CF_3SF_5 was recently identified in Earth's atmosphere and is a potential greenhouse gas. If the carbon atom and sulfur atom are bonded together, what is the Lewis structure of this compound? Do both central atoms follow the octet rule?

9.97 A compound related to the new greenhouse gas in the previous exercise is $FCSF_3$, where again the carbon and sulfur atoms are bonded together and act as central atoms. What is the Lewis structure of this compound? Do both central atoms follow the octet rule?

9.98 ▲ In the gas phase, the oxide N_2O_5 has a structure with an N–O–N core, with the other four oxygen atoms in terminal positions. In contrast, in the solid phase, the stable form is $[NO_2]^+[NO_3]^-$. Draw one Lewis structure of the molecular form (with N–O–N single bonds) and all possible resonance structures of both ions observed in the solid. Remember that second-period elements never exceed an octet.

9.99 The molecule nitrosyl chloride, NOCl, has a skeleton structure of O–N–Cl. Two resonance forms can be written; write them both. Use the formal charge stability rules to predict which form is more stable.

9.100 Draw the Lewis structure and calculate the energy needed to break all of the bonds in 1 mol of CH_3NH_2.

9.101 Draw the Lewis structure of BrNO. Which is the more polar bond in the molecule?

9.102 Phosgene, Cl_2CO, is an extremely toxic gas that can be prepared by the reaction of CO with Cl_2. Using data from Table 9.4, calculate the approximate enthalpy change for this reaction.

9.103 Calculate an approximate enthalpy change (Table 9.4) for the following reaction:

$$HCN(g) + 2H_2(g) \rightarrow H_3CNH_2(g)$$

Cumulative Exercises

9.104 Match the following three lattice energies with the three compounds LiF, KCl, and CsI. Explain your answer.

715 kJ/mol 582 kJ/mol 1036 kJ/mol

9.105 ▲ Draw the Lewis structures of N_2O and NO_2. Based on these structures, predict which has the shorter N–O bond. Does either of these molecules contain unpaired electrons?

9.106 The reaction of XeF_6 with a limited amount of H_2O yields $XeOF_4$ and HF. Write the equation for this reaction; then draw the Lewis structures of the expanded valence shell molecules in the equation.

9.107 ▲ The compound ClF_3 reacts with solid uranium to produce the volatile compounds UF_6 and ClF. ClF_3 can be used to separate uranium from plutonium, because plutonium reacts with ClF_3 to form the nonvolatile compounds PuF_4 and ClF.
(a) Write equations for these two reactions.
(b) Draw the Lewis structure of ClF_3. Is there anything unusual about the Lewis structure of ClF_3?
(c) If a mixture of uranium and plutonium reacts with excess ClF_3 to produce 43.5 g UF_6 and 22.1 g PuF_4, what were the masses of the uranium and plutonium in the starting mixture? Use 244 g/mol for the molar mass of Pu.

9.108 ▲ Natural gas is mostly methane, CH_4. Homes that are heated by natural gas obtain the energy from the combustion reaction

$$CH_4(g) + O_2(g) \rightarrow H_2O(g) + CO_2(g)$$

(a) Balance the reaction.
(b) Is this a redox reaction? Defend your answer.
(c) Determine the enthalpy change of this reaction using the enthalpy of formation data in the Appendix G.
(d) Determine the enthalpy change of this reaction using the bond energy data in Table 9.4. Compare your answer here with your answer in (c) and comment. Which answer do you expect is more accurate?
(e) Which of the bonds in the reactants and products are polar? Which are nonpolar?
(f) Most utility companies sell natural gas in units of 100 cubic feet (abbreviated ccf), where 1 ccf = 2831 L. If the natural gas is at standard temperature and pressure, how much energy is given off by the combustion of 1 ccf natural gas?
(g) If natural gas burns in limited oxygen, carbon monoxide is the product instead of carbon dioxide:

$$CH_4(g) + O_2(g) \rightarrow H_2O(g) + CO(g)$$

Balance this reaction.
(h) What are the differences in the C–O bonds in CO and CO_2? Assign formal charges to the atoms in each.
(i) Determine the energy change of the CO-containing reaction using data in Appendix G and the data in Table 9.4. Are your answers closer to each other than your answers in (c) and (d)?
(j) Exactly 1 mol CH_4 was burned and 600.0 kJ was given off. How many grams of CO and CO_2 were produced?

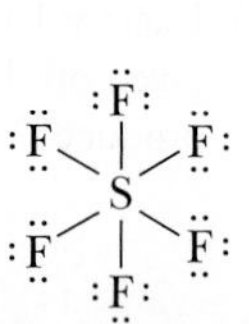

Over the years, chemists have developed experiments that allow the determination of the shapes of most molecules and polyatomic ions. These experiments have shown that molecules have many different three-dimensional shapes, and that the shapes of molecules greatly influence their physical and chemical properties. Scientists were able to exploit the properties of one of these molecules, sulfur hexafluoride, to help develop strategies to combat terrorists who threaten mass destruction with poison gas.

Sulfur hexafluoride, SF_6, is quite interesting because of its highly symmetric shape. As shown here, the six fluorine atoms are bonded around the sulfur atom in a regular geometric arrangement termed *octahedral.* The Lewis structure of this molecule indicates that it has greater than an octet of electrons around the central sulfur. Given this unusual electron count, one might expect that SF_6 would be very reactive, but the opposite is true. It is a colorless gas that is nontoxic and very nonreactive. SF_6 is so stable that it is used as an insulating gas for high-voltage electrical equipment and can be heated to 500 °C without decomposition.

The stability of SF_6 made it ideal to assist in a study important to the war on terror. On March 20, 1995, members of a Japanese terrorist group carried out a chemical attack in Tokyo, Japan. In five separate, but coordinated acts, terrorists

Molecular Structure and Bonding Theories

10

CHAPTER CONTENTS

OWL Online homework for this chapter may be assigned in OWL.

Look for the green colored vertical bar throughout this chapter, for integrated references to this chapter introduction.

released deadly sarin gas on several lines of the Tokyo Metro. The gas quickly spread through the subway cars, killing 12 people and injuring thousands more. It was the deadliest attack on Japanese soil since World War II.

Sarin prevents the nervous system from working properly. Exposure to even small amounts of sarin can lead to suffocation from paralysis of the muscles used in breathing. Discovered in 1938, sarin is still one of the deadliest chemical agents in existence.

Thankfully, chemical attacks such as the sarin incident are rare. However, governments across the world continue to analyze various scenarios of possible terrorist chemical attacks. In late March and early April of 2007, the English government conducted two full-scale tests to determine how a toxic gas would spread through London's subway system, which carries about 3.5 million riders each work day. For the tests to work effectively, scientists needed to choose a gas that was safe, nonreactive, and would move through the system about the same way as toxic gases such as sarin. After studying these factors, the scientists chose sulfur hexafluoride. The molar mass of SF_6 (146 g/mol) is close to that of sarin (140 g/mol), so it will diffuse at nearly the same rate, and is an effective, but safe, surrogate for sarin. ▮

Kyodo/Landov

Aftermath of the March 20, 1995, sarin terrorist attack on the Metro subway in Tokyo, Japan.

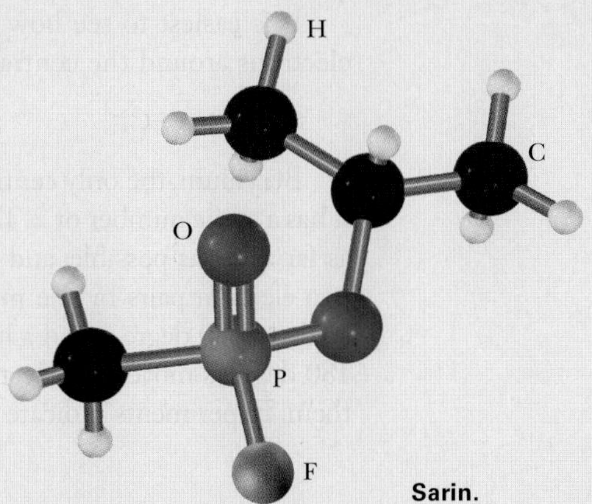

Sarin.

On the basis of many experimentally determined molecular structures, chemists have developed models that enable us to predict molecular shapes fairly accurately. These results are important because the arrangement of the atoms in a molecule has a strong influence on its physical properties and chemical reactivity. A simple model based on Lewis structures is extremely useful for predicting the shapes of many molecules and is the first model presented in this chapter. It is followed by two other theories, valence bond theory and molecular orbital theory, that explain bonding in more detail, describing it as combinations of the atomic orbitals of the individual atoms.

10.1 Valence-Shell Electron-Pair Repulsion Model

OBJECTIVES

- ☐ Explain the assumptions of the valence-shell electron-pair repulsion model
- ☐ Determine the steric number from the Lewis structure
- ☐ Use the steric number to assign the bonded-atom lone-pair arrangement of each central atom
- ☐ Distinguish between the bonded-atom lone-pair arrangement and molecular shape
- ☐ Predict the shapes of molecules from the valence-shell electron-pair repulsion model
- ☐ Determine the locations of lone pairs in molecules with a steric number of 5 or 6

The **valence-shell electron-pair repulsion (VSEPR)** model predicts the shapes of most molecules from their Lewis structures. The main premise of the model is that the electron pairs about an atom repel each other.

Because they repel each other, electron pairs are oriented as far apart as possible.

VSEPR Rule 1: A molecule has a shape that minimizes electrostatic repulsions between valence-shell electron pairs. Minimum repulsion results when the electron pairs are as far apart as possible.

The VSEPR model predicts the shape around each *central atom,* an atom in a molecule that is bonded to more than one other atom. Because the shape is determined by electrostatic repulsions among the electrons, we need to consider the electrons in both the bonds and lone pairs as shown in a proper Lewis structure. To simplify the process of determining the shape, we define the **steric number** as the number of *lone pairs* on the central atom *plus* the number of *atoms bonded to it.*

$$\text{Steric number} = (\text{number of lone pairs on central atom}) + (\text{number of atoms bonded to central atom})$$

The steric number is used to determine the **bonded-atom lone-pair arrangement,** which is the shape that maximizes the distances between regions of electron density about a central atom. Figure 10.1 shows the geometric arrangements that keep the bonded atoms and electron pairs the maximum distance apart. These arrangements are named for the geometric solids that are formed by lines drawn to connect the bonded atoms and/or electron pairs.

It is easiest to see how the VSEPR model works when there are only two pairs of electrons around the central atom, as in $BeCl_2$. The Lewis structure of $BeCl_2$ is

:C̤̈l— Be —C̤̈l:

Beryllium, the only central atom in $BeCl_2$, has two bonded atoms and no lone pairs, so it has a steric number of 2. The repulsion of the electron pairs in the two bonds drives them as far apart as possible, and for two electron pairs, a 180-degree orientation distances the two electron pairs by the maximum amount. The model thus predicts a Cl–Be–Cl bond angle of 180 degrees and a linear shape. Any reduction of the Cl–Be–Cl angle to less than 180 degrees moves the electron pairs closer together and increases the repulsion between them. Experiments indicate that $BeCl_2$ is a linear molecule.

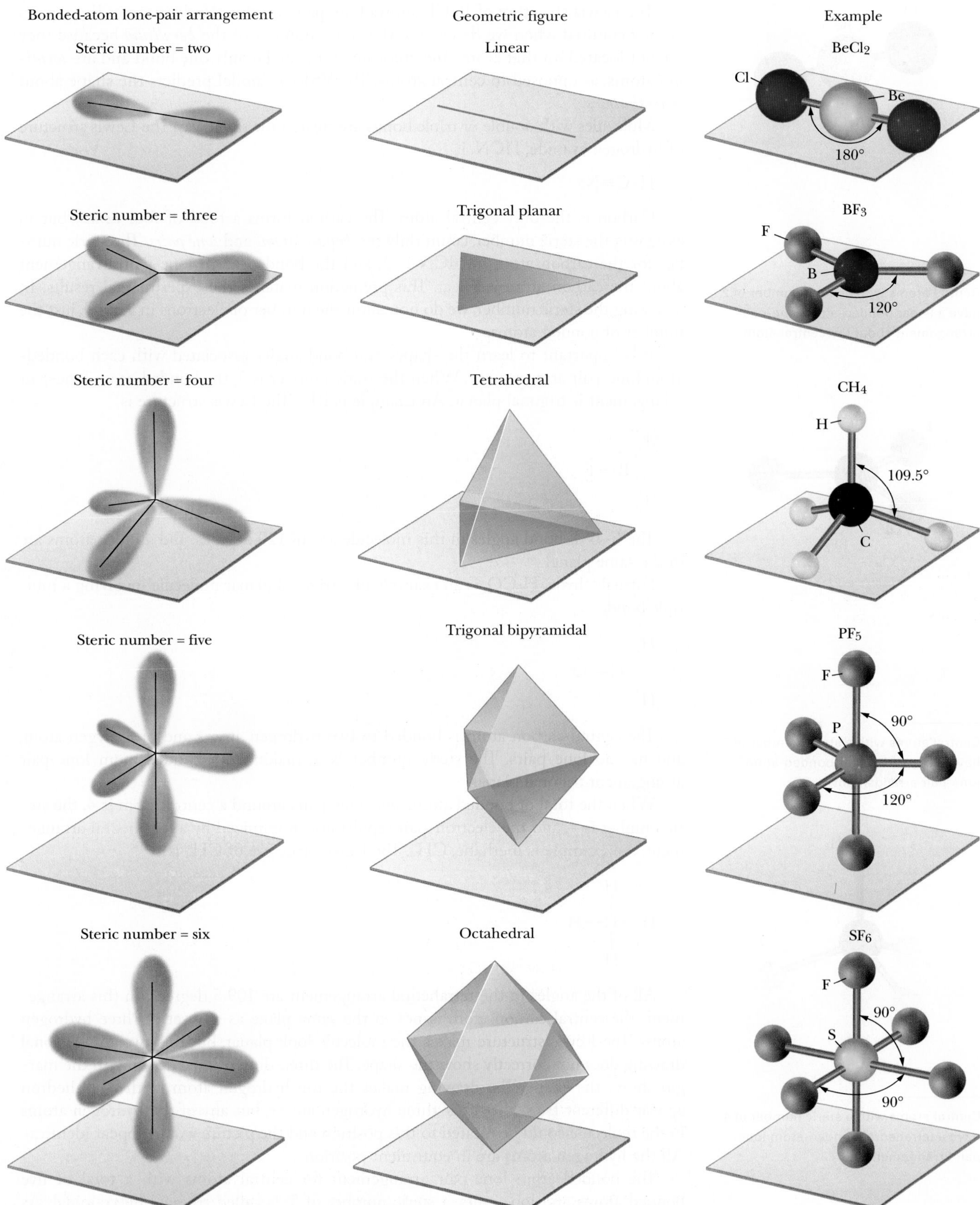

Figure 10.1 Geometric arrangements expected for different steric numbers.

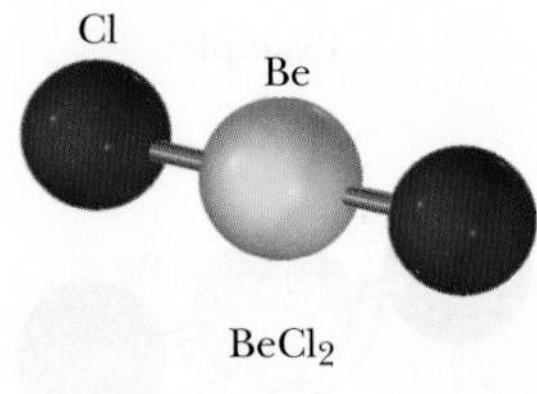

The Lewis structure of $BeCl_2$ shows lone pairs on the *chlorine* atoms. These pairs are *not* counted when we determine the steric number of the *beryllium* because they are not located on that atom. The chlorine atoms make only one bond and are *terminal* atoms, as opposed to central atoms. The VSEPR model predicts the shape about *central atoms.*

Molecules with double or triple bonds are treated the same way. The Lewis structure of hydrogen cyanide, HCN, is

H–C≡N:

Carbon is the only central atom. The carbon forms a total of four *bonds,* but in assigning the steric number, count only the *bonded atoms* and *lone pairs.* The steric number for the carbon atom in HCN is 2, and the bonded-atom lone-pair arrangement about the carbon atom is linear. This prediction matches the experimental results. In assigning the steric number, we do not count the number of electrons in a bond, just the number of bonded atoms.

Central atoms with a steric number of 2 have a linear bonded-atom lone-pair arrangement about the central atom.

It is important to learn the shapes and bond angles associated with each bonded-atom lone-pair arrangement. When the steric number is 3, the bonded-atom lone-pair arrangement is trigonal planar. An example is BF_3. The Lewis structure is

```
:F:
   \
    B—F:
   /
:F:
```

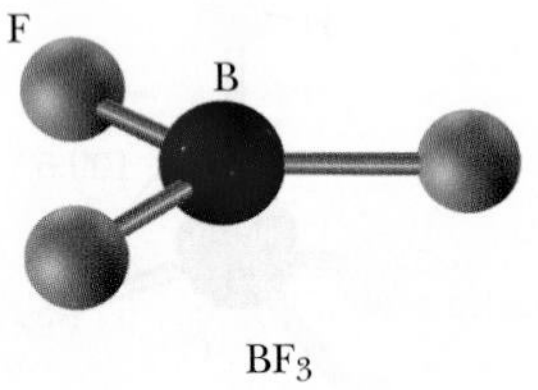

The F–B–F bond angles in this molecule are all 120 degrees, and all four atoms are in the same plane.

Formaldehyde, H_2CO, is an example of a trigonal planar molecule involving a multiple bond.

```
H
 \
  C=O:
 /
H
```

The central carbon atom is bonded to two hydrogen atoms and one oxygen atom, and has no lone pairs. The steric number is 3, making the bonded-atom lone-pair arrangement trigonal planar.

Central atoms with a steric number of 3 have a trigonal planar bonded-atom lone-pair arrangement.

When the total of bonded atoms plus lone pairs around a central atom is 4, the steric number is 4, and the electron-pair repulsion is minimized by a tetrahedral arrangement. An example is methane, CH_4. The Lewis structure of CH_4 is

```
    H
    |
H—C—H
    |
    H
```

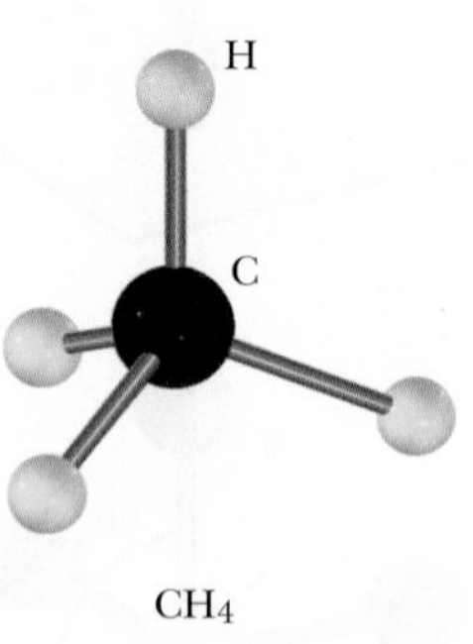

All of the angles in the tetrahedral arrangement are 109.5 degrees. In this arrangement, the central carbon atom is not in the same plane as any set of three hydrogen atoms. The Lewis structure makes the molecule look planar, but this two-dimensional drawing does not correctly show the shape. The three-dimensional drawing in the margin shows the shape. The drawing makes the top hydrogen atom in the tetrahedron appear different from the other three hydrogen atoms, but any of the hydrogen atoms in the molecule could be rotated to this position and the picture would appear identical. All the hydrogen atoms are in equivalent positions.

Central atoms with a steric number of 4 have a tetrahedral bonded-atom lone-pair arrangement.

The bonded-atom lone-pair arrangement for central atoms with a total of five bonded atoms and lone pairs, a steric number of 5, is called trigonal bipyramidal. As shown for PF_5, this shape can be considered the combination of a linear [F(1)–P–F(5)] arrangement perpendicular to a trigonal planar [P, F(2), F(3), F(4)] arrangement. In contrast with the other bonded-atom lone-pair arrangements, not every location is equivalent, and the trigonal bipyramid has two different environments in which to place bonded atoms.

The F(2), F(3), and F(4) atoms (which are equivalent) are in the *equatorial* positions (they form an equilateral triangle) and the F(1) and F(5) atoms [which are equivalent to each other but not to F(2), F(3), and F(4)] are in the *axial* positions (they are perpendicular to the plane). The bonds of the axial fluorine atoms make a 90-degree angle with the bonds of the equatorial fluorine atoms. The bonds to the equatorial fluorine atoms make an angle of 120 degrees with each other. The F(1)–P–F(5) angle is, of course, 180 degrees. The axial and equatorial positions are different, and the axial fluorine atoms have different environments from the equatorial ones. For example, the P–F bond distances for F(1) and F(5) are equal to each other but are longer than the three equal P–F distances for F(2), F(3), and F(4).

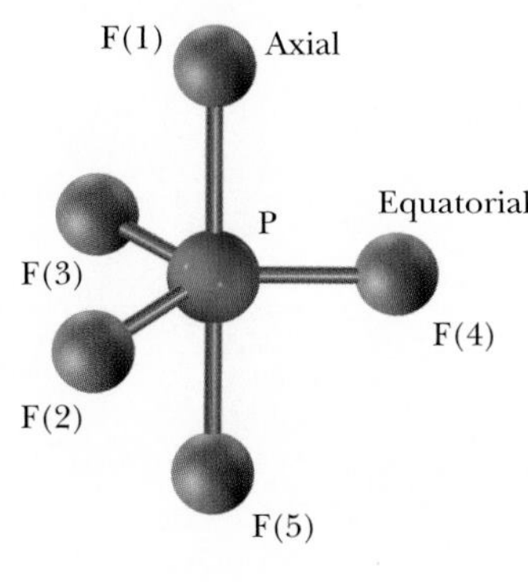

PF_5

Central atoms with a steric number of five have a trigonal bipyramidal bonded-atom lone-pair arrangement.

The bonded-atom lone-pair arrangement for a central atom with a total of six bonded atoms and lone pairs, a steric number of 6, is octahedral. An example is SF_6, the molecule discussed in the chapter introduction for its stability. The bond angles in this case are 90 or 180 degrees. In this geometry, all six positions are equivalent.

Central atoms with a steric number of six have an octahedral bonded-atom lone-pair arrangement.

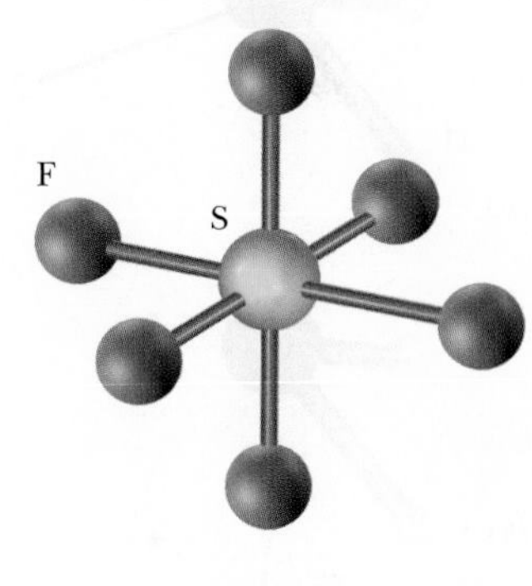

SF_6

EXAMPLE 10.1 Bonded-Atom Lone-Pair Arrangement

What is the bonded-atom lone-pair arrangement about the central carbon atom, and what are the angles between the bonds in the molecules (a) CCl_4 and (b) CO_2?

Strategy Determine the steric number from the Lewis structure, then use that number to assign the bonded-atom lone-pair arrangement.

Solution

(a) First, write the Lewis structure for CCl_4.

```
       ..
      :Cl:
       |
  ..   |   ..
 :Cl—  C —Cl:
  ..   |   ..
       |
      :Cl:
       ..
```

The central carbon atom has four bonded atoms and no lone pairs. There are lone pairs on the chlorine atoms, *but we count only lone pairs on the central atom,* so the steric number is 4. The bonded-atom lone-pair arrangement is tetrahedral. The Cl–C–Cl bond angles are all 109.5 degrees.

(b) The Lewis structure of CO_2 is

```
 ..     ..
 O=C=O
 ..     ..
```

The central carbon atom is bonded to two oxygen atoms and has no lone pairs. The steric number is 2, and the bonded-atom lone-pair arrangement is linear. Each C–O double bond is treated as a single unit. The O–C–O bond angle is 180 degrees.

Understanding

What is the bonded-atom lone-pair arrangement for phosphorus, and what are the Cl–P–Cl bond angles in PCl_5?

Answer The steric number is 5, so the bonded-atom lone-pair arrangement is trigonal bipyramidal. There are three bond angles: 90 degrees from the axial to the equatorial, 120 degrees among the equatorial, and 180 degrees between the two axial bonds.

Central Atoms That Have Lone Pairs

In all of the examples presented thus far, the central atom has had no lone pairs of electrons. An example of a molecule with lone pairs on a central atom is water. The Lewis structure of water is

```
H—Ö:
  |
  H
```

Because the oxygen atom has two lone pairs and two bonded atoms, it has a steric number of 4. The *bonded-atom lone-pair arrangement* is tetrahedral, and the predicted H–O–H angle is 109.5 degrees. When central atoms have lone pairs, the *molecular shape* is not the same as the bonded-atom lone-pair arrangement. The **molecular shape** is the geometric arrangement of the *atoms* in a species. The lone pairs influence the molecular shape but are not part of it, because there are no atoms at the locations of the lone pairs. The molecular shape of water is bent or V-shaped (Figure 10.2).

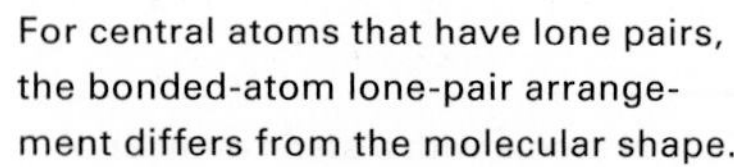

For central atoms that have lone pairs, the bonded-atom lone-pair arrangement differs from the molecular shape.

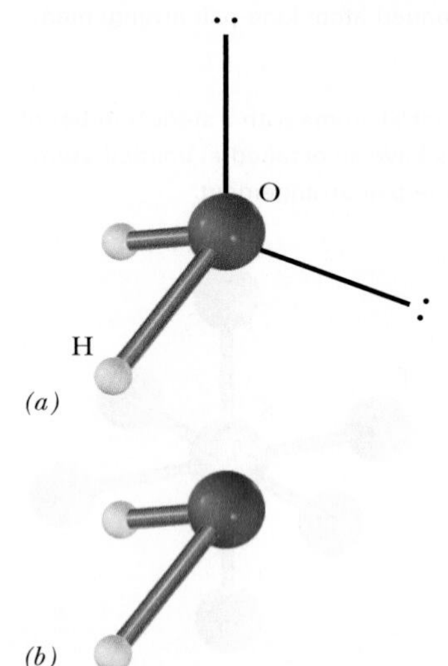

Figure 10.2 **The bonded-atom lone-pair arrangement and molecular shape of water.** *(a)* The bonded-atom lone-pair arrangement of water is tetrahedral, with two lone pairs and two bonding pairs. *(b)* The molecular shape of water is bent or V-shaped.

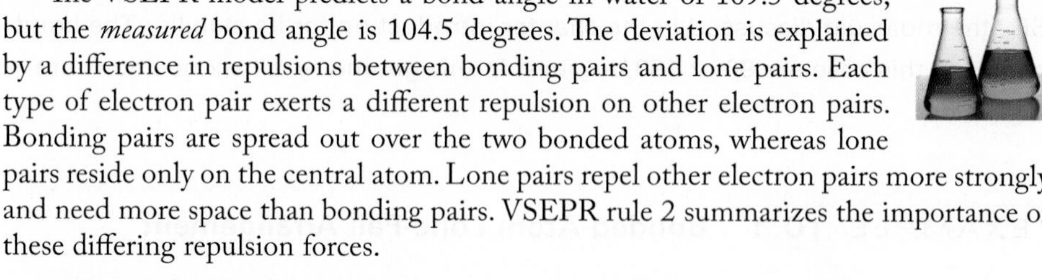

The VSEPR model predicts a bond angle in water of 109.5 degrees, but the *measured* bond angle is 104.5 degrees. The deviation is explained by a difference in repulsions between bonding pairs and lone pairs. Each type of electron pair exerts a different repulsion on other electron pairs. Bonding pairs are spread out over the two bonded atoms, whereas lone pairs reside only on the central atom. Lone pairs repel other electron pairs more strongly and need more space than bonding pairs. VSEPR rule 2 summarizes the importance of these differing repulsion forces.

VSEPR Rule 2: Forces between electron pairs vary as

lone-pair–lone-pair repulsion > lone-pair–bonding-pair repulsion
> bonding-pair–bonding-pair repulsion

Applying this rule to water, the lone-pair–lone-pair repulsions are the largest, so the lone-pair separation increases, forcing the bonding pairs closer together (Figure 10.3).

An analogous effect is observed for ammonia, NH_3. The Lewis structure shows that the steric number of nitrogen is 4.

```
H—N̈—H
  |
  H
```

The bonded-atom lone-pair arrangement is tetrahedral, as predicted by the VSEPR model, and the expected bond angles are again 109.5 degrees. In this case, the larger lone-pair–bonding-pair repulsions cause a small decrease in the H–N–H bond angles (Figure 10.4). The measured angles are 107.3 degrees.

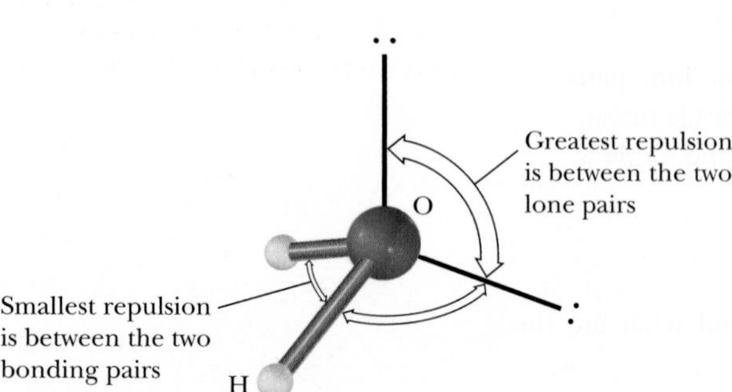

Figure 10.3 **Repulsion of electron pairs in the water molecule.** Lone-pair–lone-pair repulsion is greater than lone-pair–bonding pair and bonding-pair–bonding-pair repulsions, causing the H–O–H angle in H_2O to be 104.5 degrees, which is smaller than the 109.5-degree tetrahedral angle.

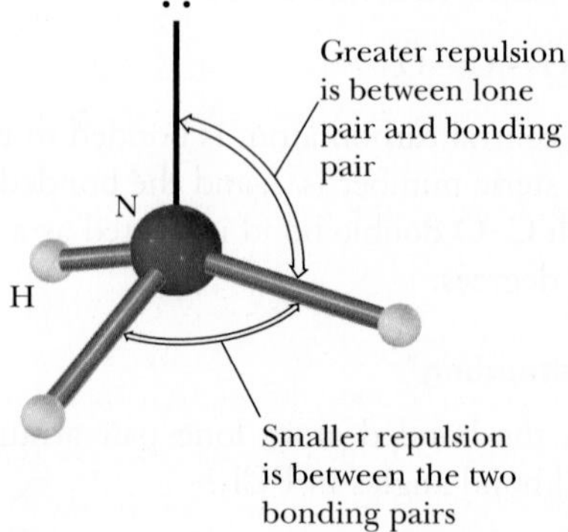

Figure 10.4 **Repulsion of electron pairs in ammonia.** Lone-pair–bonding-pair repulsion is greater than bonding-pair–bonding-pair repulsion, causing the H–N–H angles in NH_3 to be 107.3 degrees, which is slightly smaller than the 109.5-degree tetrahedral angle.

Although the bonded-atom lone-pair arrangement of NH_3 is tetrahedral, the *molecular shape* is trigonal pyramidal (Figure 10.5). Note how this shape differs from that of BF_3. The bonded-atom lone-pair arrangement for BF_3 is trigonal planar (steric number = 3), and all four atoms are in the same plane. In NH_3, the nitrogen is not in the plane of the hydrogen atoms. The difference in shape between BF_3 and NH_3 is caused by the lone pair on nitrogen.

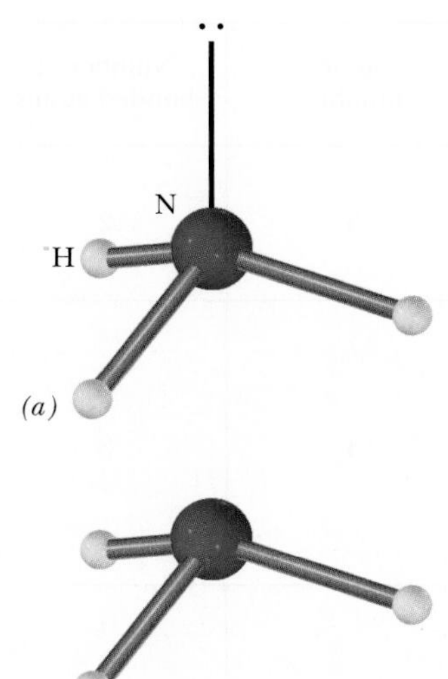

Figure 10.5 The bonded-atom lone-pair arrangement and molecular shape of ammonia. *(a)* The bonded-atom lone-pair arrangement in ammonia is tetrahedral, with one lone pair and three bonding pairs. *(b)* The molecular shape of ammonia is trigonal pyramidal.

Figure 10.6 shows the shapes around central atoms that have both bonding and lone pairs of electrons. For central atoms that have steric numbers of 3 or 4, the lone pairs can be placed in any location because all positions in these shapes are the same. In contrast, for the trigonal bipyramid, there are two types of positions: axial and equatorial. For example, the Lewis structure for SF_4 shows four bonding pairs and one lone pair around the central sulfur atom. The steric number is 5.

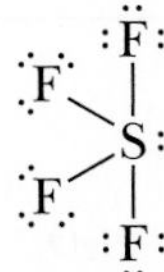

The bonded-atom lone-pair arrangement is trigonal bipyramidal. Four of the positions in the trigonal bipyramid are occupied by fluorine atoms, and the fifth by the lone pair. The lone pair should be placed to minimize the lone-pair–bonding-pair repulsions. Two molecular shapes are possible, one in which the lone pair is in an axial position (A) and another in which the lone pair is in an equatorial position (B).

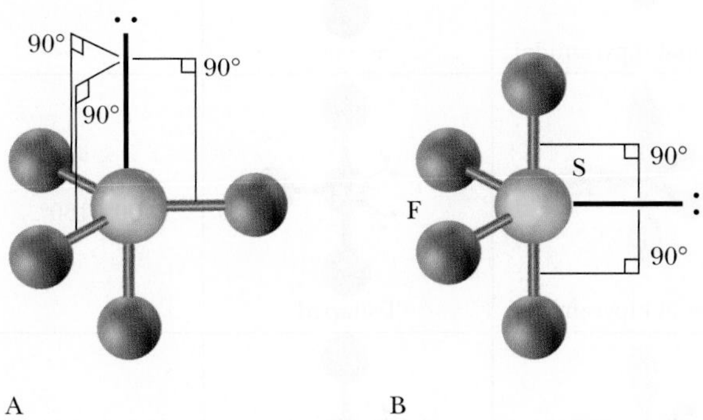

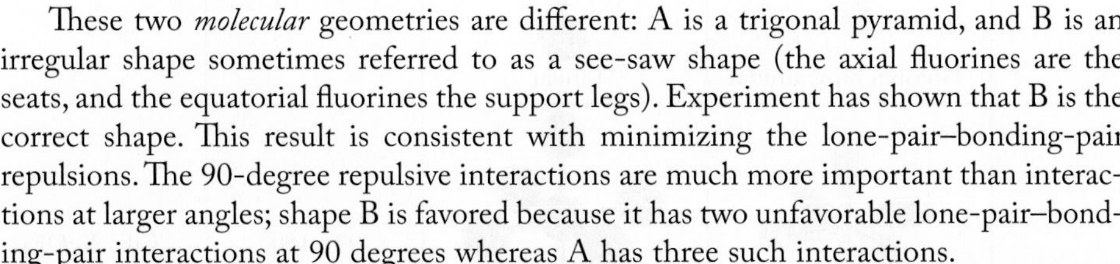

These two *molecular* geometries are different: A is a trigonal pyramid, and B is an irregular shape sometimes referred to as a see-saw shape (the axial fluorines are the seats, and the equatorial fluorines the support legs). Experiment has shown that B is the correct shape. This result is consistent with minimizing the lone-pair–bonding-pair repulsions. The 90-degree repulsive interactions are much more important than interactions at larger angles; shape B is favored because it has two unfavorable lone-pair–bonding-pair interactions at 90 degrees whereas A has three such interactions.

In a trigonal bipyramidal bonded-atom lone-pair arrangement, the structure that minimizes the number of 90-degree lone-pair interactions is favored—lone pairs always go to the equatorial position.

If a molecule with a steric number of 5 has more than one lone pair, all of the lone pairs are placed into equatorial sites to minimize lone-pair–lone-pair repulsion. For example, in the structure of ClF_3, both lone pairs are in the equatorial positions, leading to a T-shaped molecule. In this shape, the lone pairs are well separated at 120 degrees, and the number of 90-degree lone-pair–bonding-pair interactions is minimized. The structure that minimizes 90-degree interactions is always favored.

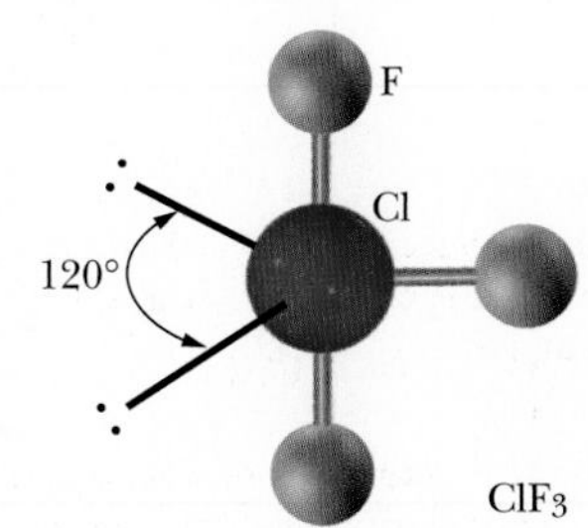

One other entry in Figure 10.6 should be considered. Xenon tetrafluoride, XeF_4, has four bonding pairs and two lone pairs, so the steric number is 6 and the bonded-atom lone-pair arrangement is octahedral. The first lone pair can go in any of the six equivalent positions in the octahedron, but the second can be either adjacent to the first (C) or away from it (D). In C, the lone pairs are at an angle of 90 degrees, whereas in D, they are at 180 degrees.

Steric number	Number of bonded atoms	Number of lone pairs	Bonded-atom lone-pair arrangement	Molecular shape	Bond angles	Examples
3	2	1	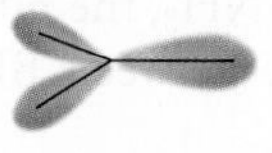 Trigonal planar	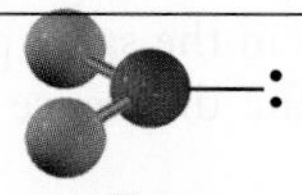 Bent	120°	SO_2, $SnCl_2$
4	3	1	 Tetrahedral	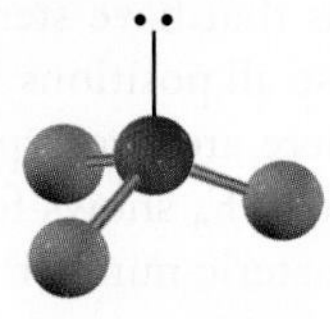 Trigonal pyramidal	109°	NH_3
4	2	2	 Tetrahedral	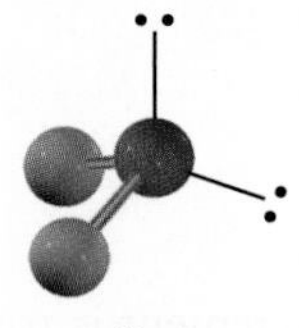 Bent	109°	H_2O
5	4	1	 Trigonal bipyramidal	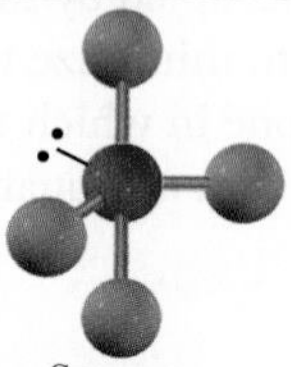 See-saw	90°, 120°, 180°	SF_4, XeO_2F_2
5	3	2	 Trigonal bipyramidal	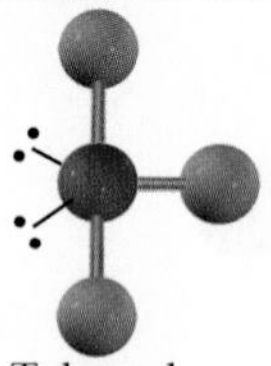 T-shaped	90°, 180°	ClF_3
5	2	3	 Trigonal bipyramidal	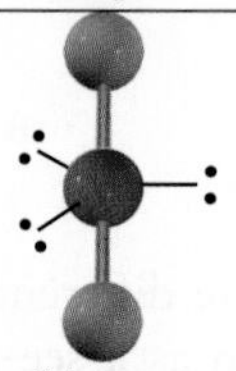 Linear	180°	XeF_2, I_3^-
6	5	1	 Octahedral	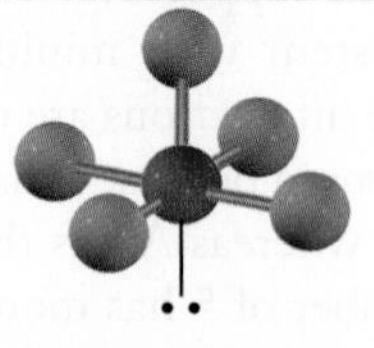 Square pyramidal	90°, 180°	ClF_5, $XeOF_4$
6	4	2	 Octahedral	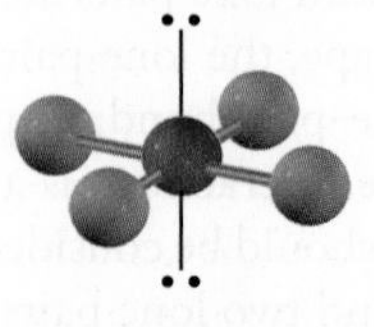 Square planar	90°, 180°	XeF_4, ICl_4^-

Figure 10.6 Molecular shapes expected for molecules with lone pairs on the central atom.

The more stable structure D minimizes the lone-pair–lone-pair repulsion. This predicted square planar molecular shape is observed experimentally.

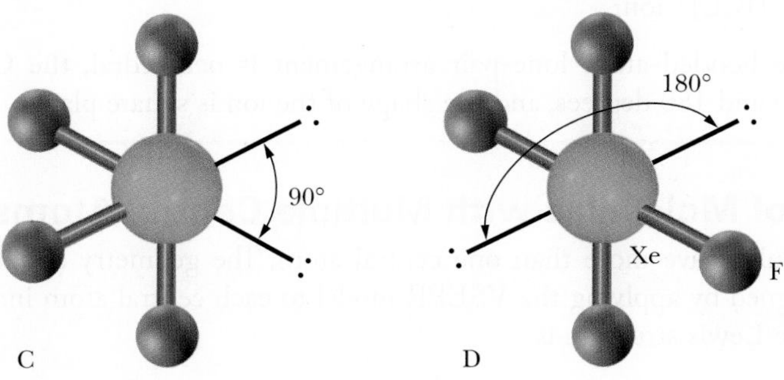

EXAMPLE 10.2 Shapes of Molecules

What are the bonded-atom lone-pair arrangements, the bond angles, and the shapes of (a) BrF_5, (b) ClNO, and (c) CO_3^{2-}?

Strategy Obtain the steric numbers of each central atom from the Lewis structures and use them to assign the bonded-atom lone-pair arrangement, which determines the bond angles. The shapes are determined from the positions of the bonded atoms; the positions of the lone pairs on central atoms are not considered in describing the *shapes*, even though they are used to determine the bonded-atom lone-pair arrangements.

Solution

(a) The Lewis structure is

The central bromine atom has five bonded atoms and one lone pair for a steric number of 6; the bonded-atom lone-pair arrangement is octahedral, and the F–Br–F bond angles are 90 and 180 degrees. The molecular shape is square pyramidal.

(b) The Lewis structure is

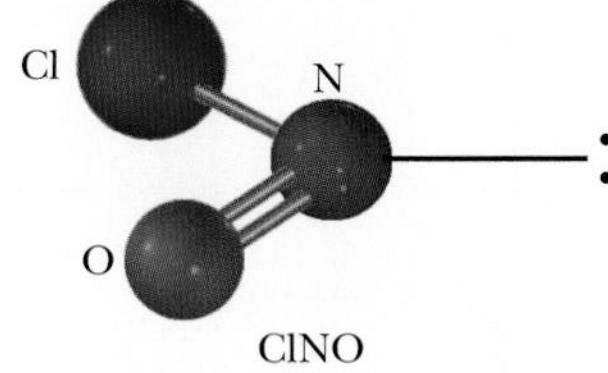

The central nitrogen atom has a steric number of 3; the bonded-atom lone-pair arrangement is trigonal planar. The Cl–N–O bond angle is about 120 degrees, and the molecule is V-shaped or bent.

(c) There are three resonance structures for CO_3^{2-}:

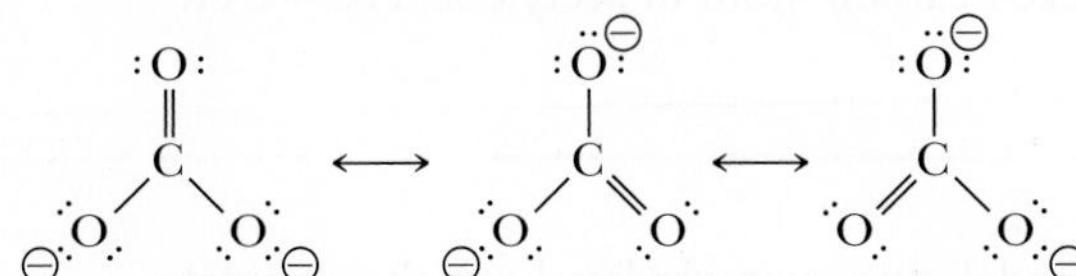

Any of the three resonance structures can be used to determine the bonded-atom lone-pair arrangement, the bond angles, and the shape of the ion. The central carbon atom has a steric number of 3; the bonded-atom lone-pair arrangement is trigonal planar. The O–C–O bond angles are 120 degrees, and the shape is trigonal planar. Because no lone pairs are present on the central atom, the bonded-atom lone-pair arrangement and the shape are the same.

Understanding

What are the bonded-atom lone-pair arrangement, the shape, and the Cl–I–Cl bond angles in the $[ICl_4]^-$ ion?

Answer The bonded-atom lone-pair arrangement is octahedral, the Cl–I–Cl bond angles are 90 and 180 degrees, and the shape of the ion is square planar.

Shapes of Molecules with Multiple Central Atoms

Most molecules have more than one central atom. The geometry about each central atom is assigned by applying the VSEPR model to each central atom individually. For CH_3CN, the Lewis structure is

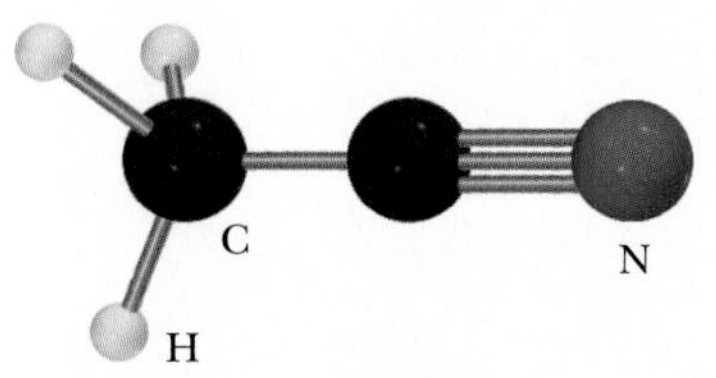

Figure 10.7 Structure of CH_3CN.

```
      H
      |
  H—C—C≡N:
      |(1) (2)
      H
```

The geometry about each central atom is determined separately by applying the VSEPR model.

The C(1) atom has a steric number of 4, so its bonded-atom lone-pair arrangement is tetrahedral. The H–C(1)–H and H–C(1)–C(2) bond angles should be about 109 degrees. The C(2) atom has a steric number of 2, so its bonded-atom lone-pair arrangement is linear. The C(1)–C(2)–N bond angle is therefore 180 degrees. Figure 10.7 is a drawing of this structure.

EXAMPLE 10.3 Shapes of Molecules

What are the bond angles around each carbon atom in ethylene, C_2H_4?

Strategy Use the Lewis structure to assign separately the steric number of both central atoms. These steric numbers determine the bonded-atom lone-pair arrangements, each of which has characteristic bond angles.

Solution
First, write the Lewis structure:

```
H     H
 \   /
  C=C
 /   \
H     H
```

Each carbon makes the same number and type of bonds, and has a steric number of 3; the bonded-atom lone-pair arrangement is trigonal planar. The geometry around each carbon atom is trigonal planar, so the H–C–H and H–C–C bond angles are approximately 120 degrees.

Understanding

What is the geometry about each carbon atom in acetylene, HC≡CH?

Answer Both are linear.

Note that the VSEPR model does not predict how the geometry around one central atom will be oriented with respect to others in the molecule. For example, is ethylene completely planar, or are the planes of the two CH_2 groups about the carbon atoms perpendicular or at some other angle (Figure 10.8)? The VSEPR model does not answer this question. The experimentally determined shape of ethylene is planar, the form shown in Figure 10.8*a*. In Section 10.4, we develop a bonding model that explains the experimentally observed shape of ethylene.

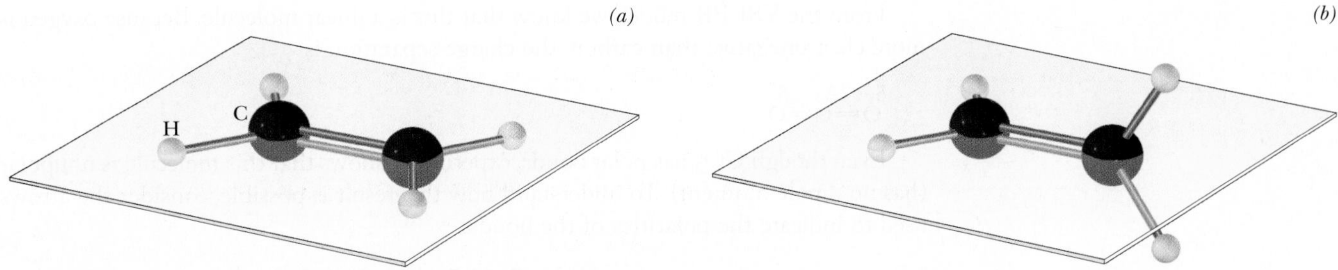

Figure 10.8 Two possible shapes of C_2H_4. The overall shape of ethylene can be *(a)* planar or *(b)* nonplanar. The valence-shell electron-pair repulsion model does not predict whether there is a preference for the planar or nonplanar arrangement.

OBJECTIVES REVIEW *Can you:*

- ☑ explain the assumptions of the VSEPR model?
- ☑ determine the steric number from the Lewis structure?
- ☑ use the steric number to assign the bonded-atom lone-pair arrangement of each central atom?
- ☑ distinguish between the bonded-atom lone-pair arrangement and molecular shape?
- ☑ predict the shapes of molecules from the VSEPR model?
- ☑ determine the locations of lone pairs in molecules with a steric number of 5 or 6?

10.2 Polarity of Molecules

OBJECTIVE

- ☐ Predict the polarity of a molecule from bond polarities and molecular shape

Knowing the structure and shape of a molecule allows us to predict many of its physical and chemical properties. One such physical property is the polarity of a molecule. **Polar molecules,** those that contain an unequal distribution of charge, generally interact with other polar molecules, which helps explain why sugar dissolves in water whereas oil does not. Section 9.4 described the experiment that determines the *dipole moment* of molecules by placing them in an electric field. Polar molecules will orient to maximize electrostatic attractions, an effect that can be measured experimentally. Only diatomic molecules were considered in Section 9.4 because the shape of a molecule must be known before we can predict whether the molecule is polar. The VSEPR model provides this information.

The degree of polarity is measured by a *dipole moment,* the magnitude of the separated charges times the distance between them. The difference in electronegativity between bonded atoms is used to predict the polarity of each bond, known as the bond dipole. For diatomic molecules, such as HF and HBr, the bond dipole is equal to the dipole moment for each. These dipoles are represented by an arrow pointing toward the more electronegative atom.

$$\overset{\longmapsto}{\mathrm{H{-}F}} \qquad \overset{\mapsto}{\mathrm{H{-}Br}}$$

The HF bond is more polar, so the arrow is longer. The arrows, which point toward the negatively charged end of the bond, are used to help emphasize that bond dipoles are vector quantities; that is, they have both direction and magnitude.

Predicting the dipole moment of more complicated molecules requires knowledge of the molecular shape in addition to the bond dipoles. Consider the molecule CO_2. The Lewis structure is

$$\mathrm{\ddot{\underset{\cdot\cdot}{O}}{=}C{=}\ddot{\underset{\cdot\cdot}{O}}}$$

From the VSEPR model, we know that this is a linear molecule. Because oxygen is more electronegative than carbon, the charge separation is

$$\overset{\delta-}{O}=\overset{\delta+}{C}=\overset{\delta-}{O}$$

Even though CO_2 has polar bonds, experiment shows that this molecule is nonpolar (has no dipole moment). To understand how this result is possible, consider the arrows used to indicate the polarities of the bonds.

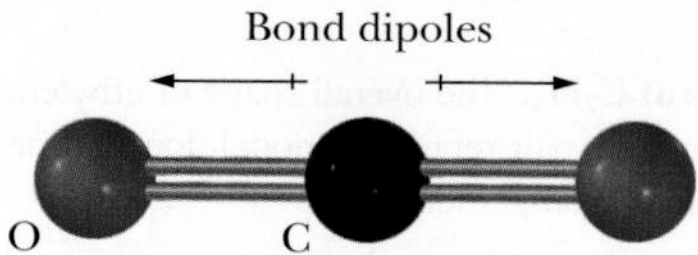

No dipole moment

Both bond polarity and molecular shape must be known to predict the polarity of a molecule.

The dipole moment of the molecule is the vector sum of the individual bond dipoles. The vectors used to represent the bond dipoles in CO_2 are of equal length. Because CO_2 has a linear geometry, the vectors point in exactly opposite directions. When the two bond dipoles are added, their vector sum is zero. We predict that CO_2 is a nonpolar molecule that contains polar bonds, consistent with experiment.

It is useful to consider two similar molecules to help clarify these ideas. First, consider OCSe. This molecule has a Lewis structure analogous to that of CO_2, but the two bonds have different polarities. In this case, the electronegativity of carbon is greater than selenium, and the C–O bond dipole is as before.

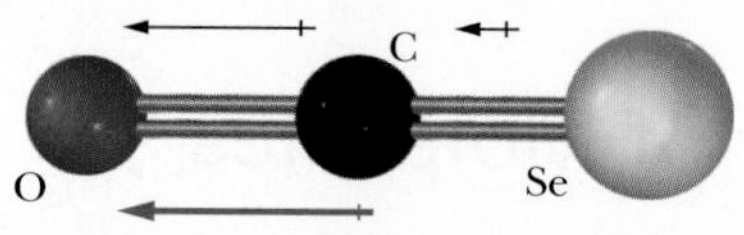

Dipole moment

The molecule is linear, the same as CO_2, but OCSe is a polar molecule because the bond dipoles are of unequal magnitude and are in the same direction; they do not cancel—they add.

Second, consider OF_2. The Lewis structure is

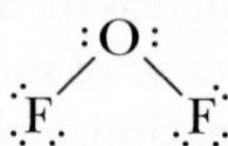

The steric number is 4 and the bonded-atom lone-pair arrangement is tetrahedral; the F–O–F bond angle is about 109 degrees. The fluorine atoms are more electronegative than oxygen, so the charge separation is

$$\overset{\delta-}{F}-\overset{\delta+}{O}-\overset{\delta-}{F}$$

The two O–F bond dipoles are of equal magnitude, but they do not cancel because they do not point in opposite directions. They sum as

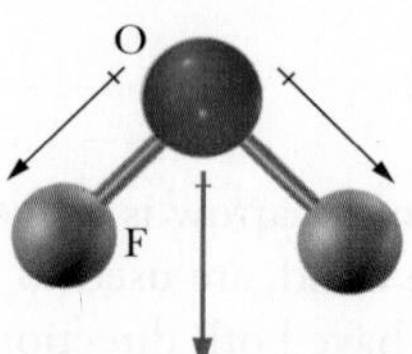

Dipole moment

Thus, OF_2 is a polar molecule; it differs from CO_2 because the lone pairs on oxygen produce a V-shaped molecule. *Molecules with lone pairs of electrons on a central atom generally are polar.* One of the few exceptions to this generalization is XeF_4, as discussed in Example 10.4d.

Two other examples of nonpolar molecules with polar bonds are BCl_3 and CCl_4. The cancellation of the bond dipoles is not as clear as with CO_2, but in the trigonal planar geometry of BCl_3 and the tetrahedral geometry of CCl_4, the vectors representing the bond dipoles are of the same length and also sum to zero.

A molecule with polar bonds is nonpolar if its geometry causes the bond polarities to sum to zero.

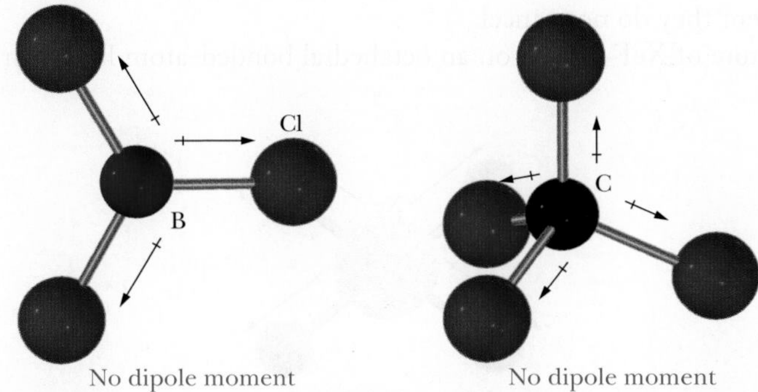

Although BCl_3 and CCl_4 are nonpolar, if an atom of another element replaces one of the chloride atoms in BCl_3 or CCl_4, the new molecule is polar. For example, both BCl_2F and $HCCl_3$ are polar molecules.

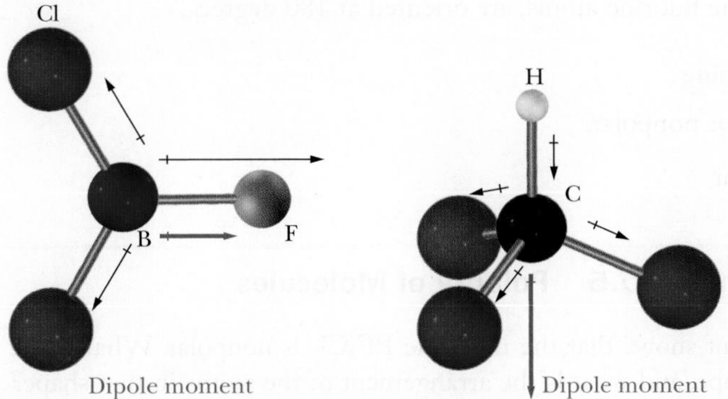

Only identical bond dipoles sum to zero for these geometries. *Molecules are nonpolar when there are no lone pairs on the central atom and all of the atoms bonded to the central atom are identical.* Most molecules are polar.

EXAMPLE 10.4 Polarity of Molecules

Predict which of the following molecules are polar and which are nonpolar: (a) CH_4, (b) HCN, (c) H_2O, (d) XeF_4.

Strategy Write the Lewis structure of each, determine the steric number, and assign the bonded-atom lone-pair arrangement. With only a few exceptions, a molecule will be polar unless all the terminal atoms are the same.

Solution

(a) As shown in Figure 10.1, CH_4 is a tetrahedral molecule. Because all vertices of the tetrahedron are filled with the same atom type (hydrogen), the four bond dipoles (which point toward carbon) are equal and sum to zero because of the geometry. The molecule is nonpolar.

(b) The structure of HCN is

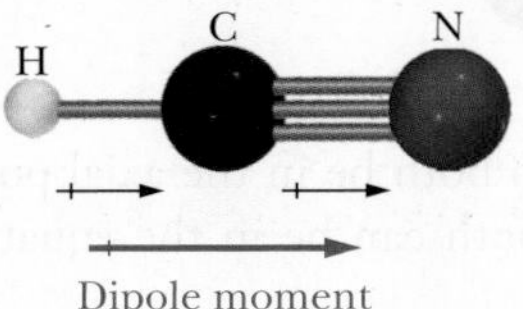

This linear molecule is polar, with a dipole moment that is the vector sum of the two bond dipoles.

(c) The oxygen atom in water has two lone pairs and is bonded to two hydrogen atoms; the bonded-atom lone-pair arrangement is tetrahedral, and the molecule is V-shaped. The molecule is polar. The bond dipoles are equal, but because the molecule is bent they do not cancel.

(d) The structure of XeF_4, based on an octahedral bonded-atom lone-pair arrangement, is

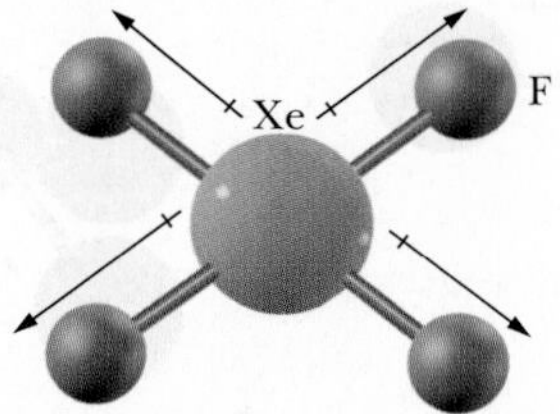

No dipole moment

The bond dipoles of the four fluorine atoms cancel because there are two pairs oriented at 180 degrees, and the molecule is nonpolar. XeF_4 is a rare example of a nonpolar molecule with lone pairs on the central atom, because the lone pairs, as well as the fluorine atoms, are oriented at 180 degrees.

Understanding

Is PF_3 polar or nonpolar?

Answer Polar

EXAMPLE 10.5 Polarity of Molecules

An experiment shows that the molecule PF_2Cl_3 is nonpolar. What is the molecular shape, and what is the arrangement of the atoms in that shape?

Strategy Write the Lewis structure and determine the bonded-atom lone-pair arrangement. Place the two types of atoms in positions such that the bond dipoles will sum to zero.

Solution

The Lewis structure of this molecule has five bonded atoms and no lone pairs about phosphorus, so the bonded-atom lone-pair arrangement is trigonal bipyramidal. Because the axial and equatorial sites are different, there are three possible arrangements of the fluorine and chlorine atoms:

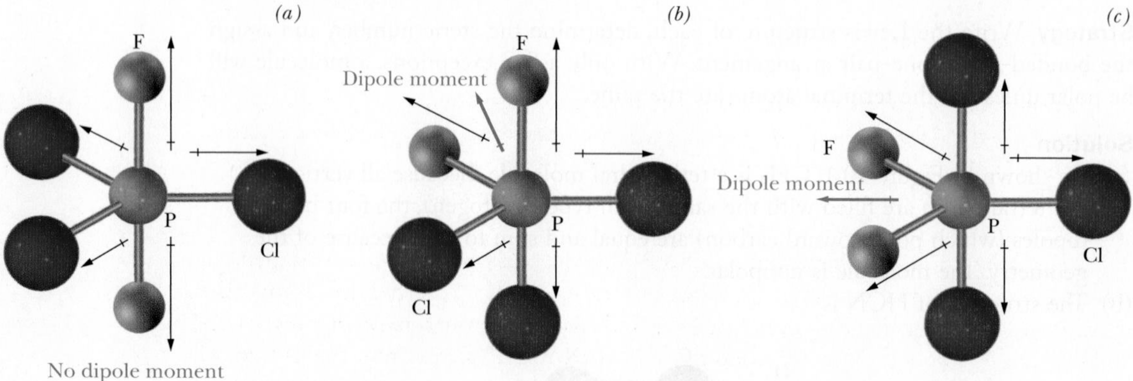

The two fluorine atoms can both be in the axial positions *(a)*, there can be one in each kind of position *(b)*, or both can be in the equatorial positions *(c)*. For *(a)*, the

bond dipoles of the axial P–F bonds cancel (this is the same as linear geometry), and the bond dipoles of the equatorial P–Cl bonds also cancel (this is a trigonal planar geometry). The molecule is nonpolar in this arrangement. The bond dipoles do not cancel in the arrangements shown in *(b)* and *(c)*. In *(b)*, neither the axial nor equatorial bonds have equal bond dipoles that could cancel. In *(c)*, the dipole moments of the axial P–Cl bonds cancel, but the bond dipoles in the equatorial plane do not cancel. Thus, the arrangement in *(a)* is correct because the experiment showed that the molecule is nonpolar.

Understanding

If the hypothetical molecule $SF_2Cl_2Br_2$ is nonpolar, what is its shape? What is the arrangement of the atoms?

Answer Octahedral, with the arrangement

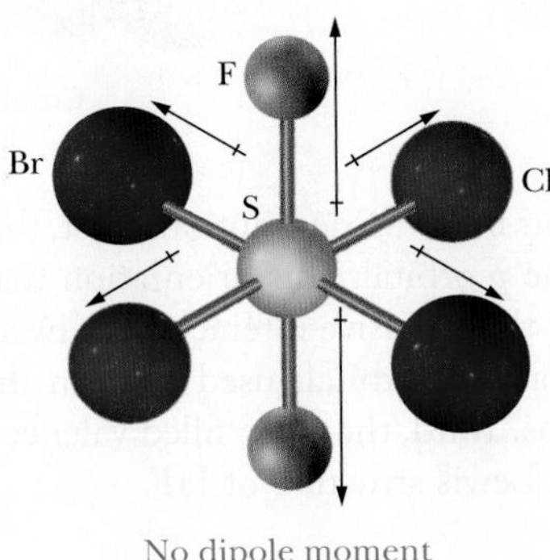

No dipole moment

OBJECTIVE REVIEW *Can you:*

☑ predict the polarity of a molecule from bond polarities and molecular shape?

10.3 Valence Bond Theory

OBJECTIVES

☐ Identify the orbitals used to form the bonds in any specific molecule

☐ Assign the hybrid orbitals used by a central atom from the Lewis structure and steric number

The VSEPR model provides useful predictions of the shapes of molecules by starting with Lewis structures and assuming that repulsion between electron pairs is minimized. A more detailed bonding description would identify the orbitals used by the atoms in making the chemical bonds. **Valence bond theory** describes covalent bonds as being formed by atoms sharing valence electrons in overlapping valence orbitals.

Section 9.3 described the orbitals used in forming the bond in H_2. As shown in Figure 10.9, the 1*s* orbitals of the two hydrogen atoms, each containing one electron, overlap to form the bond. The electron density concentrates between the two hydrogen nuclei. Each hydrogen atom provides one valence orbital and one valence electron to form the bond.

Valence bond theory states that bonds form by the overlap of valence orbitals. Each bond is made from electrons in a valence orbital on each of the bonded atoms. The bonding is based on the *orbitals available to make bonds* and the *number of valence electrons* that the atoms provide for bonding. The bond in HF forms from the overlap of a 1*s* orbital on the hydrogen atom, containing one electron, with a fluorine valence orbital. The energy-level diagram, introduced in Section 7.5, of fluorine is needed to determine which orbital on fluorine forms the bond. Only the energy levels of the

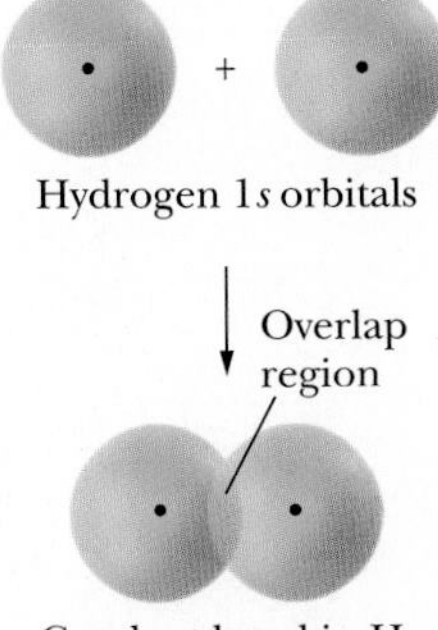

Figure 10.9 Orbitals used for the bond in H_2. The bond in H_2 is formed by overlap of a 1*s* orbital on each hydrogen atom, each containing one electron.

valence electrons of fluorine are considered because the core electrons are close to the nucleus and are unavailable for bond formation.

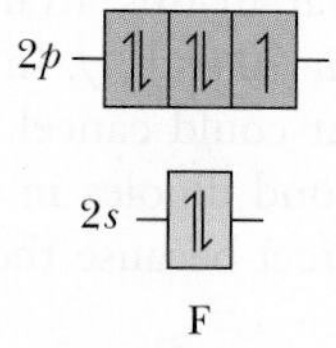

$2p$ ⥮ ⥮ ↿

$2s$ ⥮

F

Three of the valence orbitals are filled, but one of the $2p$ orbitals contains a single electron. This partially filled orbital overlaps with the partially filled hydrogen atom $1s$ orbital to form the covalent bond (Figure 10.10).

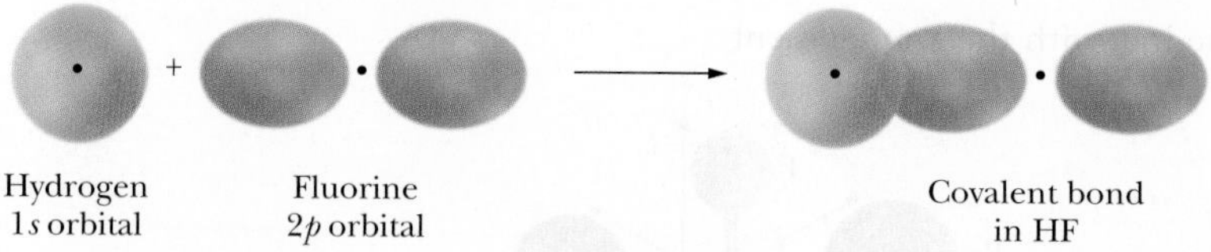

Figure 10.10 Covalent bond in HF. The bond in HF is formed from overlap of a $1s$ orbital on the hydrogen atom with a $2p$ orbital on fluorine.

Hydrogen $1s$ orbital + Fluorine $2p$ orbital → Covalent bond in HF

Note three features of this bonding description. First, the orientation of the bond is along the axis of the fluorine p orbital. This orientation maximizes the overlap of the orbitals whereas minimizing the repulsive interaction between the nuclei of the bonded atoms. Second, each of the atomic orbitals used to form the bond is occupied by one electron in the separate atoms. Third, the three filled valence orbitals on fluorine are the lone pairs on fluorine in the Lewis structure of HF.

$$\mathrm{H{-}\ddot{\underset{..}{F}}:}$$

In valence bond theory, bonds are formed by atoms sharing two electrons in overlapping atomic orbitals.

This valence bond theory description of the bonding in HF is really just an extension of Lewis structures in which we consider the orbitals that are used to form the covalent bond and those that contain the lone pairs. The difference is that now the specific orbitals forming the bond are identified.

EXAMPLE 10.6 Orbital Description of Covalent Bonds

Identify the atomic orbitals that are used to form the covalent bond in Cl_2 and the atomic orbitals that are occupied by the lone pairs.

Strategy Determine the occupancy of the valence orbitals. The bonds are made by orbitals that contain one electron.

Solution

The energy-level diagram for the valence electrons of chlorine is

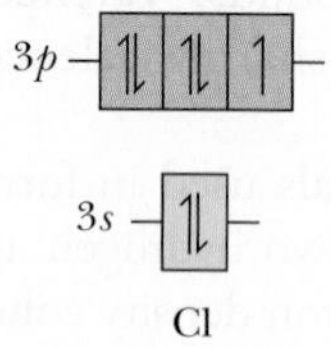

$3p$ ⥮ ⥮ ↿

$3s$ ⥮

Cl

Each chlorine atom has a single $3p$ orbital containing one electron. The covalent bond is formed from the overlap of these half-filled $3p$ orbitals on each of the chlorine atoms. The three lone pairs on each chlorine atom in the Lewis structure are those paired electrons in the $3s$ and the two filled $3p$ valence orbitals in the separate chlorine atoms.

Understanding

Identify the atomic orbitals that form the covalent bond in HCl and the atomic orbitals that hold the lone pairs.

Answer The bond forms from the overlap of the 1*s* orbital on hydrogen with a 3*p* orbital on chlorine. The three lone pairs on chlorine are in the 3*s* and two of the 3*p* valence orbitals.

Hybridization of Atomic Orbitals

Experiment shows that the *s*, *p*, and *d* valence orbitals and the ground-state arrangement of electrons in atoms do not accurately describe the bonding in all molecules. If we consider ammonia, the valence orbitals available to form bonds for the nitrogen atom are the 2*s* and three 2*p* orbitals. There is a single electron in each of the 2*p* orbitals. These electrons are well suited to form the three bonds with the single electron in each of the 1*s* orbitals of the hydrogen atoms.

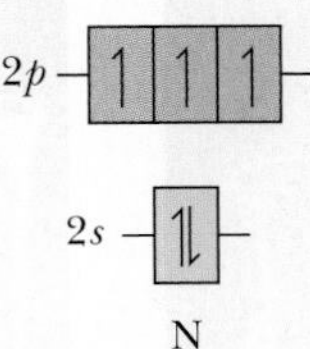

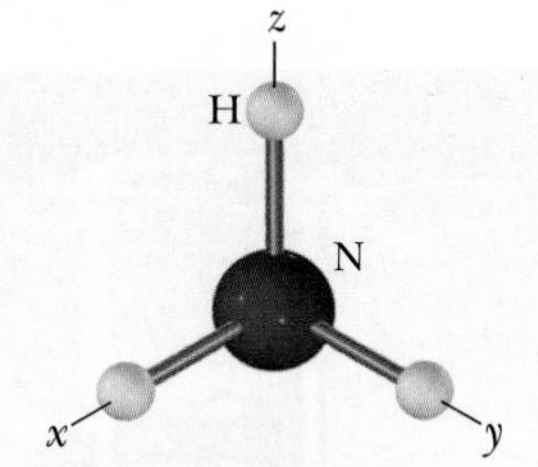

Figure 10.11 Bonding of three hydrogen atoms with three *p* orbitals. If three bonds were made by the three *p* orbitals of a central atom, the bond angles would all be 90 degrees.

However, these three *p* orbitals are oriented at 90-degree angles with respect to each other; if they formed bonds directly, there would be three 90-degree H–N–H bond angles (Figure 10.11). Experimental results do not agree with this prediction: The H–N–H bond angles in ammonia, NH_3, are all 107 degrees, and the three bonds are identical in strength.

The bond angles in ammonia and many other species can be explained if you remember that each orbital is a mathematical expression that describes the electron as a wave. Two or more of these orbitals that describe the electrons can be mathematically mixed (or averaged) to produce an equal number of orbitals that have different shapes and orientations. We leave the mathematics of the mixing for more advanced courses, but you can use the idea to understand molecular shape. **Hybrid orbitals** are orbitals obtained by mixing two or more atomic orbitals on the same central atom. The new hybrid orbitals have different shapes and directional properties from the orbitals used in constructing them. However, the total number of orbitals is the same before and after hybridization. For example, mixing one 2*s* orbital with one 2*p* orbital yields two new hybrid orbitals.

The formation of hybrid orbitals is based on the mathematical combination of the orbitals and is not very different from the averaging of numbers. A simple analogy is the mixing of different colors of liquids to obtain a new color. For example, we can make a green liquid by mixing a yellow liquid with a blue liquid (Figure 10.12). If we place the new mixture back into the original beakers, we still have *two* equal volumes of liquids that contain everything that was in the original two beakers, but the color is different. If we want a different color, we can mix two beakers of blue with one of yellow, producing *three* beakers of liquid of the new color. The mixing of orbitals is analogous; the energy and directional characteristics of the new hybrid orbitals are decided by the type and number of atomic orbitals used in the mixing. The number of new hybrid orbitals is the same as the number of orbitals used to form them.

Hybrid orbitals explain the shapes of many molecules.

sp Hybrid Orbitals

The mixing of one *s* orbital with one *p* orbital leads to the formation of two new orbitals designated *sp* hybrid orbitals. One *sp* hybrid orbital is formed by adding equal amounts of the *p* orbital and the *s* orbital. The other *sp* hybrid orbital is formed by subtracting the *p* orbital from the *s* orbital. The results of this mixing can be seen pictorially in Figure 10.13. Remember that the sign of the amplitude of the orbital in one lobe of a *p* orbital is the opposite of the sign in the other lobe (indicated by shading in Figure 10.13). Addition of the *p* orbital to the *s* orbital *(a)* reinforces the wave amplitude

HYSIL
100ml

HYSIL
100ml

HYSIL
100ml

HYSIL
100ml

Figure 10.12 Mixing liquids as an analogy to the formation of hybrid orbitals. The mathematical mixing of atomic orbitals to form hybrid orbitals is similar to mixing different colors of liquids to obtain a new color. After the liquids are mixed, there is still the same number of beakers of liquid, but the color has changed. The final colors of the mixtures are different because different amounts of the colored liquids were mixed in the two rows.

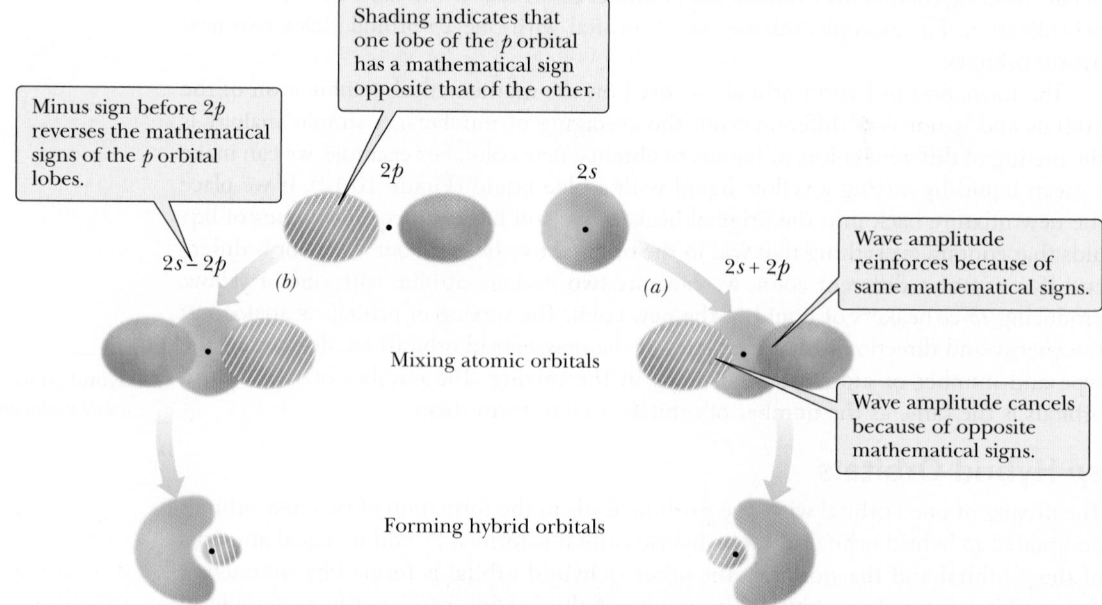
Shading indicates that one lobe of the *p* orbital has a mathematical sign opposite that of the other.
Minus sign before 2*p* reverses the mathematical signs of the *p* orbital lobes.
2*p*
2*s*
2*s* – 2*p*
(*b*)
(*a*)
2*s* + 2*p*
Wave amplitude reinforces because of same mathematical signs.
Mixing atomic orbitals
Wave amplitude cancels because of opposite mathematical signs.
Forming hybrid orbitals

Figure 10.13 *sp* hybrid orbitals. The *sp* hybrid orbitals are formed from the addition *(a)* and subtraction *(b)* of one *s* and one *p* atomic orbital. *(a)* The amplitudes of the orbitals interact constructively on the right side of the nucleus; *(b)* the amplitudes interact constructively on the left side.

on the right side of the nucleus, where both orbitals have a positive sign of the amplitude. On the left side, they have opposite signs and mainly cancel each other, forming an *sp* hybrid orbital that has one large lobe to the right and one small lobe to the left.

The other *sp* hybrid orbital is formed by subtracting the *p* orbital from the *s* orbital (see Figure 10.13*b*). The subtraction reverses the signs of the amplitude in the two lobes of the *p* orbital, causing reinforcement on the left side of the nucleus. Both hybrid orbitals have the same shape, but their large lobes are oriented at 180 degrees with respect to each other. In future drawings, for clarity, the hybrid orbitals will be shown as an elongated shape with the smaller lobes omitted.

The two *sp* hybrid orbitals are oriented 180 degrees from each other.

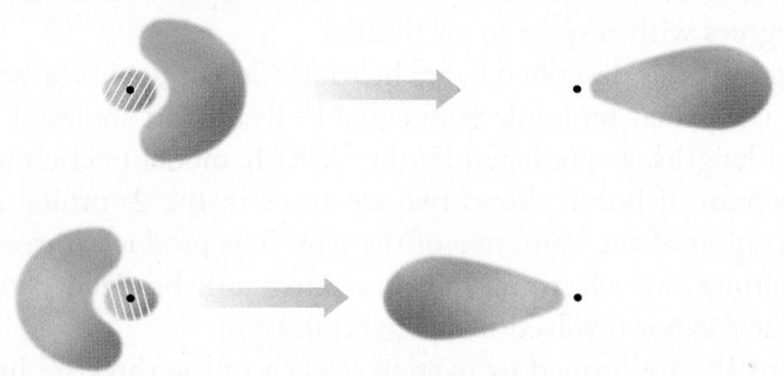

Beryllium chloride, $BeCl_2$, is an example of a molecule in which the bonding is best described by *sp* hybrid orbitals on the beryllium atom. The bonding in $BeCl_2$ can be pictured by first considering the energy-level diagram of beryllium.

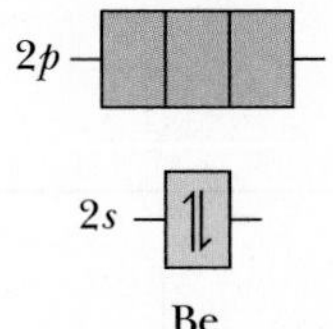

Beryllium in its ground state has no unpaired electrons and could not make any electron pair bonds, so clearly some change in the arrangement of the electrons or orbitals is needed. VSEPR theory predicts a linear shape for a beryllium atom because the steric number is 2, a prediction verified by experiment. Thus, the beryllium makes two equivalent Be–Cl bonds that are oriented 180 degrees from each other. The observed bonding is explained by the formation of two *sp* hybrid orbitals on the beryllium atom because the large lobes of these two orbitals are oriented at 180 degrees with respect to each other. The energy of the two new hybrid orbitals is the average of the energies of the *s* and *p* orbitals from which they formed. Following Hund's rule, one electron is placed in each of the *sp* hybrid orbitals with the same spin.

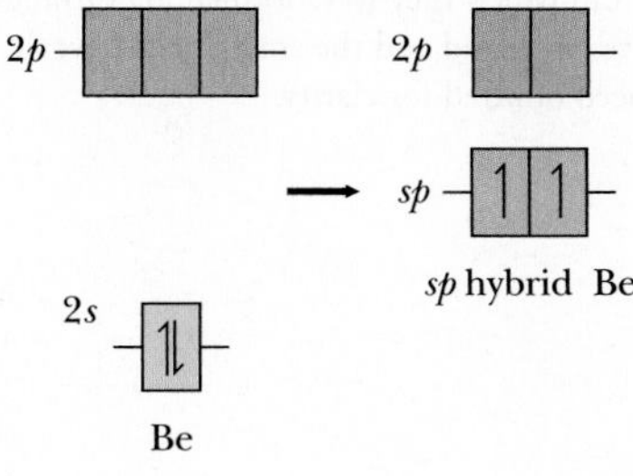

The two bonds in $BeCl_2$ are made by overlap of a 3*p* orbital on each chlorine with an *sp* hybrid orbital on beryllium (Figure 10.14). The two hybrid orbitals on beryllium each have one large lobe directed at the chlorines. These hybrid orbitals on the beryllium effectively overlap with the chlorine 3*p* orbitals making strong bonds. The two unhybridized *p* orbitals on beryllium remain vacant in the molecule. Note that the state of two electrons in the *sp* hybrid orbitals is higher in energy than the $2s^2$ ground-state configuration, but the energy required to place electrons in hybrid orbitals is much less than that released when the two Be–Cl bonds form.

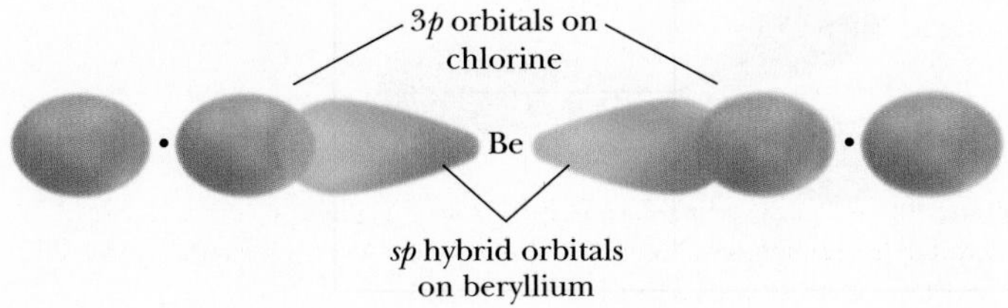

Figure 10.14 Bonding in $BeCl_2$. The bonds in $BeCl_2$ are made from the overlap of two *sp* hybrid orbitals on the beryllium atom with 3*p* orbitals on the two chlorine atoms.

It is important to remember that two equivalent *sp* hybrid orbitals form bonds with a bond angle of 180 degrees. *The sp hybrid orbitals describe the bonding on central atoms that have 180-degree bond angles*. The steric number of 2 indicates *sp hybrid orbitals.*

sp^2 Hybrid Orbitals

The combination of one *s* orbital with two *p* orbitals leads to the formation of three sp^2 hybrid orbitals. Each of these orbitals has one large and one small lobe, similar to those of *sp* hybrid orbitals. They lie in the same plane (the plane containing the two *p* orbitals from which they formed) and point toward the corners of an equilateral triangle (Figure 10.15). The three hybrid orbitals have the same shape, but their large lobes are oriented at 120 degrees with respect to each other.

The three sp^2 hybrid orbitals are oriented 120 degrees from each other.

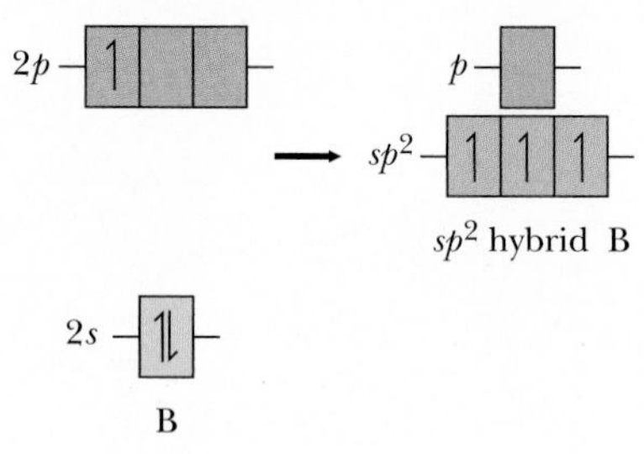

The bonding in BF_3 is described by sp^2 hybridization of the boron atom. Experiment shows that BF_3 is a planar molecule with equal F–B–F bond angles of 120 degrees and equal B–F bond lengths, as predicted by the VSEPR model (steric number = 3). The energy-level diagram of boron shows two electrons in the 2*s* orbital and one in a 2*p* orbital. Hybridization of the *s* and two of the *p* orbitals produces three sp^2 hybrid orbitals, each containing one electron. The remaining unhybridized *p* orbital on boron remains vacant and is not involved in the hybridization.

The bonds in BF_3 are formed by overlap of each of the three sp^2 hybrid orbitals on the boron atom with a 2*p* orbital on each of the fluorine atoms (Figure 10.16).

The sp^2 hybrid orbitals produce bonds that make 120-degree angles with each other. *The sp^2 hybrid orbitals describe the bonding on central atoms that have 120-degree bond angles and a steric number of 3.*

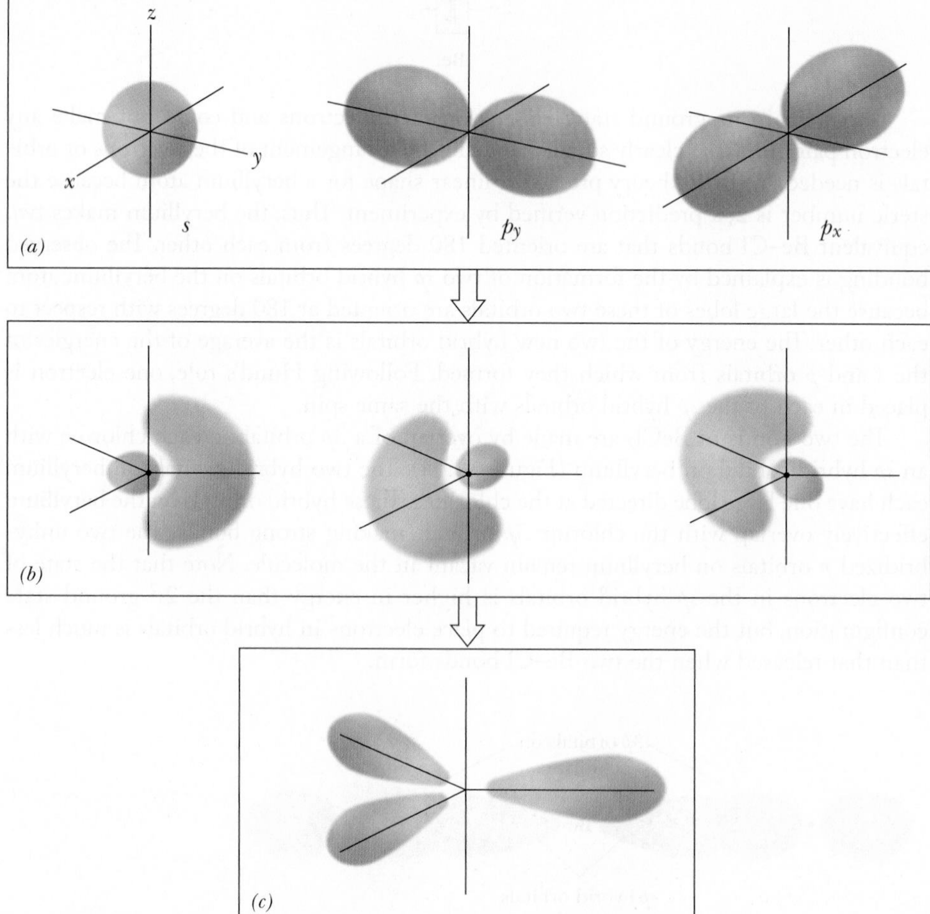

Figure 10.15 **sp^2 hybrid orbitals.** The mixing of an *s* and two *p* orbitals *(a)* leads to the formation of three sp^2 hybrid orbitals *(b)*. They point toward the corners of an equilateral triangle and are located in the same plane as the two *p* orbitals from which they were formed. *(c)* Orbitals are elongated and the small lobes have been omitted for clarity.

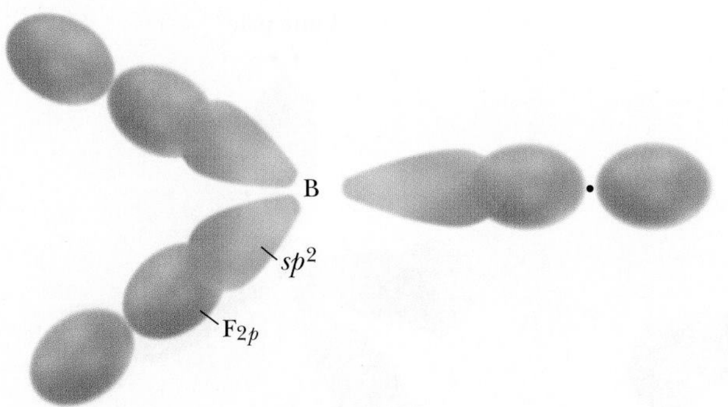

Figure 10.16 Bonding in BF_3. The bonds in BF_3 are made from the overlap of three sp^2 hybrid orbitals on the boron atom, each with a $2p$ orbital on one of the three fluorine atoms.

sp^3 Hybrid Orbitals

The mixing of one *s* and three *p* orbitals yields four sp^3 hybrid orbitals. Again, each of these orbitals has a large lobe and a small lobe. The large lobes point at the corners of a tetrahedron (Figure 10.17). The four hybrid orbitals have the same shape, but their large lobes are oriented at 109.5 degrees with respect to each other.

The four sp^3 hybrid orbitals are oriented 109.5 degrees from each other.

Methane, CH_4, is a molecule in which the central atom forms four identical bonds, a fact that cannot be explained by one *s* and three *p* orbitals. Instead, they mix and form

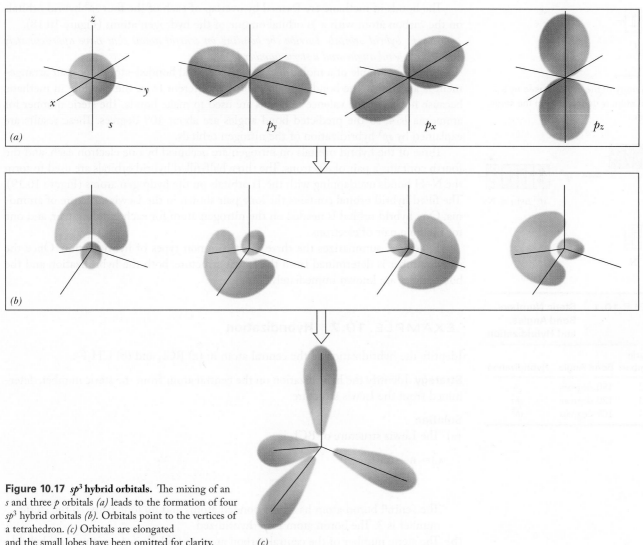

Figure 10.17 sp^3 hybrid orbitals. The mixing of an *s* and three *p* orbitals *(a)* leads to the formation of four sp^3 hybrid orbitals *(b)*. Orbitals point to the vertices of a tetrahedron. *(c)* Orbitals are elongated and the small lobes have been omitted for clarity.

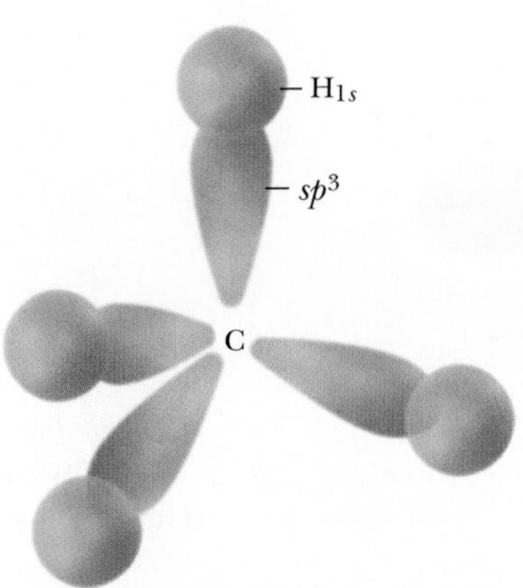

Figure 10.18 **Bonding in CH_4.** The bonds in CH_4 are made from the overlap of four sp^3 hybrid orbitals on the carbon atom with 1*s* orbitals on the four hydrogen atoms.

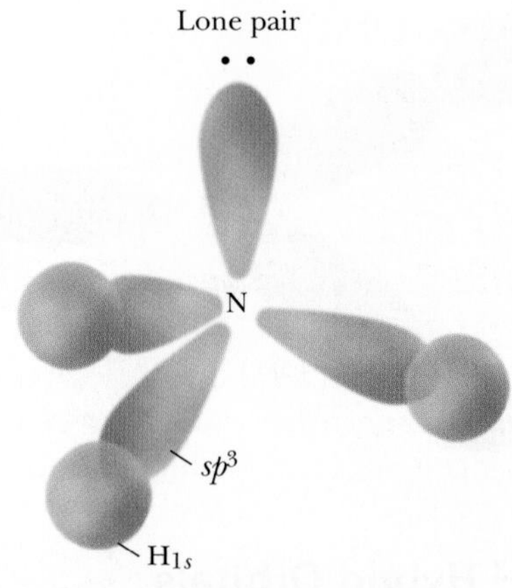

Figure 10.19 **Bonding in ammonia.** The bonds in NH_3 are made from the overlap of three sp^3 hybrid orbitals on the nitrogen atom with 1*s* orbitals on the hydrogen atoms. The lone pair on nitrogen occupies the remaining sp^3 orbital.

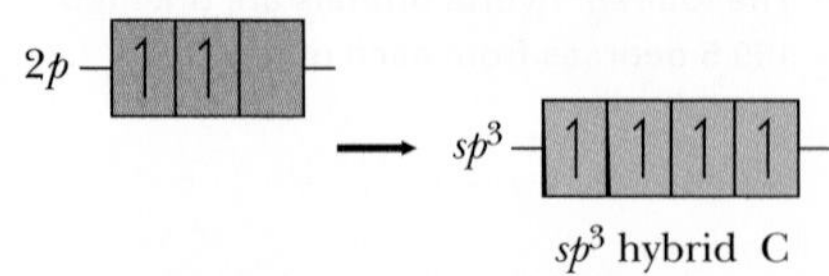

The identity of the orbitals used by a central atom is deduced from the steric number.

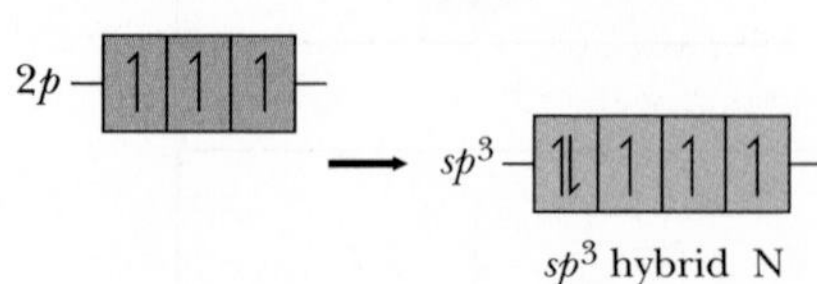

four sp^3 hybrid orbitals, all identical and directed at angles of 109.5 degrees. Each hybrid orbital contains one of the four valence electrons on the carbon atom.

The bonds in methane are formed by overlap of each of the four sp^3 hybrid orbitals on the carbon atom with a 1*s* orbital on one of the hydrogen atoms (Figure 10.18).

The sp^3 hybrid orbitals describe the bonding on central atoms that have approximately 109-degree bond angles and a steric number of 4.

Another example of a molecule with a tetrahedral bonded-atom lone-pair arrangement is ammonia. The bonding in ammonia is different from the bonding in methane because not all of the valence electrons are used to make bonds. The steric number for ammonia is 4, so the predicted bond angles are about 109 degrees. These results are explained by sp^3 hybridization of the nitrogen orbitals.

Three of the hybrid orbitals on nitrogen are occupied by one electron each, and the fourth contains a pair of electrons. The three half-filled hybrid orbitals are used to form the N–H bonds overlapping with the 1*s* orbitals on the hydrogen atoms (Figure 10.19). The filled hybrid orbital contains the lone pair shown in the Lewis structure of ammonia. One hybrid orbital is needed on the nitrogen atom for each bonding pair, and one for the lone pair of electrons.

Table 10.1 summarizes the three most common types of hybridization. Once the steric number is determined from the Lewis structure, both the hybridization and the bond angles are known immediately.

TABLE 10.1 Steric Numbers, Bond Angles, and Hybridization

Steric Number	Bond Angle	Hybridization
2	180 degrees	sp
3	120 degrees	sp^2
4	109 degrees	sp^3

EXAMPLE 10.7 Hybridization

Identify the hybridization of the central atom in (a) BCl_3 and (b) CH_2F_2.

Strategy Identify the hybridization on the central atom from the steric number, determined from the Lewis structure.

Solution

(a) The Lewis structure of BCl_3 is

:C̤̈l—B—C̤̈l:
|
:C̤l:

The central boron atom has three bonded atoms and no lone pairs, so the steric number is 3. The boron atom is sp^2 hybridized.

(b) The steric number of the central carbon atom in CH_2F_2 is 4, so it uses sp^3 hybrid orbitals.

Understanding

What hybrid oxygen orbitals form the bonds in OF_2?

Answer sp^3 hybrid orbitals

Hybridization Involving *d* Orbitals

In the valence bond model, the number of hybrid orbitals on a central atom must equal the number of bonded atoms plus lone electron pairs, or the steric number. The hybridization arrangements in Table 10.1 use only the *s* and three *p* valence orbitals, so the maximum steric number is 4. For molecules that have central atoms with steric numbers of 5 or 6, such as the phosphorus atom in PF_5, more than four atomic orbitals are needed to make the hybrid orbitals. The additional orbitals needed are taken from the *d* subshell. The bonding is described by sp^3d hybridization to explain the trigonal bipyramidal shape of PF_5 (steric number = 5). Mixing the appropriate *d* orbital with the *s* and *p* orbitals generates five hybrid orbitals pointing toward the vertices of a trigonal bipyramid (Figure 10.20*a*).

Molecules with central atoms that have a steric number of 6, such as the sulfur atom in SF_6 (the octahedral molecule outlined in the introduction to this chapter), have bonding that is described by sp^3d^2 hybridization. The six sp^3d^2 hybrid orbitals point toward the vertices of an octahedron (see Figure 10.20*b*), accounting for the known structure.

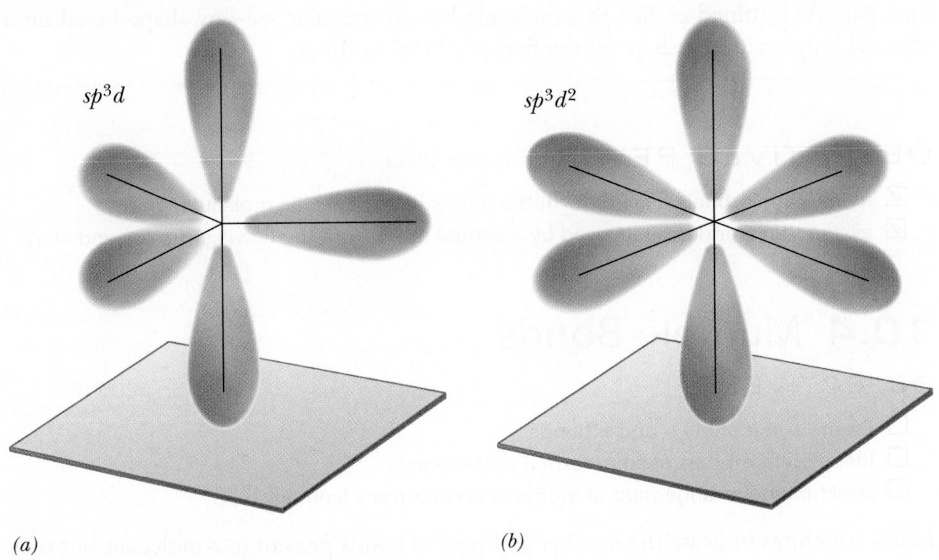

Figure 10.20 Hybridization involving *d* orbitals. *(a)* Hybridization of one *s*, three *p*, and one *d* orbital leads to the formation of five sp^3d hybrid orbitals. The five orbitals point to the vertices of a trigonal bipyramid. *(b)* Hybridization of one *s*, three *p*, and two *d* orbitals leads to the formation of six sp^3d^2 hybrid orbitals. The six orbitals point to the vertices of an octahedron.

EXAMPLE 10.8 Hybridization

Identify the hybridization of the central atom and the orbitals used for each bond in IF_5.

Strategy The steric number, determined from the Lewis structure, defines the hybrid orbitals needed for a central atom. Use the electron configuration of the peripheral atoms to determine which orbital is partially filled.

Solution

The steric number of the central iodine atom in IF_5 is 6, so it uses sp^3d^2 hybrid orbitals. The electron configuration of fluorine is $1s^22s^22p^5$—there is one $2p$ orbital containing one electron. Each bond is formed from the overlap of an sp^3d^2 hybrid orbital on iodine with a $2p$ orbital on a fluorine atom. The remaining hybrid orbital holds a lone pair (Figure 10.21).

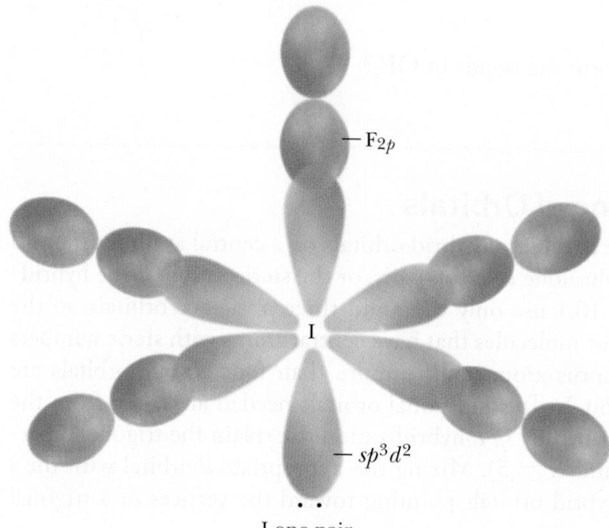

Figure 10.21 Bonding in IF_5.

Understanding

Describe the bonding in SF_4.

Answer The four S–F bonds are made from the overlap of an sp^3d orbital on sulfur with a $2p$ orbital on each fluorine atom. The remaining hybrid orbital on sulfur contains a lone pair. As outlined earlier, this molecule has an irregular see-saw shape based on a trigonal bipyramid, the shape of the five sp^3d hybrid orbitals.

OBJECTIVES REVIEW *Can you:*

- ☑ identify the orbitals used to form the bonds in any specific molecule?
- ☑ assign the hybrid orbitals used by a central atom from the Lewis structure and steric number?

10.4 Multiple Bonds

OBJECTIVES

- ☐ Define and identify σ and π bonds
- ☐ Identify the orbitals used to form σ and π bonds
- ☐ Describe the arrangement of atoms in *cis* and *trans* isomers

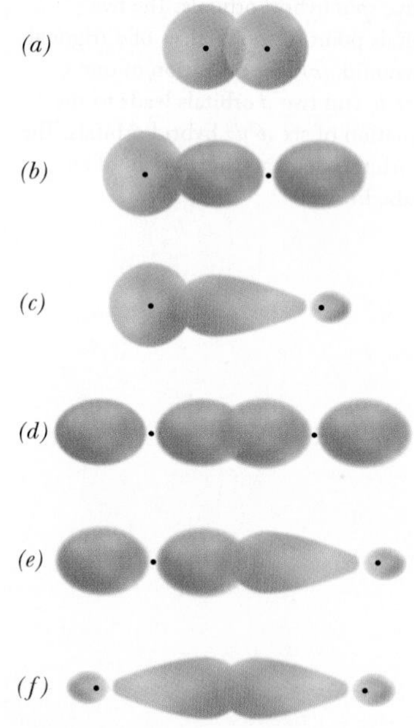

Lewis structures indicate the number and type of bonds present in a molecule, but they do not tell us where the electrons in those bonds are oriented with respect to the shape of the molecule. In a **sigma (σ) bond,** the shared pair of electrons is symmetric about the line joining the two nuclei of the bonded atoms. All single bonds are sigma bonds, as is one of the bonds in double or triple bonds. Sigma bonds form from the overlap of *s* or *p* orbitals, or hybrid orbitals oriented along the bond (Figure 10.22). The one large lobe of hybrid orbitals gives particularly good overlap in the formation of σ bonds; *all hybrid orbitals make only σ bonds.*

Figure 10.22 Sigma bonds. *(a)* Overlap of *s* orbitals. *(b)* Overlap of *s* and *p* orbitals. *(c)* Overlap of *s* and hybrid orbitals. *(d)* Head-on overlap of *p* orbitals. *(e)* Overlap of *p* and hybrid orbitals. *(f)* Overlap of hybrid orbitals.

Sigma bonds are formed by orbitals directed toward each other. Hybrid orbitals make only σ bonds.

Double Bonds

The chemistry of compounds with double and triple bonds is different from the chemistry of compounds containing only single bonds. Here, the development of valence bond theory will be expanded to include multiple bonds. These models help predict much of the reactivity of carbon compounds, compounds essential for our life and health.

Many molecules have double bonds; an important example is ethylene, C_2H_4.

H   H
C=C
H   H

The bonded-atom lone-pair arrangement at each carbon atom is trigonal planar, indicating that the hybridization is sp^2. The energy-level diagram for an sp^2-hybridized carbon atom that will make four bonds is

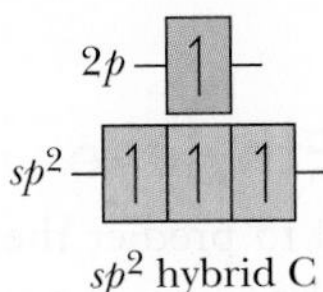

sp^2 hybrid C

Note that one of the four valence electrons is placed in each orbital so that the carbon atom can make four bonds.

There are three sp^2 hybrid orbitals and one p orbital, each containing one electron, on each carbon atom. Two of the hybrid orbitals on each carbon atom overlap with the $1s$ orbitals on the hydrogen atoms to make four of the six bonds shown in the Lewis structure. One of the two C–C bonds is formed from the overlap of two hybrid orbitals (Figure 10.23). These are all σ bonds; note that hybrid orbitals only form sigma bonds.

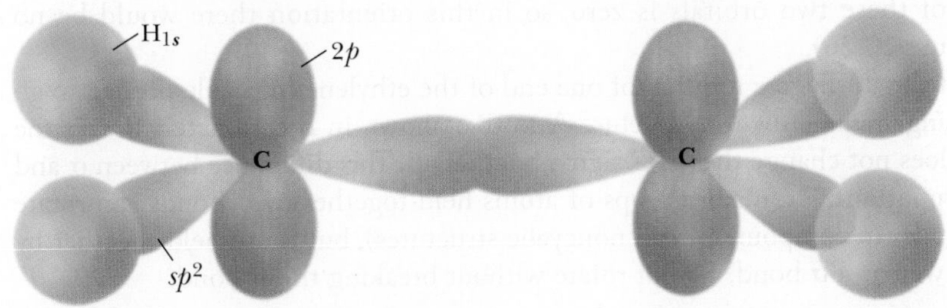

Figure 10.23 Sigma bonds in ethylene. The C–H sigma bonds in ethylene are formed from overlap of sp^2 hybrid orbitals on the carbon atoms with $1s$ orbitals on the hydrogen atoms. The C–C sigma bond is formed from overlap of carbon sp^2 hybrid orbitals.

In addition to the sp^2 hybrid orbitals that are used to form these five σ bonds, there remains one valence p orbital containing a single electron on each carbon atom. Because the hybrid orbitals are in the same plane as the two p orbitals used to make them, the remaining unhybridized p orbitals are perpendicular to the plane of the hybridized orbitals. These two p orbitals, each containing one electron, can only overlap *sideways* (Figure 10.24) to make the second bond of the double bond shown in the Lewis structure. This type of bond, called a **pi (π) bond,** places electron density above and below the line joining the bonded atoms.

The sideways overlap of two p orbitals, each containing one electron, forms one π bond. Half of the electron density in the bond is above the line joining the bonded atoms and half is below. Pi bonds have electron density in two regions of space because the p orbitals that make them have electron density above and below the nuclei. Figure 10.25 shows the complete bonding in ethylene, which includes the five σ bonds and the π bond. Two of these six bonds, one σ bond and one π bond, make up the double bond between the carbon atoms that is shown in the Lewis structure of ethylene.

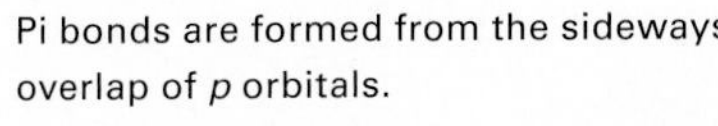

Pi bonds are formed from the sideways overlap of p orbitals.

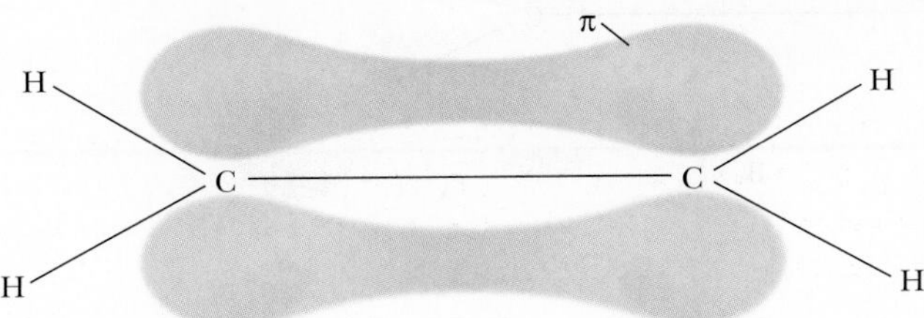

Figure 10.24 Pi bond. The sideways overlap of two p orbitals forms a pi bond. The electron density in this one π bond is concentrated above and below the axis joining the bonded atoms.

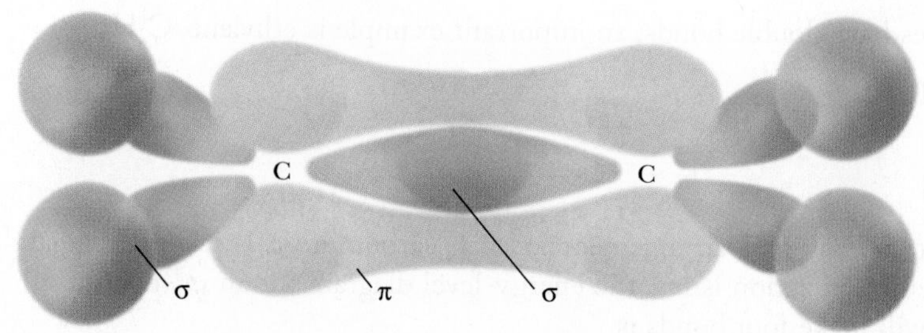

Figure 10.25 Total bonding picture of ethylene. Ethylene contains five σ bonds and one π bond.

Molecular Geometry of Ethylene

Earlier, the VSEPR model was used to predict the geometry about each carbon atom in ethylene. We concluded that the bonding of each carbon atom was trigonal planar, but the VSEPR model cannot predict how the planes of the two CH_2 groups are oriented with respect to each other.

Experiment shows that ethylene is planar. The description of the bonding orbitals shown in Figure 10.25 predicts that all six of the atoms in the molecule will be in the same plane. The molecule is planar because the *p* orbitals that overlap to form the π bond are perpendicular to the σ-bonding plane. The maximum overlap of these *p* orbitals occurs if the molecule is planar. When the two planes are oriented perpendicularly, the overlap of these two orbitals is zero, so in this orientation there would be no π bond (Figure 10.26).

Note that the 90-degree rotation of one end of the ethylene molecule needed to go from the arrangement shown in structure A to that shown in B completely breaks the π bond but does not change the C–C sigma bond at all. This difference between σ and π bonds is important. Atoms or groups of atoms held together by σ bonds can rotate about the bonds (in compounds with noncyclic structures), but those held together by a π bond, as well as a σ bond, cannot rotate without breaking the π bond.

Rotation around π bonds does not readily occur. Rotation can occur between atoms that are bonded with only a σ bond.

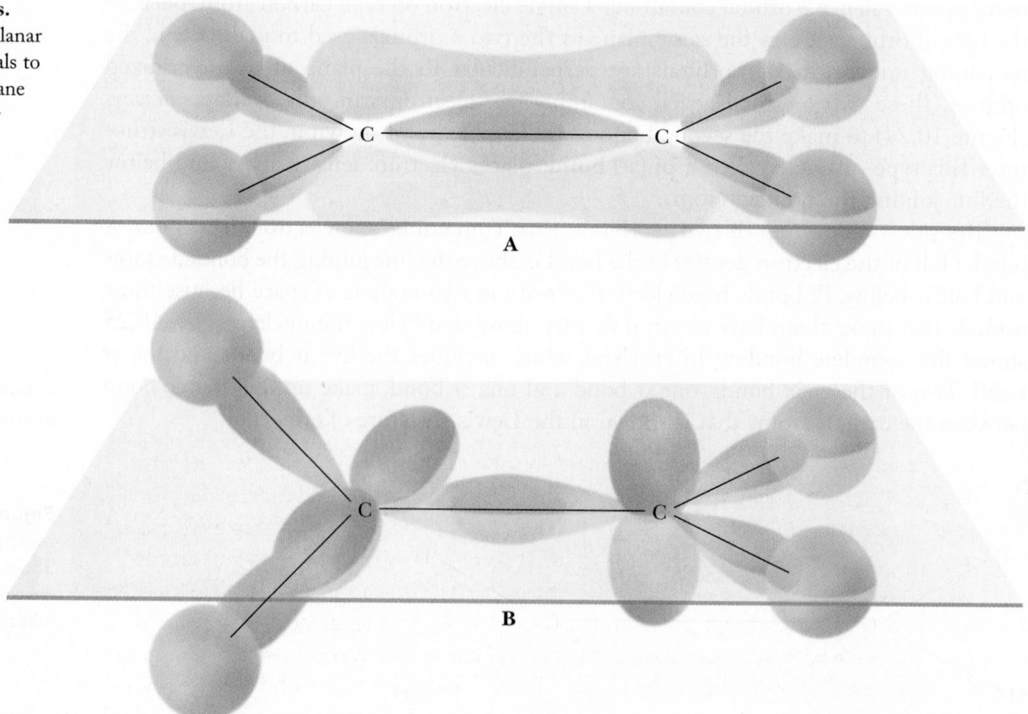

Figure 10.26 Overlap of *p* orbitals. Orientation A shows ethylene as a planar molecule with overlap of the *p* orbitals to form the π bond. Rotation of the plane of the CH_2 group on the left side by 90 degrees yields B, in which there is no overlap of the *p* orbitals and the π bond is broken. Orientation A is the correct shape.

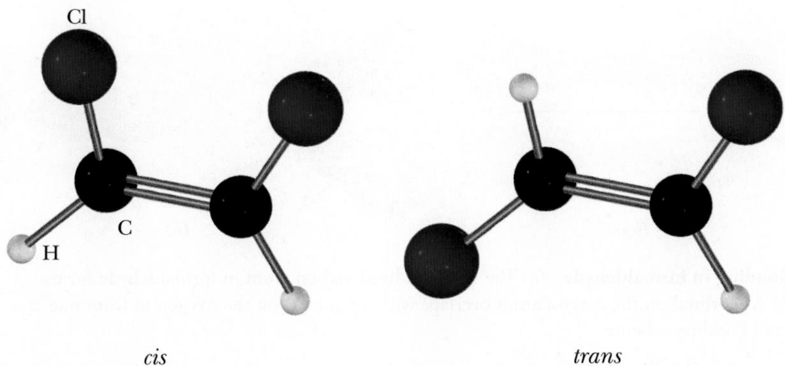

Figure 10.27 ***Cis*** **and** ***trans*** **isomers of HClC=CHCl.**

Isomers

Chemists have a lot of additional experimental evidence to show that the π bond must be broken to allow rotation about a double bond. If two of the hydrogen atoms in ethylene are replaced with chlorine atoms, three new compounds with the formula $C_2H_2Cl_2$ can form. Two of these compounds have HClC═CHCl structures (1,2-dichloroethylene). Each compound has different physical and chemical properties. For example, one compound boils at 47.5 °C, whereas the other boils at 60.3 °C. Structurally, the two forms differ in the orientation of the chlorine and hydrogen atoms in these planar molecules (Figure 10.27).

The two forms, called **isomers,** are compounds with the same molecular formula but different structures. In the form labeled *cis,* the chlorine atoms are both on the same side of the planar molecule; in the *trans* form, they are on opposite sides. The *cis* and *trans* isomers of the molecules shown in Figure 10.27 do not easily interconvert. The interconversion of the isomers requires the rotation of one H–C–Cl group about the C–C bond. Such a rotation must pass through 90 degrees where the overlap of the two *p* orbitals that form the π bond is zero. The energy needed to break the π bond is estimated to be about 263 kJ/mol. This considerable energy barrier prevents rotation about the double bond under normal conditions, explaining the existence of the two isomers. In contrast with π bonds, rotation about single bonds (σ bonds) does not change the overlap of the atomic orbitals, so rotation about single bonds is a low-energy process.

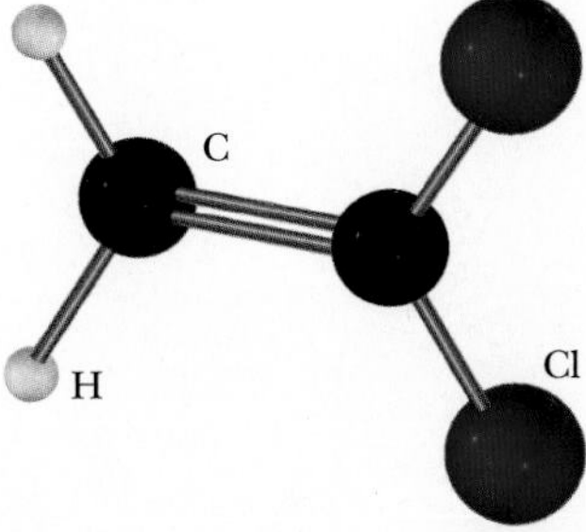

Figure 10.28 Third isomer of $C_2H_2Cl_2$. The third isomer of $C_2H_2Cl_2$, $H_2C{=}CCl_2$, has two hydrogen atoms bonded to one carbon atom and two chlorine atoms bonded to the other.

Compounds with the same molecular formula but different structures are called isomers.

Figure 10.28 shows the third isomer of those in Figure 10.27. This isomer has a $H_2C{=}CCl_2$ structure (1,1-dichloroethylene, boiling point = 32 °C). To convert either of the isomers in Figure 10.27 to this compound, σ bonds must be broken and reformed in a different way. *Cis* and *trans* isomers do not exist for this form, because rotation about the C═C bond yields the same compound. Isomers are considered in greater detail in Chapter 19.

Bonding in Formaldehyde

Formaldehyde, CH_2O, is another compound that contains a double bond.

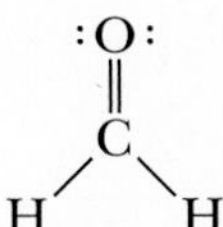

The bonded-atom lone-pair arrangement at carbon is trigonal planar; the hybridization is sp^2. As in ethylene, two of the hybrid orbitals on carbon overlap with 1*s* orbitals on the hydrogen atoms to form the C–H sigma bonds. The third sp^2 hybrid orbital points at the oxygen atom to make a σ bond between the carbon and oxygen atoms.

Next, consider the orbitals available for the oxygen atom to form bonds. The oxygen has two filled orbitals and two 2*p* orbitals each occupied by one electron.

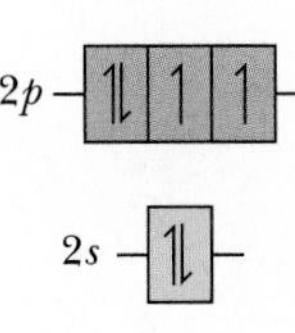

One of these *p* orbitals is directed at the carbon atom and overlaps with the remaining sp^2 orbital on carbon to form the C–O σ bond. The second C–O bond is a π bond formed from the overlap of the remaining *p* orbital on carbon with the second *p* orbital

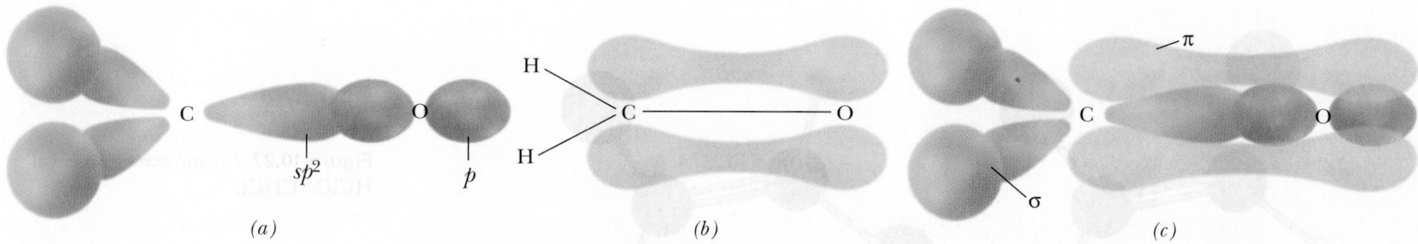

Figure 10.29 Bonding in formaldehyde. *(a)* The sp^2 hybridized carbon atom in formaldehyde forms three σ bonds. *(b)* A *p* orbital on the carbon atom overlaps with a *p* orbital on the oxygen to form one π bond. *(c)* The overall bonding scheme.

on the oxygen atom that is occupied by one electron (Figure 10.29). Because oxygen is a terminal atom, hybridization is not needed to explain its bonding.

Triple Bonds

The Lewis structure of acetylene, C_2H_2, shows that it contains a triple bond.

H–C≡C–H

The linear geometry of each carbon atom indicates *sp* hybridization; the energy-level diagram for an *sp*-hybridized carbon atom that forms four bonds is

$2p$ ⎯[1][1]⎯

sp ⎯[1][1]⎯

sp hybrid C

The *sp* hybrid orbitals form two C–H sigma bonds by overlap with the 1*s* orbitals on the two hydrogen atoms. The first of the three C–C bonds, the σ bond, results from the overlap of two *sp* hybrid orbitals, one on each carbon atom (Figure 10.30*a*).

In addition to the hybrid orbitals, there are two singly occupied *p* orbitals on each of the carbon atoms. Two π bonds are formed from the sideways overlaps of these orbitals

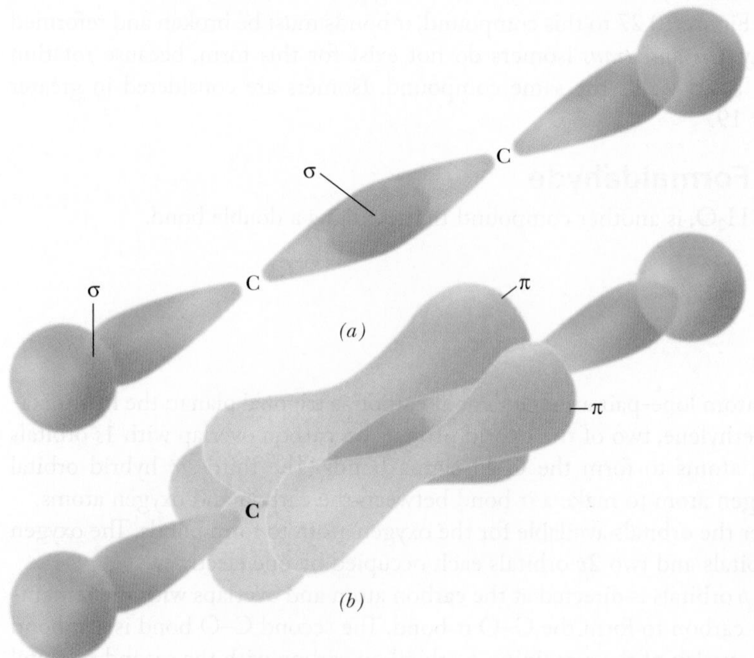

Figure 10.30 Bonding in acetylene. *(a)* Sigma bonds in acetylene. *(b)* Sigma and π bonds in acetylene.

(see Figure 10.30*b*). The triple bond in acetylene consists of a σ bond (formed by the overlap of *sp* hybrid orbitals) and two π bonds (formed from the sideways overlap of *p* orbitals). *One bond in a double or triple bond is a* σ *bond and the other bonds are* π *bonds.*

Summary of Bonding

1. When atoms are connected by a single bond, the bond is a σ bond. Sigma bonds are formed from the overlap of *s* or *p* orbitals, or hybrid orbitals oriented along the axis of the bond.
2. When atoms are connected by a multiple bond, one bond is a σ bond and the second and third bonds are π bonds, formed by the sideways overlap of *p* orbitals.

Bonding in Benzene

Hybrid orbitals explain the bonding in molecules for which more than one resonance form can be written. An interesting example is benzene, C_6H_6. Experimental measurements of the structure show that benzene is planar, all C–C–C and C–C–H bond angles are 120 degrees, and all C–C bond distances are equal. In Section 9.6, the two equivalent resonance forms of benzene were shown as well as the single structure with a dashed circle that is frequently used to represent the bonding.

The arrangement about each of the carbon atoms is trigonal planar; the hybridization for each is sp^2. Two of the hybrid orbitals on each carbon atom overlap to make C–C σ bonds, and the third makes the C–H σ bond (Figure 10.31).

Each sp^2-hybridized carbon atom has one electron in a *p* valence orbital that is perpendicular to the plane of the σ bonds (Figure 10.32*a*). These *p* orbitals can overlap to form π bonds in two different arrangements that correspond to the two resonance forms shown earlier, but the best representation analogous to the "dashed circle" structure is equal overlap of the six *p* orbitals to form a large π-electron cloud that includes all six π electrons (see Figure 10.32*b*).

The π bonds in benzene are formed by the sideways overlap of six carbon *p* orbitals.

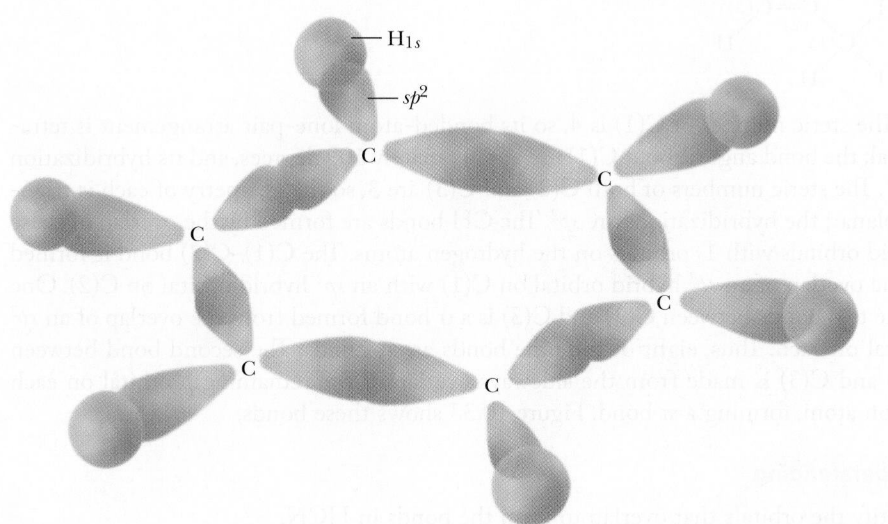

Figure 10.31 Sigma bonds of benzene. Each carbon atom in benzene is sp^2 hybridized, using one hybrid orbital to form its C–H bond and the other two hybrid orbitals to form its two C–C σ bonds.

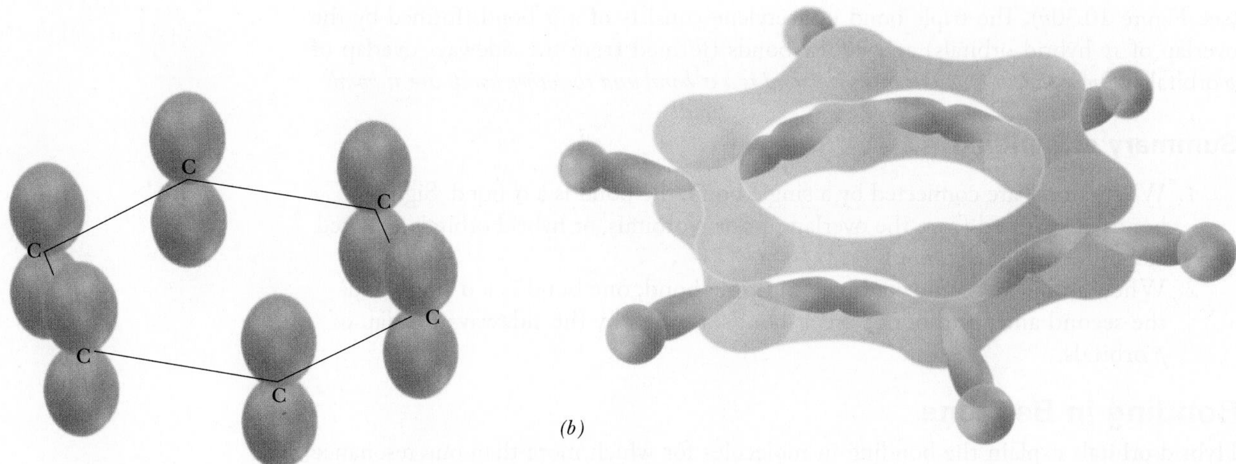

Figure 10.32 Pi bonding in benzene. *(a)* The *p* orbital on each carbon atom contains one electron. *(b)* These six orbitals overlap in benzene to form a large π bond above and below the plane of the sigma bonds.

In summary, the C–C sigma bonds in benzene are formed from the overlap of sp^2 hybrid orbitals on adjacent carbon atoms, and the C–H bonds are formed from the overlap of sp^2 hybrid orbitals on carbon with 1*s* orbitals on the hydrogen atoms. The three π bonds are made from the sideways overlap of *p* orbitals. The three π bonds can be formed by two different combinations of overlap of the *p* orbitals, but the best representation is a large, delocalized π cloud that contains all six electrons.

EXAMPLE 10.9 Bonding Descriptions

Describe the bonds in propylene, $CH_3CH{=}CH_2$. Identify the hybridization of each central atom and the type (σ or π) of each bond.

Strategy Use the steric numbers determined from the Lewis structure to assign the hybridization of each central atom. The hybrid orbitals will form the sigma bonds; any remaining *p* orbitals that contain one electron will form pi bonds.

Solution

The Lewis structure is

```
    H        H
     \      /
 H    C==C(3)
  \  / (2) \
   C(1)     H
  /  \
 H    H
```

The steric number of C(1) is 4, so its bonded-atom lone-pair arrangement is tetrahedral; the bond angles about C(1) are approximately 109 degrees, and its hybridization is sp^3. The steric numbers of both C(2) and C(3) are 3, so the geometry of each is trigonal planar; the hybridizations are sp^2. The CH bonds are formed by the overlap of these hybrid orbitals with 1*s* orbitals on the hydrogen atoms. The C(1)–C(2) bond is formed by the overlap of an sp^3 hybrid orbital on C(1) with an sp^2 hybrid orbital on C(2). One of the two bonds between C(2) and C(3) is a σ bond formed from the overlap of an sp^2 orbital on each. Thus, eight of the nine bonds are σ bonds. The second bond between C(2) and C(3) is made from the sideways overlap of the remaining *p* orbital on each carbon atom, forming a π bond. Figure 10.33 shows these bonds.

Understanding

Identify the orbitals that overlap to form the bonds in HCN.

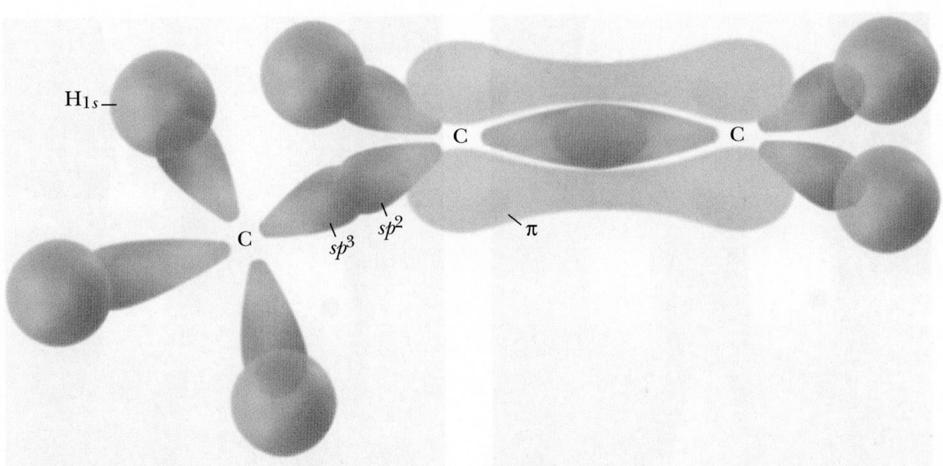

Figure 10.33 Bonding in propylene. Sigma and π bonds in propylene.

Answer The *sp*-hybridized carbon atom forms two σ bonds, one by overlapping with a 1*s* orbital on the hydrogen atom and a second by overlapping with a *p* orbital on the nitrogen atom. The carbon and nitrogen atoms form two π bonds from the sideways overlap of the two unhybridized *p* orbitals on each atom.

Summary of Bonding and Structure Models

In this chapter and in Chapter 9, we developed models and theories that describe how covalent bonds form and how the arrangement of valence electrons can determine the shapes of compounds. To describe the bonding and shape of a compound, we must know the skeleton structure and calculate the number of valence electrons. Generally, the Lewis structure describes the bonding quite well and can be used with the VSEPR model to predict the geometry about the central atoms of the molecule. The polarity of the molecule can be determined from the electronegativities of the elements and the shape of the molecule. Valence bond theory allows the determination of which orbitals are used to make each bond and to hold the lone pairs. Frequently, the shape of a molecule can be explained only if hybrid orbitals are used to form the bonds. Valence bond theory also introduces σ and π bonds, which are necessary to account for some properties of molecules that contain multiple bonds.

OBJECTIVES REVIEW *Can you:*

- ☑ define and identify σ and π bonds?
- ☑ identify the orbitals used to form σ and π bonds?
- ☑ describe the arrangement of atoms in *cis* and *trans* isomers?

10.5 Molecular Orbitals: Homonuclear Diatomic Molecules

OBJECTIVES

- ☐ Write the molecular orbital diagrams and electron configurations for homonuclear diatomic molecules and ions of the first- and second-period elements
- ☐ Determine the bond order and the number of unpaired electrons in diatomic species from the molecular orbital diagrams

Lewis structures and valence bond theory explain many known experimental results such as the shape of ethylene and the existence of *cis* and *trans* isomers of HClC=CHCl. A surprisingly simple molecule for which the Lewis structure and valence bond theory give an incorrect bonding description is O_2. The Lewis structure of O_2 is

$$\ddot{\underset{..}{O}}=\ddot{\underset{..}{O}}$$

(a)

(b)

Figure 10.34 Behavior of liquid nitrogen and oxygen in a magnetic field. *(a)* Compounds such as liquid nitrogen that contain only paired electrons are not attracted by a magnetic field; the poured liquid just flows by the magnet. *(b)* In contrast, liquid oxygen remains suspended between the poles of a magnet, indicating that it is paramagnetic and contains unpaired electrons.

The experimentally determined O–O bond distance in this molecule is consistent with a double bond, as predicted by the theory, but measurements show that O_2 is paramagnetic (attracted into a magnetic field) (Figure 10.34) and contains two unpaired electrons. The Lewis structure of O_2 does not predict two unpaired electrons; all of the electrons are present as pairs.

To explain this result and others, a more comprehensive theory has been developed. Lewis structures and valence bond theory are similar in that a bond forms when two adjacent atoms share a pair of electrons. A different approach describes the bonding using orbitals that have the valence electrons shared among all of the atoms in the molecule rather than shared between only two atoms. **Molecular orbital theory** is a model that combines atomic orbitals to form new orbitals that are shared over the entire molecule. Molecular orbital theory successfully explains certain characteristics of molecules that cannot be explained with valence bond theory.

An atomic orbital is located on one atom, whereas a **molecular orbital** extends over the entire molecule. Molecular orbitals are described by mathematically combining the valence atomic orbitals of all the atoms in a molecule. The mathematical operation used to form molecular orbitals is similar to that used to form hybrid orbitals. The difference is that hybrid orbitals are formed from combinations of valence orbitals on the *same* atom, whereas molecular orbitals are formed from combinations of orbitals on *different* atoms. In both theories, the number of orbitals that form must equal the number of atomic orbitals used to make them. As with atomic orbitals, we are interested in both the energy and the shape of the electron cloud for each molecular orbital.

Molecular orbitals are formed from combinations of atomic orbitals on different atoms.

The Hydrogen Molecule

The simplest molecule to consider is H_2, because only two 1*s* orbitals are available to form the molecular orbitals. The molecular orbitals are formed by the addition and subtraction of the two atomic orbitals (Figure 10.35).

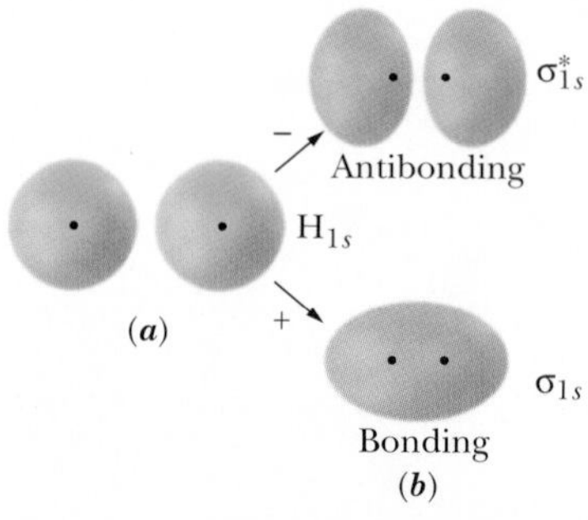

Figure 10.35 Molecular orbitals for H_2. The hydrogen atom 1*s* atomic orbitals *(a)* are added and subtracted to form bonding and antibonding molecular orbitals *(b)*.

Adding the two atomic orbitals produces one of the two new molecular orbitals. The reinforcement of the two orbitals creates a large amplitude (large electron probability) directly between the two nuclei. The new orbital is a **bonding molecular orbital,** one that concentrates the electron density between the atoms in the molecule. An electron in a bonding orbital is located mainly between the two nuclei and is attracted by both, so the molecular orbital is lower in energy than the atomic orbitals from which it formed. The bonding molecular orbital in hydrogen is labeled σ_{1s} because it is symmetric about the line joining the two nuclei (the definition of a σ bond) and is formed from 1*s* atomic orbitals.

Subtraction of the 1*s* atomic orbital on one hydrogen atom from that on the other produces the second molecular orbital. The opposite signs of the orbitals cause the electron probability between the nuclei, where the orbitals overlap, to cancel. An **antibonding molecular orbital** is one that *reduces* the electron density in the region between the atoms in the molecule. The lower electron density between the two nuclei produces a less stable (higher energy) orbital than the separate atomic orbitals from which it is made. An asterisk designates an antibonding molecular orbital, as in σ^*_{1s}. The increase in energy of an electron in an antibonding orbital is approximately the same as the decrease in energy of an electron in the corresponding bonding molecular orbital. Although less stable than the atomic orbitals from which it is formed, an antibonding orbital can hold two electrons just like any orbital.

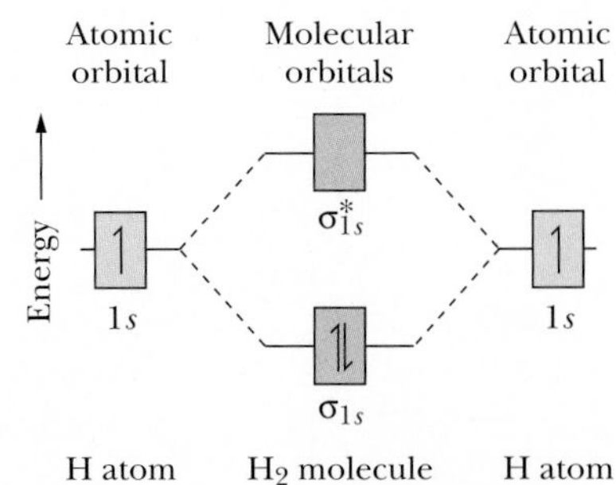

Figure 10.36 Molecular orbital diagram of H_2. The diagram for H_2 shows the starting atomic orbitals on the outside for reference and the molecular orbitals in the middle. The two electrons in the H_2 molecule are in the bonding molecular orbital.

The relative energies of the new molecular orbitals are frequently shown with the starting atomic orbitals, as in Figure 10.36. The bonding orbital is lower and the antibonding orbital higher in energy than the starting atomic orbitals. Note that the number of molecular orbitals is the same as the number of atomic orbitals used to form them. The filling of molecular orbitals follows the same procedure as the filling of atomic orbitals, using all the electrons in the molecule. The aufbau principle, the Pauli exclusion principle, and Hund's rule all apply. The two electrons in H_2 are in the σ_{1s} orbital with opposite spins. The molecular orbital electron configuration for the molecule H_2 is represented by $(\sigma_{1s})^2$. Note that the electron configurations of molecules are similar to those we have been writing for atoms, such as $1s^22s^1$ for Li, except now the orbital is delocalized over two or more atoms rather than just located on one atom.

Molecular orbital theory defines *bond order* by the equation

$$\text{Bond order} = \frac{1}{2}\,(\text{number of electrons in bonding orbitals} - \text{number of electrons in antibonding orbitals})$$

The bond order is one-half the difference of the number of electrons in bonding orbitals and the number of electrons in antibonding orbitals.

The bond order for H_2 is $\frac{1}{2}\ (2 - 0) = 1$, the same as that found using the Lewis structure.

The He_2 Molecule

The importance of the antibonding orbital is evident if the hypothetical molecule He_2 is considered. The molecular energy-level diagram for He_2 is qualitatively the same as that for H_2. A total of four electrons, two from the valence orbital of each helium atom, must now occupy the molecular orbitals. The electron configuration is $(\sigma_{1s})^2(\sigma^*_{1s})^2$ (Figure 10.37). Remember that the 1*s* atomic orbitals shown on the two sides of the diagram have combined to form the molecular orbitals. They are on the diagram to show the energies of the orbitals on the isolated atoms relative to those of the molecular orbitals and to show which atomic orbitals form the molecular orbitals.

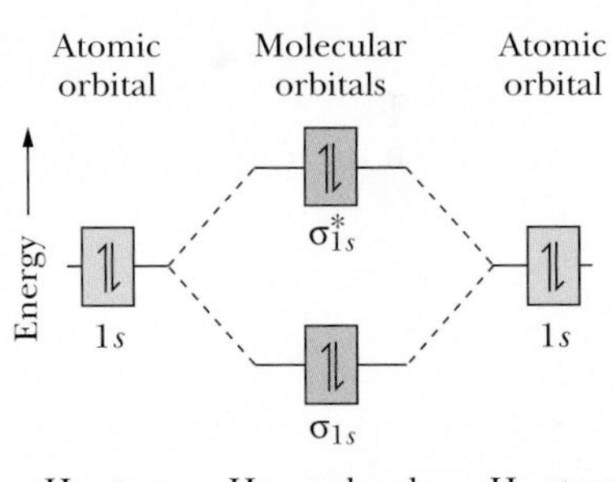

Figure 10.37 Molecular orbital diagram of He_2. The four electrons in He_2 fill both the bonding and antibonding molecular orbitals.

The bond order in the He_2 molecule is $\frac{1}{2}(2 - 2) = 0$. All of the energy that was gained (in comparison with the isolated atoms) by filling the bonding orbital is lost in filling the antibonding orbital. In fact, the completely filled bonding and antibonding orbitals are slightly less stable than the filled atomic orbitals from which they arise. Molecular orbital theory predicts that the isolated atoms are a little more stable than He_2, so this molecule will not form. We already know that atoms of He are stable and unreactive, and do not combine with each other.

A molecule is less stable than its isolated atoms if the number of electrons in bonding orbitals is equal to the number of electrons in antibonding orbitals.

EXAMPLE 10.10 Bonding in He_2^+

Write the electron configuration for He_2^+; then calculate the bond order. Predict whether this species will be stable.

Strategy Write the molecular orbital diagram, place the correct number of electrons in the diagram, and then calculate the bond order. Any molecule with a positive bond order should be stable.

Solution

The molecular orbital diagram is the same as for H_2 and He_2. The electron configuration is $(\sigma_{1s})^2(\sigma^*_{1s})^1$, and the bond order is $\frac{1}{2}(2 - 1) = \frac{1}{2}$. The He–He bond in this cation would be weak, but the presence of a fractional bond order shows that this species should exist. The He_2^+ ion has been detected experimentally and has the properties predicted by molecular orbital theory.

Understanding

Write the electron configuration for H_2^+, and calculate the bond order. Will this species be stable?

Answer $(\sigma_{1s})^1$, bond order $= \frac{1}{2}$; the species is predicted to be stable.

Second-Period Diatomic Molecules

A **homonuclear diatomic molecule** contains two atoms of the same element. Several of the nonmetallic elements in the second period exist as homonuclear diatomic molecules—N_2, O_2, and F_2. Molecular orbital theory is useful in describing the bonding in these well-known molecules and in other, less common molecules such as Li_2, B_2, and C_2. In the second period, the valence shell consists of the $2s$ and $2p$ atomic orbitals. Because of the low energy of the inner $1s$ orbitals and electrons, they will not be included in this discussion of the bonding of molecules made from second-period elements. Because each atom has four valence atomic orbitals (s, p_x, p_y, and p_z), there are a total of eight atomic orbitals used to construct the molecular orbital diagram of these diatomic molecules.

Second-period diatomic molecules form σ_{2s} and σ^*_{2s} molecular orbitals from the $2s$ orbitals on the two atoms, analogous to those formed from the $1s$ orbitals for H_2. The $2p$ orbitals combine to yield two different types of molecular orbitals. The first set of molecular orbitals arises from the head-on overlap of a p orbital on one atom with a p orbital on the other atom, oriented along the axis joining the nuclei (Figure 10.38). The

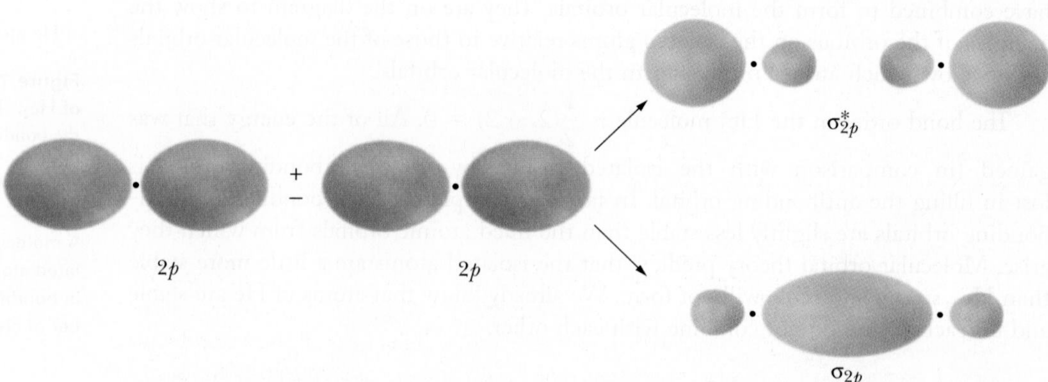

Figure 10.38 Sigma molecular orbitals from p atomic orbitals. The combination of the p orbitals oriented along the axis joining the nuclei yields a σ_{2p} bonding molecular orbital when the signs of the orbitals reinforce and a σ^*_{2p} antibonding molecular orbital when the signs of the orbitals interfere. Both orbitals can hold two electrons.

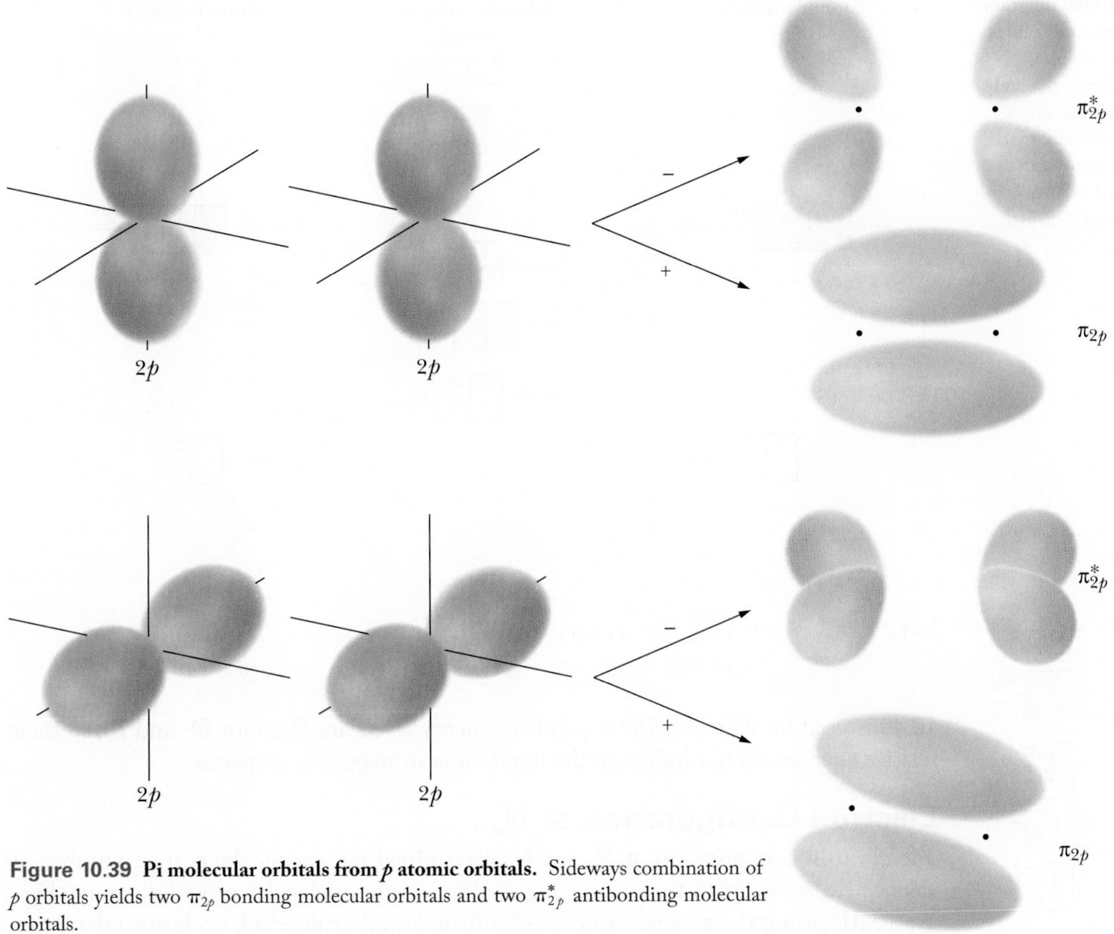

Figure 10.39 Pi molecular orbitals from *p* atomic orbitals. Sideways combination of *p* orbitals yields two π_{2p} bonding molecular orbitals and two π^*_{2p} antibonding molecular orbitals.

combination yields a σ_{2p} bonding molecular orbital when the signs of the overlapping lobes are the same, and a σ^*_{2p} antibonding molecular orbital when the signs of the overlapping lobes are opposite. As always, the combination of two atomic orbitals forms two molecular orbitals.

The remaining $2p$ orbitals, two on each atom, are perpendicular to the internuclear axis and combine as shown in Figure 10.39. Each pair of *p* orbitals combines to form one bonding and one antibonding molecular orbital. These are π molecular orbitals because the region of overlap lies on opposite sides of the line joining the nuclei. The two *p* orbitals on each atom combine with those on the other atom to form two bonding π_{2p} molecular orbitals and two antibonding π^*_{2p} molecular orbitals. The two π_{2p} molecular orbitals are perpendicular to each other and have *exactly the same* energies. The two π^*_{2p} molecular orbitals are also perpendicular to each other and have the same energies, which are higher than the energies of the π_{2p} orbitals. Overall, the four *p*-type atomic orbitals form two π-bonding molecular orbitals at one energy and two π-antibonding molecular orbitals at a higher energy.

Figure 10.40 is the complete molecular orbital diagram for a second-period diatomic molecule. The relative spacing of the molecular orbitals in this diagram is complicated; the order shown is an average of the experimentally determined values for all the second-row elements. Both the bonding and antibonding molecular orbitals that arise from the combination of the $2s$ atomic orbitals are lower in energy than the molecular orbitals arising from combination of the $2p$ atomic orbitals. Also, the π_{2p} orbitals are lower in energy than the σ_{2p} orbitals, at least for the diatomic molecules Li_2 through N_2. Evidence exists that σ_{2p} is lower in energy than π_{2p} for O_2 and F_2, but for simplicity, the diagram in Figure

Figure 10.40 Molecular orbital diagram for the second-period diatomic molecules. This diagram is correct for Li_2 through N_2. In O_2 and F_2, the energy order of σ_{2p} and π_{2p} are reversed, but because these are filled orbitals for the two molecules, this inversion does not affect the bond order or magnetic properties. Only the valence orbitals are shown in a molecular orbital diagram.

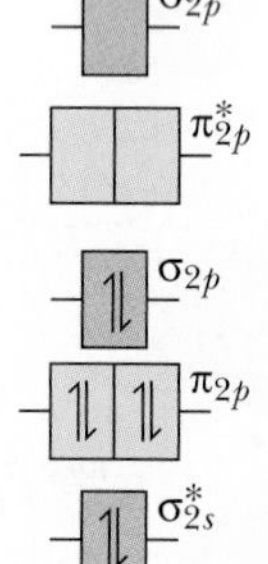

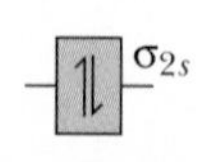

Figure 10.41 Molecular orbital diagram for N_2. N_2 has 10 valence electrons. Molecular orbital theory predicts a bond order of 3 and no unpaired electrons.

10.40 is used for all cases. The σ_{2p} and π_{2p} energy levels are filled for O_2 and F_2, so their relative energies do not influence the bond order or magnetic properties.

Electron Configuration of N_2

The electron configuration of N_2 can be determined by placing the correct number of valence electrons (10, 5 for each nitrogen atom) in the molecular orbital diagram (Figure 10.41), using the general principles (aufbau, Hund's rule, Pauli exclusion) that were introduced with atomic orbitals.

The electron configuration is $(\sigma_{2s})^2(\sigma_{2s}^*)^2(\pi_{2p})^4(\sigma_{2p})^2$. Eight valence electrons are in bonding molecular orbitals and two in antibonding orbitals, so the predicted bond order is

$$\text{Bond order} = \frac{1}{2}(8 - 2) = 3$$

The four electrons in σ_{2s} and σ_{2s}^* orbitals contribute essentially no net bonding. The two filled π_{2p} orbitals and the filled σ_{2p} orbital yield the three bonds. This bonding description is consistent with the bond order predicted by the Lewis structure (:N≡N:). The lone pairs on the nitrogen atoms correspond to the electron pairs in the filled σ_{2s} and σ_{2s}^* molecular orbitals.

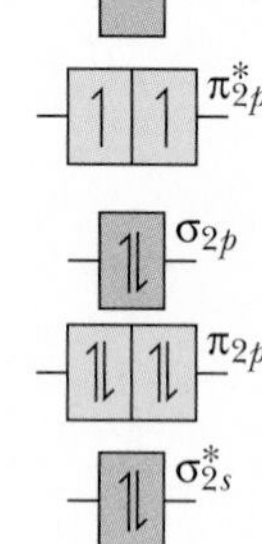

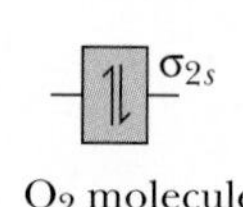

Figure 10.42 Molecular orbital diagram for O_2. O_2 has 12 valence electrons. Molecular orbital theory predicts a bond order of 2 and two unpaired electrons.

Electron Configuration of O_2

As outlined earlier, valence bond theory predicts that O_2 contains no unpaired electrons, contrary to the experimental result, which shows that it is paramagnetic and contains two unpaired electrons. In contrast, molecular orbital theory correctly predicts this property. Figure 10.42 is the molecular orbital diagram for O_2.

The energy-level diagram now has a total of 12 electrons with two electrons occupying the degenerate (equal-energy) π_{2p}^* levels. Following Hund's rule, these electrons are placed in separate orbitals with parallel spins. The predicted bond order is

$$\text{Bond order} = \frac{1}{2}(8 - 4) = 2$$

Thus, molecular orbital theory correctly accounts for the experimental observations of a double bond and the presence of *two unpaired electrons* in O_2.

Molecular orbital theory correctly predicts that O_2 will have two unpaired electrons.

EXAMPLE 10.11 Molecular Orbitals

Write the molecular orbital diagram and the electron configuration for (a) Be_2 and (b) B_2. Predict the bond order and the number of unpaired electrons for each molecule.

Strategy Use the diagram in Figure 10.40, adding electrons as appropriate to the lowest energy orbitals.

Solution

(a) Be_2 has four valence electrons. Figure 10.43 is the molecular orbital diagram. The electron configuration is $(\sigma_{2s})^2(\sigma_{2s}^*)^2$. The bond order is $\frac{1}{2}(2 - 2) = 0$. Because the bond order is zero, this molecule is not expected to be stable.

(b) B_2 has six valence electrons. Figure 10.44 is the molecular orbital diagram. The electron configuration is $(\sigma_{2s})^2(\sigma_{2s}^*)^2(\pi_{2p})^2$. The bond order is $\frac{1}{2}(4 - 2) = 1$, and the molecule has two unpaired electrons. Experiments have confirmed both of these predictions, although the molecule B_2 exists only at high temperature and low pressure because other elemental forms of boron are more stable at room temperature and pressure. In this molecule, like in O_2, the Lewis structure, B≡B, does not agree with the results of experimental measurements. Neither the bond order nor the paramagnetism is correctly predicted by the Lewis structure for B_2.

Understanding

Use molecular orbital theory to predict the bond order and the number of unpaired electrons for C_2.

Answer Bond order = 2; no unpaired electrons

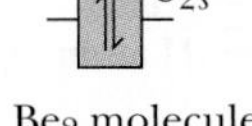

Figure 10.43 Molecular orbital diagram for Be_2. Be_2 has four valence electrons. Molecular orbital theory predicts a bond order of zero, consistent with the fact that Be_2 has not been observed.

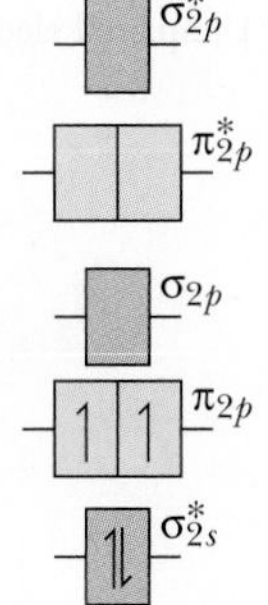

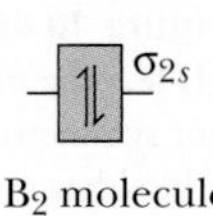

Be₂ molecule is replaced below:

Figure 10.44 Molecular orbital diagram for B_2. B_2 has six valence electrons. Molecular orbital theory correctly predicts a bond order of 1 and two unpaired electrons.

Summary of Second-Row Homonuclear Diatomic Molecules

Table 10.2 summarizes the electron configurations of the second-period homonuclear diatomic molecules, together with important physical data.

OBJECTIVES REVIEW *Can you:*

- ☑ write the molecular orbital diagrams and electron configurations for homonuclear diatomic molecules and ions of the first- and second-period elements?
- ☑ determine the bond order and the number of unpaired electrons in diatomic species from the molecular orbital diagrams?

TABLE 10.2 Molecular Orbital Electron Configurations, Bond Orders, and Physical Data for Second-Period Homonuclear Diatomic Molecules

Predicted by Molecular Orbital Theory			Experimental Measurements	
Species	Electron Configuration	Bond Order	Bond Energy (kJ/mol)	Number of Unpaired Electrons
Li_2	$(\sigma_{2s})^2$	1	105	0
Be_2	$(\sigma_{2s})^2(\sigma_{2s}^*)^2$	0	Unstable	
B_2	$(\sigma_{2s})^2(\sigma_{2s}^*)^2(\pi_{2p})^2$	1	290	2
C_2	$(\sigma_{2s})^2(\sigma_{2s}^*)^2(\pi_{2p})^4$	2	620	0
N_2	$(\sigma_{2s})^2(\sigma_{2s}^*)^2(\pi_{2p})^4(\sigma_{2p})^2$	3	946	0
O_2	$(\sigma_{2s})^2(\sigma_{2s}^*)^2(\pi_{2p})^4(\sigma_{2p})^2(\pi_{2p}^*)^2$	2	498	2
F_2	$(\sigma_{2s})^2(\sigma_{2s}^*)^2(\pi_{2p})^4(\sigma_{2p})^2(\pi_{2p}^*)^4$	1	159	0

10.6 Heteronuclear Diatomic Molecules and Delocalized Molecular Orbitals

OBJECTIVES

- ☐ Construct the molecular orbital diagram for heteronuclear diatomic molecules
- ☐ Describe the formation of delocalized molecular orbitals

In the preceding section, we constructed molecular orbital diagrams for homonuclear diatomic molecules and ions. A **heteronuclear diatomic molecule** contains one atom of each of two different elements. The molecular orbital diagrams for heteronuclear diatomics are similar to those for homonuclear molecules when the valence orbitals on one atom are fairly close in energy to the valence orbitals on the other. The diagrams are different if the orbitals are not close in energy.

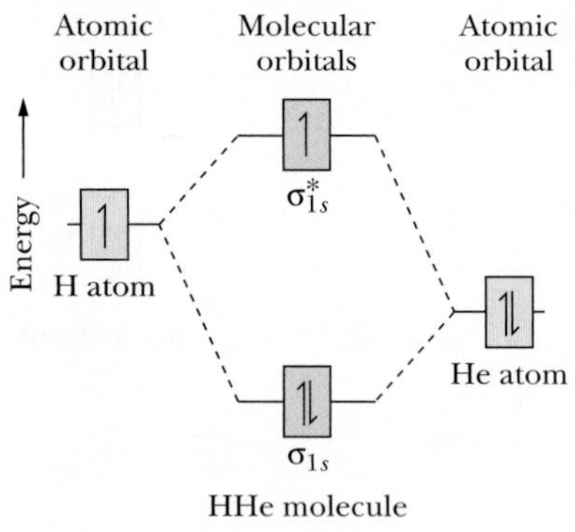

Figure 10.45 Molecular orbital diagram for HHe. The atomic orbitals that form the molecular orbitals for the HHe molecule differ in energy. The molecule has 3 electrons, a bond order of 0.5, and 1 unpaired electron.

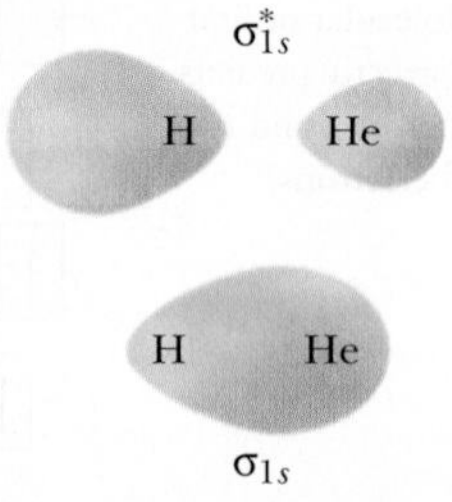

Figure 10.46 Molecular orbitals for HHe. The molecular orbitals of HHe are not symmetric. The bonding molecular orbital has more electron density near the helium atom, and the antibonding orbital has more electron density near the hydrogen atom.

Molecular orbitals for heteronuclear molecules are not symmetric.

Heteronuclear diatomic molecules formed from elements that are close to each other in the periodic table have molecular orbital diagrams that resemble those of homonuclear diatomic molecules.

The HHe Molecule

The simplest heteronuclear diatomic molecule is HHe. The molecular orbitals for HHe are formed from the 1*s* orbitals of the H and He atoms. The energy of the 1*s* orbital on the helium atom is lower than that of the hydrogen atom, as shown in Figure 10.45. We first noted this energy difference in Chapter 8, which presents trends in ionization energies within a period. Because the effective nuclear charge, Z_{eff}, increases from left to right, the atomic orbitals become more stable (lower in energy) from left to right in any period. Two molecular orbitals, σ_{1s} and σ^*_{1s}, form. The bonding orbital is lower in energy than the helium 1*s* atomic orbital, and the antibonding orbital is higher in energy than the hydrogen 1*s* orbital. Following the aufbau procedure, two of the three electrons are placed in the σ_{1s} orbital and one in the σ^*_{1s} orbital. The bond order is 0.5.

In Figure 10.36, the molecular orbital diagram of H_2, each atomic orbital contributes equally to the bonding and antibonding molecular orbitals (the mathematical mixing uses the orbitals equally). In the case of the HHe orbitals, the helium atom 1*s* orbital contributes a larger fraction to the bonding molecular orbital than does the hydrogen atom 1*s* orbital. The hydrogen atom orbital contributes more to the σ^*_{1s} orbital. In general, a molecular orbital more closely resembles the atomic orbital that is closer in energy to that molecular orbital. Because of these different contributions, the molecular orbitals in HHe are not symmetric as they are in H_2. The amplitude of the σ_{1s} orbital is greater around the helium atom, and the amplitude of the σ^*_{1s} orbital is greater around the hydrogen atom (Figure 10.46).

Second-Row Heteronuclear Diatomic Molecules

Carbon monoxide, CO, and nitrogen monoxide, NO, are two common compounds of second-row elements that exist as heteronuclear diatomic molecules. The molecular orbital energy-level diagrams for these molecules are similar to those for the homonuclear diatomic molecules. In both cases, the same valence atomic orbitals (one 2*s* and three 2*p*) are available to form molecular orbitals. The molecular orbital diagram for the heteronuclear molecules differs from the homonuclear cases because the energies of the atomic orbitals on the two atoms are not the same. For any given pair of elements, the one closer to the right side of the period has orbitals of lower energy. A diagram similar to that in Figure 10.40 is used for heteronuclear diatomic molecules such as NO that are formed from nearby elements in the periodic table, but the diagram is modified to indicate that the atomic orbitals that form the molecular orbitals are of different energies (Figure 10.47).

The electron configuration of NO is $(\sigma_{2s})^2(\sigma^*_{2s})^2(\pi_{2p})^4(\sigma_{2p})^2(\pi^*_{2p})^1$. The 11 valence electrons of NO fill the molecular orbitals through the σ_{2p} orbital completely and place one electron in the π^*_{2p} orbitals. The bond order is $\frac{1}{2}(8 - 3) = 2.5$, and one

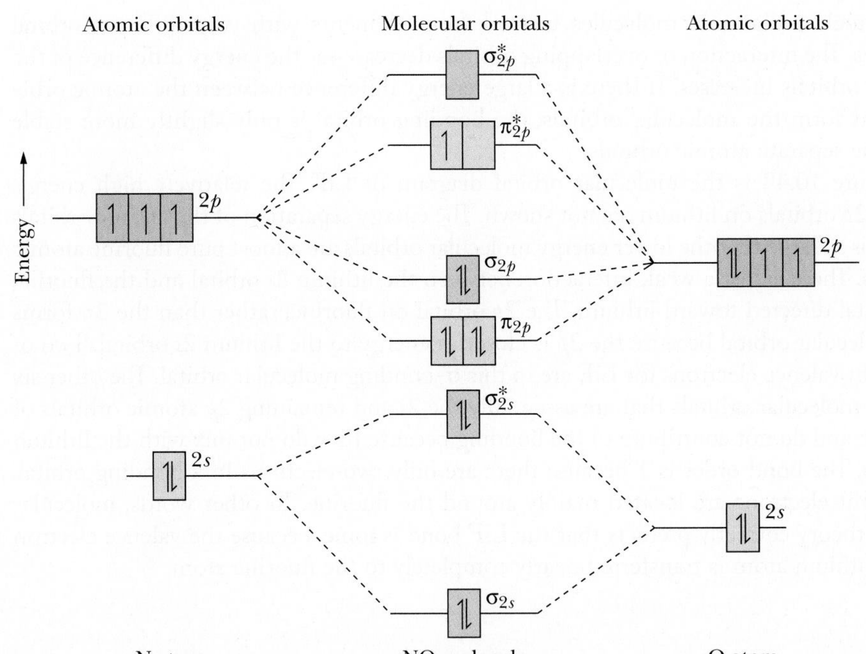

Figure 10.47 Molecular orbital diagram for nitric oxide (NO). The molecular orbital diagram for NO predicts a bond order of 2.5 and predicts that the molecule is paramagnetic with one unpaired electron. These predictions are verified by experimental measurements.

unpaired electron is expected. Experiment shows that both of these predictions are correct.

It is interesting to compare NO with NO^+. The molecular orbital configuration for NO^+ is $(\sigma_{2s})^2(\sigma_{2s}^*)^2(\pi_{2p})^4(\sigma_{2p})^2$. The NO^+ has one fewer electron than NO, increasing the bond order to 3 because the additional electron in NO occupies an antibonding orbital. It has been shown experimentally that the bond length in NO^+ is 9 pm shorter than that in NO, indicating that the bond order is greater in NO^+.

EXAMPLE 10.12 Molecular Orbitals

Write the molecular orbital electron configuration and give the bond order of CN^-.

Strategy Use the diagram in Figure 10.47, adding electrons as appropriate to the lowest energy orbitals.

Solution

The cyanide anion has 10 valence electrons. The electron configuration is $(\sigma_{2s})^2(\sigma_{2s}^*)^2(\pi_{2p})^4(\sigma_{2p})^2$. The bond order is $\frac{1}{2}(8 - 2) = 3$. This ion is isoelectronic with N_2 and NO^+.

Understanding

Write the molecular orbital electron configuration and give the bond order of NO^-. Is the species diamagnetic or paramagnetic?

Answer $(\sigma_{2s})^2(\sigma_{2s}^*)^2(\pi_{2p})^4(\sigma_{2p})^2(\pi_{2p}^*)^2$; bond order is 2; the ion is paramagnetic with two unpaired electrons.

Molecular Orbital Diagram for LiF

The molecular orbital diagram in Figure 10.47 is correct only for molecules formed by second-row elements close together in the second period because the energies of the atomic orbitals of the two elements are fairly similar. Different diagrams are needed for

Solid LiF. LiF, a hard crystalline solid, has the physical properties expected for an ionic compound.

heteronuclear diatomic molecules formed from elements with very different orbital energies. The interaction of overlapping orbitals decreases as the energy difference of the atomic orbitals increases. If there is a large energy difference between the atomic orbitals that form the molecular orbitals, the bonding orbital is only slightly more stable than the separate atomic orbitals.

Figure 10.48 is the molecular orbital diagram of LiF. The relatively high energy, empty *2p* orbitals on lithium are not shown. The energy separation of the atomic orbitals in LiF is so large that the lower energy molecular orbitals are almost pure fluorine atomic orbitals. There is just a weak interaction between the lithium *2s* orbital and the fluorine *2p* orbital directed toward lithium. The *2p* orbital on fluorine, rather than the *2s*, forms the molecular orbital because the *2p* is closer in energy to the lithium *2s* orbital. Two of the eight valence electrons for LiF are in this σ-bonding molecular orbital. The other six occupy molecular orbitals that are essentially the *2s* and remaining *2p* atomic orbitals of fluorine and do not contribute to the bonding because they do not mix with the lithium orbitals. The bond order is 1 because there are only two electrons in a bonding orbital. The eight electrons are located mainly around the fluorine. In other words, molecular orbital theory correctly predicts that the LiF bond is ionic, because the valence electron in the lithium atom is transferred nearly completely to the fluorine atom.

PRINCIPLES OF CHEMISTRY
Atomic Orbitals Overlap to Form Delocalized Molecular Orbitals

Some of the advantages of the delocalized approach of molecular orbital theory can be demonstrated by considering BeH_2. The Lewis structure of BeH_2 is

H-Be-H

This representation of the molecule, in conjunction with the VSEPR model, proves to be useful in predicting that this molecule has a linear geometry. In forming the two equivalent sigma bonds, the *2s* orbital and one *2p* orbital of the beryllium atom combine to form two *sp* hybrid orbitals. Each bond forms by overlap of a hydrogen atom *1s* orbital and one of the two *sp* hybrid orbitals on beryllium.

The molecular orbital description is based on the known linear arrangement of the three atoms. The molecular orbitals form from the overlap of valence atomic orbitals of the beryllium atom (*2s*, $2p_x$, $2p_y$, and $2p_z$) and the two *1s* orbitals on the hydrogen atoms. The *2s* orbital on the beryllium atom overlaps equally with each of the hydrogen *1s* orbitals as shown in the following diagram. These orbitals combine to give a sigma bonding orbital and a sigma antibonding orbital. An electron pair in the bonding molecular orbital contributes equally to the bonding of both the hydrogen atoms and is, therefore, a delocalized molecular orbital.

One of the *2p* atomic orbitals of the beryllium atom is directed along the molecular axis and also overlaps with the orbitals of the two hydrogen atoms. Because the sign of the amplitude of the *p* wave function is opposite in the two lobes, for both the bonding and the antibonding orbital, the sign of the molecular wave function on the first hydrogen atom is opposite that in the second hydrogen atom, as shown. Once again a

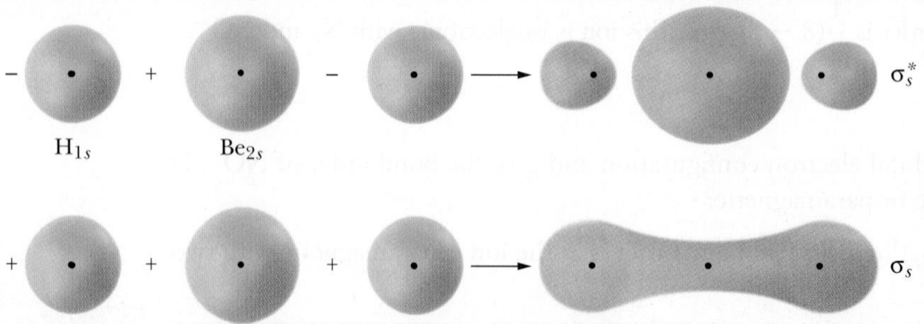

Sigma molecular orbitals from the interaction of beryllium 1*s* and hydrogen 1*s* atomic orbitals. Both bonding and antibonding orbitals are formed from the interaction of the beryllium *2s* orbital with the *1s* orbitals of the two hydrogen atoms. The new molecular orbitals are delocalized over all three atoms.

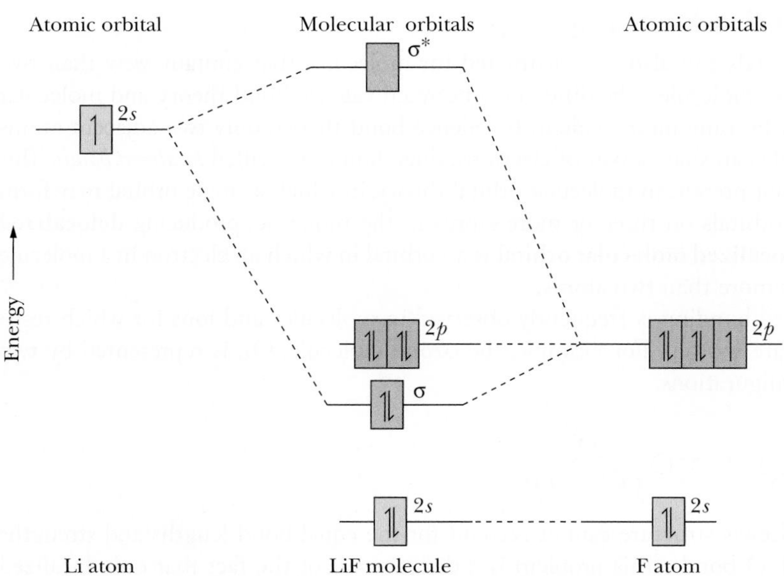

Figure 10.48 Molecular orbital diagram for LiF. The energy of the 2*s* orbital on lithium is much higher than those of the 2*s* and 2*p* orbitals on fluorine. The eight valence electrons are in molecular orbitals that are nearly the same as the atomic orbitals on fluorine.

sigma bonding orbital and a sigma antibonding orbital result from the two combinations. Note that in both of these bonding descriptions that the same two 1*s* orbitals are used; in each bonding description pictured only part of the wave function of each of the atomic orbitals contributes to each molecular orbitals.

Because the two remaining *p* orbitals of the beryllium atom are perpendicular to the Be-H bonds, there is no net overlap with the hydrogen 1*s* orbitals, so they do not contribute to the bonding in any way. The four electrons in the molecule fill the two bonding orbitals, forming overall two bonds, the same as predicted by valence bond theory. The two theories differ in that both pairs are delocalized and are involved in the bonding of each hydrogen atom in the molecular orbital model, whereas the valence bond picture assumes localized bonds. ▮

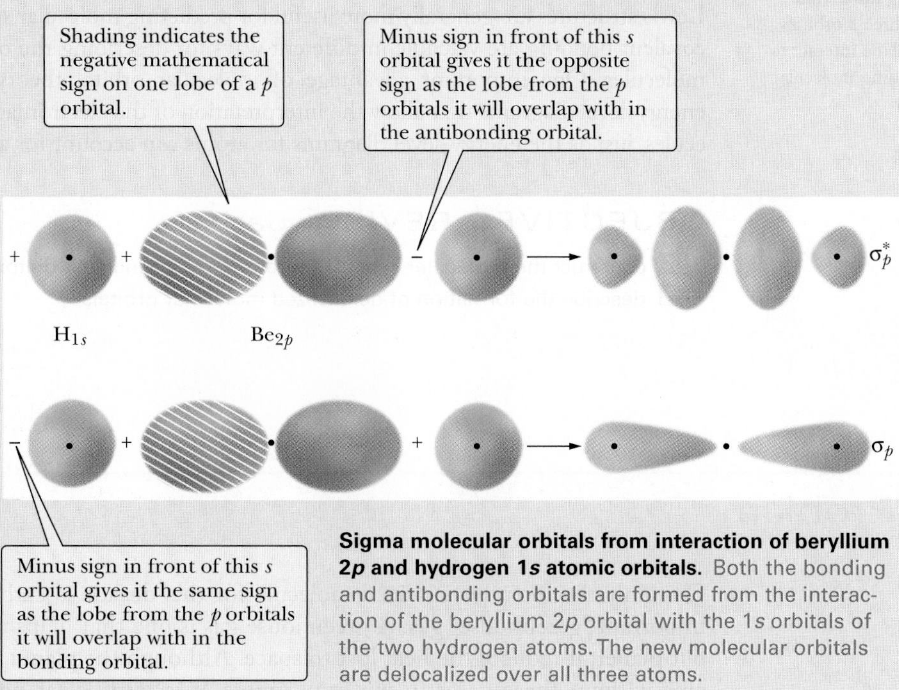

Shading indicates the negative mathematical sign on one lobe of a p orbital.
Minus sign in front of this s orbital gives it the opposite sign as the lobe from the p orbitals it will overlap with in the antibonding orbital.
+
+
−
σ_p^*
H_{1s}
Be_{2p}
−
+
+
σ_p
Minus sign in front of this s orbital gives it the same sign as the lobe from the p orbitals it will overlap with in the bonding orbital.

Sigma molecular orbitals from interaction of beryllium 2*p* and hydrogen 1*s* atomic orbitals. Both the bonding and antibonding orbitals are formed from the interaction of the beryllium 2*p* orbital with the 1*s* orbitals of the two hydrogen atoms. The new molecular orbitals are delocalized over all three atoms.

Delocalized π Bonding

Molecular orbitals can also be constructed for molecules that contain *more* than two atoms. In these molecules, the differences between valence bond theory and molecular orbital theory become more evident. In valence bond theory, only two adjacent atoms in the molecule can share a pair of electrons. Such bonds are called *localized bonds.* This limitation is not present in molecular orbital theory, in which a single orbital may form from atomic orbitals on three or more atoms in the molecule, producing **delocalized bonds.** A **delocalized molecular orbital** is an orbital in which an electron in a molecule is spread over more than two atoms.

The overlap of atomic orbitals on three or more atoms yields delocalized molecular orbitals.

Delocalized bonding is frequently observed in molecules and ions for which resonance forms are written. For example, the ozone molecule, O_3, is represented by two resonance configurations.

A single Lewis structure cannot account for the equal bond lengths and strengths of the two O–O bonds. This problem is a direct result of the fact that only localized bonds are shown in conventional Lewis diagrams. Molecular orbital theory overcomes this problem because the molecular orbitals may involve atomic orbitals on all of the atoms present.

The π bonds in the two resonance forms of O_3 are formed from three *p* orbitals, one on each oxygen atom, that are perpendicular to the molecular plane of O_3 (remember that π bonds are made from sideways overlap of *p* orbitals). The *p* orbital on the central oxygen atom overlaps with a *p* orbital on the oxygen atom on its left in the first resonance structure, and with a *p* orbital on the oxygen atom on its right in the second resonance structure. In molecular orbital theory, all three of these *p* orbitals can interact to form one large molecular orbital that is spread out over all three nuclei (Figure 10.49).

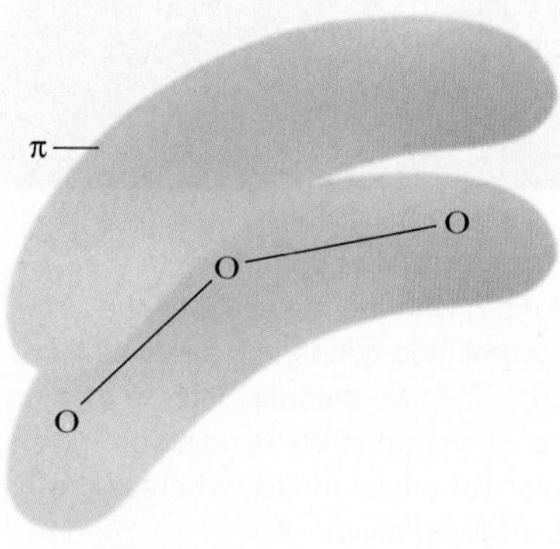

Figure 10.49 Pi bonding molecular orbitals for ozone. The three *p* orbitals perpendicular to the O_3 plane interact to form a delocalized π-bonding molecular orbital.

The complete molecular orbital treatment of O_3 and most other molecules is not appropriate at this point. The energy-level diagrams can become quite complicated, even for small molecules. Although molecular orbital theory is being used more and more frequently to describe the bonding and distribution of electrons in molecules, it does not provide a simple means of anticipating molecular geometries. Valence bond theory and Lewis structures are generally more useful for predicting molecular shapes. The models for covalent bonding are valuable in different ways for describing the observed properties of molecules. One important advantage of molecular orbital theory is that it produces energy-level diagrams that allow the interpretation of the electromagnetic spectra of molecules, just as the energy-level diagrams for atoms can account for atomic spectra.

OBJECTIVES REVIEW *Can you:*

- ☑ construct the molecular orbital diagram for heteronuclear diatomic molecules?
- ☑ describe the formation of delocalized molecular orbitals?

Summary Problem

Certainly the most popularized molecule for this decade is carbon dioxide, CO_2, an important "greenhouse" gas. A greenhouse gas is one that helps maintain the heat on our planet; it reduces the heat lost to space. Although the planet would be uninhabitable without these gases in our atmosphere, it is fairly clear now that a buildup of these gases in recent years has been too much of a good thing. The energy needs brought on by the industrial revolution have increased the amounts of CO_2 in the atmosphere contributing, together with other factors, to an increase in the average temperature of the planet.

Carbon dioxide is very stable, a fact that can be explained by its bonding. Given the O–C–O skeleton structure of CO_2, write a table that includes its Lewis structure, bonded-atom lone-pair arrangement, hybridization of the central atom, and polarity.

From the Lewis structure, the steric number of the central carbon atom is 2. The bonded-atom lone-pair arrangement and hybridization given below follow from that. The molecule is nonpolar because the identical bond dipoles of the two opposing bonds cancel each other in the linear arrangement.

Skeleton structure	Lewis structure	Bonded-atom lone-pair arrangement
O—C—O	$\ddot{O}$=C=$\ddot{O}$	linear

Hybridization of the central atom	Polarity
sp	Nonpolar

Understanding how to develop each of the answers in this type of table is important to chemists. Develop a similar table that shows the skeleton structure, Lewis structure, bonded-atom lone-pair arrangement, hybridization of each central atom, and polarity of CH_2CHOCH_3.

Skeleton structure	Lewis structure	Bonded-atom lone-pair arrangement
H_2C–CH–O–CH_3 (single bonds)	H_2C=CH–$\ddot{O}$–CH_3	CH_2, CH = trigonal planar O, CH_3 = tetrahedral

Hybridization of the central atom	Polarity
CH_2, CH = sp^2 O, CH_3 = sp^3	Polar

ETHICS IN CHEMISTRY

1. Many different bonding theories have been proposed over the years. Valence bond theory built on Lewis structures has been the most popular recently. Nevertheless, it has problems. The most obvious is the Lewis structure of O_2 shown below would predict no unpaired electrons for this important substance, but experiment shows that this molecule has two unpaired electrons. Does this failure of valence bond theory negate the theory? Does this failure clearly make molecular orbital theory the better theory because it correctly predicts that O_2 has two unpaired electrons? Explain your answer.

$\ddot{O}$=$\ddot{O}$

2. In the chapter introduction, the use of sulfur hexafluoride as a substitute for Sarin was discussed. This material did not mention that SF_6 is the most effective greenhouse gas that has been found, with a global warming potential over 20,000 times greater than CO_2. Over a recent ten year period the SF_6 concentration in the atmosphere increased steadily from 4.0 parts to 6.5 parts per trillion. What factors should be considered when SF_6 is used for an experiment? Do you think the authorities who flooded a subway system with SF_6 in 2007, as described in the chapter introduction, acted responsibly and ethically?

Chapter 10 Visual Summary

The chart shows the connections between the major topics discussed in this chapter.

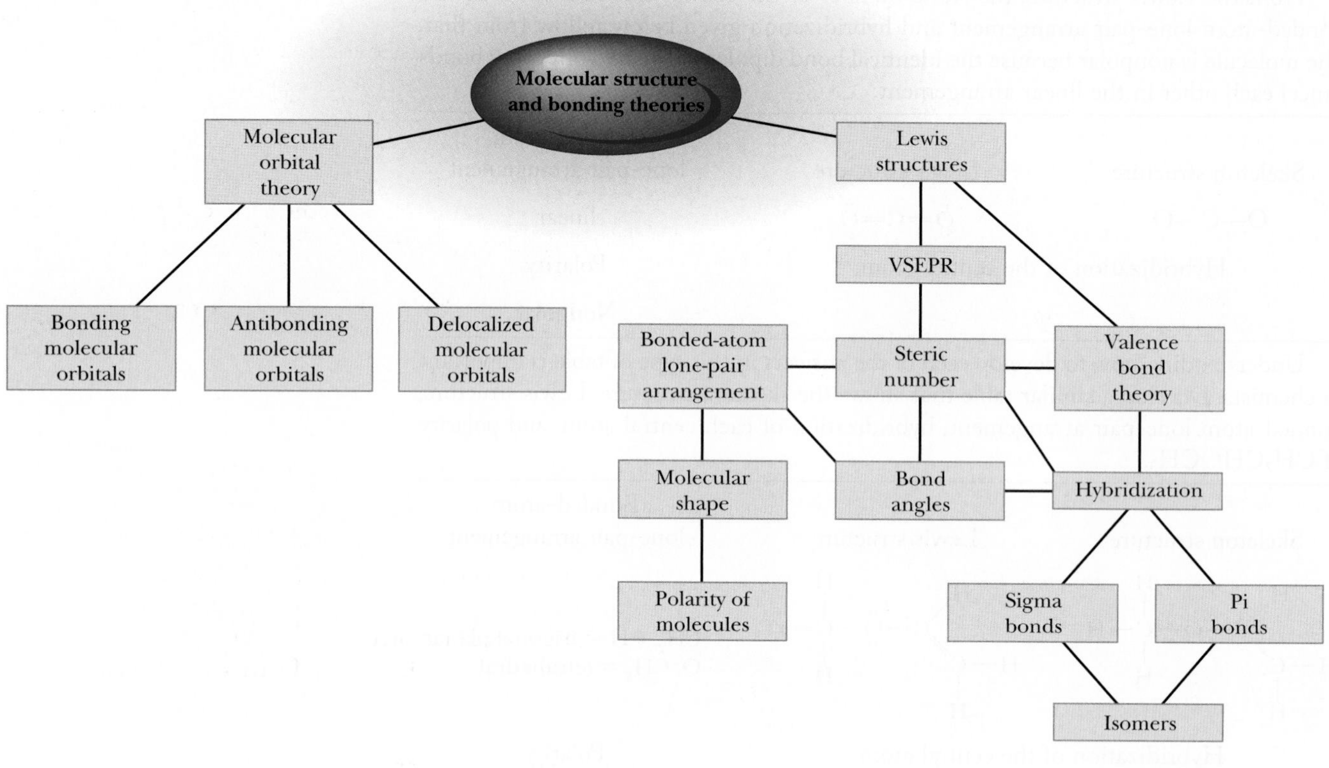

Summary

10.1 Valence-Shell Electron-Pair Repulsion Model

The *VSEPR model* predicts the shapes of many molecules and ions. In this model, we draw a Lewis structure and count the number of bonded atoms plus lone pairs for each central atom (one that is bonded to at least two other atoms). This sum is called the *steric number* and is used to determine the *bonded-atom lone-pair arrangement,* which is the geometry that maximizes the distances between the valence electron pairs. The bonded-atom lone-pair arrangements (and bond angles) for given steric numbers are linear (180 degrees) for 2; trigonal planar (120 degrees) for 3; tetrahedral (109.5 degrees) for 4; trigonal bipyramidal (90, 120, and 180 degrees) for 5; and octahedral (90 and 180 degrees) for 6. Because lone pairs are not considered part of the molecular geometry, the bonded-atom lone-pair arrangements for molecules with lone pairs on the central atom are not the same as the *molecular shapes,* but the bond angles are approximately the same in both. The observation that lone pairs exert greater repulsive forces than bonding pairs is useful for explaining details of the shapes of molecules that have lone pairs.

10.2 Polarity of Molecules

The *polarity* of a molecule can be determined from its shape and the polarities of its bonds. A molecule in which atoms of the same element occupy each vertex of the bonded-atom lone-pair arrangement is nonpolar because the individual bond dipoles sum to zero, even though the bonds may be polar. Most other molecules are polar, especially those with lone pairs on the central atom.

10.3 Valence Bond Theory

Valence bond theory is an extension of Lewis structures in which the orbitals that are used to form bonds are identified. *Hybrid orbitals,* orbitals that are mixtures of *s* and *p* and even *d* orbitals on the same atom are frequently used in this description of bonding. Use the steric number (that also determines the bond angles) to determine the correct type of hybrid orbitals formed from *s* and *p* orbitals as follows: for steric number = 2 (bonds at 180 degrees), *sp* hybrid orbitals; for steric number = 3 (bonds at 120 degrees), sp^2 hybrid orbitals; and for steric number = 4 (bonds at 109.5 degrees),

sp^3 hybrid orbitals. In addition, for steric number = 5 (bonds at 90, 120, and 180 degrees), sp^3d hybrid orbitals; for steric number = 6 (bonds at 90 and 180 degrees), sp^3d^2 hybrid orbitals.

10.4 Multiple Bonds

The bonds shown in Lewis structures are described as *sigma* or *pi bonds.* In sigma (σ) bonds, the shared pair of electrons is symmetric about the line joining the two nuclei of the bonded atoms. They form from the end-on overlap of *s*, *p*, or hybrid orbitals. Pi (π) bonds place electron density on both sides of the line joining the bonded atoms and form from the sideways overlap of *p* orbitals on different atoms. The orientation needed to form the π bond explains the planar geometry of molecules such as ethylene and its derivatives.

10.5 Molecular Orbitals: Homonuclear Diatomic Molecules and 10.6 Heteronuclear Diatomic Molecules and Delocalized Molecular Orbitals

Molecular orbital theory describes bonding as being delocalized over the entire molecule. Molecular orbitals are formed by combinations of appropriate atomic orbitals. The molecular orbitals are either *bonding molecular orbitals* or *antibonding molecular orbitals.* The electron configuration of a molecule is determined in the same manner as that of an atom, by adding the appropriate number of electrons to the molecular orbital diagram, following the aufbau procedure, the Pauli exclusion principle, and Hund's rule. The molecular orbital description of bonding accurately predicts the bond order and the number of unpaired electrons in molecules.

Download Go Chemistry concept review videos from OWL or purchase them from **www.ichapters.com**

Chapter Terms

The following terms are defined in the Glossary, Appendix I.

Section 10.1
Bonded-atom lone-pair arrangement
Molecular shape
Steric number
Valence-shell electron-pair repulsion (VSEPR)

Section 10.2
Polar molecule

Section 10.3
Hybrid orbitals
Valence bond theory

Section 10.4
Isomers
Pi (π) bond
Sigma (σ) bond

Section 10.5
Antibonding molecular orbital
Bonding molecular orbital
Homonuclear diatomic molecule
Molecular orbital
Molecular orbital theory

Section 10.6
Delocalized bond
Delocalized molecular orbital
Heteronuclear diatomic molecule

Key Equations

Steric number (10.1)

$$= (\text{number of lone pairs on central atom}) + (\text{number of atoms bonded to central atom})$$

Bond order (10.5)

$$= \frac{1}{2}(\text{number of electrons in bonding orbitals} - \text{number of electrons in antibonding orbitals})$$

Questions and Exercises

OWL Selected end of chapter Questions and Exercises may be assigned in OWL.

Blue-numbered Questions and Exercises are answered in Appendix J; questions are qualitative, are often conceptual, and include problem-solving skills.

- ■ Questions assignable in OWL
- ✎ Questions suitable for brief writing exercises
- ▲ More challenging questions

Questions

10.1 What is the basic premise of the VSEPR model?

10.2 State how the bonded-atom lone-pair arrangement is determined.

10.3 Give an example of a molecule in which a central carbon atom makes four bonds but has a trigonal planar bonded-atom lone-pair arrangement. Give an example of a molecule in which a central carbon atom makes four bonds but has a linear bonded-atom lone-pair arrangement.

10.4 How does the VSEPR model explain the fact that the measured H–O–H bond angle is 104.5 degrees, less than the predicted value of 109.5 degrees?

10.5 Draw the three possible arrangements of the fluorine atoms about the iodine atom in IF_3. Choose the shape predicted by VSEPR theory, and explain *why* this arrangement is favored.

10.6 Draw the three possible arrangements of the fluorine atoms about the xenon atom in XeF_2. Choose the shape predicted by VSEPR theory, and explain *why* this arrangement is favored.

10.7 Give an example of a nonpolar molecule that contains polar bonds. Show the polarity of the bonds with arrows, and show how these bond dipoles cancel.

10.8 Explain why SF_6 is nonpolar even though it contains polar S–F bonds.

10.9 Give the valence-bond-theory description of how chemical bonds form.

10.10 Which atomic orbitals overlap to form the bonds in HI?

10.11 Which atomic orbitals overlap to form the bonds in ClF?

10.12 Why are hybrid orbitals needed to explain the bonding in CH_4?

10.13 Identify the hybrid orbitals used by boron in BCl_3 and in BCl_4^-, the ion formed from the reaction of BCl_3 and Cl^-. Explain your choices.

10.14 Identify the hybrid orbitals used by antimony in $SbCl_5$ and in $SbCl_6^-$, the ion formed from the reaction of $SbCl_5$ and Cl^-. Explain your choices.

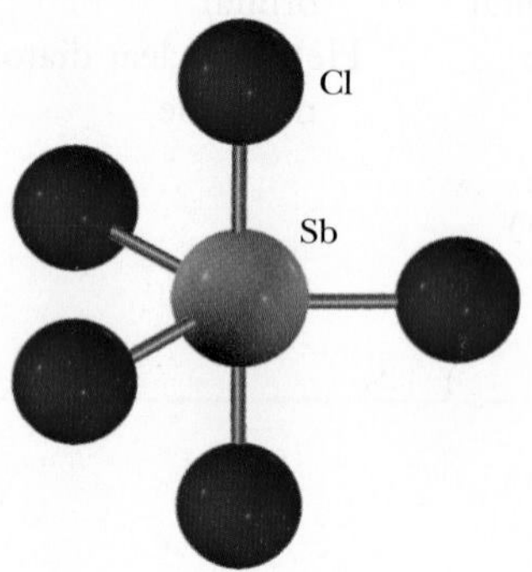

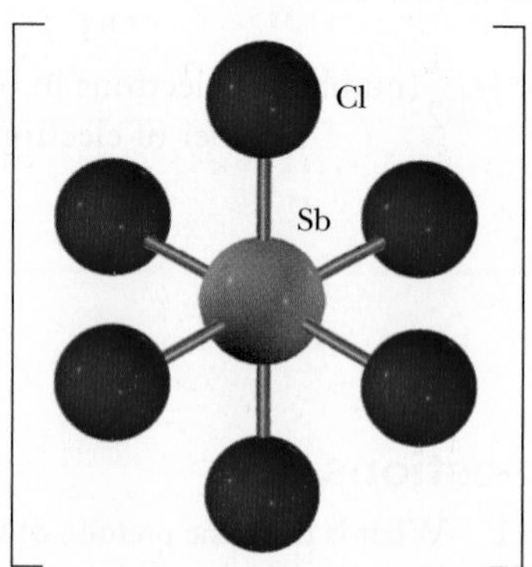

10.15 ✎ Explain why the molecular shape of HCl provides no information about the hybridization of the chlorine atom.

10.16 Make a table that shows the hybridization needed to explain bonds at angles of 180, 120, and 109.5 degrees.

10.17 Define a σ bond and a π bond. Show how *p* orbitals overlap in a σ bond and in a π bond.

10.18 Use valence bond theory to predict the planar shape of ethylene, C_2H_4.

10.19 Draw the energy-level diagram for the bonding and antibonding molecular orbitals for H_2. Indicate their relative energies with respect to the 1*s* atomic orbitals of isolated hydrogen atoms.

10.20 Draw two types of bonding molecular orbitals that can form from the overlap of 2*p* orbitals.

10.21 Compare and contrast the molecular orbital and ionic bonding descriptions of LiF.

10.22 Describe the bonding in molecular orbital terms for the delocalized π bond in O_3.

Exercises

OBJECTIVE Determine the steric number and bonded-atom lone-pair arrangement from the Lewis structure.

10.23 Give the bonded-atom lone-pair arrangement expected for a central atom that has
(a) three bonded atoms and no lone pairs.
(b) two bonded atoms and two lone pairs.
(c) four bonded atoms and no lone pairs.
(d) four bonded atoms and one lone pair.

10.24 ■ Give the bonded-atom lone-pair arrangement expected for a central atom that has
(a) two bonded atoms and one lone pair.
(b) three bonded atoms and two lone pairs.
(c) four bonded atoms and two lone pairs.
(d) five bonded atoms and one lone pair.

OBJECTIVES Predict the shapes of molecules from the valence-shell electron-pair repulsion model using the steric number, and distinguish between the bonded-atom lone-pair arrangement and molecular shape.

10.25 Use the VSEPR model to predict the shape of the following species.
(a) CF_4 (b) CS_2 (c) AsF_5
(d) F_2CO (e) NH_4^+

10.26 Use the VSEPR model to predict the shape of the following species.
(a) BeF_2 (b) SF_6 (c) SiH_4
(d) FCN (e) BeF_3^-

10.27 Give the bonded-atom lone-pair arrangement and the molecular shape of the following species.
(a) SeO_2
(b) N_2O (N is the central atom)
(c) H_3O^+
(d) IF_5
(e) SCl_4

10.28 ■ Give the bonded-atom lone-pair arrangement and the shape of the following species.
(a) XeO_2 (b) I_3^- (c) NO_2^-
(d) PCl_5 (e) $AlCl_3$

10.29 Indicate which molecule of each pair has the smaller bond angles. Explain your answer.
(a) BCl_3 or NCl_3 (b) OF_2 or SF_6

10.30 Indicate which species of each pair has the smaller bond angles. Explain your answer.
(a) SO_4^{2-} or $AlBr_3$ (b) CCl_4 or BeI_2

10.31 Indicate which species of each pair has the smaller bond angles. Explain your answer.
(a) Cl_2NH or NH_4^+ (b) SF_2 or IF_4^-

10.32 ■ Indicate which species of each pair has the smaller bond angles. Explain your answer
(a) BF_3 or $AsCl_4^+$ (b) CS_2 or $AsCl_3$

10.33 Write a Lewis structure for each of the following molecules. Indicate all of the bond angles as predicted by the VSEPR model. Deduce the skeleton structure from the way each formula is written.
(a) H_3CCCH
(b) Br_2CCH_2
(c) H_3CNH_2

10.34 Write a Lewis structure for each of the following molecules. Indicate all of the bond angles as predicted by the VSEPR model. Deduce the skeleton structure from the way each formula is written.
(a) $ClC(O)NH_2$ (oxygen bonded only to the carbon atom)
(b) $HOCH_2CH_2OH$
(c) NCCN

10.35 Write a Lewis structure for each of the following species. Indicate all of the bond angles as predicted by the VSEPR model. Deduce the skeleton structure from the way each formula is written.
(a) SO_2 (b) ClO_3^- (c) SCN^-

10.36 ■ Predict the geometry of the following species:
(a) SO_2 (b) $BeCl_2$ (c) $SeCl_4$ (d) PCl_5

10.37 Use the VSEPR model to predict the bond angles around each central atom in the following Lewis structures. Note that the drawings do not necessarily depict the bond angles correctly.

(a)
```
   H  :O:  H
   |   ‖   |
H—C — C — C—H
   |       |
   H       H
```

(b)
```
     H
     |
H—C—C≡N:
     |
     H
```

10.38 Use the VSEPR model to predict the bond angles around each central atom in the following Lewis structures (left). Note that the drawings do not necessarily depict the bond angles correctly.

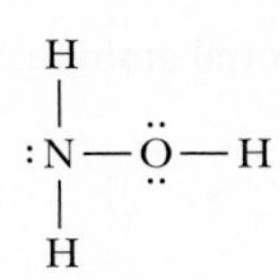

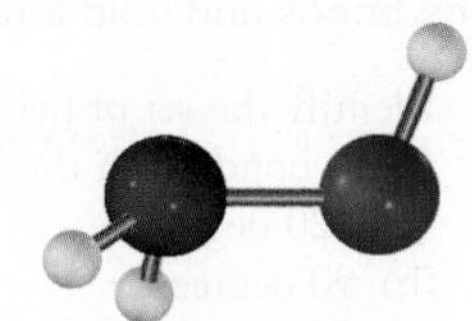

```
   H   H   H
   |   |   |
H—C — N — C—H
   |   ··  |
   H       H
```

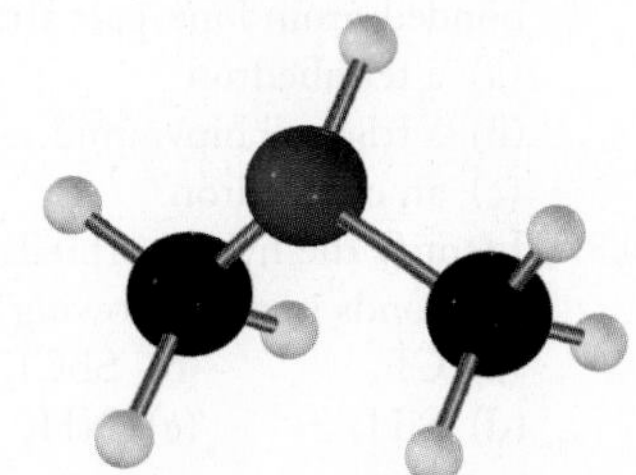

10.39 Use the VSEPR model to predict the bond angles around each central atom in the following Lewis structures. Note that the drawings do not necessarily depict the bond angles correctly.

(a)
```
   H      :C̈l:
   |   ··   |    ··
H—C — O — N — Cl:
   |   ··        ··
   H
```

(b)
```
               H
               |
H—C≡C—P—H
               ··
```

10.40 Use the VSEPR model to predict the bond angles around each central atom in the following Lewis structures (benzene rings are frequently pictured as hexagons, without the letter for the carbon atom at each vertex). Note that the drawings do not necessarily depict the bond angles correctly.

(a)
```
       H     H
  ··
H—O— (benzene ring) —H
  ··
       H     H
```

(b)
```
H   H   H   H
|   |   |   |
C = C — C = C
|           |
H           H
```

10.41 For each of the following molecules, complete the Lewis structure and use the VSEPR model to determine the bond angles around each central atom. Note that the drawings are only skeleton structures and may depict the angles incorrectly.

(a)
```
   H   O
   |   |
H—C — C—H
   |
   H
```

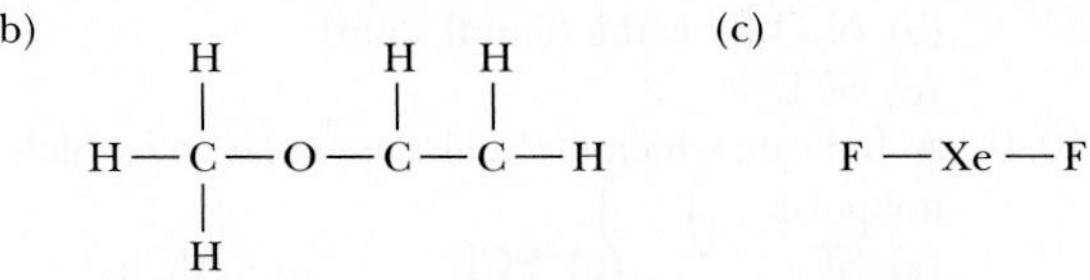

10.42 For each of the following molecules, complete the Lewis structure and use the VSEPR model to determine the bond angles around each central atom. Note that the drawings are only skeleton structures and may depict the angles incorrectly.

(a)
```
   H   H   H   H
   |   |   |   |
H—C — C — C — C—H
```

(b)
```
H       H
|       |
C—C — C—H
        |
        H
```

(c)
```
Cl
|
P—Cl
|
Cl
```

10.43 For each of the following molecules, complete the Lewis structure and use the VSEPR model to determine the bond angles around each central atom. Note that the drawings are only skeleton structures and may depict the angles incorrectly.

10.44 For each of the following molecules, complete the Lewis structure and use the VSEPR model to determine the bond angles around each central atom. Note that the drawings are only skeleton structures and may depict the angles incorrectly.

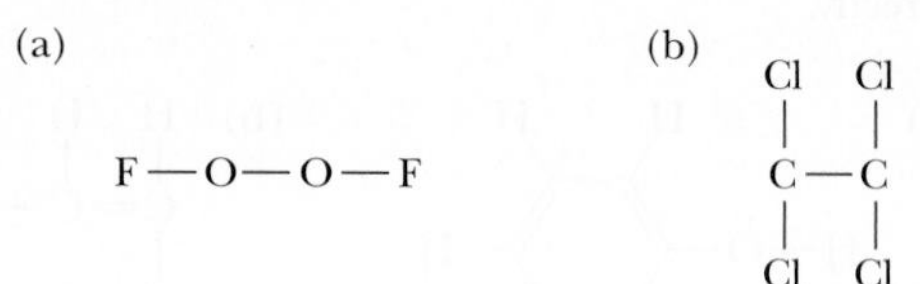

OBJECTIVE Predict the polarity of a molecule from bond polarities and molecular shape.

10.45 In Exercise 10.25, the shapes of the following molecules were determined. State whether each molecule is polar or nonpolar.
(a) CF_4 (b) CS_2 (c) AsF_5 (d) F_2CO

10.46 ■ Consider the following molecules:
(a) CH_4 (b) NH_2Cl (c) BF_3 (d) CS_2
(i) Which compound has the most polar bonds?
(ii) Which compounds in the list are *not* polar?

10.47 Indicate which molecules are polar and which are nonpolar.
(a) SeO_2
(b) N_2O (N is the central atom)
(c) SCl_4

10.48 ■ Indicate which molecules are polar and which are nonpolar.
(a) SF_2 (b) PCl_5 (c) $AlCl_3$

10.49 Indicate which of the following molecules are polar. Draw the molecular structure of each polar molecule, including the arrows that indicate the bond dipoles and the molecular dipole moment.
(a) HCN (b) I_2 (c) NO

10.50 Indicate which of the following molecules are polar. Draw the molecular structure of each polar molecule, including the arrows that indicate the bond dipoles and the molecular dipole moment.
(a) SiH_4 (b) PCl_3 (c) IF_5

10.51 Indicate which of the following molecules are polar. Draw the molecular structure of each polar molecule, including the arrows that indicate the bond dipoles and the molecular dipole moment.
(a) NF_3 (b) CBr_4 (c) BeI_2

10.52 Indicate which of the following molecules are polar. Draw the molecular structure of each polar molecule, including the arrows that indicate the bond dipoles and the molecular dipole moment.
(a) BCl_3 (b) OF_2 (c) SF_6

10.53 Following are drawings of two derivatives of acetylene. Indicate whether each is polar or nonpolar, and explain your answer.
(a) F—C≡C—F (b) H—C≡C—F

10.54 ▲ Following are drawings of two isomers of $C_6H_4Cl_2$ (benzene rings are frequently pictured as hexagons, without the letter for the carbon atom at each vertex). Indicate whether each is polar or nonpolar. Explain your answer.

(a)

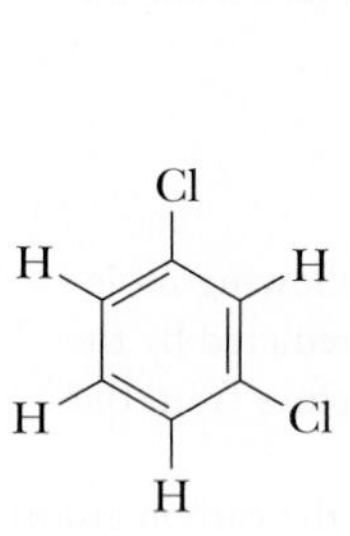

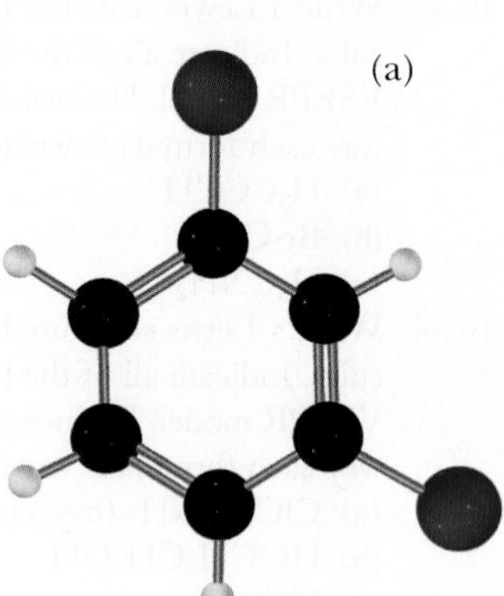

(b)

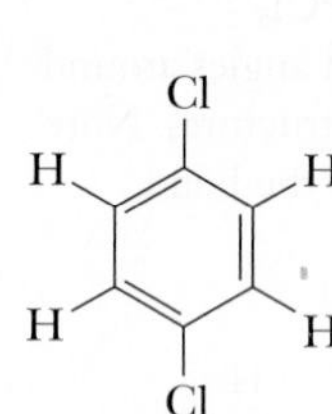

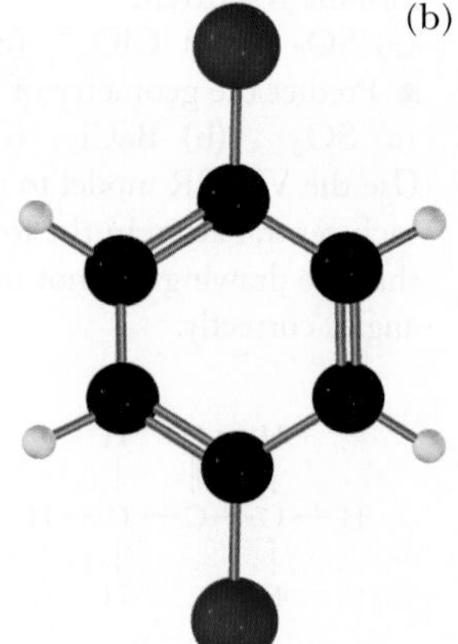

OBJECTIVE Identify the hybrid or atomic orbitals that form the bonds and hold lone pairs in any specific molecule.

10.55 Identify the set of hybrid orbitals of a central atom that forms bonds with the following angles.
(a) 120 degrees
(b) 90 degrees
(c) 180 degrees

10.56 Identify the hybridization of the central atom that has the bonded-atom lone-pair arrangement of
(a) a tetrahedron.
(b) a trigonal bipyramid.
(c) an octahedron.

10.57 Identify the hybrid orbitals on the central atom that form the bonds in the following species.
(a) CF_4 (b) $SbCl_6^-$ (c) AsF_5
(d) SiH_4 (e) NH_4^+

10.58 ■ Identify the hybrid orbitals on the central atom that form the bonds in the following species.
(a) NF_3 (b) SCl_2 (c) H_3O^+
(d) IF_5 (e) SCl_4

10.59 Identify the hybrid orbitals on the central atom that form the σ bonds in the following species.
(a) N_2O (b) $SnCl_2$
(c) I_3^- (d) SeO_2

10.60 Identify the types of hybrid orbitals on the central atom that form the σ bonds in the following molecules.
(a) ClF_3 (b) BBr_3
(c) BeF_2 (d) ONCl

10.61 Identify the hybrid orbitals on the carbon atoms that form the σ bonds in the following species.
(a) CO_3^{2-} (b) CH_2F_2 (c) H_2CO

10.62 Identify the hybrid orbitals on the carbon atoms that form the σ bonds in the following molecules.
(a) C_2H_6 (b) C_2H_4 (c) CBr_4

10.63 Identify the hybrid orbitals on the oxygen atoms that form the σ bonds in the following species.
(a) H_3O^+ (b) H_3COH (c) Cl_2O

10.64 Identify the hybrid orbitals on the nitrogen atoms that form the σ bonds in the following species.
(a) $HNCl_2$ (b) NO_3^- (c) N_2H_2

10.65 Identify all the orbitals that form the bonds and hold the lone pairs on the central atom in the following molecules.
(a) OF_2 (b) NH_3 (c) BCl_3

10.66 ■ Give the hybrid orbital set used by each of the red atoms in the following molecules.

```
     H  :O:  H
     |   ‖   |
(a) H—N—C—N—H
      ..    ..

      H  H  H
      |  ‖  |
(b) H3C—C—C—C=O:
                ..

     H  H
     |  |
(c) H—C=C—C≡N:
```

10.67 What orbitals on selenium and fluorine form the bonds in SeF_4? What orbital holds the lone pair on selenium?

10.68 Nitrous acid has the skeleton structure HONO. What are the hybrid orbitals on the nitrogen atom and the central oxygen atom?

10.69 Indicate the hybridization on each central atom in the molecules with the following Lewis structures.

```
(a)   H       :Cl:           (b)            H
      |         |                           |
    H—C—O—N—Cl:                 H—C≡C—P—H
      |  ..    ..                           ..
      H
```

10.70 ■ Indicate the hybridization on each central atom in the molecules with the following Lewis structures.

```
(a)   H :O: H             (b)     H
      |  ‖  |                     |
    H—C—C—C—H                 H—C—C≡N:
      |     |                     |
      H     H                     H
```

OBJECTIVE Identify the orbitals that form σ and π bonds.

10.71 ▲ If the z axis is defined as the bond axis, draw a picture that shows the overlap of each of the following pairs of orbitals; then indicate whether a σ or π bond forms.
(a) p_z, p_z
(b) p_y, p_y
(c) sp hybrid formed from p_z and s orbitals, p_z

10.72 ▲ If the z axis is defined as the bond axis, draw a picture that shows the overlap of each of the following pairs of orbitals; then indicate whether a σ or π bond forms.
(a) p_x, p_x
(b) s, p_z
(c) sp^2 hybrid formed from p_x, p_z, and s orbitals, s

10.73 Identify the orbitals on each of the atoms that form the bonds in H_3CCN. How many σ bonds and π bonds form?

10.74 Identify the orbitals on each of the atoms that form the bonds in propylene (shown below); then indicate whether each bond is a σ or a π bond.

```
 H       H
  \     /
   C=C      H
  /     \   /
 H        C
         / \
        H   H
```

10.75 Sketch the bonds (analogous to Figure 10.25) in H_2CNH, label the type of orbital from which each bond forms, and indicate whether the bond is a σ or a π bond.

10.76 ■ How many sigma bonds and how many pi bonds are there in each of the following molecules?

```
     H  :O:  H                          H  H  :O:
     |   ‖   |     H                    |  |   ‖
(a) H—C—C—C=C                     (b) H—C—C   C
     |             \                    |  |  /  \
     H              H                   H  H      H

    :N≡C      H
(c)     \    /                    (c) CH2CHCH2OCH2CH3
         C=C
        /   \
       H     :Cl:
```

10.77 Give the hybridization of each central atom in the following molecules.
(a) cyclohexene

```
      H  H
       \/
       C     H
      / \   /
  H—C     C—H
     ‖     |
  H—C     C—H
      \ /   \
       C     H
      / \
     H   H
```

(b) phosgene, Cl_2CO
(c) glycine, $H_2NC_{(1)}H_2C_{(2)}OOH$ (**Note:** Numbers in parentheses label each carbon atom.)

10.78 Give the hybridization of each central atom in the following molecules.
(a) CO_2
(b) H_3CCCH
(c) $H_3CC(O)H$, which has the Lewis structure

```
    H  :O:
    |   ‖
H — C — C — H
    |
    H
```

10.79 Two resonance structures can be written for NO_2^-. Indicate the hybridization on the central atom for each resonance form.

10.80 Three resonance structures can be written for N_3^-. Indicate the hybridization on the central atom for each resonance form.

10.81 Predict the hybridization at each central atom in the following molecules.

```
(a)  H   H              (b)      H
     |   |                       |
 H — C — N:               H — C — C ≡ C — H
     |   |                       |
     H   H                       H
```

10.82 Predict the hybridization at each central atom in the following molecules.

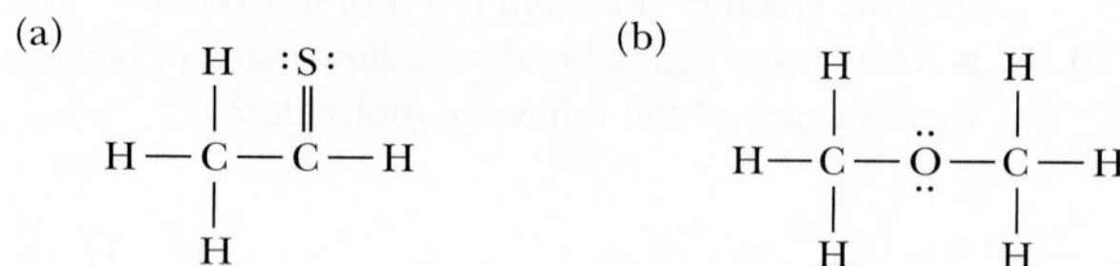

10.83 Orlon is produced from acrylonitrile, H_2CCHCN. Draw the Lewis structure of acrylonitrile, and indicate the hybridization of each central atom.

10.84 Tetrafluoroethylene, C_2F_4, is used to produce Teflon. Draw the Lewis structure of tetrafluoroethylene, and indicate the hybridization of each carbon atom.

Teflon-coated baking pan.

OBJECTIVES Write molecular orbital diagrams for homonuclear diatomic molecules and determine the bond order.

10.85 Draw the molecular orbital diagram, including the electrons, and write the electron configuration of He_2^{2+}. Give the bond order and the number of unpaired electrons, if any. Is this a stable species?

10.86 Draw the molecular orbital diagram, including the electrons, and write the electron configuration of H_2^-. Give the bond order and the number of unpaired electrons, if any. Is this a stable species?

10.87 Draw the molecular orbital diagram, including the electrons, and write the electron configuration of Li_2. Give the bond order and the number of unpaired electrons, if any. Is this a stable species?

10.88 Draw the molecular orbital diagram, including the electrons, and write the electron configuration of C_2. Give the bond order and the number of unpaired electrons, if any. Is this a stable species?

10.89 Write the molecular orbital electron configuration and determine the bond order and number of unpaired electrons for the following ions.
(a) C_2^+ (b) N_2^- (c) Be_2^-

10.90 ■ Give the electron configurations for the ions Li_2^+ and Li_2^- in molecular orbital terms. Compare the Li–Li bond order in these ions with the bond order in Li_2.

10.91 Which species, N_2 or N_2^-, has the higher bond order? Explain your answer.

10.92 Which species, O_2 or O_2^-, has the higher bond order? Explain your answer.

10.93 Use the molecular orbital diagram in Figure 10.40 to predict which species in each pair has the stronger bond.
(a) B_2 or B_2^- (b) C_2^- or C_2^+ (c) O_2^{2+} or O_2

10.94 Use the molecular orbital diagram in Figure 10.40 to predict which species in each pair has the stronger bond.
(a) F_2 or F_2^- (b) O_2^- or O_2^+ (c) C_2^{2+} or C_2

10.95 Identify two homonuclear diatomic molecules or ions with each of the following molecular orbital electron configurations. Are these species stable?
(a) $(\sigma_{2s})^2(\sigma_{2s}^*)^2(\pi_{2p})^4(\sigma_{2p})^2(\pi_{2p}^*)^3$
(b) $(\sigma_{2s})^2(\sigma_{2s}^*)^2(\pi_{2p})^4(\sigma_{2p})^2$
(c) $(\sigma_{2s})^2(\sigma_{2s}^*)^2$

10.96 Identify two homonuclear diatomic molecules or ions with each of the following molecular orbital electron configurations. Are these species stable?
(a) $(\sigma_{2s})^2(\sigma_{2s}^*)^2(\pi_{2p})^4(\sigma_{2p})^1$
(b) $(\sigma_{2s})^2(\sigma_{2s}^*)^2(\pi_{2p})^4$
(c) $(\sigma_{2s})^2(\sigma_{2s}^*)^1$

OBJECTIVES Write molecular orbital diagrams for heteronuclear diatomic molecules and determine the bond order.

10.97 Assuming that the molecular orbital diagram shown in Figure 10.40 is correct for heteronuclear diatomic molecules containing elements that are close to each other in the periodic table, write a homonuclear diatomic molecule and a heteronuclear diatomic molecule (remember that molecules are neutral) that both have the given electron configuration.
(a) $(\sigma_{2s})^2(\sigma_{2s}^*)^2(\pi_{2p})^4(\sigma_{2p})^2$
(b) $(\sigma_{2s})^2(\sigma_{2s}^*)^2(\pi_{2p})^2$

10.98 ■ The nitrosyl ion, NO^+, has an interesting chemistry.
(a) Is NO^+ diamagnetic or paramagnetic? If paramagnetic, how many unpaired electrons does it have?
(b) Assume the molecular orbital diagram for a homonuclear diatomic molecule applies to NO^+. What is the highest-energy molecular orbital occupied by electrons?
(c) What is the nitrogen–oxygen bond order?
(d) Is the N–O bond in NO^+ stronger or weaker than the bond in NO?

10.99 The molecular orbital diagram of NO shown in Figure 10.47 also applies to the following species. Write the molecular orbital electron configuration of each, indicating the bond order and the number of unpaired electrons.
(a) CN
(b) CO^-
(c) BeB^-
(d) BC^+

10.100 The molecular orbital diagram of NO shown in Figure 10.47 also applies to the following species. Write the molecular orbital electron configuration of each, indicating the bond order and the number of unpaired electrons.
(a) $LiBe^+$
(b) CO^+
(c) CN^-
(d) OF

10.101 The molecular orbital diagram of NO shown in Figure 10.47 also applies to OF^-. Draw the complete molecular orbital diagram for OF^-. What is the OF bond order?

10.102 The molecular orbital diagram of NO shown in Figure 10.47 also applies to CO. Draw the complete molecular orbital diagram for CO. What is the C–O bond order?

10.103 The delocalized bonding that describes O_3 also applies to NO_2^-. Draw the delocalized π molecular orbital for NO_2^-.

10.104 Draw the delocalized π orbital for benzene. Clearly indicate the atomic orbitals that form the molecular orbital.

Chapter Exercises

10.105 Write one Lewis structure of N_2O_5 (O_2NONO_2 skeleton structure). What are the bond angles around the central oxygen atom and the two nitrogen atoms? What is the hybridization of each?

10.106 Following are the structures of three isomers of difluorobenzene, $C_6H_4F_2$. Are any of them nonpolar?

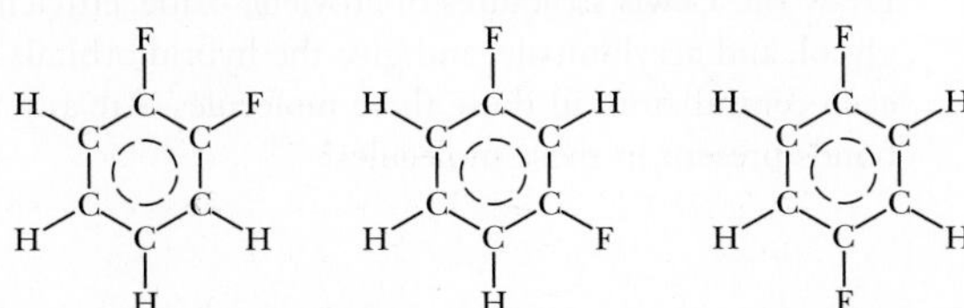

10.107 The ions ClF_2^- and ClF_2^+ have both been observed. Use the VSEPR model to predict the F–Cl–F bond angle in each.

10.108 ▲ Aspirin, or acetylsalicylic acid, has the formula $C_9H_8O_4$ and the skeleton structure

(a) Complete the Lewis structure and give the number of σ bonds and π bonds in aspirin.
(b) What is the hybridization about the CO_2H carbon atom (colored blue)?
(c) What is the hybridization about the carbon atom in the benzene-like ring that is bonded to an oxygen atom (colored red)? Also, what is the hybridization of the oxygen atom bonded to this carbon atom?

10.109 ▲ Aspartame is a compound that is 200 times sweeter than sugar and is used extensively (under the trade name NutraSweet) in diet soft drinks. The skeleton structure of the atoms in aspartame is

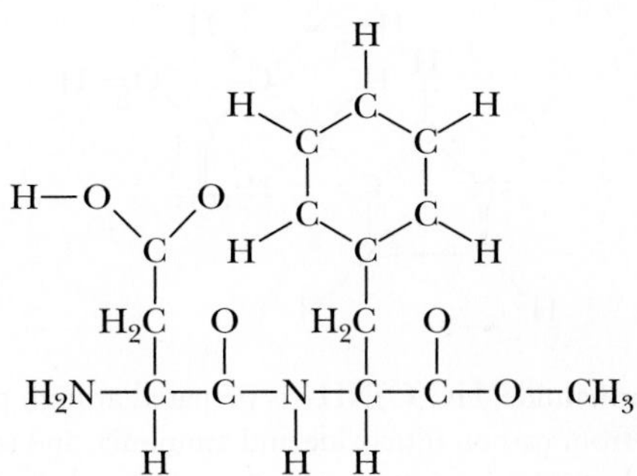

(a) Complete the Lewis structure and give the number of σ and π bonds in aspartame.
(b) What is the hybridization about each carbon atom that forms a double bond with an oxygen atom?
(c) What is the hybridization about each nitrogen atom?

10.110 ▲ Recently, the structure of an amine compound, NR_3 (R = large organic group), has been determined to have C–N–C bond angles of 119.2 degrees. It is believed that the bond angles of about 109 degrees expected from the VSEPR model are not observed because of the large substituents bonded to the nitrogen atom. Given this large bond angle, what type of orbital on the nitrogen atom makes the N–C σ bonds, and in what type of orbital is the lone pair located?

10.111 Phosgene, $COCl_2$, is a highly toxic gas that was used in combat during World War I. It is an important intermediate in the preparation of a number of organic compounds but must be handled with extreme care. Given that carbon is the central atom in phosgene, determine the Lewis structure, the bonded-atom lone-pair arrangement, the hybridization of the carbon atom, and the polarity of the molecule.

10.112 Calcium cyanamide, CaNCN, is used both to kill weeds and as a fertilizer. Give the Lewis structure of the NCN^{2-} ion and the bonded-atom lone-pair arrangement and hybridization of the carbon atom.

10.113 Histidine is an essential amino acid that the body uses to form proteins. The Lewis structure of histidine follows. What are the approximate values for bond angles 1 through 5 (indicated on the structure by blue numbers)?

[Structure of histidine with bond angles labeled 1–5]

10.114 ▲ Formamide, $HC(O)NH_2$, is prepared at high pressures from carbon monoxide and ammonia, and serves as an industrial solvent (the parentheses around the O indicate that it is bonded only to the carbon atom and that the carbon atom is also bonded to the H and the N atoms). Two resonance forms (one with formal charges) can be written for formamide. Write both resonance structures, and predict the bond angles about the carbon and nitrogen atoms for *each resonance form.* Are they the same? Describe how the experimental determination of the H–N–H bond angle could be used to indicate which resonance form is more important.

10.115 Draw the molecular orbital diagrams for NO^- and NO^+. Compare the bond orders in these two ions.

10.116 ▲ Ionization energies can be determined for molecules and atoms. Draw the molecular orbital diagrams for NO and CO, and predict which compound has the lower ionization energy.

Cumulative Exercises

10.117 Write one important resonance structure for each of the following species, and use the VSEPR model to predict the bond angles around each central atom. Also indicate the hybrid orbitals on each central atom and whether the molecule is polar or nonpolar.

(a) $H_2N{-}NO_2$ (H, H on N; N–N; O, O on N) (b) H—N—N—N

10.118 Write all important resonance structures for each of the following species, and use the VSEPR model to predict the bond angles around each central atom. Also indicate the hybrid orbitals on each central atom and whether the molecule is polar or nonpolar. Does each resonance structure use the same hybrid orbitals?

(a) $H_2C{-}N{-}N$ (b) $H{-}CH_2{-}C(O){-}NO_2$

10.119 Write all important resonance structures for each of the following species, and use the VSEPR model to determine the bond angles around each central atom. Also indicate the hybrid orbitals on each central atom. Does each resonance structure use the same hybrid orbitals?

(a) $[O{-}C{-}N]^-$ (b) $[O{-}N(O){-}O]^-$

10.120 More than 5 billion pounds of ethylene oxide, C_2H_4O, is produced annually. Ethylene oxide is used in the production of ethylene glycol, $HOCH_2CH_2OH$, the main component of antifreeze, and acrylonitrile, CH_2CHCN, used in the production of synthetic fibers and other chemicals. Ethylene oxide has an interesting cyclic structure.

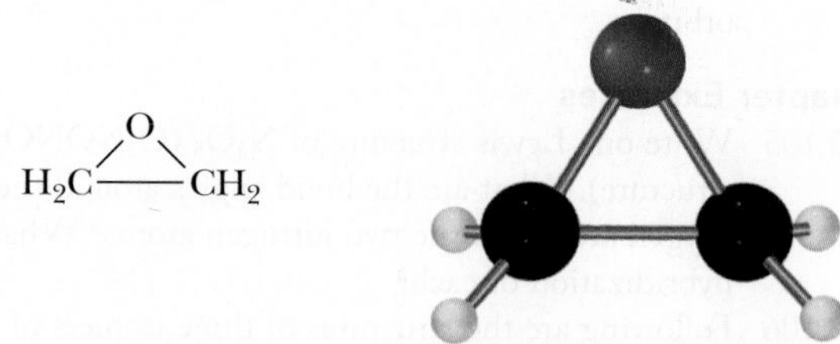

Draw the Lewis structures of ethylene oxide, ethylene glycol, and acrylonitrile, and give the hybrid orbitals on each central atom in these three molecules. Are any π bonds present in these molecules?

10.121 Vitamin A is converted by the body to retinal, a compound that is critical to human sight. The skeleton structure of vitamin A follows. Write the Lewis structure. How many sp^2 and sp^3 hybridized carbon atoms are in vitamin A?

```
        CH3   H      CH3   H      CH3   H
         |    |       |    |       |    |
 H      _C_  _C_     _C_  _C_     _C_  _C—O—H
   \   /   \/   \   /   \/   \   /   \/   |
H—C   C         C        C        C     C   H
  |   |          \       |        |     |
H—C   C—CH3       H      H        H     H
  / \ / \
 H   C   CH3
    / \
   H   H
```

10.122 ■ The reaction of calcium carbide with water is used to generate acetylene: $CaC_2 + 2H_2O \rightarrow H–C{\equiv}C–H + Ca(OH)_2$. Calcium carbide, CaC_2, contains the carbide ion, C_2^{2-}. Sketch the molecular orbital energy level diagram for the ion. How many σ and π bonds does the ion have? What is the carbon–carbon bond order? How has the bond order changed on adding electrons to C_2 to obtain C_2^{2-}? Is the C_2^{2-} ion paramagnetic?

Acetylene is used by miners to form a bright flame in headlamps.

10.123 A compound is analyzed and found to contain 54.53% carbon, 9.15% hydrogen, and 36.32% oxygen by mass. A mass spectrometry experiment shows that the molar mass is 44 g/mol. What is the molecular formula? There are two reasonable ways to draw noncyclic skeleton structures of this molecule. Draw the Lewis structure for each, indicating the bond angles and hybridization of each central atom.

10.124 The reaction of sulfur, S_8, with fluorine, F_2, yields a product with the general formula SF_x. If 4.01 g S_8 reacts with 4.76 g F_2 to yield only SF_x, what is the value of x? Draw the Lewis structure of this compound, indicating the F–S–F bond angles and the hybrid orbitals on sulfur.

10.125 Two compounds have the formula S_2F_2. Disulfur difluoride has the skeleton structure F–S–S–F, whereas thiothionyl fluoride has the skeletal structure

```
       F
      /
S — S
      \
       F
```

Determine Lewis structures for each compound.

10.126 Recently, the compound CF_3SF_5 was discovered in the atmosphere and identified as a potential greenhouse gas. Assume the carbon and sulfur atoms are both central atoms, and draw the Lewis structure for this compound. What is the hybridization of each central atom and the bond angles with the surrounding atoms?

10.127 A 1.30-g sample of C_2H_2 reacts with exactly 1.22 L H_2 gas at 27 °C and 1.01 atm of pressure to yield a compound with the formula C_2H_x. What is the value of x, and what are the orbitals on the carbon atoms that form the C–C bond(s)?

10.128 ▲ The compound cubane, C_8H_8, has an unusual structure with each carbon atom at the corner of a cube, bonded to three other carbon atoms, and a single hydrogen atom is bonded to each carbon.

(a) Draw the Lewis structure of cubane, and indicate the hybridization of the carbon atoms.

(b) Given the hybridization you assigned to each carbon atom in part a, what C–C–C bond angles are predicted by hybridization theory? What are the observed C–C–C bond angles in cubane, given the shape of the molecule? Use the difference in these values to comment on whether the theory predicted the correct hybridization.

(c) Forcing a bond angle to a value other than its naturally occurring bond angle gives rise to *bond angle strain,* an increase in energy of a compound because of the unnatural angles of the bond. The bond angle strain in cubane has been measured at 695 kJ/mol, the most of any known stable compound. What is the energy increase per C–C bond in cubane?

(d) The enthalpy for the combustion of 1 mol cubane is −4833 kJ/mol. Write the balanced chemical reaction for the combustion of cubane and, using the ΔH_f° of the other species, determine ΔH_f° [cubane].

(e) Using the table of bond energies from Chapter 9, calculate the enthalpy for the combustion of 1 mol cubane. How much does it differ from the value given in part d? What is the likely cause of the difference?

Whether uncut or brilliantly polished, the unique properties of diamond are due to its solid-state structure.

© Cengage Learning/Charles D. Winters

© E.R. Degginger/Alamy

Diamond is an unusual material. Chemically, diamond is simple: It is pure carbon. This formulation was established as early as 1772, when Antoine Lavoisier carefully burned diamond samples and showed that the combustion did not produce water but produced as much carbon dioxide as pure carbon did. (Lavoisier was rather wealthy and could afford to burn diamonds in the name of science.) However, the way the carbon atoms are bonded together gives diamond some remarkable properties.

First, diamond is the hardest known naturally occurring solid. It has the highest rank, 10, on the Mohs scale of mineral hardness, an arbitrary scale devised in 1812 by German mineralogist Frederich Mohs. The hardness of diamond is attributed to the bonding between the carbon atoms. Each carbon atom is sp^3-hybridized and covalently bonded to four other carbon atoms in an almost unending matrix. Essentially, each diamond is one large molecule! At 346 kJ/mol, the carbon-carbon bond energy is not overwhelmingly large, but the interconnected carbon atoms collectively make for a very hard substance.

Second, diamond has some unusual conductivity properties. Most materials labeled as conductors are both electrical and thermal conductors because both generally require mobile electrons. Pure diamond is an electrical insulator because the carbon-carbon bonds hold the electrons too tightly for the material to conduct electricity. Curiously, however, diamond is an excellent conductor of heat. The strongly bonded carbon atoms are efficient at moving heat from one side to the other. There is keen interest in using diamond as a

Liquids and Solids

11

CHAPTER CONTENTS

OWL Online homework for this chapter may be assigned in OWL.

Look for the green colored vertical bar throughout this chapter, for integrated references to this chapter introduction.

heat sink in the semiconductor industry, where the heat generated by smaller and smaller microprocessors can cause problems. Using diamond films to efficiently transfer the heat away from the processors would extend the processors' lifetime and ultimately allow for smaller and faster processors to be developed. Unfortunately, no one has found an effective way to make such diamond films. This research area is of intense interest in the microprocessor industry.

Third, diamond has some unusual optical properties. Pure diamond does not absorb visible light, and thus has no color (some prized diamonds do have color, but the color is due to impurities or bonding irregularities). Interestingly, diamond also does not absorb most wavelengths of infrared light down to wavelengths of 25.0 μm. Because of this property, diamond is sometimes used in infrared optics, the most famous use being an optical window on the Pioneer spacecraft that landed on Venus in 1978. Also, the velocity at which light travels through diamond depends on the wavelength, so all the different wavelengths of light thus follow slightly different paths through a diamond, separating the colors in the same way a glass prism creates a rainbow. This effect results in a spectacular sparkle in certain well-cut, gem-quality diamonds. Such a diamond is an object of beauty, as well as a solid with unique structure and bonding.

All of these properties of diamond are related to the bonding of the carbon atoms and how those bonded atoms are arranged in space. The structure of solids such as diamond is one of the topics covered in this chapter. ■

Most of the substances that are handled in the laboratory—or the kitchen, garage, or basement—are liquids or solids. This chapter is devoted to the discussion of these two states of matter. Chapter 6 presents the physical behavior of gases in some detail. Except under extreme conditions, a single relationship—the ideal gas law—describes the properties of all gases quite well. However, there is no single relationship like the ideal gas law that accounts for the properties of the liquid and solid states, so it is more difficult to make all-encompassing statements about the properties of solids and liquids.

The properties of the three states of matter—gas, liquid, and solid—are summarized in Table 11.1. The high compressibility and low density of gases are consistent with the volume of a gas being mostly empty space. In liquids and solids, called the *condensed states* (or *phases*), matter occupies a large fraction of the sample volume, causing the characteristic high densities and low compressibilities that are observed. A solid differs from a liquid in that it is rigid (it maintains a definite shape), whereas a liquid is fluid and conforms to the shape of its container.

TABLE 11.1 Characteristic Properties of Gases, Liquids, and Solids

State	Volume and Shape of Sample	Density	Compressibility
Gas	Assumes shape and volume of container	Low	Easily compressed
Liquid	Has definite volume; assumes shape of bottom of container	High	Nearly incompressible
Solid	Has definite shape and volume	High	Nearly incompressible

The high densities and low compressibilities of solids and liquids show that the particles making up these phases are quite close together. These closely packed particles have attractions that are called **intermolecular forces.** These attractions are not to be confused with the much stronger *intramolecular forces*—chemical bonds that hold atoms and ions together in compounds. When water evaporates, its intermolecular attractions are broken, but the intramolecular O–H chemical bonds remain. The intermolecular attractions between molecules are much weaker than chemical bonds.

Intermolecular forces are those *between* molecules; intramolecular forces hold atoms together in molecules.

Intermolecular forces affect the behavior of substances only when the molecules are quite close together. The deviations of real gases from ideality are caused, in part, by the intermolecular forces of attraction and become important only at high pressures, when the molecules are close together. In the solid and liquid phases, the strengths of intermolecular forces are always important in determining the properties of the substances.

This chapter presents the properties of liquids and solids that relate to the strengths of intermolecular forces and then discusses the origins of intermolecular forces. The chapter concludes with a description of solid crystalline materials.

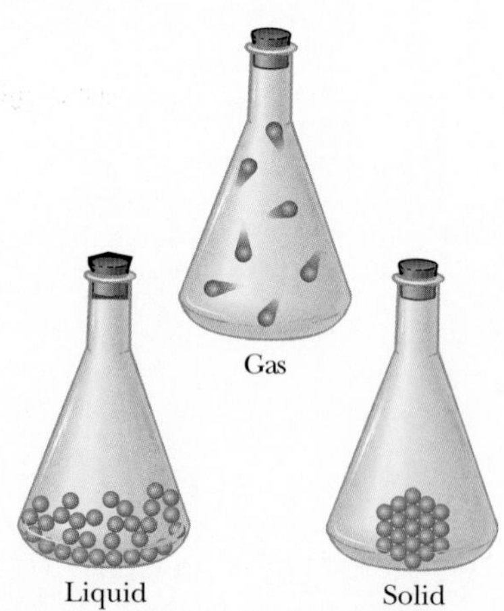

Molecular view of the states of matter. In a gas, the molecules are so far apart that each one moves independently of the others. In a liquid, the molecules are close together but can move around. In a solid, the molecules are held together in a regular arrangement.

11.1 Kinetic Molecular Theory and Intermolecular Forces

OBJECTIVE

- ☐ Relate the physical state of a substance to the strength of intermolecular forces

The physical state of any sample of matter depends on the strengths of the intermolecular attractions and the average kinetic energy of the molecules. The strengths of intermolecular attractions do not change much with the temperature, although the average kinetic energy of molecules is proportional to the absolute temperature (refer to the discussion of the kinetic molecular theory in Chapter 6). Furthermore, according to the kinetic molecular theory, a distribution of kinetic energies exists among the molecules. At any instant, some molecules have kinetic energies greater than the average and others have kinetic energies less than the average.

When the average kinetic energy is considerably greater than the attractive energy between molecules, there is no tendency for the molecules to stick together, and each molecule behaves independently of all the others. The result is the completely random behavior of molecules in the gas state. In the liquid state, the forces of attraction are large enough to keep the molecules close together but small enough that the molecules can move about. In solids, the energy of the attractions between molecules is quite large compared with the average kinetic energy of molecules at that temperature, and almost none of the molecules has enough energy to overcome the attractions. The molecules adopt an orderly arrangement that maximizes the energy of attraction, leading to the rigid characteristics of a solid. Table 11.2 summarizes the relation between the kinetic energy of molecules and intermolecular forces for the three states of matter.

The state of a substance that is stable at a given temperature depends on the strengths of intermolecular attractions. At room temperature, nitrogen is a gas, water is a liquid, and iodine is a solid. From these observations, we can conclude that the intermolecular forces in nitrogen are weaker than those in water, which, in turn, are weaker than those in iodine. At a higher temperature, both nitrogen and water are gases, whereas iodine remains a solid. The stable state of a substance is therefore temperature dependent. Several other properties of substances are closely related to the strengths of intermolecular attractions. Many of these properties are observed during phase changes, which we consider in the next section.

TABLE 11.2 Physical State and Energy of Attraction

Physical State	Relation between Energy of Attraction and Kinetic Energy of Molecules
Gas	Energy of attraction $\ll$ kinetic energy of molecules
Liquid	Energy of attraction $\cong$ kinetic energy of molecules
Solid	Energy of attraction $\gg$ kinetic energy of molecules

At a given temperature, a solid has stronger intermolecular attractions than a liquid, whereas a liquid has stronger intermolecular attractions than a gas.

EXAMPLE 11.1 Intermolecular Forces and Physical State

At room temperature, chlorine is a gas, bromine is a liquid, and iodine is a solid. Arrange these substances in order of increasing strength of intermolecular attractions.

Strategy The lower the intermolecular attractions between molecules, the more likely a substance will exist as a liquid or, in the extreme, a gas.

Solution

The physical state of a substance is determined by the energy of intermolecular attractions compared with the kinetic energy. As seen in Table 11.2, the intermolecular forces are weakest for the gas and strongest for the solid substance. Thus, the order of increasing energy of intermolecular attractions is

$$Cl_2 < Br_2 < I_2$$

Understanding

Hydrogen sulfide, H_2S, is a gas at room temperature, and water is a liquid. Which compound has the stronger intermolecular attractions?

Answer Water

OBJECTIVE REVIEW *Can you:*

☑ relate the physical state of a substance to the strength of intermolecular forces?

11.2 Phase Changes

OBJECTIVES

☐ Describe phase changes as equilibrium processes

☐ Relate the enthalpy of phase-changes and the phase change temperatures to the relative strengths of the intermolecular attractions

☐ Describe heating curves and their relation to the heat capacities and enthalpies of phase transitions

In the preceding section, we considered the role played by intermolecular forces in determining the physical state of a substance. Most substances can exist in all three states, depending on the temperature and pressure. From everyday experience, we expect a substance to change from a solid to a liquid to a gas as the temperature increases, although there are some substances (e.g., carbon dioxide) that go directly from the solid state to the gas without first forming a liquid. This section describes the behavior of a substance during a change of physical state as a dynamic equilibrium—a concept that is the theme of several of the remaining chapters of this book. As we consider phase changes, several more properties of substances related to the strengths of intermolecular attractions will be identified.

Liquid-Vapor Equilibrium

In Chapter 6, we used the kinetic molecular theory to explain the behavior of gases. The average kinetic energy of molecules is proportional to the absolute temperature of the sample, but individual molecules possess a wide range of energies. Some molecules have less energy than the average value, whereas others have more. Figure 11.1 shows the distribution of the velocities of gas molecules, called the *Maxwell–Boltzmann distribution,* for some substances at various temperatures. Note that the lighter the gas particle or the higher the temperature, the greater its spread of velocities.

The concept of distribution of energies also applies to molecules in the liquid state and is an important factor in **evaporation,** the conversion of molecules from the liquid phase to the gas phase. To escape from a liquid, a molecule must have a kinetic energy that is sufficient to overcome the forces of attraction from the other molecules in the sample. For any temperature at which the liquid state is stable, only a small fraction of the molecules possesses enough energy to evaporate, or **vaporize**—that is, have enough energy to escape from the surface of the liquid. All liquids can evaporate, with the rate of evaporation depending on the temperature of the liquid. The higher the temperature, the greater the fraction of molecules that have enough energy to evaporate, and the faster the evaporation.

Vapor Pressure

Suppose a sample of a liquid is put into an evacuated vessel (pressure = 0) held at a fixed temperature. Some of the molecules (those with enough kinetic energy) evaporate, causing an increase in the pressure within the vessel (Figure 11.2). If the temperature is held constant, the rate of evaporation is also constant (as long as the surface area of the liquid does not change), because the fraction of the molecules with enough energy to escape from the liquid does not change. However, as the concentration of the molecules

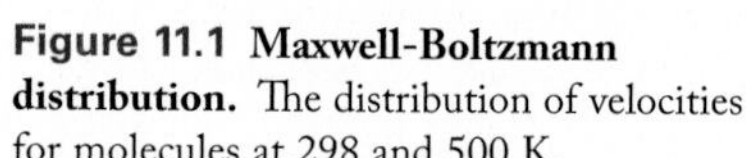

Figure 11.1 Maxwell-Boltzmann distribution. The distribution of velocities for molecules at 298 and 500 K.

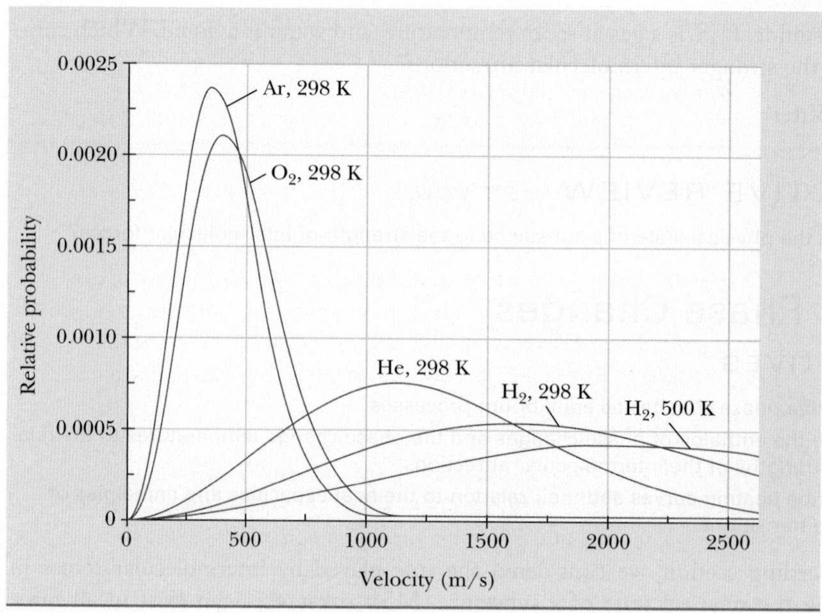

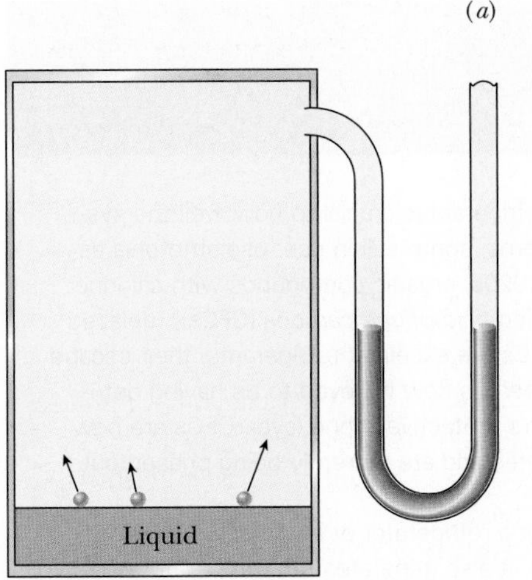

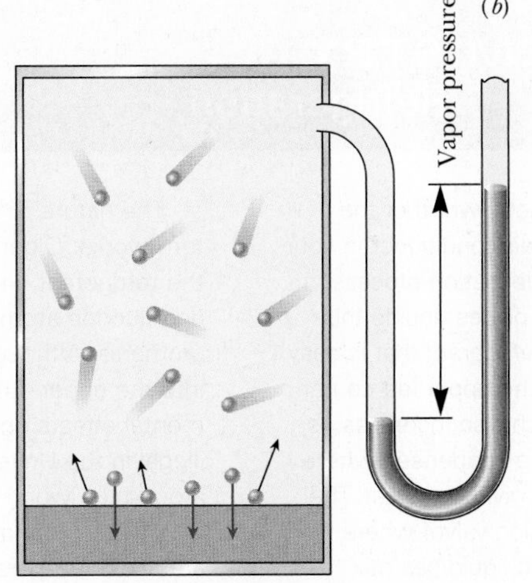

Figure 11.2 Vapor pressure. Vapor pressure of a liquid is a dynamic equilibrium. *(a)* Some of the molecules in the liquid have enough energy to escape the surface of the liquid. *(b)* The pressure in the vessel becomes constant when the rate of condensation equals the rate of evaporation.

in the gas state builds up, some of them collide with the liquid surface and rejoin it. **Condensation** is the conversion of a gas to a liquid. The greater the number of molecules in the gas state, the greater the rate of condensation, simply because there are more gaseous molecules that collide with the surface of the liquid.

Figure 11.3 shows the rates of evaporation and condensation when a liquid is placed into an empty container. After a relatively short time, the number of molecules that condense per second is equal to the number of molecules that evaporate per second. When the rate of condensation becomes equal to the rate of evaporation, the pressure in the vessel no longer changes. The constant pressure that is achieved is called the **vapor pressure** of the liquid.

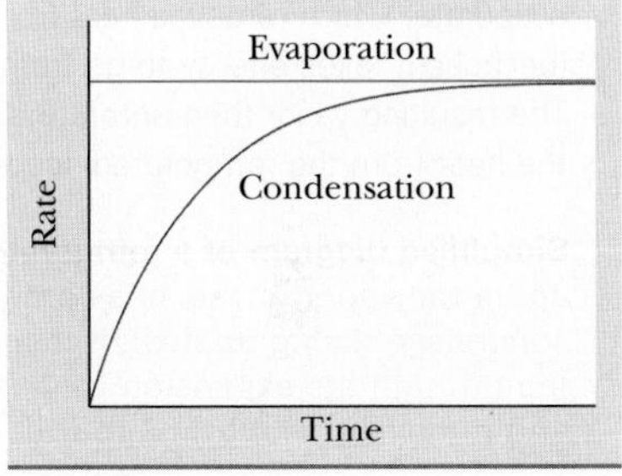

Figure 11.3 Rates of evaporation and condensation. At a constant temperature, the rate of evaporation is constant. As the number of molecules in the gas phase increases, the rate of condensation increases until it equals the rate of evaporation. At this point, a state of dynamic equilibrium has been reached.

A state of **dynamic equilibrium** is one in which two opposing changes occur at equal rates, so no net change is apparent. When considering any system at equilibrium, it is important to recognize that both opposing processes continue to occur, but because the rates are equal, no net change is observed. In a chemical equation, we indicate an equilibrium process by using two arrows, one pointing in either direction. One example is the equilibrium of liquid and gaseous diethyl ether, represented by the equation

$$(C_2H_5)_2O(\ell) \underset{\text{condensation}}{\overset{\text{vaporization}}{\rightleftharpoons}} (C_2H_5)_2O(g)$$

In this example, the two opposing processes are vaporization and condensation.

Dynamic equilibrium occurs when the rates of opposing processes are equal.

Vapor pressure changes as the temperature changes. There is a minimum kinetic energy a molecule must have to escape from the liquid, and it depends on the strength of the intermolecular attractions. As the temperature increases, the energy distribution curve shifts (see Figure 11.1), so the fraction of molecules with enough energy to evaporate increases. Therefore, the rate of evaporation and the equilibrium vapor pressure increase with increasing temperature. Figure 11.4 shows the vapor pressures of several liquids as functions of temperature. Any point on one of these lines represents a combination of temperature and pressure at which the liquid and gaseous states of the substance are in equilibrium. Each liquid has a characteristic vapor pressure curve that depends on the strength of its intermolecular attractions. Because the vapor pressure of diethyl ether is greater than that of water at all temperatures, the intermolecular attractions in diethyl ether must be weaker.

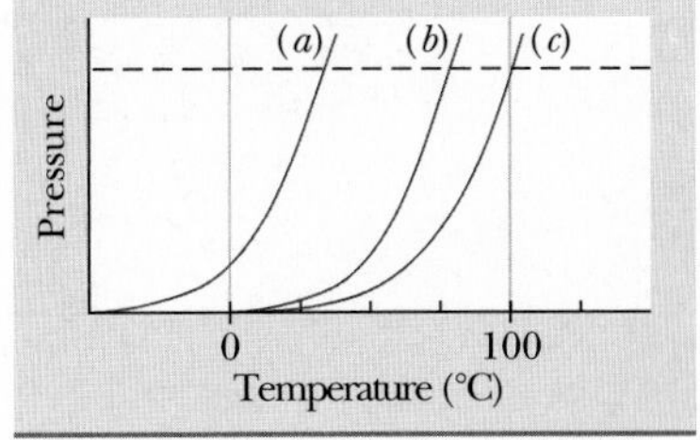

Figure 11.4 Vapor pressure curves. Vapor pressure as a function of temperature for *(a)* diethyl ether, *(b)* ethanol, and *(c)* water. The *horizontal dashed line* represents one atmosphere pressure.

Boiling Point

The **boiling point** of a liquid is the temperature at which the vapor pressure is equal to the surrounding pressure. At this and greater temperatures, the gas phase is the stable phase of the substance because the kinetic energy of the particles of the substance is high enough to overcome intermolecular forces that are holding the individual particles

PRACTICE OF CHEMISTRY
Refrigeration

Much of the refrigeration in our society, whether the refrigerator in your kitchen or the air conditioner in your car, depends on the evaporation and condensation process. Consider a long tube that is coiled in two places. Inside this tube is a small amount of liquid called a *refrigerant* that is easy to evaporate. The refrigerant is initially in the upper left corner. It is compressed by the compressor, which also increases its temperature. The hot vapor goes through a condenser, where it is cooled and it condenses into a liquid, giving up heat. The condensed liquid goes through an expansion valve, where it partially evaporates and gets cold. The cold liquid/gas mix passes through an evaporator, where warm air is blown over the coils in the evaporator. This causes the remaining liquid to evaporate, which removes heat from the refrigerated space (because it takes energy to go from the liquid to the gas phase). The resulting vapor then enters the cycle again, ridding itself of the heat from the refrigerated space in the condenser.

The nature of the refrigerant is crucial to how well the system works. Older systems, some still in use, use ammonia as the refrigerant. In the 1920s, organic compounds with chlorine and fluorine atoms, called *chlorofluorocarbons* (CFCs), replaced ammonia. Although CFCs are excellent refrigerants, their escape into the upper atmosphere is now believed to be having detrimental effects on Earth's protective ozone layer. CFCs are now illegal in the United States and are currently being phased out around the world.

Despite the fact that a refrigerator or air conditioner generates cold temperatures, it also generates hot temperatures at the same time (at the right coil, in our simplified design). In fact, the laws of thermodynamics (see Chapters 5 and 17) require that we will always give off more energy as heat than we absorb to make it cold. So do not try to cool your kitchen by opening up the refrigerator door—it may feel cold initially, but the refrigerator will generate more heat outside than cold inside! ■

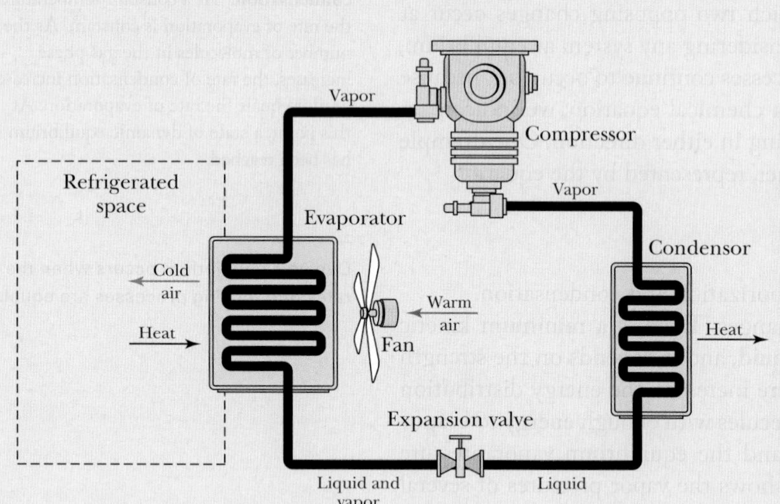

Simplified diagram of a refrigerator. The upper section contains refrigerant in the vapor phase. It is compressed, then condenses to a liquid in the condenser, giving up heat. It then cools to a liquid/vapor mix when passing through the expansion valve, evaporating further in the evaporator and removing heat from a refrigerated space. The cycle then starts again.

together. At the boiling point, bubbles of vapor can form below the surface of the liquid as heat is added to the sample. These bubbles rise and ultimately escape the liquid phase.

The boiling point is a function of the surrounding pressure—at lower surrounding pressures, a lower temperature is needed for the vapor pressure to equal that pressure. The **normal boiling point** of a liquid is the temperature at which its equilibrium vapor pressure equals one atmosphere. The horizontal dotted line in Figure 11.4 intersects each of the equilibrium vapor pressure curves at the normal boiling point of the corresponding liquid. Unless otherwise specified, a given boiling point is the normal boiling point of the substance.

The boiling point of a substance is also a useful indicator of the relative strengths of intermolecular forces. On boiling, a large increase in the distance between the individual molecules occurs, so the intermolecular attractions are almost completely broken. Therefore, the stronger the intermolecular forces of attraction, the greater the boiling point.

Enthalpy of Vaporization

When a liquid evaporates, only the higher energy molecules have enough energy to overcome the intermolecular forces of attraction in the liquid. The molecules in the liquid, therefore, have a lower average kinetic energy (a lower temperature) as a result of evaporation, unless heat is added to the sample. Thus, vaporization is an endothermic process. In nature, the cooling effect of vaporization serves to control the body temperatures of warm-blooded animals. When exercise or warm surroundings produce excess heat in the human body, the heat is removed by the evaporation of perspiration. When the humidity (partial pressure of water vapor in the air) is high, the rate of evaporation decreases, affecting comfort levels.

The **enthalpy of vaporization, ΔH_{vap},** is the enthalpy change that accompanies the conversion of one mole of a substance from the liquid state to the gaseous state at constant temperature. For water, the thermochemical equation is

$$H_2O(\ell) \rightarrow H_2O(g) \qquad \Delta H_{vap} = +44.0\text{ kJ}$$

The enthalpy of vaporization is also expressed as

$$\Delta H_{vap}[H_2O] = +44.0\text{ kJ/mol}$$

The enthalpy of vaporization of a liquid is the energy needed to separate the molecules by overcoming the intermolecular attractions. Thus, the stronger the intermolecular attractions in a substance, the greater the enthalpy of vaporization. Table 11.3 presents the boiling points and enthalpies of vaporization for several substances. The correlation of these enthalpies with the boiling points, and therefore the strengths of intermolecular attractions, are readily seen. The stronger the intermolecular forces of attraction between the molecules, the greater the boiling point and the enthalpy of vaporization.

TABLE 11.3 Boiling Points and Enthalpies of Vaporization of Selected Substances

Substance	Boiling Point (°C)	ΔH_{vap} (kJ/mol)
Argon	−185.7	6.1
Hydrogen chloride	−83.7	15.1
Carbon dioxide	−78.3	16.1
Butane	−0.6	22.3
Carbon disulfide	46.3	26.9
Water	100.0	44.0

Because condensation is the opposite of vaporization, when a condensation process occurs, the same amount of heat is involved, but now it is an exothermic process. Thus, for the condensation of one mole of water from the gas phase:

$$H_2O(g) \rightarrow H_2O(\ell) \qquad \Delta H = -44.0\text{ kJ}$$

The sign on the ΔH has changed, although the magnitude of the enthalpy change is the same as earlier.

EXAMPLE 11.2 Using Enthalpies of Vaporization

Calculate the energy necessary to boil 100.0 g carbon disulfide, CS_2, at its normal boiling point.

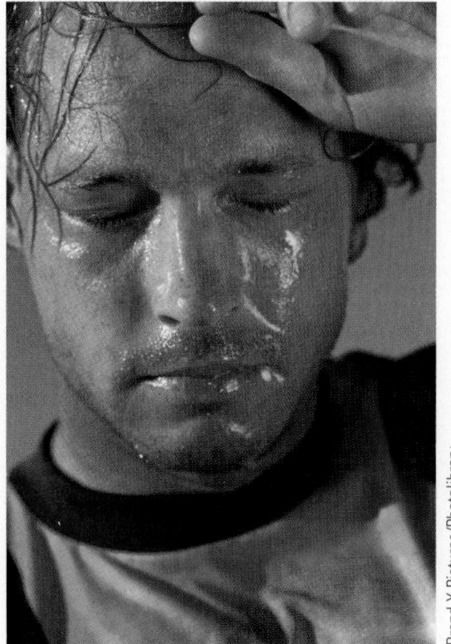

(*a*)

(*b*)

Cooling by evaporation. (*a*) Evaporation of perspiration helps cool us down. (*b*) Dogs and other animals cannot cool themselves by evaporation of sweat. Rather, they breathe rapidly, evaporating water from the lungs.

Evaporation is an endothermic process.

The stronger the intermolecular forces of attraction between the molecules, the greater the boiling point and the enthalpy of vaporization.

Strategy Determine the number of moles of carbon disulfide; then use the enthalpy of vaporization from Table 11.3 as a conversion factor to determine the energy needed.

Solution

The molar mass of CS_2 is 76.13 g/mol. The number of moles of CS_2 is:

$$100.0\ \text{g CS}_2 \times \left(\frac{1\ \text{mol CS}_2}{76.13\ \text{g}}\right) = 1.314\ \text{mol CS}_2$$

According to Table 11.3, the enthalpy of vaporization of CS_2 is 26.9 kJ/mol. Using this as a conversion factor,

$$1.314\ \text{mol CS}_2 \times \left(\frac{26.9\ \text{kJ}}{\text{mol}}\right) = 35.3\ \text{kJ}$$

Understanding

What is the enthalpy change when 75.0 g carbon dioxide condenses at its boiling point?

Answer −27.4 kJ

Critical Temperature and Pressure

At sufficiently high temperatures, substances no longer exist as liquids. The **critical temperature** is the maximum temperature at which a substance can exist in the liquid state. Above its critical temperature, no matter how high the applied pressure, a substance has only one phase that completely occupies the volume of the vessel. The **critical pressure** is the minimum pressure needed to liquefy the substance at the critical temperature. The single phase that occurs above the critical temperature is usually referred to as a gas, because it occupies the entire volume of the container, but its density at the critical pressure or higher is often comparable with those of the condensed states. The single phase that exists above the critical temperature and pressure is sometimes called a **supercritical fluid.** Like the boiling point, enthalpy of vaporization, vapor pressure, and many other properties, the critical temperature is directly related to the strength of intermolecular attractions of a substance.

Substances that have strong intermolecular forces will have high critical temperatures.

Liquid-Solid Equilibrium

The changes of a substance from liquid to solid (freezing) and from solid to liquid (melting or fusion) are also opposing changes that lead to a dynamic equilibrium. This equilibrium process for benzene, C_6H_6, is represented by the equation

$$C_6H_6(s) \underset{\text{freezing}}{\overset{\text{melting}}{\rightleftharpoons}} C_6H_6(\ell)$$

The **melting point** of a substance is the temperature at which the solid and liquid phases are in equilibrium. (We could define a normal melting point like we do a normal boiling point, but pressure changes have little effect on condensed phases unless the pressure differences are very large.) At the melting point temperature, the particles have enough kinetic energy to overcome the intermolecular forces that are keeping the particles in fixed position, but not enough energy to overcome the intermolecular forces completely and separate from each other. Associated with melting is an **enthalpy of fusion, ΔH_{fus},** the enthalpy change that occurs when one mole of solid is converted to liquid at a constant temperature. Like vaporization, fusion is an endothermic process. For any substance, the enthalpy of fusion is considerably smaller than the enthalpy of vaporization. For example, when one mole of water melts, as shown in the thermochemical equation

$$H_2O(s) \rightarrow H_2O(\ell) \qquad \Delta H_{fus} = +6.01\ \text{kJ}$$

the enthalpy of fusion is much less than the enthalpy of vaporization, +44.0 kJ. In the vaporization process, energy must be provided to separate the molecules completely. In the melting process, the molecules still remain quite close together, so only a small fraction of the attractive energy between the molecules must be provided. The energy required for melting overcomes some of the intermolecular attractions, giving the molecules greater freedom of motion but much less freedom than they would have in the vapor phase.

Solidification (or freezing) is the reverse of fusion, and as with condensation and vaporization, the enthalpy change per mole is the same as with melting, but with the opposite sign:

$$H_2O(\ell) \rightarrow H_2O(s) \qquad \Delta H_{freez} = -6.01 \text{ kJ}$$

Heating and Cooling Curves

Adding heat to a solid sample at one atmosphere of pressure usually produces the liquid phase and then the gas phase, in that order. Figure 11.5 shows the heating curve (a plot of temperature as a function of heat added) for water at one atmosphere. At the left side of the graph, the solid phase is present at −40 °C. As heat is added, the heat capacity of the solid (see Chapter 5) determines the rate of the temperature change. This part of the curve ends abruptly when the melting point is reached. At the melting point, the temperature remains constant as long as both the solid and liquid phases are present. All of the heat that is added at the melting point is used to overcome the attractive forces between the molecules in the solid and release them into the liquid state. For a well-stirred mixture of ice and water, the temperature is 0 °C, no matter how much or how little ice is present.

Once all of the solid has been converted to liquid, the temperature increases again, as determined by the heat capacity of the liquid. When the temperature reaches the normal boiling point of the liquid, it again stays constant, because the added heat is used to overcome intermolecular attractions as the molecules move far apart from each other in the vapor phase. As soon as enough heat has been added to vaporize the sample completely, the temperature increases again at a constant rate that depends on the heat capacity of the vapor. The observed lengths of the constant-temperature portions on the heating curve at the melting and boiling points are proportional to the enthalpies of fusion and vaporization, respectively.

A heating curve is a graph of the temperature of a sample versus the heat added to the sample.

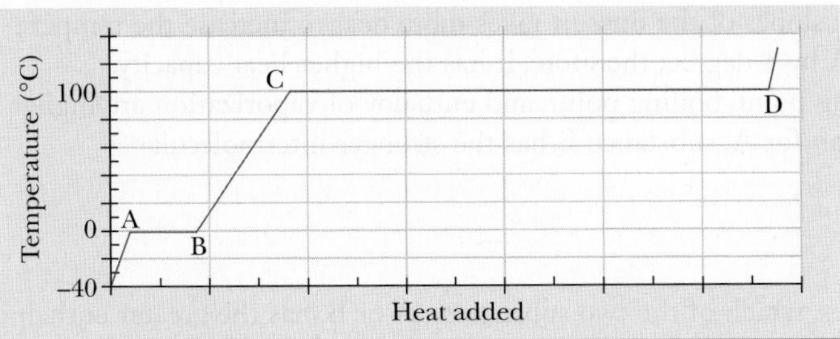

Figure 11.5 A heating curve. When heat is added to a sample of water, the temperature increases until the melting point is reached at point A. Between points A and B, the temperature remains constant until the solid water converts completely to liquid water. Addition of more heat increases the temperature until the boiling point is reached, point C. From point C to point D, the heat that is added converts the liquid into gas. Further heating increases the temperature of the gaseous sample. If the gas is cooled, the line is retraced backward as gas becomes liquid, cools, and then becomes solid.

Removing heat from a gaseous sample of this same substance retraces the heating curve. When heat is removed rapidly, the liquid can sometimes cool to temperatures less than the normal melting point. This phenomenon—the cooling of the liquid below its melting point without forming solid—is called **supercooling.** A supercooled liquid is in an unstable state, and stirring or adding a small crystal of the substance causes the rapid formation of the solid, with an abrupt increase of the temperature to the normal freezing point.

EXAMPLE 11.3 Heating Curves

The accompanying graph shows the heating curves for one mole each of substances A and B, at one atmosphere pressure.

(a) Give the melting and boiling points for each substance.
(b) Which substance has the greater enthalpy of vaporization?
(c) Which substance has a greater molar heat capacity in the liquid phase?
(d) Which substance has the stronger intermolecular forces of attraction?

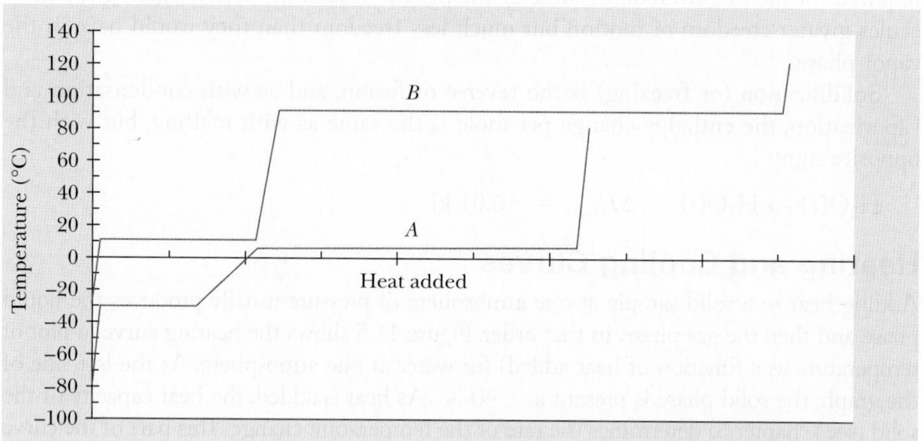

Strategy Refer to the heating curve for approximate melting and boiling points. The enthalpy of vaporization is related to the length of the horizontal vaporization line, whereas the slope of the temperature-change segments is related to the heat capacity.

Solution

(a) Each heating curve has two plateaus. The first occurs when the substance melts, and the second occurs when the substance boils. The temperatures of the first phase transition are −30 °C for substance A and +10 °C for substance B, and these are the melting points for the two substances. The boiling points are the temperatures for the second plateaus; +5 °C for A and +90 °C for B.
(b) The length of the plateau at the boiling point is proportional to the enthalpy of vaporization; therefore, substance B has the greater heat of vaporization.
(c) The heat capacity is the number of joules required to increase the temperature of the sample by 1 °C. Between the two phase transitions, each sample is in the liquid state. From the slopes of the lines, it takes more heat to increase the temperature of substance A by 1 degree; therefore, it has the higher heat capacity.
(d) Because the melting point, boiling point, and enthalpy of vaporization are higher for substance B than for A, substance B has the stronger intermolecular attractions.

Understanding

From the heating curves, which of the two substances, A or B, has the greater enthalpy of fusion?

Answer B

© Cengage Learning/Larry Cameron

Sublimation of iodine. Solid iodine has a sufficiently high vapor pressure at room temperature that the violet color of the vapor is easily seen. Volatile solids such as iodine are often purified by sublimation because many impurities will not sublime.

Solid-Gas Equilibrium

In the solid state, a few surface molecules always have sufficient kinetic energy to overcome the intermolecular attractions and escape into the gas phase. Thus, a solid has a vapor pressure, just as a liquid does. For example, the characteristic odor of mothballs comes from their main ingredient, either naphthalene or *p*-dichlorobenzene, which both have low vapor pressures. **Sublimation** is the direct conversion of a substance from the solid state to the gaseous state. The reverse of sublimation, called **deposition,** is the conversion of the gas directly to the solid state. The opposing changes of sublimation and deposition lead to a state of dynamic equilibrium at an applied pressure equal to the vapor pressure of the solid. At a pressure of 40 torr, solid iodine, I_2, is in equilibrium with the violet vapor at 97.5 °C.

$$I_2(s) \underset{\text{deposition}}{\overset{\text{sublimation}}{\rightleftharpoons}} I_2(g)$$

Carbon dioxide is a commonly encountered substance that sublimes at normal pressures around one atmosphere and is more familiarly known as dry ice. At normal pressure, it sublimes at −78 °C and is used as a convenient portable coolant. The **enthalpy of sublimation, ΔH_{sub},** is the enthalpy change for conversion of one mole of a solid to the gaseous state. The sublimation process itself is endothermic. At temperatures not much less than 0 °C, solid H_2O also has a measurable vapor pressure. Over time, ice cubes in a freezer will slowly disappear as the ice sublimes. Sublimation of solid water is also responsible for "freezer burn" of some frozen foods. Reducing the temperature of the freezer combats the sublimation of ice, as does tightly wrapping frozen food.

Figure 11.6 shows all three reversible phase transitions on an enthalpy diagram. From the enthalpy diagram and Hess's law, we see that the enthalpy of sublimation is the sum of the enthalpy of fusion and the enthalpy of vaporization when all three enthalpy changes occur at the same temperature. For most solids (such as metals and ionic solids) at normal temperatures, the strengths of the intermolecular attractions are so great that the vapor pressure is too small to measure. Increasing the temperature of these solids first causes them to melt; then the vapor pressure of the liquid increases with temperature until the substance boils. In calculating the heat absorbed or released when a sample undergoes a temperature change that also includes phase changes, the heat capacity of each phase and the enthalpy changes for the phase transitions must be used.

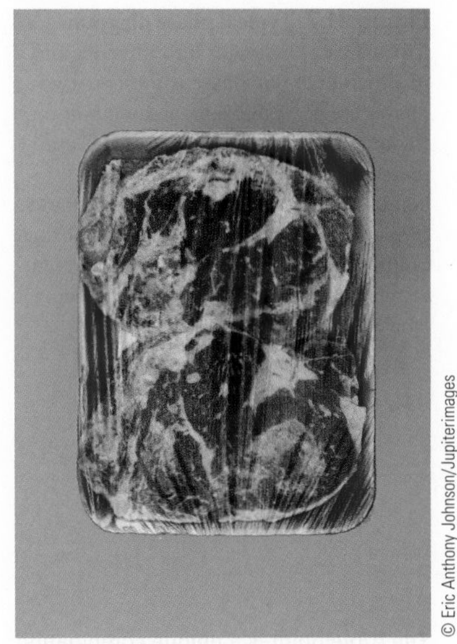

Effects of sublimation. Food that is "freezer-burned" has lost water content by sublimation. The food is still edible but usually looks unappetizing.

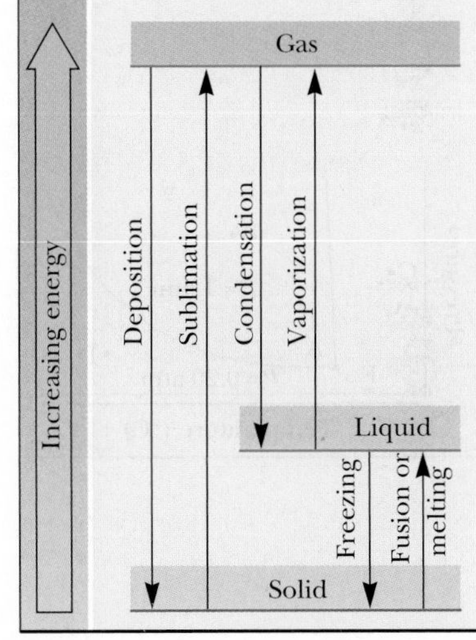

Figure 11.6 Enthalpy diagram for phase changes. All of the phase changes shown are reversible at the appropriate temperature and pressure.

OBJECTIVES REVIEW *Can you:*

- ☑ describe phase changes as equilibrium processes?
- ☑ relate the enthalpy of phase changes and the phase-change temperatures to the relative strengths of the intermolecular attractions?
- ☑ describe heating curves and their relation to the heat capacities and enthalpies of phase transitions?

11.3 Phase Diagrams

OBJECTIVES

- ☐ Use a phase diagram to identify the various phases that are stable at any particular temperature and pressure
- ☐ Relate the sign of the slope for the solid-liquid equilibrium line to the difference in the densities of the two states
- ☐ Construct from the phase diagram the main features of a heating or cooling curve at constant pressure

The physical state of a substance reflects the strength of the intermolecular attractions relative to the thermal energy of the molecules, which is determined by the temperature. Section 11.2 explained that temperature and pressure determine which phase is stable. A **phase diagram** is a graph of pressure versus temperature that shows the region of stability for each of the physical states and summarizes a great deal of information in a single picture. Figure 11.7 is a typical phase diagram. The three line segments show the combinations of pressure and temperature at which any two phases exist in equilibrium. The line that separates the liquid from the gas is the vapor pressure curve; examples for several liquids were given in Figure 11.4. The line segments in Figure 11.7 divide the diagram into three regions. Only one phase is present in each region, but on the lines the two phases are in equilibrium.

The line segments on a phase diagram represent conditions of pressure and temperature where two phases exist in equilibrium.

The **triple point** is a unique combination of pressure and temperature at which all three phases—the solid, the liquid, and the gas—exist in equilibrium. The triple point

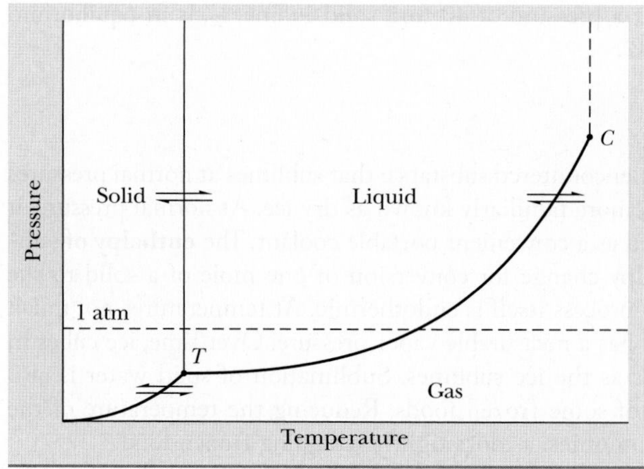

Figure 11.7 Typical phase diagram. Lines divide the graph into three regions in which only one phase is present. Two phases are at equilibrium at any point on the lines. *Dashed horizontal line* at 1 atm pressure intersects the solid-liquid equilibrium line at the melting point and intersects the liquid-gas line at the normal boiling point. All three phases are present at the triple point *(T)*. The critical point is labeled *C*.

The triple point is a unique combination of temperature and pressure where the solid, liquid, and gas phases coexist in equilibrium.

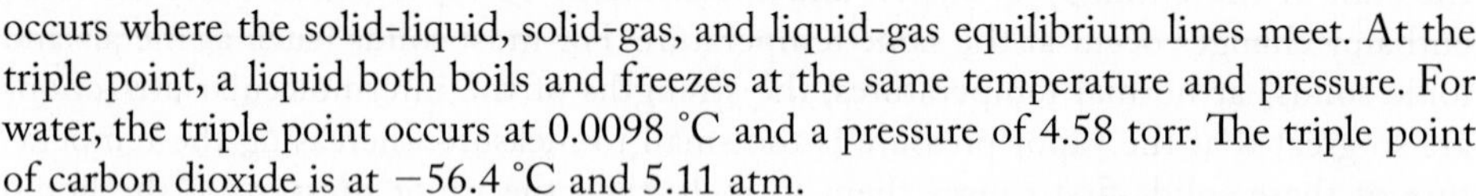

occurs where the solid-liquid, solid-gas, and liquid-gas equilibrium lines meet. At the triple point, a liquid both boils and freezes at the same temperature and pressure. For water, the triple point occurs at 0.0098 °C and a pressure of 4.58 torr. The triple point of carbon dioxide is at −56.4 °C and 5.11 atm.

The liquid-gas equilibrium line ends at the critical point. Above the critical temperature, only one phase exists. The vertical dotted line at the critical temperature does not represent a phase equilibrium. Only a single fluid phase is present at pressures greater than the critical pressure; the single phase is considered to be a liquid below the critical temperature and a gas above the critical temperature.

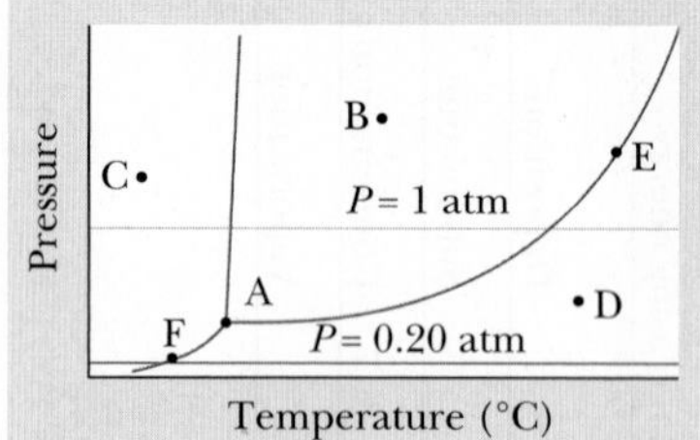

EXAMPLE 11.4 Interpreting Phase Diagrams

Using the phase diagram to the left, identify the phase or phases present at each of the lettered points A, C, and E.

Strategy Determine which phase exists in each region of the phase diagram. If a point falls on a line between two phases, both of those phases exist in equilibrium. If a point lies at the point where all three phases intersect, all three phases are in equilibrium under these conditions.

Solution

First label each of the areas with the phase that is stable in that region of the graph. You can do this easily by following any line of constant pressure that is above the triple point, such as the line at $P = 1$ atm. As the temperature increases (left to right), the stable phase changes from the solid to the liquid to the gas, in that order. We note that point A is at the intersection of all three phases—the triple point. Therefore, at A, the solid, liquid, and gas phases are all present. Point C is in the region that contains only solid. Point E is on the line that separates the liquid from the gas, so both of those phases are present in equilibrium.

Understanding

What phase or phases are present in equilibrium at points B, D, and F on the phase diagram?

Answer B: liquid only; D: gas only; F: solid and gas

The heating or cooling curve for a substance can be deduced from its phase diagram. A horizontal line at constant pressure intersects the solid-liquid line at the freezing point and intersects the liquid-vapor line at the boiling point. If the constant pressure is lower than the triple-point pressure, the substance sublimes, and the liquid phase is never observed.

EXAMPLE 11.5 Heating Curves from Phase Diagrams

Using the phase diagram in Example 11.4, sketch the heating curves expected at pressures of (a) 1 atm and (b) 0.2 atm.

Strategy Follow the proper horizontal line in the phase diagram to determine what phases occur as heat is added to the sample.

Solution

(a) A heating curve is a graph of temperature versus heat added. Starting at a temperature below the melting point of this substance at 1 atm pressure, addition of heat increases the temperature of the sample uniformly until the melting point is reached. At that temperature, continued addition of heat converts the solid to liquid at a constant temperature. Once the solid is completely melted, the temperature again increases uniformly until the boiling point is reached. The temperature remains constant while the liquid vaporizes completely, at which point the temperature increases again. For most substances, the enthalpy of vaporization is considerably greater than the enthalpy of fusion, and this observation is reflected in the heating curve. The resulting heating curve appears as line *(a)* on the graph in the margin.

(b) At a pressure of 0.2 atm, the liquid phase is never stable. Thus, the temperature of the solid increases with the addition of heat until the sublimation point (the temperature at which the constant-pressure line intersects the solid-gas equilibrium line) is reached. The temperature does not change with further addition of heat until the solid has been converted completely to gas. The temperature then increases as the added heat raises the temperature of the gas. The horizontal portion of the heating curve at 0.2 atm is approximately the same as the sum of the two horizontal portions of the heating line at 1 atm. The lengths of the horizontal portions are arbitrary, because the enthalpies of fusion and vaporization of the substance are not given. The resulting heating curve appears as line *(b)* in the graph in the margin.

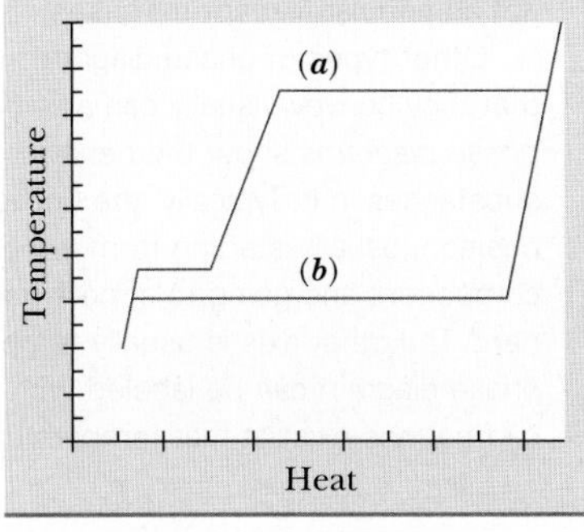

Heating curves. *(a)* Heating curve at 1 atm. *(b)* Heating curve at 0.2 atm.

The equilibrium line that separates the solid and liquid phases in Figure 11.7 is nearly vertical, in marked contrast with the solid-gas and liquid-gas equilibria. Unlike the boiling point and sublimation temperature, the melting point changes little, even when large changes in pressure occur (although the melting points of all substances do change slightly). Figure 11.8 shows the melting points of carbon dioxide and water for a large range of pressures. As the pressure increases, the melting point of carbon dioxide increases and that of water decreases.

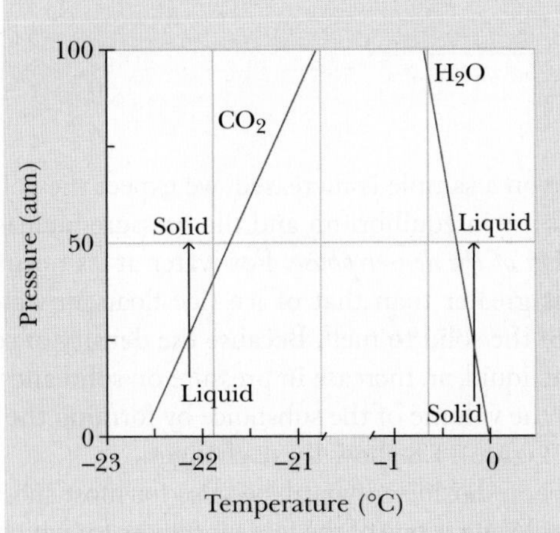

Figure 11.8 Effect of pressure on the solid-liquid equilibria for carbon dioxide and water. Lines show the change in the melting points of carbon dioxide and water over a large range of pressures. Pressure increases are indicated with vertical arrows in each phase diagram. When the pressure increases on carbon dioxide, the melting point increases. By contrast, when the pressure increases on water, the melting point decreases. Water is one of only a few substances known whose melting point behaves this way.

PRINCIPLES OF CHEMISTRY
Phase Diagrams

The phase diagrams we are considering in this chapter are fairly simple. They are called *single-component phase diagrams* because they illustrate the properties of just one substance. In addition, they are pressure-temperature phase diagrams, which give the behavior of substances at varying temperature and pressure. Volume is also something that can be varied in a system, but volume-dependent phase diagrams are not as popular (except for gases).

Other types of phase diagrams exist, and the information that they convey visually can also be important. *Two-component phase diagrams* show the behavior of a system that has two substances in it. Typically, the horizontal axis charts varying composition, usually starting from a mole fraction of zero for one component and going to a mole fraction of one for that component. The other axis is usually temperature, and regions of the phase diagram can be labeled with what substances and/or compounds exist at that temperature for that particular composition. Shown in the figure below is a phase diagram that shows the behavior of mixtures of sodium and potassium metals at varying compositions and temperatures. There is one region where the entire system is a liquid, and other regions where liquids coexist with solids (either a solid element or a solid alloy having the formula Na_2K). Unlike a one-component phase diagram, which conveys only phase information, two-component phase diagrams display some chemical information as well.

Phase diagrams can provide useful information for everyday purposes as well. Carbon steel is used in some knives; too much carbon and the knife is brittle, whereas too little carbon and the knife blade will not hold an edge. Consider also a phase diagram between H_2O and NaCl. At less than a temperature of −21 °C (which is about −6 °F), at all compositions of NaCl and H_2O, the phase diagram shows that the system is a solid. The lesson is: You cannot use salt to melt ice and snow in the wintertime if the temperature is below −21 °C! ∎

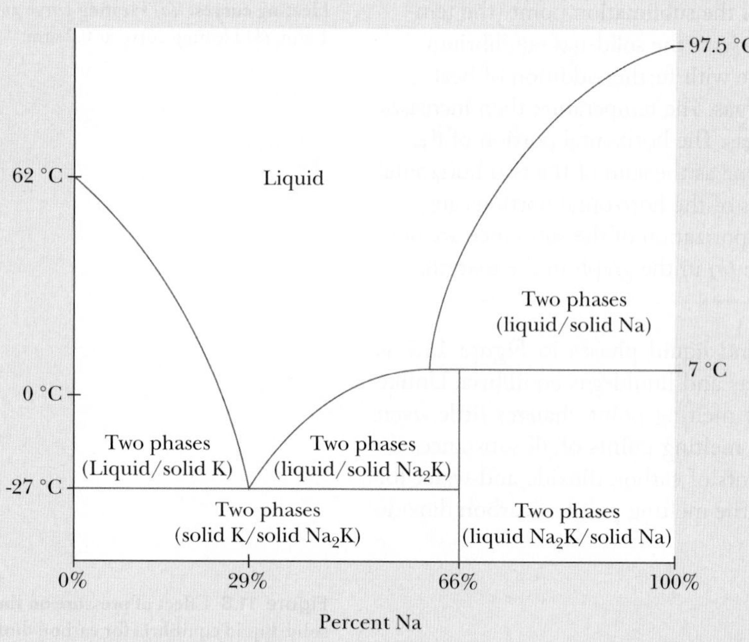

Two-component phase diagram. A two-component phase diagram, showing variations in mixtures of sodium and potassium metals.

When the pressure on a sample is increased, we expect the volume of the sample to decrease. If two phases are in equilibrium and the pressure increases, a decrease in volume results *by formation of the denser phase.* For water at its normal melting point, the density of the liquid is greater than that of ice (ice floats in water), so an increase in pressure causes some of the solid to melt. Because the density of solid carbon dioxide is greater than that of the liquid, an increase in pressure on solid and liquid carbon dioxide at equilibrium reduces the volume of the substance by forming the solid from the liquid. The vertical arrows in Figure 11.8 show these changes.

The behavior of carbon dioxide is typical, because for most substances the solid phase is denser than the liquid. Water is one of the few substances for which the solid is less dense than the liquid. Because ice melts when the pressure increases, a wire can pass through a

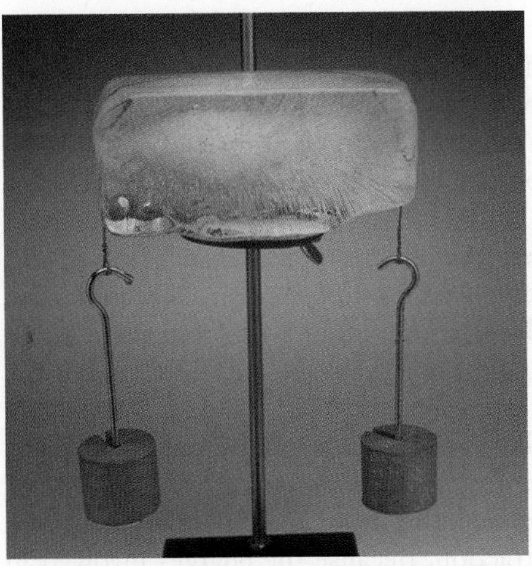
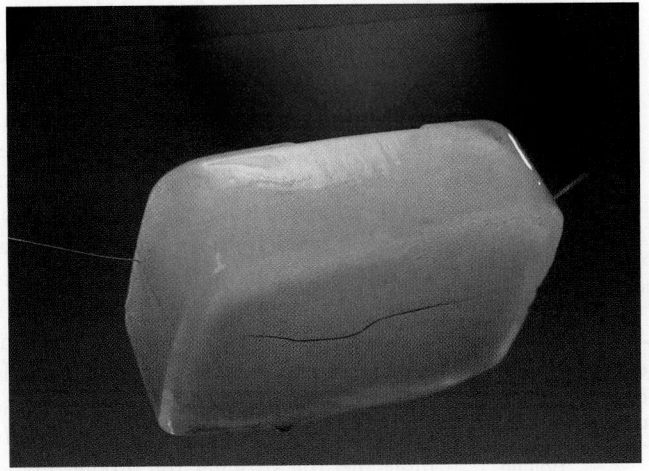

A wire that has weights attached to the ends slowly passes through a block of ice without cutting it into two pieces. The wire exerts a pressure on the ice directly under it, melting it. As the wire moves down into the ice, the liquid water refreezes above the wire, where the pressure has diminished.

block of ice without cutting it into two pieces. As the wire moves downward, the solid re-forms above the wire, because the pressure there has decreased to its original value.

The effect of pressure on solid-gas and liquid-gas equilibria can be understood using the same principle we have applied to the solid-liquid equilibrium. An equilibrium between a condensed phase and gas responds to an increase in pressure by forming the denser phase (the solid or liquid). Because there is a large difference between the densities of a gas and a condensed phase (a factor of about 1000), the effect of pressure on the equilibrium is much greater than is observed in the case of the solid-liquid equilibrium. This analysis of pressure effects on phase equilibrium is an example of Le Chatelier's principle, which is discussed in detail in Chapter 14.

Increasing the pressure on two phases in equilibrium favors the side of the equilibrium that has the denser phase.

OBJECTIVES REVIEW *Can you:*

- ☑ use a phase diagram to identify the various phases that are stable at any particular temperature and pressure?
- ☑ relate the sign of the slope for the solid-liquid equilibrium line to the difference in the densities of the two states?
- ☑ construct from the phase diagram the main features of a heating or cooling curve at constant pressure?

11.4 Intermolecular Attractions

OBJECTIVES

- ☐ List and explain the origins of the various intermolecular forces
- ☐ Use periodic trends to determine the relative strengths of intermolecular forces in different substances
- ☐ Compare the relative strengths of different kinds of intermolecular attractions

The functioning of most biologically important molecules—proteins, nucleic acids, and others—is often determined by their intermolecular forces. For example, the double-helix structure of DNA is controlled by intermolecular attractions between two long molecules. An understanding of the origins and relative strengths of different kinds of attractions between molecules is important to chemistry.

The previous sections of this chapter discussed the roles played by intermolecular forces in determining many properties of liquids, solids, and phase transitions. This section presents the origins of these forces. Several kinds of intermolecular attractions exist, but all of them depend on electrostatic attraction—the attraction between charges of opposite sign.

Dipole-Dipole Attractions

The attractions between electrical charges of opposite sign are important in nature. In Chapter 9, the stability of solid ionic compounds was shown to arise mainly from the electrostatic attraction between the ions. In covalent compounds, partial positive and negative charges on atoms are produced by the unequal sharing of electron pairs. Molecules with polar bonds (see Chapter 10) may have an overall dipole moment. Such polar molecules are attracted to one another, because the negative end of one molecule attracts the positive end of another one, in accord with Coulomb's law. **Dipole-dipole attractions** are the intermolecular forces that arise from the electrostatic attractions between the molecular dipoles. For example, the difference in the boiling points of two compounds of similar mass, Br_2 (58.8 °C) and ICl (97.4 °C), is explained by dipole-dipole attractions. The Br_2 molecule is nonpolar, but ICl is polar, and the molecules of ICl arrange themselves to maximize the attractive interactions, similar to those shown in Figure 11.9. In general, the larger the dipole moment of a molecule, the stronger the dipole-dipole attraction. Because the charges involved in dipole-dipole attractions are smaller than ionic charges, dipole-dipole attractions are much weaker than the attractions between ions. They are also much weaker than the covalent bonds between atoms.

Dipole-dipole attractions are present only in polar substances.

TABLE 11.4 Boiling Points of Some Nonpolar Substances

Substance	Molar Mass	Boiling Point (°C)
Halogens		
Fluorine (F_2)	38	−188
Chlorine (Cl_2)	71	−34.6
Bromine (Br_2)	160	58.8
Iodine (I_2)	254	184.4
Boron Halides		
BF_3	68	−101
BCl_3	117	12.5
BBr_3	250	90.1
BI_3	392	210
Group 4A Hydrides		
Methane (CH_4)	16	−184
Silane (SiH_4)	32	−111.8
Germane (GeH_4)	77	−90.0
Stannane (SnH_4)	123	−52
Noble Gases		
Helium	4	−268.9
Neon	20	−245.9
Argon	40	−185.7
Krypton	84	−152.9
Xenon	131	−107.1

London Dispersion Forces

Experiments show that nonpolar molecules also attract each other, so forces other than dipole-dipole forces must exist. Nonpolar molecular substances such as argon (Ar), nitrogen (N_2), and chlorine (Cl_2) all condense to liquids and solids at low temperatures; these experimental observations mean that attractive forces between molecules must also exist in nonpolar substances. Table 11.4 shows the boiling points of several series of related nonpolar substances. In each of these series, there is a trend of increasing boiling points and therefore of increasing strengths of intermolecular attractions, as the molar masses of the substances increase.

To gain some insight into the nature of these attractions, consider what happens when the positive end of a dipole comes close to an argon atom. The argon atom consists of a nucleus (charge = +18) surrounded by a cloud of 18 electrons; it has a spherical distribution of the electrons, so it is not polar. As the positive end of the dipole comes close to the argon atom, it attracts the outer electrons toward itself, distorting the electron cloud and causing the argon atom to have a dipole. This distortion is called an **induced dipole,** which is caused by the presence of an electrical charge close to an otherwise nonpolar molecule. A polar molecule such as HCl can induce a dipole in an argon atom by distorting the electron cloud. An electrostatic attraction exists between the ion or permanent dipole and the induced dipole. The intermolecular force in this case is called a **dipole–induced dipole attraction.**

An electrical charge close to a nonpolar molecule distorts the electron cloud of the molecule and produces an induced dipole.

In each of the preceding cases, permanent charges distort the electron cloud of a nearby nonpolar molecule. However, no permanent charges are present when all of the molecules are nonpolar. How can such molecules attract each other? Averaged over time, there is no net dipole in a nonpolar molecule, but the electrons in the nonpolar

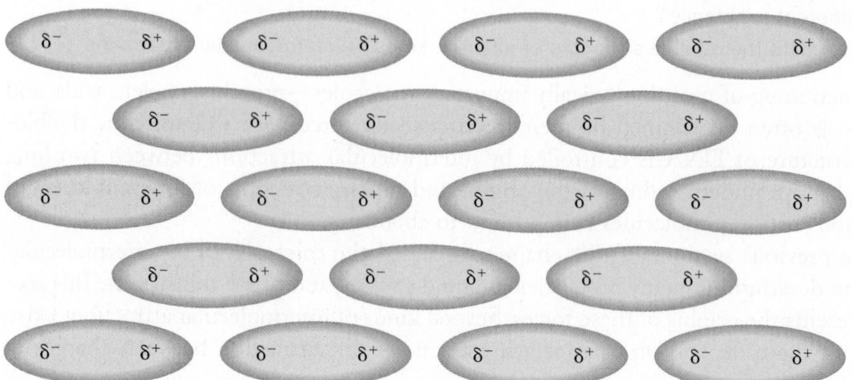

Figure 11.9 Dipole-dipole attractions. Polar molecules interact with each other so that the positive end of one molecule is close to the negative end of other molecules.

molecule are constantly in motion. If it were possible to "freeze" their positions at any instant in time, a nonsymmetric charge distribution would almost certainly exist within the molecule. An **instantaneous dipole** is the result of an unequal charge distribution within a molecule, caused by the motion of the electrons. The rapid motion of the electrons in the molecule means that this instantaneous dipole is gone or pointed in a different direction a fraction of a microsecond later. At almost any instant, though, the molecule possesses an instantaneous dipole.

Instantaneous dipoles exist in molecules as a consequence of the rapid motion of electrons.

The very small charges of an instantaneous dipole in one nonpolar molecule can induce a dipole in a nearby nonpolar molecule, causing the two molecules to attract each other (Figure 11.10). These instantaneous dipole–induced dipole attractions are called **London dispersion forces** after Fritz London (1900–1954), a German physicist who developed this model to explain the intermolecular attractions that exist between nonpolar molecules. **Polarizability** refers to the ease with which the electron cloud of a molecule can be distorted by a nearby charge. The greater the polarizability of a molecule, the greater the induced dipole and the magnitude of the electrostatic attraction. In general, polarizability increases with the size of the electron cloud. The boiling points of the series of substances shown in Table 11.4 demonstrate the effect of increasing the size of the electron cloud on the polarizability of molecules. For molecules of similar shape, the greater the number of electrons in the molecule, the more polarizable it becomes. In all four families of substances shown in Table 11.4, a regular increase in boiling points occurs as the sizes of the electron clouds increase. Because molar mass increases with the size of the electron cloud, the strengths of intermolecular attractions usually increase with increasing molar masses in related series of substances.

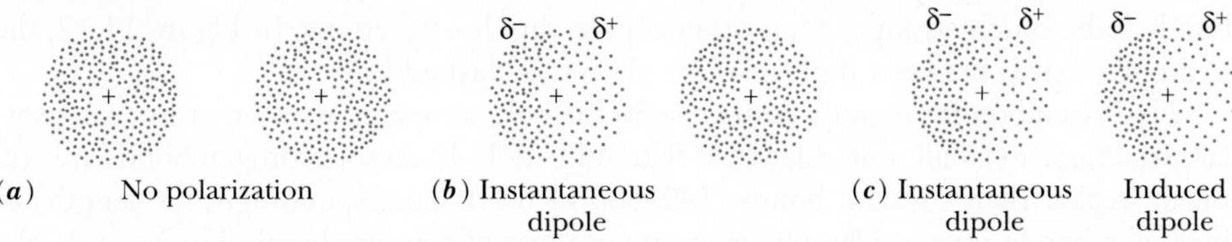

Figure 11.10 London dispersion forces. *(a)* On average, nonpolar molecules have a symmetric distribution of forces. *(b)* An instantaneous dipole can occur from the motion of the electrons. *(c)* When another molecule is nearby, the instantaneous dipole induces a dipole in its neighbor. The resulting attractions between the instantaneous dipole and the induced dipole are called *London dispersion forces.*

London dispersion forces contribute to the attractions between all molecules. Even in molecules with dipole moments, most of the energy of intermolecular attraction arises from the dispersion forces. The energy of attraction between the molecules of ICl is also caused mainly by the dispersion forces, with dipole-dipole attractions making a relatively small contribution. In comparing Br_2 and ICl, two substances with similar molecular weights and the same number of electrons, the 38.6 °C difference in boiling points (58.8 °C and 97.4 °C) is attributed to dipole-dipole attraction. However, the boiling points of both substances are more than 300 °C greater than that of H_2 (−252.8 °C)—the lightest molecule and therefore one with very weak dispersion forces. The large difference in the boiling points between bromine and hydrogen is attributed to the much larger dispersion forces of bromine.

London dispersion forces contribute to the intermolecular forces among all molecules but may be accompanied by dipole forces.

The experimentally measured boiling points of the heavier hydrogen halides (Table 11.5) increase in the order HCl < HBr < HI, which indicates that the strength of intermolecular attractions increases in the same order. For these hydrogen halides, the dipoles (and therefore the dipole-

TABLE 11.5 Boiling Points of Some Hydrogen Halides

Compound	Dipole-Dipole Forces	Dispersion Forces	Boiling Point (K)
Hydrogen chloride	Largest	Smallest	188
Hydrogen bromide	↑	↓	206
Hydrogen iodide	Smallest	Largest	237

dipole attractions) decrease with the decreasing difference between the electronegativities of the bonded atoms, HCl > HBr > HI. However, the strengths of the dispersion forces increase as the number of electrons in the molecules increase, which go as HCl < HBr < HI. (The increase in mass of the halogen in the molecule also has an impact on the boiling point.) The observed trend in the boiling points correlates with the increase in size of the dispersion forces. In most similar situations where the strengths of the London dispersion forces and dipole-dipole attractions predict different trends in boiling points, the dispersion forces dominate. All of the attractive forces discussed (dipole-dipole, dipole–induced dipole, and instantaneous dipole–induced dipole) are collectively called **van der Waals forces.**

van der Waals forces include dipole-dipole attractions and dispersion forces.

Hydrogen Bonding

The boiling points for the hydrides of the Group 4A elements (see Table 11.4) increase regularly with the strength of the dispersion forces: $CH_4 < SiH_4 < GeH_4 < SnH_4$. However, examination of the boiling points of the hydrides of Groups 5A, 6A, and 7A reveals that the first compound in each of these series has an unexpectedly high boiling point, as shown in Figure 11.11. The abnormally strong intermolecular forces of attraction that are reflected in the boiling points of ammonia (NH_3), water (H_2O), and hydrogen fluoride (HF) are attributed to a relatively strong dipole-dipole interaction called **hydrogen bonding.** Hydrogen bonding occurs when a hydrogen atom is bonded to a small, highly electronegative atom, such as nitrogen, oxygen, or fluorine. The partial positive hydrogen is attracted to the partial negative N, O, or F of another molecule. The large polarity and small size of the hydrogen atom cause hydrogen bonds to be much stronger than other dipole-dipole attractions. In Figure 11.12, the hydrogen bonds between molecules are shown as dashed lines.

Hydrogen bonding occurs when a hydrogen atom is bonded to a small, highly electronegative atom, such as nitrogen, oxygen, or fluorine. The partial positive hydrogen is attracted to the partial negative nitrogen, oxygen, or fluorine of another molecule.

Hydrogen bonds are quite strong (4–30 kJ/mol) compared with other intermolecular attractions in small molecules (less than 4 kJ/mol). Even so, hydrogen bonds are still much weaker than covalent bonds (140–600 kJ/mol). Correspondingly, the lengths of hydrogen bonds are considerably greater than those of covalent bonds. For example, the length of the O · · · · H hydrogen bond between water molecules in ice is about 177 pm, whereas the length of the covalent O–H bond in the water molecule is 99 pm.

Many of the unusual properties of water can be attributed to the strong intermolecular hydrogen bonding that occurs in that substance. For example, ice floats on water because the solid is less dense than the liquid, whereas the opposite is true for most

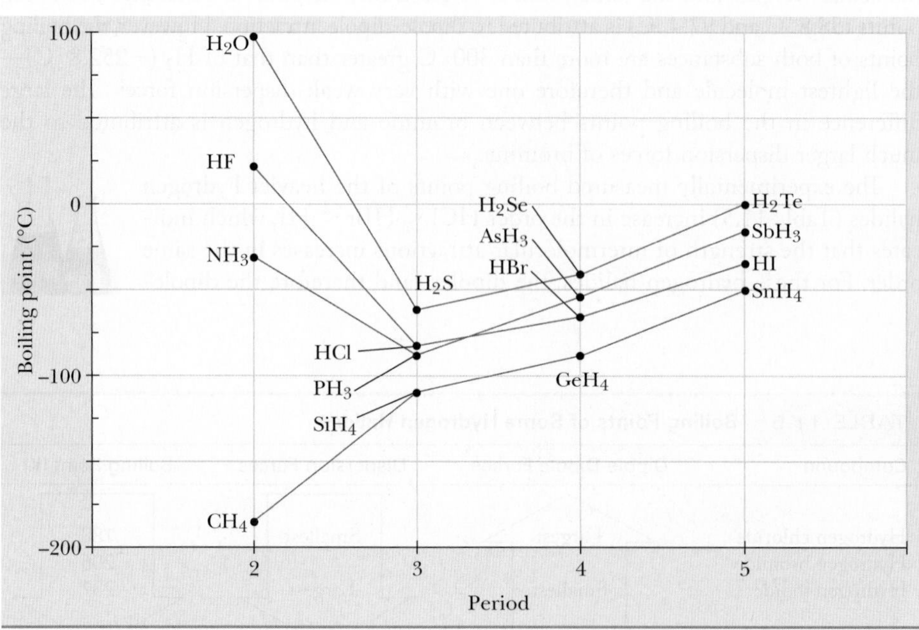

Figure 11.11 Boiling points of the hydrogen compounds of the Group 4A, 5A, 6A, and 7A elements. In all but the Group 4A compounds, the unusually high boiling point of the lightest member of the series is attributed to hydrogen bonding.

Ice floats on water because the hydrogen bonding in the solid leaves large vacant regions between the molecules. After melting, the orderly arrangement of the hydrogen bonds is destroyed, allowing the molecules to move closer together.

$$- - - - O^{\delta-} - H^{\delta+}$$
$$H^{\delta+} - O^{\delta-} - - - H^{\delta+} - O^{\delta-} - - - - -$$

Water

$$H^{\delta+} - N^{\delta-} - - - H^{\delta+} - N^{\delta-} - - - -$$

Ammonia

$$H^{\delta+} - F^{\delta-} - - - H^{\delta+} - F^{\delta-} - - - -$$

Hydrogen fluoride

Figure 11.12 Hydrogen bonds. Hydrogen bonds between molecules of H_2O, NH_3, and HF are shown as *dashed lines.*

substances. Figure 11.13 shows the arrangement of the water molecules in ice. The low density of solid water is the result of its open structure. That structure arises because the oxygen atom of each water molecule in ice is bonded to four hydrogen atoms, two by covalent bonds and two by hydrogen bonds. When ice melts, thermal energy allows some hydrogen bonds to break and re-form, allowing the water molecules to move closer together.

The fact that ice has a lower density than water has significant environmental consequences. As the temperature of the air decreases and cools a lake, a layer of ice forms

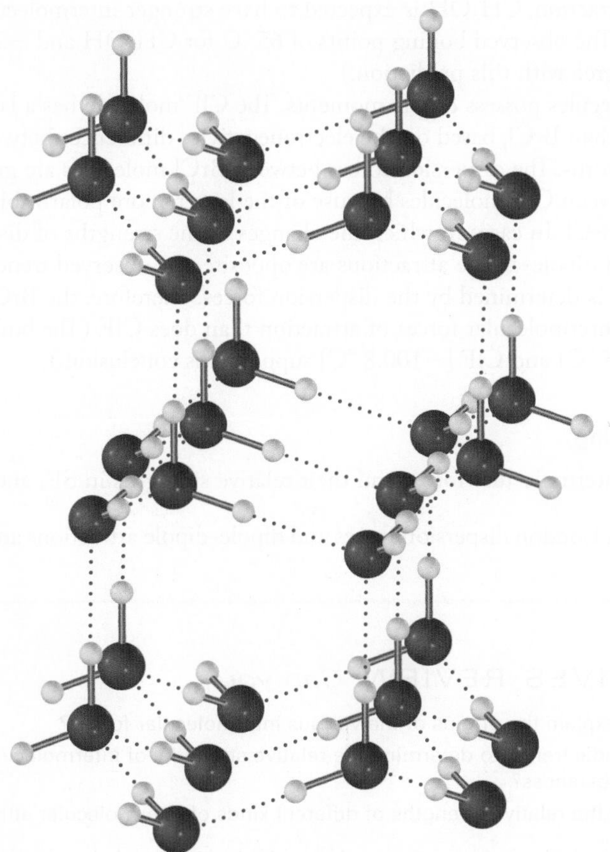

Figure 11.13 Structure of solid water. Each oxygen atom is covalently bonded to two hydrogen atoms and hydrogen-bonded to two other hydrogen atoms. Because of the hydrogen bonding, there are large vacant regions within the solid.

on the surface. Because the solid water is less dense than the liquid, it remains on the surface rather than sinking to the bottom of the lake. The surface layer of solid insulates the liquid below it, and thus provides an environment in which aquatic life can survive.

EXAMPLE 11.6 The Nature of Intermolecular Forces and Their Strengths

Identify the intermolecular forces of attraction, and predict which substance of each pair has the stronger forces of attraction.

(a) CF_4, CCl_4
(b) CH_3OH, CH_3Cl
(c) ClF, BrCl

Strategy All of the molecules will have London dispersion forces. We must determine whether a molecule is polar to determine whether dipole-dipole forces exist. If a hydrogen atom is bonded to an N, O, or F, hydrogen bonding will exist.

Solution

(a) Each of these molecules has a symmetric tetrahedral structure, so neither possesses a dipole. The only intermolecular forces that are possible in both substances are the London dispersion forces that are present in all substances. The larger number of electrons in CCl_4 means that the dispersion forces should be stronger in CCl_4 than in CF_4. (This conclusion is consistent with the observed boiling points of 76.5 °C for CCl_4 and −129 °C for CF_4.)

(b) London dispersion forces and dipole-dipole attractions are possible for both of these substances, but only the CH_3OH molecules have hydrogen-bonding attractions. Because hydrogen bonding is much stronger than other kinds of intermolecular attraction, CH_3OH is expected to have stronger intermolecular forces than CH_3Cl. (The observed boiling points of 65 °C for CH_3OH and −24 °C for CH_3Cl agree with this prediction.)

(c) Both molecules possess dipole moments. The ClF molecule has a larger dipole moment than BrCl, based on the electronegativity differences between the bonded atoms. The dispersion forces between BrCl molecules are greater than those between ClF molecules because of the larger, more polarizable electron cloud in BrCl. In cases in which the changes in the strengths of dispersion forces and dipole-dipole attractions are opposite, the observed trend of increasing attraction is determined by the dispersion forces. Therefore, the BrCl exhibits stronger intermolecular forces of attraction than does ClF. (The boiling points of BrCl [5 °C] and ClF [−100.8 °C] support this conclusion.)

Understanding

Identify the intermolecular forces and their relative strengths in SF_4 and SeF_4.

Answer Both London dispersion forces and dipole-dipole attractions are larger in SeF_4 than in SF_4.

OBJECTIVES REVIEW *Can you:*

- ☑ list and explain the origins of the various intermolecular forces?
- ☑ use periodic trends to determine the relative strengths of intermolecular forces in different substances?
- ☑ compare the relative strengths of different kinds of intermolecular attractions?

11.5 Properties of Liquids and Intermolecular Attractions

OBJECTIVES

- ☐ Relate the surface tension and viscosity of liquids to intermolecular forces
- ☐ Distinguish between cohesion and adhesion, and relate them to capillary action

The liquid state exhibits some similarities to both the solid and gaseous states. Like solids, liquids have high densities and are difficult to compress. On the molecular scale, we attribute these properties of a liquid to the intermolecular attractions holding the molecules close together so that they occupy most of the volume of a sample. Like gases, liquids are fluids and adopt the shapes of their containers. Microscopically, the fluidity of liquids occurs because the molecules can move about and lack the order found in crystalline solids.

Section 11.2 discussed the correlation of the boiling point and enthalpy of vaporization of a substance with the strengths of intermolecular forces. Some additional properties of liquids that are related to intermolecular attractions are discussed in the following section.

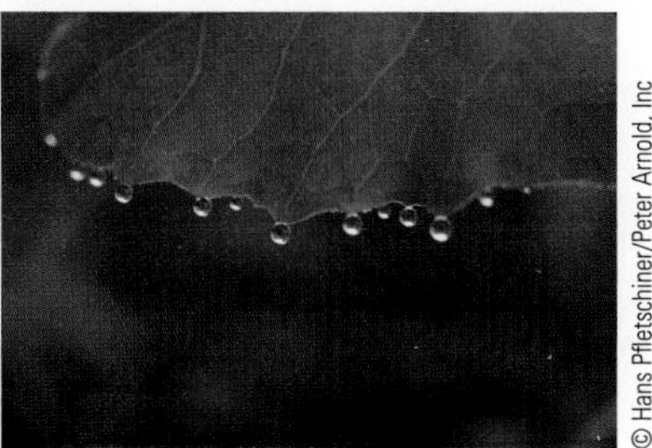

Figure 11.14 Liquid drops. The attraction of molecules for each other in a liquid produces surface tension, which causes small drops of a liquid to adopt a spherical shape.

Surface Tension

A small drop of any liquid assumes a spherical shape (Figure 11.14), a fact that can be attributed to intermolecular forces of attraction. In Figure 11.15, arrows represent intermolecular attractions for a molecule in the interior of a liquid sample and a molecule on the surface of the sample. A molecule in the interior is attracted by its neighbors in all directions. In contrast, a surface molecule has no neighbors above it, so there is a net force that attracts it toward the interior of the liquid. The unbalanced forces on the surface molecules cause a liquid to adopt a shape that has the smallest surface area possible for a fixed volume, namely, a sphere.

Increasing the surface area of a liquid requires an expenditure of energy, because the number of surface molecules increases, and each molecule on the surface has fewer neighboring molecules that attract it. **Surface tension** is the energy required to increase the surface area of a liquid and is expressed in SI units of joules per square meter (J/m^2). Liquids with strong intermolecular forces of attraction have high surface tensions.

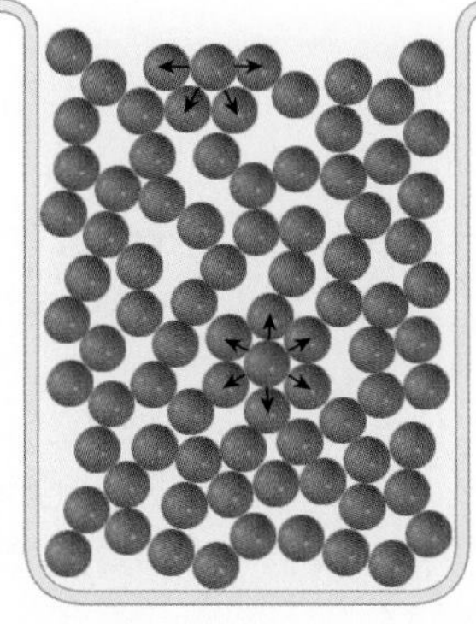

Figure 11.15 Surface tension. A molecule in the interior of a liquid is attracted by surrounding molecules equally in all directions. A molecule at the surface of a liquid has unbalanced forces of attraction toward the interior of the liquid, resulting in surface tension.

Capillary Action

Capillary action, which causes water to rise in a small-diameter glass tube (as shown in Figure 11.16*a*), is another property of liquids that results from intermolecular attractions. Two kinds of intermolecular attractions contribute to this phenomenon. **Cohesive forces** result from the intermolecular attraction of molecules for other

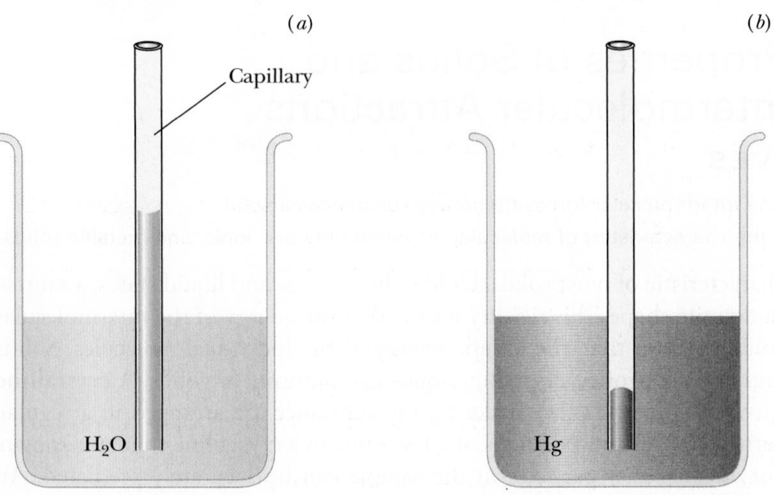

Figure 11.16 Capillary action. The meniscus, or curved surface of a liquid, results from a balance between adhesive and cohesive forces. *(a)* The curvature of the water surface results from adhesive forces that are greater than cohesive forces. The water rises in the capillary because of the unbalanced forces and has a meniscus that curves down. *(b)* Mercury has cohesive forces that are stronger than the adhesive forces toward the glass, resulting in a depression of the liquid inside the capillary and an upward curvature of the meniscus.

TABLE 11.6 Surface Tension and Viscosity of Liquids at 20 °C

Liquid	Surface Tension (J/m^2)	Viscosity ($N{\cdot}s/m^2$)
Acetone (C_3H_6O)	0.0237	0.327×10^{-3}
Chloroform ($CHCl_3$)	0.0271	0.580×10^{-3}
Benzene (C_6H_6)	0.0289	0.652×10^{-3}
Glycerol ($HOCH_2CHOHCH_2OH$)	0.0634	1.49×10^{-3}
Water (H_2O)	0.0730	1.005×10^{-3}
Mercury (Hg)	0.487	1.55×10^{-3}

molecules of the same substance. **Adhesive forces** result from the intermolecular attractions between molecules of different substances. The water rises in the capillary tube because the adhesive forces between the water and the glass are quite strong. Because the glass (largely SiO_2) has many polar sites on its surface, the adhesive attractions between these polar sites and the dipoles of the water molecules are sufficiently strong to draw the liquid up against the force of gravity. For liquid mercury (see Figure 11.16*b*), the cohesive forces are greater than the adhesive forces, so the level of the liquid inside the tube is actually depressed. The upward or downward curvature of a liquid surface is called the *meniscus.* The direction of curvature depends on the relative strengths of the adhesive and cohesive forces. Capillary action is one of the factors that contributes to the ability of plants to draw water out of the ground and is the major reason paper and cloth towels absorb water in the manner that they do.

Viscosity

Differences between syrup and water can be easily observed when they are poured. The syrup flows much more slowly than the water, reflecting the difference between the viscosities of these two liquids. **Viscosity,** the resistance of a fluid to flow, is another property that is related to the forces of intermolecular attraction. The stronger the intermolecular forces between the molecules in a liquid, the more viscous the liquid becomes. Other factors are also important in determining the viscosity of a liquid, such as the structure, size, and shape of the molecules. Although the surface tension of water is considerably greater than that of glycerol (Table 11.6), glycerol is more viscous. The higher viscosity of glycerol is attributed to the elongated shape of the polar glycerol molecules, which allows them to become entangled and slows their flow.

As intermolecular forces increase in strength, the boiling point, enthalpy of vaporization, surface tension, and viscosity generally all increase.

OBJECTIVES REVIEW *Can you:*

- ☑ relate the surface tension and viscosity of liquids to intermolecular forces?
- ☑ distinguish between cohesion and adhesion, and relate them to capillary action?

11.6 Properties of Solids and Intermolecular Attractions

OBJECTIVES

- ☐ Identify the intermolecular forces that make substances a solid
- ☐ Describe the characteristics of molecular, covalent network, ionic, and metallic solids

Rigidity is a characteristic of most solids. Unlike the gaseous and liquid states, a sample of a solid has a definite shape. This rigidity means that the energy of the intermolecular attractions is much greater than the kinetic energy of the individual molecules. Solids are classified into two categories, crystalline solids and amorphous solids. A **crystalline solid** is one in which the units that make up the substance are arranged in a regular, repeating pattern. If the relative positions of a few units in a crystalline solid are known, the locations of all the other particles in the sample can be accurately predicted. An **amorphous solid** lacks the order of a crystalline solid. Many amorphous solids consist of large molecules that cause the liquid state to become quite viscous as the temperature is reduced, and the large molecules move so slowly that they cannot arrange themselves into the pattern present in the crystalline state. Ultimately, London dispersion or dipole-dipole forces bind the individual molecules into the solid phase. Many plastics, such as polyethylene, are examples of amorphous solids; glass is another example. In this section, our consideration of the solid state is restricted to crystalline solids.

An amorphous solid does not have the regular, repeating arrangement of units that is found in crystalline solids.

We can classify crystalline solids according to the nature of the forces that hold the units together in a regular arrangement, which are usually referred to as *crystal forces.*

Molecular Solids

Molecular solids consist of atoms or small molecules held together by van der Waals forces, hydrogen bonds, or both. Physical properties (e.g., melting point and hardness) of molecular solids such as Ar, H_2O, CO_2, I_2, and $C_{20}H_{42}$ (a type of wax) vary considerably depending on the strengths of the intermolecular interactions. Covalent bonds hold the atoms together in the molecules, but only weak intermolecular attractions hold one molecule to the others. In comparison with solids held together by ionic or covalent bonds, molecular solids are usually rather soft substances. Many substances that form molecular solids exist in the liquid or gaseous states at room temperature. Most molecular solids, even those with fairly large molecules, have melting points below 300 °C.

Molecular solids such as I_2 and $C_{20}H_{42}$ are usually soft and have low melting points.

Covalent Network Solids

In a **covalent network solid,** all of the atoms are held in place by covalent bonds. Diamond, which is discussed in the chapter introduction, is an example of a covalent network solid. In the diamond structure of elemental carbon, each atom is covalently bonded to four other carbon atoms at the corners of a tetrahedron, using sp^3 hybrid orbitals, as shown in Figure 11.17*a*. Because the atoms are held together in three dimensions by strong covalent bonds, covalent network solids usually have very high melting points and are hard materials. Diamond is the hardest known substance. Cubic boron nitride (which can also exist in a graphite-like structure) and silicon carbide are also covalent network solids that are very hard and have high melting points.

Covalent network solids such as diamond and SiO_2 are usually hard and have high melting points.

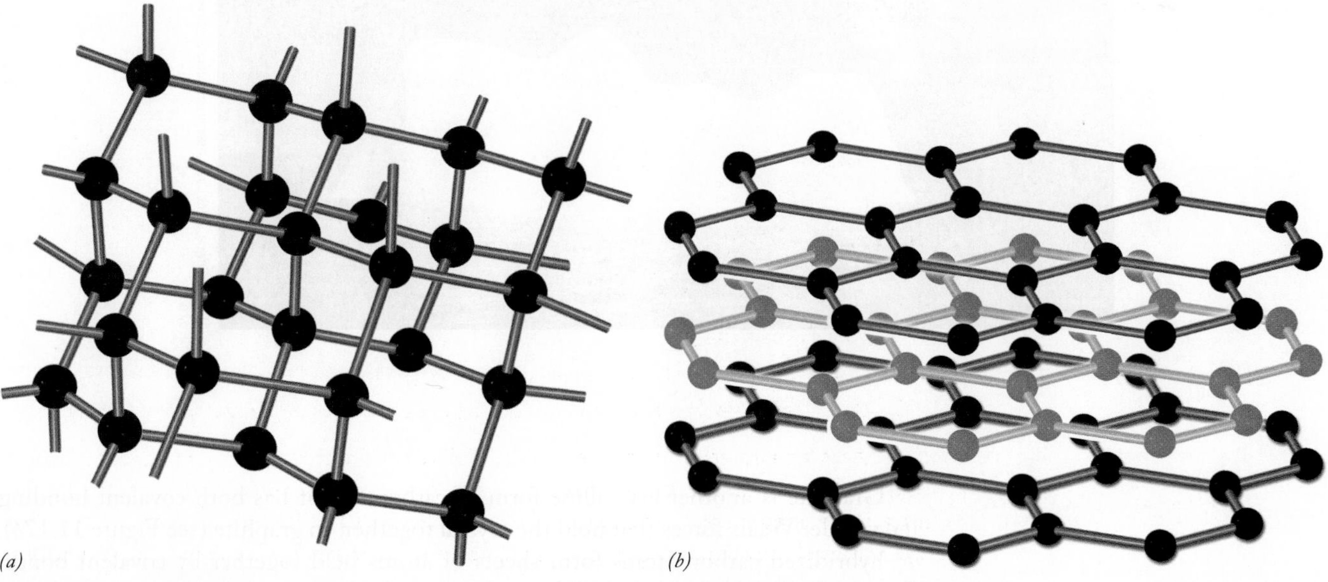

Figure 11.17 Two allotropes of carbon. *(a)* In the diamond crystal, each carbon atom is covalently bonded to four other atoms at the corners of a tetrahedron. Boron nitride and silicon carbide also form covalent network crystals with the diamond structure. *(b)* Graphite consists of layers of covalently bonded atoms of carbon, with van der Waals forces holding the layers together.

PRINCIPLES OF CHEMISTRY
The "Unusual" Properties of Water

One of the most unusual known substances is water, H_2O. Wait a minute. *Water?* An unusual substance? How can that be? It seems so, well, normal to us.

The behavior of water seems normal to us because water is so common; we are exposed to it on a daily basis. We become so familiar with how it behaves that we forget (or never realize) that water is, in fact, unusual in its properties.

What are some of the unusual properties of water?

- For such a low-molar-mass molecule (18.0 g/mol), water has an extremely high melting point and boiling point. For water, these temperatures are 0 °C and 100 °C, respectively. As a counterexample, consider methane. CH_4, with a molar mass of 16.0 g/mol, has melting and boiling points of −182 °C and −161 °C, respectively, which are hundreds of degrees lower.
- It takes a relatively large transfer of energy to increase or decrease the temperature of water. For example, consider an 8-ounce cup of coffee (which is mostly water). It takes about 95 kJ of energy to heat that cup of coffee from 0 °C to 100 °C. If the "coffee" were the same mass of aluminum instead of water, it would take only about 20 kJ. It takes almost five times as much energy to warm a given mass of water than it does the same mass of aluminum.
- Likewise, it takes a relatively large amount of energy transfer to vaporize water. It takes more than 2.2 kJ (that's *kilo*joules) to vaporize just 1 g of H_2O. It takes about half a kilojoule to vaporize 1 g of methane from liquid methane.
- Water dissolves a wide range of other substances. It does so well at dissolving substances that water is sometimes nicknamed "the universal solvent." Other liquids can dissolve substances as well, but water dissolves a wider range of substances than almost any other liquid.
- Perhaps the most unusual property of water is that the solid phase of H_2O floats on the liquid phase of H_2O; that is, the solid is less dense than the liquid. This property is unusual because most substances contract, and thus get more dense, as their temperature is decreased, with the ultimate result that the solid is *more* dense than the liquid.

What is the reason for this unusual behavior? Simply, the chemical bonding in the molecule.

In a water molecule, the hydrogen atoms are connected to the oxygen atom by a shared pair of electrons—that is, a covalent bond. It turns out, however, that this sharing is not equal: the oxygen atom attracts both electron pairs more than the individual hydrogen atoms do. This unequal sharing sets up a charge imbalance across the covalent bonds, making them polar covalent bonds. Finally, the two O–H bonds are about 104.5 degrees apart from each other, giving the water molecule a bent shape. This shape and the polar covalent bonds have a direct impact on the properties of the molecule, ultimately leading to its unusual properties. ∎

Graphite is another crystalline form of carbon, but it has both covalent bonding and van der Waals forces that hold the crystal together. In graphite (see Figure 11.17*b*), sp^2-hybridized carbon atoms form sheets of atoms held together by covalent bonds. The covalent bonds hold the solid together in only the two dimensions of the sheets, whereas the weaker van der Waals forces hold one sheet to another. In contrast with diamond, graphite is soft because the weak van der Waals attractions allow the two-dimensional sheets to slip past each other. (In fact, several graphite-based solid lubricants are available in stores.) The term **allotropes** refers to two or more molecular or crystalline forms of an element in the same physical state that exhibit different chemical and physical properties. Diamond and graphite are allotropes (other allotropic forms of carbon are discussed in Chapter 20). Sulfur, phosphorus, and tin are other elements that have allotropic forms in the solid state, and O_2 and O_3 are examples of gaseous allotropes.

Other examples of covalent network crystals are the many minerals that have structures like quartz, SiO_2. Each Si atom bonds to four oxygen atoms at the corners of a tetrahedron, and each oxygen atom bonds to two silicon atoms. Although CO_2 and SiO_2 have the same empirical formula, carbon dioxide exists at normal temperatures as gaseous triatomic molecules, whereas silicon dioxide is a network solid. In contrast with carbon, silicon forms only very weak π bonds with oxygen, so it forms four single bonds to oxygen rather than the two double bonds required in a triatomic molecule. Chapter 20 contains a more detailed discussion of the difference in bonding in these two oxides.

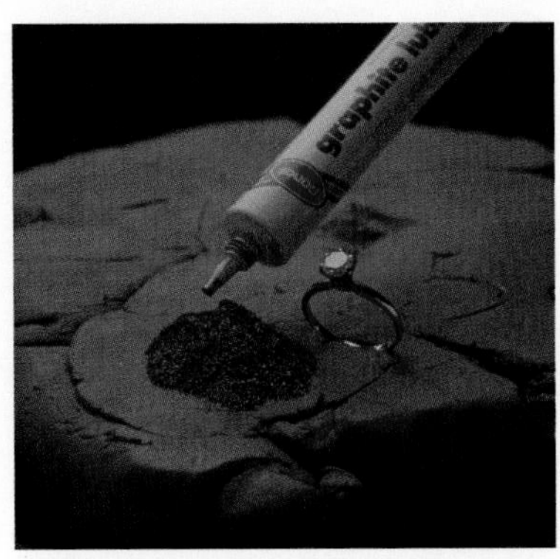

Allotropes of carbon. Because its weakly-bound layers allow for easy slippage, graphite forms the basis of several commercially-available solid lubricants, like the one pictured. Carbon can also form diamond, the hardest naturally-occurring material known.

Ionic Solids

An **ionic solid** consists of oppositely charged ions held together by electrostatic attractions that are very strong, comparable in strength to covalent bonds. Ionic compounds generally have high melting points and are relatively hard and brittle. A unique property of ionic compounds is that they do not conduct an electric current in the solid state, because the ions are held firmly in place, but are excellent electrical conductors in the liquid state, because the ions are then free to move. Many binary compounds formed by the reactions of metallic elements with nonmetals, such as NaCl and MgO, exist as ionic solids. The effects of ionic charge and size on the strengths of ionic bonds are discussed in Chapter 9.

Ionic solids are generally brittle and have high melting points.

Quartz, a crystalline form of silicon dioxide, is a covalent network solid.

Metallic Solids

Metallic solids are solids formed by metal atoms. The metallic elements form crystalline solids that exhibit many unique properties, such as high thermal conductivity, good electrical conductivity, and metallic luster. A special kind of bonding called *metallic bonding* accounts for these properties. One relatively simple model for metallic bonding is called the *electron sea model.* Because metals have relatively low ionization energies, the valence-shell electrons are easily removed and form a sea of electrons with the metal ions embedded in it. The electrons in the sea are mobile and account for the high thermal and electrical conductivity of solid metals. Chapter 20 outlines an alternative description of metallic bonding called *band theory,* based on quantum mechanics, and applies it to both metals and semiconductors. The strength of the metallic bond varies greatly, as reflected in the wide range of melting and boiling points observed for various metals. For example, mercury (Hg) boils at 356.6 °C, and rhenium (Re) has a boiling point of 5650 °C. Metals are malleable; the shape of a piece of metal can be changed by the application of a physical force, such as hammering. The malleability of metals puts them in sharp contrast with covalent network and ionic solids, which shatter when they are struck. The delocalized nature of the metallic bond allows the atoms to move without a large loss in bond energy, in contrast with network and ionic solids, in which the bonds are localized between neighboring atoms.

Table 11.7 summarizes the characteristic properties of the kinds of solids discussed in this section.

Metallic solids have luster and large thermal and electrical conductivities. A special kind of bonding, metallic bonding, accounts for these properties.

Gold leaf. Gold leaf is made by hammering out bars of the metal into very thin sheets. The dome of the State House in Boston, Massachusetts, is covered with gold leaf.

OBJECTIVES REVIEW *Can you:*

- ☑ identify the intermolecular forces that make substances a solid?
- ☑ describe the characteristics of molecular, covalent network, ionic, and metallic solids?

TABLE 11.7 Properties of Solids

	Type of Solid			
	Molecular	Covalent Network	Ionic	Metallic
Structural unit	Atoms, molecules	Atoms	Ions	Atoms
Attraction between units	Intermolecular forces	Covalent bonds	Ionic bonds	Metallic bonds
Melting points	Low-melting; often gases or liquids at room temperature	High melting	High melting	Variable, from low to very high
Characteristics	Soft	Hard and brittle	Brittle	Malleable
Electrical conductivity	Poor	Variable; depends on structure	Poor in solid, good when molten	Very high
Examples	CO_2, H_2O	SiO_2, diamond	NaCl, BaO	Mg, Fe, Hg

11.7 Structures of Crystalline Solids

OBJECTIVES

- ☐ Use the Bragg equation to relate the angle of diffraction and wavelength of x rays to the distances in a crystal
- ☐ Determine the number of ions or atoms present in a cubic unit cell of a solid
- ☐ Calculate the density of a crystalline solid from the dimensions and contents of a unit cell
- ☐ Know the relation between the cubic close packing and face-centered cubic crystal structures

Much of what is known about the bond angles and bond lengths in molecules and the sizes of atoms and ions has come from the study of the crystalline solid state. In this section, we first present the experimental measurements that provide information about a crystal. We then interpret the experimental data and use a few simple crystal structures to illustrate some of the information that is obtained from a study of the crystalline solid state.

Figure 11.18 Interference of waves. *(a)* Constructive interference occurs when two in-phase waves combine to produce a wave of greater amplitude. *(b)* Destructive interference results from the combination of two waves of equal amplitude that are exactly out of phase.

Bragg Equation

The experimental determination of crystal structures by x-ray diffraction was one of the most important discoveries of the 20th century. X rays are short-wavelength electromagnetic radiation. When a narrow beam of this radiation is directed at a crystal, each of the atoms scatters the radiation in all directions. When these scattered waves come together, they interfere with each other, either constructively or destructively. If the waves are in phase (the maximum and minimum amplitudes occur at the same places), *constructive interference* occurs and a more intense wave results. When the separate waves are exactly out of phase, *destructive interference* occurs; that is, the waves cancel each other, and no radiation is observed. Figure 11.18 illustrates constructive and destructive interferences. A narrow beam of x rays directed at a crystal is diffracted at certain angles where constructive interference occurs. Each of the dark spots on the photographic film in Figure 11.19 corresponds to one of the diffraction angles.

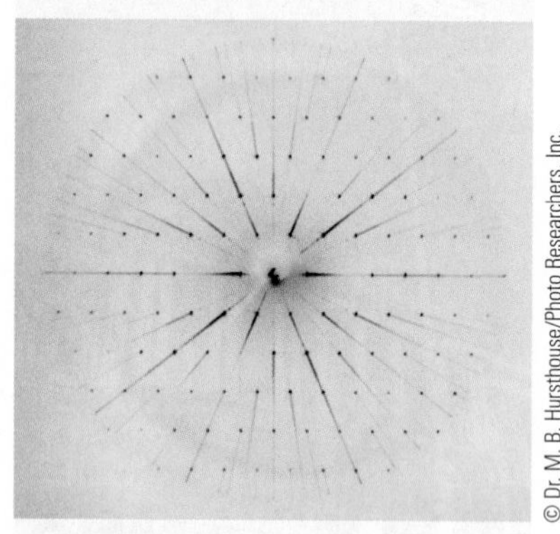

Figure 11.19 X-ray diffraction pattern. A photographic film exposed to x rays diffracted by a crystal shows many dots corresponding to diffraction angles at which constructive interference of the x rays occurs.

In 1913, William H. Bragg and his son William L. Bragg interpreted the diffraction of a narrow beam of x rays by crystals. They found that the locations of the dots on the photographic film could be predicted with a simple equation that is now called the **Bragg equation:**

$$n\lambda = 2d \sin \theta \qquad [11.1]$$

where λ is the wavelength of the x rays used, d is the distance between layers of atoms in the crystal, θ is the angle between the diffracted x rays and the layer of atoms, and n is a whole positive number called the *order*. There are many different distances between layers of atoms in a crystalline solid, so a crystal diffracts the x rays at many different angles. The Bragg equation is useful only when the spacing between layers and the wavelength of the radiation are the same order of magnitude. Example 11.7 illustrates the use of the Bragg equation.

EXAMPLE 11.7 Distances in Crystals from the Bragg Equation

A crystal is found to diffract 154 pm x rays at an angle of 19.3 degrees. Assuming that $n = 1$ in the Bragg equation, calculate the distance (in pm) between the layers of atoms that give rise to this diffraction.

Strategy Use Equation 11.1 to solve for *d*.

Solution

Solve the Bragg equation for the distance between layers of atoms:

$$d = \frac{n\lambda}{2 \sin \theta}$$

Substitution of the wavelength of the x rays and the angle of diffraction into the equation gives the distance between layers of atoms.

$$d = \frac{(1)(154 \text{ pm})}{2 \sin 19.3°} = \frac{(154 \text{ pm})}{(2)(0.3305)} = 233 \text{ pm}$$

Understanding

Using the same x-ray source, a different crystal diffracts x rays at an angle of 12.3 degrees. Using $n = 1$, calculate the distance between the layers of atoms that produce this diffraction.

Answer 361 pm

Materials can get structurally complex, so the x-ray patterns their crystals cause do not look as simple as the one shown in Figure 11.19. However, the basic concept—the Bragg equation—is the same for even complex crystals.

Crystal Structure

Although a single crystal contains an extremely large number of atomic-sized particles, it is possible to describe their arrangement based on the positions of only a few of the particles within the solid, because the particles in a crystal are arranged in a regular, repeating fashion in three dimensional space. In a crystalline solid, the particles (atoms, ions, or molecules) are arranged in a regular geometric pattern so that the attractive forces are at a maximum. X-ray diffraction data are used to define the geometric pattern within the crystal. This pattern in three-dimensional space can be described by referring to a set of mathematical points, each of which has the same environment (i.e., the same angles and distances to its nearest neighbors). Each point is called a **lattice point,** and the geometric arrangement of the lattice points is called the *crystal lattice.* The crystal lattice, together with the arrangement of physical particles with respect to the lattice points, is called the **crystal structure.**

The **unit cell** is a small, regular geometric figure that defines the repeating pattern of lattice points. A single crystal can be viewed as a large number of unit cells packed together like boxes (Figure 11.20). Because of the restrictions of geometry, only a limited number of crystal lattices are possible. These lattices can be grouped into just seven systems, defined by the relative lengths of the edges of the unit cell (*a, b,* and *c*) and the angles between the edges (α, β, and γ), which are defined in Figure 11.21. We consider only the **cubic system** of lattices, in which all three edges are equal and the angles are

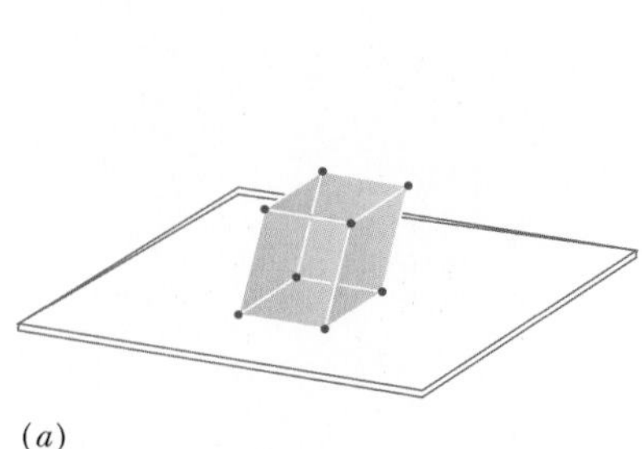
(a)

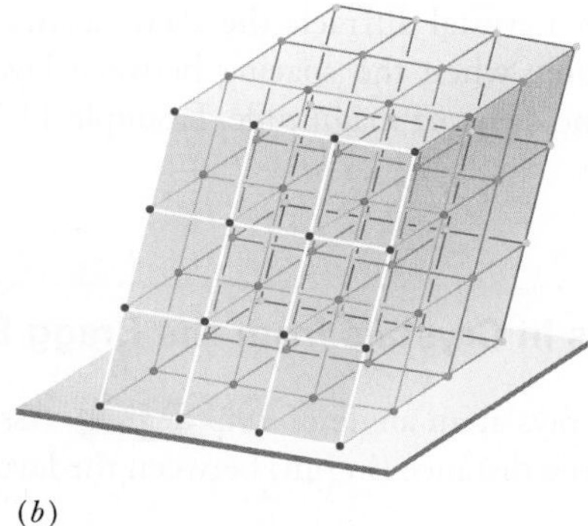
(b)

Figure 11.20 The unit cell. *Red dots* represent the lattice points, which are usually the centers of atoms or molecules. *(a)* The unit cell consists of a small number of lattice points needed to define the arrangement in the crystal. *(b)* The crystal lattice of the solid consists of unit cells packed together. In this structure, each lattice point is at a corner of eight different unit cells.

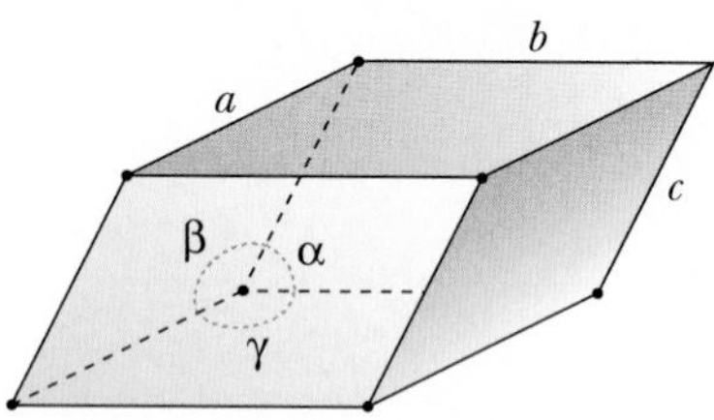

Figure 11.21 Defining the unit cell. A unit cell is defined by the lengths of the three edges—*a, b,* and *c*—and the three angles between the axes: α is the angle between *b* and *c*, β is the angle between *a* and *c*, and γ is the angle between *a* and *b*.

© Charles Falco/Photo Researchers, Inc.

Figure 11.22 Sodium chloride. Crystals of sodium chloride have a cubic shape. So does the unit cell of NaCl.

When the atoms or ions in a unit cell are counted, only a fraction of each corner, edge, or face atom is included.

all 90 degrees ($a = b = c$ and $\alpha = \beta = \gamma = 90$ degrees). Crystalline solids have planar faces and distinctive geometric shapes that are determined by the crystal systems to which they belong. For example, sodium chloride (Figure 11.22) forms crystals that are perfect cubes, and the unit cell for that compound belongs to the cubic system. In addition to the cubic shape, octahedral and other rectangular shapes are often observed in substances that crystallize with unit cells that belong to the cubic system.

There are three kinds of cubic unit cells, defined by the locations of the lattice points within them. We assume that a spherical atom occupies each lattice point in the unit cell; that is, an atom is centered at each of the eight corners of the cube. In the **simple** or **primitive cubic** unit cell, only these eight atoms are present. In the **body-centered cubic** (BCC) unit cell, an additional atom is at the center of the cube. In the **face-centered cubic** (FCC) unit cell, in addition to the eight corner atoms, an atom is at the center of each of the six faces of the cube. Figure 11.23 illustrates the three types of cubic unit cells.

Because the entire crystal consists of only repetitions of the unit cell, the intrinsic properties of a unit cell (such as elemental composition and density) must be consistent with the bulk properties of the substance. Many calculations of these properties require that we count the number of atoms or ions contained in the unit cell. In counting the number of atoms in a unit cell, we must recognize that the atoms on the corners and faces of the unit cell are shared by the adjacent cells. As shown in Figure 11.24*a*, each corner atom of a unit cell is actually part of eight cells in the lattice, so only one eighth of each corner atom is in each unit cell. The eight corner atoms in one cell therefore contribute a total of one atom $(8 \times \frac{1}{8})$ to that cell. Figure 11.24*c* shows that two unit cells share the atoms that are centered on the faces of a cubic unit cell. Therefore, one half of each face atom is contained in a single unit cell, for a total of three face atoms $(6 \times \frac{1}{2})$ in the face-centered cubic unit cell (plus one atom total from the eight corners).

Bulk properties of crystalline solids, such as the empirical formula and the density, can be found from the crystal structure and the dimensions of the unit cell. The experimentally measured densities of most metal samples differ slightly from those calculated from the unit cell because of imperfections and impurities in the crystals. Nevertheless, when the experimental and calculated densities are sufficiently close, the correct unit cell has probably been

Figure 11.23 Three cubic unit cells. The cubic unit cells are *(a)* simple cubic, *(b)* body-centered cubic (BCC), and *(c)* face-centered cubic (FCC).

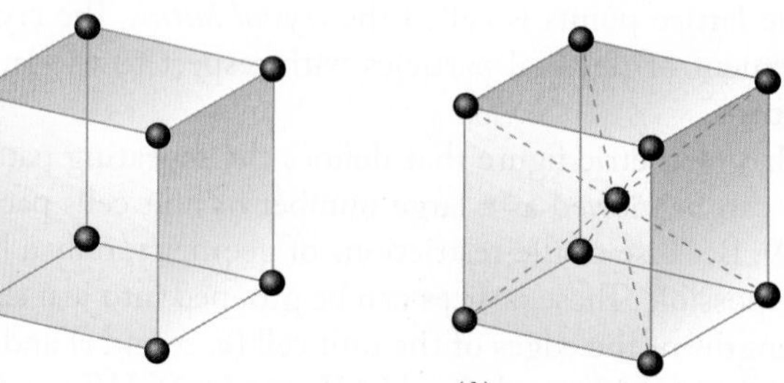
(a) (b)

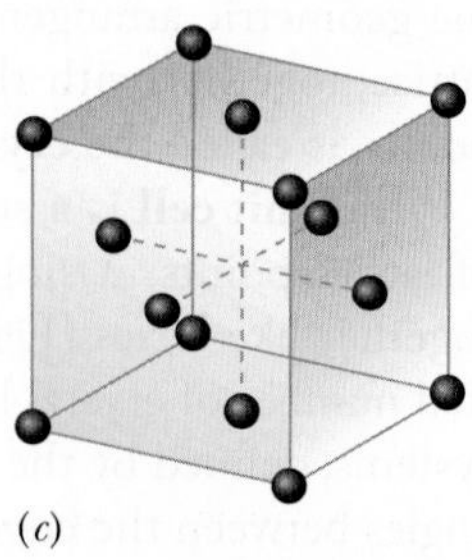
(c)

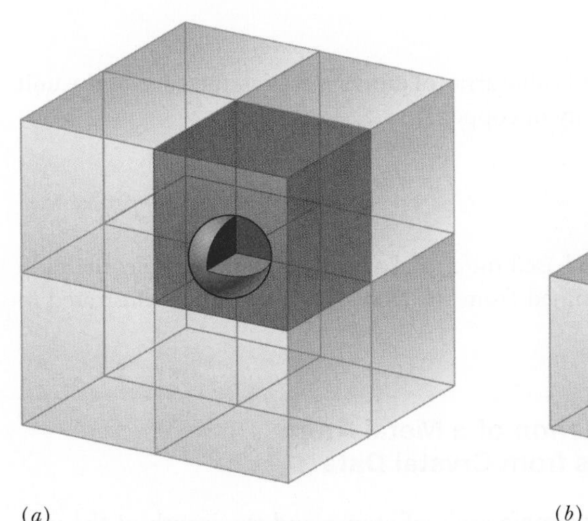

(*a*)

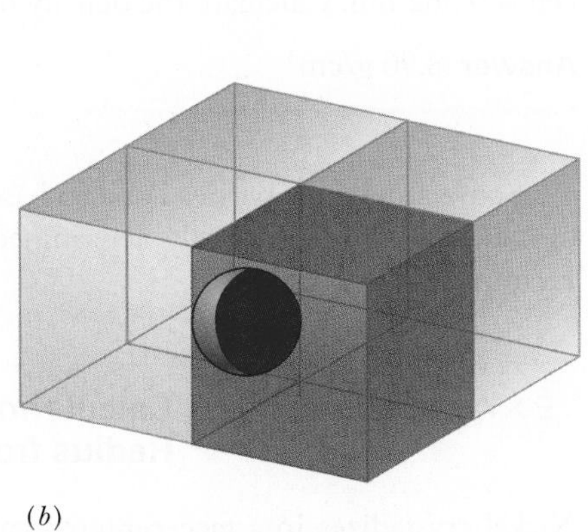

(*b*)

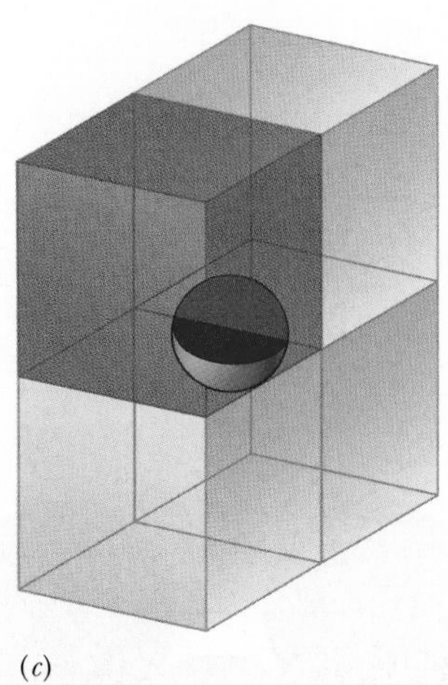

(*c*)

Figure 11.24 Sharing of atoms by adjacent unit cells. *(a)* Eight unit cells share each corner atom in any crystal lattice. *(b)* Two unit cells share a face atom. *(c)* Four unit cells share an atom located on an edge of the unit cell.

chosen. Example 11.8 illustrates the calculation of the density of a metal from its crystal structure and cell dimensions.

EXAMPLE 11.8 Calculation of Density from Crystal Data

Molybdenum crystallizes in a body-centered cubic array of atoms. The edge of the unit cell is 0.314 nm long. Calculate the density of this metal.

Strategy First, determine the net number of atoms per unit cell; then determine the mass using the atomic mass of molybdenum. Divide this mass by the volume of the cubic unit cell, knowing its edge length. Convert the units as necessary to obtain a proper unit for density.

Solution

In the body-centered cubic cell, there are atoms at the corners and in the center of the cube, so the number of atoms in the unit cell is

$$8 \text{ corners} \times \left(\frac{1 \text{ atom}}{8 \text{ corners}}\right) + 1 \text{ atom at center of cell} = 2 \text{ atoms}$$

Find the mass of the two atoms in the unit cell from the molar mass of molybdenum and Avogadro's number.

$$\text{Mass} = (2 \text{ atoms Mo}) \times \left(\frac{1 \text{ mol}}{6.022 \times 10^{23} \text{ atoms}}\right) \times \left(\frac{95.96 \text{ g}}{1 \text{ mol}}\right) = 3.187 \times 10^{-22} \text{ g Mo}$$

The volume of the cubic unit cell is the length of the edge raised to the third power. This volume should be expressed in cubic centimeters (cm^3), to obtain the units usually used for the density of solids.

$$\text{Volume} = \left(0.314 \text{ nm} \times \frac{100 \text{ cm}}{1 \times 10^9 \text{ nm}}\right)^3 = 3.10 \times 10^{-23} \text{ cm}^3$$

Finish calculating the density by combining the mass and volume.

$$\text{Density} = \frac{\text{mass}}{\text{volume}} = \frac{3.187 \times 10^{-22} \text{ g}}{3.10 \times 10^{-23} \text{ cm}^3} = 10.3 \frac{\text{g}}{\text{cm}^3}$$

The measured density of metallic molybdenum is 10.2 g/cm^3.

Understanding

Copper crystallizes in a face-centered cubic array of atoms in which the edge of the unit cell is 0.362 nm. Calculate the density of copper.

Answer 8.90 g/cm^3

Some atomic properties are also based on crystal structures. For example, the radii of metal atoms are generally determined from their crystal structures, as illustrated in Example 11.9.

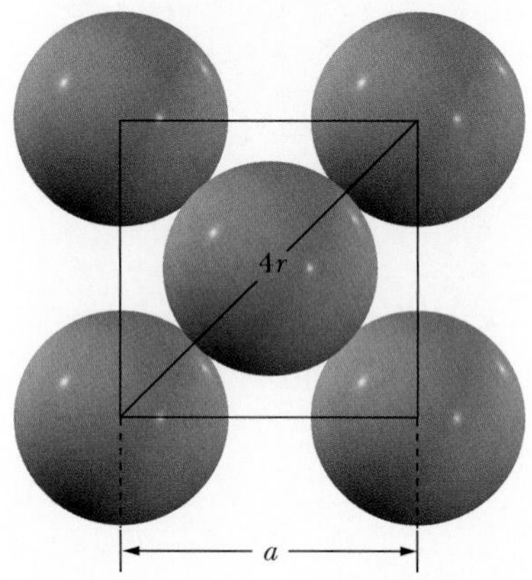

EXAMPLE 11.9 Calculation of a Metal Atom Radius from Crystal Data

Nickel crystallizes in a face-centered cubic array of atoms, and the length of the unit cell edge is 351 pm. What is the radius of the nickel atom?

Strategy On one face of this cell there are five nickel atoms—one at each corner and one in the center of the face. Since we assume that each atom is a sphere, the three atoms along the face diagonal must be in contact with each other, so the length of the face diagonal is four times the radius of a nickel atom:

$$\text{Length of diagonal} = 4 \times r$$

Because the diagonal of the face of the unit cell is the hypotenuse of a right triangle in which each of the legs is equal to the length of the edge of the unit cell *a,* we can use the Pythagorean theorem (the sum of the squares of the legs of a right triangle is equal to the square of the hypotenuse) to find the length of the diagonal in terms of the edge length.

Solution

In terms of the length of the edge of the unit cell, the diagonal of the face of the unit cell has a length of

$$\text{Length of diagonal} = \sqrt{a^2 + a^2} = a\sqrt{2}$$

Equating this expression to four atomic radii gives

$$4 \times r = a\sqrt{2}$$

We substitute the length of the cell edge and solve the equation for the radius:

$$r = \frac{(351\text{ pm})\sqrt{2}}{4} = 124\text{ pm}$$

Understanding

Use the data in Example 11.8 to calculate the atomic radius (in pm) of the molybdenum atom. In a body-centered cubic structure, the atoms along the cube diagonal are in contact, and the length of the cube diagonal is $a\sqrt{3}$.

Answer $r_{\text{Mo}} = 136$ pm

Close-Packing Structures

In Example 11.9, the atoms were viewed as spheres that were in contact with each other. In many metals and other solid monatomic elements, the atoms pack together in the most efficient manner, called **close packing,** in which empty space is a minimum. A familiar example of close packing is the stacking of oranges in a supermarket display. Figure 11.25*a* shows three layers in one close-packing arrangement. The atoms in the

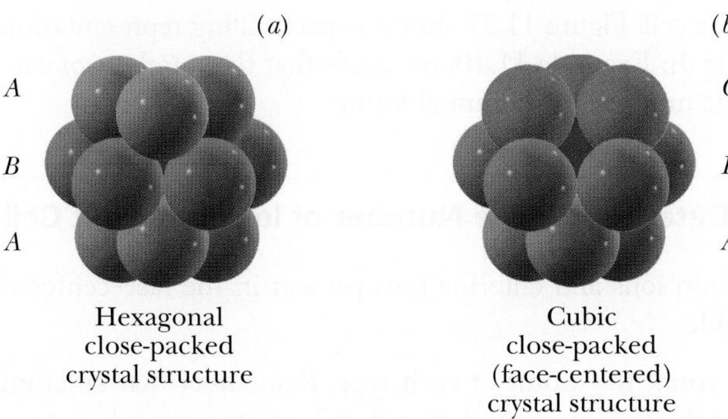

Figure 11.25 Close-packing arrays. *(a)* In hexagonal close packing, the spheres of the third layer are directly above those in the first layer. *(b)* In cubic close packing, the spheres of the third layer are not directly over the spheres in the first layer, creating a different type of close packing from hexagonal close packing.

third layer are directly above the atoms in the first layer. This kind of arrangement is called *ABA stacking,* and it produces a type of close packing called **hexagonal close packing** (hcp).

The third layer can occupy different positions, however. Figure 11.25*b* shows how the third layer can sit on top of the second layer in such as way that the atoms do not sit directly above the atoms in the first layer. This arrangement of layers is called *ABC stacking,* and it produces a type of close packing called **cubic close packing.** It can be shown that cubic close packing is the same as a face centered cubic cell. In both of the close-packing arrays, each sphere is in contact with 12 other spheres: 6 in the same layer and 3 in each of the layers above and below. The number of nearest neighbors is called the **coordination number.** The coordination number of 12 in the close packing arrays is the largest possible and results in the smallest volume of free space between spheres. The spheres occupy 74% of the total volume of the structure in both of the close-packing arrays.

The coordination number is the number of neighbors in contact with a single atom.

Ionic Crystal Structures

In ionic crystals, there are two kinds of particles: cations and anions. The unit cell is usually described in terms of the arrangement of only the cations or only the anions, with the oppositely charged ions occupying specific locations within the cell. The relative charges and sizes of the anions and cations determine the type of crystal lattice that forms. In an ionic crystal, the ratio of the number of cations to the number of anions in the unit cell must be consistent with the formula of the compound. Figure 11.26 shows a portion of a sodium chloride crystal; it identifies the unit cell as a face-centered cubic array of chloride ions, with a sodium ion at the center of each of the 12 edges of the

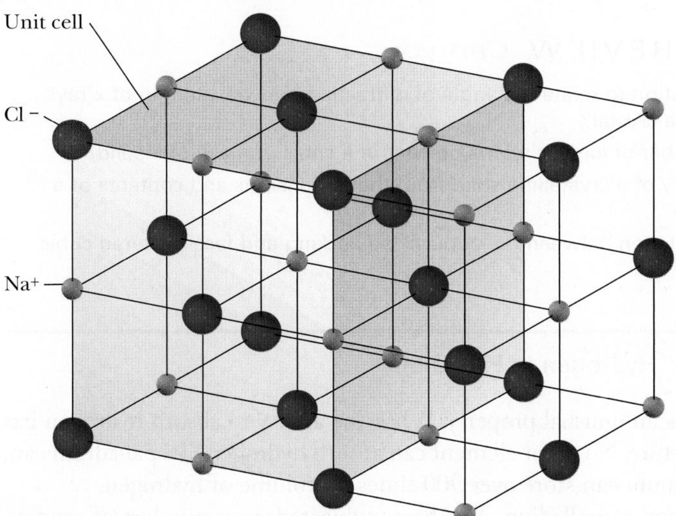

Figure 11.26 Crystal structure of NaCl. A portion of the sodium chloride crystal. The face-centered cubic unit cell is shaded and has chloride ions at the corners and in the faces. The sodium ions occupy the center and the 12 edges of the cell.

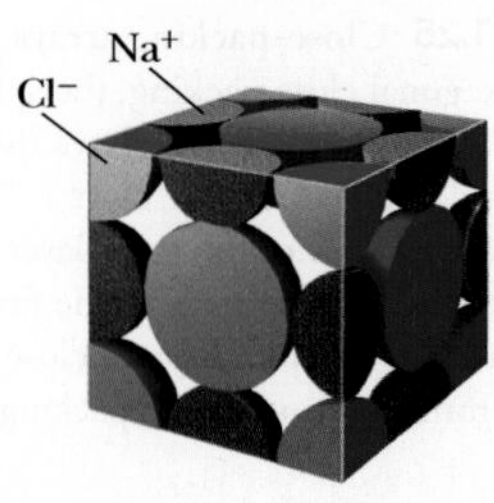

Figure 11.27 A unit cell of NaCl. A space-filling model of a single unit cell of sodium chloride. Only the portions of the ions that occupy the unit cell are shown.

cube and one in the center of the cell. Figure 11.27 shows a space-filling representation of a unit cell of sodium chloride. In Example 11.10, we verify that the numbers of cations and anions are equal in one unit cell of sodium chloride.

EXAMPLE 11.10 Determining the Number of Ions in a Unit Cell

Determine the number of sodium ions and chloride ions present in the face-centered cubic unit cell of sodium chloride.

Strategy Use Figure 11.27 to count the atoms of each type. Remember that an atom in a corner contributes one eighth of an atom to that unit cell, an atom on a face contributes only half, and an atom on an edge contributes one fourth. Recall also that there is a sodium ion in the center of the unit cell (not shown in Figure 11.27).

Solution

Only one eighth of the chloride ion on each corner is in this unit cell, and two different unit cells share each of the six ions in the centers of faces.

$$8\ \cancel{\text{corners}} \times \left(\frac{1\ \text{Cl}^-}{8\ \cancel{\text{corner}}}\right) = 1\ \text{Cl}^-$$

$$6\ \cancel{\text{faces}} \times \left(\frac{1\ \text{Cl}^-}{2\ \cancel{\text{face}}}\right) = 3\ \text{Cl}^-$$

There are $1 + 3 = 4\ Cl^-$ ions in each unit cell. One fourth of each sodium ion on each edge is present in this unit cell. In addition, there is one Na^+ ion at the center of the unit cell that is not shared with any other cell.

$$1\ \text{Na}^+ + 12\ \cancel{\text{edges}} \times \left(\frac{1\ \text{Na}^+}{4\ \cancel{\text{edge}}}\right) = 4\ \text{Na}^+$$

The unit cell of the sodium chloride contains four Na^+ ions and four Cl^- ions, giving the one-to-one ratio of cations to anions that the formula of the compound demands.

Understanding

Cesium chloride crystallizes in a simple cubic array of Cs^+ ions, with one Cl^- at the center of the unit cell. How many ions of each type are present in the unit cell?

Answer 1 Cs^+ and 1 Cl^-

OBJECTIVES REVIEW *Can you:*

- ☑ use the Bragg equation to relate the angle of diffraction and wavelength of x rays to the distances in a crystal?
- ☑ determine the number of ions or atoms present in a cubic unit cell of a solid?
- ☑ calculate the density of a crystalline solid from the dimensions and contents of a unit cell?
- ☑ demonstrate the relation between the cubic close packing and face-centered cubic crystal structures?

CASE STUDY Hydrogen in Palladium

The metal palladium has an unusual property. It has the ability to absorb hydrogen gas into its own crystal structure. No other element can absorb hydrogen like palladium can; a given volume of palladium can store over 900 times its volume of hydrogen.

This unusual behavior of palladium has been subjected to a number of studies. Experiments show that the interactions between Pd and H_2 are so strong that the H–H bond is broken, leaving individual hydrogen atoms to interact with palladium atoms. Crystalline palladium exists as a face-centered cubic lattice, and studies show that the tiny hydrogen atoms occupy the spaces within the Pd lattice; the hydrogen-containing solid still has a face-centered cubic crystal structure. Up to 70% of the spaces in the palladium lattice can be filled up with hydrogen atoms.

The absorption of hydrogen is a reversible process, as well. By either heating the palladium or reducing the surrounding pressure, hydrogen gas can be regenerated. Among other things, this allows us to produce ultrapure hydrogen. A thin palladium membrane can be placed over an opening of a container of gas. Only the hydrogen gas will diffuse through; no other gas penetrates the palladium metal like hydrogen does.

One of the possible applications of this characteristic of palladium is in fuel cells. A fuel cell is a type of battery that extracts electricity from the direct combination of a fuel and an oxidizer. A common fuel/oxidizer combination is hydrogen and oxygen, respectively. Fuel cells of this sort have been used successfully in spacecraft, in which the product of the fuel cell reaction—water—is an added bonus. However, both hydrogen and oxygen normally exist as gases, meaning that either highly compressed gas must be used to obtain enough reactant, or the gases must be cooled to very low temperatures to have a more space-efficient condensed phase. Neither option is desirable.

Hydrogen-rich palladium may be part of a solution to this dilemma. Every cubic centimeter of Pd metal can absorb 0.04 moles of H_2. Only 300 g of Pd are needed to store one mole of hydrogen; this palladium would have a volume of only 24.9 cm^3, or about one-tenth of a cup. The rapidity with which H_2 absorbs into Pd is related to the surface area of the palladium metal. Currently, scientists are working on palladium-coated ceramic beads and nanoparticles of palladium as a way to quickly absorb and desorb hydrogen.

The drawbacks, however, are also substantial. Palladium is one of the denser metals. Any useful volume of palladium that might be needed for practical purposes would be rather heavy, and fuel costs are directly related to mass. Also, palladium is one of the more expensive metals: Currently, palladium is priced at just over U.S. $180 per ounce, meaning that it would cost about $2000 to purchase enough palladium to store one mole of H_2.

Despite these drawbacks, research is ongoing at using palladium in fuel cells. In a few years, industrial demand for palladium may increase as researchers look for alternate energy sources to replace fossil fuels.

Questions

1. Palladium has a face-centered cubic unit cell that has a length of 388.9 pm. Calculate the density of palladium.
2. Based on the information in the case study, calculate the mass percentage of hydrogen in a saturated sample of H_2 in Pd.
3. Using data from the case study and your answer to Question #1 above, calculate the surface area of (a) a single sphere of 1 g of Pd metal, and (b) spheres of 1 g of Pd metal if the spheres had a radius of 5.00 nm. Recall that for a sphere, area $= 4\pi r^2$ and volume $= (4/3)\pi r^3$.

ETHICS IN CHEMISTRY

1. An advertisement for oxygenated bottled water claims that the water is "more healthy because the dissolved oxygen reduces the interactions between water molecules, allowing them to be more easily absorbed by the body!" Based on what you know about intermolecular interactions, what would be your response to such advertising claims?

2. Freeze-drying of coffee is one way to make coffee more convenient. A sample of strong coffee is brewed, then frozen. Then the frozen coffee is subjected to a vacuum, which removes the solid water, as well as other volatile substances, through sublimation. The remaining solid is more compact than brewed coffee and is easily reconstituted by adding water. Freeze-dried coffee is more expensive than regular coffee made from grounds, and it is an energy-intensive process that most coffee drinkers agree produces an inferior product. Argue whether or not the financial and energy costs are worth the convenience of freeze-dried coffee.

3. The sale of diamonds, the subject of the chapter introduction, is largely controlled by one company, De Beers. The company keeps the supply of diamonds low so that the prices are artificially high. What are the ethics involved in having a global monopoly on a chemical substance?

4. *Conflict diamonds* (or *blood diamonds*) are diamonds mined in a war zone and sold to provide funds to support one side of the war. Would you buy a low-cost diamond, knowing it was a conflict diamond?

Chapter 11 Visual Summary

The chart shows the connections between the major topics discussed in this chapter.

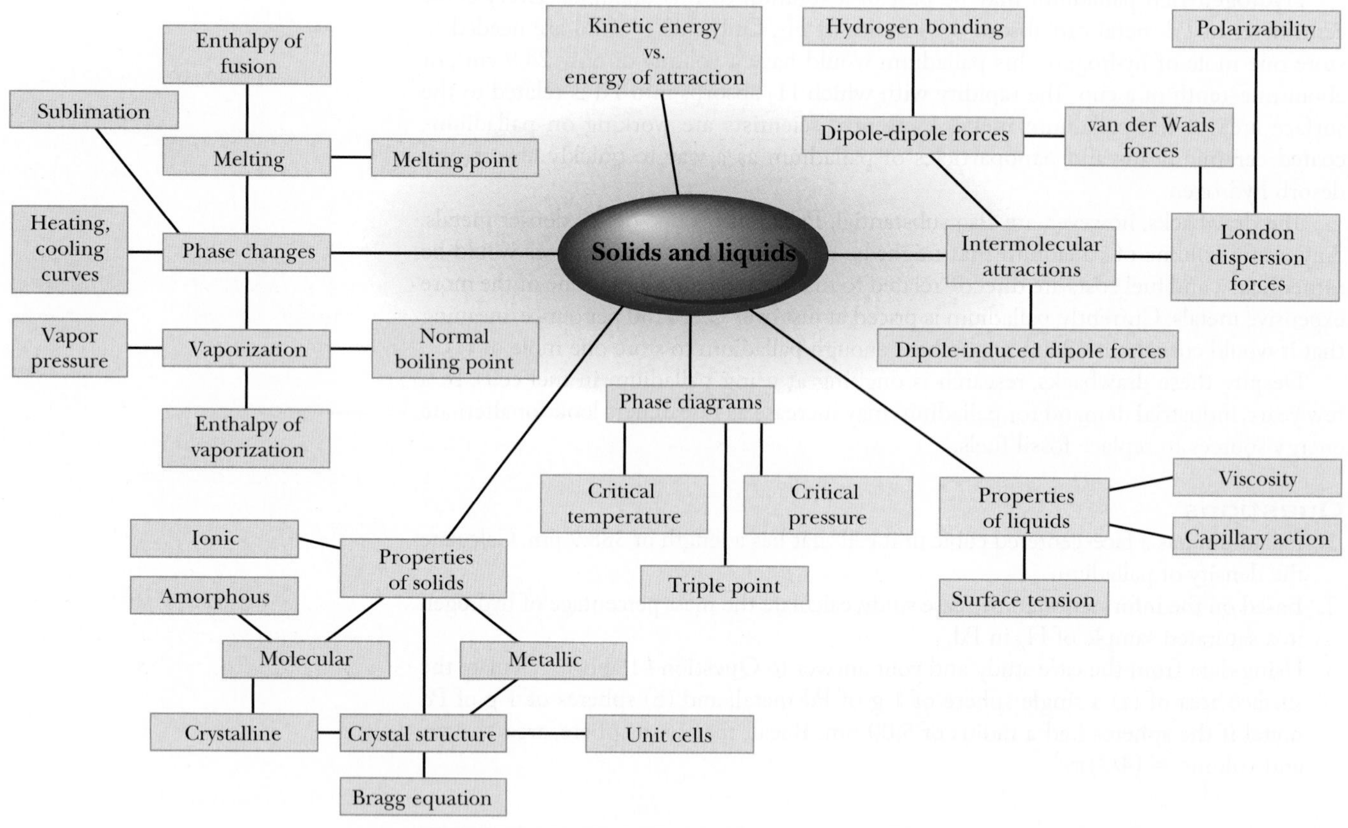

Summary

11.1 Kinetic Molecular Theory and Intermolecular Forces

Solids and liquids are condensed phases. In the solid state, the units (atoms, molecules, ions) that make up the substance are held rigidly in place and have long-range order. In the liquid state, the units are close together but have sufficient kinetic energy to move relative to each other. The strengths of *intermolecular attractions* influence and determine many of the physical properties of substances.

11.2 Phase Changes

Substances can change from one phase to another as their temperatures and pressures change. Many of the properties associated with these phase changes reflect the strength of intermolecular attractions. Phase changes are all *dynamic equilibrium* processes that occur at constant temperature and pressure. *Melting,* or fusion (solid to liquid), *evaporation,* or *vaporization* (liquid to gas), and *sublimation* (solid to gas) are all endothermic processes because energy is used to overcome the intermolecular forces of attraction. From Hess's law, the reverse processes—*freezing, condensation,* and *deposition,* respectively—are all exothermic changes. *Vapor pressure* is the partial pressure of a gas in dynamic equilibrium with its liquid or solid at a particular temperature. The vapor pressure of any substance increases with increasing temperature. The heating curve of a substance, a plot of temperature versus added heat, can be constructed from the heat capacities of each of the phases and the *enthalpies of fusion* and *vaporization.*

11.3 Phase Diagrams

A *phase diagram* is a graph of pressure versus temperature that shows the combinations of pressure and temperature at which the solid, liquid, and gas phases are in equilibrium. The equilibrium lines separate the regions in which only one of the physical states is stable. At the *triple point,* all three phases are in equilibrium with each other. When the pressure on an equilibrium mixture of two phases is increased, the system responds by forming the denser phase, explaining why the melting point of water decreases when the pressure increases. For most substances, however, the melting point increases as the pressure increases, because the solid is denser than the liquid. In sublimation and vaporization, the effect of pressure on the temperature of the equilibrium phase change is greater than for the solid-liquid equilibrium because of the much larger difference in the volumes of the phases in equilibrium.

11.4 Intermolecular Attractions

In molecular substances there are three main kinds of intermolecular attractions: *dipole-dipole attractions, London dispersion forces,* and *hydrogen bonding.* Dipole-dipole attractions and London dispersion forces are collectively called *van der Waals forces.* London dispersion forces contribute to the attraction of all atoms, molecules, and ions, and depend on the *polarizability* of the species. Hydrogen bonding is a particularly strong form of intermolecular attraction; it occurs only between molecules that contain hydrogen covalently bonded to a highly electronegative element, such as N, O, or F.

11.5 Properties of Liquids and Intermolecular Attractions

Variations in the strengths of intermolecular forces are related to the properties of molecular substances. Among the properties that may be used to indicate the relative strengths of intermolecular attractions are boiling points, melting points, enthalpies of fusion and vaporization, and the *surface tension, viscosity,* and vapor pressure of liquids. The appearance of the meniscus of a liquid in a glass tube is determined by whether the *cohesive forces* are stronger or weaker than the *adhesive forces.*

11.6 Properties of Solids and Intermolecular Attractions

The type of attractions between the particles in a crystal is one way to categorize solids. In a *molecular solid,* the attractive forces between molecules are van der Waals forces and hydrogen bonds. Molecular solids usually have low melting points. A *covalent network solid* does not contain small, discrete molecules but has a framework of covalent bonds throughout the crystal. Such a substance has a very high melting point and often is very hard. An *ionic solid* is held together in the crystal by the strong attractions that exist between the oppositely charged ions. A *metallic solid* can be viewed as metal ions held together by a "sea" of mobile electrons that extends through the entire crystal.

11.7 Structures of Crystalline Solids

The structural features of a crystalline solid are found experimentally using x-ray diffraction. The *Bragg equation* is used to find the distance of separation between layers of atoms in a crystal. The arrangement of *lattice points* in a crystalline solid can be defined by a *unit cell,* the smallest part of the crystal that can reproduce the three-dimensional arrangement of the particles by simple translations. There are three kinds of unit cells in the *cubic system*: *simple cubic, face-centered cubic,* and *body-centered cubic.* Many solids, particularly metals, are represented by *close-packing arrays,* in which each particle has a *coordination number* of 12. The composition and density of a unit cell must be the same as bulk properties of the substance. The crystal structure and unit cell dimensions are used to find the radii of atoms.

Chapter Terms

The following terms are defined in the Glossary, Appendix I.

Section 11.1
Intermolecular forces

Section 11.2
Boiling point
Condensation
Critical pressure
Critical temperature
Deposition
Dynamic equilibrium
Enthalpy of fusion
Enthalpy of sublimation
Enthalpy of vaporization
Evaporation, or vaporization
Melting point
Normal boiling point
Sublimation
Supercooling
Supercritical fluid
Vapor pressure

Section 11.3
Phase diagram
Triple point
Section 11.4
Dipole-dipole attraction
Dipole–induced dipole attraction
Hydrogen bonding
Induced dipole
Instantaneous dipole
London dispersion forces
Polarizability
van der Waals forces
Section 11.5
Adhesive forces
Capillary action
Cohesive forces
Surface tension
Viscosity
Section 11.6
Allotropes
Amorphous solid
Covalent network solid
Crystalline solid
Ionic solid
Metallic solid
Molecular solid
Section 11.7
Body-centered cubic (BCC)
Bragg equation
Close packing
Coordination number
Crystal structure
Cubic close packing
Cubic system
Face-centered cubic (FCC)
Hexagonal close packing
Lattice point
Simple (primitive) cubic
Unit cell

Key Equation

Bragg equation (11.7)

$$n\lambda = 2d \sin \theta$$

Questions and Exercises

OWL Selected end of chapter Questions and Exercises may be assigned in OWL.

Blue-numbered Questions and Exercises are answered in Appendix J; questions are qualitative, are often conceptual, and include problem-solving skills.

■ Questions assignable in OWL

✎ Questions suitable for brief writing exercises

▲ More challenging questions

Questions

11.1 The density of liquid sulfur dioxide is 1.43 g/mL, and that of the gas is 0.00293 g/mL at standard temperature and pressure. Account for the large difference between these two densities.

11.2 For any particular substance, the types of intermolecular forces in the solid, liquid, and gas phases are the same. What determines which of these phases is stable at a given temperature?

11.3 For most substances, the liquid state is less dense than the solid state. How does density affect the energy of the intermolecular attractions?

11.4 Why does water have a lower vapor pressure at 25 °C than dimethyl ether (CH_3OCH_3)? Which of these two liquids has the greater enthalpy of vaporization?

11.5 ✎ "A liquid stops evaporating when the equilibrium vapor pressure is reached." What is wrong with this statement?

11.6 Why does a perspiring body achieve greater cooling when the wind is blowing than in calm air?

11.7 ✎ How does humidity affect the efficiency of cooling by perspiration? Explain.

11.8 ✎ The enthalpy of vaporization of water (boiling point = 100 °C) is greater than that of diethyl ether (boiling point = 35 °C). Despite its smaller enthalpy of vaporization, liquid diethyl ether feels colder than water does when it evaporates from a person's skin. Explain.

11.9 Trouton's rule states that the enthalpy of vaporization for a substance divided by its normal boiling point on the Kelvin scale is approximately 88 J/K. Is this in agreement with the expected behavior of these properties as intermolecular forces change? Explain.

11.10 Each of two glasses contains 200 g of an ice-water mixture. In one glass, 10 g is ice, and in the other glass, 90 g is ice. Assuming both samples are at equilibrium, which is colder?

11.11 For a typical substance, arrange the enthalpies of sublimation, fusion, and vaporization in order of increasing value.

11.12 Even when the atmospheric temperature is above normal body temperature, sweating is effective in cooling a person. Explain how this happens, since heat flows from a warmer object to a cooler one.

11.13 There are two allotropic forms of the element tin, called *gray tin* and *white tin.* The density of gray tin is 5.75 g/cm^3, and that of white tin is 7.28 g/cm^3. Determine which allotropic form of tin is favored by high pressure.

11.14 Explain why high pressures are needed in the industrial process that makes diamond (d = 3.5 g/cm^3) from graphite (d = 2.2 g/cm^3).

11.15 Explain why the effect of hydrogen bonding on the boiling point is considerably greater for water than for ammonia and hydrogen fluoride. (*Hint:* The formation of a hydrogen bond requires both a partial positively charged hydrogen atom and unshared pairs of electrons on the electronegative atom.)

11.16 From the graph in Figure 11.11, is there any evidence that hydrogen bonding occurs for any elements other than nitrogen, oxygen, and fluorine? Explain.

11.17 Explain why the viscosity of a liquid may not be consistent with the enthalpy of vaporization and the boiling point as a measure of the strength of intermolecular forces.

11.18 ✎ Water forms beads on the surface of a newly painted car. After the car is exposed to the weather for a long time, water spreads out into a thin film on its surface. What has happened to the paint to cause this change?

11.19 State how each of the following properties changes with increasing strength of intermolecular forces.
(a) enthalpy of fusion (b) melting point
(c) surface tension (d) viscosity
(e) enthalpy of vaporization (f) boiling point

11.20 Explain why the surface of the HCl(aq) liquid in a buret is curved rather than flat.

11.21 The compounds ethanol (C_2H_5OH) and dimethyl ether (CH_3–O–CH_3) have the same molecular formula. Which is expected to have the higher surface tension? Why?

11.22 How does an amorphous solid differ from a crystalline solid?

11.23 Sometimes amorphous solids are referred to as supercooled liquids. In what ways are amorphous solids similar to liquids?

11.24 An amorphous solid can sometimes be converted to a crystalline solid by a process called *annealing*. Annealing consists of heating the substance to a temperature just below the melting point of the crystalline form and then cooling it slowly. Explain why this process helps produce a crystalline solid.

11.25 ▲ Derive the relation between the radius of the atoms and the length of the unit-cell edge for the simple cubic, body-centered cubic, and face-centered cubic cells.

11.26 ✎ Sketch what you think an edge-centered cubic unit cell looks like. How many atoms are there per unit cell for this type of crystal?

Exercises

OBJECTIVE Relate the physical state of a substance to the strength of the intermolecular forces.

11.27 At −30 °C, hydrogen sulfide (H_2S) is a gas, hydrogen telluride (H_2Te) is a liquid, and water (H_2O) is a solid. Arrange these compounds in order of increasing strength of intermolecular attractions.

11.28 ■ Rank the following in order of increasing strength of intermolecular forces in the pure substances. Which are gases at 25 °C and 1 atm?
(a) $CH_3CH_2CH_2CH_3$ (butane)
(b) CH_3OH (methanol)
(c) He

OBJECTIVE Describe phase changes.

11.29 Explain how the vapor pressure of a liquid changes with each of the following changes.
(a) The surface area of the liquid is increased from 1 to 10 cm^2 when the liquid is poured into a container with a larger diameter.
(b) Enough heat is added to the sample to increase the temperature by 5 °C.
(c) The volume of a closed container with some liquid in equilibrium with its vapor is decreased at constant temperature.

11.30 Explain how the vapor pressure of a liquid changes with each of the following changes.
(a) The surface area of the liquid decreases from 50 to 10 cm^2 when the liquid is poured into a container with a smaller diameter.
(b) The sample is cooled from 35 °C to 25 °C.
(c) The volume of a closed container with some liquid in equilibrium with its vapor is increased at constant temperature.

11.31 Isopropanol (rubbing alcohol) and methyl ethyl ether both have the same molecular formula, C_3H_8O. At 0 °C, the vapor pressure of isopropanol is 8.4 torr, and that of methyl ethyl ether is 560 torr.
(a) Which of these compounds has the stronger intermolecular attractions?
(b) Based on the answer to (a), give the relative magnitudes of three other properties of these two compounds.

11.32 Diethyl ether, a substance once used by physicians as a general anesthetic, and *n*-butanol both have the molecular formula $C_4H_{10}O$. At 25 °C, the vapor pressure of diethyl ether is 530 torr, and that of *n*-butanol is 7.1 torr.
(a) Which of these compounds has the stronger intermolecular attractions?
(b) Based on the answer to (a), give the relative magnitudes of three other properties of these two compounds.

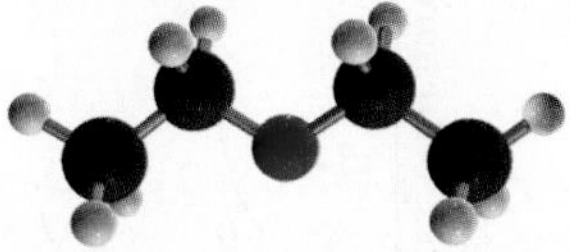

11.33 The critical temperature of butane, the fuel in cigarette lighters, is 152 °C, and that of benzene, a useful organic solvent, is 289 °C.
(a) Which of these compounds has the stronger intermolecular attractions?
(b) Based on the answer to (a), give the relative magnitudes of three other properties of these two compounds.

11.34 The critical temperature of nitrogen dioxide (NO_2) is 158 °C, and that of dinitrogen monoxide (N_2O) is 37 °C.
(a) Which of these compounds has the stronger intermolecular attractions?
(b) Based on the answer to (a), give the relative magnitudes of three other properties of these two compounds.

11.35 The enthalpy of sublimation of iodine is 60.2 kJ/mol, and its enthalpy of vaporization is 45.5 kJ/mol. What is the enthalpy of fusion of iodine?

11.36 Dichlorobenzene (mothballs) has an enthalpy of fusion of 17.2 kJ/mol and an enthalpy of sublimation of 68.3 kJ/mol. What is the enthalpy of vaporization for this compound?

11.37 How much energy is needed to melt 100.0 g $H_2O(s)$ at its melting point if its enthalpy of fusion is 6.01 kJ/mol? Is this process exothermic or endothermic?

11.38 ■ Calculate the amount of heat required to convert 70.0 g of ice at 0 °C to liquid water at 100 °C. The specific heat of liquid water is 4.184 J/mol·K.

11.39 What is the enthalpy change when a 1.00-kg block of dry ice, $CO_2(s)$, sublimes at −78 °C? The enthalpy of sublimation of $CO_2(s)$ is 26.9 kJ/mol. Is this process exothermic or endothermic?

11.40 What is the enthalpy change when 50.8 g butane condenses at its boiling point of −0.6 °C? The enthalpy of vaporization of C_4H_{10} is 22.3 kJ/mol. Is this process exothermic or endothermic?

OBJECTIVE Construct phase diagrams.

11.41 The normal melting point of dinitrogen tetroxide is −9.3 °C, and its normal boiling point is 21.0 °C. The triple point occurs at 140 torr and −10.9 °C, and the critical point is 158 °C at 100 atm. Sketch the phase diagram for dinitrogen tetroxide.

11.42 The normal melting point of iodine is 113.5 °C, and its normal boiling point is 184.3 °C. The triple point occurs at 92.3 torr and 113.4 °C, and the critical point is 512 °C at 112 atm. Sketch the phase diagram for iodine.

OBJECTIVE Use phase diagrams to identify various phases.

11.43 Use the accompanying phase diagram to do the following:
(a) Label each region of the diagram with the phase that is present.
(b) Identify the phase or phases present at each of the points A, B, C, D, and E.

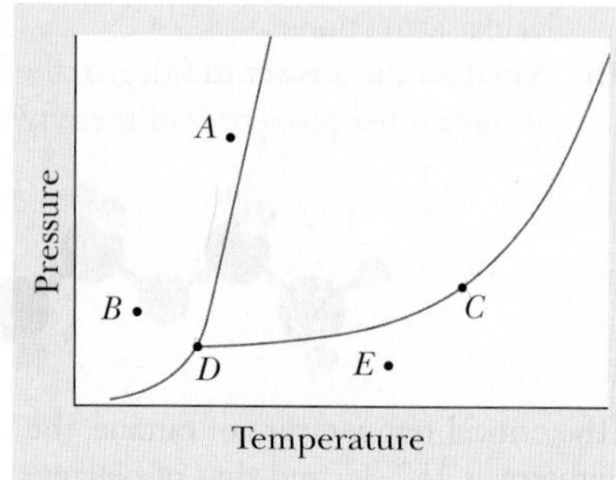

11.44 ■ Use the accompanying phase diagram to do the following:
(a) Label each region of the diagram with the phase that is present.
(b) Identify the phase or phases present at each of the points G, H, J, and K.

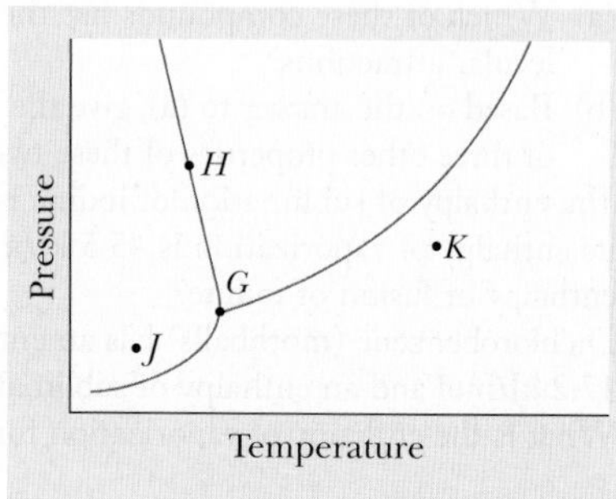

OBJECTIVE Relate phase diagram to heating or cooling curve.

11.45 Answer the following questions by using the phase diagram in Exercise 11.43.
(a) Sketch the heating curve that is expected when heat is added to the sample at constant pressure, starting at point B.
(b) Describe what happens if the pressure is lowered at constant temperature, starting at point A.
(c) What does the positive slope of the solid-liquid equilibrium line tell you about this substance?

11.46 Answer the following questions by using the phase diagram in Exercise 11.44.
(a) Sketch the heating curve that is expected when heat is added to the sample at constant pressure, starting at point J.
(b) Describe what happens if the pressure is increased at constant temperature, starting at point K.
(c) What does the negative slope of the solid-liquid equilibrium line tell you about this substance?

11.47 Describe how the equilibrium system responds to the change described in each of the following parts.
(a) Heat is added to a sample of solid in equilibrium with its vapor at constant pressure.
(b) The pressure on an equilibrium mixture of water and ice is increased abruptly.
(c) The pressure on an equilibrium mixture of a liquid and its vapor is increased at constant temperature.

11.48 Describe how the equilibrium system responds to the change described in each of the following parts.
(a) The pressure on a solid in equilibrium with its vapor is increased abruptly at constant temperature.
(b) The volume of the container holding a liquid in equilibrium with its vapor is increased at constant temperature.
(c) Heat is removed from a liquid-vapor equilibrium mixture.

OBJECTIVE List intermolecular forces.

11.49 Identify the kinds of intermolecular forces (London dispersion, dipole-dipole, hydrogen bonding) that are important in each of the following substances.
(a) propane (C_3H_8)
(b) ethylene glycol [$HO(CH_2)_2OH$]
(c) cyclohexane (C_6H_{12})
(d) phosphine oxide (PH_3O)
(e) nitrogen monoxide (NO)
(f) hydroxylamine (NH_2OH)

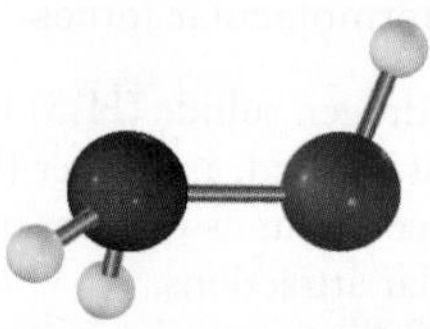

11.50 ■ Identify the kinds of intermolecular forces (London dispersion, dipole-dipole, hydrogen bonding) that are the most important in each of the following substances.
(a) methane (CH_4)
(b) methanol (CH_3OH)
(c) chloroform ($CHCl_3$)
(d) benzene (C_6H_6)
(e) ammonia (NH_3)
(f) sulfur dioxide (SO_2)

OBJECTIVE Compare relative strengths of intermolecular forces.

11.51 In each part, select the substance that has the greater boiling point, based on the relative strengths of the intermolecular attractions.
(a) C_2H_4 or CH_4 (b) Cl_2 or ClF
(c) S_2F_2 or S_2Cl_2 (d) NH_3 or PH_3
(e) CH_3I or CHI_3 (f) BBr_2I or $BBrI_2$

11.52 In each part, select the substance that has the greater boiling point, based on the relative strengths of the intermolecular attractions.
(a) C_3H_8 or CH_4 (b) I_2 or ICl
(c) H_2S or H_2Te (d) H_2Se or H_2O
(e) CH_2Cl_2 or CH_3Cl (f) NOF or NOCl

11.53 Identify the types of all the intermolecular forces that cause each of the following gases to condense to a liquid when cooled.
(a) N_2O (b) CH_4 (c) NH_3 (d) SO_2

11.54 Identify the types of all the intermolecular forces that cause each of the following gases to condense to a liquid when cooled.
(a) F_2 (b) BF_3 (c) HF (d) NO_2

OBJECTIVE Relate properties of liquids to intermolecular forces.

11.55 Select the liquid in each part that has the greater expected enthalpy of vaporization.
(a) C_3H_8 or CH_4 (b) I_2 or ICl
(c) S_2Cl_2 or S_2F_2 (d) H_2Se or H_2O
(e) CH_2Cl_2 or CH_3Cl (f) NOF or NOCl

11.56 Select the liquid in each part that has the greater expected enthalpy of vaporization.
(a) C_2H_4 or CH_4 (b) Cl_2 or ClF
(c) H_2S or H_2Te (d) NH_3 or PH_3
(e) CHI_3 or CH_3I (f) BBr_2I or $BBrI_2$

11.57 Arrange the liquids hexane (C_6H_{14}), ethanol (C_2H_5OH), and ethylene glycol [$(CH_2OH)_2$] in order of increasing expected surface tensions.

11.58 Arrange the liquids ethanol (C_2H_5OH), glycerol [$HOCH_2CH(OH)CH_2OH$], and ethylene glycol $(CH_2OH)_2$ in decreasing order of expected viscosities.

OBJECTIVE Identify intermolecular forces in solids.

11.59 Identify the kinds of forces that are most important in holding the particles together in a crystalline solid sample of each of the following substances.
(a) H_2O (b) C_6H_6 (c) $CaCl_2$
(d) SiO_2 (e) Fe

11.60 Identify the kinds of forces that are most important in holding the particles together in a crystalline solid sample of each of the following substances.
(a) Kr (b) HF (c) K_2O
(d) CO_2 (e) Zn (f) NH_3

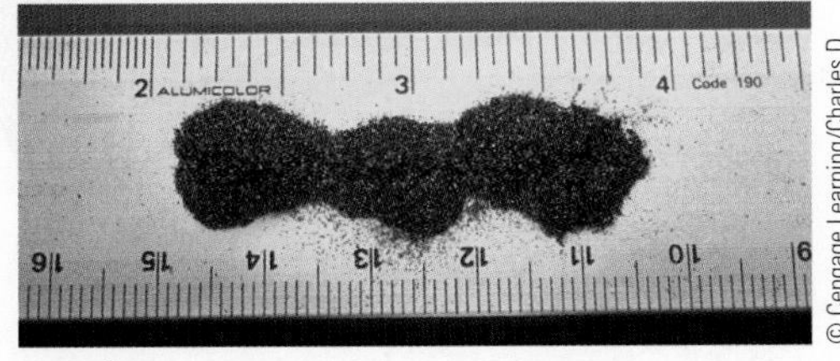

Zinc dust.

© Cengage Learning/Charles D. Winters

11.61 Arrange the following substances in order of increasing strength of crystal forces: CO_2, KCl, H_2O, N_2, CaO.

11.62 ■ Arrange the following substances in order of increasing strength of the crystal forces: He, NH_3, NO_2, NaBr, BaO.

OBJECTIVE Describe characteristics of various types of crystals.

11.63 Two of the Group 4A oxides, CO_2 and SiO_2, have very different melting points. What is the difference in their crystal forces that accounts for the very large difference in their melting points?

11.64 Silicon carbide, SiC, is a very hard, high-melting solid. What kind of crystal forces account for these properties?

OBJECTIVE Determine the number of ions or atoms in a cubic unit cell.

11.65 Ammonium chloride consists of a simple cubic array of Cl^- ions, with an NH_4^+ ion at the center of the unit cell. Calculate the number of ammonium ions and chloride ions in each unit cell.

11.66 Calcium oxide consists of a face-centered cubic array of O^{2-} ions, with Ca^{2+} ions at the center of the unit cell and along the centers of all 12 edges. Calculate the number of each ion in the unit cell.

11.67 How many atoms of tungsten are present in each unit cell of that metal, if it crystallizes in a body-centered cubic array of atoms?

11.68 Silver crystallizes in a cubic close-packing (FCC) array of atoms. How many silver atoms are present in each unit cell?

11.69 Iron crystallizes in a body-centered cubic array of atoms and has a density of 7.875 g/cm^3.
(a) What is the length of the edge of the unit cell?
(b) Assuming that the atoms are spheres in contact along the cube diagonal, what is the radius of an iron atom?

11.70 ■ Solid xenon forms crystals with a face-centered cubic unit cell that has an edge of 620 pm. Calculate the atomic radius of xenon.

11.71 In the rutile structure of TiO_2, the unit cell has a titanium ion in the center of the unit cell and a titanium ion at each of the eight corners. How many oxide ions are present in each unit cell? (Remember that the ratio of oxide ions to titanium ions must be the same as in the formula of the compound.)

11.72 Calcium fluoride (fluorite) has a unit cell with Ca^{2+} ions in a face-centered cubic arrangement. How many fluoride ions are present in each unit cell? (Remember that the ratio of fluoride ions to calcium ions must be the same as in the formula of the compound.)

OBJECTIVE Calculate the density of a crystalline solid.

11.73 Nickel crystallizes in a face-centered cubic array of atoms in which the length of the unit cell's edge is 351 pm. Calculate the density of this metal.

11.74 ■ Polonium crystallizes in a simple cubic unit cell with an edge length of 336 pm. (a) What is the mass of the unit cell? (b) What is the volume of the unit cell? (c) What is the theoretical density of Po?

11.75 Lithium hydride (LiH) has the sodium chloride structure, and the length of the edge of the unit cell is 4.086×10^{-8} cm. Calculate the density of this solid.

11.76 Cesium iodide crystallizes as a simple cubic array of iodide ions with a cesium ion in the center of the cell. The edge of this unit cell is 445 pm long. Calculate the density of CsI.

11.77 Palladium has a cubic crystal structure in which the edge of the unit cell is 389 pm long. If the density of palladium is 12.02 g/cm³, how many palladium atoms are in a unit cell? In which of the cubic unit cells does palladium crystallize?

Johnson Matthey Platinum Today www.platinum.matthey.com

11.78 Chromium has a cubic crystal structure in which the edge of the unit cell is 288 pm long. If the density of chromium is 7.20 g/cm³, how many chromium atoms are in a unit cell? In which of the cubic unit cells does chromium crystallize?

OBJECTIVE X-ray diffraction and the Bragg equation.

11.79 Which of the following materials should produce well-defined x-ray diffraction patterns?
(a) KBr (b) liquid H_2O
(c) glass (d) a sugar crystal

11.80 Which of the following materials should produce well-defined x-ray diffraction patterns?
(a) polyethylene (b) liquid ethanol
(c) CaO (d) a diamond

11.81 At what angle does first-order diffraction from layers of atoms 293 pm apart occur, using x rays with a wavelength of 154 pm?

11.82 ■ At what angle does first-order diffraction from layers of atoms 325 pm apart occur, using x rays with a wavelength of 179 pm?

11.83 What is the wavelength of x rays in which the first-order diffraction occurs at 13.4 degrees from layers of atoms spaced 232 pm apart?

11.84 What is the wavelength of x rays in which the first-order diffraction occurs at 16.5 degrees from layers of atoms spaced 315 pm apart?

Chapter Exercises

11.85 Which of the following pairs are allotropes?
(a) oxygen gas (O_2) and ozone gas (O_3)
(b) liquid oxygen (O_2) and oxygen gas (O_2)
(c) amorphous silicon dioxide and quartz (crystalline silicon dioxide)

11.86 The coordination number of uniformly sized spheres in a cubic closest-packing (FCC) array is 12. Give the coordination number of each atom in
(a) a simple cubic lattice.
(b) a body-centered cubic lattice.

11.87 ▲ Avogadro's number can be calculated quite accurately from the density of a solid and crystallographic data. Europium (molar mass = 151.94 g/mol) has a body-centered cubic array of atoms in which the edge of the unit cell is 458.27 pm long. The density of the metal is 5.243 g/cm³. Use these data to calculate Avogadro's number. (*Hint:* Calculate the atoms/cm³ from the crystallographic data and the volume of one mole of the element from the molar mass and density. Combine these values to find the atoms/mol, or Avogadro's number.)

11.88 X rays with a wavelength of 1.790×10^{-10} m are diffracted at an angle of 14.7 degrees by a crystal of KI.
(a) What is the spacing between layers in the crystal that gives rise to this angle of diffraction? (Assume $n = 1$.)
(b) KI has the same crystal structure as NaCl, and the distance calculated in part a is half the length of the cube edge. Calculate the density of KI.

11.89 The edge of the face-centered cubic unit cell of copper is 362 pm long. Assume that the copper atoms are spheres that are in contact along the diagonal of the face of the cube. Find the radius of a copper atom.

11.90 Indicate which type of bonding is expected for the following crystalline substances.
(a) N_2O (b) GeO_2 (c) $CaCl_2$
Correlate the properties below with each of these substances.
1. Melting point = −90.8 °C, liquid does not conduct electrical current.
2. Melting point = 782 °C, liquid does conduct electrical current.
3. Melting point = 1115 °C, liquid does not conduct electrical current.

11.91 Predict the types of intermolecular forces important for the following substances.
(a) He (b) CH_3OCH_3 (c) C_2ClF_5
Correlate the enthalpies of vaporization listed below with each of these substances.
1. 0.083 kJ/mol
2. 19.4 kJ/mol
3. 38.6 kJ/mol

11.92 ▲ The metallic bonding found in the element gold is such that 1.00 ounce of Au can be flattened into a sheet that is 3.00×10^2 ft². If there are 28.35 g/ounce and 30.48 cm/ft, how thin is this sheet? If a gold atom has a diameter of 358 pm, how many atoms thick is the sheet? The density of gold is 19.3 g/cm³.

© Morgan Lane Photography, 2008/Used under license from Shutterstock.com

11.93 What is the volume change when 1 mol of graphite (d = 2.26 g/cm^3) is converted to diamond (d = 3.51 g/cm^3)?

11.94 Why do free-floating liquids in an orbiting spacecraft adopt a spherical shape?

Cumulative Exercises

11.95 A 1.50-g sample of methanol (CH_3OH) is placed in an evacuated 1.00-L container at 30 °C.
(a) Calculate the pressure in the container if all of the methanol is vaporized. (Assume the ideal gas law, $PV = nRT$.)
(b) The vapor pressure of methanol at 30 °C is 158 torr. What mass of methanol actually evaporates? Is liquid in equilibrium with vapor in the vessel?

11.96 A 1.50-g sample of water is placed in an evacuated 1.00-L container at 30 °C.
(a) Calculate the pressure in the container if all of the water is vaporized. (Assume the ideal gas law, $PV = nRT$.)
(b) What is the vapor pressure of water at 30 °C? (See Table 6.3.)
(c) What mass of water actually evaporates? Is liquid in equilibrium with vapor in the vessel?

11.97 ■ ▲ The cooling process in a refrigeration unit involves reducing the pressure above a liquid called the *refrigerant.* The pressure reduction causes vaporization, which cools the remaining liquid. A common refrigerant is Freon-11 (CCl_3F), which has a boiling point of 23.8 °C, a specific heat for the liquid of 0.870 J/g·°C, and an enthalpy of vaporization of 180 J/g.
(a) How much energy must be removed from 10.0 g liquid Freon-11 to cool it from its boiling point to 0 °C?
(b) What mass of Freon-11 must vaporize to remove the amount of heat calculated in part a?

11.98 How much heat is absorbed by a 10.0-g sample of water in going from ice at −10 °C to liquid water at 95 °C? Use the data in the accompanying table.

11.99 How much heat is absorbed by a 15.0-g sample of water in going from liquid at 10 °C to steam at 105 °C and a pressure of 1.00 atm? Use the data in the accompanying table.

Some Thermal Properties of Water

Property	Value
Specific heat (J/g · °C)	
Solid	2.07
Liquid	4.18
Gas	2.01
ΔH_{fusion} (kJ/mol; at 0 °C)	6.01
$\Delta H_{vaporization}$ (kJ/mol; at 100 °C)	40.6
$\Delta H_{sublimation}$ (kJ/mol)	50.9
Melting point (°C)	0
Boiling point (°C)	100

11.100 Draw the Lewis structures of CF_4 and CF_2CCl_2. What is the hybridization of the carbon atoms in these compounds? Predict whether each is polar or nonpolar. What types of intermolecular forces do you expect for each? Which molecule do you expect will have the higher normal boiling point, and why?

11.101 A white crystalline compound that has a high melting point has the formula MCl. A 2.13-g sample dissolves in 200 mL of water, and addition of excess $AgNO_3$ gives a 4.09-g precipitate of AgCl. What is M, and what type of crystal forces do you expect for MCl?

11.102 ■ ▲ Consider a sealed flask with a movable piston that contains 5.25 L O_2 saturated with water vapor at 25 °C. The piston is depressed at constant temperature so that the gas is compressed to a volume of 2.00 L.
(a) What is the vapor pressure of water in the compressed gas mixture? (See Table 6.3.)
(b) How many grams of water condense when the gas mixture is compressed?

11.103 Two compounds, A and B, have the molecular formula C_3H_9N. One isomer has no hydrogen atoms bonded to the nitrogen atom and has a boiling point of 2.9 °C. The other isomer (boiling point of 47.8 °C) has two hydrogen atoms bonded to the nitrogen atom, and the carbon atoms and nitrogen atom are in a straight chain. Neither structure contains a ring. Draw the Lewis structure of both compounds. Compare the types of intermolecular forces for each and use them to explain the differences in the boiling point of the two compounds.

11.104 The thermite reaction is a reaction between solid iron(III) oxide and solid aluminum to make aluminum oxide and elemental iron. Usually the reactants are fine powders, to increase the contact between them. The reaction is so exothermic that the iron product is initially a liquid (melting point of iron is 1536 °C).
(a) Write a balanced chemical reaction for the thermite reaction.
(b) Identify the oxidizing agent and the reducing agent.
(c) Use the ΔH_f° data in Appendix G to determine the ΔH_{rxn} for the thermite reaction.
(d) If all of the energy given off by the reaction were absorbed by the products, what would be the temperature change of the products, assuming that their specific heats did not change with temperature? Is this temperature change enough to melt the iron? The specific heat of aluminum oxide is 79.5 J/mol·K, and the specific heat of iron is 25.2 J/mol·K.
(e) In the solid form, iron exists as a body-centered cubic unit cell with a cell parameter a of 286.64 pm. What is the density of iron?
(f) In the solid form, aluminum exists as a face-centered cubic unit cell with a cell parameter a of 404.9 pm. What is the density of aluminum?
(g) Does either iron or aluminum exist as a closed-packed crystal?
(h) At its melting point, iron has a vapor pressure of approximately 0.750 torr. If the balanced thermite reaction were to occur in molar quantities, what volume would be necessary for all of the iron to be in the vapor phase if the temperature were the melting point of iron?
(i) A similar reaction uses chromium(III) oxide instead of iron(III) oxide as a reactant, and chromium metal is a product. What is the ΔH_{rxn} for this reaction?

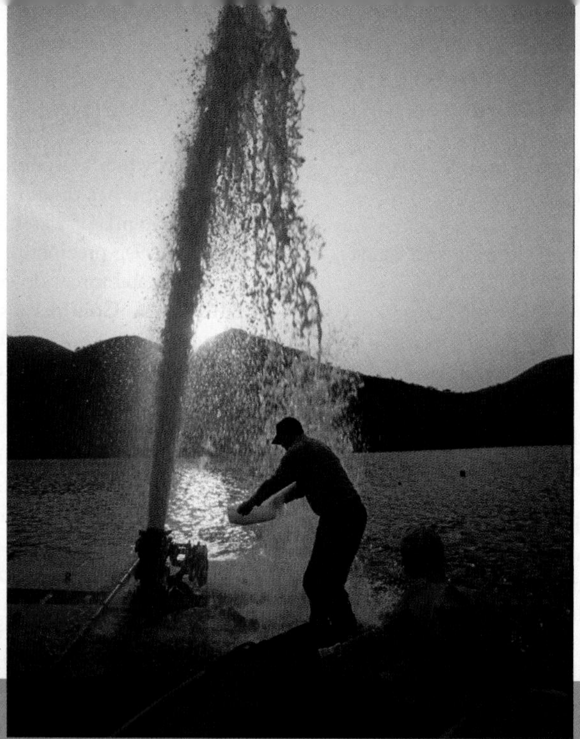

© Louise Gubb/CORBIS

Scientists collect samples from the gaseous carbon dioxide fountain that was installed in 2001 by French engineers in an international cooperative effort to relieve the amount of the gas dissolved in Lake Nyos, Cameroon. In 1986 this gas erupted, killing some 1700 local inhabitants and their livestock.

On August 26, 1986, about 1700 people and thousands of animals were found dead near the shore of Lake Nyos in northwest Cameroon, in western Africa. Although cases of widespread sudden death were not unprecedented in the area, this was by far the largest incident.

The culprit? Lake Nyos itself.

Lake Nyos is a deep lake formed inside a crater of an extinct volcano in a tectonic fault line in Cameroon. Deep in the lake, carbon dioxide gas is released by magma into the water, creating a solution of carbon dioxide. Carbon dioxide is more soluble in cold water and at the higher pressures near the bottom of the lake. Since the waters of Lake Nyos do not mix together very well, the carbon dioxide solution stayed near the bottom of the lake. Scientists hypothesize that a small earthquake or rock slide disturbed the carbon dioxide–laden water, causing the carbon dioxide gas to bubble out like spray from a shaken soda bottle. The carbon dioxide that was suddenly released blanketed the surrounding countryside, asphyxiating the people and animals that lived nearby.

Solutions

12

CHAPTER CONTENTS

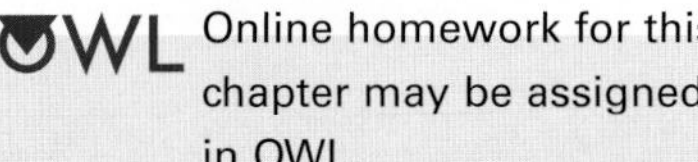
Online homework for this chapter may be assigned in OWL.

Look for the green colored vertical bar throughout this chapter, for integrated references to this chapter introduction.

Lake Nyos is one of only three lakes known that contain significant amounts of dissolved carbon dioxide; the other two lakes are Lake Monoun, also in Cameroon, and Lake Kivu in nearby Rwanda. A similar incident occurred near Lake Monoun in 1984, but the effects were much more limited so that people did not immediately connect the incident to the lake.

To prevent future catastrophes, in 1995 engineers installed some underwater pumps with pipes leading to the surface. The pumps move the water from the bottom of the lake to the top, and at the lower surface pressure the carbon dioxide escapes from the water a little at a time. Studies suggest that at least five more pipes are needed to completely eliminate the danger from Lake Nyos.

As unusual as the 1986 tragedy was, it was the result of simple chemistry: the properties of solutions. In this chapter, the properties of solutions will be introduced, and as the chapter progresses, we will show how certain properties are intimately related to the Lake Nyos tragedy. ▮

Solutions play an important role in chemistry and in the world around us. Most of the reactions that we observe occur in solution. The fuels we use, the air we breathe, and the water we drink are all solutions. The fluids in our bodies are complex solutions that distribute nutrients and oxygen throughout the body. Alloys are generally solid solutions of two or more metals with desirable properties not exhibited by the pure metals.

This chapter focuses mainly on liquid solutions because they are so important in experimental chemistry. However, understand that solute and solvent can be any phase, and a solution may be solid, liquid, or gas. Table 12.1 lists some examples of how different phases can interact to make a solution.

Particular emphasis is placed on aqueous solutions, because water is the most commonly used solvent and is important for biological systems. Chapter 4 introduced and defined several terms related to solutions, and it may be helpful to review those definitions before proceeding with this chapter.

TABLE 12.1 Solute, Solvent, and Solution Phases

Phase of Solvent	Phase of Solute	Example
Gas	Gas	Air (oxygen and other gases in nitrogen)
Liquid	Gas	Soda water (carbon dioxide in water)
Liquid	Liquid	Alcoholic beverage (ethyl alcohol in water)
Liquid	Solid	Seawater (sodium chloride and other salts in water)
Solid	Gas	Hydrogen gas absorbed in solid palladium
Solid	Liquid	Amalgams (alloys of mercury in other metals, such as silver or gold)
Solid	Solid	Steel (an alloy of carbon and metals in iron)

12.1 Solution Concentration

OBJECTIVES

- ☐ Express the concentration of a solution using several different units
- ☐ Convert between different concentration units

Many chemical processes occur in solution. Typically, we can measure a quantity of solution, usually by volume, but we also need to be able to express how much of a particular substance is present in the solution. The relative composition of a solution is expressed as a concentration. There are many different units for concentration, but all of them express the composition of the solution as the quantity of solute present in a fixed quantity of the solution or solvent. This section, after reviewing concentration units used in earlier chapters, introduces and defines some additional units.

Concentration Units

Molarity *(M)*

Chapter 4 defines *molarity* as the number of moles of solute in 1 L of solution. Molarity is the preferred concentration unit for stoichiometry calculations. This method of expressing concentrations is used extensively for the stoichiometry calculations in Chapter 4.

Mole Fraction (χ)

The mole fraction of a substance in a solution is the number of moles of that substance divided by the total number of moles of all substances present.

$$\chi_A = \frac{\text{moles of A}}{\text{moles of A} + \text{moles of B} + \text{moles of C} + \cdots}$$

When the mole fractions of all components of a solution are added together, the sum is always 1.

$$\chi_A + \chi_B + \chi_C + \cdots = 1$$

Chapter 6 introduced mole fractions to calculate the partial pressures of the components in a gaseous mixture.

Mass Percentage Composition

The concentration of solute may be expressed as a percentage of the mass of the solution with the equation

$$\text{Mass percentage solute} = \frac{\text{grams of solute}}{\text{grams of solution}} \times 100\%$$

The percentage gives the number of grams of solute present in 100 g of solution. The mass percentage concentration unit is not widely used by chemists but is still encountered frequently in pharmacies and commerce and appears on the labels of many products sold as solutions (Figure 12.1).

EXAMPLE 12.1 Mass Percentage Concentration

Determine the mass percentage concentration of a solution prepared by dissolving 5.00 g NaCl in 200 g water.

Strategy Determine the total mass of the solution and substitute the appropriate quantities into the definition of mass percentage concentration.

Solution

You need the mass of the solute and the mass of the solution to find the percentage concentration. The mass of solute is given, and the mass of the solution is the sum of the masses of the components, 205 g. The concentration is

$$\text{Mass percentage NaCl} = \frac{5.00 \text{ g NaCl}}{205 \text{ g soln}} \times 100\% = 2.44\% \text{ NaCl}$$

Understanding

A chemical analysis shows that 25.0 g of a solution contains 2.00 g glucose. Calculate the concentration in mass percentage of glucose.

Answer 8.00%

Figure 12.1 Common solutions. The concentrations of many solutions sold in grocery stores and pharmacies are expressed as percentage by mass.

Parts per Million, Parts per Billion

Closely related to mass percentage composition are two other concentration units, *parts per million* (ppm) and *parts per billion* (ppb), that are being used more widely as chemical analyses become more sensitive. Just as percent represents the grams of solute in 100 g of solution, parts per million are the grams of solute in one million (10^6) g of solution, and parts per billion are the grams of solute in 10^9 g of solution. Mathematically, these are calculated using the following equations.

$$\text{ppm} = \frac{\text{g solute}}{\text{g solution}} \times 1{,}000{,}000$$

$$\text{ppb} = \frac{\text{g solute}}{\text{g solution}} \times 1{,}000{,}000{,}000$$

Molality *(m)*

The **molality *(m)*** of a solution is defined as

$$\text{Molality } (m) = \frac{\text{moles solute}}{\text{kg solvent}}$$

Molality expresses solute concentration as mole of solute per kilogram of solvent.

Sometimes the word *molal* is used to express this unit, as in "0.65 molal." Molality is the only common unit of concentration in which the denominator expresses a quantity of *solvent* rather than *solution*. Even though their names are similar, do not confuse molality and molarity. Molarity is the moles of solute per liter of *solution*.

EXAMPLE 12.2 Molal Concentration of Solution

Determine the molality of the solution in Example 12.1 (5.00 g NaCl in 200 g water).

Strategy Determine the amount of solute and the mass of solvent before using the definition of molality.

Solution

The strategy for solving this question is

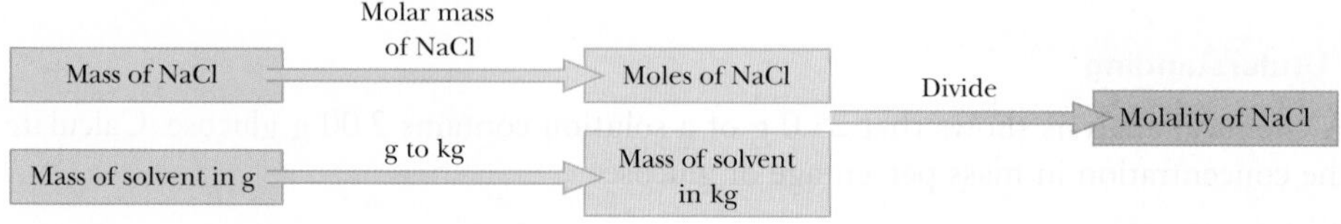

The quantity of the solute, NaCl, must be expressed in moles. Since the molar mass of NaCl is 58.44 g/mol, we have

$$\text{Amount NaCl} = 5.00 \text{ g NaCl} \times \left(\frac{1 \text{ mol}}{58.44 \text{ g}}\right) = 8.56 \times 10^{-2} \text{ mol NaCl}$$

The denominator in molality is the mass of the solvent, in kilograms. Since this mass is given as 200 g, or 0.200 kg, the molality of the solution is

$$\text{Molality NaCl} = \frac{8.56 \times 10^{-2} \text{ mol NaCl}}{0.200 \text{ kg solvent}} = 0.428\ m \text{ NaCl}$$

Understanding

What is the molality of a solution that contains 2.00 g glucose (molar mass = 180 g/mol) in 25.0 g of solvent?

Answer 0.444 molal

EXAMPLE 12.3 Calculations Involving Molality

What mass of acetic acid (CH_3COOH, molar mass = 60.05 g/mol) must be dissolved in 250 g water to produce a 0.150-*m* solution?

Strategy This problem is similar to conversions performed in Chapter 4, except that the desired molality is used as a conversion factor. The following diagram outlines the problem-solving strategy.

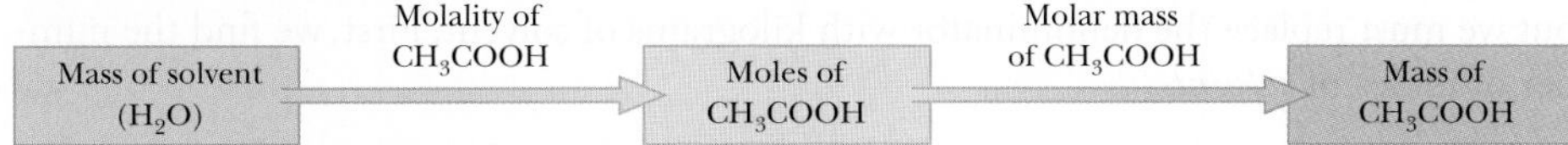

The desired molality is used as a conversion factor to find the number of moles of solute in 250 g (= 0.250 kg) of the solvent.

Solution
A 0.150-*m* solution contains 0.150 mol of solute for every kilogram of solution.

$$\text{Amount } CH_3COOH = 0.250 \text{ kg } H_2O \times \left(\frac{0.150 \text{ mol } CH_3COOH}{1 \text{ kg } H_2O}\right) = 0.0375 \text{ mol } CH_3COOH$$

The molar mass of the acetic acid is the proper conversion factor for expressing this amount of the compound, in grams.

$$\text{Mass } CH_3COOH = 0.0375 \text{ mol } CH_3COOH \times \left(\frac{60.05 \text{ g } CH_3COOH}{1 \text{ mol } CH_3COOH}\right)$$
$$= 2.25 \text{ g } CH_3COOH$$

Understanding

How many grams of water must be added to 4.00 g urea [$CO(NH_2)_2$; molar mass = 60.06 g/mol] to produce a 0.250-*m* solution of the compound?

Answer 266 g water

Conversion among Concentration Units

Often, it is necessary to convert from one concentration unit to another. All concentrations are fractions, with the quantity of the solute in the numerator and the quantity of the solution (or solvent) in the denominator. They differ only in the units used to express these quantities. Table 12.2 summarizes the units used in the numerator and denominator for the four ways of expressing concentration. In Table 12.2 and subsequent calculations, the term *moles total of solution* refers to the sum of the moles of all components.

Moles total of solution = moles solvent + moles solute

When performing conversions from one concentration unit to another, always write each unit in the form of a fraction to separate the units of the numerator and the denominator; then convert the numerator and denominator as necessary. The following examples illustrate this technique.

TABLE 12.2 Units for Concentration Conversions

Concentration Unit	Numerator Units (Solute in Each Case)	Denominator
Mass percentage	Grams	100 g of solution
Molarity	Moles	1 L of solution
Molality	Moles	1 kg of solvent
Mole fraction	Moles	1 mol total of solution

Suppose we want to calculate the molality of an NaCl solution in which we know the mole fraction of NaCl is 0.024. Start by writing the given concentration as a fraction and the units of the desired concentration.

Given concentration *Desired units*

$$\chi_{NaCl} = \frac{0.024 \text{ mol NaCl}}{1 \text{ mol total of solution}} \qquad \text{Molality} = \frac{\text{mol NaCl}}{\text{kg of solvent}}$$

To express the concentration as molality, we need not change the unit of the numerator, but we must replace the denominator with kilograms of solvent. First, we find the number of moles of *solvent.*

$$\text{Moles solvent} = \text{moles total of solution} - \text{moles solute}$$

In 1 mol total of this solution there is 0.024 mol NaCl, so the remainder is solvent, which is water:

$$\text{Moles solvent} = 1 \text{ mol total of solution} - 0.024 \text{ mol NaCl}$$
$$= 0.976 \text{ mol solvent}$$

This amount of the solvent (water) must be expressed in kilograms, so

$$\text{Mass of water} = 0.976 \cancel{\text{mol } H_2O} \times \left(\frac{18.02 \cancel{\text{g}} \text{ } H_2O}{1 \cancel{\text{mol } H_2O}}\right) \times \left(\frac{1 \text{ kg}}{1000 \cancel{\text{g}}}\right)$$
$$= 1.76 \times 10^{-2} \text{ kg } H_2O$$

Then, we find the molal concentration.

When converting from one concentration unit to another, always write the units of the numerator and denominator separately.

$$\text{Molality NaCl} = \frac{0.024 \text{ mol NaCl}}{1.76 \times 10^{-2} \text{ kg } H_2O} = 1.4 \text{ } m \text{ NaCl}$$

Examples 12.4 and 12.5 provide some additional sample calculations.

EXAMPLE 12.4 Converting Concentration Units

Hydrochloric acid is sold as a 36% aqueous solution. Express this concentration in (a) molality and (b) mole fraction.

Strategy Write each concentration unit in the form of a fraction to separate the units of the numerator and the denominator; then convert the numerator and denominator from the initial units to the final units.

Solution
Since the quantity of solvent is needed in both parts, calculate it by difference.

$$\text{Mass } H_2O = 100 \text{ g solution} - 36 \text{ g HCl} = 64 \text{ g } H_2O$$

(a) Writing the concentration units as fractions,

Given concentration *Desired units*

$$\text{Mass percentage HCl} = \frac{36 \text{ g HCl}}{100 \text{ g solution}} \qquad \text{Molality} = \frac{\text{mol HCl}}{\text{kg of solvent}}$$

The following flow diagram shows the steps involved in this conversion of units.

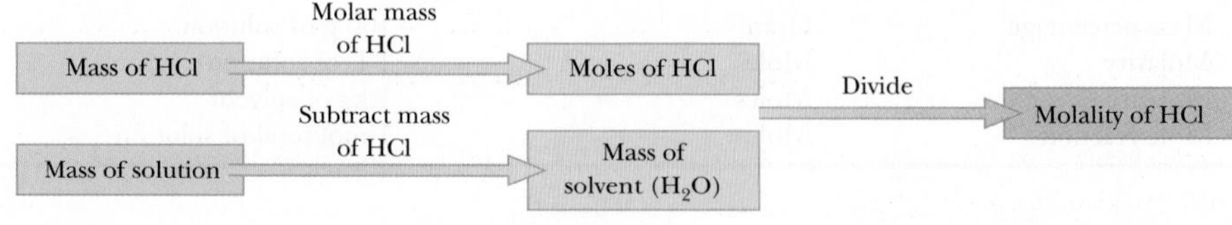

Molality expresses the quantity of solute in moles, so express the 36 g HCl in moles, using the molar mass of HCl, 36.5 g/mol.

$$\text{Moles HCl} = 36 \text{ g HCl} \times \left(\frac{1 \text{ mol}}{36.5 \text{ g HCl}}\right) = 0.99 \text{ mol HCl}$$

Since molality is the number of moles of solute per kilogram of solvent, express the quantity of water in kilograms.

$$\text{Mass } H_2O = 64 \text{ g } H_2O \times \left(\frac{1 \text{ kg}}{1000 \text{ g}}\right) = 0.064 \text{ kg } H_2O$$

Complete the conversion by combining these two numbers.

$$\text{Molality HCl} = \frac{0.99 \text{ mol HCl}}{0.064 \text{ kg } H_2O} = 15\ m \text{ HCl}$$

(b) When the concentration is expressed as the mole fraction of HCl, the desired units are

$$\chi_{HCl} = \frac{\text{moles HCl}}{\text{moles total of solution}}$$

The numerator is the number of moles of HCl, which was already calculated in part a. The denominator is the total number of moles of all substances in the solution. The following diagram shows the steps in the calculations.

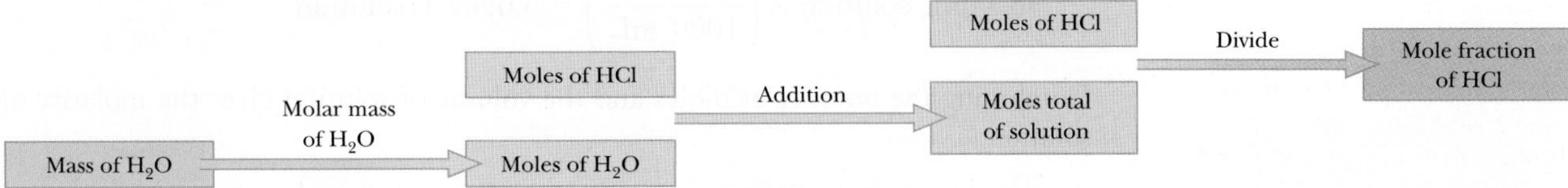

$$\text{Moles total of solution} = \text{moles HCl} + \text{moles } H_2O$$

$$\text{Moles total of solution} = 0.99 \text{ mol HCl} + 64 \text{ g } H_2O \times \left(\frac{1 \text{ mol}}{18.0 \text{ g}}\right)$$

$$\text{Moles total of solution} = 0.99 + 3.6 = 4.6 \text{ mol solution}$$

Complete the calculation by finding the quotient.

$$\text{Mole fraction HCl} = \frac{0.99 \text{ mol HCl}}{4.6 \text{ mol soln}} = 0.22$$

Understanding

Find the (a) molality and (b) mole fraction of a 24.5% solution of ammonia (NH_3) in water.

Answer (a) 19.0 *m*, (b) 0.256

For some of the conversions it is necessary to know the density of the solution. Density serves as a conversion factor between volume and mass. Any time the quantity of solution must be changed from volume to mass or vice versa, the density of the solution must be used in the calculation. The next example illustrates this process.

The density of the solution is needed to convert between units of volume and mass.

EXAMPLE 12.5 Converting Concentration Units

Determine the molarity of an aqueous solution that is 37.2% in HCl. The density of this solution is 1.034 g/mL.

Strategy Write the concentrations as fractions.

Given concentration — *Desired units*

$$\%\text{ mass HCl} = \frac{37.2\text{ g HCl}}{100\text{ g solution}} \qquad \text{Molarity of HCl} = \frac{\text{mol HCl}}{\text{L solution}}$$

The units in the numerator of percentage composition can be easily converted to moles, and we can use the density of the solution to convert from grams of solution to volume of solution.

Solution

The molar mass of HCl is 36.46 g/mol. Converting the amount of HCl in the numerator to moles:

$$\text{mol HCl} = 37.2\text{ g HCl} \times \left(\frac{1\text{ mol HCl}}{36.46\text{ g HCl}}\right) = 1.02\text{ mol HCl}$$

Using the density of the solution to convert to a volume of solution:

$$100\text{ g solution} \times \left(\frac{1\text{ mL}}{1.034\text{ g}}\right) = 96.7\text{ mL solution}$$

To determine molarity, we need to have the volume in liters, not milliliters:

$$96.7\text{ mL solution} \times \left(\frac{1\text{ L}}{1000\text{ mL}}\right) = 0.0967\text{ L solution}$$

Combining the number of moles and the volume of solution gives the molarity of the solution:

$$M = \left(\frac{1.02\text{ mol HCl}}{0.0967\text{ L solution}}\right) = 10.5\ M\text{ HCl}$$

Understanding

Find the molarity of a 7.85% aqueous ammonia solution that has a density of 0.965 g/mL.

Answer 4.77 M

Why is molality useful? Unlike molarity, the molality of a solution does not change with temperature. Typically, the volume of a solution expands as the temperature is increased, so the molarity decreases (because the volume of the solution is in the denominator of the definition of molarity). However, the mass of the solvent stays the same even if the volume expands or contracts with temperature. Thus, *molality is a temperature-independent concentration unit.* (Mass percent and mole fraction are also temperature independent.) On the other hand, molarity has the advantage of giving you stoichiometric amounts directly from the volume of the solution, which is usually easy to measure experimentally. The choice of which concentration unit to use depends on exactly how you intend to apply the concentration of the solution.

OBJECTIVES REVIEW *Can you:*

- ☑ express the concentration of a solution using several different units?
- ☑ convert between different concentration units?

12.2 Principles of Solubility

OBJECTIVES

- ☐ Define solubility and describe how to determine whether a solution is saturated, unsaturated, or supersaturated
- ☐ Describe the solution process on a molecular level
- ☐ Predict the relative solubilities of substances in different solvents based on solute-solvent interactions

Many chemical reactions do not proceed unless the reactants are dissolved in solution. Because of the major role solutions play in chemical reactions, it is helpful to understand the factors that determine whether a given substance will dissolve in a solvent. This section presents the processes that occur as a solute dissolves. These processes can be divided into two general components: an enthalpy change, caused by differences in attractive forces between molecules, and a change in the disorder of the system.

For most solutes, there is a limit to the quantity that can dissolve in a fixed volume of any given solvent. When a solid such as $CO(NH_2)_2$ (urea) is stirred with 1 L water, some of it dissolves to form a solution. If enough solute is present, we find that not all of it dissolves, but a maximum constant concentration of the solute is obtained. Adding more solute does not change this concentration further; the added solid simply remains undissolved. In this situation, a state of dynamic equilibrium has been reached, as represented in Figure 12.2. Molecules or ions continue to enter the liquid phase from the solid, but other molecules or ions of the solute leave the solution at an equal rate, analogous to the solid-liquid phase equilibrium examined in Chapter 11. The **solubility** is the concentration of solute that exists in equilibrium with an excess of that substance; that is, it is the maximum concentration that can dissolve at a particular temperature. Aqueous solubilities are typically expressed as number of grams of solute per 100 mL water, or as molar concentrations. A **saturated solution** is one that is in equilibrium with an excess of the solute. The concentration of a saturated solution is equal to the solubility. An **unsaturated solution** is one in which the concentration of the solute is less than the solubility.

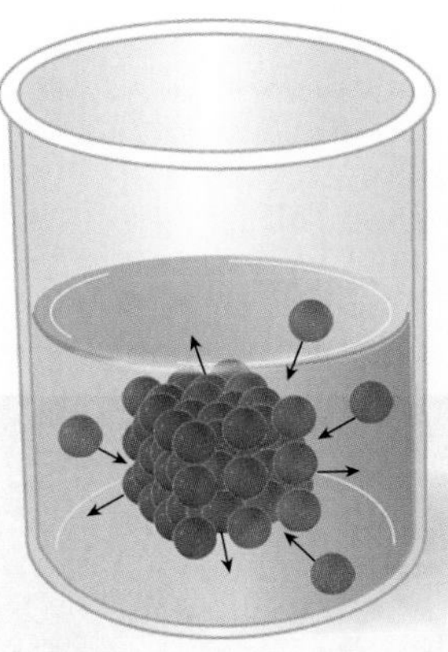

Figure 12.2 Dissolving a solid. The solubility of a solute is reached when the rate of dissolution and the rate of crystallization are equal. No net change in the concentration of the solute is observed, no matter how much excess solute is present (or how little, as long as there is some).

Under some conditions it is possible to prepare a solution called a **supersaturated solution,** in which the concentration of solute is temporarily greater than its solubility. Supersaturation is an unstable condition that is analogous to the supercooled liquids mentioned in Chapter 11.

(*a*)

(*b*)

(*c*)

Supersaturated solutions. Cooling a concentrated solution of sodium acetate often produces a supersaturated solution. Adding a seed crystal *(a)* to the supersaturated solution initiates crystallization of sodium acetate *(b),* which continues *(c)* until the concentration has decreased to that of the saturated solution.

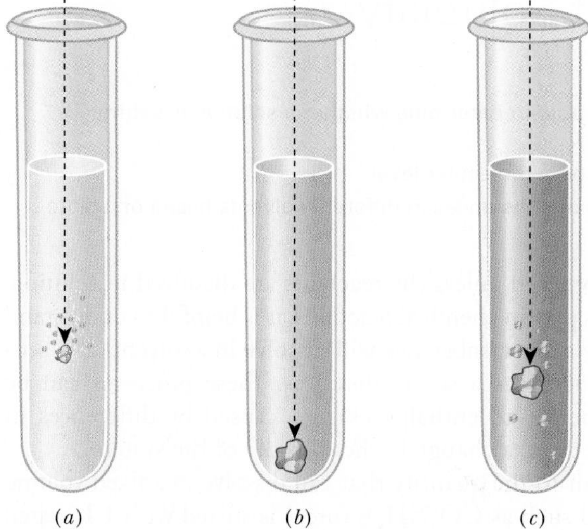

Figure 12.3 Solutions. *(a)* When pure solute is added to an unsaturated solution, it dissolves. *(b)* When solute is added to a saturated solution, no more solute dissolves. *(c)* When solute is added to a supersaturated solution, additional solid forms.

Adding a small quantity of solute to a solution is a simple way to distinguish among unsaturated, saturated, and supersaturated solutions. If the solution is unsaturated, the added solute dissolves, increasing the concentration of the solution. If the solution is saturated, the addition of solute produces no change in the concentration of the solution (although the added solute participates in the dynamic equilibrium). When the solution is supersaturated, the addition of pure solute usually causes the rapid precipitation of excess solute. The precipitation of the solute continues until the concentration decreases to its solubility limit. Figure 12.3 illustrates these three situations.

The Solution Process

Experience shows that some substances are very soluble in water, and others are quite insoluble. Sugar and alcohol readily dissolve in water, whereas sand and charcoal do not dissolve to any measurable extent. The solubilities of a single substance in different liquids also vary considerably. For example, grease does not dissolve in water, but kerosene dissolves grease stains. Figure 12.4 illustrates the differences between the solubilities of ethylene glycol and motor oil in water.

Many factors contribute to the process of dissolving, also called *dissolution.* Two of these, the enthalpy change that accompanies solute-solvent attractions and the change in disorder, are the most important and can provide some insight into the general principles of solubility.

(a)

(b)

Figure 12.4 Liquid-liquid solubilities. *(a)* Water and ethylene glycol mix in all proportions (a green dye is in the ethylene glycol for clarity). *(b)* Water and motor oil do not dissolve in each other to any great extent.

Solute-Solvent Interactions

Most spontaneous processes (those that proceed without outside forces) are accompanied by a decrease in the potential energy of the system. For example, a ball rolls down rather than up a hill. As the ball's elevation decreases, its gravitational potential energy also decreases. In general, changes accompanied by a large decrease in enthalpy are spontaneous. In considering whether a solid and liquid will form a solution, bear in mind that the change in enthalpy arises mainly from changes in the intermolecular attractions.

Three types of intermolecular forces are involved in the formation of condensed phase solutions: solute-solute, solvent-solvent, and solute-solvent interactions. In Chapter 11, we examined the nature of the intermolecular forces and emphasized the attractions among molecules of the same substance. The same forces also cause molecules of different substances to attract each other. Water molecules not only exert attractive forces on other water molecules, they also attract molecules of other substances with which they are mixed, such as alcohol. The relative strengths of these different attractive forces play an important role in determining whether two substances will form a solution.

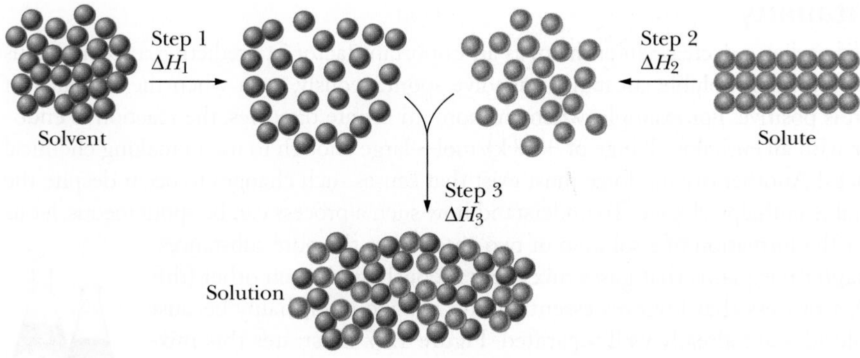

Figure 12.5 Contributions to dissolution. Pictorial representation of the factors that contribute to the enthalpy of a solution. *Step 1:* The solvent molecules move apart. *Step 2:* The solute molecules are separated. *Step 3:* The solute and solvent molecules mix. The enthalpy changes in steps 1 and 2 are both endothermic, whereas the enthalpy change in step 3 is exothermic.

When a solution forms, the molecules of the solvent must move apart to accommodate the solute molecules. Since the solvent molecules attract each other, energy must be expended to separate them. The solute molecules must also be separated from each other to enter the solution, and this process is endothermic as well. Because the solvent and solute molecules exert attractive forces on each other, energy is released when they are brought together. Figure 12.5 shows these three steps, and the energy changes are plotted in the diagrams of Figure 12.6. The overall enthalpy change that accompanies the dissolution of one mole of solute, called the **enthalpy of solution,** is simply the sum of the three individual energy changes.

The relative strengths of solute-solute, solvent-solvent, and solute-solvent attractions are important in determining whether substances will dissolve.

$$\Delta H_{soln} = \Delta H_1 + \Delta H_2 + \Delta H_3$$

Both endothermic and exothermic heats of solution are observed, as shown in Figure 12.6. When sulfuric acid dissolves in water, the enthalpy change is -74.3 kJ/mol. So much heat is released that spattering can occur when some of the water boils. In that case, the energy that is released by the exothermic solute-solvent interactions is greater than that absorbed by the endothermic processes of separating the solvent and solute molecules. In general, substances that have similar properties and thus have similar intermolecular forces have strong solute-solvent interactions and tend to form solutions. The statement "like dissolves like" is a simplification that is often used to explain observed trends in solubility.

Substances in which the intermolecular forces are similar tend to form solutions. This tendency is often expressed as "like dissolves like."

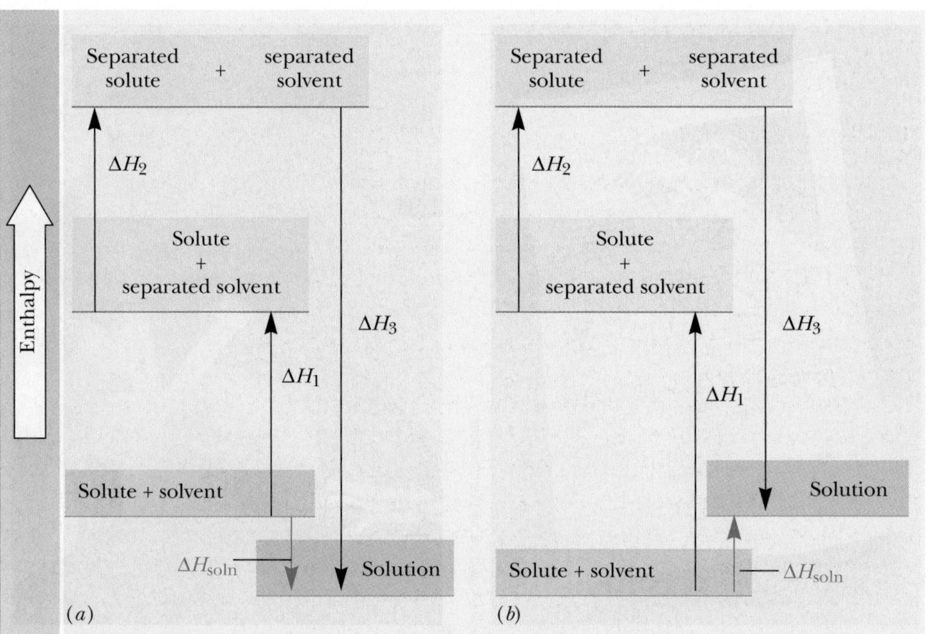

Figure 12.6 Enthalpy of solution. The enthalpy of solution can be broken into three processes: ΔH_1 = energy needed to separate the solvent molecules; ΔH_2 = energy needed to separate the solute molecules; and ΔH_3 = energy released when the solute and solvent molecules attract each other. The enthalpy of solution may be either *(a)* negative or *(b)* positive.

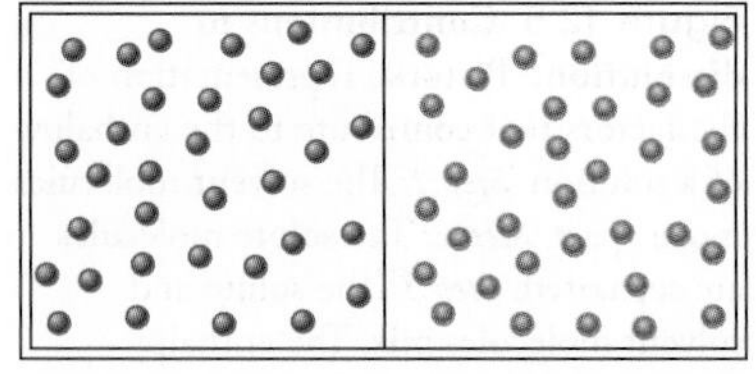

(*a*)

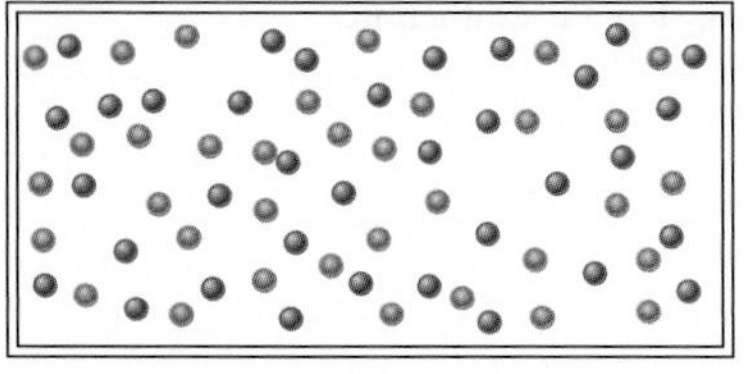

(*b*)

Figure 12.7 Mixing of gases. *(a)* Oxygen and nitrogen gases at the same pressure and temperature are separated by a partition. *(b)* When the partition is removed, there is no change in enthalpy, but the gases mix completely because of the natural tendency toward an increase in disorder.

An increase in disorder is an important driving force in the formation of solutions.

Spontaneity

As noted earlier, a decrease in enthalpy is an important factor in predicting a spontaneous change. Yet, many soluble substances dissolve spontaneously, even when the enthalpy of solution is positive. For example, when ammonium nitrate dissolves, the reaction is endothermic with an enthalpy change of +28 kJ/mol—large enough to use in making chemical cold packs! Another driving force must exist that causes such changes to occur despite the unfavorable enthalpy change. To understand how such a process can be spontaneous, let us consider the formation of a solution of two gases from the pure substances.

Chapter 6 explains that gases mix spontaneously with each other (diffusion), a process that involves essentially no change in enthalpy because the molecules are already well separated. Figure 12.7 illustrates this mixing. Suppose a container is divided by a removable partition into two compartments of equal volume. On one side of the partition is pure oxygen, and the other compartment contains pure nitrogen, with both gases at the same pressure and temperature. Removing the partition allows the molecules of the gases to mix, and in a relatively short time a uniform solution of the two gases occupies the entire volume. No matter how long we waited, we would not expect the two gases to spontaneously separate into their original arrangement.

Mixing of gases is one example of an important principle of nature: *Processes in which disorder increases tend to occur spontaneously.* The natural tendency toward disorder is one of the main driving forces in the formation of solutions. This principle is so important that it is presented more formally in Chapter 17.

An increase in disorder explains why ammonium nitrate dissolves in water even though the enthalpy of solution is sufficiently endothermic (+28 kJ/mol) that it is used in cold packs (see Figure 12.8). The ammonium nitrate is soluble because an increase in disorder occurs when the solution forms. Before dissolving, the ammonium ions and nitrate ions are in the highly ordered crystalline state. Once in solution, these ions are free to move independently of one another through the entire volume of the solution. This increase in disorder is more than enough to compensate for the unfavorable change in enthalpy.

Solubility of Molecular Compounds

Hexane (C_6H_{14}) and heptane (C_7H_{16}) are two liquid hydrocarbons. In both of these compounds, the dominant intermolecular attractions are London dispersion forces. The same dispersion forces also cause hexane and heptane molecules to attract each other. Because all

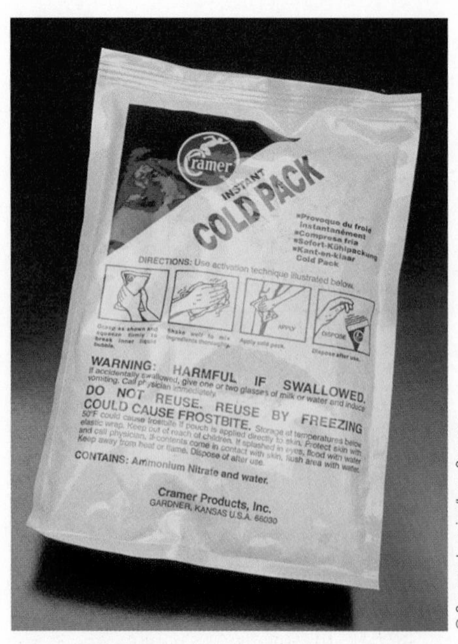

© Cengage Learning/Larry Cameron

(*a*)

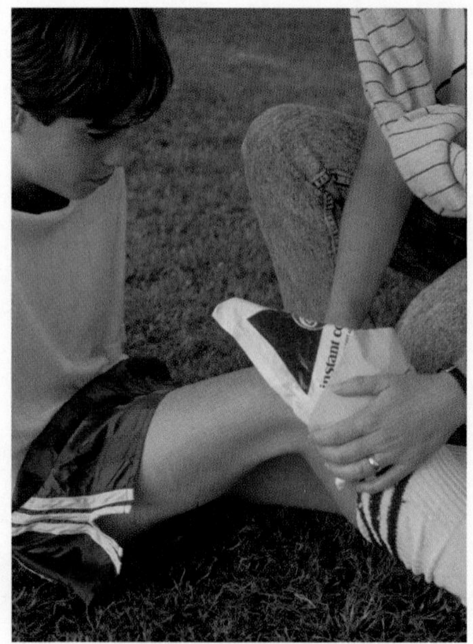

Scott Goode

(*b*)

Figure 12.8 Chemical cold packs. Cold packs such as the one shown are used to treat athletic injuries. Squeezing the plastic bag causes ammonium nitrate to mix with water, forming a solution. The endothermic heat of solution causes the pack to cool.

three of these attractions are close in energy, it is not surprising that the two liquids mix in any proportion. There is little difference in the energy of the attractions, so the increase in disorder on mixing becomes the controlling factor in the dissolution process.

Now consider the mixing of water with hexane. The intermolecular attractions among the water molecules are dominated by strong hydrogen-bonding interactions. The only interactions between water molecules and hexane molecules are the much weaker London dispersion forces. Therefore, the energy needed to break the hydrogen bonding interactions in the dissolution process is much greater than the energy released when the hexane and water molecules attract each other. In this case, the increase in the disorder of the mixture is not sufficient to overcome the unfavorable enthalpy change, so very low solubility results. The observed solubility of hexane in water is only 0.14 g/L.

Although consideration of the kinds of intermolecular attractions provides a general guide to relative solubilities, it does not help chemists make quantitative predictions with any reliability. Nevertheless, such considerations do allow us to predict the relative solubilities of a substance in two different solvents, as Example 12.6 illustrates.

EXAMPLE 12.6 Relative Solubilities

Predict the solvent in which the given compound is more soluble, and justify your prediction.

(a) Carbon tetrachloride [$CCl_4(\ell)$] in liquid water or hexane [$C_6H_{14}(\ell)$]
(b) Urea [$CO(NH_2)_2(s)$] in water or carbon tetrachloride
(c) Iodine [$I_2(s)$] in benzene [$C_6H_6(\ell)$] or water

Strategy Consider the types of intermolecular interactions in the solute and the solvents. Solubility will be greater in a solvent that has the more similar type of interaction.

Solution

(a) The following table lists the dominant intermolecular force present in each substance:

Substance	*Dominant intermolecular force*
Carbon tetrachloride	London dispersion force
Water	Hydrogen bonding
Hexane	London dispersion force

Carbon tetrachloride will be more soluble in hexane because hexane has similar dominant intermolecular forces.

(b) Again, we can construct a short table listing the dominant intermolecular force in each substance:

Substance	*Dominant intermolecular force*
Urea	Hydrogen bonding
Water	Hydrogen bonding
Carbon tetrachloride	London dispersion force

Urea will be more soluble in water because they have similar dominant intermolecular forces.

(c) Our table of forces is:

Substance	*Dominant intermolecular force*
Iodine	London dispersion force
Benzene	London dispersion force
Water	Hydrogen bonding

We would predict that iodine would be more soluble in benzene because they share the same dominant intermolecular force.

Understanding

Is methyl alcohol (CH_3OH) more soluble in water or in hexane? Explain.

Answer Water; both methyl alcohol and water have hydrogen bonding as their dominant intermolecular force

---

Solubility of Ionic Compounds in Water

Water is the most common solvent used to dissolve ionic compounds. The enthalpy changes that occur in the formation of aqueous solutions are an important factor in determining the solubilities of ionic substances. If the compound is soluble, the enthalpy of attraction between the ions in the solid must be comparable (within about 50 kJ/mol) with the enthalpy of attractions between the water molecules and the ions in the solution. The forces that hold the ionic solid together are the strong electrostatic attractions between oppositely charged ions that have energies of 400 kJ/mol or more. Because many ionic compounds are soluble, it is safe to conclude that enthalpies of interaction between solvent molecules and the ions must be approximately the same as the crystal lattice enthalpies in the solid compounds.

In solution, the polar water molecules are attracted by charged ions, as shown in Figure 12.9. Several water molecules are attracted to each ion in solution. The cations attract the negative ends of the water dipoles, whereas the positive ends of the water dipoles are drawn to the anions. Experiments indicate that the number of water molecules that surround each cation is between 4 and 10. The interaction of the ions with the water molecules is called **hydration.**

Figure 12.9 Hydration of ions. The polar water molecules are attracted to the ions in solution. The positive ends of the water dipole are attracted to the anions, whereas the cations attract the negative ends of the water dipole.

When ionic substances dissolve in water, the increase in the disorder of the solute is obvious, because the ions become free to move about. An increase in disorder also occurs when the water molecules separate to make room for the ions. At the same time, however, hydration of the ions restricts the freedom of some of the solvent molecules, decreasing their disorder. Thus, depending on the particular solute and its hydration by the water, the disorder of the solvent can either increase or decrease. Some examples exist in which the increase in order of the solvent causes the solution to have greater order than the separated components.

As seen in this section, many factors enter into the dissolution process—the change in disorder and the strengths of intermolecular attractions in the pure substances, as well as those in the mixture. In view of the complexity of the process, it is not surprising that solubilities are difficult to predict, and we must rely on experimental results or use solubility rules such as those given in Chapter 4 (which summarize experimental results).

The solubilities of ionic substances are difficult to predict; solubilities must be determined by experiment.

OBJECTIVES REVIEW *Can you:*

- ☑ define solubility and describe how to determine whether a solution is saturated, unsaturated, or supersaturated?
- ☑ describe the solution process on a molecular level?
- ☑ predict the relative solubilities of substances in different solvents based on solute-solvent interactions?

12.3 Effects of Pressure and Temperature on Solubility

OBJECTIVES

- ☐ State the effects of pressure and temperature on solubility
- ☐ Calculate the solubility of gases using Henry's law
- ☐ Explain why changes in pressure do not appreciably change the solubilities of solids and liquids

☐ Relate the sign of the enthalpy of solution to the increase or decrease of solubility with temperature

In the preceding section, we showed how the fundamental characteristics of the solute-solvent interactions influence the solubilities of substances. The solubilities of compounds also depend on the temperature and pressure. This section examines the dependence of solubility on the pressure and temperature.

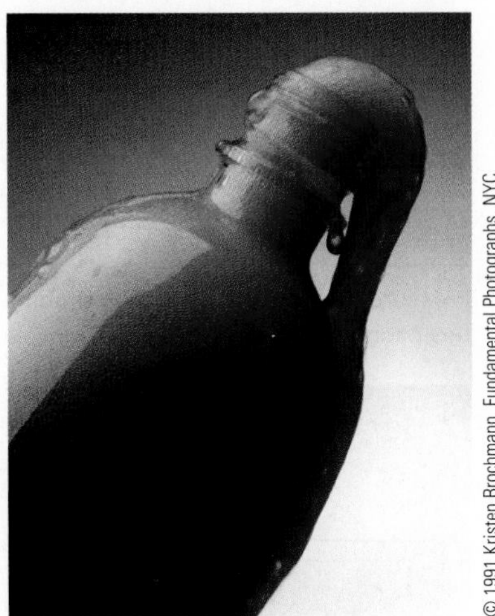

Carbonated beverages are sealed under a high pressure of carbon dioxide. When the container is opened, the sudden decrease in pressure causes bubbles of gas to form because of lower solubility of CO_2 under the new conditions. What is pictured here is similar to what happened at Lake Nyos in 1986, as recounted in the chapter introduction, though on a much smaller scale.

Effect of Pressure on Solubility

The solubility of a gas in any liquid is sensitive to pressure, just like phase equilibria are sensitive to pressure. In Section 11.3, we saw that an increase in pressure favors the denser phase (the phase that occupies a smaller volume)—the liquid, in the case of a gas-liquid equilibrium. Similarly, an increase in the pressure of a gas in contact with a saturated solution results in dissolving more gas molecules in the liquid solvent.

At pressures of a few atmospheres or less, the solubilities of gases obey **Henry's law:** The solubility of a gas is directly proportional to its partial pressure at any given temperature.

$$C = kP \qquad [12.1]$$

Here, C is the concentration of the gaseous substance in solution; k is a proportionality constant that is characteristic of the particular solute, solvent, and temperature; and P is the partial pressure of the gaseous solute in contact with the solution. The units of the constant k depend on the units used to express the concentration and the pressure.

Table 12.3 gives Henry's law constants for several gases (in units of molal/atm). These experimentally determined constants are used to calculate the solubilities of gases, as Example 12.7 illustrates.

The solubility of a gas in a liquid is directly proportional to its partial pressure.

EXAMPLE 12.7 Henry's Law Calculation

What is the molal concentration of oxygen in water at 20 °C that has been saturated with air at 1.00 atm? Assume that the mole fraction of oxygen in air is 0.21.

Strategy Use Henry's law to calculate the oxygen concentration in the water using the partial pressure of oxygen and the appropriate Henry's law constant from Table 12.3.

Solution

Before applying Henry's law, we must find the partial pressure of the oxygen in the gas phase, which is simply the mole fraction of oxygen (0.21) times the total pressure (Dalton's law of partial pressure):

$$P_{\text{oxygen}} = 0.21 \times 1.00 \text{ atm} = 0.21 \text{ atm}$$

The Henry's law constant for oxygen at 20 °C (see Table 12.3) is 1.43×10^{-3} molal/atm. Using this constant and the partial pressure of oxygen in Equation 12.1, calculate the concentration in solution.

$$C = kP = \left(1.43 \times 10^{-3} \frac{\text{molal}}{\cancel{\text{atm}}}\right) \times (0.21 \cancel{\text{atm}}) = 3.0 \times 10^{-4} \text{ molal}$$

TABLE 12.3 Henry's Law Constants for Several Gases in Water

Gas	k (molal/atm)			
	0 °C	20 °C	40 °C	60 °C
Carbon dioxide	7.60×10^{-2}	3.91×10^{-2}	2.44×10^{-2}	1.63×10^{-2}
Ethylene	1.14×10^{-2}	5.60×10^{-3}	3.43×10^{-3}	—
Helium	4.22×10^{-4}	3.87×10^{-4}	3.87×10^{-4}	4.10×10^{-4}
Nitrogen	1.03×10^{-3}	7.34×10^{-4}	5.55×10^{-4}	4.85×10^{-4}
Oxygen	2.21×10^{-3}	1.43×10^{-3}	1.02×10^{-3}	8.71×10^{-4}

Understanding

What is the molal concentration of nitrogen in this same solution? The mole fraction of nitrogen in air is 0.78.

Answer 5.7×10^{-4} molal

Pressure has little effect on the solubilities of solid and liquid solutes.

Unlike the solubilities of gases in liquids, the solubilities of liquids and solids change little with pressure. For gaseous solutes, an increase in pressure is relieved by additional gas dissolving in the liquid. When a liquid or solid dissolves in the solvent, there is little difference between the volume occupied by the solution and the sum of the volumes of the pure substances in the mixture. Thus, large changes in pressure (many atmospheres) are required to produce even small changes in the solubilities of liquids and solids in liquids.

Cengage Learning/Charles D. Winters

Figure 12.10 Change in solubility with temperature. The solubility of potassium dichromate increases when the temperature of the equilibrium system is increased.

Effect of Temperature on Solubility

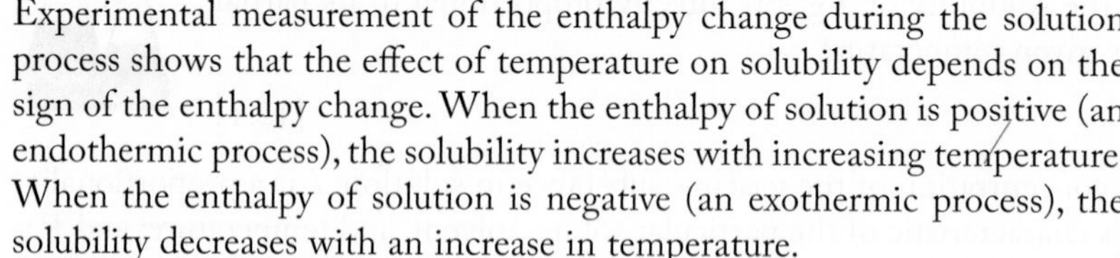

Experimental measurement of the enthalpy change during the solution process shows that the effect of temperature on solubility depends on the sign of the enthalpy change. When the enthalpy of solution is positive (an endothermic process), the solubility increases with increasing temperature. When the enthalpy of solution is negative (an exothermic process), the solubility decreases with an increase in temperature.

Figure 12.10 shows a saturated solution of potassium dichromate at two temperatures. The higher solubility of this colored salt is indicated by the more intense color of the more concentrated solution. The much greater solubility of this compound at the greater temperature is consistent with its positive (endothermic) enthalpy of solution, +66.5 kJ/mol.

As the temperature increases, the solubility increases for any substance with an endothermic enthalpy of solution and decreases for one with an exothermic enthalpy of solution.

The solubilities of most solids increase as the temperature of the solution increases. The graph in Figure 12.11 shows the solubilities of several ionic compounds as a function of temperature. In general, the more positive the enthalpy of solution, the greater the change in molar solubility with temperature. Note that the solubility of cerium(III) sulfate *decreases* as the temperature increases, consistent with its negative (exothermic) enthalpy of solution.

The solubilities of gases in water generally decrease with increasing temperature.

The enthalpy of solution for most gases in water is negative so their solubilities decrease with an increase in temperature (a behavior that may have been a factor in the Lake Nyos disaster). In the gas phase, there is almost no energy of attraction between the molecules. There are, however, attractions between the solvent and the solute molecules that result in a negative (exothermic) enthalpy of solution.

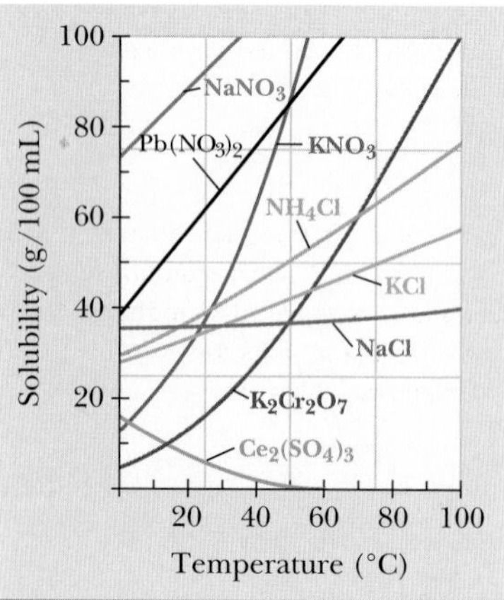

Figure 12.11 Temperature dependence of solubility. The solubilities of several ionic compounds in water are shown as a function of temperature. Most ionic compounds have greater solubility at higher temperatures.

EXAMPLE 12.8 Temperature Dependence of Solubility

The enthalpy of solution of potassium chromate (K_2CrO_4) in water is +17.4 kJ/mol. How does the solubility of this compound change when the temperature is reduced?

Strategy Determine whether the dissolution process is exothermic or endothermic, and apply the appropriate trend with decreasing temperature.

Solution

The positive sign of the enthalpy of solution means that the process is endothermic and the solubility of potassium chromate increases as the temperature increases, so decreasing the temperature reduces the solubility. Purification of solids by recrystallization depends on lower solubilities at low temperatures.

Understanding

At 1 atm pressure, the solubility of oxygen is 1.43×10^{-3} molal at 20 °C and 8.71×10^{-4} molal at 60 °C. What is the sign of the enthalpy of solution of oxygen?

Answer Negative

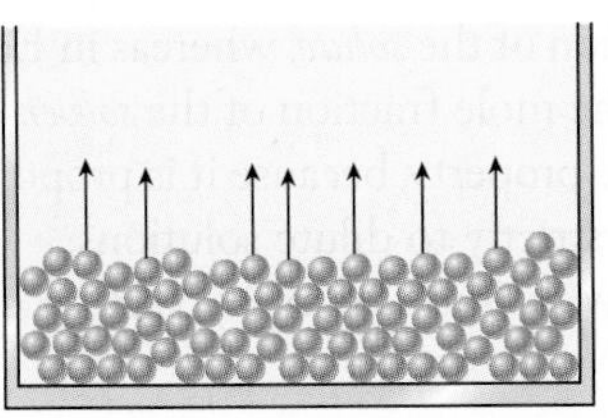

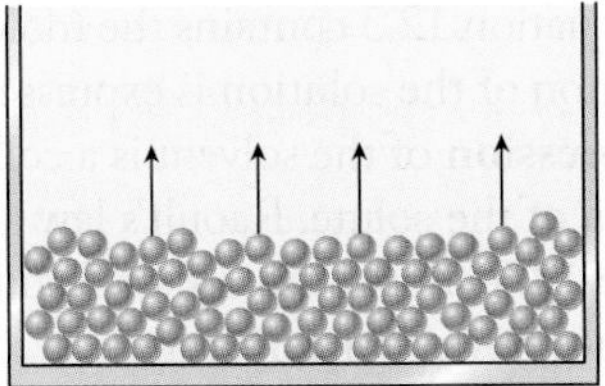

Figure 12.12 Rate of evaporation from solution. When a solution forms, the presence of the solute particles reduces the rate of evaporation of the solvent, thus decreasing the equilibrium vapor pressure.

OBJECTIVES REVIEW *Can you:*

- ☑ state the effects of pressure and temperature on solubility?
- ☑ calculate the solubility of gases using Henry's law?
- ☑ explain why changes in pressure do not appreciably change the solubilities of solids and liquids?
- ☑ relate the sign of the enthalpy of solution to the increase or decrease of solubility with temperature?

12.4 Colligative Properties of Solutions

OBJECTIVES

- ☐ List and define the colligative properties of solutions
- ☐ Relate the values of colligative properties to the concentrations of solutions
- ☐ Calculate the molar masses of solutes from measurements of colligative properties

The physical properties of a solution differ from those of the pure solvent. **Colligative properties** are those properties of solutions that change in proportion to the concentration of solute particles. They do not depend on the identity of the solute particles, only on their concentrations. Colligative properties are a means of "counting" the solute particles present in the sample and can be used to determine the molar mass of the solute. This section examines the origins and uses of several of these colligative properties.

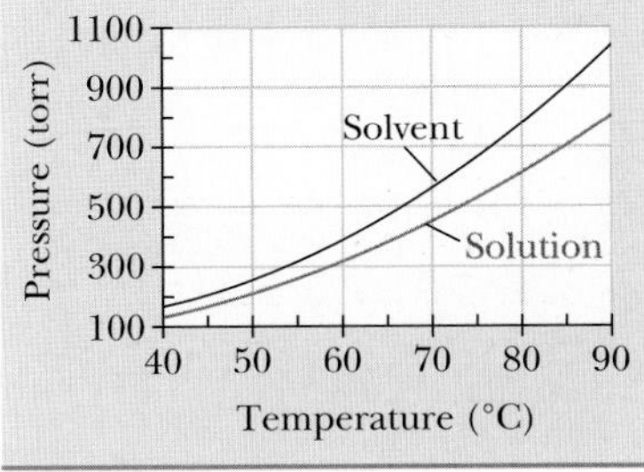

Figure 12.13 Vapor pressure of solutions as a function of temperature. The vapor pressure of benzene (upper curve) and that of a benzene solution of a nonvolatile solute (lower curve) are shown. The vapor pressure of the solvent in a solution is always lower than that of the pure solvent, at all temperatures.

Vapor-Pressure Depression of the Solvent

Experiments described in Chapter 11 showed that the equilibrium between a liquid and its vapor produces a characteristic vapor pressure for each substance, which depends on the temperature. Experimentally, it is found that adding a nonvolatile solute (one with a negligible vapor pressure of its own) to a solvent always reduces the equilibrium vapor pressure. The lowering of the vapor pressure is caused by a lesser ability of the solvent to evaporate (Figure 12.12), so equilibrium is reached with a smaller concentration (partial pressure) of the solvent in the gas phase.

Figure 12.13 shows the temperature dependence of the vapor pressure of a pure solvent and that of the same solvent that contains a nonvolatile solute. The vapor pressure of a solution is expressed quantitatively by **Raoult's law:** The vapor pressure of the solvent (P_{solv}) above a dilute solution is equal to the mole fraction of the solvent (χ_{solv}) times the vapor pressure of the pure solvent (P°_{solv}):

$$P_{solv} = \chi_{solv} P^{\circ}_{solv} \qquad [12.2]$$

Raoult's law states that the partial pressure of a substance in equilibrium with a solution is equal to its mole fraction in the solution times the vapor pressure of the pure substance.

By substituting $(1 - \chi_{solute})$ into Equation 12.2 in place of χ_{solv} and rearranging, we see that the *difference* between the vapor pressure of the pure solvent and the vapor pressure of the solution (ΔP) is proportional to the concentration of the solute:

$$P_{solv} = (1 - \chi_{solute})P^{\circ}_{solv}$$

$$P_{solv} = P^{\circ}_{solv} - \chi_{solute}P^{\circ}_{solv}$$

which rearranges to

$$\Delta P = P^{\circ}_{solv} - P_{solv} = \chi_{solute}P^{\circ}_{solv} \qquad [12.3]$$

For vapor-pressure depression and all other colligative properties, the effect is proportional to the concentration of the solute particles.

Note that Equation 12.3 contains the mole fraction of the *solute,* whereas in Equation 12.2, the composition of the solution is expressed by the mole fraction of the *solvent.* The **vapor-pressure depression** of the solvent is a colligative property, because it is proportional to the concentration of the solute. Raoult's law applies strictly to dilute solutions.

Example 12.9 shows the use of Raoult's law to find the mole fraction of solute in a solution of unknown concentration.

EXAMPLE 12.9 Finding the Solute Concentration by Vapor-Pressure Depression

At 25 °C the vapor pressure of pure benzene is 93.9 torr. A solution of a nonvolatile solute in benzene has a vapor pressure of 91.5 torr at the same temperature. What is the concentration of the solute, expressed as its mole fraction?

Strategy Determine the change in the vapor pressure of benzene. Use the vapor-pressure concentration relationship from Raoult's law and solve for mole fraction.

Solution

From the data given, the vapor-pressure depression is

$$\Delta P = 93.9 \text{ torr} - 91.5 \text{ torr} = 2.4 \text{ torr}$$

Substitute this into Raoult's law (see Equation 12.3), together with the vapor pressure of the solvent, and solve for the mole fraction of solute.

$$\Delta P = \chi_{\text{solute}} P^{\circ}_{\text{solv}}$$

$$2.4 \text{ torr} = \chi_{\text{solute}}(93.9 \text{ torr})$$

$$\chi_{\text{solute}} = \frac{2.4 \text{ torr}}{93.9 \text{ torr}} = 0.026$$

Understanding

What is the solute concentration in a benzene solution that has a vapor pressure of 90.6 torr at 25 °C?

Answer $\chi_{\text{solute}} = 0.035$

Accurate measurements of vapor pressures are required for accurate concentrations. However, these measurements are difficult to make accurately, so the depression of vapor pressure is not widely used to determine concentrations. Two other colligative properties, boiling-point elevation and freezing-point depression, are much easier to measure accurately.

Boiling-Point Elevation

Figure 12.14 shows the vapor pressure of a solvent and several dilute solutions near the normal boiling point of the solvent. At a pressure of 1 atm (760 torr), the boiling points of the solutions are all greater than those of the pure solvent. Furthermore, the graph in Figure 12.15 shows that the difference between the boiling point of each solution and that of the solvent is proportional to the concentration of the solute particles. When the **boiling-point elevation** is considered, the concentration of the solute is usually expressed as molality rather than as mole fraction of the solute. As long as the solution is fairly dilute, molality is proportional to the mole fraction of solute. However, expressing the concentration as molality makes it easier to calculate the molar mass of the solute from the boiling-point elevation.

The effect of the solute concentration on the boiling point can be written in equation form:

$$\Delta T_b = mk_b \qquad [12.4]$$

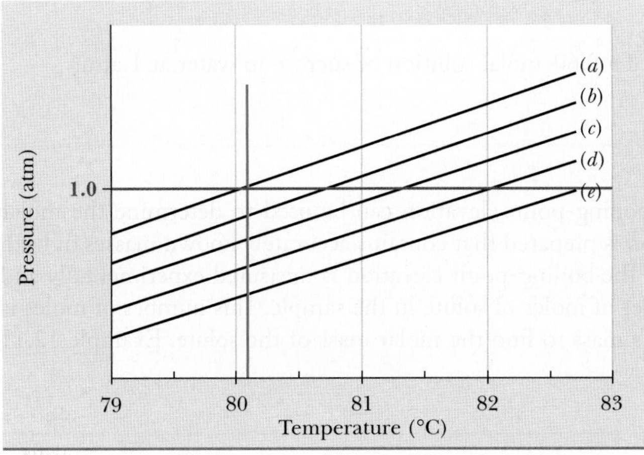

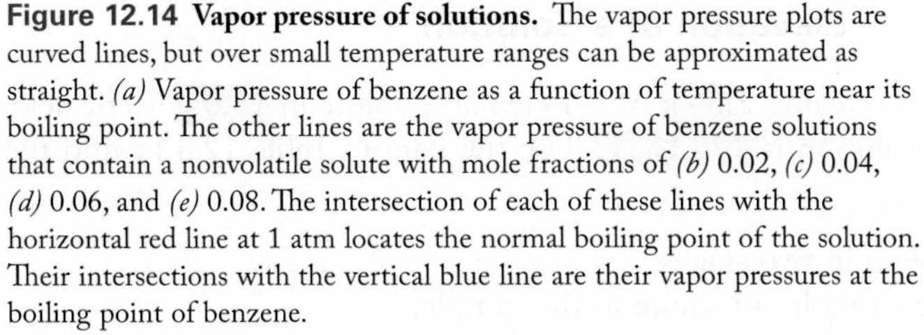
Figure 12.14 Vapor pressure of solutions. The vapor pressure plots are curved lines, but over small temperature ranges can be approximated as straight. *(a)* Vapor pressure of benzene as a function of temperature near its boiling point. The other lines are the vapor pressure of benzene solutions that contain a nonvolatile solute with mole fractions of *(b)* 0.02, *(c)* 0.04, *(d)* 0.06, and *(e)* 0.08. The intersection of each of these lines with the horizontal red line at 1 atm locates the normal boiling point of the solution. Their intersections with the vertical blue line are their vapor pressures at the boiling point of benzene.

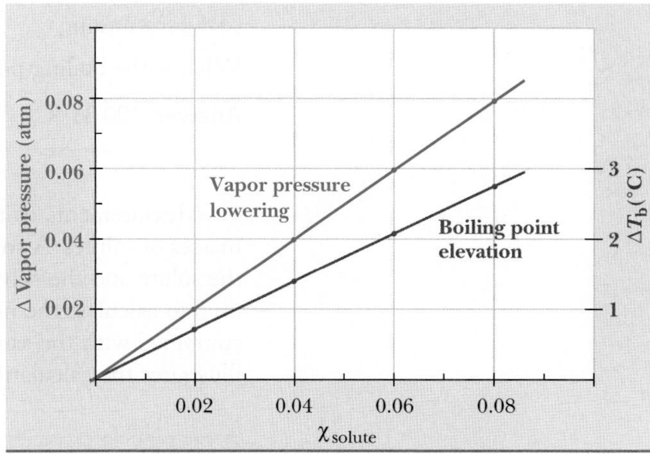

Figure 12.15 Solution vapor pressures and boiling points. As long as the solutions are dilute, both the vapor pressure depression and boiling point elevation are proportional to the concentration. Data plotted in this figure are calculated from Figure 12.14.

where ΔT_b is the increase in the boiling point (calculated as b.p.$_{\text{solution}}$ − b.p.$_{\text{solvent}}$), m is the molal concentration of the solute, and k_b is the *boiling-point elevation constant,* which depends only on the solvent used. Table 12.4 gives values of this constant for several solvents.

The proportionality constant that relates a colligative property to concentration is characteristic of the solvent but not the particular solute.

EXAMPLE 12.10 Calculating the Boiling Point of a Solution

What is the boiling point of a 0.32-molal solution of iodine (I_2) in benzene at 1 atm?

Strategy Use the value of k_b for benzene given in Table 12.4 and the molal concentration of the solute to find the *change* of the boiling point of the solution. This change must be added to the normal boiling point of the solvent.

Solution

The change in boiling point is proportional to the molal concentration of the solution:

$$\Delta T_b = mk_b$$

$$\Delta T_b = 0.32\ \cancel{m} \times 2.53\ °\text{C}/\cancel{m} = 0.81\ °\text{C}$$

The boiling point of the solution is 0.81 °C greater than that of the pure solvent, so find the boiling point of the solution by adding ΔT_b to the normal boiling point of the pure solvent.

$$T_b = 80.10\ °\text{C} + 0.81\ °\text{C} = 80.91\ °\text{C}$$

TABLE 12.4 Boiling-Point Increases and Freezing-Point Depression Constants for Solvents

Solvent	Boiling Point (°C)	k_b (°C/m)	Freezing Point (°C)	k_f (°C/m)
Acetic acid	117.90	3.07	16.60	3.90
Benzene	80.10	2.53	5.51	4.90
Naphthalene			80.2	6.8
Water	100.00	0.512	0.00	1.86

Understanding

What is the boiling point of a 0.60-molal solution of sucrose in water at 1 atm?

Answer 100.31 °C

Measurements of the boiling-point elevation can be used to determine the molar masses of solutes. A solution is prepared that contains accurately known masses of both the solute and the solvent. The boiling-point elevation is measured experimentally and used to calculate the number of moles of solute in the sample. This number of moles is combined with the sample's mass to find the molar mass of the solute. Example 12.11 illustrates the calculation.

EXAMPLE 12.11 Determining Molar Mass by Boiling-Point Elevation of a Solution

A solution is prepared by dissolving 1.00 g of a nonvolatile solute in 15.0 g acetic acid. The boiling point of this solution is 120.17 °C. Use the data in Table 12.4 to find the molar mass of the solute.

Strategy Solve this problem in two stages:

1. Calculate the number of moles of solute in the sample.

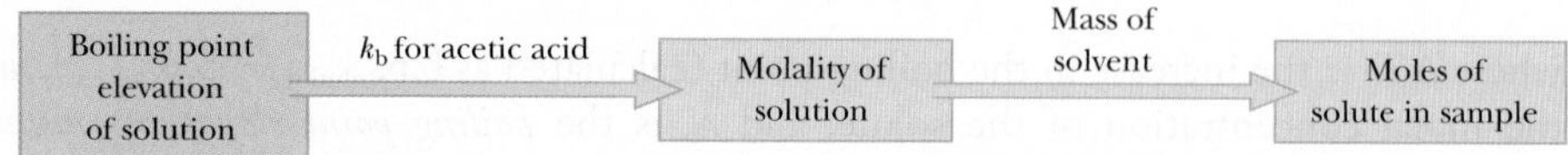

2. Calculate the molar mass by dividing the mass of the solute by the number of moles.

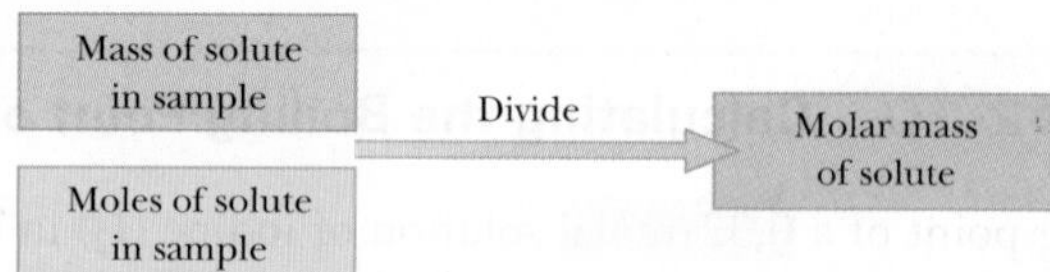

From the boiling point of the solution and the information given in Table 12.4 for the solvent, acetic acid, you can calculate the molality.

Solution

The change in the boiling point is

$$\Delta T_b = 120.17 - 117.90 = 2.27\ ^\circ\text{C}$$

Find the molal concentration of the solute with Equation 12.4, using the boiling-point elevation constant given in Table 12.4.

$$\Delta T_b = mk_b$$

$$2.27\ ^\circ\text{C} = m(3.07^\circ\text{C/molal})$$

Solving for m:

$$m = \frac{2.27\ \cancel{^\circ\text{C}}}{3.07\ \cancel{^\circ\text{C}}/\text{molal}} = 0.739\ m$$

The solution contains 0.739 mol of solute for each kilogram of solvent. Because the solution was prepared using 15.0 g of the solvent, use the molal concentration of the solution as a conversion factor to find the amount of solute present in the sample.

$$\text{Moles solute} = 15.0\ \cancel{\text{g}}\ \text{solvent} \times \left(\frac{1\ \cancel{\text{kg}}}{1000\ \cancel{\text{g}}}\right) \times \left(\frac{0.739\ \text{mol solute}}{1\ \cancel{\text{kg}}\ \text{solvent}}\right)$$

$$= 1.11 \times 10^{-2}\ \text{mol solute}$$

Because 1.00 g equals 1.11×10^{-2} mol,

$$1.00 \text{ g} = 1.11 \times 10^{-2} \text{ mol}$$

we divide both sides by 1.11×10^{-2} to get the number of grams in 1 mol:

$$\frac{1.00 \text{ g}}{1.11 \times 10^{-2}} = \frac{1.11 \times 10^{-2} \text{ mol}}{1.11 \times 10^{-2}}$$

$$90.1 \text{ g} = 1 \text{ mol}$$

Thus, the molar mass is 90.1 g/mol.

Understanding

Find the molar mass of a nonvolatile solute, if a solution of 1.20 g of the compound dissolved in 20.0 g benzene has a boiling point of 80.94 °C.

Answer 1.8×10^2 g/mol

Freezing-Point Depression

The freezing point of a solution is lower than that of the pure solvent. The decrease of the freezing point is related to the decrease of the vapor pressure of the solvent, as shown in Figure 12.16. As the concentration of the solute increases, the triple-point temperature of the solution decreases, which moves the solid-liquid equilibrium line to lower temperatures. This **freezing-point depression** is proportional to the concentration of solute particles, as long as the solution is reasonably dilute. The concentration unit used in freezing-point depression experiments is molality, as shown in Equation 12.5.

$$\Delta T_f = mk_f \qquad [12.5]$$

The freezing-point depression constant, k_f, is also a characteristic of the solvent and independent of the kind of solute particles. The freezing points and the freezing-point depression constants for some common solvents are included in Table 12.4. Note the similarity of Equations 12.5 (freezing-point depression) and 12.4 (boiling-point elevation). Note, however, that Equations 12.4 and 12.5 do not tell you explicitly the *direction* of the temperature change; rather, they only help you determine ΔT. You must remember that boiling points always go up for solutions, and freezing points always go down for solutions.

The use of freezing-point depression data is quite similar to that of boiling-point elevation data, as the following examples illustrate.

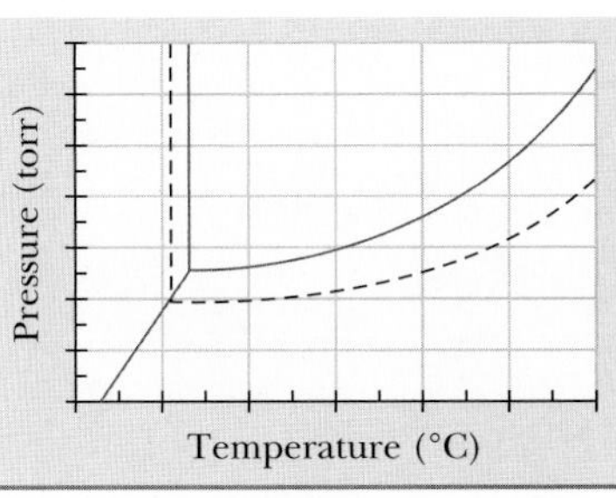

Figure 12.16 Freezing point of a solution. The phase diagram shows that the presence of a solute reduces the vapor pressure of the solvent and the triple-point temperature. As a result, the freezing point of a solution is also lower than that of the pure solvent. The solid lines show the behavior of the pure solvent, and the dotted lines show the vapor pressure and freezing point of a solution.

The solute concentration is expressed in molality for calculations with boiling-point elevation and freezing-point depression.

EXAMPLE 12.12 Determining Freezing Point of a Solution

Pure ethylene dibromide freezes at 9.80 °C. A solution is made by dissolving 0.213 g ferrocene (molecular formula $Fe(C_5H_5)_2$, molar mass = 186.04 g/mol) in 10.0 g ethylene dibromide. The freezing-point depression constant, k_f, for ethylene dibromide is 11.8 °C/molal. What is the freezing point of this solution?

Strategy To find the freezing point, use Equation 12.5, but first you must calculate the molality of the solution, m. This requires converting the quantity of ferrocene to moles and expressing the quantity of solvent in kilograms. Substitute these quantities into Equation 12.5 and solve for ΔT; then subtract this temperature change from the normal freezing point of ethylene dibromide.

Solution

First, determine the molality of the ferrocene solution:

$$\text{moles ferrocene} = 0.213 \text{ g ferrocene} \times \left(\frac{1 \text{ mol}}{186.04 \text{ g}}\right)$$

$$= 1.14 \times 10^{-3} \text{ mol ferrocene}$$

This amount of ferrocene was dissolved in 10.0 g, or 0.0100 kg, of the solvent. The molal concentration of the ferrocene in the solution is

$$\text{Molality} = \frac{1.14 \times 10^{-3}\ \text{mol ferrocene}}{0.0100\ \text{kg solvent}} = 1.14 \times 10^{-1}\ m$$

Find the freezing-point depression by using Equation 12.5 and substituting the values we have:

$$\Delta T_f = mk_f = (1.14 \times 10^{-1}\ \cancel{m})(11.8\ °\text{C}/\cancel{m})$$

$$\Delta T_f = 1.35\ °\text{C}$$

This is the *change* in the freezing point, not the actual freezing point. To find the freezing point of the solution, subtract 1.35 °C from the freezing point of the pure solvent, which is 9.80 °C:

$$T_f = 9.80\ °\text{C} - 1.35\ °\text{C}$$

$$T_f = 8.45\ °\text{C}$$

Understanding

Benzophenone has a freezing point of 49.00 °C. What is the freezing point of a 0.450-molal solution of urea in this solvent if the freezing-point depression constant is equal to 9.80 °C/*m*?

Answer 44.59 °C

EXAMPLE 12.13 Calculating Molar Mass from the Freezing-Point Depression

To identify a newly prepared substance, a scientist needs to measure its molar mass. The scientist prepares a solution by dissolving 0.350 g of the unknown compound in 5.42 g ethylene dibromide. This solution has a freezing point of 6.34 °C. Using the data from Example 12.12, find the molar mass of the solute.

Strategy Use the freezing-point depression to find the number of moles of solute in the sample.

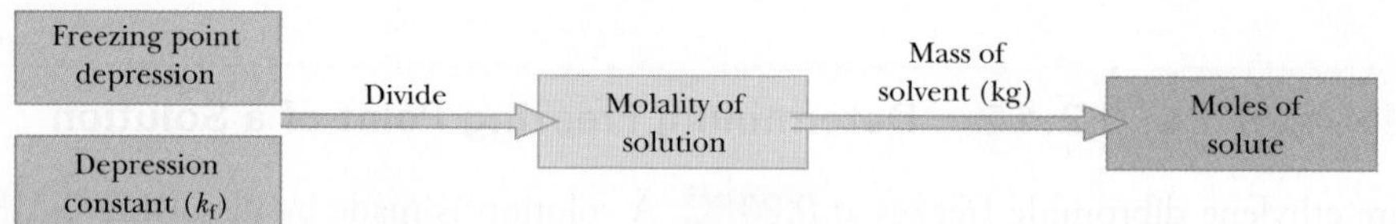

Solution

Rearranging Equation 12.5 and substituting:

$$m = \frac{\Delta T_f}{k_f} = \frac{(9.80 - 6.34)\ °\text{C}}{11.8\ °\text{C/molal}} = 0.293\ m$$

$$\text{moles solute} = 5.42\ \cancel{g}\ \text{solvent} \times \left(\frac{1\ \cancel{kg}}{1000\ \cancel{g}}\right) \times \left(\frac{0.293\ \text{mol}}{1\ \cancel{kg}\ \text{solvent}}\right)$$

$$= 1.59 \times 10^{-3}\ \text{mol solute}$$

Because 0.350 g of solute is 1.59×10^{-3} mol:

$$0.350\ \text{g} = 1.59 \times 10^{-3}\ \text{mol}$$

divide both sides of the equation by 1.59×10^{-3} to get the number of grams in 1 mol:

$$\frac{0.350 \text{ g}}{1.59 \times 10^{-3}} = \frac{1.59 \times 10^{-3} \text{ mol}}{1.59 \times 10^{-3}}$$

$220 \text{ g} = 1 \text{ mol}$

The molar mass is 220 g/mol.

Understanding

A solution of 0.134 g of a compound in 4.76 g ethylene dibromide has a freezing point of 7.62 °C. Find the molar mass of this solute.

Answer 152 g/mol

Osmotic Pressure

The final colligative property to be discussed is osmotic pressure. Thin layers of certain materials, called **semipermeable membranes,** allow only water and other small molecules to pass through them. **Osmosis** is the diffusion of a fluid through a semipermeable membrane. Animal bladders, skins of fruits and vegetables, and cellophane are examples of semipermeable membranes. Cell walls are semipermeable membranes that are crucial in biological systems; they control the transport of nutrients and waste products across cell boundaries.

Figure 12.17 shows an apparatus that uses a semipermeable membrane to separate pure water from an aqueous solution. Only the water molecules can pass through the membrane, and they move in both directions. Because the concentration of water is greater in the pure water, more water molecules strike the membrane per second on that side, and more water moves into the solution than leaves it. With time, there is a net movement of water into the solution side of the tube, the solution is diluted, and the level of liquid rises in the tube containing the solution. The different heights of liquid on either side of the semipermeable membrane cause a different pressure on the two sides of the membrane. At some point, the difference in pressure is sufficient to make the rates of passage of water equal in both directions, and a state of dynamic equilibrium is achieved. The same state of equilibrium can be achieved by applying a pressure to the solution with a piston, as shown in Figure 12.17*c*. The **osmotic pressure** of a solution is

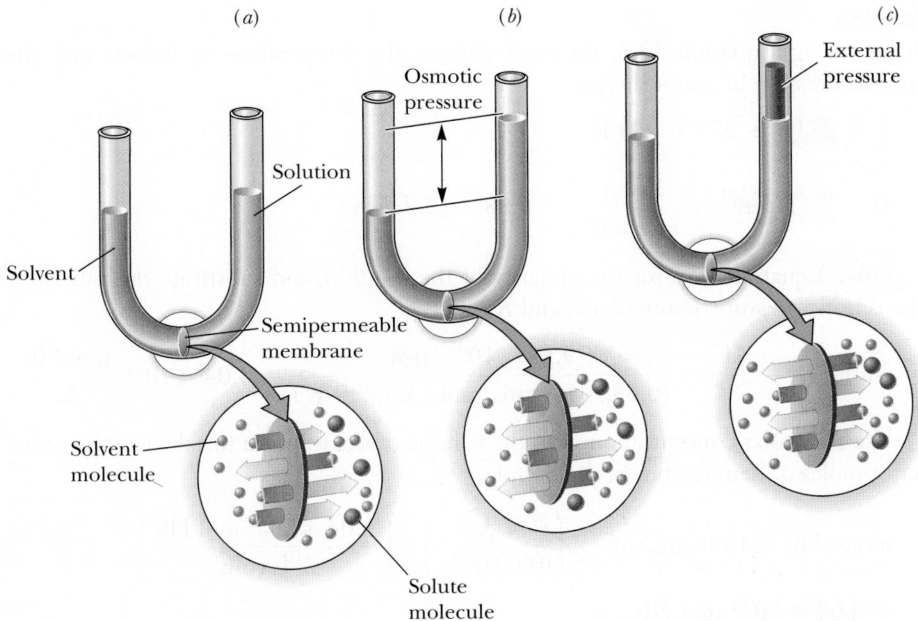

Figure 12.17 Osmotic pressure. A semipermeable membrane separates a solution from the pure solvent. *(a)* Initially, the liquid levels are equal, so there is no pressure difference. The rate of passage of water into the solution is greater. *(b)* The level of liquid on the solution side of the membrane has risen to the equilibrium height, and its weight exerts sufficient pressures that the rates of solvent transfer are equal in both directions through the membrane. *(c)* The application of an external pressure to the solution can result in the same equilibrium situation. This external pressure is equal to the osmotic pressure of the solution.

the pressure difference needed for no net transfer of solvent to occur across a semipermeable membrane that separates the solution from the pure solvent.

The osmotic pressure of a solution is proportional to the molar concentration of the solute particles in the solution. Experiments have shown that the osmotic pressure of a solution can be calculated from an equation similar to the ideal gas law:

$$\Pi = \frac{nRT}{V} = MRT \qquad [12.6]$$

where Π is the osmotic pressure, n is the number of moles of solute present, V is the volume of the solution, R is the ideal gas constant $\left(0.0821 \frac{\text{L} \cdot \text{atm}}{\text{mol} \cdot \text{K}}\right)$, and T is the temperature in kelvins. Since n/V is the molar concentration of the solute, M, the right side of equation 12.6 can be simplified as shown.

Osmotic pressure is a sensitive colligative property. The concentration of the solute is expressed in units of moles per liter (molarity).

Of all the colligative properties, osmotic pressure is the most sensitive. For example, a 0.0100 M solution of sugar in water at 25 °C has an osmotic pressure of 0.245 atm, which corresponds to a column of water 2.53 meters high. The osmotic pressure is so sensitive that it is used to measure the molar masses of very large molecules and substances that are only slightly soluble in water. Example 12.14 illustrates this procedure.

EXAMPLE 12.14 Determining Molar Mass from Osmotic Pressure

Hemoglobin is a large molecule that carries oxygen in human blood. A water solution that contains 0.263 g of hemoglobin (abbreviated here as Hb) in 10.0 mL of solution has an osmotic pressure of 7.51 torr at 25 °C. What is the molar mass of hemoglobin?

Strategy Use the measurement of the osmotic pressure to find the number of moles of solute; combine that number with the mass data given in the problem to find the molar mass.

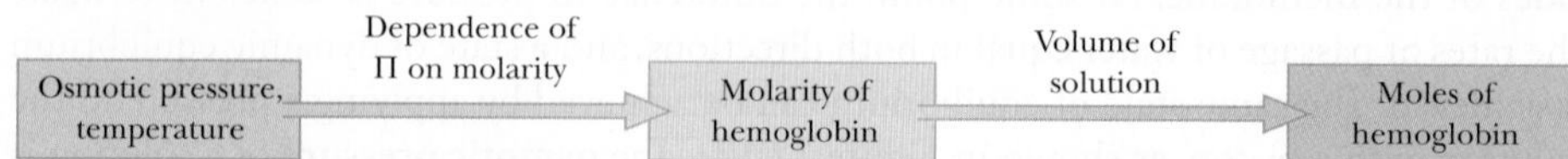

Solution

Before using Equation 12.6, we must express the temperature in kelvins and the osmotic pressure in atmospheres.

$$T = 25\ ^\circ\text{C} + 273 = 298\ \text{K}$$

$$\Pi = 7.51\ \cancel{\text{torr}} \times \left(\frac{1\ \text{atm}}{760\ \cancel{\text{torr}}}\right) = 9.88 \times 10^{-3}\ \text{atm}$$

Solve Equation 12.6 for the molarity of the solution, and substitute the values for the osmotic pressure, temperature, and R.

$$\text{Molarity} = \frac{\Pi}{RT} = \frac{9.88 \times 10^{-3}\ \cancel{\text{atm}}}{(0.0821\ \text{L} \cdot \cancel{\text{atm}}/\text{mol} \cdot \cancel{\text{K}})(298\ \cancel{\text{K}})} = 4.04 \times 10^{-4}\ \frac{\text{mol Hb}}{\text{L}}$$

Use the molar concentration with the volume of the solution to calculate the number of moles of hemoglobin in the sample.

$$\text{moles Hb} = 10.0\ \cancel{\text{mL soln}} \times \left(\frac{1\ \cancel{\text{L}}}{1000\ \cancel{\text{mL}}}\right) \times \left(\frac{4.04 \times 10^{-4}\ \text{mol Hb}}{1\ \cancel{\text{L soln}}}\right)$$

$$= 4.04 \times 10^{-6}\ \text{mol Hb}$$

PRACTICE OF CHEMISTRY
Reverse Osmosis Makes Fresh Water from Seawater

As the population of the world increases, it becomes increasingly difficult to provide fresh water suitable for drinking. This situation is particularly true in semiarid regions, such as the Middle East. Scientists have proposed various schemes to reduce the concentration of salts in seawater and water from other natural sources, making it fit for human consumption. One means of purifying water, a process called *reverse osmosis,* is the subject of continuing research.

Reverse osmosis is the movement of water from a solution on one side of a semipermeable membrane to pure water on the other side, by applying a pressure greater than the osmotic pressure to the solution. We have seen that no net movement of water across a semipermeable membrane occurs when the osmotic pressure is applied to the solution. If a pressure greater than the osmotic pressure is applied to the solution, however, net transfer of the water from the solution side to the pure water side occurs. Because the osmotic pressure of natural seawater is fairly high, one of the problems in the design of a reverse osmosis apparatus is to make a membrane that is thin enough to allow rapid movement of the water, yet is strong enough to withstand high pressures without bursting.

Reverse osmosis is used for water purification in Saudi Arabia. In 1991, this water source was threatened when the Iraqis, who were at war, dumped crude oil from wells in Kuwait into the Persian Gulf. The reverse osmosis plants were shut down, because the oil would have ruined the semipermeable membranes in the equipment.

The U.S. Navy has developed small, portable, manually operated units to desalinize seawater, for use in life rafts. Such units are capable of producing 5 L of drinkable water per hour, which is sufficient to keep several people alive. These units are replacing the bulky containers of fresh water now stored on Navy lifeboats. Backpackers can carry small reverse osmosis pumps with them to make fresh water on the trail. ▮

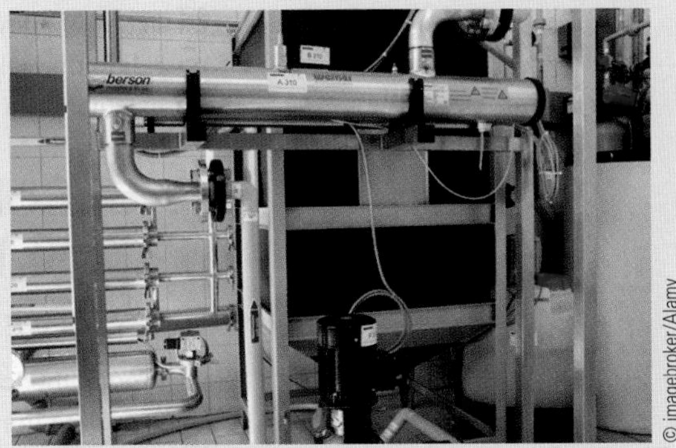

This reverse osmosis system provides the water for an island resort.

Finally, relate the mass of the sample with the number of moles of hemoglobin to find the molar mass.

$$0.263 \text{ g} = 4.04 \times 10^{-6} \text{ mol}$$

Divide both sides by 4.04×10^{-6}:

$$\frac{0.263 \text{ g}}{4.04 \times 10^{-6}} = \frac{4.04 \times 10^{-6} \text{ mol}}{4.04 \times 10^{-6}}$$

$$6.51 \times 10^{4} \text{ g} = 1 \text{ mol}$$

The molar mass of Hb is 6.51×10^4 g/mol.

Understanding

A 5.70-mg sample of a protein is dissolved in water to give 1.00 mL of solution. If the osmotic pressure of this solution is 6.52 torr at 20 °C, what is the molar mass of the protein?

Answer 1.60×10^4 g/mol

All of the colligative properties fit the following relationship:

Property = solute concentration × constant

TABLE 12.5 Concentration Units for Colligative Properties

Property	Symbol	Solute Concentration	Constant
Vapor-pressure depression	ΔP	Mole fraction	P°_{solv}
Boiling point elevation	ΔT_b	Molality	k_b
Freezing point depression	ΔT_f	Molality	k_f
Osmotic pressure	Π	Molarity	RT

One colligative property differs from another in the units in which the solute concentration is expressed. In most cases, the property is expressed as a difference between its value in the solution and its value in the pure solvent. Table 12.5 matches each of the colligative properties with the concentration units used and the usual abbreviation for the proportionality constant.

OBJECTIVES REVIEW *Can you:*

- ☑ list and define the colligative properties of solutions?
- ☑ relate the values of colligative properties to the concentrations of solutions?
- ☑ calculate the molar masses of solutes from measurements of colligative properties?

12.5 Colligative Properties of Electrolyte Solutions

OBJECTIVES

- ☐ Predict the ideal van't Hoff factor of ionic solutes
- ☐ Calculate the colligative properties for solutions of electrolytes
- ☐ Explain why colligative properties of ionic solutions vary from the predicted properties

As chemists collected data on the colligative properties of solutions, it became apparent that solutions of ionic and molecular solutes behave differently. This information provided insights into the nature of solutions, particularly the ability of some solutes to dissociate into ions. This section summarizes the results and interprets such studies.

van't Hoff Factor

Jacobus van't Hoff (1852–1911) and several other scientists noticed that the effect of some solutes on colligative properties was greater than expected. For example, a 0.01 *M* urea solution and a 0.01 *M* sucrose solution showed colligative properties of 0.01 *M* solutions. But a 0.01 *M* NaCl solution showed colligative properties of a 0.02 *M* solution, whereas a 0.01 *M* $CaCl_2$ solution showed colligative properties of a 0.03 *M* solution.

We understand now that ionic solutes separate into ions when they dissolve into solution. (Indeed, the colligative properties experiments described in the previous paragraph were valuable evidence for dissociation into ions.) For example, NaCl and $CaCl_2$ dissociate according to the following reactions:

$$NaCl(s) \xrightarrow{H_2O} Na^+(aq) + Cl^-(aq)$$

$$CaCl_2(s) \xrightarrow{H_2O} Ca^{2+}(aq) + 2Cl^-(aq)$$

So for every mole of NaCl that dissolves, two moles of ions are formed; likewise, for every mole of $CaCl_2$, three ions are formed. Because colligative properties depend only on the number of particles dissolved in solution, the colligative properties of ionic solutes will be magnified because of the increased number of dissolved particles due to ionic dissociation.

The **van't Hoff factor, *i*,** is defined by the following equation:

$$i = \frac{\text{measured colligative property}}{\text{expected value for a nonelectrolyte}}$$

TABLE 12.6 van't Hoff Factor Values for Electrolytes in Water at 0.05 Molal

Compound	ΔT_f (°C)	Measured i	Ideal i
NaCl	0.176	1.9	2.0
HIO_3	0.156	1.7	2.0
$Ca(NO_3)_2$	0.235	2.5	3.0
$MgCl_2$	0.249	2.7	3.0
$AlCl_3$	0.300	3.2	4.0

The equations for the colligative properties listed in Table 12.5 are modified by including the van't Hoff factor in the product of other terms. Thus, the equation for the boiling point elevation becomes

$$\Delta T_b = imk_b$$

For solutions that involve typical nonelectrolytes, such as urea and sucrose, the van't Hoff factor is 1. For solutions of salts and other electrolytes, i has a value greater than 1. For dilute solutions (0.01 m or less), we would expect the van't Hoff factor to equal the number of ions produced by each formula unit of the compound that dissolves: two for NaCl and $MgSO_4$, three for $Ca(NO_3)_2$, and so forth.

It is important to remember that colligative properties are proportional to the concentration of solute *particles* in solution.

Nonideal Solutions

At greater concentrations than about 0.01 m, the observed values for i tend to be smaller than the values expected. The strong electrostatic attractions between oppositely charged, closely spaced ions cause some ions to cluster together in solution and to behave as a single particle. This association can partially account for the lower values of i observed for more concentrated solutions. Table 12.6 shows some typical values of the van't Hoff factor for several electrolytes, as determined by the measured freezing points of the solutions.

EXAMPLE 12.15 van't Hoff Factor and Colligative Properties

Arrange the following aqueous solutions in order of increasing freezing point, assuming ideal behavior: 0.05 m sucrose, 0.02 m NaCl, 0.01 m $CaCl_2$, 0.03 m HCl.

Strategy The freezing point of a solution depends on the overall molal concentration of solute particles. Because several of these solutes are electrolytes, the concentration of particles is the product of the van't Hoff factor with the molal concentration of the solute.

Solution

The calculations of these products are summarized in a table, using an ideal value of the van't Hoff factor.

Compound	Present as	i	m	$i \times m$
Sucrose	Molecules	1	0.05	0.05
NaCl	$Na^+ + Cl^-$	2	0.02	0.04
$CaCl_2$	$Ca^{2+} + 2Cl^-$	3	0.01	0.03
HCl	$H^+ + Cl^-$	2	0.03	0.06

The lowest freezing point is expected for the solution with the *highest* concentration of solute particles, so the freezing points of these solutions increase in the following order:

0.03 m HCl < 0.05 m sucrose < 0.02 m NaCl < 0.01 m $CaCl_2$

Understanding

Arrange the following solutions in order of increasing osmotic pressure: 0.02 M sucrose, 0.02 M HNO_3, 0.01 M $BaCl_2$.

Answer 0.02 M sucrose < 0.01 M $BaCl_2$ < 0.02 M HNO_3

OBJECTIVES REVIEW *Can you:*

- ☑ predict the ideal van't Hoff factor of ionic solutes?
- ☑ calculate the colligative properties for solutions of electrolytes?
- ☑ explain why colligative properties of ionic solutions vary from the predicted properties?

12.6 Mixtures of Volatile Substances

OBJECTIVES

- ☐ Calculate the vapor pressure of each component and the total vapor pressure over an ideal solution
- ☐ Explain how fractional distillation works

The characteristics of solutions that contain more than one volatile substance are important in the separation and purification of many substances. For example, the separation of crude oil into components such as gasoline, diesel fuel, and asphalt depends on the different tendencies of compounds to evaporate.

In Section 12.4, we considered the vapor pressure of the solvent over a solution in which the only volatile component was the solvent. Many solutions exist in which two or more of the components are volatile—that is, have significant vapor pressures. The vapor phases in equilibrium with these solutions contain all of the volatile components. This section examines the vapor pressure and composition of the gas in equilibrium with such a mixture.

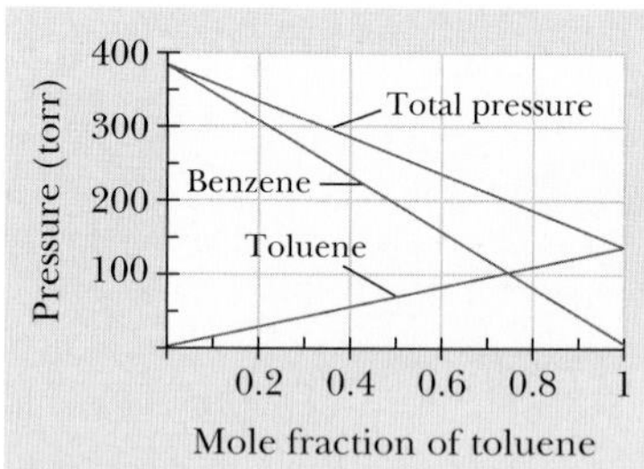

Figure 12.18 Vapor pressure of an ideal solution. The graph shows the vapor pressures of mixtures of benzene and toluene as the composition changes at a constant temperature. Benzene and toluene form an ideal solution, meaning that both liquids obey Raoult's law over the entire range of composition.

Consider a solution that contains benzene (C_6H_6) and toluene ($C_6H_5CH_3$). Both of these substances are volatile. At any given temperature, according to Raoult's law, the vapor pressure of the benzene above the solution is given by

$$P_{\text{benzene}} = \chi_{\text{benzene}} P^{\circ}_{\text{benzene}}$$

Raoult's law can also be used to calculate the partial pressure of toluene above the same solution:

$$P_{\text{toluene}} = \chi_{\text{toluene}} P^{\circ}_{\text{toluene}}$$

Figure 12.18 shows the vapor pressures of both components as the mole fraction of toluene in the solution varies from 0 to 1. At the left side of the graph, the liquid is pure benzene; on the right side, it is pure toluene. The total vapor pressure, also shown in Figure 12.18, is simply the sum of the vapor pressures of the two compounds. As expected from the equations, all three of the lines in Figure 12.18 are straight.

We can use Raoult's law to calculate the composition of the vapor above a mixture of two volatile substances. This calculation is illustrated in Example 12.16.

EXAMPLE 12.16 Composition of Vapor above a Mixture

At 60 °C, the vapor pressure of pure benzene is 384 torr, and that of pure toluene is 133 torr. A mixture is made by combining 1.20 mol toluene with 3.60 mol benzene. Find:

(a) the mole fraction of toluene in the liquid.
(b) the partial pressure of toluene above the liquid.
(c) the partial pressure of benzene above the liquid.
(d) the total vapor pressure.
(e) the mole fraction of toluene in the vapor phase.

Strategies are included with each section.

Solution

(a) Because the number of moles of each component in the mixture is given in the problem, the mole fraction of toluene is found easily.

$$\chi_{\text{toluene}} = \frac{\text{mol toluene}}{\text{mol toluene} + \text{mol benzene}}$$

$$\chi_{\text{toluene}} = \frac{1.20}{1.20 + 3.60} = 0.250$$

(b) Calculate the partial pressure of the toluene from Raoult's law.

$$P_{\text{toluene}} = \chi_{\text{toluene}} P^{\circ}_{\text{toluene}} = 0.250 \times 133 \text{ torr} = 33.2 \text{ torr}$$

(c) Raoult's law is also used to find the vapor pressure of benzene. The vapor pressure of pure benzene is given in the problem. Calculate the mole fraction of the benzene from that of toluene.

$$\chi_{\text{benzene}} = 1 - \chi_{\text{toluene}} = 1 - 0.250 = 0.750$$

$$P_{\text{benzene}} = 0.750 \times 384 \text{ torr} = 288 \text{ torr}$$

(d) The total vapor pressure above the solution is simply the sum of the vapor pressures of the two components.

$$P_{\text{total}} = 33.2 + 288 = 321 \text{ torr}$$

(e) Find the composition of the vapor from the partial pressures of the components. Recall from Chapter 6 that the partial pressures of gases in a mixture can be used to find the mole fractions of the components. Use the answers to parts b and d to calculate the mole fraction of toluene in the vapor.

$$\chi_{\text{toluene(g)}} = \frac{33.2 \text{ torr}}{321 \text{ torr}} = 0.103$$

Note that the mole fraction of the less volatile material, in this case, toluene, is lower in the vapor phase than in the liquid; thus, the mole fraction of the more volatile component, benzene, is greater in the vapor phase than in the liquid phase. In the vapor, the mole fraction of benzene is

$$\chi_{\text{benzene(g)}} = 1 - \chi_{\text{toluene(g)}} = 0.897$$

This concentration in the vapor is considerably greater than the 0.750-mol fraction of benzene found in the liquid phase.

Understanding

What is the vapor pressure, at 60 °C, of a solution with $\chi_{\text{toluene}} = 0.10$ and $\chi_{\text{benzene}} = 0.90$? Use the data given in this example.

Answer 359 torr

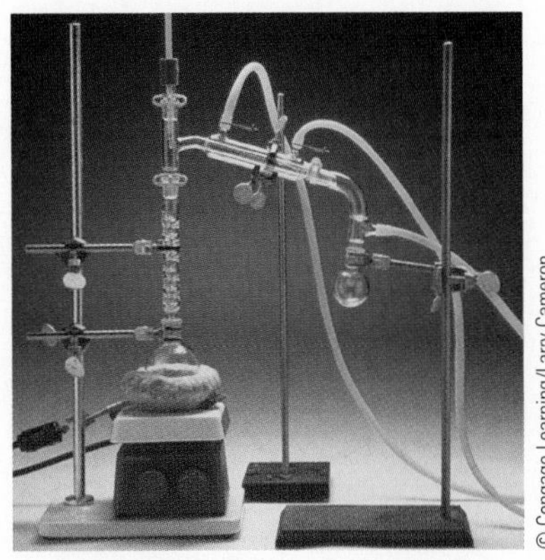

Figure 12.19 A laboratory distillation apparatus. A glass column with many indentations (Vigreaux column) is used in the laboratory for fractional distillations. The liquid repeatedly condenses and evaporates as it moves up the column. With each successive evaporation, the vapor becomes richest in the most volatile (lowest-boiling) component.

As shown in Example 12.16, the vapor in equilibrium with a mixture of two volatile substances is always richer in the more volatile component. Although the mole fraction of benzene in the liquid is 0.750, its mole fraction in the vapor is 0.897. As evaporation continues, the composition of the liquid changes, increasing the mole fraction of the less volatile (higher-boiling) substance. It is this fact that makes it possible to separate two volatile materials by a process called **fractional distillation.** (The term *distillation* is reserved for situations where a liquid and a solid are to be separated by evaporating the liquid component.)

In a fractional distillation, the liquid is repeatedly evaporated and condensed as it moves up the distillation column. Figure 12.19 shows a simple fractional distilla-

The vapor over a solution of two volatile components contains a larger fraction of the more volatile substance than does the solution.

© Dan McCoy/Rainbow

Figure 12.20 Fractional distillation of crude oil. Large-scale distillation equipment is used in the petroleum industry to separate the hydrocarbons into several fractions based on their different volatilities.

Raoult's law applies to all the volatile substances in an ideal solution.

tion apparatus used in the laboratory. Each time the vapor condenses, it produces a mixture that is richer in the more volatile component. As the mole fraction of the more volatile component increases, the boiling point of the mixture decreases, so the temperature of the column decreases from the bottom to the top. By the time the vapor reaches the top of the column, it consists of the more volatile component in a state of purity governed by the number of successive vaporizations and condensations that occurred. Very high purity can be obtained when the column is sufficiently long. Figure 12.20 shows a commercial distillation apparatus used in the petroleum industry.

In the data presented in Figure 12.18, we assumed that both of the components obeyed Raoult's law over the entire range of composition from pure benzene to pure toluene. An **ideal solution** is one that obeys Raoult's law throughout the entire range of composition. Consider a two-component mixture of compounds A and B. When the strength of the A–B attractions are close to the average of the A–A and B–B attractions, the solution behaves ideally. For two similar liquids such as benzene and toluene, this relationship is nearly true.

In most mixtures of liquids, Raoult's law is obeyed strictly by only very dilute solutions. Usually, the intermolecular forces of attraction between two different substances are stronger or weaker than the average of the A–A and B–B attractions. When the forces are unequal, we find that the straight lines in Figure 12.18 become curves. **Positive deviation** from Raoult's law means that the observed vapor pressure is greater than expected. This is illustrated in Figure 12.21, and it occurs when the A–B attractions are weaker than the average of the attractions in the pure components of the mixture. **Negative deviation** from Raoult's law (Figure 12.22) occurs when the intermolecular forces between the dissimilar molecules are stronger than the average of the intermolecular forces in the pure substances. Both positive and negative deviations from Raoult's law are observed experimentally for mixtures of different compounds.

OBJECTIVES REVIEW *Can you:*

- ☑ calculate the vapor pressure of each component and the total vapor pressure over an ideal solution?
- ☑ explain how fractional distillation works?

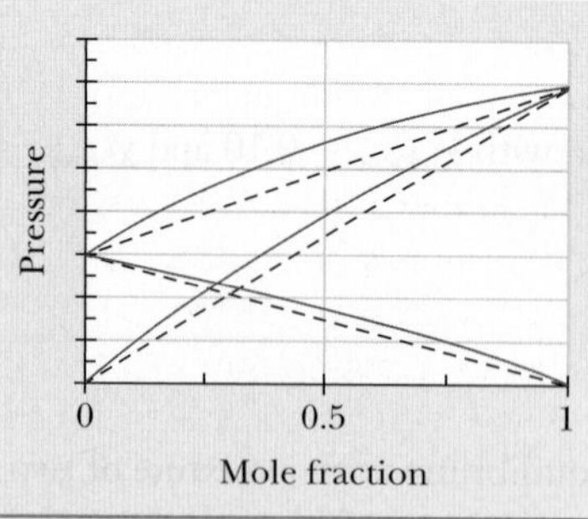

Figure 12.21 Positive deviation from Raoult's law. Graph shows the vapor pressure of each component and the total vapor pressure of the mixture as the composition of the liquid changes. *Dotted lines* represent the behavior of an ideal solution. *Solid lines* show a positive deviation from Raoult's law.

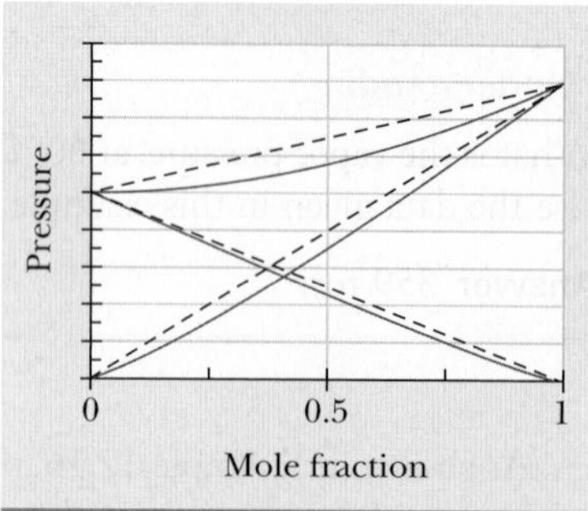

Figure 12.22 Negative deviation from Raoult's law. Graph shows the vapor pressure of each component and the total vapor pressure of the mixture as the composition of the liquid changes. *Dotted lines* represent the behavior of an ideal solution. *Solid lines* show a negative deviation from Raoult's law.

PRINCIPLES OF CHEMISTRY
Azeotropes

Deviations from ideal solution behavior sometimes produce a maximum or minimum in the vapor pressure of the solution, as shown in Figure 12.23. At a maximum or a minimum in the vapor-pressure curve for a solution, both the liquid and gas phases have exactly the same composition, and the mixture behaves as a pure substance, which is called an **azeotrope.** A fractional distillation of substances that forms an azeotrope cannot separate the mixture, because the liquid phase and gas phase will always have the same composition.

Ethanol and water form an azeotrope that has a maximum in the vapor-pressure curve where χ_{water} is 0.096. The normal boiling point of this azeotrope is 78.17 °C, slightly lower than the 78.4 °C at which pure ethanol boils. Thus, the azeotrope distills first in a fractional distillation, explaining why the distillation process can only provide ethanol that is ~95% pure. The concentrated aqueous nitric acid solution (68 mass percent HNO_3) that is sold for laboratory use is an azeotrope that boils at 120.5 °C, considerably greater than the normal boiling points of water (100 °C) and nitric acid (86 °C). Another azeotrope of some importance is the solution of hydrogen chloride in water. At 760 torr, the HCl–H_2O azeotrope boils at 108.6 °C and contains 20.22 mass percent HCl. The concentration of this azeotrope (6.11 *M*) is known with sufficient accuracy that it is used to prepare standard solutions of hydrochloric acid. ∎

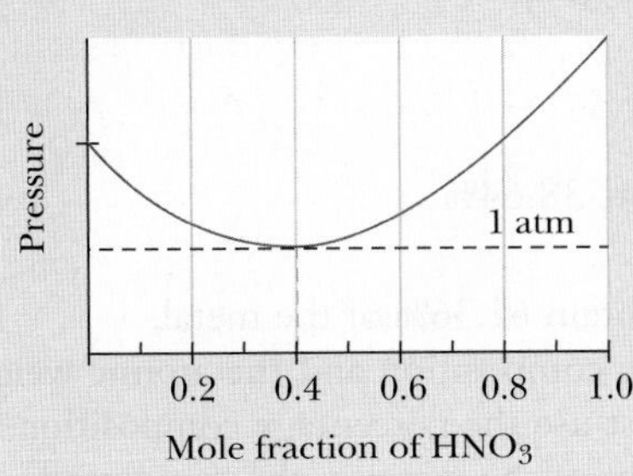

Figure 12.23 Vapor pressure of an azeotrope of nitric acid solution at constant temperature. The vapor pressure of a water solution of nitric acid at the boiling point of the azeotropic mixture (120.5 °C) is shown as a function of the composition. At the minimum in the vapor pressure curve, the mixture boils at a constant temperature and constant composition, preventing any further separation of the two substances by distillation.

CASE STUDY Determining Accurate Atomic Masses of Elements

Until relatively recently, atomic weights of elements were determined by first preparing high-purity compounds and determining the percentage composition. A Harvard chemist, Theodore W. Richards, was noted for his painstaking work that determined the accurate atomic weights of 25 elements; subsequently, in 1914, Richards became the first American to win a Nobel Prize in Chemistry. Richards was home-schooled by his mother, an author and poet, and his father, an artist, until he went to college at the age of 14. He earned a doctorate in chemistry at Harvard by the time he was 20 and was appointed to the Harvard faculty, where he remained for nearly all of his scientific career.

Williams Haynes Portrait Collection, Chemical Heritage Foundation Collections, Philadelphia, PA

T. W. Richards (1868–1928): first American winner of Nobel Prize in Chemistry.

Most of the elements were characterized using similar methods. In general, a crystal of a substance is quite pure because only one particular size of atom or ion fits into the crystalline matrix. But in practice, when compounds crystallize, the crystals might include a few impurity ions. To avoid these problems, Richards dissolved the high-purity crystal, and a higher purity crystal was grown; then the process was repeated. Some of Richards's preparations involved thousands of steps, which are called *fractional crystallizations.*

If the chloride compound were prepared, the simplest elemental analysis would be to precipitate the chloride by adding silver nitrate, forming the insoluble silver chloride. The silver chloride would be dried and weighed to determine the mass of chlorine in the original compound. Much of Richards's early work involved determining accurate atomic weights for silver and for chlorine, which were essential to his work. He also had to build laboratory apparatus, including extremely accurate and sensitive balances.

Many of the techniques we studied in this chapter were important to Richards. We can illustrate how Richards determined the atomic weight of an element with the following scenario. It is important to note that Richards would have only tens of milligrams of a compound, and that the accuracy and precision of his apparatus was not nearly as good as can be found in a modern laboratory.

Sample Experiment to Determine Atomic Weight of a New Element

Richards dissolved 0.02511 g of a metal chloride and precipitated the chloride as AgCl. The precipitate weighed 0.03922 g. (The numbers of decimal places are likely consistent with Richards's work, and the rules for significant figures are followed.) To calculate the percent of chlorine in the sample:

$$\text{Mass of sample} = 0.02511\ \text{g}$$

$$\text{Mass of AgCl precipitate} = 0.03922\ \text{g}$$

$$\text{Atomic weight of Cl} = 35.453$$

$$\text{Formula weight of AgCl} = 143.321$$

$$\text{Mass of Cl} = 0.03922\ \cancel{\text{g AgCl}} \times \left(\frac{35.453\ \text{g Cl}}{143.321\ \cancel{\text{g AgCl}}}\right)$$

$$\text{Mass of Cl} = 0.009702\ \text{g}$$

$$\text{Percentage} = \frac{0.009702\ \cancel{\text{g}}\ \text{Cl}}{0.02511\ \cancel{\text{g}}\ \text{sample}} \times 100\% = 38.64\%$$

Because the sample is 38.64% Cl, it must contain 61.36% of the metal.

On several occasions, we used the percentage composition and the atomic weights to determine an empirical formula. Now, we must use the percentage composition and empirical formula to determine the atomic weight of the metal in the compound.

Richards was an excellent chemist and would probably know that this element would likely exist as the M^{2+} or the M^{3+} ion, so the formula of the compound would be MCl_2 or MCl_3.

First, we can determine the potential atomic weights of the metal. When dealing with percentages, it is often useful to assume that we have 100 g of sample.

$$\text{Mass of Cl} = 38.64\ \text{g}$$

$$\text{Amount of Cl} = 38.64\ \cancel{\text{g}} \times \left(\frac{1\ \text{mol Cl}}{35.453\ \cancel{\text{g}}}\right) = 1.090\ \text{mol}$$

$$\text{Mass of M} = 100 - 38.64 = 61.36\ \text{g}$$

Assuming the formula is MCl_2:

$$\text{Amt of M} = 1.090\ \cancel{\text{mol Cl}} \times \left(\frac{1\ \text{mol M}}{2\ \cancel{\text{mol Cl}}}\right)$$

$$= 0.5450\ \text{mol}$$

$$0.5450\ \text{mol} = 61.36\ \text{g}$$

Divide both sides of equation by 0.5450:

$$1\ \text{mol} = 112.6\ \text{g}$$

$$\text{Formula weight of } MCl_2 = 183.5$$

Assuming the formula is MCl_3:

$$\text{Amt of M} = 1.090\ \cancel{\text{mol Cl}} \times \left(\frac{1\ \text{mol M}}{3\ \cancel{\text{mol Cl}}}\right)$$

$$= 0.3633\ \text{mol}$$

$$0.3633\ \text{mol} = 61.36\ \text{g}$$

Divide both sides of equation by 0.3633:

$$1\ \text{mol} = 168.9\ \text{g}$$

$$\text{Formula weight of } MCl_3 = 275.3$$

To determine which formula is correct, Richards might use an experiment that involved colligative properties to determine whether the compound ionized to form three or four particles. If it ionized into three particles, its formula is MCl_2; if it ionized into four particles, its formula is MCl_3.

Richards might have dissolved 0.002511 g of the compound and diluted to 10.10 mL. (He would have fabricated a quartz volume-measuring device for this work and calibrated it at the temperature of the experiment.) He measured the osmotic pressure at 15.72 °C as 65.5 torr.

The easiest way to treat the data is to calculate the osmotic pressure expected for MCl_2 and MCl_3, and determine which one matches the experiment.

Assuming the formula is MCl_2:

$$\text{mol of } MCl_2 = 0.002511\ \cancel{g} \times \left(\frac{1\text{ mol}}{183.5\ \cancel{g}}\right)$$

$$= 1.368 \times 10^{-5}\text{ mol}$$

Each mole produces three particles and the volume is 10.10 mL, so

$$\text{molarity} = 1.368 \times 10^{-5}\text{ mol} \times 3/0.01010\text{ L}$$

$$\text{molarity} = 0.004063\ M$$

$$T = 15.72 + 273.15 = 288.87\text{ K}$$

$$\Pi = MRT$$

$$= 0.004063\ M \times 0.08206\text{ L}\cdot\text{atm/mol}\cdot\cancel{K} \times 288.87\ \cancel{K}$$

$$\Pi = 0.09631\text{ atm} = 73.2\text{ torr}$$

Assuming the formula is MCl_3:

$$\text{mol of } MCl_3 = 0.002511\ \cancel{g} \times \left(\frac{1\text{ mol}}{275.3\ \cancel{g}}\right)$$

$$= 9.122 \times 10^{-6}\text{ mol}$$

Each mole produces four particles and the volume is 10.10 mL, so

$$\text{molarity} = 9.122 \times 10^{-6}\text{ mol} \times 4/0.01010\text{ L}$$

$$\text{molarity} = 0.003613\ M$$

$$T = 15.72 + 273.15 = 288.87\text{ K}$$

$$\Pi = MRT$$

$$= 0.003613\ M \times 0.08206\text{ L}\cdot\text{atm/mol}\cdot\text{K} \times 288.87\text{ K}$$

$$\Pi = 0.08564\text{ atm} = 65.09\text{ torr}$$

Because the experimental result was 65.5 torr, Richards could be confident that the compound was MCl_3, and the atomic weight of M is 168.9.

Questions

1. Identify element M.
2. If the osmotic pressure measurement had been 73.3 torr, identify element M.
3. If the formula were known to be MCl, what is the atomic weight of M?
4. What would the osmotic pressure have been if the formula for the compound were MCl?

ETHICS IN CHEMISTRY

1. After several years of stability in the lake, the government of Cameroon decides to consider repopulating the area around Lake Nyos (see chapter introduction). What factors should be considered before making any decision?

2. Some communities that have snowy and icy winters take advantage of colligative properties when they spread a salt (either NaCl or $CaCl_2$) on their roads and sidewalks. The addition of a salt reduces the melting point of the resulting solution, and the snow and ice melts, allowing for easier vehicle and foot traffic. However, large amounts of either NaCl or $CaCl_2$ can damage the plant life, and contributes to increased rusting of steel-containing objects like cars and bridges, so other communities spread sand instead. The addition of sand on the top of the snow and ice increases traction but does not help melt the snow or ice. The sand can also get quickly pulverized and blown away, contributing to visibility problems in some urban areas. Which method do you think is better? Why?

3. In 2003, the California Environmental Protection Agency proposed a limit of 4 parts per trillion (ppt) of arsenic in drinking water. However, a normal human body contains arsenic, an essential ultra-trace element, at a level of 1000 to 100,000 ppt naturally. Reducing the level of arsenic will increase the cost of doing business in California, which may cause businesses to relocate. Is it ethical to spend resources and effort to reduce the level of arsenic in drinking water to several orders of magnitude lower than currently exists in the human body?

Chapter 12 Visual Summary

The chart shows the connections between the major topics discussed in this chapter.

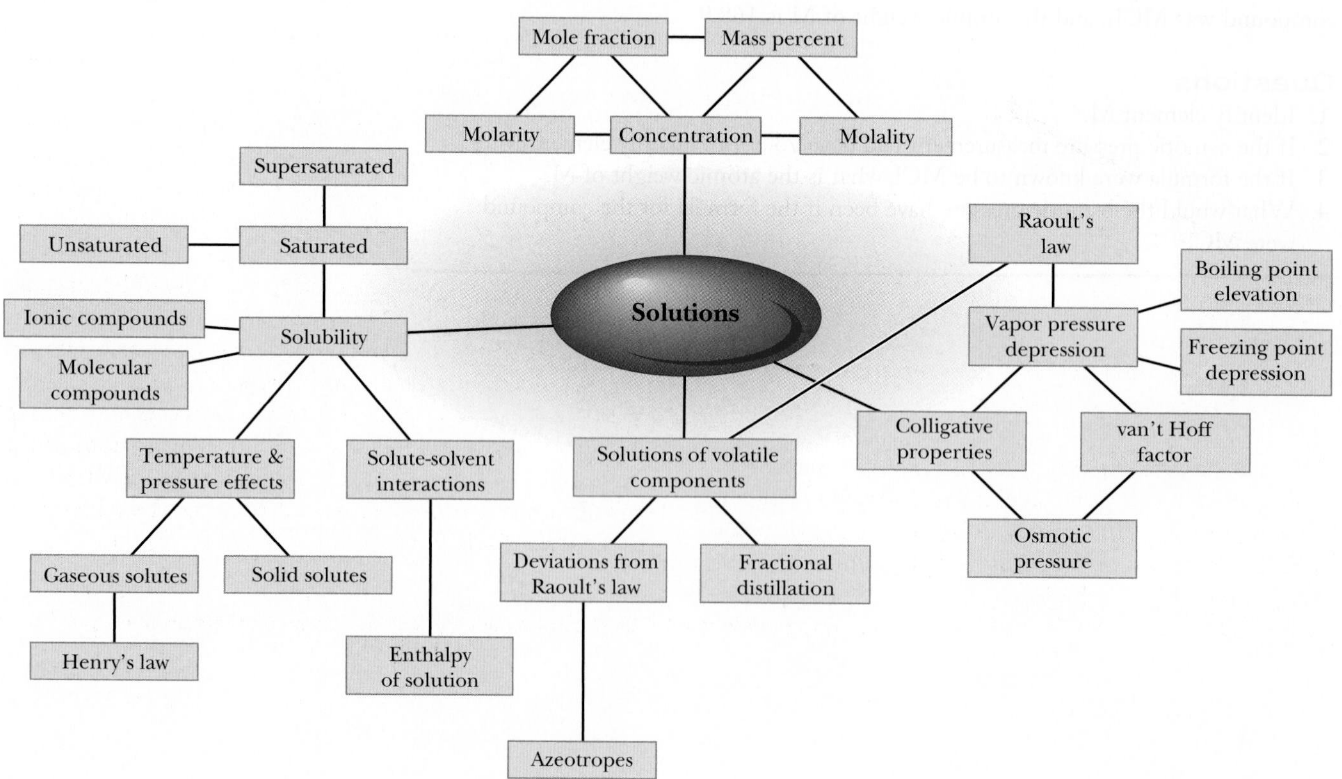

Summary

12.1 Solution Concentration

A solution is a homogeneous mixture of two or more substances and can occur in the solid, liquid, and gas phases. Solutions are described quantitatively by concentration. Some common units of concentration are mass percentage, mole fraction, *molarity*, and *molality*, defined as the moles of solute per kilogram of solvent. A *saturated* solution is one in which the dissolved and undissolved solutes are present in equilibrium. The concentration of a saturated solution is called the *solubility*, and is usually expressed in terms of the number of grams of solute dissolved per 100 mL of solvent. Any solution in which the concentration of solute is less than the solubility is *unsaturated*. A *supersaturated solution* is an unstable situation in which the concentration of the solute is greater than the solubility limit. Unsaturated, saturated, and supersaturated solutions can be distinguished by observing what happens to the solution when a small sample of the pure solute is added to it.

12.2 Principles of Solubility

The solution process is complex and depends on the strengths of solute-solute, solvent-solvent, and solute-solvent attractive forces, as well as the change in disorder that occurs on mixing. Whereas the solution process can be either endothermic or exothermic, dissolution requires that the strengths of the solute-solvent interactions be close to those of the solute-solute and solvent-solvent interactions. Although these enthalpy considerations make it possible to formulate qualitative predictions about the relative solubility of a substance in two different solvents, the dissolution process is too complex to predict the solubility of any given solute. The interaction of dipolar water molecules with ions in aqueous solution is called *hydration*. The hydration of ions is an exothermic process, comparable with the energy of attraction between oppositely charged ions in an ionic solid.

12.3 Effects of Pressure and Temperature on Solubility

Both pressure and temperature affect the solubilities of gases in liquids. The solubility of a gas increases with increasing pressure. *Henry's law* expresses the proportionality of the solubility of a gas to its partial pressure. The solubilities of most gases in water decrease with increasing temperature.

For most solids, both molecular and ionic, solubility becomes greater as the temperature of the solution increases. Solubility increases with increasing temperature when dissolution is an endothermic process (ΔH is positive) and decreases when the enthalpy of solution is negative. Pressure has almost no effect on the solubilities of liquids and solids because only a small volume change accompanies the formation of these solutions.

12.4, 12.5 Colligative Properties of Solutions and Electrolyte Solutions

Certain properties of solutions, called *colligative properties*, depend only on the concentration of solute particles, not on their identity. Four important colligative properties are *vapor-pressure depression*, *boiling-point elevation*, *freezing-point depression*, and *osmotic pressure*. All of the colligative properties obey the relationship

Colligative property = concentration (solute) × constant

where the constant is characteristic of the particular property and solvent, and the units used for the solute concentration depend on the property being measured. When the solute is an electrolyte, the *van't Hoff factor*, *i*, must be used in the equations for the colligative properties. Ideally, the van't Hoff factor is equal to the number of moles of ions produced in solution for each mole of compound dissolved. Electrostatic attractions between the ions cause the experimentally measured values of i to be somewhat less than the ideal value for all but very dilute solutions.

12.6 Mixtures of Volatile Substances

When two or more components of a liquid solution are volatile, the vapor pressure of the solution is the sum of the partial pressures of the volatile components. An *ideal solution* is one that obeys *Raoult's law* over the entire composition range of the mixture. The composition of the vapor in equilibrium with the solution is richest in the most volatile component that is present in the solution. In a fractional distillation, repeated evaporations and condensations are used to separate mixtures of two or more volatile liquids.

Most solutions of volatile substances do not behave ideally but exhibit either *positive* or *negative deviations* from Raoult's law, depending on the strength of the intermolecular attractions between the dissimilar molecules relative to those in the pure components. Many liquids form constant-boiling mixtures called *azeotropes*, in which the compositions of the vapor and liquid phases are the same. Complete separation of volatile liquids that form an azeotrope cannot be achieved by a fractional distillation.

Chapter Terms

The following terms are defined in the Glossary, Appendix I.

Section 12.1
Molality (m)
Section 12.2
Enthalpy of solution
Hydration
Saturated solution
Solubility
Supersaturated solution
Unsaturated solution
Section 12.3
Henry's law
Section 12.4
Boiling-point elevation
Colligative properties
Freezing-point depression
Osmosis
Osmotic pressure
Raoult's law
Reverse osmosis
Semipermeable membrane
Vapor-pressure depression
Section 12.5
van't Hoff factor
Section 12.6
Azeotrope
Fractional distillation
Ideal solution
Negative deviation (from Raoult's law)
Positive deviation (from Raoult's law)

Key Equations

Mole fraction (12.1)

$$\chi_A = \frac{\text{moles of A}}{\text{moles of A} + \text{moles of B} + \text{moles of C} + \cdots}$$

Mass percentage composition (12.1)

$$\text{Mass percentage solute} = \frac{\text{grams of solute}}{\text{grams of solution}} \times 100\%$$

Parts per million, parts per billion (12.1)

$$\text{ppm} = \frac{\text{g solute}}{\text{g solution}} \times 1{,}000{,}000$$

$$\text{ppb} = \frac{\text{g solute}}{\text{g solution}} \times 1{,}000{,}000{,}000$$

Molality (12.1)

$$\text{Molality } (m) = \frac{\text{mol solute}}{\text{kg solvent}}$$

Henry's law (12.3)

$$C = kP$$

Raoult's law (12.4)

$$P_{solv} = \chi_{solv} P^{\circ}_{solv}$$

Boiling-point elevation (12.4, 12.5)

$$\Delta T_b = mk_b \text{ or } \Delta T_b = imk_b$$

Freezing-point depression (12.4, 12.5)

$$\Delta T_f = mk_f \text{ or } \Delta T_f = imk_f$$

Osmotic pressure (12.4, 12.5)

$$\Pi = \frac{nRT}{V} = MRT \text{ or } \Pi = iMRT$$

Questions and Exercises

OWL Selected end of chapter Questions and Exercises may be assigned in OWL.

Blue-numbered Questions and Exercises are answered in Appendix J; questions are qualitative, are often conceptual, and include problem-solving skills.

■ Questions assignable in OWL

✎ Questions suitable for brief writing exercises

▲ More challenging questions

Questions

12.1 Concentration units all express the quantity of solute present in some quantity of solution or solvent. For molarity, mole fraction, and mass percentage, how is the quantity of solution expressed?

12.2 The concentration of a solution is often used as a conversion factor. For example, molarity is a conversion factor to find the moles of solute in a specified volume of solution. What conversions are performed with molality, mass percentage, and mole fraction?

12.3 ✎ The numeric values of the molarity and molality of a dilute solution can be very similar or quite different, depending on the density of the solvent. Explain why.

12.4 We analyzed the enthalpy change that accompanies the formation of a solution from the pure solute and solvent by considering three changes that must occur: (1) separate the solvent molecules; (2) separate the solute molecules; and (3) bring the solute particles and solvent particles together. If the enthalpy of solution is negative (an exothermic process), what can be said about the relative enthalpy changes of the three processes just listed?

12.5 A chemist adds 30 g sodium acetate to 50 mL water at room temperature. Only part of the sodium acetate dissolves. The mixture is heated with stirring, and all of the solid dissolves. With slow cooling to room temperature, no solid precipitates. Explain these observations and describe an experiment that would test your explanation.

12.6 In diluting sulfuric acid with water, you should slowly add the acid to the water while stirring. This is sometimes expressed as "Always add acid" (AAA). Give the reasons for this procedure.

12.7 Straight-chain alcohols [$CH_3(CH_2)_nOH$] that contain more than four carbon atoms have limited solubility in water. Predict how the solubility of these alcohols in water changes as the value of n in the formula increases.

12.8 ✎ Explain why opening a warm carbonated beverage results in much more frothing than is observed when the container has been refrigerated.

12.9 Why do carbonated beverages go flat if they are not stored in a tightly sealed container?

12.10 Explain why the solubilities of solids in a liquid change little with pressure.

12.11 Consider the Lake Nyos situation as discussed in the chapter introduction. Is such a situation more likely to occur with a gas that has a high Henry's law constant or a low Henry's law constant? Explain your answer.

12.12 List and define the colligative properties, and give the units used for concentrations for each.

12.13 Create a flow diagram, similar to those used in the example problems of this chapter, that outlines the determination of the molar mass of a compound from freezing-point depression measurements. Clearly indicate the data needed for this determination.

12.14 Specific samples of aqueous solutions of sucrose and urea both freeze at −0.25 °C. What other properties of these two solutions should be the same?

12.15 Compare the freezing points of 0.1-m aqueous solutions of NaCl and $CaCl_2$. Explain why one of these solutions has a lower freezing point.

12.16 ✎ Explain the difference between distillation and fractional distillation.

Exercises

OBJECTIVE Express solution concentration using several different units.

12.17 A solution contains 1.20 g benzoic acid ($C_6H_5CO_2H$) in 750.0 g water. Express the concentration of benzoic acid as
(a) mass percentage.
(b) mole fraction.
(c) molality.

12.18 A solution is prepared by dissolving 25.0 g $BaCl_2$ in 500 g water. Express the concentration of $BaCl_2$ in this solution as
(a) mass percentage.
(b) mole fraction.
(c) molality.

12.19 A solution contains 4.50 g calcium nitrate [$Ca(NO_3)_2$] in 430.0 g water. Express the concentration of $Ca(NO_3)_2$ as
(a) mass percentage.
(b) mole fraction.
(c) molality.

12.20 A solution contains 3.80 g urea [$CO(NH_2)_2$] in 125.0 g water. Express the concentration of urea as
(a) mass percentage.
(b) mole fraction.
(c) molality.

12.21 How many moles of hydrogen peroxide are present in 25.0 g of a 3.0% solution?

© Cengage Learning/Larry Cameron

12.22 ■ How many grams of sodium chloride, NaCl, are present in 35.0 g of a 3.5% solution?

12.23 Give quantitative directions for preparing 15.5 g of a 1.00% solution of boric acid (H_3BO_3). What is the molal concentration of this solution?

12.24 ■ Describe how you would prepare 465 mL of 0.355 M potassium dichromate solution.

12.25 A solution contains 12.0 g hexane (C_6H_{14}), 20.0 g octane (C_8H_{18}), and 98.0 g benzene (C_6H_6). What is the mole fraction of benzene in the solution?

12.26 A solution contains 10.0 g ethanol (C_2H_5OH), 20.0 g ethylene glycol [$C_2H_4(OH)_2$], and 90.0 g water. What is the mole fraction of water in the sample?

OBJECTIVE Convert concentration units.

12.27 What is the molality of silver nitrate ($AgNO_3$) in an aqueous 0.10% solution of that compound?

12.28 What is the molality of copper(II) bromide ($CuBr_2$) in an aqueous 0.50% solution of that compound?

12.29 What is the mole fraction of nitrous oxide (N_2O) in an aqueous 0.020-molal solution?

12.30 ■ What is the mole fraction of bromine (Br_2) in an aqueous 0.10-molal solution?

12.31 A water solution of sodium hypochlorite (NaOCl) is used as laundry bleach. The concentration of sodium hypochlorite is 0.75 m. Express this concentration as a mole fraction.

12.32 Rubbing alcohol is a water solution that contains 70% isopropanol (C_3H_7OH). What is the mole fraction of isopropanol in rubbing alcohol?

12.33 A 10.0% solution of sucrose ($C_{12}H_{22}O_{11}$) in water has a density of 1.038 g/mL. Express the concentration of the sugar as
(a) molality. (b) molarity. (c) mole fraction.

12.34 Vinegar is a 5.0% solution of acetic acid (CH_3CO_2H) in water. The density of vinegar is 1.0055 g/mL. Express the concentration of acetic acid as
(a) molality. (b) molarity. (c) mole fraction.

12.35 A 0.631 M H_3PO_4 solution in water has a density of 1.031 g/mL. Express the concentration of this solution as
(a) mass percentage. (b) mole fraction.
(c) molality.

12.36 ■ A 2.77 M NaOH solution in water has a density of 1.109 g/mL. Express the concentration of this solution as
(a) mass percentage. (b) mole fraction.
(c) molality.

12.37 Complete the following table for ammonia (NH_3) solutions in water.

Density (g/cm^3)	Molality	Molarity	Mass % NH_3	Mole Fraction
(a) 0.973			6.00	
(b) 0.936		8.80		
(c) 0.950	8.02			
(d) 0.969				0.0738

© PHOTOTAKE Inc./Alamy

12.38 Complete the following table for perchloric acid ($HClO_4$) solutions in water.

Density (g/cm^3)	Molality	Molarity	Mass % $HClO_4$	Mole Fraction
(a) 1.060			10.0	
(b) 1.011		0.2012		
(c) 1.143	2.807			
(d) 1.086				0.0284

12.39 The density of a 3.75 M aqueous sulfuric acid solution in a car battery is 1.225 g/mL. Express the concentration of the solution in molality, mole fraction H_2SO_4, and mass percentage of H_2SO_4.

12.40 Household bleach is a 5.00% solution of NaClO in water and has a density of 1.10 g/mL. Express the concentration of the solution in molality, mole fraction NaClO, and molarity of NaClO.

OBJECTIVE Predict relative solubility based on intermolecular interactions.

12.41 Using the intermolecular attractions as a guide, arrange the following solutes in order of increasing solubility in benzene (C_6H_6): hexane (C_6H_{14}), ethanol (C_2H_5OH), water.

12.42 Which pairs of liquids will be soluble in each other?
(a) H_2O and $CH_3CH_2CH_2CH_3$
(b) C_6H_6 (benzene) and CCl_4
(c) H_2O and CH_3CO_2H

12.43 Predict the relative solubility of each compound in the two solvents, on the basis of intermolecular attractions.
(a) Is Br_2 more soluble in water or in carbon tetrachloride?
(b) Is $CaCl_2$ more soluble in water or in benzene (C_6H_6)?
(c) Is chloroform ($CHCl_3$) more soluble in water or in diethyl ether [$(C_2H_5)_2O$]?
(d) Is ethylene glycol ($HOCH_2CH_2OH$) more soluble in water or in benzene (C_6H_6)?

12.44 Predict the relative solubility of each compound in the two solvents, on the basis of intermolecular attractions.
(a) Is NaCl more soluble in water or in carbon tetrachloride?
(b) Is I_2 more soluble in water or in toluene ($C_6H_5CH_3$)?
(c) Is ethanol (C_2H_5OH) more soluble in hexane or in water?
(d) Is ethylene glycol ($HOCH_2CH_2OH$) more soluble in ethanol or in benzene (C_6H_6)?

12.45 Identify the most important types of solute-solvent interactions in each of the following solutions.
(a) $(CH_3)_2CO$ in water
(b) IBr in $CHCl_3$
(c) $CaCl_2$ in water
(d) krypton in CH_3OH

12.46 Identify the most important types of solute-solvent interactions in each of the following solutions.
(a) CH_3OH in water
(b) IBr in CH_3CN
(c) KBr in water
(d) argon in water

12.47 Choose the solute of each pair that would be more soluble in hexane (C_6H_{14}). Explain your answer.
(a) $CH_3(CH_2)_{10}OH$ or $CH_3(CH_2)_2OH$
(b) $BaCl_2$ or CCl_4
(c) $Fe(C_5H_5)_2$ (a nonelectrolyte) or $FeCl_2$

12.48 Choose the solute of each pair that would be more soluble in water. Explain your answer.
(a) NaOH or CO_2
(b) $TiCl_3$ or $CHCl_3$
(c) C_3H_8 or C_3H_7OH

OBJECTIVE Use Henry's law.

12.49 The solubility of acetylene (C_2H_2) in water at 20 °C and 0.200 atm pressure is 9.38×10^{-3} molal.
(a) Calculate the Henry's law constant for this gas in units of molal/torr.
(b) How many grams of acetylene are dissolved in 1.00 kg water at 20 °C if the pressure of the gas is 300 torr?

12.50 The solubility of ethylene (C_2H_4) in water at 20 °C and 0.300 atm pressure is 1.27×10^{-4} molal.
(a) Calculate the Henry's law constant for this gas in units of molal/torr.
(b) How many grams of ethylene are dissolved in 1.00 kg water at 20 °C if the pressure of the gas is 500 torr?

12.51 The enthalpy of solution of ozone (O_3) in water is −17 kJ/mol, and its solubility at 0 °C and 1.00 atm is 0.105 g per 100 g water.
(a) What is the Henry's law constant (in molal/torr) at 0 °C?
(b) At 10 °C and 1.00 atm, would the Henry's law constant be larger, smaller, or the same as it is at 20 °C and the same pressure?
(c) Calculate the molal solubility of ozone in water at 0.500 atm and 0 °C.

12.52 The enthalpy of solution of nitrous oxide (N_2O) in water is −12.0 kJ/mol, and its solubility at 20 °C and 1.00 atm is 0.121 g per 100 g water.
(a) What is the Henry's law constant (in molal/torr) at 20 °C?
(b) At 10 °C and 1.00 atm, would the Henry's law constant be larger, smaller, or the same as it is at 20 °C and the same pressure?
(c) Calculate the molal solubility of nitrous oxide in water at 0.500 atm and 20 °C.

OBJECTIVE **Predict the pressure and temperature dependence of solubility.**

12.53 The solubility of potassium chloride in water increases from 34.7 g/100 mL at 20 °C to 56.7 g/100 mL at 100 °C. Is the enthalpy of solution for this compound endothermic or exothermic? Explain your answer.

12.54 The solubility of lead bromide in water is 0.844 g per 100 g water at 20 °C and 4.71 g per 100 g water at 100 °C. Is the dissolution of lead bromide an endothermic or exothermic process? Explain your answer.

12.55 The solubility of calcium hydroxide in water is 0.165 g per 100 g water at 20 °C and 0.128 g per 100 g water at 50 °C. Is the dissolution of calcium hydroxide an endothermic or exothermic process?

12.56 The solubilities of most gases in water decrease as the temperature increases. Are the enthalpies of solution for such gases negative or positive? Explain your answer.

12.57 From the data presented in Figure 12.11, determine which has the more positive enthalpy of solution: NaCl or NH_4Cl. Explain.

12.58 From the data presented in Figure 12.11, determine which has the more positive enthalpy of solution: NaCl or KNO_3. Explain.

12.59 At 22 °C and 1.0 atm, the enthalpy of solution of nitrogen in water is −11.0 kJ/mol, and its solubility is 6.68×10^{-4} *m*. State whether the solubility of nitrogen is greater or less than 6.68×10^{-4} *m* under each of the following conditions.
(a) 0 °C and 3.0 atm
(b) 22 °C and 0.75 atm
(c) 10 °C and 1.0 atm
(d) 50 °C and 1.0 atm
(e) 50 °C and 0.5 atm

12.60 At 25 °C and 2.0 atm, the enthalpy of solution of neon in water is −2.46 kJ/mol, and its solubility is 9.07×10^{-4} *m*. State whether the solubility of neon is greater or less than 9.07×10^{-4} *m* under each of the following conditions.
(a) 0 °C and 3.0 atm
(b) 25 °C and 1.0 atm
(c) 10 °C and 2.0 atm
(d) 50 °C and 2.0 atm
(e) 50 °C and 1.5 atm

12.61 The enthalpy of solution of nitrous oxide (N_2O) in water is −12.0 kJ/mol, and its solubility at 20 °C and 2.00 atm is 0.055 *m*. State whether the solubility of nitrous oxide is greater or less than 0.055 *m* at
(a) 2.00 atm and 0 °C.
(b) 2.00 atm and 40 °C.
(c) 1.00 atm and 20 °C.
(d) 1.00 atm and 50 °C.

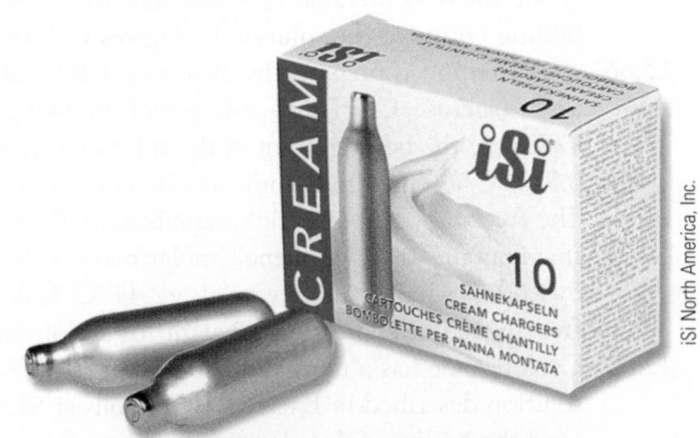

iSi North America, Inc.

© Steve Skjold/Alamy

Canned whipped cream uses nitrous oxide, N_2O, as a propellant gas.

12.62 The enthalpy of solution of ozone (O_3) in water is −17 kJ/mol, and its solubility at 25 °C and 1.00 atm is 0.0052 *m*. State whether the solubility of ozone is greater or less than 0.0052 *m* at
(a) 1.00 atm and 0 °C.
(b) 1.00 atm and 40 °C.
(c) 2.00 atm and 25 °C.
(d) 0.50 atm and 50 °C.

OBJECTIVE Relate colligative properties to concentrations of solutions.

12.63 The vapor pressure of chloroform ($CHCl_3$) is 360 torr at 40.0 °C. Find the vapor-pressure depression (in torr) produced by dissolving 10.0 g phenol (C_6H_5OH) in 95.0 g chloroform. What is the vapor pressure (in torr) of chloroform above the solution?

12.64 Cyclohexane (C_6H_{12}) has a vapor pressure of 99.0 torr at 25 °C. What is the vapor pressure (in torr) of cyclohexane above a solution of 14.0 g naphthalene ($C_{10}H_8$) in 50 g cyclohexane at 25 °C?

12.65 A solution contains 2.00 g of the nonvolatile solute urea (molar mass = 60.06 g/mol) dissolved in 25.0 g water. Using the data in Table 12.4, calculate the freezing and boiling points of the solution in degrees Celsius.

12.66 ■ A solution is prepared by dissolving 8.89 g of ordinary sugar (sucrose, $C_{12}H_{22}O_{11}$, 342 g/mol) in 34.0 g water. Calculate the boiling point of the solution using k_b as 0.512 °C/*m*. Sucrose is a nonvolatile nonelectrolyte.

12.67 The freezing point of cyclohexane is 6.50 °C. A solution that contains 0.500 g phenol (molar mass = 94.1 g/mol) in 12.0 g cyclohexane freezes at −2.44 °C. Calculate the freezing-point depression constant for cyclohexane.

12.68 Cyclohexane has a normal boiling point of 80.72 °C. The solution described in Exercise 12.67 boils at 81.94 °C. Find the boiling-point elevation constant for cyclohexane.

OBJECTIVE Calculate molar mass from measurements of colligative properties.

12.69 A 0.350-g sample of a nonvolatile compound dissolves in 12.0 g cyclohexane, producing a solution that freezes at 0.83 °C. Cyclohexane has a freezing point of 6.50 °C and a freezing-point depression constant of 20.2 °C/molal. What is the molar mass of the solute?

12.70 A 0.500-g sample of a nonvolatile, yellow crystalline solid dissolves in 15.0 g benzene, producing a solution that freezes at 5.03 °C. Use the data in Table 12.4 to find the molar mass of the yellow solid.

12.71 A solution of 1.00 g of a protein in 20.0 mL water has an osmotic pressure of 35.2 torr at 298 K. Calculate the molar mass of the protein.

12.72 ■ The molar mass of a polymer was determined by measuring the osmotic pressure, 7.6 torr, of a solution containing 5.0 g of the polymer dissolved in 1.0 L benzene. What is the molar mass of the polymer? Assume a temperature of 298.15 K.

OBJECTIVE Calculate colligative properties of electrolytes.

12.73 Arrange the following aqueous solutions in order of increasing boiling points: 0.02 *m* LiBr, 0.03 *m* sucrose, 0.03 *m* $MgSO_4$, 0.03 *m* $CaCl_2$, and 0.025 *m* $(NH_4)_2Cr_2O_7$.

12.74 Arrange the following solutions in order of decreasing osmotic pressure: 0.10 *M* urea, 0.06 *M* NaCl, 0.05 *M* $Ba(NO_3)_2$, 0.06 *M* sucrose, and 0.04 *M* $KMnO_4$.

12.75 A sample of seawater freezes at −2.01 °C. What is the total molality of solute particles? If we assume that all of the solute is NaCl, how many grams of that compound are present in 1 kg of water? (Assume an ideal value for the van't Hoff factor.)

12.76 An aqueous solution of sodium bromide freezes at −1.61 °C. What is the total molality of solute particles? How many grams of sodium bromide are present in 1 kg of water? (Assume an ideal value for the van't Hoff factor.)

12.77 A solution of 6.3 g calcium chloride in 1.20 kg water is prepared. Assuming an ideal value for the van't Hoff *i*, calculate the boiling point of this solution.

12.78 The saline solution used for intravenous injections contains 8.5 g NaCl in 1.00 kg water. Assuming an ideal value for the van't Hoff *i*, calculate the freezing point of this solution.

12.79 A 3.4-g sample of $CaCl_2$ is dissolved in water to give 500 mL of solution at 298 K. What is the osmotic pressure of this solution? (Assume an ideal value for the van't Hoff factor.)

12.80 An 8.5-g sample of NaCl is dissolved in water to give 1.00 L of solution at 298 K. What is the osmotic pressure of this solution? (Assume an ideal value for the van't Hoff factor.)

12.81 ▲ A 0.010-molar solution of sodium chloride is separated from pure water by a semipermeable membrane at 298 K. In which direction does net transport of water occur across the membrane when the applied pressure on the solution is 500 torr? (Assume an ideal value for the van't Hoff factor.)

12.82 ▲ A 0.010-molar solution of calcium chloride is separated from pure water by a semipermeable membrane at 298 K. In which direction does net transport of water occur across the membrane when the applied pressure on the solution is 500 torr? (Assume an ideal value for the van't Hoff factor. Calcium chloride is also used as a de-icer in wintry climates.)

© 1998 Larry Stepanowicz/Fundamental Photographs, NYC

12.83 A 0.029 *M* solution of potassium sulfate has an osmotic pressure of 1.79 atm at 25 °C.
(a) Calculate the van't Hoff factor, *i*, for this solution.
(b) Would the van't Hoff factor be larger, smaller, or the same for a 0.050 *M* solution of this compound?

12.84 The freezing point of a 0.031-*m* solution of copper(II) sulfate in water is −0.075 °C.
(a) Calculate the van't Hoff factor, *i*, for this solution.
(b) Would the van't Hoff factor be larger, smaller, or the same for a 0.050-*m* solution of this compound?

OBJECTIVE **Calculate the vapor pressures of solutions of volatile compounds.**

12.85 A mixture contains 15.0 g hexane (C_6H_{14}) and 20.0 g heptane (C_7H_{16}). At 40 °C, the vapor pressure of hexane is 278 torr and that of heptane is 92.3 torr. Assume that this is an ideal solution.
(a) What is the mole fraction of each of these substances in the liquid phase?
(b) What are the vapor pressures of hexane and of heptane above the solution?
(c) Find the mole fraction of each substance in the vapor phase.

12.86 A mixture contains 25.0 g cyclohexane (C_6H_{12}) and 44.0 g 2-methylpentane (C_6H_{14}). At 35 °C, the vapor pressure of cyclohexane is 150 torr and that of 2-methylpentane is 313 torr. Assume that this is an ideal solution.
(a) What is the mole fraction of each of these substances in the liquid phase?
(b) What are the vapor pressures of cyclohexane and of 2-methylpentane above the solution?
(c) Find the mole fraction of each substance in the vapor phase.

12.87 ▲ At a pressure of 760 torr, acetic acid (CH_3COOH; boiling point = 118.1 °C) and 1,1-dibromoethane ($C_2H_4Br_2$; boiling point = 109.5 °C) form an azeotropic mixture, boiling at 103.7 °C, that is 25% by mass acetic acid. At the boiling point of the azeotrope (103.7 °C), the vapor pressure of pure acetic acid is 471 torr and that of pure 1,1-dibromoethane is 637 torr.
(a) Calculate the vapor pressure of each component and the total vapor pressure at 103.7 °C if the solution had obeyed Raoult's law for both components.
(b) Compare your answer to part a with the actual vapor pressure of the azeotrope. Is the deviation of this azeotropic mixture from Raoult's law positive or negative?
(c) Compare the attractive forces between acetic acid and dibromoethane with those in the two pure substances.

12.88 ▲ At a pressure of 760 torr, formic acid (HCO_2H; boiling point = 100.7 °C) and water (H_2O; boiling point = 100.0 °C) form an azeotropic mixture, boiling at 107.1 °C, that is 77.5% by mass formic acid. At the boiling point of the azeotrope (107.1 °C), the vapor pressure of pure formic acid is 917 torr and that of pure water is 974 torr.
(a) Calculate the vapor pressure of each component and the total vapor pressure at 107.1 °C if the solution had obeyed Raoult's law for both components.
(b) Compare your answer to part a with the actual vapor pressure of the azeotrope. Is the deviation of this azeotropic mixture from Raoult's law positive or negative?
(c) Compare the attractive forces between formic acid and water with those in the two pure substances.

Chapter Exercises

12.89 Typical dinner wines are 9.00% alcohol by volume, which corresponds to 7.23% alcohol by mass. The density of the solution is 0.9877 g/mL. Express the alcohol concentration as
(a) molality.
(b) mole fraction.
(c) molarity.
(d) grams of alcohol per 100 mL.

12.90 Give quantitative directions for preparing a 0.0520-*m* aqueous solution of sodium carbonate. Assuming an ideal van't Hoff factor, what is the expected freezing point of this solution?

12.91 Predict the relative solubility of each compound in the two solvents, based on the intermolecular attractions.
(a) Is potassium iodide more soluble in water or in methylene chloride (CH_2Cl_2)?
(b) Is toluene ($C_6H_5CH_3$) more soluble in benzene (C_6H_6) or in water?
(c) Is ethylene glycol ($C_2H_4(OH)_2$) more soluble in hexane (C_6H_{14}) or in ethanol (C_2H_5OH)?

12.92 The enthalpy of solution of potassium perchlorate ($KClO_4$) is 51.0 kJ/mol. Compare the solubilities of this compound at 25 °C and at 92 °C.

12.93 At 10 °C, the solubility of CO_2 gas in water is 0.240 g per 100 mL water at a pressure of 1.00 atm. A soft drink is saturated with carbon dioxide at 4.00 atm and sealed.
(a) What mass of carbon dioxide is dissolved in a 12-oz can of this beverage (1 oz = 28.35 mL)?
(b) What volume of $CO_2(g)$, measured at standard temperature and pressure, is released when the 12-oz can is left open for several days?

12.94 The vapor pressure of trichloroethane ($C_2H_3Cl_3$) is 100 torr at 20.0 °C. What is the vapor pressure of trichloroethane, at 20.0 °C, above a solution containing 2.00 g ferrocene [$Fe(C_5H_5)_2$, which is nonvolatile] in 25.0 g trichloroethane?

12.95 A 0.325-g sample of dark red crystalline compound dissolves in 12.2 g ethylene dibromide, giving a solution that freezes at 7.97 °C. Ethylene dibromide has a normal freezing point of 9.80 °C and a freezing-point depression constant of 11.8 °C/*m*.
(a) What is the molar mass of the red solid?
(b) The compound is 32.5% iron by mass. Calculate the number of iron atoms in each molecule of this solute.

12.96 A reverse osmosis unit is used to obtain drinkable water from a source that contains 500 ppm sodium chloride (0.500 g NaCl per liter). What is the minimum pressure that must be applied across the semipermeable membrane to obtain water? (Don't forget the van't Hoff factor.)

12.97 Sketch graphs of total vapor pressure versus the mole fraction of two volatile substances that show
(a) positive deviation from Raoult's law.
(b) negative deviation from Raoult's law.
(c) a maximum boiling azeotrope.
(d) a minimum boiling azeotrope.

Cumulative Exercises

12.98 ▲ When 0.030 mol HCl dissolves in 100.0 g benzene, the solution freezes at 4.04 °C. When 0.030 mol HCl dissolves in 100.0 g water, the solution freezes at −1.07 °C. Use the data in Table 12.4 to complete the following exercises.
(a) From the freezing point, calculate the molality of HCl in the benzene solution.
(b) Use the freezing point of the aqueous solution to find the molality of HCl in the water.
(c) Offer an explanation for the different values found in parts a and b.

12.99 ▲ A 51.0-mL sample of a gas at 745 torr and 25 °C has a mass of 0.262 g. The entire gas sample dissolves in 12.0 g water, forming a solution that freezes at −0.61 °C.
(a) Calculate the molar mass of the gas using the ideal gas law.
(b) Calculate the molar mass from the freezing point of the solution, using the data in Table 12.4.
(c) Offer an explanation for the different values obtained in parts a and b.

12.100 ■ A 2.00% solution of H_2SO_4 in water freezes at −0.796 °C.
(a) Calculate the van't Hoff factor, *i*.
(b) Which of the following best represents sulfuric acid in a dilute aqueous solution: H_2SO_4, $H^+ + HSO_4^-$, or $2\ H^+ + SO_4^{2-}$?

12.101 ■ Suppose you have two aqueous solutions separated by a semipermeable membrane. One contains 5.85 g NaCl dissolved in 100 mL of solution, and the other contains 8.88 g KNO_3 dissolved in 100 mL of solution. In which direction will solvent flow: from the NaCl solution to the KNO_3 solution, or from KNO_3 to NaCl? Explain briefly.

12.102 ▲ Hemoglobin contains 0.33% Fe by mass. A 0.200-g sample of hemoglobin is dissolved in water to give 10.0 mL of solution, which has an osmotic pressure of 5.5 torr at 25 °C. How many moles of Fe atoms are present in 1 mol hemoglobin? (*Hint:* Calculate the molar mass from the osmotic pressure and find the mass of iron in one mole of the compound.)

12.103 ▲ A benzene solution and a water solution of acetic acid (CH_3CO_2H; molar mass = 60.05 g/mol) are both 0.50 mass percent acid. The freezing-point depression of the benzene solution is 0.205 °C, and that of the aqueous solution is 0.159 °C.
(a) Using the freezing-point depression constants in Table 12.4, calculate the molality of the solute in each of the two solutions.
(b) What are the van't Hoff factors for the water solution and benzene solution from these experiments?
(c) Note the difference between the van't Hoff factors, and offer an explanation for the experimental results.

12.104 ▲ A 10.00-mL sample of a 24.00% solution of ammonium bromide (NH_4Br) requires 23.41 mL of 1.200 molar silver nitrate ($AgNO_3$) to react with all of the bromide ion present.
(a) Calculate the molarity of the ammonium bromide solution.
(b) Use the molarity of the solution to find the mass of ammonium bromide in 1.000 L of this solution.
(c) From the percentage concentration and the answer to part b, find the mass of 1.000 L ammonium bromide solution.
(d) Combine the answer to part c with the volume of 1.000 L to express the density of the ammonium bromide solution (in g/mL).

12.105 A white crystalline compound is analyzed and found to have the following composition: 33.8% Na, 17.7% C, 47.0% O.
(a) Calculate the empirical formula of this compound.
(b) A solution is made by dissolving 0.500 g of the compound in 20.0 g water. This solution conducts an electric current and has a freezing point of −1.01 °C. What is the formula of the white solid? Write an equation that shows its dissociation in water solution.

12.106 In the 1986 Lake Nyos disaster (see the chapter introduction), an estimated 90 billion kilograms of CO_2 was dissolved in the lake at the time.
(a) What volume of gas is this at standard temperature and pressure?
(b) ▲ Assuming that this dissolved gas was in equilibrium with the normal partial pressure of CO_2 in the atmosphere (0.038%, or 0.29 torr), use the Henry's law constant for CO_2 in water to estimate the volume of Lake Nyos.

12.107 The ethoxide ion, $C_2H_5O^-$, is a strong base. It is similar to hydroxide, OH^-, but with the hydrogen atom in hydroxide replaced by the C_2H_5 group. When mixed with water, the salt sodium ethoxide will react with H_2O to make C_2H_5OH, ethyl alcohol, and $OH^-(aq)$ ions. An unknown mass of NaC_2H_5O was slowly added to 20.0 mL water, and the resulting solution was titrated with 1.815 M HCl solution. A total of 41.5 mL of acid was needed to neutralize the resulting solution.

(a) Write the net ionic reaction between sodium ethoxide and water.

Courtesy of M. Stading

(b) Would you expect the reaction between ethoxide ions and water to be exothermic or endothermic? Explain your answer.

(c) What mass of NaC_2H_5O was added to the water originally?

(d) At the endpoint, the system consists of two volatile liquids, water and ethanol. What is the expected equilibrium vapor pressure of this solution at 25.0 °C? $P°(H_2O) = 23.77$ torr; $P°(C_2H_5OH) = 59.02$ torr.

(e) Can fractional distillation be used to separate any of the ethanol from the water? (See the Principles of Chemistry box earlier in the chapter for a discussion of water/ethanol azeotropes.)

AP Photo/South Tyrol Museum of Archaeology press office, HO

In 1991, a glacier near the border between Italy and Austria melted enough to expose the mummified remains of a mountaineer. Discovered in Ötztal Valley, the mountaineer was named "Ötzi," but he is commonly referred to as the Ice Man. His possessions, particularly his tools, have given researchers insight into how people lived during his time. This begs the question, "When exactly *was* his time?" How old is the Ice Man? To find out, chemists used a process called *carbon-14 dating,* a conceptually simple procedure based on kinetics, the study of this chapter.

Carbon-14 (^{14}C) is a radioactive isotope of carbon that is present in all natural materials. It forms when cosmic rays strike atmospheric nitrogen. ^{14}C then decays, eventually reaching a constant concentration of 1 of every 8.48×10^{11} carbon atoms in the air. ^{14}C is incorporated into the carbon cycle of all living organisms on the planet. The incorporation of ^{14}C ceases when the living species dies and the radioactive ^{14}C in the body decays at a known rate. By measuring the concentration of ^{14}C in an archeological sample—like the Ice Man's mummified tissues and tools—chemists can calculate the age of the sample. Using such procedures, the researchers determined that Ötzi died about 5300 years ago, in a

Chemical Kinetics

13

CHAPTER CONTENTS

OWL Online homework for this chapter may be assigned in OWL.

Look for the green colored vertical bar throughout this chapter, for integrated references to this chapter introduction.

period of history called the *Bronze Age.* The Ice Man's body was remarkably well preserved, and scientists have applied forensic science to theorize how and why he died. Scientists found DNA of other people on his clothes and a wound from an arrow which leads them to hypothesize that he was mortally wounded in a skirmish. His body is preserved in a multimillion-dollar refrigerator so that as scientists develop better and more sensitive methods, Ötzi will be able to provide valuable information on the human race of 5300 years ago.

^{14}C dating is widely used by archeologists to determine the ages of samples of wood, cloth, and other organic materials. The technique is well suited for artifacts relating to human activity because the range over which ^{14}C dating is considered accurate (the last 50,000 years) spans the growth and development of human civilization. The same general method can be used with different elements to help scientists date samples older than can be measured with ^{14}C methods.

Our ability to date objects using ^{14}C depends on knowing how fast this isotope decays. The rate at which species decay is a major aspect of the study of kinetics. ▮

The study of the rates of chemical reactions is called **chemical kinetics.** The word *kinetics* is derived from the Greek *kinetikos,* meaning "putting in motion." Kinetics is inherently an experimental science; many factors influence the rate of a reaction.

Chemical reactions typically occur when reactant species strike each other and interact to form new species, called *products.* Chemists study the kinetics of reactions in an effort to determine which molecules collide, how many, under what conditions, and so on. This information is used to improve the production of materials, minimize pollution, and increase the energy efficiency of manufacturing processes.

In this chapter, we first present the experimental approaches for measuring and classifying reaction rates. We discuss the results of the experiments and the treatment of experimental data, then present a model that explains how the rates of reaction change based on factors such as concentration and temperature. This information is used to understand how catalysts increase reaction rates. A section on reaction mechanisms that focuses on molecular-level steps concludes this chapter.

13.1 Rates of Reactions

OBJECTIVES

- ☐ Relate the changes in concentration over time to the rate of reaction
- ☐ Calculate the instantaneous rate of reaction from experimental data
- ☐ Use stoichiometry to relate the rate of reaction to changes in the concentrations of reactants and products

The rates of reactions and the factors that influence those rates are crucial components of our knowledge of chemical reactions. The study of reaction rates is important to the synthesis of chemicals and the design of reactors, and provides fundamental insight into chemical reactivity as it applies to biological, geological, and industrial processes.

Rate of a Reaction

Rate is change per unit time. Although we talk about crime rates and the rate of inflation, the most common use of the word *rate* is to describe how fast something (such as a car) moves. In chemistry, the **rate of reaction** is measured in terms of a *change in concentration per unit time.*

The reaction rate is the change in concentration per unit time.

$$\text{rate} = \frac{\Delta c}{\Delta t}$$

Consider a reaction of the type

reactant $\rightarrow$ product

Chemists use square brackets to represent molar concentrations. For example, the reactant concentration is represented by [reactant]. The reactant concentration decreases during the reaction, so Δ[reactant] is negative. The product increases in concentration, and Δ[product] is positive. For this simple reaction, the rate at which the product appears is equal to the rate of disappearance of the reactant, and either can be used to define the rate of reaction. Later, we modify this definition of reaction rate for more complex reactions, because reaction stoichiometry must be included.

$$\text{rate} = \frac{\Delta[\text{product}]}{\Delta t} = \frac{-\Delta[\text{reactant}]}{\Delta t}$$

Reaction rates can be measured by following the concentrations of reactants or products. It is important to remember that *the rate of reaction is always expressed as a positive number,* regardless of which species is measured, which explains the negative sign in the preceding equation.

Instantaneous and Average Rates

The rate of reaction is not generally constant but changes over the course of the reaction, so it is important to specify the time at which the rate measurement is made.

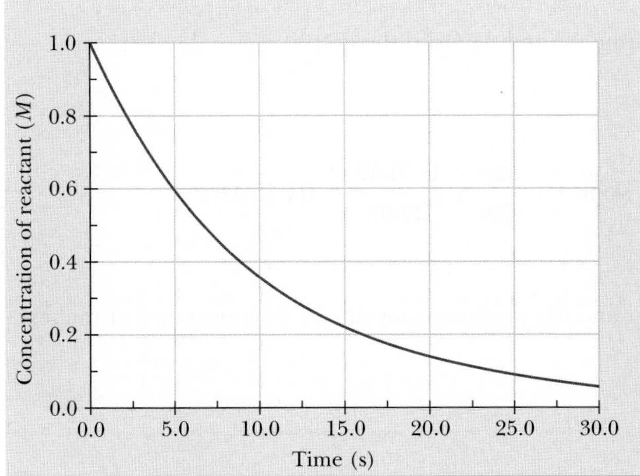

Figure 13.1 Change in concentration as a function of time. The reactant is consumed over time.

TABLE 13.1 Time Dependence of Concentration

Time (s)	Concentration (*M*)
0	1.00
5	0.61
10	0.37
15	0.22
20	0.14
25	0.082
30	0.050

Figure 13.1 is a plot of reactant concentration that decreases regularly over time, and Table 13.1 presents the same data in tabular form. Data such as these graphs and tables are always determined by experiment. Let us use these data to calculate the rate of reaction after 10 seconds has elapsed. Use the concentrations measured after 5 and 15 seconds as representative of the rate of reaction at 10 seconds. The reactant concentration decreases in this interval from 0.61 *M* (after 5 seconds) to 0.22 *M* (after 15 seconds). Because the disappearance of the reactant is observed, a negative sign is needed:

$$\text{rate} = -\frac{(0.22 - 0.61)M}{(15 - 5)\text{s}} = 0.039\ M/\text{s}$$

This calculation illustrates an important point: The most common units used to express reaction rates are *M*/s. These units are also written as (mol/L)/s or mol/(L·s) or mol L^{-1} s^{-1}. When studying the kinetics of reactions involving gases, we often use the change in pressure per unit time, typically atmospheres per second (atm/s), to express rates.

When a rate is measured over a time interval, it is called an **average rate.** In general, the average rate is not useful because it depends on the specific interval. The rate just calculated is the average rate over the 10-second interval between 5 and 15 seconds. The average rate between 0 and 20 seconds differs from the rate between 5 and 15 even though both intervals are centered on 10 seconds. As the interval decreases, the slopes of the lines approach the **instantaneous rate,** which is the slope of the tangent to the curve. The instantaneous rate is used for all chemical applications.

The instantaneous rate is the slope of the tangent to the curve of the concentration versus time graph.

Driving an automobile serves as a simple analogy with instantaneous and average rates. Explaining that it has taken 4 hours for a car to travel 200 miles, at an average rate of 50 miles per hour (mph), is not going to help defend an accused speeder whose instantaneous speed was measured at 75 mph.

Example 13.1 includes calculations over other time intervals.

EXAMPLE 13.1 Determining the Rate of Reaction

Use the data in Figure 13.1 to calculate the instantaneous rate of disappearance of the reactant at 10 seconds.

Strategy The instantaneous rate is the slope of the tangent, so place a ruler on the graph, draw the tangent, and calculate the slope. Although two people might not place the ruler in exactly the same position, slopes estimated from graphs are surprisingly accurate.

Solution

Draw the tangent and determine Δx and Δy from the graph.

$$\Delta y = -0.70\ M$$
$$\Delta x = 20 \text{ seconds}$$

$$\text{Rate of disappearance} = -\text{slope} = -\frac{\Delta y}{\Delta x} = \frac{0.70\ M}{20.0\ \text{s}} = 0.035\ M/\text{s}$$

Understanding

Use the data in Figure 13.1 to find the instantaneous rate of disappearance of the reactant at 20 seconds.

Answer rate $= 0.013\ M/\text{s}$

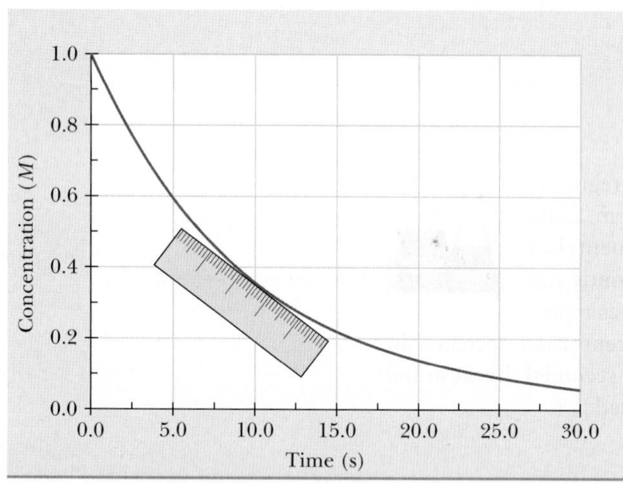

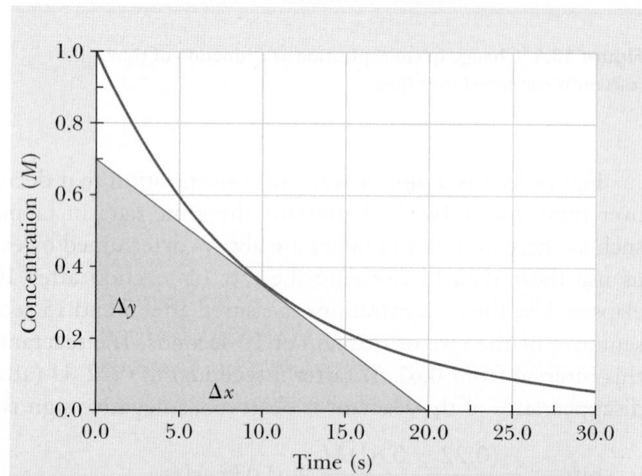

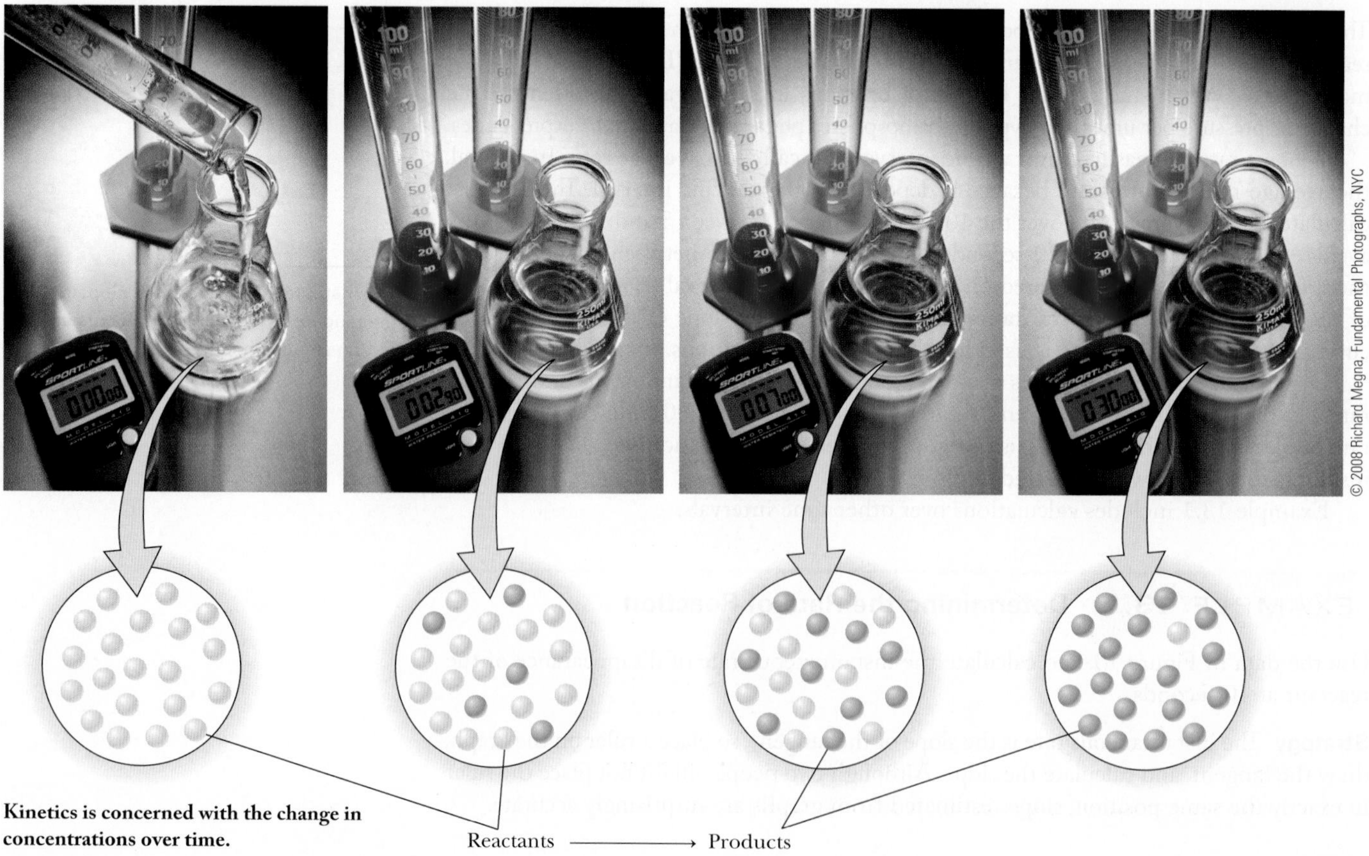

Kinetics is concerned with the change in concentrations over time.

Rate and Reaction Stoichiometry

The rates at which reactants are consumed and products are formed depend on the stoichiometry of the reaction. To illustrate the relationship between stoichiometry and rate, let us consider the decomposition of hydrogen bromide at high temperatures.

$$2HBr(g) \rightarrow H_2(g) + Br_2(g)$$

For every 2 mol of HBr that react, 1 mol of each product forms. The same 2:1 stoichiometry must apply to the *rates* as well. For example, if the concentration of HBr decreases at a rate of 0.50 *M*/s at a particular instant, the rate of change of hydrogen at that time can be calculated from the stoichiometry of the chemical equation.

$$2HBr(g) \rightarrow H_2(g) + Br_2(g)$$

$$\frac{\Delta[H_2]}{\Delta t} = \frac{-\Delta[HBr]}{\Delta t} \times \left(\frac{1 \text{ mol } H_2}{2 \text{ mol HBr}}\right)$$

$$= -(-0.50\ M/\text{s})\ \text{HBr} \times \left(\frac{1 \text{ mol } H_2}{2 \text{ mol HBr}}\right)$$

$$= 0.25\ M/\text{s}\ H_2$$

The rate at which a chemical reaction proceeds does not depend on which species is measured. Therefore, by convention, we define the *rate of reaction* as the ratio of the rate of change of any substance to its coefficient in the chemical equation. For the hydrogen bromide reaction,

The rate of reaction does not depend on which species is measured.

$$\text{Rate of reaction} = \frac{\Delta[Br_2]}{\Delta t} = \frac{\Delta[H_2]}{\Delta t} = \frac{1}{2} \times \frac{-\Delta[HBr]}{\Delta t}$$

The rate of reaction is equal to the absolute value of the rate of change of the concentration of a species times a stoichiometric coefficient that comes from the chemical equation.

Notice that because HBr changes twice as fast as Br_2 and H_2 (because the coefficient in the chemical equation is 2 for HBr), the rate of reaction is equal to half the rate of change of HBr. Example 13.2 illustrates these concepts.

EXAMPLE 13.2 Expressing the Rate of Reaction

Consider the formation of ammonia from the elements:

$$N_2(g) + 3H_2(g) \rightarrow 2NH_3(g)$$

(a) If the ammonia concentration is increasing at a rate of 0.024 *M*/s, what is the rate of reaction?
(b) What is the rate of disappearance of hydrogen?

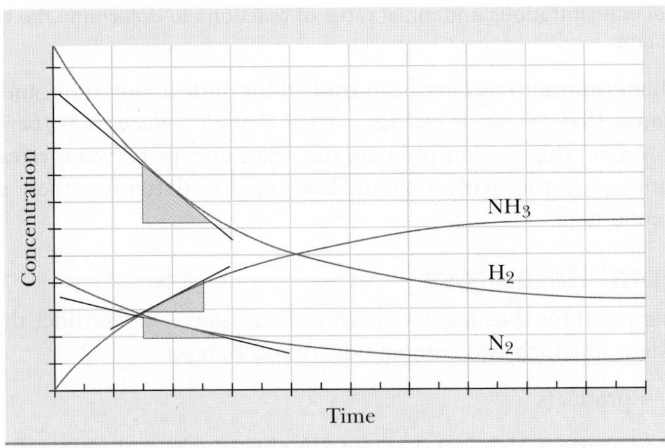

Changes in concentration during the formation of ammonia.

Strategy Use the rate of change of ammonia and the stoichiometry of the chemical equation to compute the rate of reaction. Write the chemical equation and then divide by the appropriate coefficient in the equation.

$$N_2(g) + 3H_2(g) \rightarrow 2NH_3(g)$$

Solution

(a) The rate of reaction is

$$\frac{1}{2} \times \frac{\Delta[NH_3]}{\Delta t} = \frac{1}{2} \times 0.024 \text{ mol/L} \cdot \text{s} = 0.012 \text{ mol/L} \cdot \text{s}$$

(b) The rate of disappearance of hydrogen is related to the rate of increase in the ammonia concentration by the coefficients of the chemical equation.

$$\frac{-\Delta[H_2]}{\Delta t} = \frac{\Delta[NH_3]}{\Delta t} \times \left(\frac{3 \text{ mol } H_2}{2 \text{ mol } NH_3}\right)$$

$$\frac{-\Delta[H_2]}{\Delta t} = 0.024 \cancel{\text{mol } NH_3}/\text{L} \cdot \text{s} \times \left(\frac{3 \text{ mol } H_2}{2 \cancel{\text{mol } NH_3}}\right) = 0.036 \text{ mol/L} \cdot \text{s}$$

Understanding

Under different conditions, the rate of disappearance of N_2 was 0.14 *M*/s N_2. What is the rate of reaction? What is the rate of disappearance of hydrogen?

Answer The rate of reaction is 0.14 *M*/s. The rate of disappearance of hydrogen is 0.42 *M*/s.

OBJECTIVES REVIEW *Can you:*

- ☑ relate the changes in concentration over time to the rate of reaction?
- ☑ calculate the instantaneous rate of reaction from experimental data?
- ☑ use stoichiometry to relate the rate of reaction to changes in the concentrations of reactants and products?

13.2 Relationships between Rate and Concentration

OBJECTIVES

- ☐ Define a rate law to express the dependence of the rate of reaction on the concentrations of the reactants
- ☐ Identify the reaction order from the rate law
- ☐ Use initial concentrations and initial rates of reactions to determine the rate law and rate constant

The rates of chemical reactions have been studied for quite a long time, and one important observation is that *the rate of reaction is often strongly influenced by the concentrations of the reacting species.* This section presents the relationships between rates of reaction and concentrations together with some methods used to determine these relationships from experimental data.

Experimental Rate Laws

A good starting point for learning about rates of reactions is to consider the kinetics of a chemical system in which two compounds, A and B, react:

$$aA + bB \rightarrow \text{products}$$

The investigation starts with a series of experiments in which the rate of reaction is measured as the concentrations of the reactants change. A general observation from these experiments is that the rate of reaction is not just proportional to the concentrations of the reactants, but the rate is proportional to the product of the concentrations of the reactants, *each raised to some power.* These observations can be summarized by an equation called the **rate law** that relates the rate of reaction to the concentrations of the reactants.

The rate law is the equation that relates the rate of reaction to the concentrations of the reactants.

$$\text{rate} = k[\text{A}]^x[\text{B}]^y \qquad [13.1]$$

The exponents, x and y, are the **orders of the reaction.** The order is usually a small positive integer, but occasionally it can be zero, negative, or fractional. The rate law is described as xth order in A and yth order in B. For example, if $x = 1$ and $y = 2$, then the reaction is first order in A and second order in B. The sum of the orders, termed the **overall order**, is three for this example. Notably, the orders, x and y, are not necessarily the coefficients of the balanced chemical equation but are numbers that must be determined by experiment. The proportionality constant, k, is called the specific rate constant or simply the **rate constant.**

The order is the exponent, or power, to which the concentration is raised in the equation that expresses the rate law.

EXAMPLE 13.3 Determining the Order of a Reaction

If experiments show that the reaction of

$$NO(g) + 2O_2(g) \rightarrow NO_2(g) + O_3(g)$$

follows the rate law

$$\text{rate} = k[NO][O_2],$$

what is the order in each species?

Strategy The order is the exponent to which the concentration is raised in the rate law, so we can look at the rate law, note the exponent, and state the order.

Solution

When no exponent is explicitly written on a variable, it is understood that the variable is raised to the first power. Thus, the exponent for NO is 1, so the reaction is first order in NO; it is also first order in O_2.

Understanding

If the rate law is rate $= k[NO]^{1/2}[O_2]^2$, what is the order in each species?

Answer The reaction is half order in NO and second order in O_2.

Determining the Order from Experimental Measurements of Rate and Concentration

Most often, chemists obtain experimental data and determine the rate law. The first step is to determine the order by deciding which exponent explains the observed behavior. Table 13.2 shows the dependence of rate on concentration when the value of the rate constant (k) is 1.

TABLE 13.2 Dependence of Rate on Order

	[Concentration]	Rate
First-order rate law: rate $= k[\text{conc}]^1$	1	1
	2	2
	3	3
Second-order rate law: rate $= k[\text{conc}]^2$	1	1
	2	4
	3	9
Zero-order rate law: rate $= k[\text{conc}]^0$	1	1
	2	1
	3	1

In most classes, students are given an exponent and asked to evaluate a function, such as calculating the rate in the last column of Table 13.1, but determining the rate law requires you to ask, "Which exponent explains this observation?"

Measuring the Initial Rate of Reaction

The initial rate method simplifies the determination of the relationships between rate and concentration.

One widely used method of determining the order of a reaction is called the **initial rate method.** The technique utilizes several experiments in which the initial concentrations of all substances are accurately known but are different in each experiment. The substances are mixed, and the reaction progress is followed. The time interval must be short so that the rate, which is determined from the experimentally measured change in concentration per unit time, is close to the instantaneous rate. The initial rate method correlates the initial rates of reactions with the initial concentrations.

Consider the gas-phase reaction of water with methyl chloride.

$$H_2O(g) + CH_3Cl(g) \rightarrow CH_3OH(g) + HCl(g)$$

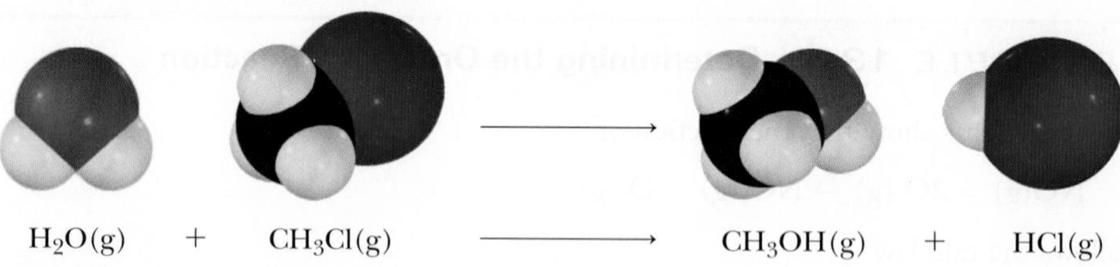

$H_2O(g)$ + $CH_3Cl(g)$ ⟶ $CH_3OH(g)$ + $HCl(g)$

We might try to measure the rate of reaction by determining the amount of HCl produced every few seconds. If we were doing a titration with sodium hydroxide, we would find that we cannot stop the reaction to collect the HCl and titrate it. Instead, we might measure a little NaOH into the container, and see how long it takes before the HCl produced by the reaction consumes the NaOH. If we add 1.80×10^{-4} mol NaOH to a 100-mL (0.100-L) container and the HCl neutralizes it in 5.00 seconds (and we know that each mole of HCl reacts with 1 mol NaOH):

$$\text{rate} = \frac{\Delta[\text{HCl}]}{\Delta t}$$

$$\text{rate} = \frac{1.8 \times 10^{-4}\ \text{mol}/0.100\ \text{L}}{5.00\ \text{s}} = 3.6 \times 10^{-4}\ (\text{mol/L})/\text{s}$$

Because the measurements are made in the early part of the reaction, the observed rate is equal to the initial rate.

A scientist has collected the following data in her laboratory. She needs to determine the rate law and evaluate the rate constant for this reaction.

	Initial Concentration *(M)*		
Experiment	H_2O	CH_3Cl	Initial Rate of Reaction (*M*/s)
1	0.010	0.030	3.6×10^{-4}
2	0.020	0.030	14.4×10^{-4}
3	0.030	0.030	32.3×10^{-4}
4	0.020	0.060	28.7×10^{-4}
5	0.020	0.090	43.2×10^{-4}

The data can be used to determine the exponents x and y in the rate law:

$$\text{rate} = k[H_2O]^x[CH_3Cl]^y$$

Experiments 1, 2, and 3 all have changing concentrations of H_2O and the same concentration of CH_3Cl. These experiments can be used to evaluate the order in H_2O. One of the best ways to recognize the dependence of rate on concentration is to express both rates and concentrations on a relative basis. Obtain the relative concentrations of water vapor and the relative rates of reaction in these three experiments by dividing each concentration by 0.010 (the smallest concentration) and each rate by 3.6×10^{-4} (the slowest rate).

Experiment	Initial Concentration (M [H_2O])	Initial Rate of Reaction (M/s)	Relative Concentration of H_2O	Relative Initial Rate of Reaction
1	0.010	3.6×10^{-4}	1.00	1.00
2	0.020	14.4×10^{-4}	2.00	4.00
3	0.030	32.3×10^{-4}	3.00	8.97

As the relative concentrations increase from 1.0 to 2.0 to 3.0, the relative rate of reaction goes from 1.0 to 4.0 to 9.0. This result indicates that the rate of reaction is proportional to the concentration of water squared (raised to the second power). *The reaction is second order in water,* so we can update the rate law to

$$\text{rate} = k[H_2O]^2[CH_3Cl]^y$$

In experiments 2, 4, and 5, only the initial concentration of CH_3Cl changes. The following table includes both the actual and relative concentrations and rates.

Experiment	Initial Concentration (M [CH_3Cl])	Initial Rate of Reaction (M/s)	Relative Concentration of CH_3Cl	Relative Initial Rate of Reaction
2	0.030	14.4×10^{-4}	1.00	1.00
4	0.060	28.7×10^{-4}	2.00	1.99
5	0.090	43.2×10^{-4}	3.00	3.00

Because the relative rate doubles and triples as the relative concentration doubles and triples, the order in methyl chloride is 1, and the rate law is

$$\text{rate} = k[H_2O]^2[CH_3Cl]$$

The final step in the solution of this problem is to evaluate the rate constant, k. We solve the rate law for k:

$$k = \frac{\text{reaction rate}}{[H_2O]^2[CH_3Cl]}$$

We can select any of the experiments; we chose the data from experiment 1.

$$k = \frac{3.6 \times 10^{-4}\ \cancel{\text{mol/L}} \cdot \text{s}}{(0.010\ \text{mol/L})^2(0.030\ \cancel{\text{mol/L}})} = 1.2 \times 10^2 \text{L}^2/\text{mol}^2 \cdot \text{s}$$

The complete rate law is

$$\text{rate} = (1.2 \times 10^2\ \text{L}^2/\text{mol}^2 \cdot \text{s})\ [H_2O]^2[CH_3Cl]$$

A good way to check our work is by calculating the rate constant from the data of a second experiment, in which *both concentrations are different.* If the rate constants found from the two experiments are the same, we can be confident that the orders in water and methyl chloride are correct; results that differ by more than experimental uncertainty indicate that an error has been made in determining the order.

Checking the rate constant verifies that the orders and numeric value of the rate constant are correct.

Once the rate law and rate constant have been determined, the rate of reaction can be calculated for any given set of reactant concentrations.

EXAMPLE 13.4 Determining the Rate Law from Initial Rate Data

When methyl bromide reacts with hydroxide ion in solution, methyl alcohol and bromide ion form.

$$CH_3Br + OH^- \rightarrow CH_3OH + Br^-$$

Determine the rate law and evaluate the rate constant from the experimental data.

	Initial Concentration (*M*)		
Experiment	$[CH_3Br]$	$[OH^-]$	Initial Rate (*M*/s)
1	0.050	0.010	2.4×10^{-3}
2	0.080	0.020	7.7×10^{-3}
3	0.080	0.010	3.8×10^{-3}

Strategy An information flow diagram helps to illustrate the problem-solving strategy. First, calculate relative concentrations and rates. Next, find experiments in which only one concentration varies, and compare the changes in concentrations and changes in rates to deduce the order of that reactant. Finally, use the rate law and the experimental data to calculate the rate constant.

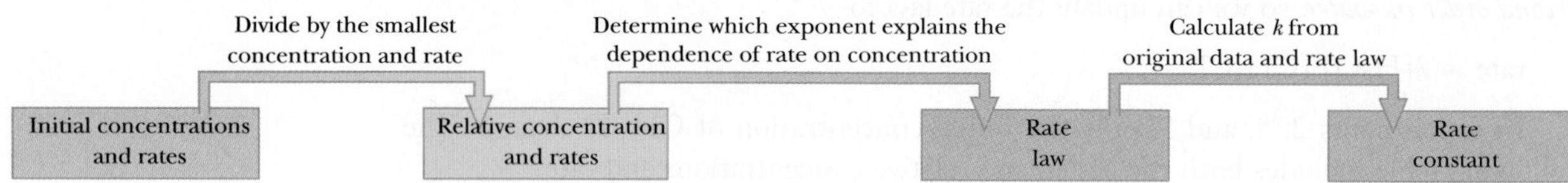

Solution

To obtain relative concentrations and rates, divide the CH_3Br concentrations by 0.050, the OH^- concentrations by 0.010, and the reaction rates by 2.4×10^{-3}.

	Relative Concentration		
Experiment	$[CH_3Br]$	$[OH^-]$	Relative Rate
1	1.0	1.0	1.0
2	1.6	2.0	3.2
3	1.6	1.0	1.6

Experiments 1 and 3 (constant hydroxide ion concentration) show that when the concentration of CH_3Br increases by a factor of 1.6, the rate increases by the same factor.

We can conclude that the reaction is first order in methyl bromide. Experiments 2 and 3 (constant concentration of methyl bromide) show that when the concentration of hydroxide doubles, so does the rate. Thus, the reaction is also first order in hydroxide ion. The rate law can now be written as

$$\text{rate} = k[CH_3Br][OH^-]$$

Experiment	Relative Concentration CH_3Br	Relative Rate
1	1.0	1.0
3	1.6	1.6

Experiment	Relative Concentration OH^-	Relative Rate
3	1.0	1.6
2	2.0	3.2

Use the original data for experiment 1 from the first table (not the relative data of the second table) to calculate the rate constant:

$$k = \frac{\text{rate}}{[CH_3Br][OH^-]}$$

$$k = \frac{2.4 \times 10^{-3} \text{ mol/L} \cdot \text{s}}{(0.050 \text{ mol/L})(0.010 \text{ mol/L})}$$

$$k = 4.8 \text{ L/mol} \cdot \text{s}$$

	Initial Concentration (*M*)		
Experiment	$[CH_3Br]$	$[OH^-]$	Initial Rate (mol/L · s)
1	0.050	0.010	2.4×10^{-3}
2	0.080	0.020	7.7×10^{-3}
3	0.080	0.010	3.8×10^{-3}

You can confirm this value using the data from experiment 2.

Understanding

The following table gives the results of a similar experiment for the reaction of *t*-butyl bromide, $(CH_3)_3CBr$, with hydroxide ion. Calculate the rate law and the rate constant.

$$(CH_3)_3CBr + OH^- \rightarrow (CH_3)_3COH + Br^-$$

	Initial Concentration (*M*)		
Experiment	$(CH_3)_3CBr$	$[OH^-]$	Initial Rate (*M*/s)
1	0.050	0.010	4.1×10^{-8}
2	0.080	0.020	6.6×10^{-8}
3	0.080	0.010	6.6×10^{-8}

Answer rate $= k[(CH_3)_3CBr]$; $k = 8.2 \times 10^{-7}\ s^{-1}$

The reactions shown in Example 13.4 are representative of some interesting organic substitution reactions. Experiments show that the first reaction (methyl bromide with hydroxide) is first order in methyl bromide and first order in hydroxide, but that when *t*-butyl bromide reacts with hydroxide (in the Understanding section), the reaction is first order in *t*-butyl bromide and zero order in hydroxide. The differences arise because the reactions proceed by different steps, as discussed in Section 13.6.

It is important to emphasize that *the rate law cannot be predicted from the reaction stoichiometry.* The rate law can be determined only by measurements of the rate of reaction. As an illustration, the three chemical systems mentioned in this section and their rate laws are:

$$CH_3Cl + H_2O \rightarrow CH_3OH + HCl \qquad \text{rate} = k[H_2O]^2[CH_3Cl]$$

$$CH_3Br + OH^- \rightarrow CH_3OH + Br^- \qquad \text{rate} = k[CH_3Br][OH^-]$$

$$(CH_3)_3CBr + OH^- \rightarrow (CH_3)_3COH + Br^- \qquad \text{rate} = k[(CH_3)_3CBr]$$

The stoichiometry of the reactions is quite similar, yet each obeys a different rate law.

The rate law must be determined by experiment and not from the coefficients of the chemical equation.

OBJECTIVES REVIEW *Can you:*

- ☑ define a rate law to express the dependence of the rate of reaction on the concentrations of the reactants?
- ☑ identify the reaction order from the rate law?
- ☑ use initial concentrations and initial rates of reactions to determine the rate law and rate constant?

13.3 Dependence of Concentrations on Time

OBJECTIVES

- ☐ Evaluate concentration-time behaviors to write a rate law
- ☐ Relate the differential and integrated forms of the rate law
- ☐ Calculate the concentration-time behavior for a first-order reaction from the rate law and the rate constant
- ☐ Relate half-life and rate constant, and calculate concentration-time behavior from the half-life of a first-order reaction
- ☐ Calculate the concentration-time behavior for a second-order reaction from the rate law and the rate constant

The experimental data shown in Figure 13.2 indicate that the rate of reaction changes over the course of the reaction. The concentrations of the reactants decrease, the concentrations of the products increase, and we observe that the rate of reaction decreases with time, eventually reaching zero.

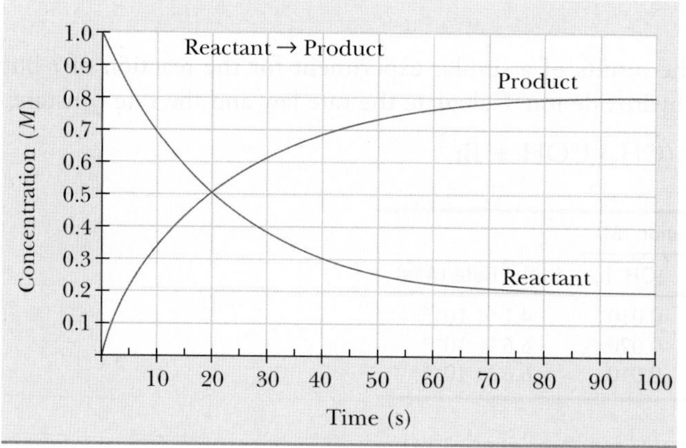

Figure 13.2 Concentration-time profile of a typical reaction. The concentration of the reactant decreases as the concentration of the product increases.

Section 13.2 used the dependence of the initial rate of reaction on the concentrations of the reactants to determine the rate law. Another way to determine the rate law is to perform experiments that measure how the concentration of a reactant changes with time during the course of a *single* experiment. Reactions of different orders behave quite differently. The units of the rate constant also depend on the order of the reaction.

If the graph of reactant concentration versus time is a straight line, then the reaction obeys zero-order kinetics.

Zero-Order Rate Laws

Some reactions show rates that are independent of the concentrations of the reactants and obey a zero-order rate law.

$$\text{rate} = k[\text{reactant}]^0$$

$$\text{rate} = k$$

The rate constant, k, has the same units as the reaction rate: M/s. One example of a zero-order reaction is the metabolism of ethyl alcohol in the body. After a person drinks alcohol, its concentration decreases at a constant rate until the alcohol is completely consumed. Experiments show that people take 1 to 2 hours to metabolize the alcohol in a serving of beer, wine, or distilled liquor, mostly depending on the size of the individual. If it takes about 2 hours for a small person to metabolize the alcohol in one drink, it takes about 4 hours to process the alcohol from two drinks, 6 hours for three drinks, and so forth. Figure 13.3 consists of graphical representations of a zero-order reaction. The graph of concentration versus time (Figure 13.3*a*) is a straight line that ends when all of the reactant has been consumed. The reaction rate (see Figure 13.3*b*) is the slope of the concentration versus time graph. It has a constant value, as indicated by the horizontal portion, and drops to zero when the reactant is completely consumed. *When the graph of concentration versus time is a straight line, the reaction is zero order.* Zero-order rate laws are common in biochemical reactions involving enzymes—an interesting class of reactions discussed later in this chapter.

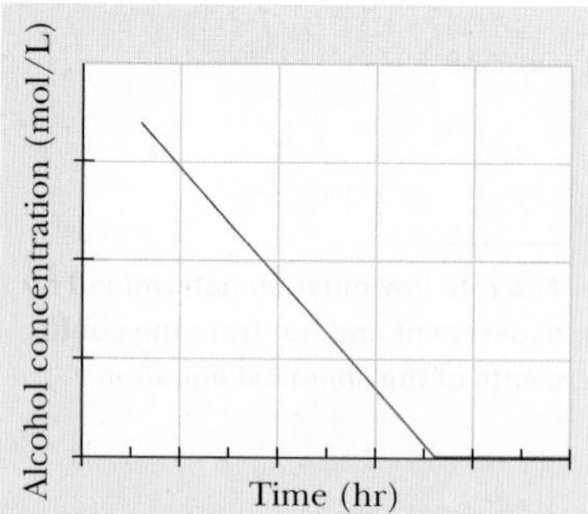

(*a*)

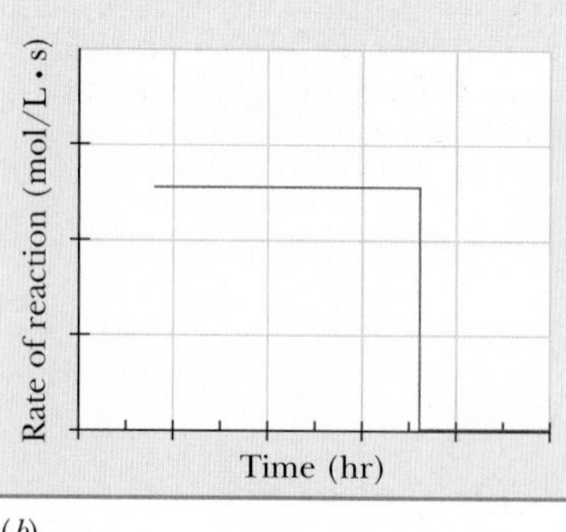

(*b*)

Figure 13.3 Metabolism of alcohol in the body. *(a)* The alcohol concentration in the bloodstream decreases linearly until all the alcohol is metabolized. *(b)* As long as any alcohol is present, the rate of reaction remains constant. One characteristic of zero-order reactions is an abrupt decline in the rate when the reactant has been consumed.

First-Order Rate Laws

When a reaction is first order in a reactant, R, the rate is proportional to the concentration of the reactant.

$$\text{R} \rightarrow \text{product}$$

$$\text{rate} = \frac{-\Delta[\text{R}]}{\Delta t} = k[\text{R}] \quad [13.2]$$

The rate constant in this case has units of s^{-1}.

Time-Dependent Behavior of Concentration

Equation 13.2 represents the rate law for a first-order reaction. This particular form of the equation is called the **differential form of the rate law** because it relates *differences* in concentration and time ($\Delta[\text{R}]/\Delta t$) to the concentrations of the species.[1] The same equation can be expressed in another way, shown in Equation 13.3. This alternative form of the equation is called the **integrated form of the rate law** and relates instantaneous concentrations, rather than *changes* in concentration to the time.

$$[R]_t = [R]_0 e^{-kt} \qquad [13.3]$$

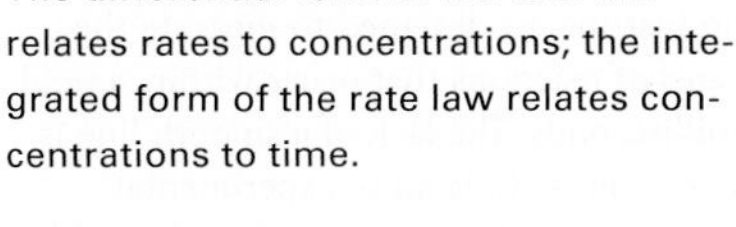

The differential form of the rate law relates rates to concentrations; the integrated form of the rate law relates concentrations to time.

The concentration of R at any time is designated by $[R]_t$; $[R]_0$ is the initial concentration when $t = 0$; and e is the base of the natural logarithms, approximately 2.718. Equation 13.3 describes an *exponential decay*. Figure 13.4 illustrates the exponential decrease of reactant concentration.

The concentration decays exponentially in systems that have first-order kinetics.

Another way to express the integrated form of the first-order rate law results when we take the natural logarithms of both sides of Equation 13.3:

$$\ln[R]_t = \ln[R]_0 - kt \qquad [13.4]$$

or

$$\ln\frac{[R]_t}{[R]_0} = -kt \qquad [13.5]$$

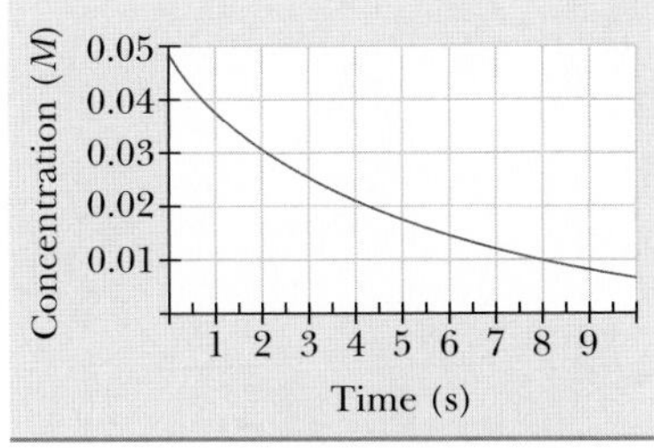

Figure 13.4 **Decrease in the concentration of a reactant in a system that shows first-order kinetics.**

These forms are often easier to use than Equation 13.3. If the system follows first-order kinetics, then a plot of $\ln[R]_t$ as a function of t is a straight line with a slope of $-k$ and an intercept of $\ln[R]_0$. The graph in Figure 13.5 shows these relationships. (Refer to the material in Appendix A if you need to review topics such as slope and intercept.)

When chemists study a new reaction to determine its rate law (in particular, the order), they often design an experiment to measure the concentration of the reactant as a function of time. After obtaining the data, they first graph the concentration as a function of time. If a straight line results, the system is zero order. If the line curves, the next step is to construct a plot of the natural logarithm of reactant concentration as a function of time. If the graph of ln[concentration] versus time is a straight line, the reaction is first order; if it is not a straight line, the investigation continues.

If a graph of ln[reactant] versus time is a straight line, then the system is described by first-order kinetics.

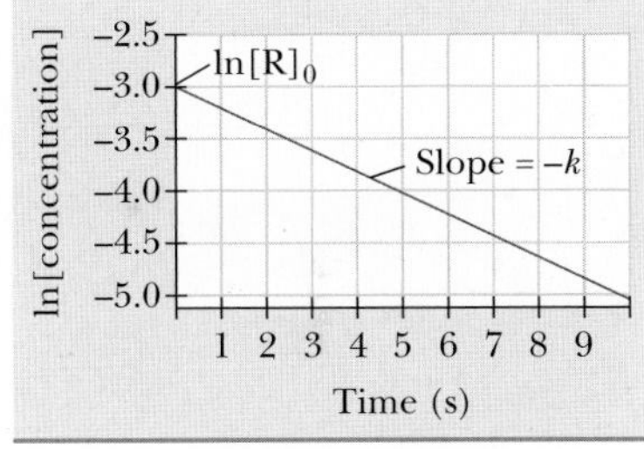

Figure 13.5 **Natural logarithm of concentration plotted against time.** The concentrations are the same as those plotted in Figure 13.4. The y-intercept equals $\ln[R]_0$ and the slope equals $-k$.

The next example shows how to determine the rate constant from concentration-time data for a first-order reaction.

EXAMPLE 13.5 Determining a First-Order Rate Law and Rate Constant

Determine the rate law (order and rate constant) from the data on the next page, obtained in a research laboratory that studies fast reactions. Assume that the reaction is

reactant $\rightarrow$ product

Strategy First, determine the rate law for the reaction using the strategy shown in the block diagram. The core of our strategy is to graph reactant concentration versus time and determine whether the relationship is a straight line. If not, graph ln[concentration] versus time and determine whether that relationship is a straight line.

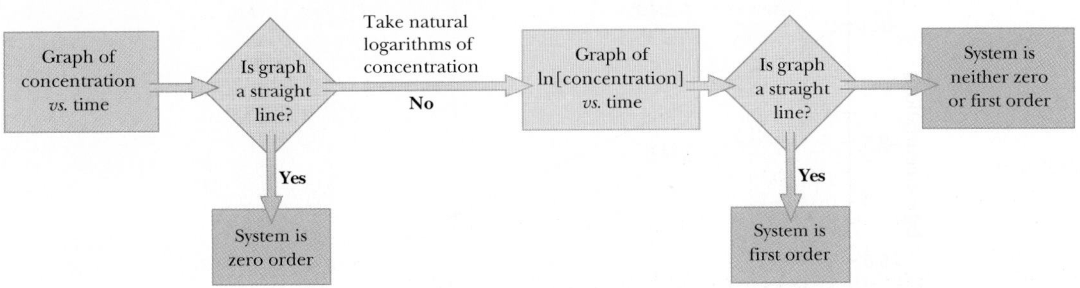

[1]Technically, the word *differential* should only be used when the differences $\Delta[R]$ and Δt are infinitesimally small.

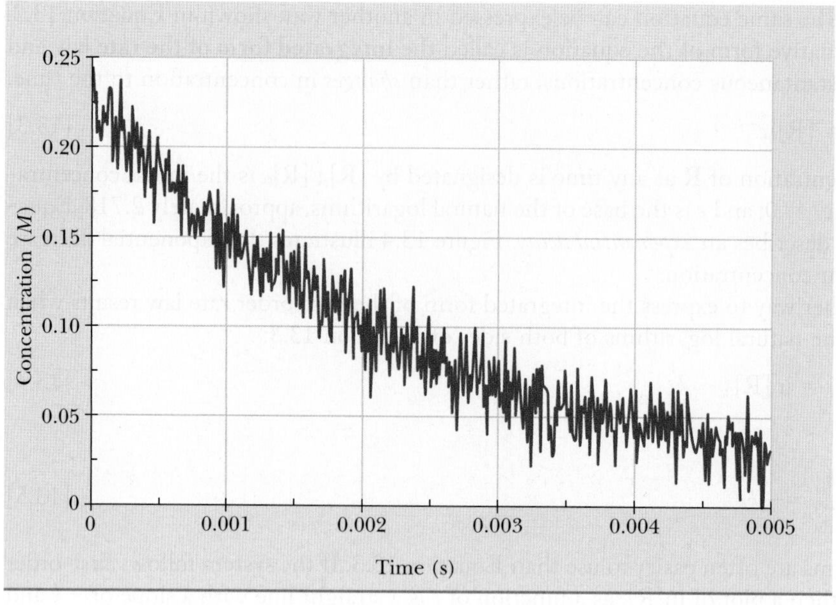

Data from a stopped-flow spectrophotometer. The data come from an instrument designed to measure the rates of reactions that occur within several milliseconds. The lack of a smooth line is due to noise (a form of experimental uncertainty) that is present together with the signal.

Solution

The graph of the original data shows that the concentration-time relationship is curved, so the reaction is not zero order. Because the data resemble those expected for first-order kinetics, start by tabulating the experimental data and the natural logarithm of the concentration. You may find it helpful to draw a smooth line through the data.

Time (s)	Concentration *(M)*	ln[concentration]
0.0	0.22	−1.51
0.001	0.15	−1.90
0.002	0.10	−2.30
0.003	0.070	−2.66
0.004	0.045	−3.10
0.005	0.030	−3.51

Prepare a graph of ln[concentration] against time. The straight-line relationship between ln[concentration] and time indicates a first-order reaction.

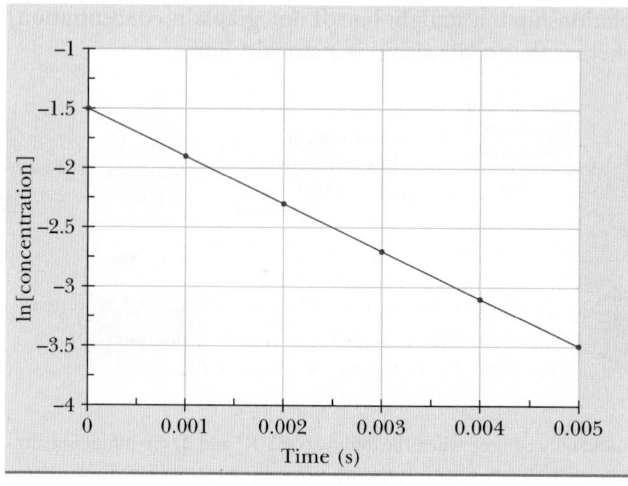

Natural logarithm of concentration as a function of time. Data show a straight-line relationship.

The rate constant is the negative of the slope, which can be measured from the graph or calculated from the tabulated data:

$$k = -\text{slope} = \frac{-3.51 - (-1.51)}{(0.0050 - 0.0)\ \text{s}}$$

$$k = 4.0 \times 10^2\ \text{s}^{-1}$$

Understanding

Graph the data in the table to determine whether the reaction is zero order, first order, or neither.

Time (s)	Concentration *(M)*
2.0	0.11
4.0	0.072
6.0	0.055
8.0	0.044
10.0	0.037

Answer Neither a plot of [R] versus t nor a plot of ln[R] versus t is a straight line. The reaction is neither zero order nor first order.

The integrated rate law can be useful. Example 13.6 illustrates how this form of rate law is used to predict the concentration of a reactant at any particular time.

Use the integrated form of the rate law to predict concentrations as a function of time.

EXAMPLE 13.6 Concentration-Time Relationships in a First-Order Reaction

The pesticide fenvalerate, $C_{25}H_{22}ClO_3N$, is a member of a class of compounds called *pyrethrins.* Fenvalerate is one of four compounds approved by New York City to combat the mosquitoes that spread West Nile virus. These compounds were originally isolated from plants (including chrysanthemums) that exhibited a tendency to repel insects. It degrades in the environment with first-order kinetics and a rate constant of $3.9 \times 10^{-7}\ \text{s}^{-1}$. An accidental discharge of 100 kg fenvalerate into a holding pond results in a fenvalerate concentration of $1.3 \times 10^{-5}\ M$. Calculate the concentration 1 month (2.6×10^6 s) after the spill.

Strategy The problem states the decay is first order and we need to determine the concentration versus time dependence. Start by writing the integrated form of the first-order rate law. Because $[R]_o$, k, and t are given, calculate $[R]_t$.

Solution
Since the decay is first order,

$$\ln[R]_t = \ln[R]_0 - kt$$

$$= \ln(1.3 \times 10^{-5}) - 3.9 \times 10^{-7} \times 2.6 \times 10^6$$

$$= -12.26$$

$$[R]_t = e^{-12.26} = 4.7 \times 10^{-6}\ M$$

The fenvalerate concentration decreased from $1.3 \times 10^{-5}\ M$ to $4.7 \times 10^{-6}\ M$ in 1 month.

Chrysanthemums. People noticed that chrysanthemums *(Pyrethrum cinerariaefolium)* resisted insect attack much more vigorously than most other plants. In 1924, chemists isolated two natural insecticides, called *pyrethrin I* and *pyrethrin II,* from chrysanthemums. These compounds are still widely used by home gardeners and organic farmers.

Understanding

What is the concentration of fenvalerate in the pond 1 year (3.15×10^7 s) after the spill?

Answer $6.0 \times 10^{-11}\ M$

Although molarity was used in the last example, the decrease in the mass of fenvalerate could have been used as well. The reason that either unit can be used becomes easier to determine whether the first-order rate equation is rearranged.

$$\frac{[R]_t}{[R]_0} = e^{-kt}$$

The units used to express $[R]_t$ and $[R]_0$ are not important as long as both quantities are expressed with the same unit.

The next example illustrates how the time needed to reach a specified quantity can be calculated.

EXAMPLE 13.7 Time-Concentration Relationships in a First-Order Reaction

Dinitrogen pentoxide decomposes to nitrogen dioxide and oxygen. When the reaction takes place in carbon tetrachloride (CCl_4), both nitrogen oxides are soluble, but the oxygen escapes as a gas.

$$N_2O_5(soln) \rightarrow 2NO_2(soln) + \frac{1}{2}O_2(g)$$

The rate of reaction ($-\Delta[N_2O_5]/\Delta t$) can be measured by monitoring the volume of oxygen gas that is produced by the reaction. Quantitative analysis of the gas generated in such an experiment shows that the rate law is first order in N_2O_5 with a rate constant of $8.1 \times 10^{-5}\ s^{-1}$ at 303 K.

If the initial mass of $[N_2O_5]$ is 0.032 g, how long does it take for the mass to decrease to 0.015 g?

Strategy This problem looks different because we are given masses, instead of concentrations, but the concentration and mass are directly proportional. Because the problem involves concentration-time behavior, use the integrated rate equation. The rate constant is given, as are the initial mass (0.032 g) and the final mass (0.015 g). We can calculate *t*.

Solution

It is probably easiest to solve with the logarithmic form of the integrated first-order rate law.

$$\ln[N_2O_5]_t = \ln[N_2O_5]_0 - kt$$

$$\ln(0.015) = \ln(0.032) - 8.1 \times 10^{-5}\ s^{-1}\ t$$

$$-4.20 = -3.44 - 8.1 \times 10^{-5}\ s^{-1}\ t$$

$$t = 9.4 \times 10^3\ s$$

It takes 9.4×10^3 seconds (about 2.6 hours) for the mass of dinitrogen pentoxide to decay from 0.032 to 0.015 g.

Understanding

In this experiment, how long does it take for the N_2O_5 to decrease from 0.015 to 0.010 g?

Answer 5.0×10^3 seconds

Half-Life

The half-life is the time needed for the concentration of a reactant to decrease to half its original value.

The rate constant, *k*, is one way to describe the speed of a reaction. A large value for *k* implies a fast reaction. Another way to describe the speed of the reaction is by the **half-life,** designated $t_{1/2}$, which is the time needed for the concentration of a reactant to decrease to half its original value. A short half-life indicates a rapid reaction.

The relationship between the rate constant and the half-life of a first-order reaction can be determined from the first-order rate law. Let $[R]_0$ equal the initial concentration, at time $t = 0$. Half-life, $t_{1/2}$, is defined as the time when $[R]_t$ is equal to $1/2[R]_0$. Substitute these values into Equation 13.3.

$$[R]_t = [R]_0 e^{-kt}$$

$$0.5[R]_0 = [R]_0 e^{-kt_{1/2}}$$

$$0.5 = e^{-kt_{1/2}}$$

Take the natural logarithm of both sides.

$$\ln(0.5) = -kt_{1/2}$$

Evaluate the logarithm and solve for $t_{1/2}$.

$$-0.693 = -kt_{1/2}$$

$$t_{1/2} = \frac{0.693}{k} \qquad [13.6]$$

Note that the *half-life of a first-order reaction is independent of the concentration of the reactant.* The time needed for the reactant to decrease to 50% of its initial concentration depends only on the rate constant. (This relationship is found only in first-order reactions.)

The half-life of a first-order reaction is independent of concentration.

Figure 13.6 shows the decrease in the concentration of the reactant in a first-order reaction. The time needed to decrease from 1.0 to 0.5 M is the same as the time needed to decrease from 0.25 to 0.125 M. A constant half-life is a mark of a first-order reaction.

Radioactive decay processes exhibit first-order kinetics. Plutonium-239, an isotope produced in nuclear reactors, has a half-life of 24,000 years. Given that the world's electrical power reactors are producing about 20,000 kg ^{239}Pu per year, we can calculate how long it would take for 1 year's total production of ^{239}Pu to decay to 0.88 kg, the minimum needed for a nuclear explosion.

First, calculate the rate constant from $t_{1/2}$:

$$t_{1/2} = \frac{0.693}{k}$$

$$24{,}000 = \frac{0.693}{k}$$

$$k = 2.9 \times 10^{-5}\ \text{yr}^{-1}$$

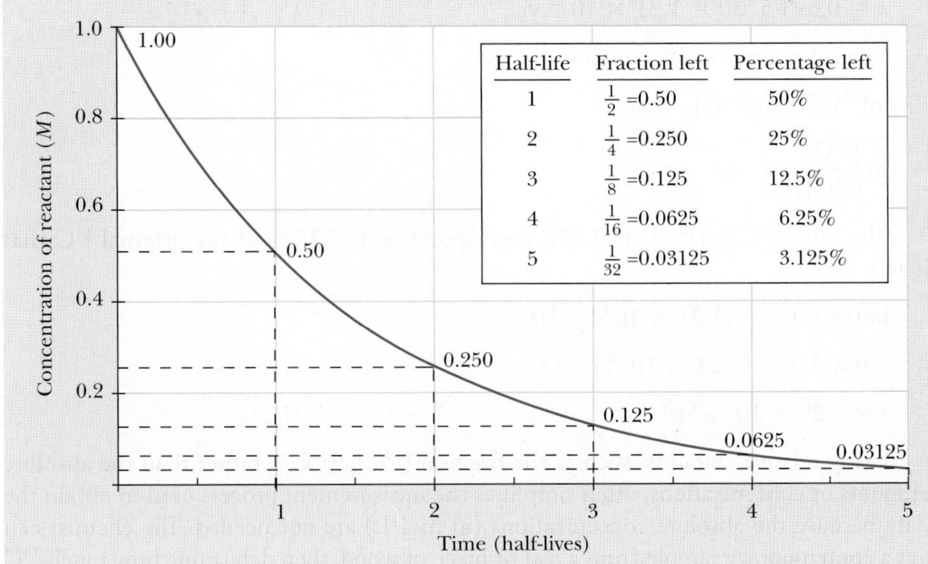

Half-life	Fraction left	Percentage left
1	$\frac{1}{2}$ =0.50	50%
2	$\frac{1}{4}$ =0.250	25%
3	$\frac{1}{8}$ =0.125	12.5%
4	$\frac{1}{16}$ =0.0625	6.25%
5	$\frac{1}{32}$ =0.03125	3.125%

Figure 13.6 Half-lives of a first-order reaction. The horizontal axis is time, expressed in half-lives rather than seconds.

Then calculate the time needed for 20,000 kg plutonium to decay to 0.88 kg. Use the integrated form of the first-order rate law:

$$\ln\frac{[Pu]}{[Pu]_0} = -kt$$

$$\ln\frac{0.88\ \text{kg}}{20{,}000\ \text{kg}} = -2.9 \times 10^{-5} t$$

$$\ln(4.4 \times 10^{-5}) = -2.9 \times 10^{-5}\ t$$

$$-10.03 = -2.9 \times 10^{-5}\ t$$

$$t = 3.5 \times 10^5 \text{ years}$$

It takes about 350,000 years for the amount of plutonium produced annually to decay to the point where there is too little to produce a nuclear bomb. This long window of vulnerability requires that the governments of the nations that produce plutonium take extreme care to ensure its security.

If we know the rate constant for the decay of an unstable isotope, and we measure the concentration of the isotope or those of its decay products, we can sometimes determine the age of a material that contains the isotope. The next example illustrates such a calculation, which is similar to the one done to determine the age of Ötzi the Ice Man, as recounted in the introduction to this chapter.

The half-life, as well as the rate constant, can be used to calculate changes in concentration over time.

EXAMPLE 13.8 Calculating a First-Order Decay

The age of Ötzi, the Ice Man, discussed in the chapter introduction, was determined by carbon-14 (^{14}C) dating. ^{14}C is a radioactive isotope with a half-life of 5730 years. A sample of carbon-containing material was found to have 52.7% of its original amount of ^{14}C. Use this information to calculate how long ago Ötzi lived.

Strategy Since we want a concentration-time relationship, use the integrated expression for a first-order decay, calculating the rate constant, k, from the half-life. The absolute concentrations are not known, but their ratio is known, because the final concentration is equal to 52.7% of the original concentration.

Solution

First, calculate k from $t_{1/2}$.

$$k = 0.693/5730 = 1.21 \times 10^{-4}\ \text{yr}^{-1}$$

Second, solve the rate equation for t.

$$\ln[^{14}C]_t = \ln[^{14}C]_0 - kt$$

$$\ln\frac{[^{14}C]_t}{[^{14}C]_0} = -kt$$

The ratio $[^{14}C]_t/[^{14}C]_0$ is 0.527, corresponding to 52.7% of the original ^{14}C that is left.

$$\ln(0.527) = -(1.21 \times 10^{-4}\ \text{y}^{-1})t$$

$$-0.641 = -(1.21 \times 10^{-4}\ \text{y}^{-1})\ t$$

$$t = 5.29 \times 10^3 \text{ years}$$

Using relative amounts, such as the fraction $[^{14}C]_t/[^{14}C]_0$, rather than the absolute amounts or concentrations, often simplifies the measurement process used to obtain the data because the absolute concentrations (in mol/L) are not needed. The chemist can get a contemporary sample from a leaf or piece of wood, then determine how much ^{14}C has decayed in the test sample. For example, if the chemist finds that 1.00 parts per trillion of the carbon in a contemporary sample is ^{14}C, and 0.85 parts per trillion in an archeological sample is ^{14}C, then the chemist could use 85% as the fraction that remains.

Understanding

A piece of wood has 25% of the ^{14}C originally present. How old is the wood?

Answer 11,000 years

Second-Order Rate Laws

Let us consider the second-order chemical reaction:

$$R \rightarrow \text{product}$$

The rate of reaction is proportional to the concentration of R raised to the second power.

$$\text{Rate} = \frac{-\Delta[R]}{\Delta t} = k[R]^2$$

The units of a second-order rate constant are typically L/mol·s. Again, be aware that zero-, first-, and second-order rate constants all have different units. In fact, if we know the units of the rate constant, we can infer the reaction order.

Concentration-Time Dependence

The rate equation can be written in its integrated form:

$$\frac{1}{[R]_t} = \frac{1}{[R]_0} + kt \qquad [13.7]$$

We can find the rate constant for a second-order reaction from a graph of 1/[R] versus t. The graph is a straight line (demonstrating a second-order reaction) with a slope equal to the rate constant, k.

If a graph of 1/[concentration] versus time is a straight line, then the system shows second-order kinetics.

Notice that the concentration-time relationship of a reaction identifies its rate law. Scientists who study kinetics nearly always start by preparing graphs of their data. The presentation that gives the best straight line identifies the order of the rate law. Table 13.3 summarizes the straight-line relationships for zero-, first-, and second-order reactions.

Half-Life

The half-life of a second-order reaction can be determined from Equation 13.7 by substituting the concentration $[R]_0/2$ and the time, $t_{1/2}$:

$$\frac{1}{[R]_0/2} = \frac{1}{[R]_0} + kt_{1/2}; \qquad t_{1/2} = \frac{1}{k[R]_0}$$

Notice that the half-life of a second-order reaction depends on the starting concentration, so it is seldom useful to measure it. The half-life of a zero-order reaction also depends on concentration.

Example 13.9 shows calculations of the rate constant of a second-order reaction.

TABLE 13.3 Concentration-Time Relationships

Order	Straight-Line Relationship	Slope of Straight Line
0	Concentration vs. time	$-k$
1	ln[concentration] vs. time	$-k$
2	1/[concentration] vs. time	$+k$

EXAMPLE 13.9 Rate Constant of a Second-Order Reaction

Nitrogen dioxide decomposes to nitrogen monoxide and oxygen in a second-order reaction.

$$NO_2(g) \rightarrow NO(g) + \frac{1}{2}O_2(g)$$

The decrease in concentration of NO_2 is shown as a function of time in the accompanying figure. Use the data to calculate the rate constant for the reaction.

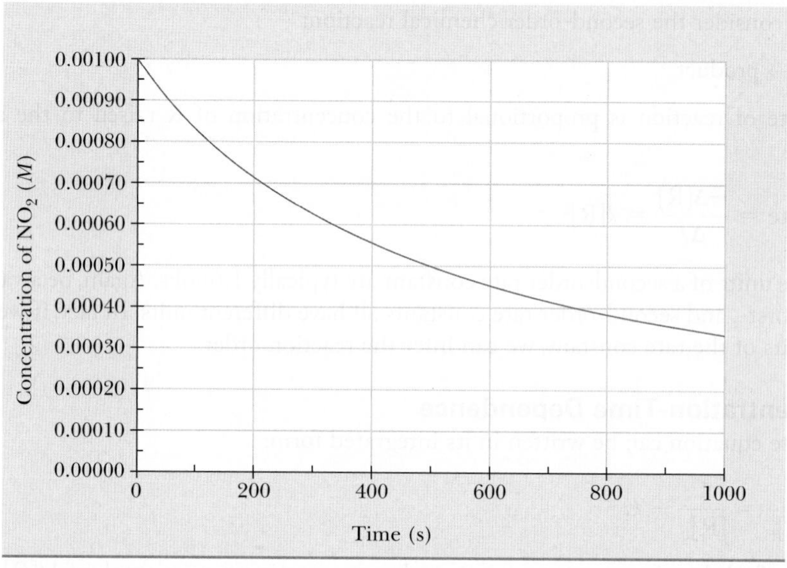

Strategy Because we are told that the reaction is second order, the graph of 1/[concentration] will be a straight line with a slope equal to the rate constant.

Solution

The data in the first two columns of the following table come from the graph. The last column in the table is the reciprocal of the concentration.

A plot of the reciprocal of concentration against time is shown below.

Time(s)	Concentration *(M)*	1/[concentration] (L/mol)
0	0.0010	1.0×10^3
200	0.00072	1.4×10^3
400	0.00056	1.8×10^3
600	0.00046	2.2×10^3
800	0.00038	2.6×10^3
1000	0.00034	2.9×10^3

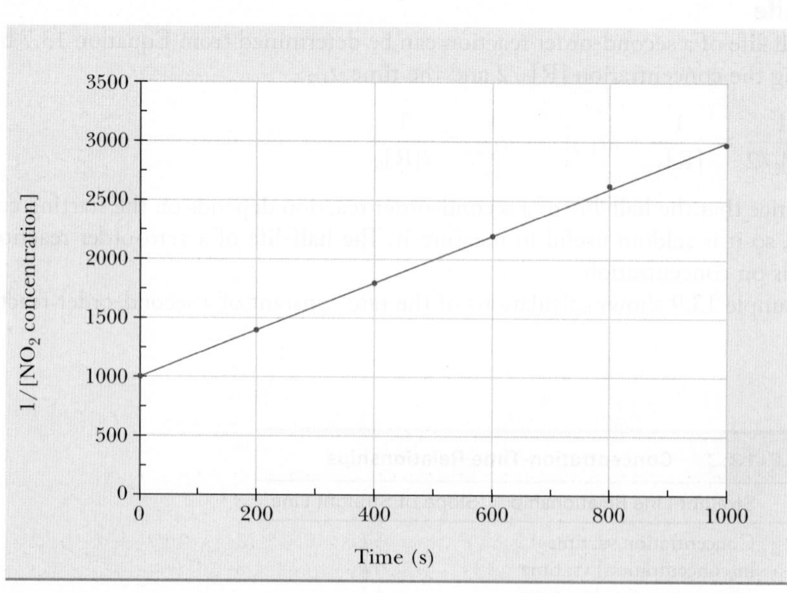

Measure the slope from the graph; the slope is equal to the second-order rate constant.

$$k = \text{slope} = \frac{2600 - 1400}{800 - 200}$$

$$k = 2.0 \text{ L/mol} \cdot \text{s}$$

Understanding

The decomposition of HI at 117 °C is measured in the laboratory.

$$2HI(g) \rightarrow H_2(g) + I_2(g)$$

What are the rate law and the rate constant?

Time (s)	Concentration of HI *(M)*
2,000	0.0088
12,000	0.0054
20,000	0.0042
30,000	0.0033

Answer rate = $k[HI]^2$, $k = 6.8 \times 10^{-3}$ L/mol · s

Summary of Rate Laws

Example 13.10 illustrates a kinetic study of a reaction. The example shows how one would go about determining the complete rate law of a reaction from experimental data.

EXAMPLE 13.10 Determining the Rate Law and Rate Constant for a Reaction

Determine the rate law and rate constant for the decomposition of the organic compound 1,3-pentadiene.

1,3-pentadiene → products

The concentration of 1,3-pentadiene was measured as a function of time. The data are as follows:

Time (s)	[1,3-pentadiene] *(M)*	ln[1,3-pentadiene]	1/[1,3-pentadiene]
0	0.480	−0.734	2.08
1000	0.179	−1.720	5.59
2000	0.110	−2.207	9.09
3000	0.0795	−2.532	12.6
4000	0.0622	−2.777	16.1
5000	0.0510	−2.976	19.6

Strategy A logic flow diagram shows the strategy. We will prepare up to three graphs: concentration versus time to check for zero-order kinetics, ln[concentration] versus time to check for first-order kinetics, and 1/[concentration] versus time to check for second-order kinetics.

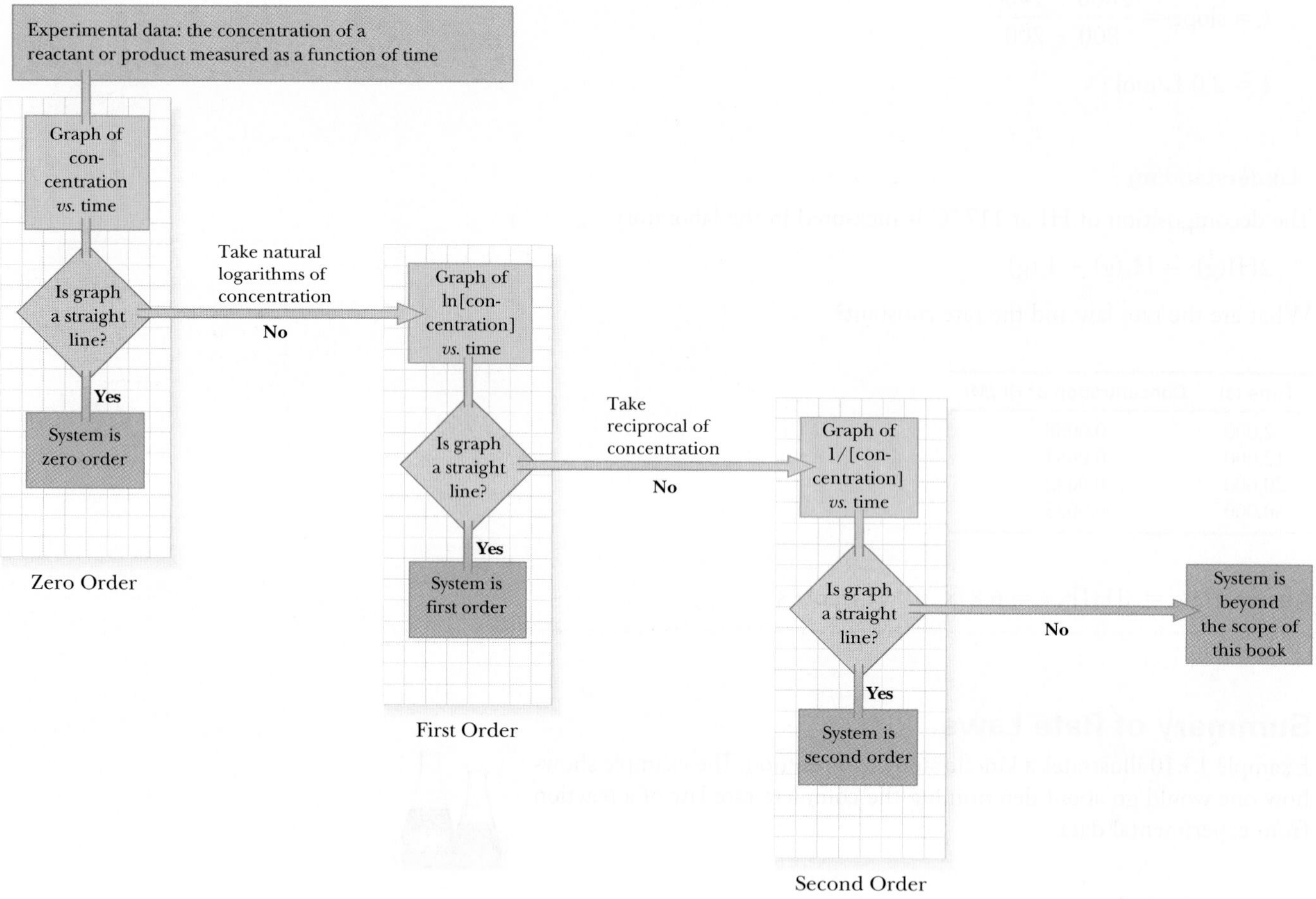

Experimental data: the concentration of a reactant or product measured as a function of time

Graph of concentration *vs.* time

Is graph a straight line?

Yes

System is zero order

Zero Order

No

Take natural logarithms of concentration

Graph of ln[concentration] *vs.* time

Is graph a straight line?

Yes

System is first order

First Order

No

Take reciprocal of concentration

Graph of 1/[concentration] *vs.* time

Is graph a straight line?

Yes

System is second order

Second Order

No

System is beyond the scope of this book

Solution

First, determine the reaction order. Begin by testing whether a plot of [1,3-pentadiene] versus time is a straight line, indicating a zero-order reaction. The original data are shown in a plot of [1,3-pentadiene] versus time; we can look at it and conclude that it is not a straight line.

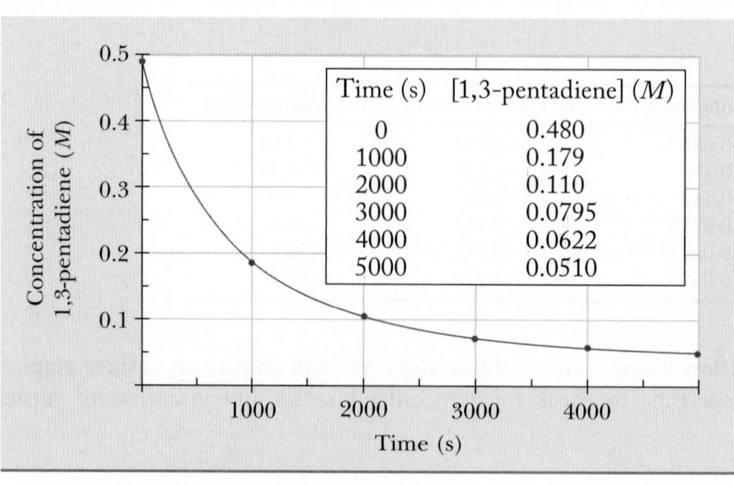

Time (s)	[1,3-pentadiene] (*M*)
0	0.480
1000	0.179
2000	0.110
3000	0.0795
4000	0.0622
5000	0.0510

Concentration of 1,3-pentadiene versus time. The lack of a straight line indicates that the reaction is not zero order.

Next, to check for first-order kinetics, plot ln[1,3-pentadiene] versus time.

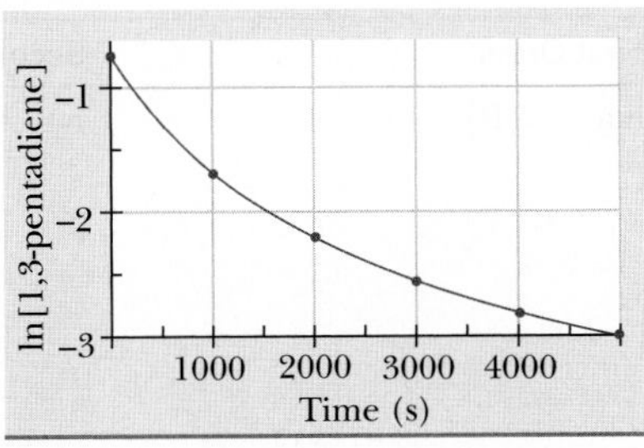

ln[1,3-pentadiene] versus time. The curvature in this graph indicates that the system does not follow first-order kinetics.

If the reaction were first order, this graph would be a straight line. The curvature indicates that the reaction is not first order. The next step is to prepare a plot of 1/[1,3-pentadiene] versus time.

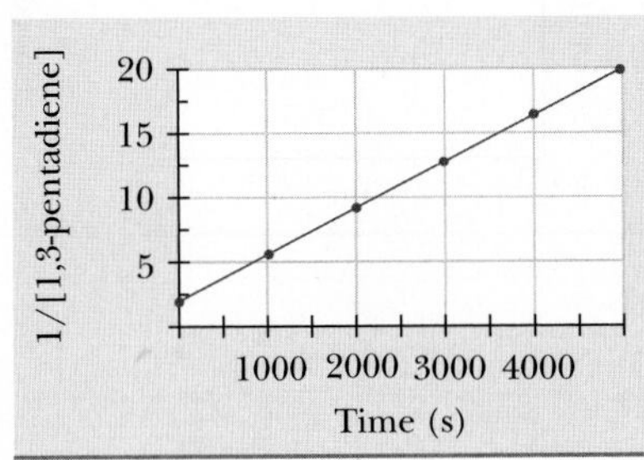

1/[1,3-pentadiene] versus time. The straight line indicates that the system follows second-order kinetics.

The straight line obtained for the last graph indicates that the reaction is second order.

rate = k[1,3-pentadiene]^2

Evaluate the rate constant from the slope of the graph.

$k = 3.5 \times 10^{-3}$ L/mol · s

Understanding

Determine the rate law and rate constant for the reaction

R → products

Time (s)	[R] (M)
0.0	0.43
2.0	0.25
4.0	0.17
6.0	0.13
8.0	0.11

Answer rate = k[R]^2; k = 0.85 L/mol · s

Table 13.4 compares rate laws and summarizes the ways that scientists evaluate experimental data to determine the rate laws.

OBJECTIVES REVIEW *Can you:*

- ☑ evaluate concentration-time behaviors to write a rate law?
- ☑ relate the differential and integrated forms of the rate law?
- ☑ calculate the concentration-time behavior for a first-order reaction from the rate law and the rate constant?
- ☑ relate half-life and rate constant, and calculate concentration-time behavior from the half-life of a first-order reaction?
- ☑ calculate the concentration-time behavior for a second-order reaction from the rate law and the rate constant?

TABLE 13.4 Comparison of Rate Laws for R → Products

	Zero Order	First Order	Second Order
Differential rate law	rate = k	rate = k[R]	rate = $k[\mathrm{R}]^2$
Concentration versus time behavior	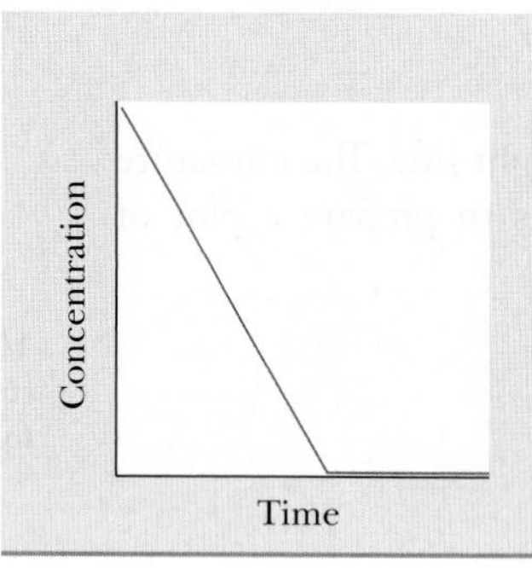	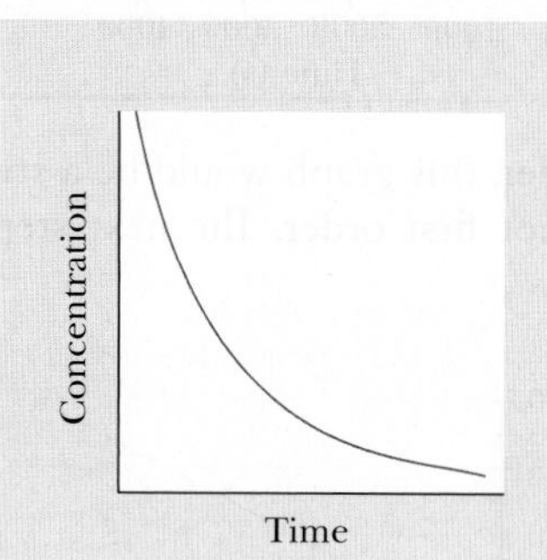	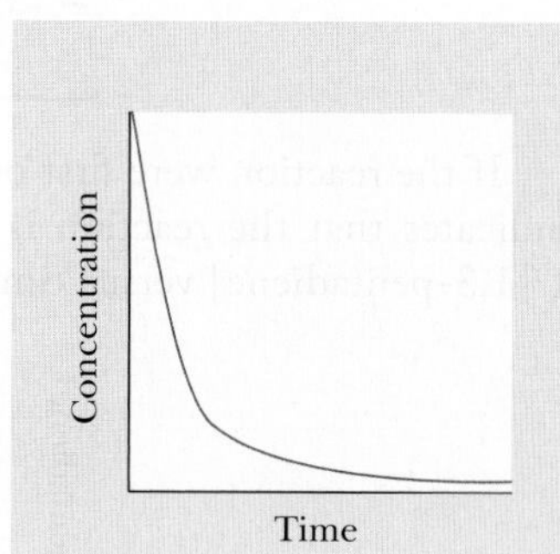
Integrated rate law	$[\mathrm{R}]_t = [\mathrm{R}]_0 - kt$	$[\mathrm{R}]_t = [\mathrm{R}]_0 e^{-kt}$ or $\ln[\mathrm{R}]_t = \ln[\mathrm{R}]_0 - kt$	$\dfrac{1}{[\mathrm{R}]_t} = \dfrac{1}{[\mathrm{R}]_0} + kt$
Straight-line plot to determine the order and rate constant	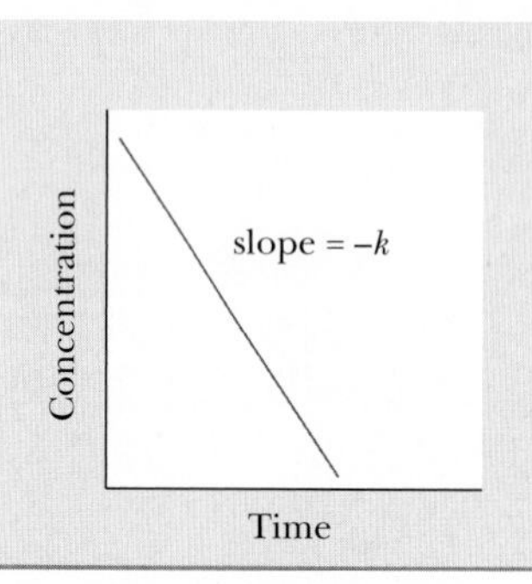	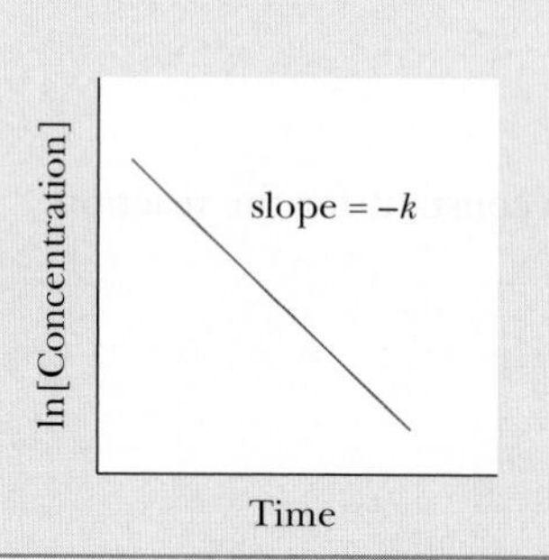	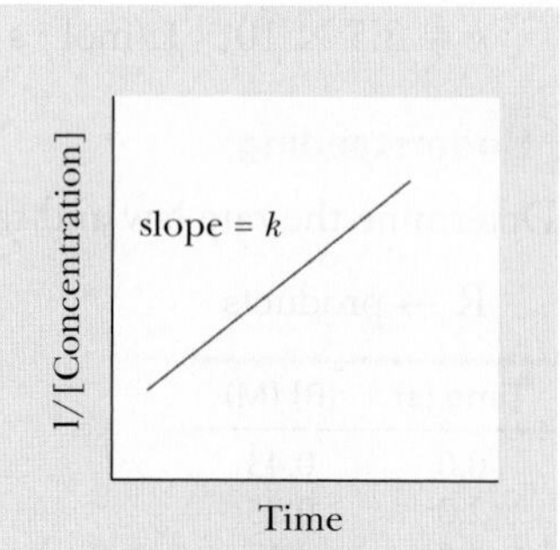
Relative rate versus concentration	[R] / rate: 1 / 1; 2 / 1; 3 / 1	[R] / rate: 1 / 1; 2 / 2; 3 / 3	[R] / rate: 1 / 1; 2 / 4; 3 / 9
Half-life	$t_{1/2} = \dfrac{[\mathrm{R}]_0}{2k}$	$t_{1/2} = \dfrac{0.693}{k}$	$t_{1/2} = \dfrac{1}{k[\mathrm{R}]_0}$
Units of k, rate constant	mol/L · s	s^{-1}	L/mol · s

13.4 Mechanisms I. Macroscopic Effects: Temperature and Energetics

OBJECTIVES

- ☐ Describe the effect of changing temperature on the rate of reaction
- ☐ Use the collision theory to relate collision frequency, activation energy, and steric factor to the rate of reaction
- ☐ Relate temperature, activation energy, and rate constant through the Arrhenius equation

Nearly all reactions go faster when the substances a ... fact that influences our everyday life, as well as chemical kinetics. We kn ... warm weather than in cold, and that higher temperatur ... Note that heating a reaction does not guarantee th ... the reaction will proceed more quickly.

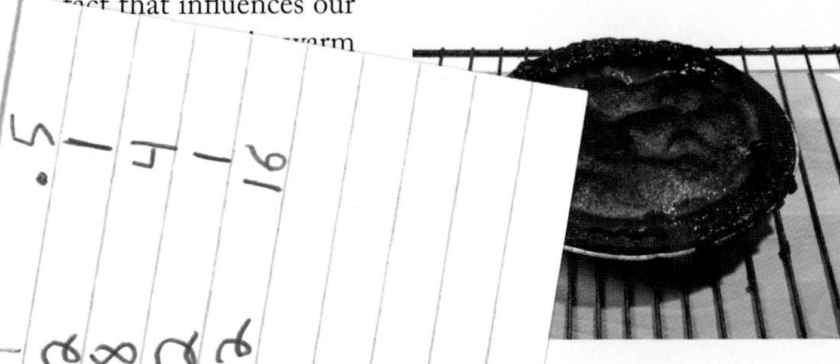

...ture and time must be correct ... desired product.

The temperature dependence of the reaction ... about how a chemical reaction proceeds. Car ... between temperature and reaction rate provid ... many chemical reactions. This section presents ... between atoms, ions, and molecules are nec ... knowing how the temperature influences the r ... collisions.

At the end of this section, we propose a ... energetics of a chemical reaction. We will us ... can be verified by measurement. If the exper ... will modify the theory appropriately.

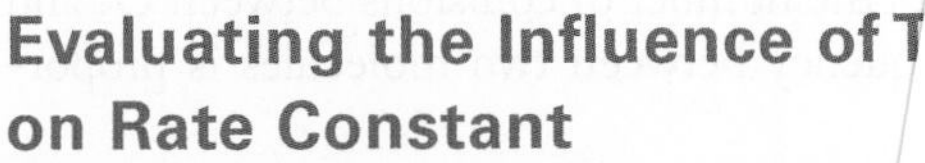

Evaluating the Influence of T... on Rate Constant

Experimental data show that the rates of r... cally with temperature. As an example, th... with ozone has been studied and found ... and second order overall.

$$NO(g) + O_3(g) \rightarrow NO_2(g) + O_2(g)$$

$$\text{rate} = k[NO][O_3]$$

Reactions proceed faster at higher ten... k, increases with temperature, as sho... order of the reaction usually does not ...

...anging the temperature does not ...ange the order of a reaction, but it ...oes increase the value of the rate ...onstant.

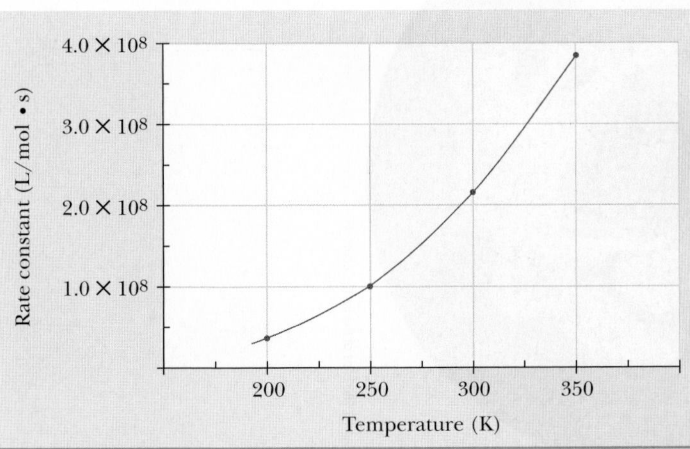

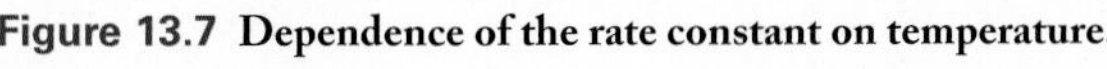

Figure 13.7 Dependence of the rate constant on temperature.

TABLE 13.5 Temperature Dependence of the Rate Constant

$NO(g) + O_3(g) \rightarrow NO_2(g) + O_2(g)$

Temperature (K)	Rate Constant (L/mol · s)
200	0.32×10^8
250	1.0×10^8
300	2.2×10^8
350	3.8×10^8

The shape of this graph is similar to those of the graphs of vapor pressure versus temperature in Figure 11.4. This similarity is more than coincidence; both the temperature dependence of vapor pressure and the temperature dependence of reaction rate are explained on the basis of molecular energy.

Two important principles explain the vapor pressure–temperature relationship: (a) the liquid molecules are in constant motion, and (b) there is a force that holds them together in the liquid state. Molecules in a liquid that have enough kinetic energy to overcome the intermolecular forces can escape to the gas phase. This process occurs more often at increased temperatures, because the hotter molecules move more rapidly and a larger fraction have enough kinetic energy to move to the gas phase.

In the following discussion, we apply a model based on molecular motion to chemical kinetics. The key parameters are the kinetic energy of the molecules and the energy changes that occur as reactants form products.

Collision Theory

Collision theory explains the rates of reactions in terms of molecular-scale collisions. A basic assumption of collision theory is that *molecules must collide to react.*

Let us continue to consider the reaction of nitrogen monoxide and ozone. Many researchers have studied this reaction, because it plays an important role in several atmospheric processes, such as the production of smog and the formation of an ozone "hole."

$$NO(g) + O_3(g) \rightarrow NO_2(g) + O_2(g)$$

The **collision frequency,** Z, is the number of molecular collisions per second. Z depends directly on the concentrations of the gases. For example, if the concentration of O_3 doubles, then the number of collisions between O_3 and NO molecules also doubles. Tripling the concentration of NO triples the number of collisions between O_3 and NO molecules. In general, the collision frequency between two molecules is proportional to the product of their concentrations.

$$\text{Collision frequency} \propto [NO][O_3]$$

An expression for the collision frequency, Z, can be written as

$$Z = Z_0[NO][O_3] \qquad [13.8]$$

where Z is the collision frequency (collisions per second between NO and O_3), and Z_0 is a proportionality constant that depends on the sizes and speed of the reacting species.

Reactions occur when molecules collide.

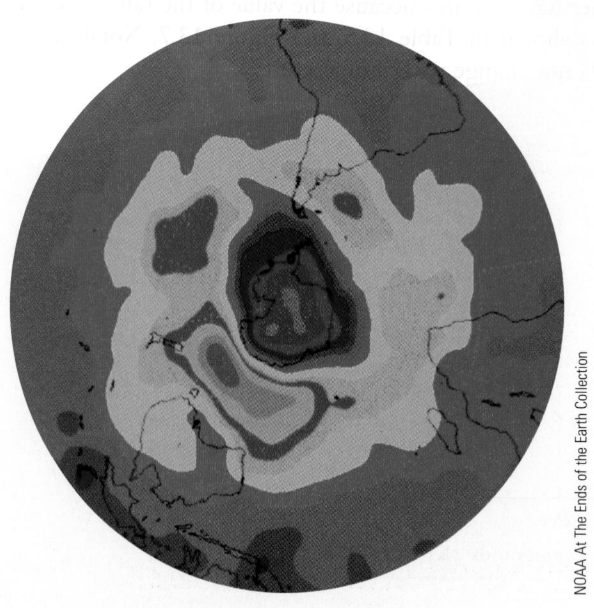

NOAA At The Ends of the Earth Collection

Ozone hole. The red area in the satellite data shows low ozone concentrations over Antarctica.

The kinetic molecular theory of gases predicts that as temperature increases, the molecules move faster and, therefore, the collision frequency increases. But the increase in collision frequency cannot account for the temperature dependence of reaction rate. We can calculate the change in collision frequency between NO and O_3 molecules as the temperature increases from 200 to 350 K and find that the calculated collision frequency increases by about 30%. When we perform the experiment, however, the observed reaction rate increases by more than 1000%. The change in rate is calculated from the experimental data summarized in Table 13.5.

The collision rate alone is insufficient to predict rates correctly.

The simple collision model also predicts collision frequencies (calculated from the kinetic theory of gases) that are much larger than the experimentally observed reaction rates. This evidence led scientists to deduce that not every collision results in a chemical reaction and to search for other factors that affect the rate of reaction.

Activation Energy

In 1888, Svante Arrhenius (1859–1927) expanded the simple collision model to include the possibility that not all collisions result in the formation of the products. According to Arrhenius, if a collision is to result in the formation of the products, then the molecules must collide with enough energy to rearrange the bonds. If the total energy of the colliding species is too small, the molecules simply bounce off each other. The **activation energy,** E_a, is the minimum collision energy required for a reaction to occur. The activation energy of a chemical reaction has an analog in the evaporation process: Only molecules with sufficient energy to overcome the intermolecular forces can escape from the liquid phase.

Only collisions with enough energy to rearrange bonds can result in the formation of products.

The Activated Complex

The energies of reactants, products, and intermediates are often displayed on an energy-level diagram. The vertical axis is potential energy; the horizontal axis, called the *reaction coordinate,* is a relative scale that begins with the reactants and ends with the products. Figure 13.8 is the energy-level diagram for the reaction of nitrogen monoxide with ozone. The overall reaction is exothermic, because the products are lower in energy than the reactants. Even so, the reactants have a "hill," or barrier, to climb before products can form. The height of the barrier is E_a, the activation energy. In the terminology of chemical kinetics, this least-stable (highest energy) arrangement of atoms is called the **activated complex,** or transition state. Because the activated complex is unstable, its concentration is extremely small and virtually undetectable.

The activation energy is the energy needed to form the activated complex from the reactants.

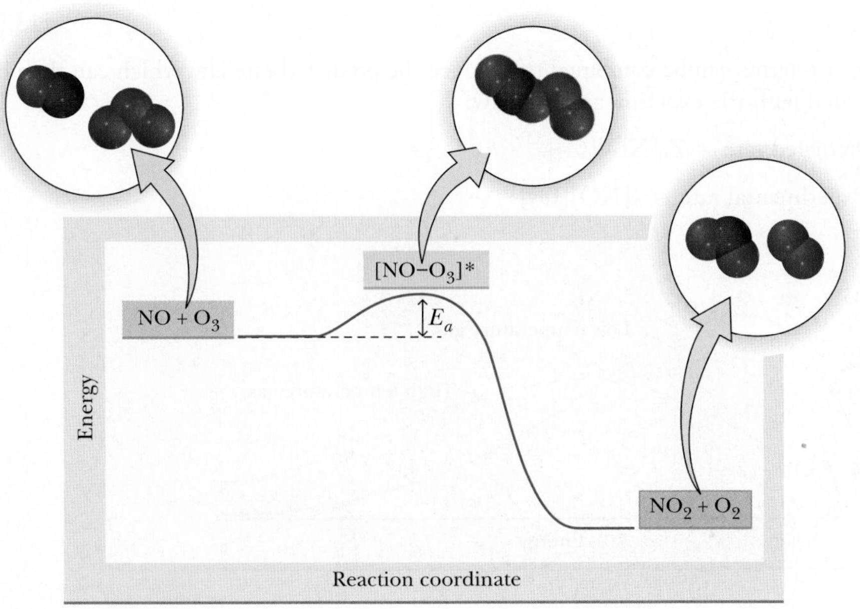

Figure 13.8 Energy-level diagram. The activation energy for this reaction has been determined to be 9.6 kJ/mol.

Another view of the activation energy is that it describes the energy needed to form the activated complex from the reactants. The instability of the activated complex is stressed by the asterisk in $[NO-O_3]^*$. The activated complex can break apart to form the products, but it can also break and re-form reactants. Reactions with high activation energies are slower than reactions with low activation energies, if all other factors are the same. The activation energies for the gas-phase reactions of NO range from 9.6 kJ/mol (for the reaction with ozone) up to 82 kJ/mol (for the reaction with Cl_2 to form NOCl).

Influence of Temperature on Kinetic Energy

The dependence of the rate of a reaction on temperature is strongly influenced by the magnitude of the activation energy. The number of molecules with kinetic energies large enough to initiate a reaction is related to the temperature, as shown in Figure 13.9. The fraction of collisions with energy in excess of E_a can be derived from the kinetic theory of gases:

$$f_r = e^{-E_a/RT} \qquad [13.9]$$

The fraction f_r is a number between 0 and 1. For example, if f_r is equal to 0.05, then the energies of 5% of the collisions exceed E_a, the activation energy. The details of the derivation are not important, but the conclusions are: f_r becomes larger (closer to 1) as T increases. It is important to recognize that the activation energy does not change with temperature, but the number of collisions exceeding E_a increases with temperature.

The number of collisions with energy that exceed E_a grows exponentially with temperature.

Equation 13.9 shows one important influence of temperature: As the temperature increases, the number of collisions with energies that exceed the activation energy grows exponentially.

According to the model, the rate of reaction should equal the rate of collision (collision frequency) times the fraction of collisions that have energies in excess of the activation energy.

$$\text{rate} = Z \times f_r$$

$$\text{rate} = (\text{collision frequency}) \times (\text{fraction exceeding } E_a)$$

The collision frequency depends on the concentrations of the colliding species, as shown by Equation 13.8.

$$Z = Z_0[NO][O_3] \qquad [13.8]$$

The fraction of collisions exceeding E_a is

$$f_r = e^{-E_a/RT} \qquad [13.9]$$

These two terms can be combined to produce the predicted rate law, which can then be compared with the experimental rate law:

$$\text{Predicted rate} = Z_0[NO][O_3]e^{-E_a/RT}$$

$$\text{Experimental rate} = k[NO][O_3]$$

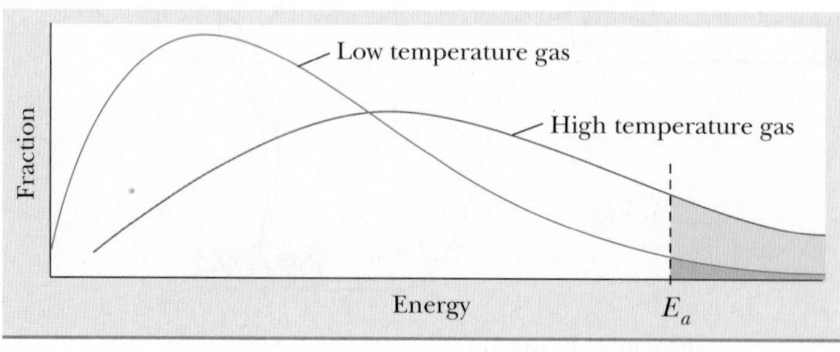

Figure 13.9 Energy distributions in gas molecules. A much larger fraction of molecules has energies in excess of E_a at high temperatures than at low temperatures.

If the model is correct, then the two expressions are equal and we can solve for k, the rate constant.

$$k = Z_0\, e^{-E_a/RT}$$

This expression can be tested by measuring rate constants at different temperatures. The conclusion is that this equation produces the correct temperature dependence of rate constants, but it predicts rates much faster than those observed in the laboratory. The model must be further refined by considering one more factor.

Steric Factor

Not all collisions with energies greater than E_a result in a reaction. Figure 13.10 depicts some possible collisions between nitrogen monoxide and ozone. Because nitrogen dioxide is a bent molecule, the orientation of the atoms in the first collision is favorable for the formation of the O–N–O bent molecule. The formation of a nitrogen-oxygen bond from the orientation shown in the bottom of the figure is unlikely. The **steric factor,** p, is a term that expresses the need for the correct orientation of reactants when they collide to form the activated complex. The steric factor (*steric* means "related to the spatial arrangement of atoms") has a value between 0 and 1.

Not all collisions with energies that exceed E_a are productive. The geometries of the collisions must also be considered.

$$\text{rate} = \underset{\text{Steric factor}}{p} \times \underset{\text{Collision frequency}}{Z_0[\mathrm{NO}][\mathrm{O_3}]} \times \underset{\text{Fraction exceeding } E_a}{e^{-E_a/RT}}$$

The steric factor and Z_0 can be collected in a single factor, which is given the symbol A and called the **pre-exponential term.**

$$\text{rate} = Ae^{-E_a/RT}[\mathrm{NO}][\mathrm{O_3}]$$

Another way to express the same result is to write the expression for the rate constant.

$$k = Ae^{-E_a/RT} \qquad [13.10]$$

Equation 13.10 is called the **Arrhenius equation.** The pre-exponential term, A, includes the steric factor, which cannot be predicted by theoretical models. A can be determined from experiment only.

The Arrhenius equation relates the rate constant to the temperature.

Arrhenius Equation

The Arrhenius equation fits the observed temperature dependence for a wide range of chemical reactions. It can be used in conjunction with experimental measurements of the rate constant at different temperatures to determine the activation energy. If we know E_a for a reaction, we can use the Arrhenius equation to predict how much faster (or slower) it will proceed as the temperature is changed. To see how to determine E_a from experimental measurements, start with Equation 13.10.

$$k = Ae^{-E_a/RT} \qquad [13.10]$$

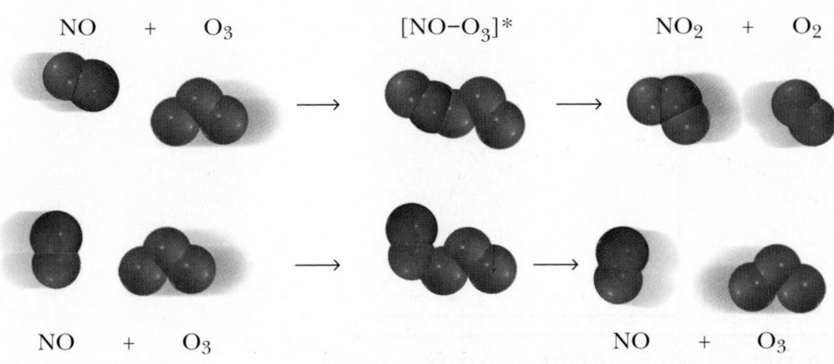

Figure 13.10 Orientation of reactants. Not every collision occurs with the reactants in the correct orientation to produce products.

Take the natural logarithms of both sides.

$$\ln k = \ln A - E_a/RT$$

Rewrite this equation to emphasize the temperature dependence:

$$\ln k = \ln A - \frac{E_a}{R}\left(\frac{1}{T}\right)$$

A plot of ln k versus $1/T$ has a slope of $-E_a/R$ and an intercept of ln A.

This method can be applied to determine the activation energy for the decomposition of NO_2.

$$2NO_2(g) \rightarrow 2NO(g) + O_2(g)$$

First, the order is determined and found to obey a second-order rate law.

$$\text{rate} = k[NO_2]^2$$

Next, the rate constant is measured at several different temperatures, summarized in Table 13.6. Figure 13.11 is a graph of ln k versus $1/T$ based on the data from Table 13.6.

Figure 13.11 is an example of an **Arrhenius plot.** The activation energy is determined by measuring the slope of the graph, which is equal to $-E_a/R$. Because we need only two experiments to calculate the slope, we can derive an equation that relates the activation energy to the rate constants at two temperatures.

$$\ln\left(\frac{k_1}{k_2}\right) = \frac{-E_a}{R}\left(\frac{1}{T_1} - \frac{1}{T_2}\right) \quad [13.11]$$

TABLE 13.6 Dependence of the Rate Constant on Temperature

$2NO_2(g) \rightarrow 2NO(g) + O_2(g)$

Temperature (K)	k (L/mol · s)	ln k	$1/T$
500	0.003	−5.8	0.00200
550	0.037	−3.30	0.00182
600	0.291	−1.234	0.00167
650	1.66	0.507	0.00154
700	7.39	2.000	0.00143

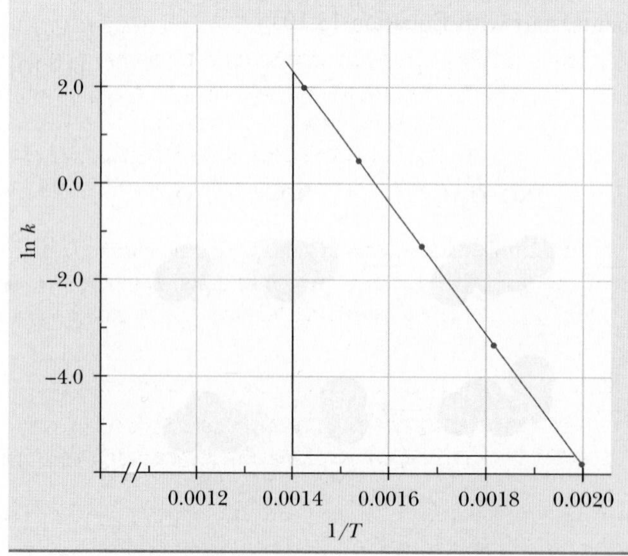

Figure 13.11 Arrhenius plot. The graph of ln k versus $1/T$. The slope is equal to $-E_a/R$.

The next example demonstrates how this equation, which is another form of the Arrhenius equation, is used to determine activation energy from rate measurements at two temperatures.

Measuring the rate constant at two different temperatures provides enough information to evaluate the activation energy.

EXAMPLE 13.11 Determining Activation Energy from the Temperature Dependence of the Rate Constant

Consider the decomposition of nitrogen dioxide.

$$2NO_2(g) \rightarrow 2NO(g) + O_2(g)$$

At 650 K, the rate constant is 1.66 L/mol·s; at 700 K, it is 7.39 L/mol·s. Use these rate constants to determine the activation energy.

Strategy Equation 13.11 relates two rate constants and two temperatures. Let subscript 1 in Equation 13.11 represent the lower temperature and corresponding rate; let subscript 2 represent the higher temperature data. Because it is conventional to express activation energy in J/mol, we will use 8.314 J/mol · K for R.

Solution

Substitute the data into Equation 13.11:

$$\ln\left(\frac{k_1}{k_2}\right) = \frac{-E_a}{R}\left(\frac{1}{T_1} - \frac{1}{T_2}\right) \qquad [13.11]$$

$$\ln\left(\frac{1.66}{7.39}\right) = \frac{-E_a}{8.314\ \text{J/mol}\cdot\text{K}}\left(\frac{1}{650\ \text{K}} - \frac{1}{700\ \text{K}}\right)$$

The units on the left side all cancel. On the right side, the kelvin units cancel, leaving joules per mole (J/mol) as the unit of measure. Solving for E_a:

$$E_a = 1.13 \times 10^5\ \text{J/mol} = 113\ \text{kJ/mol}$$

Understanding

Calculate the activation energy (in kJ/mol) for a reaction that has $k = 1.0 \times 10^8\ s^{-1}$ at 250 K and $k = 3.8 \times 10^8\ s^{-1}$ at 350 K.

Answer $E_a = 9.7$ kJ/mol

The activation energy is calculated from the dependence of the natural logarithm of rate constant (or rate) on the reciprocal of temperature. The method used in the last example uses only two points to determine the activation energy; when the rate constant is known at more than two temperatures, the activation energy can be determined from the slope of the ln k versus $1/T$ graph, as already shown. This type of data treatment has the effect of averaging several experiments, and thus minimizing experimental error, and it also allows the scientist to determine whether the result of an individual experiment lies farther from the line than might be expected from random error. When all the data in Table 13.6 are used to prepare a graph of ln k versus $1/T$, the slope of the line gives a value of 114 kJ/mol for E_a, which is close to the 113 kJ/mol calculated in Example 13.11 from just two points.

Chemists often use the rule of thumb that a "typical" reaction rate doubles with a change of 10 °C from room temperature. Most chemists remember generalizations like this because they can be handy. For example, food stored at 20 °C spoils about twice as fast as food stored at 10 °C.

EXAMPLE 13.12 Determining the Activation Energy from the Temperature Dependence of Rate

A reaction rate doubles when an investigator increases the temperature by 10 °C, from 298 to 308 K. What is the activation energy?

Strategy Start with Equation 13.11:

$$\ln\left(\frac{k_1}{k_2}\right) = \frac{-E_a}{R}\left(\frac{1}{T_1} - \frac{1}{T_2}\right)$$

To solve this problem, we need to recognize that the ratio k_1/k_2 is 1:2 because we are told that the rate doubles. Because we are given T_1 and T_2, and R is a constant that we know, we can solve for the activation energy, E_a.

Solution

Let T_1 be 298 and T_2 be 308 K. The ratio of the relative rates of reaction is the same as the ratio of the rate constants, and the ratio of the rates is 1 to 2. Thus,

$$\ln\left(\frac{1}{2}\right) = \frac{-E_a}{8.314 \text{ J/mol}\cdot\text{K}}\left(\frac{1}{298 \text{ K}} - \frac{1}{308 \text{ K}}\right)$$

$$E_a = 5.3 \times 10^4 \text{ J/mol} = 53 \text{ kJ/mol}$$

Understanding

The rate of a reaction doubles when the temperature changes by 15 °C, from 298 to 313 K. Calculate the activation energy for the reaction.

Answer $E_a = 3.6 \times 10^4$ J/mol = 36 kJ/mol

OBJECTIVES REVIEW *Can you:*

- ☑ describe the effect of changing temperature on the rate of reaction?
- ☑ use the collision theory to relate collision frequency, activation energy, and steric factor to the rate of reaction?
- ☑ relate temperature, activation energy, and rate constant through the Arrhenius equation?

13.5 Catalysis

OBJECTIVES

- ☐ Define catalysis and identify heterogeneous, homogeneous, and enzymatic catalysts
- ☐ Draw energy-level diagrams for catalyzed and uncatalyzed reactions

For the rate of a reaction to increase, the frequency of productive collisions must increase. Two different strategies each accomplish this goal. The first is to increase the collision frequency and the average energies of collisions by increasing the temperature. Although simple, this approach is not always successful, because undesirable side reactions often occur at higher temperatures.

A second approach is to make a larger fraction of the collisions productive. If a chemist can adjust conditions to reduce the activation energy or increase the steric factor, then the rate of reaction increases. One way to improve the productivity of collisions is to add a **catalyst**—a substance that increases the rate of reaction but is not consumed in the reaction. A catalyst is intimately involved in the course of the chemical reaction; it helps to make and break bonds as the reactants form products, but it does not undergo a permanent change.

A catalyst increases the rate of reaction but is not consumed.

A catalyzed reaction proceeds by a different set of steps than does an uncatalyzed reaction. The catalyzed reaction generally has lower activation energy, and thus a greater reaction rate at any given temperature. Figure 13.12 shows energy-level diagrams for the same reaction in the presence and absence of a catalyst. Some scientists use a common analogy to describe a catalyst: They say it provides a shortcut, a path with a lower energy barrier to the same product.

The energy-level diagram shows that the catalyzed reaction has lower activation energy.

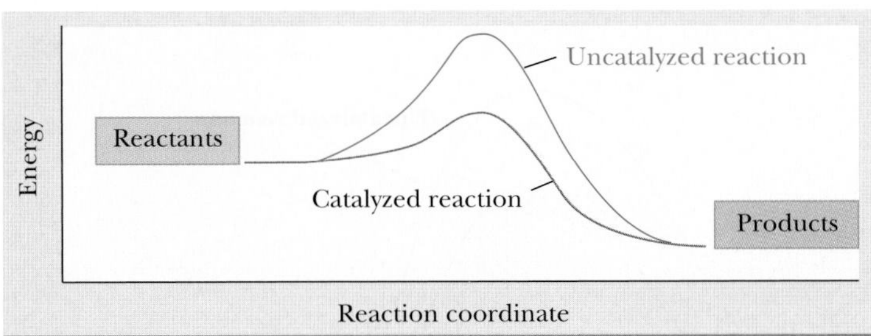

Figure 13.12 An energy-level diagram for a chemical reaction. The activation energy of the catalyzed reaction is lower than that of the uncatalyzed reaction. The catalyzed reaction is faster at any given temperature.

Homogeneous Catalysis

A **homogeneous catalyst** is one that is present in the same phase as the reactants. The bromide-catalyzed decomposition of hydrogen peroxide is an example of homogeneous catalysis.

A homogeneous catalyst is one that is in the same phase as the reactants.

$$2H_2O_2(aq) \rightarrow 2H_2O(\ell) + O_2(g)$$

Chemists believe that the catalyzed reaction occurs in two steps. First, bromide ion and the H^+ found naturally in water reacts with hydrogen peroxide to form bromine and water.

Step 1: $H_2O_2(aq) + 2Br^-(aq) + 2H^+(aq) \rightarrow Br_2(aq) + 2H_2O(\ell)$

In a second step, the bromine formed in the first step reacts with additional hydrogen peroxide to form oxygen.

Step 2: $H_2O_2(aq) + Br_2(aq) \rightarrow 2Br^-(aq) + 2H^+(aq) + O_2(g)$

The sum of the two steps gives the overall reaction.

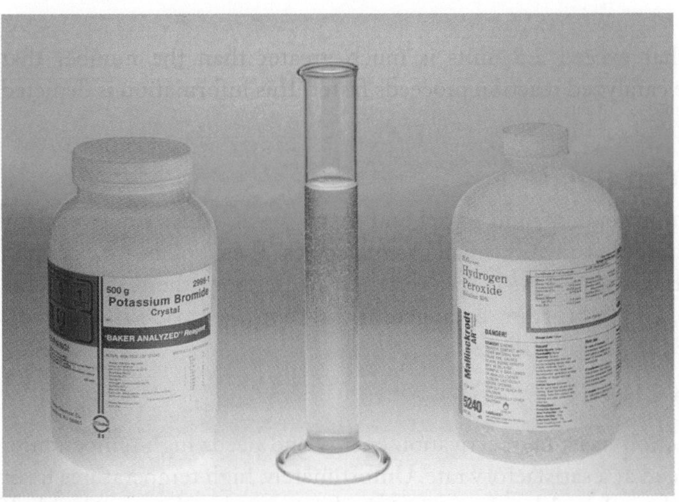

Bromide-catalyzed decomposition of hydrogen peroxide. Shortly after a bromide salt such as sodium or potassium bromide is added to a hydrogen-peroxide solution, the yellow color of bromine is seen together with bubbles of oxygen. The yellow color fades at the conclusion of the reaction.

Adding bromide ion results in violent bubbling because of the generation of oxygen gas and simultaneous appearance of the yellow color of bromine. At the end of the reaction, the bubbling stops and the yellow color disappears, because all the bromine is present then as the colorless bromide ion. Bromide ion is a catalyst because it increases the rate of reaction but is not consumed. Figure 13.13 shows the energy-level diagrams for the catalyzed and uncatalyzed reactions.

The activation energies, estimating from the scale, are about 4.0 units for the uncatalyzed reaction and 2.5 units for the catalyzed reaction. At any temperature, the

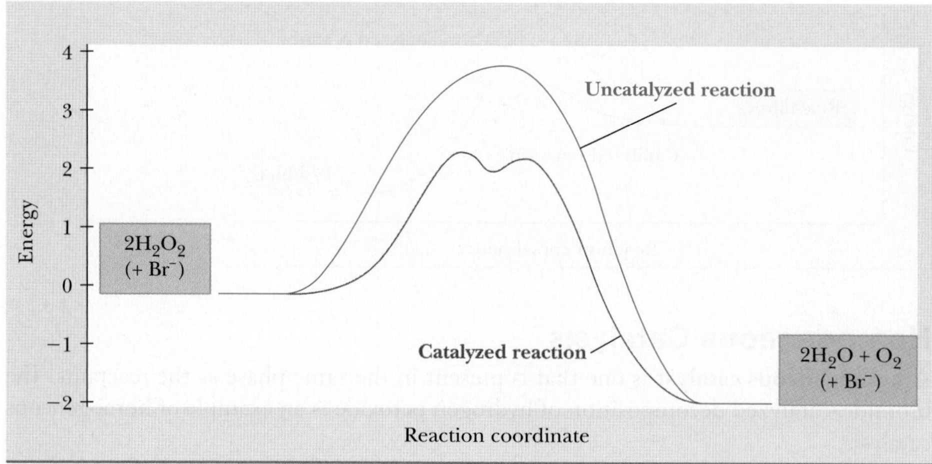

Figure 13.13 Energy-level diagram for the bromide-catalyzed decomposition of hydrogen peroxide.

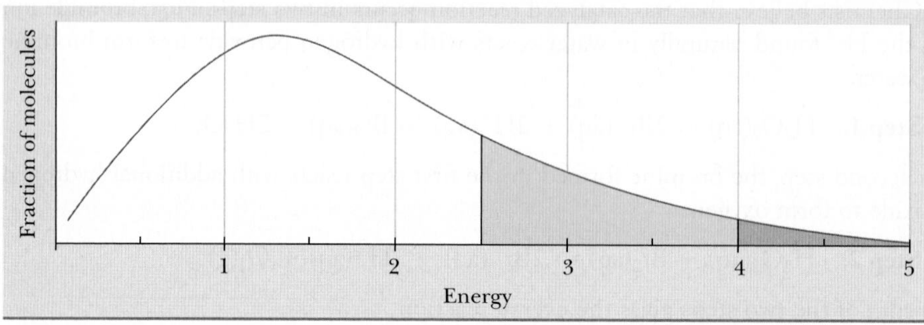

Figure 13.14 Fraction of molecules with enough energy to form the activated complex. The activation energy for the catalyzed reaction is 2.5 units. Many more collisions exceed this activation energy than exceed the activation energy of the uncatalyzed reaction, 4.0 units.

More collisions exceed the lower activation energy of a catalyzed reaction; therefore, the rate of the catalyzed reaction is larger.

number of collisions that exceed 2.5 units is much greater than the number that exceeds 4.0 units, so the catalyzed reaction proceeds faster. This information is depicted in Figure 13.14.

Heterogeneous Catalysis

A heterogeneous catalyst is one that is in a different phase from the reactants. Metals are frequently used to catalyze gas-phase reactions.

Metals such as platinum, palladium, and nickel and many metal oxides catalyze many reactions, particularly those that involve small gas molecules. These solids are examples of a **heterogeneous catalyst,** one that is in a different phase from the reactants. The formation of methanol, CH_3OH, from hydrogen and carbon monoxide is extremely slow if it is not catalyzed by metal surfaces.

$$2H_2(g) + CO(g) \rightarrow CH_3OH(g)$$

The uncatalyzed reaction has a high activation energy and needs high temperatures if the reaction is to proceed at a satisfactory rate. Unfortunately, high temperatures favor the reverse reaction, and methanol breaks up into hydrogen and carbon monoxide if heated at too high a temperature.

A better approach is to use a platinum surface to catalyze the formation of methanol from hydrogen and carbon monoxide so low temperatures can be used. We write the catalyzed reaction by naming the catalyst and placing it over the arrow:

$$2H_2(g) + CO(g) \xrightarrow{Pt} CH_3OH(g)$$

The exact way in which the catalyst takes part in the formation of methanol is not completely known. To study how heterogeneous catalysts work, scientists studied a simpler system: the metal-catalyzed hydrogen-deuterium reaction. Studying simple systems with the idea of extrapolating the knowledge to more complex systems is one important aspect of scientific research. The chemical equation for the reaction of hydrogen with deuterium is

$$H_2(g) + D_2(g) \xrightarrow{Pt} 2HD(g)$$

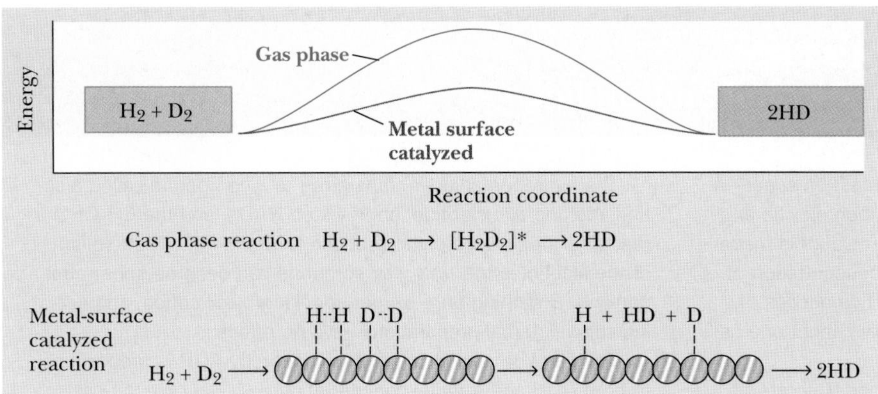

Figure 13.15 **Energy-level diagram for the reaction of hydrogen with deuterium.**

Scientists learned that when a hydrogen or deuterium molecule forms a weak bond to the metal surface, the H—H or D—D bond weakens. If a D_2 molecule bonds to the surface near an H_2 molecule, the product can form. This process is illustrated schematically in Figure 13.15. Many scientists believe that this bond-weakening process explains why so many gas-phase reactions, including the production of methanol from H_2 and CO, proceed rapidly at certain metal surfaces.

Scientists tried a variety of different metals to determine which would be the most effective catalyst for this reaction and learned that the catalyst can actually determine the nature of the product. Platinum and nickel catalysts lead to different products from the reaction of hydrogen and carbon monoxide.

$$2H_2(g) + CO(g) \xrightarrow{Pt} CH_3OH(g)$$

$$3H_2(g) + CO(g) \xrightarrow{Ni} CH_4(g) + H_2O(g)$$

The C—O bond is retained in the reaction at the platinum surface but is broken in the reaction at the nickel surface.

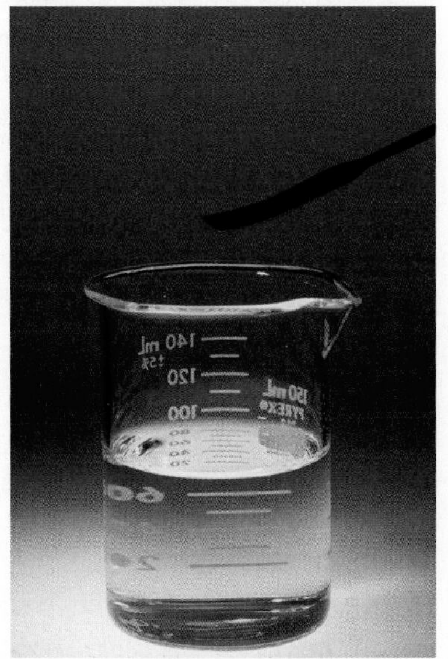

Enzyme Catalysis

Enzymes are large molecules that catalyze specific biochemical reactions. Scientists believe that some enzymes increase the rate of reaction by increasing the value of the steric factor rather than by decreasing the activation energy. These enzymes interact with the reactant molecules in a way that places them in the correct geometry to form the products.

alcohol dehyrogenase

Structure of an enzyme. Photograph of a model of an enzyme called *alcohol dehydrogenase.* Its structure was determined several years ago. Chemists are trying to match the functions and actions of enzymes with their structures, with the long-term goal of being able to predict the catalytic properties of any substance. One step is to determine the exact mechanism by which the enzyme catalyzes a particular reaction.

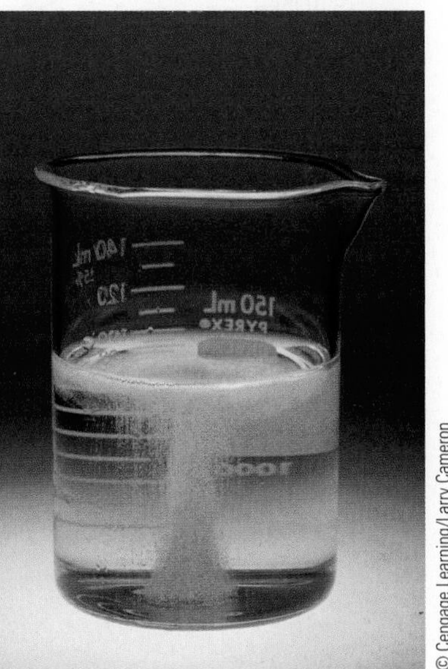

Heterogeneous catalysis. Manganese dioxide, MnO_2, catalyzes the decomposition of hydrogen peroxide. Bubbling occurs when the solid MnO_2 is added to a solution of H_2O_2.

PRACTICE OF CHEMISTRY
Alcohol and Driving

Ethanol can be produced by the action of yeast on sugar—a process called *fermentation*—in which carbon dioxide and alcohol are produced, resulting in beer or wine. Liquor is made by using a distillation process to increase the concentration of alcohol in the product. Twelve ounces of beer, five ounces of wine, and one ounce of distilled liquor all contain about one half ounce of ethyl alcohol.

The metabolism of ethanol is well understood. The rate of alcohol absorption depends on the stomach contents. If alcohol is consumed as part of a meal, then it will remain in the stomach together with the food for several hours. If alcohol is consumed on an empty stomach, it will enter the bloodstream more quickly.

Most alcohol is eventually converted to carbon dioxide and water in the liver. In the first step of the oxidation of ethanol, the alcohol is changed into acetaldehyde by an enzyme called *alcohol dehydrogenase* (ADH):

$$CH_3CH_2OH + NAD^+ \xrightarrow{ADH} CH_3CHO + NADH + H^+$$

The hydrogen anion is transferred to a compound called nicotine adenine dinucleotide, abbreviated NAD^+, and the CH_3CHO reacts further, ultimately to produce CO_2 and H_2O. Metabolism of the alcohol is the *only* way someone can become sober after drinking. Breathing pure oxygen, drinking black coffee, and eating special herbal preparations have no effect.

The reaction rate for ethanol oxidation by ADH is proportional to both ADH and ethanol concentration.

$$\text{rate} = k\,[\text{ADH}][C_2H_5OH]$$

When the concentration of ethanol is much greater than the concentration of the enzyme, the rate of reaction is limited by the concentration of the enzyme. Most enzyme-catalyzed reactions have these rate characteristics because the enzyme becomes saturated with the substrate. All the available enzyme molecules are bound to an ethanol molecule. Increasing the ethanol concentration beyond this concentration has no effect on the overall rate of acetaldehyde production.

$$\text{rate} = k'\,[\text{ADH}]$$

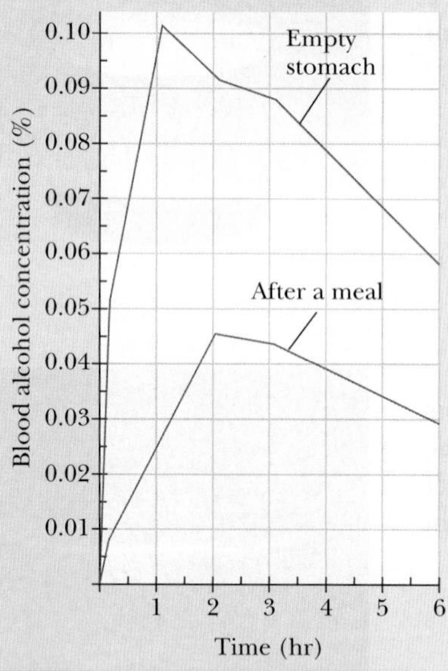

Alcohol absorption. The rate of alcohol absorption depends on several factors. The same amount of alcohol consumed on an empty stomach is absorbed more quickly than when consumed together with a meal.

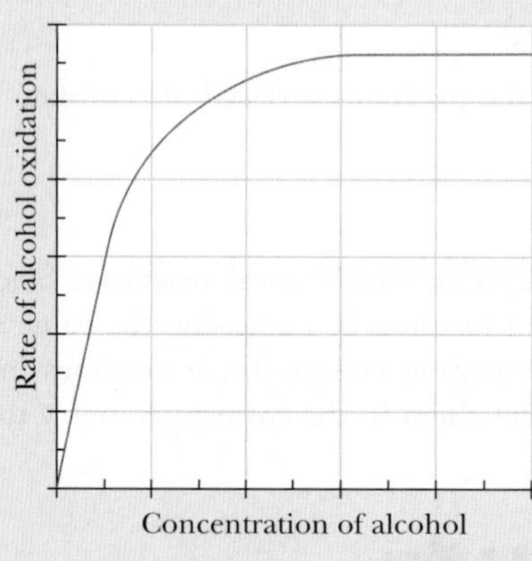

Rate of alcohol oxidation versus blood alcohol concentration. When the concentration of ethanol is much greater than that of the enzyme, as it would be after as little as one drink, the rate of reaction is a constant.

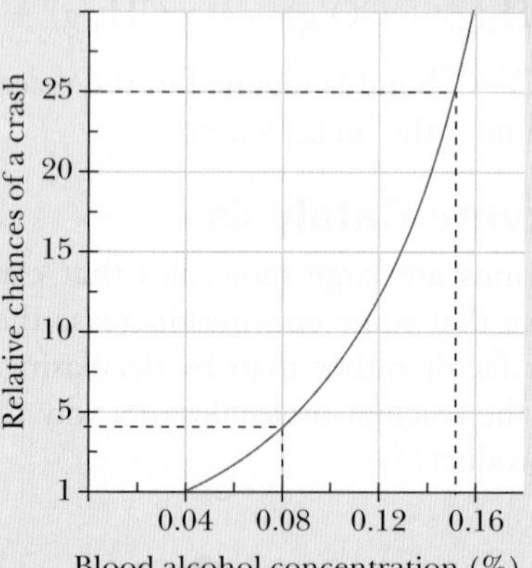

The chances of a motor-vehicle accident and blood alcohol levels. The chances of an accident are about four times greater when the blood alcohol concentration is 0.08%. The risks increase to 25 times normal at a blood-alcohol concentration of 0.15%.

People with high alcohol levels are unable to operate automobiles properly. Most people think that loss of reflexes is the main problem, but the loss of judgment and inability to avoid accidents is also an important factor. Drinking and driving simply do not mix.

Most states use a blood-alcohol test to determine whether someone operating a motor vehicle is impaired by alcohol. Some states require the courts to consider an individual impaired with blood-alcohol concentrations of 0.04%; other states define impairment and set thresholds at different levels, but because U.S. Federal policies require a maximum of 0.08% for a state to receive highway construction funds, no states have levels in excess of 0.08%.

The following nomograph can be used to determine an approximate blood-alcohol concentration from the number of drinks consumed by a person and his or her weight. Note that the predicted concentrations are corrected by subtracting the zero-order rate (0.015% per hour) at which alcohol is metabolized.

The dotted lines show that a 130-lb person who consumes six drinks on an empty stomach will have a maximum blood alcohol concentration of 0.13%. A 130-lb person who consumes four drinks together with a meal will have a maximum blood-alcohol concentration of about 0.05%. The alcohol will be metabolized at a rate of 0.015% per hour, so it will take a little longer than 3 hours for the second person's alcohol to metabolize. It is important to remember that these numbers are only estimates and that individuals respond differently to alcohol consumption. Drinking and driving simply do not mix. ▮

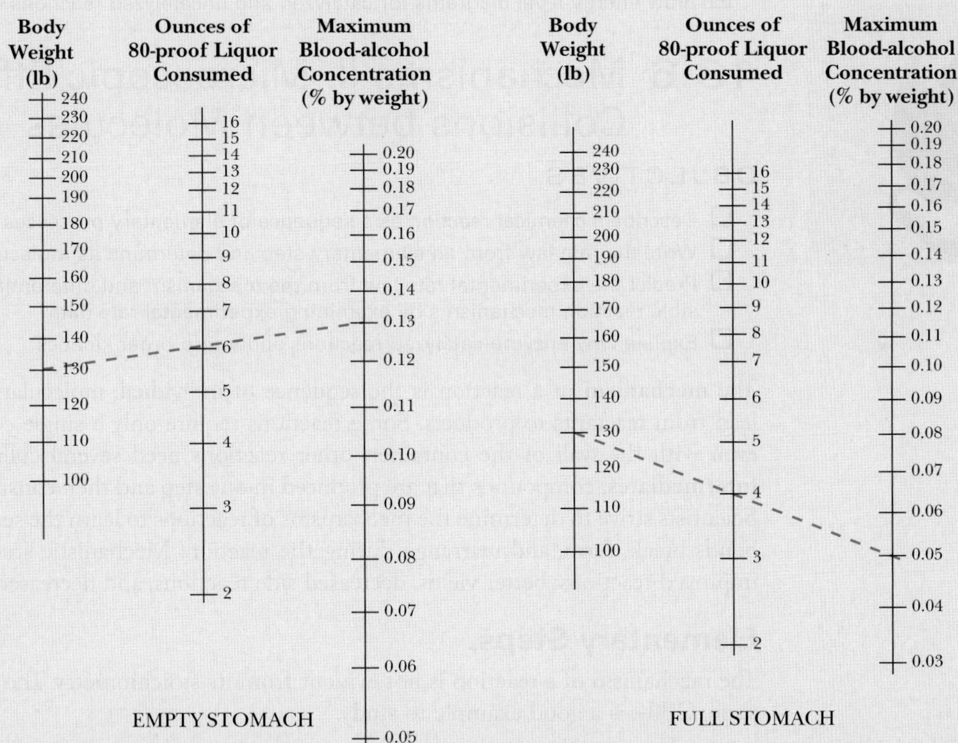

Blood-alcohol concentrations. This type of diagram is called a *nomograph.* Lay a straightedge across your weight and the amount of alcohol consumed. (One beer, one glass of wine, and one mixed drink are all equivalent.) The point at which the line crosses the column on the right is the maximum concentration of alcohol in your blood, in units of percent. Subtract 0.015% for each hour that has elapsed since the start of drinking to predict an approximate blood-alcohol concentration.

Enzymatic decomposition of hydrogen peroxide. Fresh liver contains an enzyme that catalyzes the decomposition of hydrogen peroxide. Have you noticed that the photographs in this chapter show homogeneous, heterogeneous, and enzymatic catalysts for the decomposition of hydrogen peroxide? Chemists often have several methods available for a given task and must weigh the advantages and disadvantages of each.

Enzymes are often named after their functions—that is, the reactions they catalyze. Alcohol dehydrogenase (ADH) is an enzyme that catalyzes the removal of two hydrogen atoms from ethyl alcohol, C_2H_5OH, forming acetaldehyde.

Enzymes have many practical uses. Recently, when a textile company found that their machinery was becoming clogged with the starch they use in the manufacture of a fabric, they introduced a cleaning solution that contains the enzyme amylase to digest the starch.

On a more familiar level, have you ever wondered how the liquid centers inside some candies are created? The answer is an interesting blend of kinetics and solubilities. Most liquid-filled candies are made by blending sucrose (table sugar) and water to form a paste, then coating the paste with chocolate. If these were the only ingredients, the candy would have a gritty, somewhat crystalline center. The manufacturer, however, blends an enzyme called *invertase* into the sugar-water paste. This enzyme breaks the 12-carbon sucrose molecule into two 6-carbon sugars. Fortuitously, these 6-carbon sugars are more soluble in water than the 12-carbon sugar, so the paste liquefies. By the time the candies are sold, they are filled with a sweet liquid. Although liquid-filled candies have been available for years, only recently have the details of the process been understood scientifically.

OBJECTIVES REVIEW *Can you:*

- ☑ define catalysis and identify heterogeneous, homogeneous, and enzymatic catalysts?
- ☑ draw energy-level diagrams for catalyzed and uncatalyzed reactions?

Liquid-filled candies.

13.6 Mechanisms II. Microscopic Effects: Collisions between Molecules

OBJECTIVES

- ☐ Describe a chemical reaction as a sequence of elementary processes
- ☐ Write the rate law from an elementary step and determine its molecularity
- ☐ Predict the experimental rate law from the mechanism and differentiate among possible reaction mechanisms by examining experimental rate data
- ☐ Explain why enzyme-catalyzed reactions show zero-order kinetics

The **mechanism** of a reaction is the sequence of individual, molecular-level steps that lead from reactants to products. Some reactions require only a single collision, perhaps even with the wall of the container; other reactions need several collisions and form **intermediates,** compounds that are produced in one step and then consumed in another. Scientists strive to determine the mechanisms of reactions to learn the sequence in which bonds break, form, and rearrange during the reaction. Mechanistic studies can lead to improved reactions, better yields, decreased side reactions, and decreased pollutants.

Elementary Steps

The mechanism of a reaction is not evident from its stoichiometry. The burning of propane, C_3H_8, is a good example to study.

$$C_3H_8(g) + 5O_2(g) \rightarrow 3CO_2(g) + 4H_2O(g)$$

One potential mechanism for this reaction is a single collision in which one propane molecule and five oxygen molecules strike each other simultaneously. It is extremely unlikely that six molecules will collide at the same time with enough energy and with all reactants in the proper positions to form the products. Instead, this reaction occurs through a series of **elementary reactions**—single molecular events that sum to the overall reaction. An **elementary step** is an equation that describes an actual molecular-level event. Most elementary steps involve only one or two molecules.

Elementary reactions describe the molecular collisions that add up to the overall reaction.

In the following discussion, we consider the decomposition of ozone, a reaction less complex than the burning of propane. The overall reaction that describes the decomposition of ozone is

$$2O_3(g) \rightarrow 3O_2(g)$$

One proposed mechanism involves two steps.

$O_3 \rightarrow O_2 + O$	Step 1
$O + O_3 \rightarrow 2O_2$	Step 2
$2O_3 \rightarrow 3O_2$	Overall reaction

In this mechanism, atomic oxygen is a reaction intermediate. It is produced in the first step, consumed in the second, and is never observed among the products of the overall reaction. A single O appears on each side of the elementary chemical equations, so we can remove it from the net equation, just as we remove a spectator ion from a net ionic reaction. The two steps sum to the overall stoichiometry of the reaction.

A mechanism often has more than one elementary step.

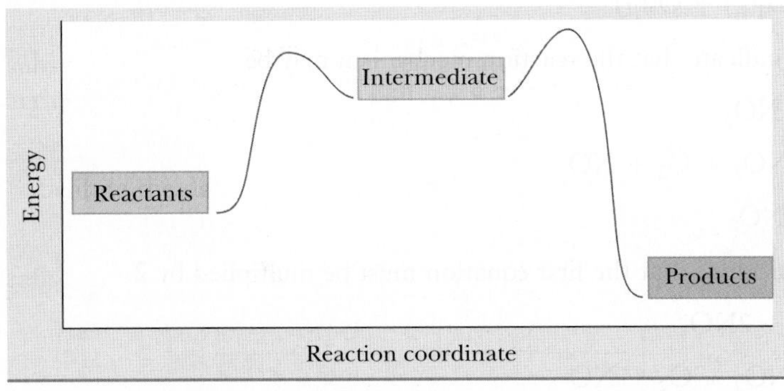

An energy level diagram for a reaction involving an unstable intermediate. The diagram shows two peaks that represent the formation of activated complexes. The valley between represents the intermediate. The unstable intermediate is more stable than the activated complex but less stable than the reactants or products.

An intermediate differs from the activated complex. Whereas the activated complex occurs at the maximum in the energy-level diagram, an intermediate is in a shallow minimum.

The next example illustrates the kinetics of a two-step process.

EXAMPLE 13.13 Evaluating Elementary Steps

The gas-phase reaction of nitrogen dioxide with fluorine proceeds to form nitrosyl fluoride, NO_2F.

$$2NO_2(g) + F_2(g) \rightarrow 2NO_2F(g)$$

Determine whether the following mechanism provides the correct overall stoichiometry, and identify intermediates, if any.

$NO_2(g) + F_2(g) \rightarrow NO_2F(g) + F(g)$	Step 1
$NO_2(g) + F(g) \rightarrow NO_2F(g)$	Step 2

Strategy Sum the proposed elementary steps to determine whether these steps produce the overall stoichiometry.

Solution

$NO_2(g) + F_2(g) \rightarrow NO_2F(g) + F(g)$	Step 1
$NO_2(g) + F(g) \rightarrow NO_2F(g)$	Step 2
$2NO_2(g) + F_2(g) + \cancel{F(g)} \rightarrow 2NO_2F(g) + \cancel{F(g)}$	
$2NO_2(g) + F_2(g) \rightarrow 2NO_2F(g)$	

We conclude that the two steps can be summed to provide the overall reaction. Atomic fluorine, F, is an intermediate.

Understanding

Evaluate the proposed mechanism (all substances are gases) to determine the overall stoichiometry. Identify intermediates.

$NO_2 + NO_2 \rightarrow NO_3 + NO$ Step 1

$NO_3 + CO \rightarrow NO_2 + CO_2$ Step 2

Answer The overall stoichiometry is $NO_2 + CO \rightarrow NO + CO_2$. Nitrogen trioxide, NO_3, is an intermediate.

In certain reactions, the mechanism is quite complicated. The gas-phase decomposition of dinitrogen pentoxide is an example of a three-step mechanism.

$$2N_2O_5(g) \rightarrow 4NO_2(g) + O_2(g)$$

Experimental results indicate that the reaction mechanism may be

$$N_2O_5 \rightarrow NO_2 + NO_3$$

$$NO_2 + NO_3 \rightarrow NO_2 + O_2 + NO$$

$$NO + NO_3 \rightarrow 2NO_2$$

The three steps can be added, but the first equation must be multiplied by 2.

$$2N_2O_5 \rightarrow 2NO_2 + 2NO_3$$

$$NO_2 + NO_3 \rightarrow NO_2 + O_2 + NO$$

$$NO + NO_3 \rightarrow 2NO_2$$

$$2N_2O_5 + NO_2 + \cancel{2NO_3} + \cancel{NO} \rightarrow 5NO_2 + \cancel{2NO_3} + O_2 + \cancel{NO}$$

$$2N_2O_5 \rightarrow 4NO_2 + O_2$$

Rate Laws for Elementary Reactions

The number and identity of colliding species determine the rate of reaction. Because an elementary step describes a molecular collision, the rate law for an *elementary step* (unlike that of the overall reaction) can be written directly from the stoichiometry of that step.

$iA + jB \rightarrow$ products (elementary step)

Rate of elementary step $= k[A]^i[B]^j$

A rate law can be written from the stoichiometry of an elementary step, but not from the overall reaction stoichiometry.

In any elementary step, the relationship between rate and concentration is determined from the number of reactant species because the reaction rate is proportional to the concentrations of the colliding species. The rate law for an overall reaction cannot be determined from the overall stoichiometry because the overall reaction generally consists of several elementary steps.

The molecularity refers to the number of species that collide in a single elementary step.

The number of species involved in a single elementary step is called its **molecularity.** When an elementary step involves the spontaneous decomposition of a single molecule, it is called a **unimolecular** step. *The kinetics of a unimolecular step is described by a first-order rate law.*

If the elementary step involves the collision of two species, the step is **bimolecular** and is described by *second-order kinetics.* An elementary step involving the collision of three species is **termolecular,** and the process shows *third-order kinetics.* Consideration of simple probability leads to the conclusion that collisions involving four (or more) species are extremely rare—a scientist would never propose such an elementary step. In fact, even termolecular reactions are extremely uncommon.

Example 13.14 illustrates how to determine the rate law of an elementary step.

EXAMPLE 13.14 Determining the Rate Law for an Elementary Step

Write the expected rate law and molecularity for each of the following elementary reactions in the gas phase.

(a) $HCl \rightarrow H + Cl$
(b) $NO_2 + NO_2 \rightarrow N_2O_4$
(c) $NO + NO_2 + O_2 \rightarrow NO_2 + NO_3$

Strategy The rate law for an elementary reaction is derived directly from its stoichiometry. The molecularity is the sum of the exponents of the concentrations in the elementary step.

Solution

(a) rate $= k[HCl]$ Unimolecular
(b) rate $= k[NO_2]^2$ Bimolecular
(c) rate $= k[NO][NO_2][O_2]$ Termolecular

Understanding

Write the rate law and expected molecularity for the overall reaction:

$$CH_4 + 2O_2 \rightarrow CO_2 + 2H_2O$$

Answer A rate law and molecularity cannot be written from an overall reaction.

Rate-Limiting Steps

Most reactions involve more than one elementary step. When one step is much slower than any of the others, the overall rate is determined by the slowest step, which is called the **rate-limiting step.** Rate-limiting processes are not unique to chemical kinetics. Perhaps you have driven down a highway that has been narrowed to one lane by construction work or one that has a toll booth. If one car can pass through the construction area or toll booth every 10 seconds, then the overall rate of travel is limited to six cars per minute. It really does not matter what the speed limit is, or how many lanes of traffic are available on the other side of the area under construction. The slowest step limits the overall flow of traffic. The same is true for reaction mechanisms. *Experimentally determined rate laws do not provide information for steps that occur after the rate-limiting step.*

© Paul White Aerial views/Alamy

Reaction products cannot form at rates faster than that of the slowest elementary step.

Consider the reaction of NO with F_2 to form ONF:

$$2NO(g) + F_2(g) \rightarrow 2ONF(g)$$

We might be tempted to assume that the reaction occurs in a single step by collision of two molecules of NO with one of F_2. Because the elementary step would be termolecular, the rate law would be third order.

$$\text{rate} = k[NO]^2[F_2]$$

Laboratory experiments, however, show that the rate law is second order.

$$\text{rate} = k[NO][F_2]$$

These results tell us that the rate-limiting step is *not* termolecular but bimolecular. The termolecular mechanism must be rejected because it does not agree with the experimental data. An alternative two-step mechanism has been proposed.

$NO + F_2 \rightarrow ONF + F$ $\quad$ $\text{rate}_1 = k_1[NO][F_2]$ $\quad$ Slow

$NO + F \rightarrow ONF$ $\quad$ $\text{rate}_2 = k_2[NO][F]$ $\quad$ Fast

Note that the sum of these two steps provides the correct overall stoichiometry. As with any multistep mechanism, the slowest elementary step—in this case, the first step—limits the rate of the overall reaction. The rate-limiting step is bimolecular, involving a collision between NO and F_2, which is consistent with the observed second-order rate law.

The chemistry of the proposed mechanism makes sense to an experienced chemist. The atomic fluorine (produced in the first step) is known from other experiments to be highly reactive and short-lived, so the second step is likely to be faster than the first. The proposed mechanism is consistent with the experimental data and with the known chemistry of atomic fluorine. Last, notice that scientists say, "The proposed mechanism is *consistent* with the experimental data." Several mechanisms may be consistent with the observed data, and selecting among them often requires many additional experiments and a great deal of knowledge about chemical reactions.

Example 13.15 illustrates another mechanistic analysis.

EXAMPLE 13.15 Evaluating a Proposed Mechanism

Nitrogen dioxide reacts with carbon monoxide to form carbon dioxide and nitrogen monoxide. The overall stoichiometry is

$$NO_2 + CO \rightarrow NO + CO_2$$

The rate law, found by experiment, is second order:

$$\text{rate} = k[NO_2]^2$$

Evaluate the following mechanism to determine whether it is consistent with experiment.

$$NO_2 + NO_2 \rightarrow NO_3 + NO \qquad \text{Slow}$$

$$NO_3 + CO \rightarrow NO_2 + CO_2 \qquad \text{Fast}$$

Strategy The rate of reaction will be limited by the rate of the slow step.

Solution

From the slow step, we deduce that the rate law is

$$\text{rate} = k[NO_2]^2$$

which agrees with the experimental rate law. When we examine the stoichiometry, we see that the two steps can be summed to provide the overall reaction. The two-step mechanism is *consistent* with both the experimental rate law and the stoichiometry. We cannot be sure, however, that the proposed mechanism is correct.

Understanding

For the same reaction,

$$NO_2 + CO \rightarrow NO + CO_2$$

is the following two-step mechanism the correct description?

$$NO_2 + NO_2 \rightarrow N_2O_4 \qquad \text{Slow}$$

$$N_2O_4 + CO \rightarrow NO + NO_2 + CO_2 \qquad \text{Fast}$$

Answer The mechanism is consistent with the experimental data because it sums to the overall reaction, and the molecularity of the rate-limiting step agrees with the experimental rate law. We cannot say that the mechanism is correct, just that it is *consistent* with experimental results. Recent experiments have found the presence of NO_3 as a short-lived intermediate, which tends to support the first mechanism presented.

The preceding examples help to clarify why scientists study chemical kinetics. Kinetic studies are necessary to identify the species that collide during the rate-limiting step, which is important information about the mechanism of a reaction. Chemists use these studies as they try to find ways to speed up a slow reaction or tame a reaction that is dangerously fast, as well as to understand how chemical reactions occur.

When scientists analyze a proposed mechanism to determine whether it is plausible, they focus on two factors. First, the sum of all the steps in the mechanism must provide the observed stoichiometry. Second, the rate-limiting step must be consistent with the observed rate law. If one or more mechanisms pass these tests, other factors, including reasonable intermediates and analogy with known reactions, help to determine which mechanism is more likely.

Complex Reaction Mechanisms

A mechanism in which the rate-limiting step is *not* the first step adds another detail. We can consider a general chemical reaction:

$$R \rightarrow P$$

The proposed mechanism has two elementary steps:

$$R \rightarrow \text{intermediates} \qquad \text{rate}_1 = k_1[R]$$

$$\text{intermediates} \rightarrow P \qquad \text{rate}_2 = k_2[\text{intermediates}]$$

If the first step limits the rate of reaction, then the rate law for the overall reaction is first order in [R]. If the second step is slower, the reaction rate depends on the intermediates. Because intermediates are often unstable and their concentrations are difficult to measure, *rate laws are not written in terms of an intermediate.* The solution is to express the concentration of the intermediate in terms of the stable species, if possible. These methods are illustrated in the next section.

Rate laws do not include the concentrations of intermediates.

Reactions with a Rapid and Reversible Step

Many multistep reactions contain rapid and reversible steps before the rate-limiting step. The general category of rapid and reversible reactions includes phase transitions such as the melting of ice to form liquid water (see discussion in Chapter 11).

The reaction of nitrogen monoxide with hydrogen contains a fast reversible step. The overall reaction is

$$2NO(g) + 2H_2(g) \rightarrow N_2(g) + 2H_2O(g)$$

Experiments show that a third-order rate law describes this reaction:

$$\text{rate} = k[NO]^2[H_2]$$

The following mechanism has been proposed.

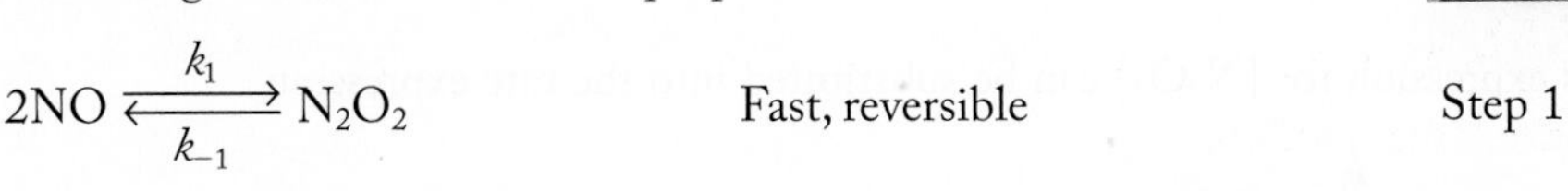

$$2NO \underset{k_{-1}}{\overset{k_1}{\rightleftharpoons}} N_2O_2 \qquad \text{Fast, reversible} \qquad \text{Step 1}$$

$$N_2O_2 + H_2 \xrightarrow{k_2} N_2O + H_2O \qquad \text{Slow} \qquad \text{rate} = k_2[N_2O_2][H_2] \qquad \text{Step 2}$$

$$N_2O + H_2 \xrightarrow{k_3} N_2 + H_2O \qquad \text{Fast} \qquad \text{rate} = k_3[N_2O][H_2] \qquad \text{Step 3}$$

The first step is to determine whether the proposed mechanism has the correct stoichiometry. Inspection of the chemical equations for the mechanism indicates that the stoichiometry of the mechanism is consistent with the stoichiometry of the reaction.

The next step is to determine whether the mechanism predicts a rate law that is consistent with the experimental results. The predicted rate of reaction is limited by the rate of the slowest step, step 2.

$$\text{rate} = k_2[N_2O_2][H_2]$$

This rate expression poses a problem because the concentration of the intermediate, N_2O_2, is present in extremely low concentration and for a short time—its concentration cannot be measured. Chemists must formulate a rate law in terms of species that have measurable concentrations. A rate law that includes unmeasurable species is of no use because it cannot be tested by experiment. Fortunately, there is a solution.

The first step is a fast reversible reaction. It is written with two arrows to show that it has both a forward component, described by the rate constant k_1, and a reverse component, described by k_{-1}.

$$2NO \underset{k_{-1}}{\overset{k_1}{\rightleftharpoons}} N_2O_2$$

The observation that the concentration of N_2O_2 is too small to measure provides some important information: The rate of formation of N_2O_2 is equal to the rate at which it decomposes. N_2O_2 cannot decompose any faster than it forms; to do so would violate the law of conservation of mass. If it formed faster than it decomposed, a measurable concentration would build up over time, but experiment shows no concentration buildup. These observations lead to the conclusion that N_2O_2 forms at the same rate as it reacts. We can analyze this step by writing the expression for the rate at which N_2O_2 forms and setting it equal to the rate at which it reacts.

First, write the rate expression for the formation of N_2O_2.

$$\text{rate of formation of } N_2O_2 = k_1[NO]^2$$

The rate of decomposition involves two possible reactions: the reverses of step 1 and step 2 both describe the decomposition of N_2O_2.

$$\text{rate of decomposition of } N_2O_2 = k_{-1}[N_2O_2] + k_2[N_2O_2][H_2]$$

In the proposed mechanism, step 2 is rate limiting and presumed to be much slower than decomposition by the reverse of step 1. Mathematically, k_{-1} is much larger than $k_2[H_2]$, so we can neglect the second term in the rate expression above. Thus,

$$\text{rate of decomposition of } N_2O_2 = k_{-1}[N_2O_2]$$

Now, set the rate of formation of N_2O_2 equal to the rate of decomposition.

$$k_1[NO]^2 = k_{-1}[N_2O_2]$$

Solve for $[N_2O_2]$.

$$[N_2O_2] = \frac{k_1}{k_{-1}}[NO]^2$$

This expression for $[N_2O_2]$ can be substituted into the rate expression.

$$\text{rate} = k_2\frac{k_1}{k_{-1}}[NO]^2[H_2]$$

We can collect all the rate constants into one term and write the rate expression with a single constant. Let $k = k_2\dfrac{k_1}{k_{-1}}$.

$$\text{rate} = k[NO]^2[H_2]$$

This result is consistent with the observed kinetics, so the proposed mechanism is consistent with both the observed rate law and the reaction stoichiometry.

It is logical to ask why a scientist might choose this mechanism instead of a termolecular collision between two molecules of NO and one of H_2. The answer, in short, is that three-body collisions are quite rare. The proposed mechanism can be tested by experiments designed to determine whether N_2O_2 is produced during the reaction. Such measurements have not yet been accomplished.

Enzyme Metabolism

Alcohol dehydrogenase (ADH) is an enzyme that catalyzes the oxidation of ethyl alcohol to acetaldehyde and other species.

$$CH_3CH_2OH \xrightarrow{ADH} CH_3CHO + \text{other products}$$

The reaction is thought to follow the **Michaelis–Menten mechanism,** which is illustrated in the following equations. Let E represent the enzyme and S represent the *substrate,* which is the compound on which the enzyme acts. In this particular example, E is ADH and S is ethanol. The first step is a rapid reversible one in which E and S form a complex.

$$E + S \underset{k_{-1}}{\overset{k_1}{\rightleftharpoons}} ES \qquad \text{Rapid, reversible} \qquad \text{Step 1}$$

The Michaelis–Menten mechanism involves a fast, reversible step to form the enzyme-substrate complex, followed by a slow step in which the complex forms the products.

The second step is the formation of the product and enzyme from the enzyme-substrate complex. This second step limits the rate of reaction.

$$ES \xrightarrow{k_2} E + P \qquad \text{rate} = k_2[ES] \qquad \text{Step 2}$$

Note that the enzyme is "recycled." It is a catalyst and not consumed in the reaction; therefore, it can act on another substrate molecule.

Let us define C_E as the total concentration of the enzyme. The enzyme is present in two forms: free enzyme, E, and bound enzyme, ES.

$$C_E = [E] + [ES]$$

Under normal conditions, the substrate is present in much greater concentration than is the enzyme, and nearly all the enzyme is bound to the substrate.

$$C_E \approx [ES]$$

Now, we can rewrite step 2.

$$\text{rate} = k_2[ES] = k_2C_E$$

The rate of reaction, which is governed by the slow second step, is zero order in substrate. This observation is true in general and explains why many biological processes, such as the metabolism of alcohol, are zero order in substrate.

OBJECTIVES REVIEW *Can you:*

- ☑ describe a chemical reaction as a sequence of elementary processes?
- ☑ write the rate law from an elementary step and determine its molecularity?
- ☑ predict the experimental rate law from the mechanism and differentiate among possible reaction mechanisms by examining experimental rate data?
- ☑ explain why enzyme-catalyzed reactions show zero-order kinetics?

CASE STUDY Hydrogen-Iodine Reaction

A reaction that has been well characterized is the formation of hydrogen iodide from the elements:

$$H_2(g) + I_2(g) \rightarrow 2HI(g)$$

Chemistry students can study this reaction by measuring the rate of consumption of iodine, a deep purple gas, as a function of time and as a function of the hydrogen concentration. An instrument called a *spectrophotometer* measures the absorbance, a property related to the absorption of light by the colored iodine gas, that is proportional to the iodine concentration. The students set up the reaction so that there is an overwhelming excess of hydrogen. As the reaction proceeds to consume $H_2(g)$ and $I_2(g)$, the HI forms, but the concentration of hydrogen changes little. In effect, the hydrogen

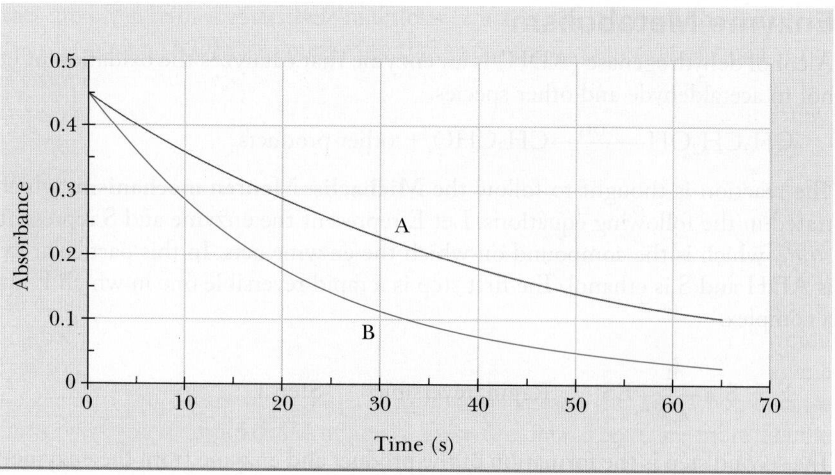

Figure 13.16 Absorbance of $I_2(g)$ as a function of time. Experiments A and B represent different concentrations of hydrogen.

TABLE 13.7 Starting Conditions for the Measurement of $H_2(g) + I_2(g) \rightarrow 2HI(g)$

	Experiment A	Experiment B
Initial pressure of H_2	100 torr	200 torr
Initial pressure of I_2	1 torr	1 torr

TABLE 13.8 Absorbance of I_2 as a Function of Time

Time (s)	Experiment A Initial P_{H_2} 100 torr	Experiment B Initial P_{H_2} 200 torr
0	0.45	0.45
10	0.36	0.28
20	0.28	0.18
30	0.22	0.11
40	0.18	0.07
50	0.14	0.04
60	0.11	0.03

concentration remains constant, so the change in reaction rate is due to the change in $I_2(g)$ concentration. The students repeat the experiment but start with a different concentration of $H_2(g)$. The experimental conditions are shown in Table 13.7 and the experimental results in Figure 13.16.

The students read the data shown in Table 13.8 from the data in Figure 13.16.

They analyze the data to determine the dependence of rate on the iodine concentration, because the hydrogen concentration remains nearly constant during the course of this experiment. Because the absorbance-time relationship is curved (see Figure 13.16), they take the natural log of the absorbance producing the data shown in Table 13.9.

The graph of ln(Absorbance) in Figure 13.17 is a straight line, indicating that the rate law is first order in iodine:

$$\text{rate} = k[H_2]^x\,[I_2]$$

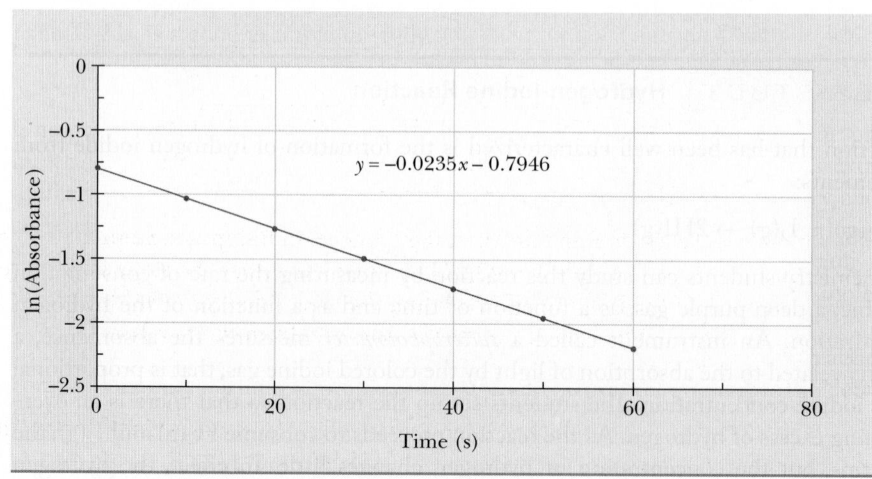

Figure 13.17 Plot of ln(Absorbance) versus time for Experiment A.

TABLE 13.9 Natural Log versus Time

Time (s)	Abs	ln(Abs)
0	0.45	−0.80
10	0.36	−1.02
20	0.28	−1.27
30	0.22	−1.51
40	0.18	−1.71
50	0.14	−1.97
60	0.11	−2.21

To determine the order in hydrogen, the students use the initial rate method. They place a ruler to measure the initial slope of the data in Figure 13.16 and determine that in going from experiment A to experiment B, the concentration of hydrogen doubles and the rate of reaction doubles.

	Experiment A	Experiment B
Initial concentration of hydrogen	100 torr	200 torr
Initial rate (−slope)	0.0104	0.0207

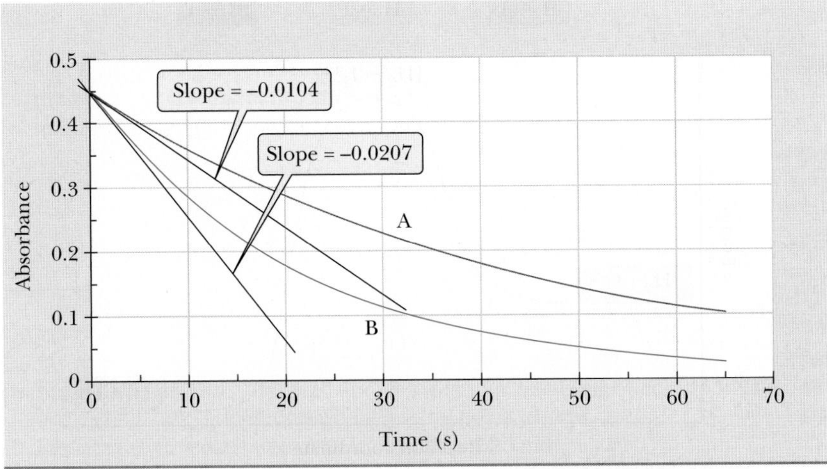

The students conclude the rate law is first order in hydrogen.

Rate $= k[H_2]\,[I_2]$

They repeat the experiment at different temperatures and obtain the rate constant at several different temperatures as shown in Table 13.10.

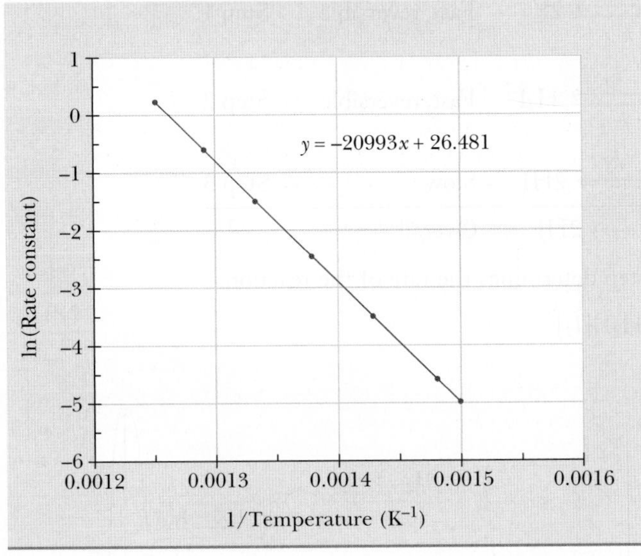

TABLE 13.10 Effect of Temperature on the Rate Constant

Temperature (K)	Rate Constant (L/mol · s)
667	6.79×10^{-3}
675	9.86×10^{-3}
700	2.99×10^{-2}
725	8.43×10^{-2}
750	2.21×10^{-1}
775	5.46×10^{-1}
800	1.27

They graph ln(k) versus $1/T$ and find a straight-line relationship, and from the graph determine that the slope is equal to −20,993, so they know that the slope is equal to $-E_a/R$. They conclude their report by calculating the value of E_a as 175 kJ/mol.

The students are asked to predict the mechanism, and from the rate law

rate $= k[H_2][I_2]$

they conclude that the mechanism is a bimolecular collision—a molecule of hydrogen strikes a molecule of iodine and two molecules of HI are produced.

Proposed Mechanism

For years, scientists assumed that the reaction proceeded in one bimolecular step, because a simple two-body collision could explain the kinetics and the stoichiometry.

Mechanism 1

$$H_2(g) + I_2(g) \rightarrow 2HI(g) \qquad \text{rate} = k[H_2][I_2] \tag{13.12}$$

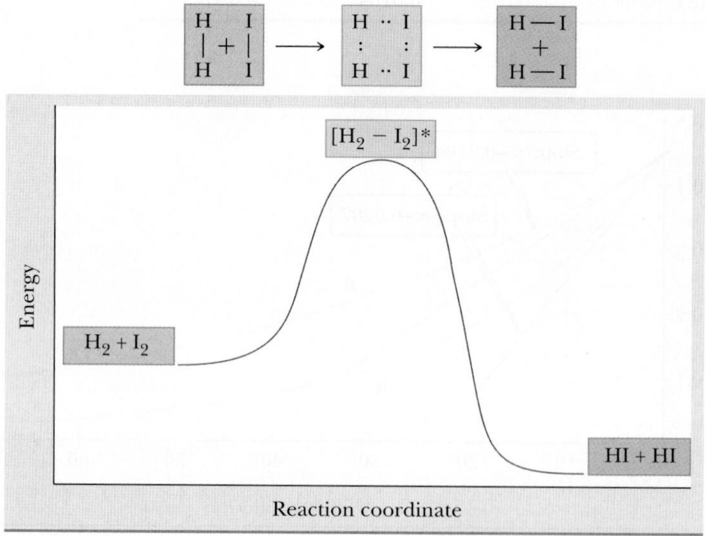

Energy level diagram for one-step bimolecular mechanism. Mechanism I.

An alternative mechanism could also be proposed.

Mechanism 2

$I_2 \underset{k_{-1}}{\overset{k_1}{\rightleftharpoons}} 2I$	Fast, reversible	Step 1
$I + H_2 \underset{k_{-2}}{\overset{k_2}{\rightleftharpoons}} H_2I$	Fast, reversible	Step 2
$H_2I + I \xrightarrow{k_3} 2HI$	Slow	Step 3
$H_2 + I_2 \longrightarrow 2HI$	Overall	

The slow step determines the rate of the reaction.

$$\text{rate} = k_3[H_2I][I_2] \tag{13.13}$$

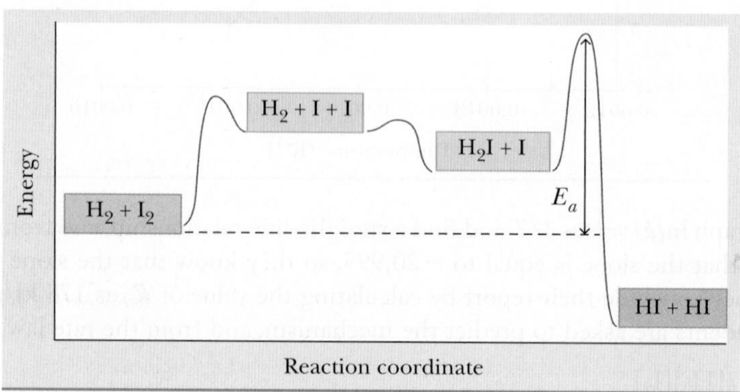

Energy level diagram for two-step bimolecular mechanism. Mechanism II.

This equation is not particularly useful, because both H_2I and I are intermediates. If we recognize that the intermediates do not build to appreciable concentrations, then we can conclude that the rates of the forward and reverse reaction in step 1 are equal. The same logic works for step 2. This logical assumption and some algebra lead to

$$\text{rate} = k[H_2][I_2] \qquad [13.14]$$

We can see that both proposed mechanisms have rate laws that are first order in hydrogen and iodine (and thus consistent with the experimental rate law), and both are consistent with the stoichiometry. Further experimentation is needed to rule out either mechanism.

Ultraviolet light causes I_2 to dissociate and form iodine atoms. This light has little effect on the I_2 concentration—the presence of ultraviolet (UV) light might cause a change from 0.0100 to 0.0099 *M*. If the reaction proceeds via mechanism I, it might proceed slightly slower in the presence of UV light, because the concentration of I_2 decreases a small amount. On the other hand, the presence of UV light causes the concentration of the intermediate, I, to increase from an undetectable concentration to 1.0×10^{-4} *M*. When the $H_2{-}I_2$ reaction mixture was irradiated, the rate increased dramatically and in proportion to the intensity of UV radiation used. The results of the experiments using UV light are consistent with a multistep mechanism involving iodine atoms as an intermediate.

Although the UV experiments support mechanism II, the simple bimolecular collision might also occur at the same time. This example provides a good idea of the difficulties in deducing reaction mechanisms.

Questions

1. Could the $H_2(g) + I_2(g) \rightleftharpoons 2HI(g)$ reaction be studied without instruments to measure the light absorption of $I_2(g)$? How?
2. Do you think that students near a window and far from a window would get the same values for rate constants and activation energies?
3. Could you use this reaction, or something similar, to measure the amount of UV radiation to warn sunbathers that they risk skin damage after a particular exposure?

ETHICS IN CHEMISTRY

1. You work for a firm that manufactures textile fibers. Your company used to dominate the market, but a competitor is making inroads into your client base because they can manufacture the fibers less expensively. Your manager asks you to investigate using a catalyst to increase the reaction yield. The competitor is reputedly using dimethylmercury, $(CH_3)_2Hg$, as a catalyst.
(a) Can a catalyst increase the yield?
(b) Can a catalyst decrease the cost?
(c) Search the Internet for information about dimethylmercury. Researchers at Dartmouth College have studied it extensively.
(d) What other information needs to be communicated to upper management before adopting a dimethylmercury catalyst?

2. You are the Quality Assurance Manager for a large company that uses chromium in its processes. Your company had some releases of chromium(VI) in the past and agreed to analyze its chromium emissions and follow the Environmental Protection Agency guidelines. Your company must limit its chromium discharge to 1 hour/day and show that the Cr(VI) emission is no more than 1 kg/day and total chromium (all oxidation states) no more than 2 kg/day. If the company exceeds its total maximum daily load, then it pays huge fines and penalties, and the owners and operators may go to jail. Once each day, the company releases a mixture of mostly Cr(VI) with a little Cr(III), but some Cr(VI) is reduced to Cr(III) in the environment. Your recent data show that your emission today will contain 1.2 kg Cr(VI) and 0.2 kg Cr(III), which is technically over the limit, but you have performed experiments and determined that the Cr(VI) is reduced to Cr(III) with a half-life of 12 hours. Management reads the Environmental Protection Agency statute literally—total release in a day of 24 hours means that you should calculate the amounts of Cr(III) and Cr(VI) 24 hours after the release. They also point out that they have a good contract for chrome-plating military hardware, and if they close for even one day, they will lose the contract, which will result in the layoff of 450 employees.

(a) Calculate the quantities of Cr(III) and Cr(VI) present in the environment after 24 hours, if the half-life of Cr(VI) is 12 hours. Are you willing to certify that the Cr release is within limits? Explain your logic.

(b) What if the half-life of Cr(VI) is much shorter or longer? Consider half-lives of 1 second, 1 hour, and 1 week, and explain if you would authorize the release under these conditions.

Chapter 13 Visual Summary

The chart shows the connections between the major topics discussed in this chapter.

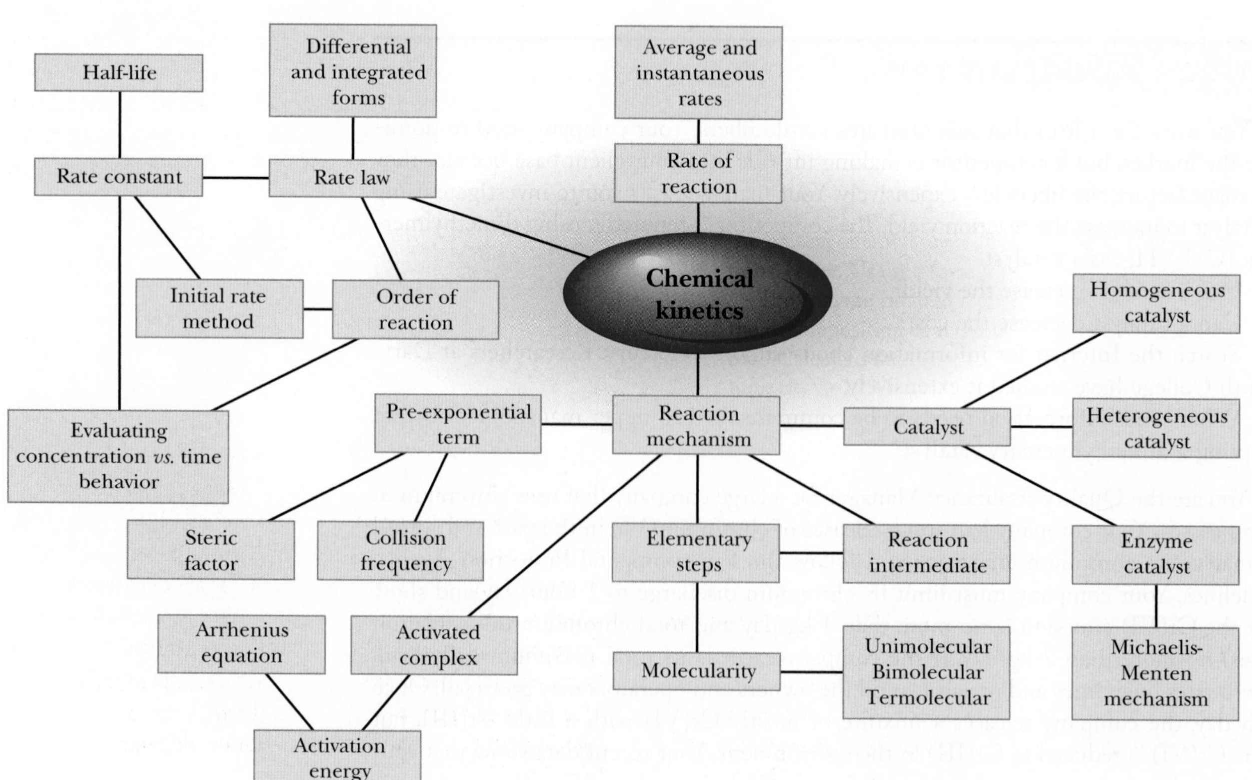

Summary

13.1 Rates of Reaction

The *rate of a reaction,* the increase in concentration of a product (or disappearance of a reactant) per unit time, is determined by evaluation of experimental data. The rate of reaction can be measured over a specified time interval, and the *instantaneous rate* can be determined at a particular time. The rate of a reaction does not depend on which species is measured.

13.2 Relationships between Rate and Concentration

The *rate law* for the reaction

$$A + B \rightarrow \text{products}$$

is of the form

$$\text{rate} = k[A]^x[B]^y$$

and is determined from experiment. The exponents x and y are called the *orders* of the reaction with respect to A and B. Because concentration influences the rate of reaction, using initial rates and initial concentrations often simplifies the determination of the rate law. The initial concentrations of reactants are varied, and the initial rates are measured to determine the relationship between rate and concentration.

13.3 Dependence of Concentrations on Time

The rate law can also be determined by measuring the change in the concentration of a reactant as a function of time. Zero-order, first-order, and second-order reactions all show different concentration-time relationships. Plots of concentration, ln[concentration], and 1/[concentration] versus time are straight lines for zero-, first-, and second-order reactions, respectively. The units of the *rate constant* differ for each reaction order as well. The *half-life* is another measure of the rate of reaction. For a first-order reaction, the half-life is related only to the rate constant. For zero- and second-order reactions, the half-lives depend on both the rate constant and the concentration, and are not widely used.

13.4 Mechanisms I. Macroscopic Effects: Temperature and Energetics

Temperature strongly influences the rate of reaction. The effects of temperature are consistent with a model that requires that the reactants collide with at least a minimum energy for a reaction to occur. The *activation energy* can be determined by measuring the rates of reaction at two or more different temperatures. The rate of reaction is a function of the *collision frequency,* the activation energy, and a steric factor that includes the geometry of the molecular collisions.

If an energy-level diagram is drawn for a particular reaction, the least stable/higher energy arrangement of atoms formed as the reactant molecules collide to form the product is the *activated complex.*

13.5 Catalysis

Catalysts are substances that alter the path of the reaction, generally by lowering the activation energy. Catalysts are not consumed in the formation of the product. Catalysts may be *homogeneous* (the catalyst is in the same phase as the reactants) or *heterogeneous. Enzymes* are biological molecules that act as catalysts.

13.6 Mechanisms II. Microscopic Effects: Collisions between Molecules

The *mechanism* of a reaction can be divided into a series of *elementary steps.* The *molecularity* describes the number of species that collide in any particular elementary step. If one step is much slower than any of the others, the slow step limits the rate of reaction. Reaction mechanisms must be consistent with both the experimentally determined rate law and the observed stoichiometry. Often, two or more mechanisms can be proposed for a reaction, each of which is consistent with experimental data.

Chapter Terms

The following terms are defined in the Glossary, Appendix I.

Chemical kinetics

Section 13.1

Average rate
Instantaneous rate
Rate
Rate of reaction

Section 13.2

Initial rate method
Overall order
Order of the reaction
Rate constant
Rate law

Section 13.3

Differential form of the rate law
Half-life
Integrated form of the rate law

Section 13.4

Activated complex
Activation energy, E_a
Arrhenius equation
Arrhenius plot
Collision frequency, Z
Pre-exponential term, A
Steric factor, p

Section 13.5

Catalyst
Enzyme
Heterogeneous catalyst
Homogeneous catalyst

Section 13.6

Bimolecular
Elementary reaction
Elementary step
Intermediate
Mechanism
Michaelis–Menten mechanism
Molecularity
Rate-limiting step
Termolecular
Unimolecular

Key Equations

rate = $k[A]^x[B]^y$ (13.2)

rate = $\frac{-\Delta[R]}{\Delta t} = k[R]$ (13.3)

$[R]_t = [R]_0 e^{-kt}$ (13.3)

$\ln[R]_t = \ln[R]_0 - kt$ (13.3)

$\ln\frac{[R]_t}{[R]_0} = -kt$ (13.3)

For a first-order reaction (13.3)

$t_{1/2} = \frac{0.693}{k}$

$\frac{1}{[R]_t} = \frac{1}{[R]_0} + kt$ (13.3)

$k = Ae^{-E_a/RT}$ (13.4)

$\ln\left(\frac{k_1}{k_2}\right) = \frac{-E_a}{R}\left(\frac{1}{T_1} - \frac{1}{T_2}\right)$ (13.4)

Questions and Exercises

OWL Selected end of chapter Questions and Exercises may be assigned in OWL.

Blue-numbered Questions and Exercises are answered in Appendix J; questions are qualitative, are often conceptual, and include problem-solving skills.

■ Questions assignable in OWL

✎ Questions suitable for brief writing exercises

▲ More challenging questions

Questions

13.1 Define rate law, rate constant, and reaction order.

13.2 What are the units for the zero-, first-, and second-order rate constants?

13.3 What is the difference between the integrated and differential forms of the rate law?

13.4 ✎ Explain how to use the method of initial rates to determine the rate law for

$$CH_3Br + OH^- \rightarrow CH_3OH + Br^-$$

13.5 ✎ Explain why half-lives are not normally used to describe reactions other than first order.

13.6 Derive an expression for the half-life of a reaction with a half-order rate law from the following integrated rate law:

$$[R]_t^{1/2} = [R]_0^{1/2} - \frac{1}{2}kt$$

13.7 Derive an expression for the half-life of a reaction with the following third-order rate law:

$$\frac{1}{[R]_t^2} - \frac{1}{[R]_0^2} = 2kt$$

13.8 List the factors that affect the rate of reaction.

13.9 The number of collisions per second that have energies exceeding the activation energy can be determined. Explain how this determination is or is not useful in predicting the rate of a chemical reaction.

13.10 Sketch the energy-level diagram for an endothermic reaction with low activation energy and for an exothermic reaction with high activation energy.

13.11 Explain the effect of doubling the concentration of one reactant on the collision frequency.

13.12 ✎ Define activation energy, and explain how it influences the rate of reaction.

13.13 ✎ Explain why a collision must exceed a minimum energy for a product to form.

13.14 ■ What is meant by the mechanism of a reaction? How does the mechanism relate to the order of the reaction?

13.15 Explain the influence of temperature on the rate of an uncatalyzed endothermic reaction and a catalyzed (lower-E_a) endothermic reaction.

13.16 Explain how temperature influences the rate of an uncatalyzed exothermic reaction and a catalyzed (lower-E_a) exothermic reaction.

13.17 Draw energy-level diagrams for catalyzed and uncatalyzed one-step endothermic reactions. Label the activation energy for each path.

13.18 Draw energy-level diagrams for catalyzed and uncatalyzed one-step exothermic reactions. Label the activation energy for each path.

13.19 ✎ Define an elementary step and explain why equations for elementary reactions can be used to predict the rate law, but the overall reaction stoichiometry cannot.

13.20 Explain why enzymatic reactions are zero order in the substrate.

Exercises

OBJECTIVES Relate the changes in concentration over time to the rate of reaction. Calculate the instantaneous rate of reaction from experimental data.

13.21 Oxalic acid can decompose to formic acid and carbon dioxide:

$$HOOC{-}COOH(g) \rightarrow HCOOH(g) + CO_2(g)$$

A graph of the concentration of oxalic acid as a function of time follows.

(a) Write an expression for the rate of reaction in terms of a changing concentration.
(b) Calculate the average rate of reaction between 10 and 30 seconds.
(c) Calculate the instantaneous rate of reaction after 20 seconds.
(d) Calculate the initial rate of reaction.
(e) Calculate the instantaneous rate of formation of CO_2 40 seconds after the start of the reaction.

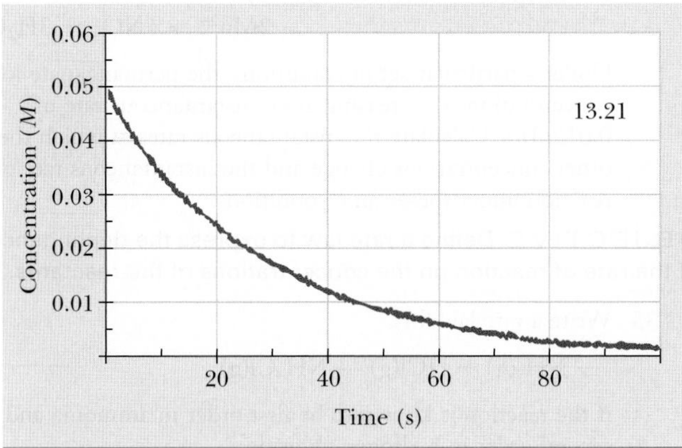

13.22 ■ Cyclobutane can decompose to form ethylene:

$$C_4H_8(g) \rightarrow 2C_2H_4(g)$$

The cyclobutane concentration can be measured as a function of time by mass spectrometry (a graph follows).

(a) Write an expression for the rate of reaction in terms of a changing concentration.
(b) Calculate the average rate of reaction between 10 and 30 seconds.
(c) Calculate the instantaneous rate of reaction after 20 seconds.
(d) Calculate the initial rate of reaction.
(e) Calculate the instantaneous rate of formation of ethylene 40 seconds after the start of the reaction.

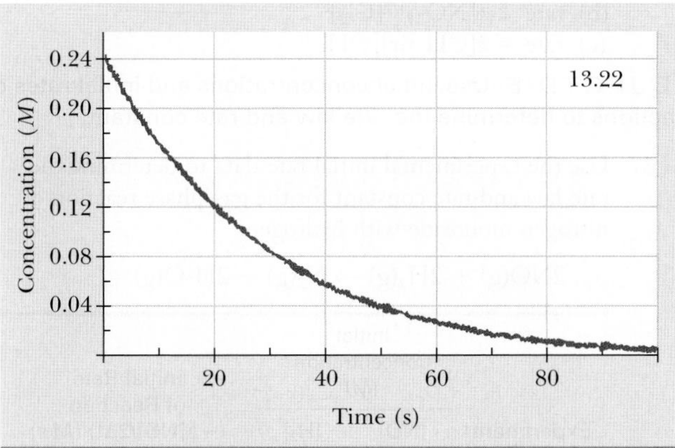

13.23 Nitrogen monoxide reacts with chlorine to form nitrosyl chloride.

$$NO(g) + \frac{1}{2}Cl_2(g) \rightarrow NOCl(g)$$

The figure shows the increase in nitrosyl chloride concentration under appropriate experimental conditions. The concentration of nitrosyl chloride actually starts at zero, although this fact may be difficult to see in the figure.

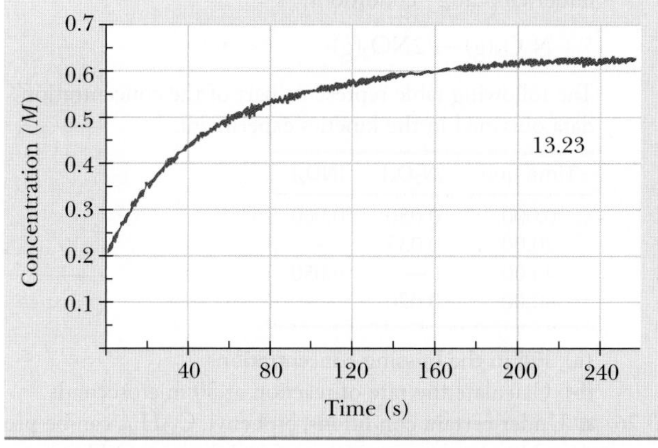

(a) Write an expression for the rate of reaction in terms of a changing concentration.
(b) Calculate the average rate of reaction between 40 and 120 seconds.
(c) Calculate the instantaneous rate of reaction after 80 seconds.
(d) Calculate the instantaneous rate of consumption of chlorine 60 seconds after the start of the reaction.

13.24 Hydrogen iodide forms from hydrogen and iodine:

$$H_2(g) + I_2(g) \rightarrow 2HI(g)$$

The following figure shows the increase in hydrogen iodide concentration under appropriate experimental conditions.

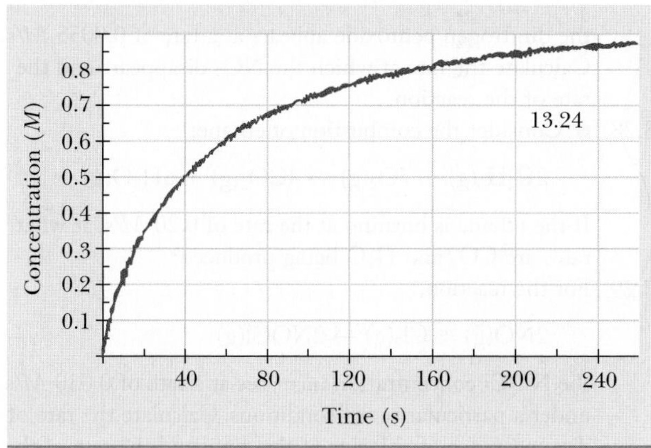

(a) Write an expression for the rate of reaction in terms of a changing concentration.
(b) Calculate the average rate of reaction between 20 and 60 seconds.
(c) Calculate the instantaneous rate of reaction after 40 seconds.
(d) Calculate the initial rate of reaction.
(e) Calculate the instantaneous rate of consumption of hydrogen 60 seconds after the start of the reaction.

OBJECTIVE Use stoichiometry to relate the rate of reaction to changes in the concentrations of reactants and products.

13.25 Dinitrogen tetroxide decomposes to nitrogen dioxide under laboratory conditions.

$$N_2O_4(g) \rightarrow 2NO_2(g)$$

The following table represents part of the concentration data obtained in the kinetics experiment.

Time (μs)	$[N_2O_4]$	$[NO_2]$
0.000	0.050	0.000
20.00	0.033	—
40.00	—	0.050
60.00	0.020	—

(a) Fill in the missing concentrations.
(b) Calculate the rate of reaction at 30 microseconds.

13.26 ■ Under certain conditions, biphenyl, $C_{12}H_{10}$, can be produced by the decomposition of cyclohexane, C_6H_{12}:

$$2C_6H_{12} \rightarrow C_{12}H_{10} + 7H_2$$

The following table represents part of the concentration data obtained in the kinetics experiment.

Time (s)	$[C_6H_{12}]$	$[C_{12}H_{10}]$	$[H_2]$
0.0	0.200	0.000	0.000
1.00	0.159	0.021	—
2.00	0.132	—	—
3.00	—	0.044	—

(a) Fill in the missing concentrations.
(b) Calculate the rate of reaction at 1.5 seconds.

13.27 For the reaction,

$$2NO_2(g) + O_3(g) \rightarrow N_2O_5(g) + O_2(g)$$

the dinitrogen pentoxide appears at a rate of 0.0055 *M*/s. Calculate the rate at which the NO_2 disappears and the rate of the reaction.

13.28 ■ Consider the combustion of ethane:

$$2C_2H_6(g) + 7O_2(g) \rightarrow 4CO_2(g) + 6H_2O(g)$$

If the ethane is burning at the rate of 0.20 *M*/s, at what rates are CO_2 and H_2O being produced?

13.29 For the reaction,

$$2NO(g) + Cl_2(g) \rightarrow 2NOCl(g)$$

the NOCl concentration increases at a rate of 0.030 *M*/s under a particular set of conditions. Calculate the rate of disappearance of chlorine at this time and the rate of the reaction.

13.30 For the reaction,

$$3N_2O(g) + C_2H_2(g) \rightarrow 3N_2(g) + 2CO(g) + H_2O(g)$$

water is produced at the rate of 0.10 *M*/s. Calculate the rates of production of the other species and the rate of the reaction.

13.31 The chromium(III) species reacts with hydrogen peroxide in aqueous solution to form the chromate ion.

$$2CrO_2^- + 3H_2O_2 + 2OH^- \rightarrow 2CrO_4^{2-} + 4H_2O$$

Under particular experimental conditions, the chromate ion, CrO_4^{2-}, is produced at an instantaneous rate of 0.0050 *M*/s. Calculate the instantaneous rates at which the other species change concentration and the instantaneous rate of reaction under these same conditions.

13.32 In aqueous solution, the permanganate ion reacts with nitrous acid to form Mn^{2+} and nitrate ions.

$$2MnO_4^- + 5HNO_2 + H^+ \rightarrow 2Mn^{2+} + 5NO_3^- + 3H_2O$$

Under a particular set of conditions, the permanganate ion concentration is decreasing at an instantaneous rate of 0.012 *M*/s. Calculate the instantaneous rates at which the other concentrations change and the instantaneous rate of reaction under these same conditions.

OBJECTIVE Define a rate law to express the dependence of the rate of reaction on the concentrations of the reactants.

13.33 Write a rate law for

$$NH_3(g) + HCl(g) \rightarrow NH_4Cl(g)$$

if the reaction is known to be first order in ammonia and second order in hydrogen chloride.

13.34 Write a rate law for

$$NO_3(g) + O_2(g) \rightarrow NO_2(g) + O_3(g)$$

if measurements show the reaction is first order in nitrogen trioxide and second order in oxygen.

OBJECTIVE Identify the reaction order from the rate law.

13.35 What is the order in each reactant and the overall order for a reaction that has the following rate law?
(a) rate = $k[NO][NO_2]^2$
(b) rate = $k[O_2]^{1/2}[Cl_2]$
(c) rate = $k[HClO]^2[OH^-]$

13.36 What is the order in each reactant and the overall order for a reaction that has the following rate law?
(a) rate = $k[O_3][NO]^2$
(b) rate = $k[NO_2]^{1/2}[Cl_2]^2$
(c) rate = $k[CH_3Br][OH^-]^0$

OBJECTIVE Use initial concentrations and initial rates of reactions to determine the rate law and rate constant.

13.37 Use the experimental initial rate data to determine the rate law and rate constant for the gas-phase reaction of nitrogen monoxide with hydrogen.

$$2NO(g) + 2H_2(g) \rightarrow N_2(g) + 2H_2O(g)$$

Experiment	Initial Concentration (*M*) [NO]	$[H_2]$	Initial Rate of Reaction ($-\Delta[NO]/2\Delta t$) (*M*/s)
1	0.10	0.10	3.8
2	0.10	0.20	7.7
3	0.10	0.30	11.4
4	0.20	0.10	15.3
5	0.30	0.10	34.2

13.38 ■ Rate data were obtained at 25 °C for the following reaction. What is the rate-law expression for this reaction?

$$A + 2B \rightarrow C + 2D$$

Experiment	Initial [A] (*M*)	Initial [B] (*M*)	Initial Rate of Formation of C (*M*/min)
1	0.10	0.10	3.0×10^{-4}
2	0.30	0.30	9.0×10^{-4}
3	0.10	0.30	3.0×10^{-4}
4	0.20	0.40	6.0×10^{-4}

13.39 Use the following experimental initial rate data to determine the rate law and rate constant for the gas-phase reaction of dinitrogen monoxide and water.

$$N_2O(g) + H_2O(g) \rightarrow 2NO(g) + H_2(g)$$

Experiment	Initial Concentration (*M*)		Initial Rate of Reaction
	$[N_2O]$	$[H_2O]$	$(-\Delta[H_2O]/\Delta t)$ (*M*/s)
1	0.12	0.10	0.051
2	0.12	0.20	0.100
3	0.25	0.30	0.313

13.40 Use the following experimental initial rate data to determine the rate law and rate constant for the reaction of hydrogen iodide with ethyl iodide.

$$HI(g) + C_2H_5I(g) \rightarrow C_2H_6(g) + I_2(g)$$

Experiment	Initial Concentration (*M*)		Initial Rate of Reaction
	[HI]	$[C_2H_5I]$	$(\Delta[I_2]/\Delta t)$ (*M*/s)
1	0.053	0.23	3.7×10^{-5}
2	0.106	0.23	7.4×10^{-5}
3	0.106	0.46	14.8×10^{-5}

13.41 In a study of the gas-phase reaction of nitrogen dioxide with ozone, the initial rate method is used to evaluate the reaction at 15 °C. Determine the rate law and the rate constant from the data.

$$NO_2(g) + O_3(g) \rightarrow NO_3(g) + O_2(g)$$

Experiment	Initial Concentration (*M*)		Initial Rate of Reaction
	$[NO_2]$	$[O_3]$	$(\Delta[O_2]/\Delta t)$ (*M*/s)
1	2.0×10^{-6}	2.0×10^{-6}	2.1×10^{-7}
2	3.0×10^{-6}	2.0×10^{-6}	3.1×10^{-7}
3	4.0×10^{-6}	3.0×10^{-6}	6.2×10^{-7}
4	4.0×10^{-6}	4.0×10^{-6}	8.3×10^{-7}

13.42 A chemist studies the kinetics of the gas-phase reaction of phosphine with diborane at 0 °C. Determine the rate law and the rate constant (use torr as the concentration unit) from the data.

$$PH_3(g) + B_2H_6(g) \rightarrow PH_3BH_3(g) + BH_3(g)$$

Experiment	Initial Concentration (torr)		Initial Rate of Reaction
	$[PH_3]$	$[B_2H_6]$	$(-\Delta[PH_3]/\Delta t)$ (torr/s)
1	1.2	1.2	1.9×10^{-3}
2	3.0	1.2	4.7×10^{-3}
3	4.0	1.2	6.4×10^{-3}
4	4.0	3.0	1.6×10^{-2}

OBJECTIVE Evaluate concentration-time behaviors to write a rate law.

For Exercises 13.43 to 13.46, assume that the chemical reaction is

reactants → products

13.43 Determine the rate constant and order from the concentration-time dependence.

Time (s)	[Reactant] (*M*)
0	0.250
1	0.216
2	0.182
3	0.148
4	0.114
5	0.080

13.44 Determine the rate constant and order from the concentration-time dependence.

Time (s)	[Reactant] (*M*)
0	0.0451
2	0.0421
5	0.0376
9	0.0316
15	0.0226

13.45 Determine the rate constant and order from the concentration-time dependence.

Time (s)	[Reactant] (*M*)
0.001	0.220
0.002	0.140
0.003	0.080
0.004	0.050
0.005	0.030

13.46 ■ Determine the rate constant and order from the concentration-time dependence.

Time (s)	[Reactant] (*M*)
0	0.0350
10	0.0223
20	0.0142
50	0.0037
70	0.0015

13.47 Nitrosyl chloride decomposes to nitrogen monoxide and chlorine at increased temperatures. Determine the rate constant and order from the concentration-time dependence.

$$NOCl(g) \rightarrow NO(g) + \frac{1}{2}Cl_2(g)$$

Time (s)	[NOCl] (*M*)
0	0.100
30	0.064
60	0.047
100	0.035
200	0.021
300	0.015
400	0.012

13.48 Determine the rate constant and order by analyzing the concentration-time dependence.

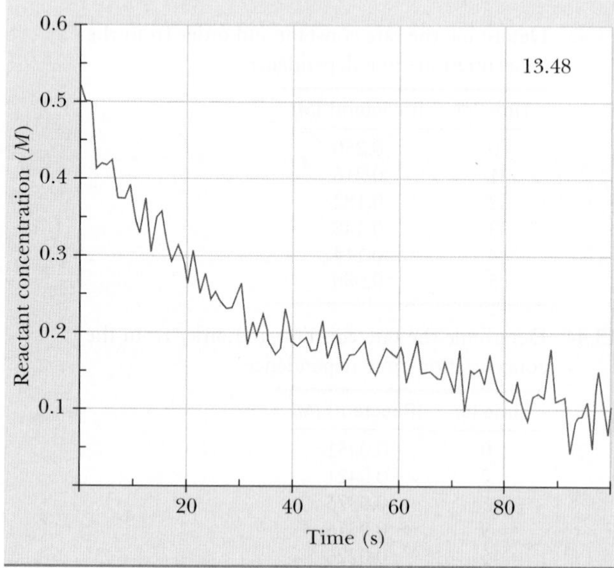

OBJECTIVE Calculate the concentration-time behavior for a first-order reaction from the rate law and the rate constant.

13.49 When formic acid is heated, it decomposes to hydrogen and carbon dioxide in a first-order decay.

$$HCOOH(g) \rightarrow CO_2(g) + H_2(g)$$

At 550 °C, the half-life of formic acid is 24.5 minutes.
(a) What is the rate constant, and what are its units?
(b) How many seconds are needed for formic acid, initially 0.15 *M*, to decrease to 0.015 *M*?

13.50 ■ The initial concentration of the reactant in a first-order reaction A → products is 0.64 *M* and the half-life is 30 seconds.
(a) Calculate the concentration of the reactant 60 seconds after initiation of the reaction.
(b) How long would it take for the concentration of the reactant to decrease to one-eighth its initial value?
(c) How long would it take for the concentration of the reactant to decrease to 0.040 mol L^{-1}?

13.51 The half-life of tritium, ^3H, is 12.26 years. Tritium is the radioactive isotope of hydrogen.
(a) What is the rate constant for the radioactive decay of tritium, in y^{-1} and s^{-1}?
(b) What percentage of the original tritium is left after 61.3 years?

13.52 The half-life of uranium-235, the major radioactive component of naturally occurring uranium, is 7.04×10^8 years.
(a) What is the rate constant for the radioactive decay of uranium-235, in y^{-1} and s^{-1}?
(b) What percentage of original uranium-235 is left after 4.5×10^9 years?

OBJECTIVE Relate half-life and rate constant, and calculate concentration-time behavior from the half-life of a first-order reaction.

13.53 Calculate the half-life of a first-order reaction if the concentration of the reactant decreases from 0.012 to 0.0082 *M* in 66.2 seconds.

13.54 ■ The hypothetical compound A decomposes in a first-order reaction that has a half-life of 2.3×10^2 seconds at 450 °C. If the initial concentration of A is 4.32×10^{-2} *M*, how long will it take for the concentration of A to decline to 3.75×10^{-3} *M*?

13.55 Calculate the half-life of a first-order reaction if the concentration of the reactant decreases from 1.02×10^{-3} *M* to 7.4×10^{-4} *M* in 116.7 seconds. How long does it take for the reactant concentration to decrease from 7.4×10^{-4} *M* to 2.0×10^{-4} *M*?

13.56 ■ ▲ Calculate the half-life of a first-order reaction if the concentration of the reactant is 0.0451 *M* at 30.5 seconds after the reaction starts and is 0.0321 *M* at 45.0 seconds after the reaction starts. How many seconds after the start of the reaction does it take for the reactant concentration to decrease to 0.0100 *M*?

OBJECTIVE Calculate the concentration-time behavior for a second-order reaction from the rate law and the rate constant.

13.57 ▲ The decomposition of ozone is a second-order reaction with a rate constant of 30.6 $atm^{-1} s^{-1}$ at 95 °C.

$$2O_3(g) \rightarrow 3O_2(g)$$

If ozone is originally present at a partial pressure of 21 torr, calculate the length of time needed for the ozone pressure to decrease to 1.0 torr.

13.58 ▲ Consider the second-order decomposition of nitrosyl chloride.

$$2NOCl(g) \rightarrow 2NO(g) + Cl_2(g)$$

At 450 K, the rate constant is 15.4 $atm^{-1} s^{-1}$.
(a) How much time is needed for NOCl originally at a partial pressure of 44 torr to decay to 22 torr?
(b) How much time is needed for NOCl originally at a concentration of 0.0044 *M* to decay to 0.0022 *M*?

OBJECTIVE Relate temperature, activation energy, and rate constant through the Arrhenius equation.

13.59 The following data were obtained by studying the change in rate constant as a function of temperature.

Rate Constant (L/mol · s)	Temperature (K)
0.36×10^6	500
3.7×10^6	550
27×10^6	600

Calculate the activation energy and the pre-exponential term. You may use a graphic method if you have a spreadsheet or graphing calculator, or an algebraic method if you do not.

13.60 The decomposition of formic acid (see Exercise 13.49) is measured at several temperatures. The temperature dependence of the first-order rate constant is:

T (K)	k (s^{-1})
800	0.00027
825	0.00049
850	0.00086
875	0.00143
900	0.00234
925	0.00372

Calculate the activation energy, in kilojoules, and the pre-exponential term. You may use a graphic method if you have a spreadsheet or graphing calculator, or an algebraic method if you do not.

13.61 A reaction rate doubles when the temperature increases from 25 °C to 40 °C. What is the activation energy?

13.62 ■ What is the activation energy for a reaction if its rate constant is found to triple when the temperature is increased from 600 to 610 K?

13.63 The activation energy for the isomerization of cyclopropane to propene is 274 kJ/mol. By what factor does the rate of reaction increase as the temperature increases from 500 °C to 550 °C?

13.64 The activation energy for the isomerization of cyclopropane to propene is 274 kJ/mol. By what factor does the rate of this reaction increase as the temperature increases from 250 °C to 280 °C?

13.65 Consider the results of an experiment in which nitrogen dioxide reacts with ozone at two different temperatures, 13 °C and 29 °C.

$$NO_2(g) + O_3(g) \rightarrow NO_3(g) + O_2(g)$$

If the activation energy is 29 kJ/mol, by what factor does the rate constant increase with this temperature change?

13.66 The activation energy for the decomposition of cyclobutane (C_4H_8) to ethylene (C_2H_4) is 261 kJ/mol. If the system produces ethylene at the rate of 0.043 g/s at 500 °C, what is the rate if the temperature increases to 600 °C?

OBJECTIVE Draw energy-level diagrams for catalyzed and uncatalyzed reactions.

13.67 A catalyst decreases the activation energy of a particular exothermic reaction by 15 kJ/mol, from 40 to 25 kJ/mol. Assuming that the reaction is exothermic, that the mechanism has only one step, and that the products differ from the reactants by 40 kJ, sketch approximate energy-level diagrams for the catalyzed and uncatalyzed reactions.

13.68 A catalyst decreases the activation energy of a particular endothermic reaction by 50 kJ/mol, from 140 to 90 kJ/mol. Assuming that the reaction is endothermic, that the mechanism has only one step, and that the products differ from the reactants by 20 kJ, sketch approximate energy-level diagrams for the catalyzed and uncatalyzed reactions.

OBJECTIVE Describe a chemical reaction as a sequence of elementary processes.

13.69 Sum the following elementary steps to determine the overall stoichiometry of the gas-phase reaction.

$$NO_2 + NO_2 \rightarrow NO_3 + NO$$

$$NO_3 + CO \rightarrow NO_2 + CO_2$$

13.70 Sum the following elementary steps to determine the overall stoichiometry of the reaction.

$$Cl_2 \rightarrow 2Cl$$

$$Cl + CO \rightarrow COCl$$

$$COCl + Cl \rightarrow COCl_2$$

13.71 Sum the following elementary steps to determine the overall stoichiometry of this hypothetical reaction.

$$NO \rightarrow N + O$$

$$O_3 + O \rightarrow 2O_2$$

$$O_2 + N \rightarrow NO_2$$

13.72 Sum the following elementary steps to determine the overall stoichiometry of the hypothetical reaction.

$$Cl_2 \rightarrow Cl^+ + Cl^-$$

$$Cl^- + H_2O \rightarrow HCl + OH^-$$

$$Cl^+ + OH^- \rightarrow HCl + O$$

OBJECTIVE Write the rate law from an elementary step and determine its molecularity.

13.73 Write the rate law and the molecularity for each of the following elementary reactions.
(a) $HCl \rightarrow H + Cl$
(b) $H_2 + Cl \rightarrow HCl + H$
(c) $2NO_2 \rightarrow N_2O_4$

13.74 ■ Assuming that each reaction is elementary, predict the rate law and molecularity.
(a) $NO(g) + NO_3(g) \rightarrow 2NO_2(g)$
(b) $O(g) + O_3(g) \rightarrow 2O_2(g)$
(c) $(CH_3)_3CBr(aq) \rightarrow (CH_3)_3C^+(aq) + Br^-(aq)$
(d) $2\ HI(g) \rightarrow H_2(g) + I_2(g)$

13.75 Write the rate law and the molecularity for each of the following elementary reactions.
(a) $C_2H_5Cl \rightarrow C_2H_4 + HCl$
(b) $NO + O_3 \rightarrow NO_2 + O_2$
(c) $HI + C_2H_5I \rightarrow C_2H_6 + I_2$

13.76 Write the rate law and the molecularity for each of the following elementary reactions.
(a) $NO + NO_2Cl \rightarrow NO_2 + NOCl$
(b) $NO_2 + SO_2 \rightarrow NO + SO_3$
(c) $N_2O_4 \rightarrow 2NO_2$

OBJECTIVE Predict the experimental rate law from the mechanism and differentiate among possible reaction mechanisms by examining experimental rate data.

13.77 Nitryl chloride, NO_2Cl, decomposes to NO_2 and Cl_2 with first-order kinetics. The following mechanism has been proposed. Identify the rate-limiting step.

$$NO_2Cl \rightarrow NO_2 + Cl$$

$$NO_2Cl + Cl \rightarrow NO_2 + Cl_2$$

13.78 Nitrogen dioxide can react with ozone to form dinitrogen pentoxide and oxygen.

$$2NO_2(g) + O_3(g) \rightarrow N_2O_5(g) + O_2(g)$$

$$\text{rate} = k[NO_2][O_3]$$

A two-step mechanism has been proposed. Identify the rate-limiting step.

$$NO_2 + O_3 \rightarrow NO_3 + O_2$$

$$NO_3 + NO_2 \rightarrow N_2O_5$$

13.79 Nitrogen dioxide reacts with carbon monoxide to form carbon dioxide and nitrogen monoxide.

$$NO_2(g) + CO(g) \rightarrow CO_2(g) + NO(g)$$

Two mechanisms are proposed:

Mechanism I (one step):

$$NO_2(g) + CO(g) \rightarrow CO_2(g) + NO(g)$$

Mechanism II (two steps):

$$NO_2(g) + NO_2(g) \rightarrow NO_3(g) + NO(g) \quad \text{Slow}$$

$$NO_3(g) + CO(g) \rightarrow NO_2(g) + CO_2(g) \quad \text{Fast}$$

Write the rate law expected for each mechanism.

13.80 Nitrogen monoxide reacts with ozone.

$$NO(g) + O_3(g) \rightarrow NO_2(g) + O_2(g)$$

Two mechanisms have been proposed:

Mechanism I (one step):

$$NO(g) + O_3(g) \rightarrow NO_2(g) + O_2(g)$$

Mechanism II (two steps):

$$O_3(g) \rightarrow O_2(g) + O(g) \quad \text{Slow}$$

$$NO(g) + O(g) \rightarrow NO_2(g) \quad \text{Fast}$$

Write the rate law that is expected for each mechanism.

13.81 The gas-phase reaction of nitrogen monoxide with chlorine proceeds to form nitrosyl chloride.

$$2NO(g) + Cl_2(g) \rightarrow 2NOCl(g)$$

$$\text{rate} = k[NO]^2[Cl_2]$$

Evaluate the following proposed mechanism to determine whether it is consistent with the experimental results, and identify intermediates, if any.

$$2NO \underset{k_{-1}}{\overset{k_1}{\rightleftharpoons}} N_2O_2 \quad \text{Fast, reversible}$$

$$N_2O_2(g) + Cl_2(g) \rightarrow 2NOCl(g) \quad \text{Slow (rate-limiting) step}$$

13.82 Evaluate each of the following proposed mechanisms to determine whether it is consistent with the experimentally-determined stochiometry and rate law, and identify intermediates, if any.

$$2NO_2 + O_3 \rightarrow N_2O_5 + O_2$$

$$\text{rate} = k[NO_2][O_3]$$

(a) $2NO_2 \underset{k_{-1}}{\overset{k_1}{\rightleftharpoons}} N_2O_4$ Fast, reversible

$N_2O_4 + O_3 \rightarrow N_2O_5 + O_2$ Slow

(b) $NO_2 + O_3 \rightarrow NO_3 + O_2$ Slow

$NO_3 + NO_2 \rightarrow N_2O_5$ Fast

Chapter Exercises

13.83 Define the rate of reaction in terms of changing concentrations for

$$aA + bB \rightarrow cC + dD$$

13.84 ■ ▲ A catalyst reduces the activation energy of a reaction from 215 to 206 kJ. By what factor would you expect the reaction-rate constant to increase at 25 °C? Assume that the pre-exponential term (A) is the same for both reactions. (*Hint:* Use the formula $\ln k = \ln A - E_a/RT$.)

13.85 ▲ The enzyme catalase reduces the activation energy for the decomposition of hydrogen peroxide from 72 to 28 kJ/mol. Calculate the factor by which the rate of reaction increases at 298 K, assuming that everything else is unchanged.

$$H_2O_2(aq) \rightarrow H_2O(\ell) + \frac{1}{2}O_2(g)$$

13.86 Kice and Bowers studied the decomposition of *p*-toluene sulfinic acid (*p*TSA) (*J Am Chem Soc* 84:605, 1962). They found the stoichiometry to be

$$3\ p\text{TSA} \rightarrow \text{products}$$

The scientists removed a portion of the reaction mixture and performed a titration at periodic intervals to determine the concentration of *p*TSA. Their results are shown in the table. Determine the rate law and reaction order.

Time (s)	[*p*TSA], *M*
0	0.100
900	0.0863
1800	0.0752
2700	0.064
3600	0.0568
7200	0.0387
10800	0.0297
18000	0.0196

13.87 The decomposition of hydrogen peroxide is catalyzed by horseradish peroxidase, an enzyme isolated from the vegetable.

$$H_2O_2(aq) \rightarrow \frac{1}{2}O_2(g) + H_2O(\ell)$$

The hydrogen peroxide concentration is measured as a function of time to produce the following data for the catalyzed reaction.

Time (s)	$[H_2O_2]$ *(M)*
0	0.0334
10	0.0300
20	0.0283
30	0.0249
40	0.0198
50	0.0164
60	0.0130

(a) What is the order?
(b) What is the rate law for the decomposition?

13.88 Use the following experimental data to determine the rate law and rate constant for formation of phosgene.

$$CO + Cl_2 \rightarrow COCl_2$$

	Initial Concentration *(M)*		Initial Rate of Reaction
Experiment	[CO]	$[Cl_2]$	*(M/s)*
1	0.053	0.23	3.7×10^{-5}
2	0.106	0.23	7.4×10^{-5}
3	0.106	0.46	10.4×10^{-5}

13.89 The reactant in a first-order reaction decreases in concentration from 0.451 to 0.235 *M* in 131 seconds. How long does it take to decrease from 0.235 to 0.100 *M*?

13.90 The following data were obtained for the decomposition of nitrogen dioxide.

$$2NO_2(g) \rightarrow 2NO(g) + O_2(g)$$

	Pressure of NO_2 (torr)	
Time (s)	310 K	315 K
0	24.0	24.0
1	18.1	15.2
2	13.7	9.7
3	10.3	6.1
4	7.8	3.9
5	5.9	2.5
6	4.5	1.6
7	3.4	1.0
8	2.6	0.6
9	1.9	0.4
10	1.5	0.3

(a) What is the reaction order?
(b) What is the rate constant at each temperature?
(c) What is the activation energy?

13.91 Determine the order and rate constant by analyzing the concentration-time graph. You may want to use a pencil to draw a smooth line through the data.

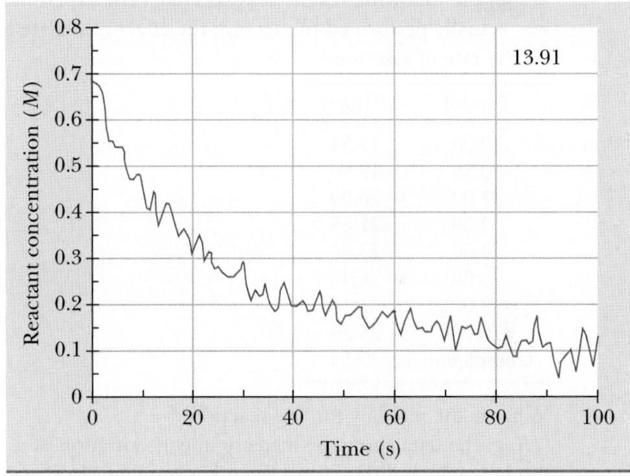

13.92 ▲ Reaction A has an activation energy of 30 kJ/mol; reaction B has an activation energy of 40 kJ/mol. The ratio of their rates is called *R*.

$$R = \frac{\text{rate of reaction A}}{\text{rate of reaction B}}$$

If the two reactions proceed at the same rate at 25 °C, what is the value of *R* at 35 °C?

13.93 ▲ Two reactions have activation energies of 45 and 40 kJ/mol, respectively. Which reaction shows the greater increase in rate with an increase in temperature?

13.94 ▲ The following experimental data were obtained in a study of the kinetics of the gas-phase formation of nitrosyl bromide at 791 K. Determine the rate law and the rate constant from the data.

$$2NO + Br_2 \rightarrow 2NOBr$$

	Initial Pressure (atm)		
Exponent	$[P_{NO}]$	$[P_{Br_2}]$	Initial Rate of Reaction $(-1/2\ \Delta[P_{NO}]/\Delta t)$ (atm/s)
1	1.1×10^{-5}	1.2×10^{-5}	0.37
2	1.9×10^{-5}	1.3×10^{-5}	0.69
3	3.5×10^{-5}	1.2×10^{-5}	1.2
4	4.0×10^{-5}	3.0×10^{-6}	3.4

(a) What is the rate law for the reaction?
(b) Calculate the rate constant, using atm as the concentration unit.

13.95 ▲ Nitramide decomposes to water and dinitrogen monoxide.

$$H_2NNO_2(aq) \rightarrow H_2O(\ell) + N_2O(g)$$

This reaction was studied by J. N. Brønsted in 1924 as part of research into the fundamental nature of acids and bases. If 1.00 L of 0.440 *M* nitramide is placed in a reactor at 20 °C, the following results are expected. (The experiment was actually performed to measure the effect of malate ion on the rate of reaction.)

T (min)	*P* (torr)
0.00	17.54
0.50	18.98
1.00	20.09
2.00	21.65
4.00	23.46
6.00	24.48
8.00	25.14
10.00	25.59
Completion	27.54

What is the rate law for this reaction?
(*Hint:* The data show the increase in concentration of a product, which differs from the other problems in this book. The problem looks more familiar if you create a decay curve. Note that the initial point represents a small amount of product, and thus a large amount of reactant. The final point represents a large amount of product and no reactant. The problem looks like any other kinetics problem if you first calculate ([final pressure − current pressure] to see the data as a decay curve).

Graph (final pressure − current pressure), rather than the current pressure, as a function of time to see the data. If this is not a straight line, you can graph ln(final pressure − current pressure) and 1/(final pressure − current pressure) to determine the order of the reaction.

Scientists frequently transform their data to a familiar form. These operations make it easier to interpret the data.

13.96 When methyl bromide reacts with hydroxide ion, methyl alcohol and bromide ion form.

$$CH_3Br + OH^- \rightarrow CH_3OH + Br^-$$

Consider the two mechanisms that follow, and write the expected rate law for each.

(a) A two-step mechanism with the rate limited by the dissociation of methyl bromide:

$$CH_3Br \xrightarrow{\text{Slow}} H_3C^+ + Br^-$$

$$H_3C^+ + OH^- \xrightarrow{\text{Fast}} CH_3OH$$

(b) Formation of a transition state followed by fast rearrangement:

$$OH^- + CH_3Br \rightarrow CH_3OH + Br^-$$

Cumulative Exercises

13.97 ▲ Ethyl chloride decomposes to form ethylene and hydrogen chloride at 437 K.

$$C_2H_5Cl(g) \rightarrow C_2H_4(g) + HCl(g)$$

The reaction takes place in a 4.0-L container and is monitored by measuring the time needed for the hydrogen chloride to react with a known amount of base.

Initial Pressure of C_2H_5Cl (torr)	Time Needed for Reactions (s)	Mass of NaOH (g)
131	11.4	0.0321
160	11.6	0.0399
172	9.1	0.0336

(a) What is the rate law?
(b) What is the rate constant using moles per liter for concentration and seconds for time?

13.98 ■ When formic acid is heated, it decomposes to hydrogen and carbon dioxide in a first-order decay.

$$HCOOH(g) \rightarrow CO_2(g) + H_2(g)$$

The rate of reaction is monitored by measuring the total pressure in the reaction container.

Time (s)	*P* (torr)
0	220
50	324
100	379
150	408
200	423
250	431
300	435

Calculate the rate constant and half-life, in seconds, for the reaction. At the start of the reaction (time = 0), only formic acid is present. (*Hint:* Find the partial pressure of formic acid [use Dalton's law of partial pressure and the reaction stoichiometry to find P_{HCOOH} at each time]).

13.99 In 1926, Hinshelwood and Green studied the reaction of nitrogen monoxide and hydrogen.

$$2NO(g) + 2H_2(g) \rightarrow N_2(g) + 2H_2O(g)$$

They measured the rate of reaction as a function of pressure at 1099 K.

P_{H_2} (torr)	P_{NO} (torr)	Initial Rate (torr/s)
289	400	0.160
205	400	0.110
147	400	0.079
400	359	0.150
400	300	0.103
400	152	0.025

(a) What is the rate law for the reaction?

(b) Use the data from the first experiment to calculate the rate constant at 1099 K, using torr as the concentration unit.

(c) The relative rate of the reaction changed with temperature, as the following data show.

T (K)	*k*, relative
956	1.00
984	2.34
1024	5.15
1061	10.9
1099	18.8

Calculate the activation energy (in kJ/mol) for the reaction.

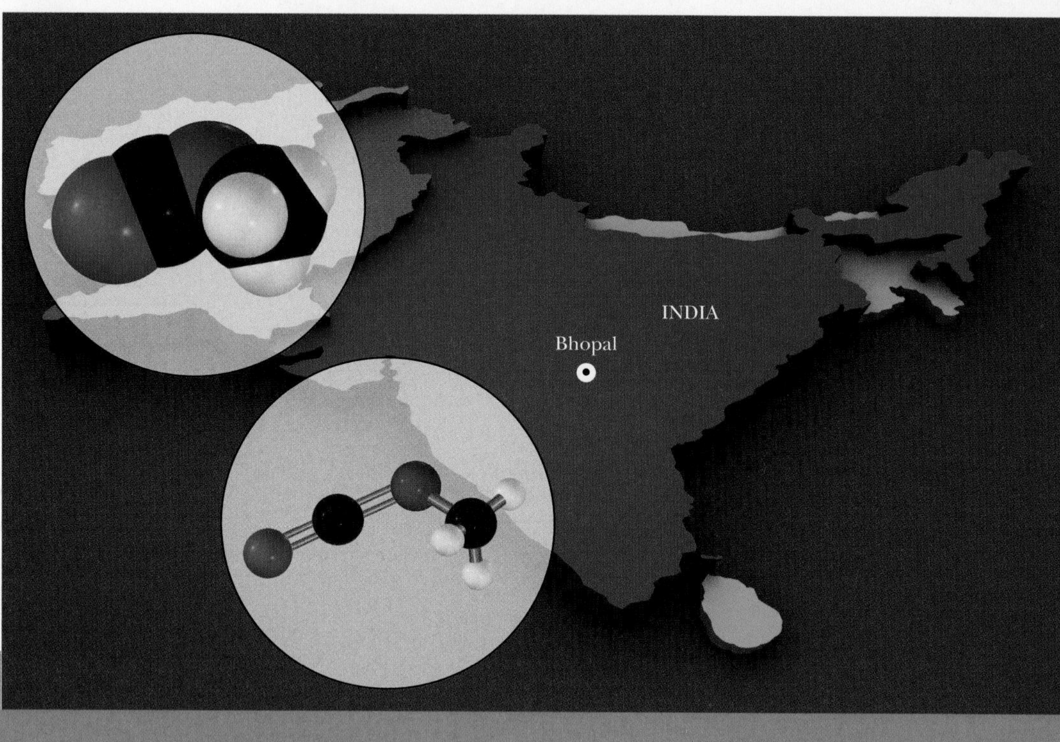

Methyl isocyanate. $CH_3-N=C=O$.

Chemistry does not always progress with prize-winning ideas. Scientists often learn as much from mistakes as from success. The weak glue on Post-it Note Pads, for example, was invented by a 3M researcher trying to synthesize a *strong* adhesive. When mistakes involve the loss of lives, however, chemists investigate carefully to try to understand the causes and, more importantly, to avoid making the same mistakes again.

In 1984, a chemical production facility in Bhopal, India, accidentally released a gas cloud of methyl isocyanate (CH_3NCO), a compound that is extremely toxic even in low concentrations. The heavier-than-air methyl isocyanate vapors cascaded into the homes of the townspeople during the early morning hours of December 3, 1984. When the sun rose, thousands of people were dead, and hundreds of thousands were severely ill, many of whom later died from complications arising from tissue damage from breathing methyl isocyanate.

The immediate cause of death was asphyxiation from the toxic gas, but much larger questions loomed. For example, how could the safety systems of the chemical plant have been compromised to allow the release of 20,000 to 40,000 kg of the gas? Why weren't the safety measures sufficient? Were there unexpected chemical reactions occurring at the plant? Who or what was responsible?

The chemistry behind the incident has been well studied and tells us a lot about what happened, but nothing about why it happened. There are (at least) two sides to the story. The company maintains that its plant was safe, that all the required safeguards were in place, and that the release was deliberate sabotage by an employee. The company maintains that no one can anticipate and guard against every possible form of employee misconduct. Its critics disagree, stating that a large number of safety systems were inoperative, and if the plant had been operating with modern systems, the accident would not have happened. They accused the U.S.-owned facility of callous disregard for the lives of its Indian neighbors and demanded the extradition of the company's president to stand charges in India. The chemical company paid $470 million to the Indian government to settle the case in 1989.

The end product of the plant was an important pesticide called *carbaryl* ($C_{12}H_{11}NO_2$). The plant made phosgene ($COCl_2$) on-site and trucked in methylamine (CH_3NH_2) to form methyl carbamoyl chloride. The solvent used was chloroform ($CHCl_3$).

$$\underset{\text{phosgene}}{COCl_2(g)} + \underset{\text{methylamine}}{CH_3NH_2(g)} \xrightarrow{CHCl_3} HCl(g) + \underset{\text{methyl carbamoyl chloride}}{CH_3NH\text{–}CO\text{–}Cl(\ell)}$$

Chemical Equilibrium

14

CHAPTER CONTENTS

OWL Online homework for this chapter may be assigned in OWL.

Look for the green colored vertical bar throughout this chapter, for integrated references to this chapter introduction.

The methyl carbamoyl chloride reacts further to eliminate another unit of HCl forming methyl isocyanate.

$$\underset{\text{methyl carbamoyl chloride}}{CH_3NH\text{–}CO\text{–}Cl(\ell)} \xrightarrow{CHCl_3} HCl(g) + \underset{\text{methyl isocyanate}}{CH_3\text{–}N{=}C{=}O(\ell)}$$

The designers of the plant knew that methyl isocyanate is volatile, flammable, and reacts violently with water, and they tried to build in safeguards. The methyl isocyanate was stored in a holding tank, as always, but inexplicably high temperatures and large pressures built up the day of the accident. The tank, like nearly all similar tanks, has a pressure-relief mechanism to allow some of the contents to escape to avoid a tank explosion. A number of safety measures failed: The operators did not refrigerate the tank, although they could have done so by pressing a button; a gas scrubber, which contained a substance that could have neutralized at least some of the toxic gas, was on standby; and the flare tower, used to burn gases that escape, was being repaired.

Almost immediately, chemists realized that the likely cause for the high temperatures and pressures was the contamination of the methyl isocyanate with water, which reacts with the methyl isocyanate to produce methylamine and carbon dioxide gas.

$$H_2O(\ell) + CH_3\text{–}N{=}C{=}O(\ell) \rightarrow CH_3NH_2(g) + CO_2(g)$$

As the pressure built up inside the holding tank, the relief valve opened, allowing the methyl isocyanate to escape into the surrounding area.

Chemists took samples from the tank and found seven products that originated from the methyl isocyanate. They were able to set up small-scale reactions in their laboratory and found that if the tank held 84.4% methyl isocyanate, 12.0% chloroform, and 3.6% water, they could reproduce the results of the explosion.

The company used the data to show that such a large volume of water had to be deliberately added to the tank because its design prevented inadvertent introduction of even small amounts of water. Its critics interpreted the data to show that the company was even more careless than they had originally charged.

The chemical reactions in the Bhopal plant formed unanticipated products, but the effects of high temperatures and pressures on the reaction can be predicted using the principles discussed in this chapter. ■

Photo by Pablo Bartholomew/Liaison

Photo by Robert Nickelsberg//Time Life Pictures/Getty Images

Bhopal, India. After the tragic deaths of people and animals, the area around the factory was deserted. Many residents felt the settlement was unjust.

Chemists have been studying chemical reactions for centuries. One of the first steps in any study is to identify the products and measure the amounts formed. The reactions presented so far in this book are said to "go to completion," and the amounts of products are determined directly from the amount of the limiting reactant.

Many reactions, however, do not go to completion and cannot be described simply by the phrase "will proceed" or "will not proceed." These types of reactions achieve a condition of equilibrium—a state of balance between opposing processes. At equilibrium, the tendency of the reactants to form products is balanced by the tendency of the products to form reactants, so a mixture of reactants and products results. We first discussed equilibrium processes with phase changes. For example, at 0 °C and 1 atm pressure, the liquid and solid forms of water are in equilibrium.

Knowledge of equilibrium and the factors that affect it enable us to calculate the extent of reaction, the amounts of products formed, and the amounts of reactants that remain. Many important industrial processes are equilibrium reactions, conducted under conditions that provide the largest yield of product at the lowest cost.

This chapter introduces chemical equilibrium primarily with homogeneous gas-phase reactions. The factors that influence the equilibrium, the response of a system at equilibrium to external changes, and calculations of the equilibrium amounts of products and reactants are presented. Heterogeneous equilibria are also discussed in the sections that cover gas-solid equilibria and the solubility of ionic compounds.

14.1 Equilibrium Constant

OBJECTIVES

- ☐ Describe equilibrium systems and write the equilibrium constant expression for any chemical reaction
- ☐ Evaluate the equilibrium constant from experimental data
- ☐ Relate the expression for the equilibrium constant to the form of the balanced equation
- ☐ Convert between equilibrium constants in which the concentrations of gases are expressed in moles per liter (mol/L or *M*) and those in terms of partial pressures expressed in atmospheres (atm).

Centuries of laboratory observation show that some reactions do not go to completion, so the amounts of products formed or reactants consumed cannot always be predicted from the stoichiometry alone. For example, when nitrogen and hydrogen are mixed, ammonia forms.

$$N_2(g) + 3H_2(g) \rightarrow 2NH_3(g)$$

The amount of ammonia that forms is less than would be expected from a stoichiometry calculation, because the reverse reaction also occurs.

$$2NH_3(g) \rightarrow N_2(g) + 3H_2(g)$$

The concentrations of reactants and products follow a pattern that led to the development of a mathematical model that relates the equilibrium concentrations of reactants and products.

Equilibrium Systems

The nitrogen-hydrogen-ammonia system reaches a balance when the forward reaction produces as much ammonia from N_2 and H_2 per second as is consumed in the reverse reaction. At that point, the reaction has achieved **equilibrium** (plural: *equilibria*), a state in which the tendency of the reactants to form products is balanced by the tendency of the products to form reactants. At equilibrium, the concentrations of nitrogen, hydrogen, and ammonia remain constant. Rates of phase-change processes were mentioned in Chapter 11, and rates of chemical reactions were discussed in detail in Chapter 13. Systems at equilibrium are depicted with a double reaction arrow to indicate that both the forward and reverse reactions occur.

Equilibrium is the balance between forward and reverse reactions.

$$N_2(g) + 3H_2(g) \rightleftharpoons 2NH_3(g)$$

Notably, when a reaction is at equilibrium, *the reaction has not stopped.* Instead, the forward reaction produces products at the same rate that the reverse reaction consumes them. Chemical equilibria are *dynamic* systems, as opposed to *static* systems in which processes stop.

It is also important to distinguish between a reaction that is at equilibrium and one that is simply proceeding too slowly to exhibit observable changes. The chemist can test a reaction to determine whether it has reached equilibrium by adding any of the compounds involved in the reaction. If some nitrogen is added to an equilibrium mixture of nitrogen, hydrogen, and ammonia, then additional ammonia forms. If some ammonia is added, then the reaction proceeds in the reverse direction, forming some nitrogen and hydrogen. When the concentrations stop changing, an equilibrium state has again been reached. Chemists generally verify that a system is at equilibrium by performing this type of test.

Chemists must be careful not to apply equilibrium methods to reactions that proceed too slowly.

Many of the most interesting chemical reactions are equilibrium systems. One example is the reaction of atmospheric nitrogen with hydrogen to produce ammonia; this reaction is the first step in the production of ammonium nitrate, which is used for products ranging from fertilizers to explosives. Without this process, many believe that Germany would not have entered World War I since it would have been unable to produce much important weaponry. In this section, we present the expressions that describe reactions at equilibrium together with a description of the factors that determine the direction in which a chemical reaction proceeds.

Consider the formation of dinitrogen tetroxide from nitrogen dioxide, a reaction that proceeds at room temperature.

$$2NO_2(g) \rightleftharpoons N_2O_4(g)$$

Figure 14.1 shows this reaction. The reaction starts with 0.0200 mol nitrogen dioxide, NO_2, sealed in a 1.0-L flask. The deep brown color of NO_2 begins to fade as the colorless dinitrogen tetroxide, N_2O_4, forms. The concentrations of NO_2 and N_2O_4 change as shown. Eventually, when the reaction reaches equilibrium, the concentrations of NO_2 and N_2O_4 cease changing.

A second experiment is performed, one that starts with 0.0100 mol N_2O_4 in the flask. The results are shown in Figure 14.2.

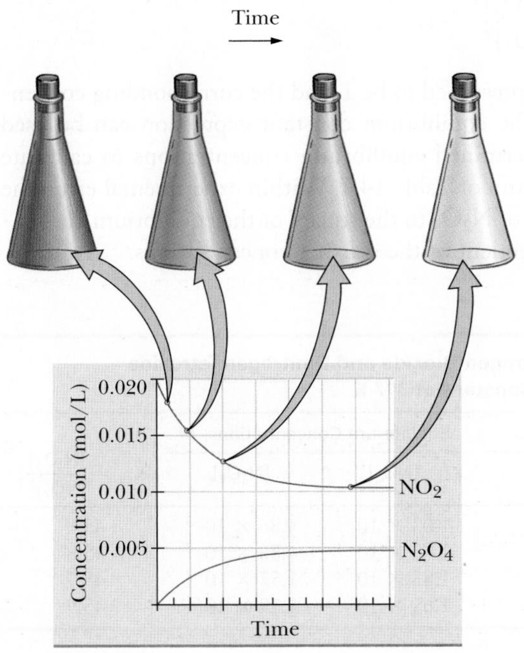

Figure 14.1 A reaction reaches equilibrium.

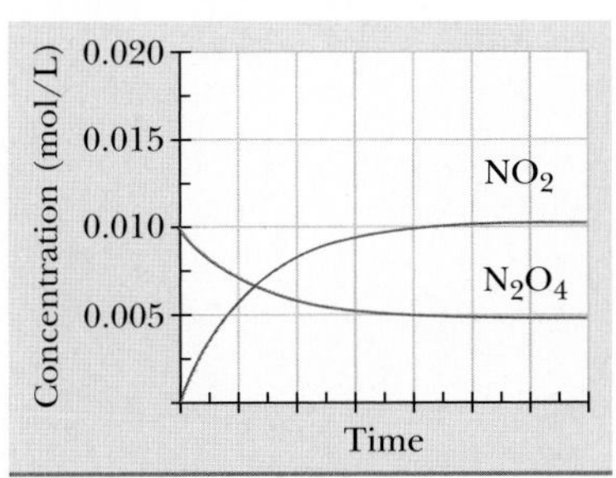

Figure 14.2 A reaction reaches equilibrium. The reaction is the same as shown in Figure 14.1, but the starting conditions are different.

The experimentally determined equilibrium concentrations at the end of each experiment are the same. The reaction produces the same equilibrium concentrations whether it is initiated with reactants, with the equivalent amount of products, or with any mixture in which the total mass (NO_2 + N_2O_4) is the same.

The experiment can be repeated with different starting concentrations. Table 14.1 shows the results of several different experiments. Chemists use square brackets, such as $[NO_2]$, to indicate molar concentrations, a convention that is followed in Table 14.1.

The relationship between the equilibrium concentrations of NO_2 and N_2O_4 is not obvious. Chemists studied many equilibrium reactions over a period of years and have proposed numerous explanations for individual reactions. No general treatment was reached until 1864, when two Norwegian scientists, Cato Guldberg (1836–1902) and Peter Waage (1833–1900), summarized the experimental results with the **law of mass action.** We can write the conclusions in modern terms: For any reaction at equilibrium,

$$aA + bB \rightleftharpoons cC + dD$$

an **equilibrium constant expression** can be written:

$$\frac{[C]^c[D]^d}{[A]^a[B]^b} = K_{eq} \qquad [14.1]$$

where the square brackets indicate the equilibrium concentrations of species A, B, C, and D (in mol/L), and the lowercase *a, b, c,* and *d* represent the stoichiometric coefficients in the balanced equation. For a given reaction, K_{eq} is a constant at any particular temperature. This mathematical relationship accurately predicts the experimentally observed relationships between the equilibrium concentrations of reactants and products.

An expression for an equilibrium constant can be written for any chemical equation. Products and their coefficients are in the numerator; reactants and their coefficients are in the denominator.

The term on the right side of Equation 14.1, K_{eq}, is called the **equilibrium constant.** If the concentrations, when substituted into the equilibrium constant expression, do not equal the equilibrium constant, a reaction occurs—and concentrations change—until the equilibrium constant is reached. This expression can be used to predict the equilibrium concentrations of species from any starting concentrations.

The equilibrium constant expression for the formation of dinitrogen tetroxide from nitrogen dioxide can be written as follows:

$$2NO_2(g) \rightleftharpoons N_2O_4(g) \quad K_{eq} = \frac{[N_2O_4]}{[NO_2]^2}$$

When a coefficient is not shown, it is presumed to be 1, and the corresponding concentration is raised to the first power. The equilibrium constant expression can be used together with the experimentally determined equilibrium concentrations to calculate the values of K_{eq} given in the last column of Table 14.1. Within experimental error, the ratio of the equilibrium concentration of N_2O_4 to the square of the equilibrium concentration of NO_2 is constant and independent of the starting concentrations.

TABLE 14.1 Concentrations of Nitrogen Dioxide and Dinitrogen Tetroxide and the Equilibrium Constant at 317 K

	Initial Concentrations, *M*		Equilibrium Concentrations, *M*		
Experiment	$[NO_2]$	$[N_2O_4]$	$[NO_2]$	$[N_2O_4]$	$K_{eq} = \frac{[N_2O_4]}{[NO_2]^2}$
1	2.00×10^{-2}	0.00	1.03×10^{-2}	4.86×10^{-3}	45.8
2	0.00	1.00×10^{-2}	1.03×10^{-2}	4.86×10^{-3}	45.8
3	3.00×10^{-2}	1.00×10^{-2}	1.85×10^{-2}	1.57×10^{-2}	45.9
4	4.00×10^{-2}	0.00	1.61×10^{-2}	1.19×10^{-2}	45.9

EXAMPLE 14.1 Write the Equilibrium Constant Expression

Write the equilibrium constant expression for

$N_2(g) + O_2(g) \rightleftharpoons 2NO(g)$

Strategy The concentrations of the products appear in the numerator, the reactants in the denominator, all raised to powers that correspond to their coefficients in the balanced equation.

Solution

The concentration of NO appears in the numerator, raised to the power 2, its coefficient in the chemical equation. The concentrations of the reactants, N_2 and O_2, appear in the denominator. Each concentration is raised to the power 1, the coefficients in the equation.

$$K_{eq} = \frac{[NO]^2}{[N_2][O_2]}$$

Understanding

Write the equilibrium constant expression for

$2SO_3(g) \rightleftharpoons 2SO_2(g) + O_2(g)$

Answer

$$K_{eq} = \frac{[SO_2]^2[O_2]}{[SO_3]^2}$$

The equilibrium constant is an experimentally determined quantity that provides information about the concentrations of reactants and products in an equilibrium mixture. In general, when K_{eq} is large (much greater than 1), the reaction favors the formation of products. One way to determine the value of K_{eq} is to measure the concentrations of all substances present in an equilibrium mixture and substitute those values into the equilibrium constant expression. For example, a scientist studying the decomposition of sulfur trioxide might perform a chemical analysis on the equilibrium mixture of gases:

$2SO_3(g) \rightleftharpoons 2SO_2(g) + O_2(g)$

The data from the chemical analysis help determine the equilibrium constant. If the results are that $[SO_2] = 0.44\ M$, $[O_2] = 0.22\ M$, and $[SO_3] = 0.11\ M$, the numerical value of the equilibrium constant is 3.5 at that particular temperature.

$$K_{eq} = \frac{[SO_2]^2[O_2]}{[SO_3]^2} = \frac{(0.44)^2(0.22)}{(0.11)^2} = 3.5$$

It is important to remember that the equilibrium constant is an experimental result and that the numerical value of K_{eq} depends on temperature. An equilibrium constant cannot be used to describe a reaction mixture unless the reaction is at the same temperature at which K_{eq} was determined.

The equilibrium constant is determined by experiment. Remember that temperature influences K_{eq}.

Relating K_{eq} to the Form of the Chemical Equation

The equilibrium constant expression and its numerical value refer to a particular chemical equation. Because equations can be balanced in a number of ways, we must know the coefficients used in the particular chemical equation. We can continue to consider the formation of nitrogen monoxide from nitrogen and oxygen to see how the coefficients of the equation influence the equilibrium constant.

$$N_2(g) + O_2(g) \rightleftharpoons 2NO(g)$$

$$K_1 = \frac{[NO]^2}{[N_2][O_2]}$$

Had the equation been written

$$\frac{1}{2}N_2(g) + \frac{1}{2}O_2(g) \rightleftharpoons NO(g)$$

the equilibrium constant expression would be

$$K_2 = \frac{[NO]}{[N_2]^{1/2}[O_2]^{1/2}}$$

Comparing the two equilibrium constant expressions, we see that $K_1 = (K_2)^2$. If each coefficient in the chemical equation is doubled, the equilibrium constant for the new reaction is the square of that for the old reaction. If the coefficients are tripled, the equilibrium constant is cubed. If we generate a new equation by multiplying the coefficients by a constant factor, *n*, then we can calculate a new equilibrium constant by raising the old equilibrium constant to the *n*th power.

The value for K_{eq} refers to a specific chemical equation.

If the reaction is written in the reverse direction, the equilibrium constant is the reciprocal (negative first power) of the equilibrium constant for the forward reaction.

$$A \rightleftharpoons B \qquad K_{forward} = \frac{[B]}{[A]}$$

$$B \rightleftharpoons A \qquad K_{reverse} = \frac{[A]}{[B]}$$

$$K_{forward} = (K_{reverse})^{-1} = \frac{1}{K_{reverse}}$$

Example 14.2 illustrates how K_{eq} varies with different forms of the chemical equation.

EXAMPLE 14.2 Dependence of K_{eq} on the Form of the Chemical Equation

The equilibrium system of hydrogen, nitrogen, and ammonia can be written in several different ways:

(1) $N_2(g) + 3H_2(g) \rightleftharpoons 2NH_3(g)$

(2) $\frac{1}{2}N_2(g) + \frac{3}{2}H_2(g) \rightleftharpoons NH_3(g)$

(3) $2NH_3(g) \rightleftharpoons N_2(g) + 3H_2(g)$

The equilibrium constant, K_{eq}, for reaction 1 is 0.19 at 532 °C. Write the equilibrium constant expression and calculate K_{eq} for reactions 2 and 3 at the same temperature.

Strategy First, write the equilibrium constant expression for each equation, then examine the concentration terms to determine the relationship between equilibrium expressions. Use these relationships to calculate the relationships between equilibrium constants.

Solution
The expressions for the equilibrium constants are:

$$K_1 = \frac{[NH_3]^2}{[N_2][H_2]^3} = 0.19$$

$$K_2 = \frac{[NH_3]}{[N_2]^{1/2}[H_2]^{3/2}}$$

$$K_3 = \frac{[N_2][H_2]^3}{[NH_3]^2}$$

Examine the concentration terms in the expressions for K_1 and K_2 to show that

$K_1 = (K_2)^2$

Check this equation by squaring each term in the equilibrium constant expression for K_2 and verifying that K_2^2 is equal to K_1. Thus,

$K_2^2 = K_1 = 0.19$

To calculate K_2, take the square root:

$K_2 = 0.44$

Note that when the stoichiometry is halved, the new K is the square root ($\frac{1}{2}$ power) of the old K.

The third equation is the reverse of the first, so the new K_3 is the old K_1 raised to the -1 power.

$$K_3 = (K_1)^{-1} = \frac{1}{K_1} = \frac{1}{0.19}$$

$K_3 = 5.3$

All the values of K are valid and contain equivalent information. Each value of K applies to a specific form of the equation.

Understanding

Use the preceding data to calculate K_{eq} at 532 °C for

$$NH_3(g) \rightleftharpoons \frac{1}{2}N_2(g) + \frac{3}{2}H_2(g)$$

Answer $K_{eq} = 2.3$

When chemical equations are added to yield a new equation, K_{eq} for the new reaction is determined by multiplying the equilibrium constants of the component equations. We can demonstrate this property by calculating the equilibrium constant for the formation of $NO_3(g)$:

(1) $2NO_2(g) \rightleftharpoons N_2O_4(g)$	K_1
(2) $N_2O_4(g) + O_2(g) \rightleftharpoons 2NO_3(g)$	K_2
(3) $2NO_2(g) + O_2(g) \rightleftharpoons 2NO_3(g)$	K_3

Reaction 3 is obtained by adding reaction 1 to reaction 2. Likewise, K_3 is the product of K_1 and K_2.

$K_3 = K_1K_2$

We can verify this relationship by substituting the concentration expression for each of the three equilibrium constants:

$$K_3 = K_1 \times K_2$$

$$= \frac{[\cancel{N_2O_4}]}{[NO_2]^2} \times \frac{[NO_3]^2}{[\cancel{N_2O_4}][O_2]}$$

$$K_3 = \frac{[NO_3]^2}{[NO_2]^2[O_2]}$$

Relationships between Pressure and Concentration

Thus far, the concentrations used for each substance in the equilibrium constant expressions have units of moles per liter (mol/L). These units are appropriate for solutes in solution, but the concentration of a gas is generally expressed in terms of its partial pressure. Either concentration unit (atm or mol/L) can be used for equilibrium calculations of gas-phase equilibria.

The starting material used in the Bhopal factory described in the chapter introduction was phosgene, $COCl_2$, a highly toxic gas. It is known to dissociate to form carbon monoxide and chlorine:

$$COCl_2(g) \rightleftharpoons CO(g) + Cl_2(g)$$

Two equilibrium constants can be defined for this reaction:

$$K_c = \frac{[CO][Cl_2]}{[COCl_2]}$$

$$K_p = \frac{P_{CO}P_{Cl_2}}{P_{COCl_2}}$$

The subscript describes the type of equilibrium constant: c is used for concentrations (expressed in mol/L), and p is used for pressures (expressed in atm). We use K_{eq} for equilibrium constants in general or when the differences between K_c and K_p are not important to the discussion. The units of moles per liter are understood for terms appearing in K_c, and atmospheres are understood for terms appearing in K_p, so units are omitted from reported values of K_c and K_p.

Express the concentration of solutes in solution in moles per liter. Express the concentration of a gas in atmospheres, unless K_c is specified.

If concentrations are provided, K_c is easier to use. If the problem makes use of partial-pressure data, then K_p is generally more convenient. But if the equilibrium constant is in one set of units and the concentrations are in the other, then we must convert either the concentrations or the equilibrium constant to match.

The relationship between the partial pressure of a gas and its molar concentration comes from the ideal gas law:

$$PV = nRT$$

$$P = (n/V)RT$$

Notice that n/V is the molar concentration—the units are moles per liter. To convert concentration in moles per liter to pressure in atmospheres, multiply by RT.

$$P = [n/V] \times RT$$

It is important to use 0.08206 L·atm/mol·K for R and express the temperature in kelvins.

To convert pressure in atmospheres to concentration in moles per liter, divide by RT.

$$[n/V] = P/RT$$

There is a second way to solve these problems. Instead of converting molar concentration to pressure, we can interconvert K_p and K_c using Equation 14.2, which is derived in the Principles of Chemistry section on the next page.

$$K_p = K_c(RT)^{\Delta n} \qquad [14.2]$$

where Δn is the change in the number of moles of gas:

Δn = total number of moles of gas on the product side

− total number of moles of gas on the reactant side

PRINCIPLES OF CHEMISTRY
Deriving the Relationship between K_p and K_c

The interconversion between K_p and K_c is

$$K_p = K_c\,(RT)^{\Delta n}$$

We can prove this by writing K_p for the gas-phase reaction

$$aA + bB \rightleftharpoons cC + dD$$

$$K_p = \frac{P_C^c P_D^d}{P_A^a P_B^b}$$

For the pressure of any species, substitute $[n/V] \times RT$ for P.

$$K_p = \frac{\left(\frac{n_C}{V} \times RT\right)^c \left(\frac{n_D}{V} \times RT\right)^d}{\left(\frac{n_A}{V} \times RT\right)^a \left(\frac{n_B}{V} \times RT\right)^b}$$

We can substitute square brackets for n/V

$$K_p = \frac{([C] \times RT)^c([D] \times RT)^d}{([A] \times RT)^a([B] \times RT)^b}$$

and collect all the RT terms.

$$K_p = \frac{[C]^c[D]^d}{[A]^a[B]^b} \times \frac{(RT)^{c+d}}{(RT)^{a+b}}$$

We can recognize that the first term is equal to K_c

$$K_p = K_c \times (RT)^{c+d-a-b}$$

and the second term is $(RT)^{\Delta n}$, so

$$K_p = K_c(RT)^{\Delta n}$$ ∎

When carbon monoxide and chlorine react to form phosgene, as in the pesticide plant in Bhopal, India, Δn is −1.0.

$$CO(g) + Cl_2(g) \rightleftharpoons COCl_2(g)$$

In other reactions, Δn can be positive, negative, or zero and can be a fraction. Example 14.3 illustrates conversions between K_c and K_p.

EXAMPLE 14.3 Convert between K_p and K_c

Consider the equilibrium

$$PCl_5(g) \rightleftharpoons PCl_3(g) + Cl_2(g)$$

If the numerical value of K_p is 0.74 at 499 K, calculate K_c.

Strategy Compute Δn from the chemical equation, then rearrange the relationship $K_p = K_c(RT)^{\Delta n}$ to solve for K_c. Be sure to use 0.08206 L·atm/mol·K for R and express the temperature in kelvins.

Solution
First, calculate Δn. There are 2 mol of gas on the product side of the chemical equation, and there is 1 mol of gaseous reactant, so Δn is +1 in this case.

Rearrange Equation 14.2 to calculate K_c.

$$K_p = K_c\,(RT)^{\Delta n}$$

$$K_c = \frac{K_p}{(RT)^{\Delta n}}$$

Now substitute the numerical values for K_p, R, and T in the equation.

$$K_c = \frac{K_p}{(RT)^{\Delta n}}$$

$$K_c = \frac{0.74}{(0.08206 \times 499)^{+1}}$$

$$K_c = 1.8 \times 10^{-2}$$

Understanding

K_p for the formation of 2 mol ammonia from nitrogen and hydrogen is 2.8×10^{-9} at 298 K. Calculate K_c for

$$N_2(g) + 3H_2(g) \rightleftharpoons 2NH_3(g)$$

Answer $K_c = 1.7 \times 10^{-6}$

OBJECTIVES REVIEW *Can you:*

- ☑ describe equilibrium systems and write the equilibrium constant expression for any chemical reaction?
- ☑ evaluate the equilibrium constant from experimental data?
- ☑ relate the expression for the equilibrium constant to the form of the balanced equation?
- ☑ convert between equilibrium constants in which the concentrations of gases are expressed in moles per liter and those in terms of partial pressures expressed in atmospheres?

14.2 Reaction Quotient

OBJECTIVES

- ☐ Write the expression for Q, the reaction quotient, and contrast it with the expression for K_{eq}
- ☐ Compare Q and K_{eq} to determine the direction in which a reaction proceeds when a system is not at equilibrium

The equilibrium constant can be used to calculate the concentrations of reactants and products in a system at equilibrium, but that is not its only use. We can also determine whether a given mixture of reactants and products will form more products, will form more reactants, or is at equilibrium. These topics and their applications to chemical systems are presented in this section.

The law of mass action not only describes equilibrium systems but also provides important information about systems not yet at equilibrium. The **reaction quotient, Q,** has the same algebraic form as K_{eq}, but the *current concentrations,* not specifically the equilibrium concentrations, are used in the calculation. Comparing Q with K_{eq} enables us to predict in which direction a reaction will proceed to achieve equilibrium.

If we examine the general chemical reaction

$$aA + bB \rightleftharpoons cC + dD$$

and use the *current* concentrations of a reaction rather than equilibrium concentrations, then the expression for Q is

$$Q = \frac{[C]^c[D]^d}{[A]^a[B]^b}$$

The concentration expression for the reaction quotient, Q, is identical to that for K_{eq}, but may contain concentrations for a mixture that is not at equilibrium.

Determining the Direction of Reaction

The numerical value of Q tells us the direction in which a reaction must proceed to reach equilibrium. *The concentrations of products and reactants change to bring Q closer in value to K_{eq}.*

If Q is less than K_{eq}, the reaction proceeds to the right to increase the concentrations of the products and decrease the concentrations of the reactants. This change increases Q and brings it closer in value to K_{eq}. Mathematically, the numerator gets larger (because the concentrations of the products increase) and the denominator gets smaller. When Q is greater than K_{eq}, the reaction proceeds to the left, to form reactants.

If $Q < K_{eq}$, the reaction proceeds toward products. If $Q > K_{eq}$, the reaction proceeds to form reactants. At equilibrium, $Q = K_{eq}$.

A mixture of nitrogen, hydrogen, and ammonia can be studied to see how the initial concentrations influence the direction of the reaction. Some N_2, H_2, and NH_3 are mixed at 532 °C, forming a mixture with the following initial concentrations:

$$[NH_3] = 0.10\ M$$
$$[H_2] = 0.20\ M$$
$$[N_2] = 0.30\ M$$

We want to determine whether this system will form more ammonia, or if it will react to consume ammonia and form more hydrogen and nitrogen. The strategy is to calculate the reaction quotient, Q, and compare it with K_{eq}. Since the concentration units are moles per liter, we use K_c for K_{eq}. The value of K_c is known to be 0.19 at 532 °C when the chemical equation is written

$$N_2(g) + 3H_2(g) \rightleftharpoons 2NH_3(g)$$

Substitute the starting concentration of each substance to evaluate the expression for Q.

$$Q = \frac{[NH_3]^2}{[N_2][H_2]^3} = \frac{(0.10)^2}{(0.30)(0.20)^3} = 4.2$$

When Q is greater than K_{eq}, the chemical system changes to form more of the reactants. As additional N_2 and H_2 are produced, Q decreases until it is equal to K_{eq}.

Another way to compare Q to K_c is with a number line, shown in Figure 14.3.

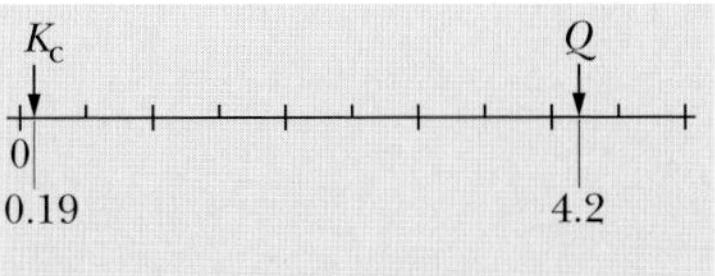

Figure 14.3 Number line representation of K_c and Q.

The symbols for Q and K_c are placed on a scale at positions related to their numerical values. In this example, Q is to the right of K_c, so the system will react to move Q left, toward K_c, by forming additional reactants. The direction in which Q moves on the number line is the direction in which the reaction proceeds.

Table 14.2 illustrates the same general approach and includes values of Q for several different mixtures of NO_2 and N_2O_4. The chemical equation is

$$2NO_2(g) \rightleftharpoons N_2O_4(g) \qquad K_c = 0.45 \text{ at } 135\ °C$$

Comparing Q with K_c enables us to predict the direction in which the system responds. Note that any equilibrium mixture must contain some of each species, so if any species is missing, as on lines 1 and 5 of the table, then the reaction proceeds to form the missing substance.

The number line in Figure 14.4, on the next page, is a graphical presentation of the data in Table 14.2. The reaction proceeds to bring Q closer to K_c. The reaction mixtures in experiments 1 and 2 will proceed to the right, to form more N_2O_4. The reaction mixtures in experiments 4 and 5 will proceed to the left, forming more NO_2. The mixture in experiment 3 is at equilibrium.

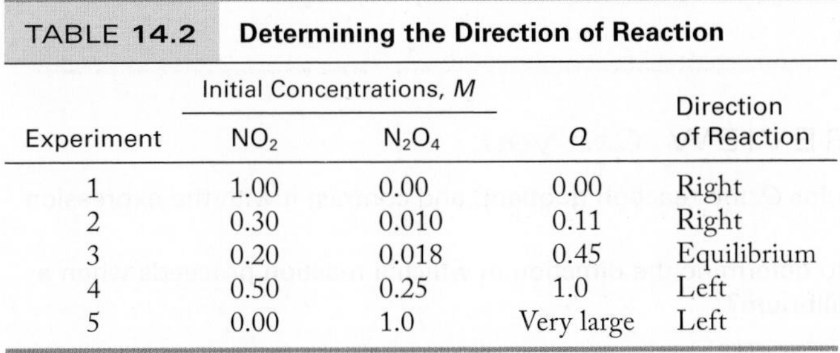

TABLE 14.2 **Determining the Direction of Reaction**

Experiment	Initial Concentrations, M NO_2	N_2O_4	Q	Direction of Reaction
1	1.00	0.00	0.00	Right
2	0.30	0.010	0.11	Right
3	0.20	0.018	0.45	Equilibrium
4	0.50	0.25	1.0	Left
5	0.00	1.0	Very large	Left

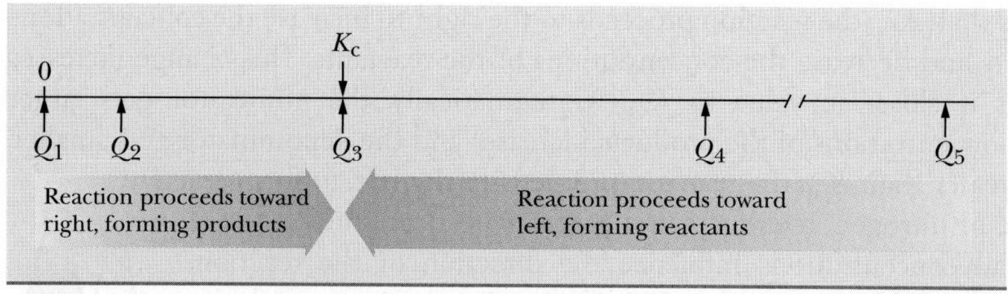

Figure 14.4 Number line representation of data presented in Table 14.2.

EXAMPLE 14.4 Determining the Direction of Reaction

A scientist mixes 0.50 mol NO_2 with 0.30 mol N_2O_4 in a 2.0-L flask at 418 K. At this temperature, K_c is 0.32 for the reaction of 2 mol NO_2 to form 1 mol N_2O_4. Does a reaction occur to form more NO_2 or more N_2O_4, or is the system at equilibrium?

Strategy Write the chemical equation and the expression for Q. Evaluate Q from the given concentrations and compare with K_{eq} to determine the direction of reaction.

Solution

The chemical equation and expression for Q is

$$2NO_2(g) \rightleftharpoons N_2O_4(g)$$

$$Q = \frac{[N_2O_4]}{[NO_2]^2}$$

Next, calculate the initial *concentration* of each species.

$$[NO_2] = 0.50 \text{ mol}/2.0 \text{ L} = 0.25\ M$$

$$[N_2O_4] = 0.30 \text{ mol}/2.0 \text{ L} = 0.15\ M$$

Last, calculate Q and compare it with K_c.

$$Q = \frac{0.15}{(0.25)^2} = 2.4$$

Because Q is *greater* than K_c (0.32), the reaction proceeds to the left, and more NO_2 forms.

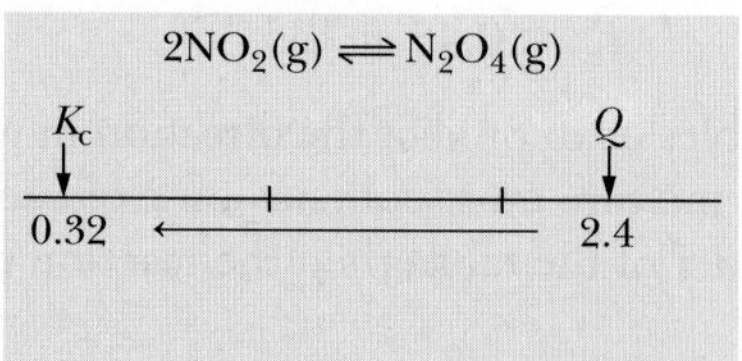

Understanding

A scientist mixes 0.24 mol NO_2 with 0.080 mol N_2O_4 in a 2.0-L flask at 418 K. Does a reaction occur to form more NO_2, more N_2O_4, or is the system at equilibrium?

Answer Q is 2.8, greater than K_c, so the reaction proceeds to the left, forming more NO_2.

OBJECTIVES REVIEW *Can you:*

- ☑ write the expression for Q, the reaction quotient, and contrast it with the expression for K_{eq}?
- ☑ compare Q and K_{eq} to determine the direction in which a reaction proceeds when a system is not at equilibrium?

14.3 Le Chatelier's Principle

OBJECTIVES

- ☐ Predict the response of an equilibrium system to changes in conditions by applying Le Chatelier's principle
- ☐ Determine how changes in temperature influence the equilibrium system

If a system at equilibrium is disturbed by changing the temperature or the concentration of reactants or products, the system will react in response to the change. This section discusses how changes in these factors influence the composition of the equilibrium mixture and the value of the equilibrium constant.

Le Chatelier's Principle

The composition of a system at equilibrium can change if the concentrations or the partial pressures of any of the reactants or products change. Henri Louis Le Chatelier (1850–1936) was first to describe qualitatively how these changes influence a chemical reaction at equilibrium. In 1884, he summarized his observations of chemical equilibria:

> *Every system in a stable chemical equilibrium submitted to the influence of an exterior force which tends to cause variation in either its temperature or its condensation (pressure, concentration, or number of molecules in the unit of volume) . . . can undergo only those interior modifications which, if they occur alone, would produce a change of temperature, or of concentration, of a sign contrary to that resulting from the exterior force.*[1]

Photo by Boyer/Roger Viollet/Getty Images

Henri Louis Le Chatelier (1850–1936).

Le Chatelier's principle can be restated in modern language: *Any change to a chemical reaction at equilibrium causes the reaction to proceed in the direction that reduces the effect of the change.*

Changes in factors such as concentration, pressure, and temperature cause a reaction to proceed in the direction that reduces the impact of the change. Consider the production of ammonia from the elements:

$$N_2(g) + 3H_2(g) \rightleftharpoons 2NH_3(g)$$

If hydrogen were added to an equilibrium mixture of nitrogen, hydrogen, and ammonia, then the hydrogen concentration would increase and the system is no longer at equilibrium. Le Chatelier's principle predicts that a reaction will occur to reduce the change. Because the hydrogen concentration increased, the system reacts to decrease the hydrogen (and nitrogen) concentration by forming additional ammonia. On the other hand, if hydrogen had been removed, the system would react to produce more hydrogen (and nitrogen). Figure 14.5, on the next page, illustrates these changes.

Chemists exploit Le Chatelier's principle to increase the yield of a reaction. For example, if we devise a way to remove ammonia as it forms, then the reaction will proceed until either the nitrogen or the hydrogen is exhausted. One way to remove ammonia is based on the observation that ammonia is easy to liquefy at modest pressures, but nitrogen and hydrogen are not. A reactor can be designed to operate at a moderate pressure and separate the liquid from the gases. As the liquid product (ammonia) is removed, the system responds by producing more. The process continues to produce ammonia until the nitrogen or hydrogen is consumed. Efficient production of ammonia is an important issue because ammonia is an important industrial product, with uses ranging from fertilizer to rocket fuels.

If the products of a reaction can be removed from the reaction mixture, the system responds by producing additional products.

Changes in Concentration or Partial Pressure

When a chemical system is at equilibrium, the reaction quotient Q is equal to K_{eq}. If the concentration of a species increases, then Q and K_{eq} are no longer equal, and a reaction proceeds in the direction that consumes the added substance. Scientists often speak

[1]H. M. Leicester and H. S. Klickstein. *A Source Book in Chemistry.* Cambridge, MA: Harvard University Press, 1963, p. 481.

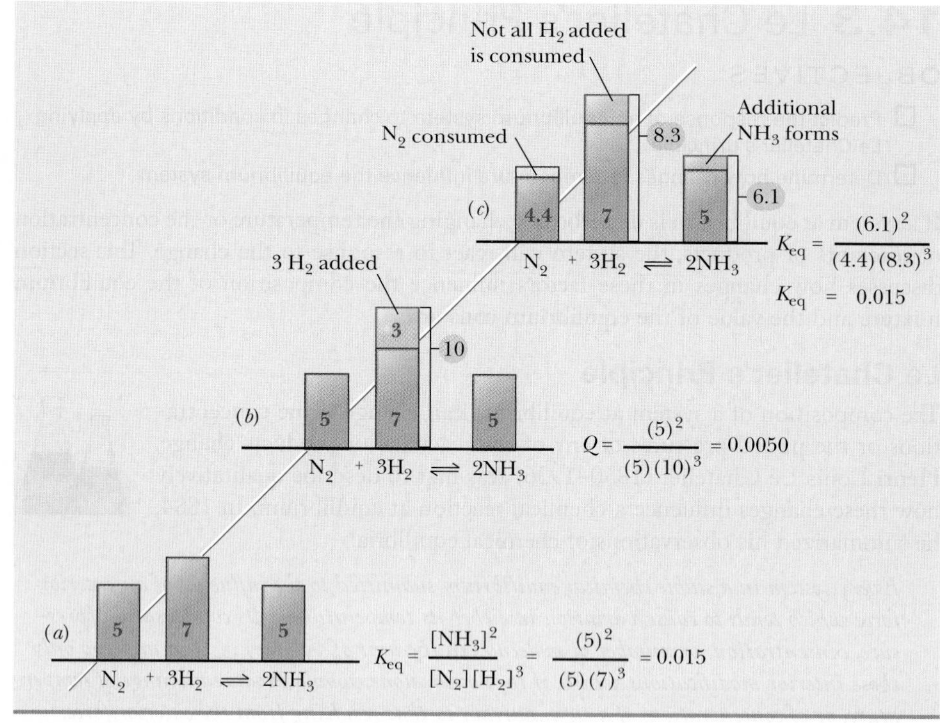

Figure 14.5 Le Chatelier's principle: The system responds to reduce the change. *(a)* Nitrogen, hydrogen, and ammonia are at equilibrium. *(b)* The hydrogen concentration is increased. *(c)* The system reacts to consume some of the added hydrogen. This reaction also decreases the concentration of nitrogen and increases the concentration of ammonia. The systems shown in parts *(a)* and *(c)* are at equilibrium; the system in *(b)* is not.

colloquially of the "shift in equilibrium" as concentrations change. The equilibrium constant certainly does not change, but the composition changes to produce additional products or reactants until equilibrium is restored. This section discusses how an equilibrium system responds to changes in concentration, pressure, and volume. Le Chatelier's principle is actually a second way to predict the direction of change because we already know we can calculate Q and compare with K_{eq}. Both methods produce the same results.

The Bhopal plant made phosgene, $COCl_2$, on-site. The reaction of carbon monoxide and chlorine to form phosgene, like any equilibrium system, shows the effects of changes in concentration.

$$Cl_2(g) + CO(g) \rightleftharpoons COCl_2(g)$$

Let us predict the direction of reaction that will occur if carbon monoxide is added.

The first way we might look at the problem is to use Le Chatelier's principle and state that the reaction will proceed to consume some of the added carbon monoxide. Because carbon monoxide is a reactant, the reaction will form some additional product in response to this change.

$$Cl_2(g) + CO(g) \Rightarrow COCl_2(g)$$

We use the symbol $\Rightarrow$ to indicate that the reaction proceeds to the right, to consume the added carbon monoxide, but the equilibrium constant does not change.

A second way to approach this problem is to see how increasing [CO] changes Q. First, write the equilibrium constant expression from the chemical equation.

$$K_{eq} = \frac{[COCl_2]}{[CO][Cl_2]}$$

K_{eq} is calculated from the concentrations of all species measured after the system reaches equilibrium. If carbon monoxide is added to the equilibrium mixture, then Q is smaller than K_{eq} and the reaction proceeds toward the right, to form more phosgene and consume some of the added CO.

The system responds to change by reducing the effect of the change.

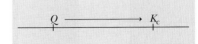

Example 14.5 demonstrates how a system responds when the equilibrium concentrations are disturbed.

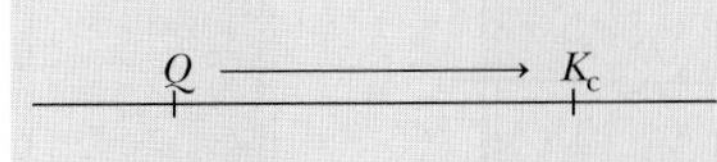

EXAMPLE 14.5 Predicting the Direction of Reaction When the System Is Disturbed

In which direction does the reaction proceed when sulfur dioxide is added to an equilibrium mixture of oxygen, sulfur dioxide, and sulfur trioxide?

$$SO_2(g) + \frac{1}{2}O_2(g) \rightleftharpoons SO_3(g)$$

Strategy Le Chatelier's principle predicts that the reaction will change in the direction that minimizes the change.

Solution

If the concentration of SO_2 increases, then the reaction will proceed in the direction that decreases the concentration of the added substance. SO_3 will form as SO_2 is consumed.

$$SO_2(g) + \tfrac{1}{2}O_2(g) \Longrightarrow SO_3(g)$$

An alternative to applying Le Chatelier's principle involves comparing Q and K_{eq}. The equilibrium constant expression is

$$K_c = \frac{[SO_3]}{[SO_2][O_2]^{1/2}}$$

If SO_2 is added to the system, Q becomes smaller than K_c, and the reaction proceeds to the right.

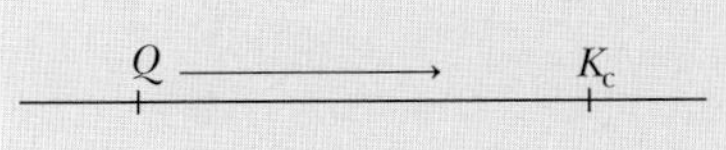

This chemical system is quite important. Sulfur is a common impurity in many fossil fuels, particularly coal. When these fuels are burned, they produce sulfur dioxide. Coals are classified as "low-sulfur" if the sulfur content is 0.6% to 1%. High-sulfur coals contain up to 4% sulfur. In the atmosphere, some of the SO_2 forms SO_3 and ultimately H_2SO_4, an important component of acid rain.

Understanding

In which direction does the reaction proceed when oxygen is removed from an equilibrium mixture of oxygen, sulfur dioxide, and sulfur trioxide?

$$SO_2(g) + \frac{1}{2}O_2(g) \rightleftharpoons SO_3(g)$$

Answer To the left, to produce more O_2 and SO_2.

Corbis/Photolibrary

Coal. Coal is a widely used fuel in the power generation industry. The United States has a 300-year supply of coal, although the supply of low-sulfur coal is much smaller.

Sometimes an inert or nonreactive material is added to a reaction container. As long as these other substances do not react or affect the partial pressures of the reactants or products, *the pressures of materials other than the reactants or products have no effect on the equilibrium.* The SO_3-SO_2 equilibrium can be studied in the presence or absence of other gases, and the results are the same as long as the other gases do not participate in the reaction.

Changes in the partial pressures of gases have the same effects as changes in concentration, because pressure is just another measure of concentration. A change

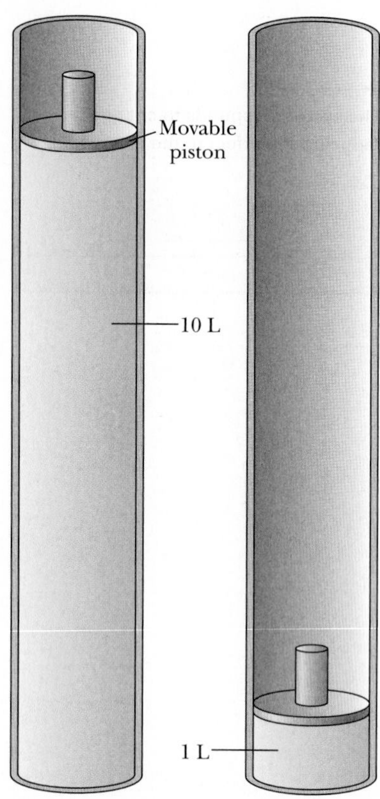

The volume decreases from 10 to 1 L when the external pressure increases.

in the partial pressures of the reacting species can be achieved either by adding (or removing) some of the reactants or products, or by changing the volume of the reaction vessel. The next example demonstrates the effect of changing the volume of the container.

EXAMPLE 14.6 Response of an Equilibrium System to Changes in Volume

Some PCl_5 is placed in a 10.0-L reaction cylinder. The temperature is increased to 500 K, and the following reaction reaches equilibrium.

$$PCl_5(g) \rightleftharpoons PCl_3(g) + Cl_2(g)$$

In which direction does the system react if the external pressure increases, causing the volume to decrease to 1.0 L whereas the temperature is kept constant at 500 K?

Strategy Two ways exist to determine the direction of the reaction. The first was is to calculate Q and compare K_{eq}, but we will use Le Chatelier's principle instead: The system will respond to minimize the effect of the change. The effect of the change will be evaluated by looking at how changing the volume changes the concentrations.

Solution

When the volume is reduced, the concentrations (partial pressures) of the species increase because the same numbers of moles are now in a smaller volume. The system responds to reduce the effect of this change by decreasing the number of moles of gases. We look at the chemical equation and see that the reactant side of the equation has 1 mol gas (PCl_5), but the product side has 2 mol of gases PCl_3 and Cl_2.

$$PCl_5(g) \rightleftharpoons PCl_3(g) + Cl_2(g)$$

1 mol gas on reactant side $\rightleftharpoons$ **2** mol gas on product side

Increasing the external pressure, which decreases the volume of the system, favors the direction that reduces the number of moles of gas. In this case, decreasing the volume causes the reaction to form more reactants.

$$PCl_5(g) \rightleftharpoons PCl_3(g) + Cl_2(g)$$

Understanding

Ammonia is formed by the reaction of nitrogen and hydrogen, all in the gas phase ($N_2 + 3H_2 \rightleftharpoons 2NH_3$). After the system reaches equilibrium, the volume of the container is decreased. In which direction does the reaction proceed?

Answer The system produces additional NH_3.

Chemists predict the response of a system to changes in volume or pressure by examining the numbers of moles of gas on the reactant and product sides of a chemical equation, because the volume occupied by liquids or solids is generally negligible compared with that occupied by gases. We can show the volume of the liquid is negligible compared to the gas by examining liquid and gaseous water at 100 °C and 1 atm: The volume of 1 mol liquid water is 19 mL, whereas 1 mol water vapor occupies about 30,000 mL—the volume of the gas is about 1600 times greater than that of the liquid. If a reaction has the same number of moles of gas on both sides, then changes in volume or pressure do not cause any net reaction. We define Δn as the change in the number of moles of gases (number of moles of product gases − number of moles of reactant gases) and can arrive at some qualitative conclusions, shown in Table 14.3.

TABLE 14.3 Relationship between Change in Number of Moles of Gases in a Reaction and Response of the System to Changes in Volume and Pressure

Example*	Δn	Decrease in Volume (or increase in external pressure) Favors Formation of	Increase in Volume (or decrease in external pressure) Favors Formation of
$CaO(s) + 3C(s) \rightleftharpoons CaC_2(s) + CO(g)$ $SO_3(g) \rightleftharpoons SO_2(g) + \frac{1}{2}O_2(g)$	+1 +0.5	Reactants	Products
$CO_2(g) + NaOH(s) \rightleftharpoons NaHCO_3(s)$ $2H_2(g) + O_2(g) \rightleftharpoons 2H_2O(\ell)$ $CO(g) + Cl_2(g) \rightleftharpoons COCl_2(g)$	−1 −3 −1	Products	Reactants
$SO_2(g) + NO_2(g) \rightleftharpoons SO_3(g) + NO(g)$ $H_2(g) + I_2(g) \rightleftharpoons 2HI(g)$	0 0	No effect	No effect

*Gas-phase reactants or products shown in blue type.

Changes in Temperature

Le Chatelier's principle also predicts how changing the temperature affects an equilibrium system. Heat is a "product" in an exothermic reaction, so adding heat causes the reaction to proceed to the left to consume the added heat; additional reactants form, and product is consumed. Heating an endothermic reaction causes the system to form additional products. *Changing the temperature of a reaction changes the value of K_{eq} in the direction predicted by Le Chatelier's principle.*

The influence of temperature can be seen by studying the formation of sulfur trioxide:

$$2SO_2(g) + O_2(g) \rightleftharpoons 2SO_3(g) \qquad \Delta H = -198 \text{ kJ}$$

When this reaction is studied at laboratory temperatures, this exothermic reaction proceeds toward the formation of SO_3. At high temperatures, such as those found in a furnace, the equilibrium constant becomes much less than 1, and sulfur trioxide decomposes to sulfur dioxide and oxygen.

Increasing the temperature of an exothermic reaction decreases K_{eq}, so more reactants form; increasing the temperature of an endothermic reaction increases K_{eq}, so more products form.

These results are consistent with that of Le Chatelier's principle. The formation of sulfur trioxide is an exothermic reaction; when heated, the system forms additional reactants as a reaction occurs to consume the added heat.

The numerical value of K_{eq} changes with temperature, so it is important to specify the temperature when describing a system at equilibrium. The accompanying figure shows the effect of temperature on the equilibrium constant for

$$2SO_2(g) + O_2(g) \rightleftharpoons 2SO_3(g)$$

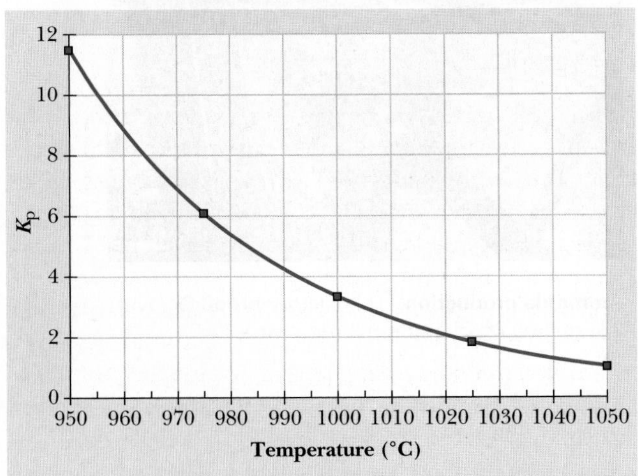

Influence of temperature on equilibrium constant. Increasing the temperature decreases K_p for the exothermic reaction: $2SO_2(g) + O_2(g) \rightleftharpoons 2SO_3(g)$.

PRACTICE OF CHEMISTRY
The Haber Process for the Production of Ammonia

Natural fertilizers such as manure and ground bones have been used since ancient times. Scientific study from the 19th century established that three elements are needed for plant growth: potassium (K), phosphorus (P), and nitrogen (N). Farmers could generally find minerals that contained potassium and phosphorus locally, but not nitrogen. One source, Chile saltpeter ($NaNO_3$), accounted for more than 60% of the world's supply of nitrogen fertilizer for most of the 19th century.

As the world population increased, so did the use of fertilizers. Because most plants cannot use atmospheric nitrogen directly, scientists searched for a chemical method to convert atmospheric N_2 to usable nitrogen-containing compounds. Three methods were investigated in the 20th century, but only one, the Haber process, was practical. The Haber process is still in use.

$$3H_2 + N_2 \rightleftharpoons 2NH_3 \qquad \Delta H° = -92 \text{ kJ/mol}$$

According to Le Chatelier's principle, increasing the pressure causes the equilibrium to favor additional ammonia because there are more gas molecules on the left side of the equation. Unfortunately, the synthesis of ammonia is slow, so the reaction needs to be heated to increase its speed (see Chapter 13). Heating the reaction indeed speeds it up, but because the reaction is exothermic, increasing the temperature will decrease the equilibrium concentration of ammonia.

Temperature (°C)	K_{eq}
25	6.1×10^5
250	6.9×10^{-2}
300	1.1×10^{-2}
400	6.1×10^{-4}
500	7.2×10^{-5}

The ammonia synthesis is not unique—many reactions require detailed study to determine the conditions that will safely produce the maximum amount of product with the minimum costs of materials and energy.

The equilibrium constant has been measured as a function of temperature and varies from 6.1×10^5 near room temperature to 7.2×10^{-5} at 500 °C.

In 1909, German chemist Fritz Haber developed a synthesis that used a temperature of 500 °C and high pressure (about 250 atm). He used a porous iron catalyst to produce ammonia, with a yield of approximately 15%. Scientists now know that even higher pressures (750 atm) give nearly complete product at 200 °C, but even modern ammonia plants use conditions quite similar to Haber's—building a plant to safely withstand 750 atm is just not cost-effective with current building materials.

In modern industrial production, the reaction never reaches equilibrium because the gases leaving the reactor are cooled, ammonia liquefies and is removed, and the reaction shifts to produce more ammonia in accordance with Le Chatelier's principle. The unreacted hydrogen and nitrogen are recycled.

Ammonia is used for fertilizer either directly or as nitrates, in industrial synthesis of compounds such as nylon, in metallurgy, and in synthesizing pharmaceuticals. One of its most important uses during Haber's time was for the production of explosives, many of which are organic nitrates such as trinitrotoluene and nitroglycerin. ■

Doug Martin/Photo Researchers, Inc.

Ammonia fertilizer. Ammonia can be applied directly into the ground as fertilizer.

Bloomberg News/Landov

Ammonia production. This factory produces over 350 million kg of ammonia per year.

Example 14.7 illustrates how to predict the direction in which the equilibrium constant changes with a change in temperature (quantitative calculations are deferred until Chapter 17).

EXAMPLE 14.7 Influence of Temperature Changes on Equilibria

How does an increase in temperature influence each of the following equilibria?

(a) $H_2(g) + I_2(g) \rightleftharpoons 2HI(g)$ $\Delta H = +52$ kJ
(b) $N_2(g) + 3H_2(g) \rightleftharpoons 2NH_3(g)$ $\Delta H = -92$ kJ

Strategy The reaction will shift in the direction that minimizes the change. Heat can be considered a product in an exothermic reaction and a reactant in an endothermic reaction.

Solution

(a) The reaction is *endothermic;* because heat is a reactant, increasing the temperature shifts the equilibrium toward products. The equilibrium constant increases as the temperature is increased so more products form.
(b) The reaction is *exothermic;* because heat is a product, increasing the temperature shifts the equilibrium toward reactants. The equilibrium constant decreases with increasing temperature so more reactants form.

Understanding

ΔH is −108 kJ for the formation of phosgene at the normal reaction temperature in Bhopal.

$$CO(g) + Cl_2(g) \rightleftharpoons COCl_2(g)$$

In which direction does the system react if temperature is *increased?*

Answer The reaction shifts to the left, forming more CO and Cl_2. Because higher temperature favors the left (reactant) side, one of the consequences of increased temperature is that the pressure builds up, because there are two moles of gases on the left and only one on the right.

OBJECTIVES REVIEW *Can you:*

- ☑ predict the response of an equilibrium system to changes in conditions by applying Le Chatelier's principle?
- ☑ determine how changes in temperature influence the equilibrium system?

14.4 Equilibrium Calculations

OBJECTIVES

- ☐ Use a systematic method, the iCe table, to solve chemical equilibria.
- ☐ Calculate equilibrium constants from experimental data and stoichiometric relationships.
- ☐ Calculate the equilibrium concentrations of species in a chemical reaction.

We found that we can determine the direction in which a reaction proceeds by calculating Q from the experimental data and comparing it with K_{eq}, but this is only one use for the equilibrium constant. Once the value of K_{eq} is known, we can use it to find the composition of any mixture of reactants and products at equilibrium. This section presents a systematic approach to help solve equilibrium problems, starting with the determination of the equilibrium constant from experimental data.

In general, equilibrium problems can be divided into two general types. In one type, the concentrations of the species are known, and the value for the equilibrium constant is determined; in the other, the value for the equilibrium constant is known, and the concentrations of the species are determined. The same general strategy works for both types of problems. We use the following five-step procedure:

1. Write the balanced chemical equation.
2. Fill in a table, which we call the iCe table, with the concentrations of the various species.
3. Write the algebraic expression for the equilibrium constant.
4. Substitute the information from the iCe table into the algebraic expression.
5. Solve the expression for the unknown quantity (or quantities).

This approach is used for essentially all equilibrium problems, as demonstrated by the next several examples.

Determining the Equilibrium Constant from Experimental Data

Although chemists use several ways to determine K_{eq}, the most fundamental is to measure the concentrations of the substances in a system that is at equilibrium, a process that was illustrated in Section 14.1. Frequently, the concentration of one species is measured, and the concentrations of the others are determined from stoichiometric relationships. Many chemists believe that equilibrium problems are best solved by systematically constructing a table of concentrations. The table has a column for each of the species that appears in the reaction. It contains rows for the initial concentration (i), change caused by the reaction (C), and equilibrium concentration (e). We call it the iCe table. The upper case letter C serves to emphasize that the changes depend on the reaction stoichiometry.

The systematic approach simplifies equilibrium calculations.

The iCe table can be used to calculate the equilibrium constant from experimental measurements of concentration.

The following example illustrates the construction and use of the iCe table.

EXAMPLE 14.8 Determining the Equilibrium Constant

Exactly 0.00200 mol hydrogen iodide is placed in a 5.00-L flask, and the temperature is increased to 600 K. Some of the HI decomposes, forming hydrogen and the violet-colored iodine gas:

$$2HI(g) \rightleftharpoons H_2(g) + I_2(g)$$

After the system reaches equilibrium, the concentration of iodine is determined by measuring the absorption of radiation (a quantitative measurement of the intensity of the purple color) and is found to be 3.8×10^{-5} M. Calculate K_c for this system at 600 K.

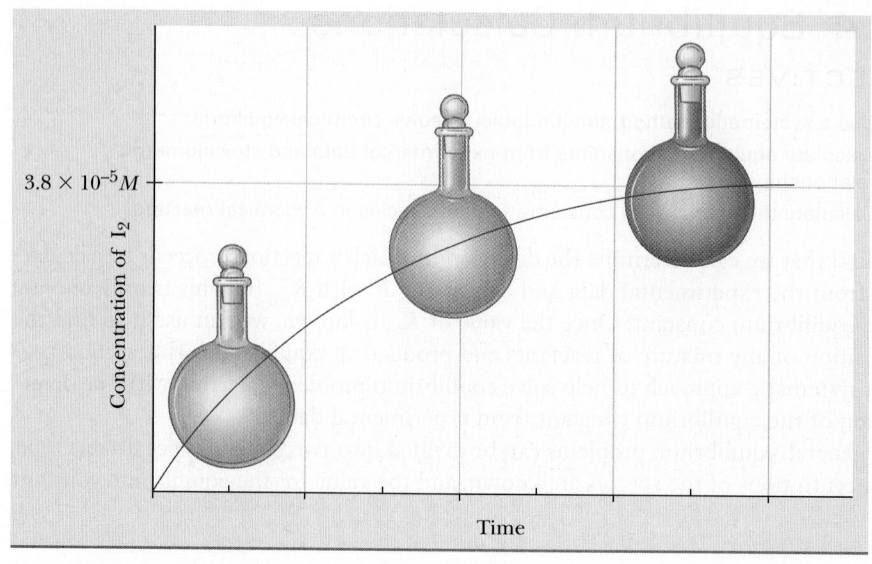

Concentration changes with time. Hydrogen iodide decomposes forming hydrogen and the violet colored iodine. See Example 14.8.

Strategy Write the balanced equation, the iCe table, the algebraic expression for K_c, substitute values from the iCe table, and solve.

Solution

First, write the balanced equation.

$$2HI(g) \rightleftharpoons H_2(g) + I_2(g)$$

Next, construct the iCe table, filling in as much data as possible. *The iCe table requires molar concentrations,* so remember that the initial concentration of HI is 0.00200 mol/ 5.00 L = 4.00×10^{-4} *M*. The equilibrium concentration of I_2 was given as 3.8×10^{-5} *M*, so this information is entered in the equilibrium (e) row.

	$2HI(g)$	$\rightleftharpoons$	$H_2(g)$	+	$I_2(g)$
initial concentration, *M*	4.00×10^{-4}		0		0
Change in concentration, *M*					
equilibrium concentration, *M*					3.8×10^{-5}

1. Write the chemical equation.

Because the equilibrium concentration is the sum of the initial concentration and the change in concentration, calculate the change by difference. The process works when both the initial and equilibrium concentrations are known, as they are for I_2. The experimental results tell us that the concentration of I_2 increased from an initial concentration of zero to 3.8×10^{-5} *M*. From this information, we know that the change in concentration is $+3.8 \times 10^{-5}$ *M*, and we write this number in the change (C) row of the table.

The stoichiometry of the chemical reaction relates the change in concentration of one species to the others. One mole of hydrogen is created for each mole of iodine, so we can fill in the change in concentration of hydrogen, which will be the same as the change in iodine. Furthermore, the stoichiometry indicates that for each mole of I_2 formed, two moles of HI are consumed, so we can determine that the change in the concentration of HI is a decrease of $2 \times (3.8 \times 10^{-5})$, or 7.6×10^{-5} *M*. Knowledge of the initial concentration of HI, 4.00×10^{-4} *M*, and of the change, -7.6×10^{-5} *M*, allows us to determine that the equilibrium concentration of HI is $(4.00 \times 10^{-4}) - (7.6 \times 10^{-5})$ *M* = 3.24×10^{-4} *M*. Now our iCe table looks like this:

	$2HI(g)$	$\rightleftharpoons$	$H_2(g)$	+	$I_2(g)$
initial concentration, *M*	4.00×10^{-4}		0		0
Change in concentration, *M*	-7.6×10^{-5}		$+3.8 \times 10^{-5}$		$+3.8 \times 10^{-5}$
equilibrium concentration, *M*	3.24×10^{-4}		3.8×10^{-5}		3.8×10^{-5}

2. Fill in the iCe table.

Next, write the algebraic expression for the equilibrium constant.

$$K_c = \frac{[H_2][I_2]}{[HI]^2}$$

3. Write the expression for *K*.

Substitute the equilibrium concentrations into the algebraic expression.

$$K_c = \frac{[3.8 \times 10^{-5}][3.8 \times 10^{-5}]}{[3.24 \times 10^{-4}]^2}$$

4. Substitute.

$$K_c = 0.014$$

5. Solve.

Understanding

A researcher places 0.0400 mol sulfuryl chloride in a 4.00-L reactor. The temperature is increased to 100 °C, and some of the sulfuryl chloride decomposes to sulfur dioxide and chlorine.

$$SO_2Cl_2(g) \rightleftharpoons SO_2(g) + Cl_2(g)$$

At equilibrium, the concentration of chlorine is found to be 3.9×10^{-3} *M*. Calculate K_c for this reaction.

Answer $K_c = 2.5 \times 10^{-3}$

Calculating the Concentrations of Species in a System at Equilibrium

Chemists are frequently asked to determine the amount of product formed in an equilibrium reaction. Typically, the starting amounts of the reactants are known, and the equilibrium constant has been evaluated in previous experiments. The iCe table provides a systematic way to solve equilibrium problems—a template to help organize the thought process. Because the concentrations of the products are unknown, these concentrations are represented by variables. If we can express the equilibrium concentrations (fill in the equilibrium [e] row in the iCe table) in terms of one unknown quantity, then the problem can be reduced to an algebraic equation that can be solved.

The systematic approach allows us to apply the same methods to all equilibrium problems.

Remember that the C in iCe reminds us that changes involve reaction stoichiometry—the stoichiometric coefficients are needed to determine the changes in the concentrations of all substances. The values in the change (C) row must be in the same proportions as the coefficients in the chemical equation.

The next example illustrates how this approach provides a guide to determining the concentrations in an equilibrium system.

EXAMPLE 14.9 Calculating Equilibrium Concentrations

Calculate the equilibrium concentrations of hydrogen and iodine that result when 0.050 mol HI is sealed in a 2.00-L reaction vessel and heated to 700 °C. At this temperature, K_c is 2.2×10^{-2} for

$$2HI(g) \rightleftharpoons H_2(g) + I_2(g)$$

Strategy Write the chemical equation, iCe table, and algebraic expression for K_{eq}; then substitute numerical values, and solve.

Solution

First, write the chemical equation. Next, begin the iCe table by writing the initial concentrations on line i (initial concentration) of the table. The initial concentration of HI is 0.050 mol/2.00 L = 0.025 *M*. The initial concentrations of H_2 and I_2 are zero. The change in concentration is unknown; we will define $+y$ as the change in concentration of H_2, with the plus sign indicating that the concentration of H_2 is increasing. The coefficients of the equation tell us that when *one* mole of H_2 forms, *one* mole of I_2 forms and *two* moles of HI are consumed. Thus, the change in concentration of I_2 is $+y$, whereas the change in concentration of HI is $-2y$ (note the minus sign indicates that the concentration of HI is decreasing). Write these changes in concentration in the change (C) row. Calculate the equilibrium concentration of each substance by summing the initial concentration and the change.

1. Write the chemical equation.

2. Fill in the iCe table.

	2HI(g)	⇌	H_2(g)	+	I_2(g)
initial concentration, *M*	0.025		0.00		0.00
Change in concentration, *M*	$-2y$		$+y$		$+y$
equilibrium concentration, *M*	$0.025 - 2y$		y		y

3. Write the expression for *K*.

Write the algebraic expression for the equilibrium constant.

$$K_c = \frac{[H_2][I_2]}{[HI]^2}$$

Substitute the numerical value for the equilibrium constant and the concentrations from line e (equilibrium).

$$2.2 \times 10^{-2} = \frac{(y)(y)}{(0.025 - 2y)^2} = \left(\frac{y}{(0.025 - 2y)}\right)^2$$

4. Substitute.

Last, solve the equation.

$$2.2 \times 10^{-2} = \frac{y^2}{(0.025 - 2y)^2} = \left(\frac{y}{(0.025 - 2y)}\right)^2$$

You can solve this particular equation by taking the square roots of both sides,

$$\frac{y}{0.025 - 2y} = 0.148$$

5. Solve.

rearranging,

$$y = 0.148 \times (0.025 - 2y) = 0.0037 - 0.296y$$

combining terms,

$$1.296y = 0.0037$$

and solving for y.

$$y = 2.9 \times 10^{-3}$$

We are not at our final answer; we have solved only for y. Now we need to go back to the e line in the iCe table, substitute 2.8×10^{-3} for y, and determine the equilibrium concentrations of all species. The equilibrium concentrations of the species are

$$[HI] = 0.025 - 2y = 1.9 \times 10^{-2}\ M$$

$$[H_2] = [I_2] = y = 2.9 \times 10^{-3}\ M$$

Not all equilibrium systems can be solved by this particular mathematical method (taking the square root of both sides). Other methods are shown later.

It is important to check your work at this stage. You can check the answer by substituting the results of the calculation into the equilibrium constant expression to determine whether they reproduce the equilibrium constant.

$$K_c = \frac{[H_2][I_2]}{[HI]^2} = \frac{(2.9 \times 10^{-3})(2.9 \times 10^{-3})}{(1.9 \times 10^{-2})^2} = 2.3 \times 10^{-2}$$

If the concentrations had not produced the given value for the equilibrium constant, within a reasonable allowance for rounding errors we would know that an error had been made.

Understanding

At very high temperatures, K_c is 0.200 for

$$N_2(g) + O_2(g) \rightleftharpoons 2NO(g)$$

Calculate the equilibrium concentrations of all species if the reaction starts with 5.00×10^{-4} mol NO in a 10.0-L container.

Answer $[N_2] = [O_2] = 2.04 \times 10^{-5}\ M$; $[NO] = 9.1 \times 10^{-6}\ M$

Sometimes additional information is needed to calculate the initial concentrations. Consider the reaction that occurs when a 1.00-L container of 0.600 M PCl_3 is connected to a 2.00-L container of 0.150 M Cl_2. The reaction is

$$PCl_3(g) + Cl_2(g) \rightleftharpoons PCl_5(g)$$

The reaction vessel is the combination of the two containers; it has a volume of 3.00 L. The concentration of $PCl_3(g)$ in the reaction vessel is *not* 0.600 M, because the volume changed from 1.00 to 3.00 L. We must calculate the initial concentrations that appear on the first row of the iCe table. This type of calculation is similar to the dilution problems in Section 4.2.

The concentration of PCl_3 is equal to the number of moles divided by the total volume. The number of moles is calculated from the initial volume and concentration—1.00 L of 0.600 M PCl_3, in this case. The total volume is 3.00 L, the volume of the interconnected 1.00- and 2.00-L containers.

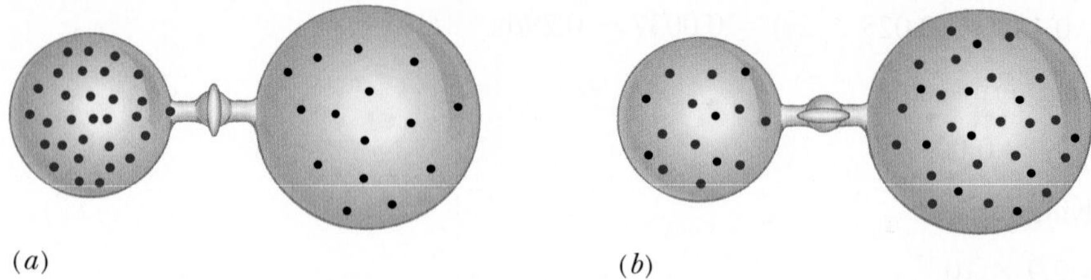

Changes in volume. When the valve is opened to mix the reactants, the reaction volume is the sum of the volumes in the original container. *(a)* System before mixing: 0.600 M PCl_3 in the left vessel and 0.150 M Cl_2 in the right vessel. *(b)* After mixing, the concentration of each gas has decreased because the volume increased.

$$\text{Moles of } PCl_3 = 1.00\ \cancel{L} \times \left(\frac{0.600 \text{ mol } PCl_3}{\cancel{L}}\right) = 0.600 \text{ mol } PCl_3$$

$$\text{Concentration } PCl_3 = \frac{0.600 \text{ mol } PCl_3}{3.00 \text{ L total volume}} = 0.200\ M$$

The concentration of the Cl_2 is calculated in a similar fashion. The steps are "chained" together in the following equation:

$$[Cl_2] = \frac{2.00\ \cancel{L} \times \left(\frac{0.150 \text{ mol } Cl_2}{L}\right)}{3.00\ \cancel{L} \text{ total volume}} = 0.100\ M$$

The next example continues this calculation, but the mathematical method used to solve the expression for the equilibrium constant differs from the one used in Example 14.9.

EXAMPLE 14.10 Calculating Equilibrium Concentrations

Phosphorus trichloride and chlorine react to form phosphorus pentachloride. At 544 K, K_c is 1.60 for

$$PCl_3(g) + Cl_2(g) \rightleftharpoons PCl_5(g)$$

Calculate the concentration of chlorine when 1.00 L of 0.600 M PCl_3 is added to 2.00 L of 0.150 M Cl_2 and allowed to reach equilibrium at 544 K.

Strategy Write the chemical equation, iCe table, algebraic and numerical expression for K_{eq}, and solve.

Solution

The initial concentrations of PCl_3 and Cl_2 were calculated above as 0.200 M PCl_3 and 0.100 M Cl_2. Notice that it is not necessary for PCl_3 and Cl_2 to be present in stoichiometric amounts.

	PCl_3	+	Cl_2	$\rightleftharpoons$	PCl_5
initial concentration, M	0.200		0.100		0.00
Change in concentration, M	$-y$		$-y$		$+y$
equilibrium concentration, M	$0.200 - y$		$0.100 - y$		y

1. Write the chemical equation.

2. Fill in the iCe table.

Write the equilibrium constant expression, then substitute the equilibrium concentrations found on row e (equilibrium) of the table.

$$K_c = \frac{[PCl_5]}{[PCl_3][Cl_2]}$$

3. Write the expression for K.

$$1.60 = \frac{y}{(0.200 - y)(0.100 - y)}$$

4. Substitute.

Rearrange and combine terms.

$$1.60y^2 - 1.48y + 3.20 \times 10^{-2} = 0$$

5. Solve.

This equation is a *quadratic equation* (one that contains a y^2 term) and can be solved with the *quadratic formula.* Any quadratic equation in the form

$$ay^2 + by + c = 0$$

has the roots

$$y = \frac{-b \pm \sqrt{b^2 - 4ac}}{2a}$$

In this problem,

$$y = \frac{+1.48 \pm \sqrt{1.48^2 - 4(1.60)(3.2 \times 10^{-2})}}{2(1.60)}$$

$$= \frac{+1.48 \pm \sqrt{1.986}}{3.20}$$

$$= \frac{+1.48 \pm 1.409}{3.20}$$

$$y = 0.903 \text{ or } 0.0222$$

Mathematically, a quadratic equation has two roots, but a chemical system has only one physically reasonable answer. The last line of the iCe table indicates that $[Cl_2] = 0.100 - y$. In this case, we reject $y = 0.903$, because it predicts a negative concentration for chlorine, which is a physical impossibility. We accept $y = 0.0222$.

Check the answer by calculating the concentrations of all the species and substituting these back into the expression for the equilibrium constant to be sure that the results are consistent.

$$[PCl_5] = y = 0.0222\ M$$

$$[Cl_2] = 0.100 - y = 0.100 - 0.0222 = 0.078\ M$$

$$[PCl_3] = 0.200 - y = 0.200 - 0.0222 = 0.178\ M$$

$$K_c = \frac{[PCl_5]}{[PCl_3][Cl_2]} = \frac{0.0222}{(0.178)(0.078)} = 1.60$$

This result is in agreement with the original value of 1.60, so we can be confident that the problem has been solved correctly.

Understanding

K_c is 0.12 at 1000 K for the dissociation of phosgene:

$$COCl_2(g) \rightleftharpoons CO(g) + Cl_2(g)$$

Calculate the equilibrium concentrations of all species if 2.00 mol $COCl_2$ is placed in a 5.00-L reactor at 1000 K.

Answer $[CO] = [Cl_2] = 0.17\ M$; $[COCl_2] = 0.23\ M$

Equilibrium problems occasionally produce more complicated mathematical equations, but even the most complex problems can be solved. Using the systematic approach and the iCe table helps reduce a difficult problem to smaller, more manageable pieces. The key process is writing the algebraic equation needed to solve for the concentrations of the species at equilibrium. The next problem emphasizes this process.

EXAMPLE 14.11 Calculating Equilibrium Concentrations

A researcher studies an industrial process by placing 0.030 mol sulfuryl chloride, a powerful chemical oxidizer, in a 100-L reactor together with 2.0 mol SO_2 and 1.0 mol Cl_2 at 173 °C. At this temperature, K_p is 3.0 for

$$SO_2Cl_2(g) \rightleftharpoons SO_2(g) + Cl_2(g)$$

Write the iCe table and derive a polynomial algebraic expression needed to calculate the equilibrium concentrations of all species.

Strategy Write the chemical equation and iCe table, expressing concentrations in units of moles per liter. We will calculate Q and compare with K_c, which we will compute from K_p to determine the direction of reaction, which often simplifies writing the iCe table. We will use algebra to develop the polynomial equation.

Solution

Write the chemical equation and blank iCe table. The initial concentrations are

$$[SO_2Cl_2] = 0.030 \text{ mol}/100 \text{ L} = 0.00030\ M$$

$$[SO_2] = 2.0 \text{ mol}/100 \text{ L} = 0.020\ M$$

$$[Cl_2] = 1.0 \text{ mol}/100 \text{ L} = 0.010\ M$$

Next, calculate Q and compare it with K_c. Knowing the direction of reaction helps you write the iCe table.

$$Q = \frac{[SO_2][Cl_2]}{[SO_2Cl_2]} = \frac{(0.020)(0.010)}{0.00030} = 0.67$$

Because K_p is given, we need to convert to K_c by using Equation 14.2. The change in the number of moles of gases, Δn, is +1 for this reaction. Remember to express the temperature as 446 K rather than 173 °C.

$$K_c = \frac{K_p}{(RT)^{\Delta n}} = \frac{3.0}{(0.08206 \times 446)} = 0.082$$

Because Q is greater than K_c, the reaction proceeds to the left, forming more SO_2Cl_2. We define y as the change in the concentration of SO_2Cl_2; the stoichiometric coefficients of the chemical reaction tell us that when y mol SO_2Cl_2 is formed, y mol each of Cl_2 and SO_2 is consumed. Write these data in the change (C) row of the table, with the proper signs. Finally, calculate the equilibrium concentration by summing the initial concentration and the change in concentration for each column.

	SO_2Cl_2	$\rightleftharpoons$	SO_2	+	Cl_2
initial concentration, M	0.00030		0.020		0.010
Change in concentration, M	$+y$		$-y$		$-y$
equilibrium concentration, M	$0.00030 + y$		$0.020 - y$		$0.010 - y$

1. Write the chemical equation.

2. Fill in the iCe table.

Write the algebraic expression for the equilibrium constant.

$$K_c = \frac{[SO_2][Cl_2]}{[SO_2Cl_2]}$$

3. Write the expression for *K*.

Substitute the concentrations from row e (equilibrium) into the equation.

$$K_c = \frac{(0.020 - y)(0.010 - y)}{(0.00030 + y)} = 0.082$$

4. Substitute.

The expression for the equilibrium constant can be reduced to a polynomial expression.

$$K_c = \frac{y^2 - 0.030y + 2.0 \times 10^{-4}}{(0.00030 + y)} = 0.082$$

Gather like terms to write the polynomial expression.

$$y^2 - 0.112y + 1.75 \times 10^{-4} = 0$$

5. Solve.

If a numerical answer were required, we would solve this equation using the quadratic formula, and we would want to check our results. One simple check would be to compare the calculated concentrations with the original concentrations. Because Q was greater than K_c, we expect to find more SO_2Cl_2 and less SO_2 and Cl_2.

Understanding

Write the iCe table and the expression for the equilibrium constant needed to solve for the concentrations of the species when 0.010 M SO_2, 0.050 M O_2, and 0.0020 M SO_3 react. At 1009 K, K_c is 2.0 for

$$2SO_2(g) + O_2(g) \rightleftharpoons 2SO_3(g)$$

Answer Define y as the change in the O_2 concentration.

	$2SO_2$	+	O_2	$\rightleftharpoons$	$2SO_3$
initial concentration, M	0.010		0.050		0.0020
Change in concentration, M	$-2y$		$-y$		$+2y$
equilibrium concentration, M	$0.010 - 2y$		$0.050 - y$		$0.0020 + 2y$

$$K_c = \frac{[SO_3]^2}{[SO_2]^2[O_2]}$$

$$2.0 = \frac{(0.0020 + 2y)^2}{(0.010 - 2y)^2(0.050 - y)}$$

This expression reduces to a polynomial equation that includes y^3 terms, or a *cubic* equation—not all equilibrium problems generate quadratic equations. Appendix A describes some methods of solving cubic and other polynomial equations.

OBJECTIVES REVIEW *Can you:*

☑ use a systematic method, the iCe table, to solve chemical equilibria?
☑ calculate equilibrium constants from experimental data and stoichiometric relationships?
☑ calculate the equilibrium concentrations of species in a chemical reaction?

14.5 Heterogeneous Equilibria

OBJECTIVE

☐ Write equilibrium constant expressions for heterogeneous equilibria

Heterogeneous equilibrium systems, in which the substances are in more than one phase, are treated in much the same manner as the homogeneous equilibria already described. This section presents some of the concepts and calculations that are used to characterize heterogeneous equilibria. These types of equilibria are important to study because chemists generally strive to make the product in a different phase than the reactants. If two solutions are mixed and the product is in solution, separation and isolation is difficult, but if the product is a gas or a solid, separating it from reactants is much easier.

Expressing the Concentrations of Solids and Pure Liquids

We use the equilibrium constant and the law of mass action to calculate the equilibrium concentrations of the reactants and products at equilibrium. But the concentrations of some species, such as pure solids and liquids, never change. For example, 1 L sodium chloride (solid) weighs 2.2 kg; it contains 37 mol. The *concentration* of NaCl(s) is 37 mol/L. Further, 2 L NaCl(s) contains twice the amount but also twice the volume, and thus the *concentration* remains constant. Solids are excluded from equilibrium calculations because their concentrations are constant.

Solid sodium chloride. A liter of sodium chloride has a mass of 2.2 kg and contains 37 moles of NaCl.

The same argument holds for pure liquids—the concentration of a pure liquid is a constant, related to its density and molar mass, and does not vary during the course of a chemical reaction. The concentration of a pure solid or liquid is independent of the amount present.[2]

The concentrations of pure solids and liquids do not appear in the expression for the equilibrium constant.

An interesting heterogeneous equilibrium system results when calcium carbonate (limestone) is heated in a closed vessel to form calcium oxide (quicklime) and carbon

[2]In more exact treatments, concentrations are expressed in terms of *activities.* The activity of a substance is a dimensionless ratio of the concentration of the substance to its concentration in the standard state. The concentrations of solids and pure liquids do not change, so this ratio is 1, and the concentrations of these substances do not appear in the expression for the equilibrium constant. The standard state for a solute is the 1.0 *M* solution, so its activity is numerically equal to its molar concentration. The activity of a gas is the ratio of its partial pressure to the 1.0-atm standard-state pressure, so its activity is numerically equal to its partial pressure, expressed in atmospheres.

dioxide. This process is used on an industrial scale to make calcium oxide, which is used in a number of areas including metallurgy, waste treatment, and cement production.

$$CaCO_3(s) \rightleftharpoons CaO(s) + CO_2(g)$$

The equilibrium constant expression is

$$K_c' = \frac{[CaO][CO_2]}{[CaCO_3]}$$

K_c' is used for the equilibrium constant because, as discussed later, this equation is just an intermediate step. The equation can be rewritten as

$$K_c' = \frac{[CaO]}{[CaCO_3]} \times [CO_2]$$

The constant K_c' and the concentrations of $CaCO_3$ and CaO do not change, so we can combine these three constants into a new one, K_c.

$$K_c = [CO_2]$$

If the equilibrium constant is written in terms of pressure, then

$$K_p = P_{CO_2}$$

Example 14.12 shows expressions for the equilibrium constants of some heterogeneous systems.

EXAMPLE 14.12 Equilibrium Constant Expressions for Heterogeneous Equilibria

Write the expressions for both K_p and K_c for the following reactions:

(a) $NaOH(s) + CO_2(g) \rightleftharpoons NaHCO_3(s)$
(b) $KOH(s) + SO_3(g) \rightleftharpoons KHSO_4(s)$
(c) $NH_4Cl(s) \rightleftharpoons HCl(g) + NH_3(g)$

Strategy Remember that pure solids and liquids do not appear in the expression for the equilibrium constant.

Solution

(a) $K_p = 1/P_{CO_2}$ $\quad K_c = 1/[CO_2]$
(b) $K_p = 1/P_{SO_3}$ $\quad K_c = 1/[SO_3]$
(c) $K_p = P_{HCl}P_{NH_3}$ $\quad K_c = [HCl][NH_3]$

Understanding

Write the expressions for K_p and K_c for

$$C(s) + H_2O(g) \rightleftharpoons CO(g) + H_2(g)$$

Answer $K_p = \dfrac{P_{CO}P_{H_2}}{P_{H_2O}}$ $\quad K_c = \dfrac{[CO][H_2]}{[H_2O]}$

Equilibria of Gases with Solids and Liquids

Many equilibria, including the phase changes studied in Chapter 11, involve gases in equilibrium with liquids and solids. The evaporation of water is typical of such systems. The first step in studying these important equilibria is to write a chemical equation and the expression for the equilibrium constant.

$$H_2O(\ell) \rightleftharpoons H_2O(g)$$

$$K_p = P_{H_2O(g)}$$

The liquid water has a constant concentration, and it does not appear as a separate term in the expression for the equilibrium constant.

The vapor pressure does not depend on the amount of water, as long as some liquid water is present. Recall from Chapter 11 that the vaporization of water is an endothermic process; this information and Le Chatelier's principle tell us that as temperature increases, both the equilibrium constant and the vapor pressure increase. Experimental measurements show that the vapor pressure is relatively small at low temperatures (17.5 torr or 2.3×10^{-2} atm at 20 °C) but increases to 760 torr or 1.0 atm at 100 °C.

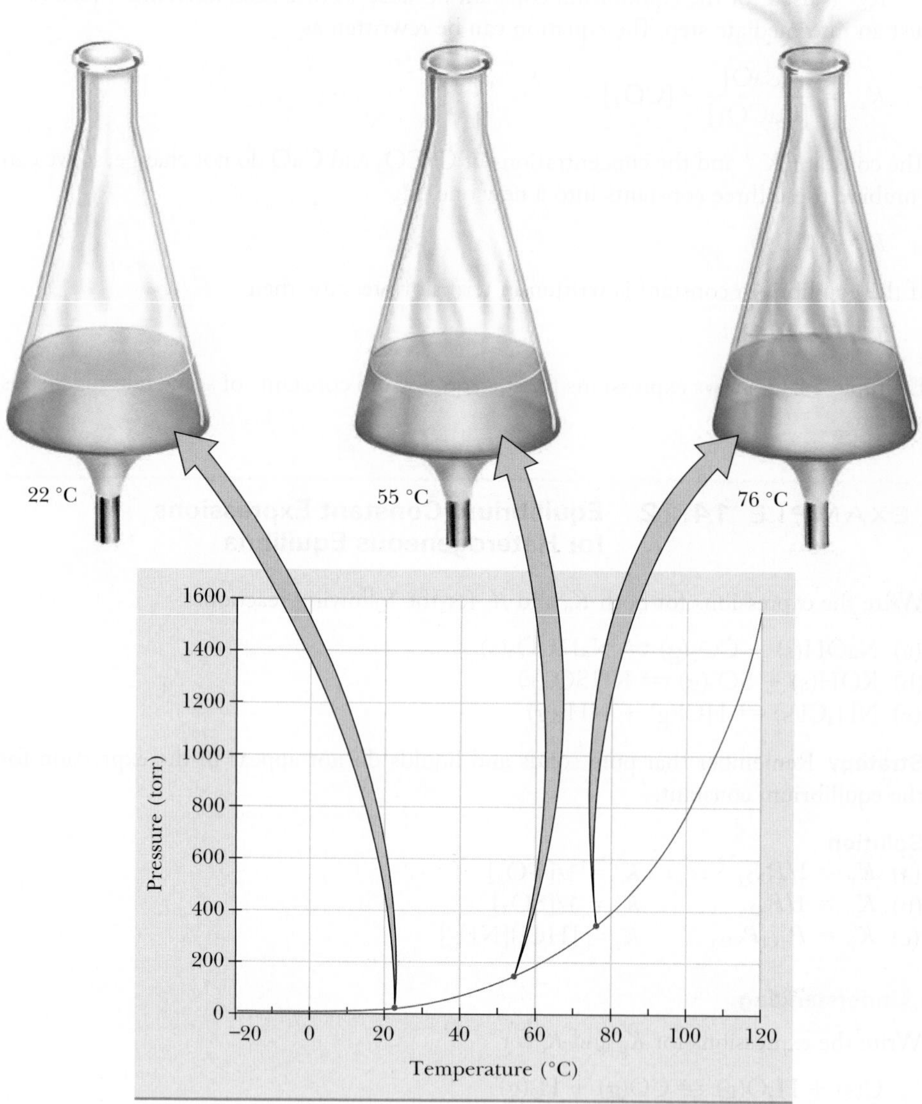

Heating water. The concentration (or pressure) of water vapor increases as the liquid is heated.

Gas-solid equilibria are similar. One example is the equilibrium established as calcium carbonate is heated to form calcium oxide and carbon dioxide:

$$CaCO_3(s) \rightleftharpoons CaO(s) + CO_2(g)$$

The equilibrium constant expression is

$$K_p = P_{CO_2}$$

Although influenced by temperature, the pressure of CO_2 is independent of the amount of $CaCO_3$ and CaO present, *as long as some of each solid is present.* Experimental

PRACTICE OF CHEMISTRY
Analyzing the Bhopal Accident

The chemists who investigated the reaction that occurred in Bhopal concluded that water reacted with the methyl isocyanate to form methylamine and carbon dioxide.

$$H_2O(\ell) + CH_3NCO(\ell) \rightleftharpoons NH_2CH_3(g) + CO_2(g)$$

Although reaction with water consumes the highly toxic methyl isocyanate, this reaction releases two moles of gas from two moles of liquid. As the reaction proceeded, the pressure built up inside the vessel. The reaction is exothermic and the extra heat also acted to increase the pressure of the gases, as predicted by the ideal gas law. Engineers always consider the possibility that a tank might explode in the design of the plant. They took the normal precaution, which is to include a pressure-relief valve. When the pressure reached a specified level, the valve opened to release some of the contents. The Bhopal design included a neutralization system and an incineration system, but tragically, both were off-line on the night of the accident.

The pressure-relief valve sent some of the contents of the vessel, including the highly toxic and volatile methyl isocyanate, into the town of Bhopal, with enormously tragic consequences. ∎

results show that the reaction is endothermic, so Le Chatelier's principle predicts that heating the system favors the formation of additional CO_2 and CaO.

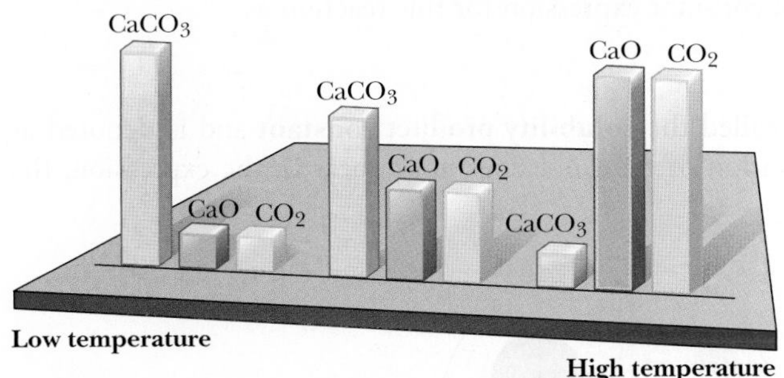

Heating calcium carbonate. As calcium carbonate is heated, more calcium oxide and carbon dioxide are produced.

OBJECTIVE REVIEW *Can you:*

☑ write equilibrium constant expressions for heterogeneous equilibria?

14.6 Solubility Equilibria

OBJECTIVES

☐ Write the expression for the solubility product constant

☐ Calculate K_{sp} from experimental data

☐ Calculate solubility of slightly soluble salts from K_{sp}

Many important chemical reactions result in the formation of a solid product from reactants in solution. These reactions vary from the synthesis of pharmaceuticals to the recovery of precious metals from waste streams. Chemists utilize precipitation reactions for practical reasons—it is easy to separate the solid product from the solution mixture.

This section presents **solubility equilibria,** reactions that involve the dissolution and formation of a solid from solution. Precipitation reactions are extensions of the heterogeneous equilibria discussed in the previous section; the equilibrium between a solute in a solution and its solid form is similar to the equilibrium between gases and solids.

One important precipitation reaction is the classic test used to determine whether silver ions are present in a solution. A chemist might

Stalagmites and stalactites. These natural geological features found in many caves form as dissolved minerals precipitate from solution.

monitor a silver recovery process by adding a few drops of dilute hydrochloric acid solution to the process solution. The formation of a white solid (Figure 14.6) indicates the presence of silver.

$$AgNO_3(aq) + HCl(aq) \rightarrow AgCl(s) + HNO_3(aq)$$

The nature of the reactants in solution is usually emphasized by the net ionic equation. Remember that soluble ionic compounds such as $AgNO_3$ and strong electrolytes such as HCl are present as ions in solution. There are no $AgNO_3$ particles in solution, but rather Ag^+ and NO_3^- ions. The equilibrium between the species in solution and the solid is represented by the net ionic equation.

$$Ag^+(aq) + Cl^-(aq) \rightleftharpoons AgCl(s)$$

The study of solubility equilibria allows us to predict many of the quantitative details of the reaction, including the amount of precipitate formed and the minimum concentration of chloride necessary to form a precipitate.

Solubility Product Constant

For historical reasons, we write an equilibrium that involves a precipitation reaction as the dissolving of a solid (dissociation into ions), as opposed to the formation of a solid.

$$AgCl(s) \rightleftharpoons Ag^+(aq) + Cl^-(aq)$$

We can write the equilibrium constant expression for this reaction as

$$K_{sp} = [Ag^+][Cl^-]$$

This equilibrium constant is called the **solubility product constant** and is denoted as K_{sp}. Notice that the concentration of the solid does not appear in the expression; the

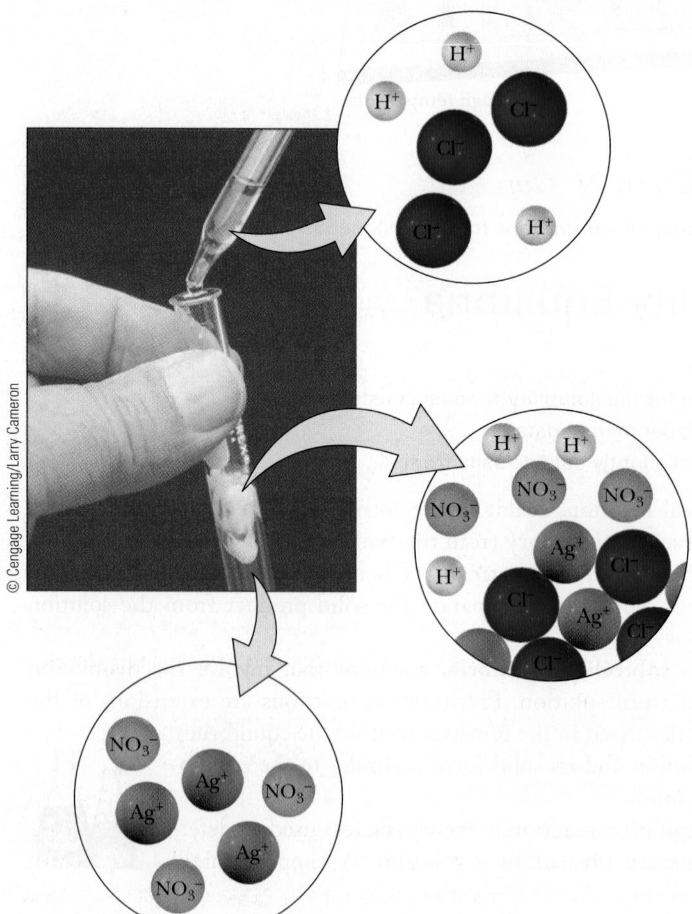

Figure 14.6 Formation of AgCl(s). A characteristic white precipitate forms when a few drops of a hydrochloric acid solution are added to a solution that contains some silver nitrate. The formation of a white precipitate does not prove conclusively that the solution contains any silver ions. Mercury and lead cations also form white chloride precipitates, so additional testing is necessary to confirm the presence of silver ions.

concentration of a solid is a constant, as discussed in Section 14.5. The solubility product constant provides a simple and effective model for predicting the concentrations of ions in equilibrium with a solid. It is not perfect because it neglects electrostatic attractions between cations and anions. These effects were mentioned in Chapter 12 in the discussion of the reasons for fractional values of the van't Hoff *i* parameter.

The solubility product is the equilibrium constant that describes a solid dissolving to produce ions in solution.

For example, when magnesium bromide, $MgBr_2$, dissolves in water, we expect just Mg^{2+} and Br^- ions. But experiments show that some $MgBr^+$ is also found in solution. These types of ions form as a result of an electrostatic attraction, called *ion pairing*, and occur mainly in concentrated solutions. For the most part, these effects are not discussed further in this textbook. Example 14.13 presents some examples of other solubility equilibria.

EXAMPLE 14.13 Expressions for the Solubility Product Constant

Write the chemical equation and expression for the solubility product constant for each of the following compounds:

(a) $Mg(OH)_2$
(b) $Ca_3(PO_4)_2$

Strategy Write the chemical equation for the solid dissolving; then write the expression for K_{sp}.

The solid does not appear in the solubility product expression.

Solution

(a) $Mg(OH)_2(s) \rightleftharpoons Mg^{2+}(aq) + 2OH^-(aq)$
$K_{sp} = [Mg^{2+}][OH^-]^2$
(b) $Ca_3(PO_4)_2(s) \rightleftharpoons 3Ca^{2+}(aq) + 2PO_4^{3-}(aq)$
$K_{sp} = [Ca^{2+}]^3[PO_4^{3-}]^2$

Understanding

Write the solubility product expression for iron(III) hydroxide.

Answer $K_{sp} = [Fe^{3+}][OH^-]^3$

The numerical value of the solubility product constant, like all equilibrium constants, is determined experimentally. A table of solubility product constants appears in Appendix F, and Table 14.4 repeats some commonly used solubility product constants.

The values of some solubility product constants are determined from experiments in which the solubilities of compounds are measured. Solubility is defined in Chapter 12; it is the concentration of solute that exists in equilibrium with an excess of that substance (typically measured in mol/L or g/100 mL). Chemists sometimes use experimental measurements of the concentrations of the dissolved species to determine the solubility product constant, but they do so with care because of effects such as ion pairing.

TABLE 14.4 Solubility Product Constants of Selected Compounds at 25 °C

Compound	Formula	K_{sp}
Barium iodate	$Ba(IO_3)_2$	4.0×10^{-9}
Barium sulfate	$BaSO_4$	1.1×10^{-10}
Calcium fluoride	CaF_2	3.5×10^{-11}
Calcium hydroxide	$Ca(OH)_2$	5.0×10^{-6}
Calcium phosphate	$Ca_3(PO_4)_2$	2.1×10^{-33}
Cerium iodate	$Ce(IO_3)_3$	3.2×10^{-10}
Copper(II) hydroxide	$Cu(OH)_2$	1.6×10^{-19}
Lanthanum hydroxide	$La(OH)_3$	1.0×10^{-19}
Lanthanum iodate	$La(IO_3)_3$	7.5×10^{-12}
Lead(II) chloride	$PbCl_2$	1.7×10^{-5}
Lead(II) iodate	$Pb(IO_3)_2$	3.7×10^{-13}
Magnesium hydroxide	$Mg(OH)_2$	5.6×10^{-12}
Mercury(II) iodide	HgI_2	2.9×10^{-29}
Silver chloride	$AgCl$	1.8×10^{-10}
Silver iodate	$AgIO_3$	3.2×10^{-8}
Silver iodide	AgI	8.5×10^{-17}
Silver sulfate	Ag_2SO_4	1.2×10^{-5}

EXAMPLE 14.14 Calculating the Solubility Product Constant

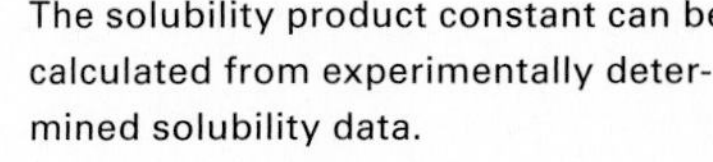

The solubility product constant can be calculated from experimentally determined solubility data.

When lead iodate, $Pb(IO_3)_2$, is added to water, a small amount dissolves. If measurements at 25 °C show that the Pb^{2+} concentration is 4.5×10^{-5} *M*, calculate the value of K_{sp} for $Pb(IO_3)_2$.

Strategy Solubility equilibria are solved by the same approach used for all equilibria. We will write the chemical equation, iCe table, and algebraic expression for the equilibrium constant; substitute numerical values from the iCe table; and solve.

1. Write the chemical equation.

Solution

The compound that dissolves dissociates into a cation and an anion, although occasionally more than one cation or anion form.

2. Fill in the iCe table.

	$Pb(IO_3)_2$	$\rightleftharpoons$	Pb^{2+}	+	$2IO_3^-$
initial concentration, M	Solid		0		0
Change in concentration, M	-4.5×10^{-5}		$+4.5 \times 10^{-5}$		$+2(4.5 \times 10^{-5})$
equilibrium concentration, M	Solid		4.5×10^{-5}		9.0×10^{-5}

3. Write the expression for K.

4. Substitute.

5. Solve.

Notice that the change in iodate concentration is twice the change in lead concentration, because the chemical equation tells us that two iodate ions form for each lead ion.

Next, write the expression for the solubility product constant and evaluate it.

$$K_{sp} = [Pb^{2+}][IO_3^-]^2$$

$$K_{sp} = (4.5 \times 10^{-5})(9.0 \times 10^{-5})^2$$

$$K_{sp} = 3.6 \times 10^{-13}$$

Note that the concentration of iodate ion is twice the concentration of the lead ion, *and* it is raised to the second power in the solubility product expression.

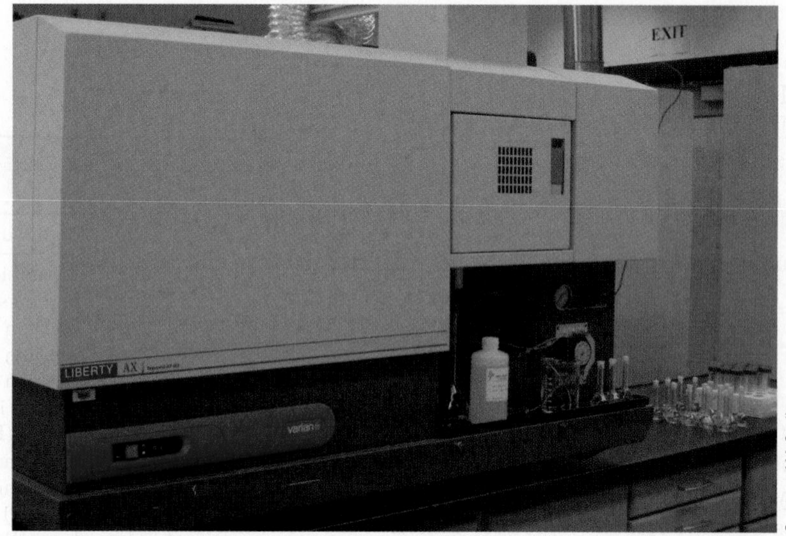

Measuring lead concentrations. Many laboratories use this type of instrument to measure the concentration of lead in water. The measurement of lead concentrations of 4.5×10^{-5} M, the solubility of lead iodate, is well within the reach of modern instruments such as the inductively coupled plasma emission spectrometer shown here. Chemists use instruments that can detect lead at the 10^{-16} M concentration level. These were originally special instruments, hand-built by individuals for their laboratories. Many of these laboratories measured levels of lead in air, dust, and soil to learn about the distribution and concentration of lead in the environment. The development of techniques that could provide accurate measurements of low lead concentrations resulted in a great deal of knowledge about the environmental chemistry of lead and its effects on humans.

Understanding

When mercury(II) iodide dissolves in water, the concentration of mercury(II) ions is found to be 2×10^{-10} M. Calculate the solubility product constant for HgI_2. The chemical equation is

$$HgI_2(s) \rightleftharpoons Hg^{2+}(aq) + 2I^-(aq)$$

Answer $K_{sp} = 3 \times 10^{-29}$

Calculating the solubility product constant from the solubility of the substance, rather than the concentration of one of the ions, is accomplished by the same general approach. Example 14.15 illustrates this method.

EXAMPLE 14.15 Calculating the Solubility Product from Solubility

Calculate the solubility product constant for silver sulfate, given that the experimentally measured solubility is 0.44 g/100 mL. (Many solubility tables, such as those published in the *Handbook of Chemistry and Physics,* express solubility in grams per 100 mL of water, not moles per liter.)

Strategy Write the chemical equation, convert the given concentration to moles per liter and fill in the iCe table, write the algebraic and numerical expressions for K_{sp}, then solve.

Solution

First, write the chemical equation

$$Ag_2SO_4(s) \rightleftharpoons 2Ag^+(aq) + SO_4^{2-}(aq)$$

1. Write the chemical equation.

Next, convert the solubility of silver sulfate from grams per 100 mL (g/100 mL) to moles per liter (mol/L). It may be simplest to convert from grams per 100 mL to moles per liter in two steps (be careful to avoid round-off errors). To simplify the calculation, we assume that the volume of solution is 100 mL. The abbreviation *s* is used to represent the molar solubility.

$$s = \frac{0.44 \text{ g } Ag_2SO_4}{100 \text{ mL}} \times \left(\frac{1000 \text{ mL}}{\text{L}}\right) = \frac{4.4 \text{ g } Ag_2SO_4}{\text{L}}$$

$$s = \frac{4.4 \text{ g } Ag_2SO_4}{\text{L}} \times \left(\frac{1 \text{ mol } Ag_2SO_4}{311.8 \text{ g } Ag_2SO_4}\right) = 1.4 \times 10^{-2} M\ Ag_2SO_4$$

2a. Convert concentrations to moles per liter (mol/L).

Use the iCe table to calculate the equilibrium concentrations of $Ag^+(aq)$ and $SO_4^{2-}(aq)$.

	Ag_2SO_4	$\rightleftharpoons$	$2Ag^+$	+	SO_4^{2-}
initial concentration, *M*	Solid		0.0		0.0
Change in concentration, *M*	$-s$		$+2s$		$+s$

Substitute 1.4×10^{-2} *M* for *s* on the change (C) line.

	Ag_2SO_4	$\rightleftharpoons$	$2Ag^+$	+	SO_4^{2-}
initial concentration, *M*	Solid		0.0		0.0
Change in concentration, *M*	-1.4×10^{-2}		$+2(1.4 \times 10^{-2})$		$+1.4 \times 10^{-2}$
equilibrium concentration, *M*	Solid		2.8×10^{-2}		1.4×10^{-2}

2b. Fill in the iCe table.

Write the algebraic expression for *K*.

$$K_{sp} = [Ag^+]^2[SO_4^{2-}]$$

3. Write the expression for *K*.

Substitute the concentrations into the algebraic expression for the solubility product constant.

$$K_{sp} = [Ag^+]^2[SO_4^{2-}] = (2.8 \times 10^{-2})^2(1.4 \times 10^{-2})$$

4. Substitute.

$$K_{sp} = 1.1 \times 10^{-5}$$

5. Solve.

Understanding

The solubility of $La(IO_3)_3$ is 7.3×10^{-4} *M* at 25 °C. Calculate the solubility product constant.

Answer $K_{sp} = 7.7 \times 10^{-12}$

Solubility Calculations

If we know the solubility product constant that describes the equilibrium between a substance and its ions in solution, we can predict the solubility of the substance. As in other equilibrium problems, we first set up a table and enter the known data. In this particular case, the solubility—generally designated by the letter *s*—is unknown. We determine it by solving the solubility product constant expression for *s*. Example 14.16 shows the procedure.

EXAMPLE 14.16 Calculating the Solubility from the Solubility Product Constant

Given the value of K_{sp} (see Table 14.4), calculate the solubility of barium iodate, $Ba(IO_3)_2$.

Strategy Write the chemical equation, iCe table, the expression for K_{sp}, and solve.

Solution

Write the chemical equation for the dissolution of $Ba(IO_3)_2$.

1. Write the chemical equation.

$$Ba(IO_3)_2(s) \rightleftharpoons Ba^{2+}(aq) + 2IO_3^-(aq)$$

Start the iCe table with the initial concentrations. In this case, $Ba(IO_3)_2$ is a solid, and the initial concentrations of Ba^{2+} and IO_3^- are zero. Next, determine the changes. We will define *s* as the number of moles of solid that dissolve in 1 L of solution. If *s* moles of $Ba(IO_3)_2$ dissolve, then *s* moles of Ba^{2+} form and 2*s* moles of IO_3^- form; we write this information on the change (C) line. Determine the equilibrium concentrations by summing the initial concentration and the change in concentration.

2. Fill in the iCe table.

	$Ba(IO_3)_2(s)$	$\rightleftharpoons$	$Ba^{2+}(aq)$	+	$2IO_3^-(aq)$
initial concentration, *M*	Solid		0		0
Change in concentration, *M*	$-s$		$+s$		$+2s$
equilibrium concentration, *M*	Solid		s		$2s$

Note that the stoichiometry tells us that when *s* mol $Ba(IO_3)_2(s)$ dissolves, *s* mol $Ba^{2+}(aq)$ and 2*s* mol $IO_3^-(aq)$ are produced.

After the iCe table is complete, write the algebraic expression for the equilibrium constant, K_{sp}.

3. Write the expression for *K*.

$$K_{sp} = [Ba^{2+}][IO_3^-]^2$$

Now, substitute the solubility product constant (4.0×10^{-9} from Table 14.4) and the values for the concentrations of Ba^{2+} and IO_3^- from the equilibrium (e) line of the iCe table.

4. Substitute.

$$4.0 \times 10^{-9} = (s)(2s)^2 = 4s^3$$

Now, solve. Some calculators do not have a specific button that takes a cube root, but almost all have some way to obtain cube roots. The y^x key and logarithms (see Appendix A) are two ways.

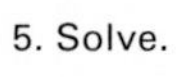

5. Solve.

$$s = 1.0 \times 10^{-3}\ M$$

Notice that we defined *s* as the number of moles of solute that dissolve in a liter of solution, which is the solubility of the solid.

Understanding

Calculate the solubilities of $AgIO_3$ and $La(IO_3)_3$ from their solubility product constants.

Answer For $AgIO_3$, $s = 1.8 \times 10^{-4}\ M$; for $La(IO_3)_3$, $s = 7.2 \times 10^{-4}\ M$

TABLE 14.5 Solubilities of Selected Iodates

Compound	K_{sp}	Solubility, *M*
$AgIO_3$	3.2×10^{-8}	1.8×10^{-4}
$Ba(IO_3)_2$	4.0×10^{-9}	1.0×10^{-3}
$La(IO_3)_3$	7.5×10^{-12}	7.3×10^{-4}

TABLE 14.6 Relationship between Solubility Product Constant, K_{sp}, and Molar Solubility, *s*

Compound	K_{sp} Expression	Solubility Cation	Solubility Anion	Expression
AgCl	$K_{sp} = [Ag^+][Cl^-]$	s	s	$K_{sp} = (s)(s) = s^2$
$CaSO_4$	$K_{sp} = [Ca^{2+}][SO_4^{2-}]$	s	s	$K_{sp} = (s)(s) = s^2$
Ag_2SO_4	$K_{sp} = [Ag^+]^2[SO_4^{2-}]$	$2s$	s	$K_{sp} = (2s)^2(s) = 4s^3$
PbI_2	$K_{sp} = [Pb^{2+}][I^-]^2$	s	$2s$	$K_{sp} = (s)(2s)^2 = 4s^3$
LaF_3	$K_{sp} = [La^{3+}][F^-]^3$	s	$3s$	$K_{sp} = (s)(3s)^3 = 27s^4$
$Ca_3(PO_4)_2$	$K_{sp} = [Ca^{2+}]^3[PO_4^{3-}]^2$	$3s$	$2s$	$K_{sp} = (3s)^3(2s)^2 = 108s^5$

Notice that there is no simple proportionality between the value of K_{sp} and the solubility of the compound, as illustrated in Table 14.5. Although the solubility product constant for $Ba(IO_3)_2$ is smaller than that for $AgIO_3$, the solubility of $Ba(IO_3)_2$ is larger. The solubility depends on both the reaction stoichiometry and the solubility product.

Table 14.6 illustrates the relationship between the expression for the solubility product constant and the solubility. You need not memorize the table, but note that the expression that relates K_{sp} to solubility comes from two sources. Both the stoichiometry (coefficients of the balanced equation) and the definition of K_{sp} contribute to the expression. When a compound dissolves, the concentration of each ion is proportional to the number of ions of that kind in the formula of the compound.

The solubility depends on both K_{sp} and the stoichiometry.

OBJECTIVES REVIEW *Can you:*

☑ write the expression for the solubility product constant?
☑ calculate K_{sp} from experimental data?
☑ calculate solubility of slightly soluble salts from K_{sp}?

14.7 Solubility and the Common Ion Effect

OBJECTIVES

☐ Predict the solubility of a solid in a solution that contains a common ion
☐ Use approximations to calculate roots of polynomial equations
☐ Decide whether a precipitate will form under a particular set of conditions

This section considers how other species in solution influence the solubility of a particular substance. In addition, the process of solving polynomial equations by numerical approximation is presented.

Common Ion Effect

Frequently, we need to calculate the solubility of a precipitate in a solution that already contains one of the ions that compose the precipitate. Such problems illustrate the **common ion effect,** a term used to describe the effect of adding a solute to a solution that contains an ion in common. In a precipitation reaction, the common ion effect predicts decreased solubility of a precipitate.

The solubility of a precipitate is lower in a solution that contains an ion in common with the substance.

EXAMPLE 14.17 Identifying Common Ions

Which systems (precipitates A or B and solutions 1–4) illustrate the common ion effect?

Precipitates	Solutions
A. Mercury(I) chloride	1. Barium chloride
B. Barium sulfate	2. Lead nitrate
	3. Lithium sulfate
	4. Potassium bromide

Strategy Look at the precipitate, identify the cation and anion, and look to see whether either is present in the solution.

Solution

A. Mercury(I) chloride has an ion in common with solution 1 (chloride).

B. Barium sulfate has an ion in common with solutions 1 (barium) and 3 (sulfate).

Understanding

Which of the solutions will show a common ion effect when the precipitate is lead bromide?

Answer Solutions 2 and 4

The common ion effect can be explained by Le Chatelier's principle. Compare the solubility of silver chloride in water with its solubility in a sodium chloride solution; chloride is the common ion.

$$AgCl(s) \rightleftharpoons Ag^+(aq) + Cl^-(aq)$$

If the solution initially contains the maximum amount of silver chloride that can dissolve, and the concentration of $Cl^-(aq)$ is increased by adding some sodium chloride, then Le Chatelier's principle predicts that the system will react to form additional AgCl(s), decreasing the solubility of silver chloride.

$$AgCl(s) \Leftarrow\!\rightleftharpoons Ag^+(aq) + Cl^-(aq)$$

A second approach, calculating Q and comparing it with K_{sp} on a number line, also leads to the conclusion that silver chloride is less soluble in a sodium chloride solution than in water. The reaction quotient is

$$Q = [Ag^+][Cl^-]$$

If chloride is added to a system at equilibrium, Q becomes greater than K_{sp}, and a reaction occurs to consume some of the added chloride, producing additional AgCl(s). Comparing Q with K_{sp} allows us to predict the direction of reaction.

$$AgCl(s) \rightleftharpoons Ag^+(aq) + Cl^-(aq)$$

Q

No solid present | Ions form solid

K_{sp}
Equilibrium

Number line representation of the common ion effect. Adding chlorine ion increases the reaction quotient and the reaction proceeds to the left, forming more AgCl(s).

Quantitative treatment of the common ion effect uses the same methods that we use for the other quantitative equilibrium calculations—we write the chemical equation, iCe table, and algebraic expression for K_{sp}; substitute numerical values into the expression; and solve. The technique is illustrated in Example 14.18.

EXAMPLE 14.18 Common Ion Effect

Complete the iCe table, write the expression for the solubility product constant needed to calculate the solubility of silver iodide ($K_{sp} = 8.5 \times 10^{-17}$) in a 0.10 M sodium iodide solution, and write the polynomial form of the equilibrium constant expression.

Strategy The same five-step approach works for this problem.

Solution

In this problem, the initial concentration of the iodide ion is 0.10 *M*, from the 0.10 *M* NaI, and *s* is the solubility of AgI (in mol/L).

	AgI(s)	⇌	$Ag^+(aq)$	+	$I^-(aq)$
initial concentration, *M*	solid		0.0		0.10
Change in concentration, *M*	$-s$		$+s$		$+s$
equilibrium concentration, *M*	solid		s		$0.10 + s$

1. Write the chemical equation.

2. Fill in the iCe table.

Write the algebraic expression for the solubility product constant.

$K_{sp} = [Ag^+][I^-]$

3. Write the expression for *K*.

Substitute the equilibrium concentrations into the solubility product expression,

$8.5 \times 10^{-17} = (s)(0.10 + s)$

4. Substitute.

and reduce to a quadratic equation:

$s^2 + 0.10s - 8.5 \times 10^{-17} = 0$

5. Solve.

Remember, this example asked for the expression for the solubility product and the resulting polynomial, not a numerical solution.

Understanding

Write (a) the equilibrium constant expression needed to determine the solubility of AgCl ($K_{sp} = 1.8 \times 10^{-10}$) in 1.0×10^{-5} *M* KCl and (b) the resulting polynomial.

Answer

(a) $1.8 \times 10^{-10} = s(1.0 \times 10^{-5} + s)$
(b) $s^2 + 1.0 \times 10^{-5}s - 1.8 \times 10^{-10} = 0$

Numerical Approximations

The quadratic formula is not always the best method of solving quadratic equations. Consider the problem that arises when we are asked to calculate the solubility of silver iodide ($K_{sp} = 8.5 \times 10^{-17}$) in 0.10 *M* sodium iodide solution (see Example 14.18). The expression for the equilibrium constant is

$$8.5 \times 10^{-17} = (s)(0.10 + s)$$

which can be expressed as a quadratic equation:

$$s^2 + 0.10s - 8.5 \times 10^{-17} = 0$$

If the quadratic formula is used to solve this quadratic equation, a major technical problem arises. The round-off error in the calculation becomes significant because the terms under the square root sign differ by many orders of magnitude. You may wish to try solving the quadratic equation to test your calculator.

$$s = \frac{-0.10 \pm \sqrt{0.10^2 - 4(1)(-8.5 \times 10^{-17})}}{2(1)}$$

Formulating an approximation is an alternative way to find the roots of a polynomial equation. (Other methods, including a form of the quadratic formula that does not require 17 digits of accuracy, are presented in Appendix A.) Consider the expression for the equilibrium constant before reducing to a quadratic equation:

$$8.5 \times 10^{-17} = (s)(0.10 + s)$$

Note that the solubility of silver iodide, which has a K_{sp} of 8.5×10^{-17}, will be much less than 0.10 *M*. If *s* is much less than 0.10, then the sum, $0.10 + s$, is approximately equal to 0.10. We can express this approximation in equation form:

If $s << 0.10$ (where $<<$ means "much less than"), then

$$8.5 \times 10^{-17} \approx (s)(0.10)$$

where $\approx$ means "very nearly equal to." This equation is easily solved:

$$s = 8.5 \times 10^{-16}\ M$$

The result is based on an approximation, so its validity must be checked. We assumed that s was much less than 0.1, and since 8.5×10^{-16} is much less than 0.10, our approximation is valid. Arithmetically, $0.10 + 8.5 \times 10^{-16}$ does not differ significantly from 0.10.

We must still establish quantitatively what is meant by "much smaller." When solving equilibrium problems, chemists generally neglect the addition or subtraction of one term if the smaller is less than 5% of the larger term. Although the approximation method limits the accuracy of the calculations to about 5%, that level is a reasonable expectation for the overall accuracy of a solubility product calculation, because other potential sources of error limit the accuracy to about this magnitude.

If the approximation fails, either solve the quadratic equation or use one of the alternative methods (including a method of *successive approximations*) discussed in Appendix A and in Chapter 15. The limited solubilities of most precipitates make the simplifying assumption valid for most cases.

Approximations can be used to solve the quadratic equation.

EXAMPLE 14.19 Calculating the Solubility of a Solid in Solution with a Common Ion

What is the solubility of calcium hydroxide ($K_{sp} = 5.0 \times 10^{-6}$) in 0.080 M sodium hydroxide solution?

Strategy Write the chemical equation, the iCe table, and the algebraic expression; substitute; and solve.

Solution

Set up the iCe table, substitute the equilibrium concentrations into the expression for the equilibrium constant, and solve for the solubility.

1. Write the chemical equation.

2. Fill in the iCe table.

	$Ca(OH)_2(s)$	$\rightleftharpoons$	$Ca^{2+}(aq)$	+	$2OH^-(aq)$
initial concentration, M	Solid		0.0		0.080
Change in concentration, M	$-s$		$+s$		$+2s$
equilibrium concentration, M	Solid		s		$0.080 + 2s$

3. Write the expression for K.

4. Substitute.

5. Solve.

$$K_{sp} = [Ca^{2+}][OH^-]^2$$

$$5.0 \times 10^{-6} = (s)(0.080 + 2s)^2$$

We make the approximation that $2s << 0.080$; if this approximation is correct, then

$$5.0 \times 10^{-6} \approx (s)(0.080)^2$$

$$s = 7.8 \times 10^{-4}\ M$$

Now check the assumption. Is $2s << 0.080$? Compare 5% of 0.080 M (4.0×10^{-3} M) and $2s$ (1.6×10^{-3} M) to see that $2s$ is much less than 0.080. We conclude that the assumption is valid and accept the solubility of 7.8×10^{-4} M as the answer.

Understanding

Calculate the solubility of CaF_2 in 0.025 M NaF.

Answer $5.6 \times 10^{-8}\ M$

If you make a numerical approximation, you must check its validity.

Formation of a Precipitate

Chemists must often predict the results of a reaction; therefore, it is important to know whether a precipitate will form when two solutions are mixed. This section presents the relationship between the concentrations of the species and the formation of a precipitate.

When two solutions are mixed, the mixture, which is not yet at equilibrium, is described by a reaction quotient, Q. If, for example, a solution that contains calcium ions is mixed with a solution that contains phosphate ions, the reaction is described by the chemical equilibrium:

$$Ca_3(PO_4)_2(s) \rightleftharpoons 3Ca^{2+}(aq) + 2PO_4^{3-}(aq)$$

Even though the reaction under study is the formation of calcium phosphate, it is traditional to write the expression for the dissolution of the solid, the reverse of the reaction that actually occurs, because that is how values for the equilibria are tabulated. The conclusions do not depend on writing the reaction in a particular direction.

The reaction quotient is calculated from the concentration data and the following equation:

When Q exceeds K_{sp}, a precipitate will form.

$$Q = [Ca^{2+}]^3[PO_4^{3-}]^2$$

The reaction quotient, Q, is compared with K_{sp} to determine the direction of reaction. If Q is greater than K_{sp}, the system will react to form solid; if the concentrations of Ca^{2+} and PO_4^{3-} are small, then Q is less than K_{sp}, no solid forms, and equilibrium cannot be established.

We can examine the dissolving of a solid with the aid of a number line. In Figure 14.7, Q_1 is smaller than K_{sp}; when Q is less than K_{sp}, a reaction occurs toward the right and some additional solid will dissolve, forming more ions. Q_2 is greater than K_{sp}, so additional precipitation will occur—the reaction proceeds toward the left. Example 14.20 illustrates the calculations used to determine whether a precipitate will form.

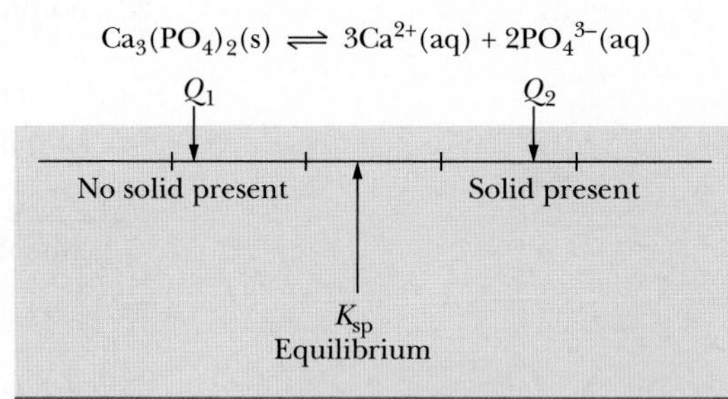

Figure 14.7 Number line representation of solubility.

EXAMPLE 14.20 Formation of a Precipitate

A chemist mixes 200 mL of 0.010 M $Pb(NO_3)_2$ with 100 mL of 0.0050 M NaCl. Will lead(II) chloride precipitate?

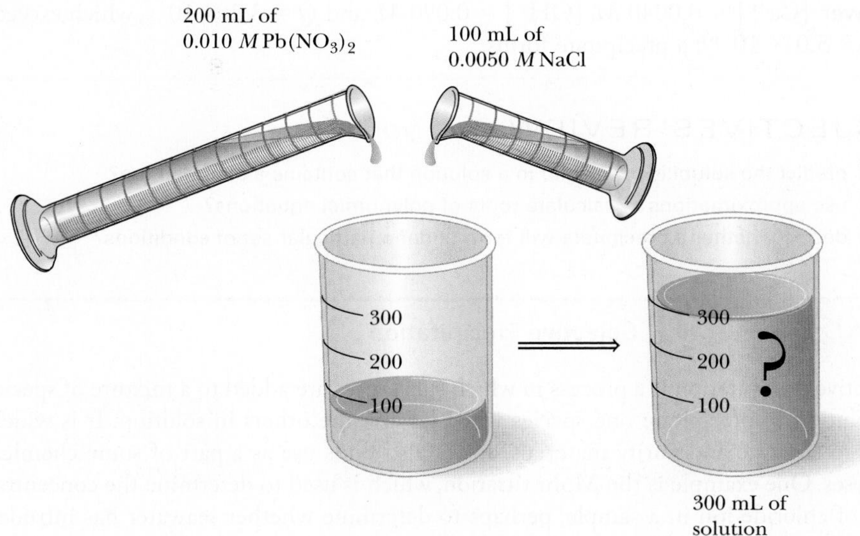

Strategy Determine Q and compare it with K_{sp} to determine whether lead chloride will form.

Solution

First, write the chemical equation.

$$PbCl_2(s) \rightleftharpoons Pb^{2+}(aq) + 2Cl^-(aq)$$

To determine Q, the concentrations of $Pb^{2+}(aq)$ and $Cl^-(aq)$ must first be calculated. The concentrations are not 0.010 and 0.0050 M because the original solutions are diluted by the mixing process. The concentrations of Pb^{2+} and Cl^- are found from calculations similar to those illustrated in Chapter 4, where dilution problems were presented. To determine the concentrations after the two solutions are mixed, calculate the number of moles of solute and divide by the total volume of solution to get:

$$[Pb^{2+}] = 6.7 \times 10^{-3}\ M$$

$$[Cl^-] = 1.7 \times 10^{-3}\ M$$

$$Q = [Pb^{2+}][Cl^-]^2$$

$$Q = [6.7 \times 10^{-3}][1.7 \times 10^{-3}]^2$$

$$Q = 1.9 \times 10^{-8}$$

The numerical value for K_{sp} is 1.7×10^{-5}. Because the reaction quotient is less than K_{sp}, no precipitate will form. Only when Q is greater than K_{sp} will precipitation occur.

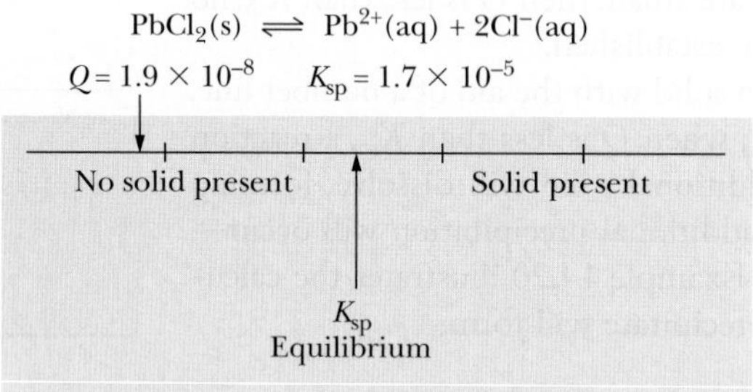

Number line representation of solubility.

Understanding

If 200 mL of 0.010 M $CaCl_2(aq)$ is mixed with 300 mL of 0.150 M NaOH(aq), will $Ca(OH)_2$ precipitate?

Answer $[Ca^{2+}] = 0.0040\ M$, $[OH^-] = 0.090\ M$, and $Q = 3.2 \times 10^{-5}$, which exceeds K_{sp} ($= 5.0 \times 10^{-6}$); a precipitate forms.

OBJECTIVES REVIEW *Can you:*

- ☑ predict the solubility of a solid in a solution that contains a common ion?
- ☑ use approximations to calculate roots of polynomial equations?
- ☑ decide whether a precipitate will form under a particular set of conditions?

CASE STUDY Selective Precipitation

Selective precipitation is a process in which substances are added to a mixture of species in solution to precipitate one species while leaving the others in solution. It is widely used to separate and purify materials, and it also finds use as a part of some chemical analyses. One example is the Mohr titration, which is used to determine the concentration of chloride ion in a sample, perhaps to determine whether seawater has intruded into an irrigation well. The procedure is a titration in which standard silver nitrate solution is added to precipitate the chloride as silver chloride.

Chemists arrange conditions so that the silver solution precipitates the chloride first, and after the chloride precipitation is complete, the excess silver reacts with the indicator. In this case, the indicator is potassium chromate and the indicator reaction is the formation of another solid silver compound, the red silver chromate. The titration ends at the first permanent appearance of the red solid.

Ideally, the concentration of chromate should be chosen so that the precipitation of silver chromate starts just after the chloride has been precipitated. At the equivalence point, the only source of silver ion is the precipitate. We can calculate the concentration of silver from the solubility product constant for silver chloride. Next, we calculate the chromate concentration needed to form a precipitate, given the silver concentration calculated from the solubility product. These coupled calculations ensure that when the chloride precipitation finishes, the silver chromate begins to precipitate.

The proper concentration of chromate ion can be calculated from the following chemical equilibria:

$$AgCl(s) \rightleftharpoons Ag^+(aq) + Cl^-(aq) \qquad K_{sp} = 1.8 \times 10^{-10}$$

$$Ag_2CrO_4(s) \rightleftharpoons 2Ag^+(aq) + CrO_4^{2-}(aq) \qquad K_{sp} = 1.1 \times 10^{-12}$$

First, calculate the concentration of silver ion at the equivalence point from the K_{sp} expression.

	AgCl(s)	$\rightleftharpoons$	$Ag^+(aq)$	+	$Cl^-(aq)$
initial concentration, M	Solid		0.0		0.0
Change in concentration, M	$-s$		$+s$		$+s$
equilibrium concentration, M	Solid		s		s

Substitute the expressions for the concentrations of silver and chloride ions into the solubility-product expression:

$$K_{sp} = [Ag^+][Cl^-]$$

$$1.8 \times 10^{-10} = s^2$$

$$s = 1.3 \times 10^{-5}\ M$$

Because we know the concentration of the silver ion at the equivalence point, $1.3 \times 10^{-5}\ M$, we next calculate the concentration of chromate ion needed to precipitate silver chromate when the silver concentration is $1.3 \times 10^{-5}\ M$.

The solubility product constant for silver chromate is $K_{sp} = 1.1 \times 10^{-12}$, so

$$K_{sp} = [Ag^+]^2[CrO_4^{2-}]$$

$$[CrO_4^{2-}] = \frac{K_{sp}}{[Ag^+]^2}$$

$$= \frac{1.1 \times 10^{-12}}{(1.3 \times 10^{-5})^2}$$

$$= 0.007\ M$$

If the chromate indicator is $0.007\ M$, the formation of the red precipitate starts as soon as the concentration of silver increases to more than $1.3 \times 10^{-5}\ M$, the point at which all the chloride has been consumed.

Other factors actually require that the chromate concentration be somewhat less than $0.007\ M$; a concentration of $0.005\ M$ is generally used. The decreased concentration of chromate requires the addition of a little excess silver solution before a color change occurs. This volume is only 0.03 mL beyond the point at which all the chloride is precipitated if we are titrating with $0.1\ M$ silver ion solution, so it does not introduce a substantial error.

Notice that even though the solubility product of silver chromate is smaller than K_{sp} for silver chloride, the silver chromate is actually more soluble. Because silver chromate

The Mohr titration. The solution is initially yellow, the color of chromate ion, as the white AgCl forms. Near the endpoint, the red silver chromate forms but it dissipates with swirling. The endpoint is the first permanent red color.

dissociates into three ions and silver chloride into two, we cannot readily compare solubilities from just the values of the solubility product constants.

Questions

1. The chloride content of a solid sample was determined by Mohr titration. A 0.4851-g sample was dissolved in 20 mL water. Potassium chromate indicator was added and the end-point was reached after 36.82 mL of standard 0.109 M $AgNO_3$ solution. Calculate the percentage of chloride in the sample.
2. A blank titration was performed. Describe how the blank should be prepared.
3. If the blank required 0.04 mL, calculate the correct percentage of chloride from the data given in Question 1.

ETHICS IN CHEMISTRY

1. You are the Environmental Quality Manager for a large company that uses a patented remediation process to neutralize a potential pollutant.

$$\text{remediator} + \text{pollutant} \rightleftharpoons \text{inert substance}$$

The process is at equilibrium, and under terms of your discharge permit, your average daily discharge must be less than 5 mg/L. Your division measures the input pollutant concentration twice a day, and calculates the concentration of the remediated pollutant from the concentrations, temperature, and the equilibrium constant. Historically, the process reduces the concentration below 2 mg/L, so it can be safely discharged. This particular day, you find the information shown in the table below. You are surprised to see that the lowest value is 1.99 mg/L and dismayed to see that the highest value is out of specifications at 7.97 mg/L.

	Calculated Pollutant Concentration after Remediation
Lowest value over 24 hours	1.99 mg/L
Highest value over 24 hours	7.97 mg/L
Average of lowest and highest values in 24 hours	4.98 mg/L

The plant manager looks at the data for the average and sees that it is less than 5 mg/L. In fact, he notices that the greatest value was 7.97 mg/L, and if the release were at that level for half the time, the average release would be half that amount, or 3.98 mg/L, which is within specifications. He signs the form to discharge the waste, but two signatures are needed. Will you countersign the authorization form? Explain your logic.

2. The U.S. company that owned the Bhopal factory discussed in the introduction to this chapter paid $470 million in 1989 as full and final compensation for its role in the accident. The money earned interest while the Bhopal Gas Tragedy Relief and Rehabilitation Department evaluated more than a million damage claims. By October 2003, 554,895 claims for injuries and 15,310 claims for death were approved. The average compensation was $2,200. Had the accident occurred in the United States, the likely award would have been at least 20 times greater, based on previous settlements in similar situations.

Should U.S. companies that have accidents outside the United States be required to reimburse victims at the local rate or the rate in the United States? What about a foreign-owned factory that has an accident in the United States? Should it be required to pay at the local (U.S.) rate or the rate in its home country?

3. The lawyers representing the Bhopal victims cited evidence that safety violations had occurred. Should the settlement amounts be greater if safety violations are present?

Chapter 14 Visual Summary

The chart shows the connections between the major topics discussed in this chapter.

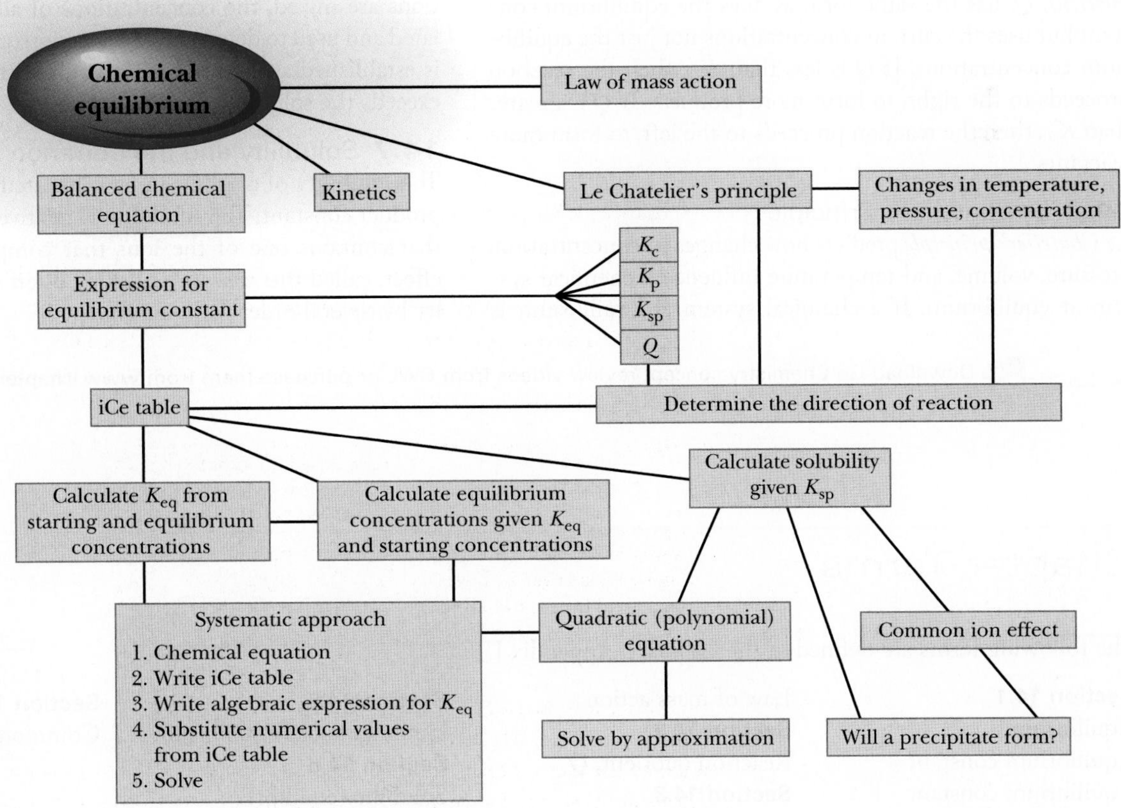

Summary

14.1 Equilibrium Constant

In many reactions, the chemical system reaches a state of *equilibrium* in which the reactants and products are present in unchanging concentrations. The *law of mass action* provides the needed mathematical basis for the evaluation of chemical equilibria. This law states that for any chemical reaction,

$$aA + bB \rightleftharpoons cC + dD$$

an equilibrium constant expression can be written that describes the equilibrium concentrations of the substances in the reaction.

$$K_{eq} = \frac{[C]^c[D]^d}{[A]^a[B]^b}$$

The numerical value of K_{eq} depends on the coefficients in the balanced equation, but if K_{eq} is known for one set of coefficients, K_{eq} can be determined for any other set of coefficients that also balance the equation. Two *equilibrium constants,* K_c and K_p, are used for systems in which concentrations are given in moles per liter and those in which concentrations are expressed as pressures in atmospheres. The relationship between K_c and K_p is derived from the ideal gas law.

14.2 Reaction Quotient

The expression for the equilibrium constant can be used to determine the direction of spontaneous reaction. The *reaction quotient,* Q, has the same form as does the equilibrium constant but uses the current concentrations not just the equilibrium concentrations. If Q is less than K_{eq}, then the reaction proceeds to the right, to form more products. If Q is greater than K_{sp}, then the reaction proceeds to the left, to form more reactants.

14.3 Le Chatelier's Principle

Le Chatelier's principle predicts how changes in concentration, pressure, volume, and temperature influence a chemical system at equilibrium. If a chemical system at equilibrium is disturbed, it reacts to reduce the effect of the disturbance. If, for example, the concentration of one of the substances is increased, then the reaction proceeds in the direction that consumes the added substance. Le Chatelier's principle also predicts how changes in temperature influence the reaction.

14.4 Equilibrium Calculations

The chemical equation, initial composition, and numerical value for K_{eq} are generally needed to calculate the concentrations of species in an equilibrium system. A systematic approach to equilibrium problems provides a template (the iCe table) with which to solve this type of problem.

14.5 Heterogeneous Equilibria

In a *heterogeneous equilibrium,* where the reaction mixture contains more than one phase, the concentrations of pure liquids and solids do not change, so they do not appear in the expression for the equilibrium constant. Equilibrium expressions contain only the concentrations of the species that change concentration as a result of the chemical reaction.

14.6 Solubility Equilibria

The solubility product expression is used to describe the process of dissolving a sparingly soluble ionic solid. The numerical value of K_{sp} can be used to determine the solubility of a solid; conversely, the solubility can be used to evaluate K_{sp}.

The expression for the *solubility product constant* is also used to determine whether a precipitate will form. If two solutions are mixed, the concentrations of all ions can be calculated and used to determine the reaction quotient. Equilibrium is established and a solid forms, only if the reaction quotient exceeds the solubility product constant.

14.7 Solubility and the Common Ion Effect

The solubility of a solid can be calculated from the solubility product constant. The solubility is relatively low in a solution that contains one of the ions that compose the solid. This effect, called the *common ion effect,* often reduces the solubility by several orders of magnitude.

Download Go Chemistry concept review videos from OWL or purchase them from **www.ichapters.com**

Chapter Terms

The following terms are defined in the Glossary, Appendix I.

Section 14.1
Equilibrium
Equilibrium constant
Equilibrium constant expression
Law of mass action

Section 14.2
Reaction quotient, Q

Section 14.3
Le Chatelier's principle

Section 14.5
Heterogeneous equilibrium

Section 14.6
Solubility equilibria
Solubility product constant

Section 14.7
Common ion effect

Questions and Exercises

OWL Selected end of chapter Questions and Exercises may be assigned in OWL.

Blue-numbered Questions and Exercises are answered in Appendix J; questions are qualitative, are often conceptual, and include problem-solving skills.

■ Questions assignable in OWL

✎ Questions suitable for brief writing exercises

▲ More challenging questions

Questions

14.1 How can you determine whether a system has reached equilibrium?

14.2 ✎ Describe a nonchemical system that is in equilibrium, and explain how the principles of equilibrium apply to the system.

14.3 Describe a nonchemical system that is *not* in equilibrium, and explain why equilibrium has not been achieved.

14.4 Sunlight strikes the upper atmosphere, creating ozone, which eventually decomposes to O_2. Is the sunlight-ozone system in equilibrium? Explain your answer.

14.5 Does a 1.5-V battery represent a system that is at equilibrium? If not, describe the equilibrium status of a battery.

14.6 Describe a chemical system in which K_p is equal to K_c. Generalize to define the conditions for which K_p and K_c are always unequal.

14.7 Compare Q with K_{eq}. Are they always different in value?

14.8 ✎ Does Le Chatelier's principle apply to nonchemical equilibrium systems? Describe the changes from equilibrium that are observed as a person squeezes a partially filled balloon.

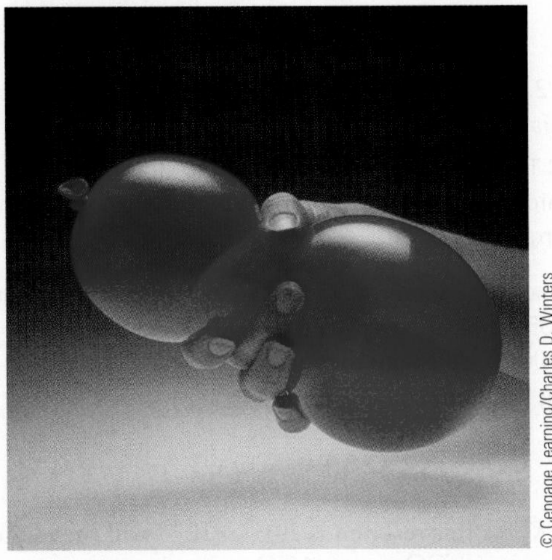
© Cengage Learning/Charles D. Winters

14.9 Under what circumstances do changes in the volume of a gaseous system *not* change the equilibrium constant?

14.10 Compare how changes in temperature influence K_{eq} and Q.

14.11 Explain why terms for pure liquids and solids do not appear in the expression for the equilibrium constant.

14.12 Temperature influences solubility. Does temperature have the same effect on all substances? Justify your answer. (*Hint:* Consider Le Chatelier's principle.)

Exercises

OBJECTIVE **Describe equilibrium systems and write the equilibrium constant expression for any chemical reaction.**

14.13 Write the expression for the equilibrium constant (K_c) for the following:

(a) $PCl_5(g) \rightleftharpoons PCl_3(g) + Cl_2(g)$
(b) $2NO_2(g) \rightleftharpoons 2NO(g) + O_2(g)$
(c) $2SO_3(g) \rightleftharpoons 2SO_2(g) + O_2(g)$
(d) $H_2(g) + I_2(g) \rightleftharpoons 2HI(g)$

14.14 Write the expression for the equilibrium constant (K_c) for the following:

(a) $2H_2O(g) \rightleftharpoons 2H_2(g) + O_2(g)$
(b) $2HCl(g) \rightleftharpoons H_2(g) + Cl_2(g)$
(c) $CO(g) + Cl_2(g) \rightleftharpoons COCl_2(g)$
(d) $2CO(g) + O_2(g) \rightleftharpoons 2CO_2(g)$

14.15 Write the expression for the equilibrium constant (K_p) for the following:

(a) $HCl(g) + \frac{1}{4}O_2(g) \rightleftharpoons \frac{1}{2}Cl_2(g) + \frac{1}{2}H_2O(g)$
(b) $\frac{1}{2}N_2O_4(g) \rightleftharpoons NO_2(g)$
(c) $N_2O_4(g) \rightleftharpoons N_2(g) + 2O_2(g)$
(d) $\frac{1}{2}O_2(g) + SO_2(g) \rightleftharpoons SO_3(g)$

14.16 Write the expression for the equilibrium constant (K_p) for the following:

(a) $Cl_2(g) + H_2O(g) \rightleftharpoons 2HCl(g) + \frac{1}{2}O_2(g)$
(b) $2NO_2(g) \rightleftharpoons N_2O_4(g)$
(c) $3O_2(g) \rightleftharpoons 2O_3(g)$
(d) $CO_2(g) \rightleftharpoons CO(g) + \frac{1}{2}O_2(g)$

OBJECTIVE **Evaluate the equilibrium constant from experimental data.**

14.17 Calculate the equilibrium constant from experimental measurements.

(a) Nitrogen dioxide dissociates into nitrogen monoxide and oxygen.

$$2NO_2(g) \rightleftharpoons 2NO(g) + O_2(g)$$

When equilibrium is reached, the concentrations are as follows:

$[NO_2] = 0.021\ M$

$[NO] = 0.0042\ M$

$[O_2] = 0.0043\ M$

Calculate K_c

(b) Dinitrogen tetroxide dissociates into nitrogen and oxygen:

$$N_2O_4(g) \rightleftharpoons N_2(g) + 2O_2(g)$$

When equilibrium is reached, the partial pressures are as follows:

$$P_{N_2O_4} = 0.431 \text{ atm}$$
$$P_{N_2} = 0.0867 \text{ atm}$$
$$P_{O_2} = 0.00868 \text{ atm}$$

Calculate K_p.

14.18 ■ An equilibrium mixture contains 3.00 mol CO, 2.00 mol Cl_2, and 9.00 mol $COCl_2$ in a 50-L reaction flask at 800 K. Calculate the value of the equilibrium constant K_c for the reaction

$$CO(g) + Cl_2(g) \rightleftharpoons COCl_2(g)$$

at this temperature.

OBJECTIVE Relate the expression for the equilibrium constant to the form of the balanced equation.

14.19 At 2000 K, experiments show that the equilibrium constant for the formation of water is 1.6×10^{10}.

$$2H_2(g) + O_2(g) \rightleftharpoons 2H_2O(g)$$

Calculate the equilibrium constant at the same temperature for

$$H_2(g) + \frac{1}{2}O_2(g) \rightleftharpoons H_2O(g)$$

14.20 At 500 K, the equilibrium constant is 155 for

$$H_2(g) + I_2(g) \rightleftharpoons 2HI(g)$$

Calculate the equilibrium constant for

$$\frac{1}{2}H_2(g) + \frac{1}{2}I_2(g) \rightleftharpoons HI(g)$$

14.21 At 77 °C, K_p is 1.7×10^4 for the formation of phosphorus pentachloride from phosphorus trichloride and chlorine.

$$PCl_3(g) + Cl_2(g) \rightleftharpoons PCl_5(g)$$

Calculate K_p for

$$PCl_5(g) \rightleftharpoons PCl_3(g) + Cl_2(g)$$

14.22 ■ Consider the following equilibria involving $SO_2(g)$ and their corresponding equilibrium constants.

$$SO_2(g) + \frac{1}{2}O_2(g) \rightleftharpoons SO_3(g) \quad K_1$$

$$2SO_3(g) \rightleftharpoons 2SO_2(g) + O_2(g) \quad K_2$$

Which of the following expressions relates K_1 to K_2?

(a) $K_2 = K_1^2$
(b) $K_2^2 = K_1$
(c) $K_2 = K_1$
(d) $K_2 = 1/K_1$
(e) $K_2 = 1/K_1^2$

14.23 K_c at 137 °C is 4.42 for

$$NO(g) + \frac{1}{2}Br_2(g) \rightleftharpoons NOBr(g)$$

Calculate K_c at 137 °C for

$$2NOBr(g) \rightleftharpoons 2NO(g) + Br_2(g)$$

14.24 K_c at 1400 K is 3.6×10^{-6} for

$$N_2(g) + O_2(g) \rightleftharpoons 2NO(g)$$

Calculate K_c at 1400 K for

$$2NO(g) \rightleftharpoons N_2(g) + O_2(g)$$

OBJECTIVE Convert between equilibrium constants in which the concentrations of gases are expressed in moles per liter and those in terms of partial pressures expressed in atmospheres.

14.25 Nitrosyl bromide is formed from nitrogen monoxide and bromine:

$$NO(g) + \frac{1}{2}Br_2(g) \rightleftharpoons NOBr(g)$$

K_p for this reaction is 116 at 25 °C. Calculate K_c at this temperature.

14.26 At 3000 K, carbon dioxide dissociates:

$$CO_2(g) \rightleftharpoons CO(g) + \frac{1}{2}O_2(g)$$

If K_p for this reaction is 2.48, calculate K_c.

14.27 Sulfur dioxide reacts with chlorine when sealed in a reactor at increased temperature. At 227 °C, $K_c = 20.9$. Calculate K_p at this same temperature.

$$SO_2(g) + Cl_2(g) \rightleftharpoons SO_2Cl_2(g)$$

14.28 ■ The value of K_c for the reaction

$$N_2(g) + 3H_2(g) \rightleftharpoons 2NH_3(g)$$

is 2.00 at 400 °C. Find the value of K_p for this reaction at this temperature.

OBJECTIVE Compare Q and K_{eq} to determine the direction in which a reaction proceeds when the system is not at equilibrium.

14.29 In an experiment, 4.95 mol CO_2, 0.050 mol CO, and 0.050 mol O_2 are placed in a 5.0-L reaction vessel at 1400 K. Calculate the reaction quotient, Q, for the following reaction:

$$CO(g) + \frac{1}{2}O_2(g) \rightleftharpoons CO_2(g)$$

If K_c for this reaction is 1.05×10^{-5}, will the mixture form more CO or more CO_2, or is the system at equilibrium? Draw a number line and place Q and K_c in the appropriate places.

14.30 The sulfur dioxide–oxygen–sulfur trioxide equilibrium is of interest to scientists who study acid rain.

$$2SO_2(g) + O_2(g) \rightleftharpoons 2SO_3(g)$$

The concentrations at the beginning of an experiment are $[SO_2] = 0.015$ *M*; $[O_2] = 0.012$ *M*; $[SO_3] = 1.45$ *M*; $K_c = 5.0 \times 10^6$ at 700 K. Calculate the reaction quotient, *Q*, and draw a number line with *Q* and K_c in the appropriate places. Will the system form more SO_2 or more SO_3, or is the system at equilibrium?

© Ted Spiegel/CORBIS

Acid rain damage. Sulfur oxides in the atmosphere are one of the most important sources of acid rain. See Exercise 14.30.

14.31 A 20.0-L flask contains 1.0 mol $PCl_5(g)$ and 2.0 mol each of $PCl_3(g)$ and $Cl_2(g)$. K_c at 425 °C is 4.0 for

$$PCl_5(g) \rightleftharpoons PCl_3(g) + Cl_2(g)$$

Calculate the reaction quotient, *Q*, and draw a number line with *Q* and K_c in the appropriate places. In which direction will the reaction proceed?

14.32 ■ Consider the equilibrium at 25 °C.

$$2SO_3(g) \rightleftharpoons 2SO_2(g) + O_2(g) \qquad K_c = 3.58 \times 10^{-3}$$

Suppose that 0.15 mol $SO_3(g)$, 0.015 mol $SO_2(g)$, and 0.0075 mol $O_2(g)$ are placed into a 10.0-L flask at 25 °C.
(a) Is the system at equilibrium?
(b) If the system is not at equilibrium, in which direction must the reaction proceed to reach equilibrium? Explain your answer.

14.33 The equilibrium constant for the decomposition of hydrogen iodide is 0.010 at 307 °C.

$$2HI(g) \rightleftharpoons H_2(g) + I_2(g)$$

Determine the direction of the reaction if the following amounts (in millimoles, where 1 mmol = 1.0×10^{-3} mol) of each compound are placed in a 1.0-L container.

	HI(g)	H_2(g)	I_2(g)	Direction
(a)	0.005	0.020	0.010	
(b)	0.020	0.20	0.20	
(c)	1.05	0.10	0.090	
(d)	2.00	0.050	0.080	

14.34 ■ The equilibrium constant for the water-gas shift reaction is 5.0 at 400 °C.

$$CO(g) + H_2O(g) \rightleftharpoons CO_2(g) + H_2(g)$$

Determine the direction of the reaction if the following amounts (in moles) of each compound are placed in a 1.0-L container.

	CO(g)	H_2O(g)	CO_2(g)	H_2(g)	Direction
(a)	0.50	0.40	0.80	0.90	
(b)	0.01	0.02	0.03	0.04	
(c)	1.22	1.22	2.78	2.78	
(d)	0.61	1.22	1.39	2.39	

OBJECTIVES **Predict the response of an equilibrium system to changes in conditions by applying Le Chatelier's principle. Determine how changes in temperature influence the equilibrium system.**

14.35 Some sulfur trioxide is sealed in a container and allowed to equilibrate at a particular temperature. The reaction is endothermic.

$$SO_3(g) \rightleftharpoons SO_2(g) + \frac{1}{2}O_2(g)$$

In which direction will the reaction proceed
(a) if more sulfur trioxide is added to the system?
(b) if oxygen is removed from the system?
(c) if the volume of the container is increased?
(d) if the temperature is increased?
(e) if argon is added to the container to increase the total pressure at constant volume?

14.36 ■ Consider the system

$$4NH_3(g) + 3O_2(g) \rightleftharpoons 2N_2(g) + 6H_2O(\ell) \qquad \Delta H = -1530.4 \text{ kJ}$$

(a) How will the concentration of ammonia at equilibrium be affected by
(1) removing $O_2(g)$?
(2) adding $N_2(g)$?
(3) adding water?
(4) expanding the container?
(5) increasing the temperature?
(b) Which of the above factors will increase the value of *K*? Which will decrease it?

OBJECTIVE **Use a systematic method, the iCe table, to solve chemical equilibria.**

14.37 Write the iCe table for the reaction and initial concentrations given in
(a) Exercise 14.29.
(b) Exercise 14.30.

14.38 Write the iCe table for the reaction and initial concentrations given in
(a) Exercise 14.31.
(b) Exercise 14.32.

14.39 Consider 1.0 mol each of hydrogen and iodine sealed in a 2.0-L flask at 1200 K.

$$H_2(g) + I_2(g) \rightleftharpoons 2HI(g)$$

The equilibrium constant, K_c, for this reaction is 4.6.
(a) Complete the iCe table, using y to represent the equilibrium concentration of HI.
(b) Write the expression for the equilibrium constant, K_c.

14.40 ■ If 1.0 mol each of SO_2 and NO_2 are sealed in a 1.0-L flask at 1500 K, they react to form SO_3 and NO:

$$SO_2(g) + NO_2(g) \rightleftharpoons SO_3(g) + NO(g)$$

The equilibrium constant, K_c, for this reaction is 1.98.
(a) Complete the iCe table, using y to represent the equilibrium concentration of SO_3.
(b) Write the algebraic expression for the equilibrium constant, K_c.
(c) Write the numerical expression for the equilibrium constant, K_c.

14.41 Exactly 0.500 mol each of sulfur trioxide and nitrogen monoxide are sealed in a 20.0-L flask.

$$SO_3(g) + NO(g) \rightleftharpoons SO_2(g) + NO_2(g)$$

K_c is 0.50 at 1500 K.
(a) Complete the iCe table.
(b) Write the algebraic expression for the equilibrium constant, K_c.
(c) Write the numerical expression for the equilibrium constant, K_c.

14.42 Exactly 2.0 mol each of carbon monoxide and water are sealed in a 4.0-L flask at 1100 K.

$$CO(g) + H_2O(g) \rightleftharpoons CO_2(g) + H_2(g)$$

The equilibrium constant, K_c, is 0.55 for this reaction.
(a) Complete the iCe table.
(b) Write the expression for the equilibrium constant, K_c.
(c) Write the polynomial form of the expression for the equilibrium constant.

OBJECTIVE Calculate equilibrium constants from experimental data and stoichiometric relationships.

14.43 When ammonia is placed in a reactor and the temperature is increased to 745 °C, some of the ammonia decomposes to nitrogen and hydrogen. The initial concentration of ammonia was 0.0240 *M*. After equilibrium is attained, the concentration of ammonia is 0.0040 *M*. Calculate K_c at 745 °C for

$$2NH_3(g) \rightleftharpoons N_2(g) + 3H_2(g)$$

14.44 ■ Chemists have conducted studies of the high temperature reaction of sulfur dioxide with oxygen in which the reactor initially contained 0.0076 *M* SO_2, 0.00360 *M* O_2, and no SO_3. After equilibrium was achieved, the SO_2 concentration decreased to 0.00320 *M*. Calculate K_c at this temperature for

$$2SO_2(g) + O_2(g) \rightleftharpoons 2SO_3(g)$$

14.45 Scientists have studied the decomposition of hydrogen iodide since the beginning of the 20th century. Some hydrogen iodide was placed in a reactor, which was then heated to 322 °C. The reaction was initiated with 1.25 atm hydrogen iodide; after equilibrium was attained, the partial pressure of HI had decreased to 1.05 atm. Calculate K_p at this temperature for

$$2HI(g) \rightleftharpoons H_2(g) + I_2(g)$$

14.46 ▲ Chemists studied the formation of phosgene, $COCl_2$, by sealing 0.96 atm carbon monoxide and 1.02 atm Cl_2 in a reactor at 682 K. The pressure declined smoothly from a total pressure of 1.98 atm to 1.22 atm as the system reached equilibrium. Calculate K_p for the following reaction:

$$CO(g) + Cl_2(g) \rightleftharpoons COCl_2(g)$$

***Gassed*, John Singer Sargent (1856-1925).** Sargent's 7-ft × 20-ft painting shows soldiers blinded by poison gas led to hospital tents with men on the ground waiting for treatment. It is thought that this particular painting depicts the effects of mustard gas although mixtures of chlorine and phosgene were also widely used in World War I.

OBJECTIVE Calculate the equilibrium concentrations of species in a chemical reaction.

14.47 A scientist seals 1.00 mol sulfur trioxide in a 1.00-L container, and the temperature is increased to 950 °C. Some SO_3 decomposes, forming sulfur dioxide and oxygen:

$$2SO_3(g) \rightleftharpoons 2SO_2(g) + O_2(g)$$

At equilibrium, experiments show that 0.50 mol sulfur trioxide is left.
(a) Calculate the concentrations of all species.
(b) Calculate K_c.

14.48 A 5.0-L reaction flask initially contains 0.030 mol sulfuryl chloride at 177 °C. Sulfur dioxide and chlorine form:

$$SO_2Cl_2(g) \rightleftharpoons SO_2(g) + Cl_2(g)$$

After equilibrium is established, chemical analysis shows that 0.0010 mol sulfuryl chloride remains.
(a) Calculate the concentrations of all species.
(b) Calculate K_c.

14.49 A 20.0-L container initially contains 2.00 mol NO_2, which reacts to form some dinitrogen tetroxide.

$$2NO_2(g) \rightleftharpoons N_2O_4(g)$$

At equilibrium, chemical analysis reveals that the concentration of $NO_2(g)$ is 0.010 *M*.
(a) Calculate the number of moles of $NO_2(g)$.
(b) Calculate the number of moles of $N_2O_4(g)$ formed and its concentration.
(c) Calculate K_c.

14.50 A 100-L reaction container is charged with 0.50 mol NOBr, which decomposes at 120 °C.

$$NOBr(g) \rightleftharpoons NO(g) + \frac{1}{2}Br_2(g)$$

The equilibrium concentration of bromine is 2.0×10^{-3} *M*.
(a) Calculate the concentrations of all species.
(b) Calculate K_c.

14.51 A 2.00-L container at 463 K contains 0.500 mol phosgene. The equilibrium constant, K_c, is 4.93×10^{-3} for

$$COCl_2(g) \rightleftharpoons CO(g) + Cl_2(g)$$

Calculate the concentrations of all species after the system reaches equilibrium.

14.52 Consider 0.200 mol phosphorus pentachloride sealed in a 2.0-L container at 620 K. The equilibrium constant, K_c, is 0.60 for

$$PCl_5(g) \rightleftharpoons PCl_3(g) + Cl_2(g)$$

Calculate the concentrations of all species after equilibrium has been reached.

14.53 Exactly 0.400 mol each of carbon monoxide and chlorine are sealed in a 2.00-L container at 919 °C. K_c is 7.52 for

$$CO(g) + Cl_2(g) \rightleftharpoons COCl_2(g)$$

Calculate the concentrations of all species after equilibrium has been reached.

14.54 ■ Consider a system that initially contains 0.100 mol each of phosphorus trichloride and chlorine sealed in a 10.0-L container. The temperature is increased to 291 °C where the equilibrium constant, K_c, is 8.18 for

$$PCl_3(g) + Cl_2(g) \rightleftharpoons PCl_5(g)$$

Calculate the concentrations of all species after the system reaches equilibrium.

14.55 Calculate the concentrations of all species formed when 1.00 mol sulfur dioxide and 2.00 mol chlorine are sealed in a 100.0-L reactor. The temperature is increased to 400 K, where K_c is 89.3 for

$$SO_2(g) + Cl_2(g) \rightleftharpoons SO_2Cl_2(g)$$

Calculate the equilibrium concentrations of all species.

14.56 Consider 1.00 mol carbon monoxide and 2.00 mol chlorine sealed in a 3.00-L container at 476 °C. The equilibrium constant, K_c, is 2.5 for

$$CO(g) + Cl_2(g) \rightleftharpoons COCl_2(g)$$

Calculate the equilibrium concentrations of all species.

14.57 Exactly 2.00 atm carbon monoxide and 3.00 atm hydrogen are placed in a reaction vessel. K_p is 5.6 for

$$CO(g) + H_2(g) \rightleftharpoons CH_2O(g)$$

Calculate the equilibrium pressures of all species.

14.58 ▲ Consider a 10.0-L vessel that contains 0.15 mol phosphorus trichloride, 0.20 mol chlorine, and 0.25 mol phosphorus pentachloride at 332 °C. K_p is 2.5 for

$$PCl_3(g) + Cl_2 \rightleftharpoons PCl_5(g)$$

Calculate the equilibrium pressures of all species.

14.59 ▲ Exactly 4 mol sulfur trioxide is sealed in a 5.0-L container at 1529 K. K_p is 1150 for

$$2SO_3(g) \rightleftharpoons 2SO_2(g) + O_2(g)$$

Calculate the concentrations of all species at equilibrium. (*Hint:* This reaction goes essentially to completion.)

14.60 ▲ Consider the formation of hydrogen fluoride (HF):

$$H_2(g) + F_2(g) \rightleftharpoons 2HF(g)$$

If a 2.0-L nickel reaction container (glass cannot be used because it reacts with HF) filled with 0.010 *M* H_2 is connected to a 5.0-L container of 0.020 *M* F_2, and K_p is 7.8×10^{14}, calculate the concentrations of all species at equilibrium. (*Hint:* This reaction goes essentially to completion.)

OBJECTIVE Write equilibrium constant expressions for heterogeneous equilibria.

14.61 Write the expression for the equilibrium constant and calculate the partial pressure of $CO_2(g)$, given that K_p is 0.12 (at 1000 K) for

$$CaCO_3(s) \rightleftharpoons CaO(s) + CO_2(g)$$

14.62 Write the expression for the equilibrium constant and calculate the partial pressure of $CO_2(g)$, given that K_p is 0.25 (at 427 °C) for

$$NaHCO_3(s) \rightleftharpoons NaOH(s) + CO_2(g)$$

14.63 Write the expression for the equilibrium constant and calculate the partial pressure of $SO_3(g)$, given K_p is 0.74 (at 2100 K) for

$$CaSO_4(s) \rightleftharpoons CaO(s) + SO_3(g)$$

14.64 Write the expression for the equilibrium constant and calculate the partial pressure of $CO_2(g)$, given K_p is 1.25 (at 1500 K) for

$$Na_2CO_3(s) \rightleftharpoons Na_2O(s) + CO_2(g)$$

OBJECTIVE Write the expression for the solubility product constant.

14.65 Write the expression for the solubility product for the dissolution of
(a) magnesium fluoride.
(b) calcium phosphate.
(c) aluminum carbonate.
(d) lanthanum fluoride.

14.66 Write the expression for the solubility product for the dissolution of
(a) barium sulfate.
(b) silver acetate.
(c) copper(I) carbonate.
(d) gold(III) chloride.

OBJECTIVE Calculate K_{sp} from experimental data.

14.67 Silver iodide is sprayed from airplanes by modern "rainmakers" in an attempt to coax rain from promising cloud formations. The silver iodide crystals provide "seeds," that is, sites for condensation of water. Silver iodide satisfies two requirements needed to form water drops. First, the crystals are quite small, and second, the solubility in water is extremely low. If the solubility of silver iodide in water is 9×10^{-9} M, calculate K_{sp} for AgI.

AP Photo/Charlie Riedel

Cloud seeding. The wing-mounted silver iodide generator seeding clouds above the skies over western Kansas.

14.68 The solubility of silver iodate, $AgIO_3$, is 1.8×10^{-4} M. Calculate the solubility product constant.

14.69 Given the following data, calculate the solubility product constant.
(a) The solubility of barium chromate, $BaCrO_4$, is 1.1×10^{-5} M.
(b) The solubility of cesium permanganate is 0.22 g/100 mL.
(c) The solubility of silver phosphate, Ag_3PO_4, is 4.4×10^{-5} M.

14.70 Given the following data, calculate the solubility product constant.
(a) The solubility of silver sulfate, Ag_2SO_4, is 1.0×10^{-2} M.
(b) The solubility of potassium iodate, KIO_3, is 43 g/L.
(c) The solubility of cadmium fluoride is 0.12 M.

OBJECTIVE Calculate solubility of slightly soluble salts from K_{sp}.

14.71 The solubility product constant for silver tungstate, Ag_2WO_4, is 5.5×10^{-12}. Calculate the solubility of this compound in water.

14.72 ■ The solubility product constant for copper(II) iodate, $Cu(IO_3)_2$, is 7.4×10^{-8}. Calculate the solubility of this compound in water.

14.73 Even though barium is toxic, a suspension of barium sulfate is administered to patients who need x rays of the gastrointestinal tract. The barium "milkshake" is safe to drink because the solubility of barium sulfate is so low. Calculate the solubility of barium sulfate in grams per liter using the data in Table 14.4.

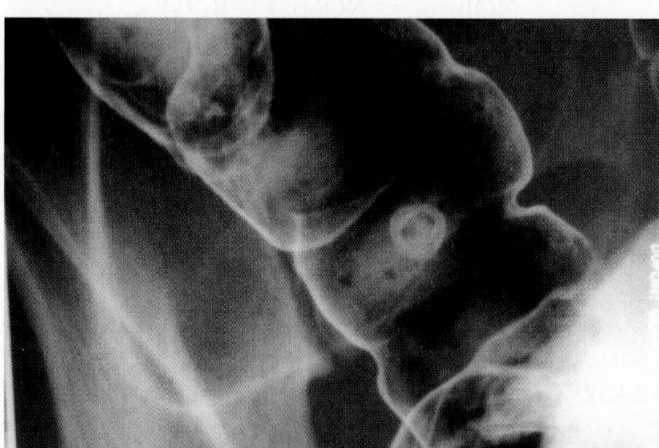
CMSP

Barium sulfate is x-ray opaque. The ingestion of the insoluble barium sulfate allows the doctors to detect a polyp in the colon of this patient.

14.74 Lead poisoning has been a hazard for centuries. Some scholars believe that the decline of the Roman Empire can be traced, in part, to high levels of lead in water from containers and pipes, and from wine that was stored in lead-glazed containers. If we presume that the typical Roman water supply was saturated with lead carbonate, $PbCO_3$ ($K_{sp} = 7.4 \times 10^{-14}$), how much lead will a Roman ingest in a year if he or she drinks 1 L/day from the container?

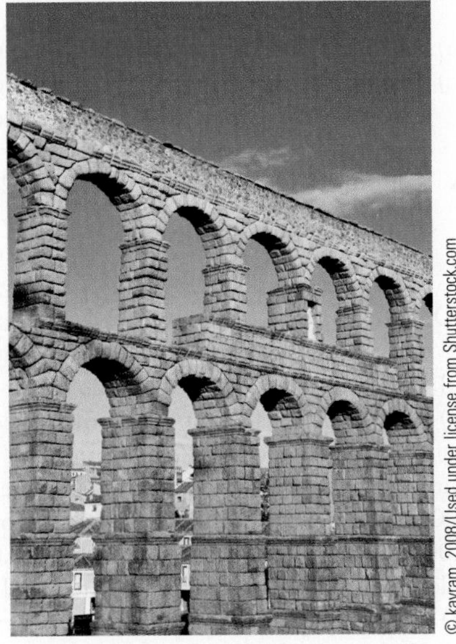

© kavram, 2008/Used under license from Shutterstock.com

Roman aqueduct.

OBJECTIVES Predict the solubility of a solid in a solution that contains a common ion. Use approximations to calculate roots of polynomial equations.

14.75 Calculate the solubility of barium sulfate ($K_{sp} = 1.1 \times 10^{-10}$) in
(a) water.
(b) a 0.10 M barium chloride solution.

14.76 Calculate the solubility of copper(II) iodate, $Cu(IO_3)_2$ ($K_{sp} = 7.4 \times 10^{-8}$), in
(a) water.
(b) a 0.10 M copper(II) nitrate solution.

14.77 Calculate the solubility of lead fluoride, PbF_2 ($K_{sp} = 3.3 \times 10^{-8}$), in
(a) water.
(b) a 0.050 M potassium fluoride solution.

14.78 ■ Calculate the solubility of zinc carbonate, $ZnCO_3$ ($K_{sp} = 1.5 \times 10^{-10}$), in
(a) water.
(b) 0.050 M $Zn(NO_3)_2$.
(c) 0.050 M K_2CO_3.

OBJECTIVE Determine whether a precipitate will form under a particular set of conditions.

14.79 Use the solubility product constant from Appendix F to determine whether a precipitate will form if 10 mL 0.0010 M $AgNO_3$ is added to 10 mL 0.0010 M Na_2SO_4.

14.80 Use the solubility product constant from Appendix F to determine whether a precipitate will form if 20.0 mL of 1.0×10^{-6} M magnesium chloride is added to 80.0 mL of 1.0×10^{-6} M potassium fluoride.

14.81 Use the solubility product constant from Appendix F to determine whether a precipitate will form if 10.0 mL of 1.0×10^{-6} M iron(II) chloride is added to 20.0 mL of 3.0×10^{-4} M barium hydroxide.

14.82 ■ Use the solubility product constant from Appendix F to determine whether a precipitate will form if 25.0 mL of 0.010 M NaOH is added to 75.0 mL of a 0.10 M solution of magnesium chloride?

14.83 ▲ Some barium chloride is added to a solution that contains both K_2SO_4 (0.050 M) and Na_3PO_4 (0.020 M).
(a) Which begins to precipitate first: the barium sulfate or the barium phosphate?
(b) The concentration of the first anion species to precipitate, either the sulfate or phosphate, decreases as the precipitate forms. What is the concentration of the first species when the second begins to precipitate?

14.84 ▲ Some trisodium phosphate, Na_3PO_4, is added to a solution that contains 0.0020 M aluminum nitrate and 0.0040 M calcium chloride.
(a) Which begins to precipitate first: the aluminum phosphate or the calcium phosphate?
(b) The concentration of the first ion to precipitate, either Al^{3+} or Ca^{2+}, decreases as the precipitate forms. What is the concentration of the first species when the second one begins to precipitate?

Chapter Exercises

14.85 A scientist seals some PCl_5 in a 20.0-L flask. After equilibrium is attained, chemical analysis shows that the flask contains 0.10 mol $PCl_5(g)$ and 0.20 mol each of $PCl_3(g)$ and $Cl_2(g)$. Calculate the equilibrium constant at the reaction temperature for

$$PCl_5(g) \rightleftharpoons PCl_3(g) + Cl_2(g)$$

14.86 To evaluate the equilibrium constant for

$$2NO_2(g) \rightleftharpoons N_2O_4(g)$$

a scientist seals 0.200 mol nitrogen dioxide in a 2.5-L container. At equilibrium, the reaction vessel is found to contain 0.15 mol nitrogen dioxide and 0.025 mol dinitrogen tetroxide. Calculate the equilibrium constant for the reaction at this particular temperature.

14.87 A total of 0.010 mol sulfur trioxide is sealed in a 2.0-L container. The temperature is increased to 832 °C, and some SO_3 decomposes, forming sulfur dioxide and oxygen:

$$2SO_3(g) \rightleftharpoons 2SO_2(g) + O_2(g)$$

Chemical analysis of the equilibrium mixture at 832 °C finds 0.0040 mol oxygen.
(a) Calculate the concentrations of all species.
(b) Calculate K_c.
(c) Calculate K_p.

14.88 Exactly 0.030 mol phosphorus pentachloride is sealed in a 500-mL container at 538 °C. Phosphorus trichloride and chlorine are formed:

$$PCl_5(g) \rightleftharpoons PCl_3(g) + Cl_2(g)$$

After equilibrium is established, chemical analysis shows that 0.0100 mol phosphorus pentachloride is present, the rest having reacted.
(a) Calculate the concentrations of all species.
(b) Calculate K_c at 538 °C.
(c) Calculate K_p at 538 °C.

14.89 Exactly 0.010 mol hydrogen iodide is sealed in a 5.0-L container. The temperature is increased to 3000 K. At this temperature, K_p is 0.050 for

$$2HI(g) \rightleftharpoons H_2(g) + I_2(g)$$

Calculate the equilibrium concentrations of all species.

14.90 Find the concentration of silver necessary to begin precipitation of AgCl from a solution in which the Cl^- concentration is 7.4×10^{-4} *M*.

Cumulative Exercises

14.91 Exactly 10.0 mL of a 0.0502 *M* KCl solution is added to 20.0 mL of 0.0259 *M* $AgNO_3$.
(a) Write the net ionic equation for the reaction.
(b) Calculate the mass of the insoluble product formed.
(c) Calculate the concentrations of the anions in the equilibrium solution.

14.92 At 3000 K, carbon dioxide dissociates

$$CO_2(g) \rightleftharpoons CO(g) + \frac{1}{2}O_2(g)$$

K_p for this reaction is 2.48. Calculate K_c for

$$2CO(g) + O_2(g) \rightleftharpoons 2CO_2(g)$$

14.93 5.0 mmol $SO_2(g)$ and 5.0 mmol $O_2(g)$ are sealed in a 1.0-L container. The mixture is heated for several hours and sulfur trioxide forms.

$$2SO_2(g) + O_2(g) \rightleftharpoons 2SO_3(g)$$

Chemical analysis of the reaction mixture shows that 2.0 mmol $SO_3(g)$ forms. The experiment is repeated, except with twice as much (10.0 mmol) $SO_2(g)$ and $O_2(g)$.
(a) *Estimate* the amount of $SO_3(g)$ formed in the second experiment.
(b) Calculate K_c from the results of the first experiment.
(c) Define *y* as the change in the oxygen concentration, and write the polynomial equation needed to solve for *y* in the second experiment.

14.94 ▲ Sulfur dioxide reacts with chlorine at 227 °C:

$$SO_2(g) + Cl_2(g) \rightleftharpoons SO_2Cl_2(g)$$

K_p for this reaction is 5.1×10^{-2}. Initially, 1.00 g each of SO_2 and Cl_2 are placed in a 1.00-L reaction vessel. After 15 minutes, the concentration of SO_2Cl_2 is 45.5 μg/mL.
(a) Has the system reached equilibrium?
(b) If the system is not at equilibrium, calculate the mass of SO_2Cl_2 expected at equilibrium.

14.95 Nitrogen, hydrogen, and ammonia are in equilibrium in a 1000-L reactor at 550 K. The concentration of N_2 is 0.00485 *M*, H_2 is 0.022 *M*, and NH_3 is 0.0016 *M*. The volume of the container is halved, to 500 L.
(a) Calculate the equilibrium constant.
(b) Define *y* as the change in the concentration of nitrogen in the 500-L container. Is *y* positive or negative?
(c) Write the iCe table and express the equilibrium constant in a polynomial in terms of *y*.

14.96 The concentration of barium in a saturated solution of barium sulfate at a particular temperature is 1.2 μg/mL. Calculate K_{sp} at this temperature.

14.97 According to the Resource Conservation and Recovery Act (RCRA), waste material is classified as toxic and must be handled as hazardous if the lead concentration exceeds 5 mg/L. By adding chloride ion, the lead ion will precipitate as $PbCl_2$, which can be separated from the liquid portion. Once the lead has been removed, the rest of the waste can be sent to a conventional waste treatment facility. How many grams of sodium chloride must be added to 500 L of a waste solution to reduce the concentration of the Pb^{2+} ion from 10 to 5 mg/L?

14.98 Will a precipitate form if 20.0 mL of 10 μg/mL solution of barium ion is mixed with 25.0 mL of 0.050 *M* potassium sulfate, K_2SO_4?
(a) Write the net ionic equation for the reaction.
(b) Calculate the mass of precipitate formed, if any.
(c) Calculate the concentrations of all species in solution.

14.99 ▲ You are in the silver recovery business. You receive 55-gal drums of waste silver solution from photography laboratories, metal plating operations, and other industrial companies. Currently, you have a drum of waste from a silver-plating factory with the silver concentration at 1 oz/gal. (Precious metals are measured in troy ounces, which are 31.103 g. Other conversion factors are in Appendix C.) You precipitate the silver as the chloride, which you sell for \$2.00/g. The sodium chloride, which is the source of the chloride ion, costs \$0.46/kg.
(a) Calculate the mass of sodium chloride needed to react completely with the silver present in one drum of waste.
(b) If more sodium chloride is added, more silver chloride precipitates, in accordance with the common ion effect. Eventually, the value of the silver precipitated is lower than the cost of the sodium chloride added. Determine the mass of silver left in the drum after the sodium chloride (computed in part a) was added by performing a solubility product calculation. Determine the monetary value of this quantity of silver.
(c) Silver is one of eight metals covered by the Resource Conservation and Recovery Act (RCRA). Express the concentration of silver present after the precipitation reaction in terms of g/mL. The legal limit for silver-containing discharges is 5 μg/mL. Is the residual silver concentration above or below this level? If the silver concentration is above the limit, calculate how much additional sodium chloride must be added to one drum to use the common ion effect to reduce it to 5.0 μg/mL.

Uranium Enrichment Facility? This satellite photograph shows a potential enrichment plant said to be operating in violation of United Nations' sanctions.

Although most people think of acids and bases, the topics of this chapter, as acrid and corrosive liquids, these compounds have an amazingly large range of properties.

Acids include:

- aspirin—used to fight pain
- gibberellins—used to regulate plant growth
- cocaine—a topical anesthetic used principally to numb the eye
- lactic acid—formed as a by-product of muscle activity
- amino acids—building blocks of proteins

One interesting acid is hydrofluoric acid. This acid is so reactive that it dissolves glass! The dissolution process is interesting because it is driven by Le Chatelier's principle, exactly as described in Chapter 14. The principal compound in glass is silicon dioxide, which reacts with hydrofluoric acid to produce water and silicon tetrafluoride:

$$4HF(aq) + SiO_2(s) \rightleftharpoons 2H_2O(\ell) + SiF_4(g)$$

Silicon tetrafluoride, like many fluorides, is volatile. It boils at −95 °C, so it is a gas under normal temperatures and pressures. As silicon tetrafluoride gas escapes from its container, Le Chatelier's principle predicts that more SiF_4 gas will form, and the reaction will continue until either the HF or the SiO_2 is completely consumed. This reaction is used to dissolve many silicate-containing minerals to determine whether they contain elements such as gold, silver, copper, and uranium in quantities high enough to make their recovery economically viable.

The high volatility of the SiF_4 is related to the weakness of the intermolecular forces between the molecules. This phenomenon is not specific to silicon tetrafluoride, but to fluorides in general. The forces between adjacent molecules are quite small, so it takes little energy to break the molecules apart as the solid becomes a liquid and then a gas. This phenomenon has been exploited in many important industrial processes. For example, uranium isotopes can be separated as the hexafluorides. UF_6 is a gas at about 65 °C. The gas is placed in a cylinder, put in a centrifuge, and spun rapidly. The heavier $^{238}UF_6$ moves to the outside and the lighter $^{235}UF_6$ stays near the center, where it is removed to a second centrifuge to increase the concentration of the fissionable $^{235}UF_6$ even further.

Solutions of Acids and Bases

15

CHAPTER CONTENTS

OWL Online homework for this chapter may be assigned in OWL.

Look for the green colored vertical bar throughout this chapter, for integrated references to this chapter introduction.

Hydrofluoric acid is used to make cryolite, sodium aluminum fluoride, Na_3AlF_6, which is the electrolyte used in the production of aluminum. Aluminum oxide ores dissolve in molten cryolite, and the aluminum is produced by an electrochemical reaction (see discussion in Chapter 18).

The volatility of fluorides was not lost on the chemists of the 16th and 17th centuries. They were able to produce HF by reaction of sulfuric acid with calcium fluoride (fluorite, a fairly common mineral). In fact, as mentioned in the introduction to Chapter 8, the name "fluorine" comes from the Latin *fluere,* meaning "to flow." Early chemists coined an expression, "Fluorine adds wings to the elements," that refers to the volatility of many fluorides.

If HF dissolves glass, what kinds of containers can be used to store it? When HF is used to dissolve minerals, chemists generally use Teflon beakers or, if extensive heating is required, crucibles made from platinum. Currently, hydrofluoric acid is shipped in plastic containers, but these containers were not available to chemists 300 years ago. The chemists used beeswax containers, a practice that continued into the 20th century.

Chemists have to be careful not to spill hydrofluoric acid on their skin. HF burns show little pain initially, but they cannot be ignored and require immediate treatment. Within several hours, the acid reacts with calcium and magnesium in the tissues and bones, forming insoluble CaF_2 and MgF_2; therefore, a spill can result in tissue and bone damage, even when the skin is not badly burned. First responders use calcium gluconate gel and apply it to the skin exposed to HF; calcium injections are called for in severe cases. ▮

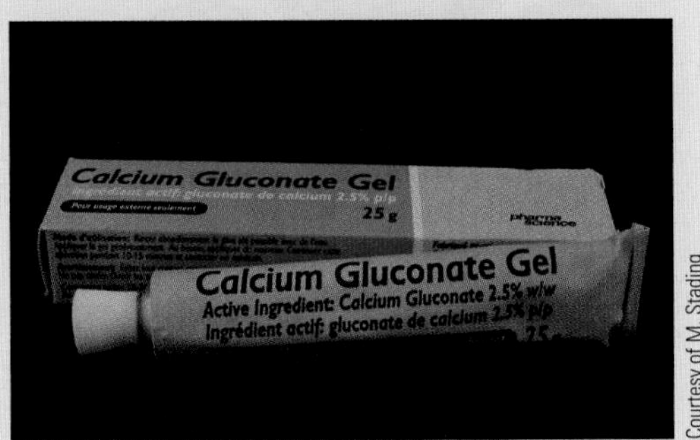

Courtesy of M. Stading

Calcium gluconate gel.

Courtesy of M. Stading

Beeswax bottle. This container, about 100 years old, held aluminum fluoride, which produced hydrofluoric acid on contact with the moisture in air.

One important characteristic of matter is based on the properties of the solution produced when the material dissolves in water. A fairly simple test helps distinguish between two different types of solutions: acids and bases. Solutions of acids taste sour, and those of bases taste bitter. Lemon juice is sour because of the presence of citric acid; soap is bitter because it is a base. Notably, tasting is neither a safe nor a sure way to identify any chemical, although it was widely used in the past.

All acids share common properties, and so do all bases. One example is the effect that acids exert on many natural compounds, causing them to change color (see Figure 15.1). Chemists have performed many experiments to determine whether acids have unique structural features that are responsible for their behaviors. All of these early experiments involved solutions of acids and bases in water. The results of those experiments led the Swedish chemist Svante Arrhenius (1859–1927) to propose a model of an acid as a substance that increases the concentration of hydrogen ions when dissolved in water. The Arrhenius model of a base is a substance that increases the concentration of hydroxide ions when dissolved in water. These descriptions are the ones we have used in this textbook beginning in Chapter 3.

In this chapter, we present other models for describing acids and bases: the Brønsted–Lowry model and the Lewis model. These models extend the definitions of acids and bases. In addition, we explain equilibria that involve acids and bases, show how to calculate concentrations of the species in an equilibrium system, and examine some relationships between structures and the acid-base properties of molecules.

15.1 Brønsted–Lowry Acid-Base Systems

OBJECTIVES

- ☐ Define Brønsted–Lowry acids and bases
- ☐ Differentiate between Brønsted–Lowry and Arrhenius acids and bases
- ☐ Identify conjugate acid-base pairs
- ☐ Identify the proton transfer in Brønsted–Lowry acid-base reactions

It is worthwhile to review some of the terms used to describe acids and bases, terms that were first introduced in Chapter 3. A hydrogen ion, H^+, is also referred to as a proton. It is important to remember that, in solution, the hydrogen ion is always associated with the solvent; "proton" does not refer to the nuclear particle. The designations $H^+(aq)$ and H_3O^+ emphasize the role of the solvent. Although H^+ is called a *hydrogen ion* and H_3O^+ is called a *hydronium ion,* the terms are interchangeable when used to describe solutions in which water is the solvent.

"H_3O^+," "$H^+(aq)$," "H^+," "hydrogen ion," "hydronium ion," and "proton" are all used interchangeably to describe the same species in water solutions.

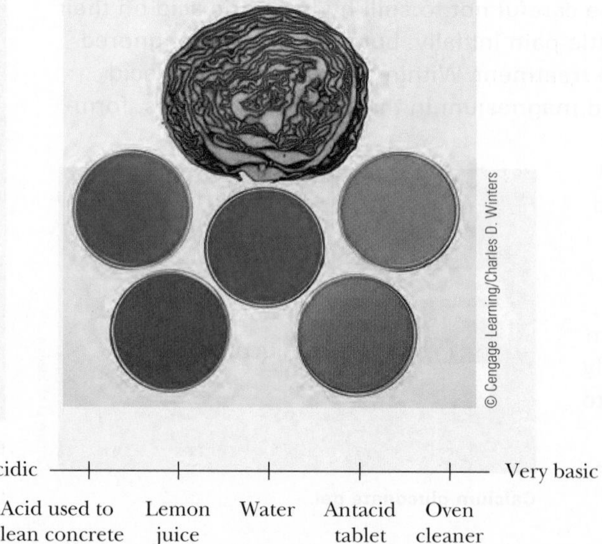

Figure 15.1 Cabbage juice. The juice of red cabbage changes color when acids or bases are added.

Scientists have successfully used the Arrhenius definition of acids and bases to predict the results of many chemical reactions. In the 1920s, two scientists, Johannes Brønsted (1879–1947) and Thomas Lowry (1874–1936), independently noted the importance of acid-base behavior in systems other than water and recognized the limitations of the Arrhenius model. The Arrhenius definition of an acid—a substance that increases the hydrogen ion concentration in aqueous solution—although correct, is quite limited. The Arrhenius definition applies only to aqueous solutions and does not describe the behaviors of substances in other solvents or in gas-phase reactions. Brønsted and Lowry recognized that the basis for an acid-base reaction is the transfer of a proton (H^+ ion) from one species to another, and they proposed a model based on this concept. A **Brønsted–Lowry acid** is defined as any species that acts as a proton donor. A **Brønsted–Lowry base** is defined as any species that acts as a proton acceptor. This section discusses the Brønsted–Lowry definitions of acids and bases, and applies them to characterize several chemical reactions.

An Arrhenius acid increases the hydrogen ion concentration when it dissolves in water. An Arrhenius base increases the hydroxide ion concentration.

A Brønsted–Lowry acid is a proton donor; a Brønsted–Lowry base is a proton acceptor.

The Brønsted–Lowry model is an extension of the Arrhenius model. In water, species that are Brønsted–Lowry acids are also Arrhenius acids, and vice versa. However, the Arrhenius model does not include gas-phase reactions such as the formation of ammonium chloride:

$$\mathrm{H{-}\overset{\displaystyle H}{\underset{\displaystyle H}{N}}{:}} + \mathrm{H{-}\ddot{\underset{..}{Cl}}{:}} \longrightarrow \left[\mathrm{H{-}\overset{\displaystyle H}{\underset{\displaystyle H}{N}}{-}H}\right]^+ \; \mathrm{:\ddot{\underset{..}{Cl}}:^-}$$

$$NH_3(g) + HCl(g) \longrightarrow NH_4Cl(s)$$

The hydrogen-chlorine bond in the hydrogen chloride molecule breaks, with both shared electrons remaining with the chlorine. The hydrogen ion transfers to the ammonia molecule, resulting in the formation of ammonium chloride, an ionic solid. You may have noticed a haze on the glassware (and maybe even the windows) in your laboratory. This film is likely to be ammonium chloride formed by the gas-phase reaction of HCl and NH_3 that are often present in the air of chemistry laboratories. The preceding equation shows that ammonia accepts a proton, so ammonia is a base in the Brønsted–Lowry scheme. The hydrogen chloride donates a proton, so it is the Brønsted–Lowry acid.

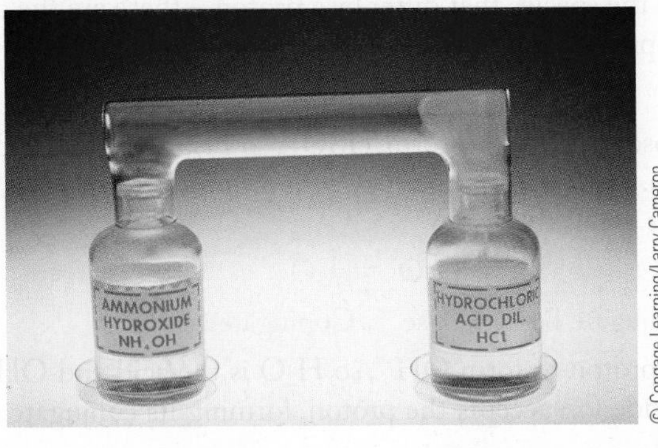

Gas-phase acid-base reaction. The white cloud visible on the right is composed of small $NH_4Cl(s)$ particles formed by the reaction of HCl(g) and $NH_3(g)$. Some containers of aqueous ammonia are still labeled NH_4OH, although $NH_3(aq)$ is preferred.

Conjugate Acid-Base Pairs

When hydrogen fluoride dissolves in water, it transfers a proton to the water, forming a hydrogen ion and a fluoride ion, F^-; thus, HF is an acid.

$$HF(aq) + H_2O(\ell) \rightarrow H_3O^+(aq) + F^-(aq)$$

In addition, a proton in the solution can form a bond with F^-, producing HF. The fluoride ion accepts a proton and behaves as a base.

$$F^-(aq) + H_3O^+(aq) \rightarrow HF(aq) + H_2O(\ell)$$

TABLE 15.1 Conjugate Acid-Base Pairs

Acid		Base	
Species	Formula	Formula	Species
Hydrogen chloride	HCl	Cl^-	Chloride ion
Sulfuric acid	H_2SO_4	HSO_4^-	Hydrogen sulfate ion
Hydrogen sulfate ion	HSO_4^-	SO_4^{2-}	Sulfate ion
Acetic acid	CH_3COOH	CH_3COO^-	Acetate ion
Ammonium ion	NH_4^+	NH_3	Ammonia
Hydronium ion	H_3O^+	H_2O	Water
Hydrogen ion	H^+		
Water	H_2O	OH^-	Hydroxide ion

Hydrogen fluoride and the fluoride ion are typical of acids and bases in general. When a Brønsted–Lowry acid transfers a proton, it also forms a base, because the remaining species can accept a proton. The two species are a **conjugate acid-base pair.** They are related by the loss and gain of a proton. In a conjugate acid-base pair, the acid form of the pair is protonated, whereas the base form has lost the proton. Every Brønsted–Lowry acid has a conjugate base, and every base has a conjugate acid. Table 15.1 lists several common conjugate acid-base pairs.

Conjugate acid-base pairs differ by a proton, which is present in the acid form and missing in the base form.

Notice that water appears in both the acid and base columns. It can behave either as a proton donor (an acid) or as a proton acceptor (a base), depending on the species with which it reacts. A substance that can act either as an acid or as a base is said to be **amphoteric.** The amphoteric behavior of water is important to an understanding of the properties of acids and bases in aqueous chemistry, and is discussed in detail later. One other species in Table 15.1, the hydrogen sulfate ion, HSO_4^-, is also an amphoteric species.

Water is an amphoteric substance because it can act as an acid or as a base.

EXAMPLE 15.1 Identifying Conjugate Acid-Base Pairs

Identify the conjugate pairs.

(a) $H_2SO_4(aq) + H_2O(\ell) \rightarrow HSO_4^-(aq) + H_3O^+(aq)$
(b) $H_2O(\ell) + F^-(aq) \rightleftharpoons OH^-(aq) + HF(aq)$

Strategy Identify the species that differ by a proton—these are the conjugate pairs. The acid form has the proton; the base form lacks the proton.

Solution

(a) The H_2SO_4 loses a proton to form HSO_4^-, so H_2SO_4 is the acid and HSO_4^- is its conjugate base. H_2O is a base, accepting a proton to form H_3O^+. The conjugate acid-base pairs are as follows:

H_2SO_4	HSO_4^-	H_2O	H_3O^+
Acid	Conjugate base	Base	Conjugate acid

(b) Water loses a proton to form OH^-, so H_2O is the acid and OH^- is its conjugate base. The fluoride ion accepts the proton, forming its conjugate acid, HF.

H_2O	OH^-	F^-	HF
Acid	Conjugate base	Base	Conjugate acid

Understanding

Identify the conjugate acid-base pairs in

$$SO_4^{2-}(aq) + HCl(aq) \rightarrow HSO_4^-(aq) + Cl^-(aq)$$

Answer

HCl	Cl^-	SO_4^{2-}	HSO_4^-
Acid	Conjugate base	Base	Conjugate acid

TABLE 15.2 Representative Acid-Base Reactions

Reaction	Acid Form of Species 1	+	Base Form of Species 2	→	Base Form of Species 1	+	Acid Form of Species 2
1	HF	+	NH_3	→	F^-	+	NH_4^+
2	HCl	+	H_2O	→	Cl^-	+	H_3O^+
3	H_2O	+	NH_2^-	→	OH^-	+	NH_3

Reactions of Acids and Bases

Acids and bases react quickly with each other and reach equilibrium at speeds that are often limited by how fast the solution is stirred. When an Arrhenius acid reacts with an Arrhenius base, the products are water and a salt.

acid + base → water + salt

The Brønsted–Lowry model, as you might guess, is not limited to this type of reaction. *An acid transfers a proton to a base to form the conjugate base of the original acid and the conjugate acid of the original base.*

Row 1 in Table 15.2 is the equation for the reaction of hydrofluoric acid with ammonia. The conjugate base of hydrofluoric acid is the fluoride ion, F^-; the conjugate acid of ammonia is the ammonium ion, NH_4^+.

Rows 2 and 3 in Table 15.2 illustrate the amphoteric behavior of water. In row 2, water acts as a Brønsted–Lowry base (defined as a proton acceptor) because it accepts a hydrogen ion from HCl to form the hydronium ion, H_3O^+.

$$HCl + H_2O \longrightarrow Cl^- + H_3O^+$$

A Brønsted–Lowry acid-base reaction involves a transfer of a proton from the acid to the base.

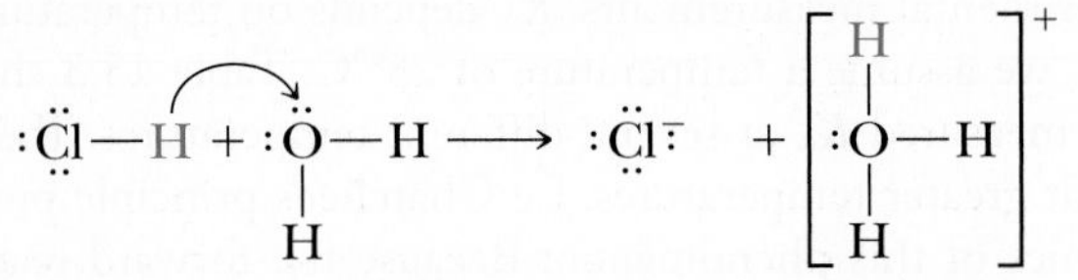

In row 3, water acts as an acid. It donates a proton to the amide ion (NH_2^-), forming ammonia. In this example, the amide ion is the base (proton acceptor) and water is the acid (proton donor).

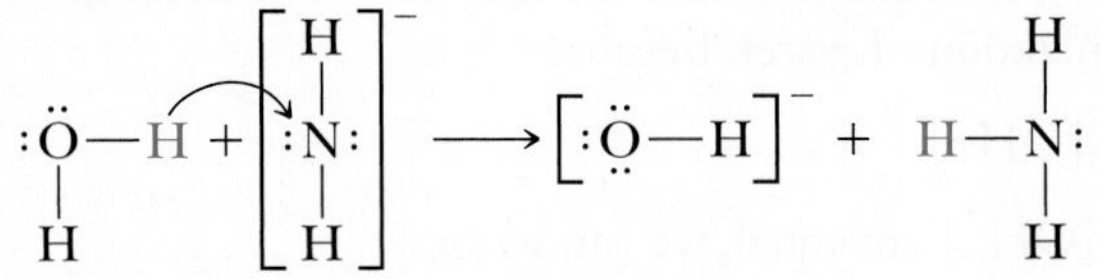

$$H_2O + NH_2^- \longrightarrow OH^- + NH_3$$

OBJECTIVES REVIEW *Can you:*

- ☑ define Brønsted–Lowry acids and bases?
- ☑ differentiate between Brønsted–Lowry and Arrhenius acids and bases?
- ☑ identify conjugate acid-base pairs?
- ☑ identify the proton transfer in Brønsted–Lowry acid-base reactions?

15.2 Autoionization of Water

OBJECTIVES

- ☐ Relate hydrogen ion concentration to hydroxide ion concentration in aqueous solutions
- ☐ Define pH and use it to express concentrations
- ☐ Convert between hydrogen ion concentrations, hydroxide ion concentrations, pH, and pOH

TABLE 15.3 Values of the Equilibrium Constant for the Autoionization of Water at Several Temperatures

Temperature (°C)	$K_w = [H_3O^+][OH^-]$
10	0.292×10^{-14}
25	1.008×10^{-14}
50	5.474×10^{-14}

Water has the ability to act as either an acid or a base. This section develops the relationship between the concentrations of the hydrogen ions and the hydroxide ions in aqueous solutions of acids and bases.

A water molecule can donate a proton (forming OH^-) or accept a proton (forming H_3O^+), depending on the experimental conditions. Water can even react with itself.

$$H_2O(\ell) + H_2O(\ell) \rightleftharpoons H_3O^+(aq) + OH^-(aq)$$

This equation illustrates the **autoionization of water,** which occurs to a small extent. The equilibrium constant for this reaction is

$$K' = \frac{[H_3O^+][OH^-]}{[H_2O]^2}$$

Water is treated like any other pure liquid; its concentration does not appear as a separate term in the expression for the equilibrium constant. The ionization of water is so important to the study of aqueous equilibria that the equilibrium constant is given the special symbol K_w.

$$K_w = [H_3O^+][OH^-]$$

The value of K_w is determined by measuring the concentrations of H_3O^+ and OH^-. In pure water, the only source of H_3O^+ and OH^- is the autoionization reaction, so only one species must be measured because the stoichiometry requires that $[H_3O^+]$ equals $[OH^-]$.

At 25 °C, the value for K_w is remarkably close to an even number, 1.0×10^{-14}. This result is an easy-to-remember coincidence, because K_w, like all equilibrium constants, is the result of experimental measurements. K_w depends on temperature; if the temperature is not stated, we assume a temperature of 25 °C. Table 15.3 shows the results of experiments that measured K_w at several different temperatures. These results indicate that K_w is larger at greater temperatures. Le Chatelier's principle provides insight into the thermodynamics of this phenomenon: Because the forward reaction is favored at higher temperatures, the ionization of water must be endothermic.

The equilibrium constant for the ionization of water is 1.0×10^{-14} at 25 °C.

Calculating Hydrogen and Hydroxide Ion Concentrations

If we know the concentration of hydrogen ion in a water solution, the concentration of hydroxide ion can be calculated from the expression for K_w, and vice versa. In pure water, $[H_3O^+]$ and $[OH^-]$ are equal, because the only source of hydrogen ions and hydroxide ions is the autoionization of water. Because

$$K_w = [H_3O^+][OH^-]$$

and $[H_3O^+]$ and $[OH^-]$ are equal, we can write

$$K_w = [H_3O^+]^2$$

At 25 °C, $K_w = 1.0 \times 10^{-14}$, so

$$[H_3O^+]^2 = 1.0 \times 10^{-14}$$

$$[H_3O^+] = 1.0 \times 10^{-7}\,M = [OH^-]$$

Water. Nearly every natural system is aqueous in nature—the majority of our planet is covered by water.

The last calculation indicates that *the hydrogen ion concentration in pure water at 25 °C is* 1.0×10^{-7} M. (Interestingly, if you measure the hydrogen ion concentration of water, it will probably be much greater. Water that contacts air dissolves small amounts of carbon dioxide, forming carbonic acid, H_2CO_3. The hydrogen ion concentration of water that is saturated with air is about $2 \times 10^{-6}\,M$.)

If an acid or a base is dissolved in the water, the concentrations of hydroxide and hydrogen ions are no longer equal, but the equilibrium constant, K_w, still applies to the system. The concentration of one of the ions can be calculated from the equilibrium expression, if the concentration of the other is known. Example 15.2 details this type of calculation.

The product of the hydrogen ion and hydroxide ion concentrations in any solution is equal to K_w.

EXAMPLE 15.2 Calculating the Concentration of Hydroxide Ion from the Concentration of Hydrogen Ion

An acid is added to water so that the hydrogen ion concentration is 0.25 *M*. Calculate the hydroxide ion concentration.

Strategy The product of the concentrations of the hydrogen and hydroxide ions is equal to K_w. Substitute the known values of K_w (from Table 15.3) and the concentration of hydrogen ion to calculate $[OH^-]$.

Solution

$$K_w = [H_3O^+][OH^-]$$

$$1.0 \times 10^{-14} = (0.25)\,[OH^-]$$

$$[OH^-] = \frac{1.0 \times 10^{-14}}{0.25}$$

$$[OH^-] = 4.0 \times 10^{-14}\ M$$

Notice that when the hydrogen ion concentration is large, the hydroxide ion concentration is small. The *product* of the two, however, is equal to a constant, K_w, which depends only on temperature.

Understanding

Calculate the hydrogen ion concentration in a solution whose hydroxide ion concentration is $2.0 \times 10^{-5}\ M$.

Answer $[H_3O^+] = 5.0 \times 10^{-10}\ M$

Concentration Scales

In any aqueous solution, the concentration of either the hydrogen ion or the hydroxide ion is small. After all, the product of the two must be 1.0×10^{-14} (at 25 °C). Powers of 10 are awkward to use, so a logarithmic notation, the **pH scale,** has been devised. The notation pH comes from the French *puissance d'hydrogène,* translated as "hydrogen power." The pH is defined by

$$pH = -\log_{10}[H_3O^+]$$

and the relationship between concentration and pH is

$$[H_3O^+] = 10^{-pH}$$

We denote common (base 10) logarithms with "log" instead of "$\log_{10}$," simply for convenience. Natural logarithms are denoted by "ln" so that they will not be confused with common logarithms. Measurements of pH are commonly expressed to two decimal places, because practical problems limit the accuracy of most pH measurements to two places. The two decimal places in pH imply that there are only two significant figures in the H_3O^+ concentration. For example, in a solution with a pH of 10.77, the 10 serves only to locate the decimal point and does not contribute to the number of significant figures. Example 15.3 shows the calculation of pH from hydrogen ion concentration.

pH is defined as $-\log[H_3O^+]$, and $[H_3O^+] = 10^{-pH}$.

EXAMPLE 15.3 Calculating pH from Concentration

Calculate the pH of the following solutions:

(a) $0.050\ M\ H_3O^+$
(b) $0.12\ M\ OH^-$
(c) $2.4\ M\ H_3O^+$

Strategy If the concentration of hydrogen ion is not given, calculate it from the concentration of hydroxide ion using the expression for K_w. To calculate the pH, take the $-\log$ of $[H_3O^+]$ and express it to two decimal places.

Solution

(a) Given: $[H_3O^+] = 0.050\ M$

Substitute this H_3O^+ concentration into the equation for pH.

$$pH = -\log(0.050) = -(-1.30) = 1.30$$

(b) Given: $[OH^-] = 0.12\ M$

First, use the expression for K_w to calculate the H_3O^+ concentration from the OH^- concentration.

$$[H_3O^+] = \frac{K_w}{[OH^-]} = \frac{1.0 \times 10^{-14}}{0.12} = 8.3 \times 10^{-14}$$

Next, calculate the pH.

$$pH = -\log(8.3 \times 10^{-14}) = 13.08$$

(c) Given: $[H_3O^+] = 2.4\ M$

$$pH = -\log(2.4) = -0.38$$

Very high concentrations of acids produce negative values of pH.

© Cengage Learning/Charles D. Winters

Measuring the pH of orange juice.

Understanding

The hydrogen ion concentration of orange juice is $3.6 \times 10^{-4}\ M$. Calculate the pH.

Answer pH = 3.44

If the pH is known, then the molarity of H_3O^+ is determined by taking the antilogarithm of the negative of the pH:

$$[H_3O^+] = 10^{-pH}$$

With many popular calculators, you can enter a number, then press the "10^x" or "inverse logarithm" button to perform this operation. Other calculators require you to press an "inverse" or a "2nd" key, then the "log" key, to perform this operation. Appendix A contains instructions for performing logarithmic operations on a calculator.

EXAMPLE 15.4 Calculating the Concentration of Hydrogen Ion from pH

Calculate the hydrogen ion concentration in a solution that has

(a) pH = 3.50.
(b) pH = 12.56.

Strategy Converting from pH to $[H_3O^+]$ involves raising 10 to the $-$pH power.

Solution

(a) pH = 3.50
$[H_3O^+] = 10^{-3.50}$
$[H_3O^+] = 3.2 \times 10^{-4}\ M$

(b) pH = 12.56
$[H_3O^+] = 10^{-12.56}$
$[H_3O^+] = 2.8 \times 10^{-13}\ M$

Understanding

Calculate the hydrogen ion concentration in a solution that has a pH of 4.76.

Answer $[H_3O^+] = 1.7 \times 10^{-5}\ M$

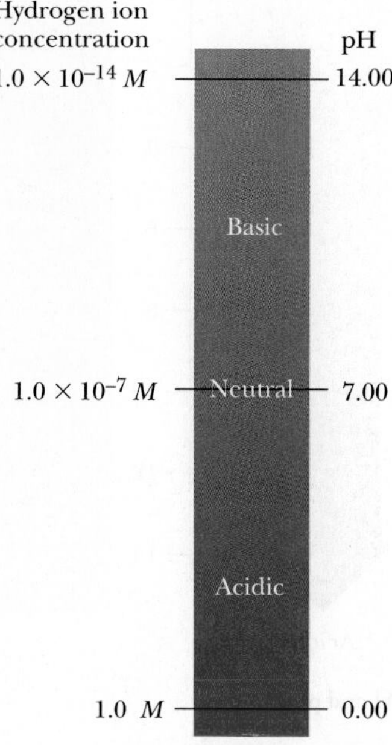

Figure 15.2 pH scale and acidity.

It has become common practice among chemists to use the p-notation for any equilibrium constant or concentration that is very small. Some examples are as follows:

$$pOH = -\log[OH^-]$$

$$pCl = -\log[Cl^-]$$

$$pK_w = -\log(K_w) = 14.00 \text{ at } 25\ ^\circ C$$

The hydrogen ion concentration in pure water at 25 °C is $1.0 \times 10^{-7}\ M$; therefore, *the pH of pure water is 7.00.* An acidic solution has a hydrogen ion concentration that is greater than that of pure water, making the pH of an acidic solution less than 7. Alternatively, if some source of hydroxide ion is added to pure water, the solution is basic and the hydrogen ion concentration decreases to less than $1.0 \times 10^{-7}\ M$, so the pH is greater than 7. Figure 15.2 summarizes these facts.

Most pH measuring devices have calibration markings over the 0- to 14-pH range, as shown in Figure 15.3, but the pH of a solution can be less than 0 or greater than 14. Solutions in which the hydrogen ion concentration exceeds 1 M have a negative pH; solutions in which the hydroxide ion concentration exceeds 1 M have a pH greater than 14.

Relationship between pH and pOH

We know that the hydrogen ion concentration can be calculated from the hydroxide ion concentration, and vice versa. The key relationship is the equilibrium constant expression that describes the ionization of water:

$$[H_3O^+][OH^-] = K_w$$

An expression can be derived for pH and pOH by taking the negative logarithm of both sides.

$$-\log([H_3O^+][OH^-]) = -\log(K_w)$$

Remember that the logarithm of a product can be written as the sum of two logarithms, so

$$-\log[H_3O^+] - \log[OH^-] = -\log(K_w)$$

Figure 15.3 pH measurements. There are many different ways to measure the pH of a solution.

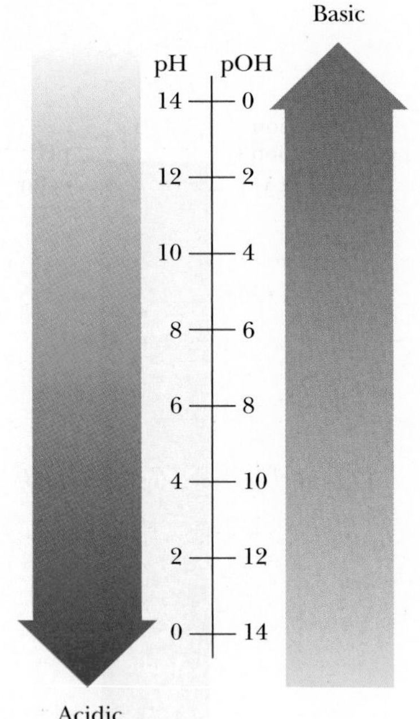

pH and pOH.

$[H_3O^+][OH^-] = 1.0 \times 10^{-14}$
$pH + pOH = 14.00$
at 25 °C

TABLE 15.4 pH Values of Some Common Substances

Substance	pH	Substance	pH
1 *M* HCl	0.0	Milk	6.9
Human stomach acid	1.7	Pure water	7.0
Lemon juice	2.2	Blood	7.4
Vinegar	2.9	Seawater	8.5
Carbonated beverage	3.0	Household detergent	9.2
Wine	3.5	Milk of magnesia	10.5
Tomato juice	4.1	Household ammonia	11.9
Black coffee	5.0	Trisodium phosphate solution	12.5
Urine	6.0	1 *M* NaOH	14.0

We have defined the p-function as −log, so we can rewrite

$$pH + pOH = pK_w$$

Because the numerical value for pK_w is 14.00 at 25 °C, the last equation can be rewritten as

$$pH + pOH = 14.00$$

Just as we know that the product of the hydrogen ion and hydroxide ion concentrations is 1.0×10^{-14}, *the sum of the pH and pOH must always be 14.00* (at 25 °C).

A review of Table 15.4 shows why categorizing matter as acids or bases is a powerful tool. Most foods are acidic, which give them a taste that our tongues discern clearly. Many scientists have noted that our taste buds evolved to help differentiate foods from poisons. Humans discern sweet, salty, sour (acidic), and bitter (basic). Most foods are sweet or sour; poisons are generally bitter.

OBJECTIVES REVIEW *Can you:*

- ☑ relate hydrogen ion concentration to hydroxide ion concentration in aqueous solutions?
- ☑ define pH and use it to express concentrations?
- ☑ convert between hydrogen ion concentrations, hydroxide ion concentrations, pH, and pOH?

15.3 Strong Acids and Bases

OBJECTIVES

- ☐ Define strong acids and bases
- ☐ List the species that are strong acids and bases
- ☐ Calculate the concentrations of species, the pH, and the pOH in solutions of strong acids and bases

Experiments show that compounds can be divided into three groups based on the degree to which they ionize when dissolved in water. Nonelectrolytes do not ionize when they dissolve. Compounds that ionize or dissociate completely in water are termed **strong electrolytes;** those that ionize or dissociate only partially are **weak electrolytes.** Much of the chemistry that occurs in aqueous solution is related to reactions of strong and weak electrolytes. This section presents the acid-base chemistry of strong electrolytes, emphasizing the relationships between their concentrations and the pH of their solutions.

Strong Acids

When chemists speak of a **strong acid,** they mean one that ionizes completely in solution. An example of a chemical equation that describes a strong acid dissolving in water is

$$HCl(g) + H_2O(\ell) \rightarrow H_3O^+(aq) + Cl^-(aq)$$

Experiment shows that this reaction goes to completion, and the equilibrium concentrations of H_3O^+ and Cl^- are equal to the starting concentration of HCl. The concentration of nonionized HCl is so small that it cannot be detected. The chemical equation has a single arrow pointing to the right, which indicates that this reaction goes to completion. A weak acid such as HF does not ionize completely in solution but rather proceeds to an equilibrium.

$$HF(aq) + H_2O(\ell) \rightleftharpoons H_3O^+(aq) + F^-(aq)$$

A strong acid ionizes completely in solution; a weak acid does not.

TABLE 15.5	Ionization of Strong Acids
Hydrochloric acid	
HCl	$+ H_2O \rightarrow H_3O^+ + Cl^-$
Hydrobromic acid	
HBr	$+ H_2O \rightarrow H_3O^+ + Br^-$
Hydroiodic acid	
HI	$+ H_2O \rightarrow H_3O^+ + I^-$
Nitric acid	
HNO_3	$+ H_2O \rightarrow H_3O^+ + NO_3^-$
Perchloric acid	
$HClO_4$	$+ H_2O \rightarrow H_3O^+ + ClO_4^-$
Sulfuric acid	
H_2SO_4	$+ H_2O \rightarrow H_3O^+ + HSO_4^-$

It may appear appropriate to use the terms *strong* and *weak* to refer to the chemical reactivity of an acid, but it is an error to think that a strong acid is more reactive than a weak one. Hydrofluoric acid, HF, is a weak acid, but it can dissolve glass! Chemists use the terms *strong* and *weak* to refer only to the degree of ionization, not the chemical reactivity or corrosiveness.

Only six strong acids are commonly encountered; they are listed in Table 15.5. You should remember these six strong acids. Assume that all the other acids you encounter in this text are weak unless you are told otherwise.

The strength of an acid depends on several factors that are discussed in more detail in Section 15.8. The principal factors are the strength of the bond between the proton and the conjugate base and the stability of the ions formed. When the bond is weak and the conjugate base is stable, then the acid is relatively strong because a weak bond and a stable anion favor the loss of a proton.

Only six common acids are strong; the others are weak.

Solutions of Strong Acids

When a strong acid dissolves in water, the concentrations of the species in the solution can be calculated from the chemical equation and the starting concentrations of the species. Because the acid ionizes completely, these calculations do not require knowledge of an equilibrium constant.

The hydrogen ion concentration is equal to the concentration of a strong acid solution.

EXAMPLE 15.5 Calculating Hydrogen Ion Concentration and pH of Solutions of Strong Acids

Calculate the hydrogen ion concentration and pH of the following:

(a) 0.010 *M* HNO_3

(b) a solution prepared by diluting 10.0 mL of 0.50 *M* $HClO_4$ to 50.0 mL

(c) a solution prepared by adding 9.66 g HCl(g) to some water and then diluting the solution to 500.0 mL

Strategy Remember that the ionization of a strong acid is complete. Use the starting quantities to calculate the amounts (number of moles) of hydrogen ion, and then calculate the concentration from the amount and the volume of solution.

Solution

(a) The ionization of nitric acid is complete:

$$HNO_3(aq) + H_2O(\ell) \rightarrow H_3O^+(aq) + NO_3^-(aq)$$

Therefore, a solution that is 0.010 *M* in HNO_3 produces 0.010 *M* hydrogen ion. The pH is −log (0.010) = 2.00.

(b) A flow diagram can be used to describe the processes.

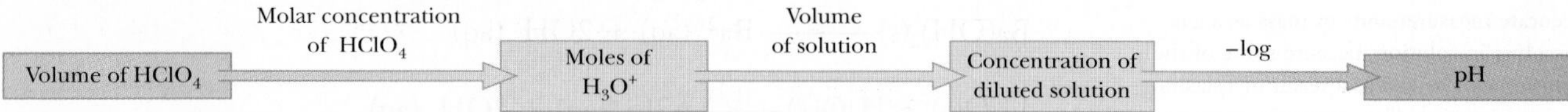

Because $HClO_4$ is a strong acid, each mole of $HClO_4$ provides a mole of H_3O^+. This problem includes a dilution step solved as in Chapter 4.

Because perchloric acid is one of the strong acids, we first calculate the number of moles of H_3O^+ from the concentration and volume of perchloric acid that is used.

$$\text{Amount } H_3O^+ = 0.0100 \text{ L soln} \times \left(\frac{0.50 \text{ mol } H_3O^+}{\text{L soln}}\right)$$

$$= 5.0 \times 10^{-3} \text{ mol } H_3O^+$$

Then divide by the total volume of the solution to determine the concentration.

$$\text{Concentration of } H_3O^+ \text{ in final solution} = \left(\frac{5.0 \times 10^{-3} \text{ mol}}{0.0500 \text{ L}}\right) = 0.10\ M$$

Finally, calculate the pH.

$$\text{pH} = -\log (0.10) = 1.00$$

(c) In this problem, 9.66 g HCl is added to some water; then more water is added until the volume of solution reaches exactly 500 mL.

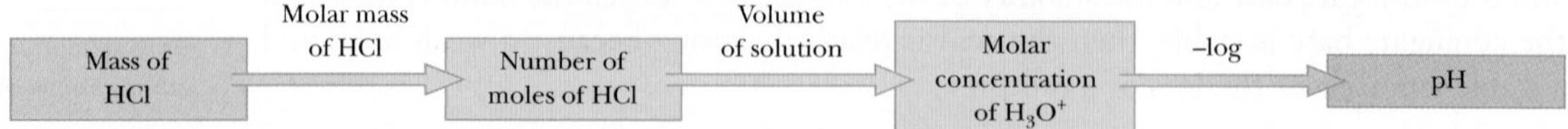

First, calculate the number of moles of HCl added to solution. Each mole of HCl yields a mole of H_3O^+, because HCl is a strong acid and dissociates completely in water. Then use the amount of H_3O^+ and the final volume to calculate the concentration.

$$\text{Amount of HCl} = 9.66 \text{ g HCl} \times \left(\frac{1 \text{ mol HCl}}{36.46 \text{ g HCl}}\right) = 0.265 \text{ mol HCl}$$

$$[H_3O^+] = \frac{0.265 \text{ mol HCl}}{0.500 \text{ L solution}} = 0.530\ M$$

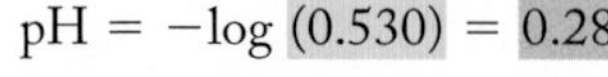

$$\text{pH} = -\log (0.530) = 0.28$$

Understanding

Calculate the pH of a solution made by dissolving 1.00 g HI in enough water to make 250 mL of solution.

Answer pH = 1.50

© Cengage Learning/Larry Cameron

Dissolving a gas in solution. One method of making a quantitative solution from a gaseous solute is to place the container of solvent on a balance and bubble the gas into it until the mass increases by the appropriate amount. In practice, it is quite difficult to perform accurate measurements of mass as a gas dissolves in solution, because some of the solution can be lost as a result of splashing and evaporation. Consequently, a titration is often used to determine the exact concentration of a solution prepared when a gas is dissolved.

Strong Bases

The **strong bases** are the soluble compounds that quantitatively produce hydroxide ion when dissolved in water. The most common strong bases are Group 1A and 2A oxides and hydroxides.

$$\text{NaOH(s)} \xrightarrow{H_2O} \text{Na}^+\text{(aq)} + \text{OH}^-\text{(aq)}$$

$$\text{Ba(OH)}_2\text{(s)} \xrightarrow{H_2O} \text{Ba}^{2+}\text{(aq)} + 2\text{OH}^-\text{(aq)}$$

$$\text{Li}_2\text{O(s)} + \text{H}_2\text{O}(\ell) \xrightarrow{H_2O} 2\text{Li}^+\text{(aq)} + 2\text{OH}^-\text{(aq)}$$

The following example illustrates how to calculate the pH of a solution that contains a strong base.

EXAMPLE 15.6 Calculating Hydroxide Ion Concentration and pH of Solutions of Strong Bases

Calculate the hydroxide ion concentration, the pOH, and the pH of the solution made when 1.00 g barium hydroxide dissolves in enough water to produce 500.0 mL of solution.

Strategy First, calculate the amount of barium hydroxide from the mass and molar mass. Next, calculate the amount of hydroxide ion from the coefficients of the chemical equation; then calculate the concentration and pH from the amount of hydroxide ion and the volume of the solution.

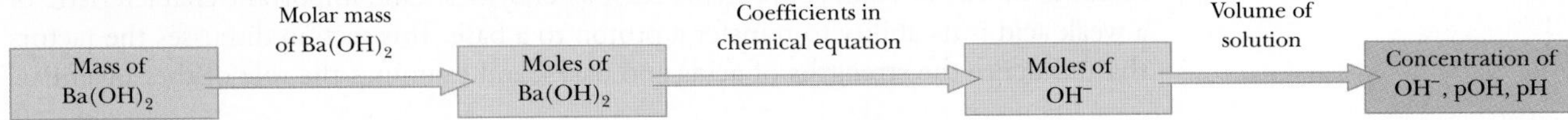

Solution

$$\text{Amount of } Ba(OH)_2 = 1.00 \text{ g } Ba(OH)_2 \times \left(\frac{1 \text{ mol } Ba(OH)_2}{171.3 \text{ g } Ba(OH)_2}\right)$$

$$= 5.84 \times 10^{-3} \text{ mol } Ba(OH)_2$$

Use the chemical equation to determine the number of moles of hydroxide ion produced from 5.84×10^{-3} mol $Ba(OH)_2$.

$$Ba(OH)_2 \xrightarrow{H_2O} Ba^{2+} + 2OH^-$$

$$\text{Amount of hydroxide} = 5.84 \times 10^{-3} \text{ mol } Ba(OH)_2 \times \left(\frac{2 \text{ mol } OH^-}{1 \text{ mol } Ba(OH)_2}\right) = 1.17 \times 10^{-2} \text{ mol } OH^-$$

Calculate the concentration of hydroxide ion from the amount and volume.

$$[OH^-] = \frac{1.17 \times 10^{-2} \text{ mol } OH^-}{0.500 \text{ L}} = 2.34 \times 10^{-2} M \text{ } OH^-$$

Last, calculate the pOH and pH.

$$\text{pOH} = -\log(2.34 \times 10^{-2}) = 1.63$$

$$\text{pH} = 14.00 - \text{pOH} = 14.00 - 1.63 = 12.37$$

Understanding

Calculate the pH when 0.010 g calcium hydroxide is dissolved in enough water to make 100.0 mL of solution.

Answer pH = 11.43

Most hydroxides of metals other than those from Groups 1A and 2A are not very soluble. The solubility product relationship is used to calculate the hydroxide ion concentrations in these solutions. Solutions of bases that are soluble but do not ionize completely, such as ammonia, are described by the equilibrium relationships discussed in the next section.

OBJECTIVES REVIEW *Can you:*

- ☑ define strong acids and bases?
- ☑ list the species that are strong acids and bases?
- ☑ calculate the concentrations of species, the pH, and the pOH in solutions of strong acids and bases?

15.4 Qualitative Aspects of Weak Acids and Weak Bases

OBJECTIVES

- ☐ Define weak acids and bases
- ☐ Relate K_a to the competition of different bases for protons

Many important compounds are weak acids or weak bases. A **weak acid** or a **weak base** is one that does not ionize completely when dissolved in water. For example, nearly all organic acids and bases are weak, and these compounds are crucial in numerous processes that occur in living systems. Amino acids are weak organic acids that are the building blocks of complex proteins such as enzymes. One important characteristic of a weak acid is its ability to transfer a proton to a base. This section discusses the factors that influence the strengths of acids and bases, and examines the role of the solvent.

We generally write the chemical equation for the ionization of a weak acid as a Brønsted–Lowry acid-base reaction, with water accepting the proton, and thus acting as the base.

$$HF(aq) + H_2O(\ell) \rightleftharpoons H_3O^+(aq) + F^-(aq)$$

The concentration of the solvent is a constant, so it does not appear in the equilibrium constant expression:

$$K_a = \frac{[H_3O^+][F^-]}{[HF]}$$

The subscript "a" in K_a is a reminder that the constant describes the ionization of an acid to form H_3O^+ and the conjugate base.

We can write a similar equation for the reaction of a weak base, such as ammonia, with water. Water functions as a Brønsted–Lowry acid in this reaction.

$$NH_3(aq) + H_2O(\ell) \rightleftharpoons NH_4^+(aq) + OH^-(aq)$$

$$K_b = \frac{[NH_4^+][OH^-]}{[NH_3]}$$

Weak acids and bases do not ionize completely in solution.

The subscript "b" indicates that the equilibrium constant is for the reaction of a base with water. As usual, the expression for the equilibrium constant does not include the concentration of the solvent.

Competition for Protons

Whenever an acid reacts with a base, the products are the conjugate acid of the base and the conjugate base of the acid. Consider an acid, HA, transferring a proton to water.

$$HA(aq) + H_2O(\ell) \rightleftharpoons H_3O^+(aq) + A^-(aq)$$

If K_a is very large—much, much greater than 1—the acid is strong. If K_a is small, the acid is weak.

The acid (HA) transfers the proton to the base (H_2O) to form the conjugate acid (H_3O^+) and A^-, the conjugate base of HA. This reaction goes essentially to completion when K_a is very large. Weak acids have small values of K_a, which indicates partial ionization.

The relative strength of an acid results from a competition for protons between the solvent and the conjugate base. Except for a few clearly noted instances, our discussion of acid-base reactions is limited to aqueous solutions.

We can compare strong and weak acids by examining the behavior of HCl and HF in aqueous solution. When HCl ionizes, the proton may be bonded either to a water molecule or to a chloride ion.

$$HCl(aq) + H_2O(\ell) \rightarrow H_3O^+(aq) + Cl^-(aq)$$

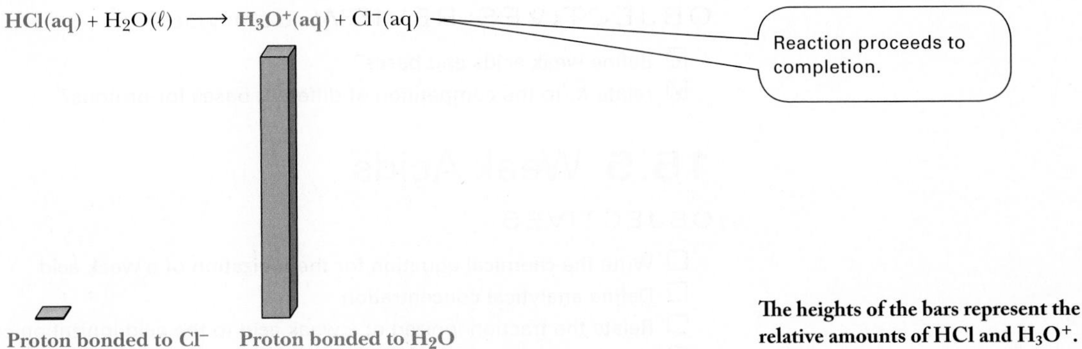

The heights of the bars represent the relative amounts of HCl and H_3O^+.

The chloride ion is a much weaker base than H_2O, so the proton preferentially bonds to the water; this reaction produces the stoichiometric amount of H_3O^+.

Weak acids, in contrast, have conjugate bases that are relatively strong. Consider the ionization of HF.

$$HF(aq) + H_2O(\ell) \rightleftharpoons H_3O^+(aq) + F^-(aq)$$

Reaction reaches equilibrium.

$HF(aq) + H_2O(\ell) \rightleftharpoons H_3O^+(aq) + F^-(aq)$

Proton bonded to F^- Proton bonded to H_2O

The heights of the bars represent the relative amounts of HF and H_3O^+.

Any individual proton bonds to a fluoride ion part of the time and to a water molecule the rest of the time. The chemical equation for the ionization of HF describes the competition of two bases, the fluoride ion and water, for the proton. Fewer protons bind to water because the fluoride ion has a stronger attraction for the proton; therefore, HF(aq) is the predominant acid species in solution at equilibrium.

The strong acids are more effective proton donors than the weak acids; the conjugate bases of strong acids are poor proton acceptors, and thus very weak bases. The conjugate bases of the strong acids, which include the chloride, bromide, iodide, hydrogen sulfate, and nitrate ions, are such weak bases that they are considered spectator ions. Spectator ions are not included in the net ionic equation for an acid-base reaction.

Strong acids have weak conjugate bases.

Influence of the Solvent

In an acid-base reaction, a proton is transferred from the stronger acid to a base, forming the weaker acid. One consequence is that *the hydrogen ion is the strongest acid that can exist in water.* Acids that are stronger than H_3O^+ quantitatively transfer their protons to the water to form H_3O^+. We cannot measure any differences in the acidities of the six strong acids in Table 15.5 in water, because they all ionize completely. This phenomenon is called the **leveling effect**—the solvent makes the strong acids appear equal, or level, in acidity. Strong bases are leveled in a similar manner. All strong bases react stoichiometrically to form hydroxide ion, which is the strongest base that can exist in water.

OBJECTIVES REVIEW *Can you:*

☑ define weak acids and bases?
☑ relate K_a to the competition of different bases for protons?

15.5 Weak Acids

OBJECTIVES

☐ Write the chemical equation for the ionization of a weak acid
☐ Define analytical concentration
☐ Relate the fraction ionized of a weak acid to the acid ionization constant
☐ Calculate acid ionization constants from experimental data
☐ Calculate the concentrations of the species present in a weak acid solution
☐ Calculate percentage ionization from K_a and concentration

The concepts and methods used in Chapter 14 to study gaseous equilibria can be applied to solve equilibria among weak acids. The systematic approach presented in Chapter 14 simplifies the process. This section presents some experimental methods to determine K_a. Also presented are the methods used to calculate the concentrations of species in a solution of a weak acid, given its concentration and the K_a value.

Expressing the Concentration of an Acid

When a weak acid dissolves in water, some of the acid molecules transfer a proton to the water. Scientists are careful when they describe the acid concentration of such a solution. For example, in a 0.010 *M* solution of acetic acid, the actual concentration of the acetic acid molecules is less than 0.010 *M* because some have lost protons to form hydrogen ions and acetate ions.

When a weak acid dissolves in water, the products are hydrogen ion and the conjugate base.

$$CH_3COOH(aq) + H_2O(\ell) \rightleftharpoons H_3O^+(aq) + CH_3COO^-(aq)$$

When speaking of the concentration of the acid, we need to specify clearly the species to which we refer. First, we may be speaking of the *nonionized acetic acid* left in the solution. To calculate the concentration of the nonionized acid, we must know the starting concentration and the quantity that has ionized. The difference between the two is the concentration of nonionized acid. The other concentration we may speak of is the *total acetic acid* concentration. The term **analytical concentration** is used to describe the concentration of all the forms of the acid, both the protonated (acetic acid) and the conjugate base, or unprotonated form (acetate ion). The symbol C_{HA} denotes the analytical, or total, concentration of the weak acid HA. If the analytical concentration of the solution is 0.010 *M,* the true acetic acid concentration, represented by $[CH_3COOH]$, is somewhat lower, because

$$C_{CH_3COOH} = 0.010\ M = [CH_3COOH] + [CH_3COO^-]$$

When a solution is described as 0.010 *M* acetic acid, this value refers to the analytical, or total, concentration. This solution could be made by dissolving 0.010 mol acetic acid and diluting to 1.00 L. Although some acetic acid ionizes, the analytical concentration of this solution is 0.010 *M.* The analytical concentration is the most common measure used to describe a solution in which some of the substances are partially or completely ionized.

The analytical concentration is the sum of the concentrations of the nonionized acid and all its conjugate base forms.

Determining K_a for Weak Acids

In addition to the analytical concentration, the ionization constant must be known to calculate the concentrations of the species in a solution of a weak acid. The chemical equation for the ionization of a weak acid is as follows:

$$HA(aq) + H_2O(\ell) \rightleftharpoons H_3O^+(aq) + A^-(aq)$$

The numerical value of K_a for this equilibrium comes from experimental measurements of concentrations.

$$K_a = \frac{[H_3O^+][A^-]}{[HA]}$$

One way to determine K_a for the acid is to measure the electrical conductivity of the solution. The conductivity is proportional to the concentrations of the ions formed in the ionization process. The degree of ionization is often expressed as a percentage. If a 0.0150 *M* HF solution is 14% ionized, then H_3O^+ accounts for 14% of the analytical concentration, and the nonionized HF accounts for 86% of the analytical concentration.

$$HF(aq) + H_2O(\ell) \rightleftharpoons H_3O^+(aq) + F^-(aq)$$

$$[HF] = 86\% \times 0.0150\ M = 0.013\ M$$

$$[H_3O^+] = [F^-] = 14\% \times 0.0150\ M = 0.0021\ M$$

The percentage ionized can be used to determine the ionization constant, as shown in Example 15.7.

The fraction ionized is equal to the ratio of the concentration of the charged (ionized) species divided by the analytical concentration.

EXAMPLE 15.7 Calculating K_a for a Weak Acid

Picric acid, a weak acid, is dissolved in water to prepare a 0.100 *M* solution. Conductivity measurements at a particular temperature indicate that the picric acid is 82% ionized. Calculate K_a and pK_a.

Strategy Use the fraction ionized to calculate the concentrations of the species in solution. Substitute these concentrations into the expression for K_a to determine the value of the equilibrium constant.

Solution

Let HA represent picric acid.

First, write the chemical equation and expression for K_a.

$$HA(aq) + H_2O(\ell) \rightleftharpoons H_3O^+(aq) + A^-(aq) \qquad K_a = \frac{[A^-][H_3O^+]}{[HA]}$$

The picric acid is 82% ionized, so

$$[A^-] = [H_3O^+] = 0.82 \times 0.100\ M = 0.082\ M$$

It is 18% nonionized, so

$$[HA] = 0.018\ M$$

Substitute these values into the equilibrium expression:

$$K_a = \frac{[A^-][H_3O^+]}{[HA]} = \frac{(0.082)(0.082)}{0.018} = 0.37$$

$$pK_a = -\log(0.37) = 0.43$$

Understanding

Calculate K_a for hydrazoic acid, HN_3, if a 0.050 *M* solution is 1.93% ionized.

Answer $K_a = 1.9 \times 10^{-5}$

A second way to determine the ionization constant for a weak acid is from experimental measurements of pH. A pH meter such as that shown in Figure 15.3 can be used to measure the pH of a solution that contains a known concentration of the weak acid. These measurements are generally straightforward to perform and interpret, and are widely used to determine values of K_a. Example 15.8 illustrates how to calculate the ionization constant from these measurements. Like most equilibrium problems, the solution consists of five steps:

1. Write the balanced chemical equation.
2. Create the iCe table.
3. Write the algebraic expression for the equilibrium constant.
4. Substitute the numerical values from the iCe table.
5. Solve for the unknown.

EXAMPLE 15.8 Determining K_a from pH

Laboratory measurements show that a 0.100 *M* solution of chlorobenzoic acid has a pH of 2.50. Calculate K_a.

Strategy Use the five-step approach and the iCe table. Start by converting the pH to the concentration of hydrogen ion, writing the chemical equation, and filling in the known quantities in the iCe table.

Solution

Calculate the hydrogen ion concentration from the pH.

$$[H_3O^+] = 10^{-pH} = 10^{-2.50} = 3.2 \times 10^{-3}\ M$$

Substitute the known quantities into the iCe table.

1. Write the chemical equation.

2. Fill in the iCe table.

	$ClC_6H_4COOH + H_2O$	$\rightleftharpoons$	H_3O^+	+	$ClC_6H_4COO^-$
initial, *M*	0.100		0		0
Change, *M*					
equilibrium, *M*			3.2×10^{-3}		

The stoichiometry of the chemical equation tells us that when $3.2 \times 10^{-3}\ M\ H_3O^+$ forms, the same concentration of chlorobenzoate ion ($ClC_6H_4COO^-$) forms and the concentration of chlorobenzoic acid decreases by the same amount. Place this information in the change (C) row of the table, then determine the equilibrium concentrations by adding the initial (i) and change (C) rows of the table.

	$ClC_6H_4COOH + H_2O$	$\rightleftharpoons$	H_3O^+	+	$ClC_6H_4COO^-$
initial, *M*	0.100		0		0
Change, *M*	-3.2×10^{-3}		$+3.2 \times 10^{-3}$		$+3.2 \times 10^{-3}$
equilibrium, *M*	9.7×10^{-2}		3.2×10^{-3}		3.2×10^{-3}

Substitute the equilibrium values in the expression for the equilibrium constant.

3. Write the algebraic expression for *K*.

$$K_a = \frac{[H_3O^+][ClC_6H_4COO^-]}{[ClC_6H_4COOH]}$$

4. Substitute values from iCe table.

$$K_a = \frac{(3.2 \times 10^{-3})(3.2 \times 10^{-3})}{9.7 \times 10^{-2}}$$

5. Solve.

$$K_a = 1.1 \times 10^{-4}$$

Understanding

The pH of a 0.100 *M* acetic acid solution is 2.88. Calculate K_a.

Answer 1.8×10^{-5}

The ionization constant for a weak acid is calculated from experimental measurements and the five-step approach using the iCe table.

Experimental methods such as determining the fraction ionized from electrical conductivity or measuring pH provide the data needed to determine acid ionization

TABLE 15.6 Weak Acid Ionization Constants

Acid	Formula	Conjugate Base	K_a at 298 K
Hydrogen sulfate ion	HSO_4^-	SO_4^{2-}	1.0×10^{-2}
Chlorous	$HClO_2$	ClO_2^-	1.1×10^{-2}
Phosphoric	H_3PO_4	$H_2PO_4^-$	7.5×10^{-3}
p-Chlorobenzoic	*p*-ClC_6H_4COOH	*p*-$ClC_6H_4COO^-$	1.1×10^{-4}
Nitrous	HNO_2	NO_2^-	5.6×10^{-4}
Hydrofluoric	HF	F^-	6.3×10^{-4}
Formic	$HCOOH$	$HCOO^-$	1.8×10^{-4}
Benzoic	C_6H_5COOH	$C_6H_5COO^-$	6.3×10^{-5}
Acetic	CH_3COOH	CH_3COO^-	1.8×10^{-5}
Hypochlorous	$HOCl$	OCl^-	4.0×10^{-8}
Hydrocyanic	HCN	CN^-	6.2×10^{-10}
Phenol	C_6H_5OH	$C_6H_5O^-$	1.0×10^{-10}

constants for many acids. Table 15.6 presents ionization constants for several weak acids. A more complete listing appears in Appendix F.

Concentrations of Species in Solutions of Weak Acids

The tabulated value of the acid ionization constant, together with the analytical concentration of the acid, provides the information needed to calculate the concentrations of the species in a solution of weak acid. These calculations enable us to determine important facts such as the pH of a particular solution. The calculations are widely used to predict properties of many important solutions; therefore, it is important to master them.

If the analytical concentration and ionization constant are known the equilibrium concentrations of species can be calculated with the iCe table in a five-step approach.

Determining the Concentrations of Species in a Weak Acid Solution

The tabular approach of the iCe table provides a framework for the systematic solution of weak acid problems. This method is illustrated by setting up the equations needed to calculate the concentrations of the species in a 0.100 M nitrous acid solution given the value of 5.6×10^{-4} for K_a (from Table 15.6).

Chemical equation → (Prepare iCe table) → Expression for equilibrium constant → (Solve expression) → Desired quantities

First, write the chemical equation and the iCe table. The initial (i) row contains the starting concentrations. The initial concentration of nitrous acid is 0.100 M, the analytical concentration. The starting concentrations of H_3O^+ and NO_2^- are essentially zero, because this problem "starts" with nitrous acid that has not yet ionized, and we can ignore the 1×10^{-7} M concentration of H_3O^+ that comes from the autoionization of water.

	$HNO_2 + H_2O$	$\rightleftharpoons$ H_3O^+	+ NO_2^-
initial concentration, M	0.100	0	0
Change in concentration, M			
equilibrium concentration, M			

1. Write the chemical equation.

We will define y as the increase in the concentration of the hydrogen ion. The stoichiometry of the dissociation tells us that the decrease in nitrous acid concentration and the increase in the nitrite ion concentration are $-y$ and $+y$, respectively.

	$HNO_2 + H_2O$	$\rightleftharpoons$ H_3O^+	+ NO_2^-
initial concentration, M	0.100	0	0
Change in concentration, M	$-y$	$+y$	$+y$
equilibrium concentration, M			

The equilibrium concentration is the sum of the starting concentration and the change in concentration in each column.

2. Fill in the iCe table.

	HNO_2 + H_2O	$\rightleftharpoons$ H_3O^+	+ NO_2^-
initial concentration, M	0.100	0	0
Change in concentration, M	$-y$	$+y$	$+y$
equilibrium concentration, M	$0.100 - y$	y	y

The algebraic expression for the equilibrium constant is

3. Write the algebraic expression for the equilibrium constant.

$$K_a = \frac{[H_3O^+][NO_2^-]}{[HNO_2]}$$

We substitute the equilibrium concentrations from the bottom line of the iCe table into the equilibrium constant expression. The numerical value for K_a comes from Table 15.6.

4. Substitute the numerical values from the iCe table.

$$5.6 \times 10^{-4} = \frac{(y)(y)}{0.100 - y}$$

You can reduce this equation to the polynomial form and solve it with the aid of the quadratic equations, although another method is presented in the next section.

Method of Successive Approximation

The method of successive approximation is an extension of the approximation methods introduced in Section 14.7, where the solubility of a precipitate in a solution containing a common ion was illustrated. The calculation of the concentration of the species and the pH of a 0.100 M solution of nitrous acid, $K_a = 5.6 \times 10^{-4}$, continues the last problem and illustrates the method of successive approximation.

The expression for the equilibrium constant is

$$5.6 \times 10^{-4} = \frac{(y)(y)}{0.100 - y}$$

This expression can be expanded into a quadratic equation and solved by the quadratic formula. Many modern calculators have built-in functions to solve expressions like these but this expression can be solved by approximation on almost any calculator.

First, assume that y is negligible with respect to 0.100. If this assumption is true, then $0.100 - y$ is about the same as 0.100. The results of this first calculation are designated as y_1, where the subscript "1" indicates the first approximation.

$$5.6 \times 10^{-4} \approx \frac{(y_1)(y_1)}{0.100}$$

$$y_1^2 = 5.6 \times 10^{-4} \times 0.100$$

$$y_1 = 0.00748$$

Because we made an approximation, it is imperative that we check it for accuracy. Is 0.00748 negligible with respect to 0.100? Scientists use a 5% rule of thumb in this circumstance.

5% of 0.100 = 0.05 × 0.100 = 0.0050

Is 0.00748 less than 0.0050? No, it is not.

When the approximation fails, as it does here, there are two choices. First, we can always solve the equation by the quadratic formula. Second, we can make an additional approximation, hence the name "successive approximations." In the first approximation, we initially assume that y is 0, but find it is 0.00748 based on our assumption. The second approximation assumes that y is 0.00748, and we calculate a new estimate for y, designated y_2, based on *this* value.

$$5.6 \times 10^{-4} = \frac{(y_2)(y_2)}{0.100 - y_1}$$

$$5.6 \times 10^{-4} = \frac{(y_2)(y_2)}{0.100 - 0.00748}$$

$$y_2 = 0.00720$$

Again, check the approximation. This time, determine whether y_2 is within 5% of y_1. "Within 5%" is understood to mean in the interval between 95% of y_1 and 105% of y_1. Calculate these values and determine whether y_2 is in this range.

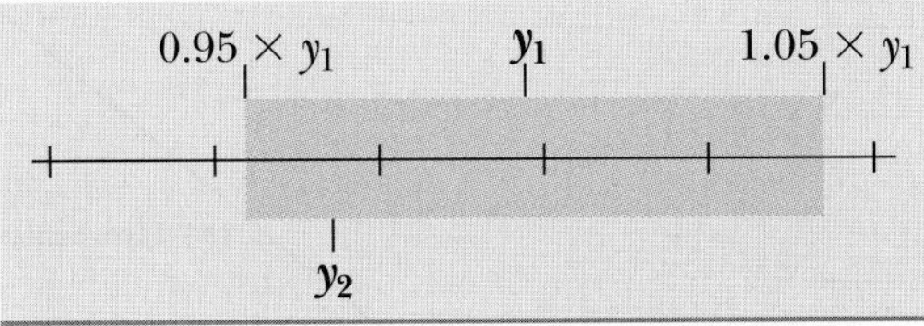

We see that y_2 is within 5% of y_1, so we are reasonably sure that the approximation is valid.

Substitute the value for y into the last row of the iCe table.

	$HNO_2 + H_2O$	$\rightleftharpoons$	H_3O^+	+	NO_2^-
equilibrium	$0.100 - y$		y		y
concentration, M	0.0928		0.00720		0.00720

Express the final results with two significant figures, which is the precision of K_a.

$[HNO_2] = 0.093\ M$

$[H_3O^+] = [NO_2^-] = 0.0072\ M$

$\text{pH} = -\log(0.00720) = 2.14$

It is always wise to check the problem by substituting the values for the concentrations of the species into the expression for the equilibrium constant.

$$K_a = \frac{[H_3O^+][NO_2^-]}{[HNO_2]}$$

$$K_a = \frac{(0.00720)(0.00720)}{0.093}$$

$$K_a = 5.6 \times 10^{-4}$$

The value for K_a calculated from the concentrations agrees with the value used in the computation, so we can be reasonably certain that we solved the problem correctly.

The approximation method limits the answer to about 5% of the mathematically exact answer, but other effects, such as uncertainty in K_a, generally limit the accuracy of this calculation to about 5% in any case.

Table 15.7 summarizes the method of successive approximation.

Fraction Ionized in Solution

Electrical conductivity measurements can be used to determine the concentration of ions in a solution. The electrical conductivity is directly proportional to the concentrations of the ions, so the concentration-conductivity relationships for strong and weak acids are quite different. Some typical experimental data appear in Figure 15.4.

The conductivity of any solution is directly proportional to the concentration of the ions. The conductivity of a strong acid solution is directly proportional to its analytical concentration, because strong acids dissociate completely. The graph of conductivity as a function of concentration is curved when the acid is weak.

TABLE 15.7 Method of Successive Approximation

Trial	Assumed y	Expression Solved	Calculated y	Change*
1	0.000000	$y_1 = \sqrt{5.6 \times 10^{-4} \times 0.100}$	0.00748	7.4%
2	0.00748	$y_2 = \sqrt{5.6 \times 10^{-4} \times (0.100 - 0.00748)}$	0.00720	3.9%
3	0.00720	$y_3 = \sqrt{5.6 \times 10^{-4} \times (0.100 - 0.00720)}$	0.00721	0.1%
4	0.00721	$y_4 = \sqrt{5.6 \times 10^{-4} \times (0.100 - 0.00721)}$	0.00721	0.00%

*The change in the first case is the difference between 0.0100 and the calculated value. The other changes are the differences between the successive values.

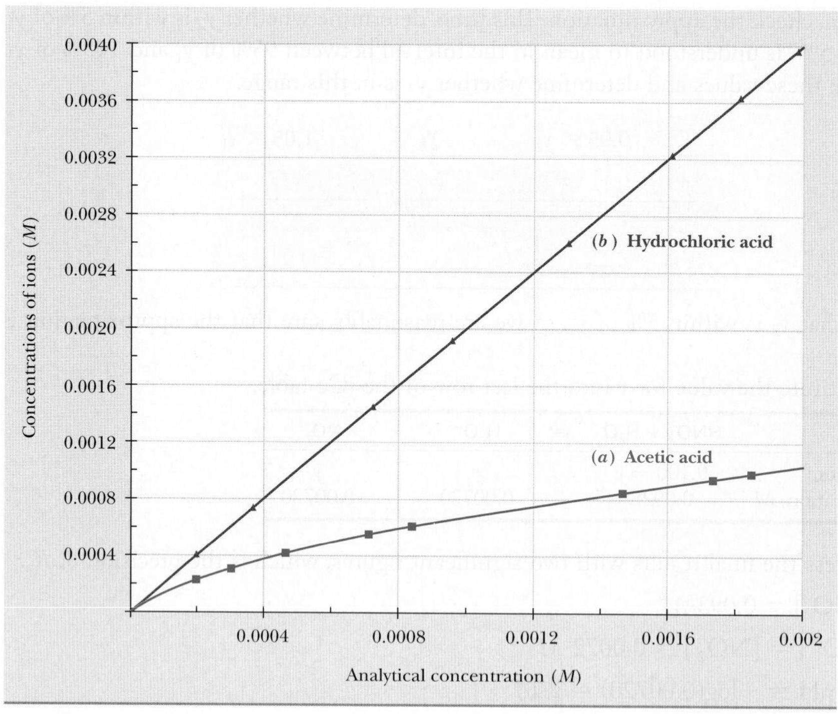

Figure 15.4 Concentration of ions as a function of concentration. Conductivity measurements provide the data to determine the fraction ionized for solutions of *(a)* Acetic acid, $K_a = 1.8 \times 10^{-5}$. *(b)* Hydrochloric acid.

The fraction ionized of a weak acid can be calculated from the analytical concentration and the ionization constant.

The fraction ionized of a weak acid is not a constant but depends on the value of K_a *and* the analytical concentration of the solution. For two different acids of the same concentration, the fraction ionized is larger for the stronger acid (larger value of K_a), as seen in Figure 15.4. For two solutions of the same acid, the fraction ionized is larger in the solution in which the concentration is lower. Example 15.9 illustrates this effect.

EXAMPLE 15.9 Calculating the pH and Fraction Ionized

Calculate the pH and the fraction ionized (in %) in a 0.100 M solution of the weak acid naphthol, for which K_a is 1.7×10^{-10}.

Strategy Write the chemical equation, iCe table, algebraic expression for K_a, numerical expression for K_a, and solve.

Solution

1. Write the balanced chemical equation.

2. Fill in the iCe table.

	HA	+	H_2O	$\rightleftharpoons$	H_3O^+	+	A^-
initial concentration, M	0.100				0		0
Change in concentration, M	$-y$				$+y$		$+y$
equilibrium concentration, M	$0.100 - y$				y		y

3. Write the algebraic expression for the equilibrium constant.

$$K_a = \frac{[H_3O^+][A^-]}{[HA]}$$

4. Substitute the numerical values from the iCe table.

$$1.7 \times 10^{-10} = \frac{y^2}{0.100 - y}$$

5. Solve for the unknown.

If $y << 0.100$, then

$$\frac{y^2}{0.100} \approx 1.7 \times 10^{-10}$$

$$y = 4.1 \times 10^{-6}$$

If you make an approximation, you must check the assumption.

Check the assumption: Is $4.1 \times 10^{-6}\ M << 0.1$? Yes, it is, so the approximation is valid.

$[H_3O^+] = y = 4.1 \times 10^{-6}$ $\quad$ $pH = -\log(4.1 \times 10^{-6}) = 5.39$

PRACTICE OF CHEMISTRY
pH and Plant Color

The hydrangea plant *(hydrangea macrophylla)* is a Japanese native that has been popular in gardens across the world for over 150 years. It is one of many plants that has a different color blossoms in acidic and alkaline soils—gardeners delight in being able to change the colors of their hydrangeas from pink to blue by adjusting the pH.

Scientists learned that the pH alone could not explain the color changes and found that the colors are related to the bioavailability of aluminum. When the soil is basic, aluminum is tied up as the highly insoluble aluminum hydroxide and unavailable to the plant, and the blossoms are pink. Gardeners can drench the soil around the plant with a solution of calcium hydroxide (hydrated lime, about 1 tablespoon per gallon) if they desire pink flowers. When the soil is acidic, which can be achieved by adding sulfur, or even small amounts of aluminum sulfate (1 tablespoon of alum per gallon of water is recommended), the hydrangeas are blue. ∎

(a)

(b)

pH and plant color. *(a)* Hydrangeas are blue in acidic soil and *(b)* pink in alkaline soils.

The fraction ionized is equal to the ratio of the concentration of the ionized acid to the analytical concentration times 100%:

$$\text{Fraction ionized} = \frac{[A^-]}{C_{HA}} = \frac{4.1 \times 10^{-6}}{0.100} \times 100\% = 0.0041\%$$

Understanding

Calculate the percentage ionized in solutions that are 0.0100 and 0.0010 *M* naphthol.

Answer 0.013%, 0.041%

The data from Example 15.9 and the Understanding section can be summarized in a table:

Analytical Concentration *(M)*	$[H_3O^+] = [A^-]$ *(M)*	pH	Fraction Ionized (%)
0.100	4.1×10^{-6}	5.39	0.0041
0.0100	1.3×10^{-6}	5.89	0.013
0.00100	4.1×10^{-7}	6.39	0.041

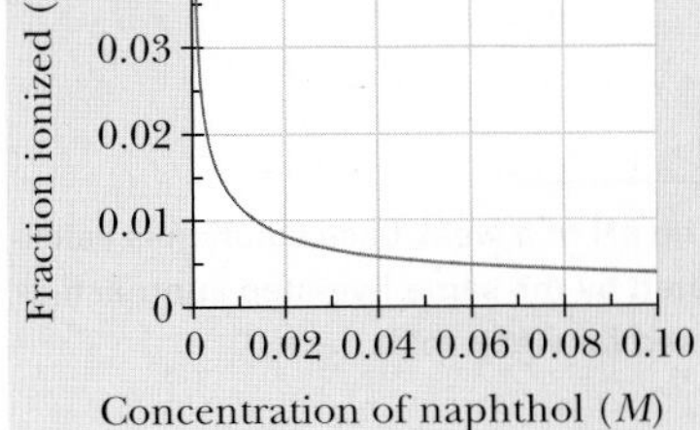

The accompanying graph shows the percentage of naphthol ionized as a function of concentration. Did you notice that you did not need to know the formula or structure of the weak acid naphthol?

OBJECTIVES REVIEW *Can you:*

- ☑ write the chemical equation for the ionization of a weak acid?
- ☑ define analytical concentration?
- ☑ relate the fraction ionized of a weak acid to the acid ionization constant?

☑ calculate acid ionization constants from experimental data?
☑ calculate the concentrations of the species present in a weak acid solution?
☑ calculate percentage ionization from K_a and concentration?

15.6 Solutions of Weak Bases and Salts

OBJECTIVES

☐ Write the chemical reaction that occurs when a weak base dissolves in water
☐ Calculate the concentrations of species present in a solution of a weak base
☐ Relate K_b for a base to K_a of its conjugate acid
☐ Calculate the pH of a salt solution

In this section, the methods used for determining the concentrations of species in solutions of weak acids are extended to solutions of weak bases. We develop the relationship between K_a of an acid and K_b of its conjugate base and use it to calculate the pH of salt solutions.

Solutions of Weak Bases

The reaction of a weak base with the solvent in aqueous solution can be written as

$$B(aq) + H_2O(\ell) \rightleftharpoons BH^+(aq) + OH^-(aq)$$

The expression for the equilibrium constant is written as usual.

$$K_b = \frac{[BH^+][OH^-]}{[B]}$$

The base, B, is represented as a neutral species in the preceding chemical equation; it can also have a negative charge, but rarely does a base have a positive charge. Some examples of reactions of weak bases are as follows:

A weak base reacts with water to produce hydroxide ion and the conjugate acid.

$$NH_3(aq) + H_2O(\ell) \rightleftharpoons NH_4^+(aq) + OH^-(aq)$$

$$HOCH_2CH_2NH_2 + H_2O(\ell) \rightleftharpoons HOCH_2CH_2NH_3{}^+ + OH^-(aq)$$

$$CN^-(aq) + H_2O(\ell) \rightleftharpoons HCN(aq) + OH^-(aq)$$

$$IO_3^-(aq) + H_2O(\ell) \rightleftharpoons HIO_3(aq) + OH^-(aq)$$

Table 15.8 lists values of K_b for several neutral bases. A more complete listing appears in Appendix F.

TABLE 15.8 Weak-Base Ionization Constants

Weak Base			Conjugate Acid	
Name	Formula	K_b at 298 K	Formula	Name
Ammonia	NH_3	1.8×10^{-5}	NH_4^+	Ammonium ion
Ethanolamine	$HOCH_2CH_2NH_2$	3.2×10^{-10}	$HOCH_2CH_2NH_3^+$	Ethanolammonium ion
Hydrazine	N_2H_4	1.3×10^{-6}	$N_2H_5^+$	Hydrazinium ion
Hydroxylamine	NH_2OH	8.7×10^{-9}	NH_3OH^+	Hydroxylammonium ion
Pyridine	C_5H_5N	1.7×10^{-9}	$C_5H_5NH^+$	Pyridinium ion

Weak bases actually react with water, removing a proton and leaving OH^- in solution. Another term used to describe this reaction is "hydrolysis," and you may run across listings of "hydrolysis constants" in other chemistry books.

The pH of a weak base solution is calculated by the same five-step approach as used for weak acids.

Calculations for solutions of weak bases are similar to those performed for weak acids. The following example illustrates one such calculation.

EXAMPLE 15.10 Calculating the pH of a Solution of a Weak Base

Calculate the pH of household ammonia, which is a 1.44 *M* aqueous solution of NH_3. The numerical value of K_b is 1.8×10^{-5} at 25 °C.

Strategy Write the chemical equation, the iCe table, and the algebraic expression for K_b; substitute equilibrium concentrations from the equilibrium (e) line of the table into the expression; and solve.

The logic flow diagram shows the strategy:

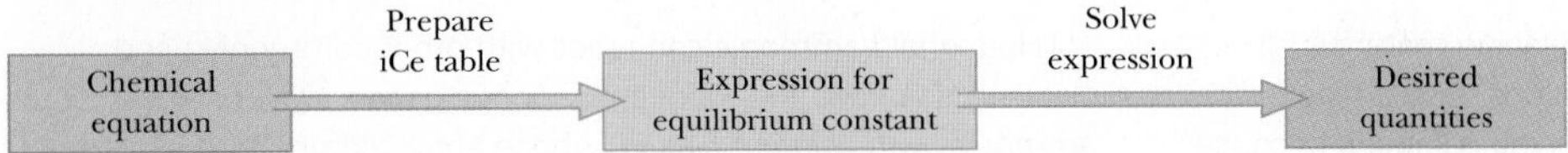

Solution

Write the chemical equation and the iCe table. Just as we ignored the $[H^+]$ from the autoionization of water in weak acid calculations, we ignore the OH^- from the autoionization of water for weak base calculations, and simply assume the starting concentration is zero.

	NH_3	+	H_2O	⇌	NH_4^+	+	OH^-
initial, M	1.44				0		0
Change, M	$-y$				$+y$		$+y$
equilibrium, M	$1.44 - y$				y		y

1. Write the balanced chemical equation.

2. Fill in the iCe table.

$$K_a = \frac{[NH_4^+][OH^-]}{[NH_3]}$$

3. Write the algebraic expression for the equilibrium constant.

$$K_b = \frac{(y)(y)}{1.44 - y} = 1.8 \times 10^{-5}$$

4. Substitute the numerical values from the iCe table.

The resulting equation can be solved by approximation. If $y << 1.44$, then we can write

$$\frac{y^2}{1.44} \approx 1.8 \times 10^{-5}$$

$$y^2 = 2.59 \times 10^{-3}$$

$$y = 5.1 \times 10^{-3} = [OH^-]$$

5. Solve for the unknown.

Check the approximation. Is $5.1 \times 10^{-3} << 1.44$? Yes, it is, so we accept the approximation.

$$[OH^-] = y = 5.1 \times 10^{-3}\ M$$

$$pOH = 2.29$$

$$pH = 14.00 - 2.29 = 11.71$$

Notice how basic the ammonia solution is. Even though ammonia is a relatively weak base, the solution is quite alkaline; its pH is much greater than 7.

Understanding

Calculate the pH of 0.20 M pyridine. Use Table 15.8 to find K_b.

Answer pH = 9.26

Solutions of Salts

When a salt dissolves in water, we introduce anions and cations into solution. Depending on the ions, the resulting solution will be acidic, basic, or neutral. Experiments show that a solution of ammonium chloride is acidic, a sodium chloride solution is neutral, and a sodium fluoride solution is basic.

PRACTICE OF CHEMISTRY
Ammonia Solutions: Good for Cleaning but Do Not Mix with Bleach

Many widely available household cleaning materials contain ammonia. Ammonia solutions have a high pH (see Example 15.10) and make good cleaning solutions because they react with water-insoluble oils, greases, and fats to form water-soluble compounds that are easy to rinse away. Although all alkaline solutions react with greases, ammonia is superior to most other alkaline cleaning materials because ammonia itself is a gas. If the final rinsing step leaves traces of the cleaning solution, the water evaporates and so does the ammonia. If we had used a different substance to make the solution basic—sodium hydroxide (lye), sodium carbonate (washing soda), and trisodium phosphate (TSP) are widely used—the alkaline ingredient is a solid, and when the water evaporates, streaks of alkaline material are left behind.

Household ammonia can react with other commonly used cleaning solutions to produce poisonous gases. *Never mix ammonia with chlorine bleach,* whose active component is sodium hypochlorite. If this happens, one or more of the following chemical reaction could occur:

$$3NaOCl(aq) + NH_3(aq) \rightarrow 3NaOH(aq) + NCl_3(\ell)$$

$$NaOCl(aq) + 2NH_3(aq) \rightarrow NaCl(aq) + H_2O(\ell) + N_2H_4(\ell)$$

Each of these reactions produces a hazardous and toxic substance: chlorine (Cl_2), trichloroamine (NCl_3), or hydrazine (N_2H_4).

In fact, household cleaners contain many of the same compounds that we use in the chemistry laboratory, and they should be treated with the same respect. Ammonia (and many other cleaning solutions) should be used only in well-ventilated areas. ▮

PARSONS
AMMONIA
CLOROX
REGULAR
CAUTION: DO NOT MIX WITH OTHER HOUSEHOLD PRODUCTS SUCH AS CHLORINE-TYPE BLEACHES, AUTOMATIC TOILET BOWL, WALL OR TILE CLEANERS. Avoid contact with eyes and prolonged contact with skin. Do not take internally. Avoid inhalation of vapors. Use in well ventilated area.
KEEP OUT OF REACH OF CHILDREN.
FIRST AID:
EYES—Flush 10-15 minutes with water. Call a physician.
SKIN—Flush thoroughly with water.
INTERNAL—Immediately give large amounts of milk or water. Do not induce vomiting. Call a physician immediately.
wash skin thoroughly with water.
Physical and chemical hazards.
Clorox bleach contains a strong oxidizer. Always flush drains before and after use. Do not use or mix with other household chemicals, such as toilet bowl cleaners, rust removers, acid or products containing ammonia. To do so will release hazardous gases. Prolonged contact with metal may cause pitting or discoloration.

Household cleaning products. Even household chemicals should be treated with respect.

To calculate the pH of a salt solution, we must know the values of K_a and K_b for each of the species in solution. Most tables, including those in this textbook, present the ionization constant for only one form of the conjugate acid-base pairs, usually the neutral form. For example, K_a for HF is tabulated, but K_b for F^- is not. The relationship between K_a and K_b for conjugate acid-base pairs is developed in the next section.

Strengths of Weak Conjugate Acid-Base Pairs

To develop the relationship between K_a and K_b, we might consider the hydrofluoric acid system. Fluoride ion is the conjugate base of hydrofluoric acid and reacts with water just as any other base does. Fluoride ions can be added to a solution without using any HF; sodium fluoride is a good source of fluoride ions. The sodium fluoride dissociates completely into sodium ions and fluoride ions.

Sodium ions are spectator ions and appear on both sides of the complete ionic equation.

$$\cancel{Na^+(aq)} + F^-(aq) + H_2O(\ell) \rightleftharpoons HF(aq) + OH^-(aq) + \cancel{Na^+(aq)}$$

The net ionic equation for the reaction of fluoride ion with water is

$$F^-(aq) + H_2O(\ell) \rightleftharpoons HF(aq) + OH^-(aq)$$

This chemical equation tells us that the base, fluoride ion, reacts with water to produce hydroxide ion in aqueous solution. The equilibrium constant, K_b, for the fluoride ion can be determined experimentally, but it can also be derived from K_a for hydrofluoric acid by the following calculations.

First, write the ionization reaction of hydrofluoric acid and the expression for K_a.

$$HF(aq) + H_2O(\ell) \rightleftharpoons F^-(aq) + H_3O^+(aq)$$

$$K_a = \frac{[F^-][H_3O^+]}{[HF]}$$

Next, write the reaction of the conjugate base with water and the expression for K_b.

$$F^-(aq) + H_2O(\ell) \rightleftharpoons HF(aq) + OH^-(aq)$$

$$K_b = \frac{[HF][OH^-]}{[F^-]}$$

Last, write the product of K_a and K_b.

$$K_aK_b = \frac{\cancel{[F^-]}[H_3O^+]}{\cancel{[HF]}} \times \frac{\cancel{[HF]}[OH^-]}{\cancel{[F^-]}}$$

After canceling concentrations, we find

$$K_aK_b = [H_3O^+][OH^-]$$

or

$$K_aK_b = K_w$$

or, using p-notation,

$$pK_a + pK_b = pK_w$$

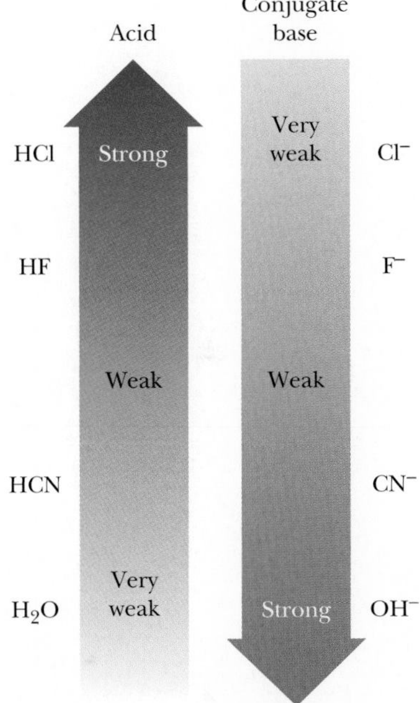

Strengths of acids and their conjugate bases. Stronger acids have weaker conjugate bases.

This equation is useful and valid for any conjugate acid-base pair in water. If we know either K_a or K_b, we can calculate the other. This equation also summarizes an important experimental observation: *the stronger an acid, the weaker its conjugate base.* It is important to remember that an acidity scale is not just two extremes, strong and weak, but a continuum. Although the strongest acids have conjugate bases that are so weak that they are spectator ions, the conjugate bases of most weak acids are themselves weak bases. The next example illustrates one way to use the relationship between K_a and K_b.

The product of the equilibrium constants of a conjugate acid-base pair is equal to K_w. Stronger acids have weaker conjugate bases, and vice versa.

The conjugate base of a strong acid is a spectator ion.

EXAMPLE 15.11 Calculating K_a and K_b for a Conjugate Acid-Base Pair

Use the information in Tables 15.6 and 15.8 to calculate K_b for the formate ion at 25 °C.

Strategy Because formate ion is the conjugate base of formic acid, use Table 15.6 to find K_a for formic acid; then calculate K_b for the formate ion from the relationship between K_a and K_b.

Solution

$$K_aK_b = K_w$$

$$K_b = \frac{K_w}{K_a}$$

$$K_b = \frac{1.0 \times 10^{-14}}{1.8 \times 10^{-4}} = 5.6 \times 10^{-11}$$

Understanding

Calculate K_a for the ammonium ion at 25 °C.

Answer K_a for ammonium $= 5.6 \times 10^{-10}$

When K_a or K_b is known, we can rank species in order of strength. Example 15.12 illustrates this process.

EXAMPLE 15.12 Ranking Bases in Order of Strength

Rank the following bases in order of relative strength, from weakest to strongest: formate ion, cyanide ion, and acetate ion.

Strategy This problem can be solved without calculations, because Table 15.6 lists values of K_a for each acid. The acids can be ranked in order of K_a, and the conjugate bases will be in the opposite order, because the stronger an acid, the weaker its conjugate base.

Solution

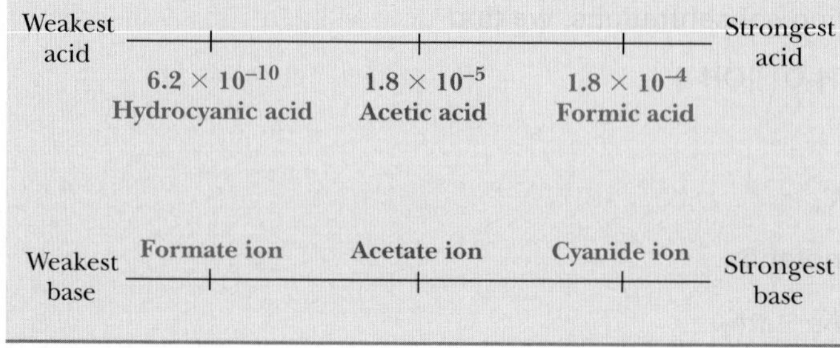

Understanding

Rank the following bases in order of strength, from weakest to strongest: nitrite ion, fluoride ion, and benzoate ion.

Answer Nitrite ion, fluoride ion, benzoate ion

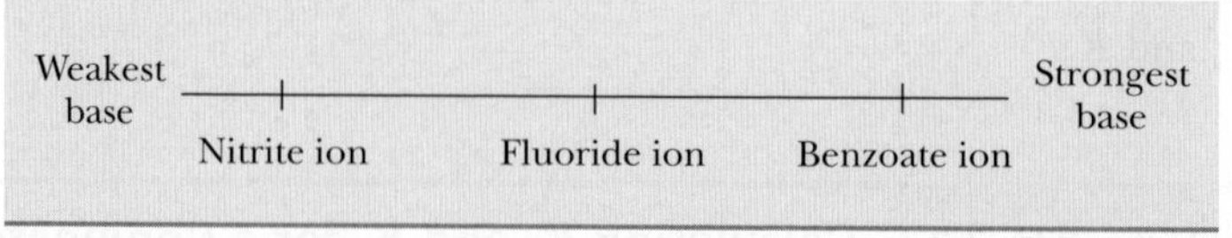

Conjugate Partners of Strong Acids and Bases

The conjugate base of a strong acid is very weak and has little tendency to remove a proton from water under ordinary circumstances.

$$Cl^-(aq) + H_2O(\ell) \not\longrightarrow HCl(aq) + OH^-(aq)$$

In fact, the reverse of this reaction goes to completion, because even the most sensitive instruments cannot detect any nonionized HCl(aq) in the solution. *The conjugate base of a strong acid has no net effect on the pH of a solution.* Examples of these very weak bases are Cl^-, NO_3^-, and ClO_4^-, the anions of strong acids. In any acid-base equilibrium, the anion of a strong acid is a spectator ion, one that does not influence the pH of the solution.

The cations of strong bases likewise lack acidic behavior in water. The strong bases, ionic compounds such as NaOH and KOH, dissociate completely to form the stoichiometric amount of hydroxide ion. Sodium and potassium ions do not affect the pH of a solution. When considering the acid-base properties of solutions, we treat these ions as spectator ions and do not include them in net ionic equations.

The spectator ions can be remembered; they go hand in hand with the strong acids and bases. The anions of the strong acids, Cl^-, Br^-, I^-, NO_3^-, and ClO_4^-, are spectator ions. The cations associated with the strong bases are also spectator ions. These include Li^+, Na^+, K^+, Rb^+, Cs^+, Ca^{2+}, Sr^{2+}, and Ba^{2+}.

Anions from strong acids and cations from strong bases are spectator ions.

pH of a Solution of a Salt

We can now consider the pH of a solution made by dissolving a salt in water. Salts dissociate completely in solutions and the ions may have acid-base properties. The acidity of the final solution depends on the relative values of K_a of the cation and K_b of the anion.

If the salt contains just spectator ions, the salt does not affect the pH of the solution. An example is sodium chloride, NaCl. When dissolved in water, neither Na^+ nor Cl^- has acid-base properties, because both HCl and NaOH are strong electrolytes. The pH of the solution is determined by the autoionization of water.

A salt that contains a spectator cation and an anion from a weak acid produces a basic solution when dissolved. An example is sodium fluoride, NaF. When sodium fluoride dissolves in water, it dissociates completely:

$$NaF(aq) \rightarrow Na^+(aq) + F^-(aq)$$

Na^+(aq) is a spectator ion

The sodium cation does not have any acidic properties. The anion is the fluoride ion, which is the conjugate base of hydrofluoric acid. HF is a weak acid, so its conjugate base affects the pH of the solution:

$$F^-(aq) + H_2O(\ell) \rightleftharpoons HF(aq) + OH^-(aq)$$

The formation of additional OH^-(aq) ions makes the resulting solution basic.

A salt that contains a spectator anion and the cation of a weak base produces an acidic solution. When NH_4Cl dissolves in water, it produces NH_4^+ and Cl^-.

$$NH_4Cl(aq) \rightarrow NH_4^+(aq) + Cl^-(aq)$$

Spectator ion

The solution is acidic because the ammonium ion is the conjugate acid of the weak base ammonia, whereas Cl^- is a spectator ion without acid-base properties. The ammonium ion donates a proton to water to form H_3O^+(aq):

$$NH_4^+(aq) + H_2O(\ell) \rightleftharpoons H_3O^+(aq) + NH_3(aq)$$

The formation of additional H_3O^+(aq) makes the resulting solution slightly acidic. If a salt contains both conjugate acids and bases of respective weak bases and acids, a more detailed calculation is necessary to determine whether the solution is acidic or basic. Table 15.9 presents qualitative descriptions of several salt solutions.

The pH of a solution of a salt can be calculated because we can determine K_a for an acid from K_b of its conjugate base, and vice versa. Example 15.13 illustrates these calculations.

The pH of a salt solution is determined from the values of K_a and K_b for the species in solution.

TABLE 15.9 Solutions of Salts

Example	Source of Cation	Source of Anion	Type of Solution
$KCl(aq) \rightarrow K^+ + Cl^-$	Strong base (KOH)	Strong acid (HCl)	Neutral
$NaF(aq) \rightarrow Na^+ + F^-$	Strong base (NaOH)	Weak acid (HF)	Basic
$NH_4NO_3(aq) \rightarrow NH_4^+ + NO_3^-$	Weak base (NH_3)	Strong acid (HNO_3)	Acidic
$NH_4F(aq) \rightarrow NH_4^+ + F^-$	Weak base (NH_3)	Weak acid (HF)	Calculations needed

EXAMPLE 15.13 Calculating the pH of a Salt Solution

Calculate the pH of the following solutions.

(a) 0.10 *M* sodium nitrate
(b) 0.050 *M* KF

Strategy Look at the ions formed when the salt dissolves. If both are spectator ions, the pH will be set by the autoionization of water. If one is a weak acid or base, then treat the problem accordingly.

Solution

(a) Sodium nitrate dissociates completely into sodium ions and nitrate ions:

$$NaNO_3(aq) \rightarrow Na^+(aq) + NO_3^-(aq)$$

The sodium ion is a spectator ion and does not influence the pH. The nitrate ion is the conjugate base of nitric acid, which is strong, so nitrate ion is also a spectator ion. The pH is determined by the autoionization of water and is 7.00.

(b) First, write the chemical equation for the dissociation of potassium cyanide.

$$KF(aq) \rightarrow K^+(aq) + F^-(aq)$$

The potassium ion is a spectator ion, but F^- is a weak base; its conjugate acid, HF, is weak. We must first determine K_b for F^-. From Table 15.6, we see that K_a for HF is 6.3×10^{-4}, so

Calculating the pH of solutions of salts uses the same approach as do solutions of weak acids or bases.

$$K_b = \frac{K_w}{K_a \text{ for HF}}$$

$$K_b = \frac{1.0 \times 10^{-14}}{6.3 \times 10^{-4}} = 1.6 \times 10^{-11}$$

We use the five-step approach and the iCe table for the reaction of the base, F^-.

1. Write the balanced chemical equation.

2. Create the iCe table.

	$F^-(aq)$	+	H_2O	$\rightleftharpoons$	HF(aq)	+	$OH^-(aq)$
initial, *M*	0.050				0		0
Change, *M*	$-y$				$+y$		$+y$
equilibrium, *M*	$0.050 - y$				y		y

3. Write the algebraic expression for the equilibrium constant.

$$K_b = \frac{[HF][OH^-]}{[F^-]}$$

4. Substitute the numerical values from the iCe table.

$$1.6 \times 10^{-11} = \frac{(y)(y)}{0.050 - y}$$

→ **If** $y << 0.050$, then

$$1.6 \times 10^{-11} \approx \frac{(y)(y)}{0.050}$$

$$y^2 = 8.0 \times 10^{-13}$$

$$y = 8.9 \times 10^{-7}$$

5. Solve for the unknown.

→ **Check** the assumption. Is $8.9 \times 10^{-7} << 0.050$? Yes, it is, so the assumption is valid.

$$[OH^-] = y = 8.9 \times 10^{-7}\ M$$

$$pOH = -\log[OH^-] = -\log(8.9 \times 10^{-7}) = 6.05$$

$$pH = 14.00 - pOH = 7.95$$

Understanding

Calculate the pH of a 0.010 M sodium nitrite solution.

Answer pH = 7.62

EXAMPLE 15.14 Calculating the pH of a Solution of a Salt of a Weak Base

Set up the equations needed to calculate the pH of a solution that is 0.050 M ammonium ion.

Strategy The strategy is detailed in the steps listed in the margin.

Solution

$$NH_4^+(aq) + H_2O(\ell) \rightleftharpoons H_3O^+(aq) + NH_3(aq)$$

$$K_a = \frac{K_w}{K_b} = \frac{1.0 \times 10^{-14}}{1.8 \times 10^{-5}} = 5.6 \times 10^{-10}$$

1. Calculate K_a from K_b for ammonia (see Table 15.8).

	NH_4^+	+	H_2O	$\rightleftharpoons$	H_3O^+	+	NH_3
initial, M	0.050				0		0
Change, M	$-y$				$+y$		$+y$
equilibrium, M	$0.050 - y$				y		y

2. Fill in the iCe table.

$$K_a = \frac{[H_3O^+][NH_3]}{[NH_4^+]}$$

3. Write the algebraic expression for K_a.

$$5.6 \times 10^{-10} = \frac{(y)(y)}{0.050 - y}$$

4. Substitute into the equilibrium constant expression.

If the problem had asked to calculate the pH, we would solve the equation and find that $[H_3O^+] = 5.3 \times 10^{-6}$ and pH = 5.28.

5. Solve.

Understanding

Calculate the pH of 0.020 M pyridinium chloride.

Answer pH = 3.46

If a salt lacks any spectator ions (both the anion and cation are derived from weak electrolytes), we can tell whether the solution is acidic or basic by comparing K_a with K_b. An exact solution, however, requires extensive calculations and is not presented here. On a qualitative basis, if $K_a > K_b$, then the solution is acidic; if $K_b > K_a$, then the solution is basic.

Consider an ammonium cyanide solution.

$$NH_4CN \xrightarrow{H_2O} NH_4^+(aq) + CN^-(aq)$$

Neither ammonium ion nor cyanide ion is listed in the tables of weak acids and bases. Ammonium ion is a weak acid, the conjugate acid of ammonia, and is described by an acid ionization constant, K_a.

$$NH_4^+(aq) + H_2O(\ell) \rightleftharpoons H_3O^+(aq) + NH_3(aq)$$

We find K_b for ammonia equal to 1.8×10^{-5} in Table 15.8, and we compute $K_a = K_w/K_b = 1.0 \times 10^{-14}/1.8 \times 10^{-5} = 5.6 \times 10^{-10}$.

Cyanide ion is a weak base, the conjugate base of HCN.

$$CN^-(aq) + H_2O(\ell) \rightleftharpoons OH^-(aq) + HCN(aq)$$

We look up K_a for HCN in Table 15.6 and calculate K_b for cyanide ion from

$$K_b = K_w/K_a = 1.0 \times 10^{-14}/6.2 \times 10^{-10} = 1.6 \times 10^{-5}$$

Last, compare K_a and K_b for the species in solution. Because K_a for ammonium ion is 5.6×10^{-10} and K_b for cyanide ion is 1.6×10^{-5}, the solution is basic.

OBJECTIVES REVIEW *Can you:*

- ☑ write the chemical reaction that occurs when a weak base dissolves in water?
- ☑ calculate the concentrations of species present in a solution of a weak base?
- ☑ relate K_b for a base to K_a of its conjugate acid?
- ☑ calculate the pH of a salt solution?

15.7 Mixtures of Acids

OBJECTIVES

- ☐ Calculate the pH of a solution that contains both strong and weak acids
- ☐ Determine the pH of a solution that contains a mixture of weak acids (or bases) with different ionization constants

When there are several different acids in solution, calculating the concentrations of the various species in solution may seem complicated, but as long as the strengths of the acids are quite different, we need consider only the strongest. The words *quite different* generally mean values of K_a that differ by a factor of 100. For example, the pH of a solution of HBr (strong) and HF (weak) can be calculated by considering only the HBr. *The contribution by the weaker acid toward the pH of a solution is generally negligible in comparison with that of a stronger acid.* Similarly, in a solution of weak acids of different strengths, only the strongest one is important in determining the pH of the solution.

Weaker acids can be neglected with respect to stronger acids when calculating the pH.

The pH of a solution that contains formic acid ($K_a = 1.8 \times 10^{-4}$) and phenol ($K_a = 1.0 \times 10^{-10}$) is determined by calculating the pH of a formic acid solution and disregarding the phenol. If a solution lacks either strong acids or weak acids, then the very weakest acid, water, determines the pH of the resulting solution. Figure 15.5 is a flow chart to help you determine how to treat mixtures of acids in water solution.

A mixture of strong and weak acids can be examined qualitatively by applying Le Chatelier's principle. It is simplest to start with a weak acid and see how the presence of a strong acid affects the equilibrium. The ionization of a weak acid can be written as

$$HA(aq) + H_2O(\ell) \rightleftharpoons H_3O^+(aq) + A^-(aq)$$

When a strong acid is added, the H_3O^+ concentration increases, and the position of equilibrium shifts to the left, partially consuming the added H_3O^+.

$$HA(aq) + H_2O(\ell) \Longleftarrow H_3O^+(aq) + A^-(aq)$$

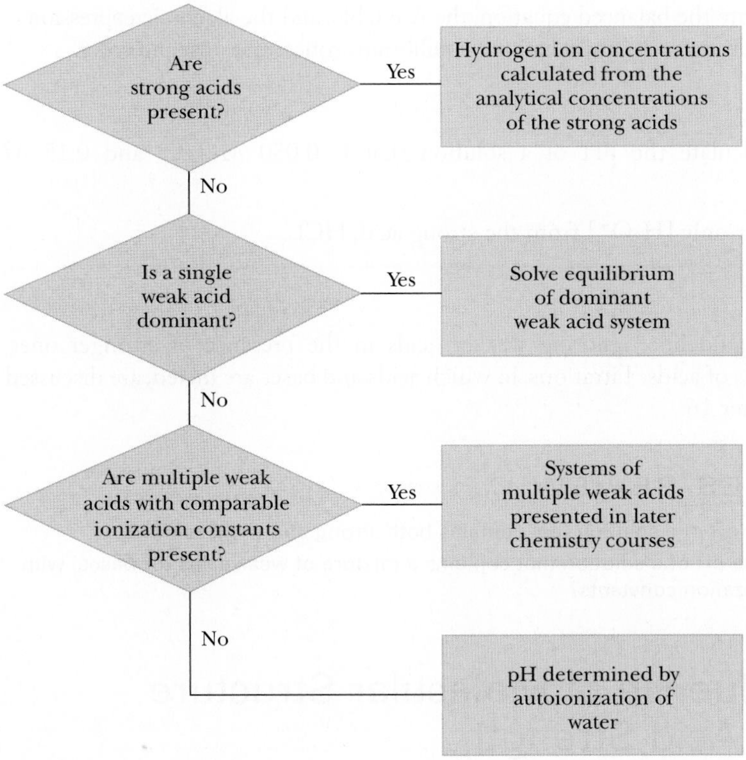

Are strong acids present? — Yes → Hydrogen ion concentrations calculated from the analytical concentrations of the strong acids

No ↓

Is a single weak acid dominant? — Yes → Solve equilibrium of dominant weak acid system

No ↓

Are multiple weak acids with comparable ionization constants present? — Yes → Systems of multiple weak acids presented in later chemistry courses

No → pH determined by autoionization of water

Figure 15.5 Calculating the pH in systems of mixed acids.

This process is similar to the effect of a common ion on the solubility of a precipitate—the solubility decreases when one of the ions in the precipitate is added to the solution. The same holds true for acid-base systems. A weak acid produces fewer protons in the presence of a strong acid, and a weak base produces fewer hydroxide ions in the presence of a strong base.

Like the solubility calculation, the calculation of pH is often simplified if a strong acid is present. Example 15.15 illustrates the calculation of pH in solutions that contain strong and weak acids and bases, as well as a solution that contains two weak acids of different strengths.

EXAMPLE 15.15 Calculating the pH of Mixtures of Acids or Bases

Explain how to calculate the pH of the following solutions:

(a) a solution that is 0.25 M KOH and 1.00 M ammonia
(b) a solution that is 0.40 M HCl, 0.20 M HBr, 0.10 M HCOOH, and 0.20 M HF
(c) a solution that is 0.80 M HCOOH and 0.50 M HOCl

Strategy If strong acids or bases are present, use their concentrations to calculate the pH. If a single weak acid or weak base dominates (is much stronger than the other weak acids or bases), calculate the pH based on its concentration, ignoring the weaker electrolytes.

Solution

(a) The hydroxide ions contributed by the weak base can be ignored in comparison with those from the strong base. Treat the problem as a solution of 0.25 M KOH.

(b) Calculate the pH from the concentrations of the strong acids.

$$[H_3O^+] = C_{HCl} + C_{HBr} = 0.40 + 0.20 = 0.60\ M$$

(c) Formic acid, with $K_a = 1.8 \times 10^{-4}$, is much stronger than hypochlorous acid ($K_a = 4.0 \times 10^{-8}$). We can consider this to be just a solution of formic acid and solve for the pH as we would for any other solution of a weak acid. We use the five-step approach: Write the balanced equation, the iCe table, and the algebraic expression for the equilibrium constant; substitute equilibrium concentrations; and solve.

Understanding

How do you calculate the pH of a solution that is 0.050 *M* HCl and 0.15 *M* HCOOH?

Answer Consider only $[H_3O^+]$ from the strong acid, HCl.

Our general guideline, ignoring weaker acids in the presence of stronger ones, applies to mixtures of acids. Titrations, in which acids and bases are mixed, are discussed in detail in Chapter 16.

OBJECTIVES REVIEW *Can you:*

- ☑ calculate the pH of a solution that contains both strong and weak acids?
- ☑ determine the pH of a solution that contains a mixture of weak acids (or bases) with different ionization constants?

15.8 Influence of Molecular Structure on Acid Strength

OBJECTIVES

- ☐ Relate the acid ionization constants of a series of related binary acids to their structure and bonding
- ☐ Define oxyacid and list several common oxyacids
- ☐ Explain how fundamental properties such as size and electronegativity affect the strengths of acids

Some acids are strong and others are weak, but we have not yet explored the fundamental reasons for these properties. This section relates the influence of structure and bonding to the strengths of acids. The ionization of an acid is a complex process influenced by the strength of the bond that holds the proton, the bond polarity, changes in the strengths of other bonds that accompany the loss of the proton, and the solvation of the ions produced in the reaction. Here we consider two types of acids—the binary hydrides and the oxyacids—and examine the factors that influence the relative strengths of these acids.

Binary Acids

A **binary acid** is an acidic compound composed of hydrogen and one other element, nearly always a nonmetal. Binary hydrides have the general formula H_nA, and the acidity of binary acids is related to the HA bond strength.

When an acid ionizes, the anion retains the electrons in the H–A bond, and the hydrogen ion shares a lone pair of electrons from a solvent molecule.

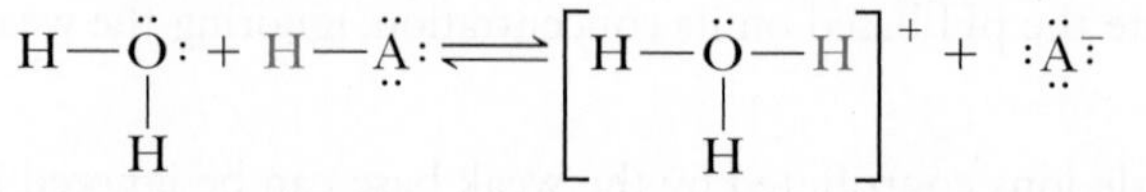

$$\mathrm{H{-}\ddot{O}{:}(H)} + \mathrm{H{-}\ddot{A}{:}} \rightleftharpoons \left[\mathrm{H{-}\ddot{O}{-}H\,(H)}\right]^+ + \mathrm{:\ddot{A}:^-}$$

$$H_2O + HA \rightleftharpoons H_3O^+ + A^-$$

The extent of reaction depends on the relative stabilities of the undissociated acid and the ions in solution. The key factors are the strength of the H–A bond and the stability of the A^- ion in solution. A strong H–A bond is difficult to break, and an unstable A^- anion is difficult to form.

Bond Strengths

The strengths of the bonds in the series of hydrogen halides is HF > HCl > HBr > HI. This experimentally observed order is consistent with the strengths of the acids.

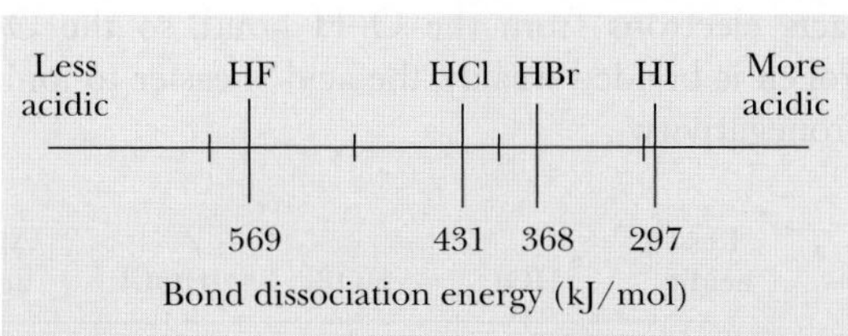

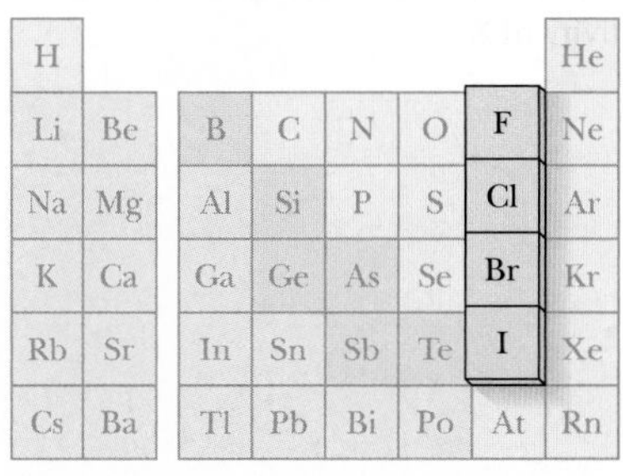

The H–A bond strengths predict the trend in acidity down a group of binary acids.

The hydrogen-fluorine covalent bond is very strong, the hydrogen-chlorine is weaker, and so forth. This trend in the strength of the bonds is due mainly to increasing size effects. The bond in HF is short, and the overlap of the 1*s* orbital on the hydrogen with the 2*p* orbital on the fluorine is substantial. Iodine is much larger than fluorine, so the distance between the atoms increases and the overlap of orbitals decreases.

The same arguments hold in other groups as well. *Within any group in the periodic table, the acidity of the hydrides increases from top to bottom.* Trends among the Group 6A hydrides follow the same logic.

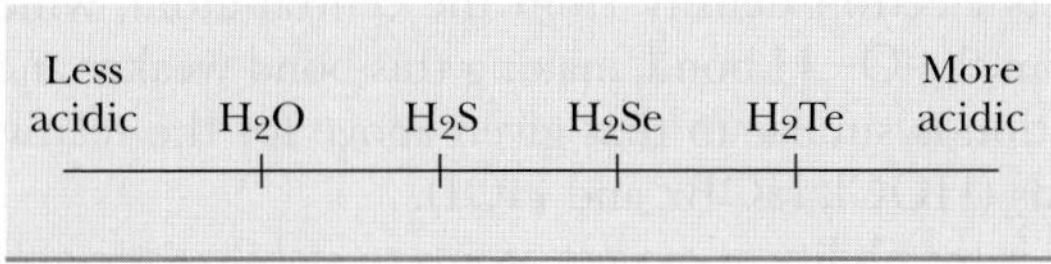

Stability of the Anion

A second factor influencing the ionization of HA is the ability of the A atom to accept additional negative charge, forming the conjugate base, A^-. *The more electronegative the atom, the more easily it can accommodate additional electron density on the atom.* A more electronegative atom results in a stronger acid.

Examine the changes in acidity of the nonmetal hydrides from left to right in the periodic chart, starting with the elements in the second period.

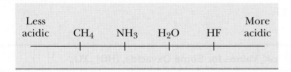

Less acidic CH_4 NH_3 H_2O HF More acidic

In this series of compounds, methane exhibits no acidic properties, ammonia behaves as an acid only in solvents much more basic than water, and hydrofluoric acid is a stronger acid than water. *Within any row of the periodic chart, the acidities of the binary nonmetal hydrides follow the trend expected on the basis of electronegativity and increase from left to right.*

Changes in electronegativities predict the trend in acidity of the binary nonmetal hydrides across a period.

Sometimes electronegativity differences and bond strengths predict opposite trends. Experimental evidence indicates that electronegativity trends dominate across a period, and bond strength is more important down a group.

Oxyacids

Nonmetal hydrides are not the only (or even the most common) compounds that exhibit acidic properties. There are many compounds, called **oxyacids,** that contain hydrogen, oxygen, and a third element. The third element is a nonmetal, such as nitrogen in HNO_2 and HNO_3, or a transition metal in a high oxidation state, such as Cr in H_2CrO_4. Oxyacids also include the organic acids such as acetic acid, CH_3COOH. The hydrogen atoms that ionize in oxyacids are always bonded to an oxygen atom, which, in turn, is bonded to the third element.

Oxyacids contain hydrogen, oxygen, and a third element. The acidic hydrogen is bonded to oxygen.

The strengths of the oxyacids HOX increase with increasing electronegativity of X.

In a series of oxyacids that have the same general structure, *the acidity increases as the electronegativity of the third element increases.* This effect is illustrated by the hypohalous acids, which have the general formula H–O–X, where X = Cl, Br, and I.

The electronegative halogen (X) attracts electrons in the O–X bond. The oxygen atom, in turn, attracts electrons from the O–H bond, so the O–H bond becomes weaker. When hydrogen is bonded weakly, the acid is easier to ionize, so acid strength increases with electronegativity.

	Less acidic	HOI	HOBr	HOCl	More acidic
K_a		3.2×10^{-11}	2.8×10^{-9}	4.0×10^{-8}	
Electronegativity		2.5	2.8	3.0	

The strengths of the oxyacids $(HO)_mXO_n$ increase with increasing n, the number of oxygen atoms that are not bound to hydrogen atoms.

Many oxyacids have additional oxygen atoms bound to the third element; the general formula of an oxyacid is $(HO)_mXO_n$. Experiments show that the oxyacid ionization constants depend mainly on n, the number of oxygen atoms not bonded to hydrogen atoms. Table 15.10 has several oxyacids grouped in this manner, together with their K_a and pK_a values.

Two factors explain these observations. First, oxygen is very electronegative and attracts electron density from the central atom, which, in turn, attracts electron density from the O−H bond, making this bond weaker and easier to ionize in water. This explanation is similar to that given above for the relative acid strengths of the hypohalous acids (HOCl, HOBr, and HOI).

A second factor is the ability of oxygen atoms to stabilize the anion (conjugate base) formed when the acid ionizes. The larger the numbers of such oxygen atoms, the more stable the anion, because the negative charge is spread out over more atoms. The oxyacids of chlorine, for example, are HClO, $HClO_2$, $HClO_3$, and $HClO_4$; each forms a conjugate base with a charge of 1−. In the hypochlorite ion, ClO^-, this charge resides mainly on the single oxygen atom, whereas in the perchlorate anion, ClO_4^-, each oxygen atom has a charge of approximately $\frac{1}{4}$ −; four equivalent resonance structures can be drawn for perchlorate ion, each with a −1 formal charge on one of the oxygen atoms. In an ionization reaction, the anion is a product, and the more stable the product, the more favored the reaction. The influence of the number of oxygen atoms on pK_a can be seen in Table 15.11.

TABLE 15.10 pK_a Values for Some Oxyacids, $(HO)_mXO_n$

n	Name	Formula	K_a	pK_a
0: Very weak	Hypoiodous acid	$(HO)I$	3.2×10^{-11}	10.5
	Arsenious acid	$(HO)_3As$	5.1×10^{-10}	9.3
	Hypobromous acid	$(HO)Br$	2.8×10^{-9}	8.6
	Telluric acid	$(HO)_6Te$	2.0×10^{-8}	7.7
	Hypochlorous acid	$(HO)Cl$	4.0×10^{-8}	7.5
1: Weak acids	Selenious acid	$(HO)_2SeO$	2.4×10^{-3}	2.6
	Arsenic acid	$(HO)_3AsO$	5.5×10^{-3}	2.3
	Phosphoric acid	$(HO)_3PO$	7.5×10^{-3}	2.1
	Chlorous acid	$(HO)ClO$	1.1×10^{-2}	2.0
	Sulfurous acid	$(HO)_2SO$	1.4×10^{-2}	1.9
	Periodic acid	$(HO)_5IO$	2.3×10^{-2}	1.6
2: Strong acids	Nitric acid	$(HO)NO_2$	(10)	(−1)
	Selenic acid	$(HO)_2SeO_2$	(1000)	(−3)
	Chloric acid	$(HO)ClO_2$	(1000)	(−3)
	Sulfuric acid	$(HO)_2SO_2$	(1000)	(−3)
3: Very strong	Perchloric acid	$(HO)ClO_3$	(10^{10})	(−10)

The pK_a values in parentheses are for acids that ionize completely in water. K_a for these values are estimated from measurements made in a more acidic solvent system.

TABLE 15.11 Acid Ionization Constants for the Oxyacids of Chlorine

Name	Formula		K_a	pK_a
Perchloric acid	$HClO_4$	$(HO)ClO_3$	(10^{10})	(-10)
Chloric acid	$HClO_3$	$(HO)ClO_2$	(10^3)	(-3)
Chlorous acid	$HClO_2$	$(HO)ClO$	1.0×10^{-2}	2.0
Hypochlorous acid	$HClO$	$(HO)Cl$	3.0×10^{-8}	7.4

OBJECTIVES REVIEW *Can you:*

- ☑ relate the acid ionization constants of a series of related binary acids to their structure and bonding?
- ☑ define oxyacid and list several common oxyacids?
- ☑ explain how fundamental properties such as size and electronegativity affect the strengths of acids?

15.9 Lewis Acids and Bases

OBJECTIVES

- ☐ Define Lewis acids and bases
- ☐ Identify Lewis acids, Lewis bases, and their reaction products

An important way in which scientific knowledge advances is by modifying and extending a model to include unstudied species. The Arrhenius definition of a base was extended by Brønsted and Lowry, who defined a base as a proton acceptor, including species other than OH^-. In 1932, G. N. Lewis expanded the definitions of acids and bases even further. Lewis noted that a common feature of all Brønsted–Lowry bases is the presence of an unshared pair of electrons, and he defined a base as a substance that can donate a pair of electrons, now called a **Lewis base.** Rather than limiting the definition of an acid to a proton donor, he defined an acid as any substance that can accept a pair of electrons—now called a **Lewis acid.** His intent was to describe phenomena *other* than proton transfer by the well-understood model used for Brønsted–Lowry acids. Table 15.12 summarizes these definitions.

A Lewis acid is an electron-pair acceptor. A Lewis base is an electron-pair donor.

Characteristics of Lewis Acid-Base Reactions

The Lewis model includes the Brønsted–Lowry acids and bases plus many other species, including some atoms and ions. To compare the two acid-base models, consider first a typical Brønsted–Lowry acid-base reaction, the reaction of NH_3 with HCl in aqueous solution:

$$NH_3(aq) + HCl(aq) \rightarrow NH_4^+(aq) + Cl^-(aq)$$

TABLE 15.12 Definitions of Acids and Bases

	Arrhenius	Brønsted–Lowry	Lewis
Acid	A substance that increases H_3O^+ concentration when dissolved in water	A proton donor	An electron pair acceptor
Example	$HF(aq) + H_2O(\ell) \rightleftharpoons H_3O^+ + F^-(aq)$	$HCl(g) + NaNH_2(s) \rightarrow NaCl(s) + NH_3(g)$	$Cd^{2+}(aq) + 4Cl^-(aq) \rightleftharpoons CdCl_4^{2-}(aq)$
Base	A substance that increases OH^- concentration when dissolved in water	A proton acceptor	An electron-pair donor
Example	$Na_2O(s) + H_2O(\ell) \rightarrow 2OH^-(aq) + 2Na^+(aq)$	$NH_3(g) + HCl(g) \rightarrow NH_4Cl(s)$	$6H_2O(\ell) + Fe^{3+} \rightleftharpoons Fe(H_2O)_6^{3+}$
Acid-base reaction	$H_3O^+(aq) + OH^-(aq) \rightarrow 2H_2O(\ell)$	Any proton transfer reaction including those above	Any electron-pair donation, including those above
	acid + base → salt + water	$acid_1 + base_2 \rightarrow acid_2 + base_1$	acid + base → adduct

PRINCIPLES OF CHEMISTRY
Superacids

In recent years, compounds called *superacids* have been used in the synthesis of materials that require exceptionally acidic conditions. Superacids are extremely strong acids, even stronger than sulfuric and nitric acids. The superacids take advantage of the fact that Lewis acid-base reactions can increase the strengths of Brønsted acids. For example, fluorosulfonic acid, HSO_3F, is a strong acid that ionizes completely in water. It is the solvent in the superacid system. In pure fluorosulfonic acid, the Brønsted–Lowry acidity is determined by the autoionization equilibrium.

$$2HSO_3F \rightleftharpoons H_2SO_3F^+ + SO_3F^-$$

Any compound that increases the concentration of $H_2SO_3F^+$ also increases the acidity of this solvent. Similarly, Le Chatelier's principle tells us that any reaction that decreases the concentration of SO_3F^-, the conjugate base in the autoionization equilibrium, also increases the acidity. Adding antimony pentafluoride, SbF_5, increases the acidity by a factor of 10,000 by forming the SbF_5–SO_3F^- adduct.

$$SbF_5 + SO_3F^- \rightarrow (SbF_5)(SO_3F)^-$$

The SbF_5–HSO_3F mixture has such a great acid strength that it is popularly referred to as "magic acid." Its pK_a is about −20 and is generally considered one of the strongest acids known. ∎

A proton is transferred from the acid (HCl) to the base (NH_3) in a Brønsted–Lowry acid-base reaction. This reaction is also classified as an acid-base reaction in the Lewis definition, because the ammonia molecule donates its unshared electron pair to the hydrogen ion in forming the ammonium ion.

$$NH_3 + HCl \longrightarrow NH_4^+ + Cl^-$$

$$H_3N\!: + H{-}\ddot{\underset{..}{Cl}}\!: \longrightarrow H_3N\!: + H^+ + :\ddot{\underset{..}{Cl}}:^- \longrightarrow [H_4N]^+ + :\ddot{\underset{..}{Cl}}:^-$$

A reaction that is not considered an acid-base reaction in the Brønsted–Lowry system but is a Lewis acid-base reaction is the reaction of BF_3 with NH_3 to form BF_3NH_3. The product of a Lewis acid-base reaction is called an **adduct,** derived from the Latin *adductus,* meaning "addition," because it forms by an addition reaction.

$$BF_3 + :NH_3 \longrightarrow F_3B{-}NH_3$$

$$BF_3 + NH_3 \longrightarrow BF_3NH_3$$

A Lewis acid must have an empty orbital available to accept the pair of electrons donated by the base.

Lewis recognized that the basis of many chemical reactions is the formation of bonds when empty orbitals on one species are filled by electron pairs from another. A covalent bond in which both electrons come from one atom is called a **coordinate covalent bond.** A coordinate covalent bond is indistinguishable from any other covalent bond. A covalent bond occurs when two atoms share two electrons and the source of the electrons is not important to the properties of the bond.

Reactions between Lewis Acids and Bases

Metal ions can be Lewis acids, because they have vacant valence-shell orbitals and can form coordinate covalent bonds with Lewis bases. Experimental evidence shows that, in water solution, many metal ions form such bonds, often with several water molecules. Most metal ions can accept four or six pairs of electrons from Lewis bases.

$$Fe^{3+} + 6H_2O(\ell) \rightarrow Fe(H_2O)_6^{3+}$$

PRINCIPLES OF CHEMISTRY
Calculating the pH of Very Dilute Acids

Calculate the pH of 5 × 10^{-8} *M* HCl.

If you are asked to calculate the pH of 5.00 × 10^{-8} *M* HCl, what do you do? If you follow the guidelines for calculating the pH of a strong acid you would take the −log of the concentration:

$pH = -\log[H_3O^+]$

$pH = -\log(5.00 \times 10^{-8})$

$pH = 7.30$

Unfortunately, this answer is incorrect, and in some ways embarrassingly so. Would you want to tell your instructor that when you add hydrochloric acid to water it becomes basic? Ultradilute solutions are found in many important natural phenomena, including molecular biology, and they are at the core of the study of acid rain.

The reason for the error is that there are *two sources* of hydrogen ion, HCl and H_2O:

$HCl(aq) + H_2O \rightarrow H_3O^+ + Cl^-$

$2H_2O \rightleftharpoons H_3O^+ + OH^-$

You may decide that you can add the *two sources* of hydrogen ion: the H_3O^+ produced by HCl (5.00 × 10^{-8} *M*) to that of water (1.00 × 10^{-7} *M*) to get $[H_3O^+]$ = 1.5 × 10^{-7} *M* and pH = 6.82. Unfortunately, Le Chatelier's principle tells us that the water equilibrium will shift in response to adding H_3O^+.

To solve problems in which we cannot neglect the contribution of water, that is, very dilute solutions of acids and bases, we use a systematic approach that includes all possible sources of H_3O^+. Most chemists refer to this process as the *exact treatment of equilibria.*

1. **Write all the reactions that occur.**

   $HCl(aq) + H_2O(\ell) \rightarrow H_3O^+(aq) + Cl^-(aq)$

   $2H_2O(\ell) \rightleftharpoons H_3O^+(aq) + OH^-(aq)$

2. **Count the number of species with unknown concentrations.**

   $[Cl^-]$, $[H_3O^+]$, $[OH^-]$

   The concentrations of H_2O and HCl are not on the list because the concentration of water does not change, and the equilibrium concentration of HCl is zero because it dissociates completely.

3. **Write algebraic equations equal in number to the number of unknowns.** The equations will consist of a mass balance equation, a charge balance equation, and some number of equilibrium constant equations.

   Mass balance. We know the analytical concentration of the acid is 5.00 × 10^{-8} *M*. In the case of a strong acid such as HCl, the acid dissociates completely, so the mass balance relationship is

   $5.00 \times 10^{-8}\ M = [Cl^-]$

   Charge balance. The sum of the positive charges must equal the sum of the negative charges.

   $[H_3O^+] = [Cl^-] + [OH^-]$

   Equilibrium constant. Write the expression for the autoionization of water.

   $K_w = [H_3O^+][OH^-]$ $K_w = 1.0 \times 10^{-14}$ at 25 °C

4. **Solve for $[H_3O^+]$.**

   We have three equations and three unknowns. Start with the charge-balance expression:

   $[H_3O^+] = [Cl^-] + [OH^-]$

   Look at the mass-balance expression and substitute for $[Cl^-]$.

   $[H_3O^+] = 5.0 \times 10^{-8} + [OH^-]$

   Look at the K_w expression and substitute for $[OH^-]$.

   $[H_3O^+] = 5.0 \times 10^{-8} + 1.0 \times 10^{-14}/[H_3O^+]$

   Rearrange.

   $0 = [H_3O^+]^2 - 5.0 \times 10^{-8} \times [H_3O^+] - 1.0 \times 10^{-14}$

   Solve this quadratic equation by the quadratic formula. There are two roots:

   $[H_3O^+] = -7.81 \times 10^{-8}\ M$

   $[H_3O^+] = 1.28 \times 10^{-7}\ M$

   We reject the root that predicts the negative concentration and accept 1.28 × 10^{-7} *M* for $[H_3O^+]$.

   $pH = 6.89$

5. **Evaluate the result.**

   Is a pH of 6.89 reasonable for 5.00 × 10^{-8} *M* HCl. Yes, it is. We would guess that the solution should be very slightly acidic. The results can be summarized in a table as follows:

Method Used	$[H_3O^+]$ (*M*)	pH
Exact method	1.28 × 10^{-7}	6.89
−log(HCl)	5.00 × 10^{-8}	7.30
−log(HCl + H_2O)	1.50 × 10^{-7}	6.82

This same method can be used for even more complicated equilibria, but those topics are discussed in other chemistry courses.

Questions

1. Calculate the pH of a solution that is 2.00 × 10^{-7} *M* HNO_3.
2. A high-school student is asked to calculate the pH of 1.00 × 10^{-9} *M* HI. She calculates 9.00 but writes a note to her teacher that says, "I know something is wrong because the solution can't be basic after I add HI." How much credit (percentage) would you award her? How much would you award the student who writes, "The pH is 9.00," without any explanatory note?
3. Do you think you will need the exact treatment to calculate the pH of 1.5 × 10^{-6} *M* HBr? Explain your reasoning. ∎

Lewis bases other than water can react with metal ions. The deep blue color that results when ammonia is added to a solution of copper(II) ions is a result of the formation of the $Cu(NH_3)_4^{2+}$ ion. The interesting chemistry of these compounds, called *coordination chemistry,* is discussed in more detail in Chapter 19.

OBJECTIVES REVIEW *Can you:*

☑ define Lewis acids and bases?

☑ identify Lewis acids, Lewis bases, and their reaction products?

CASE STUDY Chemists Identify Substance Found in Raid on Drug Lab

The Drug Enforcement Task Force receives unassailable information of a drug-manufacturing operation. After obtaining a search warrant, the task force enters an establishment that purports to manufacture ginger beer but is suspected of processing cocaine on the side. One of the members of the task force is a forensic chemist who is familiar with the methods used to make illicit drugs. She later identifies nearly all the chemicals seized as being useful for drug manufacture, but the label has fallen from a bottle of a white crystalline substance. She can read only "------- Acid."

The chemist has a strategy to determine the identity of the acid. She determines that it is not very soluble in water, which means that it is an organic (carboxylic) acid, rather than a mineral acid. She next uses an instrument to determine the amounts of carbon, hydrogen, and oxygen in the sample, and obtains the following results:

C: 68.8%

H: 5.00%

O: 26.5%

She notices that the numbers do not add to 100% due to experimental error but proceeds to calculate the empirical formula of the compound. Assuming that the sample weighs 100 g, the chemist has determined the amounts of each element as follows:

$$\text{C: Amount C} = 100 \text{ g of sample} \times \frac{68.8 \text{ g C}}{100 \text{ g sample}} \times \frac{1 \text{ mol C}}{12.01 \text{ g C}} = 5.73 \text{ mol C}$$

$$\text{H: Amount H} = 100 \text{ g of sample} \times \frac{5.00 \text{ g H}}{100 \text{ g sample}} \times \frac{1 \text{ mol H}}{1.008 \text{ g H}} = 4.96 \text{ mol H}$$

$$\text{O: Amount O} = 100 \text{ g of sample} \times \frac{26.5 \text{ g O}}{100 \text{ g sample}} \times \frac{1 \text{ mol O}}{16.00 \text{ g O}} = 1.66 \text{ mol O}$$

She divides each coefficient by 1.66, the smallest number to obtain the empirical formula for the compound, $C_{3.47}H_{2.99}O$, and realizes that she must multiply each coefficient by 2 to get the empirical formula: $C_{6.94}H_{5.98}O_2$. The chemist has a good understanding of the amounts of errors expected and determines that the empirical formula is $C_7H_6O_2$.

Because it is possible for the compound to be $C_7H_6O_2$ (122.1 g/mol), or $C_{14}H_{12}O_4$ (244.2 g/mol) or any other multiple of $C_7H_6O_2$, the chemist needs an approximate molecular weight. The chemist measures the osmotic pressure of a solution of the compound and determines that the molecular weight of the compound is in the range of 115 to 145 g/mol.

The empirical formula, $C_7H_6O_2$, has a molar mass of 122.1 g/mol, so the colligative properties indicate that the molecular formula contains one unit, so the molecular formula is the same as the empirical formula, $C_7H_6O_2$.

To help identify the compound, she decides to measure the pH of a solution of known concentration and calculate K_a, so she dissolves 0.100 g in about 90 mL water and then adds water until she has exactly 100 mL of solution. The pH of this solution is 3.16. The concentration of the weak acid is

$$\text{Amount of acid} = 0.100 \text{ g} \times \frac{1 \text{ mol acid}}{122.1 \text{ g}} = 8.19 \times 10^{-4} \text{ mol}$$

$$\text{Volume of solution} = 100 \text{ mL} = 0.100 \text{ L}$$

$$\text{Concentration of acid} = \frac{\text{amount of acid}}{\text{volume of solution}} = \frac{8.19 \times 10^{-4} \text{ mol}}{0.100 \text{ L}} = 8.19 \times 10^{-3}\ M$$

To calculate K_a, she uses the iCe table, like in Example 15.8, and the equilibrium concentration of H_3O^+, which she calculates from the pH.

$$[H_3O^+] = 10^{-pH} = 10^{-3.16} = 6.92 \times 10^{-4} M$$

	$C_7H_6O_2$	+	H_2O	$\rightleftharpoons$	H_3O^+	+	$C_7H_5O_2^-$
initial, M	8.19×10^{-3}				0		0
Change, M	$-y$				$+y$		$+y$
equilibrium, M					6.92×10^{-4}		

Because y is equal to 6.92×10^{-4}, she can complete the equilibrium (e) line of the table.

equilibrium, M	7.50×10^{-3}				6.92×10^{-4}		6.92×10^{-4}

$$K_a = \frac{[H_3O^+][C_7H_5O_2^-]}{[C_7H_6O_2]} = \frac{6.92 \times 10^{-4} \times 6.92 \times 10^{-4}}{7.50 \times 10^{-3}}$$

$$K_a = 6.4 \times 10^{-5}$$

The forensic chemist now looks up at a table of acid ionization constants, such as that of Appendix F. She finds an entry for benzoic acid, formula C_6H_5COOH, molar mass of 122.13 g/mol, with $K_a = 6.3 \times 10^{-5}$.

Questions

1. Use the Internet to determine the uses of benzoic acid. Is it associated with the drug trade or with ginger beer manufacture?
2. This treatment assumes that benzoic acid donates only one proton when it ionizes, but some acids can donate two or more protons. (Polyprotic acids are covered in Chapter 16.) How would the results change if benzoic acid donated two protons?

ETHICS IN CHEMISTRY

Derek is an environmental chemist who has been hired by a coalition of groups—civic, private, environmental, and industrial—to investigate the death of fish in the local lakes. The small town has one major industry, which employs about half the people, and a large community of retirees who moved there for the clean environment, small-town atmosphere, and good fishing. Derek finds incontrovertible evidence that industrial pollution is killing some of the weedy plants near the lake. Their tissues have a high level of the pollutant. Derek confirms that fish are dying, and that industrial pollution can be found in their tissues, but the concentrations are small and not easy to measure. Because the scientific measurements have some ambiguity, the connection between the death of the fish and the industrial pollution is not nearly as strong as the connection between the pollution and the plants. He speaks with his friend Meredith and tells her his conclusions. Meredith urges Derek to go public, saying that if they are going to be able to make a difference, Derek will have to present the results unambiguously and not mention the uncertainties. Meredith says Derek can state that industrial pollution is damaging the lakes, and the damage may be permanent unless immediate steps are taken to eliminate the pollution source. Derek feels an obligation to be objective and stay neutral, but he also recognizes that time is running out for many of the species in the lake system. If pollution accumulates unabated, the plant life will die. The coalition calls a press conference and introduces Derek as an expert.

Dead fish washed up on a lake shore.

Questions

1. Does a scientist have a responsibility to be objective and stay neutral in a debate?
2. Should Derek disavow the environmental group?
3. What should Derek say at the press conference?

Chapter 15 Visual Summary

The chart shows the connections between the major topics discussed in this chapter.

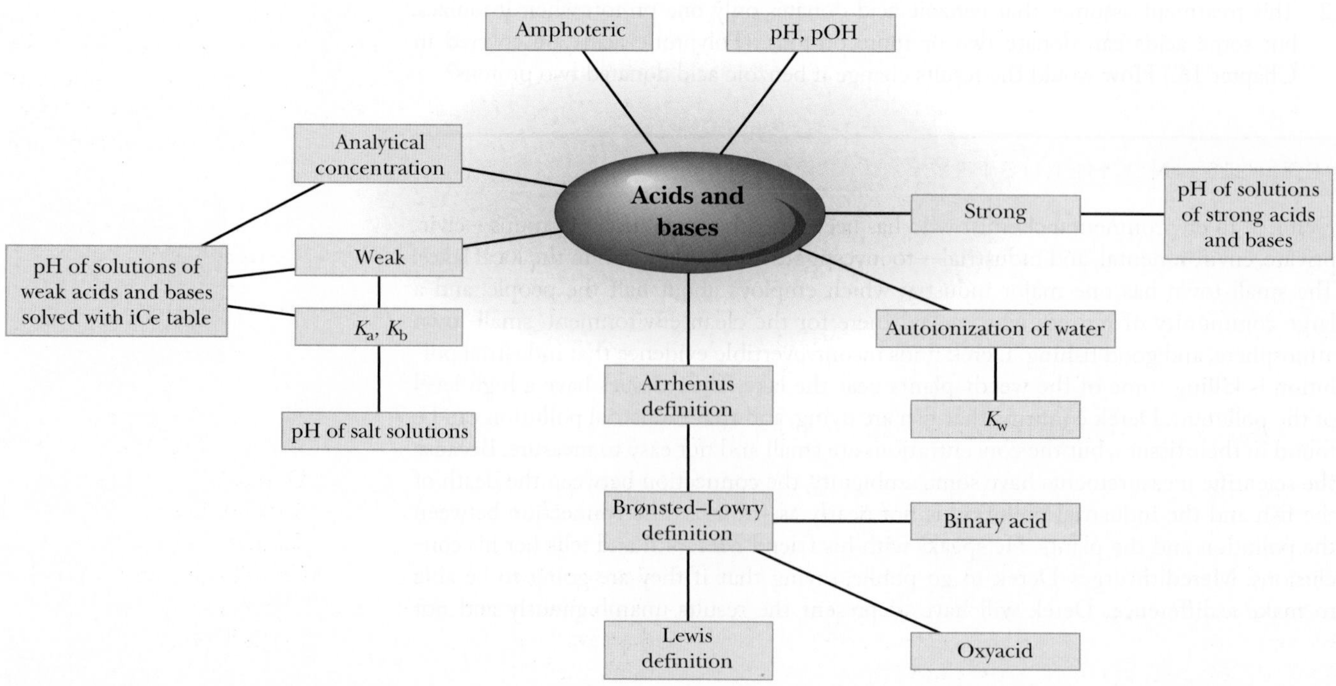

Summary

15.1 Brønsted–Lowry Acid-Base Systems

The Arrhenius definitions limited the number of species that could be classified as acids or bases. Brønsted and Lowry extended those definitions as they noted the common feature of proton transfer in acid-base reactions. They defined an acid as a proton donor and a base as a proton acceptor. Furthermore, after a compound has donated a proton, the substance that remains is capable of accepting a proton, so it is a base. The two species, differing in composition by the presence or absence of a proton, constitute a *conjugate acid-base pair.*

15.2 Autoionization of Water

The most common solvent is water, which undergoes an *autoionization* reaction

$$2H_2O(\ell) \rightleftharpoons H_3O^+(aq) + OH^-(aq)$$

The equilibrium is described numerically by the equilibrium constant, K_w:

$$K_w = [H_3O^+][OH^-] = 1.0 \times 10^{-14} \text{ at } 25\ ^\circ\text{C}$$

The product of the concentrations of the hydrogen and hydroxide ions is a constant that depends only on temperature. Concentrations are frequently quite small, and the p-notation is used to describe such solutions.

15.3 Strong Acids and Bases

When a *strong acid* or *base* dissolves in water, it ionizes completely, providing the stoichiometric amount of hydrogen ion or hydroxide ion.

15.4 Qualitative Aspects of Weak Acids and Weak Bases

Weak acids and *bases* do not ionize completely, and the acid or base ionization constant is used to calculate the concentrations of species in solution. The ionization constants are determined experimentally, and the equilibrium constants for a conjugate acid-base pair are related by

$$K_aK_b = K_w$$

This equation means that the stronger an acid, the weaker its conjugate base.

15.5 Weak Acids

Determining the concentrations of species and calculating the pH of solutions of weak acids require the *analytical concentration* and the value for the acid ionization constant, K_a. The methods are similar to those used for other equilibrium calculations. A five-step approach is applied: write the chemical equation, iCe table, and expression for K_a; substitute values from the iCe table into the expression for K_a; and solve.

15.6 Solutions of Weak Bases and Salts

The same methods used to determine the pH of a weak acid solution can be extended to bases and salts. The organized approach of the iCe table provides a template for solving this type of problem.

15.7 Mixtures of Acids

When a solution contains a mixture of acids, the concentration of hydrogen ions is determined by the strongest acid as long as its ionization constant is much greater than that of the weaker acids.

15.8 Influence of Molecular Structure on Acid Strength

The ability of a molecule to donate a proton is influenced by the strength of the bond holding the proton in the acid and the electronegativity of the atoms in the conjugate base. When the acid molecule donates a proton, the resulting conjugate base has a negative charge. The ionization is enhanced if the negatively charged base is more stable. Acids with strongly electronegative central atoms in the conjugate bases, or polyatomic species with several oxygen atoms, are stronger than those with less electronegative central atoms in the conjugate bases.

15.9 Lewis Acids and Bases

Acid-base behavior can be generalized further. G. N. Lewis extended the Brønsted–Lowry model by noting that the most general feature of a base is the presence of an unshared electron pair, and that of an acid is the presence of an empty orbital that can accept the electron pair. In the Lewis definitions of acid and base, an acid is an electron-pair acceptor and a base is an electron-pair donor.

Chapter Terms

The following terms are defined in the Glossary, Appendix I.

Section 15.1
Amphoteric
Brønsted–Lowry acid
Brønsted–Lowry base
Conjugate acid-base pair

Section 15.2
Autoionization of water
pH scale

Section 15.3
Strong acid
Strong base
Strong electrolyte
Weak electrolyte

Section 15.4
Leveling effect
Weak acid
Weak base

Section 15.5
Analytical concentration

Section 15.8
Binary acid
Oxyacid

Section 15.9
Adduct
Coordinate covalent bond
Lewis acid
Lewis base

Key Equations

$K_w = [H_3O^+][OH^-]$ (15.2)

$pH = -\log[H_3O^+]$ (15.2)

$[H_3O^+] = 10^{-pH}$ (15.2)

$pH + pOH = pK_w$ (15.2)

$pH + pOH = 14.00$ (15.2)

$K_aK_b = K_w$ (15.6)

$pK_a + pK_b = pK_w$ (15.6)

Questions and Exercises

OWL Selected end of chapter Questions and Exercises may be assigned in OWL.

Blue-numbered Questions and Exercises are answered in Appendix J; questions are qualitative, are often conceptual, and include problem-solving skills.

■ Questions assignable in OWL

✎ Questions suitable for brief writing exercises

▲ More challenging questions

Questions

15.1 ✎ Compare and contrast Brønsted–Lowry and Arrhenius acids.

15.2 Can a compound be an Arrhenius base and not a Brønsted–Lowry base? Explain your answer.

15.3 Water is not the only solvent that undergoes autoionization. Write the equation for the autoionization of acetic acid.

15.4 Write two Brønsted–Lowry acid-base reactions and show how they represent proton-transfer reactions.

15.5 Define pH and explain why pH, rather than molarity, is used as a concentration measure of H_3O^+.

15.6 ■ List the strong acids and bases. Why are they called "strong"?

15.7 Define a weak acid.

15.8 ✎ Compare and/or contrast strong and weak acids.

15.9 In acidic solvents, such as concentrated acetic acid, some of the acids considered strong (in water) behave as weak acids. Explain why HCl behaves as a weak acid and $HClO_4$ behaves as a strong acid in acetic acid solvent.

15.10 Is the conjugate base of a strong acid always a spectator ion? Explain.

15.11 ✎ What is the relationship between weak bases and their conjugate acids?

15.12 Define analytical concentration and give an example for HNO_3 and for CH_3COOH solutions.

15.13 Why have chemists not tabulated the fraction ionized for different acids? Such a table would make problems such as calculating the pH of an acid solution quite simple.

15.14 ▲ You are asked to design an experiment to determine the percentage of ionization of HCl in aqueous solution at 4 °C. You accurately dissolve 4.1349 g HCl in 1001.34 g water. Would you set up an experiment to measure the concentration of HCl(aq), H_3O^+(aq), or Cl^-(aq)? Justify your answer.

15.15 In Section 15.6, a base is stated to have a neutral or negative charge. Although positively charged bases exist, they are not common. Explain why positively charged bases are rare.

15.16 ▲ If K_b for ammonia is 1.8×10^{-5}, calculate K_a for its conjugate acid, NH_4^+. Note that NH_3 is the conjugate acid of the amide ion, NH_2^-. Can K_a and K_b for the $NH_3|NH_2^-$ pair be calculated from the preceding data and the value of K_w?

15.17 What are the expected trends in acidity of binary acids, going diagonally to the lower right ("southeast") on the periodic chart from carbon? Is the same trend observed if the starting point is Si?

15.18 ▲ Element 85, astatine (At), is a radioactive halogen that is not present in appreciable amounts in nature. The acid HAt can be prepared and compared with the other hydrogen halides. Explain why you expect HAt to be stronger or weaker than HI.

15.19 Define oxyacid and give examples from among the strong acids.

15.20 Define Lewis acids and bases, and compare with Brønsted–Lowry acids and bases.

15.21 Propose an experiment to determine whether a coordinate covalent bond is different from other covalent bonds. Use

$$:NH_3 + H_2O \rightleftharpoons NH_4^+ + OH^-$$

for a concrete example.

15.22 Compare strong and weak acids and bases.

(a) How are a strong acid and a weak acid similar? How are they different?

(b) How are a strong base and a weak base similar? How are they different?

Exercises

OBJECTIVE Identify conjugate acid-base pairs.

15.23 ■ Write the formula and name of the conjugate acid of the following substances. (The information in Tables 15.6 and 15.8 may be helpful.)
(a) hydrogen sulfate ion (b) water
(c) ammonia (d) pyridine

15.24 Write the formula and name for the conjugate base of the following substances. (The information in Tables 15.6 and 15.8 may be helpful.)
(a) nitric acid (b) hydrogen carbonate ion
(c) water (d) hydrogen chloride

15.25 Write the formula and name for the conjugate base of the following substances. (The information in Tables 15.6 and 15.8 may be helpful.)
(a) HCN (b) HSO_4^- (c) $H_2PO_3^-$ (d) HCO_3^-

15.26 Write the formula and name for the conjugate acid of the following substances. (The information in Tables 15.6 and 15.8 may be helpful.)
(a) N_2H_4 (b) NO_2^- (c) ClO_4^- (d) I^-

15.27 ■ For each of the following reactions, identify the Brønsted–Lowry acids and bases. What are the conjugate acid/base pairs?
(a) $CN^- + H_2O \rightleftharpoons HCN + OH^-$
(b) $HCO_3^- + H_3O^+ \rightleftharpoons H_2CO_3 + H_2O$
(c) $CH_3COOH + HS^- \rightleftharpoons CH_3COO^- + H_2S$

15.28 ■ Write the formula, and give the name of the conjugate acid of each of the following bases.
(a) NH_3 (b) HCO_3^- (c) Br^-

OBJECTIVE Identify the proton transfer in Brønsted-Lowry acid-base reactions.

15.29 ■ The following species react in aqueous solution. Predict the products, identify the acids and bases (and their conjugate species), and show the proton transfer in the acid-base reactions.
(a) NH_3 and CH_3COOH
(b) $N_2H_5^+$ and CO_3^{2-}
(c) H_3O^+ and OH^-
(d) HSO_4^- and $HCOO^-$

15.30 The following species react in aqueous solution. Predict the products, identify the acids and bases (and their conjugate species), and show the proton transfer in the acid-base reactions.
(a) ammonia and hydrochloric acid
(b) hydrogen carbonate ion and nitric acid
(c) formic acid and cyanide ion
(d) acetate ion and water

15.31 ■ What are the products of each of the following acid-base reactions? Indicate the acid and its conjugate base, and the base and its conjugate acid.
(a) $HClO_4 + H_2O \rightarrow$
(b) $NH_4^+ + H_2O \rightarrow$
(c) $HCO_3^- + OH^- \rightarrow$

15.32 ■ Write an equation to describe the proton transfer that occurs when each of these acids is added to water.
(a) HCO_3^- (b) HCl (c) CH_3COOH (d) HCN

OBJECTIVE Relate hydrogen ion concentration to hydroxide ion concentration in aqueous solutions.

15.33 Determine the hydrogen ion or hydroxide ion concentration in each of the following solutions, as appropriate.
(a) a solution in which $[H_3O^+] = 4.5 \times 10^{-4}$ *M*
(b) a solution in which $[OH^-] = 8.33 \times 10^{-5}$ *M*

15.34 Determine the hydrogen ion or hydroxide ion concentration in each of the following solutions, as appropriate.
(a) a solution in which $[H_3O^+] = 9.02 \times 10^{-10}$ *M*
(b) a solution in which $[OH^-] = 1.06 \times 10^{-11}$ *M*

15.35 The concentration of hydrogen ions in human blood is approximately 4.0×10^{-8} *M*. What is the hydroxide ion concentration in blood?

15.36 The hydroxide ion concentrations in wines actually range from 7.4×10^{-12} *M* to 1.6×10^{-10} *M*. What is the range of hydrogen ion concentrations in wine?

OBJECTIVE Convert between hydrogen ion concentrations, hydroxide ion concentrations, pH, and pOH.

15.37 Fill in the following table, and indicate whether the solution is acidic, basic, or neutral.

	pH	$[H_3O^+]$, *M*
(a)	2.34	
(b)		1.04×10^{-13}
(c)	−1.09	
(d)		2.12×10^{-11}
(e)		7.40×10^{-2}
(f)	13.41	
(g)		7.07×10^{-5}
(h)	9.80	
(i)		0.505

15.38 Fill in the following table, and indicate whether the solution is acidic, basic, or neutral.

	pH	$[H_3O^+]$, *M*
(a)	2.00	
(b)		1.04×10^{-3}
(c)	9.84	
(d)		2.00×10^{-1}
(e)		9.40×10^{-9}
(f)	11.34	
(g)		4.57×10^{-4}
(h)	4.51	
(i)		6.65×10^{-15}

15.39 Fill in the following table, and indicate whether the solution is acidic, basic, or neutral.

	pH	$[H_3O^+]$, *M*	pOH	$[OH^-]$, *M*
(a)	−1.04			
(b)			0.34	
(c)		1.98×10^{-7}		
(d)				4.42×10^{-2}

15.40 Fill in the following table, and indicate whether the solution is acidic, basic, or neutral.

	pH	$[H_3O^+]$, M	pOH	$[OH^-]$, M
(a)	10.34			
(b)			10.34	
(c)		0.412		
(d)				11.2×10^{-12}

15.41 What are the concentrations of hydrogen ion and hydroxide ion in each of the following? (See Table 15.4.)
(a) vinegar (b) stomach acid
(c) coffee (d) milk

15.42 What are the concentrations of hydrogen ion and hydroxide ion in each of the following? (See Table 15.4.)
(a) lemon juice (b) wine
(c) blood (d) household ammonia

OBJECTIVE Calculate the concentrations of species, the pH, and the pOH in solutions of strong acids and bases.

15.43 Calculate the pH and pOH of the following solutions.
(a) 0.050 M HCl (b) 0.024 M KOH
(c) 0.014 M $HClO_4$ (d) 1.05 M NaOH

15.44 Calculate the pH and pOH of the following solutions.
(a) 0.51 M CsOH (b) 0.0040 M HI
(c) 0.13 M LiOH (d) 0.66 M $HClO_4$

15.45 Calculate the pH and pOH of the following solutions.
(a) 0.94 M HBr (b) 0.042 M $Sr(OH)_2$
(c) 0.00033 M HCl (d) 0.88 M RbOH

15.46 Calculate the pH and pOH of the following solutions.
(a) 0.0045 M $Ba(OH)_2$ (b) 0.080 M HI
(c) 0.030 M $Sr(OH)_2$ (d) 12.3 M HNO_3

15.47 ■ A saturated solution of milk of magnesia, $Mg(OH)_2$, has a pH of 10.52. What is the hydronium ion concentration of the solution? What is the hydroxide ion concentration? Is the solution acidic or basic?

15.48 ■ Find $[OH^-]$ and the pH of the following solutions.
(a) 0.25 g barium hydroxide, $Ba(OH)_2$, dissolved in enough water to make 0.655 L of solution
(b) A 3.00 L solution of KOH is prepared by diluting 300.0 mL 0.149 M KOH with water.

OBJECTIVE Write the chemical equation for the ionization of a weak acid.

15.49 Write the chemical equation for the ionization of the following weak acids. Assume only one hydrogen ionizes in all cases.
(a) hydrazoic acid, HN_3
(b) citric acid, $H_2C_6H_6O_7$
(c) squaric acid, $H_2C_4O_4$

15.50 Write the chemical equation for the ionization of the following weak acids. Assume only one hydrogen ionizes in all cases.
(a) malic acid, $H_2C_4H_4O_5$
(b) maleic acid, $H_2C_4H_2O_4$
(c) malonic acid, $H_2C_3H_2O_4$

OBJECTIVE Relate the fraction of a weak acid ionized to the acid ionization constant.

15.51 HCN, a deadly gas that smells like bitter almonds, is formed by the reaction of H_2SO_4 and KCN. It was used in some states to execute criminals in the gas chamber. Measurements performed (carefully) on a 0.0050 M solution of HCN indicate that it is 0.035% ionized. Calculate K_a.

15.52 A solution is prepared by dissolving 0.121 g uric acid, $C_5H_3N_4O_3H$ (molar mass = 168 g/mol), and diluting to make exactly 10 mL of solution. Each uric acid molecule has only one hydrogen ion that dissociates. Conductivity measurements indicate that the acid is 4.2% ionized. Calculate K_a for uric acid, a compound that plays an important role in gout.

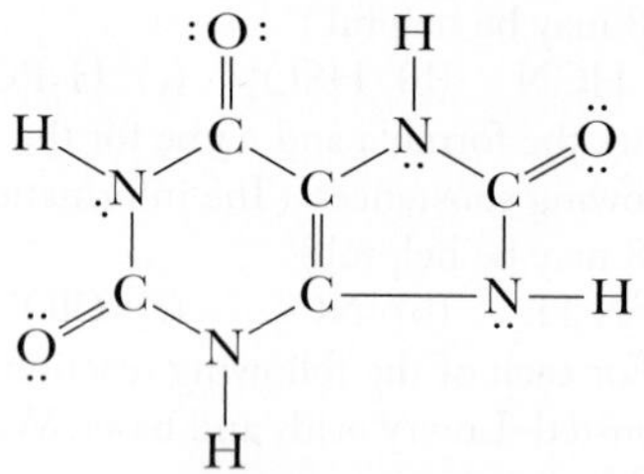

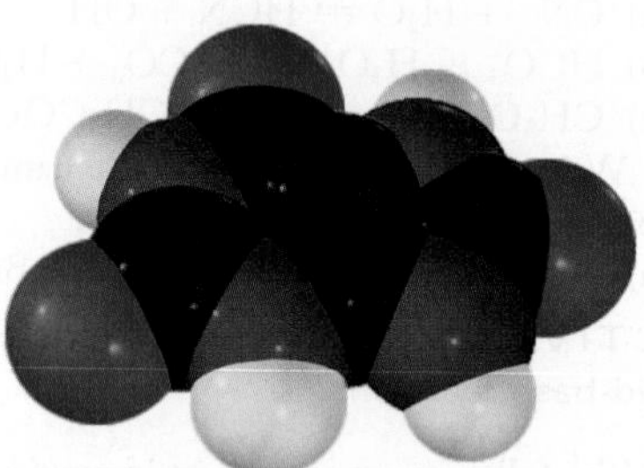

Uric acid. Uric acid crystals accumulate in the joints of people who suffer from gout. The most common joint affected is in the big toe.

15.53 Measurements of conductivity of solutions of two acids, A and B, produced the following data. Characterize each acid as strong or weak.

Concentration of Acid (g/L)	Conductivity	
	Solution A	Solution B
0.480	0.0059	0.0067
0.850	0.0078	0.0120
1.220	0.0093	0.0170

15.54 Assuming that the conductivity of an acid solution is proportional to the concentration of H_3O^+, sketch plots of conductivity versus concentration for HCl and HF over the 0- to 0.020 M concentration range.

OBJECTIVE Calculate acid ionization constants from experimental data.

15.55 Consider the solution formed when 50.0 mg butyric acid, C_3H_7COOH, a bad-smelling organic acid (formed when butter turns rancid), is dissolved in water to make 1.00 mL of solution. The pH of the solution is 2.52. Calculate K_a and pK_a for butyric acid.

15.56 ■ When 1.00 g thiamine hydrochloride (also called vitamin B_1 hydrochloride) is dissolved in water and then diluted to exactly 10.00 mL, the pH of the resulting solution is 4.50. The formula weight of thiamine hydrochloride is 337.28. Calculate K_a and pK_a for this acid.

15.57 If a 0.0100 M solution of caproic acid, thought to be at least partially responsible for the unique (and generally considered foul) smell of goats, has a pH of 3.43, what is K_a and pK_a?

15.58 Lactic acid, $CH_3CH(OH)COOH$, forms in muscles as a by-product of their contraction. If the pH of a 0.0376 M solution is 2.66, what is K_a and pK_a for lactic acid?

Lactic acid. The muscle pain after a strenuous workout was thought to be due to lactic acid buildup, but recent science suggests there are some other important factors.

15.59 ■ The pH of a 0.10 M solution of propanoic acid, CH_3CH_2COOH, a weak organic acid, is measured at equilibrium and found to be 2.93 at 25 °C. Calculate the K_a of propanoic acid.

15.60 ■ A 0.10 M solution of chloroacetic acid, $ClCH_2COOH$, has a pH of 1.95. Calculate K_a for the acid.

OBJECTIVE Calculate the concentrations of the species present in a weak acid solution.

15.61 Write the iCe table and set up the equation needed to solve for the concentration of the hydrogen ion in the following solutions.
(a) 0.20 M C_6H_5COOH (b) 1.50 M HCOOH
(c) 0.0055 M HCN (d) 0.075 M HNO_2

15.62 Write the iCe table and set up the equation needed to solve for the concentration of the hydrogen ion in the following solutions.
(a) 1.25 M HOCl (b) 0.80 M HF
(c) 0.14 M CH_3COOH (d) 0.25 M HCOOH

15.63 ▲ Use the K_a values in Table 15.6 to calculate the pH of the following solutions.
(a) 0.33 M HNO_2 (b) 0.016 M phenol, C_6H_5OH
(c) 0.25 M HF (d) 0.010 M HCOOH

15.64 ■ ▲ Use the K_a values in Table 15.6 to calculate the pH of the following solutions.
(a) 0.050 M HI (b) 0.85 M HF
(c) 0.15 M CH_3COOH (d) 0.017 M C_6H_5COOH

OBJECTIVE Calculate percentage ionization from K_a and concentration.

15.65 What is the fraction of acid ionized in each acid in Exercise 15.61?

15.66 What is the fraction of acid ionized in each acid in Exercise 15.62?

15.67 What is the fraction of acid ionized in each acid in Exercise 15.63?

15.68 What is the fraction of acid ionized in each acid in Exercise 15.64?

OBJECTIVE Write the chemical equation for the ionization of a weak base.

15.69 Write the chemical equation for the ionization of caffeine, a weak base. The chemical formula of caffeine is $C_8H_{10}N_4O_2$.

15.70 Like many narcotic drugs, cocaine is a weak base. Its chemical formula is $C_{17}H_{21}NO_4$. Write the chemical equation for the ionization of this weak base in aqueous solution.

OBJECTIVE Calculate the concentrations of species present in a solution of a weak base.

15.71 Hydrazine, N_2H_4, is weak base with $K_b = 1.3 \times 10^{-6}$. Fill in the iCe table and write the equation needed to solve for the concentration of hydroxide ion in a 0.10 M solution.

15.72 ■ Hydroxylamine, NH_2OH, is a weak base with $K_b = 8.7 \times 10^{-9}$. Fill in the iCe table and write the equation needed to solve for the concentration of hydroxide ion in a 0.10 M solution.

15.73 ▲ Coniine (2-propylpiperidine) is a weak base. It has the formula $C_8H_{17}N$. Calculate the pH of a 0.500 *M* solution of coniine (pK_b = 3.1).

Coniine is extracted from the plant *Conium maculatum,* also called *hemlock.* This harmless-looking relative of the carrot produces a deadly poison that killed the Greek philosopher Socrates in 399 B.C.

Socrates was a gadfly. He set about demonstrating that many prominent Athenians were more concerned with their own self-interest than with the needs of the society as a whole. He was charged with impiety, corrupting the youth, and disturbing the society. Socrates defended himself and was found guilty by the other Athenians. They asked him to recommend his own punishment; he recommended that he be compensated for his work with young people because he had no other source of income. This suggestion angered his peers, who sentenced him to death. Socrates was not well guarded, and the Athenians hoped he would escape. But he felt a moral obligation to follow the edict of the state. So he drank hemlock and died.

Three of Plato's dialogues speak of the events surrounding the death of Socrates. *Apology* depicts the trial, *Crito* includes his reasons for choosing death, and *Phaedro* comprises his musings as the coniine took effect.

Death of Socrates.

15.74 Morphine, $C_{17}H_{19}O_3N$, is a weak base with K_b = 1.6×10^{-6}. It is a prescription drug used to deaden pain; the average dose is 10 mg. Calculate the pH of a 0.0010 *M* solution.

15.75 ■ Calculate the $[OH^-]$ and the pH of a 0.024 *M* methylamine solution; $K_b = 4.2 \times 10^{-4}$.

15.76 ■ A hypothetical weak base has $K_b = 5.0 \times 10^{-4}$. Calculate the equilibrium concentrations of the base, its conjugate acid, and OH^- in a 0.15 *M* solution of the base.

OBJECTIVE Relate K_b for a base to K_a of its conjugate acid.

15.77 Write the chemical equation and use the data in Tables 15.6 and 15.8 to calculate the base ionization constant for the following ions.
(a) formate ion (b) nitrite ion

15.78 Write the chemical equation and use the data in Tables 15.6 and 15.8 to calculate the base ionization constant for the following ions.
(a) chlorite ion (b) fluoride ion

15.79 Write the chemical equation and use the data in Tables 15.6 and 15.8 to calculate the acid ionization constant for the following ions.
(a) hydroxylammonium ion
(b) ammonium ion

15.80 Write the chemical equation and use the data in Tables 15.6 and 15.8 to calculate the acid ionization constant for the following ions.
(a) pyridinium ion (b) hydrazinium ion

15.81 ■ Find the value of K_b for the conjugate base of the following organic acids.
(a) picric acid used in the manufacture of explosives; $K_a = 0.16$
(b) trichloroacetic acid used in the treatment of warts; $K_a = 0.20$

15.82 ■ Consider sodium acrylate, $NaC_3H_3O_2$. K_a for acrylic acid (its conjugate acid) is 5.5×10^{-5}.
(a) Write a balanced net ionic equation for the reaction that makes aqueous solutions of sodium acrylate basic.
(b) Calculate K_b for the reaction in (a).
(c) Find the pH of a solution prepared by dissolving 1.61 g $NaC_3H_3O_2$ in enough water to make 835 mL of solution.

15.83 Rank the following species in order of increasing acidity: NH_4^+, H_2O, HF, HSO_4^-.

15.84 ■ Rank the following species in order of increasing acidity: HF, HCl, NH_4^+, NH_3.

15.85 Rank the following species in order of increasing acidity: CH_3COOH, H_2O, HCOOH, F^-.

15.86 Rank the following species in order of increasing acidity: HCl, NH_3, HF, Na^+.

OBJECTIVE Calculate the pH of a salt solution.

15.87 Choose from among the labels strongly acidic, weakly acidic, neutral, weakly basic, and strongly basic to estimate the pH of the following solutions.
(a) 0.050 *M* NaF (b) 0.100 *M* KCl
(c) 0.080 *M* NH_4Br

15.88 Choose from among the labels strongly acidic, weakly acidic, neutral, weakly basic, and strongly basic to estimate the pH of the following solutions.
(a) 0.150 *M* $NaHSO_4$ (b) 0.050 *M* Na_3PO_4
(c) 0.100 *M* KBr

15.89 Write the iCe table and the equation needed to calculate the hydrogen ion concentration in 0.060 *M* pyridinium iodide.

15.90 Write the iCe table and the equation needed to calculate the hydrogen ion concentration in 1.5 *M* ammonium chloride.

15.91 Calculate the pH of each of the following solutions.
(a) 0.010 *M* sodium acetate
(b) 0.125 *M* ammonium nitrate
(c) 0.400 *M* potassium chlorite

15.92 ■ Calculate the pH of each of the following solutions.
(a) 0.25 *M* potassium nitrite
(b) 0.50 *M* sodium formate
(c) 0.015 *M* sodium fluoride

15.93 ■ State whether 1 *M* solutions of the following salts in water would be acidic, basic, or neutral.
(a) $FeCl_3$ (b) BaI_2 (c) NH_4NO_2
(d) Na_2HPO_4 (e) K_3PO_4

15.94 ■ State whether solutions of the following salts in water would be acidic, basic, or neutral.
(a) 0.1 M NH_3 (b) 0.1 M Na_2CO_3
(c) 0.1 M NaCl (d) 0.1 M CH_3CO_2H
(e) 0.1 M NH_4Cl (f) 0.1 M $NaCH_3CO_2$
(g) 0.1 M $NH_4CH_3CO_2$

OBJECTIVE Calculate the pH of a solution that contains both strong and weak acids.

15.95 Explain how to calculate the pH of a solution that is 0.20 M CH_3COOH and 0.050 M HI.

15.96 Explain how to calculate the pH of a solution that is 0.050 M HCl and 0.15 M HF.

OBJECTIVE Determine the pH of a solution that contains a mixture of weak acids (or bases) with different ionization constants.

15.97 Explain how to calculate the pH of a solution that is 0.10 M acetic acid and 0.20 M HCN.

15.98 Explain how to calculate the pH of a solution that is 0.050 M formic acid and 0.050 M phenol.

OBJECTIVE Relate the acid ionization constants of a series of related acids to their structure and bonding.

15.99 Hypofluorous acid, HOF, is known, but fluorous acid, HOFO, is not. Which acid would you expect to be stronger?

15.100 Without referring to a table in this chapter, match the acid with its acid ionization constant. You should give the formulas of these acids.

arsenious acid	$>>100$
chlorous acid	1.0×10^{-2}
selenic acid	6.3×10^{-10}

OBJECTIVE Explain how fundamental properties such as size and electronegativity affect the strengths of acids.

15.101 Which of each pair of acids is stronger? Why?
(a) GeH_4, AsH_3 (b) HNO_2, HNO_3

15.102 Which of each pair of acids is stronger? Why?
(a) H_3AsO_3, H_3AsO_4 (b) PH_3, H_2S

15.103 Which of each pair of acids is stronger? Why?
(a) $HClO_3$, $HClO_4$ (b) H_2S, H_2Se

15.104 Which of each pair of acids is stronger? Why?
(a) HClO, $HClO_2$ (b) H_2S, H_2O

OBJECTIVE Identify Lewis acids, Lewis bases, and their reaction products.

15.105 State whether each of the following reactions is an acid-base reaction, according to the definitions of Arrhenius, Brønsted–Lowry, and Lewis.
(a) $HCl(aq) + NH_3(aq) \rightarrow NH_4Cl(aq)$
(b) $SO_2(g) + NaOH(s) \rightarrow NaHSO_3(s)$

15.106 State whether each of the following reactions is an acid-base reaction, according to the definitions of Arrhenius, Brønsted–Lowry, and Lewis.
(a) $HCl(aq) + H_2O(\ell) \rightarrow H_3O^+(aq) + Cl^-(aq)$
(b) $Zn(OH)_3^-(aq) + OH^-(aq) \rightleftharpoons Zn(OH)_4^{2-}(aq)$

15.107 State whether each of the following reactions is an acid-base reaction, according to the definitions of Arrhenius, Brønsted–Lowry, and Lewis.
(a) $LiH(s) + H_2O(\ell) \rightarrow LiOH(aq) + H_2(g)$
(b) $HSO_4^-(aq) + F^-(aq) \rightleftharpoons HF(aq) + SO_4^{2-}$

15.108 State whether each of the following reactions is an acid-base reaction, according to the definitions of Arrhenius, Brønsted–Lowry, and Lewis.
(a) $CO_2(g) + LiOH(s) \rightarrow LiHCO_3(s)$
(b) $SO_2(g) + H_2O(g) \rightleftharpoons H_2SO_3(g)$

15.109 ■ Decide whether each of the following substances should be classified as a Lewis acid or a Lewis base.
(a) BCl_3 (*Hint:* Draw the electron dot structure.)
(b) H_2NNH_2, hydrazine (*Hint:* Draw the electron dot structure.)
(c) the reactants in $Ag^+ + 2\ NH_3 \rightleftharpoons [Ag(NH_3)_2]^+$

15.110 ■ Identify the Lewis acid and the Lewis base in each reaction.
(a) $I_2(s) + I^-(aq) \rightarrow I_3^-(aq)$
(b) $SO_2(g) + BF_3(g) \rightarrow O_2SBF_3(s)$
(c) $Au^+(aq) + 2CN^-(aq) \rightarrow [Au(CN)_2]^-(aq)$
(d) $CO_2(g) + H_2O(\ell) \rightarrow H_2CO_3(aq)$

Chapter Exercises

15.111 Choose from among the labels strongly acidic, weakly acidic, neutral, weakly basic, and strongly basic to estimate the pH of the following solutions.
(a) 0.250 M HBr (b) 0.50 M HF
(c) 0.020 M $Ba(OH)_2$ (d) 0.44 M NH_3

15.112 ■ Choose from among the labels strongly acidic, weakly acidic, neutral, weakly basic, and strongly basic to estimate the pH of the following solutions.
(a) 0.30 M NH_4Cl (b) 0.25 M Na_3PO_4
(c) 0.080 M HI (d) 0.12 M LiI

15.113 Choose from among the labels strongly acidic, weakly acidic, neutral, weakly basic, and strongly basic to estimate the pH of the following solutions.
(a) 0.45 M NaCl (b) 0.18 M BaF_2
(c) 0.25 M $KHSO_4$ (d) 0.33 M $NaNO_2$

15.114 Choose from among the labels strongly acidic, weakly acidic, neutral, weakly basic, and strongly basic to estimate the pH of the following solutions.
(a) 0.30 M NH_4Cl (b) 0.15 M N_2H_5Cl
(c) 0.50 M KNO_3 (d) 0.50 M HCOONa

15.115 ▲ Calculate the fraction of benzoic acid, a useful food preservative, which is ionized in 0.010- and 0.0010 M solutions.

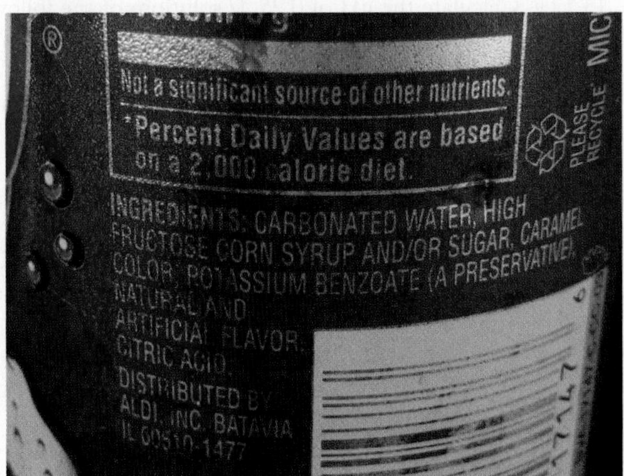

Courtesy of M. Stading

Benzoic acid. This soft drink contains potassium benzoate, the conjugate base of benzoic acid.

15.116 Phenol (C_6H_5OH, also called *carbolic acid*) has a pK_a of 9.89. It is used to preserve body tissues and is quite toxic. Calculate the fraction of the acid ionized in 0.010 M and 0.0010 M phenol.

15.117 ▲ Calculate the volume of 0.083 M HNO_2 that must be dissolved to make 1.00 L of a solution with a pH of 4.75.

15.118 ▲ ■ Calculate the volume of 0.10 M acetic acid needed to prepare 5.0 L of acetic acid solution with a pH of 4.00.

15.119 Calculate the mass of benzoic acid, C_6H_5COOH, which must be dissolved to prepare 1.00 L of a solution with a pH of 3.50.

15.120 Calculate the volume of 14.3 M HCl that must be used to prepare 100.0 L of a solution with a pH of 3.50.

15.121 ▲ A solution is made by diluting 25.0 mL of concentrated HCl (37% by weight; density = 1.19 g/mL) to exactly 500 mL. Calculate the pH of the resulting solution.

15.122 ▲ Liquid HF undergoes an autoionization reaction:

$$2HF \rightleftharpoons H_2F^+ + F^-$$

(a) Is KF an acid or a base in this solvent?
(b) Perchloric acid, $HClO_4$, is a strong acid in liquid HF. Write the chemical equation for the ionization reaction.
(c) Ammonia is a strong base in this solvent. Write the chemical equation for the ionization reaction.
(d) Write the net ionic equation for the neutralization of perchloric acid with ammonia in this solvent.

15.123 ▲ Pure liquid ammonia ionizes in a manner similar to that of water.
(a) Write the equilibrium for the autoionization of liquid ammonia.
(b) Identify the conjugate acid form and the base form of the solvent.
(c) Is $NaNH_2$ an acid or a base in this solvent?
(d) Is ammonium bromide an acid or a base in this solvent?

15.124 Calculate the pH of 0.050 M solutions of the following solutes.
(a) benzoic acid (b) sodium benzoate

15.125 Calculate the pH of 0.25 M solutions of the following solutes.
(a) hydrofluoric acid (b) potassium fluoride

15.126 Determine whether each of the following reactions favors the reactants, products, or neither.
(a) $HCl(aq) + NH_3(aq) \rightleftharpoons NH_4Cl(aq)$
(b) $HNO_3(aq) + NaOH(aq) \rightleftharpoons NaNO_3(aq) + H_2O(\ell)$
(c) $2KCl(aq) + Ba(OH)_2(aq) \rightleftharpoons BaCl_2(aq) + 2KOH(aq)$
(d) $HSO_4^-(aq) + NH_3(aq) \rightleftharpoons NH_4^+(aq) + SO_4^-(aq)$

15.127 Determine whether each of the following reactions favors the reactants, products, or neither.
(a) $NaOH(aq) + KCl(aq) \rightleftharpoons NaCl(aq) + KOH(aq)$
(b) $HF(aq) + NaOH(aq) \rightleftharpoons NaF(aq) + H_2O(\ell)$
(c) $CH_3COONa(aq) + H_2O(\ell) \rightleftharpoons CH_3COOH(aq) + NaOH(aq)$
(d) $NH_4^+(aq) + H_2O(\ell) \rightleftharpoons NH_3(aq) + H_3O^+(aq)$

Cumulative Exercises

15.128 ▲ ■ A solution is made by diluting 10.0 mL of concentrated ammonia (28% by weight; density = 0.90 g/mL) to exactly 1 L. Calculate the pH of the solution.

15.129 An aqueous solution contains formic acid and formate ion. Determine the direction in which the pH will change if each of the following chemicals is added to the solution.
(a) HCl (b) $NaHSO_4$ (c) CH_3COONa
(d) KBr (e) H_2O

15.130 A solution is made by dissolving 15.0 g sodium hydroxide in approximately 450 mL water. The solution becomes quite warm, but after it is allowed to return to room temperature, water is added to bring the volume to 500.0 mL of solution.
(a) Calculate the pH and pOH in the final solution.
(b) Why would we wait for it to return to room temperature?
(c) If the mass of the water used to initially dissolve the sodium hydroxide were exactly 450 g and the temperature of the water increased by 8.865 °C, how much heat was given off by the dissolution of 15.0 g of solute? Assume the specific heat of the solution is 4.184 J/g · K. What is the molar heat change for the dissolution of sodium hydroxide (known as the enthalpy of solution, ΔH_{sol})?

15.131 Calculate the pH of a solution prepared by adding 10.0 g sodium benzoate to 100 mL of 0.10 M KOH.

15.132 Calculate the pH of a solution prepared by adding exactly 10.0 mL of a 14.8 M KOH solution to 200 mL of water, then adding water until the volume of solution is exactly 250 mL.

15.133 ■ Calculate the pH of a solution prepared by mixing 10 mL of 1.0 M NaOH with 100 mL of 0.10 M ammonia.

15.134 When perchloric acid ionizes, it makes the perchlorate ion, ClO_4^-. Draw the Lewis electron dot symbol for the perchlorate ion.

15.135 Picric acid, or 2,4,6-trinitrophenol, has the following structure:

Picric acid

© Michael Viard/Peter Arnold, Inc.

Picric acid. Picric acid forms extremely explosive metal picrates. This substance was formerly common in chemistry and biology laboratories, and when old bottles with corroded metal caps are found in stockrooms, they must be removed by the bomb squad.

Not only is it an acid, it is also an explosive.

(a) Based on what you know of oxyacids, which hydrogen ionizes from the picric acid molecule?

(b) Picric acid has a K_a of 0.380. What is the pH of a 0.100 M solution of picric acid? You will have to use the quadratic formula.

(c) ▲ When picric acid explodes anaerobically (that is, without oxygen), it forms carbon monoxide, water, nitrogen, and solid carbon. Write the balanced chemical equation for the decomposition of picric acid.

(d) ▲ Would substitution of the nitro (NO_2) groups in the molecule with amino (NH_2) likely increase or decrease the K_a of the compound? Explain your answer.

15.136 ✎ Acids in the news.

(a) Chemists often talk about the United States senator who addressed the public about acid rain and said: "I too abhor the effects of acid rain and I pledge the United States Senate will fight to reduce acidity of rain until pH is zero!" Is battling acid rain by reducing the pH a good idea?

The story would be particularly good if it were true, but thankfully it is a myth.

(b) "Anger: an acid that can do more harm to the vessel in which it is stored than to anything on which it is poured" is a quotation attributed to Seneca, a Roman philosopher, in the mid-1st century A.D. Do you think Seneca was referring to a strong acid, a weak acid, or something else?

(c) Use the Internet and other resources to find the origin and common meaning of the phrase *acid test*.

Photo by Andrew J. Russell/National Archives/Time Life Pictures/Getty Images

Civil War Munitions Train.

As the Union forces moved through the Sumter, South Carolina area in early 1865, the Confederate Army burned the railroad bridges around Sumter, leaving a fully loaded munitions train hidden in the swamp. The Union commander heard rumors about the unguarded train, dispatched troops who found the munitions train 3 miles into the swamp, and blasted 3 locomotives and 35 cars loaded with military supplies.

Recently, historians found several small glass vials near old train tracks outside of Sumter when excavating the scene of the Civil War explosion. The vials were about 2 inches (50 mm) long and about 1/2 inch (13 mm) in diameter, and contained a clear liquid. What was it?

Courtesy of M. Stading

Glass vial containing unknown liquid.

The historians brought the vials to a chemist to identify the contents. The vials were packed in cotton, indicating importance (or explosiveness). The historians speculated that the contents might be chloroform (an anesthetic), an extract of opium, or nitroglycerin. Perhaps the vials might be leveling devices used to help aim a cannon.

The possibility of nitroglycerin, a widely used 19th-century explosive, was remote. Nitroglycerin is every bit as hazardous as people fear—small vibrations can cause it to explode. Still, extreme caution was used when one of the vials was opened.

The contents, a water-clear, viscous, oily fluid, were transferred to a screw-top container. The first step in identifying the sample was to determine whether it was water soluble, which would give some clues to its composition. If the sample did not dissolve in water, it would likely be an organic compound, and that information would direct the chemical analysis. The chemist observed that the sample dissolved when a drop was added to water in a test tube, and also noticed that the solution became so hot that it became uncomfortable to hold.

The acidity was checked next, and the solution was found to be extremely acidic, indicating that the unknown was probably a concentrated acid. Several precipitation tests were performed that indicated the dominant species in solution was the sulfate ion, indicating that the sample was sulfuric acid, H_2SO_4. As

Reactions between Acids and Bases

16

CHAPTER CONTENTS

OWL Online homework for this chapter may be assigned in OWL.

Look for the green colored vertical bar throughout this chapter, for integrated references to this chapter introduction.

additional confirmation, it is well known that sulfuric acid gets very hot when added to water. One of the important steps used to identify the material was a titration. The titration, which is described in the Case Study at the end of this chapter, showed that the sample was high-purity sulfuric acid.

It is not as easy to determine the intended use of the vials of sulfuric acid as the chemical composition of the contents. Some research was needed, and the historians found the diary of a Confederate torpedoman that explained the military use of sulfuric acid. The Confederate Army was known to have effective mines they used on land and at sea. According to the diary, soldiers filled a wooden barrel with gunpowder, then they drilled several holes in the barrel. They stuffed sugar and another substance that we know now was potassium chlorate into the holes together with the vials of sulfuric acid, and covered the vial with a thin piece of lead foil. When a ship hit the mine, the vial would break and the sulfuric acid would react with sugar and potassium-chlorate mixture in a rapid exothermic reaction releasing enough heat to ignite the gunpowder.

$$C_{12}H_{22}O_{11}(s) \xrightarrow{H_2SO_4} 12C(s) + 11H_2O(g)$$

See the Case Study at the end of the chapter for a description of the chemistry used to determine the contents of the vial.

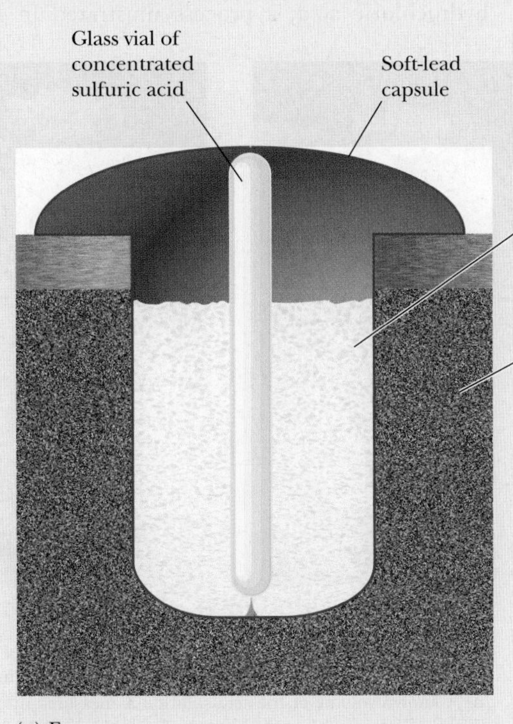

(*a*) Fuse

(*b*) Mine with several fuses

Confederate land mines used sulfuric acid in their triggers.

The reaction of an acid with a base is one of the most important classes of chemical reactions. Your life literally depends on such reactions because the human body uses them to neutralize the chemicals produced each time the heart beats. At this point in our study of chemistry, we know what products form when we mix an acid and a base, but we do not know how to predict the pH of this seemingly complex mixture of water with strong or weak acids and bases.

This chapter extends the methods used in earlier chapters to describe the chemistry of individual solutions of acids and bases to include mixtures of acids and bases. We then discuss indicators, the substances that change color and signal the end of an acid-base reaction, and polyprotic acids (acids that can donate more than one proton to a base). Last, we describe the influence of acids and bases on other equilibria. One important example of this effect is the dependence of the solubility of certain salts on the pH of the solution.

16.1 Titrations of Strong Acids and Bases

OBJECTIVES

- ☐ Describe the chemistry of a titration
- ☐ Calculate the volume of titrant needed to reach the equivalence point
- ☐ Identify regions of the titration curve in which the analyte or titrant is in excess
- ☐ Express amounts of analyte and titrant in units of millimoles

Each day, chemists perform thousands of analyses called *titrations.* A **titration** is a method used to determine the concentration of a substance, called the **analyte,** by adding another substance, called the **titrant,** which reacts in a known manner with the analyte. The titrant is generally chosen so that its reaction with the analyte goes to completion.

A titration is a chemical reaction used to measure the quantity of a substance.

analyte + titrant → products

Titrations are widely used to determine the amount or concentration of a substance as part of a quantitative analysis, and acid-base titrations are among the most important. Learning how the pH of the reaction mixture changes as the titrant is added is essential for understanding acid-base titrations. Stoichiometry calculations for titrations were introduced in Chapter 4. We review them briefly here.

Consider how to determine the concentration of a solution of hydrochloric acid, a process illustrated in Figure 16.1. First, place a

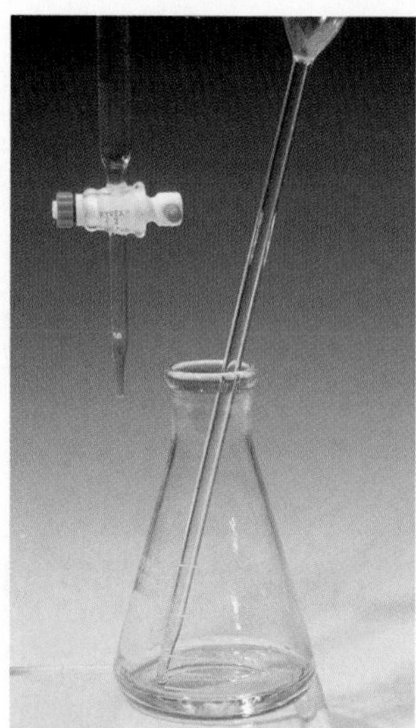
(*a*)

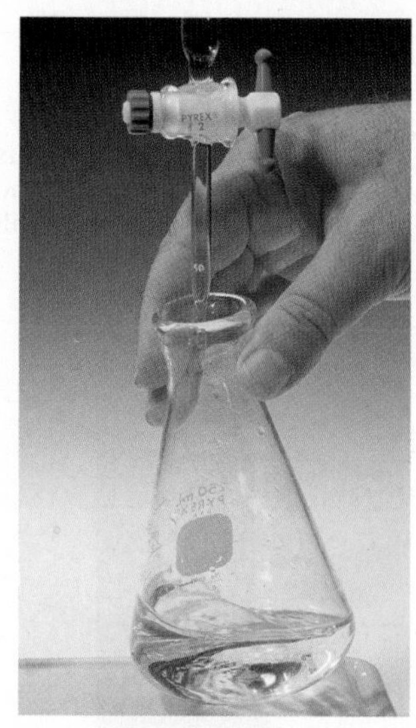
(*b*)

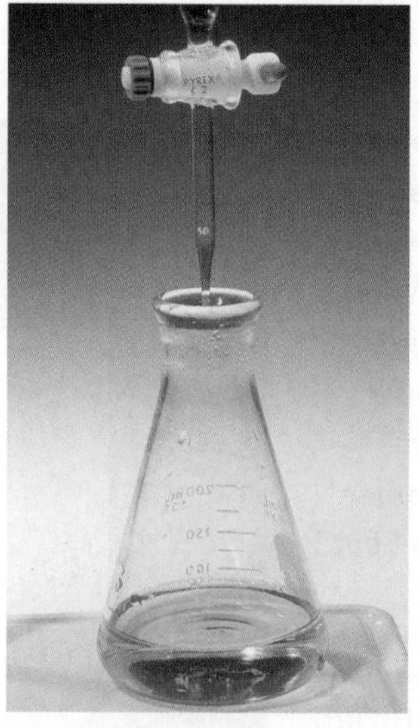
(*c*)

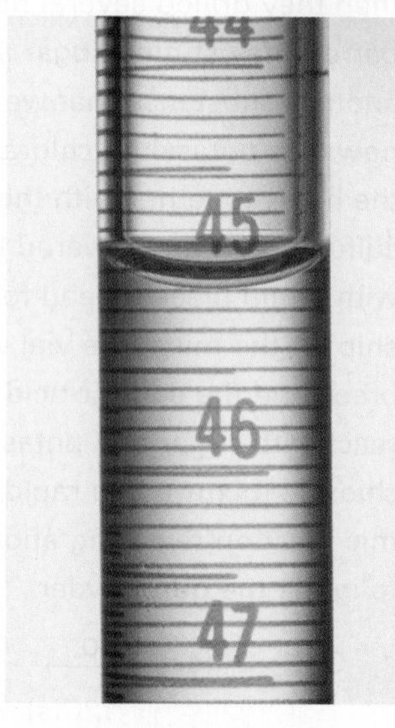

(*d*)

Figure 16.1 Using a titration to determine the concentration of a solution of hydrochloric acid. *(a)* A known volume of the acid solution is measured into a flask. *(b)* Standard base is added from a buret. *(c)* The endpoint of the titration is indicated by a color change. *(d)* The volume of base solution needed to react with the acid is recorded.

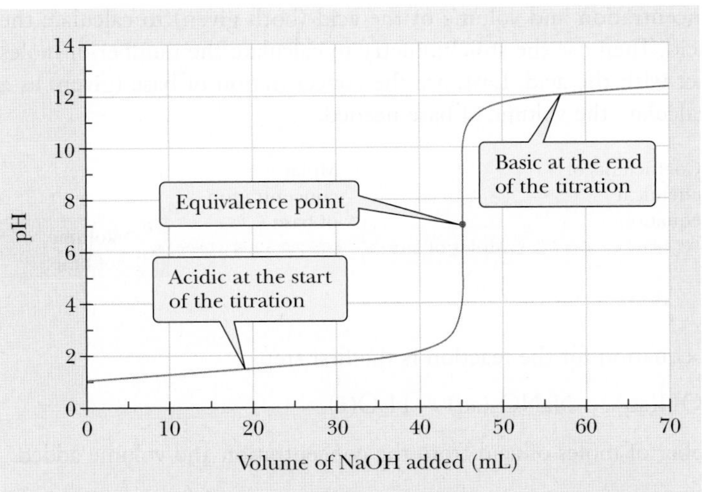

Figure 16.2 Titration of nitric acid with sodium hydroxide.

known volume of the HCl solution in a flask. Next, add an indicator that changes color when the hydrochloric acid is consumed. Phenolphthalein is a compound with such properties—it is colorless in acid solution and pink in alkaline solution. A *standard solution* of base (one of accurately known concentration) is added until the color change occurs—that is, the *endpoint* is reached. Record the volume of standard base needed to reach the endpoint, and then calculate the concentration of the original acid by the procedures presented in Chapter 4.

A titration measures the concentration or amount of the analyte in a particular sample.

Shapes of Strong Acid–Strong Base Titration Curves

Instead of using an indicator to signal the end of a titration, a pH meter can be used to record the pH during the course of the titration. The resulting graph of the pH of the solution as titrant is added is called a **titration curve.** Figure 16.2 shows the results observed for the titration of 50.00 mL of 0.0884 *M* nitric acid with 0.0980 *M* NaOH.

Some general conclusions about titration curves can be drawn from this experiment:

1. At the beginning of the titration curve, the solution is acidic, the pH is low, and pH changes are small until about 44 mL of base has been added. The exact volume depends on the specific experiment.
2. The pH changes quite abruptly from acidic to basic in the 44 to 46-mL region.
3. Beyond 46 mL, the solution is basic and the pH changes are small.

These features are common to the titrations of all strong acids with a strong base. Chemists use the titration curve to determine the volume of base required to exactly neutralize the acid present in the sample. This point in the titration is called the *equivalence point,* which occurs when exactly one mole of base has been added for each mole of acid present in the sample.

The chemical reaction that occurs in this titration is

$$HNO_3(aq) + NaOH(aq) \rightarrow NaNO_3(aq) + H_2O(\ell)$$

At the equivalence point, all of the acid and base have reacted to produce a sodium nitrate solution. Recall from Chapter 15 that a salt containing the cation of a strong base and the anion of a strong acid does not affect the pH. At the equivalence point, the pH is that of water, 7.00. From the titration curve, we see that it takes 45.1 mL of the NaOH solution to reach a pH of 7.00. Let us confirm that 45.1 mL is the expected equivalence point with a stoichiometry calculation based on the data given. Example 16.1 illustrates this calculation.

EXAMPLE 16.1 Calculating the Equivalence Point in a Titration

Calculate the equivalence-point volume in the titration of 50.00 mL of 0.0884 *M* nitric acid with 0.0980 *M* NaOH. The titration curve for this system appears in Figure 16.2.

Strategy Use the concentration and volume of the acid (both given) to calculate the number of moles of acid. Then use the stoichiometry to calculate the number of moles of base needed to react with the acid. Last, use the concentration of base (given) as a conversion factor to calculate the volume of base needed.

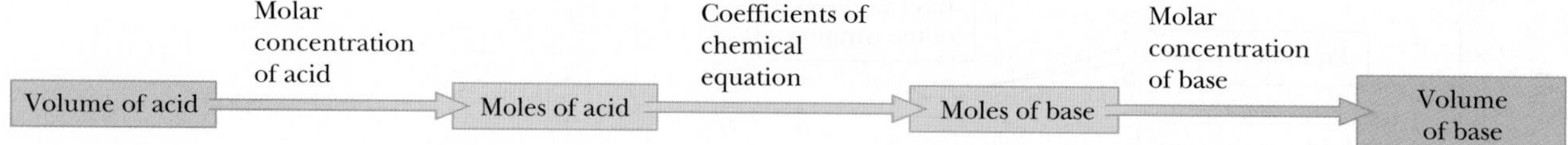

Solution

Writing the chemical equation for the reaction is the first step.

$$HNO_3(aq) + NaOH(aq) \rightarrow NaNO_3(aq) + H_2O(\ell)$$

Calculate the number of moles of acid from the concentration and volume added.

$$\text{Amount of } HNO_3 = 0.0500 \cancel{\text{ L of soln}} \times \left(\frac{0.0884 \text{ mol}}{1 \cancel{\text{ L of soln}}}\right) = 4.42 \times 10^{-3} \text{ mol}$$

Determine the amount of base from the amount of acid and the stoichiometry of the chemical equation.

$$\text{Amount of NaOH} = 4.42 \times 10^{-3} \cancel{\text{ mol } HNO_3} \times \left(\frac{1 \text{ mol NaOH}}{1 \cancel{\text{ mol } HNO_3}}\right) = 4.42 \times 10^{-3} \text{ mol}$$

Last, calculate the volume of NaOH from the number of moles and the concentration of NaOH.

$$\text{Volume NaOH} = 4.42 \times 10^{-3} \cancel{\text{ mol NaOH}} \times \left(\frac{1 \text{ L NaOH}}{0.0980 \cancel{\text{ mol NaOH}}}\right) = 0.0451 \text{ L} = 45.1 \text{ mL}$$

The results of this calculation agree with the experimentally determined equivalence point for the titration curve shown in Figure 16.2.

Understanding

What is the equivalence point in the titration of 5.00 mL of 0.0110 *M* $Ba(OH)_2$ with 0.0251 *M* HCl?

Answer 4.38 mL (Did you remember the 2:1 stoichiometry?)

Experiment shows that the pH changes abruptly from acidic to basic at the equivalence point. Consider the net ionic equation that describes the titration of a strong acid with strong base.

$$H_3O^+ + OH^- \rightarrow 2H_2O$$

In the titration of an acid with a base, initially the acid is in excess, and the pH is low. After the acid has been neutralized and excess titrant has been added, hydroxide ion is in excess, and the pH is high.

Before the equivalence point, H_3O^+ is in excess; therefore, the solution is acidic, and the pH is much less than 7. Once the equivalence point has been passed, OH^- is present in excess; therefore, the solution is basic, and the pH is much greater than 7. In every acid-base titration, the pH changes abruptly near the equivalence point. This feature is widely used to locate the equivalence point from an experimentally determined titration curve.

Numbers of a more convenient size are obtained if the amounts of materials are expressed in millimoles rather than moles. Converting moles to millimoles is shown next, and calculations in the following sections generally use millimoles.

Units of Millimoles

A titration calculation is basically a stoichiometry calculation. As in all other stoichiometry calculations, amounts of products and reactants must be expressed in units of moles. Examples in Chapter 4 show that the molarity of the solution serves as a conversion factor in such calculations. For example, the number of moles of HCl in 20.00 mL of a 0.125 M HCl solution is

$$\text{Amount of HCl} = 0.02000\ \cancel{\text{L}} \times \left(\frac{0.125\ \text{mol HCl}}{1\ \cancel{\text{L}}}\right) = 2.50 \times 10^{-3}\ \text{mol HCl}$$

The volume and concentration used in this sample conversion are typical of those that occur in titrations. For titrations, the **millimole** (1 **mmol** = 10^{-3} mol) is a more convenient-sized unit than the mole for expressing the amounts of reactants.

$$\text{Amount of HCl} = 2.50 \times \cancel{10^{-3}\ \text{mol HCl}} \times \left(\frac{1\ \text{mmol HCl}}{1 \times \cancel{10^{-3}\ \text{mol HCl}}}\right) = 2.50\ \text{mmol}$$

Most titrations involve substances that are present in millimole amounts.

Laboratory Reagents. Most laboratories use milliliter volumes of liquids and solution concentrations in molarity.

Millimoles are obtained when we start with the volume of solution in milliliters and use the molarity as a conversion factor. This relationship is true because the unit of molarity can be interpreted as millimoles per milliliter (mmol/mL), as well as moles per liter (mol/L). (Remember that 1 mmol = 1×10^{-3} mol and 1 mL = 1×10^{-3} L.)

Let us calculate the number of millimoles of HCl in 20.00 mL of a 0.125 M solution from the volume and molarity.

$$\text{Amount of HCl} = 20.00\ \cancel{\text{mL}} \times \left(\frac{0.125\ \text{mmol HCl}}{\cancel{\text{mL}}}\right) = 2.50\ \text{mmol HCl}$$

When the concentration is expressed in molarity and the volume in liters, we calculate moles of solute. When the volume is in milliliters, we calculate millimoles. Both can be used in stoichiometry problems; sometimes one is more convenient than the other. For most titrations, including the remaining examples in this chapter, it is more convenient to express amounts in millimoles.

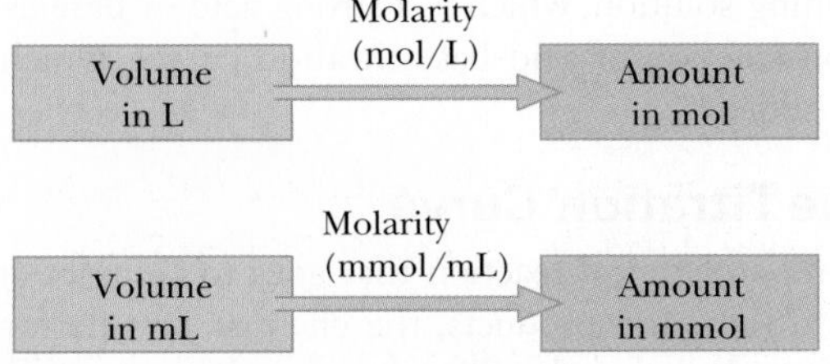

The molar concentration is used to convert between volume and amount. When the volume is expressed in milliliters, the amount is expressed in millimoles.

EXAMPLE 16.2 Calculating Millimoles

Calculate the number of millimoles of HCl needed to neutralize 10.0 mL of 0.15 M $Ba(OH)_2$.

Strategy First, write the chemical equation for the neutralization. Calculate the amount of barium hydroxide, and use the stoichiometric relationship to calculate the amount of HCl.

Solution

The chemical equation is

$$Ba(OH)_2(aq) + 2HCl(aq) \rightarrow BaCl_2(aq) + 2H_2O(\ell)$$

The amount of barium hydroxide is

$$\text{Amount of } Ba(OH)_2 = 10.0 \cancel{\text{mL}} \times \left(\frac{0.15 \text{ mmol}}{1 \cancel{\text{mL}}}\right) = 1.5 \text{ mmol } Ba(OH)_2$$

The amount of HCl is related to the amount of $Ba(OH)_2$ by the coefficients in the chemical equation.

$$\text{Amount of HCl} = 1.5 \cancel{\text{mmol } Ba(OH)_2} \times \left(\frac{2 \text{ mmol HCl}}{1 \cancel{\text{mmol } Ba(OH)_2}}\right) = 3.0 \text{ mmol HCl}$$

Understanding

Calculate the amount of $La(OH)_3(s)$ needed to neutralize 50.0 mL of 0.25 *M* HNO_3.

Answer 4.2 mmol $La(OH)_3(s)$

OBJECTIVES REVIEW *Can you:*

- ☑ describe the chemistry of a titration?
- ☑ calculate the volume of titrant needed to reach the equivalence point?
- ☑ identify regions of the titration curve in which the analyte or titrant is in excess?
- ☑ express amounts of analyte and titrant in units of millimoles?

16.2 Titration Curves of Strong Acids and Bases

OBJECTIVES

- ☐ Calculate the concentrations of all species present during the titration of a strong acid with a strong base
- ☐ Graph the titration curve
- ☐ Correlate the shape of the titration curve to the titration stoichiometry
- ☐ Estimate the pH of mixtures of strong acids and bases

Calculating the concentrations of species present in the course of a strong acid-base titration is really not that different from the calculations presented earlier. These calculations are broken into two steps. The first step uses stoichiometry to compute how much analyte and titrant remain after the reaction. The second calculation is to compute the pH of the remaining solution, which is a strong acid or base. Much of this chapter is devoted to the construction of acid-base titration curves, which show how the pH changes as titrant is added.

Calculating the Titration Curve

A titration is based on a chemical reaction that goes to completion. To determine the amounts of analyte, titrant, and products, the chemist first performs a stoichiometry calculation that starts with the total millimoles of each reactant. The concentrations of the species in solution are calculated from the numbers of millimoles and the total volume of solution. Finally, the H_3O^+ concentration and the pH are calculated. The results of these calculations are generally presented in graphic form, as a titration curve.

In the titration of an unknown concentration of a strong acid (the analyte) with a strong base (the titrant), we would begin by placing a measured volume of the acid in a flask and adding the base from a burette. Both reactants are strong electrolytes, so the net ionic equation for the titration reaction is

$$H_3O^+(aq) + OH^-(aq) \rightarrow 2H_2O(\ell)$$

Before the equivalence point, H_3O^+ is in excess and the pH is found directly from its concentration. Beyond the equivalence point, OH^- is in excess, and we calculate $[H_3O^+]$ and pH from the hydroxide ion concentration and the autoionization equilibrium of water. The pH at the equivalence point requires the same calculation as is used for the pH of water.

Titration volumes are expressed in a cumulative manner based on the volume of the titrant. If we performed a titration in the laboratory, we might measure the pH after adding 5.00 mL of titrant. We might make a second measurement after adding an additional 7.00 mL of titrant. We would describe these measurements as the pH after 5.00 and 12.00 mL of titrant are added. The pH after 12.00 mL is the same whether we add 5.00, then 7.00 mL; 7.00, then 5.00 mL; 10.00, then 2.00 mL; or any combination that adds to 12.00 mL. *Titration calculations always reference the total volume of titrant.*

A titration curve is the graph of pH versus the total volume of titrant.

A tabular approach, which we call the *sRfc table,* is helpful for organizing the stoichiometry calculations. We place the starting number of millimoles of each reactant on the s line (s for "starting"). The changes in the number of millimoles go on the R line (R for "Reaction"), and the final number of millimoles on the f line (f for "final"). We obtain the c line ("concentrations") by dividing the millimoles of each species on the f line by the total volume of solution (in mL). The chemical equation in the sRfc table is printed in color to help differentiate it from the equilibrium relationships of the iCe table.

We will fill out the sRfc table for the reaction of 15.0 mL of 0.100 M H_3O^+ with 10.0 mL of 0.100 M OH^-. On the first line, write the chemical equation.

			H_3O^+	+	OH^-	→	$2H_2O$

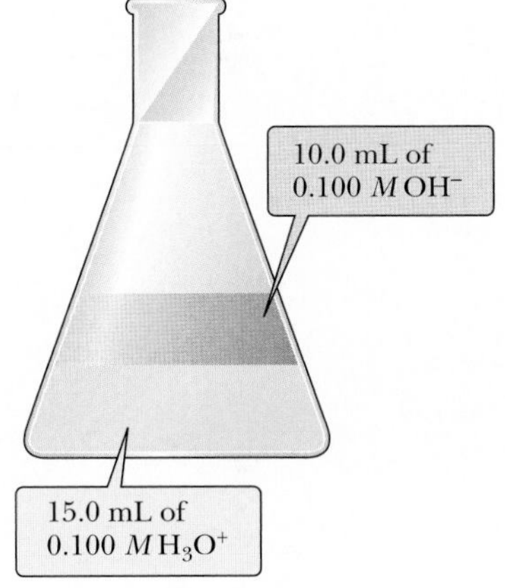

On the second line, fill in the starting information. For this example, the starting amount of H_3O^+ is 15.0 mL × 0.100 M = 1.50 mmol H_3O^+. The amount of OH^- is 10.0 mL × 0.100 M = 1.00 mmol.

			H_3O^+	+	OH^-	→	$2H_2O$
s	Fill in the "starting" information	mmol	1.50		1.00		Excess

On the R line, write the number of millimoles that react. The uppercase R serves to remind us that the stoichiometry must be considered. For this reaction, the stoichiometry is 1:1.

The reaction goes to completion, so either H_3O^+ or OH^- will react completely. We always choose the smaller of the two, as with any other limiting reactant problem.

			H_3O^+	+	OH^-	→	$2H_2O$
s		mmol	1.50		1.00		Excess
R	The "Reaction" line uses OH^- as the limiting reactant	mmol	−1.00		−1.00		+2 × 1.00

The f line shows the final amounts. The entries on this line are computed by adding the starting amounts to the amounts that react.

			H_3O^+	+	OH^-	→	$2H_2O$
s		mmol	1.50		1.00		Excess
R		mmol	−1.00		−1.00		+2 × 1.00
f	The "final" number of mmol is s + R	mmol	0.50		0.0		Excess

We can look at the f line and see that strong acid remains after the addition of the base, and predict that the solution will be strongly acidic. The pH is likely to be in the range of 1 to 3. If we are asked to estimate the pH, we do so by evaluating the species present on the f line.

When we need to calculate the pH, we add a c line, which shows the final concentrations of each species. The concentration is the number of millimoles of each (the f line) divided by the total volume of the solution. In this particular case, we added 10.0 mL of 0.100 M OH^- to 15.0 mL of 0.100 M H_3O^+, so the total volume of solution is 25.0 mL. The picture of the titration flask on the previous page shows these volumes.

The volume of solution is the volume of the sample plus the volume of the titrant.

			H_3O^+	+	OH^-	→	$2H_2O$
s		mmol	1.50		1.00		Excess
R		mmol	−1. 00		−1.000		+2 × 1.000
f		mmol	0.50		0.0		Excess
c	The concentration is the amount (mmol) divided by the total volume (mL)	M	0.020		0.0		Excess

The pH is $-\log[H_3O^+] = -\log(0.020) = 1.70$.

Example 16.3 uses an sRfc table to help compute the titration curve.

EXAMPLE 16.3 Titration of a Strong Acid with a Strong Base

Calculate the pH in the titration of 20.00 mL of 0.125 M HCl after the addition of (a) 0, (b) 2.00, (c) 10.00, and (d) 20.00 mL of 0.250 M NaOH.

Strategy First, calculate the amounts (mmol) of acid and base. The analyte and the titrant (acid and base) react with each other, and this reaction goes to completion. Use the stoichiometry of the reaction to fill out the sRfc table after each addition of titrant. We need to determine which species remain after the titration.

The logic flow diagram is shown below.

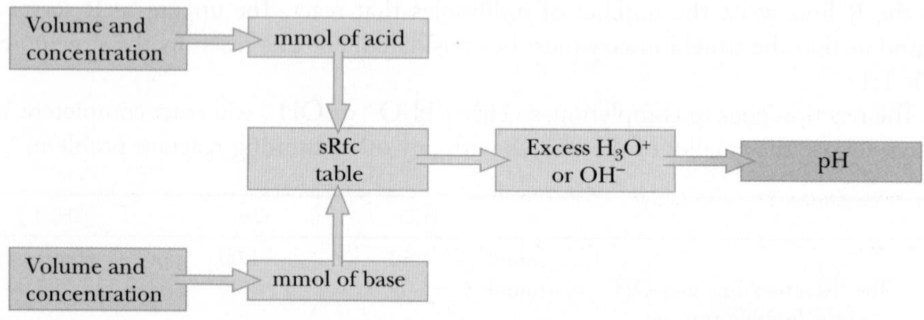

Solution

(a) **0 mL of 0.250 M NaOH added to 20.00 mL of 0.125 M HCl.** The system is 20.00 mL of 0.125 M HCl at this point. Because HCl is a strong acid, the H_3O^+ concentration is 0.125 M, and the pH is $-\log(0.125) = 0.90$.

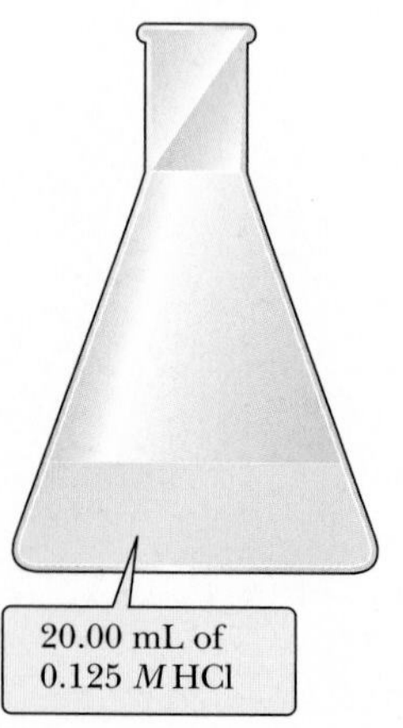

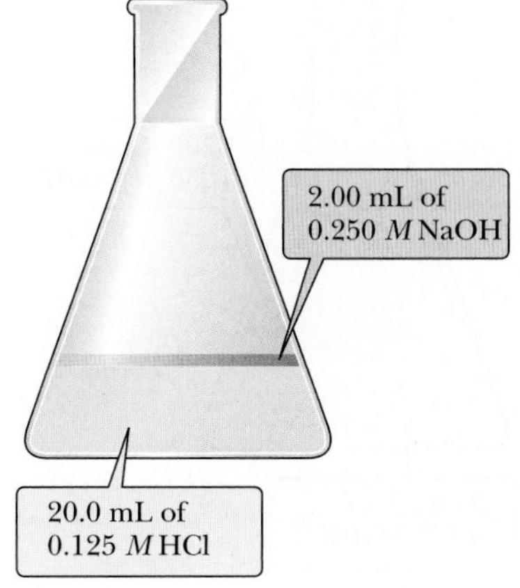

(b) **2.00 mL of 0.250 *M* NaOH added to 20.00 mL of 0.125 *M* HCl.** The sRfc table provides a template for solving this problem. Start by writing the chemical equation above a blank table. It is simplest to use the net ionic equation. The first line on the table contains the starting amounts—numbers of millimoles—of the reactants. It is important to remember that the titration calculation is really a stoichiometry problem, and amounts must be expressed in moles (or millimoles). The chemical equation and the net ionic equation for this titration are

$$HCl(aq) + NaOH(aq) \rightarrow H_2O(\ell) + NaCl(aq)$$

$$H_3O^+ + OH^- \rightarrow H_2O$$

The source of the H_3O^+ is the HCl solution, and we calculate the amount of H_3O^+ from the volume and concentration of the acid solution.

$$\text{Amount of } H_3O^+ = 20.00 \text{ mL} \times \left(\frac{0.125 \text{ mmol } H_3O^+}{\text{mL}}\right) = 2.50 \text{ mmol } H_3O^+$$

Write this amount, 2.50 mmol, on the s line under H_3O^+. Find the number of millimoles of OH^- from the volume (2.00 mL) and concentration (0.250 *M*) of the NaOH solution.

$$\text{Amount of } OH^- = 2.00 \text{ mL} \times \left(\frac{0.250 \text{ mmol } OH^-}{\text{mL}}\right) = 0.500 \text{ mmol}$$

			H_3O^+	+	OH^-	→	$2H_2O$
s	Fill in the "starting" information	mmol	2.50		0.500		Excess
R	The "Reaction" line uses OH^- as the limiting reactant.	mmol	−0.500		−0.500		+2 × 0.500
f	The "final" number of mmol is s + R.	mmol	2.00		0.0		Excess

We can look at the f line and conclude the solution is strongly acidic and estimate the pH will be quite low. Obtain the concentrations needed by dividing the amount of each species on the f line by the total volume of the solution, which, in this case, is

$$V_T = 20.00 \text{ mL} + 2.00 \text{ mL} = 22.00 \text{ mL}$$

Place the concentrations on the c line of the table. The complete table for our stoichiometry calculation follows.

	H_3O^+	+	OH^-	→	$2H_2O$
starting (mmol)	2.50		0.500		Excess
Reaction (mmol)	−0.500		−0.500		+1.00
final (mmol)	2.00		0		Excess
concentration (*M*)	0.0909		0		Excess

Because the goal is to find the pH of the solution, we need to calculate the pH from the H_3O^+ concentration.

$$pH = -\log[H_3O^+] = -\log(0.0909) = 1.04$$

The pH at any volume before the equivalence point in the titration of a strong acid with a strong base is calculated by the same methods as those already shown. It is helpful to determine the equivalence point in a titration before starting your calculations, just as a double check. In this problem, the titration of 20.0 mL of 0.125 *M* HCl, the equivalence point occurs when 10.00 mL of 0.250 *M* NaOH has been added.

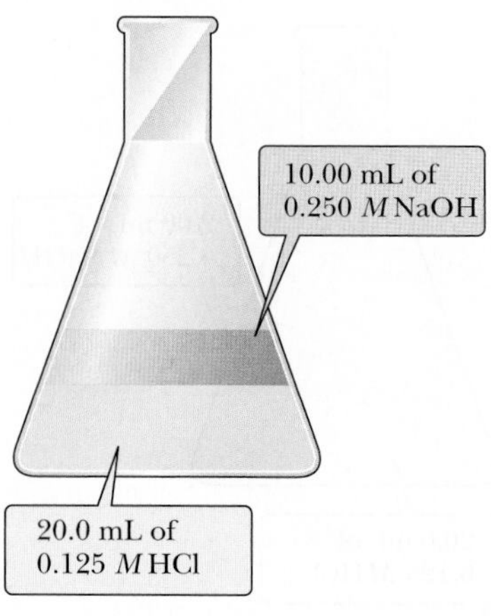

(c) **10.00 mL of 0.250 *M* NaOH added to 20.00 mL of 0.125 *M* HCl**. Again, we start with the initial amounts of H_3O^+ and OH^- that have been placed in the titration vessel. These are calculated from the volumes and concentrations of HCl (20.00 mL of a 0.125 *M* solution) and sodium hydroxide (10.00 mL of a 0.250 *M* solution).

$$\text{Amount of } H_3O^+ = 20.00 \text{ mL} \times \left(\frac{0.125 \text{ mmol } H_3O^+}{\text{mL}}\right) = 2.50 \text{ mmol } H_3O^+$$

$$\text{Amount of } [OH^-] = 10.00 \text{ mL} \times \left(\frac{0.250 \text{ mmol } OH^-}{\text{mL}}\right) = 2.50 \text{ mmol } OH^-$$

Place these values on the s line, and complete the sRfc table.

	H_3O^+	+	OH^-	→	$2H_2O$
starting (mmol)	2.50		2.50		Excess
Reaction (mmol)	−2.50		−2.50		+5.00
final (mmol)	0		0		Excess
concentration *(M)*	0		0		Excess

Because both the H_3O^+ and OH^- ions have been completely consumed, the titration is at the equivalence point. Water is the only source of H_3O^+ and OH^-, so we can calculate the hydrogen ion concentration from the autoionization equilibrium.

$$2H_2O \rightleftharpoons H_3O^+ + OH^- \qquad K_w = 1.0 \times 10^{-14}$$

This equilibrium calculation was presented in Chapter 15, with the result

$$[H_3O^+] = [OH^-] = 1.0 \times 10^{-7}\ M$$

$$pH = -\log[H_3O^+] = 7.00$$

In any strong acid-strong base titration, the pH at the equivalence point is 7.00.

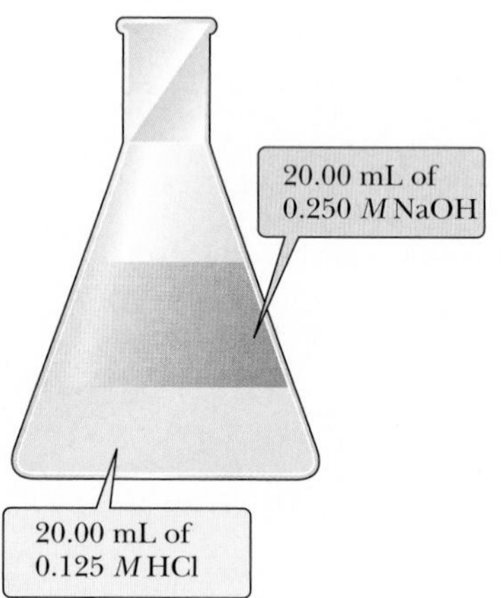

(d) **20.00 mL of 0.250 *M* NaOH added to 20.00 mL of 0.125 *M* HCl.** The starting amount of HCl is 2.50 mmol, as in parts b and c. The amount of hydroxide ion in the s line of the table is

$$\text{Amount of } OH^- = 20.00 \text{ mL} \times \left(\frac{0.250 \text{ mmol } OH^-}{\text{mL}}\right) = 5.00 \text{ mmol } OH^-$$

Complete the stoichiometry calculation using the sRfc table.

	H_3O^+	+	OH^-	→	$2H_2O$
s (mmol)	2.50		5.00		Excess
R (mmol)	−2.50		−2.50		+5.00
f (mmol)	0		2.50		Excess
c *(M)*	0		0.0625		Excess

Looking at the f line, we can predict that the solution will be quite basic because OH^- is present. The concentration of OH^- on the last line results from dividing the millimoles on the f line by the total volume of the solution, 40.00 mL. Determine the pH from the excess hydroxide ion present.

$$pOH = -\log(0.0625) = 1.20$$

$$pH = 14.00 - pOH = 14.00 - 1.20 = 12.80$$

Understanding

Calculate the pH in the titration of 20.00 mL of 0.125 *M* HCl with 0.250 *M* NaOH after adding 9.60, 10.40, and 12.00 mL of NaOH.

Answer

Volume of NaOH (mL)	pH
9.60	2.47
10.40	11.52
12.00	12.19

Note how a small volume of titrant causes a large change in pH in the vicinity of the equivalence point, which occurs after 10.00 mL of NaOH is added.

Graphing the pH as a Function of Volume of Titrant Added

Table 16.1 shows data for several points on the titration curve of 0.125 *M* HCl with 0.250 *M* NaOH. The colored shading indicates which species, hydrogen ion or hydroxide ion, is present in excess. In the region where H_3O^+ is in excess, the hydroxide ion concentration is calculated from $[OH^-] = K_w/[H_3O^+]$ and vice versa.

Figure 16.3 shows the titration curve (pH as a function of titrant volume) for the data in Table 16.1. The point at which the pH changes most rapidly is the **inflection point,** and it occurs at the equivalence point.

Figure 16.4 shows how to measure the pH during the course of a titration by using an indicator that changes color as a function of pH. Indicators are discussed in detail in Section 16.6.

Table 16.2 shows the results of calculations for the titration of 50.0 mL of 0.500 *M* KOH, a strong base, with 1.00 *M* HCl. The pH is initially high and decreases rapidly

TABLE 16.1 **Titration of 20.00 mL of 0.125 *M* HCl with 0.250 *M* NaOH**

Volume of NaOH Added (mL)	$[H_3O^+]$ (*M*)	$[OH^-]$ (*M*)	pH
0.00	0.125	8.0×10^{-14}	0.90
2.00	0.0909	1.1×10^{-13}	1.04
5.00	0.0500	2.0×10^{-13}	1.30
9.60	3.4×10^{-3}	3.0×10^{-12}	2.47
10.00	1.0×10^{-7}	1.0×10^{-7}	7.00
10.40	3.0×10^{-12}	3.3×10^{-3}	11.52
11.00	1.2×10^{-12}	8.1×10^{-3}	11.91
12.00	6.2×10^{-13}	0.016	12.19
20.00	1.6×10^{-13}	0.062	12.80

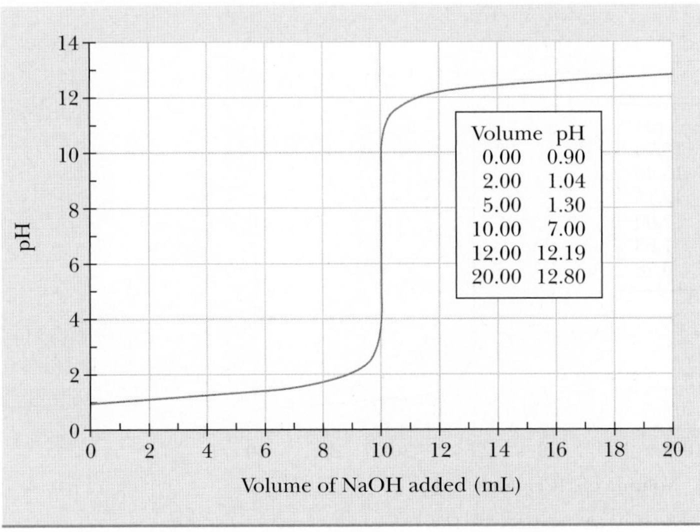

Figure 16.3 pH as a function of the volume of added base.

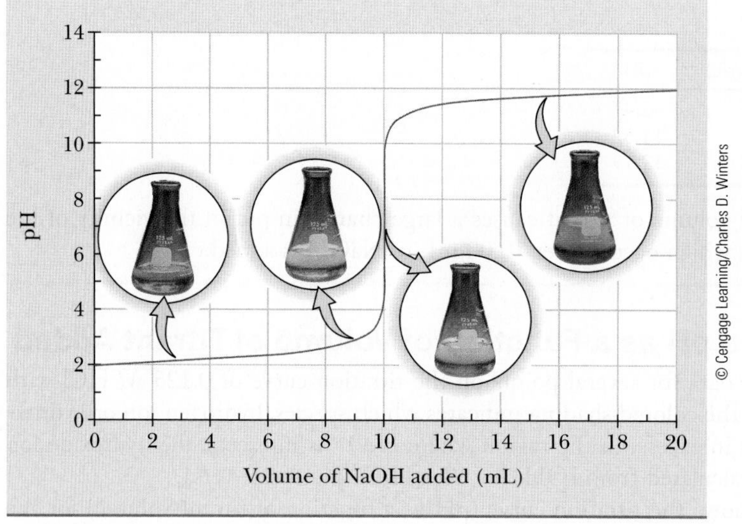

Figure 16.4 Acid-base titration curve. The titration can be evaluated by adding an indicator. Indicators change color as a function of pH and are described later in this chapter. Even though the acid here is not the same as in Figure 16.3, the general shape of the curves are similar.

TABLE 16.2 Titration of 50.0 mL of 0.500 *M* KOH with 1.00 *M* HCl

	Amount of Excess					
Volume of HCl (mL) Added	HCl (mmol)	KOH (mmol)	Total Volume (mL)	$[H_3O^+]$ *(M)*	$[OH^-]$ *(M)*	pH
0	0	25.0	50.0		0.500	13.70
10.0	0	15.0	60.0		0.250	13.40
24.0	0	1.0	74.0		0.014	12.15
25.0	0	0	75.0	1.0×10^{-7}	1.0×10^{-7}	7.00
26.0	1.0	0	76.0	0.013		1.88
40.0	15.0	0	90.0	0.167		0.78

at the equivalence point. Figure 16.5 is a plot of the titration curve. This curve is inverted from that of a strong acid (see Figure 16.3). The pH during the titration of a strong base is calculated in the same general manner as that of a strong-acid titration curve. The solution starts quite basic, with a region of excess hydroxide ion. An equivalence point exists at which $[H_3O^+]$ and $[OH^-]$ are equal, and beyond the equivalence point H_3O^+ is in excess and the solution is quite acidic.

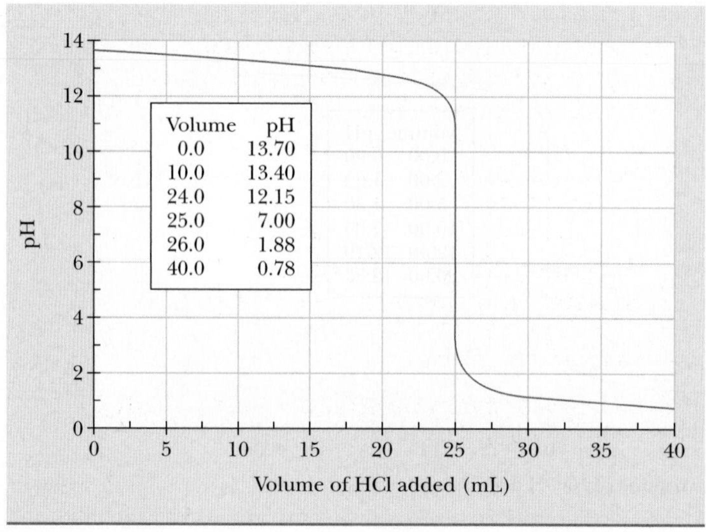

Figure 16.5 Titration curve for the titration of a strong base with strong acid. The titration curve of 50.00 mL of 0.500 *M* KOH with 1.00 *M* HCl.

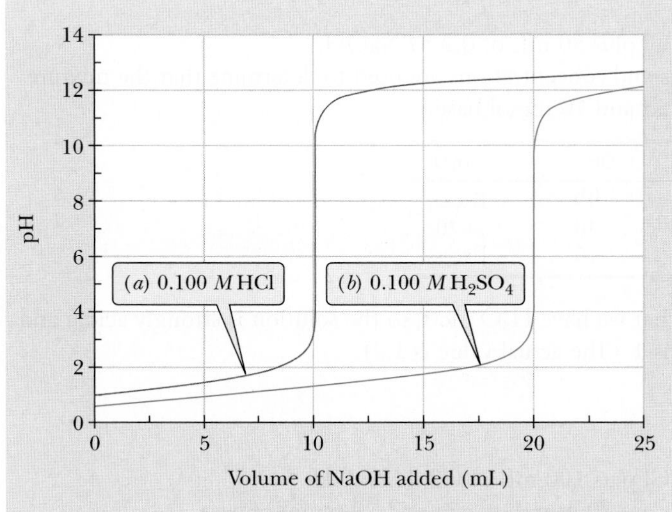

Figure 16.6 Acid-base titration curves. All start with 10.0 mL of the analyte. *(a)* 0.100 *M* HCl titrated with 0.100 *M* NaOH. *(b)* 0.100 *M* H_2SO_4 titrated with 0.100 *M* NaOH.

Titration curves calculated from concentrations and volumes generally agree well with curves that are measured in the laboratory. Some differences may exist in the pH measured and the pH calculated, but the calculated position of the inflection point is always extremely accurate.

The titration of a strong acid with a strong base shows a sharp change in pH at the equivalence point.

Influence of Stoichiometry on the Titration Curve

Most of the examples presented so far have been for systems that show 1:1 stoichiometry. Figure 16.6 shows titration curves for systems with different stoichiometries. The net ionic equation for all the systems shown in Figure 16.6 is

$$H_3O^+(aq) + OH^-(aq) \rightarrow 2H_2O(\ell)$$

One mole of HCl contributes 1 mol H_3O^+ to a titration, and 1 mol H_2SO_4 contributes 2 mol H_3O^+. These differences are reflected in the titration curves in Figure 16.6.

The titration stoichiometry, concentrations, and volumes all influence the equivalence point volume.

Estimating the pH of Mixtures of Acids and Bases

In many situations, chemists do not need an exact calculation of the pH—an estimate will suffice. When a strong acid and strong base are mixed, the net ionic equation is

$$H_3O^+(aq) + OH^-(aq) \rightarrow 2H_2O(\ell)$$

We can estimate the pH of a mixture by filling in the first three lines of the sRfc table, which we call an sRf table, and considering the substances that appear on the f line. If H_3O^+ is present, then the solution will be strongly acidic, and the pH will be in the range of 0 to 2. We can use 1 as an estimate of the pH of a strongly acidic solution. If OH^- is present, the solution will be strongly basic, probably in the range from 12 to 14, so we will use 13 as an estimate. If we do not have an excess of H_3O^+ or OH^- present, then the pH will be determined by the autoionization of water and will be 7.

Solution	Estimate of pH
Strongly acidic	1
Neutral	7
Strongly basic	13

EXAMPLE 16.4 Estimate the pH of Mixtures of Acids and Bases

Estimate the pH of a solution formed by mixing

(a) 100 mL of 0.2 *M* HCl with 50 mL of 0.2 *M* NaOH
(b) 100 mL of 0.2 *M* HCl with 100 mL of 0.2 *M* NaOH

Strategy Write the net ionic equation, calculate numbers of millimoles of each species, and fill in the sRf table. Look at the species on the f line to estimate the pH.

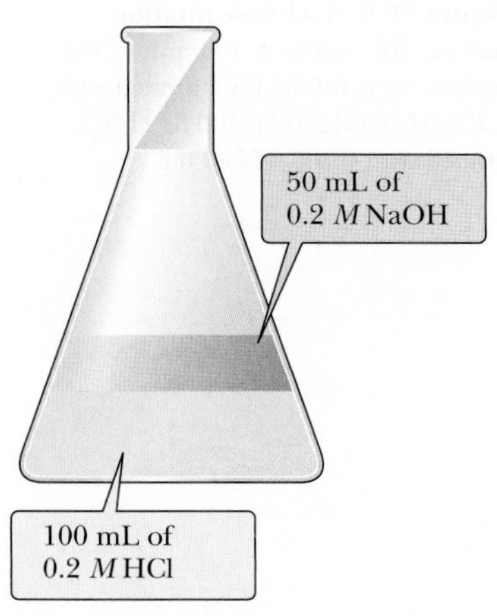

Solution

(a) 100 mL of 0.2 *M* HCl plus 50 mL of 0.2 *M* NaOH

The starting amounts and concentrations are used to determine that the mixture contains 20 mmol acid and 10 mmol base.

	H_3O^+	+	OH^-	→	$2H_2O$
s (mmol)	20		10		Excess
R (mmol)	−10		−10		+20
f (mmol)	10		0		Excess

The f line indicates that we have H_3O^+ left, so the solution is strongly acidic and we estimate the pH is 1. (The actual value is 1.2).

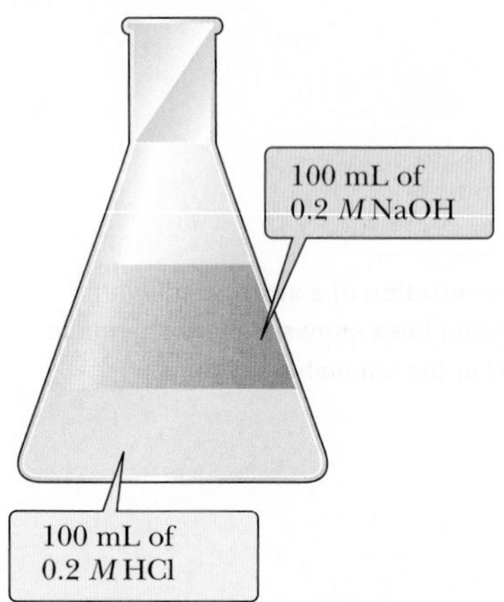

(b) 100 mL of 0.2 *M* HCl plus 100 mL of 0.2 *M* NaOH

The starting amounts are 20 mmol of acid and 20 mmol of base.

	H_3O^+	+	OH^-	→	$2H_2O$
s (mmol)	20		20		Excess
R (mmol)	−20		−20		+40
f (mmol)	0		0		Excess

The f line indicates that there is no excess H_3O^+ from the HCl or any excess OH^- from the NaOH. The pH is 7.00.

Understanding

Estimate the pH that results when 100 mL of 0.2 *M* HCl is mixed with 150 mL of 0.2 *M* NaOH.

Answer 13

OBJECTIVES REVIEW *Can you:*

- ☑ calculate the concentrations of all species present during the titration of a strong acid with a strong base?
- ☑ graph the titration curve?
- ☑ correlate the shape of the titration curve to the titration stoichiometry?
- ☑ estimate the pH of mixtures of strong acids and bases?

16.3 Buffers

OBJECTIVES

- ☐ Describe the function and composition of a pH buffer
- ☐ Calculate the pH of a buffer solution from the concentrations of the weak acid and its conjugate base
- ☐ Identify the circumstances in which the Henderson–Hasselbalch approximation fails
- ☐ Calculate the pH of a solution from the amounts of acids and bases
- ☐ Determine the change in the pH when strong acid or base is added to a buffer

The pH of blood in healthy individuals varies only over the narrow range of 7.35 to 7.45. A person with a blood pH outside of this range is at risk because the enzymatic reactions crucial to life begin to fail when the pH is outside this range. Our bodies have a system to keep the pH of blood constant, regardless of food intake, level of exercise, and stress of chemistry examinations.

Many chemical reactions require a constant or nearly constant pH to proceed. Most often, these reactions are performed in a chemical system that is designed to keep the pH constant. Such a system is called a **buffer,** a solution that resists changes in pH when hydrogen ions or hydroxide ions are added to it. A buffer system must contain both a base to react with added hydrogen ions and an acid to react with added hydroxide ions. Not every possible mixture of acid and base can form a buffer. If we mix HCl and NaOH, for example, we find that they react with each other to form water and NaCl.

A buffer resists changes in pH.

A buffer solution is prepared from a weak acid and its conjugate base (or a weak base and its conjugate acid). Let us consider a buffer made from a weak acid, HA, and its conjugate base, A^-.

If hydrogen ions are added to our buffer, they react with the base:

$$H_3O^+ + A^- \rightarrow H_2O + HA$$

The equilibrium constant for this reaction is quite large; a large equilibrium constant means that the reaction will proceed nearly to completion. The reaction between the added hydrogen ions and the weak base effectively removes the added hydrogen ions from the solution, so we find that adding hydrogen ions to a buffer solution generally does not cause a large change in the pH of the solution.

Similarly, the weak acid in the buffer reacts with any added hydroxide ions:

$$OH^- + HA \rightarrow H_2O + A^-$$

This reaction also goes to completion. The weak acid in the buffer removes excess hydroxide ions from the solution, so they, too, will not cause a large change in the pH of the solution.

The reaction of the buffer with added H_3O^+ or OH^- goes to completion.

The equilibrium constants for the neutralizations of both added hydrogen ions and added hydroxide ions are large, so these reactions go to completion and the system tends to keep the pH nearly constant; thus, it is a buffer system.

Calculating the pH of a Buffer Solution

Let us see how we can prepare a buffer system using acetic acid as the starting material. The chemical reaction for the ionization of acetic acid is

$$CH_3COOH(aq) + H_2O(\ell) \rightleftharpoons CH_3COO^-(aq) + H_3O^+(aq)$$

The conjugate acid-base pair is acetic acid, CH_3COOH, and the acetate ion, CH_3COO^-. To make a buffer solution, we need both of these species in solution. An *effective* buffer requires both species in substantial concentrations, which will be demonstrated later in this section. Dissolving acetic acid in water does not produce many acetate ions because acetic acid is weak, and only a small fraction ionizes to form acetate ion.

We will have to add the base, acetate ion, and a good way is by using a soluble salt that has acetate as the anion. Sodium acetate is a good choice. It is soluble, and it dissociates completely to form sodium ions and acetate ions.

$$NaCH_3COO(aq) \rightarrow Na^+(aq) + CH_3COO^-(aq)$$

A solution containing both acetic acid and sodium acetate is a buffer system—it contains a weak acid and its conjugate base.

A buffer system contains a weak acid and its conjugate base or a weak base and its conjugate acid.

We start to study this buffer system by writing the chemical equation for the weak acid ionization:

$$CH_3COOH(aq) + H_2O(\ell) \rightleftharpoons H_3O^+(aq) + CH_3COO^-(aq)$$

Note that in the buffer solution, the H_3O^+ concentration is *not* equal to the CH_3COO^- concentration. If the dissociation of acetic acid were the only source of the acetate ion, then $[H_3O^+]$ would equal $[CH_3COO^-]$, but the buffer solution contains additional acetate ion from the sodium acetate. Fortunately, the expression for the equilibrium constant does not depend on the source of acetate ion.

$$K_a = \frac{[H_3O^+][CH_3COO^-]}{[CH_3COOH]}$$

Upon rearrangement, the following equation is obtained.

$$[H_3O^+] = K_a \frac{[CH_3COOH]}{[CH_3COO^-]}$$

By taking the negative logarithm of both sides to convert to pH, we obtain the relationship between the pH of the solution and the concentrations of the weak acid and its conjugate base:

$$pH = pK_a + \log\frac{[CH_3COO^-]}{[CH_3COOH]}$$

Note the concentration of the base is in the numerator; the relationship is not difficult to remember because we know that the pH will increase as the concentration of the base increases.

To use this equation, we need to know the *equilibrium* concentrations of acetic acid and acetate ion in the buffer solution. For convenience, we would like to express the pH in terms of the starting, or analytical, concentrations.

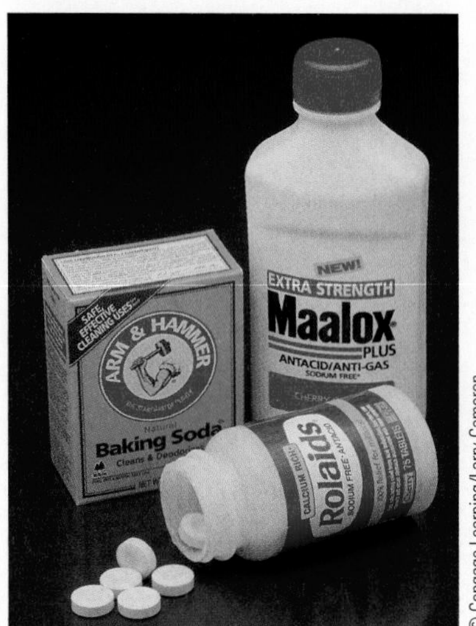

© Cengage Learning/Larry Cameron

Some common buffers. These products are all used to neutralize stomach acid; a buffer is preferred to a base for safety reasons—ingesting too much base could be fatal.

In general, we prepare a buffer solution so that the concentrations of the acid and its conjugate base are large compared with the concentrations of H_3O^+ and OH^-. A pH 5 buffer solution might have 0.1 *M* concentrations of the weak acid and its conjugate base, compared with 10^{-5} *M* H_3O^+ and 10^{-9} *M* OH^- in a pH 5 unbuffered solution. The equilibrium concentration of the acid, $[CH_3COOH]$, differs from the starting or analytical concentration, C_{CH_3COOH}, by the concentration of H_3O^+, and this difference is quite small compared with the starting concentrations involved. As long as the analytical concentrations are relatively large, we can use them in place of the equilibrium concentrations.

$$pH = pK_a + \log\frac{C_b}{C_a} \qquad [16.1]$$

where C_a is the analytical concentration of the acid, and C_b is the analytical concentration of the conjugate base.

Lawrence Henderson and Karl Hasselbalch, two chemists of the early 20th century, first described—almost simultaneously—the relationship shown in Equation 16.1. This equation is known as the **Henderson–Hasselbalch equation.**

Notice what happens when the concentrations of the acid and conjugate base are equal. When C_a equals C_b, their ratio is 1, and the log of 1 is zero; the Henderson–Hasselbalch equation predicts that the pH of the solution will equal pK_a.

$$pH = pK_a + \log(1) = pK_a$$

We usually prepare buffer solutions with approximately equal amounts of acid and base (so that they can neutralize added hydrogen ions or hydroxide ions with equal effectiveness), so the pH of a buffer solution is generally limited to the vicinity of pK_a for the acid. Most scientists use the rule of thumb that buffers are most effective in the pH range of $pK_a \pm 1$.

The Henderson–Hasselbalch equation can be used when the concentrations of the weak acid and base are large compared with the concentrations of H_3O^+ and OH^-, and K_a and K_b.

It is important to remember that the Henderson–Hasselbalch equation is an *approximation*. It works when the concentrations C_a and C_b are reasonably high, meaning large compared with $[H_3O^+]$ or $[OH^-]$. Because $[H_3O^+]$ or $[OH^-]$ is related to K_a and K_b, it can be shown that the equations should be used only when the concentrations of both the weak acid and its conjugate base are at least 100 times K_a or K_b. When the concentrations are low in comparison with K_a, the approximation is not valid.

Many scientists view the Henderson–Hasselbalch equation as consisting of two components, noting that the pH of the solution is determined primarily by the pK_a of the weak acid. This primary effect is modified by a secondary term that takes into account the concentrations of the acid and base that are present in solution. Sometimes the secondary term is called an *environmental* term because the primary term is modified by the chemical environment. Many important phenomena can be divided into a fundamental primary effect and a secondary environmental effect.

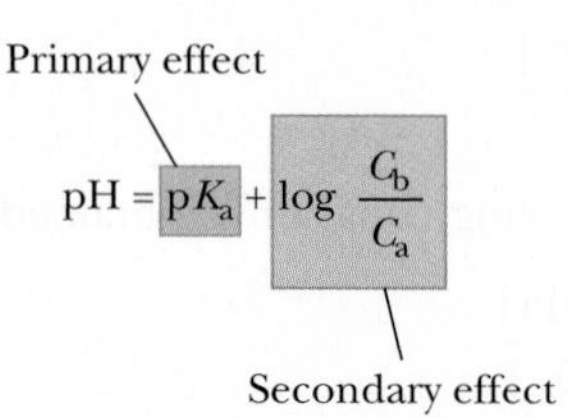

The Henderson–Hasselbalch equation and the acid ionization constant expression contain equivalent information. When you calculate the pH of a buffer, you can use whichever is most convenient. We can demonstrate how to calculate the pH of a buffer from the acid ionization expression.

$$HA + H_2O \rightleftharpoons H_3O^+ + A^-$$

$$K_a = \frac{[H_3O^+][A^-]}{[HA]}$$

$$[H_3O^+] = K_a \frac{[HA]}{[A^-]}$$

When C_a and C_b are large, then we can make some approximations. Under these conditions, $C_a \approx [HA]$ and $C_b \approx [A^-]$, so we can substitute the analytical concentrations for the equilibrium concentrations.

$$[H_3O^+] = K_a \frac{C_a}{C_b} \qquad [16.2]$$

Either the Henderson–Hasselbalch form (see Equation 16.1) or the expression for the ionization of the acid (Equation 16.2) can be used to solve for the pH of a buffer.

EXAMPLE 16.5 Calculating the pH of a Buffer Solution

Calculate the pH of a solution that is 0.40 *M* sodium acetate and 0.20 *M* acetic acid. K_a for acetic acid is 1.8×10^{-5}.

Strategy The solution contains a weak acid and its conjugate base, so it is a buffer. Use the Henderson–Hasselbalch equation to solve for the pH of a buffer system. The analytical concentrations of the acid, base, and the value of K_a are given.

Solution
First, write the Henderson–Hasselbalch expression

$$pH = pK_a + \log \frac{C_b}{C_a}$$

Remember that pK_a is the negative log of K_a, and substitute the analytical concentrations into the Henderson–Hasselbalch expression.

$$pH = -\log(1.8 \times 10^{-5}) + \log \frac{0.40}{0.20}$$

$$pH = -(-4.74) + (0.30)$$

$$pH = 5.04$$

Understanding

Calculate the pH of a solution that is 0.50 *M* hydrofluoric acid and 0.10 *M* sodium fluoride; K_a for HF is 6.3×10^{-4}.

Answer pH 2.50

Another way to solve Example 16.5 is to use the expression for the ionization of acetic acid, write the iCe table, and solve the expression for H_3O^+, just like our previous equilibrium problems. We define y as the concentration of H_3O^+ formed.

1. Write the chemical equation.

	$CH_3COOH + H_2O$	$\rightleftharpoons$	H_3O^+	+	CH_3COO^-
i (M)	0.20				0.40
C (M)	$-y$		$+y$		$+y$
e (M)	$0.20 - y$		y		$0.40 + y$

2. Fill in the iCe table.

3. Write the expression for K.

$$K_a = \frac{[H_3O^+][CH_3COO^-]}{[CH_3COOH]}$$

Substitute the numerical value for K_a ($K_a = 1.8 \times 10^{-5}$) and the equilibrium concentrations for $[CH_3COO^-]$ and $[CH_3COOH]$.

4. Substitute.

$$1.8 \times 10^{-5} = \frac{(y)(0.40 + y)}{(0.20 - y)}$$

This equation can be expanded into a quadratic equation and solved by the quadratic formula, but numerical approximation may work more quickly. First, rearrange the equation.

5. Solve.

$$y = 1.8 \times 10^{-5} \times \left(\frac{0.20 - y}{0.40 + y}\right)$$

If we can assume that $y << 0.20$, then we can neglect y when added or subtracted to 0.20 and 0.40.

$$y \approx 1.8 \times 10^{-5} \times \left(\frac{0.20}{0.40}\right)$$

$$y = [H_3O^+] = 9.0 \times 10^{-6} \qquad pH = 5.05$$

Because we made an approximation, we check: Is $y << 0.20$? Yes, 9.0×10^{-6} is much less than 0.20.

Notice that, aside from rounding issues, solving the exact equation with a mathematical approximation is the same as using the Henderson–Hasselbalch equation. Both approaches assume that C_a and C_b are much larger than $[H_3O^+]$ and $[OH^-]$.

The ratio of *concentrations* of the weak acid and conjugate base is used in the Henderson–Hasselbalch relationship. The concentration ratio is the same as the ratio of the numbers of moles of acid and base. We represent the analytical concentrations as

$$C_a = n_a/V$$

$$C_b = n_b/V$$

where n_a is the number of moles of acid, and V is the total volume of the buffer solution. We can substitute n_a/V for C_a and n_b/V for C_b in the Henderson–Hasselbalch equation.

$$pH = pK_a + \log\frac{n_b/V}{n_a/V} = pK_a + \log\frac{n_b}{n_a}$$

As long as the concentration of the acid and base in the buffer are much greater than the concentrations of H_3O^+ and OH^-, the pH of a buffer can be calculated from the number of moles of acid, n_a, and the number of moles of base, n_b. The next example illustrates this concept.

In a buffer solution, the concentration ratio C_b/C_a equals the mole ratio n_b/n_a.

EXAMPLE 16.6 Determining the Amounts of Acid and Base Needed to Prepare a pH Buffer

Calculate the mass of ammonium chloride that must be added to 500.0 mL of 0.32 M NH_3 to prepare a pH 8.50 buffer; K_b for NH_3 is 1.8×10^{-5}.

Strategy We need to calculate the amount of the acid (NH_4^+ ions from NH_4Cl) that we must add to prepare an ammonium/ammonia buffer of pH 8.50. It will be easiest if we use the Henderson–Hasselbalch equation in terms of the number of moles of acid and base.

$$\text{pH} = \text{p}K_a + \log\frac{C_b}{C_a} \quad \text{or} \quad \text{pH} = \text{p}K_a + \log\frac{n_b}{n_a}$$

We know the pH, the value for K_b, and the volume and concentration of the base. To solve for n_a, we will also need to know K_a (and pK_a), but we can calculate these values from K_b.

Solution

The value for K_b for NH_3 is given in the problem, and we can calculate K_a for the acid from K_b of the base.

$$K_a = \frac{K_w}{K_b} = \frac{1.0 \times 10^{-14}}{1.8 \times 10^{-5}} = 5.6 \times 10^{-10}$$

$$\text{p}K_a = -\log(5.6 \times 10^{-10}) = 9.25$$

Next, calculate n_b from the data given: 500.0 mL of 0.32 M NH_3.

$$n_b = 0.5000 \cancel{\text{L}} \times \left(\frac{0.32 \text{ mol NH}_3}{1 \cancel{\text{L}}}\right) = 0.16 \text{ mol NH}_3$$

Substitute these values into the Henderson–Hasselbalch equation to calculate the number of moles of acid.

$$\text{pH} = \text{p}K_a + \log\frac{n_b}{n_a}$$

$$8.50 = 9.25 + \log\frac{0.16}{n_a}$$

$$\log\frac{0.16}{n_a} = 8.50 - 9.25 = -0.75$$

$$\frac{0.16}{n_a} = 10^{-0.75} = 0.18$$

$$n_a = 0.89 \text{ mol}$$

Last, calculate the mass of ammonium chloride, the source of the acid, from the number of moles.

$$\text{Mass NH}_4\text{Cl} = 0.89 \cancel{\text{mol NH}_4\text{Cl}} \times \left(\frac{53.49 \text{ g NH}_4\text{Cl}}{\cancel{\text{mol NH}_4\text{Cl}}}\right) = 48 \text{ g NH}_4\text{Cl}$$

Understanding

How many moles of sodium benzoate must be added to 1 L of a 0.022 M solution of benzoic acid (pK_a = 4.19) to prepare a pH 4.50 buffer?

Answer 0.045 mol

Preparing and Using Buffer Solutions

A buffer solution is prepared by mixing a weak acid and its conjugate base in the correct proportions, as determined by calculations similar to the one illustrated in Example 16.6. These calculations are approximations, so after the buffer is prepared, its pH is measured and adjusted to the exact value by adding some concentrated strong acid or base.

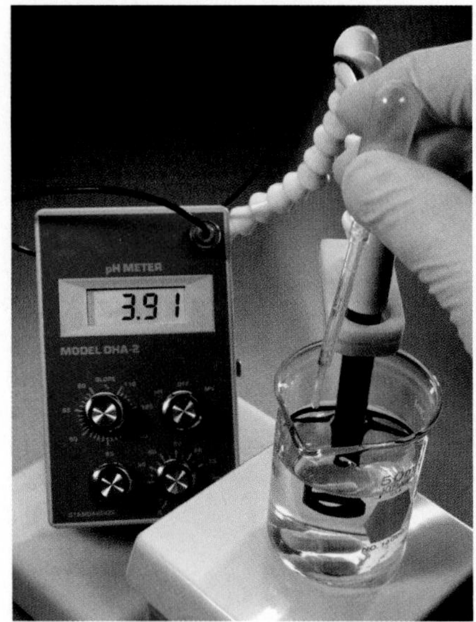

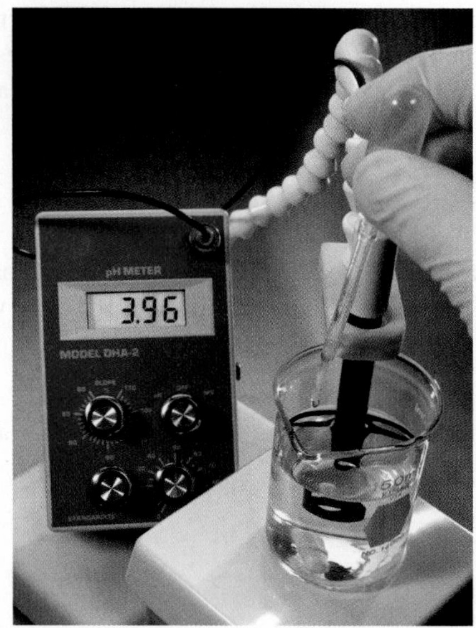

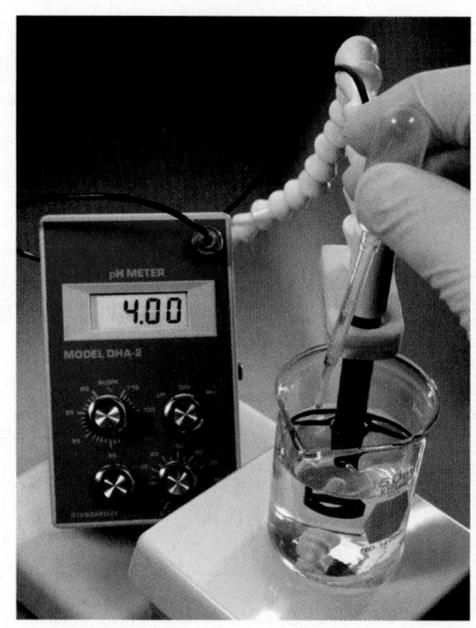

Preparing a pH buffer solution. The pH of a buffer is adjusted to the exact value by adding small volumes of concentrated solutions of strong acid or base.

Determining the Response of a Buffer to Added Acid or Base

Concentrated buffers are generally more effective than dilute buffers. The effectiveness of a buffer is measured by the **buffer capacity,** defined as the amount of strong acid or base needed to change the pH of 1 L of buffer by one unit. The buffer capacity is a quantitative measure of the ability of the buffer to resist changes in pH and is directly proportional to the concentrations. A buffer with concentrations of 0.30 M HF and 0.30 M F^- has a capacity three times that of a buffer with concentrations of 0.10 M HF and 0.10 M F^-. Note that the pH before adding strong acid or base is the same in both solutions because the ratio C_{F^-}/C_{HF} is the same; only the buffer capacity differs.

The buffer capacity is a measure of the amount of acid or base that can be effectively neutralized by the buffer.

The next example shows how to calculate the pH that results when some strong acid or base is added to a buffer system.

EXAMPLE 16.7 The Change in pH When an Acid Is Added to Unbuffered and Buffered Systems

Calculate the change in pH observed when 1.50 mL of 0.0670 M H_3O^+ is added to (a) 100.0 mL of an unbuffered HCl solution of pH 4.74 and (b) 100.0 mL of a pH 4.74 buffer that is 0.120 M acetic acid and 0.120 M sodium acetate (the pK_a for acetic acid is 4.74).

(a) Calculate the change in pH observed when 1.5 mL of 0.0670 M H_3O^+ is added to 100.0 mL of a pH 4.74 HCl solution.

Strategy

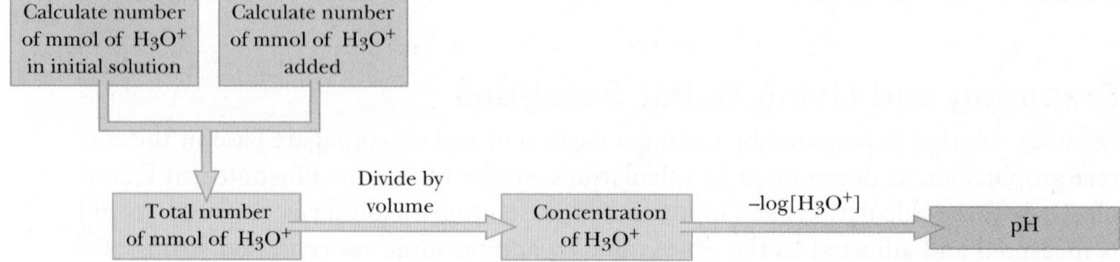

The system is unbuffered and the acid comes from two sources, the initial 100.0 mL of pH 4.74 HCl and the added 1.50 mL of 0.0679 M H_3O^+. Express amounts in moles, for both sources, add together, and divide by volume to find $[H_3O^+]$. Compute the pH from the negative log.

Solution

Consider first the 100.0 mL of HCl solution at pH 4.74. Calculate the initial H_3O^+ concentration from the pH.

$$[H_3O^+] = 10^{-4.74}\ M = 1.8 \times 10^{-5}\ M$$

Calculate the initial amount of H_3O^+ in 100 mL of this solution from the volume and concentration of HCl.

$$\text{Amount HCl} = 100.0\ \text{mL of solution} \times \left(\frac{1.8 \times 10^{-5}\ \text{mmol}}{\text{mL of solution}}\right) = 1.8 \times 10^{-3}\ \text{mmol}$$

Next, calculate the number of mmol of H_3O^+ from the additional 1.50 mL of 0.0670 M H_3O^+ that is added.

$$\text{Amount } H_3O^+ \text{ added} = 1.50\ \text{mL of solution} \times \left(\frac{0.0670\ \text{mmol}}{\text{mL of solution}}\right) = 1.00 \times 10^{-1}\ \text{mmol}$$

The *total* acid in solution is the sum of the two sources.

From the 100 mL of pH 4.74 acid	1.8×10^{-3} mmol
+ From the added 1.50 mL of 0.0670 M acid	1.00×10^{-1} mmol
Total number of moles of H_3O^+	1.02×10^{-1} mmol

Now calculate the concentration of the H_3O^+ from the total number of millimoles and the volume of the solution, which is the original 100.0 mL plus the additional 1.50 mL, or 101.5 mL.

$$[H_3O^+] = \left(\frac{1.02 \times 10^{-1}\ \text{mmol}}{101.5\ \text{mL of solution}}\right) = 1.00 \times 10^{-3}\ M$$

$$\text{pH} = 3.00$$

The pH of the solution decreases by 1.74 pH units when 1.50 mL of 0.0670 M acid is added.

(b) Calculate the change in pH observed when 1.50 mL of 0.0670 M H_3O^+ is added to 100.0 mL of a pH 4.74 buffer that is 0.120 M acetic acid and 0.120 M sodium acetate (the pK_a for acetic acid is 4.74).

Strategy This problem is similar to a titration problem such as those presented in Section 16.2. We are adding H_3O^+, which will react with the weak base form of the buffer. As a result of this reaction, which goes to completion, the number of moles of the base will decrease and the number of moles of the conjugate acid will increase. The pH will be computed by using the Henderson–Hasselbalch equation.

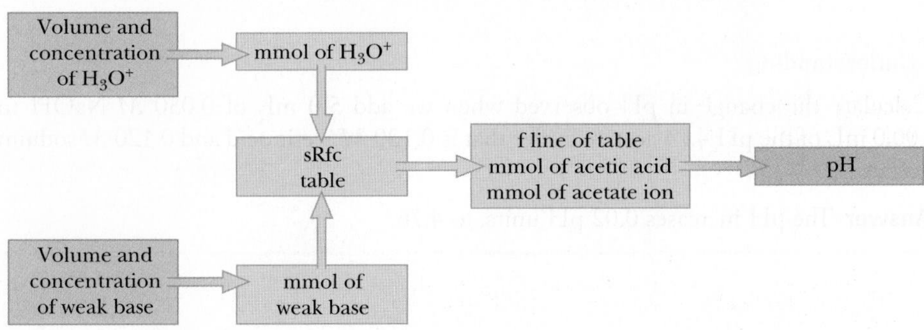

Solution

When acid is added to a buffer, the base in the buffer reacts to consume the additional acid. In the acetic acid/acetate ion buffer, the acetate ion reacts with the added acid.

$$H_3O^+ + CH_3COO^- \rightarrow CH_3COOH + H_2O$$

Adding acid to base is, of course, a neutralization reaction, and we use the techniques of stoichiometry to calculate the amounts of substances involved. We then use the results of the neutralization reaction to calculate the pH.

We need the chemical equation and the starting amounts of the substances involved. Calculate the amounts from the volumes and concentrations. The added acid is 1.50 mL and its concentration is 0.670 M.

$$\text{Moles } H_3O^+ = 1.50\ \cancel{\text{mL}} \times \left(0.0670\ \frac{\text{mmol}}{\cancel{\text{mL}}}\right) = 0.100\ \text{mmol}$$

The volume of weak base is 100.0 mL and its concentration is 0.120 M.

$$\text{Moles } CH_3COO^- = 100.0\ \cancel{\text{mL}} \times \left(0.120\ \frac{\text{mmol}}{\cancel{\text{mL}}}\right) = 12.0\ \text{mmol}$$

The volume of acetic acid is 100.0 mL and its concentration is also 0.120 M.

$$\text{Moles } CH_3COOH = 100.0\ \cancel{\text{mL}} \times \left(0.120\ \frac{\text{mmol}}{\cancel{\text{mL}}}\right) = 12.0\ \text{mmol}$$

We now have the information needed to fill in the sRfc table.

	H_3O^+	+	$CH_3COO^-(aq)$	→	$CH_3COOH(aq)$	+	$H_2O(\ell)$
starting, mmol	0.100		12.0		12.0		
Reacts, mmol	−0.100		−0.100		+0.100		
final, mmol	0.0		11.9		12.1		
concentration, mmol/mL							

We do not need to fill in the c line because we can use the amount, in millimoles, together with the Henderson–Hasselbalch equation to calculate the pH.

$$pH = pK_a + \log\frac{n_b}{n_a} = 4.74 + \log\frac{11.9}{12.1}$$

$$pH = 4.73$$

System	pH before Adding Acid	pH after Adding Acid	Change in pH
pH 4.74 HCl	4.74	3.00	−1.74
pH 4.74 acetic acid/ sodium acetate buffer	4.74	4.73	−0.01

Understanding

Calculate the change in pH observed when we add 5.0 mL of 0.050 M NaOH to 100.0 mL of the pH 4.74 acetate buffer that is 0.120 M acetic acid and 0.120 M sodium acetate.

Answer The pH increases 0.02 pH units, to 4.76.

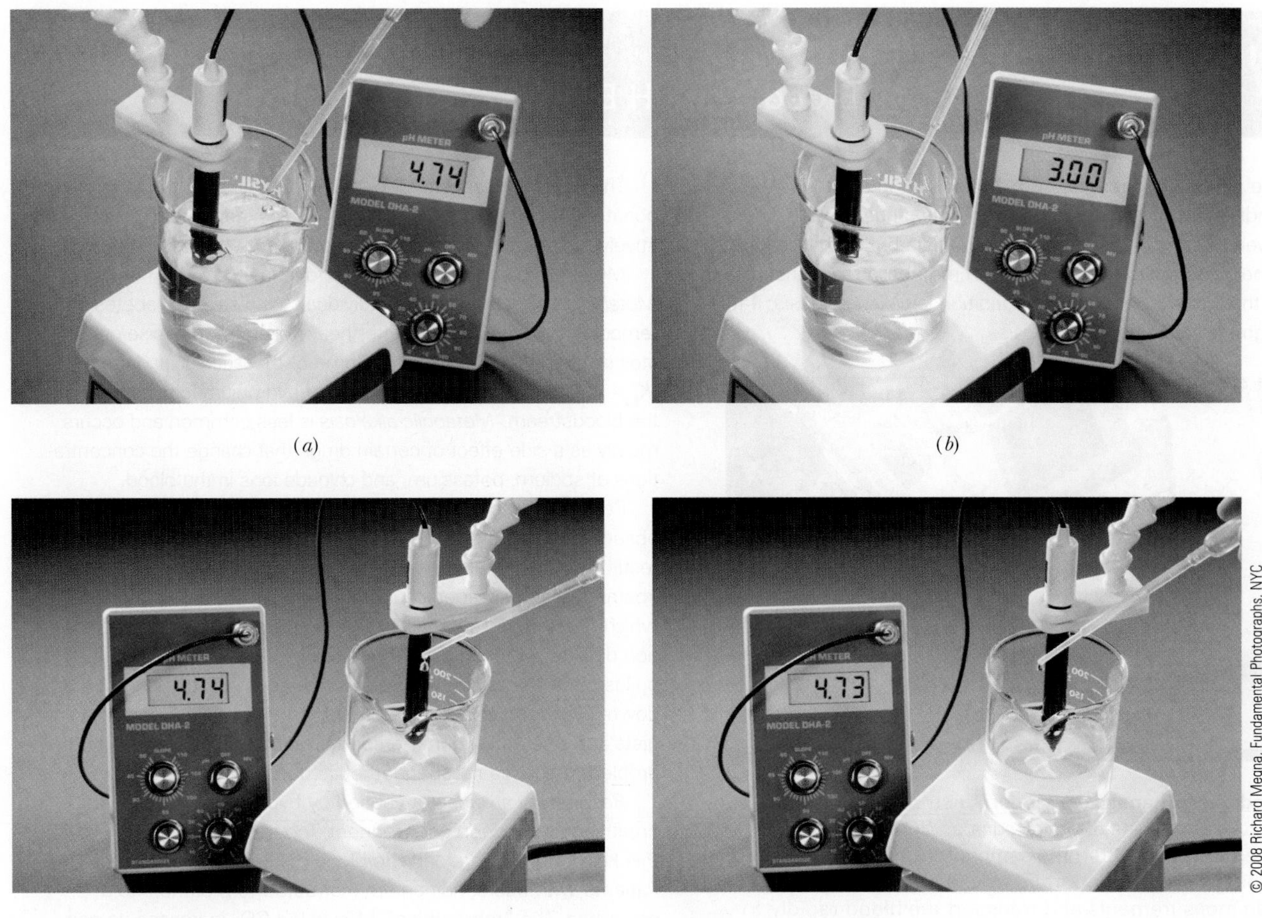

(a) (b) (c) (d)

Adding acid to unbuffered and buffered solutions. When a small amount of acid (1.50 mL of 0.0670 *M* strong acid) is added to pH 4.74 HCl *(a)*, the pH decreases to 3.00 *(b)*. When the same amount of acid is added to a pH 4.74 acetic acid/acetate ion buffer *(c)*, the pH decreases by 0.01 to 4.73 *(d)*.

Buffer solutions have wide applications, but one of the most important is maintaining a constant pH for reactions that involve enzymes. Enzymes are complex proteins that are involved in many biological functions. They help to metabolize food, store energy, and even store the compounds that are needed to contract muscles, including our heart muscle. Most enzymes stop working if the pH changes very much. Since some enzyme reactions generate hydrogen or hydroxide ions, it is important that these ions be neutralized immediately, or the reaction products might "poison" or retard the reaction.

OBJECTIVES REVIEW *Can you:*

- ☑ describe the function and composition of a pH buffer?
- ☑ calculate the pH of a buffer solution from the concentrations of the weak acid and its conjugate base?
- ☑ identify the circumstances in which the Henderson–Hasselbalch approximation fails?
- ☑ calculate the pH of a solution from the amounts of acids and bases?
- ☑ determine the change in the pH when strong acid or base is added to a buffer?

PRINCIPLES OF CHEMISTRY
Blood as a pH Buffer

The pH of blood ranges from 7.35 to 7.45 in healthy individuals; if it varies much from this range, severe illness and even death can result. Blood contains a buffer that reacts with the acids and bases produced in metabolic reactions. If the pH of the blood is too low, the condition is called *acidosis;* if it is too high, the condition is *alkalosis.*

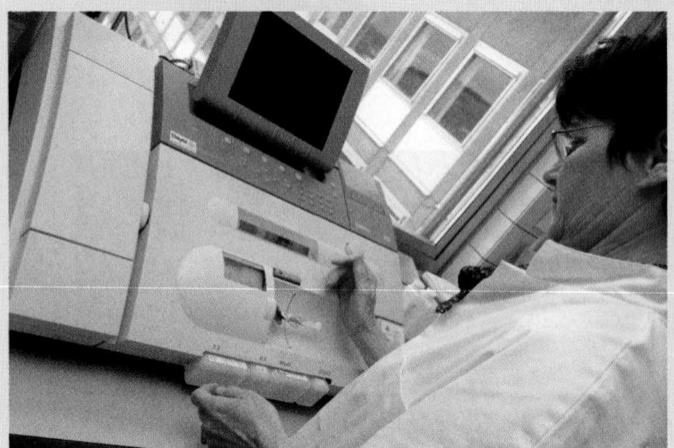
CC Studio/Photo Researchers, Inc.

Blood gas/pH analyzer. Modern instruments measure the concentrations of gases such as CO_2 and O_2 dissolved in blood. Often the same instrument measures the blood pH. It is important to use arterial blood for these measurements and transport the blood rapidly to the laboratory for measurement.

The principal buffer system in blood is the hydrogen carbonate (bicarbonate) buffer. The hydrogen carbonate ion is formed from carbon dioxide in the lungs. Carbon dioxide, which is a Lewis acid, dissolves in water to form carbonic acid, an Arrhenius acid.

$$CO_2(g) + H_2O(\ell) \rightarrow H_2CO_3(aq)$$

Carbonic acid is weak ($K_{a1} = 4.3 \times 10^{-7}$) and partially ionizes to form hydronium ion and hydrogen carbonate ion.

$$H_2CO_3(aq) + H_2O(\ell) \rightleftharpoons H_3O^+(aq) + HCO_3^-(aq)$$

The hydrogen carbonate ion is amphoteric and reacts with excess hydronium ion (either generated or ingested).

$$HCO_3^-(aq) + H_3O^+(aq) \rightleftharpoons H_2CO_3(aq) + H_2O(\ell)$$
$$K_{eq} = 1/K_{a1} = 2.3 \times 10^6$$

The hydrogen carbonate ion also reacts with excess hydroxide ion.

$$HCO_3^-(aq) + OH^-(aq) \rightleftharpoons CO_3^{2-}(aq) + H_2O(\ell)$$
$$K_{eq} = K_{a2}/K_w = 5.6 \times 10^3$$

The large values of the equilibrium constants indicate that these acid-base reactions go virtually to completion.

The buffering capacity depends on the concentration of bicarbonate ion, usually about 0.027 *M*. The concentration is kept relatively constant by chemical reactions in the kidneys and lungs.

Metabolic processes are known to influence the pH of blood. *Metabolic acidosis* is seen in individuals with severe diabetes and temporarily in all individuals after heavy exercise. Exercise generates lactic acid in the muscles. Some of the lactic acid ionizes ($K_a = 8.4 \times 10^{-4}$), and a large influx of hydronium ions appears in the bloodstream. *Metabolic alkalosis* is less common and occurs mostly as a side effect of certain drugs that change the concentrations of sodium, potassium, and chloride ions in the blood.

Respiratory acidosis and alkalosis are more commonly observed than are the metabolic conditions. These conditions result from changes in breathing patterns. Lung diseases and obstructed air passages cause hypoventilation, a condition in which breathing is so shallow that it reduces the amount of carbon dioxide removed from the blood by respiration. When CO_2 builds up in the blood, the acidity increases and the pH goes down: hypoventilation causes *respiratory acidosis.* Anesthesiologists and operating room personnel must be made aware of this problem, because most anesthetics depress respiration.

Respiratory alkalosis is caused by hyperventilation, in which breathing is too deep and frequent. Too much CO_2 is expelled in this kind of breathing, and the blood pH becomes alkaline. Patients with this condition are instructed to breathe into a paper bag. The "rebreathing" of exhaled CO_2 increases its concentration in the blood, making the blood more acidic. Hyperventilation often results from hysteria or anxiety, possibly induced by stress. ▮

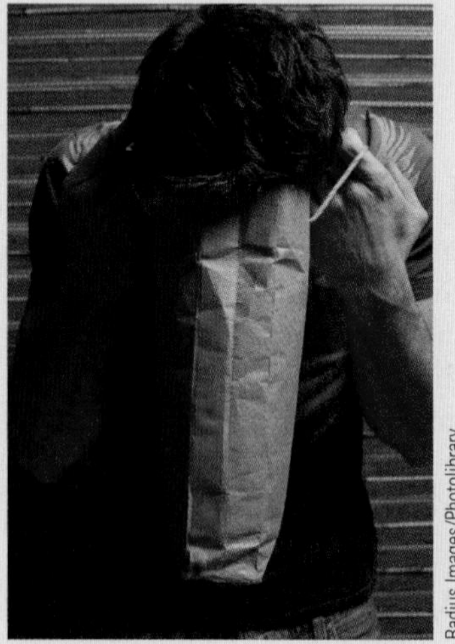
Radius Images/Photolibrary

Hyperventilation. The patient should be instructed to relax and to rebreathe air from a paper bag. Rebreathing increases the concentration of CO_2 and mitigates the respiratory alkalosis caused by hyperventilation.

16.4 Titrations of Weak Acids and Bases: Qualitative Aspects

OBJECTIVES

- ☐ Separate the titration curve for a weak acid into regions in which a single equilibrium dominates
- ☐ Estimate the pH of mixtures of a weak acid and strong base or weak base and strong acid

The titration curve of a weak acid with strong base, although similar to that of a strong-acid titration curve, exhibits some subtle but important differences. This section presents the reasons for these differences together with some ways to estimate the pH at certain points in the titration to sketch a titration curve.

Dividing the Titration Curve into Regions and Estimating the pH

Figure 16.7 shows the titration curve for 10.0 mL of 0.100 *M* HCOOH (formic acid, a weak acid with $K_a = 1.8 \times 10^{-4}$) with 0.100 *M* NaOH. The best way to study this titration curve is to divide it into segments and identify the acid-base equilibria in each region. After identifying the dominant equilibrium, we can estimate the pH.

The weak-acid titration curve is divided into four regions, each associated with a different chemical system.

The chemical equation for the titration is

$$HCOOH(aq) + OH^-(aq) \rightarrow HCOO^-(aq) + H_2O(\ell)$$

(a) The first point to consider (labeled *(a)* in Figure 16.7) is the system before any hydroxide ion is added. At this point, the system is a *weak acid solution,* and the pH is low. We can expect the pH to be 2 to 4 and will use 3 as the estimate of the pH of a weak acid solution. Weak acid solutions were discussed in Section 15.5.

The initial equilibrium in the titration, before any base is added, is a weak acid solution. A reasonable estimate of the pH is 3.

(b) Next, consider the system after some hydroxide ion has been added, but not enough to neutralize completely all the weak acid in this example, we will assume we started with 1 mmol of the weak acid and have added 0.5 mmol of hydroxide.

	HCOOH	+	OH^-	→	$HCOO^-$	+	H_2O
s (mmol)	1.00		0.50				Excess
R (mmol)	−0.50		−0.50		+0.50		+0.50
f (mmol)	**0.50**		0		**0.50**		Excess

Before the equivalence point, all the hydroxide ions are consumed; there is some weak acid, HCOOH, left, and some formate ion, $HCOO^-$, has formed. This text often uses boldface type in the sRfc tables to emphasize the most

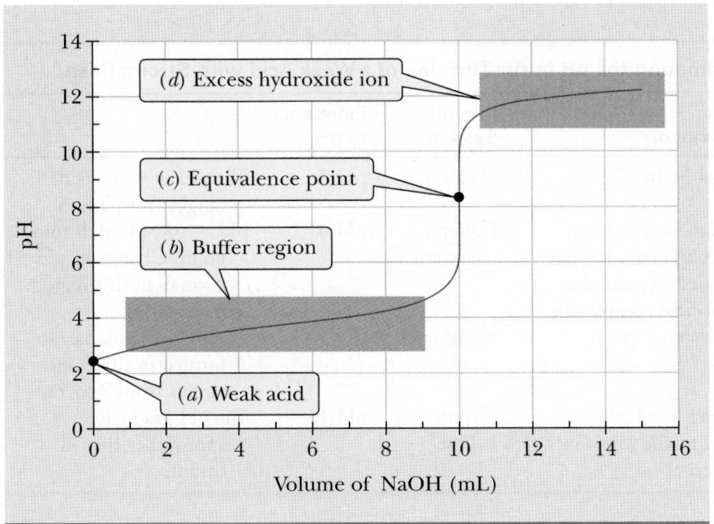

Figure 16.7 Titration curve for a weak acid with a strong base. Graph shows the titration of 10.0 mL of 0.100 *M* formic acid with 0.100 *M* sodium hydroxide. The dominant equilibria are labeled.

important species. This mixture of a weak acid and its conjugate base is a *buffer.* Half way to the equivalence point, $n_b/n_a = 1$, so from the Henderson-Hasselbalch equation pH = pK_a at this point.

At 1/11 to the equivalence point volume (9%), $n_b/n_a = 0.1$, and pH = pK_a − 1. Finally, at 10/11 of the equivalence point volume (91%), $n_b/n_a = 10$, and pH = pK_a +1. Therefore, from 9% to 91% neutralization of the weak acid, the pH changes from pK_a − 1 to pK_a + 1, a range of 2 pH units. We generally consider the 9–91% region the buffer region, where the pH remains reasonably constant.

(c) At the equivalence point (see Figure 16.7*c*), the number of moles of hydroxide added is equal to the number of moles of acid present in the original sample. We assumed that we started with 1 mmol of the weak acid, so the equivalence point is reached after we add 1 mmol of hydroxide ion. The reaction goes to completion, producing formate ion, $HCOO^-$, the conjugate base of formic acid. At the equivalence point, the formate ion is the only species that affects the pH.

	$HCOOH$	+	OH^-	→	$HCOO^-$	+	H_2O
s (mmol)	1.00		1.00				Excess
R (mmol)	−1.00		−1.00		+1.00		+1.00
f (mmol)	0		0		**1.00**		Excess

Like any solution of a weak base, the pH is greater than 7. *It is important to remember that the pH at the equivalence point in the titration of a weak acid is not 7.* Just as in the titration of a strong acid, the pH changes rapidly in the region around the equivalence point. The pH of a weak base solution formed in the titration of a typical weak acid is in the range of 9 to 11, so 10 is a reasonable estimate. Solutions of weak bases are discussed in Section 15.6.

At the equivalence point of a weak acid-strong base titration, the system is a weak base solution, and the pH is about 10.

(d) Adding hydroxide ion after all the acid has been consumed (adding 1.5 mmol of hydroxide ion to 1 mmol of weak acid in this particular example) produces a solution of a strong base plus a weak base (see Figure 16.7*d*).

	$HCOOH$	+	OH^-	→	$HCOO^-$	+	H_2O
s (mmol)	1.00		1.50				Excess
R (mmol)	−1.00		−1.00		+1.00		+1.00
f (mmol)	0		**0.50**		1.00		Excess

We can ignore the contribution of the weak base in the presence of the strong base, and we can use 13 as the estimate of the pH of a strong-base solution. Methods to calculate the pH of solutions of strong bases were discussed in Section 15.3.

Table 16.3 summarizes these preceding observations.

TABLE 16.3 Estimating the pH in the Titration of a Weak Acid with Strong Base

Region in Figure 16.7	Description	Chemical System	Estimate of pH	
(a)	Initial point	Weak acid	pH 3	The pH of weak acids ranges from 2 to 5.
(b)	Region in which not enough base has been added to neutralize the acid completely	Buffer system	pH 4	pH = pK_a (3–6) at midpoint. Change in pH is small in this region.
(c)	Equivalence point	Weak base	pH 10	The pH of weak bases formed in titrations ranges from 8 to 11.
(d)	Region in which hydroxide ion is in excess	Strong base	pH 13	The pH gradually approaches that of titrant.

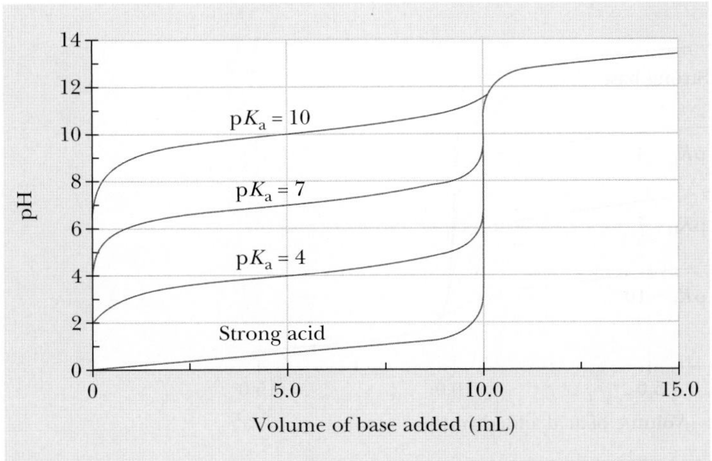

Figure 16.8 Titration curves for acids of different strengths. Titration curves for 10.00 mL of 1.00 *M* acids with 1.00 *M* strong base. The stoichiometry is presumed to be 1:1; that is, all acids are monoprotic. The change in pH at the equivalence point is largest for the strongest acids.

Sketching the Titration Curve

The exact shape of an acid-base titration curve is determined by the concentrations and the strengths of the acid and base in the titration, calculations that are presented in the next section. Figure 16.8 shows calculated titration curves for a strong acid and several weak acids with $pK_a = 4$, 7, and 10, all titrated with a strong base.

First, compare the curve of the strong acid with that of the acid with $pK_a = 4$. Notice several points:

1. *The amount of base needed to neutralize an acid depends only on the amount of acid present, not whether it is strong or weak.* The amount of base needed to neutralize 10.0 mmol of a strong acid, such as HCl, is identical to the amount needed to neutralize 10.0 mmol of a weak acid, such as HF. Each of these acids requires 10.0 mmol of base to reach the equivalence point. The stoichiometry, however, does influence the equivalence point. Although it takes 10.0 mmol KOH to neutralize 10.0 mmol hydrochloric or hydrofluoric acid,

$$HCl(aq) + KOH(aq) \rightarrow H_2O(\ell) + KCl(aq)$$

$$HF(aq) + KOH(aq) \rightarrow H_2O(\ell) + KF(aq)$$

It takes only 5.0 mmol $Ba(OH)_2$, because the stoichiometry is 2:1.

$$2HCl(aq) + Ba(OH)_2(aq) \rightarrow 2H_2O(\ell) + BaCl_2(aq)$$

$$2HF(aq) + Ba(OH)_2(aq) \rightarrow 2H_2O(\ell) + BaF_2(s)$$

2. *The titration curves for all the acids are indistinguishable beyond the equivalence point.* The excess hydroxide ion, not the strength of the acid, determines the pH of the solution.
3. *The pH halfway to the equivalence point in the titration of a weak acid is equal to pK_a.* The rapid change in pH near the equivalence point is smaller for weaker acids than for stronger acids because the abrupt change starts from about 1 pH unit greater than pK_a.
4. *The pH at the equivalence point is not 7.0,* unless both the acid and base are strong. (It is more accurate to say that the pH at the equivalence point is 7.0 only if the acid and base are equal in strength.) As the strength of the acid decreases, the pH at the equivalence point increases.

Figure 16.9 shows curves for the titrations of several bases of different strengths. Notice the similarity of these curves to the curves for acids.

Table 16.4 lists guidelines for estimating pH.

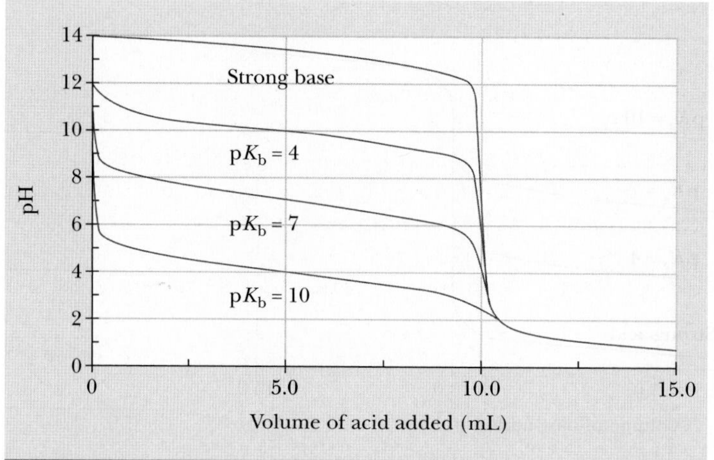

Figure 16.9 Titration curves for bases of different strengths. Stronger bases give sharper inflections. A very weak base (such as the one shown with $pK_b = 10$) does not show any inflection at all.

TABLE 16.4 Estimating pH in a Titration

Titration of an Acid with Strong Base		Titration of a Base with Strong Acid	
System	pH	System	pH
Strong acid	1	Strong base	13
Weak acid	3	Weak base	11
Acidic buffer	4	Basic buffer	10
Neutral	7	Neutral	7
Weak base	11	Weak acid	3
Strong base	13	Strong acid	1

OBJECTIVES REVIEW *Can you:*

☑ separate the titration curve for a weak acid into regions in which a single equilibrium dominates?

☑ estimate the pH of mixtures of a weak acid and strong base or weak base and strong acid?

16.5 Titrations of Weak Acids and Bases: Quantitative Aspects

OBJECTIVES

☐ Calculate the pH in the titration of a weak acid with strong base

☐ Calculate the pH in the titration of a weak base with strong acid

The titration of a weak acid with strong base is treated in the same way as is the titration of a strong acid, by using an sRfc table to determine which species are important in the equilibrium calculation. The calculation of pH is a little different, however, because the conjugate acid-base pairs formed in the titration reaction influence the final pH.

We use a two-part process to determine the pH during the course of the titration of a weak acid or base. The first step is to consider the stoichiometry of the titration reaction, which goes to completion, to find the concentrations of species before considering any equilibria. When we evaluate the results, we know whether we have a strong or weak acid, a strong or weak base, or a buffer solution. After identifying the dominant species in solution, we calculate the pH of the solution. No new equilibrium calculations are involved when we consider titrations of weak acids or bases. The solutions generated in the course of the titration are solutions of strong or weak acids or bases or buffer solutions; we have already illustrated calculations of the pH of all these types of solutions.

To analyze a titration, first use stoichiometry, then consider equilibria.

Calculating the Titration Curve for a Weak Acid with Strong Base

Example 16.8 illustrates the methods used to calculate the pH during the titration of 20.0 mL of 0.500 *M* HCOOH (formic acid, $K_a = 1.8 \times 10^{-4}$) with 0.500 *M* NaOH. We calculate the pH after 0, 10.0, 20.0, and 30.0 mL of base are added, but each step is preceded by a qualitative estimate of the pH, conducted by the methods described in the previous section. You should always have an estimate in mind so that you can check for any major mathematical errors in the calculation.

The information flow diagram for calculating a weak acid titration curve is as follows:

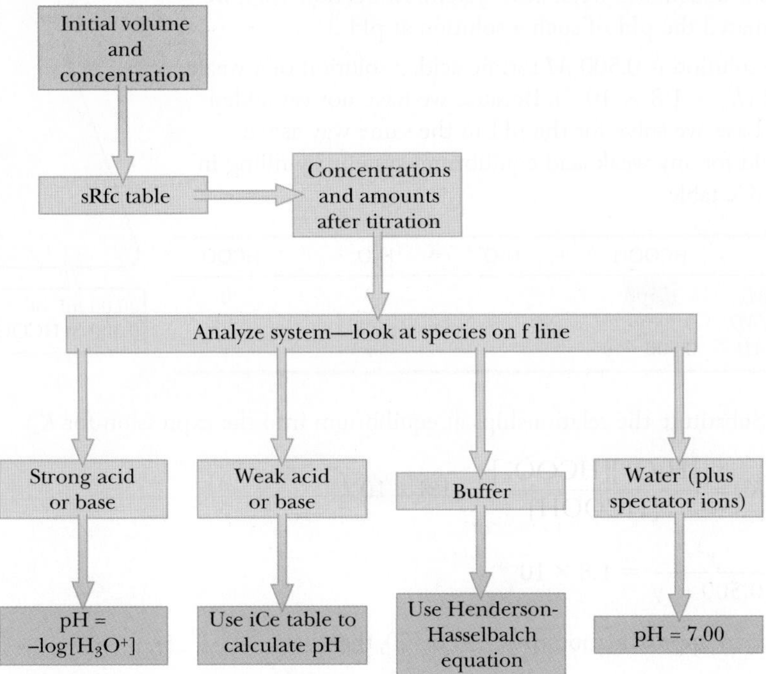

EXAMPLE 16.8 Calculating a Weak Acid Titration Curve

Calculate the pH during the titration of 20.0 mL of 0.500 *M* formic acid ($K_a = 1.8 \times 10^{-4}$) with 0.500 *M* NaOH. Calculate the pH after 0, 10.0, 20.0, and 30.0 mL of NaOH have been added, and sketch the titration curve.

Strategy First, write the chemical equation for the titration and calculate the equivalence point. We will use an sRfc table to determine which species are present after adding titrant. Once we know which species are present, we will determine whether the system contains a strong acid or base, a weak acid or base, a buffer, or just water (including spectator ions). We already know how to calculate the pH of each of these systems.

Solution

Write the chemical equation to calculate the equivalence point before beginning to calculate the pH of the solution. If we know the equivalence point, we can estimate the pH and determine whether the result of the calculation is reasonable.

The chemical equation for the titration is

$$HCOOH(aq) + OH^-(aq) \rightarrow HCOO^-(aq) + H_2O(\ell)$$

To calculate the equivalence point, determine the number of moles of acid present and the volume of base needed to provide the equivalent amount of base.

$$\text{Millimoles acid} = 20.00\ \cancel{\text{mL HCOOH}} \times \left(\frac{0.500\ \text{mmol HCOOH}}{\cancel{\text{mL HCOOH}}}\right) = 10.00\ \text{mmol}$$

The amount of hydroxide needed to neutralize this amount of acid is 10.00 mmol, from the 1:1 stoichiometry of the chemical equation. The volume of NaOH needed to reach the equivalence point is

$$\text{Volume NaOH solution} = 10.00\ \text{mmol NaOH} \times \left(\frac{1\ \text{mL NaOH}}{0.500\ \text{mmol NaOH}}\right) = 20.00\ \text{mL}$$

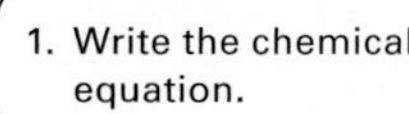

The initial point in this titration is a weak acid solution. A reasonable estimate of the pH is 3. Calculate the exact pH by using an iCe table.

Now, begin the calculations.

0 mL of 0.500 *M* NaOH added to 20.00 mL of 0.500 *M* formic acid: The initial point in the titration, before any base is added, is a weak acid system. In Section 16.4, we estimated the pH of such a solution at pH 3.

The solution is 0.500 *M* formic acid, a solution of a weak acid ($K_a = 1.8 \times 10^{-4}$). Because we have not yet added any base, we solve for the pH in the same way as we would for any weak acid equilibrium system, by filling in the iCe table.

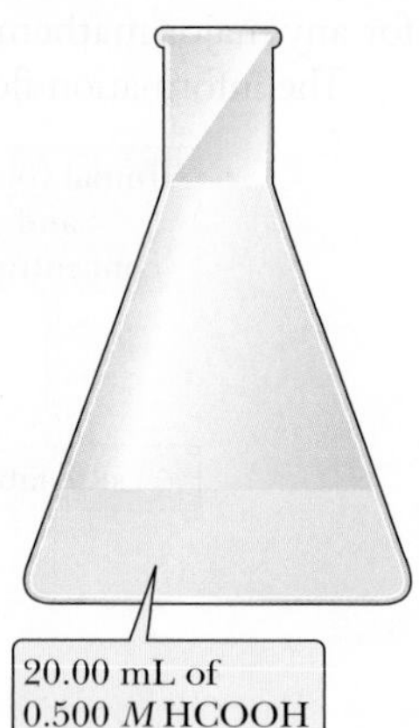

1. Write the chemical equation.

2. Fill in the iCe table.

	HCOOH	+	H_2O	⇌	H_3O^+	+	$HCOO^-$
i (*M*)	0.500				0		0
C (*M*)	$-y$				$+y$		$+y$
e (*M*)	$0.500 - y$				y		y

3. Write the expression for *K*.

Substitute the relationships at equilibrium into the expression for K_a:

$$K_a = \frac{[H_3O^+][HCOO^-]}{[HCOOH]} = 1.8 \times 10^{-4}$$

4. Substitute.

$$\frac{y^2}{0.500 - y} = 1.8 \times 10^{-4}$$

5. Solve.

Solve by approximation. If $y << 0.500$, then

$$y^2 \approx 1.8 \times 10^{-4} \times 0.500 = 9.0 \times 10^{-5}$$

$$y = 9.5 \times 10^{-3}$$

After we check that the assumption is valid and that additional calculations are not needed, we can write

$$[H_3O^+] = 9.5 \times 10^{-3}\ M \quad \text{and} \quad \text{pH} = 2.02$$

Our estimate was 3, so this result is reasonable.

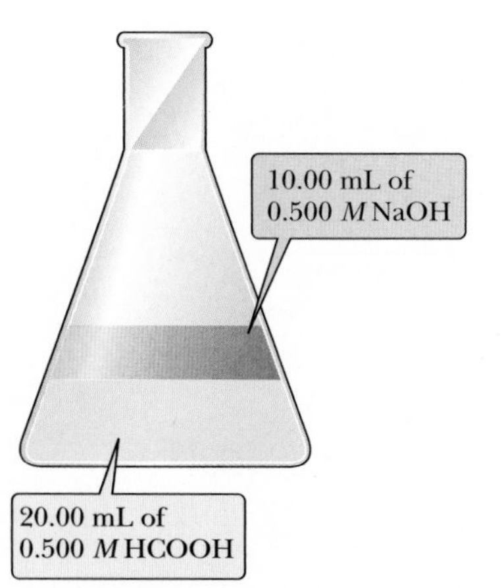

10.00 mL (total) of 0.500 *M* NaOH added to 20.00 mL of 0.500 *M* formic acid: On a qualitative basis, we would estimate the pH to be in the region of pK_a ± 1. Because pK_a for formic acid is 3.74, we estimate the pH to be between 2.7 and 4.7. To calculate the pH of the system, first prepare the sRfc table for the titration reaction. The titration is the addition of strong base to a weak acid, a reaction that goes to completion. We start with 20.00 mL of 0.500 *M* formic acid, a total of 10.0 mmol.

We have added 10.00 mL of 0.500 *M* base, or 5.00 mmol.

Look at the f line. A weak acid plus conjugate base indicates a buffer. The pH will be about 4.

	HCOOH	+	OH^-	→	$HCOO^-$	+	H_2O
s (mmol)	10.00		5.00		0		Excess
R (mmol)	−5.00		−5.00		+5.00		+5.00
f (mmol)	**5.00**		0		**5.00**		Excess

The second step is to examine the results of the titration and estimate the pH. Look at the results in the f line of the table. Is there any strong acid or strong base? No, there is none, because all of the strong base was consumed. Next, determine whether there is any weak acid or base. Both are present. Some formic acid (a weak acid) and some formate ion, its conjugate base, are present. This solution is a buffer, and we can use 4 as the estimate of its pH.

To calculate the exact pH, use the Henderson–Hasselbalch equation. We use the number of millimoles of HCOOH and $HCOO^-$ from the last line of the sRf table

After some base has been added, but not enough to reach the equivalence point, the system is a buffer. A reasonable estimate of the pH is 4.

$$\mathrm{pH} = \mathrm{p}K_a + \log\frac{n_b}{n_a} = \mathrm{p}K_a + \log\frac{5.00}{5.00}$$

$$= \mathrm{p}K_a + 0$$

$$= -\log(1.8 \times 10^{-4})$$

$$= 3.74$$

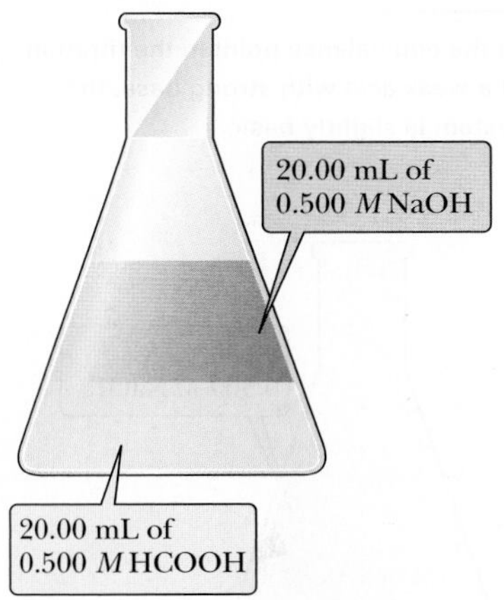

Note that when the concentrations of the acid and its conjugate base are equal, pH is equal to $\mathrm{p}K_a$. This situation is general: *The pH is equal to the* $\mathrm{p}K_a$ *halfway to the equivalence point in the titration of a weak acid with a strong base.*

20.00 mL (total) of 0.500 *M* NaOH added to 20.00 mL of 0.500 *M* formic acid: The titration has reached the equivalence point. The formic acid has been entirely converted to formate ion, and we would estimate the pH in the region of 8 to 10. But even if you did not recognize this fact, you would still get the correct answer in a calculation. Start the exact calculation by writing the titration reaction, and evaluating the amounts and concentrations of species. Calculate the numbers of millimoles of formic acid and hydroxide ion from their concentrations and volumes.

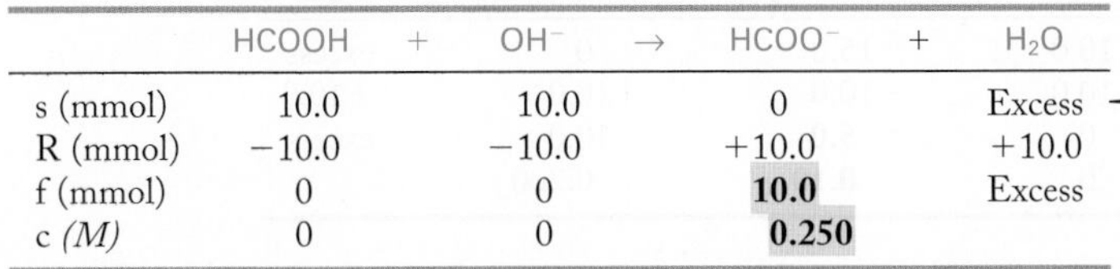

	HCOOH	+	OH^-	→	$HCOO^-$	+	H_2O
s (mmol)	10.0		10.0		0		Excess
R (mmol)	−10.0		−10.0		+10.0		+10.0
f (mmol)	0		0		**10.0**		Excess
c (*M*)	0		0		**0.250**		

Look at the f line. We have a weak base, the formate ion. The pH is about 10.

The data on the c line comes from the 10 mmol $HCOO^-$ and the *total* volume of 40.0 mL.

Both reactants have been completely consumed so the titration has reached the equivalence point. Examine the results of the titration to determine what type of calculation is needed. There is no strong acid or strong base. There is no weak acid (the formic acid has been consumed), but the titration has produced a weak base, formate ion. We estimate the pH is 10. The exact solution for this kind of equilibrium was discussed in Chapter 15 when we used the iCe table to help calculate the pH of a weak base.

	$HCOO^-$	+	H_2O	⇌	HCOOH	+	OH^-
i (*M*)	0.250				0		0
C (*M*)	$-y$				$+y$		$+y$
e (*M*)	$0.250 - y$				y		y

1. Write the chemical equation.

2. Fill in the iCe table.

Write the expression for K_b, and substitute the relationships from the bottom line of the iCe table into the expression.

3. Write the expression for *K*.

$$K_b = \frac{[\mathrm{HCOOH}][\mathrm{OH^-}]}{[\mathrm{HCOO^-}]} = \frac{y^2}{0.250 - y}$$

4. Substitute.

We calculate K_b from

$$K_b = \frac{K_w}{K_a} = \frac{1.0 \times 10^{-14}}{1.8 \times 10^{-4}} = 5.6 \times 10^{-11}$$

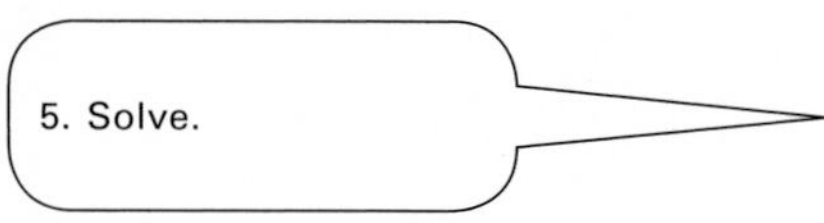

Finally, solve

$$\frac{y^2}{0.250 - y} = 5.6 \times 10^{-11}$$

This expression is most easily solved by approximation. If $y << 0.250$, then

$$y^2 \approx 5.6 \times 10^{-11} \times 0.250$$

$$y = 3.7 \times 10^{-6}$$

After we check the approximation, we can write

$$y = [OH^-] = 3.7 \times 10^{-6}\ M$$

$$pOH = 5.43;\ pH = 8.57$$

At the equivalence point in the titration of a weak acid with strong base, the system is slightly basic.

Note that in the titration of a weak acid with a strong base, the solution at the equivalence point is alkaline, not neutral.

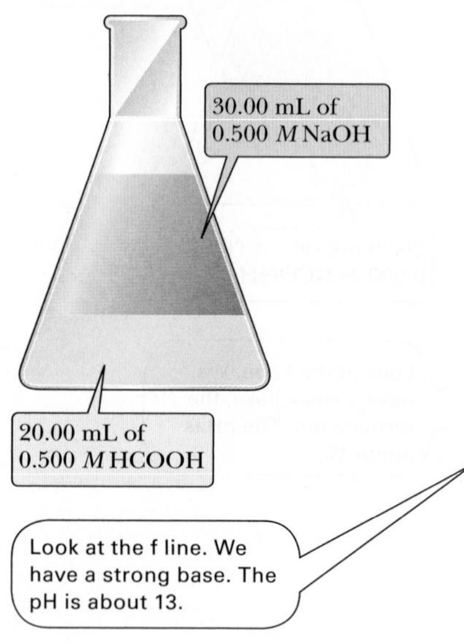

30.00 mL (total) of 0.500 *M* NaOH added to 20.00 mL of 0.500 *M* formic acid: At this point, hydroxide ion is present in excess, and we can estimate the pH as approaching 13.7, the pH of 0.500 *M* NaOH.

Again, start the calculation with the titration reaction, and assume it goes to completion. We have added 30.00 mL of 0.500 *M* NaOH (15.0 mmol NaOH), so the total volume of the solution is 50.00 mL. This volume and the number of millimoles of excess hydroxide ion are used to calculate the concentration of hydroxide ion on the c line.

	HCOOH	+	OH^-	→	$HCOO^-$	+	H_2O
s (mmol)	10.0		15.0		0		excess
R (mmol)	−10.0		−10.0		+10.0		+10.0
f (mmol)	0		**5.0**		10.0		excess
c (*M*)	0		**0.10**		0.200		

Look at the f line. We have a strong base. The pH is about 13.

We have an excess of strong base, which determines the pH of the solution.

$$[OH^-] = 0.10\ M$$

$$pOH = 1.00;\ pH = 13.00$$

This calculation is identical to the one performed in the titration of a strong acid with a strong base. After we reach the equivalence point, we have excess strong base. It does not matter whether we started with strong acid or weak acid—the acid has been consumed.

Beyond the equivalence point, OH^- is in excess.

The Titration Curve: Figure 16.10 shows the formic acid titration curve. Notice that we can divide the titration curve into four zones, depending on the type of calculation performed.

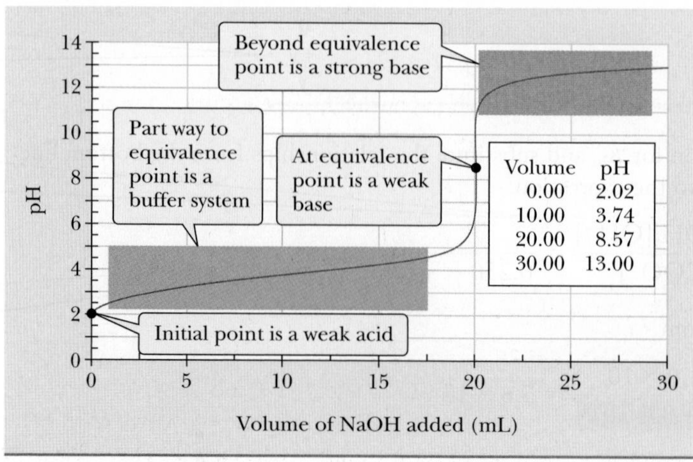

Figure 16.10 Titration curve. The titration of 20.0 mL of 0.500 *M* formic acid with 0.500 *M* sodium hydroxide.

Understanding

Calculate the pH values after the addition of 0, 10.00, 20.00, and 30.00 mL of 0.500 *M* NaOH to 20.00 mL of 0.500 *M* nitrous acid ($K_a = 5.6 \times 10^{-4}$).

Answer 0 mL, pH = 1.78; 10.00 mL, pH = 3.25; 20.00 mL (equivalence point), pH = 8.32; 30.00 mL, pH = 13.00

Titration Curves for Solutions of Weak Bases with Strong Acids

The titration curve for a solution of a weak base with strong acid is calculated by similar methods. Figure 16.11 shows the results of calculating the pH values in the titration of the 20.00 mL of 0.500 *M* methylamine ($K_b = 3.7 \times 10^{-4}$) with 0.500 *M* hydrochloric acid.

The titration curve of a weak base has an initial point at which the system is a weak base solution followed by a buffer region, then an equivalence point at which the system is a weak acid solution, and finally a region in which strong acid is in excess. When calculating the pH in the buffer region, you may use the Henderson–Hasselbalch equation.

$$\text{pH} = \text{p}K_a + \log\frac{C_b}{C_a} = \text{p}K_a + \log\frac{n_b}{n_a}$$

Take care to use K_a for the conjugate acid of methylamine (methylammonium ion), rather than K_b for methylamine, when calculating the pH with this equation.

The titration curve for a weak base is divided into four regions, and calculations are quite similar to those used for the weak acid titration.

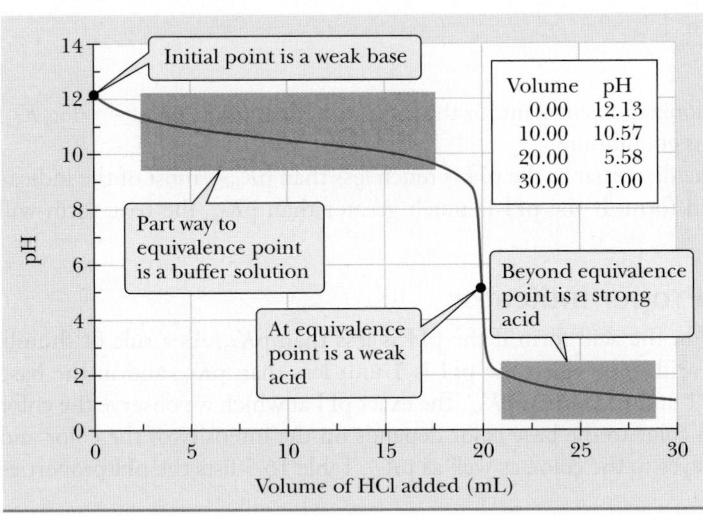

Figure 16.11 Titration curve for a weak base. The titration of 20.00 mL of 0.500 *M* methylamine with 0.500 *M* hydrochloric acid.

OBJECTIVES REVIEW *Can you:*

☑ calculate the pH in the titration of a weak acid with strong base?
☑ calculate the pH in the titration of a weak base with strong acid?

16.6 Indicators

OBJECTIVES

☐ Describe indicators by their acid-base chemistry
☐ Choose an indicator that is appropriate for a particular titration

The **indicator** is the substance that changes color to signal the end of the titration. We mentioned indicators earlier, when we compared the definitions of equivalence point and endpoint. The *equivalence point* in a titration occurs when an acid and a base are present in stoichiometrically equivalent amounts, whereas the *endpoint* is the point in the titration at which a color change occurs. When the indicator is chosen properly, the endpoint and equivalence point occur simultaneously, within experimental error.

Properties of Indicators

The indicator should have several attributes:

1. The indicator must change color as a function of pH. The change should be abrupt rather than drawn out, and the reaction that causes the change in color must occur rapidly. An indicator that requires several minutes to change color is not practical.
2. The color change should be easily discerned by the eye. A dramatic color change—for example, from red to green or from colorless to blue—is easy to detect. A subtle change—for example, from blue-gray to gray-green—is difficult to detect and reproduce.
3. The indicator must not perturb the solution. It is important for the indicator to have an intense color so that its color change does not consume a significant amount of titrant or reactant.

Most common indicators are weak acids or bases. We can represent an indicator molecule in the acid form as HIn, and in the base (deprotonated) form as In^-. The relationships between the two forms of the indicator are the same as those of any other weak acid-base conjugate pair, but the indicator is special in that the acid form and base form have different colors, even in very dilute solutions. The indicator can be described by the same equilibrium expressions as any other weak acid:

$$HIn + H_2O \rightleftharpoons H_3O^+ + In^-$$

$$K_{In} = \frac{[H_3O^+][In^-]}{[HIn]}$$

where K_{In} is the acid ionization constant for the indicator. Most often, $pK_{In}(= -\log K_{In})$ is used to describe this equilibrium.

Indicators are weak conjugate acid-base pairs in which the acid and base forms are different colors.

If the solution is acidic—that is, the pH is much less than pK_{In}—most of the indicator will be in the acid form. If the pH is much greater than pK_{In}, the base form will dominate.

Choosing the Proper Indicator

The indicator will be in the acid form if the pH is less than pK_{In}. As a rule of thumb, indicators are in the acid form when the pH is 1 unit less than pK_{In}, and in the base form when the pH is 1 unit more than pK_{In}. The exact pH at which we observe the color change from the acid color to the base color depends on the intensity of the color and the sensitivity of our eyes to the color, as well as pK_{In}. Table 16.5 lists the pH properties of several indicators.

Figure 16.12 shows titration curves for two acids, hydrochloric and acetic, with strong base and the pH range for the color changes of methyl red, which changes color between about 4.2 and 6.3, and phenolphthalein, which changes colors between 8.3 and

TABLE 16.5 Properties of Several Indicators

Name	Acid Color	Base Color	pH Range	pK_{In}
Thymol blue*	Red	Yellow	1.2–2.8	1.6
Methyl orange	Red	Yellow	3.1–4.4	3.5
Methyl red	Red	Yellow	4.2–6.3	5.0
Bromthymol blue	Yellow	Blue	6.2–7.6	7.3
Phenolphthalein	Clear	Pink	8.3–10.0	8.7
Thymol blue*	Yellow	Blue	8.0–9.6	9.2

*Thymol blue is a diprotic weak acid and has two color changes.

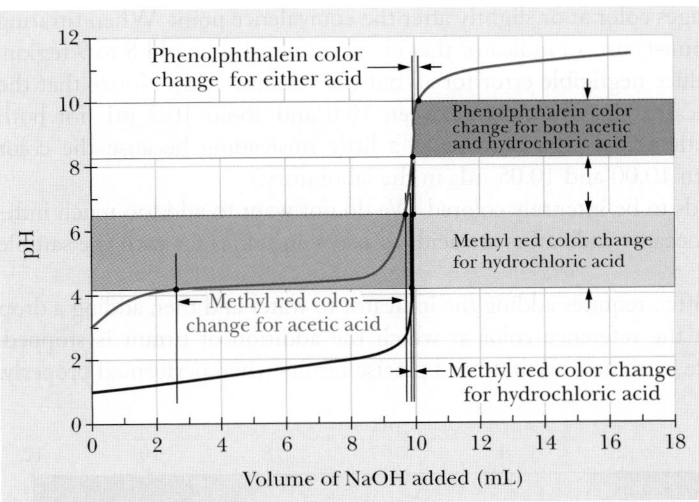

Figure 16.12 Titration curves for strong and weak acids. The color changes for two indicators methyl red (red to yellow at pH 4.2–6.4) and phenolphthalein (colorless to red pH 8.2–10) are shown along with titration curves for the strong acid HCl (blue) and the weak acetic acid (red). The indicator may not change color exactly at the equivalence point and a substantial error can occur if you choose an indicator that changes color before the equivalence point.

10.0. The error can be quite large in the titration of a weak acid if we use an indicator that changes color *before* the equivalence point. The figure shows that the methyl red begins to change color after about 2.2 mL and finishes changing color after about 9.7 mL, but the actual equivalence point occurs at 10 mL. To avoid large errors, choose

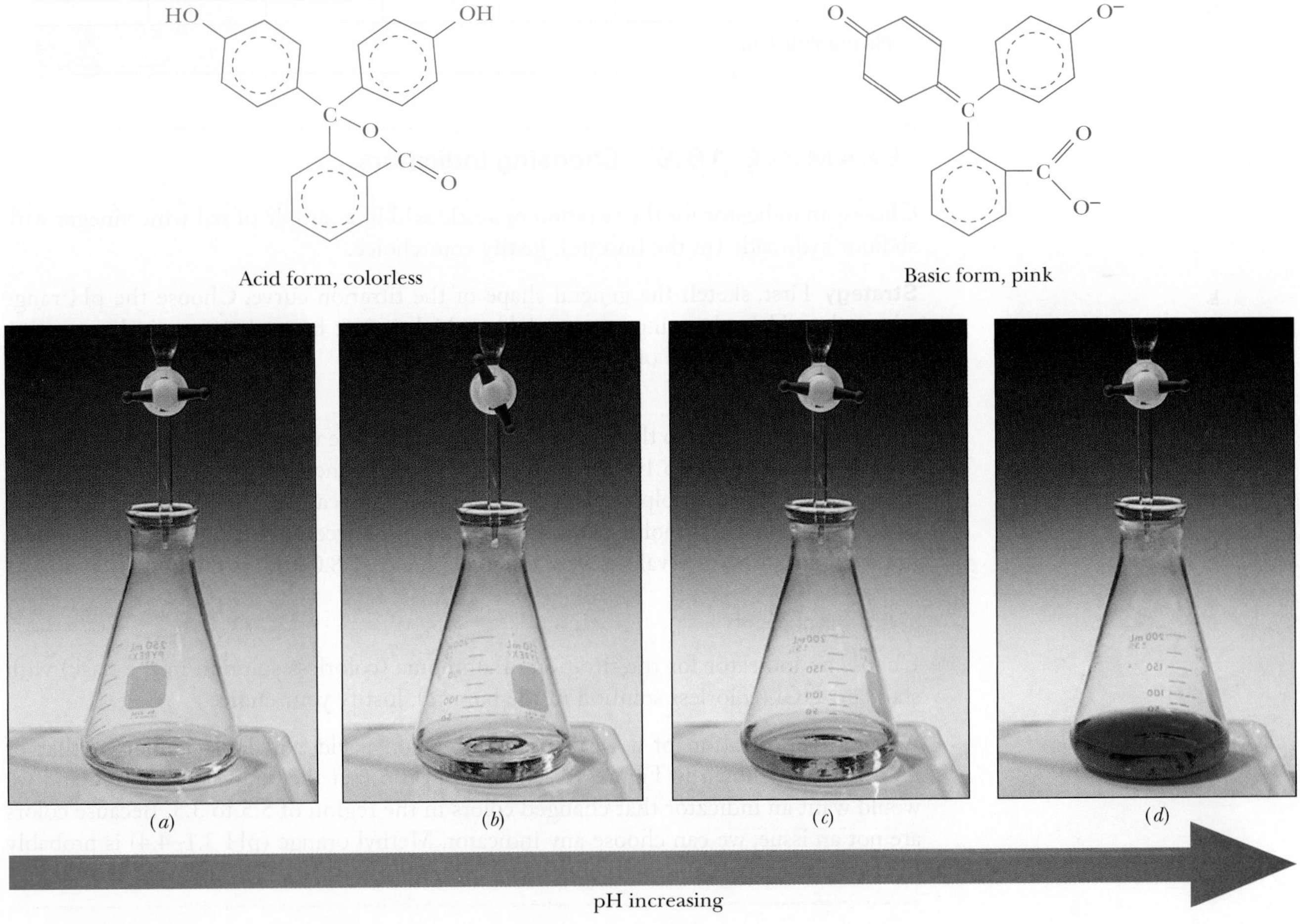

Acid form, colorless

Basic form, pink

(*a*) (*b*) (*c*) (*d*)

pH increasing

Phenolphthalein. Drawings show the acid and base forms of phenolphthalein, an indicator commonly used for the titration of an acid with strong base. *(a)* The acidic solution is initially clear. *(b)* When base is added, the solution turns pink momentarily but disappears with swirling. *(c)* The first permanent pink indicates the endpoint. *(d)* The solution is vividly colored beyond the equivalence point, where base is in excess.

an indicator that changes color at or slightly after the equivalence point. When titrating an unknown acid, chemists use an indicator that changes color in the pH 8 to 9 region; this choice will introduce negligible error for all but the weakest acids. Notice that the phenolphthalein indicator changes color between 10.0 and about 10.2 mL for both hydrochloric and acetic acids. (The drawing is a little misleading because the color change occurs between 10.00 and 10.05 mL in the laboratory.)

The indicator needs to be intensely colored. We do not want to add too much indicator to the solution because indicators are acids or bases and do react with the sample or titrant.

Indicators should be chosen to change colors slightly after the equivalence point.

Good technique often requires adding the indicator to water and then adding a drop of titrant to establish the reference color at which the addition of titrant is stopped. Titrations can produce extremely accurate and precise results when performed properly.

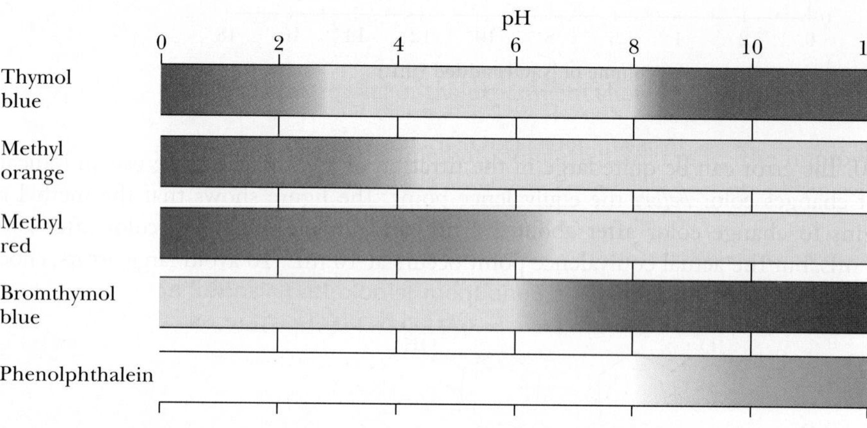

EXAMPLE 16.9 Choosing Indicators

Choose an indicator for the titration of acetic acid in a sample of red wine vinegar with sodium hydroxide (in the burette). Justify your choice.

Strategy First, sketch the general shape of the titration curve. Choose the pH range where the pH is changing most quickly, which is just beyond the equivalence point. Last, think of the colors of the solution and indicator.

Solution

The titration is similar to that shown in Figure 16.10. We want an indicator that changes color over the range 8 to 10. The red color of the red wine vinegar solution may make it difficult to see the phenolphthalein color change, so we can eliminate that indicator. We can eliminate bromthymol blue because it changes between pH 6.2 and 7.6, which does not encompass the interval we want. Thymol blue (pH 8.0–9.6) is the best choice.

Understanding

Choose an indicator for the titration of ammonia (colorless solution in the flask) with standard HCl (colorless solution in the burette). Justify your choice.

Answer The titration of a weak base with strong acid will look similar to that of methylamine (shown in Figure 16.11). The pH is 5.6 at the equivalence point, and we would want an indicator that changed colors in the region of 5.5 to 3.5. Because colors are not an issue, we can choose any indicator. Methyl orange (pH 3.1–4.4) is probably the best choice, but you could choose methyl red (pH 4.2–6.3) as well.

OBJECTIVES REVIEW *Can you:*

- ☑ describe indicators by their acid-base chemistry?
- ☑ choose an indicator that is appropriate for a particular titration?

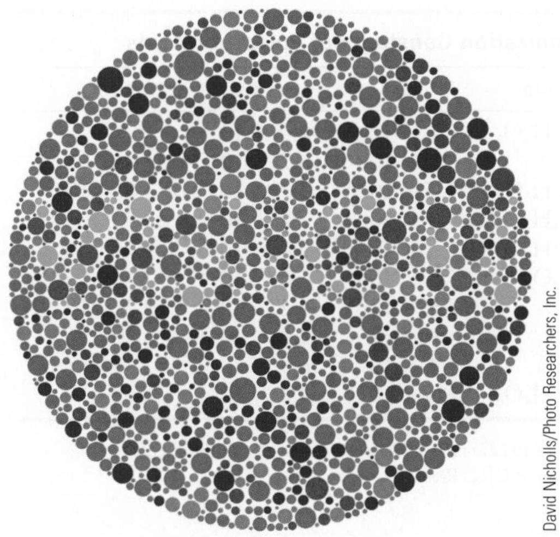
David Nicholls/Photo Researchers, Inc.

Color blindness test. Computer artwork of an Ishihara color-test card used to check for red-green color blindness. The card is comprised of red and orange dots on a background of green dots. A person with normal vision will be able to see the word "vision." A person with red deficiency (protanopia) will only see the orange parts of the word, while those with green deficiency (deuteranopia) will only see the red parts of the word. Color-blind chemists should not despair because an electronic device such as a pH meter can be used to determine the equivalence point in a titration. Several researchers published plans for a pH meter with a computerized voice that even allows vision-impaired students to acquire meaningful laboratory experiences.

16.7 Polyprotic Acid Solutions

OBJECTIVES

- ☐ Write chemical equations and expressions for the equilibrium constants for the dissociation of polyprotic acids
- ☐ Calculate the pH of solutions that contain polyprotic acids
- ☐ Estimate the pH of solutions of amphoteric species

Some acids, called **polyprotic acids,** provide more than one proton when they ionize. Some common examples include phosphoric acid, H_3PO_4, which is found in many soft drinks, and sulfuric acid, H_2SO_4, a widely used industrial chemical. The pH of a polyprotic acid solution is calculated by the same methods we have illustrated throughout this chapter and in Chapter 15.

Consider the first ionization of the diprotic oxalic acid, $H_2C_2O_4$:

$$H_2C_2O_4 + H_2O \rightleftharpoons H_3O^+ + HC_2O_4^-$$

We can write the expression for the equilibrium constant for the loss of the first proton.

$$K_{a1} = \frac{[H_3O^+][HC_2O_4^-]}{[H_2C_2O_4]}$$

K_{a1} describes the loss of the first proton.

The subscript 1 on K_{a1} signifies that the equilibrium describes the loss of the first proton.

Similarly, we can write the expression for the equilibrium constant for the loss of the second proton.

$$HC_2O_4^- + H_2O \rightleftharpoons H_3O^+ + C_2O_4^{2-}$$

$$K_{a2} = \frac{[H_3O^+][C_2O_4^{2-}]}{[HC_2O_4^-]}$$

K_{a2} describes the loss of the second proton.

Experimental measurements show that the values for K_{a1} and K_{a2} are 5.6×10^{-2} and 1.6×10^{-4}, respectively. Note that K_{a1} is always larger than K_{a2}. This situation is normal and is explained on the basis of electrostatic attraction. After the acid loses one proton, it forms the negatively charged $HC_2O_4^-$ species. Removing a second proton is more difficult because of the electrostatic attraction of the positive proton and negatively charged species. If the acid were triprotic, it would still be harder to remove the third proton.

Table 16.6 lists some common polyprotic acids together with their ionization constants.

TABLE 16.6 Ionization Constants of Polyprotic Acids

Name	Formula	K_{a1}	K_{a2}	pK_{a3}
Ascorbic	$H_2C_6H_6O_6$	9.1×10^{-5}	2.0×10^{-12}	
Carbonic	H_2CO_3	4.5×10^{-7}	4.7×10^{-11}	
Citric	$H_3C_6H_5O_7$	7.4×10^{-4}	1.7×10^{-5}	4.0×10^{-7}
Malic	$H_2C_4H_4O_5$	4.0×10^{-4}	7.8×10^{-6}	
Malonic	$H_2C_3H_2O_4$	1.6×10^{-2}	2.0×10^{-6}	
Oxalic	$H_2C_2O_4$	5.6×10^{-2}	1.6×10^{-4}	
Phosphoric	H_3PO_4	7.5×10^{-3}	6.2×10^{-8}	2.2×10^{-13}
Sulfuric	H_2SO_4	Strong	1.0×10^{-2}	
Sulfurous	H_2SO_3	1.4×10^{-2}	6.3×10^{-8}	
Tartaric	$H_2C_4H_4O_6$	1.0×10^{-3}	4.6×10^{-5}	

Values were taken from the *CRC Handbook of Chemistry and Physics,* 88th ed., © 2007–2008, by permission of CRC Press.

Calculating the Concentrations of Species in Solutions of Polyprotic Acids

When successive K_a values for an acid differ by a factor of 1000 or more, nearly all of the acid molecules lose the most easily ionized proton before any lose the second. The equilibria of polyprotic acids can be treated as a series of steps, or successive equilibria, in a manner analogous to that of a mixture of strong and weak acids. We approach polyprotic acids in the same manner as we do mixed acid systems, disregarding the ionization of the weaker species with respect to the stronger. Example 16.10 illustrates this approach.

If $K_{a1} >> K_{a2}$, then the second ionization constant can be ignored and the system treated like a monoprotic weak acid.

EXAMPLE 16.10 Concentrations of Species in a Polyprotic Acid Solution

Calculate the concentrations of all species in 0.500 *M* sulfurous acid.

Strategy Look at the values for K_{a1} and K_{a2}. If K_{a1} is much greater than K_{a2}, then K_{a1} will dominate the equilibrium, and calculations are similar to those of a monoprotic weak acid with K_a equal to K_{a1}.

Solution

Obtain the values for K_{a1} and K_{a2} from Table 16.6. Because $K_{a1} >> K_{a2}$, the dominant equilibrium is the first ionization:

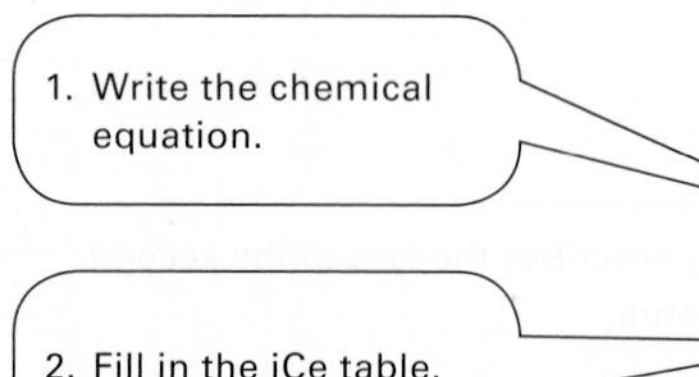
1. Write the chemical equation.

$$H_2SO_3 + H_2O \rightleftharpoons H_3O^+ + HSO_3^-$$

$$K_{a1} = 1.4 \times 10^{-2}$$

2. Fill in the iCe table.

	H_2SO_3	+	H_2O	$\rightleftharpoons$	H_3O^+	+	HSO_3^-
i (*M*)	0.500				0		0
C (*M*)	$-y$				$+y$		$+y$
e (*M*)	$0.500 - y$				y		y

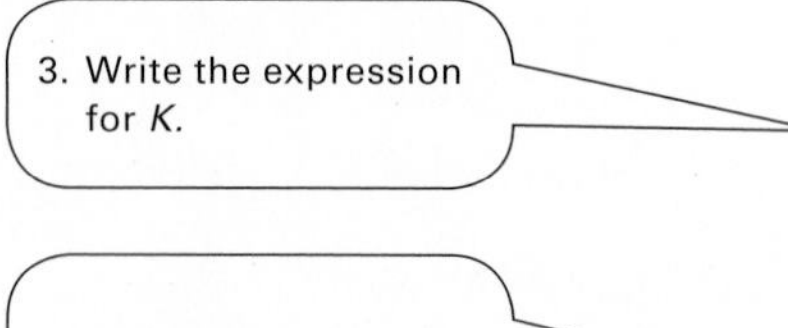
3. Write the expression for *K*.

Write the algebraic expression for the equilibrium constant.

$$K_{a1} = \frac{[H_3O^+][HSO_3^-]}{[H_2SO_3]}$$

4. Substitute.

Then substitute the information from the bottom line of the iCe table.

$$1.4 \times 10^{-2} = \frac{y^2}{0.500 - y}$$

Solve this expression, either by the quadratic formula or by the method of successive approximations, to get

$y = 7.7 \times 10^{-2}$ — 5. Solve.

Use the iCe table to calculate the concentrations of all the species present.

$[H_3O^+] = y = 7.7 \times 10^{-2}\ M$

$[HSO_3^-] = y = 7.7 \times 10^{-2}\ M$

$[H_2SO_3] = 0.500 - y = 0.423\ M$

The preceding calculations are identical to those done for any other weak acid.

To solve for the concentration of SO_3^{2-}, use these concentrations and the equilibrium expression for the ionization of the second proton. This computation will also enable us to determine the effect of ignoring the second equilibrium in the preceding calculation.

1. Write the chemical equation.

	$HSO_3^- + H_2O$	$\rightleftharpoons$	H_3O^+	+	SO_3^{2-}
i (M)	7.7×10^{-2}		7.7×10^{-2}		0
C (M)	$-z$		$+z$		$+z$
e (M)	$7.7 \times 10^{-2} - z$		$7.7 \times 10^{-2} + z$		z

2. Fill in the iCe table.

$$K_{a2} = \frac{[H_3O^+][SO_3^{2-}]}{[HSO_3^-]} = 6.3 \times 10^{-8}$$

3. Write expression for K, in this case, K_{a2}.

$$6.3 \times 10^{-8} = \frac{(7.7 \times 10^{-2} + z)(z)}{7.7 \times 10^{-2} - z}$$

4. Substitute.

If $z << 7.7 \times 10^{-2}$, then

$z = [SO_3^{2-}] = 6.3 \times 10^{-8}\ M$

5. Solve.

(The approximation is justified.)

Including the second ionization equilibrium does not change the concentrations of the other species in solutions very much. The concentration of hydrogen ion produced in the second ionization, $6.3 \times 10^{-8}\ M$, is negligible with respect to the $7.7 \times 10^{-2}\ M$ produced in the first ionization.

Understanding

Calculate the pH of 0.033 *M* carbonic acid. (This concentration corresponds to a solution that is saturated with CO_2 at 1 atm.)

Answer pH = 3.91

Amphoteric Species

Substances that have the properties of acids and bases are called **amphoteric,** from the Greek *amphoteros,* meaning "both." This section describes amphoteric species and their reactions, and presents methods used to estimate the pH of solutions containing them.

The hydrogen carbonate ion is typical of amphoteric species; Table 16.7 summarizes its behavior.

Look for a moment at the last line of the table. It is important to notice that the acid behavior of the hydrogen carbonate ion is described by the second acid ionization constant, K_{a2}. The behavior as a base is described by K_b, which comes from the first ionization constant, K_{a1}.

You can estimate the pH of a solution of an amphoteric species by comparing K_a with K_b. For the hydrogen carbonate ion, K_b is larger than K_a, so we predict that a

Greek *amphora*. The Greeks called a container with handles for both hands an *amphora*. This *amphora* (*ca.* 600-550 B.C.) shows runners in a race.

TABLE 16.7 Amphoteric Behavior of the Hydrogen Carbonate Ion

	Hydrogen Carbonate as an Acid	Hydrogen Carbonate as a Base
Chemical equation	$HCO_3^- + H_2O \rightleftharpoons H_3O^+ + CO_3^{2-}$	$HCO_3^- + H_2O \rightleftharpoons OH^- + H_2CO_3$
Expression for equilibrium constant	$K_a = \frac{[H_3O^+][CO_3^{2-}]}{[HCO_3^-]}$	$K_b = \frac{[OH^-][H_2CO_3]}{[HCO_3^-]}$
Value for equilibrium constant	$K_a = 4.7 \times 10^{-11}$	$K_b = 2.2 \times 10^{-8}$
Source of equilibrium constant	K_{a2} from Table 16.6	$K_b = K_w/K_{a1}$ K_{a1} from Table 16.6

solution of sodium hydrogen carbonate is slightly alkaline. Laboratory measurements show that the pH of a 0.10 *M* solution is 8.30.

EXAMPLE 16.11 Amphoteric Acid-Base Systems

Write the equilibria of 0.10 *M* sodium hydrogen sulfite, and determine whether its pH is greater than or less than 7.

Strategy First, write the chemical reactions in which the amphoteric species acts as an acid, and the reaction in which it acts as a base. Next, write the expression for the equilibrium constant for each reaction, K_a and K_b. Determine the numerical values of K_a and K_b, and compare them to determine which equilibrium dominates.

Solution

Sodium hydrogen sulfite dissociates into sodium ions and the amphoteric hydrogen sulfite ion.

$$NaHSO_3 \rightarrow Na^+ + HSO_3^-$$

The following table illustrates the amphoteric nature of the hydrogen sulfite ion. The solution is acidic because the HSO_3^- ion is a stronger acid than it is a base.

Amphoteric Reactions of the Hydrogen Sulfite Ion

Hydrogen Sulfite as an Acid	Hydrogen Sulfite as a Base
$HSO_3^- + H_2O \rightleftharpoons H_3O^+ + SO_3^{2-}$	$HSO_3^- + H_2O \rightleftharpoons OH^- + H_2SO_3$
$K_a = K_{a2} = 6.3 \times 10^{-8}$	$K_b = K_w/K_{a1} = 7.1 \times 10^{-13}$

Understanding

Calculate the equilibrium constants for a solution that is 0.10 *M* sodium hydrogen ascorbate, and determine whether the solution is acidic or basic.

Answer $K_a = 2.0 \times 10^{-12}$, $K_b = 1.1 \times 10^{-10}$; slightly basic

To estimate the pH of an amphoteric substance, compare K_a and K_b. If K_a is larger than K_b, the solution will be acidic, and vice versa.

OBJECTIVES REVIEW *Can you:*

- ☑ write chemical equations and expressions for the equilibrium constants for the dissociation of polyprotic acids?
- ☑ calculate the pH of solutions that contain polyprotic acids?
- ☑ estimate the pH of solutions of amphoteric species?

16.8 Factors That Influence Solubility

OBJECTIVES

- ☐ Determine how pH influences the solubility of precipitates
- ☐ Determine the effect of complex formation on the solubility of a precipitate

The solubility of a salt is influenced by the acid-base properties of the anions and cations from which it is composed. If the anion or cation reacts with hydrogen or hydroxide ion, then the solubility is influenced by pH. These effects allow chemists to control reactions, adjusting the conditions to produce the largest possible amounts of products.

The two important equilibria, the solubility equilibrium and the acid-base reaction, occur at the same time, so we refer to them as *simultaneous equilibria*. We do not cover the exact computation of simultaneous equilibria in this textbook, but computations are not necessary to understand the nature of the effect of pH on solubility.

Salts of Anions of Weak Acids

When a salt contains an anion of a weak acid, then the acid-base properties of the anion influence the solubility. We can consider the effect of pH on the solubility of barium fluoride, BaF_2. The fluoride anion is the conjugate base of the weak acid HF. When trying to predict the solubility of BaF_2 in solution, we need to consider two steps: Equation 16.3 shows the reaction for the solubility of the salt, and Equation 16.4 shows the reaction of the product ions with hydrogen ion.

$$BaF_2(s) \rightleftharpoons Ba^{2+}(aq) + 2F^-(aq) \qquad [16.3]$$

The $Ba^{2+}(aq)$ ion does not react with H_3O^+, but the $F^-(aq)$ ion does.

$$F^-(aq) + H_3O^+ \rightleftharpoons HF(aq) + H_2O(aq) \qquad [16.4]$$

We can evaluate the influence of pH qualitatively from Le Chatelier's principle. High concentrations of hydrogen ion (low pH) force the equilibrium (shown in Equation 16.4) to the right, decreasing the concentration of the fluoride ion. Additional barium fluoride will dissolve to replace the fluoride ion consumed by reaction with the acid, so the solubility of barium fluoride will increase when acid is added.

Increasing the acidity of a solution increases the solubility of any salt whose anion is a weak base.

Even in pure water, when barium fluoride dissolves, some of the fluoride ions react with hydrogen ions to form HF(aq). Each time a fluoride ion is converted to a molecule of HF, some additional BaF_2 dissociates to keep the $[Ba^{2+}][F^-]^2$ ion product equal to K_{sp}. Experiments show that *salts in which one of the ions reacts with hydrogen ion or hydroxide ion are generally more soluble than predicted by simple* K_{sp} *calculations.*

EXAMPLE 16.12 Solubility of Salts

Predict the effect of adding nitric acid to

(a) the solubility of calcium fluoride.
(b) the solubility of silver chloride.

Strategy Write the expression for the reaction that occurs as the solid dissolves, and determine whether any of the species might react with nitric acid.

Solution

(a) $CaF_2(s) \rightleftharpoons Ca^{2+} + 2F^-$

The fluoride ion will react with the added H_3O^+ from the nitric acid to form HF.

$$F^- + H_3O^+ \rightleftharpoons HF + H_2O$$

Because the F^- concentration is reduced by the second reaction, the principle of Le Chatelier tells us that the system will react to produce more; thus, more CaF_2 will dissolve.

(b) $AgCl(s) \rightleftharpoons Ag^+ + Cl^-$

Neither Ag^+ nor Cl^- will react with nitric acid (H_3O^+ and NO_3^-), so there is no effect. The potential products are HCl, a strong acid, and $AgNO_3$; both dissociate completely.

Understanding

Predict the effect of adding hydrochloric acid on

(a) the solubility of calcium nitrate.
(b) the solubility of calcium hydroxide.

Answer (a) $Ca(NO_3)_2(s) \rightarrow Ca^{2+} + 2NO_3^-$. The product ions do not react with H_3O^+ or Cl^-; therefore, the solubility is unchanged.

(b) $Ca(OH)_2 \rightarrow Ca^{2+} + 2OH^-$. The added H_3O^+ will react with the hydroxide ion; therefore, the system will react to produce more hydroxide ion. The solubility of calcium hydroxide will increase.

Salts of Transition-Metal Cations

The chemical nature of the cation also influences the solubility of the precipitate. If the cation is a transition-metal ion, it can react with substances that donate electrons. We call the species that donates electrons a **ligand;** the resulting species is a **complex** (or complex ion, if it is charged). Many common bases, such as ammonia, water, and hydroxide ions, can act as ligands and form complexes; in fact, the formation of a complex is a Lewis acid-base reaction. (Ligands are electron-pair donors, so they are Lewis bases.)

Additional discussion of complex formation is deferred to Chapter 19, but the influence of complex formation on solubility can be described by Le Chatelier's principle. When the cation forms a complex, the concentration of the free (uncomplexed) metal ion decreases, and additional precipitate dissolves until the ion product is again equal to K_{sp}. Consider the reaction of silver chloride with ammonia to form a complex ion.

$$AgCl(s) \rightleftharpoons Ag^+(aq) + Cl^-(aq)$$

$$Ag^+(aq) + 2NH_3(aq) \rightleftharpoons Ag(NH_3)_2^+(aq)$$

The solubility of a precipitate increases if a complex forms by reaction with another species in the solution.

The equilibrium constant for the formation of the complex ion, called the diamminesilver(I) ion, is about 10^8, so the concentration of $Ag^+(aq)$ is much less in the presence of ammonia. The decrease in the concentration of the silver ion causes more of the AgCl(s) to dissolve, in accordance with Le Chatelier's principle.

Complex formation plays an important role in many chemical analyses and syntheses. One important example is the qualitative test to determine whether there is silver in solution. The test is a two-step process in which the first step is to add some chloride to an unknown sample to precipitate the silver.

$$Ag^+(aq) + Cl^-(aq) \rightarrow AgCl(s)$$

(a)

(b)

Adding ammonia to silver chloride. *(a)* Silver chloride forms as solutions containing chloride and silver ions mix. *(b)* Adding ammonia dissolves the AgCl(s). After sufficient ammonia is added, the solution is clear.

If silver is present, we observe the white silver chloride. This result is not unique to silver—mercury, lead, and thallium also form insoluble white chlorides. To determine whether the precipitate is silver chloride and not one of the others requires a second step in which we add ammonia. If the white precipitate is silver chloride, the diamminesilver complex ion forms, and the precipitate dissolves.

$$AgCl(s) + 2NH_3(aq) \rightleftharpoons Ag(NH_3)_2^+(aq) + Cl^-(aq)$$

Other metals, including Cu^{2+}, Zn^{2+}, and Ni^{2+}, also form complexes with ammonia, but they do not form insoluble chlorides, so the test still works in their presence. Only silver forms an insoluble white chloride that dissolves when you add ammonia.

Ammonia forms complexes with many transition metals. Photograph shows the aqueous solutions and ammonia complexes of nickel(II) and copper(II) ions.

Solubility of Amphoteric Species

Many amphoteric species are oxides and hydroxides that react with both acids and bases. These materials form complex ions with several (often four) hydroxide ions. Aluminum hydroxide is a typical amphoteric substance. Solid aluminum hydroxide dissolves in strongly basic solutions as well as in acids.

$$Al(OH)_3(s) + OH^-(aq) \rightleftharpoons Al(OH)_4^-(aq)$$

$$Al(OH)_3(s) + 3H_3O^+(aq) \rightleftharpoons Al^{3+}(aq) + 6H_2O$$

The amphoteric nature of many species provides us with a method of separating substances from mixtures. For example, iron and calcium can be separated from an aluminum ore as part of a purification process. The ore can be added to a strongly alkaline solution. The aluminum goes into solution as the $Al(OH)_4^-$ species, but the iron and calcium form insoluble hydroxides that do not dissolve in basic solution.

OBJECTIVES REVIEW *Can you:*

- ☑ determine how pH influences the solubility of precipitates?
- ☑ determine the effect of complex formation on the solubility of a precipitate?

CASE STUDY Acid-Base Chemistry and Titrations Help Solve a Mystery

Please reread the introduction to this chapter for the background on the discovery of a vial of clear liquid. In summary, a small amount of a clear liquid likely from the Civil War was discovered, and a chemist was asked to identify the liquid.

The vial was received packed in natural cotton inside a glass bottle that had a corked top. The cotton was surprisingly different from the "cotton" balls people buy at the drugstore, which are often polyester. The old cotton had seeds and dirt, and had turned yellow with age.

To open the vial, scientists chilled the contents, and a skilled glassblower snapped off the tip. The contents were placed in a small glass container with a screw-cap closure. During the transfer of the contents, the unknown liquid was observed to be quite viscous, flowing like oil rather than water. The glass vial was cleaned carefully, dried, and then some observations and measurements were made on the glass. It was a lead-based glass ("lead crystal"), which was typical of the Civil War period.

Chemists do not have instruments that automatically determine the composition of unknown samples—testing these samples requires some knowledge of chemistry. The first test was to determine whether it would dissolve in water. If it did not dissolve, the scientist was prepared to tentatively classify it as a molecular compound and to use appropriate techniques for its identification.

Surprisingly, it dissolved in water, and the chemist observed that the solution got very hot. Not many liquids get hot when added to water, but the scientist knew from first-hand experience that sulfuric acid had a large exothermic heat of solution (Section 12.3). A test with litmus paper indicated that the solution was acidic. A follow-up test with a pH meter found the solution to be extremely acidic, with a pH that was less than 1.

Qualitative Tests and Observations
- Soluble in water
- Heat of solution exothermic
- Strongly acidic
- Forms precipitate with barium chloride
- Tests negative for phosphate

The chemist added a drop of barium chloride to the solution, and a white precipitate was observed. If you review the solubility rules in Chapter 4, you will see that barium sulfate, carbonate, and phosphate are insoluble; thus, you can tentatively identify the unknown solution as sulfuric acid, carbonic acid, or phosphoric acid. The scientist quickly ruled out carbonic acid because carbonic acid is quite weak and its pH would not be as low as observed. To distinguish between H_2SO_4 and H_3PO_4, the scientist performed a chemical test specific for phosphate ion, and the results were negative.

The chemist had a good qualitative identification of the substance in the vial as sulfuric acid, but its concentration was still a mystery. More than one way exists to measure concentration—the first way that occurs to most students is to calculate the concentration from the pH. Unfortunately, there are some serious problems with this method. Measuring the pH should not be used for concentrated solutions of acids or bases because the pH meter is known to be inaccurate at the extremes of its range. Even if the pH were accurate to within ±0.05 pH unit, that uncertainty is fairly substantial. We can quantify the uncertainty with a calculation.

If the true value of the pH of a solution is 0.50, then the range is 0.45 to 0.55. Let us calculate the H_3O^+ ion concentration for each.

Error in Concentration Because of Error in pH

	pH	$[H_3O^+]$	Error in $[H_3O^+]$
Low	0.45	0.35	+9.4%
Correct	0.50	0.32	—
High	0.55	0.28	−13%

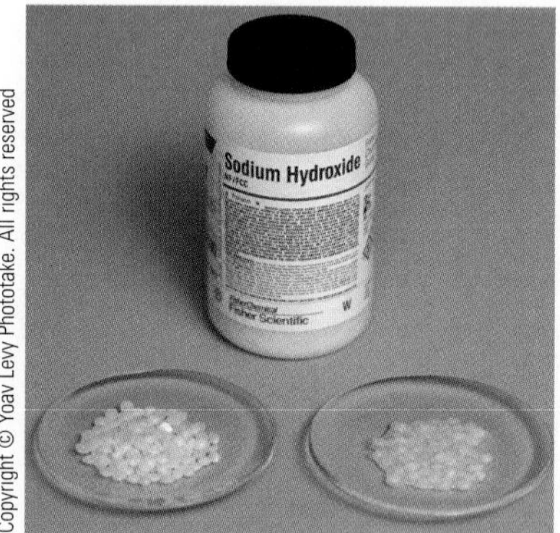

Sodium hydroxide. The dish on the left shows fresh NaOH pellets. The dish on the rights shows pellets that have been exposed to air for about one minute.

The 9–13% error is actually quite optimistic, considering that this is just one of several different errors associated with using a pH meter with a concentrated solution.

The chemist knew that the best method to determine the concentration of sulfuric acid was to titrate it with standard sodium hydroxide. Although making a standard sodium hydroxide solution sounds simple, the chemist needed to consider that sodium hydroxide solid reacts with the CO_2 and water in air, thus changing its concentration. Fortunately, procedures have been published that avoid these problems. The chemist followed these procedures and determined that the concentration of the standard NaOH solution was 0.111 *M*.

The chemist took exactly 0.100 mL of the unknown acid and added it to some water in an Erlenmeyer flask together with a drop of phenolphthalein indicator. He filled the burette with 0.111 *M* NaOH. Placing the acid in the flask and base in the burette was a conscious decision because he knew that the NaOH could react with atmospheric CO_2 if the NaOH were in the Erlenmeyer flask, it would be constantly swirled and aerated, increasing the error because of CO_2 absorption.

Digital micropipet. This picture shows a digital micropipet that can be programmed to deliver volumes between 0.010 and 1.000 mL (10-1000 µL) with a precision of ±0.001 mL. The entire solution is contained in a disposable tip, so even reactive solutions like sulfuric acid can be measured without worry that they will contact the dispensing mechanism and ruin it.

It took 31.20 mL NaOH to neutralize the sulfuric acid from the vial. To calculate the molarity of the sulfuric acid, you must first calculate the amount of NaOH used.

$$\text{Amount NaOH} = 31.20 \text{ mL} \times 0.111\ M = 3.46 \text{ mmol}$$

Because 2 mol NaOH is needed to neutralize 1 mol sulfuric acid,

$$H_2SO_4(aq) + 2NaOH(aq) \rightarrow Na_2SO_4(aq) + 2\ H_2O(\ell)$$

$$\text{Amount } H_2SO_4 = 3.46 \text{ mmol NaOH} \times \left(\frac{1 \text{ mmol } H_2SO_4}{2 \text{ mmol NaOH}}\right) = 1.73 \text{ mmol } H_2SO_4$$

We started the titration with 0.100 mL of the sulfuric acid from the vial, so the concentration of the sulfuric acid is

$$\text{Concentration} = \frac{1.73 \text{ mmol } H_2SO_4}{0.100 \text{ mL}} = 17.3\ M$$

The scientist repeated the experiment two more times and recorded the average value for the molarity of the liquid in the vial as 17.3 *M* for the sulfuric acid in the vial.

The chemist used some contemporary high-purity sulfuric acid to determine its molarity. The modern sulfuric acid was determined to be 17.9 *M*, so the chemist concluded that the liquid in the vial was relatively high-purity concentrated sulfuric acid. The historians used these results to form the conclusions described in the introduction to this chapter.

Questions

1. Had the solution been identified as nitric acid, calculate its concentration from the titration data.
2. Is this concentration plausible for nitric acid? The Internet has a number of sources of information, but a good place to start is Appendix F.
3. Consider the following hypothesis: The content of the vial is known to be one of these five compounds: sulfuric acid, hydrochloric acid, ether, vegetable oil, or acetic acid. How would each of these solutions behave when subjected to the qualitative tests described in the Case Study?

ETHICS IN CHEMISTRY

You are the Quality Assurance manager of a company that produces dried pasta for sale. Since 1988, your company has added folic acid to the pasta, in accordance with U.S. Food and Drug Administration (FDA) requirements. Folic acid is a synthetic form of vitamin B, and the FDA mandates a folic acid concentration between 0.43 and 1.4 μg folic acid per pound of pasta. Clinical results associate folic acid at these levels with a substantial decline of a particular birth defect (a neural tube defect) in infants. The chemist repeated the titration of a sample of dried pasta a total of four times with the following results:

Trial	Result (μg folic acid per lb of pasta)
1	0.33
2	0.44
3	0.46
4	0.45

The chemist looked at the first result and rejected it, stating it was an obvious outlier and the first result of a titration is "always wrong." She averaged the last three titrations to determine that the average amount of folic acid in the pasta was 0.45 μg/lb of pasta. Was the chemist correct in rejecting the first result? Explain. (Searching the Internet for information on "outlier rejection" may help in answering this question.)

Chapter 16 Visual Summary

The chart shows the connections between the major topics discussed in this chapter.

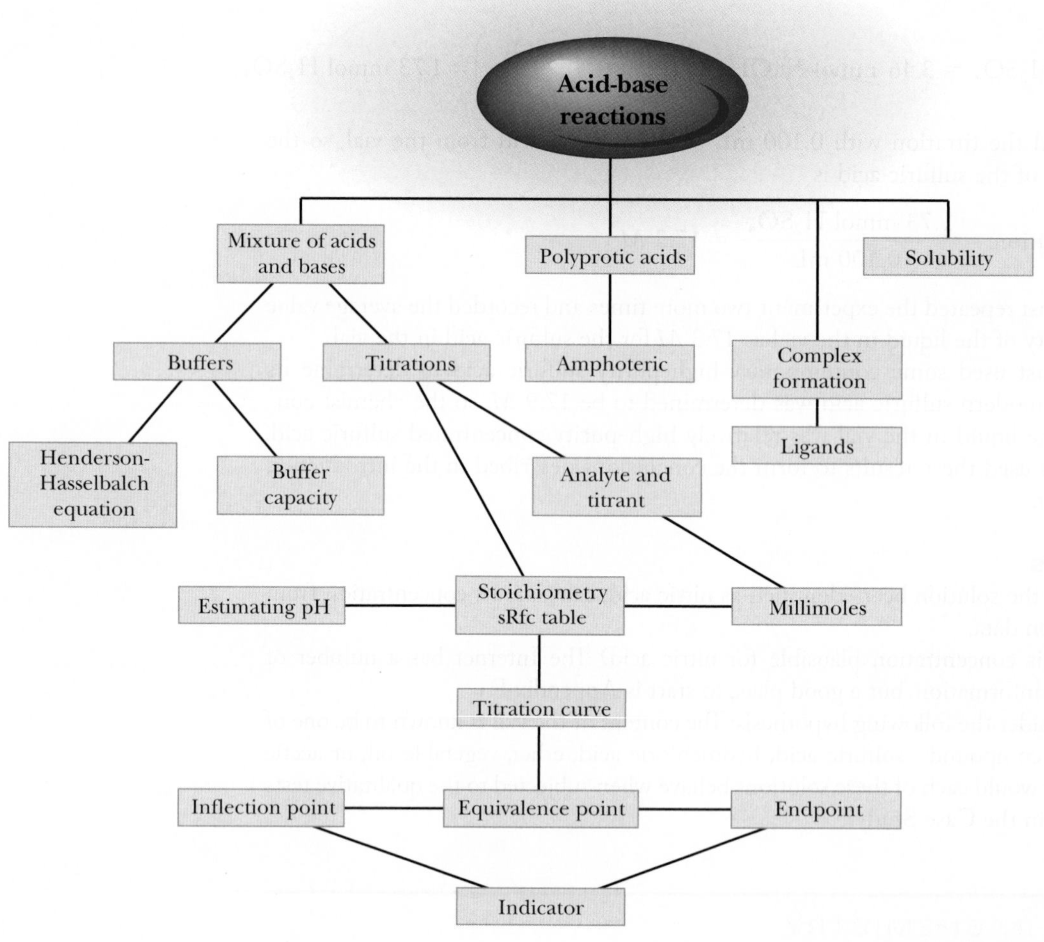

Summary

16.1 Titrations of Strong Acids and Bases

A *titration* is the stepwise addition of a reactant, generally in solution, that reacts with the *analyte* in the sample to determine the quantity of analyte present. A common titration is the neutralization of a strong acid with a strong base, and vice versa.

16.2 Titration Curves of Strong Acids and Bases

The *titration curve* is a graph of the pH during the course of a titration. The shape of the titration curve of a strong acid and a strong base can be estimated from some basics: The pH of the solution is 7 at the equivalence point, quite low before the equivalence point, and high afterward.

The concentrations of the species in the solution can be determined at any point during the titration by stoichiometry calculations. These calculations (the sRfc table) are used to determine the concentrations of the species that are present after the titration reaction goes to completion. The pH can be estimated by determining whether the acid or base is in excess, or whether equivalent amounts of both are present.

16.3 Buffers

A chemical system that contains a weak acid and its conjugate base resists changes in pH; such a solution is called a *buffer.* The acid reacts to consume added hydroxide ion; its conjugate base reacts with added hydrogen ion. The pH of a buffer is generally calculated by a relationship that chemists named Henderson and Hasselbalch first described:

$$\text{pH} = \text{p}K_a + \log\frac{C_b}{C_a}$$

where C_b and C_a are the analytical concentrations of the base and acid conjugate pairs.

16.4 Titrations of Weak Acids and Bases: Qualitative Aspects

The titration curve for a weak acid can also be estimated quickly. The curve is divided into four regions: an initial point, at which the system is a weak acid; a buffer region; the equivalence point, at which the system is a weak base; and strong base after the equivalence point.

16.5 Titrations of Weak Acids and Bases: Quantitative Aspects

Computing the pH during the course of the titration of a weak acid with a strong base requires two calculations. First, the *analyte* and *titrant* are assumed to react completely. The results of a stoichiometry calculation (the sRfc table) are used to determine the nature of the system (strong or weak acid or base, buffer, or water). Second, the pH of the resulting solution is calculated as for any other solution of weak acids, bases, or buffers. The pH at the equivalence point is greater than 7 for the titration of a weak acid with strong base and depends on the value of the ionization constant of the acid.

16.6 Indicators

An *indicator* is often added to the titration flask so that a color change occurs when all of the analyte has been consumed. The point at which the indicator changes color is called the *endpoint*. If the indicator is chosen to change color at the pH of the *equivalence point,* then the *endpoint* and the *equivalence point* coincide. If the indicator changes color at a pH that is substantially different from the equivalence-point pH, errors may result. Because the titrant is generally a strong acid or strong base, the pH changes quite sharply just beyond the equivalence point, and it is logical to select an indicator that changes color in this pH region to minimize errors.

16.7 Polyprotic Acid Solutions

Systems of *polyprotic acids* can sometimes be treated like monoprotic acids. If the first ionization constant is much larger than the second, then the second can be ignored. If the ionization constants are similar in magnitude, then this approach fails and simultaneous equilibria must be considered.

Solutions of *amphoteric* species can be evaluated qualitatively by examination of the magnitudes of the acid and base dissociation constants. If the acid ionization constant, K_a, is larger than K_b for the same ion, the solution is acidic and vice versa.

16.8 Factors That Influence Solubility

The solubilities of many compounds depend on the pH of the solution in which they dissolve. If the anion in a salt is the conjugate base of a weak acid, its solubility increases in acidic solutions. Consider the solubility of MA:

$$MA(s) \rightleftharpoons M^+(aq) + A^-(aq)$$

If the pH is adjusted so that A^- reacts to form HA, the solubility of the salt increases, because more solid dissolves to compensate for the decrease in anion concentration as the weak acid forms. Similarly, if the cation can form a *complex,* the solubility of the solid increases as the complex forms.

Download Go Chemistry concept review videos from OWL or purchase them from **www.ichapters.com**

Chapter Terms

The following terms are defined in the Glossary, Appendix I.

Section 16.1
Analyte
Millimole
Titrant
Titration
Titration curve

Section 16.2
Inflection point

Section 16.3
Buffer
Buffer capacity
Henderson–Hasselbalch equation

Section 16.6
Indicator

Section 16.7
Amphoteric
Polyprotic acid

Section 16.8
Complex
Ligand

Key Equations

$$pH = pK_a + \log\frac{C_b}{C_a} \quad (16.3)$$

$$[H_3O^+] = K_a\frac{C_a}{C_b} \quad (16.3)$$

$$pH = pK_a + \log\frac{n_b}{n_a} \quad (16.3)$$

Questions and Exercises

OWL Selected end of chapter Questions and Exercises may be assigned in OWL.

Blue-numbered Questions and Exercises are answered in Appendix J; questions are qualitative, are often conceptual, and include problem-solving skills.

- ■ Questions assignable in OWL
- ✎ Questions suitable for brief writing exercises
- ▲ More challenging questions

Questions

16.1 Define titration, analyte, and titrant.

16.2 ✎ A high-school student needs to standardize some hydrochloric acid for a project. The concentration is about 0.2 *M*. The student has several hundred milliliters of standard base (0.100 *M*), a 50-mL burette, and some phenolphthalein. Write a set of instructions that the student can use to standardize the HCl solution.

16.3 Sketch a titration curve for the titration of potassium hydroxide with HCl, both 0.100 *M*. Identify three regions in which a particular chemical species or system dominates the acid-base equilibria.

16.4 ✎ Describe the shape of a titration curve. Sketch, without calculation, a titration curve for 0.1 *M* weak acid with strong base. Label the points or regions at which the equilibrium is a strong acid or base, a weak acid or base, and a buffer.

16.5 Examine each of the following solutions and decide whether it is a buffer system. Justify your answer.
(a) 0.100 *M* ammonia and 0.100 *M* ammonium nitrate
(b) 0.100 *M* ammonia and 0.100 *M* acetic acid
(c) 0.100 *M* acetic acid and 0.100 *M* ammonium nitrate

16.6 Explain why the Henderson–Hasselbalch equation fails to predict the pH of dilute buffer solutions.

16.7 ✎ You work for a company that purchases buffers in dry form, in small envelopes similar to those in which sugar is packaged. Laboratory technicians make buffers by adding one package to 500 mL distilled water. You are asked to design an experimental plan to test the effectiveness of these buffers. Write a report describing your experiment(s) and proposing actions to be taken should your experiments determine that the effectiveness is too low or too high for the intended application within your company.

16.8 Sketch a titration curve for the titration of acetic acid (CH_3COOH) with sodium hydroxide (NaOH), both 0.100 *M*. Identify four regions in which a particular chemical system dominates the acid-base equilibria.

16.9 ✎ Preparing solutions of most acid-base indicators is more complicated than just dissolving some of the powder in water. Typical instructions may be to dissolve about 0.1 g of indicator in 50 mL alcohol; then add distilled water so that the total volume is about 100 mL. (More specific instructions are found in many standard reference books.) Explain why the mass of the indicator and the amounts of alcohol and water are important. Also explain why alcohol must be used. Why can't pure distilled water be substituted?

16.10 Several years ago, a chemistry student decided to check the color chart that came with bromthymol blue, the indicator pictured in Section 16.6, to make sure that it was accurate enough for measuring the pH in an expensive aquarium. The student put the sensor of a pH meter, known to be accurate, in a small beaker of water and measured the pH at 6.50. He added some bromthymol blue, and the color showed that the solution was fairly acidic, with a pH less than 5. When he rechecked the pH of the solution with the pH meter, it read 4.90. Explain this odd behavior. Does this experiment rule out using bromthymol blue as an indicator in fish tanks?

16.11 The polyprotic phosphoric acid is found in many soft drinks. Is the polyprotic nature of the acid important in this application? Why do you think manufacturers have chosen phosphoric acid in preference to other acids?

Courtesy of M. Stading

Soft drinks contain phosphoric acid.

16.12 List five insoluble compounds that are more soluble in acidic solution than in neutral solution. List five compounds that are not influenced by the acidity of the solution.

Exercises

OBJECTIVE Calculate the volume of titrant needed to reach the equivalence point.

16.13 Calculate how much 0.100 *M* HCl is needed to react completely with
(a) 10.0 mL of 0.150 *M* KOH.
(b) 250.0 mL of 0.00520 *M* $Ba(OH)_2$.
(c) 100.0 mL of 0.100 *M* ammonia.

16.14 Calculate how much 0.100 *M* NaOH is needed to react completely with
(a) 45.00 mL of 0.0500 *M* HCl.
(b) 5.00 mL of 0.350 *M* H_2SO_4 (forming Na_2SO_4).
(c) 10.00 mL of 0.100 *M* acetic acid.

OBJECTIVE Express amounts of analyte and titrant in units of millimoles.

16.15 How many millimoles of KOH are needed to neutralize completely 35.1 mL of 0.101 *M* nitric acid?

16.16 ■ How many millimoles of HCl are needed to neutralize completely 50.0 mL of 0.0233 *M* sodium hydroxide?

16.17 Write the chemical equation for the reaction of H_2SO_4 with lithium hydroxide, forming Li_2SO_4. Calculate the number of millimoles of H_2SO_4 needed to react completely with 25.0 mL of 0.244 *M* LiOH, forming Li_2SO_4.

16.18 Write the chemical equation for the neutralization of H_2SO_4 with $La(OH)_3$, forming $La_2(SO_4)_3$. Calculate the number of millimoles of H_2SO_4 needed to neutralize 0.8457 g $La(OH)_3$(s), forming $La_2(SO_4)_3$.

OBJECTIVES Calculate the concentrations of all species present during the titration of a strong acid with a strong base. Graph the titration curve.

16.19 Calculate the pH during the titration of 100.0 mL of 0.200 *M* HCl with 0.400 *M* NaOH after 0, 25.00, 50.00, and 75.00 mL NaOH have been added. Sketch the titration curve.

16.20 ■ Calculate the pH during the titration of 50.00 mL of 0.250 *M* HNO_3 with 0.500 *M* KOH after 0, 12.50, 25.00, and 40.00 mL KOH have been added. Sketch the titration curve.

16.21 Calculate the pH during the titration of 1.00 mL of 0.240 *M* LiOH with 0.200 *M* HNO_3 after 0, 0.25, 0.50, 1.20, and 1.50 mL nitric acid have been added. Sketch the titration curve.

16.22 Calculate the pH during the titration of 50.00 mL of 0.100 *M* NaOH with 0.100 *M* HNO_3 after 0, 25.00, 50.00, and 75.00 mL nitric acid have been added. Sketch the titration curve.

OBJECTIVE Correlate the shape of the titration curve to the titration stoichiometry.

16.23 Calculate the pH during the titration of 1.00 mL of 0.240 *M* $Ba(OH)_2$ with 0.200 *M* HNO_3 after 0, 0.50, 1.00, 2.40, and 3.00 mL nitric acid have been added. Graph the titration curve and compare with the curve obtained in Exercise 16.21.

16.24 Calculate the pH during the titration of 50.00 mL of 0.100 *M* $Sr(OH)_2$ with 0.100 *M* HNO_3 after 0, 50.00, 100.00, and 150.00 mL nitric acid have been added. Graph the titration curve and compare with the titration curve obtained in Exercise 16.22.

OBJECTIVE Estimate the pH of mixtures of strong acids and bases.

16.25 Estimate the pH that results when the following two solutions are mixed.
(a) 50 mL of 0.1 *M* HCl and 50 mL of 0.2 *M* NaOH
(b) 100 mL of 0.1 *M* HCl and 50 mL of 0.2 *M* NaOH
(c) 150 mL of 0.1 *M* HCl and 50 mL of 0.2 *M* $Ba(OH)_2$
(d) 200 mL of 0.1 *M* HCl and 50 mL of 0.2 *M* $Ba(OH)_2$

16.26 Estimate the pH that results when the following two solutions are mixed.
(a) 50 mL of 0.2 *M* HNO_3 and 50 mL of 0.1 *M* KOH
(b) 100 mL of 0.2 *M* HNO_3 and 150 mL of 0.2 *M* KOH
(c) 150 mL of 0.2 *M* HNO_3 and 150 mL of 0.2 *M* KOH
(d) 200 mL of 0.2 *M* HNO_3 and 200 mL of 0.2 *M* $Sr(OH)_2$

16.27 Estimate the pH that results when the following two solutions are mixed.
(a) 50 mL of 0.3 *M* $HClO_4$ and 50 mL of 0.4 *M* KOH
(b) 100 mL of 0.3 *M* $HClO_4$ and 50 mL of 0.4 *M* NaOH
(c) 150 mL of 0.3 *M* $HClO_4$ and 100 mL of 0.3 *M* $Ba(OH)_2$
(d) 200 mL of 0.3 *M* $HClO_4$ and 100 mL of 0.3 *M* $Ba(OH)_2$

16.28 Estimate the pH that results when the following two solutions are mixed.
(a) 50 mL of 0.4 *M* HBr and 50 mL 0.2 *M* NaOH
(b) 100 mL of 0.4 *M* HBr and 50 mL 0.2 *M* NaOH
(c) 150 mL of 0.4 *M* HBr and 100 mL 0.4 *M* $Ba(OH)_2$
(d) 200 mL of 0.4 *M* HBr and 100 mL 0.4 *M* $Ba(OH)_2$

OBJECTIVE Calculate the pH of a buffer solution from the concentrations of the weak acid and its conjugate base.

16.29 Calculate the pH of solutions that are
(a) 0.25 *M* formic acid and 0.40 *M* sodium formate.
(b) 0.50 *M* benzoic acid and 0.15 *M* sodium benzoate.

16.30 ■ Calculate the pH of solutions that are
(a) 0.20 *M* acetic acid and 0.50 *M* sodium acetate.
(b) 0.25 *M* hydrofluoric acid and 0.10 *M* potassium fluoride.
(c) 0.0250 mol sodium nitrite, $NaNO_2$, in 250.0 mL of 0.0410 *M* nitrous acid, HNO_2.

16.31 Calculate the pH that results when the following solutions are mixed.
(a) 25.0 mL of 0.250 *M* HF and 20.0 mL of 0.360 *M* NaF
(b) 20.0 mL of 0.144 *M* NH_3 and 10.0 mL of 0.152 *M* NH_4Cl

16.32 Calculate the pH that results when the following solutions are mixed.
(a) 10.00 mL of 0.500 *M* sodium acetate and 20.00 mL of 0.350 *M* acetic acid
(b) 350.0 mL of 0.150 *M* pyridinium chloride and 650.0 mL of 0.450 *M* pyridine

16.33 ▲ A buffer is made by dissolving 0.0500 mol potassium acetate and 0.0500 mol acetic acid in some water and adding water until the volume is a little less than a liter. The pH is adjusted to 5.00 by adding small amounts of concentrated acid (HCl) and base (NaOH) as needed, and the solution is then diluted to exactly 1.00 L. What are the equilibrium molar concentrations of acetic acid and acetate ion in the solution?

16.34 ▲ Saccharin is an artificial sweetener that is also a weak acid. It has the formula $C_7H_5NSO_3$, and its pK_a is 11.68. A 12-oz (350-mL) can of diet cola contains 3.0 mg saccharin and has a pH of 4.50. What are the equilibrium molar concentrations of saccharin and the saccharide ion?

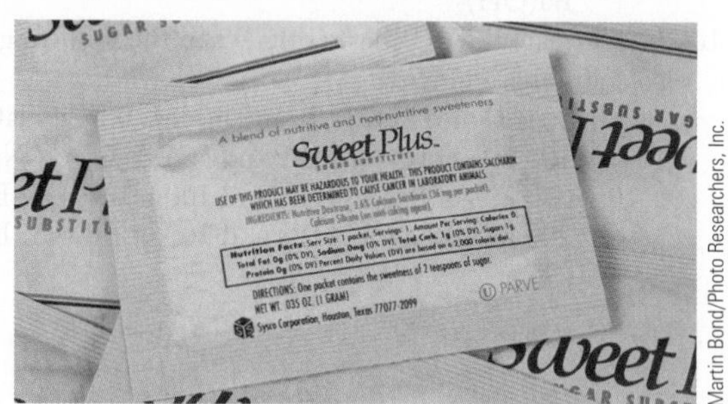

Saccharin. This sweetener contains saccharin, a weak acid. Saccarin is a sugar substitute for people who must restrict their intake of sugar.

OBJECTIVE Calculate the pH of a solution from the amounts of acids and bases.

16.35 Calculate the pH of a solution made by
(a) adding 10.0 g sodium benzoate to 3.00 g benzoic acid and dissolving in water to make 1.00 L of solution.
(b) adding 25.0 g sodium acetate to 9.0 g acetic acid and dissolving in water to make 500 mL of solution.

16.36 ■ Calculate the pH of a solution made by
(a) adding 12.5 g sodium nitrite to 6.0 g nitrous acid and adding water until the final volume is 1.00 L.
(b) adding 45.0 g formic acid to 20.0 g sodium formate and adding enough water to make 1.00 L of buffer.
(c) adding 15.00 g sodium acetate and 12.50 g acetic acid to enough water to make 0.500 L of solution.

16.37 Calculate the pH of a solution made by
(a) adding 15.45 g potassium fluoride to 100.0 mL of 0.850 *M* HF.
(b) adding 45.00 g ammonium chloride to 250.0 mL of 0.455 *M* ammonia.

16.38 Calculate the pH of a solution made by
(a) adding 30.0 g sodium formate to 300 mL of 0.30 *M* formic acid.
(b) adding 30.0 g sodium acetate to 300 mL of 0.30 *M* acetic acid.

16.39 Calculate the pH that results when the following solutions are mixed.
(a) 100.0 mL of 0.800 *M* formic acid and 200.0 mL of 0.100 *M* sodium formate
(b) 300.0 mL of 0.350 *M* ammonia and 200.0 mL of 0.150 *M* ammonium chloride

16.40 How many grams of sodium acetate must be added to 400.0 mL of 0.500 *M* acetic acid to prepare a pH 4.35 buffer?

16.41 What volume of 0.500 *M* HF must be added to 750 mL of 0.200 *M* sodium fluoride to prepare a buffer of pH 3.95?

16.42 ▲ How many grams of ammonium chloride must be added to 500 mL of 0.137 *M* ammonia to prepare a pH 9.80 buffer?

OBJECTIVE Determine the change in the pH when a strong acid or base is added to a buffer.

16.43 ■ A buffer solution that is 0.100 *M* acetate ion and 0.100 *M* acetic acid is prepared.
(a) Calculate the initial pH, final pH, and change in pH when 1.00 mL of 1.00 *M* NaOH is added to 100.0 mL of the buffer.
(b) Calculate the initial pH, final pH, and change in pH when 1.00 mL of 1.00 *M* NaOH is added to 100.0 mL pure (pH 7.00) water.

16.44 A buffer solution that is 0.100 *M* acetate and 0.200 *M* acetic acid is prepared.

(a) Calculate the initial pH, final pH, and change in pH when 1.00 mL of 0.100 *M* HCl is added to 100.0 mL of the buffer.

(b) Calculate the initial pH, final pH, and change in pH when 1.00 mL of 0.100 *M* HCl is added to 100.0 mL pure (pH 7.00) water.

16.45 ▲ Calculate the minimum concentrations of formic acid and sodium formate that are needed to prepare 500.0 mL of a pH 3.80 buffer whose pH will not change by more than 0.10 unit if 1.00 mL of 0.100 *M* strong acid or strong base is added.

16.46 ▲ Calculate the minimum concentrations of acetic acid and sodium acetate that are needed to prepare 100 mL of a pH 4.50 buffer whose pH will not change by more than 0.05 unit if 1.00 mL of 0.100 *M* strong acid or strong base is added.

OBJECTIVE Estimate the pH of mixtures of a weak acid and strong base or weak base and strong acid.

16.47 Estimate the pH that results when the following two solutions are mixed.

(a) 50 mL of 0.2 *M* HCOOH and 50 mL of 0.1 *M* KOH

(b) 100 mL of 0.2 *M* HCOOH and 150 mL of 0.2 *M* KOH

(c) 150 mL of 0.2 *M* HCOOH and 150 mL of 0.2 *M* KOH

(d) 200 mL of 0.2 *M* HCOOH and 200 mL of 0.2 *M* $Sr(OH)_2$

16.48 ■ Estimate the pH that results when the following two solutions are mixed.

(a) 50 mL of 0.1 *M* HF and 50 mL of 0.2 *M* NaOH

(b) 100 mL of 0.1 *M* HF and 50 mL of 0.2 *M* NaOH

(c) 150 mL of 0.1 *M* HF and 50 mL of 0.2 *M* $Ba(OH)_2$

(d) 200 mL of 0.1 *M* HF and 50 mL of 0.2 *M* $Ba(OH)_2$

16.49 Estimate the pH that results when the following two solutions are mixed.

(a) 50 mL of 0.3 *M* CH_3COOH and 50 mL of 0.4 *M* KOH

(b) 100 mL of 0.3 *M* CH_3COOH and 50 mL of 0.4 *M* NaOH

(c) 150 mL of 0.3 *M* CH_3COOH and 100 mL of 0.3 *M* $Ba(OH)_2$

(d) 200 mL of 0.3 *M* CH_3COOH and 100 mL of 0.3 *M* $Ba(OH)_2$

16.50 Estimate the pH that results when the following two solutions are mixed.

(a) 50 mL of 0.4 *M* HCl and 100 mL 0.2 *M* NH_3

(b) 100 mL of 0.4 *M* HCl and 100 mL 0.2 *M* NH_3

(c) 150 mL of 0.4 *M* HCl and 200 mL 0.2 *M* NH_3

(d) 200 mL of 0.4 *M* HCl and 100 mL 0.4 *M* NH_3

OBJECTIVE Calculate the pH in the titration of a weak acid with a strong base.

16.51 Calculate the pH during the titration of 25.00 mL of 0.400 *M* acetic acid with 0.500 *M* NaOH after 0, 10.00, 20.00, and 25.00 mL of base have been added. Sketch the titration curve, and label the four regions of importance.

16.52 ■ Calculate the pH during the titration of 30.00 mL of 0.150 *M* benzoic acid with 0.150 *M* NaOH after 0, 10.00, 30.00, and 40.00 mL of base have been added. Sketch the titration curve, and label the four regions of importance.

16.53 Sketch the curve for the titration of 100 mL of a 0.10 *M* weak acid ($K_a = 1.0 \times 10^{-4}$) with a 0.20 *M* strong base. On the same axes, sketch the titration curve for the same volume and concentration of HCl.

16.54 ▲ Calculate the pH during the titration of 10.00 mL of 0.400 *M* hypochlorous acid with 0.500 *M* KOH after the addition of 0%, 50%, 95%, 100%, and 105% of the base needed to reach the equivalence point. Graph the titration curve (pH vs. volume KOH), and label the four regions of importance.

OBJECTIVE Calculate the pH in the titration of a weak base with a strong acid.

16.55 Calculate the pH during the titration of 30.00 mL of 0.200 *M* pyridine with 0.200 *M* HCl after 0, 15.00, 30.00, and 40.00 mL acid have been added. Sketch the titration curve, and label the four regions of importance.

16.56 Calculate the pH in the titration of 50.00 mL of 0.100 *M* ammonia with 0.100 *M* HCl after 0, 25.00, 50.00, and 75.00 mL acid have been added. Sketch the titration curve, and label the four regions of importance.

16.57 Sketch a titration curve for the reaction of 50 mL of a 0.10 *M* weak base ($K_b = 1.0 \times 10^{-5}$) with 0.20 *M* strong acid. On the same axes, sketch the titration curve for the same volume and concentration of NaOH.

16.58 ▲ Calculate the pH during the titration of 100.0 mL of 0.230 *M* hydrofluoric acid with 0.500 *M* NaOH after the addition of 0%, 50%, 95%, 100%, and 105% of the base needed to reach the equivalence point. Graph the titration curve (pH vs. volume of NaOH), and label the four regions of importance.

OBJECTIVE Choose an indicator that is appropriate for a particular titration.

16.59 Choose an appropriate compound from Table 16.5 to serve as an indicator for the titration of a particular weak acid (in the flask) with base (in the buret), given that the pH at the equivalence point is

(a) 7.5 (b) 9.0 (c) 10.5

16.60 ■ Consider all acid-base indicators discussed in this chapter. Which of these indicators would be suitable for the titration of each of these?

(a) NaOH with $HClO_4$

(b) acetic acid with KOH

(c) NH_3 solution with HBr

(d) KOH with HNO_3

Explain your choices.

16.61 A chemist is developing a titration analysis for lactic acid. Lactic acid is a monoprotic acid with $K_a = 8.4 \times 10^{-4}$. Calculate the pH at the equivalence point of a titration of 100 mL of 0.100 *M* lactic acid with 0.500 *M* NaOH. Suggest an indicator from Table 16.4, and explain why you chose it.

16.62 Chloropropionic acid, $ClCH_2CH_2COOH$, is a weak monoprotic acid with $K_a = 7.94 \times 10^{-5}$. Calculate the pH at the equivalence point in a titration of 10.00 mL of 0.100 *M* chloropropionic acid with 0.100 *M* KOH. Choose an indicator from Table 16.4 for the titration. Explain your choice.

16.63 A 25.0-mL sample of 1.44 *M* NH_3 is titrated with 1.50 *M* HCl. Calculate the pH at the equivalence point. Choose an indicator from Table 16.4, and justify your choice.

16.64 Exactly 50 mL of a 0.0500 *M* solution of ethylamine, a base with $K_b = 1.1 \times 10^{-6}$, is titrated with 0.100 *M* HNO_3. What is the pH at the equivalence point? Suggest a good indicator from Table 16.4 for this titration, and justify your selection.

OBJECTIVE Write chemical equations and expressions for the equilibrium constants for the dissociation of polyprotic acids.

16.65 Write the chemical equilibria and expressions for the equilibrium constants for the ionizations of the following polyprotic acids.
(a) oxalic acid (b) sulfurous acid

Oxalic acid is a weak acid found in some plants. Oxalic acid can upset the stomach or in high enough concentrations cause serious illness. People with pets that nibble leaves are warned not to keep dieffenbachia plants around the house.

16.66 Write the chemical equilibrium and expression for the equilibrium constants for the ionization of
(a) tartaric acid. (b) malic acid.

OBJECTIVE Calculate the pH of solutions that contain polyprotic acids.

16.67 ▲ Calculate the pH of 0.010 *M* ascorbic acid.

16.68 ▲ Calculate the pH of 0.050 *M* phosphoric acid.

OBJECTIVE Estimate the pH of solutions of amphoteric species.

16.69 State whether each of the following solutions is acidic, basic, or neutral.
(a) sodium hydrogen oxalate
(b) potassium hydrogen malonate

16.70 State whether each of the following solutions is acidic, basic, or neutral.
(a) disodium hydrogen citrate
(b) potassium dihydrogen citrate

16.71 State whether each of the following solutions is acidic, basic, or neutral.
(a) potassium dihydrogen phosphate
(b) potassium hydrogen carbonate

16.72 State whether each of the following solutions is acidic, basic, or neutral.
(a) disodium hydrogen phosphate
(b) potassium hydrogen tartrate

OBJECTIVES Determine how pH influences the solubility of precipitates. Determine the effect of complex formation on the solubility of a precipitate.

16.73 Does adding the second compound increase, decrease, or have no effect on the solubility of the first compound?
(a) $Ca(CH_3COO)_2$ and HCl (b) MgF_2 and HCl

16.74 Does adding the second compound increase, decrease, or have no effect on the solubility of the first compound?
(a) AgCl and NH_3 (b) $PbCl_2$ and $Pb(NO_3)_2$

16.75 Does adding the second compound increase, decrease, or have no effect on the solubility of the first compound?
(a) aluminum hydroxide and NaOH
(b) magnesium phosphate and HNO_3

16.76 ■ Which compound in each pair is more soluble in water than is predicted by a calculation from K_{sp}?
(a) AgI or Ag_2CO_3
(b) $PbCO_3$ or $PbCl_2$
(c) AgCl or AgCN

Chapter Exercises

16.77 A pipet was used to measure 10.00 mL of a sulfuric acid solution into a titration flask. It took 31.77 mL of 0.102 *M* NaOH to neutralize the sulfuric acid completely. Calculate the concentration of the sulfuric acid solution. Assume that the reaction is

$$H_2SO_4(aq) + 2NaOH(aq) \rightarrow Na_2SO_4(aq) + 2H_2O(\ell)$$

16.78 ▲ You have a pH buffer made from 0.010 M acetic acid and 0.020 M sodium acetate. To buffer a biological reaction, you add 5.0 mL of this buffer to 1.0 L of a solution that contains the system of interest.
(a) Calculate the pH of the buffered biological system.
(b) You find that the concentration of sodium ion is too high for your experiment. Propose a possible solution.
(c) A worker suggests that the pH of the buffer may not be what you calculated. She says the Henderson–Hasselbalch equation fails under these conditions. Propose an experiment or a calculation to determine whether she is correct.

16.79 Calculate the pH of each of the following solutions.
(a) 1.00 mL of 0.150 M formic acid plus 2.00 mL of 0.100 M sodium hydroxide
(b) 25.00 mL of 0.250 M ammonia plus 5.00 mL of 0.100 M hydroiodic acid
(c) 5.00 mL of 0.200 M barium hydroxide plus 50.00 mL of 0.400 M hydrobromic acid

16.80 Calculate the pH of each of the following solutions.
(a) 10.0 mL of 0.300 M hydrofluoric acid plus 30.0 mL of 0.100 M sodium hydroxide
(b) 100.0 mL of 0.250 M ammonia plus 50.0 mL of 0.100 M hydrochloric acid
(c) 25.0 mL of 0.200 M sulfuric acid plus 50.0 mL of 0.400 M sodium hydroxide

16.81 Write the chemical equation and the expression for the equilibrium constant, and calculate K_b for the reaction of each of the following ions as a base.
(a) sulfate ion
(b) citrate ion

16.82 Calculate the concentration of hydroxide ion in the titration of 20.0 mL of 0.102 M NaOH with 0.207 M HCl after 0, 5, 10, 15, and 20 mL HCl are added. Graph the molar concentration of hydroxide ion (not pH or pOH) as a function of volume.

16.83 Write the chemical equation and the expression for the equilibrium constant, and calculate K_b for the reaction of each of the following ions as a base.
(a) malonate ion (b) carbonate ion

16.84 Phenolphthalein is a commonly used indicator that is colorless in the acidic form (pH < 8.3) and pink in the base form (pH > 10.0). It is a weak acid with a pK_a of 8.7. What fraction is in the acid form when the acid color is apparent? What fraction is in the base form when the base color is apparent?

16.85 The indicator methyl red is a weak acid with a pK_{In} of 5.00. Calculate the pH values at which the indicator will be 1%, 5%, 95%, and 99% in the acid form.

16.86 Use Table 16.5 as a source of data about methyl red. What fraction of the indicator is in the acid form when the acid color is observed? What fraction is in the base form when the base color is observed?

16.87 Determine the dominant acid-base equilibrium that results when each of the following pairs of solutions is mixed. Indicate the equilibrium by writing 1 for a strong acid, 3 for a weak acid, 4 for an acidic buffer, 7 for a neutral solution, 9 for a basic buffer, 10 for a weak base, and 13 for a strong base.
(a) 25.0 mL of 0.50 M NaOH + 10.0 mL of 2.00 M HCl
(b) 20.0 mL of 0.25 M HCOOH + 10.0 mL of 0.50 M KOH
(c) 100 mL of 0.20 M CH_3COOH + 50 mL of 0.10 M NaOH
(d) 25.0 mL of 0.15 M $Ba(OH)_2$ + 15.0 mL of 0.25 M H_2SO_4

16.88 ■ Determine the dominant acid-base equilibrium that results when each of the following pairs of solutions is mixed. Indicate the equilibrium by writing 1 for a strong acid, 3 for a weak acid, 4 for an acidic buffer, 7 for a neutral solution, 10 for a basic buffer, 11 for a weak base, and 13 for a strong base.
(a) 10.0 mL of 0.15 M NaOH + 15.0 mL of 0.10 M HNO_3
(b) 25.0 mL of 0.10 M HCl + 10.0 mL of 0.25 M NH_3
(c) 50.0 mL of 0.050 M NaOH + 50.0 mL of 0.10 M NH_3
(d) 50.0 mL of 0.10 M NH_3 + 50.0 mL of 0.05 M HCl

16.89 Determine the dominant acid-base equilibrium that results when each of the following pairs of solutions is mixed. Indicate the equilibrium by writing 1 for a strong acid, 3 for a weak acid, 4 for an acidic buffer, 7 for a neutral solution, 9 for a basic buffer, 10 for a weak base, and 13 for a strong base.
(a) 15.0 mL of 0.20 M NH_3 + 10.0 mL of 0.40 M HCl
(b) 5.00 mL of 0.20 M H_2SO_4 + 5.00 mL of 0.20 M NaOH
(c) 10.0 mL of 0.10 M NH_3 + 5.00 mL of 0.20 M HCl

The graphs shown in Exercises 16.90 through 16.93 are titration curves for 10.0 mL of 0.100 M acid with 0.100 M base. The identity of the acid is unknown, but the titration curves enable the chemist to rule out certain possibilities. Use the data to identify the unknown acids from the titration curves. The acids are as follows:

Acid	pK_{a1}	pK_{a2}	pK_{a3}
Citric	3.13	4.77	6.40
Oxalic	1.25	3.80	
Malic	3.40	5.11	
Phthalic	2.95	5.41	

16.90 ▲ Identify the acid from its titration curve.

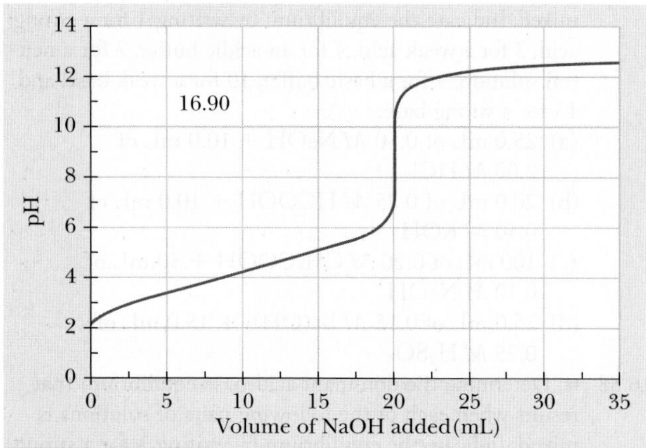

16.91 ▲ Identify the acid from its titration curve.

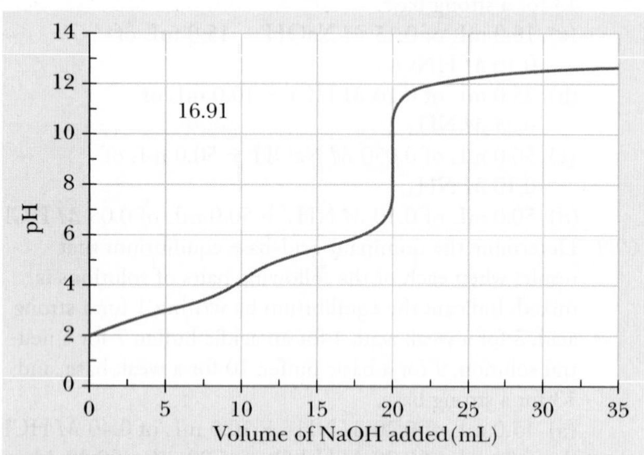

16.92 ▲ Identify the acid from its titration curve.

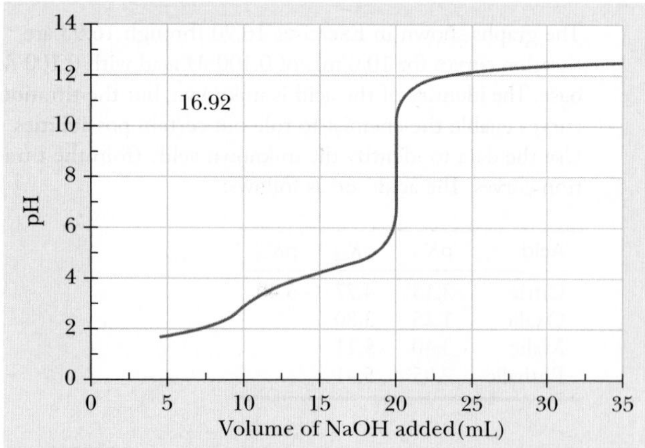

16.93 ▲ Identify the acid from its titration curve.

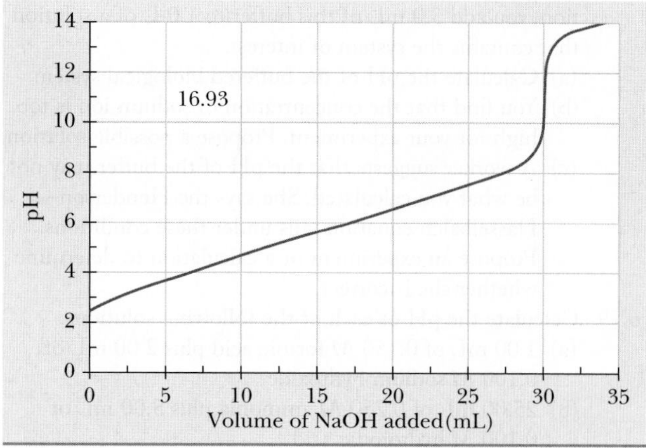

Cumulative Exercises

16.94 ▲ Calculate the volume of concentrated HCl (37% by weight, density 1.19 g/mL) that must be added to 500 mL of 0.10 *M* ammonia to make a buffer at pH 9.25.

16.95 ▲ What is the concentration of ammonium ion in a pH 9.00 solution that forms when concentrated NaOH is added to 0.100 *M* NH_4Cl?

16.96 What is the pH of a solution that is saturated with iron(II) hydroxide? (*Hint:* Find K_{sp} in Appendix F.)

16.97 A classical test for nitrogen in plant material involves adding some compounds to the plant material to produce ammonia (NH_3, a base) from the nitrogen. The solution is then heated to drive off the ammonia. The ammonia passes through a container of HCl, where it reacts. After all the ammonia has been absorbed, there is still an excess of HCl in the container. The amount of HCl remaining after reaction with the ammonia can be determined by titration. The difference between the amount of HCl put in the container and the amount determined by titration represents the amount that was neutralized by the ammonia.

Exactly 21.34 g of plant material is weighed into the reactor. All the nitrogen is converted to ammonia and collected in 100.0 mL of an HCl solution. If the initial HCl solution concentration is 0.121 *M* and the final solution requires 34.22 mL of 0.118 *M* NaOH to neutralize, calculate the percentage of nitrogen in the plant material.

16.98 A monoprotic organic acid that has a molar mass of 176.1 g/mol is synthesized. Unfortunately, the acid produced is not completely pure. In addition, it is not soluble in water. A chemist weighs a 1.8451-g sample of the impure acid and adds it to 100.0 mL of 0.1050 *M* NaOH. The acid is soluble in the NaOH solution and reacts to consume most of the NaOH. The amount of excess NaOH is determined by titration: It takes 3.28 mL of 0.0970 *M* HCl to neutralize the excess NaOH. What is the purity of the original acid, in percent?

16.99 A scientist has synthesized a diprotic organic acid, H_2A, with a molar mass of 124.0 g/mol. The acid must be neutralized (forming the potassium salt) for an important experiment. Calculate the volume of 0.221 *M* KOH that is needed to neutralize 24.93 g of the acid, forming K_2A.

16.100 Exactly 1.2451 g of a solid white acid is dissolved in water and completely neutralized by the addition of 36.69 mL of 0.404 *M* NaOH. Calculate the molar mass of the acid, assuming it to be a monoprotic acid. If additional experiments indicate that the acid is diprotic, what is its molar mass?

16.101 What is a good indicator to use in the titration of a weak acid with $K_a = 1.5 \times 10^{-2}$?

16.102 ▲ Concentrated hydrochloric acid is 38% HCl by weight and has a density of 1.19 g/mL. A solution is prepared by measuring 83 mL of the concentrated HCl, adding it to water, and diluting to 1.00 L.

(a) Calculate the approximate molarity of this solution from the volume, percentage composition, and density.

(b) The exact concentration is determined by titration. A 25.00-mL portion of the solution is titrated with 1.04 *M* NaOH. Phenolphthalein changes color after adding 23.88 mL of the base. What is the concentration of the HCl solution? How does the approximate concentration calculated in part a compare with the exact concentration?

16.103 ▲ A bottle of concentrated hydroiodic acid is 57% HI by weight and has a density of 1.70 g/mL. A solution of this strong and corrosive acid is made by adding exactly 10.0 mL to some water and diluting to 250.0 mL. If the information on the label is correct, what volume of 0.988 *M* NaOH is needed to neutralize the HI solution? Suggest an indicator for the titration.

© Philip Lange, 2008/Used under license from Shutterstock.com

Gasoline. Gasoline is produced from crude oil in large refining plants such as this one, which can produce more than 11 million gallons of gasoline per day. Limitations in gasoline production are usually not the extraction of crude oil from the ground, but rather that refineries are at capacity, producing as much gasoline as they can.

Most of us are probably familiar with the chore of filling up our cars with gasoline. But have you ever wondered how gasoline got to be such a major automotive fuel?

It was not always this way. When hydrocarbons (compounds composed of the elements carbon and hydrogen) were being examined for their fuel potential, it was to generate light in lamps. Oil-based lamps have been known for thousands of years, typically burning naturally occurring vegetable or animal oils (such as olive oil or whale oil). In the early 19th century, coal gas (a mixture of hydrogen, methane, carbon monoxide, and other gases) derived from heating coal came into widespread use because the gaseous fuel could be piped anywhere, even into homes. A liquid mixture of alcohol and turpentine called *camphene* also grew in popularity; by 1850, more than 90 million gallons of camphene had been produced. However, during the U.S. Civil War, the government imposed a heavy tax on alcohol, forcing users to consider less expensive alternatives.

Edwin Drake discovered oil in Titusville, Pennsylvania, in 1859. John Rockefeller set up a system to pipe the oil to Cleveland, Ohio, where he built factories to separate the "crude" oil into more useful, separate mixtures. The most useful mixture was kerosene, a mixture of various hydrocarbon molecules that have anywhere from 12 to 15 carbon atoms. This mixture was liquid at normal temperatures and had good properties for liquid-fueled lamps. By 1900, 500 million gallons of kerosene were used every year.

By the early 1900s, auto builders realized that kerosene was not volatile enough to work well in internal combustion engines and started using gasoline instead. Gasoline (or "petrol" to a large part of the world) is composed of hydrocarbons containing 5 to 12 carbon atoms, and it evaporates much more easily than kerosene, providing a better fuel for an automobile engine.

Chemical Thermodynamics

17

CHAPTER CONTENTS

OWL Online homework for this chapter may be assigned in OWL.

Look for the green colored vertical bar throughout this chapter, for integrated references to this chapter introduction.

By 1910, the demand for gasoline was 160 million gallons; by 1920, more than 3 billion gallons of gasoline were needed to supply the ever-increasing population of automobiles. As of 2006, in the United States alone, almost 138 billion gallons of gasoline per year were needed. Virtually all of this gasoline is obtained from crude oil. Gasoline makes up 10% to 30% of crude oil; this percentage can be increased by "cracking" or heating crude oil in the absence of air and the presence of a catalyst.

Gasoline does not contain just hydrocarbons. Many substances are added to gasoline for different purposes. Some additives keep the engine systems cleaner by minimizing the amount of carbonaceous (carbon-containing) deposits inside the engine. Others, such as ethanol, are added to minimize the amount of carbon monoxide in the engine exhaust, thereby minimizing air pollution. Still other additives are used because they minimize "knocking," a premature combustion of the gasoline/air mixture that reduces engine efficiency. Historically, a compound called *tetraethyl lead* was widely used as an antiknock additive. However, concerns about the detrimental effect of lead on the environment and the population led to a complete phase-out of tetraethyl lead. Other compounds, such as methanol, xylene, and MMT (a manganese complex, approved in the United States but banned in California), are also used to minimize knocking.

© Bettmann/CORBIS

The Revolution begins. Although the development of the automobile has a long history with contributions by many people, it was in 1893 that J. F. Duryea's Motor Wagon (shown here) became the first standardized automobile. It remained in production until the 1920s.

In this chapter, we study more closely how energy is obtained from chemical reactions. As we go through this material, you may notice how many of these topics relate to the chemistry going on inside your car's engine.

Our lives are completely dependent on our sources of energy. Our bodies require fuel, and so does our society. We need energy from food to breathe, grow, and learn, and our global society requires energy to maintain civilization as we know it. Fuels include milk, bread, and meat, as well as wood, coal, and oil; all of them provide energy when consumed in chemical reactions. The study of the energetics of chemical reactions is called **chemical thermodynamics.**

The word *thermodynamics* is derived from the Greek words *thermes,* meaning "heat," and *dynamikos,* meaning "strength." The word *dynamic* makes us think of movement, so the term *thermodynamics* is used to describe heat transfer. The word applies well to chemical systems, because heat is generally produced or absorbed in the course of a chemical reaction.

Chapter 5 introduced some of the basic concepts of thermodynamics. We considered only chemical reactions at constant pressure and temperature in that chapter, and showed that under these conditions, the chemical energy could be expressed by the enthalpy of the system. This chapter introduces additional measures of chemical energy and relates them to the enthalpy of the system.

One of the major needs of chemists is to predict whether a reaction will proceed by itself. The word **spontaneous** describes a process that can occur without outside intervention. Our everyday experience tells us that many processes occur in a particular direction spontaneously. We know that young organisms grow old; we know that dropped objects fall. If we were shown a movie of broken glass leaping together to form a flask, we would logically assume that the film was being shown in reverse, because glass fragments simply do not re-form spontaneously into an unbroken container. If a chemical reaction is spontaneous in one direction, then under the same conditions, the reverse reaction is not spontaneous. This statement does not mean that the reverse reaction *cannot* proceed, only that it may require different conditions. Changing the temperature, pressure, or concentration can sometimes make the reverse reaction occur. Water, for example, does not form ice spontaneously at room temperature and pressure, but does so at a lower temperature.

Thermodynamics provides tools to help us predict whether a process occurs spontaneously. We will have to develop a few more ideas, however, before we can apply these thermodynamic tools.

Energy from the reaction of hydrogen with oxygen. The main engines of the space shuttle consume liquid hydrogen and oxygen at a rate of 4000 L (1000 gal) per second. The space-shuttle launch requires about 2×10^{10} kJ of energy (about one-sixth from the hydrogen/oxygen engines and the rest from solid-fuel rockets that use aluminum and ammonium perchlorate in a rubberized matrix). If this energy were converted to electricity, it would heat and light a city of 1 million people for almost a day.

17.1 Work and Heat

OBJECTIVES

- ☐ Define heat, work, and energy
- ☐ Calculate work from pressure-volume relationships

It is important to know the heat absorbed or evolved in a chemical reaction. For example, chemists use this information not only to adjust conditions and thereby avoid explosions, but also to maximize efficiency. Thermodynamics provides the tools needed to evaluate the heat evolved by a reaction, to predict the maximum energy that can be produced, and to determine whether a proposed reaction is feasible. This section discusses the heat evolved or absorbed during a reaction and the models used to predict the quantity of that heat.

First, let us repeat some definitions from Chapter 5. The **system** is the part of the universe of interest. It could be a beaker full of solution, the turbines in a generator, or a patient in a hospital. Although scientists have identified several categories of systems, the systems discussed in this book are generally **closed systems,** that is, those in which the exchange of energy with the surroundings is allowed, but not the exchange of matter. In a discussion of chemical reactions, the atoms of the elements that appear in the chemical equation generally are taken as the system.

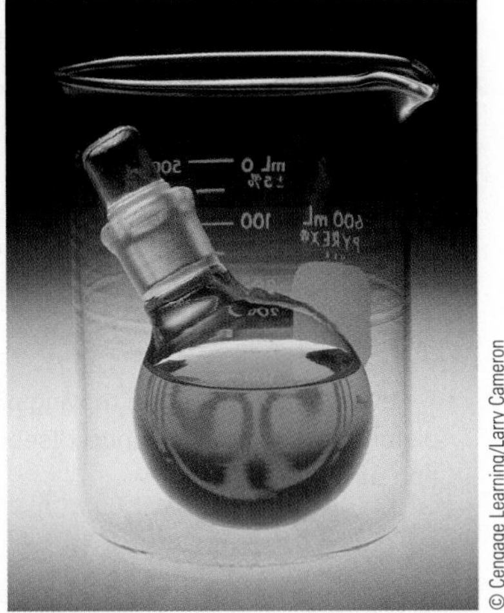

Closed system. A capped flask is a closed system.

The **surroundings** are the rest of the universe, everything that is not part of the system. If the system is defined as the particular group of atoms that undergo chemical change in a beaker in the laboratory, then the surroundings include the beaker, the laboratory bench, the chemistry building, and so on.

The **state** of the system is the set of experimental conditions needed to describe its properties completely, including temperature, pressure, and the amounts and phases of substances. Many thermodynamic properties, called **state functions,** depend only on the state of the system and not on the manner in which the system arrives at the state. Changes in state functions depend only on the initial and final states of the system, not on the path by which the change occurs.

Recall that one way to state the first law of thermodynamics is that the change of energy of a system, ΔE, is a combination of the heat, q, and the work, w:

$$\Delta E = q + w$$

Previously, we introduced ways to determine the heat that accompanies a process. Now we turn to an expanded discussion of work.

Work and heat are two ways to transfer energy between the system and surroundings.

Work

Chapter 5 introduced **work** as a force acting through a distance (e.g., lifting a weight against the force of gravity). Mathematically, work is calculated as the product of force and distance.

$$w = \text{force} \times \text{distance}$$

Units of force are newtons (kg·m/s²), and distance is measured in meters (m). When we use these units in the equation for work, we obtain the units of work.

$$\text{Units of work} = \text{N} \cdot \text{m} = \left(\frac{\text{kg} \cdot \text{m}}{\text{s}^2}\right)(\text{m}) = \left(\frac{\text{kg} \cdot \text{m}^2}{\text{s}^2}\right) \equiv \text{J}$$

Thus, work has the same units as energy, joules.

Recall the sign convention for heat that was provided in Chapter 5; heat absorbed by the system is positive ($q > 0$), and heat released by the system is negative ($q < 0$). (By way of an analogy, think about the balance in a checking account. If money goes in, the change in the balance is positive; if money goes out, the change in the balance is negative.) The same sign convention is used for work. Work is positive ($w > 0$) when work is absorbed by, or *done on,* the system, and it is negative ($w < 0$) when the work is released, or *done by,* the system on the surroundings. A positive sign for work means

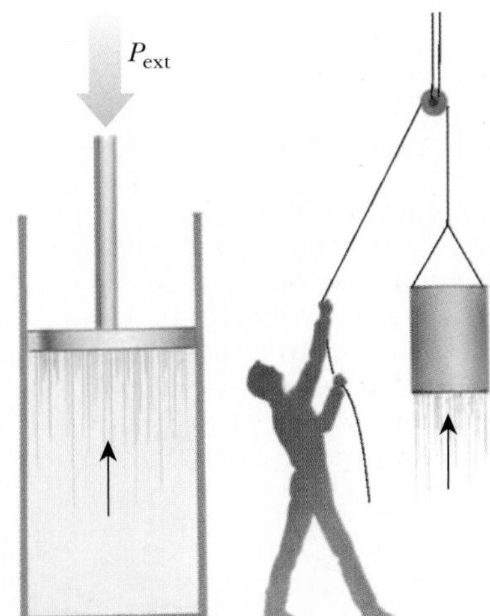

Work. A gas (the system) that expands against an external pressure is analogous to a person (the system) lifting a weight against gravity. In both cases, energy is transferred to the surroundings.

PRINCIPLES OF CHEMISTRY
Pressure-Volume Work

We can go from the original definition of work, force × distance, to the definition of *PV* work given in Equation 17.1. Pressure, *P*, is defined as force, *F*, divided by area, *A*:

$$P = \frac{F}{A}$$

This relationship can be rearranged to show that force is pressure × area:

$$F = P \times A$$

In a system that has a moving piston, let us assume that the piston moves some distance ℓ (see diagram). From the definition of work,

$$w = F \times \ell = P \times A \times \ell$$

where we have substituted for force, *F*. The product of the area of the piston, *A*, and the length of the displacement, ℓ, is the volume change of the chamber of the piston, ΔV:

$$w = P \times \Delta V$$

All that remains now is the sign convention. When work is done by the system, it loses energy. Work is done by the system when the system size expands; that is, the change in volume, ΔV, is positive (because the volume is increasing). Therefore, we need to add a negative sign to the equation so that when the volume change is positive, the overall value of the work is negative (so that we recognize that the total energy of the system is going down). Thus, we get

$$w = -P \times \Delta V$$

as our final expression. ∎

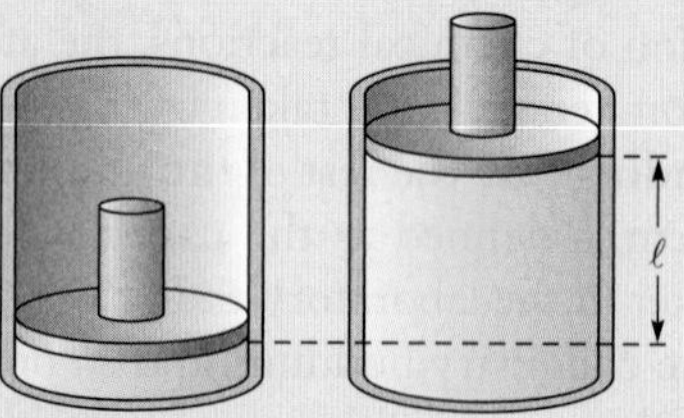

Pressure-volume work. Pressure-volume work is done when a system, such as this piston, expands or contracts.

that the energy of the system is increased, whereas a negative sign for work means that the energy of the system decreases.

Although there is only one form of heat, there are different forms of work, so there are different ways to calculate amounts of work. For instance, a system that contains a gas can perform work as the gas expands against an opposing pressure from the surroundings. This type of work is called *PV* work because a change in volume (against an external pressure) is the source of the work. An automobile engine, in which the gases formed by the combustion of gasoline (see the introduction to this chapter) push against a piston, is powered by *PV* work. The amount of work done by the expansion of a gas against a constant pressure is

$$w = -P_{ext}\Delta V \quad [17.1]$$

where P_{ext} is the external pressure and ΔV is the change in volume of the system.

A gas expanding against external pressure performs work.

When a gas expands (so that ΔV is positive), w is negative because the system transfers energy to the surroundings, so the total energy of the system decreases. When a system containing gas contracts (so that ΔV is negative), w is positive because work is done on the system as the surroundings compress it, and the energy of the system increases. When properly used, Equation 17.1 produces the correct sign for work, because a sign is associated with ΔV.

The most common units that chemists use to express the pressures and volumes of gases are atmospheres and liters, respectively (see Chapter 6). Using these units, we find that $P\Delta V$ has the units L·atm, an uncommon unit in which to express work and energy. The relationship between L·atm and joules, the preferred unit of work, is

$$1 \text{ L·atm} = 101.3 \text{ J}$$

which is used to make conversion factors in Examples 17.1 and 17.2.

PV work can be expressed in L·atm or in joules, although joules is more common.

EXAMPLE 17.1 Pressure-Volume Work

What is the work if a piston having an initial volume of 4.50 L expands to 7.65 L against a constant external pressure of 1.02 atm? Is work done on the system or by the system?

Strategy First, determine ΔV, knowing that $\Delta V = V_{\text{final}} - V_{\text{initial}}$. Then substitute quantities into Equation 17.1 and convert the final numerical answer to units of joules.

Solution
The change in volume is

$$\Delta V = 7.65 \text{ L} - 4.50 \text{ L} = 3.15 \text{ L}$$

Substituting quantities into Equation 17.1:

$$w = -(1.02 \text{ atm})(3.15 \text{ L}) = -3.21 \text{ L·atm}$$

Now converting into units of joules:

$$w = -3.21 \cancel{\text{L}} \cdot \cancel{\text{atm}} \times \left(\frac{101.3 \text{ J}}{1 \cancel{\text{L}} \cdot \cancel{\text{atm}}}\right) = -325 \text{ J}$$

Because the work is negative, the system is losing energy, so work is being done *by* the system on the surroundings.

Understanding

What is the work if a piston having an initial volume of 4.50 L contracts to 1.77 L against a constant external pressure of 1.34 atm? Is work done on the system or *by* the system?

Answer $w = +371$ J; work done *on* the system

EXAMPLE 17.2 Pressure-Volume Work

A cylinder contains 45.0 L of an ideal gas at a pressure of 140 atm. If the gas expands at a constant temperature against an opposing pressure of 0.970 atm, how much work (in joules) is done? The final pressure of the gas is 0.970 atm.

Strategy Use Boyle's law to determine the final volume; then use the definition of *PV* work to determine work.

Solution
The final volume of the gas sample is needed. Using $P_1V_1 = P_2V_2$ to determine the final volume:

Initial conditions: $V_1 = 45.0 \text{ L}$ $\quad P_1 = 140 \text{ atm}$

Final conditions: $V_2 = ?$ $\quad P_2 = P_{\text{ext}} = 0.970 \text{ atm}$

$$P_1V_1 = P_2V_2$$

$$V_2 = \frac{P_1V_1}{P_2} = \frac{(140 \cancel{\text{atm}})(45.0 \text{ L})}{(0.970 \cancel{\text{atm}})} = 6.49 \times 10^3 \text{ L}$$

We now have the data to calculate the work:

$$w = -P_{ext}\Delta V = -0.970 \text{ atm } (6.49 \times 10^3 \text{ L} - 45.0 \text{ L})$$

$$w = -6.25 \times 10^3 \text{ L·atm}$$

Convert the work to joules:

$$w = -6.25 \times 10^3 \cancel{\text{L}} \cdot \cancel{\text{atm}} \times \left(\frac{101.3 \text{ J}}{\cancel{\text{L}} \cdot \cancel{\text{atm}}}\right) = -6.33 \times 10^5 \text{ J} = -633 \text{ kJ}$$

The negative sign indicates that work is done *by* the system on the surroundings.

Understanding

Calculate the amount of work (in joules) done on the system when 6.30×10^3 L of an ideal gas (the system) initially at 1.00 atm is compressed at constant temperature and a pressure of 140 atm to a final volume of 45.0 L. The opposing pressure is the constant 140 atm applied to the system.

Answer $w = +8.76 \times 10^5$ L·atm $= +8.87 \times 10^4$ kJ. The positive sign indicates that work is done *on* the system to compress it.

If the system:
- performs work *on* the surroundings, the sign of *w* is negative.
- is worked on *by* the surroundings, the sign of *w* is positive.
- *transfers* heat to the surroundings, the sign of *q* is negative.
- *absorbs* heat from the surroundings, the sign of *q* is positive.

Heat

Heat is a second way in which energy is transferred between the system and the surroundings. Heat is a transfer of thermal energy between the system and the surroundings. Recall that thermal energy is the energy possessed by matter in the form of random motion of the particles. Notice that heat implies a *random* motion, whereas work is a *directed* motion (which includes electrons moving through a wire, as well as a gas expanding against a pressure). The measurement of heat by calorimetry is described in Chapter 5.

OBJECTIVES REVIEW *Can you:*

☑ define heat, work, and energy?
☑ calculate work from pressure-volume relationships?

17.2 The First Law of Thermodynamics

OBJECTIVES

☐ Describe how heat, work, and energy are related by the first law of thermodynamics
☐ Relate internal energy and enthalpy

Centuries of observations led to the **first law of thermodynamics,** also known as the law of conservation of energy: *Energy can be neither created nor destroyed.* Scientists cannot explain *why* the first law is observed; we can only marvel that nature is so simple and direct. The total energy of a system is called the **internal energy, *E*,** which is a state function. In a completely **isolated system,** in which matter and energy cannot enter or leave, the first law of thermodynamics can be written as

The first law of thermodynamics states: In an isolated system, the total energy of the system is constant.

$$\Delta E = 0$$

(In science, when a quantity does not change, it is said to be "conserved." Hence the first law of thermodynamics is often called the *law of conservation of energy.*) In a closed system, however, energy can enter or leave the system (but matter cannot), but *only as work or heat.* Thus, when an energy change occurs in a closed system, the first law of thermodynamics can be written as

$$\Delta E = q + w \qquad [17.2]$$

Heat and work are the only known ways in which energy enters or leaves the system. Equation 17.2 is considered another way to state the first law of thermodynamics.

Note two important details. First, thermodynamics is concerned with the *change* in energy. In general, there is no way to evaluate absolute energies; thermodynamicists cannot say, "The energy decreases from 2700 J to 2400 J," but can say only, "The energy decreases by 300 J in this change." Second, ΔE is a *state function* and is independent of the path followed between the initial and final states.

The heat and work, q and w, however, depend on the path, so conditions can be adjusted to give more heat at the expense of work, and *vice versa.* Heat and work are not state functions; that is, they are path dependent. We can demonstrate this easily, using work as an example.

Consider a sample of gas with an initial volume of 1.00 L and a pressure of 10.0 atm. Assume that this gas expands against a pressure of 1.00 atm until it reaches a final volume of 10.0 L, at which point the internal and external pressures are equal (Figure 17.1*a*). The work performed by this system is

$$w = -P_{\text{ext}}\Delta V = -(1.00\ \cancel{\text{atm}}) \times (10.0\ \cancel{\text{L}} - 1.00\ \cancel{\text{L}}) \times \left(\frac{101.3\ \text{J}}{1\ \cancel{\text{L}} \cdot \cancel{\text{atm}}}\right)$$

$$w = -912\ \text{J}$$

Now let us perform the expansion in two steps (Figure 17.1*b*), first an expansion against a pressure of 2.00 atm until the system reaches a volume of 5.00 L, then an expansion against a pressure of 1.00 atm until the volume of the system reaches 10.0 L. (You can use Boyle's law to verify that the final pressure will equal the external pressure on each step.) For the first step, we have

$$w_1 = -P_{\text{ext}}\Delta V = -(2.00\ \cancel{\text{atm}}) \times (5.00\ \cancel{\text{L}} - 1.00\ \cancel{\text{L}}) \times \left(\frac{101.3\ \text{J}}{1\ \cancel{\text{L}} \cdot \cancel{\text{atm}}}\right)$$

$$w_1 = -810\ \text{J}$$

For the second step, we have

$$w_2 = -P_{\text{ext}}\Delta V = -(1.00\ \cancel{\text{atm}}) \times (10.0\ \cancel{\text{L}} - 5.00\ \cancel{\text{L}}) \times \left(\frac{101.3\ \text{J}}{1\ \cancel{\text{L}} \cdot \cancel{\text{atm}}}\right)$$

$$w_2 = -507\ \text{J}$$

The total work is the sum of the works from each step:

$$w = w_1 + w_2 = -810\ \text{J} + (-507\ \text{J})$$

$$w = -1317\ \text{J}$$

We got more work out of the two-step expansion than the one-step expansion, despite the fact that the initial and final conditions are the same for both processes. *Work is path dependent.* So, too, is heat. But ΔE would be the same for both of these processes because E is a state function.

What are the implications of these conclusions? First, because w and q vary according to the conditions of a process, they cannot be used to determine whether a process

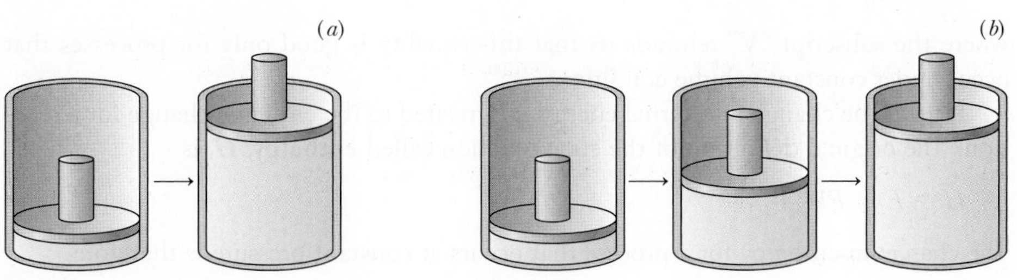

Figure 17.1 Work is path-dependent. *(a)* A one-step expansion performs a certain amount of work. *(b)* A two-step expansion performs a different amount of work than a one-step expansion, even though the initial and final states of the system are the same.

will be spontaneous. Second, processes can be performed in various ways to maximize the amount of work or heat. If, for example, heat is desired from a process, it should be necessary to design the process so that the maximum amount of heat will be provided.

EXAMPLE 17.3 Calculating Work, Heat, and ΔE

A sample of gas in a balloon is compressed from 4.50 L to 1.00 L by an external pressure of 1.25 atm. At the same time, 60.0 J of heat is generated. Calculate ΔE.

Strategy Calculate work using P_{ext} and ΔV, and determine the proper sign on heat. Then use the mathematical statement of the first law of thermodynamics to determine ΔE.

Solution

The change in volume is $1.00\text{ L} - 4.50\text{ L} = -3.50\text{ L}$. First, calculate work:

$$w = -P_{ext} \cdot \Delta V = -(1.25\ \cancel{\text{atm}}) \times (-3.50\ \cancel{\text{L}}) \times \left(\frac{101.3\text{ J}}{1\ \cancel{\text{L}} \cdot \cancel{\text{atm}}}\right)$$

$$w = 443\text{ J}$$

Because the system is generating heat, it is losing energy to the surroundings, so heat is negative: $q = -60.0\text{ J}$

The combination of the work and heat gives ΔE:

$$\Delta E = q + w = -60.0\text{ J} + 443\text{ J}$$

$$\Delta E = 383\text{ J}$$

Thus, despite the fact that some energy is leaving the gas in the balloon as heat, the overall energy of the gas in the balloon increases.

Understanding

A 7.56-g sample of gas is in a balloon that has a volume of 10.5 L. Under an external pressure of 1.05 atm, the balloon expands to a volume of 15.00 L. Then the gas is heated from 0.0 °C to 25.0 °C. If the specific heat of the gas is 0.909 J/g·°C, what are work, heat, and ΔE for the overall process?

Answer $w = -479\text{ J}$, $q = 172\text{ J}$, $\Delta E = -307\text{ J}$

Energy and Enthalpy

Chapter 5 relates the heat of a reaction at constant pressure and temperature to the change in enthalpy, ΔH, for the reaction. One way to write this is

$$\Delta H = q_P$$

where the subscript "P" reminds us that this relationship is good only for a process occurring at constant pressure. On the other hand, the change in internal energy, ΔE, is the heat absorbed by the system when the volume is constant, because when ΔV is zero, the work, w, equals zero:

$$\Delta E = q + w = q + 0 \text{ (at constant volume)}$$

$$\Delta E = q_V$$

where the subscript "V" reminds us that this equality is good only for processes that occur under constant volume conditions.

How is the change in internal energy, ΔE, related to the enthalpy change for a reaction? The original definition of the state function called **enthalpy, H,** is

$$H = E + PV$$

The change in enthalpy for a process that occurs at constant pressure is therefore

$$\Delta H = \Delta E + P\Delta V \qquad [17.3]$$

The change in enthalpy differs from the change in internal energy by the *PV* work when the change occurs at constant pressure rather than constant volume. Recall from Chapter 5 that the change in enthalpy, ΔH, is equal to the heat absorbed by the system under conditions of constant pressure when only *PV* work is done—the conditions of a typical laboratory reaction.

ΔE is the heat of reaction if the process occurs at constant volume. ΔH is the heat change at constant pressure, when the only work is *PV* work.

For reactions that involve only condensed phases (solids and liquids) at constant pressure, the change in volume is negligibly small, so ΔH and ΔE are essentially the same. A greater change in volume occurs when the reaction is accompanied by a change in the number of moles of gas. For example, in the combustion of 1 mol methane,

$$CH_4(g) + 2O_2(g) \rightarrow CO_2(g) + 2H_2O(\ell)$$

the change in the number of moles of gas is

$$\Delta n = (n_{\text{final}} - n_{\text{initial}}) = 1 \text{ mol } CO_2 - (1 \text{ mol } CH_4 + 2 \text{ mol } O_2) = -2 \text{ mol gas}$$

At normal pressures around 1 atm, we can use the ideal gas law to evaluate the $P\Delta V$ term in Equation 17.3.

$$P\Delta V = (\Delta n)RT \qquad [17.4]$$

Thus, we can rewrite Equation 17.3 as $\Delta H = \Delta E + (\Delta n)RT$. Example 17.4 illustrates the calculation of the change in enthalpy from the change in internal energy.

EXAMPLE 17.4 Calculating ΔH from ΔE

ΔE is -2.21×10^3 kJ/mol for the combustion of propane.

$$C_3H_8(g) + 5O_2(g) \rightarrow 3CO_2(g) + 4H_2O(\ell)$$

Calculate ΔH for this reaction at 25 °C and 1.00 atm.

Strategy Use Equations 17.3 and 17.4 to determine ΔH. The change in the number of moles of gas is determined by taking the number of moles of gaseous products and subtracting the number of moles of gaseous reactants.

Solution

In this example, the change in the number of moles of gas is

$$\Delta n = 3 \text{ mol } CO_2(g) - [1 \text{ mol } C_3H_8(g) + 5 \text{ mol } O_2(g)] = -3 \text{ mol gas}$$

Because we want units of energy, for R we must use 8.314 J/mol·K.

$$P\Delta V = (\Delta n)RT = (-3 \text{ mol})(8.314 \text{ J/mol·K})(298 \text{ K}) = -7.43 \times 10^3 \text{ J} = -7.43 \text{ kJ}$$

Substitution of the values for $P\Delta V$ and ΔE into Equation 17.3 produces the desired change in enthalpy:

$$\Delta H = \Delta E + P\Delta V = -2.21 \times 10^3 \text{ kJ} - 7.43 \text{ kJ} = -2.22 \times 10^3 \text{ kJ}$$

Even the fairly large change in the number of moles of gas produced only a small difference between ΔH and ΔE for this very exothermic reaction. For other reactions with smaller values of ΔE, however, the difference can be quite significant.

Understanding

What is ΔH at 25 °C and 1.00 atm for the combustion of 1 mol ethane if $\Delta E = -1553$ kJ?

$$C_2H_6(g) + \frac{7}{2}O_2(g) \rightarrow 2CO_2(g) + 3H_2O(\ell) \qquad \Delta E = -1553 \text{ kJ}$$

Answer $\Delta H = -1559$ kJ

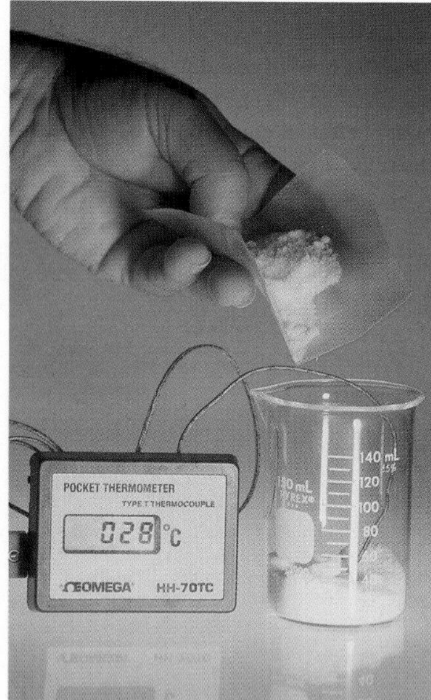

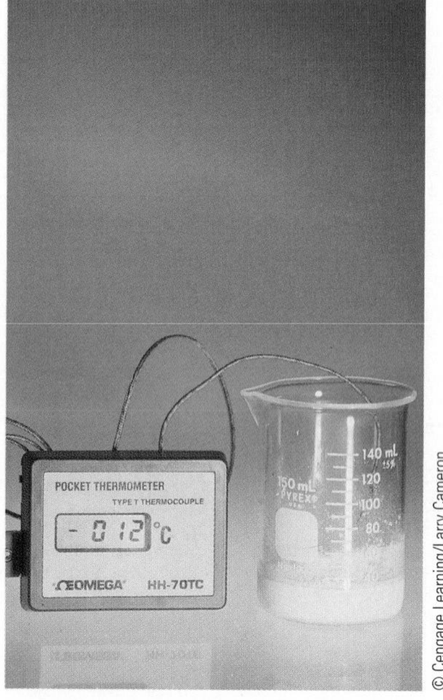

Figure 17.2 Mixing barium hydroxide and ammonium chloride. The system becomes quite cold as the reaction proceeds spontaneously. Thus, giving off energy is not a strict criterion for a spontaneous process.

Knowing the ΔE and ΔH of a process is useful, but how do they relate to the spontaneity of that process? Unfortunately, they do not. Initially it was assumed that all spontaneous processes had a negative ΔH, but there are some spontaneous processes known that have a positive ΔH (e.g., the dissolving of some salts). Another quantity is needed before we can address the spontaneity issue.

OBJECTIVES REVIEW *Can you:*

- ☑ describe how heat, work, and energy are related by the first law of thermodynamics?
- ☑ relate internal energy and enthalpy?

17.3 Entropy and Spontaneity

OBJECTIVES

- ☐ Define entropy and understand its statistical nature
- ☐ Predict the sign of entropy changes for phase changes
- ☐ Apply the second law of thermodynamics to chemical systems
- ☐ Recognize that absolute entropies can be measured because the third law of thermodynamics defines a zero point

The change in enthalpy of a process often *seems* to predict whether that process is spontaneous. Everyday experience reveals that many spontaneous processes are accompanied by a release of energy. A dropped object falling, the combustion of hydrocarbons such as methane, and the return of electrons in excited-state atoms to the ground state are a few of the many spontaneous processes that are accompanied by a release of energy to the surroundings.

However, the enthalpy change of a process does not, by itself, predict spontaneity. Some examples of endothermic processes are spontaneous. If barium hydroxide octahydrate is mixed with ammonium chloride, the two white solids form an exceptionally cold liquid solution.

$$Ba(OH)_2{\cdot}8H_2O(s) + 2NH_4Cl(s) \rightarrow BaCl_2(aq) + 2NH_3(g) + 10H_2O(\ell)$$

The products become cold enough to freeze water. Figure 17.2 illustrates the temperature changes measured for this spontaneous, endothermic reaction.

Reactions not only tend to form products with stronger bonds, they tend to form more disordered products. Disorder plays a measurably important role in predicting whether a reaction will be spontaneous. **Entropy, *S*,** is the thermodynamic state function that describes the amount of disorder. Disorder, chaos, and randomness all describe a state of high entropy. Experience indicates that disorder (entropy) increases spontaneously. For example, one's bedroom gets disorganized rather easily, but it does not clean itself up spontaneously!

Entropy, *S*, is a thermodynamic state function that is related to the disorder of the system.

The change in entropy is the sole driving force for some processes. Consider the ideal gas in the apparatus shown in Figure 17.3. When the stopcock is opened, the gas moves spontaneously to occupy both chambers. No energy change is needed to fill the empty chamber, so the change in internal energy is zero. No work is involved (the expansion of a gas against an opposing pressure of 0 atm produces no work), so ΔH is also zero. The process is driven solely by the change in disorder. In the final state, the system occupies a larger volume, and there are more possible locations for the gas particles, so the disorder has increased.

Entropy as a Measure of Randomness

In the late 19th century, Ludwig Boltzmann (1844–1906) showed that the entropy of a system is related to the number of different ways in which its components can be arranged. Boltzmann's work can be illustrated with a deck of playing cards. If a shuffled (randomized) deck has only two cards, there is a 50% chance that the deck

Practical aspects of entropy. Many everyday systems tend toward disorder.

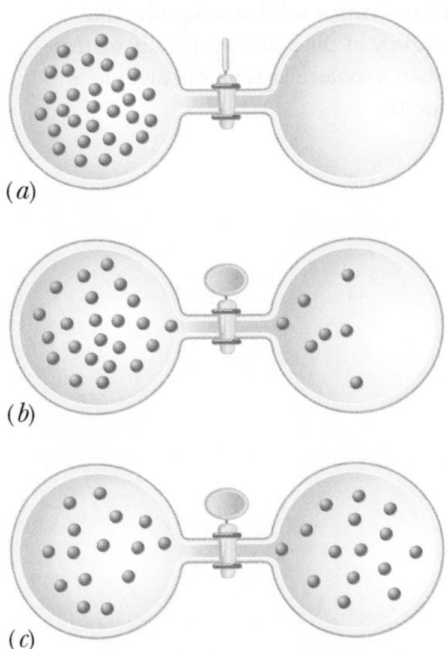

Figure 17.3 Expansion of a gas into a vacuum. *(a)* All the gas is in the container on the left. *(b)* When the valve is opened, some of the gas moves into the empty container on the right. *(c)* At equilibrium, the gas is present at the same concentration in both containers, and randomness has increased.

will be in order, arranged from low to high. If the deck has 13 cards (perhaps all the hearts), the chance of finding it in numerical order is small, about 1 in 6 billion times. If the deck has 52 cards, the chance of its being in order is still smaller (about 1 in 10^{68}).

We describe chemical systems by using similar logic. The arrangement of atoms in a crystal shows a high degree of order; the entropy is low. On the other hand, there are numerous ways in which atoms can be arranged in a gas, so the entropy of a sample of a gas is much greater than that of the same sample in solid form.

Entropy, S, is a state function, so changes in entropy, ΔS, are independent of the path and are defined as the final entropy of a system minus its initial entropy:

$$\Delta S = S_{\text{final}} - S_{\text{initial}}$$

Changes in entropy are related to changes in the randomness of the system. Most changes in entropy can be predicted from a few general concepts.

1. *The entropy of a substance increases when the solid forms a liquid and when the liquid forms a gas.* In Chapter 11, we saw that the location of every unit in a crystal can be predicted from the locations of the few units that are present in a unit cell. Relatively little randomness exists in the solid state. When the solid melts, the molecules are no longer regularly arranged, so randomness is increased and the entropy of the liquid is greater. Finally, when the liquid vaporizes, forming a gas, a large increase in the randomness of the sample occurs. Thus, the gas has much higher entropy than the liquid.

$$S_{\text{solid}} < S_{\text{liquid}} < S_{\text{gas}}$$

In general, the entropy change in the transition from liquid to gas is considerably greater than that in the change from solid to liquid.

2. *Entropy generally increases when a molecular solid or liquid dissolves in a solvent.*

$$C_{12}H_{22}O_{11}(s) \xrightarrow{H_2O} C_{12}H_{22}O_{11}(aq)$$

When sucrose dissolves in water, the sucrose molecules change from a highly ordered arrangement in the crystal to a more random state dissolved in solution. At the same time, the order of the water molecules may increase because of the hydration of the solute particles. When most molecular solutes dissolve, the increase in disorder of the solute is greater than the

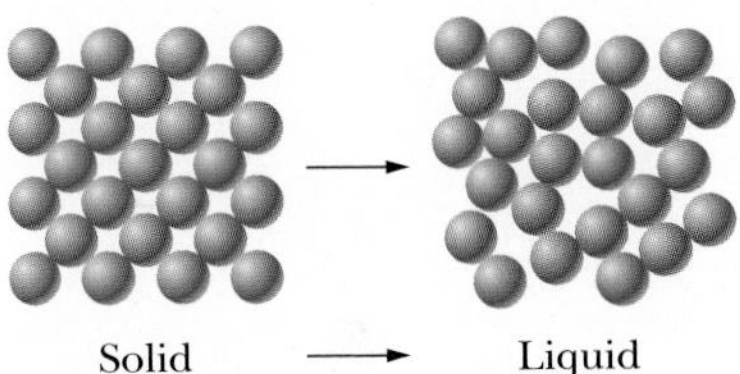

Solids and liquids. The liquid state of a substance is more random than the solid, so $S_{\text{liquid}} > S_{\text{solid}}$.

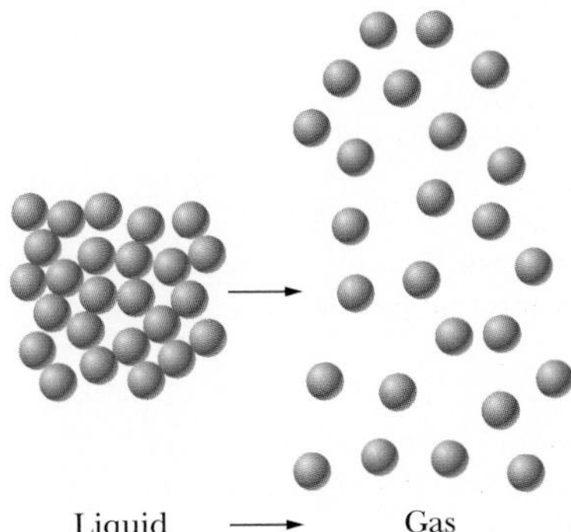

Liquids and gases. A gas is much more random than the liquid, so $S_{\text{gas}} > S_{\text{liquid}}$.

Changes in phase are the largest factors influencing changes in entropy.

Dissolving a solid in a liquid. The entropy of the system generally increases when a molecular solid dissolves in a liquid.

decrease in disorder of the solvent. Conversely, for some ionic solids, particularly those with highly charged ions, the increase in the order of the solvent because of hydration of the ions is sometimes great enough to produce a net *decrease* in entropy.

3. *Entropy decreases when a gas dissolves in a solvent.*

$$CO_2(g) \xrightarrow{H_2O} CO_2(aq)$$

When a gas such as carbon dioxide dissolves in a solvent, the solute goes from the gas phase to the liquid phase, so it becomes substantially less random, mainly because the dissolved molecules are confined to a smaller volume. The change in entropy of the solvent is small because it is a liquid in the initial and final states. Overall, the entropy decreases.

4. *Entropy increases as the temperature increases.* According to kinetic molecular theory, the kinetic energy of the particles in a sample increases as the temperature increases. The disorder increases as the motion of the particles increases.

Quite often, chemists can predict the entropy change by examining the chemical equation and determining whether the reactant side or the product side contains more moles of gas. Consider the following reaction, which is the combustion of a component of gasoline (see the introduction to this chapter):

$$C_7H_{16}(\ell) + 11O_2(g) \rightarrow 7CO_2(g) + 8H_2O(\ell)$$

Eleven moles of gas are on the reactant side and only 7 mol of gas in the products. A decrease in disorder is predicted, and the entropy of the system does decrease during this reaction.

Increasing entropy is often associated with increasing volume of the system. This general rule is consistent with the examples we have presented. In addition, the entropy of a dilute solution is greater than that of a more concentrated one. The flow of water through a semipermeable membrane in osmosis (see Chapter 12), which results in the dilution of the solution, is driven by the positive change in entropy.

The Second Law of Thermodynamics

One goal throughout this chapter has been to find a criterion to predict whether a reaction will be spontaneous. The **second law of thermodynamics** provides such a criterion: *For a spontaneous process, the entropy of the universe will increase.* In thermodynamics, we separate the universe into the system and the surroundings. Because the system and surroundings may each experience an entropy change, the second law can be written as

$$\Delta S_{\text{univ}} = \Delta S_{\text{sys}} + \Delta S_{\text{surr}} > 0 \text{ for a spontaneous process}$$

This remarkable statement has been tested against experimental observations and is entirely consistent with laboratory results. Many examples of spontaneous chemical reactions are accompanied by a decrease in entropy of the system—for example, the combustion of nearly all hydrocarbons. In such cases, there is a larger positive change in the entropy of the surroundings that produces an increase in the entropy of the universe. If the change in entropy of the system is negative, there *must* be a larger positive change in the entropy of the surroundings for the process to be spontaneous.

In any spontaneous process, the entropy of the universe increases.

Another way to consider the second law of thermodynamics is that for an isolated system, a spontaneous change always occurs with an increase in the entropy of the system. Isolated systems can be approximated in the laboratory, and the validity of the second law of thermodynamics is unquestioned. The expansion of the gas in Figure 17.2 is a direct consequence of the second law of thermodynamics. If a reaction in an isolated system causes the entropy to increase, the reaction proceeds spontaneously. If $\Delta S < 0$, then the reverse reaction is spontaneous. If the change in the entropy is zero, the system is at equilibrium—that is, it is not spontaneous in either direction. We will discuss the equilibrium condition at length later.

The Third Law of Thermodynamics

The entropy of a substance, unlike its energy or enthalpy, can be measured on an absolute basis. Although we can measure only *changes* in E and H (ΔE or ΔH), we can measure an absolute value of S, the entropy, because there is a perfectly ordered system that can serve as a reference. The **third law of thermodynamics** gives us a reference point: *The entropy of any pure crystalline substance at a temperature of 0 K is zero.* The randomness of a pure crystal at 0 K is at its minimum and is set equal to zero.

The entropy of any perfect crystalline substance at 0 K is zero.

As the temperature is increased from 0 K, the motion of the particles in a substance increases, increasing the entropy of the substance. The change in entropy is proportional to the added energy, but the proportionality constant depends on temperature. A transfer of 10 kJ of heat has a greater effect on the entropy at a low temperature than at a high temperature. For example, a change of 10 K in temperature from 300 to 310 K increases the disorder more than the change from 3000 to 3010 K. The proportionality constant is $1/T$, so the entropy change is given by the equation

$$\Delta S = \frac{q}{T} \qquad [17.5]$$

Temperature must be expressed in kelvin units. Thus, entropy has units of joules per kelvin (J/K).

When a phase change occurs—for example, a solid melts—the entropy changes abruptly because of the greater freedom of motion that the molecules have in the liquid phase. When a liquid boils, the entropy of the system increases because the gas molecules occupy a larger volume and move about more randomly in space. At any temperature greater than 0 K, the entropy of all substances is positive. Entropy changes for phase changes are easy to calculate because phase changes are accompanied by a characteristic energy change, either the heat of fusion or the heat of vaporization (see Chapter 11). Example 17.5 demonstrates calculating ΔS for a phase change.

EXAMPLE 17.5 Calculating ΔS for a Phase Change

Determine the entropy change when 1 mol H_2O boils at its normal boiling point of 100.0 °C. The heat of vaporization of water is 44.0 kJ/mol.

Strategy We will use Equation 17.5, remembering that when the temperature variable is used in this equation, it must be in kelvin units.

Solution

Because we have 1 mol H_2O, the heat involved in the process is simply 44.0 kJ, or 4.40×10^4 J. The normal boiling point of 100.0 °C is 373 K. Substituting into Equation 17.5:

$$\Delta S = \frac{q}{T} = \frac{4.40 \times 10^4 \text{ J}}{373 \text{ K}} = 118 \text{ J/K}$$

Understanding

Determine the entropy change when 1 mol H_2O freezes at its normal freezing point of 0.0 °C. The heat of fusion of water is 6.01 kJ/mol.

Answer −22.0 J/K

Appendix G lists standard molar entropies for many materials (in units of J/mol·K). The table contains the entropy for each substance in its standard state at 298 K. Standard entropies are designated by the symbol S°_{298}. Unlike the standard enthalpies of formation listed in Appendix G, the standard entropies of the most stable forms of the elements are not zero. The data in the appendix can be used to calculate entropy changes for many reactions, because S is a state function and Hess's law applies:

$$\Delta S_{\text{reaction}} = \Sigma nS^\circ[\text{products}] - \Sigma mS^\circ[\text{reactants}]$$

This is analogous to our treatment of heats of formation, ΔH_f, presented in Chapter 5. Example 17.6 shows how an entropy change for a reaction is calculated from the data in Appendix G, and demonstrates the second law of thermodynamics.

EXAMPLE 17.6 Calculating Entropy Changes

Calculate the standard entropy change when 1 mol of propane burns in oxygen.

$$C_3H_8(g) + 5O_2(g) \rightarrow 3CO_2(g) + 4H_2O(\ell)$$

Strategy Use the standard molar entropies from Appendix G. For each product and reactant, do not forget to multiply the molar entropy by the number of moles of each substance in the balanced chemical equation.

Solution

$$\Delta S° = \{3S°[CO_2(g)] + 4S°[H_2O(\ell)]\} - \{S°[C_3H_8(g)] + 5S°[O_2(g)]\}$$

$$\Delta S° = \{3\ \cancel{mol}\ (213.63\ J/\cancel{mol}{\cdot}K) + 4\ \cancel{mol}\ (69.91\ J/\cancel{mol}{\cdot}K)\} - \{1\ \cancel{mol}\ (269.91\ J/\cancel{mol}{\cdot}K) + 5\ \cancel{mol}\ (205.03\ J/\cancel{mol}{\cdot}K)\}$$

$$\Delta S° = -374.53\ J/K$$

The entropy decreases primarily because the reactants include 6 mol of gases and the products have only 3 mol.

Understanding

Calculate the standard entropy change for the decomposition of hydrogen peroxide into water and oxygen. The reaction is:

$$2H_2O_2(\ell) \rightarrow 2H_2O(\ell) + O_2(g)$$

Answer $\Delta S° = 125.7$ J/K

OBJECTIVES REVIEW *Can you:*

- ☑ define entropy and understand its statistical nature?
- ☑ predict the sign of entropy changes for phase changes?
- ☑ apply the second law of thermodynamics to chemical systems?
- ☑ recognize that absolute entropies can be measured because the third law of thermodynamics defines a zero point?

17.4 Gibbs Free Energy

OBJECTIVES

- ☐ Define Gibbs free energy and relate the sign of a Gibbs free-energy change to the direction of spontaneous reaction
- ☐ Predict the influence of temperature on Gibbs free-energy

The first and second laws of thermodynamics establish that both energy and entropy are important considerations for chemical reactions. Although the second law of thermodynamics addresses the issue of spontaneity, it focuses on an isolated system—which many of our systems are not! Is there any way to find a spontaneity test that is more generally applicable and, therefore, more useful?

Yes, there is. Yale chemist J. W. Gibbs (1839–1903) found a state function of the system that was directly related to spontaneity. Gibbs realized that the second law can be rewritten so that spontaneity is expressed by a single state function of the *system.* He defined a new state function, now called the *Gibbs free energy.* We introduce the Gibbs free energy and examine its relation to spontaneity in this section.

Gibbs Free Energy

The **Gibbs free energy,**[1] G, of the system, is defined as

$$G = H - TS$$

For conditions of constant pressure and temperature, the change in the Gibbs free energy of the system is given by the equation

$$\Delta G = \Delta H - T\Delta S \qquad [17.6]$$

[1]The International Union of Pure and Applied Chemistry (IUPAC) recommends that this state function be called *Gibbs energy,* without the word *free.*

TABLE 17.1	Gibbs Free Energy Changes and the Direction of Spontaneous Reaction
ΔG	Spontaneous Reaction
Less than zero	Forward reaction is spontaneous
Zero	At equilibrium
Greater than zero	Reverse reaction is spontaneous

Because all the quantities on the right side of Equation 17.6 are state functions, ΔG is also a state function. Like enthalpy and internal energy, the absolute values of G cannot be measured. We can, however, measure *changes* in G. We use Equation 17.6 to calculate the change in the Gibbs free energy from the changes in enthalpy and entropy at constant temperature and pressure.

Gibbs realized that *free energy change is a quantitative indicator of spontaneity for all systems.* The only restrictions are that the process must occur at constant pressure and temperature. For any spontaneous reaction, $\Delta G < 0$. Note that when $\Delta G > 0$, the *reverse* reaction is spontaneous. For a system at equilibrium, the change in the Gibbs free energy is zero. Table 17.1 summarizes the direction of spontaneous reaction.

For any spontaneous change, ΔG is negative. At equilibrium, $\Delta G = 0$.

Just as the change in the Gibbs free energy is related to the enthalpy and entropy, a similar relationship holds for the **standard Gibbs free energy of formation,** the Gibbs free energy change during the formation of one mole of a substance in its standard state from its constituent elements in their standard states.

$$\Delta G_f^\circ = \Delta H_f^\circ - T\Delta S_f^\circ \qquad [17.7]$$

When we use this equation to calculate a standard Gibbs free energy of formation, ΔS_f° must be calculated from the tabulated absolute entropies. Most tabulations of thermodynamic state functions, including Appendix G, provide the standard Gibbs free energies of formation. Changes in the Gibbs free energy of a reaction can be calculated from values of ΔH and ΔS by using $\Delta G = \Delta H - T\Delta S$ or by using the Gibbs free energies of formation of the products and reactants—as a state function, Hess's law applies to Gibbs free energies, and the usual "products minus reactants" scheme can also be used with ΔG_f°:

The Gibbs free energy change in a reaction can be calculated from tables of ΔG_f° using Hess's law: $\Delta G^\circ = \Sigma n\,\Delta G_f^\circ$ [products] $- \Sigma m\,\Delta G_f^\circ$ [reactants].

$$\Delta G_{\text{rxn}}^\circ = \Sigma n\Delta G_f^\circ[\text{products}] - \Sigma m\Delta G_f^\circ[\text{reactants}] \qquad [17.8]$$

where n and m are the number of moles of the products and reactants, respectively, in the balanced chemical equation.

Like enthalpy, the standard Gibbs free energy of formation for an element in its standard state is zero. Example 17.7 illustrates the use of tabulated values of standard Gibbs free energies of formation to determine whether a reaction is spontaneous.

EXAMPLE 17.7 Determination of ΔG° for a Reaction

Calculate ΔG° and determine whether the following reaction will take place spontaneously under standard-state conditions at 298 K.

$$H_2S(g) + 2H_2O(\ell) \rightarrow SO_2(g) + 3H_2(g)$$

Strategy Use the standard Gibbs free energies of formation from Appendix G and the coefficients of the balanced chemical reaction to calculate ΔG° for the reaction.

Solution

$$\Delta G_{\text{rxn}}^\circ = \Sigma n\Delta G_f^\circ[\text{products}] - \Sigma m\Delta G_f^\circ[\text{reactants}]$$

$$\Delta G_{\text{rxn}}^\circ = \{\Delta G_f^\circ[SO_2(g)] + 3\Delta G_f^\circ[H_2(g)]\} - \{\Delta G_f^\circ[H_2S(g)] + 2\Delta G_f^\circ[H_2O(\ell)]\}$$

$$\Delta G_{\text{rxn}}^\circ = \{1\text{ mol }(-300.19\text{ kJ/mol}) + 3\text{ mol }(0\text{ kJ/mol})\} - \{1\text{ mol }(-33.56\text{ kJ/mol}) + 2\text{ mol }(-237.18\text{ kJ/mol})\}$$

$$\Delta G_{\text{rxn}}^\circ = +207.73\text{ kJ}$$

Because ΔG°_{rxn} is positive, this reaction is *not* spontaneous at 298 K. Note that we cannot simply look at a chemical equation and predict whether a reaction is spontaneous. Calculations are required.

Understanding

Propene, C_3H_6, is proposed as a starting material in the production of butyraldehyde, C_4H_8O.

$$C_3H_6(g) + CO(g) + H_2(g) \rightarrow C_4H_8O(\ell)$$

Calculate ΔG°_{rxn} to determine whether this reaction is spontaneous under standard-state conditions at 298 K.

Answer $\Delta G^\circ = -44.8$ kJ. The reaction is spontaneous.

Influence of Temperature on Gibbs Free Energy

The Gibbs free-energy change of a process is strongly influenced by temperature. We know that a decrease in enthalpy and an increase in entropy both favor spontaneous change. These observations are consistent with Equation 17.6

$$\Delta G = \Delta H - T\Delta S \quad [17.6]$$

because a negative ΔH and a positive ΔS both make negative contributions to ΔG.

Temperature affects ΔG primarily through the $T\Delta S$ term. The effect of temperature on ΔH and ΔS is generally quite small, so any change in ΔG from a temperature change arises from the change in the $-T\Delta S$ term because of the change in temperature. At low temperatures, the sign of ΔG is dominated by the sign of ΔH; as temperature increases, the $T\Delta S$ contribution becomes increasingly important. Whether an increase in temperature favors spontaneity depends on the sign of ΔS. Because the enthalpy and entropy terms can be positive or negative, any chemical reaction falls into one of four general classes, shown in Table 17.2.

Temperature influences the free energy change primarily through the $T\Delta S$ term.

The sign of ΔH determines the sign of ΔG at low temperature, and the sign of the entropy change determines the sign of ΔG at high temperature. When the enthalpy and entropy changes have opposite signs, the two terms on the right side of Equation 17.6 have the same sign, so the direction of spontaneity does not change with temperature.

Example 17.8 illustrates an important concept, the influence of temperature on a reaction. If the thermodynamics of a reaction are understood, it is sometimes possible to change the temperature to achieve spontaneity. This example also assumes that the numerical values of ΔH°_f and ΔS° are not strongly influenced by temperature. If more exact values are needed, the temperature dependence of ΔH°_f and ΔS° must be determined by experiment.

TABLE 17.2 Influence of Temperature on the Direction of Spontaneous Reaction

ΔH	ΔS	Temperature	ΔG	Spontaneous Direction
−	+	All	−	Forward
−	−	Low	−	Forward
		High	+	Reverse
+	+	Low	+	Reverse
		High	−	Forward
+	−	All	+	Reverse

EXAMPLE 17.8 Influence of Temperature on Reaction Spontaneity

Consider the production of carbon dioxide and methane from carbon monoxide and hydrogen.

$$2CO(g) + 2H_2(g) \rightarrow CO_2(g) + CH_4(g)$$

(a) Use the following data (from Appendix G) to calculate $\Delta H°$, $\Delta S°$, and $\Delta G°$ for the reaction at 298 K. All values in the table are at 298 K.

	CO(g)	H_2(g)	CO_2(g)	CH_4(g)
$\Delta H_f°$ (kJ/mol)	−110.52	0	−393.51	−74.81
$S°$ (J/mol·K)	197.56	130.57	213.63	186.15

(b) Use the values calculated in part a to estimate $\Delta G°$ at 1000 K, assuming that $\Delta H°$ and $\Delta S°$ of the reaction do not change much with temperature.

Strategy Use Hess's law to determine $\Delta H°$ and $\Delta S°$; then calculate $\Delta G°$.

Solution

(a) Use the data provided in the table to find the $\Delta H°$ and $\Delta S°$ at 298 K.

$$\Delta H°_{rxn} = \{\Delta H_f°[CO_2(g)] + \Delta H_f°[CH_4(g)]\} - \{2\Delta H_f°[CO(g)] + 2\Delta H_f°[H_2(g)]\}$$

$$\Delta H°_{rxn} = \{1 \text{ mol } (-393.51 \text{ kJ/mol}) + 1 \text{ mol } (-74.81 \text{ kJ/mol})\} - \{2 \text{ mol } (-110.52 \text{ kJ/mol}) + 2 \text{ mol } (0 \text{ kJ/mol})\}$$

$$\Delta H°_{rxn} = -247.28 \text{ kJ}$$

Then use the standard entropies to calculate $\Delta S°$.

$$\Delta S°_{rxn} = \{S°[CO_2(g)] + S°[CH_4(g)]\} - \{2S°[CO(g)] + 2S°[H_2(g)]\}$$

$$\Delta S°_{rxn} = \{1 \text{ mol } (213.63 \text{ J/mol·K}) + 1 \text{ mol } (186.15 \text{ J/mol·K})\} - \{2 \text{ mol } (197.56 \text{ J/mol·K}) + 2 \text{ mol } (130.57 \text{ J/mol·K})\}$$

$$\Delta S°_{rxn} = -256.48 \text{ J/K}$$

Calculate the standard free energy change from these values, using the equation

$$\Delta G°_{rxn} = \Delta H°_{rxn} - T\Delta S°_{rxn}$$

A unit conversion is needed so that the energy units of both terms on the right side of the equation are the same.

$$\Delta G°_{rxn} = -247.28 \text{ kJ} - 298\cancel{\text{K}} \times \left(-256.48 \frac{\cancel{\text{J}}}{\cancel{\text{K}}}\right) \times \left(\frac{1 \text{ kJ}}{1000 \cancel{\text{J}}}\right)$$

$$\Delta G°_{rxn} = -170.85 \text{ kJ}$$

The negative sign of the standard Gibbs free-energy change tells us that the reaction is spontaneous at 298 K.

(b) Because we assume that the enthalpy and entropy do not change with temperature, use the standard enthalpy and entropy changes from part a to calculate the standard Gibbs free-energy change at 1000 K. We use the same equation that was used in part a, with a temperature of 1000 K rather than 298 K.

$$\Delta G° = \Delta H° - T\Delta S° = -247.28 \text{ kJ} - 1000 \cancel{\text{K}} \times \left(-256.48 \frac{\cancel{\text{J}}}{\cancel{\text{K}}}\right) \times \left(\frac{1 \text{ kJ}}{1000 \cancel{\text{J}}}\right)$$

$$\Delta G° = +9.20 \text{ kJ}$$

At 1000 K, the reaction is no longer spontaneous at standard-state conditions. The result of this calculation is not extremely accurate, because over the rather large temperature change from 298 to 1000 K, the assumption that ΔS° does not change is not accurate.

Understanding

Calculate ΔG° at 25 °C and at 300 °C for the synthesis of ammonia under standard conditions to determine whether the reaction is spontaneous at those temperatures.

$$N_2(g) + 3H_2(g) \rightarrow 2NH_3(g)$$

Answer $\Delta G^\circ = -33.0$ kJ at 298 K. The reaction is spontaneous. $\Delta G^\circ = 21.5$ kJ at 573 K, so the reaction is not spontaneous at the greater temperature.

Example 17.8 points out a dilemma that we often face. It turns out that both of the reactions described in this example proceed so slowly at 298 K that they are impractical for synthesis. The temperature can be increased to speed up the reaction, but then it

Effect of temperature on ΔG.

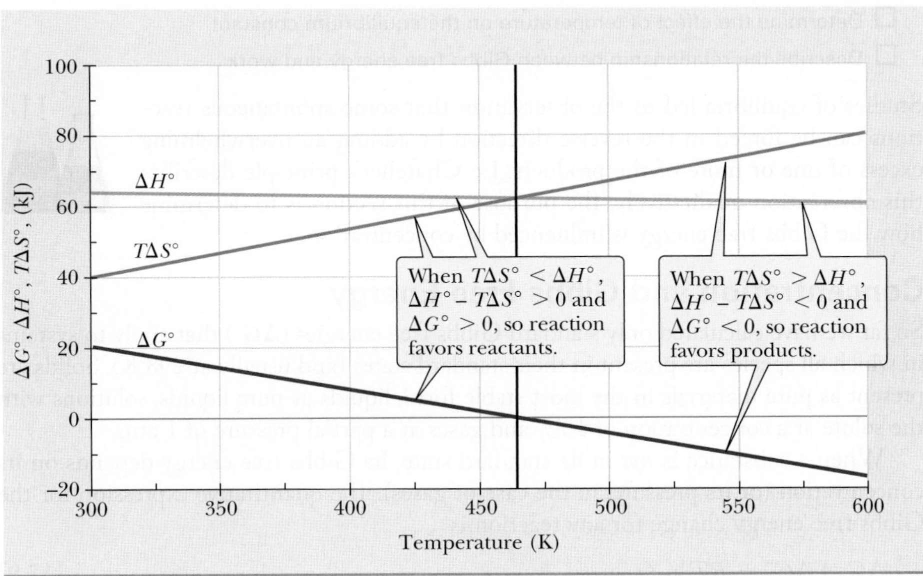

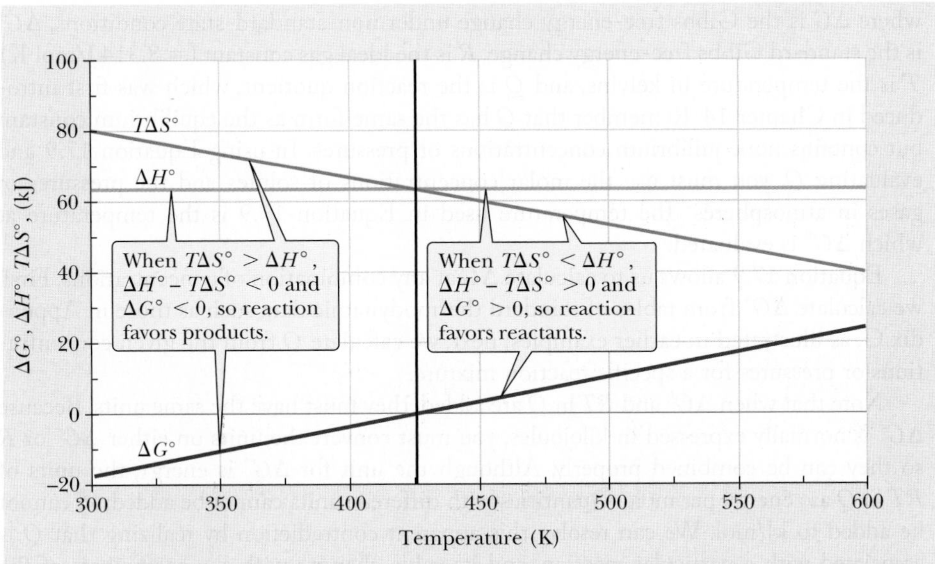

becomes less favorable, as shown by the more positive ΔG° at higher temperature. Often a compromise temperature must be found at which the rate and spontaneity of the reaction are both acceptable. Sometimes, as in the case of the formation of ammonia, a catalyst is also needed to achieve satisfactory rates at a temperature at which the reaction proceeds spontaneously.

OBJECTIVES REVIEW *Can you:*

- ☑ define Gibbs free energy and relate the sign of a Gibbs free-energy change to the direction of spontaneous reaction?
- ☑ predict the influence of temperature on Gibbs free energy?

17.5 Gibbs Free Energy and the Equilibrium Constant

OBJECTIVES

- ☐ Determine the effect of concentration on Gibbs free energy
- ☐ Calculate the standard Gibbs free-energy change from the equilibrium constant, and vice versa
- ☐ Determine the effect of temperature on the equilibrium constant
- ☐ Describe the relationship between Gibbs free energy and work

Studies of equilibria led to the observation that some spontaneous reactions can be forced in the reverse direction by adding an overwhelming excess of one or more of the products. Le Chatelier's principle describes this observation qualitatively. The purpose of this section is to determine how the Gibbs free energy is influenced by concentration.

The standard state is a pure liquid or a pure solid in its most stable form, a 1 *M* solution or a gas at 1 atm pressure. The temperature of the standard state must be specified; 298 K is frequently used as a reference temperature.

Concentration and Gibbs Free Energy

So far we have calculated only standard Gibbs free energies (ΔG°) that apply to systems in which all species are present in their standard states (and usually at 298 K). Solids are present as pure materials in the most stable form, liquids as pure liquids, solutions with the solute at a concentration of 1 *M*, and gases at a partial pressure of 1 atm.

When a substance is *not* in its standard state, its Gibbs free energy depends on its concentration (or its pressure, in the case of gases). The quantitative expression for the Gibbs free energy change for any reaction is

$$\Delta G = \Delta G^\circ + RT \ln Q \qquad [17.9]$$

where ΔG is the Gibbs free-energy change under non-standard-state conditions, ΔG° is the standard Gibbs free-energy change, R is the ideal gas constant (= 8.314 J/mol·K), T is the temperature in kelvins, and Q is the reaction quotient, which was first introduced in Chapter 14. Remember that Q has the same form as the equilibrium constant but contains nonequilibrium concentrations or pressures. In using Equation 17.9 and evaluating Q, you must use the molar concentrations of solutes and the pressures of gases in atmospheres. The temperature used in Equation 17.9 is the temperature at which ΔG° is evaluated.

Concentration influences the Gibbs free energy change, because $\Delta G = \Delta G^\circ + RT \ln Q$.

Equation 17.9 allows us to calculate ΔG at any combination of concentrations. First, we calculate ΔG° from tables of standard thermodynamic data such as those in Appendix G, as illustrated in earlier examples; next, we calculate Q from the given concentrations or pressures for a specific reaction mixture.

Note that when ΔG° and $RT \ln Q$ are added, they must have the same units. Because ΔG° is normally expressed in kilojoules, you must convert the units on either ΔG° or R so they can be combined properly. Although the unit for ΔG° is energy, the units of $RT \ln Q$ are energy per mole. Quantities with different units cannot be added; kJ cannot be added to kJ/mol. We can resolve this apparent contradiction by realizing that Q is associated with a particular reaction, and its value changes with the coefficients of the chemical equation. We understand that RT is multiplied by the appropriate number of moles when we compute $RT \ln Q$. For this reason, we omit the "per mole" units of $RT \ln Q$. Example 17.9 illustrates the influence of concentration on the Gibbs free energy.

EXAMPLE 17.9 Gibbs Free-Energy Change of a Reaction at Nonstandard Conditions

Calculate the Gibbs free-energy change for the reaction of nitrogen monoxide and bromine to form nitrosyl bromide at 298 K under two sets of conditions.

$$2NO(g) + Br_2(\ell) \rightarrow 2NOBr(g)$$

(a) The partial pressure of each gas is 1.0 atm.
(b) The partial pressure of NO is 0.10 atm, and the partial pressure of NOBr is 2.0 atm.

Solution

(a) **Strategy** If the partial pressure of each gas is 1.0 atm, we have standard-state conditions for the gases. Bromine is present as a liquid, which is its standard state. Under these conditions, ΔG equals $\Delta G°$, which we calculate from the standard Gibbs free energies of formation in Appendix G.

$$\Delta G° = \{2\Delta G_f° [NOBr(g)]\} - \{2\Delta G_f° [NO(g)] + \Delta G_f°[Br_2(\ell)]\}$$

$$\Delta G° = \{2 \text{ mol } (82.4 \text{ kJ/mol})\} - \{2 \text{ mol } (86.55 \text{ kJ/mol}) + 1 \text{ mol } (0 \text{ kJ/mol})\}$$

$$\Delta G° = -8.3 \text{ kJ}$$

Under standard conditions, the reaction to form NOBr(g) is spontaneous.

(b) **Strategy** The given partial pressure now represents *non*-standard-state conditions. When the substances are not present at standard-state conditions, the reaction quotient must be calculated.

$$2NO(g) + Br_2(\ell) \rightarrow 2NOBr(g)$$

The bromine is present as the liquid, so its concentration is not included in Q:

$$Q = \frac{P_{NOBr}^2}{P_{NO}^2} = \frac{2.0^2}{0.1^2} = 4.0 \times 10^2$$

Calculate ΔG from $\Delta G°$ and Q, after first writing -8.3 kJ as -8300 J:

$$\Delta G = \Delta G° + RT \ln Q$$

$$\Delta G = -8300 \text{ J} + (8.314 \text{ J/K})(298 \text{ K}) \ln (4.0 \times 10^2)$$

$$\Delta G = -8300 \text{ J} + 14{,}800 \text{ kJ}$$

$$\Delta G = +6500 \text{ J} = +6.5 \text{ kJ}$$

Increasing the concentration of the products and decreasing the concentration of a reactant made a reaction that is spontaneous in the forward direction (part a) become spontaneous in the reverse direction (as is predicted by Le Chatelier's principle).

Understanding

Calculate the Gibbs free-energy change for the reaction of carbon dioxide and ammonia to form urea at 298 K, when all gases are present at 0.10-atm pressures.

$$CO_2(g) + 2NH_3(g) \rightarrow CO(NH_2)_2(s) + H_2O(\ell)$$

Answer $\Delta G° = -7.30$ kJ; $Q = 1.0 \times 10^3$; $\Delta G = +9.8$ kJ. The signs on $\Delta G°$ and ΔG show that this reaction is spontaneous in the standard state, but not when the reactants are present at the lower pressures.

It is worth reiterating one point: It is the ΔG of a process that determines spontaneity, not the ΔG°. ΔG° is the free-energy change of a process under standard conditions. A process may or may not be spontaneous if all reactants and products are present under standard conditions. Calculating the ΔG value, which uses the specific conditions of a system, will determine whether a process is spontaneous under those particular conditions.

Equilibrium Constant and the Gibbs Free Energy

In Section 17.4, we mentioned that ΔG must equal zero at equilibrium, because the reaction does not proceed spontaneously in either direction. This fact can be used to find a valuable relationship between ΔG° and the equilibrium constant K_{eq}. Starting with Equation 17.9:

$$\Delta G = \Delta G^\circ + RT \ln Q$$

we know that at equilibrium, $Q = K_{eq}$ and $\Delta G = 0$, so

$$0 = \Delta G^\circ + RT \ln K_{eq}$$

Rearranging gives

$$\Delta G^\circ = -RT \ln K_{eq} \qquad [17.10]$$

Equation 17.10 relates the standard Gibbs free-energy change to the temperature and equilibrium constant. *Be sure that ΔG° is evaluated at the temperature used in this equation.* If we rearrange the equation, we can calculate the equilibrium constant from the standard Gibbs free-energy change.

$$K_{eq} = e^{-\Delta G^\circ/RT} \qquad [17.11]$$

These equations provide the relationships between the thermodynamic measure of equilibrium and the value of the equilibrium constant. Notice particularly that the temperature is an important parameter (to be discussed later in this section). The following example shows how thermodynamic data can be used to calculate an equilibrium constant.

$\Delta G^\circ = -RT \ln K_{eq}$

$K_{eq} = e^{-\Delta G^\circ/RT}$

EXAMPLE 17.10 Calculating the Gibbs Free-Energy Change from the Equilibrium Constant

The equilibrium constant for the following reaction is 1.0×10^{14} at 298 K.

$$H_3O^+(aq) + OH^-(aq) \rightarrow 2H_2O(\ell)$$

Calculate the ΔG° for the reaction.

Strategy We know that $\Delta G^\circ = -RT \ln K_{eq}$, and we have values for K_{eq} and T. Use the proper value and units for R so the units of ΔG° are joules or kilojoules.

Solution

Using $R = 8.314$ J/K and $T = 298$ K, we substitute:

$$\Delta G^\circ = -(8.314\ \text{J/}\cancel{\text{K}})(298\ \cancel{\text{K}})[\ln(1.0 \times 10^{14})]$$

The natural logarithm of 1.0×10^{14} is 32.24, and the kelvin units cancel. We have:

$$\Delta G^\circ = -(8.314\ \text{J})(298)(32.24)$$

$$\Delta G^\circ = -79{,}900\ \text{J} = -79.9\ \text{kJ}$$

Understanding

The K_{eq} for the formation of formaldehyde from water and carbon monoxide is 8.6×10^{-7}.

$$H_2(g) + CO(g) \rightarrow CH_2O(g)$$

Calculate the $\Delta G°$ at 298 K.

Answer $\Delta G° = +34.6$ kJ

EXAMPLE **17.11** **Calculating the Equilibrium Constant from the Gibbs Free-Energy Change**

The standard Gibbs free-energy change for the following reaction is +55.69 kJ.

$$AgCl(s) \rightarrow Ag^+(aq) + Cl^-(aq)$$

Calculate the equilibrium constant for this reaction at 350 K.

Strategy Because we know the $\Delta G°$ and the temperature, we can use Equation 17.11 to determine the equilibrium constant.

Solution

We should convert the $\Delta G°$ to units of J: $\Delta G° = +55{,}690$ J. Using Equation 17.11, with 8.314 J/K as the value for R:

$$K_{eq} = e^{-\Delta G°/RT} = e^{-(55690\ J)/(8.314\ J/K)(350\ K)}$$

All of the units cancel in the exponent. We get:

$$K_{eq} = e^{-19.138...}$$

$$K_{eq} = 4.88 \times 10^{-9}$$

Understanding

The Gibbs free-energy change for the following reaction is +163.2 kJ.

$$3/2\ O_2(g) \rightarrow O_3(g)$$

Calculate the equilibrium constant for this reaction at 298 K.

Answer $K_{eq} = 2.47 \times 10^{-29}$

At this point, you should be able to determine $\Delta G°$ from K_{eq} (Equation 17.10) or K_{eq} from $\Delta G°$ (Equation 17.11).

Both the sign of the Gibbs free-energy change and a comparison of the reaction quotient with the equilibrium constant can be used to determine the direction of spontaneous reaction. If the Gibbs free-energy change is negative, the reaction is spontaneous in the forward direction; if the Gibbs free-energy change is positive, the reaction is spontaneous in the reverse direction. Likewise, if the reaction quotient Q is less than K_{eq}, the forward reaction occurs spontaneously; if Q is greater than K_{eq}, the reverse reaction occurs spontaneously. As the equilibrium state is approached, ΔG approaches zero and Q approaches K_{eq}. The sign of ΔG and comparison of Q with K_{eq} are quantitative ways to consider the effect of changing concentrations, whereas Le Chatelier's principle predicts the effect of changing concentration on a qualitative basis.

If ΔG is less than zero or if Q is less than K_{eq}, the forward reaction proceeds spontaneously. If ΔG is greater than zero or if Q is greater than K_{eq}, the reverse reaction proceeds spontaneously.

Temperature and the Equilibrium Constant

Chemical reactions differ widely in their responses to changing temperature. Under any specific conditions, the direction of spontaneity is determined by the sign of ΔG. However, the equilibrium constant itself depends on temperature, as we noted in Equation 17.10:

$$\Delta G^\circ = -RT \ln K_{eq}$$

The change of K_{eq} with temperature can be quite large, because the standard free-energy change is also temperature dependent.

$$\Delta G^\circ = \Delta H^\circ - T\Delta S^\circ$$

Substituting this relationship into Equation 17.10 and rearranging, we obtain

$$\Delta H^\circ - T\Delta S^\circ = -RT \ln K_{eq}$$

$$\ln K_{eq} = \frac{\Delta S^\circ}{R} - \frac{\Delta H^\circ}{RT} \qquad [17.12]$$

Ignoring the small temperature dependencies of ΔH° and ΔS°, we see that the effect of temperature on K_{eq} depends on the sign of ΔH°. For an exothermic reaction, the negative sign of ΔH° causes the equilibrium constant to decrease with increasing temperature. Increasing the temperature of an endothermic reaction (one in which ΔH° is positive) causes the equilibrium constant to get larger, so increased temperature favors the products. In summary, the influence of temperature on the equilibrium constant is exactly as predicted by Le Chatelier's principle: Changing the temperature of a reaction shifts the equilibrium in a way consistent with whether heat is a reactant (endothermic) or a product (exothermic). Table 17.3 summarizes the effect of temperature on the equilibrium constant.

Thermodynamics provides a quantitative relationship between the equilibrium constant and the enthalpy change for a reaction.

TABLE 17.3 Influence of Temperature on the Equilibrium Constant

Sign of ΔH°	Change in Temperature	Change in K_{eq}
+ (endothermic; heat is a "reactant")	Increase	Increase
	Decrease	Decrease
− (exothermic; heat is a "product")	Increase	Decrease
	Decrease	Increase

The measurement of an equilibrium constant as a function of temperature can be used to find the value of both ΔG° and ΔH°. If the value of an equilibrium constant is measured at two temperatures, from Equation 17.12 we can obtain the following relationship:

$$\ln \frac{K_1}{K_2} = \frac{\Delta H^\circ}{R}\left(\frac{1}{T_2} - \frac{1}{T_1}\right) \qquad [17.13]$$

where K_1 and K_2 are the values of the equilibrium constant at T_1 and T_2, respectively. Equation 17.13 is known as the *Clausius–Clapeyron equation.* Example 17.12 illustrates the use of this equation to find the standard enthalpy of vaporization of a liquid.

EXAMPLE 17.12 Calculating ΔG° and ΔH° from Equilibrium Measurements

The vapor pressure of water at 25 °C is 23.77 torr, increasing to 42.20 torr at 35 °C. Calculate the standard free energy and standard enthalpy changes at 25 °C for the vaporization of water.

$$H_2O(\ell) \rightleftharpoons H_2O(g)$$

Assume that ΔH° does not change over this temperature range.

PRACTICE OF CHEMISTRY
Ice Skating

Ice is one of the few substances that we can skate on—after all, we don't skate on wood or cement floors! What are the properties of ice that would allow us to slide across it when wearing special boots with metal blades on the bottom?

There are two schools of thought. One thought is that the friction of the blade on the ice generates enough heat to melt a thin layer of ice into liquid water, and what a person is actually skating on is slippery wet ice. The other thought is the pressure of the blade on the ice actually causes the ice to melt, and the skater slides on the slippery ice.

This second hypothesis is actually based on a modified form of the Clausius–Clapeyron equation called the *Clapeyron equation:*

$$\frac{\Delta P}{\Delta T} = \frac{\Delta S}{\Delta V}$$

Here, ΔP is a pressure change, ΔT is a temperature change, ΔS is the entropy change of a process, and ΔV is the volume change of that process. For example, suppose we consider the melting of ice as the process. We know the ΔS for melting ice, as well as ΔV for that process, so we can use the Clapeyron equation to predict how the melting point of water will change (ΔT) if we change the surrounding pressure (ΔP).

Water is an unusual substance, however; when it melts, it *decreases* its volume (which is why ice floats on water). Thus, ΔV is negative for the melting of solid H_2O. Because the entropy of H_2O increases when it goes from the solid to the liquid state, ΔS is positive for the melting of ice. Thus, the right side of the Clapeyron equation is negative overall (positive divided by negative yields negative). That means that the left side of the equation must be negative as well. So if the pressure on the ice increases (ΔP is positive), the melting point temperature must decrease (ΔT must be negative). This is the science behind the second hypothesis of ice skating. It also suggests why we can skate on ice but not other materials: because ice decreases its volume when the pressure is increased, it melts. Virtually every other substance increases in volume when going from solid to liquid, so increasing the surrounding pressure would actually increase its melting point.

The question remains, is the second hypothesis, supported scientifically by the Clapeyron equation, the correct one? Probably not. A 50-kg person wearing skates with blades 20 cm by 1 mm would change the melting point by only 0.2 °C. If it were any colder than −0.2 °C (31.7 °F), not enough pressure would be available to melt the ice. Other factors, such as frictional heat, are doubtless more important in ice skating. ▮

Group of ice skaters.

Strategy First, determine the Gibbs free-energy change of the process; then use the Clausius–Clapeyron equation to determine the enthalpy change.

Solution

First, we must calculate the standard Gibbs free-energy change:

$$\Delta G^\circ = -RT \ln K_{eq}$$

Because the reactant is a pure liquid, the expression for K_{eq} is simply

$$K_{eq} = P_{H_2O}$$

so the equilibrium constant is simply the vapor pressure of water in atmospheres, 23.77/760 = 0.03128 atm. The temperature must be expressed in kelvins: 298 K. Substitution, using units of kilojoules directly, gives

$$\Delta G^\circ = -(8.314 \times 10^{-3} \text{ kJ/K})(298 \text{ K}) \ln (0.03128)$$

Omitting the units for clarity until the end:

$$\Delta G^\circ = -8.314 \times 10^{-3} \times 298 \times (-3.4648) = 8.58 \text{ kJ}$$

To calculate the standard enthalpy change, we substitute the data into the equation:

$$\ln \frac{K_1}{K_2} = \frac{\Delta H^\circ}{R}\left(\frac{1}{T_2} - \frac{1}{T_1}\right)$$

Because the equation includes the *ratio* of the two equilibrium constants, it does not matter if we express the equilibrium vapor pressure in atmosphere units or in torr units. Here we will simply use the pressure values in torr. Substituting (and leaving off the units for clarity):

$$\ln \frac{23.77}{42.20} = \frac{\Delta H^\circ}{8.314}\left(\frac{1}{308} - \frac{1}{298}\right)$$

$$\Delta H^\circ = 43{,}800 \text{ J} = 43.8 \text{ kJ}$$

You can verify that you get the same answer if you convert the equilibrium vapor pressure into atmospheres. Our answer is slightly different from the value calculated from standard enthalpies of formation ($\Delta H^\circ_{rxn} = 44.0$ kJ) because of its change with temperature from 298 to 308 K.

Understanding

The vapor pressure of methyl alcohol is 100 torr at 21.2 °C and 400 torr at 49.9 °C. What is the standard enthalpy of vaporization of methyl alcohol?

Answer 38.2 kJ

Gibbs Free Energy and Useful Work

The initial introduction to thermodynamics in Chapter 5, and this chapter's introduction, described the importance of chemical reactions in meeting the energy needs of society. The efficiency of a process is measured by the amount of useful work produced, such as turning the turbines in an electrical power plant, compared with the total chemical energy released in a reaction. In a typical coal-fueled electrical generating plant, the overall efficiency is about 34%. Higher efficiencies are possible, but the laws of thermodynamics place a theoretical limit on the maximum amount of work that can be accomplished in a spontaneous process. When a spontaneous process takes place at constant pressure and temperature, *the change in free energy, ΔG, is the maximum useful work that can be performed.*

$$w_{max} = \Delta G \qquad [17.14]$$

This relationship is the reason G is called the *free* energy—it equals the energy that is free to perform useful work. To consider the conservation of energy by improvement in the efficiency of processes, it is necessary to know this theoretical limit imposed on us by nature. When a process is spontaneous in the reverse direction (ΔG is positive), then Equation 17.14 gives the minimum amount of work needed to cause the change. We can think of maximizing the Gibbs free energy as the goal for useful work in the efficient utilization of our energy resources.

The change in the Gibbs free energy for a reaction is a theoretical limit for the maximum useful work that can be obtained under those specific conditions.

OBJECTIVES REVIEW *Can you:*

- ☑ determine the effect of concentration on the Gibbs free energy?
- ☑ calculate the standard Gibbs free-energy change from the equilibrium constant, and vice versa?
- ☑ determine the effect of temperature on the equilibrium constant?
- ☑ describe the relationship between Gibbs free energy and work?

CASE STUDY Enthalpy of Formation of Buckminsterfullerene

When a new compound is synthesized, its thermodynamic properties are among the more useful properties to study. In 1990, when macroscopic samples of buckminsterfullerene, C_{60}, were first synthesized, its basic thermodynamic properties were targets of intense interest. Chief among these properties was its enthalpy of formation, ΔH_f. Scientists need to know this value to predict how C_{60} behaves energetically in chemical reactions.

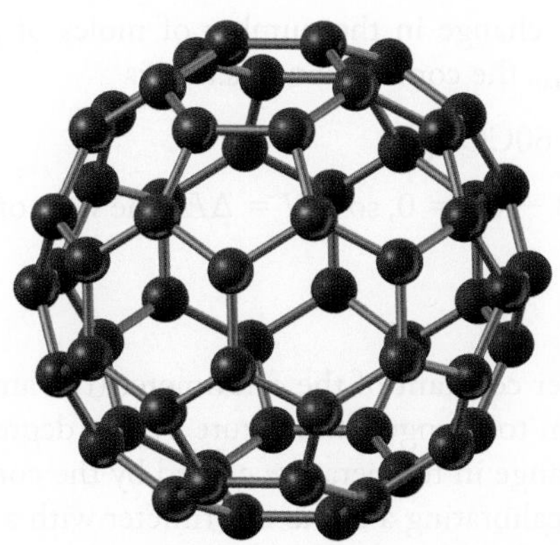

The thermodynamic properties of buckminsterfullerene, C_{60}, were of intense interest after it was first synthesized in large quantities in 1990.

Enthalpies of formation of most compounds are not measured directly, however. More typically, enthalpies of combustion are measured. Using Hess's law, these combustion energies are combined with the enthalpies of known compounds and the enthalpy of formation calculated algebraically.

Combustion energies are usually measured in a *bomb calorimeter,* a sturdy metal container that can be immersed in water. A small sample of compound is embedded with a thin platinum wire and placed in the calorimeter, which is sealed and immersed in the water. Pure oxygen gas is flushed through the calorimeter chamber, then pressurized to about 20 atm. A current is passed through the wire, which heats to red-hot and initiates combustion of the sample. The heat given off warms the calorimeter and the water surrounding it. The experimental temperature change of the water allows us to determine the energy given off by the combusting sample.

A bomb calorimeter.

Because the calorimeter is rigid, there is no volume change of the system, so work equals zero. This means that, according to the first law of thermodynamics,

$$\Delta E = q$$

in a bomb calorimeter. Therefore, the heat measured is equal to the change in internal energy, not enthalpy. We have to convert to ΔH using Equation 17.3:

$$\Delta H = \Delta E + P\Delta V$$

According to the ideal gas law, $PV = nRT$, so $P\Delta V = (\Delta n)RT$, and we can rewrite Equation 17.3 as

$$\Delta H = \Delta E + (\Delta n)RT$$

where Δn represents the change in the number of moles of gas in the course of the chemical reaction. For C_{60}, the combustion reaction is

$$C_{60}(s) + 60O_2(g) \rightarrow 60CO_2(g)$$

In this case, $\Delta n = 60 - 60 = 0$, so $\Delta H = \Delta E$. The heat of the combustion process is easily calculated as

$$q = C\Delta T$$

where C is the calorimeter constant of the calorimeter (the amount of energy required for the calorimeter system to change temperature by one degree), and ΔT is the experimentally determined change in temperature caused by the combustion of the sample.

A scientist began by calibrating a bomb calorimeter with a given mass of water. He used benzoic acid, C_6H_5COOH, which can be obtained in a highly purified form and has an enthalpy of combustion of 26,434.0 joules per gram. Approximately 1 g benzoic acid was compressed into a pellet and weighed to the nearest 0.1 mg. About 10 cm of Pt wire was pressed into the pellet, and the sample was strung on the calorimeter electrodes. The calorimeter was assembled, submerged into water, flushed with pure oxygen, then pressurized to about 20 atm of O_2. A pulse of current was sent through the wire, and the benzoic acid combusted rapidly in the high-pressure oxygen atmosphere. The scientist measured the change in temperature of the water. Using four trials performed the same way, the scientist calculated that it takes 16,775.9 J to increase the temperature of the calorimeter/water system exactly 1 °C. Thus, the calorimeter constant C is 16,775.9 J/°C, or 16,775.9 J/K.

The scientist then repeated the experiments using samples of C_{60}. Precisely weighted samples of about 150 mg pure C_{60} was mixed with sufficient high-grade graphite to yield an easily handled sample size of about 1 g. Similar experimental procedures were followed, and temperature changes of the calorimeter/water system were measured as the carbon mixtures were combusted. Knowing the precise masses of C_{60} and graphite in each sample and the known enthalpy of combustion of pure graphite, the scientist calculated a combustion energy of C_{60} as 36,123.7 J/g. Combustion energies are exothermic processes, so we can also state this as $\Delta H_{comb} = -36{,}123.7$ J/g. The molar mass of C_{60} is 720.66 g, so the molar enthalpy of combustion of C_{60} is (720.66 g/mol)(−36,123.7 J/g) = -2.6033×10^7 J/mol.

The scientist then used Hess's law to determine the enthalpy of formation of buckminsterfullerene, using the known enthalpy of formation of $CO_2(g)$, which is −393,510 J/mol:

$$60CO_2(g) \rightarrow C_{60}(s) + 60O_2(g) \qquad \Delta H = +2.6033 \times 10^7 \text{ J}$$

$$60[C(s) + O_2(g) \rightarrow CO_2(g)] \qquad \Delta H = 60[-393{,}510 \text{ J}]$$

$$60C(s) \rightarrow C_{60}(s) \qquad \Delta H \equiv \Delta H_f[C_{60}] = +2{,}422{,}400 \text{ J}$$

Thus, the calculated enthalpy of formation of buckminsterfullerene is +2422.4 kJ/mol.

Data for this case study were taken from the literature (W. V. Steele, et al. *Journal of Physical Chemistry,* 1992; volume 96: pp. 4731–4773). Students are encouraged to look up this article and learn more details about how the enthalpy of formation of buckminsterfullerene was determined.

Questions

1. Why is the bomb calorimeter flushed with oxygen before initiating the combustion reaction?
2. Why can the units on the calorimeter constant be changed to joules per degrees Kelvin if the temperature change is measured in degrees Celsius?
3. What other correction factors might have to be considered to determine a more accurate value of ΔH_f for C_{60}?

ETHICS IN CHEMISTRY

1. A great deal of attention has been given lately to the use of hydrogen as a fuel. Because it would react with oxygen to make water, some people advocate hydrogen as a clean fuel. However, other people point out that little hydrogen exists as elemental hydrogen on this planet; most of it would have to be generated by electrolysis of water, cracking of hydrocarbons, or other methods that require energy. Thus, for every joule of energy we get directly from hydrogen, we expend more energy (and increase entropy) actually generating the clean fuel itself. Which is more important: saving energy by using hydrocarbon fuels or (maybe) creating less pollution by using hydrogen as a fuel?

2. The U.S. Patent and Trademark Office does not require a working model of a new invention to consider an application for a patent except in one case: a perpetual motion machine. Because the Patent Office has never found a device that produces more energy than it consumes (an essential feature of a perpetual motion machine), they refuse to consider patents for these devices without working models. Is this policy ethical? On the one hand, nonworking machines waste time and taxpayers' dollars; on the other hand, it discourages out-of-the-box thinkers who would like protection for their ideas.

3. Suppose the scientist measuring the enthalpy of combustion of C_{60} had only enough buckminsterfullerene for a single experimental trial. Would it be ethical to publish the results of a single trial?

Chapter 17 Visual Summary

The chart shows the connections between the the major topics discussed in this chapter.

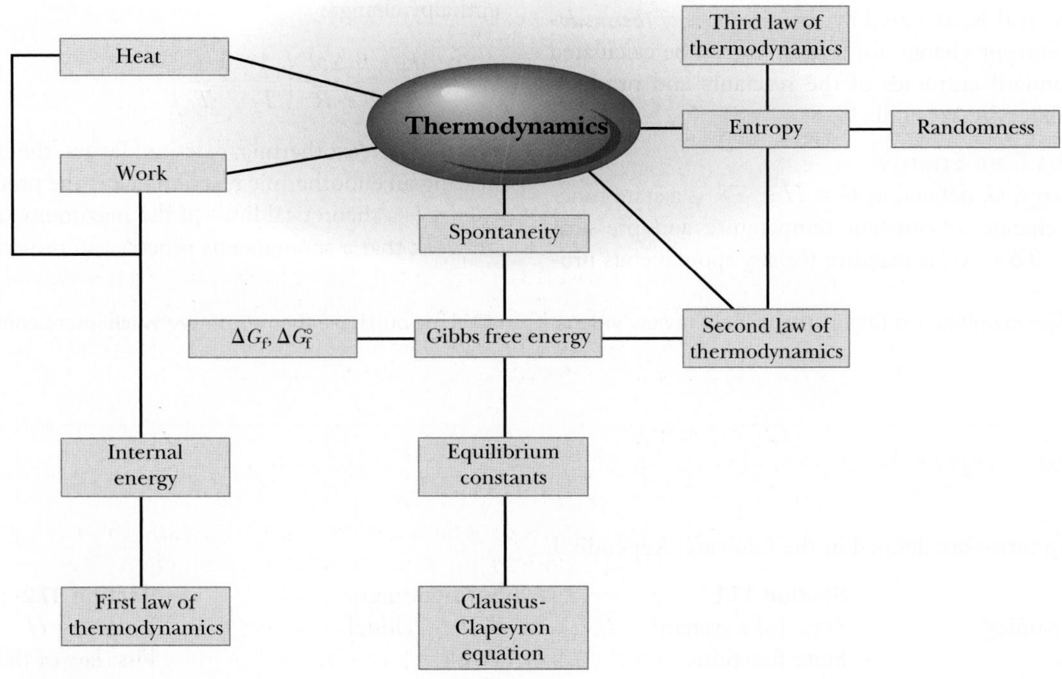

Summary

17.1 Work and Heat and 17.2 The First Law of Thermodynamics

Both heat and work are forms of energy transfer. The *internal energy, E,* of the system includes both kinetic and potential energy. Changes in the internal energy obey the *first law of thermodynamics,* $\Delta E = q + w$, where q is the heat absorbed by the system, and w is the work done on the system. If the system performs work on the surroundings, w is negative; if the surroundings do work on the system, then w is positive. Similarly, if the system absorbs heat from the surroundings, q is positive; if the surroundings absorb heat from the system, q is negative. An example of a system that does work is a gas expanding against a constant opposing pressure. Under these circumstances, $w = -P\Delta V$. If a process occurs without a change in the volume of the system, then $P\Delta V$ is zero; the heat absorbed at constant volume is ΔE. If the system is allowed to perform only PV work at constant pressure, then the heat absorbed by the system is ΔH, the change in enthalpy. Enthalpy is defined as $H = E + PV$, and it is a state function.

17.3 Entropy and Spontaneity

Reactions are driven by both changes in enthalpy and changes in *entropy,* a measure of randomness. The largest factor that influences entropy changes is a change in phase, because gases are more disordered than liquids, and liquids are more disordered than solids. Generally, if there are more moles of gases on the product side of the equation, entropy increases as the reaction proceeds. The *second law of thermodynamics* states that, for any *spontaneous* process, the entropy of the universe increases. Although absolute energy and enthalpy cannot be measured, entropy can be, because there is a reference point. The entropy of a pure crystalline substance is zero at 0 K, as stated by the *third law of thermodynamics.* The entropy change for a reaction can be calculated from the standard entropies of the reactants and products that are tabulated in Appendix G.

17.4 Gibbs Free Energy

Gibbs free energy, G, defined as $G = H - TS$, is a state function. For a change at constant temperature and pressure, $\Delta G = \Delta H - T\Delta S$. ΔG is negative for any spontaneous process and zero when the system is at equilibrium. Changes in ΔG° can be calculated from tables of ΔG_f° for the products and reactants. The change in the Gibbs free energy is influenced mainly by temperature, through the $T\Delta S$ term. Unlike enthalpy, Gibbs free energy changes significantly with a change in either temperature or pressure.

17.5 Gibbs Free Energy and the Equilibrium Constant

Concentration influences the Gibbs free-energy change, because

$$\Delta G = \Delta G^\circ + RT\ln Q$$

If Q, the reaction quotient, is less than K_{eq}, then the forward reaction proceeds spontaneously to equilibrium. When the system reaches equilibrium, then $\Delta G = 0$ and $Q = K_{eq}$, so ΔG° is related to the equilibrium constant through the equations

$$\Delta G^\circ = -RT\ln K_{eq}$$

$$K_{eq} = e^{-\Delta G^\circ/RT}$$

The effect of temperature on the equilibrium constant can be predicted qualitatively by Le Chatelier's principle and quantitatively by the equation

$$\ln K_{eq} = \frac{\Delta S^\circ}{R} - \frac{\Delta H^\circ}{RT}$$

which is based on the relationship of the equilibrium constant to the standard Gibbs free-energy change. The Clausius–Clapeyron equation relates the values of the equilibrium constants at two different temperatures to the standard enthalpy change:

$$\ln\frac{K_1}{K_2} = \frac{\Delta H^\circ}{R}\left(\frac{1}{T_2} - \frac{1}{T_1}\right)$$

Heating an exothermic reaction favors the reactants, and heating an endothermic reaction favors the products. The free energy is a theoretical limit of the maximum amount of useful work that a spontaneous process can provide.

Download Go Chemistry concept review videos from OWL or purchase them from **www.ichapters.com**

Chapter Terms

The following terms are defined in the Glossary, Appendix I.

Chemical thermodynamics
Spontaneous

Section 17.1
State (of a system)
State functions
Surroundings
System, closed
Work

Section 17.2
Enthalpy, H
First law of thermodynamics

Internal energy, E
System, isolated
Section 17.3
Entropy, S
Second law of thermodynamics
Third law of thermodynamics
Section 17.4
Gibbs free energy, G
Standard Gibbs free energy of formation

Key Equations

Work (17.1)

$$w = -P\Delta V$$

First law of thermodynamics (17.2)

$$\Delta E = q + w$$

Enthalpy and heat (17.2)

$$\Delta H = q_P$$

Internal energy and heat (17.2)

$$\Delta E = q_V$$

Internal energy and enthalpy (17.2)

$$\Delta H = \Delta E + P\Delta V$$

Entropy change (17.3)

$$\Delta S = S_{\text{final}} - S_{\text{initial}}$$

Second law of thermodynamics (17.3)

$\Delta S > 0$ for a spontaneous change in an isolated system

Third law of thermodynamics (17.3)

$S = 0$ for a perfect crystal at 0 K

Entropy change (17.3)

$$\Delta S = \frac{q}{T}$$

Gibbs free energy change (17.4)

$$\Delta G = \Delta H - T\Delta S$$

Gibbs free energy change (17.4)

$$\Delta G^\circ_{\text{rxn}} = \Sigma n\Delta G^\circ_f[\text{products}] - \Sigma m\Delta G^\circ_f[\text{reactants}]$$

Concentration dependence of ΔG (17.5)

$$\Delta G = \Delta G^\circ + RT\ln Q$$

ΔG and equilibrium constant (17.5)

$$\Delta G^\circ = -RT\ln K_{eq} \qquad K_{eq} = e^{-\Delta G^\circ/RT}$$

Temperature dependence of equilibrium constant (17.5)

$$\ln K_{eq} = \frac{\Delta S^\circ}{R} - \frac{\Delta H^\circ}{RT}$$

Clausius–Clapeyron equation (17.5)

$$\ln\frac{K_1}{K_2} = \frac{\Delta H^\circ}{R}\left(\frac{1}{T_2} - \frac{1}{T_1}\right)$$

Questions and Exercises

OWL Selected end of chapter Questions and Exercises may be assigned in OWL.

Blue-numbered Questions and Exercises are answered in Appendix J; questions are qualitative, are often conceptual, and include problem-solving skills.

■ Questions assignable in OWL

✎ Questions suitable for brief writing exercises

▲ More challenging questions

Questions

17.1 How is the sign of w, work, defined? How does it relate to the total energy of the system?

17.2 How is the sign of q, heat, defined? How does it relate to the total energy of the system?

17.3 Identify the sign of the work when a fuel-oxygen mixture (the system) burns, propelling an automobile (part of the surroundings).

17.4 What is the sign of the work when a refrigerator compresses a gas (the system) to a liquid during the refrigeration cycle?

17.5 When a rocket is launched, the burning gases are the source of the motion. If the system is the rocket (including fuel), what is the sign of the work?

17.6 A 125-L cylinder contains an ideal gas at a pressure of 100 atm. If the gas is allowed to expand against an opposing pressure of 0.0 atm at constant temperature, how much work (in kJ) is done? Explain your answer.

17.7 State the first law of thermodynamics in words and in equation form. Define all symbols used in the equation.

17.8 State the conditions under which the heat absorbed by the system is equal to the change in enthalpy.

17.9 ✎ Explain the difference between internal energy and enthalpy.

17.10 ✎ Explain why absolute enthalpies and energies cannot be measured, and only changes can be determined.

17.11 ✎ Explain why absolute entropies *can* be measured.

17.12 Under what conditions is the entropy of a substance equal to zero?

17.13 ✎ Explain why the entropy, $S°$, of an element in its standard state is not equal to zero despite the fact that ΔH_f° and ΔG_f° are equal to zero.

17.14 When most soluble ionic compounds dissolve in water, the enthalpy of solution is positive (the process is endothermic). What conclusion(s) can be made about the entropy change that accompanies the dissolution of these substances?

17.15 ✎ A colleague states, "Since there is no detectable chemical change when methane (CH_4) and oxygen mix, the reaction of these two substances is not spontaneous." Explain what is wrong with this statement.

17.16 When ice forms from liquid water at 0 °C and 1 atm pressure, ΔG is zero, because the change takes place under equilibrium conditions. Explain what the signs on the enthalpy change and entropy change must be.

17.17 ✎ Explain how the sign of the free energy as a criterion for spontaneity is a direct result of the second law of thermodynamics.

17.18 Give the relation between the change in free energy for a chemical reaction and the
(a) equilibrium constant.
(b) maximum useful work that can be obtained.
(c) free energy when reactants and products are present under standard conditions.

17.19 The free energy for a reaction decreases as temperature increases. Explain how this observation is used to determine the sign of either $\Delta H°$ or $\Delta S°$.

17.20 The equilibrium constant for a reaction decreases as temperature increases. Explain how this observation is used to determine the sign of either $\Delta H°$ or $\Delta S°$.

17.21 When solid sodium acetate crystallizes from a supersaturated solution, can you accurately predict the sign of ΔH for the crystallization? Why or why not?

17.22 When NaCl dissolves in water, can you accurately predict the sign of ΔH for the dissolution of the soluble salt? Why or why not?

Exercises

OBJECTIVE Relate heat, work, and energy.

17.23 What is the sign of w for the following processes if they occur at constant pressure? Consider only PV work from gases and assume that all gases behave ideally.
(a) $2NaHCO_3(s) + H_2SO_4(aq) \rightarrow Na_2SO_4(aq) + 2H_2O(\ell) + 2CO_2(g)$
(b) $HCl(g) + NH_3(g) \rightarrow NH_4Cl(s)$

17.24 ■ What is the sign of w for the following processes if they occur at constant pressure? Consider only PV work from gases, and assume that all gases behave ideally.
(a) $Fe_2S_3(s) + 6HNO_3(aq) \rightarrow 2Fe(NO_3)_3(aq) + 3H_2S(g)$
(b) $C_3H_8(g) + 5O_2(g) \rightarrow 3CO_2(g) + 4H_2O(\ell)$

Propane. Many regions use propane, C_3H_8, as an energy source.

OBJECTIVE Calculate work from pressure-volume relationships.

17.25 Calculate w for the following reactions that occur at 298 K and 1 atm pressure. Consider only PV work from the change in volume of gas, and assume that the gases are ideal and the chemical equation represents amounts in moles.
(a) $CO_2(g) + NaOH(s) \rightarrow NaHCO_3(s)$
(b) $3O_2(g) \rightarrow 2O_3(g)$

17.26 Calculate w for the following reactions that occur at 298 K and 1 atm pressure. Consider only PV work from the change in volume of gas, and assume that the gases are ideal and the chemical equation represents amounts in moles.
(a) $2K(s) + 2H_2O(\ell) \rightarrow 2KOH(s) + H_2(g)$
(b) $2Fe_2O_3(s) + 3C(s) \rightarrow 4Fe(s) + 3CO_2(g)$

17.27 Calculate w for the following reactions that occur at 298 K and 1 atm pressure. Consider only PV work from the change in volume of gas, and assume that the gases are ideal and the chemical equation represents amounts in moles.
(a) $Fe(s) + 5CO(g) \rightarrow Fe(CO)_5(g)$
(b) $6CO_2(g) + 6H_2O(\ell) \rightarrow C_6H_{12}O_6(s) + 6O_2(g)$

17.28 Calculate w for the following reactions that occur at 298 K and 1 atm pressure. Consider only PV work from the change in volume of gas, and assume that the gases are ideal and the chemical equation represents amounts in moles.
(a) $Fe_2O_3(s) + 2Al(s) \rightarrow 2Fe(s) + Al_2O_3(s)$
(b) $2H_2(g) + O_2(g) \rightarrow 2H_2O(\ell)$

17.29 How much work is done if a balloon expands from 1.05 to 13.8 L against a constant external pressure of 1.08 atm?

17.30 ■ Calculate the work performed if a balloon contracts from 12.90 L to 788 mL because of an external pressure of 3.70 atm.

17.31 A piston initially contains 0.400 L of air at 0.985 atm. What work is done if the piston contracts against a constant external pressure of 2.77 atm? The contraction will stop when the internal pressure equals the external pressure. Use Boyle's law to determine the final volume.

17.32 A piston initially contains 688 mL of gas at 1.22 atm. What work is done if the piston expands against a constant external pressure of 733 torr? The expansion will stop when the internal pressure equals the external pressure. Use Boyle's law to determine the final volume.

17.33 A 220-L cylinder contains an ideal gas at a pressure of 150 atm. If the gas is allowed to expand against a constant opposing pressure of 1.0 atm, how much work is done? The expansion will stop when the internal pressure equals the external pressure. Use Boyle's law to determine the final volume.

17.34 A balloon contains 2.0 L helium at 1.10 atm. Calculate the work done if the gas expands against a constant atmospheric pressure of 754 torr. The expansion will stop when the internal pressure equals the external pressure. Use Boyle's law to determine the final volume.

OBJECTIVE **Relate heat, work, energy, and the first law of thermodynamics.**

17.35 For a process, $w = -987$ J and $q = 555$ J. What is ΔE for this process?

17.36 For a process, $w = 34$ J and $q = -109$ J. What is ΔE for this process?

17.37 Calculate w for a process in which $q = 98$ J and $\Delta E = -100$ J.

17.38 ■ Calculate q for a process in which $w = 0$ J and $\Delta E = 150$ J.

17.39 A reaction between a solid and a liquid produces 4.5 L of a gas at 0.94 atm and absorbs 4.35 kJ of heat. Determine q, w, and ΔE for the reaction. (Assume an initial volume of 0.)

17.40 A reaction at 1.02 atm consumes 3.5 L of a gas and gives off 2.71 kJ of heat. Determine q, w, and ΔE for the reaction.

17.41 ▲ When an ideal gas expands at constant temperature (a condition referred to as *isothermal*), ΔE is zero because the internal energy of an ideal gas depends only on temperature, not volume. Consider 1.00 L of a gas initially at 9.00 atm and 15 °C.

(a) Calculate q and w if the gas sample expands isothermally against an opposing pressure of 1.00 atm to a final volume of 9.00 L.

(b) Calculate q and w if the gas expands isothermally, first against an opposing pressure of 3.00 atm (to an intermediate volume of 3.00 L) and then against an opposing pressure of 1.00 atm (to a final volume of 9.00 L).

(c) Calculate q and w of a three-step isothermal expansion (first to 3.00 atm and 3.00 L, then to 2.00 atm and 4.50 L, and finally to 1.00 atm and 9.00 L). Compare with the two- and one-step expansions.

(d) Based on your results for parts a to c, suggest how this expansion might be carried out so that the maximum amount of work would be done on the surroundings.

17.42 ▲ When an ideal gas is compressed at constant temperature (isothermal conditions), ΔE is zero. Consider 9.00 L of a gas that is initially at 1.00 atm and 25 °C.

(a) Calculate q and w if the gas sample is compressed isothermally at a constant pressure of 9.00 atm to a final volume of 1.00 L.

(b) Calculate q and w if the gas is compressed isothermally in two steps: first at a constant pressure of 3.00 atm (to an intermediate volume of 3.00 L) and then at a pressure of 9.00 atm (and a final volume of 1.00 L).

(c) Calculate q and w of a three-step isothermal compression (first to 2.00 atm and 4.50 L, then to 3.00 atm and 3.00 L, and finally 9.00 atm and 1.00 L). Compare with the two- and one-step compressions.

(d) Based on your results for parts a to c, suggest how this compression might be carried out so that the minimum amount of work would be done on the system.

OBJECTIVE **Distinguish between internal energy and enthalpy.**

17.43 Explain why ΔH and ΔE are so similar in value for processes that do not involve gases.

17.44 Explain why ΔH and ΔE can be quite different in value for processes that involve gases.

17.45 The products of the combustion reaction of glycerin, $C_3H_8O_3$, are gaseous carbon dioxide and liquid water. Write the balanced chemical equation for the combustion of 1 mol glycerin, and calculate ΔH for this reaction at 298 K if burning 1.240 g glycerin has a ΔE of -54.6 kJ.

17.46 The products of the combustion reaction of *n*-propanol, C_3H_7OH, are gaseous carbon dioxide and liquid water. Determine the balanced chemical equation for the combustion of *n*-propanol, and calculate ΔH for this reaction at 298 K if burning 2.09 g *n*-propanol has a ΔE of -44.08 kJ.

17.47 Calculate ΔH_f° for glycerin, $C_3H_8O_3$, at 298 K using the ΔH found in Exercise 17.45.

17.48 ■ Calculate ΔH_f° for *n*-propanol, C_3H_7OH, at 298 K using the ΔH found in Exercise 17.46.

OBJECTIVE **Define entropy and examine its statistical nature.**

17.49 What is the sign of the entropy change for each of the following processes? The system is underlined.

(a) A <u>plate</u> is dropped on the floor and shatters.

(b) A shuffled <u>deck of cards</u> is reordered from aces to deuces.

(c) <u>Iron</u> rusts into iron oxide.

(d) A <u>wooden fence</u> rots.

Does this photo suggest high entropy or low entropy?

17.50 ■ For each process, tell whether the entropy change of the system is positive or negative.
(a) A glassblower heats glass (the system) to its softening temperature.
(b) A teaspoon of sugar dissolves in a cup of coffee. (The system consists of both sugar and coffee.)
(c) Calcium carbonate precipitates out of water in a cave to form stalactites and stalagmites. (Consider only the calcium carbonate to be the system.)

OBJECTIVE Apply the second law of thermodynamics to chemical systems.

17.51 Calculate the entropy change for the following processes.
(a) 1.00 mol $H_2O(s)$ melts at 0 °C. $\Delta H_{fus} = 6.01$ kJ/mol.
(b) 2.00 mol $C_6H_6(\ell)$ vaporizes at 80.0 °C. $\Delta H_{vap} =$ 30.7 kJ/mol.

17.52 ■ Calculate the entropy change for the following processes.
(a) 2.00 mol $NH_3(\ell)$ vaporizes at −33.0 °C. $\Delta H_{vap} =$ 23.35 kJ/mol.
(b) 1.00 mol $C_2H_5OH(s)$ melts at −114 °C. $\Delta H_{fus} =$ 5.0 kJ/mol.

17.53 Use data from Appendix G to calculate the standard entropy change for the following chemical reactions.
(a) $CO(g) + 2H_2(g) \rightarrow CH_3OH(\ell)$
(b) $3H_2(g) + N_2(g) \rightarrow 2NH_3(g)$
(c) $2C_2H_2(g) + 5O_2(g) \rightarrow 4CO_2(g) + 2H_2O(\ell)$
(d) $2C(s) + O_2(g) \rightarrow 2CO(g)$

17.54 Use data from Appendix G to calculate the standard entropy change for the following chemical reactions.
(a) $C(s) + H_2O(g) \rightarrow CO(g) + H_2(g)$
(b) $2NO_2(g) \rightarrow 2NO(g) + O_2(g)$
(c) $NaCl(s) \rightarrow Na^+(aq) + Cl^-(aq)$
(d) $C_6H_{12}O_6(s) + 6O_2(g) \rightarrow 6CO_2(g) + 6H_2O(\ell)$

17.55 Use the data in Appendix G to calculate the standard entropy change for

$$H_2(g) + CuO(s) \rightarrow H_2O(\ell) + Cu(s)$$

17.56 Use the data in Appendix G to calculate the standard entropy change for

$$2Al(s) + Fe_2O_3(s) \rightarrow Al_2O_3(s) + 2Fe(s)$$

OBJECTIVE Define Gibbs free energy and relate the sign of a Gibbs free-energy change to the direction of spontaneous reaction.

17.57 Calculate $\Delta G°$ for the following reactions and state whether each reaction is spontaneous under standard conditions at 298 K.
(a) $Fe_2O_3(s) + 2Al(s) \rightarrow Al_2O_3(s) + 2Fe(s)$
(b) $CO(g) + 2H_2(g) \rightarrow CH_3OH(\ell)$

17.58 Calculate $\Delta G°$ for the following reactions and state whether each reaction is spontaneous under standard conditions at 298 K.
(a) $N_2(g) + 2H_2(g) \rightarrow N_2H_4(\ell)$
(b) $2H_2O_2(\ell) \rightarrow 2H_2O(\ell) + O_2(g)$

17.59 Calculate $\Delta G°$ for the following reactions and state whether each reaction is spontaneous under standard conditions at 298 K.
(a) $2Na(s) + H_2SO_4(\ell) \rightarrow Na_2SO_4(s) + H_2(g)$
(b) $Cu(s) + H_2SO_4(\ell) \rightarrow CuSO_4(s) + H_2(g)$

17.60 Calculate $\Delta G°$ for the following reactions and state whether each reaction is spontaneous under standard conditions at 298 K.
(a) $2NO(g) + O_2(g) \rightarrow 2NO_2(g)$
(b) $CO(g) + Cl_2(g) \rightarrow COCl_2(g)$

17.61 Calculate ΔG for the following reactions two different ways: (1) use Hess's law and the standard Gibbs free energies of formation, and (2) use $\Delta G = \Delta H - T\Delta S$. Compare the two values and judge whether you get the same ΔG_{rxn} either way. Assume $T = 298$ K.
(a) $Cl_2(g) + 2HBr(g) \rightarrow Br_2(\ell) + 2HCl(g)$
(b) $Fe_2O_3(s) + 3SO_3(\ell) \rightarrow Fe_2(SO_4)_3(s)$

17.62 Calculate ΔG for the following reactions two different ways: (1) use Hess's law and the standard Gibbs free energies of formation, and (2) use $\Delta G = \Delta H - T\Delta S$. Compare the two values and judge whether you get the same ΔG_{rxn} either way. Assume $T = 298$ K.
(a) $2SO_2(g) + O_2(g) + 2H_2O(\ell) \rightarrow 2H_2SO_4(\ell)$
(b) $H^+(aq) + OH^-(aq) \rightarrow H_2O(\ell)$

17.63 Calculate $\Delta H°$, $\Delta S°$, and $\Delta G°$ for each of the following reactions at 298 K. State whether the direction of spontaneous reaction is consistent with the sign of the enthalpy change, the entropy change, or both. Use Appendix G for data.
(a) $H_2(g) + \frac{1}{2}O_2(g) \rightarrow H_2O(\ell)$
(b) $CO(g) + 2H_2(g) \rightarrow CH_3OH(\ell)$

17.64 Calculate $\Delta H°$, $\Delta S°$, and $\Delta G°$ for each of the following reactions at 298 K. State whether the direction of spontaneous reaction is consistent with the sign of the enthalpy change, the entropy change, or both. Use Appendix G for data.
(a) $2H_2O(g) + 2Cl_2(g) \rightarrow 4HCl(g) + O_2(g)$
(b) $2CO_2(g) + 4H_2O(\ell) \rightarrow 2CH_3OH(\ell) + 3O_2(g)$

17.65 Calculate $\Delta H°$, $\Delta S°$, and $\Delta G°$ for each of the following reactions. State whether the direction of spontaneous reaction is consistent with the sign of the enthalpy change, the entropy change, or both. Use Appendix G for data.
(a) $CH_3COOH(\ell) + NaOH(s) \rightarrow Na^+(aq) + CH_3COO^-(aq) + H_2O(\ell)$
(b) $AgNO_3(s) + Cl^-(aq) \rightarrow AgCl(s) + NO_3^-(aq)$

17.66 ■ Use standard entropies and heats of formation to calculate $\Delta G_f°$ at 25°C for
(a) cadmium(II) chloride(s).
(b) methyl alcohol, $CH_3OH(\ell)$.
(c) copper(I) sulfide(s).

OBJECTIVE Predict the influence of temperature on free energy.

In Exercises 17.67 to 17.82, use data from Appendix G. Assume that $\Delta H°$ and $\Delta S°$ do not vary with temperature, and that the reactants and products are present in their standard states.

17.67 What is the sign of the standard Gibbs free-energy change at low temperatures and at high temperatures for the combustion of acetaldehyde?

$$CH_3CHO(\ell) + \tfrac{5}{2}O_2(g) \rightarrow 2CO_2(g) + 2H_2O(\ell)$$

17.68 What is the sign of the standard Gibbs free-energy change at low temperatures and at high temperatures for the formation of hydrogen sulfide from the elements?

$$H_2(g) + \tfrac{1}{8}S_8(s) \rightarrow H_2S(g)$$

17.69 What is the sign of the standard Gibbs free-energy change at low temperatures and at high temperatures for the synthesis of ammonia?

$$3H_2(g) + N_2(g) \rightarrow 2NH_3(g)$$

17.70 What is the sign of the standard Gibbs free-energy change at low temperatures and at high temperatures for the decomposition of phosgene?

$$COCl_2(g) \rightarrow CO(g) + Cl_2(g)$$

17.71 Predict the temperature at which the reaction in Exercise 17.67 comes to equilibrium. Consider the equation $\Delta G = \Delta H - T\Delta S$. At some value of T, ΔG equals zero and the reaction is at equilibrium. Set $\Delta G = 0$, substitute for ΔH° and ΔS° (assuming that they do not vary much with temperature) and solve for T.

17.72 Predict the temperature at which the reaction in Exercise 17.68 comes to equilibrium. (See Exercise 17.71 for the strategy.)

17.73 Predict the temperature at which the reaction in Exercise 17.69 comes to equilibrium. (See Exercise 17.71 for the strategy.)

17.74 Predict the temperature at which the reaction in Exercise 17.70 comes to equilibrium. (See Exercise 17.71 for the strategy.)

17.75 Calculate ΔG° at 400 and 600 K for the following reactions.
(a) $2NO(g) + Br_2(g) \rightarrow 2NOBr(g)$
(b) $2NH_3(g) \rightarrow N_2H_4(\ell) + H_2(g)$

Ammonia, NH_3, is used to make hydrazine, N_2H_4, which can be used as a fuel.

17.76 ■ Calculate ΔG° at 300 and 390 K for the following reactions.
(a) $BaCO_3(s) \rightarrow BaO(s) + CO_2(g)$
(b) $CH_3COOH(\ell) \rightarrow CH_4(g) + CO_2(g)$

17.77 Suppose you are looking for a chemical reaction that is spontaneous at low temperatures but proceeds in the reverse direction at high temperatures. What are the signs of ΔH° and ΔS° for such a reaction?

17.78 Suppose you are looking for a chemical reaction that is spontaneous at high temperatures but proceeds in the reverse direction at low temperatures. What are the signs of ΔH° and ΔS° for such a reaction?

17.79 Identify which of the following statements are incorrect and change them so that they are true. The statements refer to the formation of 1 mol methanol (CH_3OH) from carbon monoxide and hydrogen (all at 1 atm pressure and in the gas phase):

$$CO(g) + 2H_2(g) \rightarrow CH_3OH(g)$$

(a) The direction of spontaneous reaction depends entirely on ΔH°.
(b) The reaction is spontaneous at 298 K.
(c) The reaction proceeds spontaneously at all temperatures greater than 298 K.
(d) The reaction proceeds spontaneously at all temperatures less than 298 K.
(e) The equilibrium constant increases with increasing temperature.
(f) The reaction proceeds quickly at any spontaneous temperature.

17.80 ■ Decide whether each of the following statements is true or false. If false, rewrite it to make it true.
(a) The entropy of a substance increases on going from the liquid to the vapor state at any temperature.
(b) An exothermic reaction will always be spontaneous.
(c) Reactions with a positive ΔH° and a positive ΔS° can never be product favored.
(d) If ΔG° for a reaction is negative, the reaction will have an equilibrium constant greater than 1.

17.81 Determine whether the vaporization of methanol is spontaneous at 80 °C and 1 atm. Use the thermodynamic data in Appendix G. State any assumptions you make.

$$CH_3OH(\ell) \rightarrow CH_3OH(g)$$

17.82 Determine whether the condensation of nitromethane is spontaneous at 40 °C and 1 atm. Use the thermodynamic data in Appendix G. State any assumptions you make.

$$CH_3NO_2(g) \rightarrow CH_3NO_2(\ell)$$

OBJECTIVE Determine the effect of concentration on free energy.

17.83 At 298 K, $\Delta G^\circ = -70.52$ kJ for the reaction

$$2NO(g) + O_2(g) \rightleftharpoons 2NO_2(g)$$

(a) Calculate ΔG at the same temperature when $P_{NO} = 1.0 \times 10^{-4}$ atm, $P_{O_2} = 2.0 \times 10^{-3}$ atm, and $P_{NO_2} = 0.30$ atm.
(b) Under the conditions in part a, in which direction is the reaction spontaneous?

17.84 At 298 K, $\Delta G^\circ = -6.36$ kJ for the reaction

$$2N_2O(g) + 3O_2(g) \rightleftharpoons 2N_2O_4(g)$$

(a) Calculate ΔG at the same temperature when $P_{N_2O} = 4.0 \times 10^{-2}$ atm, $P_{O_2} = 4.2 \times 10^{-3}$ atm, and $P_{N_2O_4} = 0.40$ atm.
(b) Under the conditions in part a, in which direction is the reaction spontaneous?

17.85 At 298 K, $\Delta G^\circ = +27.4$ kJ for the reaction

$$PbCl_2(s) \rightleftharpoons Pb^{2+}(aq) + 2Cl^-(aq)$$

(a) Calculate ΔG at the same temperature when $[Pb^{2+}] = 4.0 \times 10^{-4}$ M and $[Cl^-] = 2.5 \times 10^{-3}$ M.
(b) Under the conditions in part a, in which direction is the reaction spontaneous?

17.86 At 298 K, $\Delta G^\circ = +11.51$ kJ for the reaction

$$CaF_2(s) \rightleftharpoons Ca^{2+}(aq) + 2F^-(aq)$$

(a) Calculate ΔG at the same temperature when $[Ca^{2+}] = 3.5 \times 10^{-2}$ M and $[F^-] = 2.3 \times 10^{-3}$ M.
(b) Under the conditions in part a, in which direction is the reaction spontaneous?

OBJECTIVE **Calculate standard free-energy change from the equilibrium constant, and vice versa.**

For Exercises 17.87 to 17.102, assume that ΔH° and ΔS° do not change with temperature.

17.87 Calculate the normal boiling point of methanol (CH_3OH). Use the thermodynamic data in Appendix G. Compare your answer with the experimentally measured boiling point.

17.88 Calculate the normal boiling point of nitromethane (CH_3NO_2). Use the thermodynamic data in Appendix G. Compare your answer with the experimentally measured boiling point.

17.89 For each reaction, an equilibrium constant at 298 K is given. Calculate ΔG° for each reaction.
(a) $H^+(aq) + OH^-(aq) \rightleftharpoons H_2O$ $K_c = 1.0 \times 10^{-14}$
(b) $CaSO_4(s) \rightleftharpoons Ca^{2+}(aq) + SO_4^{2-}(aq)$ $K_c = 7.1 \times 10^{-5}$
(c) $HIO_3(aq) \rightleftharpoons H^+(aq) + IO_3^-(aq)$ $K_c = 1.7 \times 10^{-1}$

17.90 ■ For each reaction, an equilibrium constant at 298 K is given. Calculate ΔG° for each reaction.
(a) $Br_2(\ell) + H_2(g) \rightleftharpoons 2HBr(g)$ $K_P = 4.4 \times 10^{18}$
(b) $H_2O(\ell) \rightleftharpoons H_2O(g)$ $K_P = 3.17 \times 10^{-2}$
(c) $N_2(g) + 3H_2(g) \rightleftharpoons 2NH_3(g)$ $K_c = 3.5 \times 10^8$

OBJECTIVE **Determine the effect of temperature on the equilibrium constant.**

17.91 Use the standard Gibbs free-energy change to calculate the value of the equilibrium constant for the reaction $PCl_5(g) \rightleftharpoons PCl_3(g) + Cl_2(g)$.
(a) at 25 °C. (b) at 250 °C. (c) at −50 °C.

17.92 Use the data in Appendix G to calculate the value of the equilibrium constant for the reaction $2SO_2(g) + O_2(g) \rightarrow 2SO_3(g)$.
(a) at 25 °C. (b) at 250 °C. (c) at −50 °C.

17.93 Suppose you have an endothermic reaction with $\Delta H^\circ = +15$ kJ and a ΔS° of +150 J/K. Calculate ΔG° and K_{eq} at 10, 100, and 1000 K.

17.94 Suppose you have an endothermic reaction with $\Delta H^\circ = +15$ kJ and a ΔS° of −150 J/K. Calculate ΔG° and K_{eq} at 10, 100, and 1000 K.

17.95 Suppose you have an exothermic reaction with $\Delta H^\circ = -15$ kJ and a ΔS° of +150 J/K. Calculate ΔG° and K_{eq} at 10, 100, and 1000 K.

17.96 Suppose you have an exothermic reaction with $\Delta H^\circ = -15$ kJ and a ΔS° of −150 J/K. Calculate ΔG° and K_{eq} at 10, 100, and 1000 K.

17.97 Calculate ΔG° and ΔG at 303 °C for the following equation.

$$CO(g, 2\text{ atm}) + Cl_2(g, 1\text{ atm}) \rightleftharpoons COCl_2(g, 0.1\text{ atm})$$

17.98 ■ Calculate ΔG° and ΔG at 37 °C for the following equation.

$$N_2O(g, 1\text{ atm}) + H_2(g, 0.4\text{ atm}) \rightleftharpoons N_2(g, 1\text{ atm}) + H_2O(\ell)$$

17.99 State whether increasing temperature increases or decreases the value of the equilibrium constant for the following reactions.
(a) $N_2O(g) + H_2(g) \rightleftharpoons N_2(g) + H_2O(\ell)$
(b) $CO(g) + Cl_2(g) \rightleftharpoons COCl_2(g)$
(c) $CO(g) + H_2O(g) \rightleftharpoons HCOOH(g)$
(d) $PCl_5(g) \rightleftharpoons PCl_3(g) + Cl_2(g)$
(e) $2SO_2(g) + O_2(g) \rightleftharpoons 2SO_3(g)$

17.100 State whether increasing temperature increases or decreases the value of the equilibrium constant for the following reactions.
(a) $2NO_2(g) \rightleftharpoons N_2O_4(g)$
(b) $2CO(g) + O_2(g) \rightleftharpoons 2CO_2(g)$
(c) $2NO(g) + Br_2(g) \rightleftharpoons 2NOBr(g)$
(d) $2HBr(g) \rightleftharpoons H_2(g) + Br_2(g)$
(e) C(s, graphite) $\rightleftharpoons$ C(s, diamond)

17.101 Calculate the vapor pressure of each of the following at the given temperature. (*Hint:* The vapor pressure of each substance equals 760 torr at its normal boiling point temperature.)
(a) $CS_2(\ell)$ at 5 °C
(b) $H_2O(\ell)$ at 50 °C
(c) $C_6H_6(\ell)$ at 45 °C

17.102 Calculate the vapor pressure of each of the following at the given temperature. (*Hint:* The vapor pressure of each substance equals 760 torr at its normal boiling point temperature.)
(a) $CH_3OH(\ell)$ at 58 °C
(b) $C_2H_5OH(\ell)$ at 29 °C
(c) $Hg(\ell)$ at 45 °C

Chapter Exercises

17.103 ▲ A 220-ft³ sample of gas at standard temperature and pressure is compressed into a cylinder, where it exerts pressure of 2000 psi. Calculate the work (in J) performed when this gas expands isothermally against an opposing pressure of 1.0 atm. (The amount of work that can be done is equivalent to the destructive force of about 1/4 lb of dynamite, giving you an idea of how potentially destructive compressed gas cylinders can be if improperly handled!)

17.104 ▲ What is the sign of the standard Gibbs free-energy change at low temperatures and at high temperatures for the explosive decomposition of TNT? Use your knowledge of TNT and the chemical equation, particularly the phases, to answer this question. (Thermodynamic data for TNT are not in Appendix G.)

$$2C_7H_5N_3O_6(s) \rightarrow 3N_2(g) + 5H_2O(\ell) + 7C(s) + 7CO(g)$$

17.105 ▲ The equilibrium constant for the formation of phosgene is measured at two different temperatures.

$$CO(g) + Cl_2(g) \rightarrow COCl_2(g)$$

At 506 °C, $K_{eq} = 1.3$; at 530 °C, $K_{eq} = 0.78$. Calculate ΔH° and ΔS° for this reaction. Under standard-state conditions, over what temperature range is the reaction spontaneous?

17.106 ■ Elemental boron, in the form of thin fibers, can be made by reducing a boron halide with H_2.

$$BCl_3(g) + 3/2\ H_2(g) \rightarrow B(s) + 3HCl(g)$$

Calculate ΔH°, ΔS°, and ΔG° at 25 °C for this reaction. Is the reaction predicted to be product favored at equilibrium at 25 °C? If so, is it enthalpy driven or entropy driven?

17.107 Calculate the standard Gibbs free-energy change when SO_3 forms from SO_2 and O_2 at 298 K. Why is sulfur trioxide an important substance to study? (*Hint:* What happens when it combines with water?)

17.108 ■ The *thermite reaction* is

$$2Al(s) + Fe_2O_3(s) \rightarrow Al_2O_3(s) + 2Fe(s)$$

(a) Calculate $\Delta G°$ for this reaction.
(b) Calculate K_{eq} for this reaction.
Assume $T = 298$ K. You may have to do some mathematical manipulations to get your final numerical answer.

© Phil Degginger/Alamy

The thermite reaction (see Exercise 17.108).

17.109 Chemists and engineers who design nuclear power plants have to worry about high-temperature reactions because it is possible for water to decompose.
(a) Under what conditions does this reaction occur spontaneously?

$$2H_2O(g) \rightarrow 2H_2(g) + O_2(g)$$

(b) Under conditions where the decomposition of water is spontaneous, do nuclear engineers have to worry about an oxygen/hydrogen explosion? Justify your answer.

17.110 Another type of thermite reaction (see Exercise 17.108) uses Cr_2O_3 instead of Fe_2O_3:

$$2Al(s) + Cr_2O_3(s) \rightarrow Al_2O_3(s) + 2Cr(s)$$

(a) Calculate $\Delta H°$, $\Delta S°$, and $\Delta G°$ for this reaction.
(b) On the assumption that the energy released goes to warming up the products, which reaction generates the greatest temperature, this one or the thermite reaction in Exercise 17.108? (*Hint:* You will need the heat capacities of the products of each reaction to answer.)

17.111 ▲ The reaction of carbon with metal oxide ores is used to isolate metals whenever possible, because carbon is an inexpensive reactant. You are asked to determine the feasibility of using carbon to prepare Ca from lime, which has the formula CaO. The reaction would be

$$2CaO(s) + C(s, graphite) \rightarrow 2Ca(s) + CO_2(g)$$

An analysis of the economics shows that the process is practical if the reaction occurs at less than 1500 °C.
(a) Is this reaction spontaneous under standard conditions at 298 K? Use the data in Appendix G.
(b) Estimate the minimum temperature at which this reaction proceeds under standard-state conditions.
(c) In the course of your investigation, you discover that the normal boiling point of calcium is 1484 °C. Estimate the minimum temperature at which this process becomes spontaneous under standard conditions with gaseous calcium as a product.
(d) Can Le Chatelier's principle be used to decrease the temperature needed to produce calcium by reaction of CaO with graphite? Explain your reasoning.

17.112 ▲ A cycloalkane is a hydrocarbon that contains a ring of carbon atoms with two hydrogen atoms bonded to each carbon. The standard enthalpy changes for the combustion of several gaseous cycloalkanes to form gaseous water and carbon dioxide have been measured and are summarized in the table below.

$$C_nH_{2n}(g) + \frac{3}{2}nO_2(g) \rightarrow nCO_2(g) + nH_2O(g)$$

Cycloalkane	Formula	Standard Enthalpy of Combustion, kJ/mol	C–C–C Bond Angle
Cyclopropane	$C_3H_6(g)$	−1957.7	60°
Cyclobutane	$C_4H_8(g)$	−2567.6	88°
Cyclopentane	$C_5H_{10}(g)$	−3097.6	108°
Cyclohexane	$C_6H_{12}(g)$	−3685.5	109°

Because all reactants and products are in the gaseous state, bond energies may be used to estimate the enthalpy changes for these reactions.
(a) The formulas of the related noncyclic alkanes are C_3H_8, C_4H_{10}, C_5H_{12}, and C_6H_{14}. What are the oxidation numbers of the carbon atoms in the cycloalkane and noncyclic alkanes? Can you explain the trend?
(b) Draw Lewis structures of the four cycloalkanes. According to VSEPR, what should the bond angles be?
(c) Combine the measured enthalpies of combustion with the bond energies (see Table 9.4) for C=O, O–H, O_2, and C–H to calculate the average bond energy of the C–C bond in each of these cycloalkanes.
(d) The general formula for the cycloalkanes is $(CH_2)_n$. Determine the value of n for each cycloalkane in the table, and determine the amount of energy given off per CH_2 unit for each compound. What trend do you see?
(e) Compare the observed angles given in the table with those found in part b and offer an explanation for the trend in the C–C bond energies calculated in part c.

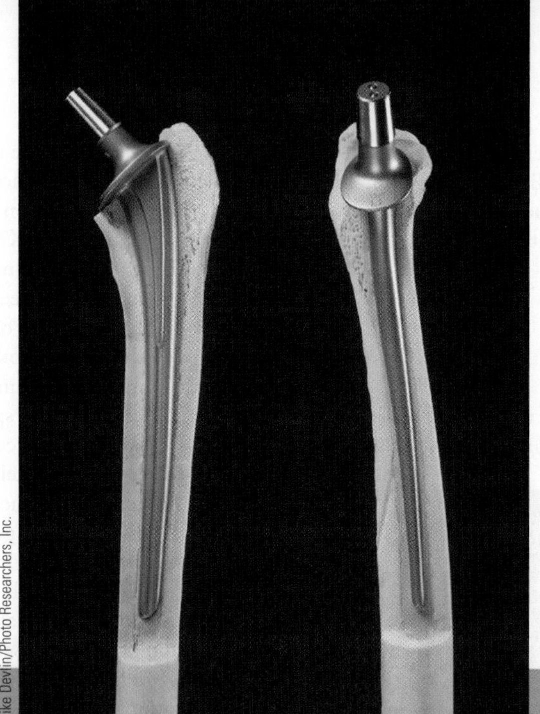

Artificial implants. The femur end of an artificial hip joint is composed of a titanium alloy that resists corrosion when exposed to body fluids.

Environmental corrosion is the ultimate cause for up to 5% of the replacement construction costs in the United States annually, costing U.S. taxpayers more than $100 billion per year. Most corrosion results from electrochemical processes, which are the subject of this chapter.

Some medical doctors need to be aware of the effects of corrosion. Surgeons who implant artificial joints, stents, or other devices that come in contact with living tissue must be concerned with how those devices may be corroded by the living environment. Thus, the corrosion of biomaterials, materials used to construct body implants, is an area of constant concern.

Some areas of the body are acidic. Biomaterials that come into contact with urine or stomach contents must be resistant to relatively low pH values. Biomaterials that come into contact with blood or intracellular fluid face a different, pernicious chemical: the chloride ion. Aqueous Cl^- ions can be destructive to the protective oxide coatings on metal surfaces, promoting additional corrosion. Corroding artificial body parts can lead to additional major surgeries, as in the case of an artificial hip joint, or serious

Electrochemistry

18

CHAPTER CONTENTS

OWL Online homework for this chapter may be assigned in OWL.

Look for the green colored vertical bar throughout this chapter, for integrated references to this chapter introduction.

cardiovascular consequences, as in the case of a malfunctioning stent that is keeping an artery open.

A variety of materials are used for making body implants that are corrosion resistant. Vascular stents, used to keep open narrowing blood vessels, are made of surgical-grade stainless steel. This type of stainless steel, although mostly iron, also contains about 16% chromium, 12% nickel, and other elements that make it resistant to corrosion. But even stainless steel is not a perfect material, because some people experience the development of allergies to the trace amount of nickel that slowly leeches out from the stent. Titanium alloys, which are also resistant to corrosion, are used for replacement hips. Gold, which is almost impervious to corrosion, is used in dental work. The metal in some replacement parts can be coated with hydroxyapatite, a major component of tooth enamel that is fairly corrosion resistant.

Ironically, a current topic of research is in the area of biodegradable inserts, which are designed to slowly decompose in the body—in a sense, intentionally corroding replacement parts! ▮

This person has gold teeth replacing his original ones.

Much of our current understanding of the structure of matter comes from studying the relationships between chemical reactions and electricity. These relationships have important practical applications: the power for flashlights, radios, calculators, and starter motors for automobiles comes from chemical reactions that produce electricity directly. Electricity helps produce many of the materials essential to our society, such as aluminum and chlorine. **Electrochemistry** is the study of the relations between chemical reactions and electricity.

The basis for all electrochemical processes is the transfer of electrons from one substance to another. We must examine electron transfer, or oxidation-reduction reactions, in more detail before beginning our study of electrochemical processes.

18.1 Oxidation Numbers

OBJECTIVES

- ☐ Define oxidation, reduction, and redox reactions
- ☐ Assign oxidation numbers to atoms in chemical species

Chapter 3 defines oxidation-reduction (or "redox") reactions as reactions in which some of the atoms change oxidation numbers. One example given is the reaction

$$\underset{0}{4Na} + \underset{0}{O_2} \rightarrow 2\underset{+1}{Na_2}\underset{-2}{O}$$

where the oxidation numbers are given below each species. In this example, in which uncombined elements are converted into an ionic compound, the changes in oxidation number are quite obvious. In reactions that involve only covalently bonded substances, the changes in oxidation numbers are not as obvious. For example, consider the following complex reaction:

$$5H_2C_2O_4(aq) + 6H_3O^+(aq) + 2MnO_4^-(aq) \rightarrow 10CO_2(g) + 2Mn^{2+} + 14H_2O(\ell)$$

It is difficult to determine by looking whether this is a redox reaction. A more formal way of identifying oxidation-reduction reactions is needed.

Assigning Oxidation Numbers

To help determine whether any reaction is an oxidation-reduction reaction, chemists have devised a bookkeeping concept called *oxidation numbers* (this is discussed briefly in Chapter 3). In this chapter, we examine this concept in further detail.

Certain rules, which should be applied in order, exist for assigning oxidation numbers to atoms. Here, we expand on the rules first presented in Chapter 3 and give several examples:

Rule 1: The oxidation state for the atoms in their elemental state is zero; for example, the oxidation number of each atom in the H_2 molecule is 0.

Rule 2: The oxidation number of a monatomic ion is the electrical charge on the ion. The oxidation number of Na^+ is +1, whereas the oxidation number of the sulfide ion, S^{2-}, is −2. Note the convention for representing oxidation numbers: the sign comes first, then the magnitude.

Rule 3: Fluorine always has an oxidation number of −1 in its compounds. The other halogens (Group 7A) have an oxidation state of −1 unless they are combined with a more electronegative halogen or oxygen (in which case, rule 6 applies).

Rule 4: In most cases, the oxidation number of oxygen is −2, except when it is bonded to fluorine (where it may be +1 or +2), and in substances that contain an O–O bond (peroxides), where it has an oxidation state of −1. For example, in H_2O, the oxygen atom has an oxidation number of −2, whereas in H_2O_2 (hydrogen peroxide), each oxygen atom has an oxidation number of −1.

Rule 5: In most cases, hydrogen has an oxidation number of +1, except when it is combined with a less electronegative element (the metallic elements and boron), in which case, its oxidation number is −1. For example, in H_2O, the hydrogen atoms have an oxidation number of +1, whereas in sodium hydride, NaH, the oxidation number of H is −1.

Rule 6: The sum of the oxidation states of all the atoms in a compound or ion is equal to the charge on the compound or ion. For example, in sulfur dioxide, SO_2, the sum of the oxidation numbers of the atoms must add to zero. Because the oxidation number of each oxygen is −2 (rule 4), the oxidation number of S is determined as follows:

$$\text{Oxidation number of S} + 2(-2) = 0$$

$$\text{Oxidation number of S} - 4 = 0$$

$$\text{Oxidation number of S} = +4$$

However, in the sulfate ion, SO_4^{2-}, the oxidation number of each oxygen is −2, so the oxidation number of S is determined as follows:

$$\text{Oxidation number of S} + 4(-2) = -2$$

$$\text{Oxidation number of S} - 8 = -2$$

$$\text{Oxidation number of S} = +6$$

This method of assigning oxidation numbers to atoms is simply a kind of "electron bookkeeping." Do not try to interpret oxidation numbers as representing the actual charge that exists on every atom in the formula of a compound or ion.

EXAMPLE 18.1 Assigning Oxidation Numbers by Rules

Assign the oxidation numbers to all elements in (a) K_2S, (b) NH_3, (c) BaO_2, (d) $Cr_2O_7^{2-}$, and (e) Br_2.

Strategy Apply the rules listed earlier, in order, for each element in the formula of the substance.

Solution

(a) From rule 2, the oxidation number of potassium is +1. Also using rule 2, the oxidation number of S must be −2. Note that the sum of oxidation numbers of all species in this formula is zero, as required by rule 6.

(b) Applying rule 5, we find the oxidation number of H is +1; thus, the oxidation number of N is −3 (rule 6).

(c) The oxidation number of Ba is +2 (rule 2), and the oxidation number of the oxygen atoms must be −1 (rule 6). This assignment suggests that this compound is a peroxide that contains an O–O bond in the anion.

(d) The oxidation number of oxygen is usually −2 (rule 4); then from rule #6:

$$\text{Sum of oxidation numbers of atoms} = \text{charge on ion}$$

$$2(\text{oxidation number of Cr}) + 7(-2) = -2$$

Solving, we get +6 for the oxidation number of Cr.

(e) Br_2 is elemental bromine, which normally exists as a diatomic molecule. By rule 1, the oxidation number of the atoms in this molecule is 0.

Understanding

Assign the oxidation numbers to the elements in ClO_3^-.

Answer O is −2, and Cl is +5.

We will now repeat the definitions of oxidation and reduction from Chapter 3. If the oxidation number of an element increases in a chemical reaction, it is oxidized. If its

An oxidation-reduction, or redox, reaction is a reaction in which some of the elements undergo a change in oxidation number.

oxidation number decreases, it is reduced. A redox reaction can be defined as a reaction in which some of the elements undergo a change in oxidation number. If you check all of the examples of redox reactions presented in this chapter so far, you will see that they all satisfy this definition of a redox reaction.

EXAMPLE 18.2 Determining Oxidation Numbers in a Redox Reaction

Determine the oxidation numbers of the elements in the following reaction, and determine whether it is a redox reaction. If so, what is being oxidized, and what is being reduced?

$$5H_2C_2O_4(aq) + 6H_3O^+(aq) + 2MnO_4^-(aq) \rightarrow 10CO_2(g) + 2Mn^{2+}(aq) + 14H_2O(\ell)$$

Strategy Apply the rules for assigning oxidation numbers and look for elements that are changing oxidation numbers. Any element increasing its oxidation number is being oxidized, whereas any element decreasing its oxidation number is being reduced.

Solution

By applying the earlier rules, the following oxidation numbers shown under the elements can be assigned:

$$5\underset{+1}{H_2}\underset{+3}{C_2}\underset{-2}{O_4}(aq) + 6\underset{+1}{H_3}\underset{-2}{O^+}(aq) + 2\underset{+7}{Mn}\underset{-2}{O_4^-}(aq) \rightarrow 10\underset{+4}{C}\underset{-2}{O_2}(g) + 2\underset{+2}{Mn^{2+}}(aq) + 14\underset{+1}{H_2}\underset{-2}{O}(\ell)$$

We can see from the numbers that the oxidation number of carbon is going from +3 to +4; therefore, it is being oxidized. The oxidation number of manganese is going from +7 to +2; therefore, it is being reduced. So this is, indeed, a redox reaction.

Understanding

Is the following reaction a redox reaction? If so, what is being oxidized, and what is being reduced?

$$2NaHCO_3(s) \rightarrow Na_2CO_3(s) + CO_2(g) + H_2O(\ell)$$

Answer No, it is not a redox reaction. All oxidation numbers stay the same.

More Definitions

The term *oxidation* refers to a reaction in which the oxidation number of an atom increases. Historically, it referred to reactions in which oxygen combined with another substance. A reaction that removed oxygen from a compound was called a *reduction,* as in the extraction of metals from their ores. Thus, the equation

$$4Fe(s) + 3O_2(g) \rightarrow 2Fe_2O_3(s)$$

describes the oxidation of iron by oxygen, whereas the reaction

$$Fe_2O_3(s) + 3C(s) \rightarrow 2Fe(s) + 3CO(g)$$

describes the reduction of iron oxide by carbon.

In the first reaction, the change of elemental iron to Fe^{3+} involves the loss of three electrons by the iron atom. A more general definition of **oxidation** is the *loss* of electrons by an element or other chemical species; many reactions that do not involve oxygen as a reactant can be classified as oxidation reactions. For example, all of the following equations represent reactions that involve the oxidation of calcium.

$$2Ca(s) + O_2(g) \rightarrow 2CaO(s)$$

$$Ca(s) + Cl_2(g) \rightarrow CaCl_2(s)$$

$$Ca(s) + 2HBr(aq) \rightarrow CaBr_2(aq) + H_2(g)$$

In each of these reactions, the loss of two electrons by calcium produces Ca^{2+} ions.

(*a*)

(*b*)

© Cengage Learning/Charles D. Winters

Nelson Morris/Photo Researchers, Inc

Oxidation of iron. *(a)* The oxidation of iron (forming rust) is generally undesirable because it weakens the iron structure. *(b)* Oxidation can also form a protective coating on a metal surface. This sculpture was created from an iron alloy that is designed to oxidize, but rather than flaking off, the oxide forms an adhering coat that protects the metal from further oxidation.

Conversely, **reduction** is defined as the *gain* of electrons by any chemical species. In the three equations above, the elements oxygen, chlorine, and hydrogen are reduced. Chemical reactions that involve the oxidation of one substance must also involve a second substance that is reduced. **Oxidation-reduction reactions,** or **redox reactions,** are those in which electrons are transferred from one species to another. In a redox reaction, at least one species must be oxidized, and at least one substance is reduced.

Oxidation involves the loss of electrons, and reduction involves the gain of electrons.

Redox reactions are often divided into two half-reactions, to emphasize that the overall reaction is the combination of oxidation and reduction processes. In a **half-reaction,** just the oxidation or the reduction is given, showing the electrons explicitly. The oxidation half-reaction has electrons on the product side of the equation; the electrons are reactants in the reduction half-reaction. In the three previous chemical equations, metallic calcium is oxidized to Ca^{2+} ions. The oxidation half-reaction in all of these equations is

$$Ca \rightarrow Ca^{2+} + 2e^-$$

The reduction reaction in each of these equations is different. The half-reactions for the reductions are

$$O_2 + 4e^- \rightarrow 2O^{2-}$$

$$Cl_2 + 2e^- \rightarrow 2Cl^-$$

$$2HBr + 2e^- \rightarrow 2\ Br^- + H_2$$

As in any chemical equation, the numbers of atoms of each element must be balanced in all half-reactions, and the overall charges on each side must be the same.

The oxidizing agent is reduced; the reducing agent is oxidized.

In an oxidation-reduction reaction, the **oxidizing agent** accepts electrons from another species; therefore, the oxidizing agent is reduced in a redox reaction. The **reducing agent** supplies the electrons transferred to a second species; because the reducing agent loses electrons, it is oxidized.

In everyday life, we encounter many oxidation-reduction reactions; life itself depends on the energy produced by a number of redox reactions that occur in each living cell. Household bleach oxidizes stains, producing either soluble products that are removed by washing or colorless products, rather than an annoying spot. Hydrogen peroxide oxidizes the pigments in hair and lightens it, possibly in preparation for coloring. Iodine acts as a disinfectant because it kills germs by oxidation. Redox reactions are used to separate important metals such as aluminum from their ores. The number of redox reactions are extremely large, and their uses vary widely.

© VStoc/Alamy

Oxidation with hydrogen peroxide. Hydrogen peroxide is an oxidizer used to bleach hair or prepare it for additional coloring.

OBJECTIVES REVIEW *Can you:*

- ☑ define oxidation, reduction, and redox reactions?
- ☑ assign the oxidation numbers to atoms in chemical species?

18.2 Balancing Oxidation-Reduction Reactions

OBJECTIVES

- ☐ Balance oxidation-reduction reactions using the half-reaction method
- ☐ Balance redox reactions in both acidic and basic solutions

Many redox reactions are simple enough that the equation can be balanced by inspection. For example:

$$2Na(s) + Cl_2(g) \rightarrow 2NaCl(s)$$

$$C(s) + O_2(g) \rightarrow CO_2(g)$$

$$Zn(s) + Cu^{2+}(aq) \rightarrow Zn^{2+}(aq) + Cu(s)$$

Many redox equations can be balanced by inspection.

Balancing other redox equations can become quite complicated; therefore, a systematic approach is used. The **half-reaction method** of balancing redox reactions emphasizes the fact that redox reactions can be separated into an **oxidation half-reaction,** which shows only the oxidation process, and a **reduction half-reaction,** which shows the reduction process. For the overall balanced reaction, the two half-reactions are combined so that the electrons in the two half-reactions cancel. The half-reaction method is particularly useful for reactions in aqueous solution, where $H_2O(\ell)$ or $H^+(aq)$ or $OH^-(aq)$ may be participating in the reaction.

An oxidation half-reaction shows only the oxidation process. A reduction half-reaction shows only the reduction process. The two half-reactions are combined to make a balanced redox equation.

The half-reaction method for balancing redox equations is a systematic sequence of steps, which we will illustrate by balancing the following reaction:

$$Fe^{2+}(aq) + Cr_2O_7^{2-}(aq) \rightarrow Cr^{3+}(aq) + Fe^{3+}(aq)$$

1. *Determine which species change oxidation number.* You may wish to assign oxidation numbers, but it is not absolutely necessary at this point.

   In this example, Fe changes oxidation number (it goes from +2 to +3), and so does Cr, which changes oxidation number as it goes from $Cr_2O_7^{2-}$ to Cr^{3+} (it goes from +6 to +3).

2. *Write a chemical expression* for each species that changes oxidation number. This expression, also called a **skeleton reaction,** shows only the species involved in the oxidation or reduction.

   The chromium skeleton reaction is

   $$Cr_2O_7^{2-} \rightarrow Cr^{3+}$$

   The iron skeleton reaction is

   $$Fe^{2+} \rightarrow Fe^{3+}$$

3. *Balance the number of atoms* of the element that changes oxidation number by using a coefficient.

   To balance Cr, use a 2 on the product side.

   $$Cr_2O_7^{2-} \rightarrow 2Cr^{3+}$$

   The Fe skeleton reaction is already balanced.

   $$Fe^{2+} \rightarrow Fe^{3+}$$

4. *Balance oxygen by adding H_2O* to the appropriate side as needed.

   First, consider

   $$Cr_2O_7^{2-} \rightarrow 2Cr^{3+}$$

Seven atoms of oxygen are on the left side, and none is on the right side; therefore, we need to add seven molecules of H_2O to the right to balance the oxygen.

$$Cr_2O_7^{2-} \rightarrow 2Cr^{3+} + 7H_2O$$

Oxygen in the iron reaction is already balanced.

$$Fe^{2+} \rightarrow Fe^{3+}$$

5. *Balance hydrogen by adding H^+* to the appropriate side as needed.

Look at the Cr reaction.

$$Cr_2O_7^{2-} \rightarrow 2Cr^{3+} + 7H_2O$$

Fourteen atoms of H are on the right side, and none is on the left side; therefore, we need to add 14 atoms of H^+ to the left side.

$$14H^+ + Cr_2O_7^{2-} \rightarrow 2Cr^{3+} + 7H_2O$$

Hydrogen in the iron reaction is already balanced.

$$Fe^{2+} \rightarrow Fe^{3+}$$

6. *Balance the overall charge by adding electrons* to the appropriate side as needed.
   This procedure results in two balanced half-reactions, in which the electrons are shown explicitly as a reactant (the reduction half-reaction) and one as a product (the oxidation half-reaction).
   The half-reaction is balanced with regard to the masses of all the species and the charge.

The charges in the Cr reaction are:

$$14H^+ + Cr_2O_7^{2-} \rightarrow 2Cr^{3+} + 7H_2O$$

+14	−2	+6
Total +12		+6

We need 6 electrons (each charge −1) on the reactant side to balance the charges on both sides.

$$6e^- + 14H^+ + Cr_2O_7^{2-} \rightarrow 2Cr^{3+} + 7H_2O$$

This last reaction is a balanced half-reaction for the reduction of $Cr_2O_7^{2-}$.

To balance the charges in the Fe reaction, add one electron to the product side.

$$Fe^{2+} \rightarrow Fe^{3+} + e^-$$

This expression is a balanced half-reaction for the oxidation of Fe^{2+}.

7. *Add the two half-reactions,* multiplying by integers so that the number of electrons in the oxidation is equal to the number of electrons in the reduction. Sometimes, the same species appears on both sides of the equation, and you can subtract the appropriate amount from both sides.

We have a six-electron reduction:

$$6e^- + 14H^+ + Cr_2O_7^{2-} \rightarrow 2Cr^{3+} + 7H_2O$$

So we will have to multiply each species in the one-electron oxidation by 6.

$$6 \times (Fe^{2+} \rightarrow Fe^{3+} + e^-)$$

Or

$$6Fe^{2+} \rightarrow 6Fe^{3+} + \cancel{6e^-}$$

$$\cancel{6e^-} + 14H^+ + Cr_2O_7^{2-} \rightarrow 2Cr^{3+} + 7H_2O$$

$$6Fe^{2+} + 14H^+ + Cr_2O_7^{2-} \rightarrow 2Cr^{3+} + 7H_2O + 6Fe^{3+}$$

8. *Check.* Always check your work.

Number of Fe on each side = **6✓**	$\mathbf{6Fe^{2+}} + 14H^+ + Cr_2O_7^{2-} \rightarrow 2Cr^{3+} + 7H_2O + \mathbf{6Fe^{3+}}$
Number of Cr on each side = **2✓**	$6Fe^{2+} + 14H^+ + \mathbf{Cr_2}O_7^{2-} \rightarrow \mathbf{2Cr^{3+}} + 7H_2O + 6Fe^{3+}$
Number of O on each side = **7✓**	$6Fe^{2+} + 14H^+ + Cr_2\mathbf{O_7^{2-}} \rightarrow 2Cr^{3+} + 7H_2\mathbf{O} + 6Fe^{3+}$
Number of H on each side = **14✓**	$6Fe^{2+} + \mathbf{14H^+} + Cr_2O_7^{2-} \rightarrow 2Cr^{3+} + \mathbf{7H_2}O + 6Fe^{3+}$
Total charge on left = total charge on right = **+24✓**	$+12 + 14 - 2 = +6 + 18$

This eight-step systematic approach works for all reactions that are balanced in acidic solution, in which we can assume that we have available as much H^+ and H_2O as the reaction needs. Let us balance another redox reaction, also in acidic solution.

$$Cl_2(g) + NO_3^-(aq) \rightarrow ClO^-(aq) + NO_2(g)$$

1. Determine which species change oxidation number.	Chlorine and nitrogen are changing oxidation numbers.
2. Write the skeleton reactions.	Chlorine: $Cl_2 \rightarrow ClO^-$ Nitrogen: $NO_3^- \rightarrow NO_2$
3. Balance the atoms changing oxidation number.	$Cl_2 \rightarrow 2ClO^-$ $NO_3^- \rightarrow NO_2$
4. Balance oxygen by adding H_2O.	$2H_2O + Cl_2 \rightarrow 2ClO^-$ $NO_3^- \rightarrow NO_2 + H_2O$
5. Balance hydrogen by adding H^+.	$2H_2O + Cl_2 \rightarrow 2ClO^- + 4H^+$ $2H^+ + NO_3^- \rightarrow NO_2 + H_2O$
6. Add electrons to balance charge.	$2H_2O + Cl_2 \rightarrow 2ClO^- + 4H^+ + 2e^-$ $e^- + 2H^+ + NO_3^- \rightarrow NO_2 + H_2O$
7. Add the two half-reactions. In this case, the second reaction is multiplied by 2.	$2H_2O + Cl_2 \rightarrow 2ClO^- + 4H^+ + 2e^-$ $2e^- + 4H^+ + 2NO_3^- \rightarrow 2NO_2 + 2H_2O$ $2e^- + 4H^+ + 2NO_3^- + 2H_2O + Cl_2 \rightarrow 2ClO^- + 4H^+ + 2e^- + 2NO_2 + 2H_2O$ which simplifies to $2NO_3^- + Cl_2 \rightarrow 2ClO^- + 2NO_2$

8. Check.

Number of N on each side = **2✓**	$\mathbf{2N}O_3^- + Cl_2 \rightarrow 2ClO^- + \mathbf{2N}O_2$
Number of Cl on each side = **2✓**	$2NO_3^- + \mathbf{Cl_2} \rightarrow \mathbf{2Cl}O^- + 2NO_2$
Number of O on each side = **6✓**	$2N\mathbf{O_3^-} + Cl_2 \rightarrow 2Cl\mathbf{O^-} + 2N\mathbf{O_2}$
Total charge on left = total charge on right = **−2✓**	$-2 = -2$

Example 18.3 provides some additional examples of the half-reaction method for balancing redox reactions.

The half-reaction method for balancing redox equations provides a systematic way to obtain the correct equation, even for complex reactions.

EXAMPLE 18.3 Half-Reaction Method of Balancing Redox Equations

Complete and balance the following oxidation-reduction reactions.

(a) $Cu(s) + NO_3^-(aq) \rightarrow Cu^{2+}(aq) + NO(g)$
(b) $H_2O_2(aq) + Cr_2O_7^{2-}(aq) \rightarrow Cr^{3+}(aq) + O_2(g)$

Strategy Use the half-reaction method to balance redox reactions.

Solution

(a) 1. First, determine the oxidation numbers of the atoms to see which are changing.

$$\underset{0}{Cu(s)} + \underset{+5\ -2}{NO_3^-(aq)} \rightarrow \underset{+2}{Cu^{2+}(aq)} + \underset{+2-2}{NO(g)}$$

The copper and nitrogen are changing oxidation numbers.

2. The skeleton reactions are:

$$Cu(s) \rightarrow Cu^{2+}(aq)$$
$$NO_3^-(aq) \rightarrow NO(g)$$

3. Balance the atoms changing oxidation number: They are already balanced, so no change is needed.

4. Use $H_2O(\ell)$ to balance oxygen atoms.

$$Cu(s) \rightarrow Cu^{2+}(aq)^- \quad \text{(no change)}$$
$$NO_3^-(aq) \rightarrow NO(g) + 2H_2O(\ell)$$

5. Add $H^+(aq)$ to balance hydrogen atoms.

$$Cu(s) \rightarrow Cu^{2+}(aq)^- \quad \text{(no change)}$$
$$4H^+(aq) + NO_3^-(aq) \rightarrow NO(g) + 2H_2O(\ell)$$

6. Add electrons to balance the overall charge on each side of the half-reactions.

$$Cu(s) \rightarrow Cu^{2+}(aq) + 2e^-$$
$$3e^- + 4H^+(aq) + NO_3^-(aq) \rightarrow NO(g) + 2H_2O(\ell)$$

These half-reactions are now balanced. With two electrons on the product side and three electrons on the reactant side, we must multiply both half-reactions to get to the least common multiple of six electrons.

7. Multiply and combine.

$$3 \times [Cu(s) \rightarrow Cu^{2+}(aq) + 2e^-]$$
$$2 \times [3e^- + 4H^+(aq) + NO_3^-(aq) \rightarrow NO(g) + 2H_2O(\ell)]$$

$$6e^- + 8H^+(aq) + 2NO_3^-(aq) + 3Cu(s) \rightarrow 2NO(g) + 4H_2O(\ell) + 3Cu^{2+}(aq) + 6e^-$$

The electrons are the only species that cancel, leading to the final balanced reaction:

$$8H^+(aq) + 2NO_3^-(aq) + 3Cu(s) \rightarrow 2NO(g) + 4H_2O(\ell) + 3Cu^{2+}(aq)$$

8. Check.

Item	Reactant Side	Product Side	Balanced?
H	8	8	Yes
N	2	2	Yes
O	6	6	Yes
Cu	3	3	Yes
Charge	+6	+6	Yes

This redox reaction is balanced.

(b) 1. The oxidation numbers are

$$\underset{+1\ -1}{H_2O_2(aq)} + \underset{+6\ -2}{Cr_2O_7^{2-}(aq)} \rightarrow \underset{+3}{Cr^{3+}(aq)} + \underset{0}{O_2(g)}$$

Oxygen and chromium are changing oxidation numbers.

2. The skeleton reactions are

$$H_2O_2(aq) \rightarrow O_2(g)$$
$$Cr_2O_7^{2-}(aq) \rightarrow Cr^{3+}(aq)$$

3. Balance the atoms changing oxidation number: They are already balanced in the oxygen half-reaction, but not the chromium half-reaction. We need a coefficient of 2 on the $Cr^{3+}(aq)$ to balance the chromium atoms.

$$H_2O_2(aq) \rightarrow O_2(g) \qquad \text{(no change)}$$
$$Cr_2O_7^{2-}(aq) \rightarrow 2Cr^{3+}(aq)$$

4. Use $H_2O(\ell)$ to balance oxygen atoms.

$$H_2O_2(aq) \rightarrow O_2(g) \qquad \text{(no change)}$$
$$Cr_2O_7^{2-}(aq) \rightarrow 2Cr^{3+}(aq) + 7H_2O(\ell)$$

5. Add $H^+(aq)$ to balance hydrogen atoms.

$$H_2O_2(aq) \rightarrow O_2(g) + 2H^+(aq)$$
$$14H^+(aq) + Cr_2O_7^{2-}(aq) \rightarrow 2Cr^{3+}(aq) + 7H_2O(\ell)$$

6. Add electrons to balance the overall charge on each side.

$$H_2O_2(aq) \rightarrow O_2(g) + 2H^+(aq) + 2e^-$$
$$6e^- + 14H^+(aq) + Cr_2O_7^{2-}(aq) \rightarrow 2Cr^{3+}(aq) + 7H_2O(\ell)$$

These half-reactions are now balanced. With two electrons on the product side and six electrons on the reactant side, we must multiply the oxygen half-reaction to get to the least common multiple of six electrons.

7. Multiply and combine.

$$3 \times [H_2O_2(aq) \rightarrow O_2(g) + 2e^- + 2H^+(aq)]$$
$$6e^- + 14H^+(aq) + Cr_2O_7^{2-}(aq) \rightarrow 2Cr^{3+}(aq) + 7H_2O(\ell)$$
$$6e^- + 14H^+(aq) + Cr_2O_7^{2-}(aq) + 3H_2O_2(aq) \rightarrow 2Cr^{3+}(aq) + 7H_2O(\ell) + 3O_2(g) + 6e^- + 6H^+(aq)$$

The electrons and six hydrogen ions cancel, leading to the final balanced reaction:

$$8H^+(aq) + Cr_2O_7^{2-}(aq) + 3H_2O_2(aq) \rightarrow 2Cr^{3+}(aq) + 7H_2O(\ell) + 3O_2(g)$$

8. Check.

Item	Reactant Side	Product Side	Balanced?
H	14	14	Yes
Cr	2	2	Yes
O	13	13	Yes
Charge	+6	+6	Yes

This redox reaction is balanced.

Understanding

Balance the following redox reaction.

$$V^{3+}(aq) + Ce^{4+}(aq) \rightarrow VO_2^{+}(aq) + Ce^{3+}(aq)$$

Answer $V^{3+}(aq) + 2Ce^{4+}(aq) + 2H_2O(\ell) \rightarrow VO_2^{+}(aq) + 2Ce^{3+}(aq) + 4H^{+}(aq)$

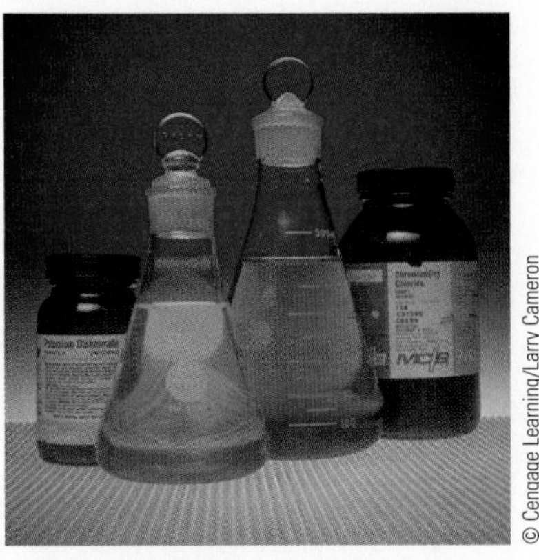

Chromium with different oxidation numbers. Chromium(VI) is orange, whereas chromium(III) is green. In fact, because chromium compounds make compounds having a wide variety of colors, the name of the element comes from the Greek word *chroma,* meaning "color."

Balancing Redox Reactions in Basic Solution

In all the previous examples, the redox reactions were assumed to take place in acidic solution, because $H^{+}(aq)$ ions were used to balance the hydrogen atoms. When a reaction occurs in a basic solution, hydrogen ions should not be shown in the final equation, because any hydrogen ions react immediately with the excess hydroxide ions present in the base, forming water. The equations for oxidation-reduction reactions that occur in basic solution can be balanced using the half-reaction method with only slight modification. Follow the steps given earlier. Then, if H^{+} appears in the final equation, add a sufficient number of hydroxide ions to both sides of the final reaction, combining them with hydrogen atoms to make water molecules. For example, hypochlorite ion oxidizes $CrO(s)$ to CrO_4^{2-} in basic solution, forming chloride ions. The two half-reactions obtained from the half-reaction method outlined earlier are as follows:

$$CrO(s) + 3H_2O \rightarrow CrO_4^{2-} + 6H^{+} + 4e^{-}$$

$$ClO^{-} + 2H^{+} + 2e^{-} \rightarrow Cl^{-} + H_2O$$

Multiply the reduction half-reaction by 2 to cancel all of the electrons, and add the two half-reactions to get the balanced equation.

$$CrO(s) + 2ClO^{-} + H_2O \rightarrow CrO_4^{2-} + 2Cl^{-} + 2H^{+}$$

In basic solution, we add two hydroxide ions to both sides of the reaction. On the product side, these two hydroxide ions combine with the hydrogen ions to make water.

$$CrO + 2ClO^{-} + H_2O + 2OH^{-} \rightarrow CrO_4^{2-} + 2Cl^{-} + \underbrace{2H^{+} + 2OH^{-}}_{2H_2O}$$

We write the final reaction in terms of H_2O, and note that we can remove one H_2O molecule from each side of the reaction.

$$CrO(s) + 2ClO^{-} + 2OH^{-} \rightarrow CrO_4^{2-} + 2Cl^{-} + H_2O$$

The combination of hydrogen ions with hydroxide ions can also be performed on the half-reactions separately. We illustrate using the half-reactions in this example. For the chromium half-reaction:

$$\begin{array}{l} CrO(s) + 3H_2O \rightarrow CrO_4^{2-} + 6H^{+} + 4e^{-} \\ \qquad\quad + 6OH^{-} \qquad\qquad +6OH^{-} \\ \hline CrO(s) + 6OH^{-} + 3H_2O \rightarrow CrO_4^{2-} + 6H_2O + 4e^{-} \end{array}$$

We cancel three water molecules from both sides to get

$$CrO(s) + 6OH^- \rightarrow CrO_4^{2-} + 3H_2O + 4e^-$$

For the chlorine half-reaction,

$$\begin{array}{l} ClO^- + 2H^+ + 2e^- \rightarrow Cl^- + H_2O \\ \quad + 2OH^- \qquad\qquad\quad + 2OH^- \\ \hline ClO^- + 2H_2O + 2e^- \rightarrow Cl^- + H_2O + 2OH^- \end{array}$$

One H_2O molecule cancels from both sides, giving us an overall half-reaction of

$$ClO^- + H_2O + 2e^- \rightarrow Cl^- + 2OH^-$$

H^+ ions cannot appear in the equation for a reaction that occurs in basic solution.

When these two half-reactions are combined, exactly the same balanced equation is obtained. When half-reactions are tabulated (see Section 18.4), the presence of OH^- or H^+ in the half-reaction shows whether the reaction occurs in acidic or basic solution.

EXAMPLE 18.4 Balancing Redox Equations in Basic Solution

Balance the following redox reaction that occurs in basic solution.

$$MnO_4^- + Br^- \rightarrow MnO_2(s) + BrO_3^-$$

Strategy Follow the steps for balancing a redox reaction, but add $OH^-(aq)$ ions to the balanced half-reactions to convert it to a basic solution.

Solution

1. The oxidation numbers are

$$\underset{+7\ -2}{MnO_4^-} + \underset{-1}{Br^-} \longrightarrow \underset{+4\ -2}{MnO_2} + \underset{+5\ -2}{BrO_3^-}$$

Bromine and manganese are changing oxidation numbers.

2. The skeleton reactions are

$$Br^- \rightarrow BrO_3^-$$

$$MnO_4^- \rightarrow MnO_2$$

3. Balance the atoms changing oxidation number: They are already balanced in both half-reactions.
4. Use H_2O to balance oxygen atoms.

$$3H_2O + Br^- \rightarrow BrO_3^-$$

$$MnO_4^- \rightarrow MnO_2 + 2H_2O$$

5. Add $H^+(aq)$ to balance hydrogen atoms.

$$3H_2O + Br^- \rightarrow BrO_3^- + 6H^+$$

$$4H^+ + MnO_4^- \rightarrow MnO_2 + 2H_2O$$

6. Add electrons to balance the charge on both sides of the half-reactions.

$$3H_2O + Br^- \rightarrow BrO_3^- + 6H^+ + 6e^-$$

$$3e^- + 4H^+ + MnO_4^- \rightarrow MnO_2 + 2H_2O$$

These half-reactions are now balanced. However, because this reaction is occurring in basic solution, we should not have H^+ as a reactant or product. Let us add six OH^- ions to each side of the first half-reaction and four OH^- ions to each side of the second half-reaction.

$$6OH^- + 3H_2O + Br^- \rightarrow BrO_3^- + 6e^- + 6H^+ + 6OH^-$$

$$4OH^- + 4H^+ + 3e^- + MnO_4^- \rightarrow MnO_2 + 2H_2O + 4OH^-$$

Combining the H^+ and OH^- ions into H_2O molecules,

$$6OH^- + 3H_2O + Br^- \rightarrow BrO_3^- + 6e^- + 6H_2O$$

$$4H_2O + 3e^- + MnO_4^- \rightarrow MnO_2 + 2H_2O + 4OH^-$$

With six electrons on the product side and three electrons on the reactant side, we must multiply the reduction half-reaction by 2 to get six electrons.

7. Multiply and combine.

$$6OH^- + 3H_2O(\ell) + Br^- \rightarrow BrO_3^- + 6e^- + 6H_2O$$

$$2 \times [4H_2O + 3e^- + MnO_4^- \rightarrow MnO_2 + 2H_2O + 4OH^-]$$

$$6OH^- + 8H_2O + 6e^- + 2MnO_4^- + 3H_2O + Br^- \rightarrow BrO_3^- + 6e^- + 6H_2O + 2MnO_2 + 4H_2O + 8OH^-$$

The electrons cancel, and we combine the 11 water molecules as reactants and 10 water molecules as products, and then cancel 10 H_2O from each side. We can also cancel 6 OH^- ions from each side. We get the following as the final balanced chemical reaction in basic solution:

$$H_2O + 2MnO_4^- + Br^- \rightarrow BrO_3^- + 2MnO_2 + 2OH^-$$

8. Check.

Item	Reactant Side	Product Side	Balanced?
H	2	2	Yes
Mn	2	2	Yes
O	9	9	Yes
Br	1	1	Yes
Charge	−3	−3	Yes

This redox reaction is balanced.

Understanding

Balance the following reaction, which occurs in basic solution.

$$CrO_2^-(aq) + BrO_4^-(aq) \rightarrow BrO_3^-(aq) + CrO_4^{2-}(aq)$$

Answer

$$2CrO_2^-(aq) + 3BrO_4^-(aq) + 2OH^-(aq) \rightarrow 3BrO_3^-(aq) + 2CrO_4^{2-}(aq) + H_2O(\ell)$$

OBJECTIVES REVIEW *Can you:*

- ☑ balance oxidation-reduction reactions using the half-reaction method?
- ☑ balance redox reactions in both acidic and basic solutions?

18.3 Voltaic Cells

OBJECTIVES

- ☐ Identify the components of a voltaic cell
- ☐ Write half-cell reactions and the overall reaction from a diagram of a voltaic cell
- ☐ Identify the direction of flow of electrons and ions through a salt bridge in a voltaic cell
- ☐ Sketch half-cells that involve metal/metal ion, metal ion/metal ion, and gas/ion redox processes

A **voltaic cell,** also called a **galvanic cell,** is an apparatus that produces electrical energy directly from the chemical energy released in a redox reaction. A battery is a familiar example of this type of cell, but there are other, less obvious examples as well. This section describes some of the basic features of voltaic cells, emphasizing their relationship to redox reactions.

(a)

(b)

(c)

Figure 18.1 **Zn/Cu^{2+} reaction.** *(a)* Metallic zinc is placed in a solution of copper(II) sulfate. *(b)* As the spontaneous reaction occurs, metallic copper forms on the zinc strip. *(c)* The blue color of the Cu(II) ions disappears as all of the copper deposits on the zinc strip. Simultaneously, $Zn^{2+}(aq)$ ions, which are colorless, enter the solution. Although this is a redox reaction, this setup provides no useful electrical work.

If someone places a strip of metallic zinc in a solution of copper(II) sulfate, a spontaneous chemical reaction occurs, as shown in Figure 18.1. The zinc dissolves, and the blue color of copper ions in the aqueous solution gradually fades as finely divided red solid—copper metal—forms at the surface of the zinc strip. The reaction is

$$Zn(s) + Cu^{2+}(aq) \rightarrow Zn^{2+}(aq) + Cu(s)$$

Although electrons are transferred in this redox reaction, the reaction does not perform any work. The energy change that occurs is manifested as heat.

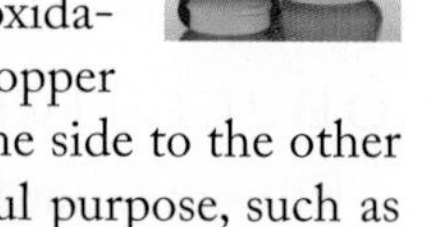

If we separate the two half reactions, as shown in Figure 18.2, and connect the two sides with a wire so that electrons can transfer from one side to the other, we can extract useful work from the redox reaction. On the left side, zinc metal and dissolved zinc ions can participate in an oxidation reaction, and on the right side, copper metal and dissolved copper ions can participate in a reduction reaction. Electrons travel from one side to the other through the wire, and this electricity can be utilized for some useful purpose, such as lighting a lamp or running a motor. Figure 18.2 is an example of a voltaic cell.

All batteries are voltaic cells, but voltaic cells have other uses as well. For example, they can be configured as sensors to measure concentrations of species in solution, and voltaic cell measurements are an important source of thermodynamic data such as free energies and equilibrium constants.

In the voltaic cell shown in Figure 18.2, the strips of zinc and copper metal are called **electrodes.** They provide electrical contacts through which the electrons leave and enter the solutions. The complete voltaic cell consists of one container in which the oxidation half-reaction occurs and a second container in which the reduction half-reaction takes place. Each container with its own half-reaction is called a **half-cell.** By convention, the half-cell for the oxidation half-reaction is drawn on the left; the electrode at which oxidation occurs is called the **anode.** Because the half-reaction in the anode produces the electrons, it is also labeled the negative electrode. The half-reaction in the negative electrode is

$$Zn(s) \rightarrow Zn^{2+}(aq) + 2e^-$$

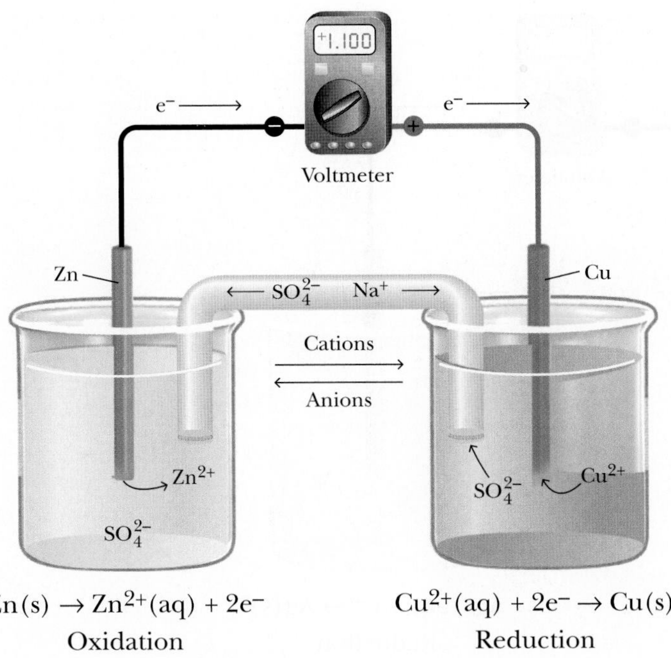

Figure 18.2 **Zinc/copper voltaic cell.**

The half-cell on the right, where the reduction reaction occurs, is called the **cathode.** It is the positive electrode. The half-reaction in the cathode is

$$Cu^{2+}(aq) + 2e^- \rightarrow Cu(s)$$

During the course of the overall reaction, electrons transfer from the negative electrode to the positive electrode.

Another detail is necessary to complete the voltaic cell. As zinc loses electrons, the contents of that half-cell would gain a net positive charge, and the contents of the other half-cell would acquire a negative change. Left alone, the charge imbalance would halt the progress of the reaction. In practice, a charge imbalance is avoided by connecting the two solutions with a **salt bridge,** a tube that contains an electrolyte solution. Figure 18.2 shows a salt bridge filled with a sodium sulfate solution. As the reaction proceeds, sulfate anions migrate into the zinc half-cell to balance the increasing positive charge that accompanies the zinc half-reaction. At the same time, sodium ions migrate into the copper half-cell to counteract the charges of the electrons that are coming into the half-cell. In this way, electrical neutrality is maintained, and a charge imbalance does not occur.

A voltaic cell is composed of two half-cells, one containing the oxidation reaction and one containing the reduction reaction. The two half-cells are connected by a salt bridge that maintains electrical neutrality.

EXAMPLE 18.5 Analyzing Voltaic Cells

Consider the cell shown in Figure 18.3. One half-cell consists of silver metal in a silver nitrate solution, and the other half-cell has a piece of copper metal immersed in a copper(II) nitrate solution. A salt bridge that contains a sodium nitrate solution connects the two half-cells. The measured voltage is positive and the electrons flow in the direction shown in the diagram.

(a) Write the half-reaction that occurs in each half-cell.
(b) Write the equation for the overall chemical change that takes place.
(c) Identify the anode and the cathode.
(d) Indicate which half-cell is the negative electrode and which half-cell is the positive electrode.
(e) Indicate the direction the nitrate ions flow through the salt bridge.

Figure 18.3 Copper/silver voltaic cell.

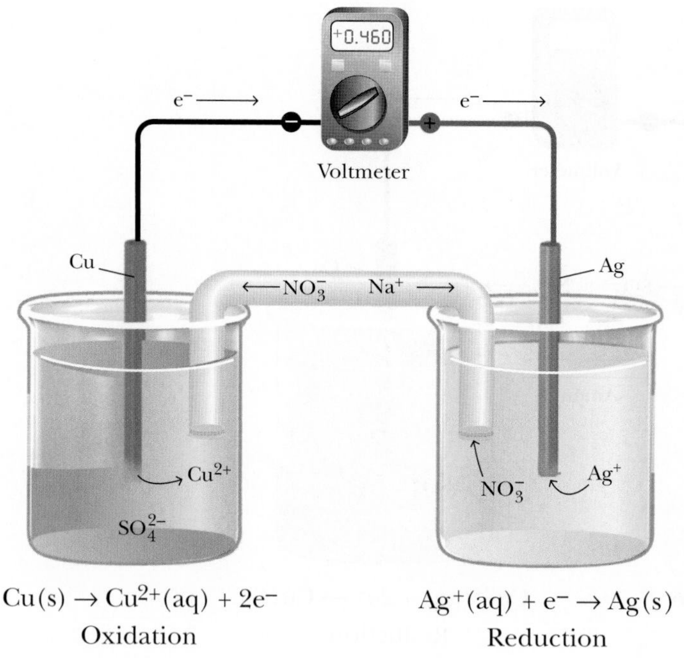

Strategy Use the definitions and conventions introduced in the text to identify the various parts of the voltaic cell and their functions.

Solution

(a) Because, by convention, the oxidation half-cell is portrayed on the left side, the copper half-cell must contain the oxidation half-reaction, which is

$$Cu(s) \rightarrow Cu^{2+}(aq) + 2e^-$$

The reduction occurs at the electrode on the right, so the silver half-cell contains the reduction half-reaction, which is

$$Ag^+(aq) + e^- \rightarrow Ag(s)$$

(b) The overall reaction must be a balanced chemical equation with no excess electrons, so multiply the silver half-reaction by 2 before adding it to the copper half-reaction so that the electrons cancel. The net ionic equation for the chemical reaction that takes place in this voltaic cell is

$$Cu(s) + 2Ag^+(aq) \rightarrow Cu^{2+}(aq) + 2Ag(s)$$

(c) By definition, the anode is the electrode at which oxidation occurs, which is the copper half-cell in this cell.

(d) The oxidation half-cell is the negative cell (the copper electrode).

(e) As the Cu is oxidized to Cu^{2+}, the net charge in the left cell would go up. Negative ions move from the salt bridge into the left cell. Sodium cations move in the opposite direction, supplying the positive charge that is lost as Ag^+ is reduced to Ag.

Understanding

Draw and label the parts of a voltaic cell that is based on the following reaction.

$$2Al(s) + 3Cu^{2+}(aq) \rightarrow 2Al^{3+}(aq) + 3Cu(s)$$

Determine which half-cell is the positive electrode, which is the negative electrode, and indicate the direction of electron flow.

Answer The drawing is left to the student, but it should resemble Figure 18.3. The electron flow is from the aluminum half-cell to the copper half-cell, making the aluminum half-cell the negative electrode and the copper half-cell the positive electrode.

Other Types of Electrodes and Half-Cells

All of the half-cells described earlier consist of a metal and one of its salts in solution. The metal serves as the electrode in these examples. A number of other redox half-reactions do not involve a metal as one of the species in the half-reaction. For example, consider the half-reaction

$$Sn^{4+}(aq) + 2e^- \rightarrow Sn^{2+}(aq)$$

A half-cell can be constructed that uses this half-reaction (or its reverse) by placing a solid conductor of electrons into the solution. The material used to make the electrode in such a half-cell must not be easily oxidized; it is referred to as an **inert electrode.** The $Sn^{4+}(aq)$ and $Sn^{2+}(aq)$ ions in the solution can transfer electrons to or from this inert electrical conductor. Metals that are difficult to oxidize, such as gold or platinum, are often chosen as inert electrodes. Graphite is another material that is used for inert electrodes, because it is a good electrical conductor and does not readily react with either oxidizing or reducing agents in aqueous solutions. Figure 18.4 is a schematic diagram for a typical half-cell involving two ions (Sn^{2+} and Sn^{4+}) in aqueous solution.

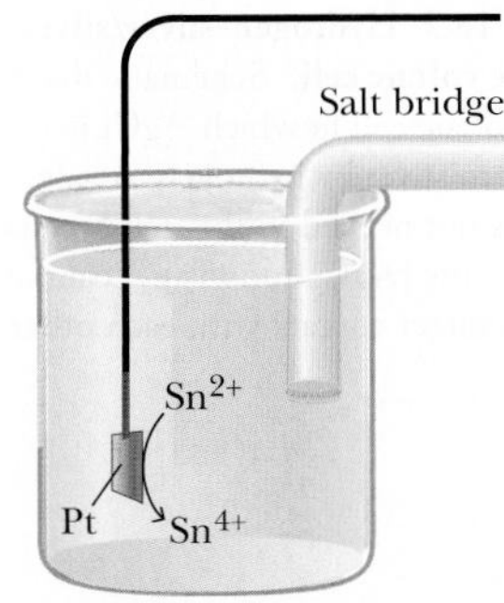

$$Sn^{2+}(aq) \rightarrow Sn^{4+}(aq) + 2e^-$$

Figure 18.4 Sn^{4+}/Sn^{2+} half-cell. A half-cell in which Sn^{2+} ions are oxidized to Sn^{4+} at a platinum surface. A Sn^{2+} ion in the solution releases two electrons to the metallic conductor, forming a Sn^{4+} ion in solution.

When neither the oxidized nor reduced species in a half-reaction is a solid electrical conductor, an inert electrode is used to provide an electrical contact to the solution.

When one of the species in the half-reaction is a gas, an inert electrode is used for electrical contact. Three such half-reactions are

$$2H^+(aq) + 2e^- \rightarrow H_2(g)$$

$$Cl_2(g) \rightarrow 2Cl^-(aq) + 2e^-$$

$$2H^+(aq) + NO_3^-(aq) + e^- \rightarrow NO_2(g) + H_2O(\ell)$$

In these half-cells, the gas is bubbled over the surface of the inert electrode that is in contact with a solution containing the ions participating in the reaction. The hydrogen electrode, shown in Figure 18.5, is an example.

Another kind of electrode involves a metal and a slightly soluble salt of that metal. An example of this type of electrode is the calomel electrode. Mercury(I) chloride (Hg_2Cl_2, which has the common name *calomel*) is placed in electrical contact with liquid mercury. The solution in the half-cell contains a soluble chloride salt such as potassium chloride, and a platinum wire is used as an inert electrical contact to the liquid mercury. The reduction half-reaction for this electrode is

$$Hg_2Cl_2(s) + 2e^- \rightarrow 2Hg(\ell) + 2Cl^-(aq)$$

Figure 18.6 is a schematic diagram of a calomel electrode.

In some voltaic cells, a salt bridge is not necessary to prevent the direct reaction between the oxidizing and reducing agents. Figure 18.7 shows one such example, in which the overall chemical reaction is

$$2AgCl(s) + H_2(g) \rightarrow 2Ag(s) + 2H^+(aq) + 2Cl^-(aq)$$

The flow of electrons occurs from the hydrogen electrode to the silver/silver chloride electrode. Because neither the silver chloride nor the gaseous hydrogen is present in the solution phase, they cannot react directly because they are separated physically.

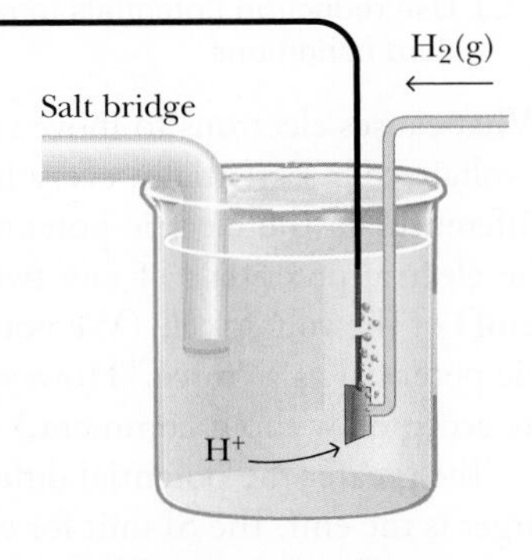

$$2H^+(aq) + 2e^- \rightarrow H_2(g)$$

Figure 18.5 Hydrogen electrode. Gaseous hydrogen is bubbled over a platinum surface that is in contact with a nitric acid solution.

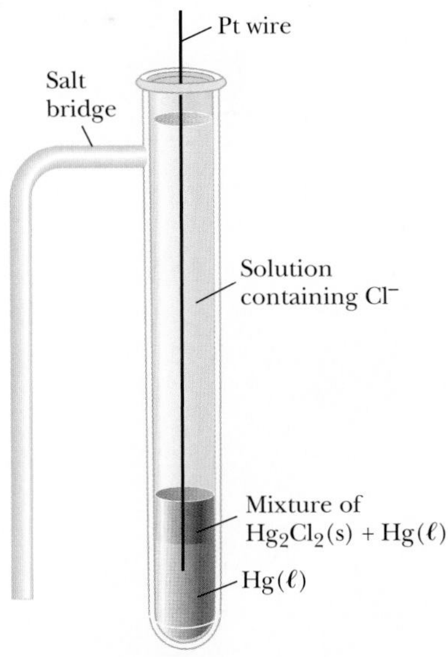

$$Hg_2Cl_2(s) + 2e^- \rightarrow 2Hg(\ell) + 2Cl^-(aq)$$

Figure 18.6 Calomel electrode. Depending on whether the calomel electrode is the anode or the cathode, mercury(I) chloride and mercury are the oxidized and reduced species, respectively.

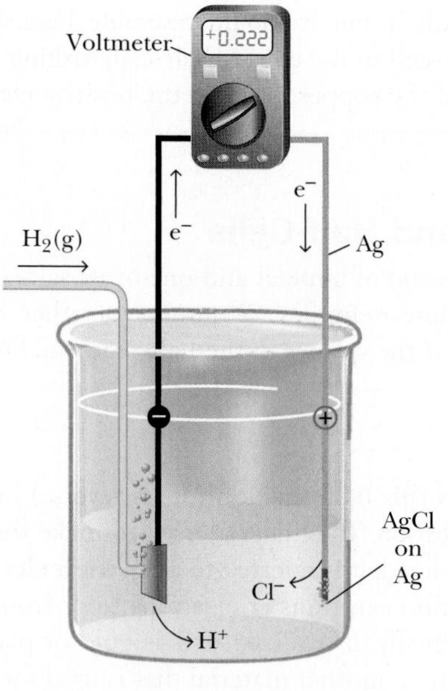

Figure 18.7 Hydrogen-silver/silver chloride voltaic cell. Schematic diagram of the voltaic cell in which AgCl is reduced to silver by hydrogen. A salt bridge is not needed in this cell because the reactants [AgCl and $H_2(g)$] cannot come in direct contact with each other.

When neither of the reactants in a voltaic cell is present in solution, a salt bridge does not have to be used in the voltaic cell.

Oxidation $H_2(g) \rightarrow 2H^+(aq) + 2e^-$

$2AgCl(s) + 2e^- \rightarrow 2Ag(s) + 2Cl^-(aq)$ Reduction

$$H_2(g) + 2AgCl(s) \rightarrow 2Ag(s) + 2Cl^-(aq) + 2H^+(aq)$$

OBJECTIVES REVIEW *Can you:*

- ☑ identify the components of a voltaic cell?
- ☑ write half-cell reactions and the overall reaction from a diagram of a voltaic cell?
- ☑ identify the direction of flow of electrons and ions through a salt bridge in a voltaic cell?
- ☑ sketch half-cells that involve metal/metal ion, metal ion/metal ion, and gas/ion redox processes?

18.4 Potentials of Voltaic Cells

OBJECTIVES

- ☐ Relate cell potential to a spontaneous reaction
- ☐ Calculate the standard potential of a voltaic cell by combining two half-reactions
- ☐ Use reduction potentials to predict the spontaneity of chemical reactions under standard conditions

What causes electrons to move from the negative electrode to the positive electrode in a voltaic cell? Essentially, every half-cell has a characteristic electric potential, and the differences in the electric potential drive the spontaneous reactions. The difference in the electric potentials of any two half-cells is referred to as the **electromotive force (emf)** of the voltaic cell. (We note the inconsistency in referring to a difference in electric potential as a "force." However, such terms are so ingrained in chemistry that they are accepted without comment.)

The greater the potential difference between the electrons at the two electrodes, the larger is the emf. The SI unit for emf is the **volt** (V). A difference of 1 volt in emf causes a charge of 1 coulomb (C) to acquire an energy of 1 joule (J).

$$1\ V = 1\ \text{joule/coulomb} = 1\ J/C$$

Chemists measure the emf of a voltaic cell with a voltmeter. The electric potential difference between the two electrodes of a voltaic cell is commonly referred to as the **potential** of the cell and is designated E. Because the cell potential is measured in units of volts, the term cell voltage is used interchangeably with cell potential or emf. The **standard potential** of a voltaic cell, E°, refers to the voltage when all of the reactants and products in the redox reaction are in their standard states—that is, solids, liquids, and gases in the pure state at 1 atm pressure, and solutes are present at a concentration of 1 *M*. The cell shown in Figure 18.2 has the reaction.

$$Zn(s) + Cu^{2+}(aq, 1\ M) \rightarrow Cu(s) + Zn^{2+}(aq, 1\ M)$$

Under standard conditions, this cell has a potential of +1.10 V. *Whenever the cell reaction occurs spontaneously, the voltage of the cell is positive.* The spontaneous flow of electrons always occurs from the negative electrode to the positive electrode. For the zinc-copper cell, the positive sign of the potential means that the electrons flow spontaneously from the zinc electrode to the copper electrode. The copper electrode is at a positive voltage with respect to the zinc electrode.

When a voltaic cell voltage is positive, the reaction proceeds spontaneously as written.

Standard Potentials for Half-Reactions

Experimental measurements record only *differences* in the potentials of two half-reactions—absolute potentials cannot be measured. This fact should not be a surprise, because chemists cannot measure absolute values of other thermodynamic functions including enthalpy *(H)* and free energy *(G)*. However, it is possible to assign standard potentials to half-reactions by arbitrarily defining the standard potential of one particular half-reaction. Chemists have chosen the reduction of hydrogen ions to hydrogen gas as a reference by assigning it a standard potential of exactly 0 V.

$$2H^+(aq, 1\ M) + 2e^- \rightarrow H_2(g, 1\ atm)$$

The standard potential of this reduction, under standard conditions and at 25 °C, is exactly 0 V *by definition.*

We can construct a voltaic cell using the standard hydrogen half-cell in combination with any other standard half-cell and measure its voltage. Because we have defined the potential of the standard hydrogen half-cell to be zero, the experimentally observed voltage of the cell is attributed to the reaction in the second half-cell. We can list the voltages of the half-reactions in tables and use them to determine the voltage of any combination of half-reactions. In tabulating these **electrode potentials,** it is conventional to write all of the half-reactions as reduction reactions; these electrode potentials are called **standard reduction potentials.** Table 18.1 lists standard reduction potentials at 25 °C for several common half-reactions. In tables of standard potentials, it is common to arrange the reduction half-reactions in order of decreasing potential. Another list of standard reduction potentials is given in Appendix H in alphabetical order of the element undergoing reduction.

In a table of standard reduction potentials, the potentials are measured with respect to the standard hydrogen electrode.

Using Standard Reduction Potentials

The sign of the standard potential for any redox reaction tells us the direction in which that reaction proceeds spontaneously under standard conditions. When given a particular redox reaction, we can divide it into a reduction half-reaction and an oxidation half-reaction, look up the potentials for each, and add them together. (The oxidation half-reaction will appear in the table as a reduction, so we have to look for the *reverse* of the oxidation reaction.) If the resulting potential is positive, the reaction proceeds spontaneously as written. If the potential is negative, then the reverse reaction is spontaneous. Thus, *we can predict the direction of spontaneous reaction for any redox process* under standard conditions from a table of standard reduction potentials such as Table 18.1. These techniques are quite useful in predicting chemical reactions. When asked questions such as "Will hydrogen ions dissolve copper?" we can calculate the potential for

$$Cu(s) + 2H^+(aq) \rightarrow Cu^{2+}(aq) + H_2(g)$$

to determine whether the reaction is spontaneous.

TABLE 18.1 Standard Reduction Potentials at 25 °C

Reduction half-reaction		$E°$(V)
$F_2(g) + 2e^-$	$\rightarrow 2F^-(aq)$	2.87
$Ce^{4+}(aq) + e^-$	$\rightarrow Ce^{3+}(aq)$	1.61
$MnO_4^-(aq) + 8H^+(aq) + 5e^-$	$\rightarrow Mn^{2+}(aq) + 4H_2O(\ell)$	1.51
$Cl_2(g) + 2e^-$	$\rightarrow 2Cl^-(aq)$	1.36
$O_2(g) + 4H^+(aq) + 4e^-$	$\rightarrow 2H_2O(\ell)$	1.23
$Br_2(\ell) + 2e^-$	$\rightarrow 2Br^-(aq)$	1.06
$NO_3^-(aq) + 4H^+(aq) + 3e^-$	$\rightarrow NO(g) + 2H_2O(\ell)$	0.96
$Ag^+(aq) + e^-$	$\rightarrow Ag(s)$	0.80
$Fe^{3+}(aq) + e^-$	$\rightarrow Fe^{2+}(aq)$	0.77
$I_2(s) + 2e^-$	$\rightarrow 2I^-(aq)$	0.54
$Cu^{2+}(aq) + 2e^-$	$\rightarrow Cu(s)$	0.34
$AgCl(s) + e^-$	$\rightarrow Ag(s) + Cl^-(aq)$	0.222
$Sn^{4+}(aq) + 2e^-$	$\rightarrow Sn^{2+}(aq)$	0.15
$2H^+(aq) + 2e^-$	$\rightarrow H_2(g)$	0.000
$Pb^{2+}(aq) + 2e^-$	$\rightarrow Pb(s)$	−0.126
$Ni^{2+}(aq) + 2e^-$	$\rightarrow Ni(s)$	−0.25
$Cr^{3+}(aq) + e^-$	$\rightarrow Cr^{2+}(aq)$	−0.41
$Fe^{2+}(aq) + 2e^-$	$\rightarrow Fe(s)$	−0.44
$Zn^{2+}(aq) + 2e^-$	$\rightarrow Zn(s)$	−0.76
$Ba^{2+}(aq) + 2e^-$	$\rightarrow Ba(s)$	−1.57
$Al^{3+}(aq) + 3e^-$	$\rightarrow Al(s)$	−1.66
$Mg^{2+}(aq) + 2e^-$	$\rightarrow Mg(s)$	−2.37
$Na^+(aq) + e^-$	$\rightarrow Na(s)$	−2.714
$Li^+(aq) + e^-$	$\rightarrow Li(s)$	−3.045

Let us determine whether copper reacts spontaneously with hydrogen ions under standard conditions. The strategy is to first identify the two half-reactions that comprise the overall reaction, much like we did when doing the half-reaction method of balancing redox equations. Next, write the cell half-reactions and their standard potentials. The oxidation half-reaction must be written in the reverse direction from the half-reaction that appears in the table of standard reduction potentials. *When we write a reaction in reverse, we change the sign of the corresponding potential.* Last, add the half-reactions and add the half-cell potentials. A positive standard potential indicates that the reaction under study proceeds spontaneously under standard conditions.

When we write a reaction in reverse, we change the sign of its corresponding potential.

We need to locate the two half-reactions that comprise the reaction of interest. From Table 18.1, they are

$$2H^+(aq) + 2e^- \rightarrow H_2(g) \qquad E° = 0.000\ V$$

$$Cu^{2+}(aq) + 2e^- \rightarrow Cu(s) \qquad E° = 0.34\ V$$

In the reaction of interest, copper metal is oxidized, so the half-reaction for copper must be reversed, and its potential has its sign changed. This oxidation is added to the other half-reaction, the reduction of hydrogen. Note that in this case, the electrons cancel.

$$2H^+(aq) + 2e^- \rightarrow H_2(g) \qquad E° = 0.000\ V$$

$$Cu(s)^- \rightarrow Cu^{2+}(aq) + 2e^- \qquad E° = -0.34\ V$$

$$Cu(s) + 2H^+(aq) \rightarrow Cu^{2+}(aq) + H_2(g) \qquad E° = -0.34\ V$$

The overall reaction is the desired reaction, and its standard cell potential is negative, indicating that this reaction is not spontaneous. Copper is quite resistant to attack by acid under standard conditions. The lack of reactivity is one important reason that copper is widely used for water pipes.

Copper in acid. Unlike its behavior toward other metals, the hydrogen ions in dilute acid do not dissolve copper metal. We can predict this behavior by determining the $E°$ of the reaction between Cu metal and H^+.

Any two half-cells—one written as a reduction and one as an oxidation—can be combined and their potentials added to calculate the voltage of a standard voltaic cell. Let us calculate the standard potential to determine whether the following reaction is spontaneous.

$$Ni(s) + 2Ag^+(aq) \rightarrow Ni^{2+}(aq) + 2Ag(s)$$

From Table 18.1, the two half-reactions are

$$Ag^+(aq) + e^- \rightarrow Ag(s) \qquad E° = +0.80\ V$$
$$Ni^{2+}(aq) + 2\ e^- \rightarrow Ni(s) \qquad E° = -0.25\ V$$

To determine the overall cell reaction, reverse the nickel half-reaction, change the sign of its standard potential, and add it to the silver reaction. For the electrons to cancel, we have to multiply the coefficients in the silver reaction by 2. However, when we do this, *we do not multiply the $E°$ by 2.* The standard potential of a half-reaction is an intensive property and therefore amount independent; it does not change if half-reactions are multiplied.

Standard potentials are not multiplied if a multiple of a half-reaction is used to balance the number of electrons.

$$2Ag^+(aq) + 2e^- \rightarrow 2Ag(s) \qquad E° = +0.80\ V$$
$$Ni(s) \rightarrow Ni^{2+}(aq) + 2e^- \qquad E° = +0.25\ V$$
$$Ni(s) + 2Ag^+(aq) \rightarrow Ni^{2+}(aq) + 2Ag(s) \qquad E° = +1.05\ V$$

This reaction, which describes the system under study, has a positive cell potential, indicating it is spontaneous.

The reason for not multiplying the voltage of the half-reaction can be understood by considering the two half-reactions.

$$Ag^+ + e^- \rightarrow Ag(s)$$
$$2Ag^+ + 2e^- \rightarrow 2Ag(s)$$

The energy change in the second half-reaction is twice that in the first one, but the charge transferred is also doubled, from 1 mol of electronic charge to 2 mol. A voltage is the energy released *per coulomb of charge transferred,* so the energy change per charge transferred is identical for both half-reactions. The energy change and the charge transferred are both extensive properties (see Chapter 1), whereas the cell voltage is an intensive property. The additivity of potentials is true only when the resulting equation is a balanced redox reaction. Other situations are not considered in this textbook.

The voltage of a cell is independent of the amount of material in the redox reaction. These common voltaic cells, or batteries, are different sizes, but all have voltages of about 1.5 V.

EXAMPLE 18.6 Using Standard Potentials to Predict Spontaneity

Calculate the standard potential and state the direction in which the reaction proceeds spontaneously for

$$2Cr^{3+}(aq) + 2Br^-(aq) \rightarrow 2Cr^{2+}(aq) + Br_2(\ell)$$

Strategy Find the two half-reactions in Table 18.1, reversing one of them to make it an oxidation process. Combine the $E°$s for the two half-reactions to determine the voltage of a voltaic cell having the above reaction.

Solution

From Table 18.1,

$$Br_2(\ell) + 2e^- \rightarrow 2Br^-(aq) \qquad E° = 1.06\ V$$
$$Cr^{3+}(aq) + e^- \rightarrow Cr^{2+}(aq) \qquad E° = -0.41\ V$$

The top half-reaction must be reversed, and the sign of the standard potential changed and then added to the bottom reaction to match the given reaction. The bottom reaction must be doubled to allow the electrons to cancel.

$$2Br^-(aq) \rightarrow Br_2(\ell) + 2e^- \qquad E° = -1.06\ V$$

$$2Cr^{3+}(aq) + 2e^- \rightarrow 2Cr^{2+}(aq) \qquad E° = -0.41\ V$$

$$2Cr^{3+}(aq) + 2Br^-(aq) \rightarrow 2Cr^{2+}(aq) + Br_2(\ell) \qquad E° = -1.47\ V$$

The negative sign indicates that this redox reaction is not spontaneous, and that the spontaneous reaction is the one that occurs in the reverse direction.

Understanding

Calculate the standard potential of the following reaction, and state whether the reaction is spontaneous as written.

$$3Ag(s) + NO_3^-(aq) + 4H^+(aq) \rightarrow NO(g) + 2H_2O(\ell) + 3Ag^+(aq)$$

Answer +0.16 V; spontaneous as written

Many important questions about chemical reactivity can be answered by observing the relationship between two half-reactions in Table 18.1. The table serves as a kind of *activity series.* Because the table lists reactions in order of their reduction potential, these reactions are also in order of chemical reactivity. When two half-cells are combined, the one that is higher on the list proceeds spontaneously as a reduction and the lower one proceeds as an oxidation, under standard conditions. The better oxidizing and reducing agents are also identified.

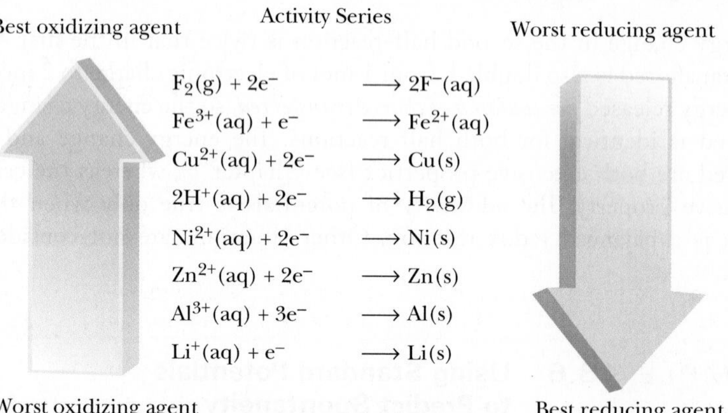

Many chemical questions can be answered easily when considering the half-reactions as an activity series. If a chemical process requires the reduction of $Ni^{2+}(aq)$ to Ni metal, a chemist looks at Table 18.1 to locate the reduction of Ni^{2+} (−0.25 V). The chemist can pick a reducing agent (on the right, or product, side of the half-reaction) from any of the half-reactions below the nickel half-reaction. The zinc reaction would work well. Because it appears below nickel, it proceeds spontaneously as an oxidation. The net reaction is the sum of the two half-reactions, the nickel written as a reduction and the zinc as an oxidation.

$$Ni^{2+}(aq) + 2e^- \rightarrow Ni(s)$$

$$Zn(s) \rightarrow Zn^{2+}(aq) + 2e^-$$

$$Ni^{2+}(aq) + Zn(s) \rightarrow Zn^{2+}(aq) + Ni(s)$$

Example 18.7 illustrates some of these techniques.

EXAMPLE 18.7 Predicting Reactions from Standard Potentials

Use Table 18.1 to

(a) list metals that are and are not oxidized by $H^+(aq)$ under standard conditions.
(b) find an oxidizing agent that will oxidize copper metal.

Strategy Consult Table 18.1 as an activity series: The higher half-reaction will be the reduction process of a spontaneous chemical reaction.

Solution

(a) The metallic species above hydrogen—copper and silver—will be spontaneously reduced by the hydrogen ion/hydrogen gas couple. These metals are not attacked by H^+ under standard conditions. The metals below hydrogen—Pb, Ni, Fe, Zn, Ba, Al, Mg, Na, and Li—will react with H^+ to form the metal cation and $H_2(g)$.
(b) Any of the half-reactions above copper will oxidize copper metal to Cu^{2+}. For example, Fe^{3+} will oxidize copper. The spontaneous reaction is

$$2Fe^{3+}(aq) + Cu(s) \rightarrow 2Fe^{2+}(aq) + Cu^{2+}(aq)$$

Understanding

Most disinfectants kill bacteria by oxidizing them. Which substance is the better oxidant, from the point of standard potentials: Cl_2 or I_2?

Answer Cl_2 is higher on the list, so it is reduced more easily than is I_2. Chlorine is the more powerful oxidant.

OBJECTIVES REVIEW *Can you:*

- ☑ relate cell potential to a spontaneous reaction?
- ☑ calculate the standard potential of a voltaic cell by combining two half-reactions?
- ☑ use reduction potentials to predict the spontaneity of chemical reactions under standard conditions?

18.5 Cell Potentials, ΔG, and K_{eq}

OBJECTIVES

- ☐ Relate cell potential, free energy, and the equilibrium constant
- ☐ Calculate the equilibrium constant for a reaction from the standard potential of a voltaic cell

The spontaneity of a redox reaction is related to the sign of the potential of the voltaic cell based on the reaction. If the potential is positive, under standard conditions, the reaction proceeds spontaneously as written. Chapter 17 explains that any spontaneous reaction has a negative value for the change in free energy. This section develops and uses the relationships among free energy changes, equilibrium constants, and standard cell potentials.

Both a positive cell potential and a negative free energy change indicate that a reaction is spontaneous under the given conditions.

Because both the sign of the cell potential and that of the free energy are able to indicate the spontaneity of reactions, there should be a simple relationship between these two quantities. Indeed, there is such a relationship; it is

$$\Delta G = -nFE \qquad [18.1]$$

where n is the number of moles of electrons transferred in the redox equation, and F is the **Faraday constant,** the negative electrical charge on one mole of electrons. The value of the Faraday constant (to five significant figures) is

$$F = 96{,}485 \frac{\text{coulombs}}{\text{mol of e}^-}$$

Three significant figures are sufficient for our purposes; therefore, a value of 96,500 C/mol is used in calculations.

Equation 18.1 is applicable under any conditions. If standard conditions apply, Equation 18.1 becomes

$$\Delta G^\circ = -nFE^\circ \quad [18.2]$$

where E° is the **standard cell potential.** The units of n are moles, F has units of coulombs per mol (C/mol), and E has units of volts, which are joules per coulomb (J/C). Thus, the free energy change, when calculated from the cell potential, will have units of joules. The next example illustrates some of these calculations.

EXAMPLE 18.8 Standard Free Energy Change from Standard Cell Potential

Calculate the standard free energy change for the reaction

$$2Fe^{3+}(aq) + 2I^-(aq) \rightarrow 2Fe^{2+}(aq) + I_2(s)$$

Strategy Use Table 18.1 to determine the cell potential; then use Equation 18.2 to calculate the ΔG°. You will also have to determine the number of moles of electrons, n, that are transferred in this reaction.

Solution

From the table of standard reduction potentials, the two half-reactions involved in this equation are

$$2Fe^{3+}(aq) + 2e^- \rightarrow 2Fe^{2+}(aq) \qquad E^\circ = +0.77\text{ V}$$

$$2I^-(aq) \rightarrow I_2(s) + 2e^- \qquad E^\circ = -0.54\text{ V}$$

so the voltage for the oxidation of iodide by iron(III) is

$$2Fe^{3+}(aq) + 2I^-(aq) \rightarrow 2Fe^{2+}(aq) + I_2(s) \qquad E^\circ = +0.77 - 0.54 = +0.23\text{ V}$$

Two electrons are transferred, so $n = 2$, and the cell voltage is +0.23 V. Substitute into Equation 18.2:

$$\Delta G^\circ = -nFE^\circ = -(2\text{ mol e}^-)\left(96{,}500\ \frac{\text{C}}{\text{mol e}^-}\right)(+0.23\text{ V})$$

$$\Delta G^\circ = -44{,}000\text{ C·V} = -44{,}000\text{ J} = -44\text{ kJ}$$

This result indicates that an aqueous solution of iron(III) iodide cannot be prepared because Fe^{3+} and I^- will react spontaneously.

Understanding

From the standard reduction potentials in Table 18.1, find the standard free energy change for the following reaction:

$$Ni(s) + Cl_2(g) \rightarrow Ni^{2+}(aq) + 2Cl^-(aq)$$

Answer $\Delta G^\circ = -311$ kJ

Relation of E° to K_{eq}

As was shown in Section 17.5, the equilibrium constant for a chemical reaction is related to the standard free energy change by the relationship.

$$\Delta G^\circ = -RT \ln K_{eq} \quad [17.10]$$

Combining Equation 17.10 with Equation 18.2 provides a direct relationship between the standard cell potential and the equilibrium constant for any redox reaction.

$$-nFE^\circ = -RT \ln K_{eq}$$

$$E^\circ = \frac{RT}{nF} \ln K_{eq} = \frac{2.303RT}{nF} \log K_{eq} \quad [18.3]$$

The value of the equilibrium constant for a redox reaction is related to the standard potential of the reaction by the equation

$$E^\circ = \frac{RT}{nF} \ln K_{eq} = \frac{2.303RT}{nF} \log K_{eq}.$$

As shown in Equation 18.3, we can substitute 2.303 log K for ln K. At 25 °C (298 K), the value of $2.303RT/F$ is 0.0591 V·mol. Therefore, Equation 18.3 becomes

$$E^\circ = \frac{0.0591}{n} \log K_{eq} \quad [18.4]$$

EXAMPLE 18.9 Calculate *K* from *E*°

What is the equilibrium constant for the following reaction?

$$Sn^{2+}(aq) + Ni(s) \rightleftharpoons Sn(s) + Ni^{2+}(aq) \qquad E^\circ = +0.11\ V$$

Strategy Determine the number of electrons transferred, then use Equation 18.4.

Solution

The standard voltage of the reaction is given, and n is 2. Rearrange Equation 18.4 to solve for log K_{eq}, and substitute the values of n and E°.

$$\log K_{eq} = \frac{nE^\circ}{0.0591} = \frac{2 \times 0.11}{0.0591} = 3.7$$

$$K_{eq} = 10^{3.7} = 5 \times 10^3$$

Understanding

Find the equilibrium constant for the following reaction.

$$Sn^{4+}(aq) + U^{4+}(aq) + 2H_2O(\ell) \rightleftharpoons UO_2^{2+}(aq) + Sn^{2+}(aq) + 4H^+(aq)$$

E° for the reaction is −0.176 V.

Answer $K_{eq} = 1.1 \times 10^{-6}$

OBJECTIVES REVIEW *Can you:*

- ☑ relate cell potential, free energy, and the equilibrium constant?
- ☑ calculate the equilibrium constant for a reaction from the standard potential of a voltaic cell?

18.6 Dependence of Voltage on Concentration: The Nernst Equation

OBJECTIVE

- ☐ Use the Nernst equation to find the voltage of cells under nonstandard conditions of concentration

Just as the free energy change for a reaction depends on concentration, the voltage also changes when concentrations of reactants and products change.

Chapter 17 shows that

$$\Delta G = \Delta G^\circ + RT \ln Q \qquad [17.9]$$

where Q is the reaction quotient. (Remember, the reaction quotient has the same form as the equilibrium constant of the chemical reaction but contains nonequilibrium values for the concentrations.) When we replace the free energy changes in Equation 17.9 with $-nFE$ and $-nFE^\circ$, respectively, we get

$$-nFE = -nFE^\circ + RT \ln Q$$

The equation obtained by dividing both sides by $-nF$ is called the **Nernst equation.**

$$E = E^\circ - \frac{RT}{nF} \ln Q \quad \text{or} \quad E = E^\circ - \frac{2.303RT}{nF} \log Q \qquad [18.5]$$

Because the value of 2.303 RT/F is 0.0591 at 25 °C, the Nernst equation can also be written as

$$E = E^\circ - \frac{0.0591}{n} \log Q \qquad [18.6]$$

The Nernst equation is used to calculate the cell potential under conditions other than the standard state.

Equation 18.6 shows how the voltage of any cell changes when the concentrations of reactants and products differ from the standard state. When using Equation 18.6, the value of Q must be calculated expressing the concentrations of solutes in solution as molarity, and the concentrations of gases using their partial pressures, expressed in atmospheres. We illustrate the use of the Nernst equation in Example 18.10.

EXAMPLE 18.10 Cell Voltage and Concentration

A voltaic cell consists of a half-cell of iron ions in a solution with $[Fe^{2+}] = 2.0\ M$ and $[Fe^{3+}] = 0.75\ M$, and a half-cell of copper metal immersed in a solution containing Cu^{2+} at a concentration of $3.6 \times 10^{-4}\ M$. What is the voltage of this cell?

Strategy First, determine E° for the spontaneous reaction and the number of electrons transferred. Then, substitute the nonstandard concentrations into the expression for Q and solve the Nernst equation for E.

Solution

The cell reaction is the oxidation of copper metal by Fe^{3+} ions.

$$2Fe^{3+}(aq) + Cu(s) \rightarrow 2Fe^{2+}(aq) + Cu^{2+}(aq)$$

Two moles of electrons are transferred from copper to iron(III) ions. From the standard reduction potentials (see Table 18.1), calculate the standard cell potential.

$$E^\circ = 0.77 - 0.34 = +0.43\ \text{V}$$

The reaction quotient, Q, is the concentrations of products divided by the concentrations of reactants, each raised to the power of their coefficients in the equation. In this cell reaction, the concentration expression for Q is

$$Q = \frac{[Cu^{2+}][Fe^{2+}]^2}{[Fe^{3+}]^2}$$

When this expression for Q is substituted into the Nernst equation, we obtain

$$E = E^\circ - \frac{0.0591}{n} \log \frac{[Cu^{2+}][Fe^{2+}]^2}{[Fe^{3+}]^2}$$

The equation involves the transfer of two electrons ($n = 2$); therefore, we substitute for n and the concentrations, and solve.

$$E = 0.43\ \text{V} - \frac{0.0591\ \text{V}}{2} \log \frac{(3.4 \times 10^{-4})(2.0)^2}{(0.75)^2} = 0.51\ \text{V}$$

The nonstandard concentrations of the ions have increased the voltage by a factor of almost 20%.

Understanding

What is the voltage of the iron ion-copper cell when the $[Fe^{2+}] = 1.55\ M$, $[Fe^{3+}] = 0.066\ M$, and $[Cu^{2+}] = 0.500\ M$?

Answer $E = 0.36$ V

Concentration Cells

So far, we have considered voltaic cells that have different half-reactions for the oxidation and reaction processes. Can we have a voltaic cell with the same reaction as both the oxidation and reduction half-reactions? Not under standard conditions, because the $E°$s would cancel and the overall voltage of the voltaic cell would be zero. But if the conditions of the half-cells were different, then there would be a nonzero voltage between the half-cells because of the $\log Q$ term in the Nernst equation. Because the concentrations of the species involved must be different so that $\log Q$ is nonzero, such voltaic cells are called **concentration cells.**

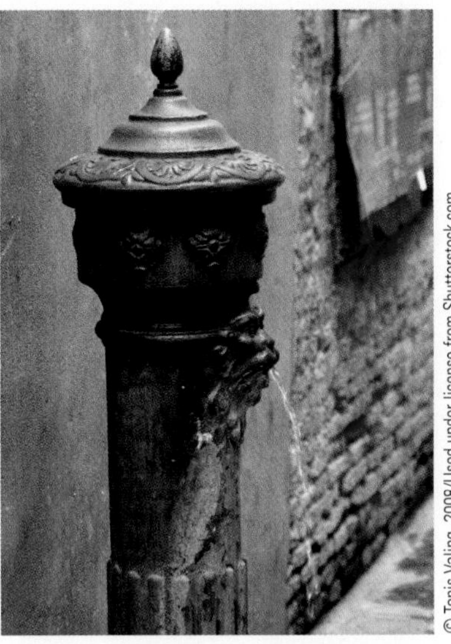

Although H^+ ions do not corrode copper metal, the Fe^{3+} ions in water can.

PRINCIPLES OF CHEMISTRY
Rusting Automobiles

Many people get the feeling that their car, clean and pristine, turns into a pile of rust overnight. Although this is a bit of an exaggeration, these people may be surprised to know that, under the right conditions, a car may turn rusty rather quickly.

Corrosion (see Section 18.9) of iron occurs when metallic iron is exposed to water and oxygen. The following redox reaction occurs:

$$2Fe(s) + O_2(g) + 4H^+(aq) \rightarrow 2Fe^{2+}(aq) + 2H_2O(\ell)$$

The Fe^{2+} ions produced react further with oxygen and water.

$$4Fe^{2+}(aq) + O_2(g) + (4 + 2x)H_2O(\ell) \rightarrow 2Fe_2O_3{\cdot}xH_2O + 8H^+(aq)$$

$Fe_2O_3{\cdot}xH_2O$, hydrated iron(III) oxide, is what we call rust. Unlike some oxides that adhere to the parent metal, iron(III) oxide flakes off, and eventually the automobile will rust away.

The problem of rusting automobiles starts with the production of a small amount of Fe^{2+}. When this happens, small sections of the automobile turn into concentration cells, which accelerate the formation of Fe^{2+} and, ultimately, rust. This process is exacerbated in the winter in regions that use salt (either NaCl or $CaCl_2$) to melt ice and snow on the roads. The dissolved salt provides ions that allow electrons to transfer, promoting the rusting process.

How can this process be stopped? Ultimately, it cannot. The oxidation of iron is thermodynamically favorable; the reactions above have a large positive standard voltage, and hence a very negative ΔG. Certain steps can be taken, however, to minimize the effect of these concentration cells on automobiles. Paint covers the metal and keeps it from being exposed to water and the oxygen in the air. Any scratches or other loss of paint should be repainted so the exposed metal will not start a concentration cell. Waxing a car not only adds to its visual appeal, but the wax repels water, protecting the metal underneath. Finally, washing your car—especially in the winter if you live in a salt-using community—removes ions that serve as conductors of electricity and promoters of rusting.

In the past, some battery-operated electronic antirusting devices have been marketed that claimed they could stop the rusting process on cars. Unfortunately, a lone battery does not have enough current to protect your entire car. You would be better off washing your car regularly and fixing any nicks in the paint. ∎

Rusting automobiles. Because exposed metal can set up concentration cells all over an automobile's body, a car can rust rather quickly, especially if there is an electrolyte present to help with the transfer of electrons.

Consider an example: one half-cell having $[Fe^{2+}]$ of 0.10 M and another half-cell having $[Fe^{2+}]$ of 0.05 M. Elemental iron electrodes are present in both half-cells, so the half-reactions are

$$Fe \rightarrow Fe^{2+}(aq, 0.05\ M) + 2e^- \qquad \text{oxidation}$$

$$Fe^{2+}(aq, 0.10\ M) + 2e^- \rightarrow Fe \qquad \text{reduction}$$

$$Fe^{2+}(aq, 0.10\ M) \rightarrow Fe^{2+}(aq, 0.05\ M) \qquad \text{overall}$$

Although $E°$ for this reaction is zero (because, under standard conditions, there is no net reaction), there is a nonzero value for Q for this reaction. It is

$$Q = \frac{[Fe^{2+}(\text{oxidation})]}{[Fe^{2+}(\text{reduction})]} = \frac{0.05}{0.10}$$

At 25 °C (298 K), the Nernst equation is

$$E = E° - \frac{0.0591\ V}{n}\log Q = 0.000\ V - \frac{0.0591\ V}{2}\log\frac{0.05}{0.10}$$

$$E = 0.000\ V + 0.009\ V$$

$$E = +0.009\ V$$

Thus, the presence of dissimilar concentrations of the same ion is enough to produce a nonzero voltage and a spontaneous reaction.

OBJECTIVE REVIEW *Can you:*

☑ use the Nernst equation to find the voltage of cells under nonstandard conditions of concentration?

18.7 Applications of Voltaic Cells

OBJECTIVES

☐ Describe some of the applications of voltaic cells in chemistry
☐ Discuss the application of voltaic cells as portable energy sources
☐ Describe the chemical reactions that occur in the more common commercial cells
☐ Describe the advantages and disadvantages of fuel cells as an alternative to combustion of hydrocarbon fuels

Voltaic cells have wide applications in society. You may be surprised to realize how many voltaic cells are nearby as you read this section.

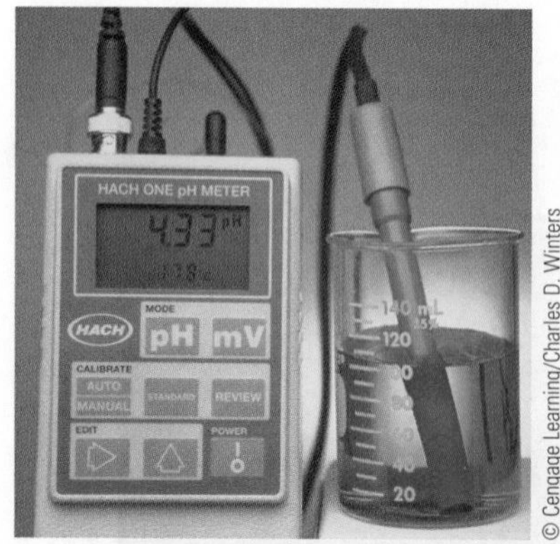

Figure 18.8 pH meter. A pH meter provides a direct reading of the pH of a solution, by measuring the voltage of a voltaic cell. The voltage changes by 0.0591 V for a change of 1 unit of pH.

Measurement of the Concentrations of Ions in Solution

The concentration dependence of cell voltages provides a convenient means of measuring the concentrations of species in solution. Perhaps the most important solution species we measure on a regular basis is the hydronium ion, $H_3O^+(aq)$. A **pH meter** (Figure 18.8) uses cell voltages to determine concentrations of hydronium ions in solution. The probe that is connected to the meter consists of a reference electrode (an electrode that has a known voltage) and an electrode that obeys the Nernst equation for hydrogen ions. The meter itself is a voltmeter, with a scale marked in pH units, rather than volts. A change of 1 pH unit causes the potential of the electrode to change by 0.0591 V at 298 K. The potential of the cell is related to the hydrogen ion concentration by the equation

$$E = k - 0.0591 \log [H^+] = k + 0.0591\ \text{pH}$$

where k is a constant that depends on the particular electrodes used in the measurement. For many pH electrodes, the value of k often changes slightly with time. In an experiment using a pH meter, the electrodes are first placed in a standard buffer solution of

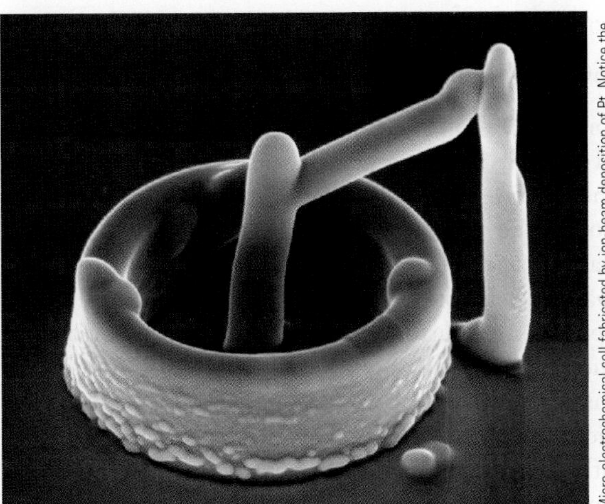

Micro-electrochemical cell fabricated by ion-beam deposition of Pt. Notice the horizontal branch connecting the inner and outer electrodes. Nanofabrication by G.C.Gazzadi, S3 Center (Italy).

Figure 18.9 Tiny voltaic cells. Micrograph of one of the smallest electrodes produced. It was designed to measure ion concentrations in a single nerve cell. The distance between the two vertical electrodes is about 1000 nm.

known pH, and the meter is adjusted to give the correct reading for the pH. The voltage produced when the electrode is placed in a test solution then provides an accurate measurement of the pH.

Although pH electrodes may be the most common type of what are called *ion selective electrodes,* other voltaic cells exist that can determine the concentration of other ions as well. Some of these voltaic cells are extremely small and are designed to probe the contents of individual animal cells (Figure 18.9).

The potential of an electrode provides a rapid means of measuring the concentration of a species in solution.

Batteries

By far the most common use of voltaic cells is the **battery.** Batteries are convenient sources of portable energy, and all of them are based on chemical reactions. Some batteries are single voltaic cells; others are several voltaic cells connected together to generate a different (usually higher) voltage.

Batteries can be described as primary batteries or secondary batteries. In a **primary battery,** the chemical reaction is essentially irreversible, so the battery is used once, then discarded. In a secondary battery, the chemical reaction can be reversed (typically by the application of electricity), so the battery can be reused, sometimes hundreds of times.

Alessandro Volta invented the first battery in 1800, using alternating zinc and copper plates sandwiched between cardboard soaked in brine. This so-called wet cell was not safe and was rather chemically corrosive. In 1866, George Leclanché introduced the **dry cell,** which is essentially the same as a modern dry cell battery. The cell (Figure 18.10) consists of a zinc case that serves as the negative electrode, and a carbon rod in the center that serves as an inert positive electrode. The space between the electrodes contains a moist paste of MnO_2, NH_4Cl, and carbon. The electrode reactions are quite complex but are generally represented by

$$Zn(s) \rightarrow Zn^{2+}(aq) + 2e^-$$

$$2MnO_2(s) + 2NH_4^+(aq) + 2e^- \rightarrow Mn_2O_3(s) + 2NH_3(aq) + H_2O(\ell)$$

The Leclanché cell produces a voltage of about 1.56 V when it is new. As the cell is used, the concentrations of the soluble products (Zn^{2+} and NH_3) increase, causing the voltage produced by the cell to decrease gradually.

An alkaline version (Figure 18.11) of this same cell replaces NH_4Cl with KOH as the electrolyte. Under alkaline conditions, the half-reactions can be represented by

$$Zn(s) + 2OH^-(aq) \rightarrow ZnO(s) + H_2O(\ell) + 2e^-$$

$$2MnO_2(s) + H_2O(\ell) + 2e^- \rightarrow Mn_2O_3(s) + 2OH^-(aq)$$

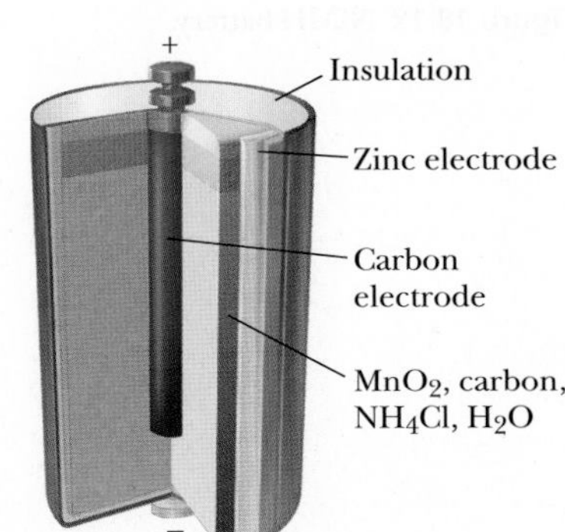

Figure 18.10 Leclanché dry cell. The initial voltage of the cell is about 1.56 V, but it decreases as the cell discharges. Most dry cell batteries are this type.

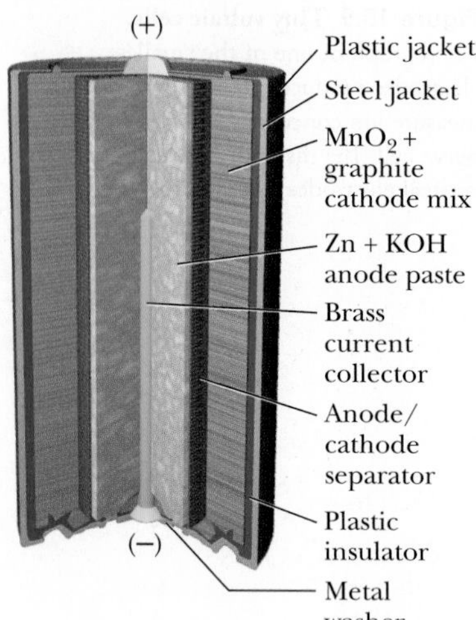

Figure 18.11 Alkaline battery. The alkaline cell has a voltage of 1.54 V that remains constant throughout the useful life of the cell.

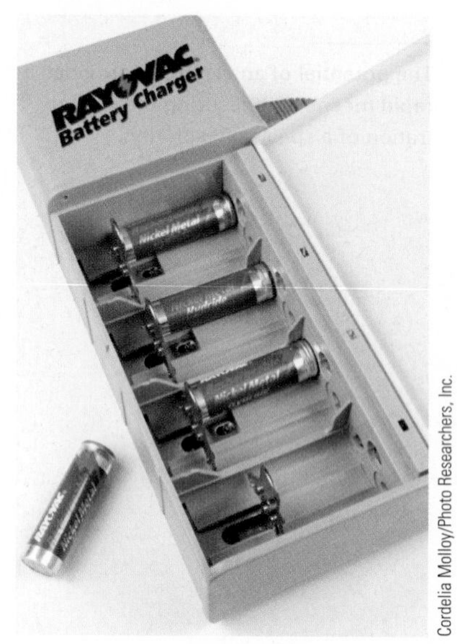

Cordelia Molloy/Photo Researchers, Inc.

Figure 18.12 NiMH battery.

The overall cell reaction is

$$Zn(s) + 2MnO_2(s) \rightarrow ZnO(s) + Mn_2O_3(s)$$

Unlike the previous version of the dry cell, which has acidic contents, the alkaline cell keeps a constant voltage as it discharges. Because the reactants and products are all solids, the value of the reaction quotient in the Nernst equation is 1, so the cell voltage does not change. Although they are more expensive to produce, alkaline cells have a longer useful lifetime than do acid cells. In both the acid and alkaline cell, the outer zinc case is consumed. Toward the end of the useful life of the Leclanché cells, holes in the casing may develop, exposing the surroundings to the rather corrosive contents. Modern designs of alkaline batteries no longer use the zinc electrode as the outer casing, so potential problems with leakage are avoided.

Batteries based on the Leclanché cell cannot be recharged; they are primary batteries. For reasons of economics and safety (disposal of a large number of batteries can be environmentally problematic), **secondary batteries,** batteries whose reactions can be reversed and thus are reusable, are gaining in popularity. One of the first truly successful secondary batteries for everyday use was the nickel-cadmium, or NiCad, battery. The NiCad battery is a dry cell that can be recharged, and for that reason has become popular for use in battery-operated tools that require moderately large power. The products of the cell reaction adhere to the electrodes, and the cell reaction can be reversed by forcing electricity through the cell ("recharging"). The electrode reactions are

$$Cd(s) + 2OH^-(aq) \rightarrow Cd(OH)_2(s) + 2e^-$$

$$NiO(OH)(s) + H_2O(\ell) + e^- \rightarrow Ni(OH)_2(s) + OH^-(aq)$$

$$\text{Overall reaction: } Cd(s) + 2NiO(OH)(s) + 2H_2O(\ell) \rightarrow Cd(OH)_2(s) + 2Ni(OH)_2(s)$$

The overall reaction can be reversed because all of the reactants and products are in the condensed phase, and the substances involved all adhere to the electrodes.

Another type of secondary battery that is gaining in popularity is the nickel-metal-hydride (NiMH) battery (Figure 18.12). It uses the same nickel reaction as the NiCad battery, but a metal alloy that absorbs hydrogen is used in place of the cadmium half-reaction. Because cadmium is toxic, this electrode has advantages from an environmental perspective. The half reaction is

$$MH(s) + OH^-(aq) \rightarrow M(s) + H_2O(\ell) + e^-$$

where MH represents the metal alloy that has absorbed hydrogen. NiMH batteries are being increasingly used in power tools and cell phones.

Lead Storage Battery

The lead storage battery (Figure 18.13) has been used in automobiles since 1915 to provide the energy to operate the starter motor. The most common batteries produce 12 V, with six individual cells that are connected. In each cell, the oxidation reaction involves the conversion of lead to lead sulfate, and the reduction reaction produces lead sulfate from lead dioxide. The electrolyte solution in the cell is sulfuric acid. The half-reactions are

$$Pb(s) + HSO_4^- \rightarrow PbSO_4(s) + H^+ + 2e^-$$

$$PbO_2(s) + HSO_4^- + 3H^+ + 2e^- \rightarrow PbSO_4(s) + 2H_2O$$

The overall cell reaction is

$$Pb(s) + PbO_2(s) + 2HSO_4^- + 2H^+ \rightarrow 2PbSO_4(s) + 2H_2O \qquad E° = 2.041 \text{ V}$$

A salt bridge is not needed to connect the oxidation and reduction half-reactions because they both involve solids that are not in direct contact with each other

(Figure 8.13). Without a salt bridge, the internal resistance of the cell is low, allowing the large currents needed for an energy supply. The Nernst equation for the cell is

$$E = E^\circ - \frac{0.0591}{2}\log\frac{1}{[H^+]^2[HSO_4^-]^2}$$

As electrical energy is drawn from the cell, the voltage decreases because the sulfuric acid is consumed as it produces lead sulfate. The electrodes are designed so the lead sulfate formed adheres to the surface; this feature allows the battery to be recharged. In an automobile, some of the mechanical energy produced by the engine is converted into an electric current, which reverses the cell reaction. The battery can usually be discharged and recharged several thousand times before a cell fails.

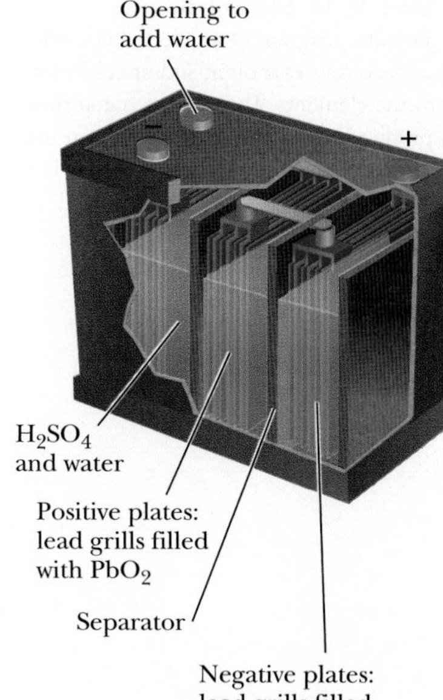

Figure 18.13 Lead storage battery. The design of the lead storage cell keeps the solid products of the reaction in contact with the electrodes so the cell can be recharged by reversing the reaction.

Fuel Cells

A **fuel cell** is a voltaic cell in which the reactants are supplied continuously, and generally the products of the cell reaction are removed continuously. The major source of electrical energy in the world is the combustion reaction of fuels (natural gas, petroleum products, and coal), all of which are oxidation-reduction processes. A great deal of research has been done in recent years to develop practical and economical fuel cells. The U.S. space program developed a fuel cell using the combustion reaction of hydrogen. The electrode reactions are

$$H_2(g) + 2OH^-(aq) \rightarrow 2H_2O(\ell) + 2e^-$$

$$O_2(g) + 2H_2O(\ell) + 4e^- \rightarrow 4OH^-(aq)$$

Power plants now burn hydrocarbon fuels to produce steam for turbines that turn electrical generators. A fuel cell that consumes hydrocarbons and oxygen could convert a much larger fraction of the available chemical energy into electricity than is obtained by combustion. Although such cells have been available for a number of years, they are still too expensive to compete economically with the current methods for energy production.

Pasquale Sorrentino/Photo Researchers, Inc.

A fuel cell.

OBJECTIVES REVIEW *Can you:*

- ☑ describe some of the applications of voltaic cells in chemistry?
- ☑ discuss the application of voltaic cells as portable energy sources?
- ☑ describe the chemical reactions that occur in the more common commercial cells?
- ☑ describe the advantages and disadvantages of fuel cells as an alternative to combustion of hydrocarbon fuels?

18.8 Electrolysis

OBJECTIVES

- ☐ Describe the operation of an electrolytic cell
- ☐ Calculate the amount of material that can be produced from an electrolytic cell
- ☐ List several applications of electrolytic cells

Voltaic cells produce electrical energy from spontaneous oxidation-reduction reactions. It is also possible to use electrical energy to force a chemical reaction to occur, by a process called *electrolysis.* **Electrolysis** is the passage of an electric current through an electrolyte, causing an otherwise nonspontaneous oxidation-reduction reaction to occur. For example, passing a current through a molten mixture of anhydrous HF with some dissolved KHF_2 decomposes the HF into the elements.

$$2HF(\ell) \rightarrow H_2(g) + F_2(g)$$

Without the consumption of electrical energy, the reverse chemical reaction would be spontaneous.

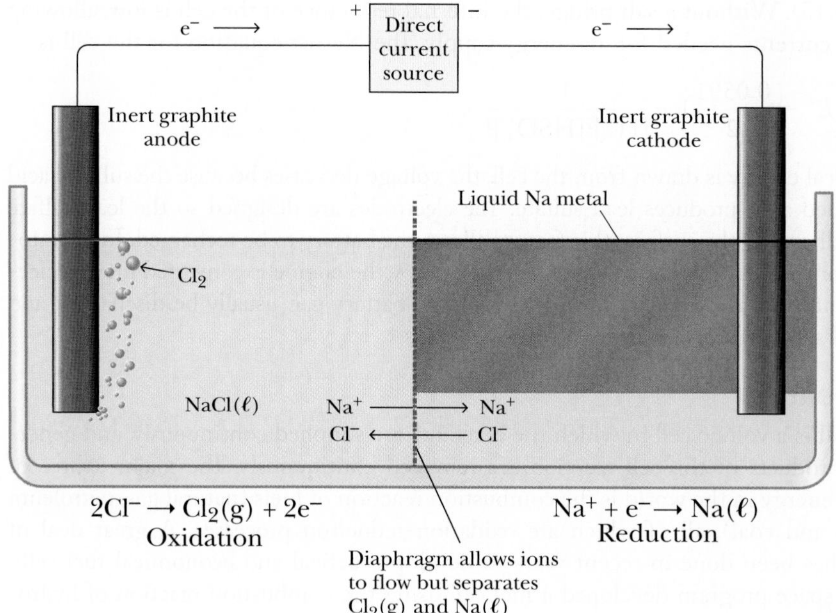

Figure 18.14 Electrolytic cell. Schematic diagram of an electrolytic cell that decomposes molten sodium chloride into the elements. The sodium metal that is produced is molten, and it floats on top of the molten sodium chloride, where it is removed.

An **electrolytic cell** consists of two electrodes in a molten salt or an electrolyte solution. A battery or other voltage source is attached across the two electrodes, as shown in Figure 18.14. The battery serves as an electron pump, drawing electrons in at the positive electrode and forcing them out at the negative electrode. A reduction half-reaction occurs at the electrode that is attached to the negative terminal of the battery, using the electrons provided by the battery. The electrons that enter the battery at its positive terminal must be obtained from an oxidation half-reaction in an electrolytic cell. In electrolytic cells, as with voltaic cells, oxidation occurs at the anode, whereas reduction occurs at the cathode. Thus, in the case of the electrolysis of molten sodium chloride in Figure 18.14, the processes at the two electrodes are

$$2Cl^-(\ell) \rightarrow Cl_2(g) + 2e^- \qquad \text{(anode reaction)}$$

$$Na^+(\ell) + e^- \rightarrow Na(\ell) \qquad \text{(cathode reaction)}$$

$$\text{Overall reaction: } 2Na^+(\ell) + 2Cl^-(\ell) \rightarrow 2Na(\ell) + Cl_2(g)$$

The E° of this reaction is about -4.2 V, so the electrolysis of molten NaCl requires us to apply at least $+4.2$ V to force the reaction to occur. (The reason that the E° is not exactly as predicted by the voltages of the half-reactions in Table 18.1 is because those half-reactions occur in aqueous solution, not the molten salt. The overall voltages are, however, similar.)

Electrolysis in Aqueous Solutions

Electrolysis of aqueous electrolyte solutions is complicated by the fact that, in addition to the species dissolved in solution, water can be oxidized and reduced.

$$2H_2O(\ell) + 2e^- \rightarrow H_2(g) + 2OH^-(aq) \qquad \text{(reduction)}$$

$$2H_2O(\ell) \rightarrow 4H^+(aq) + O_2(g) + 4e^- \qquad \text{(oxidation)}$$

How do we determine which oxidation and reduction reaction will proceed? Actually, there are two simple rules:

- The oxidation half-reaction with the most positive voltage will occur first.
- The reduction half-reaction with the most positive voltage will occur first.

In applying these rules, you should write the oxidation half-reactions *as oxidation reactions,* and keep in mind that “most positive” can also mean “least negative.”

For example, in the electrolysis of aqueous sodium fluoride, the fluoride ion or water can be oxidized:

$$2F^-(aq) \rightarrow F_2(g) + 2e^- \qquad E^\circ = -2.87\ V$$

$$2H_2O(\ell) \rightarrow 4H^+(aq) + O_2(g) + 4e^- \qquad E^\circ = -1.23\ V$$

The more positive voltage is the oxidation of water (it is less negative by 1.64 V), so this is the half-reaction that occurs in the electrolysis.

Two reductions are also possible in the electrolysis of aqueous sodium fluoride. Either the sodium cation, Na^+, or water might be reduced.

$$Na^+(aq) + e^- \rightarrow Na(s) \qquad E^\circ = -2.71\ V$$

$$2H_2O(\ell) + 2e^- \rightarrow H_2(g) + 2OH^-(aq) \qquad E^\circ = -0.83\ V$$

Water will be reduced instead of sodium ions because its reduction potential is more positive (less negative). The net chemical reaction that occurs in the electrolysis of aqueous sodium fluoride is the decomposition of water into its elements.

$$2H_2O(\ell) \rightarrow 2H_2(g) + O_2(g)$$

The species that react in the electrolysis of an aqueous solution are those that have the most positive potentials.

The sodium fluoride undergoes no permanent chemical change—it simply serves as an electrolyte that allows large electrical currents to flow through the solution.

For kinetic reasons, sometimes reactions whose voltages are close to those of water will occur instead of water being oxidized or reduced. In solutions that have high concentrations of $Cl^-(aq)$ ion, for example, the possible oxidation reactions are

$$2Cl^-(aq) \rightarrow Cl_2(g) + 2\ e^- \qquad E^\circ = -1.36\ V$$

$$2H_2O(\ell) \rightarrow 4H^+(aq) + O_2(g) + 4\ e^- \qquad E^\circ = -1.23\ V$$

The more positive potential for the oxidation of water indicates that gaseous oxygen should be the product formed at the anode. However, in reality, $Cl_2(g)$ is produced. Thus, for reactions whose potentials are close to those for water, what actually happens must be determined by experiment.

EXAMPLE 18.11 Predicting Electrolysis Products in Water Solution

Using standard potentials, predict the electrolysis reaction that occurs, and the standard potential of the electrolysis reaction, for a solution of nickel perchlorate using inert electrodes.

Strategy Determine what oxidation and reduction half-reactions might occur, including those involving the solvent. Whichever half-reaction has the most positive potential should be the one that occurs.

Solution

Appendix H shows no reactions in which perchlorate or Ni^{2+} ions are oxidized, so the oxidation half-reaction must be the oxidation of water.

$$2H_2O(\ell) \rightarrow 4H^+(aq) + O_2(g) + 4e^- \qquad E^\circ = -1.23\ V$$

At the cathode, either water ($E^\circ = -0.83$ V) or Ni^{2+} ($E^\circ = -0.28$ V) could be reduced.

$$Ni^{2+}(aq) + 2e^- \rightarrow Ni(s) \qquad E^\circ = -0.28\ V$$

$$2H_2O(\ell) + 2e^- \rightarrow H_2(g) + 2OH^-(aq) \qquad E^\circ = -0.83\ V$$

Because the potential for reducing nickel ions is more positive, the cathode reaction is the reduction of $Ni^{2+}(aq)$. The overall cell reaction expected in this electrolysis is

$$2Ni^{2+}(aq) + 2H_2O(\ell) \rightarrow 2Ni(s) + 4H^+(aq) + O_2(g)$$

PRACTICE OF CHEMISTRY
Overvoltage

Not all electrolysis reactions proceed exactly at the potential calculated from standard reduction potentials because of a phenomenon analogous to the activation energy of a reaction (see Chapter 13). For an electrochemical reaction to proceed at a noticeable rate, the applied voltage must exceed the predicted voltage plus some additional voltage called the *overvoltage*. Overvoltage must be determined experimentally, and it is generally high for reactions that generate gases.

The overvoltages for both the oxidation and reduction of water are particularly high, about 0.40 V. The overvoltage is probably caused by slow electron transfer between the electrode surface and the water molecules. Because of this overvoltage, the potential needed to oxidize water to oxygen is about −1.6 V, which is considerably more negative than the potential needed to oxidize chloride ions to elemental chlorine. Thus, electrolysis of a sodium chloride solution yields chlorine gas and hydroxide ions.

$$2Cl^-(aq) + 2H_2O(\ell) \xrightarrow{\text{electrolysis}} H_2(g) + Cl_2(g) + 2OH^-(aq)$$

Most elemental chlorine is generated in this manner. Because one of the other chemical products is sodium hydroxide (the sodium coming from sodium chloride), this electrochemical process is called the *chloralkali process*.

The products of electrochemical reactions cannot be predicted with unfailing accuracy because overvoltages can be determined only by experiment. Often, electrolysis products are

Understanding

Predict the oxidation-reduction reaction that occurs in the electrolysis of a solution of $ZnBr_2$.

Answer $2H_2O(\ell) + 2Br^-(aq) \rightarrow 2OH^-(aq) + Br_2(\ell)$

Quantitative Aspects of Electrolysis

When electrolysis is performed, the amount of electricity that is used determines the quantity of products formed. Consider the nickel produced by electrolysis of a nickel sulfate solution (see Example 18.11). The half-reaction at the cathode is

$$Ni^{2+}(aq) + 2e^- \rightarrow Ni(s)$$

One mole of nickel metal (58.69 g) is produced for each 2 mol of electrons that pass through the solution. Experimentally, we measure the electric current in amperes, and the length of time the current flows, rather than the number of moles of electrons. The SI unit of charge is the coulomb, which is 1 ampere-second. Therefore the total electrical charge, Q, in electrolysis is simply the product of the current, I, and the time, t.

$$\text{Charge (coulombs)} = \text{current (amperes)} \times \text{time (seconds)}$$

$$Q = I \times t$$

The total charge in an electrolysis serves as the basis for stoichiometry calculations. In performing these stoichiometry calculations, it is convenient to replace the unit of ampere with its equivalent of coulombs/second. For example, if a constant current of 0.200 A were passed through a nickel sulfate solution for 30 minutes, the total charge is the current times the time, or

$$\text{Total charge} = 30.0 \text{ min} \times \left(\frac{60 \text{ s}}{1 \text{ min}}\right) \times \left(\frac{0.200 \text{ C}}{\text{s}}\right) = 3.60 \times 10^2 \text{ C}$$

The Faraday constant, 96,500 C/mol e^-, is the charge on 1 mol of electrons. The total charge from above is converted to the number of moles of electrons, using the Faraday constant.

$$\text{mol } e^- = 3.60 \times 10^2 \not{C} \times \left(\frac{1 \text{ mol } e^-}{96{,}500 \not{C}}\right) = 3.73 \times 10^{-3} \text{ mol } e^-$$

From the coefficients in the half-reaction, one mole of nickel is deposited at the cathode for each two moles of electrons, so the mass of nickel produced is

$$\text{Mass Ni} = 3.73 \times 10^{-3} \text{ mol } e^- \times \left(\frac{1 \text{ mol Ni}}{2 \text{ mol } e^-}\right) \times \left(\frac{58.69 \text{ g Ni}}{1 \text{ mol Ni}}\right) = 0.109 \text{ g Ni}$$

This procedure is how we determine how much material is produced in the course of an electrolysis process.

The product of current and time gives the total amount of charge in an electrochemical process. The Faraday constant is used to convert the total charge into moles of electrons.

EXAMPLE 18.12 Stoichiometry of Electrolysis

A constant current of 0.500 A passes through a silver nitrate solution for 90.0 minutes. What mass of silver metal is deposited at the anode?

Strategy First, determine the half-reaction involved. Then, determine the total charge of the process and relate that to the stoichiometry of the half-reaction to determine the mass of silver deposited. The flow diagram below shows the specific steps.

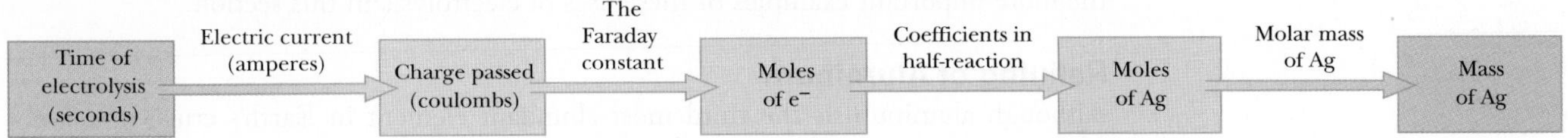

Solution

The half-reaction for the reduction of silver ions is

$$Ag^+(aq) + e^- \rightarrow Ag(s)$$

The product of the time and the current is the total charge used, which we want to express as the equivalent number of moles of electrons.

$$\text{mol } e^- = 90.0 \text{ min} \times \left(\frac{60 \text{ s}}{1 \text{ min}}\right) \times \left(\frac{0.500 \text{ C}}{\text{s}}\right) \times \left(\frac{1 \text{ mol } e^-}{96{,}500 \text{ C}}\right)$$

$$\text{mol } e^- = 2.80 \times 10^{-2} \text{ mol } e^-$$

We complete the calculation by converting to the number of moles of silver, using the stoichiometry of the half-reaction, and then converting to mass.

$$\text{Mass Ag} = 2.80 \times 10^{-2} \text{ mol } e^- \times \left(\frac{1 \text{ mol Ag}}{1 \text{ mol } e^-}\right) \times \left(\frac{107.87 \text{ g Ag}}{1 \text{ mol Ag}}\right)$$

$$\text{Mass Ag} = 3.02 \text{ g}$$

Understanding

The anode reaction in this electrolysis cell is the oxidation of water to $O_2(g)$. What volume of $O_2(g)$, measured at standard temperature and pressure was produced?

Answer 0.157 L O_2

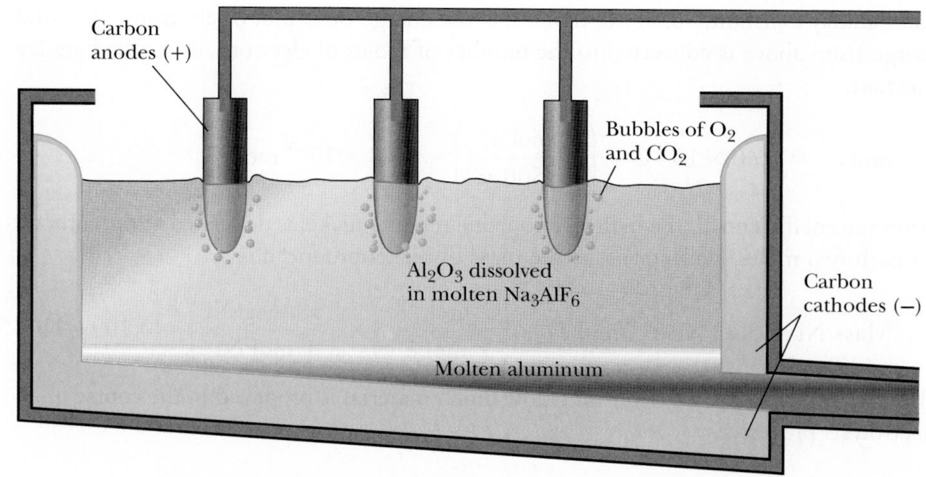

Figure 18.15 Electrolysis to produce aluminum. Aluminum is manufactured in the Hall-Héroult process by the electrolysis of molten salts at carbon electrodes. The anode reaction produces oxygen, part of which reacts at the high temperatures with the carbon in the electrode forming CO_2.

Industrial Applications of Electrolysis

Several industrial processes use electrolysis to isolate different elements from the naturally occurring ores. In fact, electrolysis is the only practical method to isolate certain highly reactive elements. Metals such as copper and aluminum are separated from other substances by electrolysis. In addition, electroplating places decorative and protective layers of metals such as silver and gold, as well as other materials. We describe some of the more important examples of these uses of electrolysis in this section.

Refining of Aluminum

Although aluminum is the third most abundant element in Earth's crust, it was not until 1886 that Charles Hall and Paul Héroult independently developed a practical means of isolating the metal from its compounds. Before the discovery of the Hall–Héroult process, aluminum was made by the reduction of aluminum chloride with sodium metal. In the mid-19th century, aluminum was considered a precious metal and was used mainly in jewelry.

Aluminum metal has a low density and is resistant to corrosion. The resistance of aluminum to corrosion, despite its very positive voltage for Al^{3+} formation ($E° = +1.66$ V), is attributed to the formation of a very thin transparent layer of aluminum oxide that strongly adheres on the surface of the metal and protects it from further oxidation by air. The low density of the metal makes it a desirable structural material in airplanes and automobiles where low weight is important.

In the Hall–Héroult process for manufacturing aluminum, aluminum oxide from the ore is mixed with cryolite, Na_3AlF_6, to produce a mixture that melts at about 980 °C, considerably lower than the melting point of the oxide alone. Carbon electrodes are used in the electrolysis cell, which operates at about 4.5 V. Molten aluminum is formed at the cathode, and oxygen is the principal product formed at the anode, together with some carbon dioxide. Figure 18.15 shows a schematic diagram of the electrolysis cell needed for the Hall–Héroult process.

Production of Other Elements

Some elements are so reactive that reaction with electrons from an electrical current is the only feasible way to produce them. Sodium, potassium, fluorine, and chlorine are four elements that are produced electrolytically. Because of the chemical reactivity of many of the elements produced electrolytically, special electrolytic cells are necessary. Figure 18.16 shows an electrolytic cell used to generate fluorine. Although copper is not refined from its ore electrolytically, copper metal can be purified or recycled using electrolytic techniques, which are called **electrorefining.**

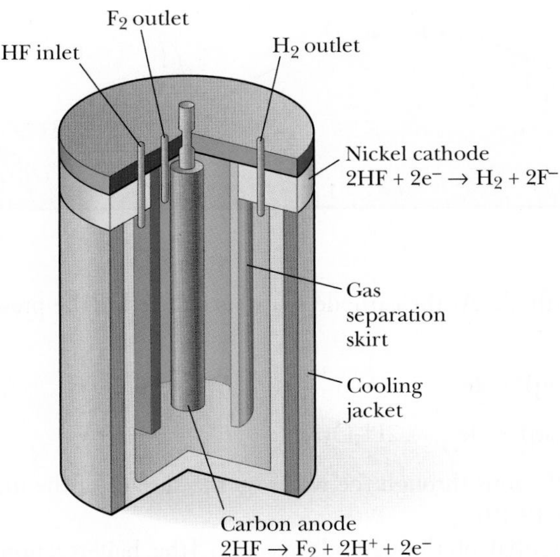

Figure 18.16 Electrolytic cell to produce fluorine. The electrolysis cell is used to isolate fluorine from an HF-KF electrolyte. A barrier that keeps the H_2 separated from the F_2 is needed to avoid the explosive reaction of these two gases.

Electroplating

The electrolytic deposition of a thin metal film on the surface of a metallic object is called **electroplating.** Chromium is often plated onto iron and steel surfaces to improve the appearance and protect the object from corrosion. Objects are often plated with thin layers of silver or gold for appearance. In the electroplating process, the object being plated is used as the cathode, and the source of the metal in the coating is usually in the form of a complex metal ion in solution (Figure 18.17).

Figure 18.17 In electroplating, decorative and protective films of metals are deposited on the surface of objects by electrolysis.

OBJECTIVES REVIEW *Can you:*

- ☑ Describe the operation of an electrolytic cell?
- ☑ calculate the amount of material that can be produced from an electrolytic cell?
- ☑ list several applications of electrolytic cells?

18.9 Corrosion

OBJECTIVES

- ☐ Describe corrosion as an electrochemical process
- ☐ Explain how acidity and electrolyte concentration contributes to corrosion
- ☐ Explain the electrochemistry of anodic and cathodic protection

Corrosion is the oxidation of a metal to produce compounds of the metal through interaction with its environment. The most familiar and most costly example of corrosion is the rusting of iron and its alloys. Rust is a mixture of hydrated forms of iron(III) oxide. The green coating common on bronze statues (called a *patina*) is a familiar sight and is another example of corrosion. The green color is caused by copper(II) compounds formed from the corrosion of the copper in bronze (Figure 18.18). In addition to being unsightly, some corrosion processes lead to severe safety concerns by weakening structures made from metals. There is also a high cost for replacing certain items made largely from metals, such as bridges. For these reasons, chemists and other scientists have devoted great effort to understanding the causes of corrosion and developing methods to control it.

Corrosion is complex, but it can be understood if it is viewed as an electrochemical process. The formation of rust, $Fe_2O_3{\cdot}xH_2O$, requires the presence of oxygen and water. One part of a piece of iron serves as the anode in an electrochemical cell. The iron is oxidized to Fe^{2+}, and the electrons released flow through the metal to a region that

Figure 18.18 The Statue of Liberty. The green color of the Statue of Liberty is caused by the oxidation of the copper to copper(II) compounds, primarily oxides and carbonates. Before the recent restoration of the statue, corrosion had created holes in the outer surface.

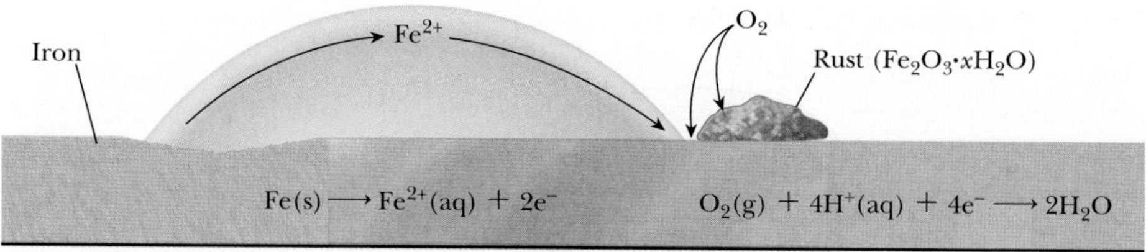

Figure 18.19 Rusting of iron. Iron in contact with water forms a negative electrode where the metal is oxidized to $Fe^{2+}(aq)$. The metal in contact with oxygen and water acts as a positive electrode where the oxygen is reduced. Ultimately, the $Fe^{2+}(aq)$ ions are oxidized to Fe^{3+}, which in the presence of oxygen and water forms $Fe_2O_3{\cdot}xH_2O$, which is called rust.

functions as the cathode. At the cathode, $O_2(g)$ is reduced in the presence of water. The half-reactions are

$$Fe(s) \rightarrow Fe^{2+}(aq) + 2e^- \qquad E^\circ = +0.44 \text{ V}$$

$$O_2(g) + 4H^+(aq) + 4e^- \rightarrow 2H_2O(\ell) \qquad E^\circ = +1.23 \text{ V}$$

The migration of the ions through the water on the surface of the metal completes the voltaic cell (Figure 18.19).

The actual potential of the cathode reaction [the half-reaction involving $O_2(g)$, above] depends on the acidity of the water solution, as the presence of H^+ ions as reactants would indicate. As the concentration of hydrogen ions increases in the water solution, the electrode potential increases, which can be confirmed by the Nernst equation. Sulfur oxides in the atmosphere dissolve to produce acidic solutions and tend to enhance the corrosion of metals. Not only is the potential of the electrochemical cell increased, but protective oxide films are more soluble in the acidic solutions.

The Fe^{2+} ions formed at the anode are further oxidized to hydrates of iron(III) oxide when they migrate to the surface of the water, where oxygen is available.

$$4Fe^{2+}(aq) + O_2(g) + (4 + 2x)H_2O(\ell) \rightarrow 2Fe_2O_3{\cdot}xH_2O + 8H^+(aq)$$

The concentration of electrolytes in the aqueous solution also affects the rate at which metals corrode. The presence of ions increases the electrical conductivity of the solution, and the larger currents that result speed the rate of deterioration of the metal. The chloride ion is particularly prone to promote corrosion, as mentioned in the introduction to this chapter. Because seawater has a high salt concentration, rusting and corrosion of metals occurs much more rapidly in waterfront locations. High electrolyte concentrations cause the bodies of automobiles to rust out more quickly in areas where salt is used on roads for the melting of snow and ice.

Corrosion. A break in the painted surface of a car allows the iron to come into contact with water and oxygen, both of which are necessary for corrosion to occur.

Protection from Corrosion

Corrosion limits the useful lifetime of many products, so a great deal of effort has been expended to find ways to inhibit or prevent corrosion. Because most corrosion requires the presence of water and oxygen in direct contact with the metal, one obvious way to reduce it is to apply an impervious coating to the metal surface. The painting of automobiles, in addition to improving appearance, reduces rusting. Plating of iron with chromium also protects auto parts from rusting. Although the voltage for the half-reaction

$$Cr(s) \rightarrow Cr^{2+}(aq) + 2e^- \qquad E° = +0.91\ V$$

indicates that chromium should be quite reactive, it forms a thin surface film of oxide that protects the chromium underneath from further oxidation. Even a small break in the chromium plating, however, allows the iron alloys underneath to corrode.

Another popular method for inhibiting corrosion is a process called **anodic protection,** in which the metal is intentionally oxidized under carefully controlled conditions to form a thin, adhering layer of oxide on the surface of the metal. The treatment of iron with aqueous sodium chromate forms a layer of Fe(III) and Cr(III) oxides that protects the iron from contact with oxygen and water.

$$2Fe(s) + 2Na_2CrO_4(aq) + 2H_2O(\ell) \rightarrow Fe_2O_3(s) + Cr_2O_3(s) + 4NaOH(aq)$$

A protective oxide layer can also be produced by electrolytic oxidation. *Anodized aluminum* is coated with an impervious layer of aluminum oxide by an electrolytic process.

Anodic protection. This aluminum drinkware has been anodized to provide a protective, corrosion-resistant surface.

Another way to protect metals from corrosion is to force the metal to behave as a cathode in the electrochemical cell. In **cathodic protection,** a second, more reactive metal is placed in electrical contact with the metal object being protected from corrosion. The more reactive metal will behave as the anode in the electrochemical cell, thus forcing the other metal to function as the cathode. In such a situation, the other metal will corrode preferentially. Iron that has been coated with a layer of zinc is called *galvanized iron* (Figure 18.20). Even if the zinc coating is broken to expose the iron, the iron does not oxidize as long as the more reactive zinc metal is present. The zinc is referred to as the **sacrificial anode.** Bars of magnesium are often attached to ocean vessels to serve as a sacrificial anode and prevent the rusting of the iron hulls of the vessel. Sacrificial anodes are attached to water, gas, and oil pipelines to minimize the potential catastrophic results of a corroded pipeline. Water heaters have magnesium bars in them to keep the iron body of the water heater from corroding, potentially avoiding a minor flood.

OBJECTIVES REVIEW *Can you:*

- ☑ describe corrosion as an electrochemical process?
- ☑ explain how acidity and electrolyte concentration contributes to corrosion?
- ☑ explain the electrochemistry of anodic and cathodic protection?

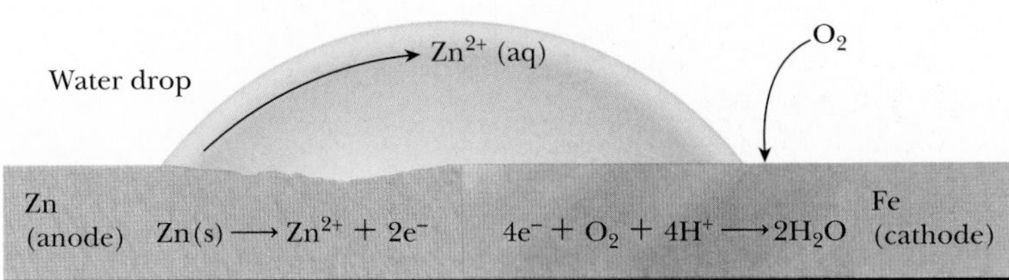

Figure 18.20 Cathodic protection. Iron that is in contact with a more reactive metal does not rust. Galvanized iron has a coating of zinc. The oxidation of the zinc to Zn^{2+} takes place more readily, so the iron does not rust.

CASE STUDY Cold Fusion

On March 23, 1989, two chemists from the University of Utah held a press conference and made a startling claim: They had been able to induce a nuclear fusion reaction in an electrochemical cell.

Fusion is the nuclear process in which small nuclei, like hydrogen, combine to make larger nuclei. In the process, copious amounts of energy are released. The Sun generates its energy by fusion; nuclear bombs that generate fusion reactions are the most destructive known weapons. However, the ability to create a controlled fusion reaction, so that useful energy could be extracted, had been unachievable to date. Thus, the announcement at the press conference generated an enormous amount of interest.

Chemists Stanley Pons and Martin Fleischmann claimed to have performed fusion in a tabletop system consisting of a jar of deuterium oxide (or "heavy water"; as mentioned in Chapter 2, deuterium is the isotope of hydrogen with a proton and a neutron in its nucleus) with palladium electrodes. A current was passed through the jar using deuterated potassium hydroxide as an electrolyte. Pons and Fleischmann claimed that they measured an increase in heat that could not be explained by a chemical reaction. They claimed that the only process that could generate such energy was nuclear fusion, called *cold fusion* because it occurs at relatively low temperatures. Aware that a fellow researcher at nearby Brigham Young University was also working on electrolyzing deuterium oxide solutions, Pons and Fleischmann apparently used a press conference to establish the priority of their discovery, rather than waiting for their research to be published in the peer-reviewed scientific literature. The manuscript that they prepared for publication was faxed all over the world before it was evaluated by other experts in the field, but it was eventually published in a journal on electroanalytical chemistry. The article gave more details, and it included not only claims of excess energy production, but production of gamma rays characteristic of nuclear fusion, as well as generation of neutrons and tritium, the heaviest and radioactive isotope of hydrogen.

Although interest in Pons and Fleischmann's claims was intense, reaction was split. Many scientists, including many chemists, were thrilled that the two electrochemists were able to succeed after so many years of attempts, many of them quite expensive, had failed. Other scientists were skeptical of the claims, stating that no electrochemical process could force nuclei close enough to fuse. The State of Utah appropriated $5 million to establish an institute to study cold fusion, and researchers all over the world rushed to confirm Pons and Fleischmann's findings.

Most researchers were not able to replicate the claims in Pons and Fleischmann's article. Although Pons and Fleischmann initially claimed that there was a "secret" to the experiment—perhaps to better maintain control of their own discovery—later they admitted that the description of the experiment in their article should allow others to reproduce their results. Several experiments to date have shown an increase in heat, but the magnitude of the heat produced makes it clear that the process is chemical, not nuclear. (Palladium has the unusual property of physically absorbing hydrogen, and the energy changes produced when hydrogen is absorbed could be responsible for any heat generated.) Many scientists criticized the original experiment, citing poor experimental setup, not enough neutron production to support a fusion process, lower density of deuterium than needed for fusion, and other factors. Some of these issues, such as experimental design, could be fixed; others could not. After several years and multiple experiments by numerous investigators, most of the scientific community now considers the original claims unsupported by the evidence.

Why did Pons and Fleischmann make such grandiose claims? Why did they announce their conclusions by press conference rather than through the normal peer-review process? We may never know for certain, but it is likely that one of the reasons is the desire for fame. Scientists, after all, are human beings; many humans crave the attention that would accompany a major scientific breakthrough. Achieving controlled nuclear fusion has been the subject of an enormous scientific effort; the thought of making it happen, and in such a simple apparatus, would tempt many scientists to swerve from the established scientific method.

Some people use the cold fusion story to argue that science is never *certain* about anything, that science is always changing its mind (figuratively speaking), that what we believed in the past was wrong, that science (and presumably scientists) cannot be trusted. Other people—and most scientists—point to the cold fusion issue as an example of how well science works. Science is constantly testing and retesting, designing experiments, and checking facts. Ultimately, science is self-correcting, and in the long run, results in a more accurate picture of how the universe operates. Science is the best process that humanity has developed to help us learn about our universe. Perhaps "bad science" is what happens when people try to thwart the normal scientific process.

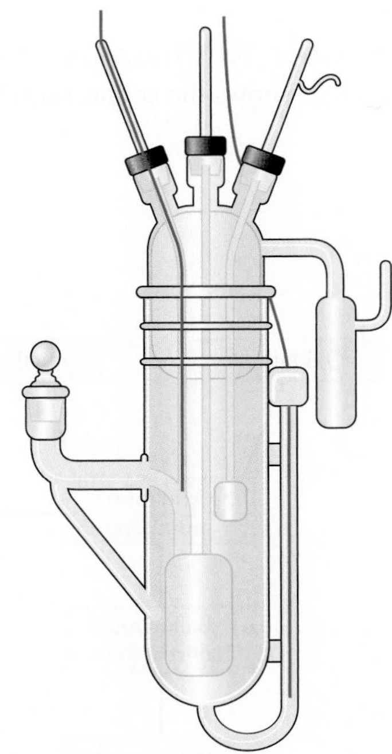

The "cold fusion" cell. Chemists Pons and Fleischmann claimed that a simple electrochemical cell such as this, with palladium electrodes and deuterium oxide, could produce nuclear fusion. Virtually every experiment that tried to replicate their claims failed. Electrochemical cold fusion is widely considered to be discredited.

Questions

1. What would be the expected electrolysis products of the electrochemical cell that Pons and Fleischmann constructed?
2. Based on this case study, why was palladium necessary as an electrode material? Would the same results occur if the electrode were some other inert material, such as carbon?
3. Does the cold fusion episode demonstrate the positive side of the scientific method, or the negative side? What would your arguments be in support of your choice?

ETHICS IN CHEMISTRY

1. Although it may be cheaper and mechanically sound to do so, plumbers should never solder a copper pipe directly to an iron one, despite the fact that both copper and iron may be cheaper than certain plastic pipe. Can you explain why?

2. Consider the Case Study. Was it ethical for Pons and Fleischmann to announce their research results by press conference, rather than submitting them to a normal "review-by-experts" evaluation that most research undergoes?

3. Consider the Case Study. Would Pons and Fleischmann's press conference announcement be more or less ethical (see question 2) if their results had been replicated and electrochemical cold fusion turned into a major new energy source?

Chapter 18 Visual Summary

The chart shows the connections between the major topics discussed in this chapter.

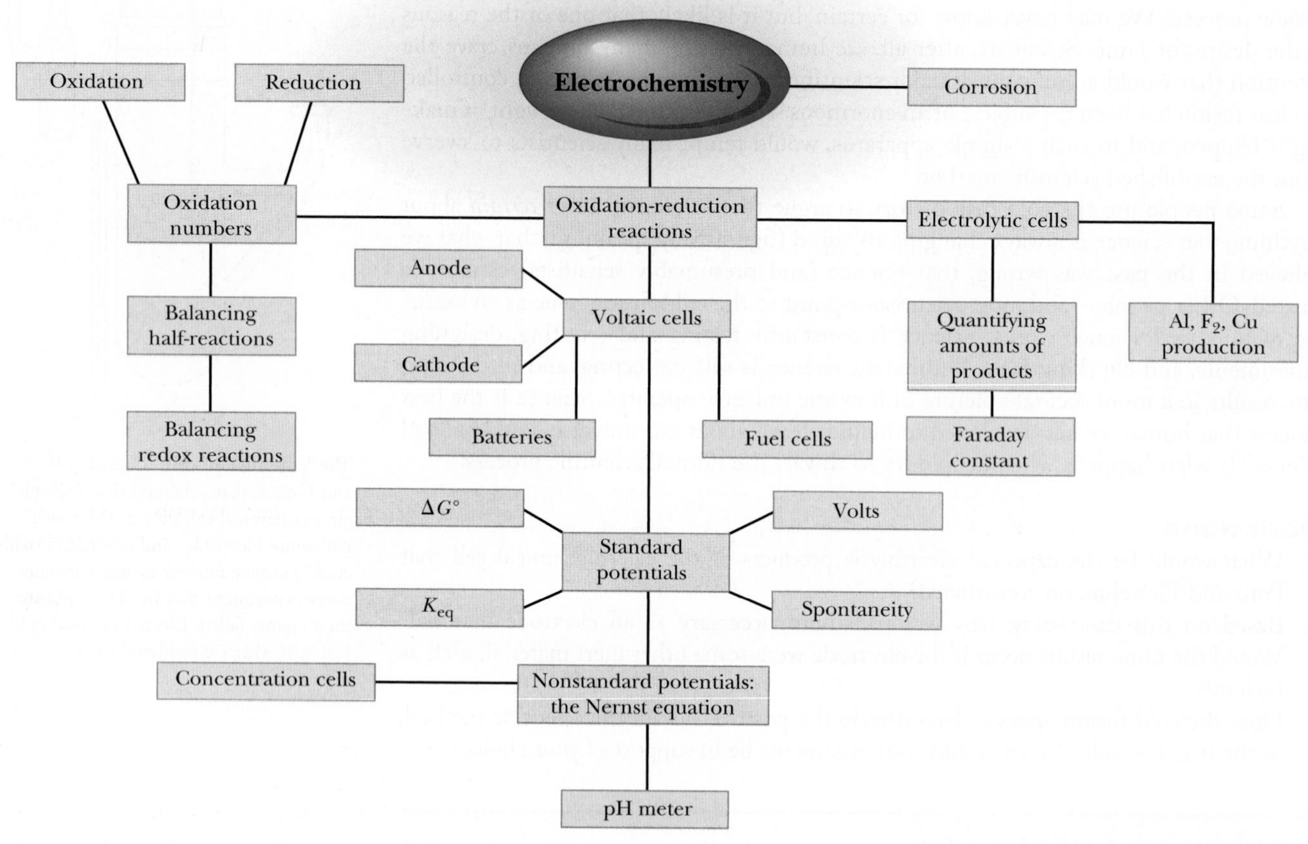

Summary

18.1 Oxidation Numbers

Electrochemistry is the study of the relationship between chemical reactions and electricity. All electrochemical processes involve *oxidation-reduction* (or *redox*) reactions (electron transfer reactions). In a redox reaction, the substance that loses electrons is *oxidized,* and it is called the *reducing agent.* The substance that gains electrons is *reduced,* and it is called the *oxidizing agent.* Assigning *oxidation numbers* to the elements in substances helps us recognize redox processes. The oxidation number is determined by using the rules given in Section 18.1.

18.2 Balancing Oxidation-Reduction Reactions

The *half-reaction method* for balancing oxidation-reduction reactions first divides the chemical expression into two *half-reactions,* each of which is balanced separately. The resulting oxidation and reduction half-reactions are then combined, after multiplying each by constants that make the number of electrons released in the oxidation equal to the number consumed by the reduction. The half-reaction method may be used for redox reactions that occur in either acidic or basic solutions.

18.3 Voltaic Cells

The chemical energy released in a spontaneous redox reaction can be directly converted into electrical energy using a *voltaic cell.* A voltaic cell consists of two *half-cells,* which are often connected through a *salt bridge* to maintain a charge balance. The word *electrode* is used to refer to the solid electrical conductor in a half-cell.

18.4 Potentials of Voltaic Cells

The *electromotive force* (emf), measured in volts, is an intensive property that measures the electrical driving force moving the electrons from the negative to the positive electrode of a voltaic cell. The voltage of a cell is an indication of the electric potential difference between the negative electrode and the positive electrode. The additivity of cell potentials allows us to assign a voltage to half-reactions, by defining the standard voltage of the hydrogen half-reaction as zero. The *standard reduction potentials* of half-reactions are tabulated and may be used to calculate the voltage of any redox reaction under standard conditions. When the voltage of a redox reaction is positive, the reaction proceeds spontaneously as written. A negative voltage shows that the reverse reaction is spontaneous.

18.5 Cell Potentials, ΔG, and K_{eq}

The standard free energy change for a redox reaction is related to the standard potential by the equation

$$\Delta G^\circ = -nFE^\circ$$

where n is the number of electrons transferred, and F is the *Faraday constant,* which is the charge on 1 mol of electrons (96,500 C). The standard potential is also related to the equilibrium constant for the cell reaction by the relationship

$$\log K_{eq} = nFE^\circ/2.303\ RT$$

Thus, standard cell potentials provide a means of determining free energy changes and equilibrium constants.

18.6 Dependence of Voltage on Concentration: The Nernst Equation

The potentials of redox reaction depend on the concentrations of the reactant and products. The concentration dependence of potentials is given by the *Nernst equation:*

$$E = E^\circ - \frac{2.303RT}{nF}\log Q = E^\circ - \frac{0.0591}{n}\log Q$$

where Q is the reaction quotient. The factor $2.303RT/F$ has a value of 0.0591 V at 25 °C. The Nernst equation also allows us to determine the voltages of concentration cells, in which the oxidation and reduction half-reactions are the same but ionic concentrations differ.

18.7 Applications of Voltaic Cells

Voltaic cells and *batteries* provide portable energy sources to power radios, watches, flashlights, cordless tools, calculators, among other items. Among the more important of these cells and batteries are Leclanché's *dry cell,* lead storage cell, and a variety of other cells, such as the nickel-cadmium cell and the nickel-metal hydride cell. *Fuel cells* convert the chemical energy of combustion reactions directly into electrical energy.

18.8 Electrolysis

Although voltaic cells produce electrical energy from chemical reactions, *electrolytic cells* consume electrical energy to force nonspontaneous chemical reactions to occur. In the electrolysis of aqueous solutions, the oxidation and reduction of water, as well as species in solution, must be considered in determining the expected products.

The amount of products formed in electrolysis is proportional to the total charge (current × time) that is passed through the electrolyte. The stoichiometry of the half-reaction can then be used to determine the amount of products formed.

Many very reactive elements (e.g., fluorine, chlorine, aluminum) are isolated from their compounds by electrolysis. In addition, the purification and recycling of some metals often involve electrolysis methods. The production of protective and decorative coatings by *electroplating* is another commercial application of electrolysis.

18.9 Corrosion

Corrosion of metals is an electrochemical process, in which oxygen combines with the metal. Corrosion is reduced by *anodic protection,* which produces a protective oxide film on the surface of the metal. In *cathodic protection,* a more reactive metal is used as a *sacrificial anode* to prevent oxidation of the protected metallic object.

Download Go Chemistry concept review videos from OWL or purchase them from **www.ichapters.com**

Chapter Terms

The following terms are defined in the Glossary, Appendix I.

Electrochemistry

Section 18.1

Half-reaction
Oxidizing agent
Oxidation
Oxidation-reduction reaction
Redox reaction
Reducing agent
Reduction

Section 18.2

Half-reaction method
Oxidation half-reaction
Reduction half-reaction
Skeleton reaction

Section 18.3

Anode
Cathode
Electrode
Half-cell
Inert electrode
Salt bridge
Voltaic cell (or galvanic cell)

Section 18.4

Electrode potentials
Electromotive force (emf)
Cell potential
Potential
Standard potential
Standard reduction potentials
Volt (V)

Section 18.5

Faraday constant, F
Standard cell potential, E°

Section 18.6

Concentration cells
Nernst equation

Section 18.7
Battery
Dry cell
Fuel cell
pH meter
Primary battery
Secondary battery
Section 18.8
Electrolysis
Electrolytic cell
Electroplating
Electrorefining
Section 18.9
Anodic protection
Cathodic protection
Corrosion
Sacrificial anode

Key Equations

Cell potential and free energy (18.5)

$$\Delta G = -nFE;\ \Delta G^\circ = -nFE^\circ$$

Relation of E° to K_{eq} (18.5)

$$E^\circ = \frac{RT}{nF}\ln K_{eq} = \frac{2.303RT}{nF}\log K_{eq} = \frac{0.0591}{n}\log K_{eq}$$

Nernst equation (18.6)

$$E = E^\circ - \frac{RT}{nF}\ln Q = E^\circ - \frac{2.303RT}{nF}\log Q = E^\circ - \frac{0.0591}{n}\log Q$$

Questions and Exercises

OWL Selected end of chapter Questions and Exercises may be assigned in OWL.

Blue-numbered Questions and Exercises are answered in Appendix J; questions are qualitative, are often conceptual, and include problem-solving skills.

■ Questions assignable in OWL

✎ Questions suitable for brief writing exercises

▲ More challenging questions

Questions

18.1 ✎ Describe oxidation and reduction. Compare the electron transfer in a redox reaction with the electron donation in a Lewis acid-base reaction.

18.2 List the halogens in order of increasing oxidizing power.

18.3 Which is a better reducing agent: zinc or mercury?

18.4 List four species that can oxidize Fe^{2+} to Fe^{3+}.

18.5 List three species that can reduce Al^{3+} to Al.

18.6 In a "dead" battery, the chemical reaction has come to equilibrium. What are the values of ΔG and E for a dead battery?

18.7 What is the difference between a battery and a fuel cell?

18.8 What are the differences between anodic and cathodic protection from corrosion?

Exercises

OBJECTIVE Assign oxidation numbers to atoms in chemical species.

18.9 Assign the oxidation numbers of all atoms in the following species.
(a) ClO_3^- (b) PF_3 (c) CO

18.10 ■ Assign the oxidation numbers of all atoms in the following species.
(a) N_2 (b) $B(OH)_3$ (c) IF_4^-

18.11 Assign the oxidation numbers of all atoms in the following ions.
(a) NO_3^- (b) NO_2^- (c) NH_4^+

18.12 ■ Assign the oxidation numbers of all atoms in the following species.
(a) Br_2 (b) CO_3^{2-} (c) CO_2

18.13 Assign the oxidation numbers of all atoms in the following compounds.
(a) ZrO_2 (b) FeO (c) $Ca(NO_3)_2$

18.14 Assign the oxidation numbers of all atoms in the following species.
(a) PF_5 (b) Na_2CrO_4 (c) NO_2^-

18.15 Assign the oxidation numbers of all atoms in the following species.
(a) BaO_2 (b) F_2 (c) Sn^{2+}

18.16 Assign the oxidation numbers of all atoms in the following species.
(a) $KMnO_4$ (b) H_2O (c) Cl_2

18.17 Assign the oxidation numbers of all atoms in the following species.
(a) NO_2 (b) CrO_2^- (c) $Co(NO_3)_3$

18.18 Assign the oxidation numbers of all atoms in the following species.
(a) $CaCO_3$ (b) $HBrO_4$ (c) Fe^{3+}

18.19 Assign the oxidation numbers of all atoms in the following compounds.
(a) KHF_2 (b) H_2Se
(c) NaO_2 (d) C_2H_6

18.20 Assign the oxidation numbers of all atoms in the following species.
(a) NO (b) BO_2^-
(c) $Cr(NO_3)_3$ (d) CH_3OH

OBJECTIVE Balance oxidation-reduction reactions.

18.21 Balance the following reactions, and specify which species is oxidized and which is reduced.
(a) $H_2 + O_2 \rightarrow H_2O$
(b) $Fe + O_2 \rightarrow Fe_2O_3$
(c) $Al_2O_3 + C \rightarrow Al + CO_2$

18.22 Balance the following reactions, and specify which species is oxidized and which is reduced.
(a) $Fe_2O_3 + H_2 \rightarrow Fe + H_2O$
(b) $CuCl_2 + Na \rightarrow NaCl + Cu$
(c) $C + O_2 \rightarrow CO_2$

18.23 Balance the following reactions, and specify which species is oxidized and which is reduced.
(a) $Na + FeCl_3 \rightarrow Fe + NaCl$
(b) $SnCl_2 + FeCl_3 \rightarrow SnCl_4 + FeCl_2$
(c) $CO + Cr_2O_3 \rightarrow Cr + CO_2$

18.24 Balance the following reactions, and specify which species is oxidized and which is reduced.
(a) $Na + Hg_2Cl_2 \rightarrow NaCl + Hg$
(b) $HCl + Zn \rightarrow ZnCl_2 + H_2$
(c) $H_2 + CO_2 \rightarrow CO + H_2O$

OBJECTIVE Balance redox reactions using the half-reaction method in acidic and basic solutions.

18.25 Complete and balance each half-reaction in acid solution, and identify it as an oxidation or a reduction.
(a) $Cr^{3+}(aq) \rightarrow Cr(s)$
(b) $I^-(aq) \rightarrow I_2(aq)$
(c) $NO_2^-(aq) \rightarrow NO_3^-(aq)$

18.26 ■ Write balanced equations for the following half reactions. Specify whether each is an oxidation or reduction.
(a) $H_2O_2(aq) \rightarrow O_2(g)$
(b) $H_2C_2O_4(aq) \rightarrow CO_2(g)$
(c) $NO_3^-(aq) \rightarrow NO(g)$

18.27 Complete and balance each half-reaction in acid solution, and identify it as an oxidation or a reduction.
(a) $UO_2^{2+}(aq) \rightarrow U^{4+}(aq)$
(b) $Zn(s) \rightarrow Zn^{2+}(aq)$
(c) $IO_3^-(aq) \rightarrow I^-(aq)$

18.28 Complete and balance each half-reaction in acid solution, and identify it as an oxidation or a reduction.
(a) $N_2O_4(g) \rightarrow NO_3^-(aq)$
(b) $Mn^{3+}(aq) \rightarrow MnO_4^-(aq)$
(c) $HOCl(aq) \rightarrow ClO_3^-(aq)$

18.29 Balance each of the following redox reactions in acid solution.
(a) $Sn(s) + Fe^{3+}(aq) \rightarrow Sn^{2+}(aq) + Fe^{2+}(aq)$
(b) $HAsO_3^{2-}(aq) + I_2(aq) \rightarrow H_2AsO_4^-(aq) + I^-(aq)$
(c) $Cu(s) + Ag^+(aq) \rightarrow Cu^{2+}(aq) + Ag(s)$

18.30 Balance each of the following redox reactions in acid solution.
(a) $MnO_4^-(aq) + H_2C_2O_4(aq) \rightarrow Mn^{2+}(aq) + CO_2(g)$
(b) $Cl_2(g) + Br^-(aq) \rightarrow Cl^-(aq) + Br_2(\ell)$
(c) $Cu(s) + NO_3^-(aq) \rightarrow NO(g) + Cu^{2+}(aq)$

18.31 Balance each of the following redox reactions in acid solution.
(a) $Fe(s) + Ag^+(aq) \rightarrow Ag(s) + Fe^{2+}(aq)$
(b) $I_2(aq) + S_2O_3^{2-}(aq) \rightarrow I^-(aq) + S_4O_6^{2-}(aq)$
(c) $MnO_4^-(aq) + Fe^{2+}(aq) \rightarrow Fe^{3+}(aq) + Mn^{2+}(aq)$

18.32 Balance each of the following redox reactions in acid solution.
(a) $Zn(s) + NO_3^-(aq) \rightarrow Zn^{2+}(aq) + N_2(g)$
(b) $IO_3^-(aq) + I^-(aq) \rightarrow I_2(aq)$
(c) $Ce^{4+}(aq) + Cl^-(aq) \rightarrow Cl_2(aq) + Ce^{3+}(aq)$

18.33 Balance each of the following redox reactions in basic solution.
(a) $Al(s) + ClO^-(aq) \rightarrow Al(OH)_4^-(aq) + Cl^-(aq)$
(b) $MnO_4^-(aq) + SO_3^{2-}(aq) \rightarrow MnO_2(s) + SO_4^{2-}(aq)$
(c) $Zn(s) + NO_3^-(aq) \rightarrow Zn(OH)_4^{2-}(aq) + NH_3(aq)$

18.34 Balance each of the following redox reactions in basic solution.
(a) $ClO^-(aq) + CrO_2^-(aq) \rightarrow Cl^-(aq) + CrO_4^{2-}(aq)$
(b) $Br_2(aq) \rightarrow Br^-(aq) + BrO_3^-(aq)$
(c) $H_2O_2(aq) + N_2H_4(aq) \rightarrow N_2(g) + H_2O(\ell)$

18.35 Balance each of the following redox reactions in basic solution.
(a) $Cl_2(aq) \rightarrow Cl^-(aq) + ClO_3^-(aq)$
(b) $MnO_4^-(aq) + I^-(aq) \rightarrow IO_3^-(aq) + MnO_2(s)$
(c) $ClO_3^-(aq) + CN^-(aq) \rightarrow Cl^-(aq) + CNO^-(aq)$

18.36 Balance each of the following redox reactions in basic solution.
(a) $PH_3(g) + CrO_4^{2-}(aq) \rightarrow CrO_2^-(aq) + P_4(s)$
(b) $F_2(g) + H_2O(\ell) \rightarrow F^-(aq) + O_2(g)$
(c) $H_2O_2(aq) + Cr(OH)_3(s) \rightarrow CrO_4^{2-}(aq)$

18.37 Why is the following balanced reaction not a proper redox reaction?

$$Fe^{2+}(aq) + 2Br^-(aq) \rightarrow Fe^{3+}(aq) + Br_2(\ell)$$

18.38 Why is the following balanced reaction not a proper redox reaction?

$$O_2(g) + 2H_2O(\ell) + I_2(s) \rightarrow 4OH^-(aq) + 2I^-(aq)$$

OBJECTIVES Identify the components of a voltaic cell. Write overall cell reaction. Identify direction of flow of electrons.

18.39 A voltaic cell is based on the reaction

$$Pb(s) + 2Ag^+(aq) \rightarrow Pb^{2+}(aq) + 2Ag(s)$$

Voltage measurements show that the Ag electrode is positive. Sketch the cell, and label the anode and cathode, the positive and negative electrodes, the direction of electron flow in the external circuit, and the direction of flow of cations and anions through the salt bridge. Write the half-reaction that occurs at each electrode.

18.40 A voltaic cell is based on the reaction

$$Zn(s) + Ni^{2+}(aq) \rightarrow Zn^{2+}(aq) + Ni(s)$$

Voltage measurements show that the Ni electrode is positive. Sketch the cell, and label the anode and cathode, the positive and negative electrodes, the direction of electron flow in the external circuit, and the direction of flow of cations and anions through the salt bridge. Write the half-reaction that occurs at each electrode.

18.41 A platinum wire is in contact with a mixture of mercury and solid mercury(I) chloride (Hg_2Cl_2) in a beaker containing 1 *M* KCl solution. A salt bridge connects this half-cell to a beaker that contains a copper electrode immersed in 1 *M* $CuSO_4$ solution. Voltage measurements show that the copper electrode is positive.
(a) Write balanced half-reactions for the two electrodes.
(b) Write the equation for the spontaneous cell reaction.
(c) In which direction do electrons flow in the external circuit?
(d) Would direct reaction occur if both the Hg/Hg_2Cl_2 and copper electrodes were placed in a container holding an aqueous solution that is 1 *M* $CuSO_4$ and 1 *M* KCl?

18.42 Two electrodes are immersed in a 1 *M* HBr solution. One of the electrodes is a silver wire coated with a deposit of AgBr(s). The second electrode is a platinum wire in electrical contact with a mixture of metallic mercury and Hg_2Br_2(s). Voltage measurements show that the Pt electrode is positive.
(a) Write balanced half-reactions for the two electrodes.
(b) Write the equation for the spontaneous cell reaction.
(c) In which direction do electrons flow in the external circuit?
(d) Why is a salt bridge unnecessary in this cell?

OBJECTIVE **Calculate the potential of a standard voltaic cell; then relate cell potential to a spontaneous reaction.**

18.43 For each of the reactions, calculate $E°$ from the table of standard potentials, and state whether the reaction is spontaneous as written or spontaneous in the reverse direction under standard conditions.
(a) $Cu^{2+}(aq) + Ni(s) \rightarrow Cu(s) + Ni^{2+}(aq)$
(b) $2Ag(s) + Cl_2(g) \rightarrow 2AgCl(s)$
(c) $Cl_2(g) + 2I^-(aq) \rightarrow 2Cl^-(aq) + I_2(s)$

18.44 For each of the reactions, calculate $E°$ from the table of standard potentials, and state whether the reaction is spontaneous as written or spontaneous in the reverse direction under standard conditions.
(a) $Zn(s) + Fe^{2+}(aq) \rightarrow Zn^{2+}(aq) + Fe(s)$
(b) $AgCl(s) + Fe^{2+}(aq) \rightarrow Ag(s) + Fe^{3+}(aq) + Cl^-(aq)$
(c) $Br_2(\ell) + 2Cl^-(aq) \rightarrow Cl_2(g) + 2Br^-(aq)$

18.45 Use the data from the table of standard reduction potentials in Appendix H to calculate the standard potential of the cell based on each of the following reactions. In each case, state whether the reaction proceeds spontaneously as written or spontaneously in the reverse direction under standard-state conditions.
(a) $H_2(g) + Cl_2(g) \rightarrow 2H^+(aq) + 2Cl^-(aq)$
(b) $Al^{3+}(aq) + 3Cr^{2+}(aq) \rightarrow Al(s) + 3Cr^{3+}(aq)$
(c) $Fe^{2+}(aq) + Ag^+(aq) \rightarrow Fe^{3+}(aq) + Ag(s)$

18.46 Use the data from the table of standard reduction potentials in Appendix H to calculate the standard potential of the cell based on each of the following reactions. In each case, state whether the reaction proceeds spontaneously as written or spontaneously in the reverse direction under standard-state conditions.
(a) $Ce^{4+}(aq) + Cu^+(aq) \rightarrow Cu^{2+}(aq) + Ce^{3+}(aq)$
(b) $Sn^{2+}(aq) + 2Fe^{3+}(aq) \rightarrow Sn^{4+}(aq) + 2Fe^{2+}(aq)$
(c) $Ni(s) + Br_2(\ell) \rightarrow Ni^{2+}(aq) + 2Br^-(aq)$

18.47 The standard potential for the cell reaction

$$UO_2^{2+}(aq) + Pb(s) + 4H^+(aq) \rightarrow U^{4+}(aq) + Pb^{2+}(aq) + 2H_2O(\ell)$$

is $E° = +0.460$ V. Use the tabulated standard potential of the lead half-reaction to find the standard reduction potential of the uranium half-reaction.

18.48 The standard potential of the cell reaction

$$Ag^+(aq) + Eu^{2+}(aq) \rightarrow Ag(s) + Eu^{3+}(aq)$$

is $E° = +1.23$ V. Use the tabulated standard potential of the silver half-reaction to find the standard reduction potential for the europium half-reaction.

18.49 A half-cell that consists of a copper wire in a 1.00 *M* $Cu(NO_3)_2$ solution is connected by a salt bridge to a solution that is 1.00 *M* in both Pu^{3+} and Pu^{4+}, and contains an inert metal electrode. The voltage of the cell is 0.642 V, with the copper as the negative electrode.
(a) Write the half-reactions and the overall equation for the spontaneous chemical reaction.
(b) Use the standard potential of the copper half-reaction, with the voltage of the cell, to calculate the standard reduction potential for the plutonium half-reaction.

18.50 A half-cell that consists of a silver wire in a 1.00 *M* $AgNO_3$ solution is connected by a salt bridge to a 1.00 *M* thallium(I) acetate solution that contains a metallic Tl electrode. The voltage of the cell is 1.136 V, with the silver as the positive electrode.
(a) Write the half-reactions and the overall chemical equation for the spontaneous reaction.
(b) Use the standard potential of the silver half-reaction, with the voltage of the cell, to calculate the standard reduction potential for the thallium half-reaction.

18.51 Use the data in Appendix H and assume standard conditions when answering the following questions.
(a) Which is the better oxidant: H_2O_2 or MnO_4^-?
(b) Which is a better reducing agent: Cu(s) or Ag(s)?

18.52 Use the data in Appendix H and assume standard conditions when answering the following questions.
(a) Which is the better oxidant: $Cr_2O_7^{2-}$ or Ce^{4+}?
(b) Which is the better reducing agent: Sn^{2+} or Fe^{2+}?

18.53 Use the standard reduction potentials in Table 18.1 to find
(a) a metal ion that reduces Ni^{2+}.
(b) a metal ion that can oxidize Cu.
(c) a metal ion that is reduced by Cr^{2+} but not H_2.

18.54 ■ Use the standard reduction potentials in Table 18.1 to find
(a) a reducing agent that will reduce Cu^{2+} but not Pb^{2+}.
(b) an oxidizing agent that will react with Cu but not Fe^{2+}.
(c) a metal ion that can reduce Fe^{3+} to Fe^{2+}.

OBJECTIVE Relate cell potential, free energy, and the equilibrium constant.

18.55 Write an expression and determine a value for K_{eq} for each voltaic cell in Exercise 18.43.

18.56 Write an expression and determine a value for K_{eq} for each voltaic cell in Exercise 18.44.

18.57 What is $\Delta G°$ for the oxidation of metallic iron by dichromate ($Cr_2O_7^{2-}$) in acidic solution assuming that the iron is oxidized to Fe^{2+}? Use the data in Appendix H.

18.58 What is $\Delta G°$ for the oxidation of metallic copper by permanganate (MnO_4^-) in acidic solution assuming that the copper is oxidized to Cu^{2+}? Use the data in Appendix H.

18.59 The standard potential of the half-reaction

$$2D^+(aq) + 2e^- \rightarrow D_2(g)$$

(where D = deuterium, or ^2H) is −0.013 V. Determine $\Delta G°$ and K_{eq} for the reaction

$$H_2(g) + 2D^+(aq) \rightarrow 2H^+(aq) + D_2(g)$$

In a mixture of hydrogen and deuterium, which isotope more favors its elemental form under standard conditions?

18.60 *Disproportionation* is a type of redox reaction in which the same species is simultaneously oxidized and reduced. One species that undergoes disproportionation is $Cu^+(aq)$.

$$2Cu^+(aq) \rightarrow Cu(s) + Cu^{2+}(aq)$$

If the half-reactions are

$$Cu^{2+}(aq) + e^- \rightarrow Cu^+(aq) \qquad E° = 0.153\ V$$

$$Cu^+(aq) + e^- \rightarrow Cu(s) \qquad E° = 0.521\ V$$

what are $E°$, $\Delta G°$, and K_{eq} for the overall reaction?

OBJECTIVE Use the Nernst equation to find the voltage of cells under nonstandard conditions.

18.61 Calculate the potential for each of the voltaic cells in Exercise 18.43 when the concentrations of the soluble species and gas pressures are as follows:
(a) $[Cu^{2+}] = 0.050\ M$, $[Ni^{2+}] = 1.40\ M$
(b) $P_{Cl_2} = 320$ torr
(c) $[I^-] = 0.0010\ M$, $P_{Cl_2} = 0.300$ atm, $[Cl^-] = 0.60\ M$

18.62 Calculate the potential for each of the voltaic cells in Exercise 18.44 when the concentrations of the soluble species and gas pressures are as follows:
(a) $[Fe^{2+}] = 0.050\ M$, $[Zn^{2+}] = 1.0 \times 10^{-3}\ M$
(b) $[Fe^{2+}] = 0.20\ M$, $[Fe^{3+}] = 0.010\ M$, $[Cl^-] = 4.0 \times 10^{-3}\ M$
(c) $[Br^-] = 3.5 \times 10^{-3}\ M$, $[Cl^-] = 0.10\ M$, $P_{Cl_2} = 0.50$ atm

18.63 A voltaic cell consists of a lead electrode and a reference electrode with a constant potential. This cell has a voltage of 53 mV when the lead electrode is placed in a 0.100 *M* $Pb(NO_3)_2$ solution (the lead electrode is positive). What voltage is measured when the lead electrode is placed in a saturated lead chloride solution, in which $[Pb^{2+}]$ is $1.6 \times 10^{-2}\ M$?

18.64 ■ Consider the voltaic cell

$$2Ag^+(aq) + Cd(s) \rightarrow 2Ag(s) + Cd^{2+}(aq)$$

operating at 298 K.
(a) What is the $E°_{cell}$ for this cell?
(b) If $[Cd^{2+}] = 2.0$ M and $[Ag^+] = 0.25\ M$, what is E_{cell}?
(c) If $E_{cell} = 1.25$ V and $[Cd^{2+}] = 0.100\ M$, what is $[Ag^+]$?

18.65 The voltaic cell in Exercise 18.63 produced a voltage of 0.010 V in an unknown test solution. What is the concentration of lead ions in the unknown solution?

18.66 Using the voltaic cell in Exercise 18.64, a voltage of 0.425 V was measured when the cell was placed in a solution of unknown ion concentrations. What is the ratio $[Cd^{2+}]/[Ag^+]^2$ in this solution?

18.67 Calculate the value of the solubility product constant for $Cd(OH)_2$ from the half-cell potentials.

$$Cd^{2+}(aq) + 2e^- \rightarrow Cd(s) \qquad E° = -0.403\ V$$

$$Cd(OH)_2(s) + 2e^- \rightarrow Cd(s) + 2OH^-(aq) \qquad E° = -0.83\ V$$

18.68 Calculate the value of the solubility product constant for $PbSO_4$ from the half-cell potentials.

$$PbSO_4(s) + 2e^- \rightarrow Pb(s) + SO_4^{2-}(aq) \qquad E° = -0.356\ V$$

$$Pb^{2+}(aq) + 2e^- \rightarrow Pb(s) \qquad E° = -0.126\ V$$

18.69 What is the voltage of a concentration cell of Fe^{2+} ions where the concentrations are 0.0025 and 0.750 *M*? What is the spontaneous reaction?

18.70 What is the voltage of a concentration cell of Cl^- ions where the concentrations are 1.045 and 0.085 *M*? What is the spontaneous reaction?

OBJECTIVE Review some applications of voltaic cells; then describe the chemical reactions that occur in commercial batteries.

18.71 (a) Use standard reduction potentials to calculate the potential of a silver-zinc cell that uses a basic electrolyte. The silver oxide is reduced from Ag_2O to Ag, and zinc metal is oxidized to $Zn(OH)_2$.
(b) Does the potential of this cell change as the cell is discharged? Explain.
(c) Would this cell make a good battery? Explain your answer.

18.72 (a) Use standard reduction potentials to calculate the potential of a nickel-cadmium cell that uses a basic electrolyte. The nickel in NiO(OH)(s) is reduced to $Ni(OH)_2(s)$, and cadmium metal is oxidized to $Cd(OH)_2(s)$.
(b) Will the potential of this cell change as the cell is discharged? Explain.
(c) Would this cell make a good battery? Explain your answer.

18.73 A possible reaction for a fuel cell is

$$C_3H_8(g) + 5O_2(g) \rightarrow 3CO_2(g) + 4H_2O(\ell)$$

(a) Write the oxidation and reduction half-reactions that occur, assuming a basic electrolyte.
(b) Use standard free energies of formation to calculate ΔG° for a propane-oxygen cell. From the standard free energy change, calculate the potential this cell could produce under standard conditions.
(c) Calculate the number of kilojoules of electrical energy produced when 1 g propane, C_3H_8, is consumed.

18.74 ■ A possible reaction for a fuel cell is

$$CH_4(g) + 2O_2(g) \rightarrow CO_2(g) + 2H_2O(\ell)$$

(a) Write the oxidation and reduction half-reactions that occur, assuming a basic electrolyte.
(b) Use standard free energies of formation from Appendix G to calculate ΔG° for a methane-oxygen cell. From the standard free energy change, calculate the potential this cell could produce under standard conditions.
(c) Calculate the number of kilojoules of electrical energy produced when 1 g methane, CH_4, is consumed.

OBJECTIVE Describe the operation of an electrolytic cell.

18.75 Write the half-reactions and the balanced chemical equations for the reactions that occur in the electrolysis of
(a) a zinc chloride aqueous solution, using zinc electrodes.
(b) a calcium bromide solution, using inert electrodes.
(c) a sodium iodide solution, using inert electrodes.

18.76 Write the half-reaction and the chemical equations for the reactions that occur in the electrolysis of
(a) molten $CaCl_2$, using inert electrodes.
(b) a saturated solution of magnesium sulfate, using inert electrodes.
(c) the electrolysis cell represented in this diagram.

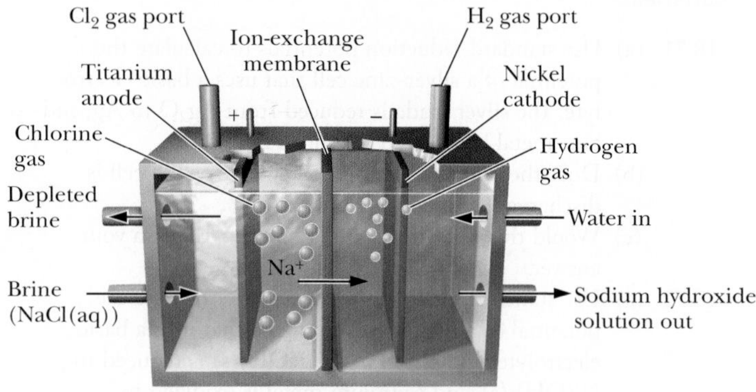

Electrolysis of brine

18.77 A solution contains the ions H^+, Ag^+, Pb^{2+}, and Ba^{2+}, each at a concentration of 1.0 *M*.
(a) Which of these ions would be reduced first at the cathode during an electrolysis?
(b) After the first ion has been completely removed by electrolysis, which is the second ion to be reduced?
(c) Which, if any, of these ions cannot be reduced by the electrolysis of the aqueous solution?

18.78 A solution contains the ions H^+, Cu^{2+}, Ca^{2+}, and Ni^{2+}, each at a concentration of 1.0 *M*.
(a) Which of these ions would be reduced first at the cathode during an electrolysis?
(b) After the first ion has been completely removed by electrolysis, which is the second ion to be reduced?
(c) Which, if any, of these ions cannot be reduced by the electrolysis of the aqueous solution?

18.79 The commercial production of fluorine uses a mixture of molten potassium fluoride and anhydrous hydrogen fluoride as the electrolyte. The products of the electrolysis are hydrogen and fluorine.
(a) Why is the potassium fluoride necessary, because it is not involved in the redox reaction that occurs?
(b) What products would form if the hydrogen fluoride were not present?

18.80 The commercial production of magnesium is accomplished by electrolysis of molten $MgCl_2$.
(a) Why is electrolysis of an aqueous solution of $MgCl_2$ not used in this process?
(b) Write the anode and cathode half-reaction in the electrolysis of molten $MgCl_2$.

OBJECTIVE Calculate the amount of material produced by an electrolytic cell.

18.81 How many coulombs of charge are needed to accomplish each of the following conversions by electrolysis?
(a) Produce 0.50 mol Al by electrolysis of Al_2O_3.
(b) Reduce all of the Cu^{2+} in 100 mL of 0.20 *M* $Cu(NO_3)_2$.
(c) Make 10.0 g Cl_2 by electrolysis of molten NaCl.
(d) Deposit 0.32 g silver from an aqueous $AgNO_3$ solution.

18.82 How many coulombs of charge are needed to accomplish each of the following conversions by electrolysis?
(a) Form 0.50 mol Ca from $CaCl_2$.
(b) Produce 3.0 g Al by electrolysis of Al_2O_3.
(c) Form 0.52 g O_2 by electrolysis of an aqueous Na_2SO_4 solution.
(d) Make 1.0 L of gaseous H_2 at standard temperature and pressure by electrolysis of water.

18.83 Find the mass of hydrogen produced by electrolysis of hydrochloric acid for 59.0 minutes, using a current of 0.500 A.

18.84 What mass of cadmium is deposited from a $CdCl_2$ solution by passing a current of 1.50 A for 38.0 minutes?

18.85 How many minutes are needed to deposit 15.0 g copper from a Cu^{2+} solution, using a current of 3.00 A?

18.86 ■ How long would it take to electroplate a metal surface with 0.500 g nickel metal from a solution of Ni^{2+} with a current of 4.00 A?

OBJECTIVE Describe corrosion as an electrochemical process; then explain the electrochemistry of anodic and cathodic protection.

18.87 Like zinc, sodium is a rather active metal. Would it be possible to use metallic sodium for cathodic protection of the iron hull of an ocean vessel? Explain.

18.88 The electrochemical processes that occur in the corrosion of iron are represented by the half-reactions

$$Fe(s) \rightarrow Fe^{2+}(aq) + 2e^-$$

$$O_2(g) + 4H^+(aq) + 4e^- \rightarrow 2H_2O(\ell)$$

(a) Write the overall reaction.
(b) What is the standard potential for the overall chemical reaction?
(c) Natural water has a pH of about 5.9, and air is 0.21 mol fraction oxygen. If the concentration of iron(II) in the water is 5×10^{-5} *M*, what is the potential of the corrosion reaction in the presence of air and natural water at 1 atm pressure and 298 K?

18.89 An aluminum bulkhead in a swimming pool collapsed. Stainless steel (mostly iron) braces were bolted to the aluminum to strengthen the bulkhead. Within a few months, the bulkhead collapsed again and showed extreme corrosion of the aluminum close to the steel bolts. Explain the electrochemical processes that occurred in the reinforced bulkhead.

18.90 Magnesium is used in the cathodic protection of metal coffins that are guaranteed to last a century without corroding. If the average current produced by the electrochemical cell is 2.0 mA, how much magnesium (in pounds) is needed to protect the coffin for 100 years?

Chapter Exercises

18.91 Assign the oxidation states of all elements in each of the following:
(a) CaC_2O_4 (b) $Ba(ClO_4)_2$ (c) Tl^{3+}

18.92 Sphalerite is the naturally occurring mineral zinc sulfide, from which zinc metal is extracted. The ore is heated in oxygen to form zinc oxide and sulfur dioxide, followed by the reaction of metal oxide with elemental carbon to form CO.
(a) Write a balanced equation for each of these reactions.
(b) For each reaction, identify the element that is oxidized and the one that is reduced.

18.93 What is the reduction potential of the hydrogen electrode at 298 K if the pressure of gaseous hydrogen is 2.5 atm in a solution of pH 6.00?

18.94 Use the standard reduction potentials in Appendix H to answer the following questions.
(a) What products, if any, are formed when $KClO_3$ and KCl are mixed in an acid solution?
(b) Which is a stronger oxidizing agent in acid solution: Fe^{3+} or Cr^{3+}?
(c) Which is a stronger reducing agent: $Fe(CN)_6^{4-}$ or Fe^{2+}?

18.95 Write the chemical equation and calculate the standard potentials for each of the following, using the data in Appendix H.
(a) Fe^{3+} is added to $H_2SO_3(aq)$.
(b) Iron(II) is titrated with $K_2Cr_2O_7$.
(c) Nitrous acid (HNO_2) is added to $Fe^{2+}(aq)$.

18.96 The standard free energy change at 25 °C, $\Delta G°$, is equal to −34.3 kJ for

$$2Fe(CN)_6^{3-}(aq) + 2I^-(aq) \rightarrow 2Fe(CN)_6^{4-}(aq) + I_2(s)$$

Calculate the standard potential for this reaction.

18.97 The equilibrium constant at 25 °C is 1.58×10^2 for

$$2VO^{2+}(aq) + Br_2(\ell) + 2H_2O(\ell) \rightarrow 2VO_2^{2+}(aq) + 2Br^-(aq) + 4H^+(aq)$$

Calculate $\Delta G°$ and $E°$ for this reaction.

18.98 Calculate the potential of the half-reaction

$$Fe^{3+} + e^- \rightarrow Fe^{2+}$$

when the concentrations in solution are $[Fe^{3+}] = 0.033$ *M* and $[Fe^{2+}] = 0.0025$ *M*, and the temperature is 298 K.

18.99 Another type of battery is the alkaline zinc-mercury cell, in which the cell reaction is

$$Zn(s) + HgO(s) \rightarrow Hg(\ell) + ZnO(s) \quad E° = +1.35 \text{ V}$$

(a) What is the standard free energy change for this reaction?
(b) The standard free energy change in a voltaic cell is the maximum electrical energy that the cell can produce. If the reaction in a zinc-mercury cell consumes 1.00 g mercury oxide, what is the standard free energy change?
(c) For how many hours could a mercury cell produce a 10-mA current if the limiting reactant is 3.50 g mercury oxide?

Cumulative Exercises

18.100 In the analytical technique called *electrogravimetry*, electrolysis is used to separate the analyte from a solution by depositing it on an inert electrode. The electrode is weighed before and after the experiment to find the mass of analyte deposited. A 0.122-g sample of a copper-zinc alloy was treated with concentrated sulfuric acid to produce a solution containing copper(II) and zinc(II) sulfates. The platinum cathode used in the electrolysis of this solution increased in mass by 0.073 g after exhaustive electrolysis.
(a) Which metal was deposited on the cathode during the electrolysis? Write the balanced equation for the electrolysis reaction.
(b) What was the mass percentages of copper and zinc in the alloy sample?

18.101 ▲ At 298 K, the solubility product constant for PbC_2O_4 is 8.5×10^{-10}, and the standard reduction potential of the Pb^{2+}(aq) to Pb(s) is −0.126 V.

(a) Find the standard potential of the half-reaction

$$PbC_2O_4(s) + 2e^- \rightarrow Pb(s) + C_2O_4^{2-}(aq)$$

(*Hint:* The desired half-reaction is the sum of the equations for the solubility product and the reduction of Pb^{2+}. Find ΔG° for these two reactions and add them to find ΔG° for their sum. Convert the ΔG° to the potential of the desired half-reaction.)

(b) Calculate the potential of the Pb/PbC_2O_4 electrode in a 0.025 *M* solution of $Na_2C_2O_4$.

18.102 ▲ At 298 K, the solubility product constant for $Pb(IO_3)_2$ is 2.6×10^{-13}, and the standard reduction potential of the Pb^{2+}(aq) to Pb(s) is −0.126 V.

(a) Find the standard potential of the half-reaction

$$Pb(IO_3)_2(s) + 2e^- \rightarrow Pb(s) + 2IO_3^-(aq)$$

(*Hint:* The desired half-reaction is the sum of the equations for the solubility product and the reduction of Pb^{2+}. Find ΔG° for these two reactions, and add them to find ΔG° for their sum. Convert the ΔG° to the potential of the desired half-reaction.)

(b) Calculate the potential of the $Pb/Pb(IO_3)_2$ electrode in a 3.5×10^{-3} *M* solution of $NaIO_3$.

18.103 The acid-base titration curves discussed in Chapter 16 can be determined using a pH meter, which measures the potential of a cell made up of a reference electrode and an indicating electrode that responds to the hydrogen ion concentration in solution. Assume that the cell potential, in volts, follows the equation

$$E_{cell} = k - 0.059 \text{ pH}$$

The potential of the cell is 135 mV at the start of a titration of a 0.032 *M* HCl solution with NaOH(aq). What is the potential of this cell when the equivalence point is reached?

18.104 ■ Calculate the rate of oxygen gas production at standard temperature and pressure, in units of milliliters per minute (mL/min), by the electrolysis of water at a 0.250-A current.

18.105 An electrolytic cell produces aluminum from Al_2O_3 at the rate of 10 kg/day. Assuming a yield of 100%,

(a) how many moles of electrons must pass through the cell in one day?

(b) how many coulombs are passing through the cell?

(c) how many moles of oxygen (O_2) are being produced simultaneously?

18.106 Cardiac pacemakers use a lithium-iodine battery that is based on the reaction

$$Li(s) + I_2(s) \rightarrow 2LiI(s)$$

(a) What are the E°, ΔG°, and K_{eq} for this reaction?

(b) Given the phases of the reactants and products, what do you think happens to the voltage of this battery as the reaction proceeds? (*Hint:* Write the Nernst equation for this reaction.)

18.107 ▲ An object with a surface area of 100 cm^2 is gold plated. The source of the gold is a solution of $Au(CN)_4^-$. How many minutes does it take to cover the object with gold to a thickness of 0.0020 mm, using a current of 0.500 A? The density of gold is 19.3 g/cm^3.

AP Photo/Charles Rex Arbogast

Gold is often used to plate objects to give them an attractive finish. The thickness of the gold plating can be as little as a few micrometers.

18.108 ▲ At 298 K, the solubility product constant for solid $Ba(IO_3)_2$ is 1.5×10^{-9}. Use the standard reduction potential of Ba^{2+}(aq) to find the standard potential for the half-reaction

$$Ba(IO_3)_2(s) + 2e^- \rightarrow Ba(s) + 2IO_3^-(aq)$$

(*Hint:* Find ΔG° for both the solubility equilibrium and the reduction half-reaction for Ba^{2+}, and add the reactions. Use the ΔG° for the sum reaction to find E°.)

18.109 Another type of fuel cell uses CH_4 as the fuel and O_2 as the oxidizer. Without having the half-reactions for the fuel cell reaction, use other data available to predict what the E° for this fuel cell would be.

18.110 ▲ Consider the standard reduction potentials of cesium and lithium.

$$Cs^+(aq) + e^- \rightarrow Cs(s) \qquad E° = -3.026\ V$$

$$Li^+(aq) + e^- \rightarrow Li(s) \qquad E° = -3.095\ V$$

The periodic trends in the properties of the element indicate that fluorine is the most chemically reactive nonmetal, so perhaps it is not surprising that the standard reduction potential of fluorine has the highest positive value for a nonmetallic element. However, periodic properties of the elements also indicate that cesium should be the most reactive metal. Comparison of the voltage of the cesium half-reaction with that of lithium shows that the standard reduction potential of lithium is less negative than that of cesium, indicating that lithium is a better oxidizer than is cesium.

(a) Calculate the standard cell voltages of the voltaic cells based on the reaction

$$2M(s) + F_2(g) \rightarrow 2M^+(aq) + 2F^-(aq)$$

where M is Cs and Li.

(b) Assuming that the pressure of $F_2(g)$ stays at 1.00 atm, what concentration does $Li^+(aq)$ have to be for the voltage of the Li/F_2 voltaic cell to equal the standard voltage of the Cs/F_2 voltaic cell?

(c) Can you suggest a reason why the standard reduction potential of lithium is lower than that of cesium, even though periodic trends indicate that cesium is the more reactive metal?

(d) Calculate $\Delta G°$ for both the Li/F_2 and the Cs/F_2 voltaic cells from their $E°$s. Compare this with the Gibbs' free energies of formation of 2 mol LiF and CsF. Can you explain the difference?

(e) Given the fact that alkali metals react rather violently with water, it would be unlikely that any voltaic cell can be constructed using Li(s) or Cs(s) in the presence of water. A more likely scenario is that the voltaic cell would have no solvent, so that the voltaic cell reaction would be

$$2M(s) + F_2(g) \rightarrow 2MF(xtal)$$

where M is Cs or Li. What would be the $E°$s of the two different voltaic cells if this were the reaction? (*Hint:* See your answer to part d.)

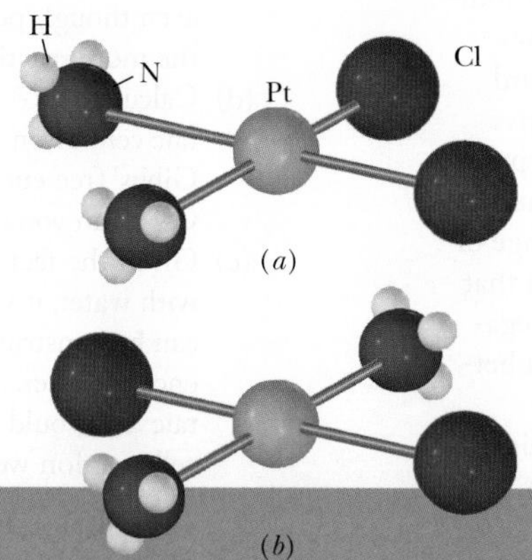

Isomers of $[Pt(NH_3)_2Cl_2]$. The structures of *(a)* the *cis* isomer and *(b)* the *trans* isomer.

One of the more effective compounds for fighting cancer has the common name *cisplatin* and has the formula *cis*-$[Pt(NH_3)_2Cl_2]$. Cisplatin was first described in 1845, but it took more than 100 years for scientists to realize that it was effective in fighting cancer. Like many important advances, some scientists coupled an observation—that bacteria did not multiply in the region of a platinum electrode—with a careful analysis that showed cisplatin formed from the electrode material and was responsible for the inhibition of the bacteria. The discovery that this compound was useful in chemotherapy came as a surprise because compounds of the heavier transition metals are generally toxic. It has been known for a long time that certain lighter transition metals, such as the iron in hemoglobin and the cobalt in vitamin B_{12}, are absolutely necessary for many living systems, but the heavier metals tend to disrupt biological systems, and are thus toxic.

Cisplatin has been shown to interact with the DNA of cancer cells and help prevent fast replication. Like many chemotherapy drugs, cisplatin does have

Transition Metals, Coordination Chemistry, and Metallurgy

19

OWL Online homework for this chapter may be assigned in OWL.

Look for the green colored vertical bar throughout this chapter, for integrated references to this chapter introduction.

major side effects. One of them is neurotoxicity wherein many patients experience numbness in their feet and hands. Because cisplatin can improve the chance of eliminating the cancer, other pharmaceuticals that can minimize the side effects are administered along with cisplatin.

The structure of cisplatin, together with its *trans* isomer, are pictured. The two isomers are different in the arrangement of the groups bonded to platinum (these groups are called *ligands*) in the square planar structure of this compound. As in Chapter 10, which describes *cis* and *trans* isomers of alkenes, the like ligands about the platinum in the *cis* isomer are adjacent, whereas they are opposite in the *trans* isomer. As with the alkenes, the physical and chemical properties of the two isomers are different; for example, the *trans* isomer of cisplatin is toxic and has no anticancer value.

The preparation of the *cis* isomer, free of the *trans* isomer, is a classic experiment in inorganic chemistry. The reaction of 1 mol $[PtCl_4]^{2-}$ with 2 mol NH_3 yields the *cis* complex, whereas the reaction of 1 mol $[Pt(NH_3)_4]^{2+}$ and 2 mol Cl^- yields the *trans* isomer. ▮

Most of the elements are metals, and many of them have played key roles in the development of civilization. In our modern society, metals are used as electrical conductors, structural materials, and catalysts. The representative elements sodium and other Group 1A elements are used as strong reducing agents, aluminum and magnesium are used in lightweight alloys, and lead is used in the manufacture of storage batteries. Chapter 8 presents the chemistry of the Group 1A and 2A metals.

The transition elements and their compounds are important in all aspects of our society. Iron and its alloys are widely used structural materials. Copper is used in electrical wiring; silver compounds are used in photographic processes; titanium is used in lightweight alloys, and its compounds are used as paint pigments; gold has long been a standard of value and beauty; and platinum is an important catalyst. In addition, the transition elements play vital roles in the chemistry of biological systems. Iron, copper, cobalt, zinc, molybdenum, and many other metals are essential for many biological functions.

Lewis bases bond to transition metals and form metal complexes with interesting and varied properties, and many of the compounds have beautiful colors. These properties can be understood by examining the structure and bonding of metal complexes. A good reason to understand the characteristics of transition-metal complexes is their importance in many biological systems; most notable is hemoglobin, an iron complex found in the red blood cells that transports oxygen.

With a few exceptions (Cu, Au, Ag, Ru, Rh, Pd, Ir, and Pt), all of the metals occur in nature as compounds; therefore, extensive processing is needed to prepare the free elements. Section 19.5 presents methods used to isolate metals from their naturally occurring ores.

19.1 Properties of the Transition Elements

OBJECTIVES

- ☐ Define which elements are transition metals and locate them on the periodic table
- ☐ Relate the trends in physical properties (melting points, boiling points, and hardness) and atomic properties (atomic radius and ionization energy) of the transition elements to their relative positions in the periodic table
- ☐ Describe the effect of the lanthanide contraction on the sizes and ionization energies of the transition elements

Transition elements have partially filled *d* or *f* orbitals in the metal or one of its oxidation states; they include all of the elements in Groups 3B through 1B, and the lanthanides and actinides.

The **transition elements** are characterized by the presence of a partially-filled *d* or *f* subshell in the metal atom or one of its oxidation states. The *d*-block transition elements, the ones that are primarily covered in this chapter, are found in the center of the periodic table, in the nine columns from Group 3B through 1B (3 through 11). The elements in Group 2B (12), Zn, Cd, and Hg, are not transition metals as we have defined them. Even though they are found in the *d* block of the periodic table, their *d* orbitals are always completely filled. The *f*-block transition elements comprise the 2 rows of 14 elements across the bottom of the periodic table. Elements in the first of these two rows are called the *lanthanides* and those in the second the *actinides;* all of these elements with partially-filled *f* orbitals are frequently referred to as the *inner transition elements.*

Figure 19.1 Vanadium compounds. All of the oxidation states of vanadium are colored. (right to left) V^{2+} (violet), V^{3+} (green), VO^{2+} (blue), and VO_2^+ (yellow).

Several properties distinguish the transition elements from the representative elements. Unlike the representative elements, all of the transition elements have high melting points. With the exception of the Group 3B elements, each of the transition elements can have more than one stable oxidation state in its compounds. Transition-metal compounds are generally colored, and the color depends on the oxidation state of the metal, as well as on the other elements in the compound. For example, Figure 19.1 shows solutions that contain compounds of vanadium in four different oxidation states, each of a different color. This section examines several of the properties of the transition metals and relates them to the atomic properties.

TABLE 19.1 Melting and Boiling Points of the Transition Elements (°C)

		Fourth Period			Fifth Period			Sixth Period		
Group No.		Element	m.p.	b.p.	Element	m.p.	b.p.	Element	m.p.	b.p.
3B	(3)	Sc	1541	2831	Y	1522	3338	La	921	3457
4B	(4)	Ti	1668	3287	Zr	1852	4409	Hf	2227	4602
5B	(5)	V	1910	3407	Nb	2468	4742	Ta	3017	5458
6B	(6)	Cr	1907	2672	Mo	2617	4639	W	3422	5555
7B	(7)	Mn	1244	2061	Tc	2152	4265	Re	3180	5596
8B	(8)	Fe	1535	2861	Ru	2034	4150	Os	3033	5012
	(9)	Co	1495	2927	Rh	1966	3695	Ir	2446	4428
	(10)	Ni	1453	2913	Pd	1552	2970	Pt	1772	3827
1B	(11)	Cu	1083	2567	Ag	962	2162	Au	1064	2856

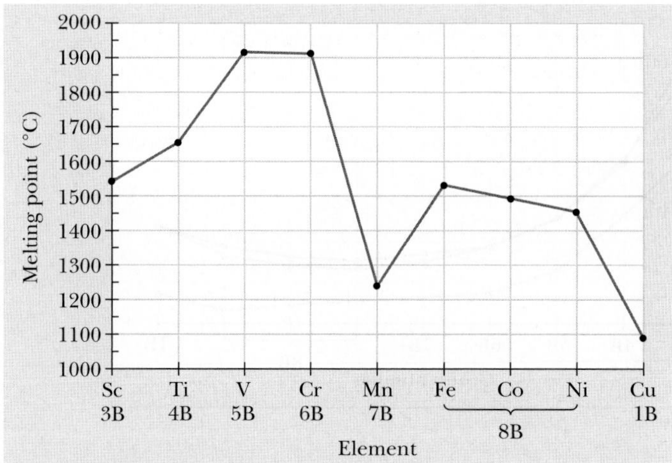

Figure 19.2 Melting points of the fourth-period transition metals. The melting points of the transition metals are highest in Groups 5B and 6B.

Melting Points and Boiling Points

Table 19.1 lists the melting points and boiling points for the transition elements, and the melting points for the fourth-period transition elements are shown graphically in Figure 19.2. These properties reach a maximum in Groups 5B and 6B, suggesting that the metallic bonds are strongest in those groups. If we assume that only the $(n-1)d$ and ns orbitals are used by the transition metals in the formation of highly delocalized metallic bonds, then a metal atom can form a maximum of one bond for each singly occupied s or d orbital. The number of electrons available on each atom to form bonds increases from three for the Group 3B (3) elements to six in Group 6B (6). In Group 7B (7), one of the six available orbitals is occupied by two electrons and is unavailable to form a metal-metal bond, which results in fewer bonds per metal atom. With each successive group after that, the number of doubly occupied orbitals increases by one, causing a further reduction in the number of bonds each metal atom may form. Consistent with this simple model, there is a large drop in the melting point from the Group 6B elements to the 7B elements in all the periods.

Several other properties, such as heats of fusion and hardness, also reflect the strength of the metallic bonding in the transition elements.

Atomic and Ionic Radii

The trends in the chemical properties of the transition metals in each group parallel the changes in atomic and ionic radii. Table 19.2 gives the atomic radius of each transition element and Figure 19.3 graphs these trends. In the fourth period, the atomic radius decreases fairly rapidly through the element chromium, in Group 6B. After chromium, the radius decreases much more slowly and actually increases from nickel to copper. This pattern is

TABLE 19.2 Atomic Radii (pm) and Densities (g/cm³) of the Transition Metals

Group		Fourth Period			Fifth Period			Sixth Period		
		Element	Radius	Density	Element	Radius	Density	Element	Radius	Density
3B	(3)	Sc	162	3.00	Y	180	4.50	La	187	6.17
4B	(4)	Ti	147	4.50	Zr	160	6.51	Hf	159	13.28
5B	(5)	V	134	6.11	Nb	146	8.57	Ta	146	16.65
6B	(6)	Cr	128	7.14	Mo	139	10.28	W	139	19.30
7B	(7)	Mn	127	7.43	Tc	136	11.5	Re	137	21.00
8B	(8)	Fe	126	7.87	Ru	134	12.41	Os	135	22.57
	(9)	Co	125	8.90	Rh	134	12.39	Ir	136	22.61
	(10)	Ni	124	8.91	Pd	137	12.0	Pt	139	21.41
1B	(11)	Cu	128	8.95	Ag	144	10.49	Au	144	19.32

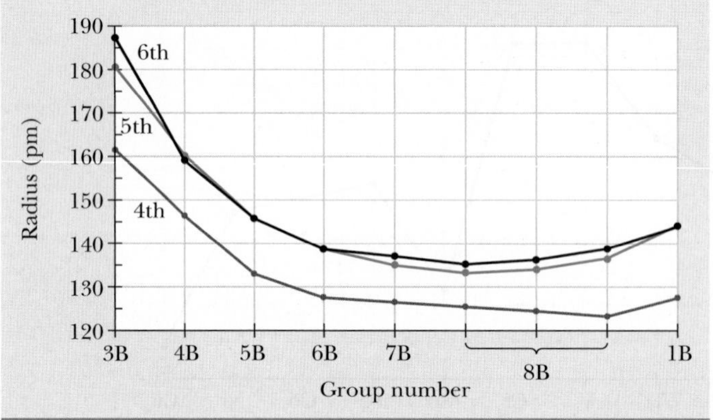

Figure 19.3 Atomic radii of the transition elements. In the three transition series, the atomic radii decrease from left to right, with a small increase near the end of the series. The atoms in each group of the fifth and sixth periods have nearly identical radii because of the lanthanide contraction.

repeated for the elements in the fifth and sixth periods. The general trend of decreasing radius from left to right within a period is the same as that observed in the representative elements, but the changes in size are not as great for the transition elements.

It is the outermost electrons (those in orbitals with the highest principal quantum number) and the effect of nuclear charge that determine the sizes of the atoms (see Section 8.2). The outer electrons in transition-metal elements are always in an *ns* subshell. Along any transition-metal series, the number of electrons in the $(n - 1)d$ level increases as the nuclear charge increases. These *d* electrons in the transition elements are closer to the nucleus; they are effective in shielding the outer *s* electrons (the ones that determine the size of the metal atom) from the nuclear charge. Therefore, the attractive effect of the nuclear charge on the outer *s* electrons increases quite slowly across the transition elements, resulting in a relatively small decrease in the radius of the atoms. The strength of the metallic bonding also contributes to the atomic size. Through Group 6B, the metallic bonds increase in strength, reducing the distance between nearest neighbors in the solid, contributing to the decrease in the radius. Toward the end of the *d* block, the decrease in metal-metal bonding causes the radii to increase slightly.

The radii of transition metals decrease slowly from left to right through Group 8B and then increase slightly.

Figure 19.3 also shows trends in the atomic radii of the elements for each of the transition series. As expected, the radii of the elements in each group increase from the fourth to the fifth period. However, with the exception of group 3B, the radii of the fifth- and sixth-period elements in each group are nearly identical. Between the elements lanthanum and hafnium are the 14 elements (the lanthanides) in which the valence electrons are in the 4*f* subshell. The electrons in the filled 4*f* subshell do not completely shield the outer electrons, causing a small increase in the Z_{eff} and a decrease in the atomic radii. The contraction in size over the entire 14 elements is sufficient to reduce the sizes of the sixth-period transition-metal atoms to nearly the same radii as those in the fifth period. This effect is known as the **lanthanide contraction,** the small decrease in the radii of the lanthanides as the 4*f* subshell is filled. This decrease in size of the elements just before the sixth-period transition metals is about the same as the expected increase in size expected for this row, causing the transition elements of the fifth and sixth periods to have nearly identical radii within each group.

As expected, the radii of the positively-charged transition metal ions are smaller than the radii of the parent metal atoms, and for a given charge they decrease slightly from left to right within any period. The ionic radii of the fifth- and sixth-row transition elements, just like the atomic radii, are almost the same within any group because of the *lanthanide contraction.*

The lanthanide contraction causes the sizes and chemical properties of the fifth- and sixth-period transition elements to be similar.

The trends in the chemical properties of the transition metals in each group parallel the changes in the radii within each group. There is usually a much larger difference between the chemical behavior of the first and second elements in a transition-metal group than is observed between the second and third elements. Because of the lanthanide contraction, zirconium and hafnium in Group 4B have nearly the same atomic radii, and their chemistries are nearly identical but quite different from that of titanium. The chemical behaviors of zirconium and hafnium are so similar that the two elements are difficult to separate.

Oxidation States and Ionization Energies

With the exception of the Group 3B elements, each of the transition metals has at least two stable oxidation states. Figure 19.4 shows the full range of oxidation states for the transition elements. The maximum positive oxidation state is equal to the group number for Groups 3B through 7B (Mn), and also for the Group 8B elements in the fifth and sixth periods (Ru and Os). Note that the group number for these metals equals the total number of valence-shell electrons [the *ns* and $(n-1)d$ electrons] in the element. To the right of the elements Mn, Ru, and Os in each period, the maximum positive oxidation state observed for each successive element decreases. Generally, the highest oxidation state of each of the transition elements is observed only when the metals are combined with the very electronegative elements oxygen and fluorine.

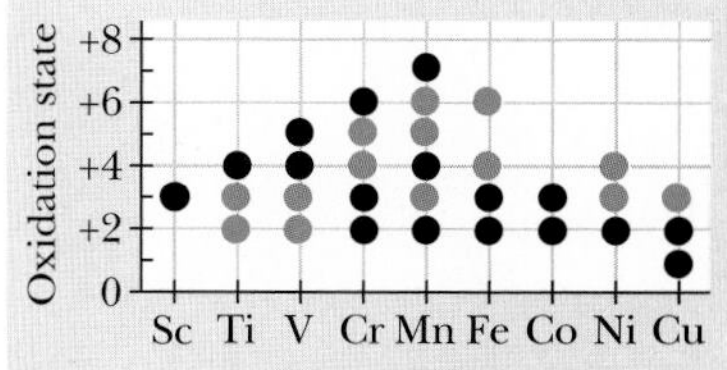

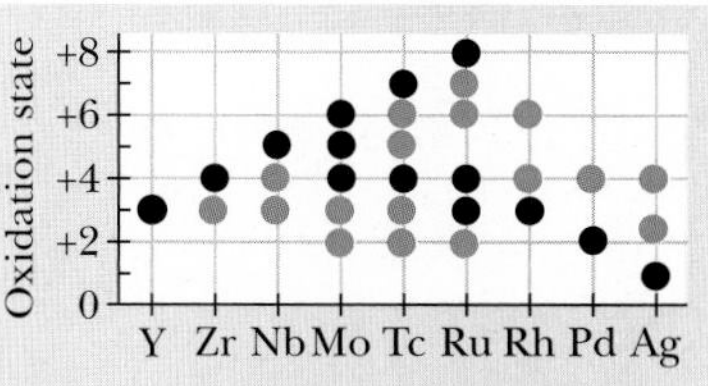

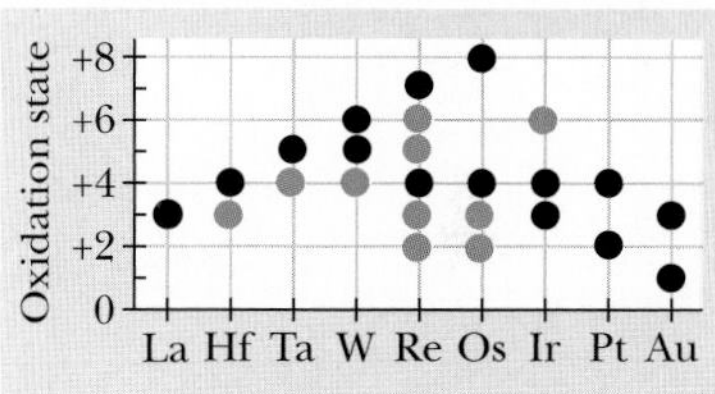

Figure 19.4 Stable oxidation states of the transition metals. The stable nonzero oxidation states of the transition metals are shown. The more common oxidation states are indicated by black circles.

Most oxidation states of the transition metals are characterized by partially-filled *d* orbitals. The partially-filled *d* orbitals in these compounds cause them to exhibit a wide range of colors and lead to interesting magnetic properties. We consider both of these topics in detail later in this chapter. The large number of oxidation states exhibited by the transition elements leads to an extensive oxidation-reduction chemistry.

The first, second, and third ionization energies for the fourth-period transition metals are shown graphically in Figure 19.5. Each of these lines exhibits a general increase from left to right. As can be seen in Figure 19.6, the ionization energies of elements within any group in the fourth and fifth period are similar and may increase or decrease depending on the group. In Group 5B and beyond a large increase in the ionization energy is observed from the fifth to the sixth period that is at least partially explained by the lanthanide contraction.

Sixth-period transition metals have larger ionization energies than those of fourth- and fifth-period transition metals.

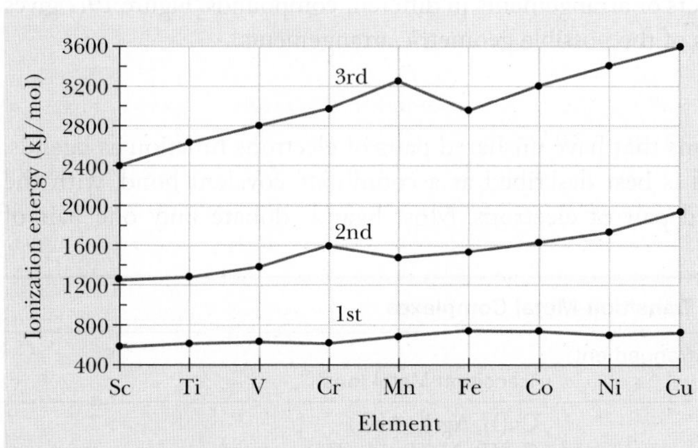

Figure 19.5 Ionization energies of transition elements. The first, second, and third ionization energies of the fourth-period transition metals.

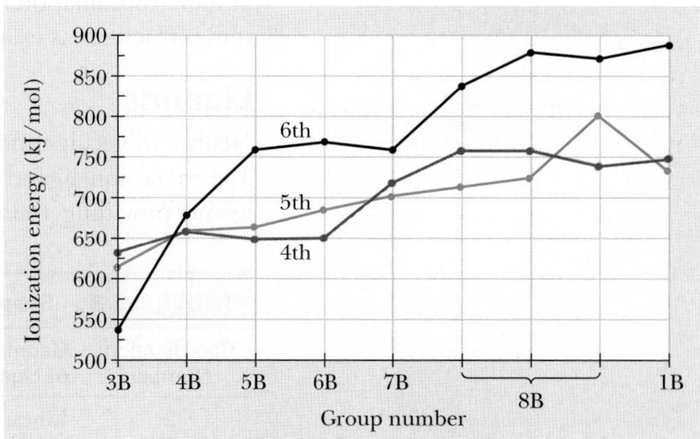

Figure 19.6 First ionization energies. The ionization energies of elements in the fourth and fifth periods are roughly the same, but the ionization energies of the sixth-period elements are higher.

OBJECTIVES REVIEW *Can you:*

- ☑ define which elements are transition metals and locate them on the periodic table?
- ☑ relate the trends in physical properties (melting points, boiling points, and hardness) and atomic properties (atomic radius and ionization energy) of the transition elements to their relative positions in the periodic table?
- ☑ describe the effect of the lanthanide contraction on the sizes and ionization energies of the transition elements?

19.2 Coordination Compounds: Structure and Nomenclature

OBJECTIVES

- ☐ Identify the geometric arrangements of complexes containing two, four, and six ligands
- ☐ Use the standard conventions for writing the names and formulas of coordination compounds

Much of the chemistry of the transition metals is that of compounds formed from metal cations acting as Lewis acids, forming covalent bonds with molecules or ions that donate unshared pairs of electrons. These Lewis acid-base adducts play an important role in developing photographic film, electroplating, the function of metals in enzymes, and the oxygen transport systems in living systems. The study of these compounds is known as *coordination chemistry.*

Each transition-metal ion in a complex is a Lewis acid capable of forming bonds with a number of Lewis bases. In the terminology of coordination chemistry, the Lewis bases bonded to the metal ion are **ligands,** and they may be either neutral molecules or anions. For a species to behave as a ligand, it must have an unshared pair of electrons that can be donated to an empty orbital of the metal ion, forming a coordinate covalent bond. The atom in the ligand that bonds to the metal, one that contains a lone pair, is called the **donor atom.** The **coordination number** is the number of donor atoms that are bonded to the metal ion. A **coordination compound** or **complex** is one that contains a metal ion bound to ligands by Lewis acid-base interactions.

Bonds in coordination complexes are made from electron pairs donated to the central metal ion. The Lewis base that donates the electron pair is called a *ligand.*

Coordination Number

The number of bonds between the ligands and a metal ion in a coordination complex ranges from two to nine, depending on the nature of the particular metal ion and the ligands that are present. The most common coordination numbers are four and six. Table 19.3 shows some typical coordination numbers for several transition-metal ions and the geometric arrangement of the ligands for each coordination number. Both tetrahedral and square-planar shapes are found for a coordination number of four. Several of the transition-metal ions exhibit two or more coordination numbers or arrangements in different compounds. Figure 19.7 gives some structures as examples of the possible geometric arrangements.

Ligands

Neutral molecules and anions that have unshared pairs of electrons function as ligands. The metal ion-ligand bond is best described as a coordinate covalent bond, with the ligand providing the shared pair of electrons. Most ligands donate only one pair of

TABLE 19.3 **Shapes of Transition-Metal Complexes**

Coordination Number	Geometric Arrangement of Ligands	Transition Metal Ions
2	Linear	Cu(I), Ag(I), Au(I)
4	Tetrahedral	Co(II), Ni(II), Mn(II)
4	Square planar	Cu(II), Ni(II), Pt(II), Au(III)
6	Octahedral	Fe(II), Fe(III), Cr(III), Co(II), Co(III), Ni(II), Mn(II), Mn(III), Ti(III), Pt(IV)

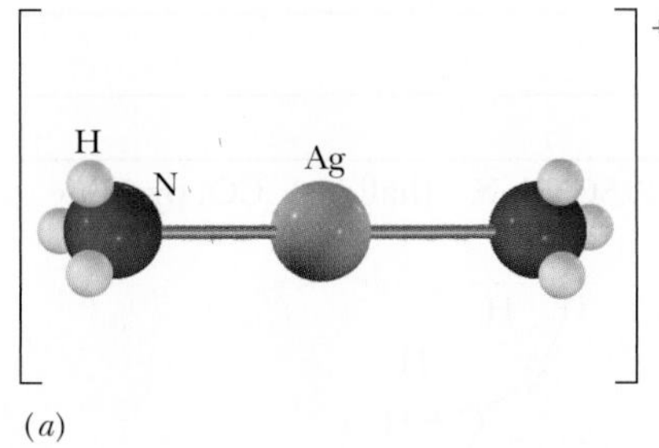

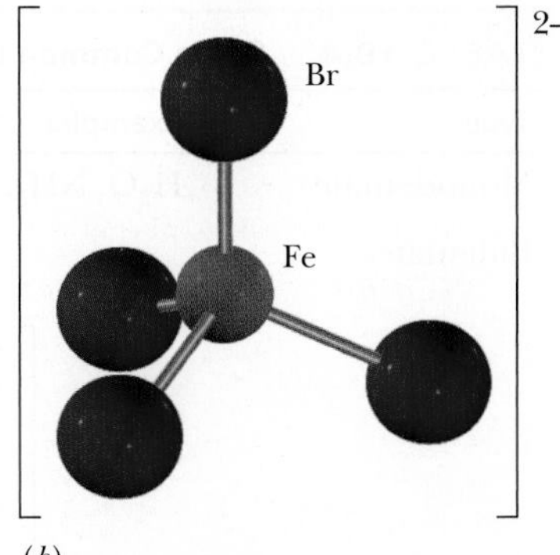

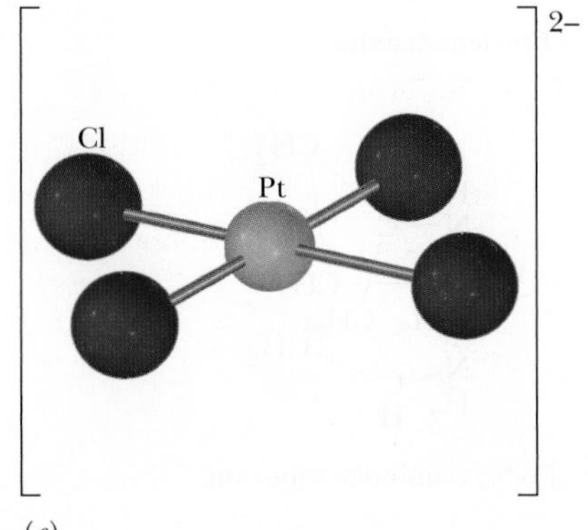

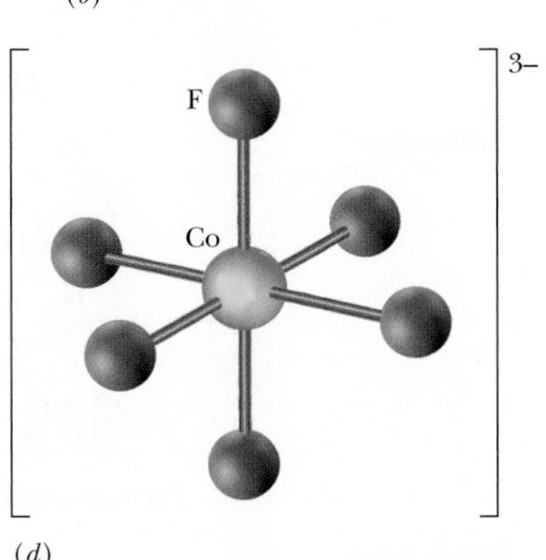

Figure 19.7 Structural formulas of complexes. Structural formulas for common geometric arrangements found in complexes: *(a)* linear complex, *(b)* tetrahedral complex, *(c)* square planar complex, and *(d)* octahedral complex.

electrons to the metal atom (have one donor atom) and are called **monodentate ligands.** The use of "dentate" in describing ligands is derived from the Latin word for teeth. Table 19.4 gives examples of monodentate ligands.

Some molecules and ions have more than one donor atom that can form bonds to the same metal atom. These ligands are called **chelating ligands,** and the complexes formed are called **chelates.** The word *chelate* comes from the Greek word for claw (to the originator of the term, a chelating ligand must have looked like a crab using both of its claws to attach itself to the metal ion). Ligands that have two donor atoms that simultaneously bond to metal ions are called *bidentate;* those with three donor atoms are *tridentate;* and so forth. As a group, chelating ligands are referred to as **polydentate ligands.** Examples of several polydentate ligands are also shown in Table 19.4. In the hexadentate ligand ethylenediaminetetraacetate ion ($EDTA^{4-}$), two of the bonds to the metal form using electron pairs from the nitrogen atoms, and four of the bonds to the metal form using electron pairs from the oxygen atoms. $EDTA^{4-}$ forms especially stable complexes with most metal ions. The addition of $EDTA^{4-}$ to many food products increases shelf life, because it combines with transition-metal ions that catalyze the decomposition of food. It is also used as a water softener, mostly to remove calcium and magnesium ions that are responsible for making water "hard."

A chelate is a coordination compound in which two or more donor atoms in a single, multidentate ligand share a pair of electrons with a metal ion.

Formulas of Coordination Compounds

Before the development of the modern view of coordination compounds, the formulas of coordination compounds were written in forms such as $CoCl_3{\cdot}6NH_3$, showing that the compound contained six molecules of ammonia for each metal atom or ion present. Today, we enclose the metal ion and its coordinated ligands in square brackets, with other ions of the compound outside the brackets. Thus, the formula of the cobalt(III) chloride compound containing six ammonia ligands is written as $[Co(NH_3)_6]Cl_3$. In this compound, the six ammonia molecules are covalently bonded to the cobalt ion to form a complex cation. The three chloride ions are not coordinated to the cobalt and are referred to as *counterions;* as with all ionic compounds the charges on the ions must balance. When a complex is an anion, its formula is enclosed in brackets preceded by the cations that are present—for example, $K_3[Fe(CN)_6]$. Here the six cyanide ions are covalently bonded to the Fe^{3+} ion,

Atoms joined to the metal by a Lewis acid-base interaction are said to be "coordinated."

TABLE 19.4 Some Common Ligands

Type	Examples
Monodentate	H_2O, NH_3, CN^-, NO_2^-, OH^-, SCN^-, X^- (halides), CO, pyridine
Bidentate	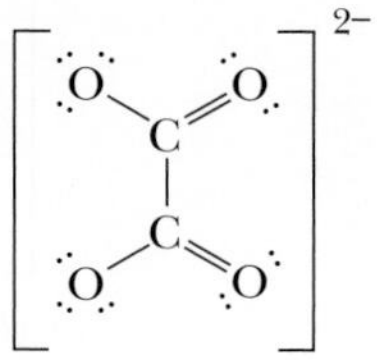 Oxalate; 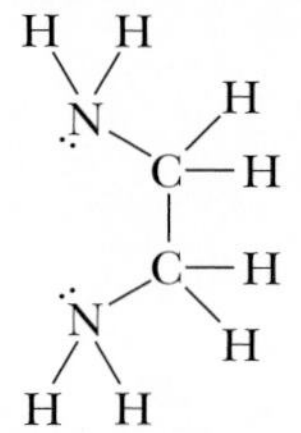 Ethylenediamine
Tridentate	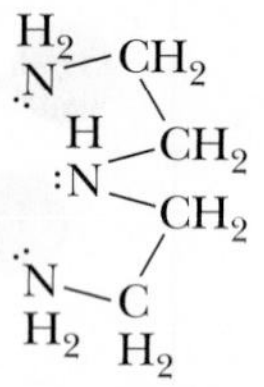 Diethylenetriamine; 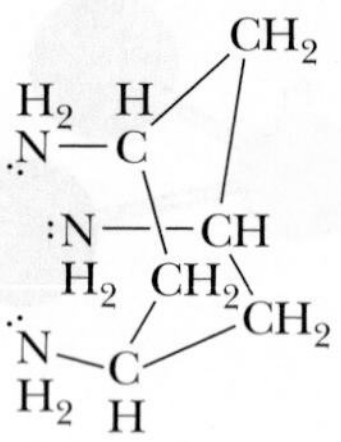 1,3,5 Triaminocyclohexane
Polydentate	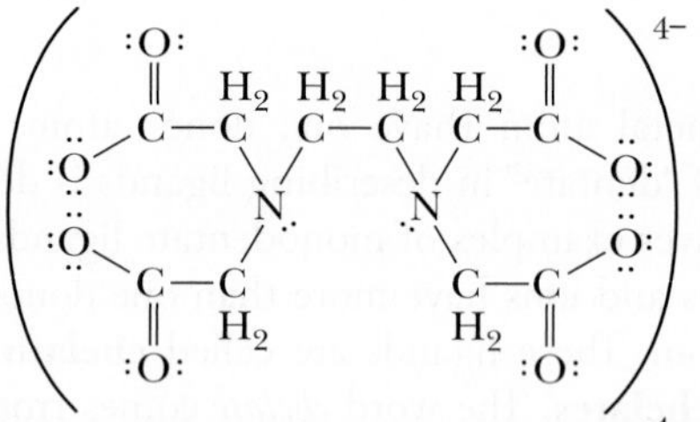 Ethylenediaminetetraacetate ion, $EDTA^{4-}$

Oxalate

Ethylenediamine

Diethylenetriamine

1,3,5 Triaminocyclohexane

Ethylenediaminetetraacetate ion, $EDTA^{4-}$

In the formula of a coordination compound, square brackets enclose the formula of the coordination complex.

forming the complex anion $[Fe(CN)_6]^{3-}$; the three potassium ions are the counterions. Some metal complexes are neutral, such as the Mn^{2+} complex $[Mn(Cl)_2(NH_3)_4]$; in this case, the two chloride ligands exactly balance the charge of the metal.

EXAMPLE 19.1 Writing the Formulas of Coordination Compounds

Write the formula for each of the following coordination compounds.

(a) $CoCl_3{\cdot}5NH_3$, in which only one of the chloride ions is coordinated to the metal
(b) $CoCl_3{\cdot}4NH_3$, in which two chloride ions are coordinated to the metal
(c) The potassium salt of a complex containing chromium(III) coordinated to five CN^- ions and one CO molecule (note that the CO ligand is a neutral molecule, carbon monoxide)

Strategy Determine which ligands are coordinated to the metal and which are counterions. Place the coordinated ligands and the metal in square brackets.

Solution

(a) The complex ion consists of the Co^{3+} ion coordinated to the five ammonia molecules and one Cl^- ion. The complex ion is enclosed in square brackets; because it is a cation, the two uncoordinated chloride anions are outside the brackets. The formula is

$$[Co(NH_3)_5Cl]Cl_2$$

(b) In this compound, the metal, four ammonia molecules, and two of the Cl^- ions constitute the complex ion, which is enclosed in square brackets. The formula is

$$[Co(NH_3)_4Cl_2]Cl$$

(c) The complex ion, enclosed in square brackets, has a charge of 2−, which is the sum of the charges on the chromium ion and five CN^- ions. There must be two K^+ ions in the compound, which precede the formula of the complex anion. The formula is

$$K_2[Cr(CN)_5CO]$$

Understanding

Write the formula of the coordination compound that contains platinum(II) coordinated to four ammonia molecules and platinum(IV) coordinated to six chloride ions.

Answer $[Pt(NH_3)_4][PtCl_6]$

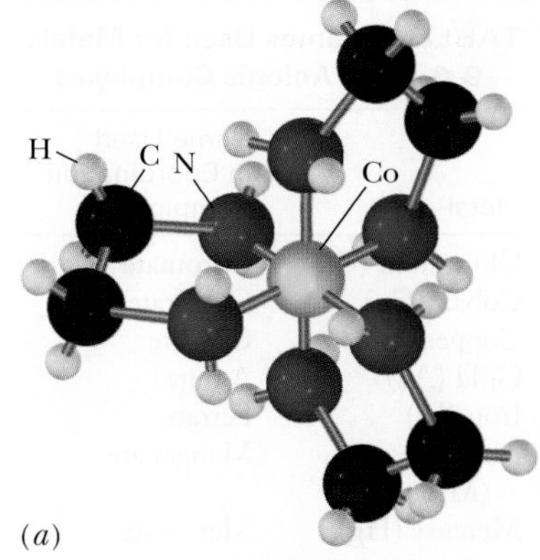

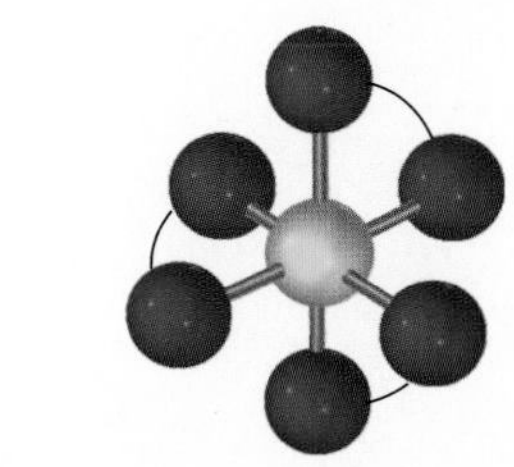

Figure 19.8 Representation of chelating ligands. *(a)* The structural formula of $[Co(en)_3]^{3+}$ showing all of the atoms in the ethylenediamine (en) ligand. *(b)* The same structure but representing the en ligand with a *curved line* connecting the donor atoms.

In structural formulas, a bidentate ligand, such as ethylenediamine (Figure 19.8*a*, abbreviated "en"), is usually represented as a curved line connecting the two donor atoms, as shown in Figure 19.8*b*. This convention greatly simplifies the representation of these structures. Curved lines connecting the donor atoms are also used to represent any polydentate ligand.

Naming Coordination Compounds

The naming of compounds that contain metal complexes is a part of the International Union of Pure and Applied Chemistry (IUPAC) system of nomenclature. The following rules are an expansion of the simple nomenclature presented in Section 2.8.

1. In naming an ionic compound, the cation name comes first, followed by the name of the anion as a separate word. The same rule is used to name any ionic compound, regardless of whether it contains a complex ion.
2. In naming a coordination complex, the ligands come first in alphabetical order, and the metal in the complex is named last, followed by the oxidation state of the metal as a Roman numeral in parentheses. The name of the complex is written as one word.
3. The names of the anionic ligands are changed to end in the letter *o*. In naming the neutral ligands, the name of the molecule is usually unchanged. Four common neutral ligands are exceptions to this rule: "aqua" is used for water, "ammine" is used for ammonia, "nitrosyl" is used for NO, and "carbonyl" is used for carbon monoxide. Table 19.5 lists the names of some commonly encountered ligands.
4. The number of times each ligand occurs in a complex is indicated by a prefix: (1) mono- (usually omitted), (2) di-, (3) tri-, (4) tetra-, (5) penta-, or (6) hexa-. When there are numeric prefixes anywhere in the name of a ligand, its name is usually enclosed in parentheses, and the numerical prefixes are changed to bis-, tris-, tetrakis-, and pentakis- for two, three, four, and five ligands, respectively.
5. For cationic and neutral complexes, the name of the metal is used. When the complex is an anion, the ending *-ate* is attached to the name of the metal, replacing the endings *-um* or *-ium* if present in the name of the metal. For a metal whose symbol is based on the Latin name for the element, the Latin stem is used in naming anionic complexes. (Mercury is one exception to this rule—"mercurate" is used rather than "hydrargentate.") (See Table 19.6.)

TABLE 19.5 Names of Some Common Ligands

Ligand	Name Used in Coordination Complex
Anions	
Bromide, Br^-	bromo
Carbonate, CO_3^{2-}	carbonato
Chloride, Cl^-	chloro
Cyanide, CN^-	cyano
Hydroxide, OH^-	hydroxo
Nitrite, NO_2^-	nitrito, nitro
Oxalate, $C_2O_4^{2-}$	oxalato
Oxide, O^{2-}	oxo
Sulfate, SO_4^{2-}	sulfato
Thiocyanate, SCN^-	thiocyanato
Neutral molecules	
Ammonia, NH_3	ammine
Carbon monoxide, CO	carbonyl
Ethylenediamine, $NH_2CH_2CH_2NH_2$	ethylenediamine (en)
Nitrogen monoxide, NO	nitrosyl
Pyridine, C_5H_5N	pyridine
Water, H_2O	aqua

TABLE 19.6 Names Used for Metals in Anionic Complexes

Metal	Name Used in Coordination Complex
Chromium (Cr)	Chromate
Cobalt (Co)	Cobaltate
Copper (Cu)	Cuprate
Gold (Au)	Aurate
Iron (Fe)	Ferrate
Manganese (Mn)	Manganate
Mercury (Hg)	Mercurate

Consider the following examples:

$[Ni(NH_3)(H_2O)_4Cl]Cl_2$ Name: amminetetraaquachloronickel(III) chloride

$Na_3[MnCl_6]$ Name: sodium hexachloromanganate(III)

Examples 19.2 and 19.3 illustrate additional application of these rules.

EXAMPLE 19.2 Naming of Coordination Compounds

Give the IUPAC name for each of the following compounds.

(a) $K_3[Fe(CN)_6]$
(b) $[Co(NH_3)_5Br]Br_2$
(c) $[Cr(NH_3)_2(en)Cl_2]Cl$ (en = ethylenediamine)
(d) $Ni(CO)_4$

Strategy Cationic counterions will be named before the complex, anionic ones after. The net charge of the complex is determined from the charges on the counterions outside the brackets. The oxidation state of the transition metal is determined from this charge and the charges on the ligands bonded to it. The ligands are then named in alphabetical order and the metal name will end in *-ate* if the complex is an anion.

Solution

(a) Because of the 1+ charge of each of the potassium counterions (they are outside the bracket and will be named first), the complex has a 3− charge and is named second with an *-ate* ending on the metal. The charge of each of the six cyanide ions is 1−, so the oxidation state of the iron (called *ferrate* in Table 19.6) is +3 ($6CN^-$ + Fe^{3+} yields the overall charge of 3−). The name of the compound is potassium hexacyanoferrate(III). The number of potassium counterions is not specified; it can be determined from the charges of the ligands and the oxidation state of the metal.

(b) Because of the 1− charge of each of the bromide counterions, the charge of the complex is 2+. In the complex, the charge of the bromide ion ligand is 1− and ammonia is a neutral molecule, so the cobalt is in the +3 oxidation state. The complex is named first because it is the cation in the compound. Using "ammine" for the ammonia ligands and ordering them alphabetically based on the name of the ligand, not the prefix, the name of the compound is pentaamminebromocobalt(III) bromide.

(c) Because ammonia and ethylenediamine are both neutral ligands and the charge of a chloride ion is 1−, the chromium is in the +3 oxidation state. Naming the ligands in alphabetical order with the use of numerical prefixes gives the name diamminedichloro(ethylenediamine)chromium(III) chloride.

(d) The metal complex is neutral and the carbon monoxide is a neutral ligand, so the nickel is in the 0 oxidation state. The IUPAC name is tetracarbonylnickel(0).

Understanding

Give the name of $K_2[RuCl_5(H_2O)]$.

Answer potassium aquapentachlororuthenate(III)

EXAMPLE 19.3 Formulas from the Names of Coordination Compounds

Write the formula of each of the following compounds.

(a) sodium aquapentacyanocobaltate(III)
(b) dichlorobis(ethylenediamine)chromium(III) nitrate
(c) triamminetrichlorocobalt(III)
(d) ammonium diaquadioxalatoferrate(II)

Strategy Use the charge of each ligand, the number of ligands in the complex, and the given oxidation state of the metal to determine the overall charge of the complex. Write the formula of the complex, using the symbol of the metal followed by the formulas of each ligand, using subscripts to show the number of ligands present. Positively charged counterions precede the complex in the formula and negative counterions follow it.

Solution

(a) The complex anion, written in brackets, is made up of one water, five CN^- ligands, and one cobalt in the 3+ oxidation state. It has a net charge of $5(1-) + (3+) = 2-$, from the sum of the charges on the cyanide ions and the oxidation state of the cobalt. There must be two sodium cations to produce a neutral compound and they are written in the formula before the complex. The formula is therefore $Na_2[Co(CN)_5(H_2O)]$.

(b) From the charge of each of the two chloride ions (1−) and the oxidation state of the chromium (+3), the complex cation has a charge of 1+, which must be balanced by a single NO_3^- ion in the compound. The formula of the complex is enclosed in square brackets to give the formula $[CrCl_2(en)_2]NO_3$. The common abbreviation "en" for the ethylenediamine ligand is used in this formula.

(c) The complex is neutral, containing three chloride ions, three molecules of ammonia, and a Co^{3+} ion. The formula is written as $[Co(NH_3)_3Cl_3]$.

(d) The anion consists of two oxalate ions ($C_2O_4^{2-}$), two water molecules, and an Fe^{2+} ion. The net charge of the complex is thus 2−, which combines with two NH_4^+ ions to give a neutral compound. The formula is $(NH_4)_2[Fe(C_2O_4)_2(H_2O)_2]$.

Understanding

What is the formula of hexaamminecobalt(III) hexacyanoferrate(III)?

Answer $[Co(NH_3)_6][Fe(CN)_6]$

OBJECTIVES REVIEW *Can you:*

- ☑ identify the geometric arrangements for complexes containing two, four, and six ligands?
- ☑ use the standard conventions for writing the names and formulas of coordination compounds?

19.3 Isomers

OBJECTIVES

- ☐ Identify and distinguish structural isomers and stereoisomers
- ☐ Recognize and classify structural isomers as coordination isomers or linkage isomers
- ☐ Identify examples of geometric isomers and optical isomers
- ☐ Use the terminology introduced for describing the different isomers (*cis* and *trans*, *mer* and *fac*, chiral, enantiomers, racemic mixture, optical isomers)

Isomers are different compounds that have the same chemical formula (see Section 10.4). Experimentally, isomers are recognized as different compounds because they differ in one or more physical or chemical properties. Coordination compounds exhibit a wide variety of isomers, as a result of the fairly large number of atoms present in the species. Figure 19.9 outlines the categories of isomerism that are used to classify coordination compounds.

Structural Isomers

Structural isomers contain the same numbers and kinds of atoms but differ in the bonding. Coordination compounds have two kinds of structural isomerism: coordination isomerism and linkage isomerism.

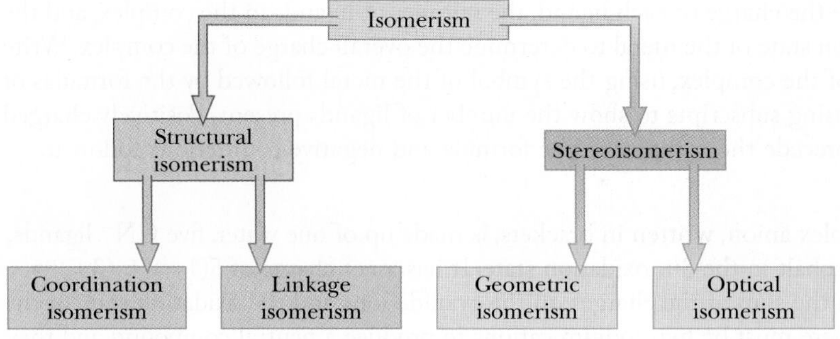

Figure 19.9 Categories of isomerism. Block diagram shows how the various kinds of isomerism exhibited by coordination compounds are related.

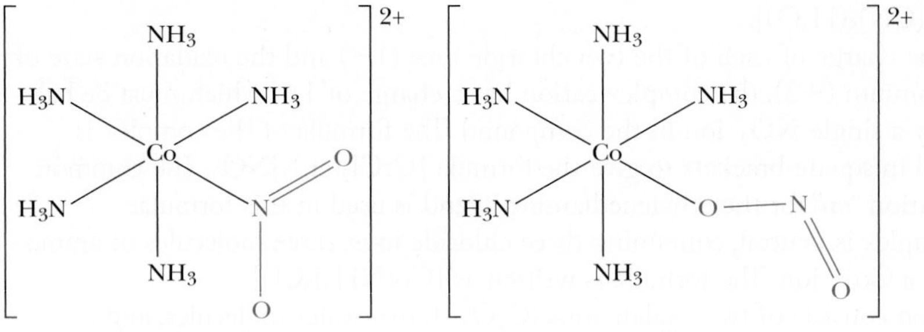

Figure 19.10 Linkage isomers of $[Co(NH_3)_5NO_2]^{2+}$. Two linkage isomers of the NO_2^- ligand are possible. *(a)* The attachment to the metal ion is through the nitrogen atom; this complex is called the *nitro complex.* *(b)* The bond is formed by donation of the election pair of one of the oxygen atoms of the ligand; this complex is called the *nitrito complex.*

Coordination isomers contain the same numbers and kinds of atoms but have different ligands directly bonded to the metal. Coordination isomers can form because anions can function either as counterions or as ligands. The compounds $[Co(NH_3)_4Cl_2]Br$ and $[Co(NH_3)_4ClBr]Cl$ are *coordination* isomers because the bromide ion in the second compound is coordinated to the cobalt, replacing one of the chloride ions coordinated to the cobalt in the first compound. Coordination isomerism can also occur in ionic compounds that contain coordination complexes as both the cation and anion, such as the compounds $[Cr(NH_3)_6][Co(C_2O_4)_3]$ and $[Co(NH_3)_6][Cr(C_2O_4)_3]$. These two compounds have exactly the same atoms present, but the oxalate ions and ammonia molecules are coordinated to different metal ions in the two compounds.

A second type of structural isomerism is **linkage isomerism,** in which the same ligand is coordinated to the metal through a different donor atom. The most frequently encountered example of this occurs with the ligand NO_2^-, the nitrite ion. All three of the atoms in this anion have unshared pairs of electrons that can form bonds to a metal ion center. Figure 19.10 shows the structures of two linkage isomers with the nitrite ion as a ligand. In one isomer, the metal is bonded to the nitrogen atom, and in the other isomer, the metal is bonded to an oxygen atom. In the case of NO_2^-, the ligand is named differently, depending on the atom that forms the bond to the metal ion. The N-bonded ligand is called *nitro* and the O-bonded ligand is called *nitrito.* The traditional names, not part of the IUPAC nomenclature system (see later), of the complexes shown in Figure 19.10 are (a) pentaamminenitrocobalt(III) and (b) pentaamminenitritocobalt(III).

Structural isomers, both coordination and linkage, have different connectivities.

Another ligand that exhibits linkage isomerism is the thiocyanate ion, NCS^-. The ion can coordinate to a metal through either the nitrogen or the sulfur atom. In the IUPAC nomenclature system, linkage isomers are distinguished by the symbol of the coordinated atom preceding the name of the ligand. The name of the complex $[Pt(NH_3)_3SCN]^+$ in which the sulfur atom is bonded to the platinum is called triammine-S-thiocyanatoplatinum(II). If the thiocyanate ion were bonded through the nitrogen atom, this complex would be called triammine-N-thiocyanatoplatinum(II). The IUPAC name for the complex shown in Figure 19.10*a* is pentaammine-N-nitritocobalt(III) and in Figure 19.10*b* is pentaammine-O-nitritocobalt(III).

Stereoisomers

Stereoisomers have the same bonds but differ in the arrangement of the atoms in space. There are two categories of stereoisomers: geometric isomers and optical isomers.

Geometric Isomers

Geometric isomers have the same numbers and kinds of bonds but differ in the relative positions of the atoms. The compound diamminedichloroplatinum(II) [Pt(NH_3)$_2$$Cl_2$] discussed in the introduction to this chapter is a square-planar complex. Two geometric isomers are possible, and both have been made, that differ in the Cl–Pt–Cl bond angle. The prefix *cis* identifies the isomer in which the two like ligands are on the same side of the metal atom, and *trans* identifies the isomer in which the like ligands are on opposite sides of the metal. In the *cis* isomer, the Cl–Pt–Cl angle is 90 degrees, and in the *trans* isomer, it is 180 degrees. This change in the arrangement of the ligands about the platinum atom changes the properties of the compounds markedly. Table 19.7 gives some of the properties of the two geometric isomers. As outlined, *cis*-diamminedichloroplatinum(II) is a highly effective antitumor agent, the *trans* isomer shows virtually no therapeutic activity. Note that, although *cis-trans* isomers are found for many square-planar complexes, isomers of this type are not possible in the other regular four-coordinate geometry, the tetrahedron. In a tetrahedron, all of the bond angles are 109.5 degrees, so it is not possible to arrange the ligands with different bond angles as it is in square-planar geometry.

TABLE 19.7 **Some Properties of Geometric Isomers of Diamminedichloroplatinum(II)**

Property	*Cis* Isomer	*Trans* Isomer
Density	3.738 g/cm^3	3.746 g/cm^3
Color	Bright yellow	Pale yellow
Solubility	0.2523 g/100 g H_2O	0.0366 g/100 g H_2O
Synthesis	$[PtCl_4]^{2-} + 2NH_3 \rightarrow [Pt(NH_3)_2Cl_2] + 2Cl^-$	$[Pt(NH_3)_4]^{2+} + 2Cl^- \rightarrow [Pt(NH_3)_2Cl_2] + 2NH_3$

Geometric isomerism is also observed in octahedral complexes that have the general formulas MA_4B_2, and MA_3B_3, where A and B represent different ligands. Figure 19.11 illustrates examples of these kinds of geometric isomers. In the MA_4B_2 formula, the *cis* and *trans* designations are used. Although *cis* and *trans* are sometimes used to name the

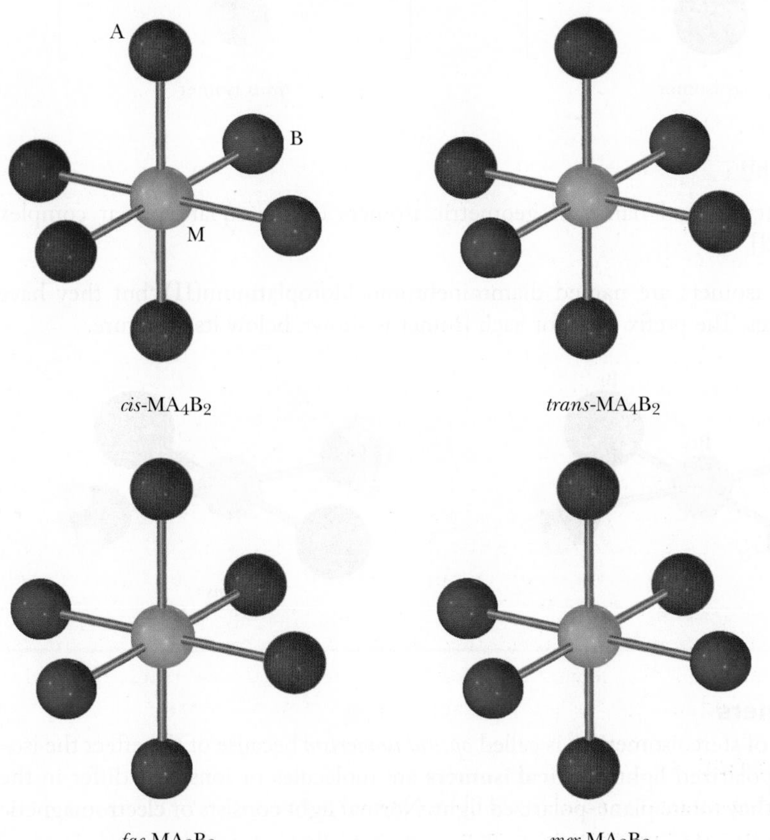

Figure 19.11 Geometric isomers in octahedral complexes. (top) *Cis* and *trans* isomers of an octahedral complex having the general formula MA_4B_2. (bottom) *Fac* and *mer* isomers of an octahedral complex having the general formula MA_3B_3.

Geometric isomers contain the same bonded atoms but differ in the arrangement of the ligands around the metal ion.

two isomers of the type MA_3B_3, these isomers are usually identified as *fac* (facial) and *mer* (meridional) isomers, respectively. In the *fac* isomer, the three like ligands are at the corners of one of the triangular faces of the octahedron, whereas in the *mer* isomer, two of the like ligands are at opposite vertices of the octahedron.

EXAMPLE 19.4 Geometric Isomers of Coordination Complexes

How many geometric isomers exist for the octahedral complex $[Ru(H_2O)_2Cl_4]^-$? Name each isomer.

Strategy Determine the arrangements of the ligands that will form different complexes. Be careful that each "new" possibility is not the same as an earlier example in which the complex is simply rotated.

Solution

The complex has two water molecules and four chloride ligands. Complexes of the MA_4B_2 family have two geometric isomers. In one, the O–Ru–O angle is 90 degrees; in the other, this angle is 180 degrees. The isomer with both water molecules on the same side of the metal (a 90-degree angle) is *cis*-diaquatetrachlororuthenate(III), and the other isomer is *trans*-diaquatetrachlororuthenate(III).

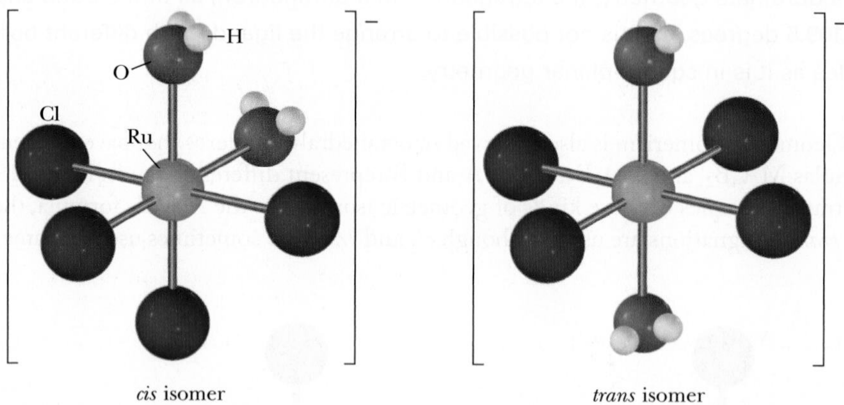

Understanding

Show the structure and name all geometric isomers of the square planar complex $[Pt(NH_3)_2BrCl]$.

Answer Both isomers are named diamminebromochloroplatinum(II), but they have different prefixes. The prefix used for each isomer is shown below its structure.

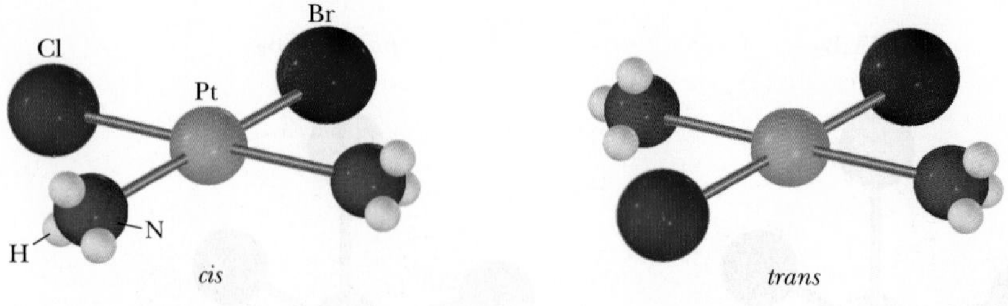

Optical Isomers

A second class of stereoisomerism is called *optical isomerism* because of the effect the isomers have on polarized light. **Optical isomers** are molecules or ions that differ in the ways in which they rotate plane-polarized light. Normal light consists of electromagnetic

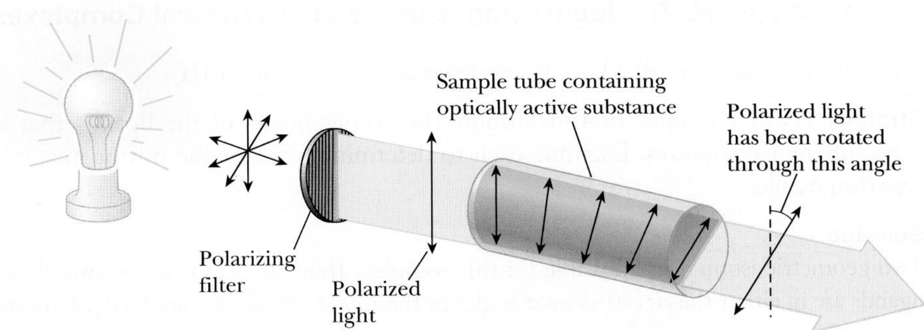

Figure 19.12 Rotation of polarized light. The electric field in unpolarized light (left) oscillates in all directions perpendicular to the direction in which the light travels. Only light that is oscillating in a single plane is transmitted by a polarizing filter. When plane-polarized light passes through a sample containing an optical isomer, the plane of polarization is rotated.

radiation in which the electric field oscillates randomly in all directions perpendicular to the direction of the beam. In **polarized light,** the electric field oscillates in a single plane, as shown in the middle of Figure 19.12. When polarized light passes through a sample of an optical isomer, the plane of polarization of the light is rotated, as shown in the right side of Figure 19.12. Molecules or ions that rotate plane-polarized light are said to be *optically active.*

Enantiomeric isomers are mirror-image compounds that cannot be superimposed on each other; they rotate plane-polarized light in opposite directions.

Chiral molecules or **ions** are those with mirror-image structures that cannot be superimposed. A familiar example of a pair of objects with nonsuperimposable mirror images is a person's right and left hands. The *reflection* of a left hand in a mirror is identical to the right hand (see Figure 19.13), but the two hands are not the same. No matter how a right hand is twisted and turned, it can never be superimposed on the left hand.

Enantiomers are a special class of chiral molecules that are nonsuperimposable mirror images of *each other.* Enantiomers rotate plane-polarized light in exactly opposite directions. Common coordination complexes that exist as enantiomers have octahedral geometry containing two or three bidentate ligands, such as ethylenediamine, or have tetrahedral geometry with two bidentate ligands. For example, the mirror-image relationship of the complex ion tris(ethylenediamine)cobalt(III) is shown in Figure 19.14. No matter how the mirror images are turned, they cannot be superimposed. One enantiomer rotates plane-polarized light clockwise, and its mirror image rotates the light counterclockwise by the same amount.

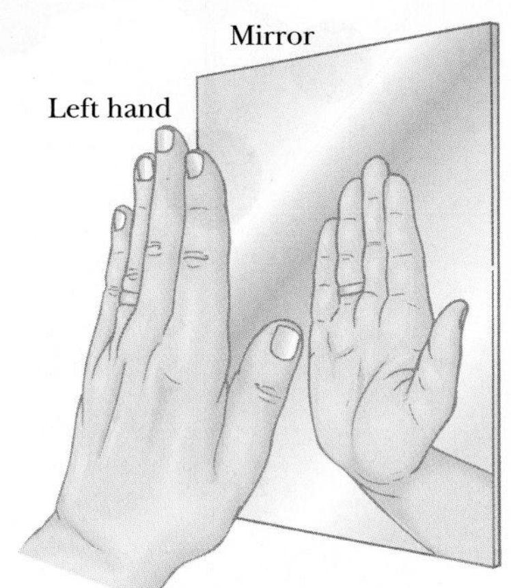

Figure 19.13 Nonsuperimposable mirror images. The mirror image of a left hand is the same as a right hand; a right hand and a left hand cannot be superimposed.

The synthesis of a compound that exists as enantiomers usually results in a mixture of equal quantities of both isomers, called a **racemic mixture.** A racemic mixture produces no net rotation of plane-polarized light, because the presence of equal numbers of molecules of the two enantiomers results in cancellation of the rotations.

An important class of chiral compounds consists of compounds that contain carbon atoms bonded to four different substituents. As discussed in Chapter 22, many biologically important molecules contain large numbers of chiral carbon atoms.

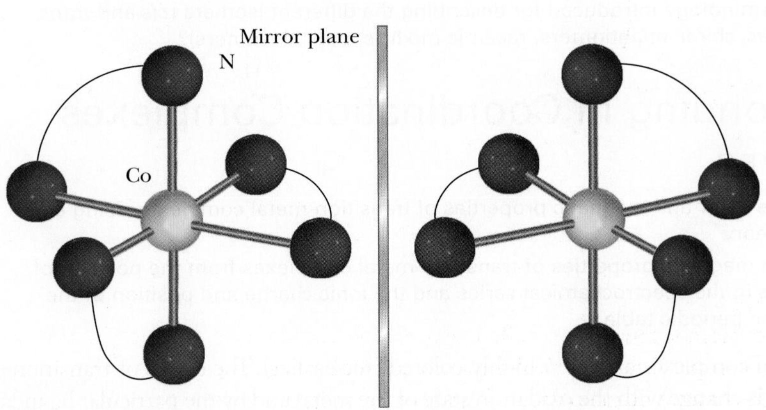

Figure 19.14 Enantiomers of $[Co(en)_3]^{3+}$. The mirror images of the tris(ethylenediamine)cobalt(III) ion cannot be superimposed; therefore, they constitute a pair of enantiomers. The lines connecting pairs of nitrogen atoms bonded to the cobalt represent the CH_2CH_2 part of the ethylenediamine ligands.

EXAMPLE 19.5 Identifying Isomers of Octahedral Complexes

Draw all the isomers of dichlorobis(ethylenediamine)chromium(III).

Strategy As in Example 19.4, determine the arrangements of the ligands that will form different complexes. Examine each to determine whether the mirror images are superimposable.

Solution
Two geometric isomers are possible for this complex. They differ in that the two chloride ligands are in either the *cis* (90-degree angle) or the *trans* (180-degree angle) configuration. Comparison of the mirror images of these geometric isomers shows that the *cis* complex exists as two enantiomers. The mirror image is not superimposable on the original object. In contrast, the mirror image of the *trans* isomer is identical to the original object, so there are no optical isomers for this geometry. The three possible isomers for this ion follow.

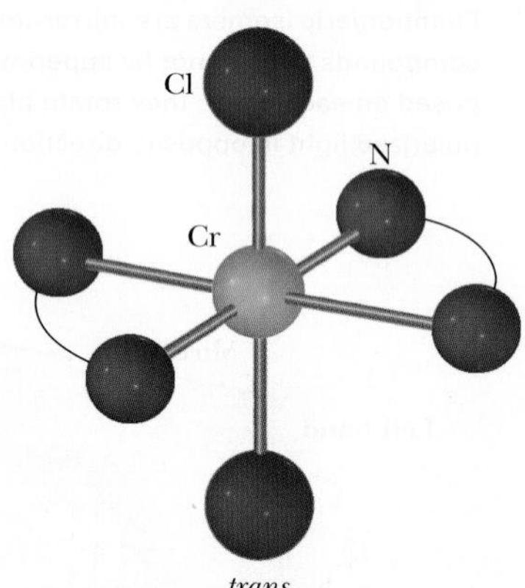

trans

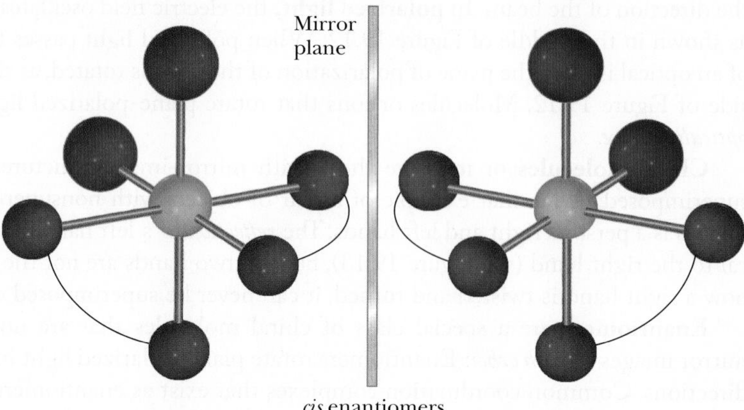

cis enantiomers

Understanding

Are optical isomers possible for a square-planar complex of platinum(II), in which four different ligands are coordinated to the metal?

Answer No, although there are a number of geometrical isomers, because each isomer is planar, the mirror image can always be rotated to superimpose on the initial structure.

OBJECTIVES REVIEW *Can you:*

- ☑ identify and distinguish structural isomers and stereoisomers?
- ☑ recognize and classify structural isomers as coordination isomers or linkage isomers?
- ☑ identify examples of geometric isomers and optical isomers?
- ☑ use the terminology introduced for describing the different isomers (*cis* and *trans*, *mer* and *fac*, chiral, enantiomers, racemic mixture, optical isomers)?

Figure 19.15 Nickel(II) compounds. The colors of transition metal complexes change in the presence of different ligands. (left to right) Nickel(II) is coordinated to water, dimethylgloxime, and ammonia.

19.4 Bonding in Coordination Complexes

OBJECTIVES

- ☐ Explain the color and magnetic properties of transition-metal complexes using crystal field theory
- ☐ Predict the magnetic properties of transition-metal complexes from the position of the ligands in the spectrochemical series and the ionic charge and position of the metal in the periodic table

Transition-metal complexes are often highly colored (see earlier). The colors of transition-metal compounds change with the oxidation state of the metal and by the particular ligands that are bound to the metal ion (Figure 19.15). The magnetic properties of transition-metal

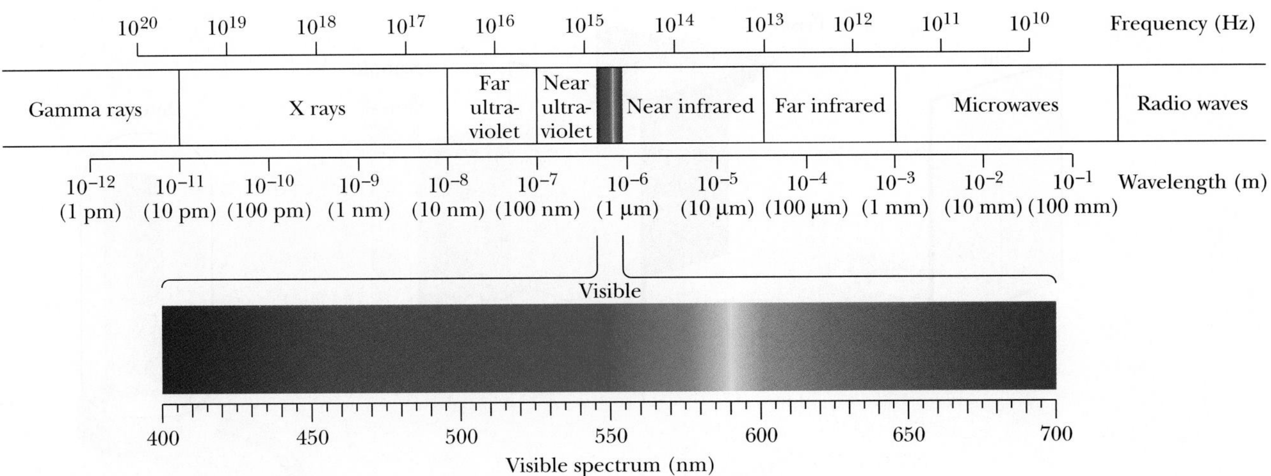

Figure 19.16 Visible spectrum of light. Electromagnetic radiation with wavelengths from 400 to 700 nm is called *visible light*. These wavelengths are the only ones detected by the human eye.

compounds are also unusual when compared with the magnetic properties of compounds of the representative elements. An explanation of these properties can be found in the bonding description outlined in this section, a description that is somewhat unique to this class of compounds.

A successful bonding model for transition-metal complexes must account for their colors and magnetic properties. Currently, several such models exist, but none is entirely satisfactory. This section presents the crystal field model because of its simplicity and its ability to account easily for the colors and magnetic properties of transition-metal complexes. Before we begin our discussion of crystal field theory, a brief review of color and magnetism is in order.

Color

We discussed the interaction of electromagnetic radiation (light) with matter in earlier chapters. Our eyes detect only a small range of wavelengths of the electromagnetic spectrum from about 400 to 700 nm. The colors of visible light are shown in the spectrum in Figure 19.16.

The energy of the photons at any given wavelength is given by Planck's relationship:

$$E = h\nu = hc/\lambda$$

The energy of a photon of red light ($\lambda = 700$ nm) is lower than the energy of a photon of violet light ($\lambda = 400$ nm). When all wavelengths of visible light strike our eyes, we observe white light. A sample that absorbs all wavelengths of visible light equally appears black. If a sample of matter absorbs part of the visible light, only the remaining wavelengths strike the eye, and the object appears colored. When only yellow light strikes the eye, we see a yellow color. A yellow color is also observed if all wavelengths except violet light are present. The colors violet and yellow are referred to as complementary. Complementary colors appear opposite each other on an artist's color wheel, such as the one in Figure 19.17.

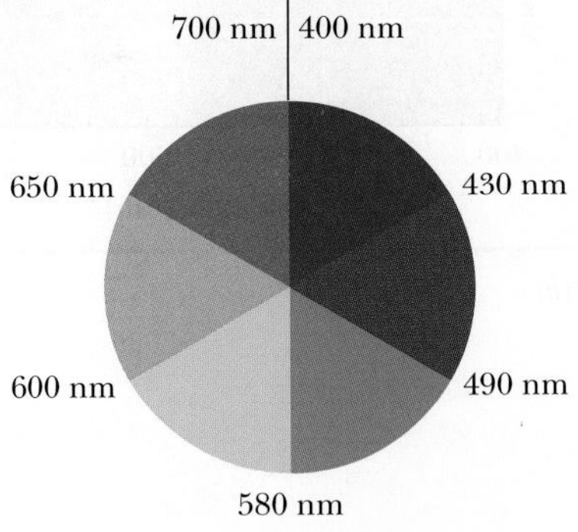

Figure 19.17 Artist's color wheel. The artist's color wheel arranges the six colors in order of wavelength around a circle. Complementary colors are located opposite each other. The presence of yellow light or the absence of violet light (the complementary color of yellow) both appear yellow to the human eye.

Absorption Spectra

The observed colors of most transition-metal complexes arise because the sample is absorbing the light of the complementary color. These observed colors are related to the allowed energy states of electrons in the coordination complexes. A spectrophotometer (Figure 19.18*a*) records an accurate measurement of the light absorbed by a sample. In a spectrophotometer, white light passes through a prism (or diffraction grating) to separate the individual wavelengths of light. Each of the wavelengths is then passed through

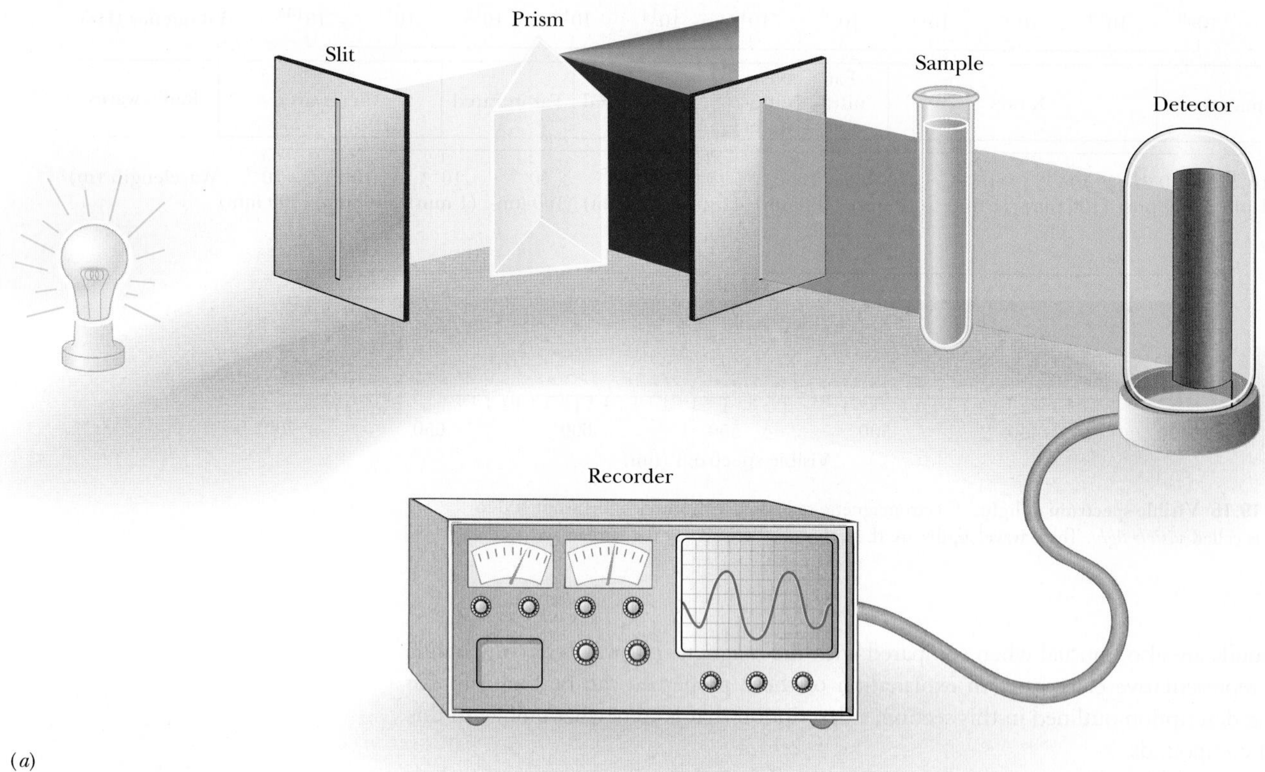

(*a*)

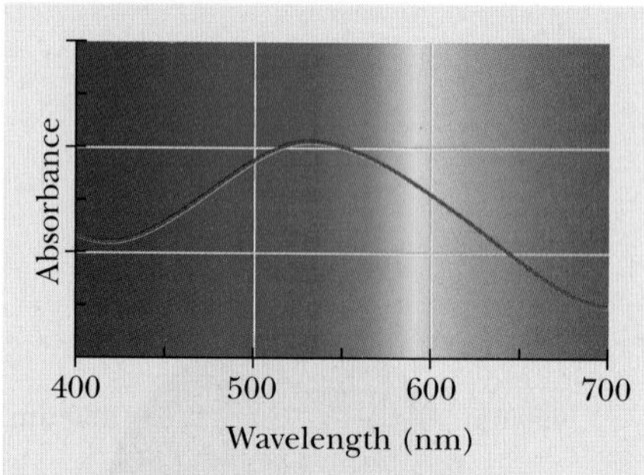

(*b*)

Figure 19.18 Absorption spectrum. *(a)* Spectrophotometer used to record an absorption spectrum. White light passes through a prism, separating the individual wavelengths of light. Then each wavelength of light passes through the sample. A detector is used to determine which wavelengths are absorbed by the sample. *(b)* The output of the spectrophotometer is a graph of absorbance versus wavelength.

the sample, and a detector is used to determine which wavelengths are absorbed by the sample. The results are displayed as an **absorption spectrum**—a graph of the quantity of light absorbed by the sample as a function of wavelength.

Figure 19.18*b* shows the absorption spectrum of $Ti(H_2O)_6^{3+}$. This complex absorbs light in the yellow-green region. The eye detects the absence of yellow-green light as a red-violet color, the complementary color shown in Figure 19.17.

Magnetic Properties of Coordination Compounds

All matter is affected by the presence of an external magnetic field. Matter that has no unpaired electrons is slightly repelled by a magnetic field. Such matter is called **diamagnetic.** In matter that has unpaired electrons, the unpaired electrons act as tiny magnets themselves. In the presence of an external magnetic field, the matter is attracted to the field. Such matter is called **paramagnetic.** For paramagnetic substances, precise measurements of the force of attraction between a sample of matter and an external magnetic field can experimentally determine the number of unpaired electrons in the atoms of the sample.

Although paramagnetism is relatively rare in compounds of representative elements, it is quite commonly observed in transition-metal compounds. In addition, different complexes of the same transition-metal ion can have different numbers of unpaired electrons. For example, the $[Co(NH_3)_6]^{3+}$ ion contains no unpaired electrons, although there are four unpaired electrons in the $[CoF_6]^{3-}$ complex. Any successful model that explains the properties of transition-metal complexes must adequately account for this kind of magnetic behavior, as well as the absorption spectrum.

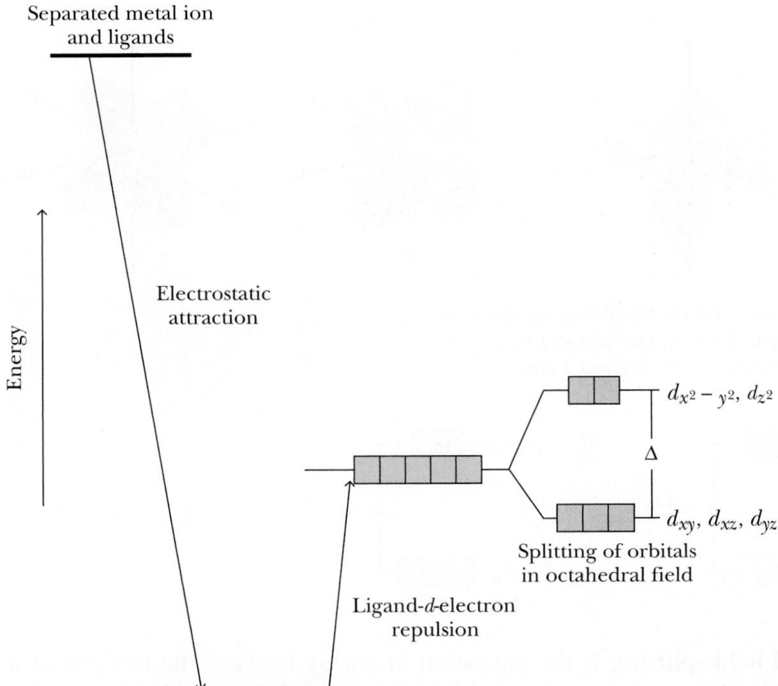

Figure 19.19 Energy changes in complex ion formation. (left) The large decrease in energy results from the attraction of the negative charge of the ligands by the positive transition-metal ion. (center) The repulsion of the *d* electrons of the metal by the negative charges of the ligands causes these orbitals to increase in energy. (right) The separation of the *d* orbitals into twofold and threefold degenerate sets by the octahedral arrangement of the ligands is shown.

Crystal Field Theory

In the early 1930s, physicists explained the colors and the magnetic properties of crystalline solids that contain transition-metal ions by using a purely ionic model called *crystal field theory.* It was more than 20 years later that chemists recognized the usefulness of this model for explaining the properties of transition-metal complexes in solution.

Crystal field theory assumes that the interaction between the ligands and the metal ion in a complex is electrostatic. When the ligands are anions, such as chloride or cyanide, a strong attractive force is exerted on them by the positive charge of the metal ion. Neutral ligands such as water and ammonia are polar molecules, and the negative end of the dipole is strongly attracted by the positive charge of the metal ion. In either case, the electrostatic forces cause a decrease in the energy of the system as the ligands are drawn close to the metal ion. At the same time, the negative charges of the ligands repel each other, so they stay as far apart as possible. We can describe an octahedral arrangement by having the six ligands approach the metal ion along the *x, y,* and *z* axes in a Cartesian coordinate system. The left side of Figure 19.19 shows the decrease in the energy that results from the attraction of the negative ligands by the positive metal ion.

The valence-shell electrons in transition-metal ions occupy the *d* subshell (the *s* electrons have been lost in any transition-metal ion with a charge of 2+ or greater). Although there is a net positive charge on the metal ion attracting the ligands, the negative charges of the ligands repel electrons located in the valence *d* orbitals, causing them to be less stable (of higher energy) in the complex than they are in the absence of the ligands. The destabilizing of the *d* electrons by the charges of the ligands is shown as an increase in energy in the center of Figure 19.19. However, the electrons in the different *d* orbitals do not experience the same repulsions by the ligands. Figure 19.20 shows the contours for the five *d* orbitals. Three of these orbitals (d_{xy}, d_{xz}, and d_{yz}) have the lobes of high-electron density directed diagonally between the *x* and *y, x* and *z,* and *y* and *z* axes, respectively, and away from the locations of the negative charges of the ligands. The other two *d* orbitals, $d_{x^2-y^2}$ and d_{z^2} are directed along the Cartesian axes and consequently point directly at the negative charges of the ligands. As a result, the five *d* orbitals are no longer degenerate (equal in energy) in the presence of the six ligands—three of them are lower in energy than the other two. The separation of the *d* orbitals into a doubly degenerate and a triply degenerate set is shown in the right side of Figure 19.19.

When a transition-metal ion is surrounded by ligands, some *d* orbitals point at the ligands and others point between them, so the *d* orbitals are no longer equal in energy.

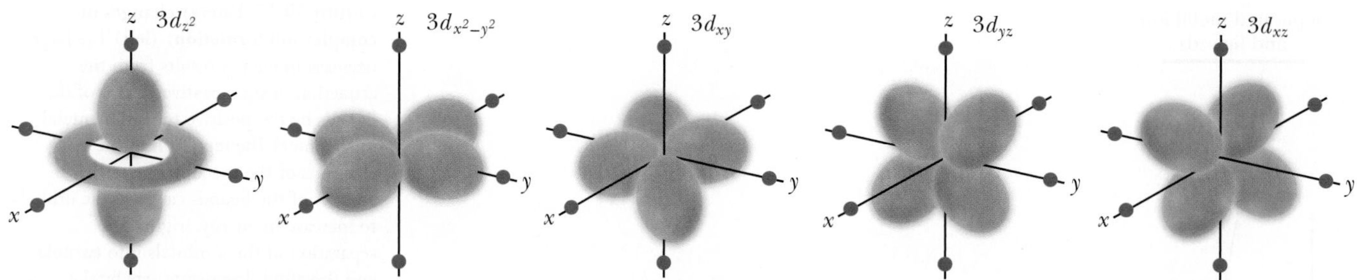

Figure 19.20 Degeneracy of the *d* orbitals. Electron density in three of the *d* orbitals (d_{xy}, d_{yz}, and d_{xz}) is farther from the locations of the ligands than is the electron density in the other two orbitals ($d_{x^2-y^2}$ and d_{z^2}). The locations of the negatively charged ligands are shown as *red dots* on the *x*, *y*, and *z* axes.

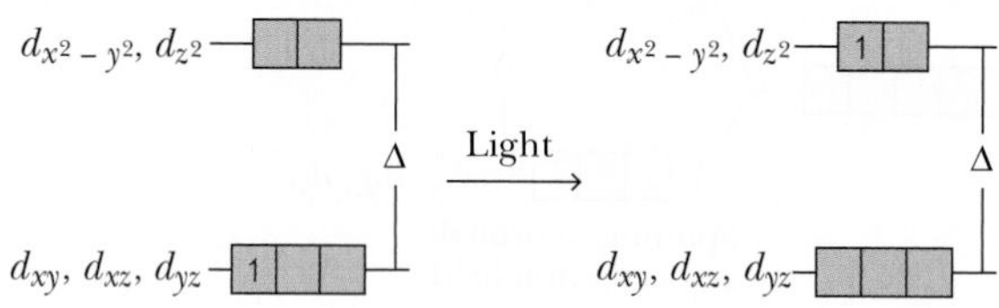

Figure 19.21 Crystal field energy-level diagram. The energies of the *d* orbitals in the $Ti(H_2O)_6^{3+}$ ion are shown. The absorption of light with an energy of Δ excites the electron from the ground-state configuration on the left into a higher-energy *d* orbital, producing the excited state shown on the right.

The **crystal field splitting** is the separation in energy between the two sets of *d* orbitals, caused by the unequal repulsion of the *d* electrons of the metal by the negative charges of the ligands that are arranged octahedrally about the central metal ion. The crystal field splitting is represented by Δ, the Greek capital letter delta. Figure 19.21 shows how this energy separation is usually represented in orbital energy-level diagrams for a metal ion in an octahedral crystal field. The energy separation of the *d* orbitals by the crystal field offers a simple explanation of the colors and magnetic properties of transition-metal complexes.

The color of a transition-metal ion is caused by the absorption of visible light when an electron moves from one of the lower-energy *d* orbitals to one of the higher-energy *d* orbitals.

Visible Spectra of Complex Ions

The size of Δ, the energy difference of the *d* orbitals, for most transition-metal complexes is within the energy range of visible light (2.8×10^{-19} to 5.0×10^{-19} J per photon, or 170 to 300 kJ/mol). The absorption of light that moves an electron from one of the lower-energy *d* orbitals to one of the higher-energy *d* orbitals produces the colors observed in transition-metal complexes. Figure 19.21 represents the electron transition for the single *d* electron in the red-violet $[Ti(H_2O)_6]^{3+}$ ion. This absorption of energy is a direct measure of the size of Δ for $[Ti(H_2O)_6]^{3+}$ and is responsible for the absorption spectrum shown in Figure 19.18*b*.

When more than one electron occupies the *d* orbitals of the metal, the relation between the value of Δ and the energy of the absorption bands in the visible spectrum is more complicated. For many of these situations, more than one absorption band is observed. Even in these more complicated situations, it is possible to determine Δ from the absorption spectra.

Factors That Affect Crystal Field Splitting

The value of the crystal field splitting, Δ, is calculated from the observed spectrum of a transition-metal complex. The spectra of many transition-metal complexes have been observed, and several generalizations about the resulting values of Δ have been made. As the ligands around a given metal ion change, the size of Δ changes. Table 19.8 shows the value of Δ for three metal ions with five different ligands. As the ligands change from Cl^- to CN^- down the table, the size of Δ increases. It is found that regardless of the metal ion used, as the ligands change, the size of Δ increases in the following order:

$$I^- < Br^- < Cl^- < F^- < H_2O < NH_3 < en < NO_2^- < CN^- < CO$$

Colors of solid transition-metal compounds. Crystal field theory accounts for the colors of gemstones. The red color of rubies is due to chromium(III), and sapphires are blue because of iron(II), iron(III), and titanium(IV).

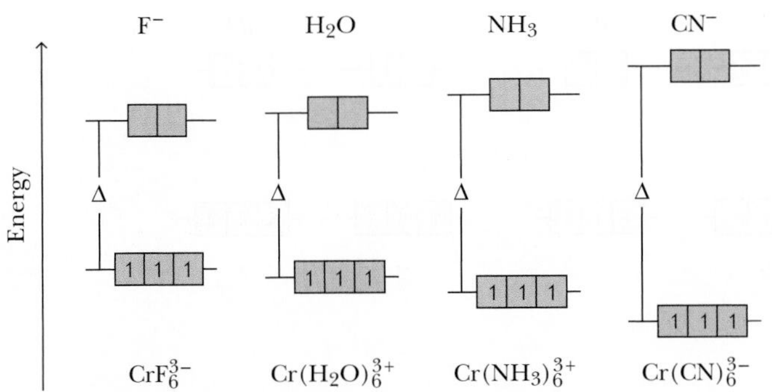

Figure 19.22 Spectrochemical series. The effect on Δ of changing ligands for a series of Cr(III) complexes. In the energy-level diagrams, Δ increases from left to right as the ligands change from fluoride to water to ammonia to cyanide.

This arrangement of ligands in order of increasing Δ is called the **spectrochemical series.** More complete spectrochemical series that contain many additional ions and molecules have been compiled. Figure 19.22 shows the change of Δ with different ligands for a series of chromium(III) complexes.

The charge of the metal ion also influences the size of Δ. In general, Δ is greater for a metal ion in a higher oxidation state. Thus, in the water complexes of Fe^{2+} and Fe^{3+}, the value of Δ increases 40% from 120 to 168 kJ/mol. Furthermore, within any periodic group; the size of Δ increases from the 3*d* to the 4*d* to the 5*d* transition series. Table 19.8 presents a comparison of the values of Δ for the Group 8B(9) complexes of Co^{3+}, Rh^{3+}, and Ir^{3+} with several different ligands.

The size of Δ depends on the particular ligand in a consistent manner; it increases as the charge of the metal ion increases and increases within a group from top to bottom.

TABLE 19.8 Δ (kJ/mol) for Selected Complexes

	Metal Ion		
Ligand	Co^{3+}	Rh^{3+}	Ir^{3+}
Cl^-	—	243	299
H_2O	218	323	—
NH_3	274	408	490
en	278	414	495
CN^-	401	544	—

EXAMPLE 19.6 Comparisons of the Size of Δ

For each pair of complexes listed, select the one that has the larger value of Δ.

(a) $[Ti(H_2O)_6]^{3+}$ or $[Ti(CN)_6]^{3-}$
(b) $[Cr(NH_3)_6]^{3+}$ or $[Mo(NH_3)_6]^{3+}$

Strategy Consult the spectrochemical series, and also consider the charge on the metal and the row in which the metal is located on the periodic table.

Solution

(a) The cyanide ligands are further along the spectrochemical series. The oxidation state of titanium is the same (+3) in the two complexes; therefore, based on the spectrochemical series, the $[Ti(CN)_6]^{3-}$ complex has the greater value of Δ.
(b) The ligands and oxidation state of the metals are constant, but Δ is greater for a metal with 4*d* valence orbitals than for one with 3*d*. $[Mo(NH_3)_6]^{3+}$ has the greater value of Δ.

Understanding

From the pair of complexes $[Cr(NH_3)_6]^{2+}$ and $[Cr(NH_3)_6]^{3+}$, select the one that has the larger value of Δ.

Answer The metal and ligands are constant, but the oxidation state of the metal is higher for $[Cr(NH_3)_6]^{3+}$, causing a larger Δ.

Electron Configurations of Complexes

The valence-shell electron configuration of a transition-metal atom in its ground state is usually $ns^2(n-1)d^m$, where *n* is the value of the principal quantum number, and *m* is the number of electrons in the occupied *d* subshell. When a transition element ionizes, *the* ns *electrons are lost before any electrons are removed from the* (n − 1)d *subshell* (see Section 8.1). Thus, in a transition-metal ion, *all* of the valence-shell electrons are in the $(n-1)d$ subshell. The electrons in the *d* orbitals of a complex ion obey Hund's rule of maximum spin; they remain unpaired, with one electron in each orbital, as long as possible. The d^1, d^2, and

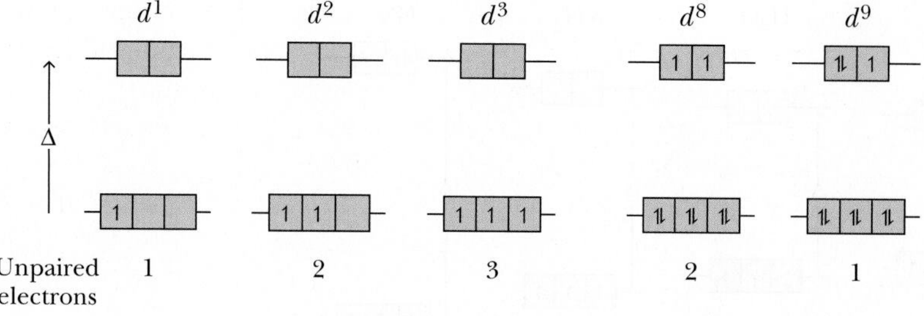

Figure 19.23 Orbital diagrams for complexes. The electron configurations of octahedral complexes containing one, two, three, eight, and nine d electrons are the same, regardless of the size of Δ. The number of unpaired electrons is given at the bottom of the figure.

All of the valence-shell electrons in a transition-metal ion occupy the d orbitals.

d^3 configurations have one, two, and three unpaired electrons, respectively, as shown in Figure 19.23. For the d^3 electron configuration in a complex ion, each d orbital in the lower-energy degenerate set contains a single unpaired electron.

For octahedral complexes of the d^1, d^2, and d^3 configurations, we find experimentally that the number of unpaired electrons shown in Figure 19.23 is correct. Experimentally, octahedral complexes with a d^4 configuration are found to have either 2 or 4 unpaired electrons. Those complexes with small Δ (low metal–ion charge and ligands early in the spectrochemical series) have 4 unpaired electrons, while those with large Δ (large metal–ion charge; the 5d and 6d transition metals; ligands farther down in the spectrochemical series) have only 2 unpaired electrons. These experimental results can be explained by realizing that the addition of the fourth electron into the lower-energy orbitals produces a pair of electrons. An energy price is paid for this because the electrons that share the same orbital are close together and repel each other more strongly than do the electrons that occupy different orbitals. The **electron pairing energy,** P, is the additional energy required for two electrons to occupy the same orbital, compared with the occupation of separate degenerate orbitals. It is this pairing energy that causes the electrons in a degenerate level to occupy separate orbitals whenever possible (Hund's rule).

Instead of forming a pair of electrons, the fourth electron could be located in one of the two d orbitals at an energy Δ above the others. The two possible arrangements of the four electrons in the d orbitals are shown on the left side of Figure 19.24. The relative sizes of Δ and P determine which of these configurations is preferred in a given complex. If $P > \Delta$, the fourth electron enters the higher-energy d orbital, and the complex contains four unpaired electrons. When $P < \Delta$, a pair of electrons is located in the lower-energy d orbitals, and the complex has only two unpaired electrons. Depending on the size of Δ (which depends on the nature of the ligands), both of these situations are observed for d^4 metal complexes; the $[MnF_6]^{3-}$ complex contains four unpaired electrons, but the $[Mn(CN)_6]^{3-}$ complex has only two unpaired electrons. A **high-spin**

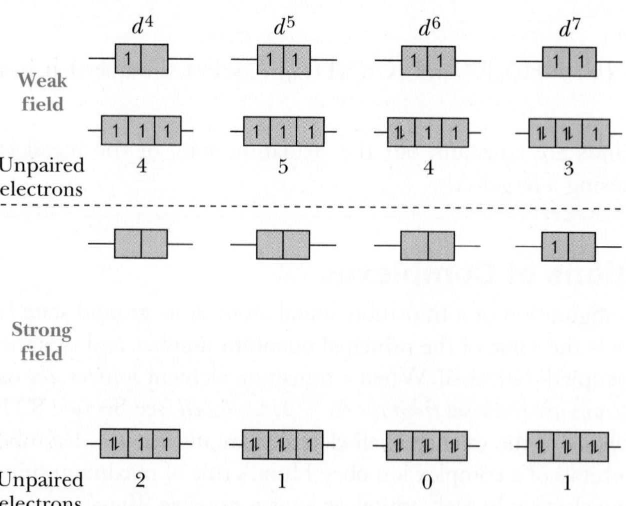

Figure 19.24 d electron configurations in octahedral complexes. Two electron configurations are possible for octahedral complexes that contain four to seven d electrons. (top row) Weak-field (high-spin) complexes. (bottom row) Strong-field (low-spin) complexes.

or **weak-field** complex occurs when $P > \Delta$. A **low-spin** or **strong-field** complex occurs when $P < \Delta$. Note that a small value of Δ means that the ligand field is weak and results in a high spin state. A large Δ produces a strong ligand field and a low-spin state.

Two spin states are possible for the d^4 to d^7 electron configurations, depending on the strength of the crystal field. Figure 19.24 shows the different spin states and orbital occupation for all of these configurations. Once eight or more electrons are present in the *d* subshell (see Figure 19.23), the number of unpaired electrons is the same regardless of the strength of the crystal field.

The factors that influence the size of Δ determine the number of unpaired electrons in complexes that have four through seven *d* electrons. For example, the $[CoF_6]^{3-}$ ion contains four unpaired electrons, whereas $[Co(NH_3)_6]^{3+}$ contains no unpaired electrons, because ammonia is higher in the spectrochemical series, and Δ is larger than the pairing energy. Because Δ increases with the charge on the metal ion, low-spin complexes are more common for complexes of 3+ metal ions than for 2+ ions. Because Δ is larger for the transition elements in the 4*d* and 5*d* series than for the 3*d* series, nearly all complexes of the heavier transition metals are low spin.

When the crystal field splitting (Δ) is small the electrons occupy the higher energy *d* orbitals singly to form a high-spin complex, before pairing the electrons in the lower-energy orbitals.

There are two possible spin states for octahedral complexes that contain four, five, six, or seven *d* electrons.

EXAMPLE 19.7 Magnetic Properties of Transition-Metal Complexes

Predict whether the following octahedral complexes are high spin or low spin, and given your prediction, give the number of unpaired electrons in each.

(a) $[FeF_6]^{3-}$
(b) $[Cr(CN)_6]^{4-}$
(c) $[Mn(H_2O)_6]^{2+}$
(d) $[RhCl_6]^{3-}$

Strategy Use a combination of the location in the spectrochemical series of the ligands, the charge on the metal, and the row in which the metal is located on the periodic table to predict the magnitude of Δ. If Δ is large, the complex will be low spin; if it is small, then it will be high spin.

Solution

(a) The fluoride ion produces a weak crystal field (see the spectrochemical series); consequently, we expect a weak-field complex for a 3*d* transition metal ion. The Fe^{3+} ion has a d^5 electron configuration, so the high-spin complex should contain five unpaired electrons (Figure 19.24).

(b) This is a complex of Cr^{2+}, which has the d^4 configuration. From its position in the spectrochemical series, the CN^- ion produces one of the strongest crystal field splittings, so a low-spin complex is expected. All four of the electrons are in the lower-energy *d* orbitals, so there are two unpaired electrons.

(c) Water is a moderately weak-field ligand. With the low ionic charge of the Mn^{2+} ion, a high-spin complex is anticipated. Five *d* electrons are in the complex, and each singly occupies a *d* orbital, so all five electrons are unpaired.

(d) The chloride ligand occurs early in the spectrochemical series. However, Rh is a metal in the 4*d* transition series, so all of its complexes are expected to be low spin. The d^6 configuration of Rh^{3+} forms a complex in which all of the electrons are paired in the lower-energy orbitals.

Understanding

Two complexes of iron(II), $[Fe(H_2O)_6]^{2+}$ and $[Fe(CN)_6]^{4-}$, have different numbers of unpaired electrons. How many unpaired electrons are present in each of the complexes?

Answer $[Fe(H_2O)_6]^{2+}$ has four unpaired electrons, and $[Fe(CN)_6]^{4-}$ has no unpaired electrons.

Complexes of Other Shapes

The crystal field theory also explains the colors and magnetic properties of tetrahedral and square-planar complexes, the other two arrangements most commonly observed in transition-metal complexes.

Let us first examine a tetrahedral complex of a transition-metal ion. One way of visualizing this arrangement of ligands about the metal is to place the metal at the center of a cube with the Cartesian axes passing through the centers of the cube's faces. The four ligands then are located at diagonally opposite corners of the cube, as shown in Figure 19.25. With this orientation, three of the *d* orbitals (d_{xy}, d_{xz}, and d_{yz}) have the lobes of electron density directed at the centers of the 12 edges of the cube (four lobes for each of the *d* orbitals). The other two *d* orbitals ($d_{x^2-y^2}$ and d_{z^2}) are directed at the centers of the faces of the cube. The centers of the edges of the cube are closer to the ligands ($a/2$) than are the centers of the faces ($a/\sqrt{2}$). As a result, electrons in three of the *d* orbitals experience stronger repulsions than do those in the other two. Just as in the case of an octahedral complex, the difference in the repulsion energy causes the *d* orbitals to split into a threefold degenerate set and a twofold degenerate set. In the case of the tetrahedral complex, however, the stronger repulsions occur with the electrons in the set of three orbitals, so these are of higher energy. The result of this analysis is that the crystal field energy-level diagram for a tetrahedral complex is inverted from that for an octahedral complex, with two orbitals at lower energy and three orbitals at higher energy, as shown in Figure 19.25*b*.

A quantitative treatment of the ligand field produced by a tetrahedral arrangement of four ligands about a metal ion shows that the crystal field splitting, Δ, is 4/9 of the Δ for an octahedral complex with the same ligands and metal ion. Experimental measurements confirm that Δ for tetrahedral complexes is about half that observed in similar octahedral complexes. As a consequence of the much smaller Δ in the tetrahedral arrangement, *nearly all tetrahedral complexes are high spin.* Generally, it is safe to assume that a tetrahedral arrangement means that a weak-field complex forms.

The crystal field energy-level diagram for a metal surrounded by ligands in a tetrahedral arrangement is inverted from that produced by an octahedral arrangement and Δ is much smaller.

Square-planar geometry is observed almost exclusively in complexes where the metal ion has a d^8 electron configuration. In these d^8 complexes, the electrons are invariably paired, suggesting a strong-field configuration. The square-planar geometry is usually visualized as a distortion of an octahedral complex, produced by removing the two ligands along the *z* axis. As the negative ligands are removed from the *z* axis, the electrons in the d_{z^2} orbital become more stable than those in the $d_{x^2-y^2}$ orbital because of the reduced electrostatic repulsion. The electrons in the d_{yz} and d_{xz} orbitals also experience a greater reduction in the repulsions (as the ligands on the *z* axis are removed) than do those in the d_{xy} orbital. Figure 19.26 includes the resulting energy-level diagram for the *d* electrons in a square-planar complex. In this arrangement, four of the *d* orbitals are much lower in energy than the fifth one. Metal ions that contain eight *d* electrons favor a square-planar arrangement, because there are just enough electrons to fill the four lower-energy orbitals and they have no unpaired electrons.

The crystal field theory can be applied to any coordination number and geometry.

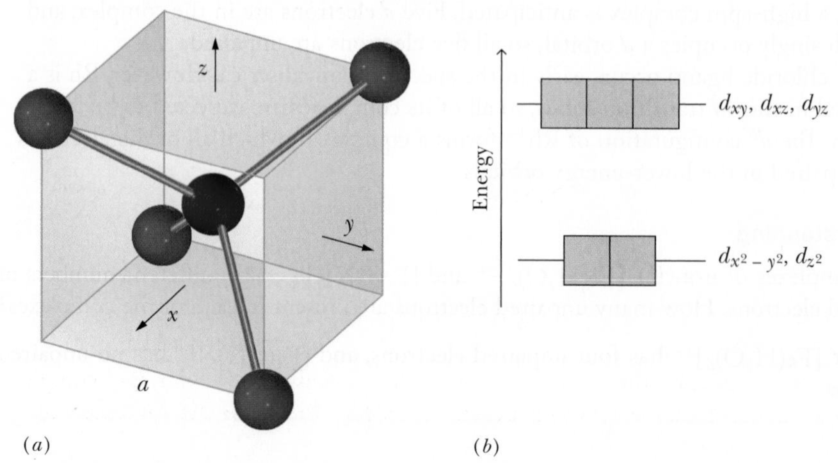

Figure 19.25 Tetrahedral complexes. *(a)* A tetrahedral complex can be represented by placing the ligands at alternating corners of a cube, with the metal ion at the center of the cube. The length of the edge of the cube is *a*. The *x*, *y*, and *z* axes pass through the centers of the cube's faces. The electrons in the three *d* orbitals (d_{xy}, d_{yz}, and d_{xz}) with lobes directed at the centers of the cube's edges are closer to the ligands than are those in the other two orbitals ($d_{x^2-y^2}$ and d_{z^2}), which point at the centers of the cube's faces. *(b)* Unequal repulsions cause splitting of the orbitals into a twofold degenerate set and a higher-energy threefold degenerate set.

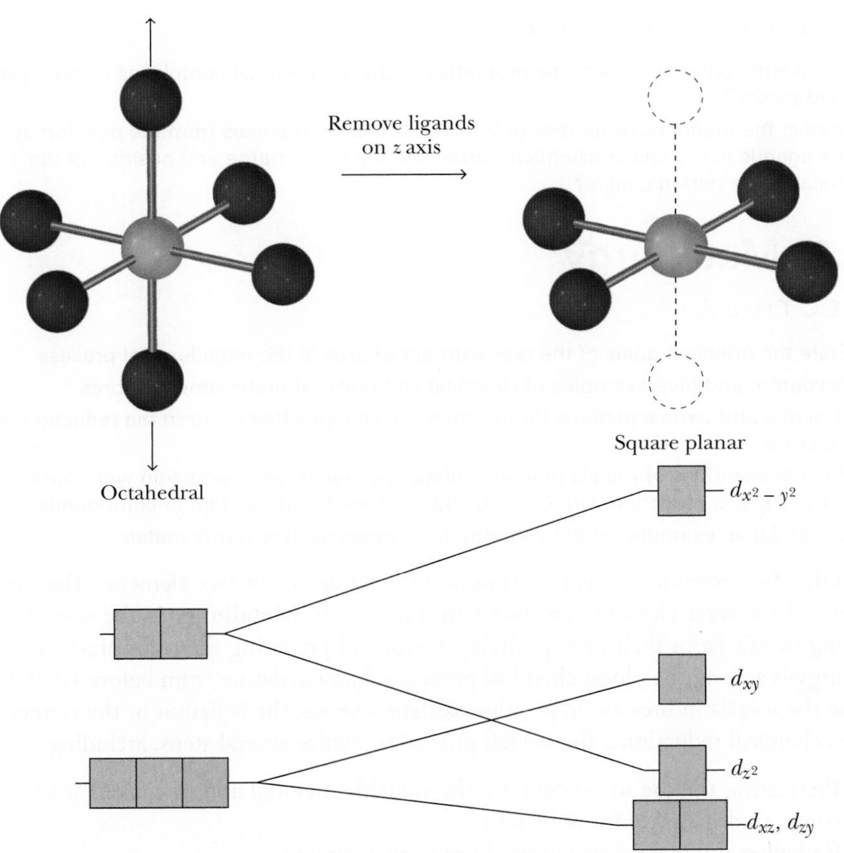

Figure 19.26 Crystal field diagram for a square complex. The change in the relative energies of the *d* orbitals is shown as an octahedral complex is converted into a square-planar complex by removing the two ligands along the *z* axis.

EXAMPLE 19.8 Electron Configurations of Four-Coordinate Complexes

Predict the geometry, the *d* orbital energy-level diagram, and the number of unpaired electrons in (a) $[Ni(CN)_4]^{2-}$ and (b) $[FeBr_4]^-$.

Strategy Based on the *d* electron configuration of the metal ion, decide whether the four-coordinate complex is tetrahedral or square planar. Use the appropriate diagram from Figure 19.25 or 19.26 to write the orbital energy-level diagram, and determine the number of unpaired electrons.

Solution

(a) Nickel is in the +2 oxidation state and, therefore, has a d^8 electron configuration. Furthermore, the cyanide ion is a strong-field ligand. Under these conditions, a square-planar complex is expected. The energy-level diagram for the *d* orbitals is that on the right in Figure 19.26, with a pair of electrons in all of the orbitals except $d_{x^2-y^2}$. No unpaired electrons are in the complex, so it is diamagnetic.

(b) This complex forms from $Fe^{3+}(d^5)$ and Br^- ions. The small size of the 3+ metal ion and the relatively large bromide ligands favor the formation of a tetrahedral complex. The *d* orbital energy-level diagram expected for a tetrahedral complex follows.

[↑1 ↑1 ↑1] (d_{xy}, d_{xz}, d_{yz})

Δ

[↑1 ↑1] $(d_{x^2-y^2}, d_{z^2})$

The crystal field splitting for tetrahedral complexes is small, so the high-spin complex is expected. Therefore, one electron is present in each of the *d* orbitals, for a total of five unpaired electrons.

OBJECTIVES REVIEW *Can you:*

- ☑ explain the color and magnetic properties of transition-metal complexes using crystal field theory?
- ☑ predict the magnetic properties of transition-metal complexes from the position of the ligands in the spectrochemical series and the ionic charge and position of the metal in the periodic table?

19.5 Metallurgy

OBJECTIVES

- ☐ State the principal goals of the pretreatment of ores in the metallurgical process
- ☐ Recognize and give examples of chemical and physical pretreatment of ores
- ☐ Describe and write equations for the chemical changes that occur in the reduction of iron ores
- ☐ Provide examples of metals that are isolated by electrolysis, reduction with more active metals, reduction with carbon, and thermal decomposition of compounds
- ☐ List and give examples of the common techniques used to purify metals

With only a few exceptions, metals do not occur in nature as the free elements. The common ores for several elements are listed in Table 19.9. **Metallurgy** is the science of extracting metals from their ores, purifying them, and preparing them for practical use. Metallurgy is among the oldest chemical processes known, dating from before 1000 B.C. Because the metals in ores are in positive oxidation states, the isolation of the elements involves chemical reductions. The overall processes involve several steps, including:

1. Pretreating the ore to concentrate the valuable material and to convert it to compounds suitable for reduction
2. Reducing the metal concentrated ore to the element
3. Purifying the metal

Pretreatment of Ores

The ores that are found in nature are usually complex mixtures of several minerals that interfere with the isolation of the desired metal. In the first step of purification, the ore is pulverized. Differences in physical and chemical properties then make it possible to concentrate the desired minerals before attempting to extract the metal.

Panning for gold. Separation of gold from other materials in the ore takes advantage of the high density of the metal.

Concentration

Several physical processes are used to concentrate ores. A procedure used in the recovery of gold takes advantage of the high density of the metal. When gold ore (actually metallic gold suspended in some mineral) is stirred under a flow of water, the less dense materials, often silicates, are suspended and carried away with the water, whereas the metallic gold settles to the bottom of the container. The prospectors who panned for gold in the 1849 gold rush in California used this method.

Another process, often used to concentrate sulfide ores (e.g., PbS and ZnS), is called *flotation.* The ore is added to a mixture of water and oil with other additives. The resulting

TABLE 19.9 **Compositions of Some Typical Metallic Ores**

Type of Ore	Examples
Oxide	Fe_2O_3, Fe_3O_4, Al_2O_3, SnO_2
Sulfide	PbS, ZnS, FeS_2, HgS, Cu_2S
Chloride	NaCl, KCl
Carbonate	$FeCO_3$, $CaCO_3$, $MgCO_3$
Sulfate	$BaSO_4$, $CaSO_4{\cdot}2H_2O$
Silicate	$Be_3Al_2Si_6O_{18}$, $Al_2(Si_2O_5)(OH)_4$
Free metal	Cu, Au, Ag, Pt, Pd, Rh, Ir, Ru

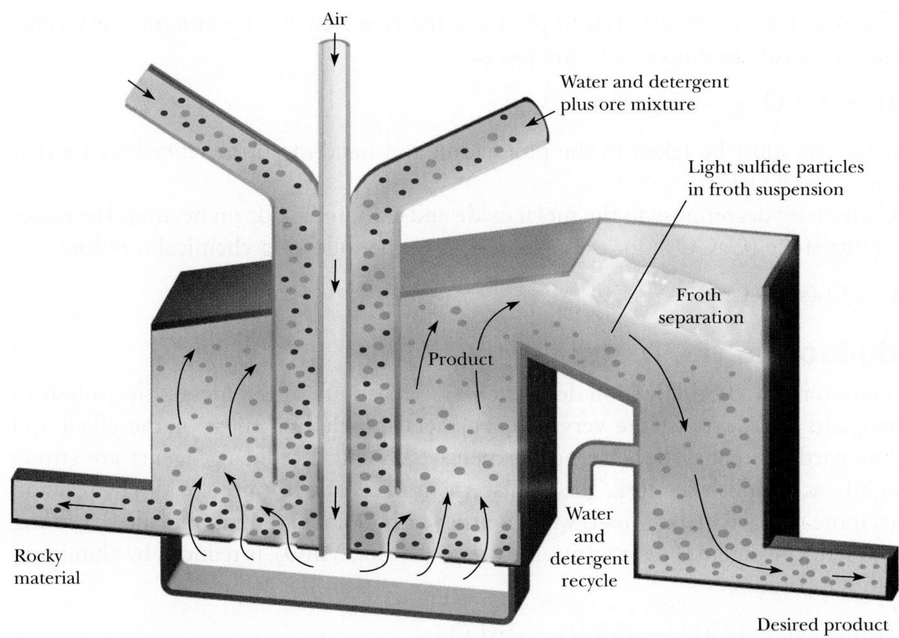

Ore concentration by flotation. Schematic diagram of the flotation process used to concentrate sulfide ores. The oil adheres to the metal sulfides and forms a foam with the air bubbles, which carry the desired compounds to the surface.

mixture is then stirred while a stream of air is bubbled through it. The metal sulfides become coated with the oil and are carried to the surface as a foam, whereas the unwanted material, called *gangue,* settles to the bottom of the container; it is later discarded.

Flotation is an example of concentrating ores by using differences in physical properties.

Chemical processes for concentrating ores vary greatly because they depend on the properties of the compounds containing the desired metals. As an example, the principal ore of aluminum is bauxite (hydrated aluminum oxide, $Al_2O_3{\cdot}xH_2O$), which usually contains iron(III) oxide as a major impurity. Treatment of the ore with aqueous sodium hydroxide dissolves the amphoteric Al_2O_3, leaving the other metal oxides as solids.

$$Al_2O_3(s) + 2OH^-(aq) + 3H_2O(\ell) \rightarrow 2[Al(OH)_4]^-(aq)$$

After the solution is separated from the solid residue, acid or carbon dioxide is added to precipitate $Al(OH)_3$, which is heated to produce the purified oxide.

Aluminum ores are concentrated by the use of the differences in chemical properties of aluminum oxide and iron(III) oxide.

$$[Al(OH)_4]^-(aq) + H^+(aq) \rightarrow Al(OH)_3(s) + H_2O(\ell)$$

$$2Al(OH)_3(s) \xrightarrow{\text{Heat}} Al_2O_3(s) + 3H_2O(g)$$

Roasting

Another process in the pretreatment of some ores, called **roasting,** consists of heating the mineral at a temperature below its melting point, usually in the presence of air, to convert the ore to a chemical form more suitable for the reduction step. Dehydration of aluminum hydroxide (shown in the preceding paragraph) could be classified as roasting. More commonly, the roasting step converts sulfide and carbonate ores to oxides. Galena (PbS), pyrite (FeS_2), and sphalerite (ZnS) are all converted to the metal oxides by roasting in air.

Sulfide ores are roasted to form the metal oxides, which can often be reduced by carbon, a relatively inexpensive reducing agent.

$$2ZnS(s) + 3O_2(g) \rightarrow 2ZnO(s) + 2SO_2(g)$$

$$2PbS(s) + 3O_2(g) \rightarrow 2PbO(s) + 2SO_2(g)$$

$$3FeS_2(s) + 8O_2(g) \rightarrow Fe_3O_4(s) + 6SO_2(g)$$

The conversion from sulfides to oxides allows the use of carbon in the later reduction to the metal, because the carbon compound that forms, carbon dioxide, is much more stable than carbon disulfide. In modern operations, the SO_2 is trapped and used to manufacture sulfuric acid. Any SO_2 that is vented to the atmosphere is a serious pollutant.

The roasting of cinnabar (HgS) produces the free element, because mercury compounds are easily decomposed to the metal.

$$HgS(s) + O_2(g) \rightarrow Hg(\ell) + SO_2(g)$$

Extreme care must be taken in the production and handling of mercury because that element is very toxic.

Carbonates decompose to the metal oxide and carbon dioxide on heating. The roasting of limestone ($CaCO_3$), for example, results in the following chemical reaction:

$$CaCO_3(s) \rightarrow CaO(s) + CO_2(g)$$

Reduction to the Metal

The chosen method of reduction depends on the reactivity of the metal. Electrolysis of molten salts must reduce the very reactive metals, including most of the alkali and alkaline-earth elements, because no inexpensive chemical reducing agents are strong enough to accomplish the task. The displacement of metals from the oxides by reaction with a more reactive metal is used when neither carbon nor carbon monoxide is a strong enough reducing agent. For example, the ore pyrolusite, MnO_2, is reduced by aluminum at high temperature.

$$3MnO_2(s) + 4Al(\ell) \rightarrow 3Mn(\ell) + 2Al_2O_3(s)$$

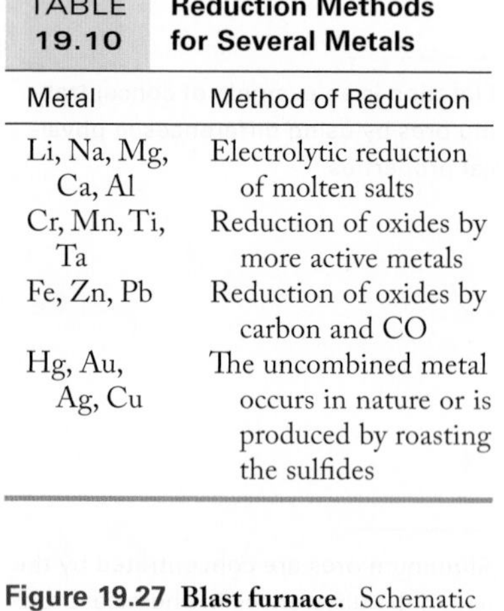

TABLE 19.10 Reduction Methods for Several Metals

Metal	Method of Reduction
Li, Na, Mg, Ca, Al	Electrolytic reduction of molten salts
Cr, Mn, Ti, Ta	Reduction of oxides by more active metals
Fe, Zn, Pb	Reduction of oxides by carbon and CO
Hg, Au, Ag, Cu	The uncombined metal occurs in nature or is produced by roasting the sulfides

Whenever possible, carbon is used as the reducing agent, because it is fairly abundant and inexpensive. The least reactive metals occur in nature as the uncombined elements or are released during the roasting of the ore. Table 19.10 shows the reduction methods for isolating several of the common metals.

Iron is by far the most widely produced metal, and it is typically isolated from oxide ores by using carbon as the reducing agent. The reduction is carried out in a blast furnace (Figure 19.27). The iron ore, mixed with coke (mainly carbon) and limestone ($CaCO_3$), is added continuously to the top of the furnace. Hot air is fed into the furnace at the bottom. Several chemical reactions are involved in the overall process of reducing the ore to iron metal. The oxygen in the air combines with the carbon, forming carbon monoxide. This exothermic reaction maintains the high temperatures needed in the furnace.

$$2C(s) + O_2(g) \rightarrow 2CO(g) \qquad \Delta H^\circ = -221.0 \text{ kJ}$$

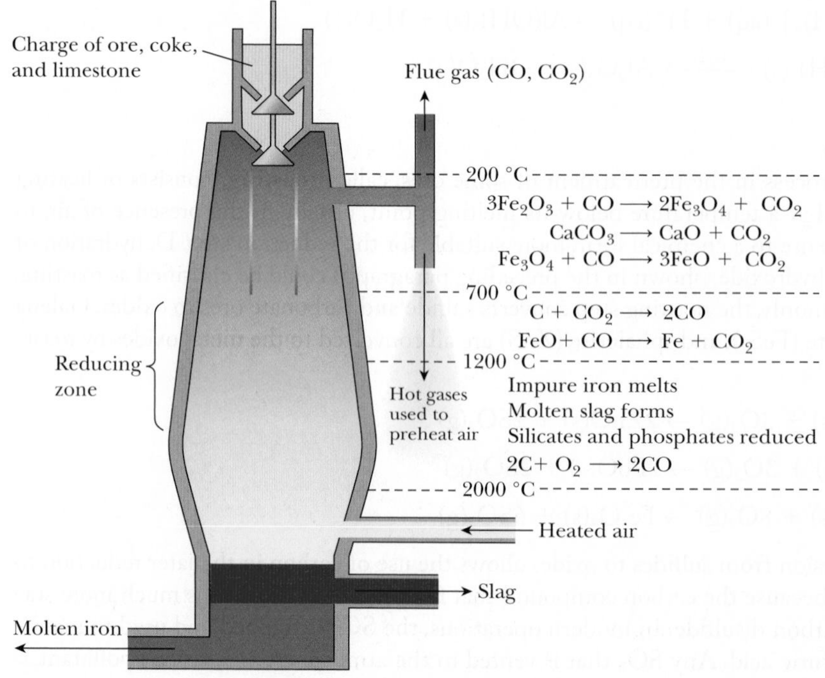

Figure 19.27 Blast furnace. Schematic diagram of the blast furnace. The iron ore, coke, and calcium carbonate are added at the top of the furnace, and molten iron and slag are removed separately at the bottom.

Carbon monoxide forms rather than carbon dioxide because oxygen is the limiting reactant. The hot gases rise through the furnace, mixing with the ore. Carbon monoxide is the principal reducing agent, leading to the production of molten iron.

$$Fe_2O_3(s) + 3CO(g) \rightarrow 2Fe(\ell) + 3CO_2(g)$$

The limestone decomposes at the high temperatures in the furnace to form carbon dioxide and calcium oxide. The calcium oxide combines with impurities in the ore, mostly aluminum oxide and silicon dioxide, to form molten silicates and aluminates.

$$CaCO_3(s) \rightarrow CaO(s) + CO_2(g)$$

$$CaO(s) + SiO_2(s) \rightarrow CaSiO_3(\ell)$$

$$CaO(s) + Al_2O_3(s) \rightarrow Ca(AlO_2)_2(\ell)$$

The mixture of calcium silicate and calcium aluminate, called *slag,* is a liquid at the temperatures inside the blast furnace. The liquid iron and slag fall to the bottom of the furnace, forming two separate layers (see Figure 19.27). These are drawn off separately, and the iron is cast into bars, called *pig iron.* The pig iron is impure, containing up to 5% carbon and silicon, manganese, and several other impurities, so this crude material must be further purified before it can be used in modern applications.

Carbon or CO cannot reduce aluminum ores. The Hall process uses electrolytic reduction to produce elemental aluminum. The Hall electrolysis process is described in Section 18.11. Because of its cost, electrolysis is used only to produce the more reactive metals that cannot be isolated from their ores with carbon or other inexpensive chemical reducing agents.

Purifying Metals

The method used to purify a metal depends on the chemical properties of the particular metal, the nature of the impurities, and the degree of purity needed. Many metals, particularly iron, have more desirable physical and chemical properties when mixed with other elements. An *alloy* is a mixture of two or more elements that has the properties of a metal. Thus, in the production of steel, an alloy of iron, the complete removal of chromium and carbon from the crude pig iron is not desired. Some of the more common techniques used to purify metals are described briefly in the following paragraphs.

Distillation

Several metals are sufficiently volatile that they can be purified by fractional distillation. Mercury, zinc, and magnesium are all refined (purified) by distillation. When purifying reactive metals, the absence of oxygen is essential during the high-temperature distillation to avoid forming metal oxides, so the separation is carried out under an inert atmosphere or in a vacuum. Vacuum distillation offers a second advantage—it reduces the temperature needed to distill the metal.

The Mond process for purifying nickel is an interesting example that uses volatility as a method of separation. The impure nickel is treated with carbon monoxide at about 70 °C to form the compound $Ni(CO)_4$, tetracarbonylnickel(0), which has a boiling point of 43 °C, and the $Ni(CO)_4$ gas is separated from the solid impurities in the nickel. After separation, $Ni(CO)_4$ is heated to a higher temperature at which the nickel compound decomposes, yielding the purified solid metal.

$$Ni(CO)_4(g) \xrightarrow{200\ ^\circ C} Ni(s) + 4CO(g)$$

The carbon monoxide released in this decomposition is recycled through the reactor.

Electrolysis

In the electrorefining of copper, impure copper is dissolved at the anode and pure copper is deposited at the cathode. Impurities form sludge at the bottom of the cell. Other metals that are purified by electrorefining are cobalt and lead.

Electrolysis for purification. Plates of copper produced in the electrorefining of the metal are shown. Impure copper is dissolved at the anode and deposited in much higher purity at the cathode.

Purification of metals is achieved by separations based on either chemical or physical properties.

Zone Refining

When extremely high purity is needed, the process of zone refining can be used. Section 20.4 details this technique for the purification of the nonmetal silicon. Because zone refining is a relatively expensive process and is effective in removing only small amounts of impurities, it is used when very high purity (99.99%+) of the element is essential.

Refining of Iron and Manufacture of Steel

Before iron is used, the pig iron produced in the blast furnace must be further refined to remove most of the nonmetal impurities, which include silicon, sulfur, phosphorus, and relatively large amounts of carbon. The *basic oxygen process* is the most widely used process for manufacturing steel from the impure pig iron. In a typical operation, the furnace is charged with 75 tons of molten iron and 10 tons of limestone. Oxygen gas, sometimes diluted with argon, is forced into the bottom of the furnace. The nonmetal impurities are rapidly converted to oxides in exothermic reactions. The heat released by these oxidations maintains the temperature of the mixture well above the melting point of the iron. The oxides of silicon and phosphorus combine with calcium oxide to form a slag. The entire process takes about 20 minutes; the furnace is then tilted to pour the molten metal into molds. Some of the important chemical reactions involved include:

$$C + O_2 \rightarrow CO_2$$

$$S + O_2 \rightarrow SO_2$$

$$Si + O_2 \rightarrow SiO_2$$

$$CaO + SiO_2 \rightarrow CaSiO_3$$

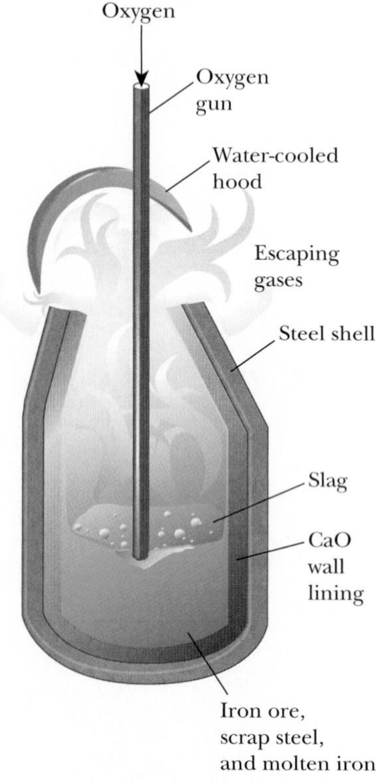

Oxygen furnace. At high temperatures, reaction with oxygen removes most of the nonmetal impurities in crude iron.

Purified molten iron is poured into molds.

OBJECTIVES REVIEW *Can you:*

- ☑ state the principal goals of the pretreatment of ores in the metallurgical process?
- ☑ recognize and give examples of chemical and physical pretreatment of ores?
- ☑ describe and write equations for the chemical changes that occur in the reduction of iron ores?
- ☑ provide examples of metals that are isolated by electrolysis, reduction with more active metals, reduction with carbon, and thermal decomposition of compounds?
- ☑ list and give examples of the common techniques used to purify metals?

CASE STUDY Shape of 4-Coordinate Complexes

An interesting area of research is the synthesis and determination of the shapes and structures of new transition-metal complexes. In the laboratory, you have just prepared a new four-coordinate compound of nickel(II), $[Ni(SCN)_2(H_2O)_2]$. How would you determine the arrangement of the donor atoms about the nickel center?

The best way to approach the problem is to do a literature search of similar compounds of nickel(II) that have been made by others and compare the properties of your new compound with those that have known structures. The search turns up an interesting result: Some of the known complexes are tetrahedral, whereas others are square planar. How will you tell which one of these shapes is correct for your new compound? The key is to compare the color and magnetism of your new compound with the known complexes. First, you realize that the *d*-orbital splitting will be different in the two shapes. Shown below are the arrangements of the *d* orbitals for the two geometries containing the eight *d* electrons in a Ni(II) complex.

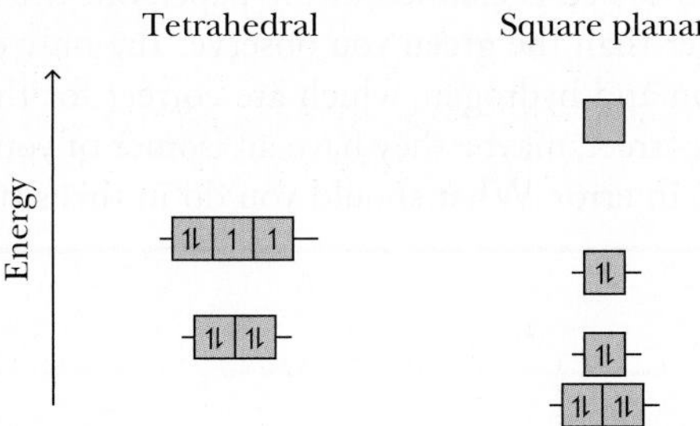

Prior research had shown that the split of the orbitals in the tetrahedral case is small and less than the pairing energy, so the eight *d* electrons available for Ni(II) would be arranged as shown, leading to a prediction of two unpaired electrons. In contrast, given that in a square-planar arrangement, the $d_{x^2-y^2}$ orbital is much higher in energy than the other four orbitals, all eight electrons arrange themselves as shown, leaving the $d_{x^2-y^2}$ orbital empty and producing a species that has no unpaired electrons. Measurements previously made on the compounds that you found in your literature search confirm these predictions. Now you need to measure whether the compound does (is paramagnetic) or does not (is diamagnetic) contain unpaired electrons. Also, given that the *d* electrons are arranged differently in the two possibilities, you should also expect the color of the complexes in the two shapes to be different. Your new compound is green, and an investigation of the colors of the known compounds indicated that many of the tetrahedral compounds are green, whereas the predominate color for the square-planar complexes is red. So when the green compound is found experimentally to have two unpaired electrons, there is no doubt that it has tetrahedral geometry.

It also needs to be determined if any isomers are possible for your new compound. First, consider geometrical isomers. If the compound had been square planar, you could have had either a *cis* or *trans* arrangement of the ligands, but in a tetrahedral arrangement, isomers are not possible. Build a model and convince yourself of that! One could also have linkage isomerization because the SCN^- ligand can be both N- and S-bound. Although a number of experiments can be used to determine which is correct, an experienced inorganic chemist would predict that it will be N-bound because results show that most first-row transition-metal complexes of this ligand are bonded that way. But you should prove this by experiment; one good experiment is to grow a crystal of the material and do an x ray crystal structure. This analysis shows the location of all the atoms. If N-bound as expected, the name would be diaquadi-N-thiocyanatonickel(II). The x ray experiment will also definitively determine the shape.

ETHICS IN CHEMISTRY

1. You have just finished the complete analysis outlined in the Case Study that completely characterizes your new nickel complex. The next day, an editor of *Inorganic Chemistry,* an important research journal in your field, asks you to review a paper she has just been sent to determine whether it reports high-quality new science and should be published. This review process is an important step in the research publication process. In the paper, the authors also report synthesizing your compound but have not done the magnetic study or the x ray crystallography study that you have done. They predict the tetrahedral shape based on the color and do not even consider linkage isomerization. What should you do? You could suggest to the editor that she reject the paper, saying they should do those studies, and then rush your paper out reporting your data, possibly getting your paper published first. You could simply tell the editor the situation, but that might lead to your not being able to publish your more complete data at all. This outcome is negative because your job depends on the number and quality of your research publications. Other possible actions are available as well. What would you do?

2. A second scenario that could arise from the paper you were sent to review is that the compound you synthesized is claimed in the paper, but the authors report that the compound is blue, rather than the green you observe. The only other data they present are percentage of carbon and hydrogen, which are correct for the formula of the compound. If that data are correct, maybe they have an isomer of your compound, or maybe their analytical data are in error. What should you do in this situation?

Chapter 19 Visual Summary

The chart shows the connections between the major topics discussed in this chapter.

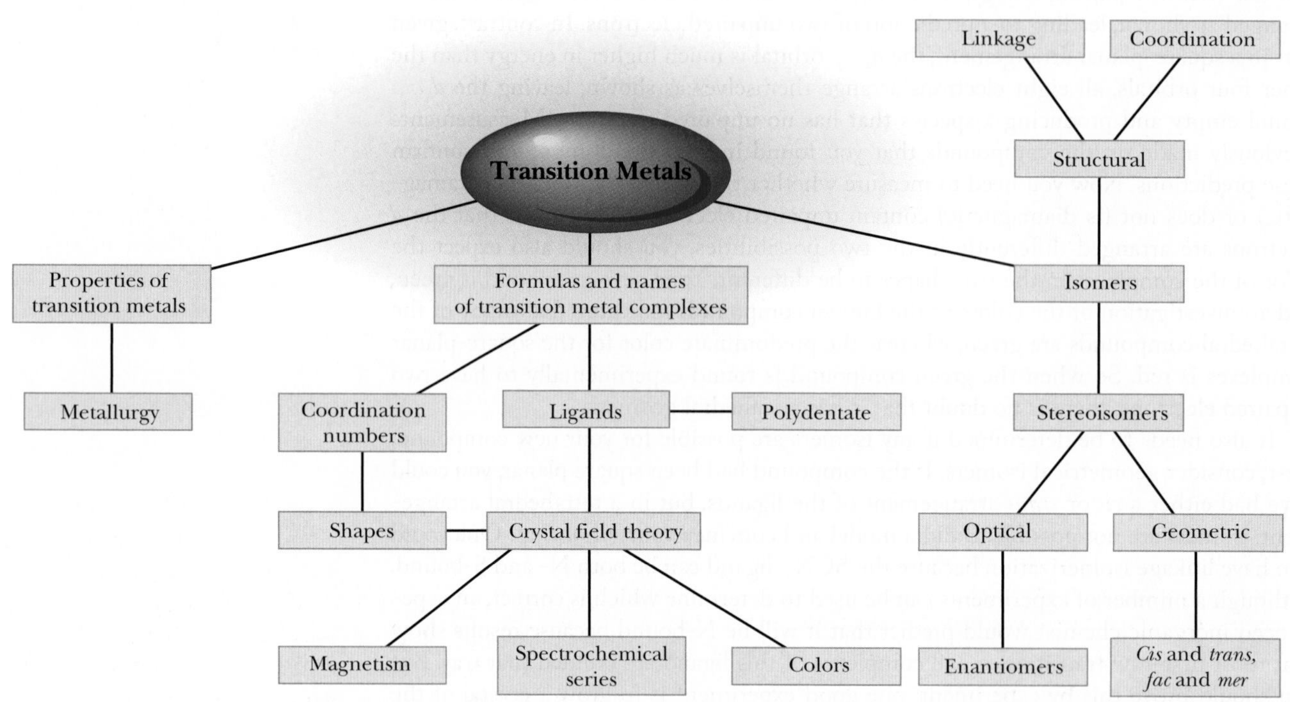

Summary

19.1 Properties of the Transition Elements

Most of the elements are metals. The *transition elements* are those metals that have partially-filled *d* orbitals in at least one oxidation state and generally exhibit more than one stable oxidation state. High melting and boiling points are characteristic of the transition metals and indicate the presence of very strong metallic bonds. The assumption that the *ns* and $(n - 1)d$ electrons and orbitals are used to form highly delocalized bonds in these metals is consistent with the increase in melting points within each period up to Group 6B, followed by decreasing melting points through the remainder of the transition series. A small increase occurs in the effective nuclear charge within each period as the atomic number increases and electrons are added to the *d* subshell, causing a small decrease in atomic radius from left to right across the transition elements. The decrease in radius is smaller for the transition metals than for the representative elements because the outer electrons in each transition-metal period are always in the same *ns* level. The *lanthanide contraction* causes the sizes of the transition-metal atoms in the sixth period to be nearly the same as those of the elements directly above them in the fifth period. Consequently, there is usually a much greater difference between the chemical behaviors of the first and second elements in a transition-metal group than between the second and third elements in a group.

All of the transition elements exhibit stable oxidation states of +2, +3, or both, as well as higher and sometimes lower states. The maximum oxidation state achieved by the transition elements is equal to the number of *s* and *d* electrons present in the valence shell through manganese (+7) in the first transition series and through ruthenium and osmium (+8) in the second and third rows. Following the maximum in positive oxidation state in Group 7B or 8B, the highest observed oxidation state decreases through the remaining groups. The lanthanide contraction is also evident in the ionization energies of the elements in each group of the transition elements. The first and second elements in each group have similar ionization energies, whereas the ionization energy of the third element in the group is markedly greater.

19.2 Coordination Compounds: Structure and Nomenclature

A large part of the chemistry of the transition elements involves *coordination complexes.* In many compounds, transition-metal ions behave as Lewis acids, forming coordinate covalent bonds with several *ligands*—molecules or ions that donate unshared electron pairs to the metal ion. The most commonly observed *coordination numbers* in transition-metal complexes and their geometric shapes are two (linear), four (square planar or tetrahedral), and six (octahedral). Ligands are classified as *monodentate,* bidentate, tridentate, and so forth by the number of *donor atoms* (one, two, three, ...) in the ligand that bond to a metal ion. *Chelating ligands* have two or more donor atoms (are *polydentate*) that coordinate to the same metal atom. In the formula of a coordination compound, the metal ion and its coordinated ligands are enclosed in square brackets.

19.3 Isomers

Several kinds of isomers are possible in coordination compounds. These fall into two main categories: *structural isomers* and *stereoisomers.* Structural isomers are further subdivided into *coordination isomers,* which differ in the ligands coordinated to the metal, and *linkage isomers,* in which the same ligand is bonded to the metal ion through different donor atoms. *Geometric isomers,* a kind of stereoisomer, contain the same ligands with different geometric arrangements about the metal ion. Stereoisomers also include *optical isomers,* which rotate the plane of *polarized light.* This behavior is characteristic of *chiral molecules,* which have nonsuperimposable mirror-image structures. The nonsuperimposable mirror images are called *enantiomers,* and they rotate the plane of polarized light in opposite directions. A *racemic mixture* is an equimolar mixture of enantiomers that produces no net rotation of plane-polarized light.

19.4 Bonding in Coordination Complexes

Crystal field theory is a model for the bonding in coordination complexes. It explains the magnetic properties and visible spectra that are so characteristic of these species. The arrangement of ligands surrounding the transition-metal ion in a regular geometric pattern removes the degeneracy of the partially-filled *d* orbitals. In an octahedral complex, three of the *d* orbitals have a lower energy than the other two *d* orbitals. The energy difference between the two sets of *d* orbitals is called the *crystal field splitting,* Δ, and it corresponds to the energy of light in the visible region of the electromagnetic spectrum. The absorption of light in the visible spectrum promotes an electron from a lower-energy *d* orbital to a higher one, producing the observed color of coordination compounds. The *spectrochemical series* arranges ligands in order of the values of Δ they produce. When the ligands produce a small splitting of the *d* orbitals, the complexes are called *weak-field. Strong-field complexes* have a large energy separation of the *d* orbitals. When the ligands produce a strong field, the electrons in the complex form pairs in the lower-energy *d* orbitals before any enter the higher-energy *d* orbitals. The electrons in a weak-field complex singly occupy all five *d* orbitals before pairing occurs in the lower-energy *d* orbitals. Weak-field complexes are called *high-spin,* and strong-field complexes are called *low-spin.* Crystal field theory is also successful in accounting for the spectral and magnetic properties of other arrangements of ligands, such as tetrahedral and square-planar complexes.

19.5 Metallurgy

The isolation of the metals from naturally occurring compounds has been an important factor in the development of our modern technological society. *Metallurgy,* the science of extracting metals from their ores, purifying them, and preparing them for practical use, can involve chemical reactions in all the stages of pretreatment, concentration, reduction, and purification. Concentration of ores is accomplished by both physical and chemical means. The *roasting* of ores converts naturally occurring compounds to chemical forms more suitable for reduction.

The isolation of metals from their compounds often involves a chemical reduction. The agent used for reduction depends on the reactivity of the metal being isolated. Very reactive metals, such as sodium and aluminum, are prepared by electrolysis. With other metals, including chromium and titanium, displacement by more reactive metals is used to isolate the metals from ores. For economic reasons, carbon and carbon monoxide are the reducing agents of choice in the isolation of metals such as iron and zinc.

The final purification of metals is highly dependent on the chemical and physical properties of the metal and the impurities present, as well as the purity needed in the final application. Often alloys, a mixture of two or more elements that has the properties of a metal, have more desirable properties than pure metals. Distillation, electrolysis, and zone refining are three of the processes used for purifying metals. In the oxygen furnace, used to purify pig iron, several chemical reactions occur.

Chapter Terms

The following terms are defined in the Glossary, Appendix I.

Section 19.1
Lanthanide contraction
Transition elements

Section 19.2
Chelate
Chelating ligand
Coordination compound or complex
Coordination number
Donor atom
Ligand
Monodentate ligand
Polydentate ligand

Section 19.3
Chiral molecule or ion
Coordination isomers
Enantiomers
Geometric isomers
Linkage isomerism
Optical isomers
Polarized light
Racemic mixture
Stereoisomers
Structural isomers

Section 19.4
Absorption spectrum
Crystal field splitting, Δ
Diamagnetic
Electron pairing energy, *P*
High-spin (weak-field) complex
Low-spin (strong-field) complex
Paramagnetic
Spectrochemical series

Section 19.5
Metallurgy
Roasting

Questions and Exercises

OWL Selected end of chapter Questions and Exercises may be assigned in OWL.

Blue-numbered Questions and Exercises are answered in Appendix J; questions are qualitative, are often conceptual, and include problem-solving skills.

■ Questions assignable in OWL

✎ Questions suitable for brief writing exercises

▲ More challenging questions

Questions

19.1 What distinguishes a transition element from a representative element?

19.2 Define the transition elements.

19.3 ✎ The ratio of the density of tantalum to that of niobium is 1.94, which is nearly identical to the 1.95 ratio of their atomic weights. Explain how this similarity is a result of the lanthanide contraction.

19.4 ✎ Why do the atomic radii of the transition elements within a period decrease more rapidly from Group 3B through 6B than through the rest of the transition elements in that period?

19.5 Use the information in Figure 19.5 to explain why the +3 oxidation state becomes less common for the elements near the end of the transition series.

19.6 Name and sketch the four most important shapes of transition-metal complexes.

19.7 Only the Group 1B transition elements form simple compounds in which the oxidation state of the metal is +1. For all of the other transition elements, the lowest positive oxidation state is +2. What common feature in the electron configuration of the transition elements contributes to this fact?

19.8 ✎ List and describe three methods used to purify metals once they have been reduced.

19.9 List two goals in the pretreatment of ores. Give an example of each.

19.10 Write equations for the principal reactions involved in the reduction of Fe_2O_3 in a blast furnace.

Exercises

OBJECTIVE Explain which elements are transition metals.

19.11 Which elements appear in the *d* block of the periodic table but do not meet the definition of a transition element?

19.12 Is actinium ($Z = 89$) a transition element? Explain.

19.13 Which of the following elements are transition metals?
(a) Fe (b) Ba (c) Hg

19.14 Which of the following elements are transition metals?
(a) Mo (b) La (c) Pd

19.15 Which element in the fourth period has one or more $3d$ electrons in the free element but none in any of its common oxidation states?

19.16 Which element in the fourth period has the greatest number of unpaired electrons?

OBJECTIVE Explain the trends in physical and atomic properties of the transition elements.

19.17 In each part, select the transition element that has the higher melting point and explain why.
(a) Cr or Co (b) Ti or Hf
(c) Nb or V (d) Y or W

19.18 In each part, select the transition element that has the higher melting point and explain why.
(a) Cr or Cu (b) Fe or Os
(c) Cr or V (d) La or W

19.19 Based on the general trends in metallic bonding, explain which transition metal in the fourth period should have the highest heat of fusion?

19.20 Use the melting point data in Table 19.1 to predict which element in the transition-metal series in the fifth period is hardest.

19.21 Arrange the following transition-metal atoms in order of decreasing atomic radius: V, Co, Nb, W.

19.22 ■ Arrange the following transition-metal atoms in order of decreasing atomic radius: Fe, Mo, Hf, Ta.

19.23 Determine the maximum positive oxidation state expected for each of the following.
(a) Ti (b) W (c) Ta (d) Re

19.24 Determine the maximum positive oxidation state expected for each of the following.
(a) Cr (b) Zr (c) Y (d) Tc

19.25 In each part, explain why one element has the higher first ionization energy.
(a) Ti or Mn (b) V or Ta
(c) Ru or Rh (d) Mo or Os

19.26 In each part, explain why one element has the higher first ionization energy.
(a) Zr or Tc (b) Mo or W
(c) Fe or Pt (d) Mn or Co

OBJECTIVE Use the standard conventions for writing the names and formulas of coordination compounds.

19.27 Write the formula for each of the following coordination compounds.
(a) chromium(III) chloride, in which one Cl^- and five water molecules are coordinated to the metal
(b) $CrCl_3{\cdot}4NH_3$, which contains two coordinated chloride ions
(c) the potassium salt of the coordination complex containing six CN^- ions and Fe(III)

19.28 Write the formula for each of the following coordination compounds.
(a) a coordination compound containing two complex ions of Co(III), one with six CN^- ions and the other with three ethylenediamine molecules (you may use the abbreviation "en" for the ethylenediamine molecule)
(b) a coordination compound of platinum(II) nitrate, in which four ammonia molecules are coordinated to the transition-metal ion
(c) the sodium salt of the complex formed from Rh(III), five Cl^- ions, and one water molecule

19.29 Name each of the following compounds.
(a) $[Pt(NH_3)_2Cl_2]$
(b) $[Co(en)_2(NO_2)_2]NO_3$ (en = ethylenediamine)
(c) $K_3[RhCl_6]$
(d) $[Pt(NH_3)_4][PtCl_4]$
(e) $Cr(CO)_6$

19.30 ■ Give the name or formula for each ion or compound, as appropriate.
(a) tetraaquadichlorochromium(III) chloride
(b) $[Cr(NH_3)_5SO_4]Cl$
(c) sodium tetrachlorocobaltate(II)

19.31 Write the formula of each of the following ions or compounds.
(a) pentaaquachlorochromium(III) chloride
(b) tetraamminedinitrorhodium(III) bromide
(c) dichlorobis(ethylenediamine)ruthenium(III)
(d) diaquatetrachlororhodate(III)
(e) triamminetribromoplatinum(IV)

19.32 Write the formula of each of the following ions or compounds.
(a) hexaaquachromium(III) hexacyanoferrate(III)
(b) bromochlorobis(ethylenediamine)cobalt(III)
(c) carbonylpentacyanocobaltate(III)
(d) (diethylenetriamine)trinitrochromium(III) (abbreviate the neutral ligand with "dien")
(e) pentaaquathiocyanatoiron(III)

OBJECTIVE Identify isomers in coordination compounds.

19.33 Draw the structures and name the geometric isomers of tetraaquadibromochromium(III).

19.34 Draw the structures and name the geometric isomers of triamminetrichlorocobalt(III).

19.35 Use the following list of complexes to answer the questions. There may be more than one correct choice for each answer.
(1) $[Co(NH_3)_3Cl_3]$
(2) $[Co(en)_2Br_2]^+$
(3) $[Cr(H_2O)_2Cl_2Br_2]^-$
(4) $[Pt(NH_3)_3SCN]^+$
(5) $[Cr(C_2O_4)_3]^{3-}$
(a) Which complex has three isomers, two of which are enantiomers?
(b) Which complex may have linkage isomers?
(c) Which complexes cannot form optically active isomers?
(d) Identify the complex that has the greatest number of possible isomers. Show the structures for all of the isomers.

19.36 ■ In which of the following complexes are geometric isomers possible? If isomers are possible, draw their structures and label them as *cis* or *trans,* or as *fac* or *mer.*
(a) $[Co(H_2O)_4Cl_2]^+$ (b) $[Co(NH_3)_3F_3]$
(c) $[Pt(NH_3)Br_3]^-$ (d) $[Co(en)_2(NH_3)Cl]^{2+}$

19.37 Draw the structure of the following complexes.
(a) *mer*-triamminetribromorhodium(III)
(b) *trans*-dinitrobromochloroplatinate(II)

19.38 Draw the structure of the following complexes.
(a) *fac*-bromotrichloro(ethylenediamine)cobalt(III)
(b) *cis*-dibromotetrachloroferrate(II)

19.39 Identify which of the following ligands could display linkage isomerism: N_3^-, SCN^-, NO_2^-, NCO^-, ethylenediamine.

19.40 What structural feature is used to determine whether a compound can exist as optical isomers?

19.41 What is a racemic mixture?

19.42 What physical property is different for two enantiomers?

OBJECTIVE Predict the magnetic properties of transition-metal complexes using crystal field theory.

19.43 From each pair of complexes, select the one that has the greater crystal field splitting.
(a) $[Co(NH_3)_6]^{3+}$ or $[Co(CN)_6]^{3-}$
(b) $[Cr(H_2O)_6]^{2+}$ or $[Cr(H_2O)_6]^{3+}$
(c) $[Fe(H_2O)_6]^{2+}$ or $[Ru(H_2O)_6]^{2+}$
(d) $[CrF_6]^{3-}$ or $[Cr(H_2O)_6]^{3+}$

19.44 ■ From each pair of complexes, select the one that has the greater crystal field splitting.
(a) $[Rh(NH_3)_6]^{3+}$ or $[Rh(CN)_6]^{3-}$
(b) $[Fe(H_2O)_6]^{2+}$ or $[Fe(H_2O)_6]^{3+}$
(c) $[Co(H_2O)_6]^{3+}$ or $[Rh(H_2O)_6]^{3+}$
(d) $[TiF_6]^{3-}$ or $[Ti(H_2O)_6]^{3+}$

19.45 For each d electron configuration, state the number of unpaired electrons expected in octahedral complexes. Give an example complex for each case. (Two answers are possible for some of these cases.)
(a) d^2 (b) d^4 (c) d^6 (d) d^8

19.46 ■ For a tetrahedral complex of a metal in the first transition series, which of the following statements concerning energies of the 3d orbitals is correct?
(a) The five d orbitals have the same energy.
(b) The $d_{x^2-y^2}$ and d_{z^2} orbitals are higher in energy than the d_{xz}, d_{yz}, and d_{xy} orbitals.
(c) The d_{xz}, d_{yz}, and d_{xy} orbitals are higher in energy than the $d_{x^2-y^2}$ and d_{z^2} orbitals.
(d) The d orbitals all have different energies.

19.47 For each of the following octahedral complexes, give the number of unpaired electrons expected.
(a) $[CrCl_6]^{3-}$
(b) $[Co(CN)_5(H_2O)]^{2-}$
(c) $[Mn(H_2O)_6]^{2+}$
(d) $[Rh(H_2O)_6]^{3+}$
(e) $[V(H_2O)_6]^{3+}$

19.48 ■ For the low-spin complex $[Co(en)(NH_3)_2Cl_2]ClO_4$, identify the following:
(a) the coordination number of cobalt.
(b) the coordination geometry for cobalt.
(c) the oxidation number of cobalt.
(d) the number of unpaired electrons.
(e) whether the complex is diamagnetic or paramagnetic.

19.49 Ni(II) forms the complex $[NiBr_4]^{2-}$, which contains two unpaired electrons. Which of the four coordinate geometries discussed in this chapter is most likely for this complex?

19.50 Pt(II) forms the complex $[PtCl_4]^{2-}$, which contains no unpaired electrons. Which of the four coordinate geometries discussed in this chapter is most likely for this complex?

19.51 Each of the following pairs of complexes contains identical numbers of d electrons, but in each pair, one complex is high-spin and the other is low-spin. In each case, predict which is high-spin and which is low-spin, and give the number of unpaired electrons present in each complex.
(a) $[Cr(H_2O)_6]^{2+}$ and $[Mn(CN)_6]^{3-}$
(b) $[Fe(H_2O)_6]^{2+}$ and $[Ru(H_2O)_6]^{2+}$
(c) $[Co(H_2O)_6]^{2+}$ and $[Co(CN)_5(H_2O)]^{3-}$

19.52 Recently, researchers found a low-spin tetrahedral complex of cobalt(III) in which the ligands are large hydrocarbon groups bonded to the metal through carbon atoms. Construct a crystal field energy-level diagram containing the metal d electrons of this complex. How many unpaired electrons are expected in this complex?

19.53 Give the number of unpaired electrons and the geometry expected for each of the following four-coordinate complexes.
(a) $[Au(CN)_4]^-$ (b) $[CoCl_4]^{2-}$ (c) $[Pd(NH_3)_4]^{2+}$

19.54 ■ The complex $[Mn(H_2O)_6]^{2+}$ has five unpaired electrons, whereas $[Mn(CN)_6]^{4-}$ has only one. Using the ligand field model, depict the electron configuration for each ion. What can you conclude about the effects of the different ligands on the magnitude of Δ?

OBJECTIVE Describe the chemical changes that occur in the reduction of metal ores.

19.55 For each of the following compounds, give the preferred method of reduction for isolating the metal.
(a) LiCl (b) Fe_2O_3

19.56 For each of the following compounds, give the preferred method of reduction for isolating the metal.
(a) Al_2O_3 (b) HgS

Chapter Exercises

19.57 Show that the high-spin complex for the d^5 electron configuration is favored when $P > \Delta$.

19.58 It has been shown experimentally that the compound $[Co(NH_3)_4Br_2]^+$ exists as two (and only two) isomers. Assuming that the compound is octahedral, draw the two isomers. Another possible shape of the compound is a trigonal prism arrangement. Show why the existence of only two isomers rules out the trigonal prism shape. (*Hint:* Draw all possible isomers for the trigonal prism.)

19.59 Use the spectrochemical series, the oxidation state of the metal, and the geometry of the complex to predict the number of unpaired electrons in $[FeCl_4(H_2O)_2]^{2-}$ and $[FeCl_4]^{2-}$ (tetrahedral). Name each complex and draw all possible isomers for each species. Does each possible isomer have the same number of unpaired electrons?

19.60 The ion $[Fe(H_2O)_6]^{3+}$ is a weak Brønsted–Lowry acid in water. Write the equilibrium for this process. Discuss the number of unpaired electrons in the $[Fe(H_2O)_6]^{3+}$ ion and the iron product of the equilibrium reaction. Is it possible to determine the number of unpaired electrons without an experiment?

19.61 The compound $[PdCl_4]^{2-}$ is diamagnetic. Discuss whether this information is sufficient to determine the geometry of the ion.

19.62 ■ A compound is analyzed and found to consist of Co^{3+}, $3Cl^-$, and $4H_2O$ groups. If this compound is shown by electrochemical measurements to consist of ions with 1− and 1+ charges, write the formula and name the compound. How many unpaired electrons, if any, do you expect for this compound?

19.63 Write all the possible isomers of dicyanobis(ethylenediamine)iron(II).

Cumulative Exercises

19.64 Write the equation for the roasting of ZnS. What volume of SO_2 gas at standard temperature and pressure is produced by the roasting of 1000 g ZnS?

19.65 Although copper metal does not dissolve in HCl, it does dissolve in hot, concentrated sulfuric acid. Write the equation for this reaction, and explain this difference in reactivity. What mass of copper can be dissolved by 125 mL of 10.4 molar H_2SO_4?

19.66 Ethylenediamine (en) is a bidentate ligand, and the ligand diethylenetriamine (dien) is the tridentate analog of en, $NH_2CH_2CH_2NHCH_2CH_2NH_2$ (the donor atoms are in color). Draw the Lewis structure of dien and all possible geometric isomers of $[Co(dien)_2]^{3+}$.

19.67 ▲ A brown-yellow complex of vanadium is analyzed and found to contain 49.75% V, 15.62% O, and 34.63% Cl. The substance boils at 127 °C and 1 atm of pressure. What is the empirical formula of the compound? How many electrons are in the vanadium 3*d* orbitals? Predict whether the bonding is mainly covalent or ionic.

19.68 ▲ The oxide MnO_2 reacts with limited amounts of carbon, C(s), to produce Mn(s) and CO(g). Write the balanced equation, and calculate the amount of carbon needed to produce 22.8 L CO(g) at 1.00 atm and 298 K in a reaction with excess MnO_2. Calculate $\Delta G°$, $\Delta H°$, and $\Delta S°$ at 298 K for the reaction. Estimate the minimum temperature needed to make this reaction spontaneous at standard-state conditions ($\Delta G° < 0$).

19.69 ▲ The introduction to this chapter describes the important anticancer agent cisplatin, *cis*-$[Pt(NH_3)_2Cl_2]$. As outlined earlier, the reaction of 1 mol $[PtCl_4]^{2-}$ with 2 mol NH_3 yields this *cis* complex, whereas the reaction of 1 mol $[Pt(NH_3)_4]^{2+}$ and 2 mol Cl^- yields the *trans* isomer. The difference in these reactions has to do with the difference in the *rate* of reaction. Substitution occurs more rapidly for a ligand *trans* to a chloride ion than one *trans* to ammonia. Draw the structure of both $[PtCl_4]^{2-}$ and $[Pt(NH_3)_4]^{2+}$, and sequentially do the substitution reactions for each stepwise, one ligand at a time, always substituting a ligand *trans* to a Cl^- to see how the difference in rates produces the different isomers.

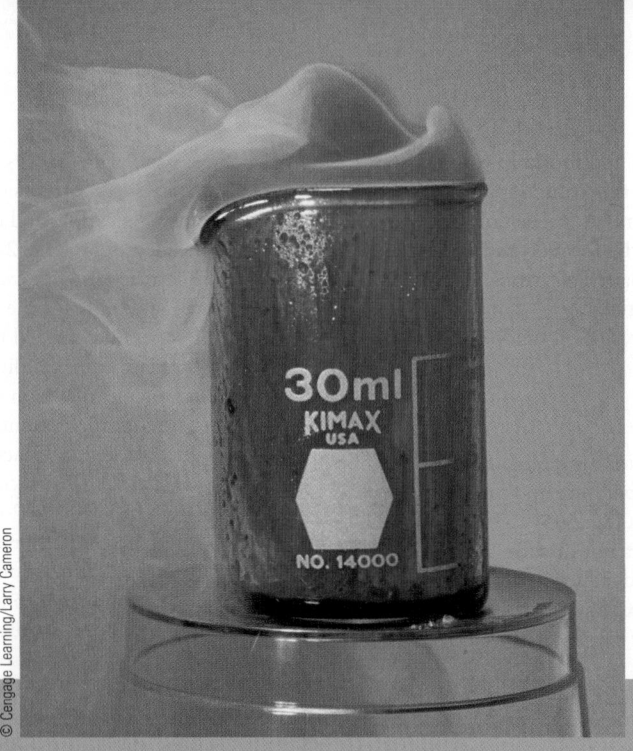

© Cengage Learning/Larry Cameron

An important part of the development of the chemistry of any element or compound is to determine how it reacts with other elements or compounds. Thus, whenever a new compound is made, how it reacts with other substances is closely studied. Although an experienced chemist can frequently predict what the results will be, sometimes the reactions can be unexpected. The photograph shows the results of mixing liquid bromine with gallium metal.

As you can see, a rapid reaction takes place that is so exothermic that the beaker gets very hot and some of the liquid bromine vaporizes. Because reaction rates generally increase as temperature increases, the rate of this reaction

20

Chemistry of Hydrogen, Elements in Groups 3A through 6A, and the Noble Gases

CHAPTER CONTENTS

OWL Online homework for this chapter may be assigned in OWL.

Look for the green colored vertical bar throughout this chapter, for integrated references to this chapter introduction.

increases as the temperature increases. If large amounts of the materials had been used, the high temperature and large reaction rate could have caused a dangerous situation, especially if the reaction had not been properly carried out in a fume hood. The reaction ends, of course, when one of the reactants is consumed. We can conclude from the experiment that the reaction of gallium with bromine is rapid and exothermic. The chemists would then study the properties of any new compounds formed in the reaction to be able to write the balanced equation. Also, the products of the reaction may have interesting and possibly useful properties. ■

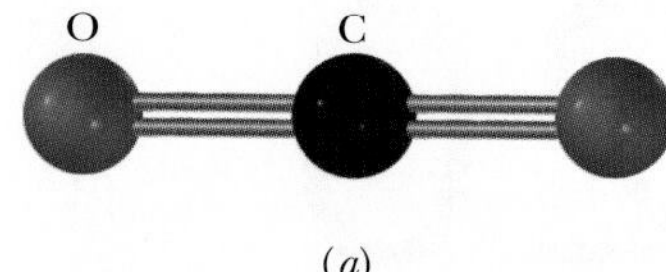

(a)

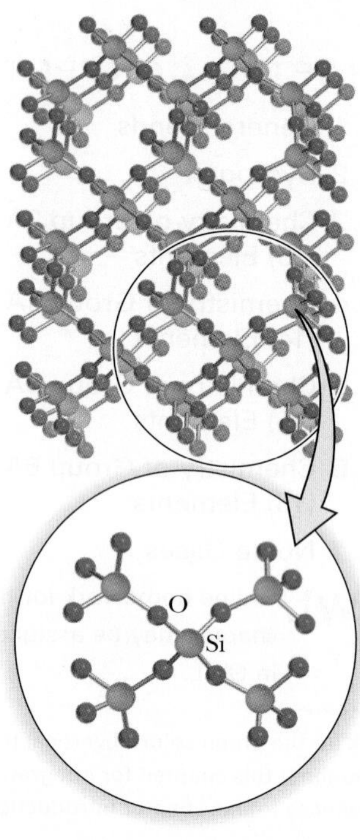

(b)

Figure 20.1 Structures of CO_2 and SiO_2. *(a)* Carbon dioxide is a triatomic molecule containing two σ and two π bonds. *(b)* Silicon dioxide exists as a covalent network solid with each silicon (green) forming σ bonds to four oxygen atoms.

The atomic properties, such as electronegativity, atomic and ionic radii, and ionization energy, influence the chemistry of the representative elements. This chapter surveys the properties and chemical reactivity of hydrogen, the elements in Groups 3A through 6A, and the noble gases. Chapter 8 presents a similar survey of Groups 1A, 2A, and 7A.

20.1 General Trends

OBJECTIVE

- ☐ Discuss how and why the properties of the second-period elements are different from those of other elements in their groups

The nonmetallic elements, with the exception of hydrogen, are located in the upper right portion of the periodic table. The chemistry of these elements is controlled by relatively high electronegativities and ionization energies. The nonmetals form ionic compounds with metals and form covalent compounds with each other. Within any group, there is usually a large difference between the chemistry of the second-period element and that of the remaining members of the group. High electronegativity is particularly important for the nonmetallic elements of the second period. Fluorine is the most electronegative element, and oxygen is second. The nonmetallic elements of the second period have the smallest radii in their respective groups (helium is the smallest noble gas).

An important consequence of the small sizes of the elements in the second period is the tendency to form strong π bonds from the sideways overlap of *p* orbitals. Elements of the third and higher periods are too large to form strong π bonds from extensive *p*-orbital overlap. Because of the weakness of the π bond, these heavier elements tend to form two σ bonds rather than one σ bond and one π bond. This point is demonstrated by a comparison of the structures of CO_2 and SiO_2. Carbon dioxide is a molecular compound with strong π bonding, whereas SiO_2 is a covalent network solid in which each silicon atom forms four σ bonds to four bridging oxygen atoms in an extended array (Figure 20.1). The silicon in SiO_2 does not form π bonds as observed for carbon in CO_2, because only second-period elements form strong π bonds. The formation of four σ bonds is a more stable bonding arrangement for this third-period element.

The difference in the importance of π bonding is also noticeable in the elemental forms of nitrogen and phosphorus. The stable form of nitrogen is N_2, a molecule containing a triple bond (one σ and two π bonds). The simplest form of phosphorus is P_4, in which the atoms are arranged at the vertices of a tetrahedron. In this arrangement, each phosphorus atom forms three σ bonds. The other stable forms of elemental phosphorus also contain only σ bonds.

Another major distinction of the second-period elements is that they do not form compounds in which the Lewis structures would place more than eight electrons around a central atom. For example, in Group 6A, only one type of compound is formed from the combination of one oxygen atom with fluorine atoms, OF_2, but three compounds (two have expanded valence shells) can form from the combination of one sulfur atom with fluorine atoms, SF_2, SF_4, and SF_6 (Figure 20.2). As a Group 6A element from the second period, oxygen has six valence electrons and four valence orbitals, and thus can share only two additional electrons with fluorine atoms. The oxygen atom in OF_2 is sp^3 hybridized. The valence shell of sulfur is the $n = 3$ level, and the valence orbitals can include the 3*d* as well as the 3*s* and 3*p* orbitals. The six valence electrons can be used to form up to six electron-pair bonds. In SF_2, the sulfur uses sp^3 hybrid orbitals; in SF_4, sp^3d hybrid orbitals; and in SF_6, sp^3d^2 hybrid orbitals. As outlined in Chapter 9, recent calculations have questioned the extent to which *d* orbitals are used in expanded valence shell molecules. Chemists continue to debate this issue, but there is no doubt that the chemistry of the third-row element sulfur is quite different from the second-row element oxygen.

The chemistry of the second-period elements is different from that of other elements in their groups because of their relatively small size and high electronegativity and the availability of only four valence orbitals.

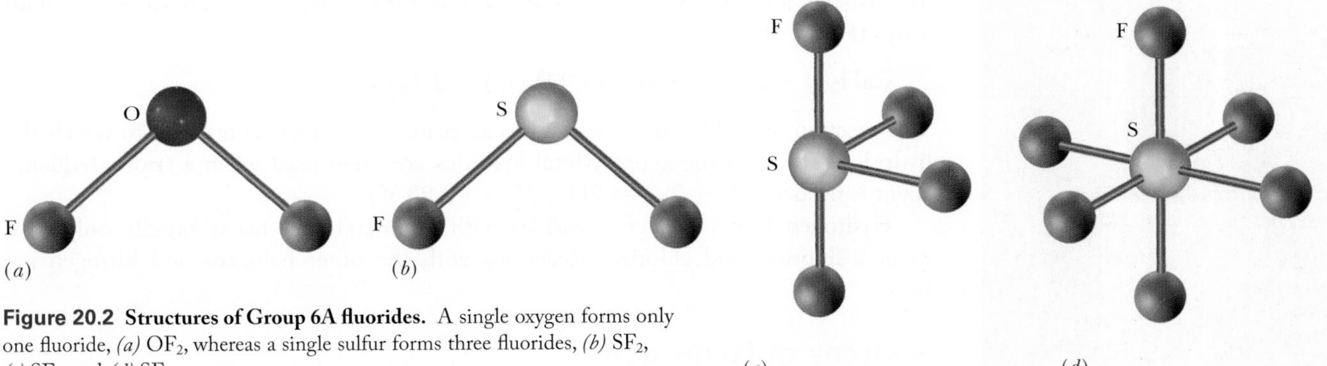

Figure 20.2 Structures of Group 6A fluorides. A single oxygen forms only one fluoride, *(a)* OF_2, whereas a single sulfur forms three fluorides, *(b)* SF_2, *(c)* SF_4, and *(d)* SF_6.

Another important general trend in the elements of Groups 3A through 6A is that metallic character increases down the group. This trend is expected from the observed decreases in ionization energies and electronegativities down a group. Metals tend to form cations, a property favored by low ionization energies.

OBJECTIVE REVIEW *Can you:*

- ☑ discuss how and why the properties of the second-period elements are different from those of other elements in their groups?

20.2 Hydrogen

OBJECTIVE

- ☐ Describe the properties, sources, and important uses of hydrogen

Although hydrogen is generally listed in Group 1A (1) on a periodic table, it should probably be considered in a group by itself. It is not surprising that the lightest element has unique properties. Hydrogen can lose an electron to form a proton [$H^+(aq)$ in water]; in combination with electropositive metals, it can gain an electron to form the hydride ion, H^-, which has the electron configuration of the noble gas helium. With an electronegativity near the middle of the scale (2.1), hydrogen can also form strong covalent bonds with the other nonmetallic elements.

Hydrogen is the most abundant element in the universe. It is the nuclear fuel consumed by the sun in its production of energy. In contrast, hydrogen makes up only 0.87% of the mass of Earth's crust. Hydrogen has three isotopes: ^1H (99.985% abundant) and ^2H (frequently called *deuterium,* 0.015% abundant) are stable, and ^3H (frequently called *tritium*) is radioactive and rare. The molecular form of the element is H_2, a tasteless and odorless gas. Because it is nonpolar and has a small, nonpolarizable electron cloud, H_2 has weak intermolecular forces and boils at -253 °C. The gas has a low density compared with air, and it was used to lift several transatlantic passenger airships during the early 1930s. Unfortunately, one of the H_2 filled airships, the *Hindenburg,* caught fire on landing at Lakehurst, New Jersey, in 1937 (Figure 20.3). The *Hindenburg* caught fire because hydrogen reacts rapidly when mixed with oxygen in the presence of a source of ignition such as a spark. Hydrogen gas and reactions that produce it must be handled carefully.

Figure 20.3 The *Hindenburg* fire. The hydrogen gas in the airship *Hindenburg* caught fire while docking at Lakehurst, New Jersey, in 1937.

Although the famous fire of the *Hindenburg* would indicate that hydrogen is very reactive, reactions of H_2 with most substances at room temperature are relatively difficult to initiate, mainly because of the strong H–H bond (with a bond energy of 436 kJ/mol). Hydrogen gas does react with the highly electropositive Group 1A metals and calcium, strontium, and barium to form ionic hydrides.

$$2Na(s) + H_2(g) \rightarrow 2NaH(s)$$

$$Ba(s) + H_2(g) \rightarrow BaH_2(s)$$

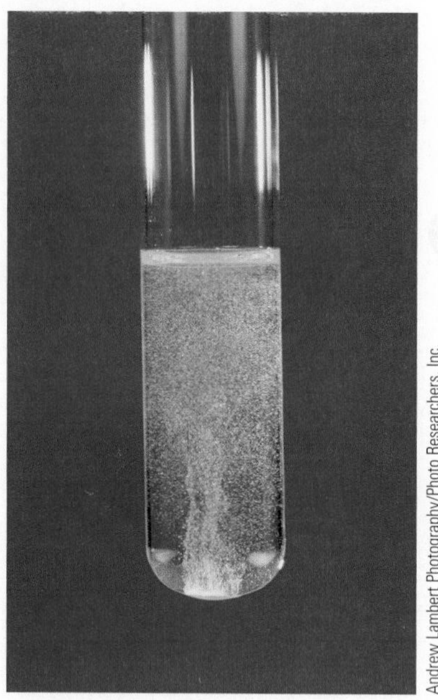

Figure 20.4 Reaction of calcium hydride and water. Calcium hydride reacts vigorously with water, producing hydrogen gas.

Hydrogen can be produced by reactions of either hydrocarbons (mainly CH_4) or carbon with water.

Figure 20.5 Reaction of zinc and HCl. Zinc metal reacts with HCl(aq) to produce hydrogen gas.

The hydride ion, H^-, in these salts is a strong Lewis base and reacts vigorously with water (Figure 20.4).

$$CaH_2(s) + 2H_2O(\ell) \rightarrow Ca(OH)_2(aq) + 2H_2(g)$$

This reaction of CaH_2 can be viewed as an oxidation-reduction reaction in which the hydride is the reducing agent. Metal hydrides are often used when a strong reducing agent is needed ($H_2 + 2e^- \rightarrow 2H^-$; $E^\circ = -2.23$ V).

Hydrogen forms covalent hydrides with nonmetals but reacts rapidly only with oxygen, fluorine, and chlorine. Reactions with the other halogens and nitrogen are slow.

Sources of Hydrogen

Chemists generally prepare small amounts of hydrogen by using the reaction of hydrochloric or sulfuric acid and zinc (Figure 20.5). However, these reactants are too expensive for industrial use.

$$Zn(s) + 2HCl(aq) \rightarrow H_2(g) + ZnCl_2(aq)$$

Hydrogen is an important industrial chemical. Currently, the major commercial source of hydrogen is the high-temperature reaction of methane and steam.

$$CH_4(g) + H_2O(g) \rightarrow 3H_2(g) + CO(g)$$

Hydrogen also forms in the reaction of red-hot carbon (from coal) with steam.

$$C(s) + H_2O(g) \rightarrow H_2(g) + CO(g)$$

The equimolar mixture of hydrogen and carbon monoxide formed in this reaction is known as *synthesis gas* (also known previously as water gas). It was used extensively as a fuel in the late 19th and early 20th centuries, in much the same way that natural gas is used today. Natural gas is safer, because the carbon monoxide in synthesis gas is toxic.

The carbon monoxide produced in either of these reactions can react further with water to produce additional hydrogen.

$$CO(g) + H_2O(g) \rightarrow H_2(g) + CO_2(g)$$

This reaction is known as the **water-gas shift reaction.** Note that in the synthesis of hydrogen from either methane or carbon followed by the water-gas shift reaction, large amounts of carbon dioxide are produced. Carbon dioxide, although not poisonous, is a "greenhouse" gas that is involved in global warming (see Section 5.5). It is likely that any new plants built in the future that use coal to produce hydrogen using the two above reactions would include technology to capture and sequester underground the CO_2 produced.

Hydrogen is also formed by the electrolysis of water, the other product being oxygen. Unfortunately, this clean method of preparing hydrogen is expensive because of the high cost of electricity. If a method could be developed to produce hydrogen inexpensively from water (sunlight or wind would be good sources of the energy), it would be an extremely clean form of energy because the only product of its combustion is water. A large-scale economic system based on hydrogen as an energy source is still a hope for the future. In addition to the problems associated with producing the hydrogen (no "natural" source of hydrogen exists like there is for natural gas or oil), it is difficult to store. It has a low molecular mass so it diffuses readily through anything but the strongest containers, and its low boiling point makes it difficult to store in the liquid state the way natural gas can be stored.

Uses of Hydrogen

The largest single commercial use of hydrogen is in the synthesis of ammonia by the Haber process.

$$N_2(g) + 3H_2(g) \rightleftharpoons 2NH_3(g)$$

More than 20 million tons of ammonia are prepared in the United States annually. Most of the ammonia is used as fertilizer, either directly or after conversion to other compounds.

Hydrogen is used in the syntheses of ammonia and methanol.

Hydrogen is also used in the synthesis of methanol.

$$2H_2(g) + CO(g) \rightarrow CH_3OH(\ell)$$

Methanol is used as a solvent and an additive in gasoline. Recently, a process was developed to convert methanol to gasoline, which is a mixture of hydrocarbons. Thus, coal, a source of carbon, can be converted to water gas and then to methanol, and the methanol can then be converted to gasoline. The synthesis of methanol requires a 2:1 ratio of H_2 to CO, and the additional H_2 needed can be obtained from the water-gas shift reaction. Because coal is abundant in the United States, gasoline from this process could replace petroleum as an energy source. With the increased cost of crude oil, gasoline made from coal is now potentially competitive with gasoline refined from crude oil, but there are also a number of environmental problems (buildup of CO_2 gas in the atmosphere and problems with coal mining). Nevertheless, the conversion of coal to a liquid hydrocarbon fuel may be important in the future.

Hydrogen is also used to hydrogenate some of the double bonds in vegetable oils.

$$\sim\overset{H}{\underset{}{C}}=\overset{H}{\underset{}{C}}\sim + H_2 \longrightarrow \sim\overset{H}{\underset{H}{C}}-\overset{H}{\underset{H}{C}}\sim$$

The reaction converts liquid oils to solid cooking fats such as margarine. Fats with double bonds are known as *unsaturated fats* and are generally liquids, whereas the solid fats are mostly saturated. Although solid fats were widely used for fried foods because they are more stable at high temperatures, it has been shown that consuming saturated fats can cause health problems.

OBJECTIVE REVIEW *Can you:*

☑ describe the properties, sources, and important uses of hydrogen?

20.3 Chemistry of Group 3A (13) Elements

OBJECTIVES

☐ Discuss the inert pair effect

☐ Describe the isolation, purification, and fundamental chemistry of boron and aluminum

H							He
Li	Be	B	C	N	O	F	Ne
Na	Mg	Al	Si	P	S	Cl	Ar
K	Ca	Ga	Ge	As	Se	Br	Kr
Rb	Sr	In	Sn	Sb	Te	I	Xe
Cs	Ba	Tl	Pb	Bi	Po	At	Rn

Group 3A (13) elements have the valence electron configuration ns^2np^1. They generally form compounds in which the element has an oxidation number of +3, but heavier members of the family, especially thallium, also form many compounds in the +1 oxidation state. The trend for the heavier members of the group to use only the *p* valence electron(s) in forming compounds is general and is observed in Groups 4A and 5A as well. This tendency for the heavier members of Groups 3A to 5A not to use the pair of valence *s* electrons for bonding is called the **inert pair effect.** The origins of the inert pair effect are complicated, but it does not arise from simple differences in ionization energies. The main reason for the effect appears to be that these large elements form weaker bonds; thus the energy needed to use the *s* electrons in bonding is not returned by making the bonds.

The heavier members of Groups 3A to 5A form some compounds in which the pair of valence *s* electrons is not used for bonding.

All of the elements of the group are metals with the exception of boron, which is a metalloid. Group 3A elements generally occur in nature as oxides. In contrast with Groups 1A and 2A, these elements become less reactive toward the bottom of the group.

Science VU/Visuals Unlimited, Inc.

(*a*)

(*b*)

Figure 20.6 Borax. *(a)* View of a borax mine in the Mojave Desert. *(b)* Borax has a variety of uses in the home.

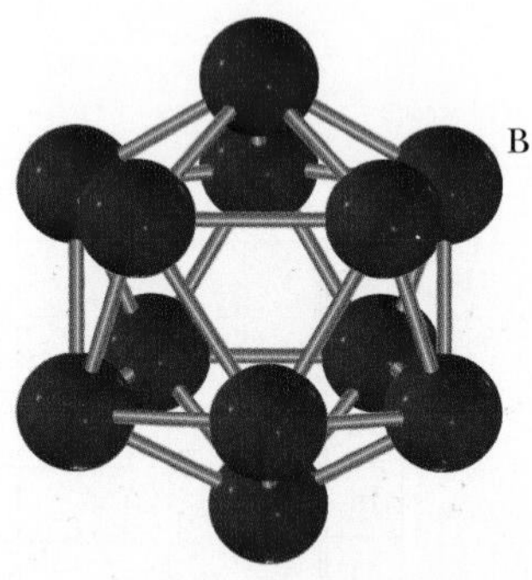

Figure 20.7 An icosahedron of boron atoms. The elemental forms of boron contain icosahedral arrays of atoms.

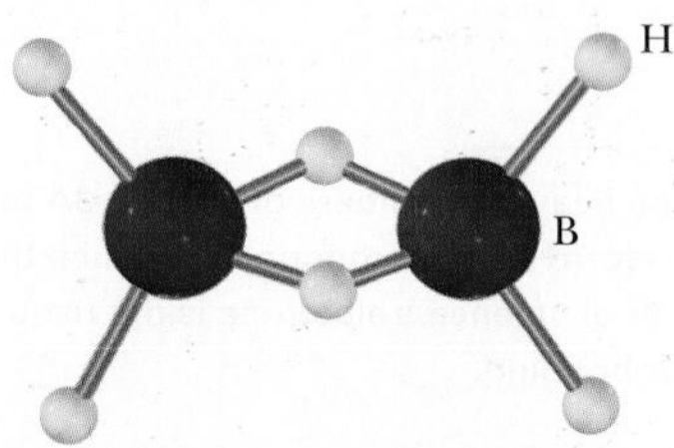

Figure 20.8 Structure of B_2H_6. In the structure of B_2H_6 two of the hydrogen atoms bridge the boron atoms and four form normal terminal bonds.

Boron

Boron occurs in nature in combination with oxygen. Although the element is extremely rare in Earth's crust, the boron-containing mineral borax is found in high concentrations in the Mojave Desert in California (Figure 20.6). The formula of borax was generally written as $Na_2B_4O_7{\cdot}10H_2O$ until its structure was found to be $Na_2B_4O_5(OH)_4{\cdot}8H_2O$. It has been mined for years for use as a water softener (it precipitates Ca^{2+} and Mg^{2+}) and in cleaning products it forms weakly basic solutions of pH $\approx$ 9.

The pure element is difficult to prepare. A form of low-purity boron can be prepared by the reduction of B_2O_3 with magnesium.

$$B_2O_3(s) + 3Mg(s) \rightarrow 2B(s) + 3MgO(s)$$

High-purity boron can be prepared by the high-temperature reduction of BBr_3 by hydrogen in the presence of a solid catalyst.

$$2BBr_3(g) + 3H_2(g) \rightarrow 2B(s) + 6HBr(g)$$

Boron exists in a number of allotropic forms that all contain an unusual arrangement of the boron atoms, an **icosahedron,** which is a regular polyhedron with 20 faces and 12 vertices (Figure 20.7). The icosahedra are connected differently in the various allotropic forms, but all have extended bonding arrangements between the icosahedra. As a result of this stable arrangement, several of the elemental forms of boron produce very hard crystals.

Treatment of purified borax with sulfuric acid produces boric acid, H_3BO_3. Its formula is frequently written as $B(OH)_3$ to reflect its molecular structure. In the solid, it exists as sheets containing trigonal planar $B(OH)_3$ groups, held together by hydrogen bonds to oxygen atoms in other boric acid molecules. The electron-deficient, sp^2-hybridized boron atom in boric acid acts as an acid that reacts with water to form weakly acidic solutions containing the $B(OH)_4^-$ ion. These solutions have antiseptic qualities and are used as eyewashes.

$$B(OH)_3 + 2H_2O \rightleftharpoons B(OH)_4^- + H_3O^+ \qquad K_a = 7.3 \times 10^{-10}$$

Heating boric acid causes its dehydration to B_2O_3. This oxide is used extensively in the manufacture of borosilicate glass (see Section 20.4) and in the preparation of elemental boron.

The boron trihalides are interesting nonpolar molecules in which the boron atom is also sp^2 hybridized. Although only six electrons are about the boron atom in these BX_3 derivatives, it is believed that the lone pairs on the halogen atoms interact with the empty p orbital on boron to help stabilize the compounds. The boron trihalides also act as Lewis acids to form adducts with neutral donor molecules, such as NH_3.

$$BCl_3 + {:}NH_3 \rightarrow Cl_3B - NH_3$$

In addition, they can react with anionic donors to form ions such as BF_4^-. In these adducts, the boron atom uses sp^3 hybrid orbitals that are directed at the corners of a tetrahedron.

$$BF_3(aq) + HF(aq) \rightarrow H^+(aq) + BF_4^-(aq)$$

Another important example of a tetrahedral anion of boron is BH_4^-. Sodium borohydride, $NaBH_4$, is a reducing agent used in industry and in research laboratories.

Boron Hydrides

Boron forms an extremely interesting series of binary compounds with hydrogen. The simplest member of the series might be expected to be BH_3, in which boron uses each of its three valence electrons to form a σ bond with each of the three hydrogen atoms. At high temperatures, BH_3 can be observed in the gaseous state by mass spectrometry, but it dimerizes at room temperature to form diborane, B_2H_6. Diborane is prepared by the reaction of $NaBH_4$ with I_2 and has the interesting structure shown in Figure 20.8.

$$2NaBH_4 + I_2 \rightarrow B_2H_6 + 2NaI + H_2$$

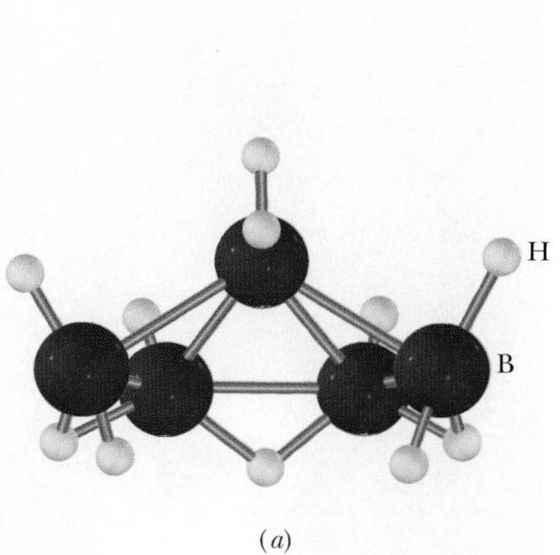

(a)

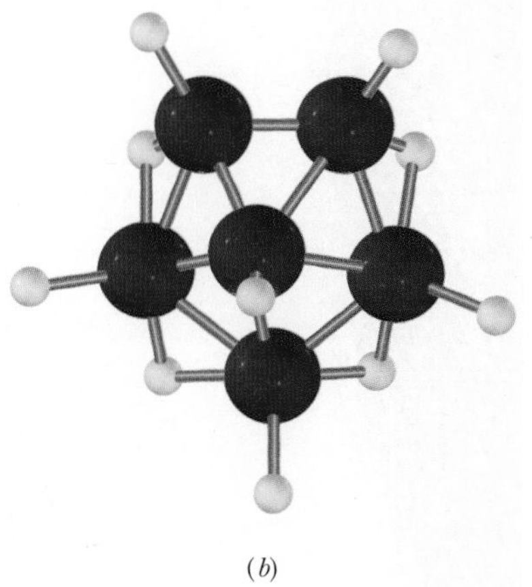
(b)

Figure 20.9 Structures of B_5H_{11} and B_6H_{10}. Both *(a)* B_5H_{11} and *(b)* B_6H_{10} contain three-center bonds, and the basic structures can be viewed as part of the icosahedron in Figure 20.7.

If BH_3 were a monomer, it would have an empty valence orbital and only six valence electrons about boron. To use all four valence orbitals and attain an octet of electrons, it forms the dimer. The bonding in this dimer can be explained as two sp^3-hybridized boron atoms, each of which forms two normal two-center, two-electron bonds with two of the hydrogen atoms and two three-center, two-electron bonds with the other boron atom and the bridging hydrogen atoms. In a **three-center, two-electron bond,** three orbitals (one from each atom) overlap to form an orbital that contains two electrons. Like normal two-center bonds, this three-center bond contains two electrons. In B_2H_6, each boron atom uses all four valence orbitals and attains an octet of electrons.

Delocalized three-center bonds are used to describe the bonding in the boron hydrides.

Because three-center bonding allows boron to attain an octet of electrons, it often occurs in boron compounds. Figure 20.9 shows the structures of two other boron hydrides, B_5H_{11} and B_6H_{10}. In these compounds, three-center bonds can also be formed by three adjacent boron atoms. Interestingly, the basic arrangement of the boron atoms in these two compounds, as in most of the boron hydrides, consists of pieces of the icosahedron pictured in Figure 20.7. For example, the arrangement of the boron atoms in B_6H_{10} is the same as that of the top six atoms in the icosahedron in Figure 20.7.

These boron hydrides ignite spontaneously in air, giving a green flame (Figure 20.10). The boron hydrides were considered as possible rocket fuels because of their high enthalpy of combustion (shown in the following equation), but they are expensive to produce, the reactions are hard to control, and the resulting B_2O_3 would cause damage to the engines.

$$B_2H_6(g) + 3O_2(g) \rightarrow B_2O_3(s) + 3H_2O(g) \qquad \Delta H = -2034 \text{ kJ}$$

Aluminum

Aluminum is the most abundant metal and the third most abundant element in Earth's crust. Metallic aluminum, generally in alloys with silicon and copper or magnesium, is an important structural material in aircraft because of its low density. Aluminum protects itself from corrosion by forming a thin, strongly adhering, protective coating of the inert oxide Al_2O_3.

$$4Al(s) + 3O_2(g) \rightarrow 2Al_2O_3(s)$$

The formation of Al_2O_3 from the elements is so exothermic that powdered aluminum is used as a solid rocket fuel. Aluminum is also a good conductor of electricity and is used in overhead power lines because it is less dense than the more highly conducting copper.

Aluminum is isolated from the ore bauxite, a hydrated oxide ($Al_2O_3{\cdot}xH_2O$), by electrolysis, in a process that Charles Hall developed just after his graduation from Oberlin College (see Section 18.8). Bauxite ores contain small amounts of Fe_2O_3 and SiO_2 that must be separated before electrolysis. The basis for this separation, known as the Bayer process, takes advantage of the amphoteric properties of Al_2O_3. The Al_2O_3

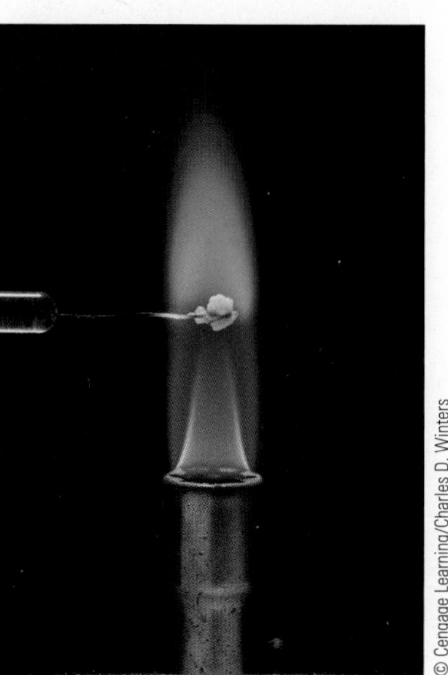

Figure 20.10 Flame test for boron compounds. Compounds containing boron burn with a green flame.

(*a*)

(*b*)

(*c*)

Figure 20.11 Thermite reaction. The reaction of aluminum metal with iron(III) oxide is extremely exothermic, producing molten iron. *(a)* Aluminum metal and iron oxide are placed in a clay pot, and the reaction is initiated by addition of a piece of burning magnesium. *(b)* Once the reaction is started, it is rapid and exothermic. *(c)* The violence and heat of the reaction break the clay pot that has initially held the reactants, and the molten iron produced melts through the sheet of iron placed below the pot.

and SiO_2 are dissolved in a strong base, and the solid Fe_2O_3 is removed by filtration. The solution is then acidified, and the aluminum oxide precipitates, leaving the silicon in solution as silicates (silicates are discussed in Section 20.4).

Aluminum reacts with transition-metal oxides in extremely exothermic reactions. The reaction with iron(III) oxide is known as the *thermite reaction* (Figure 20.11).

$$2Al(s) + Fe_2O_3(s) \rightarrow Al_2O_3(s) + 2Fe(\ell) \qquad \Delta H = -852 \text{ kJ}$$

The thermite reaction is so exothermic that it produces molten iron that can be used to weld iron and steel.

Aluminum oxide, also known as alumina, has many important industrial uses. It exists in a number of different forms. The α-alumina form is a very hard substance known as corundum, which is used as an abrasive. Several gemstones, including rubies (Figure 20.12), are clear crystals of α-alumina that contain metal ion impurities.

Figure 20.12 Aluminum oxides. Bauxite is a white hydrated ore of aluminum oxide from which aluminum metal is isolated. A ruby is α-alumina colored red by Cr^{3+} impurities.

A less dense and more reactive form of Al_2O_3 is γ-alumina, which is used as a support in chromatographic separations and as a heterogeneous catalyst or catalyst support for many chemical reactions. Because heterogeneous catalysis occurs on the surface of a solid, a high surface-to-volume ratio makes γ-alumina desirable in these applications.

Mixing Al_2O_3 with sulfuric acid produces aluminum sulfate, which is used to strengthen paper.

$$Al_2O_3(s) + 3H_2SO_4(aq) \rightarrow Al_2(SO_4)_3(aq) + 3H_2O(\ell)$$

Alumina also dissolves in base to form the aluminate anion.

$$Al_2O_3(s) + 2OH^-(aq) + 3H_2O(\ell) \rightarrow 2[Al(OH)_4]^-(aq)$$

A mixture of $Al_2(SO_4)_3$ and $Na[Al(OH)_4]$ produces insoluble $Al(OH)_3$, which on precipitation is used to remove impurities from water. The gelatinous $Al(OH)_3$ adsorbs dissolved impurities (adsorption is the process by which a substance adheres to the surface of a solid) and carries small, suspended solid particles with it as it precipitates.

Another aluminum compound familiar to many people is the antiperspirant "aluminum chlorohydrate." This compound is actually aluminum hydroxychloride, $Al_2(OH)_5Cl{\cdot}2H_2O$. It is an astringent; that is, it contracts the pores of the skin.

Aluminum chloride, $AlCl_3$, has a solid-phase structure in which aluminum is six-coordinate. It sublimes at increased temperatures to form molecular Al_2Cl_6. The bromide Al_2Br_6 and the iodide Al_2I_6 are molecular in the solid phase, as well as in the gas phase. The dimers have structures similar to that of B_2H_6 (Figure 20.13).

Although the structures of Al_2Cl_6 and B_2H_6 are similar, the two-electron, three-center bonding description used for B_2H_6 is not needed for Al_2Cl_6. The bonding in Al_2Cl_6 can be viewed simply as a Lewis acid-Lewis base interaction of a lone pair on the bridging chlorine atoms of each $AlCl_3$ unit with the empty orbital on each aluminum atom. This type of bonding interaction is common in metal halides.

Figure 20.13 Structure of Al_2Cl_6. Al_2Cl_6 is a dimer in the gas phase.

Gallium, Indium, and Thallium

Gallium and indium are not abundant but are becoming increasingly important because gallium arsenide (GaAs) and indium phosphide (InP) are useful semiconductor materials. Metallic gallium is unique in having a large liquid range, from 30 °C to 2403 °C, and it will melt in your hand (Figure 20.14).

All three elements are isolated as by-products of the refining of other metals. Gallium is recovered from the refining of aluminum, indium from the purification of zinc, and thallium from the smelting of lead.

OBJECTIVES REVIEW *Can you:*

- ☑ discuss the inert pair effect?
- ☑ describe the isolation, purification, and fundamental chemistry of boron and aluminum?

20.4 Chemistry of Group 4A (14) Elements

OBJECTIVES

- ☐ Describe the bonding and properties of three allotropic forms of carbon
- ☐ Discuss the occurrence and chemistry of silicon
- ☐ Describe the properties of semiconductors and how doping influences their properties.

H							He
Li	Be	B	C	N	O	F	Ne
Na	Mg	Al	Si	P	S	Cl	Ar
K	Ca	Ga	Ge	As	Se	Br	Kr
Rb	Sr	In	Sn	Sb	Te	I	Xe
Cs	Ba	Tl	Pb	Bi	Po	At	Rn

Group 4A (14) elements have the electron configuration ns^2np^2. With four valence electrons and four valence orbitals, these elements generally attain an octet by forming four covalent bonds. As with the Group 3A elements, the heavier members of the group, tin and especially lead, exhibit the *inert pair effect* and form compounds in the

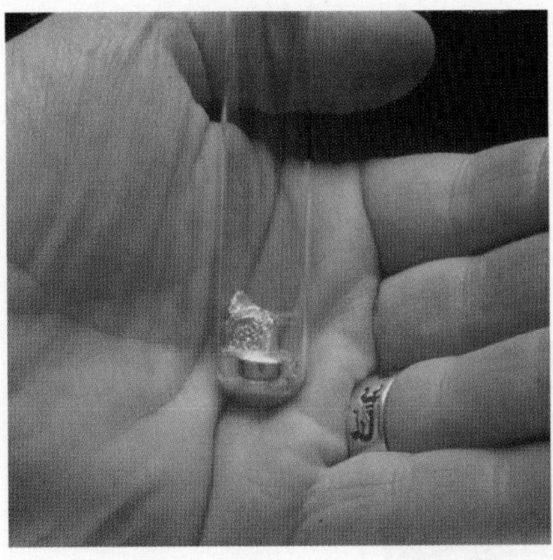

(*a*)

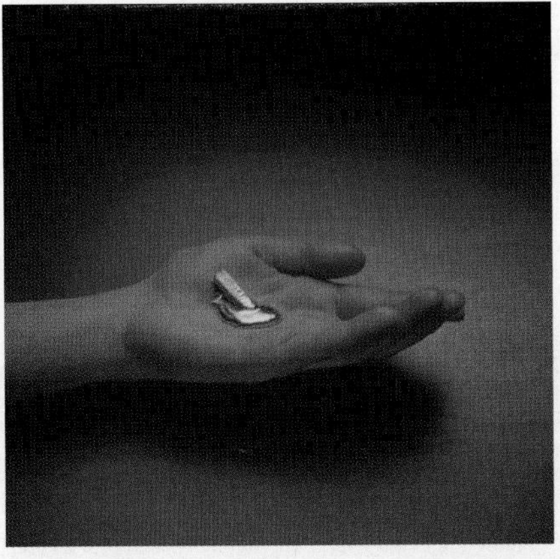

(*b*)

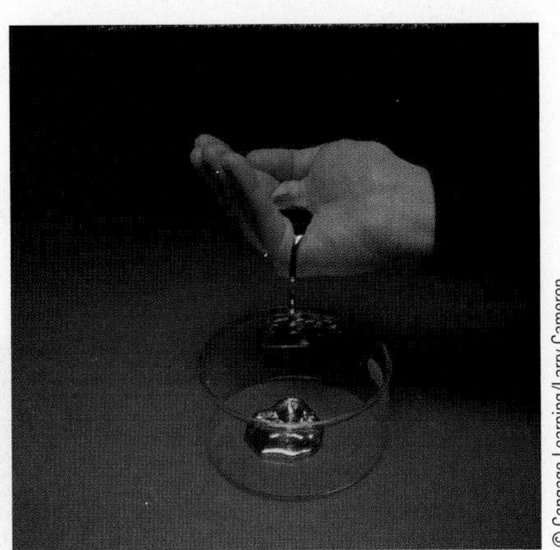

(*c*)

Figure 20.14 Gallium. Gallium metal *(a)* warmed to body temperature *(b)* melts *(c)*.

Philip Wallick/Photolibrary

© Cengage Learning/Charles D. Winters

Figure 20.15 Uses of graphite. Graphite composites are used to make Stealth fighter planes and sports equipment because they are strong and very lightweight.

+2 oxidation state, as well as in the +4 oxidation state. The elements span the entire range of properties from carbon, a typical nonmetal, to metallic lead. Silicon and germanium are metalloids. Tin exists in two allotropic forms, white and gray tin. White tin is a metal and is the form used in plating "tin" cans; gray tin is a nonmetallic form that is quite brittle and is not an electrical conductor.

Carbon

Carbon is distributed widely in Earth's crust, mostly as the calcium and magnesium salts of the carbonate ion, CO_3^{2-}. The matter that makes up living organisms contains a high percentage of carbon, as do the fossil fuels—oil, coal, and natural gas. Millions of compounds of carbon have been isolated from plants and animals or synthesized in laboratories. Chapter 22 is an introduction to the chemistry of carbon compounds.

Carbon is also found as the free element. Graphite, composed of sheets of sp^2-hybridized carbon atoms (see Figure 11.17*b*), is the stable form of the element at room temperature and pressure. Graphite has a high melting point and is used to make molds for casting metals. It is a reasonable conductor of electricity and is used as an electrode material in many industrial electrolytic processes, such as the production of aluminum. It is also the "lead" in lead pencils (no elemental lead is present). Recently, high-strength, lightweight materials have been prepared from graphite fibers mixed with plastics. These "composite" materials have uses ranging from the shell of the Stealth fighter plane to high-quality sports equipment (Figure 20.15).

Charcoal and carbon black are finely divided forms of graphite. Charcoal (particularly "activated charcoal," formed by heating charcoal with steam or CO_2) has a large surface area per unit volume. It is used in the purification of water and other liquids and gases because it efficiently adsorbs impurities (Figure 20.16). Carbon black and other amorphous forms of carbon are used to reinforce rubber and are the reason tires are black.

© 1998 Paul Silverman/Fundamental Photographs, NYC

Figure 20.16 Activated charcoal. Activated charcoal is used as part of this water purification system because it adsorbs impurities.

PRACTICE OF CHEMISTRY
Buckminsterfullerene: Tough, Pliable, and Full of Potential

Buckminsterfullerene is a spherical cluster of 60 carbon atoms arranged in a series of five- and six-membered rings to form a soccer-ball shape (see Figure 20.17). The unusual name comes from the American architect Buckminster Fuller, who designed geodesic dome structures with a similar shape.

Buckminsterfullerene, frequently called *buckyball,* is an allotrope of carbon. Until its serendipitous discovery at Rice University in 1985 by Harold W. Kroto (visiting from Sussex University), Richard Smalley, and their colleagues, only two common allotropes of carbon were known to exist: graphite and diamond. Buckyballs can be made by vaporizing graphite rods in a helium atmosphere, using a high-current electric arc. It is a black, powdery material that can be dissolved in solvents such as benzene, forming deep magenta solutions. In addition to C_{60}, a whole series of additional allotropes of carbon called *fullerenes,* such as C_{70}, also form in the soot formed by vaporizing the rods. Buckytubes can also form, which have interesting properties that give them the potential to be used in important new devices.

Buckminsterfullerene is a surprisingly tough and resilient molecule. It can be accelerated to 15,000 miles per hour and slammed against steel surfaces without suffering damage, a property unknown in other molecular particles. Buckyballs can be compressed to less than 70% of their initial volume without destroying the carbon cage. Buckminsterfullerene is also stable toward heating; one method of purification is to sublime the crude material at 600 °C under vacuum. Unlike diamond and graphite, which exist as covalent network solids, buckminsterfullerene is a discrete molecule. ∎

Buckminsterfullerene. Buckminsterfullerene dissolves in benzene, forming a deep magenta solution.

Diamond is an extremely hard allotrope of carbon formed at high pressures and temperatures, in which the carbon atoms are sp^3-hybridized (see Figure 11.17*a*). Diamond is the more thermodynamically stable form at high pressures because diamond is denser than graphite. In fact, graphite can be converted to diamond by applying high pressures and temperatures. Because diamonds are extremely hard, they are used in cutting and drilling tools. Larger, more nearly perfect crystals of diamonds are valued for their beauty.

Graphite and diamond are two allotropes of carbon in which the carbon atoms are sp^2- and sp^3-hybridized, respectively.

A whole series of interesting new allotropes of carbon have recently been isolated from the carbon dust formed by heating graphite at high temperatures. The main allotrope, with the formula C_{60}, is named *buckminsterfullerene.* Its shape is similar to the surface of a soccer ball (Figure 20.17), and its nickname is "buckyball." Buckyball is just the first of a whole series of new allotropes of carbon, called *fullerenes,* that can be formed by heating graphite. Scientists are studying the properties and potential uses of these unusual species.

Buckyball, C_{60}, is an allotrope of carbon whose shape is similar to the surface of a soccer ball. It is the parent compound of a whole series of fullerenes.

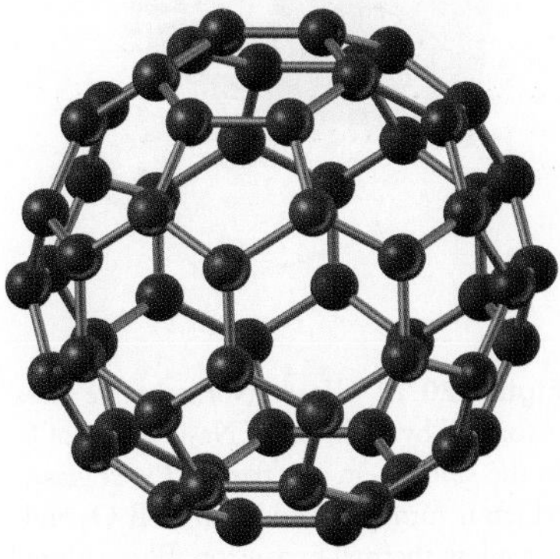

Figure 20.17 Buckminsterfullerene. Buckminsterfullerene is an allotrope of carbon that was discovered in 1985.

Silicon

Silicon is the second most abundant element on Earth. It is not found as the free element in nature but in minerals called **silicates,** compounds containing silicon together with oxygen and various metals. Many different types of silicates exist with different ratios of silicon to oxygen, but in all forms the silicon atoms are in the +4 oxidation state and are tetrahedrally bonded to oxygen atoms (Figure 20.18).

Silicon is also found as silica, SiO_2, which is the main component of common beach sand. We have already seen (see Figure 20.1) that silica is a covalent network solid with silicon in the center of a tetrahedral arrangement of oxygen atoms. It does not form discrete molecules analogous to CO_2, because third-period elements do not form strong π bonds. Silica is used to prepare glass, an amorphous solid (see Section 11.6). The silica is melted and the liquid melt is cooled rapidly to make a glass known as fused silica. The rapid cooling prevents the formation of crystalline silica. Additives such as Na_2CO_3,

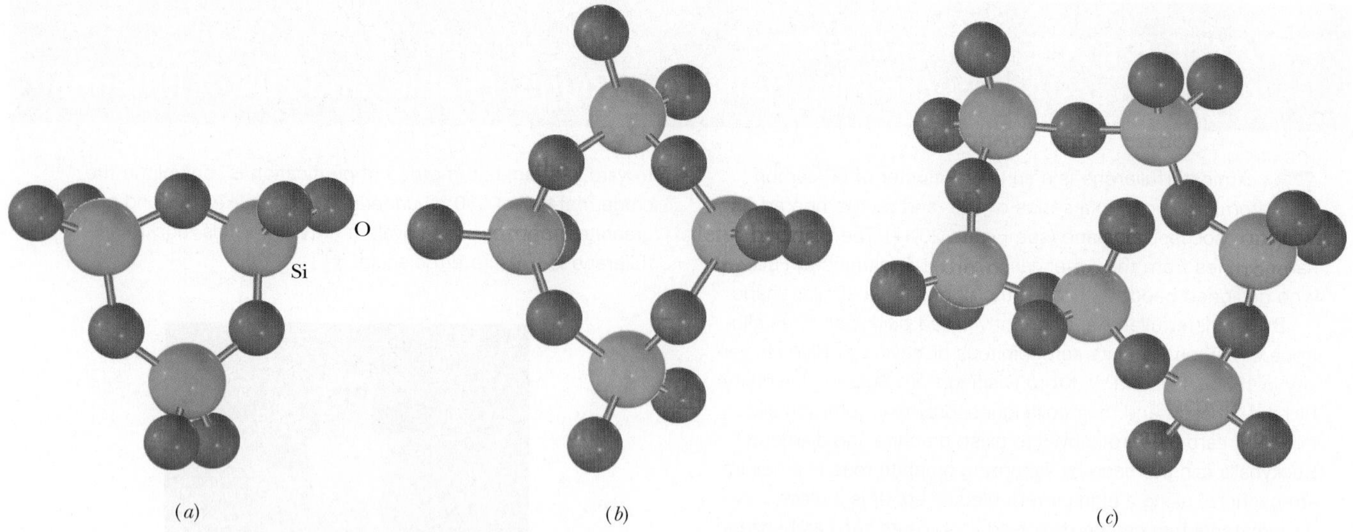

Figure 20.18 Silicates. Structures of three silicate anions: *(a)* $Si_3O_9^{6-}$, *(b)* $Si_4O_{12}^{8-}$, and *(c)* $Si_6O_{18}^{12-}$.

Silicon is found in nature in combination with oxygen.

B_2O_3 (to form Pyrex glass), and K_2O (to make an especially hard glass) are mixed into the melt to change the appearance and properties of the glass (Figure 20.19).

Glass is just one example of a group of materials known as **ceramics**—nonmetallic solid materials that are hard, resistant to heat, and chemically inert. Most ceramics are formed from silicates. Clays, which are mainly aluminosilicates (e.g., kaolinite, $Al_2Si_2O_5[OH]_4$), have been used for more than 5000 years to make pottery and dishes.

Glass, clays, and cement are important materials based on silicon compounds.

Cement is also mainly an aluminosilicate material. It is formed by heating a mixture of clay and limestone ($CaCO_3$) in a kiln at 1400 to 1600 °C to produce small lumps called *clinkers,* which are ground with some gypsum ($CaSO_4{\cdot}2H_2O$) into a powder. Concrete is formed by adding water to a mixture of cement and sand or gravel. This mixture slowly hardens (sets) through a complicated series of reactions that are not fully understood. As with glass, the properties of cement can be varied by using different additives and heating procedures.

Another important silicate is *asbestos* (Figure 20.20). Asbestos is the term for a family of silicates with fibrous properties. Asbestos is a good heat insulator and does not burn, so it has been used extensively to insulate buildings and ships. Unfortunately, research has shown that several forms of asbestos can be dangerous to lung tissue, and use of asbestos as an insulating material has been banned. The asbestos in buildings

Figure 20.19 Glass. Normal plate glass is formed by addition of Na_2O and CaO to the silica melt. The flask is Pyrex glass, which is formed by addition of B_2O_3 and has a low thermal expansion. Blue-colored glass is produced by addition of cobalt(II) compounds.

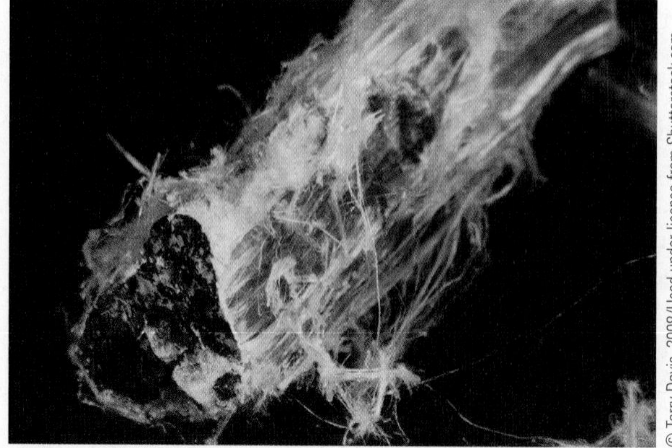

Figure 20.20 Chrysotile asbestos.

PRACTICE OF CHEMISTRY
Properties of Glass Changed by Additives; Hubble Space Telescope Requires Ultrastable Mirrors

The volume of a solid changes little when it is heated, but small changes can be significant. Consider what happens when a glass container is heated and then rapidly cooled. Because glass is a poor conductor of heat, temperature differences cause different parts of the object to change volume at different rates. These changes lead to stress, which often results in the glass breaking into pieces.

Some glasses are specially formulated to change volume only slightly with temperature. The beakers and flasks used in chemistry laboratories that are made of borosilicate glass are examples of such materials. These containers can be heated with a Bunsen burner and then placed on a cold laboratory bench without shattering. One common brand of borosilicate glass is Pyrex, a trademark of Corning Glass Company Incorporated.

Pure silica glass (silica, SiO_2) has even better thermal properties than borosilicate. It can be heated to more than 1000 K and then placed in liquid nitrogen at 77 K without shattering. Although silica glass has good thermal properties, compounds such as sodium oxide (soda ash, Na_2O) and calcium oxide (lime, CaO) are added to silica to reduce its melting point. The lower melting point decreases the cost of producing glass objects, because the ovens used to melt the glass last longer at the lower temperatures and the energy costs are also lower. However, these additives make the glass less tolerant of rapid changes in temperature.

Materials scientists characterize glasses by their coefficients of linear expansion; the most common units are parts per million per kelvin (ppm/K), the relative change in length (in ppm, or microinches of change per inch of material) per degree change in temperature. The following table lists some common glasses and their coefficients of thermal expansion.

Glass	Compounds Added to Silica	Coefficient of Linear Expansion (ppm/K)
Soda lime	Na_2O, CaO	5
Borosilicate	Na_2O, B_2O_3	3.25
Silica glass		0.8

One group of scientists concerned with the thermal properties of glasses is astronomers. Telescope lenses and mirrors are fabricated from dimensionally stable glasses that do not change size or shape greatly as the temperature changes. A mirror or lens that changed shape would render the telescope useless.

The Hubble space telescope is an orbiting observatory launched in 1989. For a telescope in space, the effects of extreme temperature changes are important. The primary mirror is 94 in. (2.4 m) in diameter and was ground to a particular shape. The mirror was designed to be within 10 nm (4×10^{-7} in.) of the specific shape *while in the microgravity environment of space.* (Correcting for forces because of Earth's gravity during the fabrication process was a difficult and time-consuming procedure.) Even small fluctuations in shape cannot be tolerated in the glass of the mirrors. The mirrors were constructed from ultra-low expansion glass and thermostated in the telescope.

Unfortunately, the primary mirror was originally flawed; its shape was slightly too flat. Five pairs of error-correcting mirrors, each about the size of a postage stamp, were fitted to the telescope by astronauts from the space shuttle. The telescope now operates as planned, but new repairs are needed to other parts for this to remain true. ▮

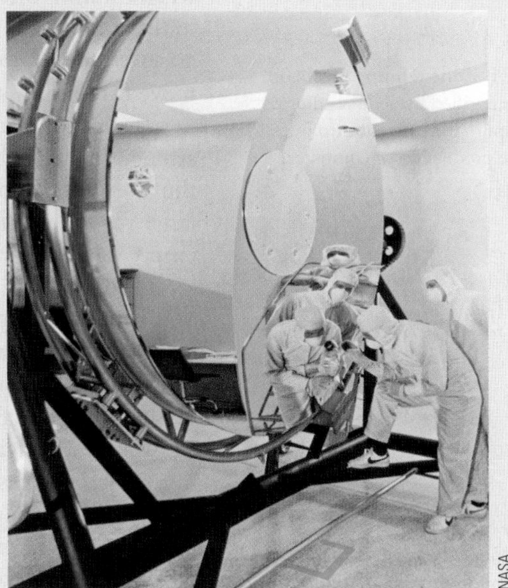
NASA

Mirror for Hubble space telescope.

in the United States is generally replaced by specially trained workers when the buildings are renovated.

Preparations of Silicon

Thousands of tons of silicon are used in alloys of iron and aluminum, in silicone polymers, and (in highly purified form) in solid-state electronic components. In the preparation of the element, SiO_2 is reduced by carbon in an electric-arc furnace. In this procedure, the SiO_2 must be kept in excess to prevent formation of SiC. Highly purified

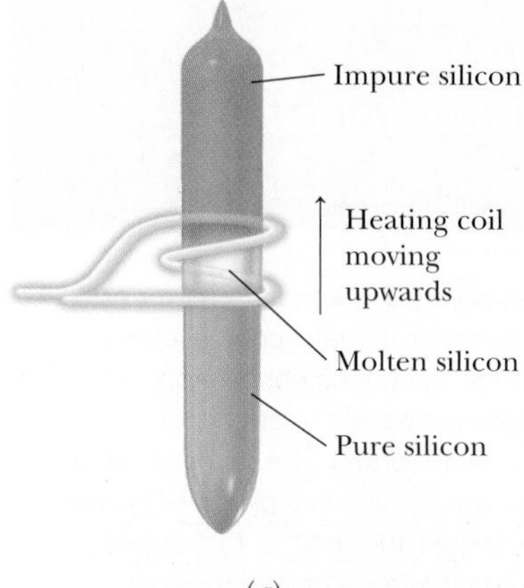

Figure 20.21 Zone refining of silicon. *(a)* Silicon is purified by zone refining. *(b)* The very pure rods are cut into wafers for use in the production of computer chips.

silicon is prepared by treating this silicon with chlorine gas to form the volatile tetrachloride. Silicon tetrachloride can be purified by repeated distillation, followed by reduction with magnesium or hydrogen to recover the element.

$$SiO_2(\ell) + 2C(s) \xrightarrow{3000\ ^\circ C} Si(\ell) + 2CO(g)$$

$$Si(s) + 2Cl_2(g) \rightarrow SiCl_4(\ell)$$

$$SiCl_4(g) + 2Mg(s) \rightarrow Si(s) + 2MgCl_2(s)$$

Rods of silicon are then further purified by zone refining. In this process, a thin band at one end of a silicon rod is melted, and the heat source is slowly moved toward the other end. As the heat source is moved, the impurities stay in the molten silicon zone, leaving behind high-purity silicon (Figure 20.21).

Semiconductors

The bonding in solid metals or metalloids can be described as arising from the overlap of many orbitals to form energy bands that extend throughout the solid. For example, the 3*s* orbitals in a piece of sodium that contains one mole of atoms overlap to form a band of one mole of orbitals (in the terms of molecular orbital theory, a half mole of bonding orbitals and a half mole of antibonding orbitals) closely spaced in energy. Because for sodium, one mole of electrons is available to fill this band, the band (called a 3*s* band because it arises from the overlap of 3*s* atomic orbitals) is half filled (Figure 20.22*a*). Sodium can conduct an electrical current because electrons can move easily through the empty but low-energy orbitals of the 3*s* band. A **conduction band** is a partially or completely empty band of orbitals that can conduct an electrical current. Magnesium is also a conductor, even though this 3*s* band by itself would be filled (one mole of magnesium atoms has two moles of electrons). In magnesium, the band that arises from the 3*p* orbitals overlaps the 3*s* band (see Figure 20.22*b*), and electrons can move through this mixed 3*p*/3*s* band, which is now the conduction band.

The properties of a nonmetallic solid such as diamond are quite different. The strong, directed C–C bonds in diamond cause a large energy gap (band gap) between the filled band and the conduction band. Diamond is an insulator; the energy gap is too large for the electrons to move into the conduction band (see Figure 20.22*c*), so no empty low-energy orbitals are available to conduct an electric current.

Pure silicon has the same structure as diamond, but the solid is not nearly as hard as diamond because the Si–Si bonds are much weaker. Because of the weaker bonds, the energy gap between the filled band and the empty band is much smaller in silicon than in diamond (see Figure 20.22*d*). In this situation, a few electrons can cross the band gap because of thermal energy, making silicon a weak conductor of electricity.

The conductivity of silicon can be increased by adding trace quantities of a selected impurity, a process called *doping*. In one type of doping, a small quantity of an element that has five valence electrons (one more than silicon), such as phosphorus, is added to pure silicon. The extra electrons enter the conduction band, increasing the conductivity of the solid (Figure 20.23*a*). The doping atoms add negative-charge carriers (electrons) to the silicon, so the result is called an *n-type semiconductor.*

Pure silicon and doped silicon are used in the production of semiconductor devices.

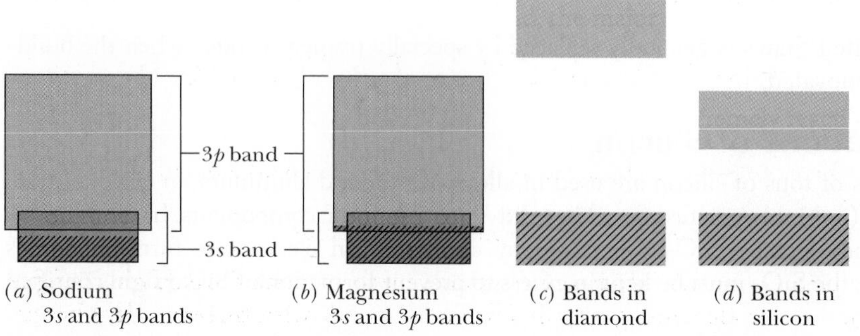

Figure 20.22 Bands of orbitals. *(a)* The overlap of 3*s* atomic orbitals in sodium forms a band of orbitals (orange) that are half filled (orbitals occupied by electrons are indicated by hatching). *(b)* The 3*s* band in magnesium overlaps the 3*p* band (blue). *(c)* In diamond, the filled band is well separated in energy from the conduction band. *(d)* In silicon, the filled band is close in energy to the conduction band.

In contrast, silicon can be doped with an element that has three valence electrons (one less than silicon), such as boron. Each of these atoms leaves a "hole" in the lower-energy band, again increasing the conductivity (see Figure 20.23*b*) by allowing an electron to move into the "hole" and leaving a "hole" at its previous location. Because the entity that appears to move is a positive "hole" (the absence of an electron), this kind of doped silicon is called a *p-type semiconductor.* Note that the extreme sensitivity of the electrical conductivity of silicon and other semiconductors to even trace amounts of impurities makes it important that these materials be carefully purified.

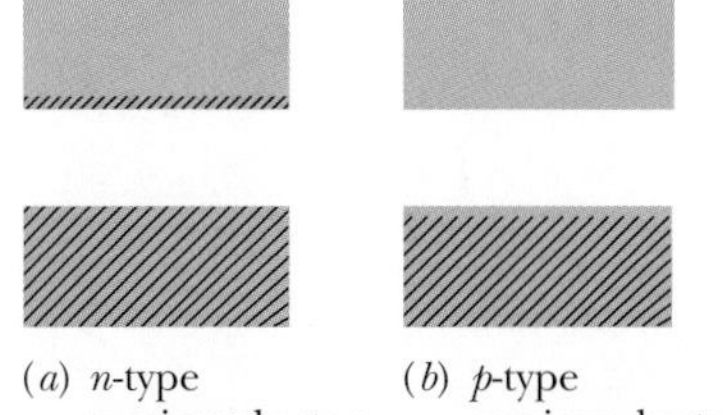

(*a*) *n*-type semiconductor (*b*) *p*-type semiconductor

Figure 20.23 Doping of semiconductors. *(a)* An *n*-doped semiconductor is formed by the substitution of impurity atoms that contain more valence electrons (hatching indicates orbitals that are occupied by electrons). *(b)* A *p*-doped semiconductor is formed by the substitution of impurity atoms that contain fewer valence electrons than are present in the pure substance.

Germanium, Tin, and Lead

Germanium was once important in the semiconductor industry, but it has given way to silicon, which retains its desirable properties better at higher temperatures. Germanium is recovered from the flue dust produced in the processing of zinc ore, and it may be purified by zone refining. It forms gray-white crystals that have the same structure as silicon, and it has similar properties.

Both tin and lead were among the first metals isolated by humans. Tin is found as the oxide in the mineral cassiterite, SnO_2 (Figure 20.24*a*). The metal can easily be isolated by reduction with charcoal.

$$SnO_2(s) + C(s) \rightarrow Sn(s) + CO_2(g)$$

The metallic form of tin is soft and is used in low-melting alloys such as solder and pewter. Tin became important in the development of early civilizations, when it was alloyed with copper to form bronze. Bronze is much harder than either copper or tin. Tin is also used as a coating for other metals such as iron because it does not react with air and water. An interesting use of tin is in the production of "plate" glass. The molten glass is poured onto the surface of molten tin; on cooling, the glass surface is so smooth that it does not need to be polished.

The Romans used lead extensively to make eating utensils and for plumbing. Lead occurs naturally as lead(II) sulfide, PbS, in the mineral galena (see Figure 20.24*b*). Practically all common lead compounds contain the element in the +2 oxidation state. The pure metal is obtained by conversion of the sulfide to the oxide by reaction with oxygen (roasting), followed by reduction of the oxide with carbon and carbon monoxide.

$$2PbS(s) + 3O_2(g) \rightarrow 2PbO(s) + 2SO_2(g)$$

$$PbO(s) + C(s) \rightarrow Pb(\ell) + CO(g)$$

$$PbO(s) + CO(g) \rightarrow Pb(\ell) + CO_2(g)$$

The main use of lead today is in the electrodes in lead storage batteries. In addition, lead shields are used to absorb high-energy radiation such as X-rays. Lead was used extensively as an antiknock additive to gasoline in the form of tetraethyllead, $(C_2H_5)_4Pb$, but this use has been curtailed because of harmful environmental and physiologic effects of the metal. The use of lead compounds in paints has been significantly reduced for the same reasons.

(*a*)

(*b*)

Figure 20.24 Tin and lead minerals. Tin is found in the mineral cassiterite *(a)* and lead in galena *(b)*.

OBJECTIVES REVIEW *Can you:*

- ☑ describe the bonding and properties of three allotropic forms of carbon?
- ☑ discuss the occurrence and chemistry of silicon?
- ☑ describe the properties of semiconductors and how doping influences their properties?

20.5 Chemistry of Group 5A (15) Elements

OBJECTIVES

- ☐ Discuss the properties of the compounds of nitrogen and phosphorus with hydrogen and oxygen
- ☐ Compare the elemental forms of nitrogen and phosphorus

H							He
Li	Be	B	C	**N**	O	F	Ne
Na	Mg	Al	Si	**P**	S	Cl	Ar
K	Ca	Ga	Ge	**As**	Se	Br	Kr
Rb	Sr	In	Sn	**Sb**	Te	I	Xe
Cs	Ba	Tl	Pb	**Bi**	Po	At	Rn

The elements in Group 5A (15) have five valence electrons with the electron configuration ns^2np^3. These elements generally form compounds with three covalent bonds, leaving one lone pair on the central atom. The first three elements in the group can also gain three electrons to form 3− anions in compounds with the highly electropositive Group 1A and 2A metals. The properties of the elements in Group 5A range from the nonmetals nitrogen and phosphorus to bismuth, an element that is a metal. As with other groups, the chemistry of the lightest member, nitrogen, is unique because it can form compounds with strong multiple bonds. In contrast with nitrogen, the heavier members of the group form numerous species with an expanded valence shell such as PF_5 and AsF_6^-.

Nitrogen

Nitrogen (boiling point, −196 °C) comprises 78% (by volume) of the atmosphere, and it is easily isolated by fractional distillation of liquid air. The liquid is used extensively as a low-temperature coolant, and the gas is used to protect foods and reactive chemicals from oxidation by oxygen in the air. Nitrogen is second only to sulfuric acid in the quantity produced by the chemical industry.

Nitrogen is relatively nonreactive because it exists as nonpolar diatomic molecules that contain a strong triple bond (945 kJ/mol). However, it reacts with the active metals lithium and magnesium to form ionic nitrides.

$$6Li(s) + N_2(g) \rightarrow 2Li_3N(s)$$

$$3Mg(s) + N_2(g) \rightarrow Mg_3N_2(s)$$

Ammonia

Nitrogen also reacts with hydrogen gas to form ammonia, but only under special conditions. This reaction is known as nitrogen *fixation,* the combination of the element with another element. Nitrogen is fixed by bacteria that live in the roots of leguminous plants such as soybeans and alfalfa. Plants, as well as all other living organisms, need nitrogen compounds. Frequently, farmers grow alfalfa and soybeans in rotation (in a given field) with crops such as corn and wheat that do not fix nitrogen. Other natural sources of fixed nitrogen are guano (the excrement of bats, found concentrated in caves, and of seabirds, found on isolated islands) and the minerals saltpeter, KNO_3, and Chile saltpeter, $NaNO_3$. In modern farming, the nitrogen needed for the growth of crops such as corn and wheat is frequently supplied through the addition of commercial fertilizers.

At the turn of the 20th century, a tremendous demand existed for fertilizer and for nitrates used to prepare explosives such as TNT. Searching for a way to meet this need, the German chemist Fritz Haber conducted considerable research and concluded that the direct synthesis of ammonia from its elements was practical.

$$N_2(g) + 3H_2(g) \rightleftharpoons 2NH_3(g) \qquad \Delta H^\circ = -92 \text{ kJ}$$

The reaction requires high pressures (200 atm or more) and high temperatures for the efficient production of ammonia from N_2 and H_2. Le Chatelier's principle predicts that high pressures will favor formation of product because four volumes of reactant gas are converted to two volumes of product. This reaction has very high activation energy because of the strong bond in N_2, and the **Haber process** is performed at increased temperatures (380 °C–450 °C) in the presence of an iron catalyst. The high temperatures are unfavorable for the position of the equilibrium but are necessary for the reaction to proceed at a reasonable rate. The temperature selected for the reaction is a compromise between the rate and the position of the equilibrium.

Ammonia, NH_3, is prepared from its elements at high temperatures and pressures.

Ammonia is a colorless gas that condenses at −33 °C. The gas has a pungent odor, which is the smell in "smelling salts." It dissolves in water, forming a solution that is often called *ammonium hydroxide,* although a discrete substance with the formula NH_4OH has never been isolated. Like water, liquid ammonia undergoes autoionization, but the equilibrium constant at −50 °C is only about 10^{-33}.

$$2NH_3 \rightleftharpoons NH_4^+ + NH_2^- \qquad K \approx 10^{-33}$$

Bloomberg News/Landov

Ammonia production. The Haber process is used to produce ammonia from elemental nitrogen and hydrogen.

The other important hydride of nitrogen is hydrazine, N_2H_4. Hydrazine is a liquid with a variety of industrial uses. One use of hydrazine is to remove oxygen from water that is used in boilers of electrical generating plants.

$$N_2H_4(aq) + O_2(aq) \rightarrow N_2(g) + 2H_2O(\ell)$$

Dissolved oxygen causes rapid deterioration of metal pipes at the high temperatures and pressures used in boilers. The reaction of derivatives of hydrazine with N_2O_4 is quite violent, and this reaction was used to propel the Apollo lunar lander.

Nitrogen Oxides

Nitrogen forms numerous oxides, six of which are listed in Table 20.1.

Dinitrogen monoxide (N_2O, sometimes called *nitrous oxide*) is a nontoxic gas that is used as an analgesic (laughing gas). An interesting commercial use of dinitrogen oxide is as the propellant in cans of whipped cream. It is prepared by the thermal decomposition of ammonium nitrate.

$$NH_4NO_3(s) \xrightarrow{200\ °C} N_2O(g) + 2H_2O(g)$$

Nitrogen monoxide (NO, sometimes called *nitric oxide*) is a gas at room temperature. It has an odd number of electrons and is an example of a molecule that does not satisfy the octet rule. In the solid state, NO dimerizes to form N_2O_2, thus pairing all the electrons.

Nitrogen monoxide reacts readily with oxygen to form nitrogen dioxide.

$$2NO(g) + O_2(g) \rightarrow 2NO_2(g)$$

Nitrogen monoxide is a major contributor to pollution in the lower atmosphere. It forms in the engines of automobiles and jet-propelled planes by the direct combination of $O_2(g)$ and $N_2(g)$. When formed by jets in the upper atmosphere, it can lead to the destruction of ozone, O_3.

$$NO(g) + O_3(g) \rightarrow NO_2(g) + O_2(g)$$

$$NO_2(g) + O_3(g) \rightarrow NO(g) + 2O_2(g)$$

The NO is regenerated in these reactions (it is a catalyst), so each NO molecule can destroy a large amount of ozone.

Nitrogen dioxide, NO_2, is also an odd-electron species; the odd electron is located mainly on the nitrogen atom. In the gas phase, NO_2 is in equilibrium with dinitrogen

TABLE 20.1 Nitrogen Oxides

Name	Formula
Dinitrogen monoxide	N_2O
Nitrogen monoxide	NO
Dinitrogen trioxide	N_2O_3
Nitrogen dioxide	NO_2
Dinitrogen tetroxide	N_2O_4
Dinitrogen pentoxide	N_2O_5

Figure 20.25 Nitrogen dioxide. Copper metal reacts with nitric acid, producing brown NO_2 gas.

tetroxide, N_2O_4, and exists as N_2O_4 in the solid state. Nitrogen dioxide is formed in the reactions of many metals, such as copper, with concentrated nitric acid (Figure 20.25).

$$Cu(s) + 4HNO_3(aq) \rightarrow Cu(NO_3)_2(aq) + 2H_2O(\ell) + 2NO_2(g)$$

Nitrogen dioxide is a red-brown gas that is toxic. It reacts with water to form nitric acid, HNO_3, and nitrous acid, HNO_2, which can further react to produce more nitric acid, water, and NO.

$$2NO_2(g) + H_2O(\ell) \rightarrow HNO_3(aq) + HNO_2(aq)$$

The nitrous acid is then easily oxidized to nitric acid by oxygen. The reaction of oxygen with NO to form NO_2 and the reaction of NO_2 with water make both of these odd-electron molecules major contributors to acid rain.

Dinitrogen oxide is a nonreactive gas, whereas both nitrogen monoxide and nitrogen dioxide are reactive, toxic gases that contribute to air pollution and acid rain.

Nitric Acid

Nitric acid is an important industrial product; more than 8 million tons are produced yearly. It can be formed in the laboratory by the reaction of $NaNO_3$ and sulfuric acid.

$$2NaNO_3(s) + H_2SO_4(aq) \rightarrow 2HNO_3(aq) + Na_2SO_4(aq)$$

Commercially, most of this acid is produced by the **Ostwald process.** In the first step of this process, ammonia (formed by the Haber process) is oxidized by oxygen over a platinum catalyst.

$$4NH_3(g) + 5O_2(g) \xrightarrow{Pt} 4NO(g) + 6H_2O(g)$$

In the second step, the NO is oxidized with $O_2(g)$ to give $NO_2(g)$, as described earlier. In the final step, the $NO_2(g)$ is mixed with water to form nitric acid. The NO that is also formed in this reaction is recycled through the process.

$$2NO(g) + O_2(g) \rightarrow 2NO_2(g)$$

$$3NO_2(g) + H_2O(\ell) \rightarrow 2HNO_3(aq) + NO(g)$$

Nitric acid is an unstable, colorless liquid usually distributed as a 70% solution in water. It frequently has a yellow color because of the presence of NO_2 formed on exposure to light. Nitric acid is a strong oxidizing agent, as well as a strong acid. It oxidizes all metals except the noble metals gold, iridium, platinum, and rhodium. The largest use of HNO_3 is in the production of ammonium nitrate, NH_4NO_3, by reaction with ammonia.

$$NH_3(g) + HNO_3(aq) \rightarrow NH_4NO_3(aq)$$

The ammonium nitrate is used widely as a fertilizer.

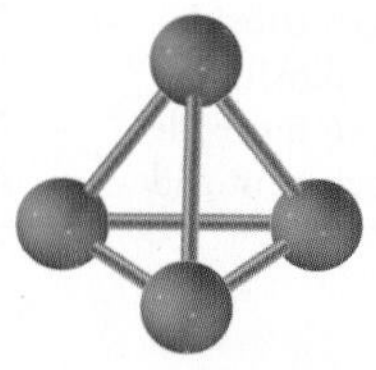

Figure 20.26 Structure of white phosphorus. White phosphorus contains tetrahedral P_4 molecules in which each phosphorus atom makes three sigma bonds.

Phosphorus

Phosphorus is found in many minerals in the form of the tetrahedral PO_4^{3-} ion or related species. The element exists in a number of allotropic forms, none of which is analogous to N_2 because of the tendency for elements of the third period to form σ rather than π bonds. The most common form is white phosphorus, P_4, which is prepared by the reaction of calcium phosphate with coke and sand at high temperatures.

$$2Ca_3(PO_4)_2(s) + 6SiO_2(s) + 10C(s) \rightarrow P_4(g) + 6CaSiO_3(\ell) + 10CO(g)$$

The gaseous P_4 condenses from this reaction as white phosphorus, which has a tetrahedral arrangement of phosphorus atoms (Figure 20.26). White phosphorus is a toxic material that burns when exposed to the air. Consequently, this form is usually stored under water to protect it from air. Heating white phosphorus in the absence of air converts it to a second allotropic form called *red phosphorus,* which is stable in air. This form is believed to have a polymeric structure of linked P_4 units.

White phosphorus consists of reactive P_4 molecules, whereas red phosphorus is a less reactive polymer of P_4 units.

Industrially, most of the white phosphorus is oxidized to P_4O_{10}, which is mixed with water to give phosphoric acid.

$$P_4(s) + 5O_2(g) \rightarrow P_4O_{10}(s)$$

$$P_4O_{10}(s) + 6H_2O(\ell) \rightarrow 4H_3PO_4(aq)$$

The second reaction is quite rapid and complete, making $P_4O_{10}(s)$ a useful drying agent for gases and liquids. Pure H_3PO_4 is a solid that melts at 42 °C. It is very **hygroscopic**—that is, it absorbs water vapor from the air—and is generally sold as a water solution. It is used in very dilute solution (0.01–0.05%) to give a tart taste to carbonated drinks. Phosphoric acid can also be prepared by the reaction of $Ca_3(PO_4)_2$ with sulfuric acid.

$$Ca_3(PO_4)_2(s) + 3H_2SO_4(aq) \rightarrow 2H_3PO_4(aq) + 3CaSO_4(s)$$

Another oxide of phosphorus, P_4O_6, can be prepared by the controlled oxidation of white phosphorus. Hydrolysis of P_4O_6 produces phosphorous acid, H_3PO_3:

$$P_4(s) + 3O_2(g) \rightarrow P_4O_6(s)$$

$$P_4O_6(s) + 6H_2O(\ell) \rightarrow 4H_3PO_3(aq)$$

Phosphorus is essential to life. In biological systems, it is generally found in the form of phosphates, compounds that contain the PO_4^{3-} ion. Calcium phosphates are major constituents of human bones and teeth. Phosphorus is important for the growth of plants and, together with nitrogen, is a component of fertilizers. The minerals found in nature, such as $Ca_5(PO_4)_3F$ (fluorapatite), are insoluble in water and are converted to more soluble compounds for use as fertilizers by treatment with sulfuric acid or phosphoric acid.

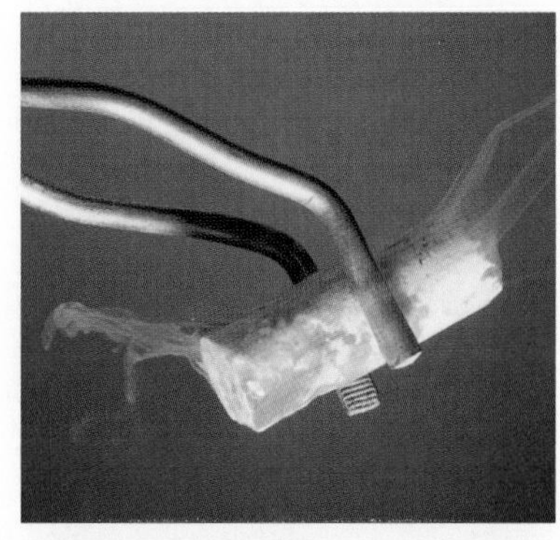

(a)

(b)

White and red phosphorus. *(a)* White phosphorus reacts with air. *(b)* White phosphorus is stored under water to prevent this reaction, but red phosphorus is stable in air.

Nitrogen and phosphorus compounds are produced by industry for use as fertilizers.

Phosphine

The stable hydride of phosphorus, PH_3, is called *phosphine.* This highly poisonous gas boils at −88 °C. Its boiling point is 55 °C lower than that of ammonia, because no hydrogen bonding forces exist between molecules of PH_3.

The structure of PH_3 differs from that of ammonia in that the H–P–H bond angles are only 93.7 degrees, compared with the 107.3-degree angles in NH_3 (Figure 20.27). Clearly, the valence shell electron-pair repulsion model does not accurately predict the bond angles in PH_3. According to valence bond theory, the nearly 90-degree H–P–H bond angles suggest that the bonds are formed from almost pure 3*p* orbitals on the phosphorus, leaving the 3*s* orbital to accommodate the lone pair of electrons.

Phosphine is prepared from the reaction between white phosphorus and aqueous NaOH.

$$P_4(s) + 3NaOH(aq) + 3H_2O(\ell) \rightarrow 3NaH_2PO_2(aq) + PH_3(g)$$

Despite its toxicity, phosphine is an important starting material in the manufacture of a major flame-proofing material used on cotton cloth.

Arsenic, Antimony, and Bismuth

The heavier Group 5A elements occur in nature as sulfides—As_2S_3, Sb_2S_3, and Bi_2S_3. Arsenic and antimony are metalloids, and their chemistry resembles that of phosphorus. The +3 oxidation state becomes increasingly more stable than +5 for the heavier elements in the group. For example, burning the elements in excess air leads to the formation of As_4O_6 and Sb_4O_6 rather than the P_4O_{10} formed by phosphorus. As_4O_{10} can be prepared, but only when strong oxidizing agents are used. The oxide of bismuth, Bi_2O_3, is basic and will dissolve in acid solution, as is expected for the oxide of a metal.

$$Bi_2O_3(s) + 2H^+(aq) \rightarrow 2BiO^+(aq) + H_2O$$

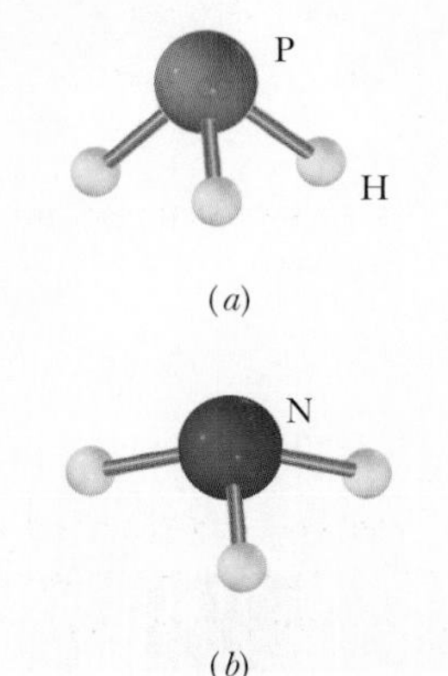

Figure 20.27 Structures of NH_3 and PH_3. The bond angles in phosphine *(a)* are much smaller than those in ammonia *(b)*.

OBJECTIVES REVIEW *Can you:*

- ☑ discuss the properties of the compounds of nitrogen and phosphorus with hydrogen and oxygen?
- ☑ compare the elemental forms of nitrogen and phosphorus?

20.6 Chemistry of Group 6A (16) Elements

OBJECTIVES

- ☐ Describe the allotropic forms of oxygen and sulfur and their properties
- ☐ Know the chemistry of compounds containing both oxygen and sulfur

The elements in Group 6A (16) have six valence electrons with the electron configuration ns^2np^4. These elements generally form two covalent bonds and have two lone pairs on the central atom. As in the other groups, the heavier elements also form expanded valence shell compounds, such as SF_4 and SF_6.

Oxygen

Oxygen is the most abundant element on our planet; it makes up about half of Earth's crust and is present in both air (21% O_2 by volume) and water. On Earth's surface, oxygen exists mostly in combination with a variety of other elements. It occurs mainly in water and combined with silicon in silica and the silicates.

The most common allotrope of oxygen is O_2. The liquid has a light blue color and boils at −183 °C. The other important allotrope is ozone, O_3. Ozone is a reactive gas that causes the pungent odor sometimes noticed during electrical storms. The solid and liquid phases (boiling point, −112 °C) of ozone are unstable and decompose explosively. Although ozone is a pollutant in the lower atmosphere, its presence in the upper atmosphere is extremely important because it absorbs much of the ultraviolet light coming from the sun. Exposure to ultraviolet light can increase the incidence of skin cancer. A number of synthetic compounds, mainly the chlorofluorocarbons (Freons, or CFCs), were consuming ozone at high altitude in a complicated series of reactions, such as these of trichlorofluoromethane, $CFCl_3$:

The stable allotrope of oxygen is O_2(g), but a second allotrope, O_3(g), is important in the upper atmosphere.

$$CFCl_3(g) \xrightarrow{\text{Ultraviolet light}} CFCl_2(g) + Cl(g)$$

$$Cl(g) + O_3(g) \rightarrow ClO(g) + O_2(g)$$

$$O_3(g) \rightarrow O(g) + O_2(g)$$

$$ClO(g) + O(g) \rightarrow Cl(g) + O_2(g)$$

Note that the last three reactions constitute a cycle, showing that Cl atoms are a catalyst, so each molecule of $CFCl_3$ can destroy many ozone molecules. Fortunately, the use of most of these CFCs has been reduced substantially.

Molecular oxygen can be prepared on a small scale by heating potassium chlorate, using MnO_2 as a catalyst.

$$2KClO_3(s) \xrightarrow{MnO_2,\ 150\ °C} 2KCl(s) + 3O_2(g)$$

On an industrial scale, oxygen, like molecular nitrogen, is recovered from the fractional distillation of liquid air. The largest industrial consumption of O_2 is in the production of steel (see Section 19.5). Oxygen is also used in the oxidation of hydrocarbons, in the treatment of wastewater, in medicine, and in rocket engines.

Reaction of metals with oxygen. At high temperatures, iron powder reacts with the oxygen present in air.

Most elements react directly with molecular oxygen. All elements except helium, neon, argon, and possibly krypton form binary compounds with oxygen. Oxygen forms strong covalent bonds to most of the nonmetals and many of the transition metals. In its ionic compounds, the small size and 2− charge of the oxide ion produce stable ionic structures.

Changes in the electronegativity of the elements across the periodic table lead to dramatic changes in the properties of the binary compounds of oxygen. Oxides of the metals on the left side of the table are generally high-melting ionic solids and react to form basic solutions in water. Oxides of the other metals and metalloids are also generally solids, but they are less ionic and may be amphoteric in water. Oxides of the nonmetals are generally covalent compounds that form acidic solutions by reaction with water.

In addition to water, hydrogen forms a second compound with oxygen: hydrogen peroxide, H_2O_2. Hydrogen peroxide has a freezing point close to that of water, −0.41 °C, but boils at 150 °C and has a much higher density, 1.4 g/mL. Figure 20.28 shows its structure.

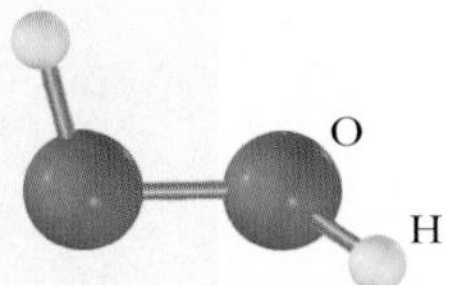

Figure 20.28 Structure of hydrogen peroxide.

Hydrogen peroxide is unstable, particularly when pure, and its decomposition is catalyzed by many metals and even by glass.

$$2H_2O_2(\ell) \rightarrow 2H_2O(\ell) + O_2(g)$$

The 2% to 3% solutions of hydrogen peroxide that are sold as a germicide are safe to handle, but solutions with concentrations of more than 20% can be dangerous. Note from the decomposition reaction that when hydrogen peroxide is used as a germicide it gives off $O_2(g)$. The reaction causing this "foaming" is catalyzed by the iron enzymes in the blood and helps clean the wound. Hydrogen peroxide is also a bleach that is used extensively in industry and for hair coloring. An important reason for its widespread use is that the products of the reaction, $O_2(g)$ and $H_2O(\ell)$, are nonpollutants.

Sulfur

Sulfur is abundant in nature. It occurs as the free element, as well as in sulfides and sulfates. Sulfur deposits are generally found underground and are brought to the surface by the **Frasch process.** In this process, the sulfur is melted with superheated steam and pushed to the surface by air pressure (Figures 20.29 and 20.30).

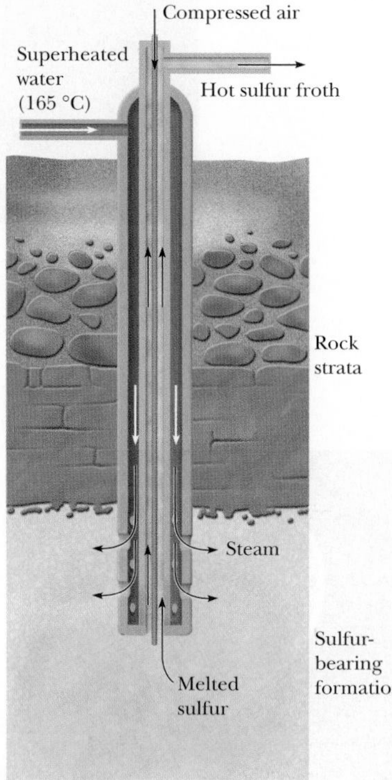

Figure 20.29 Frasch process. Underground deposits of sulfur are melted with hot steam and forced to the surface with compressed air.

Another major source of sulfur (and sulfur compounds) is its production as a byproduct in the purification of fossil fuels. Both crude oil and natural gas are contaminated by H_2S and other sulfur-containing compounds that must be removed before the fuels are burned, to prevent air pollution. As a pollution control measure, sulfur dioxide is removed from the flue gas that is produced when coal is burned. Sulfur dioxide is also recovered from the roasting of metal sulfide ores in the refining of many metals.

$$2PbS(s) + 3O_2(g) \rightarrow 2PbO(s) + 2SO_2(g)$$

Sulfur exists in many allotropic forms. The yellow solid that forms around some volcanic steam vents in the earth consists of orthorhombic sulfur, a form that contains molecules of S_8 rings (Figure 20.31).

When orthorhombic sulfur is heated above its melting point, 113 °C, the rings break and the short chains link together to form longer chains. Depending on the temperature and the rate of heating or cooling, many different forms of sulfur can occur. If molten sulfur is cooled rapidly—for example, by pouring it into water (Figure 20.32)—the result is a rubbery form called *plastic sulfur,* which contains long chains of sulfur atoms. At room temperature, plastic sulfur reverts to orthorhombic sulfur.

The most common allotrope of sulfur is S_8, which has a cyclic structure.

Figure 20.30 Sulfur. A hot mixture of sulfur and water is pushed to the surface in the Frasch process.

(*a*) (*b*)

Figure 20.31 Orthorhombic sulfur. *(a)* Sulfur deposits can be found around natural steam vents in the earth. *(b)* This allotropic form consists of eight-member rings of sulfur atoms.

Figure 20.32 Plastic sulfur. A rubbery form of sulfur called *plastic sulfur,* which contains long chains of sulfur atoms, forms on rapid cooling of hot sulfur.

The most important hydride of sulfur is hydrogen sulfide, H_2S. Hydrogen sulfide has a lower boiling point (−60 °C) than water because of the lack of strong hydrogen bonding forces. It is extremely toxic and has the odor of rotten eggs. Although easily detected by its characteristic smell at low concentrations, it can actually deaden the olfactory nerve at greater concentrations. Volcanic activity emits large volumes of H_2S, and at these high temperatures it reacts with oxygen to produce SO_2.

$$2H_2S(g) + 3O_2(g) \rightarrow 2SO_2(g) + 2H_2O(g)$$

Compounds of Oxygen and Sulfur

The most common oxide of sulfur is sulfur dioxide, SO_2. Sulfur dioxide is a toxic gas (boiling point, −10 °C) that has a choking odor. It is produced industrially on a large scale by the burning of sulfur and as a by-product in the roasting of sulfide ores. Although the situation has improved greatly in recent years, some of the SO_2 that is produced, mainly from the burning of coal, is not trapped and is emitted into the air. The SO_2 from this source, together with SO_2 produced naturally, reacts with water to contribute to acid rain.

$$SO_2(g) + H_2O(\ell) \rightarrow H_2SO_3(aq)$$

Much of this SO_2 that is produced (both industrially and in nature) is converted to sulfur trioxide, SO_3, by reaction with oxygen.

$$2SO_2(g) + O_2(g) \xrightarrow{\text{Catalyst}} 2SO_3(g)$$

This reaction is slow but is catalyzed by a variety of solids, such as dust particles in the presence of sunlight, and by metal ions, such as Fe^{3+}, dissolved in water. The SO_3 thus produced reacts with water to form sulfuric acid, the major contributor to acid rain.

$$SO_3(g) + H_2O(\ell) \rightarrow H_2SO_4(aq)$$

The SO_3 is a volatile liquid (boiling point, 45 °C) that is extremely reactive. It has a triangular structure in the gas phase, but in the solid or liquid phase, it exists in various polymeric forms, such as the cyclic trimer S_3O_9 (Figure 20.33).

Sulfuric acid, H_2SO_4, is produced commercially in larger quantity than any other compound. The main industrial method of preparing sulfuric acid is the **contact**

process. In this process, the reaction of SO_2 with O_2 is catalyzed by V_2O_5; the SO_3 product is trapped by dissolution in concentrated H_2SO_4, which is then diluted by mixing with additional water. The resulting acid is generally sold as a solution that is 96%–98% H_2SO_4 by weight.

Sulfuric acid is a strong, corrosive acid that must be handled carefully. Dilution of the concentrated acid with water produces a large amount of heat. When diluting sulfuric acid, remember the rule to add acid to water, and do it slowly, while stirring. If you make the mistake of adding water to sulfuric acid, the heat that is released can cause droplets of the water to boil and spatter the concentrated acid out of its container. Even when you perform the dilution properly, you should wear protective clothing and a full face shield. The affinity of concentrated sulfuric acid for water is so high that it is frequently used to remove water from gases.

A dramatic demonstration of the ability of concentrated H_2SO_4 to act as a dehydrating agent is its reaction with sugar to form carbon and hydrated sulfuric acid (Figure 20.34).

$$C_{12}H_{22}O_{11}(s) + 11H_2SO_4(\ell) \rightarrow 12C(s) + 11H_2SO_4{\cdot}H_2O(\ell)$$

In addition to being a strong acid and dehydrating agent, H_2SO_4 is an oxidizing agent. Although it has only mild oxidizing properties at room temperature, it is much more reactive at higher temperatures.

The main use of sulfuric acid is in the conversion of insoluble phosphorus minerals to soluble phosphate fertilizers, as outlined in Section 20.5. It is also used in the refining of petroleum and in the synthesis of many chemicals.

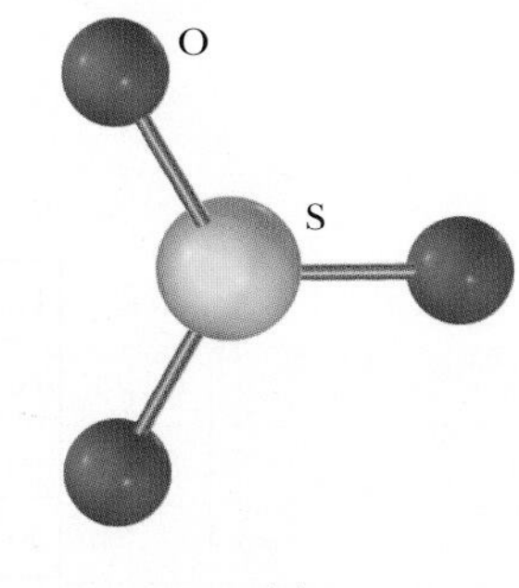

(*a*)

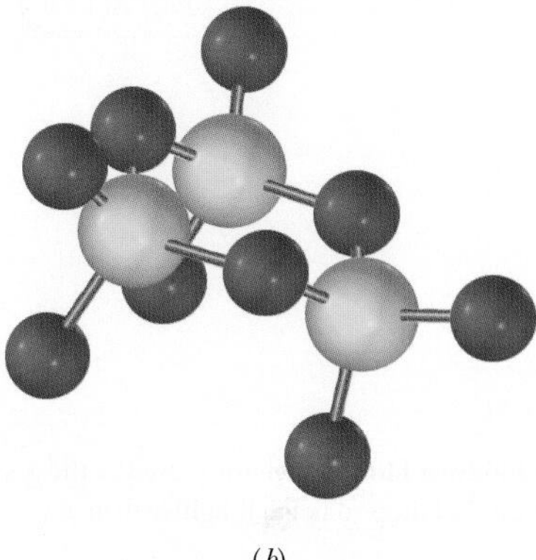

(*b*)

Figure 20.33 **SO_3 and S_3O_9.** *(a)* SO_3 has a trigonal planar structure. *(b)* S_3O_9 is a cyclic trimer.

More sulfuric acid is produced commercially than any other compound. It is used as an acid, an oxidizing agent, and a dehydrating agent.

Selenium and Tellurium

Selenium and tellurium are rare elements whose chemistry resembles that of sulfur. Selenium exists in a number of allotropic forms, including an Se_8 ring form. Another form of selenium and the only crystalline form of tellurium are composed of spiral chains of atoms. The last element of the family, polonium, is radioactive and exists in only trace amounts. Selenium is recovered as a by-product in the roasting of certain ores. It is used in the manufacture of red glass. Although it is toxic, selenium has been found as a trace element in the human body. This is an interesting case in which trace amounts of an element are beneficial, whereas larger amounts are toxic.

(*a*)

(*b*)

(*c*)

Figure 20.34 **Reaction of H_2SO_4 and sugar.** *(a)* Sulfuric acid reacts with sugar, *(b)* removing the water and *(c)* leaving only a column of carbon.

OBJECTIVES REVIEW *Can you:*

- ☑ describe the allotropic forms of oxygen and sulfur and their properties?
- ☑ know the chemistry of compounds containing both oxygen and sulfur?

H							He
Li	Be	B	C	N	O	F	Ne
Na	Mg	Al	Si	P	S	Cl	Ar
K	Ca	Ga	Ge	As	Se	Br	Kr
Rb	Sr	In	Sn	Sb	Te	I	Xe
Cs	Ba	Tl	Pb	Bi	Po	At	Rn

20.7 Noble Gases

OBJECTIVE

- ☐ Describe the properties and chemical behavior of the noble-gas elements

The elements in Group 8A (18) have a noble-gas electron configuration, ns^2np^6 ($1s^2$ for helium). As expected, they are unreactive and exist as monatomic gases. Because of their lack of reactivity, these gases were once called the *inert gases,* but since the early 1960s, compounds of krypton and especially xenon have been prepared, with an unstable argon compound recently reported. Although no known compounds of helium or neon exist, these gases have a number of important uses.

Helium, the second most abundant element in the universe, is found in very low concentrations in our atmosphere because the gravitational pull of Earth is too weak to prevent it from escaping into outer space. It is found in considerable concentrations in the United States in certain natural gas deposits, where it forms from the α-particle decay (He^{2+}) of radioactive elements. Helium is much less dense than air and is used to float balloons and lighter-than-air ships.

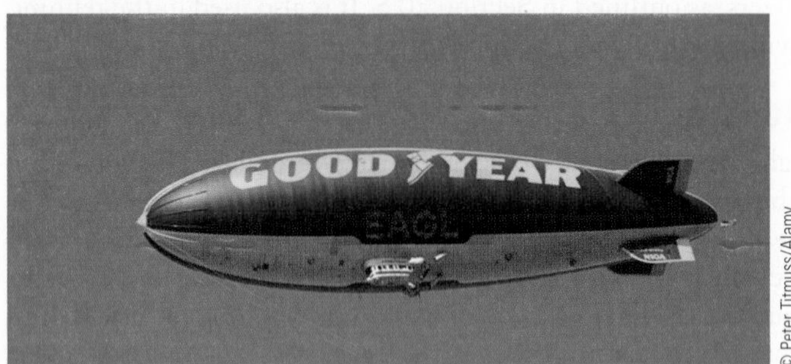

© Peter Titmuss/Alamy

Goodyear blimp. Helium is used as the gas inside a blimp to make it lighter than air.

Liquid helium, with a boiling point of 4 K, is used as a coolant for low-temperature experiments and for devices such as the superconducting magnets used in magnetic resonance imaging, an important diagnostic technique in the health field (Figure 20.35). The use of helium in magnets for magnetic resonance imaging has increased to such a degree that there is considerable concern whether enough helium will be produced in the next few years to meet demand.

A helium-oxygen mixture is often used instead of a nitrogen-oxygen mixture for deep-sea diving and in spacecraft. At the high pressures encountered during a deep dive, the concentration of dissolved nitrogen in the blood increases and can cause narcosis. This problem is avoided by using helium rather than nitrogen as the diluting gas for oxygen.

Neon is used as the gas in "neon" lights. The red color of the light can be changed by mixing in some argon or mercury vapors. Argon is the third most abundant gas in

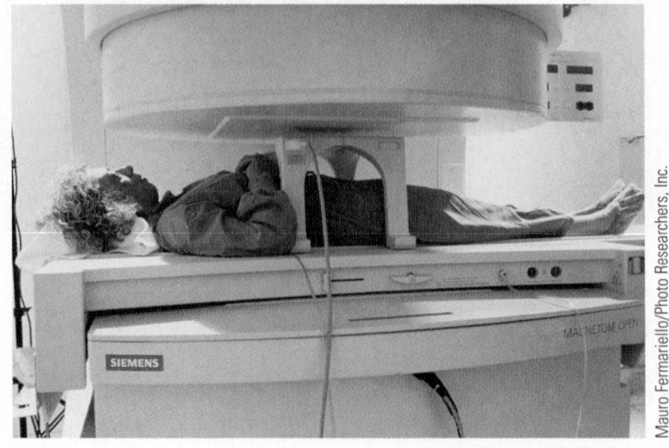

Mauro Fermariello/Photo Researchers, Inc.

Figure 20.35 Magnetic resonance imaging (MRI). MRI is an important technique for medical diagnosis. The superconducting magnet used in this equipment is cooled by liquid helium.

dry air (0.93%); together with neon, krypton, and xenon, it is produced by the distillation of liquid air. Argon has a higher density than air and is used as the inert atmosphere in electric lightbulbs and in welding to protect metals from oxygen.

Krypton and Xenon

In 1962, Neil Bartlett prepared the compound $[O_2^+][PtF_6^-]$ from the reaction of $O_2(g)$ with PtF_6. Realizing that the ionization energy of xenon is similar to that of O_2, he conducted a similar reaction with xenon and isolated $[Xe^+][PtF_6^-]$ (now known to have a more complex formula), the first compound of an "inert gas." Although his decision to carry out this reaction may seem obvious in retrospect, at the time it was extraordinary for a scientist to even try a reaction with an element thought to be completely nonreactive. Bartlett's success demonstrates that it is important to question and test accepted scientific theories. Soon after this discovery, workers at the Argonne National Laboratory demonstrated that fluorine will also oxidize xenon at increased temperatures, leading to the covalent compound XeF_4.

$$Xe(g) + 2F_2(g) \xrightarrow{400\ °C} XeF_4(s)$$

With careful control of the conditions, XeF_2 and XeF_6 can also be prepared. A number of oxides (XeO_3, an explosive compound, and XeO_4) and mixed fluorooxides (XeF_2O_3 and XeF_4O_2) have also been prepared. Sodium perxenate, Na_4XeO_6, forms in the reaction of XeO_3 with O_3 in aqueous NaOH solution. Both XeO_3 and Na_4XeO_6 are strong oxidizing agents. Although krypton is not as reactive, KrF_2 and KrF_4 have been prepared.

Although a few compounds of krypton and xenon are known, the Group 8A elements are very unreactive.

Radon

All isotopes of radon are radioactive and are difficult to work with in the laboratory. Radon has received considerable publicity in recent years because it forms in the radioactive decay of an isotope of uranium that occurs in natural deposits in many parts of North America. Because radon is a gaseous element, it can escape from soil into air. In well-insulated houses with little ventilation, the gas can build up to levels that constitute a considerable health hazard. Because it is impossible to remove the source of the radon, the problem is best solved by ventilation. In most cases, a small exhaust fan in the basement is sufficient to prevent buildup of the gas. Houses in areas that are known to have this problem should be tested for radon (Figure 20.36).

Blair Seitz/Photo Researchers, Inc.

Figure 20.36 Radon testing. Several testing devices are available to determine whether radon is building up in the basement of a home.

OBJECTIVE REVIEW *Can you:*

☑ describe the properties and chemical behavior of the noble-gas elements?

Summary Problem

Describe the bonding in BH_3. What is the hybridization at the boron atom? Is the boron really electron deficient?

Boron is in the $n = 2$ row and contains three valence electrons. Valence shell electron-pair repulsion theory predicts that the shape is trigonal planar with bond angles of 120 degrees. Its Lewis structure is as follows:

```
        H
       /
  H—B
       \
        H
```

Because the boron contains only three valence electrons, it can make only three normal electron-pair bonds with hydrogen and is electron deficient. The hybridization at the boron atom is sp^2, the hybridization that yields bond angles at 120 degrees. As outlined earlier, this electron deficient compound dimerizes to form B_2H_6, in which the boron is not electron deficient because of the two three-center, two-electron bonds.

ETHICS IN CHEMISTRY

1. You are building a new plant to produce ammonia using the Haber process. You are considering two ways to produce the hydrogen needed in the process. One is the use of the high-temperature reaction of methane and steam, and the other is the reaction of red-hot carbon (from coal) with steam. In both cases, you will use the water-gas shift reaction to produce more hydrogen from the CO produced in the first reactions. Your economic analysis shows that the methane reaction is a little cheaper, but your plant is located in a coal-rich area with many unemployed workers. The mayor of the local town is pushing you to use coal and has hinted at tax breaks if you should you choose coal as the source of the hydrogen. Is it ethical to push the local mayor hard for tax breaks (which could be viewed as a "bribe"), possibly raising taxes on the local homeowners?

2. The mayor makes a good offer of tax breaks, making the coal method the cheaper hydrogen source. But the local environmentalists are urging you to use the methane process because it is the "cleaner" source of hydrogen. What will you do?

Chapter 20 Visual Summary

The chart shows the connections between the major topics discussed in this chapter.

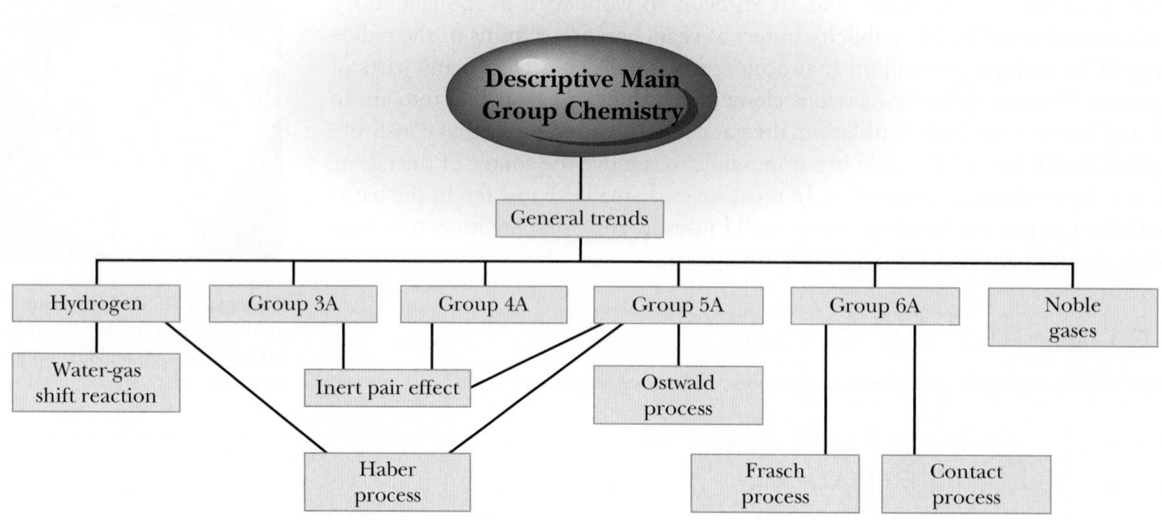

Summary

20.1 General Trends

The chemistry of the nonmetallic elements of the second period is dominated by their small atomic size and relatively high electronegativity. These second-period elements readily form both σ and π bonds, whereas elements in the later periods tend to form only σ bonds because π overlap of *p* orbitals in these larger elements produces relatively weak bonds. For example, the elemental form of nitrogen is N_2, a molecule with one σ and two π bonds, whereas the simplest form of phosphorus is P_4, in which each phosphorus atom makes three σ bonds. The metallic character of the elements increases down each group, as does the stability of the lower oxidation states.

20.2 Hydrogen

Hydrogen is generally listed in Group 1A on a periodic table but is really in a group by itself. Hydrogen can lose an electron to form a proton, $H^+(aq)$ in water, or gain an electron to form the hydride ion (H^-). It also forms strong covalent bonds with other nonmetals. Hydrogen is produced on a large scale by the high-temperature reaction of methane and water, and by the reaction of carbon and water. It combines with nitrogen to form ammonia in the *Haber process.* It also can be used to prepare methanol by reaction with carbon monoxide.

20.3 Chemistry of Group 3A (13) Elements

Group 3A elements generally form compounds in which the element has an oxidation number of +3. Heavier members of the family, especially thallium, also exist in the +1 oxidation state in some compounds. The trend for the heavier members of a group not to use the pair of valence *s* electrons in the formation of bonds is observed in groups 4A and 5A as well, and is called the *inert pair effect.* The elemental forms of boron contain an unusual arrangement of the boron atoms, an *icosahedron.* The types of compounds formed by boron are dominated by the fact that boron has only three valence electrons. Electron-deficient compounds such as boric acid and the boron halides react with neutral and anionic Lewis bases to form compounds with an octet of electrons on boron. In the boron hydrides, the boron atoms make four bonds by forming three-center, two-electron bonds.

Because of their low density, aluminum alloys that contain silicon, copper, and magnesium are used extensively as structural materials. Aluminum metal is protected from corrosion by the formation of a thin, strongly adhering protective coating of inert aluminum oxide, Al_2O_3. Aluminum oxide exists in multiple forms and has many important industrial uses. Corundum, or α-alumina, is a very hard substance that is used as an abrasive. Crystals of α-alumina that contain metal ion impurities are well-known gems, such as rubies and sapphires. Another form, γ-alumina, is used as a support in chromatographic separations and as a catalyst or catalyst support for many chemical reactions. Gallium and indium are becoming increasingly important because gallium arsenide (GaAs) and indium phosphide (InP) are useful semiconductor materials.

20.4 Chemistry of Group 4A (14) Elements

Elements in Group 4A have four valence electrons and generally attain an octet by forming four covalent bonds. The heavier members of the group, tin and especially lead, exhibit the inert pair effect and form the +2 oxidation state in some compounds. Millions of compounds of carbon have been isolated from living organisms or synthesized in laboratories. Graphite and diamond are both elemental forms of carbon. The former is composed of layers consisting of sp^2-hybridized carbon atoms, and the latter is an extremely hard allotrope in which the carbon atoms are sp^3 hybridized. Silicon is the second most abundant element in Earth's crust and is found in nature in minerals called *silicates* and *silica,* SiO_2. In these minerals, the silicon atom is surrounded by a tetrahedral arrangement of the oxygen atoms. Commercial glass is formed by melting then slowly cooling a mixture of silica and various additives that modify the properties of the glass. Silicon is used in common alloys and to make silicone polymers. Pure silicon is used in the fabrication of solid-state electronic components. Both tin and lead were among the first metals isolated.

20.5 Chemistry of Group 5A (15) Elements

Elements in Group 5A have five valence electrons and generally form three covalent bonds. Elemental nitrogen is easily isolated by fractional distillation of liquid air and is used extensively by the chemical industry as a low-temperature coolant, as an inert gas to protect materials from reactions with oxygen in the air, and for the synthesis of ammonia by the Haber process. Hydrazine (N_2H_4) is another compound of nitrogen and hydrogen. Nitrogen forms a series of oxides that have interesting properties and structures. Two of these oxides, NO and NO_2, are major sources of air pollution and acid rain. Nitric acid, HNO_3, is made from ammonia by reaction with oxygen and water in the *Ostwald process.* The major compounds prepared from phosphorus are phosphoric acid, H_3PO_4, used in carbonated drinks, and various phosphates (compounds of PO_4^{3-}), used as fertilizers.

20.6 Chemistry of Group 6A (16) Elements

Group 6A elements, especially oxygen, generally form two covalent bonds and have two lone pairs on the central atom. Oxygen is the most abundant element on Earth; it is found in the atmosphere, in water, and in Earth's crust as oxygen compounds of silicon such as silica and the silicates. Oxygen combines with nearly every other element on the periodic table. It forms ionic oxides with metals that form basic solutions in water, and covalent oxides with nonmetals that form acidic solutions in water. Sulfur is found as the free element in nature and is recovered by the *Frasch process.* It is also isolated as H_2S or SO_2 from contaminants in crude oil and natural gas or from the combustion of coal. Commercially, most of the sulfur is converted to sulfuric acid, H_2SO_4, the world's largest volume industrial compound, by the *contact process.*

20.7 Noble Gases

The elements in Group 8A are called the *noble gases* because they are not very reactive and exist as monatomic gases. Although these elements are nonreactive, compounds of both krypton and xenon are known. The compounds of xenon include three fluorides, XeF_2, XeF_4, and XeF_6, and a number of oxides and mixed fluorooxides.

Download Go Chemistry concept review videos from OWL or purchase them from **www.ichapters.com**

Chapter Terms

The following terms are defined in the Glossary, Appendix I.

Section 20.2
Water-gas shift reaction

Section 20.3
Icosahedron
Inert pair effect
Three-center, two-electron bond

Section 20.4
Ceramics
Conduction band
Silicates

Section 20.5
Haber process
Hygroscopic
Ostwald process

Section 20.6
Contact process
Frasch process

Questions and Exercises

OWL Selected end of chapter Questions and Exercises may be assigned in OWL.

Blue-numbered Questions and Exercises are answered in Appendix J; questions are qualitative, are often conceptual, and include problem-solving skills.

■ Questions assignable in OWL

✎ Questions suitable for brief writing exercises

▲ More challenging questions

Questions

20.1 ✎ Discuss the factors that cause the chemistry of the elements in the second period to be different from that of the elements in the same group in later periods.

20.2 Why do elements of the second period form stronger π bonds than elements of the third period? Give a specific example of a structural contrast between elemental forms from a single group that can be explained by this difference.

20.3 Compare the electronegativities and ionization energies of metals and nonmetals.

20.4 Why does sulfur form expanded valence shell compounds such as SF_6, whereas oxygen does not?

20.5 Three different bonding modes occur for hydrogen. Describe them and give a specific example of each.

20.6 Describe the bonding of the hydrogen in
(a) KH (b) HCl (c) H_2

20.7 Why does H_2 have such a low boiling point (-253 °C)?

20.8 List the symbols of the three isotopes of hydrogen and the approximate abundance of each.

20.9 ✎ Why is helium rather than hydrogen currently used as the gas in blimps?

20.10 What is the main difference between saturated and unsaturated cooking oils?

20.11 What is the inert pair effect? How does it affect the chemistry of Group 3A?

20.12 Explain why the $+1$ oxidation state is more stable for thallium than for aluminum.

20.13 Classify the Group 3A elements as nonmetals, metals, or metalloids.

20.14 What unusual structural feature is found in the elemental forms of boron?

20.15 Explain why aluminum, a reactive metal, can be used in airplanes and on the exteriors of houses, where it is exposed to the oxygen in the air.

20.16 Explain why Group 4A elements, especially carbon and silicon, are ideally suited to making four electron-pair bonds.

20.17 State three commercial uses for graphite.

20.18 What is meant by the term *adsorption,* and why does activated charcoal have high adsorption qualities?

20.19 How is silicon purified for use in the electronics industry?

20.20 How does doping silicon with phosphorus change its conducting properties?

20.21 What mineral is mined for the production of lead? Describe the process for obtaining the metal from this mineral.

20.22 ✎ Describe the Haber process for the synthesis of ammonia. Be sure to comment on the positive and negative effects that Le Chatelier's principle has on the production of ammonia by this process.

20.23 ✎ Even though phosphates are found widely in most soils, fertilizers are used to supply additional phosphates. What is the problem with the "natural phosphates," and how is this problem overcome in commercial fertilizers?

20.24 Briefly outline the important physical and chemical properties of the two main allotropes of oxygen.

20.25 Classify as acidic, basic, or amphoteric the oxides of the metals, the nonmetals, and the metalloids.

20.26 Why are elements in Group 8A expected to be monomeric and relatively nonreactive?

20.27 What realization led Bartlett to prepare the first compound of a noble gas?

20.28 What is the main source of radon in homes?

Exercises

OBJECTIVE **Discuss how and why the properties of the second-period elements are different from those of other elements in their groups.**

20.29 Of the following pairs of elements, which element is more likely to be able to form a double bond with carbon?
(a) nitrogen or phosphorus (b) oxygen or sulfur

20.30 ■ Of the following pairs of elements, which element is more likely to be able to form a double bond with oxygen?
(a) carbon or silicon (b) oxygen or sulfur

20.31 From the elements nitrogen, silicon, and gallium, pick the ones with the most and the least metallic properties. Explain your choices.

20.32 From the elements silicon, germanium, and tin, pick the ones with the most and the least metallic properties. Explain your choices.

OBJECTIVE **Describe the properties, sources, and important uses of hydrogen.**

20.33 Write the equation for the reaction of NaH and water. What mass of NaH is needed to prepare 1.00 L hydrogen gas at 25 °C and 1.00 atm pressure?

20.34 ■ Write the equation for the reaction of zinc metal with hydrochloric acid. What mass of zinc metal is needed to prepare 1.00 L hydrogen gas at 25 °C and 1.00 atm pressure, assuming excess HCl?

20.35 Give two important industrial preparations for H_2.

20.36 What is the most important industrial use of H_2? Write the equation for this use.

20.37 Write the equation for the water-gas shift reaction.

20.38 Write a series of equations that shows how coal plus water can be converted to methanol. Be careful to balance all equations.

OBJECTIVE **Describe the isolation, purification, and fundamental chemistry of boron and aluminum.**

20.39 Draw the structure of B_2H_6, and describe the bonding in this molecule. What is the hybridization at the boron atoms?

20.40 Describe the bonding in BCl_3. What is the hybridization at the boron atom? Is the boron really electron deficient?

20.41 What is a three-center, two-electron bond?

20.42 How does a three-center, two-electron bond differ from a normal two-center, two-electron bond?

20.43 Because of the high reactivity of the boron hydrides with oxygen, they were considered as possible solid rocket fuels. Write the equation for the reaction of oxygen with $B_{10}H_{14}$ (a solid at room temperature).

20.44 ■ Write the equation and describe the changes in hybridization of the boron atom in the reaction between BCl_3 and NH_3.

20.45 Describe the Hall process for the production of aluminum from the mineral bauxite. How is energy saved by recycling aluminum rather than preparing it by the Hall process?

20.46 What is the thermite reaction, and why can it be used to weld steel?

20.47 Describe the composition of a ruby.

20.48 How is γ-alumina used to purify water?

20.49 Draw the structure of Al_2Cl_6. Compare the bonding in Al_2Cl_6 and B_2H_6.

20.50 The oxide Ga_2O_3 is amphoteric. Write an equation for its reactions, if any, with HCl and NaOH.

OBJECTIVE Discuss the occurrence and chemistry of silicon.

20.51 What are hybridizations of silicon and carbon in SiO_2 and CO_2?

20.52 ■ Draw the structure of silica, SiO_2, and compare it with the structure of CO_2. Why are the structures so different?

20.53 Describe a *p*-type semiconductor based on silicon.

20.54 Describe an *n*-type semiconductor based on silicon.

20.55 What is the hybridization of silicon in $SiCl_4$? Is this compound polar or nonpolar?

20.56 What is the hybridization of silicon in silicates?

OBJECTIVE Discuss the properties of the compounds of nitrogen and phosphorus with hydrogen and oxygen.

20.57 What are the structures of the most common allotropic forms of nitrogen and phosphorus? Explain why they are so different.

20.58 Write the Lewis structure of P_4.

20.59 Nitrogen is the substance isolated in second-largest quantity in the chemical industry. How is it isolated?

20.60 What is meant by the *fixation* of nitrogen with hydrogen?

20.61 Write the Lewis structures of NO_2 and N_2O_3. What is the hybridization of the nitrogen atoms in each compound?

20.62 Write the Lewis structures of N_2O and N_2O_4. What is the hybridization of the nitrogen atoms in each?

20.63 Write an equation for each of the following reactions.
(a) reaction between magnesium and nitrogen
(b) preparation of P_4O_{10}
(c) reaction of nitrogen dioxide with water

20.64 Write an equation for each of the following reactions.
(a) preparation of dinitrogen oxide
(b) reaction of hydrazine with oxygen
(c) reaction between P_4O_{10} and water

20.65 Discuss the structure of $(NO_2)_x$ in both the gas and solid phases.

20.66 Write two reactions that show how NO gas can catalytically decompose large amounts of ozone, O_3.

20.67 How is nitric acid prepared by the Ostwald process?

20.68 ■ The largest use of nitric acid is in the production of ammonium nitrate. Write the equation for this process.

OBJECTIVES Describe the allotropic forms of oxygen and sulfur and their properties, and know the chemistry of compounds containing both oxygen and sulfur.

20.69 Indicate which elements form binary compounds with oxygen.

20.70 Describe two allotropic forms of sulfur.

20.71 Write equations for the industrial preparation of sulfuric acid from sulfur.

20.72 ■ Identify the three main types of reactions that sulfuric acid undergoes.

OBJECTIVE Describe the properties and chemical behavior of the noble-gas elements.

20.73 Draw the Lewis structures and assign the shapes of XeF_2 and XeO_3. Is either of these compounds polar?

20.74 ■ Draw the Lewis structures and assign the shapes of XeF_4 and XeO_4. Is either of these compounds polar?

Cumulative Exercises

20.75 Write the equation for the oxidation of ammonia to nitrogen monoxide (the first step in the Ostwald process). What mass of ammonia is needed to produce 25 kg nitrogen monoxide?

20.76 Write the equation for the preparation of ammonium nitrate from NH_3. What mass of ammonia is needed to produce 5.22 kg ammonium nitrate?

20.77 Compare the boiling points of NH_3 and PH_3, and explain the difference.

20.78 ■ Given that the H–P–H angles in phosphine are about 90 degrees, what orbitals on the phosphorus are used to make the H–P bonds?

20.79 Oxygen can be prepared in the laboratory by heating $KClO_3$ in the presence of MnO_2. Write the equation; then determine the mass of $KClO_3$ needed to produce 0.50 L O_2 gas at 27 °C and 755 torr pressure.

20.80 Write the equation for the roasting of lead sulfide. What volume of SO_2 gas measured at standard temperature and pressure is produced from the roasting of 1.0×10^2 g lead sulfide?

20.81 As outlined in this chapter, hydrazine is used to remove oxygen from water used in boilers of electrical generating plants. If the water contains 2.0×10^{-8} g O_2 per gram of water, what mass of hydrazine is needed to remove the oxygen in 10 tons of water?

20.82 What is the most important crystal force that holds each of the following substances together?
(a) Pb (b) SiO_2 (c) P_4 (d) buckyballs

20.83 Write the Lewis structure and use valence shell electron-pair repulsion theory to predict the shape of XeF_4O_2.

20.84 Draw all possible resonance forms of H_2SO_4 and HNO_3. Explain why resonance forms that do not show formal charges can be written for H_2SO_4 but not HNO_3.

20.85 Write the Lewis structure and molecular orbital diagram for NO. What bond order and number of unpaired electrons does each predict for NO?

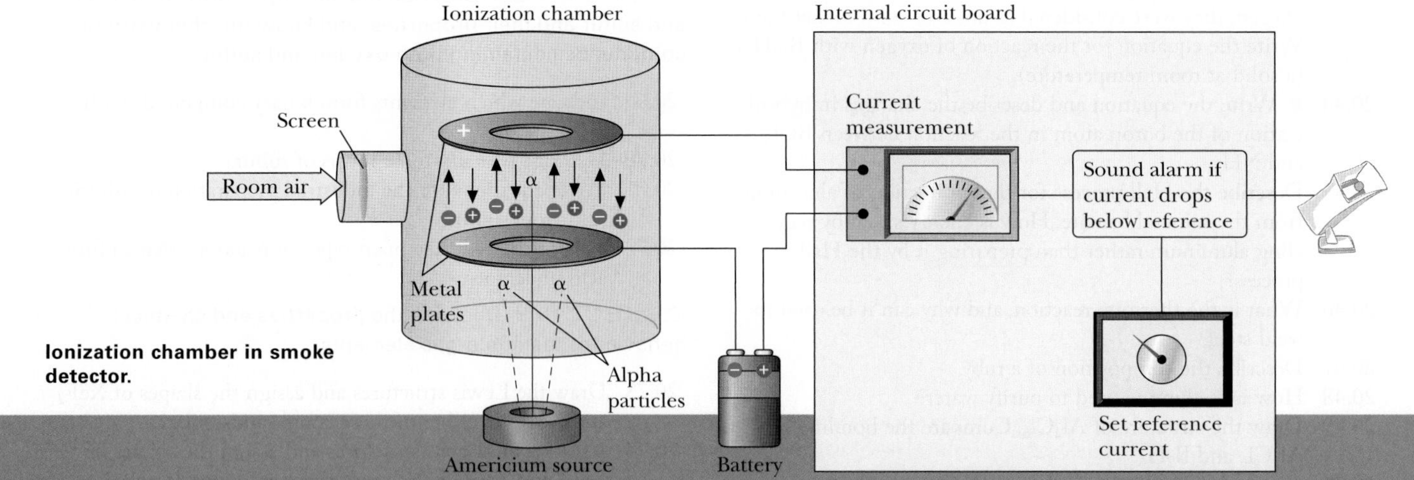

Ionization chamber in smoke detector.

Thousands of lives are saved every year because of smoke detectors. According to the National Fire Protection Association, since the late 1970s, properly installed and maintained smoke alarms have contributed to an almost 50% decrease in deaths caused by fires. Two different types of smoke detectors exist, each of which uses a different principle of chemistry in their operation. Photoelectric smoke detectors use properties of light (see Chapter 7); ionization smoke detectors utilize nuclear chemistry, the topic of this chapter. Some manufacturers also make hybrid models that contain both photoelectric and ionization technology to increase the performance of the detectors.

Ionization smoke detectors use the radioactive element americium to detect smoke particles that generally precede a full-blown fire. In 1944, at the University of Chicago, Glenn Seaborg and colleagues synthesized americium. The most common isotope, Am-241, emits alpha particles (see Section 2.2) together with gamma radiation. The basic operating principle of the detector is that the alpha particles flow steadily into the detector's chamber and the presence of smoke interrupts the alpha particle flow. A great enough flow disruption activates the alarm.

The associated figures show the construction of a smoke detector. A microscopic amount of Am-241 oxide (AmO_2) is mixed with gold and formed into a thin film. The gold film is encapsulated in palladium foil, which serves to prevent americium from entering the environment. However, the palladium foil is thin enough to pass the alpha particles.

A mass of 0.0003 g Am-241 is placed in each detector. The alpha particles enter the ionization chamber, composed of two oppositely charged plates that are attached to a battery through a screen that keeps debris from the chamber.

Nuclear Chemistry

21

CHAPTER CONTENTS

OWL Online homework for this chapter may be assigned in OWL.

Look for the green colored vertical bar throughout this chapter, for integrated references to this chapter introduction.

The alpha particles emitted from the Am-241 ionize the air between the plates, leading to an electrical current. Alpha particles are very efficient at creating ions, thus the term "ionizing radiation" is often applied to alpha particles.

When smoke drifts between the two plates, the smoke particles interact with the ions to decrease the current, causing the smoke detector to alert everyone of a possible threat to their lives. This peaceful application of nuclear radiation has saved countless lives by providing extra time to escape from a fire.

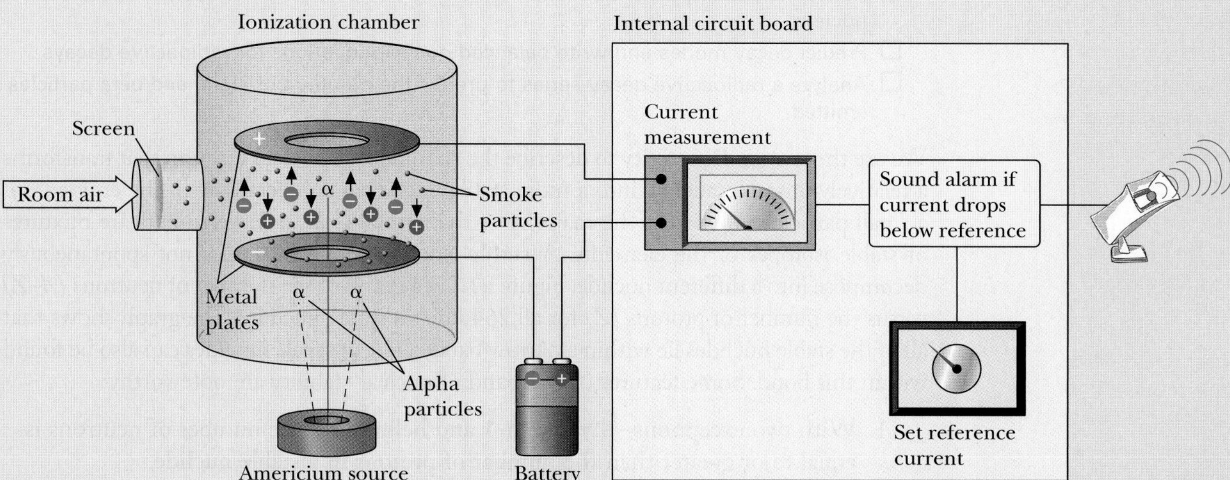

Ionization current decreases in the presence of smoke particles.

Throughout the preceding chapters, the focus has been on chemical reactions, in which changes in the arrangements of the electrons in ions, atoms, and molecules cause chemical changes. In normal chemical reactions, the nuclei of the atoms do not change.

Under other circumstances, however, nuclei do change. Changes in the nuclei of atoms have become increasingly important; therefore, no survey of chemistry is complete without considering nuclear chemistry. This chapter discusses the processes, both natural and induced, that involve changes in the nuclei of atoms. We present important applications of nuclear processes and the new problems created for society in the nuclear age. We emphasize the role of chemistry in the uses of nuclear processes, in their discoveries, and in possible solutions to problems arising from these processes.

A brief review of the nucleus of the atom is necessary before discussing nuclear changes. The nucleus of an atom consists of protons and neutrons held together in a small fraction of the volume occupied by the atom. This chapter uses the word **nucleon** to describe the particles in the nucleus—a proton or a neutron. The radius of an atom is about 10^{-8} cm, whereas that of a nucleus is between 10^{-13} and 10^{-12} cm. The density of nuclear matter is great, about 10^{13} to 10^{14} g/cm^3. A pea-sized object made of nuclear matter would have a mass of more than 1 million *tons!*

Most of the naturally occurring elements are mixtures of several isotopes; that is, atoms with the same number of protons but different numbers of neutrons. The term **nuclide** refers to the nucleus of a particular isotope. The atomic number, *Z,* is the number of protons in a nucleus. The mass number, *A,* is the number of nucleons, or the sum of the number of protons and neutrons in a particular nuclide. The symbol of a nuclide of an atom is represented by

$$^{A}_{Z}\mathrm{X}$$

where X is the chemical symbol for the element (H, He, Fe); the subscript, *Z,* is the atomic number; and the superscript, *A,* is the mass number. Using this notation, we label the three naturally occurring isotopes of oxygen (oxygen-16, oxygen-17, and oxygen-18) with the symbols $^{16}_{8}\mathrm{O}$, $^{17}_{8}\mathrm{O}$, and $^{18}_{8}\mathrm{O}$, respectively.

21.1 Nuclear Stability and Radioactivity

OBJECTIVES

- ☐ Relate the stability of a nuclide to the numbers of protons and neutrons in the nucleus
- ☐ Predict decay modes and write balanced nuclear equations for radioactive decays
- ☐ Analyze a radioactive decay series to predict the numbers of alpha and beta particles emitted

We use the term **radioactivity** to describe the spontaneous nuclear reaction that transforms a relatively unstable nuclide into a more stable nuclide and generally with the emission of a small particle and energy. The majority of the naturally occurring elements are mixtures of stable isotopes of the elements. A **stable isotope** is one that does not spontaneously decompose into a different nuclide. Figure 21.1 is a graph of the number of neutrons *(A–Z)* versus the number of protons *(Z)* for all 264 known stable nuclides. The graph shows that all of the stable nuclides lie within a narrow band; a few unstable nuclides can also be found within this bond. Some features of this band of nuclear stability are noteworthy:

1. With two exceptions—hydrogen-1 and helium-3—the number of neutrons is equal to or greater than the number of protons in a stable nuclide.
2. In the elements with low atomic numbers, the numbers of neutrons and protons in stable nuclei are nearly equal. Above an atomic number of about 20, the ratio of neutrons to protons gradually increases to a maximum of about 1.6:1.
3. Nuclear stability is greater for nuclides that contain even numbers of protons, neutrons, or both (see Table 21.1). The only stable odd-odd nuclei occur in the four lightest elements with odd atomic numbers (H, Li, B, and N), and each stable nuclide has the same number of protons and neutrons.

TABLE 21.1 Numbers of Protons and Neutrons in the Stable Nuclides

Number of Protons	Number of Neutrons	Number of Stable Nuclides	Examples
Even	Even	157	${}^{4}_{2}He$, ${}^{24}_{12}Mg$
Even	Odd	53	${}^{47}_{22}Ti$, ${}^{67}_{30}Zn$
Odd	Even	50	${}^{35}_{17}Cl$, ${}^{63}_{29}Cu$
Odd	Odd	4	${}^{2}_{1}H$, ${}^{14}_{7}N$

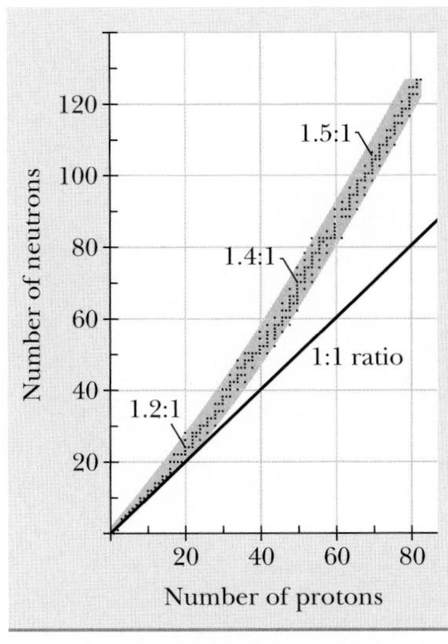

Figure 21.1 Zone of stability of nuclides. Dots represent the combinations of neutrons and protons in the known stable isotopes. As the number of protons increases, the ratio of neutrons to protons needed to produce a stable nuclide also increases.

4. Certain numbers of protons and neutrons confer unusual stability to the nuclides. These numbers, called **magic numbers,** are 2, 8, 20, 26, 28, 50, 82, and 126. There is an excellent correlation between the magic numbers for the nucleus and the unusual electronic stability of the noble gases (2, 10, 18, 36, 54, and 86 electrons).
5. The zone of stability marked in Figure 21.1 contains all of the stable nuclides. However, not all of the nuclides within this band are stable. For example, argon ($Z = 18$) has stable isotopes with mass numbers of 36, 38, and 40, whereas isotopes with mass numbers of 37 and 39 are unstable.
6. None of the elements beyond bismuth ($Z = 83$) have any stable isotopes. Two other elements—technetium, Tc ($Z = 43$), and promethium, Pm ($Z = 61$)—also have no stable isotopes.

These and other observations concerning the properties of atomic nuclei have helped nuclear physicists develop a quantum theory for nuclear particles. There are specific allowed energy states that the nucleons may have, which account for the magic numbers and the emissions that are produced by unstable nuclides.

The stability of a particular isotope depends on the numbers of protons and neutrons.

Types of Radioactivity

Early researchers observed three kinds of emissions from radioactive isotopes, represented by the first three letters of the Greek alphabet: alpha (α), beta (β), and gamma (γ). Quantitative studies of nuclear radiation show that the **alpha particles** are high-energy helium-4 nuclei, **beta particles** consist of high-energy electrons that originate in the nucleus, and **gamma rays** are very-short-wavelength (high-frequency, thus high-energy) electromagnetic radiation. The origin of the gamma rays is similar to that of the light observed in an atomic emission spectrum. An atom in an excited electronic state can return to the ground state by emission of a photon of light (see Section 7.2).

SPL/Photo Researchers, Inc.

Cloud chamber. When radiation passes through a supersaturated gas, the ionization produced causes small droplets of liquid to form, leaving a visible trail. The white lines are the trails left by alpha particles. In this late 1920 photograph by physicist Patrick Blackett shows alpha particles streaming upward into a chamber filled with helium atoms. One particle collides with a helium atom and the 90-degree difference between the two tracks indicates that the mass of the alpha is equal to the mass of the helium atom. The cloud chamber has been used to study radiation since C. T. R. Wilson built the original one in 1911.

TABLE 21.2 Nuclear Particles and Those Involved in Radioactive Decays

Particle	Charge	Mass	Symbol(s)
Proton	+1	1	${}^{1}_{1}H$, ${}^{1}_{1}p$
Neutron	0	1	${}^{1}_{0}n$
Alpha particle	+2	4	${}^{4}_{2}He$, ${}^{4}_{2}\alpha$
Beta particle	−1	0	${}^{0}_{-1}e$, ${}^{0}_{-1}\beta$
Gamma ray	0	0	${}^{0}_{0}\gamma$
Positron	+1	0	${}^{0}_{+1}e$, ${}^{0}_{+1}\beta$

Similarly, when a radioactive nuclide decomposes, it leaves some of the nucleons in excited nuclear states. Gamma rays carry away the energy released as the nucleons return to the ground-state nuclear configuration.

Table 21.2 identifies the particles that make up the nucleus and the kinds of emission observed when unstable nuclides decompose. Included are the mass numbers and atomic numbers used to describe these particles when writing nuclear equations.

A **nuclear equation** describes a process in which nuclides undergo change. A nuclear equation must be balanced, just as a chemical equation must. In a nuclear equation, the sum of the mass numbers (mass balance) and the sum of the atomic numbers (charge balance) on the two sides of the equation must be equal. The complete symbols, including the Z and A for each particle, are used. Examples of the kinds of emission observed in radioactive decay processes follow.

Nuclear equations are balanced when the sums of both the mass numbers and atomic numbers are equal on both sides of the equation.

1. *Alpha decay:* Uranium-238 emits an alpha particle to form thorium-234. The nuclear equation is

$$ {}^{238}_{92}U \longrightarrow {}^{234}_{90}Th + {}^{4}_{2}He $$

The 2+ charge of the alpha particle was not labeled explicitly in this equation. A nuclear charge balance is always required; however, it is not necessary to complicate the nuclear equations with ionic charges, because in nuclear equations there is no need to keep track of the charges on the particles. The equation shows that the neutral uranium atom contains 92 electrons and the neutral thorium atom contains 90 electrons, so the atom loses two electrons when the nuclear decay occurs. The two electrons produced with the thorium atom are eventually captured by an alpha particle when it slows down, forming a neutral helium-4 atom.

2. *Beta decay:* Carbon-14 is a radioactive isotope formed in the outer fringes of the atmosphere by a nuclear reaction of nitrogen-14 with neutrons in the upper atmosphere (see Section 21.3). Carbon-14 emits a beta particle when it decomposes.

$$ {}^{14}_{6}C \longrightarrow {}^{14}_{7}N + {}^{0}_{-1}\beta $$

Note that we use 0 and −1, respectively, for the mass and charge of the beta particle, which are relative to the values of 1 possessed by the proton for each quantity. Beta emission is observed when a neutron in the original nuclide converts to a proton and a beta particle in the product nucleus.

$$ {}^{1}_{0}n \longrightarrow {}^{1}_{1}p + {}^{0}_{-1}\beta $$

It is important to recognize that the nucleus does not contain electrons, even though some unstable nuclides emit them (as beta particles). The beta particle is created by the decomposition of a neutron.

3. *Positron emission* (${}^{0}_{+1}\beta$): A **positron** is a particle that has properties identical to those of the electron except that its charge is positive. Although positrons are not found in the nucleus, they are emitted by some nuclei.

$$ {}^{1}_{1}p \longrightarrow {}^{1}_{0}n + {}^{0}_{+1}\beta $$

Sodium-22 is a radioactive isotope that decomposes by positron emission. The nuclear equation for this decay is

$$^{22}_{11}\text{Na} \longrightarrow ^{22}_{10}\text{Ne} + ^{0}_{+1}\beta$$

4. *Electron capture:* Some unstable nuclides can capture one of the electrons in the atom, usually from the 1*s* subshell, a process that converts one of the protons in the nucleus to a neutron.

$$^{1}_{1}\text{p} + ^{0}_{-1}\text{e} \longrightarrow ^{1}_{0}\text{n}$$

An example of electron capture is the decay of ^{44}Ti.

$$^{44}_{22}\text{Ti} + ^{0}_{-1}\text{e} \longrightarrow ^{44}_{21}\text{Sc}$$

When **electron capture** occurs, the product atom is formed in an excited *electronic* state, because the capture of an electron leaves a vacancy in the 1*s* subshell. The atom returns to its ground electronic state by the emission of x rays. The observed electromagnetic radiation from electron capture is designated x rays, rather than gamma rays, because it arises from the electronic transitions *outside* the nucleus. Generally, gamma radiation has higher energy than x rays, but there is some overlap of the energy ranges of these two sources of electromagnetic radiation.

Both positron emission and electron capture reduce the atomic number of the atom by 1, leaving the mass number unchanged. Positron emission is more common with elements of low *Z*, and electron capture occurs more frequently with elements of high *Z*. Some unstable isotopes decay by both processes.

EXAMPLE 21.1 Balancing Nuclear Equations for Radioactive Decays

Write nuclear equations for the following radioactive decays:

(a) beta decay of radium-228
(b) alpha decay of polonium-210
(c) electron capture by lanthanum-137

Strategy The numbers of nucleons (protons plus neutrons) are conserved. If we know the starting nuclide and the particle emitted or captured, the atomic number and mass number of the product nuclide can be computed by difference.

Solution

(a) From the periodic table, we find that the atomic number of Ra is 88. The beta particle has $A = 0$ and $Z = -1$. We can write part of the nuclear equation as

$$^{228}_{88}\text{Ra} \longrightarrow ? + ^{0}_{-1}\beta$$

The missing nuclide must have $A = 228$ and $Z = 89$ to balance the atomic numbers and mass numbers, respectively. The element with $Z = 89$ is actinium, so the complete equation is

$$^{228}_{88}\text{Ra} \longrightarrow ^{228}_{89}\text{Ac} + ^{0}_{-1}\beta$$

(b) Polonium has an atomic number of 84, and *Z* for the alpha particle is 2; therefore, the product nucleus has $Z = 82$, which is the element lead, Pb. The product nucleus must have a mass number that is 4 less than the unstable polonium nuclide; therefore, $A = 206$. The nuclear equation is

$$^{210}_{84}\text{Po} \longrightarrow ^{206}_{82}\text{Pb} + ^{4}_{2}\alpha \qquad \text{or} \qquad ^{210}_{84}\text{Po} \longrightarrow ^{206}_{82}\text{Pb} + ^{4}_{2}\text{He}$$

(c) Lanthanum is element number 57, so we can write the partial equation as

$$^{137}_{57}\text{La} + ^{0}_{-1}\text{e} \longrightarrow ?$$

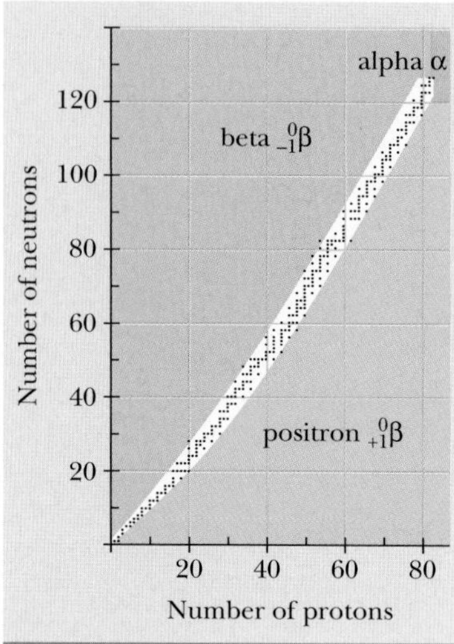

Modes of decay. The type of radioactive decay depends on the location of the unstable nuclide relative to the zone of stability.

The product nucleus has $A = 137 + 0 = 137$ and $Z = 57 - 1 = 56$. The element with an atomic number of 56 is Ba; therefore, the complete equation is

$$ {}_{57}^{137}\text{La} + {}_{-1}^{0}\text{e} \longrightarrow {}_{56}^{137}\text{Ba} $$

Understanding

Write the nuclear equation that forms tin-117 by a positron emission.

Answer ${}_{51}^{117}\text{Sb} \longrightarrow {}_{50}^{117}\text{Sn} + {}_{+1}^{0}\beta$

Predicting Decay Modes

Example 21.1 states what kind of radioactive decay occurred. However, it is possible to predict the mode of decay for a radioactive element from its position relative to the zone of stability shown in Figure 21.1. The unstable heavy elements, with $Z > 83$, generally decay by alpha emission. When elements have too many neutrons (or too few protons) they decay by beta emission that reduces the neutron-to-proton ratio in the nucleus. This type of decay occurs for isotopes that are above and to the left of the zone of stability shown in Figure 21.1. Both positron emission and electron capture occur with nuclides in which the neutron-to-proton ratio is low, the zone to the right of the zone of stability. Figure 21.2 shows an information flow diagram that is useful in predicting the decay mode of a particular isotope.

The position of a radioactive isotope relative to the zone of stability can be determined by comparing its mass number with the rounded atomic mass of the element found on the periodic table. If the atomic mass of the nuclide is greater than the average atomic mass from the periodic table, the nuclide is too heavy (has too many neutrons) to be stable, and a neutron-to-proton decay is observed with the emission of a beta particle. If the atomic mass of the nuclide is less than the average atomic mass from the periodic table, the nuclide is light, and positron emission or electron capture is observed as a proton is converted to a neutron. Example 21.2 illustrates the application of these guidelines.

Decay modes can be predicted by examining the atomic number and then comparing the mass number of the isotope with the mass found in the periodic table.

Figure 21.2 Evaluating decay modes of unstable nuclides.

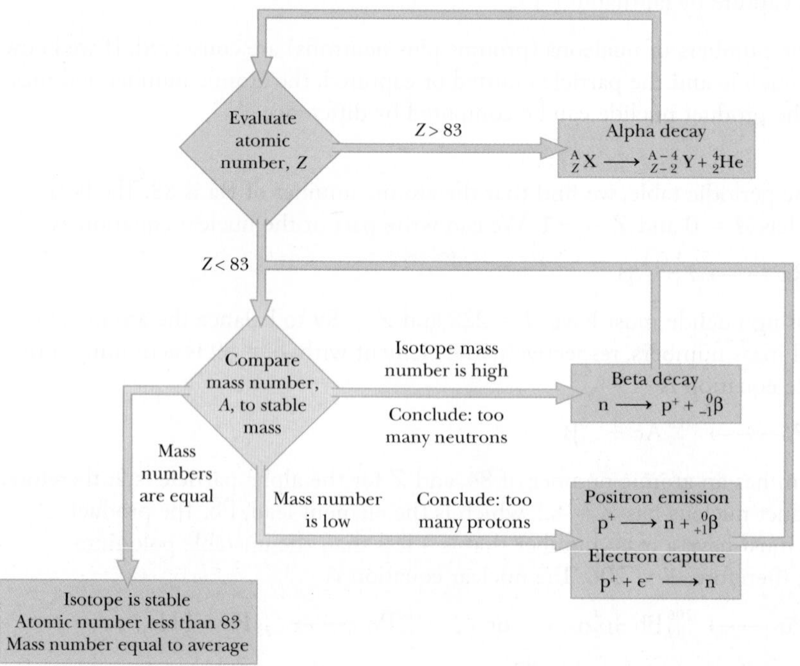

EXAMPLE 21.2 Predicting Radioactive Decays

Predict the kind of decay expected, and write the nuclear equation for each of the following radioactive nuclides.

(a) ^{234}U (b) ^{77}As (c) ^{26}Al

Strategy If the atomic number is greater than 83, alpha emission is observed. If the atomic number is lower, compare the mass number with the atomic mass shown in the periodic table; if the mass number is high, there are too many neutrons and one decays to a proton plus a beta particle. If the mass number is lower than the average from the periodic table, there are too few neutrons or too many protons. A proton converts to a neutron-plus positron particle, or electron capture occurs in this situation.

Solution

(a) The atomic number of uranium is 92, which is higher than 83, so it is likely that this nuclide will emit an alpha particle. The equation is

$$^{234}_{92}U \longrightarrow ^{230}_{90}Th + ^{4}_{2}\alpha$$

(b) We obtain the atomic mass of natural arsenic, 74.92, from the periodic table. The mass number of 77 for the isotope in question is greater than the atomic mass of the stable element, placing it above the zone of stability, so we expect a beta decay to occur. The equation for this decay is

$$^{77}_{33}As \longrightarrow ^{77}_{34}Se + ^{0}_{-1}\beta$$

(c) The aluminum-26 nuclide has a mass number smaller than the atomic mass of the element, 26.98, so it contains too many protons (or too few neutrons) for stability. Positron emission and electron capture are the modes of decay that convert a proton in the nucleus to a neutron, we expect one or both of these modes of decay.

$$^{26}_{13}Al \longrightarrow ^{26}_{12}Mg + ^{0}_{+1}\beta$$

$$^{26}_{13}Al + ^{0}_{-1}e \longrightarrow ^{26}_{12}Mg$$

Experimentally, aluminum-26 decays by both electron capture and positron emission.

Understanding

Predict the kind of decay expected, and write the nuclear equation for the radioactive isotope $^{72}_{30}Zn$.

Answer The mass number of 72 is greater than the stable mass of 65.39, so this nuclide undergoes beta decay.

$$^{72}_{30}Zn \longrightarrow ^{72}_{31}Ga + ^{0}_{-1}\beta$$

Radioactive Series

In a number of cases, the radioactive decay of a nuclide produces another unstable nuclide that also undergoes decay. With the heavy elements, several decays occur in sequence until a stable nuclide is produced. One such series of decays begins with ^{238}U, finally producing the stable isotope ^{206}Pb after several alpha and beta decays have occurred. Figure 21.3 shows the ^{238}U series. The number of alpha and beta decays that occur in such a series can be determined if the parent nuclide and the final stable isotope are known, as shown in Example 21.3.

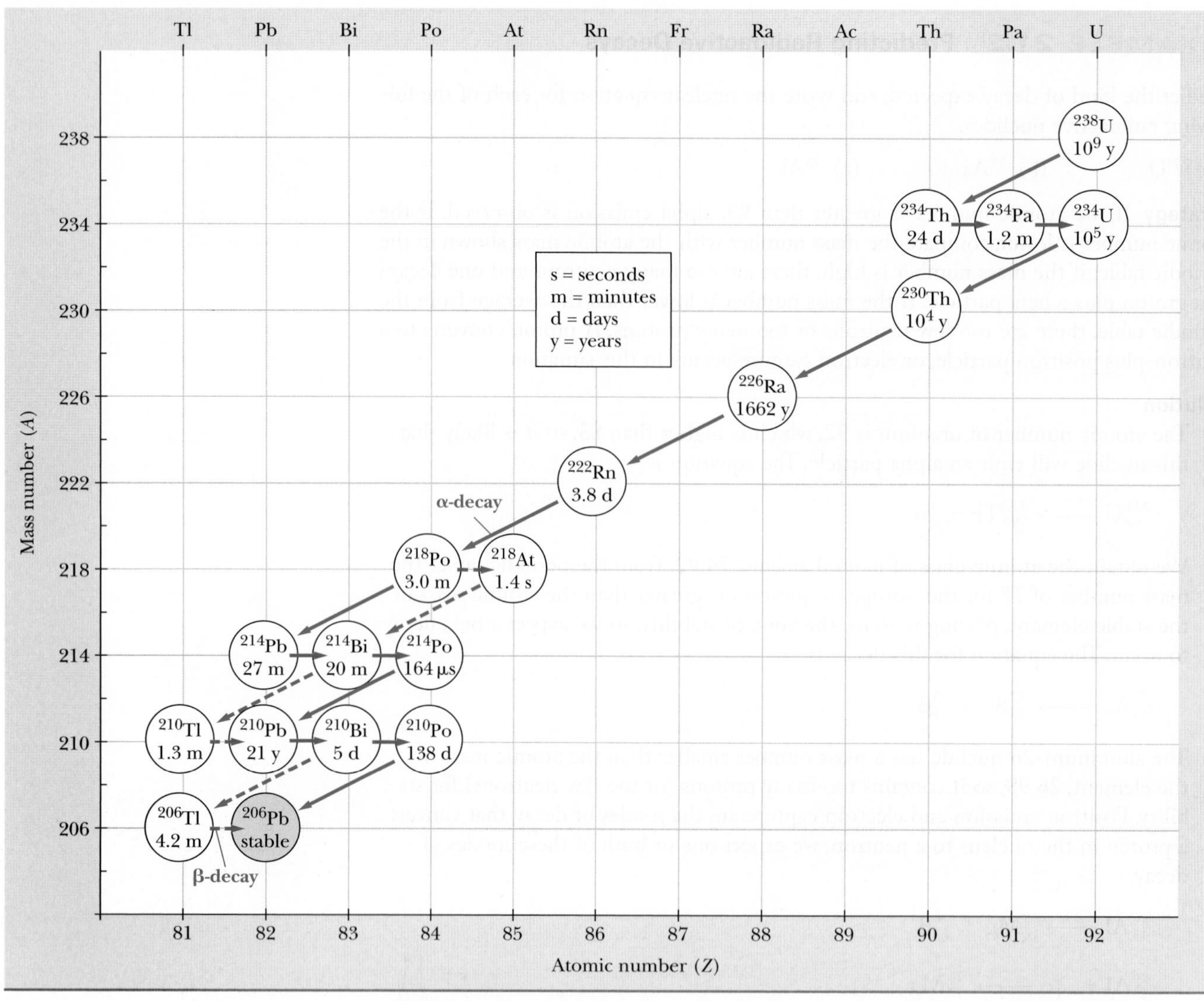

Figure 21.3 Radioactive decay series. The radioactive decay series shown starts at uranium-238 and concludes with lead-206. The half-life of each atom appears in its *circle. Blue arrows* are α decays; *red arrows* represent β decays. *Dashed arrows* indicate secondary decay paths. Note that the slowest decay rate is that of ^{238}U, so nearly all of the uranium that has decayed will be present in the sample as ^{206}Pb.

EXAMPLE 21.3 Analysis of a Decay Series

Determine the numbers of alpha and beta decays needed to change ^{238}U into ^{206}Pb.

Strategy The mass number decreases by 4 for each alpha particle emitted, whereas a beta decay does not change the mass number. The atomic number decreases by 2 for each alpha particle emitted and increases by 1 for each beta particle.

Solution

In the decay of ^{238}U into ^{206}Pb, the change in mass number is

$$\Delta A = 206 - 238 = -32$$

A total of 8 (32/4) alpha particles must be emitted to balance the mass numbers in the decay series.

Once the number of alpha particles is known, we turn our attention to the change in Z. In the present series,

$$\Delta Z = 82 - 92 = -10$$

Each alpha particle emitted must reduce the atomic number by 2, whereas a beta emission increases the atomic number by 1. This information can be expressed in the following equation:

$$\Delta Z = \text{number of } \beta - (2 \times \text{number of } \alpha)$$

$$-10 = \text{number of } \beta - (2 \times 8)$$

$$\text{number of } \beta = 16 - 10 = 6$$

The nuclear equation for the overall change is

$$^{238}_{92}\text{U} \longrightarrow {}^{206}_{82}\text{Pb} + 8\,^{4}_{2}\alpha + 6\,^{0}_{-1}\beta$$

Of course, it is impossible to determine from the given data the order in which the eight alpha and six beta decays occur. Counting the number of alpha and beta decays in Figure 21.3 confirms that our solution is correct.

Understanding

The decay of ^{241}Am in smoke detectors terminates in ^{209}Bi. Find the number of alpha and beta decays that occurred in this transformation.

Answer $8\,^{4}_{2}\alpha$ and $4\,^{0}_{-1}\beta$

Detecting Radioactivity

Nuclear radiation can be detected in a number of ways. All of these methods depend on the ability of high-energy particles to ionize atoms and molecules. The exposure of photographic film by radioactive materials was the earliest means of detecting radioactivity, and it is still used by some people in the nuclear industry. The workers wear film badges that are developed periodically so that their exposure to radiation can be monitored. Film badges are particularly useful for exposure to gamma and x rays; other systems are used where appropriate.

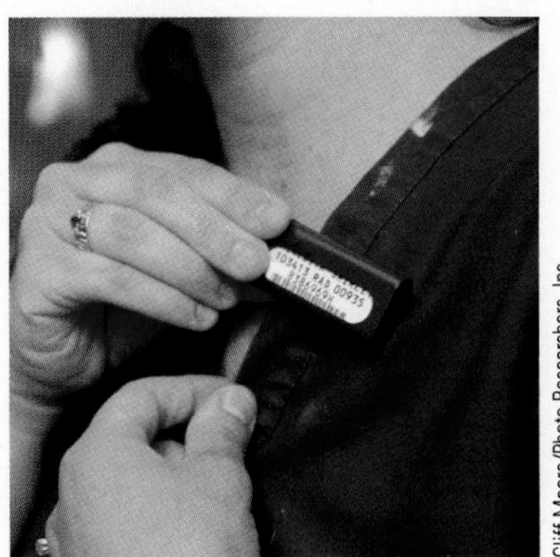

Cliff Moore/Photo Researchers, Inc

A film badge. Workers exposed to gamma radiation and x rays wear film badges. The badges are periodically developed to determine the amount of radiation exposure of the worker.

The Geiger counter, shown schematically in Figure 21.4, is an electrical device for measuring radiation. Most of the portable devices used by radiation-safety officers work on the same principle. A Geiger counter consists of a metal tube with a wire in the center. The tube is filled with a gas at low pressure, and a high voltage is applied between the center wire and the outer tube. Alpha and beta particles enter the tube through a thin mica window and ionize the gas inside the tube. The ions and the electrons cause the gas to conduct electricity for a brief instant, and a pulse of electric current is amplified and detected. One electrical pulse is produced for each alpha or beta particle that enters the tube, but even the thinnest of windows absorb nearly all alpha and some beta particles. Although most gamma radiation passes through the window, the ability of the gamma to ionize the gas in the Geiger counter is small, so much of the gamma radiation is not detected.

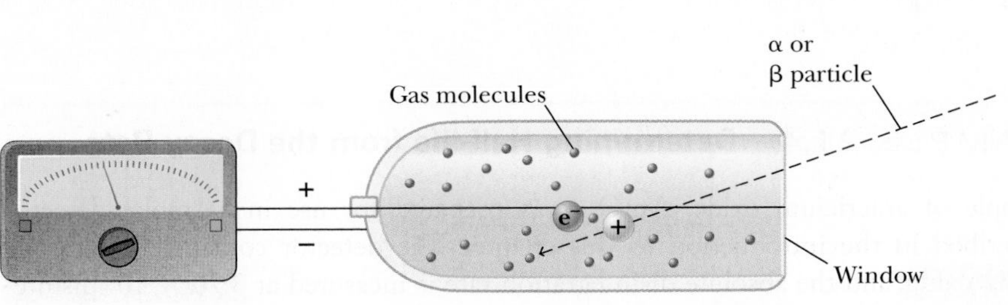

Figure 21.4 Schematic representation of a Geiger counter. Radiation enters the Geiger tube and produces electrons and ions in the gas. A pulse of current is produced for each particle of radiation that enters the tube.

Erwin Reguindin

Figure 21.5 Scintillation counter. Each radioactive particle causes a burst of light, which is converted to an electrical signal that is amplified. The electrical signal is often proportional to the energy of the radiation.

Some substances emit visible light as a result of the ionization caused by radioactivity. In a scintillation counter (Figure 21.5), each particle of ionizing radiation produces a small burst of light, which is converted to an electrical signal by the photon detector that is based on the photoelectric effect (see Section 7.1).

OBJECTIVES REVIEW *Can you:*

- ☑ relate the stability of a nuclide to the numbers of protons and neutrons in the nucleus?
- ☑ predict decay modes and write balanced nuclear equations for radioactive decays?
- ☑ analyze a radioactive decay series to predict the numbers of alpha and beta particles emitted?

21.2 Rates of Radioactive Decays

OBJECTIVES

- ☐ Determine the half-life of a reaction from the absolute decay rate and the number of atoms present
- ☐ Determine the half-life of a reaction from changes in the decay rate
- ☐ Determine the ages of samples of matter from radioactive decay data

One of the characteristics used to identify radioactive nuclides is the rate of decay. Several of the uses of radioactive isotopes of the elements also depend on how rapidly an unstable nuclide emits radiation. Finally, the hazards of radioactive wastes that are frequently discussed in the popular press are closely related to the lifetime of the unstable isotopes. For all these reasons, it is important to understand the laws that dictate the disintegration of the radioactive isotopes of the elements.

Measuring the Half-Lives of Radioactive Materials

The spontaneous decay of any unstable nucleus obeys a first-order rate law.

$$\text{Rate} = \frac{-\Delta N}{\Delta t} = kN \qquad [21.1]$$

The rate is the number of disintegrations per unit time, k is the decay constant, and N is the number of radioactive nuclei present in the sample. This relationship is identical to the rate law for a first-order chemical process, with one important difference. The constant k in a nuclear decay does not change with the temperature, whereas the specific rate constant for a chemical reaction does change. Therefore, it is not necessary to control the temperature when measuring the rate of a radioactive decay.

The rate constant for a nuclear decay can be computed directly from the decay rate and the number of nuclei.

When describing radioactive isotopes, we commonly express the rate of decay in terms of the half-life (the length of time it takes for half of the atoms in the sample to undergo nuclear disintegration) instead of the specific rate constant k. As detailed in Chapter 13, the half-life of a first-order process is related to the rate constant by the equation

$$t_{1/2} = \frac{0.693}{k} \qquad [21.2]$$

EXAMPLE 21.4 Determining Half-life from the Decay Rate

A sample of americium oxide, AmO_2, was prepared for use in a smoke detector, as described in the introduction to this chapter. The detector contains 0.000330 g Am-242 oxide, and the absolute disintegration rate is measured at 3.70×10^7 disintegrations/s. Calculate the half-life of ^{241}Am, and express the answer in years.

Strategy Equation 21.1 relates the decay rate to the number of atoms present, which can be calculated from the mass of AmO_2 and Avogadro's number.

$$\text{Rate} = kN$$

Equation 21.2 relates the half-life to the rate constant

$$t_{1/2} = 0.693/k$$

Solution

First, calculate the number of atoms of Am-241. The molar mass of $^{241}AmO_2$ is 273 g/mol.

$$\text{Amount of } ^{241}\text{Am} = 0.000330 \text{ g} \times \frac{1 \text{ mol}}{273 \text{ g}} = 1.21 \times 10^{-6} \text{ mol}$$

$$\text{Atoms } ^{241}\text{Am} = 1.21 \times 10^{-6} \text{ mol} \left(\frac{6.02 \times 10^{23} \text{ atoms } ^{241}\text{Am}}{\text{mol}}\right)$$

$$= 7.28 \times 10^{17} \text{ atoms } ^{241}\text{Am}$$

Use Equation 21.1 to find k from the number of atoms and the given decay rate, 3.70×10^7 disintegrations/s.

$$\text{Rate} = kN$$

$$3.70 \times 10^7 \frac{\text{atoms}}{\text{s}} = k \times 7.28 \times 10^{17} \text{ atoms}$$

$$k = \frac{3.70 \times 10^7 \text{ atoms/s}}{7.28 \times 10^{17} \text{ atoms}} = 5.08 \times 10^{-11} \text{s}^{-1}$$

Obtain the half-life from the rate constant by using Equation 21.2.

$$t_{1/2} = \frac{0.693}{5.08 \times 10^{-11} \text{s}^{-1}} = 1.36 \times 10^{10} \text{ s}$$

Finally, convert the half-life to years.

$$t_{1/2} = 1.36 \times 10^{10} \text{ s} \times \left(\frac{1 \text{ min}}{60 \text{ s}}\right) \times \left(\frac{1 \text{ hr}}{60 \text{ min}}\right) \times \left(\frac{1 \text{ day}}{24 \text{ hr}}\right) \times \left(\frac{1 \text{ yr}}{365.25 \text{ day}}\right) = 431 \text{ yr}$$

Although people should replace the batteries in their smoke detectors once a year, the ^{241}Am sources will last a lifetime.

Understanding

A sample that contains 4.50×10^{14} atoms of a radioactive isotope decays at a rate of 503 disintegrations/min. What is the half-life (in years) of this isotope?

Answer 1.18×10^6 years

The half-lives for known radioactive nuclides cover a very wide range, from fractions of a second up to more than 10^{15} years. If the rate constant is large enough that the rate changes a measurable amount over a period, the integrated rate law can be used to determine the half-life for the radioactive decay.

$$\ln\left(\frac{N}{N_0}\right) = -kt = -\frac{0.693}{t_{1/2}}t$$

where N is the number of radioactive atoms present at time t, when N_0 atoms were present at time $t = 0$. Because the rate of decay is proportional to the number

The half-life and rate constant for a nuclear decay can be determined measuring the decay rate at different times.

of atoms present, we can use rates, as well as numbers of atoms in the following equation:

$$\ln\left(\frac{N}{N_0}\right) = \ln\left(\frac{R}{R_0}\right) = -kt = -\frac{0.693}{t_{1/2}}t \qquad [21.3]$$

Suppose a sample containing iodine-131 has a decay rate of 3153 disintegrations/min at $t = 0$. The decay rate of this sample 52.5 hours later is 2613 disintegrations/min. What is the half-life of ^{131}I? The solution to this problem uses Equation 21.3.

$$\ln\left(\frac{R}{R_0}\right) = -kt$$

$$\ln\left(\frac{2613}{3153}\right) = -k \times 52.5 \text{ h}$$

Substitute for k using the relationship between k and $t_{1/2}$

$$k = 0.693/t_{1/2}$$

$$\ln\left(\frac{2613}{3153}\right) = -\frac{0.693}{t_{1/2}} \times 52.5 \text{ h}$$

and solve for $t_{1/2}$.

$$t_{1/2} = -\frac{0.693 \times 52.5 \text{ h}}{\ln(0.8287)} = 194 \text{ h}$$

This type of calculation is shown in Example 21.5.

EXAMPLE 21.5 Finding Half-life from Changes in the Decay Rate

The nuclide ^{31}Si is a radioactive isotope that disintegrates by beta decay.

$$^{31}_{14}\text{Si} \longrightarrow {}^{31}_{15}\text{P} + {}^{0}_{-1}\beta$$

A sample of silicon-31 is found to produce 251 disintegrations/s. Exactly 3.00 hours later, the same sample produces 113 disintegrations/s. What is the half-life of ^{31}Si?

Strategy The relative rate of decay is proportional to the number of atoms present, so we can use Equation 21.3 to calculate the half-life as follows:

$$\ln\left(\frac{N}{N_0}\right) = -kt = -\left(\frac{0.693}{t_{1/2}}\right) \times t$$

Solution

Because the disintegration rate at any time is proportional to the number of radioactive atoms,

$$\frac{N}{N_0} = \frac{R}{R_0} = \frac{113}{251} = 0.450$$

Substitute this value and the time, 3.00 hours, into Equation 21.3 and solve for $t_{1/2}$.

$$\ln(0.450) = -\left(\frac{0.693}{t_{1/2}}\right) \times 3.00 \text{ h}$$

$$t_{1/2} = \frac{-0.693 \times 3.00 \text{ h}}{-0.798} = 2.60 \text{ h}$$

This value for the half-life is reasonable because slightly more than half of the atoms decayed in 3.00 hours.

Understanding

The radioactive decay of a sample containing ^{41}Ar produces 555 disintegrations/min. The rate decreases to 314 disintegrations/min exactly 90.0 minutes later. Calculate the half-life of ^{41}Ar.

Answer 110 minutes

The integrated rate law cannot be used to determine the half-life of a very long-lived isotope such as ^{36}Cl, which has a half-life of 3.1×10^5 years. The change in the decay rate of such an isotope is too small to measure accurately in any reasonable period. Instead, we would use Equation 21.1, as illustrated in Example 21.4.

Dating Artifacts by Radioactivity

Archaeologists and geologists use their knowledge of the rates of radioactive decays to determine the ages of objects. One of the techniques for determining the age of a sample, radiocarbon dating, is useful with organic matter. Atmospheric carbon dioxide contains a very small amount of carbon-14 (^{14}C), a beta emitter with a half-life of 5730 years. The ^{14}C forms by reaction between nitrogen and neutrons that are produced by collisions between cosmic rays and gas molecules in the upper atmosphere. If we assume that the ^{14}C content of the atmosphere has remained constant for the past several thousand years, then we can determine the age of a sample.

The ^{14}C in a sample of atmospheric carbon dioxide produces about 15.3 disintegrations per minute per gram of carbon. Plants and animals incorporate this radioactive isotope within their cells during their lifetimes. When an organism dies, it no longer exchanges carbon with the atmosphere and the amount of ^{14}C decreases because of the radioactive decay. The known initial rate of decay, 15.3 disintegrations per minute per gram of carbon, and the current rate of decay allow us to determine the age of the sample, as shown in Example 21.6.

The age of a sample can be determined from the decay rate of a radioactive isotope and the amount of the isotope present.

EXAMPLE 21.6 Radiocarbon Dating

An artifact found in an ancient Egyptian tomb produced 11.8 disintegrations of ^{14}C per minute per gram of carbon in the sample. Estimate the age of this sample, assuming that its original radioactivity was 15.3 disintegrations per minute per gram of carbon.

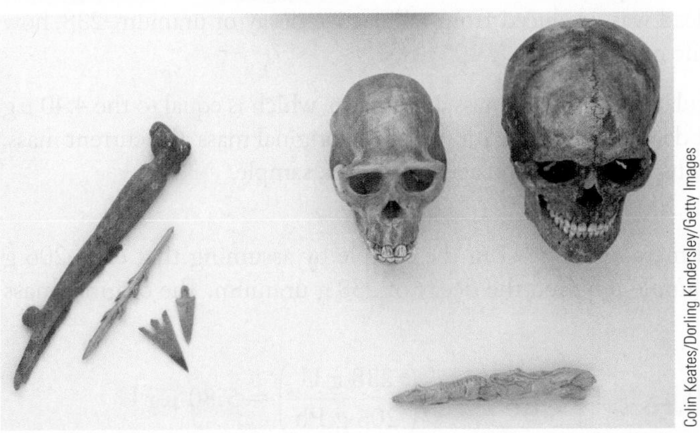

Colin Keates/Dorling Kindersley/Getty Images

Dating artifacts. The human (and chimpanzee) skulls can be dated by C-14 measurements because they contain carbon. The flint (stone) tools and artifacts found buried next to these skulls are dated by their association with the skulls—any other dating method would provide the dates and ages of the stones themselves, rather than the dates that they were used as tools.

Strategy Equation 21.3 relates the rate of decay at two different times to the rate constant and time:

$$\ln\left(\frac{N}{N_0}\right) = \ln\left(\frac{R}{R_0}\right) = -kt = -\frac{0.693}{t_{1/2}}t$$

Solution

The rate of decay at time zero is 15.3 disintegrations/min; at the current time, the rate is 11.8 disintegrations/min; and the half-life of ^{14}C is 5730 years.

$$\ln\left(\frac{11.8}{15.3}\right) = -\left(\frac{0.693}{5730\ \text{yr}}\right) \times t$$

$$t = \frac{-5730\ \text{yr} \times \ln(0.771)}{0.693} = 2.15 \times 10^3\ \text{yr}$$

Understanding

What is the age of a bone found by an archaeologist if its ^{14}C activity was 2.9 disintegrations per minute per gram of carbon?

Answer 1.4×10^4 years

The accuracy with which the age of a sample can be predicted with ^{14}C dating depends on the 5730-year half-life and on fluctuations in the abundance of ^{14}C in the environment from year to year. Carbon-14 dating of samples less than 500 years old is impractical because of the uncertainty in its abundance and the small fraction of the isotope that has decayed. For samples older than 50,000 years, an accurate age determination is impossible because nearly all of the ^{14}C has decayed.

Other radioactive elements provide ways to determine the ages of very old samples. Scientists have estimated the ages of some uranium ore deposits by another dating technique. We have seen (see Example 21.3) that ^{238}U decays and finally produces ^{206}Pb, a stable isotope. The slowest decay in this series is that of ^{238}U, with a half-life of 4.51×10^9 years (see Figure 21.3), so nearly all of the uranium that has decayed is present in the sample as ^{206}Pb. Assuming that no lead was present in the sample when the deposit formed, the $^{206}Pb/^{238}U$ ratio can be used to determine the age of the mineral deposit. With the use of this technique, a number of corrections must be made and tests run to verify the assumptions. Example 21.7 illustrates the use of this method.

EXAMPLE 21.7 Using Pb/U Ratio for Dating

The mass spectrum of a rock sample shows that it contains 4.40 μg ^{238}U and 1.21 μg ^{206}Pb. Assuming that all of the lead was produced from radioactive decay of uranium-238, how old is the rock? The half-life of ^{238}U is 4.5×10^9 years.

Strategy We need to calculate the original mass of uranium, which is equal to the 4.40 μg present, plus the mass that decayed into 1.21 μg ^{206}Pb. The original mass, the current mass, and the half-life will allow us to calculate the age of the rock sample.

Solution

We calculate the original mass of the ^{238}U in the sample by assuming that each 206 g lead now present in the sample required the decay of 238 g uranium. The original mass of uranium-238 was

$$\text{Original mass} = 4.40\ \mu\text{g U} + 1.21\ \mu\text{g Pb} \times \left(\frac{238\ \text{g U}}{206\ \text{g Pb}}\right) = 5.80\ \mu\text{g U}$$

Thus, during the lifetime of the rock, the original 5.80 μg of the uranium decreased to 4.40 μg today. Using these quantities in the integrated rate law (Equation 21.3), together with the half-life of ^{238}U, we can solve for the age of the rock.

$$\ln\left(\frac{4.40}{5.80}\right) = -\frac{0.693}{4.5 \times 10^9 \text{ yr}}t$$

$$t = -\frac{4.5 \times 10^9 \text{ yr}}{0.693}(-0.276) = 1.8 \times 10^9 \text{ yr}$$

Understanding

The ratio of ^{40}Ar/^{40}K can also be used for the dating of minerals, because ^{40}Ar is a stable nuclide.

$$^{40}_{19}\text{K} \longrightarrow {}^{40}_{18}\text{Ar} + {}^{0}_{+1}\beta \qquad t_{1/2} = 1.28 \times 10^9 \text{ years}$$

Analysis of a rock showed that the atom ratio of ^{40}Ar/^{40}K was 0.44. What is the age of the sample?

Answer 6.7×10^8 years

OBJECTIVES REVIEW *Can you:*

- ☑ determine the half-life of a reaction from the absolute decay rate and the number of atoms present?
- ☑ determine the half-life of a reaction from changes in the decay rate?
- ☑ determine the ages of samples of matter from radioactive decay data?

21.3 Induced Nuclear Reactions

OBJECTIVE

- ☐ Write the nuclear equations that describe transmutations

Scientists who observed natural nuclear reactions soon realized that it might be possible to create new isotopes in their own laboratories. In 1919, Ernest Rutherford showed that bombarding nitrogen with alpha particles produced protons. The nuclear reaction that occurs is

$$^{14}_{7}\text{N} + {}^{4}_{2}\text{He} \longrightarrow {}^{17}_{8}\text{O} + {}^{1}_{1}\text{p}$$

James Chadwick discovered the neutron in 1939 when he bombarded beryllium with alpha particles.

$$^{9}_{4}\text{Be} + {}^{4}_{2}\alpha \longrightarrow {}^{12}_{6}\text{C} + {}^{1}_{0}\text{n}$$

© Hulton-Deutsch Collection/CORBIS

(a)

Argonne National Laboratory photo

(b)

Nuclear Chemistry Research Laboratories. *(a)* Laboratory of Ernest Rutherford, 1926. *(b)* Aerial view of the Advanced Photon Source, one of the resources available to researchers at Argonne National Laboratory.

Nuclear reactions can be induced by bombarding target nuclei with high-energy particles.

These two reactions are examples of **induced nuclear reactions** or **nuclear transmutations,** reactions in which two particles or nuclei produce elements or isotopes that are different from the reactant species.

Although the first transmutation of elements involved the reactions of alpha particles with other nuclei, these reactions are generally unfavorable. Reaction does not occur until the nucleus and the alpha particle are within 10^{-12} cm of each other. To reach this small distance, the alpha particle must have a very high energy so that it can overcome the electrostatic repulsion by the nucleus, because both it and the nucleus have positive charges. The energy required to get a single alpha particle close enough to the beryllium nucleus to react is about 10^{-12} J/particle or 6×10^8 kJ/mol, an enormous energy.

For elements with high atomic numbers, much higher energies are needed to induce nuclear reactions by bombardment with positively charged particles. Several machines have been developed that accelerate protons, alpha particles, and other small nuclei to these high energies. In one of these machines, the cyclotron, an alternating voltage is

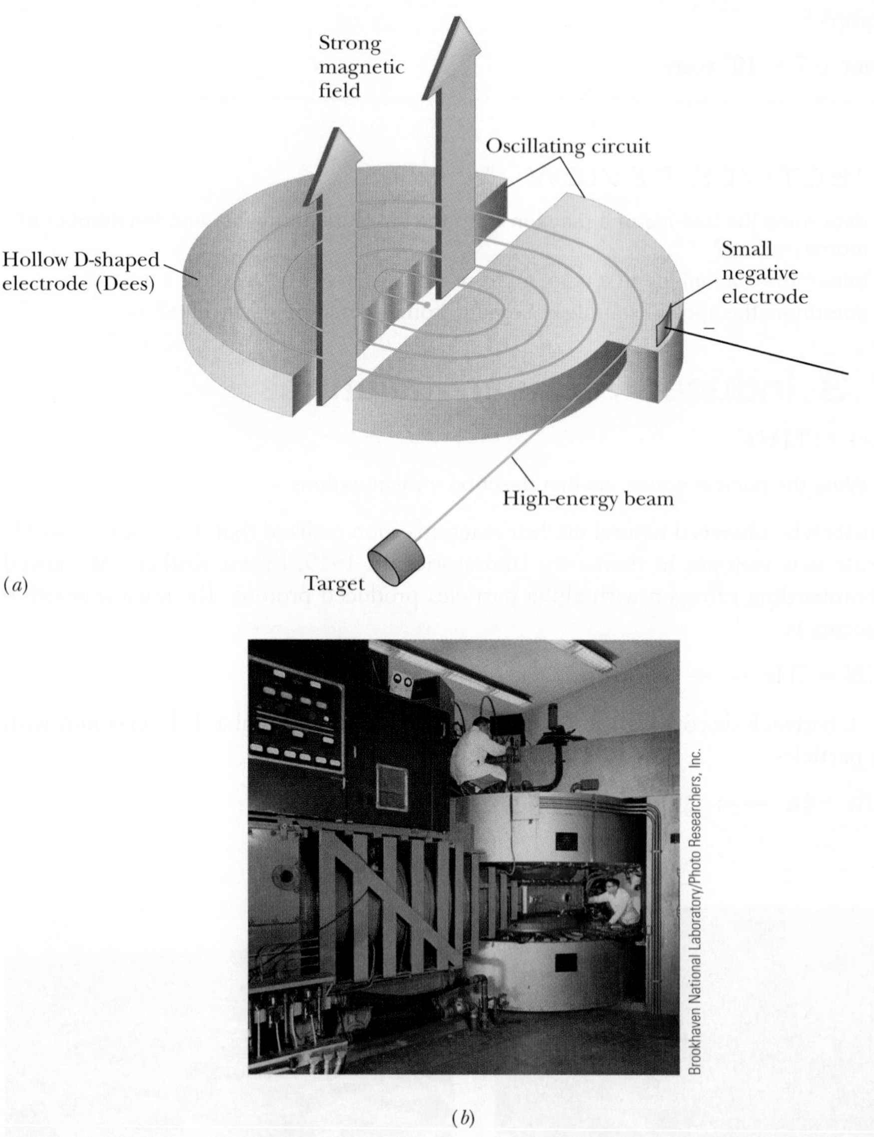

Figure 21.6 Cyclotron. The cyclotron is used to accelerate protons and other small, positively charged particles to very high energies. An alternating voltage is applied to the two electrodes and a large magnetic field is used to produce the spiral path of the accelerated particles. *(a)* Schematic diagram of the cyclotron. The small negative electrode deflects the beam of protons to the exit port. *(b)* 60-inch diameter cyclotron. This device was built and operated at the Brookhaven National Laboratory, Long Island, NY.

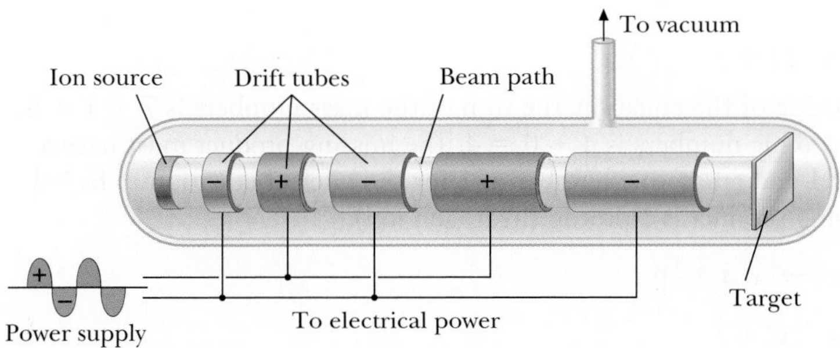

Figure 21.7 Linear accelerator. Positive ions are introduced into the first tube. By the time these ions leave the first tube, the voltage has been reversed and they are accelerated further between the first and second tubes. Very high ion energies are achieved by repeating this process several times. The consumption of electrical power by a linear accelerator is so great that its use is often restricted to nonpeak hours.

applied between two hollow D-shaped electrodes. The particles are accelerated by the electric field when they are in the gap between the two electrodes. A magnetic field perpendicular to the electrodes causes the ions to follow a spiral of increasing radius as their velocities increase. Figure 21.6*a* is a schematic diagram for a cyclotron.

The linear accelerator, another device used to accelerate positive ions, is shown schematically in Figure 21.7. Positive ions are introduced into the first tube. The voltage accelerates the ions between the first and second tubes. By the time these ions leave the second tube, the voltage has been reversed, and they are accelerated further between the second and third tubes. Each successive tube is longer than the previous one, so the particle arrives at the next gap just in time to coincide with the maximum accelerating voltage. In 1958, scientists at the Lawrence Radiation Laboratory in California used the linear accelerator to produce the first sample of nobelium ($Z = 102$) by bombarding a sample of curium with high-energy ^{12}C nuclei. The reaction that occurred was

Accelerators produce charged particles that are used to initiate nuclear reactions.

$$^{246}_{96}\text{Cm} + ^{12}_{6}\text{C} \longrightarrow ^{254}_{102}\text{No} + 4^{1}_{0}\text{n}$$

When neutrons are used as the bombarding particles, there is no electrostatic repulsion barrier, so nuclear transmutations occur at much lower energies. Essentially all nuclides react with neutrons, so most radioactive elements are produced by the reactions of neutrons with nuclei. Some examples of neutron-induced nuclear reactions are as follows:

$$^{35}_{17}\text{Cl} + ^{1}_{0}\text{n} \longrightarrow ^{36}_{17}\text{Cl} + ^{0}_{0}\gamma$$

$$^{10}_{5}\text{B} + ^{1}_{0}\text{n} \longrightarrow ^{7}_{3}\text{Li} + ^{4}_{2}\alpha$$

$$^{14}_{7}\text{N} + ^{1}_{0}\text{n} \longrightarrow ^{14}_{6}\text{C} + ^{1}_{1}\text{p}$$

The last reaction is quite important because it is the natural source of the ^{14}C that is used in radiocarbon dating (see Section 21.2). Example 21.8 illustrates the use of nuclear equations to describe the reactions of small particles with other nuclei.

Low-energy neutrons react with most nuclides.

EXAMPLE 21.8 Nuclear Equations for Transmutations

In each part, write a complete, balanced nuclear equation.

(a) A neutron transforms ^{7}Be into ^{7}Li.
(b) A neutron reacts with ^{17}O to form ^{14}C.
(c) ^{69}Ga reacts with a neutron and emits a gamma ray.
(d) ^{238}U is bombarded with ^{16}O ions, forming ^{250}Fm.

Strategy The sum of the atomic numbers and the sum of the mass numbers on the product side must equal those on the reactant side. The atomic number and mass number of the missing particle is assigned by difference.

Solution

(a) $^{7}_{4}\text{Be} + ^{1}_{0}\text{n} \longrightarrow ^{7}_{3}\text{Li} + ?$

On the reactant side of the equation, the sum of the mass numbers is 7 + 1 = 8, and that of the atomic numbers is 4 + 0 = 4. The missing product must have a mass number of 1 and an atomic number of 1 for the nuclear equation to be balanced. The missing product is a proton (hydrogen ion).

$$^{7}_{4}\text{Be} + ^{1}_{0}\text{n} \longrightarrow ^{7}_{3}\text{Li} + ^{1}_{1}\text{p}$$

(b) $^{17}_{8}\text{O} + ^{1}_{0}\text{n} \longrightarrow ^{14}_{6}\text{C} + ?$

To balance the mass numbers in the equation, the missing product must have a mass number of 17 + 1 − 14 = 4. The atomic number is 8 + 0 − 6 = 2. The particle with A = 4 and Z = 2 is an alpha particle, so the complete equation is

$$^{17}_{8}\text{O} + ^{1}_{0}\text{n} \longrightarrow ^{14}_{6}\text{C} + ^{4}_{2}\alpha$$

It is also correct to write

$$^{17}_{8}\text{O} + ^{1}_{0}\text{n} \longrightarrow ^{14}_{6}\text{C} + ^{4}_{2}\text{He}$$

(c) $^{69}_{31}\text{Ga} + ^{1}_{0}\text{n} \longrightarrow ? + ^{0}_{0}\gamma$

Both the mass number and the atomic number are equal to zero for a gamma ray, so the product isotope has the same atomic number as the reacting nucleus and is a gallium atom (Z = 31); the mass number increases by 1 (Z = 70). The balanced nuclear equation is

$$^{69}_{31}\text{Ga} + ^{1}_{0}\text{n} \longrightarrow ^{70}_{31}\text{Ga} + ^{0}_{0}\gamma$$

(d) $^{238}_{92}\text{U} + ^{16}_{8}\text{O} \longrightarrow ^{250}_{100}\text{Fm} + ?$

Balancing the mass numbers in the partial equation shows that four mass units are missing from the product side, whereas the atomic numbers are balanced. Because the neutron is the only particle that has mass but no charge, four neutrons must be produced in this reaction.

$$^{238}_{92}\text{U} + ^{16}_{8}\text{O} \longrightarrow ^{250}_{100}\text{Fm} + 4\,^{1}_{0}\text{n}$$

Understanding

Write the balanced nuclear reaction for the bombardment of ^{56}Fe with protons to form ^{56}Co.

Answer $^{56}_{26}\text{Fe} + ^{1}_{1}\text{p} \longrightarrow ^{56}_{27}\text{Co} + ^{1}_{0}\text{n}$

OBJECTIVE REVIEW *Can you:*

☑ write the nuclear equations that describe transmutations?

21.4 Nuclear Binding Energy

OBJECTIVES

☐ Calculate the mass defect of a nuclide
☐ Calculate the nuclear binding energy from the mass defect
☐ Express the binding energy in megaelectron volts per nucleon (MeV/nucleon)

The energy released by nuclear weapons and in nuclear power plants is testimony to the tremendous energies that are involved in the formation and disintegration of the nuclei of atoms. Einstein's theory of relativity shows that mass is equivalent to energy through the following equation:

$$E = mc^2 \qquad [21.4]$$

where c is the speed of light (2.9979 × 10^{8} m/s), m is the mass in kilograms, and E is the energy in joules. According to Einstein's equation, for any exothermic process, the

TABLE 21.3 Masses of the Hydrogen-1 Atom and the Neutron

		Mass	
Particle	Symbol	Atomic Mass Units (u)	Grams (g)
Hydrogen-1 atom*	${}^{1}_{1}H$	1.007825	1.67353×10^{-24}
Neutron	${}^{1}_{0}n$	1.008665	1.67496×10^{-24}

*Because the measured mass of an atom includes Z electrons, the mass of the hydrogen-1 atom (one proton and one electron) must be used to calculate the mass defect.

products have a smaller mass than the reactants. For any chemical change, the predicted difference in mass is much smaller than the uncertainty in the measurement of the mass. For example, consider the combustion of methane:

$$CH_4(g) + 2O_2(g) \rightarrow CO_2(g) + 2H_2O(\ell) \qquad \Delta H^\circ = -890 \text{ kJ}$$

The mass change we calculate is less than 1×10^{-8} g when 80 g of reactants is converted to the products. Thus, although violated in theory, the law of conservation of mass for any ordinary chemical change is valid within experimental accuracy.

When nuclear changes take place, the energy changes are much larger, and measurable differences in mass occur. The **nuclear binding energy** is the amount of energy required to keep the protons and neutrons bound together in any nuclide, overcoming the coulombic repulsions among the positive charges. The source of this energy is the conversion of some of the mass of the nucleons (Equation 21.4). Suppose we measure the mass of an atom and then compare it with the masses of an equivalent number of atoms of hydrogen-1 (a proton and electron) and neutrons (Table 21.3). In every case (except the hydrogen-1 atom itself), the mass of the atom is less than the mass of the individual particles in it. The difference is known as the **mass defect,** and because of the large energies involved in nuclear reactions, this mass defect is easily determined. Example 21.9 demonstrates this type of calculation.

Calculate the mass defect of an isotope by the difference between the sum of the masses of the individual particles and the measured mass of the atom.

EXAMPLE 21.9 Calculate the Mass Defect of a Nucleus

Calculate the mass defect of ${}^{4}He$. The atomic mass of the atom is 4.002602 atomic mass units (u).

Strategy Use the information in Table 21.3, to calculate the atomic masses of the nucleons; then calculate the mass defect by the difference from its atomic mass of 4.002602 u.

Solution

The ${}^{4}He$ atom has two protons (plus electrons) and two neutrons.

$$\text{mass of 2 } {}^{1}H = 2 \times 1.007825 = 2.015650 \text{ u}$$

$$+ \text{ mass of 2 } {}^{1}n = 2 \times 1.008665 = 2.017330 \text{ u}$$

$$\text{sum of nucleon masses} = 4.032980 \text{ u}$$

$$- \text{ mass of } {}^{4}He = 4.002602 \text{ u}$$

$$\text{difference (mass defect)} = 0.030378 \text{ u}$$

Understanding

Calculate the mass defect for ${}^{16}O$. The mass of the atom is 15.99502 u.

Answer 0.1369 u

The nuclear binding energy, ΔE, can be calculated from the mass defect:

$$\Delta E = \Delta mc^2 \qquad [21.5]$$

Calculate the nuclear binding energy from the mass defect.

where Δm is the mass defect. To obtain joules as the appropriate unit for the nuclear binding energy, you must express the mass defect in units of kilograms and the speed of light in units of meters per second. This process is illustrated in Example 21.10.

EXAMPLE 21.10 Calculating Nuclear Binding Energy from Mass Defect

What is the energy change when 1 mol ^{4}He is made from fundamental particles?

Strategy Use the results of the calculation of Example 21.9 and Equation 21.5:

$$\Delta E = \Delta mc^2$$

Remember to express the mass defect in kilograms and the speed of light in meters per second.

Solution

The mass defect of 1 mol ^{4}He is 0.030378 u; therefore, 1 mol ^{4}He will have a mass defect of 0.030378 g, or 3.0378×10^{-5} kg. Substitute this mass defect in Equation 21.5:

$$\Delta E = (3.0378 \times 10^{-5}\ \text{kg/mol}) \times (2.9979 \times 10^8\ \text{m/s})^2 = 2.7302 \times 10^{12}\ \text{J/mol}$$

This energy is roughly 10 million times the energy of a typical covalent bond. A mass of 54 *tons* of natural gas would have to be burned to produce the energy that is released when 1 mol (4 g) helium is formed from the nucleons.

Understanding

The mass defect for ^{16}O is 0.1369 u. Express the nuclear binding energy for this nucleus in kilojoules per mole.

Answer 1.230×10^{10} kJ/mol

Nuclear binding energies are extremely large, so a different energy unit, the mega-electron volt (MeV), traditionally has been used to express these energies.

$$1\ \text{MeV} = 1.602 \times 10^{-13}\ \text{J}$$

A convenient equivalency for the direct conversion of atomic mass units to megaelectron volts can be obtained by first using Equation 21.4 to calculate the energy equivalent of 1 amu, which is $1.6605402 \times 10^{-27}$ kg.

$$1\ \text{amu} = mc^2 = 1.6605 \times 10^{-27}\ \text{kg}\ (2.9979 \times 10^8\ \text{m/s})^2 = 1.4924 \times 10^{-10}\ \text{J}$$

Then use the definition of the megaelectron volt to convert the energy units.

$$1\ \text{u} = 1.4924 \times 10^{-10}\ \cancel{\text{J}} \times \left(\frac{1\ \text{MeV}}{1.602 \times 10^{-13}\ \cancel{\text{J}}}\right) = 931.5\ \text{MeV}$$

Expressing the binding energy in megaelectron volts per nucleon (MeV/nucleon) allows different atoms and isotopes to be compared.

From the mass defect, the nuclear binding energy for the ^{4}He nucleus, expressed in megaelectron volts, is

$$\Delta E = 0.030378\ \cancel{\text{u}} \times \left(\frac{931.5\ \text{MeV}}{1\ \cancel{\text{u}}}\right) = 28.30\ \text{MeV}$$

In any nuclear transformation, the number of nucleons (protons plus neutrons) does not change.

Although it may appear that isotopes with large binding energies are more stable, we have to recognize that binding energies also increase with the number of nucleons. In any nuclear change, the number of nucleons is conserved, so scientists generally compare the binding energy per nucleon.

EXAMPLE 21.11 Calculating Binding Energy per Nucleon

The atomic mass of ^{40}Ca is 39.96259 u. What is the binding energy per nucleon (in MeV) in this atom?

Strategy Calculate the mass defect from the sum of the masses of the particles (see Table 21.3) and from the atomic mass of the atom. Convert the mass defect into binding energy in megaelectron volts and divide by the number of nucleons.

Solution

We know that the nucleus contains 20 neutrons and 20 protons (plus 20 electrons) from the atomic and mass numbers of this nuclide. The mass defect is

$$\Delta m = (20 \times 1.007825) + (20 \times 1.008665) - 39.96259 = 0.36721 \text{ u}$$

Find the binding energy per nucleon by converting this mass defect to megaelectron volts and dividing by the 40 nucleons that compose the nucleus.

$$\text{Binding energy} = \left(\frac{0.36721 \not{u}}{40 \text{ nucleons}}\right) \times \left(\frac{931.5 \text{ MeV}}{1 \not{u}}\right) = 8.551 \text{ MeV/nucleon}$$

Understanding

What is the binding energy per nucleon (in MeV) for ^{16}O, which has a mass defect of 0.1369 u?

Answer 7.970 MeV/nucleon

OBJECTIVES REVIEW *Can you:*

- ☑ calculate the mass defect of a nuclide?
- ☑ calculate the nuclear binding energy from the mass defect?
- ☑ express the binding energy in megaelectron volts per nucleon (MeV/nucleon)?

21.5 Fission and Fusion

OBJECTIVES

- ☐ Recognize that the stability of nuclei varies, and nuclear reactions that form stable nuclei release large amounts of energy
- ☐ Distinguish among critical, subcritical, and supercritical conditions for fission
- ☐ Recognize that fission requires a critical mass of material arranged in a critical geometry
- ☐ Explain why fission products are highly radioactive
- ☐ Describe a typical fusion reaction with a nuclear equation

Because the number of nucleons does not change in a nuclear reaction, the binding energy per nucleon is a measure of nuclear stability. Figure 21.8, a graph of the binding energy per nucleon versus mass number, shows that the most stable nuclei are in the region of ^{56}Fe.

Fission Reactions

The binding curve in Figure 21.8 suggests that the nucleus is a possible source of great energy, if nuclei of intermediate mass, such as ^{56}Fe, are formed from either lighter or heavier nuclides. This section describes the nuclear processes that take advantage of these changes in binding energy per nucleon to release very large quantities of energy.

Nuclear reactions that form stable nuclei such as ^{56}Fe can release large amounts of energy.

In 1938, Otto Hahn and Fritz Strassman detected the presence of radioactive barium among the products formed when uranium was bombarded with neutrons. Other

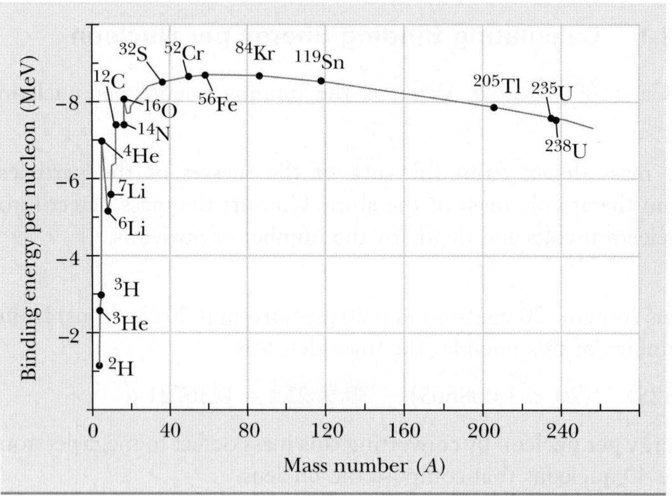

Figure 21.8 Binding energy per nucleon. The binding energy per nucleon is shown as a function of the mass number of the nuclides. The nuclei at the top of the curve are the most stable.

AP Photo/File

Nuclear explosion. Photograph shows the first nuclear explosion, a test in Alamagordo, New Mexico, on July 16, 1945.

studies led to the conclusion that ^{235}U had captured a neutron to form ^{236}U, which split into two smaller nuclei according to the nuclear equation

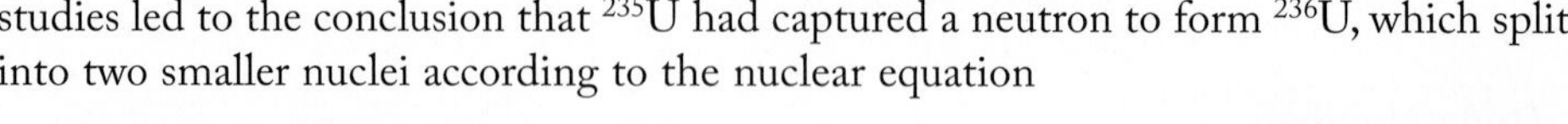

$$^{236}_{92}U \longrightarrow\ ^{141}_{56}Ba + ^{92}_{36}Kr + 3^{1}_{0}n \qquad [21.6]$$

This experiment was the first example of **nuclear fission,** the splitting of a heavy nucleus into two nuclei of comparable size. The energy released in this fission is about 6.8×10^7 kJ/g uranium-236.

On August 2, 1939, Albert Einstein, at the request of several other scientists, sent a letter to President Franklin Roosevelt. In that letter he urged that the United States begin research to develop a bomb based on the energy released from nuclear fission, so that Germany would not develop one first. As a result of Einstein's letter, the Manhattan Project was begun, with the purpose of developing the atomic bomb. The success of this top-secret research effort was demonstrated when the first fission bomb was tested on July 16, 1945, in New Mexico. On August 6 and 9 of that year, the United States used similar bombs against Japan. They are the only two nuclear devices that have ever been used in war.

When fission of a heavy element occurs, many products form. More than 370 product nuclides, with mass numbers ranging from 72 to 161, have been observed from the neutron-induced fission of ^{235}U. Figure 21.9 shows a fission yield curve for this reaction. For reasons not fully understood, fission is unsymmetric, with the mass number of one product nuclide being about 1.4 times the mass number of the other nuclide. The determination of fission yield curves required the chemical separation of the products, by procedures not greatly different from those used in the qualitative analysis exercises in most general chemistry laboratories.

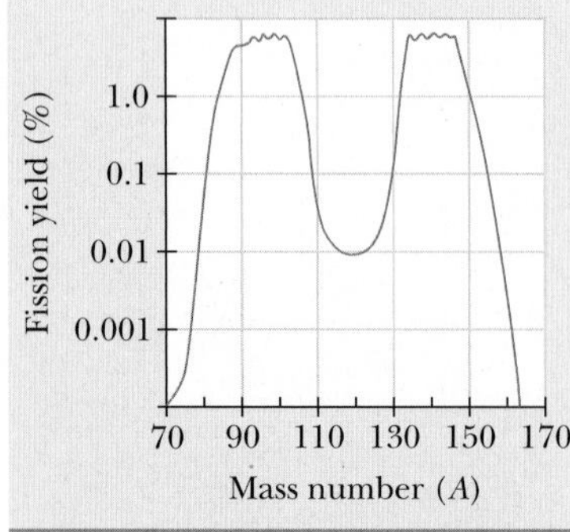

Figure 21.9 Fission yield curve. The distribution of fission products from the neutron-induced fission of ^{235}U. Most of the fission products fall in the two mass ranges of 84 to 105 and 130 to 150.

Each fission event releases about 200 MeV, equivalent to 80 million kJ/g ^{235}U. By way of contrast, the energy released by the detonation of TNT, a powerful chemical explosive, is 16 kJ/g; therefore, a nuclear fission produces about 5 *million* times as much energy per gram as does a chemical reaction. Often, the energies of nuclear explosives are expressed in megatons (millions of tons) of chemical explosives.

On average, the nuclear fission of a single ^{235}U nucleus produces 2.5 neutrons, and each may induce another fission reaction if it is absorbed by another ^{235}U atom. Thus, a single fission event may start a **nuclear chain reaction,** leading to many more nuclear fission events. If each of the neutrons formed initiates another fission, then a rapidly expanding sequence of events occurs, as illustrated in Figure 21.10.

The diagram in Figure 21.10 assumes that every neutron produced by fission induces another fission. If the size of the sample is small, many of the neutrons escape from the sample before reacting with another nucleus to initiate another fission. The chain reaction is said to be *critical* when, on average, exactly one of the neutrons

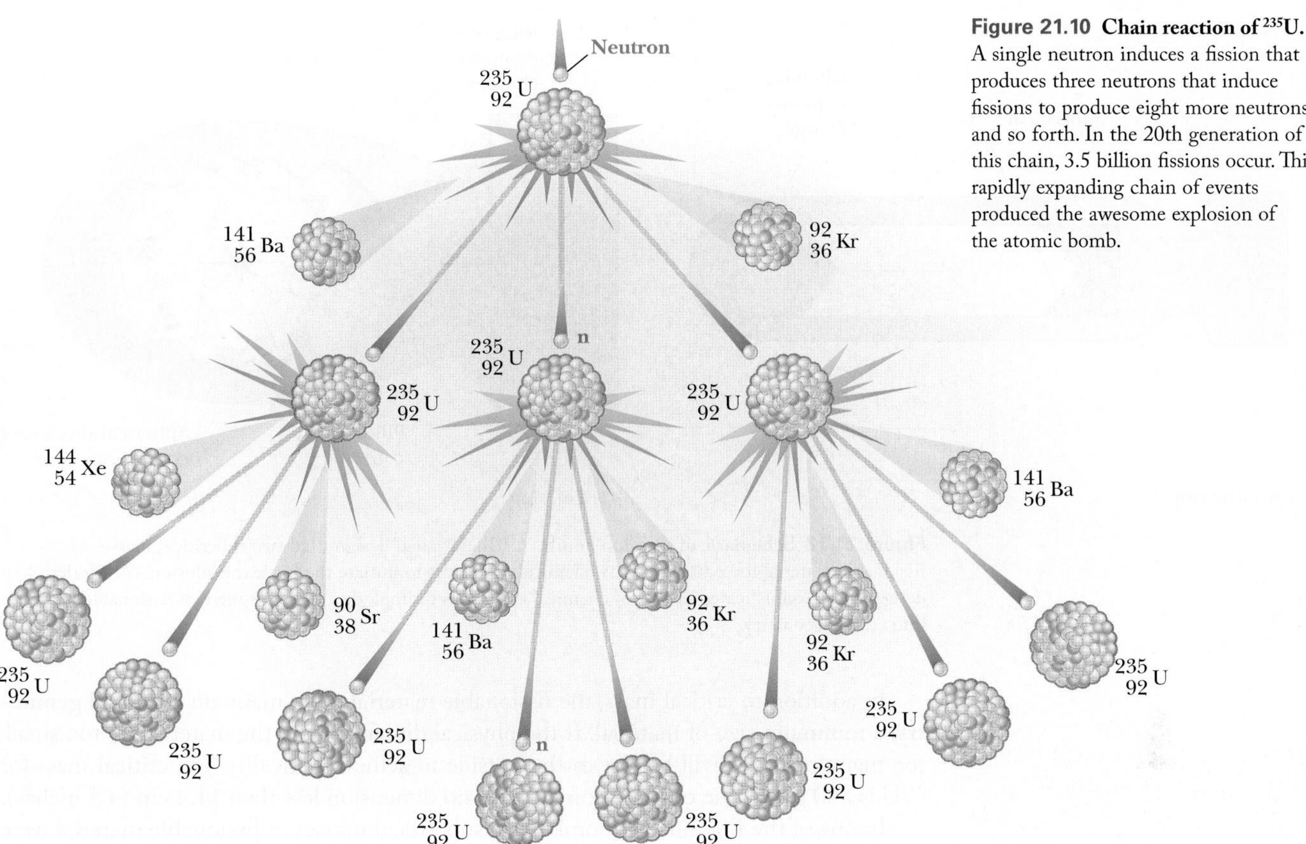

Figure 21.10 Chain reaction of ^{235}U. A single neutron induces a fission that produces three neutrons that induce fissions to produce eight more neutrons, and so forth. In the 20th generation of this chain, 3.5 billion fissions occur. This rapidly expanding chain of events produced the awesome explosion of the atomic bomb.

produced by a fission is absorbed to induce a second fission, thus causing a self-sustaining reaction. The reaction is *supercritical* if, on average, each fission event induces more than one fission, and it is *subcritical* if each fission event causes less than one fission. The **critical mass** is the minimum mass of fissionable matter needed for the chain reaction to be critical. Figure 21.11 shows the conditions of subcritical and critical nuclear fission schematically.

Type of Nuclear Reaction	*Number of Fissions Produced per Neutron*
Subcritical	Less than 1
Critical	1
Supercritical	More than 1

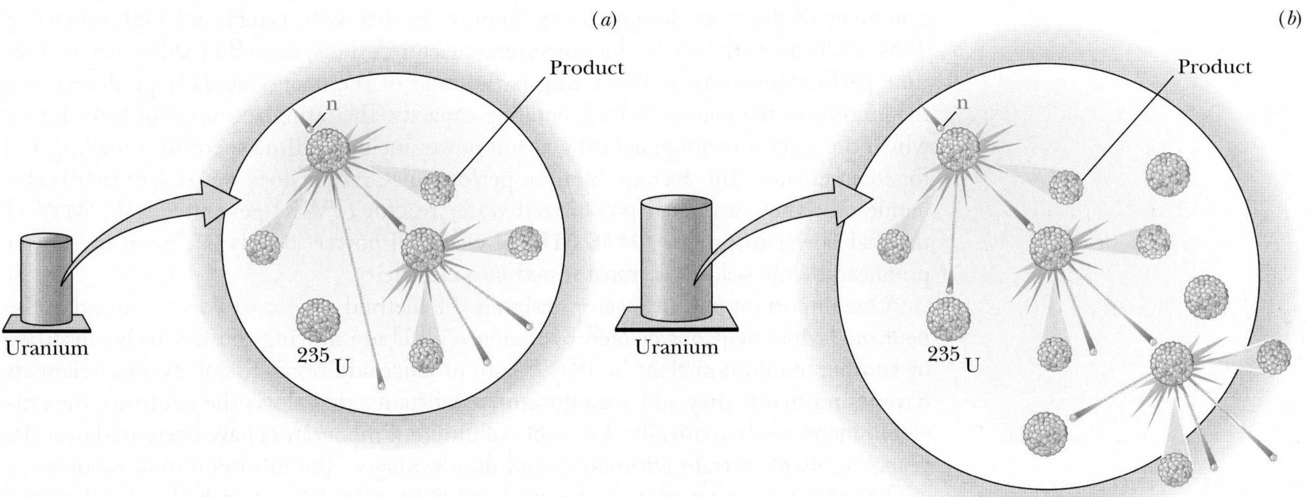

Figure 21.11 Subcritical and critical nuclear fission. *(a)* When most of the neutrons produced escape from the sample, the chain reaction is subcritical. *(b)* The chain reaction is critical when, on average, each fission induces one additional fission.

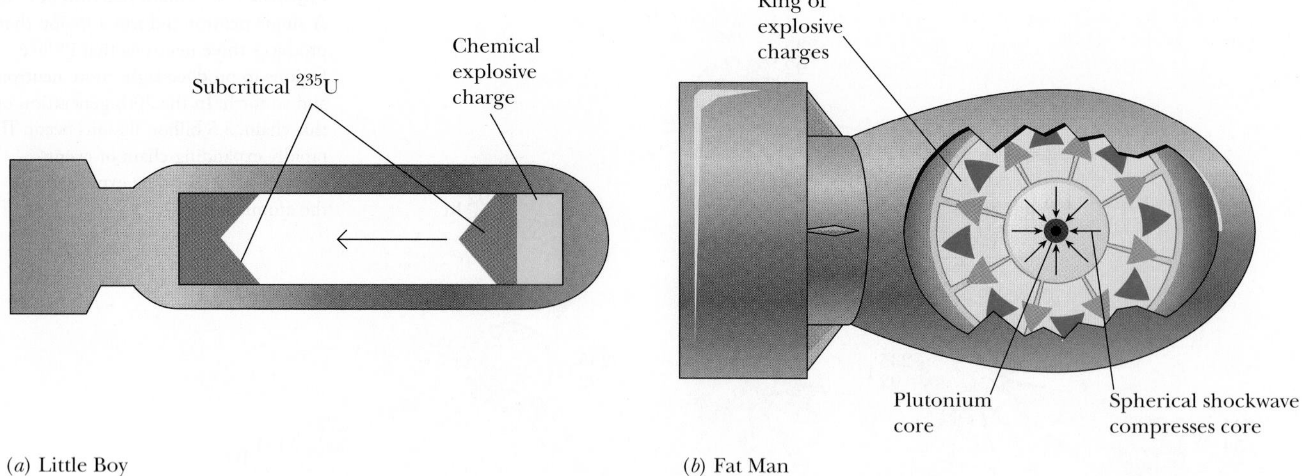

Figure 21.12 Schematic of a fission bomb. *(a)* The original design used two subcritical masses of fissionable material forced together by chemical explosives to initiate the nuclear explosion. *(b)* Modern design uses a sophisticated design of chemical explosives to implode a hollow sphere of fissionable material into critical geometry.

In addition to critical mass, the fissionable material must maintain a **critical geometry,** a minimum size of material. If the physical dimensions of the material are too small, too many neutrons will be lost to the outside to achieve criticality. The critical mass for ^{235}U is 880 g, and the critical geometry has no dimension less than 11.4 cm (4.5 inches).

In one of the first nuclear bombs, two subcritical masses of fissionable material were forced together to form a supercritical mass (see Figure 21.12*a*) and a supercritical geometry by a chemical explosive. In later bombs, carefully spaced explosive charges have been used to implode a hollow sphere of fissile material (^{239}Pu is preferred) to the critical geometry as shown in Figure 21.12*b*.

Both critical mass and critical geometry are required for fission.

Nuclear Power Reactors

In a nuclear power reactor, the chain reaction is carefully controlled so that it remains barely critical at a constant power level. In 2005, there were 104 power reactors in 31 states in the United States producing about 20% of the nation's power. Worldwide, there are about 441 nuclear reactors (26 more under construction) in operation in 32 countries; most use the fission of ^{235}U as the basis for generating power. Scientists have used a number of different designs to maintain criticality with safety and high efficiency. Table 21.4 shows the efficiencies of several reactor designs, ranked by their average lifetime performance rating. This rating is the ratio of the energy actually produced over the lifetime of the reactor to its generating capacity. The rating accounts for time during which the reactor is not generating at full power including time spent for refueling and for maintenance. The average lifetime performance rating does not reflect thermodynamic efficiency: a typical pressurized water reactor (PWR) generates 3411 MW of thermal power to produce 1148 MW of electrical power so only 33.7% of the power produced by the reactor is transformed into electricity.

One important part of reactor design is the method chosen to slow the speed of the neutrons. Most neutrons ejected by fission of ^{235}U are moving too fast to be absorbed by another uranium nucleus, so they do not produce additional fission events. Scientists have found that if they add a **moderator,** a substance that slows the neutrons, the efficiency increases dramatically. A number of different moderators have been used over the years, each with certain advantages and disadvantages. The most common moderators are heavy water (enriched in deuterium) and light water (non-enriched, natural water). In addition to choosing different moderators, the designers choose different levels of enrichment in the uranium fuel, ranging from natural uranium (which is 0.7% ^{235}U) to 5%-enriched ^{235}U. The fuel is nearly always present as pellets of uranium dioxide, UO_2,

TABLE 21.4 **Average Lifetime Performance Ratings**

Reactor Type	Designation	Average Age (yr)	ALPR*	Number	Countries
Pressurized water reactor	PWR	19.2	73%	215	Worldwide
Boiling water reactor	BWR	21.8	71%	90	Worldwide
Canadian deuterium uranium	CANDU	18.0	71%	29	Canada, India, three others
Large-power boiling reactor	RBMK	20.4	65%	16	Former Soviet Union
Magnesium natural uranium oxide	MAGNOX	34.6	58%	8	England, France
Advanced gas cooled reactor	AGR	19.2	61%	14	England (replacing MAGNOX)
Other		13.5	60%	69	

*ALPR is the average lifetime performance rating, defined as the percentage of electricity produced/rated capacity.

although a few reactors use other fuel. These UO_2 pellets are packed into rods that are generally made of zirconium alloys. The rods are slender, so they cannot become critical until a relatively large number (several hundred) are loaded in a precise geometry in the reactor core. The core geometry is an important safety factor. Finally, reactor engineers choose different operating temperatures and methods to cool the reactor and generate steam. Some of these choices are summarized in Table 21.5.

Figure 21.13 shows a diagram of a PWR. This nuclear reactor design was an extension of the design first used in nuclear submarines. PWR reactor fuels are enriched to about 3.5% in ^{235}U. A typical nuclear reactor that supplies 1000 MW of electricity consumes nearly a ton of uranium fuel in 1 year of operation. The PWR uses light water as a coolant and moderator. The reactor cooling system has water that is in contact with the fuel; the water reaches about 334 °C. The water is not allowed to boil, even at this extreme temperature, by keeping the pressure quite high, approximately 150 atm or 2200 psi. A thick steel jacket surrounds the reactor cooling loop to operate at pressures this high, and the fuel must be clad with a high-temperature zirconium alloy to keep from melting.

The pressurized water passes through a heat exchanger to make steam that drives the turbine blades of the generator, transferring some of its heat in the process. The generator that is driven by pressurized cooling water provides the name for this design: a pressurized water reactor (PWR). The boiling water reactor is largely the same, but the reactor cooling water is allowed to expand into steam to drive the turbine directly.

TABLE 21.5 **Reactor Designs**

Reactor Type	Designation	Steam Generator	Moderator	Coolant	Fuel (UO_2)	Cladding
Pressurized water reactor	PWR	Heat exchanger	Light water	Light water	2–5% enriched	Zirconium
Boiling water reactor	BWR	Direct	Light water	Light water	2–3% enriched	Zirconium
Canadian deuterium uranium	CANDU	Heat exchanger	Heavy water	Heavy water	Natural	Zirconium
Large-power boiling reactor	RBMK Reactor Bolshoi Mochnosti Kipyashiy	Direct	Graphite	Light water	2% enriched	Zirconium/niobium
Magnesium natural uranium oxide	MAGNOX	Heat exchanger	Graphite	Carbon dioxide	Natural	Mg
Advanced gas cooled reactor	AGR	Heat exchanger	Graphite	Carbon dioxide	2–3% enriched	Stainless steel

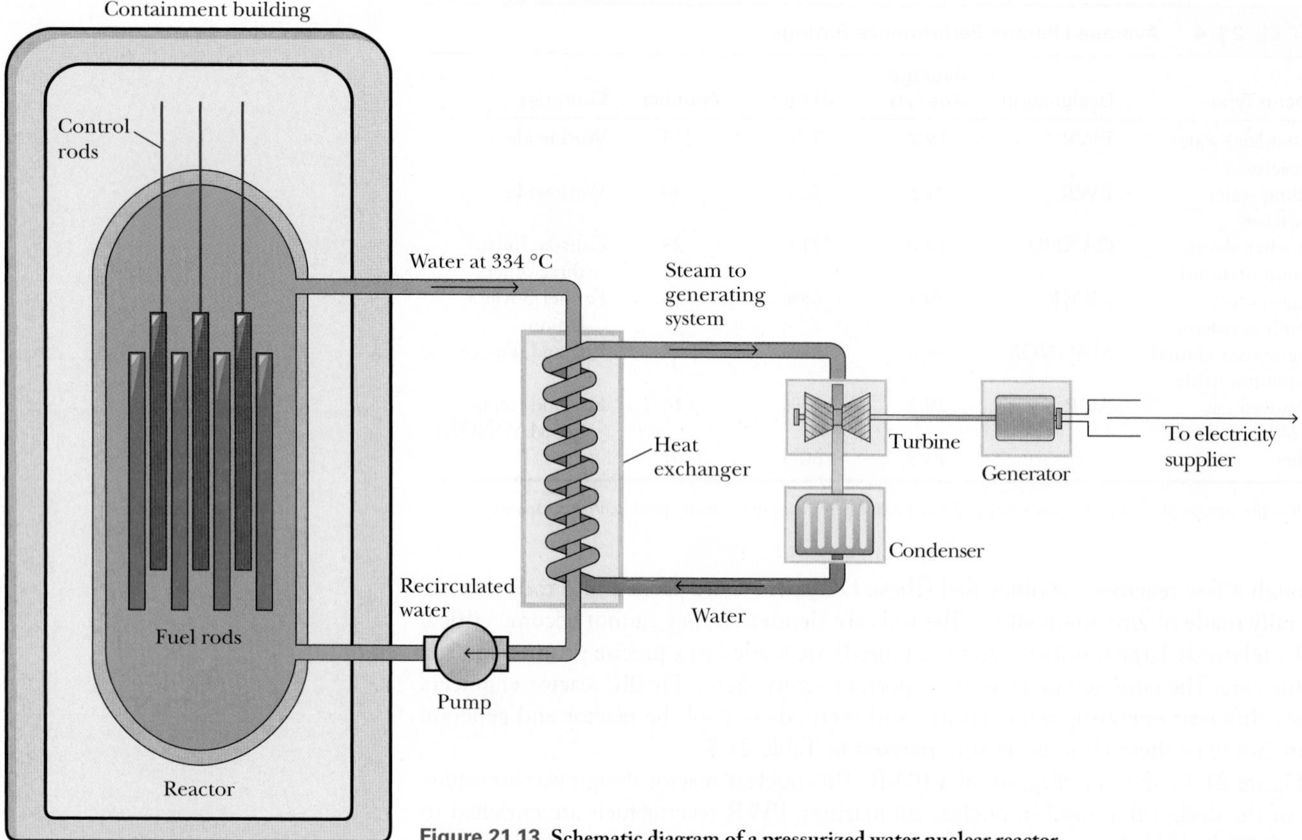

Figure 21.13 Schematic diagram of a pressurized water nuclear reactor.

Nuclear Power and Safety

It may appear that fission reactors offer an abundant supply of the energy needed by modern society. Compared with the generation of electricity by burning fossil fuels, nuclear power plants are quite clean. They do not introduce ash, smoke, sulfur oxides, nitrogen oxides, and carbon dioxide into the atmosphere as conventional power plants do. As discussed in previous chapters, these waste products from conventional fuels cause pollution of the environment and make substantial contributions to such problems as acid rain and global warming through the greenhouse effect.

Nuclear power generation, however, has several undesirable aspects. The nuclides produced in fission reactions lie above the region of stability and are highly radioactive themselves. Equation 21.6 shows that two prominent products from the fission of ^{235}U are ^{141}Ba and ^{92}Kr. These isotopes are radioactive and decay by the following routes, each step of which has a characteristic half-life:

$$^{92}_{36}Kr \xrightarrow{3.0s} {}^{92}_{37}Rb \xrightarrow{5.3s} {}^{92}_{38}Sr \xrightarrow{2.7h} {}^{92}_{39}Y \xrightarrow{3.5h} {}^{92}_{40}Zr$$

$$^{141}_{56}Ba \xrightarrow{18m} {}^{141}_{57}La \xrightarrow{3.9h} {}^{141}_{58}Ce \xrightarrow{33d} {}^{141}_{59}Pr$$

Gamma-ray emission accompanies all of these beta decays; therefore, fission products are extremely hazardous. The nuclear decay continues and after about a decade, the principal nuclides are ^{90}Sr with a half-life of 28.1 years and ^{137}Cs with a half-life of 30.2 years; thus, fission products from nuclear reactors remain dangerously radioactive for years and even centuries. Safe storage of radioactive products from reactors is a long-term problem that must be solved. Currently, the U.S. Department of Energy has begun to immobilize radioactive waste from the atomic weapons program by mixing it with other materials to form a glass. The molten glass is poured into thick-walled stainless-steel canisters that are welded shut and eventually buried deep underground in a specially designed facility. No such program has been developed for U.S. power reactor waste.

Many people fear disastrous accidents from the operation of nuclear reactors. These accidents have the potential to release radioactive nuclides into the surroundings. Reactors are designed with a number of safety features, but one of the most important is to use the cooling fluid as the moderator. If cooling fluid is lost, then the reactor becomes subcritical because the absorption of neutrons decreases when they are no longer moderated. In addition, control rods made from neutron-absorbing materials such as cadmium can be lowered into the reactor to reduce the neutrons to a subcritical flux; and last, neutron-absorbing elements such as gadolinium can be added to the reactor cooling water. The design of every reactor in Table 21.5 has inherent safety features that convinced the designers and the regulatory authorities that a reactor accident is highly unlikely, if not impossible.

Fission products are highly radioactive because the nuclides produced by fission lie far from the zone of stability. They produce beta and gamma radiation for many decades.

Two reactor accidents have received wide publicity. In 1979, a cooling system failure led to the release of radioactive gases into the atmosphere at the PWR Three Mile Island reactor in Pennsylvania. A second and more serious accident occurred in an RBMK reactor near Chernobyl in the former Soviet Union in 1986. Subsequent investigation of both of these incidents revealed that they had resulted from avoidable operator errors. Concerns about the safety of nuclear reactors and the disposal of their radioactive waste have deterred the construction of additional nuclear power reactors in the United States for the past several decades although several are being planned for the future.

All means of meeting the demand for power in modern technological society involve risks in the form of environmental contamination and potential loss of life. Policy makers must carefully and rationally consider nuclear power, fossil fuels, solar power, and other energy sources in planning for future energy needs.

Nuclear Fusion

The binding energy curve in Figure 21.8 suggests that **nuclear fusion,** the combination of two light nuclei to form a larger one, is also a source of energy. Fusion reactions are the source of the energy emitted by the Sun and other stars. The Sun is about 73% hydrogen and 26% helium, and some of the fusion reactions believed to occur constantly in the Sun are

$$^{1}_{1}\text{H} + {}^{1}_{1}\text{H} \longrightarrow {}^{2}_{1}\text{H} + {}^{0}_{+1}\beta \qquad \Delta E = -9.9 \times 10^{7}\ \text{kJ/mol}$$

$$^{1}_{1}\text{H} + {}^{2}_{1}\text{H} \longrightarrow {}^{3}_{2}\text{He} \qquad \Delta E = -5.2 \times 10^{8}\ \text{kJ/mol}$$

$$^{3}_{2}\text{He} + {}^{1}_{1}\text{H} \longrightarrow {}^{4}_{2}\text{He} + {}^{0}_{1}\text{n} \qquad \Delta E = -1.9 \times 10^{9}\ \text{kJ/mol}$$

Several equations indicate that the fusion of small nuclei such as hydrogen or helium results in the production of large quantities of energy.

Fusion is a much cleaner nuclear process than fission because it does not directly form radioactive products.

Scientists estimate that temperatures in the range of 10^6 to 10^7 K are needed before fusion reactions can occur. At these high temperatures, all matter exists as a **plasma**—a gaseous mixture of ions, free electrons, and atoms. In the fusion bomb, or hydrogen bomb, a fission explosion increases the temperature of light nuclei and initiates the fusion reaction. This reaction leads to the name *thermonuclear* for the fusion bomb.

An explosive release of nuclear energy is of no use for generation of electrical power; therefore, if fusion is to become a useful source of energy, some means of controlled release of the nuclear energy must be developed. Investigations related to the development of a controlled fusion reactor have been in progress for more than 50 years. The basic problem is to create a high-pressure, high-temperature environment for a time long enough to initiate fusion. Two different technologies have emerged. One method, called *magnetic confinement,* uses strong magnetic fields to confine a plasma at high pressures and temperatures in excess of 10^7 K. The second method, called *inertial confinement,* pumps energy into a plasma so fast that the plasma cannot expand quickly enough to dissipate the energy.

AP Photo/Claude Paris

Model of ITER.

The People's Republic of China, the European Union, Japan, the Republic of Korea, the Russian Federation, and the United States of America are jointly participating in the design of an experimental international thermonuclear reactor (ITER) based on magnetic confinement. The original project, begun in 1988, was to design and build a 1000-MW reactor. The design stage took 10 years, and the estimated cost of the reactor was \$5.5 billion. The governments funding the program rejected this design and cost, and asked for a downscaled design. The engineers finalized a design for a 500-MW reactor that would cost \$2.8 billion to build several years ago. Two sites, one in France and one in Japan, have been discussed, and support is evenly split, so a final decision has not been made to date. The construction of ITER will take about 10 years, followed by an estimated 20 years of operation to generate data for a second-generation fusion project, called DEMO, which could be constructed from 2020 to 2030 and operational until 2050. The DEMO project would be a prototype reactor/generator station and is expected to produce reasonable amounts of electrical power. A commercial fusion power plant based on DEMO could be in operation by 2050.

The United States has invested heavily in the National Ignition Facility (NIF) that will use 192 high-powered laser beams in an inertial confinement system. The beams will interact with a deuterium and tritium target material that is just a few millimeters in diameter to achieve fusion. The pressures at the target are estimated to be six times the pressure at the center of the Sun. The laser beams are powerful but very short in duration. During the 10-ns laser burst, the power at the target will be about 1000 times the electrical-generating capacity of the United States.

The \$4 billion project is scheduled for completion in 2009 with ignition (fusion energy) expected in 2010. The facility will support high-energy density plasmas that will be used by scientists and engineers to study fusion ignition, to study astrophysics, and to develop new materials. The NIF also has a major role in national security because it will allow scientists to produce the energies achieved by nuclear weapons without actually firing them. Scientists hope to learn the effects of aging on weapons, and through these experiments, to study the interaction of radiation with matter.

OBJECTIVES REVIEW *Can you:*

- ☑ recognize that the stability of nuclei varies, and nuclear reactions that form stable nuclei release large amounts of energy?
- ☑ distinguish among critical, subcritical, and supercritical conditions for fission?
- ☑ recognize that fission requires a critical mass of material arranged in a critical geometry?
- ☑ explain why fission products are highly radioactive?
- ☑ describe a typical fusion reaction with a nuclear equation?

(a)

(b)

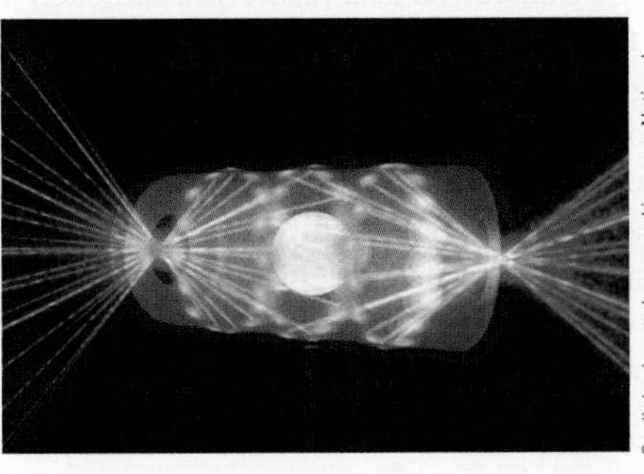
(c)

Credit is given to Lawrence Livermore National Security, LLC, Lawrence Livermore National Laboratory, and the Department of Energy under whose auspices this work was performed.

National Ignition Facility. *(a)* Laser and target room of the national ignition facility is the size of three football fields. *(b)* The target chamber is 10 m in diameter and weights 287,000 lb. The high-power laser beams will come to focus within 10 μm tolerances in this chamber. *(c)* Artists drawing of the NIF target. 192 laser beams are focused on the inside of a gold-plated cylindrical can the size of a pencil eraser. High intensity x rays are created from this interaction which then bathe the spherical target, rapidly heating it and causing it to compress and ignite the fusion reaction.

21.6 Biological Effects of Radiation and Medical Applications

OBJECTIVES

- ☐ Distinguish among radioactivity, radiation dose, and effective radiation dose
- ☐ Describe the biological effect of radon
- ☐ Explain the advantages of nuclear medicine procedures over conventional x-ray diagnostics
- ☐ Describe the factors that influence the effect of radiation on the human body

The effect of radiation on the functions of biological systems, especially the human body, is a major concern. The analysis of the effects of radiation is complicated by several factors. The different kinds of radiation—alpha particles, beta particles, gamma rays, x rays, and neutrons—differ in their interactions with matter. Furthermore, the radiation effects depend on whether the source of the radiation is inside or outside the body, or whether the entire body or only part of it is exposed to the radiation. For these reasons, the established radiation safety standards involve many qualitative judgments and assumptions.

Some of the important terms and units used in measurements of radioactivity are shown in Table 21.6 and explained further in the following paragraphs.

TABLE 21.6 Units of Radioactivity and Radiation Dose

Unit	Name	Abbreviation	Definition or Conversion
Radioactivity			
SI unit	Becquerel	Bq	1 disintegration per second
Common unit	Curie	Ci	3.7×10^{10} disintegrations per second
Radiation Dose	rad	rad	Quantity of radiation that transfers 1×10^{-2} J energy per kilogram of matter
Effective Radiation Dose			
SI	Sievert	Sv	1 Sv = 100 rem
Common	rem	rem	1 rad of beta or gamma produces a dose of 1 rem 1 rem = 0.01 Sv

Radioactivity is any spontaneous nuclear reaction that transforms a relatively unstable nuclide into a more stable nuclide and a small particle plus energy. The SI unit for radioactivity is the *becquerel* (Bq), defined as an activity of 1 disintegration/s. The *curie* is a common alternative to the becquerel, and was originally defined as the radioactivity of 1 g ^{226}Ra, the isotope first studied in detail by Pierre and Marie Curie. It is now defined as 3.7×10^{10} Bq.

The biological damage produced by radiation originating outside the body depends on the depth of penetration of the particles and their energies, as well as on the number of particles. Alpha particles are generally stopped by the skin and produce little internal damage. Beta radiation penetrates only about 1 cm below the skin. Gamma rays and x rays are the most penetrating forms of radiation, with a range of 30 cm or more, and can therefore produce damage throughout the interior of the body.

The biological damage caused by radiation is the result of ionization of molecules as energy is transferred to matter. The physical effects of radiation on matter depend on the **radiation dose,** that is, the amount of energy transferred as a result of a radioactive decay to a target. The unit of radiation dose is the **rad,** defined as the quantity of radiation that transfers 1×10^{-2} J energy per kilogram of matter. The damage to biological tissues per rad differs with the kind of radiation. For example, 1 rad of alpha radiation is more damaging than 1 rad of beta particles. In measuring the biological effects of radiation, it is necessary to multiply the radiation dose by a quality factor, Q, which depends on the type of radiation and other factors. The value of Q is approximately 1 for beta and gamma radiation, and 10 for alpha particles. The **effective radiation dose** is the product of the quality factor times the radiation dose.

Radioactivity is a measure of the number of disintegrations per second; radiation dose measures the amount of energy absorbed by the target material, and effective radiation dose measures the biological effect of the absorbed energy.

$$\text{Effective dose (in rems)} = Q \times \text{dose (in rads)}$$

The unit for effective dose is the rem, an abbreviation for "roentgen equivalent man." Note that Q is quite variable and depends on the rate of the dosage, the kind of exposed tissue, and the total dose received.

People are exposed to a number of sources of radiation; most are surprised to learn that the largest sources of exposure are natural sources. Table 21.7 presents the sources of radiation and their contributions to the average effective dose of the U.S. population. These data are shown graphically in Figure 21.14 as percentages of the average total exposure.

These figures are approximations because they depend a great deal on geographical location, altitude, and occupational exposure. For example, the cosmic ray exposure of a person on a commercial jetliner is well above the average dose from that source at sea level. The exposure to radon depends on the concentration of uranium in the ground, which varies considerably from one location to another.

TABLE 21.7 Estimated Annual Effective Dose of Radiation in U.S. Population

Source	Effective Dose (millirem)
Natural Sources	
Cosmic rays	28
Terrestrial	28
In the body	39
Inhaled radon	~200
Human Activity	
Occupational	0.9
Nuclear fuel	0.05
Consumer products	5–13
Environmental sources	0.06
Medical	
Diagnostic x rays	39
Nuclear medicine	14
Total	360

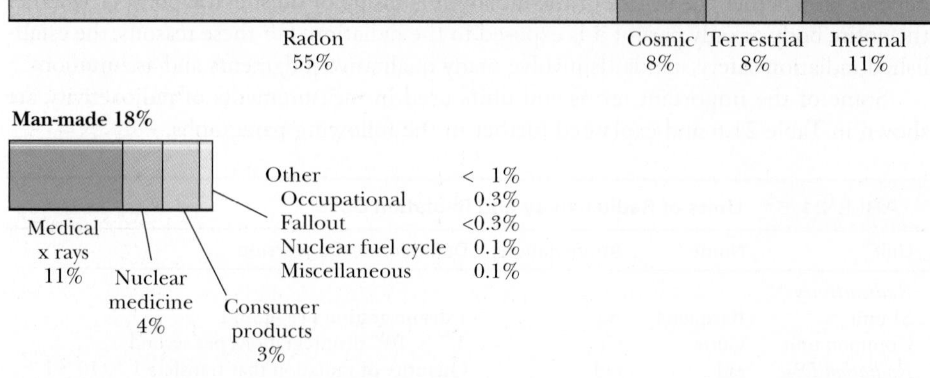

Figure 21.14 Average radiation exposure. Sources of average radiation exposure for the U.S. population, expressed in percentages. Note that the exposure from natural sources far exceeds that from man-made origins.

Radon

As Figure 21.14 shows, a majority of the average radiation exposure in the United States comes from radon, the heaviest of the noble gases. Radon-222 is produced by the decay of natural ^{238}U found in rocks such as granite. Because radon is a noble gas, it escapes from the ground easily and enters the atmosphere. It undergoes alpha decay with a 3.82-day half-life, producing ^{218}Po. If this decay occurs while the radon is in a person's lungs, the ^{218}Po (Group VIA) remains in the tissue. Within a few days, the ^{218}Po that is trapped within the body emits two more alpha particles and two beta particles, causing further radiation damage to the person's lungs. Home radon test kits (see Figure 20.36) are sold in some parts of the United States because of the great public awareness and the potential dangers of radon accumulation. Because radon generally enters homes through the basement, one of the most effective means of eliminating the radiation danger is to add ventilation fans in the basement.

The principal biological effect of radon is due to the alpha and beta particles emitted by the ^{218}Po nuclide formed when the radon decays.

Nuclear Medicine

Nuclear medicine is used more than 10 million times a year in the United States. Not only can radioisotopes be used for diagnosis, they can also be used for treatment.

When a patient has an overactive thyroid gland, he or she is often given a dose of ^{131}I, a beta emitter. Iodine concentrates in the thyroid gland, and the beta radiation from the 131-iodine isotope reduces the amount of hormone produced by the thyroid gland.

Gamma-Radiation Scans

Diagnostic procedures take advantage of the ability of gamma radiation to penetrate the organs of the body. Radiopharmaceuticals are synthesized with isotopes that emit gamma radiation coupled to organic molecules that are taken up specifically in target organs. Although x rays can be used to image an organ, nuclear medicine goes further because the uptake of the radiopharmaceutical is related to the function of the organ. Thus, nuclear medicine provides functional information, as well as the size and shape of the organ. Modern scanners use gamma-ray cameras that take measurements at thousands of different locations in conjunction with sophisticated computer programs that reconstruct the information into three-dimensional images in a process called *tomography*. Computerized tomography is often abbreviated CT and referred to as a CT scan.

Nuclear medicine provides information about the size, shape, and functionality of an organ.

The most widely used radioisotope is ^{99m}Tc. The *m* indicates a metastable nuclear energy state. This isotope emits a gamma ray with energy of 141 keV, which is energetic enough to pass through the body but not energetic enough to pass thorough lead shielding. The decay of ^{99m}Tc does not produce any alpha or beta particles; thus, the effect to the patient is as small as possible. Also, ^{99m}Tc has a half-life of 6 hours—long enough to allow for the preparation of the radiopharmaceutical but short enough to avoid harmful effects to the patient. The radiation dosage from nuclear medicine is comparable with that received from medical x rays.

Technetium-99*m* is made available to hospitals and nuclear medicine clinics in the form of ^{99}Mo. The clinic receives a molybdate solution (MoO_4^{2-}) in which the molybdenum is ^{99}Mo. The molybdenum decays with a half-life of 66 hours to $^{99m}TcO_4^-$, the pertechnetate ion. The 66-hour half-life is good for overnight shipping to the hospital. When it arrives, the molybdate and pertechnetate ions are separated by chromatography on an alumina column (colloquially known as a "technetium cow" because it is "milked" for the product) supplied with the radiopharmaceutical.

Technetium is a transition metal with a variety of stable oxidation states that makes it an excellent species to use to react with organic molecules. For example, heart imaging is done with a compound formed by technetium(I) and six isonitrile ligands: $[Tc(C{\equiv}N{-}R)_6]^+$. Other compounds have been developed for mapping the brain, lungs, and other organs. More than 100 different nuclear medicine diagnoses are available to clinicians.

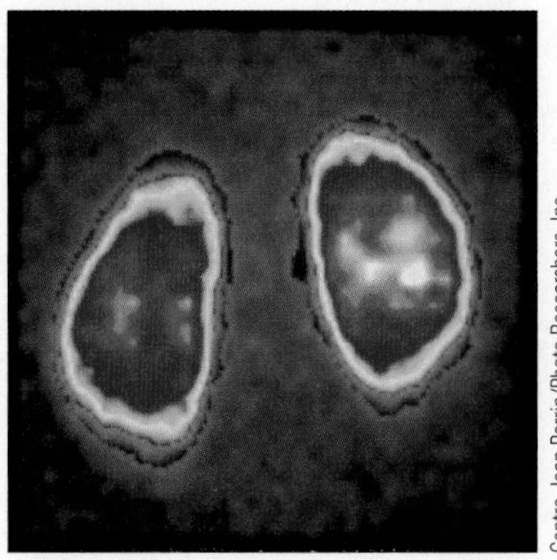

Centre Jean Perrin/Photo Researchers, Inc.

Gamma camera scan of human kidneys. Dimercaptosuccinic acid (DMSA), a compound that is taken up by the renal tubules in the kidneys, is labeled with ^{99m}Tc. The colors are generated by computer and provide information about the ability of the kidneys to function.

PRINCIPLES OF CHEMISTRY
Exposure and Contamination

A well-known scientist often begins a discussion of radioactivity with a quiz for his audience. He points out that modern laboratories have sensitive methods to detect radioactivity, and he mentions that Marie Curie's laboratory notebook is so heavily contaminated that it is kept in a lead container for purposes of public safety. He then asks how long will radioactivity be detected in your body after you are exposed to enough radiation (a mixture of alpha, beta, and gamma) that would fog a roll of film in about 1 second.

Most people guess years, but the right answer is, *you cannot be made radioactive by exposure to radiation.* (Neutron exposure is an exception because some radioactive nuclides can form by collision of a neutron and a stable nuclide. Neutron exposure is extremely rare.) Radiation is like light; when you turn it off, it goes away.

Even the science press often confuses exposure to radiation with contamination by radioactive materials. If radioactive materials are ingested, inhaled, or absorbed through skin or open wounds, then you are *contaminated* and your body will be radioactive. Radiation workers can be tested for contamination by beta and gamma emitters by using a Geiger counter or personal radiation portal monitor. Alpha contamination is detected by bioassay—that is, analysis of body fluids.

All radiation that enters the body, regardless of source, can damage cells, with ensuing medical consequences. Exposure to radiation (absorbed dose) ends when the exposure stops. Contamination ends either when the radionuclide decays with its particular half-life, or when it is eliminated through normal bodily functions. The biological half-life of water is about a week, but water-soluble metals have half-lives of several months. Additionally, heavy metals such as lead and plutonium accumulate in bones where lead has a half-life of about 10 years and plutonium about five times longer.

Fortunately, chemists have developed materials called *chelates* that bind to metals such as plutonium and uranium, and increase the elimination rate. These compounds are discussed in Chapter 19. Chelation therapy is also used for cases of heavy metal (lead, arsenic) poisoning. ▮

Photo by Krafft Angerer/Getty Images

Personal portal monitor.

Positron Emission Tomography Scans

A number of diagnostic tests have made use of positron emission tomography (PET) scans. These scans use ^{18}F or ^{11}C compounds to map the function of a number of organs. Most PET scans are performed with a fluorinated sugar (fluorodeoxyglucose) that contains ^{18}F. This sugar is taken up by a number of different organs, but the highest

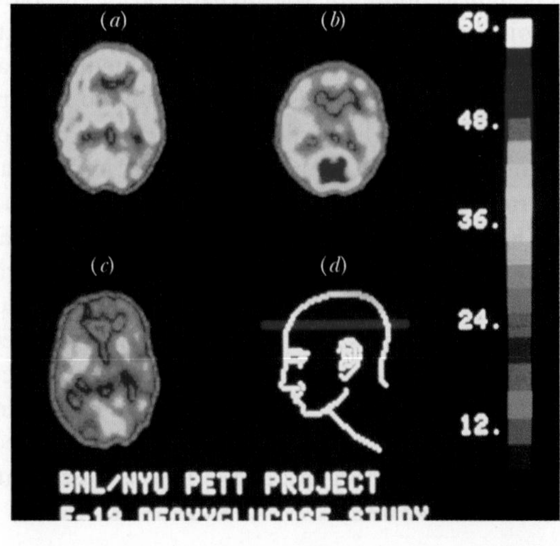

Science Source/Photo Researchers, Inc.

PET scans of brain. ^{18}F deoxyglucose provides the positrons for this scan. *(a)* Image of the brain of a normal individual. *(b)* An individual suffering from schizophrenia. *(c)* An individual with depression. *(d)* The location that is sampled.

concentrations are found in the organs that are subdividing most quickly, and cancer cells are among the fastest multiplying cells; therefore, PET scans are frequently used to diagnose cancer and to monitor changes during treatment. PET scans are particularly effective in diagnosing lung, head and neck, colorectal, esophageal, lymphoma, melanoma, breast, thyroid, cervical, pancreatic, and brain cancers.

Gamma imaging agents synthesized from ^{99m}Tc target a specific organ. Positron-emitting agents are not organ-specific; they accumulate in areas of rapid cell division, such as cancers.

OBJECTIVES REVIEW *Can you:*

- ☑ distinguish among radioactivity, radiation dose, and effective radiation dose?
- ☑ describe the biological effect of radon?
- ☑ explain the advantages of nuclear medicine procedures over conventional x-ray diagnostics?
- ☑ describe the factors that influence the effect of radiation on the human body?

CASE STUDY Nuclear Forensics

Nuclear terrorism sadly has progressed from science fiction and mystery novels to the real world. Two cases have been documented in which radioactive isotopes were used in a deliberate attempt to murder someone.

In 2000, a German man who worked at a nuclear reprocessing facility attempted to kill his former wife with plutonium that he stole. He did not steal a large amount, just some cleaning rags and a small quantity of liquid waste, but that was enough to contaminate his ex-wife and her mother; fortunately, both survived. Two apartments had to be decontaminated at a cost of several million dollars, and the man was sent to prison.

In 2006, a Russian man was killed by administration of ^{210}Po. Alexander Litvinenko, a former security officer who became a critic of Russia, died 3 weeks after ingesting ^{210}Po. Litvinenko, a vocal critic of Vladimir Putin, President of the Russian Federation (2000–2008), had been granted political asylum in London. On November 1, he met with two former Russian State Security (FSB, formerly known as KGB) officers while doing research for a book he was writing. He died on November 23.

AP Photo/Alistair Fuller

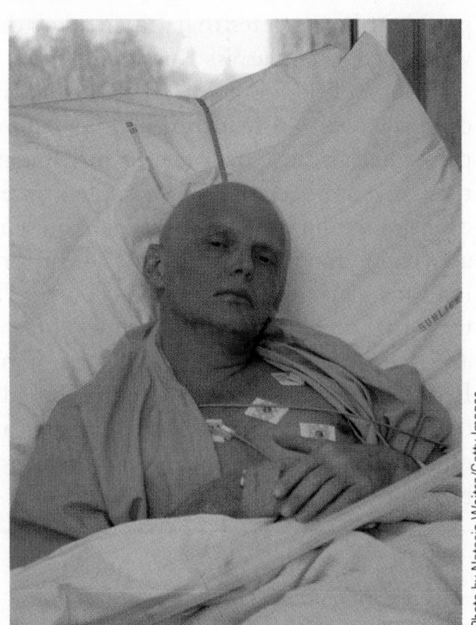

Photo by Natasja Weitsz/Getty Images

^{210}Po is essentially the perfect radioactive poison. It is an alpha emitter, but unlike nearly every other alpha emitter, it is nearly undetectable. When most alpha-emitting nuclides decay, they also emit a high-energy gamma ray in concert with the alpha particle, and the gamma radiation can be detected by sensors such as the Geiger

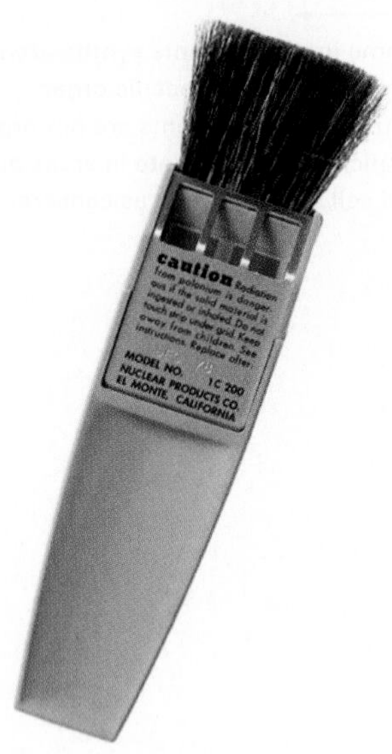

D. Hurst/Alamy

Antistatic brush. This type of brush is widely used by photographers who must deal with dust and static when developing pictures from negatives.

counter. This technology is used to safeguard stores of plutonium and enriched uranium. But ^{210}Po is essentially unique because only 1 of every 10,000 decays produces a gamma ray. Because alpha particles have low-penetrating power, an aqueous solution of ^{210}Po can be prepared and put into a glass vial. No airport has equipment sensitive enough to detect ^{210}Po inside a vial, so it could easily be transported across international borders.

^{210}Po is synthesized in reactors by bombarding ^{209}Bi with neutrons.

$$^{1}_{0}n + ^{209}_{83}Bi \longrightarrow ^{210}_{83}Bi \longrightarrow ^{210}_{84}Po + ^{0}_{-1}\beta$$

The ^{210}Po decays with a half-life of 138 days via alpha emission to the stable ^{206}Pb isotope.

$$^{210}_{84}Po \longrightarrow ^{206}_{82}Pb + ^{4}_{2}\alpha$$

Russian nuclear reactors produce about 85 g/yr, all sold to the United States to be used to manufacture commercial devices such as antistatic brushes.

The radiation from ^{210}Po alpha is weak and cannot penetrate through paper, but if inhaled, ingested, or absorbed through cuts in the skin, polonium enters the bloodstream and the alpha particles immediately begin to damage the cells.

Unknown to the perpetrators, technology had advanced to the point that investigators could trace the ^{210}Po, finding it in restaurants, a car, and an airplane, as well as Litvinenko's apartment. Everything he touched for the first 3 days after ingesting the poison was contaminated, presumably from perspiration and body oils that were transferred from Litvinenko to his surroundings. His apartment could not be occupied for 6 months, and his car could never be made safe for use.

Investigators believe they found traces from the two Russians who handled the substance, and authorities inferred that these traces were consistent with carrying the poison and administering it in Litvinenko's tea. As part of their investigation, the British government analyzed dishes and eating utensils in all the places Litvinenko visited and took swabs (wipes) from surfaces that he might have touched. Traces were found on the airliner that flew the two Russians to London. The British government asked for the extradition of several Russians implicated in Litvinenko's death, but the Russian Federation refused, stating their constitution did not allow for extradition. Former president Putin states that Litvinenko's death was the work of anti-Russian forces and was aimed to embarrass Russia.

Much of the information about the amounts of ^{210}Po is circumstantial because the diagnosis was not made until a day before Litvinenko's death. A full autopsy was not performed because of concerns for the safety of the medical team. In addition, scientists have not been granted complete access to the data because of the criminal nature of the radiation exposure. Litvinenko's symptoms were consistent with ingesting radioactivity of about 2×10^9 Bq (2 GBq or 50 mCi). This quantity of radiation would indicate ingestion of a mass of 1 μg ^{210}Po, a mass that is approximately 200 times the lethal dose (and 1 billion times the amount of ^{210}Po found in products such as the antistatic brush pictured here). Other corroboration comes from a statement from the British authorities that the coffin should not be opened for 22 years, although even that length of time is disputed. Perhaps as science develops even more sensitive techniques and radiation safety methods improve, the first nuclear murderers will be brought to justice.

ETHICS IN CHEMISTRY

1. The power company that supplies power to the town in which your school is located proposes to build a nuclear power plant about 15 miles upriver from your town. The school and townspeople unite to express their disapproval, stating that the multiple dangers from a nuclear accident far outweigh any small decrease in the cost of electricity. They are particularly concerned that the river, which the power company proposes to use for cooling, might become contaminated with radioactive isotopes. The power company does not mount a fight because a better option has become available. The Yonkalla Indian Nation has invited the company to build the power plant on its reservation, about 150 miles upriver from your town. The Indian leadership states that bringing construction to the reservation will allow the Yonkalla community to escape from a nearly inexorable life of poverty. The college and townspeople are extremely upset, because the power company has not addressed the danger to their water or any of the other points they raised.

a. List the arguments that the townspeople will likely raise.
b. List the arguments the Yonkalla Nation will likely raise.
c. What are the major points that the power company will raise.
d. As a scientist and as a citizen of the town, explain what you believe the power company should do and justify your explanation.

2. Although solar panels are used for exploration of the inner solar system, the sunlight is too faint to provide enough energy to explore the outer planets. NASA proposes to send a spacecraft to Saturn using a radioisotope thermoelectric generator (RTG). The isotope ^{238}Pu decays (via alpha) with a half-life of 88 years. The nuclear reaction produces heat that is used as a source of electricity. NASA says that the RTG is the only viable option. Opponents say the risks of even a one in a million chance of a crash scattering plutonium dust all over the beaches of Florida are far too large to warrant putting 72 lb of plutonium in a space vehicle.

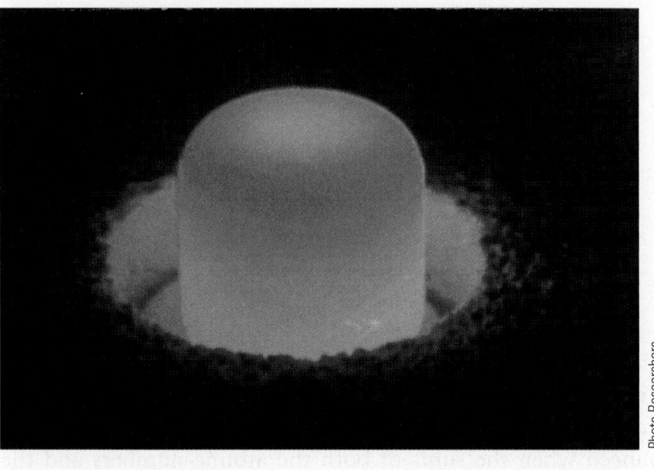

Photo Researchers

238-Plutonium dioxide pellet prepared for the *Cassini-Huygens* space mission.

a. Compare the risks with the rewards; then decide your position on launching an RTG and justify it.
b. Use the Internet to research the *Cassini-Huygens* mission to Saturn; then determine whether the outcomes justified the risk.
c. If you search the Web, you will find hundreds of Websites that present positions for, against, and neutral toward the proposition of launching a plutonium source into space. Examine the best two arguments for and against launching the RTG, and state how you would counter these arguments.

Chapter 21 Visual Summary

The chart shows the connections between the major topics discussed in this chapter.

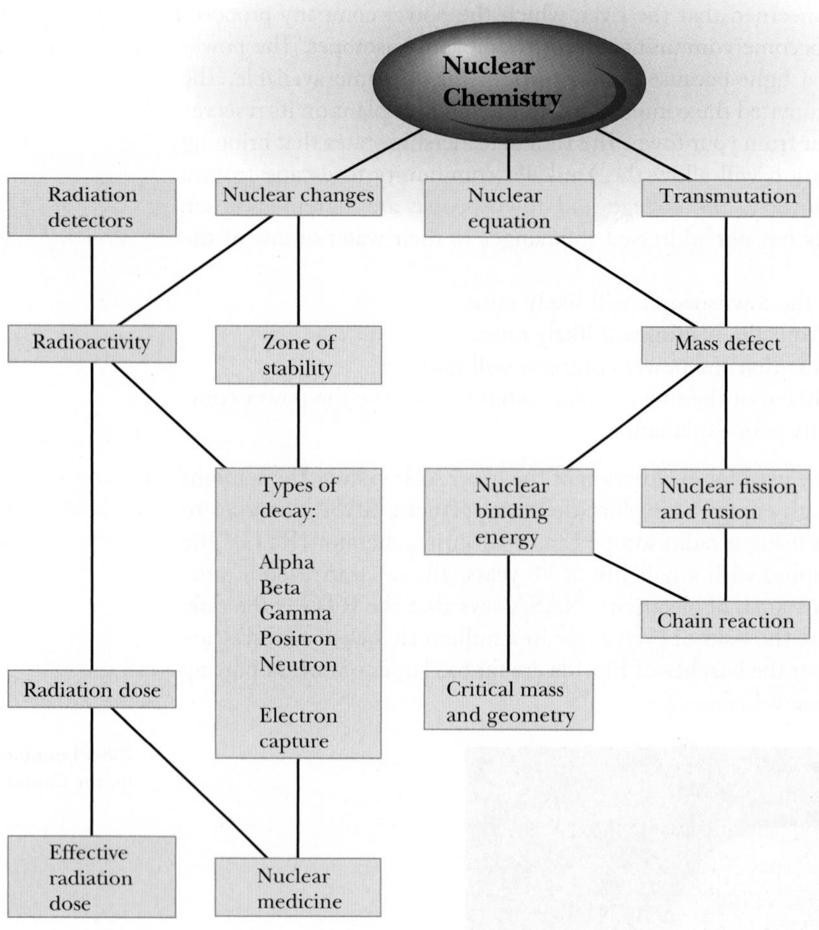

Summary

21.1 Nuclear Stability and Radioactivity

Although many isotopes of the elements are stable, there are many more that are radioactive. Because radioactive decays form product nuclides that are closer to the zone of stability, it is often possible to predict the type of decay a given nuclide undergoes. With few exceptions, the unstable *nuclides* are transformed into more stable ones by emitting an *alpha particle,* a *beta particle,* a *gamma ray,* or a *positron.* An alternative to positron decay is *electron capture,* in which the nucleus absorbs one of the electrons of the atom, converting a proton to a neutron inside the nucleus. Factors that contribute to the stability of nuclei are the neutron-to-proton ratio, *magic numbers,* even numbers of protons, and even numbers of neutrons. Any radioactive decay, as well as other nuclear changes, can be described by a *nuclear* equation, which is balanced when the sums of both the atomic numbers and the mass numbers on the reactant and product sides of the equation are the same.

Alpha decay is generally observed for the heavy elements, those with $Z > 83$, although beta decays are also observed. The type of decay expected for lighter elements is predicted by comparing the mass number of the unstable nuclide with the atomic mass of the element found in the periodic table. Among the most common means of detecting and measuring radioactivity are photographic film, Geiger counters, and scintillation counters.

21.2 Rates of Radioactive Decays

Many radioactive elements form products that are also radioactive; therefore, the heavy elements undergo several decays before a stable nuclide is produced. The required numbers of alpha and beta decays can be calculated from the mass and atomic numbers of the starting nuclide and those of the stable product. The rate of decay of radioisotopes follows first-order kinetics. The rate is usually expressed as a half-life, which is determined from the absolute decay rate (number of disintegrations per second and number of atoms) or changes in the decay rate with time. These techniques are applied to determine the ages of artifacts.

21.3 Induced Nuclear Reactions

Bombarding nuclei with high-energy ions induces nuclear transformations and creates some new isotopes. This technique produced the first samples of many of the elements beyond uranium in the periodic table. Neutron bombardment has been used to make many other radioisotopes. Nuclear equations are used to describe these transformations.

21.4 Nuclear Binding Energy

The energy changes that accompany nuclear transformations are large enough to produce measurable changes in mass in accordance with Einstein's relation, $\Delta E = \Delta mc^2$. The *mass defect,* which results from the *nuclear binding energy,* is the difference between the mass of the constituent hydrogen-1 and neutrons and that of the isotope. The energy released in nuclear reactions is about 10 million times larger than the energy of a chemical change involving the same amount of matter.

21.5 Fission and Fusion

Comparison of the nuclear binding energies per nucleon indicates that *nuclear fission* and *nuclear fusion* can release large amounts of energy. Many heavy nuclides, ^{235}U and ^{239}Pu in particular, undergo fission when the nucleus absorbs a neutron. Fission reactions are usually *chain reactions,* which lead to devastating explosions when they proceed under supercritical conditions. These conditions include both a *critical mass* and *critical geometry.* When the fission process is controlled at barely critical conditions in a nuclear reactor, it is a useful source of energy. Nuclear reactors are now used to generate a significant fraction of the world's electrical energy. Concerns about the safety of nuclear reactors and the disposal of the radioactive waste they generate have deterred the construction of additional nuclear power reactors in the United States, but planning for construction of several new reactors is progressing.

Although nuclear fusion has been achieved as an uncontrolled explosion, the development of a controlled fusion reactor that generates useful amounts of energy is still several decades away. The containment of the reacting nuclides at the pressures and temperatures needed (several million kelvins) for fusion to occur has proved to be a formidable task.

21.6 Biological Effects of Radiation and Medical Applications

Radiation damage to biological tissue is caused mainly by the ionization of molecules as the radioactive particles give up their energy. The SI unit for radioactivity is the *becquerel,* and the energy transferred to matter, called *radiation dose,* is measured in *rads*. An estimate of radiation damage to biological tissue must also include a quality factor, *Q*, for different kinds of radiation. The effective radiation dose (measured in rems) is the product of the quality factor and the radiation dose (in rads). Current estimates indicate that most of the exposure of the U.S. population to radiation is from natural sources, particularly radon.

Nuclear medicine uses radioactive molecules to diagnose and sometimes treat diseases. The accumulation of the radiopharmaceuticals in target organs is a measure of the function of the organ, information that cannot be obtained from x ray images alone.

Download Go Chemistry concept review videos from OWL or purchase them from **www.ichapters.com**

Chapter Terms

The following terms are defined in the Glossary, Appendix I.

Nucleon
Nuclide

Section 21.1
Alpha particle
Beta particle
Electron capture
Gamma ray
Magic numbers
Nuclear equation
Positron
Radioactivity
Stable isotope

Section 21.3
Induced nuclear reaction or nuclear transmutation

Section 21.4
Mass defect
Nuclear binding energy

Section 21.5
Critical mass
Critical geometry
Moderator
Nuclear chain reaction
Nuclear fission
Nuclear fusion
Plasma

Section 21.6
Effective radiation dose
Rad
Radiation dose
Radioactivity

Key Equations

$$\text{Rate} = \frac{-\Delta N}{\Delta t} = kN \quad (21.2)$$

$$t_{1/2} = \frac{0.693}{k} \quad (21.2)$$

$$\ln\left(\frac{N}{N_0}\right) = \ln\left(\frac{R}{R_0}\right) = kt = -\frac{0.693}{t_{1/2}}t \quad (21.2)$$

$$E = mc^2 \quad (21.4)$$

$$\Delta E = \Delta mc^2 \quad (21.4)$$

Questions and Exercises

OWL Selected end of chapter Questions and Exercises may be assigned in OWL.

Blue-numbered Questions and Exercises are answered in Appendix J; questions are qualitative, are often conceptual, and include problem-solving skills.

■ Questions assignable in OWL

✎ Questions suitable for brief writing exercises

▲ More challenging questions

Questions

21.1 A nuclide that lies above the band of stability shown in Figure 21.1 has too large a neutron-to-proton ratio to be stable. Which of the natural decay processes for such a nuclide will move it toward the band of stability?

21.2 Does the neutron-to-proton ratio increase, decrease, or remain unchanged when each of the following radioactive decays occurs?
(a) beta decay (b) positron decay
(c) gamma decay (d) alpha decay

21.3 A nuclide that lies below the band of stability shown in Figure 21.1 has too small a neutron-to-proton ratio to be stable. Which of the natural decay processes for such a nuclide will move it toward the band of stability?

21.4 Why are the high-energy photons that accompany an electron capture decay called x rays rather than gamma rays?

21.5 Most alpha and beta decays are accompanied by the emission of gamma rays as well. Why?

21.6 What property of the products of radioactive decay is used to detect and measure the presence of the radiation?

21.7 Would carbon-14 dating be useful for determining whether a sample of paper was 20 or 50 years old?

21.8 Why are charged particles with very high energies needed to cause nuclear reactions with the heavier elements?

21.9 Why are low-energy neutrons able to react with nearly all nuclides?

21.10 ✎ What are the roles of the moderator and the control rods in a nuclear reactor?

21.11 ✎ Compare the general penetrating abilities of alpha, beta, and gamma radiation. Why are alpha particles produced inside the body particularly dangerous?

21.12 Explain why most fission products formed in a nuclear reactor decay by beta emission rather than undergoing another kind of nuclear decay.

Photo by Jack Fletcher/National Geographic/Getty Images

Fuel rod manufacturing facility. The worker shown here, working with enriched uranium fuel pellets, is wearing plastic coveralls. The cuffs and sleeves are taped to form a barrier that prevents small particles from contaminating his skin.

21.13 ✎ A person's exposure to radiation can depend greatly on occupation. List several occupations that may result in an exposure to radiation greater than the average exposure of the U.S. population. Explain your choices.

21.14 ✎ In 1999, workers were preparing a shipment of uranium solutions from a nuclear fuel fabrication plant in Japan. The workers decided to consolidate the solutions in a precipitation vat before placing them into the shipment containers. The procedure worked well for the first six solutions, but the seventh batch contained 18.8% ^{235}U. While two workers were adding a seventh batch of uranium solution to the tank, a critical excursion occurred. The two workers, together with a third worker nearby, observed a blue flash and fled; simultaneously, gamma-radiation detectors went off in the building and two adjacent buildings, prompting all workers to evacuate. The two workers who had been pouring were transported to the hospital and began to show signs of radiation sickness during the ambulance ride. They had absorbed fatal doses in the range of 600 to 2000 rem. Explain why the criticality occurred and how it could have been avoided.

21.15 Nuclear fuel rods are fabricated in a standard manufacturing facility, but after the same rods are removed from the reactor, they are hazardous to approach. Explain.

21.16 Write the fusion reaction that occurs in the Sun. Why are researchers trying to use fusion as a power source? What are the advantages and disadvantages compared with nuclear fission?

Exercises

OBJECTIVE Relate the stability of a nuclide to the numbers of protons and neutrons in the nucleus.

21.17 Fill in the missing entries in the table below.

Symbol	Z	A	Number of Protons	Number of Neutrons
$^{40}_{20}Ca$	—	—	—	—
—	15	31	—	—
—	—	—	50	68
—	—	239	93	—

21.18 ■ Fill in the missing entries in the table below.

Symbol	Z	A	Number of Protons	Number of Neutrons
$^{23}_{11}Na$	—	—	—	—
—	45	103	—	—
—	—	—	32	38
—	—	234	90	—

21.19 Calculate the neutron-to-proton ratio (to two decimal places) for each of the following stable isotopes.
(a) $^{12}_{6}C$ (b) $^{40}_{20}Ca$ (c) $^{90}_{40}Zr$
(d) $^{138}_{56}Ba$ (e) $^{208}_{82}Pb$

21.20 Calculate the neutron-to-proton ratio (to two decimal places) for each of the following stable isotopes.
(a) $^{17}_{8}O$ (b) $^{39}_{19}K$ (c) $^{75}_{33}As$
(d) $^{118}_{50}Sn$ (e) $^{196}_{78}Pt$

21.21 Which of the following isotopes lie(s) within the band of nuclear stability shown in Figure 21.1?
(a) $^{17}_{8}O$ (b) $^{93}_{38}Sr$ (c) $^{67}_{30}Zn$
(d) $^{233}_{92}U$ (e) $^{28}_{12}Mg$

21.22 Which of the following isotopes lie(s) within the band of nuclear stability shown in Figure 21.1?
(a) $^{16}_{8}O$ (b) $^{37}_{17}Cl$ (c) $^{68}_{29}Cu$
(d) $^{11}_{6}C$ (e) $^{88}_{41}Nb$

OBJECTIVE Predict decay modes and write balanced nuclear equations for radioactive decays.

21.23 Write the balanced nuclear equation for each of the following nuclear decays. Show the mass and atomic numbers for each of the particles in the equation.
(a) alpha decay of bismuth-201
(b) positron decay of iridium-184
(c) electron capture of lanthanum-135
(d) beta decay of bromine-80

21.24 Write the balanced nuclear equation for each of the following nuclear decays. Show the mass and atomic numbers for each of the particles in the equation.
(a) beta decay of carbon-14
(b) alpha decay of thorium-234
(c) positron decay of potassium-40
(d) electron capture decay of lead-198

21.25 Identify the missing particles by balancing the mass and atomic numbers in each of the following nuclear decay equations.
(a) $\_\_\_\_ \longrightarrow {}^{223}_{88}Ra + {}^{4}_{2}\alpha$
(b) ${}^{22}_{11}Na + {}^{0}_{-1}e \longrightarrow \_\_\_\_$
(c) ${}^{223}_{90}Th \longrightarrow \_\_\_\_ + {}^{4}_{2}\alpha$
(d) ${}^{72}_{31}Ga \longrightarrow {}^{72}_{32}Ge + \_\_\_\_$
(e) ${}^{60}_{29}Cu \longrightarrow {}^{60}_{28}Ni + \_\_\_\_$

21.26 ■ Identify the missing particles by balancing the mass and atomic numbers in each of the following nuclear decay equations.
(a) ${}^{9}_{4}Be + ? \longrightarrow {}^{6}_{3}Li + {}^{4}_{2}\alpha$
(b) $? + {}^{1}_{0}n \longrightarrow {}^{24}_{11}Na + {}^{4}_{2}\alpha$
(c) ${}^{40}_{20}Ca + ? \longrightarrow {}^{40}_{19}K + {}^{1}_{1}H$
(d) ${}^{241}_{95}Be + {}^{4}_{2}\alpha \longrightarrow {}^{243}_{97}Bk + ?$
(e) ${}^{246}_{96}Cm + {}^{12}_{6}C \longrightarrow 4{}^{1}_{0}n + ?$
(f) ${}^{238}_{92}U + ? \longrightarrow {}^{249}_{100}Fm + 5{}^{1}_{0}n$

21.27 Identify the missing particles by balancing the mass and atomic numbers in each of the following nuclear decay equations.
(a) ${}^{67}_{31}Ga + {}^{0}_{-1}e \longrightarrow \_\_\_\_$
(b) $\_\_\_\_ \longrightarrow {}^{211}_{85}At + {}^{4}_{2}\alpha$
(c) ${}^{67}_{29}Cu \longrightarrow {}^{67}_{30}Zn + \_\_\_\_$
(d) $\_\_\_\_ \longrightarrow {}^{124}_{53}I + {}^{0}_{1}\beta$
(e) ${}^{233}_{92}U \longrightarrow \_\_\_\_ + {}^{4}_{2}\alpha$

21.28 ■ Identify the missing particles by balancing the mass and atomic numbers in each of the following nuclear decay equations.
(a) ${}^{242}_{94}Pu \longrightarrow {}^{4}_{2}\alpha + \_\_\_\_$
(b) $\_\_\_\_ \longrightarrow {}^{32}_{16}S + {}^{0}_{-1}\beta$
(c) ${}^{252}_{98}Cf + \_\_\_\_ \longrightarrow 3\,{}^{1}_{0}n + {}^{259}_{103}Lr$
(d) ${}^{55}_{26}Fe + \_\_\_\_ \longrightarrow {}^{55}_{25}Mn$
(e) ${}^{15}_{8}O \longrightarrow \_\_\_\_ + {}^{0}_{+1}\beta$

21.29 Radioactive decay always moves the nuclide toward the band of stability. Based on the mode of decay for each of the following radioactive nuclides, state whether it lies above or below the band of stability.
(a) ^{32}Si, beta decay
(b) ^{44}Ti, electron capture
(c) ^{52}Mn, positron decay

21.30 Radioactive decay always moves the nuclide toward the band of stability. Based on the mode of decay for each of the following radioactive nuclides, state whether it lies above or below the band of stability.
(a) ^{116}Ag, beta decay
(b) ^{86}Zr, electron capture
(c) ^{70}Se, positron decay

21.31 For each of the following radioactive nuclides, predict the mode of decay from its location in the periodic table and by comparing the mass number with the atomic mass of the element in the periodic table. Write a nuclear equation for the decay.
(a) ^{117}Sb (b) ^{83}Se (c) ^{221}Ac (d) ^{42}Ar

21.32 ■ Predict the probable mode of decay for each of the following radioactive isotopes, and write an equation to show the products of decay.
(a) ^{54}Mn (b) ^{241}Am (c) ^{110}Ag (d) ^{197}Hg

21.33 For each of the following radioactive nuclides, predict the mode of decay from its location in the periodic table and by comparing the mass number with the atomic mass of the element in the periodic table. Write a nuclear equation for the decay.
(a) ^{76}Br (b) ^{84}Br (c) ^{109}Pd (d) ^{241}Am

21.34 ■ Write a nuclear equation for the type of decay each of these unstable isotopes is most likely to undergo.
(a) ^{19}Ne (b) ^{230}Th (c) ^{92}Br (d) ^{212}Po

OBJECTIVE Analyze a radioactive decay series to predict the numbers of alpha and beta particles emitted.

21.35 The decay of ^{238}Pu terminates in ^{206}Pb. Determine the number of alpha and beta decays that occurred in this transformation.

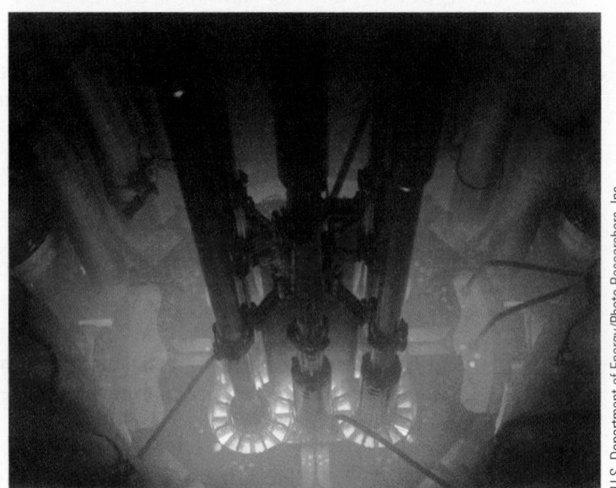

U.S. Department of Energy/Photo Researchers, Inc.

Nuclear reactor core. The blue glow is from the interaction of charged particles with the cooling water.

21.36 ■ A decay series begins with ^{235}U, finally producing the stable isotope ^{207}Pb. Find the numbers of alpha and beta decays that occurred in this transformation.

OBJECTIVE Determine the half-life of a reaction from the absolute decay rate and the number of atoms present.

21.37 A sample that contains 3.75×10^{13} atoms of a beta emitter has a disintegration rate of 382 disintegrations/min. What is the half-life (in years) for this isotope?

21.38 By mass spectral analysis, a sample of strontium is known to contain 2.64×10^{10} atoms of ^{90}Sr as the only radioactive element. The absolute disintegration rate of this sample is measured as 1238 disintegrations/min.
(a) Calculate the half-life (in years) of ^{90}Sr.
(b) How long will it take for the disintegration rate of this sample to decline to 1000 disintegrations/min?

OBJECTIVE Determine the half-life of a reaction from changes in the decay rate.

21.39 The radioactive decay of a sample containing ^{121}Te produced 865 disintegrations/min. Exactly 7.00 days later, the rate of decay was found to be 650 disintegrations/min. Calculate the half-life, in days, for the decay of ^{121}Te.

21.40 ■ ^{130}I decays by emission of beta particles to form stable ^{130}Xe. A sample of iodine-130 was recorded as having 1245 disintegrations per second at 11:00 A.M. on June 23. At 9:32 A.M. on June 24, the same sample produced 350 disintegrations/s. What is the half-life, in hours, for the beta decay of ^{130}I?

OBJECTIVE Determine the ages of samples of matter from radioactive decay data.

21.41 A sample of uranium ore contains 6.73 mg ^{238}U and 3.22 mg ^{206}Pb. Assuming that all of the ^{206}Pb arose from decay of the ^{238}U and that the half-life of ^{238}U is 4.51×10^9 years, determine the age of the ore.

21.42 A tree was cut down and used to make a statue 2300 years ago. What fraction of the ^{14}C that was present originally remains today? (*Note:* $t_{1/2}$ of ^{14}C is 5730 years.)

21.43 The ratio of ^{87}Rb to ^{87}Sr can be used for the dating of geological samples, assuming that all of the strontium came from the beta decay of rubidium ($t_{1/2} = 5.0 \times 10^{10}$ years). What is the age of a rock sample if $^{87}Sr/^{87}Rb = 0.051$?

21.44 ▲ In a living organism, the decay of ^{14}C produces 15.3 disintegrations per minute per gram of carbon. The half-life of ^{14}C is 5730 years. What percentage of the atoms of carbon in the biosphere is ^{14}C?

OBJECTIVE Write the nuclear equations that describe transmutations.

21.45 Complete and balance the following nuclear equations.
(a) $^{54}_{26}Fe + ^{4}_{2}\alpha \longrightarrow 2^{1}_{1}p +$ ____
(b) ____ $+ ^{1}_{0}n \longrightarrow ^{24}_{11}Na + ^{4}_{2}\alpha$
(c) $^{238}_{92}U + ^{16}_{8}O \longrightarrow$ ____ $+ 5^{1}_{0}n$
(d) $^{96}_{42}Mo +$ ____ $\longrightarrow ^{97}_{43}Tc + ^{1}_{0}n$
(e) $^{250}_{98}Cf + ^{11}_{5}B \longrightarrow$ ____ $+ 5^{1}_{0}n$

21.46 Complete and balance the following nuclear equations.
(a) $^{249}_{98}Cf + ^{10}_{5}B \longrightarrow ^{257}_{103}Lr +$ ____
(b) $^{14}_{7}N + ^{1}_{1}p \longrightarrow ^{4}_{2}\alpha +$ ____
(c) $^{238}_{92}U + ^{1}_{0}n \longrightarrow$ ____ $+ ^{0}_{-1}\beta$
(d) $^{6}_{3}Li + ^{1}_{0}n \longrightarrow ^{4}_{2}\alpha +$ ____
(e) ____ $+ ^{4}_{2}\alpha \longrightarrow ^{12}_{6}C + ^{1}_{0}n$

OBJECTIVE Calculate the mass defect of a nuclide. Calculate the nuclear binding energy from the mass defect. Express the binding energy in megaelectron volts per nucleon (MeV/nucleon).

21.47 The molar mass of ^{31}P is 30.9738 g/mol.
(a) Use the masses of hydrogen-1 and the neutron given in Table 21.3 to calculate the mass defect of this nuclide in grams per mole.
(b) Express the total nuclear binding energy, in megaelectron volts, for this nuclide.
(c) Calculate the binding energy per nucleon for ^{31}P.

21.48 ■ The actual mass of a ^{108}Pd atom is 107.90389 u.
(a) Calculate the mass defect in atomic mass units per atom (u/atom) and in grams per mole (g/mol) for this isotope.
(b) What is the nuclear binding energy in kilojoules per mole (kJ/mol) for this isotope?

21.49 The atomic masses of three isotopes of aluminum are ^{26}Al 25.9869 ^{27}Al 26.9815 ^{28}Al 27.9819
(a) Calculate the binding energy per nucleon, in megaelectron volts, for each of these isotopes of aluminum.
(b) Two of these nuclides are radioactive, and the third one is stable. Based on your answer to part (a), which isotopes are unstable?

21.50 The atomic masses of three isotopes of phosphorus are ^{30}P 29.9783 ^{31}P 30.9738 ^{32}P 31.9739
(a) Calculate the binding energy per nucleon, in megaelectron volts, for each of these isotopes of phosphorus.
(b) Two of these nuclides are radioactive, and the third one is stable. Based on your answer to part a, which isotopes are unstable?

OBJECTIVE Recognize that the stability of nuclei varies, and nuclear reactions that form stable nuclei release great amounts of energy.

21.51 When ^{239}Pu is used in a nuclear reactor, one of the fission events that occurs is

$$^{1}_{0}\text{n} + \underset{239.052\ \text{u}}{^{239}_{94}\text{Pu}} \longrightarrow \underset{97.913\ \text{u}}{^{98}_{40}\text{Zr}} + \underset{138.919\ \text{u}}{^{139}_{54}\text{Xe}} + 3^{1}_{0}\text{n}$$

The atomic mass of each atom is given below its symbol in the equation.
(a) Calculate the change in mass, expressed in atomic mass units, that accompanies this fission event.
(b) Convert the mass loss to joules per fission.
(c) Find the energy released when 1.00 g plutonium undergoes this particular fission.

21.52 When ^{239}Pu is used in a nuclear reactor, one of the fission events that occurs is

$$^{1}_{0}\text{n} + \underset{239.052\ \text{u}}{^{239}_{94}\text{Pu}} \longrightarrow \underset{95.916\ \text{u}}{^{96}_{39}\text{Y}} + \underset{139.917\ \text{u}}{^{140}_{55}\text{Cs}} + 4^{1}_{0}\text{n}$$

The atomic mass of each atom is given below its symbol in the equation.
(a) Calculate the change in mass, expressed in atomic mass units, that accompanies this fission event.
(b) Convert the mass loss to joules per fission.
(c) Find the energy released when 1.00 g plutonium undergoes this particular fission.

21.53 A product of the fission of ^{235}U is ^{137}Cs, which undergoes beta decay with a half-life of 30.17 years. What fraction of the ^{137}Cs that was formed in the detonation of the first fission bombs in August 1945 will still be present in the environment 100 years later?

21.54 One of the fission products that causes major concern is ^{90}Sr, because it is incorporated into milk and other high-calcium foods. ^{90}Sr undergoes beta decay with a half-life of 28.1 years. What fraction of the ^{90}Sr that was formed in the detonation of the first fission bombs in August 1945 will still be present in the environment in August 2015?

Chapter Exercises

21.55 Cadmium is used in the control rods of fission reactors because it absorbs neutrons efficiently. Most of the neutrons are absorbed by ^{113}Cd to produce ^{114}Cd and a gamma ray. The atomic masses of these two isotopes of cadmium are 112.9044 u for ^{113}Cd and 113.9034 u for ^{114}Cd. Calculate the energy of the gamma ray emitted in this nuclear transformation, expressing your answer in megaelectron volts. What is the wavelength of the gamma ray?

21.56 ■ The $^{14}_{6}$C activity of an artifact from a burial site was 8.6/min per gram carbon. The half-life of $^{14}_{6}$C is 5730 years, and the current $^{14}_{6}$C activity is 15.3 disintegrations per minute per gram of carbon. How old is the artifact?

21.57 One of the promising reactions for a controlled fusion reactor consumes deuterium (^{2}H) and tritium (^{3}H) in the reaction

$$\underset{2.0140}{^{2}_{1}\text{H}} + \underset{3.0161}{^{3}_{1}\text{H}} \longrightarrow \underset{4.0026}{^{4}_{2}\text{He}} + \underset{1.008665}{^{1}_{0}\text{n}}$$

The atomic mass of each particle is given below the equation. Calculate the energy released by this fusion reaction per gram of helium formed, and compare it with the energy generated by the fission of 1 g ^{235}U (8×10^{7} kJ/g).

21.58 ^{128}I decays by both beta and positron decay. What are the product nuclides from each of these decays?

Cumulative Exercises

21.59 One kilogram of high-grade coal produces about 2.8×10^{4} kJ energy when it is burned. Fission of 1 mol ^{235}U releases 1.9×10^{10} kJ.
(a) Calculate the number of metric tons (1 metric ton = 1000 kg) of coal needed to produce the same amount of energy as the fission of 1 kg uranium.
(b) How many metric tons of sulfur dioxide (a major source of acid rain) are produced from the burning of the coal in part a, if the coal is 0.90% by mass sulfur?

21.60 ■ Collision of an electron and a positron results in formation of two gamma rays. In the process, their masses are converted completely into energy.
(a) Calculate the energy evolved from the annihilation of an electron and a positron, in kilojoules per mole.
(b) Using Planck's equation (Equation 7.2), determine the frequency of the gamma rays emitted in this process.

21.61 Gaseous diffusion is used to enrich natural uranium in ^{235}U. Use Graham's law (see Section 6.7) to calculate the ratio of enrichment of ^{235}U by a single diffusion of UF_6. The atomic masses of the two principal isotopes of uranium are 235.0493 and 238.0508 for ^{235}U and ^{238}U, respectively, and that of fluorine is 18.9984.

21.62 ■ How many disintegrations per second occur in a basement that is 40 × 40 × 10 feet if the radiation level from radon is the maximum allowed, 4×10^{-12} Ci/L?

21.63 Natural potassium is 0.0117% radioactive ^{40}K, which decays by positron emission with a half-life of 1.26×10^{9} years. If an average banana has 6.00×10^{-1} g potassium in it, what is the radioactivity of a banana in becquerels?

21.64 ■ A 100.0 g sample of water containing tritium, $^{3}_{1}$H, emits 2.89×10^{3} beta particles per second. Tritium has a half-life of 12.3 years. What percentage of all the hydrogen atoms in the water sample is tritium?

21.65 The human body also contains potassium. If a 70-kg person has an average of 140 g potassium in his or her body, what is the radioactivity of this person, in becquerel, coming from ^{40}K? See Exercise 21.63 for necessary data.

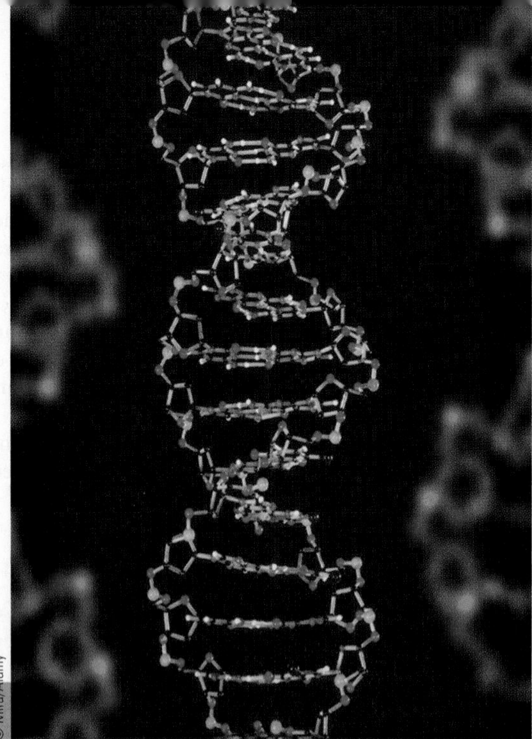

Structure of DNA.

One of the fundamental questions of science is how species inherit traits from their parents. Years of studies have established that genetic information is carried by deoxyribonucleic acids (DNA). One of the most interesting stories in science is the account of how James Watson and Francis Crick unraveled the famous double-helix structure of DNA. They were an unlikely pair working in 1953 at Cavendish Laboratory at Cambridge University. Although a number of workers were trying to determine the structure of DNA, most notably Linus Pauling, Watson, and Crick interpreted many facts correctly to lead them to the structure. One important piece of information they had was the knowledge that DNA was a helix, information learned from the work of Rosalind Franklin, a brilliant scientist in the field of x-ray crystallography. She also critiqued the first (and incorrect) model put forward by Watson and Crick, and provided data that led to the double-helix model.

Organic Chemistry and Biochemistry

22

CHAPTER CONTENTS

OWL Online homework for this chapter may be assigned in OWL.

Look for the green colored vertical bar throughout this chapter, for integrated references to this chapter introduction.

In the end, one key to their success was that Watson and Crick built large three-dimensional molecular models out of metal plates that allowed them to determine that adenine and thymine bases were held together by hydrogen bonds, as were guanine and cytosine bases. These bonds held the two strands of DNA together in the highly organized double strand.

Genetic information is preserved as the helix unwinds, and each strand creates a new matching strand, forming two DNA units identical to the first, held in place by hydrogen bonds. For their discovery, Watson and Crick received the 1962 Nobel Prize for Physiology and Medicine, shared with another important worker in the field, Maurice Wilkins. Sadly, Franklin had died at age 37 in 1958, and the Nobel Prize is not awarded posthumously. ■

Organic chemistry is the study of carbon-containing compounds. Compounds that contain carbon atoms have a rich and diverse chemistry, and well over 30 million different organic compounds have been characterized. Originally, it was believed that living systems had a "vital force" that produced special compounds. Substances produced by living systems were called *organic* and were differentiated from all others (called *inorganic*). However, in 1828, the German chemist Friedrich Wöhler demonstrated that urea, an organic compound first isolated from urine, could be prepared by the reaction of three inorganic compounds, lead(II) cyanate, ammonia, and water.

$$Pb(OCN)_2 + 2NH_3 + 2H_2O \rightarrow \underset{\text{Urea}}{2(NH_2)_2CO} + Pb(OH)_2$$

The name *organic* is now used for carbon-containing compounds in which carbon forms bonds to itself, to hydrogen, and to other nonmetals such as nitrogen, oxygen, and sulfur. The many carbon-containing minerals, such as carbonate salts, are not considered organic compounds. **Biochemistry** is the study of the chemical systems in living organisms.

Why is the chemistry of carbon so extensive? The primary reason is that carbon atoms make strong bonds to each other, forming chains or rings—a process known as **catenation.** Carbon atoms are ideally suited to form four strong covalent bonds because they have four valence orbitals and four valence electrons. In addition to forming strong carbon-carbon bonds, carbon atoms make strong bonds to hydrogen, nitrogen, oxygen, and sulfur. Also, carbon readily forms multiple bonds with nitrogen, oxygen, and other carbon atoms.

22.1 Alkanes

OBJECTIVES

- ☐ Identify and name linear and cyclic alkanes
- ☐ Write the structural isomers of alkanes
- ☐ Name the important substituent groups
- ☐ Describe the shapes of the two important conformers of cyclohexane
- ☐ Write some reactions of alkanes

Alkanes, also known as saturated hydrocarbons, are hydrocarbons that contain no multiple bonds or rings.

Hydrocarbons are compounds that contain only the elements hydrogen and carbon. An **alkane** is a hydrocarbon that contains no multiple bonds or rings. Alkanes are also known as **saturated hydrocarbons** because each carbon atom makes bonds to four other atoms, the maximum number possible. The simplest hydrocarbon is methane, CH_4. In the terms of valence bond theory, each C–H bond in this tetrahedral molecule forms from the overlap of an sp^3-hybridized orbital on the carbon atom with the 1*s* orbital on a hydrogen atom (Figure 22.1).

The alkane with two carbon atoms is ethane, C_2H_6. The carbon atoms in ethane, like those in all alkanes, are sp^3 hybridized. Three of these hybrid orbitals on each carbon

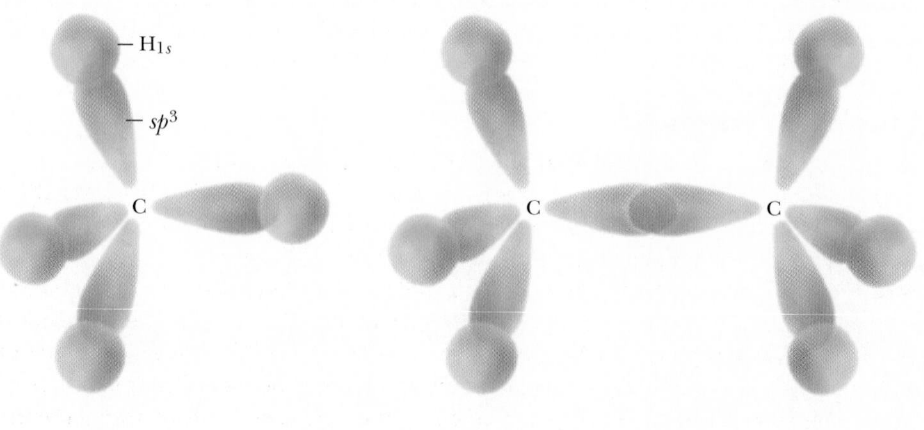

Figure 22.1 Bonding in ethane and methane. Carbon atoms form covalent bonds from sp^3-hybridized orbitals in *(a)* methane and *(b)* ethane.

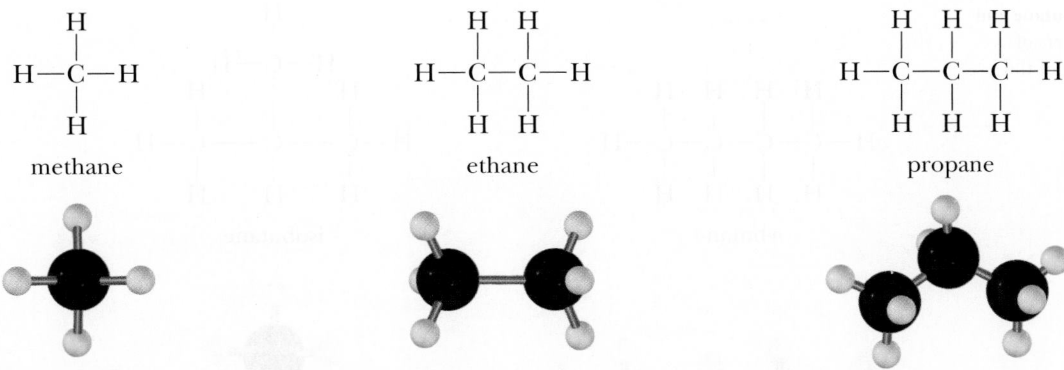

Figure 22.2 Structures of methane, ethane, and propane. The tetrahedral geometry about the carbon atoms in alkanes makes the actual arrangement of the atoms in space (computer-generated figures; bottom) different from those shown by structural formulas (top).

TABLE 22.1 The First 10 Straight-Chain Alkanes

Name	Formula	Normal Boiling Point (°C)	Alkyl Group	Formula
Methane	CH_4	−162	Methyl	CH_3–
Ethane	C_2H_6	−89	Ethyl	C_2H_5–
Propane	C_3H_8	−42	Propyl	C_3H_7–
n-Butane	C_4H_{10}	0	Butyl	C_4H_9–
n-Pentane	C_5H_{12}	36	Pentyl	C_5H_{11}–
n-Hexane	C_6H_{14}	69	Hexyl	C_6H_{13}–
n-Heptane	C_7H_{16}	98	Heptyl	C_7H_{15}–
n-Octane	C_8H_{18}	126	Octyl	C_8H_{17}–
n-Nonane	C_9H_{20}	151	Nonyl	C_9H_{19}–
n-Decane	$C_{10}H_{22}$	174	Decyl	$C_{10}H_{21}$–

atom overlap with 1*s* orbitals on the hydrogen atoms, and the fourth forms the C–C bond. The third member of the family is propane, C_3H_8.

A number of ways are possible to draw the structures of these molecules (Figure 22.2). The commonly used structural formulas, although convenient to write, do not properly show the shapes of these molecules. When you look at the molecular formula or a structural formula of an alkane, remember that the geometry around each carbon atom is tetrahedral, so the carbon chain in propane forms a bent, or V, shape.

Note that from methane to ethane and from ethane to propane, the chemical formula increases by one CH_2 group. In general, the alkanes have the formula C_nH_{2n+2}, where $n = 1, 2, \ldots$. Table 22.1 lists the names of the first 10 straight-chain members (isomers with all of the carbon atoms in a continuous chain) of the alkane family, together with their boiling points. As expected, the boiling points increase with increasing molecular weight, because increasing the chain length increases the polarizability of the alkane (see Section 11.4), leading to stronger London dispersion forces.

Structural Isomers

There are two hydrocarbons with the formula C_4H_{10}. The carbon atoms can be arranged in an unbranched chain (called a *straight chain* even though it is puckered) or a branched chain (Figure 22.3). The straight-chain isomer is called *normal butane* (*n*-butane), and the branched isomer is called *isobutane.* These two alkanes are *structural isomers,* molecules that have the same molecular formula but differ in the connectivity or arrangement of the bonds. They have different physical and chemical properties. For example, the boiling point of isobutane (−12 °C) is lower than that of *n*-butane (0 °C). The boiling points are different because the more linear shape of *n*-butane allows the chains to line up next to each other and exert slightly stronger intermolecular forces. In comparison, the more spherical shape of isobutane limits the contact between molecules, so its intermolecular forces are slightly weaker.

Figure 22.3 Structures of *n*-butane and isobutane. Two structural isomers of butane differ in the connectivity of the carbon chain.

n-butane

isobutane

Alkanes heavier than propane have many structural isomers because of different ways to connect the carbon atoms.

The number of possible structural isomers for an alkane increases rapidly as the number of carbon atoms increases. Example 22.1 shows the isomers of C_5H_{12}, pentane.

EXAMPLE 22.1 Isomers of Pentane

Draw structural formulas of all of the possible structural isomers of the pentanes, C_5H_{12}.

Strategy Determine the unique ways the chain can be arranged.

Solution

One isomer, *n*-pentane, has all of the carbon atoms in a linear chain. A second isomer, isopentane, forms when the chain branches once at the second carbon atom in the chain. A third isomer, neopentane, forms when the chain branches twice at the second carbon atom. No other connectivity patterns can be drawn in which each carbon atom has four bonds and each hydrogen atom has one bond; therefore, only these three isomers are possible.

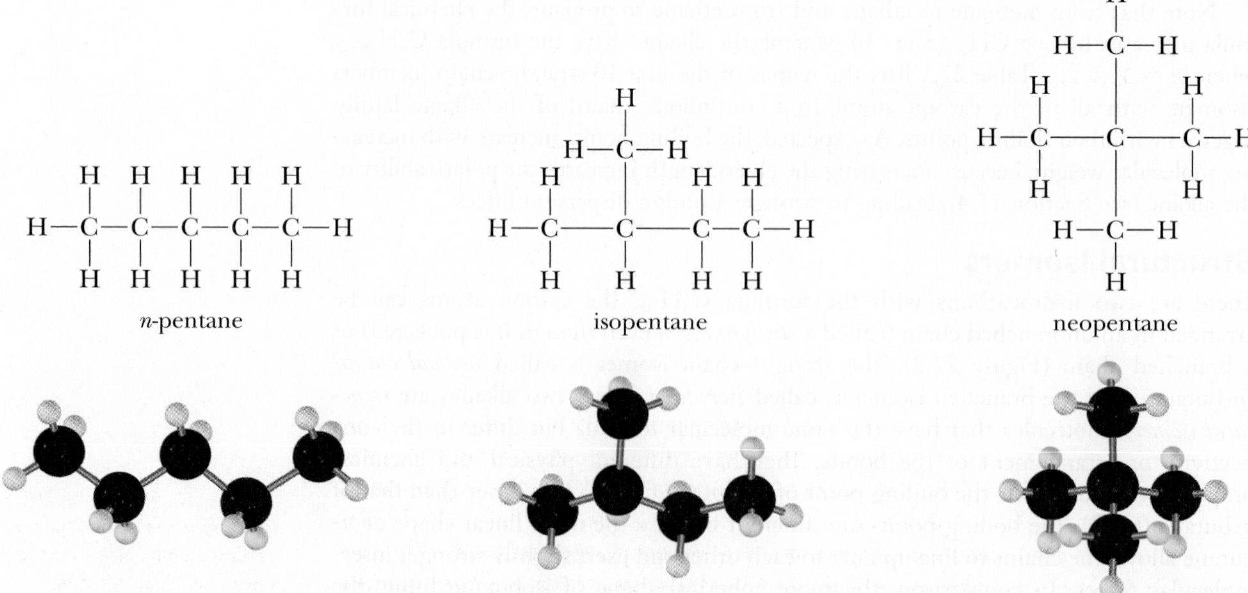

Note that each isomer has a different pattern of connections among the carbon atoms. Simply changing the position of a group by rotating the drawing or rotating about a single bond does not lead to additional isomers. The following drawings are two different representations of isopentane. Although they look different, they both have the same connectivity, and thus are different representations of the same isomer.

```
         H                          H
         |                          |
       H—C—H                      H—C—H
     H   |   H   H            H   H   |
     |   |   |   |            |   |   |
   H—C———C———C———C—H        H—C———C———C—H
     |   |   |   |            |   |   |
     H   H   H   H            H   H   |
                                    H—C—H
                                      |
                                      H
```

Alkane Nomenclature

Given the large number of organic compounds, the International Union of Pure and Applied Chemistry (IUPAC) has established an extensive system for naming them. A few rules guide the naming of alkanes. Some of the names used earlier are not systematic, but rather are based on tradition.

1. Alkanes are named by using the suffix *-ane* with the appropriate prefix (*eth-* for two, *prop-* for three, *but-* for four; after that, numerical prefixes are used: *pent-* for five, *hex-* for six, and so forth, as shown in Table 22.1) to name the longest continuous chain of carbon atoms. For example, the alkane below is named as an octane.

```
   CH3CH2CH2CHCH2CH2CH2CH3
            |
           CH3
```

2. Carbon chains that branch from the longest continuous chain are named as alkyl groups. An **alkyl group** is an alkane from which one hydrogen atom has been removed. It is named with the prefix followed by *-yl.* For example, the $-CH_3$ group formed by removing a hydrogen atom from methane is named *methyl.* Alkyl groups are known as *substituents* of the longest chain. Table 22.1 contains the names of straight-chain alkyl substituents, and Table 22.2 lists other types of important substituents.
3. The positions of the substituents are designated by numbering the carbon atoms in the longest chain (the one used in step 1). We start the numbering at the end of the chain that minimizes the numbers assigned to the carbon atoms to which the substituents are bonded. For example, if we number the octane shown in step 1 from left to right, the methyl group is attached at position 4 (**A**); if we number it from right to left, the methyl group would be attached at position 5 (**B**). The numbering scheme in **A** is correct because it minimizes the position number of the substituent.

```
   1   2   3   4 5   6   7   8            8   7   6   5 4   3   2   1
   CH3CH2CH2CHCH2CH2CH2CH3                CH3CH2CH2CHCH2CH2CH2CH3
            |                                      |
           CH3                                    CH3

            A                                      B
```

The position of the substituent is indicated by a number, followed by a hyphen, before its name. This number and substituent name are placed before the name of the longest chain. The preceding compound is 4-methyloctane. In compounds with more than one substituent, the substituents are listed in alphabetical order. If there is more than one substituent of the same type, the prefixes *di-* (2), *tri-* (3), and *tetra-* (4) are used, and the position numbers are separated by commas.

An alkane is named starting with the prefix that designates the number of carbon atoms in the longest chain, followed by the suffix *-ane*.

TABLE 22.2 Table Substituents

Name	Formula	Name	Formula
Fluoro	$-F$	Phenyl	$-C_6H_5$ (benzene ring)
Chloro	$-Cl$	Vinyl	$-CH{=}CH_2$
Bromo	$-Br$	Isopropyl	$-C(CH_3)(CH_3)-H$
Iodo	$-I$	Isobutyl	$-CH_2-C(CH_3)(CH_3)-H$
Nitro	$-NO_2$	*sec*-Butyl	$-C(CH_3)(CH_2-CH_3)-H$
Amino	$-NH_2$	*tert*-Butyl	$-C(CH_3)(CH_3)-CH_3$

EXAMPLE 22.2 Naming Substituted Alkanes

Name the following compounds.

(a) $CH_3CH_2CH_2CHCH_2CH_3$ with CH_3 on the fourth carbon (the CH)

(b) $CH_3CHCH_2CHCH_2CH_2CH_3$ with Br on the second carbon and CH_3 on the fourth carbon

(c) $CH_3CHCH_2CCH_2CH_2CH_2CH_3$ with NO_2 on the second carbon and two CH_3 groups (one above, one below) on the fourth carbon

Strategy Follow the steps outlined earlier in the order presented.

Solution

(a) The longest chain contains six carbon atoms, so it is a hexane.

$$\overset{6}{C}H_3\overset{5}{C}H_2\overset{4}{C}H_2\overset{3}{C}H\overset{2}{C}H_2\overset{1}{C}H_3 \quad (CH_3 \text{ on C-3})$$

In this case, numbering from right to left yields a lower number for the methyl substituent, the 3 position. The name is 3-methylhexane.

(b) The longest chain is a heptane.

$$\overset{1}{C}H_3\overset{2}{C}H\overset{3}{C}H_2\overset{4}{C}H\overset{5}{C}H_2\overset{6}{C}H_2\overset{7}{C}H_3 \quad (Br \text{ on C-2}, CH_3 \text{ on C-4})$$

The substituents are at positions 2 and 4. The name is 2-bromo-4-methylheptane.

(c) The longest chain is an octane. The nitro substituent is at the 2 position, and the two methyl substituents are both at the 4 position. Numbering the chain in the other direction would give higher position numbers, 5 and 7. The name is 4,4-dimethyl-2-nitrooctane.

Understanding

Name the following substituted alkane.

$$CH_3CH_2\overset{\displaystyle Cl}{\underset{\displaystyle Cl}{C}}CH_2CH_2CH_2CH_3$$

Answer 3,3-dichloroheptane

Cycloalkanes

A **cycloalkane** is a saturated hydrocarbon that contains a ring of carbon atoms. The general formula of cycloalkanes that contain one ring and no substituents is C_nH_{2n}, where $n = 3, 4, \ldots$. The first four cycloalkanes are shown in Figure 22.4. These compounds are frequently drawn as polygons (as shown here), with each corner representing a carbon atom that has the appropriate number of hydrogen atoms attached to make four bonds.

We know that a carbon atom making bonds to four other atoms generally has a tetrahedral shape with approximately 109-degree bond angles. In cyclopropane and cyclobutane, the small rings force C–C–C bond angles that are much smaller than this value. The resulting *ring strain* causes these two molecules to be less stable and consequently more reactive than other hydrocarbons. The carbon-carbon bond angles in larger rings can be approximately 109 degrees only if the rings are not planar. The carbon atoms in cyclohexane can achieve approximately 109-degree bond angles by two arrangements, a "chair" form and a "boat" form, as shown in Figure 22.5. These two arrangements are called **conformers,** different arrangements of atoms caused by rotations about single bonds. The two conformers of cyclohexane can interconvert by rotations about the

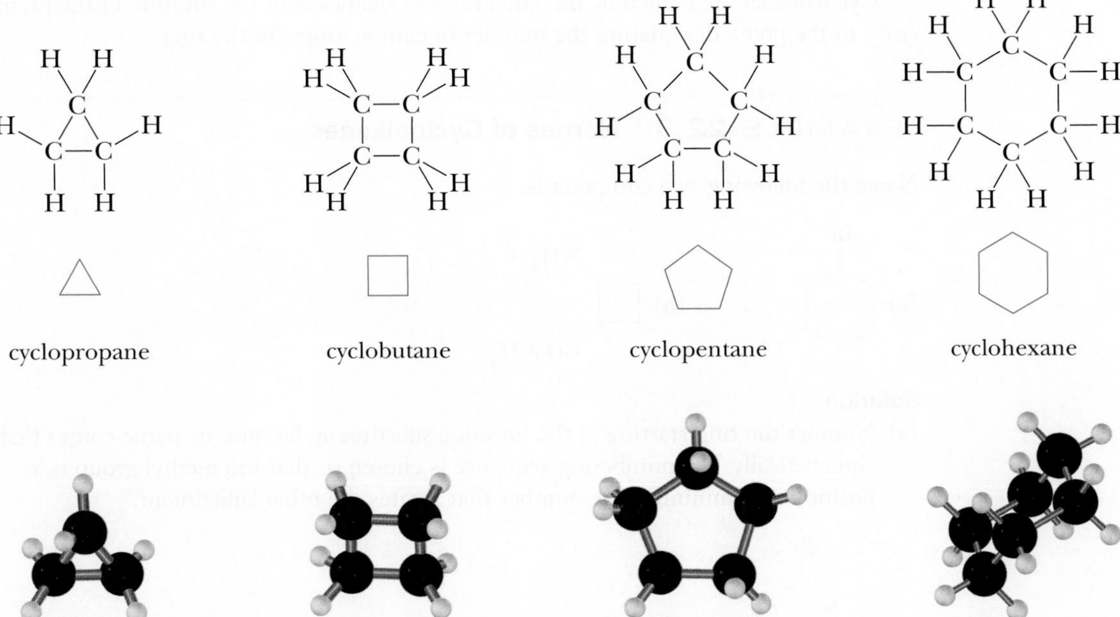

Figure 22.4 Cycloalkanes. The structural formula of a cycloalkane (top) is frequently written as the polygon of the carbon atom framework (center). The computer-generated figures (bottom) represent the spatial arrangements of the atoms.

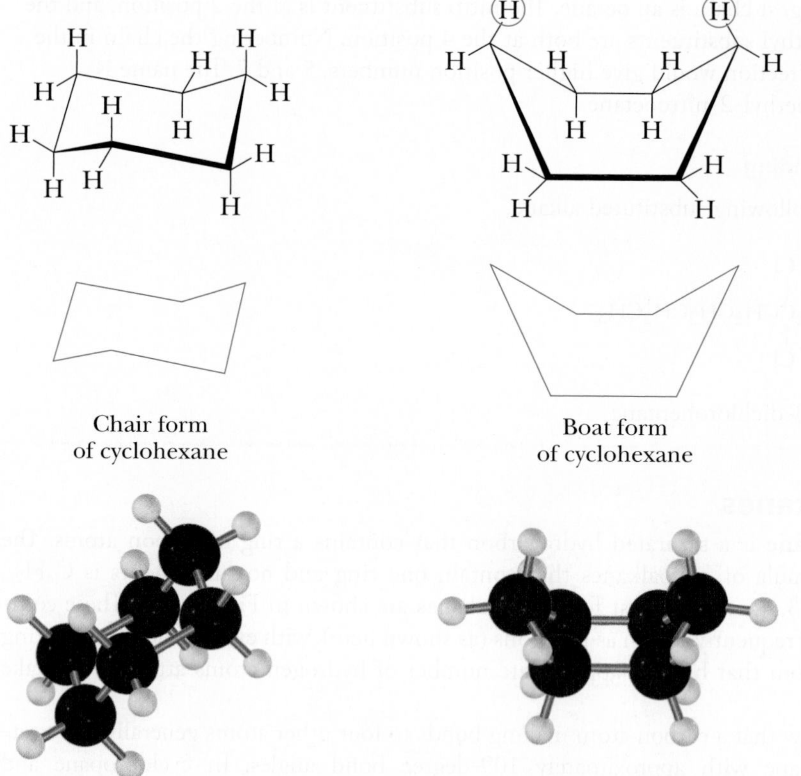

Figure 22.5 Chair and boat forms of cyclohexane. The chair form of cyclohexane is favored over the boat form because it has fewer repulsive interactions between the hydrogen atoms. The line drawings (top) are frequently written as just the bent polygons of the carbon framework (center). The computer-generated structures (bottom) represent the spatial arrangements of the atoms.

C–C bonds without breaking them, and they are in rapid equilibrium with each other at room temperature. The chair arrangement is favored, because in the boat form, several pairs of the hydrogen atoms are fairly close to each other, especially the pair at the inside top of the structure (circled in red in the top drawing of Figure 22.5). The closeness of these hydrogen atoms causes repulsions that make the boat form higher in energy than the chair form. The properties of many biological molecules are determined by the orientations of the six-member rings.

Cyclohexane has two stable conformers. The chair form is more stable than the boat form.

Cycloalkanes are named by the same rules as alkanes with the addition of the prefix *cyclo-* to the prefix designating the number of carbon atoms in the ring.

EXAMPLE 22.3 Names of Cycloalkanes

Name the following two compounds.

(a) Br, CH_3 (b) NH_2, CH_2CH_3

Solution

(a) Number the ring starting at the bromine substituent, because its name comes first alphabetically. The numbering sequence is chosen so that the methyl group is at position 3 to minimize the number that locates the other substituent.

Br, 1, 2, 3, 4, 5, 6, CH_3

The name is 1-bromo-3-methylcyclohexane.

(b) The substituents are at the 1 and 2 positions. The name is 1-amino-2-ethylcyclobutane.

NH_2

4 1
3 2

CH_2CH_3

Understanding

Name the following compound.

CH_3

CH_3

Answer 1,2-dimethylcyclopentane

Reactions of Alkanes

In general, alkanes are not very reactive, because only relatively strong C–C (348 kJ/mol) and C–H (414 kJ/mol) bonds are present. Despite this stability, alkanes are important fuels, because at high temperatures they do react with oxygen in a combustion reaction. We know that the reaction of a hydrocarbon with oxygen is highly exothermic and yields carbon dioxide and water. For example, methane is the main component of the natural gas that heats many homes.

$$CH_4(g) + 2O_2(g) \rightarrow CO_2(g) + 2H_2O(\ell) \qquad \Delta H = -890 \text{ kJ}$$

Another important fuel, gasoline, is a complex mixture of compounds. The main components of gasoline are alkanes that contain 5 to 12 carbon atoms. The combustion of gasoline in an automobile engine is carried out under conditions that produce some CO, as well as CO_2.

Alkanes also react with halogens in the presence of light or an appropriate catalyst, leading to replacement of one or more of the hydrogen atoms—a **substitution reaction.**

A substitution reaction replaces one atom or group of atoms with a different atom or group of atoms.

$$CH_3CH_3 + Cl_2 \xrightarrow{\text{light}} CH_3CH_2Cl + HCl$$

$$CH_3CH_2Cl + Cl_2 \xrightarrow{\text{light}} CH_3CHCl_2 \quad \text{or} \quad CH_2ClCH_2Cl + HCl$$

In the second reaction, two possible isomers can form: 1,1-dichloroethane or 1,2-dichloroethane. Chlorinated hydrocarbons such as these are important as cleaners (dry-cleaning solvents), as insecticides, and as intermediates in the production of numerous chemical products. Their widespread syntheses and uses are controversial because of environmental problems.

The mechanism of these substitution reactions is complex. Light is needed to provide the energy to break the bond in Cl_2.

$$Cl_2 \xrightarrow{\text{light}} 2Cl$$

The chlorine atoms have an unpaired electron and are reactive enough to break C–H bonds. Two products form in the second reaction, because these highly reactive chlorine atoms react with any of the C–H bonds.

OBJECTIVES REVIEW *Can you:*

- ☑ identify and name linear and cyclic alkanes?
- ☑ write the structural isomers of alkanes?
- ☑ name the important substituent groups?
- ☑ describe the shapes of the two important conformers of cyclohexane?
- ☑ write some reactions of alkanes?

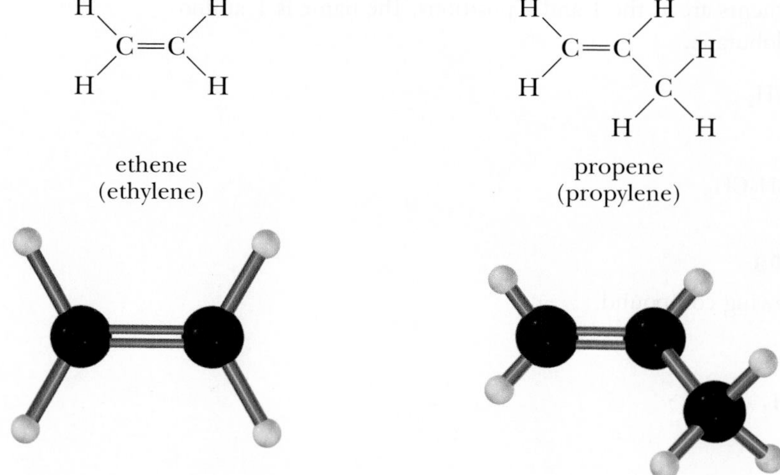

Figure 22.6 Structures of ethene and propene.

22.2 Alkenes, Alkynes, and Aromatic Compounds

OBJECTIVES

- ☐ Identify and name unsaturated compounds
- ☐ List the possible isomers of alkenes, alkynes, and aromatic compounds
- ☐ Write some reactions of alkenes, alkynes, and aromatic compounds

Alkenes

An **unsaturated hydrocarbon** contains one or more multiple carbon-carbon bonds. **Alkenes** are unsaturated hydrocarbons that contain carbon-carbon double bonds. The general formula of an alkene containing one double bond is C_nH_{2n}, where $n = 2, 3 \ldots$. The two simplest alkenes are ethene, C_2H_4, and propene, C_3H_6 (Figure 22.6). The older common names for these compounds are ethylene and propylene. Each carbon atom involved in forming the double bond uses sp^2-hybridized orbitals to form three σ bonds and a p orbital to form a π bond (Figure 22.7).

Alkenes are named by the same system as alkanes, but use the suffix *-ene* instead of *-ane*. The position of the double bond is indicated by the number in the chain of the first carbon atom involved in the multiple bonding. The chain is numbered so as to minimize the number given to the double bond, and this numbering takes precedence over the number of any substituent. The second member of the alkene family, C_3H_6, is called *propene* and not "1-propene" because only one location of the double bond is possible in this molecule.

An alkene contains at least one carbon-carbon double bond. It is named in the same way as an alkane, by replacing the *-ane* suffix with *-ene*.

The third member of the alkene family, C_4H_8, has four possible isomers (Figure 22.8). Two isomers differ in the position of the double bond in the carbon chain. The double

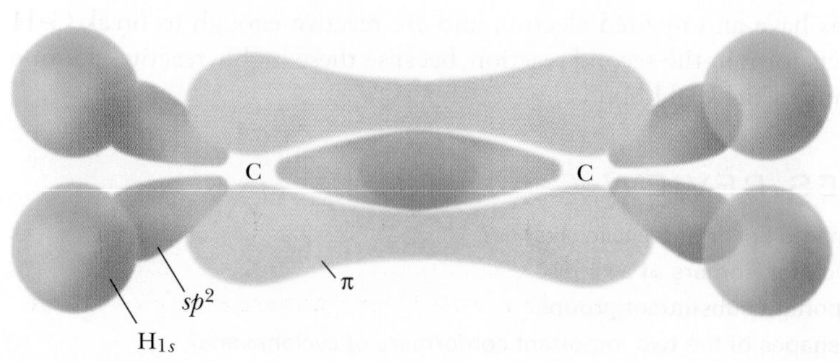

Figure 22.7 Bonding in ethene. The σ bonds in ethene form from sp^2-hybridized orbitals on the carbon atoms, and the π bond forms from the sideways overlap of a p orbital on each carbon atom.

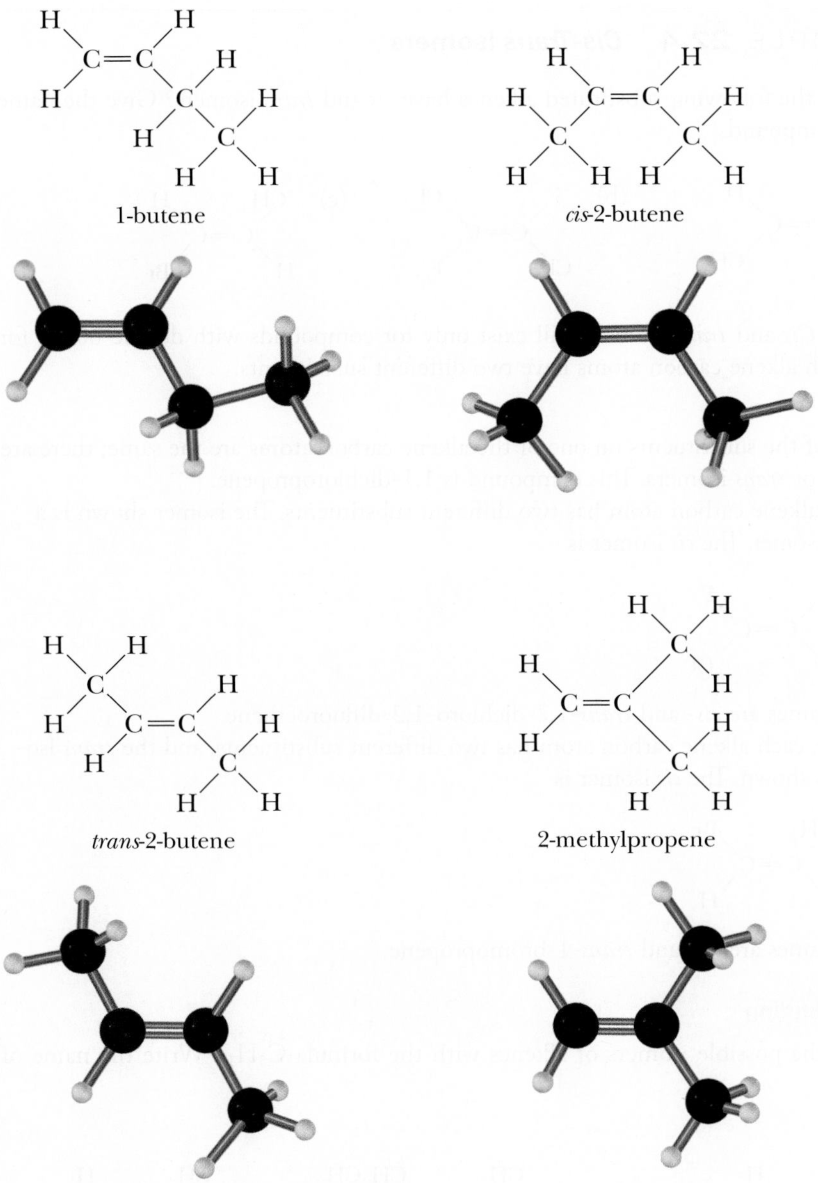

1-butene

cis-2-butene

trans-2-butene

2-methylpropene

Figure 22.8 Structures of the four alkene isomers of C_4H_8.

bond can be between the first two carbon atoms in the chain, $CH_2{=}CHCH_2CH_3$ (1-butene), or between the second and third carbon atoms, $CH_3CH{=}CHCH_3$ (2-butene). The presence of the double bond in this position also leads to the possibility of *geometric isomers,* isomers with the same connectivity and kind of bonds between atoms but with different arrangements of atoms in space (see Section 19.3), for 2-butene. These isomers, the *cis* isomer with both methyl groups on the same side of the double bond and the *trans* isomer with the methyl groups on opposite sides, exist because rotation about a double bond does not occur because of the presence of the π bond.

Geometric *cis* and *trans* isomers exist for alkenes in which both alkene carbon atoms have two different substituents.

The final alkene isomer, with the formula C_4H_8, is the branched isomer 2-methylpropene, $CH_2{=}C(CH_3)_2$. *Cis* and *trans* isomers of 2-methylpropene are not possible. *Cis* and *trans* isomers of any alkene are not possible if one of the carbon atoms involved in the double bond has two substituents that are identical. The 1-butene, the two geometric isomers of 2-butene, and 2-methylpropene are examples of structural isomers.

EXAMPLE 22.4 *Cis-Trans* Isomers

Which of the following substituted alkenes have *cis* and *trans* isomers? Give the name of each compound.

(a) 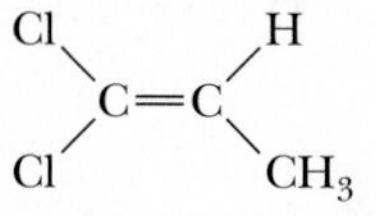 Cl, Cl on one carbon; H, CH_3 on the other:

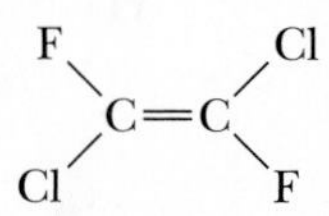

Cl   H
C=C
Cl CH_3

(b)

F   Cl
C=C
Cl   F

(c)

CH_3 H
C=C
H   Br

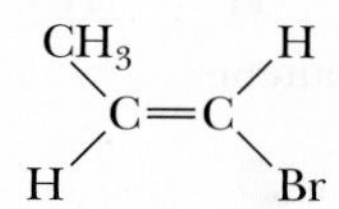

Strategy *Cis* and *trans* isomers will exist only for compounds with double bonds for which both alkene carbon atoms have two different substituents.

Solution

(a) Two of the substituents on one of the alkene carbon atoms are the same; there are no *cis* or *trans* isomers. This compound is 1,1-dichloropropene.

(b) Each alkene carbon atom has two different substituents. The isomer shown is a *trans* isomer. The *cis* isomer is

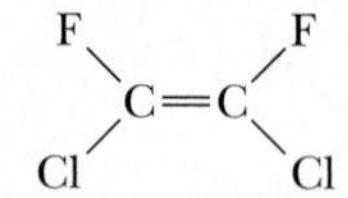

F   F
C=C
Cl   Cl

The names are *cis*- and *trans*-1,2-dichloro-1,2-difluoroethene.

(c) Again, each alkene carbon atom has two different substituents, and the *trans* isomer is shown. The *cis* isomer is

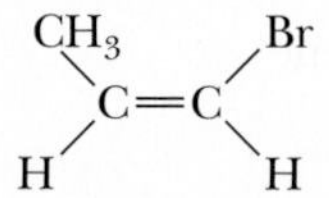

CH_3 Br
C=C
H   H

The names are *cis*- and *trans*-1-bromopropene.

Understanding

Draw all the possible isomers of alkenes with the formula C_5H_{10}. Write the name of each.

Answer

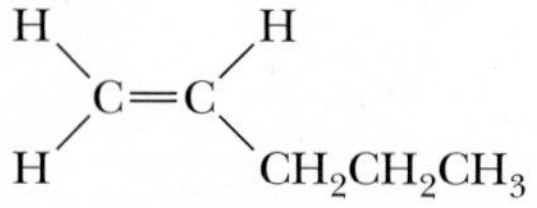

H   H
C=C
H $CH_2CH_2CH_3$

1-pentene

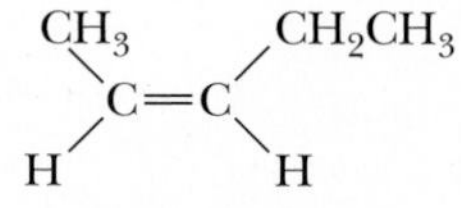

CH_3 CH_2CH_3
C=C
H   H

cis-2-pentene

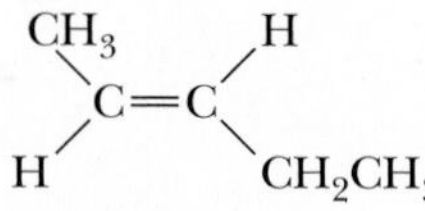

CH_3 H
C=C
H CH_2CH_3

trans-2-pentene

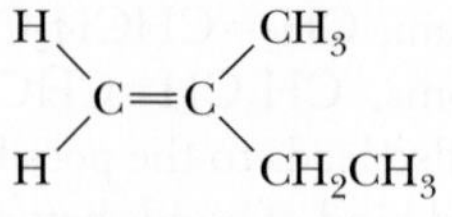

H CH_3
C=C
H CH_2CH_3

2-methyl-1-butene

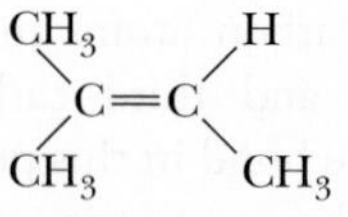

CH_3 H
C=C
CH_3 CH_3

2-methyl-2-butene

H   H
C=C
H $CH(CH_3)_2$

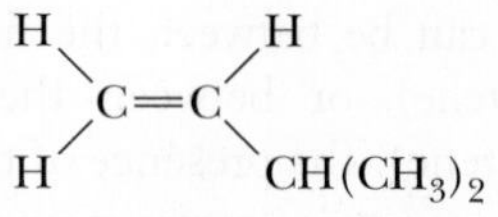

3-methyl-1-butene

Alkenes can be prepared from alkanes by the high-temperature reaction known as thermal cracking, which is an important industrial process.

$$CH_3(CH_2)_nCH_3 \xrightarrow[\text{Steam}]{900\ ^\circ\text{C}} H_2 + CH_4 + CH_2{=}CH_2 + CH_2{=}CHCH_3 + CH_2{=}CHCH_2CH_3, \ldots$$
$$n = 0\text{–}6$$

The thermodynamics of this reaction are dominated by entropy. The fragmentation of larger molecules into many smaller molecules makes the entropy term very favorable (i.e., gives it a large positive value). The high temperature used in the reaction makes the $T\Delta S$ term in $\Delta G = \Delta H - T\Delta S$ larger than the ΔH term; thus, ΔG has a favorable negative value. Manufacturers use the main products of this reaction, ethene, propene, and butene, to prepare products such as ethylene glycol (antifreeze) and plastics.

Alkynes

Alkynes are unsaturated hydrocarbons that contain carbon-carbon triple bonds. The general formula of an alkyne with one triple bond is C_nH_{2n-2}. Alkynes are named as alkenes are, but an *-yne* suffix is used instead of *-ene*. Figure 22.9 shows the alkynes for $n = 2$ to 4.

Alkynes are named in the same way as alkenes, except that an *-yne* suffix is used instead of *-ene*. Like alkenes, alkynes have structural isomers; however, because the –C≡C– bonding is linear, there are no *cis* or *trans* isomers.

Each alkyne carbon atom uses *sp*-hybridized orbitals to form two σ bonds and two *p* orbitals to form two π bonds. Because of the linear arrangement of the –C≡C– group, there are no *cis* or *trans* isomers for alkynes.

The simplest alkyne is ethyne, which is more commonly called by its older name, acetylene. Acetylene was once used as a starting material for the synthesis of a variety of compounds, such as acetic acid, but better routes to these compounds that use ethene as the starting material have been developed. The industrial preparation of acetylene is based on the reaction of methane at extremely high temperature, another reaction that is helped by a large and favorable $T\Delta S$ term.

$$2CH_4 \xrightarrow[\text{Steam}]{1200\ ^\circ C} HC{\equiv}CH + 3H_2$$

In the laboratory, acetylene is prepared from the reaction of calcium carbide (CaC_2) and water (Figure 22.10).

$$CaC_2(s) + 2H_2O(\ell) \rightarrow C_2H_2(g) + Ca(OH)_2(aq)$$

Acetylene is one of the few hydrocarbons with a positive standard enthalpy of formation. The reverse reaction, the decomposition reaction into hydrogen and carbon, is thus thermodynamically favorable.

$$HC{\equiv}CH(g) \rightarrow H_2(g) + 2C(s) \qquad \Delta H = -227\ \text{kJ/mol}$$

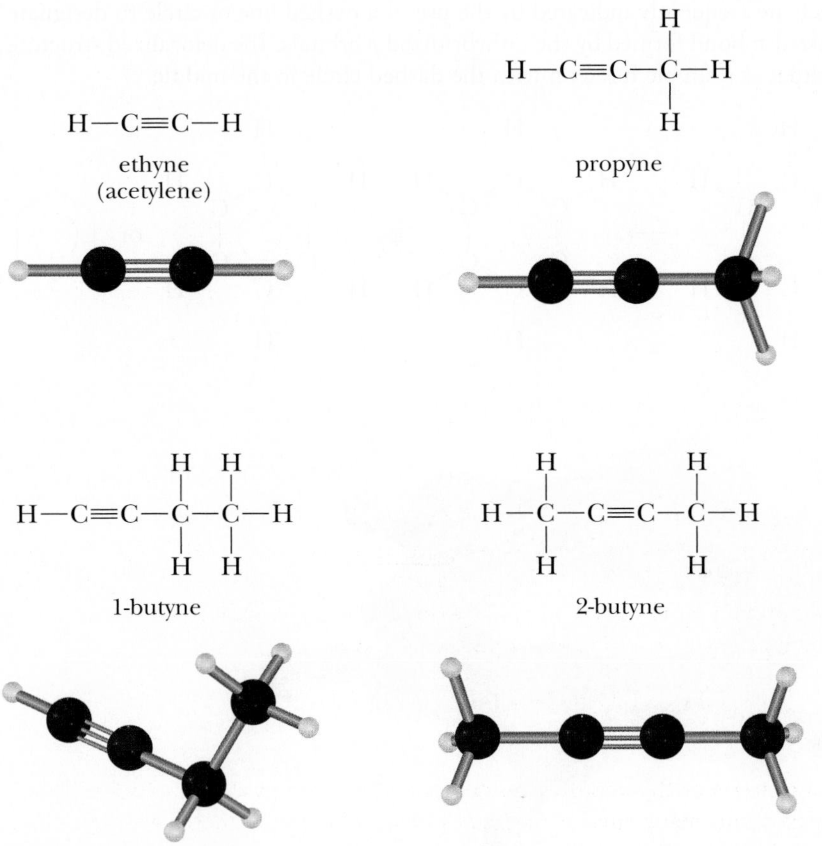

H—C≡C—H

ethyne
(acetylene)

H—C≡C—CH₃ (structural formula)

propyne

H—C≡C—CH₂—CH₃ (structural formula)

1-butyne

H₃C—C≡C—CH₃ (structural formula)

2-butyne

Figure 22.9 **Structures of alkynes.**

Figure 22.10 **Preparation of acetylene.** Acetylene gas is prepared from the reaction of calcium carbide and water. The gas will burn in air if ignited.

Figure 22.11 Oxyacetylene torch. The reaction of acetylene and oxygen produces very high temperatures and is used to weld metals.

The π bonds make alkenes and alkynes more reactive than alkanes.

Acetylene can decompose explosively under certain conditions, especially when it is present as a liquid. The combustion of acetylene and oxygen produces a flame that is hot enough to be used for welding (Figure 22.11).

Addition Reactions of Alkenes and Alkynes

Unsaturated hydrocarbon compounds are more reactive than saturated hydrocarbons because of the double or triple bonds. One common type of reaction is an **addition reaction,** the combination of two or more substances to form one new substance. Small molecules such as hydrogen and HCl can add across the multiple bond in unsaturated hydrocarbons. For example, hydrogen can add across the double bond of an alkene to form an alkane. The hybridization of each carbon atom changes from sp^2 to sp^3, and one hydrogen atom attaches to each carbon atom.

$$CH_2{=}CH_2 + H_2 \xrightarrow{\text{Catalyst}} CH_3CH_3$$

This reaction, known as a *hydrogenation reaction,* is generally catalyzed by a metal such as nickel. The reaction is used to convert liquid vegetable oils that contain double bonds into margarine or solid cooking fats. In general, the more saturated oils have higher melting points. The fixed arrangement of the double bonds in unsaturated oils prevents them from packing together as well as the more saturated oils. The better packing in the more saturated oils leads to stronger intermolecular forces, and thus to higher melting points.

Halogens also add across double and triple bonds. Either 1 or 2 mol halogen can be added per mole of alkyne, depending on the conditions of the reaction.

$$HC{\equiv}CH \xrightarrow{Br_2} CHBr{=}CHBr \xrightarrow{Br_2} CHBr_2CHBr_2$$

Aromatic Hydrocarbons

Aromatic hydrocarbons contain one or more benzene rings. Remember from Section 10.4 that benzene, C_6H_6, has a planar cyclic structure in which all of the carbon atoms are sp^2 hybridized (Figure 22.12). The bonding can be represented by two resonance forms, which are frequently indicated by the use of a dashed line or circle to designate the delocalized π bond formed by the unhybridized p orbitals. The delocalized structure is often written as a simple hexagon with the dashed circle in the middle.

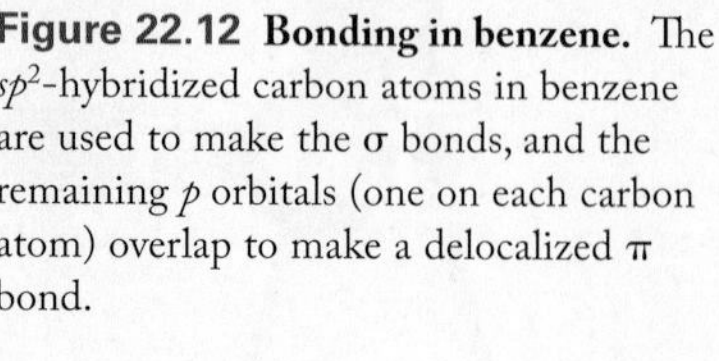

Figure 22.12 Bonding in benzene. The sp^2-hybridized carbon atoms in benzene are used to make the σ bonds, and the remaining p orbitals (one on each carbon atom) overlap to make a delocalized π bond.

Many compounds can be prepared by replacing one or more of the hydrogen atoms in benzene with some other group or groups. These compounds are named as substituted benzenes, although many aromatic compounds also have nonsystematic names. Toluene, a common solvent, has one methyl substituent on its ring. The systematic name, methylbenzene, is rarely used.

toluene

A second substituent on the benzene ring can occupy any of three different positions. Counting the carbon atom in the ring with the first substituent as position 1, a second substituent at position 2 is named *ortho (o)*, one at position 3 is named *meta (m)*, and one at position 4 is named *para (p)*. (There are two *ortho* and *meta* positions, but as always in naming compounds, the numbers are kept as low as possible.) The three isomers of dimethyl-substituted benzenes are known as xylenes, and they can also be named as substituted benzenes (Figure 22.13). Both toluene and the xylenes are used as industrial solvents and in paints. They are much less toxic than benzene.

Many aromatic compounds with three or more substituents are also known. These are named as substituted benzenes, with numbers used to indicate the position of the substituents; they may sometimes be named from the nonsystematic names toluene or

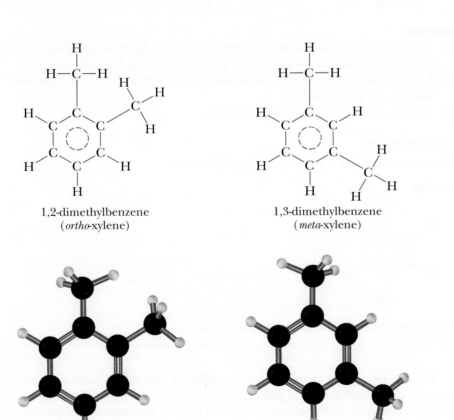

1,2-dimethylbenzene
(*ortho*-xylene)

1,3-dimethylbenzene
(*meta*-xylene)

1,4-dimethylbenzene
(*para*-xylene)

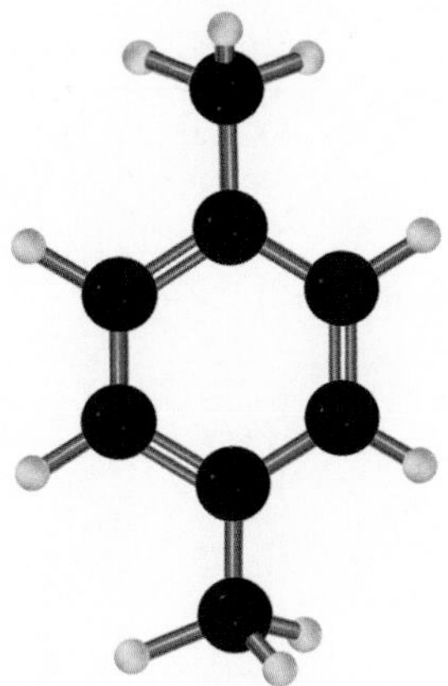

Figure 22.13 Structures of the xylenes.

xylene. The ring is numbered to minimize the sum of the numbers for the alphabetically listed substituents. Three examples are:

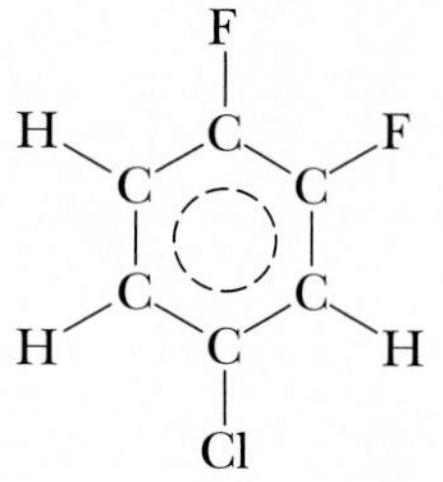

4-chloro-1,2-difluorobenzene

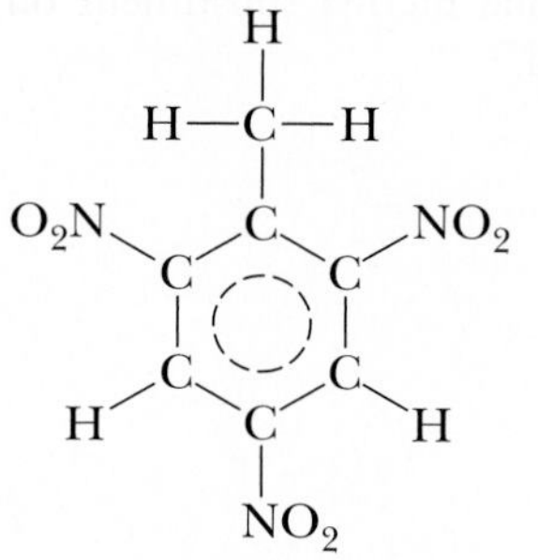

2,4,6-trinitrotoluene (TNT)

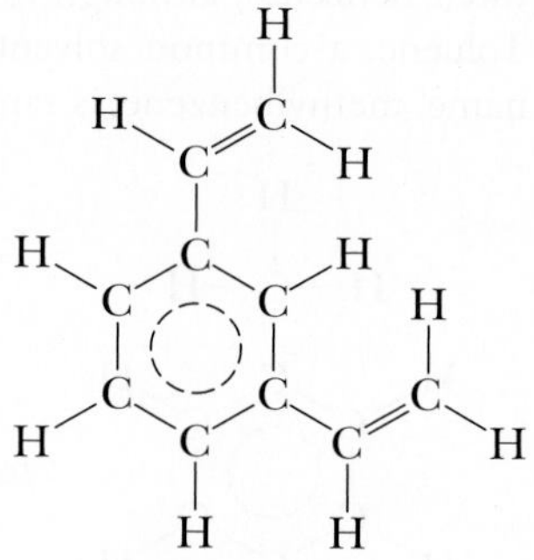

1,3-divinylbenzene

EXAMPLE 22.5 Naming Substituted Benzenes

Name the following compounds.

(a)

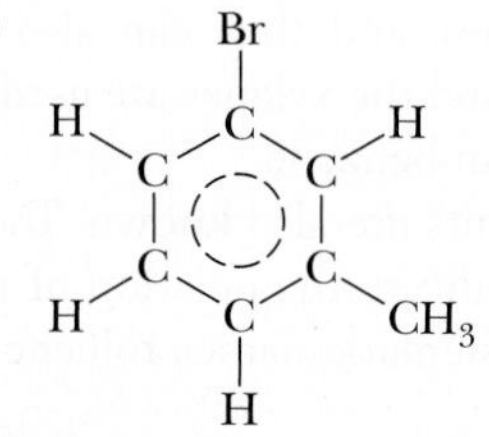

(b)

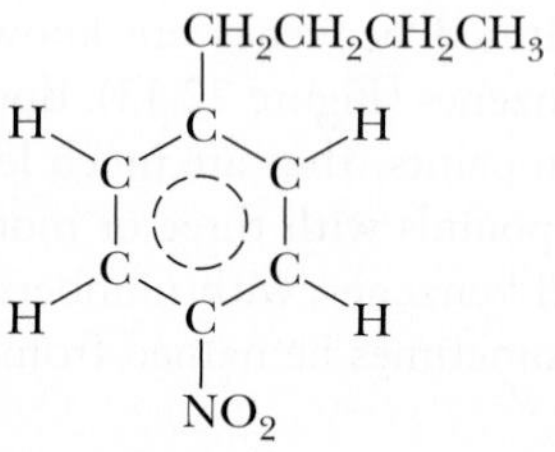

Solution

(a) The ring is numbered starting with the first group to be listed alphabetically, the bromine-substituted carbon. Thus, its systematic name is 1-bromo-3-methylbenzene. The compound can also be named as a substituted toluene: *m*-bromotoluene or 3-bromotoluene.

(b) This compound is 1-butyl-4-nitrobenzene.

Understanding

Draw the structure of 1,2-dibromo-4-methylbenzene.

Answer

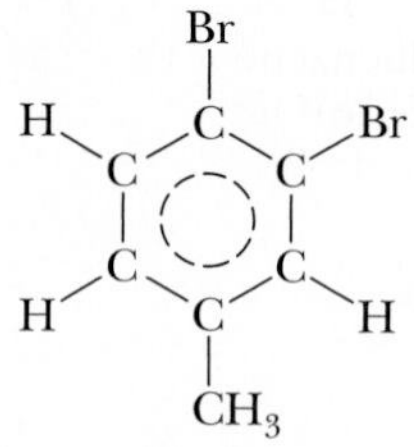

This compound could also be named 3,4-dibromotoluene.

Benzene rings can fuse together at two adjacent positions to form more complex aromatic compounds known as polycyclic aromatic hydrocarbons (Figure 22.14). The polycyclic aromatic hydrocarbons are planar because the aromatic rings are fused together with sp^2-hybridized carbon atoms. Many of these polycyclic hydrocarbons and their derivatives are formed by heating soft coal in the absence of oxygen. Coke (mainly carbon), used as the reducing agent in the steel industry, is also formed in this process.

naphthalene anthracene phenanthrene

Figure 22.14 Fused-ring aromatic hydrocarbons.

The reactions of aromatic hydrocarbons differ from those of other unsaturated hydrocarbons in that addition reactions are not favored. An addition reaction would destroy the delocalized π bonding of the aromatic ring. Instead, aromatic compounds undergo substitution reactions, frequently in the presence of a catalyst, as shown in the following reaction.

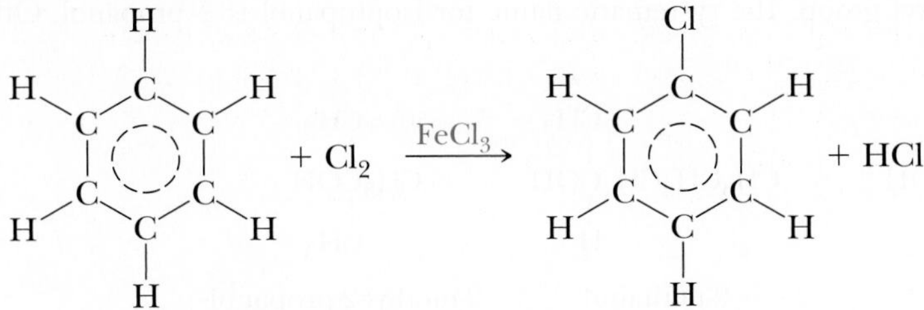

Aromatic compounds generally undergo substitution reactions rather than addition reactions.

OBJECTIVES REVIEW *Can you:*

- ☑ identify and name unsaturated compounds?
- ☑ list the possible isomers of alkenes, alkynes, and aromatic compounds?
- ☑ write some reactions of alkenes, alkynes, and aromatic compounds?

22.3 Functional Groups

OBJECTIVES

- ☐ Identify the important functional groups
- ☐ Name organic compounds that contain functional groups
- ☐ Identify chiral carbon atoms

Organic compounds are frequently categorized by certain groups of atoms that have characteristic chemical properties. A **functional group** is an atom or small group of atoms in a molecule that undergoes characteristic reactions. The C=C and C≡C parts of alkanes and alkynes are functional groups. For example, propene and 1-butene would be expected to react similarly because both contain the carbon-carbon double bond functional group. This section focuses on functional groups that contain oxygen or nitrogen atoms.

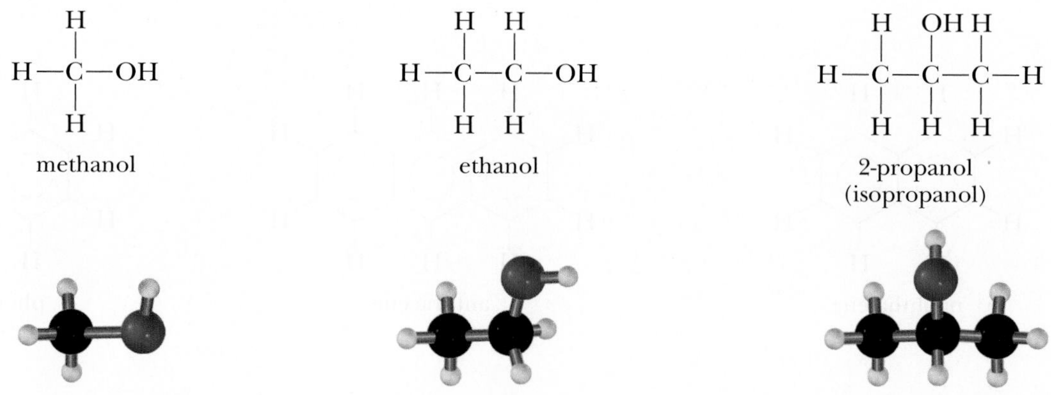

$$\begin{array}{c} H \\ | \\ H-C-OH \\ | \\ H \end{array} \qquad \begin{array}{c} H\ \ H \\ |\ \ \ | \\ H-C-C-OH \\ |\ \ \ | \\ H\ \ H \end{array} \qquad \begin{array}{c} H\ \ OH\ H \\ |\ \ \ \ |\ \ \ \ | \\ H-C-C-C-H \\ |\ \ \ \ |\ \ \ \ | \\ H\ \ \ H\ \ \ H \end{array}$$

methanol — ethanol — 2-propanol (isopropanol)

Figure 22.15 Structures of methanol, ethanol, and isopropanol.

Alcohols

An **alcohol** contains the $-OH$ (hydroxyl) functional group. Alcohols can be thought of as being derived from water by the replacement of one of the hydrogen atoms with an alkyl or substituted alkyl group. Like water, alcohols have strong intermolecular hydrogen-bonding forces. As a result, alcohols tend to have higher boiling points than do hydrocarbons of approximately equal molecular weight. For example, the boiling point of butane is 0 °C, whereas the boiling point for an alcohol of similar molecular weight, $CH_3CH_2CH_2OH$, is 97 °C. Despite the similarity of this functional group to the hydroxide ion, OH^-, alcohols are not bases.

Alcohols that contain fairly short alkyl substituents mix freely with water, but as the hydrocarbon portion increases in size, the water solubility decreases significantly. The three most common alcohols are methanol, ethanol, and isopropanol (Figure 22.15).

The name of an alcohol is based on the longest carbon chain to which the $-OH$ group is attached, with the suffix *-ol* added. The chain is numbered beginning at the end nearer the hydroxyl group. The systematic name for isopropanol is 2-propanol. Other examples are:

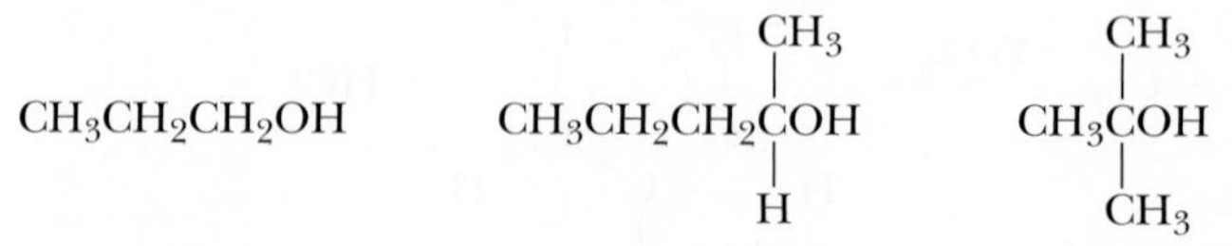

$$CH_3CH_2CH_2OH \qquad CH_3CH_2CH_2\overset{\displaystyle CH_3}{\underset{\displaystyle H}{C}}OH \qquad CH_3\overset{\displaystyle CH_3}{\underset{\displaystyle CH_3}{C}}OH$$

1-propanol (primary) — 2-pentanol (secondary) — 2-methyl-2-propanol (tertiary)

Alcohols are classified as *primary* when the $-OH$ group is bonded to a carbon atom that is bonded to no more than one other carbon atom, *secondary* when the $-OH$ group is bonded to a carbon atom that is bonded to two other carbon atoms, and *tertiary* when the $-OH$ group is bonded to a carbon atom that is bonded to three other carbon atoms. It is important to distinguish these three types of alcohols because they react differently.

Alcohols contain the –OH functional group and are named with the suffix *-ol*.

EXAMPLE 22.6 Naming Alcohols

Name each of the following alcohols, and indicate whether it is primary, secondary, or tertiary.

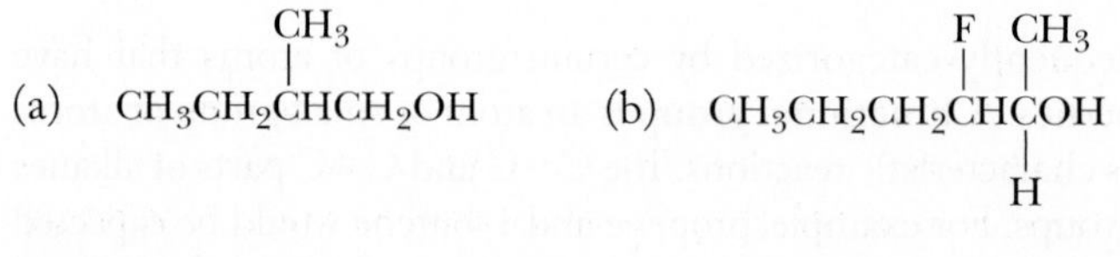

(a) $CH_3CH_2\overset{\displaystyle CH_3}{C}HCH_2OH$ (b) $CH_3CH_2CH_2\overset{\displaystyle F}{C}H\overset{\displaystyle CH_3}{\underset{\displaystyle H}{C}}OH$

Solution

(a) The longest chain containing the hydroxyl group has four carbon atoms. The hydroxyl group is at the 1 position, and the additional methyl group that is not in this chain is at the 2 position. The name is 2-methyl-1-butanol. This compound is a primary alcohol.

(b) All six of the carbon atoms are in the chain. The hydroxyl group is at the 2 position, and the fluorine atom is at the 3 position. The name is 3-fluoro-2-hexanol. This compound is a secondary alcohol.

Understanding

Name the following alcohol, and indicate whether it is primary, secondary, or tertiary.

$$\begin{array}{c} \quad\quad\quad CH_3 \\ \quad\quad\quad | \\ CH_3CH_2COH \\ \quad\quad\quad | \\ \quad\quad\quad CH_2CH_3 \end{array}$$

Answer 3-methyl-3-pentanol; tertiary

Ethanol, often referred to as grain alcohol, is the alcohol in beer, wine, and other alcoholic beverages. It is formed by the fermentation of sugars and starches.

$$\underset{\text{sugar}}{C_6H_{12}O_6} \xrightarrow{\text{Yeast}} \underset{\text{ethanol}}{2CH_3CH_2OH} + 2CO_2$$

Oxygen must be excluded in this reaction or acetic acid, CH_3CO_2H, will also form. The carbon dioxide gas produced in the reaction helps to protect the alcohol from the oxygen in air in the early stages of fermentation of grapes (Figure 22.16).

© Cephas Picture Library/Alamy

Figure 22.16 Fermentation of grapes. The fermentation of grapes is so vigorous in its early stages that it produces large amounts of CO_2 gas.

Only until recently has anything but a small fraction of the ethanol used industrially been produced by fermentation. The first major exception occurred in Brazil, where they prepare ethanol from sugarcane to supply much of the country's energy needs. In the United States, there is now considerable effort to produce ethanol from corn, with the hope in the future that other, less valuable organic materials such as the corn stalks and switchgrass will be used. Much of the ethanol used industrially is prepared by the acid-catalyzed addition reaction of ethene and water.

$$CH_2{=}CH_2 + H_2O \xrightarrow{H_2SO_4} CH_3CH_2OH$$

Ethanol is used as an intermediate for the production of other compounds and as a solvent. It is increasingly being used as an additive (5–10%) in gasoline (gasohol) and as a fuel. A product sold mainly in the midwest United States is E-85, which is a blend of 85 percent ethanol and 15 percent unleaded gasoline. The addition of the gasoline helps cars start in cold weather.

Methanol is another important alcohol. It was once prepared by heating wood in the absence of air (to prevent combustion), and thus became known as "wood alcohol." Methanol is toxic to humans, causing blindness and death when ingested. A small amount of methanol or other alcohol is sometimes added to ethanol intended for commercial use, to prevent its consumption. This toxic mixture is known as *denatured alcohol,* and it should not be ingested.

Methanol is prepared industrially by the zinc oxide catalyzed reaction of carbon monoxide and hydrogen.

$$CO + 2H_2 \xrightarrow{400\ ^\circ\text{C, ZnO}} CH_3OH$$

Despite the simplicity of this reaction, chemists continue to devote a considerable amount of research—some quite successful—to reducing the necessary temperature. Methanol is widely used as an industrial solvent. Recently, there has been interest in using it as a starting material for a variety of compounds because carbon monoxide and hydrogen can be prepared from coal and water. The increased use of methanol as a starting material would reduce the demand for crude oil, the present source of many products in the chemical industry.

The compound ethylene glycol, CH_2OHCH_2OH, is sold as antifreeze. This alcohol is added to automobile radiators to prevent the water in the cooling system from freezing at low temperatures and bursting the engine block. It is a *diol,* a compound that contains two hydroxyl groups.

Phenols

A **phenol** is a compound in which a hydroxyl group is substituted for a hydrogen atom on an aromatic ring. The name *phenol* is also used for the parent compound of this type, C_6H_5OH. Phenols are considered a separate class of compounds from alcohols because they are much more acidic. For example, the reaction of phenol and sodium hydroxide goes to completion, whereas the same reaction with cyclohexanol does not occur to any appreciable extent.

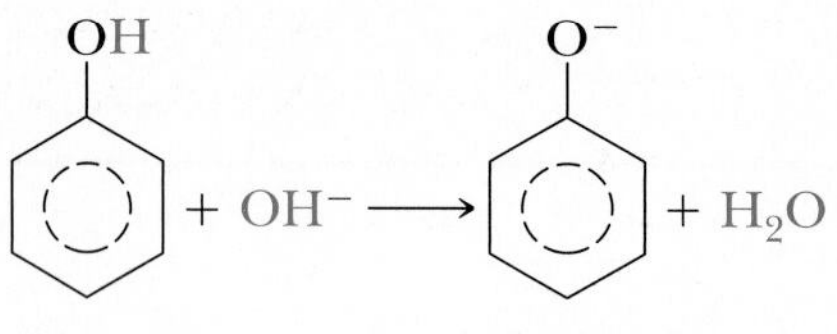

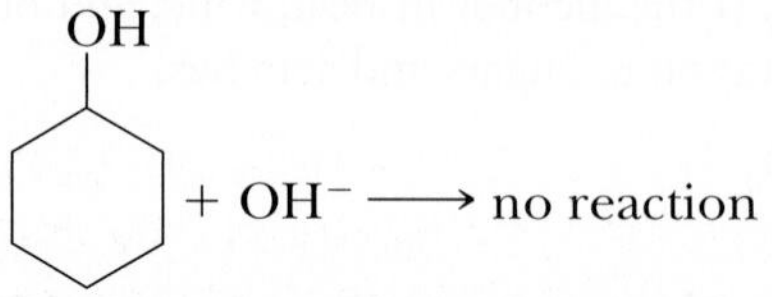

$$C_6H_5OH + OH^- \longrightarrow C_6H_5O^- + H_2O$$

$$C_6H_{11}OH + OH^- \longrightarrow \text{no reaction}$$

Many phenols have industrial applications; phenol itself is used in the manufacture of resins and dyes, and 4-methylphenol (*p*-cresol) has been used as a disinfectant.

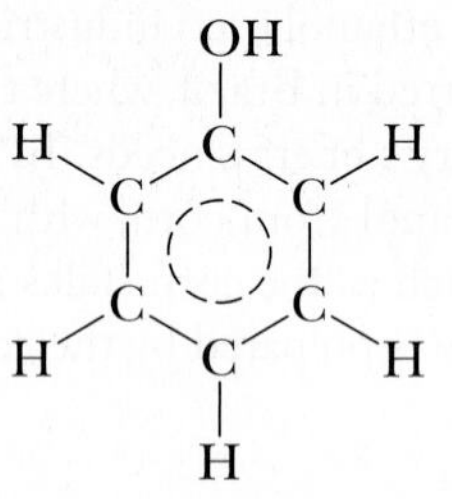

phenol

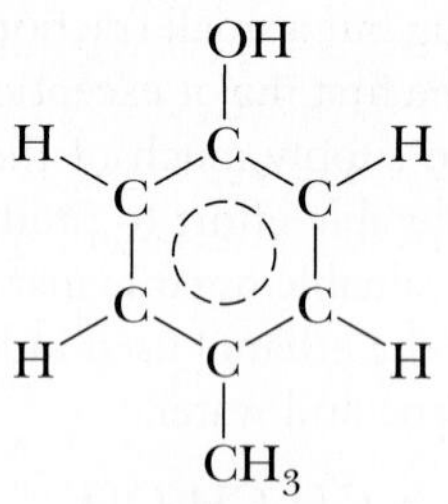

4-methylphenol
(*p*-cresol)

Ethers

An **ether** contains a C–O–C functional group. An ether can be viewed as being formed from water by the replacement of both hydrogen atoms with carbon substituents. Unlike water and alcohols, the ether functional group cannot form hydrogen bonds to itself (no hydrogen atoms with highly polarized bonds are present). Because ethers lack the ability to form hydrogen bonds, they have lower boiling points than do alcohols of similar molecular weight.

Most simple ethers are named by giving the names for the two substituents on the oxygen atom, followed by the word *ether.* Diethyl ether, CH_3CH_2–O–CH_2CH_3, is used extensively and is the "ether" that was once used as a general anesthetic. More recently, divinyl ether (CH_2=CH–O–CH=CH_2) and methyl propyl ether (CH_3–O–$CH_2CH_2CH_3$) have been used because they have fewer side effects. The boiling point of diethyl ether is quite low, 35 °C, only a few degrees above room temperature. This low boiling point means that a room in which diethyl ether is being used contains a substantial buildup of the highly flammable vapors, and care must be taken to avoid introducing any source of a spark or flame. A potential danger in storing ethers is that they react slowly with oxygen to form peroxides (compounds with a –O–O– functional group). Peroxides have a tendency to explode violently.

Ethers contain the C–O–C functional group and are named by the substituents, followed by the word *ether.*

A **condensation reaction** is a reaction that joins two molecules while producing a small molecule such as water. An ether can be prepared by an acid-catalyzed condensation reaction of two alcohol molecules.

$$CH_3CH_2OH + HOCH_2CH_3 \xrightarrow{H_2SO_4} CH_3CH_2OCH_2CH_3 + H_2O$$

Aldehydes and Ketones

The C=O functional group is called the **carbonyl group.** An **aldehyde** contains the

$$-\overset{\overset{\displaystyle O}{\|}}{C}-H$$

functional group (Figure 22.17), where the R is an alkyl or aryl substituent. The systematic names for aldehydes are based on the name of the longest carbon chain containing the C=O group, with the suffix *-al* added. The carbonyl carbon atom is counted in the base name for the chain.

Aldehydes contain the $-\overset{\overset{\displaystyle O}{\|}}{C}-H$ functional group and are named with the suffix *-al.*

A **ketone** contains the

$$-\overset{\overset{\displaystyle O}{\|}}{C}-R$$

functional group, where the R group is an alkyl or aryl substituent (Figure 22.18). Ketones are frequently named by giving the names of the two substituents followed by the word *ketone,* similar to the naming of ethers. However, the systematic names for ketones are based on the name of the longest carbon chain containing the C=O group, with the suffix *-one* added. A numeral indicates the position of the carbonyl group. Propanone, the simplest ketone, has the carbonyl group at the 2 position and is better known by its common name, acetone. Acetone is widely used as a solvent. The carbonyl group makes acetone polar, so it is soluble in water, as well as being a good solvent for many organic substances. Methyl ethyl ketone is also used extensively as a solvent and a gasoline additive. Both of these ketones have the important property of being relatively nontoxic compounds.

Ketones contain the $-\overset{\overset{\displaystyle O}{\|}}{C}-R$ functional group and are named with the suffix *-one.*

$$H-\overset{\overset{\displaystyle O}{\|}}{C}-H$$

methanal
(formaldehyde)

$$H-\underset{\underset{\displaystyle H}{|}}{\overset{\overset{\displaystyle H}{|}}{C}}-\overset{\overset{\displaystyle O}{\|}}{C}-H$$

ethanal
(acetaldehyde)

Figure 22.17 Aldehydes.

$$H-\underset{\underset{\displaystyle H}{|}}{\overset{\overset{\displaystyle H}{|}}{C}}-\overset{\overset{\displaystyle O}{\|}}{C}-\underset{\underset{\displaystyle H}{|}}{\overset{\overset{\displaystyle H}{|}}{C}}-H$$

propanone
(acetone)

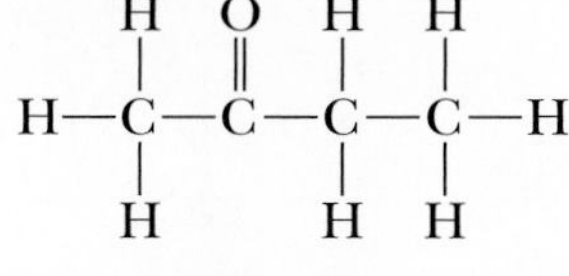

2-butanone
(methyl ethyl ketone)

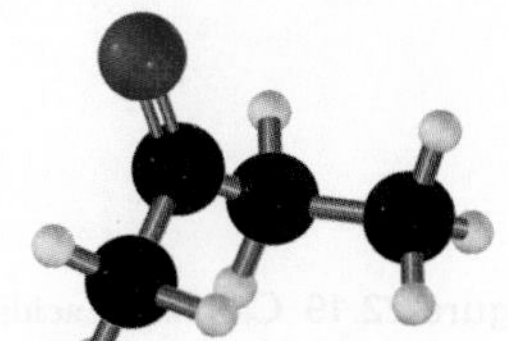

Figure 22.18 Ketones.

Aldehydes and ketones may be prepared by the controlled oxidation of alcohols. The oxidation of a primary alcohol yields an aldehyde, and the oxidation of a secondary alcohol yields a ketone.

$$CH_3CH_2OH + \tfrac{1}{2}O_2 \longrightarrow CH_3-\overset{\overset{\displaystyle O}{\|}}{C}-H + H_2O$$

$$CH_3-\underset{\underset{\displaystyle H}{|}}{\overset{\overset{\displaystyle OH}{|}}{C}}-CH_3 + \tfrac{1}{2}O_2 \longrightarrow CH_3-\overset{\overset{\displaystyle O}{\|}}{C}-CH_3 + H_2O$$

Carboxylic Acids and Esters

A **carboxylic acid** contains the

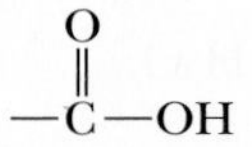

$$-\overset{\overset{\displaystyle O}{\|}}{C}-OH$$

(or CO_2H) functional group, the carboxyl group. The carboxyl group is the combination of a carbonyl and a hydroxyl group. Compounds that contain the carboxyl group are acids because the electron-withdrawing carbonyl group makes the O–H bond more polar than in alcohols, so the ionization of a proton is more likely. Ionization of the proton leaves the **carboxylate group,** $-CO_2^-$, an anion that is stabilized by two resonance structures.

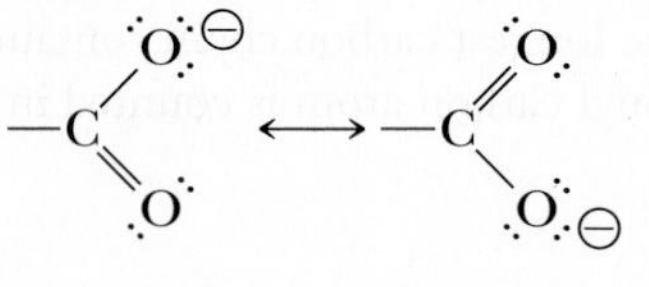

$$-C(=O)-O^{\ominus} \longleftrightarrow {}^{\ominus}O-C(=O)-$$

Still, carboxylic acids are fairly weak acids compared with H_2SO_4 or HCl.

The systematic method for naming carboxylic acids is based on the name of the longest chain attached to the carboxylic acid group, with the suffix *-oic* and the word *acid.* The carboxyl carbon atom is counted as part of the base name for the chain.

Carboxylic acids contain the $-\overset{\overset{\displaystyle O}{\|}}{C}-OH$ functional group and are named with the suffix *-oic,* followed by "acid."

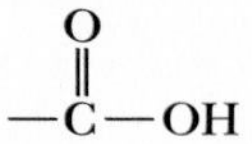

Carboxylic acids have been known for centuries. The more common ones have nonsystematic names based on their natural sources (Figure 22.19). The simplest acid, methanoic acid, was first isolated from ants, so its common name is formic acid (from the Latin *formica,* meaning "ant"). Ethanoic acid, known as acetic acid, is responsible for the sour taste of vinegar. Butanoic acid, also called *butyric acid,* is the compound that gives rancid butter its unpleasant odor.

Carboxylic acids are prepared from the oxidation of primary alcohols or aldehydes. A more powerful oxidizing agent is used in these reactions than is used in the oxidation of alcohols to form aldehydes and ketones. Milder oxidizing agents in the presence of a catalyst can also be used.

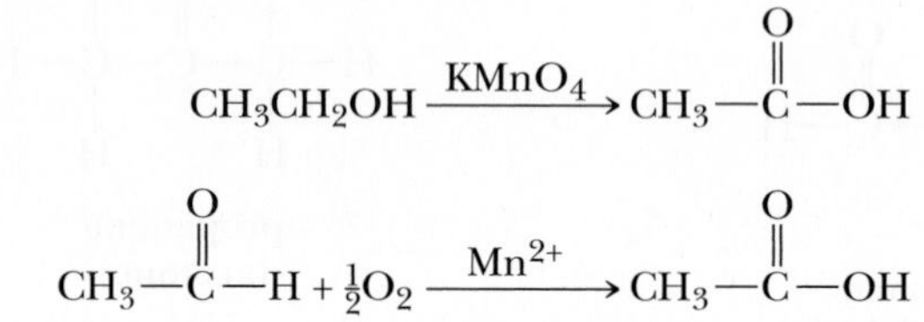

$$CH_3CH_2OH \xrightarrow{KMnO_4} CH_3-\overset{\overset{\displaystyle O}{\|}}{C}-OH$$

$$CH_3-\overset{\overset{\displaystyle O}{\|}}{C}-H + \tfrac{1}{2}O_2 \xrightarrow{Mn^{2+}} CH_3-\overset{\overset{\displaystyle O}{\|}}{C}-OH$$

Figure 22.19 Carboxylic acids.

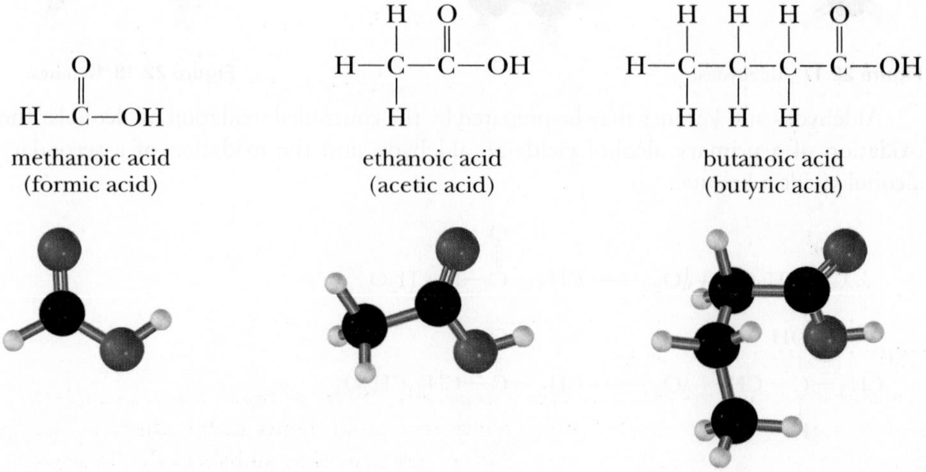

$H-\overset{\overset{\displaystyle O}{\|}}{C}-OH$ methanoic acid (formic acid)

$CH_3-\overset{\overset{\displaystyle O}{\|}}{C}-OH$ ethanoic acid (acetic acid)

$CH_3CH_2CH_2-\overset{\overset{\displaystyle O}{\|}}{C}-OH$ butanoic acid (butyric acid)

An **ester** contains the

$$-\overset{\overset{\displaystyle O}{\|}}{C}-OR$$

functional group, where R is an organic substituent such as methyl or phenyl. Esters are prepared by the acid-catalyzed condensation of a carboxylic acid and an alcohol. The OR portion of the ester originates with the alcohol.

Esters contain the $-\overset{\overset{\displaystyle O}{\|}}{C}-OR$ functional group.

$$CH_3\overset{\overset{\displaystyle O}{\|}}{C}-OH + HOCH_3 \longrightarrow CH_3\overset{\overset{\displaystyle O}{\|}}{C}-OCH_3 + H_2O$$

An ester is named by combining the name of the parent alcohol with the name of the parent acid, modified by the replacement of *-ic* with the suffix *-ate* and dropping the word "acid." The systematic name for the ester shown in the preceding equation is methyl ethanoate. It is more commonly called *methyl acetate.* In general, esters formed from the $CH_3C(O)O$– group are called *acetates.* An important ester for industrial applications is vinyl acetate.

$$H_3C-\overset{\overset{\displaystyle O}{\|}}{C}-OCH{=}CH_2$$

vinyl acetate

In contrast with many carboxylic acids, esters generally have pleasant odors; they also contribute to the flavors of fruits.

EXAMPLE 22.7 Names of Organic Compounds with Functional Groups

Name the following compounds.

(a) $CH_3CH_2\underset{\underset{\displaystyle CH_2CH_3}{|}}{CH}\overset{\overset{\displaystyle O}{\|}}{C}-OH$ (b) $CH_3CH_2\overset{\overset{\displaystyle O}{\|}}{C}CH_2Br$ (c) C_6H_5–O–CH_3

Solution

(a) The carboxyl functional group is part of a four-carbon chain (regardless of which CH_3CH_2– group is counted, both chains are the same length). A five-carbon atom chain is also present, but it does not contain the carbon atom of the carboxyl group. Therefore, the name is 2-ethylbutanoic acid. Note that it is not necessary to indicate the position of the acid group because it must always be at position 1, the first carbon atom of the chain on which the name is based.
(b) This compound is a four carbon ketone, with the bromide group at position 1, and the carbonyl group at position 2. The name is 1-bromo-2-butanone.
(c) This compound has a benzene substituent (a phenyl group) and a methyl group connected to an oxygen atom. It is an ether: methyl phenyl ether.

Understanding

Draw the structure of phenyl propanoate.

Answer $CH_3CH_2\overset{\overset{\displaystyle O}{\|}}{C}-OC_6H_5$

Amines and Amides

An **amine** is a derivative of ammonia in which one or more hydrogen atoms are replaced by an organic substituent (R). Amines are related to ammonia in the same way that alcohols and ethers are related to water. Amines are named by alphabetically listing the names of the substituents attached to the nitrogen atom, followed by the suffix *-amine.* Amines with the formula RNH_2 are *primary amines* (methylamine and phenylamine), R_2NH are *secondary amines* (diethylamine), and R_3N are *tertiary amines* (triethylamine). Primary and secondary amines, like ammonia and alcohols, can form intermolecular hydrogen bonds, and thus generally have fairly high boiling points.

CH_3NH_2
methylamine

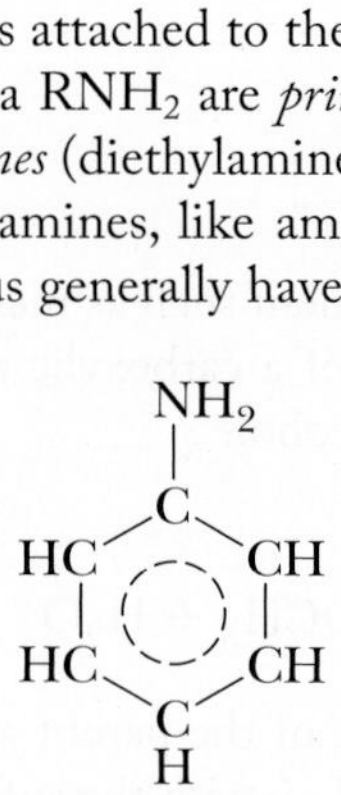

$C_6H_5NH_2$ (benzene ring: C bearing NH_2, HC, CH, HC, CH, C–H)
phenylamine

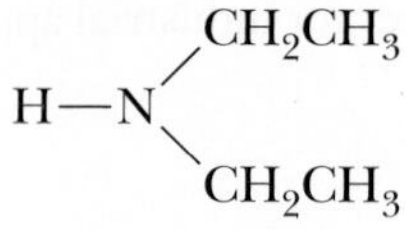

$H{-}N(CH_2CH_3)_2$
diethylamine

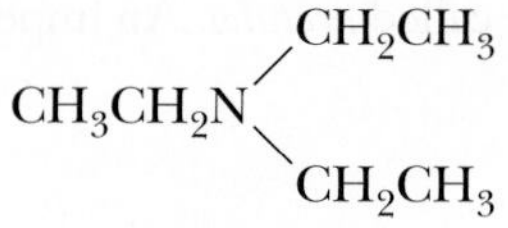

$CH_3CH_2N(CH_2CH_3)_2$
triethylamine

Amines are derivatives of ammonia.

An **amide** contains the

$$-\overset{\overset{\displaystyle O}{\|}}{C}-NR_2$$

functional group. An amide forms in a condensation reaction of a primary or secondary amine or ammonia with a carboxylic acid.

$$CH_3\overset{\overset{\displaystyle O}{\|}}{C}-OH + HNR_2 \longrightarrow CH_3\overset{\overset{\displaystyle O}{\|}}{C}-NR_2 + H_2O$$

Amides contain the $-\overset{\overset{\displaystyle O}{\|}}{C}-NR_2$ functional group.

The amide functional group is an important link in the backbones of protein molecules (see Section 22.5).

Review of Functional Groups

Table 22.3 lists the important functional groups covered in the preceding two sections.

Amino Acids and Chirality at Carbon

An **amino acid** contains a carboxyl group ($-CO_2H$) and an amine group ($-NH_2$). Glycine is the simplest amino acid.

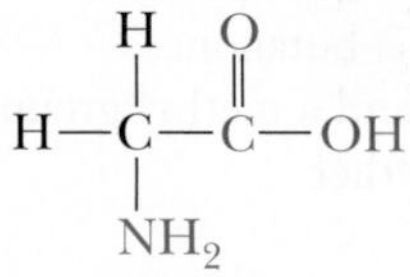

$$H-\underset{\underset{\displaystyle NH_2}{|}}{\overset{\overset{\displaystyle H}{|}}{C}}-\overset{\overset{\displaystyle O}{\|}}{C}-OH \qquad CH_3-\underset{\underset{\displaystyle NH_2}{|}}{\overset{\overset{\displaystyle H}{|}}{C}}-\overset{\overset{\displaystyle O}{\|}}{C}-OH$$

glycine 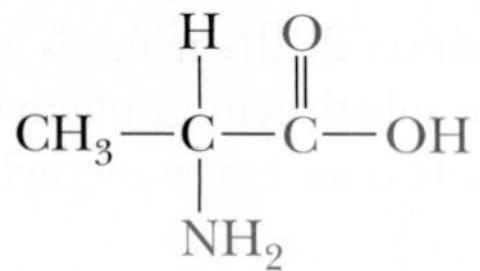 — alanine

Glycine and alanine are called α*-amino acids,* indicating that the carboxylic acid and amine functional groups are attached to the same carbon atom. α-Amino acids are the building blocks of proteins.

As outlined in Section 19.3, *chiral* molecules or ions are those whose mirror-image structures cannot be superimposed on the original structure. A carbon atom that is bonded to four different substituents is chiral. As with most α-amino acids, alanine is chiral because the central carbon atom is bonded to four different groups—in this case, $-H$, $-CH_3$, $-NH_2$, and $-COOH$. Chiral carbon atoms are called *asymmetric* centers. As shown in Figure 22.20, the mirror image of alanine is not superimposable on the

TABLE 22.3 Important Functional Groups

Functional Group	Name	Example
C=C	Alkene	$H_2C{=}CH_2$ (ethylene or ethene)
C≡C	Alkyne	HC≡CH (acetylene or ethyne)
–OH	Alcohol (attached to alkyl group)	CH_3CH_2OH (ethanol)
–OH	Phenol (attached to aryl group)	C_6H_5OH (phenol)
C–O–C	Ether	$CH_3CH_2OCH_2CH_3$ (diethyl ether)
$-\overset{O}{\overset{\|}{C}}-H$	Aldehyde	$CH_3\overset{O}{\overset{\|}{C}}-H$ (acetaldehyde or ethanal)
$-\overset{O}{\overset{\|}{C}}-R$ (R = alkyl, aryl)	Ketone	$CH_3\overset{O}{\overset{\|}{C}}CH_3$ (acetone or propanone)
$-\overset{O}{\overset{\|}{C}}-OH$	Carboxylic acid	$CH_3\overset{O}{\overset{\|}{C}}-OH$ (acetic acid or ethanoic acid)
$-\overset{O}{\overset{\|}{C}}-OR$ (R = alkyl, aryl)	Ester	$CH_3\overset{O}{\overset{\|}{C}}-O-CH_2CH_3$ (ethyl acetate or ethyl ethanoate)
$-NR_2$ (R = H, alkyl, aryl)	Amine	$(CH_3CH_2)_2NH$ (diethylamine)
$-\overset{O}{\overset{\|}{C}}-NR_2$ (R = H, alkyl, aryl)	Amide	$CH_3\overset{O}{\overset{\|}{C}}-NH_2$ (acetamide or ethanamide)

original. The form on the left is designated L-alanine, and that on the right D-alanine, to distinguish the arrangements of the groups.

Alanine mirror images. Drawings of L-alanine (left) and D-alanine (right).

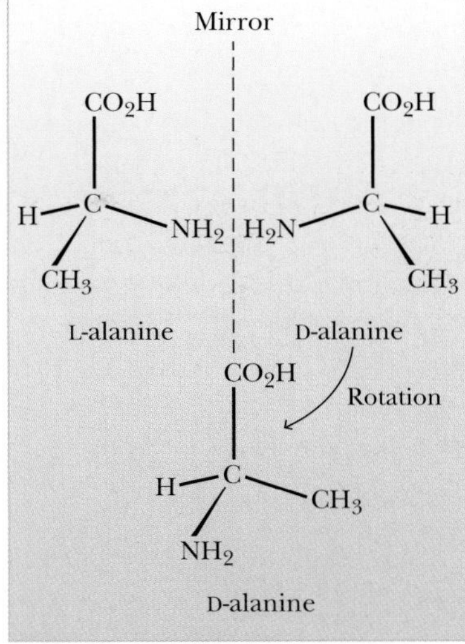

Figure 22.20 D and L isomers of alanine. L-Alanine and its mirror image, D-alanine, are not superimposable. Rotation of the structure on the right, so that the orientation of the C–H bond matches that of the structure on the left, places the CH_3 and NH_2 substituents in reversed positions.

Although the difference between the two forms of alanine and other molecules that contain a chiral carbon atom may seem small, it is crucial in biological reactions. Nearly all the amino acids that occur in living organisms are L-amino acids. The D-amino acids are not biologically equivalent to L-amino acids of the same formula.

OBJECTIVES REVIEW *Can you:*

- ☑ identify the important functional groups?
- ☑ name organic compounds that contain functional groups?
- ☑ identify chiral carbon atoms?

A carbon atom that is bonded to four different substituents is a chiral center.

PRACTICE OF CHEMISTRY
The Unique Chemical Structure of Soap Enables It to Dissolve Oil into Water

It is difficult to wash an oil spot out of clothing with plain water, because oil is a hydrocarbon that does not dissolve in water. Oil and water actually repel one another, so that oil adheres even more strongly to clothing in the presence of water. The addition of soap or detergent to water changes the situation; soapy water can dissolve oil from clothing and rinse it away. What is special about the structure of soaps that makes them effective cleaning agents for oils and greases?

Most soaps are soluble sodium or potassium salts of carboxylic acids. The most common commercial soap is sodium stearate, $Na[C_{17}H_{35}CO_2]$. It dissolves in water, forming the sodium and stearate ions. Even though most of the stearate ion is a hydrocarbon chain, it dissolves in water because of the carboxylate group. The carboxylate end is called *hydrophilic* (water loving), and the hydrocarbon tail is called *hydrophobic* (water fearing).

It is the long hydrocarbon chains of the stearate anions that dissolve the oils and greases. If water containing dissolved soap is mixed with oil, the hydrocarbon chains strongly attract the oil, whereas the ionic ends keep the soap dissolved into water. The oil spot is broken up into small droplets and dispersed into the water. The "tails" of many soap anions are needed to remove each oil droplet. ■

The hydrocarbon chain is hydrophobic. It penetrates the oil deposit and pulls small droplets into the water. The hydrophilic end of the soap keeps the droplets suspended in the water.

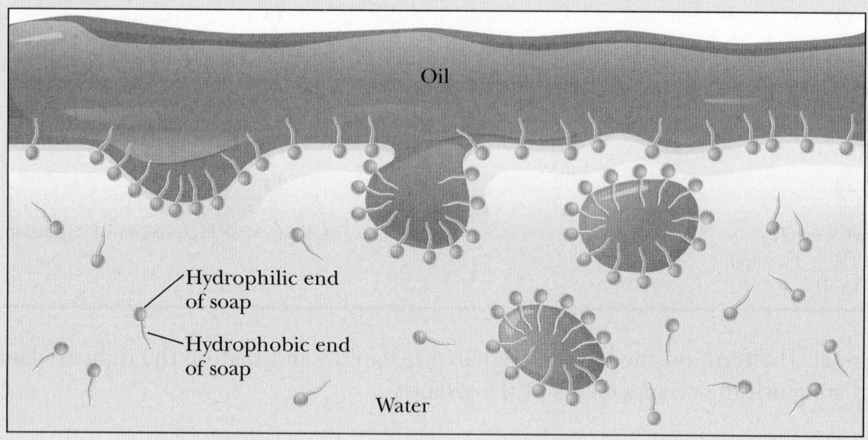

22.4 Organic Polymers

OBJECTIVES

- ☐ Describe two common methods for the formation of polymers
- ☐ Identify the monomeric unit of a polymer

Most of the organic molecules considered so far in this chapter are composed of a relatively small number of atoms, 2 to 40. In contrast, many materials that we deal with every day, such as the clear plastic over food or the tires on your car, are made of organic molecules that are composed of many thousands of atoms. Also, the molecules that nature uses, such as cellulose and DNA, are made of large numbers of atoms. Polymers are an important class of these large molecules.

A **polymer** is a large molecule formed by the repeated bonding together of many smaller units *(monomers)*. Organic polymers are formed by joining together many small organic molecules. They may be synthetic polymers prepared by the chemical industry or natural polymers made by living systems. The natural polymers essential to biological systems are discussed in sections 22.5 through 22.7.

Chain-Growth Polymers

A **chain-growth polymer** (also called an **addition polymer**) is a polymer chain formed from monomeric units with no loss of atoms. The simplest are **homopolymers,** polymers formed from the combination of many units of a single monomer compound. An example of a homopolymer is polyethylene, formed by addition reactions of numerous ethylene molecules in the presence of a catalyst.

$$n\mathrm{CH_2{=}CH_2} \xrightarrow{\text{catalyst}} \begin{array}{c} \mathrm{H\ \ H\ \ H\ \ H\ \ H\ \ H} \\ -\mathrm{C{-}C{-}C{-}C{-}C{-}C}- \\ \mathrm{H\ \ H\ \ H\ \ H\ \ H\ \ H} \end{array} \ \text{or}\ \left(\begin{array}{c} \mathrm{H\ \ H} \\ -\mathrm{C{-}C}- \\ \mathrm{H\ \ H} \end{array} \right)_n$$

polyethylene

Chain-growth or addition polymers form from monomeric units with no loss of atoms.

In this case, the repeating unit of the polymer is simply a CH_2CH_2 group.

In the preparation of polyethylene, not all of the chains are completely linear, as shown in the preceding equation. Depending on the conditions of the reaction, some branching can occur.

$$\begin{array}{c} \mathrm{H\ \ H\ \ H\ \ H\ \ H} \\ -\mathrm{C{-}C{-}C{-}C{-}C}- \\ \mathrm{H\ \ H\ \ \ |\ \ \ H\ \ H} \\ \mathrm{CH_2} \\ | \\ \mathrm{CH_3} \end{array}$$

Two forms of polyethylene are available commercially. Low-density polyethylene has some branching in the chains, whereas in high-density polyethylene, the chains are nearly all linear (the more uniform linear chains can pack better, producing a denser solid). The high-density polymer is harder and has greater strength, but the low-density polymer is more transparent. Low-density polyethylene is used as a film to wrap food and other products, and the high-density polymer is used in tubes, toys, and bottles (Figure 22.21).

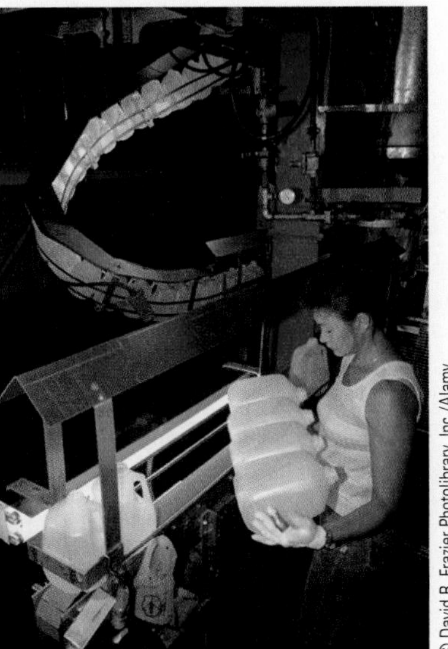

© David R. Frazier Photolibrary, Inc./Alamy

Figure 22.21 Polyethylene bottles. Polyethylene is used to make containers for milk.

The properties of the polymer can be further altered by starting with substituted alkenes. The chain-growth polymerization of propene (common name propylene) yields polypropylene, with the repeating unit of a CH_2CHCH_3 group.

$$n\mathrm{CH_2{=}CHCH_3} \xrightarrow{\text{catalyst}} \left(\begin{array}{c} \mathrm{H\ \ \ H} \\ -\mathrm{C{-}C}- \\ \mathrm{H\ \ CH_3} \end{array} \right)_n$$

polypropylene

The formation of a polymer from a substituted ethylene is complicated by the fact that different arrangements of the substituents along the chain are possible, in addition to differences in the branching of the chains. Depending on the conditions of the polypropylene polymerization reaction, products ranging from a soft, rubbery material to a hard solid can form. Much research has been conducted in this area with the aim of producing materials that have a variety of desirable physical properties.

The chain-growth polymerization of tetrafluoroethylene produces the polymer known as Teflon.

$$n\mathrm{CF_2{=}CF_2} \xrightarrow{\text{catalyst}} \left(\begin{array}{c} \mathrm{F\ \ F} \\ -\mathrm{C{-}C}- \\ \mathrm{F\ \ F} \end{array} \right)_n$$

Teflon

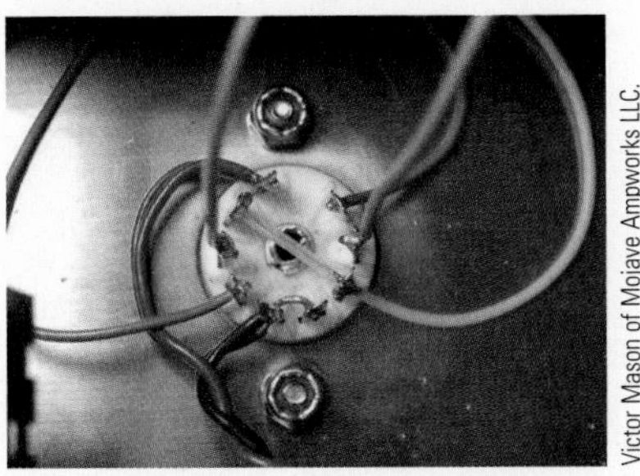

Victor Mason of Mojave Ampworks LLC.

Figure 22.22 Teflon-insulated wires. Teflon (four different colors here) covers the copper wires. Teflon is a good electrical and thermal insulator, so a much thinner layer of insulation can be used. The thinner Teflon-insulated wires are frequently used when old buildings are rewired (e.g., for computer networks), because a conduit can hold more Teflon-insulated wires than wires with conventional insulation.

Teflon is an extremely unreactive material that is used for nonsticking cooking utensils, for coatings on valves, and as insulation around wires (Figure 22.22).

Natural and Synthetic Rubbers

Natural rubber is a polymer secreted as a liquid by rubber trees. It is a polymer of isoprene (2-methyl-1,3-butadiene), $CH_2{=}C(CH_3){-}CH{=}CH_2$. In the polymerization of a monomer containing two double bonds (a *diene*), the backbone of the polymer can have either a *cis* or a *trans* arrangement about the double bonds.

$$\left(\begin{array}{ccccc} CH_3 & & H \quad CH_3 & & H \\ & C{=}C & & C{=}C & \\ -CH_2 & & CH_2{-}CH_2 & & CH_2- \end{array}\right)_n$$

natural rubber (all *cis*)

$$\left(\begin{array}{ccccc} -CH_2 & & H \quad CH_3 & & CH_2- \\ & C{=}C & & C{=}C & \\ CH_3 & & CH_2{-}CH_2 & & H \end{array}\right)_n$$

gutta-percha (all *trans*)

The isomer generally found in nature has a *cis* configuration in the backbone of the chain, which leads to a material that is elastic. The *trans* isomer is hard and nonelastic, but it had a few specialty uses, such as in hockey pucks and the covers of golf balls, before being replaced by modern synthetic polymers. Both isoprene isomers can also be produced synthetically.

Natural rubber is soft and can be pulled apart easily. Charles Goodyear developed a process known as *vulcanization*, which improves the properties of rubber. In vulcanization, the polymer is heated with a small amount of sulfur, and the polymer chains become interconnected with C–S–S–C bridges (Figure 22.23). These bridges prevent the individual chains from sliding over each other, leading to a harder and more elastic polymer. This bridging of the individual chains in the polymer is known as *cross-linking*.

Soft polymers can be converted to more useful hard materials by cross-linking the polymer chains.

The difficulty of importing natural rubber during World War II spurred the development of synthetic polymers. An example of a process developed at that time is the conversion of $CH_2{=}CCl{-}CH{=}CH_2$ to the polymer with *trans* double bonds to produce neoprene, a compound with characteristics somewhat similar to those of natural rubbers.

$$2n\,CH_2{=}\overset{\displaystyle Cl}{\overset{|}{C}}{-}CH{=}CH_2 \longrightarrow \left(\begin{array}{ccccc} -CH_2 & & H \quad Cl & & CH_2- \\ & C{=}C & & C{=}C & \\ Cl & & CH_2{-}CH_2 & & H \end{array}\right)_n$$

neoprene

Figure 22.23 Vulcanized rubber. Cross-linking the polymer chains *(a)* prevents them from coming apart. This process converts soft natural rubber *(b)* to a harder material *(c)*.

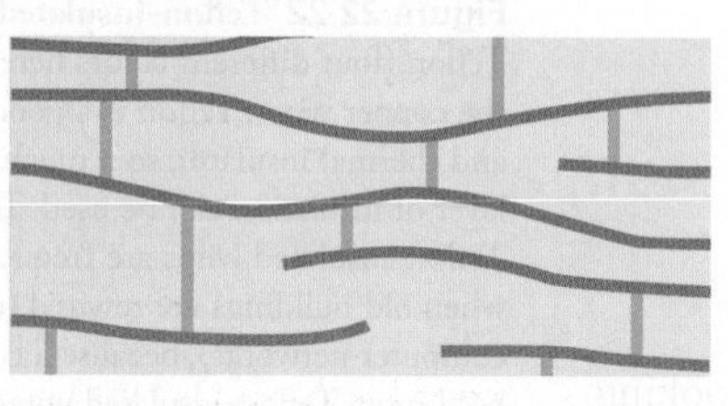

(a)

(b)

(c)

Copolymers

The chain-growth polymers discussed so far are homopolymers, which are formed from one type of monomer. **Copolymers** are polymers formed from the combination of any units of two or more types of monomers. The plastic film Saran, most familiar as a wrap for food, is made from chloroethene (vinyl chloride) and 1,1-dichloroethene.

$$2n\text{CH}_2{=}\text{CH(Cl)} + 2n\text{CH}_2{=}\text{CCl}_2 \longrightarrow \left(\text{CH}_2{-}\text{CH(Cl)}{-}\text{CH}_2{-}\text{CCl}_2{-}\text{CH}_2{-}\text{CH(Cl)}{-}\text{CH}_2{-}\text{CCl}_2 \right)_n$$

Saran

In the formation of Saran shown here, the two groups that form the polymer alternate in the backbone of the chain. In practice, some imperfections in the alternation occur.

Step-Growth or Condensation Polymers

Step-growth or condensation polymers form from monomeric units with the elimination of small molecules.

A **step-growth** or **condensation polymer** is a polymer formed by a reaction that eliminates a small molecule each time a monomer is linked to the polymer chain. Nylon 66 is an example; it results from the condensation of hexamethylendiamine and adipic acid. In this polymerization reaction, an *amide* linkage forms with the elimination of water.

$$n\text{HOOC(CH}_2)_4\text{COOH} + n\text{H}_2\text{N(CH}_2)_6\text{NH}_2 \longrightarrow$$

adipic acid hexamethylendiamine

$$\left(\text{C(=O)(CH}_2)_4\text{C(=O)}{-}\text{NH(CH}_2)_6\text{NH} \right)_n + 2n\text{H}_2\text{O}$$

Nylon 66

Formation of Nylon 66. Nylon forms at the interface of hexamethylendiamine (dissolved in water) and a derivative of adipic acid (adipyl chloride, dissolved in hexane).

Polyesters form from the condensation of diacids and dialcohols. An example is Dacron, an extremely strong but lightweight material that is used in clothing and as a tire cord. It has also been used in surgery to help support weak or damaged arteries. Dacron is produced from terephthalic acid and ethylene glycol.

$$n\text{HOOC}{-}\text{C}_6\text{H}_4{-}\text{COOH} + n\text{HOCH}_2\text{CH}_2\text{OH} \longrightarrow$$

terephthalic acid ethylene glycol

$$\left(\text{C(=O)}{-}\text{C}_6\text{H}_4{-}\text{C(=O)}{-}\text{O}{-}\text{CH}_2\text{CH}_2\text{O} \right)_n + 2n\text{H}_2\text{O}$$

Dacron

Dacron. Dacron was used to cover the wings and body of the Gossamer Albatross, the first human-powered plane to cross the English Channel. Dacron was chosen because it is lightweight and durable.

OBJECTIVES REVIEW *Can you:*

- ☑ describe two common methods for the formation of polymers?
- ☑ identify the monomeric unit of a polymer?

22.5 Proteins

OBJECTIVE

- ☐ Describe the primary, secondary, tertiary, and quaternary structures of proteins

A variety of polymeric compounds is synthesized by living systems. **Proteins,** polymers of amino acids, are one of three important classes of polymers found in living systems. Proteins account for about 15% of the mass of the human body and perform many functions. Some proteins are used for structural purposes, such as in the skin, muscles, and cartilage. Others can act as catalysts for important biological reactions, as hormones to regulate reactions, and as antibodies to help fight disease.

Proteins form as a result of the step-growth polymerization of α-amino acids. Because each amino acid contains both an amine group and an acid group, amino acids can polymerize by forming amide linkages. This method of polymerization is similar to the formation of nylon except that nature may use any of 20 different amino acids, each containing a different R group (see Table 22.4), to form the chains in proteins.

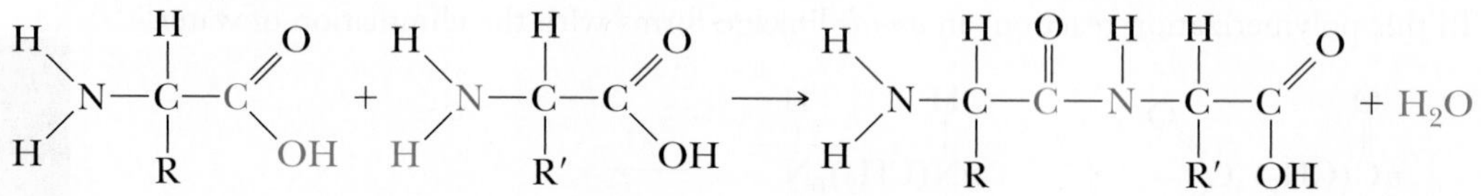

$$H_2N-CH(R)-C(=O)-OH + H_2N-CH(R')-C(=O)-OH \longrightarrow H_2N-CH(R)-C(=O)-NH-CH(R')-C(=O)-OH + H_2O$$

The new molecule formed from two or more amino acids is called a **peptide,** and the new amide linkage is called a *peptide bond. Polypeptides* contain many amino acids, and extremely long ones are termed proteins.

With 20 different amino acids available to form a polymer, nature has the potential to synthesize a wide variety of proteins. One way to classify the various amino acids is to note that some have polar R groups, such as the alcohol group in serine, whereas others have nonpolar groups, such as the methyl group in alanine. Another useful classification involves whether the R group is acidic, basic, or neutral. The types of R groups and the order in which they occur in proteins greatly influence the properties of the proteins.

Protein Structure

The primary structure of a protein is the sequence of amino acids in the polypeptide chain.

The structures of proteins are complex and can be described at four different levels. The **primary structure** is the sequence of amino acids in the polypeptide chain. It can be represented by a picture of all the atoms or by the use of the abbreviations in Table 22.4 (see Figure 22.24).

When we write the primary structure using the three-letter abbreviations, we assume that the amino-terminal end of each amino acid, and thus of the whole peptide, is on the left and the carboxylic acid end is on the right. This convention is important because, for example, the dipeptide Ser-Val is a different compound from Val-Ser.

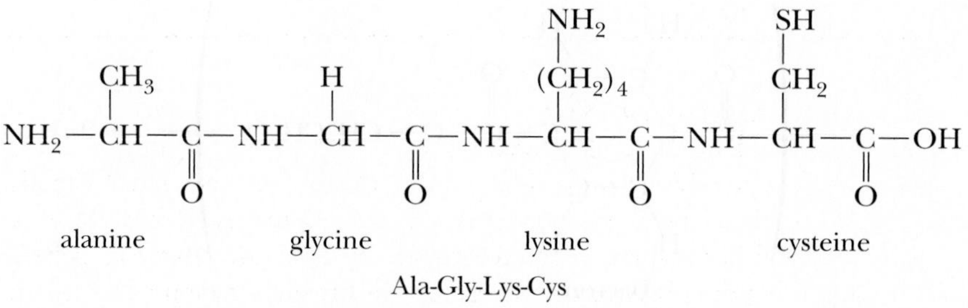

$$NH_2-CH(CH_3)-C(=O)-NH-CH(H)-C(=O)-NH-CH((CH_2)_4NH_2)-C(=O)-NH-CH(CH_2SH)-C(=O)-OH$$

alanine glycine lysine cysteine

Ala-Gly-Lys-Cys

Figure 22.24 Primary chain of a polypeptide. The primary sequence of a peptide is the order of amino acids in the polypeptide chain. The structure can be written using the structural formula or the shorthand notation for each amino acid.

TABLE 22.4 The 20 α-Amino Acids in Protein

Nonpolar R Groups

Glycine (Gly)

Valine (Val)

Alanine (Ala)

Isoleucine (Ile)

Proline (Pro)

Tryptophan (Trp)

Phenylalanine (Phe)

Methionine (Met)

Leucine (Leu)

Polar R Groups

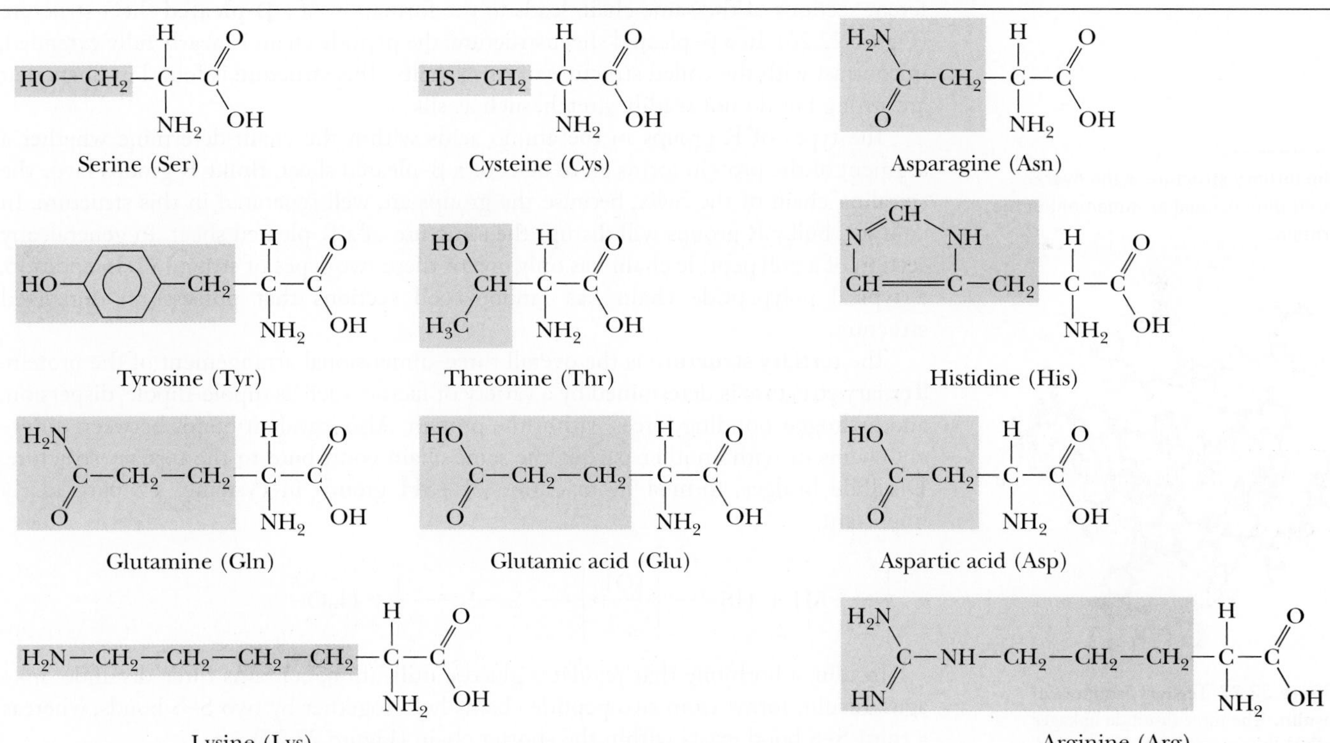

Serine (Ser)

Cysteine (Cys)

Asparagine (Asn)

Tyrosine (Tyr)

Threonine (Thr)

Histidine (His)

Glutamine (Gln)

Glutamic acid (Glu)

Aspartic acid (Asp)

Lysine (Lys)

Arginine (Arg)

The protein chains can assume different shapes. The **secondary structure** is the shape of the polypeptide chain, which is determined by the type of hydrogen bonds made by the amide portion of the chains. The hydrogen atoms of the polar N–H bonds can form hydrogen bonds to the amide oxygen atoms in the same chain or to those in other chains. An **α-helix** is the spiral structure adopted by a peptide chain that is held together by hydrogen bonding between amide groups in the same region of the chain

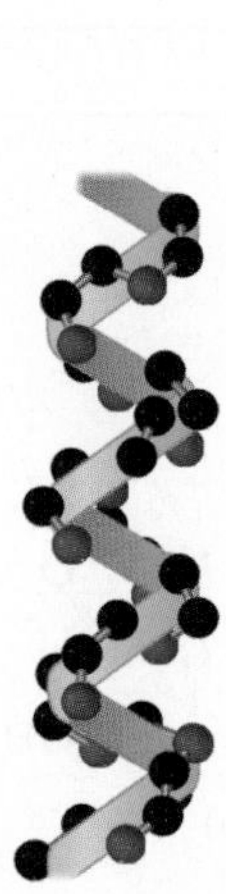

Figure 22.25 α-Helix. Projection of an α-helix structure containing only the atoms in the polypeptide chain shows the spiral clearly.

Figure 22.26 β-Pleated sheet. Hydrogen bonding between different sections of a polypeptide chain or different chains leads to β-pleated sheets. Only a part of the extended sheet is shown.

The secondary structure is the shape of the polypeptide chain.

(Figure 22.25). The coiled structure of an α helix is the main arrangement in structural proteins that need to stretch, such as those in hair and skin.

Hydrogen bonding between amide groups in different polypeptide chains or in different sections of the same chain leads to the formation of a **β-pleated** sheet structure (Figure 22.26). In a β-pleated sheet structure, the peptide chain is nearly fully extended, in contrast with the coiled structure of the α-helix. This structure is found in fibers that are strong but do not readily stretch, such as silk.

The tertiary structure is the overall three-dimensional arrangement of the protein.

The types of R groups in the amino acids within the chain determine whether a segment of the protein forms an α-helix or a β-pleated sheet. Bulky R groups favor the spiraling chain of the helix, because the groups are well separated in this structure. In contrast, bulky R groups will disrupt the structure of a β-pleated sheet. In general, any section of a polypeptide chain has only one of these two types of structures. In addition, a typical polypeptide chain has random coil sections that possess no organized structure.

The **tertiary structure** is the overall three-dimensional arrangement of the protein. Tertiary structure is determined by a variety of factors such as dipole-dipole, dispersion, and hydrogen-bonding forces within the protein. Also, covalent bonds between different chains or with another part of the same chain contribute to the tertiary structure. Disulfide bridges, formed by oxidation of –SH groups in cysteine, are particularly important.

$$\vert\!\text{—SH} + \text{HS—}\!\vert \xrightarrow{[O]} \vert\!\text{—S—S—}\!\vert + H_2O$$

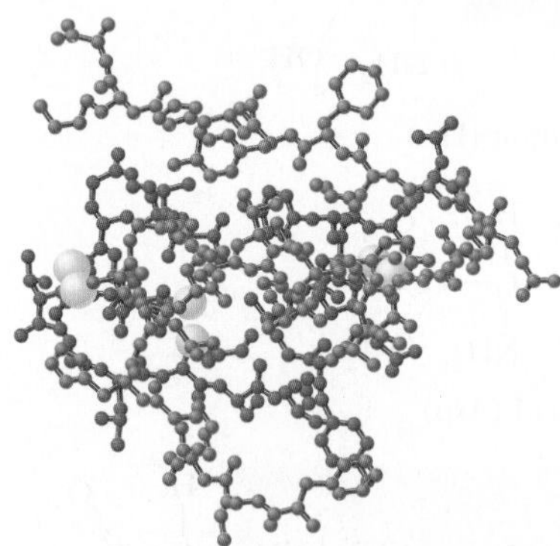

Figure 22.27 Tertiary structure of insulin. The three disulfide linkages (yellow) are a controlling factor in the arrangement of the tertiary structure of insulin. Insulin is formed from two peptide chains held together by two S–S bonds. A third S–S bond is within one chain.

Insulin, a hormone that regulates glucose utilization, contains three disulfide linkages. Insulin forms from two peptide chains held together by two S–S bonds, whereas a third S–S bond exists within the shorter chain (Figure 22.27).

Finally, some proteins have **quaternary structure**—the orientations of different polypeptide chains with respect to each other. The classic example is hemoglobin, which consists of four polypeptide chains (Figure 22.28).

The quaternary structure is the relative orientations of polypeptide chains within the protein.

Denaturation is the loss of the structural organization of a protein. Mild heating or the presence of metal ions such as mercury or lead causes many proteins to lose part or all of their structure. One example is the changes that take place when an egg is boiled. On heating, the protein albumin is converted from a soluble protein (clear egg white) to an insoluble denatured protein (a white opaque gel). Heavy metals

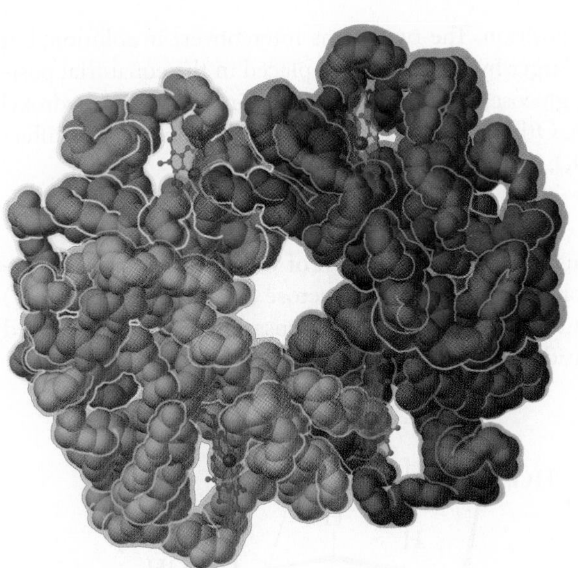

Figure 22.28 Hemoglobin. The structure of hemoglobin consists of four interacting polypeptide chains, shown here as two green and two blue regions. Each polypeptide chain coordinates an iron ion (large red ball in the center of the red areas) that is used for the transport of oxygen.

disrupt protein structure by breaking the disulfide bridges and bonding to the sulfide functional groups.

OBJECTIVE REVIEW *Can you:*

☑ describe the primary, secondary, tertiary, and quaternary structures of proteins?

22.6 Carbohydrates

OBJECTIVE

☐ Describe the structures of carbohydrates

Carbohydrates, polyhydroxyl aldehydes or ketones, or substances that react with water to yield such compounds, are the second important class of polymers in living systems. The name arises from the fact that most carbohydrates have the empirical formula $C_n(H_2O)_m$. A wide variety of carbohydrates exists, including sugars, starches, and cellulose. Sugar and starch are common food sources, and cellulose is the main structural material of plants. Carbohydrates form in plants from water and CO_2 in a process catalyzed by chlorophyll and activated by sunlight.

$$m\text{H}_2\text{O} + n\text{CO}_2 \xrightarrow{\text{Chlorophyll, sunlight}} \text{C}_n(\text{H}_2\text{O})_m + n\text{O}_2$$

This overall reaction, known as photosynthesis, is endothermic; the energy that is consumed is supplied by sunlight.

Monosaccharides

Monosaccharides are the basic units of carbohydrates, having the general formula $(CH_2O)_x$, where $x = 3$ to 8. Monosaccharides are simple sugars such as glucose, $C_6H_{12}O_6$ (Figure 22.29).

As shown in Figure 22.29, glucose exists in water in two different cyclic forms. Both are six-member rings that adopt the chair conformation, with two locations for the hydrogen atoms and hydroxyl groups on the ring carbon atoms. The bonds to these groups can be parallel to the rough plane formed by the carbon atoms in the ring (equatorial) or perpendicular to this plane (axial). Bulky functional groups are generally located in the equatorial sites to reduce steric interactions. Groups placed in the axial sites are close to other atoms located in axial sites in the molecule, but steric interactions are less between the equatorial positions. The two forms of glucose differ in that the β form has all of the hydroxyl groups in equatorial positions, whereas the α form has the

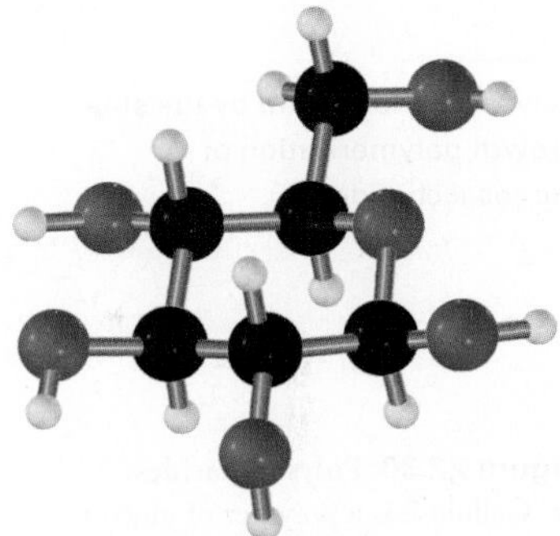

β-glucose

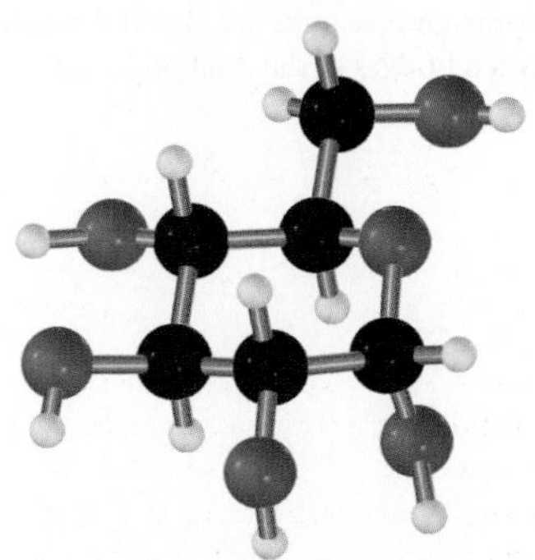

α-glucose

Figure 22.29 α- and β-glucose. The two isomers of glucose differ by the arrangement of the –OH and –H substituents at the ring carbon atom pictured at the bottom right (C1).

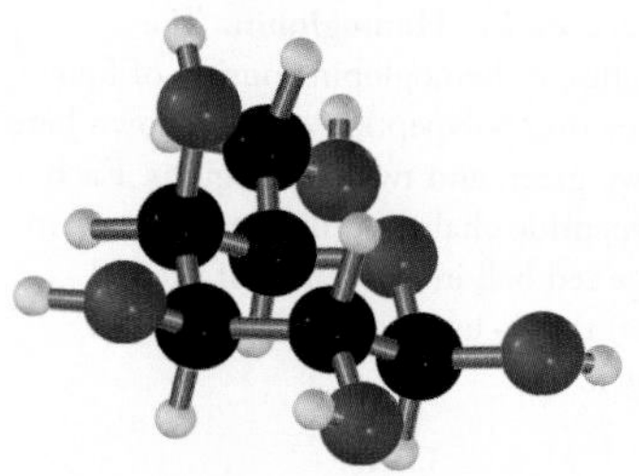

β-Galactose.

hydroxyl group at C1 in an axial position. The two forms interconvert in solution, but the β form is favored because the larger hydroxyl group is placed in the equatorial position. There are other isomers of glucose, such as β-galactose, in which the hydroxyl group at C4 is in an axial position. Other monosaccharides are based on rings of different sizes, such as ribose ($C_5H_{10}O_5$), which has a five-member ring.

Disaccharides

Two monosaccharides can condense with the elimination of water to yield a **disaccharide.** For example, the combination of α-glucose with fructose (a component of honey) yields sucrose, the main constituent of cane sugar. This condensation reaction is reversed in digestion, and the monosaccharides formed can then be used as an energy source.

α-glucose + fructose → sucrose + H_2O

Polysaccharides

Polysaccharides form by the step-growth polymerization of monosaccharides.

A **polysaccharide** is a step-growth polymer of monosaccharides. Well-known polysaccharides are cellulose and starch, both polymers of glucose. Cellulose contains long chains of β-glucose units joined as shown in Figure 22.30*a*.

The strands of cellulose can form strong hydrogen bonds, leading to strong fibers such as those found in cotton. Cellulose is the main structural component of plants. Humans cannot digest cellulose, although cows and some other species can. Thus, cellulose can be

(*a*)

(*b*)

Figure 22.30 Polysaccharides. *(a)* Cellulose is a polymer of glucose. The glucose units are joined through oxygen bridges at the equatorial positions. *(b)* Starch is also a polymer of glucose, but the glucose units are joined through oxygen bridges at the axial positions.

a source of glucose for those animals, but it passes through the digestive systems of humans essentially unchanged (cellulose in food is popularly known as "fiber").

On the other hand, starch (also a polymer of glucose) is digested by humans and is an important source of energy for them. In starch, the polymer linkages are through the axial positions (more than one type of starch exists). The structure of starch is not as suitable for hydrogen-bonding interactions; thus, it does not form strong fibers as cellulose does.

OBJECTIVE REVIEW *Can you:*

☑ describe the structures of carbohydrates?

22.7 Nucleic Acids

OBJECTIVE

☐ Explain the three different units of nucleic acids and describe how they are arranged

Nucleic acids are the third important type of polymer found in living systems. Nucleic acids are responsible for directing the syntheses of the various proteins necessary for the existence of each species. The building blocks of these polymers are nucleotides. A **nucleotide** is composed of three units: a five-carbon sugar, a base (a cyclic amine), and a phosphate group.

1. *Sugar:* The sugars ribose and deoxyribose are used to make nucleotides.

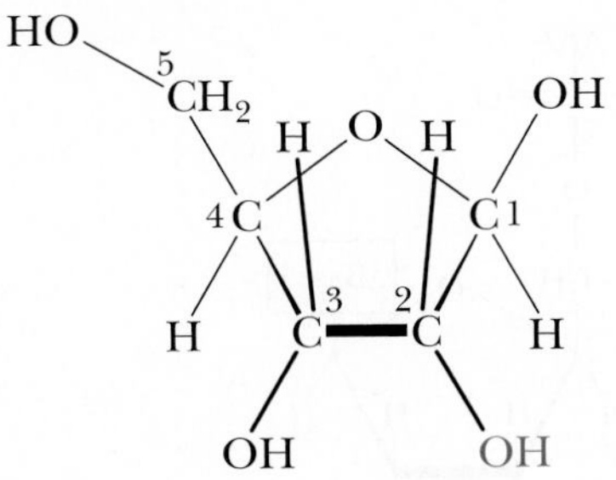

ribose

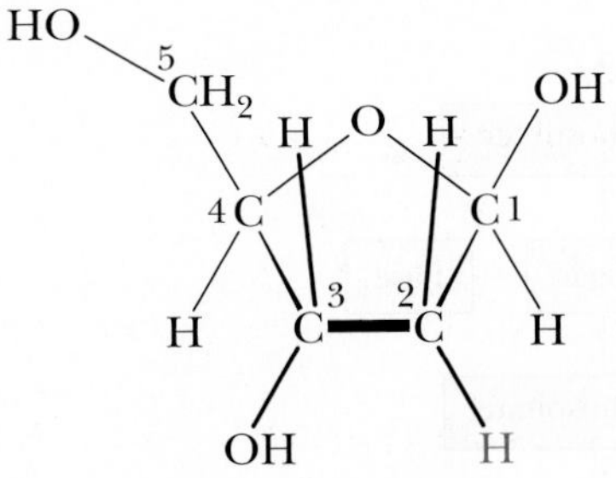

deoxyribose

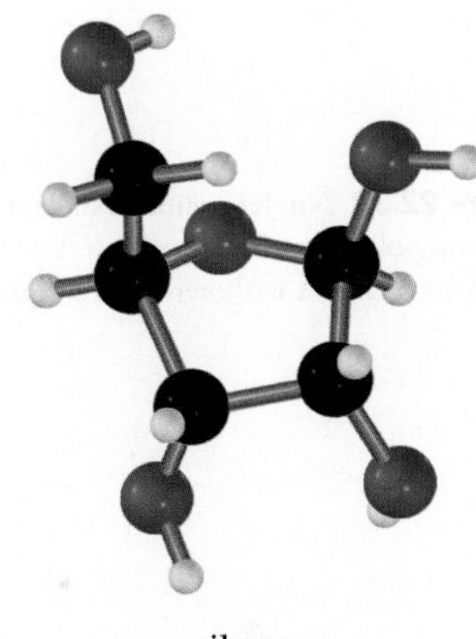

ribose

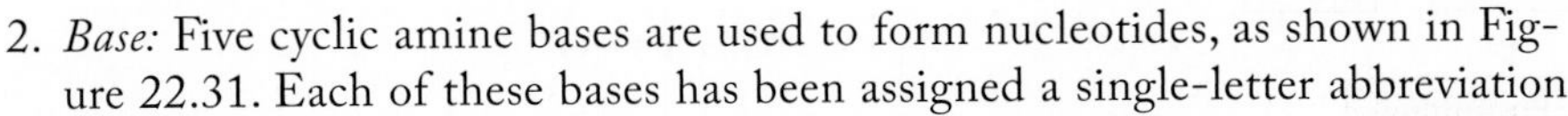

2. *Base:* Five cyclic amine bases are used to form nucleotides, as shown in Figure 22.31. Each of these bases has been assigned a single-letter abbreviation

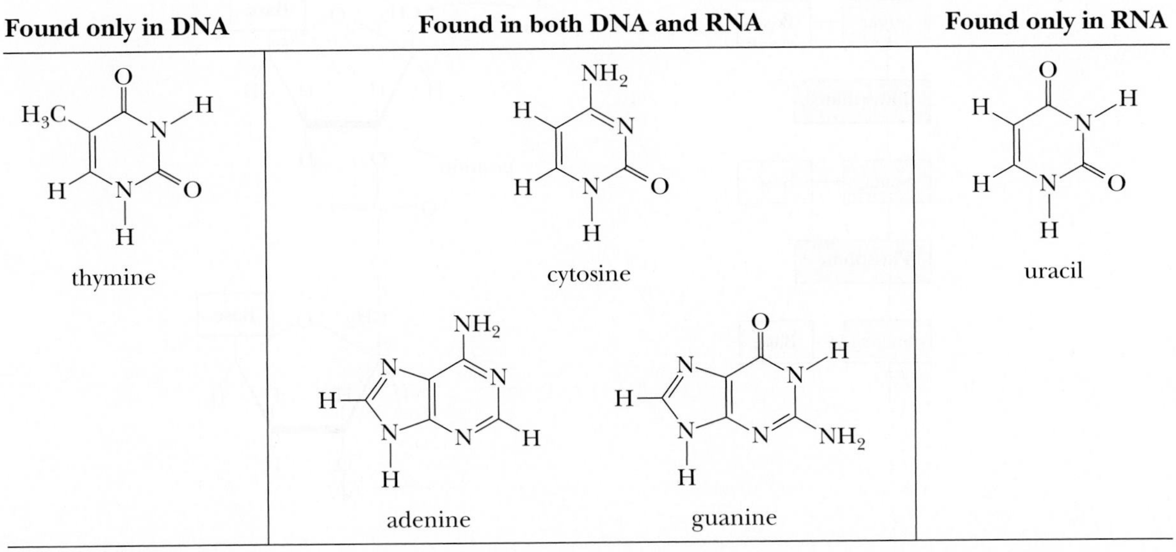

Found only in DNA	Found in both DNA and RNA	Found only in RNA
thymine	cytosine	uracil
	adenine, guanine	

Figure 22.31 The five bases in DNA and RNA.

(the first letter of its name), much like the three-letter abbreviations given to the amino acids used to form proteins.

3. *Phosphate:* The third component is derived from phosphoric acid, H_3PO_4.

The nucleotide is formed by condensation reactions of the phosphoric acid, sugar, and the base.

A nucleotide is composed of a five-carbon sugar, a base, and a phosphate group.

A **nucleic acid** consists of a chain of nucleotides. The phosphates bond to carbon 3 of one sugar and to carbon 5 of the next, forming a sugar-phosphate backbone (Figure 22.32). The base is attached at carbon 1 of the sugar.

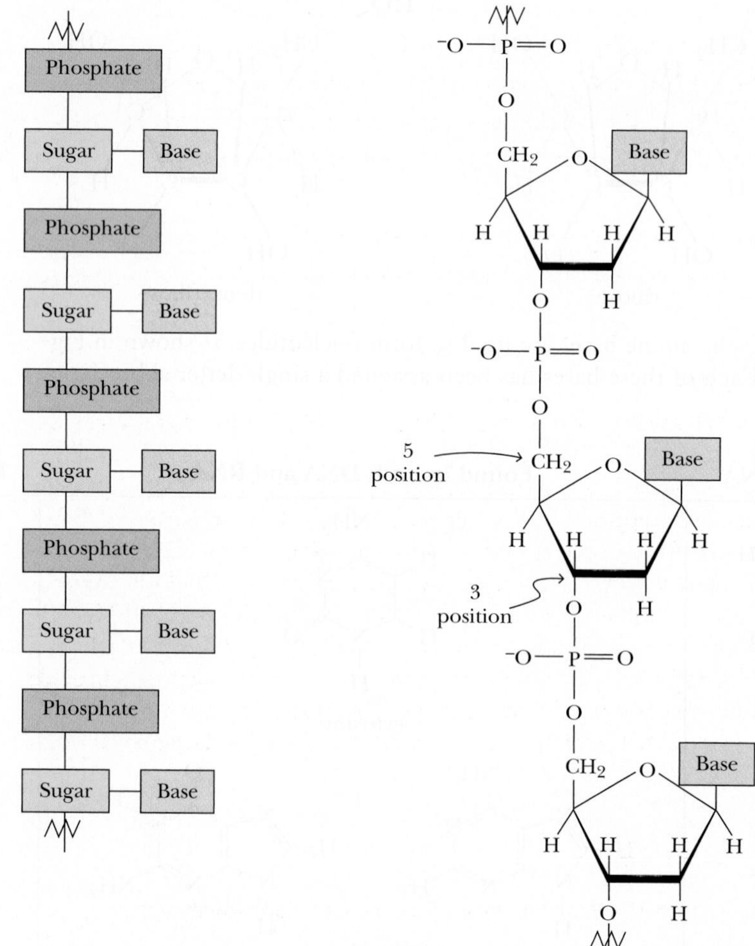

Figure 22.32 Nucleic acids. Nucleic acids are polymers of nucleotides. A fragment of DNA is shown on the right.

Two types of nucleic acids exist. **Deoxyribonucleic acids (DNA)** are polymers of nucleotide units located inside the chromosomes. DNA is the molecule that stores genetic information. **Ribonucleic acids (RNA)** are polymers of nucleotide units located outside the chromosomes. They transfer genetic information from DNA and direct the syntheses of proteins. The sugar in RNA is ribose, and that in DNA is deoxyribose. The bases adenine, guanine, and cytosine are found in both DNA and RNA, but uracil occurs only in RNA and thymine only in DNA.

The primary structure of a nucleic acid is determined by the sequence of its bases. It is the ordering of the bases that stores information in DNA and RNA. Each three-base sequence signals the incorporation of a particular amino acid into a protein and is called a *codon*. For example, the sequence UCU calls for the incorporation of serine.

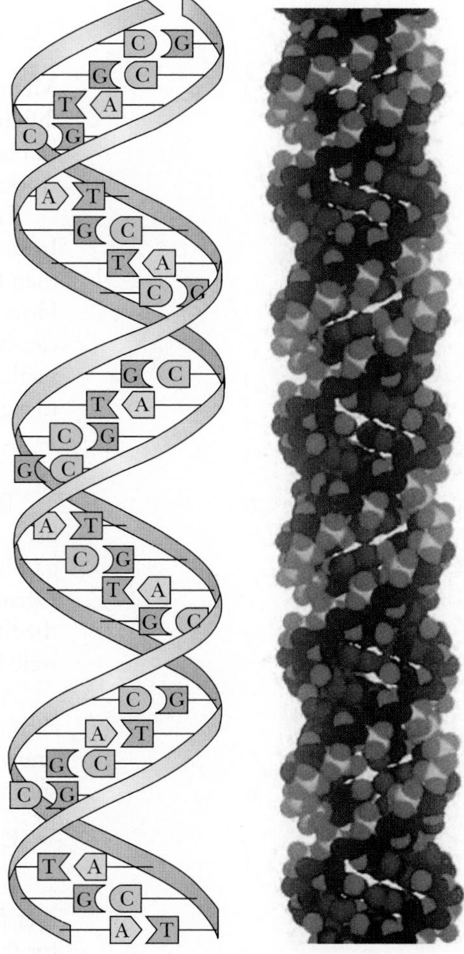

Figure 22.33 Structure of DNA. The two strands of DNA are held together by specific adenine-thymine (A-T) and guanine-cytosine (G-C) hydrogen-bonding interactions.

Secondary Structure of DNA

Careful analysis of the makeup of DNA reveals that any sample of DNA has the same molar amounts of adenine base and thymine base, and the same amounts of guanine base and cytosine base, but the ratio of the two pairs of bases varies considerably from one type of DNA to another. Using this information and available information from x-ray crystallographic studies, as outlined in the introduction to this chapter, Watson and Crick proposed in 1953 that the secondary structure of DNA is a double helix of two entwined nucleic acid strands (Figure 22.33). This double helix is held together by very specific hydrogen-bonding interactions in which the adenine bases (A) on one strand interact only with thymine bases (T) on the second strand, and the guanine bases (G) interact *only* with cytosine bases (C) (see Figure 22.33). This base pairing explains why the A/T and G/C ratios are 1. It also explains how cell division produces two identical cells. The double helix unwinds, and each strand creates a new matching strand to form two new *identical* double helices.

DNA is composed of two nucleic acid chains held in a double helix by specific hydrogen-bonding interactions.

Protein Synthesis

Nucleic acids control the assembly of proteins. The types of proteins that are synthesized determine the factors that distinguish species and differentiate individuals of each species. Protein synthesis does not take place in the nucleus of the cell where the DNA is, but in the cytoplasm. For the genetic information stored in the DNA to be transferred, a molecule of RNA (called *messenger RNA*) is synthesized by the DNA, much in the same way as DNA reproduces, except that in RNA the sugar is ribose and uracil replaces thymine as the base that interacts with adenine. The messenger RNA travels to the cytoplasm, where, with the help of two other types of RNA—*transfer* RNA and *ribosomal* RNA—it directs the order in which amino acids are incorporated into newly synthesized proteins. This order is determined by the order of the bases in the RNA, which, in turn, was determined by the order of bases in DNA.

OBJECTIVE REVIEW *Can you:*

- ☑ explain the three different units of nucleic acids and describe how they are arranged

CASE STUDY Methanol Fuel from Coal

Methanol, CH_3OH, is an attractive future fuel that can be produced from a number of different feedstocks. Currently, natural gas is the main feedstock for methanol, and oil can be used as well, but as the costs of these materials increase dramatically, other feedstocks become economically feasible.

The most obvious alternative is coal, which is readily available in much of the world. In fact, coal has been used to produce methanol for many years; it was just more expensive than from oil or natural gas until the recent increases in the prices of these two sources. How can methanol be made from coal? Even though it is a very complex material, coal can be written as C_n (that is as mostly carbon), when considering it as a source for methanol production. An attractive well-known reaction to convert the coal from a solid to a gaseous mixture that could be used to produce methanol is its reaction with water.

$$C_n + nH_2O \rightarrow nCO + nH_2$$

This mixture of CO and H_2 is known as "synthesis gas" because the technology has been worked out to convert it to many different compounds, including methanol and gasoline itself. Although an attractive fuel in countries such as the United States and Germany where coal is abundant, there are many problems as well. Any combustion of the fuels produced from this reaction will produce additional CO_2 in the atmosphere, a well-known greenhouse gas.

A solution to this problem is to make methanol from biomass, thus recycling the carbon rather than releasing carbon presently stored in coal. Processes that do this are known, but a technology that many are working on now would use carbon dioxide, the substance most likely responsible for climate change, as the feedstock.

$$CO_2 + 3H_2 \rightarrow CH_3OH + H_2O$$

In this case, one needs to make H_2 from another source, preferably from water, but that type of process is currently expensive. An important reason to study chemistry is for the new scientists using this book to learn the basic knowledge needed to successfully develop the technology to produce hydrogen in an economical and pollution-free method!

ETHICS IN CHEMISTRY

You have just prepared a new molecule using coal as the main starting material that when added to gasoline at 5% levels improves gas mileage. Although not expensive to make, tests show that the compound has a small probability to impact harmfully on the development of unborn babies if the fumes were breathed by pregnant women pumping gas. If you are the government regulator that needs to decide whether to test this new compound on a large scale, which issue is more important: the reduction of the consumption of gasoline or the slight chance that the development of unborn babies could be impacted?

Chapter 22 Visual Summary

The chart shows the connections between the major topics discussed in this chapter.

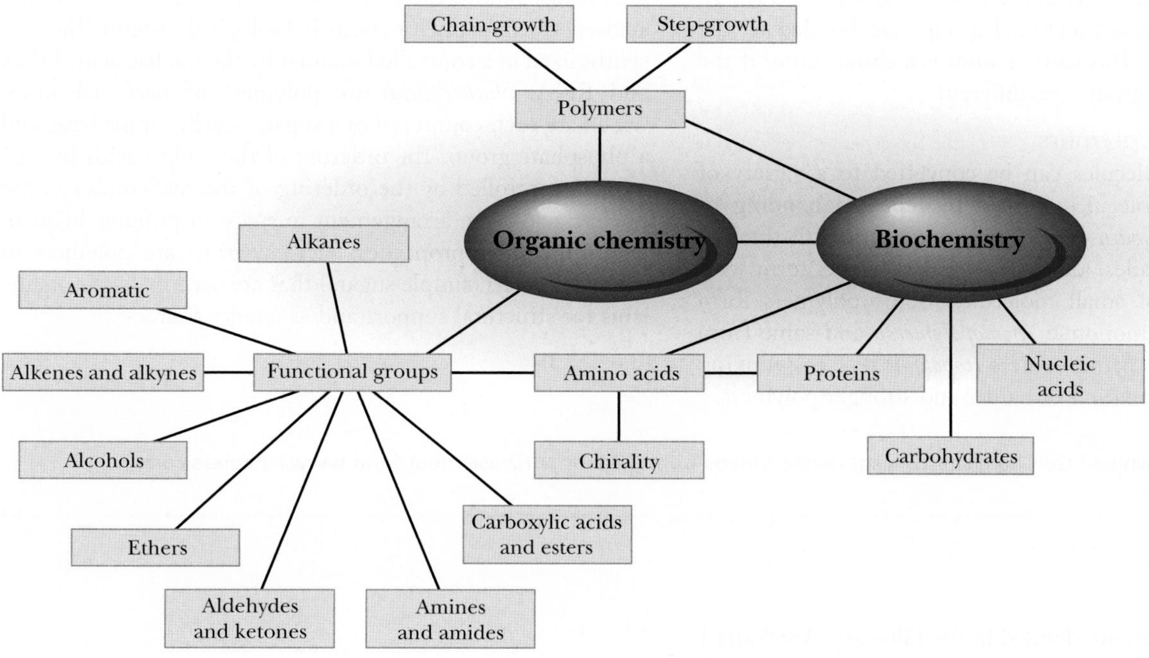

Summary

22.1 Alkanes and 22.2 Alkenes, Alkynes, and Aromatic Compounds

The study of carbon-containing compounds is called *organic chemistry.* The simplest organic compounds are *hydrocarbons,* compounds that contain only the elements hydrogen and carbon. If hydrocarbons have no multiple bonds, they are *saturated* and are known as *alkanes.* Saturated hydrocarbons that contain rings are known as *cycloalkanes.* Hydrocarbons that contain double bonds are *alkenes,* and those that contain triple bonds are known as *alkynes.* The IUPAC has established a series of rules for naming these compounds and their substituted derivatives, although many nonsystematic names are still in use. Most hydrocarbons exist as two or more isomers. Alkanes are not very reactive, but the π bonds in alkenes and alkynes provide good sites for *addition reactions. Aromatic hydrocarbons* are also unsaturated but rarely undergo addition reactions because of the stability of benzene (aromatic) rings. Aromatic hydrocarbons undergo *substitution reactions.*

22.3 Functional Groups

Functional groups occur in many organic compounds. Important groups that contain elements in addition to carbon and hydrogen are found in the following:

Alcohols and	–OH
Phenols	(aromatic alcohols)
Ethers	C–O–C
Aldehydes	$-\overset{\displaystyle O}{\overset{\|\|}{C}}-H$
Ketones	$-\overset{\displaystyle O}{\overset{\|\|}{C}}-R$
Carboxylic acids	$-\overset{\displaystyle O}{\overset{\|\|}{C}}-OH$
Esters	$-\overset{\displaystyle O}{\overset{\|\|}{C}}-OR$
Amines	$-NR_2$
Amides	$-\overset{\displaystyle O}{\overset{\|\|}{C}}-NR_2$

The properties of organic molecules that contain these functional groups are determined mainly by the degree of unsaturation, the polarity, and whether hydrogen bonding by the functional group is possible. An *amino acid* contains a carboxyl group ($-CO_2H$) and an amino group ($-NH_2$). In α-amino acids, these functional groups are bonded to the same carbon atom. This carbon atom is a chiral center if the other two bonded groups are different.

22.4 Organic Polymers

Small organic molecules can be converted to a variety of *polymers,* large molecules formed by repeated bonding of monomer units. *Chain-growth* polymers form with the loss of no other molecules, and *step-growth* polymers form with the elimination of small molecules. Some polymers form from one type of monomer *(homopolymers),* and some from two or more types of monomers *(copolymers).* Cross-linking can lead to the formation of harder and stronger polymers.

22.5 Proteins, 22.6 Carbohydrates, and 22.7 Nucleic Acids

Organic molecules can be either formed in living systems or synthesized. The study of the chemical systems in living organisms is *biochemistry. Proteins,* polymers of α-amino acids, perform many functions in biological systems. They are synthesized in a controlled manner by the nucleic acids DNA and RNA. *Nucleic acids* are polymers of *nucleotide* units, which are each composed of a sugar, a cyclic amine base, and a phosphate group. The ordering of the amino acids in proteins is controlled by the ordering of the nucleotides in the nucleic acids. The arrangement in space of proteins helps to determine their properties. *Polysaccharides* are polymers of *monosaccharides* (simple sugars) that are used by living organisms for structural support and as energy sources.

Download Go Chemistry concept review videos from OWL or purchase them from **www.ichapters.com**

Chapter Terms

The following terms are defined in the Glossary, Appendix I.

Biochemistry
Catenation
Organic chemistry

Section 22.1
Alkane
Alkyl group
Conformers
Cycloalkane
Hydrocarbon
Saturated hydrocarbon
Substitution reaction

Section 22.2
Addition reaction
Alkene
Alkyne
Aromatic hydrocarbon
Aryl group
Unsaturated hydrocarbon

Section 22.3
Alcohol
Aldehyde
Amide
Amine
Amino acid
Carbonyl group
Carboxylate group
Carboxylic acid
Condensation reaction
Ester
Ether
Functional group
Ketone
Phenol

Section 22.4
Chain-growth polymer (addition polymer)
Condensation polymer
Copolymer
Homopolymer
Polyester
Polymer
Step-growth polymer (condensation polymer)

Section 22.5
α-Helix
β-Pleated sheet
Denaturation
Peptide
Primary structure
Protein
Quaternary structure
Secondary structure
Tertiary structure

Section 22.6
Carbohydrate
Disaccharide
Monosaccharide
Polysaccharide

Section 22.7
Deoxyribonucleic acid (DNA)
Nucleic acid
Nucleotide
Ribonucleic acid (RNA)

Questions and Exercises

OWL Selected end of chapter Questions and Exercises may be assigned in OWL.

Blue-numbered Questions and Exercises are answered in Appendix J; questions are qualitative, are often conceptual, and include problem-solving skills.

- ■ Questions assignable in OWL
- ✎ Questions suitable for brief writing exercises
- ▲ More challenging questions

Questions

22.1 ✎ Explain the differences between organic chemistry and biochemistry.

22.2 ✎ Explain what is special about the element carbon that causes it to have such an extensive chemistry.

22.3 Why are alkanes also called *saturated hydrocarbons?* What is the hybridization of the carbon atoms in alkanes?

22.4 Draw the boat and chair configurations of 1,1,3,3-tetramethylcyclohexane. Which is more stable? Why?

22.5 What are the hybridized orbitals on the carbon atoms that form the σ bonds in ethylene and acetylene?

22.6 Explain why *cis* and *trans* isomers exist for 1,2-dibromoethene but not for 1,1-dibromoethene.

22.7 Both carboxylic acids and alcohols contain the –OH functional group, but only one of them is acidic in water. Which one is acidic? Why?

22.8 Explain why ethyl alcohol has a higher boiling point than methyl ethyl ether even though the latter has the greater molecular weight.

22.9 Explain what is meant by cross-linking a polymer. How does cross-linking affect the properties of the polymer?

22.10 Describe the four levels of structure of a protein.

22.11 Describe the bonding in an α-helix and in a β-pleated sheet. To what type of structure within the protein do these arrangements refer?

22.12 Describe the difference between cellulose and starch. How are they treated differently by the human digestive system?

22.13 Discuss the functions of DNA and RNA.

22.14 Describe the three components of a nucleotide. Be specific about the components used in DNA and those in RNA.

22.15 Describe the Watson and Crick model of DNA. How does this model account for the experimental fact that the A/T and G/C ratios are 1?

22.16 Describe how both DNA and RNA are used in the synthesis of a specific protein.

Exercises

OBJECTIVE Name and write structural isomers and conformers of alkanes.

22.17 Which of the following molecules are noncyclic alkanes?
(a) C_8H_{16} (b) C_6H_{14} (c) C_4H_9F (d) C_9H_{20}

22.18 Which of the following molecules are noncyclic alkanes?
(a) C_5H_{12} (b) C_6H_{12} (c) C_3H_6 (d) C_4H_9Cl

22.19 Write the molecular formula of a noncyclic alkane that contains eight carbon atoms.

22.20 Write the molecular formula of a noncyclic alkane that contains 12 carbon atoms.

22.21 Show a structural formula for the straight-chain isomer of C_5H_{12}. Name the alkyl group formed by the removal of hydrogen atom from one of the terminal carbon atoms.

22.22 Show the structural formula for C_3H_8. Name the alkyl group formed by the removal of a hydrogen atom from the center carbon atom.

22.23 Draw the structural formulas for the isomers of hexane, and name each isomer.

22.24 ■ Draw structural formulas for possible isomers of the dichlorinated propane, $C_3H_6Cl_2$. Name each compound.

22.25 Write the structural formula for each of the following alkanes.
(a) 2-methylhexane
(b) 3,3-dichloroheptane
(c) 2-methyl-3-phenyloctane
(d) 1,1-diethylcyclohexane

22.26 Write the structural formula for each of the following alkanes.
(a) 1-bromobutane
(b) 2-methyl-3-nitropentane
(c) 2,2-dimethylhexane
(d) 1-chloro-1-methylcyclopentane

22.27 Name the following compounds.

(a) $CH_2-CH-CH_2-CH_2-CH_3$ (F on CH_2; CH_3 on CH)

(b) $CH_3-CH-CH_2-CH_2-CH_3$ (CH_2CH_3 on CH)

(c)
$$CH_3-\underset{CH_3}{\overset{CH_3}{C}}-CH_2-CH_3$$

(d) Cyclobutane ring bearing Cl and CH_2-CH_3 on the same carbon

22.28 Name the following compounds.

(a) $CH_3-CH-CH_2-CH-CH_2-CH_3$ (CH_3 on first CH; Br on second CH)

(b) Cyclohexane ring bearing CH_3

(c) $CH_3-CH-CH-CH_2-CH_2-CH_2-CH_2-CH_3$ (NH_2 on first CH; CH_3 on second CH)

(d)
$$CH_3-CH_2-\underset{CH_2-CH_3}{\overset{CH_2-CH_2-CH_3}{CH}}$$

OBJECTIVES Identify and name unsaturated compounds, list the possible isomers, and write some reactions of alkenes, alkynes, and aromatic compounds.

22.29 Write the general formulas for a noncyclic alkene and a noncyclic alkyne.

22.30 Show an example of an alkene that can exist as both *cis* and *trans* isomers and another that cannot.

22.31 Name the following unsaturated compounds.

(a) 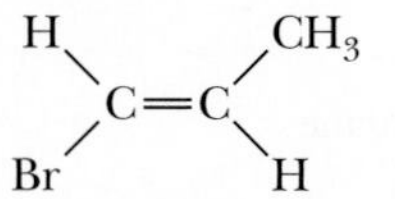
H $\quad$ CH_3
$C{=}C$
Br $\quad$ H

(b) 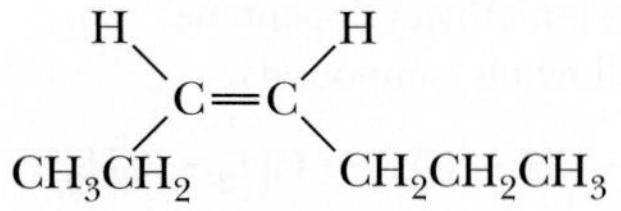
H $\quad$ H
$C{=}C$
CH_3CH_2 $\quad$ $CH_2CH_2CH_3$

(c) $CH_3C{\equiv}CCH_2CH_2F$

(d) 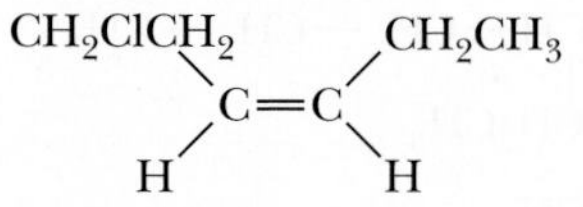
CH_2ClCH_2 $\quad$ CH_2CH_3
$C{=}C$
H $\quad$ H

22.32 Name the following unsaturated compounds.

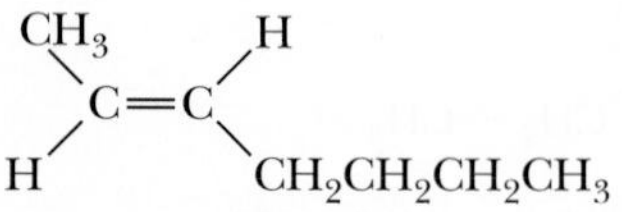

(a) 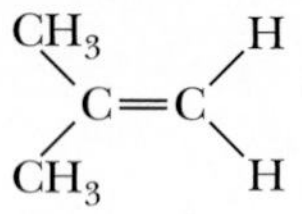
CH_3 $\quad$ H
$C{=}C$
CH_3 $\quad$ H

(b)
CH_3 $\quad$ H
$C{=}C$
H $\quad$ $CH_2CH_2CH_2CH_3$

(c)
$CH_3CH_2CH_2$ $\quad$ F
$C{=}C$
H $\quad$ F

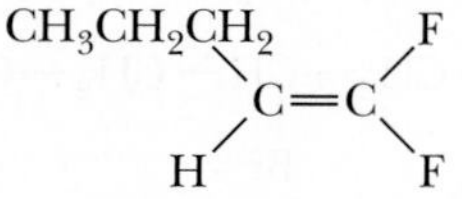

(d) $CH_3C{\equiv}CH_2CH_2Br$

22.33 Which compound exists as *cis* and *trans* isomers?

Br $\quad$ H
$C{=}C$
Br $\quad$ Cl

Cl $\quad$ Br
$C{=}C$
Br $\quad$ H

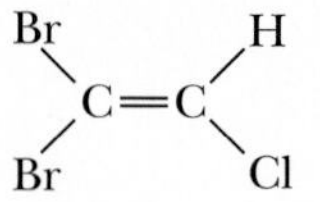

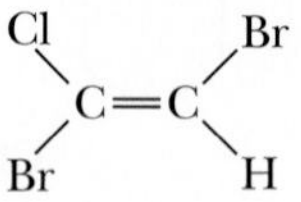

22.34 Which compound exists as *cis* and *trans* isomers?

H $\quad$ CH_3
$C{=}C$
H $\quad$ CH_3

H $\quad$ CH_3
$C{=}C$
CH_3 $\quad$ H

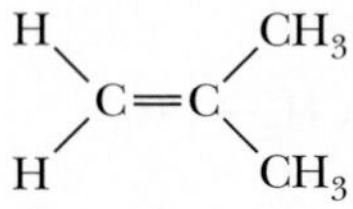

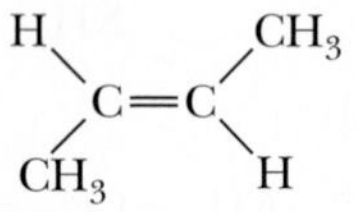

22.35 Draw all possible isomers for the substituted alkene C_3H_5F.

22.36 ■ Draw the structure, and give the systematic name for the products of the following reactions:
(a) $CH_3CH{=}CH_2 + Br_2 \rightarrow$
(b) $CH_3CH_2CH{=}CHCH_3 + H_2 \rightarrow$

22.37 Write the structural formula for each of the following unsaturated compounds.
(a) 2-bromo-1-hexene
(b) *cis*-4-nitro-2-pentene
(c) 1,2-dichloro-3-hexyne
(d) 1,1-difluoro-3-chloro-1-heptene

22.38 Write the structural formula for each of the following unsaturated compounds.
(a) 2,3-dimethyl-1-pentene
(b) 1-methylcyclohexene
(c) *cis*-1-chloro-2-butene
(d) 4-methyl-1-hexene

22.39 Write the structural formula of propyne, and name the product of its reaction with two molecules of Cl_2.

22.40 Write the structural formula of 2-hexene, and name the product of its reaction with Br_2.

22.41 Complete the following equations.

(a) $CH_2{=}CH(CH_3) + H_2 \longrightarrow$

(b) 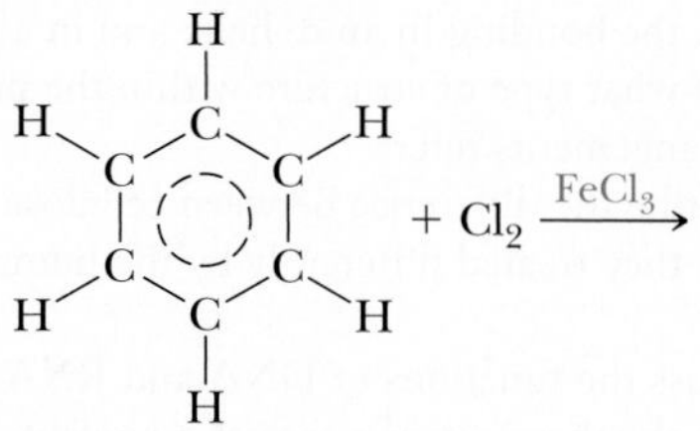 C_6H_6 (benzene ring, each C bearing an H) $+ Cl_2 \xrightarrow{FeCl_3}$

22.42 ■ Which of the following compounds can exist as *cis* and *trans* isomers? Draw and label them.
(a) 2,3-dimethyl-2-butene
(b) 2-chloro-2-butene
(c) dichlorobenzene

22.43 Draw the three isomers of dimethylbenzene.

22.44 Draw all the isomers of trichlorobenzene.

OBJECTIVE Identify and name the important functional groups in organic compounds.

22.45 Identify the class of organic compounds (ester, ether, ketone, etc.) to which each of the following belongs

(a) 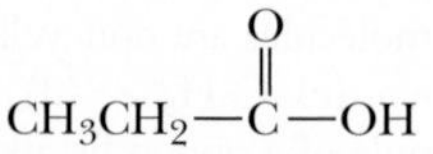 $CH_3CH_2{-}\overset{\overset{\large O}{\|}}{C}{-}OH$

(b) 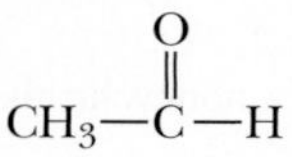 $CH_3{-}\overset{\overset{\large O}{\|}}{C}{-}H$

(c) $H_3C{-}O{-}CH_3$

22.46 ■ Identify the class of organic compounds (ester, ether, ketone, etc.) to which each of the following belongs.

(a) 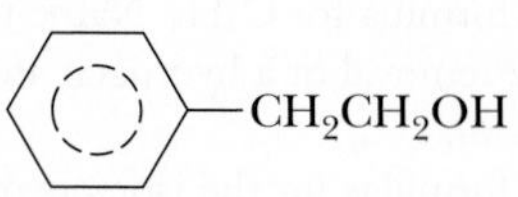 $C_6H_5{-}CH_2CH_2OH$

(b) $C_6H_5{-}O{-}C_6H_5$

(c) $C_6H_5{-}C_6H_4{-}OH$

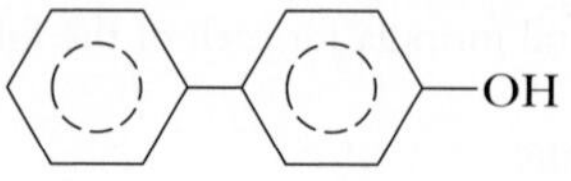

(d) 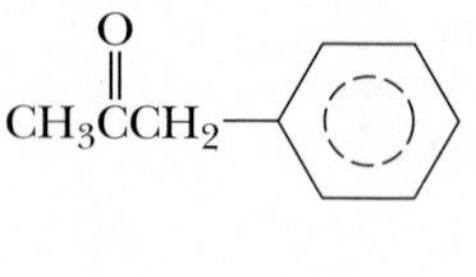 $CH_3\overset{\overset{\large O}{\|}}{C}CH_2{-}C_6H_5$

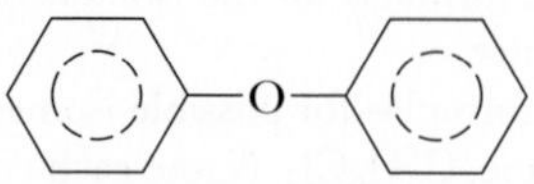

22.47 State which compound in each of the following pairs you expect to have the higher boiling point. Explain your answer.
(a) $CH_3CH_2OCH_2CH_3$ or $CH_3CH_2\overset{\overset{OH}{|}}{C}{=}O$
(b) $CH_3CH_2CH_2CH_2NH_2$ or $CH_3CH_2OCH_3$
(c) $CH_3CH_2C{\equiv}CF$ or $CH_2{=}CH_2$

22.48 State which compound in each of the following pairs you expect to have the higher boiling point. Explain your answer.
(a) CH_3OCH_3 or $CH_3CH_2CH_2CH{=}CH_2$
(b) $CH_3\overset{\overset{O}{\|}}{C}NH_2$ or $CH_3C{\equiv}CCH_3$
(c) $CH_3CH_2CH_2CH_2OH$ or $CH_3\overset{\overset{O}{\|}}{C}OCH_3$

22.49 Write the structural formula and name the organic product expected from the acid-catalyzed condensation reaction of CH_3OH.

22.50 Write the structural formula and name the organic product expected from the mild air oxidation of CH_3OH.

22.51 Write the structural formula and name the organic product expected from the acid-catalyzed condensation reaction of CH_3CO_2H and CH_3CH_2OH.

22.52 ■ Aldehydes and carboxylic acids are formed by oxidation of primary alcohols, and ketones are formed when secondary alcohols are oxidized. Give the name and formula for the alcohol that, when oxidized, gives the following products.
(a) $CH_3CH_2CH_2CHO$
(b) 2-hexanone

22.53 Draw the structure of each of the following molecules.
(a) 1-butanol
(b) 3-methyl-2-pentanone
(c) methyl acetate
(d) ethylphenylamine

22.54 Draw the structure of each of the following molecules.
(a) ethyl vinyl ether
(b) 2-bromopropanal
(c) pentanoic acid
(d) 3-fluorophenol

22.55 Name the following compounds.
(a) $FCH_2CH_2CH_2OH$
(b) $CH_3CH_2\overset{\overset{O}{\|}}{C}OH$
(c) $CH_3\overset{\overset{O}{\|}}{C}O\overset{\overset{CH_3}{|}}{C}HCH_3$
(d) $CH_3CH_2CH_2NH_2$

22.56 Name the following compounds.
(a) $CH_3CH_2CH{=}CH_2$
(b) $CH_3CH_2CH_2\overset{\overset{O}{\|}}{C}OCH_2CH_3$
(c) $CH_3CH_2O\overset{\overset{CH_3}{|}}{C}HCH_3$
(d) $CH_3CH_2CH_2CH_2\overset{\overset{O}{\|}}{C}NH_2$

OBJECTIVE **Identify chiral carbon atoms.**

22.57 Indicate which of the following compounds have a chiral or asymmetric center. Mark the chiral center in those molecules.
(a) $CH_3CH_2CH_2CH_2CH_3$
(b) $H{-}\underset{\underset{NH_2}{|}}{\overset{\overset{H}{|}}{C}}{-}\overset{\overset{O}{\|}}{C}{-}OH$
(c) $CH_3\underset{\underset{NH_2}{|}}{\overset{\overset{H}{|}}{C}}{-}\overset{\overset{O}{\|}}{C}{-}OH$

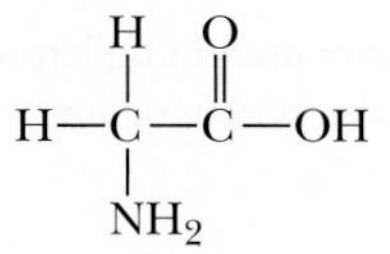

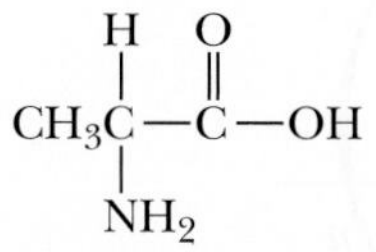

22.58 ■ Locate the chiral carbon(s), if any, in the following molecules.
(a) $CH_3{-}\underset{\underset{OH}{|}}{\overset{\overset{H}{|}}{C}}{-}\underset{\underset{OH}{|}}{\overset{\overset{H}{|}}{C}}{-}H$
(b) $H{-}\underset{\underset{H}{|}}{C}{=}\underset{\underset{H}{|}}{C}{-}CH_2{-}OH$
(c) $CH_3{-}\underset{\underset{Cl}{|}}{\overset{\overset{Cl}{|}}{C}}{-}\underset{\underset{H}{|}}{\overset{\overset{F}{|}}{C}}{-}Cl$

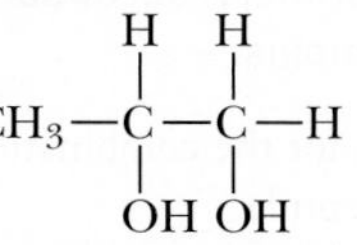

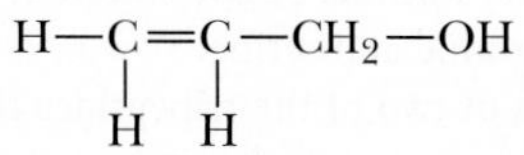

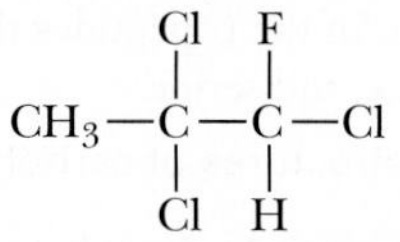

OBJECTIVES **Describe two common methods for the formation of polymers and identify the monomeric unit of a polymer.**

22.59 Describe and give a specific example of the formation of
(a) a chain-growth polymer.
(b) a homopolymer.

22.60 Describe and give a specific example of the formation of
(a) a step-growth polymer.
(b) a copolymer.

22.61 Polyvinyl chloride, a chemically resistant polymer used in house siding and floor tiles, is a chain-growth polymer of chloroethene (vinyl chloride). Draw the repeating unit of polyvinyl chloride.

22.62 ■ Polystyrene, a polymer used for thermal insulation (e.g., coolers) and toys, is a chain-growth polymer of phenylethylene (styrene). Draw the repeating unit of polystyrene.

22.63 Viton is a strong, flexible chain-growth copolymer used in gaskets. It is formed from 1,1-difluoroethylene and hexafluoropropene. Draw the repeating unit.

22.64 Tires are frequently made from SBR, the trade name of the chain-growth copolymer made from styrene ($CH_2{=}CHC_6H_5$) and butadiene ($CH_2{=}CHCH{=}CH_2$). Draw the repeating unit.

22.65 Polyacrylonitrile, a chain-growth polymer used in carpets, is formulated as follows. What monomer is used to make polyacrylonitrile?

$$\left(\!-\mathrm{C(H)(H)}-\mathrm{C(H)(CN)}-\!\right)_n$$

22.66 Kevlar, a step-growth polymer used in bulletproof vests, is formulated as follows. What monomeric unit is used to make Kevlar?

$$\left(\!-\mathrm{N(H)(C_6H_4)C(H)}-\!\right)_n$$

OBJECTIVE Describe the primary, secondary, tertiary, and quaternary structures of proteins.

22.67 Write the general reaction for the combination of two amino acids to form a dipeptide.

22.68 ■ Give the structural formula of two different dipeptides formed between arginine and serine.

22.69 Draw the structures of two of the tripeptides that can be formed from alanine, glycine, and cysteine.

22.70 Draw the structures of two of the tripeptides that can be formed from valine, alanine, and serine.

OBJECTIVE Describe the structures of carbohydrates.

22.71 Draw the two forms of glucose, clearly indicating how they differ.

22.72 Draw the structure of β-galactose. How does it differ from glucose?

22.73 Draw a disaccharide formed from β-galactose and ribose.

22.74 ■ Mannose has the same molecular formula as glucose and the same geometry except at carbon-2, where the H and OH groups are interchanged. Draw the structures of α- and β-mannose.

OBJECTIVE Know the three different units of nucleic acids and describe how they are arranged.

22.75 Draw and name the five principal bases used in DNA and RNA. Be specific about the bases used in DNA and those used in RNA.

22.76 Draw the nucleotides that contain the following bases and sugars.
(a) Adenine and ribose
(b) Cytosine and deoxyribose
(c) Uracil and ribose

Chapter Exercises

22.77 Write the structural formula for each of the following molecules.
(a) *cis*-2-bromo-3-hexene
(b) 2-nitrophenol

22.78 Indicate the functional groups in each of the following compounds.

(a) aspirin

$C_6H_4(COOH)(O{-}C(=O){-}CH_3)$

(b) estrone

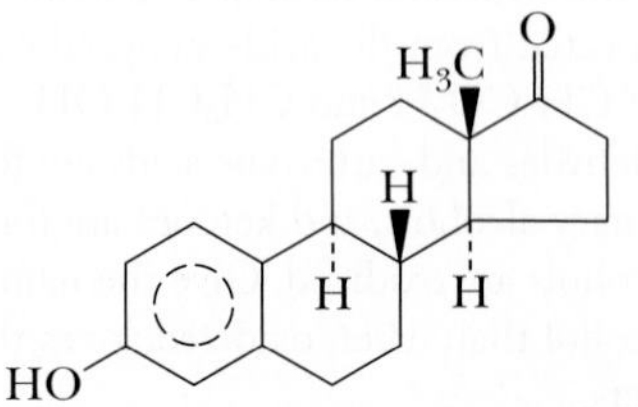

(c) butacetin, an analgesic (pain-killing) agent

$(CH_3)_3CO{-}C_6H_4{-}NHC(=O)CH_3$

22.79 Explain why ethanol dissolves freely in water, whereas its isomer, dimethyl ether, is only slightly soluble in water.

22.80 Draw the structure of an α-amino acid with a chiral center.

22.81 Lexan is a step-growth polymer with high-impact strength formed from the following monomers with the elimination of phenol. Draw the copolymer.

$HO{-}C_6H_4{-}C(CH_3)_2{-}C_6H_4{-}OH \qquad (C_6H_5O)_2CO$

22.82 Write the structures of two monomers that you think might be used to make an interesting copolymer. Draw the repeating unit of the copolymer.

22.83 Explain the correlation between the three-letter base pairs in a codon and the order of α-amino acids in proteins.

22.84 How many different tripeptides can be formed that contain one glycine, one valine, and one alanine?

Cumulative Exercises

22.85 Write the reaction for bromination of propene. What mass of bromine is needed to completely react with 22.1 g propene?

22.86 Write the product of the reaction of equimolar amounts of chlorine gas and 1-butyne. Are *cis* and *trans* isomers possible in the product? How does the hybridization of the alkyne carbon atoms change in the transition from reactant to product?

22.87 Describe the most important intermolecular force between molecules of

(a) 2-methyl-2-propanol.

(b) 2-butanone.

22.88 A compound with the formula C_5H_8 could have either of the following structures. If the addition of an equimolar amount of molecular hydrogen yields only a single pure compound, which structure is likely to be correct?

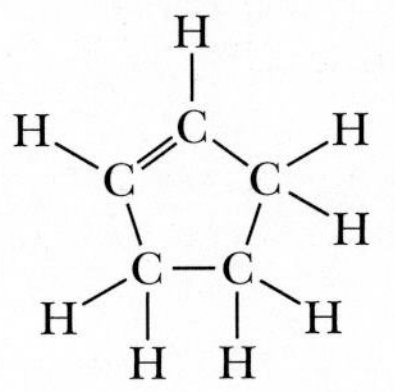

```
        H
        |
  H    C     H
   \ //  \  /
    C      C
    \     / \
     C - C    H
   / |   | \
  H  H   H  H
```

$CH_2{=}CHCH{=}CHCH_3$

APPENDIX A

Math Procedures

The study of chemistry requires certain mathematical skills. This appendix briefly reviews some of the mathematical operations that are used in this textbook, although it is not intended to replace a mathematics textbook.

A.1 Electronic Calculators

A modern scientific calculator is an important tool that will help you in your study of chemistry. Your calculator should be able to express numbers in exponential notation (explained in the next section), and perform operations such as natural and base-10 logarithms, antilogarithms (raising e or 10 to a power), roots, and raising a number to any power. The order of many operations change from one calculator to another, so you may have to consult your instruction book from time to time.

To use a calculator efficiently, you should learn to chain calculations together. When evaluating an expression such as

$$\frac{(1.202 \times 0.850) - 0.0307}{0.576},$$

it is easiest to multiply 1.202 by 0.850 and leave the result (1.0217) on the calculator display, subtract 0.0307 (to get 0.991), and divide by 0.576 to get 1.7204861, the final result. On most algebraic notation calculators, it is not necessary to press the [=] key after the first multiplication to obtain 1.0217 on the display. The preceding calculation can be performed by these keystrokes:

1.202 [×] .85 [−] .0307 [=] [÷] .576 [=]

	↑	↑	↑
Display reads	1.0217	0.991	1.7204861

On most algebraic calculators, if the [=] is not pressed before the [÷], only 0.0307 will be divided by 0.576, and the final result will be incorrect (0.96840…). (*Note:* The calculator does not determine the correct number of significant figures. You must truncate the answer as necessary, according to the rules described in Chapter 1.)

A.2 Exponential Notation

The numbers of science range from very large to very small. Exponential notation (or scientific notation) frequently is used to express a number as a product of a digit term and an exponential term. The digit term is a number between 1 and 10. The exponential term represents 10 raised to a whole number power. For example, the number 2468 is expressed as 2.468×1000 or 2.468×10^3 in exponential notation. Some other examples are as follows:

$10000 = 1 \times 10^4$ $\qquad$ $13579 = 1.3579 \times 10^4$

$1000 = 1 \times 10^3$ $\qquad$ $1357 = 1.357 \times 10^3$

$100 = 1 \times 10^2$ $\qquad$ $135 = 1.35 \times 10^2$

$10 = 1 \times 10^1$ $\qquad$ $13 = 1.3 \times 10^1$

$1 = 1 \times 10^0$

$0.1 = 1 \times 10^{-1}$ $\qquad$ $0.13579 = 1.3579 \times 10^{-1}$

$0.01 = 1 \times 10^{-2}$ $\qquad$ $0.01357 = 1.357 \times 10^{-2}$

$0.001 = 1 \times 10^{-3}$ $\qquad$ $0.00135 = 1.35 \times 10^{-3}$

$0.0001 = 1 \times 10^{-4}$ $\qquad$ $0.00013 = 1.3 \times 10^{-4}$

The exponential term locates the decimal point. The exponent is the number of places that the decimal point is shifted while going from the original number to the digit term. A positive exponent indicates that the decimal point is shifted to the left; a negative exponent indicates that the decimal point is shifted to the right. Most calculators have a mode in which any number entered or calculated is displayed in exponential notation. In this mode, the display shows the digit part of the number on the left, followed by a gap and the power of 10 at the right side of the display.

Addition and Subtraction

To add or subtract two numbers, convert the numbers to the same power of 10 and add or subtract the digit terms.

$$
\begin{aligned}
1.23 \times 10^{-2} + 4.5 \times 10^{-3} &= 1.23 \times 10^{-2} + 0.45 \times 10^{-2} \\
&= 1.68 \times 10^{-2}
\end{aligned}
$$

Multiplication

Multiply the digit terms and add the exponents. Shift the decimal place, if necessary, so the result is written as a digit term between 1 and 10 times the exponential term. The exponents are simply added in this calculation.

$$
\begin{aligned}
1.23 \times 10^{-2} \times 4.5 \times 10^{-3} &= (1.23)(4.5) \times 10^{-2+(-3)} \\
&= 5.535 \times 10^{-5} \\
&= 5.5 \times 10^{-5}
\end{aligned}
$$

Modern calculators perform calculations with high accuracy and can display numbers with more digits than can be justified by the data. The concept of significant digits is discussed in Chapter 1.

Division

Divide the digit terms and subtract the exponents.

$$
\frac{4.03 \times 10^{2}}{1.24 \times 10^{-3}} = \frac{4.03}{1.24} \times 10^{2-(-3)} = 3.25 \times 10^{5}
$$

Powers and Roots

To raise a number to a power, raise the digit term to the power and multiply the exponent by the power.

$$
(4.0 \times 10^{-2})^{3} = (4.0)^{3} \times 10^{(-2\times 3)} = 64 \times 10^{-6} = 6.4 \times 10^{-5}
$$

Many calculators have an $\boxed{x^y}$ (or $\boxed{y^x}$) operation. If you enter 4.0×10^{-2}, press $\boxed{x^y}$, then enter 3 and press $\boxed{=}$; you should see 6.4×10^{-5}.

A root is a fractional power: the square root is equivalent to raising a number to the 1/2 power, a cube root is the 1/3 power, and so forth. Many calculators have a separate button for the square root operation but not for other roots. Treating a root as a fractional power will work, as will using logarithms, which are discussed in the next section.

A.3 Logarithms

A logarithm is the power to which you must raise a base number to obtain the desired number. In this text, two types of logarithms are used: common logarithms (log), for which the base is 10, and natural logarithms (ln), for which the base is e (2.71828...). Important logarithmic relationships are as follows:

$$\log x = y, \text{ where } x = 10^{y}$$

$$\ln x = z, \text{ where } x = e^{z}$$

The common logarithms and natural logarithms are related by the equation

$$\ln x = (\ln 10)(\log x) \approx 2.303 \log x$$

The common logarithm of a number that is a power of 10 is always a whole number.

Number	Exponential Notation	Common Logarithm
100	10^2	2
10	10^1	1
1	10^0	0
0.1	10^{-1}	−1
0.01	10^{-2}	−2

The logarithm of a number that is not an integral power of 10 is found by using a calculator or table of logarithms. The common logarithm of 45 is 1.65. We can use an estimate to check this value; because 45 is partway between 10 and 100, we expect the logarithm to be partway between 1 and 2. An alternative way of expressing the relationship is

$$\log 45 = 1.65 \text{ or } 45 = 10^{1.65}.$$

When taking the logarithm of a number, the number of decimal places in the result is equal to the number of significant digits in the original quantity.

To find the log on the calculator, enter 45; then depress the [log] key to obtain 1.65. Alternatively, some calculators require that you press the [log] key then enter 45. No one set of instructions can cover all calculators.

The second kind of logarithm is the natural logarithm. The base of natural logarithms is $e \approx 2.7183$. Natural logarithms make some operations more convenient. To find the natural logarithm of 45, enter 45; then depress the [ln] key to obtain 3.81.

$$\ln 45 = 3.81 \text{ or } 45 = e^{3.81}$$

Antilogarithms

Frequently, the logarithm is given, and the equivalent number must be found. For example, we might have to determine the number whose common logarithm is −7.82. To calculate the antilogarithm, we must raise 10 (the base) to the −7.82 power. On most calculators, enter −7.82, then depress the [10^x] button (often labeled *inv log*) to obtain 1.5136×10^{-8}. Note that there are two decimal places in the logarithm because there are two significant figures in the number.

$$10^{-7.82} = 1.5 \times 10^{-8}$$

The process of finding the antilogarithm of a natural logarithm is called *exponentiation* because it involves the exponential function. To find the natural antilogarithm of −7.82, first enter −7.82 and then depress [e^x] (often called *exp* or *inv ln*) to obtain 4.0162×10^{-4}.

$$e^{-7.82} = 4.0 \times 10^{-4}$$

Different calculators have different labels on their keys, so refer to your user's manual for the proper key to press for each function.

Operations Using Logarithms

Logarithms are exponents so adding logarithms is the same as multiplying the numbers that they represent. Subtracting corresponds to division. The type of the logarithm, natural or common, does not matter in these operations.

$$\log(xy) = \log(x) + \log(y)$$

$$\log(x/y) = \log(x) - \log(y)$$

$$\log(x^y) = y \log(x)$$

$$\log \sqrt[y]{x} = \log(x^{1/y}) = (1/y) \log(x)$$

Consider raising 4.300 to the fifth power. We will find the logarithm, multiply by the power, and then take the antilogarithm.

$\log 4.300 = 0.63347$

$\log 4.300^5 = 5 \times \log 4.300 = 5 \times 0.63347 = 3.16735$

$10^{3.16735} = 1470 = 4.300^5$

Similarly, the cube root of 4.300 can be found by taking the logarithm, dividing by 3, and then taking the antilogarithm. Either common or natural logarithms can be used; natural logarithms are used just to offer a comparison with the last calculation.

$\ln 4.300 = 1.45866$

$\ln 4.300^{1/3} = 1/3 \times \ln 4.300 = 1/3 \times 1.45862 = 0.4862$

$e^{0.4862} = 1.626 = 4.300^{1/3}$

A.4 Graphs

In many situations, the value of one quantity depends on the value of another. It is often easier to understand the relationship between two quantities by looking at a graph that displays the value of one quantity against the value of the other. Graphs also help us to predict trends and to answer "what-if" problems.

Let us consider the relationship between the amount of some product formed in a reaction and the temperature of the reaction. The experiment is designed so that we change the temperature and then measure the amount of product at that temperature. The horizontal axis (x-axis) is used for the quantity that can be controlled or adjusted; in this example temperature is placed on the horizontal axis. The vertical axis (y-axis) is used for the quantity that responds to changes in the quantity on the x-axis—in this case, the amount of product formed. We can say that the amount of product depends on temperature; often, the quantity on the y-axis is called the *dependent variable,* and the quantity on the x-axis is called the *independent variable.* The experimental data are placed on the graph and connected with a smooth line.

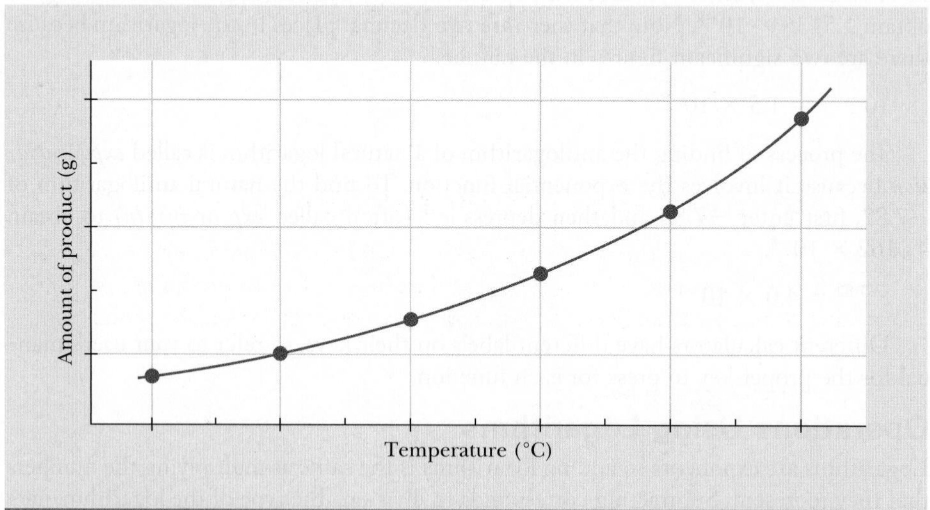

Figure A.1

The line connecting the points on a graph can assume many different shapes, but one of the most important is the straight line. A straight-line or linear relationship can be described by an equation such as

$y = 3x + 2.$

This relationship predicts the value of y for a given value of x. For example, when x is equal to 4, then y will be 14. The general equation of a straight line is

$$y = mx + b.$$

The quantity b is called the y-intercept, and it provides the value of y when x is equal to 0. The quantity m is called the *slope* of the line, and it provides information about the degree of slant. Lines with positive values of m slant up toward the right, a line with a value of zero for the slope is level, and lines with negative values of the slope slant down toward the right. The slope is evaluated by measuring the change in the dependent variable (Δy) for a given change in the independent variable (Δx). The change in any variable is represented by using the Greek letter delta, Δ, which should be understood to mean (final value − initial value).

$$m = \frac{\Delta y}{\Delta x}$$

The slope of a line representing a measured quantity must have units. If, for example, the x-axis is time in seconds and the y-axis is temperature in degrees Celsius, then the slope will have units of degrees Celsius per second (°C/s).

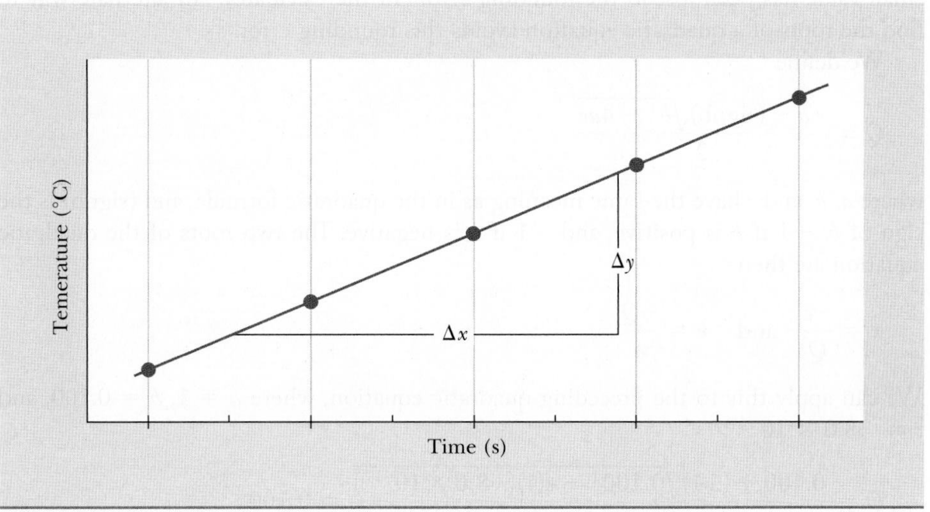

Figure A.2

A.5 Solving Polynomial Equations

The equation of a straight line is only one mathematical model that is used to describe data. Sometimes the equation that best fits the data has x raised to the second power (a quadratic equation), the third power (a cubic equation), or even higher. The general form of the relationship is

$$y = \beta_0 + \beta_1 x + \beta_2 x^2 + \beta_3 x^3 + \ldots$$

This relationship is called a *polynomial equation.* In this section, some of the methods used to solve these equations are presented. When we speak of "solving" an equation, we generally mean finding a *root,* which is a value of x for which y is equal to zero. The number of roots of a polynomial equation cannot be larger than the largest power of x in the equation but may be smaller.

Quadratic Equations

If the highest power of x found in the equation is 2, then the equation is called a quadratic equation. It has the general form

$$y = ax^2 + bx + c$$

where a, b, and c are constants. To find the roots of the equation, we need to find the values of x for which y is zero.

$$ax^2 + bx + c = 0$$

A formula, called the *quadratic formula,* enables us to determine the roots of the quadratic equation:

$$x = \frac{-b \pm \sqrt{b^2 - 4ac}}{2a}.$$

A quadratic equation has two roots, one calculated from the addition and the other from the subtraction process. Both represent values of x for which y is equal to zero; the text (particularly Chapters 14 through 16) includes ways to determine which of the two values should be used in a particular situation.

There is an alternate solution to the quadratic formula that has some advantages.[1] For example, if the quadratic formula and a calculator are used to solve the equation

$$x^2 + 0.100x - 8.0 \times 10^{-17} = 0,$$

the roots found are 0 and −0.100. The smaller root is not exactly equal to 0, but its very small value has been lost in the rounding errors of the calculator. An alternate way to find the roots of a quadratic equation avoids this rounding error.

We define

$$Q = \frac{-b + (\text{signb})\sqrt{b^2 - 4ac}}{2}$$

where a, b, and c have the same meaning as in the quadratic formula, and (signb) is the sign of b: +1 if b is positive, and −1 if b is negative. The two roots of the quadratic equation are then

$$x = \frac{-c}{Q} \quad \text{and} \quad x = \frac{-Q}{a}.$$

We can apply this to the preceding quadratic equation, where $a = 1$, $b = 0.100$, and $c = -8.0 \times 10^{-17}$:

$$Q = \frac{0.100 + (+1)\sqrt{0.100^2 - 4(1)(-8.0 \times 10^{-17})}}{2} = 0.100.$$

The two roots are

$$x = \frac{-c}{Q} = 8.0 \times 10^{-16} \quad \text{and} \quad x = \frac{-Q}{a} = -0.100.$$

Chemical situations exist in which the small number is the physically meaningful solution. By using this alternate solution to the quadratic equation, the desired solution is not lost in rounding errors.

Other ways can be used to solve for the root of a quadratic equation. These methods, which are applicable to any polynomial equation, are discussed in the next section.

Higher-Order Polynomial Equations

Several ways exist to determine the roots of a higher-order polynomial equation. We use a quadratic and a cubic (third-order) equation as examples, but the methods discussed can be applied to any polynomial.

[1]Brown, R. J. C. (1990). *J. Chem. Ed.*, *67*(5), 409.

One way to estimate the root is by graphing the function and analyzing the graph to find the root. Remember that a root is the value of x for which $y = 0$.

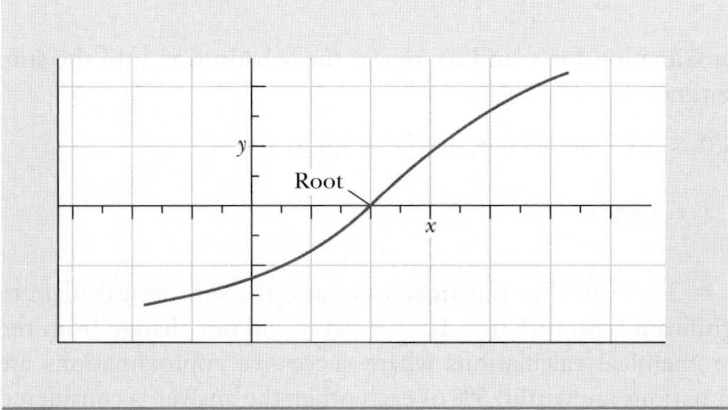

Figure A.3

Exact Solutions

When the equation is simple, an exact, or analytical, solution can be found. We have presented the formula for the roots of a quadratic equation; the formula to find the roots of higher-order equations is much more complex. Furthermore, mathematicians can prove that for equations of order 5 and higher, no formula can be used. In general, the roots of equations of order greater than 2 are found by numerical methods, as described in the following section.

Numerical Solutions

The root of an equation is a number. In the study of chemistry, the method used to find the root of an equation is not important; what is important is the number itself and its interpretation. All numerical methods are similar in that they require an initial estimate of the root, and they require a test for the desired accuracy of the approximation to the root.

Successive Approximation

Often, quadratic equations can be solved without using the quadratic formula by a method called *successive approximation.* We can illustrate the method by using an equation from the solubility equilibrium considered in Chapter 14. We will solve a cubic equation (a quadratic would be solved in similar fashion) by successive approximation.

$$1.0 \times 10^{-8} = s(5.0 \times 10^{-3} + 2s)^2$$

From the origin of this equation there is reason to believe that $2s$ is quite a bit smaller than 5.0×10^{-3}, so we make the approximation that s in the sum is zero. In using successive approximations, the approximate value is substituted only in sums and differences. With this approximation, the equation becomes

$$1.0 \times 10^{-8} = s(5.0 \times 10^{-3})^2,$$

which is easily solved for s.

$$s = \frac{1.0 \times 10^{-8}}{(5.0 \times 10^{-3})^2} = 4.0 \times 10^{-4}$$

This is a better approximation for s than is zero, so we substitute it into the sum in the original equation.

$$1.0 \times 10^{-8} = s(5.0 \times 10^{-3} + 2s)^2$$

$$1.0 \times 10^{-8} = s(5.0 \times 10^{-3} + 2[4.0 \times 10^{-4}])^2 = s(5.8 \times 10^{-3})^2$$

We now solve this equation for the third approximation for s:

$$s = \frac{1.0 \times 10^{-8}}{(5.8 \times 10^{-3})^2} = 3.0 \times 10^{-4}$$

This value is a better approximation for s, and we repeat the substitution into the sum in the equation one more time.

$$5.0 \times 10^{-3} + 2s = 5.0 \times 10^{-3} + 2(3.0 \times 10^{-4}) = 5.6 \times 10^{-3}$$

$$s = \frac{1.0 \times 10^{-8}}{(5.6 \times 10^{-3})^2} = 3.2 \times 10^{-4}$$

We accept this value of $s = 3.2 \times 10^{-4}$ as the answer because the sum we calculate to the correct number of significant figures ($5.0 \times 10^{-3} + 2s$) would not change from the previous iteration. In the chemical calculations where successive approximations are used, if successive approximations are within 5% of each other, the answer is considered satisfactory.

Interval Halving

When it is known that the desired root of a polynomial lies between two values, it can be approached by a method of halving. In an acid ionization equilibrium problem, we obtain the equation

$$1.3 \times 10^{-3} = \frac{x^2}{0.0500 - x}.$$

From the chemistry involved, we know that the desired value of x must lie between 0 and 0.0500. First, expand the equation into the standard form of a quadratic equation.

$$f(x) = x^2 + 1.3 \times 10^{-3}\, x - 6.5 \times 10^{-5} = 0$$

We evaluate $f(x)$ for the two limits (0 and 0.0500) and for the midpoint of these two values.

$$f(0) = -6.5 \times 10^{-5}$$

$$f(0.0500) = 2.5 \times 10^{-3}$$

$$f(0.0250) = 5.9 \times 10^{-4}$$

If the sign of the function changes between $x = 0$ and $x = 0.0250$, the desired root must lie in that interval (remember that the function is equal to zero when x is equal to the root), so the midpoint of the two numbers, 0.0125, is used for the next evaluation of the function.

$$f(0.0125) = 2.6 \times 10^{-4}$$

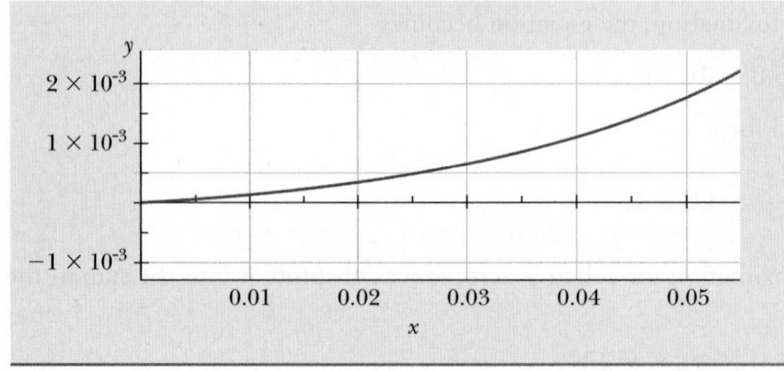

Figure A.4

This value of x is also larger than the root because the function has a positive value. The following table summarizes the rest of the procedure.

Lower x	Upper x	Middle x	$f(x)$ (at middle x)	Root between
0.00	5.00×10^{-2}	2.50×10^{-2}	5.9×10^{-4}	lower and middle
0.00	2.50×10^{-2}	1.25×10^{-2}	1.1×10^{-4}	lower and middle
0.00	1.25×10^{-2}	6.25×10^{-3}	-1.8×10^{-5}	middle and upper
6.25×10^{-3}	1.25×10^{-2}	9.38×10^{-3}	3.5×10^{-5}	lower and middle
6.25×10^{-3}	9.38×10^{-3}	7.82×10^{-3}	6.3×10^{-6}	lower and middle
6.25×10^{-3}	7.82×10^{-3}	7.04×10^{-3}	-6.3×10^{-6}	middle and upper
7.04×10^{-3}	7.82×10^{-3}	7.43×10^{-3}	-1.4×10^{-7}	middle and upper

We can accept 7.43×10^{-3} as the root. Note how close the function is to zero, $f(x) = -1.4 \times 10^{-7}$, when x is equal to 7.43×10^{-3}.

Newton–Raphson Method

Probably the fastest method for finding the roots of a polynomial equation is the Newton–Raphson method. We will illustrate this method by finding a root (a value of x for which $y = 0$) of the equation

$$y = 4x^3 - 3x^2 + 2x - 1.$$

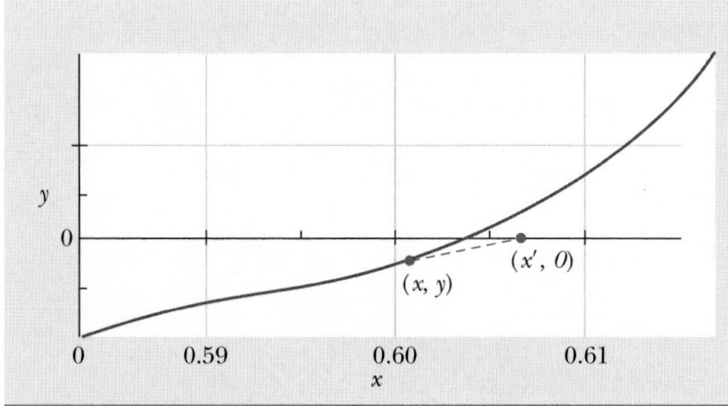

Figure A.5

The graph of this equation shows that there is a root at about 0.6, so we will use that value as our first estimate for x. The value of y at $x = 0.6$ is

$$y = 4(0.6)^3 - 3(0.6)^2 + 2(0.6) - 1 = -0.016.$$

To find the second estimate for x, determine the slope of the function at the first estimate and draw a straight line with that slope from the first estimate to the x-axis. The point at which the extrapolated line crosses the x-axis is x', the second estimate for the root.

Notice that the line of slope m passes through points with coordinates (x, y) and $(x', 0)$.

$$\text{Slope} = m = \frac{\Delta y}{\Delta x} = \frac{0 - y}{x' - x}$$

This relationship can be rearranged to solve for x':

$$x' = x - \frac{y}{m}.$$

These relationships can be applied to the particular example. In this particular example, $x = 0.6$, and $y = -0.016$. We will find the slope by evaluating the derivative of the function. (If you do not know how to obtain the derivative, the slope is found by graphing the function and drawing the tangent.)

$$\text{Slope} = m = 12x^2 - 6x + 2$$
$$= 12(0.6)^2 - 6(0.6) + 2 = 2.72$$

Now, solve for x':

$$x' = x - \frac{y}{m} = 0.6 - \frac{-0.016}{2.72} = 0.606$$

This is our second estimate of the root of the function. The value of the function when $x = 0.606$ is $y = 4.72 \times 10^{-4}$, which is much closer to zero than the previous estimate.

We can repeat this procedure until the estimates converge to the desired accuracy. (In fact, in this example, the third estimate is the same as the second one to three significant figures.) The Newton–Raphson method is easily automated (programmed into a computer or calculator), and it is one of the best overall methods of solving for roots of polynomial equations.

APPENDIX B

Selected Physical Constants

Quantity	Symbol	Value	Units
Atmosphere (standard)	atm	1.01325×10^{5}	Pa
Atomic mass unit	u	1.660539×10^{-27}	kg
Avogadro's number	N_A	6.02214×10^{23}	particles/mol
Bohr radius	a_0	5.2918×10^{-11}	m
Boltzmann constant	k	1.38065×10^{-23}	J/K
Charge-to-mass ratio of electron	e/m_e	-1.758820×10^{11}	C/kg
Elementary charge	e	1.602176×10^{-19}	C
Electron mass	m_e	9.10938×10^{-31}	kg
Faraday constant	F	9.64853×10^{4}	C/mol e^-
Gas constant	R	0.082061	L·atm/mol·K
		8.3145	J/mol·K
Neutron mass	m_n	1.674927×10^{-27}	kg
Planck's constant	h	6.62607×10^{-34}	J·s
Proton mass	m_p	1.672622×10^{-27}	kg
Rydberg constant	R_H	1.09737×10^{7}	m^{-1}
		1.09737×10^{5}	cm^{-1}
Speed of light (in vacuum)	c	2.99792458×10^{8}	m/s
		1.862824×10^{5}	mi/s

The uncertainty is less than ±1 in the last place shown.

CODATA, 2006.

APPENDIX C

Unit Conversion Factors

The metric system was implemented by the French National Assembly in 1790 and has been modified many times. The International System of Units, or *le Système International d'Unités* (SI), represents an extension of the metric system and is widely used by scientists. It was adopted by the 11th General Conference of Weights and Measures in 1960 and has been modified since that time.

The SI defines seven base units and other units based on these seven. The seven base units are shown in Table C.1.

TABLE C.1 SI Base Units

Physical Quantity	Name of Unit	Symbol
Length	meter	m
Mass	kilogram	kg
Time	second	s
Temperature	kelvin	K
Amount of substance	mole	mol
Electric current	ampere	A
Luminous intensity	candela	cd

Decimal fractions and multiples of metric and SI units are designated by the prefixes listed in Table C.2. Those most commonly used in chemistry are in boldface.

TABLE C.2 Metric and SI Prefixes

Factor	Prefix	Symbol	Factor	Prefix	Symbol
10^{24}	yotta	Y	10^{-1}	deci	d
10^{21}	zetta	Z	10^{-2}	**centi**	c
10^{18}	exa	E	10^{-3}	**milli**	m
10^{15}	peta	P	10^{-6}	**micro**	μ
10^{12}	tera	T	10^{-9}	**nano**	n
10^{9}	giga	G	10^{-12}	**pico**	p
10^{6}	mega	M	10^{-15}	femto	f
10^{3}	**kilo**	k	10^{-18}	atto	a
10^{2}	hecto	h	10^{-21}	zepto	z
10^{1}	deka	da	10^{-24}	yocto	y

All other units are derived from the seven base units. Table C.3 shows important derived units.

TABLE C.3 Derived SI Units

Physical Quantity	Name of Unit	Symbol	Definition
Area	square meter	m^2	
Volume	cubic meter	m^3	
Density	kilogram per cubic meter	kg/m^3	
Force	newton	N	$kg \cdot m/s^2$
Pressure	pascal	Pa	N/m^2
Energy	joule	J	$kg \cdot m^2/s^2$
Electric charge	coulomb	C	$A \cdot s$
Electric potential difference	volt	V	J/C

Important Equalities

Many different units of measure are used. The following table gives important equalities among these units.

Common Units of Mass and Weight

1 pound = 453.59 grams = 0.45359 kilogram
1 kilogram = 1000 grams = 2.205 pounds
1 gram = 1000 milligrams = 6.022×10^{23} atomic mass units
1 atomic mass unit = 1.660540×10^{-27} kilograms
1 short ton = 2000 pounds = 907.2 kilograms
1 metric ton = 1000 kilograms = 2205 pounds

Common Units of Length

1 inch = 2.54 centimeters (exactly) = 0.0254 meter
1 foot = 12 inches = 30.48 centimeters
1 mile = 5280 feet = 1.609 kilometers
1 yard = 36 inches = 0.9144 meter
1 meter = 100 centimeters = 39.37 inches = 3.281 feet = 1.094 yards
1 kilometer = 1000 meters = 1094 yards = 0.6214 mile
1 picometer = 100 Ångstrom = 1.0×10^{-10} centimeter = 1.0×10^{-12} meter
= 3.937×10^{-11} inch

Common Units of Volume

1 quart = 0.94635 liter
1 liter = 1 cubic decimeter = 1000 cubic centimeters = 0.001 cubic meter
1 milliliter = 1 cubic centimeter = 0.001 liter = 1.0567×10^{-3} quart
1 cubic foot = 28.317 liters = 29.922 quarts = 7.480 gallons

Common Units of Pressure

1 atmosphere = 760 torr = 1.01325×10^{5} pascals = 14.70 pounds per square inch
1 bar = 10^{5} pascals
1 torr = 1 millimeter of mercury
1 pascal = 1 $kg/m{\cdot}s^2$ = 1 N/m^2

Common Units of Energy

1 thermochemical calorie = 4.184 joules = 4.129×10^{-2} liter-atmospheres
= 2.611×10^{19} electron volts
1 electron volt = 1.6022×10^{-19} joules = 96.488 kJ/mol
1 liter-atmosphere = 24.217 calories = 101.325 joules
1 British thermal unit = 1055.06 joules = 252.2 calories

APPENDIX D

Names of Ions

TABLE D.1 Common Ions

Cations	Anions
1+	**1−**
Ammonium (NH_4^+)	Acetate (CH_3COO^-)
Cesium (Cs^+)	Bromide (Br^-)
Copper(I) (Cu^+)	Chlorate (ClO_3^-)
Hydrogen (H^+, H_3O^+)	Chloride (Cl^-)
Lithium (Li^+)	Chlorite (ClO_2^-)
Potassium (K^+)	Cyanide (CN^-)
Silver (Ag^+)	Dihydrogen phosphate ($H_2PO_4^-$)
Sodium (Na^+)	Fluoride (F^-)
Thallium(I) (Tl^+)	Hydride (H^-)
	Hydrogen carbonate (HCO_3^-)
	Hydrogen sulfate (HSO_4^-)
	Hydrogen sulfite (HSO_3^-)
	Hydroxide (OH^-)
	Hypochlorite (ClO^-)
	Iodide (I^-)
	Nitrate (NO_3^-)
	Nitrite (NO_2^-)
	Perchlorate (ClO_4^-)
	Permanganate (MnO_4^-)
	Thiocyanate (SCN^-)
2+	**2−**
Barium (Ba^{2+})	Carbonate (CO_3^{2-})
Cadmium (Cd^{2+})	Chromate (CrO_4^{2-})
Calcium (Ca^{2+})	Dichromate ($Cr_2O_7^{2-}$)
Chromium(II) (Cr^{2+})	Hydrogen phosphate (HPO_4^{2-})
Cobalt(II) (Co^{2+})	Oxalate ($C_2O_4^{2-}$)
Copper(II) (Cu^{2+})	Oxide (O^{2-})
Iron(II) (Fe^{2+})	Peroxide (O_2^{2-})
Lead(II) (Pb^{2+})	Sulfide (S^{2-})
Magnesium (Mg^{2+})	Sulfate (SO_4^{2-})
Manganese(II) (Mn^{2+})	Sulfite (SO_3^{2-})
Mercury(I) (Hg_2^{2+})	
Mercury(II) (Hg^{2+})	
Strontium (Sr^{2+})	
Nickel (Ni^{2+})	
Tin(II) (Sn^{2+})	
Zinc (Zn^{2+})	
3+	**3−**
Aluminum (Al^{3+})	Arsenate (AsO_4^{3-})
Chromium(III) (Cr^{3+})	Phosphate (PO_4^{3-})
Cobalt(III) (Co^{3+})	
Iron(III) (Fe^{3+})	
Lanthanum (La^{3+})	
Thallium(III) (Tl^{3+})	
Titanium(III) (Ti^{3+})	
Vanadium(III) (V^{3+})	
Higher	
Manganese(IV) (Mn^{4+})	
Tin(IV) (Sn^{4+})	
Lead(IV) (Pb^{4+})	Several transition metals with high oxidation states are nearly always present as anions Chromium(VI) (CrO_4^{2-}, $Cr_2O_7^{2-}$) Manganese(VII) (MnO_4^-)

TABLE D.2 Oxyanions and Their Corresponding Oxyacids

Oxyanions		Oxyacids	
CO_3^{2-}	Carbonate ion	H_2CO_3	Carbonic acid
NO_2^-	Nitrite ion	HNO_2	Nitrous acid
NO_3^-	Nitrate ion	HNO_3	Nitric acid
PO_4^{3-}	Phosphate ion	H_3PO_4	Phosphoric acid
SO_3^{2-}	Sulfite ion	H_2SO_3	Sulfurous acid
SO_4^{2-}	Sulfate ion	H_2SO_4	Sulfuric acid
ClO^-	Hypochlorite ion	$HClO$	Hypochlorous acid
ClO_2^-	Chlorite ion	$HClO_2$	Chlorous acid
ClO_3^-	Chlorate ion	$HClO_3$	Chloric acid
ClO_4^-	Perchlorate ion	$HClO_4$	Perchloric acid

APPENDIX E

Properties of Water

Physical Properties of Water

Molar mass	18.01528 g/mol
Melting point (1 atm)	0.00 °C
Boiling point (1 atm)	100.00 °C
Triple-point temperature	0.01 °C
Triple-point pressure	4.58 torr
Triple-point density (ℓ)	0.99978 g/mL
Triple-point density (g)	4.885 g/L
Critical temperature	374.1 °C
Critical pressure	218.3 atm
Critical density	0.322 g/mL
Maximum density	0.99995 g/mL
Temperature of maximum density	4.0 °C
Enthalpy of fusion (at 0 °C)	6.01 kJ/mol
Entropy of fusion (at 0 °C)	22.0 J/mol · K
Enthalpy of vaporization (at 25 °C)	44.0 kJ/mol
Entropy of vaporization (at 25 °C)	118.8 J/mol · K
Freezing point depression constant	1.86 °C/molal
Boiling point elevation constant	0.512 °C/molal

Lide, D. R. (Ed.). (2000–2001). *CRC Handbook of Chemistry and Physics* (82nd ed.). CRC Press: Boca Raton, FL.

Dependence of Water Density, Vapor Pressure, and pK_w (−log of ionization constant) on Temperature

T, °C	Density, g/mL	Vapor Pressure, torr	pK_w
0	0.99984	4.585	14.9435
10	0.99970	9.212	14.5346
20	0.99821	17.542	14.1669
25	0.99707	23.769	13.9965
30	0.99565	31.844	13.8330
40	0.99222	55.365	13.5348
50	0.98803	92.588	13.2617
60	0.98320	149.50	13.0171
70	0.97778	233.84	
80	0.97182	355.33	
90	0.96535	525.92	
100	0.95840	760.00	

Adapted from *CRC Handbook of Chemistry and Physics*. CRC Press: Boca Raton, FL, 1991.

APPENDIX F

Solubility Product, Acid, and Base Constants

Solubility Product Constants at 25 °C

Substance	Formula	Solubility Product
Aluminum phosphate	$AlPO_4$	9.8×10^{-21}
Barium carbonate	$BaCO_3$	2.6×10^{-9}
Barium chromate	$BaCrO_4$	1.2×10^{-10}
Barium fluoride	BaF_2	1.8×10^{-7}
Barium iodate	$Ba(IO_3)_2$	4.0×10^{-9}
Barium phosphate	$Ba_3(PO_4)_2$	6.0×10^{-39}
Barium sulfate	$BaSO_4$	1.1×10^{-10}
Bismuth arsenate	$BiAsO_4$	4.4×10^{-10}
Cadmium arsenate	$Cd_3(AsO_4)_2$	2.2×10^{-33}
Cadmium carbonate	$CdCO_3$	1.0×10^{-12}
Cadmium cyanide	$Cd(CN)_2$	1.0×10^{-8}
Cadmium fluoride	CdF_2	6.4×10^{-3}
Cadmium hydroxide	$Cd(OH)_2$	7.2×10^{-15}
Cadmium iodate	$Cd(IO_3)_2$	2.5×10^{-8}
Cadmium phosphate	$Cd_3(PO_4)_2$	2.5×10^{-33}
Calcium carbonate	$CaCO_3$	3.4×10^{-9}
Calcium fluoride	CaF_2	3.5×10^{-11}
Calcium hydroxide	$Ca(OH)_2$	5.0×10^{-6}
Calcium iodate	$Ca(IO_3)_2$	6.5×10^{-6}
Calcium phosphate	$Ca_3(PO_4)_2$	2.1×10^{-33}
Calcium sulfate	$CaSO_4$	4.9×10^{-5}
Cerium(III) iodate	$Ce(IO_3)_3$	3.2×10^{-10}
Cobalt(II) arsenate	$Co_3(AsO_4)_2$	6.8×10^{-29}
Cobalt(II) phosphate	$Co_3(PO_4)_2$	2.1×10^{-35}
Copper(I) bromide	$CuBr$	6.3×10^{-9}
Copper(I) chloride	$CuCl$	1.7×10^{-7}
Copper(I) iodide	CuI	1.3×10^{-12}
Copper(I) thiocyanate	$CuSCN$	1.8×10^{-13}
Copper(II) arsenate	$Cu_3(AsO_4)_2$	7.9×10^{-36}
Copper(II) hydroxide	$Cu(OH)_2$	1.6×10^{-19}
Copper(II) iodate	$Cu(IO_3)_2$	7.4×10^{-8}
Copper(II) oxalate	CuC_2O_4	4.4×10^{-10}
Copper(II) phosphate	$Cu_3(PO_4)_2$	1.4×10^{-37}
Iron(II) carbonate	$FeCO_3$	3.1×10^{-11}
Iron(II) fluoride	FeF_2	2.4×10^{-6}
Iron(II) hydroxide	$Fe(OH)_2$	4.9×10^{-17}
Iron(III) hydroxide	$Fe(OH)_3$	2.8×10^{-39}
Iron(III) phosphate	$FePO_4$	9.9×10^{-16}
Lanthanum hydroxide	$La(OH)_3$	1.0×10^{-19}
Lanthanum iodate	$La(IO_3)_3$	7.5×10^{-12}
Lead(II) bromide	$PbBr_2$	6.6×10^{-6}
Lead(II) carbonate	$PbCO_3$	7.4×10^{-14}
Lead(II) chloride	$PbCl_2$	1.7×10^{-5}
Lead(II) fluoride	PbF_2	3.3×10^{-8}
Lead(II) hydroxide	$Pb(OH)_2$	1.4×10^{-20}
Lead(II) iodate	$Pb(IO_3)_2$	3.7×10^{-13}
Lead(II) iodide	PbI_2	9.8×10^{-9}
Lead(II) oxalate	PbC_2O_4	8.5×10^{-10}
Lead(II) sulfate	$PbSO_4$	2.5×10^{-8}
Lead(II) thiocyanate	$Pb(SCN)_2$	2.1×10^{-5}
Lithium carbonate	Li_2CO_3	8.2×10^{-4}

Lide, D. R. (Ed.). (2000–2001). *CRC Handbook of Chemistry and Physics* (82nd ed.). CRC Press: Boca Raton, FL.

Continued

Solubility Product Constants at 25 °C *Continued*

Substance	Formula	Solubility Product
Magnesium carbonate	$MgCO_3$	6.8×10^{-6}
Magnesium fluoride	MgF_2	5.2×10^{-11}
Magnesium hydroxide	$Mg(OH)_2$	5.6×10^{-12}
Magnesium phosphate	$Mg_3(PO_4)_2$	1.0×10^{-24}
Manganese(II) carbonate	$MnCO_3$	2.2×10^{-11}
Manganese(II) hydroxide	$Mn(OH)_2$	2.1×10^{-13}
Manganese(II) iodate	$Mn(IO_3)_2$	4.4×10^{-7}
Mercury(I) bromide	Hg_2Br_2	6.4×10^{-23}
Mercury(I) carbonate	Hg_2CO_3	3.6×10^{-17}
Mercury(I) chloride	Hg_2Cl_2	1.4×10^{-18}
Mercury(I) fluoride	Hg_2F_2	3.1×10^{-6}
Mercury(I) iodide	Hg_2I_2	5.2×10^{-29}
Mercury(I) oxalate	$Hg_2C_2O_4$	1.8×10^{-13}
Mercury(I) sulfate	Hg_2SO_4	6.5×10^{-7}
Mercury(I) thiocyanate	$Hg_2(SCN)_2$	3.2×10^{-20}
Mercury(II) hydroxide	$Hg(OH)_2$	3.1×10^{-26}
Mercury(II) iodide	HgI_2	2.9×10^{-29}
Nickel(II) carbonate	$NiCO_3$	1.4×10^{-7}
Nickel(II) hydroxide	$Ni(OH)_2$	5.5×10^{-16}
Nickel(II) iodate	$Ni(IO_3)_2$	4.7×10^{-5}
Nickel(II) phosphate	$Ni_3(PO_4)_2$	4.7×10^{-32}
Palladium(II) thiocyanate	$Pd(SCN)_2$	4.4×10^{-23}
Potassium hexachloroplatinate(IV)	$K_2[PtCl_6]$	7.5×10^{-6}
Potassium perchlorate	$KClO_4$	1.1×10^{-2}
Silver acetate	$AgCH_3COO$	1.9×10^{-3}
Silver arsenate	Ag_3AsO_4	1.0×10^{-22}
Silver bromate	$AgBrO_3$	5.4×10^{-5}
Silver bromide	$AgBr$	5.4×10^{-13}
Silver carbonate	Ag_2CO_3	8.5×10^{-12}
Silver chloride	$AgCl$	1.8×10^{-10}
Silver chromate	Ag_2CrO_4	1.1×10^{-12}
Silver cyanide	$AgCN$	6.0×10^{-17}
Silver dichromate	$Ag_2Cr_2O_7$	3.0×10^{-7}
Silver iodate	$AgIO_3$	3.2×10^{-8}
Silver iodide	AgI	8.5×10^{-17}
Silver oxalate	$Ag_2C_2O_4$	5.4×10^{-12}
Silver phosphate	Ag_3PO_4	8.9×10^{-17}
Silver sulfate	Ag_2SO_4	1.2×10^{-5}
Silver sulfite	Ag_2SO_3	1.5×10^{-14}
Silver thiocyanate	$AgSCN$	1.0×10^{-12}
Strontium arsenate	$Sr_3(AsO_4)_2$	4.3×10^{-19}
Strontium carbonate	$SrCO_3$	5.6×10^{-10}
Strontium fluoride	SrF_2	4.3×10^{-9}
Strontium iodate	$Sr(IO_3)_2$	1.1×10^{-7}
Strontium sulfate	$SrSO_4$	3.4×10^{-7}
Tin(II) hydroxide	$Sn(OH)_2$	5.5×10^{-27}
Zinc arsenate	$Zn_3(AsO_4)_2$	2.8×10^{-28}
Zinc carbonate	$ZnCO_3$	1.5×10^{-10}
Zinc fluoride	ZnF_2	3.0×10^{-2}
Zinc hydroxide	$Zn(OH)_2$	3.0×10^{-17}
Zinc iodate	$Zn(IO_3)_2$	4.1×10^{-6}

Selected Acids and Their Properties

Properties of Some Common, Commercially Available Acids

Name	Formula	Approximate Weight Percent	Approximate Molarity	Density of Concentrated Acid (g/mL)	Volume Needed to Prepare 1.0 L of a 1 M Solution (mL)
Acetic	CH_3COOH	99.8	17.4	1.05	57.6
Hydrochloric	HCl	37.2	12.1	1.19	82.6
Hydrofluoric	HF	49.0	28.9	1.18	34.6
Nitric	HNO_3	70.4	15.9	1.42	62.9
Perchloric	$HClO_4$	70.5	11.7	1.67	85.5
Phosphoric	H_3PO_4	85.5	14.8	1.7	67.7
Sulfuric	H_2SO_4	96.0	18.0	1.84	55.6

Lide, D. R. (Ed.). (2000–2001). *CRC Handbook of Chemistry and Physics* (82nd ed.). CRC Press: Boca Raton, FL.

Safety Concerns

All acids, particularly when present in concentrated form, are potentially dangerous. They can all cause burns to skin. Always wear goggles, gloves, face shield, and an acid-resistant apron when handling concentrated solutions; even dilute acid solutions need to be treated with respect. (Goggles, gloves, and an apron are always appropriate when handling chemicals.) When you prepare an acid solution, remember to add acid to water. Adding water to acid can cause localized heating that spatters the acid out of its container. You should also stir while you are adding the acid.

Particular safety problems associated with the behavior of some acids are as follows:

HCl Strong fumes, handle in hood only.

HF Dissolves glass. Use polymer beakers and bottles. Avoid skin contact.

HNO_3 Strong fumes, handle in hood only. Powerful oxidant. Avoid contact with organic materials.

$HClO_4$ Powerful oxidant. Can form explosive compounds with organic materials.

H_2SO_4 Reacts violently with water.

Selected Bases and Their Properties

Properties of Some Common, Commercially Available Bases

Name	Formula	Approximate Weight (Percent)	Approximate Molarity	Density (g/mL)	Volume Needed to Prepare 1.0 L of a 1 M Solution (mL)
Ammonia	NH_3	28	14.5	0.88	69.0
Sodium hydroxide	$NaOH$	47	17.6	1.50	56.7
Potassium hydroxide	KOH	45	11.7	1.46	85.5

Lide, D. R. (Ed.). (2000–2001). *CRC Handbook of Chemistry and Physics* (82nd ed.). CRC Press: Boca Raton, FL.

Safety Concerns

All bases, particularly when present in concentrated form, are potentially dangerous. They are especially hazardous if spilled into eyes. Bases require extensive washing, particularly of the eyes; at least 15 minutes of washing is needed, even for dilute solutions of bases. Always wear goggles, gloves, and an apron when handling concentrated solutions.

Ionization Constants of Selected Weak Acids at 25 °C

Acid	Formula	Molar Mass (g/mol)	K_{a1}	K_{a2}	K_{a3}
Acetic	CH_3COOH	60.05	1.8×10^{-5}		
Arsenic	H_3AsO_4	141.94	5.5×10^{-3}	1.7×10^{-7}	5.1×10^{-12}
Arsenious	H_3AsO_3	125.94	5.1×10^{-10}		
Ascorbic	$H_2C_6H_6O_6$	176.13	9.1×10^{-5}	2.0×10^{-12}	
Benzoic	C_6H_5COOH	122.13	6.3×10^{-5}		
Carbonic	H_2CO_3	62.03	4.5×10^{-7}	4.7×10^{-11}	
Chlorobenzoic	$C_6H_4ClCOOH$	156.58	1.1×10^{-4}		
Chlorous	$HClO_2$	68.46	1.1×10^{-2}		
Citric	$H_3C_6H_5O_7$	192.13	7.4×10^{-4}	1.7×10^{-5}	4.0×10^{-7}
Formic	$HCOOH$	46.03	1.8×10^{-4}		
Hydrocyanic	HCN	27.03	6.2×10^{-10}		
Hydrofluoric	HF	20.01	6.3×10^{-4}		
Hypobromous	$HOBr$	96.91	2.8×10^{-9}		
Hypochlorous	$HOCl$	52.46	4.0×10^{-8}		
Hypoiodous	HOI	143.91	3.2×10^{-11}		
Iodic	HIO_3	175.91	1.7×10^{-1}		
Malic	$H_2C_4H_4O_5$	134.09	4.0×10^{-4}	7.8×10^{-6}	
Malonic	$H_2C_3H_2O_4$	104.06	1.4×10^{-3}	2.0×10^{-6}	
Nitrous	HNO_2	47.02	5.6×10^{-4}		
Oxalic	$H_2C_2O_4$	90.04	5.6×10^{-2}	1.6×10^{-4}	
Periodic	HIO_4	191.94	2.3×10^{-2}		
Phenol	C_6H_5OH	94.11	1.0×10^{-10}		
Phosphoric	H_3PO_4	98.00	7.5×10^{-3}	6.2×10^{-8}	2.2×10^{-13}
Selenious	H_2SeO_3	128.97	2.4×10^{-3}	4.8×10^{-9}	
Sulfuric	H_2SO_4	98.08	Strong	1.0×10^{-2}	
Sulfurous	H_2SO_3	82.08	1.4×10^{-2}	6.3×10^{-8}	
Tartaric	$H_2C_4H_4O_6$	150.09	1.0×10^{-3}	4.6×10^{-5}	

Lide, D. R. (Ed.). (2000–2001). *CRC Handbook of Chemistry and Physics* (82nd ed.). CRC Press: Boca Raton, FL.

Dissociation Constants of Selected Weak Bases at 25 °C

Base	Formula	K_b
Ammonia	NH_3	1.8×10^{-5}
Ethylamine	$C_2H_5NH_2$	1.1×10^{-6}
Hydrazine	N_2H_4	1.3×10^{-6}
Hydroxylamine	NH_2OH	8.7×10^{-9}
Morphine	$C_{17}H_{19}NO_3$	1.6×10^{-6}
Pyridine	C_5H_5N	1.7×10^{-9}
Strychnine	$C_{21}H_{22}N_2O_2$	1.8×10^{-6}
Urea	$(NH_2)_2CO$	1.3×10^{-14}

Lide, D. R. (Ed.). (2000–2001). *CRC Handbook of Chemistry and Physics* (82nd ed.). CRC Press: Boca Raton, FL.

Thermodynamic Constants for Selected Compounds

APPENDIX G

This table lists standard enthalpies of formation ΔH_f°, standard entropies S°, and standard free energies of formation ΔG_f° for a variety of substances all at 25 °C (298.15 K) and 1 atm. The entries in the table are ordered by group number in the periodic table.

Note that the solution-phase entropies are not absolute entropies but are measured relative to the arbitrary standard that S° for H_3O^+(aq) is equal to 0. It is for this reason that some of them are negative.

Most of the thermodynamic data in these tables were taken from the *NBS Tables of Chemical Thermodynamic Properties* (1982). The data for organic compounds C_nH_m ($n > 2$) were taken from the *Handbook of Chemistry and Physics* (1981) (D. R. Lide, [Ed.]. [2000–2001]. *CRC Handbook of Chemistry and Physics,* [82nd ed]. CRC Press: Boca Raton, FL.)

	Substance	ΔH_f° (25 °C), kJ/mol	S° (25 °C), J/mol·K	ΔG_f° (25 °C), kJ/mol
	H(g)	217.96	114.60	203.26
	H_2(g)	0	130.57	0
	H^+(aq)	0	0	0
	H_3O^+(aq)	−285.83	69.91	−237.18
1A	Li(s)	0	29.12	0
	Li(g)	159.37	138.66	126.69
	Li^+(aq)	−278.49	13.4	−293.31
	LiH(s)	−90.54	20.01	−68.37
	Li_2O(s)	−597.94	37.57	−561.20
	LiF(s)	−615.97	35.65	−587.73
	LiCl(s)	−408.61	59.33	−384.39
	LiBr(s)	−351.21	74.27	−342.00
	LiI(s)	−270.41	86.78	−270.29
	Na(s)	0	51.21	0
	Na(g)	107.32	153.60	76.79
	Na^+(aq)	−240.12	59.0	−261.90
	Na_2O(s)	−414.22	75.06	−375.48
	NaOH(s)	−425.61	64.46	−379.53
	NaF(s)	−573.65	51.46	−543.51
	NaCl(s)	−411.15	72.13	−384.15
	NaBr(s)	−361.06	86.82	−348.98
	NaI(s)	−287.78	98.53	−286.06
	$NaNO_3$(s)	−467.85	116.52	−367.07
	Na_2S(s)	−364.8	83.7	−349.8
	Na_2SO_4(s)	−1387.08	149.58	−1270.23
	$NaHSO_4$(s)	−1125.5	113.0	−992.9
	Na_2CO_3(s)	−1130.68	134.98	−1044.49
	$NaHCO_3$(s)	−950.81	101.7	−851.1
	K(s)	0	64.18	0
	K(g)	89.24	160.23	60.62
	K^+(aq)	−252.38	102.5	−283.27
	KO_2(s)	−284.93	116.7	−239.4
	K_2O_2(s)	−494.1	102.1	−425.1
	KOH(s)	−424.76	78.9	−379.11
	KF(s)	−567.27	66.57	−537.77
	KCl(s)	−436.75	82.59	−409.16
	$KClO_3$(s)	−397.73	143.1	−296.25
	KBr(s)	−393.80	95.90	−380.66
	KI(s)	−327.90	106.32	−324.89
	$KMnO_4$(s)	−837.2	171.71	−737.7

	Substance	ΔH_f°(25 °C), kJ/mol	S° (25 °C), J/mol·K	ΔG_f°(25 °C), kJ/mol
	K_2CrO_4(s)	−1403.7	200.12	−1295.8
	$K_2Cr_2O_7$(s)	−2061.5	291.2	−1881.9
	Rb(s)	0	76.78	0
	Rb(g)	80.88	169.98	53.09
	Rb^+(aq)	−251.17	121.50	−283.98
	RbCl(s)	−435.35	95.90	−407.82
	RbBr(s)	−394.59	109.96	−381.79
	RbI(s)	−333.80	118.41	−328.86
	Cs(s)	0	85.23	0
	Cs(g)	76.06	175.49	49.15
	Cs^+(aq)	−258.28	133.05	−292.02
	CsF(s)	−553.5	92.80	−525.5
	CsCl(s)	−443.04	101.17	−414.55
	CsBr(s)	−405.81	113.05	−391.41
	CsI(s)	−346.60	123.05	−340.58
2A	Be(s)	0	9.50	0
	Be(g)	324.3	136.16	286.6
	BeO(s)	−609.6	14.14	−580.3
	Mg(s)	0	32.68	0
	Mg(g)	147.70	148.54	113.13
	Mg^{2+}(aq)	−466.85	−138.1	−454.8
	MgO(s)	−601.70	26.94	−569.45
	$MgCl_2$(s)	−641.32	89.62	−591.82
	$MgSO_4$(s)	−1284.9	91.6	−1170.7
	Ca(s)	0	41.42	0
	Ca(g)	178.2	154.77	144.33
	Ca^{2+}(aq)	−542.83	−53.1	−553.58
	CaH_2(s)	−186.2	42	−147.2
	CaO(s)	−635.09	39.75	−604.05
	CaS(s)	−482.4	56.5	−477.4
	$Ca(OH)_2$(s)	−986.09	83.39	−898.56
	CaF_2(s)	−1219.6	68.87	−1167.3
	$CaCl_2$(s)	−795.8	104.6	−748.1
	$CaBr_2$(s)	−682.8	130	−663.6
	CaI_2(s)	−533.5	142	−528.9
	$Ca(NO_3)_2$(s)	−938.39	193.3	−743.20
	CaC_2(s)	−59.8	69.96	−64.9
	$CaCO_3$(s, calcite)	−1206.92	92.9	−1128.84
	$CaCO_3$(s, aragonite)	−1207.13	88.7	−1127.80
	$CaSO_4$(s)	−1434.11	106.9	−1321.86
	$CaSiO_3$(s)	−1634.94	81.92	−1549.66
	$CaMg(CO_3)_2$(s)	−2326.3	155.18	−2163.4
	Sr(s)	0	52.3	0
	Sr(g)	164.4	164.51	130.9
	Sr^{2+}(aq)	−545.80	−32.6	−559.48
	$SrCl_2$(s)	−828.9	114.85	−781.1
	$SrCO_3$(s)	−1220.0	97.1	−1140.1
	Ba(s)	0	62.8	0
	Ba(g)	180	170.24	146
	Ba^{2+}(aq)	−537.64	9.6	−560.77
	BaO(s)	−582.0	70.3	−552.3
	$BaCl_2$(s)	−858.6	123.68	−810.4
	$BaCO_3$(s)	−1216.3	112.1	−1137.6
	$BaSO_4$(s)	−1473.2	132.2	−1362.3
3B	Sc(s)	0	34.64	0
	Sc(g)	377.8	174.68	336.06
	Sc^{3+}(aq)	−614.2	−255	−586.6
4B	Ti(s)	0	30.63	0
	Ti(g)	469.9	180.19	425.1
	TiO_2(s)	−944.7	50.33	−889.5
	$TiCl_4$(ℓ)	−804.2	252.3	−737.2
	TiCl4(g)	−763.2	354.8	−726.8

	Substance	ΔH_f°(25 °C), kJ/mol	S° (25 °C), J/mol·K	ΔG_f°(25 °C), kJ/mol
6B	Cr(s)	0	23.77	0
	Cr(g)	396.6	174.39	351.8
	Cr_2O_3(s)	−1139.7	81.2	−1058.1
	CrO_4^{2-}(aq)	−881.15	50.21	−727.75
	$Cr_2O_7^{2-}$(aq)	−1490.3	261.9	−1301.1
	W(s)	0	32.64	0
	W(g)	849.4	173.84	807.1
	WO_2(s)	−589.69	50.54	−533.92
	WO_3(s)	−842.87	75.90	−764.08
7B	Mn(s)	0	32.01	0
	Mn(g)	280.7	238.5	173.59
	Mn^{2+}(aq)	−220.75	−73.6	−228.1
	MnO(s)	−385.22	59.71	−362.92
	MnO_2(s)	−520.03	53.05	−465.17
	MnO_4^-(aq)	−541.4	191.2	−447.2
8B	Fe(s)	0	27.28	0
	Fe(g)	416.3	180.38	370.7
	Fe^{2+}(aq)	−89.1	−137.7	−78.9
	Fe^{3+}(aq)	−48.5	−315.9	−4.7
	FeO(s)	−272	—	—
	Fe_2O_3(s, hematite)	−824.2	87.40	−742.2
	Fe_3O_4(s, magnetite)	−1118.4	146.4	−1015.5
	$Fe(OH)_3$(s)	−569.0	88	−486.6
	FeS(s)	−100.0	60.29	−100.4
	$FeCO_3$(s)	−740.57	93.1	−666.72
	$Fe(CN)_6^{3-}$(aq)	561.9	270.3	729.4
	$Fe(CN)_6^{4-}$(aq)	455.6	95.0	695.1
	$Fe_2(SO_4)_3$	−2383.00	307.46	−2262.7
	Co(s)	0	30.04	0
	Co(g)	424.7	179.41	380.3
	Co^{2+}(aq)	−58.2	−113	−54.4
	Co^{3+}(aq)	92	−305	134
	CoO(s)	−237.94	52.97	−214.22
	$CoCl_2$(s)	−312.5	109.16	−269.8
	Ni(s)	0	29.87	0
	Ni(g)	429.7	182.08	384.5
	Ni^{2+}(aq)	−54.0	−128.9	−45.6
	NiO(s)	−239.7	37.99	−211.7
	Pt(s)	0	41.63	0
	Pt(g)	565.3	192.30	520.5
	$PtCl_6^{2-}$(aq)	−668.2	219.7	−482.7
1B	Cu(s)	0	33.15	0
	Cu(g)	338.32	166.27	298.61
	Cu^+(aq)	71.67	40.6	49.98
	Cu^{2+}(aq)	64.77	−99.6	65.49
	CuO(s)	−157.3	42.63	−129.7
	Cu_2O(s)	−168.6	93.14	−146.0
	Cu_2S(s)	−79.5	120.9	−86.2
	CuCl(s)	−137.2	86.2	−119.88
	$CuCl_2$(s)	−220.1	108.07	−175.7
	$CuSO_4$(s)	−771.36	109	−661.9
	$Cu(NH_3)_4^{2+}$(aq)	−348.5	273.6	−111.07
	Ag(s)	0	42.55	0
	Ag(g)	284.55	172.89	245.68
	Ag^+(aq)	105.58	72.68	77.11
	AgCl(s)	−127.07	96.2	−109.81
	$AgNO_3$(s)	−124.39	140.92	−33.48
	$Ag(NH_3)_2^+$(aq)	−111.29	245.2	−17.12
	Au(s)	0	47.40	0
	Au(g)	366.1	180.39	326.3

	Substance	ΔH_f°(25 °C), kJ/mol	S° (25 °C), J/mol·K	ΔG_f°(25 °C), kJ/mol
2B	$Zn(s)$	0	41.63	0
	$Zn(g)$	130.73	160.87	95.18
	$Zn^{2+}(aq)$	−153.89	−112.1	−147.06
	$ZnO(s)$	−348.28	43.64	−318.32
	ZnS(s, sphalerite)	−205.98	57.7	−201.29
	$ZnCl_2(s)$	−415.05	111.46	−369.43
	$ZnSO_4(s)$	−982.8	110.5	−871.5
	$Zn(NH_3)_4^{2+}(aq)$	−533.5	301	−301.9
	$CdCl_2(s)$	−391.5	115.3	−343.9
	$Hg(\ell)$	0	76.02	0
	$Hg(g)$	61.32	174.85	31.85
	$HgO(s)$	−90.83	70.29	−58.56
	$HgCl_2(s)$	−224.3	146.0	−178.6
	$Hg_2Cl_2(s)$	−265.22	192.5	−210.78
3A	$B(s)$	0	5.86	0
	$B(g)$	562.7	153.34	518.8
	$B_2H_6(g)$	35.6	232.00	86.6
	$B_5H_9(g)$	73.2	275.81	174.9
	$B_2O_3(s)$	−1272.77	53.97	−1193.70
	$H_3BO_3(s)$	−1094.33	88.83	−969.02
	$BF_3(g)$	−1137.00	254.01	−1120.35
	BF_4^-	−1574.9	180	−1486.9
	$BCl_3(g)$	−403.76	289.99	−388.74
	$BBr_3(g)$	−205.64	324.13	−232.47
	$Al(s)$	0	28.33	0
	$Al(g)$	326.4	164.43	285.7
	$Al^{3+}(aq)$	−531	−321.7	−485
	$Al(OH)_3(s)$	−1287	85.4	−1150
	$Al_2O_3(s)$	−1675.7	50.92	−1582.3
	$AlCl_3(s)$	−704.2	110.67	−628.8
	$Ga(s)$	0	40.88	0
	$Ga(g)$	277.0	168.95	238.9
	$Tl(s)$	0	64.18	0
	$Tl(g)$	182.21	180.85	147.44
4A	C(s, graphite)	0	5.74	0
	C(s, diamond)	1.895	2.377	2.900
	$C(g)$	716.682	157.99	671.29
	$CH_4(g)$	−74.81	186.15	−50.75
	$C_2H_2(g)$	226.73	200.83	209.20
	$C_2H_4(g)$	52.26	219.45	68.12
	$C_2H_6(g)$	−84.68	229.49	−32.89
	C_3H_6(g, propene)	20.41	267	62.74
	C_3H_6(g, cyclopropane)	53.3	237.4	104.39
	$C_3H_8(g)$	−103.85	269.91	−23.49
	n-$C_4H_{10}(g)$	−124.73	310.03	−15.71
	i-$C_4H_{10}(g)$	−131.60	294.64	−17.97
	C_4H_8O(ℓ, butyraldehyde)	−238.7	246.8	−119.2
	n-$C_5H_{12}(g)$	−146.44	348.40	−8.20
	$C_6H_6(g)$	82.93	269.2	129.66
	$C_6H_6(\ell)$	49.03	172.8	124.50
	$C_7H_5N_3O_6(s)$	−63.2	284.9	230.0
	$CO(g)$	−110.52	197.56	−137.15
	$CO_2(g)$	−393.51	213.63	−394.36
	$CO_2(aq)$	−413.80	117.6	−385.98
	$CS_2(\ell)$	89.70	151.34	65.27
	$CS_2(g)$	117.36	237.73	67.15
	$H_2CO_3(aq)$	−699.65	187.4	−623.08
	$HCO_3^-(aq)$	−691.99	91.2	−586.77
	$CO_3^{2-}(aq)$	−677.14	−56.9	−527.81
	$HCOOH(\ell)$	−424.72	128.95	−361.42
	$HCOOH(g)$	−378.61	248.74	−351.00

Substance	ΔH_f°(25 °C), kJ/mol	S° (25 °C), J/mol·K	ΔG_f°(25 °C), kJ/mol
$HCOOH(aq)$	−425.43	163	−372.3
$HCOO^-(aq)$	−425.55	92	−351.0
$CH_2O(g)$	−108.57	218.66	−102.55
$CH_3OH(\ell)$	−238.66	126.8	−166.35
$CH_3OH(g)$	−200.66	239.70	−162.01
$CH_3OH(aq)$	−245.93	133.1	−175.31
$H_2C_2O_4(s)$	−827.2	—	—
$HC_2O_4^-(aq)$	−818.4	149.4	−698.34
$C_2O_4^{2-}(aq)$	−825.1	45.6	−673.9
$CH_3COOH(\ell)$	−484.5	159.8	−390.0
$CH_3COOH(g)$	−432.25	282.4	−374.1
$CH_3COOH(aq)$	−485.76	178.7	−396.46
$CH_3COO^-(aq)$	−486.01	86.6	−369.31
$CH_3CHO(\ell)$	−192.30	160.2	−128.12
$C_2H_5OH(\ell)$	−277.69	160.7	−174.89
$C_2H_5OH(g)$	−235.10	282.59	−168.57
$C_2H_5OH(aq)$	−288.3	148.5	−181.64
$(CH_3)_2CO(\ell)$	−248.1	200.4	155.4
$CH_3OCH_3(g)$	−184.05	266.27	−112.67
$C_6H_{12}O_6(s)$	−1268	212	−910
$CF_4(g)$	−925	261.50	−879
$CCl_4(\ell)$	−135.44	216.40	−65.28
$CCl_4(g)$	−102.9	309.74	−60.62
$CHCl_3(g)$	−103.14	295.60	−70.37
$COCl_2(g)$	−218.8	283.53	−204.6
$CH_2Cl_2(g)$	−92.47	270.12	−65.90
$CH_3Cl(g)$	−80.83	234.47	−57.40
$CBr_4(s)$	79	357.94	67
$CH_3I(\ell)$	−15.5	163.2	13.4
$HCN(g)$	135.1	201.67	124.7
$HCN(aq)$	107.1	124.7	119.7
$CN^-(aq)$	150.6	94.1	172.4
$CH_3NH_2(g)$	−22.97	243.30	32.09
$CH_3NO_2(\ell)$	−113.1	171.76	−14.52
$CH_3NO_2(g)$	−74.73	274.42	−6.91
$CO(NH_2)_2(s)$	−333.51	104.49	−197.44
$Si(s)$	0	18.83	0
$Si(g)$	455.6	167.86	411.3
$SiC(s)$	−65.3	16.61	−62.8
SiO_2(s, quartz)	−910.94	41.84	−856.67
SiO_2(s, cristobalite)	−909.48	42.68	−855.43
$Ge(s)$	0	31.09	0
$Ge(g)$	376.6	335.9	167.79
Sn(s, white)	0	51.55	0
Sn(s, gray)	−2.09	44.14	0.13
$Sn(g)$	302.1	168.38	267.3
$SnO(s)$	−285.8	56.5	−256.9
$SnO_2(s)$	−580.7	52.3	−519.6
$Sn(OH)_2(s)$	−561.1	155	−491.7
$Pb(s)$	0	64.81	0
$Pb(g)$	195.0	161.9	175.26
$Pb^{2+}(aq)$	−1.7	10.5	−24.43
$PbBr_2(s)$	−277.02	161.5	−260.41
$PbCl_2(s)$	−359.20	136.40	−314.01
$PbCO_3(s)$	−699.1	130.96	−625.5
$PbF_2(s)$	−663.16	121.34	−619.65
$PbO(s)$	−218.99	66.5	−188.95
$PbO_2(s)$	−277.4	68.6	−217.36
$PbS(s)$	−100.4	91.2	−98.7
$PbI_2(s)$	−175.48	174.85	−173.64
$PbSO_4(s)$	−919.94	148.57	−813.21

	Substance	ΔH_f°(25 °C), kJ/mol	S° (25 °C), J/mol·K	ΔG_f°(25 °C), kJ/mol
5A	$N_2(g)$	0	191.50	0
	$N(g)$	472.70	153.19	455.58
	$NH_3(g)$	−46.11	192.34	−16.48
	$NH_3(aq)$	−80.29	111.3	−26.50
	$NH_4^+(aq)$	−132.51	113.4	−79.31
	$N_2H_4(\ell)$	50.63	121.21	149.24
	$N_2H_4(aq)$	34.31	138	128.1
	$NO(g)$	90.25	210.65	86.55
	$NO_2(g)$	33.18	239.95	51.29
	$NO_2^-(aq)$	−104.6	123.0	−32.2
	$NO_3^-(aq)$	−205.0	146.4	−108.74
	$N_2O(g)$	82.05	219.74	104.18
	$N_2O_4(g)$	9.16	304.18	97.82
	$N_2O_5(s)$	−43.1	178.2	113.8
	$NOBr(g)$	82.1	273.5	82.4
	$HNO_2(g)$	−79.5	254.0	−46.0
	$HNO_3(\ell)$	−174.10	155.49	−80.76
	$NH_4NO_3(s)$	−365.56	151.08	−184.02
	$NH_4Cl(s)$	−314.43	94.6	−202.97
	$(NH_4)_2SO_4(s)$	−1180.85	220.1	−901.90
	P(s, white, $\frac{1}{4}$ P_4)	0	41.09	0
	$P(g)$	314.64	163.08	278.28
	$P_2(g)$	144.3	218.02	103.7
	$P_4(g)$	58.91	279.87	24.47
	$PH_3(g)$	5.4	210.12	13.4
	$H_3PO_4(s)$	−1279.0	110.50	−1119.2
	$H_3PO_4(aq)$	−1288.34	158.2	−1142.54
	$H_2PO_4^-(aq)$	−1296.29	90.4	−1130.28
	$HPO_4^{2-}(aq)$	−1292.14	−33.5	−1089.15
	$PO_4^{3-}(aq)$	−1277.4	−222	−1018.7
	$POCl_3(g)$	−592.7	324.6	−545.2
	$PCl_3(g)$	−287.0	311.67	−267.8
	$PCl_5(g)$	−374.9	364.47	−305.0
	As(s, gray)	0	35.1	0
	$As(g)$	302.5	174.10	261.0
	$As_2(g)$	222.2	239.3	171.9
	$As_4(g)$	143.9	314	92.4
	$AsH_3(g)$	66.44	222.67	68.91
	$As_4O_6(s)$	−1313.94	214.2	−1152.53
	Sb(s)	0	45.69	0
	Sb(g)	262.3	180.16	222.1
	Bi(s)	0	56.74	0
	Bi(g)	207.1	186.90	168.2
6A	$O_2(g)$	0	205.03	0
	O(g)	249.17	160.95	231.76
	$O_3(g)$	142.7	238.82	163.2
	$OH^-(aq)$	−229.99	−10.75	−157.24
	$H_2O(\ell)$	−285.83	69.91	−237.18
	$H_2O(g)$	−241.82	188.72	−228.59
	$H_2O_2(\ell)$	−187.78	109.6	−120.42
	$H_2O_2(aq)$	−191.17	143.9	−134.03
	S(s, rhombic, $\frac{1}{8}$ S_8)	0	31.80	0
	S(s, monoclinic)	0.30	32.6	0.096
	S(g)	278.80	167.71	238.28
	$S_8(g)$	102.30	430.87	49.66
	$S^{2-}(aq)$	33.1	−14.6	85.8
	$H_2S(g)$	−20.63	205.68	−33.56
	$H_2S(aq)$	−39.7	121	−27.83
	$HS^-(aq)$	−17.6	62.8	12.08
	SO(g)	6.26	221.84	−19.87
	$SO_2(g)$	−296.83	248.11	−300.19
	$SO_3(g)$	−395.72	256.65	−371.08
	$SO_3(\ell)$	−438	95.6	−368

	Substance	ΔH_f° (25 °C), kJ/mol	S° (25 °C), J/mol·K	ΔG_f° (25 °C), kJ/mol
	H_2SO_3(aq)	−608.81	232.2	−537.81
	HSO_3^-(aq)	−626.22	139.7	−527.73
	SO_3^{2-}(aq)	−635.5	−29	−486.5
	H_2SO_4(ℓ)	−813.99	156.90	−690.10
	HSO_4^-(aq)	−887.34	131.8	−755.91
	SO_4^{2-}(aq)	−909.27	20.1	−744.53
	SF_6(g)	−1209	291.71	−1105.4
	Se(s, black)	0	42.44	0
	Se(g)	227.07	176.61	187.06
7A	F_2(g)	0	202.67	0
	F(g)	78.99	158.64	61.94
	F^-(aq)	−332.63	−13.8	−278.79
	HF(g)	−271.1	173.67	−273.2
	HF(aq)	−320.08	88.7	−296.82
	Cl_2(g)	0	222.96	0
	Cl(g)	121.68	165.09	105.71
	Cl^-(aq)	−167.16	56.5	−131.23
	HCl(g)	−92.31	186.80	−95.30
	ClO^-(aq)	−107.1	42	−36.8
	ClO_2(g)	102.5	256.73	120.5
	ClO_2^-(aq)	−66.5	101.3	17.2
	ClO_3^-(aq)	−103.97	162.3	−7.95
	ClO_4^-(aq)	−129.33	182.0	−8.52
	Cl_2O(g)	80.3	266.10	97.9
	HClO(aq)	−120.9	142	−79.9
	ClF_3(g)	−163.2	281.50	−123.0
	Br_2(ℓ)	0	152.23	0
	Br_2(g)	30.91	245.35	3.14
	Br_2(aq)	−2.59	130.5	3.93
	Br(g)	111.88	174.91	82.41
	Br^-(aq)	−121.55	82.4	−103.96
	HBr(g)	−36.40	198.59	−53.43
	BrO_3^-(aq)	−67.07	161.71	18.60
	I_2(s)	0	116.14	0
	I_2(g)	62.44	260.58	19.36
	I_2(aq)	22.6	137.2	16.40
	I(g)	106.84	180.68	70.28
	I^-(aq)	−55.19	111.3	−51.57
	I_3^-(aq)	−51.5	239.3	−51.4
	HI(g)	26.48	206.48	1.72
	ICl(g)	17.78	247.44	−5.44
	IBr(g)	40.84	258.66	3.71
8A	He(g)	0	126.04	0
	Ne(g)	0	146.22	0
	Ar(g)	0	154.73	0
	Kr(g)	0	163.97	0
	Xe(g)	0	169.57	0
	XeF_4(s)	−261.5	—	—

APPENDIX H

Standard Reduction Potentials at 25 °C

Half-reaction	$E°$, V
$Ag^+(aq) + e^- \rightarrow Ag(s)$	+0.800
$AgBr(s) + e^- \rightarrow Ag(s) + Br^-(aq)$	+0.095
$AgCl(s) + e^- \rightarrow Ag(s) + Cl^-(aq)$	+0.222
$Ag_2CrO_4(s) + 2e^- \rightarrow 2Ag(s) + CrO_4^{2-}(aq)$	+0.446
$AgI(s) + e^- \rightarrow Ag(s) + I^-(aq)$	−0.151
$Ag_2O(s) + H_2O(\ell) + 2e^- \rightarrow 2Ag(s) + 2OH^-(aq)$	0.342
$Al^{3+}(aq) + 3e^- \rightarrow Al(s)$	−1.66
$H_3AsO_4(aq) + 2H^+(aq) + 2e^- \rightarrow H_3AsO_3(aq) + H_2O(\ell)$	+0.559
$Ba^{2+}(aq) + 2e^- \rightarrow Ba(s)$	−2.90
$Br_2(\ell) + 2e^- \rightarrow 2Br^-(aq)$	+1.06
$BrO_3^-(aq) + 6H^+(aq) + 5e^- \rightarrow \frac{1}{2}Br_2(\ell) + 3H_2O(\ell)$	+1.52
$Ca^{2+}(aq) + 2e^- \rightarrow Ca(s)$	−2.87
$2CO_2(g) + 2H^+(aq) + 2e^- \rightarrow H_2C_2O_4(aq)$	−0.49
$Cd^{2+}(aq) + 2e^- \rightarrow Cd(s)$	−0.403
$Cd(OH)_2(s) + 2e^- \rightarrow Cd(s) + 2OH^-(aq)$	−0.83
$Ce^{4+}(aq) + e^- \rightarrow Ce^{3+}(aq)$	+1.61
$Cl_2(g) + 2e^- \rightarrow 2Cl^-(aq)$	+1.36
$HClO(aq) + H^+(aq) + 2e^- \rightarrow Cl^-(aq) + H_2O(\ell)$	+1.63
$ClO^-(aq) + H_2O(\ell) + 2e^- \rightarrow Cl^-(aq) + 2OH^-(aq)$	+0.89
$ClO_3^-(aq) + 6H^+(aq) + 5e^- \rightarrow \frac{1}{2}Cl_2(g) + 3H_2O(\ell)$	+1.47
$Co^{2+}(aq) + 2e^- \rightarrow Co(s)$	−0.28
$Co^{3+}(aq) + e^- \rightarrow Co^{2+}(aq)$	+1.83
$Cr^{3+}(aq) + e^- \rightarrow Cr^{2+}(aq)$	−0.41
$Cr^{2+}(aq) + 2e^- \rightarrow Cr(s)$	−0.91
$Cr_2O_7^{2-}(aq) + 14H^+(aq) + 6e^- \rightarrow 2Cr^{3+}(aq) + 7H_2O(\ell)$	+1.33
$CrO_4^{2-}(aq) + 4H_2O(\ell) + 3e^- \rightarrow Cr(OH)_3(s) + 5OH^-(aq)$	−0.13
$Cu^{2+}(aq) + 2e^- \rightarrow Cu(s)$	+0.34
$Cu^{2+}(aq) + e^- \rightarrow Cu^+(aq)$	+0.153
$Cu^+(aq) + e^- \rightarrow Cu(s)$	+0.521
$F_2(g) + 2e^- \rightarrow 2F^-(aq)$	+2.87
$Fe^{2+}(aq) + 2e^- \rightarrow Fe(s)$	−0.44
$Fe^{3+}(aq) + e^- \rightarrow Fe^{2+}(aq)$	+0.771
$Fe(CN)_6^{3-}(aq) + e^- \rightarrow Fe(CN)_6^{4-}(aq)$	+0.358
$2H^+(aq) + 2e^- \rightarrow H_2(g)$	0.000
$H_2(g) + 2e^- \rightarrow 2H^-(aq)$	−2.23
$2H_2O(\ell) + 2e^- \rightarrow H_2(g) + 2OH^-(aq)$	−0.83
$HO_2^-(aq) + H_2O(\ell) + 2e^- \rightarrow 3OH^-(aq)$	+0.88
$H_2O_2(aq) + 2H^+(aq) + 2e^- \rightarrow 2H_2O(\ell)$	+1.776
$Hg^{2+}(aq) + 2e^- \rightarrow Hg(\ell)$	+0.85
$2Hg^{2+}(aq) + 2e^- \rightarrow Hg_2^{2+}(aq)$	+0.92
$Hg_2Br_2(s) + 2e^- \rightarrow 2Hg(\ell) + 2Br^-(aq)$	+0.1396
$Hg_2Cl_2(s) + 2e^- \rightarrow 2Hg(\ell) + 2Cl^-(aq)$	+0.2682
$I_2(g) + 2e^- \rightarrow 2I^-(aq)$	+0.54
$IO_3^-(aq) + 6H^+(aq) + 5e^- \rightarrow \frac{1}{2}I_2(aq) + 3H_2O(\ell)$	+1.195
$K^+(aq) + e^- \rightarrow K(s)$	−2.93
$Li^+(aq) + e^- \rightarrow Li(s)$	−3.045
$Mg^{2+}(aq) + 2e^- \rightarrow Mg(s)$	−2.37
$Mn^{2+}(aq) + 2e^- \rightarrow Mn(s)$	−1.18
$MnO_4^-(aq) + 8H^+(aq) + 5e^- \rightarrow Mn^{2+}(aq) + 4H_2O(\ell)$	+1.51
$MnO_4^-(aq) + 2H_2O(\ell) + 3e^- \rightarrow MnO_2(s) + 4OH^-(aq)$	+0.59
$HNO_2(aq) + H^+(aq) + e^- \rightarrow NO(g) + H_2O(\ell)$	+0.983
$N_2(g) + 2H_2O(\ell) + 4H^+(aq) + 2e^- \rightarrow 2NH_3OH^+(aq)$	−1.87
$N_2(g) + 4H_2O(\ell) + 2e^- \rightarrow 2NH_2OH(aq) + 2OH^-(aq)$	−3.17
$NO_3^-(aq) + 4H^+(aq) + 3e^- \rightarrow NO(g) + 2H_2O(\ell)$	+0.96

Half-reaction	$E°$, V
$Na^+(aq) + e^- \rightarrow Na(s)$	−2.714
$Ni^{2+}(aq) + 2e^- \rightarrow Ni(s)$	−0.25
$NiO(OH)(s) + H_2O(\ell) + e^- \rightarrow Ni(OH)_2(s) + OH^-(aq)$	+0.52
$O_2(g) + 2H_2O(\ell) + 4e^- \rightarrow 4OH^-(aq)$	+0.40
$O_2(g) + 4H^+(aq) + 4e^- \rightarrow 2H_2O(\ell)$	+1.23
$O_2(g) + 2H^+(aq) + 2e^- \rightarrow H_2O_2(aq)$	+0.68
$Pb^{2+}(aq) + 2e^- \rightarrow Pb(s)$	−0.126
$PbSO_4(s) + H^+(aq) + 2e^- \rightarrow Pb(s) + HSO_4^-(aq)$	−0.356
$PbO_2(s) + HSO_4^-(aq) + 3H^+(aq) + 2e^- \rightarrow PbSO_4(s) + 2H_2O(\ell)$	+1.685
$S(s) + 2H^+(aq) + 2e^- \rightarrow H_2S(g)$	+0.14
$H_2SO_3(aq) + 4H^+(aq) + 4e^- \rightarrow S(s) + 3H_2O(\ell)$	+0.45
$HSO_4^-(aq) + 3H^+(aq) + 2e^- \rightarrow H_2SO_3(s) + H_2O(\ell)$	+0.17
$Sn^{2+}(aq) + 2e^- \rightarrow Sn(s)$	−0.14
$Sn^{4+}(aq) + 2e^- \rightarrow Sn^{2+}(aq)$	+0.15
$VO_2^+(aq) + 2H^+(aq) + e^- \rightarrow VO^{2+}(aq) + H_2O(\ell)$	+1.00
$Zn^{2+}(aq) + 2e^- \rightarrow Zn(s)$	−0.76
$Zn(OH)_2(s) + 2e^- \rightarrow Zn(s) + 2OH^-$	−1.25

APPENDIX I

Glossary

α-Helix—the spiral structure adopted by a peptide chain that is held together by hydrogen bonding between amide groups in the same region of the chain. *(22.5)*

Absorption spectrum—a graph of the quantity of light a sample absorbs as a function of wavelength. *(19.4)*

Accuracy—the agreement of the measured value to a true or accurately known value of the same quantity. *(1.3)*

Acid—a substance that provides the hydrogen cation in water solution (see also *Lewis acids* and *Brønsted–Lowry acids*). *(3.1)*

Actinides—the elements in period 7 after actinium (Ac); thorium (Th) to lawrencium (Lr). *(2.5)*

Activated complex—an unstable arrangement of atoms at the highest energy point on the reaction coordinate. *(13.4)*

Activation energy (E_a)—the minimum collision energy required for a reaction to occur; equal to the difference in energy between the reactants and the activated complex. *(13.4)*

Actual yield—the quantity of product isolated when a chemical reaction occurs. *(3.5)*

Addition polymer—a polymer chain formed from monomeric units with no loss of atoms. *(22.4)*

Addition reaction—the combination of two or more substances to form one new substance. *(22.2)*

Adduct—the product of a Lewis acid–base addition reaction. *(15.9)*

Adhesive forces—the attractions that molecules of one substance exert on those of a different substance. *(11.5)*

Alcohol—an organic compound that contains the —OH (hydroxyl) functional group. *(22.3)*

Aldehyde—an organic compound that contains the $-\overset{\text{O}}{\overset{\|}{\text{C}}}-\text{H}$ functional group. *(22.3)*

Alkali metals—the metallic elements in Group 1A (1): Li, Na, K, Rb, Cs, Fr. *(2.5)*

Alkaline earth metals—the elements in Group 2A (2): Be, Mg, Ca, Sr, Ba, Ra. *(2.5)*

Alkane—a hydrocarbon that contains no multiple bonds or rings. *(22.1)*

Alkene—an unsaturated hydrocarbon that contains one or more carbon–carbon double bonds. *(22.2)*

*The number in parentheses is the number of the section in which the term is defined.

Alkyl group—an alkane from which one hydrogen atom has been removed. *(22.1)*

Alkyne—an unsaturated hydrocarbon that contains one or more carbon–carbon triple bonds. *(22.2)*

Allotropes—structurally distinct forms of an element in the same physical state that exhibit different chemical and physical properties. *(11.6)*

Alloy—a mixture of a metal and one or more additional elements, often a second metal. *(1.2)*

Alpha particle—a high-energy helium nucleus produced by the radioactive decay of some atoms. *(21.1)*

Amide—an organic compound that contains the —C(O)NR_2 functional group, where R is an organic substituent or hydrogen. *(22.3)*

Amine—a derivative of ammonia in which one or more of the hydrogen atoms is replaced with an organic substituent (R). Amines with the formula RNH_2 are primary amines, R_2NH are secondary amines, and R_3N are tertiary amines. *(22.3)*

Amino acid—an organic compound that contains both a carboxyl group (—CO_2H) and an amino group (—NH_2). *(22.3)*

Amorphous solid—a solid that lacks the long-range order of a crystalline solid. *(11.6)*

Amphoteric—a type of species that can act as either an acid or a base. *(15.1, 16.7)*

Amplitude—the height of a wave. *(7.1)*

Analyte—a substance whose concentration is being determined. *(16.1)*

Analytical concentration—the total concentration of all protonated and deprotonated forms of the same species. *(15.5)*

Angular momentum quantum number, ℓ—the quantum number that describes the shape of the electron probability function in an atom. The allowed values of this quantum number are zero and all positive whole numbers up to $(n - 1)$. *(7.4)*

Anion—a negatively charged ion. *(2.3)*

Anode—the electrode in a cell at which oxidation occurs. *(18.3)*

Anodic protection—reducing corrosion by the intentional oxidation of a metal under carefully controlled conditions, to form a thin, adhering layer of oxide on the surface of the metal. The protective layer can be generated either chemically or electrochemically. *(18.9)*

Antibonding molecular orbital—an orbital that reduces the electron density in the region between the atoms in the molecule. *(10.5)*

Aqueous solution—solutions with water as the solvent. *(4.1)*

Aromatic hydrocarbon—a compound that contains one or more benzene rings. *(22.2)*

Arrhenius acid—a species that increases the hydrogen ion concentration when dissolved in water. *(15.0)*

Arrhenius base—a species that increases the hydroxide ion concentration when dissolved in water. *(15.0)*

Arrhenius equation—the relation between rate constant and temperature. *(13.4)*

Arrhenius plot—a graph of the natural logarithm of rate constant against 1/temperature (in kelvin). *(13.4)*

Aryl group—an aromatic ring that is being considered as a substituent. *(22.2)*

Atom—the smallest unit of an element that has all the properties of that element. *(2.1)*

Atomic mass (atomic weight)—the mass in atomic mass units of one atom of an element. It is an average mass that reflects the natural isotopic distribution of the element. *(2.4)*

Atomic mass unit (u)—the base unit of a mass scale that defines one unit (u) as one-twelfth the mass of a single atom of ^{12}C. *(2.4)*

Atomic number—the number of protons in the nucleus of an atom. It has the symbol Z. *(2.3)*

Atomic orbital—a wave function of an electron in an atom that has assigned values for all three of the quantum numbers: n, ℓ, and m_ℓ. *(7.4)*

Atomic radius—half the distance between adjacent atoms of the same element in a molecule. *(8.3)*

Aufbau principle—the process by which multielectron atoms have their electrons assigned to shells and subshells. *(7.6)*

Autoionization of water—the equilibrium that describes the ionization of water. *(15.2)*

$$2H_2O(\ell) \rightleftharpoons H_3O^+(aq) + OH^-(aq)$$

$$K_w = [H_3O^+][OH^-]$$

Average rate—the rate measured over an interval of time. *(13.1)*

Avogadro's law—at constant pressure and temperature, the volume of a gas sample is proportional to the number of moles of gas present (V = constant $\times$ n). *(6.2)*

Avogadro's number—the number of units in 1 mol, 6.022×10^{23} units/mol. *(3.2)*

Azeotrope—a solution that has the composition of the constant boiling mixture in which the compositions of the vapor and liquid phases are identical. *(12.6)*

β-Pleated sheet—the flattened structure adopted by a protein that is held together by hydrogen bonding between amide groups in different chains or in different sections of the same chain. *(22.5)*

Base—a substance that provides hydroxide anion in water (see also *Lewis base* and *Brønsted–Lowry bases*). *(3.1)*

Base unit—any of the defined quantities in the SI. *(1.4)*

Battery—a voltaic cell that is used as a portable source of energy, which comes from chemical reactions. Batteries can also consist of two or more voltaic cells with the negative electrode of each cell connected to the positive electrode of the next cell. *(18.7)*

Beta particle—a high-energy electron emitted by the nucleus of some radioactive atoms. *(21.1)*

Bimolecular—an elementary step that involves the collision of two species. *(13.6)*

Binary acid—an acid compound composed of hydrogen and one other element, nearly always a nonmetal. *(15.8)*

Binary compound—a compound composed of only two elements. *(2.8)*

Biochemistry—the study of the chemical systems in living organisms. *(22.0)*

Body-centered cubic (BCC)—the cubic structure with identical atoms at the corners and the center of the unit cell. *(11.7)*

Boiling point—the temperature at which the vapor pressure of a liquid is equal to the applied pressure. *(11.2)*

Boiling-point elevation—a colligative property of solutions described by the equation

$$\Delta T_b = mk_b,$$

where ΔT_b = (boiling point of solution − boiling point of solvent), m is the molal concentration of the solute particles, and k_b is the boiling-point elevation constant characteristic of the solvent. *(12.4)*

Bond dissociation energy or bond energy—the energy required to break one mole of bonds in a gaseous species. *(9.8)*

Bond length—the distance between the nuclei of two bonded atoms in a molecule. *(9.3)*

Bond order—the number of electron pairs that are shared between two atoms. *(9.3)*

Bonded-atom lone-pair arrangement—the orientation of valence-shell electron pairs that maximizes the distances between regions of electron density about a central atom. *(10.1)*

Bonding molecular orbital—an orbital that concentrates the electron density between the atoms in the molecule. *(10.5)*

Bonding pairs—pairs of electrons shared between two atoms. *(9.3)*

Boyle's law—at constant temperature, the volume of a gas sample is inversely proportional to the pressure ($P \times V = constant$). *(6.2)*

Bragg equation—the relation between the angle of diffraction of x rays and the distance between layers of atoms in a crystalline solid:

$$n\lambda = 2d\sin\theta,$$

where λ is the wavelength of the x rays used, d is the distance between layers of atoms in the crystal, θ is the angle between the x-ray beam and the layers of atoms, and n is a positive whole number called the *order.* *(11.7)*

Brønsted–Lowry acid—a species that can donate a proton. *(15.1)*

Brønsted–Lowry base—a species that can accept a proton. *(15.1)*

Buffer—a system that resists changes in pH. Most buffers are solutions that contain a weak acid–base conjugate pair. *(16.3)*

Buffer capacity—the amount of strong acid or base needed to change the pH of 1 L buffer solution by 1 unit. *(16.3)*

Buret—a device calibrated to measure the volume of added liquid in a titration. *(4.4)*

Calorimeter—the device used to measure heat. *(5.3)*

Calorimetry—the measurement of the heat absorbed or released when a chemical or physical change occurs. *(5.3)*

Capillary action—the change in the height of a liquid in a capillary caused by differences in cohesive and adhesive forces. *(11.5)*

Carbohydrate—a polyhydroxyl aldehyde or ketone, or substances that react with water to yield such a compound. *(22.6)*

Carbonyl group—the C═O functional group. *(22.3)*

Carboxylate group—the $—CO_2^-$ functional group. *(22.3)*

Carboxylic acid—an organic compound that contains the $—CO_2H$ functional group, a carboxyl group. *(22.3)*

Catalyst—a substance that increases the rate of reaction but is not consumed in the reaction. *(13.5)*

Catenation—the formation of chains or rings of like atoms. *(22.0)*

Cathode—the electrode in a cell at which reduction occurs. *(18.3)*

Cathodic protection—protection of a metallic object from corrosion by placing it in electrical contact with a second, more reactive metal. *(18.9)*

Cation—a positively charged ion. *(2.3)*

Cell potential—the voltage difference between the two electrodes in a voltaic cell. *(18.4)*

Central atom—an atom bonded to two or more other atoms. *(9.3)*

Ceramics—nonmetallic, solid materials that are hard, resistant to heat, and chemically inert. *(20.4)*

Chain-growth polymer (addition polymer)—a polymer formed from monomeric units with no loss of atoms. *(22.4)*

Change in enthalpy, ΔH—the heat absorbed or given off by the system under constant pressure. *(5.1)*

Charles's law—at constant pressure, the volume of a sample of gas is proportional to the absolute temperature ($V = constant \times T$). *(6.2)*

Chelate—a coordination complex that contains one or more chelating ligands. *(19.2)*

Chelating ligand—a ligand that simultaneously bonds to the same metal ion through two or more different atoms. *(19.2)*

Chemical bonds—the forces that hold the atoms together in substances. *(9.0)*

Chemical change—a process in which one or more new substances are produced. *(1.2)*

Chemical energy—a form of potential energy derived from the forces that hold the atoms together. *(5.1)*

Chemical equation—an equation that describes the identities and relative amounts of reactants and products in a chemical reaction. *(3.1)*

Chemical kinetics—the study of the rates of chemical reactions. *(13.0)*

Chemical nomenclature—the organized system for the naming of substances. *(2.8)*

Chemical property—the tendency to react and form new substances. *(1.2)*

Chemical thermodynamics—the study of the energetics of chemical reactions. *(17.0)*

Chemistry—the study of matter and its interactions with other matter and with energy. *(1.1)*

Chiral molecule or ion—a molecule or ion that has a mirror image structure that cannot be superimposed on the original. *(19.3)*

Close packing—the most efficient packing of spheres, which has the smallest empty space (26%). Each sphere is in contact with 12 other spheres. *(11.7)*

Closed system—see **System, closed.**

Coefficient—the number of units of each substance in the chemical equation. *(3.1)*

Cohesive forces—the attraction of molecules for other molecules of the same substance. *(11.5)*

Colligative properties—those properties of a solution that are proportional to the concentration of solute particles. *(12.4)*

Collision frequency (Z)—the number of collisions per second. *(13.4)*

Combined gas law—an equation that describes the relation of the volume (V), pressure (P), and temperature (T) of a gas sample undergoing change *(6.2)*:

$$\frac{P_1V_1}{T_1} = \frac{P_2V_2}{T_2}$$

Combustion analysis—a method that determines the quantity of carbon and hydrogen in a sample of an organic compound. *(3.3)*

Combustion reaction—the process of burning. Examples are the reactions of organic compounds with excess oxygen to yield carbon dioxide and water. *(3.1)*

Common ion effect—the term used to describe the effect of adding a solute to a solution that contains an ion in common. *(14.7)*

Complete ionic equation—the equation that shows separately all species, both ions and molecules, as they are present in solution. *(4.1)*

Complex—the species formed by the Lewis acid–base reaction of a metal atom or ion with ligands. *(16.8)*

Compound—a substance that can be decomposed into its elements by chemical processes. *(1.2)*

Concentration—the amount of solute in a given quantity of that solution. *(4.2)*

Condensation—the conversion of a gas to a liquid. *(11.2)*

Condensation polymer—a polymer formed by a reaction that eliminates a small molecule each time a monomer is linked to the polymer chain. *(22.4)*

Condensation reaction—a reaction that joins two molecules with the elimination of a small molecule such as water. *(22.3)*

Condensed phase—the solid and liquid states of matter. Any phase that is very resistant to volume changes. *(6.1)*

Conduction band—a partially or completely empty band of orbitals in solids that results in high electrical conductivity. *(20.4)*

Conformers—different arrangements of atoms in a molecule that are caused by rotations about single bonds. *(22.1)*

Conjugate acid–base pair—two species that differ by the presence or absence of a proton. The protonated species is the acid; the deprotonated species is the base. *(15.1)*

Conservation of energy (law)—a law stating that the total energy of the universe—the system plus the surroundings—is constant during a chemical or physical change. *(5.1)*

Contact process—a process for the production of sulfuric acid from sulfur dioxide, oxygen, and water. *(20.6)*

Continuous spectrum—a spectrum in which all wavelengths of light are present. This spectrum is characteristic of the light emitted by a heated solid. *(7.2)*

Conversion factor—a fraction in which the numerator and denominator express the same quantity in different units. *(1.4)*

Coordinate covalent bond—a covalent bond in which both electrons of the bonding pair have come from one atom. *(15.9)*

Coordination compound or complex—a species that contains a metal ion bound to ligands by Lewis acid–base interactions. *(19.2)*

Coordination isomers—compounds that contain the same numbers and kinds of atoms but have different Lewis bases directly bonded to the metal ion; for example, $[Co(NH_3)_4Cl_2]Br$ and $[Co(NH_3)_4ClBr]Cl$ are coordination isomers. *(19.3)*

Coordination number—the number of nearest neighbors an atom, ion, or molecule has in a crystalline solid. The largest possible coordination number of uniform-sized spheres is 12, and it occurs in both of the closest packing arrays *(11.7)*; also the number of donor atoms bonded to a single metal ion in a coordination complex. *(19.2)*

Copolymer—a polymer formed from the combination of many units of more than one type of monomer. *(22.4)*

Core electrons—The inner shell electrons that are not in the valence shell. *(8.1)*

Corrosion—the oxidation of a metal to produce compounds of the metal through interaction with its environment. *(18.9)*

Covalent bond—the bond that arises from atoms sharing electron pairs. *(9.3)*

Covalent network solid—a solid that consists of atoms held together in the crystal by covalent bonds. *(11.6)*

Critical geometry—the minimum size a fissionable material must maintain for a chain reaction to be critical. *(21.5)*

Critical mass—the minimum mass of the sample of fissionable matter needed for the chain reaction to be self-sustained; that is, on average, at least one neutron formed by a fission is used to induce a second fission. *(21.5)*

Critical pressure—the minimum pressure needed to cause the liquid state to exist up to the critical temperature. *(11.2)*

Critical temperature—the maximum temperature at which a substance can exist in the liquid phase. *(11.2)*

Crystal—See **crystalline solid.** *(2.7)*

Crystal field splitting, Δ—the separation in energy between the *d* orbitals caused by the electric field produced by the arrangement of ligands about the central metal ion. *(19.4)*

Crystal structure—the geometric arrangement of particles (atoms, ions, or molecules) in a crystalline solid. *(11.7)*

Crystalline solid—a solid in which the units that make up the substance are arranged in a regular repeating pattern. *(11.6)*

Cubic close packing—the close packing array with an *ABCABC*... stacking pattern of the layers. *(11.7)*

Cubic system—the crystal system in which $a = b = c$, and $\alpha = \beta = \gamma = 90°$. *(11.7)*

Cycloalkane—a saturated hydrocarbon that contains a ring of carbon atoms. *(22.1)*

***d*-Block elements**—the section of the periodic table containing the elements whose *d* subshell is being filled with electrons. *(8.1)*

Dalton's law of partial pressure—the total pressure of a mixture of gases is the sum of the partial pressure of the component gases. *(6.5)*

Degenerate orbitals—all orbitals in an atom that have identically the same energy. In the hydrogen atom and one-electron ions, all wave functions that belong to the same principal shell form a degenerate set. In many-electron atoms, all orbitals in the same subshell are degenerate. *(7.5)*

Delocalized bond—a bond in a molecule that is spread over more than two atoms. *(10.6)*

Delocalized molecular orbital—a molecular orbital that involves atomic orbitals on more than two atoms. *(10.6)*

Denaturation—the loss of structural organization of a protein. *(22.5)*

Density—the ratio of mass to volume. *(1.4)*

Deoxyribonucleic acid (DNA)—a polymer of nucleotide units located inside the chromosomes. DNA is the molecule that stores genetic information. *(22.7)*

Deposition—the direct conversion of a gas to a solid. Deposition is the reverse of sublimation. *(11.2)*

Derived unit—a unit that is composed of combinations of base units. *(1.4)*

Diamagnetic matter—matter that is repelled by a magnetic field. All electrons are paired in diamagnetic materials. *(19.4)*

Diatomic molecule—a molecule that is composed of two atoms. *(2.6)*

Differential form of the rate law—an expression that relates the ratio of *differences* in concentration and time ($\Delta[A]/\Delta t$) to the concentrations of the species. *(13.3)*

Diffusion—the mixing of particles caused by motion. *(6.7)*

Dipole moment—the magnitudes of separated charges times the distance between the charges. *(9.4, 10.2)*

Dipole–dipole attractions—the intramolecular forces that arise from electrostatic attractions between molecular dipoles. *(11.4)*

Dipole-induced dipole attraction—the intramolecular force that arises between a polar molecule and an induced dipole moment in a nonpolar molecule. *(11.4)*

Disaccharide—the product from the condensation of two monosaccharides with the elimination of water. *(22.6)*

Dissociation—the separation of an ionic solid into individual cations and anions when dissolved in a solvent (generally water). *(2.9)*

Distillation—the separation of components based on differences in volatility. It consists of two steps: (1) the mixture is heated to convert the liquid into a vapor, its gaseous form; and (2) the vapor is condensed to a liquid in a different container. *(1.4)*

Donor atom—the atom in the ligand that donates an unshared pair of electrons to the metal forming a bond. *(19.2)*

Double bond—the bond formed by sharing two pairs of electrons between two atoms. *(9.3)*

Dry cell—a cell that consists of a zinc case that serves as the negative electrode, and a carbon rod in the center that serves as an inert positive electrode. *(18.7)*

Dynamic equilibrium—a situation in which two opposing changes occur at equal rates, so no *net* change is apparent. *(11.2)*

Effective nuclear charge—the weighted average of the nuclear charge that influences any particular electron in an atom, after correcting for the effect of shielding and inter-electronic repulsions. *(7.5, 8.3)*

Effective radiation dose—the product of the quality factor times the number of rads of radiation dose. *(21.6)*

Effusion—the passage of a gas through a small hole into an evacuated space. *(6.7)*

Electrochemistry—the study of the relation between chemical reactions and electricity. *(18.0)*

Electrode—(1) a metal or other electrical conductor that connects an electrochemical cell to the external circuit; (2) a half-cell. *(18.3)*

Electrode potentials—the voltages assigned to the individual half-cells in an electrochemical cell. *(18.4)*

Electrolysis—an otherwise nonspontaneous oxidation-reduction reaction that is caused by the passage of an electric current. *(18.8)*

Electrolyte—a substance that separates into ions when it dissolves in water. *(2.9)*

Electrolytic cell—an apparatus that consists of two electrodes immersed in a molten salt or electrolyte solution, which is used to perform electrolysis. *(18.8)*

Electromagnetic radiation—oscillating electric and magnetic fields that are both perpendicular to each other and to the direction of motion. *(7.1)*

Electromotive force (emf)—the electrical driving force that pushes the electrons released in the oxidation half-cell (the negative electrode) through the external circuit into the reduction half-cell (the positive electrode). The emf is an intensive property of a cell and is measured in volts. *(18.4)*

Electron—a small particle that has a mass of 9.11×10^{-31} kg and a charge of -1.602×10^{-19} coulombs. The relative charge is $1-$, and the relative mass is approximately zero (< 0.0006, compared to 1 for both the proton and neutron). *(2.2)*

Electron affinity—the energy change that accompanies the addition of an electron to a gaseous atom or ion. *(8.5)*

Electron capture—a mode of decay for some unstable nuclides in which the nucleus captures an electron, converting a proton in the nucleus into a neutron. *(21.1)*

Electron configuration—a notation to describe the number of electrons in each subshell of an atom or ion; for example, $1s^22s^22p^2$ is the electron configuration of the carbon atom. *(7.6)*

Electron configuration, abbreviated—an electron configuration that uses the configurations of the noble gases as shorthand. *(7.7)*

Electron configuration, anomalous—a ground-state electron configuration that does not strictly follow the expected order of filling of electron subshells. *(7.7)*

Electron pair (paired electrons)—two electrons with opposite spins that occupy the same orbital. *(7.6)*

Electron pairing energy (*P*)—the additional energy required for two electrons to occupy the same orbital, compared with the two electrons singly occupying separate degenerate orbitals. *(19.4)*

Electron shielding—the reduction of the effect of nuclear charge by inner electrons on the electrons in multielectron atoms. *(7.5)*

Electron spin quantum number, m_s—the quantum number that represents one of the two allowed spin states of an electron. The allowed values of m_s are $+\frac{1}{2}$ and $-\frac{1}{2}$ only. *(7.4)*

Electron-deficient molecule—a molecule for which the Lewis structure has fewer than eight electrons around any atom. *(9.7)*

Electronegativity—a measure of the ability of an atom to attract the shared electrons in a chemical bond. *(9.4)*

Electroplating—the electrolytic deposition of a thin uniform metal film on the surface of an object. It is possible to plate metal on any electrically conducting object. *(18.8)*

Electrorefining—an electrolytic process that converts an impure metal anode into a pure sample of the metal at the cathode. *(18.8)*

Element—a substance that cannot be decomposed into a simpler substance by normal chemical means. *(1.2)*

Elementary reactions—single molecular events that are summed to provide the overall reaction. *(13.6)*

Elementary step—an equation that describes a molecular-level interaction or collision. *(13.6)*

Empirical formula—gives the relative numbers of different atoms or ions in a substance, using the smallest whole numbers for subscripts. *(2.7)*

Enantiomers—molecules or ions that are nonsuperimposable mirror images of each other. *(19.3)*

Endothermic processes—reactions that absorb heat from the surroundings. *(5.1)*

Endpoint—the point in a titration at which the indicator changes color. *(4.4, 16.1)*

Enthalpy (*H*)—a measure of the total energy of the system at a given pressure and temperature *(5.2)*; heat content, $H = E + PV$. *(17.2)*

Enthalpy change (ΔH)—the change in the enthalpy that accompanies a physical or chemical process. *(5.2)*

Enthalpy of combustion—the enthalpy change that accompanies a combustion reaction. *(5.5)*

Enthalpy of formation—See **Standard enthalpy of formation.** *(5.5)*

Enthalpy of fusion—the enthalpy change that accompanies the conversion of one mole of a solid to the liquid phase. *(11.2)*

Enthalpy of solution—the enthalpy change that accompanies the dissolution of one mole of a solute. *(12.2)*

Enthalpy of sublimation—the enthalpy change for the conversion of one mole of a solid to the gaseous state. *(11.2)*

Enthalpy of vaporization—the enthalpy change that accompanies the conversion of one mole of a liquid to the gas phase. *(11.2)*

Entropy (*S*)—the property of a system that describes the amount of disorder. It is a state function. *(17.3)*

Enzyme—a biological compound that catalyzes a specific biochemical reaction. *(13.5)*

Equilibrium—a state in which the tendency of reactants to form products is balanced by the tendency of products to form reactants. *(14.1)*

Equilibrium constant—an experimentally measured value that describes the equilibrium condition of a reaction. See **law of mass action.** *(14.1)*

Equilibrium constant expression—the mathematical equation that relates the equilibrium constant to a term that contains the concentrations of reactants and products as determined by the law of mass action. *(14.1)*

Equivalence point—the point in a titration at which the reactants are present in stoichiometrically equivalent amounts. *(4.4, 16.1)*

Error—the difference between the measured result and the true value. *(1.3)*

Ester—an organic compound that contains the —CO_2R functional group, where R is an organic substituent. *(22.3)*

Ether—an organic compound that contains a C—O—C functional group. *(22.3)*

Evaporation (or vaporization)—the escape of molecules from the liquid phase into the gas phase. *(11.2)*

Excited state—a species in which one or more electrons occupy orbitals that leave lower-energy orbitals partially or completely vacant. *(7.6)*

Exothermic processes—reactions that release heat to the surroundings. *(5.1)*

Expanded valence-shell molecule—a molecule with more than eight electrons about an atom in a Lewis structure. *(9.7)*

Exponential notation (scientific notation)—a quantity expressed as the product of a number between 1 and 10 multiplied by a power of 10. *(1.3)*

Extensive property—a property that depends on the size of the specific sample that is under observation. Examples are mass and volume. *(1.2)*

***f*-Block elements**—the section of the periodic table containing sequence of the elements whose *f* subshell is being filled with electrons. *(8.1)*

Face-centered cubic (FCC)—the cubic unit cell with identical atoms at the corners and in the center of each of the six square faces. The FCC and the cubic close packing arrays are identical. *(11.7)*

Faraday constant (*F*)—the negative electrical charge on one mole of electrons, 96,485 C. *(18.5)*

First law of thermodynamics—energy can be neither created nor destroyed. In equation form: $\Delta E = q + w$. *(17.2)*

Formal charge—a charge assigned to each atom in a Lewis structure, obtained by assuming the shared electrons are divided equally between the bonded atoms. *(9.5)*

Formation reaction—a chemical reaction that makes one mole of a substance from its constituent elements in their standard states. *(5.5)*

Formula mass—the sum of the atomic masses of atoms in a formula. *(2.7)*

Fraction ionized, α—the concentration of the ionized form divided by the analytical concentration. *(15.5)*

Fractional distillation—a distillation process by which two or more volatile liquids are separated from each other. *(12.6)*

Frasch process—a method for melting underground deposits of sulfur and bringing them to the surface. *(20.6)*

Freezing-point depression—a colligative property of solutions described by the equation

$$\Delta T_f = mk_f,$$

where ΔT_f = (freezing point of solvent − freezing point of solution), m is the molal concentration of the solute particles, and k_f is the freezing-point depression constant characteristic of the solvent. *(12.4)*

Frequency (ν)—the number of waves that pass a fixed point in 1 second. The SI unit for frequency is s^{-1} and has been given the name hertz. *(7.1)*

Fuel cell—a voltaic cell in which the reactants are continuously supplied, and generally the products of the redox reaction are continuously removed. *(18.7)*

Functional group—an atom or small group of atoms in a molecule that undergoes characteristic reactions. *(22.3)*

Galvanic cell—see **Voltaic cell.**

Gamma ray—high-energy electromagnetic radiation originating from the radioactive decay of unstable nuclides. *(21.1)*

Gas—a fluid that has no definite shape or volume. *(6.1)*

Geometric isomers—stereoisomers that have the same number and kind of bonds but differ in the relative positions of the atoms. *(19.3)*

Gibbs free energy—$G = H - TS$. It is a state function. *(17.4)*

Graham's law—the rate of effusion of gases is inversely proportional to the square root of the molar mass. *(6.7)*

Gravimetric analysis—the selective precipitation of one component of a solution, followed by isolation and weighing of the precipitate, to determine the amount of that component. *(4.4)*

Ground state—the state of the atom or molecule in which the electron configuration has the lowest possible energy. *(7.2)*

Group—a column of the periodic table. *(2.5)*

Haber process—the direct synthesis of ammonia from nitrogen and hydrogen. *(20.5)*

Half-cell—the compartment in a voltaic cell in which either the reduction or oxidation half-reaction occurs. *(18.3)*

Half-life—the time needed for the concentration of a reactant to decrease to half its original value. *(13.3)*

Half-reaction—an equation describing either the oxidation or the reduction portion of a redox reaction, with the electrons explicitly shown. An oxidation half-reaction has electrons on the product side of the equation. The electrons are on the reactant side of the equation in a reduction half-reaction. *(18.2)*

Halogens—the elements in Group 7A (17): F, Cl, Br, I, At. *(2.5)*

Heat (*q*)—a form of energy transfer that produces a change in the thermal energy of matter. *(5.1)*

Heat capacity (*C*)—the quantity of heat required to increase the temperature of a sample by 1 kelvin (or 1 °C). *(5.3)*

Heisenberg uncertainty principle—it is not possible to know simultaneously and precisely both the position and momentum of a particle. *(7.3)*

Henderson–Hasselbalch equation—the relationship generally used to calculate the pH of a buffer solution. *(16.3)*

$$\text{pH} = \text{p}K_a + \log\frac{C_b}{C_a}$$

Henry's law—the solubility of a gas is directly proportional to its pressure above the solution at any given temperature:

$$C = k \times P,$$

where C is the concentration of the gaseous compound in solution, k is a proportionality constant that is characteristic of the particular solute and temperature, and P is the partial pressure of the solute in contact with the solution. *(12.3)*

Hertz (Hz)—the SI unit of frequency; 1 Hz = 1 s^{-1}. *(7.1)*

Hess's law—the change in enthalpy for an equation obtained by adding two or more thermochemical equations is the sum of the enthalpy changes of the equations that have been added. *(5.4)*

Heterogeneous catalyst—a catalyst that is in a different phase from that containing the reactants. *(13.5)*

Heterogeneous equilibrium—a chemical equilibrium in which the reactants and the products are present in more than one phase. *(14.5)*

Heterogeneous mixture—a mixture in which different parts have different properties and composition. *(1.2)*

Heteronuclear diatomic molecule—a molecule formed by one atom of each of two different elements. *(10.6)*

Hexagonal close packing—the close packing array with an *ABABA* . . . stacking pattern of the layers. *(11.7)*

High-spin (weak-field) complex—a coordination complex in which each of the five *d* orbitals is occupied by a single electron before two electrons occupy one of the low-energy orbitals. High-spin complexes are observed when $P > \Delta$. *(19.4)*

Homogeneous catalyst—a catalyst that is in the same phase as the reactants. *(13.5)*

Homogeneous mixture—a mixture in which all parts of the sample exhibit identical properties. *(1.2)*

Homonuclear diatomic molecule—a molecule formed by two atoms of the same element. *(10.5)*

Homopolymer—a polymer formed from the combination of many units of a single monomer compound. *(22.4)*

Hund's rule—in filling a set of degenerate orbitals, each orbital is occupied by one electron, all with identical spins, before two electrons are placed in the same orbital. *(7.6)*

Hybrid orbitals—orbitals obtained by mixing two or more atomic orbitals on the same atom. *(10.3)*

Hydration—the interaction of ions with water molecules. *(12.2)*

Hydrocarbon—a compound that contains only the elements hydrogen and carbon. *(2.8, 22.1)*

Hydrogen bonding—a particularly strong intermolecular attraction between a hydrogen atom that is bonded to a highly electronegative atom and a lone pair of electrons on N, O, or F. *(11.4)*

Hygroscopic—having the ability to absorb water vapor from the air. *(20.5)*

Hypothesis—a possible explanation for observed results. *(1.1)*

Icosahedron—a regular polyhedron with 20 faces and 12 vertices. *(20.3)*

Ideal gas law—the equation that describes the state of a gas, $PV = nRT$, where P = pressure, V = volume, n = number of moles, R = ideal gas constant (0.08206 L·atm/mol·K), and T = temperature. *(6.3)*

Ideal solution—a solution that obeys Raoult's law throughout the entire range of composition. *(12.6)*

Indicator—a compound that changes color at the endpoint of a titration. The indicator should be chosen so that the endpoint coincides as nearly as possible with the equivalence point. *(4.4, 16.6)*

Induced dipole—a dipole caused by the presence of an electrical charge close to an otherwise nonpolar molecule. *(11.4)*

Induced nuclear reaction or nuclear transmutation—a reaction in which two particles or nuclei produce elements or isotopes that are different from the reactant species. *(21.3)*

Inert electrode—a solid electrical conductor, usually an unreactive metal or graphite, that is neither oxidized nor reduced in the reaction of a voltaic cell. It serves only as a means of electrical contact to the solution. *(18.3)*

Inert pair effect—the tendency for the heavier members of Groups 3A to 5A not to use the valence *s* electrons for bonding. *(20.3)*

Inflection point—the point in a titration curve at which the pH changes most rapidly. *(16.2)*

Initial rate method—determination of the rate law by measuring the initial rate of reaction in several experiments in which the initial concentrations of the reactants are varied. *(13.2)*

Inner transition metals—the lanthanides and the actinides. The inner transition elements are placed at the bottom of the periodic table. *(2.5)*

Instantaneous dipole—a dipole moment that results from unequal charge distribution within a molecule caused by the motion of the electrons. *(11.4)*

Instantaneous rate—the slope of the tangent to the concentration-time graph at any given point, rather than an average over a time interval. *(13.1)*

Integrated form of the rate law—equation that provides concentration-time relations. *(13.3)*

Intensive property—a property that is identical in any sample of the substance; examples are color, density, and melting point. *(1.2)*

Interhalogens—compounds formed from two different halogens. *(8.6)*

Intermediate—a compound that is produced in one step of the reaction mechanism and then consumed in a subsequent one. Intermediates are not found among the reactants or products. *(13.6)*

Intermolecular forces—the attractive forces that exist between molecules. *(11.1)*

Internal energy (*E*)—the total energy of the system. *(17.2)*

Ion—a charged particle formed by the addition or removal of electrons from an atom or group of atoms. *(2.3)*

Ionic bonding—the bonding that results from the electrostatic attraction between positively charged cations and negatively charged anions. *(9.2)*

Ionic compound—a compound composed of cations and anions joined to form a neutral species. *(2.7)*

Ionic radius—the measure of the size of an ion in an ionic solid. *(8.3)*

Ionic solid—a solid that consists of oppositely charged ions that are held together by electrostatic attractions. *(11.6)*

Ionization—a process that forms ions; the term is also used to describe the separation of a molecular compound into individual cations and anions when dissolved (generally in water). *(3.1)*

Ionization energy—the energy required to remove the highest-energy electron from a gaseous atom or ion in its electronic ground state. *(8.4)*

Isoelectronic series—a group of atoms and ions that have the same number of electrons. *(8.2)*

Isolated system—see **System, isolated.**

Isomers—different compounds with the same molecular formula but with different structural formulas. *(10.4)*

Isotopes—atoms of the same element that have different numbers of neutrons. *(2.3)*

Isotopic mass—the atomic mass of a particular isotope of any element. *(2.4)*

Joule (J)—the SI unit of heat, work, and energy, defined as

$$1\ \text{J} = 1\ \text{kg·m}^2/\text{s}^2.$$

In comparison with the calorie, 1 cal = 4.184 J. *(5.1)*

Ketone—an organic compound that contains the $-\overset{\displaystyle O}{\overset{\|}{C}}-R$ functional group, where the R group is an alkyl or aryl substituents. *(22.3)*

Kinetic energy—the energy that matter possesses because of its motion. *(5.1)*

Kinetic molecular theory—a model that describes the behavior of gas particles at the atomic or molecular level. *(6.6)*

Lanthanide contraction—the small decrease in the radii of the lanthanides as the 4*f* subshell fills, causing the transition elements of the fifth and sixth periods to have nearly identical radii within each group in Groups 4B and beyond. *(19.1)*

Lanthanides—the elements in period 6 between lanthanum (La) and hafnium (Hf): cerium (Ce) through lutetium (Lu). *(2.5)*

Lattice energy—the energy required to separate one mole of an ionic solid into isolated gaseous ions. *(9.2)*

Lattice point—the point in space that has the same geometric environment as every other lattice point in a crystal lattice. *(11.7)*

Law—a statement or equation that summarizes a large number of observations. *(1.1)*

Law of conservation of energy—the total energy of the universe—the system plus the surroundings—is constant during a chemical or physical change. *(5.1)*

Law of conservation of mass—there is no detectable loss or gain in mass when a chemical reaction occurs. *(2.1)*

Law of constant composition—all samples of a pure substance contain the same elements in the same proportions by mass. *(2.1)*

Law of mass action—for any reaction $a\text{A} + b\text{B} \rightleftharpoons c\text{C} + d\text{D}$, the equilibrium constant can be calculated from measured equilibrium concentrations. *(14.1)*

$$K_{\text{eq}} = \frac{[\text{C}]^c[\text{D}]^d}{[\text{A}]^a[B]^b}$$

Law of multiple proportions—if two elements unite to form more than one compound, the masses of one element (in each compound) that combine with a fixed mass of the second element are in a ratio of small whole numbers. *(2.1)*

Le Chatelier's principle—a change to a system at equilibrium causes a shift in the position of the equilibrium that reduces the effect of the change. *(14.3)*

Leveling effect—the effect that makes strong acids (or bases) appear equal in strength in a given solvent because they ionize or dissociate completely. *(15.4)*

Lewis acid—an electron-pair acceptor. *(15.9)*

Lewis base—an electron-pair donor. *(15.9)*

Lewis electron-dot symbol—the symbol of the element with one dot to represent each valence electron. *(9.1)*

Lewis structure—a representation of covalent bonding, using Lewis symbols, that shows shared electrons as dots or lines between atoms and unshared electrons as dots. *(9.3)*

Ligand—an anion or molecule that functions as a Lewis base in forming one or more coordinate covalent bonds to metal ions. *(16.8, 19.2)*

Limiting reactant—the reactant that is completely consumed in a chemical reaction. The amounts of products that can form and the amounts of the other reactants that can be consumed are determined by the amount of the limiting reactant present. *(3.5)*

Line spectrum—a spectrum that contains light at discrete wavelengths separated by regions where no light is emitted. Line spectra are produced by gaseous atoms of the elements. *(7.2)*

Linkage isomers—compounds in which the same ligand is coordinated to the metal ion through one of two possible donor atoms; for example, NO_2^- may form a bond to the metal ion through one of the oxygen atoms (nitrito) or the nitrogen atom (nitro). *(19.3)*

Liquid—a fluid that has a fixed volume but no definite shape. *(6.1)*

London dispersion forces—the instantaneous dipole-induced dipole attractions that explain the attractions between nonpolar molecules, but contribute to the attractions between all molecules. *(11.4)*

Lone or nonbonding pairs—pairs of electrons that are not shared. *(9.3)*

Low-spin (strong-field) complex—a coordination complex in which the lower-energy *d* orbitals are completely filled with two electrons each, before electrons enter the higher-energy *d* orbitals. Low-spin complexes are observed when $P < \Delta$. *(19.4)*

Magic numbers—certain numbers of protons and neutrons that confer unusual stability to a nuclide. These numbers are 2, 8, 20, 26, 28, 50, 82, and 126. *(21.1)*

Magnetic quantum number, m_ℓ—the quantum number that describes the orientation of an electron wave function in an atom. The allowed values of this quantum number depend on the value of the ℓ quantum number and may have all integer values from $-\ell$ to $+\ell$. *(7.4)*

Main-group elements—See **Representative elements.** *(2.5)*

Mass—a measure of the quantity of matter in a sample or object. *(1.2)*

Mass defect—the difference between the mass of the atom and the mass of the equivalent number of ^1H atoms and neutrons. *(21.4)*

Mass number—the total number of protons and neutrons in an atom. It has the symbol *A*. *(2.3)*

Mass percent solute—a concentration unit sometimes used to express the concentration of a solution. *(12.1)*

$$\text{Mass percentage solute} = \frac{\text{mass solute}}{\text{mass solution}} \times 100\%$$

Matter—anything that has mass and occupies space. *(1.2)*

Mechanism—the sequence of molecular-level steps that lead from reactants to products. *(13.6)*

Melting point—the temperature at which a solid substance changes to the liquid phase. *(11.2)*

Metal—a material that has luster and is a good conductor of electricity. Metallic elements are located in the center and left-hand side of the periodic table. *(2.5)*

Metallic solid—a solid formed by metal atoms. In metals, the outer electrons occupy highly delocalized orbitals that extend throughout the crystal and account for their electrical and thermal conductivity. *(11.6)*

Metalloid—an element with properties between those of a metal and a nonmetal. The elements along the dividing line between metals and nonmetals in the periodic table. *(2.5)*

Metallurgy—the science of extracting metals from their ores, purifying, and preparing them for practical use. *(19.5)*

Michaelis–Menten mechanism—an enzyme catalysis mechanism in which an enzyme-substrate complex is formed in a reversible step, followed by a rate-limiting formation of products from the complex. *(13.6)*

Millimole—1×10^{-3} mol. *(16.1)*

Mixture—two or more substances that can be separated by taking advantage of different physical properties of the substances. *(1.2)*

Moderator—a substance used in nuclear reactors that slows the speed of neutrons. *(21.5)*

Molality (*m*)—the unit of concentration of a solute expressed as the moles of solute present in one kilogram of solvent. *(12.1)*

Molar mass—the mass in grams of one mole of any substance; numerically the same as the molecular or formula mass. *(3.2)*

Molarity (*M*)—the concentration unit defined as the number of moles of solute in one liter of solution. *(4.2)*

Mole—the amount of substance that contains as many entities as there are atoms in exactly 12 g of ^{12}C (the isotope of carbon that has 6 neutrons). *(3.2)*

Mole fraction—the number of moles of one component of a homogeneous mixture divided by the total number of moles of all substances present in the mixture. *(6.5)*

Molecular compound—atoms of two or more elements joined so strongly they behave as a single particle. *(2.6)*

Molecular formula—gives the number of every type of atom in a molecule. *(2.6)*

Molecular mass—the sum of the atomic masses of all atoms present in the molecular formula of a molecule. *(2.6)*

Molecular orbital—a wave function of an electron in a molecule. *(10.5)*

Molecular orbital theory—a model that treats bonding as delocalized over the entire molecule. *(10.5)*

Molecular shape—a description of the positions of the atoms, not the lone pairs. *(10.1)*

Molecular solid—a solid that consists of small covalently bonded molecules, held together by van der Waals forces or hydrogen bonding. *(11.6)*

Molecularity—the number of reactant species that are involved in a single elementary step. *(13.6)*

Molecule—a combination of atoms joined so strongly that they behave as a single particle. *(2.6)*

Monatomic ion—an ion formed by the loss or gain of electrons by a single atom. *(2.7)*

Monodentate ligand—a ligand that donates one pair of electrons to the metal in a coordination complex. *(19.2)*

Monosaccharide—the basic unit of carbohydrates, having the general formula $(CH_2O)_x$, where $x = 3$ to 8. *(22.6)*

Negative deviations (from Raoult's law)—the observed vapor pressure is less than expected for an ideal solution and occurs when the attractions between dissimilar molecules are stronger than the average of the attractions in the pure components of the mixture. *(12.6)*

Nernst equation—the equation that describes the dependence of cell voltage on the concentrations of reactants and products. *(18.6)*

$$E = E^\circ - \frac{RT}{nF}\ln Q = E^\circ - \frac{2.303RT}{nF}\log Q$$

Net ionic equation—the equation that shows only those species in the reaction that undergo change. *(4.1)*

Neutralization—the reaction of an acid and a base to yield water and a salt. *(3.1)*

Neutron—a small particle that has a mass of 1.675×10^{-27} kg and no charge. Its relative mass is 1. *(2.2)*

Noble gases—the elements of Group 8A. They were formerly known as the "inert gases": He, Ne, Ar, Kr, Xe, Rn. *(2.5)*

Nonelectrolyte—water and compounds that dissolve in water as neutral molecules. *(2.9, 4.1)*

Nonmetal—a material that lacks the characteristics of a metal. Nonmetallic elements are in the top right part of the periodic table. *(2.5)*

Normal boiling point—the boiling point at a pressure of 1 atm. *(11.2)*

Normal melting point—the temperature at which the solid and liquid phases are in equilibrium at 1 atm pressure. Melting points change little with pressure. *(11.2)*

Nuclear binding energy—the energy of attraction among the protons and neutrons in a nuclide. This energy is calculated from the mass defect by using

$$\Delta E = \Delta mc^2,$$

where Δm is the mass defect of the nucleus in kilograms per mole, c is the speed of light, 3.00×10^8 m/s, and ΔE is in joules per mole. *(21.4)*

Nuclear chain reaction—a fission event, initiated by the absorption of a neutron, that produces more neutrons than are consumed, allowing a self-sustained nuclear reaction. *(21.5)*

Nuclear equation—describes the changes in a nuclear reaction in shorthand notation. Nuclear equations are balanced by conservation of atomic number and mass number of the particles on the reactant and product sides of the equation. *(21.1)*

Nuclear fission—the splitting of a heavy nucleus into two nuclei of comparable size. *(21.5)*

Nuclear fusion—the combination of two light nuclides to form a heavier one. *(21.5)*

Nuclear transmutation—a nuclear reaction that involves two or more particles or nuclei as reactants and produces elements or isotopes that are different from the reactants. *(21.3)*

Nucleic acid—a chain of nucleotides. *(22.7)*

Nucleon—a particle that is present in the nucleus of an atom; both protons and neutrons are called *nucleons*. *(21.0)*

Nucleotide—the building block of DNA and RNA; each nucleotide is composed of three units: a five-carbon sugar, a base (a cyclic amine), and a phosphate group. *(22.7)*

Nucleus—the small, heavy, positively charged core of an atom. *(2.2)*

Nuclide—the nucleus of a particular isotope. *(21.0)*

Number of significant digits—the number of digits from the first nonzero digit to the first digit that is uncertain in the quantity. *(1.3)*

Octet rule—each atom in a molecule shares electrons until it is surrounded by eight valence electrons. The octet rule is most important for elements in the second period. *(9.3)*

Optical isomers—stereoisomers that rotate the plane of polarized light. *(19.3)*

Orbital diagram—a drawing that represents the spins of electrons as "up" and "down" arrows placed in boxes that represent the orbitals. *(7.6)*

Order of the reaction—the power to which the concentration is raised in the rate law. *(13.2)*

Organic chemistry—the study of carbon-containing compounds. *(22.0)*

Organic compound—a compound made up of carbon atoms in combination with other elements such as hydrogen, oxygen, and nitrogen. *(2.8, 3.1)*

Osmosis—the diffusion of a fluid through a semipermeable membrane. *(12.4)*

Osmotic pressure—the pressure difference needed to prevent net transport of solvent across a semipermeable membrane that separates a solution from the pure solvent. This is one of the most sensitive colligative properties and is described by the equation

$$\Pi = MRT,$$

where Π is the osmotic pressure, M is the molar concentration of solute particles, R is the ideal gas constant (0.08206 L·atm/mol·K), and T is the temperature in kelvins. *(12.4)*

Ostwald process—a multistep process for the production of nitric acid from ammonia, oxygen, and water. *(20.5)*

Overall equation—the equation that shows all of the reactants and products in undissociated form. *(4.1)*

Overall order—the sum of the orders for all substances that appear in the rate law. *(13.2)*

Oxidation—the loss of electrons by an element, compound, or ion. *(3.1, 18.1)*

Oxidation half-reaction—the part of an oxidation-reduction reaction that shows only the oxidation process. *(18.2)*

Oxidation numbers (states)—the numerical equivalent of the charge of monatomic ions, or the charge an atom would possess if the shared electrons in each covalent bond were assigned to the more electronegative atom. *(3.1, 18.1)*

Oxidation-reduction (redox) reaction—a reaction in which electrons are transferred from one species to another. *(3.1, 18.1)*

Oxidizing agent—the reactant that is reduced in a redox reaction. *(18.1)*

Oxyacid—a molecule of the general formula $(OH)_mXO_n$ that has at least one hydrogen atom attached to an oxygen atom. *(9.7, 15.8)*

***p*-Block elements**—the section of the periodic table that contains the elements whose *p* subshell is being filled with electrons, Groups 3A to 8A. *(8.1)*

Paired electron—see **Electron pair.** *(7.6)*

Paramagnetic matter—matter that is attracted by a magnetic field. Only substances that contain unpaired electron spins are paramagnetic. *(19.4)*

Partial pressure—the pressure exerted by each gas in a mixture of gases, if it alone occupied the container at the same temperature. *(6.5)*

Pauli exclusion principle—no two electrons in the same atom can have the same set of all four quantum numbers. *(7.6)*

Peptide—a small polymer of amino acids. *(22.5)*

Percent yield—the actual yield divided by the theoretical yield and multiplied by 100%. *(3.5)*

Period—a row of the periodic table. *(2.5)*

Periodic table—an arrangement of the elements (shown by their symbols) into rows and columns so that elements with similar chemical properties are placed in vertical columns. *(2.5)*

pH meter—a voltmeter that is scaled to read the pH of a solution from the measured potential of a voltaic cell. *(18.6)*

pH scale—equal to $-\log[H_3O^+]$. *(15.2)*

Phase diagram—a graph of pressure versus temperature that shows the regions of stability for each phase (solid, liquid, and gas). The lines between the phases represent conditions of temperature and pressure at which two or more phases are in equilibrium. *(11.3)*

Phenol—an organic compound in which a hydroxyl group is substituted for a hydrogen atom on an aromatic ring. The parent compound of this type, C_6H_5OH, is called *phenol.* *(22.3)*

Photoelectric effect—the ejection of electrons from a solid by the absorption of a photon of light. *(7.1)*

Photon—a particle of light that possesses an energy of $h\nu$. *(7.1)*

Physical change—a change that occurs without a change in the composition of the substance. *(1.2)*

Physical property—one that can be observed without changing the substances present in the sample. *(1.2)*

Pi (π) bond—a bond that places electron density on opposite sides of the line joining the bonded atoms. It is formed from sideways overlap of *p* orbitals. *(10.4)*

Pipet—a device calibrated either to deliver or to contain a specific volume of liquid; it is used to measure accurately a fixed volume of solution. *(4.2)*

Planck's constant (*h*)—the proportionality constant that relates a quantum of energy to the frequency of the radiation absorbed or emitted. $h = 6.626 \times 10^{-34}$ J·s. *(7.1)*

Planck's equation—$E = h\nu$ *(7.1)*

Plasma—an electrically neutral gas at high temperature that consists of highly charged ions and electrons. *(21.5)*

Polar bond—a covalent bond in which the bonding electrons are not equally shared by the two atoms. *(9.4)*

Polar molecule—a molecule that contains an unequal distribution of charge and thus has a dipole moment. *(10.2)*

Polarizability—the ease with which the electron cloud of a molecule can be distorted. *(11.4)*

Polarized light—electromagnetic radiation that has the electric field oscillating in a single plane. *(19.3)*

Polyatomic ion—charged species made up of more than one atom. *(2.7)*

Polydentate ligand—a ligand with more than one atom that forms a coordinate covalent bond with a metal ion. Numerical prefixes (bi-, tri-, tetra-, and so forth) are combined with "dentate" to designate the number of donor atoms in the ligand. *(19.2)*

Polyester—a polymer formed from the condensation of a dicarboxylic acid and a dialcohol. *(22.4)*

Polymer—a large molecule formed by the repeated bonding together of many smaller units (monomers). *(22.4)*

Polyprotic acid—an acid that can ionize to produce more than one proton per molecule. *(16.7)*

Polysaccharide—a step-growth polymer of monosaccharides. *(22.6)*

Positive deviations (from Raoult's law)—the observed vapor pressure is greater than expected for an ideal solution, and occurs when the attractions between dissimilar molecules are weaker than the average of the attractions in the pure components of the mixture. *(12.6)*

Positron—a positively charged electron that is produced in the radioactive decay of some unstable isotopes. *(21.1)*

Potential—another term for the voltage of an electrochemical cell or half-cell. *(18.4)*

Potential energy—the energy derived from the position or condition of matter. *(5.1)*

Precipitation reaction—the formation of an insoluble product or products from reactants in solution. *(4.1)*

Precision—agreement among repeated measurements. *(1.3)*

Pre-exponential term (*A*)—the product of the steric factor and collision frequency, collected in a single term. *(13.4)*

Pressure—the force per unit area exerted on a surface. *(6.1)*

Primary battery—a battery in which the chemical reaction is essentially irreversible. *(18.7)*

Primary structure—the sequence of amino acids in the polypeptide chain. *(22.5)*

Principal quantum number (*n*)—the quantum number that contains information about the distance of an electron from the nucleus. It may have any positive integer values, and it affects the energy of the electron. The value determines the energy of an electron in the hydrogen atom. *(7.4)*

Principal shell—the set of all atomic orbitals that have the same value of *n*. *(7.4)*

Product—a substance that is formed in a chemical reaction. *(3.1)*

Property—anything that can be observed or measured. *(1.2)*

Protein—a polymer of amino acids. *(22.5)*

Proton—a small particle that has a mass of 1.673×10^{-27} kg and a charge of $+1.602 \times 10^{-19}$ coulombs. The absolute charge is the same as that of the electron but opposite in sign. The relative charge is 1+, and the relative mass is 1. *(2.2)*

Pseudo-noble gas electron configurations—electron configurations of the type [noble gas] $(n-1)d^{10}$. *(8.4)*

Quantum—the smallest quantity of energy that is absorbed or emitted by matter as electromagnetic radiation. *(7.1)*

Quantum mechanics—the most successful theory to date that explains the behavior of electrons in atoms and molecules. *(7.3)*

Quantum numbers—numbers that describe the characteristics of wave functions and are analogous to coordinates that describe the location of a particle. In the hydrogen atom, four quantum numbers are needed to describe the wave function of the electron. *(7.3)*

Quaternary structure—the orientation of different polypeptide chains with respect to each other. *(22.5)*

Racemic mixture—an equimolar mixture of enantiomers that produces no net rotation of the plane of polarized light. *(19.3)*

Rad—the unit of radiation dose, defined as the quantity of radiation that transfers 1×10^{-2} joule per kilogram of matter. *(21.6)*

Radiation dose—the amount of energy transferred to a target as a result of a radioactive decay. *(21.6)*

Radioactivity—the spontaneous nuclear reaction that transforms a relatively unstable nucleus into a more stable nucleus and a small particle and energy. *(21.1)*

Range—the difference between the largest and smallest measured values. *(1.3)*

Raoult's law—the vapor pressure (*P*) of the solvent over a dilute solution is equal to the mole fraction of the solvent (χ) times the vapor pressure of the pure solvent ($P°$):

$$P = \chi_{solvent}P°$$

A more useful form of this law is

$$\Delta P = \chi_{solute}P°,$$

where ΔP is $(P° - P_{solution})$. *(12.4)*

Rate—change per unit time. *(13.1)*

Rate constant (*k*)—the proportionality constant in the rate law, equal to the rate of reaction when the reactants are present at 1.0 *M* concentrations. *(13.2)*

Rate law—the relation between rate and concentrations. It is of the form: rate $= k[A]^x[B]^y$. *(13.2)*

Rate of reaction—the ratio of the rate of change of any substance to its coefficient in the chemical equation. *(13.1)*

Rate-limiting step—a step that is much slower than any of the others in the mechanism. *(13.6)*

Reactant—a substance that is consumed in a chemical reaction. *(3.1)*

Reaction quotient (*Q*)—the result of a calculation based on the same expression as the equilibrium constant, except that the concentrations are not restricted to those observed after the system reaches equilibrium. Initial concentrations are often used. *(14.2)*

Redox reaction—an abbreviation for an oxidation-reduction reaction. *(18.1)*

Reducing agent—the reactant that is oxidized in a redox reaction. *(18.1)*

Reduction—the gain of electrons by an element, compound, or ion. *(3.1, 18.1)*

Reduction half-reaction—the part of a redox-reaction that shows only the reduction process. *(18.2)*

Representative elements (main-group elements)—the elements in Groups labeled A (1A–8A) or Groups 1 and 2 and 13 through 18 on the periodic table. *(2.5)*

Resonance structures—structures that differ only in the distribution of the valence electrons; the skeleton structure does *not* change, only the placement of the electrons. *(9.6)*

Reverse osmosis—the transport of water across a semipermeable membrane from a solution to pure water on the other side of the membrane, by applying a pressure greater than the osmotic pressure to the solution. *(12.4)*

Ribonucleic acid (RNA)—a polymer of nucleotide units located outside the chromosomes. RNA transmits genetic information and directs the assembly of proteins. *(22.7)*

Roasting—pretreatment of ore by heating the material below its melting point, usually in the presence of air, to convert the ore into a chemical form more suitable for the reduction step. *(19.5)*

Root-mean-square (rms) speed (u_{rms})—the square root of the average squared speeds of a collection of particles. *(6.6)*

Rydberg equation—the equation that predicts the wavelengths of the lines in the hydrogen atom spectrum,

$$\frac{1}{\lambda} = R_H\left(\frac{1}{n_A^2} - \frac{1}{n_B^2}\right),$$

where n_A and n_B are whole positive numbers, with $n_A < n_B$, and R_H is a constant, called the *Rydberg constant,* which has a value of $1.097 \times 10^7\ m^{-1}$. *(7.2)*

***s*-Block elements**—the section of the periodic table that contains the elements whose *s* subshell is being filled with electrons, Groups 1A and 2A. *(8.1)*

Sacrificial anode—a piece of active metal placed in electrical contact with a less reactive metal. *(18.9)*

Salt—a compound made up of the cation from a base and the anion from an acid. *(3.1)*

Salt bridge—an electrolyte medium that allows the transport of ions between the two half-cells of a voltaic cell. *(18.3)*

Saturated hydrocarbon—a hydrocarbon with no multiple bonds. *(22.1)*

Saturated solution—a solution that is in equilibrium with an excess of the solute. *(12.2)*

Science—the systematic knowledge of the natural universe. *(1.1)*

Scientific method—investigations that are guided by theory and past experiments. *(1.1)*

Scientific notation—see **Exponential notation.** *(1.3)*

Second law of thermodynamics—in any spontaneous process, the entropy of the universe will increase. *(17.3)*

Secondary battery—a battery in which the chemical reaction can be reversed. *(18.7)*

Secondary structure—the shape of the polypeptide chain. *(22.5)*

Semiconductor—a weak conductor of electricity. *(2.5)*

Semipermeable membrane—thin films of materials that allow only water and other small molecules to pass through them. Animal bladders, the skins of fruits and vegetables, and cellophane are examples of semipermeable membranes. *(12.4)*

Shell—the group of orbitals that have the same principal quantum number *n*. *(7.4)*

Sigma (σ) bond—a bond in which the shared pair of electrons is concentrated along the line joining the nuclei of the bonded atoms. *(10.4)*

Significant figures (significant digits)—all digits in a measurement from the first nonzero digit through the first digit that is uncertain. *(1.3)*

Silicates—compounds that contain silicon combined with oxygen and various metals. *(20.4)*

Simple (primitive) cubic—the cubic structure with identical atoms located only at the corners of the unit cell. *(11.7)*

Single bond—the bond formed by sharing one pair of electrons between two atoms. *(9.3)*

Skeleton reaction—a chemical reaction that shows only the species oxidized and reduced in the reaction. *(18.2)*

Skeleton structure—the drawing that shows which atoms are bonded to each other in a molecule. *(9.3)*

Solid—state of matter with a fixed shape and volume. *(6.1)*

Solubility—the concentration of solute that exists in equilibrium with an excess of that substance. *(4.1, 12.2)*

Solubility equilibrium—an equilibrium that describes a solid dissolving in solution or a solid forming from the solution. *(14.6)*

Solubility-product constant—the equilibrium constant for the dissolving of a sparingly soluble salt. *(14.6)*

Solute—any substance being dissolved to form a solution. Solutes are usually present in lesser quantity than the solvent. *(4.1)*

Solution—another name for a homogeneous mixture. *(1.2)*

Solvent—the substance that has the same physical state as the solution. It is generally the component present in largest quantity; in aqueous solutions, the solvent is water. *(4.1)*

Specific heat (C_s)—the heat needed to increase the temperature of one gram of a substance by one kelvin; it has the units of J/g·K (or J/g·°C). *(5.3)*

Spectator ions—ions present in solution that do not undergo change. *(4.1)*

Spectrochemical series—the arrangement of ligands in order of increasing size of the crystal field splitting that they cause. *(19.4)*

Spectrum—a graph of the intensity of light as a function of the wavelength or frequency. *(7.2)*

Spin quantum number—see **Electron spin quantum number.** *(7.4)*

Spontaneous—able to occur without outside intervention. *(17.0)*

Stable isotope—an atom with a nuclide that does not spontaneously decompose into one or more different nuclides. *(21.1)*

Standard cell potential, E°_{cell}—the voltage of a voltaic cell when all of the reactants and products are in their standard states (pure solids, liquids, or gases at 1 atm pressure; solutes at 1 *M* concentration). *(18.4)*

Standard enthalpy of formation (ΔH_f°)—the change in enthalpy when one mole of a substance in its standard state is formed from the most stable forms of the elements in their standard states. The change in enthalpy is positive if heat is absorbed and negative if heat is released by the system. *(5.5)*

Standard Gibbs free energy of formation—the free energy change during the formation of one mole of a substance in its standard state from the most stable form of the elements in their standard states. *(17.4)*

Standard reduction potential—the standard voltage of a reduction half-reaction based on a voltage of zero for the standard hydrogen electrode. *(18.4)*

Standard solution—a solution of accurately known concentration. *(4.4)*

Standard state—a substance as the pure liquid, solid, or gas, at the designated temperature, usually 298 K, and 1 atm pressure. *(5.5)*

Standard temperature and pressure (STP)—the conditions of 273 K (or 0 °C) and 1.00 atm. *(6.3)*

State (of a system)—the set of experimental conditions needed to describe its properties completely, including temperature, pressure, and the amounts and phases of substances. *(17.1)*

State function—a thermodynamic property that depends only on the state of the system and not the manner in which the system arrived at the state. *(5.4, 17.1)*

Step-growth polymer (condensation polymer)—a polymer formed by a reaction that eliminates a small molecule each time a monomer is linked to the polymer chain. *(22.4)*

Stereoisomers—molecules or ions that have the same bonds but are different in the arrangement of the atoms in space. *(19.3)*

Steric factor (*p*)—a term in the predicted rate law that expresses the need for the correct orientation of reactants when they collide to form the activated complex. *(13.4)*

Steric number—the number of lone pairs on a central atom plus the number of atoms bonded to it. *(10.1)*

Stoichiometry—the study of quantitative relationships involving substances and their reactions. *(3.1)*

Strong acid—an acid that ionizes completely in solution. *(15.3)*

Strong base—one of the soluble metal oxides or hydroxides that dissociates completely in solution. *(15.3)*

Strong electrolytes—substances that dissociate or ionize completely in solution. *(4.1, 15.3)*

Strong-field complex—see **Low-spin complex.** *(19.6)*

Structural formula—a formula that indicates how the atoms are connected in the molecule. *(2.6)*

Structural isomers—compounds that contain the same numbers and kinds of atoms but differ in the bonds that are present. *(19.3)*

Sublimation—the direct conversion of a solid to a gas without first changing to a liquid. *(11.2)*

Subshell—all the possible atomic orbitals that have the same values for both the n and ℓ quantum numbers. Each principal shell consists of n subshells. *(7.4)*

Substance—matter that cannot be separated into component parts by a physical process. *(1.2)*

Substitution reaction—a reaction in which one atom or group of atoms in a molecule is replaced by a different atom or group of atoms. *(22.1)*

Supercooling—reducing the temperature of a liquid below its freezing point without forming a solid. This is an unstable condition. *(11.2)*

Supercritical fluid—the single fluid phase that exists above the critical temperature and pressure. *(11.2)*

Supersaturated solution—a solution in which the concentration of the solute is temporarily greater than its solubility. This is an unstable condition. *(12.2)*

Surface tension—the quantity of energy required to increase the surface area of a liquid; it has SI units of joules per square meter. *(11.5)*

Surroundings—all matter other than the system in chemical thermodynamics. *(5.1, 17.1)*

Symbol—an abbreviation for an element that consists of one or two letters usually related to the name of the element. *(1.2)*

System—the matter of interest in a chemical reaction. *(5.1)*

System, closed—system in which the exchange of energy with the surroundings is allowed, but not the exchange of matter. *(17.1)*

System, isolated—system in which matter and energy cannot enter or leave. *(17.2)*

Termolecular—an elementary step that involves the collision of three species. *(13.6)*

Tertiary structure—the overall three-dimensional arrangement of the protein. *(22.5)*

Theoretical yield—the maximum quantity of product that can be obtained in a chemical reaction based on the amounts of starting materials. *(3.4)*

Theory—an explanation of the laws of nature. *(1.1)*

Thermochemical energy-level diagram—a diagram that illustrates the energy change that accompanies a physical or chemical change. *(5.4)*

Thermochemical equation—a chemical equation that includes the change in enthalpy. *(5.2)*

Thermochemistry—the study of the relation between heat and chemical reactions. *(5.1)*

Third law of thermodynamics—the entropy of a perfect crystalline substance is zero at a temperature of absolute zero. *(17.3)*

Titrant—the substance added to react with the analyte. *(16.1)*

Titration—a procedure for the determination of the quantity of one substance by the addition of a measured amount of a second substance. *(4.4, 16.1)*

Titration curve—a graph of pH as a function of the added volume of titrant. *(16.1)*

Transition elements—those elements characterized by having partially filled *d* orbitals in the metal or at least one of its oxidation states. These elements are found in the center of the periodic chart, in the nine columns from Group 3B through 1B (3–11). *(2.5, 19.2)*

Transition metals—See **transition elements.** *(2.5)*

Triple bond—the bond formed by sharing three pairs of electrons between two atoms. *(9.3)*

Triple point—the unique combination of pressure and temperature at which all three phases (solid, liquid, and gas) exist in equilibrium. *(11.3)*

Uncertainty—related to precision. Measurements of high precision have small uncertainty. *(1.3)*

Unimolecular—an elementary step that involves the spontaneous decomposition of a single molecule. *(13.6)*

Unit—standard used for quantitative comparison between measurements of the same type of quantity. *(1.4)*

Unit cell—a small geometric figure needed to define the pattern of all lattice points in the entire crystal. *(11.7)*

Unpaired electron—an electron that is the only one occupying an orbital. *(7.6)*

Unsaturated hydrocarbon—a hydrocarbon that contains one or more multiple carbon-carbon bonds. *(22.2)*

Unsaturated solution—a solution in which the concentration of the solute is less than its solubility. *(12.2)*

Valence bond theory—a theory that describes bonds as being formed by atoms sharing valence electrons in overlapping valence orbitals. *(10.3)*

Valence electrons—electrons that occupy the valence orbitals. *(8.1)*

Valence orbitals—those orbitals in the atom of the highest occupied principal level, and the orbitals of partially filled subshells of lower principal quantum number. *(8.1)*

Valence-shell electron-pair repulsion (VSEPR)—a model that predicts the shapes of molecules that is based on the premise that electron pairs on a central atom repel each other so they are as far apart as possible. *(10.1)*

van der Waals equation—a description of the behavior of real gases, based on correcting the ideal gas law for particle size and attractive forces:

$$\left(P + \frac{an^2}{V^2}\right)(V - nb) = nRT,$$

where a and b are constants experimentally determined for each gas. *(6.8)*

van der Waals forces—the weak intermolecular forces between small molecules. The term includes both dipole-dipole attractions and London dispersion forces. *(11.4)*

van't Hoff factor (*i*)—the ratio of the measured colligative property to that expected for a nonelectrolyte.

$$i = \frac{\text{measured colligative property}}{\text{expected value for a nonelectrolyte}}$$

In dilute solutions of strong electrolytes, the van't Hoff factor approaches the number of ions in solution for each formula unit of the compound that dissolves. *(12.5)*

Vapor pressure—the partial pressure of a substance that is in equilibrium with one of the condensed phases of that substance. *(11.2)*

Vapor-pressure lowering—the difference between the vapor pressure of a pure solvent and that of the solvent in a solution; it is a colligative property (see Raoult's law). *(12.4)*

Vaporization—the same as evaporation. *(11.2)*

Viscosity—the resistance of a fluid to flow; it is one of the properties that is related to the strength of intermolecular attraction. *(11.5)*

Visible light—electromagnetic radiation in the wavelength range from 400 to 700 nm. Light in this wavelength range is seen by the human eye. *(7.1)*

Volt (V)—the SI unit for electromotive force, which is defined as the potential difference required to increase the energy of a charge of 1 coulomb by 1 joule: 1 V = 1 joule/coulomb = 1 J/C. *(18.4)*

Voltaic cell—an apparatus that converts the chemical energy produced by a redox reaction directly into electrical energy. This is also called a galvanic cell. *(18.3)*

Volumetric analysis—a quantitative determination in which the volume of a solution or substance is measured. *(4.4)*

Volumetric flask—a container calibrated to hold an accurately known volume of liquid. *(4.2)*

Water-gas shift reaction—the reaction of carbon monoxide and steam to produce hydrogen and carbon dioxide. *(20.2)*

Wave—a periodic disturbance in a medium or in space that is described by specifying its amplitude, speed, wavelength, and frequency. *(7.1)*

Wave function (ψ)—the equation of the wave that describes the location and energy of a particle in the wave model of matter. *(7.3)*

Wavelength (λ)—the distance from one peak of a wave to the next. In the SI system, wavelength is measured in meters. *(7.1)*

Weak acid—an acid that partially ionizes in solution. *(15.4)*

Weak base—a base that partially ionizes in solution. *(15.4)*

Weak electrolytes—substances that only partially dissociate or ionize in solution. *(4.1, 15.3)*

Weak-field complex—see **High-spin complex.** *(19.6)*

Weight—the force of attraction between two objects. The weight of an object will change from one location to another, but the mass is always the same. *(1.2)*

Work—the product of force times distance. *(5.1, 17.1)*

Answers to Selected Exercises

APPENDIX J

Chapter 1

1.1 Science is knowledge, especially that gained by experience. Fields of science include chemistry, biology, and physics. Fields that are not science include astrology, art, and music.

1.3 An athletic team might test the effects of mealtime or contents on performance. They could divide the team randomly into a test group and a control group, making sure one group did not have all the best players. They could then test a hypothesis such as "eating fruit before a meet helps improve performance." The coaches and scientists could monitor performance and use the data to determine whether any particular meal plan produced better results.

1.5 The explanation of the extinction of dinosaurs is a theory, but only if substantiated by experimental results, such as the presence of meteoritic elements at certain locations in the earth's crust.

1.7 Matter is anything that has mass and occupies space. Mass is a measure of the quantity of matter. Weight is the force that results when a sample of matter is accelerated (usually by gravity).

1.9 Homogeneous mixtures have the same composition throughout; examples include sugar water, 14-karat gold, and clean air. Heterogeneous mixtures have a variable composition; examples include a bag of mixed nuts, oil and vinegar salad dressing, and smoke.

1.11 Deciding whether a solution is homogeneous or heterogeneous depends on the fineness of the measurement. Traditionally, the eye or an optical microscope was used, so the differences on the scale of the wavelength of light, hundreds of nanometers, was used as a standard. But now, microscopes are much more powerful, so a scientist would want to qualify by saying, "The solution is heterogeneous when examined by a microscope that can perform measurements with 1-nanometer resolution."

1.13 An alloy is the metal that results from the homogeneous mixture of two or more substances, where at least one of the substances is a metal.

1.15 American football fans complain not about the accuracy of the markers, but of the judgment of the official who decides exactly where the ball was located.

1.17 Division operations will frequently cause a calculator to display many digits that are not significant.

1.19 A dozen eggs weigh 1.5 lbs.

1.21 (a) 0.1 mm (b) ~20,000 km (c) 10^{16} kg
(d) ~1 L (e) 10^6 kg

1.23 SI length (meters to kilometers): $\left(\frac{1\text{ km}}{10^3\text{ m}}\right)$

SI mass English (grams to pounds): $\left(\frac{1\text{ lb}}{453.6\text{ g}}\right)$

1.25 (a) intensive, physical (b) intensive, physical
(c) intensive, chemical (d) extensive, physical
(e) intensive, physical

1.27 (a) physical (b) physical
(c) chemical (d) chemical

1.29 (a) chemical (b) physical
(c) physical (d) chemical

1.31 (a) chemical (b) physical
(c) physical (d) chemical

1.33 Chemical: "... corrosive gas that reacts with practically all substances. Finely divided metals, glass, ceramics, carbon, and even water burn in fluorine with a bright flame. Small amounts of compounds of this element in drinking water and toothpaste prevent dental cavities. . . . Fluorine is one of the few elements that forms compounds with the element xenon."
Physical: "Fluorine is a pale-yellow . . . gas . . . The free element has a melting point of −219.6 °C and boils at −188.1 °C."

1.35 Chemical: "... reacts with water to form sodium hydroxide and hydrogen gas. It is stored under oil because it reacts with air."
Physical: "Sodium is a soft, silver-colored metal. . . . Sodium melts at 98 °C, which is relatively low for a metal."

1.37 (a) Pure, clean air is a homogeneous mixture. Normally, air is a heterogeneous mixture containing dust and other finely divided particulate matter suspended in the gases that make up air.
(b) compound
(c) homogeneous mixture
(d) element

1.39 (a) element
(b) heterogeneous mixture
(c) homogeneous mixture
(d) heterogeneous mixture

1.41 (a) and (d) are solutions.

1.43 (a) not precise (b) accurate and precise
(c) not precise (d) not accurate

1.45 (a) 5 (b) 1
(c) 3 (d) 3

1.47 (a) 1 (b) 5
(c) 6 (d) 2

1.49 (a) 9.65×10^4 J/C (b) 3.00 g/cm^3
(c) 6×10^{-2} mL (d) 6.6×10^{-34} kg

1.51 (a) 5.62 mL (b) 37.8 °C
(c) 762.2 mmHg

1.53 (a) 2.16×10^2 (b) 3.7×10^{-2}
(c) 0.224 (d) 0.4

1.55 (a) 13.56 (b) 11
(c) 3.14×10^4 (d) 0.0475

1.57 (a) 0.377 (b) 7.5
(c) 132.0

1.59 1.01×10^{242}

1.61 (a) kilograms (b) meters
(c) kelvins (d) seconds

1.63 $\left(\frac{10^6\ \mu m}{1\ m}\right)$ or $\left(\frac{1\ m}{10^6\ \mu m}\right)$ or $\left(\frac{10^{-6}\ m}{1\ \mu m}\right)$ or $\left(\frac{1\ \mu m}{10^{-6}\ m}\right)$

1.65 $\left(\frac{10^6\ mL}{1\ kL}\right)$ or $\left(\frac{1\ kL}{10^6\ mL}\right)$ or $\left(\frac{10^{-6}\ kL}{1\ mL}\right)$ or $\left(\frac{1\ mL}{10^{-6}\ kL}\right)$

1.67 $\left(\frac{0.911\ ft/s}{1\ km/hr}\right)$

1.69 761 miles/hr

1.71 (a) 5.88×10^{12} mi
(b) 4.13×10^{16} or 41,300,000,000,000,000

1.73 (a) 0.0639 m; 63.9 mm; 6.39×10^7 nm
(b) 0.0550 dm^3; 55.0 mL; 0.0550 L; $5.50 \times 10^{-5}\ m^3$
(c) 2.31×10^4 mg; 0.0231 kg
(d) 37.0 °C; 310.2 K

1.75 0.9321 mi

1.77 9.51 qt

1.79 75 cL

1.81 $T_F\ (T_K - 273.15\ K) \times \frac{9\ °F}{5\ K} + 32\ °F$

1.83 (a) −268.94 °C, −452.09 °F (b) 204 °C

1.85 1474 °F, 1074 K

1.87 2.84 L

1.89 1.58 kg

1.91 7.99 g/cm^3

1.93 50 m^2

1.95 A substance can exist in two phases under certain conditions. For example, at 0 °C both water liquid and solid (ice) can be present.

1.97 124 ft^2

1.99 19.1 g/mL

1.101 (a) trial 1: 3.1×10^8 m/s; trial 2: 3.0×10^8 m/s
(b) The largest uncertainty is in the time measurement. Try to make the peak sharper and decrease the noise to get better results.

1.103 38.6 °C, 311.8 K

1.105 49.5 kg

Chapter 2

2.1 (a) All samples of a pure substance have the same composition.
(b) Atoms do not change mass in chemical reactions.

2.3 Atoms are the simplest chemical component of matter. A substance in which each atom has the same chemical properties is called an *element.* Each different type of atom represents a different element, such as carbon, oxygen, or sodium. Molecules are species formed by the combination of two or more atoms. O_2 is a molecule where the atoms are of the same element, and H_2O is a molecule where different types of atoms combine chemically to form a compound.

2.5 Rutherford used a uranium source to direct alpha particles through a thin metal target. He had detectors of alpha particles surrounding the metal target, and the results were that most of the alpha particles went through the metal foil with no deflection, but a few of the alpha particles were deflected, some through large angles.

2.7 Because aluminum nuclei are much lighter and have less charge than gold, collisions between alpha particles and aluminum nuclei would result in less scattering of alpha particles.

2.9 (a) atomic number: the number of protons in the nucleus
(b) mass number: total number of protons and neutrons in the nucleus
(c) isotopes: atoms with the same atomic number but different mass numbers

2.11 The atomic mass of C is the weighted average of the isotopic masses of ^{12}C and ^{13}C. The isotopic mass is measured quite accurately, but the relative abundance is known to no more than four significant figures, which limits the precision of the atomic mass.

2.13 8S means eight sulfur atoms; S_8 means one sulfur molecule that contains 8S atoms.

2.15 N_2O_4: one molecule of dinitrogen tetroxide is composed of two nitrogen atoms and four oxygen atoms bonded together.

2.17 Sulfur dioxide is made up of discrete molecules, each composed of one sulfur atom and two oxygen atoms. Calcium chloride is an ionic compound made up of Ca^{2+} ions and Cl^- ions in a 1:2 ratio in a regular three-dimensional crystal lattice held together by attractions between oppositely charged particles.

2.19 The name *nitrogen monoxide* should be used to distinguish this compound from other oxides of nitrogen.

2.21 Ionic compounds are usually formed by metals combined with nonmetals. Molecular compounds are usually formed by nonmetals or metalloids combined with other nonmetals.

2.23 Ionic compounds consist of large numbers of positive and negative ions, held together by strong attractions among the charges. Each particle in a molecular compound consists of a relatively small number of atoms. Small molecules do not attract each other strongly.

2.25

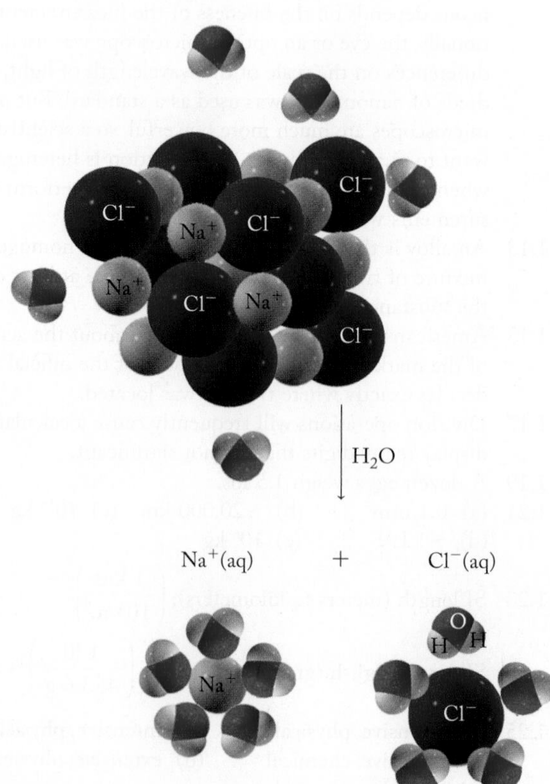

2.27 A group is all the elements in any column, and a period is any horizontal row of the periodic table.

2.29 (a) $^{37}_{17}Cl$ (b) $^{40}_{20}Ca$

2.31 (a) $^{15}_{7}N$ (b) $^{70}_{31}Ga$ (c) $^{40}_{18}Ar$

2.33 (a) 33p, 46n (b) 23p, 28n (c) 52p, 76n

2.35 $^{88}_{38}Sr^{2+}$

2.37 (a) $^{16}_{8}O^{2-}$ (b) $^{79}_{34}Se^{2-}$ (c) $^{59}_{28}Ni^{2+}$

2.39 (a) $^{25}_{12}Mg^{2+}$ (b) $^{27}_{13}Al^{3+}$
(c) $^{29}_{14}Si$ (d) $^{79}_{35}Br^{-}$

2.41

Symbol	$^{23}_{11}Na^{+}$	$^{40}_{20}Ca^{2+}$	$^{81}_{35}Br^{-}$	$^{128}_{52}Te^{2-}$
Atomic number	11	20	35	52
Mass number	23	40	81	128
Charge	1+	2+	1−	2−
Number of protons	11	20	35	52
Number of electrons	10	18	36	54
Number of neutrons	12	20	46	76

2.43 69.72 Ga, gallium
2.45 85.47
2.47 (a) 24.30 (b) ^{24}Mg, ^{25}Mg, ^{26}Mg
2.49 107.9, Ag
2.51 (a) ^{121}Sb, ^{123}Sb (b) 57.5% ^{121}Sb, 42.5% ^{123}Sb
2.53 (a) Rb, rubidium (b) Ru, ruthenium
(c) I, iodine (d) Xe, xenon
2.55 (a) Zr, zirconium (b) S, sulfur
(c) I, iodine (d) Mg, magnesium
2.57 (a) representative (b) transition metal
(c) representative (d) inner transition metal
2.59 (a) representative (b) transition metal
(c) representative (d) inner transition metal
2.61 (a) Na, sodium (b) Cl, chlorine
(c) Ra, radium (d) Ne, neon
2.63 (a) 6 (b) 5 (c) 14
(d) 32 (e) 3
2.65 Ra, Sr; these two elements are in the same group
2.67 Pb, Sn; these two elements are both in Group 14 and are both metals. Carbon is also in Group 14, but it is a nonmental.
2.69 (a) N_2H_2 (b) SO_3
2.71

S
F

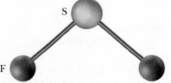

2.73 (a) 70.091 u (b) 100.91115 u (c) 76.0117 u
2.75 294.307 u
2.77 (a) I^- (b) Mg^{2+} (c) O^{2-} (d) Na^+
2.79 (a) CaS (b) Mg_3N_2 (c) FeF_2
2.81 (a) $CaCl_2$ (b) Rb_2S (c) Li_3N (d) Y_2Se_3

2.83 (a) OH^- (b) ClO_3^- (c) MnO_4^-
2.85 (a) HSO_4^- (b) CN^- (c) $H_2PO_4^-$
2.87 (a) $Mg(NO_2)_2$ (b) Li_3PO_4
(c) $Ba(CN)_2$ (d) $(NH_4)_2SO_4$
2.89 (a) $Sr(NO_3)_2$ (b) NaH_2PO_4
(c) $KClO_4$ (d) $LiHSO_4$
2.91 (a) 174.260 u (b) 169.8731 u (c) 53.4912 u
2.93 (a) lithium iodide (b) magnesium nitride
(c) sodium phosphate (d) barium perchlorate
2.95 (a) cobalt(III) chloride (b) iron(II) sulfate
(c) copper(II) oxide
2.97 (a) Mn_2S_3 (b) $Fe(CN)_2$
(c) K_2S (d) $HgCl_2$
2.99 (a) HCl, hydrochloric acid
(b) HNO_2, nitrous acid
(c) $HClO_4$, perchloric acid
2.101 (a) phosphoric acid (b) sulfurous acid
(c) hydrotelluric acid
2.103 KCl
2.105 (a) SF_4 (b) NCl_3
(c) N_2O_5 (d) ClF_3
2.107 (a) phosphorus pentabromide
(b) selenium dioxide
(c) diboron tetrachloride
(d) disulfur dichloride
2.109 (a) *n*-pentane (b) 3-ethylhexane
2.111 (a) *n*-hexane (b) 3-chloroheptane
2.113 The two elements in LiCl are a metal and nonmetal, so it is an ionic compound that will be a solid. Many ionic compounds dissolve in water. The two elements in CO_2 are nonmetals; it is a molecular compound that is a gas.
2.115 Beaker c: Na_2SO_3 is ionic and dissociates in water into two Na^+ for every SO_3^{2-}.
2.117 (a) conductive; Fe^{3+}, Cl^-
(b) nonconductive
(c) conductive; NH_4^+, Br^-
(d) conductive; Na^+, ClO_4^-
(e) nonconductive
2.119 Ca_3N_2
2.121 magnesium hydroxide
2.123 (a) $^{35}_{17}Cl^-$ (b) $^{39}_{19}K^+$
2.125 (a) nitrogen monoxide, molecular
(b) yttrium(III) sulfate, ionic
(c) sodium oxide, ionic
(d) nitrogen tribromide, molecular

2.127

Symbol	$^{70}_{31}Ga^{3+}$	$^{103}_{45}Rh^{3+}$	$^{114}_{49}In^{+}$	$^{28}_{14}Si^{2-}$
Atomic number	31	45	49	14
Mass number	70	103	114	28
Charge	3+	3+	1+	2−
Number of protons	31	45	49	14
Number of electrons	28	42	48	16
Number of neutrons	39	58	65	14

2.129 (a) $CaCl_2$, calcium chloride
(b) CO_2, carbon dioxide
(c) Fe_2O_3, iron(III) oxide

2.131 (a) ${}^{23}_{11}Na^+$ (b) ${}^{121}_{51}Sb^{3+}$ (c) ${}^{84}_{36}Kr$

2.133 Addition often yields a sum with more significant figures than the addends.

2.135 15.0

Chapter 3

3.1 In addition to identifying the reactants and products, a balanced equation also describes the relative molar quantities in the reaction.

3.3

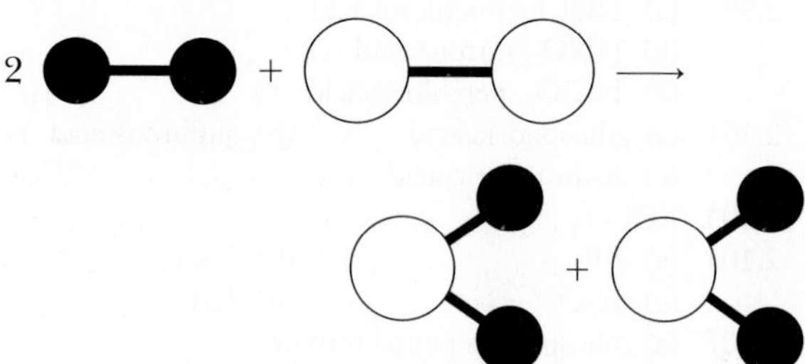

3.5 Mole: amount of matter containing the same number of particles as there are atoms in exactly 12 g ^{12}C.

3.7 Quantities of reactants and products in the laboratory are closer to the order of molar quantities than to molecular quantities.

3.9 [number of atoms] — Avogadro's number → [moles]

3.11 Weigh the sample, then completely react the hydrocarbon with oxygen and pass the products through chambers that selectively absorb carbon dioxide and water. Measure the mass of carbon dioxide and water formed, and calculate %C and %H.

3.13 The molar mass must be known; the ratio of the molecular molar mass to the empirical molar mass is the multiplier that must be used on the empirical formula to obtain the molecular formula.

3.15

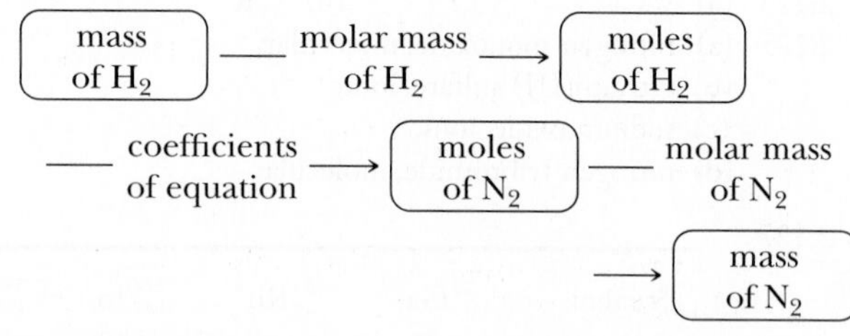

3.17 Convert the masses of each reactant to moles of product. The reactant that produces the least amount of product is the limiting reactant.

3.19 (a) $2CO + O_2 \rightarrow 2CO_2$
(b) carbon dioxide

3.21 (a) $C_5H_{12} + 8O_2 \rightarrow 5CO_2 + 6H_2O$
(b) $4NH_3 + 3O_2 \rightarrow 2N_2 + 6H_2O$
(c) $2KOH + H_2SO_4 \rightarrow K_2SO_4 + 2H_2O$

3.23 (a) $2N_2H_4 + N_2O_4 \rightarrow 3N_2 + 4H_2O$
(b) $2F_2 + 2H_2O \rightarrow 4HF + O_2$
(c) $Na_2O + H_2O \rightarrow 2NaOH$

3.25 (a) $HClO_4(\ell) \xrightarrow{H_2O} H^+(aq) + ClO_4^-(aq)$
(b) $NaNO_3(s) \xrightarrow{H_2O} Na^+(aq) + NO_3^-(aq)$

3.27 (a) $HNO_3(\ell) \xrightarrow{H_2O} H^+(aq) + NO_3^-(aq)$
(b) $K_2CO_3(s) \xrightarrow{H_2O} 2K^+(aq) + CO_3^{2-}(aq)$

3.29 (a) $2NaOH + H_2SO_4 \rightarrow Na_2SO_4 + 2H_2O$
(b) $Ca(OH)_2 + 2HCl \rightarrow CaCl_2 + 2H_2O$
(c) $LiOH + HNO_3 \rightarrow LiNO_3 + H_2O$

3.31 (a) $C_6H_{12} + 9O_2 \rightarrow 6CO_2 + 6H_2O$
(b) $C_4H_8 + 6O_2 \rightarrow 4CO_2 + 4H_2O$
(c) $2C_2H_4O + 5O_2 \rightarrow 4CO_2 + 4H_2O$
(d) $2C_4H_6O_2 + 9O_2 \rightarrow 8CO_2 + 6H_2O$

3.33 (a) $2C_6H_{10} + 17O_2 \rightarrow 12CO_2 + 10H_2O$
(b) $Be(OH)_2 + 2HNO_3 \rightarrow Be(NO_3)_2 + 2H_2O$

3.35 $CS_2 + 3O_2 \rightarrow 2SO_2 + CO_2$

3.37 $(CH_3)_2CO + 4O_2 \rightarrow 3CO_2 + 3H_2O$

3.39 $S_8 + 4Cl_2 \rightarrow 4S_2Cl_2$

3.41 (a) N_2: 0 (b) NaBr: +1 −1 (c) Na_2SO_4: +1 +6 −2
(d) HNO_3: +1 +5 −2 (e) PCl_5: +5 −1 (f) CH_2O: 0 +1 −2

3.43 −1

3.45 $N_2H_4 + 3O_2 \rightarrow 2NO_2 + 2H_2O$
−2 +1 | 0 | +4 −2 | +1 −2
N in N_2H_4 oxidized, O in O_2 reduced

3.47 $Zn + 2HCl \rightarrow ZnCl_2 + H_2$
0 | +1 −1 | +2 −1 | 0
Zn oxidized, H in HCl reduced

3.49 $6NO + 4NH_3 \rightarrow 5N_2 + 6H_2O$
+2 −2 | −3 +1 | 0 | +1 −2
N in NH_3 oxidized, N in NO reduced

3.51 (a) 8.67×10^{23} atoms
(b) 5.88×10^{24} atoms
(c) 6.0×10^{22} atoms

3.53 (a) 5.97×10^{25} molecules
(b) 7.35×10^{23} molecules
(c) 1.38×10^{25} molecules
(d) 1.3×10^{21} molecules

3.55 (a) 5.71 mol (b) 0.0184 mol
(c) 9.25×10^6 mol (d) 2.76 mol

3.57 (a) 39.9971 g/mol (b) 28.054 g/mol
(c) 58.3197 g/mol

3.59 (a) 225.19 g/mol (b) 194.1903 g/mol
(c) 169.393 g/mol

3.61 (a) 0.147 mol (b) 4.4×10^2 g
(c) 0.186 mol

3.63 (a) 0.183 mol (b) 1.72 g
(c) 3.71×10^{22} molecules

3.65 (a) 0.013 mol (b) 0.039 mol
(c) 0.0383 mol

3.67 (a) 1.41 mol (b) 1.70×10^{24} atoms

3.69 (a) 161 g (b) 2.11×10^{24} molecules
(c) 2.11×10^{24} atoms N, 4.22×10^{24} atoms O

3.71 0.016 mol

3.73 (a) 399.443 g/mol (b) 21 g
(c) 0.816 mol (d) 1.1×10^3 C atoms

3.75 CH_3OH, 1.3 mol H

3.77 (a) 85.63% C, 14.37% H
(b) 52.93%C, 5.92% H, 41.15% N
(c) 69.94% Fe, 30.06% O
3.79 62.05% C, 10.41% H, 27.55% O
3.81 32.37% Na, 22.58% S, 45.05% O
3.83 (a) 37.833 g/mol
(b) 60.77% Na, 28.58% B, 10.66% H
3.85 12.79% Fe, 87.21% I; the formula is not correct
3.87 (a) 2.1 g C (b) 1.9 g C
(c) 4.86 g C
3.89 (a) 1.18 g C (b) 1.89 g C
(c) 6.99×10^{-3} g C
3.91 0.491 g C, 45.9% C; 0.114 g H, 10.7% H; 0.465 g O, 43.5% O
3.93 1.38 g C, 44.4% C; 0.232 g H, 7.46% H; 1.50 g N, 48.2% N
3.95 CH_2
3.97 C_2H_3N
3.99 $FeCl_2$
3.101 $TiCl_4$
3.103 C_4H_8O
3.105 $PtN_2H_6Cl_2$
3.107 $C_3H_4O_2$
3.109 $C_4H_6NO_2$
3.111 $C_3H_6O_3$
3.113 (a) $C_6H_{12}O_3$ (b) $C_{12}H_{16}N_4O_{12}$
3.115 $C_9H_{18}O_3$
3.117 $C_2H_4O_2$
3.119 (a) $2C_3H_6 + 9O_2 \rightarrow 6CO_2 + 6H_2O$
(b) 7.69 g CO_2
3.121 3.25 g Cl_2
3.123 118 g Al
3.125 0.97 g Li_2O
3.127 (a) $3H_2 + N_2 \rightarrow 2NH_3$ (b) H_2
3.129 5.6 g NH_3
3.131 0.862 g CO_2
3.133 21%
3.135 70.6%
3.137 37.4%
3.139 (a) 45.1% (b) 9.6 g NaOH remains
3.141 (a) O_2 (b) 28.7 g NH_3
3.143 Fe
3.145 Ca_3N_2: 81.10% Ca, 18.90% N
3.147 572 tons
3.149 $CaSO_4 \cdot \frac{1}{2}H_2O$ 6.2057% water, $CaSO_4 \cdot 2H_2O$ 20.927% water
3.151 76 g CO_2
3.153 9.49×10^{23} atoms N
3.155

Name	Empirical Formula	Molar Mass (g/mol)	Molecular Formula
Dimethyl sulfoxide	C_2H_6SO	78	C_2H_6SO
Cyclopropane	CH_2	42	C_3H_6
Tryptamine	C_5H_6N	160	$C_{10}H_{12}N_2$
Lactose	$C_{12}H_{22}O_{11}$	342	$C_{12}H_{22}O_{11}$

3.157 (a) 40.7 g $K[PtCl_3(C_2H_4)]$
(b) C_2H_4 is in excess, 9.4 g C_2H_4 remains
3.159 214 g NO
3.161 $C_6H_6 + Cl_2 \rightarrow C_6H_5Cl + HCl$

Chapter 4

4.1 Sugar is the solute; water is the solvent.
4.3 Add NaCl; a precipitate indicates the presence of Hg_2^{2+}.
4.5 $CH_3CH_2OOH(\ell) + H_2O(\ell) \rightleftharpoons CH_3CH_2OO^-(aq) + H_3O^+(aq)$
4.7 Two moles of Li^+ ions are present for each mole of Li_2SO_4 present in solution.
4.9 Partially fill the volumetric flask with water before adding the acid; then add additional water to the line.
4.11 In a dilution problem, the moles in the dilute solution are the same as the moles in the concentrated solution, but in a titration problem, the coefficients of the equation can change the numbers of moles of the two substances involved.
4.13 Assuming the Cl is present as Cl^-, it can be precipitated as AgCl by adding $AgNO_3$. The AgCl precipitate can be separated from solution by filtration, dried, and weighed. The amount of Cl present in the sample can be obtained from the mass of AgCl precipitate using g Cl = g precipitate × (g Cl/g AgCl), and percentage Cl can be calculated using % = (g Cl/g sample) × 100%.
4.15 (a) dissolves
(b) does not dissolve
(c) dissolves
(d) dissolves
4.17 (a) dissolves
(b) dissolves
(c) dissolves
(d) does not dissolve
4.19 (a) $Mg^{2+}(aq) + 2OH^-(aq) \rightarrow Mg(OH)_2(s)$
(b) no reaction
(c) $Mg(s) + 2HCl(aq) \rightarrow MgCl_2(aq) + H_2(g)$
4.21 (a) $H^+(aq) + OH^-(aq) \rightarrow H_2O(\ell)$
(b) $Ag^+(aq) + Cl^-(aq) \rightarrow AgCl(s)$
(c) $Ba^{2+}(aq) + CO_3^{2-}(aq) \rightarrow BaCO_3(s)$
4.23 $2AgNO_3(aq) + CaCl_2(aq) \rightarrow 2AgCl(s) + Ca(NO_3)_2(aq)$
$2Ag^+(aq) + 2NO_3^-(aq) + Ca^{2+}(aq) + 2Cl^-(aq) \rightarrow 2AgCl(s) + Ca^{2+}(aq) + 2NO_3^-(aq)$
$Ag^+(aq) + Cl^-(aq) \rightarrow AgCl(s)$
4.25 $(NH_4)_3PO_4(aq) + 3AgNO_3(aq) \rightarrow Ag_3PO_4(s) + 3NH_4NO_3(aq)$
$3NH_4^+(aq) + PO_4^{3-}(aq) + 3Ag^+(aq) + 3NO_3^-(aq) \rightarrow Ag_3PO_4(s) + 3NH_4^+(aq) + 3NO_3^-(aq)$
$3Ag^+(aq) + PO_4^{3-}(aq) \rightarrow Ag_3PO_4(s)$
4.27 Pb^{2+}
4.29 Mg^{2+}
4.31 overall: $Pb(NO_3)_2(aq) + 2NaCl(aq) \rightarrow PbCl_2(s) + 2NaNO_3(aq)$
total ionic: $Pb^{2+}(aq) + 2NO_3^-(aq) + 2Na^+(aq) + 2Cl^-(aq) \rightarrow PbCl_2(s) + 2NO_3^-(aq) + 2Na^+(aq)$
net ionic: $Pb^{2+}(aq) + 2Cl^-(aq) \rightarrow PbCl_2(s)$
4.33 0.587 *M*
4.35 0.00848 *M*
4.37 0.49 L
4.39 0.112 L
4.41 0.028 *M*
4.43 (a) 0.044 *M* (b) 0.018 *M*
4.45 (a) 252 g (b) 1.3×10^2 g
4.47 51.0 g

4.49 20.8 g
4.51 0.0259 *M*; 0.0259 *M* Sr^{2+}, 0.0518 *M* Cl^-
4.53 0.0290 *M*; 0.0290 *M* Mg^{2+}, 0.0580 *M* NO_3^-
4.55 19.4 g
4.57 Transfer 600 mL stock solution to a 2-L volumetric flask and add water to fill the flask to the calibration mark.
4.59 (a) 0.10 mol (b) 3.2 mol
4.61 (a) 0.281 mol (b) 0.00113 mol
4.63 (a) 3.7×10^{-3} mol K_2SO_4 (b) 8.3×10^{-2} mol NaCl
4.65 0.54 L
4.67 0.049 g
4.69 25.0 g
4.71 0.55 L
4.73 0.221 *M*
4.75 0.93 g
4.77 6.73 g
4.79 $BaSO_4$, 4.90 g
4.81 118.8 mL
4.83 $(NH_4)_2SO_4(aq) + BaCl_2(aq) \rightarrow BaSO_4(s) + 2NH_4Cl(aq)$
17 g
4.85 overall: $AgNO_3(aq) + NaBr(aq) \rightarrow AgBr(s) + NaNO_3(aq)$
complete ionic: $Ag^+(aq) + NO_3^-(aq) + Na^+(aq) + Br^-(aq) \rightarrow AgBr(s) + Na^+(aq) + NO_3^-(aq)$
net ionic: $Ag^+(aq) + Br^-(aq) \rightarrow AgBr(s)$
mass: 21.4 g
4.87 0.0205 *M*
4.89 0.735 *M*
4.91 (a) 15 mL (b) 43.0 mL
4.93 0.00664 *M*
4.95 0.00168 *M*
4.97 0.032 *M*
4.99 $CaCO_3$, 0.150 *M*
4.101 34.0%
4.103 90.9%
4.105 0.0039 *M*
4.107 $Be(OH)_2$ will precipitate.
4.109 54.8 mL
4.111 0.0172 *M*
4.113 49.8 g
4.115 100 mL
4.117 21.5%
4.119 $BaSO_4$, 0.432 g
4.121 0.299%
4.123 25.1 g

Chapter 5

5.1 The enthalpy of a substance is a function of its physical state.
5.3 The enthalpy of formation is the enthalpy of reaction when 1 mol of a substance is formed from the most stable form of its constituent elements in their standard states. All enthalpies of formation are enthalpies of reaction, but not all enthalpies of reaction are enthalpies of formation.
5.5 The specific heats of the contents of a calorimeter affect the temperature change that accompanies the chemical or physical change that occurs in the calorimeter.
5.7 Heat is the flow of energy from one object to another that causes a change in the temperature of the object. It has units of joules. Heat is a form of energy.
5.9 The system is the portion of the universe of interest. The surroundings are everything else.
5.11 An endothermic reaction absorbs energy from the surroundings into the system.
5.13 The heat of a process is equal to the enthalpy change for the process if the process occurs under conditions of constant pressure.
5.15 The law of conservation of energy states that the total energy of the universe—the system plus the surroundings—is constant during a chemical or physical change.
5.17

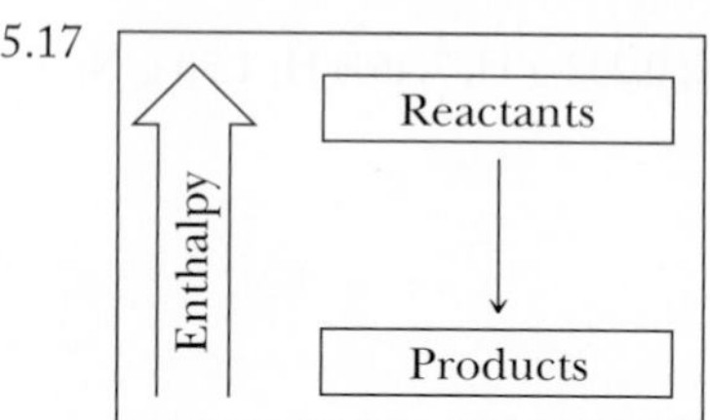

5.19

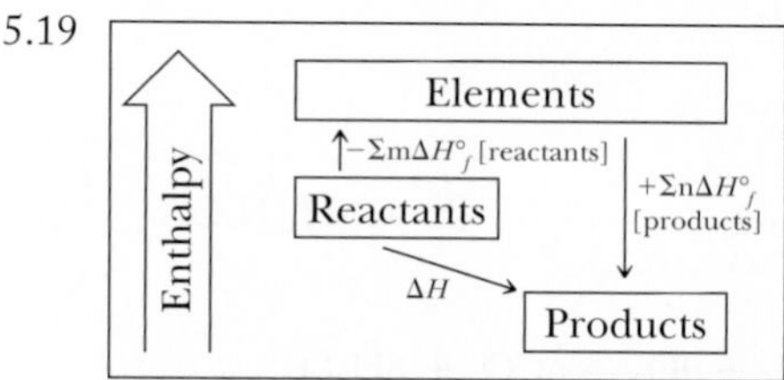

5.21 Methane gives off more energy per gram, 55.6 kJ/g versus 47.9 kJ/g, than octane. The difference is 7.7 kJ/g.
5.23 The calorimeter and its contents are the only parts of the surroundings that are used to calculate the ΔH of reaction because the energy change of the reaction is confined to the inside of the calorimeter by the insulating layer.
5.25 Hess's law leads to Equation 5.11 because when a general chemical reaction is broken down into the sum of its formation reactions, all of the products of a reaction are represented by the forward formation reaction, whereas all of the reactants of the reaction are represented by the reverse of the formation reaction.
5.27 The table of standard enthalpies of formation does not include the common forms of the elements because their enthalpies of reaction are equal to zero.
5.29 The two factors are the mass and the specific heat of the material in the system.
5.31 7770 cal; exothermic
5.33 (a) energy is released
(b) 1 mol C; 1mol O_2; 1 mol CO_2
(c) −98.3 kJ
5.35 (a) exothermic
(b) 1 mol CH_4; 2 mol O_2; 1 mol CO_2; 2 mol H_2O
(c) −55.5 kJ
5.37 (a) −1240 kJ (b) heat is released
5.39 −789 kJ
5.41 (a) $2C_8H_{18}(\ell) + 25O_2(g) \rightarrow 16CO_2(g) + 18H_2O(\ell)$; $\Delta H = -10{,}900$ kJ
(b) −478 kJ
5.43 (a) negative (b) -1.14×10^3 kJ

5.45 +14.3 kJ
5.47 −379 kJ
5.49 −59.3 kJ
5.51 0.0440 g
5.53 34.7 kJ
5.55 0.211 kJ
5.57 (a) 830 J (b) 830 J (c) 0.232 J/g·°C
5.59 44.4 °C
5.61 −2.2 kJ; endothermic
5.63 (a) −9.12 kJ; exothermic
(b) $Mg(s) + 2HCl(aq) \rightarrow MgCl_2(aq) + H_2(g)$; $\Delta H = -472$ kJ

5.65 (a)

$CO(g) + H_2(g)$

↑ $\Delta H = +131.3$ kJ

$C(s) + H_2O(g)$

(b)

$CO(g) + H_2O(g)$

↓ $\Delta H = -41.2$ kJ

$CO_2(g) + H_2(g)$

(c)

$2SO_2(g) + O_2(g)$

↓ $\Delta H = -197.8$ kJ

$2SO_3(g)$

5.67 −312 kJ
5.69 −348.0 kJ
5.71 −220.1 kJ
5.73

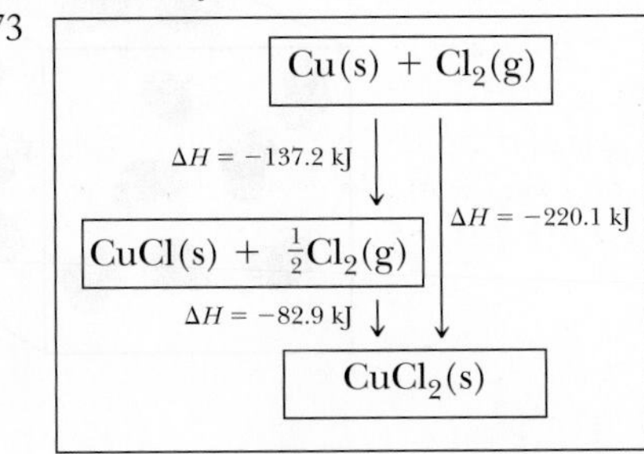

5.75

$Cu(s) + Cl_2(g)$

↑ $\Delta H = 137.2$ kJ ↑ $\Delta H = 220.1$ kJ

$CuCl(s) + \frac{1}{2}Cl_2(g)$

↑ $\Delta H = 82.9$ kJ

$CuCl_2(s)$

5.77 (a) $\frac{1}{2} H_2(g) + \frac{1}{2} Br_2(\ell) \rightarrow HBr(g)$
(b) $H_2(g) + S(s) + 2O_2(g) \rightarrow H_2SO_4(\ell)$
(c) $3/2\ O_2(g) \rightarrow O_3(g)$
(d) $Na(s) + \frac{1}{2} H_2(g) + S(s) + 2O_2(g) \rightarrow NaHSO_4(s)$

5.79 (a) −74 kJ, exothermic
(b) −2878.46 kJ, exothermic
5.81 (a) 131.69 kJ, endothermic
(b) −132.44 kJ, exothermic
(c) −176.45 kJ, exothermic
5.83 -5.9×10^2 kJ
5.85 (a) $C_8H_{18}(\ell) + 25/2\ O_2(g) \rightarrow 8CO_2(g) + 9H_2O(\ell)$; $\Delta H = -5456.6$ kJ
(b) −264.0 kJ/mol
5.87 (a) 2.9 metric tons
(b) 52 kg
(c) In the short term, burning coal will be much worse, but the long-term effect of treating and storing 1 g radioactive waste is difficult to assess.
5.89 −172 kJ
5.91 68.1 g
5.93 2.42 J/g·°C
5.95 (a) 222 J/°C (b) +40.6 kJ/mol
5.97 31.6 kJ/mol
5.99 (a) $2C_3H_6(g) + 9O_2(g) \rightarrow 6CO_2(g) + 6H_2O(\ell)$; $\Delta H = -4182$ kJ
(b) $\Delta H_f = +53.0$ kJ/mol
5.101 (a) CH_2
(b) $\Delta H_{vap} = 28.3$ kJ/mol; approximate molar mass = 72.8 g/mol
(c) molecular formula = C_5H_{10}
5.103 (a) \$58.06/mol (b) 36.6 kJ
5.105 (a) 54.3 g/mol (b) MCl_3 (c) M = Mn
5.107 6.42×10^4 kJ is released.

Chapter 6

6.1 Gases and liquids both take on the form of their containers, but gases expand to occupy the volume of the container, whereas liquids retain a constant volume.

6.3 A barometer is an inverted, closed tube filled with mercury. The downward force caused by the weight of the mercury in the tube is counterbalanced by the air pressure. Therefore, the height of the column of mercury is directly proportional to the atmospheric pressure.

A manometer works on a similar principle, except that one end of the tube is connected to a sealed sample of gas; therefore, the height of the column of mercury is directly proportional to the difference in pressure between the sample and the atmosphere.

6.5 atmosphere: a unit of pressure equal to the normal atmospheric pressure at sea level
torr (or mm Hg): a unit of pressure equal to 1/760 of 1 atmosphere
pascal: the SI unit of pressure, equal to $1\ kg/m \times s^2$

6.7 Measure the volume of a fixed sample of gas at fixed pressure as a function of temperature. A plot of V versus T will yield a straight line with an x-intercept equal to absolute zero. Because all gases condense into liquids or solids above absolute zero, the value is determined indirectly.

6.9 $PV = nRT$ (ideal gas law), where P = pressure, V = volume, n = moles, R = ideal gas law constant, and T = temperature

$n = \frac{m}{M}$, where n = moles, m = mass and M = molar mass

substituting for n gives $PV = \frac{m}{M}RT$

$D = \frac{m}{V}$, where D = density, m = mass and V = volume

solving for density gives $D = \frac{m}{V} = \frac{PM}{RT}$

6.11 At low temperatures, the attractive forces of particles are significant enough to cause deviations from the ideal gas law. At extremely low temperatures, gases near their condensation point where the attractive forces dominate the kinetic energy and liquid formation occurs.

6.13
1. Gases are made up of small particles that are in constant and random motion.
2. Collisions between gas molecules with each other or the walls of the container are elastic. Attractive and repulsive forces between molecules are negligible.
3. Gas particles are small compared with the average distance that separates them.
4. The average kinetic energy of a gas is directly proportional to the absolute temperature.

6.15

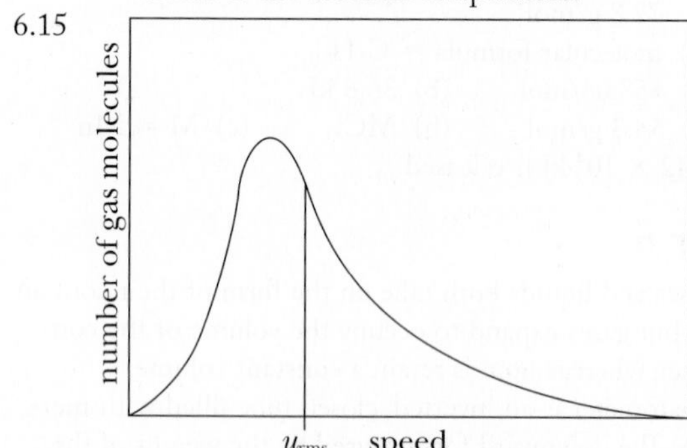

6.17 At high pressures, the gas molecules are close together. This has the effect of (a) making the volume of the molecule significant and (b) making the attractive forces between gas molecules significant.

6.19 (a) 0.439 atm (b) 0.03892 atm (c) 1.8×10^3 torr

6.21 (a) 318 K (b) 245 K (c) 503 K

6.23 2.52 atm

6.25 52.8 mL

6.27 87.2 mL

6.29 474 mL

6.31 4.38×10^4 ft^3

6.33 5 (enough gas for 5.5)

6.35

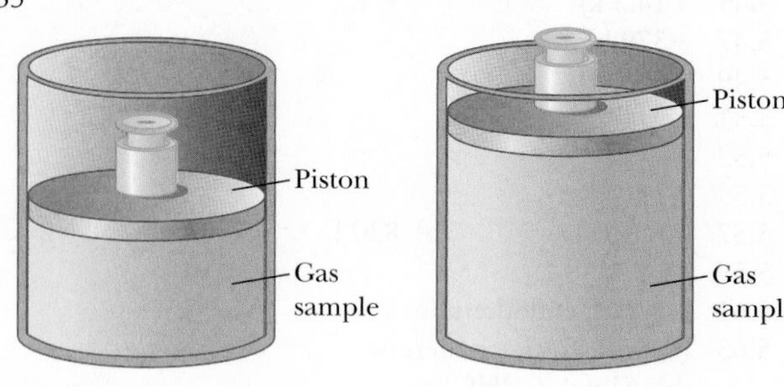

6.37 0.0388 mol

6.39 12.2 mol

6.41 2.02×10^3 L

6.43 0.558 atm

6.45 6.97×10^{23} molecules

6.47 6.4 atm

6.49 52.7 g/mol

6.51 147 g/mol

6.53 1.79 g/L

6.55 1.79 g/L

6.57 67.0 g/L

6.59 30.4 mL

6.61 1.31 L

6.63 0.444 L

6.65 0.713 L

6.67 12.6 L

6.69 2.44 L SO_2; 3.66 L O_2

6.71 2 L NH_3 formed, 2 L N_2 remains

6.73

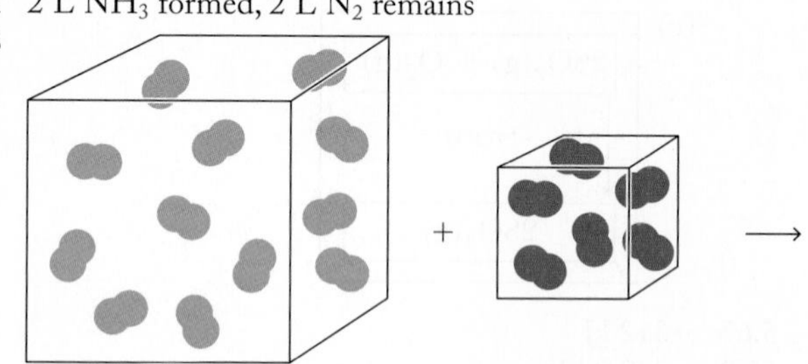

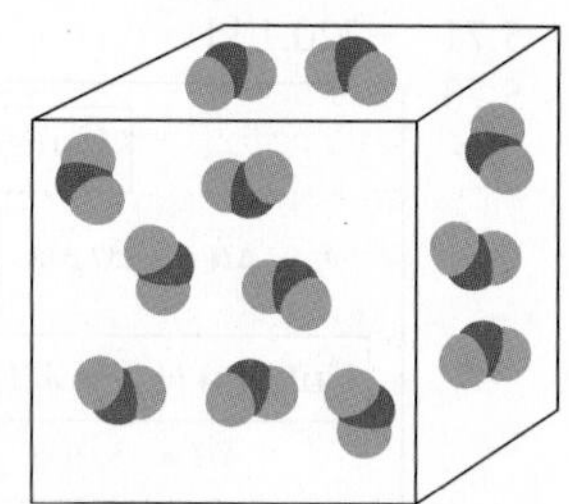

6.75 5.55 atm

6.77 6.06 atm

6.79 17.6 atm

6.81 P_{H_2} = 5.60 atm, 7.13 atm total

6.83 1.76 atm

6.85 $P_{Ne} = 0.74$ atm; $P_{N_2} = 1.1$ atm, $P_{O_2} = 0.74$ atm
6.87 730 torr
6.89 44.0 mL
6.91 $P_{left} = 0.985$ atm
$P_{right} = 0.493$ atm
$P_{open} = 0.739$ atm
6.93 $P_{CO_2} = 87.9$ atm, $P_{N_2} = 3.1$ atm, $P_{H_2O} = 0.18$ atm
6.95 $O_2 < N_2 < Ne$
6.97 Ar < Ne (25°) < Ne (100°)
6.99 679 m/s
6.101 28.0 g/mol
6.103 $\frac{He}{Ne} = 2.245$
6.105 $\frac{He}{Ar} = 3.16$
6.107 116 g/mol
6.109 70.2 g/mol
6.111 (a) O_2 at 30° is further from its boiling point
(b) N_2 is further from its boiling point
(c) Ar at 1 atm; the molecules are more widely separated
6.113 (a) CO_2 at 0.05 atm; the molecules are more widely separated
(b) Ne is further from its boiling point
(c) SO_2 at 50° C is further from its boiling point
6.115 208 atm, ideally; 218 atm by van der Waals equation
6.117 167 atm, ideally; 174 atm by van der Waals equation
6.119 pressure declines 5 psi
6.121 64 atm
6.123 5.0×10^3 L
6.125 6.04 g
6.127 786 g
6.129 (a) B (b) B
6.131 31.6 mL
6.133 4.38 atm
6.135 24.5%

Chapter 7

7.1 Speed, frequency, and wavelength are the same. Only amplitude is different. The energy of the photons from the two light sources is the same, but the total energy from the brighter light source is twice that of the other.
7.3 Nearly all atoms are in the ground state, so when a hydrogen atom absorbs energy, it goes in one step from its ground-state energy to a higher energy level. When an excited atom loses energy, it can do so in a single large step or in a sequence of smaller steps from any one of many excited energy levels.
7.5 If de Broglie's hypothesis of the wave nature of matter were true, then the electron orbit could be viewed as a standing wave, for which the circumference must be some integer multiple of the wavelength.
7.7 The spin quantum number has no effect on the energy of the H atom.
7.9 The 2*s* electron experiences a higher effective nuclear charge because its radial charge density penetrates the 1*s* orbital better.
7.11 The nuclear charge of a Be atom is 4+. The effective nuclear charge is less because of shielding. The electrons with $n = 2$ contribute little to the shielding of 1*s* electrons, so Z_{eff} is larger for the 1*s* electrons than for the 2*s* electrons.
7.13 The uncertainty of the position of the electron is large compared with the size of an atom, but the uncertainty is negligible compared with the size of a baseball.
7.15

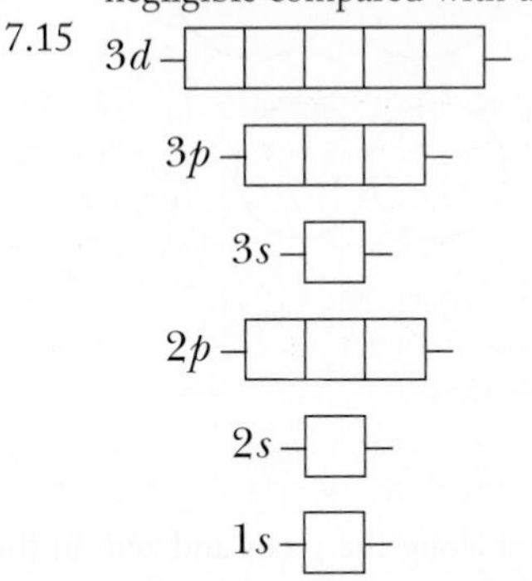

7.17 Electrons occupy orbitals in a subshell singly, with parallel spin, until each orbital is half full, before pairing up. Carbon has only two electrons in the 2*p* subshell, and they are unpaired, in accordance with Hund's rule. Oxygen has four electrons in the 2*p* subshell; the first three half fill the three 2*p* orbitals, and the fourth pairs with the first, leaving two electrons unpaired.
7.19 1.056×10^{15} s^{-1}, ultraviolet
7.21 5.54×10^{-7} m, visible
7.23 517 m = 5.17×10^{11} nm
7.25 1.21 pm
7.27 1.4×10^7 s^{-1}
7.29 (a) 3.73×10^{-19} J (b) 2.68×10^{18} photons/s
7.31 1.48×10^3 kJ
7.33 3.37×10^{-19} J
7.35 (a) 259 nm, near ultraviolet (b) no
7.37 6.2×10^{15} electrons are ejected, 6.2×10^{15} photons are absorbed per second
7.39 102.6 nm, ultraviolet
7.41 2.3 nm
7.43 332 pm, which equals the circumference of the first Bohr orbit
7.45 (a) 1.66×10^{-34} m (b) 3.58×10^{-38} m
(c) 1.47×10^{-10} m
7.47 2.65×10^3 m/s
7.49 (a) 6*p* (b) 3*s*
(c) 5*d* (d) 4*s*
(e) not allowed, ℓ cannot be greater than $n - 1$
7.51 (a) $n = 3, \ell = 1$ (b) $n = 5, \ell = 2$
(c) $n = 7, \ell = 0$ (d) $n = 4, \ell = 3$
(e) $n = 2, \ell = 0$

7.53 (a) $2p$
(b) not allowed, ℓ cannot be greater than $n - 1$
(c) $3s$
(d) not allowed, $|m_\ell|$ must be less than or equal to ℓ
(e) $3d$
(f) $5s$

7.55 (a) 4 (b) 5
(c) 2 (d) $n = 3, \ell = 1, m_\ell = -1, 0,$ or $+1$

7.57 (a)

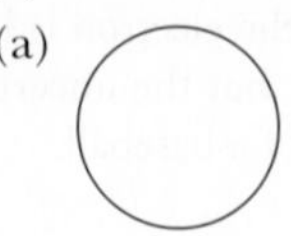

(b)

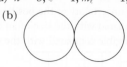

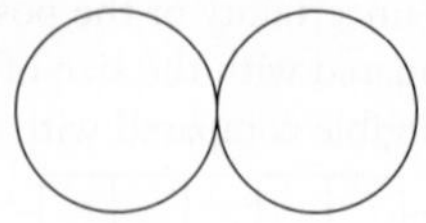

(c)

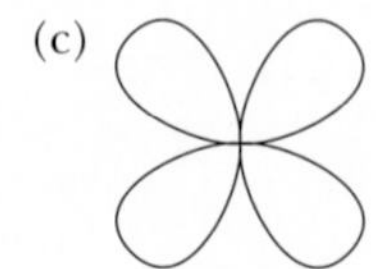

(d)

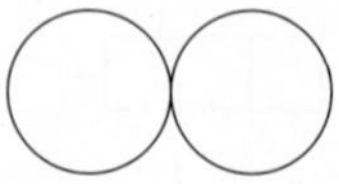

(e)

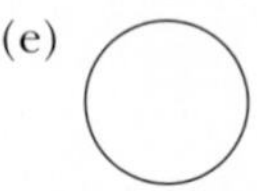

7.59 The probability is greatest along the y axis and zero in the xz plane.

7.61 s

7.63

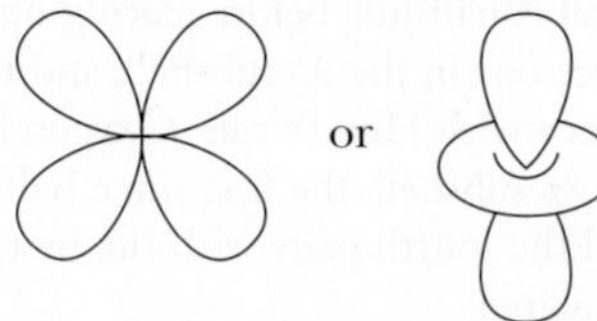

7.65 $2s < 3s = 3p_x = 3p_y < 4s = 4p_z = 4d_{xy}$

7.67 11.4 nm, far ultraviolet

7.69 (a) $2s < 2p < 3p < 3d < 5p$
(b) $1s < 2s < 2p < 3s < 3d < 4d$
(c) $1s < 2s < 2p < 3s < 3p < 3d < 4p$

7.71 (a) C $1s^2\,2s^2\,2p^2$ and O $1s^2\,2s^2\,2p^4$
(b) N $1s^2\,2s^2\,2p^3$
(c) Li $1s^2\,2s^1$ and Be $1s^2\,2s^2$

7.73

(a) C: $2p$ [↑][↑][ ]; $2s$ [↑↓]; $1s$ [↑↓]
O: $2p$ [↑↓][↑][↑]; $2s$ [↑↓]; $1s$ [↑↓]

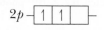

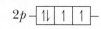

(b) N: $2p$ [↑][↑][↑]; $2s$ [↑↓]; $1s$ [↑↓]

(c) Li: $2s$ [↑]; $1s$ [↑↓]
Be: $2s$ [↑↓]; $1s$ [↑↓]

7.75 Ne and Pd have no unpaired electrons.

7.77 (a) $1s$ (b) $4p$ (c) $2p$ (d) $4f$

7.79 2; 1; 1, 0 or −1; +1/2 or −1/2

7.81 2, 0, 0, +1/2

7.83 (a) 10 (b) 2 (c) 8 (d) 50

7.85 (a) O ground state

$2s$ [↑↓]

(b) Be excited state, $1s$ [↑↓]
(c) P ground state

$2s$ [↑]

(d) Li excited state, $1s$ [↑↓]

7.87 0.672 m

7.89 $\nu = 6.08 \times 10^{14}$ Hz, $E = 4.03 \times 10^{-19}$ J

7.91 1875 nm and 1282 nm

7.93 $\Delta x = 1.8 \times 10^{-37}$ m; this uncertainty is too small to be significant

7.95 $-1313/n^2$ kJ/mol

7.97 (a) Li, $1s^2\,2s^1$ $2s$ [↑]

$1s$ [↑↓]

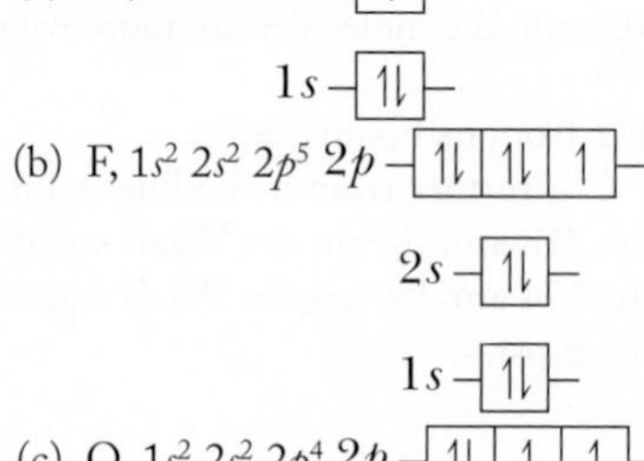

(b) F, $1s^2\,2s^2\,2p^5$ $2p$ [↑↓][↑↓][↑]

$2s$ [↑↓]

$1s$ [↑↓]

(c) O, $1s^2\,2s^2\,2p^4$ $2p$ [↑↓][↑][↑]

$2s$ [↑↓]

$1s$ [↑↓]

(d) Ga, $1s^2\,2s^2\,2p^6\,3s^2\,3p^6\,4s^2\,3d^{10}\,4p^1$

$4p$ [↑][ ][ ]

$3d$ [↑↓][↑↓][↑↓][↑↓][↑↓]

$4s$ [↑↓]

$3p$ [↑↓][↑↓][↑↓]

$3s$ [↑↓]

$2p$ [↑↓][↑↓][↑↓]

$2s$ [↑↓]

$1s$ [↑↓]

7.99 (a) $\Delta E\ (3,2) = 3.03 \times 10^{-19}$ J,
$\Delta E\ (32,33) = 1.27 \times 10^{-22}$ J
(b) 2.18×10^{-18} J
(c) The energy levels get closer together as n increases, until the electron is removed from the atom.
(d) This energy corresponds to the ionization energy of H.

7.101 (a) B, ground state
(b) C, ground state
(c) S, excited state; ground state is

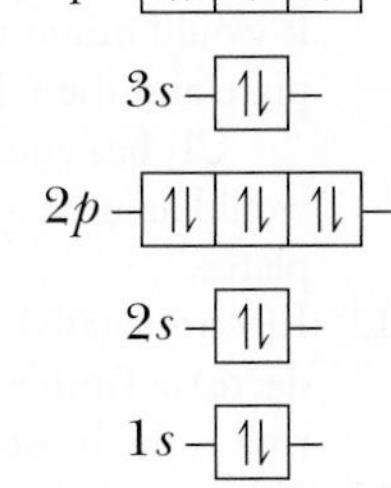

(d) impossible state; cannot have three electrons in one orbital

7.103 (a) 7.27×10^6 m/s
(b) 3.97×10^3 m/s; $u_{rms} = 2.73 \times 10^3$ m/s

7.105 $R = 1.07 \times 10^{-16}$ J; species is N^{6+}

Chapter 8

8.1 An isoelectronic series is a group of atoms and ions that have the same electron configuration. An example would be O^{2-}, F^-, Ne, and Na^+.

8.3 See Figure 8.10 for an example of how your graph should look. Across the period, the atomic radius generally decreases. Down Group 1A, the radius increases.

8.5 S is larger than O because it has one additional occupied shell.

8.7 An atom has as many ionization energies as it has electrons. This is because it requires a different amount of energy to remove each electron.

8.9 Although ionization energies generally increase as you move across the periodic table, the first ionization energy of aluminum is less than that of magnesium because the electron being removed from an aluminum atom is in the $3p$ subshell, which is slightly less bound to the atom than is a $3s$ electron.

8.11 For manganese, iron, and cobalt, the first electrons being removed are in the $4s$ subshell, and the shielding they experience by the increasing numbers of $3d$ electrons largely cancels out the increase in nuclear charge. In gallium, germanium, and arsenic, the $3d$ subshell is filled and provides the same amount of shielding for each of the elements, so the effect of increasing nuclear charge is more noticeable.

8.13 The second ionization energy of boron will be significantly higher than 800 kJ/mol because the second electron is coming out of the $2s$ subshell. The third ionization energy will be larger than the second because, whereas it is also a $2s$ electron, the resulting ion has a larger charge, making it more difficult to remove the next electron. The next ionization energy will again be significantly higher than the previous one because this fourth electron is coming from the $1s$ subshell.

8.15 The ionization energy of silicon is greater than that of aluminum because the $3p$ electron removed from silicon experiences a larger effective nuclear charge than the $3p$ electron removed from aluminum.

8.17 Large exothermic electron affinities are expected in Group 7A, the halogen elements, because those atoms preferentially form 1− ions.

8.19 In Li, the added electron is completing the $2s$ subshell, which is a slightly energetically favorable process. In Be, an added electron goes into the previously empty $2p$ orbital, which is not as energetically favorable. On the other hand, the first ionization energy of Be is significantly larger than that of Li, because in removing an electron, the Be atom is going from a filled $2s$ subshell to an unfilled $2s$ subshell; in Li, the $2s$ subshell is already unfilled and does not experience the stability that accompanies filled subshells.

8.21 The elements in Group 2A are all solid, metallic elements with a shiny appearance.

8.23 (a) Reactivity increases as one goes down the group.
(b) Reactivity increases as one goes down the group.
(c) Reactivity decreases as one goes down the group.

8.25 (a) $1s^2\ 2s^2\ 2p^6\ 3s^2\ 3p^3$
(b) $1s^2\ 2s^2\ 2p^6\ 3s^2\ 3p^6\ 4s^2\ 3d^{10}\ 4p^6\ 5s^2$
(c) $1s^2\ 2s^2\ 2p^6\ 3s^2\ 3p^6\ 4s^2\ 3d^{10}\ 4p^6\ 5s^2\ 4d^{10}\ 5p^6\ 6s^2\ 4f^6$
(d) $1s^2\ 2s^2\ 2p^6\ 3s^2\ 3p^6\ 4s^2\ 3d^{10}\ 4p^6\ 5s^2\ 4d^{10}\ 5p^6\ 6s^2\ 4f^{14}\ 5d^{10}\ 6p^6\ 7s^2$

8.27 (a) f block (b) d block
(c) p block (d) d block

8.29 (a) [Ne] $3s^2\ 3p^6$ (b) [Ar] $3d^5$ (c) [Ar] $4s^2\ 3d^{10}$

8.31 (a) [Ne] $3s^2\ 3p^5$ (b) [Ar] $3d^6$ (c) [Ar] $3d^6$

8.33 (a) Si^+ (b) Mg^+ (c) Ga^+

8.35 (a) Se^{2-} (b) N^{2-} (c) S^{2-}

8.37 Fe^{3+}: [Ar] $3d^5$; Cr^{3+}: [Ar] $3d^3$

8.39 As^-, Ge^{2-}, Br^+, Kr^{2+}

8.41 (a) Ne (b) Cr (c) V (d) Cd

8.43 Ca^{2+}

8.45 (a) Na, which has one electron in the $n = 3$ shell
(b) O^{2-}, which has a lower effective nuclear charge
(c) Ni^{2+}, which has more electrons

8.47 (a) Na, which has a lower effective nuclear charge
(b) B, which has a lower effective nuclear charge
(c) Be^{2+}, which has more electrons

8.49 (a) O < B < Li (b) N < C < Si
(c) S < As < Sn

8.51 (a) Be^{2+} < Be < Li (b) Cl < S < S^{2-}
(c) O < S < Si

8.53 Li, because it has the smallest effective nuclear charge

8.55 $1s^2 2s^2 2p^6 3s^2 3p^3$, which is P

8.57 (a) Cl, which has the larger effective nuclear charge
(b) Na, whose valence shell is smaller
(c) F^-, which has a larger effective nuclear charge

8.59 (a) Cl, whose valence shell is smaller
(b) F, which has a larger effective nuclear charge
(c) Al^{3+}, which has more protons in its nucleus

8.61 (a) O^{2-} < O < F (b) Si < C < N
(c) Sr < Ru < Te

8.63 (a) $N^{3-} < N < Ne$ (b) $Si < P < Cl$
(c) $Ga < Se < O$

8.65 Aluminum, because it has a larger effective nuclear charge

8.67 Magnesium, because the third electron must be removed from an inner shell orbital

8.69 The second ionization energy of Li is so much higher than the first that the formation of a 2+ ion is energetically unfavorable.

8.71 [Kr] $5s^2\ 4d^{10}\ 5p^5$. The third electron must be removed from the $5p$ subshell, which requires much more energy than removing the first two electrons.

8.73 (a) Cl, which has a higher effective nuclear charge
(b) O, which has a larger effective nuclear charge
(c) F, which has a smaller valence shell

8.75 (a) F, which has a larger effective nuclear charge
(b) Cl, which has a larger effective nuclear charge
(c) Br, which has a larger effective nuclear charge

8.77 (a) $2Na + O_2 \rightarrow Na_2O_2$
(b) no reaction
(c) $2Na + Cl_2 \rightarrow 2NaCl$
(d) $2Na + 2H_2O \rightarrow 2NaOH + H_2$

8.79 (a) $Ba + O_2 \rightarrow BaO_2$
(b) $Ba + 2H_2O \rightarrow Ba(OH)_2 + H_2$

8.81 Reactivity increases from Li to Cs because the valence shell gets larger.

8.83 $2K + 2H_2O \rightarrow 2KOH + H_2$; $Ca + 2H_2O \rightarrow Ca(OH)_2 + H_2$. The potassium reaction is faster because potassium is more chemically active than calcium.

8.85 (a) size decreases
(b) ionization energy increases
(c) electron affinity increases (becomes more positive)

8.87 $1s^22s^22p^63s^23p^6$

8.89 $1s^22s^22p^63s^2$ is Mg, 160 pm; $[Ar]4s^2$ is Ca, 231 pm; $1s^22s^22p^63s^23p^5$ is Cl, 99 pm

8.91 [Kr] $4d^8$; the configuration of Pd does not influence the electron configuration of Pd^{2+}.

8.93 2

8.95 (a) $1s^22s^22p^1$ (b) $1s^22s^22p^1$ (c) $1s^22s^22p^5$

8.97 Ga^{4+}; the fourth ionization energy of Ga is too high for the ion to form

8.99 (a) $O < C < Li$ (b) $Li < C < O$
(c) $C < O < Li$ (d) $Li < C = O$

8.101 $1s^2\ 2s^2\ 2p^6\ 3s^2\ 3p^6$

8.103 $2Na + 2H_2O \rightarrow 2NaOH + H_2$; 1.83 g

8.105 3.47 g Fe, 4.41 g Cl_2

8.107 N; $1s^2\ 2s^2\ 2p^3$; [He] 2s: [↑↓] 2p: [↑|↑|↑]

Chapter 9

9.1 A Lewis electron-dot symbol is a notation for showing the valence electron arrangement for an atom, ion, or molecule.

9.3 (a) $\cdot Be \cdot + \cdot\ddot{O}\cdot \longrightarrow Be^{2+} + :\ddot{O}:^{2-}$
(b) $\cdot\dot{Y}\cdot + 3:\ddot{Cl}\cdot \longrightarrow Y^{3+} + 3:\ddot{Cl}:^{-}$

9.5 (a) Na^+ is smaller than K^+, so the ionic bond is shorter and, therefore, stronger.
(b) Mg^{2+} has a larger charge than Na^+, resulting in a larger coulombic force.

9.7 Na does not form 2+ ions because the second ionization energy for Na is too high because it removes a core electron.

9.9 When forming covalent compounds, atoms tend to gain enough electrons through sharing to completely occupy the outer s and p subshells. The rule applies to most 2nd period elements because they have no outer d subshell that could contribute to bonding.

9.11 HCl has unequal charge distribution in the HCl bond, so it would orient the H (positive end) toward the negative plate and the Cl (negative end) toward the positive plate.
Cl_2 has equal charge distribution in its bond, so it would assume a random orientation relative to the charged plates.

9.13 Electronegativities tend to increase from left to right and decrease from top to bottom of the periodic chart. Fluorine is the most electronegative element.

9.15 The formal charge is the charge on the atom when shared electrons are divided equally between the bonded atoms.

9.17 Structures that minimize the amount of formal charge found on each atom are more likely to be correct than structures that place large amounts of formal charge on atoms. Also, structures that have adjacent atoms with formal charges of the same sign are much less favorable.

9.19 Radicals are atoms with one or more unpaired electrons. They violate the octet rule because they do not have completely filled s and p subshells.

9.21 Bond dissociation energy is the energy required to break all the bonds in a mole of gaseous molecules. The enthalpy of reaction can be approximated by the sum of the bond dissociation energies of all the reactants minus the sum of the bond dissociation energies of all the products.

9.23 (a) $\dot{Na}$ (b) $:\ddot{F}\cdot$ (c) $:\ddot{O}:^{2-}$ (d) Mg^{2+}

9.25 (a) $\dot{Li}$ (b) $:\ddot{S}:^{2-}$ (c) $\cdot Mg\cdot$ (d) $:\ddot{Br}\cdot$

9.27 (a) Li_2O, Na_2S—Li_2O will have the greater lattice energy because the ions are smaller
(b) KCl, MgF_2—MgF_2 will have the greater lattice energy because the Mg ion has a larger charge

9.29 LiCl, because Cl^- is smaller than I^-

9.31 (a) $KBr < NaBr < NaCl$
(b) $CaCl_2 < CaO < MgO$
(c) $LiF < BeF_2 < BeO$

9.33 (a) CF_4 (b) NI_3 (c) Cl_2O

9.35 (a) H—$\ddot{S}$—H; lone pairs; bonding pairs
(b) H—C(=$\ddot{O}$:)—H; lone pairs; bonding pairs
(c) $:\ddot{F}$—$\ddot{P}$(—$:\ddot{F}:$)—$\ddot{F}:$; bonding pairs; lone pairs; lone pairs

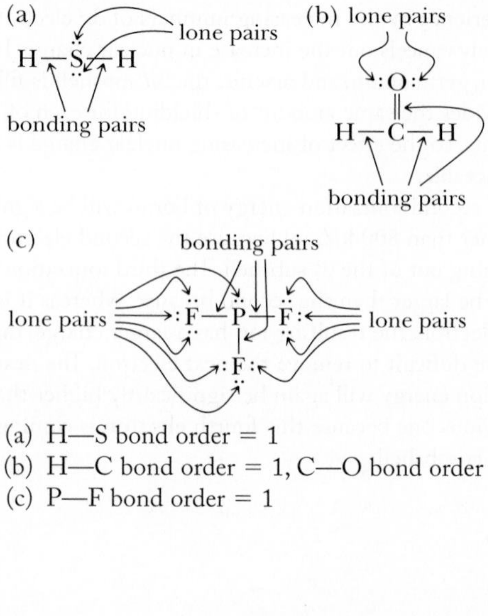

9.37 (a) H—S bond order = 1
(b) H—C bond order = 1, C—O bond order = 2
(c) P—F bond order = 1

9.39 (a) H—As(H)—H (with one lone pair on As) (b) :Cl—F: (with lone pairs) (c) F—C(F)(F)—O—H (with lone pairs)

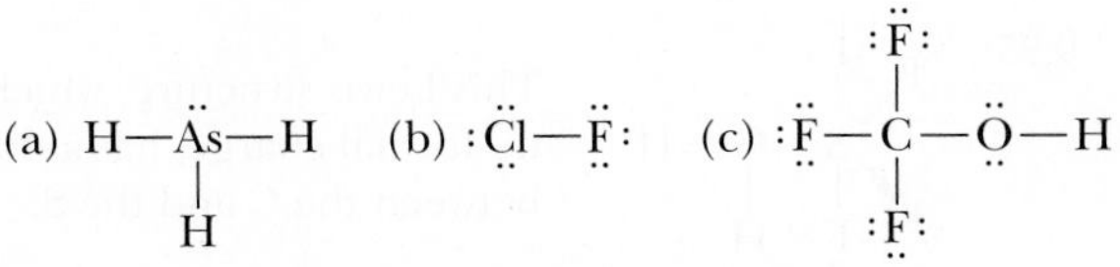

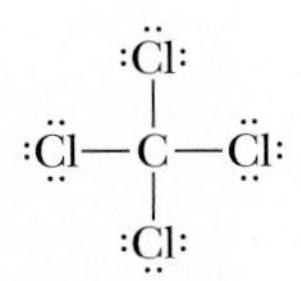

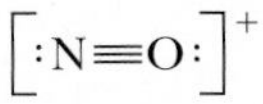

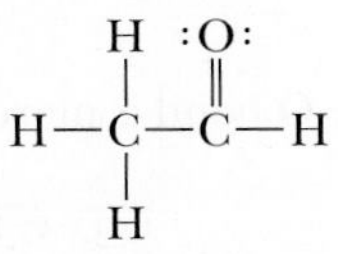

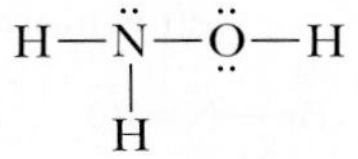

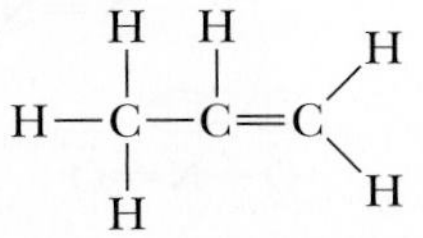

9.41 (a) CCl₄: Cl—C(Cl)(Cl)—Cl (with lone pairs on each Cl) (b) [:N≡O:]⁺

(c) [Cl—B(Cl)(Cl)—Cl]⁻ (with lone pairs on each Cl)

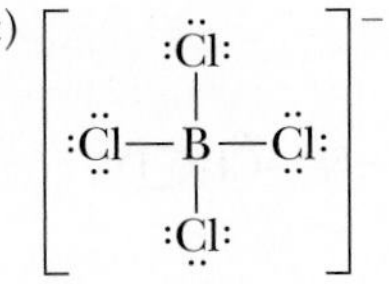

9.43 (a) H—C(H)(H)—C(=O)—H (b) H—N(H)—O—H

(c) H—C(H)(H)—C(H)=C(H)H (d) H—N=N—H

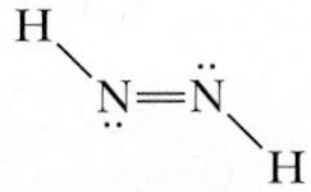

9.45 (a) H—C≡C—H (b) H—O—Cl:

(c) H₂C=C(H)—C≡N: (d) H—O—O—H

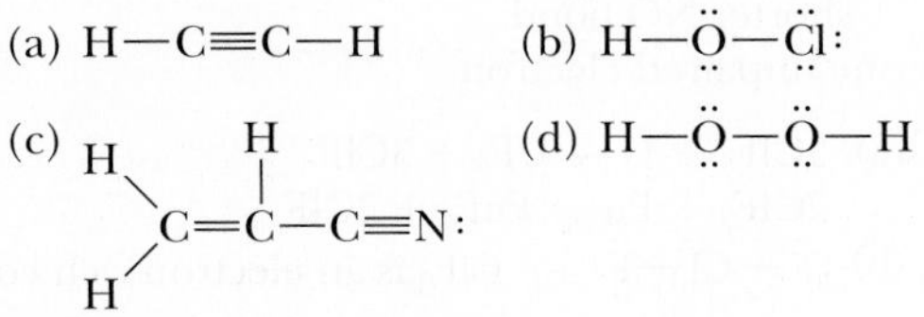

9.47 (a) Br (b) Cl (c) N

9.49 (a) I < Br < Cl (b) Ca < Ga < Br
(c) K < Ge < O

9.51 (a) polar covalent (b) ionic
(c) nonpolar covalent (d) ionic

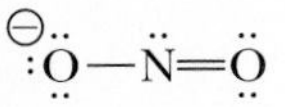

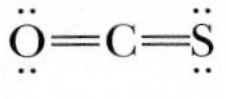

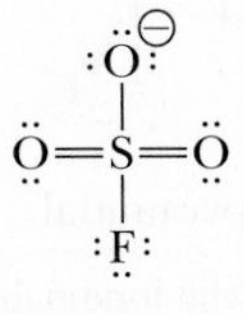

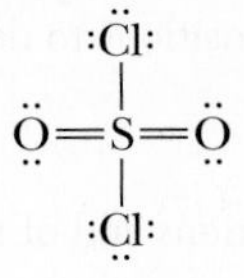

9.53 (a) C—O (arrow →) (b) Ge—C (arrow →) (c) O—H (arrow ←) (d) B—C (arrow →)

9.55 Br—F (arrow →)

9.57 (a) ⊖:O—N=O (b) O=C=S

(c) O=S(=O)=O

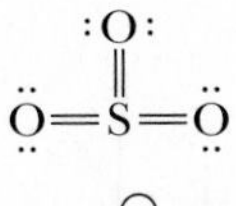

9.59 (a) O⊖—S(=O)(=O)—F (b) H—C≡N:

(c) Cl—S(=O)(=O)—Cl

9.61 H₂C=O and (−2)C=O(+2)(H)H

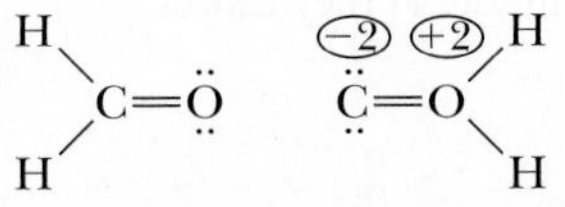

The structure on the left is better because it minimizes formal charge.

9.63 H—C(H)=N—H and H—C(H)(−1)=N(+1)—H

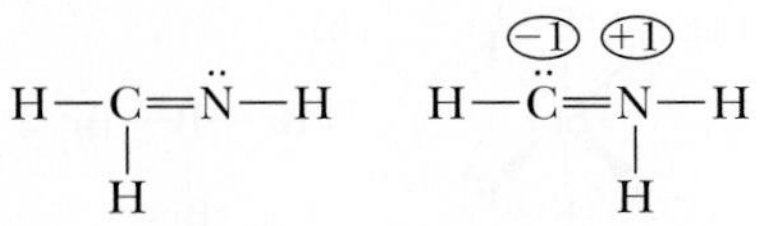

The structure on the left is better because it minimizes formal charge.

9.65 (a) O=N—O:(−1) and (−1):O—N=O

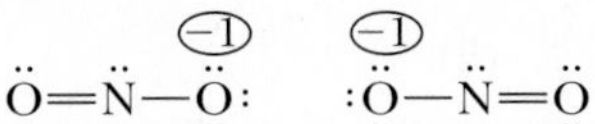

Both structures are equally important

(b) :Cl—C≡N: (+1)Cl=C=N(−1) (+2):Cl≡C—N:(−2)

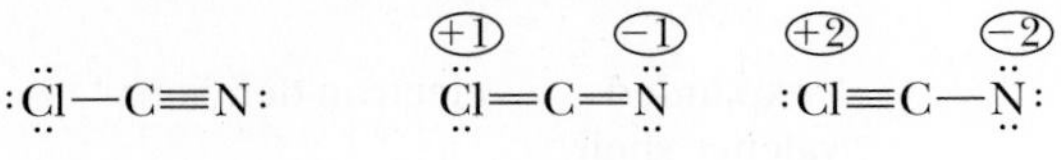

This structure is most important [:Cl—C≡N:]

9.67 (a) H₂C=N(+1)=N:(−1) and H₂C(−1)—N(+1)≡N:

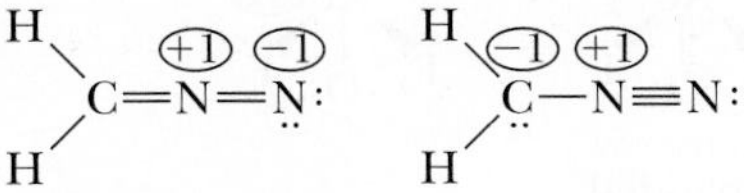

This structure is more important [H₂C=N=N:]

(b) :O=C=N:(−1) (+1):O≡C—N:(−2) (−1):O—C≡N:

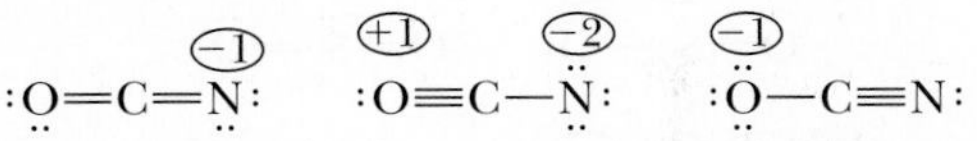

This structure is most important [(−1):O—C≡N:]

9.69 (a) (−1)N=N(+1)=N(−1) (+1):N≡N—N:(−2) (−2):N—N(+1)≡N:

This structure is most important [(−1)N=N(+1)=N(−1)]

(b) O=C(—O:(−1))—O:(−1) (−1):O—C(=O)—O:(−1) (−1):O—C(—O:(−1))=O

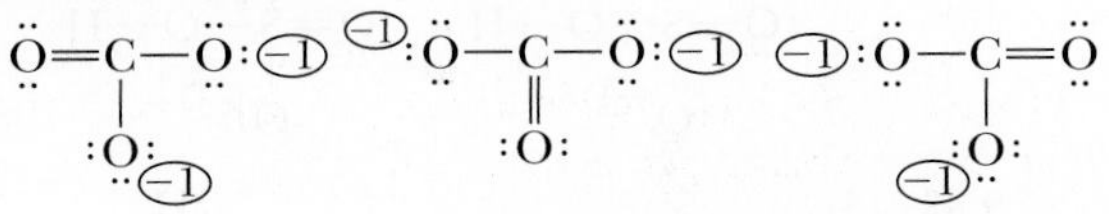

All three structures are equally important

9.71 Two Kekulé structures of $C_6H_5CH_3$ (toluene), with alternating double bonds in the ring.

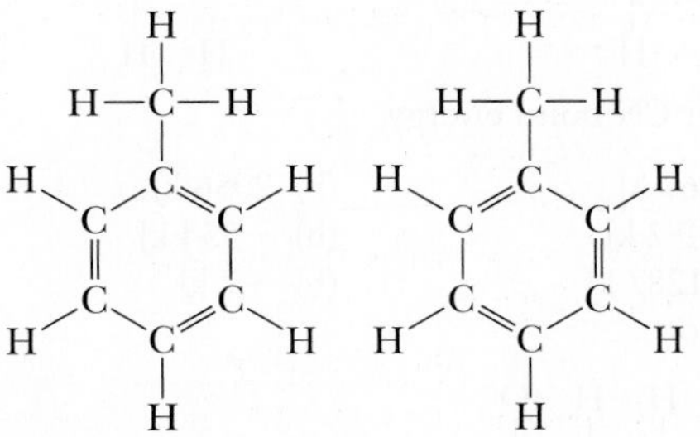

Both structures are equally important

9.73 H—C(H)(H)—N=C=O and H—C(H)(H)—N(+1)≡C—O:(−1)

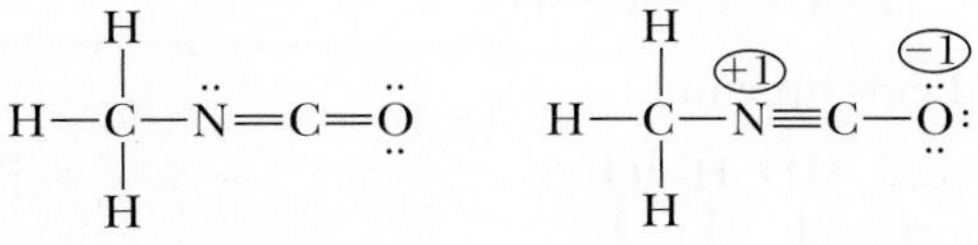

This structure is most important [H—C(H)(H)—N=C=O]

H—C(H)(H)—N(−1)—C≡O:(+1)

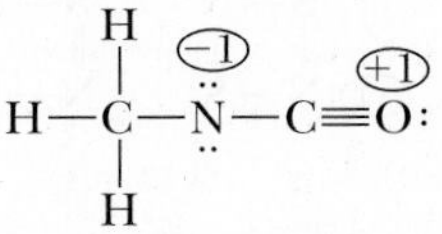

9.75 (a)

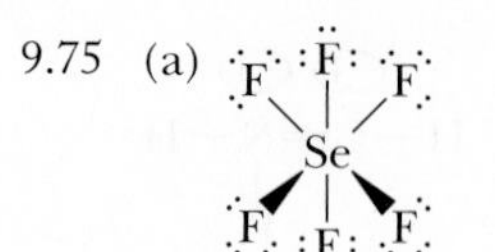

expanded valence shell

(b) :Br—B—Br: with :Br: bonded to B

electron-deficient

(c) O=N(+1)—O:(−1)

odd-electron

9.77 (a) 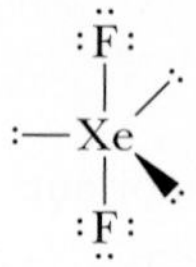 XeF_4 (Xe with :F: above and below)

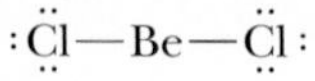

expanded valence shell

(b) :Cl—Be—Cl:

electron-deficient

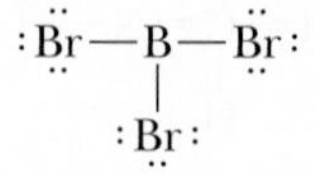

(c)

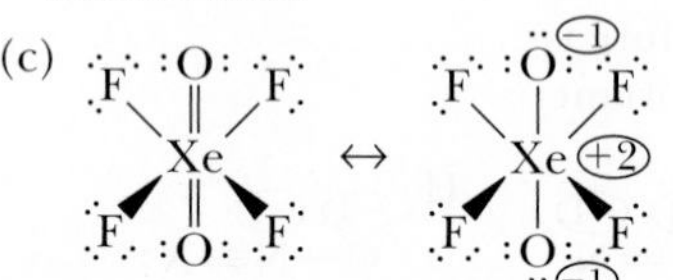

expanded valence shell

9.79 (a) 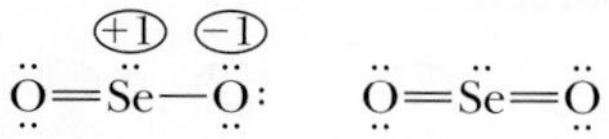 O=Se(+1)—O:(−1) O=Se=O

(b) 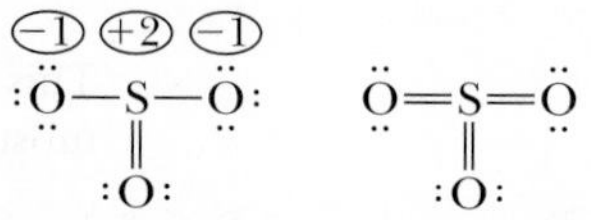 :O(−1)—S(+2)—O:(−1) with :O: double-bonded to S O=S=O with :O: double-bonded to S

9.81 (a) :O(−1)—Cl(+1)—O:(−1) :O(−1)—Cl=O

(b) 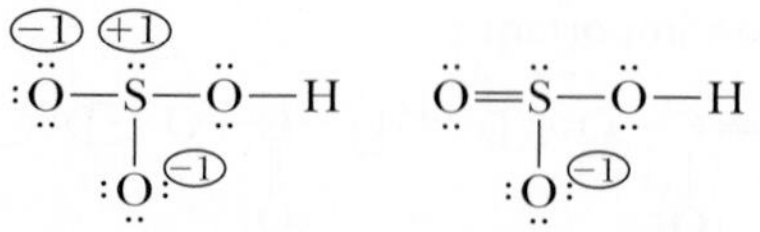 :O(−1)—S(+1)—O—H with :O:(−1) bonded to S O=S—O—H with :O:(−1) bonded to S

9.83 H—C=N—H (H on C) H—C—N—H (H on C; H, H below)

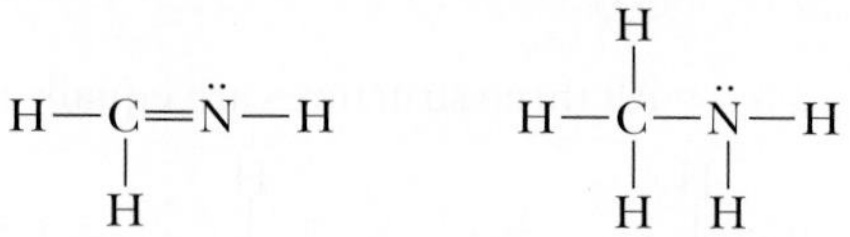

greater CN bond energy

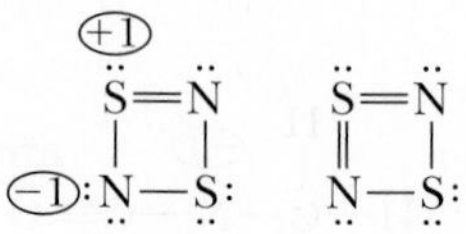

9.85 (a) 1167 kJ (b) 2056 kJ

9.87 (a) −482 kJ (b) −554 kJ

9.89 (a) −1287 kJ (b) +7 kJ

9.91 skeleton:

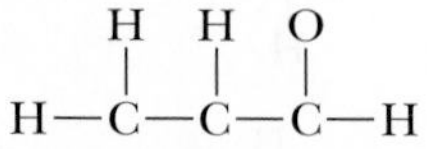

H—C—C—C—H (H, H, O above)

Lewis structure:

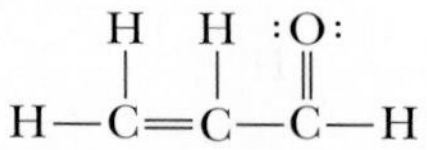

H—C=C—C—H (H, H, :O: above; C=O double bond)

9.93 S=N(+1) above :N(−1)—S: S=N above N—S: (ring structures)

9.95 SF_3—C—H with S=C double bond; H below C

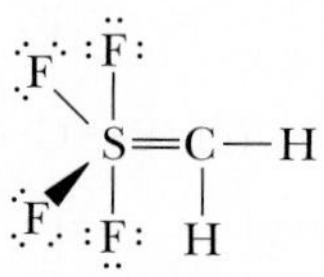

This Lewis structure, which shows no formal charge, has a double bond between the C and the S.

9.97 :F—C≡S—F: with :F: above and below S

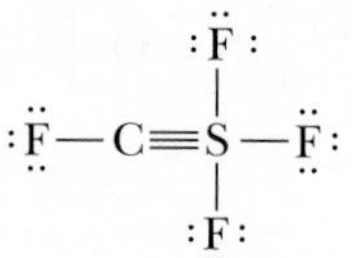

Carbon follows the octet rule; sulfur does not

9.99 :O=N—Cl: (−1):O—N=Cl:(+1)

The form on the left is more stable.

9.101 :Br—N=O The N—O bond is more polar

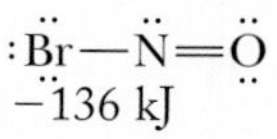

9.103 −136 kJ

9.105 N≡N(+1)—O:(−1) N(−1)=N(+1)=O

O=N(+1)—O:(−1) :O(−1)—N(+1)=O

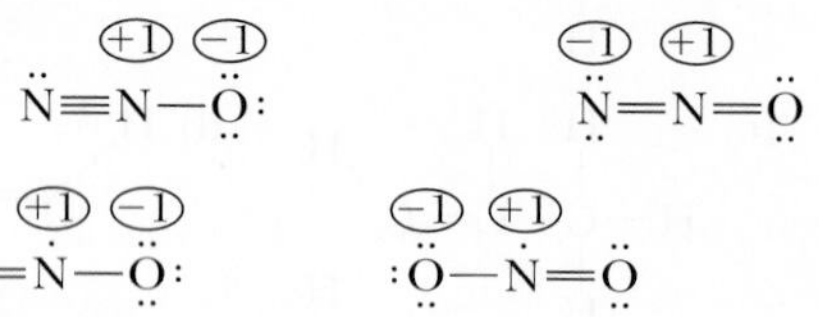

shorter NO bond
one unpaired electron

9.107 (a) $3ClF_3 + U \rightarrow UF_6 + 3ClF$
$2ClF_3 + Pu \rightarrow PuF_4 + 2ClF$

(b) 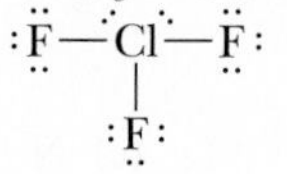 :F—Cl—F: with :F: below Cl ClF_3 is an electron-rich compound

(c) 29.4 g U, 16.9 g Pu

Chapter 10

10.1 Bonding pairs and lone pairs of electrons in a molecule repel each other, so they move into the configuration that maximizes separation and minimizes repulsion.

10.3

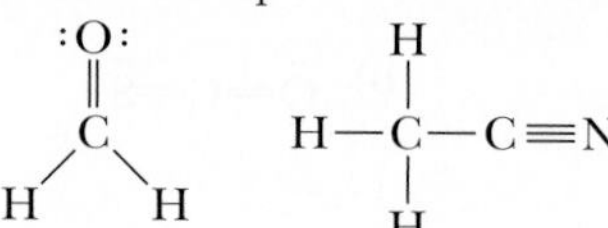

10.5

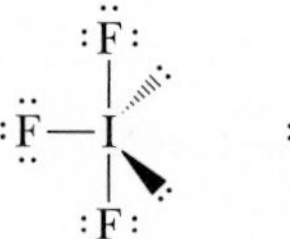

:F—I with F, F (trigonal planar)

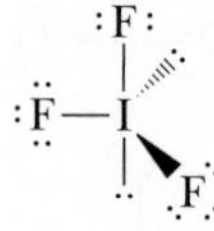

T-shape trigonal planar pyramidal

The molecule is T-shaped because the lone pairs of electrons prefer to occupy equatorial positions to decrease repulsion.

10.7 O=C=O (bond dipole arrows pointing outward)

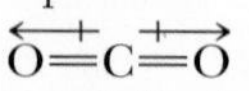

Bond dipoles are in opposite directions and of the same magnitude, so they cancel.

10.9 Atomic orbitals overlap to allow electron pairs to be shared.

10.11 A chlorine 3*p* orbital overlaps with a fluorine 2*p* orbital.

10.13 The B in BCl_3 is sp^2 hybridized because B is trigonal planar. The B in BCl_4^- is sp^3 hybridized because B is tetrahedral.

10.15 Hybridization cannot be assigned because there are no bond angles on which to base the hybridization.

10.17 A sigma (σ) bond places the shared pair of electrons symmetrically about the line joining the two nuclei of the bonded atoms. A pi (π) bond places electron density above and below the line joining the bonded atoms.

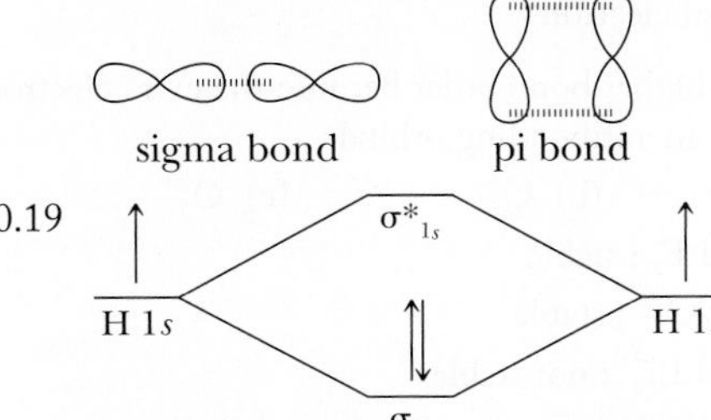

10.21 The molecular orbital description has a σ bond formed by the overlap of the Li 2*s* orbital with an F 2*p* orbital. This orbital will be mainly F 2*p* in character, with the shared pair of electrons residing predominately on the F. The ionic bonding description has Li losing one electron to F, thus forming an Li^+ ion and an F^- ion, which will be attracted by strong coulomb forces.

10.23 (a) trigonal planar (b) tetrahedral (c) tetrahedral (d) trigonal bipyramid

10.25 (a) tetrahedral (b) linear (c) trigonal bipyramid (d) trigonal planar (e) tetrahedral

10.27 (a) trigonal planar, bent
(b) linear, linear
(c) tetrahedral, trigonal pyramid
(d) octahedral, square pyramid
(e) trigonal bipyramid, see-saw

10.29 (a) NCl_3; N has a lone pair
(b) SF_6; octahedral geometry has 90-degree angles

10.31 (a) Cl_2NH; lone pair on N constrains bond angles
(b) IF_4^-; square planar geometry has 90-degree angles

10.33 (a) H–C(H)(H)–C≡C–H

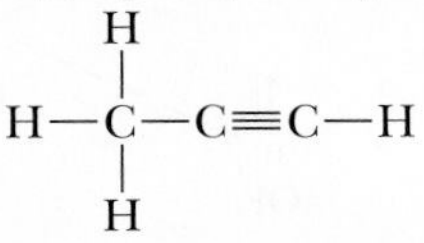

109.5° around first C, 180° around others

(b) (Br)(Br)C=C(H)(H)

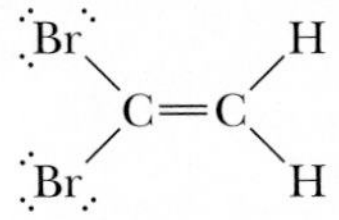

120° around each C

(c) H–C(H)(H)–N̈(H)–H

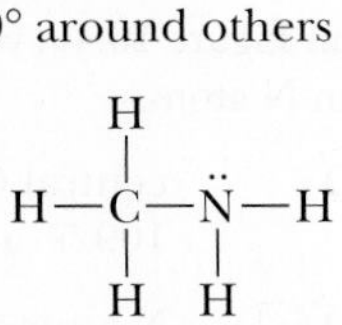

109.5° around C and N

10.35 (a) :Ö=S̈=Ö:

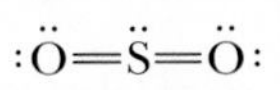

120°

(b) $[$:Ö–Cl(–:Ö:)–Ö: $]^-$

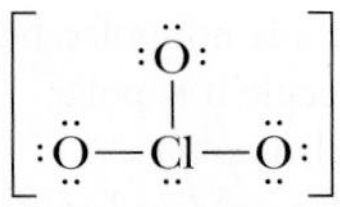

approximately 109°

(c) $[$:S̈=C=N̈:$]^-$

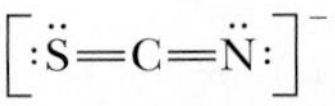

180°

10.37 (a) 109.5 degrees around each end C, 120 degrees around center C
(b) 109.5 degrees around first C, 180 degrees around center C

10.39 (a) 109.5 degrees around C, O, and N
(b) 180 degrees around each C, 109.5 degrees around P

10.41 (a) H–C(H)(H)–C(=:O:)–H

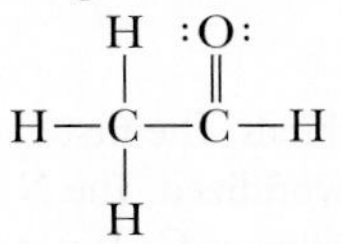

109.5° around first C, 120° around second

(b) H–C(H)(H)–Ö–C(H)=C(H)–H

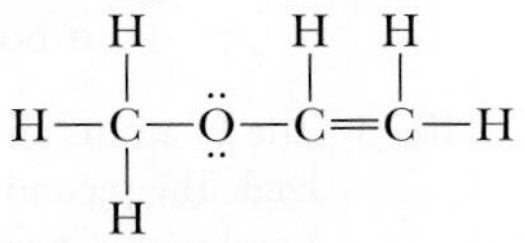

109.5° around first C and O, 120° around second and third C

(c) :F̈–Xe–F̈:

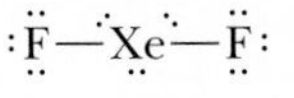

180°

10.43 (a) H–C(H)(H)–S̈–H

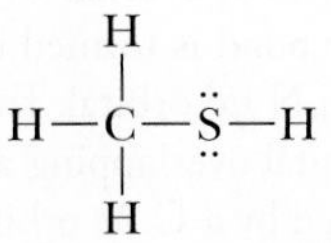

all angles are 109.5°

(b) H–N̈=N̈–H

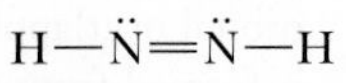

all angles are 120°

10.45 (a) nonpolar (b) nonpolar (c) nonpolar (d) polar

10.47 (a) polar (b) polar (c) polar

10.49 (a) polar

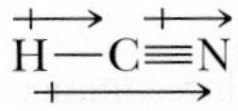

H–C≡N:

molecular dipole is vector sum of bond dipoles

(b) nonpolar

(c) polar

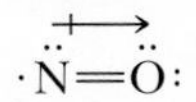

·N̈=Ö:

molecular dipole is same as bond dipole

10.51 (a) polar

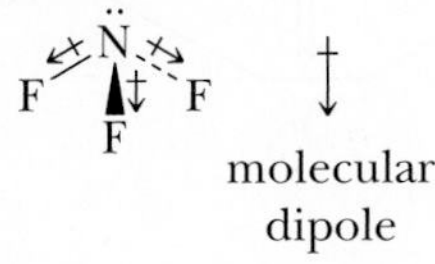

molecular dipole

(b) nonpolar
(c) nonpolar

10.53 Molecule a is nonpolar, because the C—F dipoles cancel out. Molecule b is polar.

10.55 (a) sp^2 (also, sp^3d) (b) sp^3d^2 (also, sp^3d) (c) sp (also, sp^3d and sp^3d^2)

10.57 (a) sp^3 (b) sp^3d^2 (c) sp^3d (d) sp^3 (e) sp^3

10.59 (a) sp (b) sp^2 (c) sp^3d (d) sp^2

10.61 (a) sp^2 (b) sp^3 (c) sp^2

10.63 (a) sp^3 (b) sp^3 (c) sp^3

10.65 (a) sp^3 (b) sp^3 (c) sp^2

10.67 Se is sp^3d hybridized, with a lone pair in an equatorial sp^3d hybrid orbital. F uses a $2p$ orbital.

10.69 (a) C, sp^3 O, sp^3 N, sp^3 (b) C, sp P, sp^3

10.71 (a)

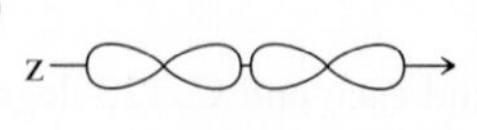

σ bond

(b)

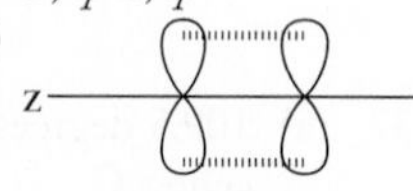

π bond

(c)

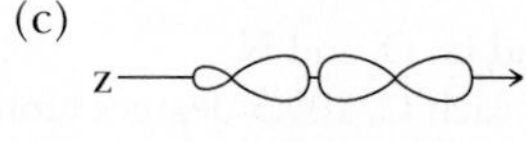

σ bond

10.73 The H atoms all use $1s$ orbitals. The first C is sp^3 hybridized. The second C is sp hybridized. The N makes a σ bond with a p orbital pointing at C. Two p orbitals on this C and two p orbitals on N overlap to form two π bonds. There are five σ bonds and two π bonds.

10.75 H, C—N, H, H

H—C bonds are formed by H $1s$ orbitals overlapping C sp^2 orbitals. The C—N σ bond is formed by a C sp^2 orbital overlapping with an N sp^2 orbital. The N—H bond is formed from a sp^2 N orbital overlapping an H $1s$ orbital. The C—N π bond is formed by a C $2p$ orbital overlapping with an N $2p$ orbital.

10.77 (a) The double bonded atoms are sp^2; the other carbons are sp^3.
(b) sp^2
(c) All central atoms are sp^3 except C_2, which is sp^2.

10.79 :Ö—N̈=Ö: (−1 on the left O) :Ö=N̈—Ö: (−1 on the right O)

The N is sp^2 hybridized in each resonance structure.

10.81 (a) C and N are each sp^3.
(b) The first C is sp^3, and the other C atoms are sp.

10.83 H₂C=C(H)—C≡N:

The first two C atoms are sp^2, and the third is sp.

10.85

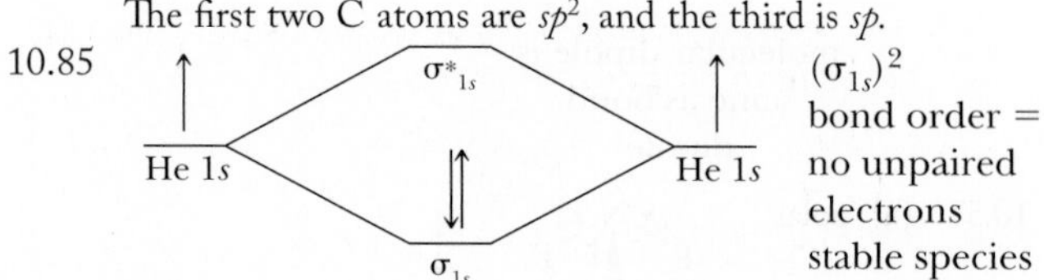

$(\sigma_{1s})^2$
bond order = 1
no unpaired electrons
stable species

10.87

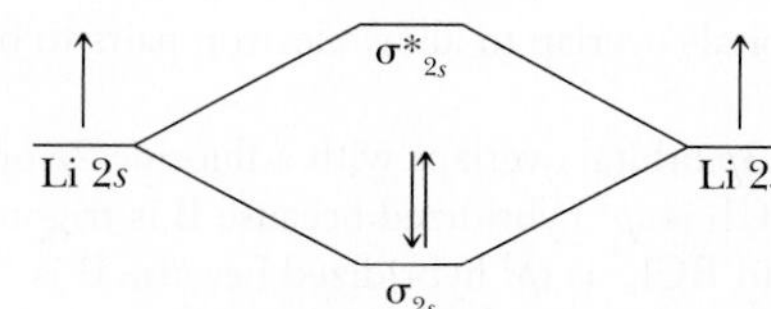

$(\sigma_{2s})^2$
bond order = 1
no unpaired electrons
stable species

10.89 (a) $(\sigma_{2s})^2(\sigma^*_{2s})^2\ (\pi_{2p})^3$; bond order is 1.5 with one unpaired electron
(b) $(\sigma_{2s})^2(\sigma^*_{2s})^2\ (\pi_{2p})^4\ (\sigma_{2p})^2(\pi^*_{2p})^1$; bond order is 2.5 with one unpaired electron
(c) $(\sigma_{2s})^2(\sigma^*_{2s})^2\ (\pi_{2p})^1$; bond order is 0.5 with one unpaired electron

10.91 N_2 has the higher bond order because the extra electron in N_2^- is in an antibonding orbital.

10.93 (a) B_2^- (b) C_2^- (c) O_2^{2+}

10.95 (a) O_2^- and F_2^+; stable
(b) N_2 and O_2^{2+}; stable
(c) Be_2 and Li_2^{2-}; not stable

10.97 (a) N_2 and CO (b) B_2 and BeC

10.99 (a) $(\sigma_{2s})^2(\sigma^*_{2s})^2\ (\pi_{2p})^4(\sigma_{2p})^1$; bond order is 2.5 with one unpaired electron
(b) $(\sigma_{2s})^2(\sigma^*_{2s})^2\ (\pi_{2p})^4\ (\sigma_{2p})^2(\pi^*_{2p})^1$; bond order is 2.5 with one unpaired electron
(c) $(\sigma_{2s})^2(\sigma^*_{2s})^2\ (\pi_{2p})^2$; bond order is 1 with two unpaired electrons
(d) $(\sigma_{2s})^2(\sigma^*_{2s})^2\ (\pi_{2p})^2$; bond order is 1 with two unpaired electrons

10.101

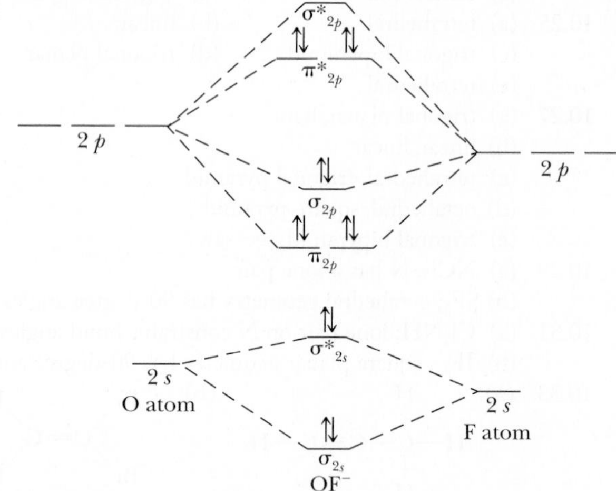

bond order is 1

10.103 Similar to the drawing in Figure 10.49, with the central O atom replaced with an N atom.

10.105

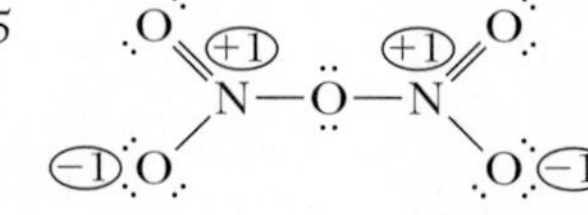

central O is sp^3 with 109.5° angles

N atoms are sp^2 with 120° angles

10.107 ClF_2^- is linear (180 degrees); ClF_2^+ is bent (109.5 degrees)

10.109

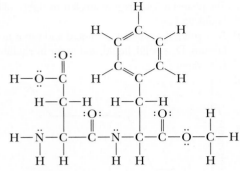

(a) 39 σ bonds, 6 π bonds
(b) sp^2
(c) sp^3

10.111 trigonal planar, sp^2, polar

10.113 angle 1: 120° angle 2: 109.5° angle 3: 109.5°
angle 4: 120° angle 5: 109.5°

10.115

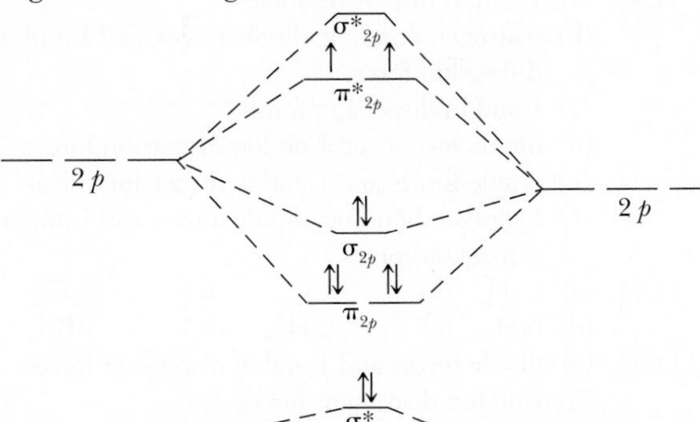

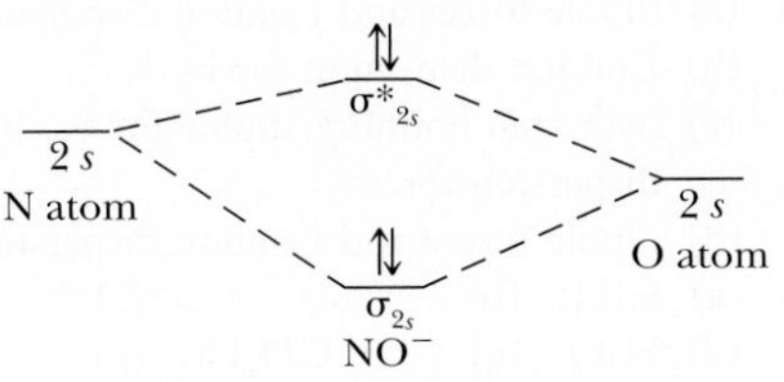

NO^- has a bond order of 2

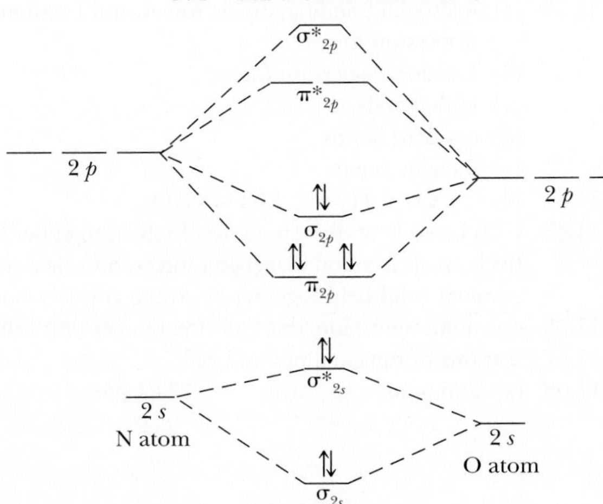

NO^+ has a bond order of 3

10.117 (a)

first N is 109.5°, sp^3
second N is 120°, sp^2
molecule is polar

(b)

first N is 120°, sp^2
second N is 180°, sp
molecule is polar

10.119 (a)

bond angle is 180°, C is sp in all three resonance structures

(b)

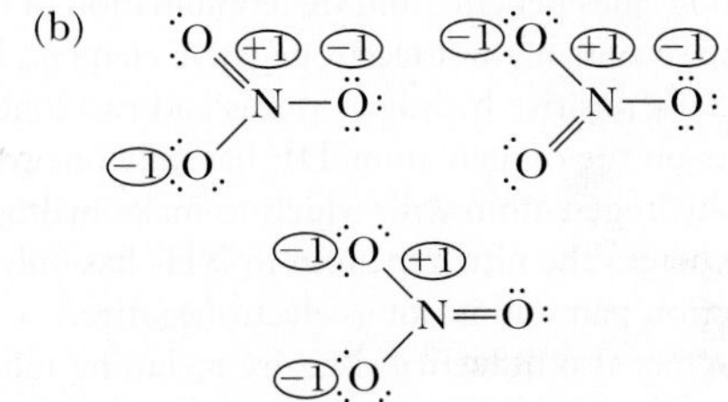

bond angle is 120°, N is sp^2 in all three resonance structures

10.121

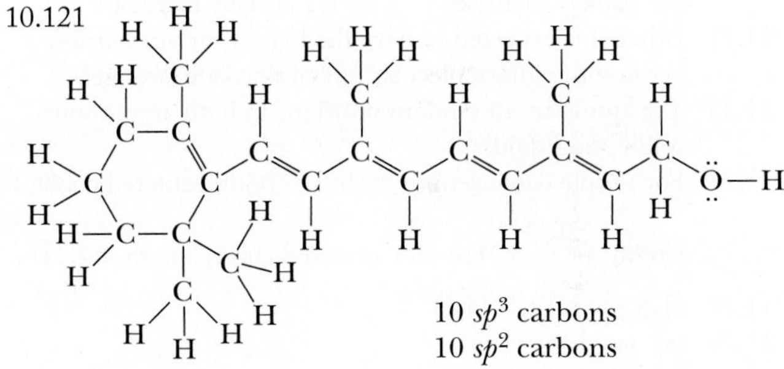

10 sp^3 carbons
10 sp^2 carbons

10.123 C_2H_4O

first C, sp^3, 109.5°
second C, sp^2, 120°

both carbons sp^2, 120°
O, sp^3, 109.5°

10.125

10.127 x = 4; carbon uses sp^2 orbitals to form the σ bond and p orbitals to form the π bond

Chapter 11

11.1 The molecules are much closer together in the liquid phase than in the vapor phase.

11.3 In the liquid state, the molecules are farther apart. The larger the distance, the smaller the energy of attraction.

11.5 When the equilibrium vapor pressure is reached, the liquid is in dynamic equilibrium with its vapor. The liquid continues to evaporate at a rate equal to the rate at which the vapor is condensing.

11.7 When the humidity is high, the liquid water is closer to equilibrium with its vapor, so the vapor condenses more rapidly, resulting in a net decrease in the overall rate of evaporation. Therefore, less heat is absorbed.

11.9 Trouton's rule can be stated as $\Delta H_{vap} = T_b \times 88$ J/K, which is consistent with the idea that both ΔH_{vap} and T_b should increase as intermolecular forces increase.

11.11 $\Delta H_{fus} < \Delta H_{vap} < \Delta H_{sub}$

11.13 white tin

11.15 Water molecules benefit from the combination of oxygen being the second-highest electronegative element, having two partially positive hydrogen atoms and two lone electron pairs on the oxygen atom. HF has only one partially positive hydrogen atom with which to make hydrogen bonds, whereas the nitrogen atom in NH_3 has only one lone electron pair and is not as electronegative.

11.17 Other factors also influence viscosity including molecular size, shape, and structure.

11.19 (a) usually increases (b) usually increases
(c) usually increases (d) may increase
(e) usually increases (f) usually increases

11.21 Ethanol is expected to have the higher surface tension because the intermolecular interactions are stronger.

11.23 The atoms are in random positions in both amorphous solids and liquids.

11.25 For simple cubic, length $= 2r$. For body-centered cubic, length $= \dfrac{2\sqrt{3}}{3}r$. For face-centered cubic, length $= 2\sqrt{2}r$.

11.27 $H_2S < H_2Te < H_2O$

11.29 (a) no change
(b) vapor pressure increases
(c) no change

11.31 (a) isopropyl alcohol
(b) ΔH_{vap}, melting point, and boiling point are all greater for isopropyl alcohol than for methyl ethyl ether

11.33 (a) benzene
(b) ΔH_{vap}, melting point, and boiling point are all greater for benzene than for butane.

11.35 14.7 kJ/mol

11.37 33.4 kJ; endothermic

11.39 611 kJ; endothermic

11.41

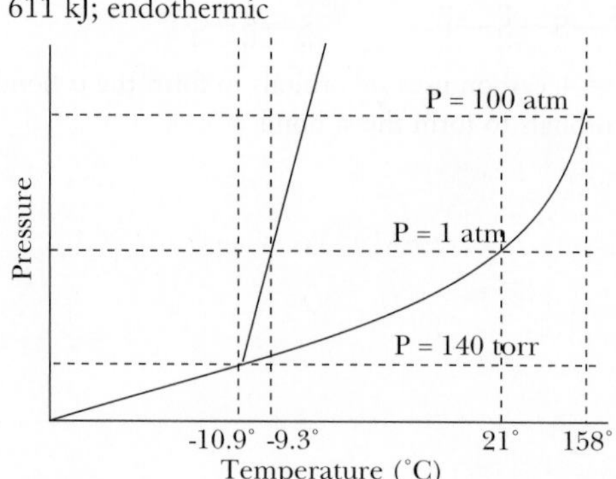

11.43 (a) The phases in each region are, left to right, solid, liquid, and gas.
(b) A = solid, B = solid, C = liquid and vapor in equilibrium, D = solid, liquid, and vapor in equilibrium, E = vapor

11.45 (a)

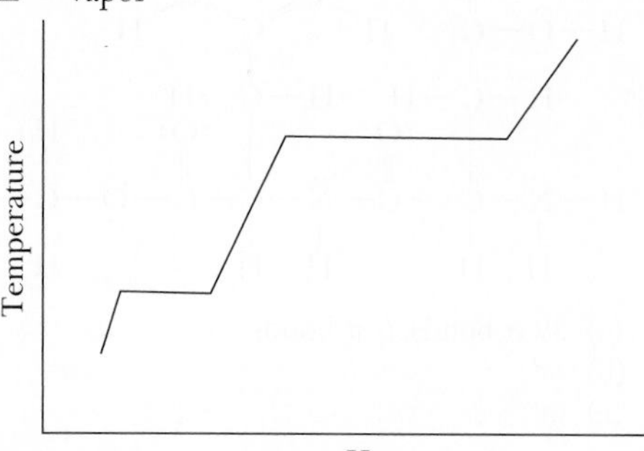

(b) solid melts to liquid, liquid evaporates
(c) solid is denser than liquid

11.47 (a) additional solid will sublime
(b) the ice will melt
(c) the vapor will condense into a liquid

11.49 (a) London dispersion forces
(b) hydrogen bonding, dipole forces, and London dispersion forces
(c) London dispersion forces
(d) dipole forces and London dispersion forces
(e) dipole forces and London dispersion forces
(f) hydrogen bonding, dipole forces, and London dispersion forces

11.51 (a) C_2H_4 (b) Cl_2 (c) S_2Cl_2
(d) NH_3 (e) CHI_3 (f) $BBrI_2$

11.53 (a) dipole forces and London dispersion forces
(b) London dispersion forces
(c) hydrogen bonding, dipole forces, and London dispersion forces
(d) dipole forces and London dispersion forces

11.55 (a) C_3H_8 (b) I_2 (c) S_2Cl_2
(d) H_2O (e) CH_2Cl_2 (f) NOCl

11.57 $C_6H_{14} < C_2H_5OH < (CH_2OH)_2$

11.59 (a) hydrogen bonding, dipole forces, and London dispersion forces
(b) London dispersion forces
(c) ionic bonds
(d) covalent bonds
(e) metallic bonds

11.61 $N_2 < CO_2 < H_2O < KCl < CaO$

11.63 CO_2 consists of discrete molecules held together by relatively weak London dispersion forces. SiO_2 is a network covalent solid held together by strong covalent bonds.

11.65 one ammonium ion, one chloride ion per unit cell

11.67 2 atoms of tungsten per unit cell

11.69 (a) 286.6 pm (b) 124.1 pm

11.71 four oxide ions per unit cell
11.73 9.01 g/cm^3
11.75 0.774 g/cm^3
11.77 four atoms per unit cell; face-centered cubic
11.79 KBr (a) and a sugar crystal (d)
11.81 15.2 degrees
11.83 $\lambda = 108$ pm
11.85 (a) only
11.87 6.023×10^{23} atoms/mol
11.89 128 pm
11.91 (a) He: London dispersion forces, $\Delta H_{vap} = 0.083$ kJ/mol
(b) CH_3OCH_3: dipole-dipole forces, London dispersion forces, $\Delta H_{vap} = 19.4$ kJ/mol
(c) C_2ClF_5: dipole-dipole forces, London dispersion forces, $\Delta H_{vap} = 38.6$ kJ/mol
11.93 change in volume $= -1.89$ cm^3
11.95 (a) $P = 1.16$ atm
(b) 0.268 g; yes, there is liquid-vapor equilibrium
11.97 (a) 207 J (b) 1.15 g
11.99 39.6 kJ
11.101 M = potassium, ionic bonding
11.103 $CH_3{-}\ddot{N}(CH_3){-}CH_3$ $\quad$ $CH_3{-}CH_2{-}CH_2{-}\ddot{N}H_2$

A, b.p. 2.9°C $\quad$ B, b.p. 47.8°

Compound A has dipole-dipole bonding intermolecular forces, but Compound B has hydrogen bonding, which is a stronger intermolecular force. That is why Compound B has a higher boiling point than Compound A.

Chapter 12

12.1 Molarity expresses the volume of solution in liters. Mole fraction expresses the amount of solution in moles. Mass percentage expresses the amount of solution as total mass.
12.3 For a dilute solution, the mass of solvent is approximately equal to the volume of solution times the density of the solution. Therefore, molality will be approximately equal to molarity when the density of both the solvent and the solution are close to 1.
12.5 The solution is supercooled. Addition of a seed crystal of sodium acetate will cause the excess solute to precipitate.
12.7 Solubility decreases with increasing values of n.
12.9 The dissolved CO_2 escapes from the container until there is none left in solution.
12.11 The situation at Lake Nyos is more likely to occur with a gas with a low Henry's law constant, because the formation of a supersaturated solution of the gas is unstable and the excess gas may escape catastrophically given the opportunity.

12.13

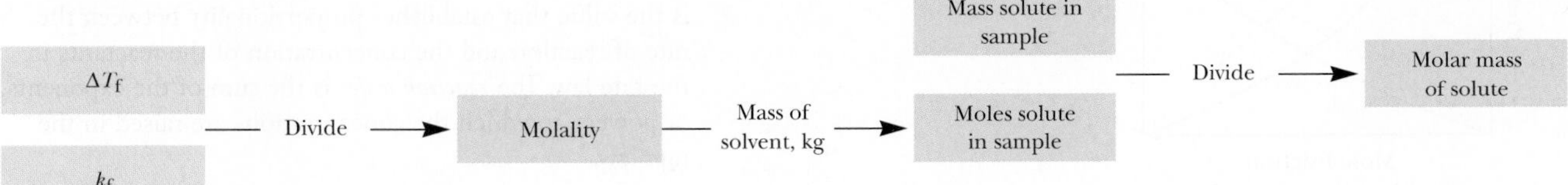

12.15 The freezing point of the NaCl solution is -0.4 °C. The freezing point of the $CaCl_2$ solution is -0.6 °C. The $CaCl_2$ solution has a lower freezing point because each mole of $CaCl_2$ forms three moles of particles in solution whereas each mole of NaCl forms only two moles of particles in solution.
12.17 (a) 0.160% (b) 0.000236 (c) 0.0131 mol/kg
12.19 (a) 1.04% (b) 0.00115 (c) 0.0638 mol/kg
12.21 0.022 mol
12.23 dissolve 0.155 g boric acid in 15.3 g water; 0.164 molal
12.25 0.801
12.27 0.0059 m
12.29 3.6×10^{-4}
12.31 0.013
12.33 (a) 0.325 m (b) 0.303 M (c) 5.81×10^{-3}
12.35 (a) 6.00% (b) 0.0116 (c) 0.651 m
12.37

	Molality	Molarity	Mass%	Mole Fraction
(a)	3.76 m	3.44 M	**6.00%**	0.0635
(b)	11.2 m	**8.80 M**	16.0%	0.168
(c)	**8.02 m**	6.68 M	12.0%	0.126
(d)	4.42 m	3.97 M	7.00%	**0.0738**

12.39 molality = 4.38 m, mole fraction = 0.0730, mass percentage = 30.0%
12.41 water < ethanol < hexane
12.43 (a) carbon tetrachloride (b) water
(c) diethyl ether (d) water
12.45 (a) hydrogen bonding
(b) dipole dipole and London dispersion forces
(c) ion-dipole forces
(d) London dispersion forces
12.47 (a) $CH_3(CH_2)_{10}OH$; less polar
(b) CCl_4; nonpolar
(c) $Fe(C_5H_5)_2$; nonpolar
12.49 (a) 6.17×10^{-5} molal/torr (b) 0.482 g C_2H_2
12.51 (a) 2.88×10^{-5} molal/torr (b) larger
(c) 1.09×10^{-2} molal
12.53 The enthalpy of solution is endothermic because the solubility increases with temperature.
12.55 exothermic
12.57 NH_4Cl has the more positive enthalpy of solution because its solubility increases faster with increased temperature.
12.59 (a) greater (b) less (c) greater
(d) less (e) less
12.61 (a) greater (b) less
(c) less (d) less
12.63 The vapor pressure is lowered by 42.4 torr. The vapor pressure is 318 torr.

12.65 $T_f = -2.48$ °C, $T_b = 100.682$ °C
12.67 $k_f = 20.2$ °C/molal
12.69 104 g/mol
12.71 2.64×10^4 g/mol
12.73 0.03 *m* sucrose < 0.02 *m* LiBr < 0.03 *m* $MgSO_4$ < 0.025 *m* $(NH_4)_2Cr_2O_7$ < 0.03 *m* $CaCl_2$
12.75 1.08 *m* in solute particles, 31.6 g NaCl
12.77 100.074 °C
12.79 4.5 atm
12.81 $\Pi = 3.7 \times 10^2$ torr, so solvent flows from solution to pure water (reverse osmosis)
12.83 (a) $i = 2.5$ (b) smaller
12.85 (a) mole fraction of hexane = 0.465; mole fraction of heptane = 0.535
(b) vapor pressure of hexane = 129 torr; vapor pressure of heptane = 49.4 torr
(c) mole fraction of hexane = 0.723; mole fraction of heptane = 0.277
12.87 (a) vapor pressure of acetic acid = 241 torr; vapor pressure of 1,1-dibromoethane = 311 torr; total vapor pressure = 552 torr
(b) actual vapor pressure = 760 torr; positive deviation
(c) the attractive forces between the dissimilar molecules are weaker than between the similar molecules
12.89 (a) 1.69 molal (b) 0.0296
(c) 1.55 *M* (d) 7.14 g/100 mL
12.91 (a) water (b) benzene
(c) ethanol
12.93 (a) 3.27 g (b) 1.66 L CO_2
12.95 (a) 172 g/mol (b) 1 Fe atom per molecule

12.97 (a)

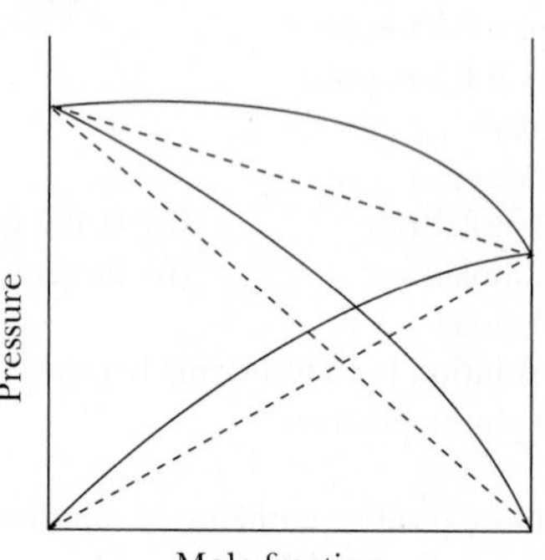

Positive deviation

(b)

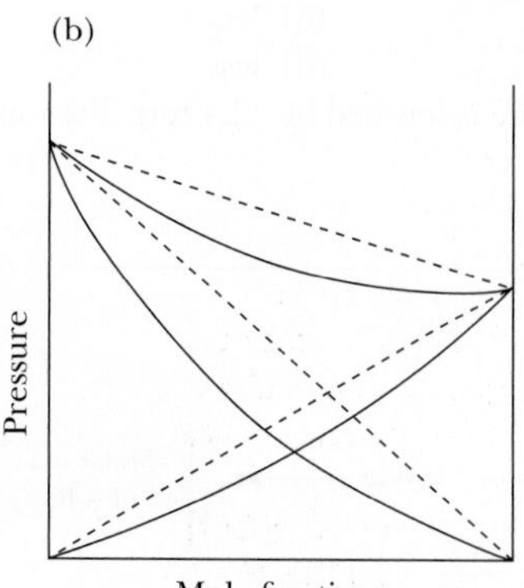

Negative deviation

(c)

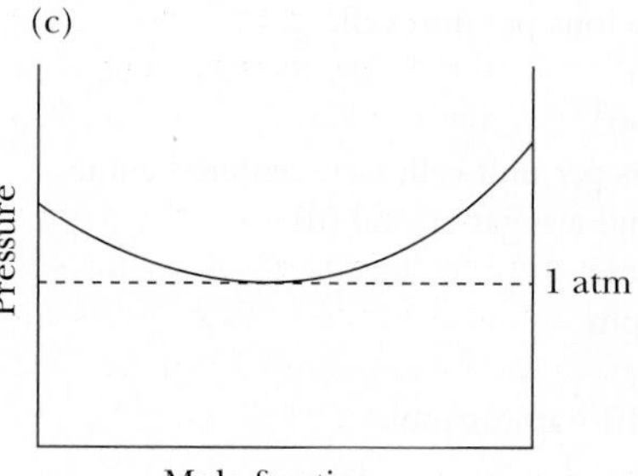

Maximum boiling azeotrope

(d)

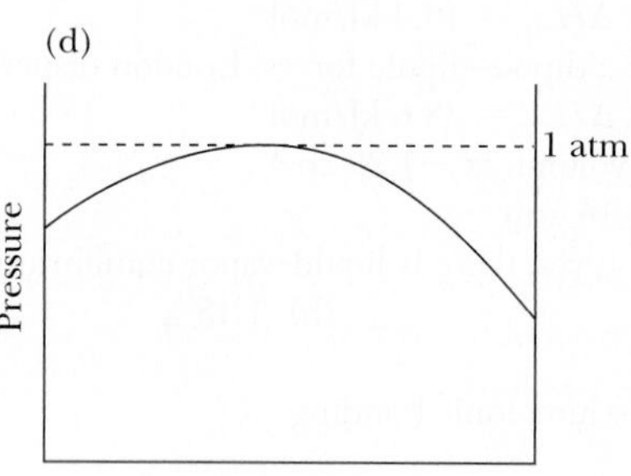

Minimum boiling azeotrope

12.99 (a) 128 g/mol
(b) 67 g/mol
(c) The gas dissociates in water to form two ions. From the molar mass, the gas may be HI: $HI(g) \rightarrow H^+(aq) + I^-(aq)$.
12.101 Solvent will flow from the KNO_3 solution to the NaCl solution. The concentration of the NaCl solution is 1.00 *m*, whereas the concentration of the KNO_3 solution is 0.878 *m*. Solvent will move from the more dilute solution to the more concentrated solution.
12.103 (a) 0.0418 molal in benzene, 0.0855 molal in water
(b) $i = 1$ in water, $i = 0.5$ in benzene
(c) acetic acid molecules dimerize (associate in pairs) in benzene because of hydrogen bonding
12.105 (a) $NaCO_2$
(b) $Na_2C_2O_4$; $Na_2C_2O_4 \rightarrow 2Na^+(aq) + C_2O_4^{2-}(aq)$
12.107 (a) $C_2H_5O^-(aq) + H_2O(\ell) \rightarrow C_2H_5OH(aq) + OH^-(aq)$
(b) exothermic; the combination of ethanol and hydroxide ion is more energetically stable than water and the ethoxide ion
(c) 5.13 g
(d) 24.53 torr
(e) The mole fraction of ethanol in the resulting solution is 0.0216, so only an azeotrope is distilled.

Chapter 13

13.1 The *rate law* is the equation that relates the rate of a reaction to the concentration of the reactants. The *rate constant* is the value that establishes proportionality between the rate of reaction and the concentration of the reactants in the rate law. The *reaction order* is the sum of the exponents, or powers, to which the concentrations are raised in the rate law.

13.3 The integral rate law gives concentration as a function of time. The differential rate law gives the rate of reaction (changes in concentration per unit time) as a function of concentration.

13.5 Half-lives for first-order reactions do not change with changes in the starting concentrations of the reactants, whereas half-lives for the other order reactions do change with changes in the starting concentrations. This means the half-lives for other orders depend on the starting concentrations and even vary from one half-life to another for a given reaction because the starting concentration for each successive half-life is half of what it was for the just completed half-life.

13.7 $t_{1/2} = \dfrac{1}{2k[\mathrm{R}]_0^2}$

13.9 For a collision to result in the formation of the activated complex, the reactants must be in the proper orientation; thus, not all collisions whose energy exceeds E_a can result in the formation of the activated complex.

13.11 Doubling the concentration of a reactant doubles the number of collisions per time, which doubles the collision frequency.

13.13 Only collisions with sufficient energy to rearrange bonds can result in the formation of products.

13.15 At any temperature, the catalyzed reaction will be faster than the uncatalyzed reaction. As temperature increases, the rate of the uncatalyzed reaction will increase faster than the rate of the catalyzed reaction.

13.17

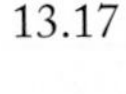

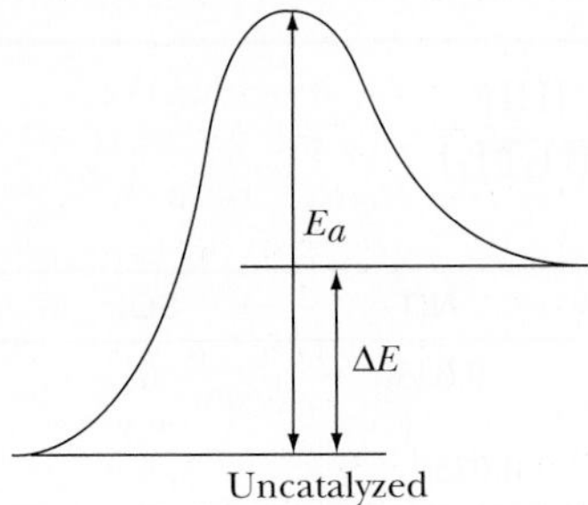

Uncatalyzed

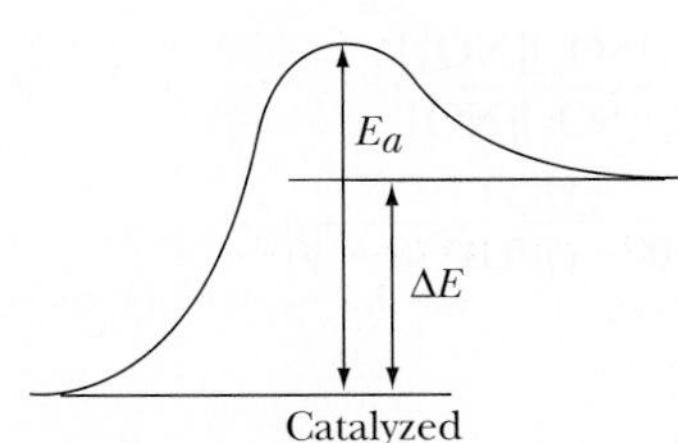

Catalyzed

13.19 An elementary step is a step that describes a single molecular interaction or collision. You can write a rate law if you know the identity and number of species that collide in an elementary step. The rate of reaction will be limited by the slowest elementary step. You cannot write a rate law from the overall stoichiometry because the balanced equation does not describe a molecular level collision, but rather the sum of many.

13.21 (a) rate $= -\dfrac{\Delta[\mathrm{C_2H_2O_4}]}{\Delta t}$
(b) 9.0×10^{-4} *M*/s
(c) 8.0×10^{-4} *M*/s
(d) 2.5×10^{-3} *M*/s
(e) 4.0×10^{-4} *M*/s

13.23 (a) $-\dfrac{\Delta[\mathrm{NO}]}{\Delta t} = -\dfrac{2\Delta[\mathrm{Cl_2}]}{\Delta t} = \dfrac{\Delta[\mathrm{NOCl}]}{\Delta t}$
(b) rate = 0.0016 *M*/s
(c) rate = 0.0013 *M*/s
(d) rate = 0.0010 *M*/s

13.25 (a) $[\mathrm{NO_2}] = 0.034\ M$, $[\mathrm{N_2O_4}] = 0.025\ M$, $[\mathrm{NO_2}] = 0.060\ M$
(b) rate $= 1.4 \times 10^{-3}\ M/\mu\mathrm{s}$

13.27 $-\dfrac{\Delta[\mathrm{NO_2}]}{\Delta t} = 0.011\ M/\mathrm{s}$, rate $= 0.0055\ M/\mathrm{s}$

13.29 0.015 *M*/s, 0.015 *M*/s

13.31 $-\dfrac{\Delta[\mathrm{H_2O_2}]}{\Delta t} = 0.0075\ M/\mathrm{s}$
$-\dfrac{\Delta[\mathrm{OH^-}]}{\Delta t} = 0.0050\ M/\mathrm{s}$,
$-\dfrac{\Delta[\mathrm{H_2O}]}{\Delta t} = 0.010\ M/\mathrm{s}$
$-\dfrac{\Delta[\mathrm{CrO_2^-}]}{\Delta t} = 0.0050\ M/\mathrm{s}$
reaction rate = 0.0025 *M*/s

13.33 rate $= k[\mathrm{NH_3}][\mathrm{HCl}]^2$

13.35 (a) 1 for NO, 2 for NO_2, 3 for the reaction
(b) 1/2 for O_2, 1 for Cl_2, 1.5 for the reaction
(c) 2 for HClO, 1 for OH^-, 3 for the reaction

13.37 rate $= (3.8 \times 10^3\ \mathrm{L^2/mol^2{\cdot}s})[\mathrm{NO}]^2[\mathrm{H_2}]$

13.39 rate $= (4.2\ \mathrm{L/mol{\cdot}s})[\mathrm{N_2O}][\mathrm{H_2O}]$

13.41 rate $= (5.2 \times 10^4\ \mathrm{L/mol{\cdot}s})[\mathrm{NO_2}][\mathrm{O_3}]$

13.43 zero order, $k = 0.034\ M/\mathrm{s}$

13.45 first order, $k = 503\ \mathrm{s^{-1}}$

13.47 second order, $k = 0.185\ \mathrm{L/mol{\cdot}s}$

13.49 (a) $k = 0.0283\ \mathrm{min^{-1}}$ (b) $t = 81.4$ min

13.51 (a) $0.05653\ \mathrm{y^{-1}}$, $1.797 \times 10^{-9}\ \mathrm{s^{-1}}$
(b) 3.13% of the tritium remains

13.53 120 s

13.55 $t_{1/2} = 2.6 \times 10^2$ s, $t = 4.8 \times 10^2$ s

13.57 24 s

13.59 $E_a = 108$ kJ/mol, $A = 7.8 \times 10^{16}$

13.61 35.8 kJ/mol

13.63 factor of 13

13.65 factor of 1.9

13.67

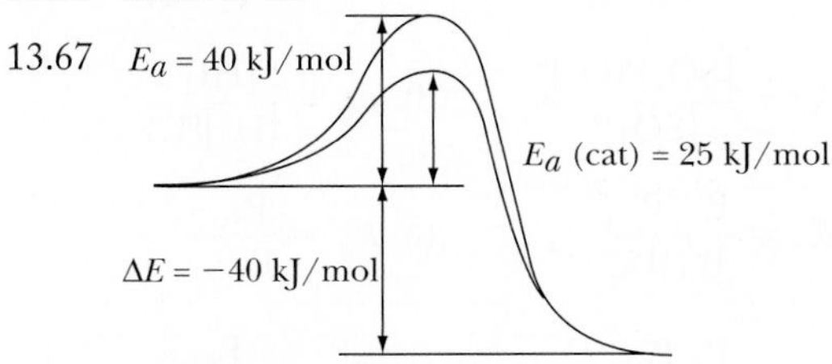

13.69 $\mathrm{NO_2 + CO \rightarrow NO + CO_2}$

13.71 $\mathrm{NO + O_3 \rightarrow NO_2 + O_2}$

13.73 (a) rate $= k\,[\mathrm{HCl}]$, unimolecular
(b) rate $= k\,[\mathrm{H_2}][\mathrm{Cl}]$, bimolecular
(c) rate $= k\,[\mathrm{NO_2}]^2$, bimolecular

13.75 (a) rate $= k\,[\mathrm{C_2H_5Cl}]$, unimolecular
(b) rate $= k\,[\mathrm{NO}][\mathrm{O_3}]$, bimolecular
(c) rate $= k\,[\mathrm{HI}][\mathrm{C_2H_5I}]$, bimolecular

13.77 step 1

13.79 by mechanism I: rate $= k\,[\mathrm{NO_2}][\mathrm{CO}]$
by mechanism II: rate $= k\,[\mathrm{NO_2}]^2$

13.81 mechanism is consistent, N_2O_2 is an intermediate

13.83 rate $= -\dfrac{\Delta[\mathrm{A}]}{a\Delta t} = -\dfrac{\Delta[\mathrm{B}]}{b\Delta t} = \dfrac{\Delta[\mathrm{C}]}{c\Delta t} = \dfrac{\Delta[\mathrm{D}]}{d\Delta t}$

13.85 factor of 5×10^7
13.87 zero order, $k = 3.5 \times 10^4$ M/s, rate = 3.5×10^4 M/s
13.89 172 s
13.91 second order, $k = 0.078$ L/mol·s
13.93 The reaction with the larger E_a will have a larger increase in rate with increasing temperature.
13.95 second order, rate = $k\,[H_2NNO_2]^2$
$k = 0.041$ torr^{-1} s^{-1}
13.97 (a) first order, rate = $k\,[C_2H_5Cl]$
(b) $k = 3.8 \times 10^{-3}$ s^{-1}
13.99 (a) rate = $k\,[H_2][NO]^2$
(b) $k = 3.46 \times 10^{-9}$ torr^{-2} s^{-1}
(c) E_a = 175 kJ/mol

Chapter 14

14.1 A system is at equilibrium when the tendency of the reactants to form products is balanced by the tendency of the products to form reactants. At this point, the concentrations of the reactants and products do not appear to change with time.
14.3 A vehicle rolling down a hill is not at equilibrium because its position and momentum are changing. It will be at equilibrium when it coasts to a stop.
14.5 A completely discharged battery is in its equilibrium state. A battery is not at equilibrium as long as it is holding a charge.
14.7 Q and K_{eq} have the same form but K_{eq} calculations require equilibrium concentrations whereas Q is not so restricted. At equilibrium $Q = K_{eq}$.
14.9 Changes in volume will not affect the equilibrium of a gaseous system when the number of moles of gaseous reactants is equal to the number of moles of gaseous products.
14.11 The molar concentration of a pure liquid or solid is a constant (proportional to its density). Because its concentration does not change in a chemical reaction, it is omitted from the expression for the equilibrium constant.

14.13 (a) $K_c = \dfrac{[PCl_3][Cl_2]}{[PCl_5]}$ (b) $K_c = \dfrac{[NO]^2[O_2]}{[NO_2]^2}$

(c) $K_c = \dfrac{[SO_2]^2[O_2]}{[SO_3]^2}$ (d) $K_c = \dfrac{[HI]^2}{[H_2][I_2]}$

14.15 (a) $K_p = \dfrac{P_{Cl_2}^{1/2}P_{H_2O}^{1/2}}{P_{HCl}P_{O_2}^{1/4}}$ (b) $K_p = \dfrac{P_{NO_2}}{P_{N_2O_4}^{1/2}}$

(c) $K_p = \dfrac{P_{N_2}P_{O_2}^2}{P_{N_2O_4}}$ (d) $K_p = \dfrac{P_{SO_3}}{P_{O_2}^{1/2}P_{SO_2}}$

14.17 (a) 1.7×10^{-4}
(b) 1.52×10^{-5}
14.19 $K = 1.3 \times 10^5$
14.21 $K_P = 5.9 \times 10^{-5}$
14.23 $K_c = 0.0512$
14.25 $K_c = 574$
14.27 $K_p = 0.509$
14.29 $Q = 9.9 \times 10^2$, more CO will form

K ⟵ Q

14.31 $Q = 0.20$, reaction proceeds to the right

Q ⟶ K

14.33 (a) $Q = 8$, reaction proceeds to the left
(b) $Q = 1 \times 10^2$, reaction proceeds to the left
(c) $Q = 8 \times 10^{-3}$, reaction proceeds to the right
(d) $Q = 1 \times 10^{-3}$, reaction proceeds to the right
14.35 (a) reaction proceeds to the right
(b) reaction proceeds to the right
(c) reaction proceeds to the right
(d) reaction proceeds to the right
(e) no change
14.37 (a)

	$CO(g)$	+	$\frac{1}{2}O_2(g)$	⇌	$CO_2(g)$
i	0.01		0.01		0.99
C	$+y$		$+\frac{1}{2}y$		$-y$
E	$0.01 + y$		$0.01 + \frac{1}{2}y$		$0.99 - y$

(b)

	$2SO_2(g)$	+	$O_2(g)$	⇌	$2SO_3(g)$
i	0.015		0.012		1.45
C	$+2y$		$+y$		$-2y$
E	$0.015 + 2y$		$0.012 + y$		$1.45 - 2y$

14.39 (a)

	H_2	+	I_2	⇌	2 HI
i	0.50		0.50		0
C	$-y$		$-y$		$+2y$
e	$0.50 - y$		$0.50 - y$		$2y$

(b) $K_c = 4.6 = \dfrac{[HI]^2}{[H_2][I_2]}$

14.41 (a)

	SO_3	+	NO	⇌	SO_2	+	NO_2
i	0.0250		0.0250		0		0
C	$-y$		$-y$		$+y$		$+y$
E	$0.0250 - y$		$0.0250 - y$		y		y

(b) $K_c = 0.50 = \dfrac{[SO_2][NO_2]}{[SO_3][NO]}$

(c) $K_c = \dfrac{(y)(y)}{(0.0250 - y)(0.0250 - y)}$

14.43 $K_c = 0.017$
14.45 $K_p = 9.1 \times 10^{-3}$
14.47 $[SO_3] = 0.50$ M, $[SO_2] = 0.50$ M, $[O_2] = 0.25$;
$K_c = 0.25$
14.49 (a) $n(NO_2) = 0.20$ mol
(b) $n(N_2O_4) = 0.90$ mol, 0.045 M
(c) $K_c = 4.5 \times 10^2$
14.51 $[COCl_2] = 0.217$ M, $[CO] = 0.0327$ M,
$[Cl_2] = 0.0327$ M
14.53 $[CO] = 0.110$ M, $[Cl_2] = 0.110$ M,
$[COCl_2] = 0.0904$ M
14.55 $[SO_2] = 0.00438$ M, $[Cl_2] = 0.0144$ M,
$[SO_2Cl_2] = 0.00562$ M

14.57 $P_{CO} = 0.25$ atm, $P_{H_2} = 1.25$ atm, $P_{CH_2O} = 1.75$ atm
14.59 $[SO_3] = 0.13\ M$, $[SO_2] = 0.67\ M$, $[O_2] = 0.33\ M$
14.61 $K_p = P_{CO_2} = 0.12$ atm
14.63 $K_p = P_{SO_3} = 0.74$ atm
14.65 (a) $K_{sp} = [Mg^{2+}][F^-]^2$
(b) $K_{sp} = [Ca^{2+}]^3[PO_4^{3-}]^2$
(c) $K_{sp} = [Al^{3+}]^2[CO_3^{2-}]^3$
(d) $K_{sp} = [La^{3+}][F^-]^3$
14.67 $K_{sp} = 8 \times 10^{-17}$
14.69 (a) $K_{sp} = 1.2 \times 10^{-10}$
(b) $K_{sp} = 7.6 \times 10^{-5}$
(c) $K_{sp} = 1.0 \times 10^{-16}$
14.71 $1.1 \times 10^{-4}\ M$
14.73 2.2×10^{-3} g/L
14.75 (a) $9.3 \times 10^{-6}\ M$
(b) $8.7 \times 10^{-10}\ M$
14.77 (a) $5.6 \times 10^{-3}\ M$
(b) $2.8 \times 10^{-4}\ M$
14.79 $Q = 1.2 \times 10^{-10}$; no precipitate will form
14.81 $Q = 5.3 \times 10^{-14}$; a precipitate will form
14.83 (a) the phosphate precipitates first
(b) $[PO_4^{3-}] = 1.1 \times 10^{-6}\ M$ when sulfate begins to precipitate
14.85 $K_c = 0.020$
14.87 (a) $[SO_3] = 0.0010\ M$, $[SO_2] = 0.0040\ M$, $[O_2] = 0.0020\ M$
(b) $K_c = 0.032$
(c) $K_p = 2.9$
14.89 $[HI] = 0.0014\ M$, $[H_2] = 3.1 \times 10^{-4}\ M$, $[I_2] = 3.1 \times 10^{-4}\ M$
14.91 (a) $Ag^+(aq) + Cl^-(aq) \rightarrow AgCl(s)$
(b) 0.0719 g
(c) $[NO_3^-] = 0.0173\ M$, $[Cl^-] = 3.4 \times 10^{-7}\ M$
14.93 (a) $[SO_3]$ increases by a factor of $\sqrt{2}$, to approximately $2.8 \times 10^{-3}\ M$.
(b) $K_c = 1.1 \times 10^2$
(c) $1.1 \times 10^2 = \dfrac{(2y)^2}{(0.010 - 2y)^2(0.010 - y)}$
14.95 (a) $K_c = 50$
(b) y is negative (more NH_3 will form if the volume is decreased)
(c)

	N_2	+	$3H_2$	$\rightleftharpoons$	$2NH_3$
i	0.00970		0.044		0.0032
C	$-y$		$-3y$		$+2y$
e	$0.00970 - y$		$0.044 - 3y$		$0.0032 + 2y$

$$K_c = 50 = \frac{(0.0032 + 2y)^2}{(0.00970 - y)(0.044 - 3y)^3}$$

14.97 24,000 g NaCl
14.99 (a) 927 g NaCl
(b) 0.30 g, \$0.60
(c) 1.45 μg/mL; no treatment is needed

Chapter 15

15.1 A Brønsted–Lowry acid is a proton donor, and an Arrhenius acid causes the formation of hydrogen ions in water. All Arrhenius acids are Brønsted–Lowry acids, but not all Brønsted–Lowry acids are Arrhenius acids.
15.3 $CH_3COOH + CH_3COOH \rightleftharpoons CH_3COOH_2^+ + CH_3COO^-$
15.5 The pH is defined as the negative log of the molar concentration of hydronium ion, or $pH = -\log[H_3O^+]$. Because the molar concentration of hydronium ion is usually very small (typical values are on the order of 10^{-3} to $10^{-10}\ M$), the pH scale was developed to make working with these values easier.
15.7 A weak acid is one that does not ionize completely when dissolved in water.
15.9 In strongly acidic solvents, strong acids are not able to fully ionize. In this system, $HClO_4$ still behaves as a strong acid, because the additional oxygens pull electron density away from the H—Cl bond, making it weaker and more easily ionized. HCl is not as strong (because it has no oxygen atoms to attract electrons) and does not fully ionize.
15.11 The conjugate acid of a weak base is an acid. The strength of the acid can vary from weak to strong.
15.13 The fraction of an acid that is ionized varies with the starting concentration of the acid.
15.15 Positive bases are rare because they tend to repel positively charged protons.
15.17 Acids become stronger going "southeast" from C or Si.
15.19 Oxyacids are compounds that contain hydrogen, oxygen, and a third element, which is a nonmetal. Of the six strong acids, sulfuric acid (H_2SO_4), perchloric acid ($HClO_4$), and nitric acid (HNO_3) are oxyacids.
15.21 Determining the molecular geometry of NH_4^+ is one approach that could be used to determine whether all four N—H bonds are equivalent or whether the coordinate covalent $N—H^+$ bond is different.
15.23 (a) H_2SO_4: sulfuric acid
(b) H_3O^+: hydrogen ion
(c) NH_4^+: ammonium ion
(d) $C_5H_5NH^+$: pyridinium ion
15.25 (a) CN^-: cyanide ion
(b) SO_4^{2-}: sulfate ion
(c) HPO_3^{2-} monohydrogen phosphite ion
(d) CO_3^{2-}: carbonate ion
15.27 (a) CN^-(base) – HCN(acid) and H_2O(acid) – OH^-(base)
(b) HCO_3^- (base) – H_2CO_3(acid) and H_3O^+(acid) – H_2O(base)
(c) CH_3COOH(acid) – CH_3COO^-(base) and HS^-(acid) – S^{2-}(base)

15.30 (a) $NH_3 + HCl \longrightarrow NH_4^+ + Cl^-$
base acid acid base

(b) $HCO_3^{2-} + HNO_3 \longrightarrow H_2CO_3 + NO_3^-$
base acid acid base

(c) $HCOOH + CN^- \longrightarrow HCN + HCOO^-$
acid base acid base

(d) $CH_3COO^- + H_2O \longrightarrow CH_3COOH + OH^-$

15.33 (a) 2.2×10^{-11} *M* OH^-
(b) 1.20×10^{-10} *M* H_3O^+

15.35 2.5×10^{-7} *M* OH^-

15.37
(a) 2.34 0.0046 acidic
(b) 12.98 1.04×10^{-13} basic
(c) −1.09 12.0 acidic
(d) 10.67 2.12×10^{-11} basic
(e) 1.13 7.40×10^{-2} acidic
(f) 13.41 3.9×10^{-14} basic
(g) 4.15 7.07×10^{-5} acidic
(h) 9.80 1.6×10^{-10} basic
(i) 0.30 0.505 acidic

15.39
(a) −1.04 11 15.04 9.1×10^{-16} acidic
(b) 13.66 2.2×10^{-14} 0.34 0.46 basic
(c) 6.70 1.98×10^{-7} 7.30 5.05×10^{-8} acidic
(d) 12.65 2.2×10^{-13} 1.35 4.4×10^{-2} basic

15.41

	pH	[H⁺]	[OH⁻]
(a)	2.9	1×10^{-3}	1×10^{-11}
(b)	1.7	2×10^{-2}	5×10^{-13}
(c)	5.0	1×10^{-5}	1×10^{-9}
(d)	6.9	1×10^{-7}	1×10^{-7}

15.43 (a) pH: 1.30; pOH: 12.70
(b) pH: 12.38; pOH: 1.62
(c) pH: 1.85; pOH: 12.15
(d) pH: 14.02; pOH: −0.02

15.45 (a) pH: 0.03; pOH: 13.97
(b) pH: 12.92; pOH: 1.08
(c) pH: 3.48; pOH: 10.52
(d) pH: 13.94; pOH: 0.06

15.49 (a) $HN_3(aq) + H_2O(\ell) \rightleftharpoons H_3O^+(aq) + N_3^-(aq)$
(b) $H_2C_6H_6O_7(aq) + H_2O(\ell) \rightleftharpoons H_3O^+(aq) + HC_6H_6O_7^-(aq)$
(c) $H_2C_4O_4(aq) + H_2O(\ell) \rightleftharpoons H_3O^+(aq) + HC_4O_4^-(aq)$

15.51 7.2×10^{-10}

15.53 A is a weak acid, B is a strong acid

15.55 $pK_a = 4.79$

15.57 $K_a = 1.43 \times 10^{-5}$, $pK_a = 4.85$

15.61 (a)

	C_6H_5COOH	+	H_2O	→	H_3O^+	+	$C_6H_5COO^-$
i	0.20				0		0
C	$-x$				$+x$		$+x$
E	$0.20 - x$				x		x

$$K_a = 6.3 \times 10^{-5} = \frac{(x)(x)}{(0.20 - x)}$$

(b)

	HCOOH	+	H_2O	→	H_3O^+	+	$HCOO^-$
i	1.50				0		0
C	$-x$				$+x$		$+x$
E	$1.50 - x$				x		x

$$K_a = 1.8 \times 10^{-4} = \frac{(x)(x)}{(1.50 - x)}$$

(c)

	HCN	+	H_2O	→	H_3O^+	+	CN^-
i	0.0055				0		0
C	$-x$				$+x$		$+x$
E	$0.0055 - x$				x		x

$$K_a = 6.2 \times 10^{-10} = \frac{(x)(x)}{(0.0055 - x)}$$

(d)

	HNO_2	+	H_2O	→	H_3O^+	+	NO_2^-
i	0.075				0		0
C	$-x$				$+x$		$+x$
E	$0.075 - x$				x		x

$$K_a = 5.6 \times 10^{-4} = \frac{(x)(x)}{(0.075 - x)}$$

15.63 (a) 1.87 (b) 5.90 (c) 1.90 (d) 2.90

15.65 (a) 1.8% (b) 1.1% (c) 0.034% (d) 8.3%

15.67 (a) 4.1% (b) 7.9×10^{-3}% (c) 4.9% (d) 12%

15.69 $C_8H_{10}N_4O_2(aq) + H_2O(\ell) \rightleftharpoons HC_8H_{10}N_4O_2^+(aq) + OH^-(aq)$

15.71

	N_2H_4	+	H_2O	→	$N_2H_5^+$	+	OH^-
i	0.10				0		0
C	$-x$				$+x$		$+x$
E	$0.10 - x$				x		x

$$K_b = 1.3 \times 10^{-6} = \frac{(x)(x)}{(0.10 - x)}$$

15.73 pH 12.30

15.77 (a) $HCOO^- + H_2O \rightleftharpoons HCOOH + OH^-$

$$K_b = \frac{K_w}{K_a} = \frac{1.0 \times 10^{-14}}{1.8 \times 10^{-4}} = 5.6 \times 10^{-11}$$

(b) $NO_2^- + H_2O \rightleftharpoons HNO_2 + OH^-$

$$K_b = \frac{K_w}{K_a} = \frac{1.0 \times 10^{-14}}{5.6 \times 10^{-4}} = 1.8 \times 10^{-11}$$

15.79 (a) $NH_3OH^+ + H_2O \rightleftharpoons NH_2OH + H_3O^+$

$$K_a = \frac{K_w}{K_b} = \frac{1.0 \times 10^{-14}}{8.7 \times 10^{-9}} = 1.1 \times 10^{-7}$$

(b) $NH_4^+ + H_2O \rightleftharpoons NH_3 + H_3O^+$

$$K_a = \frac{K_w}{K_b} = \frac{1.0 \times 10^{-14}}{1.8 \times 10^{-5}} = 5.6 \times 10^{-10}$$

15.83 $H_2O < NH_4^+ < HF < HSO_4^-$

15.85 $F^- < H_2O < CH_3COOH < HCOOH$

15.87 (a) weakly basic (b) neutral (c) weakly acidic

15.89

	$C_5H_5NH^+$	+	H_2O	→	C_5H_5N	+	H_3O^+
i	0.060				0		0
C	$-x$				$+x$		$+x$
E	$0.060 - x$				x		x

$$K_a = \frac{K_w}{K_b} = 5.9 \times 10^{-6} = \frac{(x)(x)}{(0.060 - x)}$$

15.91 (a) pH 8.37 (b) pH 5.08 (c) pH 7.78

15.95 HI is a strong acid, so the effect of CH_3COOH will be negligible. Calculate $[H_3O^+]$ based on the concentration of HI.

15.97 K_a for acetic acid is so much larger than for HCN that $[H_3O^+]$ can be calculated based only on the acetic acid concentration.

15.99 HOFO

15.101 (a) AsH_3; within a period, acidity increases with increasing electronegativity
(b) HNO_3; the acidity of an oxyacid increases with increasing number of O atoms not bonded to H

15.103 (a) $HClO_4$; the acidity of an oxyacid increases with increasing number of O atoms not bonded to H
(b) H_2Se; within a period, acidity increases with increasing electronegativity.

15.105 (a) Arrhenius, Brønsted–Lowry, Lewis
(b) Lewis

15.107 (a) Lewis
(b) Arrhenius, Brønsted–Lowry, Lewis

15.111 (a) strongly acidic (b) weakly acidic
(c) strongly basic (d) weakly basic

15.113 (a) neutral (b) weakly basic
(c) weakly acidic (d) weakly basic

15.115 7.7%, 22%

15.117 0.22 mL

15.119 0.23 g

15.121 0.22

15.123 (a) $NH_3 + NH_3 \rightleftharpoons NH_4^+ + NH_2^-$
(b) NH_4^+ is the conjugate acid of NH_3, and NH_2^- is the conjugate base of NH_3.
(c) $NaNH_2$ is a source of NH_2^- and, therefore, acts as a base.
(d) NH_4Br is a source of NH_4^+ and, therefore, acts as an acid.

15.125 (a) pH_{HF} 1.91 (b) pH_{KF} 8.30

15.127 (a) neither (b) products
(c) reactants (d) reactants

15.129 (a) pH will be lower (b) pH will be lower
(c) pH will increase (d) no change
(e) no change

15.131 pH 13.00

15.135 (a) The hydrogen bonded to the oxygen ionizes.
(b) 1.09
(c) $2C_6H_3N_3O_7 \rightarrow 11CO + 3H_2O + 3N_2 + C$
(d) Decrease. Amino groups do not pull as much of the electron density between the oxygen and hydrogen away from hydrogen as nitro groups do.

Chapter 16

16.1 A *titration* is a procedure for the determination of the quantity of one substance by the addition of a measured amount of a second substance. The *analyte* is a substance whose concentration is being determined. The *titrant* is the substance added to react with the analyte.

16.3

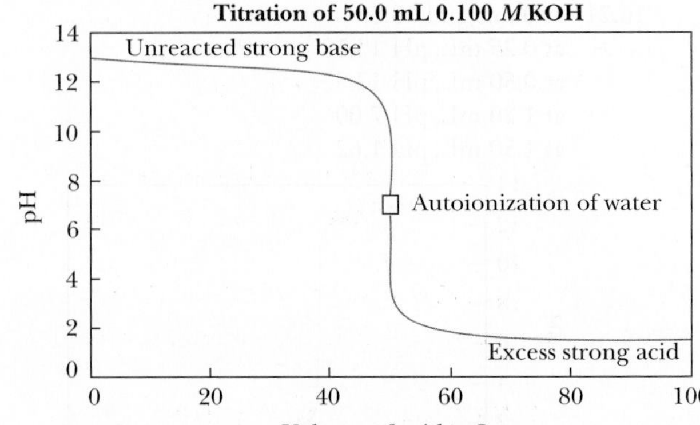

16.5 (a) This is a buffer system, consisting of a weak base (NH_3) and its conjugate acid (NH_4^+).
(b) This is a buffer system consisting of a weak acid and a weak base that are not a conjugate acid–base pair. It has little buffer capacity and would not be a good choice for a buffer system.
(c) This is not a buffer because it contains two weak acids and no weak base.

16.7 Prepare the buffer solution according to directions. Titrate a sample with 0.10 *M* HCl, recording pH versus volume acid. Perform a similar titration with NaOH. The capacity of the buffer can be judged by evaluating how flat the pH curve is and at what point the pH makes a drastic change. If the buffer capacity is too low, add a second package of the buffer mix. If the capacity is higher than needed, the buffer may be diluted with additional water to reduce the cost.

16.9 The mass of indicator is important for two reasons. If too little indicator is used, the color may be too faint to be detected. If too much is used, the volume of titrant needed to react with the indicator may become significant. The amount of alcohol specified is important because most acid–base indicators are insoluble or only sparingly soluble in water.

16.11 Phosphoric acid is a weak acid that can be used to buffer the soft drink. The manufacturer must consider other factors such as the taste of the acid and its effect on health, in addition to its K_a values.

16.13 (a) 15.0 mL (b) 26.0 mL (c) 100.0 mL

16.15 1.16 mmol

16.17 $H_2SO_4 + 2LiOH \rightarrow Li_2SO_4 + 2H_2O$; 3.05 mmol

16.19 at 0 mL, pH 0.60
at 12.50 mL, pH 1.00
at 25.00 mL, pH 7.00
at 40.00 mL, pH 12.92

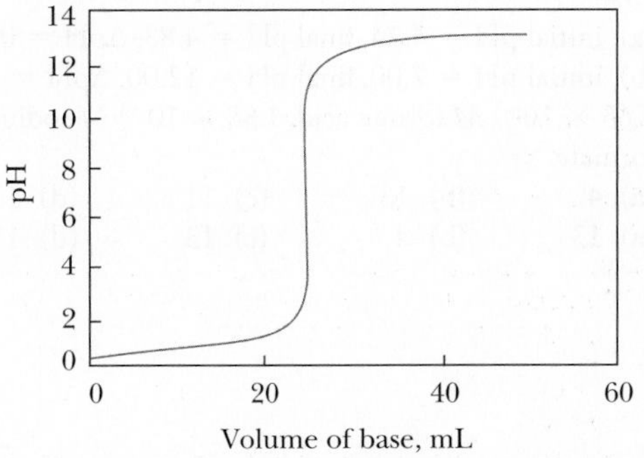

pH

Volume of base, mL

16.21 at 0 mL, pH 13.38
at 0.25 mL, pH 13.18
at 0.50 mL, pH 12.97
at 1.20 mL, pH 7.00
at 1.50 mL, pH 1.62

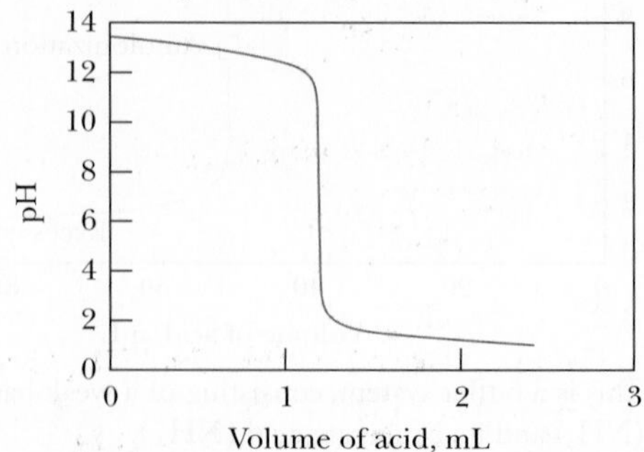

16.23

Volume HNO_3	pH
0.00 mL	13.68
0.50 mL	13.40
1.00 mL	13.15
2.40 mL	7.00
3.00 mL	1.52

The curve in Question 16.21 has the same general form, but the equivalence point occurs at half the volume the equivalence point occurs in Question 16.23.

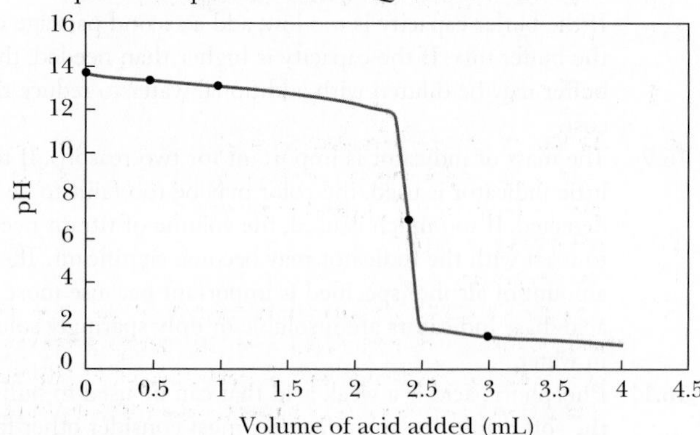

16.25 (a) 13 (b) 7 (c) 13 (d) 7
16.27 (a) 13 (b) 1 (c) 13 (d) 7
16.29 (a) 3.70 (b) 3.68
16.31 (a) 3.26 (b) 9.53
16.33 $[CH_3CO^-] = 0.064\ M$; $[CH_3COOH] = 0.036\ M$
16.35 (a) pH 4.65 (b) pH 5.05
16.37 (a) 3.70 (b) 8.39
16.39 (a) 3.14 (b) 9.80
16.41 53.4
16.43 (a) initial pH = 4.74, final pH = 4.83, ΔpH = 0.09
(b) initial pH = 7.00, final pH = 12.00, ΔpH = 5.00
16.45 $1.65 \times 10^{-3}\ M$ formic acid, $1.88 \times 10^{-3}\ M$ sodium formate
16.47 (a) 4 (b) 13 (c) 11 (d) 13
16.49 (a) 13 (b) 4 (c) 13 (d) 11

16.51

Volume NaOH	pH
0.00 mL	2.57
10.00 mL	4.74
20.00 mL	9.04
25.00 mL	12.70

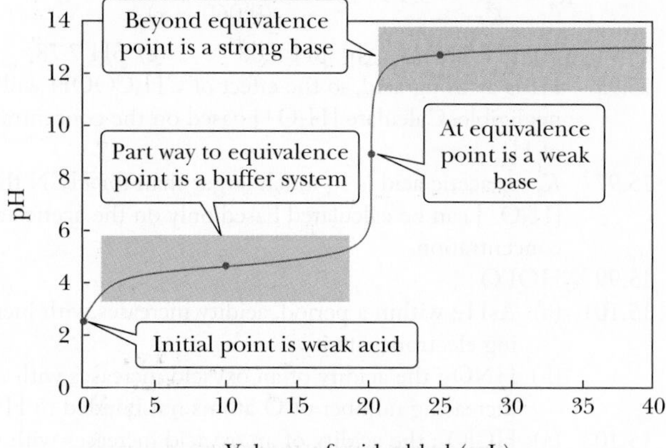

16.53

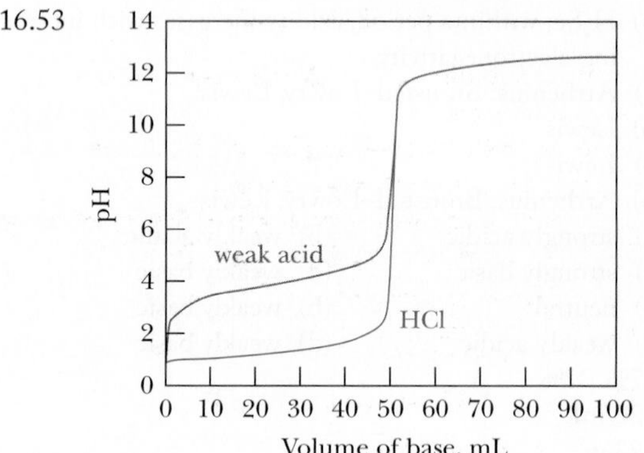

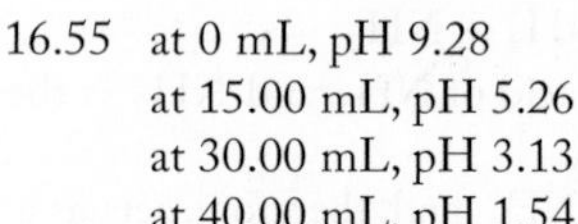

16.55 at 0 mL, pH 9.28
at 15.00 mL, pH 5.26
at 30.00 mL, pH 3.13
at 40.00 mL, pH 1.54

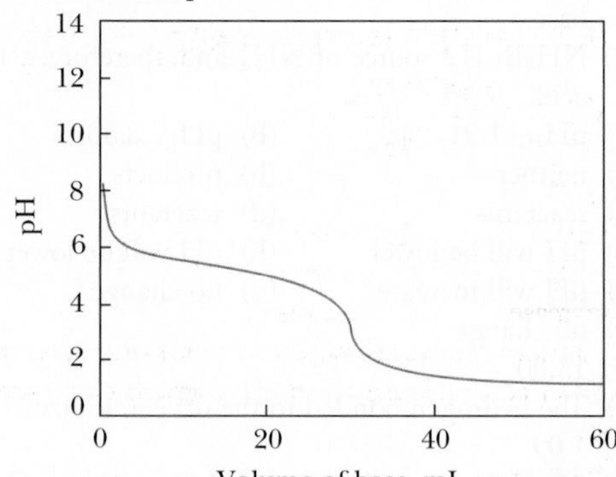

16.57

Strong Acid (mL)	Weak Base (pH)	Strong Base (pH)
0.0	11.00	13.00
10.0	9.18	12.70
20.0	8.40	12.15
24.5	7.31	11.13
25.0	5.09	7.00
25.5	2.88	2.88
40.0	1.48	1.48

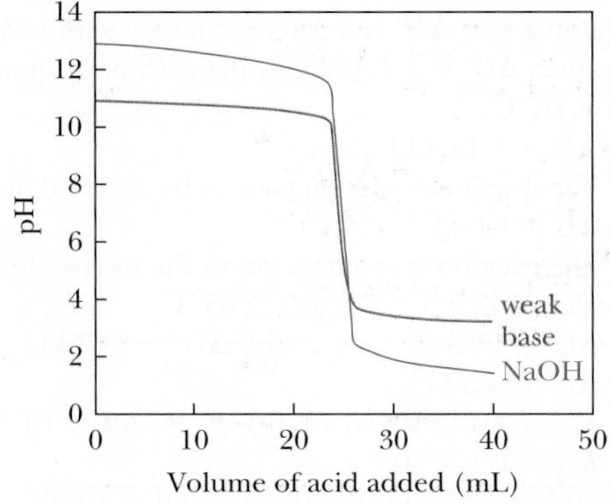

16.59 (a) bromthymol blue (b) phenolphthalein
(c) phenolphthalein

16.61 pH at the endpoint is 8.00, so phenolphthalein or bromthymol blue are both good choices.

16.63 pH at the endpoint is 4.69, so methyl red is a good choice.

16.65 (a) $H_2C_2O_4 + H_2O \rightleftharpoons H_3O^+ + HC_2O_4^-$

$$K_{a1} = \frac{[H_3O^+][HC_2O_4^-]}{[H_2C_2O_4]}$$

$HC_2O_4^- + H_2O \rightleftharpoons H_3O^+ + C_2O_4^{2-}$

$$K_{a2} = \frac{[H_3O^+][C_2O_4^{2-}]}{[HC_2O_4^-]}$$

(b) $H_2SO_3 + H_2O \rightleftharpoons H_3O^+ + HSO_3^-$

$$K_{a1} = \frac{[H_3O^+][HSO_3^-]}{[H_2SO_3]}$$

$HSO_3^- + H_2O \rightleftharpoons H_3O^+ + SO_3^{2-}$

$$K_{a2} = \frac{[H_3O^+][SO_3^{2-}]}{[HSO_3^-]}$$

16.67 pH 3.02

16.69 (a) acidic (b) acidic

16.71 (a) acidic (b) basic

16.73 (a) increase (b) increase

16.75 (a) Cannot predict. Solubility decreases due to common ion effect and increases from formation of complex ion.
(b) increases

16.77 0.162 *M*

16.79 (a) 12.22 (b) 10.32 (c) 0.49

16.81 (a) $SO_4^{2-} + H_2O \rightleftharpoons HSO_4^- + OH^-$

$$K_b = \frac{[HSO_4^-][OH^-]}{[SO_4^{2-}]} = 1.0 \times 10^{-12}$$

(b) $C_6H_5O_7^{3-} + H_2O \rightleftharpoons HC_6H_5O_7^{2-} + OH^-$

$$K_b = \frac{[HC_6H_5O_7^{2-}][OH^-]}{[C_6H_5O_7^{3-}]} = 2.5 \times 10^{-8}$$

16.83 (a) $C_3H_2O_4^{2-}(aq) + H_2O(\ell) \rightleftharpoons HC_3H_2O_4^-(aq) + OH^-(aq)$

$$K_b = \frac{[HC_3H_2O_4^-][OH^-]}{[C_3H_2O_4^{2-}]}$$

$K_b = 5 \times 10^{-9}$

(b) $CO_3^{2-}(aq) + H_2O(\ell) \rightleftharpoons HCO_3^-(aq) + OH^-(aq)$

$$K_b = \frac{[HCO_3^-][OH^-]}{[HCO_3^-]}$$

$K_b = 2.1 \times 10^{-4}$

16.85 1%: pH 7.00; 5%: pH 6.28; 95%: 3.72; 99%: pH 3.00

16.87 (a) $HCl + H_2O \rightarrow H_3O^+ + Cl^-$ 1
(b) $HCOOH + OH^- \rightarrow HCOO^- + H_3O^+$ 4
(c) $CH_3COOH + OH^- \rightarrow CH_3COO^- + H_3O^+$ 4
(d) $SO_4^{2-} + H_2O \rightleftharpoons HSO_4^- + OH^-$ 11

16.89 (a) $HCl + H_2O \rightarrow H_3O^+ + Cl^-$pH 1
(b) $HSO_4^- + H_2O \rightleftharpoons SO_4^{2-} + H_3O^+$ 3
(c) $NH_4^+ + H_2O \rightleftharpoons NH_3 + H_3O^+$ 3

16.91 phthalic acid

16.93 citric acid

16.95 0.064 *M* NH_4^+

16.97 0.529% N

16.99 1.82 L

16.101 either bromthymol blue or phenolphthalein

16.103 77 mL

Chapter 17

17.1 Work is positive for work done on the system, negative for work done by the system on the surroundings. Work is part of the energy change of the system.

17.3 w is negative

17.5 w is negative

17.7 Energy can neither be created nor destroyed. $\Delta E = q + w$, where ΔE is the change in the internal energy of the system under study, q is the heat absorbed or emitted by the system, and w is the work performed on or by the system.

17.9 $H = E + PV$. $\Delta H = \Delta E + P\Delta V$ for a change at constant pressure. $\Delta E = q + w$. In a system where the only work is that of expansion or contraction, $\Delta E = q$ for a constant volume system, whereas $\Delta H = q$ for a constant pressure system.

17.11 The entropy of a pure crystalline substance at absolute zero is known to be zero; therefore, entropy at any non-zero temperature can be determined by measuring the ΔS as the temperature increases.

17.13 Entropies of elements are zero only when the element is in a crystalline form at absolute zero. Because most calculations use a benchmark temperature of 25 °C, not absolute zero, the entropies of elements are not zero.

17.15 Thermodynamics can predict whether a reaction is spontaneous, but it cannot predict the kinetics (rate) of the reaction because thermodynamics is concerned with state functions, whereas kinetics is dependent on reaction pathways.

17.17 $\Delta G = -T\Delta S_{univ}$. By the second law of thermodynamics, $\Delta S_{univ} > 0$ for a spontaneous process. Because $T \geq 0$, $\Delta G < 0$ for a spontaneous process.

17.19 Because the entropy term in the expression for ΔG is $-T\Delta S$, if the free energy decreases as temperature increases, then the ΔS must be positive.

17.21 You cannot accurately predict the sign on ΔH. All you know is that the entropy is decreasing, so the $T\Delta S$ term in ΔG is negative. ΔH may also be negative, or it may be slightly positive but overwhelmed by the $T\Delta S$ term so that the overall value of ΔG is negative.

17.23 (a) w is negative (b) w is positive

17.25 (a) 24.5 L·atm = 2.48 kJ (b) 24.5 L·atm = 2.48 kJ

17.27 (a) 98.0 L·atm = 9.93 kJ (b) 0

17.29 −13.8 L·atm = −1.39 kJ

17.31 0.715 L·atm = 72.4 J

17.33 -3.3×10^3 kJ

17.35 −432 J

17.37 −198 J

17.39 q = 4.35 kJ; w = −430 J; ΔE = 4.31 kJ

17.41 (a) w = −0.810 kJ, q = +0.810 kJ
(b) w = −1.22 kJ, q = +1.22 kJ
(c) w = −1.37 kJ, q = +1.37 kJ
(d) expand against a slowly decreasing pressure

17.43 ΔH and ΔE will be similar because the $P\Delta V$ term is small for processes that do not involve gases.

17.45 $C_3H_8O_3 + 7/2O_2(g) \rightarrow 3CO_2(g) + 4H_2O(\ell)$; $\Delta H = -54.6$ kJ

17.47 $\Delta H_f^\circ = 1.73 \times 10^3$ kJ/mol

17.49 (a) positive (b) negative
(c) positive (d) positive

17.51 (a) 22.0 J/K (b) 174 J/K

17.53 (a) ΔS° = −331.9 J/K (b) ΔS° = −198.53 J/K
(c) ΔS° = −432.47 J/K (d) ΔS° = 178.61 J/K

17.55 ΔS° = −70.14 J/K

17.57 (a) ΔG° = −840.1 kJ, spontaneous
(b) ΔG° = −29.20 kJ, spontaneous

17.59 (a) ΔG° = −580.13 kJ, spontaneous
(b) ΔG° = 28.20 kJ, not spontaneous

17.61 (a) (1) −83.74 kJ (2) −83.72 kJ
(b) (1) −416.5 kJ (2) −425 kJ

17.63 (a) ΔH° = −285.83 kJ; ΔS° = −163.18 J/K; ΔG° = −237.2 kJ; direction of spontaneous reaction is consistent with the sign on ΔH°, not ΔS°
(b) ΔH° = −128.14 kJ; ΔS° = −331.9 J/K; ΔG° = −29.2 kJ; direction of spontaneous reaction is consistent with the sign on ΔH°, not ΔS°

17.65 (a) ΔH° = −101.8 kJ; ΔS° = −8.8 J/K; ΔG° = −99.2 kJ; direction of spontaneous reaction is consistent with the sign on ΔH°, not ΔS°
(b) ΔH° = −40.5 kJ; ΔS° = +45.2 J/K; ΔG° = −54.0 kJ; direction of spontaneous reaction is consistent with the signs on ΔH° and ΔS°

17.67 ΔG° is negative at low T and positive at high T.

17.69 ΔG° is negative at low T and positive at high T.

17.71 11,000 K

17.73 465 K

17.75 (a) ΔG°(400 K) = +0.7 kJ, ΔG°(600 K) = +24.6 kJ
(b) ΔG°(400 K) = +196.0 kJ, ΔG°(600 K) = +222.6 kJ

17.77 ΔH° and ΔS° should both be negative.

17.79 (a) false; the direction depends on temperature
(b) true; ΔG° = −24.88 kJ
(c) false; the reaction is spontaneous at temperatures below 411 K
(d) true
(e) false; the reaction is exothermic, so K_{eq} decreases with increased temperature
(f) false; we cannot say anything about the speed of reaction

17.81 Assuming that ΔH° and ΔS° do not vary with temperature, then ΔG° = −1.85 kJ, so the reaction is spontaneous at 80 °C.

17.83 (a) ΔG = −15.4 kJ
(b) The reaction is spontaneous in the forward direction.

17.85 (a) ΔG = 8.0 kJ
(b) The reaction is spontaneous in the reverse direction.

17.87 T_b(calc) = 63.4 °C, T_b(expt) = 65 °C

17.89 (a) ΔG° = 79.9 kJ (b) ΔG° = 23.7 kJ
(c) ΔG° = 4.39 kJ

17.91 (a) $K_{eq} = 3.0 \times 10^{-7}$ (b) $K_{eq} = 1.9 \times 10^{-4}$
(c) $K_{eq} = 1.9 \times 10^{-9}$

17.93 ΔG(10 K) = 13.5 kJ, K_{eq}(10 K) = 3.0×10^{-71};
ΔG(100 K) = 0.0 kJ, K_{eq}(100 K) = 1.0;
ΔG(1000 K) = −135 kJ, K_{eq}(1000 K) = 1.1×10^7

17.95 ΔG(10 K) = −16.5 kJ, K_{eq}(10 K) = 1.5×10^{86};
ΔG(100 K) = −30.0 kJ, K_{eq}(100 K) = 4.7×10^{15};
ΔG(1000 K) = −165 kJ, K_{eq}(1000 K) = 4.2×10^8

17.97 At 303 °C, ΔG° = −82.6 kJ, ΔG = −96.9 kJ

17.99 (a) K_{eq} decreases with increasing temperature.
(b) K_{eq} decreases with increasing temperature.
(c) K_{eq} decreases with increasing temperature.
(d) K_{eq} increases with increasing temperature.
(e) K_{eq} decreases with increasing temperature.

17.101 (a) 0.207 atm (b) 0.111 atm (c) 0.293 atm

17.103 -6.26×10^5 J of work is done.

17.105 $\Delta H = -1.1 \times 10^2$ kJ, $\Delta S = -1.4 \times 10^2$ J/K. The reaction is spontaneous at temperatures less than 780 K.

17.107 ΔG° = −70.89 kJ for every mole of SO_3 made. SO_3 combines with water to make H_2SO_4, which is a component of acid rain.

17.109 (a) The reaction becomes spontaneous at about 5450 K.
(b) Concern may be warranted, because at these high temperatures the hydrogen/oxygen mixture may explode spontaneously.

17.111 (a) With a ΔG° of 813.74 kJ, the reaction is not spontaneous at room temperature.
(b) The reaction becomes spontaneous at about 3850 K.
(c) ΔH° = 1233.07 kJ, ΔS° = 437.93 J/K, so the reaction becomes spontaneous at about 2816 K.
(d) The reaction can be driven forward at temperatures greater than 1484 °C by removing Ca vapor as it forms.

Chapter 18

18.1 Oxidation is the loss of electrons. Reduction is the gain of electrons. A redox reaction is an electron-transfer reaction in which one or more electrons are transferred from the species being oxidized (the reducing agent) to the species being reduced (the oxidizing agent). The difference between a redox reaction and a Lewis acid–base reaction is that the redox reaction involves a change in the oxidation numbers of some elements.

18.3 Zinc is the better reducing agent.

18.5 Mg, Na, Li

18.7 A battery contains the chemicals it needs to generate electricity. A fuel cell depends on a constant flow of reactants from some external source (the atmosphere, a tank of fuel, among others) to generate electricity.

18.9 (a) Cl = +5, O = −2 (b) P = +3, Cl = −1
(c) C = +2, O = −2

18.11 (a) N = +5, O = −2 (b) N = +3, O = −2
(c) N = −3, H = +1

18.13 (a) Zr = +4, O = −2 (b) Fe = +2, O = −2
(c) Ca = +2, N = +5, O = −2

18.15 (a) Ba = +2, O = −1 (b) F = 0
(c) Sn = +2

18.17 (a) N = +4, O = −2 (b) Cr = +3, O = −2
(c) Co = +3, N = +5, O = −2

18.19 (a) K = +1, H = +1, F = −1
(b) H = +1, Se = −2
(c) Na = +1, O = −1/2
(d) C = −3, H = +1

18.21 (a) $2H_2 + O_2 \rightarrow 2H_2O$ H_2 is oxidized, O is reduced
(b) $4Fe + 3O_2 \rightarrow 2Fe_2O_3$ Fe is oxidized, O is reduced
(c) $2Al_2O_3 + 3C \rightarrow 4Al + 3CO_2$ Al is reduced, C is oxidized, O is unchanged

18.23 (a) $3Na + FeCl_3 \rightarrow Fe + 3NaCl$ Na is oxidized, Fe is reduced
(b) $SnCl_2 + 2FeCl_3 \rightarrow SnCl_4 + 2FeCl_2$ Sn is oxidized, Fe is reduced
(c) $3CO + Cr_2O_3 \rightarrow 2Cr + 3CO_2$ C is oxidized, Cr is reduced

18.25 (a) $Cr^{3+}(aq) + 3e^- \rightarrow Cr(s)$, reduction
(b) $2I^-(aq) \rightarrow I_2 + 2e^-$, oxidation
(c) $NO_2^-(aq) + H_2O(\ell) \rightarrow NO_3^-(aq) + 2H^+(aq) + 2e^-$, oxidation

18.27 (a) $UO_2^{2+}(aq) + 4H^+(aq) + 2e^- \rightarrow U^{4+}(aq) + 2H_2O(\ell)$, reduction
(b) $Zn(s) \rightarrow Zn^{2+}(aq) + 2e^-$, oxidation
(c) $IO_3^-(aq) + 6H^+(aq) + 6e^- \rightarrow I^-(aq) + 3H_2O(\ell)$, reduction

18.29 (a) $Sn(s) + 2Fe^{3+}(aq) \rightarrow Sn^{2+} + 2Fe^{2+}$
(b) $HAsO_3^{2-}(aq) + I_2(s) + H_2O(\ell) \rightarrow H_2AsO_4^-(aq) + 2I^-(aq) + H^+(aq)$
(c) $Cu(s) + 2Ag^+(aq) \rightarrow Cu^{2+}(aq) + 2Ag(s)$

18.31 (a) $Fe(s) + 2Ag^+(aq) \rightarrow 2Ag(s) + Fe^{2+}(aq)$
(b) $I_2(s) + 2S_2O_3^{2-}(aq) \rightarrow 2I^-(aq) + S_4O_6^{2-}(aq)$
(c) $MnO_4^-(aq) + 5Fe^{2+}(aq) + 8H^+(aq) \rightarrow 5Fe^{3+} + Mn^{2+}(aq) + 4H_2O(\ell)$

18.33 (a) $2Al(s) + 3ClO^-(aq) + 3H_2O(\ell) + 2OH^-(aq) \rightarrow 2Al(OH)_4^-(aq) + 3Cl^-(aq)$
(b) $2MnO_4^-(aq) + 3SO_3^{2-}(aq) + H_2O(\ell) \rightarrow 2MnO_2(s) + 3SO_4^{2-}(aq) + 2OH^-(aq)$
(c) $4Zn(s) + NO_3^-(aq) + 7OH^-(aq) + 6H_2O(\ell) \rightarrow 4Zn(OH)_4^{2-}(aq) + NH_3(aq)$

18.35 (a) $3Cl_2(aq) + 6OH^-(aq) \rightarrow 5Cl^-(aq) + ClO_3^-(aq) + 3H_2O(\ell)$
(b) $2MnO_4^-(aq) + I^-(aq) + H_2O(\ell) \rightarrow IO_3^-(aq) + 2MnO_2(s) + 2OH^-(aq)$
(c) $ClO_3^-(aq) + 3CN^-(aq) \rightarrow Cl^-(aq) + 3CNO^-(aq)$

18.37 The charges are not balanced in the proposed reaction. The proposed reaction is also two oxidations.

18.39

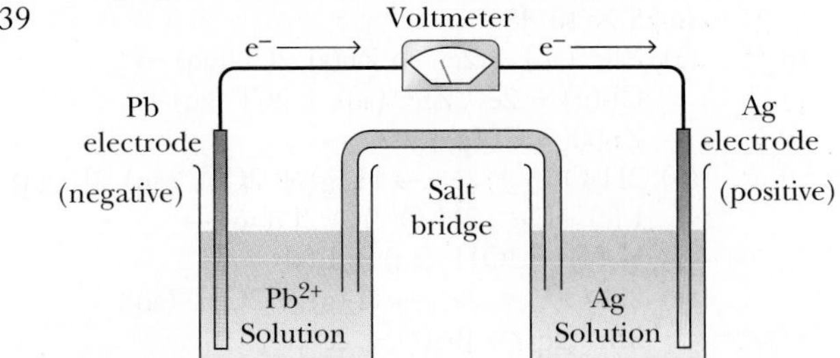

Half reactions: $Pb(s) \rightarrow Pb^{2+}(aq) + 2e^-$, $Ag^+(aq) + e^- \rightarrow Ag(s)$

18.41 (a) $2Hg(\ell) + 2Cl^-(aq) \rightarrow Hg_2Cl_2(s) + 2e^-$, $Cu^{2+}(aq) + 2e^- \rightarrow Cu(s)$
(b) $2Hg(\ell) + 2Cl^-(aq) + Cu^{2+}(aq) \rightarrow Hg_2Cl_2(s) + Cu(s)$
(c) Electrons flow from the Pt electrode to the Cu electrode.
(d) yes

18.43 (a) $E° = +0.59$ V, spontaneous as written
(b) $E° = +2.16$ V, spontaneous as written
(c) $E° = +0.82$ V, spontaneous as written

18.45 (a) $E° = +1.36$ V, spontaneous in the forward direction
(b) $E° = -1.25$ V, spontaneous in the reverse direction
(c) $E° = +0.029$ V, spontaneous in the forward direction

18.47 $E°(\text{red}) = +0.334$ V

18.49 (a) $Cu(s) \rightarrow Cu^{2+}(aq) + 2e^-$, $Pu^{4+}(aq) + e^- \rightarrow Pu^{3+}(aq)$, $Cu(s) + 2Pu^{4+}(aq) \rightarrow Cu^{2+}(aq) + 2Pu^{3+}(aq)$
(b) $E°(\text{red}) = +0.98$ V

18.51 (a) H_2O_2 is the better oxidant.
(b) Cu is the better reducing agent.

18.53 (a) Cr^{2+} (b) Fe^{3+}, Ag^+, or Ce^{4+}
(c) Ni^{2+} or Pb^{2+}

18.55 (a) $K_{eq} = \dfrac{[Ni^{2+}]}{[Cu^{2+}]} = 9 \times 10^{19}$

(b) $K_{eq} = \dfrac{1}{P_{Cl_2}} = 3.7 \times 10^{38}$

(c) $K_{eq} = \dfrac{[Cl^-]^2}{[I^-]^2} = 5.5 \times 10^{27}$

18.57 $\Delta G° = -1025$ kJ

18.59 $\Delta G° = +2.51$ kJ, $K_{eq} = 0.363$, H_2 is favored

18.61 (a) 0.55 V (b) 2.15 V (c) 0.64 V

18.63 0.030 V
18.65 $[Pb^{2+}] = 0.0034\ M$
18.67 $K_{sp} = 2.8 \times 10^{-15}$
18.69 $E = 0.073$ V, $Fe(s) + Fe^{2+}(aq, 0.750\ M) \rightarrow$
$Fe(s) + Fe^{2+}(aq, 0.0025\ M)$
18.71 (a) $E = +1.59$ V
(b) No, $[OH^-]$ is constant, and all other species are condensed phases.
(c) Yes; in fact, such cells are currently on the market.
18.73 (a) $C_3H_8(g) + 20OH^-(aq) \rightarrow$
$3CO_2(g) + 14H_2O(\ell) + 20e^-$,
$5O_2(g) + 10H_2O(\ell) + 20e^- \rightarrow 20OH^-(aq)$
(b) $\Delta G^\circ = -2108.3$ kJ, $E^\circ = 1.09$ V
(c) 5×10^4 J
18.75 (a) $Zn^{2+}(aq) + 2e^- \rightarrow Zn(s), 2Cl^-(aq) \rightarrow$
$Cl_2(g) + 2e^-, Zn^{2+}(aq) + 2Cl^-(aq) \rightarrow$
$Zn(s) + Cl_2(g)$
(b) $2H_2O(\ell) + 2e^- \rightarrow H_2(g) + 2OH^-(aq), 2I^-(aq) \rightarrow$
$I_2(s) + 2e^-, 2H_2O(\ell) + 2I^-(aq) \rightarrow$
$H_2(g) + 2OH^-(aq) + I_2(s)$
(c) $2H_2O(\ell) + 2e^- \rightarrow H_2(g) + 2OH^-(aq)$,
$2Br^-(aq) \rightarrow Br_2(\ell) + 2e^-$,
$2H_2O(\ell) + 2Br^-(aq) \rightarrow H_2(g) + 2OH^-(aq) +$
$Br_2(\ell)$
18.77 (a) Ag^+ (b) Pb^{2+} (c) Ba^{2+}
18.79 (a) To increase the conductivity, because HF is a weak electrolyte.
(b) K and F_2
18.81 (a) 1.4×10^5 C (b) 3.9×10^3 C
(c) 2.72×10^4 C (d) 2.9×10^2 C
18.83 0.0185 g
18.85 253 minutes
18.87 No, because Na reacts with water.
18.89 Because Al has a more positive oxidation potential than Fe, the aluminum acts like a sacrificial anode, protecting the iron from corrosion.
18.91 (a) Ca = +2, C = +3, O = −2
(b) Ba = +2, Cl = +7, O = −2
(c) Tl = +3
18.93 $E = -0.37$ V
18.95 (a) $H_2SO_3(aq) + H_2O(\ell) + 2Fe^{3+}(aq) \rightarrow$
$HSO_4^-(aq) + 3H^+(aq) + 2Fe^{2+}(aq), E^\circ = 0.60$ V
(b) $6Fe^{2+}(aq) + Cr_2O_7^{2-}(aq) + 14H^+(aq) \rightarrow$
$6Fe^{3+}(aq) + 2Cr^{3+}(aq) + 7H_2O(\ell), E^\circ = 0.56$ V
(c) $HNO_2(aq) + H^+(aq) + Fe^{2+}(aq) \rightarrow$
$NO(g) + H_2O(\ell) + Fe^{3+}(aq), E^\circ = 0.212$ V
18.97 $\Delta G^\circ = -12.5$ kJ, $E^\circ = 0.0650$ V
18.99 (a) $\Delta G^\circ = -261$ kJ (b) $\Delta G = -1.20$ kJ
(c) 86.6 hours
18.101 (a) $E^\circ = -0.394$ V (b) $E = -0.347$ V
18.103 $E = -0.19$ V
18.105 (a) 1112 mol e^- (b) 1.24×10^3 amps
(c) 278 mol O_2
18.107 19 minutes
18.109 $E^\circ = 1.06$ V

Chapter 19

19.1 Transition elements have partially occupied *d* subshells.
19.3 Because of the lanthanide contraction, Nb and Ta have the same atomic radius, 146 pm. However, the atomic mass of Ta, 180.95 u, is nearly twice that of Nb, 92.91 u. Because Ta and Nb are both body-centered cubic, the ratio of the densities of the solids will be nearly equal to that of the atoms.
19.5 The third ionization energy is much higher toward the end of the transition series.
19.7 The atoms of elements in Group 1B have just one electron in the outermost *s* subshell, whereas the atoms of most other transition elements have two electrons in their outermost *s* subshell.
19.9 Concentrate the mineral and remove impurities. This can be done physically, such as in panning for gold or flotation of sulfide ores, or chemically, such as the treatment of bauxite with base.
Convert the mineral into a form more suitable for reduction. This is usually accomplished by roasting.
19.11 Zn, Cd, Hg
19.13 Fe is a transition metal
19.15 Sc
19.17 (a) Cr (b) Hf (c) Nb (d) W
19.19 Based on the trend, Cr should have the greatest heat of fusion, but actually, V is higher.
19.21 Nb > W > V > Co
19.23 (a) +4 (b) +6 (c) +5 (d) +7
19.25 (a) Mn (b) Ta (c) Rh (d) Os
19.27 (a) $[Cr(H_2O)_5Cl]Cl_2$
(b) $[Cr(NH_3)_4Cl_2]Cl$
(c) $K_3[Fe(CN)_6]$
19.29 (a) diamminedichloroplatinum(II)
(b) bis(ethylenediamine)dinitrocobalt(III) nitrate
(c) potassium hexachlororhodate(III)
(d) tetrammineplatinum(II) tetrachloroplatinate(II)
(e) hexacarbonylchromium(0)
19.31 (a) $[Cr(H_2O)_5Cl]Cl_2$
(b) $[Rh(NH_3)_4(NO_2)_2]Br$
(c) $[RuCl_2(en)_2]^+$
(d) $[Rh(H_2O)_2Cl_4]^-$
(e) $[Pt(NH_3)_3Br_3]^+$
19.33

[Cr(Br)(Br)(H_2O)(H_2O)(H_2O)(H_2O)]$^+$

cis-tetraaquadibromochromium(III)

[Cr(Br)(Br)(H_2O)(H_2O)(H_2O)(H_2O)]$^+$

trans-tetraaquadibromochromium(III)

19.35 (a) $[Co(en)_2Br_2]^+$
(b) $[Pt(NH_3)_3SCN]^+$
(c) $[Co(NH_3)_3Cl_3]$ and $[Pt(NH_3)_3SCN]^+$
(d) $[Cr(H_2O)_2Cl_2Br_2]^-$

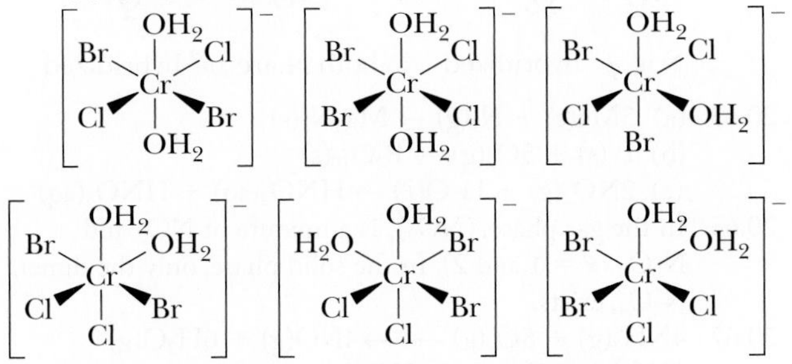

19.37 (a) 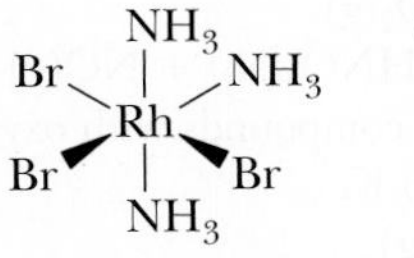 (b)

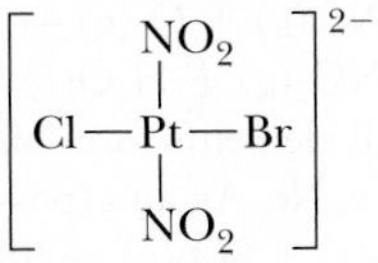

19.39 SCN^-, NO_2^-, NCO^-
19.41 a mixture containing equal quantities of two enantiomers
19.43 (a) $[Co(CN)_6]^{3-}$
(b) $[Cr(H_2O)_6]^{3+}$
(c) $[Ru(H_2O)_6]^{2+}$
(d) $[Cr(H_2O)_6]^{3+}$
19.45 (a) 2 unpaired electrons $[V(H_2O)_6]^{3+}$
(b) 4 unpaired electrons $[CrF_6]^{4-}$ or 2 $[Cr(CN)_6]^{4-}$
(c) 4 unpaired electrons $[Fe(H_2O)_6]^{2+}$ or 0 $[Co(CN)_6]^{3-}$
(d) 2 unpaired electrons $[Ni(H_2O)_6]^{2+}$
19.47 (a) 3 unpaired electrons
(b) 0 unpaired electrons
(c) 5 unpaired electrons
(d) 0 unpaired electrons
(e) 2 unpaired electrons
19.49 tetrahedral
19.51 (a) $[Cr(H_2O)_6]^{2+}$ is high spin with 4 unpaired electrons
$[Mn(CN)_6]^{3-}$ is low spin with 2 unpaired electrons
(b) $[Fe(H_2O)_6]^{2+}$ is high spin with 4 unpaired electrons
$[Ru(H_2O)_6]^{2+}$ is low spin with 0 unpaired electrons
(c) $[Co(H_2O)_6]^{2+}$ is high spin with 3 unpaired electrons
$[Co(CN)_5H_2O]^{3-}$ is low spin with 1 unpaired electron
19.53 (a) square planar, no unpaired electrons
(b) tetrahedral, three unpaired electrons
(c) square planar, no unpaired electrons
19.55 (a) electrolytic reduction of molten salt
(b) reduction by C and CO
19.57

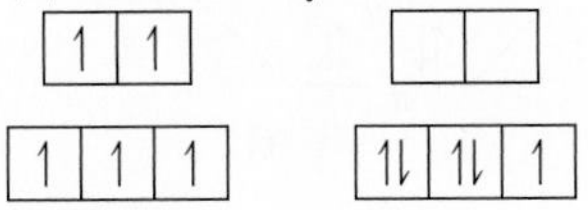

high spin
$\Delta < P$

low spin
$\Delta > P$

19.59 $[FeCl_4(H_2O)_2]^{2-}$, d^6, high spin, both isomers have 4 unpaired electrons

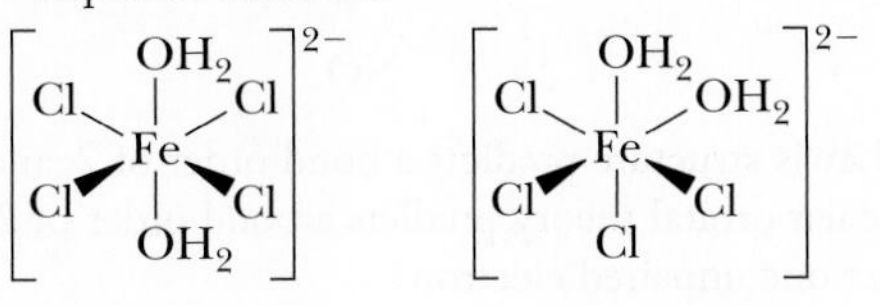

trans-diaquatetra-chloroferrate(II)

cis-diaquatetra-chloroferrate(II)

$[FeCl_4]^{2-}$ Fe^{2+}, d^6 high spin, 4 unpaired electrons

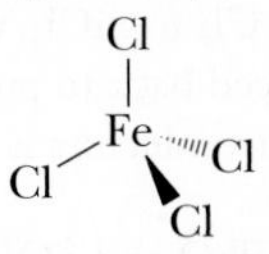

tetrachloroferrate(II)

19.61 Paladium is d^8 in $[PdCl_4]^{2-}$. If the complex were tetrahedral, it would be paramagnetic (two unpaired electrons) regardless of the magnitude of Δ. Therefore, the complex must be low-spin square planar.

19.63

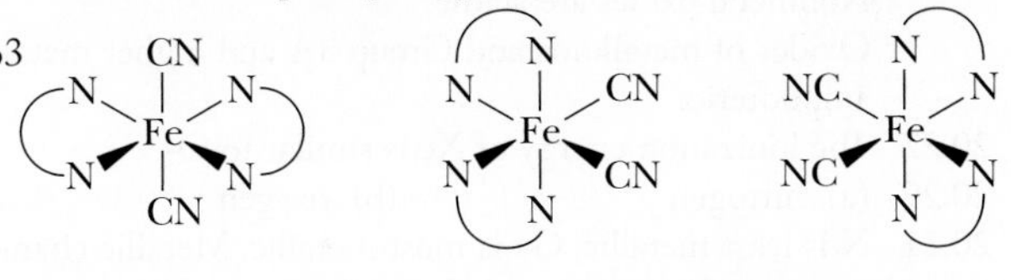

enantiomers

19.65 $Cu + 2H_2SO_4 \rightarrow CuSO_4 + H_2SO_3 + H_2O$
41.3 g Cu
H_2SO_4 reacts with copper because the sulfate anion is an oxidizing agent
19.67 VOCl, two electrons are in the $3d$ orbitals. The low melting point indicates that the bonding is mainly covalent.
19.69

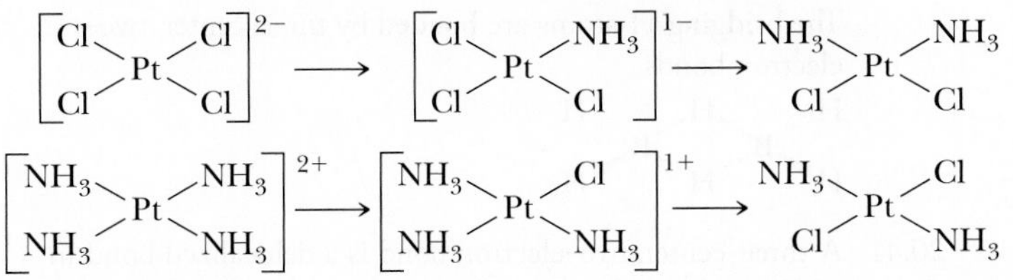

Chapter 20

20.1 The second period elements are smaller and have higher electronegativities than the elements of the later periods. Because there is no $2d$ subshell, the second period elements cannot form expanded valence compounds.
20.3 Electronegativities and ionization energies are high for nonmetals and low for metals.
20.5 Hydrogen can gain an electron to form H^-, as in NaH. Hydrogen can lose an electron to form H^+ in water. Hydrogen can share its electron with another atom, with the shared electrons being evenly distributed between the two bonded atoms, as in H_2, or with the shared electrons being more strongly attracted to one of the bonded atoms, causing a buildup of positive charge on one atom and negative charge on the other, as in HF.
20.7 H_2 is a very small nonpolar molecule, so the molecules are held together by very weak London dispersion forces.
20.9 He is not flammable but, like H_2, has a much lower density than air.
20.11 The inert pair effect is the tendency for the heavier metals of Groups 3A, 4A, and 5A not to use the outer shell s electrons in bonding. It causes the Group 3A metals to form +1 cations.
20.13 B is a metalloid; Al, Ga, In, and Tl are metals.
20.15 A protective coating of aluminum oxide, Al_2O_3, forms on the aluminum.
20.17 graphite electrodes, carbon black, carbon fiber for composites

20.19 SiO_2 is reduced to Si with C in an electric arc. The crude Si is converted by reaction with Cl_2 to $SiCl_4$, which is purified by distillation and reduced back to pure Si with Mg or H_2. Ultrapure Si for semiconductors is obtained by zone refining.

20.21 PbS (galena) is roasted to form PbO and SO_2. The PbO is then reduced with C to form Pb and CO.

20.23 Minerals that contain P are generally insoluble in water. They are converted into soluble forms by treatment with H_2SO_4 or H_3PO_4.

20.25 Group 1A and 2A metal oxides are basic.
Nonmetal oxides are acidic.
Oxides of metalloids and Group 3A and higher metals are amphoteric.

20.27 The ionization energy of Xe is similar to O_2.

20.29 (a) nitrogen (b) oxygen

20.31 N is least metallic, Ga is most metallic. Metallic character increases from right to left and from top to bottom on the periodic chart.

20.33 $NaH(s) + H_2O(\ell) \rightarrow NaOH(aq) + H_2(g)$
0.981 g NaH

20.35 $CH_4(g) + H_2O(g) \rightarrow 3H_2(g) + CO(g)$
$C(s) + H_2O(g) \rightarrow H_2(g) + CO(g)$

20.37 $CO(g) + H_2O(g) \rightarrow H_2(g) + CO_2(g)$

20.39 B atoms are sp^3 hybridized. The terminal H atoms are bonded by conventional two-center, two-electron bonds. The bridging H atoms are bonded by three-center, two-electron bonds.

[Structure of B_2H_6: two B atoms bridged by two H atoms, each B bearing two terminal H atoms]

20.41 A three-center, two-electron bond is a delocalized bond in which atomic orbitals on three atoms overlap to form an orbital containing two electrons, centered on the three bonded atoms.

20.43 $2B_{10}H_{14}(s) + 22O_2(g) \rightarrow 10B_2O_3(s) + 14H_2O(g)$

20.45 $Al_2O_3(s) + 2OH^-(aq) + 3H_2O(\ell) \rightarrow 2Al(OH)_4^-(aq)$ to remove insoluble impurities
$Al(OH)_4^-(aq) + H^+(aq) \rightarrow Al(OH)_3(aq) + H_2O(\ell)$
$2Al(OH)_3(s) \xrightarrow{heat} Al_2O_3(s) + 3H_2O(\ell)$

20.47 Ruby is α-alumina (Al_2O_3, corundum) with chromium impurities

20.49 Although the structures of Al_2Cl_6 and B_2H_6 are similar, all the bonds in Al_2Cl_6 are conventional two-center, two-electron bonds, whereas the bridging H atoms in B_2H_6 are bonded by three-center, two-electron bonds.

[Structure of Al_2Cl_6: two Al atoms bridged by two Cl atoms, each Al bearing two terminal Cl atoms]

20.51 Si is sp^3 hybridized, C is sp hybridized

20.53 A p-type Si semiconductor has Group 3A impurities (such as Ga), which causes vacancies in the conduction band.

20.55 Si is sp^3 hybridized, so the molecule is tetrahedral and nonpolar.

20.57 Nitrogen exists as $N_2(g)$. Phosphorus exists as $P_4(s)$. P does not form diatomic molecules because it is too large to form strong π bonds.

20.59 Nitrogen is isolated by fractional distillation of liquified air.

20.61 [Lewis structures of NO_2 and N_2O_4 with formal charges]

N is sp^2 hybridized both N are sp^2 hybridized

20.63 (a) $3Mg(s) + N_2(g) \rightarrow Mg_3N_2(s)$
(b) $P_4(s) + 5O_2(g) \rightarrow P_4O_{10}(s)$
(c) $2NO_2(g) + H_2O(\ell) \rightarrow HNO_2(aq) + HNO_3(aq)$

20.65 In the gas phase, $(NO_2)_x$ is a mixture of NO_2 and N_2O_4 ($x = 1$ and 2). In the solid phase, only the dimer, N_2O_4, exists.

20.67 $4NH_3(g) + 5O_2(g) \xrightarrow{Pt} 4NO(g) + 6H_2O(g)$
$2NO(g) + O_2(g) \rightarrow 2NO_2(g)$
$3NO_2(g) + H_2O(\ell) \rightarrow 2HNO_3(aq) + NO(g)$

20.69 All elements form binary compounds with oxygen except He, Ne, Ar, and (possibly) Kr.

20.71 $S_8(s) + 8O_2(g) \rightarrow 8SO_2(g)$
$2SO_2(g) + O_2(g) \xrightarrow{V_2O_5} 2SO_3(g)$
$SO_3(g) + H_2O(\ell) \xrightarrow{H_2SO_4} H_2SO_4(aq)$

20.73 [Lewis structures of XeF_2 and XeO_3]

linear, non-polar tetrahedral, polar

20.75 $4NH_3(g) + 5O_2(g) \rightarrow 4NO(g) + 6H_2O(g)$
14 kg NH_3

20.77 NH_3 boils at −33 °C. PH_3 boils at −88 °C. NH_3 has the higher boiling point because of hydrogen bonding.

20.79 $2KClO_3(s) \xrightarrow{MnO_2} 2KCl(s) + 3O_2(g)$
1.6 g $KClO_3$

20.81 0.18 g N_2H_4

20.83 [Lewis structure of XeO_2F_4]

octahedral

20.85 $\cdot\ddot{N}=\ddot{O}$

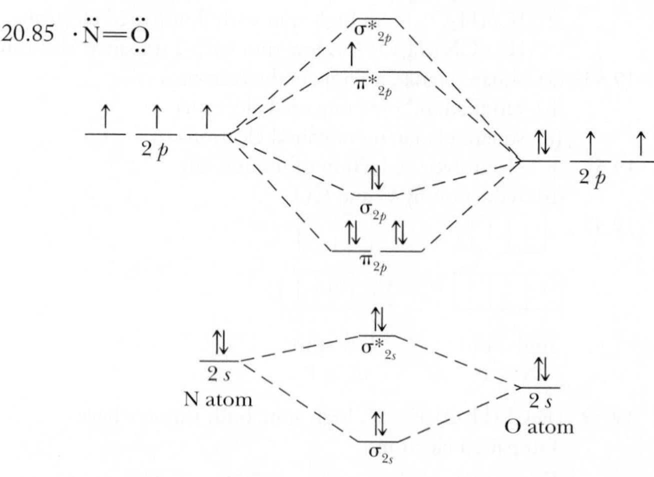

The Lewis structure predicts a bond order of 2, and molecular orbital theory predicts a bond order of 2.5; both predict one unpaired electron.

Chapter 21

21.1 beta decay

21.3 electron capture, positron emission

21.5 Alpha and beta decays leave the remaining nucleons in an excited nuclear state, which returns to the ground state by gamma emission.

21.7 ^{14}C dating is not accurate enough to date an object 20–50 years old.

21.9 Because neutrons are electrically neutral, there is no electrostatic repulsion between the neutron and the target nucleus.

21.11 Alpha particles are the least penetrating; gamma rays are the most penetrating. Alpha particles formed within the body are dangerous because they are more likely to interact with living tissue than to escape to the outside.

21.13 People working in nuclear medicine and nuclear power plants near the proximity of nuclear processes, and those mining and processing uranium are likely to receive a greater exposure to radiation than the average exposure of the U.S. population. Pilots and flight attendants on commercial airplanes also are likely to receive greater than average exposure to radiation because the presence of cosmic radiation is greater at higher elevations than it is at sea level.

21.15 The rods contain highly radioactive fission products.

21.17

Symbol	Z	A	Number of Protons	Number of Neutrons
$^{40}_{20}Ca$	**20**	**40**	**20**	**20**
$^{31}_{15}\mathbf{P}$	15	31	**15**	**16**
$^{118}_{50}\mathbf{Sn}$	**50**	**118**	50	68
$^{239}_{93}\mathbf{Np}$	**93**	239	93	**146**

21.19 (a) 1.00 (b) 1.00 (c) 1.25 (d) 1.46 (e) 1.54

21.21 (a) within the band of stability (b) above the band of stability (c) within the band of stability (d) beyond the band of stability (e) above the band of stability

21.23 (a) $^{201}_{83}Bi \rightarrow {}^{197}_{81}Tl + {}^{4}_{2}\alpha$

(b) $^{184}_{77}Ir \rightarrow {}^{184}_{76}Os + {}^{0}_{+1}\beta$

(c) $^{135}_{57}La + {}^{0}_{-1}e \rightarrow {}^{135}_{56}Ba$

(d) $^{80}_{35}Br \rightarrow {}^{80}_{36}Kr + {}^{0}_{-1}\beta$

21.25 (a) $^{227}_{90}Th$ (b) $^{22}_{10}Ne$ (c) $^{219}_{88}Ra$ (d) $^{0}_{-1}\beta$ (e) $^{0}_{+1}\beta$

21.27 (a) $^{67}_{30}Zn$ (b) $^{215}_{87}Fr$ (c) $^{0}_{-1}\beta$ (d) $^{124}_{54}Xe$ (e) $^{229}_{90}Th$

21.29 (a) above (b) below (c) below

21.31 (a) positron emission or electron capture, $^{117}_{51}Sb \rightarrow {}^{117}_{50}Sn + {}^{0}_{+1}\beta$ or $^{117}_{51}Sb + {}^{0}_{-1}e \rightarrow {}^{117}_{50}Sn$

(b) beta emission, $^{83}_{34}Se \rightarrow {}^{83}_{35}Br + {}^{0}_{-1}\beta$

(c) alpha emission, $^{221}_{89}Ac \rightarrow {}^{219}_{89}Fr + {}^{4}_{2}\alpha$

(d) beta emission, $^{42}_{18}Ar \rightarrow {}^{42}_{19}K + {}^{0}_{-1}\beta$

21.33 (a) $^{76}_{35}Br \rightarrow {}^{76}_{34}Se + {}^{0}_{+1}\beta$ or $^{76}_{35}Br + {}^{0}_{-1}e \rightarrow {}^{76}_{34}Se$

(b) $^{84}_{35}Br \rightarrow {}^{84}_{36}Kr + {}^{0}_{-1}\beta$

(c) $^{109}_{46}Pd \rightarrow {}^{109}_{47}Ag + {}^{0}_{-1}\beta$

(d) $^{241}_{95}Am \rightarrow {}^{237}_{93}Np + {}^{4}_{2}\alpha$

21.35 eight alpha decays and four beta decays

21.37 $t_{1/2} = 1.29 \times 10^5$ years

21.39 $t_{1/2} = 17.0$ days

21.41 2.86×10^9 years

21.43 3.6×10^9 years

21.45 (a) $^{56}_{26}Fe$ (b) $^{27}_{13}Al$ (c) $^{249}_{100}Fm$ (d) $^{2}_{1}H$ (e) $^{256}_{103}Lr$

21.47 (a) 0.2822 g (b) 262.9 MeV (c) 8.481 MeV/nucleon

21.49 (a) ^{26}Al: 8.151 MeV/nucleon, ^{27}Al: 8.332 MeV/nucleon, ^{28}Al:8.310 MeV/nucleon (b) ^{27}Al is most stable

21.51 (a) 0.203 u (b) 3.03×10^{-11} J (c) 7.63×10^{10} J

21.53 0.10

21.55 E = 9.0 MeV, $\lambda = 1.4 \times 10^{-13}$ m

21.57 4.22×10^8 kJ/g He, about 5 times that of 1 g ^{235}U

21.59 (a) 2.9×10^3 metric tons coal (b) 52 metric tons SO_2

21.61 1.0043

21.63 18 Bq

21.65 4300 Bq

Chapter 22

22.1 Organic chemistry is the study of carbon-containing compounds. Biochemistry is the study of the chemistry of systems in living organisms.

22.3 Alkanes are called *saturated hydrocarbons* because they contain the maximum possible number of hydrogen atoms per molecule. The carbon atoms are sp^3 hybridized.

22.5 Carbon atoms are sp^2 hybridized in ethylene and sp hybridized in acetylene.

22.7 Carboxylic acids are acidic in water because of the electron-withdrawing power of the carbonyl group.

22.9 Cross-linking is the creation of bonds between chains of the polymer. These links strengthen the polymer and prevent it from crystallizing with age.

22.11 An α-helix is a spiral structure formed by hydrogen bonding between amide groups within the same region of a protein chain. A β-pleated sheet is formed by hydrogen bonding between amide groups in different chains or different sections of the same chain. These structural features are secondary structures.

22.13 DNA stores genetic information. RNA transfers information and directs the synthesis of proteins.

22.15 DNA is made of complementary strands of a sugar-phosphate polymer with base side chains. Because of the structure of the bases, strong hydrogen bonding occurs only between the adenine-thymine pairs and the guanine-cytosine pairs, making each of their ratios 1:1. The double helix of DNA is made up of two strands, in which each base on one strand is hydrogen bonded to its complement on the other.

22.17 C_6H_{14} and C_9H_{20} are noncyclic alkanes.

22.19 C_8H_{18}

22.21

```
    H  H  H  H  H
    |  |  |  |  |
 H—C—C—C—C—C—H    pentyl group
    |  |  |  |  |
    H  H  H  H  H
```

22.23

```
   H  H  H  H  H  H
   |  |  |  |  |  |
H—C—C—C—C—C—C—H
   |  |  |  |  |  |
   H  H  H  H  H  H
```

n-hexane

```
   H  H  H  H  H
   |  |  |  |  |
H—C—C—C—C—C—H
   |  |  |  |  |
   H  |  H  H  H
   H—C—H
      |
      H
```

2-methylpentane

```
   H  H  H  H  H
   |  |  |  |  |
H—C—C—C—C—C—H
   |  |  |  |  |
   H  H  |  H  H
      H—C—H
         |
         H
```

3-methylpentane

```
         H
         |
      H—C—H
   H  |  H  H
   |  |  |  |
H—C—C—C—C—H
   |  |  |  |
   H  |  H  H
      H—C—H
         |
         H
```

2,2-dimethylbutane

```
         H
         |
      H—C—H
   H  |  H  H
   |  |  |  |
H—C—C—C—C—H
   |  |  |  |
   H  H  |  H
      H—C—H
         |
         H
```

2,3-dimethylbutane

22.25 (a)

```
    H  H  H  H  H  H
    |  |  |  |  |  |
 H—C—C—C—C—C—C—H
    |  |  |  |  |  |
    H  |  H  H  H  H
    H—C—H
       |
       H
```

(b)

```
    H  H  Cl H  H  H  H
    |  |  |  |  |  |  |
 H—C—C—C—C—C—C—C—H
    |  |  |  |  |  |  |
    H  H  Cl H  H  H  H
```

(c)

```
          (C6H5)
    H  H  |  H  H  H  H  H
    |  |  |  |  |  |  |  |
 H—C—C—C—C—C—C—C—C—H
    |  |  |  |  |  |  |  |
    H  |  H  H  H  H  H  H
    H—C—H
       |
       H
```

(d)

```
 H3C—CH2   CH2—CH3
         \ /
     H2C    CH2
      |      |
     H2C    CH2
        \  /
        CH2
```

22.27 (a) 1-fluoro-2-methylpentane
(b) 3-methylhexane
(c) 2,2-dimethylbutane
(d) 1-chloro-2-ethylcyclobutane

22.29 C_nH_{2n}, C_nH_{2n-2}

22.31 (a) *trans*-1-bromopropene
(b) *cis*-3-heptene
(c) 5-fluoro-2-pentyne
(d) *cis*-1-chloro-3-hexene

22.33

```
 Cl      H
   \    /
    C=C
   /    \
 Br      Br
```

cis-1-chloro-1,2-dibromoethene

```
 Cl      Br
   \    /
    C=C
   /    \
 Br      H
```

trans-1-chloro-1,2-dibromoethene

22.35

```
 F      CH3        H      CH3
  \    /            \    /
   C=C               C=C
  /    \            /    \
 H      H          F      H
```

cis-1-fluoropropene *trans*-1-fluoropropene

```
 H      CH3        H      CH2F
  \    /            \    /
   C=C               C=C
  /    \            /    \
 H      F          H      H
```

2-fluoropropene 3-fluoropropene

22.37 (a)

```
             H  H  H  H
             |  |  |  |
 H—C=C—C—C—C—C—H
    |  |  |  |  |  |
    H  Br H  H  H  H
```

(b)

```
    H        NO2 H
    |         |  |
 H—C         C—C—H
    |  \     / |  |
    H   C=C   H  H
    H         H
```

(c)

```
    H  H        H  H
    |  |        |  |
 H—C—C—C≡C—C—C—H
    |  |        |  |
    Cl Cl       H  H
```

(d)

```
 F          H  H  H  H  H
  \         |  |  |  |  |
   C=C—C—C—C—C—C—H
  /     |  |  |  |  |  |
 F      H  Cl H  H  H  H
```

22.39

```
             H
             |
 H—C≡C—C—H + 2Cl2 ⟶
             |
             H
```

1,1,2,2-tetrachloropropane

22.41 (a) $CH_3CH_2CH_3$ (b) $C_6H_5Cl + HCl$

22.43 [three benzene rings: ortho-, meta-, and para-dimethylbenzene, each with two CH_3 groups]

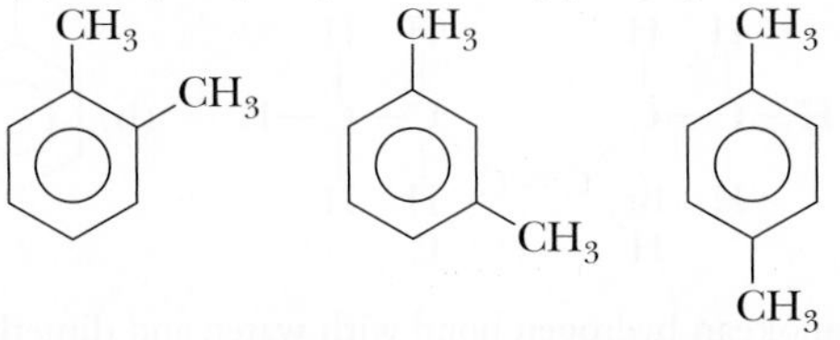

22.45 (a) carboxylic acid (b) aldehyde
(c) ether

22.47 (a) CH_3CH_2COOH; hydrogen bonding
(b) $CH_3CH_2CH_2CH_2NH_2$; hydrogen bonding
(c) $CH_3CH_2C{\equiv}CF$; larger molecule, stronger dispersion force

22.49 CH_3OCH_3, dimethyl ether

22.51 $H_3C{-}C({=}O){-}O{-}CH_2{-}CH_3$ ethyl acetate

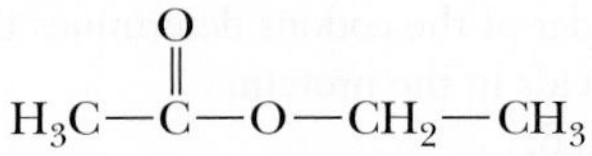

22.53 (a) $H{-}O{-}CH_2{-}CH_2{-}CH_2{-}CH_2{-}H$ (all H shown)

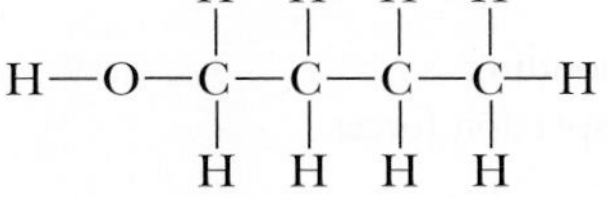

(b) $H_3C{-}C({=}O){-}CH(CH_3){-}CH_2{-}CH_3$

(c) $H_3C{-}C({=}O){-}O{-}CH_3$

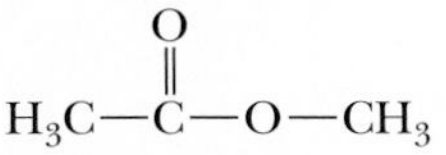

(d) $C_6H_5{-}N(H){-}CH_2{-}CH_3$

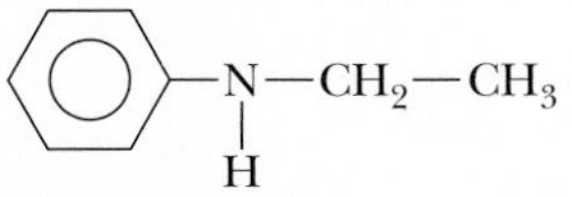

22.55 (a) 3-fluoro-1-propanol (b) propanoic acid
(c) isopropyl acetate (d) *n*-propylamine

22.57 (a) no chiral center
(b) no chiral center
(c) $H_3C{-}C^{*}H(NH_2){-}C({=}O){-}O{-}H$

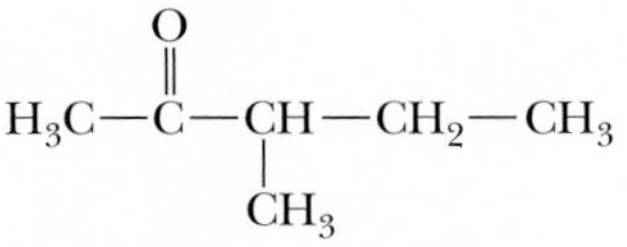

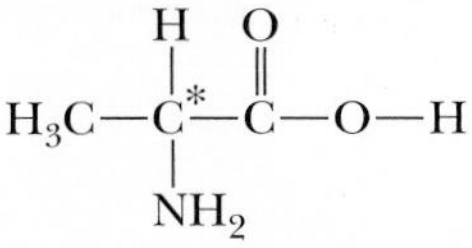

22.59 (a) A chain-growth polymer (or addition polymer) is a polymer formed with no loss of atoms.

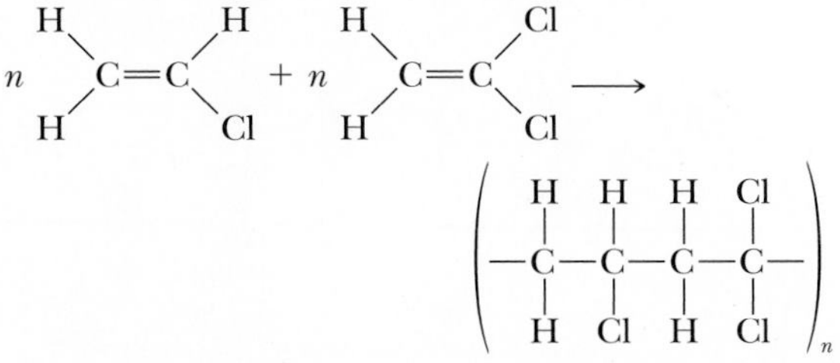

$n\ H_2C{=}CHCl + n\ H_2C{=}CCl_2 \longrightarrow {-}(CH_2{-}CHCl{-}CH_2{-}CCl_2){-}_n$

(b) A homopolymer is a chain-growth polymer in which all the monomeric units are the same.

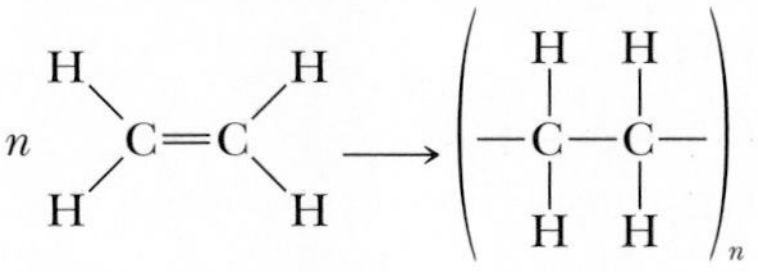

$n\ H_2C{=}CH_2 \longrightarrow {-}(CH_2{-}CH_2){-}_n$

22.61 ${-}(CH_2{-}CHCl){-}_n$

22.63 ${-}(CF_2{-}CH_2{-}CF_2{-}CF(CF_3)){-}_n$

22.65 $HC(H){=}C(H){-}C{\equiv}N$ (H₂C=CH–C≡N)

22.67 $H{-}C(R)(NH_2){-}C({=}O){-}OH + H{-}C(R')(NH_2){-}C({=}O){-}OH \longrightarrow$
$H{-}C(R)(NH_2){-}C({=}O){-}N(H){-}C(R')(H){-}C({=}O){-}OH + H_2O$

22.69 $H_3C{-}CH(NH_2){-}C({=}O){-}NH{-}CH_2{-}C({=}O){-}NH{-}CH(CH_2SH){-}C({=}O){-}OH$

$H_3C{-}CH(NH_2){-}C({=}O){-}NH{-}CH(CH_2SH){-}C({=}O){-}NH{-}CH_2{-}C({=}O){-}OH$

22.71 [ring structure of glucose with CH_2OH, HO, OH, H substituents]

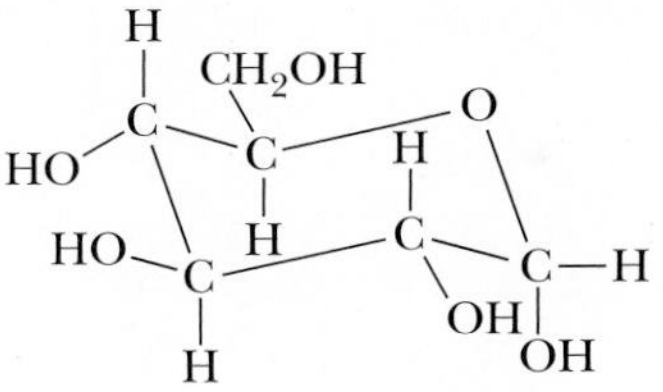

α-glucose

[ring structure of glucose with CH_2OH, HO, OH, H substituents]

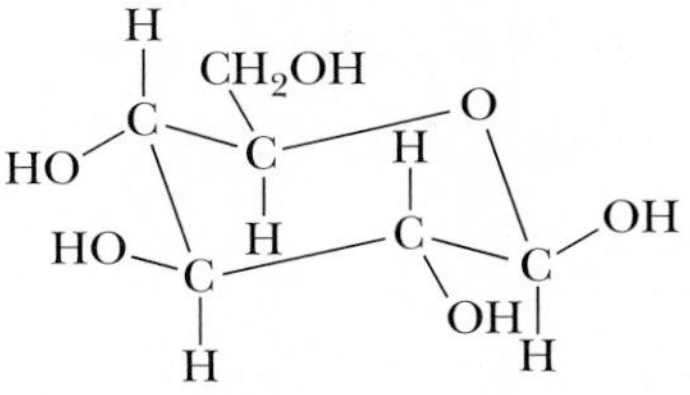

β-glucose

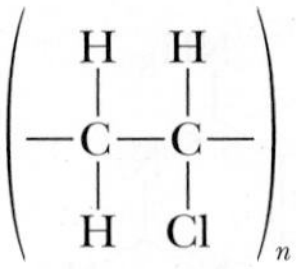

22.73 [glucose ring joined through O–CH_2 to a five-membered ring with OH, HO, H substituents]

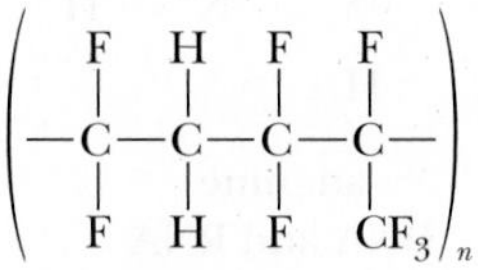

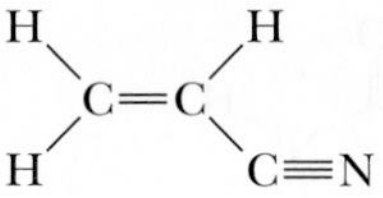

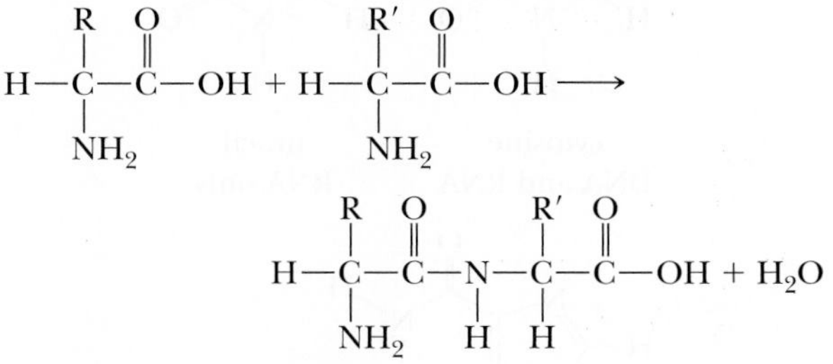

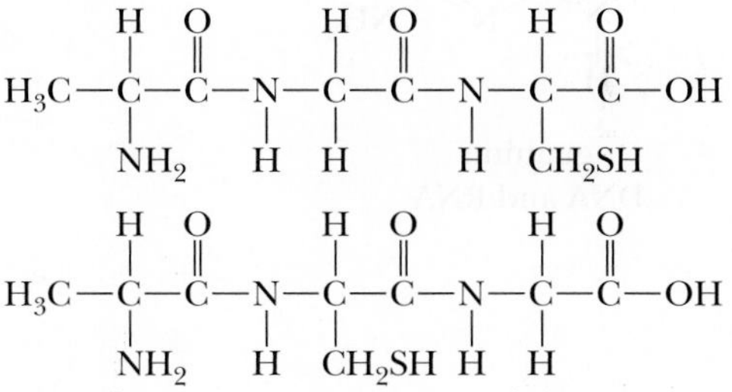

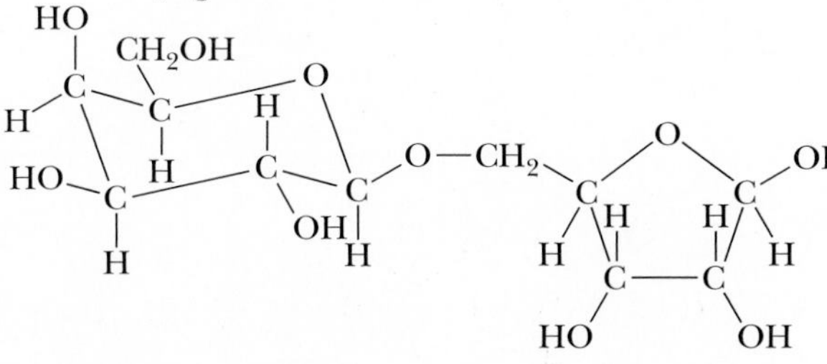

22.75

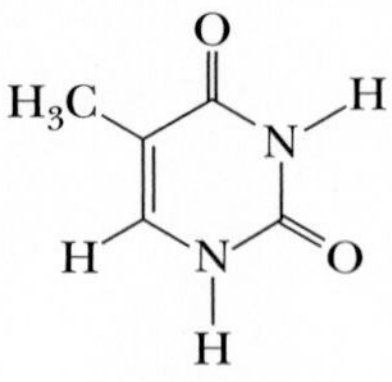

thymine
DNA only

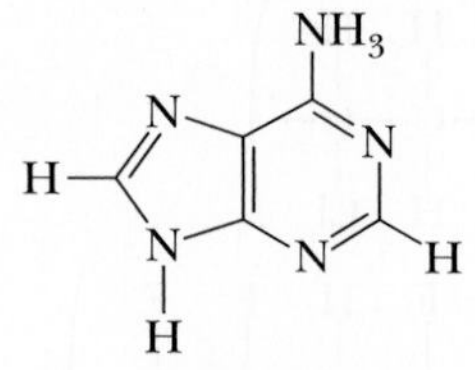

adenine
DNA and RNA

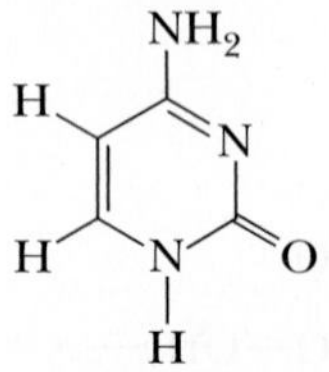

cytosine
DNA and RNA

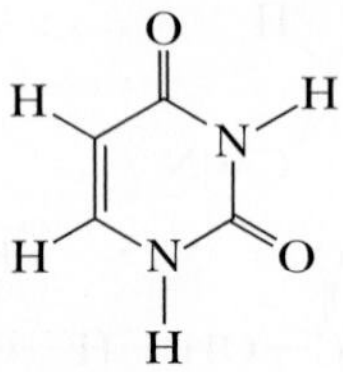

uracil
RNA only

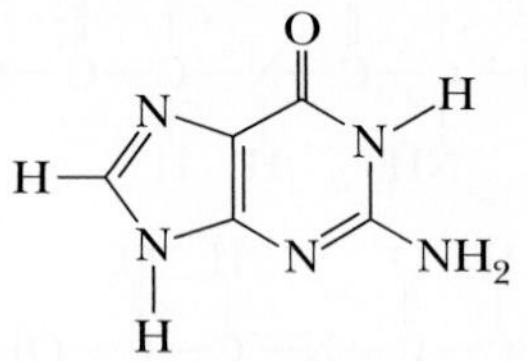

guanine
DNA and RNA

22.77 (a) 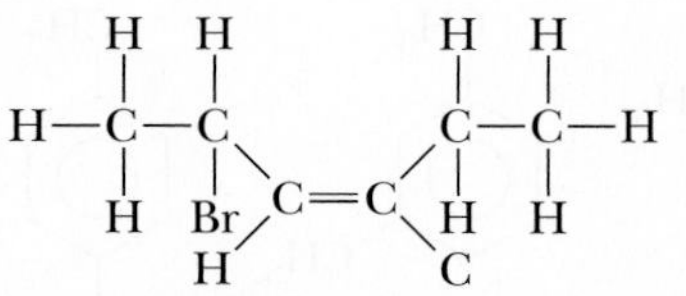 (b)

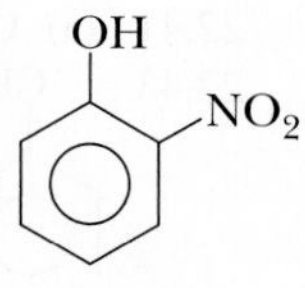

22.79 Ethanol can hydrogen bond with water, and dimethyl ether cannot.

22.81

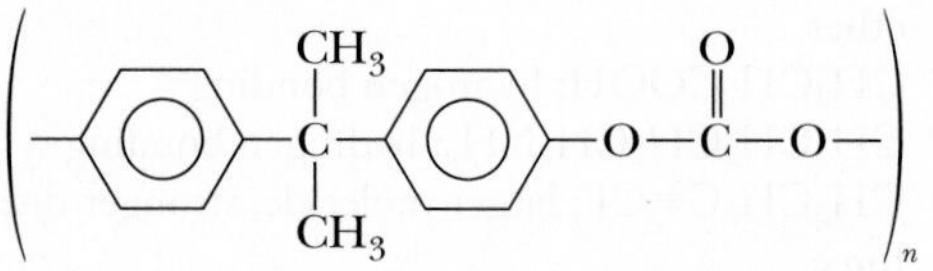

22.83 Each amino acid is represented by one or more three-letter codons. The order of the codons determines the order of the amino acids in the protein.

22.85 $C_3H_6 + Br_2 \rightarrow C_3H_6Br_2$
83.9 g

22.87 (a) hydrogen bonding
(b) London dispersion forces

Index

Note: Figures, tables, and footnotes are indicated by *italics,* "t", and "n" respectively.

A

B

C

D

E

H

Q

R

S

T